# CHILTON.

# DAIMLERCHRYSLER
## DIAGNOSTIC SERVICE
## 2006 EDITION

THOMSON
DELMAR LEARNING

Australia • Canada • Mexico • Singapore • Spain • United Kingdom • United States

# CHILTON®

## DAIMLERCHRYSLER
### DIAGNOSTIC SERVICE
### 2006 Edition

**Vice President,**
**Technology Professional Business Unit:**
Gregory L. Clayton

**Publisher,**
**Professional Business Unit:**
David Koontz

**Production Director:**
Mary Ellen Black

**Marketing Director:**
Beth A. Lutz

**Marketing Specialist:**
Brian McGrath

**Marketing Coordinator:**
Marissa Mariella

**Marketing Assistant:**
Jennifer Stall

**Sr. Production Editor:**
Elizabeth Hough

**Editorial Assistant:**
Christine Wade

**Editor:**
Dennis Bailey

**Publishing Coordinator:**
Paula Baillie

**Cover Design:**
Melinda Possinger

**ISBN: 1-4180-2118-0**

## NOTICE TO THE READER

# TABLE OF CONTENTS

# USING THIS INFORMATION

## Organization

To find where a particular model section or procedure is located, look in the Table of Contents. Main topics are listed with the page number on which they may be found. Following the main topics is an alphabetical listing of all of the procedures within the section and their page numbers.

## Manufacturer and Model Coverage

This product covers 1990-2006 DaimlerChrysler models that are produced in sufficient quantities to warrant coverage, and which have technical content available from the vehicle manufacturers before our publication date. Although this information is as complete as possible at the time of publication, some manufacturers may make changes which cannot be included here. While striving for total accuracy, the publisher cannot assume responsibility for any errors, changes, or omissions that may occur in the compilation of this data.

## Part Numbers & Special Tools

Part numbers and special tools are recommended by the publisher and vehicle manufacturer to perform specific jobs. Before substituting any part or tool for the one recommended, you must be completely satisfied that neither your personal safety, nor the performance of the vehicle will be endangered.

# ACKNOWLEDGEMENT

The publisher would like to express appreciation to DaimlerChrysler for its assistance in producing this product.

# PRECAUTIONS

Before servicing any vehicle, please be sure to read all of the following precautions, which deal with personal safety, prevention of component damage, and important points to take into consideration when servicing a motor vehicle:

• Always wear safety glasses or goggles when drilling, cutting, grinding or prying.

• Steel-toed work shoes should be worn when working with heavy parts. Pockets should not be used for carrying tools. A slip or fall can drive a screwdriver into your body.

• Work surfaces, including tools and the floor should be kept clean of grease, oil or other slippery material.

• When working around moving parts, don't wear loose clothing. Long hair should be tied back under a hat or cap, or in a hair net.

• Always use tools only for the purpose for which they were designed. Never pry with a screwdriver.

• Keep a fire extinguisher and first aid kit handy.

• Always properly support the vehicle with approved stands or lift.

• Always have adequate ventilation when working with chemicals or hazardous material.

• Carbon monoxide is colorless, odorless and dangerous. If it is necessary to operate the engine with vehicle in a closed area such as a garage, always use an exhaust collector to vent the exhaust gases outside the closed area.

• When draining coolant, keep in mind that small children and some pets are attracted by ethylene glycol antifreeze, and are quite likely to drink any left in an open container, or in puddles on the ground. This will prove fatal in sufficient quantity. Always drain the coolant into a sealable container.

• To avoid personal injury, do not remove the coolant pressure relief cap while the engine is operating or hot. The cooling system is under pressure; steam and hot liquid can come out forcefully when the cap is loosened slightly. Failure to follow these instructions may result in personal injury. The coolant must be recovered in a suitable, clean container for reuse. If the coolant is contaminated it must be recycled or disposed of correctly.

• When carrying out maintenance on the starting system be aware that heavy gauge leads are connected directly to the battery. Make sure the protective caps are in place when maintenance is completed. Failure to follow these instructions may result in personal injury.

• Do not remove any part of the engine emission control system. Operating the engine without the engine emission control system will reduce fuel economy and engine ventilation. This will weaken engine performance and shorten engine life. It is also a violation of Federal law.

• Due to environmental concerns, when the air conditioning system is drained, the refrigerant must be collected using refrigerant recovery/recycling equipment. Federal law requires that refrigerant be recovered into appropriate recovery equipment and the process be conducted by qualified technicians who have been certified by an approved organization, such as MACS, ASI, etc. Use of a recovery machine dedicated to the appropriate refrigerant is necessary to reduce the possibility of oil and refrigerant incompatibility concerns. Refer to the instructions provided by the equipment manufacturer when removing refrigerant from or charging the air conditioning system.

• Always disconnect the battery ground when working on or around the electrical system.

• Batteries contain sulfuric acid. Avoid contact with skin, eyes, or clothing. Also, shield your eyes when working near batteries to protect against possible splashing of the acid solution. In case of acid contact with skin or eyes, flush immediately with water for a minimum of 15 minutes and get prompt medical attention. If acid is swallowed, call a physician immediately. Failure to follow these instructions may result in personal injury.

• Batteries normally produce explosive gases. Therefore, do not allow flames, sparks or lighted substances to come near the battery. When charging or working near a battery, always shield your face and protect your eyes. Always provide ventilation. Failure to follow these instructions may result in personal injury.

• When lifting a battery, excessive pressure on the end walls could cause acid to spew through the vent caps, resulting in personal injury, damage to the vehicle or battery. Lift with a battery carrier or with your hands on opposite corners. Failure to follow these instructions may result in personal injury.

• Observe all applicable safety precautions when working around fuel. Whenever servicing the fuel system, always work in a well-ventilated area. Do not allow fuel spray or vapors to come in contact with a spark, open flame, or excessive heat (a hot drop light, for example). Keep a dry chemical fire extinguisher near the work area. Always keep fuel in a container specifically designed for fuel storage; also, always properly seal fuel containers to avoid the possibility of fire or explosion. Do not smoke or carry lighted tobacco or open flame of any type when working on or near any fuel-related components.

• Fuel injection systems often remain pressurized, even after the engine has been turned OFF. The fuel system pressure must be relieved before disconnecting any fuel lines. Failure to do so may result in fire and/or personal injury.

• The evaporative emissions system contains fuel vapor and condensed fuel vapor. Although not present in large quantities, it still presents the danger of explosion or fire. Disconnect the battery ground cable from the battery to minimize the possibility of an electrical spark occurring, possibly causing a fire or explosion if fuel vapor or liquid fuel is present in the area. Failure to follow these instructions can result in personal injury.

• The EPA warns that prolonged contact with used engine oil may cause a number of skin disorders, including cancer! You should make every effort to minimize your exposure to used engine oil. Protective gloves should be worn when changing oil. Wash your hands and any other exposed skin areas as soon as possible after exposure to used engine oil. Soap and water, or waterless hand cleaner should be used.

• Some vehicles are equipped with an air bag system, often referred to as a Supplemental Restraint System (SRS) or Supplemental Inflatable Restraint (SIR) system. The system must be disabled before performing service on or around system components, steering column, instrument panel components, wiring and sensors. Failure to follow safety and disabling procedures could result in accidental air bag deployment, possible personal injury and unnecessary system repairs.

• Always wear safety goggles when working with, or around, the air bag system. When carrying a non-deployed air bag, be sure the bag and trim cover are pointed away from your body. When placing a non-deployed air bag on a work surface, always face the bag and trim cover upward, away from the surface. This will reduce the motion of the module if it is accidentally deployed.

• Electronic modules are sensitive to electrical charges. The ABS module can be damaged if exposed to these charges.

• Brake pads and shoes may contain asbestos, which has been determined to be a cancer-causing agent. Never clean brake surfaces with compressed air. Avoid inhaling brake dust. Clean all brake surfaces with a commercially available brake cleaning fluid.

• When replacing brake pads, shoes, discs or drums, replace them as complete axle sets.

• When servicing drum brakes, disassemble and assemble one side at a time, leaving the remaining side intact for reference.

• Brake fluid often contains polyglycol ethers and polyglycols. Avoid contact with the eyes and wash your hands thoroughly after handling brake fluid. If you do get brake fluid in your eyes, flush your eyes with clean, running water for 15 minutes. If eye irritation persists, or if you have taken brake fluid internally, immediately seek medical assistance.

• Clean, high quality brake fluid from a sealed container is essential to the safe and proper operation of the brake system. You should always buy the correct type of brake fluid for your vehicle. If the brake fluid becomes contaminated, completely flush the system with new fluid. Never reuse any brake fluid. Any brake fluid that is removed from the system should be discarded. Also, do not allow any brake fluid to come in contact with a painted or plastic surface; it will damage the paint.

• Never operate the engine without the proper amount and type of engine oil; doing so will result in severe engine damage.

• Timing belt maintenance is extremely important! Many models utilize an interference-type, non-freewheeling engine. If the timing belt breaks, the valves in the cylinder head may strike the pistons, causing potentially serious (also time-consuming and expensive) engine damage.

• Disconnecting the negative battery cable on some vehicles may interfere with the functions of the on-board computer system(s) and may require the computer to undergo a relearning process once the negative battery cable is reconnected.

• Steering and suspension fasteners are critical parts because they affect performance of vital components and systems and their failure can result in major service expense. They must be replaced with the same grade or part number or an equivalent part if replacement is necessary. Do not use a replacement part of lesser quality or substitute design. Torque values must be used as specified during reassembly to ensure proper retention of these parts.

## Contents

## OBD II Vehicle Applications

DAIMLERCHRYSLER & DODGE CARS

### Avenger (FJ Body Code) & Sebring (JX Body Code)
1996-2000

| | |
|---|---|
| 2.0L 4-Cyl. MPI | VIN C |
| 2.0L 4-Cyl. MPI | VIN Y |
| 2.4L 4-Cyl. MPI | VIN X |
| 2.5L V6 MPI | VIN H, N |

### Eagle (FJ Body Code)
1995-98 Talon

| | |
|---|---|
| 2.0L 4-Cyl. MPI | VIN Y |
| 2.0L 4-Cyl. Turbo MPI | VIN F |

### Crossfire (ZH Body Code)
2004-05

| | |
|---|---|
| 3.2L V6 | VIN L |
| 3.2L V6 Supercharged | VIN N |

### Breeze, Cirrus & Stratus (JA Body Code)
1996-2000

| | |
|---|---|
| 2.0L 4-Cyl. MPI | VIN C |
| 2.4L 4-Cyl. MPI | VIN X |
| 2.5L V6 MPI | VIN H |

### Concorde, Intrepid, LHS, New Yorker, Vision, 300M (LH Body Code)
1996-2005

| | |
|---|---|
| 2.7L V6 MPI | VIN R, U, V |
| 3.2L V6 MPI | VIN J |
| 3.3L V6 MPI | VIN T |
| 3.5L V6 MPI | VIN F, G, K, M, V |

### Magnum, 300 (LX Body Code)
2004-05

| | |
|---|---|
| 2.7L V6 MPI | VIN R |
| 3.5L V6 HO MPI | VIN G |
| 5.7L V8 Hemi MPI | VIN H |
| 6.1L V8 Hemi SMPI | VIN W |

### Neon (PL Body Code)
1995-2003

| | |
|---|---|
| 2.0L 4-Cyl. MPI | VIN C, F |
| 2.0L 4-Cyl. MPI | VIN Y |
| 2.4L 4-Cyl. H.O. Turbo | VIN S |

### Pacifica (CS-Body Code)
2005

| | |
|---|---|
| 3.5L V6 MPI | VIN 4 |
| 3.8L V6 SMPI | VIN L |

### Prowler (PL-Body Code) & Viper (SR & ZB Body Code)
1997-2001

| | |
|---|---|
| 3.5L V6 MPI | VIN F |
| 3.5L V6 MPI | VIN G |

1996-2003

| | |
|---|---|
| 8.0L V10 MPI | VIN E |

2004-05

| | |
|---|---|
| 8.3L V10 SFI | VIN Z |

**PT Cruiser (PT Body Code)**

2001-05

    1.6L 4-Cyl. SMPI .................................................................................................................. VIN F

    2.2L 4-Cyl. Turbo Diesel ...................................................................................................... VIN U

    2.4L 4-Cyl. SMPI ............................................................................................................. VIN B, X

    2.4L 4-Cyl. Turbo Gas ................................................................................................ VIN E, G, S, 8

**Sebring, Stratus (JR Body Code: Sedan & Convertible; ST Body Code: Coupe)**

2001-05

    2.0L 4-Cyl. MPI .................................................................................................................. VIN Y

    2.4L 4-Cyl. MPI ............................................................................................................ VIN G, J, S, X

    2.7L V6 MPI ............................................................................................................... VIN R, T, U

    3.0L V6 MPI ....................................................................................................................... VIN H

DAIMLERCHRYSLER & DODGE MINIVANS

**Caravan, Town & Country, Voyager (NS & RS Body Codes)**

1996-2005

| | |
|---|---|
| 2.4L 4-Cyl. MPI | VIN B, X |
| 2.5L 4-Cyl. Turbo Diesel | VIN 7 |
| 2.8L 4-Cyl. Turbo Diesel | VIN 5 |
| 3.0L V6 MPI | VIN 3 |
| 3.3L V6 MPI/SMPI | VIN R |
| 3.3L V6 MPI (Flex Fuel) | VIN E, G |
| 3.8L V6 MPI/SMPI | VIN L |

DODGE TRUCKS & VANS

**Ram Truck, Ram Van & Ram Wagon (B, R & DR Body Codes)**

1996-2003 (B & R Body Codes)

| | |
|---|---|
| 3.9L V6 MPI | VIN X |
| 5.2L V8 MPI | VIN 2, T & Y |
| 5.9L V8 MPI LD | VIN Z |

2004-05 (DR Body Code)

| | |
|---|---|
| 3.7L V6 MPI | VIN K |
| 4.7L V8 MPI | VIN N |
| 5.7L V8 SMPI | VIN D |
| 5.9L 6-Cyl. Turbo Diesel | VIN 6, C |
| 8.3L 10-Cyl. SFI Gasoline | VIN H |

**Dakota (N, ND Body Codes) & Durango (DN, HB Body Codes)**

1996-2005

| | |
|---|---|
| 3.7L V6 MPI | VIN K |
| 4.7L V8 MPI | VIN N |
| 4.7L V8 MPI H.O. | VIN J |
| 5.2L V8 MPI | VIN T & Y |
| 5.7L V8 SMPI | VIN D |
| 5.9L V8 MPI LD | VIN 5 & Z |
| 5.9L I6 Diesel | VIN 6 & 7 |
| 8.0L V10 MPI | VIN W |

<u>JEEP MODELS</u>

**Cherokee (XJ Body Code) & Grand Cherokee (ZJ Body Code)**

1996-2003

2.5L 4-Cyl. MPI ...................................................................................................................................... VIN P

4.0L 6-Cyl. MPI ...................................................................................................................................... VIN S

4.7L V8 MPI .......................................................................................................................................... VIN N

5.2L V8 MPI .......................................................................................................................................... VIN Y

5.9L V8 MPI .......................................................................................................................................... VIN Z

**Grand Cherokee (WK Body Code)**

2004

2.7L 4-Cyl. Turbo Diesel Direct Injection ............................................................................................. VIN A

4.0L 6-Cyl. MPI ...................................................................................................................................... VIN S

4.7L V8 MPI .......................................................................................................................................... VIN N

4.7L V8 H.O. ......................................................................................................................................... VIN J

2005

3.0L 6-Cyl. Turbo Diesel ....................................................................................................................... VIN M

3.7L 6-Cyl. MPI ..................................................................................................................................... VIN K

4.7L V8 MPI .......................................................................................................................................... VIN N

5.7L V8 HEMI Multi Displacement ....................................................................................................... VIN 2

**Liberty (KJ Body Code)**

2002-05

2.4L 4-Cyl. MPI ..................................................................................................................................... VIN 1

2.5L 4-Cyl. Diesel ................................................................................................................................. VIN 7

2.8L 4-Cyl. Diesel ................................................................................................................................. VIN 5

3.7L V6 MPI .......................................................................................................................................... VIN K

**Wrangler (TJ-Body Code)**

1997-2005

2.4L 4-Cyl. MPI ..................................................................................................................................... VIN 1

2.5L 4-Cyl. MPI ..................................................................................................................................... VIN P

4.0L 6-Cyl. MPI ..................................................................................................................................... VIN S

## NOTES & CAUTIONS

Before servicing any vehicle, please be sure to read all of the following precautions, which deal with personal safety, prevention of component damage, and important points to take into consideration when servicing a motor vehicle:

- Observe all applicable safety precautions when working around fuel. Whenever servicing the fuel system, always work in a well-ventilated area. Do NOT allow fuel spray or vapors to come in contact with a spark, open flame, or excessive heat (a hot drop light, for example). Keep a dry chemical fire extinguisher near the work area. Always keep fuel in a container specifically designed for fuel storage; also, always properly seal fuel containers to avoid the possibility of fire or explosion. Refer to the additional fuel system precautions later in this section.

- Fuel injection systems often remain pressurized, even after the engine has been turned **OFF**. The fuel system pressure must be relieved before disconnecting any fuel lines. Failure to do so may result in fire and/or personal injury.

- Brake fluid often contains Polyglycol Ethers and Polyglycols. Avoid contact with the eyes and wash your hands thoroughly after handling brake fluid. If you do get brake fluid in your eyes, flush your eyes with clean, running water for 15 minutes. If eye irritation persists, or if you have taken brake fluid internally, IMMEDIATELY seek medical assistance.

- The EPA warns that prolonged contact with used engine oil may cause a number of skin disorders, including cancer. You should make every effort to minimize your exposure to used engine oil. Protective gloves should be worn when changing oil. Wash your hands and any other exposed skin areas as soon as possible after exposure to used engine oil. Soap and water, or waterless hand cleaner should be used.

- The air bag system must be disabled (negative battery cable disconnected and/or air bag system main fuse removed) for at least 30 seconds before performing service on or around system components, steering column, instrument panel components, wiring and sensors. Failure to follow safety and disabling procedures could result in accidental air bag deployment, possible personal injury and unnecessary system repairs.

- Always wear safety goggles when working with, or around, the air bag system. When carrying a non-deployed air bag, be sure the bag and trim cover are pointed away from your body. When placing a non-deployed air bag on a work surface, always face the bag and trim cover upward, away from the surface. This will reduce the motion of the module if it is accidentally deployed. Refer to the additional air bag system precautions later in this section.

- Disconnecting the negative battery cable on some vehicles may interfere with the functions of the on-board computer system(s) and may require the computer to undergo a relearning process once the negative battery cable is reconnected.

- It is critically important to observe all instructions regarding ground disconnects, ignition switch positions, etc., in each diagnostic routine provided. Ignoring these instructions can result in false readings, damage to electronic components or circuits, or personal injury.

**Preliminary Diagnostics**

## HISTORY OF OBD SYSTEMS

Starting in 1978, several vehicle manufacturers introduced a new type of control for several vehicle systems and computer control of engine management systems. These computer-controlled systems included programs to test for problems in the engine mechanical area, electrical fault identification and tests to help diagnose the computer control system. Early attempts at diagnosis involved expensive and specialized diagnostic testers that hooked up externally to the computer in series with the wiring connector and monitored the input/output operations of the computer.

By early 1980, vehicle manufacturers had designed systems in which the onboard computer incorporated programs to monitor selected components, and to store a trouble code in its memory that could be retrieved at a later time. These trouble codes identified failure conditions that could be used to refer a technician to diagnostic repair charts or test procedures to help pinpoint the problem area.

## EVOLUTION OF DAIMLERCHRYSLER COMPUTERIZED ENGINE CONTROLS

The evolution of Computerized Engine Controls on DaimlerChrysler vehicles equipped with fuel injection is highlighted in the Graphic below.

## OBD I SYSTEM DIAGNOSTICS

One of the most important things to understand about the automotive repair industry is the fact that you have to continually learn new systems and new diagnostic routines (the test procedures designed to isolate a problem on a vehicle system). For OBD I and II systems, a diagnostic routine can be defined as a procedure (a series of steps) that you follow to find the cause of a problem, make a repair and then verify the problem is fixed.

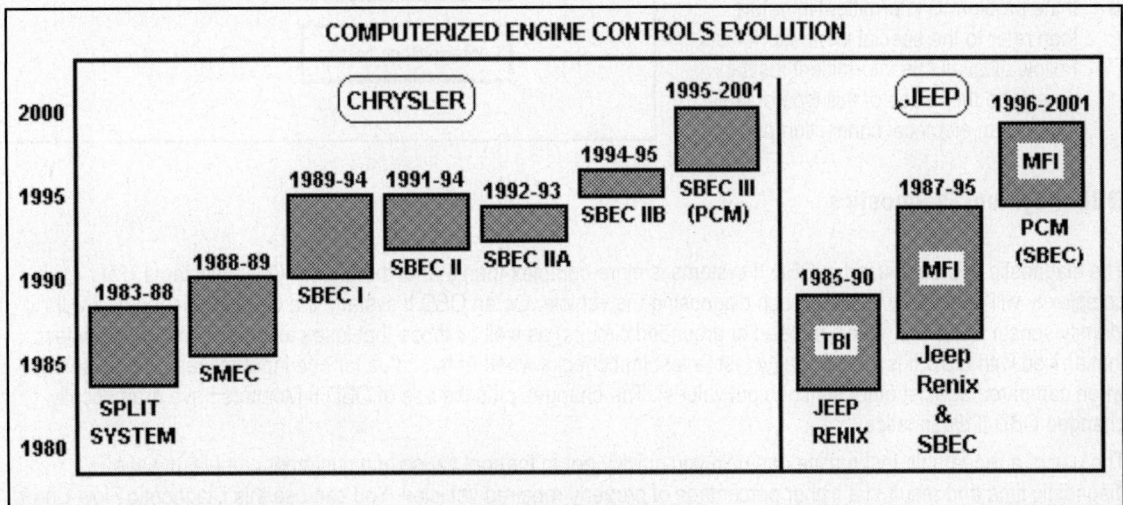

## CHANGES IN DIAGNOSTIC ROUTINES

In some cases, a new Engine Control system may be similar to an earlier system, but it can have more indepth control of vehicle emissions, input and output devices and it may include a diagnostic "monitor" embedded in the engine controller designed to run a thorough set of emission control system tests.

## OBD I Diagnostic Flowchart

The OBD I Diagnostic Flowchart on this page can be used to find the cause of problems related to Engine Control system trouble codes or driveability symptoms detected on OBD I systems. It includes a step-by-step procedure to use to repair these systems. To compare this flowchart with the one used on OBD II systems, refer to the next page.

The steps in this flow chart should be followed as described below (from top to bottom).

- Do the Pre-Computer Checks.
- Check for any trouble codes stored in memory.
- Read the trouble codes - If trouble codes are set, record them and then clear the codes.
- Start the vehicle and see if the trouble code(s) reset. If they do, then use the correct trouble code repair chart to make the repair.
- If the codes do not reset, than the problem may be intermittent in nature. In this case, refer to the test steps used to find the cause of an intermittent fault (wiggle test).
- In no trouble codes are found at the initial check, then determine if a driveability symptom is present. If so, then refer to the approriate driveability symptom repair chart to make the repair. If the first symptom chart does not isolate the cause of the condition, then go on to another driveability symptom and follow that procedure to conclusion.
- If the problem is intermittent in nature, then refer to the special intermittent tests. Follow all available intermittent tests to determine the cause of this type of fault (usually an electrical connection problem).

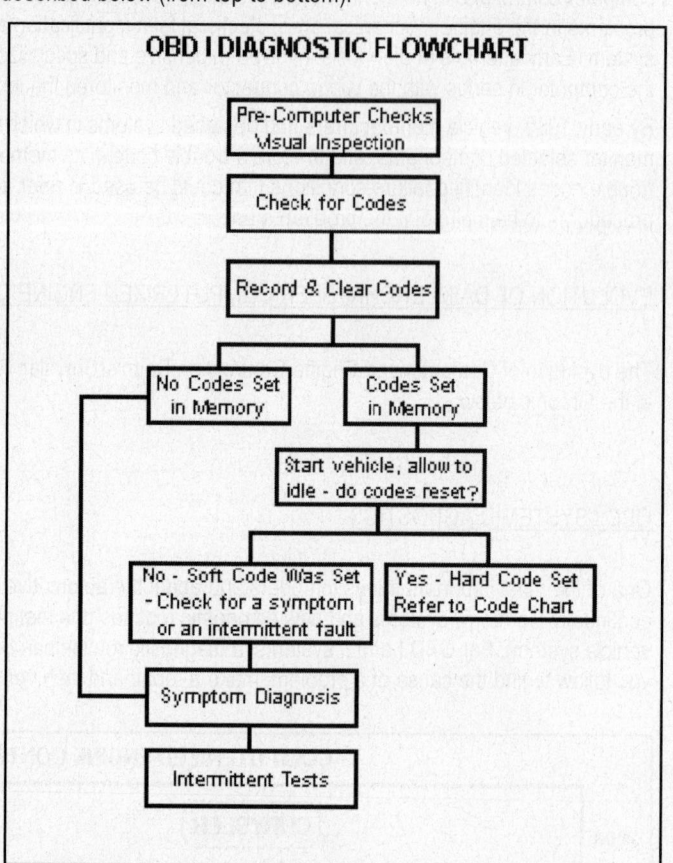

## OBD II System Diagnostics

The diagnostic approach used in OBD II systems is more complex than that of the one for OBD I systems. This complexity will effect how you approach diagnosing the vehicle. On an OBD II system, the onboard diagnostics will identify sensor faults (i.e., open, shorted or grounded circuits) as well as those that lose calibration. Another new test that arrived with OBD II is the rationality test (a test that checks whether the value for one input makes rational sense when compared against other sensor input values). The changes plus the use of OBD II Monitors have dramatically changed OBD II diagnostics.

The use of a repeatable test routine can help you quickly get to the root cause of a customer complaint, save diagnostic time and result in a higher percentage of properly repaired vehicles. You can use this Diagnostic Flow Chart to keep on track as you diagnose an Engine Control problem or a base engine fault on vehicles with OBD II.

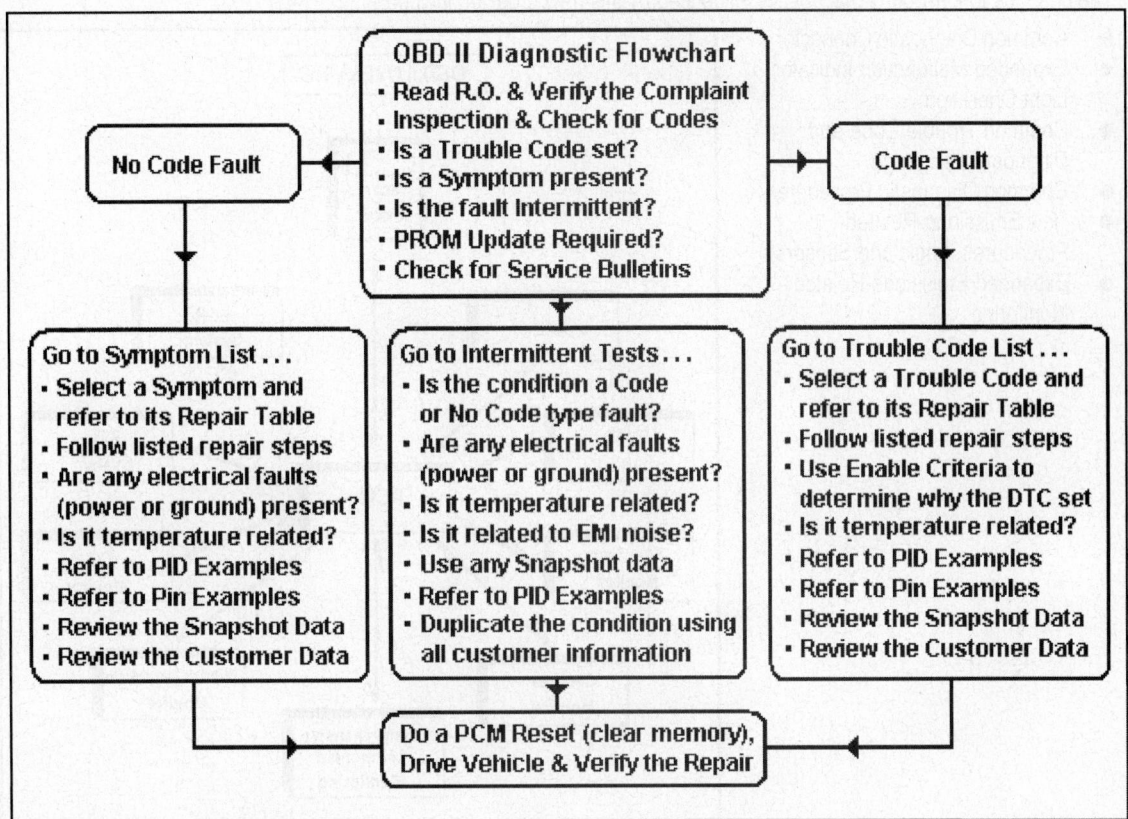

**OBD II Diagnostic Flowchart**
- Read R.O. & Verify the Complaint
- Inspection & Check for Codes
- Is a Trouble Code set?
- Is a Symptom present?
- Is the fault Intermittent?
- PROM Update Required?
- Check for Service Bulletins

**No Code Fault**

**Code Fault**

**Go to Symptom List . . .**
- Select a Symptom and refer to its Repair Table
- Follow listed repair steps
- Are any electrical faults (power or ground) present?
- Is it temperature related?
- Refer to PID Examples
- Refer to Pin Examples
- Review the Snapshot Data
- Review the Customer Data

**Go to Intermittent Tests . . .**
- Is the condition a Code or No Code type fault?
- Are any electrical faults (power or ground) present?
- Is it temperature related?
- Is it related to EMI noise?
- Use any Snapshot data
- Refer to PID Examples
- Duplicate the condition using all customer information

**Go to Trouble Code List . . .**
- Select a Trouble Code and refer to its Repair Table
- Follow listed repair steps
- Use Enable Criteria to determine why the DTC set
- Is it temperature related?
- Refer to PID Examples
- Refer to Pin Examples
- Review the Snapshot Data
- Review the Customer Data

**Do a PCM Reset (clear memory), Drive Vehicle & Verify the Repair**

## Flow Chart Steps

Here are some of the steps included in the Diagnostic Routine:

- Review the repair order and verify the customer complaint as described
- Perform a Visual Inspection of underhood or engine related items
- If the engine will not start, refer to No Start Tests
- If codes are set, refer to the trouble code list, select a code and use the repair chart
- If no codes are set, and a symptom is present, refer to the Symptom List
- Check for any related technical service bulletins (for both Code and No Code Faults)
- If the problem is intermittent in nature, refer to the special Intermittent Tests

## OBD II SYSTEM OVERVIEW

The OBD II system was developed as a step toward compliance with California and Federal regulations that set standards for vehicle emission control monitoring for all automotive manufacturers. The primary goal of this system is to detect when the degradation or failure of a component or system will cause emissions to rise by 50%. Every manufacturer must meet OBD II standards by the 1996 model year. Some manufacturers began programs that were OBD II mandated as early as 1992, but most manufacturers began an OBD II phase-in period starting in 1994.

The changes to On-Board Diagnostics influenced by this new program include:

- Common Diagnostic Connector
- Expanded Malfunction Indicator Light Operation
- Common Trouble Code and Diagnostic Language
- Common Diagnostic Procedures
- New Emissions-Related Procedures, Logic and Sensors
- Expanded Emissions-Related Monitoring

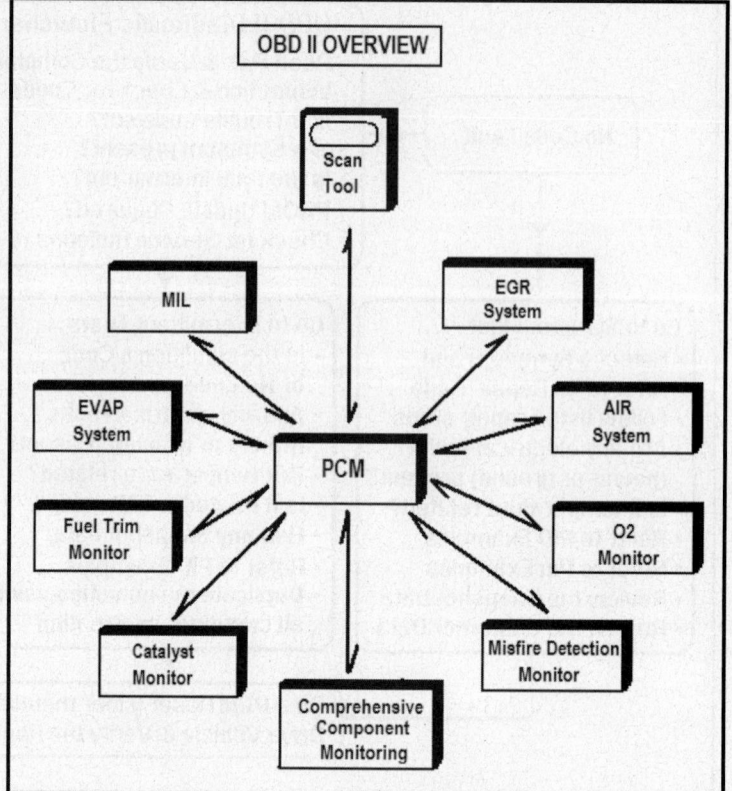

## COMMON TERMINOLOGY

OBD II introduces common terms, connectors, diagnostic language and new emissions-related monitoring procedures. The most important benefit of OBD II is that all vehicles will have a common data output system with a common connector. This allows equipment Scan Tool manufacturers to read data from every vehicle and pull codes with common names and similar descriptions of fault conditions. In the future, emissions testing will require the use of an OBD II certifiable Scan Tool.

## Diagnostic Tools & Circuit Testing

## HAND TOOLS & METER OPERATION

To effectively use this or any diagnostic information, you should have a solid understanding of how to operate required tools and test equipment.

## SCAN TOOLS

Domestic vehicle manufacturers designed their computers to have an accessible data line where a diagnostic tester could retrieve data on sensors and the status of operation for components.

These testers became known in the automotive repair industry as "Scan Tools" because they scanned the data on the computers and provided information for the technician.

The Scan Tool is your basic tool link into the on-board electronic control system of the vehicle. Scan Tools are equipped with, or have separate software cards, for each OEM needed to be diagnosed. In this case, always secure a scan tool that has the latest OEM-specific diagnostic software included. Spend some time in the scan tool user's manual to ensure you know how to properly operate the tool and how to select the necessary programs required for full and proper diagnostics.

DaimlerChrysler specifies the use of a "DRB-II or DRB-III" scan tool with its diagnostic processes. However, there are aftermarket scan tools, when equipped with the right software, that can provide proper diagnosis as well.

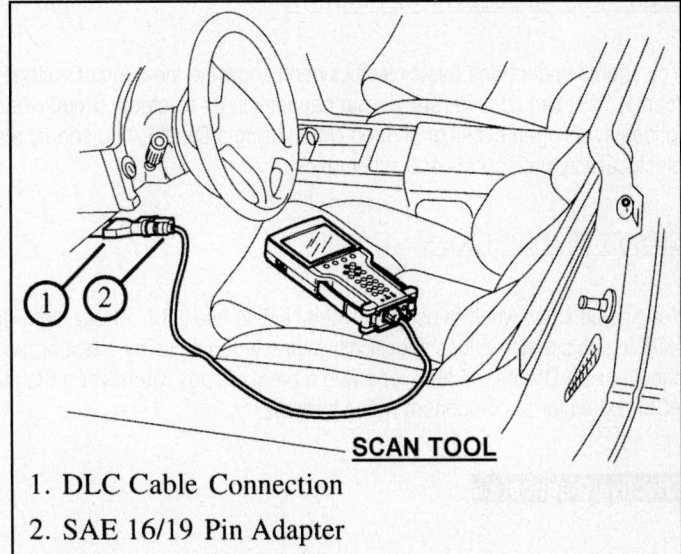

**SCAN TOOL**

1. DLC Cable Connection
2. SAE 16/19 Pin Adapter

MALFUNCTION INDICATOR LAMP

Emission regulations require that a Malfunction Indicator Lamp (MIL) be illuminated when an emissions related fault is detected and that a Diagnostic Trouble Code be stored in the vehicle controller (PCM) memory.

When the MIL is illuminated, it is an indication of a problem within one of the electronic components or circuits. When the scan tool is attached to the Data Link Connector (DLC) in the vehicle, it can access the DTCs. In some situations, without the use of a scan tool, the MIL can be activated to flash a series of long and short flashes, which correspond to the numbering of the DTC.

OBD II guidelines define *when* an emissions-related fault will cause the MIL to activate and set a Diagnostic Trouble Code (DTC). There are some DTCs that will not cause the MIL to illuminate. OBD II guidelines determine how quickly the onboard diagnostics must be able to identify a fault, set the trouble code in memory and activate the MIL (lamp).

ELECTRONIC CONTROLS

You should have a basic knowledge of electronic controls when performing test procedures to keep from making an incorrect diagnosis or damaging components. Do NOT attempt to diagnose an electronic control problem without this basic knowledge!

## ELECTRICITY & ELECTRICAL CIRCUITS

You should understand basic electricity and know the meaning of voltage (volts), current (amps), and resistance (ohms). You should understand what happens in an electrical circuit when it is open or shorted, and you should be able to identify an open circuit or shorted circuit using a DVOM. You should also be able to read and understand automotive electrical wiring diagrams and schematics.

## CIRCUIT TESTING TOOLS

You should know when to use and when NOT to use a 12-volt test light during diagnosis of electronic controls (Do NOT use this tester unless specifically instructed to do so by a test procedure). Instead of using a 12-volt test light, you should use a DVOM or Lab Scope with a breakout box whenever a diagnostic procedure calls for a measurement at a PCM connector or component wiring harness.

## Effective Diagnostics

## GETTING STARTED

If you are reasonably certain that the problem is related to a particular electronic control system, the first step is to check for any stored trouble codes in that controller.

On vehicles with more than one vehicle controller (i.e., PCM, BCM, MIC, TCM, etc.), if you are unsure whether the problem is Powertrain related, start by checking for codes in the other controllers to determine if the problem is related to another vehicle system.

If there are no codes set, and you are certain which Powertrain subsystem has a problem, you can start by checking one of the subsystems. The subsystems include the Charging, Cooling, Fuel, Ignition and Speed Control systems.

If a wiring problem is found during testing, you will need to refer to wiring diagrams in the appropriate information resource.  Using a wiring schematic can help you determine:

- Wiring circuits, circuit numbers, and wire colors
- Electrical component connector and component relationships within a circuit
- Power, ground, and splice locations within a circuit
- Related circuits connected into the circuit you are reviewing

Once you decide how to repair the vehicle, in addition to performing the repair, it is a good idea to clear any trouble codes that were set and to verify they do not reset.

An explanation of how to use the "PCM Reset" step to clear codes is included in DIAGNOSTIC TROUBLE CODES Section.

To verify a repair, you should confirm that the Check Engine Light is operational and goes out after the 4-second key-on bulb check. Then, you need to duplicate the conditions present when the customer complaint occurred or when a trouble code set; these are the actual code conditions that caused a code to set. The individual code conditions and possible causes are included in DIAGNOSTIC TROUBLE CODES Section. You can use this information to find out how to drive a vehicle for problem verification.

## Contents

## INTRODUCTION

### System Control Modules

Before attempting diagnosis of the Electronic Engine Control system, familiarize yourself with the basics of how the system is designed to operate. It consists of a central processing unit: Powertrain Control Module (PCM), Engine Control Module (ECM), Transmission Control Module (TCM) and/or the Body Control Module (BCM). These units are the "heart" of the electronic control systems on the vehicle. In some cases, these units are integral with one another, and on some applications, they are separate. As you get deeper into actual diagnostic testing, you will find out which units are used on the vehicle you are testing.

The PCM is a digital computer that contains a microprocessor. The PCM receives input signals from various sensors and switches that are referred to as PCM inputs. Based on these inputs, the PCM adjusts various engine and vehicle operations through devices that are referred to as PCM outputs. Examples of the input and output devices are shown in the graphic below.

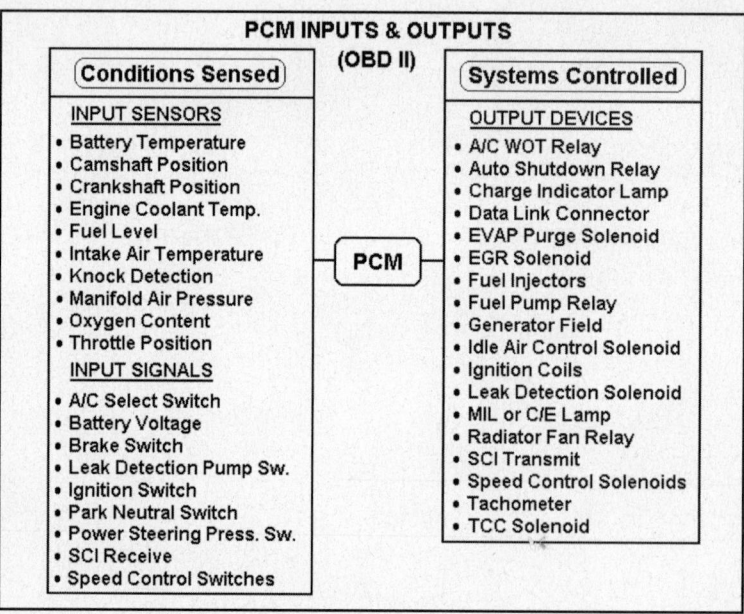

**PCM INPUTS & OUTPUTS**
**(OBD II)**

**Conditions Sensed**

INPUT SENSORS
- Battery Temperature
- Camshaft Position
- Crankshaft Position
- Engine Coolant Temp.
- Fuel Level
- Intake Air Temperature
- Knock Detection
- Manifold Air Pressure
- Oxygen Content
- Throttle Position

INPUT SIGNALS
- A/C Select Switch
- Battery Voltage
- Brake Switch
- Leak Detection Pump Sw.
- Ignition Switch
- Park Neutral Switch
- Power Steering Press. Sw.
- SCI Receive
- Speed Control Switches

**PCM**

**Systems Controlled**

OUTPUT DEVICES
- A/C WOT Relay
- Auto Shutdown Relay
- Charge Indicator Lamp
- Data Link Connector
- EVAP Purge Solenoid
- EGR Solenoid
- Fuel Injectors
- Fuel Pump Relay
- Generator Field
- Idle Air Control Solenoid
- Ignition Coils
- Leak Detection Solenoid
- MIL or C/E Lamp
- Radiator Fan Relay
- SCI Transmit
- Speed Control Solenoids
- Tachometer
- TCC Solenoid

**Input & Output Device Graphic (Example)**

### Powertrain Subsystems

A key to the diagnosis of the PCM and its subsystems is to determine which subsystems are on a vehicle. Examples of typical subsystems appear below:

- Cranking & Charging System
- Emission Control Systems
- Engine Cooling System
- Engine Air/Fuel Controls
- Exhaust System
- Ignition System
- Speed Control System
- Transaxle Controls

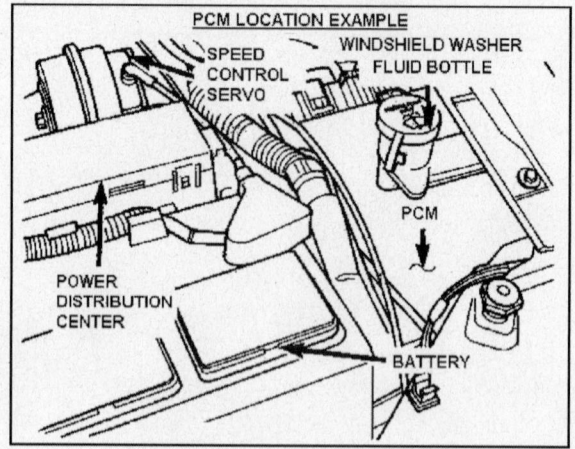

PCM LOCATION EXAMPLE

SPEED CONTROL SERVO

WINDSHIELD WASHER FLUID BOTTLE

PCM

POWER DISTRIBUTION CENTER

BATTERY

WHERE TO BEGIN

Diagnosis of engine performance or drivability problems on a vehicle with an onboard computer requires that you have a logical plan on how to approach the problem. The "Six Step Test Procedure" is designed to provide a uniform approach to repair any problems that occur in one or more of the vehicle subsystems.

The diagnostic flow built into this test procedure has been field-tested for several years at dealerships - *it is the starting point when a repair is required!*

It should be noted that a commonly overlooked part of the "Problem Resolution" step is to check for any related Technical Service Bulletins.

**Six-Step Test Procedure**
The steps outlined on this page were defined to help you determine how to perform a proper diagnosis. Refer to the flow chart that outlines the Six Step Test Procedure on the previous page as needed. The recommended steps include:

*1. VERIFY THE COMPLAINT & CHECK FOR TSBS*
To verify the customer complaint, the technician should understand the normal operation of the system. Conduct a thorough visual and operational inspection, review the service history, detect unusual sounds or odors, and gather diagnostic trouble code (DTC) information resources to achieve an effective repair.

```
┌─────────────────────────────────────────────┐
│   SIX STEP TROUBLESHOOTING PROCEDURE         │
│                                              │
│   ┌──────────────────────────────────────┐   │
│   │  VERIFICATION OF THE COMPLAINT        │   │
│   │  & CHECK FOR ANY RELATED TSBs         │   │
│   └──────────────────────────────────────┘   │
│                                              │
│   ┌──────────────────────────────────────┐   │
│   │       CHECK FOR TROUBLE CODES         │   │
│   └──────────────────────────────────────┘   │
│                                              │
│   ┌──────────────────────────────────────┐   │
│   │     SYMPTOM OR TROUBLE CODE TESTS     │   │
│   └──────────────────────────────────────┘   │
│                                              │
│   ┌──────────────────────────────────────┐   │
│   │    PROBLEM RESOLUTION & REPAIR        │   │
│   └──────────────────────────────────────┘   │
│                                              │
│   ┌──────────────────────────────────────┐   │
│   │             PCM RESET                 │   │
│   └──────────────────────────────────────┘   │
│                                              │
│   ┌──────────────────────────────────────┐   │
│   │        REPAIR VERIFICATION            │   │
│   └──────────────────────────────────────┘   │
└─────────────────────────────────────────────┘
```

This check should include videos, newsletters, and any other information in the form of TSBs or Dealer Service Bulletins. Analyze the complaint and then use the recommended Six Step Test Procedure. Utilize the wiring diagrams and theory of operation articles. Combine your own knowledge with efficient use of the available service information.

Verify the cause of any related symptoms that may or may not be supported by one or more trouble codes. There are various checks that can be performed to Engine Controls that will help verify the cause of a related symptom. This step helps to lead you in an organized diagnostic approach.

*2. CHECK FOR TROUBLE CODES OR SYMPTOMS*
Determine if the problem is a Code or a No Code Fault. Then refer to the appropriate published service diagnostic information to make the repair.

*3. PROBLEM RESOLUTION & REPAIR*
Once the problem component or circuit has been properly identified and verified using published diagnostic procedures, make any needed repairs or replacement to restore the vehicle to proper working order. If the condition has set a DTC, follow the designated repair chart to make an effective repair. If there is not a DTC set, but you can determine specific symptoms that are evident during the failure, select the symptom from the symptom tables and follow the diagnostic paths or suggestions to complete the repair or refer to the applicable component or system in service information.

### 4. SYMPTOM OR TROUBLE CODE TESTS

If the vehicle does not set a DTC and has only intermittent operating failures or concerns, to resolve an intermittent fault, perform the following steps:

- Observe trouble codes, DTC modes and freeze frame data.
- Evaluate the symptoms and conditions described by the customer.
- Use a check sheet to identify the circuit or electrical system component.
- Many Aftermarket Scan Tools and Lab Scopes have data capturing features.

### 5. PCM RESET

It is a good idea, prior to tracing any faults, to clear the DTCs, attempt to replicate the condition and see if the same DTC resets. Also, once any repairs are made, it will be necessary to clear the DTC(s) – PCM Reset – to ensure the repair has totally resolved the problem. For procedures on PCM Reset, see the DIAGNOSTIC TROUBLE CODES section.

### 6. REPAIR VERIFICATION

Once a repair is completed, the next step is to verify the vehicle operates properly and that the original symptom was corrected. Verification Tests, related to specific DTC diagnostic steps, can be used to verify a repair.

## Base Engine Tests

To determine that an engine is mechanically sound, certain tests need to be performed to verify that the correct A/F mixture enters the engine, is compressed, ignited, burnt, and then discharged out of the exhaust system. These tests can be used to help determine the mechanical condition of the engine.

To diagnose an engine-related complaint, compare the results of the Compression, Cylinder Balance, Engine Cylinder Leakage (not included) and Engine Vacuum Tests.

### ENGINE COMPRESSION TEST

The Engine Compression Test is used to determine if each cylinder is contributing its equal share of power. The compression readings of all the cylinders are recorded and then compared to each other and to the manufacturer's specification (if available).

Cylinders that have low compression readings have lost their ability to seal. It this type of problem exists, the location of the compression leak must be identified. The leak can be in any of these areas: piston, head gasket, spark plugs, and exhaust or intake valves.

The results of this test can be used to determine the overall condition of the engine and to identify any problem cylinders as well as the most likely cause of the problem.

**CAUTION:** *Prior to starting this procedure, set the parking brake, place the gear selector in P/N and block the drive wheels for safety. The battery must be fully charged.*

### COMPRESSION TEST PROCEDURE

1. Allow the engine to run until it is fully warmed up.

2. Remove the spark plugs and disable the Ignition system and the Fuel system for safety. Disconnecting the CKP sensor harness connector will disable both fuel and ignition (except on NGC vehicles).

3. Carefully block the throttle to the wide-open position.

4. Insert the compression gauge into the cylinder and tighten it firmly by hand.

5. Use a remote starter switch or ignition key and crank the engine for 3-5 complete engine cycles. If the test is interrupted for any reason, release the gauge pressure and retest. Repeat this test procedure on all cylinders and record the readings.

The lowest cylinder compression reading should not be less than 70% of the highest cylinder compression reading and no cylinder should read less than 100 psi.

### EVALUATING THE TEST RESULTS

To determine why an individual cylinder has a low compression reading, insert a small amount of engine oil (3 squirts) into the suspect cylinder. Reinstall the compression gauge and retest the cylinder and record the reading. Review the explanations below.

**Reading is higher** - If the reading is higher at this point, oil inserted into the cylinder helped to seal the piston rings against the cylinder walls. Look for worn piston rings.

**Reading did not change** - If the reading didn't change, the most likely cause of the low cylinder compression reading is the head gasket or valves.

**Low readings on companion cylinders** - If low compression readings were recorded from cylinders located next to each other, the most likely cause is a blown head gasket.

**Readings are higher than normal** - If the compression readings are higher than normal, excessive carbon may have collected on the pistons and in the exhaust areas. One way to remove the carbon is with an approved brand of "Top Engine Cleaner."

**Note:**  *Always clean spark plug threads and seat with a spark plug thread chaser and seat cleaning tool prior to reinstallation. Use anti-seize compound on aluminum heads.*

### ENGINE VACUUM TESTS

An engine vacuum test can be used to determine if each cylinder is contributing an equal share of power. Engine vacuum, defined as any pressure lower than atmospheric pressure, is produced in each cylinder during the intake stroke. If each cylinder produces an equal amount of vacuum, the measured vacuum in the intake manifold will be even during engine cranking, at idle speed, and at off-idle speeds.

Engine vacuum is measured with a vacuum gauge calibrated to show the difference between engine vacuum (the lack of pressure in the intake manifold) and atmospheric pressure. Vacuum gauge measurements are usually shown in inches of Mercury (in. Hg).

**Note:**  *In the tests described in this article, connect the vacuum gauge to an intake manifold vacuum source at a point below the throttle plate on the throttle body.*

### ENGINE CRANKING VACUUM TEST PROCEDURE

The Engine Cranking Vacuum Test can be used to verify that low engine vacuum is not the cause of a No Start, Hard Start, Starts and Dies or Rough Idle condition (symptom).

The vacuum gauge needle fluctuations that occur during engine cranking are indications of individual cylinder problems. If a cylinder produces less than normal engine vacuum, the needle will respond by fluctuating between a steady high reading (from normal cylinders) and a lower reading (from the faulty cylinder). If more than one cylinder has a low vacuum reading, the needle will fluctuate very rapidly.

1.  Prior to starting this test, set the parking brake, place the gearshift in P/N and block the drive wheels for safety. Then block the PCV valve and disable the idle air control device.

2.  Disable the fuel and/or ignition system to prevent the vehicle from starting during the test (while it is cranking).

3.  Close the throttle plate and connect a vacuum gauge to an intake manifold vacuum source. Crank the engine for three seconds (do this step at least twice).

The test results will vary due to engine design characteristics, the type of PCV valve and the position of the AIS or IAC motor and throttle plate. However, the engine vacuum should be steady between 1.0-4.0 in. Hg during normal cranking.

### ENGINE RUNNING VACUUM TEST PROCEDURE

1.  Allow the engine to run until fully warmed up. Connect a vacuum gauge to a clean intake manifold source. Connect a tachometer or Scan Tool to read engine speed.

2.  Start the engine and let the idle speed stabilize. Raise the engine speed rapidly to just over 2000 rpm. Repeat the test (3) times. Compare the idle and cruise readings.

## EVALUATING THE TEST RESULTS

If the engine wear is even, the gauge should read over 16 in. Hg and be steady. Test results can vary due to engine design and the altitude above or below sea level.

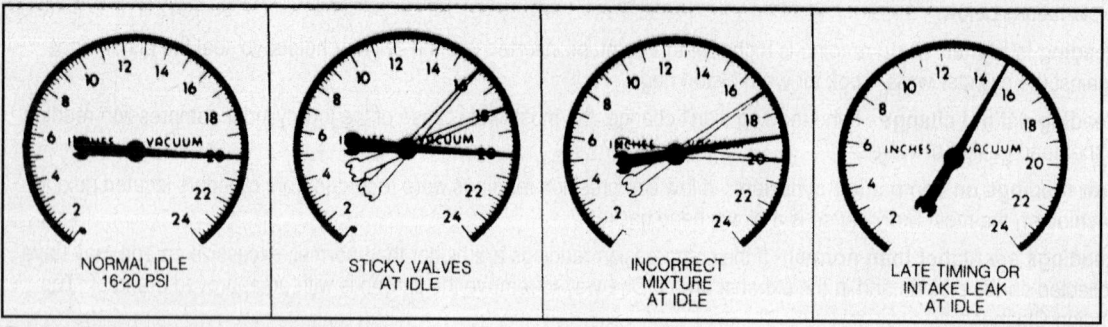

NORMAL IDLE
16-20 PSI

STICKY VALVES
AT IDLE

INCORRECT
MIXTURE
AT IDLE

LATE TIMING OR
INTAKE LEAK
AT IDLE

**Engine Running Vacuum Test Graphic**

## Ignition System Tests - Distributor

This next section gives an overview of ignition tests (with examples) for a Distributor Ignition System.

### PRELIMINARY INSPECTION

1. Perform these checks prior to connecting the Engine Analyzer:
2. Check the battery condition (verify that it can sustain a cranking voltage of 9.6v).
3. Inspect the ignition coil for signs of damage or carbon tracking at the coil tower.
4. Remove the coil wire and check for signs of corrosion on the wire or tower.
5. Test the coil wire resistance with a DVOM (it should be less than 7 k/ohm per foot).
6. Connect a *low* output spark tester to the coil wire and engine ground. Verify that the ignition coil can sustain adequate spark output while cranking for 3-6 seconds.
7. Connect the Engine Analyzer to the Ignition System, and choose Parade display. Run the engine at 2000 RPM, and note the display patterns, looking for any abnormalities.

## Ignition System Tests - Distributorless

Perform the following checks prior to connecting the Engine Analyzer:

1. Check the battery condition (verify that it can sustain a cranking voltage of 9.6v).
2. Inspect the ignition coils for signs of damage or carbon tracking at the coil towers.
3. Remove the secondary ignition wires and check for signs of corrosion.
4. Test the plug wire resistance with a DVOM (specification varies from 15-30 k/ohm).
5. Connect a *low* output spark tester to a plug wire and to engine ground. Verify that the ignition coil can sustain adequate spark output for 3-6 seconds.

### SECONDARY IGNITION SYSTEM SCOPE PATTERNS (V6 ENGINE)

1. Connect the Engine Analyzer to the ignition system.
2. Turn the scope selector to view the "Parade Display" of the ignition secondary.
3. Start the engine in Park or Neutral and slowly increase the engine speed from idle to 2000 rpm.
4. Compare actual display to the secondary ignition system examples below.

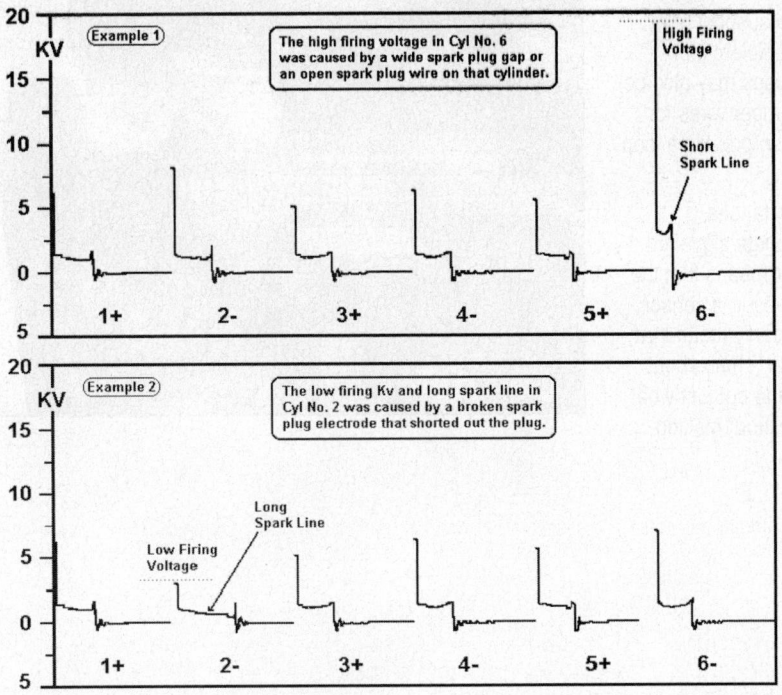

Example 1: The high firing voltage in Cyl No. 6 was caused by a wide spark plug gap or an open spark plug wire on that cylinder.

Example 2: The low firing Kv and long spark line in Cyl No. 2 was caused by a broken spark plug electrode that shorted out the plug.

### Symptom Diagnosis

To determine whether vehicle problems are identified by a set Diagnostic Trouble Code, you will first have to connect a proper scan tool to the Data Link Connector and retrieve any set codes. See DIAGNOSTIC TROUBLE CODES section for information on retrieving and reading codes.

If no codes are set, the problem must be diagnosed using only vehicle operating symptoms. A complete set of "No Code" symptoms is found in the SYMPTOM DIAGNOSIS (NO CODES) section.

Do NOT attempt to diagnose driveability symptoms without having a logical plan to use to determine which engine control system is the cause of the symptom - this plan should include a way to determine which systems do NOT have a problem! Remember, there are 2 kinds of NO CODE conditions:

● Symptom diagnosis, in which a continuous problem exists, but no DTC is set as a result. Therefore, only the operating symptoms of the vehicle can be used to pinpoint the root cause of the problem.
● Intermittent problem diagnosis, in which the problem does not occur all the time and does not set any DTCs.

Both of these NO CODE conditions are covered in the SYMPTOM DIAGNOSIS (NO CODES) section.

### Accessing Components & Circuits

Every vehicle and every diagnostic situation is different. It is a good idea to first determine the best diagnostic path to follow using flow charts, wiring diagrams, TSBs, etc. Part of choosing steps is to determine how time-consuming and effective each step will be. It may be easy to access a component or circuit in one vehicle, but difficult in another. Many circuits are integrated into a large harness and are difficult to test. Many components are inaccessible without disassembly of unrelated systems.

In the graphic, you will note that the protective covers have been removed from the PCM connectors, and any circuit can be easily identified and back probed. In other cases, PCM access is difficult, and it may be easier to access circuits at the component side of the harness.

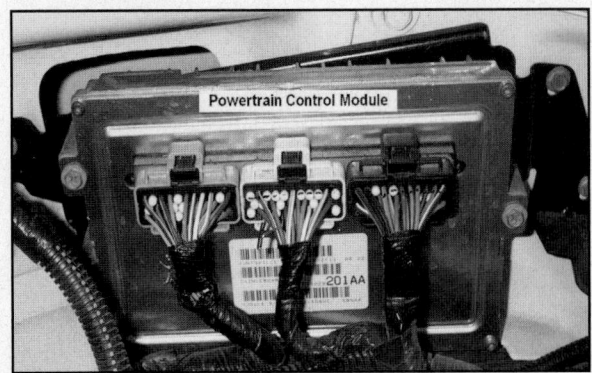

Another important point to remember is that any circuit or component controlled by a relay or fused circuit can be monitored from the appropriate fuse box.

There is generally more than one of each type of relay or fuse. Therefore, swapping a suspect relay from another system may be more efficient than testing the relay itself. Relays and fuses may also be removed and replaced with fused jumper wires for testing circuits. Jumper wires can also provide a loop for inductive amperage tests.

Choosing the easiest way has its limitations, however. Remember that an appropriate signal on a PCM controlled circuit at an actuator means that the signal at the PCM is also good. However, a sensor signal at the sensor does not necessarily mean that the PCM is receiving the same signal. Think about the direction flow through a circuit, and not just what signal is appropriate, to save time without making costly assumptions.

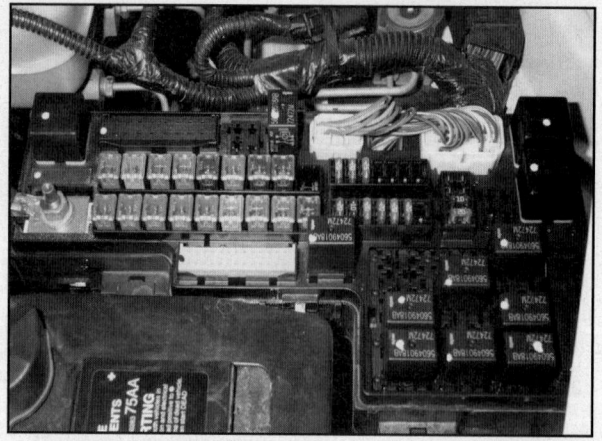

## DaimlerChrysler OBD II System Contents

## GENERAL INFORMATION

### DTC DESCRIPTIONS

The OBD II system uses a *Diagnostic Trouble Code (DTC)* identification system established by the Society of Automotive Engineers (SAE) and the EPA. The codes are a standardized set of identifiers used to help pinpoint the location of a fault within the on-board electronic controls systems.

The first letter of the DTC is used to identify the type of computer system that has failed. This letter will be followed by 4 numbers, the letters are described as shown below:

The letter 'P' indicates a Powertrain related device.
The letter 'C' indicates a Chassis related device.
The letter 'B' indicates a Body related device.
The letter 'U' indicates a Data Link or Network device code.

The first DTC number indicates a generic (P0XXX) or manufacturer (P1XXX) type code. A list of trouble codes is included in this section.

## INTRODUCTION TO ON-BOARD DIAGNOSTICS

### INTRODUCTION

The purpose of the DaimlerChrysler On-Board Diagnostic System is to provide optimum control of the engine and transmission while meeting the objectives of the OBD II regulations.

At the center of this system is the Powertrain Control Module (PCM) connected to various input and output devices through a wiring harness with two connectors with anywhere from 1 to 4 connectors, depending on controller type. The PCM receives input information from various sensors and switches, performs calculations based on data stored in long term memory, and controls output devices (actuators, relays, and solenoids).

### PCM Hardware & Software

The PCM is divided into two main parts, the system hardware and system software. Hardware components include:

- All related actuators, relays, and solenoids
- All related sensors and switches
- All interconnecting wires, connectors and terminals
- The Automatic Shutdown Relay (ASD)
- The Power Control Module (PCM)

### System Software

The software includes all the programs that make up strategies that the PCM uses for Engine Control outputs based on related inputs. Generally, these are the strategies used to control engine operation, the electronic transmission, engine idle speed, fuel control and Fail Safe circuitry (Limp-In) should any major failures occur inside the PCM.

### PCM Decision Making

As in the past, the PCM receives input signals from various sensors and switches (called PCM inputs). Based on these inputs, the PCM adjusts various engine systems by controlling its outputs. On vehicles with OBD II systems, the PCM operates software called the Task Manager to control the OBD II system.

## Task Manager

In order to perform the new strategies and emission control device tests that are part of the OBD II system, the PCM was changed to include a unique piece of software referred to as the Task Manager. Many diagnostic steps and tests required by OBD II must be performed under specific operating conditions (called enable criteria).

The software in the Task Manager organizes and prioritizes the OBD II diagnostics. The job of the Task Manager is to determine if conditions are appropriate for tests to be run, to monitor the parameters for a trip (for each test), and to record the results of each test.

The list below contains some of the tasks performed by the Task Manager:

- Sequence all OBD II Monitor tests
- Monitor the Trip and Readiness Indicators
- Control the operation of the MIL
- Trouble Code Identification
- Freeze Frame Data Storage
- Display the Similar Conditions Window

## OBD II TERMINOLOGY

In order to diagnose DaimlerChrysler vehicles equipped with an OBD II System, it is important that you understand the terms related to these test procedures. Some of these terms and their definitions are discussed in the next few articles.

### Two-Trip Detection

In many cases, an emission related system or component must fail a Monitor test more than once before it activates the MIL. The first time an OBD II Monitor detects a fault during a related trip, it sets a "pending code" in PCM memory. These codes appear when the Memory or Continuous codes are read. For a "pending code" to mature into a hard code (and illuminate the MIL), the original fault must occur for two consecutive trips (two-trip detection). However, a "pending code" can remain in the PCM for a long time before the conditions that caused the code to set reappear.

Fuel Trim and Misfire Detection trouble codes can cause the PCM to flash the MIL after <u>one</u> trip because faults in these systems can cause damage to the catalytic converter.

### Pending Code

The term "pending code" is used to describe a fault that has been detected once and is stored in memory. This type of fault has not been detected on two consecutive trips (i.e., it has not matured into a hard code).

It is possible to access a "pending code" with a Generic Scan Tool (GST) on most DaimlerChrysler vehicles. Be aware that you may not be able to read a pending code with a Generic Scan Tool on some 1995 phase-in models.

```
PENDING DTCs
ECU: $40 (Engine)
Number of Codes:1
P0703 Torque
      Converter/Brake
      Switch B Circuit
      Malfunction
```

### Similar Conditions

If a "pending code" is set because of a Fuel System or Misfire Monitor detected fault, the vehicle must meet *similar conditions* for two consecutive trips before the code matures and the PCM activates the MIL and stores the code in memory. The meaning of *similar conditions* is important when you attempt to diagnose a fault detected by the Fuel System Monitor or Misfire Detection Monitor.

To achieve *similar conditions*, the vehicle must reach the following engine running conditions simultaneously (for the first failure recorded that set the code):

- Engine speed must be within 375 rpm
- Engine load must be within 10%
- Engine warmup state must match the previous state (cold or warm)

Summary - Similar conditions are defined as conditions that match those recorded when the fault was detected and a code was set.

### Warmup Cycle

*A warmup cycle is defined as vehicle operation (after a cool-down period) when the engine temperature increases by at least 40°F and reaches at least 160°F.*

Most trouble codes are cleared from the PCM memory after 40 "warmup cycles" if the fault does not reappear.

### Cylinder Bank

The cylinder bank identifies the location of a specific component. For example, a specific group of engine cylinders may share a common control sensor and it would be identified as "Bank 1", which is the location of cylinder No. 1 while "Bank 2" identifies cylinders on the opposite cylinder head.

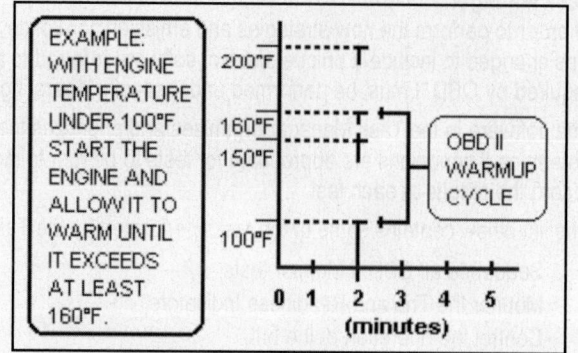

### Sensor

If sensors are numbered (Bank 1 Sensor 1, or B1 S1), they follow the convention described above. If they are identified with letters ('A', 'B', 'C'), they are manufacturer defined. If only 1 sensor is used, the letter or number may be omitted.

### Data Link Connector

DaimlerChrysler vehicles equipped with OBD II Systems use a standardized Data Link Connector (DLC). It is typically located between the left end of the instrument panel and 12 inches past vehicle centerline. The connector is mounted out of sight from the passengers, but should be easy to see from outside by a technician in a kneeling position (door open).

### DLC Features

The DLC is rectangular in design. It can accept up to 16 terminals. The DLC in the graphic to the right shows many common pin designations, but does not represent any specific vehicle.

OBD II DaimlerChrysler vehicle will have power at pin 16, and ground at pins 4 and 5. All other pins vary by year and model. All applications use the SCI circuits, but the pin assignments vary. All vehicles will utilize either the PCI bus, or the CCD bus circuits.

*NOTE: Not all DaimlerChrysler vehicle DLCs are wired the same. Always use a wiring diagram when attempting to diagnose these circuits.*

Both the DLC and Scan Tool have latching features that ensure that the Scan Tool will remain connected to the vehicle during operation.

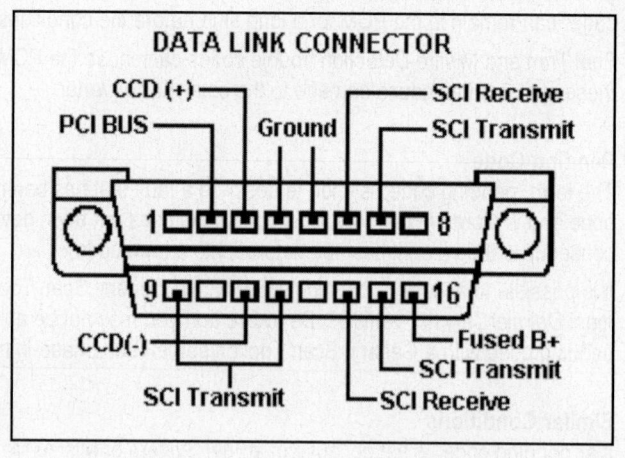

### Common uses of the Scan Tool while connected to the DLC include:

- Display the results of the most current I/M Readiness Tests
- Read and clear any diagnostic trouble codes
- Read the Serial Data from the PCM
- Perform Enhanced Diagnostic Tests (System Tests or Actuator Tests)

## DIAGNOSTIC TROUBLE CODES

### SCAN TOOL COMMUNICATIONS

For the Scan Tool to communicate with various vehicle stand-alone modules, it must be connected into the system. On OBD I systems, connect the Scan Tool to the underhood test connector. On OBD II systems, connect the Scan Tool to the DLC under the dash.

### RETRIEVING & READING DTCs

To access the PCM, connect the Scan Tool to the appropriate DLC. Turn the ignition on to allow the vehicle to identify itself to the Scan Tool. Once communication is established with the PCM, select a function (codes, data, actuator tests, system tests, etc) from the tool main menu. The Scan Tool performs these functions by communicating with the vehicle modules through the data bus circuits.

To read trouble codes without a Scan Tool, cycle the key from "off" to "on" 3 times within 5 seconds. This test can establish whether the PCM is capable of responding to a request. It can also be used if the Scan Tool will not communicate.

The Scan Tool "talks" to the vehicle controllers on separate networks. The tool communicates with the PCM via the SCI Transmit and Receive circuits, and with other modules through the CCD or PCI data bus circuits.

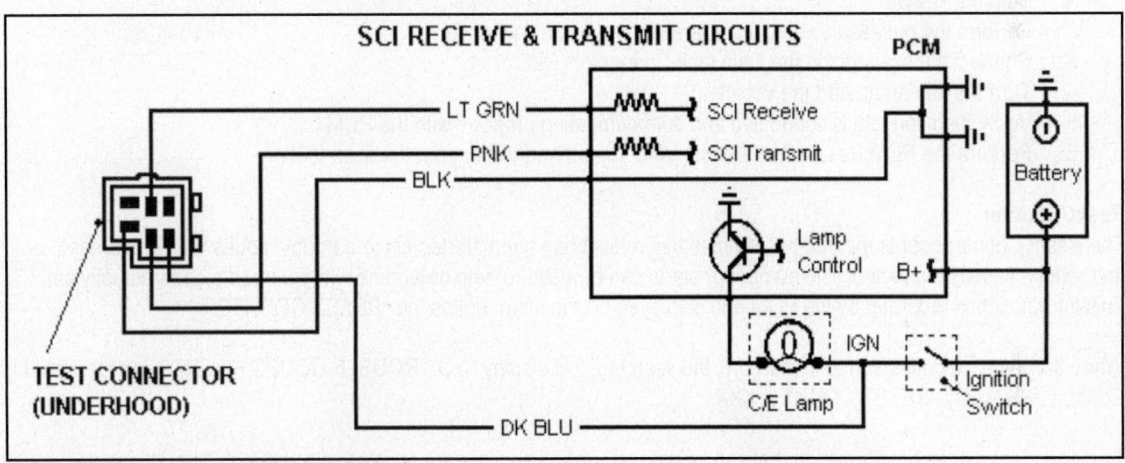

OBD I SCI Schematic (1983-95 Vehicles)

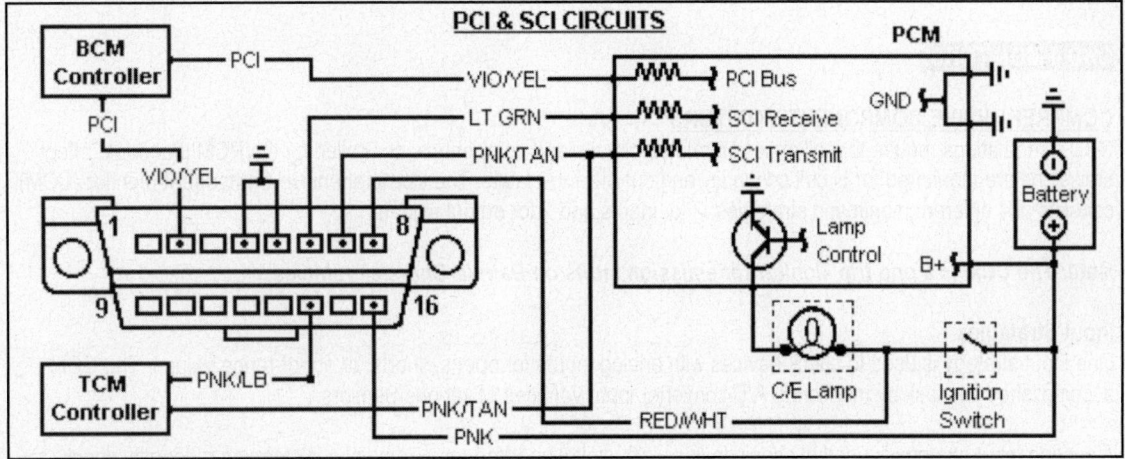

OBD II DLC Schematic (1999 LH Body)

## CLEARING DTCs

The PCM Reset function allows the technician to command the scan tool to clear all emissions-related diagnostic information from the Powertrain Control Module (PCM). Emissions related codes are stored in the PCM until all the OBD II System monitors or components have been tested to satisfy a drive cycle without any other faults occurring. Refer to the OBD II Drive Cycle article in this section for additional information on this subject. The events listed below occur after the OBD II Drive Cycle is completed and the PCM Reset procedure is performed:

- The trouble code number total will be cleared
- All trouble codes will be cleared
- All Freeze Frame data will be cleared
- All Oxygen Sensor Test data will be cleared
- All OBD II System Monitors will have their status reset

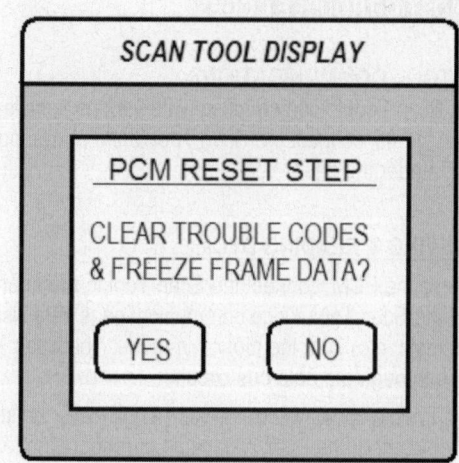

### PCM Reset Instructions

Refer to your scan tool Operating Manual for specific instructions on how to perform a PCM Reset. The list of instructions should include:

1. Turn the key off
2. Perform the necessary vehicle preparation and visual inspection
3. Connect the scan tool to the Data Link Connector (DLC.
4. Turn the key on or start the vehicle
5. Verify the scan tool is connected and communicating properly with the PCM
6. Perform the PCM Reset step, then turn the key off and disconnect the scan tool

### Reset Counter

The Reset Counter counts the number of times the vehicle has been started since a trouble codes was set, erased, or the battery was disconnected. The number of starts can be used to help determine when a trouble code actually set. This information is recorded by the PCM and displayed on the scan tool as the "RESET COUNTER."

When there are no codes stored in memory, the scan tool will display "NO TROUBLE CODES FOUND" and the Reset Counter will display "RESET COUNT = XXX."

Two important pieces of information that can help speed up a diagnosis are code conditions (including all enable criteria), and the parameter information (PID) stored in the Freeze Frame at the moment a trouble code is set and stored in memory

## OBD II MONITORS

## COMPREHENSIVE COMPONENT MONITORS

OBD II regulations require that all emission related circuits and components controlled by the PCM that could affect emissions are monitored for circuit continuity and out-of-range faults. The Comprehensive Component Monitor (CCM) consists of 4 different monitoring strategies: 2 for inputs and 2 for output signals.

*Note: The CCM is a one-trip Monitor for emission faults on DaimlerChrysler vehicles.*

### Input Strategies

One input strategy is used to check devices with analog inputs for opens, shorts, or out-of-range values. The CCM accomplishes this task by monitoring A/D converter input voltages of various Sensors.

A second input strategy is used to check devices with digital and frequency inputs by performing rationality checks. The PCM uses other Sensor readings and calculations to determine if a Sensor or switch reading is correct under existing conditions. Some tests of the CKP, CMP and VSS run continuously. Other tests run only after actuation.

**Output Strategies**

An Output Device Monitor in the PCM checks for any open and shorted circuits by observing the control voltage level of a particular device.

The control voltage is low when the device is "on" and the voltage is high when the device is "off".
Monitored outputs include the AC relay, ASD relay SS1, SS2, SS3, EGR and EVAP and O2 heater control.

## CRANKSHAFT RELEARN

Select "Crank Relearn" on the scan tool to allow the PCM to enable a fast CMP and CKP Sensor relearn procedure. The PCM relearns the crank target window spacing during Decel Fuel Shutoff mode The PCM reset step requires 3 closed throttle decelerations from 55 mph in order to "relearn" the Crank Target Spacing.

## CATALYST MONITOR

OBD II regulations require that the functionality of the Catalyst system be monitored. If the catalyst system deteriorates to a point where vehicle emissions increase by more than 1.5 times the Federal Test Procedure (FTP) standard, the MIL must be activated.

The oxygen content in a catalyst is important for efficient conversion of exhaust gases. When a lean air-fuel (A/F) ratio is present for an extended period of time, the oxygen content in the catalyst can reach a maximum. Conversely, when a rich A/F ratio is present for too long a period, the oxygen content in the catalyst can become totally depleted.

Catalyst operation is dependent on its ability to store and release oxygen needed to complete the emissions-reducing chemical reactions. As a catalyst deteriorates, its ability to store oxygen is reduced. Since the catalyst's ability to store oxygen is somewhat related to proper operation, oxygen storage can be used as an indicator of catalyst performance.

## FRONT AND REAR HEATED OXYGEN SENSORS

There are 2 Oxygen (O2) Sensors used to monitor the catalyst storage capability. One Sensor is mounted in front of the catalyst (pre-catalyst O2) and another Sensor is mounted in back of the catalyst (post-catalyst O2). By utilizing an O2 Sensor in front of the catalytic converter, and a second Sensor located behind the catalyst, the oxygen storage capability of the catalyst can be determined by comparing the voltage signals from the 2 Sensors.

## EGR SYSTEM MONITOR

Emissions of NOx increase proportionally with the temperature in the combustion chamber. The Exhaust Gas Recirculation (EGR) system is designed to circulate non-combustible exhaust gases into the manifold to dilute the A/F mixture. In this manner, the EGR system lowers the combustion temperature and reduces NOx emissions and pre-detonation (knocking).

## EVAP SYSTEM MONITOR

OBD II regulations require that all vehicles equipped with an EVAP system be monitored for component integrity, system functionality and loss of hydrocarbons. The Enhanced EVAP system test is required to detect leaks as small as 0.040" in diameter. The test must detect leaks of 0.040-0.080" in diameter on fuel tanks larger than 25 gallons.

## FUEL SYSTEM MONITOR

To comply with clean air regulations, DaimlerChrysler vehicles are equipped with catalytic converters that reduce the amount of HC, CO and NOx emissions released from the vehicle. The catalyst works best when the A/F ratio is at or near an optimum value of 14.7:1. The PCM is programmed to maintain this optimum A/F ratio.

This is accomplished by making Short Term fuel trim corrections in injector pulse width based upon the O2 Sensor input. The programmed memory in the PCM acts as a self-calibration tool to compensate for variations in engine specifications, sensor tolerances and engine fatigue over the life span of the engine.

As injector pulse width increases, the amount of fuel delivered by the fuel injector is increased. A shorter pulse width decreases the amount of fuel delivered. The PCM monitors engine load, RPM, TP Sensor, ECT Sensor signals in order to make changes to injector pulse width.

## MISFIRE MONITOR

OBD II regulations require that the OBD II system monitor the engine for misfire conditions and identify specific cylinders that experience a misfire. The PCM should also identify and set a different code to indicate if a multiple misfire condition exists. It should also be able to identify a specific cylinder that misfires in a multiple misfire condition. This monitor detects misfires related to engine mechanical, ignition or fuel system faults under positive load conditions (idle, cruise, etc.).

## REVOLUTION COUNTER

The PCM contains a program called a Revolution Counter. In the logic portion of this counter, each 1000-revolution window contains five 200-revolution windows. The PCM counts the misfires for each 200-window and carries that value over to the 1000-revolution window.

## SECONDARY AIR MONITOR

This monitor is a PCM diagnostic that monitors the Secondary AIR system for component integrity, system functionality and faults that could cause vehicle tailpipe output levels to exceed 1.5 times the FTP Standard. The Secondary AIR System Monitor is run once per trip once the required enable criteria are met.

## ON-BOARD DIAGNOSTICS

### CIRCUIT ACTUATION TEST MODE

When the Circuit Actuation Test Mode option is selected on the scan tool, you will have access to Actuator Test menus. The tests listed in the menu provide the capability to activate certain controlled output circuits to the PCM. The lists of tests available vary depending on engine and vehicle equipment. The tests may be used to find device faults that the PCM may not recognize.

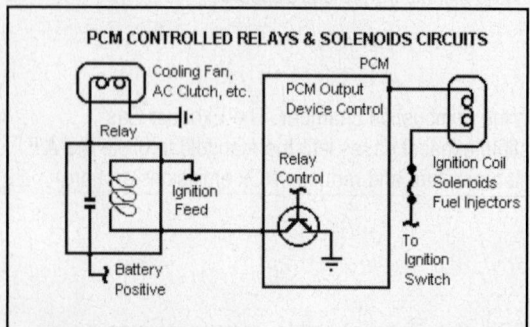

During Actuator Tests, both electrical and mechanical activity can be checked and verified. Once a certain device is selected, the PCM activates the device that allows you to listen for an audible click or observe a visual indication of correct device operation.

The scan tool Actuator Test Mode can be used to diagnose many PCM controlled devices, and to help diagnose the cause of one or more trouble codes on these vehicles.

### TESTING RELAYS, SOLENOIDS & SWITCHES WITH SCAN TOOL

Refer to the specific instructions in the scan tool Operating Manual to test a particular device or its circuit.

## STATE DISPLAY SWITCH TEST MODE

The PCM switch inputs have two recognized states: they are either in a HIGH or LOW state. Because these inputs are either high or low, the PCM may not be able to detect the difference between when a switch is in a high or low state from when a switch has a circuit problem (i.e., when a circuit is open, grounded or shorted or even if the switch is defective).

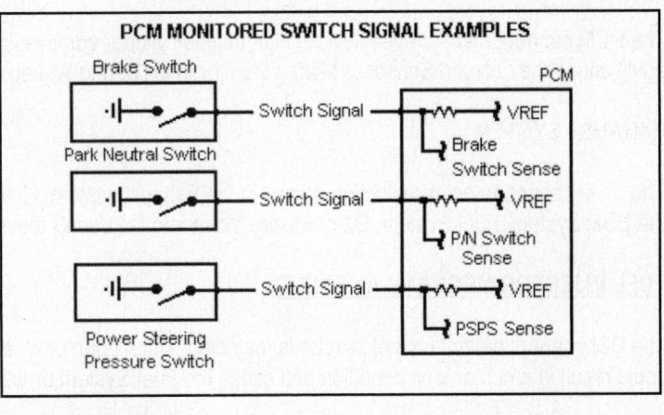

If the scan tool State Display (meaning the Switch State) shows a change from HIGH to LOW; or from LOW to HIGH when the selected switch is activated, it can be assumed that the entire switch circuit to the PCM is functioning properly. The scan tool State Display Mode can be used to diagnose these codes:

- Brake Switch: P0703
- LDP Switch: P1494
- P/N Switch: P1899
- PSP Switch: P0551

## NON-MONITORED SYSTEMS & CIRCUITS

The PCM and CCM cannot monitor all of the Base Engine systems and components for faults or conditions that might cause a driveability problem. This handicap can cause confusing diagnostic situations when a Base Engine problem occurs. If the fuel pressure is too low or high, a fuel pressure code will not be set, but a misfire or oxygen Sensor code might be set due to a lean or rich A/F condition.

## FUEL PRESSURE

On engines with fuel injection, the fuel pressure regulator controls fuel system pressure. The PCM cannot detect a restricted fuel pump inlet filter, a dirty in-line fuel filter, or a pinched fuel supply or return line. An O2 Sensor or Fuel system code might set due to a lean A/F condition.

## IGNITION SYSTEM SECONDARY

The PCM cannot detect a faulty ignition coil, fouled or worn out spark plugs, ignition wires that are cross firing, or an open spark plug wire. However, the Misfire Monitor would detect these faults during testing.

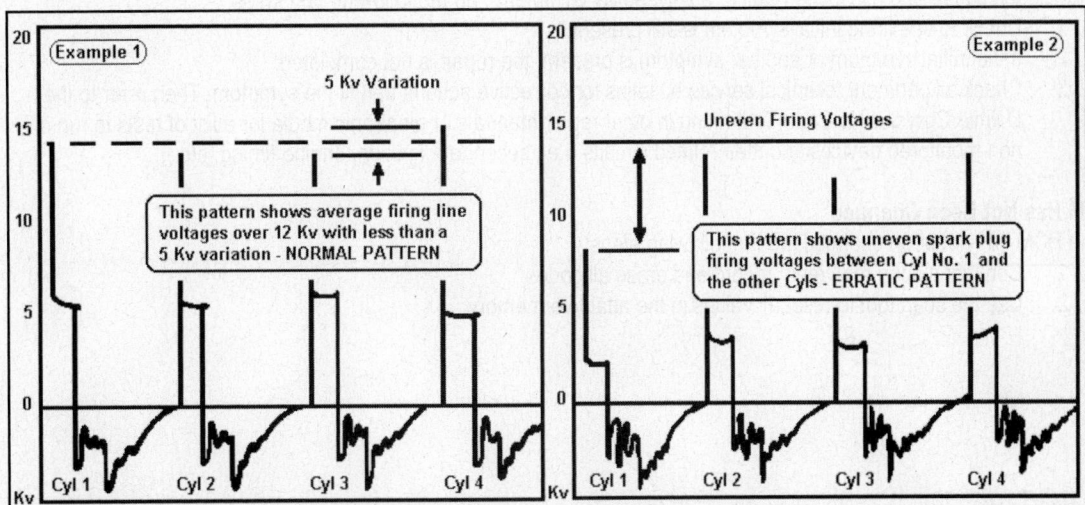

## ENGINE COMPRESSION

The PCM cannot detect uneven, low, or high engine cylinder compression. However, a fault in one of these areas could cause the Oxygen Sensor or Misfire Monitor to fail during testing.

## EXHAUST SYSTEM

The PCM cannot detect a restriction or leak in the Exhaust system. However, a fault in one of these areas could cause the EGR System, Fuel System, O2 or Misfire Monitor to fail during testing.

## FUEL INJECTOR MECHANICAL FAULT

The PCM cannot detect if a fuel injector is restricted, stuck open or closed. However, a fault in one of these areas could result in a rich or lean condition and cause the Fuel System or O2 Monitor to fail.

## STARTING THE DIAGNOSTIC PROCESS

## VISUAL INSPECTION

Verify that all engine components have been reconnected and installed properly. If this has not been done, reassemble or reconnect all components as needed. Check the oil level and coolant level. If this Road Test Verification procedure is for an OBD II System code, refer to OBD II DRIVE CYCLE in this section as needed. For previously read codes that have not been repaired, refer to the correct code repair chart and perform the repair test steps before continuing.

## ROAD TEST VERIFICATION - CODE FAULTS

Prior to proceeding with Road Test Verification for an OBD II system related trouble code, you should perform a complete visual inspection (as above).

## ROAD TEST VERIFICATION - NO CODE FAULTS

Prior to proceeding with Road Test Verification for Non-OBD II codes, inspect the vehicle to ensure that all engine components are properly connected. Reassemble and connect all components as needed.

If the Road Test Verification procedure is for a No Code Fault, refer to the OBD II Drive Cycle article in this section.

For trouble codes that are not OBD II related, continue with this Road Test Verification step. For previously read codes that have not been repaired, refer to the correct code repair test and perform the repair test steps.

### For No Code Faults
To locate the cause of a "No Code Fault or a Driveability Symptom", do the following test steps:
1. Check to see if the initial symptom is still present.
2. If the initial symptom or another symptom is present, the repair is not completed.
3. Check all pertinent technical service bulletins for corrective actions that fit the symptom. Then refer to the DaimlerChrysler No Code Test menu in other repair manuals or electronic media for a list of tests to run on non-monitored devices and their related circuits (i.e., secondary ignition, engine timing, etc.).

### PCM Has Not Been Changed
If the PCM has not been changed, do the following steps:
1. Connect a scan tool, read, record and erase all codes.
2. Use the scan tool to reset all values in the adaptive memory.

### Road Test Verification for A/C Relay Non-OBD II Codes
1. Drive the vehicle for at least 5 minutes with the A/C on.
2. During the drive cycle, go at least 40 mph and at some point, stop the vehicle and turn off the engine for 10 seconds or more.
3. Then, restart the engine and drive the vehicle while shifting through all gears.
4. Return to the service bay and verify that no codes are stored.

### Road Test Verification for Charging System Non-OBD II Codes
If the PCM has been changed, and the vehicle is equipped with factory theft alarm, perform the following:
1. Start the vehicle 20 times to activate the alarm.
2. Connect the scan tool, read, record and erase all trouble codes.
3. Start the engine, then raise and hold the engine speed at 2000 rpm for 30 seconds.

Return to idle speed, turn the engine off, then turn the key on and read the Charging System trouble codes.

## OBD II TROUBLE CODE LISTS

The first step is to read and record all trouble codes and Freeze Frame data - this information will be used during diagnosis. *If the Reset function of the PCM is done first, all codes and related data will be lost!*
Look up the appropriate trouble code in the list on the following pages. This information includes the DTC number, Code Title and Conditions (the enable criteria that indicate conditions present when a code sets).

You can use this information to determine how to drive a vehicle to validate a repair procedure is complete. The enable criteria can be used to help diagnose a trouble code related to the AIR, Catalyst, CCM (component monitor), EGR or EVAP, Fuel and Misfire Monitors. The **(1T)** and **(2T)** designators in the first column indicate whether the code requires one or two "trips" to occur before the MIL is activated.

*Note: In many cases, you will notice the same DTC for multiple applications. Be sure to check the year, make and model (and, in some cases, engine and VIN) when looking for a specific DTC. Don't assume that the first one you find is the only occurrence of that DTC.*

### OBD II Trouble Code List (P0XXX Codes)

| DTC | Trouble Code Title, Conditions & Possible Causes: |
|---|---|
| **P0016**<br>1T CCM<br>2000-05<br>300M, Acclaim, Avenger, Breeze, Caravan, Cherokee, Cirrus, Concorde, PT Cruiser, Dakota, Durango, Grand Cherokee, LeBaron, Spirit, Intrepid, LHS, Liberty, Magnum, Neon, New Yorker, Sebring, Stratus, Prowler, Pacifica, Ram, Town & Country, Viper, Vision, Voyager, Wrangler | **Crankshaft/Camshaft Timing Misalignment Conditions:**<br>Engine cranking or running; and the PCM detected the camshaft was out of phase with the crankshaft during the CCM test period.<br>**Possible Causes:**<br>Base engine problem (i.e., the camshaft timing is not correct)<br>Intermittent condition<br>CKP or CMP Sensor signal is erratic (check with lab scope)<br>Tone wheel or pulse wheel is damaged or contains debris<br>CKP or CMP Sensor, harness or connector has failed<br>PCM has failed |

**OBD II Trouble Code List (P0XXX Codes) - Continued**

| DTC | Trouble Code Title, Conditions & Possible Causes: |
|---|---|
| **P0030**<br>1T CCM<br>2000-03<br>300M, Cherokee, Concorde, Grand Cherokee, Intrepid, LHS, Wrangler | **O2 (B1 S1) Heater Circuit Fault Conditions:**<br>Engine started; system voltage over 10.6v, and the PCM detected a fault in the O2 heater element feedback sense circuit.<br>**Possible Causes:**<br>O2 assembly is damaged or it has failed<br>O2 heater control circuit is open, shorted to ground or B+<br>O2 heater ground circuit is open<br>O2 heater element is damaged or has failed<br>PCM has failed |
| **P0031**<br>1T CCM<br>2001-05<br>300M, Caravan, Cherokee, Concorde, PT Cruiser, Dakota, Durango, Grand Cherokee, Intrepid, LHS, Liberty, Magnum, Neon, Pacifica, Ram, Sebring, Stratus, Town & Country, Viper, Voyager, Wrangler | **O2 (B1 S1) Heater Circuit Low Conditions:**<br>Key on; system voltage over 10.6v; ASD relay on; O2 heater "on". The PCM detected the O2 Heater circuit is out of acceptable range low.<br>**Possible Causes:**<br>O2 assembly is damaged or it has failed<br>O2 heater element is damaged or has failed<br>O2 heater control circuit is shorted to ground<br>PCM has failed |
| **P0032**<br>1T CCM<br>2001-05<br>300M, Caravan, Cherokee, Concorde, PT Cruiser, Dakota, Durango, Grand Cherokee, Intrepid, LHS, Liberty, Magnum, Neon, Pacifica, Ram, Sebring, Stratus, Town & Country, Viper, Voyager, Wrangler | **O2 (B1 S1) Heater Circuit High Conditions:**<br>Key on; system voltage over 10.6v; ASD relay on; O2 heater "off". The PCM detected the O2 Heater circuit is out of range high.<br>**Possible Causes:**<br>O2 heater element is damaged or the heater has failed<br>O2 heater control circuit is open or it is shorted to power<br>O2 heater ground circuit is open<br>PCM has failed |
| **P0033**<br>1T CCM<br>2003-05<br>Neon, PT Cruiser, Stratus, Sebring with I4 | **Surge Valve Solenoid Circuit Fault Conditions:**<br>Key on or engine running; system voltage over 10v. The PCM detected the Actual state of the Surge Valve solenoid circuit did not match the Intended state during the CCM test.<br>**Possible Causes:**<br>Surge valve power supply is open (test power from ASD relay)<br>Surge valve solenoid circuit is open<br>Surge valve solenoid circuit is shorted to ground or power (B+)<br>Surge valve solenoid is damaged or it has failed<br>PCM has failed |
| **P0036**<br>1T CCM<br>2000-03<br>300M, Cherokee, Concorde, Intrepid, LHS, Grand Cherokee, Wrangler | **O2 (B1 S2) Heater Circuit Fault Conditions:**<br>Engine started; system voltage over 10.8v and the PCM detected a problem in the Heater Relay circuit.<br>**Possible Causes:**<br>O2 assembly is damaged or it has failed<br>O2 heater control circuit is open, shorted to ground or B+<br>O2 heater ground circuit is open<br>O2 heater element is damaged or has failed<br>PCM has failed |

**OBD II Trouble Code List (P0XXX Codes) - Continued**

| DTC | Trouble Code Title, Conditions & Possible Causes: |
|---|---|
| **P0037**<br>1T CCM<br>2001-05<br>300M, Caravan, Cherokee, Concorde, PT Cruiser, Dakota, Durango, Grand Cherokee, Intrepid, LHS, Liberty, Magnum, Neon, Pacifica, Ram, Sebring, Stratus, Town & Country, Viper, Voyager, Wrangler | **O2 (B1 S2) Heater Circuit Low Conditions:**<br>Key on; system voltage over 10.6v; ASD relay on; O2 heater "on". The PCM detected the O2 Heater circuit is out of acceptable range (i.e., below 0.0926v).<br>**Possible Causes:**<br>O2 assembly is damaged or it has failed<br>O2 heater element is damaged or has failed<br>O2 heater control circuit is shorted to ground<br>PCM has failed |
| **P0038**<br>1T CCM<br>2001-05<br>300M, Caravan, Cherokee, Concorde, PT Cruiser, Dakota, Durango, Grand Cherokee, Intrepid, LHS, Liberty, Magnum, Neon, Pacifica, Ram, Sebring, Stratus, Town & Country, Viper, Voyager, Wrangler | **O2 (B1 S2) Heater Circuit High Conditions:**<br>Key on; system voltage over 10.6v; ASD relay on; O2 heater "off". The PCM detected the O2 heater voltage is out of range high.<br>**Possible Causes:**<br>O2 Sensor failed or improper operation<br>O2 heater element is damaged or failed<br>O2 heater control circuit is open or it is shorted to power<br>O2 heater ground circuit is open<br>PCM has failed |
| **P0043**<br>1T CCM<br>2001-03<br>Ram with 3.9L V6, 5.9L V8 & 8.0L V10 engine | **O2 (B1 S3) Heater Relay Circuit Low Conditions:**<br>Engine started; system voltage over 10.6v, ECT less within the test condition value, and the PCM detected the Post Catalyst O2 heater Relay circuit voltage remained too high.<br>**Possible Causes:**<br>O2 assembly is damaged or the O2 heater element is damaged or has failed<br>O2 heater control circuit is shorted to ground<br>PCM has failed |
| **P0044**<br>1T CCM<br>2001-03<br>Truck, Van models equipped with 3.9L V6, 5.9L V8 & 8.0L V10 engine | **O2 (B1 S3) Heater Relay Circuit High Conditions:**<br>Engine started; system voltage over 10.6v, ECT less within the test condition value, and the PCM detected the Post Catalyst O2 heater Relay circuit voltage remained too high.<br>**Possible Causes:**<br>O2 heater element is damaged, or the control circuit is open or shorted to power<br>O2 heater ground circuit is open<br>PCM has failed |
| **P0045**<br>1T CCM<br>2004-05<br>Liberty with Diesel | **Boost Pressure Solenoid Excessive Current Conditions:**<br>Ignition on; ECM Boost Pressure Solenoid commanded ON. The ECM detects excessive current on the BP Solenoid Control circuit.<br>**Possible Causes:**<br>Intermittent condition<br>Boost Pressure Solenoid has failed<br>BP Solenoid control circuit is shorted to voltage<br>ECM has failed |

## OBD II Trouble Code List (P0XXX Codes) - Continued

| DTC | Trouble Code Title, Conditions & Possible Causes: |
|---|---|
| **P0045**<br>1T CCM<br>2004-05<br>Liberty with Diesel | **Boost Pressure Solenoid Open Circuit Conditions:**<br>Ignition on; ECM Boost Pressure Solenoid commanded OFF. The ECM did not detect voltage on the BP Solenoid Control circuit.<br>**Possible Causes:**<br>Intermittent condition<br>ASD relay output circuit is open<br>BP Solenoid control circuit is open or is shorted to ground<br>Boost Pressure Solenoid has failed<br>ECM has failed |
| **P0047**<br>1T CCM<br>2004-05<br>Liberty with Diesel | **Boost Pressure Solenoid Short To Ground Circuit Conditions:**<br>Ignition on; ECM Boost Pressure Solenoid commanded OFF. The ECM did not detect voltage on the BP Solenoid Control circuit.<br>**Possible Causes:**<br>Intermittent condition<br>ASD relay output circuit is open<br>BP Solenoid control circuit is open or is shorted to ground<br>Boost Pressure Solenoid has failed<br>ECM has failed |
| **P0048**<br>1T CCM<br>2004-05<br>Liberty with Diesel | **Boost Pressure Solenoid Short Circuit Conditions:**<br>Ignition on; ECM Boost Pressure Solenoid commanded ON. The ECM detects excessive current on the BP Solenoid Control circuit.<br>**Possible Causes:**<br>Intermittent condition<br>Boost Pressure Solenoid has failed<br>BP Solenoid control circuit is shorted to voltage<br>ECM has failed |
| **P0050**<br>1T CCM<br>2002-03 300M, Concorde, PT Cruiser, Intrepid, LHS, Neon, Prowler, Sebring, Stratus, Vision | **O2 (B2 S1) Heater Relay Circuit Low Conditions:**<br>Key on, system voltage over 10.6v, ASD relay on, O2 heater "on", and the PCM detected the Heater Relay circuit Actual state did not match the Desired state (low circuit).<br>**Possible Causes:**<br>O2 assembly is damaged, or the O2 heater element is damaged or has failed<br>O2 heater control circuit is shorted to ground<br>PCM has failed |
| **P0051**<br>1T CCM<br>2001-05<br>300M, Caravan, Cherokee, Concorde, PT Cruiser, Dakota, Durango, Grand Cherokee, Intrepid, LHS, Liberty, Magnum, Neon, Pacifica, Ram, Sebring, Stratus, Town & Country, Viper, Voyager, Wrangler | **O2 (B2 S1) Heater Relay Circuit Low Conditions:**<br>Key on, system voltage over 10.6v, ASD relay on, O2 heater "on", and the PCM detected the Heater Relay circuit Actual state did not match the Desired state (low circuit). 3 good trips are required to turn off the MIL.<br>**Possible Causes:**<br>O2 assembly is damaged or it has failed<br>O2 heater element is damaged or has failed<br>O2 heater control circuit is shorted to ground<br>PCM has failed |
| **P0051**<br>1T CCM<br>2001-05<br>Dakota, Durango | **O2 (B2 S1) Heater Circuit Low Conditions:**<br>Key on; system voltage over 10.6v; ASD relay on; O2 heater "on". The PCM detected the O2 Heater circuit is below acceptable range.<br>**Possible Causes:**<br>O2 assembly is damaged or it has failed<br>O2 heater control circuit is shorted to ground<br>PCM has failed |

**OBD II Trouble Code List (P0XXX Codes) - Continued**

| DTC | Trouble Code Title, Conditions & Possible Causes: |
|---|---|
| **P0052**<br>1T CCM<br>2001-05<br>300M, Caravan, Cherokee, Concorde, PT Cruiser, Dakota, Durango, Grand Cherokee, Intrepid, LHS, Liberty, Magnum, Neon, Pacifica, Ram, Sebring, Stratus, Town & Country, Viper, Voyager, Wrangler | **O2 (B2 S1) Heater Relay Circuit High Conditions:**<br>Key on, system voltage over 10.6v, ASD relay on, O2 heater "off", and the PCM detected the Heater Relay circuit Actual state did not match the Desired state (high circuit).<br>**Possible Causes:**<br>O2 heater element is damaged or the heater has failed<br>O2 heater control circuit is open or it is shorted to power<br>O2 heater ground circuit is open<br>PCM has failed |
| **P0052**<br>1T CCM<br>2001-05<br>Dakota, Durango | **O2 (B2 S1) Heater Circuit High Conditions:**<br>Key on; system voltage over 10.6v; ASD relay on; O2 heater "off". The PCM detected the O2 Heater circuit is above acceptable range.<br>**Possible Causes:**<br>O2 heater control circuit is open or is shorted to battery voltage<br>O2 heater ground circuit is open<br>O2 Sensor has failed<br>PCM has failed |
| **P0056**<br>1T CCM<br>2002-03 300M, Concorde, PT Cruiser, Intrepid, LHS, Neon, Prowler, Sebring, Stratus, Vision | **O2 (B2 S1) Heater Relay Circuit Low Conditions:**<br>Key on, system voltage over 10.6v, ASD relay on, O2 heater "on", and the PCM detected the Heater Relay circuit Actual state did not match the Desired state (low circuit).<br>**Possible Causes:**<br>O2 assembly is damaged or it has failed<br>O2 heater element is damaged or has failed<br>O2 heater control circuit is shorted to ground<br>PCM has failed |
| **P0057**<br>1T CCM<br>2001-05<br>300M, Caravan, Cherokee, Concorde, PT Cruiser, Grand Cherokee, Intrepid, LHS, Liberty, Magnum, Neon, Pacifica, Ram, Sebring, Stratus, Town & Country, Viper, Voyager, Wrangler | **O2 (B2 S2) Heater Relay Circuit Low Conditions:**<br>Key on, system voltage over 10.6v, ECT input under test condition value, and the PCM detected the Heater Relay signal was too low.<br>**Possible Causes:**<br>O2 assembly is damaged or it has failed<br>O2 heater element is damaged or has failed<br>O2 heater control circuit is shorted to ground<br>PCM has failed |
| **P0057**<br>1T CCM<br>2001-05<br>Dakota, Durango | **O2 (B2 S2) Heater Circuit Low Conditions:**<br>Key on; system voltage over 10.6v; ASD relay on; O2 heater "on". The PCM detected the O2 Heater circuit is below acceptable range.<br>**Possible Causes:**<br>O2 assembly is damaged or it has failed<br>O2 heater control circuit is shorted to ground<br>PCM has failed |

**OBD II Trouble Code List (P0XXX Codes) - Continued**

| DTC | Trouble Code Title, Conditions & Possible Causes: |
|---|---|
| **P0058** <br> 1T CCM <br> 2002-05 <br> 300M, Caravan, Concorde, PT Cruiser, Grand Cherokee, Intrepid, LHS, Liberty, Magnum, Neon, Pacifica, Ram, Sebring, Stratus, Town & Country, Viper, Voyager, Wrangler | **O2 (B2 S2) Heater Relay Circuit High Conditions:** <br> Key on, system voltage over 10.6v, ASD powered up, and O2 heater is off. ECT input under test condition value and the PCM detected the Heater Relay signal was too high. <br> **Possible Causes:** <br> O2 heater element is damaged or the heater has failed <br> O2 heater control circuit is open or it is shorted to power <br> O2 heater ground circuit is open <br> PCM has failed |
| **P0058** <br> 1T CCM <br> 2001-05 <br> Dakota, Durango | **O2 (B2 S2) Heater Circuit High Conditions:** <br> Key on; system voltage over 10.6v; ASD relay on; O2 heater "off". The PCM detected the O2 Heater circuit is above acceptable range. <br> **Possible Causes:** <br> O2 heater control circuit is open or is shorted to battery voltage <br> O2 heater ground circuit is open <br> O2 Sensor has failed <br> PCM has failed |
| **P0068** <br> 1T CCM <br> 2003-05 <br> 300M, Concorde, PT Cruiser, Dakota, Durango, Grand Cherokee, Intrepid, Liberty, Magnum, Pacifica, Ram, Sebring, Stratus, Viper, Voyager, Wrangler | **MAP Sensor/TP Sensor Correction - High Flow/Vacuum Leak Conditions:** <br> Engine started; engine speed over 2000 rpm, and the PCM detected the Manifold Air Pressure (MAP) value dropped to less than 1.5" Hg with the throttle closed during the test. <br> **Possible Causes:** <br> An engine vacuum leak present <br> High resistance in the MAP ground circuit, MAP Sensor signal or VREF (5v) circuit <br> High resistance in the TP ground, TP circuit or the TP Sensor VREF (5v) circuit <br> MAP Sensor is damaged or it has failed <br> TP Sensor is damaged or it has failed <br> PCM has failed |
| **P0068** <br> 2T CCM <br> 2003-05 <br> Caravan, LHS, Neon, Sebring Coupe, Stratus Coupe, Town & Country | **MAP Sensor/TP Sensor Correlation Conditions:** <br> Engine started; no MAP Sensor or TP Sensor DTCs are present. The PCM determines a valid range in which the TP Sensor should be, at a given rpm/engine load. The actual TP Sensor voltage is then compared to this value. If the TP Sensor voltage does not fall within the expected range within a predetermined time, an error is detected. The DTC will set after 2 trips. <br> **Possible Causes:** <br> An engine vacuum leak present <br> High resistance in the MAP ground circuit, MAP Sensor signal or VREF (5v) circuit <br> High resistance in the TP ground, TP circuit or the TP Sensor VREF (5v) circuit <br> MAP Sensor is damaged or it has failed <br> TP Sensor is damaged or it has failed <br> PCM has failed |

**OBD II Trouble Code List (P0XXX Codes) - Continued**

| DTC | Trouble Code Title, Conditions & Possible Causes: |
|---|---|
| **P0068**<br>1T CCM<br>2004-05<br>Dakota, Durango, Ram | **MAP Sensor/TP Sensor Correlation Conditions:**<br>Condition is monitored during all drive modes. This DTC will set when an unexpectedly high intake manifold airflow condition exists that can lead to increased engine speed and which then puts the Next Generation Controller into a High Airflow Protection Limiting Mode. This feature includes rpm limits whenever a TP and/or MAP Sensor limp-in fault is present. If vacuum drops below 1.5 in. Hg, with engine speed greater than 2000 rpm and closed throttle, this DTC will set.<br>**Possible Causes:**<br>An engine vacuum leak present<br>High resistance in the 5v supply circuit<br>5v supply circuit is shorted to ground<br>High resistance in the MAP signal circuit or the TP signal circuit<br>TP signal circuit is shorted to ground<br>High resistance in the Sensor ground circuit<br>MAP Sensor is damaged or it has failed<br>TP Sensor is damaged or it has failed<br>PCM has failed |
| **P0070**<br>1T CCM<br>2001-03<br>300M, Caravan, Cherokee, Concorde, PT Cruiser, Grand Cherokee, Intrepid, LHS, Liberty, Neon, Pacifica, Ram, Sebring, Stratus, Town & Country, Viper, Voyager, Wrangler | **Ambient Temperature Sensor Circuit Stuck Conditions:**<br>Engine started 4 times, 4 warm-up cycles completed, vehicle driven for 200 miles, and the PCM did not detect more than a 6°F change in the Ambient Air Temperature (AAT) Sensor signal in the test.<br>**Possible Causes:**<br>AAT Sensor signal circuit shorted to power (VREF)<br>AAT Sensor ground circuit is open<br>AAT Sensor signal circuit is open<br>AAT Sensor signal circuit is shorted to ground<br>PCM High or Low circuit is damaged or it has failed |
| **P0070**<br>1T CCM<br>2004-05<br>Liberty with Diesel | **Ambient Air Temperature Signal Voltage Too High Conditions:**<br>Ignition on; Ambient Air Temp. Sensor signal is above 4.82v.<br>**Possible Causes:**<br>Intermittent condition<br>AAT Sensor signal circuit is open or is shorted to voltage<br>AAT Sensor ground circuit is open<br>AAT Sensor has failed<br>ECM has failed |
| **P0070**<br>1T CCM<br>2004-05<br>Liberty with Diesel | **Ambient Air Temperature Signal Voltage Too Low Conditions:**<br>Ignition on; Ambient Air Temp. Sensor signal is below 0.068v.<br>**Possible Causes:**<br>Intermittent condition<br>AAT Sensor signal circuit is shorted to ground or to sensor ground<br>AAT Sensor has failed<br>ECM has failed |

**OBD II Trouble Code List (P0XXX Codes) - Continued**

| DTC | Trouble Code Title, Conditions & Possible Causes: |
|---|---|
| **P0071**<br>2T CCM<br>2001-05<br>300M, Caravan, Cherokee, Concorde, PT Cruiser, Dakota, Durango, Grand Cherokee, Intrepid, LHS, Liberty, Magnum, Neon, Pacifica, Ram, Sebring, Stratus, Town & Country, Viper, Voyager, Wrangler | **Ambient Temperature Sensor Performance Conditions:**<br>Engine "off" time over 8 hours; DTC P0072 and P0073 not set; ambient temperature more than 38ºF (4°C). The PCM determined the Ambient Air Temperature (AAT) Sensor was not within calibrated temperature of the ECT and IAT Sensor signals after a cool down period.<br>**Possible Causes:**<br>AAT Sensor circuit open, shorted to ground or VREF<br>AAT Sensor voltage below 1.0v<br>AAT signal circuit is open or is shorted to ground, to battery voltage or to Sensor ground<br>AAT Sensor ground circuit is open<br>AAT Sensor is damaged or it has failed<br>PCM High or Low circuit is damaged or it has failed |
| **P0071**<br>1T CCM<br>2004-05<br>Viper | **Ambient Temperature/Battery Temp. Sensor Performance Conditions:**<br>Ignition is on and no Battery Temperature Sensor DTCs present. After 5 warm cycles have occurred, in which coolant temperature increases to at least 160°F (71°C), and vehicle was driven at least 200 miles, if the Battery Temperature did not change more than a 7ºF (4°C), this DTC will set.<br>**Possible Causes:**<br>Battery Temp. Sensor signal circuit shorted to ground or to voltage<br>Battery Temp. Sensor has failed<br>Battery Temp. Sensor signal circuit or ground circuit has high resistance<br>AAT Sensor signal circuit is shorted to ground<br>PCM has failed |
| **P0072**<br>1T CCM<br>2002-05<br>300M, Caravan, Cherokee, Concorde, PT Cruiser, Dakota, Durango, Grand Cherokee, Intrepid, LHS, Liberty, Magnum, Neon, Pacifica, Ram, Sebring, Stratus, Town & Country, Viper, Voyager | **Ambient Temperature Sensor Circuit Low Input Conditions:**<br>Related DTCs not set, key on or engine running; system voltage over 10.5v, at least 5 warm-up cycles completed, odometer mileage change at least 196.6 miles, and the PCM detected the AAT Sensor signal was less than 0.3v (0.78v: Jeep) at PCM. 3 good trips required to turn off MIL.<br>**Possible Causes:**<br>AAT Sensor internal failure<br>AAT signal circuit shorted to ground<br>AAT signal circuit shorted to Sensor ground<br>PCM has failed |
| **P0072**<br>1T CCM<br>2004-05<br>Dakota, Durango, Liberty | **Ambient Temperature Sensor Circuit Low Input Conditions:**<br>Ignition is on. When the Ambient Temperature Sensor is less than 0.078v at the PCM for 2.8 seconds (Dakota, Durango) or less than 0.039v for 4.2 seconds (Liberty), this DTC with set. 3 good trips required to turn off MIL.<br>**Possible Causes:**<br>AAT Sensor internal failure<br>AAT signal circuit shorted to ground<br>AAT signal circuit shorted to Sensor ground<br>Front Control Module has failed |

**OBD II Trouble Code List (P0XXX Codes) - Continued**

| DTC | Trouble Code Title, Conditions & Possible Causes: |
|---|---|
| **P0073**<br>1T CCM<br>2002-05<br>300M, Caravan, Cherokee, Concorde, PT Cruiser, Dakota, Durango, Grand Cherokee, Intrepid, LHS, Liberty, Magnum, Neon, Pacifica, Ram, Sebring, Stratus, Town & Country, Viper, Voyager | **Ambient Temperature Sensor Circuit High Input Conditions:**<br>Key on or engine running; system voltage over 10.0v and the PCM detected the Ambient Air Temperature (AAT) Sensor signal was more than 4.9v for more than 2.8 seconds. Note that this code can be set due to an intermittent failure. 3 good trips are required to turn off the MIL.<br>**Possible Causes:**<br>AAT Sensor signal shorted to VREF (5v) or battery voltage<br>AAT Sensor signal circuit or Sensor ground circuit is open<br>AAT Sensor is damaged (it may be open)<br>PCM has failed |
| **P0087**<br>1T CCM<br>2004-05<br>Liberty with Diesel | **Fuel Rail Pressure Too Low Malfunction Conditions:**<br>Engine running; ECM determines that the fuel rail pressure is too low for a given engine speed.<br>**Possible Causes:**<br>Air in fuel system<br>Fuel injector problems<br>Fuel Pressure Solenoid has failed<br>Fuel Pump has malfunctioned or failed<br>Fuel system has contamination<br>Fuel system has a leak<br>Intermittent condition |
| **P0088**<br>1T CCM<br>2004-05<br>Liberty with Diesel | **Fuel Rail Pressure Too High Malfunction Conditions:**<br>Engine running; ECM detects the fuel rail pressure is above 23,000 psi.<br>**Possible Causes:**<br>Air in fuel system<br>Fuel injector problems<br>Fuel Pressure Solenoid has failed<br>Fuel Pump has malfunctioned or failed<br>Fuel system has contamination<br>Fuel system has a leak<br>Intermittent condition |
| **P0089**<br>1T CCM<br>2004-05<br>Liberty with Diesel | **Fuel Pressure Solenoid After-Run Plausibility Conditions:**<br>Engine running; ECM determines that fuel rail pressure is too low for a given engine speed.<br>**Possible Causes:**<br>Air in fuel system<br>Fuel injector problems<br>Fuel Pressure Solenoid has failed<br>Fuel Pump has malfunctioned or failed<br>Fuel system has contamination<br>Fuel system has a leak<br>Intermittent condition |
| **P0090**<br>1T CCM<br>2004-05<br>Liberty with Diesel | **Fuel Quality Solenoid Open Circuit Conditions:**<br>Ignition on; ECM Fuel Quality Solenoid commanded OFF; ECM detects an open in the Fuel Quality Solenoid circuit.<br>**Possible Causes:**<br>FQ Solenoid circuit(s) are open, shorted to ground, shorted to voltage, or shorted together<br>Intermittent condition<br>FQ Solenoid has failed<br>ECM has failed |

**OBD II Trouble Code List (P0XXX Codes) - Continued**

| DTC | Trouble Code Title, Conditions & Possible Causes: |
|---|---|
| **P0091**<br>1T CCM<br>2004-05<br>Liberty with Diesel | **Fuel Quality Solenoid Short To Ground Circuit Conditions:**<br>Ignition on; ECM Fuel Quality Solenoid commanded OFF; ECM detects short to ground on the Fuel Quality Solenoid circuit.<br>**Possible Causes:**<br>FQ Solenoid circuit(s) are open, shorted to ground, shorted to voltage, or shorted together<br>Intermittent condition<br>FQ Solenoid has failed<br>ECM has failed |
| **P0092**<br>1T CCM<br>2004-05<br>Liberty with Diesel | **Fuel Quality Solenoid Short Circuit Conditions:**<br>Ignition on; ECM Fuel Quality Solenoid commanded ON; ECM detects excessive current on the Fuel Quality Solenoid circuit.<br>**Possible Causes:**<br>FQ Solenoid circuit(s) are open, shorted to ground, shorted to voltage, or shorted together<br>Intermittent condition<br>FQ Solenoid has failed<br>ECM has failed |
| **P0100**<br>1T CCM<br>1995-98<br>Talon with Turbo | **Volume Airflow Sensor Circuit Conditions:**<br>Engine started; engine speed over 500 rpm, and the PCM detected the VAF Sensor input was 3 Hz or less (the correct reading should be somewhere between 22/48 Hz).<br>**Possible Causes:**<br>Volume Airflow Sensor signal shorted to power<br>Volume Airflow Sensor signal shorted to ground<br>Volume Airflow Sensor has failed<br>PCM has failed |
| **P0100**<br>1T CCM<br>2004-05<br>Liberty with Diesel | **MAF Sensor Signal Voltage Too Low Or Too High Conditions:**<br>Engine running between 500-5000 rpm; ECM detects the MAF Sensor signal is blow 15 kg/h for 0.5 seconds, or is above 800 kg/h for 0.5 seconds.<br>**Possible Causes:**<br>ASD relay output circuit is open<br>ECM 5v supply circuit problem<br>MAF Sensor ground circuit is open<br>MAF Sensor has failed<br>Intermittent condition<br>MAF Sensor signal circuit is open or is shorted to ground or shorted to MAF Sensor ground<br>MAF Sensor 5v supply circuit is shorted to MAF Sensor ground circuit or shorted to ground<br>MAF Sensor ground circuit is open<br>MAF Sensor 5v supply circuit is shorted to voltage<br>MAF Sensor circuit is shorted to voltage<br>ECM has failed |
| **P0101**<br>1T CCM<br>2004-05<br>Liberty with Diesel | **MAF Sensor Signal Negative Or Positive Deviation Conditions:**<br>Engine running; engine coolant temperature between (60-100C); IAT Sensor reading is steady; atmospheric pressure is below 21.8 psi (1500 hPa); boost pressure is between 10.9-34.8 psi (750-2400 hPa). ECM detects the MAF Sensor reading is below (negative) or above (positive) the calibrated value for more than 2 seconds.<br>**Possible Causes:**<br>Air filer problem<br>Air restriction in intake system<br>Air leak(s) in intake system<br>MAF Sensor has failed<br>Intermittent condition |

**OBD II Trouble Code List (P0XXX Codes) - Continued**

| DTC | Trouble Code Title, Conditions & Possible Causes: |
|---|---|
| **P0105**<br>1T CCM<br>1995-98<br>Talon with Turbo | **Barometric Pressure Sensor Circuit Conditions:**<br>Key on for less than 60 seconds or right after startup, system voltage over 8v, and PCM detected a BARO input of more than 4.5v or less than 0.2v for 4 sec's (normal is 3.2-3.8v).<br>**Possible Causes:**<br>Loss of 5-volt supply from PCM (internal failure)<br>Sensor 5-volt supply circuit is shorted, open or grounded<br>Sensor signal circuit is open or shorted to ground<br>Sensor has failed<br>PCM has failed |
| **P0105**<br>1T CCM<br>2004-05<br>Liberty with Diesel | **Inlet Pressure Sensor Signal Plausibility Conditions:**<br>Engine running below 800 rpm; no other IAT DTCs are present. The difference between Inlet Pressure Sensor signal and Atmospheric Pressure Sensor signal is 50.8 psi for 5 seconds.<br>**Possible Causes:**<br>Air filer problem<br>Air restriction in intake system<br>Intermittent condition<br>High resistance in Inlet Pressure Sensor signal circuit, ground circuit, or 5v supply circuit<br>ECM has failed |
| **P0105**<br>1T CCM<br>2004-05<br>Liberty with Diesel | **Inlet Pressure Sensor Signal Voltage Too High Conditions:**<br>Ignition on; ECM detects the Inlet Pressure Sensor signal is above 4.75v for 2 seconds.<br>**Possible Causes:**<br>Intermittent condition<br>Inlet Pressure Sensor ground circuit is open, or shorted to voltage<br>Inlet Pressure Sensor signal circuit is shorted to voltage<br>Inlet Pressure Sensor has failed<br>ECM has failed |
| **P0105**<br>1T CCM<br>2004-05<br>Liberty with Diesel | **Inlet Pressure Sensor Signal Voltage Too Low Conditions:**<br>Ignition on; ECM detects the Inlet Pressure Sensor signal is below 0.25v for 2 seconds.<br>**Possible Causes:**<br>Intermittent condition<br>Inlet Pressure Sensor 5v supply circuit problem<br>Inlet Pressure Sensor has failed<br>Inlet Pressure Sensor signal circuit is shorted to ground<br>Inlet Pressure Sensor signal circuit and ground circuit are shorted together<br>ECM has failed<br>Inlet Pressure Sensor signal circuit is open |

## OBD II Trouble Code List (P0XXX Codes) - Continued

| DTC | Trouble Code Title, Conditions & Possible Causes: |
|---|---|
| **P0106**<br>1T CCM<br>1996-2003<br>300M, Acclaim, Avenger, Breeze, Caravan, Cherokee, Cirrus, Concorde, PT Cruiser, Dakota, Durango, Grand Cherokee, LeBaron, Spirit, Intrepid, LHS, Liberty, Neon, New Yorker, Sebring, Stratus, Prowler, Pacifica, Ram, Town & Country, Viper, Vision, Voyager, Wrangler | **BARO Out-Of-Range at Key On / MAP Sensor Low Conditions:**<br>Key on for less than 350 ms; engine speed less than 255 rpm, and the PCM detected the MAP Sensor input was less than 2.196v but more than 0.019v during a 300 ms period.<br>**Possible Causes:**<br>Loss of 5-volt supply from PCM (internal failure)<br>Sensor 5-volt supply circuit is shorted, open or grounded<br>Sensor signal circuit is open or shorted to ground<br>Sensor has failed<br>PCM has failed |
| **P0107**<br>1T CCM<br>1995-98<br>Talon (Non-Turbo) | **MAP Sensor Circuit Low Input Conditions:**<br>Engine speed from 400-1500 rpm; TP Sensor input less than 1.3v, and PCM detected a MDP input less than 0.02v for 2 seconds (it should be between 0.8-2.5v).<br>**Possible Causes:**<br>Loss of 5-volt supply from PCM (internal failure)<br>Sensor 5-volt supply circuit is open or shorted to ground<br>Sensor signal circuit is shorted to ground<br>Sensor has failed<br>PCM has failed |
| **P0107**<br>1T CCM<br>1995-2003<br>300M, Acclaim, Avenger, Breeze, Caravan, Cherokee, Cirrus, Concorde, PT Cruiser, Grand Cherokee, LeBaron, Spirit, Intrepid, LHS, Neon, New Yorker, Prowler, Pacifica, Town & Country, Viper, Vision, Voyager, Wrangler | **MAP Sensor Circuit Low Input Conditions:**<br>Engine speed from 600-3500 rpm; TP Sensor input less than 1.2v, system voltage over 10.6v, and the PCM detected the MAP input was below 0.0392v (conditions met for 1.76 seconds).<br>**Possible Causes:**<br>MAP Sensor connector is damaged or shorted<br>MAP Sensor 5v supply circuit is open or it is shorted to ground<br>MAP Sensor signal circuit is shorted to ground<br>MAP Sensor has failed<br>PCM has failed |

**OBD II Trouble Code List (P0XXX Codes) - Continued**

| DTC | Trouble Code Title, Conditions & Possible Causes: |
|---|---|
| **P0107**<br>1T CCM<br>2001-03<br>Sebring, Stratus with 2.4L or 3.0L | **Barometric Pressure Sensor Circuit Low Input Conditions:**<br>Engine started, engine runtime over 2 seconds, system voltage over 10.5v and the PCM detected the BARO Sensor was 1.95v or lower (a value equal to 7.3 in. Hg or higher) for 10 seconds. Note: This value indicates the vehicle was at 15,000 feet above sea level.<br>**Possible Causes:**<br>BARO Sensor connector is damaged or shorted<br>BARO Sensor 5v supply circuit is open or it is shorted to ground<br>BARO Sensor signal circuit is shorted to ground<br>BARO Sensor has failed<br>PCM has failed |
| **P0107**<br>1T CCM<br>1996-2003<br>Cherokee, Dakota, Durango, Grand Cherokee, Liberty, Ram, Wrangler | **MAP Sensor Circuit Low Input Conditions:**<br>Engine speed from 416-1470 rpm; TP Sensor input less than 1v, and the PCM detected the MAP Sensor input was less than 2.35v at startup, or was less than 0.20v with the engine running (conditions met for 1.76 seconds).<br>**Possible Causes:**<br>Loss of 5-volt supply from PCM (internal failure)<br>Sensor 5-volt supply circuit is open or shorted to ground<br>Sensor signal circuit is shorted to ground<br>Sensor or PCM has failed |
| **P0107**<br>1T CCM<br>2004-05<br>300M, Caravan, Concorde, PT Cruiser, Dakota, Durango, Grand Cherokee, Intrepid, LHS, Liberty, Magnum, Neon, Pacifica, Ram, Sebring, Stratus, Town & Country, Viper, Voyager | **MAP Sensor Circuit Low Input Conditions:**<br>Ignition on or engine from 600-3500 rpm (416-1500 rpm: Viper); TP Sensor input less than 1.2v (exc. Ram, Viper, Jeep), less than 0.8v (Ram, Jeep) or less than 1.13v (Viper). Battery voltage greater than 10v. MAP Sensor signal voltage was less than 0.782v for 1.7 seconds (exc. Dakota, Durango, Viper, Jeep) or less than 0.08v for 3 seconds (Dakota, Durango, Jeep), less than 0.1v for 2 seconds (Viper). If equipped, ETC light will flash.<br>**Possible Causes:**<br>5-volt supply circuit is open or shorted to ground<br>MAP Sensor signal circuit is shorted to ground or shorted to Sensor ground circuit<br>MAP Sensor has failed<br>PCM 5-volt supply circuit has failed |
| **P0108**<br>1T CCM<br>1995-98<br>Talon (Non-Turbo) | **MAP Sensor Circuit High Input Conditions:**<br>Engine speed from 600-3500 rpm; TP input more than 1.3v; and the PCM detected the MAP input was more than 4.7v for 2 seconds.<br>**Possible Causes:**<br>MAP Sensor signal circuit open, or the ground circuit is open<br>MAP Sensor signal circuit is shorted to VREF (5v)<br>MAP Sensor is damaged or it has failed<br>PCM has failed |
| **P0108**<br>1T CCM<br>2004-05<br>300M, Caravan, Concorde, PT Cruiser, Dakota, Durango, Grand Cherokee, Intrepid, LHS, Liberty, Magnum, Neon, Pacifica, Ram, Sebring, Stratus, Town & Country, Viper, Voyager | **MAP Sensor Circuit High Input Conditions:**<br>Ignition on or engine speed from 600-3500 rpm (above 400 for Viper); TP Sensor input more than 1.2v, for more than 1.7 seconds; battery voltage over 10.0v. The PCM detected the MAP Sensor signal voltage (input) was over 4.92v (exc. Viper), or was greater than 4.88v at start or with engine running for 2.2 seconds (Viper).<br>**Possible Causes:**<br>MAP Sensor signal circuit is open<br>MAP Sensor ground circuit is open<br>MAP Sensor signal circuit shorted to 5-volt supply circuit or to battery voltage<br>MAP Sensor has failed or it has failed (possible open circuit)<br>MAP Sensor has failed<br>PCM has failed |

**OBD II Trouble Code List (P0XXX Codes) - Continued**

| DTC | Trouble Code Title, Conditions & Possible Causes: |
|---|---|
| **P0108**<br>1T CCM<br>2001-03<br>Sebring, Stratus with 2.4L or 3.0L | **Barometric Pressure Sensor Circuit High Input Conditions:**<br>Engine started; engine runtime over 2 seconds, system voltage over 8.0v, and the PCM detected the BARO Sensor was 4.45v or higher (a value equal to 114 kPa or 16.5" psi or higher) for 10 seconds. Note: This value indicates the vehicle is 4,000 feet below sea level.<br>**Possible Causes:**<br>MAP Sensor signal circuit open, or the ground circuit is open<br>MAP Sensor signal circuit is shorted to VREF (5v)<br>MAP Sensor is damaged or it has failed<br>PCM has failed |
| **P0108**<br>1T CCM<br>1996-2003<br>Cherokee, Dakota, Durango, Grand Cherokee, Liberty, Ram, Wrangler | **MAP Sensor Circuit High Input Conditions:**<br>Engine speed from 416-1470 rpm with the throttle closed; and the PCM detected a MAP Sensor input of more than 4.6v (conditions met for 1 second).<br>**Possible Causes:**<br>MAP Sensor signal circuit is open<br>MAP Sensor ground circuit is open<br>MAP Sensor signal circuit shorted to VREF (5v)<br>MAP Sensor has failed or it has failed (possible open circuit)<br>PCM has failed |
| **P0110**<br>1T CCM<br>1995-98<br>Talon | **IAT Sensor Circuit High or Low Input Conditions:**<br>Key on for 60 seconds or right after startup; and the PCM detected an IAT Sensor input of 4.6v or less than 0.2v for 4 seconds.<br>**Possible Causes:**<br>IAT Sensor signal circuit is open, shorted to ground or to VREF<br>IAT Sensor signal circuit has a high resistance condition<br>IAT Sensor is damaged or it has failed<br>PCM has failed |
| **P0110**<br>1T CCM<br>2002-03<br>300M, Caravan, Concorde, PT Cruiser, Dakota, Durango, Grand Cherokee, Intrepid, Neon, Prowler, Ram, Sebring, Stratus, Town & Country, Voyager, Wrangler | **IAT Sensor Circuit High or Low Input Conditions:**<br>Key on for 60 seconds or right after startup; and the PCM detected an IAT Sensor input of 4.6v or less than 0.2v for 4 seconds.<br>**Possible Causes:**<br>IAT Sensor signal circuit is open, shorted to ground or to VREF<br>IAT Sensor signal circuit has a high resistance condition<br>IAT Sensor is damaged or it has failed<br>PCM has failed |
| **P0110**<br>1T CCM<br>2004-05<br>Liberty with Diesel | **IAT Sensor Signal Voltage Too Low Or Too High Conditions:**<br>Ignition on; ECM detects the IAT Sensor signal is either below 0.45v (too low) or above 4.95v (too high).<br>**Possible Causes:**<br>Intermittent condition<br>IAT Sensor signal circuit is open or is shorted to voltage (too high)<br>IAT Sensor signal circuit is shorted to ground or to Sensor ground (too low)<br>IAT Sensor ground circuit is open (too high)<br>Boost Pressure/IAT Sensor has failed<br>ECM has failed |

**OBD II Trouble Code List (P0XXX Codes) - Continued**

| DTC | Trouble Code Title, Conditions & Possible Causes: |
|---|---|
| **P0111**<br>1T CCM<br>2001-05<br>300M, Caravan, Cherokee, Concorde, PT Cruiser, Dakota, Durango, Grand Cherokee, Intrepid, LHS, Liberty, Magnum, Neon, Pacifica, Ram, Sebring, Stratus, Town & Country, Viper, Voyager, Wrangler | **IAT Sensor Performance Conditions:**<br>DTC P0112 and P0113 not set, key on, ECT Sensor input more than 160ºF at startup, at least 5 warm-up cycles have occurred with vehicle mileage change of more than 196.6 miles, and the PCM detected the IAT input changed less than 5.4ºF during this period.<br>**Possible Causes:**<br>IAT Sensor signal circuit is open, shorted to ground or VREF<br>IAT Sensor ground circuit is open<br>IAT Sensor is damaged or it has failed<br>PCM High or Low circuit is damaged or it has failed |
| **P0111**<br>2T CCM<br>2004-05<br>Caravan, PT Cruiser, Dakota, Durango, Grand Cherokee, Liberty, Neon, Pacifica, Ram, Sebring, Stratus, Town & Country, Voyager | **IAT Sensor Performance Conditions:**<br>Engine off. After calibrated amount of cool down (over 8 hours). PCM detects IAT Sensor is not within calibrated temperature amount of ECT Sensor and AAT Sensor. Engine time off when monitored is more than 8 hours and ambient temperature is more than 38°F (4°C). 3 good trips are required to turn off the MIL.<br>**Possible Causes:**<br>IAT Sensor signal circuit is open, shorted to ground, to Sensor ground, or to battery voltage<br>IAT Sensor voltage is below 1.0v<br>IAT Sensor ground circuit is open<br>IAT Sensor is damaged or has failed<br>PCM High or Low circuit is damaged or has failed |
| **P0112**<br>1T CCM<br>1995-98<br>Talon | **IAT Sensor Circuit Low Input Conditions:**<br>Engine started; system voltage over 10.5v and the PCM detected the IAT Sensor was less than 0.157v for 2 seconds.<br>**Possible Causes:**<br>IAT Sensor signal circuit is shorted to chassis ground<br>IAT Sensor signal circuit is shorted to Sensor ground<br>IAT Sensor is damaged or it has failed (an internal short circuit)<br>PCM has failed |
| **P0112**<br>1T CCM<br>1995-2005<br>300M, Caravan, Cherokee, Concorde, PT Cruiser, Dakota, Durango, Grand Cherokee, Intrepid, LHS, Liberty, Magnum, Neon, Pacifica, Ram, Sebring, Stratus, Town & Country, Voyager, Wrangler | **IAT Sensor Circuit Low Input Conditions:**<br>Ignition on or engine started; battery voltage greater than 10v. If the PCM detected an IAT Sensor input of less than 0.1v (Neon), 0.157v (Cirrus/Sebring, Stratus, Caravan/Town & Country) or less than 0.078v (Durango, Ram, Grand Cherokee) for 3 seconds, less than 0.8v (Viper), or less than 0.5v (Liberty) this DTC will set.<br>**Possible Causes:**<br>IAT Sensor signal circuit is shorted to chassis ground<br>IAT Sensor signal circuit is shorted to Sensor ground<br>IAT Sensor is damaged or it has failed (an internal short circuit)<br>PCM has failed |

## OBD II Trouble Code List (P0XXX Codes) - Continued

| DTC | Trouble Code Title, Conditions & Possible Causes: |
|---|---|
| **P0113**<br>1T CCM<br>1995-2005<br>300M, Caravan, Cherokee, Concorde, PT Cruiser, Dakota, Durango, Grand Cherokee, Intrepid, LHS, Liberty, Magnum, Neon, Pacifica, Ram, Sebring, Stratus, Town & Country, Viper, Voyager, Wrangler | **IAT Sensor Circuit High Input Conditions:**<br>Check with ignition on or engine running; battery voltage more than 10v; The PCM detected the IAT Sensor input was over 4.90v (exc. Dakota, Durango, Ram) or 4.98v (Dakota, Durango, Ram) for 3 seconds.<br>**Possible Causes:**<br>IAT Sensor signal circuit shorted to VREF (5v)<br>IAT Sensor signal circuit is open, or the ground circuit is open<br>IAT Sensor is damaged or it has failed (an internal open circuit)<br>PCM has failed |
| **P0113**<br>1T CCM<br>1998-2003<br>Ram Trucks with 5.9L Diesel | **IAT Sensor Circuit High Input Conditions:**<br>Engine started; system voltage over 10.5v and the PCM detected the IAT Sensor input was more than 4.90v for 2 seconds.<br>**Possible Causes:**<br>IAT Sensor signal circuit shorted to VREF (5v)<br>IAT Sensor signal circuit is open, or the ground circuit is open<br>IAT Sensor is damaged or it has failed (an internal open circuit)<br>PCM has failed |
| **P0115**<br>1T CCM<br>1995-98<br>Talon | **ECT Sensor Circuit High or Low Input Conditions:**<br>Key on for 60 seconds or right after startup, and the PCM detected the ECT Sensor was more than 4.6v or less than 0.2v for 4 seconds, or with engine running, the signal went from under 1.6v to over 1.6v and remained over 1.6v for 5 minutes, or with ECT and IAT inputs over 68°F at startup, the ECT Sensor took 5 minutes to reach 122°F.<br>**Possible Causes:**<br>ECT Sensor signal circuit is open or shorted to ground<br>ECT Sensor signal circuit is shorted to VREF (5v)<br>ECT Sensor signal circuit has a high resistance condition<br>ECT Sensor is damaged or it has failed<br>PCM has failed. |
| **P0115**<br>1T CCM<br>2004-05<br>Liberty with Diesel | **ECT Sensor Signal Voltage Too Low Or Too High Conditions:**<br>Ignition on; ECM detects the ECT Sensor signal is either below 0.12v (too low) or above 4.95v (too high).<br>**Possible Causes:**<br>Intermittent condition<br>ECT Sensor signal circuit is open or is shorted to voltage (too high)<br>ECT Sensor signal circuit is shorted to ground or to Sensor ground (too low)<br>ECT Sensor ground circuit is open (too high)<br>Boost Pressure/IAT Sensor has failed<br>ECM has failed |

**OBD II Trouble Code List (P0XXX Codes) - Continued**

| DTC | Trouble Code Title, Conditions & Possible Causes: |
|---|---|
| **P0116**<br>1T CCM<br>2001-03<br>300M, Caravan, Cherokee, Concorde, PT Cruiser, Dakota, Durango, Grand Cherokee, Intrepid, LHS, Liberty, Neon, Pacifica, Ram, Sebring, Stratus, Town & Country, Viper, Voyager, Wrangler | **ECT Sensor Circuit Performance Conditions:**<br>Engine started; engine runtime over 10 minutes, and the PCM detected the ECT Sensor did not reach a calibrated level during the test period (i.e., it failed the CCM rationality test).<br>**Possible Causes:**<br>ECT Sensor signal circuit is open or it is shorted to ground<br>ECT Sensor signal circuit is shorted to VREF (5v)<br>ECT Sensor ground circuit is open<br>ECT Sensor is damaged or it has failed<br>PCM High or Low circuit is damaged or it has failed |
| **P0116**<br>2T CCM<br>2004-05<br>300M, Caravan, Concorde, PT Cruiser, Dakota, Durango, Grand Cherokee, Intrepid, LHS, Liberty, Magnum, Neon, Pacifica, Ram, Sebring, Stratus, Town & Country, Viper, Voyager | **ECT Sensor Circuit Performance Conditions:**<br>Engine off time is more than 8 hours; ambient temperature is more than 38°F (4°C); and after a calibrated amount of cool-down time, the PCM compares the ECT Sensor, IAT Sensor and AAT Sensor values; if ECT Sensor is not within a calibrated temperature amount of the other 2 Sensors, an error is detected. 3 good trips are required to turn off the MIL. If equipped, the ECT light will also illuminate when the MIL illuminates.<br>**Possible Causes:**<br>ECT Sensor signal circuit is open or it is shorted battery voltage or to ground<br>ECT Sensor signal circuit is shorted to Sensor ground<br>ECT Sensor ground circuit is open<br>ECT Sensor is damaged or it has failed<br>PCM High or Low circuit is damaged or it has failed |
| **P0117**<br>1T CCM<br>1995-2005<br>300M, Avenger, Breeze, Caravan, Cherokee, Cirrus, Concorde, PT Cruiser, Dakota, Durango, Grand Cherokee, Intrepid, LHS, Liberty, Magnum, Neon, New Yorker, Pacifica, Ram, Sebring, Stratus, Town & Country, Viper, Vision, Voyager, Wrangler | **ECT Sensor Circuit Low Input Conditions:**<br>Ignition on or engine started. The PCM detected the ECT Sensor input voltage was below 0.51v (95-03) or 0.1v (04-05, exc. Grand Cherokee, Liberty), less than 0.5v (04-05 Liberty), less than 0.78v (Grand Cherokee), or below 0.8v (Viper) for 3 seconds.<br>**Possible Causes:**<br>ECT Sensor signal circuit is shorted to chassis ground<br>ECT Sensor signal circuit is shorted to Sensor ground<br>ECT Sensor is damaged or it has failed (it may be shorted)<br>PCM has failed |

**OBD II Trouble Code List (P0XXX Codes) – Continued**

| DTC | Trouble Code Title, Conditions & Possible Causes: |
|---|---|
| **P0118**<br>1T CCM<br>1995-2005<br>300M, Avenger, Breeze, Caravan, Cherokee, Cirrus, Concorde, PT Cruiser, Dakota, Durango, Intrepid, LHS, Neon, New Yorker, Pacifica, Sebring, Stratus, Town & Country, Viper, Vision, Voyager, Wrangler | **ECT Sensor Circuit High Input Conditions:**<br>Ignition on or engine started and the PCM detected the ECT Sensor input was over 4.9v for 3 seconds. If equipped, the ETC lamp will illuminate with the MIL. 3 good trips are required to turn off the MIL.<br>**Possible Causes:**<br>ECT Sensor signal circuit is shorted to VREF (5v)<br>ECT Sensor signal circuit is open<br>ECT Sensor ground circuit is open<br>ECT Sensor is damaged or it has failed (possible open circuit)<br>PCM has failed |
| **P0120**<br>1T CCM<br>1995-98<br>Talon | **Throttle Position Sensor Circuit Conditions:**<br>Key on for 60 seconds or right after startup, closed throttle switch on; and the PCM detected a TP Sensor input of 2v or more or a TP Sensor input of 0.2v or less for 4 seconds.<br>**Possible Causes:**<br>TP Sensor 5-volt supply circuit is shorted, open or grounded<br>TP Sensor signal circuit is open or shorted to ground<br>TP Sensor is damaged or it has failed<br>PCM has failed |
| **P0120**<br>1T CCM<br>2001-03<br>PT Cruiser | **Throttle Position Sensor Circuit Conditions:**<br>Engine started; and the TCM detected an unexpected change in the throttle angle, or that the throttle angle went out-of-range abruptly.<br>**Possible Causes:**<br>TP Sensor signal circuit is open or shorted to ground<br>TP Sensor ground circuit is open<br>TP Sensor signal circuit to TCM is open or shorted to ground<br>TP Sensor is damaged or it has failed<br>PCM has failed |
| **P0121**<br>1T CCM<br>1995-2003<br>300M, Avenger, Breeze, Caravan, Cherokee, Cirrus, Concorde, PT Cruiser, Dakota, Durango, Intrepid, LHS, Neon, New Yorker, Pacifica, Sebring, Stratus, Town & Country, Viper, Vision, Voyager, Wrangler | **TP Sensor Does Not Agree With MAP Conditions:**<br>DTC P0107, P0108, P0122 and P0123 not set, engine running with throttle plate closed (at high vacuum), and the PCM detected the TP Sensor input was high when it should have been low (High Input Test), or with the engine running and the VSS input over 25 mph, throttle plate open (at low vacuum), the PCM detected the TP Sensor input was low when it should have been high, conditions met for 4 seconds.<br>**Possible Causes:**<br>MAP Sensor and/or TP Sensor VREF (5v) not present<br>MAP or TP Sensor ground circuit has high resistance<br>MAP or TP Sensor signal circuit has high resistance<br>TP Sensor is damaged or has failed<br>PCM has failed |

**OBD II Trouble Code List (P0XXX Codes) - Continued**

| DTC | Trouble Code Title, Conditions & Possible Causes: |
|---|---|
| **P0121**<br>1T CCM<br>2004-05<br>300M, Durango,<br>Grand Cherokee,<br>Liberty, Magnum,<br>Ram | **TP Sensor No. 1 Does Not Agree With MAP Conditions:**<br>Ignition is on and no MAP Sensor DTCs are set. The PCM determines the TP Sensor signals do not correlate with the MAP Sensor signal. The ECT light will illuminate. DTC P2135 should also set with this DTC.<br>**Possible Causes:**<br>TP Sensor No. 1 signal circuit shorted to battery voltage<br>Resistance in either TP Sensor No. 1 or 2 signal circuit, 5v supply circuit or TP Sensor return circuit<br>TP Sensor No. 1 signal circuit shorted to group or to TP Sensor No. 2 signal circuit<br>5v supply circuit shorted to ground<br>TP Sensor or throttle body damaged or failed<br>PCM has failed |
| **P0121**<br>1T CCM<br>2004-05<br>Viper | **TP Sensor No. 1 Voltage Does Not Agree With MAP Conditions:**<br>Engine running; no MAP Sensor or TP Sensor DTCs are present; engine speed over 1600 rpm. The PCM performs 2 separate tests. When the manifold vacuum is high, the TP Sensor should be low. When the manifold vacuum in lower, the TP Sensor signal should be high. If the proper TP Sensor voltage is not detected when the 2 condition are met, a DTC will set after 4 seconds.<br>**Possible Causes:**<br>High resistance in 5v supply circuit<br>5v supply circuit shorted to ground<br>MAP Sensor has failed<br>Resistance in MAP Sensor signal circuit, including to ground<br>TP Sensor No. 1 has failed<br>High resistance in TP Sensor 5v supply circuit<br>TP Sensor No. 1 5v supply circuit shorted to ground<br>High resistance in TP Sensor No. 1 signal circuit, including to ground<br>Resistance in Sensor ground circuit<br>PCM has failed |
| **P0121**<br>1T CCM<br>1998-2003<br>Ram with 5.9L<br>Diesel | **Accelerator Position Sensor Signal Does Not Agree With IVS:**<br>Engine running at idle speed (throttle indicating less than 15% open) for 1 second or running at off-idle with the VSS input indicating over 10 mph for 1 second, and the PCM detected the IVS indicated idle when the APPS signal indicated off-idle, or the IVS indicated off-idle when the APPS indicated the engine was at idle position.<br>**Possible Causes:**<br>APP Sensor signal circuit open<br>APP Sensor signal circuit shorted to voltage (IVS 2 circuit)<br>APP Sensor short to vehicle power<br>APP Sensor is damaged or it has failed<br>PCM has failed |
| **P0122**<br>1T CCM<br>1995-98<br>Talon | **TP Sensor Circuit Low Input Conditions:**<br>Key on, and the PCM detected the TP Sensor was less than 0.20v, or with engine speed over 1500 rpm, VSS over 20 mph, engine vacuum below 2" Hg, it was less than 0.5v for 704 ms.<br>**Possible Causes:**<br>Sensor 5-volt supply circuit open or shorted to ground<br>Sensor signal circuit is shorted to ground<br>TP Sensor is damaged or it has failed<br>PCM has failed |

## OBD II Trouble Code List (P0XXX Codes) – Continued

| DTC | Trouble Code Title, Conditions & Possible Causes: |
|---|---|
| **P0122**<br>1T CCM<br>1995-2003<br>300M, Avenger, Breeze, Caravan, Cherokee, Cirrus, Concorde, PT Cruiser, Dakota, Durango, Intrepid, LHS, Neon, New Yorker, Pacifica, Sebring, Stratus, Town & Country, Viper, Vision, Voyager, Wrangler | **TP Sensor Circuit Low Input Conditions:**<br>Key on or engine running; system voltage over 10.5v and the PCM detected the TP Sensor indicated less than 0.0978v for 700 ms.<br>**Possible Causes:**<br>TP Sensor VREF (5v) circuit is open or shorted to ground<br>TP Sensor 5v supply circuit is shorted to ground<br>TP Sensor signal circuit is shorted to ground<br>TP Sensor is damaged or it has failed<br>Possible intermittent condition<br>PCM has failed |
| **P0122**<br>1T CCM<br>1995-2003<br>Grand Cherokee, Liberty, Ram | **TP Sensor Circuit Low Input Conditions:**<br>Key on or engine running; system voltage over 10.5v and the PCM detected the TP Sensor was less than 0.10v, condition met for 1.3 seconds.<br>**Possible Causes:**<br>Sensor 5-volt supply circuit open or shorted to ground<br>Sensor signal circuit is shorted to ground<br>TP Sensor is damaged or it has failed<br>PCM has failed |
| **P0122**<br>1T CCM<br>2001-03<br>Ram with 5.9L Diesel | **Accelerator Position Sensor Circuit Low Conditions:**<br>Engine started; and the PCM detected the APPS signal was to low, or the APPS signal between the ECM and the Powertrain Control Module (PCM) was too low at any time.<br>**Possible Causes:**<br>APP Sensor 5-volt supply circuit open or shorted to ground<br>APP Sensor signal shorted to ground<br>APP Sensor is damaged or it has failed<br>PCM has failed |
| **P0122**<br>1T CCM<br>2004-05<br>300M, Caravan, Concorde, PT Cruiser, Dakota, Durango, Grand Cherokee, Intrepid, LHS, Liberty, Magnum, Neon, Pacifica, Ram, Sebring, Stratus, Town & Country, Viper, Voyager | **TP Sensor No. 1 Circuit Low Input Conditions:**<br>Key on; system voltage over 10v; and the PCM detected the TP Sensor indicated less than 0.16v for 0.7 second (300M, Magnum, Liberty), 0.078v (Pacifica, PT Cruiser), or 0.0978v (Neon, Sebring, Stratus) or 0.1v (Viper). 3 good trips are required to turn off the MIL. If equipped, the ETC light will illuminate.<br>**Possible Causes:**<br>TP Sensor sweep<br>Intermittent condition<br>5v supply circuit open or shorted to ground<br>TP Sensor No. 1 signal circuit shorted to ground or to Sensor return circuit<br>TP Sensor or throttle body damaged or has failed<br>PCM has failed |
| **P0122**<br>1T CCM<br>2004-05<br>Liberty with Diesel | **TP Sensor/APP Sensor Circuit Low Input Conditions:**<br>Engine running. If the monitored APP Sensor voltage drops below 0.078v for 0.48 second, this DTC will set.<br>**Possible Causes:**<br>TP Sensor sweep<br>Intermittent condition<br>5v supply circuit open or shorted to ground<br>TP Sensor No. 1 signal circuit shorted to ground or to Sensor return circuit<br>TP Sensor or throttle body damaged or has failed<br>PCM has failed |

## OBD II Trouble Code List (P0XXX Codes) – Continued

| DTC | Trouble Code Title, Conditions & Possible Causes: |
|---|---|
| **P0122**<br>1T CCM<br>2004-05<br>Dakota, Durango, Ram | **TP Sensor No. 1 Circuit Low Input Conditions:**<br>Key on; system voltage over 10.4v. The PCM detected the TP Sensor voltage is less than 0.0978v for 1.3 seconds (3.7L, 4.7L), or less than 0.16v for 0.7 second (5.7L). 3 good trips are required to turn off the MIL. If equipped, the ETC light will illuminate.<br>**Possible Causes:**<br>5v supply circuit open or shorted to ground<br>TP Sensor No. 1 signal circuit shorted to ground or to Sensor ground circuit<br>TP Sensor has failed<br>Throttle body is damaged<br>PCM has failed |
| **P0123**<br>1T CCM<br>1995-98<br>Talon | **TP Sensor Circuit High Input Conditions:**<br>Key on or engine running; and the PCM detected the TP Sensor input was more than 4.7v (all conditions met for 1 second).<br>• Refer to "Mini-Test" or code repair chart in other media. |
| **P0123**<br>1T CCM<br>1995-2003<br>300M, Avenger, Breeze, Caravan, Cherokee, Cirrus, Concorde, PT Cruiser, Dakota, Durango, Intrepid, LHS, Neon, New Yorker, Pacifica, Sebring, Stratus, Town & Country, Viper, Vision, Voyager, Wrangler | **TP Sensor Circuit High Input Conditions:**<br>Key on or engine running; system voltage over 10.5v and the PCM detected the TP Sensor indicated more than 4.50v for 700 ms.<br>**Possible Causes:**<br>TP Sensor sweep<br>TP Sensor ground circuit is open<br>TP Sensor signal circuit is open or is shorted to VREF (5v)<br>TP Sensor is damaged or has failed (an internal open circuit)<br>Possible intermittent condition<br>PCM has failed |
| **P0123**<br>1T CCM<br>2004-05<br>300M, Caravan, Concorde, PT Cruiser, Dakota, Durango, Grand Cherokee, Intrepid, LHS, Liberty, Magnum, Neon, Pacifica, Ram, Sebring, Stratus, Town & Country, Viper, Voyager | **TP Sensor/APPS Circuit High Input Conditions:**<br>Key on or engine running; system voltage over 10v. The PCM detected the TP Sensor indicated more than 4.47v (Neon, Sebring, Stratus, Viper, Liberty) or 4.94v (all other models) for 0.48 second. If equipped, ETC light will illuminate.<br>**Possible Causes:**<br>Related TP Sensor engine DTCs present<br>Intermittent wiring or connector problem<br>TP signal circuit is open or is shorted to battery voltage or to 5v supply circuit<br>TP sensor has failed<br>TP sensor ground circuit is open<br>PCM has failed |
| **P0123**<br>1T CCM<br>2001-03<br>Ram with 5.9L Diesel | **Accelerator Position Sensor Circuit High Conditions:**<br>Engine started; and the PCM detected the APPS signal was too high, or the APPS signal between the Engine Control Module (ECM) and the Powertrain Control Module (PCM) was too high at any time.<br>**Possible Causes:**<br>APP Sensor circuit is open, or Sensor ground circuit is open<br>APP Sensor signal shorted to power<br>APP Sensor has failed or the PCM has failed<br>PCM has failed |

**OBD II Trouble Code List (P0XXX Codes) – Continued**

| DTC | Trouble Code Title, Conditions & Possible Causes: |
|---|---|
| **P0123**<br>1T CCM<br>2003-04<br>Dakota, Durango, Ram | **Throttle Position Sensor No. 1 Circuit High Conditions:**<br>Ignition is on; battery voltage is more than 10.4v. The PCM detected the TP Sensor voltage is more than 4.47v for 1.3 seconds (3.7L, 4.7L) or more than 4.8v for 25ms (5.7L). If equipped, the ETC light will illuminate.<br>**Possible Causes:**<br>TP Sensor No. 1 circuit is open, or is shorted to battery voltage<br>TP Sensor No. 1 signal circuit is shorted to 5v supply circuit<br>TP Sensor ground circuit is open<br>TP Sensor has failed<br>Throttle body is damaged (5.7L)<br>Throttle plate is jammed against the maximum stop (5.7L)<br>PCM has failed |
| **P0123**<br>1T CCM<br>2004-05<br>Liberty with Diesel | **TP Sensor/APP Sensor Circuit High Input Conditions:**<br>Engine running. If the monitored APP Sensor voltage rises above 4.94v for 0.48 second, this DTC will set.<br>**Possible Causes:**<br>Speed Sensor ground circuit open or shorted to voltage<br>5v supply circuit shorted to voltage<br>APP Sensor has failed<br>TCM has failed<br>Intermittent wiring and connector problems |
| **P0124**<br>1T CCM<br>2004-05<br>Caravan, PT Cruiser, Neon, Pacifica, Sebring, Stratus, Town & Country | **TP Sensor/APPS Circuit Intermittent Conditions:**<br>Key on or engine running; system voltage over 10.5v. This DTC will set if the monitored TP Sensor angle between 6-120° and the degree change is greater than 5° within a period of less than 7.0ms.<br>**Possible Causes:**<br>Related TP Sensor engine DTCs present<br>TP Sensor has failed<br>Intermittent wiring or connector problem<br>PCM has failed |
| **P0124**<br>1T CCM<br>2004-05<br>Liberty with Diesel | **TP Sensor/APP Sensor Intermittent Circuit Conditions:**<br>Engine running. This DTC will set with a throttle angle between 6° and 120.6° with a 5° or higher change under 7ms. Related DTCs may be present.<br>**Possible Causes:**<br>Intermittent wiring and connector problems<br>APP Sensor has failed<br>TCM has failed |
| **P0125**<br>2T CCM<br>1995-98<br>Talon | **Closed Loop Temperature Not Reached Conditions:**<br>DTC P0117 and P0118 not set, engine speed from 2400-4000 rpm, IAT Sensor from 32-122°F, and the PCM detected the ECT Sensor value did not exceed 140°F after 10 minutes.<br>**Possible Causes:**<br>Check the operation of the thermostat (it may be stuck open)<br>Inspect for low coolant level or for an incorrect coolant mixture<br>ECT Sensor is damaged or it has failed. |

## OBD II Trouble Code List (P0XXX Codes) – Continued

| DTC | Trouble Code Title, Conditions & Possible Causes: |
|-----|---------------------------------------------------|
| **P0125**<br>2T CCM<br>1995-2003<br>300M, Avenger, Breeze, Caravan, Cherokee, Cirrus, Concorde, PT Cruiser, Dakota, Durango, Intrepid, LHS, Neon, New Yorker, Pacifica, Sebring, Stratus, Town & Country, Viper, Vision, Voyager, Wrangler | **Closed Loop Temperature Not Reached Conditions:**<br>DTC P0117 and P0118 not set, ECT Sensor between -20ºF and +21.2ºF at startup, engine runtime over 10 minutes, vehicle speed more than 28 mph, and the PCM detected the ECT Sensor did not reach 174ºF after 20 minutes of sustained engine operation.<br>**Possible Causes:**<br>Check the operation of the thermostat (it may be stuck open)<br>ECT Sensor is damaged or it is out-of-calibration<br>Inspect for low coolant level or for an incorrect coolant mixture |
| **P0125**<br>2T CCM<br>2004-05<br>300M, Caravan, Concorde, PT Cruiser, Dakota, Durango, Grand Cherokee, Intrepid, LHS, Liberty, Magnum, Neon, Pacifica, Ram, Sebring, Stratus, Town & Country, Viper, Voyager | **Closed Loop Temperature Not Reached Conditions:**<br>Engine running; battery over 10v. Engine temperature does not enable closed loop. Failure time depends on start-up coolant temperature and ambient temperature (i.e., 2 minutes for a start temperature of 50ºF (10°C), or up to 10 minutes for a vehicle with start-up temperature of -18°F (-28°C).<br>**Possible Causes:**<br>Low coolant level<br>Improper thermostat operation or thermostat failure<br>ECT has failed |
| **P0125**<br>2T CCM<br>1998-2003<br>Ram with 5.9L Diesel | **Engine Does Not Reach Closed Loop Conditions:**<br>DTC P0112, P0113, P0117, P0118, P1492 and P1493 not set, key on, IAT Sensor signal less than -80ºF, system voltage over 10.4v, and the PCM detected the ECT Sensor was more than 140ºF, or with the ECT Sensor between -40ºF and 150ºF at startup, or it detected the ECT Sensor did not change more than 6ºF after 10 minutes.<br>**Possible Causes:**<br>Check the operation of the thermostat (it may be stuck open)<br>ECT Sensor is damaged or it is out-of-calibration<br>Inspect for low coolant level or for an incorrect coolant mixture |
| **P0128**<br>1T CCM<br>2001-03<br>300M, Avenger, Breeze, Caravan, Cherokee, Cirrus, Concorde, PT Cruiser, Dakota, Durango, Intrepid, LHS, Neon, New Yorker, Pacifica, Sebring, Stratus, Town & Country, Viper, Vision, Voyager | **Thermostat Rationality Test Conditions:**<br>DTC P0117, P0118, P1492 and P1493 not set, ECT Sensor between 20ºF and 130ºF at startup, and the PCM detected the ECT Sensor did not exceed 170ºF after 10-32 minutes of sustained engine operation (the actual time depends on the ECT Sensor at startup).<br>**Possible Causes:**<br>Check the operation of the thermostat (it may be stuck open)<br>ECT Sensor is contaminated, damaged or it has failed<br>Inspect for low coolant level or for an incorrect coolant mixture |

## OBD II Trouble Code List (P0XXX Codes) – Continued

| DTC | Trouble Code Title, Conditions & Possible Causes: |
|---|---|
| **P0128**<br>1T CCM<br>2001-03<br>Sebring, Stratus with 2.4L or 3.0L | **Thermostat Rationality Test Conditions:**<br>ECT Sensor from 14ºF and 82ºF at startup with the ECT and IAT Sensor inputs within 48ºF and the IAT input less than 36ºF, volume airflow from 50-100 Hz for under 300 seconds, and the PCM detected the ECT Sensor took too long to reach 180ºF. The actual time vary with the ECT Sensor at startup (e.g., from 11-23 minutes to reach 180ºF if the ECT Sensor is over 68ºF at startup, or 20-54 minutes to reach 180ºF if the ECT Sensor is less than 68F at startup).<br>**Possible Causes:**<br>Check the operation of the thermostat (it may be stuck open)<br>Inspect for low coolant level or for an incorrect coolant mixture |
| **P0128**<br>2T CCM<br>2004-05<br>300M, Caravan, Concorde, PT Cruiser, Dakota, Durango, Grand Cherokee, Intrepid, LHS, Liberty, Magnum, Neon, Pacifica, Ram, Sebring, Stratus, Town & Country, Viper, Voyager | **Thermostat Rationality Test Conditions:**<br>With engine running after cold start. PCM predicts a coolant temperature value that it will compare to the actual coolant temperature. If the 2 coolant temperature values are not within 50ºF (10ºC) of each other, an error is detected.<br>**Possible Causes:**<br>Low coolant level<br>Thermostat has failed<br>Signal circuit shorted to battery voltage<br>ECT Sensor has failed or voltage is below 1.0v<br>Signal circuit is open, shorted to ground or shorted to Sensor ground<br>ECT Sensor ground circuit or signal circuit is open<br>PCM has failed |
| **P0128**<br>1T CCM<br>2004-05<br>Liberty with Diesel | **ECT Sensor: Engine Is Cold Too Long Conditions:**<br>Ignition on; With engine running and engine temperature is below 40C, this DTC will set.<br>**Possible Causes:**<br>Intermittent condition<br>Cooling system problems<br>ECT Sensor has failed<br>ECM has failed |
| **P0129**<br>1T CCM<br>2000-05<br>300M, Avenger, Breeze, Caravan, Cherokee, Cirrus, Concorde, PT Cruiser, Dakota, Durango, Grand Cherokee, Intrepid, LHS, Liberty, Magnum, Pacifica, Ram, Sebring, Stratus, Town & Country, Viper, Voyager | **Barometric Pressure Out-Of-Range Conditions:**<br>Engine cranking at less than 250 rpm; no CKP or CMP Sensor signals within 75ms. The PCM detected the MAP/BARO Sensor signal range was 0.04-2.2v for 300ms during testing. If equipped, the ETC lamp will be illuminated. 3 good trips are required to turn off the MIL.<br>**Possible Causes:**<br>IAC motor control low or control high circuit has failed<br>MAP Sensor VREF (5v) circuit is open or shorted to ground<br>MAP Sensor signal circuit is open or shorted to ground<br>MAP Sensor is damaged or it has failed<br>May be an intermittent condition<br>PCM has failed |

**OBD II Trouble Code List (P0XXX Codes) – Continued**

| DTC | Trouble Code Title, Conditions & Possible Causes: |
|---|---|
| **P0129**<br>1T CCM<br>2000-05<br>Neon | **Barometric Pressure Out-Of-Range Conditions:**<br>Ignition on; no CKP or CMP Sensor signals for 75ms; engine speed less than 250 rpm. The PCM detected the MAP Sensor signal range was greater than 4.9v but less than 2.28v (non-turbo) or was greater than 2.4v but less than 1.2v (turbo) for 300ms during testing.<br>**Possible Causes:**<br>IAC signal circuit is low or high<br>Intermittent condition<br>5v supply circuit is open or shorted to ground or to voltage<br>MAP Sensor signal circuit is open or shorted to ground<br>MAP Sensor is damaged or it has failed<br>PCM has failed |
| **P0130**<br>1T CCM<br>1995-98<br>Talon | **O2 (B1 S1) Circuit Fault Conditions:**<br>Engine running in closed loop, ECT Sensor over 176ºF, IAT Sensor signal over 14ºF, BARO signal more than 75 kPa, engine speed above 1200 rpm, volumetric efficiency over 25%, then after the PCM commanded the fuel injector pulse width lean and rich, it detected the O2 response time was too slow, or it detected the O2 signal was less than 0.10v for 3 minutes.<br>**Possible Causes:**<br>O2 signal circuit is open or shorted to ground<br>O2 ground circuit is open<br>O2 may be contaminated or it has failed<br>PCM has failed |
| **P0130**<br>1T CCM<br>2001<br>Sebring, Stratus with 2.4L or 3.0L | **O2 (B1 S1) Circuit Fault Conditions:**<br>Engine runtime over 3 minutes, engine speed more than 1200 rpm, volumetric efficiency more than 25%, ECT Sensor more than 180ºF, and the PCM detected the O2 signal was less than 0.2v during the 7 second test period, and the PCM detected the O2 signal remained at more than 4.5v with 5v applied to the circuit during the O2 Test period.<br>**Possible Causes:**<br>O2 signal circuit is open or shorted to ground<br>O2 ground circuit is open<br>O2 may be contaminated or it has failed<br>PCM has failed |
| **P0130**<br>1T CCM<br>1999-2000<br>Cherokee, Dakota, Durango, Grand Cherokee, Ram, Wrangler with 4.7L | **O2 (B1 S1) Circuit Fault Conditions:**<br>Key on or engine running; system voltage over 10.5v and the PCM detected the state of the O2 relay coil circuit (between the PCM and the relay) did not match the expected state.<br>**Possible Causes:**<br>Fused ignition feed (power) circuit open<br>Heater relay control circuit open<br>Heater relay has failed (an internal winding is open)<br>PCM has failed |
| **P0131**<br>1T CCM<br>1995-98<br>Talon | **O2 (B1 S1) Short to Ground Conditions:**<br>ECT Sensor No. 170ºF or more at previous key on, ECT Sensor below 90ºF and BTS signal within 27ºF of the ECT, and the PCM detected the O2 input was below 156 mV for 28 seconds.<br>**Possible Causes:**<br>O2 signal circuit is shorted to chassis or Sensor ground<br>O2 signal circuit is open (circuit reads 450 mV when open)<br>O2 ground circuit is open (circuit reads 170 mV when open)<br>O2 may be contaminated or it has failed<br>PCM has failed |

**OBD II Trouble Code List (P0XXX Codes) – Continued**

| DTC | Trouble Code Title, Conditions & Possible Causes: |
|---|---|
| **P0131**<br>1T CCM<br>1995-2003<br>300M, Avenger, Breeze, Caravan, Cherokee, Cirrus, Concorde, PT Cruiser, Dakota, Durango, Intrepid, LHS, Neon, New Yorker, Pacifica, Sebring, Stratus, Town & Country, Viper, Vision, Voyager, Wrangler | **O2 (B1 S1) Short to Ground Conditions:**<br>ECT Sensor over 170°F on previous key on, ECT Sensor under 98°F and BTS within ±27°F of ECT Sensor, and the PCM detected the O2 signal was below 156 mV for 28 seconds.<br>**Possible Causes:**<br>O2 signal circuit is shorted to chassis or Sensor ground<br>O2 ground circuit is open (circuit reads 170 mV when open)<br>O2 may be contaminated or it has failed<br>PCM has failed |
| **P0131**<br>2T CCM<br>2004-05<br>300M, Caravan, Concorde, PT Cruiser, Dakota, Durango, Grand Cherokee, Intrepid, LHS, Liberty, Magnum, Pacifica, Ram, Sebring, Stratus, Town & Country, Viper | **O2 (B1 S1) Circuit Short to Ground Conditions:**<br>Engine running; cold start. O2 Sensor signal voltage is below 2.402v for 9 seconds (exc. Durango, Grand Cherokee, Liberty, Ram, Viper), or below 2.52v for 6 seconds (Durango, Grand Cherokee, Liberty, Ram), or below 0.156v for 28 seconds (Viper).<br>**Possible Causes:**<br>O2 signal circuit is shorted to chassis or Sensor ground<br>O2 return circuit is shorted to ground or to signal circuit<br>O2 signal circuit is shorted to O2 return upstream circuit<br>O2 may be contaminated or it has failed<br>PCM has failed |
| **P0131**<br>1T CCM<br>1996-2003<br>Cherokee, Grand Cherokee, Liberty, Ram, Wrangler | **O2 (B1 S1) Circuit Short to Ground Conditions:**<br>Engine running with ECT Sensor over 146°F, followed by an engine off period in which the O2 heater Test ran and passed, then after an engine cool-down period, and a cold engine startup (ECT Sensor under 100°F and the BTS signal within 44°F of the ECT Sensor), the PCM detected the O2 signal was less than 78 mV for 5 seconds.<br>**Possible Causes:**<br>O2 signal circuit is shorted to chassis or Sensor ground<br>O2 ground circuit is open (circuit reads 170 mV when open)<br>O2 may be contaminated or it has failed<br>PCM has failed |
| **P0132**<br>1T CCM<br>1995-2003<br>300M, Avenger, Breeze, Cirrus, Concorde, PT Cruiser, Intrepid, LHS, Neon, New Yorker, Pacifica, Sebring, Stratus, Viper, Vision | **O2 (B1 S1) Circuit Short to Voltage Conditions:**<br>Engine started; engine runtime over 119 seconds, system voltage over 10.5v, ECT Sensor more than 176°F, and the PCM detected the O2 signal was more than 1.29v for 3 seconds.<br>**Possible Causes:**<br>O2 signal circuit shorted to heater B+ circuit (inspect the connector for oil or moisture inside the terminal area)<br>O2 signal circuit is open<br>PCM has failed |

## OBD II Trouble Code List (P0XXX Codes) – Continued

| DTC | Trouble Code Title, Conditions & Possible Causes: |
|---|---|
| **P0132**<br>1T CCM<br>1996-2003<br>Caravan, Cherokee, Dakota, Durango, Grand Cherokee, Ram, Town & Country, Voyager, Wrangler | **O2 (B1 S1) Circuit Short to Voltage Conditions:**<br>Engine started; engine runtime over 119 seconds, system voltage over 10.99v, O2 heater temperature over 1085ºF and the PCM detected the O2 input was over 3.70v for 1 minute.<br>**Possible Causes:**<br>O2 signal tracking (wet/oily) in connector causing a short between the signal and heater power circuits<br>O2 ground circuit is open<br>O2 signal circuit is open<br>PCM has failed |
| **P0132**<br>2T CCM<br>2004-05<br>Caravan, Concorde, PT Cruiser, Dakota, Durango, Grand Cherokee, Intrepid, LHS, Liberty, Pacifica, Sebring, Stratus, Town & Country | **O2 Sensor (B1 S1) Voltage High**<br>Engine started; engine runtime over 119 seconds. Battery voltage over 10.99v. O2 Sensor heater temperature is more than 662°F (350°C). O2 Sensor voltage is more than 3.7v for 60 seconds (Pacifica), more than 3.99v for 60 seconds (Neon, Sebring, Stratus, PT Cruiser), or more than 3.9902v for 30 seconds (Caravan/Town & Country), or more than 3.99v for 66.56 seconds (Liberty).<br>**Possible Causes:**<br>O2 Sensor signal circuit and/or return circuit shorted to voltage<br>O2 Sensor has failed<br>O2 Sensor signal or return circuit open<br>PCM has failed |
| **P0132**<br>1T CCM<br>2004-05<br>300M, Magnum, Ram | **O2 Sensor (B1 S1) Voltage High**<br>Engine started; battery voltage above 10.4v; O2 Sensor heater temperature is more than 925°F (496°C). O2 Sensor is more than 3.7v for 40 seconds. 3 good trips are required to turn the MIL off.<br>**Possible Causes:**<br>O2 Sensor signal circuit and/or return circuit shorted to voltage<br>O2 Sensor has failed<br>O2 Sensor signal or return circuit open<br>PCM has failed |
| **P0132**<br>1T CCM<br>2004-05<br>Viper | **O2 Sensor (B1 S1) Voltage High**<br>Engine running for more than 4 minute; battery voltage above 10.4v; ECT Sensor above 180°F. O2 Sensor voltage is more than 1.5v.<br>**Possible Causes:**<br>O2 Sensor has failed<br>O2 Sensor signal circuit is open or is shorted to voltage<br>O2 Sensor signal circuit is shorted to heater supply circuit<br>O2 Sensor ground circuit is open<br>O2 Sensor heater control circuit is open |
| **P0133**<br>2T O2<br>1995-98<br>Talon | **O2 (B1 S1) Slow Response Conditions:**<br>Engine started; engine runtime over 3 minutes, ECT Sensor more than 170ºF, vehicle driven to a speed over 24 mph at an engine speed of 518-864 rpm (near idle speed), A/C and PSP switches indicating off, the PCM detected the O2 signal did not cross 670 mV or the O2 switch rate was too low during the O2 Test.<br>**Possible Causes:**<br>Exhaust leak present in the exhaust manifold or exhaust pipes<br>O2 element is fuel contaminated<br>O2 element is deteriorated or it has failed |

## OBD II Trouble Code List (P0XXX Codes) – Continued

| DTC | Trouble Code Title, Conditions & Possible Causes: |
|---|---|
| **P0133**<br>2T O2<br>1996-2003<br>300M, Caravan, Cherokee, Concorde, Grand Cherokee, Intrepid, LHS, New Yorker, Ram, Town & Country, Vision, Voyager, Wrangler | **O2 (B1 S1) Slow Response Conditions:**<br>Engine runtime over 2 minutes, ECT Sensor more than 147°F, VSS over 10 mph, A/C and PSPS indicating off, then at idle speed in Drive (A/T) or Neutral (M/T), and the PCM detected the O2 signal did not switch enough times from 270-620 mV in the test period.<br>**Possible Causes:**<br>Exhaust leak present in the exhaust manifold or exhaust pipes<br>O2 element is fuel contaminated<br>O2 element is deteriorated or it has failed |
| **P0133**<br>2T O2<br>1995-2003<br>Avenger, Breeze, Cirrus, PT Cruiser, Dakota, Durango, Neon, Prowler, Sebring, Stratus, Viper | **O2 (B1 S1) Slow Response Conditions:**<br>Engine at idle speed, system voltage over 10.5v, then engine speed from 1216-1984 rpm, VSS input from 19-46 mph, BTS signal more than 44°F, BARO signal more than 22.16" Hg, MAP Sensor signal from 11.9-18.15" Hg, then back to idle speed in Drive, and the PCM detected the O2 signal did not switch enough times from below 330 mV to above 610 mV, condition met for 60 seconds.<br>**Possible Causes:**<br>Exhaust leak present in the exhaust manifold or exhaust pipes<br>O2 element is fuel contaminated<br>O2 element is deteriorated or it has failed |
| **P0133**<br>2T O2<br>2004-05<br>300M, Caravan, Concorde, PT Cruiser, Intrepid, LHS, Magnum, Pacifica, Sebring, Stratus, Town & Country | **O2 (B1 S1) Slow Response Conditions:**<br>Vehicle driven at speeds between 20-55 mph, with throttle open for minimum of 120 seconds; coolant temperature is greater than 158°F (70°C); catalytic converter temperature is greater than 1112°F (600°C). PCM compared differences (state of change) between front and rear O2 Sensors indicate difference is greater than calibrated amount.<br>**Possible Causes:**<br>Exhaust leak present in the exhaust manifold or exhaust pipes<br>O2 signal circuit or return circuit has failed<br>O2 element has failed |
| **P0133**<br>1T CCM<br>2004-05<br>Viper | **O2 Sensor (B1 S1) Slow Response Conditions:**<br>With ECT greater than 147F, after reaching a vehicle speed of 10 mph, and throttle remaining open (off idle) for 2 minutes, bring the vehicle to a stop and all to idle with transmission in Drive. If the O2 sensor signal voltage is switching from less than 0.27v to more than 0.62v and back fewer times than required, this DTC will set.<br>**Possible Causes:**<br>Exhaust leak present in the exhaust manifold or exhaust pipes<br>O2 Sensor signal circuit has resistance<br>O2 Sensor ground circuit has resistance<br>O2 Sensor has failed |
| **P0133**<br>2T O2<br>2004-05<br>Dakota, Durango, Grand Cherokee, Liberty, Ram | **O2 (B1 S1) Slow Response Conditions:**<br>Vehicle driven at speeds between 20-55 mph, with throttle open for minimum of 120 seconds; coolant temperature is greater than 158°F (70°C); catalytic converter temperature is greater than 1112°F (600°C); EVAP purge is active. The PCM detects the O2 Sensor signal voltage switches less than 16 times from lean to rich with 20 seconds during monitoring. 3 good trips are required to turn off MIL.<br>**Possible Causes:**<br>Exhaust leak present in the exhaust manifold or exhaust pipes<br>O2 signal circuit or upstream circuit has failed<br>O2 element has failed |

### OBD II Trouble Code List (P0XXX Codes) - Continued

| DTC | Trouble Code Title, Conditions & Possible Causes: |
|---|---|
| **P0134**<br>2T O2<br>1995-98<br>Talon | **O2 (B1 S1) Remains At Center Conditions:**<br>Engine runtime over 2 minutes, system voltage over 10.5v, ECT Sensor more than 170°F, and the PCM detected the O2 signal remained fixed between 350-550 mV for 1.5 minutes.<br>**Possible Causes:**<br>Exhaust leak present in exhaust manifold or exhaust pipes<br>O2 element is fuel contaminated or it is deteriorated<br>PCM has failed |
| **P0134**<br>2T O2<br>1995-2003<br>300M, Avenger, Breeze, Cirrus, Concorde, PT Cruiser, Intrepid, LHS, Neon, New Yorker, Pacifica, Prowler, Sebring, Stratus, Viper, Vision | **O2 (B1 S1) Remains At Center Conditions:**<br>Engine runtime more than 121 seconds, system voltage over 10.5v, ECT Sensor more than 150.8°F, fuel control system in closed loop mode, and the PCM detected the O2 signal remained fixed in a range between 350-580 mV, condition met for 60 seconds.<br>**Possible Causes:**<br>Exhaust leak present in exhaust manifold or exhaust pipes<br>O2 element is fuel contaminated or has deteriorated<br>O2S signal circuit or ground circuit has high resistance<br>PCM has failed |
| **P0134**<br>2T O2<br>1996-2003<br>Caravan, Cherokee, Dakota, Durango, Grand Cherokee, Liberty, Town & Country, Voyager, Wrangler | **O2 (B1 S1) Remains At Center Conditions:**<br>Engine runtime more than 121 seconds, system voltage over 10.5v, ECT Sensor more than 150.8°F, fuel control system in closed loop mode, and the PCM detected the O2 signal remained fixed in a range between 350-580 mV, condition met for 60 seconds.<br>**Possible Causes:**<br>Exhaust leak present in exhaust manifold or exhaust pipes<br>O2 element is fuel contaminated or has deteriorated<br>O2S signal circuit or ground circuit has high resistance<br>PCM has failed |
| **P0135**<br>1T CCM<br>1995-98<br>Talon | **O2 (B1 S1) Heater Circuit Conditions:**<br>ECT Sensor less than 147°F, BTS signal within ±27°F of ECT Sensor, system voltage over 10.5v, running at idle for 12 seconds, and the PCM detected the O2 signal was more than 3.0v for 30-90 seconds.<br>**Possible Causes:**<br>ASD relay output (power) circuit to the heater is open<br>O2 heater ground circuit is open or has high resistance<br>O2 heater element is damaged or it has failed<br>PCM has failed |
| **P0135**<br>1T CCM<br>2001-03<br>Sebring, Stratus with 2.4L or 3.0L | **O2 (B1 S1) Heater Circuit Conditions:**<br>Engine started; system voltage from 11-16v, ECT Sensor over 68°F, and the PCM detected the O2 heater current was less than 0.16 amps or more than 7.5 amps for 4 seconds.<br>**Possible Causes:**<br>O2 heater ground circuit open or O2 signal circuit is open<br>O2 heater element has high resistance<br>O2 heater element has failed (open or shorted)<br>MFI relay output (power) circuit to the heater is open<br>PCM has failed |

**OBD II Trouble Code List (P0XXX Codes) - Continued**

| DTC | Trouble Code Title, Conditions & Possible Causes: |
|---|---|
| **P0135**<br>1T CCM<br>1995-2003<br>300M, Avenger, Breeze, Cirrus, Concorde, PT Cruiser, Intrepid, LHS, Neon, New Yorker, Pacifica, Prowler, Sebring, Stratus, Viper, Vision | **O2 (B1 S1) Heater Circuit Conditions:**<br>ECT Sensor less than 147°F and BTS signal within 27°F of the ECT Sensor (cold engine), engine started, engine running at idle speed for over 12 seconds, system voltage over 10.5v and the PCM detected the O2 signal was more than 3.0v for 30-90 seconds.<br>**Possible Causes:**<br>ASD relay output (power) circuit to the heater is open<br>O2 heater ground circuit is open<br>O2 heater element is damaged or has high resistance<br>PCM has failed |
| **P0135**<br>1T CCM<br>1996-2003<br>Caravan, Cherokee, Dakota, Durango, Grand Cherokee, Liberty, Town & Country, Voyager, Wrangler | **O2 (B1 S1) Heater Circuit Conditions:**<br>Key off after a warm engine drive cycle, engine cool-down finished (at least 5 seconds after the key is turned off), system voltage over 10.5v, then with the ASD relay energized, the PCM detected the O2 signal rose to 0.49v or more within a 144 second period, and the initial rise of the oxygen Sensor signal was less than 1.57v.<br>**Possible Causes:**<br>O2 heater ground circuit open or O2 signal circuit is open<br>O2 heater element has high resistance<br>O2 heater element has failed (open or shorted)<br>PCM has failed |
| **P0135**<br>2T CCM<br>2004-05<br>300M, Caravan, Concorde, PT Cruiser, Intrepid, LHS, Liberty, Magnum, Pacifica, Sebring, Stratus, Town & Country | **O2 (B1 S1) Heater Circuit Conditions:**<br>Engine running and O2 heater duty cycle is greater than 0%. O2 heater temperature does not reach 959°F (575°C), exc. PT Cruiser, or 662°F (350°C), Neon and PT Cruiser, within 90 seconds (Sebring, Stratus, 45 seconds) during monitoring conditions. No Sensor output is received when the PCM powers up the Sensor heater. 3 good trips are required to turn off the MIL.<br>**Possible Causes:**<br>O2 heater ground circuit open or O2 signal circuit is open<br>O2 heater element has failed<br>PCM has failed |
| **P0135**<br>2T CCM<br>2004-05<br>Grand Cherokee, Liberty, Wrangler | **O2 (B1 S1) Heater Circuit Conditions:**<br>Engine running; O2 heater duty cycle is greater than 0%; ASD relay is energized; battery voltage is greater than 10.4v. No sensor output is received when the PCM powers up the sensor heater. The O2 heater is out of control for 128 seconds after it has reached 572°F (300°C).<br>**Possible Causes:**<br>O2 heater ground circuit open or O2 signal circuit is open<br>O2 heater element has failed<br>PCM has failed |
| **P0135**<br>1T CCM<br>2004-05<br>Viper | **O2 Sensor (B1 S1) Heater Failure Conditions:**<br>At cold start, battery voltage greater than 9v; ECT less than 147°F; Battery Temp. Sensor equal to of less than 27°F; engine at idle for at least 12 seconds. If O2 Sensor voltage is greater than 3v for 30-90 seconds, this DTC will set.<br>**Possible Causes:**<br>O2 Sensor heater operation has failed<br>O2 Sensor heater element has failed<br>02 Sensor ground circuit or control circuit is open |

**OBD II Trouble Code List (P0XXX Codes) - Continued**

| DTC | Trouble Code Title, Conditions & Possible Causes: |
|---|---|
| **P0135**<br>2T CCM<br>2004-05<br>Dakota, Durango,<br>Ram | **O2 (B1 S1) Heater Circuit Conditions:**<br>Engine running and O2 heater duty cycle is greater than 0%; battery voltage is more than 11v. No O2 sensor output signal is received when the PCM powers up the sensor heater. 3 good trips are required to turn off the MIL.<br>**Possible Causes:**<br>O2 heater ground circuit open or heater control circuit is open<br>O2 heater element has failed<br>PCM has failed |
| **P0136**<br>1T CCM<br>1995-98<br>Talon | **O2 (B1 S2) Circuit Conditions:**<br>Engine started; engine speed above 1200 rpm, ECT Sensor over 176°F, IAT Sensor over 14°F, BARO signal over 75 kPa, volumetric efficiency over 25%, Fuel Injector pulse width commanded lean and rich, and the PCM detected the O2 response time was too slow, or it detected the O2 signal was less than 0.10v for 3 minutes.<br>**Possible Causes:**<br>O2 signal circuit is open or shorted to ground<br>O2 ground circuit is open<br>O2 may be contaminated or it has failed<br>PCM has failed |
| **P0136**<br>1T CCM<br>2000-03<br>Cherokee, Dakota,<br>Durango, Grand<br>Cherokee, Liberty,<br>Ram, Wrangler | **O2 (B1 S2) Heater Circuit Low Input Conditions**<br>Key on, cold engine conditions, system voltage over 10.5v, ASD relay on, and the PCM detected the Actual state of the Heater relay circuit did not match the Intended state (i.e., the circuit status remained in a low state).<br>**Possible Causes:**<br>Fused ignition feed (power) circuit to the relay is open<br>Heater relay control circuit is open<br>Heater relay or PCM has failed |
| **P0136**<br>1T CCM<br>2004-05<br>Viper | **O2 Sensor (B1 S2) Heater Circuit Malfunction Conditions:**<br>Ignition on; battery voltage greater than 10.4v. If the state of the PCM relay control circuit, between the PCM and relay coil does not match the desired state, this DTC will set.<br>**Possible Causes:**<br>O2 Sensor heater relay has failed<br>Fused ASD relay output circuit problem<br>O2 heater relay control circuit is open or is shorted to ground<br>PCM has failed |
| **P0137**<br>1T CCM<br>1995-2003<br>300M, Avenger,<br>Breeze, Cirrus,<br>Concorde, PT<br>Cruiser, Intrepid, LHS,<br>Neon, New Yorker,<br>Pacifica, Prowler,<br>Sebring, Stratus,<br>Viper, Vision | **O2 (B1 S2) Circuit Short to Ground Conditions:**<br>Engine started; engine runtime over 119 seconds, system voltage over 10.99v, O2 heater temperature over 1085°F and the PCM detected the O2 input was over 3.70v for 1 minute.<br>**Possible Causes:**<br>O2 signal tracking in connector due to wet/oil conditions<br>O2 signal circuit is open, or the ground circuit is open<br>O2 heater supply circuit is open<br>PCM has failed |
| **P0137**<br>1T CCM<br>1996-2003<br>Cherokee, Dakota,<br>Durango, Grand<br>Cherokee, Liberty,<br>Ram, Wrangler | **O2 (B1 S2) Circuit Short to Ground Conditions:**<br>ECT Sensor over 170°F at previous key off, O2 heater Test completed, engine cool down period finished, engine started, ECT Sensor less than 98°F and BTS signal within 27°F of the ECT Sensor, and PCM detected the O2 signal was below 156 mV for 28 seconds.<br>**Possible Causes:**<br>O2 signal circuit is shorted to chassis or Sensor ground<br>O2 ground circuit is open (circuit reads 170 mV when open)<br>O2 may be contaminated or it has failed<br>PCM has failed |

**OBD II Trouble Code List (P0XXX Codes) - Continued**

| DTC | Trouble Code Title, Conditions & Possible Causes: |
|---|---|
| **P0137**<br>1T CCM<br>2004-05<br>300M, Caravan, Concorde, PT Cruiser, Dakota, Durango, Grand Cherokee, Intrepid, LHS, Liberty, Magnum, Pacifica, Ram, Sebring, Stratus, Town & Country, Grand Cherokee, Viper, Voyager, Wrangler | **O2 (B1 S2) Sensor Circuit Low Conditions:**<br>Engine running; battery voltage over 10.9v; O2 heater temperature below 484°F (251°C) or ECT above 170°F from previous key off. The PCM detected the O2 Sensor signal voltage was less than 1.5v for 3 seconds (300M, Magnum), less than 2.402v for 9 seconds (Neon, Sebring, Stratus, PT Cruiser, Caravan, Town & Country, Liberty), or 2.5194v for 3 seconds (Durango, Ram), or less than 0.156v for 28 seconds (Viper).<br>**Possible Causes:**<br>O2 return circuit is shorted to ground<br>O2 signal circuit is shorted to ground, or to O2 return circuit, or O2 heater ground circuit<br>O2 Sensor has failed<br>PCM has failed |
| **P0138**<br>1T CCM<br>1995-2003<br>300M, Avenger, Breeze, Caravan, Cirrus, Concorde, PT Cruiser, Intrepid, LHS, Neon, New Yorker, Pacifica, Sebring, Stratus, Town & Country, Viper, Vision, Voyager | **O2 (B1 S2) Circuit Short to Voltage Conditions:**<br>Engine runtime over 2 minutes, system voltage over 10.5v, ECT Sensor more than 170°F, and the PCM detected the O2 signal was more than 1.20v for over 3 seconds.<br>**Possible Causes:**<br>O2 signal tracking (wet/oily) causing a short to heater power<br>O2 signal circuit open, or ground circuit open<br>PCM has failed |
| **P0138**<br>1T CCM<br>1996-2003<br>Cherokee, Dakota, Durango, Grand Cherokee, Liberty, Ram, Wrangler | **O2 (B1 S2) Circuit Short to Voltage Conditions:**<br>Engine started; engine runtime more than 4 minutes, system voltage over 10.5v, ECT Sensor more than 180°F, and the PCM detected the O2 signal was more than 1.50v for 3 seconds.<br>**Possible Causes:**<br>O2 signal tracking (wet/oily) in connector due to short from signal to power circuit<br>O2 signal circuit is open<br>O2 ground circuit is open<br>O2 is damaged or it has failed<br>PCM has failed |
| **P0138**<br>2T CCM<br>2004-05<br>300M, Caravan, Concorde, PT Cruiser, Dakota, Durango, Grand Cherokee, Intrepid, LHS, Liberty, Magnum, Pacifica, Ram, Sebring, Stratus, Town & Country, Wrangler | **O2 (B1 S2) Sensor Voltage High Condition:**<br>Engine runtime for 119 seconds; O2 Sensor heater temperature is more than 662°F (350°C); Battery voltage more than 10.99v. O2 Sensor voltage is above 3.7v for 60 seconds (300M, Magnum, Pacifica), or 3.99v for 60 seconds (PT Cruiser, Sebring, Stratus), above 3.9902v for 30 seconds (Neon, Caravan, Dakota, Durango, Grand Cherokee, Ram, Town & Country, Voyager, Wrangler), or above 3.99v for 76.8 seconds (Liberty).<br>**Possible Causes:**<br>O2 Sensor signal circuit or return circuit shorted to voltage<br>O2 Sensor has failed<br>O2 Sensor signal circuit or return circuit open<br>PCM has failed |

**OBD II Trouble Code List (P0XXX Codes) - Continued**

| DTC | Trouble Code Title, Conditions & Possible Causes: |
|---|---|
| **P0138**<br>1T CCM<br>2004-05<br>Viper | **O2 Sensor (B1 S2) Voltage High**<br>Engine running for more than 4 minute; battery voltage above 10.4v; ECT Sensor above 180°F. O2 Sensor voltage is more than 1.5v.<br>**Possible Causes:**<br>O2 Sensor has failed<br>O2 Sensor signal circuit is open or is shorted to voltage<br>O2 Sensor signal circuit is shorted to heater supply circuit<br>O2 Sensor ground circuit is open<br>O2 Sensor heater control circuit is open |
| **P0139**<br>1T CCM<br>1996-2003<br>300M, Avenger, Breeze, Caravan, Cherokee, Cirrus, Concorde, PT Cruiser, Dakota, Durango, Intrepid, Grand Cherokee, LHS, Liberty, Neon, New Yorker, Pacifica, Ram, Sebring, Stratus, Town & Country, Viper, Vision, Voyager, Wrangler | **O2 (B1 S2) Slow Response Conditions:**<br>Engine started; vehicle driven at 20-55 mph with the throttle open for 2 minutes, ECT Sensor more than 158°F, Converter temperature over 1112°F, EVAP purge "on", and the PCM detected the O2 signal switched from rich-to-lean less than 11 times in 20 seconds.<br>**Possible Causes:**<br>Exhaust leak present in the exhaust manifold or exhaust pipes<br>O2 element is contaminated, deteriorated or it has failed<br>O2 ground or O2 signal circuit has high resistance |
| **P0139**<br>2T O2<br>2002-03<br>300M, Concorde, Grand Cherokee, Intrepid, LHS, Liberty, Wrangler | **O2 (B1 S2) Slow Response Conditions:**<br>Engine started; vehicle driven at 20-55 mph with the throttle open for 2 minutes, ECT Sensor more than 158°F (70°C), Converter Temperature over 1112°F, EVAP purge on, and the PCM detected the O2 signal switched from rich-to-lean less than 11 times in 20 seconds.<br>**Possible Causes:**<br>Exhaust leak present in the exhaust manifold or exhaust pipes<br>O2 element is contaminated, deteriorated or it has failed<br>O2 ground or O2 signal circuit has high resistance |
| **P0139**<br>2T O2<br>2002-03<br>Sebring, Stratus with 2.4L or 3.0L | **O2 (B1 S2) Slow Response Conditions:**<br>Engine started; ECT Sensor more than 169°F, front O2 active, volumetric airflow Sensor more than 4000 Hz, then vehicle speed over 18.7 mph at over 1500 rpm with the volumetric efficiency over 40%, then vehicle speed below 0.9 mph, Fuel Shutoff active, and the PCM detected the O2 signal was less than 0.78v for 38 seconds.<br>**Possible Causes:**<br>Exhaust leak present in the exhaust manifold or exhaust pipes<br>O2 element is fuel contaminated or it has failed<br>O2 ground or O2 signal circuit has high resistance |
| **P0139**<br>2T O2<br>2004-05<br>Pacifica | **O2 (B1 S2) Slow Response Conditions:**<br>Engine started; vehicle driven at 20-60 mph with the throttle open for 2 minutes; ECT Sensor more than 158°F (70°C); engine is between 1200-2000 rpm; vacuum is between 28-56 kPa. PCM-compared differences (state of change) between front and rear O2 Sensors indicate difference is greater than calibrated amount.<br>**Possible Causes:**<br>Exhaust leak present in the exhaust manifold or exhaust pipes<br>O2 element is contaminated, deteriorated or it has failed<br>O2 signal circuit or return circuit has failed |

**OBD II Trouble Code List (P0XXX Codes) - Continued**

| DTC | Trouble Code Title, Conditions & Possible Causes: |
|---|---|
| **P0139**<br>2T O2<br>2004-05<br>300M, Caravan, Concorde, PT Cruiser, Dakota, Durango, Grand Cherokee, Intrepid, LHS, Liberty, Magnum, Neon, Pacifica, Prowler Ram, Sebring, Stratus, Town & Country, Wrangler | **O2 (B1 S2) Slow Response Conditions:**<br>Engine started; vehicle driven at 20-55 mph with the throttle open for 2 minutes; ECT at more than 158°F (70°C); catalytic converter temperature is more than 1112°F (600°C); and EVAP purge is active. O2 Sensor signal voltage switches less than 16 times from lean to rich within 20 seconds during monitoring, or will compare the state of change between the front and rear O2 Sensors and if the differences are greater than a calibrated amount, the DTC will set. 3 good trips are required to turn off the MIL.<br>**Possible Causes:**<br>Exhaust leak<br>O2 element is contaminated, deteriorated or it has failed<br>O2 signal circuit or return circuit has failed |
| **P0139**<br>1T CCM<br>2004-05<br>Viper | **O2 Sensor (B1 S2) Slow Response Conditions:**<br>Start engine and allow to idle. For first part of the test, if limits are exceeded, test passes. If not, second part of test is run, checking for AAT Sensor of greater than 44°F, BARO Sensor greater than 22.13 in., Battery voltage of more than 10.5v, MAP Sensor 11.79-18.15 in., engine rpm between 1350-2200 rpm, and VSS Sensor between 50-65 rpm.<br>**Possible Causes:**<br>Exhaust leak present in the exhaust manifold or exhaust pipes<br>O2 Sensor signal circuit has resistance<br>O2 Sensor ground circuit has resistance<br>O2 Sensor has failed |
| **P0140**<br>2T O2<br>1996-2003<br>Cherokee, Dakota, Durango, Grand Cherokee, Liberty, Ram | **O2 (B1 S2) Remains At Center Conditions:**<br>Engine started; system voltage over 10.5v, ECT Sensor over 150.8°F, engine running in closed loop, and the PCM detected the O2 signal was fixed at 350-580 mV for 60 seconds.<br>**Possible Causes:**<br>Exhaust leak present in exhaust manifold or exhaust pipes<br>O2 element is fuel contaminated or has deteriorated<br>O2 signal circuit or ground circuit has high resistance<br>PCM has failed |
| **P0141**<br>1T CCM<br>1995-98<br>Talon | **O2 (B1 S2) Heater Circuit Conditions:**<br>ECT Sensor less than 147°F and BTS signal within 27°F of the ECT Sensor (cold engine), engine running at idle speed for over 12 seconds, system voltage over 10.5v and the PCM detected the O2 signal was more than 3.0v for 30-90 seconds.<br>**Possible Causes:**<br>ASD relay power supply circuit to the heater is open<br>O2 heater ground circuit or the O2 signal circuit is open<br>O2 heater element has high resistance or it has failed<br>PCM has failed |
| **P0141**<br>1T CCM<br>2001-03<br>Sebring, Stratus with 2.4L or 3.0L | **O2 (B1 S1) Heater Circuit Conditions:**<br>Engine started; system voltage from 11-16v, ECT Sensor over 68°F, and the PCM detected the O2 heater current was less than 0.16 amps or more than 7.5 amps for 4 seconds.<br>**Possible Causes:**<br>O2 heater ground circuit open or O2 signal circuit is open<br>O2 heater element has high resistance<br>O2 heater element has failed (open or shorted)<br>MFI relay output (power) circuit to the heater is open<br>PCM has failed |

**OBD II Trouble Code List (P0XXX Codes) - Continued**

| DTC | Trouble Code Title, Conditions & Possible Causes: |
|---|---|
| **P0141**<br>1T CCM<br>1996-2003<br>300M, Breeze, Caravan, Cirrus, Concorde, Intrepid, LHS, Liberty, Neon, New Yorker, Pacifica, Town & Country, Vision, Voyager, Wrangler | **O2 (B1 S2) Heater Circuit Conditions:**<br>Key off after the vehicle has been driven for at least 10 miles with the throttle open for 3 minutes, system voltage over 11v, then the O2 heater Test is enabled (i.e., the PCM energizes the ASD relay to provide power to the heater, and the PCM monitors the O2 signal voltage. If it continues to increase (instead of decreasing), the O2S Monitor Test fails.<br>**Possible Causes:**<br>ASD relay output (power) circuit to the heater is open<br>O2 heater ground circuit is open or has high resistance<br>O2 heater element has failed (open or shorted)<br>PCM has failed |
| **P0141**<br>1T CCM<br>1995-2003<br>Avenger, Cirrus, Cherokee, PT Cruiser, Dakota, Durango, Grand Cherokee, Liberty, Neon, Prowler, Ram, Sebring, Stratus, Prowler, Viper, Voyager, Wrangler | **O2 (B1 S2) Heater Circuit Conditions:**<br>ECT Sensor less than 147°F, BTS signal within ±27°F of ECT Sensor (cold engine); engine running at idle speed for over 12 seconds, system voltage over 10.5v and the PCM detected the O2 signal was more than 3.0v for 30-90 seconds.<br>**Possible Causes:**<br>ASD relay output (power) circuit to the heater is open<br>O2 heater ground circuit open or O2 signal circuit is open<br>O2 heater element has high resistance<br>O2 heater element has failed (open or shorted)<br>PCM has failed |
| **P0141**<br>2T CCM<br>2004-05<br>300M, Magnum, Neon, Sebring, Stratus, PT Cruiser, Caravan/Town & Country/Voyager, Dakota, Durango, Ram | **O2 (B1 S2) Heater Circuit Conditions:**<br>Engine running and O2 heater duty cycle is greater than 0%. O2 heater temperature does not reach 959°F (575°C), exc. 300M, Magnum, or 662°F (350°C), all other models, within 90 seconds (45 seconds for Sebring, Stratus), or no O2 Sensor output is received when the PCM attempts to power up the Sensor heater. 3 good trips are required to turn off the MIL.<br>**Possible Causes:**<br>O2 heater ground circuit open or O2 signal circuit is open<br>O2 heater element has failed<br>O2 heater ground circuit or control circuit is open<br>PCM has failed |
| **P0141**<br>1T CCM<br>2004-05<br>Viper | **O2 Sensor (B1 S2) Heater Failure Conditions:**<br>At cold start, battery voltage greater than 10.5v; ECT less than 147°F; Battery Temp. Sensor equal to of less than 27°F; engine at idle for at least 12 seconds. If O2 Sensor voltage is greater than 3v for 60-240 seconds, this DTC will set.<br>**Possible Causes:**<br>O2 Sensor heater operation has failed<br>O2 Sensor heater element has failed<br>O2 Sensor heater supply circuit or control circuit is open |
| **P0141**<br>2T CCM<br>2004-05<br>Grand Cherokee, Liberty, Wrangler | **O2 (B1 S2) Heater Circuit Conditions:**<br>Engine running; O2 heater duty cycle is greater than 0%; ASD relay is energized; battery voltage is greater than 10.4v. No sensor output is received when the PCM powers up the sensor heater. The O2 heater is out of control for 128 seconds after it has reached 662°F (350°C).<br>**Possible Causes:**<br>O2 heater ground circuit open or O2 signal circuit is open<br>O2 heater element has failed<br>PCM has failed |

## OBD II Trouble Code List (P0XXX Codes) - Continued

| DTC | Trouble Code Title, Conditions & Possible Causes: |
|---|---|
| **P0143**<br>1T CCM<br>1996-2002<br>Ram with 5.9L or 8.0L<br>(CA models) | **O2 (B1 S3) Short to Ground Conditions:**<br>ECT Sensor more than 170ºF on previous key on, ECT Sensor less than 98ºF and the BTS signal within ±59ºF of the ECT Sensor at startup (cold engine), and the PCM detected the O2 signal was less than 156 mV for 28 seconds.<br>**Possible Causes:**<br>O2 signal circuit is open or is shorted to chassis or Sensor ground<br>O2 signal circuit is open (circuit reads 450 mV when open)<br>O2 may be contaminated or it has failed |
| **P0144**<br>1T CCM<br>1996-2002<br>Ram with 5.9L or 8.0L<br>(CA models) | **O2 (B1 S3) Short to Voltage Conditions:**<br>Engine runtime over 2 minutes, ECT Sensor more than 180ºF, and the PCM detected the O2 signal was more than 1.50v.<br>**Possible Causes:**<br>O2 signal tracking in connector due to a short between the signal and power circuits<br>O2 signal circuit shorted to system power<br>O2 signal circuit open, or ground circuit open<br>O2 heater supply circuit is open<br>PCM has failed |
| **P0145**<br>2T O2<br>1996-2002<br>Ram with 5.9L or 8.0L<br>(CA models) | **O2 (B1 S3) Slow Response Conditions:**<br>Engine started; ECT Sensor more than 170ºF, VSS input over 10 mph with throttle open for 2 minutes, then back to idle speed in Drive (A/T) or Neutral (M/T), and the PCM detected the O2 signal did not switch enough times from 270-620 mV during the O2 Test.<br>**Possible Causes:**<br>Exhaust leak present in the exhaust manifold or exhaust pipes<br>O2 element is fuel contaminated or the O2 element is deteriorated<br>PCM has failed |
| **P0147**<br>1T CCM<br>1996-2002<br>Ram with 5.9L or 8.0L<br>(CA models) | **O2 (B1 S3) Heater Fault Conditions:**<br>ECT Sensor less than 147ºF, BTS signal within ±27ºF of ECT Sensor (cold engine), engine running at idle speed for over 12 seconds, system voltage over 10.5v and the PCM detected the O2 signal was more than 3.0v for 30-90 seconds.<br>**Possible Causes:**<br>ASD relay output (power) circuit to the heater is open<br>O2 heater ground circuit open or O2 signal circuit is open<br>O2 heater element has high resistance<br>O2 heater element has failed (open or shorted)<br>PCM has failed |
| **P0151**<br>1T CCM<br>1996-2003<br>Avenger, Breeze, Cirrus, Sebring, Stratus with V6 | **O2 (B2 S1) Circuit Short to Ground Conditions:**<br>Engine runtime less than 3 seconds, ECT Sensor less than 120ºF at engine startup, and the PCM detected the O2 signal indicated less than 160 mV right after engine startup.<br>**Possible Causes:**<br>O2 signal circuit is shorted to chassis or Sensor ground<br>O2 signal circuit is open (circuit reads 450 mV when open)<br>O2 ground circuit is open (circuit reads 170 mV when open)<br>O2 may be contaminated or it has failed<br>PCM has failed |
| **P0151**<br>1T CCM<br>1996-2003<br>300M, Concorde, PT Cruiser, Intrepid, LHS, Neon, Prowler, Sebring, Stratus, Vision | **O2 (B2 S1) Circuit Short to Ground Conditions:**<br>Engine runtime less than 3 seconds, ECT Sensor less than 120ºF at engine startup, and the PCM detected the O2 signal indicated less than 160 mV right after engine startup.<br>**Possible Causes:**<br>O2 signal circuit is shorted to chassis or Sensor ground<br>O2 signal circuit is open (circuit reads 450 mV when open)<br>O2 ground circuit is open (circuit reads 170 mV when open)<br>O2 may be contaminated or it has failed<br>PCM has failed |

## OBD II Trouble Code List (P0XXX Codes) - Continued

| DTC | Trouble Code Title, Conditions & Possible Causes: |
|---|---|
| **P0151**<br>1T CCM<br>1996-2003<br>Dakota, Durango, Prowler, Viper, Jeep, Ram | **O2 (B2 S1) Circuit Short to Ground Conditions:**<br>ECT Sensor more than 170ºF on previous key on, ECT Sensor less than 98ºF and the BTS signal within ±27ºF of the ECT Sensor at startup (cold engine), and the PCM detected the O2 signal was less than 156 mV for 28 seconds.<br>**Possible Causes:**<br>O2 signal circuit is shorted to chassis or Sensor ground<br>O2 signal circuit is open (circuit reads 450 mV when open)<br>O2 ground circuit is open (circuit reads 170 mV when open)<br>O2 may be contaminated or it has failed<br>PCM has failed |
| **P0151**<br>1T CCM<br>2004-05<br>300M, Dakota, Durango, Grand Cherokee, Liberty, Magnum, Pacifica, Ram, Sebring, Stratus, Wrangler | **O2 (B2 S1) Circuit Short to Ground Conditions:**<br>Engine runtime under 30 seconds, system voltage over 10.99v, O2 heater temperature below 484ºF. The PCM detected the O2 signal was below 1.5v (300M, Magnum, Sebring, Stratus) or below 2.5196v (Dakota, Durango, Grand Cherokee, Ram, Wrangler) for 3 seconds, or below 2.411v for 10 seconds (Liberty) after engine start.<br>**Possible Causes:**<br>O2 upstream circuit is shorted to ground<br>O2 signal circuit is shorted to ground or to O2 upstream return circuit<br>O2 signal circuit is shorted to the heater ground circuit<br>O2 may be contaminated or it has failed<br>PCM has failed |
| **P0151**<br>1T CCM<br>2004-05<br>Viper | **O2 (B2 S1) Circuit Voltage Too Low Conditions:**<br>Engine runtime from cold start; ECT below 98F; ECT signal above 170F from previous key off. The PCM detected the O2 signal voltage was below 0.156v for 28 seconds.<br>**Possible Causes:**<br>O2 Sensor below 0.16v<br>O2 may be contaminated or it has failed<br>O2 signal circuit is shorted to ground or to ground circuit<br>PCM has failed |
| **P0152**<br>1T CCM<br>1996-2003<br>Avenger, Breeze, Cirrus, Sebring, Stratus | **O2 (B2 S1) Circuit Shorted to Voltage Conditions:**<br>Engine runtime more than 2 minutes, ECT Sensor more than 176ºF, and the PCM detected the O2 signal was more than 1.2v, condition met for 3 seconds.<br>**Possible Causes:**<br>O2 signal tracking due to wet/oily condition in the connector<br>O2 signal circuit shorted to system power<br>O2 signal circuit open, or ground circuit open<br>PCM has failed |
| **P0152**<br>1T CCM<br>1996-2001<br>300M, Avenger, Breeze, Cirrus, Concorde, PT Cruiser, Intrepid, LHS, Neon, New Yorker, Prowler, Sebring, Stratus, Vision | **O2 (B2 S1) Circuit Shorted to Power Conditions:**<br>Engine runtime more than 2 minutes, system voltage over 10.5v, ECT Sensor more than 176ºF, and the PCM detected the O2 signal was more than 1.2v for 3 seconds.<br>**Possible Causes:**<br>O2 signal tracking in the connector due to wet/oily condition<br>O2 signal circuit is open, or the ground circuit open<br>O2 heater supply circuit is open<br>PCM has failed |

## OBD II Trouble Code List (P0XXX Codes) - Continued

| DTC | Trouble Code Title, Conditions & Possible Causes: |
|---|---|
| **P0152**<br>1T CCM<br>2002-03<br>300M, Concorde, PT Cruiser, Intrepid, LHS, Neon, Prowler, Sebring, Stratus, Vision | **O2 (B2 S1) Circuit Shorted to Power Conditions:**<br>Engine runtime over 119 seconds, system voltage over 10.99v, O2 heater temperature more than 1085ºF, and the PCM detected the O2 signal was more than 3.70v for 1 minute.<br>**Possible Causes:**<br>O2 signal tracking (wet/oily) in connector causing a short between the signal and heater power circuits<br>O2 ground circuit is open<br>O2 signal circuit is open<br>PCM has failed |
| **P0152**<br>1T CCM<br>1996-2003<br>Caravan, Cherokee, Dakota, Durango, Grand Cherokee, Liberty, Prowler, Ram, Town & Country, Wrangler, Viper, Voyager | **O2 (B2 S1) Short to Voltage Conditions:**<br>Engine runtime over 4 minutes, system voltage over 10.5v, ECT Sensor more than 180ºF, and the PCM detected the O2 signal was more than 1.50v for 3 seconds during the CCM test.<br>**Possible Causes:**<br>O2 signal tracking in connector due to wet/oily conditions<br>O2 ground circuit is open<br>O2 heater supply circuit is open<br>O2 signal circuit is open<br>O2 element is damaged or it has failed<br>PCM has failed |
| **P0152**<br>1T CCM<br>2004-05<br>300M, Caravan, Concorde, PT Cruiser, Dakota, Durango, Grand Cherokee, Intrepid, LHS, Liberty, Magnum, Neon, Pacifica, Prowler Ram, Sebring, Stratus, Town & Country, Wrangler | **O2 (B2 S1) Circuit High Conditions:**<br>O2 Sensor heater temperature is more than 925°F (496°C) on 300M, Magnum or 1085°F (350°C) on Sebring; battery voltage is more than 10.99v. O2 Sensor voltage is more than 3.7v for 30 seconds (300M, Magnum), more than 3.99v for 30 seconds (Dakota, Durango, Grand Cherokee, Ram, Wrangler), 3.99v for 60 seconds (Sebring, Stratus), or 3.99v for 66.56 seconds (Liberty). 3 good trips are required to turn off the MIL.<br>**Possible Causes:**<br>O2 signal circuit is open or is shorted to battery voltage.<br>O2 upstream return circuit is open or is shorted to battery voltage<br>O2 Sensor is damaged or has failed<br>PCM has failed |
| **P0152**<br>1T CCM<br>2004-05<br>Viper | **O2 Sensor (B2 S1) Voltage High**<br>Engine running for more than 4 minute; battery voltage above 10.4v; ECT Sensor above 180°F. O2 Sensor voltage is more than 1.5v.<br>**Possible Causes:**<br>O2 Sensor has failed<br>O2 Sensor signal circuit is open or is shorted to voltage<br>O2 Sensor signal circuit is shorted to heater supply circuit<br>O2 Sensor ground circuit is open<br>O2 Sensor heater control circuit is open |
| **P0153**<br>1T CCM<br>1996-2003<br>Avenger, Breeze, Sebring, Stratus with V6 | **O2 (B2 S1) Slow Response Conditions:**<br>Engine runtime over 3 minutes, ECT Sensor over 170°F, VSS more than 24 mph for 75 seconds, A/C and PSPS both indicating off, then back to idle in Drive or Neutral (M/T), and the PCM detected the O2 signal did not reach 670 mV, or the O2 switched from 350-550 mV too few times, condition met for 6 seconds.<br>**Possible Causes:**<br>Exhaust leak present in the exhaust manifold or exhaust pipes<br>O2 signal or ground circuit has a high resistance condition<br>O2 element is fuel contaminated<br>O2 element is deteriorated or it has failed |

**OBD II Trouble Code List (P0XXX Codes) - Continued**

| DTC | Trouble Code Title, Conditions & Possible Causes: |
|---|---|
| **P0153**<br>1T CCM<br>1996-2001<br>300M, Avenger, Breeze, Cirrus, Concorde, PT Cruiser, Intrepid, LHS, Neon, New Yorker, Prowler, Sebring, Stratus, Vision | **O2 (B2 S1) Slow Response Conditions:**<br>Engine runtime over 2 minutes, ECT Sensor more than 147°F, VSS indicating over 10 mph, then back to idle speed in Drive (A/T) or Neutral (M/T), and the PCM detected the O2 signal did not switch enough times from 270-620 mV in the O2 Monitor Test.<br>**Possible Causes:**<br>Exhaust leak present in the exhaust manifold or exhaust pipes<br>O2 signal or ground circuit has a high resistance condition<br>O2 element is fuel contaminated<br>O2 element is deteriorated or it has failed |
| **P0153**<br>1T CCM<br>2002-03<br>300M, Concorde, PT Cruiser, Intrepid, LHS, Neon, Prowler, Sebring, Stratus, Vision | **O2 (B2 S1) Slow Response Conditions:**<br>Engine started; vehicle driven at a steady speed of 20-55 mph with the throttle open for at least 2 minutes, ECT Sensor more than 158°F (70°C), Catalytic Converter temperature more than 1112°F (600°C), and the PCM detected the O2 signal switched from rich-to-lean less than 12 times within a 60 second period.<br>**Possible Causes:**<br>Exhaust leak present in the exhaust manifold or exhaust pipes<br>O2 signal or ground circuit has a high resistance condition<br>O2 element is fuel contaminated<br>O2 element is deteriorated or it has failed |
| **P0153**<br>1T CCM<br>1996-2003<br>Caravan, Cherokee, Dakota, Durango, Grand Cherokee, Liberty, Prowler, Ram, Town & Country, Wrangler, Viper, Voyager | **O2 (B2 S1) Slow Response Conditions:**<br>Engine started; ECT Sensor more than 147°F, VSS input over 10 mph with throttle open for 2 minutes, then back to idle speed in Drive (A/T) or Neutral (M/T), and the PCM detected the O2 signal did not switch enough from 270-620 mV during the O2 Test.<br>**Possible Causes:**<br>Exhaust leak present in the exhaust manifold or exhaust pipes<br>O2 signal or ground circuit has a high resistance condition<br>O2 element is fuel contaminated<br>O2 element is deteriorated or it has failed |
| **P0153**<br>2T CCM<br>2004-05<br>300M, Caravan, Concorde, PT Cruiser, Dakota, Durango, Grand Cherokee, Intrepid, LHS, Liberty, Magnum, Neon, Pacifica, Prowler Ram, Sebring, Stratus, Town & Country, Wrangler | **O2 (B2 S1) Slow Response Conditions:**<br>Engine started; vehicle driven at a steady speed of 20-55 mph with the throttle open for at least 2 minutes, ECT Sensor more than 158°F (70°C), Catalytic Converter temperature more than 1112°F (600°C), EVAP purge is active, and the PCM detected the O2 signal switched from lean to rich less than 16 times (3.5L, 3.7L, 4.7L, 5.7L) or 11 times (2.7L) within a 20 second period during monitoring. 3 good trips are required to turn off MIL.<br>**Possible Causes:**<br>Exhaust leak<br>O2 signal circuit has an open or grounded condition<br>O2 upstream return circuit has an open or grounded condition<br>O2 element is deteriorated or it has failed |
| **P0153**<br>1T CCM<br>2004-05<br>Viper | **O2 Sensor (B2 S1) Slow Response Conditions:**<br>With ECT greater than 147°F, after reaching a vehicle speed of 10 mph, and throttle remaining open (off idle) for 2 minutes, bring the vehicle to a stop and all to idle with transmission in Drive. If the O2 sensor signal voltage is switching from less than 0.27v to more than 0.62v and back fewer times than required, this DTC will set.<br>**Possible Causes:**<br>Exhaust leak present in the exhaust manifold or exhaust pipes<br>O2 Sensor signal circuit has resistance<br>O2 Sensor ground circuit has resistance<br>O2 Sensor has failed |

## OBD II Trouble Code List (P0XXX Codes) - Continued

| DTC | Trouble Code Title, Conditions & Possible Causes: |
|---|---|
| **P0154**<br>1T CCM<br>1996-2003<br>Avenger, Breeze, Sebring, Stratus with V6 | **O2 (B2 S1) Remains At Center Conditions:**<br>Engine runtime over 2 minutes, system voltage over 10.5v, ECT Sensor more than 170°F, and the PCM detected the O2 signal remained fixed between 350-550 mV for 1.5 minutes.<br>**Possible Causes:**<br>Exhaust leak present in exhaust manifold or exhaust pipes<br>O2 element is fuel contaminated or has deteriorated<br>O2S signal circuit or ground circuit has high resistance<br>PCM has failed |
| **P0154**<br>1T CCM<br>1996-2001<br>300M, Avenger, Breeze, Cirrus, Concorde, PT Cruiser, Intrepid, LHS, Neon, New Yorker, Prowler, Sebring, Stratus, Vision | **O2 (B2 S1) Remains At Center Conditions:**<br>Engine runtime over 2 minutes, system voltage over 10.5v, ECT Sensor more than 170°F, and the PCM detected the O2 signal remained fixed between 350-550 mV for 1.5 minutes.<br>**Possible Causes:**<br>Exhaust leak present in exhaust manifold or exhaust pipes<br>O2 element is fuel contaminated or has deteriorated<br>O2S signal circuit or ground circuit has high resistance<br>PCM has failed |
| **P0154**<br>1T CCM<br>2002-03<br>300M, Concorde, PT Cruiser, Intrepid, LHS, Neon, Prowler, Sebring, Stratus, Vision | **O2 (B2 S1) Remains At Center Conditions:**<br>Engine runtime more than 121 seconds, system voltage over 10.5v, ECT Sensor more than 150.8°F, fuel control system in closed loop mode, and the PCM detected the O2 signal remained fixed in a range between 350-580 mV, condition met for 60 seconds.<br>**Possible Causes:**<br>Exhaust leak present in exhaust manifold or exhaust pipes<br>O2 element is fuel contaminated or has deteriorated<br>O2S signal circuit or ground circuit has high resistance<br>PCM has failed |
| **P0154**<br>1T CCM<br>1996-2003<br>Caravan, Cherokee, Dakota, Durango, Grand Cherokee, Liberty, Prowler, Ram, Town & Country, Wrangler, Viper, Voyager | **O2 (B2 S1) Remains At Center Conditions:**<br>Engine started; ECT Sensor more than 147°F, VSS input over 10 mph with throttle open for 2 minutes, then back to idle speed in Drive (A/T) or Neutral (M/T), and the PCM detected the O2 signal did not switch enough from 270-620 mV during the O2 Test.<br>**Possible Causes:**<br>Exhaust leak present in the exhaust manifold or exhaust pipes<br>O2 signal or ground circuit has a high resistance condition<br>O2 element is fuel contaminated<br>O2 element is deteriorated or it has failed |
| **P0155**<br>1T CCM<br>1996-2003<br>Avenger, Breeze, Sebring, Stratus with V6 | **O2 (B2 S1) Heater Circuit Conditions:**<br>Key off after a warm engine drive cycle, engine cool-down finished (at least 5 seconds after the key is turned off), system voltage over 10.5v, then with the ASD relay energized, the PCM detected the O2 signal rose to 0.49v or more within a 144 second period, and the initial rise of the oxygen Sensor signal was less than 1.57v.<br>**Possible Causes:**<br>ASD relay power supply circuit to the heater is open<br>O2 heater ground circuit open or O2 signal circuit is open<br>O2 heater element has high resistance<br>O2 heater element has failed (open or shorted)<br>PCM has failed |

**OBD II Trouble Code List (P0XXX Codes) - Continued**

| DTC | Trouble Code Title, Conditions & Possible Causes: |
|---|---|
| **P0155**<br>1T CCM<br>1996-2001<br>300M, Avenger, Breeze, Cirrus, Concorde, PT Cruiser, Intrepid, LHS, Neon, New Yorker, Prowler, Sebring, Stratus, Vision | **O2 (B2 S1) Heater Circuit Conditions:**<br>Key off after a warm engine drive cycle, engine cool-down finished (at least 5 seconds after the key is turned off), system voltage over 10.5v, then with the ASD relay energized, the PCM detected the O2 signal rose to 0.49v or more within a 144 second period, and the initial rise of the oxygen Sensor signal was less than 1.57v.<br>**Possible Causes:**<br>ASD relay power supply circuit to the heater is open<br>O2 heater ground circuit open or O2 signal circuit is open<br>O2 heater element has high resistance<br>O2 heater element has failed (open or shorted)<br>PCM has failed |
| **P0155**<br>1T CCM<br>2002-03<br>300M, Concorde, PT Cruiser, Intrepid, LHS, Neon, Prowler, Sebring, Stratus, Vision | **O2 (B2 S1) Heater Circuit Conditions:**<br>Engine started; O2 heater duty cycle from 1-99%, and the PCM detected the Heater temperature did not reach 959°F in 90 seconds.<br>**Possible Causes:**<br>O2 heater control circuit is open<br>O2 heater ground circuit is open<br>O2 heater is damaged or it has failed (open or shorted)<br>PCM has failed |
| **P0155**<br>1T CCM<br>1996-2003<br>Caravan, Cherokee, Dakota, Durango, Grand Cherokee, Liberty, Prowler, Ram, Town & Country, Wrangler, Viper, Voyager | **O2 (B2 S1) Heater Circuit Conditions:**<br>ECT Sensor less than 147°F, BTS signal within ±27°F of ECT Sensor (cold engine), engine running at idle speed for over 12 seconds, system voltage over 10.5v and the PCM detected the O2 signal was more than 3.0v for 30-90 seconds.<br>**Possible Causes:**<br>ASD relay output (power) circuit to the heater is open<br>O2 heater ground circuit open or O2 signal circuit is open<br>O2 heater element has high resistance<br>O2 heater element has failed (open or shorted)<br>PCM has failed |
| **P0155**<br>2T CCM<br>2004-05<br>300M, Caravan, Concorde, PT Cruiser, Dakota, Durango, Grand Cherokee, Intrepid, LHS, Magnum, Neon, Pacifica, Prowler Ram, Sebring, Stratus, Town & Country, Wrangler | **O2 (B2 S1) Heater Circuit Conditions:**<br>Engine running and heater duty cycle is greater than 0%; battery voltage is more than 11v. O2 heater temperature does not reach 959°F (575°C) within 90 seconds, or no Sensor output is received when the PCM powers up the Sensor heater. 3 good trips required to turn off MIL.<br>**Possible Causes:**<br>O2 heater control circuit is open<br>O2 heater ground circuit is open<br>O2 heater element is damaged or has failed<br>PCM has failed |
| **P0155**<br>1T CCM<br>2004-05<br>Viper | **O2 Sensor (B2 S1) Heater Failure Conditions:**<br>At cold start, battery voltage greater than 9v; ECT less than 147°F; Battery Temp. Sensor equal to of less than 27°F; engine at idle for at least 12 seconds. If O2 Sensor voltage is greater than 3v for 60-240 seconds, this DTC will set.<br>**Possible Causes:**<br>O2 Sensor heater operation has failed<br>O2 Sensor heater element has failed<br>02 Sensor heater supply circuit or control circuit is open |

## OBD II Trouble Code List (P0XXX Codes) - Continued

| DTC | Trouble Code Title, Conditions & Possible Causes: |
|---|---|
| **P0155**<br>2T CCM<br>2004-05<br>Liberty | **O2 (B1 S1) Heater Circuit Conditions:**<br>Engine running; O2 heater duty cycle is more than 0%; ASD relay is energized; battery voltage is more than 10.4v. No sensor output is received when PCM powers up the sensor heater. The O2 heater is out of control for 128 seconds after it has reached 572°F (300°C).<br>**Possible Causes:**<br>O2 heater ground circuit open or O2 signal circuit is open<br>O2 heater element has failed<br>PCM has failed |
| **P0157**<br>1T CCM<br>1998-2003<br>Avenger, Breeze, Sebring, Stratus with V6 | **O2 (B2 S2) Short to Ground Conditions:**<br>Engine runtime less than 3 seconds, ECT Sensor No. 120°F or less at engine startup, and the PCM detected the O2 signal indicated less than 160 mV during the CCM test period.<br>**Possible Causes:**<br>O2 signal circuit is shorted to chassis or Sensor ground<br>O2 signal circuit is open (circuit reads 450 mV when open)<br>O2 ground circuit is open (circuit reads 170 mV when open)<br>O2 may be contaminated or it has failed<br>PCM has failed |
| **P0157**<br>1T CCM<br>1996-2001<br>300M, Avenger, Breeze, Cirrus, Concorde, PT Cruiser, Intrepid, LHS, Neon, New Yorker, Prowler, Sebring, Stratus, Vision | **O2 (B2 S2) Circuit Shorted to Ground Conditions:**<br>ECT Sensor more than 170°F on previous key on, ECT Sensor less than 98°F and the BTS signal within ±27°F of ECT Sensor at startup (cold engine), and the PCM detected the O2 signal was less than 156 mV for 28 seconds.<br>**Possible Causes:**<br>O2 signal circuit is shorted to chassis or Sensor ground<br>O2 may be contaminated or it has failed<br>PCM has failed |
| **P0157**<br>1T CCM<br>1996-2003<br>Caravan, Cherokee, Dakota, Durango, Grand Cherokee, Liberty, Prowler, Ram, Town & Country, Wrangler, Viper, Voyager | **O2 (B2 S2) Circuit Shorted to Ground Conditions:**<br>ECT Sensor more than 170°F at previous key off, O2 heater Test completed after shutdown, engine cool-down period completed, then engine started, ECT Sensor less than 98°F and the BTS signal within 27°F of the ECT Sensor (cold engine), and the PCM detected the O2 signal was less than 156 mV for 28 seconds during the test.<br>**Possible Causes:**<br>O2 signal circuit is shorted to chassis or Sensor ground<br>O2 may be contaminated or it has failed<br>PCM return or signal circuit is damaged or has failed |
| **P0157**<br>1T CCM<br>2002-03<br>300M, Concorde, PT Cruiser, Intrepid, LHS, Neon, Prowler, Sebring, Stratus, Vision | **O2 (B2 S2) Circuit Shorted to Ground Conditions:**<br>Engine runtime under 20 seconds, system voltage over 10.99v, O2 heater temperature below 705°F, and the PCM detected the O2 signal was less than 1.50 volts for 3 seconds.<br>**Possible Causes:**<br>O2 signal circuit is shorted to chassis or Sensor ground<br>O2 is damaged or it has failed<br>PCM has failed |

**OBD II Trouble Code List (P0XXX Codes) - Continued**

| DTC | Trouble Code Title, Conditions & Possible Causes: |
|---|---|
| **P0157**<br>1T CCM<br>2004-05<br>300M, Caravan, Concorde, PT Cruiser, Dakota, Durango, Grand Cherokee, Intrepid, LHS, Liberty, Magnum, Neon, Pacifica, Prowler Ram, Sebring, Stratus, Town & Country, Wrangler | **O2 (B2 S2) Circuit Low Conditions:**<br>Engine runtime under 30 seconds; system voltage over 10.99v; O2 heater temperature below 484ºF (251ºC); O2 Sensor signal was less than 1.5v (300M, Magnum, Viper) or less than 2.5196v (Durango, Grand Cherokee, Ram, Wrangler) for 3 seconds (28 seconds: Viper), or below 2.411v for 10 seconds (Liberty) after engine start. 3 good trips are required to turn off the MIL.<br>**Possible Causes:**<br>O2 signal circuit is shorted to chassis or Sensor ground<br>O2 is damaged or it has failed<br>PCM has failed |
| **P0158**<br>1T CCM<br>1998-2003<br>Avenger, Breeze, Sebring, Stratus with V6 | **O2 (B2 S2) Circuit Shorted to Power Conditions:**<br>Engine runtime over 2 minutes, ECT Sensor over 176ºF, and the PCM detected the O2 signal was more than 1.2v for 3 seconds.<br>**Possible Causes:**<br>O2 signal tracking in connector due to wet/oily condition<br>O2 signal circuit is open or shorted to system power<br>PCM has failed |
| **P0158**<br>1T CCM<br>1996-2001<br>300M, Avenger, Breeze, Cirrus, Concorde, PT Cruiser, Intrepid, LHS, Neon, New Yorker, Prowler, Sebring, Stratus, Vision | **O2 (B2 S2) Circuit Shorted to Power Conditions:**<br>Engine runtime over 2 minutes, system voltage over 10.5v, ECT Sensor more than 176ºF, and the PCM detected the O2 signal was more than 1.21v, condition met for 3 seconds.<br>**Possible Causes:**<br>O2 signal tracking due to wet/oily condition in the connector<br>O2 signal circuit is open or shorted to system power<br>PCM has failed |
| **P0158**<br>1T CCM<br>1996-2003<br>Caravan, Cherokee, Dakota, Durango, Grand Cherokee, Liberty, Prowler, Ram, Town & Country, Wrangler, Viper, Voyager | **O2 (B2 S2) Circuit Shorted to Ground Conditions:**<br>Engine runtime over 4 minutes, system voltage over 10.5v, ECT Sensor more than 180ºF, and the PCM detected the O2 signal was more than 1.50v, condition met for 3 seconds.<br>**Possible Causes:**<br>O2 signal tracking due to wet/oily condition in the connector<br>O2 signal circuit is open or shorted to system power<br>O2 is damaged or is has failed<br>PCM has failed |
| **P0158**<br>1T CCM<br>2002-03<br>300M, Concorde, PT Cruiser, Intrepid, LHS, Neon, Prowler, Sebring, Stratus, Vision | **O2 (B2 S2) Circuit Shorted to Power Conditions:**<br>Engine runtime over 119 seconds, system voltage over 10.99v, O2 heater temperature more than 1085ºF, and the PCM detected the O2 signal was more than 3.70v for 1 minute.<br>**Possible Causes:**<br>O2 signal tracking due to wet/oily condition in the connector<br>O2 ground circuit or the signal circuit is open<br>O2 is damaged or it has failed<br>PCM has failed |

## OBD II Trouble Code List (P0XXX Codes) - Continued

| DTC | Trouble Code Title, Conditions & Possible Causes: |
|---|---|
| **P0158**<br>1T CCM<br>2004-05<br>300M, Caravan, Concorde, PT Cruiser, Dakota, Durango, Grand Cherokee, Intrepid, LHS, Liberty, Magnum, Neon, Pacifica, Prowler Ram, Sebring, Stratus, Town & Country, Wrangler | **O2 (B2 S2) Circuit High Conditions:**<br>Engine is running; system voltage over 10.99v; O2 heater temperature more than 925°F (496°C) on all exc. Sebring, Stratus, or more than 1085°F (350°C) on Sebring, Stratus. The PCM detected the O2 signal was more than 3.70v for 30 seconds (300M, Magnum), more than 3.99v for 30 seconds (Dakota, Durango, Grand Cherokee, Ram, Wrangler), more than 3.99v for 60 seconds (Sebring, Stratus), or more than 3.99v for 76.8 seconds (Liberty). 3 good trips required to turn off MIL.<br>**Possible Causes:**<br>O2 signal circuit is open or is shorted to battery<br>O2 downstream return circuit is open or is shorted to battery<br>O2 is damaged or it has failed<br>PCM has failed |
| **P0158**<br>1T CCM<br>2004-05<br>Viper | **O2 Sensor (B2 S2) Voltage High**<br>Engine running for more than 4 minute; battery voltage above 10.4v; ECT Sensor above 180°F. O2 Sensor voltage is more than 1.5v.<br>**Possible Causes:**<br>O2 Sensor has failed<br>O2 Sensor signal circuit is open or is shorted to voltage<br>O2 Sensor signal circuit is shorted to heater supply circuit<br>O2 Sensor ground circuit is open<br>O2 Sensor heater control circuit is open |
| **P0159**<br>2T O2<br>1998-2003<br>Avenger, Breeze, Sebring, Stratus with V6 | **O2 (B2 S2) Slow Response Conditions:**<br>Engine runtime over 3 minutes, ECT Sensor over 170°F, VSS over 24 mph for 75 seconds, A/C and PSPS off, then back to idle speed in Drive or Neutral, and the PCM detected the O2 signal did not reach 670 mV, or switched from 350-550 mV too few times in 6 seconds.<br>**Possible Causes:**<br>Exhaust leak present in the exhaust manifold or exhaust pipes<br>O2 element is fuel contaminated<br>O2 element is deteriorated or it has failed |
| **P0159**<br>2T O2<br>1996-2001<br>300M, Avenger, Breeze, Cirrus, Concorde, PT Cruiser, Intrepid, LHS, Neon, New Yorker, Prowler, Sebring, Stratus, Vision | **O2 (B2 S2) Slow Response Conditions:**<br>Engine runtime over 2 minutes, ECT Sensor more than 147°F, VSS indicating over 10 mph, then back to idle speed in Drive (A/T) or Neutral (M/T), and the PCM detected the O2 signal did not switch enough times from 270-620 mV in the O2 Monitor Test.<br>**Possible Causes:**<br>Exhaust leak present in the exhaust manifold or exhaust pipes<br>O2 element is fuel contaminated<br>O2 element is deteriorated or it has failed |
| **P0159**<br>2T O2<br>2002-03<br>300M, Concorde, PT Cruiser, Intrepid, LHS, Neon, Prowler, Sebring, Stratus, Vision | **O2 (B2 S2) Slow Response Conditions:**<br>Engine speed from 1200-2000 rpm at 20-60 mph with throttle open for 2 minutes, MAP Sensor from 28-65 kPa, ECT Sensor over 158°F, Converter temperature over 1115°F, and the PCM detected the O2 signal switched from rich-to-lean less than 12 times in 60 seconds.<br>**Possible Causes:**<br>Exhaust leak present in the exhaust manifold or exhaust pipes<br>O2 element is fuel contaminated<br>O2 element is deteriorated or it has failed |

**OBD II Trouble Code List (P0XXX Codes) - Continued**

| DTC | Trouble Code Title, Conditions & Possible Causes: |
|---|---|
| **P0159**<br>2T O2<br>2004-05<br>300M, Caravan, Concorde, PT Cruiser, Dakota, Durango, Grand Cherokee, Intrepid, LHS, Liberty, Magnum, Neon, Pacifica, Prowler Ram, Sebring, Stratus, Town & Country, Wrangler | **O2 (B2 S2) Slow Response Conditions:**<br>Engine is driven at 20-55 mph with throttle open for 2 minutes; ECT Sensor over 158ºF (70°C); catalytic converter temperature over 1112ºF (600°C); EVAP purge is active. The O2 Sensor signal voltage switches less than 16 times or 11 times (2.7L exc. Liberty) from lean to rich with 20 seconds during monitoring. 3 good trips required to turn off MIL.<br>**Possible Causes:**<br>Exhaust leak<br>O2 signal circuit is open or shorted<br>O2 downstream return circuit is open or shorted<br>O2 Sensor is damaged or has failed |
| **P0159**<br>1T CCM<br>2004-05<br>Viper | **O2 Sensor (B2 S2) Slow Response Conditions:**<br>Start engine and allow to idle. For first part of the test, if limits are exceeded, test passes. If not, second part of test is run, checking for AAT Sensor of greater than 44°F, BARO Sensor greater than 22.13 in., Battery voltage of more than 10.5v, MAP Sensor 11.79-18.15 in., engine rpm between 1350-2200 rpm, and VSS Sensor between 50-65 rpm.<br>**Possible Causes:**<br>Exhaust leak present in the exhaust manifold or exhaust pipes<br>O2 Sensor signal circuit has resistance<br>O2 Sensor ground circuit has resistance<br>O2 Sensor has failed |
| **P0159**<br>2T O2<br>1996-2003<br>Caravan, Cherokee, Dakota, Durango, Grand Cherokee, Liberty, Prowler, Ram, Town & Country, Wrangler, Viper, Voyager | **O2 (B2 S2) Slow Response Conditions:**<br>Engine started; ECT Sensor more than 170ºF, VSS input over 10 mph with throttle open for 2 minutes, then back to idle speed in Drive (A/T) or Neutral (M/T), and the PCM detected the O2 signal did not switch enough times from 270-620 mV during the O2 Test.<br>**Possible Causes:**<br>Exhaust leak present in the exhaust manifold or exhaust pipes<br>O2 element is fuel contaminated<br>O2 element is deteriorated or it has failed |
| **P0160**<br>2T O2<br>1998-2003<br>Avenger, Breeze, Sebring, Stratus with V6 | **O2 (B2 S2) Slow Response Conditions:**<br>Engine runtime over 2 minutes, system voltage over 10.5v, ECT Sensor more than 170ºF, and the PCM detected the O2 signal remained fixed between 350-550 mV for 1.5 minutes.<br>**Possible Causes:**<br>Exhaust leak present in exhaust manifold or exhaust pipes<br>O2 element is fuel contaminated or has deteriorated<br>O2 signal circuit or ground circuit has high resistance<br>PCM has failed |
| **P0160**<br>2T O2<br>1996-2001<br>300M, Avenger, Breeze, Cirrus, Concorde, PT Cruiser, Intrepid, LHS, Neon, New Yorker, Prowler, Sebring, Stratus, Vision | **O2 (B2 S2) Remains at Center Conditions:**<br>Engine runtime over 2 minutes, system voltage over 10.5v, ECT Sensor more than 170ºF, and the PCM detected the O2 signal remained fixed between 350-550 mV for 1.5 minutes.<br>**Possible Causes:**<br>Exhaust leak present in exhaust manifold or exhaust pipes<br>O2 element is fuel contaminated or has deteriorated<br>O2 signal circuit or ground circuit has high resistance<br>PCM has failed |

**OBD II Trouble Code List (P0XXX Codes) - Continued**

| DTC | Trouble Code Title, Conditions & Possible Causes: |
|---|---|
| **P0160**<br>2T O2<br>2002-03<br>300M, Concorde, PT Cruiser, Intrepid, LHS, Neon, Prowler, Sebring, Stratus, Vision | **O2 (B2 S2) Remains at Center Conditions:**<br>Engine runtime over 121 seconds, system voltage over 10.5v, ECT Sensor more than 170ºF, and the PCM detected the O2 signal remained fixed between 350-550 mV for 1.5 minutes.<br>**Possible Causes:**<br>Exhaust leak present in exhaust manifold or exhaust pipes<br>O2 element is fuel contaminated or has deteriorated<br>O2 signal circuit or ground circuit has high resistance<br>PCM has failed |
| **P0161**<br>1T CCM<br>1998-2003<br>Avenger, Breeze, Sebring, Stratus with V6 | **O2 (B2 S2) Heater Circuit Conditions:**<br>Key off after a warm engine drive cycle, engine cool-down finished (at least 5 seconds after the key is turned off), system voltage over 10.5v, then with the ASD relay energized, the PCM detected the O2 signal rose to 0.49v or more within a 144 second period, and the initial rise of the oxygen Sensor signal was less than 1.57v.<br>**Possible Causes:**<br>ASD relay output (power) circuit to the heater is open<br>O2 heater ground circuit open or O2 signal circuit is open<br>O2 heater element has high resistance or it has failed<br>PCM has failed |
| **P0161**<br>1T CCM<br>1996-2001<br>300M, Avenger, Breeze, Cirrus, Concorde, PT Cruiser, Intrepid, LHS, Neon, New Yorker, Prowler, Sebring, Stratus, Vision | **O2 (B2 S2) Heater Circuit Conditions:**<br>Key off after a warm engine drive cycle, engine cool-down finished (at least 5 seconds after the key is turned off), system voltage over 10.5v, then with the ASD relay energized, the PCM detected the O2 signal rose to 0.49v or more within a 144 second period, and the initial rise of the oxygen Sensor signal was less than 1.57v.<br>**Possible Causes:**<br>ASD relay output (power) circuit to the heater is open<br>O2S heater ground circuit is open or O2S signal circuit is open<br>O2S heater element has high resistance or it has failed<br>PCM has failed |
| **P0161**<br>1T CCM<br>2002-03<br>300M, Concorde, PT Cruiser, Intrepid, LHS, Neon, Prowler, Sebring, Stratus, Vision | **O2 (B2 S2) Heater Circuit Conditions:**<br>Engine runtime over 2 min.; system voltage over 10.5v; ECT Sensor more than 170ºF. The PCM detected the O2 signal remained fixed between 350-550 mV for 1.5 minutes.<br>**Possible Causes:**<br>Exhaust leak present in exhaust manifold or exhaust pipes<br>O2 element is fuel contaminated or has deteriorated<br>O2 signal circuit or ground circuit has high resistance<br>PCM has failed |
| **P0161**<br>1T CCM<br>1996-2003<br>Caravan, Cherokee, Dakota, Durango, Grand Cherokee, Liberty, Prowler, Ram, Town & Country, Wrangler, Viper, Voyager | **O2 (B2 S2) Slow Response Conditions:**<br>ECT Sensor less than 147ºF and BTS signal within 27ºF of the ECT Sensor (cold engine), engine started, system voltage over 10.5v, engine running at idle speed for over 12 seconds, and the PCM detected the O2 signal was more than 3.0v for 30-90 seconds.<br>**Possible Causes:**<br>ASD relay output (power) circuit to the heater is open<br>O2 heater ground circuit is open or O2 signal circuit open<br>O2 heater element has high resistance, is open or shorted<br>PCM has failed |

**OBD II Trouble Code List (P0XXX Codes) - Continued**

| DTC | Trouble Code Title, Conditions & Possible Causes: |
|---|---|
| **P0161**<br>2T CCM<br>2004-05<br>300M, Magnum, Dakota, Durango, Ram | **O2 (B2 S2) Heater Performance Conditions:**<br>Engine running; heater duty cycle is greater than 0%; battery voltage is more than 11v. O2 heater temperature does not reach 662°F (350°C) within 90 seconds, or no Sensor output is received when the PCM powers up Sensor heater. 3 good trips required to turn off MIL.<br>**Possible Causes:**<br>O2 heater control circuit or ground circuit is open<br>O2 Sensor heater element is damaged or has failed<br>PCM has failed |
| **P0161**<br>1T CCM<br>2004-05<br>Viper | **O2 Sensor (B2 S2) Heater Failure Conditions:**<br>At cold start, battery voltage greater than 10.5v; ECT less than 147°F; Battery Temp. Sensor equal to of less than 27°F; engine at idle for at least 12 seconds. If O2 Sensor voltage is greater than 3v for 60-240 seconds, this DTC will set.<br>**Possible Causes:**<br>O2 Sensor heater operation has failed<br>O2 Sensor heater element has failed<br>O2 Sensor heater supply circuit or control circuit is open |
| **P0161**<br>2T CCM<br>2004-05<br>Liberty | **O2 (B1 S1) Heater Circuit Conditions:**<br>Engine running; O2 heater duty cycle is more than 0%; ASD relay is energized; battery voltage is over 10.4v. No sensor output is received when the PCM powers up the sensor heater. The O2 heater is out of control for 2 minutes after it has reached 662°F (350°C).<br>**Possible Causes:**<br>O2 heater ground circuit open or O2 signal circuit is open<br>O2 heater element or PCM has failed |
| **P0165**<br>1T CCM<br>2001-03<br>300M, Caravan, Cherokee, Concorde, PT Cruiser, Dakota, Durango, Grand Cherokee, Intrepid, LHS, Liberty, Neon, Prowler, Ram, Sebring, Stratus, Town & Country, Voyager, Wrangler | **Starter Relay Circuit Conditions:**<br>Engine cranking and the PCM did not detect the correct voltage signal from the Starter relay control circuit.<br>**Possible Causes:**<br>Starter Relay control circuit is open<br>Starter Relay Control circuit is grounded<br>PCM has failed |
| **P0168**<br>1T CCM<br>2001-03<br>Ram with 5.9L Diesel | **Decreased Performance (High Injection Pump Fuel Temperature) Conditions:**<br>Engine started. The PCM detected decreased performance from the Injector Pump due to a high injection fuel pump temperature.<br>**Possible Causes:**<br>Fuel Injection pump module has failed<br>Overflow Valve Test failed |
| **P0170**<br>2T Fuel<br>1995-98<br>Talon | **Fuel Trim Too Lean or Too Rich (Bank 1) Conditions:**<br>Engine running in closed loop; IAT Sensor signal over 14°F, BARO Sensor signal over 75 kPa. The PCM detected Long Term fuel trim value was below -12.5% (a rich A/F ratio) or over +12.5% (a lean A/F ratio), or the Short Term fuel trim value was +10% or higher, or it was -10% or lower for 10 seconds.<br>**Possible Causes:**<br>Air leaks present in the exhaust manifold or exhaust pipes<br>Air is being drawn in from leaks in gaskets or other seals<br>Fuel control Sensor is out calibration (BARO, ECT, IAT or VAF)<br>Fuel pressure too high or low, leaking or restricted fuel injector<br>O2 element is contaminated, deteriorated or it has failed |

**OBD II Trouble Code List (P0XXX Codes) - Continued**

| DTC | Trouble Code Title, Conditions & Possible Causes: |
|---|---|
| **P0171**<br>2T Fuel<br>1996-2000<br>Avenger, Breeze, Cirrus, Talon, Eagle Vision, Sebring, Stratus | **Fuel System Lean (Bank 1) Conditions:**<br>Engine running in closed loop, IAT Sensor signal over 20ºF, altitude less than 8,000 feet, and the PCM detected too large an amount of Fuel Trim correction due to a lean A/F condition.<br>**Possible Causes:**<br>Air leaks in intake manifold, exhaust pipes or exhaust manifold<br>Base engine mechanical problem causing a lean A/F condition<br>Fuel control Sensor is out of calibration (e.g., ECT, IAT or MAP)<br>Fuel delivery component fault (clogged filter, low fuel pressure)<br>O2 element is contaminated, deteriorated or it has failed<br>Vacuum hose is disconnected, broken, leaking or loose |
| **P0171**<br>2T Fuel<br>1996-2003<br>300M, Avenger, Breeze, Caravan, Cherokee, Cirrus, Concorde, PT Cruiser, Dakota, Durango, Grand Cherokee, Intrepid, LHS, Liberty, Neon, New Yorker, Prowler, Ram, Sebring, Stratus, Town & Country, Viper, Vision, Voyager, Wrangler | **Fuel System Lean (Bank 1) Conditions:**<br>Engine running in closed loop, ambient temperature over 20ºF, altitude less than 8,000 feet, and the PCM detected too large an amount of Fuel Trim correction due to a lean A/F condition.<br>**Possible Causes:**<br>Air leaks in intake manifold, exhaust pipes or exhaust manifold<br>Base engine mechanical problem causing a lean A/F condition<br>Fuel control Sensor is out of calibration (e.g., ECT, IAT or MAP)<br>Fuel delivery component fault (clogged filter, low fuel pressure)<br>O2 element is contaminated, deteriorated or it has failed<br>Vacuum hose is disconnected, broken, leaking or loose |
| **P0171**<br>2T Fuel<br>2001-03<br>Sebring, Stratus Coupe with 2.4L I4 or 3.0L V6 | **Fuel System Lean (Bank 1) Conditions:**<br>ECT Sensor between 140ºF to 212ºF, IAT Sensor less than 140ºF at startup, engine running in closed loop, volume airflow Sensor less than 88 Hz, and the PCM detected the Long Term fuel trim exceed +12.5% or the Short Term fuel trim exceeded +25% for 5 seconds.<br>**Possible Causes:**<br>Air leaks in intake manifold, exhaust pipes or exhaust manifold<br>Base engine mechanical problem causing a lean A/F condition<br>Fuel control Sensor is out of calibration (e.g., BARO or VAF)<br>Fuel delivery component fault (clogged filter, low fuel pressure)<br>O2 element is contaminated, deteriorated or it has failed<br>Vacuum hose is disconnected, broken, leaking or loose |
| **P0171**<br>2T Fuel<br>1996-2003<br>Cherokee, Dakota, Durango, Grand Cherokee, Prowler, Ram, Viper, Voyager, Wrangler | **Fuel System Lean (Bank 1) Conditions:**<br>Engine running in closed loop; IAT Sensor signal over 20ºF; altitude less than 8,000 feet. The PCM detected too large an amount of Fuel Trim correction due to a lean A/F condition.<br>**Possible Causes:**<br>Air leaks in intake manifold, exhaust pipes or exhaust manifold<br>Base engine mechanical problem causing a lean A/F condition<br>Fuel control Sensor is out of calibration (e.g., ECT, IAT or MAP)<br>Fuel delivery component fault (clogged filter, low fuel pressure)<br>O2 element is contaminated, deteriorated or it has failed<br>Vacuum hose is disconnected, broken, leaking or loose |

**OBD II Trouble Code List (P0XXX Codes) - Continued**

| DTC | Trouble Code Title, Conditions & Possible Causes: |
|---|---|
| **P0171**<br>2T Fuel<br>2004-05<br>300M, Caravan, Concorde, PT Cruiser, Dakota, Durango, Grand Cherokee, Intrepid, LHS, Liberty, Magnum, Neon, Pacifica, Prowler Ram, Sebring, Stratus, Town & Country, Wrangler | **Fuel System Lean (B1 S1) Conditions:**<br>Engine running in closed loop. AAT Sensor signal over 20ºF (-7°C). Altitude less than 8,500 feet. Fuel level greater than 15%. If PCM multiplies short-term compensation by long-term adaptive and a certain percentage is exceeded for 2 trips, a freeze frame is stored, the MIL illuminates, and a DTC is stored.<br>**Possible Causes:**<br>Restricted fuel supply line<br>Fuel pump inlet strainer plugged or fuel pump has failed<br>O2 Sensor has failed<br>O2 signal circuit or return circuit has failed<br>O2 Sensor heater operation is faulty<br>TP Sensor sweep has failed<br>MAP Sensor operation has failed<br>ECT Sensor operation has failed<br>Engine mechanical problem is present<br>Fuel is contaminated<br>Exhaust leak exists |
| **P0172**<br>2T Fuel<br>1996-2000<br>Avenger, Breeze, Cirrus, Talon, Eagle Vision, Sebring, Stratus | **Fuel System Rich (Bank 1) Conditions:**<br>Engine running in closed loop; IAT Sensor signal over 20ºF; altitude below 8,000 feet, and the PCM detected too large an amount of Fuel Trim correction due to a rich A/F condition.<br>**Possible Causes:**<br>Base engine fault (i.e., cam timing incorrect, oil level too high)<br>EVAP vapor recovery system failure (e.g., canister full of fuel)<br>Fuel control Sensor is out of calibration (e.g., ECT, IAT or MAP)<br>Fuel delivery component fault (injector leak, high fuel pressure)<br>O2 element is contaminated, deteriorated or it has failed<br>O2 heater is damaged or it has failed |
| **P0172**<br>2T Fuel<br>1996-2003<br>300M, Avenger, Breeze, Caravan, Cherokee, Cirrus, Concorde, PT Cruiser, Dakota, Durango, Grand Cherokee, Intrepid, LHS, Liberty, Neon, New Yorker, Prowler, Ram, Sebring, Stratus, Town & Country, Viper, Vision, Voyager, Wrangler | **Fuel System Rich (Bank 1) Conditions:**<br>Engine running in closed loop, IAT Sensor signal over 20ºF, altitude less than 8,000 feet, and the PCM detected too large an amount of Fuel Trim correction due to a rich A/F condition.<br>**Possible Causes:**<br>Base engine fault (i.e., cam timing incorrect, oil level too high)<br>EVAP vapor recovery system failure (e.g., canister full of fuel)<br>Fuel control Sensor is out of calibration (e.g., ECT, IAT or MAP)<br>Fuel delivery component fault (injector leak, high fuel pressure)<br>O2 element is contaminated, deteriorated or it has failed<br>O2 heater is damaged or it has failed |
| **P0172**<br>2T Fuel<br>2001-03<br>Sebring, Stratus Coupe with 2.4L I4 or 3.0L V6 | **Fuel System Rich (Bank 1) Conditions:**<br>ECT Sensor from 140-212ºF, IAT Sensor less than 140ºF at startup, engine running in closed loop, volume airflow Sensor less than 88 Hz, and the PCM detected the Long Term fuel trim exceeded -12.5% or the Short Term fuel trim exceeded -25% for 5 seconds.<br>**Possible Causes:**<br>Base engine fault (i.e., cam timing incorrect, oil level too high)<br>EVAP vapor recovery system has failed (canister full of fuel)<br>High fuel pressure (fuel pressure regulator is sticking or failed)<br>One or more injectors leaking or pressure regulator is leaking<br>O2 element is contaminated, deteriorated or it has failed |

**OBD II Trouble Code List (P0XXX Codes) - Continued**

| DTC | Trouble Code Title, Conditions & Possible Causes: |
|---|---|
| **P0172**<br>2T Fuel<br>1996-2003<br>Cherokee, Dakota, Durango, Grand Cherokee, Prowler, Ram, Viper, Voyager, Wrangler | **Fuel System Rich (Bank 1) Conditions:**<br>Engine running in closed loop, IAT Sensor signal over 20°F, altitude less than 8,000 feet, and the PCM detected too large an amount of Fuel Trim correction due to a rich A/F condition.<br>**Possible Causes:**<br>Base engine fault (i.e., cam timing incorrect, oil level too high)<br>EVAP vapor recovery system has failed (canister full of fuel)<br>Fuel injector (one or more) leaking or pressure regulator leaking<br>O2 element is deteriorated or has failed |
| **P0172**<br>2T Fuel<br>2004-05<br>300M, Caravan, Concorde, PT Cruiser, Dakota, Durango, Grand Cherokee, Intrepid, LHS, Liberty, Magnum, Neon, Pacifica, Prowler Ram, Sebring, Stratus, Town & Country, Wrangler | **Fuel System (S1 B1) Rich Conditions:**<br>Engine running in closed loop. IAT Sensor signal over 20°F (-7°C). Altitude less than 8,500 feet. PCM multiplies short-term compensation by long-term adaptive, as well as a purge fuel multiplier, and the result is below a certain value for 30 seconds over 2 trips, a freeze frame is stored. MIL illuminates and DTC is stored.<br>**Possible Causes:**<br>O2 Sensor heater or O2 Sensor has failed<br>EVAP purge solenoid failed or improper operation<br>O2 signal circuit or return circuit has failed<br>MAP Sensor has failed or circuit malfunction<br>ECT Sensor has failed or circuit malfunction<br>Engine mechanical problem<br>Fuel filter/pressure regulator has failed or needs repair<br>PCM has failed |
| **P0174**<br>2T Fuel<br>1996-2000<br>Avenger, Breeze, Eagle Vision, Sebring, Stratus with V6 | **Fuel System Lean (Bank 2) Conditions:**<br>Engine running in closed loop, IAT Sensor signal over 20°F, altitude less than 8,000 feet, and the PCM detected too large an amount of Fuel Trim correction due to a lean A/F condition.<br>**Possible Causes:**<br>Air leaks in intake manifold, exhaust pipes or exhaust manifold<br>Base engine mechanical problem causing a lean A/F condition<br>Fuel control Sensor is out of calibration (e.g., ECT, IAT or MAP)<br>Fuel delivery component fault (clogged filter, low fuel pressure)<br>O2 element is contaminated, deteriorated or it has failed<br>Vacuum hose is disconnected, broken, leaking or loose |
| **P0174**<br>2T Fuel<br>1996-2003<br>300M, Avenger, Breeze, Caravan, Cherokee, Cirrus, Concorde, PT Cruiser, Dakota, Durango, Grand Cherokee, Intrepid, LHS, Liberty, Neon, New Yorker, Prowler, Ram, Sebring, Stratus, Town & Country, Viper, Vision, Voyager, Wrangler | **Fuel System Lean (Bank 2) Conditions:**<br>Engine running in closed loop, IAT Sensor signal over 20°F, altitude less than 8,000 feet, and the PCM detected too large an amount of Fuel Trim correction due to a lean A/F condition.<br>**Possible Causes:**<br>Air leaks in intake manifold, exhaust pipes or exhaust manifold<br>Base engine mechanical problem causing a lean A/F condition<br>Fuel control Sensor is out of calibration (e.g., ECT, IAT or MAP)<br>Fuel delivery component fault (clogged filter, low fuel pressure)<br>O2 element is contaminated, deteriorated or it has failed<br>Vacuum hose is disconnected, broken, leaking or loose |

## OBD II Trouble Code List (P0XXX Codes) - Continued

| DTC | Trouble Code Title, Conditions & Possible Causes: |
|---|---|
| **P0174**<br>2T Fuel<br>2001-03<br>Sebring, Stratus<br>Coupe with 3.0L V6 | **Fuel System Lean (Bank 2) Conditions:**<br>ECT Sensor between 140ºF to 212ºF, IAT Sensor less than 140ºF at startup, engine running in closed loop, volume airflow Sensor less than 88 Hz, and the PCM detected the Long Term fuel trim exceed +12.5% or the Short Term fuel trim exceeded +25% for 5 seconds.<br>**Possible Causes:**<br>Air leaks in intake manifold, exhaust pipes or exhaust manifold<br>Base engine mechanical problem causing a lean A/F condition<br>Fuel control Sensor is out of calibration (e.g., BARO or VAF)<br>Fuel delivery component fault (clogged filter, low fuel pressure)<br>O2 element is contaminated, deteriorated or it has failed<br>Vacuum hose is disconnected, broken, leaking or loose |
| **P0174**<br>2T Fuel<br>1996-2003<br>Cherokee, Dakota, Durango, Grand Cherokee, Prowler, Ram, Viper, Voyager, Wrangler | **Fuel System Lean (Bank 2) Conditions:**<br>Engine running in closed loop, IAT Sensor signal over 20ºF, altitude less than 8,000 feet, and the PCM detected too large an amount of Fuel Trim correction due to a lean A/F condition.<br>**Possible Causes:**<br>Air leaks in intake manifold, exhaust pipes or exhaust manifold<br>Base engine mechanical problem causing a lean A/F condition<br>Fuel control Sensor is out of calibration (e.g., ECT, IAT or MAP)<br>Fuel delivery component fault (clogged filter, low fuel pressure)<br>O2 element is contaminated, deteriorated or it has failed<br>Vacuum hose is disconnected, broken, leaking or loose |
| **P0174**<br>2T Fuel<br>2004-05<br>300M, Caravan, Concorde, PT Cruiser, Dakota, Durango, Grand Cherokee, Intrepid, LHS, Liberty, Magnum, Neon, Pacifica, Prowler Ram, Sebring, Stratus, Town & Country, Viper, Voyager, Wrangler | **Fuel System (S2 B1) Lean Conditions:**<br>Engine running in closed loop. IAT Sensor signal over 20ºF (-7°C). Altitude less than 8,500 feet. PCM multiplies short-term compensation by long-term adaptive, and a certain percentage is exceeded in 2 trips, a freeze frame is stored. MIL illuminates and DTC is stored. 3 good trips required to turn off MIL.<br>**Possible Causes:**<br>Restricted fuel supply line<br>Fuel pump inlet strainer plugged<br>Fuel pump is damaged or has failed<br>O2 signal circuit or return circuit has failed<br>MAP Sensor has failed or circuit malfunction<br>ECT Sensor has failed or circuit malfunction<br>Engine mechanical problem<br>Fuel filter/pressure regulator has failed or needs repair<br>O2 Sensor has failed<br>PCM has failed |
| **P0175**<br>2T Fuel<br>1996-2000<br>Avenger, Breeze, Eagle Vision, Sebring, Stratus with V6 | **Fuel System Rich (Bank 2) Conditions:**<br>Engine running in closed loop, IAT Sensor signal over 20ºF, altitude less than 8,000 feet, and the PCM detected too large an amount of Fuel Trim correction due to a rich A/F condition.<br>**Possible Causes:**<br>Base engine fault (i.e., cam timing incorrect, oil level too high)<br>EVAP vapor recovery system failure (e.g., canister full of fuel)<br>Fuel control Sensor is out of calibration (e.g., ECT, IAT or MAP)<br>Fuel delivery component fault (injector leak, high fuel pressure)<br>O2 element is contaminated, deteriorated or it has failed<br>O2 heater is damaged or it has failed |

**OBD II Trouble Code List (P0XXX Codes) - Continued**

| DTC | Trouble Code Title, Conditions & Possible Causes: |
|---|---|
| **P0175**<br>2T Fuel<br>1996-2003<br>300M, Concorde, Intrepid, LHS, Neon, Prowler, Sebring, Stratus, Vision, Neon, PT Cruiser, Caravan/Town & Country/Voyager | **Fuel System Rich (Bank 2) Conditions:**<br>Engine running in closed loop, IAT Sensor signal over 20°F, altitude less than 8,000 feet, and the PCM detected too large an amount of Fuel Trim correction due to a rich A/F condition.<br>**Possible Causes:**<br>Base engine fault (i.e., cam timing incorrect, oil level too high)<br>EVAP vapor recovery system failure (e.g., canister full of fuel)<br>Fuel control Sensor is out of calibration (e.g., ECT, IAT or MAP)<br>Fuel delivery component fault (injector leak, high fuel pressure)<br>O2 element is contaminated, deteriorated or it has failed<br>O2 heater is damaged or it has failed |
| **P0175**<br>2T Fuel<br>2001-03<br>Sebring, Stratus Coupe with 3.0L V6 | **Fuel System Rich (Bank 2) Conditions:**<br>ECT Sensor from 140-212°F, IAT Sensor less than 140°F at startup, engine running in closed loop, volume airflow Sensor less than 88 Hz, and the PCM detected the Long Term fuel trim exceeded -12.5% or the Short Term fuel trim exceeded -25% for 5 seconds.<br>**Possible Causes:**<br>Base engine fault (i.e., cam timing incorrect, oil level too high)<br>EVAP vapor recovery system has failed (canister full of fuel)<br>High fuel pressure (fuel pressure regulator is sticking or failed)<br>One or more injectors leaking or pressure regulator is leaking<br>O2 element is contaminated, deteriorated or it has failed |
| **P0175**<br>2T Fuel<br>1996-2003<br>Cherokee, Dakota, Durango, Grand Cherokee, Prowler, Ram, Viper, Voyager, Wrangler | **Fuel System Rich (Bank 2) Conditions:**<br>Engine running in closed loop, IAT Sensor signal over 20°F, altitude less than 8,000 feet, and the PCM detected too large an amount of Fuel Trim correction due to a rich A/F condition.<br>**Possible Causes:**<br>Base engine fault (i.e., cam timing incorrect, oil level too high)<br>EVAP vapor recovery system has failed (canister full of fuel)<br>Fuel injector (one or more) leaking or pressure regulator leaking<br>O2 element is deteriorated or has failed |
| **P0175**<br>2T Fuel<br>2004-05<br>300M, Caravan, Concorde, PT Cruiser, Dakota, Durango, Grand Cherokee, Intrepid, LHS, Liberty, Magnum, Neon, Pacifica, Prowler Ram, Sebring, Stratus, Town & Country, Viper, Voyager, Wrangler | **Fuel System (S2 B1) Rich Conditions:**<br>Engine running in closed loop. IAT Sensor signal over 20°F (-7°C). Altitude less than 8,500 feet. If the PCM multiplies short-term compensation by long-term adaptive, and a purge fuel multiplier, and the result is below a certain value for 30 seconds in 2 trips, a freeze frame is stored. MIL illuminates and DTC is stored. 3 good trips required to turn off MIL.<br>**Possible Causes:**<br>Restricted fuel supply line<br>Fuel pump inlet strainer plugged<br>Fuel pump is damaged or has failed<br>O2 signal circuit or return circuit has failed<br>MAP Sensor has failed or circuit malfunction<br>ECT Sensor has failed or circuit malfunction<br>Engine mechanical problem<br>Fuel filter/pressure regulator has failed or needs repair<br>O2 Sensor has failed<br>PCM has failed |

**OBD II Trouble Code List (P0XXX Codes) - Continued**

| DTC | Trouble Code Title, Conditions & Possible Causes: |
|---|---|
| **P0176**<br>1T CCM<br>1998-2002<br>Caravan/Town &<br>Country/Voyager | **Loss Of Flexible Fuel Calibration Signal Conditions:**<br>Key on. The PCM detected the Flexible Fuel (FF) Sensor signal was too low or too high.<br>**Possible Causes:**<br>Flexible Fuel Sensor signal circuit shorted to Sensor ground<br>Flexible Fuel Sensor signal circuit is open<br>Flexible Fuel Sensor has failed<br>PCM has failed |
| **P0177**<br>1T CCM<br>1998-2003<br>Ram with 5.9L Diesel | **Water In Fuel Signal Circuit Fault Conditions:**<br>Key on. The PCM detected the Water In Fuel (WIF) Sensor signal was out of range low.<br>**Possible Causes:**<br>Water In Fuel Sensor has detected water in the fuel supply (drain the water from the fuel filter and then retest for the code)<br>Water In Fuel Sensor signal circuit shorted to Sensor ground<br>Water In Fuel Sensor signal circuit shorted to chassis ground<br>PCM has failed |
| **P0178**<br>1T CCM<br>1998-2002<br>Caravan, Town &<br>Country, Voyager | **Flexible Fuel Sensor Voltage Too Low Conditions:**<br>Key on, and the PCM detected the Flexible Fuel (FF) Sensor signal was less than 0.51v.<br>**Possible Causes:**<br>Flexible Fuel Sensor signal circuit shorted to Sensor ground<br>Flexible Fuel Sensor signal circuit is shorted to chassis ground<br>Flexible Fuel Sensor has failed<br>PCM has failed |
| **P0178**<br>1T CCM<br>1998-2003<br>Ram with 5.9L Diesel | **Water in Fuel Sensor Low Input Conditions:**<br>Key on, and the PCM detected the Water In Fuel (WIF) Sensor signal was less than 0.51v.<br>**Possible Causes:**<br>Water In Fuel Sensor signal circuit shorted to Sensor ground<br>Water In Fuel Sensor signal circuit shorted to chassis ground<br>Water In Fuel Sensor has failed<br>PCM has failed |
| **P0179**<br>1T CCM<br>1998-2002<br>Caravan, Town &<br>Country, Voyager | **Flexible Fuel Sensor Voltage Too High Conditions:**<br>Key on or engine running; and the PCM detected the Flexible Fuel (FF) Sensor signal was more than 4.96v during the test period.<br>**Possible Causes:**<br>Flexible Fuel Sensor signal circuit open or ground circuit is open<br>Flexible Fuel Sensor signal circuit shorted to vehicle power<br>Flexible Fuel Sensor has failed<br>PCM has failed |
| **P0180**<br>1T CCM<br>1996-2003<br>Ram with 5.9L Diesel | **Fuel Injection Pump Temperature Out-Of-Range Conditions:**<br>Engine started; and the PCM detected the temperature of the fuel was out of its normal operating range during the test period.<br>**Possible Causes:**<br>Fuel injection pump has failed (this code is normally caused by an internal failure of the fuel injection pump)<br>PCM has failed |
| **P0180**<br>1T CCM<br>2004-05<br>Liberty with Diesel | **Fuel Temperature Sensor Signal Voltage Too Low Or Too High Conditions:**<br>Ignition on; Fuel Temperature Sensor signal is below 0.12v or above 4.95v for 0.5 seconds.<br>**Possible Causes:**<br>Intermittent condition<br>Fuel Temp. Sensor signal circuit is open or is shorted to voltage (too high)<br>Fuel Temp. Sensor signal circuit is shorted to ground (too low)<br>Fuel Temp. Sensor signal and ground circuits are shorted together<br>Fuel Temp. Sensor has failed<br>ECM has failed |

| | |
|---|---|
| **P0181**<br>1T CCM<br>2001-03<br>Sebring, Stratus<br>Coupe with 2.4L I4 or 3.0L V6 | **Fuel Temperature Sensor Circuit Malfunction Conditions:**<br>ECT Sensor from 14-97°F and IAT Sensor within 5°F of the ECT at startup, ECT Sensor over 140°F during testing, engine running with the VSS less than 17 mph, and the PCM detected the difference between the Fuel Temperature and ECT Sensor signals was more than 27°F.<br>**Possible Causes:**<br>Fuel Temperature Sensor connector is damaged or loose<br>Fuel Temperature Sensor is damaged or it has failed<br>PCM has failed |
| **P0181**<br>1T CCM<br>1996-2003<br>Ram with 5.9L Diesel | **Fuel Injection Pump Failure Conditions:**<br>Key on or engine running; and the PCM detected the signal from the Fuel Temperature Sensor signal was too high or too low in the test. Note: The Sensor is located inside the Bosch VP44 Pump Controller.<br>**Possible Causes:**<br>Fuel Temperature Sensor has failed (this Sensor is not serviceable without replacing the VP44 pump controller unit).<br>Engine power is "de-rated" when this trouble code is set<br>PCM has failed |
| **P0182**<br>1T CCM<br>2001-03<br>Sebring, Stratus<br>Coupe with 2.4L I4 or 3.0L V6 | **Fuel Temperature Sensor Circuit Low Input Conditions:**<br>Engine started; engine runtime more than 2 seconds and the PCM detected the Fuel Temperature Sensor indicated less than 0.10v.<br>**Possible Causes:**<br>Fuel temperature Sensor connector is damaged or shorted<br>Fuel temperature Sensor circuit is shorted to ground<br>Fuel temperature Sensor is damaged or it has failed<br>PCM has failed |
| **P0182**<br>1T CCM<br>1996-2003<br>Ram with 5.2L VIN T CNG | **CNG Temperature Sensor Low Input Conditions:**<br>Key on or engine running; and the PCM detected the CNG temperature Sensor signal was less than 0.51v during the test.<br>**Possible Causes:**<br>CNG Sensor signal circuit is shorted to Sensor ground<br>CNG Sensor signal circuit is shorted to chassis ground<br>CNG Sensor has failed (it may be shorted internally)<br>PCM has failed |
| **P0183**<br>1T CCM<br>2001-03<br>Sebring, Stratus<br>Coupe with 2.4L I4 or 3.0L V6 | **Fuel Temperature Sensor Circuit High Input Conditions:**<br>Engine started; engine runtime more than 2 seconds and the PCM detected the Fuel Temperature Sensor indicated less than 4.60v.<br>**Possible Causes:**<br>Fuel temperature Sensor connector is damaged or shorted<br>Fuel temperature Sensor signal circuit is shorted to VREF (5v)<br>Fuel temperature Sensor is damaged or it has failed<br>PCM has failed |
| **P0183**<br>1T CCM<br>1996-2003<br>Ram with 5.2L VIN T CNG | **CNG Temperature Sensor High Input Conditions:**<br>Key on or engine running; and the PCM detected the CNG temperature Sensor signal was more than 4.96v during the test.<br>**Possible Causes:**<br>CNG Sensor signal circuit is open<br>CNG Sensor signal circuit is shorted to vehicle power<br>CNG Sensor has failed (it may be open internally)<br>PCM has failed |

**OBD II Trouble Code List (P0XXX Codes) - Continued**

| DTC | Trouble Code Title, Conditions & Possible Causes: |
|---|---|
| **P0190**<br>1T CCM<br>2004-05<br>Liberty with Diesel | **Fuel Pressure Sensor Signal Voltage Too Low Or Too High Conditions:**<br>Ignition on; Fuel Pressure Sensor signal is below 0.2v or above 4.8v for 0.5 second.<br>**Possible Causes:**<br>Fuel Pressure Sensor signal circuit is open or is shorted to voltage (too high)<br>Fuel Pressure Sensor signal circuit is shorted to ground (too low)<br>Fuel Pressure Sensor ground circuit is open (too high)<br>Fuel Pressure Sensor signal circuit is shorted to Sensor ground (too low)<br>Intermittent condition<br>Fuel Pressure Sensor 5v supply circuit is open (too high)<br>Fuel Pressure Sensor ground circuit is shorted to voltage (too high)<br>Fuel Pressure Sensor has failed<br>ECM has failed |
| **P0191**<br>1T CCM<br>2004-05<br>Liberty with Diesel | **Fuel Pressure Sensor After-Run Negative Or Positive Plausibility Conditions:**<br>At ignition shut off during After-Run; Fuel Pressure Sensor signal is below 0.415v or above 0.615v for 1.0 second.<br>**Possible Causes:**<br>Fuel Pressure Sensor has failed<br>Intermittent condition |
| **P0196**<br>1T CCM<br>2004-05<br>300M, Caravan, Concorde, PT Cruiser, Dakota, Durango, Grand Cherokee, Intrepid, LHS, Liberty, Magnum, Neon, Pacifica, Prowler Ram, Sebring, Stratus, Town & Country, Viper, Voyager, Wrangler | **Engine Oil Temperature Sensor Circuit Performance Conditions:**<br>Engine off time is more than 8 hours; ambient temperature is more than 38°F (4°C). If the PCM detects the engine oil temperature value is incorrect, by comparing it with other engine inputs, then the DTC will set. 3 good trips required to turn off MIL.<br>**Possible Causes:**<br>Engine oil temp signal circuit is open or is shorted to ground or to battery voltage<br>Engine oil temp Sensor ground circuit is open<br>Engine oil temp signal circuit is shorted to Sensor ground<br>Engine oil temp Sensor has failed<br>PCM has failed |
| **P0197**<br>1T CCM<br>2004-05<br>300M, Caravan, Concorde, PT Cruiser, Dakota, Durango, Grand Cherokee, Intrepid, LHS, Liberty, Magnum, Neon, Pacifica, Prowler Ram, Sebring, Stratus, Town & Country, Viper, Voyager, Wrangler | **Engine Oil Temperature Sensor Circuit Low Conditions:**<br>Ignition is on; battery voltage is more than 10.4v. The engine oil temperature Sensor circuit voltage at the PCM is less than the calibrated amount. 3 good trips required to turn off MIL.<br>**Possible Causes:**<br>Engine oil temp signal circuit is shorted to ground<br>Engine oil temp signal circuit is shorted to Sensor ground<br>Engine oil temp Sensor has failed<br>PCM has failed |

**OBD II Trouble Code List (P0XXX Codes) - Continued**

| DTC | Trouble Code Title, Conditions & Possible Causes: |
|---|---|
| **P0198**<br>1T CCM<br>2004-05<br>300M, Caravan, Concorde, PT Cruiser, Dakota, Durango, Grand Cherokee, Intrepid, LHS, Liberty, Magnum, Neon, Pacifica, Prowler Ram, Sebring, Stratus, Town & Country, Viper, Voyager, Wrangler | **Engine Oil Temperature Sensor Circuit High Conditions:**<br>Ignition is on; battery voltage is more than 10.4v. The engine oil temperature Sensor circuit voltage at the PCM is higher than the calibrated amount. 3 good trips required to turn off MIL.<br>**Possible Causes:**<br>Engine oil temp signal circuit is open or is shorted to battery voltage<br>Engine oil temp Sensor ground circuit is open<br>Engine oil temp Sensor has failed<br>PCM has failed |
| **P0201-P0204**<br>1T CCM<br>1995-98<br>Talon | **Fuel Injector 1, 2, 3 or 4 Control Conditions:**<br>Engine speed below 1000 rpm, system voltage over 12v, Actuator Test off, TP Sensor signal under 1.16v, and the PCM did not detect any injector coil surge voltage for 3 ms after Injector 1-4 was turned "off" (a surge voltage of 2v higher than system voltage is expected).<br>**Possible Causes:**<br>Fuel injector 1-4 control circuit is open or grounded<br>Fuel injector 1-4 power circuit from the ASD relay is open<br>Fuel injector 1 has failed or PCM injector 1 driver has failed |
| **P0201-P0204**<br>1T CCM<br>2004-05<br>PT Cruiser | **Fuel Injector 1, 2, 3 or 4 Control Circuit Open Conditions:**<br>Ignition on. The PCM tests the injector circuit internally for more than 27 injector pulses and has determined that the circuit is open.<br>**Possible Causes:**<br>Intermittent condition<br>Fuel injector 1-4 control circuit is open<br>Fuel injector 1-4 ASD relay output circuit is open<br>Fuel injector driver circuit is open<br>Fuel injector is clogged or has failed<br>PCM has failed |
| **P0201-P0204**<br>1T CCM<br>2004-05<br>Neon | **Fuel Injector 1, 2, 3 or 4 Control Circuit Open Conditions:**<br>Engine running at less than 3000 rpm; battery voltage is greater than 10v; the ASD relay is energized. The PCM did not detect any inductive spikes after injector turn off.<br>**Possible Causes:**<br>ASD relay output circuit problem<br>Fuel injector is clogged or has failed<br>Fuel injector control circuit is open or is shorted to ground<br>PCM has failed |
| **P0201-P0204**<br>1T CCM<br>2004-05<br>Liberty with Diesel | **Fuel Injector 1, 2, 3 or 4 Injector Circuit Load Drop Or Overcurrent High Side Or Low Side Conditions:**<br>Engine running. The ECM detects insufficient current through the injector driver when commanded ON (circuit load drop), or the ECM detects excessive current on the high side driver circuit or on the low side driver circuit.<br>**Possible Causes:**<br>ECM has failed<br>Intermittent condition<br>Fuel injector control circuit is open or is shorted to ground or to voltage<br>Fuel injector control circuits are shorted together<br>Fuel injector has failed |

**OBD II Trouble Code List (P0XXX Codes) - Continued**

| DTC | Trouble Code Title, Conditions & Possible Causes: |
|---|---|
| **P0201-P0210**<br>1T CCM<br>1995-2005<br>Ram, Viper | **Injector 1, 2, 3, 4, 5, 6, 7, 8, 9 or 10 Control Conditions:**<br>ASD relay "on", engine speed under 3000 rpm, injector pulse width under 10 ms, system voltage over 12v, and the PCM did not detect any inductive spike from the injector for 0.18 ms after it is turned off. Note: This code takes 0.64-10 seconds to set once the injector is off.<br>**Possible Causes:**<br>Fuel injector 1-10 control circuit is open or grounded<br>Fuel injector 1-10 power circuit from the ASD relay is open<br>Fuel injector 1-10 has failed<br>PCM injector 1-10 driver has failed |
| **P0201-P0206**<br>1T CCM<br>2004-05<br>300M, Caravan, Concorde, PT Cruiser, Dakota, Durango, Grand Cherokee, Intrepid, LHS, Liberty, Magnum, Neon, Pacifica, Prowler Ram, Sebring, Stratus, Town & Country, Viper, Voyager, Wrangler | **Injector 1, 2, 3, 4, 5, or 6 Control Conditions:**<br>ASD relay "on"; engine speed under 3000 rpm; battery voltage greater than 10v (11.9981v: Liberty). No inductive spike is detected after injector turns off.<br>**Possible Causes:**<br>ASD relay output circuit failure<br>Fuel injector has malfunctioned or failed<br>Fuel injector control circuit is open or shorted to ground<br>PCM has failed |
| **P0201-P0208**<br>1T CCM<br>2004-05<br>300M, Concorde, Dakota, Durango, Grand Cherokee, Intrepid, LHS, Magnum, Ram, Sebring, Stratus, Town & Country, Viper | **Injector 1, 2, 3, 4, 5, 6, 7 or 8 Control Conditions:**<br>ASD relay "on", engine speed under 3000 rpm, battery voltage greater than 10v. No inductive spike is detected after injector turn off.<br>**Possible Causes:**<br>ASD relay output circuit failure<br>Fuel injector has malfunctioned or failed<br>Fuel injector control circuit is open or shorted to ground<br>PCM has failed |
| **P0215**<br>1T CCM<br>1998-2003<br>Ram with 5.9L Diesel | **Fuel Injection Pump Control Circuit Conditions:**<br>Key on, and the PCM did not detect any power present on the Fuel Injection Pump control circuit (VP44) during the CCM test period.<br>**Possible Causes:**<br>Fuel injection pump relay control circuit open, shorted to ground<br>Fuel injection pump relay ground circuit is open<br>Fuel injection pump relay has failed<br>Engine Control Module has failed |
| **P0216**<br>1T CCM<br>1998-2003<br>Ram with 5.9L Diesel | **Fuel Injection Pump Timing Failure Conditions:**<br>No other fuel injection pump trouble codes set, engine speed more than 300 rpm, fuel command 5 ml/stroke, and the PCM detected the pump timing command was out of range.<br>Possible Causes:<br>Fuel injection pump gear not aligned properly<br>Fuel filter plugged or restricted<br>Fuel injection pump is damaged<br>Transfer pump inlet line restricted |

## OBD II Trouble Code List (P0XXX Codes) - Continued

| DTC | Trouble Code Title, Conditions & Possible Causes: |
|---|---|
| **P0217**<br>1T CCM<br>1998-2003<br>Ram with 5.9L Diesel | **Decreased Performance (Engine Overheat) Conditions:**<br>Engine started; and the PCM detected that the signal from the ECT Sensor indicated an engine overheat condition existed. Note: Measure coolant temperature (with thermal couple) and compare to the scan tool ECT PID (they should be within 10ºF of each other).<br>**Possible Causes:**<br>ECT Sensor circuit has high resistance<br>ECT Sensor is damaged or has deteriorated<br>PCM has failed |
| **P0218**<br>1T CCM<br>2002-05<br>300M, Caravan, Cherokee, Concorde, PT Cruiser, Dakota, Durango, Grand Cherokee, Intrepid, LHS, Liberty, Magnum, Neon, Pacifica, Prowler Ram, Sebring, Stratus, Town & Country, Viper, Voyager, Wrangler | **A/T High Temperature Operation Activated Conditions:**<br>Engine started; vehicle driven in gear, and the TCM indicated the Overheat shift schedule was activated (i.e., the TCM had detected a transmission oil temperature of more than 240ºF). Note: This is an informational DTC, designed to aid the technician in diagnosing shift quality complaints.<br>**Possible Causes:**<br>Engine cooling system malfunction present<br>High temperature operations activated<br>Transmission oil pump flow is too low or it is restricted |
| **P0218**<br>1T CCM<br>2004-05<br>Liberty with Diesel | **Transmission High Temperature Operation Activated Conditions:**<br>Engine running. This DTC is an informational code and does not necessarily indicate that a failure exists. It merely flags the fact that the transmission sump oil temperature reached 240°F (116°C). This temperature level can be reached when operating under a heavy load in hot weather. This causes the transmission controller to use an "overheat" shift schedule, which changes the shift patterns in an attempt to control the temperature. Customers may notice a different feeling or response under these conditions. The Owners' Manual includes an explanation of this "Over Temperature Mode" for information purposes. The DTC sets immediately when the Overheat Shift Schedule is activation with a transmission oil temperature above 240°F (116°C).<br>**Possible Causes:**<br>Sever operation: trailer towing in hot weather<br>Engine cooling system problem<br>Oil pump volume check<br>Torque converter failure<br>High temperature operations activated |
| **P0219**<br>1T CCM<br>1998-2003<br>Ram with 5.9L Diesel | **Crankshaft Position Sensor Over-Speed Signal Conditions:**<br>Engine started; and the PCM detected the engine had exceeded its over-speed value.<br>**Possible Causes:**<br>CKP Sensor or its tone wheel is damaged (incorrect engine speed reading)<br>Induction of an alternate fuel source (i.e., ether or propane)<br>Improper operating conditions (i.e., motoring downhill)<br>Turbocharger seals are leaking |

**OBD II Trouble Code List (P0XXX Codes) - Continued**

| DTC | Trouble Code Title, Conditions & Possible Causes: |
|---|---|
| **P0221**<br>1T CCM<br>2004-05<br>300M, Caravan, Concorde, PT Cruiser, Dakota, Durango, Grand Cherokee, Intrepid, LHS, Liberty, Magnum, Neon, Pacifica, Prowler Ram, Sebring, Stratus, Town & Country, Viper, Voyager, Wrangler | **Throttle Position Sensor No. 2 Performance Conditions:**<br>Ignition on; No MAP Sensor DTCs are set. TP Sensor signals Do NOT correlate to the MAP Sensor signal. If equipped, ETC light will illuminate. P2135 should also set.<br>**Possible Cause:**<br>TP Sensor No. 1 or 2 signal circuit is shorted to battery voltage or to ground<br>TP Sensor No. 1 or 2 signal circuit has high resistance<br>5v supply circuit is shorted to ground<br>TP Sensor return circuit has high resistance<br>TP Sensor No. 1 signal circuit shorted to TP Sensor No. 2 signal circuit<br>TP Sensor or throttle body damaged or has failed<br>PCM has failed |
| **P0222**<br>1T CCM<br>2004-05<br>Liberty with Diesel | **Throttle Position Sensor No. 2 Circuit Low Conditions:**<br>Ignition on; battery voltage is more than 10v. TP Sensor voltage at the PCM is less than 0.16v for 0.7 second. 3 good trips required to turn off MIL.<br>**Possible Cause:**<br>5v supply circuit is open or shorted to ground<br>TP Sensor No. 2 signal circuit shorted to ground or to Sensor return circuit<br>TP Sensor or throttle body damaged or has failed<br>Throttle plate jammed against the maximum stop<br>PCM has failed |
| **P0222**<br>1T CCM<br>1998-2003<br>Ram with 5.9L Diesel | **Idle Validation Signals Both Low Conditions:**<br>Key on, and the PCM detected that the Idle Validation Signal No. 1 and Idle Validation Signal No. 2 both indicated "no" voltage during the test.<br>**Possible Causes:**<br>APP Sensor is damaged<br>IVS No. 1 or IVS No. 2 circuit(s) is shorted to Sensor ground or to chassis ground<br>IVS No. 1 or IVS No. 2 is damaged or has failed<br>Engine Control Module has failed |
| **P0223**<br>1T CCM<br>1998-2003<br>Ram with 5.9L Diesel | **Idle Validation Signals Both High Conditions:**<br>Key on, and the PCM detected that the Idle Validation Signal No. 1 and Idle Validation Signal No. 2 both indicated high voltage during the test.<br>**Possible Causes:**<br>APP Sensor is damaged, or the Sensor ground circuit is open<br>IVS No. 1 or IVS No. 2 circuit(s) is open or shorted to system power<br>IVS No. 1 or IVS No. 2 is damaged or has failed<br>Engine Control Module has failed |
| **P0223**<br>1T CCM<br>2004-05<br>300M, Caravan, Concorde, PT Cruiser, Dakota, Durango, Grand Cherokee, Intrepid, LHS, Liberty, Magnum, Neon, Pacifica, Prowler Ram, Sebring, Stratus, Town & Country, Viper, Voyager, Wrangler | **Throttle Position Sensor No. 2 Circuit High Conditions:**<br>Ignition on; battery voltage is more than 10v. TP Sensor voltage at the PCM is more than 4.9v for 25ms. If equipped, ETC light will illuminate.<br>**Possible Cause:**<br>TP Sensor No. 2 signal circuit shorted to battery voltage or to 5v supply circuit<br>TP Sensor return circuit is open<br>TP Sensor or throttle body is damaged or has failed<br>PCM has failed |

| | |
|---|---|
| **P0224**<br>1T CCM<br>1996-2003<br>Ram with 5.9L Diesel | **Turbocharger Boost Limited Exceeded Conditions:**<br>No Intake Air Pressure Sensor codes set, engine started, engine running at a speed of more than 2200 rpm, and the PCM detected that the Turbocharger Boost Limit had been exceeded.<br>**Possible Causes:**<br>Wastegate has a mechanical problem or it has failed<br>Wastegate stuck open |
| **P0230**<br>1T CCM<br>1998-2002<br>Ram with 5.9L Diesel | **Transfer Pump (Lift Pump) Circuit Out-Of-Range Conditions:**<br>Key on, and the PCM detected the Transfer Pump signal was too high or low during testing.<br>**Possible Causes:**<br>Transfer Pump power feed circuit open, or shorted to ground<br>Transfer Pump ground circuit is open<br>Transfer Pump (internal) resistance out of range<br>Engine Control Module has failed |
| **P0232**<br>1T CCM<br>1998-2002<br>Ram with 5.9L Diesel | **Fuel Shutoff Signal Voltage To High Conditions:**<br>Key on, and the PCM detected that the Fuel Shutoff Signal was too high due a problem.<br>**Possible Causes:**<br>Transfer Pump may be damaged<br>Engine Control Module has failed |
| **P0234**<br>1T CCM<br>1998-2003<br>Ram with 5.9L Diesel | **Turbo Boost Limit Exceeded Conditions:**<br>No Intake Air Pressure Sensor codes set, engine speed over 2200 rpm, and the PCM detected the Turbo Boost exceeded its limit due to a problem in the Turbocharger Wastegate Unit.<br>**Possible Causes:**<br>Wastegate is stuck<br>Wastegate has failed (due to a mechanical failure) |
| **P0234**<br>1T CCM<br>2004-05<br>Neon with Turbo | **Turbo Overboost Performance Conditions:**<br>Engine running. The TIP Sensor signal indicates excessive boost pressure. Other related DTCs may be present.<br>**Possible Causes:**<br>Hoses and/or tubing may be disconnected or cracked<br>Vacuum supply to surge solenoid may be blocked or disconnected<br>Solenoid has failed<br>Turbocharger is malfunctioning<br>Wastegate actuator is damaged or has failed. |
| **P0234**<br>1T CCM<br>2004-05<br>Liberty with Diesel | **Turbo Boost Pressure Negative Deviation Performance Conditions:**<br>Engine running. Actual boost pressure differs from the boost pressure set point by more than 14.5 psi (1000 hPa).<br>**Possible Causes:**<br>Air filter is clogged or malfunctioning<br>Air restrictions exist in intake or boost components<br>Air leaks exist in intake or boost components<br>Boost control vacuum supply is insufficient<br>Boost pressure actuator is malfunctioning or has failed<br>Turbocharger is malfunctioning or has failed |
| **P0235**<br>1T CCM<br>2004-05<br>Liberty with Diesel | **Turbo Boost Pressure Sensor Plausibility Conditions:**<br>Engine running at less than 850 rpm; no other Boost Pressure Sensor DTCs are present; No Atmospheric Pressure DTCs are present. The Boost Pressure Sensor signal differs from the Atmospheric Pressure signal by 2.18 (150 hPa) or more for at least 2 seconds.<br>**Possible Causes:**<br>Intermittent condition<br>High resistance in Boost Pressure Sensor signal circuit, ground circuit or 5v supply circuit<br>Boost Pressure/Intake Air Temperature Sensor has failed<br>ECM has failed |

**OBD II Trouble Code List (P0XXX Codes) - Continued**

| DTC | Trouble Code Title, Conditions & Possible Causes: |
|---|---|
| **P0235**<br>1T CCM<br>2004-05<br>Liberty with Diesel | **Turbo Boost Pressure Sensor Signal Voltage Too High Or Too Low Conditions:**<br>Ignition on. The Boost Pressure Sensor signal voltage is above 4.79v for 0.5 second (too high) or is below 0.29v for 0.5 second (too low).<br>**Possible Causes:**<br>Intermittent condition<br>BP Sensor ground circuit is shorted to voltage or is open (too high)<br>BP Sensor 5v supply circuit is open (too low)<br>BP Sensor signal circuit is shorted to voltage (too high)<br>BP Sensor signal circuit is shorted to ground or is open (too low)<br>BP Sensor signal and ground circuits are shorted together (too low)<br>BP/IAT Sensor has failed<br>Poor connector terminal contact (too high)<br>ECM has failed |
| **P0236**<br>1T CCM<br>1998-2003<br>Ram with 5.9L Diesel | **MAP Sensor Voltage Too High To Long Conditions:**<br>DTC P0106, P0107 and P0108 not set, engine started and the PCM detected the Pressure Test failed because the MAP Sensor signal was too high. This occurs if the Boost Pressure Sensor indicates high when other related signals (load and speed signals) indicate low.<br>**Possible Causes:**<br>MAP Sensor signal circuit open, shorted to ground or to power<br>MAP Sensor has failed<br>Engine Control Module has failed |
| **P0237**<br>1T CCM<br>1998-2003<br>Ram with 5.9L Diesel | **MAP Sensor Voltage Too Low Conditions:**<br>Engine started, engine speed from 416-3520 rpm, system voltage over 10.5v and the PCM detected the MAP Sensor was under 0.10v.<br>**Possible Causes:**<br>Boost Pressure Test did not pass, or a MAP VREF DTC is present<br>MAP Sensor signal circuit shorted to chassis or Sensor ground<br>MAP Sensor has failed |
| **P0238**<br>1T CCM<br>1998-2003<br>Ram with 5.9L Diesel | **MAP Sensor Voltage Too High Conditions:**<br>Engine speed more than 416 rpm but less than 3520 rpm, system voltage over 10.5v and the PCM detected the MAP Sensor signal was more than 4.88v, condition met for 2 seconds.<br>**Possible Causes:**<br>Boost Pressure Test did not pass, ECT or IAT DTC present<br>MAP Sensor signal circuit open or shorted to vehicle power<br>MAP Sensor has failed<br>Engine Control Module has failed |
| **P0243**<br>1T CCM<br>2003-05<br>Neon Turbo, Sebring Sedan with 2.4L | **Wastegate Solenoid Circuit Malfunction Conditions:**<br>Key on or engine running; system voltage over 10.5v. The PCM detected an unexpected voltage condition on the Wastegate Solenoid control circuit during the CCM test period.<br>**Possible Causes:**<br>Wastegate solenoid control circuit is open or shorted to ground<br>Wastegate solenoid control circuit is shorted to system power<br>Wastegate solenoid is damaged or it has failed<br>PCM has failed |
| **P0251**<br>1T CCM<br>1998-2003<br>Ram with 5.9L Diesel | **Injector Pump Mechanical / Fuel Valve Feedback Circuit Fault Conditions:**<br>DTC P1284 not set, key on, system voltage over 12.0v, fuel command 20 ml/stroke, Fuel Delivery valve enabled, and the PCM detected the Fuel Valve feedback signal was over 9.0v.<br>**Possible Causes:**<br>Fuel injection pump relay control circuit is open or grounded<br>Fused battery power (B+) circuit or relay output circuit is open<br>Fuel injection pump relay or fuel injection pump is damaged |

## OBD II Trouble Code List (P0XXX Codes) - Continued

| DTC | Trouble Code Title, Conditions & Possible Causes: |
|---|---|
| **P0251**<br>1T CCM<br>2004-05<br>Liberty with Diesel | **Fuel Quality Solenoid Open Or Short Circuit Conditions:**<br>Ignition on; ECM detects an open or short in the Fuel Quality Solenoid circuit.<br>**Possible Causes:**<br>FQ Solenoid circuit(s) are open, shorted to ground, shorted to voltage, or shorted together<br>Intermittent condition<br>FQ Solenoid or ECM has failed |
| **P0252**<br>1T CCM<br>1998-2003<br>Ram with 5.9L Diesel | **Fuel Valve Signal Missing Conditions:**<br>Engine started; system voltage over 10.5v, Fuel Valve current feedback test completed, and the PCM detected a signal from the Fuel Metering valve that indicated the valve had moved.<br>**Possible Causes:**<br>Fuel injection pump relay control circuit is open or grounded<br>Fused battery power (B+) circuit or relay output circuit is open<br>Fuel injection pump relay or fuel injection pump is damaged |
| **P0252**<br>1T CCM<br>2004-05<br>Liberty with Diesel | **Fuel Quality Solenoid Circuit Malfunction Conditions:**<br>Engine running; ECM detects a malfunction with the Fuel Quality Solenoid circuit.<br>**Possible Causes:**<br>FQ Solenoid circuit(s) are open, shorted to ground, shorted to voltage, or shorted together<br>Intermittent condition<br>FQ Solenoid or ECM has failed |
| **P0253**<br>1T CCM<br>1998-2003<br>Ram with 5.9L Diesel | **Injector Pump Fuel Valve Current Too Low Conditions:**<br>Engine started; engine speed more than 100 rpm and the PCM detected a condition of low current (or no current) at the Fuel Metering Valve during the test period (there may be an open circuit).<br>• Refer to code repair chart in other manuals to test this code. |
| **P0253**<br>1T CCM<br>2004-05<br>Liberty with Diesel | **Fuel Quality Solenoid Short To Ground Circuit Conditions:**<br>Ignition on; ECM detects a short to ground in the Fuel Quality Solenoid circuit(s).<br>**Possible Causes:**<br>FQ Solenoid circuit(s) are open, shorted to ground, shorted to voltage, or shorted together<br>Intermittent condition<br>FQ Solenoid has failed<br>ECM has failed |
| **P0254**<br>1T CCM<br>1998-2003<br>Ram with 5.9L Diesel | **Injector Pump Fuel Valve Current Too High Conditions:**<br>Key "off", and the PCM detected system voltage (power) at the Fuel Metering Valve VP44 circuit during the test period.<br>**Possible Causes:**<br>Fuel injection pump is damaged or it has failed |
| **P0254**<br>1T CCM<br>2004-05<br>Liberty with Diesel | **Fuel Quality Solenoid Short Circuit Conditions:**<br>Ignition on; ECM detects a short in the Fuel Quality Solenoid circuit.<br>**Possible Causes:**<br>FQ Solenoid circuit(s) are open, shorted to ground, shorted to voltage, or shorted together<br>Intermittent condition<br>FQ Solenoid has failed<br>ECM has failed |
| **P0261**<br>1T CCM<br>2004-05<br>PT Cruiser | **Injector No. 1 Control Circuit Low Conditions:**<br>Ignition is on. The PCM tests the injector circuit internally for more than 27 injector pulses and has determined that the circuit is shorted to ground. MIL is illuminated.<br>**Possible Causes:**<br>Intermittent condition<br>Fuel injector has malfunctioned or failed<br>ASD relay output circuit is open<br>Fuel injector driver circuit is shorted to ground<br>PCM has failed |

**OBD II Trouble Code List (P0XXX Codes) - Continued**

| DTC | Trouble Code Title, Conditions & Possible Causes: |
|---|---|
| **P0262**<br>1T CCM<br>2004-05<br>PT Cruiser | **Injector No. 1 Control Circuit High Conditions:**<br>Ignition is on. The PCM tests the injector circuit internally for more than 27 injector pulses and has determined that the circuit is shorted to voltage. MIL is illuminated.<br>**Possible Causes:**<br>Intermittent condition<br>Fuel injector has malfunctioned or failed<br>ASD relay output circuit is open<br>Fuel injector driver circuit is shorted to ASD relay output circuit or to battery voltage<br>PCM has failed |
| **P0264**<br>1T CCM<br>2004-05<br>PT Cruiser | **Injector No. 2 Control Circuit Low Conditions:**<br>Ignition is on. The PCM tests the injector circuit internally for more than 27 injector pulses and has determined that the circuit is shorted to ground. MIL is illuminated.<br>**Possible Causes:**<br>Intermittent condition<br>Fuel injector has malfunctioned or failed<br>ASD relay output circuit is open<br>Fuel injector driver circuit is shorted to ground<br>PCM has failed |
| **P0265**<br>1T CCM<br>2004-05<br>PT Cruiser | **Injector No. 2 Control Circuit High Conditions:**<br>Ignition is on. The PCM tests the injector circuit internally for more than 27 injector pulses and has determined that the circuit is shorted to voltage. MIL is illuminated.<br>**Possible Causes:**<br>Intermittent condition<br>Fuel injector has malfunctioned or failed<br>ASD relay output circuit is open<br>Fuel injector driver circuit is shorted to ASD relay output circuit or to battery voltage<br>PCM has failed |
| **P0267**<br>1T CCM<br>2004-05<br>PT Cruiser | **Injector No. 3 Control Circuit Low Conditions:**<br>Ignition is on. The PCM tests the injector circuit internally for more than 27 injector pulses and has determined that the circuit is shorted to ground. MIL is illuminated.<br>**Possible Causes:**<br>Intermittent condition<br>Fuel injector has malfunctioned or failed<br>ASD relay output circuit is open<br>Fuel injector driver circuit is shorted to ground<br>PCM has failed |
| **P0268**<br>1T CCM<br>2004-05<br>PT Cruiser | **Injector No. 3 Control Circuit High Conditions:**<br>Ignition is on. The PCM tests the injector circuit internally for more than 27 injector pulses and has determined that the circuit is shorted to voltage. MIL is illuminated.<br>**Possible Causes:**<br>Intermittent condition<br>Fuel injector has malfunctioned or failed<br>ASD relay output circuit is open<br>Fuel injector driver circuit is shorted to ASD relay output circuit or to battery voltage<br>PCM has failed |

**OBD II Trouble Code List (P0XXX Codes) - Continued**

| DTC | Trouble Code Title, Conditions & Possible Causes: |
|---|---|
| **P0270**<br>1T CCM<br>2004-05<br>PT Cruiser | **Injector No. 4 Control Circuit Low Conditions:**<br>Ignition is on. The PCM tests the injector circuit internally for more than 27 injector pulses and has determined that the circuit is shorted to ground. MIL is illuminated.<br>**Possible Causes:**<br>Intermittent condition<br>Fuel injector has malfunctioned or failed<br>ASD relay output circuit is open<br>Fuel injector driver circuit is shorted to ground<br>PCM has failed |
| **P0271**<br>1T CCM<br>2004-05<br>PT Cruiser | **Injector No. 4 Control Circuit High Conditions:**<br>Ignition is on. The PCM tests the injector circuit internally for more than 27 injector pulses and has determined that the circuit is shorted to voltage. MIL is illuminated.<br>**Possible Causes:**<br>Intermittent condition<br>Fuel injector has malfunctioned or failed<br>ASD relay output circuit is open<br>Fuel injector driver circuit is shorted to ASD relay output circuit or to battery voltage<br>PCM has failed |
| **P0299**<br>1T CCM<br>2004-05<br>Liberty with Diesel | **Turbo Boost Pressure Positive Deviation Performance Conditions:**<br>Engine running. Actual boost pressure differs from the boost pressure set point by more than 14.5 psi (1000 hPa).<br>**Possible Causes:**<br>Air filter is clogged or malfunctioning<br>Air restrictions exist in intake or boost components<br>Air leaks exist in intake or boost components<br>Boost control vacuum supply is insufficient<br>Boost pressure actuator is malfunctioning or has failed<br>Turbocharger is malfunctioning or has failed |
| **P0300**<br>1T Catalyst<br>2T Emission<br>1995-2000<br>Talon | **Multiple Cylinder Misfire Detected Conditions:**<br>Engine started; adaptive numerator updated, engine speed less than 3000 rpm, and the PCM detected a misfire condition (1-2%) within 1000 revolutions (High Emissions) or a 2-10% misfire condition within 200 revolutions (Catalyst Damaging) in two or more cylinders.<br>Note: If the misfire is severe, the MIL will flash on/off on the 1st trip!<br>**Possible Causes:**<br>Air leak in the intake manifold, or in the EGR or EVAP system<br>Base engine mechanical fault that affects one or more cylinders<br>Erratic or interrupted CKP or CMP Sensor signals<br>Fuel delivery component fault that affects one or more cylinders (i.e., a contaminated, dirty or sticking fuel injector)<br>Ignition system problem (coil or plug) in one or more cylinders<br>Vehicle driven with low fuel pressure or while very low on fuel |

**OBD II Trouble Code List (P0XXX Codes) - Continued**

| DTC | Trouble Code Title, Conditions & Possible Causes: |
|---|---|
| **P0300**<br>1T Catalyst<br>2T Emission<br>1995-2003<br>300M, Avenger, Breeze, Caravan, Cherokee, Cirrus, Concorde, PT Cruiser, Dakota, Durango, Grand Cherokee, Intrepid, LHS, Liberty, Neon, New Yorker, Prowler, Ram, Sebring, Stratus, Town & Country, Viper, Vision, Voyager, Wrangler | **Multiple Cylinder Misfire Detected Conditions:**<br>Engine less than 3000 rpm; adaptive numerator updated. PCM detected a misfire rate of 1-2% (High Emissions 2T), or a misfire rate of 6-30% (Catalyst Damaging 1T) in 2 or more cylinders. Note: If the misfire is severe, the MIL will flash on/off on the 1st trip!<br>**Possible Causes:**<br>Air leak in the intake manifold, or in the EGR or EVAP system<br>Base engine mechanical fault that affects two or more cylinders<br>Erratic or interrupted CKP or CMP Sensor signals<br>Fuel delivery component fault that affects two or more cylinders (i.e., a contaminated, dirty or sticking fuel injector)<br>Ignition system problem (coil or plug) in two or more cylinders<br>Vehicle driven with low fuel pressure or while very low on fuel |
| **P0300**<br>1T Catalyst<br>2T Emission<br>2001-03<br>Sebring, Stratus Coupe with 2.4L I4 or 3.0L V6 | **Multiple Cylinder Misfire Detected Conditions:**<br>Engine speed from 500-6000 rpm at a steady throttle, ECT and IAT Sensors over 14ºF, BARO Sensor over 11 psi, and the PCM detected a misfire rate of 1.5% (High Emissions 2T) or a high misfire rate with Converter temperature over 1742ºF in 2 or more cylinders. Note: If the misfire is severe, the MIL will flash on/off on the 1st trip!<br>**Possible Causes:**<br>Air leak in the intake manifold, or in the EGR or EVAP systems<br>Base engine mechanical fault (e.g., a skipped timing belt)<br>Erratic or interrupted CKP or CMP Sensor signals<br>Fuel delivery component fault that affects two or more cylinders<br>Ignition system problem (coil or plug) in two or more cylinders<br>Vehicle driven with low fuel pressure or while very low on fuel<br>TSB 18-028-02 contains a repair procedure for this code |
| **P0300**<br>1T Catalyst<br>2T Emission<br>1996-2003<br>Cherokee, Dakota, Durango, Grand Cherokee, Liberty, Ram, Wrangler | **Multiple Cylinder Misfire Detected Conditions:**<br>Engine less than 3000 rpm; adaptive numerator updated. PCM detected a misfire rate of 1% (High Emissions 2T), or a misfire rate of 6-30% (Catalyst Damaging 1T) in two or more cylinders. Note: If the misfire is severe, the MIL will flash on/off on the 1st trip!<br>**Possible Causes:**<br>Air leak in the intake manifold, or in the EGR or EVAP system<br>Base engine mechanical fault that affects two or more cylinders<br>Erratic or interrupted CKP or CMP Sensor signals<br>Fuel delivery component fault that affects two or more cylinders<br>Ignition system problem (coil or plug) in two or more cylinders<br>Vehicle driven with low fuel pressure or while very low on fuel |
| **P0300**<br>1T CCM<br>2T Emission<br>1998-2003<br>Ram with 5.9L Diesel | **Multiple Cylinder Misfire Conditions:**<br>Engine less than 880 rpm; ECT Sensor above 140ºF; VSS at 0 mph; PTO inactive. PCM detected the time for crankshaft to turn 120° during firing event of one or more cylinders compared to the time for the crank to turn 120° for the previous cylinder exceeded 240 microseconds. Note: If the misfire is severe, the MIL will flash on/off on the 1st trip!<br>**Possible Causes:**<br>Air leak (manifold, PCV) that affects more than 1 cylinder<br>Base engine problem that affects more than 1 cylinder<br>CMP Sensor, Sensor wiring harness or tone wheel is damaged<br>Ignition system problem that affects more than one cylinder<br>Fuel metering problem (fuel injectors, fuel pressure) that affects more than one cylinder<br>EGR or EVAP Purge problem that affects more than 1 cylinder |

**OBD II Trouble Code List (P0XXX Codes) - Continued**

| DTC | Trouble Code Title, Conditions & Possible Causes: |
|---|---|
| **P0300**<br>2T CCM<br>2004-05<br>300M, Caravan, Concorde, Dakota, Durango, Grand Cherokee, Intrepid, LHS, Liberty, Magnum, Neon, Pacifica, Prowler Ram, Sebring, Stratus, Town & Country, Voyager, Wrangler | **Multiple Cylinder Misfire Conditions:**<br>Any time engine is running and Adaptive Numerator (Target Learning Coefficient) has been successfully updated. If more than 1.0% (Neon, Pacifica), 1.5% (300M, Magnum) or 1.8% (Sebring, Stratus, Caravan/Town & Country), 2% (Durango, Ram, Liberty), 2.5% (Durango LEV, Ram LEV, Liberty LEV), misfire rate is measured during 2 trips. 3 good trips required to turn off MIL.<br>**Possible Causes:**<br>ASD relay output circuit fault<br>Injector control circuit fault<br>Coil control circuit fault<br>Ignition wiring, coil control circuit or coil fault<br>Fuel pump inlet strainer plugged<br>Restricted fuel supply line<br>Fuel pump module is damaged<br>Fuel pressure leakdown fault<br>Fuel injector damaged or has failed<br>Engine mechanical problems exist<br>PCM has failed |
| **P0300**<br>2T CCM<br>2004-05<br>Viper | **Multiple Cylinder Misfire Conditions:**<br>Any time engine is running and Adaptive Numerator (Target Learning Coefficient) has been successfully updated. If more than 1.0% misfire rate is measured during 2 trips, or with 6-30% misfire rate during 1 trip, this DTC will set.<br>**Possible Causes:**<br>Ignition wiring, coil control circuit or coil fault<br>ASD relay output circuit fault<br>Engine mechanical problems exist<br>Injector control circuit fault<br>Ignition coil problems<br>Coil control circuit fault<br>Spark plugs fouled or burned<br>Fuel pressure problem<br>Fuel pump inlet strainer plugged<br>Restricted fuel supply line<br>Fuel pump module is damaged<br>Fuel leak down fault<br>Fuel injector damaged or has failed<br>PCM has failed |
| **P0300**<br>2T CCM<br>2004-05<br>PT Cruiser | **Multiple Cylinder Misfire Conditions:**<br>Engine is running; atmospheric pressure, fuel level, battery voltage, engine speed, engine load are above minimal predetermined specifications; the engine is not in fuel cutoff and intake air temperature and coolant temperature are within normal operating ranges; there are no DTCs relating to MAP, ECT, CMP, CKP or IAT Sensor, to Crank rationality, to Crank Sensor Learn invalid, or to System Voltage. If total misfires detected are greater than 8.5% during the first 1000 rpm, or are 40% during any 200 rpm, with no single cylinder or multiple cylinder misfire counters above 80%, this DTC will set. The MIL will flash after first trip occurrence and will be steady after second trip.<br>**Possible Causes:**<br>Fuel system problems exist<br>Erratic CMP or CKP Sensor signals<br>Ignition system problems exist<br>Engine mechanical problems exist<br>PCM has failed |

**OBD II Trouble Code List (P0XXX Codes) - Continued**

| DTC | Trouble Code Title, Conditions & Possible Causes: |
|---|---|
| **P0300**<br>1T CCM<br>2004-05<br>Liberty with Diesel | **Misfire Conditions:**<br>Engine running. The ECM detects multiple misfires from one or more cylinders.<br>**Possible Causes:**<br>Engine compression problems exist<br>Fuel injector quantity is insufficient<br>Fuel injector leaking<br>Intermittent condition |
| **P0301-P0304**<br>1T Catalyst<br>2T Emission<br>1995-98<br>Talon | **Cylinder 1, 2, 3 or 4 Misfire Detected Conditions:**<br>Engine speed less than 3000 rpm, adaptive numerator updated, and the PCM detected a misfire rate of 1-2% (High Emissions 2T), or a misfire rate of 6-30% (Catalyst Damaging 1T) in a single cylinder. Note: If the misfire is severe, the MIL will flash on/off on the 1st trip!<br>**Possible Causes:**<br>Air leak in the intake manifold, or in the EGR or PCV system<br>Base engine mechanical fault that affects only one cylinder<br>Fuel delivery component fault that affects only one cylinder (i.e., a contaminated, dirty or sticking fuel injector)<br>Ignition system problem (coil or plug) that affects one cylinder |
| **P0301-P0310**<br>1T Catalyst<br>2T Emission<br>1995-2003<br>300M, Avenger, Breeze, Caravan, Cherokee, Cirrus, Concorde, PT Cruiser, Dakota, Durango, Grand Cherokee, Intrepid, LHS, Liberty, Neon, New Yorker, Prowler, Ram, Sebring, Stratus, Town & Country, Viper, Vision, Voyager, Wrangler | **Cylinder 1-10 Misfire Detected Conditions:**<br>Engine speed less than 3000 rpm, adaptive numerator updated, and the PCM detected a misfire rate of 1-2% (High Emissions 2T), or a misfire rate of 6-30% (Catalyst Damaging 1T) in only one cylinder. Note: If the misfire is severe, the MIL will flash on/off on the 1st trip!<br>**Possible Causes:**<br>Air leak in the intake manifold, or in the EGR or EVAP system<br>Base engine mechanical fault that affects only one cylinder<br>Fuel delivery component fault that affects only one cylinder (e.g., a dirty fuel injector)<br>Ignition system problem (coil or plug) in only one cylinder |
| **P0301-P0306**<br>1T Catalyst<br>2T Emission<br>2001-03<br>Sebring, Stratus Coupe with 2.4L I4 or 3.0L V6 | **Cylinder 1-6 Misfire Detected Conditions:**<br>Engine speed from 500-6000 rpm at a steady throttle, ECT and IAT Sensors over 14°F, BARO Sensor over 76 kPa (11 psi), and the PCM detected a misfire rate of 1.5% (High Emissions 2T) or a high misfire rate with Converter temperature over 1742°F in only one cylinder. Note: If the misfire is severe, the MIL will flash on/off on the 1st trip!<br>**Possible Causes:**<br>Air leak in the intake manifold, or in the EGR or EVAP systems<br>Base engine mechanical fault (e.g., a skipped timing belt)<br>Fuel delivery component fault that affects only one cylinder (e.g., a dirty fuel injector)<br>Ignition system problem (coil or plug) in only one cylinder<br>TSB 18-028-02 contains a repair procedure for this code |

**OBD II Trouble Code List (P0XXX Codes) - Continued**

| DTC | Trouble Code Title, Conditions & Possible Causes: |
|---|---|
| **P0301-P0308**<br>1T Catalyst<br>2T Emission<br>1996-2003<br>Cherokee, Dakota,<br>Durango, Grand<br>Cherokee, Liberty,<br>Ram, Wrangler | **Cylinder 1-8 Misfire Detected Conditions:**<br>Engine speed less than 3000 rpm, adaptive numerator updated, and PCM detected a misfire rate of 1% (High Emissions 2T), or a misfire rate of 6-30% (Catalyst Damaging 1T) in a single cylinder. Note: If the misfire is severe, the MIL will flash on/off on the 1st trip!<br>**Possible Causes:**<br>Air leak in the intake manifold, or in the EGR or EVAP system<br>Base engine mechanical fault that affects only one cylinder<br>Erratic or interrupted CKP or CMP Sensor signals<br>Fuel delivery component fault that affects only one cylinder (e.g., a dirty fuel injector)<br>Ignition system problem (coil or plug) that affects only one cylinder |
| **P0301-P0306**<br>1T Catalyst<br>2T Emission<br>1998-2003<br>Ram with 5.9L Diesel | **Cylinder 1-6 Misfire Detected Conditions:**<br>Engine started; engine speed below 860 rpm, ECT Sensor over 140°F, VSS at 0 mph, PTO "off", and the PCM detected the time for the crankshaft to turn 120 degrees while firing one cylinder when compared to the time for the crankshaft to turn 120 degrees for the previous cylinder exceeded 40 ms. Note: If the misfire is severe, the MIL will flash on/off on 1st trip!<br>**Possible Causes:**<br>Base engine mechanical fault that affects only one cylinder<br>CMP Sensor, Sensor wiring harness or tone wheel is damaged<br>Fuel delivery component fault that affects only one cylinder (e.g., a dirty fuel injector) |
| **P0301-P0304**<br>2T CCM<br>2004-05<br>Neon with Turbo | **Cylinder 1-4 Misfire Detected Conditions:**<br>Any time engine is running and the Target Learning Coefficient has been successfully updated. When more than 1.0% misfire rate is measured during 2 trips.<br>**Possible Causes:**<br>Intermittent misfire<br>Base engine or ignition system problem |
| **P0301-P0304**<br>2T CCM<br>2004-05<br>PT Cruiser | **Cylinder No. 1, 2, 3 or 4 Misfire Detected Conditions:**<br>Engine is running; atmospheric pressure, fuel level, battery voltage, engine speed, engine load are above minimal predetermined specifications; the engine is not in fuel cutoff and intake air temperature and coolant temperature are within normal operating ranges; there are no DTCs relating to ECT, CMP, CKP, MAP or IAT Sensor, to Crank rationality, to Crank Sensor Learn invalid, or to System Voltage. If total misfires detected are greater than 7% during the first 1000 rpm, or are over 3% after the first 1000 rpm, or are 25% during any 200 rpm with suspect cylinder misfire counter more than 80% and no other cylinder misfire counters above 80%, this DTC will set. The MIL will flash after first trip occurrence and will be steady after second trip.<br>**Possible Causes:**<br>Fuel system problems exist<br>Erratic CMP or CKP Sensor signals<br>Ignition system problems exist<br>Engine mechanical problems exist<br>PCM has failed |
| **P0301-P0304**<br>1T CCM<br>2004-05<br>Liberty with Diesel | **Misfire Detected In Specific Cylinder Conditions:**<br>Engine running. The ECM detects multiple misfires from cylinder No. 1 (P0301), No. 2 (P0302), No. 3 (P0303) or No. 4 (P0304).<br>**Possible Causes:**<br>Engine compression problems exist<br>Fuel injector quantity is insufficient<br>Fuel injector leaking<br>Intermittent condition |

| P0301-P0306<br>2T Catalyst<br>2T CCM<br>2004-05<br>300M, Caravan, Concorde, PT Cruiser, Dakota, Durango, Grand Cherokee, Intrepid, LHS, Liberty, Neon, Prowler, Ram, Sebring, Stratus, Town & Country, Voyager, Wrangler | **Cylinder 1-6 Misfire Detected Conditions:**<br>Any time engine is running and Target Learning Coefficient (TLC) has been successfully updated, if more than 1.0% (Pacifica) or 1.8% (Sebring, Stratus, Caravan/Town & Country), 2% (Durango, Ram, Liberty), 2.5% (Durango LEV, Ram LEV, Liberty LEV), misfire rate is measured during 2 trips or with 10-30% misfire rate during one trip.<br>**Possible Causes:**<br>Intermittent misfire<br>Base engine mechanical fault that affects only 1 cylinder<br>Ignition wiring, coil control circuit or coil fault<br>ASD relay output circuit (coil or injector) problem<br>Spark plug malfunction or failure on 1 cylinder<br>CMP Sensor, Sensor wiring harness or tone wheel is damaged<br>Fuel delivery component fault that affects only 1 cylinder (e.g., a dirty fuel injector)<br>Injector or control circuit failure<br>PCM has failed |
| --- | --- |
| P0301-P0308<br>2TCatalyst<br>2004-05<br>300M, Magnum, Dakota, Durango, Ram | **Cylinder 1-8 Misfire Detected Conditions:**<br>Any time engine is running and the adaptive numerator has been successfully updated. When more than 1.5% (300M, Magnum), 2% (Durango, Ram), 2.5% (Durango LEV, Ram LEV), misfire rate is measured during 2 trips, or with 10-30% misfire during 1 trip. 3 good trips required to turn off MIL.<br>**Possible Causes:**<br>ASD relay output 2 circuit fault<br>Injector control 1 circuit fault<br>Coil control 1 circuit fault<br>Ignition wiring, spark plug, or ignition coil fault<br>Fuel pump inlet strainer plugged<br>Restricted fuel supply line<br>Fuel pump module fault<br>Fuel pressure leakdown fault<br>Fuel injector is damaged or has failed<br>Engine mechanical problems exist<br>PCM has failed |
| P0301-P0310<br>2TCatalyst<br>2004-05<br>Viper | **Cylinder 1-10 Misfire Detected Conditions:**<br>Any time engine is running and the adaptive numerator has been successfully updated. When more than 1% misfire rate is measured during 2 trips, or with 6-30% misfire during 1 trip, this DTC will set.<br>**Possible Causes:**<br>Ignition wiring, coil control circuit or coil fault<br>ASD relay output circuit fault<br>Engine mechanical problems exist<br>Injector control circuit fault<br>Ignition coil problems<br>Coil control circuit fault<br>Spark plugs fouled or burned<br>Fuel pressure problem<br>Fuel pump inlet strainer plugged<br>Restricted fuel supply line<br>Fuel pump module is damaged<br>Fuel leak down fault<br>Fuel injector damaged or has failed<br>PCM has failed |

**OBD II Trouble Code List (P0XXX Codes) - Continued**

| DTC | Trouble Code Title, Conditions & Possible Causes: |
|---|---|
| **P0315**<br>1T CCM<br>2001-05<br>300M, Caravan, Concorde, PT Cruiser, Dakota, Durango, Grand Cherokee, Intrepid, LHS, Liberty, Magnum, Neon, Pacifica, Prowler Ram, Sebring, Stratus, Town & Country, Viper, Voyager, Wrangler | **No Crankshaft Position Sensor Learned Conditions:**<br>Engine started; engine runtime more than 50 seconds under closed throttle conditions; A/C off; ECT Sensor more than 167°F (75°C). The PCM detected that one of the CKP Sensor windows had too much variance (e.g., over 2.86%) from its calibrated reference point.<br>**Possible Causes:**<br>Crankshaft tone wheel flex plate is damaged<br>Tone wheel/pulse ring may be damaged<br>Erratic CKP Sensor signals (wiring/connector problem)<br>CKP Sensor has failed<br>PCM has failed |
| **P0315**<br>1T CCM<br>2004-05<br>PT Cruiser | **No Crankshaft Position Sensor Learned Conditions:**<br>Engine started; no MAP, CKP or CMP faults detected; engine rpm is 2000-3000; engine state includes Deceleration Fuel Cutoff, Transmission in gear, and CMP/CKP synch has been achieved. When the CKP Sensor correction value is out of limits for more than 1 second, this DTC will set. MIL is illuminated.<br>**Possible Causes:**<br>Intermittent condition<br>Crankshaft tone wheel/pulse ring is damaged<br>Wiring harness problems exist<br>CKP Sensor has failed |
| **P0320**<br>1T CCM<br>1996-2003<br>300M, Avenger, Breeze, Caravan, Cherokee, Cirrus, Concorde, PT Cruiser, Dakota, Durango, Grand Cherokee, Intrepid, LHS, Liberty, Neon, New Yorker, Prowler, Ram, Sebring, Stratus, Town & Country, Viper, Vision, Voyager, Wrangler | **No Crank Reference Signal to PCM Conditions:**<br>Engine cranking, vacuum signals present; and the PCM detected from 3-8 CMP Sensor signals without detecting a CKP Sensor signal.<br>**Possible Causes:**<br>CKP Sensor signal circuit is open or grounded<br>CKP Sensor VREF circuit is open or grounded<br>Tone wheel or flex plate damage or erratic CKP Sensor signals<br>CKP Sensor is damaged or has failed<br>PCM has failed |
| **P0320**<br>1T CCM<br>1998-2002<br>Ram with 5.9L Diesel | **No Crank Reference Signal to PCM Conditions:**<br>Engine started; system voltage from 10-15v, and the PCM detected the engine speed signal indicated under 800 rpm while the engine speed signal at the ECM indicated over 1024 rpm.<br>**Possible Causes:**<br>CKP Sensor signal circuit is open or grounded<br>CKP Sensor VREF circuit is open or grounded<br>Tone wheel or flex plate damage or erratic CKP Sensor signals<br>CKP Sensor is damaged or has failed<br>PCM has failed |

**OBD II Trouble Code List (P0XXX Codes) - Continued**

| DTC | Trouble Code Title, Conditions & Possible Causes: |
|---|---|
| **P0320**<br>1T CCM<br>2004-05<br>Viper | **No Crank Reference Signal to PCM Conditions:**<br>Ignition on. If the PCM detected no signal from the CKP Sensor during engine cranking, when at least 3 CMP Sensor signals have occurred, this DTC will set.<br>**Possible Causes:**<br>CMP Sensor signal fault<br>Intermittent CKP Sensor signal fault<br>5v supply circuit is open or is shorted to voltage or to ground<br>CKP Sensor signal circuit is open, shorted to ground, shorted to voltage or to 5v supply circuit<br>Sensor ground circuit is open<br>CKP Sensor is damaged or has failed<br>PCM has failed |
| **P0325**<br>1T CCM<br>1995-98<br>Talon | **Knock Sensor No. 1 Circuit Conditions:**<br>Engine running at idle or in deceleration mode, and the PCM detected the Knock Sensor signal was below a minimum value (value depends on engine speed), or if Sensor voltage was more than 5.0v<br>**Possible Causes:**<br>Knock Sensor signal circuit open or grounded<br>Knock Sensor not tightened properly<br>Knock Sensor damaged or has failed (it may be open internally)<br>PCM has failed |
| **P0325**<br>1T CCM<br>1995-2005<br>300M, Avenger, Breeze, Caravan, Cherokee, Cirrus, Concorde, PT Cruiser, Dakota, Durango, Grand Cherokee, Intrepid, LHS, Liberty, Magnum, Neon, New Yorker, Pacifica, Prowler, Ram, Sebring, Stratus, Talon, Town & Country, Viper, Vision, Voyager, Wrangler | **Knock Sensor No. 1 Circuit Conditions:**<br>Engine running at idle or in deceleration mode, and the PCM detected the Knock Sensor signal was below a minimum value (value depends on engine speed), or if Sensor voltage was about 5.0v with engine within idle range.<br>**Possible Causes:**<br>Knock Sensor connector is damaged or shorted<br>Knock Sensor signal circuit open or grounded<br>Knock Sensor signal circuit shorted to return circuit<br>Knock Sensor return circuit is open<br>Knock Sensor not tightened properly<br>Knock Sensor damaged or has failed (it may be open internally)<br>PCM has failed |
| **P0325**<br>1T CCM<br>2004-05<br>Liberty | **Knock Sensor No. 1 Circuit Conditions:**<br>Engine running at higher than 1312 rpm; coolant temperature greater than 150°F (65°C); MAF signal greater than 250 mg/tdc; no MAF, ECT or CMP Sensor DTCs present. The Knock Sensor error program internal to the PCM is on; KS voltage was less than 0.49v and the value of the KS changes less than 0.06v every 11 or more seconds. 3 good trips required to turn off MIL.<br>**Possible Causes:**<br>Knock Sensor signal circuit open or is shorted to ground or to voltage<br>Knock Sensor return circuit is open<br>Knock Sensor signal circuit is short to KS return circuit<br>Knock Sensor damaged or has failed<br>PCM has failed |

## OBD II Trouble Code List (P0XXX Codes) - Continued

| DTC | Trouble Code Title, Conditions & Possible Causes: |
|---|---|
| **P0330**<br>1T CCM<br>2002-03<br>Liberty, Ram with 2.4L I4 or 3.7L V6 | **Knock Sensor No. 2 Circuit Conditions:**<br>Engine running at idle or in deceleration mode, and the PCM detected the Knock Sensor No. 2 signal was below a minimum value (value depends on engine speed), or was more than 5.0v.<br>**Possible Causes:**<br>Knock Sensor signal circuit open or grounded<br>Knock Sensor not tightened properly<br>Knock Sensor damaged or has failed (it may be open internally)<br>PCM has failed |
| **P0330**<br>1T CCM<br>2004-05<br>300M, Caravan, Concorde, PT Cruiser, Dakota, Durango, Grand Cherokee, Intrepid, LHS, Magnum, Neon, Pacifica, Prowler Ram, Sebring, Stratus, Town & Country, Viper, Voyager, Wrangler | **Knock Sensor No. 2 Circuit Conditions:**<br>Engine running. The Knock Sensor circuit voltage falls below a minimum value at idle or deceleration. The minimum value is from a lookup table internal to the PCM and is based on engine rpm. This DTC will also set if the Sensor voltage goes above 5v. 3 good trips required to turn off MIL.<br>**Possible Causes:**<br>Knock Sensor No. 2 signal circuit shorted to battery voltage or to KS 2 return circuit<br>Knock Sensor No. 2 signal circuit or return circuit is open<br>Knock Sensor No. 2 signal circuit is shorted to KS 2 return circuit or to ground<br>Knock Sensor damaged or has failed<br>PCM has failed |
| **P0330**<br>1T CCM<br>2004-05<br>Liberty | **Knock Sensor No. 2 Circuit Conditions:**<br>Engine running at higher than 1312 rpm; coolant temperature greater than 150°F (65°C); MAF signal greater than 250 mg/tdc; no MAF, ECT or CMP Sensor DTCs present. The Knock Sensor error program internal to the PCM is on; KS voltage was less than 0.49v and the value of the KS changes less than 0.06v every 11 or more seconds. 3 good trips required to turn off MIL.<br>**Possible Causes:**<br>Knock Sensor signal circuit open or is shorted to ground or to voltage<br>Knock Sensor return circuit is open<br>Knock Sensor signal circuit is short to KS return circuit<br>Knock Sensor damaged or has failed<br>PCM has failed |
| **P0335**<br>1T CCM<br>1995-2000<br>Avenger, Sebring, Stratus, Talon | **Crankshaft Position Sensor Circuit Conditions:**<br>Engine started and the PCM did not detect any change in the CKP Sensor input for 2 seconds.<br>**Possible Causes:**<br>CKP Sensor signal circuit is open or shorted to ground<br>CKP Sensor VREF circuit is open or shorted to ground<br>CKP Sensor is damaged or has failed<br>PCM has failed |

**OBD II Trouble Code List (P0XXX Codes) - Continued**

| DTC | Trouble Code Title, Conditions & Possible Causes: |
|---|---|
| **P0335**<br>1T CCM<br>2001-05<br>300M, Caravan, Cherokee, Cirrus, Concorde, PT Cruiser, Dakota, Durango, Grand Cherokee, Intrepid, LHS, Liberty, Magnum, Neon, New Yorker, Pacifica, Prowler, Ram, Sebring, Stratus, Talon, Town & Country, Viper, Vision, Voyager, Wrangler | **Crankshaft Position Sensor Circuit Conditions:**<br>Engine cranking with at least 8 CMP Sensor signals detected. The PCM did not detect any CKP Sensor signals for 2 seconds.<br>**Possible Causes:**<br>Intermittent CKP signal<br>CKP Sensor signal circuit is open or it is shorted to ground or voltage<br>CKP Sensor 5v supply circuit is open or shorted to ground or voltage<br>CKP Sensor ground circuit is open<br>CKP Sensor or CMP Sensor is damaged or has failed<br>PCM has failed |
| **P0335**<br>1T CCM<br>2004-05<br>PT Cruiser | **No Crankshaft Position Sensor Signal At PCM Conditions:**<br>Engine running and no CMP Sensor DTCs are present. If the PCM did not detect any CKP Sensor pulses when synchronization is attempted with the CMP signal, this DTC will set.<br>**Possible Causes:**<br>Intermittent CKP signal<br>5v supply circuit open, or shorted to ground or voltage<br>CKP Sensor signal circuit is open or shorted to ground or voltage<br>CKP Sensor signal circuit is shorted to 5v supply circuit<br>CKP Sensor ground circuit is open<br>CKP Sensor or PCM has failed |
| **P0335**<br>1T CCM<br>2004-05<br>Liberty with Diesel | **Crankshaft Position Sensor Circuit Incorrect Or Missing Signal Conditions:**<br>Engine running at less than 6000 rpm. The ECM does not receive a CKP Sensor signal or receives an incorrect signal.<br>**Possible Causes:**<br>CKP Sensor is damaged, improperly positioned or has failed<br>ECM has failed<br>Intermittent condition<br>CKP Sensor signal circuit(s) shorted to ground or shorted together<br>CKP Sensor signal circuits open or shorted to voltage |
| **P0336**<br>1T CCM<br>2004-05<br>PT Cruiser | **Crankshaft Position Sensor Incorrect Performance Conditions:**<br>Engine running and no CMP Sensor DTCs are present. If the PCM detects an incorrect amount of CKP Sensor pulses when compared to CMP Sensor pulse, this DTC will set.<br>**Possible Causes:**<br>Intermittent CMP signal<br>Wiring harness problems exist<br>Tone wheel/pulse ring damaged or has failed<br>CKP Sensor is damaged or has failed<br>CMP Sensor has failed |
| **P0336**<br>1T CCM<br>1998-2003<br>Ram with 5.9L Diesel | **Crankshaft Position Sensor Signal Conditions:**<br>Engine started; CMP Sensor signals detected, and the PCM did not detect any change in the CKP Sensor signal during the test period.<br>**Possible Causes:**<br>Check the tone or pulse ring for damage or debris collection<br>CKP Sensor signal circuit is open or shorted to ground<br>CKP Sensor VREF circuit is open or shorted to ground<br>CKP Sensor is damaged or has failed<br>PCM has failed |

## OBD II Trouble Code List (P0XXX Codes) - Continued

| DTC | Trouble Code Title, Conditions & Possible Causes: |
|---|---|
| **P0339**<br>1T CCM<br>2000-05<br>300M, Caravan, Cherokee, Cirrus, Concorde, PT Cruiser, Dakota, Durango, Grand Cherokee, Intrepid, LHS, Liberty, Magnum, Neon, New Yorker, Pacifica, Prowler, Ram, Sebring, Stratus, Talon, Town & Country, Viper, Vision, Voyager, Wrangler | **Crankshaft Position Sensor Circuit Intermittent Conditions:**<br>Engine cranking or running; CMP Sensor signals detected. The PCM detected an intermittent loss of the CKP Sensor signal. The Failure counter must reach 20 before this code will set.<br>**Possible Causes:**<br>Check the tone wheel/pulse ring for damage or debris collection<br>CKP Sensor signal circuit is open or shorted to ground<br>CKP Sensor 5v supply circuit is open or shorted to ground<br>CKP Sensor is damaged or it has failed<br>PCM has failed |
| **P0339**<br>1T CCM<br>2004-05<br>Liberty with Diesel | **Crankshaft Position Sensor Circuit Intermittent Or Missing Signal Conditions:**<br>Engine running at less than 6000 rpm. The ECM does not receive a CKP Sensor signal or receives an incorrect signal.<br>**Possible Causes:**<br>CKP Sensor is damaged, improperly positioned or has failed<br>ECM has failed<br>Intermittent condition<br>CKP Sensor signal circuit(s) shorted to ground or shorted together<br>CKP Sensor signal circuits open or shorted to voltage |
| **P0340**<br>1T CCM<br>1995-98<br>Talon | **No Camshaft Synch Signal To PCM Conditions:**<br>Engine cranking, CKP Sensor pulses detected, and the PCM did not detect any CMP Sensor pulses, condition met for 2 seconds.<br>**Possible Causes:**<br>CMP Sensor signal circuit is open or shorted to ground<br>CMP Sensor VREF circuit is open or shorted to ground<br>CMP Sensor is damaged or has failed<br>PCM has failed |
| **P0340**<br>1T CCM<br>1995-2005<br>300M, Avenger, Breeze, Caravan, Cherokee, Cirrus, Concorde, PT Cruiser, Dakota, Durango, Grand Cherokee, Intrepid, LHS, Liberty, Magnum, Neon, New Yorker, Pacifica, Prowler, Ram, Sebring, Stratus, Talon, Town & Country, Viper, Vision, Voyager, Wrangler | **No Camshaft Position Sensor Circuit Failure Conditions:**<br>Engine cranking or running, system voltage over 10v. The PCM detected CKP pulses without detecting any CMP Sensor pulses for 5 seconds or 2.5 engine revolutions.<br>**Possible Causes:**<br>CMP Sensor connector is damaged, open or it is shorted<br>CMP Sensor signal circuit is open or shorted to ground or to battery voltage or 5v supply circuit<br>CMP Sensor 5v supply circuit is open or shorted to ground or to battery voltage<br>CMP Sensor ground circuit is open<br>CMP Sensor is damaged or has failed<br>CKP Sensor is damaged or has failed<br>PCM has failed |

**OBD II Trouble Code List (P0XXX Codes) - Continued**

| DTC | Trouble Code Title, Conditions & Possible Causes: |
|---|---|
| **P0340**<br>1T CCM<br>2004-05<br>PT Cruiser | **No CMP Sensor Signal Conditions:**<br>Engine cranking or running; no CKP Sensor DTCs present. At least 10 engine revolutions have elapsed, with CKP Sensor signals present, but no CMP Sensor signal polarity change.<br>**Possible Causes:**<br>Intermittent CMP signal<br>5v supply circuit open or is shorted to ground or to voltage<br>CMP Sensor signal circuit is open or shorted to ground or to battery voltage or 5v supply circuit<br>CMP Sensor ground circuit is open<br>5v supply has failed<br>CMP Sensor is damaged or has failed |
| **P0340**<br>1T CCM<br>2004-05<br>Liberty with Diesel | **Camshaft Position Sensor Circuit Incorrect Or Missing Signal Conditions:**<br>Engine running at less than 6000 rpm. The ECM does not receive a CMP Sensor signal or receives an incorrect signal.<br>**Possible Causes:**<br>5v supply circuit is open<br>CMP Sensor signal circuit shorted to voltage<br>CMP Sensor is damaged, improperly positioned or has failed<br>ECM has failed |
| **P0341**<br>1T CCM<br>1998-2003<br>Ram with 5.9L Diesel | **Camshaft Position Sensor Signal Conditions:**<br>Engine started; system voltage from 8.0-15v, and the PCM detected an incorrect tone wheel tooth signal occurred 25 times in 8 seconds, or the PCM did not detect any CMP Sensor signals for 2 seconds.<br>**Possible Causes:**<br>CMP Sensor signal circuit is open or shorted to ground<br>CMP Sensor VREF circuit is open or shorted to ground<br>CMP Sensor is damaged or has failed<br>PCM has failed |
| **P0342**<br>1T CCM<br>1998-2003<br>Ram with 5.9L Diesel | **CMP Sensor Supply Voltage Too Low Conditions:**<br>Engine started; system voltage from 8.0-15v, and the PCM detected the CMP Sensor VREF signal was less than 4.2v for 2 seconds.<br>**Possible Causes:**<br>CMP Sensor connector is damaged, open or it is shorted<br>CMP Sensor VREF circuit is open<br>CMP Sensor VREF circuit is shorted to ground<br>CMP Sensor is damaged or has failed<br>PCM has failed |
| **P0343**<br>1T CCM<br>1998-2003<br>Ram with 5.9L Diesel | **CMP Sensor Supply Voltage Too Low Conditions:**<br>Engine cranking or engine running, system voltage over 10.5v and the PCM detected an intermittent loss of the CMP Sensor signal during the period of 2.5 complete engine revolutions. The failure counter must reach 20 before this code matures and a code is set.<br>**Possible Causes:**<br>CMP Sensor signal circuit is open (intermittent fault)<br>CMP Sensor signal circuit shorted to ground (intermittent fault)<br>CMP Sensor ground circuit is open (an intermittent fault)<br>PCM has failed |

## OBD II Trouble Code List (P0XXX Codes) - Continued

| DTC | Trouble Code Title, Conditions & Possible Causes: |
|---|---|
| **P0344**<br>1T CCM<br>2000-05<br>300M, Avenger, Breeze, Caravan, Cherokee, Cirrus, Concorde, PT Cruiser, Dakota, Durango, Grand Cherokee, Intrepid, LHS, Liberty, Magnum, Neon, New Yorker, Pacifica, Prowler, Ram, Sebring, Stratus, Talon, Town & Country, Viper, Vision, Voyager, Wrangler | **Camshaft Position Sensor Circuit Intermittent Conditions:**<br>Engine cranking or running; system voltage over 10.5v. The PCM detected an intermittent loss of the CMP Sensor signal during the period of 2.5 complete engine revolutions. The failure counter must reach 20 before this code matures and a code is set.<br>**Possible Causes:**<br>Wiring harness fault<br>5v supply circuit open or shorted to ground<br>Tone wheel/pulse ring is damaged or corroded<br>CMP Sensor has failed<br>CMP Sensor signal circuit is open, shorted to ground or battery voltage or 5v supply<br>CMP Sensor ground circuit is open<br>PCM has failed |
| **P0344**<br>1T CCM<br>2004-05<br>Liberty with Diesel | **Camshaft Position Sensor Circuit Intermittent Or Missing Signal Conditions:**<br>Engine running at less than 6000 rpm. The ECM does not receive a CMP Sensor signal or receives an intermittent signal.<br>**Possible Causes:**<br>5v supply circuit is open or is shorted to ground or to Sensor ground circuit<br>CMP Sensor signal circuit is open or is shorted to voltage or to ground<br>CMP Sensor is damaged, improperly positioned or has failed<br>ECM has failed<br>CMP Sensor ground circuit open<br>Intermittent condition |
| **P0350**<br>1T CCM<br>2001-2002<br>Ram with 5.9L Diesel | **Ignition Coil Current Too High Conditions:**<br>Engine started; and the PCM detected the Ignition Coil current level was too high.<br>**Possible Causes:**<br>Ignition Coil No. 1, 2 or 3 control circuit (the coil windings) shorted internally<br>Ignition Coil No. 1, 2 or 3 control circuit is shorted to ground<br>PCM has failed |
| **P0351, P0152, P0353**<br>1T CCM<br>1995-98<br>Talon | **Ignition Coil No. 1, 2 or 3 Primary Circuit Conditions:**<br>Engine speed less than 4000 rpm, MFI relay on, system voltage over 13v, and the PCM detected the Coil No. 1, 2 or 3 primary circuit did not achieve peak current (dwell) for 3 seconds.<br>**Possible Causes:**<br>Ignition coil 1, 2 or 3 primary "driver" circuit is open or it is grounded<br>Ignition coil 1, 2 or 3 is damaged or it has failed<br>MFI relay power circuit to ignition coil is open<br>PCM has failed |

**OBD II Trouble Code List (P0XXX Codes) - Continued**

| DTC | Trouble Code Title, Conditions & Possible Causes: |
|---|---|
| **P0351, P0152, P0153**<br>1T CCM<br>1996-2003<br>300M, Avenger, Breeze, Caravan, Cherokee, Cirrus, Concorde, PT Cruiser, Dakota, Durango, Grand Cherokee, Intrepid, LHS, Liberty, Neon, New Yorker, Prowler, Ram, Sebring, Stratus, Talon, Town & Country, Viper, Vision, Voyager, Wrangler | **Ignition Coil No. 1, 2 or 3 Primary Circuit Conditions:**<br>Engine cranking or running; ASD relay on, system voltage over 10v, Coil No. 1, 2 or 3 was not in dwell period when it was checked, and the PCM detected that the peak coil current was not reached in 2.5 ms of dwell for 3-6 seconds. The coil primary resistance is less than 2 ohms.<br>**Possible Causes:**<br>ASD relay power circuit to the ignition coil is open<br>Ignition Coil No. 1, 2 or 3 primary "driver" circuit open or grounded<br>Ignition Coil No. 1, 2 or 3 is damaged or it has failed<br>PCM has failed |
| **P0351-P0355**<br>1T CCM<br>2004-05<br>Viper | **Ignition Coil No. 1-5 Primary Circuit Conditions:**<br>Battery voltage greater than 8v during engine cranking or greater than 12v with engine running; engine rpm less than 2016 rpm; coils not in dwell period. If the PCM detected peak coil current was not reached with battery based dwell, plus 1.5ms of diagnostic offset, this DTC will set It takes less than 3 seconds during cranking, or up to 6 seconds with engine running, for peak to be achieved normally.<br>**Possible Causes:**<br>ASD relay output circuit fault<br>Ignition coil has resistance<br>Ignition coil is damaged or it has failed<br>Ignition coil control circuit is open or shorted to ground<br>PCM has failed |
| **P0351-P0358**<br>1T CCM<br>1999-2003<br>Cherokee, Dakota, Durango, Grand Cherokee, Liberty, Ram, Wrangler | **Ignition Coil No. 1-8 Primary Circuit Conditions:**<br>Engine cranking or speed less than 2012 rpm, ASD relay on, system voltage 8v or 12v, coils not in dwell period, and the PCM detected peak coil current was not reached in 2.5 ms of dwell for 3-6 seconds. The coil primary resistance is less than 2 ohms.<br>**Possible Causes:**<br>ASD relay power circuit to the ignition coil is open<br>Ignition coil primary "driver" circuit open or grounded<br>Ignition coil is damaged or it has failed<br>PCM has failed |
| **P0370**<br>1T CCM<br>1998-2003<br>Ram with 5.9L Diesel | **Fuel Injection Pump Speed Sensor Signal Lost Conditions:**<br>Engine started; and the PCM detected that it lost the Fuel Injection Pump Speed/Position Signal (due to a fault in the internal Fuel injection Pump (the pump may have failed).<br>**Possible Causes:**<br>ECM cannot control the engine speed through circuit WP44<br>Fuel Injection pump has failed due to an internal problem |
| **P0380**<br>1T CCM<br>1998-2003<br>Ram with 5.9L Diesel | **Intake Air Heater Relay No. 1 Control Circuit Conditions:**<br>Key on for 1 second, and the PCM detected the Intake Air Heater Relay No. 1 circuit was not activated for more than one second. This problem is not due to a fault in the heater element.<br>**Possible Causes:**<br>Intake Air Heater relay control circuit open or shorted to ground<br>Intake Air Heater relay ground circuit is open<br>Intake Air Heater relay is damaged or has failed<br>PCM has failed |

## OBD II Trouble Code List (P0XXX Codes) - Continued

| DTC | Trouble Code Title, Conditions & Possible Causes: |
|---|---|
| **P0381**<br>1T CCM<br>1998-2003<br>Ram with 5.9L Diesel | **Wait To Start Lamp Inoperative Conditions:**<br>Key on for 2 seconds, and the PCM detected the Wait To Start control circuit was not activated for more than two seconds. The PCM expects to detect a voltage drop of 0.5v.<br>**Possible Causes:**<br>Wait To Start lamp driver circuit shorted to ground (the lamp will be on continuously)<br>Wait To Start lamp driver circuit open (lamp will not come on)<br>Wait To Start bulb has failed<br>Wait To Start ignition feed circuit is open, or the Cluster is open<br>PCM has failed |
| **P0382**<br>1T CCM<br>1998-2003<br>Ram with 5.9L Diesel | **Intake Air Heater Relay No. 2 Control Circuit Conditions:**<br>Key on for one second; and the PCM detected the Intake Air Heater Relay 2 Signal circuit was not activated for more than one second.<br>**Possible Causes:**<br>Intake air heater relay control circuit is open or shorted to ground<br>Intake air heater relay power circuit is open<br>Intake air heater relay is damaged or has failed<br>PCM has failed |
| **P0387**<br>1T CCM<br>1998-2000<br>Ram with 5.9L Diesel | **Crankshaft Position Sensor Supply Voltage Low Conditions:**<br>Engine started and the PCM did not detect any CKP Sensor position or engine speed signals.<br>**Possible Causes:**<br>CKP Sensor signal circuit is open or shorted to ground<br>CKP Sensor is damaged or has failed<br>PCM has failed |
| **P0388**<br>1T CCM<br>1998-2000<br>Ram with 5.9L Diesel | **Crankshaft Position Sensor Supply Voltage High Conditions:**<br>Engine started and the PCM did not detect any CKP Sensor position or engine speed signals.<br>**Possible Causes:**<br>CKP Sensor ground circuit is open<br>CKP Sensor signal circuit is shorted to VREF or to system power<br>CKP Sensor is damaged or has failed<br>PCM has failed |
| **P0400**<br>2T EGR<br>1995-98<br>Talon with VIN F<br>Turbo | **EGR System Fault Conditions:**<br>Engine runtime over 3 minutes in closed loop, ECT input over 170°F, engine speed at 1952-2400 rpm, MAP input at 1.80-2.70v, TP input from 0.6-1.8v, VSS over 3 mph, EGR Test "on", and the PCM detected too little change in EGR flow with the valve cycled.<br>**Possible Causes:**<br>EGR valve source vacuum supply line open or blocked<br>EGR exhaust transfer tubes blocked in the exhaust manifold<br>EGR transducer tube to the transducer is blocked or restricted<br>EGR valve or solenoid transducer is damaged or has failed<br>PCM has failed |

**OBD II Trouble Code List (P0XXX Codes) - Continued**

| DTC | Trouble Code Title, Conditions & Possible Causes: |
|---|---|
| **P0401**<br>2T EGR<br>1995-2000<br>300M, Avenger, Breeze, Caravan, Cherokee, Cirrus, Concorde, PT Cruiser, Dakota, Durango, Grand Cherokee, Intrepid, LHS, Neon, New Yorker, Prowler, Ram, Sebring, Stratus, Talon, Town & Country, Viper, Vision, Voyager, Wrangler | **EGR System Malfunction Conditions:**<br>Engine running with throttle open under steady load in closed loop, ECT Sensor more than 180ºF, BTS signal more than 40ºF, fuel trim not operating near its limits, then the EGR valve was cycled off to on and the PCM detected very little change in the O2 signal. Note: This test is repeated up to three (3) times for each vehicle trip.<br>**Possible Causes:**<br>EGR valve source vacuum supply line open or blocked<br>EGR exhaust transfer tubes blocked in the exhaust manifold<br>EGR transducer tube to the transducer is blocked or restricted<br>EGR valve or solenoid transducer is damaged or has failed<br>PCM has failed |
| **P0401**<br>2T EGR<br>1995-2003<br>300M, Avenger, Breeze, Caravan, Cherokee, Cirrus, Concorde, PT Cruiser, Dakota, Durango, Grand Cherokee, Intrepid, LHS, Liberty, Neon, New Yorker, Prowler, Ram, Sebring, Stratus, Talon, Town & Country, Viper, Vision, Voyager, Wrangler | **EGR System Fault Conditions:**<br>Engine running with throttle open under steady load in closed loop, ECT Sensor more than 180ºF, BTS signal more than 0ºF, system voltage over 10.5v, fuel trim not operating near its limits, then the EGR valve was cycled off to on and the PCM detected very little change in the O2 signal. Note: This test is repeated up to three (3) times for each vehicle trip.<br>**Possible Causes:**<br>EGR valve control circuit open, grounded or shorted to power<br>EGR exhaust transfer tubes blocked in the exhaust manifold<br>ASD relay power circuit open to the EGR solenoid<br>EGR valve is damaged or has failed<br>PCM has failed<br>TSB 18-33-98 contains a repair procedure for this code |
| **P0401**<br>2T EGR<br>2004-05<br>300M, Caravan, Concorde, PT Cruiser, Dakota, Durango, Grand Cherokee, Intrepid, LHS, Liberty, Magnum, Neon, Pacifica, Prowler Ram, Sebring, Stratus, Town & Country, Viper, Voyager, Wrangler | **EGR System Fault Conditions:**<br>Engine running for more than 2 minutes with ECT more than 158ºF (70°C). EGR is active. Vehicle is at less than 8500 feet altitude. Ambient temperature more than 20°F (-6°C). PCM closes EGR valve while monitoring O2 Sensor signal. Once a closed EGR fueling sample has been established, PCM then ramps in EGR and additional fueling, while monitoring the O2 Sensor signal in the open state. A fueling sample is again established. The PCM then compares the 2 different O2 Sensor signal readings (fueling samples). If a larger than expected variation is detected, a soft failure is recorded. Three soft failures set a one-trip (1T) failure. After 2 failed trips (2T), a DTC is set and the MIL illuminated.<br>**Possible Causes:**<br>EGR valve is open at idle<br>EGR solenoid ground circuit is open<br>EGR solenoid control circuit is open, shorted to ground or to voltage<br>ASD relay power circuit open to the EGR solenoid<br>EGR valve or solenoid is damaged or has failed<br>PCM has failed (EGR open or EGR closed) |

**OBD II Trouble Code List (P0XXX Codes) - Continued**

| DTC | Trouble Code Title, Conditions & Possible Causes: |
|---|---|
| **P0402**<br>1T CCM<br>2004-05<br>Liberty with Diesel | **EGR Solenoid Circuit Deviation Conditions:**<br>Engine running. The ECM detects the EGR flow is less than the requested flow (negative deviation) or is greater than the requested flow (positive deviation).<br>**Possible Causes:**<br>Air filter is restricted or damaged<br>Air restrictions in intake air system or EGR system<br>Air leaks in intake air system or EGR system<br>EGR valve has malfunctioned or failed<br>Intermittent condition |
| **P0403**<br>1T CCM<br>1995-98<br>Talon | **EGR Solenoid Circuit Conditions:**<br>Engine started; system voltage over 10.5v and the PCM detected the EGR Solenoid control circuit was not in its expected state when requested to operate by the PCM. Note: The EGR solenoid resistance range is 25-35 ohms at 68°F.<br>**Possible Causes:**<br>EGR solenoid control circuit is open or shorted to ground<br>EGR solenoid power circuit is open<br>EGR solenoid is damaged or has failed<br>PCM has failed |
| **P0403**<br>1T CCM<br>1995-2005<br>300M, Avenger, Breeze, Caravan, Cherokee, Cirrus, Concorde, PT Cruiser, Dakota, Durango, Grand Cherokee, Intrepid, LHS, Liberty, Magnum, Neon, New Yorker, Pacifica, Prowler, Ram, Sebring, Stratus, Talon, Town & Country, Viper, Vision, Voyager, Wrangler | **EGR Solenoid Circuit Conditions:**<br>Engine started; system voltage over 10.5v. The EGR solenoid control circuit was not in its expected state when requested to operate by the PCM.<br>**Possible Causes:**<br>EGR solenoid ground circuit is open<br>EGR solenoid control circuit is open or shorted to ground or to voltage<br>EGR solenoid power circuit is open<br>EGR solenoid is damaged or has failed<br>PCM has failed |
| **P0403**<br>1T CCM<br>2004-05<br>Liberty with Diesel | **EGR Solenoid Circuit Excessive Current Conditions:**<br>Ignition on; ECM commands EGR Solenoid on. The ECM detects excessive current on the EGR Solenoid control circuit.<br>**Possible Causes:**<br>Intermittent condition<br>EGR Solenoid has malfunctioned<br>EGR Solenoid control circuit is shorted to voltage<br>EMC has internal short to voltage |
| **P0403**<br>1T CCM<br>2004-05<br>Liberty with Diesel | **EGR Solenoid Circuit Open Conditions:**<br>Ignition on; ECM commands EGR Solenoid off. The ECM does not detect voltage on the EGR Solenoid control circuit.<br>**Possible Causes:**<br>Intermittent condition<br>ASD Relay output circuit is open<br>EGR Solenoid control circuit is open or is shorted to ground<br>EGR Solenoid has malfunctioned<br>EMC has failed |

**OBD II Trouble Code List (P0XXX Codes) - Continued**

| DTC | Trouble Code Title, Conditions & Possible Causes: |
|---|---|
| **P0404**<br>1T CCM<br>1995-2005<br>300M, Avenger, Breeze, Caravan, Cherokee, Cirrus, Concorde, PT Cruiser, Dakota, Durango, Grand Cherokee, Intrepid, LHS, Liberty, Magnum, Neon, New Yorker, Pacifica, Prowler, Ram, Sebring, Stratus, Talon, Town & Country, Viper, Vision, Voyager, Wrangler | **EGR Position Sensor Signal Performance Conditions:**<br>Engine started; system voltage over 10.5v and the PCM detected that the EGR flow (or valve movement) was not what was expected during the test period.<br>**Possible Causes:**<br>EGR Sensor signal circuit is open or shorted to ground<br>EGR Sensor 5v supply circuit is open or has high resistance<br>EGR Sensor ground circuit is open<br>EGR solenoid control circuit has a problem<br>EGR valve actuator loose, sticking, blocked or improperly grounded<br>EGR Sensor is damaged or has failed<br>Intermittent condition<br>PCM has failed |
| **P0405**<br>1T CCM<br>1995-2005<br>300M, Avenger, Breeze, Caravan, Cherokee, Cirrus, Concorde, PT Cruiser, Dakota, Durango, Grand Cherokee, Intrepid, LHS, Liberty, Magnum, Neon, New Yorker, Pacifica, Prowler, Ram, Sebring, Stratus, Talon, Town & Country, Viper, Vision, Voyager, Wrangler | **EGR Position Sensor Circuit Low Input Conditions:**<br>Key on or engine running; system voltage over 10v. The PCM detected that the EGR Sensor signal indicated less than 0.1v.<br>**Possible Causes:**<br>EGR Sensor signal circuit is shorted to ground or open<br>EGR Sensor VREF (5v) circuit is open or shorted to ground<br>EGR Sensor is damaged (shorted internally) or it has failed<br>EGR position internal failure<br>PCM has failed |
| **P0406**<br>1T CCM<br>1995-2005<br>300M, Avenger, Breeze, Caravan, Cherokee, Cirrus, Concorde, PT Cruiser, Dakota, Durango, Grand Cherokee, Intrepid, LHS, Liberty, Magnum, Neon, New Yorker, Pacifica, Prowler, Ram, Sebring, Stratus, Talon, Town & Country, Viper, Vision, Voyager, Wrangler | **EGR Position Sensor Circuit High Input Conditions:**<br>Key on or engine running; system voltage over 10.5v. The PCM detected the EGR Sensor indicated more than 4.89v for 6 seconds.<br>**Possible Causes:**<br>Intermittent condition<br>EGR Sensor signal is shorted to VREF (5v) supply circuit or to battery voltage<br>EGR Sensor ground circuit is open<br>EGR Sensor signal circuit is open<br>EGR Sensor is damaged (it may have an internal open circuit)<br>EGR solenoid failure<br>PCM has failed |

## OBD II Trouble Code List (P0XXX Codes) - Continued

| DTC | Trouble Code Title, Conditions & Possible Causes: |
|---|---|
| **P0410**<br>2T CCM<br>2004-05<br>Sebring, Stratus | **Air Injection Air Flow Failure Conditions:**<br>Engine running; Air Injection Pump is active. If the PCM detects that there is not enough airflow entering the exhaust stream, this DTC will set.<br>**Possible Causes:**<br>Air injection system damage or failure of components<br>Exhaust 1-way valve has failed<br>Fused B+ voltage output circuit problem<br>Air injection pump relay failure<br>Air injection pump ground circuit problem<br>Air injection pump motor has failed<br>Air injection passage is blocked or damaged<br>O2 Sensor failed |
| **P0411**<br>2T AIR<br>1995-96<br>Talon with VIN Y | **Too Little or Too Much Secondary Air Conditions:**<br>Engine runtime over 11 minutes, engine speed over 700 rpm, then after the PCM shifted the A/F mixture rich by 10% for 18 seconds, it detected the O2 signal did not switch to rich, or with the AIR solenoid commanded ON, the O2 did not indicate a lean ratio.<br>**Possible Causes:**<br>AIR solenoid air injection valve is damaged or has failed<br>AIR solenoid source vacuum hoses loose or disconnected<br>AIR solenoid air injection tube is restricted or clogged<br>PCM has failed |
| **P0412**<br>1T CCM<br>1995-96<br>Talon with VIN Y | **Secondary Air Solenoid Circuit Conditions:**<br>Engine started; system voltage over 10.5v and the PCM detected an unexpected voltage state after the AIR solenoid was enabled. The AIR solenoid resistance is from 33-39 ohms at 68°F.<br>**Possible Causes:**<br>AIR solenoid control circuit is open or shorted to ground<br>AIR solenoid power circuit is open<br>AIR solenoid is damaged or has failed<br>PCM has failed |
| **P0418**<br>2T CCM<br>2004-05<br>Sebring, Stratus | **Air Pump Relay Control Circuit Failure Conditions:**<br>Ignition is on; Battery voltage is more than 10,4v. If the PCM detects that there is an open or shorted condition in the Air Pump Relay Control Circuit, this DTC will set.<br>**Possible Causes:**<br>Air injection system damage or failure of components<br>Exhaust 1-way valve has failed<br>Fused B+ voltage output circuit problem<br>Air injection pump relay failure<br>Air injection pump ground circuit problem<br>Air injection pump motor has failed<br>Air injection passage is blocked or damaged<br>O2 Sensor failed |

**OBD II Trouble Code List (P0XXX Codes) - Continued**

| DTC | Trouble Code Title, Conditions & Possible Causes: |
|---|---|
| **P0420**<br>2T Catalyst<br>1995-2005<br>300M, Avenger, Breeze, Caravan, Cherokee, Cirrus, Concorde, PT Cruiser, Dakota, Durango, Grand Cherokee, Intrepid, LHS, Liberty, Magnum, Neon, New Yorker, Pacifica, Prowler, Ram, Sebring, Stratus, Talon, Town & Country, Viper, Vision, Voyager, Wrangler | **Catalyst Efficiency Below Normal (Bank 1) Conditions:**<br>Engine speed at 1200-1700 rpm in closed loop with the throttle open for over 2 minutes, ECT Sensor more than 147°F, MAP Sensor signal from 15.0-21.0 in. Hg, and the PCM detected the switch rate of the rear O2 reached 70% of the switch rate of the front O2.<br>**Possible Causes:**<br>Air leaks in at the exhaust manifold or exhaust pipes<br>Base engine problems (high coolant or engine oil consumption)<br>Catalytic converter damaged or has failed<br>Front O2 older (aged) than the rear O2 (O2 is lazy) |
| **P0420**<br>1T Catalyst<br>2004-05<br>Dakota, Durango, Caravan, Grand Cherokee, Liberty, Neon, Ram, Sebring, Stratus, Town & Country, Voyager | **Catalyst Efficiency Below Normal (Bank 1) Conditions:**<br>Engine is running for more than 90 seconds; engine coolant is more than 158°F (70°C); vehicle speed is 20-55 mph; engine speed is 1200-1900 rpm; MAP vacuum at 15-20 in. Hg. As catalyst efficiency deteriorates, the switch rate of the downstream O2 Sensor approaches that of the upstream O2 Sensor. If at any point during the test, the switch ratio reaches a predetermined value, a counter is incremented by one.<br>**Possible Causes:**<br>Catalytic converter damaged or has failed<br>Air leaks in at the exhaust manifold or exhaust pipes<br>Base engine problems (high coolant or engine oil consumption)<br>Front O2 older (aged) than the rear O2 (O2 is lazy) |
| **P0420**<br>2T Catalyst<br>2004-05<br>PT Cruiser | **Catalyst Efficiency Below Normal (Bank 1) Conditions:**<br>Engine running more than 90 sec.; ECT above 158°F (70°C); vehicle speed 20-55 mph; engine at 1216-1952 rpm. As catalyst efficiency deteriorates, the switch rate of the downstream O2 Sensor approaches that of the upstream O2 Sensor. If at any point during the test, the switch ratio reaches a predetermined value, a counter is incremented by one.<br>**Possible Causes:**<br>Exhaust leak is present<br>Base engine mechanical problems<br>Catalytic converter damaged or has failed<br>Front O2 older (aged) than the rear O2 (O2 is lazy) |
| **P0421**<br>2T Catalyst<br>2001-03<br>Sebring, Stratus Coupe with 2.4L I4 or 3.0L V6 | **Catalyst Efficiency Fault (Bank 1) Conditions:**<br>Engine started, engine speed below 3000 rpm in closed loop with throttle open at more than 1 mph for 84 seconds, IAT Sensor over 14°F, VAF Sensor over 4000 Hz, BARO Sensor over 75 kPa, volumetric efficiency from 63-169 Hz, and the PCM detected the frequency of the front and rear O2 exceeded 0.8.<br>**Possible Causes:**<br>Air leaks in at the exhaust manifold or exhaust pipes<br>Base engine problems (high coolant or engine oil consumption)<br>Catalytic converter is damaged or it has failed<br>Front O2 older (aged) than the rear O2 (O2 is lazy) |

**OBD II Trouble Code List (P0XXX Codes) - Continued**

| DTC | Trouble Code Title, Conditions & Possible Causes: |
|---|---|
| **P0421**<br>2T Catalyst<br>2001-03<br>Sebring, Stratus<br>Coupe with 3.0L V6 | **Catalyst Efficiency Fault (Bank 1) Conditions:**<br>Engine speed below 3000 rpm with throttle open at over 1 mph for 84 seconds, IAT Sensor over 14ºF, VAF Sensor over 4000 Hz, volumetric efficiency from 63-169 Hz, BARO Sensor over 75 kPa, and the PCM detected the frequency of the front and rear O2 exceeded 0.8.<br>**Possible Causes:**<br>Air leaks in at the exhaust manifold or exhaust pipes<br>Base engine problems (high coolant or engine oil consumption)<br>Catalytic converter is damaged or it has failed<br>Front O2 older (aged) than the rear O2 (O2 is lazy) |
| **P0422**<br>2T Catalyst<br>1995-98<br>Talon | **Catalyst Efficiency Below Normal Conditions:**<br>Engine speed from 1248-2400 in closed loop with the throttle open for over 2 minutes, ECT Sensor more than 170ºF, MAP Sensor signal from 1.50-2.60v, and the PCM detected the switch rate of the rear O2 reached 70% of the switch rate of the front O2.<br>**Possible Causes:**<br>Air leaks in at the exhaust manifold or exhaust pipes<br>Base engine problems (high coolant or engine oil consumption)<br>Catalytic converter damaged or has failed<br>Front O2 older (aged) than the rear O2 (O2 is lazy) |
| **P0430**<br>1T Catalyst<br>2004-05<br>300M, Caravan, Concorde, PT Cruiser, Dakota, Durango, Grand Cherokee, Intrepid, LHS, Liberty, Magnum, Neon, Pacifica, Prowler Ram, Sebring, Stratus, Town & Country, Viper, Voyager, Wrangler | **Catalyst (2/1) Efficiency Below Normal Conditions:**<br>After engine warm-up, ECT Sensor more than 170ºF for 180 seconds of open throttle operation and over 20 mph (engine between 1200-1700 rpm and MAP vacuum between 15-20 in. Hg). As catalyst efficiency deteriorates, the switch rate of the downstream O2 Sensor approaches that of the upstream O2 Sensor. If, at any point during the test, the switch ratio reaches a predetermined value, a counter is incremented by one. 3 good trips required to turn off MIL.<br>**Possible Causes:**<br>Exhaust leaks<br>Base engine problems<br>Catalytic converter damaged or has failed<br>Front O2 older (aged) than the rear O2 (O2 is lazy) |
| **P0432**<br>2T Catalyst<br>1995-2005<br>300M, Avenger, Breeze, Caravan, Cirrus, Concorde, PT Cruiser, Grand Cherokee, Intrepid, LHS, Magnum, Neon, New Yorker, Pacifica, Prowler, Sebring, Stratus, Talon, Town & Country, Viper, Vision, Voyager | **Catalyst Efficiency Below Normal (Bank 2) Conditions:**<br>Engine speed at 1200-1700 rpm in closed loop with the throttle open for over 2 minutes, ECT Sensor more than 147ºF, MAP Sensor signal from 15.0-21.0" Hg, and the PCM detected the switch rate of the rear O2 reached 70% of the switch rate of the front O2.<br>**Possible Causes:**<br>Air leaks in at the exhaust manifold or exhaust pipes<br>Base engine problems (high coolant or engine oil consumption)<br>Catalytic converter damaged or has failed<br>Front O2 older (aged) than the rear O2 (O2 is lazy) |

**OBD II Trouble Code List (P0XXX Codes) - Continued**

| DTC | Trouble Code Title, Conditions & Possible Causes: |
|---|---|
| **P0432**<br>2T Catalyst<br>1996-2003<br>Cherokee, Dakota, Durango, Grand Cherokee, Liberty, Ram, Wrangler | **Catalyst Efficiency Fault (Bank 2) Conditions:**<br>Engine started; then vehicle driven at 1200-1700 rpm at over 20 mph with the throttle open and steady for 3 minutes, ECT Sensor over 147ºF, MAP Sensor at 15.0-21.0" Hg, and the PCM detected the switch rate of the rear O2 approached the switch rate of the front O2 during the Catalyst Monitor test period.<br>**Possible Causes:**<br>Air leaks in at the exhaust manifold or exhaust pipes<br>Base engine problems (high coolant or engine oil consumption)<br>Catalytic converter damaged or has failed<br>Front O2 older (aged) than the rear O2 (O2 is lazy) |
| **P0440**<br>2T EVAP<br>1995-98<br>Talon with 2.0L Turbo | **EVAP Purge System Fault Conditions:**<br>ECT Sensor from 40-90ºF and IAT Sensor within 10ºF of ECT at startup (cold engine), engine speed under 2048 rpm, Purge solenoid commanded ON, EVAP Stricter test completed and passed, and the PCM detected too little change in the Fuel Control system.<br>**Possible Causes:**<br>EVAP system vacuum hoses plugged, loose or disconnected<br>EVAP purge control solenoid is damaged or has failed<br>EVAP purge control valve is damaged |
| **P0440**<br>2T EVAP<br>2002-05<br>300M, Caravan, Cherokee, Cirrus, Concorde, PT Cruiser, Dakota, Durango, Grand Cherokee, Intrepid, LHS, Liberty, Magnum, Neon, New Yorker, Pacifica, Prowler, Ram, Sebring, Stratus, Talon, Town & Country, Viper, Vision, Voyager, Wrangler | **EVAP Purge System Fault Conditions:**<br>Ambient Air Temperature from 39-89ºF (4-32°C); engine running; Fuel level over 12%. The PCM detected that the NVLD switch did not close during medium/large leak test. Once this event occurs, the PCM will increase the amount of vacuum in the system that flows past the purge valve. If the NVLD switch does not close under these conditions, the PCM will set this code.<br>**Possible Causes:**<br>EVAP purge valve vacuum supply is leaking or clogged<br>EVAP purge valve is stuck closed<br>EVAP purge solenoid has failed<br>NVLD assembly (leak detection) is damaged or has failed<br>NVLD switch circuit is open or the NVLD switch has failed<br>Ground circuit is open<br>PCM has failed |
| **P0441**<br>2T EVAP<br>300M, Avenger, Breeze, Caravan, Cherokee, Cirrus, Concorde, PT Cruiser, Dakota, Durango, Grand Cherokee, Intrepid, LHS, Liberty, Neon, New Yorker, Prowler, Ram, Sebring, Stratus, Talon, Town & Country, Viper, Vision, Voyager, Wrangler | **EVAP Purge Flow Monitor Fault Conditions:**<br>Engine at idle speed in closed loop for 200 seconds, BARO Sensor signal less than 8,000 feet, ECT Sensor more than 160ºF, no Low Fuel, MAP Sensor signal less than 23.7" Hg, and the PCM did not detect any purge flow through the EVAP system during this test.<br>**Possible Causes:**<br>EVAP purge solenoid vacuum line loose, leaking or restricted<br>EVAP purge solenoid stuck leaking, stuck open or stuck closed<br>EVAP purge vacuum line to canister leaking or disconnected<br>EVAP canister leaking, damaged or has failed |

## OBD II Trouble Code List (P0XXX Codes) - Continued

| DTC | Trouble Code Title, Conditions & Possible Causes: |
|---|---|
| **P0441**<br>2T EVAP<br>2004-05<br>300M, Caravan, Concorde, PT Cruiser, Dakota, Durango, Grand Cherokee, Intrepid, LHS, Liberty, Magnum, Neon, Pacifica, Prowler Ram, Sebring, Stratus, Town & Country, Voyager, Wrangler | **EVAP Purge System Performance**<br>Check with cold start test. Engine running. Small leak test passed. The PCM activates the EVAP purge solenoid and it gradually increases to maximum flow. During flow, the PCM looks for the NVLD switch to close. If the PCM does not see the NVLD switch close at maximum flow, an error is detected.<br>Possible Causes:<br>Intermittent condition<br>EVAP purge solenoid functioning improperly<br>EVAP purge solenoid vacuum supply leaking or clogged |
| **P0441**<br>2T EVAP<br>2004-05<br>Viper | **EVAP Purge Flow Monitor Fault Conditions:**<br>Engine at idle speed in closed loop for 2 minutes; BARO Sensor signal less than 8,000 feet; ECT Sensor more than 170ºF; no Low Fuel; MAP Sensor signal less than 15.7" Hg. After having passed the Leak Detection Pump test, no airflow through the EVAP system is detected by the EVAP monitor.<br>**Possible Causes:**<br>Intermittent condition<br>EVAP purge solenoid vacuum line loose, leaking or restricted<br>EVAP purge solenoid stuck leaking, stuck open or stuck closed<br>EVAP purge vacuum line to canister leaking or disconnected<br>EVAP canister leaking, damaged or has failed |
| **P0442**<br>2T EVAP<br>1996-2003<br>300M, Avenger, Breeze, Caravan, Cherokee, Cirrus, Concorde, PT Cruiser, Dakota, Durango, Grand Cherokee, Intrepid, LHS, Liberty, Neon, New Yorker, Prowler, Ram, Sebring, Stratus, Talon, Town & Country, Viper, Vision, Voyager, Wrangler | **EVAP System Small Leak Detected Conditions:**<br>ECT Sensor from 40-90ºF and within 10ºF of the BTS input at startup (cold engine), fuel level more than 1/2 full, engine started, EVAP leak test enabled, and the PCM detected a leak greater than 0.040" but less than 0.080" somewhere in the EVAP system.<br>**Possible Causes:**<br>EVAP fuel tank or canister vapor hoses leaking or damaged<br>EVAP system component leaking, damaged or has failed<br>Fuel tank cap is loose, or the cap release pressure is incorrect<br>Leak detection pump damaged or leaking |
| **P0442**<br>2T EVAP<br>2001-03<br>Sebring, Stratus Coupe with 2.4L I4 or 3.0L V6 | **EVAP System Small Leak Detected Conditions:**<br>ECT and IAT Sensors under 86ºF at startup; engine runtime less than 12 min.; BARO Sensor over 76 kPa; engine over 1600 rpm; volumetric efficiency from 20-80%; ECT Sensor over 140ºF; PSP switch shows OFF. The test stops as FTP Sensor indicates 0.065 psi and fluctuates less than 0.094 psi, with FTP input of 1-4.0v, ECT input over 140ºF and IAT input under 41ºF with Purge and Vent solenoid closed. The test fails if the PCM detects the FTP signal changes more than 0.114 psi within 20 seconds.<br>**Possible Causes:**<br>EVAP canister vent solenoid is damaged or has failed<br>EVAP emission canister seal is damaged or leaking<br>Fuel tank cap is loose, or the cap release pressure is incorrect<br>Fuel tank, vapor line or vacuum seal is damaged or leaking |

**OBD II Trouble Code List (P0XXX Codes) - Continued**

| DTC | Trouble Code Title, Conditions & Possible Causes: |
|---|---|
| **P0442**<br>2T EVAP<br>2004-05<br>300M, Caravan, Concorde, PT Cruiser, Dakota, Durango, Grand Cherokee, Intrepid, LHS, Liberty, Magnum, Neon, Pacifica, Prowler Ram, Sebring, Stratus, Town & Country, Voyager, Wrangler | **EVAP System Medium Leak Detected Conditions:**<br>Monitor with engine running. Cold start test. Fuel level more than 12%. Ambient temperature between 39-89°F (4-32°C). Closed loop fuel system. Test runs when small leak test is maturing. The PCM activates EVAP purge solenoid to pull EVAP system into a vacuum to close the NVLD switch. Once this switch is closed, the PCM turns the EVAP purge solenoid off to seal the EVAP system. If the NVLD switch re-opens before the calibrated amount of time for a Medium leak, an error is detected.<br>**Possible Causes:**<br>Intermittent condition<br>Vacuum hoses, connections or switches have come loose or malfunctioned<br>EVAP emission system has a leak<br>EVAP purge solenoid operation has malfunctioned<br>NVLD switch operation has malfunctioned |
| **P0442**<br>2T EVAP<br>2004-05<br>Viper | **EVAP System Medium Leak Detected Conditions:**<br>Monitor immediately after cold start; battery/ambient temperature between 40-90F; coolant temperature within 10F of battery/ambient temperature. If there is a leak of 0.040-0.080 in. in the EVAP system, this DTC will set.<br>**Possible Causes:**<br>Intermittent condition<br>Vacuum hoses, connections or switches have come loose or malfunctioned<br>EVAP emission system has a leak<br>EVAP purge solenoid operation has malfunctioned |
| **P0443**<br>1T CCM<br>1995-98<br>Talon | **EVAP Purge Solenoid Circuit Fault Conditions:**<br>Engine started; system voltage from 10-16v, and after the EVAP purge solenoid was commanded ON and off, the PCM did not detect any solenoid surge voltage (system voltage +2v) for 2 seconds. Note: The solenoid resistance is from 25-35 ohms at 68°F.<br>**Possible Causes:**<br>EVAP purge solenoid control circuit open or shorted to ground<br>EVAP purge solenoid power circuit open<br>EVAP purge solenoid damaged or has failed<br>PCM has failed |
| **P0443**<br>1T CCM<br>1995-2003<br>300M, Avenger, Breeze, Caravan, Cherokee, Cirrus, Concorde, PT Cruiser, Dakota, Durango, Grand Cherokee, Intrepid, LHS, Liberty, Neon, New Yorker, Prowler, Ram, Sebring, Stratus, Talon, Town & Country, Viper, Vision, Voyager, Wrangler | **EVAP Purge Solenoid Circuit Fault Conditions:**<br>Engine started; system voltage over 10v, EVAP Purge solenoid commanded ON and off, and the PCM detected an unexpected voltage condition at the EVAP purge solenoid for 3 seconds. Note: The solenoid resistance at 68°F is from 25-35 ohms.<br>**Possible Causes:**<br>EVAP purge solenoid control circuit open or shorted to ground<br>EVAP purge solenoid power circuit open<br>EVAP purge solenoid is damaged or it has failed<br>PCM has failed |

## OBD II Trouble Code List (P0XXX Codes) - Continued

| DTC | Trouble Code Title, Conditions & Possible Causes: |
|---|---|
| **P0443**<br>1T CCM<br>2001-03<br>Sebring, Stratus<br>Coupe with 2.4L I4 or 3.0L V6 | **EVAP Purge Solenoid Circuit Fault Conditions:**<br>Engine started; system voltage from 10-16v, and after the EVAP purge solenoid was commanded ON and off, the PCM did not detect enough solenoid surge voltage (system voltage +2v) for 200 ms. Note: The solenoid resistance at 68°F is from 25-35 ohms.<br>**Possible Causes:**<br>EVAP purge solenoid control circuit open or shorted to ground<br>EVAP purge solenoid power circuit open<br>EVAP purge solenoid or PCM has failed |
| **P0443**<br>1T CCM<br>2004-05<br>300M, Caravan, Concorde, Dakota, Durango, Grand Cherokee, Intrepid, LHS, Liberty, Magnum, Neon, Pacifica, Prowler Ram, Sebring, Stratus, Town & Country, Voyager, Wrangler | **EVAP Purge Solenoid Circuit Fault Conditions:**<br>Ignition on or engine running. Battery voltage more than 10v. The PCM will set a trouble code if the actual state of the solenoid does not match the intended state.<br>**Possible Causes:**<br>EVAP purge solenoid control circuit open or shorted to ground<br>EVAP purge solenoid return circuit open or shorted to ground<br>EVAP purge solenoid is damaged or it has failed<br>PCM has failed |
| **P0443**<br>1T CCM<br>2004-05<br>Viper | **EVAP Purge Solenoid Circuit Fault Conditions:**<br>Solenoid is monitored continuously after ignition is ON and battery is more than 10.4v. System is not powering down, not in limp-in mode, and time since last solenoid activation is more than 72ms. The PCM will set this DTC if the actual state of the solenoid does not match the intended state on 2 consecutive cycles.<br>**Possible Causes:**<br>Fused ignition switch output circuit is open<br>EVAP purge solenoid control circuit open or shorted to ground<br>EVAP purge solenoid or PCM has failed |
| **P0444**<br>1T CCM<br>2004-05<br>PT Cruiser | **EVAP Purge Solenoid Circuit Open Conditions:**<br>Any time key is on, if the PCM detects an open circuit in the purge solenoid or in the circuit for more than 280ms, a one-trip fault is set and the MIL is illuminated.<br>**Possible Causes:**<br>Intermittent condition<br>12v supply circuit is open<br>EVAP purge solenoid control circuit open<br>EVAP purge solenoid is open<br>PCM has failed |
| **P0445**<br>2T EVAP<br>2004-05<br>Pacifica, Sebring, Stratus | **EVAP System Large Leak Detected Conditions:**<br>Monitor with engine running. Cold start test. Fuel level more than 12%. Ambient temperature between 39-89°F (4-32°C). Closed loop fuel system. Test runs when small leak test is maturing. The PCM activates EVAP purge solenoid to pull EVAP system into a vacuum to close the NVLD switch. Once this switch is closed, the PCM turns the EVAP purge solenoid off to seal the EVAP system. If the NVLD switch re-opens before the calibrated amount of time for a Large leak, an error is detected.<br>**Possible Causes:**<br>Intermittent condition<br>Vacuum hoses, connections or switches have come loose or malfunctioned<br>EVAP emission system has a leak<br>EVAP purge solenoid operation has malfunctioned<br>NVLD switch operation has malfunctioned |

**OBD II Trouble Code List (P0XXX Codes) - Continued**

| DTC | Trouble Code Title, Conditions & Possible Causes: |
|---|---|
| **P0446**<br>1T CCM<br>2001-03<br>Sebring, Stratus<br>Coupe with 2.4L I4 or<br>3.0L V6 | **EVAP Canister Vent Solenoid Circuit Fault Conditions:**<br>Engine started; system voltage from 10-16v, EVAP canister vent solenoid commanded ON and off, and the PCM did not detect a solenoid surge voltage (system voltage +2v) for 30 ms.<br>**Possible Causes:**<br>EVAP vent solenoid circuit open, shorted to ground or power<br>EVAP canister vent solenoid has failed (solenoid resistance is 25-35 ohms at 68ºF)<br>PCM has failed |
| **P0450**<br>1T CCM<br>1998<br>Talon | **EVAP Fuel Tank Pressure Sensor Range/Performance Conditions:**<br>Engine speed 1600 rpm or higher, volumetric efficiency at 20-80%, Purge solenoid "on", and the PCM detected the EVAP pressure Sensor input was over 4.50v, or with the solenoid "off", the Sensor signal was 0.5v, or the Sensor signal varied 0.2v over 20 times with throttle open.<br>**Possible Causes:**<br>EVAP pressure Sensor signal circuit open or shorted to ground<br>EVAP pressure Sensor is damaged or has failed<br>EVAP fuel vent valve or fuel vapor line clogged or restricted<br>PCM has failed |
| **P0450**<br>1T CCM<br>2001-03<br>Sebring, Stratus<br>Coupe with 2.4L I4 or<br>3.0L V6 | **EVAP Fuel Tank Pressure Sensor Range/Performance Conditions:**<br>Engine speed 1600 rpm or higher, volumetric efficiency at 20-80%, IAT Sensor signal more than 41ºF, and with the purge solenoid on (100% duty cycle), the PCM detected the EVAP pressure Sensor signal was more than 4.50v, or with the purge solenoid off (0%), the pressure Sensor signal was 0.5v, or the Sensor signal varied 0.2v at least 20 times with either the throttle open, or at idle with it closed.<br>**Possible Causes:**<br>EVAP pressure Sensor signal circuit open or shorted to ground<br>EVAP pressure Sensor is damaged or has failed<br>EVAP fuel vent valve or fuel vapor line clogged or restricted<br>PCM has failed |
| **P0451**<br>1T CCM<br>2001-03<br>Sebring, Stratus<br>Coupe with 2.4L I4 or<br>3.0L V6 | **EVAP Fuel Tank Pressure Sensor Performance Conditions:**<br>Engine started; engine at idle speed, throttle closed (throttle switch is "on"), and the PCM detected the fuel tank pressure differential Sensor value fluctuated over 0.20v at least 20 times, or with engine speed over 2500 rpm at over 9.5 mph with the volumetric efficiency over 55%, the Sensor value changed over 0.20v at least 20 times.<br>**Possible Causes:**<br>Fuel tank pressure differential Sensor connector is damaged<br>Fuel tank pressure differential Sensor is damaged or has failed<br>PCM has failed |
| **P0452**<br>1T CCM<br>2001-03<br>Sebring, Stratus<br>Coupe with 2.4L I4 or<br>3.0L V6 | **Fuel Tank Differential Pressure Sensor Circuit Low Input Conditions:**<br>Engine started; IAT Sensor more than 41ºF, engine speed over 1600 rpm, volumetric efficiency from 20-80%, and the PCM detected the Fuel Tank pressure differential Sensor indicated less than 0.1v.<br>**Possible Causes:**<br>Fuel tank pressure differential Sensor connector is shorted<br>Fuel tank pressure differential Sensor circuit shorted to ground<br>Fuel tank pressure differential Sensor is damaged or has failed<br>PCM has failed |

## OBD II Trouble Code List (P0XXX Codes) - Continued

| DTC | Trouble Code Title, Conditions & Possible Causes: |
|---|---|
| **P0452**<br>1T CCM<br>2002-05<br>300M, Caravan, Cherokee, Cirrus, Concorde, PT Cruiser, Dakota, Durango, Grand Cherokee, Intrepid, LHS, Liberty, Magnum, Neon, New Yorker, Pacifica, Prowler, Ram, Sebring, Stratus, Talon, Town & Country, Viper, Vision, Voyager, Wrangler | **NVLD Pressure Switch Sense Circuit Low Input Conditions:**<br>Engine started; and immediately after the engine is running. The PCM activates the NVLD solenoid to test the NVLD switch circuit. If the switch is not open, the PCM sets this code.<br>**Possible Causes:**<br>EVAP purge solenoid control circuit is shorted to ground<br>EVAP purge solenoid is leaking or it is stuck in open position<br>NVLD assembly or NVLD switch is damaged or it has failed<br>NVLD switch signal circuit is shorted to ground<br>PCM has failed |
| **P0453**<br>1T CCM<br>2001-03<br>Sebring, Stratus Coupe with 2.4L I4 or 3.0L V6 | **Fuel Tank Differential Pressure Sensor Circuit High Input Conditions:**<br>Engine speed over 1600 rpm, IAT Sensor from 41-113°F, volumetric efficiency from 20-80%, and the PCM detected the Fuel Tank pressure differential Sensor indicated less than 4.0v.<br>**Possible Causes:**<br>Fuel tank pressure differential Sensor connector is open<br>Fuel tank pressure differential Sensor circuit is open<br>Fuel tank pressure differential Sensor is damaged or has failed<br>PCM has failed |
| **P0453**<br>1T CCM<br>2002-05<br>300M, Caravan, Cherokee, Cirrus, Concorde, PT Cruiser, Dakota, Durango, Grand Cherokee, Intrepid, LHS, Liberty, Magnum, Neon, New Yorker, Pacifica, Prowler, Ram, Sebring, Stratus, Talon, Town & Country, Viper, Vision, Voyager, Wrangler | **NVLD Pressure Switch Sense Circuit High Input Conditions:**<br>Engine started; and immediately after the engine is running, the PCM activates the NVLD solenoid to test the NVLD switch circuit. If the switch does not close under these conditions, this code is set.<br>**Possible Causes:**<br>NVLD assembly ground circuit is open<br>NVLD switch signal circuit is open or shorted to power (B+) or to NVLD solenoid control circuit<br>NVLD assembly or switch is damaged or it has failed<br>PCM has failed |
| **P0455**<br>2T EVAP<br>1998<br>Talon | **EVAP Large Leak (0.80") Detected Conditions:**<br>ECT Sensor from 40-90°F and within 10°F of BTS input at startup, Fuel Level over 50%, engine running, EVAP leak test "on", and the PCM detected a leak greater than 0.080" somewhere in the EVAP system.<br>**Possible Causes:**<br>EVAP fuel tank or canister vapor hoses leaking or damaged<br>EVAP system component leaking, fuel cap loose or missing<br>Leak detection pump damaged or leaking |

**OBD II Trouble Code List (P0XXX Codes) - Continued**

| DTC | Trouble Code Title, Conditions & Possible Causes: |
|---|---|
| **P0455**<br>2T EVAP<br>2002-05<br>300M, Caravan, Cherokee, Cirrus, Concorde, PT Cruiser, Dakota, Durango, Grand Cherokee, Intrepid, LHS, Liberty, Magnum, Neon, New Yorker, Pacifica, Prowler, Ram, Sebring, Stratus, Talon, Town & Country, Viper, Vision, Voyager, Wrangler | **EVAP Large Leak Detected Conditions:**<br>Ambient Air Temperature from 39-89°F at engine startup, engine running under closed loop conditions, Fuel Level over 12%, then with the EVAP purge solenoid enabled (to pull vacuum into the system to close the NVLD switch) and the EVAP "small leak" test maturing, the PCM turns "off" the EVAP purge solenoid once the NVLD switch closes. If the NVLD switch reopens before a calibrated amount of time expires, a "large" leak in the system is detected (larger than 0.080 in.).<br>**Possible Causes:**<br>EVAP purge solenoid is damaged or it has failed<br>Fuel tank cap is damaged, missing or the wrong part number<br>NVLD switch is damaged or it has failed |
| **P0455**<br>2T EVAP<br>2001-03<br>Sebring, Stratus Coupe with 2.4L I4 or 3.0L V6 | **EVAP System Large Leak Detected Conditions:**<br>ECT and IAT Sensor signals less than 86°F at startup, engine runtime under 12 minutes, BARO Sensor over 76 kPa, engine speed over 1600 rpm, volumetric efficiency 20-80%, ECT Sensor more than 140°F, PSP switch indicating off. The test stops when the FTP Sensor indicates 451 Pa (0.065 psi) and fluctuates less than 647 Pa (0.094 psi) with FTP input of 1-4.0v, an ECT input over 140°F and an IAT input under 41°F with the purge and vent solenoid closed. The test fails if the PCM detects that the FTP input changes less than 324 Pa (0.047 psi) within 20 seconds. Test takes from 75-125 seconds (depends on fuel level, etc.).<br>**Possible Causes:**<br>EVAP emission canister seal is damaged or leaking<br>Fuel tank, vapor line or vacuum seal is damaged or leaking<br>EVAP canister vent solenoid is damaged or has failed<br>Fuel tank cap is loose, or the cap release pressure is incorrect |
| **P0455**<br>2T EVAP<br>1996-2003<br>300M, Avenger, Breeze, Caravan, Cherokee, Cirrus, Concorde, PT Cruiser, Dakota, Durango, Grand Cherokee, Intrepid, LHS, Neon, New Yorker, Prowler, Ram, Sebring, Stratus, Talon, Town & Country, Viper, Vision, Voyager, Wrangler | **EVAP Large Leak (0.80") Detected Conditions:**<br>ECT Sensor from 40-90°F and within 10°F of the BTS input at startup (cold startup), fuel level more than 1/2 full, engine started, EVAP leak test enabled, and the PCM detected a leak greater than 0.080" existed somewhere in the EVAP system.<br>**Possible Causes:**<br>EVAP fuel tank or canister vapor hoses leaking or damaged<br>EVAP system component leaking, fuel cap loose or missing<br>Leak detection pump damaged or leaking |

**OBD II Trouble Code List (P0XXX Codes) - Continued**

| DTC | Trouble Code Title, Conditions & Possible Causes: |
|---|---|
| **P0456**<br>2T EVAP<br>1998-2003<br>300M, Avenger, Breeze, Caravan, Cherokee, Cirrus, Concorde, PT Cruiser, Dakota, Durango, Grand Cherokee, Intrepid, LHS, Liberty, Neon, New Yorker, Prowler, Ram, Sebring, Stratus, Talon, Town & Country, Viper, Vision, Voyager, Wrangler | **EVAP Small Leak (0.020") Detected Conditions:**<br>ECT Sensor from 40-90°F and within 10°F of BTS Sensor at startup, Fuel Level from 15-85%, engine started, EVAP leak test enabled, and the PCM detected a leak greater than 0.020" but less than 0.040" somewhere in the EVAP system.<br>**Possible Causes:**<br>EVAP fuel tank or canister vapor hoses leaking or damaged<br>EVAP system component leaking, fuel cap loose or missing<br>Leak detection pump damaged or leaking |
| **P0456**<br>2T EVAP<br>2002-05<br>300M, Caravan, Cherokee, Cirrus, Concorde, PT Cruiser, Dakota, Durango, Grand Cherokee, Intrepid, LHS, Liberty, Magnum, Neon, New Yorker, Pacifica, Prowler, Ram, Sebring, Stratus, Talon, Town & Country, Viper, Vision, Voyager, Wrangler | **EVAP System Small Leak Detected Conditions:**<br>Ambient Air Temperature from 39-109°F at engine startup, engine running under closed loop conditions, Fuel Level below 88%, then with the EVAP system sealed, the PCM monitors the NVLD switch. If the NVLD switch does not close within a calibrated amount of time expires, a "small" leak in the EVAP system was detected.<br>**Possible Causes:**<br>Fuel tank cap is damaged, loose or the wrong part number<br>Small leak present somewhere in the EVAP system |
| **P0457**<br>2T EVAP<br>2004-05<br>300M, Caravan, Concorde, Dakota, Durango, Grand Cherokee, Intrepid, LHS, Liberty, Magnum, Neon, Pacifica, Prowler, Ram, Sebring, Stratus, Town & Country, Viper, Voyager, Wrangler | **Loose Fuel Cap Condition:**<br>Monitor with ignition on. Ambient temperature should be between 39-109°F (4-43°C). Vehicle should be in closed loop fuel system. The PCM has detected an EVAP system leak after a fuel level increase. If the NVLD switch reopens before the calibrated amount of time after a fuel tank fill, an error is detected. MIL will illuminate. Condition requires 3 good trips to turn off MIL.<br>**Possible Causes:**<br>Loose or missing fuel fill cap<br>Intermittent condition<br>NVLD system or switch malfunction<br>EVAP system leaking<br>EVAP purge solenoid malfunction |

**OBD II Trouble Code List (P0XXX Codes) - Continued**

| DTC | Trouble Code Title, Conditions & Possible Causes: |
|---|---|
| **P0458**<br>1T EVAP<br>2004-05<br>PT Cruiser | **EVAP Purge Solenoid Low Condition:**<br>Monitor with ignition on. If PCM detected an open in the purge solenoid of in the circuit for more than 280ms, a one-trip fault is set and the MIL is illuminated.<br>**Possible Causes:**<br>Intermittent condition<br>EVAP purge solenoid connector problems<br>EVAP purge solenoid control circuit is shorted to ground<br>EVAP purge solenoid leaks, stuck open, or stuck closed<br>PCM has failed |
| **P0459**<br>1T EVAP<br>2004-05<br>PT Cruiser | **EVAP Purge Solenoid High Condition:**<br>Monitor with ignition on. If PCM detected a short to ground in the purge solenoid of in the circuit for more than 280ms, a one-trip fault is set and the MIL is illuminated.<br>**Possible Causes:**<br>Intermittent condition<br>EVAP purge solenoid connector problems<br>EVAP purge solenoid control circuit is shorted to battery voltage<br>EVAP purge solenoid leaks, stuck open, or stuck closed<br>PCM has failed |
| **P0460**<br>1T CCM<br>1998-2003<br>300M, Avenger, Breeze, Caravan, Cherokee, Cirrus, Concorde, PT Cruiser, Dakota, Durango, Grand Cherokee, Intrepid, LHS, Liberty, Neon, New Yorker, Prowler, Ram, Sebring, Stratus, Talon, Town & Country, Viper, Vision, Voyager, Wrangler | **Fuel Level Sending No Change Over Miles Conditions:**<br>Engine started; fuel level less than 15% or more than 85%, and the PCM detected a Low Fuel condition that the fuel level was less than 15% for 120 miles, or that the fuel level remained at more than 85% and did not change by at least 10% after traveling over 100 miles.<br>**Possible Causes:**<br>Fuel level sending unit signal circuit open or shorted to ground<br>Fuel level sending unit ground circuit is open<br>Fuel level sensing unit is damaged or the fuel tank is damaged |
| **P0460**<br>1T CCM<br>2001-03<br>Sebring, Stratus Coupe with 2.4L I4 or 3.0L V6 | **Fuel Gauge Unit Circuit Malfunction Conditions:**<br>Engine started; then after the vehicle was driven enough miles for the fuel calculation from the fuel injector usage to reach 20 liters, the PCM detected the diversity of the amount of fuel in the fuel tank calculated from the fuel level Sensor indicated less than 2 liters.<br>**Possible Causes:**<br>Fuel gauge unit connector is damaged, open or shorted<br>Fuel gauge unit signal circuit is open or shorted to ground<br>Fuel gauge unit is damaged or the fuel tank is damaged |
| **P0460**<br>1T CCM<br>2004-05<br>Viper | **Fuel Level Sending No Change Over Miles Conditions:**<br>Engine running; fuel level less than 15% or more than 85%. The PCM detected a Low Fuel condition, less than 15% for 120 miles, or fuel level does not change by at least 4% for more than 250 miles.<br>**Possible Causes:**<br>Physically damaged, deformed or obstructed fuel tank<br>Fuel level sensor has failed |

**OBD II Trouble Code List (P0XXX Codes) - Continued**

| DTC | Trouble Code Title, Conditions & Possible Causes: |
|---|---|
| **P0460**<br>1T CCM<br>2004-05<br>Liberty with Diesel | **Fuel Level Sensor Circuit Signal Voltage Too High Or Too Low Conditions:**<br>Ignition on. The Fuel Level Sensor signal voltage is above 4.51v for 0.6 second (too high) or is below 0.19v for 0.6 second (too low).<br>**Possible Causes:**<br>Intermittent condition<br>Fuel Level Sensor signal circuit is open or is shorted to voltage (too high)<br>Fuel Level Sensor ground circuit is open (too high)<br>Fuel Level Sensor signal circuit is short to ground (too low)<br>Fuel Level Sensor signal circuit and ground circuit are shorted together (too low)<br>Fuel Level Sensor has failed<br>FCM has failed |
| **P0461**<br>2T CCM<br>2004-05<br>300M, Caravan, Concorde, Dakota, Durango, Grand Cherokee, Intrepid, LHS, Liberty, Magnum, Neon, Pacifica, Prowler, Ram, Sebring, Stratus, Town & Country, Viper, Voyager, Wrangler | **Fuel Level Sensor No. 1 Malfunction Conditions:**<br>Test No. 1: With ignition on, fuel level is compared to the previous key-down after a 20-second delay. If the PCM does not see a difference in the fuel level of more than 0.1v, the test will fail.<br>Test No. 2: The PCM monitors the fuel level with ignition on. If the PCM does not see a change in the fuel level of 0.1765 in. over a set amount of miles, the test will fail.<br>**Possible Causes:**<br>Fuel tank or internal siphon hose damage<br>Fuel level signal circuit open or shorted to ground<br>Ground circuit is open<br>Fuel level Sensor malfunction |
| **P0461**<br>1T CCM<br>2004-05<br>PT Cruiser | **Fuel Level Sensor Malfunction Conditions:**<br>With engine running and under partial load, the fuel level must be above 0%, and there must not be any fuel level Sensor electrical DTCs. If the PCM detects the fuel level change is less than 2% over 2 hours, this DTC will set.<br>**Possible Causes:**<br>Intermittent condition<br>Fuel tank or component damage<br>Fuel level Sensor signal circuit open or shorted to ground<br>Ground circuit is open<br>Fuel level Sensor malfunction |
| **P0462**<br>1T CCM<br>1998-2003<br>300M, Avenger, Breeze, Caravan, Cherokee, Cirrus, Concorde, PT Cruiser, Dakota, Durango, Grand Cherokee, Intrepid, LHS, Liberty, Neon, New Yorker, Prowler, Ram, Sebring, Stratus, Talon, Town & Country, Viper, Vision, Voyager, Wrangler | **Fuel Level Sensing Unit Low Input Conditions:**<br>Key on or engine running; system voltage over 10.5v and the PCM detected the fuel level sensing unit signal indicated less than 0.98v, condition met for 200 seconds. No MIL.<br>**Possible Causes:**<br>Fuel level sending unit signal circuit shorted to Sensor ground<br>Fuel level sending unit signal circuit shorted to chassis ground<br>Fuel level sensing unit is damaged or the fuel tank is damaged<br>BCM or PCM has failed |

**OBD II Trouble Code List (P0XXX Codes) - Continued**

| DTC | Trouble Code Title, Conditions & Possible Causes: |
|---|---|
| **P0462**<br>1T CCM<br>2004-05<br>300M, Caravan, Concorde, Dakota, Durango, Grand Cherokee, Intrepid, LHS, Liberty, Magnum, Neon, Pacifica, Prowler, Ram, Sebring, Stratus, Town & Country, Viper, Voyager, Wrangler | **Fuel Level Sensor No. 1 Low Input Conditions:**<br>Key on. Battery voltage over 10.4v. Fuel level Sensor signal goes below 0.1961v (exc. Durango, Ram) for more than 5 seconds, or below 0.4v for more than 90 seconds (Durango, Ram). DTC is recorded.<br>**Possible Causes:**<br>Intermittent condition<br>Fuel level sending unit signal circuit shorted to Sensor or chassis ground<br>Fuel level sensing unit is damaged or the fuel tank is damaged<br>Instrument cluster problem |
| **P0462**<br>1T CCM<br>2004-05<br>PT Cruiser | **Fuel Level Sending Unit Low Voltage Conditions:**<br>Key on. If the PCM detects the fuel level as more than 100% for more than 4.5 minutes, this DTC will set.<br>**Possible Causes:**<br>Intermittent condition |
| **P0463**<br>1T CCM<br>1998-2003<br>300M, Avenger, Breeze, Caravan, Cherokee, Cirrus, Concorde, PT Cruiser, Dakota, Durango, Grand Cherokee, Intrepid, LHS, Liberty, Neon, New Yorker, Prowler, Ram, Sebring, Stratus, Talon, Town & Country, Viper, Vision, Voyager, Wrangler | **Fuel Level Sending Unit High Input Conditions:**<br>Key on or engine running; system voltage over 10.5v and the PCM detected the fuel level sensing unit signal indicated more than 4.90v, condition met for 200 seconds. No MIL.<br>**Possible Causes:**<br>Fuel level sending unit signal circuit shorted to Sensor or chassis ground<br>Fuel level sensing unit is damaged or the fuel tank is damaged<br>BCM or PCM has failed |
| **P0463**<br>1T CCM<br>2004-05<br>300M, Caravan, Concorde, PT Cruiser, Dakota, Durango, Grand Cherokee, Intrepid, LHS, Liberty, Magnum, Neon, Pacifica, Prowler, Ram, Sebring, Stratus, Town & Country, Viper, Voyager, Wrangler | **Fuel Level Sensor No. 1 High Input Conditions:**<br>Key on. Battery voltage over 10.4v. If Fuel Level Sensor signal goes above 4.7v for more than 5 seconds (exc. Durango, Ram, Viper, Liberty), above 4.9v for more than 90 seconds (Durango, Ram, Viper, Liberty), this DTC is recorded.<br>**Possible Causes:**<br>Fuel level sending unit signal circuit is open or is shorted to battery voltage<br>Fuel level Sensor ground circuit is open<br>Fuel level Sensor is damaged or the fuel tank is damaged<br>Instrument cluster is faulty<br>BCM or PCM has failed |

**OBD II Trouble Code List (P0XXX Codes) - Continued**

| DTC | Trouble Code Title, Conditions & Possible Causes: |
|---|---|
| **P0480**<br>1T CCM<br>2001-03<br>300M, Caravan, Cherokee, Cirrus, Concorde, PT Cruiser, Dakota, Durango, Grand Cherokee, Intrepid, LHS, Liberty, Neon, New Yorker, Prowler, Ram, Sebring, Stratus, Town & Country, Viper, Voyager, Wrangler | **Low Speed (No. 1) Fan Relay Control Circuit Conditions:**<br>Key on or engine running; and the PCM detected an unexpected low or high voltage condition on the Low Speed Fan Relay control circuit.<br>**Possible Causes:**<br>LFAN relay power circuit is open from the relay to fused power<br>LFAN relay control circuit is open or shorted to chassis ground<br>LFAN relay is damaged or has failed<br>PCM has failed |
| **P0480**<br>1T CCM<br>2004-05<br>300M, Caravan, Concorde, PT Cruiser, Dakota, Durango, Grand Cherokee, Intrepid, LHS, Liberty, Magnum, Neon, Pacifica, Prowler, Ram, Sebring, Stratus, Town & Country, Viper, Voyager, Wrangler | **Low Speed (No.1) Fan Control Relay Circuit Conditions:**<br>Key on. Battery voltage over 10v. An open or shorted circuit is detected in the Low Speed Fan Relay control circuit system.<br>**Possible Causes:**<br>Fan relay intermittent condition<br>Ground circuit is open<br>Fused B+ output circuit malfunction<br>Fan relay control circuit is open or shorted to battery voltage or ground<br>Fan relay is damaged or has failed<br>PCM has failed |
| **P0480**<br>1T CCM<br>2004-05<br>Liberty with Diesel | **Fan No. 1 Control Circuit Excessive Current Or Short Circuit Conditions:**<br>Ignition on; Low Speed Radiator Fan Relay commanded OFF (excessive current) or commanded ON (short circuit). The ECM detects excessive current on Low Speed Radiator Fan Relay control circuit.<br>**Possible Causes:**<br>Intermittent condition<br>Low Speed Radiator Fan Relay has failed<br>Low Speed Radiator Fan control circuit is shorted to voltage<br>ECM has failed |
| **P0480**<br>1T CCM<br>2004-05<br>Liberty with Diesel | **Fan No. 1 Control Circuit Open Or Short-to-Ground Conditions:**<br>Ignition on; Low Speed Radiator Fan Relay commanded OFF. The ECM does not detect voltage on Low Speed Radiator Fan Relay control circuit.<br>**Possible Causes:**<br>Intermittent condition<br>ASD Relay output circuit is open<br>Low Speed Radiator Fan Relay has failed<br>Low Speed Radiator Fan control circuit is open or is shorted to ground<br>ECM has failed |

**OBD II Trouble Code List (P0XXX Codes) - Continued**

| DTC | Trouble Code Title, Conditions & Possible Causes: |
|---|---|
| **P0481**<br>1T CCM<br>2001-05<br>300M, Caravan, Cherokee, Cirrus, Concorde, PT Cruiser, Dakota, Durango, Grand Cherokee, Intrepid, LHS, Liberty, Magnum, Neon, New Yorker, Pacifica, Prowler, Ram, Sebring, Stratus, Town & Country, Viper, Vision, Voyager, Wrangler | **High Speed (No. 2) Fan Relay Control Circuit Conditions:**<br>Key on or engine running; and the PCM detected an unexpected low or high voltage condition (open or shorted condition) on the High Speed Fan Relay circuit.<br>**Possible Causes:**<br>HFAN relay power circuit is open from the relay to fused power<br>HFAN relay control circuit is open or shorted to chassis ground<br>Fan relay(s) failed<br>PCM has failed. |
| **P0481**<br>2T CCM<br>2004-05<br>Sebring, Stratus | **Air Injection System Malfunction Conditions:**<br>Engine running; Air Injection Pump is active for a calibrated amount of time. Once enough airflow has accumulated through the AI system, the test will begin. If the PCM detects excessive airflow, or not enough airflow through the AI system, this DTC will be set.<br>**Possible Causes:**<br>Air Injection system may be damaged, restricted or disconnected<br>ASD relay output circuit failure<br>MAF Sensor internal failure<br>MAF signal circuit open or shorted to ground or to ASD relay output circuit or to battery voltage<br>MAF signal circuit is shorted to sensor ground circuit<br>MAF Sensor ground circuit is open<br>PCM has failed. |
| **P0489**<br>1T CCM<br>2004-05<br>Liberty with Diesel | **EGR Solenoid Circuit Short-To-Ground Conditions:**<br>Ignition on; ECM commands EGR Solenoid off. The ECM does not detect voltage on the EGR Solenoid control circuit.<br>**Possible Causes:**<br>Intermittent condition<br>ASD Relay output circuit is open<br>EGR Solenoid control circuit is open or is shorted to ground<br>EGR Solenoid has malfunctioned<br>EMC has failed |
| **P0490**<br>1T CCM<br>2004-05<br>Liberty with Diesel | **EGR Solenoid Circuit Short Conditions:**<br>Ignition on; ECM commands EGR Solenoid on. The ECM detects excessive current on the EGR Solenoid control circuit.<br>**Possible Causes:**<br>Intermittent condition<br>EGR Solenoid has malfunctioned<br>EGR Solenoid control circuit is shorted to voltage<br>EMC has internal short to voltage |

**OBD II Trouble Code List (P0XXX Codes) - Continued**

| DTC | Trouble Code Title, Conditions & Possible Causes: |
|---|---|
| **P0498** 1T CCM 2001-05 300M, Caravan, Cherokee, Cirrus, Concorde, PT Cruiser, Dakota, Durango, Grand Cherokee, Intrepid, LHS, Liberty, Magnum, Neon, New Yorker, Pacifica, Prowler, Ram, Sebring, Stratus, Town & Country, Viper, Vision, Voyager, Wrangler | **NVLD Canister Vent Solenoid Circuit Low Conditions:** Key on or engine running; and the PCM detected an unexpected low voltage condition on the Natural Vacuum Leak Detection (NVLD) control circuit during the CCM test period. **Possible Causes:** NVLD canister vent solenoid control circuit is shorted to ground NVLD canister vent solenoid is damaged or it has failed PCM has failed |
| **P0499** 1T CCM 2001-05 300M, Caravan, Cherokee, Cirrus, Concorde, PT Cruiser, Dakota, Durango, Grand Cherokee, Intrepid, LHS, Liberty, Magnum, Neon, New Yorker, Pacifica, Prowler, Ram, Sebring, Stratus, Town & Country, Viper, Vision, Voyager, Wrangler | **NVLD Canister Vent Solenoid Circuit High Conditions:** Key on or engine running. The PCM detected an open or unexpected high voltage condition on the Natural Vacuum Leak Detection (NVLD) circuit. **Possible Causes:** NVLD canister vent solenoid control circuit is open or is shorted to power NVLD canister vent solenoid ground circuit is open NVLD canister vent solenoid is damaged or it has failed PCM has failed |
| **P0500** 1T CCM 1995-98 Talon | **Vehicle Speed Sensor Circuit Fault Conditions:** Engine started; engine runtime over 31 seconds, ECT Sensor more than 120°F, transaxle not in Park or Neutral, brakes not applied, engine speed over 1800 rpm, MAP Sensor less than 11" Hg, and the PCM did not detect a VSS signal from the TCM for over 11 seconds. **Possible Causes:** OSS signal circuit to the TCM is open or shorted to ground OSS is damaged or has failed VSS circuit between the PCM and TCM is open or shorted TCM has failed or speedometer pinion factor not programmed |

## OBD II Trouble Code List (P0XXX Codes) - Continued

| DTC | Trouble Code Title, Conditions & Possible Causes: |
|---|---|
| **P0500**<br>1T CCM<br>1995-2003<br>300M, Avenger, Breeze, Caravan, Cherokee, Cirrus, Concorde, PT Cruiser, Dakota, Durango, Grand Cherokee, Intrepid, LHS, Liberty, Neon, New Yorker, Prowler, Ram, Sebring, Stratus, Talon, Town & Country, Viper, Vision, Voyager, Wrangler | **Vehicle Speed Sensor Circuit Fault Conditions:**<br>Engine started; engine runtime over 31 seconds, ECT Sensor more than 180°F, transaxle not in Park or Neutral, brakes not applied, engine speed over 1800 rpm, MAP Sensor less than 11" Hg, and the PCM did not detect a VSS signal from the TCM for over 11 seconds.<br>**Possible Causes:**<br>VSS or OSS signal circuit to the PCM is open or shorted to ground<br>VSS or OSS power circuit (8v) is open or shorted to ground<br>VSS or OSS ground circuit to the PCM is open<br>Speedometer pinion or the VSS is damaged or has failed |
| **P0500**<br>2T CCM<br>1998-2002<br>Ram with 5.9L Diesel | **Vehicle Speed Sensor Circuit Fault Conditions:**<br>Engine started; engine speed from 1024-2784 rpm, engine boost more than 7 psi, and the PCM did not detect a VSS signal from the CAB module for over 15 seconds during testing.<br>**Possible Causes:**<br>CAB circuit from the CAB to PCM is open or it is shorted<br>OSS signal circuit to the TCM is open or shorted to ground<br>OSS is damaged or has failed<br>TCM or PCM has failed |
| **P0500**<br>2T CCM<br>2004-05<br>Viper | **No Vehicle Speed Signal Conditions:**<br>Engine started; engine temperature above 104°F; MAP vacuum approximately 15-16 in. Hg; engine speed is 1400-3000 rpm. If the PCM did not detect a VSS signal for more than 15 seconds on 2 consecutive trips, this DTC will set.<br>**Possible Causes:**<br>VSS speed signal circuit is short to voltage or to ground<br>VSS speed signal circuit is open between CAB and PCM<br>PCM has failed |
| **P0500**<br>1T CCM<br>2004-05<br>PT Cruiser | **No Vehicle Speed Sensor Signal Conditions:**<br>Engine started; if the PCM did not detect a VSS signal for over 6 seconds, this DTC will set.<br>**Possible Causes:**<br>Intermittent condition<br>5v supply circuit open<br>VSS circuit is open or is shorted to ground<br>VSS ground circuit is open<br>VSS has failed<br>PCM has failed |

**OBD II Trouble Code List (P0XXX Codes) - Continued**

| DTC | Trouble Code Title, Conditions & Possible Causes: |
|---|---|
| **P0501**<br>2T CCM<br>2001-05<br>300M, Caravan, Cherokee, Cirrus, Concorde, PT Cruiser, Dakota, Durango, Grand Cherokee, Intrepid, LHS, Liberty, Magnum, Neon, New Yorker, Pacifica, Prowler, Ram, Sebring, Stratus, Town & Country, Viper, Vision, Voyager, Wrangler | **Vehicle Speed Sensor Performance Conditions:**<br>Engine started; vehicle driven at over 1500 rpm for 10 seconds, gear selector not in Park or Neutral (or clutch is not depressed on M/T); brakes not applied. The PCM (or ECM: Liberty with Diesel) did not receive any VSS signals from the TCM (BCM: Liberty) for 11 seconds for 2 consecutive trips. 3 good trips required to turn off MIL.<br>**Possible Causes:**<br>Check for any ABS/RWAL or TCM codes related to the VSS<br>VSS connector is damaged, open or it is shorted<br>VSS signal is open, shorted to ground or shorted to power<br>Incorrect tire circumference<br>ABS/RWAL controller, BCM, ECM, TCM or PCM has failed |
| **P0501**<br>1T CCM<br>2001-03<br>Ram with 5.9L Diesel | **Vehicle Speed Sensor Signal Range/Performance Conditions:**<br>Engine started; vehicle driven to a speed over 20 mph for 2 seconds (as indicated by the CCD Bus signal to the PCM), ECM vehicle speed signal less than 10 mph, and the PCM did not receive a CAB vehicle speed signal during the CCM Rationality test.<br>**Possible Causes:**<br>Check for a CAB controller code related to the VSS signal<br>CAB (ABS) controller has failed<br>VSS signal is open, shorted to ground or shorted to power<br>ECM or PCM has failed |
| **P0503**<br>1T or 2T CCM<br>2004-05<br>300M, Caravan, Concorde, PT Cruiser, Dakota, Durango, Grand Cherokee, Intrepid, LHS, Liberty, Magnum, Neon, Pacifica, Prowler, Ram, Sebring, Stratus, Town & Country, Viper, Voyager, Wrangler | **Vehicle Speed Sensor No. 1 Erratic Performance Conditions:**<br>Ignition is on; battery voltage over 10v; transmission in Drive or Reverse; brakes not applied. Vehicle speed signal is erratic during road load conditions. One-trip fault for ETC vehicles; Two-trip fault for ETC vehicles. 3 good trips required to turn off MIL.<br>**Possible Causes:**<br>Check for active Bus or Communication DTCs<br>Incorrect Tire Circumference<br>PCM has failed |
| **P0504**<br>1T CCM<br>2004-05<br>Liberty with Diesel | **Brake Switch Signal Circuits Plausibility With Redundant Contact Conditions:**<br>Ignition is on. The Primary Brake Switch signal and the Secondary Brake Switch signal inputs to the ECM do not agree.<br>**Possible Causes:**<br>Intermittent condition<br>Brake Lamp Switch sense circuit is open or shorted to ground<br>Brake Switch sense circuit is open or shorted to ground<br>Brake Lamp Switch output circuit is open or shorted to voltage<br>ECM has failed |

**OBD II Trouble Code List (P0XXX Codes) - Continued**

| DTC | Trouble Code Title, Conditions & Possible Causes: |
|---|---|
| **P0505**<br>2T IAC<br>1995-98<br>Talon | **Idle Air Control Motor System Fault Conditions:**<br>IAT Sensor less than 131ºF during last key cycle, engine started, system voltage over 10v, engine running at hot idle speed, ECT Sensor over 176ºF, IAT Sensor over 14ºF, Short Term fuel trim from -8% to +8%, BARO Sensor over 76 kPa, and the PCM detected the Actual idle speed was more than 200 rpm above or 100 rpm below the Target idle speed for 10 seconds.<br>**Possible Causes:**<br>Stepper motor Coil No. 1, 2, 3 or 4 circuit open or shorted to ground<br>Stepper motor coil circuit(s) shorted to system power (B+)<br>Stepper motor is damaged or has failed<br>PCM has failed |
| **P0505**<br>2T IAC<br>1995-2005<br>300M, Avenger, Breeze, Caravan, Cherokee, Cirrus, Concorde, PT Cruiser, Dakota, Durango, Grand Cherokee, Intrepid, LHS, Liberty, Magnum, Neon, New Yorker, Prowler, Ram, Sebring, Stratus, Talon, Town & Country, Viper, Vision, Voyager, Wrangler | **Idle Air Control Motor Circuit Conditions:**<br>Engine started; system voltage over 11.5v and the PCM detected an unexpected voltage condition on one or more of the IAC motor circuits for 2.75 seconds (exc. Viper) or for 100ms (Viper), with the IAC motor active.<br>**Possible Causes:**<br>Stepper motor Coil No. 1, 2, 3 or 4 circuit open or shorted to ground<br>Stepper motor coil circuit(s) shorted to system power (B+)<br>Stepper motor is damaged or has failed<br>PCM has failed |
| **P0506**<br>1T CCM<br>2002-03<br>300M, Caravan, Cherokee, Concorde, PT Cruiser, Dakota, Durango, Grand Cherokee, Intrepid, LHS, Liberty, Neon, New Yorker, Prowler, Ram, Sebring, Stratus, Town & Country, Viper, Voyager, Wrangler | **Idle Speed Low Performance Conditions:**<br>Engine running at idle speed in closed loop, and the PCM detected the actual idle speed was not within the expected low idle speed limit when compared to the target idle speed limit.<br>**Possible Causes:**<br>Air induction system restrictions (clogged air filter, etc.)<br>Idle air control passage is clogged or dirty (clean and retest)<br>Throttle body or linkage is binding, damaged or sticking |
| **P0506**<br>1T CCM<br>2001-03<br>Sebring, Stratus Coupe with 2.4L I4 or 3.0L V6 | **Idle Control System RPM Lower Than Expected Conditions:**<br>Engine started; system voltage over 10v, ECT Sensor over 171ºF, volumetric efficiency less than 40%, BARO Sensor over 76 kPa, IAT Sensor over 14ºF, PSPS is off, 25 seconds since last test, Target IAC position over 100 steps, and the PCM detected the Actual idle speed was 100 rpm less than the Target idle speed for 12 seconds.<br>**Possible Causes:**<br>Stepper motor Coil A1, A2, B1 and B2 circuit is open<br>Stepper motor Coil A1, A2, B1 and B2 circuit shorted to ground<br>Stepper motor Coil A1, A2, B1 and B2 circuit shorted to B+<br>Stepper motor is damaged or it has failed<br>PCM has failed |

## OBD II Trouble Code List (P0XXX Codes) - Continued

| DTC | Trouble Code Title, Conditions & Possible Causes: |
|---|---|
| **P0506**<br>2T CCM<br>2004-05<br>300M, Caravan, Concorde, Intrepid, LHS, Magnum, Neon, Pacifica, Prowler, Sebring, Stratus, Town & Country, Viper, Voyager | **Idle Speed Low Performance Conditions:**<br>Engine running at idle speed in closed loop. If engine rpm does not come within a calibratable low limit of the target idle speed, a failure timer will increment. When the appropriate failure timer reaches its maximum threshold without sign of rpm trending toward control, a soft-fail is generated. When a calibratable number of the soft-fails is reached, a 1-trip fault is set. When two 1-trip faults occur in a row, the DTC is set and the MIL illuminates.<br>**Possible Causes:**<br>PCV system malfunction<br>Air induction system restrictions (clogged air filter, etc.)<br>Idle air control passage is clogged or dirty (clean and retest)<br>Air induction system malfunction<br>Throttle body or linkage is binding, damaged or sticking |
| **P0506**<br>1T CCM<br>2004-05<br>Dakota, Durango, Grand Cherokee, Liberty, Ram, Wrangler | **Idle Speed Low Performance Conditions:**<br>Engine running at idle speed; MAF is less than 250 mg/tdc; air temperature is greater than 0°F (-18°C) and less than 19°F (-7°C) enable after coolant temperature is greater than 158°F (70°C) or air temperature is greater than 19°F (-7°C); coolant temperature is between 19 to 266°F (-7 to 130°C); canister purge is less than 100% duty cycle; no DTCs are present for VSS, MAF/MAP, ECT, TPS, ECT and CKP Sensors; also no fuel system or injector related DTCs are present. If the DTC detects that engine speed is 100 rpm or more below the normal idle speed for 7 seconds, this DTC will set.<br>**Possible Causes:**<br>Air induction system restrictions<br>Throttle body or linkage is binding, damaged or sticking<br>Intermittent condition<br>PCM has failed |
| **P0507**<br>1T CCM<br>2001-03<br>300M, Caravan, Cherokee, Concorde, PT Cruiser, Dakota, Durango, Grand Cherokee, Intrepid, LHS, Liberty, Neon, New Yorker, Prowler, Ram, Sebring, Stratus, Town & Country, Viper, Voyager, Wrangler | **Idle Speed High Performance Conditions:**<br>No CKP, ECT, MAF, MAP or VSS codes set, engine started, engine running at idle speed, BTS Sensor from 0-19.4°F, ECT Sensor from 158-266°F, MAF Sensor under 250 mg, and the PCM detected the Actual speed was 200 rpm above the Target speed for 7 seconds.<br>**Possible Causes:**<br>Idle air control passage is clogged or dirty (clean and retest)<br>Throttle body or linkage is binding, damaged or sticking<br>Vacuum leaks in the engine or PCV system components<br>Intermittent condition<br>PCM has failed |
| **P0507**<br>1T CCM<br>2001-03<br>Sebring, Stratus Coupe with 2.4L I4 or 3.0L V6 | **Idle Control System RPM Higher Than Expected Conditions:**<br>Engine running in closed loop, ECT Sensor over 171°F, system voltage over 10v, volumetric efficiency less than 40%, BARO Sensor over 76 kPa, IAT Sensor over 14°F, PSPS signal "off", 25 seconds have elapsed since last test, Target IAC position at (0) steps, and the PCM detected the Actual idle speed was 200 rpm more than the Target idle speed for 12 seconds.<br>**Possible Causes:**<br>Stepper motor Coil A1 or A2 circuit is open or shorted to ground<br>Stepper motor Coil B1 or B2 circuit is open or shorted to ground<br>Stepper motor coil circuit(s) is shorted to system power (B+)<br>Stepper motor power circuit is open (test power from MFI relay)<br>Stepper motor is damaged or it has failed<br>PCM has failed |

## OBD II Trouble Code List (P0XXX Codes) - Continued

| DTC | Trouble Code Title, Conditions & Possible Causes: |
|---|---|
| **P0507**<br>2T CCM<br>2004-05<br>300M, Caravan, Concorde, Dakota, Durango, Grand Cherokee, Intrepid, LHS, Liberty, Magnum, Neon, Pacifica, Prowler, Ram, Sebring, Stratus, Town & Country, Viper, Voyager, Wrangler | **Idle Speed High Performance Conditions:**<br>Engine running at idle speed in closed loop. If engine rpm does not come within a calibratable high limit of the target idle speed, a failure timer will increment. When the appropriate failure timer reaches its maximum threshold without sign of rpm trending toward control, a soft-fail is generated. When a calibratable number of the soft-fails is reached, a 1-trip fault is set. When two 1-trip faults occur in a row, the DTC is set and the MIL illuminates.<br>**Possible Causes:**<br>PCV system malfunction<br>Air induction system restrictions (clogged air filter, etc.)<br>Idle air control passage is clogged or dirty (clean and retest)<br>Air induction system malfunction<br>Throttle body or linkage is binding, damaged or sticking |
| **P0507**<br>1T CCM<br>2004-05<br>PT Cruiser, Dakota, Durango, Grand Cherokee, Liberty, Ram, Wrangler | **Idle Speed High Performance Higher Than Expected Conditions:**<br>Engine running at idle speed; MAF is less than 250 mg/tdc; air temperature is greater than 0°F (-18°C) and less than 19°F (-7°C) enable after coolant temperature is greater than 158°F (70°C) or air temperature is greater than 19°F (-7°C); coolant temperature is between 19 to 266°F (-7 to 130°C); canister purge is less than 100% duty cycle; no DTCs are present for VSS, MAF/MAP, ECT, TPS, ECT and CKP Sensors; also no fuel system or injector related DTCs are present. If the DTC detects that engine speed is 200 rpm or more above the normal idle speed for 7 seconds, this DTC will set.<br>**Possible Causes:**<br>Air induction system restrictions (clogged air filter, etc.)<br>Vacuum leaks<br>Intermittent condition<br>Throttle body or linkage is binding, damaged or sticking<br>PCM has failed |
| **P0508**<br>2T CCM<br>2001-05<br>300M, Caravan, Cherokee, Concorde, PT Cruiser, Dakota, Durango, Grand Cherokee, Intrepid, LHS, Liberty, Magnum, Neon, New Yorker, Pacifica, Prowler, Ram, Sebring, Stratus, Town & Country, Viper, Vision, Voyager, Wrangler | **Idle Air Control Motor Sense Circuit Low Input Conditions:**<br>Engine started; system voltage over 10.5v; IAC motor operating. The PCM detected the IAC Motor Sense circuit current was less than 175mA during the CCM test period.<br>**Possible Causes:**<br>IAC motor driver circuit is open or shorted to ground<br>IAC motor sense circuit is open or shorted to ground<br>IAC motor is damaged or it has failed<br>PCM has failed |
| **P0508**<br>1 CCM<br>2004-05<br>Dakota, Durango, Grand Cherokee, Liberty, Ram, Wrangler | **Idle Air Control Motor Sense Circuit Low Input Conditions:**<br>Engine running; system voltage over 10v; IAC motor operating. The PCM senses an open or short to ground on any of the Linear Idle Air Control (LIAC) control circuits for 2.75 seconds while the IAC motor is active. 3 good trips are required to turn off the MIL.<br>**Possible Causes:**<br>IAC motor control circuit is open or shorted to ground<br>IAC motor signal circuit is open or shorted to ground<br>IAC motor is damaged or it has failed<br>PCM has failed |

## OBD II Trouble Code List (P0XXX Codes) - Continued

| DTC | Trouble Code Title, Conditions & Possible Causes: |
|---|---|
| **P0509**<br>2T CCM<br>2001-05<br>300M, Caravan, Cherokee, Concorde, PT Cruiser, Dakota, Durango, Grand Cherokee, Intrepid, LHS, Liberty, Magnum, Neon, New Yorker, Pacifica, Prowler, Ram, Sebring, Stratus, Town & Country, Viper, Vision, Voyager, Wrangler | **Idle Air Control Motor Circuit High Conditions:**<br>Engine started; system voltage over 10.5v, IAC motor activated, and the PCM detected a high voltage on one or more of the IAC motor circuits (over 980mA) during the CCM test period.<br>**Possible Causes:**<br>IAC motor driver circuit is shorted to power<br>IAC motor sense circuit is shorted to power<br>IAC motor is damaged or it has failed<br>PCM has failed |
| **P0509**<br>1 CCM<br>2004-05<br>Dakota, Durango, Grand Cherokee, Liberty, Ram, Wrangler | **Idle Air Control Motor Sense Circuit High Conditions:**<br>Engine running; system voltage over 10v; IAC motor operating. The PCM senses a short-to-power on any of the Linear Idle Air Control (LIAC) control circuits for 2.75 seconds while the IAC motor is active. 3 good trips are required to turn off the MIL.<br>**Possible Causes:**<br>IAC motor control circuit is shorted to battery power<br>IAC motor signal circuit is open or shorted to battery power<br>IAC control circuit is shorted to IAC return circuit<br>IAC motor is damaged or it has failed<br>PCM has failed |
| **P0513**<br>1T PCM<br>2001-05<br>300M, Caravan, Cherokee, Concorde, PT Cruiser, Dakota, Durango, Grand Cherokee, Intrepid, LHS, Liberty, Magnum, Neon, New Yorker, Pacifica, Prowler, Ram, Sebring, Stratus, Town & Country, Viper, Vision, Voyager, Wrangler | **Invalid SKIM Key Detected Conditions:**<br>Key on, and the PCM detected an invalid Sentry Key Immobilizer key had been inserted into the ignition key assembly.<br>**Possible Causes:**<br>Incorrect VIN in the PCM<br>No communication between the PCM and the SKIM<br>SKIM trouble codes present (check for any SKIM codes)<br>Valid SKIM key not present<br>VIN not programmed into the PCM<br>PCM has failed |

**OBD II Trouble Code List (P0XXX Codes) - Continued**

| DTC | Trouble Code Title, Conditions & Possible Causes: |
|---|---|
| **P0516**<br>1T CCM<br>2003-05<br>300M, Caravan, Concorde, PT Cruiser, Dakota, Durango, Grand Cherokee, Intrepid, LHS, Liberty, Magnum, Neon, Pacifica, Prowler, Ram, Sebring, Stratus, Town & Country, Viper, Voyager, Wrangler | **Battery Temperature Sensor Circuit Low Input Conditions:**<br>Key on or engine running; and the PCM detected a Battery Temperature Sensor signal that indicated less than 0.10v (exc. Ram, Liberty) or less than 0.039v (Ram, Liberty). 3 good trips required to turn off MIL.<br>**Possible Causes:**<br>BTS signal circuit is shorted to Sensor or chassis ground<br>BTS assembly is damaged or it has failed<br>PCM has failed |
| **P0517**<br>1T CCM<br>2003-05<br>300M, Caravan, Concorde, PT Cruiser, Dakota, Durango, Grand Cherokee, Intrepid, LHS, Liberty, Magnum, Neon, Pacifica, Prowler, Ram, Sebring, Stratus, Town & Country, Viper, Voyager, Wrangler | **Battery Temperature Sensor Circuit High Input Conditions:**<br>Key on or engine running; and the PCM detected a Battery Temperature Sensor signal that indicated more than 4.8v (exc. Ram, Liberty) or more than 4.94v (Ram, Liberty).<br>**Possible Causes:**<br>BTS signal circuit is shorted to VREF (5v)<br>BTS signal circuit is open or the BTS ground circuit is open<br>BTS assembly is damaged or it has failed<br>PCM has failed |
| **P0519**<br>2T IAC<br>2002-03<br>300M, Caravan, Cherokee, Concorde, PT Cruiser, Dakota, Durango, Grand Cherokee, Intrepid, LHS, Liberty, Neon, New Yorker, Prowler, Ram, Sebring, Stratus, Town & Country, Viper, Voyager, Wrangler | **Idle Air Performance Conditions:**<br>DTC P0106, P0107, P0108, P0121, P0122 and P0123 not set, engine started, engine running with the gear selector indicating Drive position, and the PCM detected the engine idle speed was not within 200 rpm of the high idle limit or within 100 rpm of the low idle limit when compared to the Target Idle Speed limit for 40 seconds.<br>**Possible Causes:**<br>Idle air control passage is clogged or dirty (clean and retest)<br>Throttle body or linkage is binding, damaged or sticking<br>Vacuum leaks in the engine or PCV system components |
| **P0519**<br>2T IAC<br>2002-05<br>Neon | **Idle Air Performance Conditions:**<br>DTC P0106, P0107, P0108, P0121, P0122 and P0123 not set, engine started, engine running with the gear selector indicating Drive position, and the PCM detected the engine idle speed was not within 200 rpm of the high idle limit or within 100 rpm of the low idle limit when compared to the Target Idle Speed limit for 40 seconds.<br>**Possible Causes:**<br>Idle air control passage is clogged or dirty (clean and retest)<br>Throttle body or linkage is binding, damaged or sticking<br>Vacuum leaks in the engine or PCV system components |

**OBD II Trouble Code List (P0XXX Codes) - Continued**

| DTC | Trouble Code Title, Conditions & Possible Causes: |
|---|---|
| **P0520**<br>1T CCM<br>2004-05<br>300M, Caravan, Concorde, PT Cruiser, Dakota, Durango, Grand Cherokee, Intrepid, LHS, Liberty, Magnum, Neon, Pacifica, Prowler, Ram, Sebring, Stratus, Town & Country, Viper, Voyager, Wrangler | **Engine Oil Pressure Sensor Out of Range Conditions:**<br>Key on (engine not started). The PCM detected an engine oil pressure reading out of the calibrated range.<br>**Possible Causes:**<br>Oil pressure Sensor signal circuit is open or shorted to ground<br>Oil pressure Sensor signal circuit has high resistance<br>5v supply circuit has high resistance<br>5v supply circuit is shorted to ground<br>Oil pressure Sensor ground circuit has high resistance<br>Oil pressure Sensor is damaged or has failed<br>PCM has failed |
| **P0504**<br>1T CCM<br>2004-05<br>Liberty with Diesel | **Engine Oil Pressure Sensor Circuit Too Low Or Too High Conditions:**<br>Engine running at start up. The Oil Pressure signal is below the lower limit for 8 seconds after engine startup. After the engine is running, the OP Sensor signal is above 4.8v for 0.5 seconds or it may be below 0.19v for 0.5 second.<br>**Possible Causes:**<br>5v supply circuit is open<br>Front Control Module (FCM) Oil Pressure Sensor signal circuit is shorted to ground or to voltage<br>Engine mechanical problem exists<br>Oil Pressure Sensor has failed<br>Oil Pressure Sensor signal circuit is open, shorted to voltage or to ground<br>Oil Pressure Sensor signal circuit is shorted to Sensor ground<br>Oil Pressure Sensor ground circuit is open<br>Intermittent condition |
| **P0521**<br>1T CCM<br>2004-05<br>300M, Caravan, Concorde, PT Cruiser, Dakota, Durango, Grand Cherokee, Intrepid, LHS, Liberty, Magnum, Neon, Pacifica, Prowler, Ram, Sebring, Stratus, Town & Country, Viper, Voyager, Wrangler | **Engine Oil Pressure Sensor Does Not Reach Range Conditions:**<br>Engine running. The PCM detected an engine oil pressure reading never reaches the calibrated specification when engine is at 1250 rpm.<br>**Possible Causes:**<br>Engine oil or engine mechanical fault<br>Oil pressure Sensor signal circuit is shorted to battery<br>Oil pressure Sensor signal circuit has high resistance<br>5v supply circuit has high resistance<br>5v supply circuit is shorted to ground<br>Oil pressure Sensor return circuit has high resistance<br>Oil pressure Sensor is damaged or has failed<br>PCM has failed |
| **P0522**<br>1T CCM<br>2002-03<br>Ram with 5.9L Diesel | **Engine Oil Pressure Sensor Circuit Low Input Conditions:**<br>Key on or engine running; system voltage more than 10.4v, and the PCM detected the Oil Pressure Sensor signal was less than 0.10v.<br>**Possible Causes:**<br>Check for multiple trouble codes set related to this circuit<br>Oil pressure Sensor signal circuit is open or shorted to ground<br>Oil pressure Sensor VREF circuit is open<br>Oil pressure Sensor is damaged or has failed<br>PCM has failed |

**OBD II Trouble Code List (P0XXX Codes) - Continued**

| DTC | Trouble Code Title, Conditions & Possible Causes: |
|---|---|
| **P0522**<br>1T CCM<br>1999-2003<br>300M, Avenger, Breeze, Caravan, Cherokee, Cirrus, Concorde, PT Cruiser, Dakota, Durango, Grand Cherokee, Intrepid, LHS, Liberty, Neon, New Yorker, Prowler, Ram, Sebring, Stratus, Talon, Town & Country, Viper, Vision, Voyager, Wrangler | **Engine Oil Pressure Sensor Rationality Conditions:**<br>Key on (engine not started), engine speed (rpm) indicating 0 rpm, and the PCM detected an engine oil pressure reading not within calibrated range.<br>**Possible Causes:**<br>Oil pressure Sensor signal circuit is open<br>Oil pressure Sensor signal circuit is shorted to ground<br>Oil pressure Sensor is damaged or has failed<br>PCM has failed |
| **P0522**<br>1T CCM<br>2004-05<br>300M, Caravan, Concorde, PT Cruiser, Dakota, Durango, Grand Cherokee, Intrepid, LHS, Liberty, Magnum, Neon, Pacifica, Prowler, Ram, Sebring, Stratus, Town & Country, Viper, Voyager, Wrangler | **Engine Oil Pressure Sensor Rationality Conditions:**<br>Engine running; battery voltage over 10.4v. If the PCM detected an engine oil pressure voltage reading of less than 0.1v for 0.5 second (Neon, Dakota, Durango, Ram), less than 0.3v (Viper), or less that 0.942v (Liberty), this DTC will set.<br>**Possible Causes:**<br>5v supply circuit is open or is shorted to ground<br>Oil pressure Sensor signal circuit is shorted to ground or to sensor ground<br>Oil pressure Sensor is damaged or has failed<br>PCM has failed |
| **P0523**<br>1T CCM<br>1999-2003<br>300M, Avenger, Breeze, Caravan, Cherokee, Cirrus, Concorde, PT Cruiser, Dakota, Durango, Grand Cherokee, Intrepid, LHS, Liberty, Neon, New Yorker, Prowler, Ram, Sebring, Stratus, Talon, Town & Country, Viper, Vision, Voyager, Wrangler | **Engine Oil Pressure Sensor Rationality Conditions:**<br>Key on (engine not started), engine speed (rpm) indicating 0 rpm, and the PCM detected an engine oil pressure reading.<br>**Possible Causes:**<br>Oil pressure Sensor signal circuit is open<br>Oil pressure Sensor signal circuit is shorted to ground<br>Oil pressure Sensor is damaged or has failed<br>PCM has failed |

**OBD II Trouble Code List (P0XXX Codes) - Continued**

| DTC | Trouble Code Title, Conditions & Possible Causes: |
|---|---|
| **P0523**<br>1T CCM<br>2004-05<br>300M, Caravan, Concorde, PT Cruiser, Dakota, Durango, Grand Cherokee, Intrepid, LHS, Liberty, Magnum, Neon, Pacifica, Prowler, Ram, Sebring, Stratus, Town & Country, Viper, Voyager, Wrangler | **Engine Oil Pressure Sensor Circuit High Conditions:**<br>Key on (engine not started). The PCM detected an engine oil pressure reading greater than the calibrated amount.<br>**Possible Causes:**<br>Oil pressure Sensor signal circuit is open or shorted to battery voltage or to 5v supply circuit<br>Oil pressure Sensor ground circuit is open<br>Oil pressure Sensor is damaged or has failed<br>PCM has failed |
| **P0523**<br>1T CCM<br>1998-2003<br>Ram | **Engine Oil Pressure Sensor Circuit High Input Conditions:**<br>Key on or engine running; system voltage more than 10.4v, and the PCM detected the Oil Pressure Sensor signal indicated more than 4.90v during the CCM test.<br>**Possible Causes:**<br>Oil pressure Sensor signal circuit is shorted to VREF<br>Oil pressure Sensor ground circuit is open<br>Oil pressure Sensor is damaged or has failed<br>PCM has failed |
| **P0524**<br>1T CCM<br>2004-05<br>300M, Caravan, Concorde, PT Cruiser, Dakota, Durango, Grand Cherokee, Intrepid, LHS, Liberty, Magnum, Neon, Pacifica, Prowler, Ram, Sebring, Stratus, Town & Country, Viper, Voyager, Wrangler | **Engine Oil Pressure Low Conditions:**<br>Engine running. The PCM detected that the engine oil pressure never reaches the calibrated specification to allow the MDS activation.<br>**Possible Causes:**<br>Engine oil system or engine mechanical faults<br>Oil pressure Sensor signal circuit is shorted to battery voltage or to ground<br>Oil pressure Sensor signal circuit has high resistance<br>5v supply circuit has high resistance<br>5v supply circuit is shorted to ground<br>Oil pressure Sensor return circuit has high resistance<br>Oil pressure Sensor is damaged or has failed<br>PCM has failed |
| **P0524**<br>1T CCM<br>1998-2003<br>Ram with 5.9L Diesel | **Engine Oil Pressure Sensor Signal Too Low Conditions:**<br>Key on or engine running; system voltage more than 10.4v, and the PCM detected a signal from the Oil Pressure Sensor that indicated the engine oil pressure was too low.<br>**Possible Causes:**<br>Engine oil level is very low<br>Oil pressure Sensor is damaged or has failed<br>PCM has failed |

**OBD II Trouble Code List (P0XXX Codes) - Continued**

| DTC | Trouble Code Title, Conditions & Possible Causes: |
|---|---|
| **P0530**<br>1T CCM<br>2004-05<br>Liberty with Diesel | **A/C Pressure Sensor Circuit Too Low Or Too High Conditions:**<br>Ignition on. An error occurs with the A/C Pressure CAN Bus message from the Front Control Module (FCM) to the ECM, or the signal is below 0.06v or above 4.74v for 0.6 second.<br>**Possible Causes:**<br>Intermittent condition<br>A/C Pressure Sensor signal circuit is shorted to 5v supply or to voltage<br>A/C Pressure Sensor ground circuit is shorted to voltage<br>5v supply circuit is open<br>A/C Pressure Sensor has failed<br>A/C Pressure Sensor signal circuit is open<br>A/C Pressure Sensor signal circuit is shorted to ground or to Sensor ground circuit<br>FCM to 5v supply circuit problem |
| **P0532**<br>1T CCM<br>2002-05<br>300M, Caravan, Cherokee, Concorde, PT Cruiser, Dakota, Durango, Grand Cherokee, Intrepid, LHS, Liberty, Magnum, Neon, New Yorker, Pacifica, Prowler, Ram, Sebring, Stratus, Town & Country, Viper, Vision, Voyager, Wrangler | **Air Conditioning Pressure Sensor Circuit Low Input Conditions:**<br>Engine running with the A/C relay energized. The PCM detected the signal from the A/C Pressure Sensor indicated less than 0.58v for over 2.6 seconds during the CCM test period.<br>**Possible Causes:**<br>A/C pressure Sensor 5v power supply (VREF) circuit is open or is shorted to ground<br>A/C pressure Sensor signal circuit is shorted to ground or to Sensor ground circuit<br>A/C pressure Sensor is damaged or it has failed<br>Front A/C control module damaged or has failed<br>PCM has failed |
| **P0533**<br>1T CCM<br>2002-05<br>300M, Caravan, Cherokee, Concorde, PT Cruiser, Dakota, Durango, Grand Cherokee, Intrepid, LHS, Liberty, Magnum, Neon, New Yorker, Pacifica, Prowler, Ram, Sebring, Stratus, Town & Country, Viper, Vision, Voyager, Wrangler | **Air Conditioning Pressure Sensor Circuit High Input Conditions:**<br>Engine running with the A/C relay energized, and the PCM detected the signal from the A/C Pressure Sensor indicated less than 4.92v for over 2.6 seconds during the CCM test period.<br>**Possible Causes:**<br>A/C pressure Sensor signal circuit is shorted to 5v VREF power<br>A/C pressure Sensor signal circuit or ground circuit is open<br>A/C pressure Sensor is damaged or it has failed<br>Front A/C control module is damaged or has failed<br>PCM has failed |
| **P0545**<br>1T CCM<br>1996-2002<br>Ram with V8 | **Air Conditioning Clutch Circuit Failure Conditions:**<br>Key on or engine running; and the PCM detected an unexpected voltage condition on the A/C clutch control circuit in the CCM test.<br>**Possible Causes:**<br>A/C clutch control circuit is open between the clutch and PCM<br>A/C clutch control circuit is shorted to ground<br>A/C clutch power circuit is open<br>A/C clutch or PCM has failed |

## OBD II Trouble Code List (P0XXX Codes) - Continued

| DTC | Trouble Code Title, Conditions & Possible Causes: |
|---|---|
| **P0551**<br>1T CCM<br>1995-2003<br>300M, Avenger, Breeze, Caravan, Cherokee, Cirrus, Concorde, PT Cruiser, Dakota, Durango, Grand Cherokee, Intrepid, LHS, Liberty, Neon, New Yorker, Prowler, Ram, Sebring, Stratus, Talon, Town & Country, Viper, Vision, Voyager, Wrangler | **Power Steering Pressure Switch Circuit Failure Conditions:**<br>Vehicle driven at more than 56 mph for more than 30 seconds and the PCM detected a high voltage input on the PSPS circuit (pressure exceeds 500 psi). This is a normally open switch.<br>**Possible Causes:**<br>PSPS sense circuit is shorted to ground<br>PSPS is damaged or has failed<br>PCM has failed |
| **P0551**<br>1T CCM<br>2001-03<br>Sebring, Stratus Coupe with 2.4L or 3.0L | **Power Steering Pressure Switch Circuit Failure Conditions:**<br>Engine started; vehicle driven at more than 31 mph for 4 seconds, BARO Sensor over 75 kPa, ECT Sensor more than 86°F, followed by a deceleration period to a stop with VSS at 0.93 mph, and the PCM detected the PSPS switch signal indicated "on" (test fails 10 times).<br>**Possible Causes:**<br>PSPS sense circuit is open between the switch and the PCM<br>PSPS ground circuit is open between the switch and ground<br>PSPS or PCM has failed |
| **P0551**<br>1T CCM<br>2004-05<br>300M, Caravan, Concorde, Dakota, Durango, Grand Cherokee, Intrepid, LHS, Liberty, Magnum, Pacifica, Prowler, Ram, Town & Country, Viper, Voyager, Wrangler | **Power Steering Pressure Switch Circuit Failure Conditions:**<br>Engine running and vehicle driven at more than 40 mph for 30 seconds. If the PCM detected the PSPS signal remains open for 2 consecutive trips, this DTC will set. 3 good trips required to turn off MIL.<br>**Possible Causes:**<br>PSPS sense circuit is open between the switch and the PCM<br>PSPS ground circuit is open between the switch and ground<br>PSPS is damaged or it has failed<br>PCM has failed |
| **P0551**<br>2T CCM<br>2004-05<br>Neon, Sebring, Stratus | **Power Steering Pressure Switch Circuit Failure Conditions:**<br>Engine running and vehicle driven at more than 50 mph or more; coolant temperature is above 68°F (20°C). If the PCM detected the PSPS signal remains open for 40 seconds or more, this DTC will set.<br>**Possible Causes:**<br>PSPS sense circuit is open or is shorted to ground<br>PSPS ground circuit is open<br>PSPS or PCM has failed |
| **P0551**<br>1T CCM<br>2004-05<br>PT Cruiser | **Power Steering Pressure Switch Circuit Failure Conditions:**<br>Engine running and vehicle driven at more than 30 mph for 60 seconds; no VSS DTC is present. If the PCM detected the PSPS signal remains high after 60 seconds, this DTC will set.<br>**Possible Causes:**<br>PSPS sense circuit is open or is shorted to ground<br>PSPS ground circuit is open<br>PSPS is damaged or it has failed<br>PCM has failed |

## OBD II Trouble Code List (P0XXX Codes) - Continued

| DTC | Trouble Code Title, Conditions & Possible Causes: |
|---|---|
| **P0562**<br>1T CCM<br>1996-2003<br>300M, Avenger, Breeze, Caravan, Cherokee, Cirrus, Concorde, PT Cruiser, Dakota, Durango, Grand Cherokee, Intrepid, LHS, Liberty, Magnum, Neon, New Yorker, Prowler, Ram, Sebring, Stratus, Talon, Town & Country, Viper, Vision, Voyager, Wrangler | **Battery Sense Circuit Low Input Conditions:**<br>Engine running at a speed over 380 rpm, and the PCM detected the Battery Sense circuit voltage indicated one volt less than the Charging system "goal" (11.50v) for 13.47 seconds.<br>**Possible Causes:**<br>Battery sense circuit has a high resistance condition<br>Generator ground circuit has a high resistance condition<br>Generator field ground circuit is open<br>Generator field control circuit is open or shorted to ground<br>Generator is damaged or it has failed<br>PCM has failed |
| **P0562**<br>1T CCM<br>2004-05<br>300M, Concorde, Dakota, Durango, Grand Cherokee, Intrepid, LHS, Liberty, Magnum, Pacifica, Prowler, Ram, Viper, Wrangler | **Battery Voltage Low Input Conditions:**<br>Engine running at a speed over 1000 rpm. If battery voltage is 1v less than desired voltage for a set period of time, this DTC will set. The ETC light is flashing.<br>**Possible Causes:**<br>Resistance in battery positive circuit<br>Resistance in the generator case ground<br>Generator field ground circuit is open or shorted to ground<br>Generator is damaged or it has failed<br>Ground circuit is open<br>PCM has failed |
| **P0562**<br>1T CCM<br>2004-05<br>Neon, Sebring, Stratus | **Battery Voltage Low Input Conditions:**<br>Engine running at a speed over 1150 rpm. This DTC will set if the following occur: the battery sensed voltage is 1v below the charging goal for 13.47 seconds, of less than 6.5v for 200ms; the battery voltage of the Transmission Control Relay output sense circuit(s) to the PCM is less than 10v for 15 seconds, of less than 7.2v for 200ms. The PCM senses the battery voltage, turns off the field driver, and senses the battery voltage again. If the voltages are the same, this DTC will be set. Note: This DTC generally indicates a gradually failing battery voltage output or a resistive connection to the PCM.<br>**Possible Causes:**<br>B+ circuit to relay or to PCM has high resistance<br>Generator ground circuit has high resistance<br>Generator is damaged or it has failed<br>Generator field ground circuit is open<br>Generator field control circuit is open or is shorted to ground<br>TC relay output to TCM is open or has high resistance<br>TC relay has failed<br>Intermittent wiring or connector problems<br>PCM has failed |

**OBD II Trouble Code List (P0XXX Codes) - Continued**

| DTC | Trouble Code Title, Conditions & Possible Causes: |
|---|---|
| **P0562**<br>1T CCM<br>2004-05<br>PT Cruiser, Caravan,<br>Town & Country,<br>Voyager | **Charging System Voltage Low Conditions:**<br>PCM: Engine is running for over 30 sec. If the PCM detects battery voltage is under 11.5v for more than 5 sec., this DTC will set. Other charging system DTCs may be present.<br>TCM w/NGC: Engine is running and PCM has closed the Transmission Control Relay. If battery voltage of the TC Relay output sense circuit is less than 10v for 15 seconds, this DTC will set. Note: P0562 usually indicates failing battery voltage or resistive connection to the PCM. This DTC also sets if battery voltage sensed at PCM is less than 6.5v for 200ms or when the TC Relay output circuit is less than 7.2v for 200ms.<br>**Possible Causes:**<br>Resistance is high in battery positive circuit or in generator case ground<br>Ground circuit is open or has high resistance<br>Fused B+ circuit to TC relay or to PCM has high resistance<br>Intermittent wiring or connector condition<br>TC relay output to TCM is open or has high resistance<br>TC relay has failed<br>Generator field driver circuit is open<br>ASD relay output circuit is open<br>Generator has failed<br>PCM has failed |
| **P0562**<br>1T CCM<br>1998-2003<br>Ram with 5.9L Diesel | **Charging System Voltage Low Input Conditions:**<br>Engine started; engine speed over 1000 rpm during testing, and the PCM detected the system voltage was less than 6.0v at any time.<br>**Possible Causes:**<br>Battery connections corroded (high resistance) or loose<br>Ignition system voltage circuit is open at the PCM terminals<br>Generator is damaged or has failed (output is too low)<br>PCM has failed |
| **P0563**<br>1T CCM<br>1996-2005<br>300M, Avenger,<br>Breeze, Caravan,<br>Cherokee, Cirrus,<br>Concorde, PT Cruiser,<br>Dakota, Durango,<br>Grand Cherokee,<br>Intrepid, LHS, Liberty,<br>Magnum, Neon, New<br>Yorker, Prowler, Ram,<br>Sebring, Stratus, Talon,<br>Town & Country, Viper,<br>Vision, Voyager,<br>Wrangler | **Battery Sense Circuit High Input Conditions:**<br>Engine running at a speed over 380 rpm, and the PCM detected the Battery Sense circuit voltage indicated 1v higher than the Charging system "goal" during the CCM test.<br>**Possible Causes:**<br>Generator field control circuit is shorted to system power (B+)<br>Generator is damaged or it has failed<br>PCM has failed |
| **P0563**<br>1T CCM<br>2004-05<br>300M, PT Cruiser,<br>Dakota, Durango,<br>Liberty, Magnum, Ram | **Battery (Charging System) Voltage High Input Conditions:**<br>Engine running at a speed over 1000 rpm. No other charging system codes are set. If battery voltage is 1v more than desired voltage for a set period of time, this DTC will set. 3 good trips required to turn off MIL.<br>**Possible Causes:**<br>Intermittent condition<br>Generator field ground circuit is shorted to battery voltage<br>Generator is damaged or it has failed<br>PCM has failed |

**OBD II Trouble Code List (P0XXX Codes) - Continued**

| DTC | Trouble Code Title, Conditions & Possible Causes: |
|---|---|
| **P0563**<br>1T CCM<br>1998-2003<br>Ram with 5.9L Diesel | **Charging System Voltage High Input Conditions:**<br>Engine started; engine speed over 1000 rpm during the test period, and the PCM detected the system voltage was more than 17.0v at any time during the CCM test.<br>**Possible Causes:**<br>Generator is damaged or has failed (output is too high)<br>PCM has failed |
| **P0564**<br>1T CCM<br>2004-05<br>Liberty | **Speed Control Switch No. 1 Circuit Plausibility, Signal Too High, Too Low Or Stuck Switch Conditions:**<br>Ignition is on. ECM detects Speed Control (S/C) Switch circuit signal is not in agreement with expected result.<br>**Possible Causes:**<br>ECM to S/C signal circuit is open or shorted to ground<br>ECM to S/C Sensor ground is open<br>S/C Switch signal circuit is open or shorted to voltage or to ground<br>S/C Sensor ground is open<br>S/C switch is damaged or has failed |
| **P0571**<br>1T CCM<br>2004-05<br>300M, PT Cruiser, Dakota, Durango, Liberty, Magnum, Ram | **Brake Switch No. 1 No Output Signal Conditions:**<br>Ignition is on. If output of BS 1 to PCM looks like brake is not applied, while BS 2 circuit is applied, the fault will mature in 60ms.<br>**Possible Causes:**<br>BS 1 signal open or shorted to ground<br>BS 2 signal open or shorted to ground<br>Ground circuit is open<br>Fused ignition switch output is open<br>Stop lamp switch is damaged or has failed<br>PCM has failed |
| **P0572**<br>1T CCM<br>1998-2003<br>Ram with 5.9L Diesel | **Brake Switch No. 1 Signal Circuit Failure Conditions:**<br>Vehicle driven at over 55 mph for 1 minute, and then returned to idle speed, test performed at least 10 times, and the PCM did not detect any change in the Brake Switch No. 1 signal.<br>**Possible Causes:**<br>Brake switch signal circuit to open or shorted to ground<br>Brake switch power circuit is open<br>Brake switch is damaged or has failed<br>PCM has failed |
| **P0572**<br>1T CCM<br>2004-05<br>300M, Caravan, Concorde, Dakota, Durango, Grand Cherokee, Intrepid, LHS, Liberty, Magnum, Neon, Pacifica, Prowler, Ram, Sebring, Stratus, Town & Country, Viper, Voyager, Wrangler | **Brake Switch Signal No. 1 Circuit Low Conditions:**<br>Ignition is on. When PCM recognizes that brake switch is mechanically stuck in the low/on position, a DTC will set. Three global good trips are necessary to turn off MIL.<br>**Possible Causes:**<br>Brake switch signal circuit is shorted to ground<br>BS No. 2 signal open (5.7L)<br>Brake switch is damaged or has failed<br>PCM has failed |

**OBD II Trouble Code List (P0XXX Codes) - Continued**

| DTC | Trouble Code Title, Conditions & Possible Causes: |
|---|---|
| **P0572**<br>1T CCM<br>2004-05<br>PT Cruiser | **Brake Switch Circuit Low Conditions:**<br>Engine is running and vehicle speed is above 31 mph; battery voltage is more than 9.5v; no other brake switch DTCs are present. If the PCM detects no change from the brake switch sense circuit input after coming to a complete stop, it will increment a counter. If the counter increments 10 times, the DTC will set and the MIL will illuminate.<br>**Possible Causes:**<br>Intermittent condition<br>Fused B+ circuit problem<br>Brake switch sense circuit is shorted to ground<br>Brake switch is closed<br>PCM has failed |
| **P0573**<br>1T CCM<br>1998-2003<br>Ram with 5.9L Diesel | **Brake Switch Signal No. 2 Circuit Malfunction Conditions:**<br>Engine started; engine running and the PCM did not detect any Brake Switch No. 2 signals on the CCD Bus circuit during the test.<br>**Possible Causes:**<br>Brake switch signal to the PCM is missing (circuit problem)<br>CCD Bus circuit is open or shorted to ground<br>CCD Bus circuit is shorted to system power<br>Generator field control circuit shorted to battery voltage<br>PCM has failed |
| **P0573**<br>1T CCM<br>2004-05<br>300M, Caravan, Concorde, Dakota, Durango, Grand Cherokee, Intrepid, LHS, Liberty, Magnum, Neon, Pacifica, Prowler, Ram, Sebring, Stratus, Town & Country, Viper, Voyager, Wrangler | **Brake Switch No. 1 Stuck High/Off Conditions:**<br>Ignition is on. If PCM recognizes BS No. 11 is mechanically stuck in the high/off position, this DTC will set.<br>**Possible Causes:**<br>BS No. 2 signal open or shorted to ground<br>BS No. 1 signal shorted to ground or to voltage<br>Ground circuit is open<br>Fused ignition switch output is open<br>Stop lamp switch is damaged or has failed<br>PCM has failed |
| **P0573**<br>1T CCM<br>2004-05<br>PT Cruiser | **Brake Switch Circuit High Conditions:**<br>Engine is running and vehicle speed is above 31 mph; accelerator pedal position is more than 25%; battery voltage is more than 9.5v; no other brake switch DTCs are present. If the PCM detects a high brake switch sense circuit input, it will increment a counter. If the counter increments 10 times, the DTC will set and the MIL will illuminate.<br>**Possible Causes:**<br>Intermittent condition<br>Fused B+ circuit problem<br>Brake switch sense circuit is open or is shorted to voltage<br>Ground circuit is open<br>Brake switch lamp operation has failed<br>PCM has failed |
| **P0575**<br>1T CCM<br>2001-03<br>Ram with 5.9L Diesel | **Cruise Control Switch Signal Low Input Conditions:**<br>Engine started; engine running and the PCM detected a continuous "low" voltage condition on the Cruise Control switch circuit.<br>**Possible Causes:**<br>C/C switch signal shorted to ground between switch and PCM<br>C/C clockspring is shorted to ground<br>C/C switch is damaged or has failed<br>PCM has failed |

**OBD II Trouble Code List (P0XXX Codes) - Continued**

| DTC | Trouble Code Title, Conditions & Possible Causes: |
|---|---|
| **P0576**<br>1T CCM<br>2001-03<br>Ram with 5.9L Diesel | **Cruise Control Switch Signal High Input Conditions**:<br>Engine started; engine running and the PCM detected a continuous "high" voltage condition on the Cruise Control switch circuit.<br>**Possible Causes:**<br>C/C switch circuit is open or shorted to VREF or system power<br>C/C switch or clockspring ground circuit is open<br>C/C switch is damaged or has failed<br>PCM has failed |
| **P0577**<br>1T CCM<br>2001-03<br>Ram with 5.9L Diesel | **Cruise Control Switch Signal High Input Conditions:**<br>Engine started; engine running and the PCM detected a continuous "high" voltage condition on the Cruise Control switch circuit.<br>**Possible Causes:**<br>C/C switch circuit is open or shorted to VREF or system power<br>C/C switch or clockspring ground circuit is open<br>C/C switch is damaged or has failed<br>PCM has failed |
| **P0579**<br>1T CCM<br>2002-03<br>300M, Caravan, Cherokee, Concorde, PT Cruiser, Dakota, Durango, Grand Cherokee, Intrepid, LHS, Liberty, Neon, New Yorker, Prowler, Ram, Sebring, Stratus, Town & Country, Viper, Voyager, Wrangler | **Speed Control Switch No. 1 Performance Conditions:**<br>Engine started; and the PCM detected a continuous invalid voltage condition on the Speed Control Switch No. 1 signal.<br>**Possible Causes:**<br>S/C switch signal circuit is shorted to chassis or Sensor ground<br>S/C switch signal circuit is open or shorted to system power<br>S/C switch is damaged or it has failed<br>Sensor ground circuit is open<br>PCM has failed |
| **P0579**<br>1T CCM<br>2004-05<br>Ram with 5.7L | **Speed Control Switch No. 1 Performance Conditions:**<br>Ignition on. The PCM detected the cruise switch voltage output is not out of range, but it does not equal any of the values for any of the button positions.<br>**Possible Causes:**<br>S/C switch No. 1 signal circuit is open or shorted to battery voltage<br>S/C switch No. 1 signal circuit is shorted to ground or to switch return circuit<br>S/C switch No. 1 switch return circuit is open<br>Clockspring or S/C switch is damaged or it has failed<br>PCM has failed |
| **P0580**<br>1T CCM<br>2002-05<br>300M, Caravan, Cherokee, Concorde, PT Cruiser, Dakota, Durango, Grand Cherokee, Intrepid, LHS, Liberty, Magnum, Neon, New Yorker, Pacifica, Prowler, Ram, Sebring, Stratus, Town & Country, Viper, Vision, Voyager, Wrangler | **Speed Control Switch No. 1 Circuit Low Input Conditions:**<br>Key on or engine started; system voltage over 10.0v and the PCM detected the Speed Control Switch No. 1 signal indicated less than 0.43v (exc. Liberty) or 0.60v (Liberty) for 2 minutes. 3 good trips required to turn off MIL.<br>**Possible Causes:**<br>Intermittent condition<br>S/C switch signal circuit is shorted to chassis or Sensor ground<br>S/C On/Off switch is damaged or it has failed<br>S/C Resume/Accel switch is damaged or it has failed<br>PCM has failed. |

## OBD II Trouble Code List (P0XXX Codes) - Continued

| DTC | Trouble Code Title, Conditions & Possible Causes: |
|---|---|
| **P0580**<br>1T CCM<br>2002-05<br>Dakota, Durango,<br>Grand Cherokee, Ram<br>with 5.7L | **Speed Control Switch No. 1 Circuit Low Input Conditions:**<br>Key on. The PCM detected the Speed Control Switch No. 1 signal is below the minimum acceptable voltage.<br>**Possible Causes:**<br>S/C switch No. 1 signal circuit is shorted to S/C switch return circuit<br>S/C switch No. 1 signal circuit is shorted to ground<br>Clockspring is damaged or it has failed<br>Speed control switch No. 1 has failed<br>PCM has failed |
| **P0581**<br>1T CCM<br>2002-05<br>300M, Caravan,<br>Cherokee, Concorde,<br>PT Cruiser, Dakota,<br>Durango, Grand<br>Cherokee, Intrepid,<br>LHS, Liberty, Magnum,<br>Neon, New Yorker,<br>Pacifica, Prowler, Ram,<br>Sebring, Stratus, Town<br>& Country, Viper,<br>Vision, Voyager,<br>Wrangler | **Speed Control Switch No. 1 Circuit High Input Conditions:**<br>Engine started; system voltage over 10v and the PCM detected an open or shorted condition, or above maximum acceptable S/C switch voltage, in the Speed Control Switch signal circuit.<br>**Possible Causes:**<br>S/C switch No.1 signal circuit is shorted to system power (B+)<br>S/C switch ground circuit is open<br>S/C switch signal circuit is open between PCM and clockspring<br>S/C Sensor ground circuit is open between PCM and clockspring or clockspring and S/C switch<br>Clockspring has failed<br>S/C switch (one or more) is damaged or has failed<br>PCM has failed |
| **P0582**<br>1T CCM<br>2002-05<br>300M, Caravan,<br>Cherokee, Concorde,<br>PT Cruiser, Dakota,<br>Durango, Grand<br>Cherokee, Intrepid,<br>LHS, Liberty, Magnum,<br>Neon, New Yorker,<br>Pacifica, Prowler, Ram,<br>Sebring, Stratus, Town<br>& Country, Viper,<br>Vision, Voyager,<br>Wrangler | **Speed Control Vacuum Solenoid Circuit Malfunction Conditions:**<br>Ignition on or engine started; Speed Control (S/C) system activated, and the PCM detected an open or short to voltage condition on the S/C Vacuum solenoid circuit during the CCM test period.<br>**Possible Causes:**<br>S/C supply circuit is open or is short to ground or to battery voltage<br>S/C vacuum solenoid control circuit is open<br>S/C vacuum solenoid control circuit is shorted to ground<br>S/C vacuum solenoid is damaged or has failed<br>PCM has failed |
| **P0585**<br>1T CCM<br>2004-05<br>Dakota, Durango, Ram<br>with 5.7L | **Speed Control Vacuum Solenoid Circuit Malfunction Conditions:**<br>Ignition on; Speed Control (S/C) system activated. The PCM detected the C/S switch inputs are not coherent with each other; for example, the PCM is reading C/S switch No. 1 as "Accel" and switch No. 2 as "Coast" at the same time.<br>**Possible Causes:**<br>S/C signal circuits shorted to battery voltage<br>High resistance in the S/C switch signal or return circuits<br>S/C switch No. 1 signal circuit is shorted to switch No. 2 signal circuit<br>S/C signal circuits shorted to ground<br>Clockspring is damaged or has failed<br>S/C switch(es) has failed<br>PCM has failed |

**OBD II Trouble Code List (P0XXX Codes) - Continued**

| DTC | Trouble Code Title, Conditions & Possible Causes: |
|---|---|
| **P0585**<br>1T CCM<br>2004-05<br>Liberty | **Speed Control Switch Plausibility Between Switch No. 1 & Switch No. 2**<br>**Conditions:**<br>Ignition is on. ECM detects a discrepancy between S/C switch No. 1 and No. 2 signals.<br>**Possible Causes:**<br>Intermittent condition<br>High resistance in the S/C Switch signal circuit or ground circuit<br>S/C switch is damaged or has failed<br>ECM has failed |
| **P0586**<br>1T CCM<br>2002-05<br>300M, Caravan,<br>Cherokee, Concorde,<br>PT Cruiser, Dakota,<br>Durango, Grand<br>Cherokee, Intrepid,<br>LHS, Liberty, Magnum,<br>Neon, New Yorker,<br>Pacifica, Prowler, Ram,<br>Sebring, Stratus, Town<br>& Country, Viper,<br>Vision, Voyager,<br>Wrangler | **Speed Control Vent Solenoid Circuit Malfunction Conditions:**<br>Engine started; battery voltage over 10v; Speed Control (S/C) system activated. The PCM detected an unexpected voltage condition on the S/C Vent solenoid circuit during the CCM test period.<br>**Possible Causes:**<br>S/C supply circuit is open or is short to ground<br>S/C vent solenoid control circuit is open<br>S/C vent solenoid control circuit is shorted to ground<br>S/C vent solenoid is damaged or has failed<br>PCM has failed |
| **P0589**<br>1T CCM<br>2004-05<br>Liberty | **Speed Control Switch No. 2 Plausibility, Voltage Too High Or Too Low, Or Switch Stuck Conditions:**<br>Ignition is on. ECM detects a discrepancy in the S/C switch No. 2 circuit signals.<br>**Possible Causes:**<br>ECM to S/C signal circuit is open or shorted to voltage<br>ECM to S/C Sensor ground is open<br>S/C Switch signal circuit is open or shorted to voltage or to ground<br>S/C Sensor ground is open<br>S/C switches damaged or failed |
| **P0591**<br>1T CCM<br>2004-05<br>Dakota, Durango,<br>Grand Cherokee, Ram | **Speed Control Switch No. 2 Malfunction Conditions:**<br>Ignition on; Speed Control (S/C) system activated. The PCM detected S/C switch No. 2 output voltage is not out of range, but it does not equal any of the values for any of the button positions.<br>**Possible Causes:**<br>S/C switch No. 2 signal circuit is open or is shorted to ground or to battery voltage<br>S/C return circuit is open<br>S/C switch No. 2 signal circuit is shorted to switch return circuit<br>S/C switch No. 2 has failed<br>Clockspring has failed<br>PCM has failed |
| **P0592**<br>1T CCM<br>2004-05<br>Dakota, Durango,<br>Grand Cherokee, Ram | **Speed Control Switch No. 2 Circuit Low Conditions:**<br>Ignition on; Speed Control (S/C) system activated. The PCM detected S/C switch No. 2 input voltage is below minimum acceptable voltage at the PCM.<br>**Possible Causes:**<br>S/C switch No. 2 signal circuit is shorted to ground or to switch return circuit<br>S/C switch No. 2 has failed<br>Clockspring has failed<br>PCM has failed |

**OBD II Trouble Code List (P0XXX Codes) - Continued**

| DTC | Trouble Code Title, Conditions & Possible Causes: |
|---|---|
| **P0593**<br>1T CCM<br>2004-05<br>Dakota, Durango, Grand Cherokee, Ram | **Speed Control Switch No. 2 Circuit High Conditions:**<br>Ignition on; Speed Control (S/C) system activated. The PCM detected S/C switch No. 2 input voltage is above maximum acceptable voltage at the PCM.<br>**Possible Causes:**<br>S/C switch No. 2 signal circuit shorted to voltage or open between PCM and clockspring<br>S/C switch No. 2 signal circuit is open between the clockspring and S/C switch<br>S/C switch return circuit is open between PCM and clockspring or S/C switch<br>S/C switch No. 2 or clockspring has failed<br>PCM has failed |
| **P0594**<br>1T CCM<br>2002-05<br>300M, Caravan, Cherokee, Concorde, PT Cruiser, Dakota, Durango, Grand Cherokee, Intrepid, LHS, Liberty, Magnum, Neon, New Yorker, Pacifica, Prowler, Ram, Sebring, Stratus, Town & Country, Viper, Vision, Voyager, Wrangler | **Speed Control Servo Power Circuit Malfunction Conditions:**<br>Engine started; Speed Control (S/C) system activated. The PCM detected an unexpected voltage condition on the S/C Vent solenoid circuit during the CCM test period.<br>**Possible Causes:**<br>S/C solenoid or vent solenoid has failed<br>Brake switch is damaged or it has failed<br>S/C brake switch circuit is open or it is shorted to ground<br>S/C power circuit is open or it is shorted to ground<br>PCM has failed |
| **P0600**<br>1T PCM<br>1995-2003<br>300M, Avenger, Breeze, Caravan, Cherokee, Cirrus, Concorde, PT Cruiser, Dakota, Durango, Grand Cherokee, Intrepid, LHS, Liberty, Neon, New Yorker, Prowler, Ram, Sebring, Stratus, Talon, Town & Country, Viper, Vision, Voyager, Wrangler | **PCM Internal Failure, No SPI Communications Conditions:**<br>Key on, system voltage over 10.5v and the PCM detected the initial serial data communications attempt failed 8 times in succession.<br>**Possible Causes:**<br>Turn the key off. Remove the PCM connector(s) and inspect the wiring harness connector pins and PCM pins for damaged, bent or missing pins. If any problems are located, make the repair and then perform the PCM Reset function. If this code resets after the repairs are made, the PCM has failed.<br>PCM needs to be replaced, and then reprogrammed |
| **P0600**<br>1T PCM<br>2004-05<br>300M, Caravan, Concorde, PT Cruiser, Dakota, Durango, Grand Cherokee, Intrepid, LHS, Liberty, Magnum, Neon, Pacifica, Prowler, Ram, Sebring, Stratus, Town & Country, Viper, Voyager, Wrangler | **Serial Communication Link Malfunction Conditions:**<br>Ignition on. Internal Bus communication failure is recognized between engine and transmission processors.<br>**Possible Causes:**<br>PCM or SPI failure |

**OBD II Trouble Code List (P0XXX Codes) - Continued**

| DTC | Trouble Code Title, Conditions & Possible Causes: |
|---|---|
| **P0600**<br>1T ECM<br>2004-05<br>Liberty with Diesel | **ECM Communication Error Conditions:**<br>Ignition on. ECM detects an internal failure.<br>**Possible Causes:**<br>ECM has failed<br>Intermittent condition |
| **P0601**<br>1T PCM<br>1995-2003<br>300M, Avenger, Breeze, Caravan, Cherokee, Cirrus, Concorde, PT Cruiser, Dakota, Durango, Grand Cherokee, Intrepid, LHS, Liberty, Neon, New Yorker, Prowler, Ram, Sebring, Stratus, Talon, Town & Country, Viper, Vision, Voyager, Wrangler | **PCM Random Access Memory Failure, Self-Test Failed Conditions:**<br>Key on, system voltage over 10.5v and the PCM detected the Random Access Memory (RAM) test failed in the initial Self-Test.<br>**Possible Causes:**<br>PCM needs to be replaced, and then reprogrammed |
| **P0601**<br>1T PCM<br>2004-05<br>300M, Caravan, Concorde, PT Cruiser, Dakota, Durango, Grand Cherokee, Intrepid, LHS, Liberty, Magnum, Neon, Pacifica, Prowler, Ram, Sebring, Stratus, Town & Country, Viper, Voyager, Wrangler | **PCM Internal Controller Failure Conditions:**<br>Ignition on. Internal CHECKSUM (or Bus communication) for software has failed; no communication between processors; cannot match calculated value.<br>**Possible Causes:**<br>PCM or SPI failure |
| **P0602**<br>1T PCM<br>2004-05<br>PT Cruiser, Sebring, Stratus with NGC | **Control Module Programming Error Conditions:**<br>Condition is monitored continuously. This DTC will always illuminate the MIL and is designed to signal the technician that the controller still has generic software installed.<br>**Possible Causes:**<br>Control module needs updated programming |
| **P0602**<br>1T PCM<br>2001-03<br>Ram with 5.9L Diesel | **PCM Fueling Calibration Conditions:**<br>Key on or engine starting, and the PCM detected one or more parameters was out-of-range for over one (1) second during testing.<br>**Possible Causes:**<br>PCM needs to be replaced, and then reprogrammed |
| **P0602**<br>1T ECM<br>2004-05<br>Liberty with Diesel | **ECM Invalid Code Word Conditions:**<br>Ignition on. ECM detects an internal failure.<br>**Possible Causes:**<br>ECM has failed<br>Intermittent condition |
| **P0604**<br>1T ECM<br>2004-05<br>Liberty with Diesel | **TCM Internal Problem Conditions:**<br>Ignition on. TCM detects an internal controller problem<br>**Possible Causes:**<br>TCM has failed |

**OBD II Trouble Code List (P0XXX Codes) - Continued**

| DTC | Trouble Code Title, Conditions & Possible Causes: |
|-----|---------------------------------------------------|
| **P0604**<br>1T PCM<br>2004-05<br>Caravan, PT Cruiser, Neon, Sebring, Stratus, Town & Country with NGC | **Internal TCM Error Conditions:**<br>Condition is monitored continuously. This DTC will always illuminate the MIL and is designed to signal the technician that there is a PCM internal error.<br>**Possible Causes:**<br>PCM needs to be replaced, and then reprogrammed |
| **P0604**<br>1T PCM<br>1999-2003<br>Dakota, Durango, Grand Cherokee, Ram, Wrangler | **TCM Random Access Memory Failure, Self-Test Failed Conditions:**<br>Key on, system voltage over 10.5v and the TCM detected the Random Access Memory (RAM) test failed during the initial Self-Test.<br>**Possible Causes:**<br>TCM needs to be replaced, and then reprogrammed |
| **P0605**<br>2T PCM<br>1995-98<br>Talon | **PCM Fault, SPI Communications Conditions:**<br>Key on or engine running; and the PCM detected that the serial communications inside the controller failed 8 times.<br>**Possible Causes:**<br>PCM needs to be replaced, and then reprogrammed |
| **P0605**<br>1T PCM<br>2004-05<br>Neon, Sebring, Stratus, PT Cruiser with NGC | **Internal TCM Error Conditions:**<br>Condition is monitored continuously. This DTC will always illuminate the MIL and is designed to signal the technician that there is a PCM internal error.<br>**Possible Causes:**<br>PCM needs to be replaced, and then reprogrammed |
| **P0605**<br>1T PCM<br>1999-2003<br>Dakota, Durango, Grand Cherokee, Ram, Wrangler | **TCM Random Access Memory Failure, Self-Test Failed Conditions:**<br>Key on, system voltage over 10.5v and the TCM detected the Read Only Memory (ROM) test failed during the initial Self-Test.<br>**Possible Causes:**<br>TCM needs to be replaced, and then reprogrammed |
| **P0605**<br>1T ECM<br>2004-05<br>Liberty with Diesel | **TCM Internal Problem Conditions:**<br>Ignition on. TCM detects an internal controller problem<br>**Possible Causes:**<br>TCM has failed |
| **P0606**<br>1T PCM<br>2004-05<br>300M, Dakota, Durango, Grand Cherokee, Magnum, Ram | **Engine Control Module Processor Malfunction Conditions:**<br>Engine running. When the PCM detected an internal failure to communicate with the ECM, or the CMP and CKP Sensor count periods are too short, the DTC will set. The ETC light will be flashing.<br>**Possible Causes:**<br>PCM has failed |
| **P0606**<br>1T PCM<br>2002-03<br>Ram with 5.9L Diesel | **Powertrain Control Module Malfunction Conditions:**<br>Key on or engine running; and the PCM detected an internal malfunction had occurred during the initialization step.<br>**Possible Causes:**<br>PCM needs to be replaced, and then reprogrammed |
| **P0606**<br>1T ECM<br>2004-05<br>Liberty with Diesel | **ECM CHECKSUM Error Or Deviation Error Conditions:**<br>Ignition on. ECM detects an internal failure.<br>**Possible Causes:**<br>ECM has failed<br>Intermittent condition |
| **P0607**<br>1T ECM<br>2004-05<br>Liberty with Diesel | **ECM Internal Error Conditions:**<br>Ignition on. ECM detects an internal failure.<br>**Possible Causes:**<br>ECM has failed<br>Intermittent condition |

**OBD II Trouble Code List (P0XXX Codes) - Continued**

| DTC | Trouble Code Title, Conditions & Possible Causes: |
|---|---|
| **P060B**<br>1T PCM<br>2004-05<br>300M, Dakota, Durango, Grand Cherokee, Magnum, Ram | **Engine Temperature Control A-D Ground Malfunction Conditions:**<br>When throttle motor is powered, if A to D reading does not return to ground within a set period of time from the test activation, this DTC will set. The test typically runs a couple of times per second, and is the reason why the APP2 signal spikes to ground a couple of times per second in normal running. Reprogramming the module may not always fix this fault. The ETC lamp will flash.<br>**Possible Causes:**<br>PCM need to be reprogrammed<br>PCM has failed |
| **P060D**<br>1T PCM<br>2004-05<br>Dakota, Durango, Ram | **Engine Temperature Control Level 2 Performance Conditions:**<br>When throttle motor is powered and no matured faults related to APP Sensors are present. When secondary software determines that APPS 1 and APPS 2 signals do not match for a period of time, this DTC will set. The ETC lamp will flash.<br>**Possible Causes:**<br>PCM need to be reprogrammed<br>PCM has failed |
| **P060E**<br>1T PCM<br>2004-05<br>Dakota, Durango, Ram | **Engine Temperature Control Level 2 TPS Performance Conditions:**<br>When throttle motor is powered and no matured faults related to TP Sensors are present. When secondary software determines that TP Sensor No. 1 and TP Sensor No. 2 signals do not match for a period of time, this DTC will set. The ETC lamp will flash.<br>**Possible Causes:**<br>PCM need to be reprogrammed<br>PCM has failed |
| **P060F**<br>1T PCM<br>2004-05<br>Dakota, Durango, Ram | **Engine Temperature Control Level 2 ETC Performance Conditions:**<br>When throttle motor is powered and no matured faults related to ETC Sensor is present. When secondary software determines that ETC Sensor signal is implausible for a period of time, this DTC will set. The ETC lamp will flash.<br>**Possible Causes:**<br>PCM need to be reprogrammed<br>PCM has failed |
| **P0610**<br>1T ECM<br>2004-05<br>Liberty with Diesel | **A/T Or M/T Miscoding Error Conditions:**<br>Ignition on. ECM detects an automatic transmission has been programmed as a manual transmission, or it detects a manual transmission has been programmed as an automatic transmission.<br>**Possible Causes:**<br>ECM needs to be reprogrammed<br>ECM has failed |
| **P0611**<br>1T ECM<br>2004-05<br>Liberty with Diesel | **ECM Capacitor Voltage 1 Error Conditions:**<br>Engine running; capacitor monitored during every 180 degrees of engine rotation. ECM determines that the capacitor voltage is greater than 100v.<br>**Possible Causes:**<br>ECM has failed<br>Intermittent condition |
| **P0613**<br>1T PCM<br>2004-05<br>PT Cruiser, Neon, Sebring, Stratus with NGC | **Internal TCM Error Conditions:**<br>Condition is monitored continuously. This DTC will always illuminate the MIL and is designed to signal the technician that there is a PCM internal error.<br>**Possible Causes:**<br>PCM needs to be replaced, and then reprogrammed |

## OBD II Trouble Code List (P0XXX Codes) - Continued

| DTC | Trouble Code Title, Conditions & Possible Causes: |
|---|---|
| **P0613**<br>1T PCM<br>2002-03<br>Truck, Van models | **TCM Internal Failure Conditions:**<br>Key on, system voltage over 10.5v and the TCM detected an internal failure (possible open condition in the TCM ground circuit).<br>**Possible Causes:**<br>Turn the key off. Remove the TCM connector(s) and inspect the wiring harness connector pins and TCM pins for damaged, bent or missing pins. Make repairs as needed. Then perform the TCM Reset function. If the code resets, the PCM has failed. |
| **P0613**<br>1T ECM<br>2004-05<br>Liberty with Diesel | **TCM Internal Problem Conditions:**<br>Ignition on. TCM detects an internal controller problem<br>**Possible Causes:**<br>Ground circuit is open<br>TCM has failed |
| **P0615**<br>1T ECM<br>2004-05<br>Liberty with Diesel | **Starter Relay Circuit Excessive Current Or Open Circuit Conditions:**<br>Ignition on and ECM Starter Relay commanded ON (excessive current), or commanded OFF (open circuit). ECM detects excessive current or does not detect voltage on the Starter Relay control circuit.<br>**Possible Causes:**<br>Intermittent condition<br>Starter Relay has failed<br>Ignition Switch Start output is open<br>Starter Relay control circuit is open, or is shorted to ground or to voltage<br>ECM has failed |
| **P0616**<br>1T ECM<br>2004-05<br>Liberty with Diesel | **Starter Relay Circuit Short-To-Ground Conditions:**<br>Ignition on and ECM Starter Relay commanded OFF. ECM does not detect voltage on the Starter Relay control circuit.<br>**Possible Causes:**<br>Intermittent condition<br>Starter Relay has failed<br>Ignition Switch Start output is open<br>Starter Relay control circuit is open, or is shorted to ground or to voltage<br>ECM has failed |
| **P0617**<br>1T ECM<br>2004-05<br>Liberty with Diesel | **Starter Relay Circuit Short Conditions:**<br>Ignition on and ECM Starter Relay commanded ON. ECM detects excessive current on the Starter Relay control circuit.<br>**Possible Causes:**<br>Intermittent condition<br>Starter Relay has failed<br>Ignition Switch Start output is open<br>Starter Relay control circuit is open, or is shorted to ground or to voltage<br>ECM has failed |
| **P061C**<br>1T PCM<br>2004-05<br>300M, Dakota, Durango,<br>Grand Cherokee,<br>Magnum, Ram | **Engine Temperature Control Level 2 RPM Performance Conditions:**<br>When throttle motor is powered, and no CMP or CKP electrical signal related DTCs are set, if the secondary software determines that the engine speed is implausible for a period of time, this DTC will set. The ETC lamp will flash.<br>**Possible Causes:**<br>PCM need to be reprogrammed<br>PCM has failed |

**OBD II Trouble Code List (P0XXX Codes) - Continued**

| DTC | Trouble Code Title, Conditions & Possible Causes: |
|---|---|
| **P0622**<br>1T CCM<br>1999-2005<br>300M, Avenger, Breeze, Caravan, Cherokee, Cirrus, Concorde, PT Cruiser, Dakota, Durango, Grand Cherokee, Intrepid, LHS, Liberty, Magnum, Neon, New Yorker, Prowler, Ram, Sebring, Stratus, Talon, Town & Country, Viper, Vision, Voyager, Wrangler | **Generator Field Control Circuit Malfunction Conditions:**<br>Engine running. The PCM detected the Generator Field control circuit had malfunctioned (PCM tries to regulate the generator field with no result).<br>**Possible Causes:**<br>Generator field control circuit is open or is shorted to ground<br>Generator field control circuit is shorted to system power (B+)<br>Generator field ground circuit is open<br>Generator is damaged or PCM has failed<br>PCM has failed |
| **P0625**<br>1T CCM<br>2004-05<br>PT Cruiser | **Generator Field Control Circuit Low Conditions:**<br>Engine running for more than 25 seconds. The PCM detected the Generator Field circuit is open or is shorted to ground.<br>**Possible Causes:**<br>Wiring harness intermittent problem<br>Generator field driver circuit is open or is shorted to ground<br>Generator field has malfunctioned<br>Generator field coil is open<br>PCM has failed |
| **P0626**<br>1T CCM<br>2004-05<br>PT Cruiser | **Generator Field Control Circuit High Conditions:**<br>Engine running for more than 25 seconds. The PCM detected the Generator Field circuit is shorted B+.<br>**Possible Causes:**<br>Wiring harness intermittent problem<br>Generator field circuit is shorted to voltage<br>Generator has malfunctioned |
| **P0627**<br>1T CCM<br>2002-05<br>300M, Caravan, Cherokee, Concorde, PT Cruiser, Dakota, Durango, Grand Cherokee, Intrepid, LHS, Liberty, Magnum, Neon, New Yorker, Pacifica, Prowler, Ram, Sebring, Stratus, Town & Country, Viper, Vision, Voyager, Wrangler | **Fuel Pump Relay Control Circuit Malfunction Conditions:**<br>Engine started; system voltage over 10.5v. The PCM detected an unexpected voltage condition (open or short) on the Fuel Pump relay control circuit during the CCM test period.<br>**Possible Causes:**<br>Fuel pump relay control circuit is open or is shorted to ground<br>Fuel pump relay control circuit is shorted to system power (B+)<br>Fuel pump relay power circuit (fused ignition) circuit is open<br>Fuel pump relay is damaged or it has failed<br>PCM has failed |
| **P062C**<br>1T PCM<br>2004-05<br>300M, Grand Cherokee, Dakota, Durango, Magnum, Ram | **Engine Temperature Control Level 2 MPH Performance Conditions:**<br>When throttle motor is powered, and no vehicle speed related DTCs are set, if the secondary software determines that the vehicle speed is implausible for a period of time, this DTC will set. The ETC lamp will flash.<br>**Possible Causes:**<br>PCM need to be reprogrammed<br>PCM has failed |

**OBD II Trouble Code List (P0XXX Codes) - Continued**

| DTC | Trouble Code Title, Conditions & Possible Causes: |
|---|---|
| **P0628**<br>1T CCM<br>2002-03<br>300M, Caravan, Cherokee, Concorde, PT Cruiser, Dakota, Durango, Grand Cherokee, Intrepid, LHS, Liberty, Neon, New Yorker, Prowler, Ram, Sebring, Stratus, Town & Country, Viper, Voyager, Wrangler | **Fuel Pump Relay Control Circuit Malfunction Conditions:**<br>Engine started; system voltage over 10.5v and the PCM detected an unexpected voltage condition on the Fuel Pump relay control circuit during the CCM test period.<br>**Possible Causes:**<br>Fuel pump relay circuit is open, shorted to ground or to power<br>Fuel pump relay power circuit (fused ignition) circuit is open<br>Fuel pump relay is damaged or it has failed<br>PCM has failed |
| **P0628**<br>1T CCM<br>2004-05<br>PT Cruiser | **Fuel Pump Relay Control Circuit Low Conditions:**<br>Ignition is on. If the PCM detects no voltage on the fuel pump relay control circuit for more than 3 seconds, this DTC will set.<br>**Possible Causes:**<br>Fuel pump relay intermittent operation<br>Fuel system/circuit intermittent operation<br>Fused ignition switch output circuit problem<br>Fuel pump relay has failed<br>Fuel pump relay control circuit is open or shorted to ground<br>PCM has failed |
| **P0629**<br>1T CCM<br>2004-05<br>PT Cruiser | **Fuel Pump Relay Control Circuit High Conditions:**<br>Ignition is on. If the PCM detects high voltage on the fuel pump relay control circuit for more than 3 seconds, this DTC will set.<br>**Possible Causes:**<br>Fuel pump relay intermittent operation<br>Fuel system/circuit intermittent operation<br>Fuel pump relay has failed<br>Fuel pump relay control circuit is shorted to battery voltage<br>PCM has failed |
| **P0630**<br>1T CCM<br>2002-05<br>300M, Caravan, Cherokee, Concorde, PT Cruiser, Dakota, Durango, Grand Cherokee, Intrepid, LHS, Liberty, Magnum, Neon, New Yorker, Pacifica, Prowler, Ram, Sebring, Stratus, Town & Country, Viper, Vision, Voyager, Wrangler | **VIN Not Programmed Into The PCM Conditions:**<br>Key on, and the PCM determined that the Vehicle Identification Number (VIN) had not been programmed into its memory.<br>**Possible Causes:**<br>Reprogram the correct VIN into the PCM<br>PCM has failed |

**OBD II Trouble Code List (P0XXX Codes) - Continued**

| DTC | Trouble Code Title, Conditions & Possible Causes: |
|---|---|
| **P0632**<br>1T CCM<br>2002-05<br>300M, Caravan, Cherokee, Concorde, PT Cruiser, Dakota, Durango, Grand Cherokee, Intrepid, LHS, Liberty, Magnum, Neon, New Yorker, Pacifica, Prowler, Ram, Sebring, Stratus, Town & Country, Viper, Vision, Voyager, Wrangler | **Odometer Not Programmed Into The PCM Conditions:**<br>Key on, and the PCM detected the vehicle mileage had not been programmed into memory.<br>**Possible Causes:**<br>Reprogram the correct mileage into the PCM<br>PCM has failed |
| **P0633**<br>1T CCM<br>2002-05<br>300M, Caravan, Cherokee, Concorde, PT Cruiser, Dakota, Durango, Grand Cherokee, Intrepid, LHS, Liberty, Magnum, Neon, New Yorker, Pacifica, Prowler, Ram, Sebring, Stratus, Town & Country, Viper, Vision, Voyager, Wrangler | **SKIM Key Not Programmed Into The PCM Conditions:**<br>Key on, and the PCM determined that the Security Key Immobilizer (SKIM) information had not been programmed into its memory.<br>**Possible Causes:**<br>Reprogram the SKIM key into the PCM<br>PCM has failed |
| **P0641**<br>1T ECM<br>2004-05<br>Liberty with Diesel | **Sensor Supply No. 1 Voltage Too High Or Too Low Conditions:**<br>Ignition on. ECM detects a short-to-voltage (too high) or no voltage (too low) on the Sensor Supply No. 1 circuit, which supplies 5v to the CMP Sensor and the APP Sensor No. 1.<br>**Possible Causes:**<br>Intermittent condition<br>Wiring problem in either circuit<br>APP Sensor No. 1 5v supply circuit is shorted to voltage (too high)<br>CMP Sensor 5v supply circuit is shorted to voltage (too high)<br>APP Sensor No. 1 5v supply circuit is shorted to ground (too low)<br>CMP Sensor 5v supply circuit is shorted to ground (too low)<br>APP Sensor or CMP Sensor has failed (too low)<br>ECM has failed |
| **P0642**<br>1T CCM<br>2004-05<br>300M, Grand Cherokee, Dakota, Durango, Magnum, Ram | **Sensor Reference Voltage 1 Circuit Low Conditions:**<br>Ignition is on. When the PCM recognizes the primary 5v supply circuit voltage is too low, this DTC will set. The ETC light is flashing.<br>**Possible Causes:**<br>Primary 5v supply shorted to ground<br>Sensor is shorted to ground<br>5v Sensor has failed<br>PCM has failed |
| **P0643**<br>1T CCM<br>2004-05<br>300M, Grand Cherokee, Dakota, Durango, Magnum, Ram | **Sensor Reference Voltage 1 Circuit High Conditions:**<br>Ignition is on. When the PCM recognizes the primary 5v supply circuit voltage is too high, this DTC will set. The ETC light is flashing.<br>**Possible Causes:**<br>Primary 5v supply shorted to battery voltage<br>PCM has failed |

## OBD II Trouble Code List (P0XXX Codes) - Continued

| DTC | Trouble Code Title, Conditions & Possible Causes: |
|---|---|
| **P0645**<br>1T CCM<br>2002-05<br>300M, Caravan, Cherokee, Concorde, PT Cruiser, Dakota, Durango, Grand Cherokee, Intrepid, LHS, Liberty, Magnum, Neon, New Yorker, Pacifica, Prowler, Ram, Sebring, Stratus, Town & Country, Viper, Vision, Voyager, Wrangler | **A/C Clutch Relay Circuit Malfunction Conditions:**<br>Engine started; system voltage over 10.0v, A/C switch "on". The PCM detected an unexpected voltage condition (open or shorted condition) on the A/C Clutch relay control circuit during the CCM test.<br>**Possible Causes:**<br>Internally fused ignition switch output circuit is faulty<br>A/C relay clutch control circuit is open or it is shorted to ground<br>A/C relay clutch power supply (fused ignition) circuit is open<br>A/C relay is damaged or it has failed<br>PCM has failed |
| **P0645**<br>1T ECM<br>2004-05<br>Liberty with Diesel | **A/C Clutch Relay Circuit Malfunction Conditions:**<br>Ignition on and A/C Clutch Relay commanded ON. If the ECM detects an excessive current on the A/C Clutch Relay control circuit (excessive current or short circuit), or if the A/C Clutch Relay is commanded OFF, and the ECM does not detect voltage on the A/C Clutch Relay control circuit (open circuit or short-to-ground), this DTC will set.<br>**Possible Causes:**<br>Intermittent condition<br>A/C Clutch Relay has failed<br>A/C Clutch Relay control circuit is shorted to voltage<br>Fused ASD Relay output circuit is open<br>A/C Clutch Relay control circuit is open or is shorted to ground<br>ECM has failed |
| **P0646**<br>1T CCM<br>2004-05<br>PT Cruiser | **A/C Clutch Relay Control Circuit Low Conditions:**<br>Ignition is on. If the PCM detects low voltage on the A/C clutch relay control circuit for more than 2.73 seconds, this DTC will set.<br>**Possible Causes:**<br>A/C relay intermittent operation<br>Fused B+ circuit problem<br>A/C clutch relay has failed<br>A/C clutch relay control circuit is open or is shorted to ground<br>PCM has failed |
| **P0647**<br>1T CCM<br>2004-05<br>PT Cruiser | **A/C Clutch Relay Control Circuit High Conditions:**<br>Ignition is on. If the PCM detects high voltage on the A/C clutch relay control circuit for more than 2.73 seconds, this DTC will set.<br>**Possible Causes:**<br>A/C relay intermittent operation<br>A/C system intermittent condition<br>A/C clutch relay has failed<br>A/C clutch relay control circuit is shorted to battery voltage<br>PCM has failed |

## OBD II Trouble Code List (P0XXX Codes) - Continued

| DTC | Trouble Code Title, Conditions & Possible Causes: |
|---|---|
| **P0651**<br>1T ECM<br>2004-05<br>Liberty with Diesel | **Sensor Supply No. 2 Voltage Too High Or Too Low Conditions:**<br>Ignition on. If the ECM detects a short-to-voltage (current too high), or if a low voltage (too low) is detected on the Sensor Supply No. 2 circuit, which supplies 5v to the MAF Sensor, Fuel Pressure Sensor, and Boost Pressure Sensor, this DTC will set.<br>**Possible Causes:**<br>Intermittent condition<br>Boost Pressure Sensor 5v supply circuit is shorted to voltage (too high)<br>Fuel Pressure Sensor 5v supply circuit is shorted to voltage (too high)<br>MAF Sensor 5v supply circuit is shorted to voltage (too high)<br>Boost Pressure Sensor 5v supply circuit is shorted to ground (too low)<br>Fuel Pressure Sensor 5v supply circuit is shorted to ground (too low)<br>MAF Sensor 5v supply circuit is shorted to ground (too low)<br>FP Sensor, BP Sensor or MAF Sensor has failed<br>ECM has failed |
| **P0652**<br>1T CCM<br>2004-05<br>300M, Dakota, Durango, Grand Cherokee, Magnum, Ram | **Sensor Reference Voltage 2 Circuit Low Conditions:**<br>Ignition is on. When the PCM recognizes the auxiliary 5v supply circuit voltage is too low, this DTC will set. The ETC light is flashing.<br>**Possible Causes:**<br>Auxiliary 5v supply shorted to ground<br>Sensor is shorted to ground<br>CMP Sensor has failed<br>PCM has failed |
| **P0653**<br>1T CCM<br>2004-05<br>300M, Dakota, Durango, Grand Cherokee, Magnum, Ram | **Sensor Reference Voltage 2 Circuit High Conditions:**<br>Ignition is on. When the PCM recognizes the auxiliary 5v supply circuit voltage is too high, this DTC will set. The ETC light is flashing.<br>**Possible Causes:**<br>Auxiliary 5v supply shorted to battery voltage<br>PCM has failed |
| **P0660**<br>1T CCM<br>2002-05<br>300M, Caravan, Cherokee, Concorde, PT Cruiser, Dakota, Durango, Grand Cherokee, Intrepid, LHS, Liberty, Magnum, Neon, New Yorker, Pacifica, Prowler, Ram, Sebring, Stratus, Town & Country, Viper, Vision, Voyager, Wrangler | **Manifold Tune Valve Solenoid Circuit Malfunction Conditions:**<br>Engine started; ASD relay "on", system voltage over 10.0v, and the PCM detected an unexpected voltage condition on the Manifold Tune Valve (MTV) solenoid control circuit.<br>**Possible Causes:**<br>Fused B+ circuit has failed<br>MTV solenoid/relay circuit is open or it is shorted to ground<br>MTV solenoid/relay circuit is shorted to power<br>MTV solenoid/relay ground circuit is open<br>MTV solenoid/relay is damaged or it has failed<br>PCM has failed |
| **P0660**<br>1T CCM<br>2004-05<br>300M, Magnum | **Manifold Tune Valve Solenoid Circuit Malfunction Conditions:**<br>Ignition on; ASD relay energized; battery voltage more than 10v. If the PCM senses the MTV is not at the desired state, this DTC will set.<br>**Possible Causes:**<br>MTV ground circuit is open<br>MTV control circuit is open or is shorted to ground or to battery voltage<br>MTV solenoid has failed<br>PCM has failed |

## OBD II Trouble Code List (P0XXX Codes) - Continued

| DTC | Trouble Code Title, Conditions & Possible Causes: |
|---|---|
| **P0670**<br>1T ECM<br>2004-05<br>Liberty with Diesel | **Glow Plug Controller Circuit Malfunction Conditions:**<br>Ignition on. If the ECM detects an open or shorted condition on the glow plug module signal/control circuit, this DTC will set.<br>**Possible Causes:**<br>Glow Plug Module has failed<br>Glow Plug Module signal/control circuit is open or is shorted to ground or to voltage<br>Intermittent condition<br>ECM has failed |
| **P0671-P0674**<br>1T ECM<br>2004-05<br>Liberty with Diesel | **Glow Plug Failure Or Short Circuit Conditions:**<br>Ignition on and Glow Plug Module Glow Plug commanded ON. If the ECM detects no current or excessive current on the respective Glow Plug output circuit, this DTC will set.<br>**Possible Causes:**<br>Glow Plug has failed<br>Glow Plug control circuit is open or is shorted to ground or to voltage<br>Glow Plug Module has failed<br>Intermittent condition |
| **P0683**<br>1T ECM<br>2004-05<br>Liberty with Diesel | **Glow Plug Module Signal Circuit Malfunction Conditions:**<br>Ignition on. If the ECM detects an open or shorted condition on the glow plug module signal/control circuit, this DTC will set.<br>**Possible Causes:**<br>Glow Plug Module has failed<br>Glow Plug Module signal/control circuit is open or is shorted to ground or to voltage<br>Intermittent condition<br>ECM has failed |
| **P0685**<br>1T CCM<br>2002-05<br>300M, Caravan, Cherokee, Concorde, PT Cruiser, Dakota, Durango, Grand Cherokee, Intrepid, LHS, Liberty, Magnum, Neon, New Yorker, Pacifica, Prowler, Ram, Sebring, Stratus, Town & Country, Viper, Vision, Voyager, Wrangler | **ASD Relay Control Circuit Malfunction Conditions:**<br>Key on; system voltage over 10.0v. The PCM detected an unexpected voltage condition (open or short) on the Automatic Shutdown (ASD) relay control circuit during the CCM test period (ASD actual state is not equal to the desired state). 3 good trips are required to turn off MIL. P0688 may also set.<br>**Possible Causes:**<br>Fused B+ circuit faults<br>ASD relay connector is damaged, loose or shorted<br>ASD relay control circuit is open or it is shorted to ground<br>ASD power supply (fused B+) circuit is open<br>ASD relay is damaged, has high resistance, or it has failed<br>PCM has failed |
| **P0685**<br>1T ECM<br>2004-05<br>Liberty with Diesel | **ASD Relay Control Circuit Shuts Off Malfunction Conditions:**<br>During after-run. The internal ECM timer determines that the ASD Relay has shut off before the After-Run mode of operation has been completed, or remains on too long when the After-Run mode of operation has been completed. Other DTCs may be present.<br>**Possible Causes:**<br>Intermittent condition<br>Replace ASD Relay and retest<br>ASD Relay control circuit is open intermittently (shut off too soon)<br>ASD Relay control circuit is shorted to ground intermittently (shut off too late)<br>ASD Relay control circuit is shorted to voltage (shut off too late)<br>ECM has failed |

**OBD II Trouble Code List (P0XXX Codes) - Continued**

| DTC | Trouble Code Title, Conditions & Possible Causes: |
|---|---|
| **P0686**<br>1T CCM<br>2004-05<br>PT Cruiser | **ASD Relay Control Circuit Low Conditions:**<br>Ignition is on. If the PCM detects no voltage on the ASD relay control circuit for more than 3.5 seconds, this DTC will set.<br>**Possible Causes:**<br>ASD relay intermittent operation<br>ASD system intermittent condition<br>Fused B+ circuit problem<br>ASD relay has failed<br>ASD relay control circuit is open or is shorted to ground<br>PCM has failed |
| **P0687**<br>1T CCM<br>2004-05<br>PT Cruiser | **ASD Relay Control Circuit High Conditions:**<br>Ignition is on. If the PCM detects high voltage on the ASD relay control circuit for more than 3.5 seconds, this DTC will set.<br>**Possible Causes:**<br>ASD relay intermittent operation<br>ASD system intermittent condition<br>ASD relay has failed<br>ASD relay control circuit is shorted to battery voltage<br>PCM has failed |
| **P0688**<br>1T CCM<br>2002-05<br>300M, Caravan, Cherokee, Concorde, PT Cruiser, Dakota, Durango, Grand Cherokee, Intrepid, LHS, Liberty, Magnum, Neon, New Yorker, Pacifica, Prowler, Ram, Sebring, Stratus, Town & Country, Viper, Vision, Voyager, Wrangler | **ASD Relay Sense Circuit Low Conditions:**<br>Key on, ASD relay energized, system voltage over 10.0v, and the PCM did not detect any voltage on the Automatic Shutdown (ASD) Sense circuit during the CCM test period.<br>**Possible Causes:**<br>ASD relay output circuit is open<br>ASD power supply (fused B+) circuit is open<br>ASD relay is damaged or it has failed<br>Problem in fuse/relay center<br>PCM no start condition<br>PCM has failed |
| **P0689**<br>1T CCM<br>2004-05<br>Dakota, Durango | **ASD Relay Sense Circuit Low Conditions:**<br>Key on; ASD relay energized; system voltage 9-16v. The ASD output circuit voltage drops below an acceptable value at the Front Control Module (FCM). This circuit is continuously monitored.<br>**Possible Causes:**<br>ASD power supply (fused B+) problem<br>ASD relay output circuit is open or is shorted to ground<br>ASD relay has failed<br>PCM has failed |
| **P0690**<br>1T CCM<br>2004-05<br>PT Cruiser, Dakota, Durango | **ASD Relay Sense Circuit High Conditions:**<br>Key on, ASD relay energized, system voltage over 10.0v, and the PCM (PT Cruiser) or FCM (Durango) detects high voltage on the Automatic Shutdown (ASD) Sense circuit during the CCM test period.<br>**Possible Causes:**<br>Intermittent condition<br>ASD relay output circuit is shorted to voltage<br>ASD relay is damaged or it has failed<br>PCM internal short to voltage<br>PCM has failed |

**OBD II Trouble Code List (P0XXX Codes) - Continued**

| DTC | Trouble Code Title, Conditions & Possible Causes: |
|---|---|
| **P0691**<br>1T CCM<br>2004-05<br>PT Cruiser, Dakota,<br>Durango | **Low Speed Fan (Fan No. 1) Relay Control Circuit Low Conditions:**<br>Key on; No. 1 cooling fan relay is actuated. If the PCM (PT Cruiser) or FCM (Durango) detects no voltage (open or shorted to ground) on the Radiator Fan Relay control circuit for more than 3 seconds, this DTC will set.<br>**Possible Causes:**<br>Intermittent condition<br>Fused ignition switch output circuit problems<br>Radiator fan relay has failed<br>Radiator fan control circuit is open or is shorted to ground<br>PCM has failed |
| **P0692**<br>1T CCM<br>2004-05<br>PT Cruiser, Dakota,<br>Durango | **Low Speed Fan (Fan No. 1) Relay Control Circuit High Conditions:**<br>Key on; radiator fan commanded ON. If the PCM (PT Cruiser) or FCM (Durango) detects an open or high voltage on the Radiator Fan Relay circuit for more than 3 seconds, this DTC will set.<br>**Possible Causes:**<br>Intermittent condition<br>Radiator fan relay has failed<br>Radiator fan control circuit is shorted to battery voltage<br>PCM has failed |
| **P0693**<br>1T CCM<br>2004-05<br>PT Cruiser, Dakota,<br>Durango | **High Speed Fan (Fan No. 2) Relay Control Circuit Low Conditions:**<br>Key on; fan relay is powered on. If the PCM (PT Cruiser) or FCM (Durango) detects no voltage (open or shorted) on the Radiator Fan Relay control circuit for more than 3 seconds, this DTC will set. Circuit is continuously monitored.<br>**Possible Causes:**<br>Intermittent condition<br>Fused ignition switch output circuit problems<br>Radiator fan relay has failed<br>Radiator fan control circuit is open or is shorted to ground<br>PCM has failed |
| **P0694**<br>1T CCM<br>2004-05<br>PT Cruiser, Dakota,<br>Durango | **High Speed Fan (Fan No. 2) Relay Control Circuit High Conditions:**<br>Key on; radiator fan commanded ON. If the PCM (PT Cruiser) or FCM (Durango) detects an open, short or high voltage on the Radiator Fan Relay circuit for more than 3 seconds, this DTC will set. This circuit is continuously monitored.<br>**Possible Causes:**<br>Intermittent condition<br>Radiator fan relay has failed<br>Radiator fan control circuit is open or shorted to battery voltage<br>PCM has failed |
| **P0697**<br>1T ECM<br>2004-05<br>Liberty with Diesel | **Sensor Supply No. 3 Voltage Too High Or Too Low Conditions:**<br>Ignition on. If the ECM detects a short-to-voltage (current too high), or if a low voltage (too low) is detected on the Sensor Supply No. 3 circuit, which supplies 5v to the Inlet Pressure Sensor and the APP Sensor No. 2, this DTC will set.<br>**Possible Causes:**<br>Intermittent condition<br>5v supply circuit(s) shorted to voltage<br>APP Sensor has failed (too low)<br>Inlet Pressure Sensor has failed (too low)<br>ECM has failed |

**OBD II Trouble Code List (P0XXX Codes) - Continued**

| DTC | Trouble Code Title, Conditions & Possible Causes: |
|---|---|
| **P0700**<br>2T TCM<br>1995-98<br>Talon | **Transaxle Control System Fault Conditions:**<br>Engine started and the PCM received a signal from the TCM that it had detected a fault.<br>Possible Causes:<br>The presence of this code means the TCM detected a problem<br>TCM related Sensor has solenoid is damaged or has failed<br>This code is for information only - check for other TCM codes<br>TCM or PCM has failed |
| **P0700**<br>2T TCM<br>1995-2005<br>300M, Avenger, Breeze,<br>Caravan, Cherokee,<br>Cirrus, Concorde, PT<br>Cruiser, Dakota, Durango,<br>Grand Cherokee, Intrepid,<br>LHS, Liberty, Magnum,<br>Neon, New Yorker,<br>Prowler, Ram, Sebring,<br>Stratus, Talon, Town &<br>Country, Viper, Vision,<br>Voyager, Wrangler | **Automatic Transmission Control System Malfunction Conditions:**<br>Ignition on or engine started. The PCM received a message over the CCD Bus from the Transmission Control Module (TCM) that it had detected a problem and set a trouble code in memory.<br>**Possible Causes:**<br>The presence of this code means the TCM detected a problem<br>TCM related Sensor has solenoid is damaged or has failed<br>This code is for information only - check for other TCM codes<br>TCM or PCM has failed |
| **P0700**<br>1T ECM<br>2004-05<br>Liberty with Diesel | **TCM DTC Conditions:**<br>Ignition on. If the ECM detects a CAN Bus message indicating the presences of a TCM-related DTC, this code will set.<br>**Possible Causes:**<br>Verify presence of any DTCs |
| **P0703**<br>1T CCM<br>1996-2003<br>300M, Avenger, Breeze,<br>Caravan, Cherokee,<br>Cirrus, Concorde, PT<br>Cruiser, Dakota, Durango,<br>Grand Cherokee, Intrepid,<br>LHS, Liberty, Neon, New<br>Yorker, Prowler, Ram,<br>Sebring, Stratus, Talon,<br>Town & Country, Viper,<br>Vision, Voyager, Wrangler | **A/T Brake Switch Sense Circuit Malfunction Conditions:**<br>Engine started; vehicle driven to over 20 mph with the TP Sensor over 0.02v for 6 seconds, followed by a deceleration period to 0 mph at least 16 times, and the PCM detected the Brake switch signal status was not correct during the acceleration/deceleration periods.<br>**Possible Causes:**<br>Brake switch signal circuit or the switch ground circuit is open<br>Brake switch is damaged, out of adjustment or has failed<br>PCM has failed |
| **P0703**<br>1T CCM<br>2004-05<br>300M, Dakota, Durango,<br>Grand Cherokee,<br>Magnum, Ram | **A/T Brake Switch No. 2 Performance Malfunction Conditions:**<br>Ignition is on. When the PCM recognizes brake switch No.2 voltage is not equal to applied value at the PCM when brake switch No. 1 is applied, this DTC will set. Note: This could be a normal condition; however, if this condition is seen repeatedly by the PCM, the DTC will be set. Cruise control will not work for the rest of the key cycle.<br>**Possible Causes:**<br>Fused B+ circuit malfunction<br>Brake switch output circuit is open or is shorted to battery voltage or to ground<br>Brake switch 1 signal circuit is open<br>Brake switch has failed<br>PCM has failed |

**OBD II Trouble Code List (P0XXX Codes) - Continued**

| DTC | Trouble Code Title, Conditions & Possible Causes: |
|---|---|
| **P0703**<br>1T CCM<br>2004-05<br>PT Cruiser | **Brake Lamp Switch No. 2 Performance Malfunction Conditions:**<br>Engine is running; battery voltage is more than 9.5v; no other brake switch DTCs are present. When Brake Switch 1 output is the same as Brake Switch No. 2 output for more than 25 seconds, this DTC will set.<br>**Possible Causes:**<br>Intermittent condition<br>Fused B+ circuit malfunction<br>Brake switch output circuit is open or is shorted to battery voltage or to ground<br>Ground circuit is open<br>Brake switch or PCM has failed |
| **P0705**<br>1T CCM<br>1995-98<br>Talon | **A/T Transmission Range Switch Circuit Malfunction Conditions:**<br>Key on or engine running. The PCM detected an invalid PRNDL switch signal occurred (i.e., an invalid PRNDL switch signal occurred 3 times for 100 ms) during the CCM test.<br>**Possible Causes:**<br>Manual Lever (Rooster Comb) is worn out (check the contacts)<br>TR switch signal circuit is open, shorted to ground or to power<br>TR switch is damaged or has failed<br>TCM or the PCM has failed |
| **P0705**<br>1T CCM<br>1996-2003<br>300M, Avenger, Breeze, Caravan, Cherokee, Cirrus, Concorde, PT Cruiser, Dakota, Durango, Grand Cherokee, Intrepid, LHS, Liberty, Neon, New Yorker, Prowler, Ram, Sebring, Stratus, Talon, Town & Country, Viper, Vision, Voyager, Wrangler | **A/T Check Shifter Signal Circuit Malfunction Conditions:**<br>Key on or engine running; and the PCM detected at least three occurrences of an invalid TR Sensor PRNDL signal for over 100 ms.<br>**Possible Causes:**<br>TR1, T3, T41 or T42 sense circuit is open<br>TR1, T3, T41 or T42 sense circuit is shorted to ground<br>TR1, T3, T41 or T42 Sensor circuit is shorted to system power<br>Transmission Range Sensor is damaged or has failed<br>PCM has failed |
| **P0706**<br>1T CCM<br>2004-05<br>300M, Caravan, Concorde, PT Cruiser, Dakota, Durango, Grand Cherokee, Intrepid, LHS, Liberty, Magnum, Neon, Pacifica, Prowler, Ram, Sebring, Stratus, Town & Country, Viper, Voyager, Wrangler | **A/T Check Shifter Signal Circuit Malfunction Conditions:**<br>Key on. After 3 occurrences in one ignition cycle of an invalid PRNDL DDTC which last for more than 0.1 second. NOTE: All indicator lights on the instrument cluster will illuminate boxed when the vehicle engine is not running, ignition on, or engine running in Park or Neutral if a problem exists.<br>**Possible Causes:**<br>Shifter out of adjustment<br>TRS T1, T3, T41 or T42 sense circuit is open, shorted to ground or to voltage<br>TRS Sensor has failed<br>Intermittent wiring or connector problems<br>PCM has failed |
| **P0706**<br>1T ECM<br>2004-05<br>Liberty with Diesel | **Check Shifter Signal Conditions:**<br>Ignition on. This DTC will set with 3 occurrences in one ignition start with an invaled PRNDL code, which lasts more than 0.1 second.<br>**Possible Causes:**<br>Shifter out of adjustment<br>TRS T1, T2 or T3 sense circuit is open, shorted to ground or to voltage<br>T41 or T42 sense circuit is open, shorted to ground or to voltage<br>Transmission Range Sensor (TRS) has failed<br>TCM has failed<br>Intermittent wiring and connector problems |

**OBD II Trouble Code List (P0XXX Codes) - Continued**

| DTC | Trouble Code Title, Conditions & Possible Causes: |
|---|---|
| **P0710**<br>1T CCM<br>2001-03<br>Sebring, Stratus Coupe<br>with 2.4L I4 or 3.0L V6 | **TCM Transmission Fluid Temperature Sensor Circuit Malfunction Conditions:**<br>Key on or engine running; and the PCM detected an unexpected low or high voltage condition on the TFT Sensor circuit in the CCM test.<br>**Possible Causes:**<br>TFT Sensor signal circuit is open between the Sensor and PCM<br>TFT Sensor signal circuit is shorted to ground<br>TFT Sensor is damaged or has failed (open or shorted)<br>PCM has failed |
| **P0711**<br>1T CCM<br>1996-2003<br>Cherokee, Dakota, Durango, Grand Cherokee, Ram, Wrangler | **A/T Transmission Fluid Temperature Sensor Signal - No Rise After Startup Conditions:**<br>Engine started; engine runtime over 20 minutes, and the PCM detected the TFT Sensor signal did not increase at least 16°F; or the PCM detected the TFT Sensor indicated more than 260°F with the ECT Sensor signal indicating less than 100°F during the CCM test.<br>**Possible Causes:**<br>TFT Sensor signal has an intermittent high resistance condition<br>TFT Sensor is damaged, skewed or has drifted out of range<br>PCM has failed |
| **P0711**<br>1T CCM<br>2004-05<br>300M, Caravan, Concorde, PT Cruiser, Dakota, Durango, Grand Cherokee, Intrepid, LHS, Liberty, Magnum, Neon, Pacifica, Prowler, Ram, Sebring, Stratus, Town & Country, Viper, Voyager, Wrangler | **A/T Transmission Fluid Temperature Sensor Signal - No Rise After Startup Conditions:**<br>Engine started. This DTC will set when the desired transmission temperature does not reach a normal operation temperature within a given time frame. Time is variable due to ambient temperature at cold engine start: from 35 minutes at –40°F (-40°C) to 10 minutes at 60°F (15°C).<br>**Possible Causes:**<br>Related DTCs will be present<br>Transmission temperature Sensor has failed<br>Intermittent wiring or connector problems<br>PCM has failed |
| **P0711**<br>1T ECM<br>2004-05<br>Liberty with Diesel | **Transmission Temperature Sensor Performance Conditions:**<br>Ignition on and engine running. This DTC will set when the desired transmission temperature does not reach a normal operating temperature within a given time frame. Time is variable due to ambient temperature. At 60°F (16°C) ambient temperature, warmup time is approximately 10 minutes. Related DTCs may be present.<br>**Possible Causes:**<br>Transmission Temperature Sensor (TTS) has failed<br>TCM has failed<br>Intermittent wiring and connector problems |
| **P0712**<br>1T CCM<br>1996-2003<br>Cherokee, Dakota, Durango, Grand Cherokee, Ram, Wrangler | **A/T Transmission Fluid Temperature Sensor Low Input Conditions:**<br>Engine started and the PCM detected the TFT Sensor signal was under 1.55v for 2.2 seconds.<br>**Possible Causes:**<br>TFT Sensor signal circuit is shorted to ground<br>TFT Sensor is damaged or has failed (it may be shorted)<br>PCM has failed |

**OBD II Trouble Code List (P0XXX Codes) - Continued**

| DTC | Trouble Code Title, Conditions & Possible Causes: |
|---|---|
| **P0712** 1T CCM 2004-05 300M, Caravan, Concorde, PT Cruiser, Dakota, Durango, Grand Cherokee, Intrepid, LHS, Liberty, Magnum, Neon, Pacifica, Prowler, Ram, Sebring, Stratus, Town & Country, Viper, Voyager, Wrangler | **A/T Transmission Fluid Temperature Sensor Low Input Conditions:** Engine started and the PCM detected the TFT Sensor signal was under 0.078v for 0.45 second. **Possible Causes:** Related DTCs are present TFT Sensor signal circuit is shorted to ground TFT Sensor is damaged or has failed (it may be shorted) Intermittent wiring or connector problems PCM has failed |
| **P0712** 1T ECM 2004-05 Liberty with Diesel | **Transmission Temperature Sensor Low Conditions:** Ignition on and engine running. This DTC will set when the monitored transmission temperature drops below 0.78v for 0.45 second. Related DTCs may be present. **Possible Causes:** Transmission Temperature Sensor (TTS) signal circuit is shorted to ground TTS has failed TCM has failed Intermittent wiring and connector problems |
| **P0713** 1T CCM 1996-2003 Cherokee, Dakota, Durango, Grand Cherokee, Ram, Wrangler | **A/T Transmission Fluid Temperature Sensor High Input Conditions:** Engine started and the PCM detected the TFT Sensor signal was over 3.76v for 2.2 seconds. **Possible Causes:** TFT Sensor signal circuit is open between the Sensor and PCM TFT Sensor ground circuit is open between Sensor and PCM TFT Sensor is damaged or has failed (it may be open internally) PCM has failed |
| **P0713** 1T CCM 2004-05 300M, Caravan, Concorde, PT Cruiser, Dakota, Durango, Grand Cherokee, Intrepid, LHS, Liberty, Magnum, Neon, Pacifica, Prowler, Ram, Sebring, Stratus, Town & Country, Viper, Voyager, Wrangler | **A/T Transmission Fluid Temperature Sensor High Input Conditions:** Engine started and the PCM detected the TFT Sensor signal was over 4.94v for 0.45 second. **Possible Causes:** Related DTCs are present TFT Sensor signal circuit is open or is shorted to voltage TFT Sensor is damaged or has failed (it may be shorted) Intermittent wiring or connector problems PCM has failed |
| **P0714** 1T CCM 2004-05 300M, Caravan, Concorde, PT Cruiser, Dakota, Durango, Grand Cherokee, Intrepid, LHS, Liberty, Magnum, Neon, Pacifica, Prowler, Ram, Sebring, Stratus, Town & Country, Viper, Voyager, Wrangler | **A/T Transmission Fluid Temperature Sensor Intermittent Conditions:** Engine started and the PCM detected the TFT Sensor signal was fluctuating or changes abruptly within a predetermined period of time. **Possible Causes:** Related DTCs are present TFT Sensor is damaged or has failed (it may be shorted) Intermittent wiring or connector problems PCM has failed |

**OBD II Trouble Code List (P0XXX Codes) - Continued**

| DTC | Trouble Code Title, Conditions & Possible Causes: |
|---|---|
| **P0713**<br>1T ECM<br>2004-05<br>Liberty with Diesel | **Transmission Temperature Sensor High Conditions:**<br>Ignition on and engine running. This DTC will set when the monitored Transmission Temperature Sensor signal rises above 4.94v for 0.45 second.<br>**Possible Causes:**<br>Transmission Temperature Sensor (TTS) signal circuit is open or is shorted to voltage<br>TTS has failed<br>TCM has failed<br>Intermittent wiring and connector problems |
| **P0714**<br>1T ECM<br>2004-05<br>Liberty with Diesel | **Transmission Temperature Sensor Intermittent Conditions:**<br>Ignition on and engine running. This DTC will set when the monitored Transmission Temperature Sensor voltage fluctuates or changes abruptly within a predetermined period. Related DTCs may be present.<br>**Possible Causes:**<br>TTS has failed<br>TCM has failed<br>Intermittent wiring and connector problems |
| **P0715**<br>1T CCM<br>1996-2003<br>300M, Avenger, Breeze, Caravan, Cherokee, Cirrus, Concorde, PT Cruiser, Dakota, Durango, Grand Cherokee, Intrepid, LHS, Liberty, Neon, New Yorker, Prowler, Ram, Sebring, Stratus, Talon, Town & Country, Viper, Vision, Voyager, Wrangler | **A/T Input Speed Sensor Circuit Malfunction Conditions:**<br>Vehicle driven in 1st, 2nd or 3rd gear with Output Speed Sensor (OSS) signals present, and the PCM did not detect any ISS signals less for 1 second with the vehicle in 2nd gear.<br>**Possible Causes:**<br>ISS signal (+) circuit is open, shorted to ground or to power<br>ISS signal (-) circuit is open, shorted to ground or to power<br>ISS is damaged or has failed (it may be open internally)<br>Overdrive clutch drum lugs are damaged or missing<br>PCM has failed |
| **P0715**<br>1T CCM<br>2001-03<br>Sebring, Stratus Coupe with 2.4L I4 or 3.0L V6 | **TCM Input Shaft Speed Sensor Circuit Malfunction Conditions:**<br>Engine started; and the PCM detected a signal from the TCM indicating it had detected an unexpected voltage condition on the Input Speed Sensor (ISS) signal circuit during the test.<br>**Possible Causes:**<br>ISS positive (+) circuit is open, shorted to ground or to power<br>ISS positive (-) circuit is open, shorted to ground or to power<br>ISS Sensor is damaged or it has failed |
| **P0715**<br>1T CCM<br>2004-05<br>300M, Caravan, Concorde, PT Cruiser, Dakota, Durango, Grand Cherokee, Intrepid, LHS, Liberty, Magnum, Neon, Pacifica, Prowler, Ram, Sebring, Stratus, Town & Country, Viper, Voyager, Wrangler | **TCM Input Speed Sensor Circuit Malfunction Conditions:**<br>Engine started; the transmission gear ratio is monitored continuously while the transmission is in gear. This DTC will set if there is an excessive change in the Input RPM in any gear.<br>**Possible Causes:**<br>ISS ground circuit is open or is shorted to voltage<br>ISS signal circuit is open, shorted to ground or to power<br>ISS Sensor is damaged or it has failed<br>Intermittent wiring or connector problems<br>PCM has failed |

**OBD II Trouble Code List (P0XXX Codes) - Continued**

| DTC | Trouble Code Title, Conditions & Possible Causes: |
|---|---|
| **P0719**<br>1T CCM<br>2004-05<br>PT Cruiser | **Brake Lamp Switch Circuit Low Conditions:**<br>Vehicle is driven over 31 mph; battery over 9.5v. If the PCM detects no change from Brake Lamp Switch circuit input, after a complete stop, it will increment a counter. If the counter increments 10 times, this DTC will set and the MIL will illuminate.<br>**Possible Causes:**<br>Intermittent condition<br>Fused B+ circuit problems<br>Brake lamp switch sense circuit is open or is shorted to ground<br>Ground circuit is open<br>Brake lamp switch has failed<br>PCM has failed |
| **P0720**<br>1T CCM<br>1996-2003<br>300M, Avenger, Breeze, Caravan, Cherokee, Cirrus, Concorde, PT Cruiser, Dakota, Durango, Grand Cherokee, Intrepid, LHS, Liberty, Neon, New Yorker, Prowler, Ram, Sebring, Stratus, Talon, Town & Country, Viper, Vision, Voyager, Wrangler | **A/T Output Speed Sensor - Low Output Above 15 MPH Conditions:**<br>Engine started; vehicle driven to a speed over 15 mph, and the PCM detected the OSS signal was less than 60 RPM for 2.6 seconds.<br>**Possible Causes:**<br>OSS signal (+) circuit is open, shorted to ground or to power<br>OSS signal (-) circuit is open, shorted to ground or to power<br>OSS is damaged or has failed (it may be open internally)<br>Parking pawl lugs damaged or missing<br>PCM has failed |
| **P0720**<br>1T CCM<br>2001-03<br>Sebring, Stratus Coupe with 2.4L I4 or 3.0L V6 | **TCM Output Shaft Speed Sensor Circuit Malfunction Conditions:**<br>Engine started; and the PCM detected a signal from the TCM indicating it had detected an unexpected voltage condition on the Output Speed Sensor (OSS) circuit during the test.<br>**Possible Causes:**<br>OSS positive (+) circuit is open, shorted to ground or to power<br>OSS positive (-) circuit is open, shorted to ground or to power<br>OSS Sensor or PCM has failed |
| **P0720**<br>1T CCM<br>2004-05<br>300M, Caravan, Concorde, PT Cruiser, Dakota, Durango, Grand Cherokee, Intrepid, LHS, Liberty, Magnum, Neon, Pacifica, Prowler, Ram, Sebring, Stratus, Town & Country, Viper, Voyager, Wrangler | **TCM Output Speed Sensor Circuit Malfunction Conditions:**<br>Engine started; the transmission gear ratio is monitored continuously while the transmission is in gear. This DTC will set if there is an excessive change in the Output RPM in any gear. On some models, this DTC can take up to 5 minutes of problem identification before lighting the MIL.<br>**Possible Causes:**<br>OSS ground circuit is open or is shorted to voltage or to ground<br>Speed Sensor ground circuit is open, shorted to ground or to voltage<br>OSS Sensor is damaged or it has failed<br>Intermittent wiring or connector problems<br>PCM or TCM has failed |
| **P0720**<br>1T CCM<br>1996-2003<br>Cherokee, Grand Cherokee, Liberty, Wrangler | **A/T Output Speed Sensor - Low Output Above 15 MPH Conditions:**<br>Engine started; vehicle speed over 15 mph (determined by the VSS inputs). PCM detected the OSS input was less than 60 RPM for 2.6 seconds during the CCM test.<br>**Possible Causes:**<br>OSS signal (+) circuit is open, shorted to ground or to power<br>OSS signal (-) circuit is open, shorted to ground or to power<br>OSS is damaged or has failed (it may be open internally)<br>Parking pawl lugs damaged or missing<br>PCM has failed |

**OBD II Trouble Code List (P0XXX Codes) - Continued**

| DTC | Trouble Code Title, Conditions & Possible Causes: |
|---|---|
| **P0724**<br>1T CCM<br>2004-05<br>PT Cruiser | **Brake Lamp Switch Circuit High Conditions:**<br>Vehicle is driven at speeds over 31 mph; accelerator pedal position is more than 25%; battery voltage over 9.5v. If the PCM detects a high Brake Lamp Switch circuit input, it will increment a counter. If the counter increments 10 times, this DTC will set and the MIL will illuminate.<br>**Possible Causes:**<br>Intermittent condition<br>Brake lamp switch sense circuit is shorted to battery voltage<br>Brake lamp switch has failed<br>PCM has failed |
| **P0725**<br>1T CCM<br>2001-05<br>300M, Caravan, Cherokee, Concorde, PT Cruiser, Dakota, Durango, Grand Cherokee, Intrepid, LHS, Liberty, Magnum, Neon, New Yorker, Pacifica, Prowler, Ram, Sebring, Stratus, Town & Country, Viper, Vision, Voyager, Wrangler | **A/T Engine Speed Sensor Circuit Malfunction Conditions:**<br>Engine running; and the PCM detected an Engine Speed Sensor reading of less than 390 rpm or more than 8000 rpm occurred for 2 seconds during the CCM test.<br>**Possible Causes:**<br>Check for trouble codes related to the CKP Sensor<br>CKP Sensor signal circuit open, shorted to ground or to power<br>CKP Sensor is damaged or has failed (open or shorted)<br>Intermittent wiring or connector problems<br>PCM has failed |
| **P0725**<br>1T CCM<br>2001-03<br>Sebring, Stratus Coupe with 2.4L I4 or 3.0L V6 | **TCM Engine Speed Sensor Circuit Malfunction Conditions:**<br>Engine started; and the PCM detected a signal from the TCM indicating it had detected an unexpected voltage condition on the CKP Sensor signal circuit during the CCM test.<br>**Possible Causes:**<br>CKP Sensor signal circuit open, shorted to ground or to power<br>CKP Sensor is damaged or has failed (open or shorted)<br>TCM engine speed signal circuit from the PCM has failed |
| **P0725**<br>1T ECM<br>2004-05<br>Liberty with Diesel | **Engine Speed Sensor Circuit Malfunction Conditions:**<br>Engine running. This DTC will set when the TCM senses an engine rpm less than 400 rpm, with engine running for at least 2 seconds. Engine rpm information is transferred over the communication Bus from the ECM. This DTC can take up to 5 minutes of problem identification before lighting the MIL.<br>**Possible Causes:**<br>Engine speed signal circuit is open, or is shorted to ground or to voltage<br>TCM has failed<br>ECM has failed<br>Intermittent wiring and connector problems |
| **P0731**<br>1T CCM<br>1999-2003<br>300M, Avenger, Breeze, Caravan, Cherokee, Cirrus, Concorde, PT Cruiser, Dakota, Durango, Grand Cherokee, Intrepid, LHS, Liberty, Neon, New Yorker, Prowler, Ram, Sebring, Stratus, Town & Country, Viper, Voyager, Wrangler | **A/T Additional Gear Ratio Error In First Gear Conditions:**<br>Vehicle driven to a speed of 4-37 mph at more than 500 rpm, gear selector in 1 Gear in Drive, and the PCM detected too much difference in the engine speed and output shaft speed.<br>**Possible Causes:**<br>U/D and 2/4 pressures less than 105 psi<br>U/D or L/R clutch volume indexes out of specification<br>L/R accumulator seals are damaged or L/R switch valve stuck<br>L/R switch valve stuck in first gear position<br>Transmission has an internal leak (U/D or other seals leaking) |

**OBD II Trouble Code List (P0XXX Codes) - Continued**

| DTC | Trouble Code Title, Conditions & Possible Causes: |
|---|---|
| **P0731**<br>1T CCM<br>2004-05<br>300M, Caravan, Concorde, PT Cruiser, Dakota, Durango, Grand Cherokee, Intrepid, LHS, Liberty, Magnum, Neon, Pacifica, Prowler, Ram, Sebring, Stratus, Town & Country, Viper, Voyager, Wrangler | **A/T Additional Gear Ratio Error In First Gear Conditions:**<br>The transmission gear ratio is monitored continuously while the transmission is in gear. If the ratio of the Input RPM to the Output RPM does not match the current gear ratio, this DTC will set.<br>**Possible Causes:**<br>Related DTCs will be present<br>Internal transmission mechanical problems may exist<br>Intermittent gear ratio errors are present |
| **P0732**<br>1T CCM<br>1999-2003<br>300M, Avenger, Breeze, Caravan, Cherokee, Cirrus, Concorde, PT Cruiser, Dakota, Durango, Grand Cherokee, Intrepid, LHS, Liberty, Neon, New Yorker, Prowler, Ram, Sebring, Stratus, Town & Country, Viper, Voyager, Wrangler | **A/T Additional Gear Ratio Error In Second Gear Conditions:**<br>Vehicle driven to a speed of 4-37 mph at more than 500 rpm, gear selector in 2nd Gear in Drive, and the PCM detected too much difference in the engine speed and output shaft speed.<br>**Possible Causes:**<br>U/D and L/R pressures less than 95 psi at 2000 rpm<br>U/D or 2/4 clutch volume indexes out of specification<br>L/R accumulator seals are damaged or L/R switch valve stuck<br>Transmission has an internal leak (U/D or 2/4 seals leaking)<br>L/R switch valve stuck in first gear position |
| **P0732**<br>1T CCM<br>2004-05<br>300M, Caravan, Concorde, PT Cruiser, Dakota, Durango, Grand Cherokee, Intrepid, LHS, Liberty, Magnum, Neon, Pacifica, Prowler, Ram, Sebring, Stratus, Town & Country, Viper, Voyager, Wrangler | **A/T Additional Gear Ratio Error In Second Gear Conditions:**<br>The transmission gear ratio is monitored continuously while the transmission is in gear. If the ratio of the Input RPM to the Output RPM does not match the current gear ratio, this DTC will set.<br>**Possible Causes:**<br>Related DTCs will be present<br>Transmission solenoid/pressure switch assembly has malfunctioned or failed<br>Internal transmission mechanical problems may exist<br>Intermittent gear ratio errors are present |
| **P0733**<br>1T CCM<br>1999-2003<br>300M, Avenger, Breeze, Caravan, Cherokee, Cirrus, Concorde, PT Cruiser, Dakota, Durango, Grand Cherokee, Intrepid, LHS, Liberty, Neon, New Yorker, Prowler, Ram, Sebring, Stratus, Town & Country, Viper, Voyager, Wrangler | **A/T Additional Gear Ratio Error In Third Gear Conditions:**<br>Vehicle driven to a speed of 4-37 mph at more than 500 rpm, gear selector in 3rd Gear in Drive, and the PCM detected too much difference in the engine speed and output shaft speed.<br>**Possible Causes:**<br>O/D and U/D pressures less than 75 psi<br>O/D or U/D clutch volume indexes out of specification<br>O/D accumulator seals are damaged<br>L/R switch valve stuck in first gear position<br>Transmission has an internal leak (U/D seals leaking) |

**OBD II Trouble Code List (P0XXX Codes) - Continued**

| DTC | Trouble Code Title, Conditions & Possible Causes: |
|---|---|
| **P0733**<br>1T CCM<br>2004-05<br>300M, Caravan, Concorde, PT Cruiser, Dakota, Durango, Grand Cherokee, Intrepid, LHS, Liberty, Magnum, Neon, Pacifica, Prowler, Ram, Sebring, Stratus, Town & Country, Viper, Voyager, Wrangler | **A/T Additional Gear Ratio Error In Third Gear Conditions:**<br>The transmission gear ratio is monitored continuously while the transmission is in gear. If the ratio of the Input RPM to the Output RPM does not match the current gear ratio, this DTC will set.<br>**Possible Causes:**<br>Related DTCs will be present<br>Transmission solenoid/pressure switch assembly has malfunctioned or failed<br>Internal transmission mechanical problems may exist<br>Intermittent gear ratio errors are present |
| **P0734**<br>1T CCM<br>1999-2003<br>300M, Avenger, Breeze, Caravan, Cherokee, Cirrus, Concorde, PT Cruiser, Dakota, Durango, Grand Cherokee, Intrepid, LHS, Liberty, Neon, New Yorker, Prowler, Ram, Sebring, Stratus, Town & Country, Viper, Voyager, Wrangler | **A/T Additional Gear Ratio Error In Fourth Gear Conditions:**<br>Vehicle driven to a speed of 4-37 mph at more than 500 rpm, gear selector in 4th Gear in Drive, and the PCM detected too much difference in the engine speed and output shaft speed.<br>**Possible Causes:**<br>Customer may complain of intermittent 4th gear operation<br>O/D and 2/4 pressures less than 75 psi<br>O/D or 2/4 clutch volume indexes out of specification<br>O/D accumulator seals are damaged<br>Transmission enters "limp in" mode running in 4th gear |
| **P0734**<br>1T CCM<br>2004-05<br>300M, Avenger, Breeze, Caravan, Cherokee, Cirrus, Concorde, PT Cruiser, Dakota, Durango, Grand Cherokee, Intrepid, LHS, Liberty, Magnum, Neon, New Yorker, Prowler, Ram, Sebring, Stratus, Talon, Town & Country, Viper, Vision, Voyager, Wrangler | **A/T Additional Gear Ratio Error In Fourth Gear Conditions:**<br>The transmission gear ratio is monitored continuously while the transmission is in gear. If the ratio of the Input RPM to the Output RPM does not match the current gear ratio, this DTC will set.<br>**Possible Causes:**<br>Related DTCs will be present<br>Transmission solenoid/pressure switch assembly has malfunctioned or failed<br>Internal transmission mechanical problems may exist<br>Intermittent gear ratio errors are present |
| **P0735**<br>1T CCM<br>1999-2005<br>300M, Avenger, Breeze, Caravan, Cherokee, Cirrus, Concorde, PT Cruiser, Dakota, Durango, Grand Cherokee, Intrepid, LHS, Liberty, Magnum, Neon, New Yorker, Prowler, Ram, Sebring, Stratus, Talon, Town & Country, Viper, Vision, Voyager, Wrangler | **A/T Gear Ratio Error Fourth Prime Conditions:**<br>Vehicle driven any forward Gear, and the TCM detected the ratio of the Input speed to the Output Speed did not match the current Gear Ratio (this test can take up to 5 minutes).<br>**Possible Causes:**<br>Related Gear Ratio trouble codes may be stored (note that some of these Gear Ratio trouble codes may be intermittent)<br>Transmission has internal problems or damage present |

**OBD II Trouble Code List (P0XXX Codes) - Continued**

| DTC | Trouble Code Title, Conditions & Possible Causes: |
|---|---|
| **P0736**<br>1T CCM<br>1999-2003<br>300M, Avenger, Breeze, Caravan, Cherokee, Cirrus, Concorde, PT Cruiser, Dakota, Durango, Grand Cherokee, Intrepid, LHS, Liberty, Neon, New Yorker, Prowler, Ram, Sebring, Stratus, Town & Country, Viper, Voyager, Wrangler | **A/T Additional Gear Ratio Error In Reverse Gear Conditions:**<br>Vehicle driven in Reverse Gear at more than 500 rpm, and the PCM detected the ratio of the Input speed to the Output Speed did not match the current Gear Ratio during the test.<br>**Possible Causes:**<br>Customer may complain of intermittent Reverse gear operation<br>Related Gear Ratio trouble codes may be stored<br>Transmission has internal problems or damage present |
| **P0736**<br>1T CCM<br>2004-05<br>300M, Caravan, Concorde, PT Cruiser, Dakota, Durango, Grand Cherokee, Intrepid, LHS, Liberty, Magnum, Neon, Pacifica, Prowler, Ram, Sebring, Stratus, Town & Country, Viper, Voyager, Wrangler | **A/T Additional Gear Ratio Error In Reverse Gear Conditions:**<br>The transmission gear ratio is monitored continuously while the transmission is in gear. If the ratio of the Input RPM to the Output RPM does not match the current gear ratio, this DTC will set.<br>**Possible Causes:**<br>Related DTCs will be present<br>Internal transmission mechanical problems may exist<br>Intermittent gear ratio errors are present |
| **P0740**<br>1T CCM<br>1995-2003<br>300M, Avenger, Breeze, Caravan, Cherokee, Cirrus, Concorde, PT Cruiser, Dakota, Durango, Grand Cherokee, Intrepid, LHS, Liberty, Magnum, Neon, New Yorker, Prowler, Ram, Sebring, Stratus, Talon, Town & Country, Viper, Vision, Voyager, Wrangler | **A/T Torque Converter Clutch - No RPM Drop At Lockup Conditions:**<br>Engine started; vehicle driven at over 1750 rpm at 50 mph for 20 seconds, TCC command at 100%, and the PCM detected the engine speed was not within 60 rpm of ISS speed with the TP angle less than 30 degrees. The test must fail 3 times on one trip to set a code.<br>**Possible Causes:**<br>Internal transmission component fault (Input or Reaction shaft)<br>Internal transmission leakage (valve body leaking or seal rings)<br>Converter clutch valve or clutch timing valve is sticking<br>TCC assembly has failed |
| **P0740**<br>1T CCM<br>2001-03<br>Sebring, Stratus Coupe with 2.4L I4 or 3.0L V6 | **TCM Torque Converter Clutch System Failure Conditions:**<br>Engine started; and the TCM indicating it had detected a malfunction in the TCC system.<br>**Possible Causes:**<br>TCC solenoid circuit open, shorted to ground or power<br>TCC solenoid is damaged or it has failed<br>TCC System is damaged (possible mechanical fault present)<br>PCM has failed |

**OBD II Trouble Code List (P0XXX Codes) - Continued**

| DTC | Trouble Code Title, Conditions & Possible Causes: |
|---|---|
| **P0740**<br>1T CCM<br>2004-05<br>300M, Caravan, Concorde, PT Cruiser, Dakota, Durango, Grand Cherokee, Intrepid, LHS, Liberty, Magnum, Neon, Pacifica, Prowler, Ram, Sebring, Stratus, Town & Country, Viper, Voyager, Wrangler | **A/T Torque Converter Clutch System Out of Range Conditions:**<br>The TCC is in FEMCC or PEMCC, transmission temperature is hot, engine temperature is more than 100°F (38°C), transmission input speed is more than 1750 rpm, with TPS less than 30°. The TCC is modulated by controlling the duty cycle of the L/R solenoid, until the difference between the engine and transmission input speed rpm or duty cycle is within desired range. The DTC is set after the period of 10 seconds and 3 occurrences of either: FEMCC – with slip greater than 100 rpm or PEMCC – duty cycle is more than 85%.<br>**Possible Causes:**<br>Related DTCs will be present<br>Internal transmission mechanical problems may exist<br>Intermittent gear ratio errors are present |
| **P0740**<br>1T ECM<br>2004-05<br>Liberty with Diesel | **A/T Torque Converter Clutch System Out of Range Conditions:**<br>System is monitored during Electronically Modulated Converter Clutch (EMCC) operation. Transmission must be in EMCC, with input speed of more than 1750 rpm. This DTC will set when the TCC-L/R Solenoid achieves the maximum duty cycle and cannot pull engine speed within 60 rpm of input speed. Also, it will set when the transmission is in EMCC and the engine slips TCC more than 100 rpm for 10 seconds. This DTC can take up to 5 minutes of problem identification before lighting the MIL.<br>**Possible Causes:**<br>Related DTC 0750 is present<br>Internal transmission problem<br>Transmission Solenoid/TRS Assembly has failed<br>Intermittent wiring and connector problems |
| **P0743**<br>1T CCM<br>1995-2003<br>300M, Avenger, Breeze, Caravan, Cherokee, Cirrus, Concorde, PT Cruiser, Dakota, Durango, Grand Cherokee, Intrepid, LHS, Liberty, Neon, New Yorker, Prowler, Ram, Sebring, Stratus, Talon, Town & Country, Viper, Vision, Voyager, Wrangler | **A/T Torque Converter Clutch Solenoid Circuit Failure Conditions:**<br>Key on or engine started, and the PCM detected an unexpected voltage condition on the TCC solenoid control circuit during the test.<br>**Possible Causes:**<br>TCC power circuit is open between the solenoid and Fused B+<br>TCC solenoid control circuit is open or shorted to ground<br>TCC solenoid control circuit is shorted to system power<br>TCC solenoid is damaged or has failed<br>PCM has failed |
| **P0748**<br>1T CCM<br>1996-2003<br>Cherokee, Dakota, Durango, Grand Cherokee, Liberty, Ram, Wrangler | **A/T Governor Pressure Solenoid/Transmission Relay Fault Conditions:**<br>Engine started, and the PCM detected an unexpected voltage on the GPS control circuit. This solenoid is used to control governor pressure so the transmission can determine shift points.<br>**Possible Causes:**<br>A/T transmission relay power circuit is open (test power to B+)<br>A/T transmission relay is damaged or has failed<br>Governor pressure solenoid circuit is open or shorted to ground<br>Governor pressure solenoid is damaged or has failed<br>PCM has failed |

**OBD II Trouble Code List (P0XXX Codes) - Continued**

| DTC | Trouble Code Title, Conditions & Possible Causes: |
|---|---|
| **P0750**<br>1T CCM<br>1998-2003<br>300M, Avenger, Breeze, Caravan, Cherokee, Cirrus, Concorde, PT Cruiser, Dakota, Durango, Grand Cherokee, Intrepid, LHS, Liberty, Neon, New Yorker, Prowler, Ram, Sebring, Stratus, Town & Country, Viper, Voyager, Wrangler | **A/T Low/Reverse Solenoid Circuit Failure Conditions:**<br>Key on, and the PCM detected an unexpected voltage condition on the L/R Solenoid control circuit after the solenoid was enabled and disabled (the PCM checks for an inductive spike).<br>**Possible Causes:**<br>L/R solenoid control circuit is open or shorted to ground<br>L/R solenoid control circuit is shorted to system power<br>L/R Solenoid is damaged or has failed<br>TCM or PCM has failed |
| **P0750**<br>1T CCM<br>2004-05<br>300M, Caravan, Concorde, PT Cruiser, Dakota, Durango, Grand Cherokee, Intrepid, LHS, Liberty, Magnum, Neon, Pacifica, Prowler, Ram, Sebring, Stratus, Town & Country, Viper, Voyager, Wrangler | **A/T Low/Reverse Solenoid Circuit Failure Conditions:**<br>Solenoids are tested initially at power-up, then every 10 seconds thereafter, the solenoids will also be tested immediately after a gear ratio or pressure switch error is detected. 3 consecutive solenoid continuity test failures, or one failure if test is run in response to a gear ratio or pressure switch error.<br>**Possible Causes:**<br>Related relay DTCs present<br>Transmission control relay output circuit open<br>L/R solenoid control circuit open or shorted to ground or to voltage<br>L/R solenoid/pressure switch assembly has malfunctioned or failed<br>Intermittent wiring and connectors<br>PCM has failed |
| **P0750**<br>1T CCM<br>2001-03<br>Sebring, Stratus Coupe with 2.4L I4 or 3.0L V6 | **TCM Shift Solenoid 'A' Control Circuit Failure Conditions:**<br>Engine started; and the TCM detected a problem in the Low/Reverse or Shift Solenoid 'A' Control circuit during the CCM continuous test.<br>**Possible Causes:**<br>L/R solenoid control circuit open, shorted to ground or power<br>L/R solenoid power circuit is open (test power to Fused B+)<br>L/R control solenoid is damaged or has failed<br>TCM or the PCM has failed |
| **P0751**<br>1T CCM<br>1996-2003<br>300M, Avenger, Breeze, Caravan, Cherokee, Cirrus, Concorde, PT Cruiser, Dakota, Durango, Grand Cherokee, Intrepid, LHS, Liberty, Neon, New Yorker, Prowler, Ram, Sebring, Stratus, Talon, Town & Country, Viper, Vision, Voyager, Wrangler | **A/T Overdrive Switch Pressed Low For Over 5 Minutes Conditions:**<br>Engine started; engine runtime over 10 seconds and the PCM detected the Overdrive Off Switch indicated "low" for over 5 minutes.<br>**Possible Causes:**<br>Overdrive Switch Off" circuit is shorted to ground<br>Overdrive Switch is damaged or has failed<br>PCM has failed |

**OBD II Trouble Code List (P0XXX Codes) - Continued**

| DTC | Trouble Code Title, Conditions & Possible Causes: |
|---|---|
| **P0753**<br>1T CCM<br>1996-2003<br>Cherokee, Dakota, Durango, Grand Cherokee, Liberty, Ram, Wrangler | **A/T 3-4 Solenoid/Transmission Relay Circuit Failure Conditions:**<br>Key on or engine started, and the PCM detected an unexpected voltage condition on the 3-4 Solenoid control circuit during the test.<br>**Possible Causes:**<br>3-4 Solenoid control circuit is open or shorted to ground<br>3-4 Solenoid is damaged or has failed<br>Transmission relay power circuit is open (check the fused B+)<br>Transmission relay is damaged or has failed<br>PCM has failed |
| **P0755**<br>1T CCM<br>1996-2003<br>300M, Avenger, Breeze, Caravan, Cherokee, Cirrus, Concorde, PT Cruiser, Dakota, Durango, Grand Cherokee, Intrepid, LHS, Liberty, Neon, New Yorker, Prowler, Ram, Sebring, Stratus, Talon, Town & Country, Viper, Vision, Voyager, Wrangler | **A/T 2/4 Solenoid Circuit Failure Conditions:**<br>Key on or engine running; and the PCM detected an unexpected voltage condition on the2/4 Solenoid control circuit immediately the solenoid was energized or de-energized. The PCM checks for an inductive spike when the solenoid is turned "on" and then "off".<br>**Possible Causes:**<br>2/4 Solenoid control circuit is open or shorted to ground<br>2/4 Solenoid control circuit is shorted to system power<br>2/4 Solenoid is damaged or has failed<br>TCM or PCM has failed |
| **P0755**<br>1T CCM<br>2004-05<br>300M, Caravan, Concorde, PT Cruiser, Dakota, Durango, Grand Cherokee, Intrepid, LHS, Liberty, Magnum, Neon, Pacifica, Prowler, Ram, Sebring, Stratus, Town & Country, Viper, Voyager, Wrangler | **A/T 2/4 Solenoid Circuit Failure Conditions:**<br>2/4 solenoid in monitored initially at power-up, then every 10 seconds thereafter. Also, immediately<br>after a gear ratio or pressure switch error is detected. 3 consecutive solenoid continuity test failures, or one failure if test is run in response to a gear ratio or pressure switch error.<br>**Possible Causes:**<br>Related DTCs present<br>Transmission control relay output circuit open<br>2/4 Solenoid control circuit is open or shorted to ground<br>2/4 Solenoid control circuit is shorted to system power<br>2/4 Solenoid is damaged or has failed<br>Intermittent wiring or connector problems<br>PCM has failed |
| **P0755**<br>1T CCM<br>2001-03<br>Sebring, Stratus Coupe with 2.4L I4 or 3.0L V6 | **TCM Shift Solenoid 'B' Control Circuit Failure Conditions:**<br>Engine started; and the PCM detected a signal from the TCM indicating it had detected a problem in the Underdrive or Shift Solenoid 'B' Control circuit during the CCM test.<br>**Possible Causes:**<br>U/D solenoid control circuit open, shorted to ground or power<br>U/D solenoid power circuit is open (test power to Fused B+)<br>U/D control solenoid is damaged or has failed<br>TCM or the PCM has failed |

**OBD II Trouble Code List (P0XXX Codes) - Continued**

| DTC | Trouble Code Title, Conditions & Possible Causes: |
|---|---|
| **P0755**<br>1T ECM<br>2004-05<br>Liberty with Diesel | **A/T 2C Solenoid Circuit Malfunction Conditions:**<br>System is monitored initially at powerup and every 10 seconds thereafter. It will also be tested immediately after a gear ratio or pressure switch error is detected. After 3 consecutive solenoid continuity test failures, or after one failure if a test is run in response to a gear ratio or pressure switch error, this DTC will set.<br>**Possible Causes:**<br>Related Relay DTCs are present<br>Transmission Control Relay output circuit is open<br>2C Solenoid control circuit is open or is shorted to ground or to voltage<br>Transmission Solenoid/TRS Assembly has failed<br>TCM has failed<br>Intermittent wiring and connector problems |
| **P0756**<br>1T CCM<br>1999-2003<br>Cherokee, Grand Cherokee, Liberty, Wrangler | **A/T Shift Solenoid 'B' Mechanical Failure Conditions:**<br>DTC P1746 and P1747 not set, engine started, vehicle driven at cruise speed under heavy load conditions, and the PCM detected a problem while operating the Shift Solenoid 'B'.<br>**Possible Causes:**<br>ATF level is low, or the ATF fluid is contaminated<br>A/T SSB is damaged (mechanical fault) or has failed<br>Automatic transmission has internal damage or has failed |
| **P0760**<br>1T CCM<br>1999-2003<br>300M, Avenger, Breeze, Caravan, Cherokee, Cirrus, Concorde, PT Cruiser, Dakota, Durango, Grand Cherokee, Intrepid, LHS, Liberty, Neon, New Yorker, Prowler, Ram, Sebring, Stratus, Town & Country, Viper, Voyager, Wrangler | **A/T Overdrive Solenoid Circuit Failure Conditions:**<br>Engine started, and the PCM detected an unexpected voltage condition on the Overdrive solenoid control circuit. The solenoids are tested at power-up and then ever 10 seconds.<br>**Possible Causes:**<br>O/D Solenoid control circuit is open, shorted to ground or power<br>O/D Solenoid is damaged or has failed<br>TCM relay power circuit to O/D solenoid is open (loss of B+)<br>TCM or PCM has failed |
| **P0760**<br>1T CCM<br>2004-05<br>300M, Caravan, Concorde, PT Cruiser, Dakota, Durango, Grand Cherokee, Intrepid, LHS, Liberty, Magnum, Neon, Pacifica, Prowler, Ram, Sebring, Stratus, Town & Country, Viper, Voyager, Wrangler | **A/T Overdrive Solenoid Circuit Failure Conditions:**<br>O/D solenoid in monitored initially at power-up, then every 10 seconds thereafter. Also, immediately after a gear ratio or pressure switch error is detected. 3 consecutive solenoid continuity test failures, or one failure if test is run in response to a gear ratio or pressure switch error.<br>**Possible Causes:**<br>Related DTCs present<br>Transmission control relay output circuit open<br>O/D Solenoid control circuit is open or shorted to ground<br>O/D Solenoid control circuit is shorted to system power<br>O/D Solenoid is damaged or has failed<br>PCM has failed |
| **P0760**<br>1T CCM<br>2001-03<br>Sebring, Stratus Coupe with 2.4L I4 or 3.0L V6 | **TCM Shift Solenoid 'C' Control Circuit Failure Conditions:**<br>Engine started; and the PCM detected a signal from the TCM indicating it had detected a problem in the 2nd Solenoid or Shift Solenoid 'C' control circuit during the CCM test.<br>**Possible Causes:**<br>2nd solenoid control circuit open, shorted to ground or power<br>2nd solenoid power circuit is open (test power to Fused B+)<br>2nd control solenoid is damaged or has failed<br>TCM or the PCM has failed |

**OBD II Trouble Code List (P0XXX Codes) - Continued**

| DTC | Trouble Code Title, Conditions & Possible Causes: |
|---|---|
| **P0765**<br>1T CCM<br>1999-2003<br>300M, Avenger, Breeze, Caravan, Cherokee, Cirrus, Concorde, PT Cruiser, Dakota, Durango, Grand Cherokee, Intrepid, LHS, Liberty, Neon, New Yorker, Prowler, Ram, Sebring, Stratus, Town & Country, Viper, Voyager, Wrangler | **A/T Underdrive Solenoid Circuit Malfunction Conditions:**<br>Key on or engine running; and the PCM detected an unexpected voltage condition on the U/D Solenoid control circuit immediately after a Gear Ratio or Pressure Switch error was detected (this code can set due to a single or multiple circuit faults).<br>**Possible Causes:**<br>U/D Solenoid control circuit is open or shorted to ground<br>U/D Solenoid control circuit is shorted to system power<br>U/D Solenoid is damaged or has failed<br>TCM or PCM has failed |
| **P0765**<br>1T CCM<br>2004-05<br>300M, Caravan, Concorde, PT Cruiser, Dakota, Durango, Grand Cherokee, Intrepid, LHS, Liberty, Magnum, Neon, Pacifica, Prowler, Ram, Sebring, Stratus, Town & Country, Viper, Voyager, Wrangler | **A/T Underdrive Solenoid Circuit Failure Conditions:**<br>U/D solenoid in monitored initially at power-up, then every 10 seconds thereafter. Also, immediately after a gear ratio or pressure switch error is detected. 3 consecutive solenoid continuity test failures, or one failure if test is run in response to a gear ratio or pressure switch error.<br>**Possible Causes:**<br>Related DTCs present<br>Transmission control relay output circuit open<br>U/D Solenoid control circuit is open or shorted to ground<br>U/D Solenoid control circuit is shorted to system power<br>U/D Solenoid is damaged or has failed<br>Intermittent wiring or connector problems<br>PCM has failed |
| **P0770**<br>1T ECM<br>2004-05<br>Liberty with Diesel | **A/T 4C Solenoid Circuit Malfunction Conditions:**<br>System is monitored initially at powerup and every 10 seconds thereafter. It will also be tested immediately after a gear ratio or pressure switch error is detected. After 3 consecutive solenoid continuity test failures, or after one failure if a test is run in response to a gear ratio or pressure switch error, this DTC will set.<br>**Possible Causes:**<br>Related Relay DTCs are present<br>Transmission Control Relay output circuit is open<br>4C Solenoid control circuit is open or is shorted to ground or to voltage<br>Transmission Solenoid/TRS Assembly has failed<br>TCM has failed<br>Intermittent wiring and connector problems |
| **P0781**<br>1T CCM<br>2004-05<br>PT Cruiser | **A/T Low/Reverse Pressure Switch Circuit Failure Conditions:**<br>Monitored whenever engine is running. If one of the pressure switches are open or closed at the wrong time, this DTC will set.<br>**Possible Causes:**<br>Related DTCs present<br>Loss of Prime P0944 DTC present<br>L/R pressure switch sense circuit is open or is shorted to ground or to voltage<br>L/R pressure switch has failed<br>Intermittent wiring or connector problems<br>PCM has failed |

## OBD II Trouble Code List (P0XXX Codes) - Continued

| DTC | Trouble Code Title, Conditions & Possible Causes: |
|---|---|
| **P0783**<br>1T CCM<br>1996-2003<br>300M, Avenger, Breeze, Caravan, Cherokee, Cirrus, Concorde, PT Cruiser, Dakota, Durango, Grand Cherokee, Intrepid, LHS, Liberty, Neon, New Yorker, Prowler, Ram, Sebring, Stratus, Town & Country, Viper, Voyager, Wrangler | **A/T 3-4 Shift Solenoid - No RPM Drop At Lockup Conditions:**<br>Engine started; vehicle driven to over 30 mph, and the PCM did not detect the correct amount of rpm drop after a gear change occurred.<br>**Possible Causes:**<br>ATF fluid level is burnt, contaminated or low<br>3-4 Solenoid control circuit is open or shorted (intermittent fault)<br>3-4 Solenoid is damaged or has failed<br>Transmission oil pan has excessive debris |
| **P0801**<br>1T CCM<br>1996-2005<br>Viper models | **Reverse Gear Lockout Control Circuit Malfunction Conditions:**<br>Vehicle driven at over 5 mph at an engine speed over 608 rpm, and the PCM detected an unexpected "low" or high voltage condition on the Reverse Gear Lockout solenoid circuit.<br>**Possible Causes:**<br>Reverse gear lockout solenoid circuit open or shorted to ground<br>Reverse gear lockout solenoid is damaged or has failed<br>PCM has failed |
| **P0830**<br>1T CCM<br>1999-2003<br>Ram with M/T | **Clutch Pedal Depressed Circuit Malfunction Conditions:**<br>Engine cranking or vehicle driven to over 15 mph at an engine speed of 1550-2880 rpm with the delta throttle at over 1.1v for 4 seconds, and the PCM detected an unexpected voltage condition on the Clutch Pedal switch circuit (test must fail 5 times during one trip).<br>**Possible Causes:**<br>CPP switch signal circuit is open or shorted to ground<br>CPP switch power circuit is open (test the power at the PDC)<br>Clutch pedal is damaged or has failed<br>PCM has failed |
| **P0833**<br>1T CCM<br>1999-2003<br>Ram with M/T | **Clutch Pedal Released Circuit Malfunction Conditions:**<br>Engine cranking or vehicle driven to over 15 mph at an engine speed of 1550-2880 rpm with the delta throttle at over 1.1v for 4 seconds, and the PCM detected an unexpected voltage condition on the Clutch Pedal switch circuit (test must fail 5 times during one trip).<br>**Possible Causes:**<br>CPP switch signal circuit is open or shorted to ground<br>CPP switch power circuit is open (test the power at the PDC)<br>Clutch pedal is damaged or has failed<br>PCM has failed |
| **P0833**<br>1T CCM<br>2001-05<br>PT Cruiser, Neon with M/T | **Clutch Pedal Position Switch Circuit Malfunction Conditions:**<br>Engine cranking, or vehicle speed over 25 mph at an engine speed from 1550-2880 rpm with the delta throttle over 1.1v for 4 seconds, and the PCM detected an unexpected voltage condition on the Clutch Pedal switch circuit (test must fail 5 times during one trip).<br>**Possible Causes:**<br>CPP switch signal circuit is open or shorted to ground<br>CPP switch power circuit is open (test the power at the PDC)<br>Clutch pedal is damaged or has failed<br>PCM has failed |

**OBD II Trouble Code List (P0XXX Codes) - Continued**

| DTC | Trouble Code Title, Conditions & Possible Causes: |
|---|---|
| **P0833**<br>1T CCM<br>2004-05<br>PT Cruiser | **Clutch Pedal Position Switch Circuit Malfunction Conditions:**<br>Condition is monitored with the engine running; battery voltage over 9v; vehicle speed is less than 0 mph, then more than 27 mph; MAP is greater than 90mbar; and engine speed is over 3200 rpm. If the Clutch Switch status did not change for more than 15 seconds, this DTC will set.<br>**Possible Causes:**<br>Intermittent condition<br>CPP switch signal circuit is open or shorted to ground<br>CPP switch ground circuit is open<br>Clutch pedal is damaged or has failed<br>PCM has failed |
| **P0836**<br>1T ECM<br>2004-05<br>Liberty with Diesel | **Transfer Case Position Sensor Plausibility Or Improper Voltage Signal Conditions:**<br>Ignition on. If the ECM detects a voltage signal from the Transfer Case Switch that does not fall into a valid switch position voltage range, or if the Position Sensor signal is above 4.8v or below 0.14v for 0.5 second, this DTC will set.<br>**Possible Causes:**<br>Transfer Case Position Sensor has failed<br>Intermittent wiring and/or connector problems<br>Transfer Case Position Sensor signal circuit is open, shorted to ground, shorted to voltage or shorted to Sensor ground circuit<br>ECM has failed |
| **P0841**<br>1T CCM<br>2001-03<br>300M, Caravan, Cherokee, Concorde, PT Cruiser, Dakota, Durango, Grand Cherokee, Intrepid, LHS, Liberty, Neon, New Yorker, Prowler, Ram, Sebring, Stratus, Town & Country, Viper, Voyager, Wrangler | **Low/Reverse Pressure Switch Sense Circuit Malfunction Conditions:**<br>Engine started; vehicle driven in a forward gear, and the PCM detected an unexpected voltage condition on the Low/Reverse Pressure Switch Sense circuit during the CCM test.<br>**Possible Causes:**<br>L/R pressure switch sense circuit is open or shorted to ground<br>L/R pressure switch sense circuit is shorted to power (B+)<br>L/R pressure switch is damaged or it has failed<br>TCM relay power circuit to L/R switch is open (loss of B+)<br>TCM has failed |
| **P0841**<br>1T CCM<br>2004-05<br>300M, Caravan, Concorde, PT Cruiser, Dakota, Durango, Grand Cherokee, Intrepid, LHS, Liberty, Magnum, Neon, Pacifica, Prowler, Ram, Sebring, Stratus, Town & Country, Viper, Voyager, Wrangler | **Low/Reverse Pressure Switch Sense Circuit Malfunction Conditions:**<br>Switches are monitored whenever engine is running. This DTC will set if one of the pressure switches in open or closed at the wrong time in a given gear.<br>**Possible Causes:**<br>Related DTCs present<br>Loss of Prime P0944 DTC present<br>Transmission control relay output circuit open<br>L/R switch sense circuit is open or is shorted to ground or to voltage<br>L/R pressure switch is damaged or has failed<br>Intermittent wiring or connector problems<br>PCM has failed |

**OBD II Trouble Code List (P0XXX Codes) - Continued**

| DTC | Trouble Code Title, Conditions & Possible Causes: |
|---|---|
| **P0845**<br>1T CCM<br>2001-05<br>300M, Caravan, Cherokee, Concorde, PT Cruiser, Dakota, Durango, Grand Cherokee, Intrepid, LHS, Liberty, Magnum, Neon, New Yorker, Pacifica, Prowler, Ram, Sebring, Stratus, Town & Country, Viper, Vision, Voyager, Wrangler | **A/T 2/4 Hydraulic Pressure Test Malfunction Conditions:**<br>Engine speed over 1000 rpm, then immediately after a shift event, the PCM detected a failure in one or more of the Pressure Switch circuits (i.e., it tests switches that are not operating).<br>**Possible Causes:**<br>2/4 pressure is incorrect, or internal transmission faults exist<br>2/4 pressure switch circuit is open, shorted to ground or power<br>2/4 pressure switch is damaged or it has failed<br>Transmission solenoids/TRS assembly is damaged or have failed<br>TCM relay power circuit to 2/4 switch is open (loss of B+)<br>Intermittent wiring or connector problems exist<br>PCM or TCM has failed |
| **P0845**<br>1T ECM<br>2004-05<br>Liberty with Diesel | **A/T 2C Hydraulic Pressure Test Failure Conditions:**<br>System hydraulic pressure is monitored in any forward gear with engine speed above 1000 rpm, shortly after a shift, and every minute thereafter. After a shift into a forward gear, with engine speed above 1000 rpm, the TCM momentarily turns ON element pressure to the Clutch circuits that don't have pressure, in order to identify the correct Pressure Switch closes. If the Pressure Switch does not close 2 times, the DTC will set.<br>**Possible Causes:**<br>Related Relay DTCs are present<br>Transmission Solenoid/TRS Assembly has failed<br>2C Pressure Switch sense circuit is open or is shorted to ground or to voltage<br>5v supply circuit is open or is shorted to ground<br>Pressure Sensor has poor line connection<br>Transmission Control Relay output circuit is open<br>Excessive debris in oil pan<br>Line Pressure Sensor has failed<br>Internal transmission problems<br>TCM has failed<br>Intermittent wiring and connector problems |
| **P0846**<br>1T CCM<br>2001-05<br>300M, Caravan, Cherokee, Concorde, PT Cruiser, Dakota, Durango, Grand Cherokee, Intrepid, LHS, Liberty, Magnum, Neon, New Yorker, Pacifica, Prowler, Ram, Sebring, Stratus, Town & Country, Viper, Vision, Voyager, Wrangler | **A/T 2/4 Pressure Switch Circuit Malfunction Conditions:**<br>Engine started; vehicle driven in a forward gear, and the PCM detected that the 2/4 Pressure Switch circuit indicated open or closed at the wrong time. Related relay DTCs may be present.<br>**Possible Causes:**<br>2/4 pressure is incorrect, or internal transmission faults exist<br>2/4 pressure switch circuit is open, shorted to ground or power<br>2/4 pressure switch is damaged or it has failed<br>TCM relay power circuit to L/R switch is open (loss of B+)<br>PCM/TCM has failed |

**OBD II Trouble Code List (P0XXX Codes) - Continued**

| DTC | Trouble Code Title, Conditions & Possible Causes: |
|---|---|
| **P0846**<br>1T ECM<br>2004-05<br>Liberty with Diesel | **A/T 2C Pressure Switch Sense Circuit Malfunction Conditions:**<br>Pressure Switch circuit is monitored whenever engine is running. The appropriate DTC is set if one of the Pressure Switches is open or closed at the wrong time for a given gear.<br>**Possible Causes:**<br>Related Relay DTCs are present<br>Transmission Solenoid/TRS Assembly has failed<br>2C Pressure Switch sense circuit is open or is shorted to ground or to voltage<br>2C Pressure Switch has failed<br>TCM has failed<br>Intermittent wiring and connector problems |
| **P0850**<br>2T CCM<br>2001-05<br>300M, Caravan, Cherokee, Concorde, PT Cruiser, Dakota, Durango, Grand Cherokee, Intrepid, LHS, Liberty, Magnum, Neon, New Yorker, Pacifica, Prowler, Ram, Sebring, Stratus, Town & Country, Viper, Vision, Voyager, Wrangler | **A/T Park/Neutral Switch Performance Conditions:**<br>Engine running; gearshift selector in Park, Neutral or Drive position (not "Limp-In" mode). The PCM detected an invalid Park/Neutral switch state during vehicle operation.<br>**Possible Causes:**<br>Check for any TCM related codes stored in the TCM controller<br>PCM has failed |
| **P0864**<br>1T ECM<br>2004-05<br>Liberty with Diesel | **TCM Torque Reduction Signal Error Conditions:**<br>Ignition on or engine running. If the TCM receives an improper or implausible Torque Management Request signal, this DTC will set.<br>**Possible Causes:**<br>ECM has failed<br>Torque Management Request signal circuit is open, shorted to ground, shorted to voltage<br>TCM has failed<br>Intermittent condition |
| **P0867**<br>1T CCM<br>2002-03<br>Grand Cherokee, Liberty, Ram, Wrangler | **A/T Line Pressure Malfunction Conditions:**<br>Engine started; vehicle driven, and the PCM detected the Actual Line Pressure was not within 10 psi of the Desired Line pressure.<br>**Possible Causes:**<br>Check for related trouble codes<br>Check TCM connectors for loose or damaged terminals<br>A/T line pressure is out of range<br>TCM line pressure is out of range |
| **P0868**<br>1T CCM<br>2002-03<br>Grand Cherokee, Liberty, Ram, Wrangler | **A/T Line Pressure Low Conditions:**<br>Engine started; vehicle driven, and the PCM detected the Actual Line Pressure indicated 10 psi less than the Desired Line pressure.<br>**Possible Causes:**<br>Check for related trouble codes<br>Check TCM connectors for loose or damaged terminals<br>A/T line pressure is out of range<br>TCM line pressure is out of range |

**OBD II Trouble Code List (P0XXX Codes) - Continued**

| DTC | Trouble Code Title, Conditions & Possible Causes: |
|---|---|
| **P0868**<br>1T ECM<br>2004-05<br>Liberty with Diesel | **A/T Line Pressure Low Conditions:**<br>Line pressure is monitored whenever driving in a forward gear. The TCM continuously monitors the Transducer Line Pressure output and compares it to a desired line pressure. If the actual pressure is more than 10 psi below the desired line pressure, the DTC will set in about 2.1 seconds.<br>**Possible Causes:**<br>Related Relay DTCs may be present<br>5v supply circuit is open or is shorted to ground  or to voltage<br>Poor Line Pressure Sensor connection<br>Pressure Control Solenoid control circuit is shorted to voltage<br>Internal transmission problems<br>Line Pressure Sensor has failed<br>Plugged filter<br>TCM has failed<br>Intermittent wiring and connector problems |
| **P0869**<br>1T CCM<br>2002-03<br>Grand Cherokee, Liberty, Ram, Wrangler | **A/T Line Pressure High Conditions:**<br>Engine started; vehicle driven, and the PCM detected the Actual Line Pressure indicated 10 psi more than the Desired Line pressure.<br>**Possible Causes:**<br>Check for related trouble codes<br>Check TCM connectors for loose or damaged terminals<br>A/T line pressure is out of range<br>TCM line pressure is out of range |
| **P0869**<br>1T ECM<br>2004-05<br>Liberty with Diesel | **A/T Line Pressure High Conditions:**<br>Line pressure is monitored whenever driving in a forward gear. The TCM continuously monitors the Transducer Line Pressure output and compares it to a desired line pressure. If the actual pressure is more than the highest desired line pressure ever used in the current gear, while the Pressure Control Solenoid duty cycle is at or near its maximum value (which should result in minimum line pressure), the DTC will set.<br>**Possible Causes:**<br>Related Relay DTCs may be present<br>5v supply circuit is open or is shorted to ground<br>Poor Line Pressure Sensor connection<br>Pressure Control Solenoid control circuit is open or is shorted to ground<br>Internal transmission problems (line pressure high)<br>Line Pressure Sensor has failed<br>TCM has failed<br>Intermittent wiring and connector problems |
| **P0870**<br>1T CCM<br>2002-05<br>300M, Caravan, Cherokee, Concorde, PT Cruiser, Dakota, Durango, Grand Cherokee, Intrepid, LHS, Liberty, Magnum, Neon, New Yorker, Pacifica, Prowler, Ram, Sebring, Stratus, Town & Country, Viper, Vision, Voyager, Wrangler | **A/T Hydraulic Pressure Line Malfunction Conditions:**<br>Engine started; vehicle driven at over 1000 rpm, then immediately after a shift, the PCM detected a malfunction in one or more of the Pressure Switch circuits (it detected the switch did not close twice). DTC P0944 may be present.<br>**Possible Causes:**<br>Check for related line pressure trouble codes<br>Check for related speed ratio and pressure switch codes<br>5v supply circuit is open or is shorted to ground<br>Transmission Control Relay output circuit is open<br>OD Pressure Switch sense circuit is shorted to ground or to voltage<br>Excessive debris in the oil pan<br>Line pressure Sensor connector is loose or damaged<br>Oil pressure switch is damaged or has failed<br>Intermittent wiring or connector problems exist<br>PCM or TCM has failed |

**OBD II Trouble Code List (P0XXX Codes) - Continued**

| DTC | Trouble Code Title, Conditions & Possible Causes: |
|---|---|
| **P0871**<br>1T CCM<br>2002-05<br>300M, Caravan, Cherokee, Concorde, PT Cruiser, Dakota, Durango, Grand Cherokee, Intrepid, LHS, Liberty, Magnum, Neon, New Yorker, Pacifica, Prowler, Ram, Sebring, Stratus, Town & Country, Viper, Vision, Voyager, Wrangler | **A/T O/D Pressure Switch Circuit Malfunction Conditions:**<br>Engine started; vehicle driven in a forward gear, and the PCM detected that the O/D Pressure Switch circuit indicated open or closed at the wrong time.<br>**Possible Causes:**<br>Related DTCs may be present<br>O/D pressure is incorrect, or internal transmission faults exist<br>O/D pressure switch circuit is open, shorted to ground or power<br>O/D pressure switch is damaged or it has failed<br>TCM relay power circuit to O/D switch is open (loss of B+)<br>Intermittent wiring or connector problems exist<br>PCM or TCM has failed |
| **P0875**<br>1T CCM<br>2002-05<br>Grand Cherokee, Liberty, Ram, Wrangler | **A/T U/D Hydraulic Pressure Test Malfunction Conditions:**<br>Engine speed over 1000 rpm, then immediately after a shift event, the TCM/PCM detected a fault in one or more of the Pressure Switch circuits (it detected the switch did not close twice).<br>**Possible Causes:**<br>Check for related line pressure trouble codes<br>Check for related speed ratio and pressure switch codes<br>5v supply circuit is open or shorted to ground<br>U/D Pressure switch sense circuit is open or shorted to ground or to voltage<br>Excessive debris in the oil pan<br>Line pressure Sensor connector is loose or damaged<br>U/D Oil Pressure Switch is damaged or it has failed<br>TCM has failed<br>Intermittent wiring and connector problems |
| **P0876**<br>1T CCM<br>2002-05<br>Grand Cherokee, Liberty, Ram, Wrangler | **A/T U/D Pressure Switch Sense Malfunction Conditions:**<br>Engine started; vehicle driven in a forward gear, and the PCM detected that the U/D Pressure Switch Sense circuit indicated open or closed at the wrong time during the CCM test period.<br>**Possible Causes:**<br>U/D pressure is incorrect, or internal transmission faults exist<br>U/D pressure switch circuit is open, shorted to ground or power<br>U/D pressure switch is damaged or it has failed<br>TCM relay power circuit to L/R switch is open (loss of B+)<br>TCM has failed |
| **P0884**<br>1T CCM<br>2002-03<br>Grand Cherokee, Liberty, Ram, Wrangler | **A/T U/D Pressure Switch Sense Malfunction Conditions:**<br>Engine started; vehicle driven in a forward gear, and the PCM detected that the U/D Pressure Switch Sense circuit indicated open or closed at the wrong time during the CCM test period.<br>**Possible Causes:**<br>U/D pressure is incorrect, or internal transmission faults exist<br>U/D pressure switch circuit is open, shorted to ground or power<br>U/D pressure switch is damaged or it has failed<br>TCM relay power circuit to L/R switch is open (loss of B+)<br>TCM has failed |

## OBD II Trouble Code List (P0XXX Codes) - Continued

| DTC | Trouble Code Title, Conditions & Possible Causes: |
|---|---|
| **P0884**<br>1T CCM<br>2002-05<br>300M, Caravan, Cherokee, Concorde, PT Cruiser, Dakota, Durango, Grand Cherokee, Intrepid, LHS, Liberty, Magnum, Neon, New Yorker, Pacifica, Prowler, Ram, Sebring, Stratus, Town & Country, Viper, Vision, Voyager, Wrangler | **Power-Up Automatic Transmission Speed Malfunction Conditions:**<br>Engine started, TCM relay enabled; and the TCM detected a valid forward gear PNDRL signal with the Output Speed more than 800 rpm indicating a vehicle speed of over 20 mph. Note: The TCM has separate powers and grounds specifically to its portion of the PCM.<br>**Possible Causes:**<br>TCM power supply circuit to direct battery is open<br>TCM power supply circuit to the ignition switch is open<br>TCM power ground circuit is open or the connector is loose<br>TCM has failed |
| **P0888**<br>1T CCM<br>2004-05<br>300M, Caravan, Concorde, PT Cruiser, Dakota, Durango, Grand Cherokee, Intrepid, LHS, Liberty, Magnum, Neon, Pacifica, Prowler, Ram, Sebring, Stratus, Town & Country, Viper, Voyager, Wrangler | **A/T Relay Output Malfunction Conditions:**<br>Engine started, TCM relay enabled and monitored continuously. This DTC sets when less than 3v are present at the relay output circuits at the TCM when the TCM is energizing the relay. Note: Due to the integration of the PCM and TCM, the transmission part of the PCM has its own specific power and ground circuits.<br>**Possible Causes:**<br>Fused B+ circuit is open<br>TC relay output circuit is open or is shorted to ground<br>TC relay control circuit is open or is shorted to ground<br>TC relay ground circuit is open<br>TC relay has failed<br>Intermittent wiring or connector problems exist<br>Transmission solenoid/pressure switch assembly has malfunctioned or failed<br>PCM/TCM has failed |
| **P0890**<br>1T CCM<br>2001-05<br>Concorde, Intrepid, LHS, Neon, Sebring, Stratus, PT Cruiser, Pacifica, Caravan/Town & Country/Voyager, Jeep Liberty, Dakota, Durango, Ram | **A/T TCM Switched Battery Circuit Malfunction Conditions:**<br>Ignition switch position is changed from one position to another. TCM relay "not" energized, and the TCM detected voltage present at any of the Pressure Switch input circuits. Note: Due to the integration of the PCM and TCM, the transmission part of the PCM has its own specific power and ground circuits.<br>**Possible Causes:**<br>2/4 switch circuit is shorted to system power (B+)<br>L/R switch circuit is shorted to system power (B+)<br>O/D switch circuit is shorted to system power (B+)<br>TCM switched battery circuit is damaged<br>Intermittent wiring or connector problems exist<br>PCM/TCM has failed |
| **P0891**<br>1T CCM<br>2001-05<br>300M, Caravan, Cherokee, Concorde, PT Cruiser, Dakota, Durango, Grand Cherokee, Intrepid, LHS, Liberty, Magnum, Neon, New Yorker, Pacifica, Prowler, Ram, Sebring, Stratus, Town & Country, Viper, Vision, Voyager, Wrangler | **A/T TCM Relay Always On Conditions:**<br>Key on or engine cranking; TCM relay "not" energized, and the TCM detected voltage present at the TCM output circuit during the test. Note: Due to the integration of the PCM and TCM, the transmission part of the PCM has its own specific power and ground circuits.<br>**Possible Causes:**<br>TCM relay output circuit is shorted to system power (B+)<br>TCM relay control circuit is shorted to system power (B+)<br>TCM relay is damaged or it has failed (it may be stuck closed)<br>Intermittent wiring or connector problems exist<br>PCM/TCM has failed |

**OBD II Trouble Code List (P0XXX Codes) - Continued**

| DTC | Trouble Code Title, Conditions & Possible Causes: |
|---|---|
| **P0897**<br>1T CCM<br>2001-05<br>Concorde, Intrepid, LHS, Neon, Sebring, Stratus, PT Cruiser, Pacifica, Caravan/Town & Country/Voyager, Jeep, Dakota, Durango, Ram | **A/T Transmission Fluid Burnt Or Worn Out Conditions:**<br>Engine started; vehicle driven, and immediately after a transition from full TCC lockup to partial TCC engagement (for A/C bump prevention), the TCM detected vehicle shutter during engagement.<br>**Possible Causes:**<br>Automatic transmission fluid is burnt or contaminated<br>Automatic transmission fluid is worn out |
| **P0932**<br>1T CCM<br>2002-03<br>Cherokee, Grand Cherokee, Liberty, Ram, Wrangler | **A/T Line Pressure Malfunction Conditions:**<br>Engine started; vehicle driven in any forward gear, and the TCM detected the Line Pressure Sensor signal was less than 0.20v or more than 4.75v during the CCM test period.<br>**Possible Causes:**<br>Line pressure Sensor connector is damaged or loose<br>Line pressure Sensor circuit is open or shorted to ground<br>Line pressure Sensor circuit is shorted to VREF (5v)<br>Line pressure Sensor ground circuit is open<br>Line pressure Sensor supply circuit (5v) is open or missing<br>TCM Line Pressure Low or High circuit is open |
| **P0932**<br>1T CCM<br>2004-05<br>Jeep Liberty with Diesel | **A/T Line Pressure Sensor Malfunction Conditions:**<br>Sensor is monitored continuously while driving in a forward gear. T PCM continuously monitors actual line pressure and compares it to desired line pressure. If the actual pressure is more than 25 psi higher than the desire line pressure, but is less than the highest line pressure ever used in the current gear, this DTC will set. Related DTCs may be present.<br>**Possible Causes:**<br>Poor line pressure connection<br>Poor wiring connection<br>Internal transmission problems<br>TCM has failed<br>Intermittent wiring and connector problems |
| **P0934**<br>1T CCM<br>2004-05<br>Jeep Liberty with Diesel | **A/T Line Pressure Sensor Low Conditions:**<br>Sensor is monitored continuously while engine is running and Output Speed is more than 390 rpm. This DTC will set when the Line Pressure Sensor output signal is less than 0.35v for 1.4 seconds.<br>**Possible Causes:**<br>5v supply circuit is open or is shorted to ground<br>Line Pressure Sensor signal circuit is shorted to ground<br>Line Pressure Sensor has failed<br>TCM has failed<br>Intermittent wiring and connector problems |
| **P0935**<br>1T CCM<br>2004-05<br>Jeep Liberty with Diesel | **A/T Line Pressure Sensor High Conditions:**<br>Sensor is monitored continuously while engine is running and Output Speed is more than 390 rpm and desired line pressure is less than 200 psi. This DTC will set when the Line Pressure Sensor output signal is more than 4.75v for 1.4 seconds.<br>**Possible Causes:**<br>Line Pressure Sensor ground circuit is open<br>Line Pressure Sensor signal circuit is open or is shorted to voltage<br>Line Pressure Sensor has failed<br>TCM has failed<br>Intermittent wiring and connector problems |

**OBD II Trouble Code List (P0XXX Codes) - Continued**

| DTC | Trouble Code Title, Conditions & Possible Causes: |
|---|---|
| **P0944**<br>1T CCM<br>2002-05<br>300M, Caravan, Cherokee, Concorde, PT Cruiser, Dakota, Durango, Grand Cherokee, Intrepid, LHS, Liberty, Magnum, Neon, New Yorker, Pacifica, Prowler, Ram, Sebring, Stratus, Town & Country, Viper, Vision, Voyager, Wrangler | **A/T Loss Of Prime Pressure Conditions:**<br>Engine started; vehicle driven, and immediately after a slipping condition is detected with the pressure switches "not" indicating pressure, the PCM detected a loss of prime pressure. In effect, the TCM turns "on" available elements to detect if prime pressure exists. The DTC sets if no pressure switches respond.<br>**Possible Causes:**<br>A/T pressure switch connector is damaged, loose or shorted<br>Invalid PRNDL code (shift lever position error)<br>Automatic transmission fluid level is too low<br>Transmission oil filter is clogged or severely restricted<br>Transmission oil pump is damaged or weak<br>Intermittent wiring or connector problems exist |
| **P0951**<br>1T CCM<br>2001-03<br>300M, Caravan, Cherokee, Concorde, PT Cruiser, Dakota, Durango, Grand Cherokee, Intrepid, LHS, Liberty, Neon, New Yorker, Prowler, Ram, Sebring, Stratus, Town & Country, Viper, Voyager, Wrangler | **A/T AutoStick Sensor Circuit Malfunction Conditions:**<br>Engine started; vehicle driven, transmission not in AutoStick position, and the TCM that either the Upshift or Downshift switch was closed, or if the Upshift and Downshift switches are closed at the same time.<br>**Possible Causes:**<br>AutoStick assembly is damaged or has failed<br>AutoStick connector is damaged, loose or shorted<br>Downshift switch circuit is shorted too ground<br>Upshift switch circuit is shorted to ground |
| **P0952**<br>1T CCM<br>2004-05<br>300M, Caravan, Concorde, PT Cruiser, Dakota, Durango, Grand Cherokee, Intrepid, LHS, Liberty, Magnum, Neon, Pacifica, Prowler, Ram, Sebring, Stratus, Town & Country, Viper, Voyager, Wrangler | **A/T AutoStick Sensor Circuit Malfunction Conditions:**<br>Engine started; vehicle driven, transmission not in AutoStick position, and the TCM that either the Upshift or Downshift switch was closed (below 0.3v), or if both the Upshift and Downshift switches are closed at the same time.<br>**Possible Causes:**<br>AutoStick assembly is damaged or has failed<br>Intermittent wiring or connector problems exist<br>Downshift sense or Upshift sense circuit is shorted to ground<br>PCM/TCM has failed |
| **P0953**<br>1T CCM<br>2004-05<br>300M, Caravan, Concorde, PT Cruiser, Dakota, Durango, Grand Cherokee, Intrepid, LHS, Liberty, Magnum, Neon, Pacifica, Prowler, Ram, Sebring, Stratus, Town & Country, Viper, Voyager, Wrangler | **A/T AutoStick Sensor Circuit High Conditions:**<br>The AutoStick circuit is checked every .007 second, with the ignition on and in both AutoStick and non-AutoStick modes. If the TCM detects circuit voltage rises above 4.8v, this DTC will set.<br>**Possible Causes:**<br>AutoStick assembly is damaged or has failed<br>Intermittent wiring or connector problems exist<br>Downshift sense or Upshift sense circuit is shorted to ground<br>PCM/TCM has failed |

## OBD II Trouble Code List (P0XXX Codes) - Continued

| DTC | Trouble Code Title, Conditions & Possible Causes: |
|---|---|
| **P0987**<br>1T CCM<br>2002-05<br>Cherokee, Grand Cherokee, Liberty, Ram, Wrangler | **A/T 4C Hydraulic Pressure Test Malfunction Conditions:**<br>Engine started; vehicle driven at over 1000 rpm, then immediately after a shift and every minute thereafter. After a shift into a forward gear, with engine speed more than 1000 rpm, the TCM momentarily turns on element pressure to the clutch circuits that don't have pressure to identify the correct pressure switch that closes. If the pressure switch does not close 2 times, this DTC will set. Related line pressure DTCs are present.<br>**Possible Causes:**<br>Check for related speed ratio and pressure switch codes<br>Excessive debris in the oil pan<br>5v supply circuit is open or is shorted to ground<br>Transmission Control Relay output circuit is open<br>4C Pressure Switch sense circuit is shorted to ground or to voltage<br>Line Pressure Sensor connector is loose or damaged<br>4C Line Pressure Sensor has failed<br>Transmission Solenoid/TRS Assembly has failed<br>Internal transmission problems<br>TCM has failed<br>Intermittent wiring and connector problems |
| **P0988**<br>1T CCM<br>2002-05<br>Cherokee, Grand Cherokee, Liberty, Ram, Wrangler | **A/T 4C Pressure Switch Sense Circuit Malfunction Conditions:**<br>Engine started; vehicle driven in any forward gear, and the PCM detected that the 4C Pressure Switch circuit indicated open or closed at the wrong time during the CCM test. Related Relay DTCs are present.<br>**Possible Causes:**<br>4C pressure is incorrect, or internal transmission faults exist<br>4C Pressure Switch sense circuit is open, shorted to ground or power<br>4C Pressure Switch is damaged or it has failed<br>TCM Relay power circuit to L/R switch is open (loss of B+)<br>PCM/TCM has failed<br>Intermittent wiring and connector problems |
| **P0992**<br>1T CCM<br>2001-05<br>300M, Caravan, Cherokee, Concorde, PT Cruiser, Dakota, Durango, Grand Cherokee, Intrepid, LHS, Liberty, Magnum, Neon, New Yorker, Pacifica, Prowler, Ram, Sebring, Stratus, Town & Country, Viper, Vision, Voyager, Wrangler | **A/T 2/4 & O/D Hydraulic Pressure Test Malfunction Conditions:**<br>Engine started; vehicle driven at over 1000 rpm, then immediately after a shift, the PCM detected a malfunction in one or more of the Pressure Switch circuits (it tests the switches that are not operating). If the pressure switch does not close 2 times, the DTC will set.<br>**Possible Causes:**<br>2/4 pressure switch circuit is open, shorted to ground or power<br>2/4 pressure switch is damaged or it has failed<br>O/D pressure switch circuit is open, shorted to ground or power<br>O/D pressure switch is damaged or it has failed<br>Internal transmission faults exist<br>TCM relay power circuit to 2/4 or O/D switch open (loss of B+)<br>PCM/TCM has failed |

## OBD II Trouble Code List (P1XXX Codes)

| DTC | Trouble Code Title, Conditions & Possible Causes: |
|---|---|
| **P1101**<br>1T ECM<br>2004-05<br>Liberty with Diesel | **ACM Crash Signal Received Conditions:**<br>Ignition on. If crash signal is received from Airbag Control Module, this DTC will set.<br>**Possible Causes:**<br>Clear DTC<br>Examine airbag system integrity<br>Check connections and grounds |
| **P1102**<br>1T ECM<br>2004-05<br>Liberty with Diesel | **Viscous/Cabin Heater Relay Error Conditions:**<br>Ignition on. ECM Viscous/Cabin Heater Relay is commanded on (excessive current or short circuit), or is commanded OFF (open circuit or short-to-ground). If the ECM detects excessive current or no voltage signal on the Viscous/Cabin Heater Relay control circuit, this DTC will set.<br>**Possible Causes:**<br>ASD Relay output circuit is open<br>Cabin Heater Relay has failed<br>Cabin Heater Relay control circuit is open, shorted to voltage, or to ground<br>ECM has failed |
| **P1103**<br>1T CCM<br>1995-98<br>Talon Turbo | **Turbocharger Wastegate Actuator Malfunction Conditions:**<br>Engine started, ECT Sensor signal 176°F, and the PCM detected the volumetric efficiency was more than 200% for over 1.5 seconds during the CCM Rationality test.<br>**Possible Causes:**<br>Charging pressure control system has failed<br>Turbocharger Wastegate actuator is damaged or has failed<br>Vacuum hose routing is incorrect<br>PCM has failed |
| **P1104**<br>1T CCM<br>1995-98<br>Talon Turbo | **Turbocharger Wastegate Actuator Circuit Malfunction Conditions:**<br>Engine started, system voltage over 10.5v, and the PCM did not detect any surge voltage (system voltage +2V) on the Wastegate actuator control circuit when the solenoid was cycled "on" to "off". Note: The scan tool Actuator test can be used to cycle the device.<br>**Possible Causes:**<br>Wastegate solenoid control circuit is open or shorted to ground<br>Wastegate solenoid power circuit open (power from MPI relay)<br>Wastegate solenoid is damaged or has failed<br>PCM has failed |
| **P1105**<br>1T CCM<br>1995-98<br>Talon Turbo | **Fuel Pressure Control Solenoid Circuit Malfunction Conditions:**<br>Engine started, system voltage over 10.5v, and the PCM did not detect any surge voltage (system voltage +2V) on the Fuel Pressure Solenoid control circuit when the solenoid was cycled "on" to "off". Note: The scan tool Actuator test can be used to cycle the device.<br>**Possible Causes:**<br>Fuel pressure solenoid circuit is open or shorted to ground<br>Fuel pressure solenoid power circuit open (power from relay)<br>Fuel pressure control solenoid is damaged or has failed<br>PCM has failed |
| **P1105**<br>2T CCM<br>2003-05<br>Neon, PT Cruiser Turbo | **Throttle Inlet Pressure Sensor Solenoid Circuit Malfunction Conditions:**<br>Engine started, system voltage over 10.5v, Turbo Boost mode enabled, and the PCM detected the Actual and Intended state of the Throttle Inlet Pressure Sensor solenoid did not match.<br>**Possible Causes:**<br>ASD output circuit to the TIP solenoid is open<br>Throttle inlet pressure Sensor solenoid is damaged or has failed<br>TIP solenoid control circuit is open, shorted to ground or power<br>PCM has failed |

## OBD II Trouble Code List (P1XXX Codes) – Continued

| DTC | Trouble Code Title, Conditions & Possible Causes: |
|---|---|
| **P1106**<br>2T CCM<br>2003-05<br>Neon, PT Cruiser Turbo | **Throttle Inlet Pressure Sensor Solenoid Circuit Malfunction Conditions:**<br>Engine started; battery over 10.5v; Turbo Boost mode enabled. The PCM did not detect enough difference between BARO Sensor and TIP Sensor values during boost.<br>**Possible Causes:**<br>Check the vacuum supply to the turbo surge solenoid unit<br>Inspect the hoses and tubing to the turbo charger assembly<br>Review results of Solenoid Tests (Test 1, 2, 3 and 4 results)<br>Turbocharger assembly is damaged or it has failed<br>Wastegate actuator has failed (due to a mechanical failure) |
| **P1110**<br>1T CCM<br>1998-99<br>Ram with 5.9L Diesel | **Decreased Engine Performance Due To High Intake Air Temperature Conditions:**<br>Engine started, engine running, and the PCM detected the Intake Air Temperature exceeded its normal operating range during the test.<br>**Possible Causes:**<br>IAT Sensor is contaminated, dirty or skewed<br>Base engine conditions causing the high operating temperature<br>ECM has failed |
| **P1115**<br>1T CCM<br>2002-05<br>300M, Caravan, Cherokee, Concorde, PT Cruiser, Dakota, Durango, Grand Cherokee, Intrepid, LHS, Liberty, Magnum, Neon, New Yorker, Pacifica, Prowler, Ram, Sebring, Stratus, Town & Country, Viper, Vision, Voyager, Wrangler | **General Temperature Sensor Performance Conditions:**<br>Engine "off" more than 8 hours, then engine started, ambient temperature above -10ºF; and after a calibrated amount of cool-down time, the PCM compares the values from the Ambient Air Temperature (AAT), Engine Coolant Temperature (ECT) and Intake Air Temperature (IAT) Sensors. If the PCM detects that the value of any combination of these Sensors (AAT-IAT, AAT-ECT or ECT-IAT) is less than a calibrated value, it will set this trouble code.<br>**Possible Causes:**<br>Sensor signal circuit is open or shorted to ground<br>Sensor ground circuit is open or shorted to VREF (5v)<br>One or more of the identified Sensors is out-of-calibration<br>Ambient air temperature Sensor is damaged or it has failed<br>PCM High or Low circuit is damaged or it has failed |
| **P1131**<br>1T ECM<br>2004-05<br>Liberty with Diesel | **Glow Plug Module Voltage Supply Conditions:**<br>Ignition on or engine running. If the ECM detects an improper voltage supply signal, this DTC will set.<br>**Possible Causes:**<br>Battery supply circuit is open<br>Ground circuit is open<br>Intermittent condition<br>Glow Plug Control Module has failed |
| **P1132**<br>1T ECM<br>2004-05<br>Liberty with Diesel | **Glow Plug Module Internal Fault Conditions:**<br>Ignition on or engine running. If the ECM detects an improper voltage supply signal, this DTC will set.<br>**Possible Causes:**<br>Battery supply circuit is open<br>Ground circuit is open<br>Intermittent condition<br>Glow Plug Control Module has failed |
| **P1135**<br>1T ECM<br>2004-05<br>Liberty with Diesel | **Glow Plug Module Control Circuit Fault Conditions:**<br>Ignition on or engine running. If the ECM detects a no-signal or improper current signal on the control circuit, this DTC will set.<br>**Possible Causes:**<br>ECM has failed<br>Glow Plug Module has failed<br>Glow Plug Module control circuit is open or is shorted to voltage or to ground<br>Intermittent condition |

**OBD II Trouble Code List (P1XXX Codes) – Continued**

| DTC | Trouble Code Title, Conditions & Possible Causes: |
|---|---|
| **P1135**<br>1T CCM<br>2004-05<br>PT Cruiser | **O2 (B1 S1) Heater Element Resistance Out-Of-Range Conditions:**<br>Monitored with engine running; O2 (B1 S1) Sensor has reached 98% of duty cycle at least once since cranking; vehicle speed is between 20-93 mph; catalyst temperature is between 1112-1706°F (600-930°C); battery voltage is 9-16v; and no O2 Sensor electrical DTCs are present. If the PCM determines the O2 Sensor resistance is less than 2 ohms for more than 30 seconds, this DTC will set.<br>**Possible Causes:**<br>O2 heater element has failed<br>O2 Sensor heater ground circuit is open<br>ASD relay output circuit is open<br>Intermittent condition<br>PCM has failed |
| **P1136**<br>1T CCM<br>2004-05<br>PT Cruiser | **O2 (B1 S2) Heater Element Resistance Out-Of-Range Conditions:**<br>Monitored with engine running; O2 Sensor has reached 98% of duty cycle at least once since cranking; vehicle speed is between 20-93 mph; catalyst temperature is between 1112-1706°F (600-930°C); battery voltage is 9-16v; and no O2 Sensor electrical DTCs are present. If the PCM determines the B1 S2 O2 Sensor resistance is less than 2 ohms for more than 30 seconds, this DTC will set.<br>**Possible Causes:**<br>O2 heater element has failed<br>O2 Sensor heater ground circuit is open<br>ASD relay output circuit is open<br>Intermittent condition<br>PCM has failed |
| **P1140**<br>1T ECM<br>2004-05<br>Liberty with Diesel | **Vacuum Reservoir Solenoid Open Or Short-To-Ground Conditions:**<br>Ignition on; Vacuum Reservoir Solenoid commanded OFF. If the ECM does not detect a voltage signal or a change in voltage on the control circuit, this DTC will set.<br>**Possible Causes:**<br>Intermittent condition<br>ASD Relay output circuit is open<br>Vacuum Reservoir Solenoid control circuit is open or is shorted to ground<br>Vacuum Reservoir Solenoid has failed<br>ECM has failed |
| **P1142**<br>1T ECM<br>2004-05<br>Liberty with Diesel | **Fuel Pressure Solenoid Open Or Short-To-Ground Circuit Conditions:**<br>Ignition on; ECM Fuel Pressure Solenoid commanded OFF. If the ECM does not detect a voltage signal or detects excessive current in voltage on the control circuit, this DTC will set.<br>**Possible Causes:**<br>FP Solenoid circuit(s) open, shorted to voltage, shorted to ground, or shorted together<br>Intermittent condition<br>FP Solenoid has failed<br>ECM has failed |
| **P1155**<br>1T CCM<br>2004-05<br>Liberty with Diesel | **Fuel Rail Rail Pressure Too High Malfunction Conditions:**<br>Engine running; ECM determines that the fuel rail pressure exceeds 1700 bar.<br>**Possible Causes:**<br>Air in fuel system<br>Fuel injector problems<br>Fuel Pressure Solenoid has failed<br>Fuel Pump has malfunctioned or failed<br>Fuel system has contamination<br>Fuel system has a leak<br>Intermittent condition |

**OBD II Trouble Code List (P1XXX Codes) – Continued**

| DTC | Trouble Code Title, Conditions & Possible Causes: |
|---|---|
| **P1159**<br>1T ECM<br>2004-05<br>Liberty with Diesel | **Improper Start Attempt Conditions:**<br>Engine running; vehicle drive at less than 2 mph. If the ECM detects engine speed above 100 rpm without activating the starter relay control, this DTC will set. Verify the active DTCs.<br>**Possible Causes:**<br>ECM has failed |
| **P1160**<br>1T ECM<br>2004-05<br>Liberty with Diesel | **Ignition Voltage Improper Signal Conditions:**<br>Engine running. If the ECM detects an improper ignition voltage at any time, this DTC will set.<br>**Possible Causes:**<br>ECM power and/or ground connection problems<br>ECM has failed<br>Intermittent condition |
| **P1167**<br>1T ECM<br>2004-05<br>Liberty with Diesel | **Capacitor Voltage Problem Conditions:**<br>Engine cranking or running. If the ECM detects a capacitor voltage problem during injector actuation, this DTC will set. Verify any other injector-related DTCs.<br>**Possible Causes:**<br>ECM has failed<br>Intermittent condition |
| **P1168**<br>1T ECM<br>2004-05<br>Liberty with Diesel | **ECM Communication Error Conditions:**<br>Ignition on. ECM detects an internal failure.<br>**Possible Causes:**<br>ECM has failed<br>Intermittent condition |
| **P1169**<br>1T ECM<br>2004-05<br>Liberty with Diesel | **ECM A/D Converter Error Conditions:**<br>Ignition on. ECM detects an internal failure.<br>**Possible Causes:**<br>ECM has failed<br>Intermittent condition |
| **P1187**<br>2T CCM<br>2004-05<br>Neon | **Throttle Inlet Pressure/MAP Correlation Conditions:**<br>Ignition on. The PCM compares the MAP and TIP Sensor with the ignition on. If the Sensors are not close in value, a failure is recorded. The MAP and TIP should be within the range of 56-112 kPa with ignition on. Related DTCs may be present.<br>**Possible Causes:**<br>Hoses and tubing may be cracked, disconnected or misrouted<br>TIP signal circuit is open<br>Solenoid(s) failed tests<br>TIP Sensor has failed<br>Vacuum passage may be blocked<br>5v supply circuit problem<br>MAP Sensor has failed (internal open)<br>MAP signal circuit is open or is shorted to ground<br>PCM has failed |
| **P1188**<br>2T CCM<br>2003-05<br>Neon, PT Cruiser Turbo | **Throttle Inlet Pressure Sensor Signal Range/Performance Conditions:**<br>Engine started, engine running in Turbo Boost or Non-Boost mode, and the PCM detected a significant difference between the BARO Sensor and TIP Sensor signals (i.e., the TIP Sensor cannot read the signal correctly).<br>**Possible Causes:**<br>ASD output circuit to the TIP solenoid is open<br>Throttle inlet pressure Sensor solenoid is damaged or has failed<br>TIP solenoid control circuit is open, shorted to ground or power<br>PCM has failed |

## OBD II Trouble Code List (P1XXX Codes) – Continued

| DTC | Trouble Code Title, Conditions & Possible Causes: |
|---|---|
| **P1189**<br>2T CCM<br>2003-05<br>Neon, PT Cruiser Turbo | **Throttle Inlet Pressure Sensor Circuit Low Input Conditions:**<br>Engine started, TP Sensor less than 1.2v, system voltage over 10.5v, and the PCM detected the Throttle Inlet Pressure (TIP) Sensor was less than 0.0782v for a period of 1-7 seconds.<br>**Possible Causes:**<br>TIP Sensor VREF circuit is open<br>TIP Sensor signal circuit is open<br>TIP Sensor signal circuit is shorted to chassis or Sensor ground<br>TIP Sensor is damaged or it has failed<br>PCM has failed |
| **P1190**<br>2T CCM<br>2003-05<br>Neon, PT Cruiser Turbo | **Throttle Inlet Pressure Sensor Circuit High Input Conditions:**<br>Engine started, TP Sensor less than 1.2v, system voltage over 10.5v, and the PCM detected the Throttle Inlet Pressure (TIP) Sensor was more than 4.92v for a period of 1-7 seconds.<br>**Possible Causes:**<br>TIP Sensor VREF circuit is open<br>TIP Sensor signal circuit is open<br>TIP Sensor signal circuit is shorted to chassis or Sensor ground<br>TIP Sensor is damaged or it has failed<br>PCM has failed |
| **P1192**<br>1T CCM<br>2001-03<br>300M, Caravan, Cherokee, Concorde, PT Cruiser, Dakota, Durango, Grand Cherokee, Intrepid, LHS, Liberty, Neon, New Yorker, Prowler, Ram, Sebring, Stratus, Town & Country, Viper, Voyager, Wrangler | **Intake Air Temperature Sensor Circuit Low Input Conditions:**<br>Engine started; system voltage over 10.5v and the PCM detected the IAT Sensor signal was less than 0.80v during the CCM test period.<br>**Possible Causes:**<br>IAT Sensor signal circuit is shorted to ground<br>IAT Sensor is damaged or has failed<br>PCM has failed |
| **P1193**<br>1T CCM<br>2001-03<br>300M, Caravan, Cherokee, Concorde, PT Cruiser, Dakota, Durango, Grand Cherokee, Intrepid, LHS, Liberty, Neon, New Yorker, Prowler, Ram, Sebring, Stratus, Town & Country, Viper, Voyager, Wrangler | **Intake Air Temperature Sensor Circuit High Input Conditions:**<br>Engine started; system voltage over 10.5v and the PCM detected the IAT Sensor signal was more than 4.90v during the CCM test.<br>**Possible Causes:**<br>IAT Sensor connector is damaged or it is open<br>IAT Sensor signal circuit is open between the Sensor and PCM<br>IAT Sensor ground circuit is open between the Sensor and PCM<br>IAT Sensor damaged or has failed<br>PCM has failed |

**OBD II Trouble Code List (P1XXX Codes) – Continued**

| DTC | Trouble Code Title, Conditions & Possible Causes: |
| --- | --- |
| **P1194**<br>1T CCM<br>2001-02<br>PT Cruiser | **O2 (B1 S1) Heater Performance Conditions:**<br>Key off after a warm engine drive cycle, engine cool-down finished (at least 5 seconds after the key is turned off), system voltage over 10.5v, then with the ASD relay energized, the PCM detected the O2 signal rose to 0.49v or more within a 144 second period, and the initial rise of the oxygen Sensor signal was less than 1.57v. Note: This test is done at key off.<br>**Possible Causes:**<br>ASD relay output (power) circuit to the heater is open<br>O2 heater ground circuit open or O2 signal circuit is open<br>O2 heater element has high resistance<br>O2 heater element has failed (open or shorted)<br>PCM has failed |
| **P1195**<br>1T CCM<br>1998-2005<br>Car, Jeep, SUV, Truck, Van models | **O2 (B1 S1) Circuit Insufficient Activity Conditions:**<br>Engine started, vehicle driven with the throttle open at a speed over 18 mph at light engine load for over 5 minutes, ECT Sensor more than 170ºF, and the PCM detected the O2 signal switched from 0.39v to 0.60v too few times during the Oxygen Sensor Monitor test.<br>**Possible Causes:**<br>Base engine mechanical fault affecting more than one cylinder<br>Exhaust leak present in exhaust manifold or exhaust pipes<br>O2 element fuel contamination or has deteriorated<br>O2 signal circuit or ground circuit has high resistance<br>PCM has failed |
| **P1196**<br>1T CCM<br>1998-2005<br>300M, Avenger, Breeze, Caravan, Cherokee, Cirrus, Concorde, PT Cruiser, Dakota, Durango, Grand Cherokee, Intrepid, LHS, Liberty, Magnum, Neon, New Yorker, Prowler, Ram, Sebring, Stratus, Talon, Town & Country, Viper, Vision, Voyager, Wrangler | **O2 (B2 S1) Circuit Insufficient Activity Conditions:**<br>Engine started, vehicle driven with the throttle open at a speed over 18-55 mph at light engine load for over 5 minutes, ECT Sensor more than 170ºF, and the PCM detected the O2 signal switched from 0.39v to 0.60v too few times in the Oxygen Sensor Monitor test.<br>**Possible Causes:**<br>Base engine mechanical fault affecting more than one cylinder<br>Exhaust leak present in exhaust manifold or exhaust pipes<br>O2 element fuel contamination or has deteriorated<br>O2 signal circuit or ground circuit has high resistance |
| **P1197**<br>1T CCM<br>1998-2003<br>Grand Cherokee, Ram with V8 or V10 | **O2 (B1 S2) Circuit Insufficient Activity Conditions:**<br>Engine started, vehicle driven with the throttle open at a speed over 18-55 mph at light engine load for over 5 minutes, ECT Sensor more than 170ºF, and the PCM detected the O2 signal switched from 0.39v to 0.60v too few times in the Oxygen Sensor Monitor test.<br>**Possible Causes:**<br>Base engine mechanical fault affecting more than one cylinder<br>Exhaust leak present in exhaust manifold or exhaust pipes<br>O2 element fuel contamination or has deteriorated<br>O2 signal circuit or ground circuit has high resistance |

**OBD II Trouble Code List (P1XXX Codes) – Continued**

| DTC | Trouble Code Title, Conditions & Possible Causes: |
|---|---|
| **P1250**<br>1T ECM<br>2004-05<br>Liberty with Diesel | **Vacuum Reservoir Solenoid Open Circuit Conditions:**<br>Ignition on; solenoid commanded ON. If the ECM does not detect a voltage signal on the Vacuum Reservoir Solenoid control circuit, this DTC will set.<br>**Possible Causes:**<br>Intermittent condition<br>ASD Relay output circuit is open<br>VR Solenoid control circuit is open or is shorted to ground<br>VR Solenoid or ECM has failed |
| **P1251**<br>1T ECM<br>2004-05<br>Liberty with Diesel | **Vacuum Reservoir Solenoid Short-To-Ground Circuit Conditions:**<br>Ignition on; solenoid commanded ON. If the ECM does not detect a voltage signal on the Vacuum Reservoir Solenoid control circuit, this DTC will set.<br>**Possible Causes:**<br>Intermittent condition<br>ASD Relay output circuit is open<br>VR Solenoid control circuit is open or shorted to ground<br>VR Solenoid or ECM has failed |
| **P1252**<br>1T ECM<br>2004-05<br>Liberty with Diesel | **Vacuum Reservoir Solenoid Short Circuit Conditions:**<br>Ignition on; solenoid commanded ON. If the ECM detects excessive voltage signal on the Vacuum Reservoir Solenoid control circuit, this DTC will set.<br>**Possible Causes:**<br>Intermittent condition<br>ASD Relay output circuit is open<br>VR Solenoid control circuit is open or shorted to ground<br>VR Solenoid or ECM has failed |
| **P1281**<br>2T ECT<br>1996-2005<br>300M, Avenger, Breeze, Caravan, Cherokee, Cirrus, Concorde, PT Cruiser, Dakota, Durango, Grand Cherokee, Intrepid, LHS, Liberty, Magnum, Neon, New Yorker, Prowler, Ram, Sebring, Stratus, Talon, Town & Country, Viper, Vision, Voyager, Wrangler | **Engine Is Cold Too Long Conditions:**<br>Engine started, engine runtime more than 20 minutes, and the PCM detected the engine temperature did not exceed 176°F in the period.<br>**Possible Causes:**<br>Check the operation of the thermostat (it may be stuck open)<br>ECT Sensor signal circuit has high resistance<br>ECT Sensor is damaged or it has failed<br>Inspect for low coolant level or an incorrect coolant mixture |
| **P1282**<br>1T CCM<br>1996-2005<br>300M, Avenger, Breeze, Caravan, Cherokee, Cirrus, Concorde, PT Cruiser, Dakota, Durango, Grand Cherokee, Intrepid, LHS, Liberty, Magnum, Neon, New Yorker, Prowler, Ram, Sebring, Stratus, Talon, Town & Country, Viper, Vision, Voyager, Wrangler | **Fuel Pump Relay Control Circuit Malfunction Conditions:**<br>Key on or engine started, system voltage over 10.5v, and the PCM detected an unexpected voltage condition on the Fuel Pump Relay control circuit during the CCM test period.<br>**Possible Causes:**<br>Fuel pump relay control circuit is open or shorted to ground<br>Fuel pump relay power circuit is open (test power from Ignition)<br>Fuel pump relay is damaged or has failed<br>PCM has failed |

**OBD II Trouble Code List (P1XXX Codes) – Continued**

| DTC | Trouble Code Title, Conditions & Possible Causes: |
|---|---|
| **P1283**<br>1T CCM<br>1998-2002<br>Ram with 5.9L Diesel | **Idle Select Signal Invalid Conditions:**<br>Key on and the Fuel Pump Control Module (VP44) detected an invalid Low Idle Select signal from the PCM (ECM controller).<br>**Possible Causes:**<br>Low idle select signal circuit is open or shorted to ground<br>Low idle select signal circuit is shorted to VREF or to power<br>ECM internal circuit is shorted to ground<br>ECM internal regulator output is more than 6.0v<br>Fuel injection pump is damaged or has failed |
| **P1284**<br>1T CCM<br>1998-2002<br>Ram with 5.9L Diesel | **Fuel Injection Pump Battery Voltage Out-Of-Range Conditions:**<br>Key on and the Fuel Pump Control Module (VP44) detected an invalid Low Idle Select signal from the PCM (ECM controller).<br>**Possible Causes:**<br>Fuel injection pump is damaged or has failed |
| **P1285**<br>1T CCM<br>1998-2002<br>Ram with 5.9L Diesel | **Fuel Injection Pump Controller Always On Conditions:**<br>Engine started, engine running, and the PCM detected the Fuel Injection Pump Controller was in an "always on" condition.<br>**Possible Causes:**<br>Fuel injection pump relay driver circuit shorted to system power<br>Fuel injection pump relay output circuit shorted to power<br>Fuel injection pump relay is damaged or has failed<br>ECM has failed |
| **P1286**<br>1T CCM<br>1998-2002<br>Ram with 5.9L Diesel | **Accelerator Position Sensor Supply Voltage High Input Conditions:**<br>Key on or engine running; and the PCM detected the supply voltage circuit to the Accelerator Position (APP) Sensor was too high.<br>**Possible Causes:**<br>APP Sensor supply voltage circuit is shorted to system power<br>APP Sensor signal circuit is shorted to power<br>APP Sensor ground circuit is open<br>ECM has failed |
| **P1287**<br>1T CCM<br>1998-2002<br>Ram with 5.9L Diesel | **Fuel Injection Pump Battery Voltage Out-Of-Range Conditions:**<br>Engine started and the PCM detected the Fuel Injection Pump battery voltage was too low.<br>**Possible Causes:**<br>Fuel injection pump ground circuit is open<br>Generator voltage is less than 12.0v (the Generator has failed)<br>Battery voltage is less than 8.0v (the battery is defective)<br>PCM has failed |
| **P1288**<br>1T CCM<br>1998-2003<br>Concorde, Intrepid,<br>Prowler, Sebring,<br>Stratus with 2.7L V6 | **Short Runner Valve Control Circuit Malfunction Conditions:**<br>Key on or engine cranking; and the PCM detected an unexpected voltage condition on the Short Runner Solenoid (SRV) Control circuit during the CCM test.<br>**Possible Causes:**<br>SRV control circuit is open between the solenoid and PCM<br>SRV control circuit is shorted to Sensor or chassis ground<br>SRV is damaged or has failed<br>PCM has failed |

**OBD II Trouble Code List (P1XXX Codes) – Continued**

| DTC | Trouble Code Title, Conditions & Possible Causes: |
|---|---|
| **P1289**<br>1T CCM<br>1998-2002<br>300M, Avenger, Breeze, Caravan, Cherokee, Cirrus, Concorde, PT Cruiser, Dakota, Durango, Grand Cherokee, Intrepid, LHS, Liberty, Neon, New Yorker, Prowler, Ram, Sebring, Stratus, Town & Country, Viper, Voyager, Wrangler | **Manifold Tuning Valve Control Circuit Malfunction Conditions:**<br>Key on or engine cranking; and the PCM detected an unexpected voltage condition on the Manifold Tuning Valve (MTV) control circuit.<br>**Possible Causes:**<br>MTV control circuit is open between the solenoid and PCM<br>MTV control circuit is shorted to Sensor or chassis ground<br>MTV is damaged or has failed<br>PCM has failed |
| **P1290**<br>1T CCM<br>1996-2002<br>Ram with CNG VIN T | **Certified Natural Gas Fuel System Pressure Too High Conditions:**<br>Engine started, and the PCM detected the CNG Fuel System was operating outside of its normal operating range during the CCM test.<br>**Possible Causes:**<br>CNG pressure Sensor signal is skewed<br>CNG fuel system component is damaged or has failed<br>PCM has failed |
| **P1290**<br>1T CCM<br>1998-2002<br>Caravan, Town & Country, Voyager with 3.3L CNG | **CNG Pressure Sensor Circuit High Input Conditions:**<br>Key on or engine running; and the PCM detected the Certified Natural Gas (CNG) Sensor was more than 4.96v during the test.<br>**Possible Causes:**<br>CNG Sensor signal circuit open between the Sensor and PCM<br>CNG Sensor ground circuit open between the Sensor and PCM<br>CNG Sensor is damaged or has failed (it may be open)<br>PCM has failed |
| **P1291**<br>1T CCM<br>1998-2002<br>Caravan, Town & Country, Voyager with 3.3L CNG | **CNG Pressure Sensor Circuit Low Input Conditions:**<br>Engine started and the PCM detected the Certified Natural Gas (CNG) Sensor was less than 0.49v.<br>**Possible Causes:**<br>CNG Sensor signal circuit is shorted to chassis or Sensor ground between the Sensor and PCM<br>CNG Sensor is damaged or has failed (it may be shorted)<br>PCM has failed |
| **P1291**<br>1T CCM<br>1998-2002<br>Ram with 5.9L Diesel | **No Temperature Rise Detected From The Intake Heaters Conditions:**<br>No IAT or IAH Relay codes set. Preheat function completed before startup. Post-heat function active. Engine cranking for less than 5 seconds. Engine runtime over 15 seconds. IAT Sensor from 0-66°F. BTS and IAT Sensors within 10°F of each other. Time between engine preheat period and engine run state is less than 30 seconds. PCM did not detect a temperature increase at the Intake Heaters.<br>**Possible Causes:**<br>Battery cable from No. 1 Relay to No. 1 Heater is open<br>Battery cable from No. 2 Relay to No. 1 Heater is open<br>Battery cable to the No. 1 Relay is open or has high resistance<br>Battery cable to the No. 2 Relay is open or has high resistance<br>Intake Air Heater Relay is damaged or has failed<br>PCM has failed |

**OBD II Trouble Code List (P1XXX Codes) – Continued**

| DTC | Trouble Code Title, Conditions & Possible Causes: |
|---|---|
| **P1292**<br>1T CCM<br>1996-2002<br>Ram with CNG VIN T | **Certified Natural Gas Pressure Sensor Circuit Low Input Conditions:**<br>Engine started and the PCM detected the CNG Pressure Sensor indicated less than 0.49v.<br>**Possible Causes:**<br>CNG pressure Sensor signal circuit is shorted to ground<br>CNG pressure Sensor is damaged or has failed<br>PCM has failed |
| **P1294**<br>1T CCM<br>1995-98<br>Talon | **Target Idle Speed Not Reached Conditions:**<br>DTC P0106, P0107, P0108, P0121, P0122 and P0123 not set, engine started, engine running at idle in Drive or Neutral, and the PCM detected the Actual idle speed was more than 200 rpm over or more than 100 rpm less than the Target speed for over 14 seconds.<br>**Possible Causes:**<br>Engine vacuum leak in a hose, Brake Booster or in the engine<br>IAC motor control circuits open or grounded in the wire harness<br>Throttle body dirty or restricted (trying cleaning it and retesting)<br>Throttle linkage or throttle plate not in the correct position<br>PCM has failed |
| **P1294**<br>1T CCM<br>1996-2005<br>300M, Avenger, Breeze, Caravan, Cherokee, Cirrus, Concorde, PT Cruiser, Dakota, Durango, Grand Cherokee, Intrepid, LHS, Liberty, Magnum, Neon, New Yorker, Prowler, Ram, Sebring, Stratus, Talon, Town & Country, Viper, Vision, Voyager, Wrangler | **Target Idle Speed Not Reached Conditions:**<br>DTC P0106, P0107, P0108, P0121, P0122 and P0123 not set, engine started, running at idle in Drive or Neutral, and the PCM detected the Actual idle speed was more than 200 rpm over or more than 100 rpm less than the Target speed for over 14 seconds.<br>**Possible Causes:**<br>Engine vacuum leak in a hose, brake booster or in the engine<br>IAC motor control circuits open or grounded in the wire harness<br>Throttle body dirty or restricted (trying cleaning it and retesting)<br>Throttle linkage or throttle plate not in the correct position<br>PCM has failed |
| **P1294**<br>1T CCM<br>1996-2003<br>Cherokee, Grand Cherokee, Liberty, Ram, Wrangler | **Target Idle Speed Not Reached Conditions:**<br>DTC P0106, P0107, P0108, P0121, P0122 and P0123 not set, engine started, engine running at idle in Drive or Neutral, and the PCM detected the Actual idle speed was more than 200 rpm over or more than 100 rpm less than the Target speed for over 14 seconds.<br>**Possible Causes:**<br>Engine vacuum leak in a hose, Brake Booster or in the engine<br>IAC motor control circuits open or grounded in the wire harness<br>Throttle body dirty or restricted (trying cleaning it and retesting)<br>Throttle linkage or throttle plate not in the correct position<br>PCM has failed |
| **P1295**<br>1T CCM<br>1996-2000<br>Avenger, Breeze, Cirrus, Sebring, Stratus, Talon | **5-Volt VREF Missing To Position Sensor Conditions:**<br>Engine started, vehicle driven to over 20 mph at more than 1500 rpm with MAP Sensor less than 13 kPa, and the PCM detected the TP Sensor signal was less than a specified value during the test.<br>**Possible Causes:**<br>TP Sensor VREF circuit is open between the Sensor and PCM<br>TP Sensor ground circuit is open between the Sensor and PCM<br>TP Sensor is damaged or has failed<br>PCM has failed |

**OBD II Trouble Code List (P1XXX Codes) – Continued**

| DTC | Trouble Code Title, Conditions & Possible Causes: |
|---|---|
| **P1295**<br>1T CCM<br>1998-2002<br>Ram with 5.9L Diesel | **5-Volt VREF Missing To APP Sensor Conditions:**<br>Key on or engine running; and the PCM detected the supply voltage circuit to the Accelerator Position (APP) Sensor was too low.<br>**Possible Causes:**<br>APP Sensor supply voltage circuit is open<br>ACCEL pedal position Sensor is damaged or has failed<br>APP Sensor supply circuit shorted to chassis or Sensor ground<br>PCM has failed |
| **P1296**<br>1T CCM<br>1999-2005<br>300M, Avenger, Breeze, Caravan, Cherokee, Cirrus, Concorde, PT Cruiser, Dakota, Durango, Grand Cherokee, Intrepid, LHS, Liberty, Magnum, Neon, New Yorker, Prowler, Ram, Sebring, Stratus, Talon, Town & Country, Viper, Vision, Voyager, Wrangler | **5-Volt VREF Supply Not Present Conditions:**<br>Key on, altitude indicating zero feet above seal level, then the PCM detected the MAP Sensor was near 101 kPa; or with altitude at 1200 feet above sea level, the MAP Sensor was near 88 kPa.<br>**Possible Causes:**<br>MAP Sensor VREF circuit open between the Sensor and PCM<br>MAP Sensor ground circuit open between the Sensor and PCM<br>MAP Sensor is damaged or has failed<br>PCM has failed |
| **P1297**<br>1T CCM<br>1996-2000<br>Avenger, Breeze, Cirrus, Sebring, Stratus, Talon | **No Change In MAP Signal From Start To Run Transition Conditions:**<br>Engine started, engine speed between 400 and 1200 rpm, and the PCM detected too small a difference between the BARO signal at key "on" and the engine running MAP Sensor input for 1.76 seconds.<br>**Possible Causes:**<br>Engine vacuum port to MAP Sensor clogged, dirty or restricted<br>MAP Sensor signal is skewed or the Sensor is out-of-calibration<br>MAP Sensor VREF circuit open or grounded (intermittent fault)<br>PCM has failed |
| **P1297**<br>1T CCM<br>1996-2005<br>300M, Avenger, Breeze, Caravan, Cherokee, Cirrus, Concorde, PT Cruiser, Dakota, Durango, Grand Cherokee, Intrepid, LHS, Liberty, Magnum, Neon, New Yorker, Prowler, Ram, Sebring, Stratus, Talon, Town & Country, Viper, Vision, Voyager, Wrangler | **No Change In MAP Signal From Start To Run Transition Conditions:**<br>Engine started, and with the engine speed within ±64 rpm of the Target idle speed, the PCM detected too small a difference between the BARO and MAP Sensor signals for 8.80 seconds.<br>**Possible Causes:**<br>Engine vacuum port to MAP Sensor clogged, dirty or restricted<br>MAP Sensor signal is skewed or the Sensor is out-of-calibration<br>MAP Sensor VREF circuit open or grounded (intermittent fault)<br>PCM has failed |

**OBD II Trouble Code List (P1XXX Codes) – Continued**

| DTC | Trouble Code Title, Conditions & Possible Causes: |
|---|---|
| **P1297**<br>1T CCM<br>1996-2003<br>300M, Avenger, Breeze, Caravan, Cherokee, Cirrus, Concorde, PT Cruiser, Dakota, Durango, Grand Cherokee, Intrepid, LHS, Liberty, Neon, New Yorker, Prowler, Ram, Sebring, Stratus, Talon, Town & Country, Viper, Vision, Voyager, Wrangler | **No Change In MAP Signal From Start To Run Transition Conditions:**<br>Engine started, and with the engine speed within ±64 rpm of the Target idle speed, the PCM detected too small a difference between the BARO and MAP Sensor signals for 8.80 seconds.<br>**Possible Causes:**<br>Engine vacuum port to MAP Sensor clogged, dirty or restricted<br>MAP Sensor signal is skewed or the Sensor is out-of-calibration<br>MAP Sensor VREF circuit open or grounded (intermittent fault)<br>PCM has failed |
| **P1299**<br>2T IAC<br>1996-2003<br>300M, Avenger, Breeze, Caravan, Cherokee, Cirrus, Concorde, PT Cruiser, Dakota, Durango, Grand Cherokee, Intrepid, LHS, Liberty, Neon, New Yorker, Prowler, Ram, Sebring, Stratus, Talon, Town & Country, Viper, Vision, Voyager, Wrangler | **Vacuum Leak Present With IAC Valve Fully Seated Conditions:**<br>Engine running at idle speed in closed loop, and the PCM detected the MAP Sensor signal did not correlate to the TP Sensor signal under these operating conditions during the test.<br>**Possible Causes:**<br>Leaking engine vacuum hose, brake booster, or in the engine<br>MAP Sensor is out-of-calibration or skewed<br>TP Sensor is damaged or has failed (perform a sweep test)<br>PCM has failed |
| **P1388**<br>1T CCM<br>1996-2005<br>300M, Avenger, Breeze, Caravan, Cherokee, Cirrus, Concorde, PT Cruiser, Dakota, Durango, Grand Cherokee, Intrepid, LHS, Liberty, Magnum, Neon, New Yorker, Prowler, Ram, Sebring, Stratus, Talon, Town & Country, Viper, Vision, Voyager, Wrangler | **Auto Shutdown Relay Control Circuit Malfunction Conditions:**<br>Key on or engine cranking; and the PCM detected an unexpected voltage condition on the ASD Relay Control circuit. The ASD Relay coil resistance is 95-105 ohms at 68°F.<br>**Possible Causes:**<br>ASD relay control circuit is open between the relay and PCM<br>ASD relay control circuit is shorted to ground<br>ASD relay power circuit is open (test power from Fused B+)<br>ASD relay is damaged or has failed<br>PCM has failed |
| **P1388**<br>1T CCM<br>1998-2002<br>Ram with 5.9L Diesel | **Auto Shutdown Relay Control Circuit Malfunction Conditions:**<br>Key on or engine cranking; and the PCM detected an unexpected voltage condition on the ASD Relay Control circuit. The ASD Relay coil resistance is 95-105 ohms at 68°F.<br>**Possible Causes:**<br>ASD relay control circuit is open between the relay and PCM<br>ASD relay control circuit is shorted to ground<br>ASD relay power circuit is open (test power from Fused B+)<br>ASD relay is damaged or has failed<br>PCM has failed |

**OBD II Trouble Code List (P1XXX Codes) – Continued**

| DTC | Trouble Code Title, Conditions & Possible Causes: |
|---|---|
| **P1389**<br>1T CCM<br>1996-2005<br>300M, Avenger, Breeze, Caravan, Cherokee, Cirrus, Concorde, PT Cruiser, Dakota, Durango, Grand Cherokee, Intrepid, LHS, Liberty, Magnum, Neon, New Yorker, Prowler, Ram, Sebring, Stratus, Talon, Town & Country, Viper, Vision, Voyager, Wrangler | **No Auto Shutdown Relay Output Voltage To PCM Conditions:**<br>Engine cranking; and the PCM did not detect any voltage on the ASD Relay Output circuit to the PCM during the CCM test.<br>**Possible Causes:**<br>ASD relay connector is damaged, loose or shorted<br>ASD relay output circuit is open between the relay and PCM<br>ASD relay power circuit is open (test power from Fused B+)<br>ASD relay is damaged or has failed<br>PCM has failed |
| **P1390**<br>1T CCM<br>1995-2003<br>300M, Avenger, Breeze, Caravan, Cherokee, Cirrus, Concorde, PT Cruiser, Dakota, Durango, Grand Cherokee, Intrepid, LHS, Liberty, Neon, New Yorker, Prowler, Ram, Sebring, Stratus, Talon, Town & Country, Viper, Vision, Voyager, Wrangler | **Timing Belt Skipped One Tooth Or More Conditions:**<br>Engine started, engine running, then with the Inhibit Test not active, the PCM checked the CKP and CMP Sensor alignment. If the PCM detects the CMP Sensor is offset from the CKP Sensor signal by 1 tooth, this trouble code is set. The PCM performs the Inhibit Test whenever the engine is cold, if the engine speed is outside of a given window, or if there is a large change in the MAP Sensor signal.<br>**Possible Causes:**<br>Camshaft timing is out of specifications<br>Valve timing is out of specifications |
| **P1391**<br>1T CCM<br>1995-2005<br>300M, Avenger, Breeze, Caravan, Cherokee, Cirrus, Concorde, Cruiser, Dakota, Durango, Grand Cherokee, Intrepid, LHS, Liberty, Magnum, Neon, New Yorker, Prowler, Ram, Sebring, Stratus, Talon, Town & Country, Viper, Vision, Voyager, Wrangler | **CKP Or CMP Sensor Signal Intermittent Conditions:**<br>Engine started, engine running, and after every 69-degree CKP Sensor leading edge and trailing signal edge is determined, the PCM updates this data and compares it to the true CMP Sensor port level. If the PCM detects a disagreement between these two values 20 times in succession, this trouble code is set.<br>**Possible Causes:**<br>Camshaft Sensor is not installed properly<br>Engine valve timing is not within specifications<br>Perform a CKP and CMP Sensor relearn with the scan tool<br>Tone wheel or pulse ring is damaged |
| **P1398**<br>1T CCM<br>1995-2005<br>300M, Avenger, Breeze, Caravan, Cherokee, Cirrus, Concorde, PT Cruiser, Dakota, Durango, Grand Cherokee, Intrepid, LHS, Liberty, Magnum, Neon, New Yorker, Prowler, Ram, Sebring, Stratus, Talon, Town & Country, Viper, Vision, Voyager, Wrangler | **Misfire Adaptive Numerator At Limit Conditions:**<br>Engine started; ECT Sensor under 75ºF; engine runtime over 50 sec.; A/C "OFF"; vehicle speed over 36 mph in 1st gear, or over 65 mph in high gear, followed by a closed throttle decel period. This code sets if the PCM detects one of the CKP Sensor target windows varies more than 2.86% from the reference window.<br>Background - PCM needs to learn any variation in engine machining to detect when a misfire is present. CKP Sensor has 2 40° windows that are 180° apart. The window for Cylinders 1 and 4 is the reference window. It is checked against the window for Cylinders 2 and 3. The PCM checks for any variation to make engine speed adjustments.<br>**Possible Causes:**<br>Base engine problem (i.e., low cylinder compression)<br>CKP Sensor crankshaft target variation too large<br>CKP Sensor improperly installed or the CKP Sensor has failed<br>CKP Sensor signal circuit open or shorted (intermittent fault)<br>Tone wheel or pulse ring is damaged |

**OBD II Trouble Code List (P1XXX Codes) – Continued**

| DTC | Trouble Code Title, Conditions & Possible Causes: |
|---|---|
| **P1400**<br>1T CCM<br>1995-98<br>Talon with 2.0L Turbo | **Manifold Differential Pressure Sensor Circuit Malfunction Conditions:**<br>Engine running at low to medium load, ECT Sensor over 65.4ºF, and the PCM detected the Manifold Differential Pressure (MDP) Sensor was over 4.50v or under 0.20v for 4 seconds.<br>**Possible Causes:**<br>MDP Sensor signal circuit is open or shorted to ground<br>MDP Sensor power (VREF) circuit is open<br>MDP Sensor is damaged or has failed<br>PCM has failed |
| **P1400**<br>1T CCM<br>2001-03<br>Sebring, Stratus Coupe<br>with 2.4L I4 or 3.0L V6 | **Manifold Differential Pressure Sensor Circuit Malfunction Conditions:**<br>Engine runtime over 8 minutes if the ECT Sensor is less than 32ºF at startup, ECT more than 113ºF during testing, volumetric efficiency from 30-45%, IAT Sensor over 14ºF and the PCM detected the Manifold Differential Pressure (MDP) Sensor was more than 4.60v (scan tool reads over 108 kPa) or less than 0.10v (scan tool reads under 2.4 kPa) for 2 seconds. If the volumetric efficiency is less than 30%, P1400 sets if the MDP Sensor is over 4.20v (scan tool reads over 108 kPa) for 2 seconds). If the volumetric efficiency is more than 70%, P1400 sets if the MDP Sensor is less than 1.80v (San Tool read 46 kPa) for 2 seconds.<br>**Possible Causes:**<br>MDP Sensor signal circuit is open or shorted to ground<br>MDP Sensor ground circuit is open (fault may be intermittent)<br>MDP power supply (VREF) circuit is open<br>MDP Sensor is damaged, skewed or it has failed<br>PCM has failed |
| **P1411**<br>1T CCM<br>2004-05<br>300M, Magnum | **Cylinder No. 1 Reactivation Control Performance Malfunction Conditions:**<br>This condition is monitored when transitioning from 8-cylinder to 4-cylinder mode. If the MDS solenoid fails to activate for cylinder No. 1. By actuating the solenoid, oil pressure is raised to the pair of lifters that coincide with each particular solenoid. The oil pressure pushes in the locking pins that allow the lifter to collapse, decoupling the valves and camshaft. If this does not occur, the DTC will set.<br>**Possible Causes:**<br>MDS solenoid No. 1 control is open or is shorted to ground<br>MDS solenoid ground circuit is open<br>Insufficient oil pressure acting on the lifter locking pins<br>Oil passages restricted<br>Lifter is damaged or has failed<br>MDS solenoid No. 1 has failed<br>PCM has failed |
| **P1414**<br>1T CCM<br>2004-05<br>300M, Magnum | **Cylinder No. 4 Reactivation Control Performance Malfunction Conditions:**<br>This condition is monitored when transitioning from 8-cylinder to 4-cylinder mode. If the MDS solenoid fails to activate for cylinder No. 4. By actuating the solenoid, oil pressure is raised to the pair of lifters that coincide with each particular solenoid. The oil pressure pushes in the locking pins that allow the lifter to collapse, decoupling the valves and camshaft. If this does not occur, the DTC will set.<br>**Possible Causes:**<br>MDS solenoid No. 4 control is open or is shorted to ground<br>MDS solenoid ground circuit is open<br>Insufficient oil pressure acting on the lifter locking pins<br>Oil passages restricted<br>Lifter is damaged or has failed<br>MDS solenoid No. 4 has failed<br>PCM has failed |

**OBD II Trouble Code List (P1XXX Codes) – Continued**

| DTC | Trouble Code Title, Conditions & Possible Causes: |
|---|---|
| **P1416**<br>1T CCM<br>2004-05<br>300M, Magnum | **Cylinder No. 6 Reactivation Control Performance Malfunction Conditions:**<br>This condition is monitored when transitioning from 8-cylinder to 4-cylinder mode. If the MDS solenoid fails to activate for cylinder No. 6. By actuating the solenoid, oil pressure is raised to the pair of lifters that coincide with each particular solenoid. The oil pressure pushes in the locking pins that allow the lifter to collapse, decoupling the valves and camshaft. If this does not occur, the DTC will set.<br>**Possible Causes:**<br>MDS solenoid No. 6 control is open or is shorted to ground<br>MDS solenoid ground circuit is open<br>Insufficient oil pressure acting on the lifter locking pins<br>Oil passages restricted<br>Lifter is damaged or has failed<br>MDS solenoid No. 6 has failed<br>PCM has failed |
| **P1417**<br>1T CCM<br>2004-05<br>300M, Magnum | **Cylinder No. 7 Reactivation Control Performance Malfunction Conditions:**<br>This condition is monitored when transitioning from 8-cylinder to 4-cylinder mode. If the MDS solenoid fails to activate for cylinder No. 7. By actuating the solenoid, oil pressure is raised to the pair of lifters that coincide with each particular solenoid. The oil pressure pushes in the locking pins that allow the lifter to collapse, decoupling the valves and camshaft. If this does not occur, the DTC will set.<br>**Possible Causes:**<br>MDS solenoid No. 7 control is open or is shorted to ground<br>MDS solenoid ground circuit is open<br>Insufficient oil pressure acting on the lifter locking pins<br>Oil passages restricted<br>Lifter is damaged or has failed<br>MDS solenoid No. 7 has failed<br>PCM has failed |
| **P1475**<br>1T CCM<br>1998-2002<br>Ram with 5.9L Diesel | **Auxiliary 5-Volt Supply Circuit High Input Conditions:**<br>Key on or engine running; and the PCM detected an unexpected "high" voltage condition on the Auxiliary 5-volt power circuit.<br>**Possible Causes:**<br>Auxiliary 5v supply circuit shorted to system power<br>MAP Sensor VREF circuit is shorted to ground<br>MAP Sensor is open internally<br>PCM has failed |
| **P1478**<br>1T CCM<br>2002-03<br>Caravan, Town &<br>Country, Voyager | **Battery Temperature Sensor Circuit Out-Of-Limits Conditions:**<br>Key on or engine running; and the PCM detected the Battery Temperature Sensor was under 0.1v or over 4.90v for 3.2 seconds.<br>**Possible Causes:**<br>Battery temperature Sensor is damaged or it has failed.<br>Clear the codes and retest for this trouble code. If the same code resets, the PCM will have to be replaced, as the Battery Temperature Sensor is located inside the controller. |
| **P1479**<br>1T CCM<br>1997-2002<br>Prowler | **A/T Fan Relay Circuit Malfunction Conditions:**<br>Engine started, engine running, and the PCM detected an unexpected voltage condition on the A/T Transmission Relay circuit.<br>**Possible Causes:**<br>A/T fan relay control circuit is open or shorted to ground<br>A/T fan relay power circuit is open (test power to Fused IGN)<br>A/T fan relay is damaged or has failed<br>PCM has failed |

## OBD II Trouble Code List (P1XXX Codes) – Continued

| DTC | Trouble Code Title, Conditions & Possible Causes: |
|---|---|
| **P1480**<br>1T CCM<br>2001-03<br>Cherokee, Grand Cherokee, Liberty, Ram, Wrangler | **Positive Crankcase Ventilation Solenoid Circuit Failure Conditions:**<br>Engine started and the PCM detected an unexpected voltage on the PCV Solenoid circuit.<br>**Possible Causes:**<br>PCV solenoid control circuit is open or shorted to ground<br>PCV solenoid power circuit is open to the fuse in the PDC<br>PCV solenoid is damaged or has failed<br>PCM has failed |
| **P1481**<br>1T CCM<br>2001-03<br>Cherokee, Grand Cherokee, Liberty, Ram, Wrangler | **EVAP Leak Detection Monitor Pinched Hose Detected Conditions:**<br>BTS from 40-96ºF, ECT Sensor within 10ºF of the BTS signal at startup (cold engine); engine started, and after the EVAP Leak test started, the PCM detected a no flow condition.<br>**Possible Causes:**<br>EVAP vapor hose blocked between the fuel tank and the LDP (check rollover valve)<br>EVAP ventilation solenoid is damaged or has failed<br>Purge line is loose, damaged or incorrectly routed |
| **P1486**<br>2T EVAP<br>1996-2005<br>300M, Avenger, Breeze, Caravan, Cherokee, Cirrus, Concorde, PT Cruiser, Dakota, Durango, Grand Cherokee, Intrepid, LHS, Liberty, Magnum, Neon, New Yorker, Prowler, Ram, Sebring, Stratus, Talon, Town & Country, Viper, Vision, Voyager, Wrangler | **EVAP Leak Detection Monitor Pinched Hose Detected Conditions:**<br>BTS from 40-96ºF and ECT Sensor within 20ºF of the BTS signal at startup (cold engine), engine started, and after the EVAP Leak Detection test was enabled, the PCM detected the LDP switch did not reach 3 closures (i.e., a "no flow" condition was present).<br>**Possible Causes:**<br>EVAP vapor hose blocked between the fuel tank and the LDP (i.e., in the OLFV, rollover or vapor hose)<br>EVAP canister is clogged or full of dirt or moisture<br>EVAP ventilation solenoid is damaged or has failed<br>Purge line is loose, damaged or incorrectly routed<br>PCM has failed |
| **P1487**<br>1T CCM<br>1995-98<br>Sebring, Stratus, Talon | **High Speed Radiator Fan Relay Circuit Failure Conditions:**<br>Key on or engine running; system voltage over 10.5v, and the PCM detected an unexpected voltage condition on the High Speed Radiator Fan Relay control circuit during the CCM test.<br>**Possible Causes:**<br>HFAN radiator relay control circuit is open or shorted to ground<br>HFAN radiator relay power circuit is open (test power to IGN)<br>HFAN radiator fan relay is damaged or has failed<br>PCM has failed |
| **P1488**<br>1T CCM<br>1996-2002<br>Ram with 5.9L Diesel | **Auxiliary 5-Volt Supply Circuit Low Input Conditions:**<br>Key on or engine running; and the PCM detected an unexpected "low" voltage condition on the Auxiliary 5-volt power circuit.<br>**Possible Causes:**<br>Auxiliary 5v supply circuit shorted to Sensor or chassis ground<br>Camshaft position Sensor VREF circuit is shorted to ground<br>MAP Sensor VREF circuit is shorted to ground<br>Oil Pressure Sensor VREF circuit is shorted to ground<br>PCM has failed |

**OBD II Trouble Code List (P1XXX Codes) – Continued**

| DTC | Trouble Code Title, Conditions & Possible Causes: |
|---|---|
| **P1489**<br>1T CCM<br>1995-2003<br>300M, Avenger, Breeze, Caravan, Cherokee, Cirrus, Concorde, PT Cruiser, Dakota, Durango, Grand Cherokee, Intrepid, LHS, Liberty, Neon, New Yorker, Prowler, Ram, Sebring, Stratus, Talon, Town & Country, Viper, Vision, Voyager, Wrangler | **High Speed Radiator Fan Relay Circuit Malfunction Conditions:**<br>Key on or engine running; system voltage over 10.5v, and the PCM detected an unexpected voltage condition on the High Speed Radiator Fan Relay control circuit during the CCM test.<br>**Possible Causes:**<br>HFAN radiator relay control circuit is open or shorted to ground<br>HFAN radiator relay power circuit is open (test power to IGN)<br>HFAN radiator fan relay is damaged or has failed<br>PCM has failed |
| **P1490**<br>1T CCM<br>1995-2003<br>300M, Avenger, Breeze, Caravan, Cherokee, Cirrus, Concorde, PT Cruiser, Dakota, Durango, Grand Cherokee, Intrepid, LHS, Liberty, Neon, New Yorker, Prowler, Ram, Sebring, Stratus, Talon, Town & Country, Viper, Vision, Voyager, Wrangler | **Low Speed Radiator Fan Relay Circuit Malfunction Conditions:**<br>Key on or engine running; system voltage over 10.5v, and the PCM detected an unexpected voltage condition on the Low Speed Radiator Fan Relay control circuit during the CCM test.<br>**Possible Causes:**<br>LFAN radiator relay control circuit is open or shorted to ground<br>LFAN radiator relay power circuit is open (test power to IGN)<br>LFAN radiator fan relay is damaged or has failed<br>PCM has failed |
| **P1491**<br>1T CCM<br>1995-2003<br>300M, Avenger, Breeze, Caravan, Cherokee, Cirrus, Concorde, PT Cruiser, Dakota, Durango, Grand Cherokee, Intrepid, LHS, Liberty, Neon, New Yorker, Prowler, Ram, Sebring, Stratus, Talon, Town & Country, Viper, Vision, Voyager, Wrangler | **Radiator Fan Control Relay Circuit Malfunction Conditions:**<br>Key on or engine running; system voltage over 10.5v, and the PCM detected an unexpected voltage condition on the Radiator Fan Control Relay circuit during the CCM test period.<br>**Possible Causes:**<br>Radiator fan control relay circuit is open or shorted to ground<br>Radiator fan control relay circuit is open (test power to IGN)<br>Radiator fan control relay is damaged or has failed<br>PCM has failed |
| **P1492**<br>1T CCM<br>1995-2005<br>300M, Avenger, Breeze, Caravan, Cherokee, Cirrus, Concorde, PT Cruiser, Dakota, Durango, Grand Cherokee, Intrepid, LHS, Liberty, Magnum, Neon, New Yorker, Prowler, Ram, Sebring, Stratus, Talon, Town & Country, Viper, Vision, Voyager, Wrangler | **Battery Temperature Sensor Circuit High Input Conditions:**<br>Key on or engine running; and the PCM detected the BTS signal indicated more than 4.90v for 3 seconds during the CCM test.<br>**Possible Causes:**<br>BTS signal circuit is open between the Sensor and the PCM<br>BTS ground circuit is open between the Sensor and the PCM<br>BTS (Sensor) is damaged or the PCM has failed |

**OBD II Trouble Code List (P1XXX Codes) – Continued**

| DTC | Trouble Code Title, Conditions & Possible Causes: |
|---|---|
| **P1493**<br>1T CCM<br>1995-2005<br>300M, Avenger, Breeze, Caravan, Cherokee, Cirrus, Concorde, PT Cruiser, Dakota, Durango, Grand Cherokee, Intrepid, LHS, Liberty, Magnum, Neon, New Yorker, Prowler, Ram, Sebring, Stratus, Talon, Town & Country, Viper, Vision, Voyager, Wrangler | **Battery Temperature Sensor Circuit Low Input Conditions:**<br>Key on or engine running; and the PCM detected the BTS signal indicated less than 0.30v for 3 seconds during the CCM test.<br>**Possible Causes:**<br>BTS circuit is shorted to ground between Sensor and the PCM<br>BTS (Sensor) is damaged or has failed<br>PCM has failed |
| **P1494**<br>1T CCM<br>1996-2005<br>300M, Avenger, Breeze, Caravan, Cherokee, Cirrus, Concorde, PT Cruiser, Dakota, Durango, Grand Cherokee, Intrepid, LHS, Liberty, Magnum, Neon, New Yorker, Prowler, Ram, Sebring, Stratus, Talon, Town & Country, Viper, Vision, Voyager, Wrangler | **EVAP Leak Detection Pump Switch Or Mechanical Fault Conditions:**<br>BTS from 40-96ºF and ECT Sensor within 10ºF of the BTS signal at startup (cold engine), engine started, and the PCM detected the LDP switch was not in its expected state at key "on" or engine running.<br>**Possible Causes:**<br>LDP switch signal circuit is open or shorted to ground<br>LDP switch power circuit is open (test power to Fused Ignition)<br>LDP vacuum hose is clogged, loose or restricted<br>LDP assembly is damaged or has failed (the switch has failed) |
| **P1495**<br>1T CCM<br>1996-2005<br>300M, Avenger, Breeze, Caravan, Cherokee, Cirrus, Concorde, PT Cruiser, Dakota, Durango, Grand Cherokee, Intrepid, LHS, Liberty, Magnum, Neon, New Yorker, Prowler, Ram, Sebring, Stratus, Talon, Town & Country, Viper, Vision, Voyager, Wrangler | **Leak Detection Pump Solenoid Circuit Malfunction Conditions:**<br>Engine started, ECT Sensor from 40-90ºF and within 10ºF of the Battery Temperature Sensor signal, engine running, and the PCM detected the Actual state of the Leak Detection Pump solenoid did not match the Intended state of the solenoid during the test period.<br>**Possible Causes:**<br>LDP power supply circuit from the ignition switch is open<br>LDP solenoid control circuit is open or shorted to ground<br>LDP assembly is damaged or it has failed<br>PCM has failed |

**OBD II Trouble Code List (P1XXX Codes) – Continued**

| DTC | Trouble Code Title, Conditions & Possible Causes: |
|---|---|
| **P1496**<br>1T CCM<br>1996-2003<br>300M, Avenger, Breeze, Caravan, Cherokee, Cirrus, Concorde, PT Cruiser, Dakota, Durango, Grand Cherokee, Intrepid, LHS, Liberty, Neon, New Yorker, Prowler, Ram, Sebring, Stratus, Talon, Town & Country, Viper, Vision, Voyager, Wrangler | **5-Volt VREF Supply Voltage Too Low Conditions:**<br>Key on or engine running; and the PCM detected the 5-volt VREF supply was less than 3.5v for 4 seconds during the CCM test.<br>**Possible Causes:**<br>A/C pressure Sensor has failed (a short to ground condition)<br>MAP Sensor has failed (a short to ground condition)<br>TP Sensor has failed (a short to ground condition)<br>PCM has failed |
| **P1497**<br>1T CCM<br>1996-97<br>Concorde, Intrepid, LHS, Prowler | **PCM Failure (SRI Mileage Not Stored) Conditions:**<br>Key on, and the PCM detected an unsuccessful attempt to "write" the Service Reminder Indicator (SRI) or Emission Mileage Request (EMR) mileage to an EEPROM located occurred during initialization.<br>**Possible Causes:**<br>Clear the trouble codes and retest for the same trouble code. If DTC P1697 resets, replace the PCM and then reprogram it. |
| **P1498**<br>1T CCM<br>1996-2002<br>300M, Avenger, Breeze, Caravan, Cherokee, Cirrus, Concorde, PT Cruiser, Dakota, Durango, Grand Cherokee, Intrepid, LHS, Liberty, Neon, New Yorker, Prowler, Ram, Sebring, Stratus, Talon, Town & Country, Viper, Vision, Voyager, Wrangler | **No CCD Messages Received From The TCM Conditions:**<br>Key on or engine running; and the PCM detected a failure to communicate with the TCM over the CCD data Bus circuit.<br>**Possible Causes:**<br>CCD data Bus circuit is open, shorted to ground or to power<br>TCM or the PCM has failed |
| **P1499**<br>1T CCM<br>1996-2005<br>Cherokee, Grand Cherokee, Liberty, Viper, Wrangler | **Radiator (Hydraulic) Fan Solenoid Circuit Failure Conditions:**<br>Key on or engine running; and the PCM detected an unexpected voltage condition on the Radiator Fan Solenoid Control circuit.<br>**Possible Causes:**<br>Radiator fan solenoid control circuit is open<br>Radiator fan solenoid ground circuit is open<br>Radiator fan solenoid power circuit is open<br>Radiator fan solenoid is damaged or has failed<br>PCM has failed |
| **P1500**<br>1T CCM<br>1995-98<br>Talon | **Generator 'FR' Terminal Circuit Failure Conditions:**<br>Engine started, and the PCM detected the Generator 'FR' terminal remained at more than 4.5v for over 20 seconds during the test.<br>**Possible Causes:**<br>Generator 'FR' terminal circuit is open between the Generator and the PCM terminal<br>Generator is damaged or has failed<br>PCM has failed |

**OBD II Trouble Code List (P1XXX Codes) – Continued**

| DTC | Trouble Code Title, Conditions & Possible Causes: |
|---|---|
| **P1500**<br>1T CCM<br>2001-03<br>Sebring, Stratus Coupe<br>with 2.4L I4 or 3.0L V6 | **Generator 'FR' Terminal Circuit Failure Conditions:**<br>Engine started, engine running and the PCM detected the Generator 'FR' terminal input signal indicated more than 4.50v for 20 seconds.<br>**Possible Causes:**<br>Generator 'FR' circuit is open between the Generator and PCM<br>Generator is damaged or it has failed<br>PCM has failed |
| **P1501**<br>1T CCM<br>2004-05<br>300M, Dakota, Durango,<br>Magnum, Ram | **Vehicle Speed Sensor No. 1/2 Drive Wheel Correlation Conditions:**<br>Engine is running and vehicle is moving. Speed control is learned and the speed control is trying to be activated. If the PCM recognizes the rear wheel speed is greater than the front wheel speed, this DTC will set.<br>**Possible Causes:**<br>Other active Bus or Communication DTCs<br>Incorrect tire circumference<br>PCM has failed |
| **P1502**<br>1T CCM<br>2004-05<br>300M, Dakota, Durango,<br>Magnum, Ram | **Vehicle Speed Sensor No. 1/2 Non-Drive Wheel Correlation Conditions:**<br>Engine is running and vehicle is moving; brake pedal must not be applied. If the PCM recognizes the rear wheel speed is greater than the front wheel speed, this DTC will set.<br>**Possible Causes:**<br>Other active Bus or Communication DTCs<br>Incorrect tire circumference<br>PCM has failed |
| **P1521**<br>1T CCM<br>2004-05<br>300M, Magnum | **Incorrect Engine Oil Type Conditions:**<br>Engine is running. The PCM will use oil pressure, oil temperature and other vital engine inputs to determine the engine oil viscosity. Incorrect viscosity will affect the operation of the MDS by delaying cylinder activation.<br>**Possible Causes:**<br>Incorrect engine oil type<br>Engine oil contamination<br>Engine oil has aged and is breaking down |
| **P1572**<br>1T CCM<br>2004-05<br>300M, Dakota, Durango,<br>Magnum, Dakota,<br>Durango, Ram | **Brake Switch Stuck ON Conditions:**<br>Ignition is on. The PCM recognizes that brake switch 1 is mechanically stuck in the Low/On position.<br>**Possible Causes:**<br>Brake switch 1 signal is shorted to ground<br>Brake switch 2 signal is open<br>Stop lamp switch has failed<br>PCM has failed |
| **P1573**<br>1T CCM<br>2004-05<br>300M, Dakota, Durango,<br>Magnum, Ram | **Brake Switch Stuck ON Conditions:**<br>Ignition is on. The PCM recognizes that brake switch 1 is mechanically stuck in the High/Off position.<br>**Possible Causes:**<br>Brake switch 1 signal is shorted to ground or to voltage<br>Brake switch 2 signal is open or is shorted to ground<br>Ground circuit is open<br>Fused ignition switch output is open<br>Stop lamp switch has failed<br>PCM has failed |

**OBD II Trouble Code List (P1XXX Codes) – Continued**

| DTC | Trouble Code Title, Conditions & Possible Causes: |
|---|---|
| **P1593**<br>1T CCM<br>2004-05<br>300M, Caravan, Concorde, PT Cruiser, Dakota, Durango, Grand Cherokee, Intrepid, LHS, Liberty, Magnum, Neon, Pacifica, Prowler, Ram, Sebring, Stratus, Town & Country, Viper, Voyager, Wrangler | **Speed Control Switch Stuck Operation Conditions:**<br>Ignition on. Either S/C switch is mechanically stuck in On/Off, Resume/Accel or Set position for too long.<br>**Possible Causes:**<br>Intermittent speed control switch 1/2 stuck DTC<br>S/C switches or Steering Column Control Module malfunctioning<br>S/C signal circuit open or shorted ground or to battery voltage<br>S/C switch signal circuit shorted to switch return circuit<br>S/C Sensor ground open<br>PCM has failed |
| **P1594**<br>1T CCM<br>1996-97<br>Concorde, Intrepid, LHS, Prowler | **Charging System Voltage Too High Conditions:**<br>Engine started, and the PCM detected the Charging System voltage was too high even after it tried to lower the output. Note: The Generator illuminates when this code sets.<br>**Possible Causes:**<br>Battery temperature Sensor is damaged or has failed (skewed)<br>Generator field driver circuit is shorted to ground<br>Generator has an internal short circuit condition<br>PCM has failed |
| **P1594**<br>1T CCM<br>1996-2005<br>Cherokee, Grand Cherokee, Liberty, Viper, Wrangler | **Charging System Voltage Too High Conditions:**<br>Engine started, engine running, and the PCM detected the Charging System voltage was too high even after it tried to lower the generator output by controlling the Field control circuit (Generator Lamp is on).<br>**Possible Causes:**<br>Battery temperature Sensor is damaged or has failed (skewed)<br>Generator field driver circuit is shorted to ground<br>Generator has an internal short circuit condition<br>PCM has failed |
| **P1594**<br>1T CCM<br>2004-05<br>PT Cruiser | **Charging System Voltage Too High Conditions:**<br>Engine is running at more than 380 rpm. If the battery voltage is 1v greater than the desired voltage, this DTC will set.<br>**Possible Causes:**<br>Intermittent condition<br>Generator field driver circuit is shorted to ground<br>Generator field is damaged or has failed<br>PCM has failed |
| **P1595**<br>1T CCM<br>1996-2003<br>300M, Avenger, Breeze, Caravan, Cherokee, Cirrus, Concorde, PT Cruiser, Dakota, Durango, Grand Cherokee, Intrepid, LHS, Liberty, Neon, New Yorker, Prowler, Ram, Sebring, Stratus, Talon, Town & Country, Viper, Vision, Voyager, Wrangler | **Speed Control Solenoid Circuit Failure Conditions:**<br>Engine started, vehicle driven at over 35 mph, S/C enabled with the Set switch "on", and the PCM detected it could not control the operation of the vacuum and vent control solenoids.<br>**Possible Causes:**<br>S/C power supply circuit is open (test power from Brake switch)<br>S/C vacuum solenoid control circuit open or shorted to ground<br>S/C vent solenoid control circuit is open or shorted to ground<br>S/C vacuum or vent solenoid is damaged or has failed<br>PCM has failed |

**OBD II Trouble Code List (P1XXX Codes) – Continued**

| DTC | Trouble Code Title, Conditions & Possible Causes: |
|---|---|
| **P1596**<br>1T CCM<br>1996-2003<br>300M, Avenger, Breeze, Caravan, Cherokee, Cirrus, Concorde, PT Cruiser, Dakota, Durango, Grand Cherokee, Intrepid, LHS, Liberty, Neon, New Yorker, Prowler, Ram, Sebring, Stratus, Talon, Town & Country, Viper, Vision, Voyager, Wrangler | **Speed Control Switch Continuous High Input Conditions:**<br>Key on or engine running; and the PCM detected the S/C switch was in a continuous high voltage state (over 4.70v) during the CCM test.<br>**Possible Causes:**<br>S/C On/Off switch is open<br>S/C switch (MUX switch) is open<br>S/C switch (MUX switch) is shorted to VREF or system power<br>S/C switch (MUX switch) is damaged or has failed<br>PCM has failed |
| **P1597**<br>1T CCM<br>1996-2003<br>300M, Avenger, Breeze, Caravan, Cherokee, Cirrus, Concorde, PT Cruiser, Dakota, Durango, Grand Cherokee, Intrepid, LHS, Liberty, Neon, New Yorker, Prowler, Ram, Sebring, Stratus, Talon, Town & Country, Viper, Vision, Voyager, Wrangler | **Speed Control Switch Continuous Low Input Conditions:**<br>Key on or engine running; and the PCM detected the S/C switch was in a continuous low voltage state (below 4.50v) during the CCM test.<br>**Possible Causes:**<br>S/C On/Off switch is shorted to ground<br>S/C switch (MUX switch) is shorted to ground<br>S/C switch (MUX switch) is damaged or has failed<br>PCM has failed |
| **P1598**<br>1T CCM<br>1996-2005<br>300M, Avenger, Breeze, Caravan, Cherokee, Cirrus, Concorde, PT Cruiser, Dakota, Durango, Grand Cherokee, Intrepid, LHS, Liberty, Magnum, Neon, New Yorker, Prowler, Ram, Sebring, Stratus, Talon, Town & Country, Viper, Vision, Voyager, Wrangler | **A/C Pressure Sensor Circuit High Input Conditions:**<br>Engine started, engine running, A/C Relay is "on", and the PCM detected the A/C Pressure Sensor indicated more than 4.90v.<br>**Possible Causes:**<br>A/C pressure Sensor circuit is open or shorted to VREF (5v)<br>A/C pressure Sensor ground circuit is open<br>A/C pressure Sensor is damaged or has failed<br>PCM has failed |
| **P1599**<br>1T CCM<br>1996-2005<br>300M, Avenger, Breeze, Caravan, Cherokee, Cirrus, Concorde, PT Cruiser, Dakota, Durango, Grand Cherokee, Intrepid, LHS, Liberty, Magnum, Neon, New Yorker, Prowler, Ram, Sebring, Stratus, Talon, Town & Country, Viper, Vision, Voyager, Wrangler | **A/C Pressure Sensor Circuit Low Input Conditions:**<br>Engine started, engine running, A/C Relay is "on", and the PCM detected the A/C Pressure Sensor indicated less than 0.70v.<br>**Possible Causes:**<br>A/C pressure Sensor circuit is shorted to ground<br>A/C pressure Sensor power circuit is open<br>A/C pressure Sensor is damaged or has failed<br>PCM has failed |

## OBD II Trouble Code List (P1XXX Codes) – Continued

| DTC | Trouble Code Title, Conditions & Possible Causes: |
|---|---|
| **P1602**<br>1T PCM<br>1995-98<br>Talon | **A/T Serial Communication Link Circuit Malfunction Conditions:**<br>Key on or engine running; and the PCM detected an unexpected voltage condition on the serial communication link used to communicate between it and the TCM.<br>**Possible Causes:**<br>CCD data Bus (+) circuit is open or shorted to ground<br>CCD data Bus (-) circuit is open<br>TCM has failed<br>PCM has failed |
| **P1602**<br>1T PCM<br>2001-05<br>300M, Magnum, Concorde, Intrepid, LHS, Neon, Sebring, Stratus, PT Cruiser, Pacifica, Caravan/Town & Country/Voyager, Jeep, Dakota, Durango, Ram | **PCM Not Programmed Conditions:**<br>Key on. The PCM detected that it had not been programmed.<br>**Possible Causes:**<br>Program the PCM and then retest for this same trouble code<br>PCM has failed |
| **P1603**<br>1T PCM<br>2001-05<br>300M, Caravan, Cherokee, Concorde, PT Cruiser, Dakota, Durango, Grand Cherokee, Intrepid, LHS, Liberty, Magnum, Neon, New Yorker, Pacifica, Prowler, Ram, Sebring, Stratus, Town & Country, Viper, Vision, Voyager, Wrangler | **Powertrain Control Module Internal Dual-Port Ram Communication Conditions:**<br>Key on; and the PCM detected an error message that indicated that it had not been programmed or that it was programmed properly.<br>**Possible Causes:**<br>Fused ignition switch output is missing (off-start-run circuit)<br>PCM is damaged or it has an internal failure |
| **P1603**<br>1T PCM<br>2001-03<br>Sebring, Stratus with 2.4L I4 or 3.0L V6 | **Powertrain Control Module Internal Dual-Port Ram Communication Conditions:**<br>Key on; and the PCM detected an error message that indicated that it had not been programmed or that it was programmed properly.<br>**Possible Causes:**<br>Fused ignition switch output is missing (off-start-run circuit)<br>PCM is damaged or it has an internal failure |
| **P1604**<br>1T PCM<br>2001-05<br>300M, Caravan, Cherokee, Concorde, PT Cruiser, Dakota, Durango, Grand Cherokee, Intrepid, LHS, Liberty, Magnum, Neon, New Yorker, Pacifica, Prowler, Ram, Sebring, Stratus, Town & Country, Viper, Vision, Voyager, Wrangler | **PCM Internal Dual-Port Ram Read/Write Integrity Failure Conditions:**<br>Key on; and the PCM detected an error message that indicated it had not been programmed, or it was not programmed properly.<br>**Possible Causes:**<br>Fused ignition switch output is missing (off-start-run circuit)<br>PCM is damaged or it has an internal failure |

**OBD II Trouble Code List (P1XXX Codes) – Continued**

| DTC | Trouble Code Title, Conditions & Possible Causes: |
|---|---|
| **P1607**<br>1T PCM<br>2001-05<br>300M, Caravan, Cherokee, Concorde, PT Cruiser, Dakota, Durango, Grand Cherokee, Intrepid, LHS, Liberty, Magnum, Neon, New Yorker, Pacifica, Prowler, Ram, Sebring, Stratus, Town & Country, Viper, Vision, Voyager, Wrangler | **Powertrain Control Module Internal Shutdown Timer Rationality Conditions:**<br>Cold engine startup, and after the PCM compared the coolant temperature to the shutdown time, it detected a rationality fault.<br>**Possible Causes:**<br>Fused ignition switch output is missing (off-start-run circuit)<br>PCM is damaged or it has an internal failure |
| **P1610**<br>1T PCM<br>2001-03<br>Sebring, Stratus with 2.4L I4 or 3.0L V6 | **PCM Signal Line To Immobilizer Circuit Malfunction Conditions:**<br>Key on, and the PCM detected an unexpected voltage condition on the communication line between the Immobilizer ECU and the PCM.<br>**Possible Causes:**<br>Immobilizer communication line to the PCM is open<br>Immobilizer communication line to the PCM is shored to ground<br>Immobilizer communication line to the PCM is shorted to power<br>Immobilizer ECU is damaged or it has failed<br>PCM has failed |
| **P1616**<br>1T PCM<br>2004-05<br>PT Cruiser | **Primary 5v Sensor Reference Voltage Low Conditions:**<br>Key on. If the PCM detects a voltage of less than 4.75v on the primary 5v supply circuit for at least 100ms, this DTC will set.<br>**Possible Causes:**<br>Intermittent condition<br>Primary 5v supply circuit shorted to ground<br>PCM has failed |
| **P1617**<br>1T PCM<br>2004-05<br>PT Cruiser | **Primary 5v Sensor Reference Voltage High Conditions:**<br>Key on. If the PCM detects a voltage of more than 5.25v on the primary 5v supply circuit for at least 100ms, this DTC will set.<br>**Possible Causes:**<br>Intermittent condition<br>Primary 5v supply circuit is open or is shorted to voltage<br>PCM has failed |
| **P1618**<br>1T PCM<br>2004-05<br>PT Cruiser | **Primary 5v Sensor Reference Voltage Unstable Conditions:**<br>Key on. If the PCM detects a voltage variance of more than 0.25v on the primary 5v supply circuit for more than 100ms, this DTC will set.<br>**Possible Causes:**<br>Primary 5v supply circuit open or shorted to ground or to battery voltage<br>5v Sensor has failed<br>PCM has failed |
| **P1618**<br>1T PCM<br>2004-05<br>300M, Dakota, Durango, Magnum, Ram | **Primary 5v Sensor Reference Voltage Malfunction Conditions:**<br>Key on. The PCM recognizes the primary 5v supply circuit voltage is varying too much too quickly. ETC light is flashing.<br>**Possible Causes:**<br>Primary 5v supply circuit open or shorted to ground or to battery voltage<br>5v Sensor has failed<br>PCM has failed |

**OBD II Trouble Code List (P1XXX Codes) – Continued**

| DTC | Trouble Code Title, Conditions & Possible Causes: |
|---|---|
| **P1626**<br>1T PCM<br>2004-05<br>PT Cruiser | **Secondary 5v Sensor Reference Voltage Low Conditions:**<br>Key on. If the PCM detects a voltage of less than 4.75v on the secondary 5v supply circuit for more than 100ms, this DTC will set.<br>**Possible Causes:**<br>Intermittent condition<br>5v supply circuit shorted to ground<br>5v Sensor has failed<br>PCM has failed |
| **P1627**<br>1T PCM<br>2004-05<br>PT Cruiser | **Secondary 5v Sensor Reference Voltage High Conditions:**<br>Key on. If the PCM detects a voltage of more than 5.25v on the secondary 5v supply circuit for more than 100ms, this DTC will set.<br>**Possible Causes:**<br>Intermittent condition<br>5v supply circuit is open or is shorted to voltage<br>5v Sensor has failed<br>PCM has failed |
| **P1628**<br>1T PCM<br>2004-05<br>PT Cruiser | **Secondary 5v Sensor Reference Voltage Unstable Conditions:**<br>Key on. If the PCM detects a voltage variance of more than 0.25v on the secondary 5v supply circuit for more than 100ms, this DTC will set.<br>**Possible Causes:**<br>Intermittent condition<br>ETC assembly has failed<br>5v supply circuit has high resistance<br>PCM has failed |
| **P1628**<br>1T PCM<br>2004-05<br>300M, Dakota, Durango, Magnum, Ram | **Auxiliary 5v Sensor Reference Voltage Malfunction Conditions:**<br>Key on. The PCM recognizes the auxiliary 5v supply circuit voltage is varying too much too quickly. ETC light is flashing.<br>**Possible Causes:**<br>Auxiliary 5v supply circuit open or shorted to ground or to battery voltage<br>5v Sensor has failed<br>PCM has failed |
| **P1652**<br>1T CCM<br>2001-05<br>300M, Caravan, Cherokee, Concorde, PT Cruiser, Dakota, Durango, Grand Cherokee, Intrepid, LHS, Liberty, Magnum, Neon, New Yorker, Pacifica, Prowler, Ram, Sebring, Stratus, Town & Country, Viper, Vision, Voyager, Wrangler | **Serial Communication Link Malfunction Conditions:**<br>Engine started; and after the TCM did not detect any signals on the Serial Communication Line for more than 20 seconds. Note: Due to the integration of the PCM and TCM, Bus communication between the modules is internal.<br>**Possible Causes:**<br>TCM cannot communicate with the Instrument Cluster (MIC)<br>TCM cannot communicate with the Powertrain Control Module<br>PCM/TCM is damaged or it has an internal failure |
| **P1653**<br>1T PCM<br>2004-05<br>PT Cruiser | **PCI Bus Shorted To Ground Conditions:**<br>Key on. If the PCM detects a short to ground on the PCI Bus for more than 5 seconds, this DTC will set.<br>**Possible Causes:**<br>Intermittent condition<br>Internal controller short to ground<br>PCI Bus short to ground |

**OBD II Trouble Code List (P1XXX Codes) – Continued**

| DTC | Trouble Code Title, Conditions & Possible Causes: |
|---|---|
| **P1654**<br>1T PCM<br>2004-05<br>PT Cruiser | **PCI Bus Shorted To Voltage Conditions:**<br>Key on. If the PCM detects a short to voltage on the PCI Bus for more than 5 seconds, this DTC will set.<br>**Possible Causes:**<br>Intermittent condition<br>Internal controller short to battery voltage<br>PCI Bus short to voltage |
| **P1654**<br>1T PCM<br>2004-05<br>PT Cruiser | **PCI Bus Not Available Conditions:**<br>Key on. If the PCM detects the PCI Bus is not available for more than 5 seconds, this DTC will set.<br>**Possible Causes:**<br>Intermittent condition<br>PCI Bus circuit is open<br>PCM has failed |
| **P1681**<br>1T PCM<br>2003<br>300M, Concorde, PT Cruiser, Intrepid, Neon, Prowler, Ram, Sebring, Stratus | **No Fuel Level Bus Messages Conditions:**<br>Key on, and the PCM determined that it did not receive any Fuel Level messages over the Data Bus line for 20 seconds.<br>**Possible Causes:**<br>Data Bus circuit from BCM to PCM is damaged or it has failed<br>Fuel level Bus message to the PCM is invalid<br>BCM is damaged or has failed<br>PCM unable to communicate with the Body Control Module |
| **P1682**<br>1T CCM<br>1996-2005<br>300M, Avenger, Breeze, Caravan, Cherokee, Cirrus, Concorde, PT Cruiser, Dakota, Durango, Grand Cherokee, Intrepid, LHS, Liberty, Magnum, Neon, New Yorker, Prowler, Ram, Sebring, Stratus, Talon, Town & Country, Viper, Vision, Voyager, Wrangler | **Charging System Voltage Too Low Conditions:**<br>Engine started; engine speed over 1152 rpm, and the PCM detected the Battery Sense circuit was 1.0v less than the Charging System circuit for 25 seconds during the CCM test (Generator Lamp is "on").<br>**Possible Causes:**<br>Battery positive or Fused Ignition circuit has high resistance<br>Generator drive belt out-of-adjustment or worn out<br>Generator field circuit has a high resistance condition<br>PCM has failed |
| **P1683**<br>1T CCM<br>1996-2003<br>300M, Avenger, Breeze, Caravan, Cherokee, Cirrus, Concorde, PT Cruiser, Dakota, Durango, Grand Cherokee, Intrepid, LHS, Liberty, Neon, New Yorker, Prowler, Ram, Sebring, Stratus, Talon, Town & Country, Viper, Vision, Voyager, Wrangler | **Speed Control Relay Or Driver Circuit Malfunction Conditions:**<br>Engine started, engine running with the S/C switch "on"; and the PCM detected an unusual voltage condition on the S/C Relay control circuit during the CCM test.<br>**Possible Causes:**<br>S/C power supply circuit is open or shorted to ground<br>S/C dump solenoid (servo) is damaged or has failed<br>PCM has failed |

**OBD II Trouble Code List (P1XXX Codes) – Continued**

| DTC | Trouble Code Title, Conditions & Possible Causes: |
|---|---|
| **P1684**<br>1T PCM<br>2001-05<br>300M, Avenger, Breeze, Caravan, Cherokee, Cirrus, Concorde, PT Cruiser, Dakota, Durango, Grand Cherokee, Intrepid, LHS, Liberty, Magnum, Neon, New Yorker, Prowler, Ram, Sebring, Stratus, Talon, Town & Country, Viper, Vision, Voyager, Wrangler | **Battery Has Been Disconnected Conditions:**<br>Key on, and the TCM detected that it had been disconnected from the Battery Direct (B+) circuit or its Power Ground circuit. This DTC will also set during the scan tool Quick Battery Disconnect procedure. Note: Due to the integration of the PCM and TCM, the transmission part of the PCM has its own specific power and ground circuits.<br>**Possible Causes:**<br>Quick Learn procedure was performed with scan tool<br>TCM battery direct (B+) circuit is open or disconnected<br>TCM power ground circuit is open<br>PCM/TCM was disconnected or it has been replaced |
| **P1685**<br>1T PCM<br>1996-2003<br>300M, Avenger, Breeze, Caravan, Cherokee, Cirrus, Concorde, PT Cruiser, Dakota, Durango, Grand Cherokee, Intrepid, LHS, Liberty, Neon, New Yorker, Prowler, Ram, Sebring, Stratus, Talon, Town & Country, Viper, Vision, Voyager, Wrangler | **Smart Key Immobilizer Module Invalid Key Conditions:**<br>Key on, and the PCM received a message from the Smart Key Immobilizer Module (SKIM) that an invalid key had been inserted.<br>**Possible Causes:**<br>A theft attempt may have occurred.<br>Obtain the correct key and attempt to start the vehicle<br>Do a PCM Reset function to clear the code after engine startup |
| **P1686**<br>1T PCM<br>1999-2003<br>300M, Avenger, Breeze, Caravan, Cherokee, Cirrus, Concorde, PT Cruiser, Dakota, Durango, Grand Cherokee, Intrepid, LHS, Liberty, Neon, New Yorker, Prowler, Ram, Sebring, Stratus, Town & Country, Viper, Voyager, Wrangler | **No SKIM Bus Messages Received Conditions:**<br>Engine started; and the PCM did not detect any MIC (I/P Cluster) messages over the Data Bus circuit for 20 seconds.<br>**Possible Causes:**<br>Data Bus circuit from MIC to the PCM is damaged or it is open<br>Instrument Cluster is damaged or has failed<br>PCM unable to communicate with the Instrument Cluster<br>PCM has failed |
| **P1686**<br>1T PCM<br>2004-05<br>PT Cruiser | **No SKIM Bus Message Received Conditions:**<br>Key on. If the PCM does not receive a Bus message from the SKIM when expected, this DTC will set.<br>**Possible Causes:**<br>Intermittent condition<br>SKIM/PCM has failed<br>Loss of SKIM communication link<br>PCI Bus circuit open from PCM to SKIM |

**OBD II Trouble Code List (P1XXX Codes) – Continued**

| DTC | Trouble Code Title, Conditions & Possible Causes: |
|---|---|
| **P1687**<br>1T PCM<br>1999-2005<br>300M, Avenger, Breeze, Caravan, Cherokee, Cirrus, Concorde, PT Cruiser, Dakota, Durango, Grand Cherokee, Intrepid, LHS, Liberty, Magnum, Neon, New Yorker, Prowler, Ram, Sebring, Stratus, Talon, Town & Country, Viper, Vision, Voyager, Wrangler | **No Cluster Bus Messages Conditions:**<br>Key on or engine running; and the PCM determined that it did not receive any Security Key Bus Messages over the Data Bus line for 20 seconds. This malfunction may be an intermittent problem.<br>**Possible Causes:**<br>Data Bus circuit from SKIM to PCM is damaged or it is open<br>PCM unable to communicate with the Body Control Module<br>PCM has failed, or the SKIM is damaged or has failed |
| **P1687**<br>1T PCM<br>2004-05<br>300M, Caravan, Concorde, PT Cruiser, Dakota, Durango, Grand Cherokee, Intrepid, LHS, Liberty, Magnum, Neon, Pacifica, Prowler, Ram, Sebring, Stratus, Town & Country, Viper, Voyager, Wrangler | **No Communication with MIC Conditions:**<br>Communications are monitored continuously with engine running. The DTC sets in about 25 seconds if no Bus messages are received from the MIC.<br>**Possible Causes:**<br>Other Bus problems exist<br>Intermittent wiring or connector problems exist<br>PCM has failed |
| **P1688**<br>1T CCM<br>1998-2002<br>Ram with 5.9L Diesel | **Fuel Injection Pump Internal Malfunction Conditions:**<br>Key on or engine running; and the PCM detected an unexpected "low" voltage condition on the Auxiliary 5-volt power circuit.<br>**Possible Causes:**<br>Fuel injection pump DTC counter malfunction<br>Fuel injection pump Good Trip counter malfunction<br>Fuel injection pump is damaged or has failed<br>ECM has failed |
| **P1689**<br>1T PCM<br>1998-2002<br>Ram with 5.9L Diesel | **No Communication Between ECM & Injection Pump Module Conditions:**<br>Key on, and the PCM detected the time between the CAN messages received from the Instrument Panel (I/P) module was more than 3 seconds, or it detected no messages arrived.<br>**Possible Causes:**<br>Fuel injection pump ground circuit open or has high resistance<br>Fuel injection module wiring harness is damaged or has failed<br>CAN data Bus (+) circuit is open, shorted to ground or to power<br>ECM has failed |
| **P1690**<br>1T CCM<br>1998-2002<br>Ram with 5.9L Diesel | **Injection Pump CKP Signal Different Than The CKP Signal Conditions:**<br>No CKP or CMP Sensor codes set, engine started, and the PCM detected that the CKP signal received by the Instrument Panel (I/P) module was not within its normal operating range.<br>**Possible Causes:**<br>DTC counter did not change to (0)<br>Fuel injection module wiring harness is damaged or has failed<br>Fuel injection static timing is incorrect<br>Fuel "sync" circuit is open or shorted to ground<br>ECM has failed |

**OBD II Trouble Code List (P1XXX Codes) – Continued**

| DTC | Trouble Code Title, Conditions & Possible Causes: |
|---|---|
| **P1691**<br>1T PCM<br>1998-2002<br>Ram with 5.9L Diesel | **Fuel Injection Pump Calibration Error Conditions:**<br>Key on or engine running; and the PCM detected a calibration error related to the Fuel Injection pump operation during the CCM test.<br>**Possible Causes:**<br>Fuel Injection pump error at key "on" or at startup<br>Fuel injection pump is damaged or has failed<br>ECM has failed |
| **P1692**<br>1T PCM<br>1998-2002<br>Ram with 5.9L Diesel | **Diagnostic Trouble Code Set In The Companion Module Conditions:**<br>Key on and the PCM detected a diagnostic trouble code was set in the Companion Module.<br>**Possible Causes:**<br>Companion Module detected a problem and set a trouble code in memory<br>ECM has failed |
| **P1693**<br>1T PCM<br>1998-2002<br>Ram with 5.9L Diesel | **Diagnostic Trouble Code Set In The ECM Conditions:**<br>Key on or engine running; and the PCM detected a diagnostic trouble code was set in the Electronic Control Module (ECM).<br>**Possible Causes:**<br>PCM has detected a problem and set a trouble code in memory<br>PCM has failed |
| **P1694**<br>1999-2005<br>300M, Avenger, Breeze, Caravan, Cherokee, Cirrus, Concorde, PT Cruiser, Dakota, Durango, Grand Cherokee, Intrepid, LHS, Liberty, Magnum, Neon, New Yorker, Prowler, Ram, Sebring, Stratus, Talon, Town & Country, Viper, Vision, Voyager, Wrangler | **No PCM Bus Messages Conditions:**<br>Ignition on or engine started; system voltage over 10.5v and the PCM determined that it did not receive any Bus messages for 10 seconds. Note: Due to the integration of the PCM and TCM, Bus communication between the modules is internal.<br>**Possible Causes:**<br>Data Bus circuit connector is damaged, open or it is shorted<br>Data Bus circuit to the PCM is damaged or it is open<br>PCM unable to communicate with the body control module<br>Intermittent wiring or connector problems exist<br>PCM has failed |
| **P1694**<br>1998-2002<br>Ram with 5.9L Diesel | **No PCM Bus Messages Conditions:**<br>Engine started; system voltage over 10.5v and the PCM determined that it did not receive any Bus messages for 10 seconds.<br>**Possible Causes:**<br>Data Bus circuit connector is damaged, open or it is shorted<br>Data Bus circuit to the PCM is damaged or it is open<br>PCM unable to communicate with BCM or has failed |
| **P1695**<br>1T PCM<br>1998-2005<br>300M, Avenger, Breeze, Caravan, Cherokee, Cirrus, Concorde, PT Cruiser, Dakota, Durango, Grand Cherokee, Intrepid, LHS, Liberty, Magnum, Neon, New Yorker, Prowler, Ram, Sebring, Stratus, Talon, Town & Country, Viper, Vision, Voyager, Wrangler | **No BCM Bus Messages Conditions:**<br>Engine started; system voltage over 10.5v and the TCM determined that it did not receive any BCM messages for 20 seconds.<br>**Possible Causes:**<br>Data Bus circuit from BCM to the PCM is damaged or it is open<br>BCM is damaged or has failed<br>TCM unable to communicate with the TCM<br>TCM has failed |

**OBD II Trouble Code List (P1XXX Codes) – Continued**

| DTC | Trouble Code Title, Conditions & Possible Causes: |
|---|---|
| **P1696**<br>1T PCM<br>1998-2005<br>300M, Avenger, Breeze, Caravan, Cherokee, Cirrus, Concorde, PT Cruiser, Dakota, Durango, Grand Cherokee, Intrepid, LHS, Liberty, Magnum, Neon, New Yorker, Prowler, Ram, Sebring, Stratus, Talon, Town & Country, Viper, Vision, Voyager, Wrangler | **PCM EEPROM Write Operation Denied/Invalid Conditions:**<br>Engine started or ignition on continuously. PCM detected an unsuccessful attempt to program/write to the internal EEPROM. Occurred at initialization or shutdown.<br>**Possible Causes:**<br>DRB or scan tool displays a "write" failure occurred<br>DRB or scan tool displays "write" refused a second time<br>DRB or scan tool displays SRI mileage invalid (compare the SRI mileage reading to the reading on the odometer)<br>PCM has failed |
| **P1696**<br>1T PCM<br>1998-2002<br>Ram with 5.9L Diesel | **PCM Failure (EEPROM Write Operation Denied) Conditions:**<br>Key on, and the PCM detected an unsuccessful attempt to "write" to an EEPROM location occurred during initialization.<br>**Possible Causes:**<br>Clear the trouble codes and retest for the same trouble code. If DTC P1696 resets, replace the PCM and then reprogram it. |
| **P1697**<br>1T PCM<br>1996-2005<br>300M, Avenger, Breeze, Caravan, Cherokee, Cirrus, Concorde, PT Cruiser, Dakota, Durango, Grand Cherokee, Intrepid, LHS, Liberty, Magnum, Neon, New Yorker, Prowler, Ram, Sebring, Stratus, Talon, Town & Country, Viper, Vision, Voyager, Wrangler | **PCM Failure (EMR/SRI Mileage Not Stored) Conditions:**<br>Key on, and the PCM detected an unsuccessful attempt to "write" the Service Reminder Indicator (SRI) or Emission Mileage Request (EMR) mileage to an EEPROM located occurred during initialization.<br>**Possible Causes:**<br>Clear the trouble codes and retest for the same trouble code. If DTC P1697 resets, replace the PCM and then reprogram it. |
| **P1697**<br>1T PCM<br>1996-2003<br>Cherokee, Dakota, Durango, Grand Cherokee, Liberty, Ram, Wrangler | **EMR Or SRI Mileage Not Stored Conditions:**<br>Engine started; and the PCM detected an unsuccessful attempt to "write" the Service Reminder Indicator (SRI) or Emission Mileage Request (EMR) mileage to an EEPROM located occurred during initialization or at engine shutdown.<br>**Possible Causes:**<br>DRB or scan tool displays a "write" failure occurred<br>DRB or scan tool displays "write" refused a second time<br>DRB or scan tool displays SRI mileage invalid (compare the SRI mileage reading to the reading on the odometer)<br>Clear the trouble codes and retest for the same trouble code. If DTC P1696 resets, replace the PCM and then reprogram it. |
| **P1698**<br>1T CCM<br>1995-98<br>Talon | **No CCD Messages Received From The TCM Conditions:**<br>Key on or engine running; and the PCM detected a failure to communicate with the TCM over the CCD data Bus circuit.<br>**Possible Causes:**<br>CCD data Bus circuit is open, shorted to ground or to power<br>TCM or the PCM has failed |

**OBD II Trouble Code List (P1XXX Codes) – Continued**

| DTC | Trouble Code Title, Conditions & Possible Causes: |
|---|---|
| **P1698**<br>1T PCM<br>1995-2003<br>300M, Avenger, Breeze, Caravan, Cherokee, Cirrus, Concorde, PT Cruiser, Dakota, Durango, Grand Cherokee, Intrepid, LHS, Liberty, Neon, New Yorker, Prowler, Ram, Sebring, Stratus, Talon, Town & Country, Viper, Vision, Voyager, Wrangler | **No CCD Messages Received From The TCM Conditions:**<br>Key on or engine running; and the PCM detected a failure to communicate with the TCM over the CCD data Bus circuit.<br>**Possible Causes:**<br>CCD data Bus circuit is open, shorted to ground or to power<br>TCM or the PCM has failed |
| **P1714**<br>1T CCM<br>1999-2003<br>Breeze, Cirrus, Prowler, Stratus, Sebring | **A/T Transmission Control Relay Low Battery Voltage Conditions:**<br>Engine started; and the PCM detected a "low" voltage condition on the Transmission Control Relay output circuit after the relay was energized during the CCM test.<br>**Possible Causes:**<br>TCM relay output circuit(s) to the TCM have high resistance<br>TCM relay control circuit is open or shorted to ground<br>TCM relay control power circuit is open (check the fused B+)<br>TCM relay ground circuit is open<br>TCM control relay is damaged or has failed<br>TCM has failed |
| **P1715**<br>1T CCM<br>1995-98<br>Talon with 2.0L Turbo | **A/T Pulse Generator Assembly Circuit Malfunction Conditions:**<br>Engine started; vehicle driven to over 10 mph and the PCM detected an unexpected voltage condition on the Pulse Generator circuit.<br>**Possible Causes:**<br>Pulse generator positive (+) circuit is open or shorted to ground<br>Pulse generator negative (-) circuit is open or shorted to ground<br>Pulse generator is damaged or has failed |
| **P1715**<br>1T CCM<br>2002-03<br>Jeep, Ram | **A/T Pulse Generator Assembly Circuit Malfunction Conditions:**<br>Engine started; vehicle driven to over 10 mph and the PCM detected an unexpected voltage condition on the Pulse Generator circuit.<br>**Possible Causes:**<br>Pulse generator positive (+) circuit is open or shorted to ground<br>Pulse generator negative (-) circuit is open or shorted to ground<br>Pulse generator is damaged or has failed |
| **P1715**<br>1T CCM<br>2004-05<br>Liberty with Diesel | **Restricted Port in T3 Range Conditions:**<br>Monitored whenever the PRNDL code indicates "Temp 3". This DTC will set whenever the conditions for a code P1776 are satisfied with the shifter in the "Temp 3" zone. This causes a restricted port. Related transmission DTCs are present.<br>**Possible Causes:**<br>Improper customer driving habits<br>Misadjusted shifter |

## OBD II Trouble Code List (P1XXX Codes) – Continued

| DTC | Trouble Code Title, Conditions & Possible Causes: |
|---|---|
| **P1716**<br>1T CCM<br>1999-2002<br>300M, Avenger, Breeze, Caravan, Cherokee, Cirrus, Concorde, PT Cruiser, Dakota, Durango, Grand Cherokee, Intrepid, LHS, Liberty, Neon, New Yorker, Prowler, Ram, Sebring, Stratus, Town & Country, Viper, Voyager, Wrangler | **Bus Communication Failure With The PCM Conditions:**<br>Key on, and the PCM detected it could not communicate with the TCM for over 10 seconds.<br>**Possible Causes:**<br>CCD data Bus circuit is open, shorted to ground or to power<br>Extremely low battery (system) voltage<br>TCM or the PCM has failed |
| **P1717**<br>1T CCM<br>2002-03<br>300M, Caravan, Cherokee, Concorde, PT Cruiser, Dakota, Durango, Grand Cherokee, Intrepid, LHS, Liberty, Neon, New Yorker, Prowler, Ram, Sebring, Stratus, Town & Country, Viper, Voyager, Wrangler | **Bus Communication Failure With The MIC Conditions:**<br>Key on or engine started, and the PCM detected it could not communicate with the Mechanical Instrument Cluster (MIC) for over 25 seconds during the test.<br>**Possible Causes:**<br>CCD data Bus circuit is open, shorted to ground or to power<br>Extremely low battery (system) voltage<br>MIC or the TCM has failed |
| **P1718**<br>1T CCM<br>1999-2003<br>Cherokee, Grand Cherokee, Liberty, Wrangler | **TCM Internal Malfunction Conditions:**<br>Key on or engine started, and the PCM received a signal from the TCM over the data Bus circuit that the TCM had an internal problem.<br>**Possible Causes:**<br>TCM has failed.<br>Replace the TCM<br>Perform Transmission Verification Test VER-1A |
| **P1719**<br>1T CCM<br>1996-2005<br>Sebring, Stratus Coupe, Viper | **A/T Skip Shift Solenoid Control Circuit Malfunction Conditions:**<br>Engine started; vehicle driven to a speed of 12-18 mph in 1st Gear at light to moderate engine load at an engine speed over 608 rpm, and the PCM detected an unexpected "low" or high voltage condition on the Reverse Gear Lockout solenoid circuit.<br>**Possible Causes:**<br>Skip Shift solenoid control circuit is open<br>Skip Shift solenoid control circuit shorted to ground<br>Skip Shift solenoid is damaged or has failed<br>PCM has failed |
| **P1736**<br>1T CCM<br>2002-05<br>Liberty, Ram with 3.7L V6 | **A/T Gear Ratio Error In Second Prime Conditions:**<br>Transmission gear ratio is monitored whenever the transmission is in gear. If the ratio of the Input Speed (rpm) to the Output Speed did not match the current gear ratio, this DTC will set. This DTC can take up to 5 minutes of problem identification before lighting the MIL. Related DTCs are present.<br>**Possible Causes:**<br>Internal transmission problems<br>Transmission intermittent gear ratio malfunction |

## OBD II Trouble Code List (P1XXX Codes) – Continued

| DTC | Trouble Code Title, Conditions & Possible Causes: |
|---|---|
| **P1738**<br>1T CCM<br>1999-2003<br>300M, Avenger, Breeze, Caravan, Cherokee, Cirrus, Concorde, PT Cruiser, Dakota, Durango, Grand Cherokee, Intrepid, LHS, Liberty, Neon, New Yorker, Prowler, Ram, Sebring, Stratus, Town & Country, Viper, Voyager, Wrangler | **A/T High Temperature Operation Activated Conditions:**<br>Engine started and the PCM detected the TFT Sensor was more than 240°F during the test.<br>**Possible Causes:**<br>Engine Cooling Fan System is not operating properly<br>Engine cooling fan has failed<br>ATF oil level is too high (overfilled)<br>Transmission oil cooler capacity too low, or cooler is plugged<br>TCM or the PCM has failed |
| **P1739**<br>1T CCM<br>1999-2002<br>300M, Avenger, Breeze, Caravan, Cherokee, Cirrus, Concorde, PT Cruiser, Dakota, Durango, Grand Cherokee, Intrepid, LHS, Liberty, Neon, New Yorker, Prowler, Ram, Sebring, Stratus, Town & Country, Viper, Voyager, Wrangler | **A/T Power Up Circuit At Speed Conditions:**<br>Key on or engine running; and the PCM detected an unexpected "low" voltage on the TCM power up circuit (battery direct circuit).<br>**Possible Causes:**<br>Fused Ignition power circuit is open between TCM and PCM<br>TCM Power UP circuit is open (test power from PDC or Fuse)<br>TCM Power UP circuit is grounded (test power from PDC or<br>TCM or the PCM has failed |
| **P1740**<br>2T PCM<br>1996-2003<br>Cherokee, Dakota, Durango, Grand Cherokee, Liberty, Ram, Wrangler | **TCM Internal Malfunction Conditions:**<br>Engine started; vehicle driven at over 30 mph in 3rd gear with TCC and Overdrive Clutch engaged, and PCM detected an invalid engine speed to output shaft ratio with the TCC or O/D Clutch engaged.<br>**Possible Causes:**<br>ATF fluid is burnt, contaminated or contains excessive debris<br>A/T cooler flow restriction or problems in the oil pump shaft<br>Internal transmission problem (O/D clutch seals or valve body)<br>Overdrive clutch or TCC clutch is damaged or has failed<br>Transmission valve body is leaking, damaged or has failed |
| **P1742**<br>1T CCM<br>1999-2003<br>Cherokee, Grand Cherokee, Liberty, Wrangler | **TCM Internal Malfunction Conditions:**<br>Key on or engine started, and the PCM received a signal from the TCM over the data Bus circuit that the TCM had an internal problem.<br>**Possible Causes:**<br>TCM has failed.<br>Replace the TCM<br>Perform Transmission Verification Test VER-1A |
| **P1743**<br>1T CCM<br>1998-2002<br>Cherokee, Grand Cherokee, Liberty, Wrangler | **TCM Internal Malfunction Conditions:**<br>Key on or engine started, and the PCM received a signal from the TCM over the data Bus circuit that the TCM had an internal problem.<br>**Possible Causes:**<br>TCM has failed.<br>Replace the TCM<br>Perform Transmission Verification Test VER-1A |

**OBD II Trouble Code List (P1XXX Codes) – Continued**

| DTC | Trouble Code Title, Conditions & Possible Causes: |
|---|---|
| **P1744**<br>1T CCM<br>1998-2002<br>Cherokee, Grand<br>Cherokee, Liberty,<br>Wrangler | **A/T Shift Solenoid 'A' Control Circuit Low Input Conditions:**<br>Engine started; vehicle driven in 1st or 2nd gear position, and the PCM detected an unexpected "low" voltage condition on the SSA control circuit during the CCM test.<br>**Possible Causes:**<br>A/T SSA control circuit is shorted to ground<br>A/T SSA is damaged or has failed (it may be shorted)<br>PCM has failed |
| **P1745**<br>1T CCM<br>1998-2002<br>Cherokee, Grand<br>Cherokee, Liberty,<br>Wrangler | **A/T Shift Solenoid 'A' Control Circuit High Input Conditions:**<br>Engine started; vehicle driven in 1st or 4th gear position, and the PCM detected an unexpected "high" voltage condition on the SSA control circuit during the CCM test.<br>**Possible Causes:**<br>A/T SSA control circuit is open between the solenoid and PCM<br>A/T SSA control circuit is shorted to system power (B+)<br>A/T SSA is damaged or has failed (it may be shorted)<br>PCM has failed |
| **P1746**<br>1T CCM<br>1998-2002<br>Cherokee, Grand<br>Cherokee, Liberty,<br>Wrangler | **A/T Shift Solenoid 'B' Control Circuit Low Input Conditions:**<br>Engine started; vehicle driven in 1st or 2nd gear position, and the PCM detected an unexpected "low" voltage condition on the SSB control circuit during the CCM test.<br>**Possible Causes:**<br>A/T SSB control circuit is shorted to ground<br>A/T SSB is damaged or has failed (it may be shorted)<br>PCM has failed |
| **P1747**<br>1T CCM<br>1998-2002<br>Cherokee, Grand<br>Cherokee, Liberty,<br>Wrangler | **A/T Shift Solenoid 'B' Control Circuit High Input Conditions:**<br>Engine started; vehicle driven in 3rd or 4th gear position, and the PCM detected an unexpected "high" voltage condition on the SSB control circuit during the CCM test.<br>**Possible Causes:**<br>A/T SSB control circuit is open between the solenoid and PCM or shorted to voltage<br>A/T SSB is damaged or has failed (it may be shorted)<br>PCM has failed |
| **P1748**<br>1T CCM<br>1998-2002<br>Cherokee, Grand<br>Cherokee, Liberty,<br>Wrangler | **A/T TCC Solenoid 'C' Control Circuit Low Input Conditions:**<br>Engine started; vehicle driven at cruise with the TCC engaged, and the PCM detected an unexpected "low" voltage condition for 12.5 seconds on the TCC control circuit in the test.<br>**Possible Causes:**<br>A/T TCC control circuit is shorted to ground<br>A/T TCC is damaged or has failed (it may be shorted)<br>PCM has failed |
| **P1749**<br>1T CCM<br>1998-2002<br>Cherokee, Grand<br>Cherokee, Liberty,<br>Wrangler | **A/T Shift Solenoid 'C' Control Circuit High Input Conditions:**<br>Engine started; vehicle driven at cruise with the TCC engaged, and the PCM detected an unexpected "high" voltage condition on the SSC control circuit during the CCM test.<br>**Possible Causes:**<br>A/T SSC control circuit is open between the solenoid and PCM<br>A/T SSC control circuit is shorted to system power (B+)<br>A/T SSC is damaged or has failed (it may be shorted)<br>PCM has failed |
| **P1750**<br>1T CCM<br>1995-98<br>Talon with 2.0L VIN F | **A/T Solenoid Assembly Control Circuit Malfunction Conditions:**<br>Key on or engine running; and the PCM detected an unexpected voltage condition on the A/T Solenoid Assembly Control circuit.<br>**Possible Causes:**<br>A/T converter clutch solenoid circuit open or shorted to ground<br>A/T pressure solenoid circuit is open or shorted to ground<br>A/T shift solenoid control circuit is open or shorted to ground<br>A/T solenoid assembly power circuit is open (power from relay) |

**OBD II Trouble Code List (P1XXX Codes) – Continued**

| DTC | Trouble Code Title, Conditions & Possible Causes: |
|---|---|
| **P1751**<br>1T CCM<br>2001<br>Sebring, Stratus with 2.4L<br>I4 or 3.0L V6 | **A/T Control Relay Circuit Malfunction Conditions:**<br>Engine started; and the TCM detected a malfunction in the A/T Control Relay circuit, and then sent a signal to the PCM indicating that the malfunction had occurred.<br>**Possible Causes:**<br>A/T control relay circuit is open, shorted to ground or to power<br>A/T control relay power circuit is open (test power to Fused B+)<br>A/T control relay is damaged or has failed<br>PCM has failed |
| **P1756**<br>1T CCM<br>1996-2003<br>Cherokee, Dakota, Durango, Grand Cherokee, Liberty, Ram, Wrangler | **A/T Governor Pressure Not Equal To Target At 15-20 PSI Conditions:**<br>Engine started; vehicle driven to a speed over 30 mph, and the PCM detected the Governor Pressure Sensor indicated less than 15 psi or more than 30 psi when the requested pressure was 20-25 psi for 2.2 seconds. This fault must occur 5 times on one trip to set this code.<br>**Possible Causes:**<br>Check for the presence of other A/T related trouble codes<br>Governor pressure Sensor is damaged or has failed<br>Governor pressure Sensor VREF (5v) circuit is open or shorted<br>Transmission valve body is leaking, damaged or has failed<br>PCM has failed |
| **P1757**<br>1T CCM<br>1996-2003<br>Cherokee, Dakota, Durango, Grand Cherokee, Liberty, Ram, Wrangler | **A/T Governor Pressure Above 3 PSI In Gear At 0 MPH Conditions:**<br>Engine started; vehicle driven to a speed over 30 mph, and the PCM detected the Governor Pressure Sensor was more than 3 psi with the requested pressure at 0 psi (95% duty cycle command) for 2.65 seconds. The fault must occur twice on one trip to set the code.<br>**Possible Causes:**<br>Check for the presence of other A/T related trouble codes<br>Governor pressure Sensor is damaged or has failed<br>Governor pressure Sensor VREF (5v) circuit is open or shorted<br>Transmission valve body is leaking, damaged or has failed<br>PCM has failed |
| **P1762**<br>1T CCM<br>1996-2003<br>Cherokee, Dakota, Durango, Grand Cherokee, Liberty, Ram, Wrangler | **A/T Governor Pressure Sensor Offset Volts Too High/Low Conditions:**<br>Key on or engine running. Gear selector in Park or Neutral. PCM detected an out-of-range Governor Pressure Sensor signal for 1.3 seconds. The fault must occur 3 times on one trip to set a code.<br>**Possible Causes:**<br>Governor pressure Sensor VREF (5v) circuit is open<br>Governor pressure Sensor ground circuit is open<br>P/N switch is damaged or has failed (not operating properly)<br>Transmission fluid has excessive debris or it is contaminated<br>PCM has failed |
| **P1763**<br>1T CCM<br>1996-2003<br>Cherokee, Dakota, Durango, Grand Cherokee, Liberty, Ram, Wrangler | **A/T Governor Pressure Sensor Signal Too High Conditions:**<br>Key on or engine running; and the PCM detected the Governor Pressure Sensor signal indicated more than 4.89v for 8.5 seconds.<br>**Possible Causes:**<br>Governor pressure Sensor signal circuit is open<br>Governor pressure Sensor ground circuit is open<br>Governor pressure Sensor signal circuit is shorted to VREF<br>Governor pressure Sensor is damaged or has failed<br>Transmission harness solenoid circuit has a problem<br>PCM has failed |

**OBD II Trouble Code List (P1XXX Codes) – Continued**

| DTC | Trouble Code Title, Conditions & Possible Causes: |
|---|---|
| **P1764**<br>1T CCM<br>1996-2003<br>Cherokee, Dakota, Durango, Grand Cherokee, Liberty, Ram, Wrangler | **A/T Governor Pressure Sensor Signal Too Low Conditions:**<br>Key on or engine running; and the PCM detected the Governor Pressure Sensor signal indicated less than 0.10v for 8.5 seconds.<br>**Possible Causes:**<br>Governor pressure Sensor signal circuit is shorted to ground<br>Governor pressure Sensor VREF (5v) circuit is open<br>Governor pressure Sensor is damaged or has failed<br>Transmission harness solenoid circuit has a problem<br>PCM has failed |
| **P1765**<br>1T CCM<br>1996-2003<br>Cherokee, Dakota, Durango, Grand Cherokee, Liberty, Ram, Wrangler | **A/T Transmission Relay Circuit Malfunction Conditions:**<br>Key on or engine started, and the PCM detected an unexpected voltage condition on the TCC solenoid control circuit during the test.<br>**Possible Causes:**<br>Generator power source circuit to relay control circuit is open<br>Transmission relay power circuit is open (check fuse in PCC)<br>Transmission control circuit is open or shorted to ground<br>Transmission relay is damaged or has failed<br>PCM has failed |
| **P1767**<br>1T CCM<br>1996-2003<br>300M, Avenger, Breeze, Caravan, Cherokee, Cirrus, Concorde, PT Cruiser, Dakota, Durango, Grand Cherokee, Intrepid, LHS, Liberty, Neon, New Yorker, Prowler, Ram, Sebring, Stratus, Talon, Town & Country, Viper, Vision, Voyager, Wrangler | **A/T Transmission Relay Circuit Malfunction Conditions:**<br>Key on or engine started, and the PCM detected an unexpected voltage condition on the TCC solenoid control circuit during the test.<br>**Possible Causes:**<br>Generator power source circuit to relay control circuit is open<br>Transmission relay power circuit is open (check fuse in PCC)<br>Transmission control circuit is open or shorted to ground<br>Transmission relay is damaged or has failed<br>PCM has failed |
| **P1768**<br>1T CCM<br>1996-2003<br>300M, Avenger, Breeze, Caravan, Cherokee, Cirrus, Concorde, PT Cruiser, Dakota, Durango, Grand Cherokee, Intrepid, LHS, Liberty, Neon, New Yorker, Prowler, Ram, Sebring, Stratus, Talon, Town & Country, Viper, Vision, Voyager, Wrangler | **A/T Transmission Relay Output Always Off Conditions:**<br>Engine started; and the TCM did not detect voltage on the Transmission Control Relay output circuit with the relay "on".<br>**Possible Causes:**<br>2/4 or L/R Solenoid control circuit is open<br>O/D or U/D Solenoid control circuit is open<br>Transmission control relay power circuit is open to Fused B+<br>Transmission control relay is damaged (contacts have failed)<br>TCM has failed |

**OBD II Trouble Code List (P1XXX Codes) – Continued**

| DTC | Trouble Code Title, Conditions & Possible Causes: |
|---|---|
| **P1775**<br>1T CCM<br>1999-2005<br>300M, Avenger, Breeze, Caravan, Cherokee, Cirrus, Concorde, PT Cruiser, Dakota, Durango, Grand Cherokee, Intrepid, LHS, Liberty, Magnum, Neon, New Yorker, Prowler, Ram, Sebring, Stratus, Town & Country, Viper, Voyager, Wrangler | **A/T Solenoid Switch Latched In TCC Position Conditions:**<br>Engine started; vehicle driven to over 15 mph and the TCM detected the Transmission did not shift into 1st Gear (test must fail 3 times).<br>**Possible Causes:**<br>Related DTC P0841 may be present.<br>Intermittent wiring or connector problems<br>Extremely low battery (system) voltage<br>L/R Solenoid pressure switch circuit is open or switch has failed<br>Transmission solenoid pack is damaged or has failed<br>Transmission control relay circuit is shorted to L/R solenoid<br>Valve body engine idle too high<br>Valve body solenoid switch stuck in "lockup" position<br>PCM has failed |
| **P1776**<br>2T CCM<br>1999-2005<br>300M, Avenger, Breeze, Caravan, Cherokee, Cirrus, Concorde, PT Cruiser, Dakota, Durango, Grand Cherokee, Intrepid, LHS, Liberty, Magnum, Neon, New Yorker, Pacifica, Prowler, Ram, Sebring, Stratus, Town & Country, Viper, Voyager, Wrangler | **A/T Solenoid Switch Latched In Low/Reverse Position Conditions:**<br>Engine started; vehicle driven to over 30 mph and the TCM detected the L/R switch was closed while performing partial or full PEMCC or FEMCC.<br>**Possible Causes:**<br>Related DTC P0841 may be present.<br>Intermittent wiring or connector problems<br>L/R pressure switch sense circuit is open, shorted to ground or to voltage<br>Extremely low battery (system) voltage<br>Transmission pan has debris caused by valve body damage<br>Transmission internal problems, SSV sticking, or valve body damage<br>PCM has failed |
| **P1781**<br>1T CCM<br>1999-2003<br>300M, Avenger, Breeze, Caravan, Cherokee, Cirrus, Concorde, PT Cruiser, Dakota, Durango, Grand Cherokee, Intrepid, LHS, Liberty, Neon, New Yorker, Prowler, Ram, Sebring, Stratus, Town & Country, Viper, Voyager, Wrangler | **A/T Overdrive Pressure Switch Circuit Malfunction Conditions:**<br>Engine started; engine runtime over 2 seconds, engine speed over 500 rpm, no "loss of prime" test in progress, no pressure switch mismatch detected, and the TCM detected the Overdrive pressure switch was open or closed at the wrong time in any gear position.<br>**Possible Causes:**<br>O/D pressure switch circuit is open or shorted to ground<br>O/D pressure switch circuit is shorted to the TCR solenoid pack<br>O/D pressure switch is damaged or has failed<br>O/D solenoid is damaged or has failed<br>Transmission seals are damaged or have failed |
| **P1782**<br>1T CCM<br>1999-2003<br>300M, Avenger, Breeze, Caravan, Cherokee, Cirrus, Concorde, PT Cruiser, Dakota, Durango, Grand Cherokee, Intrepid, LHS, Liberty, Neon, New Yorker, Prowler, Ram, Sebring, Stratus, Town & Country, Viper, Voyager, Wrangler | **A/T Overdrive Pressure Switch Circuit Malfunction Conditions:**<br>Engine started; and the TCM detected an unexpected voltage condition on the Overdrive Pressure Switch while driving in any gear positions during the CCM test.<br>**Possible Causes:**<br>2/4 line pressure is too high<br>2/4 pressure switch is open or shorted to ground<br>Solenoid pack is damaged or has failed<br>Transmission has internal problems or has failed<br>Valve body torque is out of specification |

**OBD II Trouble Code List (P1XXX Codes) – Continued**

| DTC | Trouble Code Title, Conditions & Possible Causes: |
|---|---|
| **P1784**<br>1T CCM<br>1999-2003<br>300M, Avenger, Breeze, Caravan, Cherokee, Cirrus, Concorde, PT Cruiser, Dakota, Durango, Grand Cherokee, Intrepid, LHS, Liberty, Neon, New Yorker, Prowler, Ram, Sebring, Stratus, Town & Country, Viper, Voyager, Wrangler | **A/T Additional L/R Pressure Switch Circuit Malfunction Conditions:**<br>Engine started; vehicle driven to a speed over 30 mph in Drive, and the PCM detected an invalid Low/Reverse Pressure switch value.<br>**Possible Causes:**<br>L/R pressure switch circuit open, shorted to ground or to power<br>L/R solenoid switch valve is stuck in "lockup" position<br>L/R solenoid is damaged or has failed<br>Solenoid pack is damaged or has failed (causing this code)<br>Valve body bolt torque is out of specification |
| **P1787**<br>1T CCM<br>1999-2002<br>Avenger, Breeze, Cirrus, Sebring, Stratus | **A/T Additional O/D Pressure Switch Circuit Malfunction Conditions:**<br>Engine started; vehicle driven to a speed over 30 mph in Drive, and the PCM detected an invalid Overdrive Pressure Switch pressure.<br>**Possible Causes:**<br>O/D pressure switch circuit is open or shorted to ground<br>O/D pressure switch circuit is shorted to TCR output circuit<br>O/D solenoid is damaged or has failed |
| **P1788**<br>1T CCM<br>1999-2003<br>300M, Avenger, Breeze, Caravan, Cherokee, Cirrus, Concorde, PT Cruiser, Dakota, Durango, Grand Cherokee, Intrepid, LHS, Liberty, Neon, New Yorker, Prowler, Ram, Sebring, Stratus, Town & Country, Viper, Voyager, Wrangler | **A/T Additional 2/4 Pressure Switch Circuit Malfunction Conditions:**<br>Engine started; vehicle driven to a speed of 4-37 mph in Drive, and the PCM detected an invalid 2/4 Pressure Switch circuit pressure.<br>**Possible Causes:**<br>2/4 pressure switch circuit is open or shorted to ground<br>2/4 pressure switch circuit is shorted to TCR output circuit<br>2/4 solenoid is damaged or has failed |
| **P1789**<br>1T CCM<br>1999-2003<br>300M, Avenger, Breeze, Caravan, Cherokee, Cirrus, Concorde, PT Cruiser, Dakota, Durango, Grand Cherokee, Intrepid, LHS, Liberty, Neon, New Yorker, Prowler, Ram, Sebring, Stratus, Town & Country, Viper, Voyager, Wrangler | **A/T Overdrive/2/4 Pressure Switch Circuit Malfunction Conditions:**<br>Engine started; vehicle driven in 1st, 2nd or 3rd Gear at an engine speed over 1000 rpm, and the TCM detected the O/D or 2/4 Pressure switch did not close (test fails twice to set a code).<br>**Possible Causes:**<br>Extremely low battery (system) voltage<br>2/4 pressure switch circuit is open or shorted to ground<br>O/D pressure switch circuit is open or shorted to ground |

**OBD II Trouble Code List (P1XXX Codes) – Continued**

| DTC | Trouble Code Title, Conditions & Possible Causes: |
|---|---|
| **P1790**<br>1T CCM<br>1999-2005<br>300M, Avenger, Breeze, Caravan, Cherokee, Cirrus, Concorde, PT Cruiser, Dakota, Durango, Grand Cherokee, Intrepid, LHS, Liberty, Magnum, Neon, New Yorker, Pacifica, Prowler, Ram, Sebring, Stratus, Town & Country, Viper, Voyager, Wrangler | **A/T Malfunction Immediately After Shift Event Conditions:**<br>Engine started; vehicle driven to a speed over 10 mph in Drive, and the TCM detected a Speed Ratio error within 1.3 seconds of a shift.<br>**Possible Causes:**<br>Transmission internal mechanical problem |
| **P1791**<br>1T CCM<br>1995-98<br>Talon | **A/T Engine Coolant Temperature Circuit Malfunction Conditions:**<br>Key on or engine running; and the PCM detected an unexpected voltage condition on the Engine Coolant Temperature (ECT) Sensor circuit shared by the PCM and TCM in the test.<br>**Possible Causes:**<br>ECT Sensor signal circuit is open or shorted to ground<br>ECT Sensor ground circuit is open between Sensor and PCM<br>ECT Sensor is damaged o has failed (It is open or shorted) |
| **P1791**<br>1T CCM<br>1999-2003<br>300M, Avenger, Breeze, Caravan, Cherokee, Cirrus, Concorde, PT Cruiser, Dakota, Durango, Grand Cherokee, Intrepid, LHS, Liberty, Neon, New Yorker, Prowler, Ram, Sebring, Stratus, Town & Country, Viper, Voyager, Wrangler | **A/T Loss Of Prime Malfunction Conditions:**<br>Engine started; vehicle driven to a speed of 4-37 mph in Drive, and the vehicle exhibited a "no drive" condition during the CCM test.<br>**Possible Causes:**<br>ATF cooler lines damaged or leaking<br>ATF fluid level incorrect or filled with debris<br>Customer complains of "no drive" condition<br>Oil pump pressure drop occurs with "no drive" condition<br>Transmission filter clogged or restricted |
| **P1792**<br>1T CCM<br>1999-2003<br>300M, Avenger, Breeze, Caravan, Cherokee, Cirrus, Concorde, PT Cruiser, Dakota, Durango, Grand Cherokee, Intrepid, LHS, Liberty, Neon, New Yorker, Prowler, Ram, Sebring, Stratus, Town & Country, Viper, Voyager, Wrangler | **Battery Was Disconnected Conditions:**<br>Key on and the TCM detected an interruption of the Battery Direct Power (B+) circuit.<br>**Possible Causes:**<br>Battery direct power (B+) circuit is open (test power to PDC)<br>Battery direct power (B+) circuit is shorted to ground (test fuse)<br>Battery was discharged to a low voltage state or disconnected<br>scan tool "quick learn" for quick disconnect function performed |

**OBD II Trouble Code List (P1XXX Codes) – Continued**

| DTC | Trouble Code Title, Conditions & Possible Causes: |
|---|---|
| **P1793**<br>1T CCM<br>1999-2003<br>300M, Avenger, Breeze, Caravan, Cherokee, Cirrus, Concorde, PT Cruiser, Dakota, Durango, Grand Cherokee, Intrepid, LHS, Liberty, Neon, New Yorker, Prowler, Ram, Sebring, Stratus, Town & Country, Viper, Voyager, Wrangler | **Torque Management Request Circuit Malfunction Conditions:**<br>Key on and the TCM did not detect any signals on the Torque Management Request circuit.<br>**Possible Causes:**<br>Extremely low battery (system) voltage<br>Torque management request circuit is open, shorted to ground or shorted to system power (i.e., the TRD circuit has failed) or the TCM has failed |
| **P1793**<br>1T CCM<br>2004-05<br>300M, Caravan, Concorde, PT Cruiser, Dakota, Durango, Grand Cherokee, Intrepid, LHS, Liberty, Magnum, Neon, Pacifica, Prowler, Ram, Sebring, Stratus, Town & Country, Viper, Voyager, Wrangler | **TCM TRD Link Communication Error Conditions:**<br>The TCM pulses the 12v TRD signal from the PCM to ground, during torque managed shifts, with the throttle angle above 54 degrees. The TRD system is also tested whenever the vehicle is stopped and the engine is at idle. This DTC is set when the TCM sends 2 subsequent torque reduction messages to the PCM and the TCM does not receive a confirmation from the PCM. Note: Due to the integrations of the PCM and TCM, Bus communication between the modules is internal. Related DTCs are present.<br>**Possible Causes:**<br>Torque Management Request (TMR) sense circuit is open or is shorted to ground or to voltage<br>PCM or TCM has failed<br>Intermittent wiring and connector problems |
| **P1794**<br>1T CCM<br>1999-2005<br>300M, Avenger, Breeze, Caravan, Cherokee, Cirrus, Concorde, PT Cruiser, Dakota, Durango, Grand Cherokee, Intrepid, LHS, Liberty, Magnum, Neon, New Yorker, Pacifica, Prowler, Ram, Sebring, Stratus, Talon, Town & Country, Viper, Vision, Voyager, Wrangler | **A/T Speed Sensor Ground Circuit Malfunction Conditions:**<br>Engine started; gear selector position indicating Neutral, and the PCM an error in the Output Speed Sensor signal during the test.<br>**Possible Causes:**<br>Extremely low battery (system) voltage<br>TCM "reset" function has just been performed |
| **P1794**<br>1T CCM<br>2004-05<br>Liberty with Diesel | **A/T Speed Sensor Ground Error Conditions:**<br>The gear ratio is monitored whenever the transmission is in gear. After a TCM reset in Neutral and a ratio for Input to Output is 1 to 2, this DTC will set. This DTC can take up to 5 minutes of problem identification to light the MIL.<br>**Possible Causes:**<br>Speed Sensor ground circuit is open or is shorted to ground or to voltage<br>TCM has failed<br>Intermittent wiring or connector problems |

**OBD II Trouble Code List (P1XXX Codes) – Continued**

| DTC | Trouble Code Title, Conditions & Possible Causes: |
|---|---|
| **P1795**<br>1T CCM<br>1999-2003<br>300M, Avenger, Breeze, Caravan, Cherokee, Cirrus, Concorde, PT Cruiser, Dakota, Durango, Grand Cherokee, Intrepid, LHS, Liberty, Neon, New Yorker, Prowler, Ram, Sebring, Stratus, Town & Country, Viper, Voyager, Wrangler | **Transmission Control Module Internal Malfunction Conditions:**<br>Engine started; gear position is Neutral, and the PCM detected the engine model stored in RAM was different than the EEPROM value.<br>**Possible Causes:**<br>Extremely low battery (system) voltage<br>Engine "starts" since set counter less than three (3) |
| **P1796**<br>1T CCM<br>1999-2003<br>300M, Avenger, Breeze, Caravan, Cherokee, Cirrus, Concorde, PT Cruiser, Dakota, Durango, Grand Cherokee, Intrepid, LHS, Liberty, Neon, New Yorker, Prowler, Ram, Sebring, Stratus, Town & Country, Viper, Voyager, Wrangler | **A/T Additional AutoStick Input Circuit Malfunction Conditions:**<br>Engine started; vehicle driven to a speed of 4-37 mph in Drive, and the PCM detected a problem on the AutoStick signal circuit.<br>**Possible Causes:**<br>AutoStick switch ground circuit is open<br>AutoStick downshift switch is stuck in open or closed position<br>AutoStick downshift Sensor circuit is open or shorted to ground<br>AutoStick switch power circuit is open (test power at Fused B+)<br>AutoStick Upshift switch is stuck in open or closed position<br>AutoStick Upshift switch Sensor is open or shorted to ground |
| **P1797**<br>1T CCM<br>1999-2003<br>300M, Avenger, Breeze, Caravan, Cherokee, Cirrus, Concorde, PT Cruiser, Dakota, Durango, Grand Cherokee, Intrepid, LHS, Liberty, Neon, New Yorker, Prowler, Ram, Sebring, Stratus, Town & Country, Viper, Voyager, Wrangler | **A/T Manual Shift Overheat Malfunction Conditions:**<br>Engine started; vehicle driven to a speed over 30 mph in Drive, and the TCM detected A Manual Shift Overheat condition was present.<br>**Possible Causes:**<br>ATF fluid level too high (transmission may be overfilled)<br>Engine Cooling System or engine cooling fan malfunction<br>Excessive drive time in low gear, or aggressive drive patterns<br>Transmission oil cooler is clogged or restricted |
| **P1797**<br>1T CCM<br>2004-05<br>300M, Concorde, PT Cruiser, Dakota, Durango, Grand Cherokee, Intrepid, LHS, Liberty, Magnum, Neon, Pacifica, Prowler, Ram, Sebring, Stratus, Viper, Wrangler | **A/T Manual Shift Overheat Malfunction Conditions:**<br>Whenever the engine is running and the transmission is in the AutoStick mode, if the engine temperature exceeds 275°F (135°C), this DTC will set. Note: Aggressive driving or driving in Low for extended periods in AutoStick mode will set this DTC.<br>**Possible Causes:**<br>ATF fluid level too high (transmission may be overfilled)<br>Engine Cooling System or engine cooling fan malfunction<br>Excessive drive time in low gear, or aggressive drive patterns<br>Transmission oil cooler is clogged or restricted |

## OBD II Trouble Code List (P1XXX Codes) – Continued

| DTC | Trouble Code Title, Conditions & Possible Causes: |
|---|---|
| **P1797**<br>1T CCM<br>2004-05<br>Caravan, Town & Country, Voyager | **Manual Shift Overheat Conditions:**<br>Engine running; transmission in AutoStick mode. If the ECT Sensor exceeds 255°F (123°C), or the transmission temperature exceeds 275°F (135°C) while in AutoStick mode, this DTC will set. Note: Aggressive driving or driving in Low for extended periods of time in AutoStick will set this DTC.<br>**Possible Causes:**<br>Aggressive driving/shift patterns<br>Driving in Low for extended periods |
| **P1798**<br>1T CCM<br>1999-2003<br>300M, Avenger, Breeze, Caravan, Cherokee, Cirrus, Concorde, PT Cruiser, Dakota, Durango, Grand Cherokee, Intrepid, LHS, Liberty, Neon, New Yorker, Prowler, Ram, Sebring, Stratus, Town & Country, Viper, Voyager, Wrangler | **A/T Transmission Fluid Is Burnt Or Contaminated Conditions:**<br>Engine started; vehicle driven to a speed over 30 mph in Drive, and the TCM detected a vehicle "shutter" condition during TCC lockup.<br>**Possible Causes:**<br>ATF is burnt, contaminated or very dirty<br>Extremely low battery (system) voltage<br>Transmission pan has debris<br>Transmission has internal damage causing the vehicle shudder |
| **P1799**<br>1T CCM<br>1999-2003<br>300M, Avenger, Breeze, Caravan, Cherokee, Cirrus, Concorde, PT Cruiser, Dakota, Durango, Grand Cherokee, Intrepid, LHS, Liberty, Neon, New Yorker, Prowler, Ram, Sebring, Stratus, Town & Country, Viper, Voyager, Wrangler | **A/T Calculated Oil Temperature In Use Malfunction Conditions:**<br>Engine started; vehicle driven to a speed over 30 mph in Drive, and the TCM detected A Manual Shift Overheat condition was present.<br>**Possible Causes:**<br>TFT Sensor signal circuit is open<br>TFT Sensor ground circuit is open<br>TFT Sensor signal circuit is shorted to ground<br>TFT Sensor signal circuit is shorted to VREF (5v)<br>TFT Sensor signal not correct for the ATF fluid temperature<br>Speed Sensor signal circuit is open or shorted to ground<br>Speed Sensor signal circuit is shorted to system power<br>TCM oil temperature Sensor high or low circuit has failed |
| **P1830**<br>1T CCM<br>2002-03<br>Jeep with 2.4L VIN 1 | **Clutch Override Relay Control Circuit Malfunction Conditions:**<br>Key on or engine running; and the PCM detected an unexpected voltage condition on the Clutch Override Relay Control circuit.<br>**Possible Causes:**<br>Clutch override relay control circuit is open<br>Clutch override relay control circuit is shorted to ground<br>Clutch override relay is damaged or it has failed<br>Fused ignition power circuit to the clutch override relay is open<br>PCM has failed |
| **P1854**<br>2T CCM<br>2003-05<br>PT Cruiser, Neon | **Throttle Inlet Pressure BARO Reading Out Of Range Conditions:**<br>Engine started. The PCM detected the BARO Sensor indicated an incorrect reading. On Neon, MAP Sensor voltage is greater than 4.9v or below 2.28v (non-Turbo) or greater than 2.4v or below 1.2v (Turbo) for 400ms.<br>**Possible Causes:**<br>Inspect the hoses and tubing to the turbo charger assembly<br>Review results of Solenoid Tests (Test 1, 2, 3 and 4 results)<br>TIP signal circuit is open (may be an intermittent fault)<br>TIP ground circuit is open (may be an intermittent fault)<br>PCM has failed |

**OBD II Trouble Code List (P1XXX Codes) – Continued**

| DTC | Trouble Code Title, Conditions & Possible Causes: |
|---|---|
| **P1861**<br>1T PCM<br>2004-05<br>Pacifica | **Siphon Line Disconnected Conditions:**<br>Ignition on. PCM compares the primary tank level with the secondary tank level. If the PCM detects the primary side is lower than the secondary side, by a calibrated amount, the DTC will set.<br>Possible Causes:<br>Damage to fuel tank<br>Fuel level signal circuit is open or shorted to ground<br>Ground circuit is open<br>Internal tank components or siphon hose damaged<br>Fuel level Sensor has failed |
| **P1899**<br>1T CCM<br>1995-2003<br>300M, Avenger, Breeze, Caravan, Cherokee, Cirrus, Concorde, PT Cruiser, Dakota, Durango, Grand Cherokee, Intrepid, LHS, Liberty, Neon, New Yorker, Prowler, Ram, Sebring, Stratus, Talon, Town & Country, Viper, Vision, Voyager, Wrangler | **A/T Park Neutral Switch Stuck in Park Position or In Gear Conditions:**<br>Key on or engine running (while not in Limp-In mode), gear selector in Park, Neutral or Drive position, and the PCM detected an invalid P/N position switch state for a given mode of vehicle operation.<br>**Possible Causes:**<br>P/N switch signal circuit is open between the switch and PCM<br>P/N switch signal circuit is shorted to Sensor or chassis ground<br>P/N switch ground circuit is open (at the switch mounting point)<br>P/N switch is stuck in P/N position, or stuck in Drive position<br>P/N switch is damaged or has failed<br>PCM has failed |

## OBD II Trouble Code List (P2XXX Codes)

| DTC | Trouble Code Title, Conditions & Possible Causes: |
|---|---|
| **P2008**<br>1T CCM<br>2002-05<br>300M, Concorde, Intrepid, LHS, Magnum, Pacifica | **Short Runner Solenoid Circuit Malfunction Conditions:**<br>Engine started. ASD relay energized. PCM detected the Short Runner solenoid circuit was not in its expected voltage state.<br>**Possible Causes:**<br>S/R solenoid control circuit is open<br>S/R solenoid control circuit is shorted to ground or power (B+)<br>S/R solenoid power supply circuit is open to the ASD relay<br>Short runner solenoid is damaged or it has failed<br>PCM has failed |
| **P2066**<br>2T CCM<br>2004-05<br>300M, Magnum, Pacifica | **Fuel Level Sensor No. 2 Malfunction Conditions:**<br>Test No. 1: With ignition on, fuel level is compared to the previous key-down after a 20-second delay. If the PCM does not see a difference in the fuel level of more than 0.1v, the test will fail.<br>Test No. 2: The PCM monitors the fuel level with ignition on. If the PCM does not see a change in the fuel level of 0.1765 in. over a set amount of miles, the test will fail.<br>**Possible Causes:**<br>Fuel tank or internal siphon hose damage<br>Fuel level signal circuit open or shorted to ground<br>Ground circuit is open<br>Fuel level Sensor malfunction |
| **P2067**<br>1T CCM<br>2004-05<br>300M, Magnum, Pacifica | **Fuel Level Sensor No. 2 Low Input Conditions:**<br>Key on. Battery voltage over 10.4v. Fuel level Sensor signal goes below 0.4v for more than 90 seconds (300M, Magnum), or below 0.1961v for more than 5 seconds (Pacifica). DTC is recorded.<br>**Possible Causes:**<br>Intermittent condition<br>Fuel level sending unit signal circuit shorted to Sensor or chassis ground<br>Fuel level sensing unit is damaged or the fuel tank is damaged<br>BCM or PCM has failed |
| **P2068**<br>1T CCM<br>2004-05<br>300M, Magnum, Pacifica | **Fuel Level Sensor No. 2 High Input Conditions:**<br>Key on. Battery voltage over 10.4v. Fuel level Sensor signal goes above 4.9v for more than 90 seconds (300M, Magnum) or above 4.7v for more than 5 seconds (Pacifica). DTC is recorded.<br>**Possible Causes:**<br>Fuel level sending unit signal circuit shorted to Sensor or chassis ground<br>Fuel level sensing unit is damaged or the fuel tank is damaged<br>Instrument cluster module faulty<br>BCM or PCM has failed |
| **P2072**<br>1T CCM<br>2004-05<br>300M, Dakota, Durango, Magnum, Ram | **Electronic Throttle Control System Malfunction Conditions:**<br>Key on. The PCM recognizes the throttle plate is stuck during extremely cold ambient temperature conditions. The throttle plate goes through a de-icing procedure, but if the throttle plate still does not move, this DTC will set. The MIL will not illuminate. The vehicle will be in the "Limp Home" mode, limiting rpm and vehicle speed.<br>**Possible Causes:**<br>Throttle plate frozen |

**OBD II Trouble Code List (P2XXX Codes) – Continued**

| DTC | Trouble Code Title, Conditions & Possible Causes: |
|---|---|
| **P2074**<br>1T CCM<br>2004-05<br>300M, Caravan, Concorde, PT Cruiser, Dakota, Durango, Grand Cherokee, Intrepid, LHS, Liberty, Magnum, Neon, Pacifica, Prowler, Ram, Sebring, Stratus, Town & Country, Viper, Voyager, Wrangler | **Manifold Pressure/Throttle Position Correlation; High Flow/Vacuum Leak Conditions:**<br>Engine running in all drive modes. The relationship between the MAP Sensor and TP Sensor exceeds a predetermined value for a given engine speed. If vacuum drops below 1.5 in. Hg with engine rpm at more than 2000 rpm at closed throttle, or if an unexpectedly high intake manifold airflow exists that can lead to increased engine speed and puts the NGC (Ram) into a "High Airflow Protection Limiting" mode; in this case, rpm limits for when a TP Sensor and/or MAP Sensor limp-in fault is present.<br>**Possible Causes:**<br>Vacuum leak in hoses or component connections<br>High resistance or resistance to ground in MAP 5v supply or signal circuit<br>MAP Sensor has failed<br>High resistance in MAP ground circuit<br>TP Sensor has failed or is improperly adjusted<br>High resistance or resistance to ground in TP Sensor 5v supply or signal circuit<br>High resistance in TP Sensor ground circuit<br>PCM has failed |
| **P2096**<br>2T CCM<br>2004-05<br>300M, Caravan, Concorde, PT Cruiser, Dakota, Durango, Grand Cherokee, Intrepid, LHS, Liberty, Magnum, Neon, Pacifica, Prowler, Ram, Sebring, Stratus, Town & Country, Viper, Voyager, Wrangler | **Downstream Fuel System 1/2 Lean Conditions:**<br>Engine running in closed loop mode. Ambient/battery temperature above 20°F (-7°C). Altitude below 8500 feet. Fuel level is more than 15%. If the PCM adds downstream short-term compensation to long-term adaptive, and a certain percentage is exceeded for 2 trips, a freeze frame is stored, the MIL illuminates and a DTC is set.<br>**Possible Causes:**<br>Exhaust leak<br>Engine mechanical problem<br>O2 Sensor has failed<br>O2 Sensor signal circuit or return circuit problem<br>Fuel contamination |
| **P2097**<br>2T CCM<br>2004-05<br>300M, Magnum, Neon, Sebring, Stratus, Pacifica, Caravan/Town & Country/Voyager, Dakota, Durango, Ram, Liberty | **Downstream Fuel System 1/2 Rich Conditions:**<br>Engine running in closed loop mode. Ambient/battery temperature above 20°F (-7°C). Altitude below 8500 feet. Fuel level is more than 15%. If the PCM adds downstream short-term compensation to long-term adaptive, and a certain percentage is exceeded for 2 trips, a freeze frame is stored, the MIL illuminates and a DTC is set.<br>**Possible Causes:**<br>Exhaust leak<br>Engine mechanical problem<br>O2 Sensor No. 1/2 has failed<br>O2 Sensor No. 1/2 signal circuit or return circuit problem<br>Fuel contamination |
| **P2098**<br>2T CCM<br>2004-05<br>300M, Caravan, Concorde, PT Cruiser, Dakota, Durango, Grand Cherokee, Intrepid, LHS, Liberty, Magnum, Neon, Pacifica, Prowler, Ram, Sebring, Stratus, Town & Country, Viper, Voyager, Wrangler | **Downstream Fuel System 2/2 Lean Conditions:**<br>Engine running in closed loop mode. Ambient/battery temperature above 20°F (-7°C). Altitude below 8500 feet. Fuel level is more than 15%. If the PCM adds downstream short-term compensation to long-term adaptive, and a certain percentage is exceeded for 2 trips, a freeze frame is stored, the MIL illuminates and a DTC is set.<br>**Possible Causes:**<br>Exhaust leak<br>Engine mechanical problem<br>O2 Sensor No. 2/2 has failed<br>O2 Sensor No. 2/2 signal circuit or return circuit problem<br>Fuel contamination |

**OBD II Trouble Code List (P2XXX Codes) – Continued**

| DTC | Trouble Code Title, Conditions & Possible Causes: |
|---|---|
| **P2099**<br>2T CCM<br>2004-05<br>300M, Caravan, Concorde, PT Cruiser, Dakota, Durango, Grand Cherokee, Intrepid, LHS, Liberty, Magnum, Neon, Pacifica, Prowler, Ram, Sebring, Stratus, Town & Country, Viper, Voyager, Wrangler | **Downstream Fuel System 2/2 Rich Conditions:**<br>Engine running in closed loop mode. Ambient/battery temperature above 20°F (-7°C). Altitude below 8500 feet. Fuel level is more than 15%. If the PCM adds downstream short-term compensation to long-term adaptive, and a certain percentage is exceeded for 2 trips, a freeze frame is stored, the MIL illuminates and a DTC is set.<br>**Possible Causes:**<br>Exhaust leak<br>Engine mechanical problem<br>O2 Sensor No. 2/2 has failed<br>O2 Sensor No. 2/2 signal circuit or return circuit problem<br>Fuel contamination |
| **P2100**<br>1T CCM<br>2004-05<br>300M, PT Cruiser, Dakota, Durango, Magnum, Ram | **Electronic Throttle Control Motor Circuit Malfunction Conditions:**<br>Ignition on and the ETC motor is not is "Limp Home" mode. When the PCM detects an internal error or a short between the ETC Motor and the ETC Motor positive circuit in the ETC Motor Driver, this DTC will set. The ETC light will be flashing.<br>**Possible Causes:**<br>Intermittent condition<br>Throttle plate or bore may have foreign object blockage<br>ETC positive circuit is open or is shorted to battery voltage, to ground, or to ETC negative circuit<br>ETC negative circuit is open or is shorted to battery voltage or to ground<br>Low battery voltage<br>ETC Motor or Throttle Body has failed<br>PCM has failed |
| **P2101**<br>1T CCM<br>2004-05<br>300M, Dakota, Durango, Magnum, Ram | **Electronic Throttle Control Motor Malfunction Conditions:**<br>With vehicle running and ETC motor is not is "Limp Home" mode, and the TPS adaptation is complete. The PCM recognizes too large of an error between the actual position of the throttle plate and the set point position. This DTC will set within 5 seconds. 3 good trips required to turn off MIL. The ETC light will be flashing.<br>**Possible Causes:**<br>Throttle body assembly may have failed<br>Low battery voltage<br>PCM has failed |
| **P2101**<br>1T CCM<br>2004-05<br>PT Cruiser | **Electronic Throttle Control Motor Malfunction Conditions:**<br>With vehicle running and ETC motor is not is "Limp Home" mode, and the TPS adaptation is complete. If the PCM recognizes the difference between the TPS set point and the TPS actual setting is 8° for more than 405ms, this DTC will set.<br>**Possible Causes:**<br>Intermittent condition<br>Electronic throttle control motor operation has failed<br>Throttle position is out of adjustment or has malfunctioned |
| **P2101**<br>1T ECM<br>2004-05<br>Liberty with Diesel | **EGR Airflow Control Valve Excessive Current Or Open Circuit Conditions:**<br>Ignition on; EGR Airflow Control Valve commanded ON (excessive current) or OFF (open circuit). If the ECM detects excessive voltage signal or no voltage signal (open) on the EGR Airflow Control Valve control circuit, this DTC will set.<br>**Possible Causes:**<br>Intermittent condition<br>ASD Relay Output circuit is open<br>EGR Airflow Control Valve has failed<br>EGR Airflow Control Valve control circuit is open, shorted to voltage or to ground<br>ECM has failed |

## OBD II Trouble Code List (P2XXX Codes) – Continued

| DTC | Trouble Code Title, Conditions & Possible Causes: |
|---|---|
| **P2107**<br>1T CCM<br>2004-05<br>300M, PT Cruiser, Dakota, Durango, Magnum, Ram | **Electronic Throttle Control Module Processor Malfunction Conditions:**<br>Ignition is on. This condition is caused by an internal PCM failure. The module will attempt to reset, so you will be able to hear the throttle relearning. If the condition is continuous, the vehicle may not be drivable. The ETC light will be flashing.<br>**Possible Causes:**<br>PCM requires reprogramming |
| **P2108**<br>1T CCM<br>2004-05<br>300M, Magnum, PT Cruiser, Dakota, Durango, Ram | **Electronic Throttle Control Module Processor Malfunction Conditions:**<br>Ignition is on. This condition is caused by an internal PCM failure. Customer may experience an extended cranking condition, with limited driving and a rough idle. This code will set within 5 seconds. The ETC light will be flashing.<br>**Possible Causes:**<br>PCM requires reprogramming |
| **P2110**<br>1T CCM<br>2004-05<br>300M, Dakota, Durango, Magnum, Ram | **Electronic Throttle Control – Forced Limited RPM Conditions:**<br>Ignition is on and ETC motor is working. When the PCM requests to limit engine speed, if the PWM is too high for 20.5 seconds and before P2118 sets. This one-trip fault will set within 5 seconds. The ETC light will be illuminated.<br>**Possible Causes:**<br>Throttle plate stuck<br>ETC positive circuit is open or is shorted to ground<br>ETC negative circuit is open or is shorted to ground<br>ETC motor has failed<br>PCM has failed |
| **P2111**<br>1T CCM<br>2004-05<br>300M, Dakota, Durango, Magnum, Ram | **Electronic Throttle Control – Forced Limited RPM Conditions:**<br>Ignition is on and battery voltage is more than 10v. If the TP Sensor does not return to "Limp Home" position at the end of this test, the DTC will set. This one-trip fault will set within 5 seconds. The ETC light will be flashing.<br>**Possible Causes:**<br>Throttle plate stuck above "Limp Home" position<br>TP Sensors 1 & 2 both read 2.5v<br>ETC positive circuit is open or is shorted to ground or to battery voltage<br>ETC negative circuit is open or is shorted to ground<br>PCM has failed |
| **P2112**<br>1T CCM<br>2004-05<br>300M, Dakota, Durango, Magnum, Ram | **Electronic Throttle Control – Unable To Open Conditions:**<br>Ignition is on and battery voltage is more than 10v. Just after the ignition is turned on, the throttle is opened and closed to test the system. If the TP Sensor does not return to "Limp Home" position at the end of this test, the DTC will set. This one-trip fault will set within 5 seconds. The ETC light will be flashing.<br>**Possible Causes:**<br>Throttle plate stuck at or below "Limp Home" position<br>ETC positive circuit is open or is shorted to ground<br>ETC negative circuit is open or is shorted to ground or to battery voltage<br>PCM has failed |
| **P2115**<br>1T CCM<br>2004-05<br>300M, Dakota, Durango, Magnum, Ram | **Accelerator Pedal Position Sensor No. 1 Minimum Stop Performance Conditions:**<br>Ignition is on. During "in-plant" mode the APP Sensors need to be checked to make sure that idle and full pedal travel can be reached on both Sensors. The test for this DTC is enabled once the test for DTC P2166 has passed. This DTC will set if the APP Sensor No. 1 has failed to achieve the required minimum value during "in-plant" testing. This one-trip fault will set within 5 seconds. The engine will only idle.<br>**Possible Causes:**<br>APP Sensors must be reprogrammed to relearn |

**OBD II Trouble Code List (P2XXX Codes) – Continued**

| DTC | Trouble Code Title, Conditions & Possible Causes: |
|---|---|
| **P2116**<br>1T CCM<br>2004-05<br>300M, Dakota, Durango, Magnum, Ram | **Accelerator Pedal Position Sensor No. 2 Minimum Stop Performance Conditions:**<br>Ignition is on. During "in-plant" mode the APP Sensors need to be checked to make sure that idle and full pedal travel can be reached on both Sensors. The test for this DTC is enabled once the test for DTC P2167 has passed. This DTC will set if the APP Sensor No. 2 has failed to achieve the required minimum value during "in-plant" testing. This one-trip fault will set within 5 seconds. The engine will only idle.<br>**Possible Causes:**<br>APP Sensors must be reprogrammed to relearn |
| **P2118**<br>1T CCM<br>2004-05<br>300M, Dakota, Durango, Magnum, Ram | **Electronic Throttle Control Motor Circuit Malfunction Conditions:**<br>Ignition is on and ETC motor is not in "limp-home" mode. When the PCM detects an internal error or short between the ETC motor and ETC motor positive circuits in the ETC motor driver. The ETC light will be flashing.<br>**Possible Causes:**<br>Throttle plate or bore malfunctions<br>ETC positive circuit is open or is shorted to ground, battery voltage or ETC negative circuit<br>ETC negative circuit is open or is shorted to ground or to battery voltage<br>ETC motor has malfunctioned<br>PCM has failed |
| **P2120**<br>1T ECM<br>2004-05<br>Liberty with Diesel | **APP Sensor No. 1 Circuit Plausibility Or Signal Voltage Too High Or Too Low Conditions:**<br>Ignition on. If the APP Sensor No. 1 and No. 2 signals do not agree (plausibility) or if Sensor No. 1 voltage signal is above 4.8v (too high) or below 0.29v (too low), this DTC will set.<br>**Possible Causes:**<br>APP Sensor has failed<br>APP Sensor No. 1 5v supply circuit is open, shorted to ground, to voltage, or shorted to Sensor ground<br>APP Sensor ground circuit or signal circuit is open<br>Intermittent condition<br>APP Sensor signal circuit is shorted to the Sensor ground circuit<br>ECM has failed |
| **P2122**<br>1T CCM<br>2004-05<br>300M, Dakota, Durango, Magnum, Ram | **Accelerator Pedal Position Sensor No. 1 Circuit Low Conditions:**<br>Ignition is on and no other APP Sensor No. 1 DTCs are present. When APP Sensor No. 1 voltage is too low, the engine will additionally idle, if the brake pedal is pressed or has failed. Acceleration rate and engine output are limited. This one-trip fault will set within 5 seconds. The ETC light will be flashing.<br>**Possible Causes:**<br>5v supply circuit is open or shorted to ground<br>APP Sensor No. 1 signal circuit is open, shorted to ground or to Sensor return circuit<br>APP Sensor No. 1 has failed<br>PCM has failed |
| **P2122**<br>1T CCM<br>2004-05<br>PT Cruiser | **Accelerator Pedal Position Sensor No. 1 Circuit Low Conditions:**<br>Ignition is on. When APP Sensor No. 1 voltage is less than 0.2444v, or the circuit is shorted to ground or open for more than 120msec, this DTC will set.<br>**Possible Causes:**<br>APP Sensor sweep<br>Intermittent condition<br>5v supply circuit is open or shorted to ground<br>APP Sensor No. 1 signal circuit is open, shorted to ground or to Sensor return circuit<br>APP Sensor No. 1 has failed<br>PCM 5v supply circuit problem<br>PCM has failed |

**OBD II Trouble Code List (P2XXX Codes) – Continued**

| DTC | Trouble Code Title, Conditions & Possible Causes: |
|---|---|
| **P2123**<br>1T CCM<br>2004-05<br>300M, Dakota, Durango, Magnum, Ram | **Accelerator Pedal Position Sensor No. 1 Circuit High Conditions:**<br>Ignition is on and no other APP Sensor No. 1 DTCs are present. When APP Sensor No. 1 voltage is too high, the engine will additionally idle, if the brake pedal is pressed or has failed. Acceleration rate and engine output are limited. This one-trip fault will set within 5 seconds. The ETC light will be flashing.<br>**Possible Causes:**<br>APP Sensor No. 1 return circuit is open<br>APP Sensor No. 1 signal circuit is shorted to either 5v supply circuit<br>APP Sensor No. 1 has failed<br>PCM has failed |
| **P2123**<br>1T CCM<br>2004-05<br>PT Cruiser | **Accelerator Pedal Position Sensor No. 1 Circuit High Conditions:**<br>Ignition is on. When APP Sensor No. 1 voltage is more than 4.8192v, because of a short to voltage, this DTC will set.<br>**Possible Causes:**<br>Intermittent condition<br>5v supply circuit is shorted to battery voltage<br>APP Sensor No. 1 signal circuit is shorted to battery voltage or to 5v circuit<br>APP Sensor sweep<br>PCM has failed |
| **P2125**<br>1T ECM<br>2004-05<br>Liberty with Diesel | **APP Sensor No. 2 Circuit Plausibility Or Signal Voltage Too High Or Too Low Conditions:**<br>Ignition on. If the APP Sensor No. 1 and No. 2 signals do not agree (plausibility) or if Sensor No. 1 voltage signal is above 2.4v (too high) or below 0.15v (too low), this DTC will set.<br>**Possible Causes:**<br>APP Sensor has failed<br>APP Sensor No. 1 5v supply circuit is open, shorted to ground, to voltage, or shorted to Sensor ground<br>APP Sensor ground circuit or signal circuit is open<br>Intermittent condition<br>APP Sensor signal circuit is shorted to the Sensor ground circuit<br>ECM has failed |
| **P2127**<br>1T CCM<br>2004-05<br>300M, Dakota, Durango, Magnum, Ram | **Accelerator Pedal Position Sensor No. 2 Circuit Low Conditions:**<br>Ignition is on and no other APP Sensor No. 2 DTCs are present. When APP Sensor No. 2 voltage is too high, the engine will additionally idle, if the brake pedal is pressed or has failed. Acceleration rate and engine output are limited. This one-trip fault will set within 5 seconds. The ETC light will be flashing.<br>**Possible Causes:**<br>5v supply circuit is open or shorted to ground<br>APP Sensor No. 2 signal circuit is open, shorted to ground or to Sensor return circuit<br>APP Sensor No. 2has failed<br>PCM has failed |
| **P2127**<br>1T CCM<br>2004-05<br>PT Cruiser | **Accelerator Pedal Position Sensor No. 2 Circuit Low Conditions:**<br>Ignition is on. When APP Sensor No. 2 voltage is less than 0.2444v, or the circuit is shorted to ground or open for more than 120msec, this DTC will set.<br>**Possible Causes:**<br>APP Sensor sweep<br>Intermittent condition<br>5v supply circuit is open or shorted to ground<br>APP Sensor No. 2 signal circuit is open, shorted to ground or to Sensor return circuit<br>APP Sensor No. 2 has failed<br>PCM 5v supply circuit problem<br>PCM has failed |

**OBD II Trouble Code List (P2XXX Codes) – Continued**

| DTC | Trouble Code Title, Conditions & Possible Causes: |
|---|---|
| **P2128**<br>1T CCM<br>2004-05<br>300M, Dakota, Durango,<br>Magnum, Ram | **Accelerator Pedal Position Sensor No. 2 Circuit High Conditions:**<br>Ignition is on and no other APP Sensor No. 2 DTCs are present. When APP Sensor No. 2 voltage is too high, the engine will additionally idle, if the brake pedal is pressed or has failed. Acceleration rate and engine output are limited. This one-trip fault will set within 5 seconds. The ETC light will be flashing.<br>**Possible Causes:**<br>APP Sensor No. 2 return circuit is open<br>APP Sensor No. 2 signal circuit is shorted to either 5v supply circuit<br>APP Sensor No. 2 has failed<br>PCM has failed |
| **P2123**<br>1T CCM<br>2004-05<br>PT Cruiser | **Accelerator Pedal Position Sensor No. 2 Circuit High Conditions:**<br>Ignition is on. When APP Sensor No. 2 voltage is more than 4.8192v, because of a short to voltage, this DTC will set.<br>**Possible Causes:**<br>Intermittent condition<br>5v supply circuit is shorted to battery voltage<br>APP Sensor No. 1 signal circuit is shorted to battery voltage or to 5v circuit<br>APP Sensor sweep<br>PCM has failed |
| **P2135**<br>1T CCM<br>2004-05<br>300M, Dakota, Durango,<br>Magnum, Ram | **Throttle Position Sensors 1 & 2 Correlation Conditions:**<br>Ignition is on and no other TP Sensor DTCs are present. The PCM recognizes that TP Sensors 1 and 2 are not coherent, this one-trip fault will set within 5 seconds. The ETC light will be illuminated.<br>**Possible Causes:**<br>TP Sensor No. 1 or 2 signal circuit is shorted to ground or to battery voltage<br>TP Sensor No. 1 or 2 signal circuit has high resistance<br>5v supply circuit has high resistance<br>5v supply circuit shorted to ground<br>TP Sensor ground circuit has high resistance<br>TP Sensor No. 1 signal circuit is shorted to Sensor No. 2 signal circuit<br>TP Sensor has failed<br>PCM has failed |
| **P2135**<br>1T CCM<br>2004-05<br>PT Cruiser | **Throttle Position Sensors 1 & 2 Voltage Correlation Conditions:**<br>Ignition is on. When the difference between TP Sensor No. 1 degrees and TP Sensor No. 2 degrees is more than 1.995 degrees, this DTC will set.<br>**Possible Causes:**<br>Intermittent condition<br>5v supply circuit has high resistance<br>TP Sensor signal circuit has high resistance<br>Ground signal circuit shows high resistance<br>TP Sensors 1 & 2 require lab scope check<br>PCM has failed |
| **P2138**<br>1T CCM<br>2004-05<br>300M, Dakota, Durango,<br>Magnum, Ram | **Accelerator Pedal Position Sensors 1 & 2 Correlation Conditions:**<br>Ignition is on and no other APP Sensor DTCs are present. The PCM recognizes that APP Sensors 1 and 2 are not coherent. Acceleration rate and engine output are limited. This one-trip fault will set within 5 seconds. The ETC light will be flashing.<br>**Possible Causes:**<br>APP Sensor No. 1 or 2 signal circuit has high resistance<br>APP Sensor No. 1 or 2 return circuit has high resistance<br>5v supply circuit has high resistance<br>APP Sensor has failed<br>PCM has failed |

## OBD II Trouble Code List (P2XXX Codes) – Continued

| DTC | Trouble Code Title, Conditions & Possible Causes: |
|---|---|
| **P2138**<br>1T CCM<br>2004-05<br>PT Cruiser | **Accelerator Pedal Position Sensors 1 & 2 Voltage Correlation Conditions:**<br>Ignition is on. DTCs P2122, 2123, 2127 and 2128 are not present. When APP Sensor No. 1 voltage is 1.7 times the APP Sensor No. 2 voltage, and this equals more than 0.2v for 120msec, this DTC will set.<br>**Possible Causes:**<br>Intermittent condition<br>5v supply circuit is open or is shorted to ground<br>APP Sensor signal circuit has high resistance<br>5v supply circuit has high resistance<br>Ground signal circuit shows high resistance<br>APP Sensors 1 & 2 require lab scope check<br>PCM has failed |
| **P2141**<br>1T ECM<br>2004-05<br>Liberty with Diesel | **EGR Airflow Control Valve Short-To-Ground Circuit Conditions:**<br>Ignition on; EGR Airflow Control Valve commanded OFF. If the ECM detects no voltage signal on the EGR Airflow Control Valve control circuit, this DTC will set.<br>**Possible Causes:**<br>Intermittent condition<br>ASD Relay Output circuit open<br>EGR Airflow Control Valve has failed<br>EGR Airflow Control Valve control circuit is open, shorted to ground<br>ECM has failed |
| **P2142**<br>1T ECM<br>2004-05<br>Liberty with Diesel | **EGR Airflow Control Valve Short Circuit Conditions:**<br>Ignition on; EGR Airflow Control Valve commanded ON. If the ECM detects excessive voltage signal on the EGR Airflow Control Valve control circuit, this DTC will set.<br>**Possible Causes:**<br>Intermittent condition<br>EGR Airflow Control Valve has failed<br>EGR Airflow Control Valve control circuit is shorted to voltage<br>ECM has failed |
| **P2147**<br>1T CCM<br>2004-05<br>Liberty with Diesel | **Fuel Injector Bank 1 Open Circuit Conditions:**<br>Engine running. ECM detects unexpected current flow through injector control circuit.<br>**Possible Causes:**<br>Intermittent condition<br>Fuel injector control circuit is open or is shorted to ground<br>Fuel injector has failed<br>ECM has failed |
| **P2148**<br>1T CCM<br>2004-05<br>Liberty with Diesel | **Fuel Injector Bank 1 Short Circuit Conditions:**<br>Engine running. The ECM detects unexpected current flow through the injector control circuit.<br>**Possible Causes:**<br>Intermittent condition<br>Fuel injector control circuit is shorted to ground or to voltage<br>Fuel injector control circuits are shorted together<br>Fuel injector has failed<br>ECM has failed |
| **P2150**<br>1T CCM<br>2004-05<br>Liberty with Diesel | **Fuel Injector Bank 2 Open Circuit Conditions:**<br>Engine running. The ECM detects unexpected current flow through the injector control circuit.<br>**Possible Causes:**<br>Intermittent condition<br>Fuel injector control circuit is open or is shorted to ground<br>Fuel injector has failed<br>ECM has failed |

**OBD II Trouble Code List (P2XXX Codes) – Continued**

| DTC | Trouble Code Title, Conditions & Possible Causes: |
|---|---|
| **P2148**<br>1T CCM<br>2004-05<br>Liberty with Diesel | **Fuel Injector Bank 2 Short Circuit Conditions:**<br>Engine running. The ECM detects unexpected current flow through the injector control circuit.<br>**Possible Causes:**<br>Intermittent condition<br>Fuel injector control circuit is shorted to ground or to voltage<br>Fuel injector control circuits are shorted together<br>Fuel injector has failed<br>ECM has failed |
| **P2161**<br>1T CCM<br>2004-05<br>300M, Dakota, Durango, Magnum, Ram | **Vehicle Speed Sensor No. 2 Erratic Conditions:**<br>Ignition is on and battery voltage is greater than 10v. Transmission is in Drive or Reverse. The PCM recognizes the VSS 2 speed signal is erratic or high. No MIL and no ETC light. The cruise control is disabled.<br>**Possible Causes:**<br>Active Bus or Communications DTCs<br>Incorrect tire circumference<br>PCM has failed |
| **P2166**<br>1T CCM<br>2004-05<br>300M, Dakota, Durango, Magnum, Ram | **Accelerator Pedal Position Sensor No. 1 Maximum Stop Performance Conditions:**<br>Ignition is on. During "in-plant" mode the APP Sensors need to be checked to make sure that idle and full pedal travel can be reached on both Sensors. This DTC will set if the APP Sensor No. 1 has failed to achieve the required maximum value during "in-plant" testing. This one-trip fault will set within 5 seconds. The engine will only idle.<br>**Possible Causes:**<br>In-Plant test failure<br>APP Sensors must be reprogrammed to relearn |
| **P2167**<br>1T CCM<br>2004-05<br>300M, Dakota, Durango, Magnum, Ram | **Accelerator Pedal Position Sensor No. 2 Maximum Stop Performance Conditions:**<br>Ignition is on. During "in-plant" mode the APP Sensors need to be checked to make sure that idle and full pedal travel can be reached on both Sensors. This DTC will set if the APP Sensor No. 2 has failed to achieve the required maximum value during "in-plant" testing. This one-trip fault will set within 5 seconds. The engine will only idle.<br>**Possible Causes:**<br>In-Plant test failure<br>APP Sensors must be reprogrammed to relearn |
| **P2172**<br>1T CCM<br>2004-05<br>300M, Dakota, Durango, Magnum, Ram | **High Airflow/Vacuum Leak Detected (Instantaneous Accumulation) Conditions:**<br>Ignition is on and engine running with no MAP Sensor DTCs present. A large vacuum leak has been detected or both of the TP Sensors have failed, based on their position being 2.5v and the calculated MAP value is less than the actual MAP, minus an Offset value. This one-trip fault will set within 5 seconds. The ETC light will flash.<br>**Possible Causes:**<br>Vacuum leak<br>5v supply circuit has high resistance or is shorted to ground<br>MAP signal circuit has high resistance or is shorted to ground<br>TP Sensor ground circuit has high resistance<br>TP Sensor signal circuit is shorted to ground<br>TP Sensor return circuit has high resistance<br>MAP Sensor has failed<br>TP Sensor has failed<br>PCM has failed |

**OBD II Trouble Code List (P2XXX Codes) – Continued**

| DTC | Trouble Code Title, Conditions & Possible Causes: |
|-----|---------------------------------------------------|
| **P2173**<br>1T CCM<br>2004-05<br>300M, Dakota, Durango, Magnum, Ram | **High Airflow/Vacuum Leak Detected (Slow Accumulation) Conditions:**<br>Ignition is on and engine running with no MAP Sensor DTCs present. A large vacuum leak has been detected or both of the TP Sensors have failed, based on their position being 2.5v and the calculated MAP value is less than the Gas Flow Adaptation value. This one-trip fault will set within 5 seconds. The ETC light will flash.<br>**Possible Causes:**<br>Vacuum leak<br>5v supply circuit has high resistance or is shorted to ground<br>MAP signal circuit has high resistance or is shorted to ground<br>TP Sensor ground circuit has high resistance<br>TP Sensor signal circuit is shorted to ground<br>TP Sensor return circuit has high resistance<br>MAP Sensor has failed<br>TP Sensor has failed<br>PCM has failed |
| **P2174**<br>1T CCM<br>2004-05<br>300M, Dakota, Durango, Magnum, Ram | **Low Airflow/Vacuum Leak Detected (Instantaneous Accumulation) Conditions:**<br>Ignition is on and engine running with no MAP Sensor DTCs present. The PCM calculated the MAP value is greater than actual MAP value, plus an Offset value. 3 good trips required to turn off MIL. The ETC light will flash.<br>**Possible Causes:**<br>Restricted air inlet system<br>5v supply circuit has high resistance or is shorted to ground<br>MAP signal circuit has high resistance or is shorted to ground<br>TP Sensor ground circuit has high resistance<br>TP Sensor signal circuit is shorted to ground<br>TP Sensor return circuit has high resistance<br>MAP Sensor has failed<br>TP Sensor has failed<br>PCM has failed |
| **P2175**<br>1T CCM<br>2004-05<br>300M, Dakota, Durango, Magnum, Ram | **Low Airflow/Vacuum Leak Detected (Slow Accumulation) Conditions:**<br>Ignition is on and engine running with no MAP Sensor DTCs present. The PCM calculated the MAP value is greater than actual MAP value, plus an Offset value. This DTC will set in 5 seconds after occurrence. 3 good trips required to turn off MIL. The ETC light will flash.<br>**Possible Causes:**<br>Restricted air inlet system<br>5v supply circuit has high resistance or is shorted to ground<br>MAP signal circuit has high resistance or is shorted to ground<br>TP Sensor ground circuit has high resistance<br>TP Sensor signal circuit is shorted to ground<br>TP Sensor return circuit has high resistance<br>MAP Sensor has failed<br>TP Sensor has failed<br>PCM has failed |

**OBD II Trouble Code List (P2XXX Codes) – Continued**

| DTC | Trouble Code Title, Conditions & Possible Causes: |
|---|---|
| **P2181**<br>2T CCM<br>2004-05<br>300M, Dakota, Durango, Magnum, Ram | **Cooling System Performance Conditions:**<br>Ignition is on and engine running with no ECT Sensor DTCs present. The PCM recognizes that the ECT has failed its self-coherence test. The coolant temperature should only change at a certain rate. If this rate is too slow or too fast, this DTC will set. 3 good trips required to turn off MIL. The ETC light will illuminate on first trip failure.<br>**Possible Causes:**<br>Low coolant level<br>ECT signal circuit is open or shorted to ground, Sensor ground, or battery voltage<br>ECT Sensor ground circuit is open<br>Thermostat has failed<br>ECT Sensor has failed<br>PCM has failed |
| **P2226**<br>1T ECM<br>2004-05<br>Liberty with Diesel | **ECM Barometric Pressure Error Conditions:**<br>Ignition on. ECM detects an internal failure.<br>**Possible Causes:**<br>ECM has failed<br>Intermittent condition |
| **P2264**<br>1T ECM<br>2004-05<br>Liberty with Diesel | **Water In Fuel Voltage Above Upper Limit Or Below Lower Limit Conditions:**<br>Ignition on. If the ECM detects high voltage (above upper limit) or low voltage (below lower limit) on the Water In Fuel Sensor signal circuit, this DTC will set.<br>**Possible Causes:**<br>Intermittent condition<br>WIF Sensor signal circuit is open or is shorted to voltage (high) or to ground (low)<br>WIF Sensor ground circuit is open<br>WIF Sensor signal and ground circuits are shorted together (low)<br>WIF Sensor has failed<br>FCM has failed (low)<br>ECM has failed (high) |
| **P2294**<br>1T ECM<br>2004-05<br>Liberty with Diesel | **Fuel Pressure Solenoid Short-To-Ground Circuit Conditions:**<br>Ignition on; ECM Fuel Pressure Solenoid commanded OFF. If the ECM detects a short-to-ground on the control circuit, this DTC will set.<br>**Possible Causes:**<br>FP Solenoid circuit(s) open, shorted to voltage, shorted to ground, or shorted together<br>Intermittent condition<br>FP Solenoid has failed<br>ECM has failed |
| **P2296**<br>1T ECM<br>2004-05<br>Liberty with Diesel | **Fuel Pressure Solenoid Short Circuit Conditions:**<br>Ignition on; ECM Fuel Pressure Solenoid commanded ON. If the ECM detects excessive current on the control circuit, this DTC will set.<br>**Possible Causes:**<br>FP Solenoid circuit(s) open, shorted to voltage, shorted to ground, or shorted together<br>Intermittent condition<br>FP Solenoid has failed<br>ECM has failed |

**OBD II Trouble Code List (P2XXX Codes) – Continued**

| DTC | Trouble Code Title, Conditions & Possible Causes: |
|---|---|
| **P2299**<br>1T CCM<br>2004-05<br>300M, Dakota, Durango, Magnum, Ram | **Brake Pedal Position/Accelerator Pedal Position Incompatible Conditions:**<br>Ignition is on and no Brake or APPS DTCs present. The PCM recognizes that a brake application following the APPS showing a fixed pedal opening. Temporary or permanent in nature. Internally, the PCM will reduce throttle opening below driver demand. This one-trip fault code will set in 5 seconds. The ETC light will illuminate and will only stay on while the DTC is active.<br>**Possible Causes:**<br>Customer pressing accelerator pedal, then pressing brake pedal and holds both down at the same time<br>Stop lamp switch has failed<br>APP Sensor has failed |
| **P2300**<br>1T CCM<br>2004-05<br>PT Cruiser | **Ignition Coil No. 1 Secondary Circuit Low Conditions:**<br>Ignition is on. If the PCM detects an open or short to ground on the Ignition Coil No. 1 control circuit for more than 15 coil change requests, it will set this DTC.<br>**Possible Causes:**<br>Ignition Coil No. 1 is damaged or it has failed<br>ASD relay output circuit problems<br>Ignition coil driver circuit is shorted to ground<br>PCM has failed |
| **P2301**<br>1T CCM<br>2004-05<br>PT Cruiser | **Ignition Coil No. 1 Secondary Circuit High Conditions:**<br>Ignition is on. If the PCM detects a short to voltage on the Ignition Coil No. 1 control circuit for more than 15 coil change requests, it will set this DTC.<br>**Possible Causes:**<br>ASD relay output circuit problems<br>Ignition coil driver circuit is open<br>Ignition Coil No. 1 is damaged or it has failed<br>Ignition coil driver circuit is shorted to ASD output circuit<br>PCM has failed<br>Intermittent condition |
| **P2302**<br>1T CCM<br>2002-05<br>300M, Caravan, Cherokee, Concorde, PT Cruiser, Dakota, Durango, Grand Cherokee, Intrepid, LHS, Liberty, Magnum, Neon, New Yorker, Pacifica, Prowler, Ram, Sebring, Stratus, Town & Country, Viper, Vision, Voyager, Wrangler | **Ignition Coil No. 1 Secondary Circuit Insufficient Ionization Conditions:**<br>Engine started; and the PCM detected the Ignition Coil No. 1 secondary "burn time" was insufficient, or it was missing.<br>**Possible Causes:**<br>Intermittent condition<br>Cylinder No. 1 spark plug or wire is damaged or it has failed<br>Ignition Coil No. 1 is damaged or it has failed<br>Ignition coil control circuit is open or shorted to ground<br>ASD relay output circuit problems<br>PCM has failed |
| **P2303**<br>1T CCM<br>2004-05<br>PT Cruiser | **Ignition Coil No. 2 Secondary Circuit Low Conditions:**<br>Ignition is on. If the PCM detects an open or short to ground on the Ignition Coil No. 2 control circuit for more than 15 coil change requests, it will set this DTC.<br>**Possible Causes:**<br>Ignition Coil No. 2 is damaged or it has failed<br>ASD relay output circuit problems<br>Ignition coil driver circuit is shorted to ground<br>PCM has failed |

**OBD II Trouble Code List (P2XXX Codes) – Continued**

| DTC | Trouble Code Title, Conditions & Possible Causes: |
|---|---|
| **P2304**<br>1T CCM<br>2004-05<br>PT Cruiser | **Ignition Coil No. 2 Secondary Circuit High Conditions:**<br>Ignition is on. If the PCM detects a short to voltage on the Ignition Coil No. 2 control circuit for more than 15 coil change requests, it will set this DTC.<br>**Possible Causes:**<br>ASD relay output circuit problems<br>Ignition coil driver circuit is open<br>Ignition Coil No. 2 is damaged or it has failed<br>Ignition coil driver circuit is shorted to ASD output circuit<br>PCM has failed<br>Intermittent condition |
| **P2305**<br>1T CCM<br>2002-05<br>300M, Caravan, Cherokee, Concorde, PT Cruiser, Dakota, Durango, Grand Cherokee, Intrepid, LHS, Liberty, Magnum, Neon, New Yorker, Pacifica, Prowler, Ram, Sebring, Stratus, Town & Country, Viper, Vision, Voyager, Wrangler | **Ignition Coil No. 2 Secondary Circuit Insufficient Ionization Conditions:**<br>Engine started; and the PCM detected the Ignition Coil No. 2 secondary "burn time" was insufficient, or it was missing.<br>**Possible Causes:**<br>Intermittent condition<br>Cylinder No. 2 spark plug or wire is damaged or it has failed<br>Ignition Coil No. 2 is damaged or it has failed<br>Ignition coil control circuit is open or shorted to ground<br>ASD relay output circuit problems<br>PCM has failed |
| **P2308**<br>1T CCM<br>2002-05<br>300M, Caravan, Cherokee, Concorde, PT Cruiser, Dakota, Durango, Grand Cherokee, Intrepid, LHS, Liberty, Magnum, Neon, New Yorker, Pacifica, Prowler, Ram, Sebring, Stratus, Town & Country, Viper, Vision, Voyager, Wrangler | **Ignition Coil No. 3 Secondary Circuit Insufficient Ionization Conditions:**<br>Engine started; and the PCM detected the Ignition Coil No. 3 secondary "burn time" was insufficient, or it was missing.<br>**Possible Causes:**<br>Intermittent condition<br>Cylinder No. 3 spark plug or wire is damaged or it has failed<br>Ignition Coil No. 3 is damaged or it has failed<br>Ignition coil control circuit is open or shorted to ground<br>ASD relay output circuit problems<br>PCM has failed |
| **P2311**<br>1T CCM<br>2002-05<br>300M, Caravan, Cherokee, Concorde, PT Cruiser, Dakota, Durango, Grand Cherokee, Intrepid, LHS, Liberty, Magnum, Neon, New Yorker, Pacifica, Prowler, Ram, Sebring, Stratus, Town & Country, Viper, Vision, Voyager, Wrangler | **Ignition Coil No. 4 Secondary Circuit Insufficient Ionization Conditions:**<br>Engine started; and the PCM detected the Ignition Coil No. 4 secondary "burn time" was insufficient, or it was missing.<br>**Possible Causes:**<br>Intermittent condition<br>Cylinder No. 4 spark plug or wire is damaged or it has failed<br>Ignition Coil No. 4 is damaged or it has failed<br>Ignition coil control circuit is open or shorted to ground<br>ASD relay output circuit problems<br>PCM has failed |

**OBD II Trouble Code List (P2XXX Codes) – Continued**

| DTC | Trouble Code Title, Conditions & Possible Causes: |
|---|---|
| **P2314**<br>1T CCM<br>2002-05<br>300M, Caravan, Cherokee, Concorde, PT Cruiser, Dakota, Durango, Grand Cherokee, Intrepid, LHS, Liberty, Magnum, Neon, New Yorker, Pacifica, Prowler, Ram, Sebring, Stratus, Town & Country, Viper, Vision, Voyager, Wrangler | **Ignition Coil No. 5 Secondary Circuit Insufficient Ionization Conditions:**<br>Engine started; and the PCM detected the Ignition Coil No. 5 secondary "burn time" was insufficient, or it was missing.<br>**Possible Causes:**<br>Intermittent condition<br>Cylinder No. 5 spark plug or wire is damaged or it has failed<br>Ignition Coil No. 5 is damaged or it has failed<br>Ignition coil control circuit is open or shorted to ground<br>ASD relay output circuit problems<br>PCM has failed |
| **P2317**<br>1T CCM<br>2002-05<br>300M, Caravan, Cherokee, Concorde, PT Cruiser, Dakota, Durango, Grand Cherokee, Intrepid, LHS, Liberty, Magnum, Neon, New Yorker, Pacifica, Prowler, Ram, Sebring, Stratus, Town & Country, Viper, Vision, Voyager, Wrangler | **Ignition Coil No. 6 Secondary Circuit Insufficient Ionization Conditions:**<br>Engine started; and the PCM detected the Ignition Coil No. 6 secondary "burn time" was insufficient, or it was missing.<br>**Possible Causes:**<br>Intermittent condition<br>Cylinder No. 6 spark plug or wire is damaged or it has failed<br>Ignition Coil No. 6 is damaged or it has failed<br>Ignition coil control circuit is open or shorted to ground<br>ASD relay output circuit problems<br>PCM has failed |
| **P2320**<br>1T CCM<br>2002-05<br>300M, Caravan, Cherokee, Concorde, PT Cruiser, Dakota, Durango, Grand Cherokee, Intrepid, LHS, Liberty, Magnum, Neon, New Yorker, Pacifica, Prowler, Ram, Sebring, Stratus, Town & Country, Viper, Vision, Voyager, Wrangler | **Ignition Coil No. 7 Secondary Circuit Insufficient Ionization Conditions:**<br>Engine started; and the PCM detected the Ignition Coil No. 7 secondary "burn time" was insufficient, or it was missing.<br>**Possible Causes:**<br>Cylinder No. 7 spark plug or wire is damaged or it has failed<br>Ignition Coil No. 7 is damaged or it has failed<br>Ignition coil control circuit is open or shorted to ground<br>PCM has failed |
| **P2323**<br>1T CCM<br>2002-05<br>300M, Caravan, Cherokee, Concorde, PT Cruiser, Dakota, Durango, Grand Cherokee, Intrepid, LHS, Liberty, Magnum, Neon, New Yorker, Pacifica, Prowler, Ram, Sebring, Stratus, Town & Country, Viper, Vision, Voyager, Wrangler | **Ignition Coil No. 8 Secondary Circuit Insufficient Ionization Conditions:**<br>Engine started; and the PCM detected the Ignition Coil No. 8 secondary "burn time" was insufficient, or it was missing.<br>**Possible Causes:**<br>Cylinder No. 8 spark plug or wire is damaged or it has failed<br>Ignition Coil No. 8 is damaged or it has failed<br>Ignition coil control circuit is open or shorted to ground<br>PCM has failed |

## OBD II Trouble Code List (P2XXX Codes) – Continued

| DTC | Trouble Code Title, Conditions & Possible Causes: |
|---|---|
| **P2431**<br>2T CCM<br>2004-05<br>Sebring, Stratus | **MAF Sensor Malfunction Conditions:**<br>Engine running; no other MAF Sensor faults are present; Air Injection system is active. If the PCM detects the MAF Sensor has an excessive amount of airflow or not enough airflow through it, this DTC will set.<br>**Possible Causes:**<br>Air Injection system may be damaged, restricted or disconnected<br>ASD relay output circuit failure<br>MAF Sensor internal failure<br>MAF signal circuit open or shorted to ground or to ASD relay output circuit or to battery voltage<br>MAF signal circuit is shorted to sensor ground circuit<br>MAF Sensor ground circuit is open<br>PCM has failed. |
| **P2432**<br>1T CCM<br>2004-05<br>Sebring, Stratus | **MAF Sensor High Conditions:**<br>Engine running between 600-3500 rpm; TP sensor voltage is less than 1.2v for more than 1.7 seconds; battery voltage is more than 10v. If the PCM detects the MAF sensor signal is greater than 4.9267v, this DTC will set.<br>**Possible Causes:**<br>MAF signal circuit open or shorted to ground or to ASD relay output circuit or to battery voltage<br>MAF Sensor internal failure<br>MAF signal circuit is shorted to sensor ground circuit<br>MAF Sensor ground circuit is open<br>PCM has failed. |
| **P2433**<br>1T CCM<br>2004-05<br>Sebring, Stratus | **MAF Sensor Low Conditions:**<br>Engine running between 600-3500 rpm; TP sensor voltage is less than 1.2v for more than 1.7 seconds; battery voltage is more than 10v. If the PCM detects the MAF sensor signal is less than 0.07829v, this DTC will set.<br>**Possible Causes:**<br>ASD relay output circuit problem<br>MAF Sensor internal failure<br>MAF signal circuit is shorted to ground or to sensor ground circuit<br>PCM has failed. |
| **P2448**<br>2T CCM<br>2004-05<br>Sebring, Stratus | **Air Injection System High Flow Conditions:**<br>Engine running; Air Injection Pump is active. If the PCM detects there is too much airflow through the AI system, this DTC will set.<br>**Possible Causes:**<br>Make a visual inspection of the AI system and components for any damage or disconnects<br>Exhaust 1-way valve has failed<br>Fused battery voltage output circuit problem<br>AI pump relay has failed<br>AI pump ground circuit problem<br>AI pump motor is damaged or has failed<br>AI passages are blocked or leaking<br>MAF sensor has an internal failure<br>MAF signal circuit is open, or is shorted to battery voltage or to ASD relay output circuit<br>MAF signal circuit is shorted to ground or to sensor ground circuit<br>MAF sensor ground circuit is open<br>PCM has failed. |

## OBD II Trouble Code List (P2XXX Codes) – Continued

| DTC | Trouble Code Title, Conditions & Possible Causes: |
| --- | --- |
| **P2503**<br>1T CCM<br>2002-05<br>300M, Caravan, Cherokee, Concorde, PT Cruiser, Dakota, Durango, Grand Cherokee, Intrepid, LHS, Liberty, Magnum, Neon, New Yorker, Pacifica, Prowler, Ram, Sebring, Stratus, Town & Country, Viper, Vision, Voyager, Wrangler | **Charging System Voltage Low Conditions:**<br>Engine started; engine speed over 1157 rpm; PCM detected the Battery Sense voltage was 1v less than the Charging system voltage "goal" for 13.47 seconds during the CCM test. The PCM senses the battery voltage turns off the field driver and then senses the battery voltage again. If the voltages are the same, the DTC is set.<br>**Possible Causes:**<br>Battery sense circuit has a high resistance condition<br>Generator ground circuit has a high resistance condition<br>Generator field ground circuit is open<br>Generator field control circuit is open or shorted to ground<br>Generator is damaged or it has failed |
| **P2525**<br>1T ECM<br>2004-05<br>Liberty with Diesel | **Vacuum Reservoir Solenoid Open Circuit Conditions:**<br>Ignition on; Vacuum Reservoir Solenoid commanded OFF. If the ECM does not detect a voltage signal on the control circuit, this DTC will set.<br>**Possible Causes:**<br>Intermittent condition<br>ASD Relay output circuit is open<br>Vacuum Reservoir Solenoid control circuit is open or is shorted to ground<br>Vacuum Reservoir Solenoid has failed<br>ECM has failed |
| **P2527**<br>1T ECM<br>2004-05<br>Liberty with Diesel | **Vacuum Reservoir Solenoid Short-To-Ground Conditions:**<br>Ignition on; Vacuum Reservoir Solenoid commanded OFF. If the ECM does not detect a voltage signal on the control circuit, this DTC will set.<br>**Possible Causes:**<br>Intermittent condition<br>ASD Relay output circuit is open<br>Vacuum Reservoir Solenoid control circuit is open or is shorted to ground<br>Vacuum Reservoir Solenoid has failed<br>ECM has failed |
| **P2700**<br>1T CCM<br>2002-05<br>Dakota, Durango, Grand Cherokee, Liberty, Ram, Wrangler with 3.7L or 4.7L V6 | **A/T L/R Inadequate Element Volume Detected Conditions:**<br>Engine started; transmission fluid temperature more than 110°F, vehicle driven, and the PCM updated the L/R volume (during a 3-1 or 2-1 Manual downshift) with the throttle angle less than 5 degrees, and it detected that the L/R volume fell below 16 during the test.<br>**Possible Causes:**<br>L/R volume clutch index is too low<br>TCM L/R volume clutch circuit is damaged or has failed |
| **P2701**<br>1T CCM<br>2002-05<br>Dakota, Durango, Grand Cherokee, Liberty, Ram, Wrangler with 3.7L or 4.7L V6 | **A/T 2C Inadequate Element Volume Detected Conditions:**<br>Engine started; transmission fluid temperature more than 110°F, vehicle driven, then after the PCM updated the 2C volume (during a 3-2 kickdown event) with the throttle angle from 10-54 degrees, the PCM detected that the 2C volume fell below 5 during the CCM test.<br>**Possible Causes:**<br>2C volume clutch index is too low<br>TCM 2C volume clutch circuit is damaged or has failed |

**OBD II Trouble Code List (P2XXX Codes) – Continued**

| DTC | Trouble Code Title, Conditions & Possible Causes: |
| --- | --- |
| **P2702**<br>1T CCM<br>2002-05<br>Dakota, Durango, Grand Cherokee, Liberty, Ram, Wrangler with 3.7L or 4.7L V6 | **A/T O/D Inadequate Element Volume Detected Conditions:**<br>Engine started; transmission fluid temperature more than 110ºF, vehicle driven, then after he PCM updated the O/D volume (during a 2-3 Upshift event) with the throttle angle from 10-54 degrees, the PCM detected that the O/D volume fell below 5 during the CCM test.<br>**Possible Causes:**<br>O/D volume clutch index is too low<br>TCM O/D volume clutch circuit is damaged or has failed |
| **P2703**<br>1T CCM<br>2002-05<br>Dakota, Durango, Jeep Liberty, Truck with 3.7L or 4.7L V6 | **A/T U/D Inadequate Element Volume Detected Conditions:**<br>Engine started; transmission fluid temperature more than 110ºF, vehicle driven, and the TCM updated the U/D volume (during a 4-3 kickdown) with the throttle angle from 10-54 degrees, and it detected that the U/D volume fell below 11 during the test.<br>**Possible Causes:**<br>U/D volume clutch index is too low<br>TCM U/D volume clutch circuit is damaged or has failed |
| **P2704**<br>1T CCM<br>2002-05<br>Dakota, Durango, Grand Cherokee, Liberty, Ram, Wrangler with 3.7L or 4.7L V6 | **A/T 4C Inadequate Element Volume Detected Conditions:**<br>Engine started; transmission fluid temperature more than 110ºF, vehicle driven, then after the TCM updated the 4C volume (during a 3-4 Upshift event) with the throttle angle from 10-54 degrees, the PCM detected that the 4C volume fell below 5 during the CCM test.<br>**Possible Causes:**<br>4C volume clutch index is too low<br>TCM 4C volume clutch circuit is damaged or has failed |
| **P2706**<br>1T CCM<br>2002-05<br>Dakota, Durango, Grand Cherokee, Liberty, Ram, Wrangler with 3.7L or 4.7L V6 | **A/T MS Solenoid Circuit Malfunction Conditions:**<br>Engine started; vehicle driven in a forward gear, and immediately after a gear ratio or pressure switch change, the TCM detected a detected a MS solenoid error. The PCM sets this code when it detects three consecutive solenoid continuity test faults; or 1 failure if the test is run in response to a gear ratio of pressure switch fault.<br>**Possible Causes:**<br>Check for a loose connector to the MS solenoid (intermittent)<br>MS solenoid control circuit is open or shorted to ground<br>MS solenoid control circuit is shorted to system power (B+)<br>MS solenoid is damaged or it has failed<br>Transmission control relay output supply circuit is open<br>TCM MS solenoid circuit is damaged or it has failed |

## OBD II Trouble Code List (P3XXX Codes)

| DTC | Trouble Code Title, Conditions & Possible Causes: |
|-----|---------------------------------------------------|
| **P3400**<br>1T CCM<br>2004-05<br>300M, Magnum | **MDS Rationality Bank 1 Conditions:**<br>Engine running and is in transition from 8 to 4-cylinder operation. The O2 Sensor readings on Bank 1 side indicate a lean condition while in the 4-cylinder mode.<br>**Possible Causes:**<br>Insufficient oil pressure acting on the lifter locking pins<br>Oil passages restricted<br>Lifter has failed<br>MDS solenoid has failed |
| **P3401**<br>1T CCM<br>2004-05<br>300M, Magnum | **MDS Solenoid 1 Circuit Malfunction Conditions:**<br>Engine running and is in transition from 8 to 4-cylinder operation. The PCM recognizes a problem with the solenoid control circuit.<br>**Possible Causes:**<br>MDS solenoid 1 control circuit is open, or is shorted to ground or to battery voltage<br>Ground circuit is open<br>MDS solenoid 1 has failed<br>PCM has failed |
| **P3402**<br>1T CCM<br>2004-05<br>300M, Magnum | **Cylinder 1 Deactivation Control Performance Conditions:**<br>Engine running and is in transition from 8 to 4-cylinder operation. The MDS fails to disengage for cylinder 1.<br>**Possible Causes:**<br>MDS solenoid 1 control circuit is shorted to voltage<br>Oil passages restricted<br>Lifter has failed<br>MDS solenoid 1 has failed<br>PCM has failed |
| **P3425**<br>1T CCM<br>2004-05<br>300M, Magnum | **MDS Solenoid 4 Circuit Malfunction Conditions:**<br>Engine running and is in transition from 8 to 4-cylinder operation. The PCM recognizes a problem with the solenoid control circuit.<br>**Possible Causes:**<br>MDS solenoid 4 control circuit is open, or is shorted to ground or to battery voltage<br>Ground circuit is open<br>MDS solenoid 4 has failed<br>PCM has failed |
| **P3426**<br>1T CCM<br>2004-05<br>300M, Magnum | **Cylinder 4 Deactivation Control Performance Conditions:**<br>Engine running and is in transition from 8 to 4-cylinder operation. The MDS fails to disengage for cylinder 4.<br>**Possible Causes:**<br>MDS solenoid 4 control circuit is shorted to voltage<br>Oil passages restricted<br>Lifter has failed<br>MDS solenoid 4 has failed<br>PCM has failed |
| **P3441**<br>1T CCM<br>2004-05<br>300M, Magnum | **MDS Solenoid 6 Circuit Malfunction Conditions:**<br>Engine running and is in transition from 8 to 4-cylinder operation. The PCM recognizes a problem with the solenoid control circuit.<br>**Possible Causes:**<br>MDS solenoid 6 control circuit is open, or is shorted to ground or to battery voltage<br>Ground circuit is open<br>MDS solenoid 6 has failed<br>PCM has failed |

**OBD II Trouble Code List (P3XXX Codes) – Continued**

| DTC | Trouble Code Title, Conditions & Possible Causes: |
|---|---|
| **P3442**<br>1T CCM<br>2004-05<br>300M, Magnum | **Cylinder 6 Deactivation Control Performance Conditions:**<br>Engine running and is in transition from 8 to 4-cylinder operation. The MDS fails to disengage for cylinder 6.<br>**Possible Causes:**<br>MDS solenoid 6 control circuit is shorted to voltage<br>Oil passages restricted<br>Lifter has failed<br>MDS solenoid 6 has failed<br>PCM has failed |
| **P3449**<br>1T CCM<br>2004-05<br>300M, Magnum | **MDS Solenoid 7 Circuit Malfunction Conditions:**<br>Engine running and is in transition from 8 to 4-cylinder operation. The PCM recognizes a problem with the solenoid control circuit.<br>**Possible Causes:**<br>MDS solenoid 7 control circuit is open, or is shorted to ground or to battery voltage<br>Ground circuit is open<br>MDS solenoid 7 has failed<br>PCM has failed |
| **P3450**<br>1T CCM<br>2004-05<br>300M, Magnum | **Cylinder 7 Deactivation Control Performance Conditions:**<br>Engine running and is in transition from 8 to 4-cylinder operation. The MDS fails to disengage for cylinder 7.<br>**Possible Causes:**<br>MDS solenoid 7 control circuit is shorted to voltage<br>Oil passages restricted<br>Lifter has failed<br>MDS solenoid 7 has failed<br>PCM has failed |
| **P3497**<br>1T CCM<br>2004-05<br>300M, Magnum | **MDS Rationality Bank 2 Conditions:**<br>Engine running and is in transition from 8 to 4-cylinder operation. The O2 Sensor readings on Bank 2 side indicate a lean condition while in the 4-cylinder mode.<br>**Possible Causes:**<br>Insufficient oil pressure acting on the lifter locking pins<br>Oil passages restricted<br>Lifter has failed<br>MDS solenoid has failed |

## OBD II Trouble Code List (UXXXX Codes)

| DTC | Trouble Code Title, Conditions & Possible Causes: |
|---|---|
| **U0001**<br>1T TCM<br>2004-05<br>300M, Magnum | **CAN C Bus Circuit Malfunction Conditions:**<br>Ignition is on and battery voltage is 9-16v. Engine is running for more than 3 seconds. The PCM loses communication over the CAN C Bus circuit. The circuit is continuously monitored.<br>**Possible Causes:**<br>CAN C Bus failure open or shorted<br>PCM has failed |
| **U0101**<br>1T TCM<br>2004-05<br>300M, Caravan, Concorde, PT Cruiser, Dakota, Durango, Grand Cherokee, Intrepid, LHS, Liberty, Magnum, Neon, Pacifica, Prowler, Ram, Sebring, Stratus, Town & Country, Viper, Voyager, Wrangler | **No TCM Bus Message Conditions:**<br>Engine running. Battery voltage more than 10v. No Bus messages are received from the TCM for 20 seconds. 2 trips required.<br>**Possible Causes:**<br>PCI Bus unable to communicate with (DRBIII) scan tool<br>Fused ignition switch output incorrect (off-run-start)<br>Intermittent condition<br>PCM has failed |
| **U0101**<br>1T TCM<br>2004-05<br>300M, Dakota, Durango, Magnum, Ram | **No TCM Bus Message Conditions:**<br>Ignition is on and battery voltage is 9-16v. Engine is running for more than 3 seconds. The PCM does not receive a Bus message from the TCM for 7 consecutive seconds. The circuit is continuously monitored.<br>**Possible Causes:**<br>CAN C Bus failure open or shorted<br>PCM has failed |
| **U0103**<br>1T TCM<br>2004-05<br>300M, Magnum | **Lost Communication With Electric Gear Shift Module Conditions:**<br>Ignition is on and battery voltage is 9-16v. Engine is running for more than 3 seconds. The PCM does not receive an Electric Gear Shift Module message over the CAN C circuit. The circuit is continuously monitored.<br>**Possible Causes:**<br>CAN C Bus failure open or shorted<br>Electric gear shift module has failed<br>PCM has failed |
| **U0121**<br>1T TCM<br>2004-05<br>300M, Magnum | **Lost Communication With ABS Module Conditions:**<br>Ignition is on and battery voltage is 9-16v. Engine is running for more than 3 seconds. The PCM does not receive an ABS message over the CAN C circuit for 7 consecutive seconds. The circuit is continuously monitored.<br>**Possible Causes:**<br>CAN C Bus failure open or shorted<br>ABS module has failed<br>PCM has failed |
| **U0140**<br>1T BCM<br>2004-05<br>Caravan, Pacifica, Sebring, Stratus, Town & Country, Voyager | **No Body Bus Message Conditions:**<br>Engine running. Battery voltage more than 10v. No Bus messages are received from the BCM for 20 seconds.<br>**Possible Causes:**<br>Communication link with BCM has failed<br>PCI Bus circuit open<br>PCM has failed |

**OBD II Trouble Code List (UXXXX Codes) – Continued**

| DTC | Trouble Code Title, Conditions & Possible Causes: |
|---|---|
| **U0141**<br>1T TCM<br>2004-05<br>300M, Magnum | **Lost Communication With Front Control Module Conditions:**<br>Ignition is on and battery voltage is 9-16v. Engine is running for more than 3 seconds. The PCM does not receive an FCM message over the CAN C circuit for 7 consecutive seconds. The circuit is continuously monitored.<br>**Possible Causes:**<br>CAN C Bus failure open or shorted<br>Front control module has failed<br>PCM has failed |
| **U0155**<br>1T MIC<br>2004-05<br>300M, Caravan, Concorde, PT Cruiser, Dakota, Durango, Grand Cherokee, Intrepid, LHS, Liberty, Magnum, Neon, Pacifica, Prowler, Ram, Sebring, Stratus, Town & Country, Viper, Voyager, Wrangler | **No Cluster Bus Message Conditions:**<br>Engine running. Battery voltage more than 10v. No Bus messages are received from the MIC (instrument cluster) for 20 seconds.<br>**Possible Causes:**<br>Communication link with instrument cluster has failed<br>Instrument cluster operation improper or has failed<br>PCM has failed |
| **U0155**<br>1T TCM<br>2004-05<br>300M, Magnum | **Lost Communication With Instrument Cluster/CCN Conditions:**<br>Ignition is on and battery voltage is 9-16v. Engine is running for more than 3 seconds. The PCM does not receive a Cluster message over the CAN C circuit. The circuit is continuously monitored.<br>**Possible Causes:**<br>CAN C Bus failure open or shorted<br>Front control module has failed<br>PCM has failed |
| **U0168**<br>1T MIC<br>2004-05<br>300M, Caravan, Concorde, PT Cruiser, Dakota, Durango, Grand Cherokee, Intrepid, LHS, Liberty, Magnum, Neon, Pacifica, Prowler, Ram, Sebring, Stratus, Town & Country, Viper, Voyager, Wrangler | **No SKIM Bus Message Conditions:**<br>Engine running or ignition on. Battery voltage more than 10v. No Bus or J1850 messages are received from the SKIM for 20 seconds.<br>**Possible Causes:**<br>Intermittent operation<br>PCI Bus circuit open or shorted from PCM to SKIM<br>Loss of communication between PCM and SKIM<br>SKIM or PCM has failed |
| **U0168**<br>1T TCM<br>2004-05<br>300M, Magnum | **Lost Communication With Vehicle Security Control Module (SKREEM/WCM) Conditions:**<br>Ignition is on and battery voltage is 9-16v. Engine is running for more than 3 seconds. Bus message not received from the SKREEM/WCM from about 2-5 seconds.<br>**Possible Causes:**<br>CAN C Bus failure open or shorted<br>SKREEM/WCM module has failed<br>PCM has failed |

## OBD II Trouble Code List (UXXXX Codes) – Continued

| DTC | Trouble Code Title, Conditions & Possible Causes: |
|---|---|
| **U110A**<br>1T TCM<br>2004-05<br>300M, Magnum | **Lost Communication With Steering Control Module (SCCM) Conditions:**<br>Ignition is on and battery voltage is 9-16v. Engine is running for more than 3 seconds. Bus message not received from the SCCM from about 2-5 seconds.<br>**Possible Causes:**<br>CAN C Bus failure open or shorted<br>SCCM module has failed<br>PCM has failed |
| **U110C**<br>1T MIC<br>2004-05<br>Caravan, Pacifica, Neon, Sebring, Stratus, Town & Country, Voyager | **No Fuel Level Bus Message Conditions:**<br>Ignition on. Battery voltage more than 10v. No fuel level Bus messages are received from the PCM for 20 seconds.<br>**Possible Causes:**<br>PCI Bus circuit open between PCM and BCM<br>Fuel level Bus message circuit failure<br>BCM has failed |
| **U110C**<br>1T TCM<br>2004-05<br>300M, Magnum | **No Fuel Level Bus Message Conditions:**<br>Ignition is on. PCM does not receive a fuel level signal from the FCM over the CAN C circuit. The circuit is constantly monitored.<br>**Possible Causes:**<br>CAN C Bus failure open or shorted<br>Front Control Module (FCM) has failed<br>PCM has failed |
| **U110E**<br>1T TCM<br>2004-05<br>300M, Magnum | **No Ambient Temperature Message Conditions:**<br>Ignition is on. PCM does not receive an ambient temperature signal over the CAN C circuit from the FCM. The circuit is constantly monitored.<br>**Possible Causes:**<br>CAN C Bus failure open or shorted<br>Front Control Module (FCM) has failed<br>PCM has failed |
| **U110F**<br>1T TCM<br>2004-05<br>300M, Magnum | **No Fuel Volume Message Conditions:**<br>Ignition is on. PCM does not receive a fuel volume signal over the CAN C circuit from the FCM. The circuit is constantly monitored.<br>**Possible Causes:**<br>CAN C Bus failure open or shorted<br>Front Control Module (FCM) has failed<br>PCM has failed |
| **U1110**<br>1T TCM<br>2004-05<br>300M, Magnum | **No Vehicle Speed Message Conditions:**<br>Ignition is on. PCM does not receive a vehicle speed signal from the ABS module or FCM (non-ABS) over the CAN C circuit.<br>**Possible Causes:**<br>CAN C Bus failure open or shorted<br>Front Control Module (FCM) has failed<br>ABS Module has failed<br>PCM has failed |
| **U1120**<br>1T TCM<br>2004-05<br>300M, Magnum | **No Wheel Distance Message Conditions:**<br>Ignition is on. PCM does not receive a wheel distance signal from the ABS module or FCM (non-ABS) over the CAN C circuit.<br>**Possible Causes:**<br>CAN C Bus failure open or shorted<br>Front Control Module (FCM) has failed<br>ABS Module has failed<br>PCM has failed |

**OBD II Trouble Code List (UXXXX Codes) – Continued**

| DTC | Trouble Code Title, Conditions & Possible Causes: |
|---|---|
| **U1403**<br>1T TCM<br>2004-05<br>300M, Magnum | **Implausible Fuel Level Signal Conditions:**<br>Ignition is on. The fuel level message that the PCM is receiving is implausible. The circuit is continuously monitored.<br>**Possible Causes:**<br>CAN B Bus failure open or shorted<br>Instrument Cluster Module has failed<br>Front Control Module has failed<br>PCM has failed |
| **U1411**<br>1T TCM<br>2004-05<br>300M, Magnum | **Implausible Fuel Volume Signal Conditions:**<br>Ignition is on. The fuel volume message that the PCM is receiving is implausible. The circuit is continuously monitored.<br>**Possible Causes:**<br>CAN B Bus failure open or shorted<br>Instrument Cluster Module has failed<br>Front Control Module has failed<br>PCM has failed |
| **U1412**<br>1T TCM<br>2004-05<br>300M, Magnum | **Implausible Vehicle Speed Signal Conditions:**<br>Ignition is on. The vehicle speed message that the PCM is receiving over the CAN C circuit from the ABS module or FCM (non-ABS) is implausible. The circuit is continuously monitored.<br>**Possible Causes:**<br>CAN C Bus failure open or shorted<br>ABS Module has failed<br>Front Control Module has failed<br>PCM has failed |
| **U1417**<br>1T TCM<br>2004-05<br>300M, Magnum | **Implausible Left Wheel Distance Signal Conditions:**<br>Ignition is on. The left wheel distance message that the PCM is receiving over the CAN C circuit from the ABS module or FCM (non-ABS) is implausible. The circuit is continuously monitored.<br>**Possible Causes:**<br>Vehicle speed Sensor fault active in ABS module<br>CAN C Bus failure open or shorted<br>ABS Module has failed<br>Front Control Module has failed<br>PCM has failed |
| **U1418**<br>1T TCM<br>2004-05<br>300M, Magnum | **Implausible Right Wheel Distance Signal Conditions:**<br>Ignition is on. The left wheel distance message that the PCM is receiving over the CAN C circuit from the ABS module or FCM (non-ABS) is implausible. The circuit is continuously monitored.<br>**Possible Causes:**<br>Vehicle speed Sensor fault active in ABS module<br>CAN C Bus failure open or shorted<br>ABS Module has failed<br>Front Control Module has failed<br>PCM has failed |

## Glossary of Terms & Acronyms

| | |
|---|---|
| (<) - Indicates less than the value | (>) - Indicates more than the value |
| A/C - Air Conditioning System | A/D - Analog to Digital Converter |
| A/F - Air Fuel Ratio | A/T - Automatic Transmission |
| ABS - Antilock Brake System | ACM - Airbag control Module |
| AIR - Secondary Air Injection System | AIS - Automatic Idle Speed |
| ASD - Automatic Shutdown Relay | AWD - All Wheel Drive |
| B+ - Battery Voltage | BARO - Barometric Pressure Sensor |
| BCM - Body Control Module | BOB - Breakout Box |
| CANP - EVAP Canister Purge Solenoid | CARB - California Air Resources Board |
| CCD - Chrysler Collision Detection Serial Data Bus | CCP - Comprehensive Component Monitor |
| CCS - Coast Clutch Solenoid | CKP - Crankshaft Position |
| CKT - Circuit | CMP - Camshaft Position |
| CNG - Certified Natural Gas | CO - Carbon Monoxide |
| CO2 - Carbon Dioxide | CTRL - Control |
| CYL - Cylinder | DI - Distributor Ignition |
| DIS - Direct Ignition System | DLC - Data Link Connector |
| DOHC - Double Overhead Cam Engine | DTC - Diagnostic Trouble Code |
| DRL - Daytime Running Lights | DVOM - Digital Volt/Ohm Meter |
| EATX - Electronic Automatic Transaxle | EBCM - Electronic Brake Control Module |
| EBTCM - Electronic Brake T/C Module | ECT - Engine Coolant Temperature |
| EEPROM - Electronic Erasable Programmable Read Only Memory | EFI - Electronic Fuel Injection |
| EGR - Exhaust Gas Recirculation | EGR Monitor - OBD II EGR Test |
| EI - Electronic Ignition System | EMCC - Electronically Modulated Converter Clutch |
| EPA - Environmental Protection Agency | EVAP - Evaporative Emission System |
| FAN - Cooling Fan (Low or High Speed) | FF - Flexible Fuel Vehicle |
| FTP - Fuel Tank Pressure | FWD - Front Wheel Drive |
| GEM - Generic Electronic Module | GND - Electrical ground connection |
| GPS - Governor Position Sensor | GVW - Gross Vehicle Weight |
| HC - Hydrocarbons | HMSL - High Mounted Stop Lamp |
| O2 (B1 S1) Signal | O2 (B1 S2) Signal |
| O2 (B2 S1) Signal | O2 (Bank 2 Sensor No. 2) Signal |
| O2 - Heated Oxygen Sensor | Hz - Hertz |
| IAC - Idle Air Control Sensor | IAT - Intake Air Temperature Sensor |
| ICM - Ignition Control Module | IGN GND - Ignition Ground |
| ISO - International Standards Organization | JTEC - Jeep/Truck Engine Control Module |
| KAM - Keep Alive Memory | KAPWR - Direct Battery Power |
| Kg/cm$^2$ - Kilograms/Cubic Centimeters | KOEC - Key On, Engine Cranking |
| KOEO - Key On, Engine Off | KOER - Key On, Engine Running |
| KS - Knock Sensor | LDP - Leak Detection Pump |
| LED - Light Emitting Diode | LONGFT - Long Term Fuel Trim |
| LOOP - Engine Operating Loop Status | LPG - Liquid Petroleum Gas |

**DaimlerChrysler OBD II GLOSSARY**

### Glossary of Terms & Acronyms

| | |
|---|---|
| MAF - Mass Airflow (Sensor) | MAP - Manifold Air Pressure |
| MFI - Multiport Fuel Injection | MIL - Malfunction Indicator Lamp |
| MPD - Manifold Pressure Differential | MPH - Miles Per Hour |
| MPI - Multiport Injection (relay) | M/T - Manual Transmission |
| MTV - Manifold Tuning Valve | Ms - Milliseconds |
| MV - Millivolt | NGV - Natural Gas Vehicles |
| NOx - Oxides of Nitrogen | NTC - Negative Temperature Coefficient |
| O2S-11 (B1 S1) Signal | O2S-21 (B2 S1) Signal |
| OBD I - On Board Diagnostics Version I | OBD II - On Board Diagnostics Version II |
| ORD - Overdrive Running Clutch | OSS - Output Speed Shaft |
| PCI - Programmable Communications Interface | PCM - Powertrain Control Module |
| PCV - Positive Crankcase Ventilation | PDC - Power Distribution Center |
| PFI - Port Fuel Injection | PID - Parameter Identification Location |
| PNP - Park Neutral Position (switch) | PSP - Power Steering Pressure (switch) |
| PWR GND - Power Ground for PCM | PWM - Pulse width Modulated (signal) |
| RAM - Random Access Memory | ROM - Read Only Memory |
| RPM - Revolutions Per Minute | RWD - Rear Wheel Drive |
| S/C - Speed Control | SBEC - Single Board Engine Controller |
| SCI - Serial Communication Interface | SFI - Sequential Fuel Injection |
| SIL - Shift Indicator Lamp | SKIM: Sentry Key Immobilizer Module |
| SMEC- Single Module Engine Controller | SOHC - Single Overhead Cam Engine |
| SRI - Service Reminder Indicator | SRS - Supplemental Restraint System |
| TAC - Thermostatic Air Cleaner | TACH - Tachometer (signal) |
| TBI - Throttle Body Injection | TCC - Torque Converter Clutch |
| TCCS - Torque Converter Clutch Solenoid | TCM - Transmission Control Module |
| TCS - Traction Control Switch | TDC - Top Dead Center |
| TPS - Throttle Position Sensor | TRS - Transmission Range Switch |
| TSB - Technical Service Bulletin | TTS - Transmission Temp. Sensor |
| Turbo - Turbo Charged | TWC - Three Way Catalyst |
| VAC - Vacuum | VAF - Volume Airflow (Sensor) |
| VECI - Vehicle Emission Control Information (Label) | VREF - Reference Voltage (from PCM) |
| VSS - Vehicle Speed Sensor | WOT - Wide Open Throttle |

## Contents

*NOTE:* *This section covers electronic engine control sensors, switches, and other key components. Where available, the coverage includes a Description and Operation explanation, Remove & Installation, and Testing procedures. Not all models and engine combination, or procedures for each, may be included; however, many like components or similar engines may contain information that is useful for your specific vehicle application.*

## DAIMLERCHRYSLER CARS: 300 & MAGNUM

<u>AMBIENT AIR TEMPERATURE (AAT) SENSOR</u>

**Connector Pinouts**

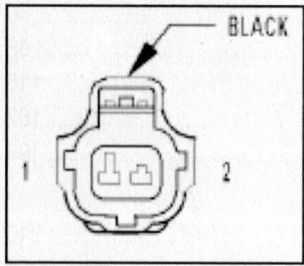

| PIN # | WIRE COLOR | CIRCUIT DESCRIPTION |
|-------|-----------|---------------------|
| 1 | VT/LG | AAT SIGNAL |
| 2 | VT/BR | SENSOR GROUND |

**2005 300 & Magnum Ambient Air Temperature Sensor**

<u>CAMSHAFT POSITION (CMP) SENSOR</u>

**Removal & Installation**

*2.7L*

1. Remove sensor. Disconnect negative battery cable.
2. Unlock and disconnect electrical connector.
3. Remove mounting bolt.

**To install:**

4. Install sensor.
5. Install mounting bolt.
6. Tighten bolt to 12 Nm (105 inch lbs.).
7. Connect electrical connector and lock.
8. Connect negative battery cable.

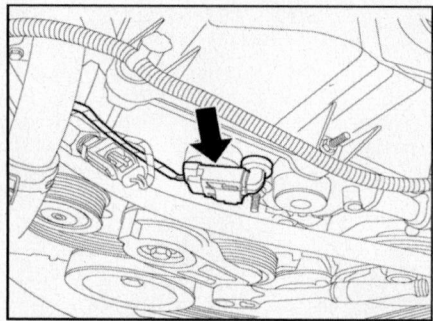

**Location of CMP Sensor on 2.7L engine**

*3.5L*

1. Disconnect negative battery cable.
2. Unlock electrical connector.
3. Remove electrical connector.
4. Remove mounting bolt.
5. Remove sensor.

**To install:**

6. Install new spacer to sensor if installing the old sensor or new sensor will have the spacer installed.
7. Install sensor and insert until sensor makes contact with tone wheel.
8. Install mounting bolt.
9. Tighten mounting bolt to 12 Nm (105 inch lbs.).
10. Install and lock the electrical connector.
11. Connect negative battery cable.

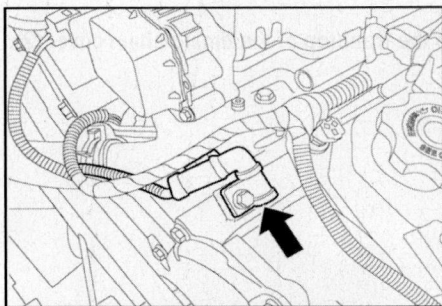

**Location of CMP Sensor on 3.5L engine**

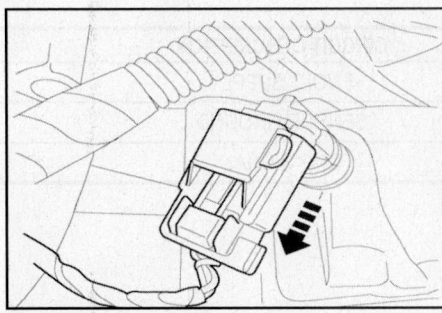

**Disconnecting electrical connector from CMP Sensor on 3.5L engine**

*5.7L & 6.1*

1. Disconnect electrical connector (4) at CMP sensor.
2. Remove sensor mounting bolt (3).
3. Carefully twist sensor from timing chain cover.
4. Check condition of sensor O-ring.

**To install:**

5. Clean out machined hole in timing chain cover.
6. Apply a small amount of engine oil to sensor O-ring.

7. Install sensor (1) into timing chain cover (2) with a slight rocking action. Do not twist sensor into position as damage to O-ring may result.

*CAUTION: Before tightening sensor mounting bolt, be sure sensor is completely flush to timing chain cover. If sensor is not flush, damage to sensor mounting tang may result.*

8. Install mounting bolt (3) and tighten to 12 Nm (105 inch lbs.)

9. Connect electrical connector (4) to sensor.

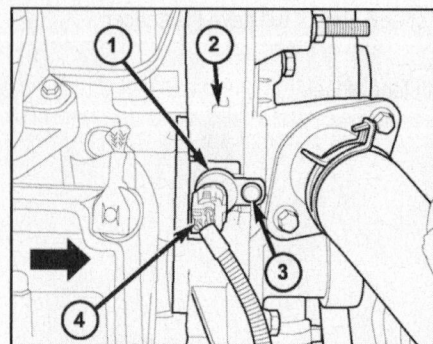

**The CMP Sensor (CMP) on the 5.7L & 6.1L V-8 engine (1) is bolted to the front/top of the timing chain cover (2).**

**Connector Pinouts**

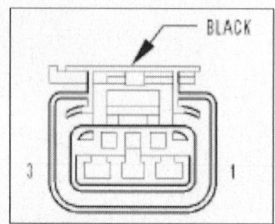

| PIN # | WIRE COLOR | CIRCUIT DESCRIPTION |
|-------|------------|---------------------|
| 1 | YL/PK | 5 VOLT SUPPLY |
| 2 | DB/DG | SENSOR GROUND |
| 3 | DB/GY | CMP SIGNAL |

**2004-05 300 2.5L & 3.7L Camshaft Position Sensor**

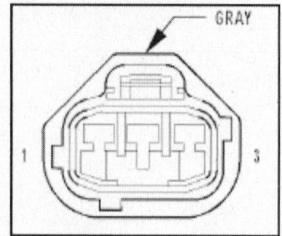

| PIN # | WIRE COLOR | CIRCUIT DESCRIPTION |
|-------|------------|---------------------|
| 1 | DB/GY | CMP SIGNAL |
| 2 | DB/DG | SENSOR GROUND |
| 3 | YL/PK | 5 VOLT SUPPLY |

**2004-05 300 & Magnum 5.7L & 6.1L Camshaft Position Sensor**

CRANKSHAFT POSITION (CKP) SENSOR

**Removal & Installation**

**2.7L & 3.5L**

1. Disconnect negative battery cable.
2. Raise vehicle and support.
3. Unlock and disconnect the electrical connector.
4. Remove mounting bolt.
5. Remove sensor.

**To install:**

6. Install sensor.
7. Install mounting bolt and tighten to 12 Nm (105 inch lbs.).
8. Connect the electrical connector and lock.
9. Lower vehicle.
10. Connect negative battery cable.

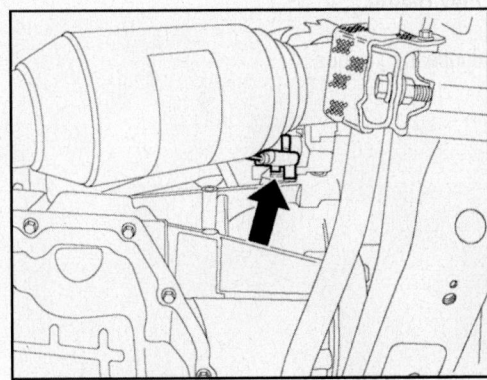

**Showing the location of the CKP Sensor on 2.7L & 3.5L engines**

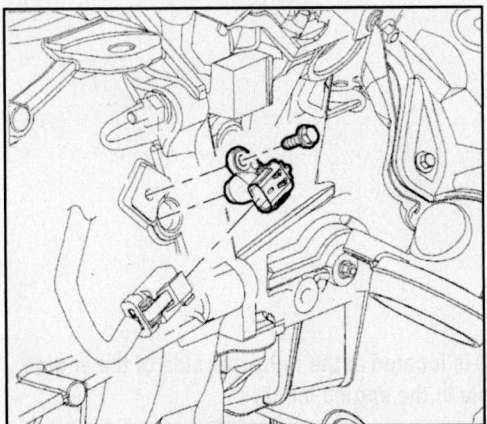

**Detach electrical connector and remove CKP Sensor on 2.7L & 3.5L engines**

**5.7L & 6.1L**

1. If equipped with AWD (All Wheel Drive) disconnect and isolate negative battery cable.
2. Raise vehicle.
3. If equipped with AWD (All Wheel Drive) remove starter motor.
4. Disconnect CKP electrical connector (2) at sensor.
5. Remove CKP mounting bolt (3).
6. Carefully twist sensor from cylinder block.
7. Remove sensor from vehicle.
8. Check condition of sensor O-ring.

**To install:**

9. Clean out machined hole in engine block.
10. Apply a small amount of engine oil to sensor O-ring.
11. Install sensor (4) into engine block with a slight rocking and twisting action.

*CAUTION: Before tightening sensor mounting bolt (3), be sure sensor is completely flush to cylinder block. If sensor is not flush, damage to sensor mounting tang may result.*

12. Install mounting bolt (3) and tighten to 12 Nm (106 inch lbs.) torque.
13. Connect electrical connector (2) to sensor.
14. Lower vehicle.

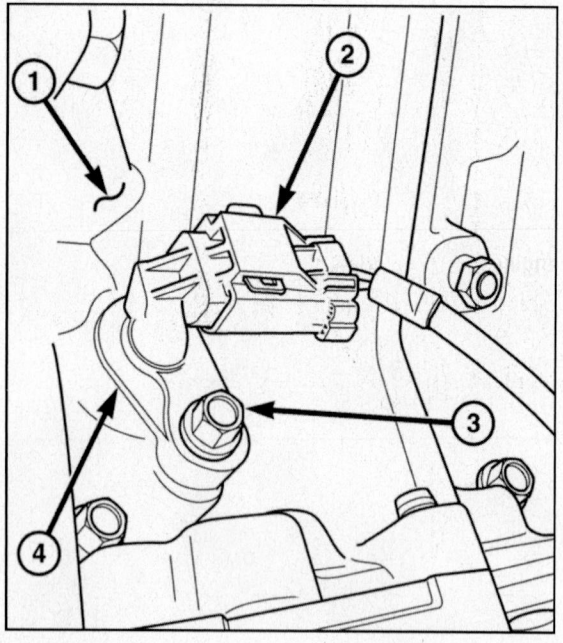

On 5.7L & 6.1L engines the Crankshaft Position (CKP) sensor (4) is located at the right-rear side of the engine cylinder block (1). It is positioned and bolted into a machined hole in the engine block.

**Connector Pinouts**

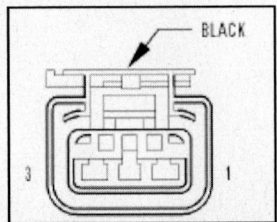

| PIN # | WIRE COLOR | CIRCUIT DESCRIPTION |
|-------|------------|---------------------|
| 1 | PK/YL | 5 VOLT SUPPLY |
| 2 | DB/DG | SENSOR GROUND |
| 3 | DB/WT | CKP SIGNAL |

**300 & Magnum Crankshaft Position Sensor (All Engines)**

DATA LINK CONNECTOR

**Connector Pinouts**

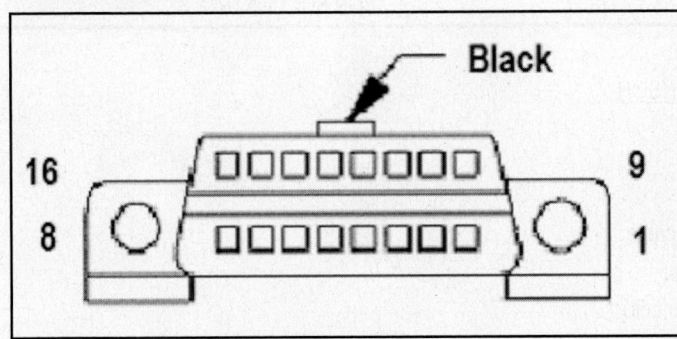

| PIN # | WIRE COLOR | CIRCUIT DESCRIPTION |
|-------|------------|---------------------|
| 1 | - | - |
| 2 | - | - |
| 3 | - | - |
| 4 | BK/TN | GROUND |
| 5 | BK/TN | GROUND |
| 6 | WT/LB | CAN C DIAGNOSTIC (+) |
| 7 | PK | SCI TRANSMIT (PCM) |
| 8 | - | - |
| 9 | WT/OR | SCI RECEIVE (TCM) |
| 10 | WT/DG (ABS) | FLASH ABS |
| 11 | - | - |
| 12 | LG | SCI RECEIVE (PCM) |
| 13 | - | - |
| 14 | WT/DB | CAN C DIAGNOSTIC (-) |
| 15 | WT/DG | SCI TRANSMIT (TCM) |
| 16 | LB/RD | FUSED B (+) |

**300 & Magnum Data Link Connector**

## EXHAUST GAS RECIRCULATION (EGR) VALVE

**Connector Pinouts**

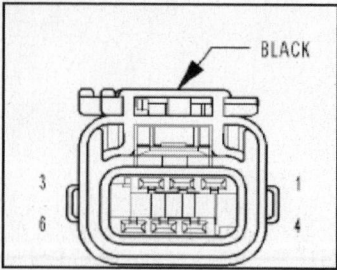

| PIN # | WIRE COLOR | CIRCUIT DESCRIPTION |
|-------|-----------|---------------------|
| 1 | DB/LG | EGR SIGNAL |
| 2 | YL/PK | 5 VOLT SUPPLY |
| 3 | DB/DG | SENSOR GROUND |
| 4 | BK/BR | GROUND |
| 5 | - | - |
| 6 | DB/VT | EGR SOL CONTROL |

**300 & Magnum EGR Valve Connector Pinout**

## ENGINE COOLANT TEMPERATURE (ECT) SENSOR

**Removal & Installation**

*2.7L & 3.5L*

1. Disconnect negative battery cable.
2. Partially drain cooling system.
3. Disconnect the electrical connector (2).
4. Remove engine coolant sensor (1) from coolant outlet tube (3).

**To install:**

5. Apply thread sealant to sensor threads.
6. Install engine coolant temperature sensor (1) into coolant outlet tube (3).
7. Tighten sensor to 28 Nm (20 ft. lbs.) torque.
8. Connect electrical connector (2) to engine coolant temperature sensor (1).

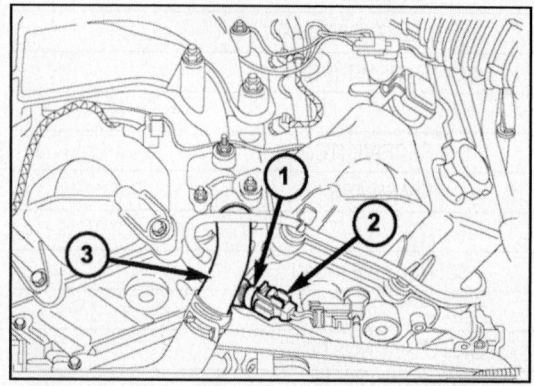

**Identifying the locations of the engine coolant sensor (1), electrical connector (2) and the coolant outlet tube (3) on the 2.7L engine**

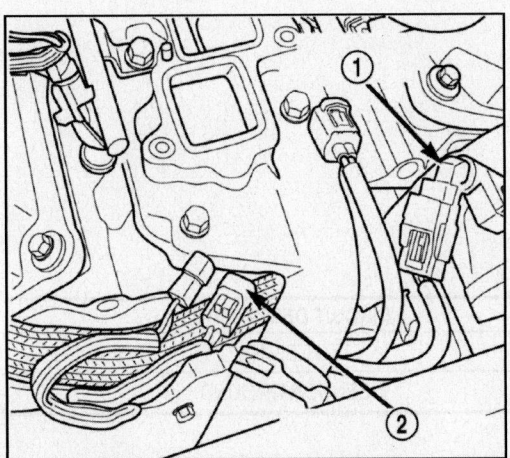

**Showing the location of the ECT Sensor (2) and the CMP Sensor (1) on the 3.5L engine**

### 5.7L/6.1L

The Engine Coolant Temperature (ECT) sensor on the 5.7L and 6.1L engines is located under the air conditioning compressor. It is installed into a water jacket at the front of the cylinder block.

**WARNING: HOT, PRESSURIZED COOLANT CAN CAUSE INJURY BY SCALDING. COOLING SYSTEM MUST BE PARTIALLY DRAINED BEFORE REMOVING THE COOLANT TEMPERATURE SENSOR.**

1. Partially drain the cooling system.
2. Remove accessory drive belt.
3. Carefully unbolt air conditioning compressor from front of engine. Do not disconnect any A/C hoses from compressor.
4. Temporarily support compressor to gain access to ECT sensor (3).
5. Disconnect electrical connector (2) from sensor (3).
6. Remove sensor (3) from cylinder block.

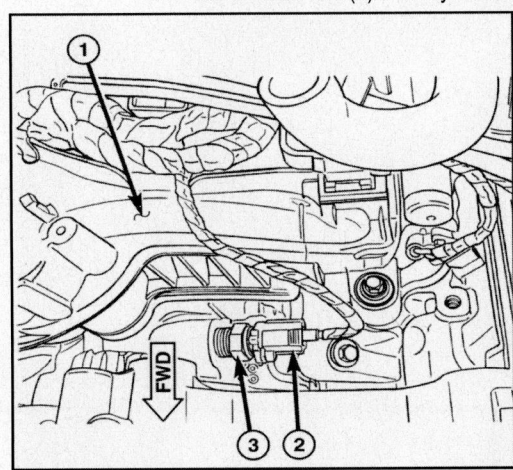

**Identifying the location of the ECT sensor (3), electrical connector (2), and the intake manifold (1) on the 5.7L & 6.1L engines**

**Connector Pinouts**

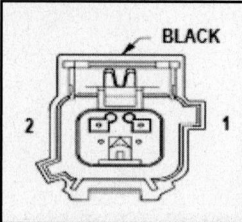

| PIN # | WIRE COLOR | CIRCUIT DESCRIPTION |
|-------|------------|---------------------|
| 1 | VT/OR | ECT SIGNAL |
| 2 | DB/DG | SENSOR GROUND |

**300 & Magnum Engine Coolant Temperature Sensor**

## ENGINE OIL TEMPERATURE (EOT) SENSOR

**Connector Pinouts**

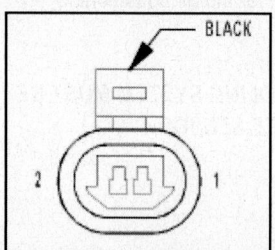

| PIN # | WIRE COLOR | CIRCUIT DESCRIPTION |
|-------|------------|---------------------|
| 1 | VT/YL | ENGINE OIL TEMPERATURE SIGNAL |
| 2 | DB/DG | SENSOR GROUND |

**300 & Magnum 5.7L Oil Temperature Sensor**

## FAN/RADIATOR CONNECTORS

**Connector Pinouts**

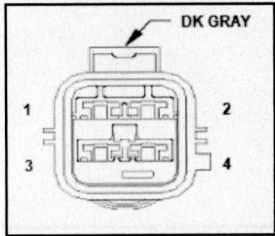

| PIN # | WIRE COLOR | CIRCUIT DESCRIPTION |
|-------|------------|---------------------|
| 1 | DG/DB | RADIATOR FAN CONTROL RELAY OUTPUT |
| 2 | DB/DG | RADIATOR FAN HIGH/LOW CONTROL FEED |
| 3 | DB/DG | RADIATOR FAN HIGH RELAY OUTPUT |
| 4 | BK/BR | GROUND |

**300 & Magnum Fan/Radiator**

## FUEL INJECTOR CONNECTORS

**Connector Pinouts**

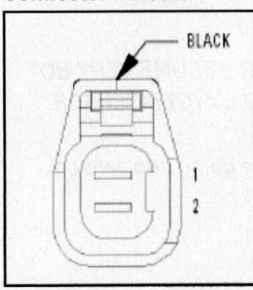

| PIN # | WIRE COLOR | CIRCUIT DESCRIPTION |
|-------|------------|---------------------|
| 1 | BR/YL | INJECTOR CONTROL NO. 1 |
| 1 | BR/DB | INJECTOR CONTROL NO. 2 |
| 1 | BR/LB | INJECTOR CONTROL NO. 3 |
| 1 | BR/TN | INJECTOR CONTROL NO. 4 |
| 1 | BR/OR | INJECTOR CONTROL NO. 5 |
| 1 | BR/VT | INJECTOR CONTROL NO. 6 |
| 1 | BR/YL | INJECTOR CONTROL NO. 7 |
| 1 | BR/LB | INJECTOR CONTROL NO. 8 |
| 2 | BR/YL | FUSED AUTO SHUT DOWN RELAY OUTPUT 2 |

**300 & Magnum Fuel Injector**

## GENERATOR CONNECTOR

**Connector Pinout**

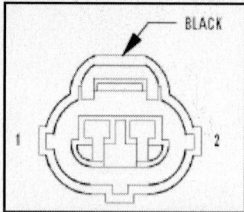

| PIN # | WIRE COLOR | CIRCUIT DESCRIPTION |
|-------|------------|---------------------|
| 1 | BK/BR | GROUND |
| 2 | BR/GY | GEN FIELD CONTROL |

**300 & Magnum Generator**

## HEATED OXYGEN SENSOR (HO2S)

**Removal & Installation**

*2.7L, 3.5L, 5.7L & 6.1L*

The engines uses two heated oxygen sensors, one in each exhaust manifold.

*CAUTION: Never apply any type of grease to the oxygen sensor electrical connector, or attempt any soldering of the sensor wiring harness.*

*WARNING: THE EXHAUST MANIFOLD, EXHAUST PIPES AND CATALYTIC CONVERTER BECOME VERY HOT DURING ENGINE Operation. ALLOW ENGINE TO COOL BEFORE REMOVING OXYGEN SENSOR.*

*CAUTION: When disconnecting sensor electrical connector, do not pull directly on wire going into sensor.*

1. Remove the negative battery cable.
2. Raise vehicle and support.
3. Disconnect the heated oxygen sensor electrical connector.
4. Use a suitable socket or a crowfoot wrench to remove oxygen sensor.

**To install:**

*NOTE: When replacing an O2 Sensor, the PCM RAM memory must be cleared, either by disconnecting the PCM C-1 connector or momentarily disconnecting the Battery negative terminal. The NGC learns the characteristics of each O2 heater element and these old values should be cleared when installing a new O2 sensor. The customer may experience driveability issues if this is not performed.*

*CAUTION: Never apply any type of grease to the oxygen sensor electrical connector, or attempt any soldering of the sensor wiring harness.*

5. After removing the sensor, the exhaust manifold threads must be cleaned with an 18 mm X 1.5 + 6E tap. If reusing the original sensor, coat the sensor threads with an anti-seize compound such as Loctite 771- 64 or equivalent. New sensors have compound on the threads and do not require an additional coating.
6. Tighten the sensor to 28 Nm (20 ft. lbs.) torque on 2.7L and 3.5L engines, and to 30 Nm (22 ft. lbs.) torque on 5.7L and 6.1L engines.
7. Connect the heated oxygen sensor electrical connector.
8. Lower vehicle. Install the negative battery cable.

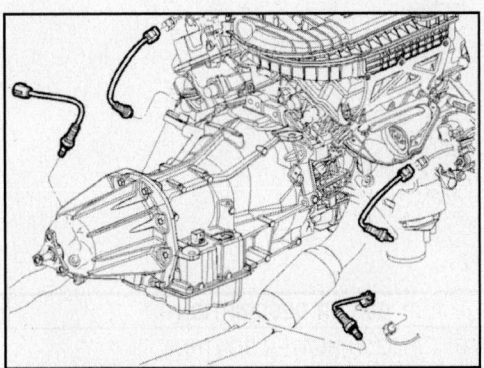

These engines use 2 heated oxygen sensors are used on each side of the exhaust system

## Connector Pinouts

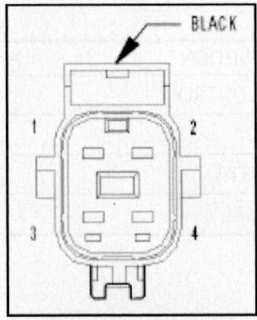

| PIN # | WIRE COLOR | CIRCUIT DESCRIPTION |
|-------|------------|---------------------|
| 1 | BR/LG | O2 1/1 HEATER CONTROL |
| 2 | BK | GROUND |
| 3 | BR/DG | O2 RETURN (UP) |
| 4 | DB/LB | O2 1/1 SIGNAL |

**300 & Magnum Left Front Oxygen Sensor Connector Pinout**

| PIN # | WIRE COLOR | CIRCUIT DESCRIPTION |
|-------|------------|---------------------|
| 1 | BR/WT | O2 1/2 HEATER CONTROL |
| 2 | BK/LB | GROUND |
| 3 | DB/DG | O2 RETURN (DOWN) |
| 4 | DB/YL | O2 1/2 SIGNAL |

**300 & Magnum Left Rear Oxygen Sensor Connector Pinout**

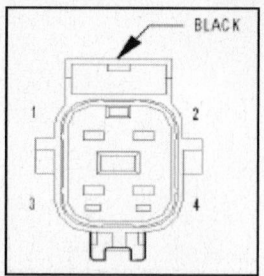

| PIN # | WIRE COLOR | CIRCUIT DESCRIPTION |
|-------|-----------|---------------------|
| 1 | BR/VT | O2 2/1 HEATER CONTROL |
| 2 | BK | GROUND |
| 3 | BR/DG | O2 RETURN (UP) |
| 4 | DB/LG | O2 2/1 SIGNAL |

**300 & Magnum Right Front Oxygen Sensor Connector Pinout**

| PIN # | WIRE COLOR | CIRCUIT DESCRIPTION |
|-------|-----------|---------------------|
| 1 | BR/GY | O2 2/2 HEATER CONTROL |
| 2 | BK/LB | GROUND |
| 3 | DB/DG | O2 RETURN (DOWN) |
| 4 | BR | O2 2/2 SIGNAL |

**300 & Magnum Right Rear Oxygen Sensor Connector Pinout**

## IGNITION COIL CONNECTORS

### Connector Pinouts

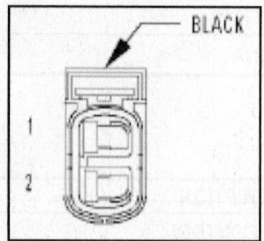

| PIN # | WIRE COLOR | CIRCUIT DESCRIPTION |
|-------|-----------|---------------------|
| 1 | BR/YL | FUSED AUTO SHUT DOWN RELAY OUTPUT 2 |
| 2 | DB/DG | COIL CONTROL NO. 1 |
| 2 | DB/TN | COIL CONTROL NO. 2 |
| 2 | DB/LG | COIL CONTROL NO. 3 |
| 2 | DB/GY | COIL CONTROL NO. 4 |
| 2 | DB/YL | COIL CONTROL NO. 5 |
| 2 | DB/OR | COIL CONTROL NO. 6 |

**300 2.7L & 3.5L Ignition Coil Connector Pinout**

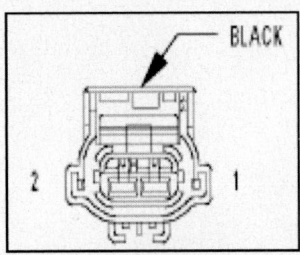

| PIN # | WIRE COLOR | CIRCUIT DESCRIPTION |
|---|---|---|
| 1 | BR/YL | FUSED AUTO SHUT DOWN RELAY OUTPUT 2 |
| 2 | DB/DG | COIL CONTROL NO. 1 |
| 2 | DB/TN | COIL CONTROL NO. 2 |
| 2 | DB/LG | COIL CONTROL NO. 3 |
| 2 | DB/GY | COIL CONTROL NO. 4 |
| 2 | DB/YL | COIL CONTROL NO. 5 |
| 2 | DB/OR | COIL CONTROL NO. 6 |
| 2 | DB/YL | COIL CONTROL NO. 7 |
| 2 | DB/YL | COIL CONTROL NO. 8 |

**300 & Magnum 5.7L Ignition Coil Connector Pinout**

## INLET AIR TEMPERATURE (IAT) SENSOR

**Connector Pinouts**

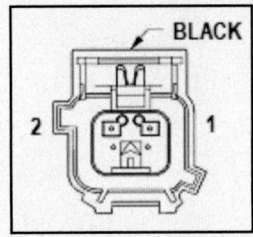

| PIN # | WIRE COLOR | CIRCUIT DESCRIPTION |
|---|---|---|
| 1 | DB/LG | IAT SIGNAL |
| 2 | DB/DG | SENSOR GROUND |

**300 & Magnum Intake Air Temperature Sensor Connector Pinout**

## INPUT SPEED SENSOR (RLE) CONNECTOR

**Connector Pinouts**

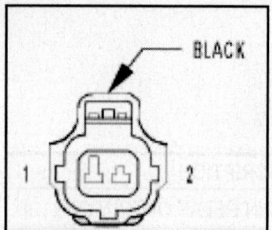

| PIN # | WIRE COLOR | CIRCUIT DESCRIPTION |
|-------|-----------|---------------------|
| 1 | DG/VT | SENSOR GROUND |
| 2 | DG/OR | INPUT SPEED SENSOR SIGNAL |

**300 & Magnum (RLE) Input Speed Sensor Connector Pinout**

## KNOCK SENSORS

**Removal & Installation**

**2.7L**

The sensors screws into the cylinder block, directly below the intake manifold.

1. Disconnect negative battery cable.
2. Remove intake manifold.
3. Remove the passenger side cylinder head.
4. Disconnect electrical connector from knock sensor.
5. Use a crowfoot socket to remove the knock sensors.

**To install:**

6. Install knock sensor. Tighten knock sensor to 10 Nm (7 ft. lbs.) torque.

*CAUTION: Over or under tightening effects knock sensor performance resulting in possible improper spark control.*

7. Attach electrical connector to knock sensor.
8. Install the passenger side cylinder head.
9. Install intake manifold.
10. Connect negative battery cable.

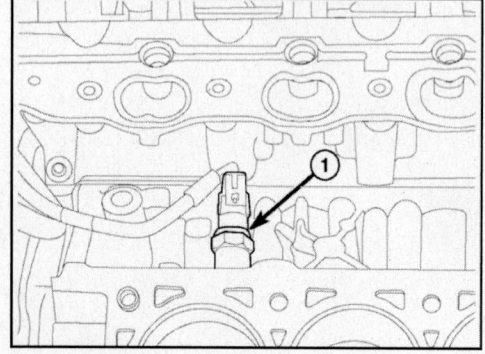

**Indicating the location of the Knock Sensor on 2.7L engine**

**3.5L**

1. Disconnect the negative battery cable.
2. Remove the upper intake manifold.
3. Disconnect the electrical connector.
4. Remove the knock sensor.

**To install:**

5. Install knock sensor. Tighten knock sensor to 20 Nm (15.2 inch lbs.) torque.

*CAUTION: Over or under tightening effects knock sensor performance, possibly causing improper spark control.*

6. Route the knock sensor wire in the proper location.
7. Install the intake manifold.
8. Connect electrical connector,
9. Connect negative battery cable.

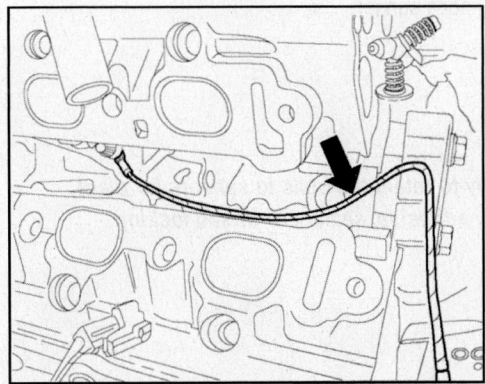

**Indicating proper Knock Sensor wire routing on 3.5L engine**

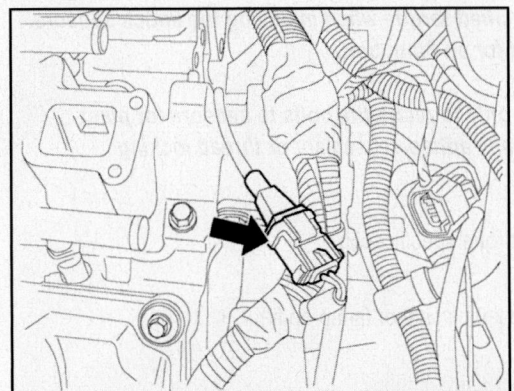

**Showing location of electrical connector for Knock Sensor on 3.5L engine**

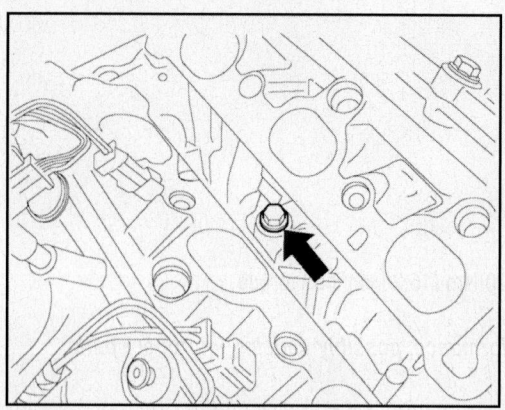

**Identifying location of Knock Sensor on 3.5L engine**

### 5.7L & 6.1L

Two sensors (1) are used. Each sensor is bolted to the outside of cylinder block below the exhaust manifold (3).

1. Raise vehicle.
2. On 6.1L engine, remove the heat shield from the knock sensor.
3. Disconnect knock sensor electrical connector (5).
4. Remove sensor mounting bolt (2).
5. Remove sensor from engine.

**CAUTION: Note foam strip on bolt threads. This foam is used only to retain the bolts to sensors for plant assembly. It is not used as a sealant. Do not apply any adhesive, sealant or thread locking compound to these bolts.**

**To install:**

6. Thoroughly clean knock sensor mounting hole.
7. Install sensor (1) into cylinder block.

**NOTE: Over or under tightening the sensor mounting bolts will affect knock sensor performance, possibly causing improper spark control. Always use the specified torque when installing the knock sensors. The torque for the knock senor bolt is relatively light for an 8mm bolt.**

**CAUTION: Note foam strip on bolt threads. This foam is used only to retain the bolts to sensors for plant assembly. It is not used as a sealant. Do not apply any adhesive, sealant or thread locking compound to these bolts.**

8. Install and tighten mounting bolt (2). Refer to torque specification.
9. Install electrical connector to sensor.
10. On 6.1L engine, install the heat shield over the knock sensor (snap on fit).

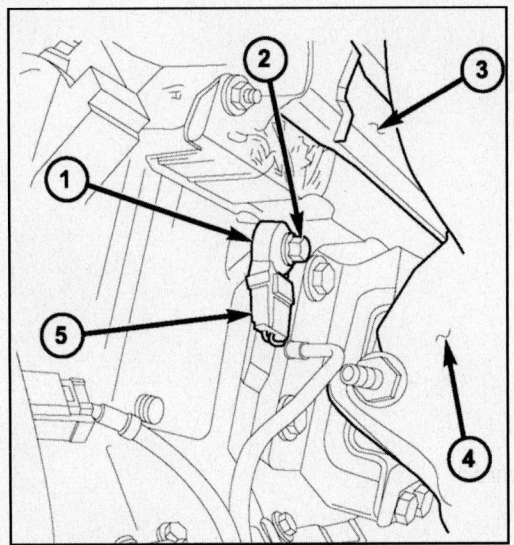

Identifying the location of the Knock Sensor (1), mounting bolt (2), exhaust manifold (3), body (4), and electrical connector (5) on the 5.7L engine

**Connector Pinouts**

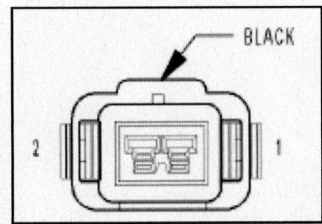

| PIN # | WIRE COLOR | CIRCUIT DESCRIPTION |
|-------|-----------|---------------------|
| 1 | BR/LG | KNOCK SENSOR NO. 1 RETURN |
| 2 | DB/YL | KNOCK SENSOR NO. 1 SIGNAL |

**300 & Magnum Knock Sensor #1 Connector Pinout**

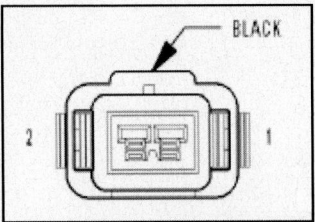

| PIN # | WIRE COLOR | CIRCUIT DESCRIPTION |
|-------|-----------|---------------------|
| 1 | WT/BR | KNOCK SENSOR NO. 2 RETURN |
| 2 | BR/WT | KNOCK SENSOR NO. 2 SIGNAL |

**300 & Magnum Knock Sensor #2 (5.7L) Connector Pinout**

MANIFOLD ABSOLUTE PRESSURE (MAP) SENSOR

**Removal & Installation**

*2.7L*

1. Unlock the electrical connector.
2. Remove electrical connector.
3. Turn sensor 1/4 turn counterclockwise.
4. Pull sensor straight up.
5. Remove sensor.
6. Inspect O-ring.

**To install:**

7. Clean MAP sensor mounting hole at intake manifold.
8. Check MAP sensor O-ring seal for cuts or tears.
9. Position sensor into intake manifold.
10. Rotate sensor 1/4 turn clockwise for installation.
11. Connect electrical connector to sensor.
12. Lock electrical connector.
13. Connect negative battery cable.

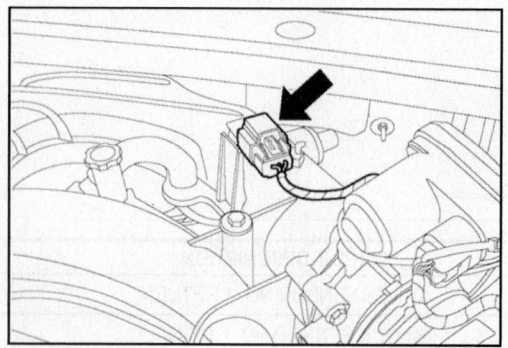

**Showing MAP Sensor location on 2.7L engine**

*3.5L*

1. Disconnect negative battery cable.
2. Unlock the electrical connector.
3. Disconnect the electrical connector.
4. Rotate sensor 1/4 turn clockwise.
5. Pull up on sensor.
6. Remove sensor.

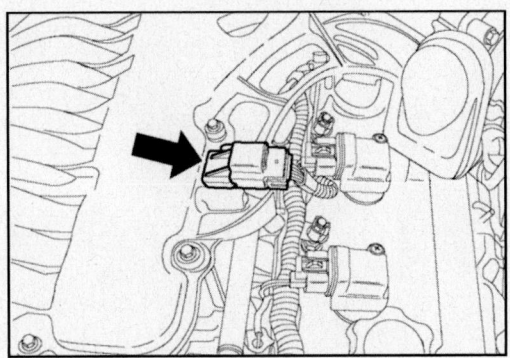

**Showing MAP Sensor location on the 3.5L engine**

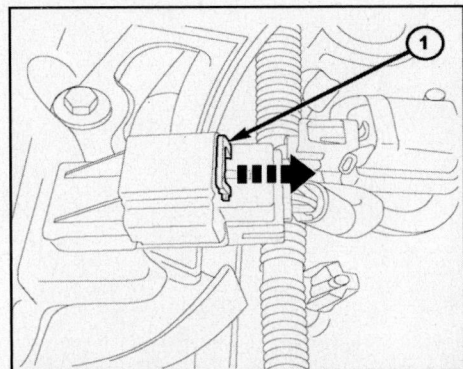

**To unlock the electrical connector from the MAP Sensor on the 3.5L engine, pull locking tab (1) straight out**

*5.7L*

The Manifold Absolute Pressure (MAP) sensor (3) is mounted into the top/rear of the intake manifold (4) near the cowl/hood seal (1).

1. Disconnect electrical connector at sensor by sliding release lock out (1). Press down on lock tab (2) for removal.
2. Rotate sensor 1/4 turn counter-clockwise for removal.
3. Check condition of sensor O-ring.

**To install:**

4. Clean MAP sensor mounting hole at intake manifold.
5. Check MAP sensor O-ring seal for cuts or tears.
6. Position sensor (3) into intake manifold.
7. Rotate sensor 1/4 turn clockwise for installation.
8. Connect electrical connector (2) to sensor (3).

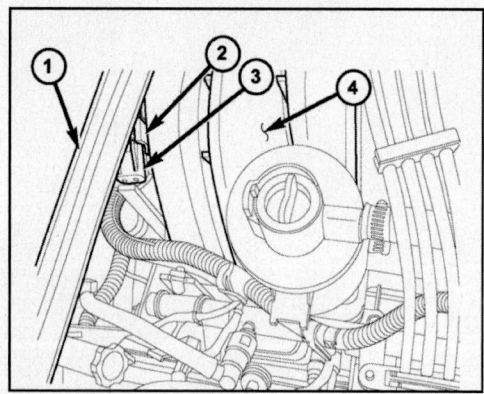

**Indicating the location of the MAP Sensor (3), cowl seal (1), lock tab (2) and intake manifold (4) on 5.7L engine**

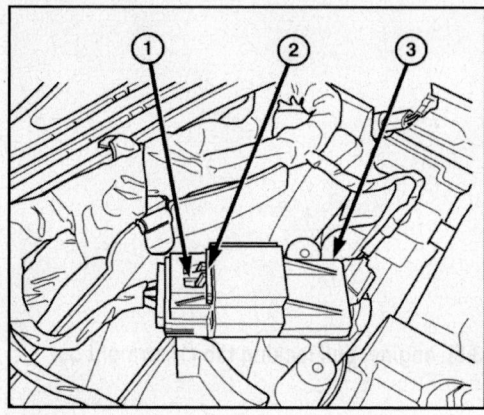

**To release the electrical connector (3), slide out the release lock (1) and press down on the lock tab (2) – 5.7L engine**

*6.1L*

1. The Manifold Absolute Pressure (MAP) sensor (2) is located at the rear of the intake manifold (1).
2. Disconnect electrical connector at sensor.
3. Remove two sensor mounting bolts (3).
4. Check condition of sensor O-ring (2).

**To install:**

5. Clean MAP sensor mounting hole at rear of intake manifold.
6. Check MAP sensor O-ring seal (2) for cuts or tears.
7. Position sensor (1) into intake manifold.
8. Install two sensor mounting bolts (3).
9. Connect electrical connector to sensor.

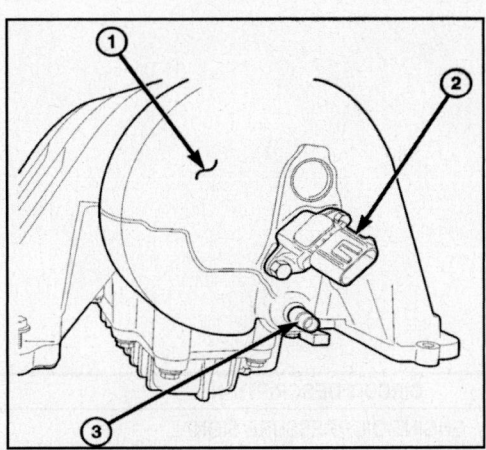

**Identifying the location of the MAP Sensor on the 6.1L engine**

**Connector Pinouts**

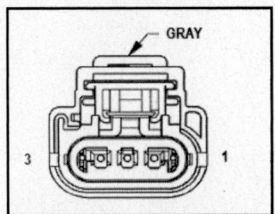

| PIN # | WIRE COLOR | CIRCUIT DESCRIPTION |
|-------|-----------|---------------------|
| 1 | VT/BR | MAP SIGNAL |
| 2 | DB/DG | SENSOR GROUND |
| 3 | YL/PK | 5 VOLT SUPPLY |

**300 & Magnum MAP Sensor Connector Pinout**

## NATURAL VACUUM LEAK DETECTOR (NLVD)

**Connector Pinouts**

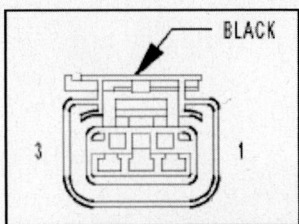

| PIN # | WIRE COLOR | CIRCUIT DESCRIPTION |
|-------|-----------|---------------------|
| 1 | BK/GY | GROUND |
| 2 | VT/WT | NVLD SWITCH SIGNAL |
| 3 | VT/LB | NVLD SOL CONTROL |

**300 & Magnum NVLD Connector Pinout**

## OIL PRESSURE (OP) SENSOR

**Connector Pinouts**

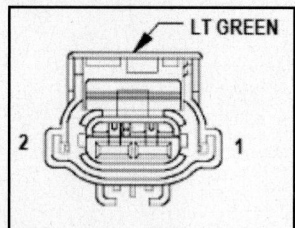

| PIN # | WIRE COLOR | CIRCUIT DESCRIPTION |
|-------|------------|---------------------|
| 1 | VT/GY | ENGINE OIL PRESSURE SIGNAL |
| 2 | - | - |

**300 2.5L, 3.7L Oil Pressure Sensor**

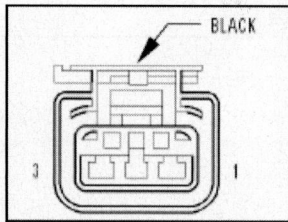

| PIN # | WIRE COLOR | CIRCUIT DESCRIPTION |
|-------|------------|---------------------|
| 1 | PK/YL | 5 VOLT SUPPLY |
| 2 | VT/GY | ENGINE OIL PRESSURE SIGNAL |
| 3 | DB/DG | SENSOR GROUND |

**300 & Magnum 5.7L Oil Pressure Sensor**

## OUTPUT SPEED SENSOR (OSS)

**Connector Pinouts**

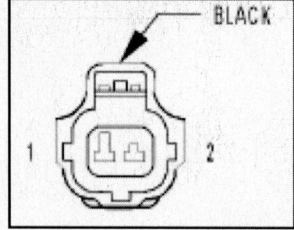

| PIN # | WIRE COLOR | CIRCUIT DESCRIPTION |
|-------|------------|---------------------|
| 1 | DG/VT | SENSOR GROUND |
| 2 | DG/BR | OUTPUT SPEED SENSOR SIGNAL |

**300 & Magnum (w/RLE) Output Speed Sensor Connector Pinout**

THROTTLE BODY CONNECTOR

**Connector Pinouts**

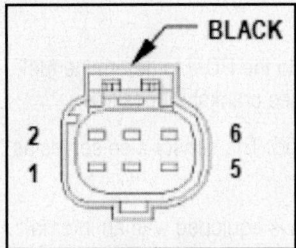

| PIN # | WIRE COLOR | CIRCUIT DESCRIPTION |
|-------|-----------|---------------------|
| 1 | BR/OR | TP SENSOR NO. 1 SIGNAL |
| 2 | PK/YL | 5 VOLT SUPPLY |
| 3 | DB/GY | ETC MOTOR (+) |
| 4 | BR/DG | TP SENSOR NO. 2 SIGNAL |
| 5 | DB/LG | ETC MOTOR (-) |
| 6 | BR/DB | TP SENSOR RETURN |

**300 & Magnum Throttle Body Connector Pinout**

# DAIMLERCHRYSLER CARS: BREEZE, CIRRUS & STRATUS

CAMSHAFT POSITION (CMP) SENSOR

## Description & Operation

The camshaft position sensor (along with the crankshaft position sensor) provides inputs to the PCM to determine fuel injection synchronization and cylinder identification. From these inputs, the PCM determines crankshaft position.

On 4-cylinder engines, the camshaft position sensor mounts to the rear of the cylinder head. The sensor also serves as a thrust plate to control endplay of the camshaft.

The 6-cylinder engines are equipped with a camshaft driven mechanical distributor, which is equipped with an internal camshaft position (fuel sync) sensor.

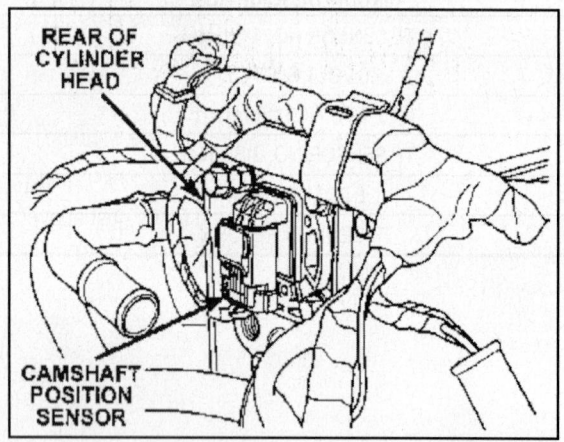

Camshaft position sensor location on 2.0L SOHC engines

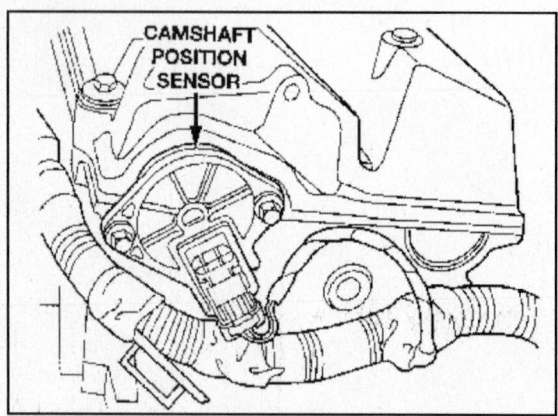

Camshaft position sensor location on 2.0L DOHC and 2.4L engines

## Testing

To test this sensor, you will need the use of an oscilloscope.

Visually check the connector, making sure it is attached properly and that all of the terminals are straight, tight and free of corrosion. The output voltage of a properly operating camshaft position sensor switches from high (5.0 volts) to low (0.3 volts). By connecting an oscilloscope to the sensor output circuit, you can view the square wave pattern produced by the voltage swing.

CRANKSHAFT POSITION (CKP) SENSOR

## Description & Operation

The PCM determines what cylinder to fire from the crankshaft position sensor input and the camshaft position sensor input. On 4-cylinder engines, the second crankshaft counterweight has two sets of four timing reference notches, including a 60° signature notch. From the crankshaft position sensor input, the PCM determines engine speed and crankshaft angle (position). On 6-cylinder engines, this sensor is a Hall effect device that detects notches in the flexplate.

The notches generate pulses from high to low in the crankshaft position sensor output voltage. When a metal portion of the notches line up with the crankshaft position sensor, the sensor output voltage goes low (less than 0.5 volts). When a notch aligns with the sensor, voltage goes high (5.0 volts). As a group of notches pass under the sensor, the output voltage switches from low (metal) to high (notch), then back to low.

If available, an oscilloscope can display the square wave patterns of each voltage pulse. From the width of the output voltage pulses, the PCM calculates engine speed. The width of the pulses represents the amount of time the output voltage stays high before switching back to low. The period of time the sensor output voltage stays high before switching back to low is referred to as pulse width. The faster the engine is operating, the smaller the pulse width on the oscilloscope.

On 4-cylinder engines, the crankshaft position sensor is mounted to the engine block behind the alternator, just above the oil filter. On 6-cylinder engines, the crankshaft position sensor is mounted on the transaxle housing, above the vehicle speed sensor.

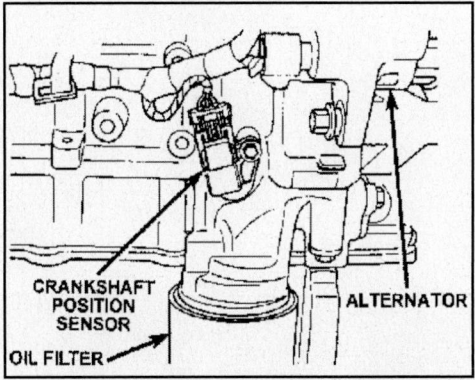

**Location of the crankshaft position (CKP) sensor on 2.0L engines**

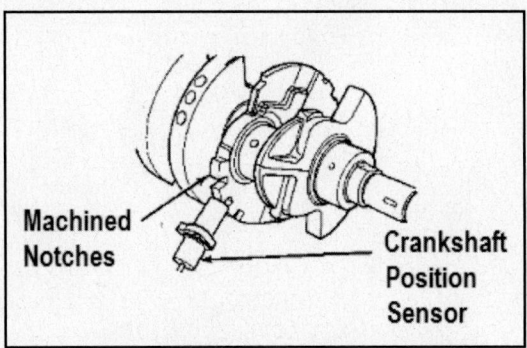

**Timing reference notches on 4-cylinder engines**

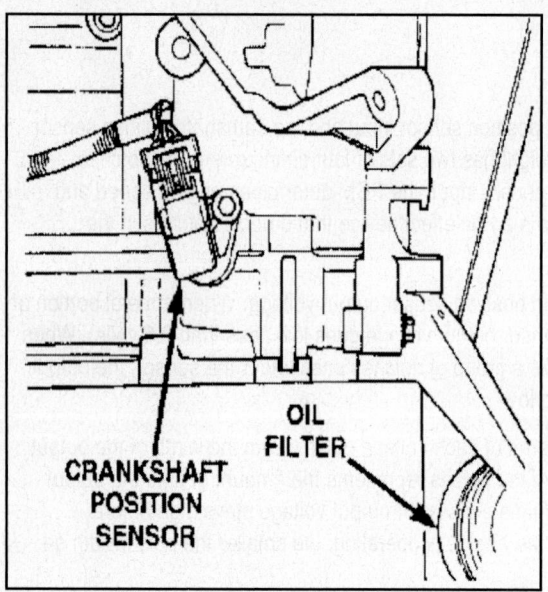

Location of the crankshaft position sensor on 2.4L engines

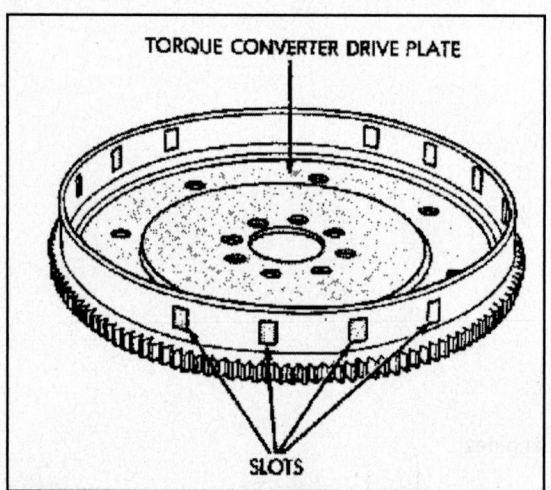

Location of the crankshaft position sensor on 2.5L engines

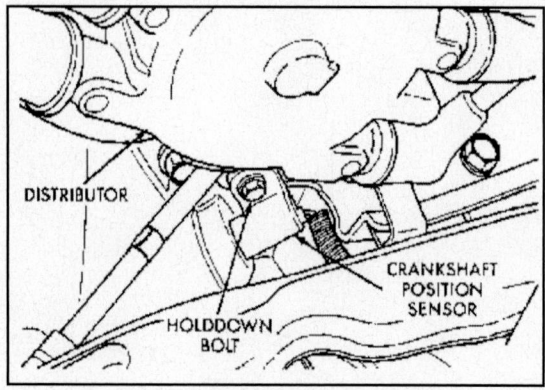

Timing reference notches on Breeze, Cirrus, Stratus with 2.5L engines

**Testing**

To test the CKP sensor, you will need the use of an oscilloscope.

Visually check the connector, making sure it is attached properly and that all of the terminals are straight, tight and free of corrosion. Also inspect the notches in the crankshaft (4-cylinder) or flywheel (6-cylinder) for damage, and replace if necessary.

The output voltage of a properly operating crankshaft position sensor switches from high (5.0 volts) to low (0.3 volts). By connecting an oscilloscope to the sensor output circuit, you can view the square wave pattern produced by the voltage swing.

# DAIMLERCHRYSLER CARS: CONCORDE, INTREPID, LHS, NEW YORKER & VISION

BATTERY TEMPERATURE SENSOR

## Description & Operation

The PCM incorporates a Battery Temperature Sensor (BTS) on its circuit board.

The PCM uses the temperature of the battery area to control the charge rate. This temperature data, along with data from monitored line voltage, is used by the PCM to vary the battery charging rate. The system voltage is higher at cold temperatures and is gradually reduced as temperature around the battery increases.

The function of the battery temperature sensor (BTS) is to enable control of the generator output based upon ambient temperature. As temperature increases, the charging rate should decrease. As temperature decreases, the charging rate should increase. The sensor functions similar to the ECT sensor with one major difference, the ambient sensor does not have a dual temperature range program. The PCM maintains the maximum output of the generator by monitoring battery voltage and controlling battery voltage to a range of 13.5-14.7 volts.

The battery temperature sensor is also used for OBD II diagnostics. Certain faults and OBD II monitors are either enabled or disabled depending upon the battery temperature sensor input (example: disable purge and EGR, enable LDP). Most OBD II monitors are disabled below 20 °F.

CAMSHAFT POSITION (CMP) SENSOR

## Description & Operation

The Camshaft Position (CMP) sensor provides cylinder identification to the PCM. The sensor generates pulses as groups of notches on the camshaft sprocket pass underneath it. The PCM keeps track of the crankshaft rotation and identifies each cylinder by the pulses generated by the notches on the camshaft sprocket. Four crankshaft pulses follow each group of camshaft pulses.

When the PCM receives 2 camshaft pulses followed by the long flat spot on the camshaft sprocket, it knows that the crankshaft timing marks for cylinder one are next (on the drive plate). When the PCM receives one camshaft pulse after the long flat spot on the sprocket, cylinder No. 2 crankshaft timing marks are next. After 3 camshaft pulses, the PCM knows cylinder 4 crankshaft timing marks follow. One camshaft pulse after the 3 pulses indicates cylinder No. 5. Two camshaft pulses after cylinder No. 5, signals cylinder No. 6. The PCM can synchronize on cylinders Nos. 1 or 4.

When metal aligns with the sensor, voltage goes low (less than 0.3 volts). When a notch aligns with the sensor, voltage spikes high (5.0 volts). As a group of notches pass under the sensor, the voltage switches from low to high then back to low again. The number of notches determines the amount of pulses. If available, an oscilloscope can display the square wave patterns of each timing event.

Top Dead Center (TDC) does not occur when notches on the camshaft sprocket pass below the sensor. TDC occurs after the camshaft pulse (or pulses) and after the 4 crankshaft pulses associated with the particular cylinder. The arrows and cylinder call in the illustration represent which cylinder the flat spot and notches identify, they do not indicate the TDC position.

The camshaft position sensor is mounted to the top of the timing case cover. The bottom of the sensor is positioned above the camshaft sprocket. The distance between the bottom of sensor and the camshaft sprocket is critical to the operation of the system.

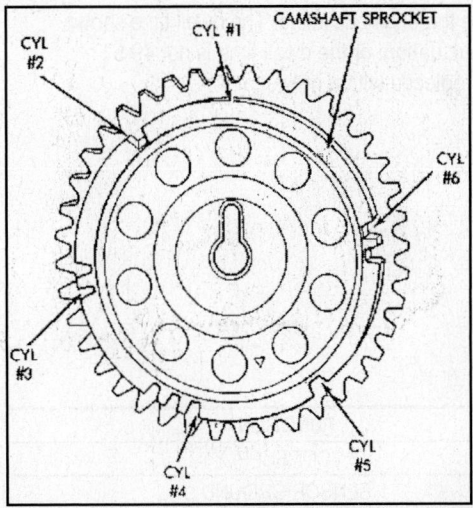

**Camshaft sprocket with gear cylinder numbering on 3.3L engine**

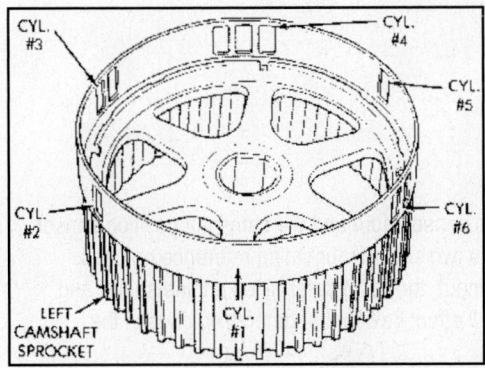

**Camshaft sprocket with gear cylinder numbering on 3.5L engine**

### Testing

Before testing any electrical component, inspect the wiring and connectors for damage. Also wiggle the connectors to ensure that they are firmly engaged. For this procedure a dwell meter, or the equivalent, will be needed.

1. Unplug the CMP sensor connector.
2. Turn the ignition ON.
3. Using a voltmeter, measure the voltage from the wiring harness connector 8-volt supply circuit (orange wire) to ground.
4. If the voltage is 8-9.5 volts, skip to the next step.
5. If the voltage measured is lower than 8 volts, or greater than 9.5 volts, the CMP sensor is not receiving the correct current to function properly. There is a problem in the wiring or related components.
6. Turn the ignition OFF.
7. Attach the CMP sensor wiring and engine wiring harness connectors back together.
8. Attach a dwell meter to the battery. Attach the lead probe of the dwell meter to the sensor signal wire (light blue wire with dark blue tracer) by back-probing the connector or by using jumper cables between the terminals.
9. Place the dwell meter out of the way of any moving components of the engine, and in a position in which it can be seen once the engine is started.
10. Turn the engine ON.

11. Watch the dwell meter for one or two minutes while the engine is idling. The dwell time shown should be a steady 49-51 percent. If there is any fluctuation, or the dwell time is not 49-51 percent the CMP sensor is defective and must be replaced with a new one.

**Connector Pinouts**

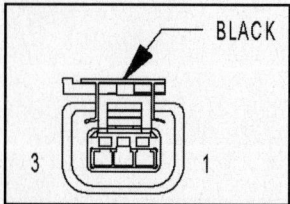

| PIN # | CIRCUIT | FUNCTION |
|-------|---------|----------|
| 1 | K6 20VT/WT | 5 VOLT SUPPLY |
| 2 | K4 20BK/LB | SENSOR GROUND |
| 3 | K44 20TN/YL | CMP SIGNAL |

**CMP Sensor Connector Pinout**

## CRANKSHAFT POSITION (CKP) SENSOR

### Description & Operation

The PCM determines what cylinder to fire from the crankshaft position sensor input and the camshaft position sensor input. On 4-cylinder engines, the second crankshaft counterweight has two sets of four timing reference notches, including a 60° signature notch. From the crankshaft position sensor input, the PCM determines engine speed and crankshaft angle (position). On 6-cylinder engines, this sensor is a Hall effect device that detects notches in the flexplate.

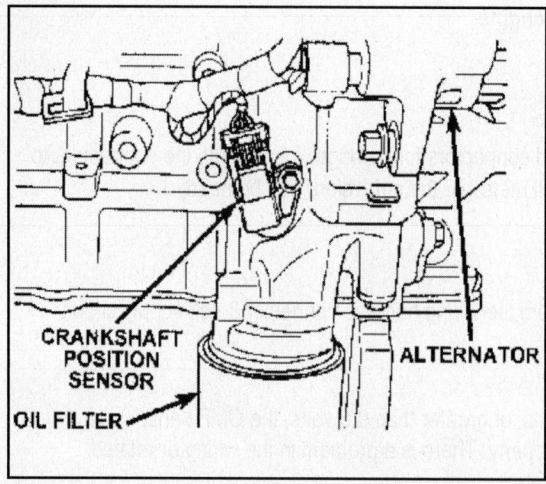

**Location of the crankshaft position sensor on 2.0L engines**

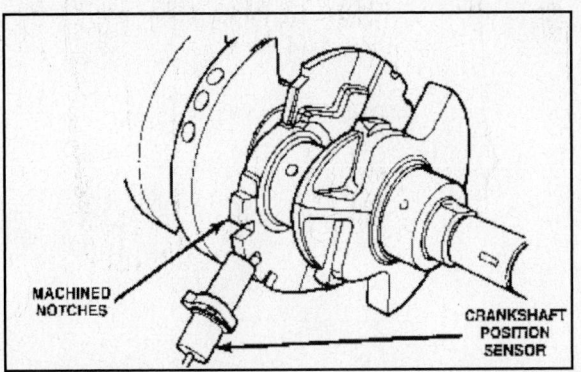

**Timing reference notches on 4-cylinder engines**

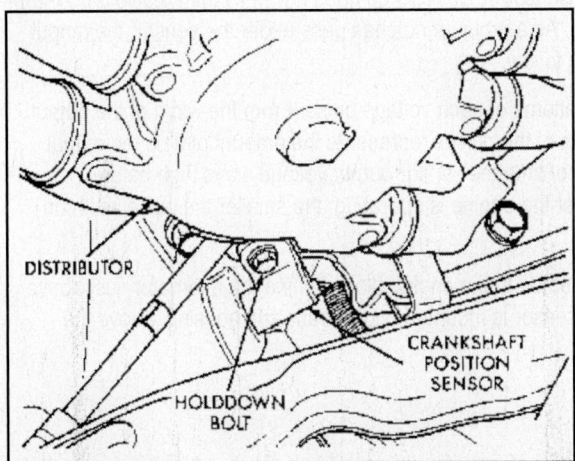

**Location of the crankshaft position sensor on 2.4L engines**

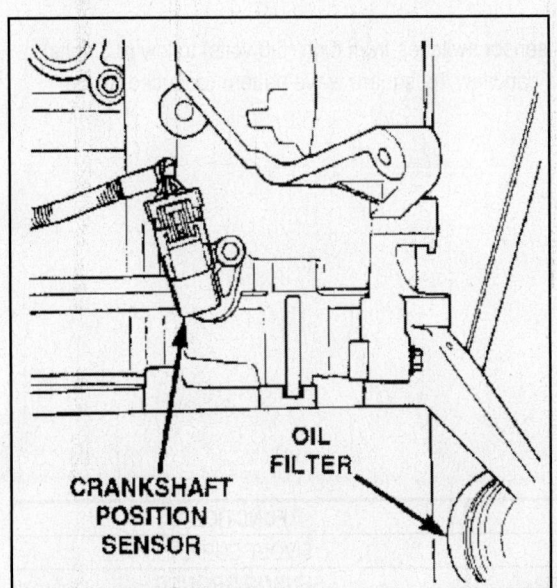

**Location of the crankshaft position sensor on 2.5L engines**

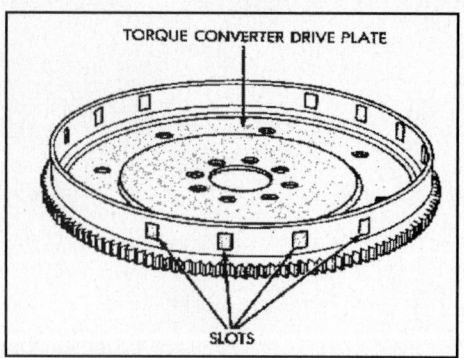

**Timing reference notches on 2.5L engines**

The notches generate pulses from high to low in the crankshaft position sensor output voltage. When a metal portion of the notches line up with the crankshaft position sensor, the sensor output voltage goes low (less than 0.5 volts). When a notch aligns with the sensor, voltage goes high (5.0 volts). As a group of notches pass under the sensor, the output voltage switches from low (metal) to high (notch), then back to low.

If available, an oscilloscope can display the square wave patterns of each voltage pulse. From the width of the output voltage pulses, the PCM calculates engine speed. The width of the pulses represents the amount of time the output voltage stays high before switching back to low. The period of time the sensor output voltage stays high before switching back to low is referred to as pulse width. The faster the engine is operating, the smaller the pulse width on the oscilloscope.

On 4-cylinder engines, the crankshaft position sensor is mounted to the engine block behind the alternator, just above the oil filter. On 6-cylinder engines, the crankshaft position sensor is mounted on the transaxle housing, above the vehicle speed sensor.

### Testing

To test this sensor, you will need the use of an oscilloscope.

Visually check the connector, making sure it is attached properly and that all of the terminals are straight, tight and free of corrosion. Also inspect the notches in the crankshaft (4-cylinder) or flywheel (6-cylinder) for damage, and replace if necessary.

The output voltage of a properly operating crankshaft position sensor switches from high (5.0 volts) to low (0.3 volts). By connecting an oscilloscope to the sensor output circuit, you can view the square wave pattern produced by the voltage swing.

### Connector Pinouts

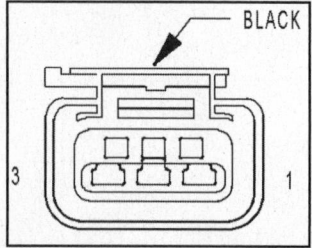

| PIN # | CIRCUIT | FUNCTION |
|-------|---------|----------|
| 1 | K6 20VT/WT | 5 VOLT SUPPLY |
| 2 | K4 20BK/LB | SENSOR GROUND |
| 3 | K24 20GY/BK | CKP SIGNAL |

**CKP Sensor**

ENGINE COOLANT TEMPERATURE (ECT) SENSOR

### Description & Operation

The sensor provides an input to the Powertrain Control Module (PCM). As coolant temperature varies, the sensor resistance changes, resulting in a different input voltage to the PCM. When the engine is cold, the PCM will demand slightly richer air-fuel mixtures and higher idle speeds until normal operating temperatures are reached.
The engine coolant sensor input also determines operation of the low and high speed cooling fans.

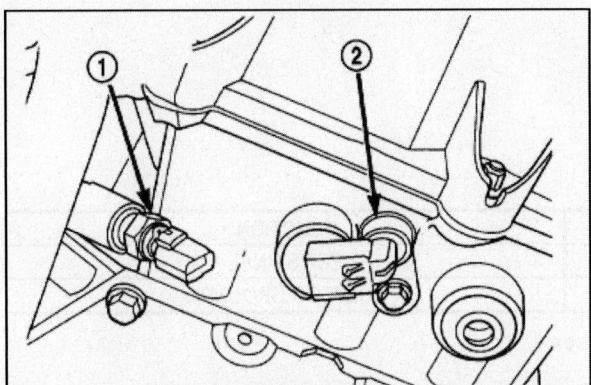

**Location of Coolant Temperature Sensor (1) near Cam Sensor (2) on 2.7L engine**

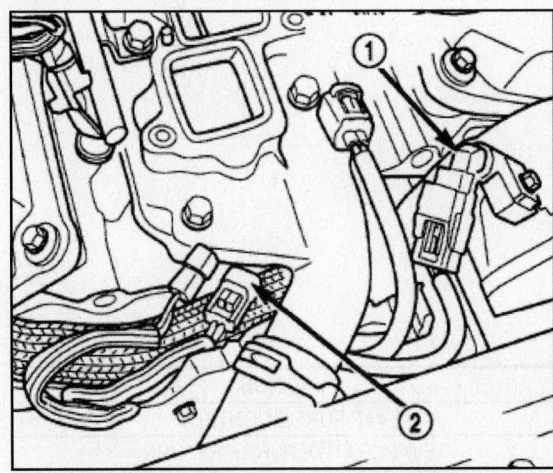

**Location of Engine Coolant Temperature Sensor (2), near Camshaft Sensor (1) on 3.2L engine**

### Removal & Installation

1. Remove the negative battery cable.
2. With the engine cold, disconnect coolant sensor electrical connector.
3. Remove sensor.

**To install:**

4. Install engine coolant temperature sensor. Tighten sensor to 28 Nm (20 ft. lbs.) torque.
5. Attach electrical connector to sensor.
6. Install the negative battery cable.

**Connector Pinouts**

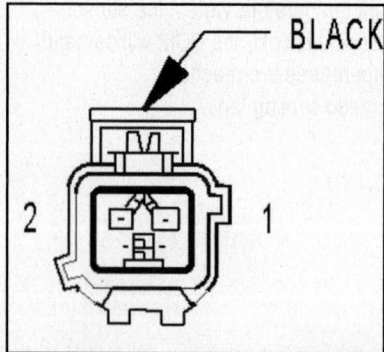

| PIN # | CIRCUIT | FUNCTION |
|:-----:|:-------:|:--------:|
| 1 | K2 20TN/BK | ECT SIGNAL |
| 2 | K4 20BK/LB | SENSOR GROUND |

**Concorde, Intrepid, LHS, New Yorker, Vision ECT Sensor**

## EVAP PURGE SOLENOID

**Connector Pinouts**

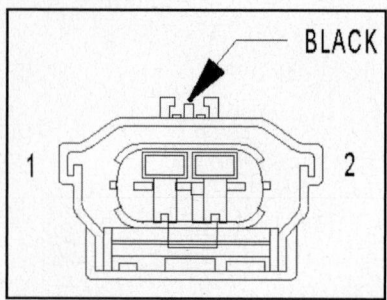

| PIN # | CIRCUIT | FUNCTION |
|:-----:|:-------:|:--------:|
| 1 | K52 18PK/BK | EVAP PURGE CONTROL |
| 2 | K108 18DG/LG | EVAPORATIVE PURGE RETURN |

**EVAP Purge Solenoid**

## FUEL LEVEL SENSOR

### Description & Operation

The fuel level sending unit is attached to the side of fuel pump module. The level sensor is a variable resistor. Its resistance changes with the amount of fuel in the tank. The float arm attached to the sensor moves as the fuel level changes. The fuel level input is used as an input for OBD II. If the fuel level is below 15% of total tank capacity several monitors are disabled. There are diagnostics for the level circuit open and shorted.

**Removal & Installation**

1. Remove fuel pump module.

2. Depress the retaining tab and remove the fuel pump/level sensor connector from the bottom of the fuel pump module electrical connector.

3. Remove wire terminal retaining clip from connector. Note the location of terminals for the level sensor wires.

4. Using special tool 7812, or an equivalent, push level sensor signal and ground terminals out of the connector.

5. Insert a screwdriver between the fuel pump module and the top of the level sensor housing. Push level sensor down slightly.

6. Slide level sensor wires through standpipe inside fuel pump module.

7. Slide level sensor out of the channel.

**To install:**

8. Insert level sensor wires in bottom of stand- pipe.

9. Wrap wires into groove in back of level sensor.

10. While feeding wires into standpipe, slide level sensor up into the channel until it snaps into place. Ensure tab at bottom of sensor locks in place.

11. Install level sensor wires in connector. Push the wires up through the connector and then pull them down until they lock in place. Ensure signal and ground wires are installed in the correct position.

12. Install retaining clip on connector.

13. Push fuel pump/level sensor connector up into bottom of fuel pump module electrical connector.

14. Install fuel pump module.

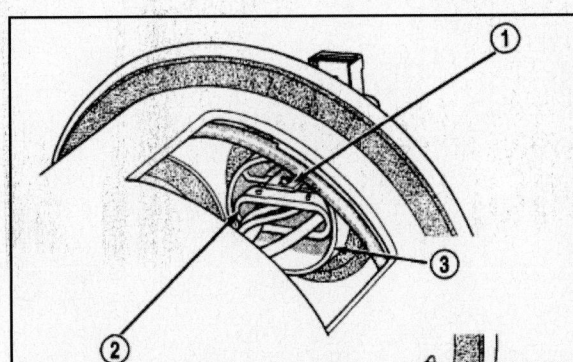

**Showing the location of the Fuel Level Sensor (2), the locating tab (1) and the bottom of the fuel pump module connector (3)**

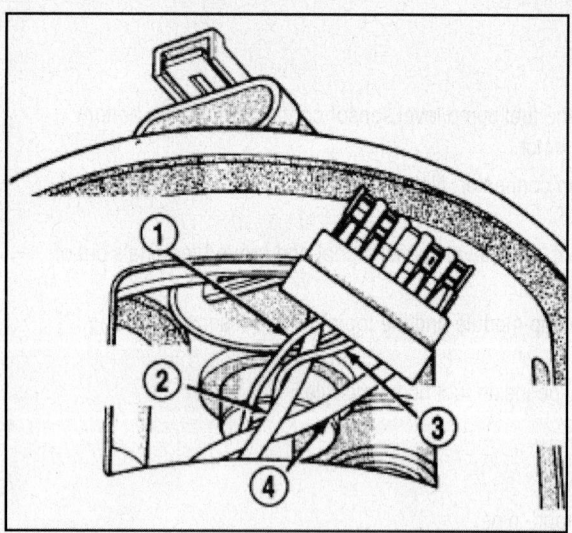

Identifying the connector wiring: 1 – Level Sensor ground; 2 – Fuel Pump; 3 – Level Sensor signal; 4 – Fuel Pump (+)

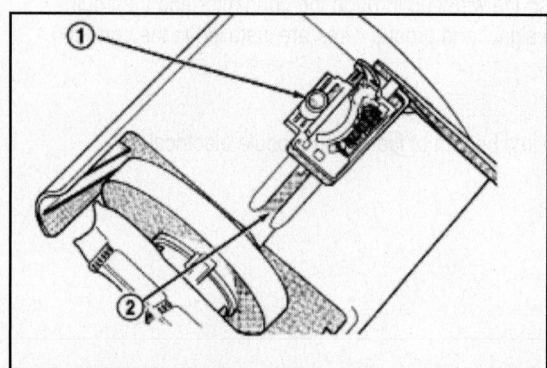

Identifying Fuel Level Sensor (1) in channel (2)

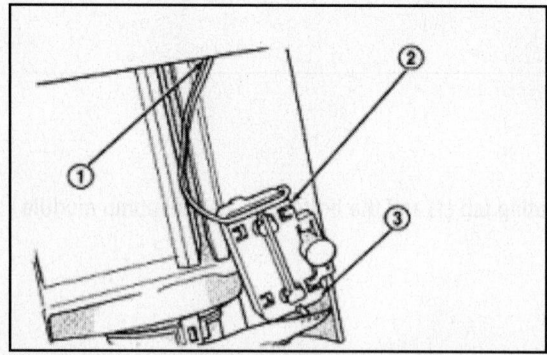

Fuel Level Sensor (3) installation:, noting groove (2) and standpipe opening (1)

## FUEL PUMP MODULE

### Connector Pinouts

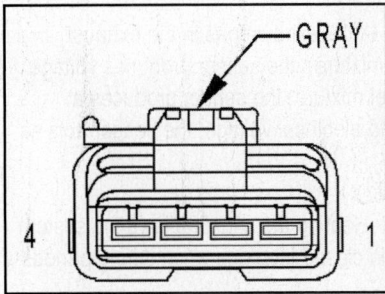

| PIN # | CIRCUIT | FUNCTION |
|-------|---------|----------|
| 1 | Z1 12BK | GROUND |
| 2 | Z1 18BK | GROUND |
| 3 | G4 18DB | FUEL LEVEL SIGNAL |
| 4 | A141 12DG/WT | FUEL PUMP RELAY OUTPUT |

**Fuel Pump Module Connector Pinout**

## HEATED OXYGEN SENSOR (HO2S)

### Description & Operation

As vehicles accumulate mileage, the catalytic converter deteriorates. The deterioration results in a less efficient catalyst. To monitor catalytic converter deterioration, the fuel injection system uses two heated oxygen sensors: One sensor upstream of the catalytic converter, one downstream of the converter. The PCM compares the reading from the sensors to calculate the catalytic converter oxygen storage capacity and converter efficiency. Also, the PCM uses the upstream heated oxygen sensor input when adjusting injector pulse width.

When the catalytic converter efficiency drops below emission standards, the PCM stores a diagnostic trouble code and illuminates the malfunction indicator lamp (MIL).

The O2S produce voltages from 0 to 1 volt, depending upon the oxygen content of the exhaust gas in the exhaust manifold. When a large amount of oxygen is present (caused by a lean air/fuel mixture), the sensor produces a low voltage. When there is a lesser amount present (rich air/fuel mixture) it produces a higher voltage. By monitoring the oxygen content and converting it to electrical voltage, the sensors act as a rich-lean switch.

The oxygen sensors are equipped with a heating element that keeps the sensors at proper operating temperature during all operating modes. Maintaining correct sensor temperature at all times allows the system to enter into closed loop operation sooner. Also, it allows the system to remain in closed loop operation during periods of extended idle.

In Closed Loop operation the PCM monitors the O2S input (along with other inputs) and adjusts the injector pulse width accordingly. During Open Loop operation the PCM ignores the O2 sensor input. The PCM adjusts injector pulse width based on preprogrammed (fixed) values and inputs from other sensors.

The Automatic Shutdown (ASD) relay supplies battery voltage to both the upstream and downstream heated oxygen sensors. The oxygen sensors are equipped with a heating element. The heating elements reduce the time required for the sensors to reach operating temperature.

### UPSTREAM OXYGEN SENSOR

The input from the upstream heated oxygen sensor tells the PCM the oxygen content of the exhaust gas. Based on this input, the PCM fine tunes the air-fuel ratio by adjusting injector pulse width.

The sensor input switches from 0 to 1 volt, depending upon the oxygen content of the exhaust gas in the exhaust manifold. When a large amount of oxygen is present (caused by a lean air-fuel mixture), the sensor produces voltage as low as 0.1 volt. When there is a lesser amount of oxygen present (rich air-fuel mixture) the sensor produces a voltage as high as 1.0 volt. By monitoring the oxygen content and converting it to electrical voltage, the sensor acts as a rich-lean switch.

The heating element in the sensor provides heat to the sensor ceramic element. Heating the sensor allows the system to enter into closed loop operation sooner. Also, it allows the system to remain in closed loop operation during periods of extended idle.

In Closed Loop, the PCM adjusts injector pulse width based on the upstream heated oxygen sensor input along with other inputs. In Open Loop, the PCM adjusts injector pulse width based on preprogrammed (fixed) values and inputs from other sensors.

### DOWNSTREAM OXYGEN SENSOR 1/2

The downstream heated oxygen sensor input is used to detect catalytic converter deterioration. As the converter deteriorates, the input from the downstream sensor begins to match the upstream sensor input except for a slight time delay. By comparing the downstream heated oxygen sensor input to the input from the upstream sensor, the PCM calculates catalytic converter efficiency. The downstream heated oxygen sensor threads into the outlet pipe at the rear of the catalytic converter. Separate controlled ground circuits are run through the PCM for the upstream O2 sensors.

This engine uses two upstream heated oxygen sensors. One oxygen sensor is threaded into the outlet flange of each exhaust manifold.

**WARNING: The exhaust manifold, exhaust pipes and catalytic converter become very hot during engine operation. Allow engine to cool before removing oxygen sensor.**

**NOTE: See CONNECTOR VIEWS & PINOUTS at the end of this section.**

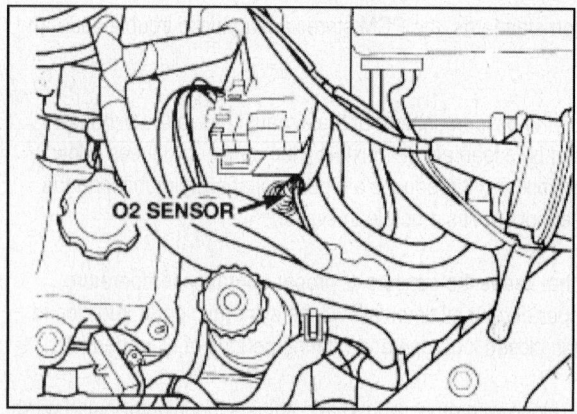

**Upstream O2 Sensor Location**

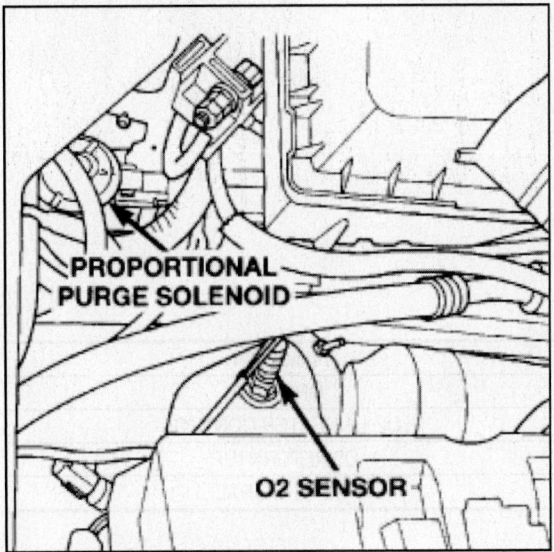

Downstream O2 Sensor Location

### Removal & Installation

*CAUTION: When disconnecting the sensor electrical connector, do not pull directly on wire going into sensor.*

1. Remove the negative battery cable.
2. Disconnect the heated oxygen sensor electrical connector.
3. Use a socket such as a crow's foot wrench to remove oxygen sensor.

**To install:**

4. After removing the sensor, the exhaust manifold threads must be cleaned with an 18 mm X 1.5 + 6E tap. If reusing the original sensor, coat the sensor threads with an anti-seize compound such as Loctite 771-64 or equivalent. New sensors have compound on the threads and do not require an additional coating. Tighten the sensor to 28 Nm (20 ft. lbs.) torque.
5. Connect the heated oxygen sensor electrical connector. Install the negative battery cable.

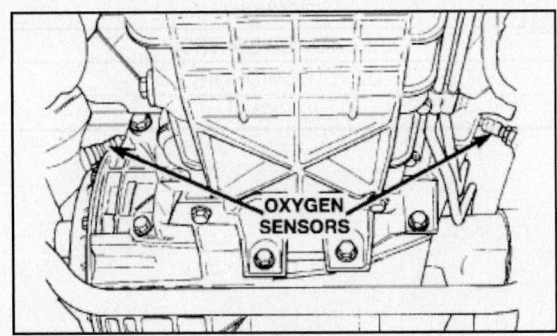

Dual Downstream O2 Sensors

### Testing

1. Use an ohmmeter to test the heating element of the oxygen sensors.
2. Disconnect the electrical connector from each oxygen sensor. The white wires in the sensor connector are the power and ground circuits for the boater.
3. Connect the ohmmeter test leads to terminals of the white wires in the heated oxygen sensor connector. Replace the heated oxygen sensor if the resistance is not between 4 and 7 ohms.

## Connector Pinouts

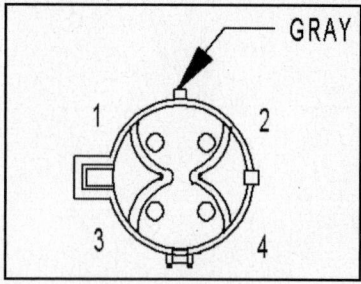

| PIN # | CIRCUIT | FUNCTION |
|---|---|---|
| 1 | Z1 18BK | GROUND |
| 2 | K99 18BR/OR | O2 1/1 HEATER CONTROL |
| 3 | K902 18BR/DG | O2 RETURN (UP) |
| 4 | K41 20BK/DG | O2 1/1 SIGNAL |

**O2 Sensor (1/1 Right Bank) Connector Pinout**

| PIN # | CIRCUIT | FUNCTION |
|---|---|---|
| 1 | Z1 18BK | GROUND |
| 2 | K199 18BR/VT | O2 1/2 HEATER CONTROL |
| 3 | K904 18DB/DG | O2 RETURN (DOWN) |
| 4 | K141 20TN/WT | O2 1/2 SIGNAL |

**O2 Sensor (1/2 Right Bank) Connector Pinout**

| PIN # | CIRCUIT | FUNCTION |
|---|---|---|
| 1 | Z1 18BK | GROUND |
| 2 | K299 18BR/WT | O2 2/1 HEATER CONTROL |
| 3 | K902 18BR/DG | O2 RETURN (UP) |
| 4 | K241 20LG/RD | O2 2/1 SIGNAL |

**O2 Sensor (2/1 Left Bank) Connector Pinout**

| PIN # | CIRCUIT | FUNCTION |
|---|---|---|
| 1 | Z1 18BK | GROUND |
| 2 | K399 18BR/GY | O2 2/2 HEATER CONTROL |
| 3 | K904 18DB/DG | O2 RETURN (DOWN) |
| 4 | K341 20PK/WT | O2 2/2 SIGNAL |

**O2 Sensor (2/2 Left Bank) Connector Pinout**

## IDLE AIR CONTROL (IAC) MOTOR

### Description & Operation

The idle air control motor (IAC) is attached to the throttle body. It is an electric stepper motor. The PCM adjusts engine idle speed through the idle air control motor to compensate for engine load, coolant temperature or barometric pressure changes. The throttle body has an air bypass passage that provides air for the engine during closed throttle idle. The idle air control motor pintle protrudes into the air bypass passage and regulates airflow through it.

The PCM adjusts engine idle speed by moving the IAC motor pintle in and out of the bypass passage. The adjustments are based on inputs the PCM receives. The inputs are from the throttle position sensor, crankshaft position sensor, coolant temperature sensor, MAP sensor, vehicle speed sensor and various switch operations (brake, park/neutral, air conditioning). When engine rpm is above idle speed, the IAC is used for the following functions:

- Off-idle dashpot
- Deceleration air flow control
- A/C compressor load control (also opens the passage slightly before the compressor is engaged so that the engine rpm does not dip down when the compressor engages)

*TARGET IDLE*

Target idle is determined by the following inputs:

- Gear position ECT Sensor
- Battery voltage
- Ambient/Battery Temperature Sensor
- VSS
- TPS
- MAP Sensor

**Removal & Installation**

4. Disconnect the negative battery cable.
5. Disconnect the IAC electrical connector.
6. Remove the IAC mounting screws.
7. Remove the IAC.

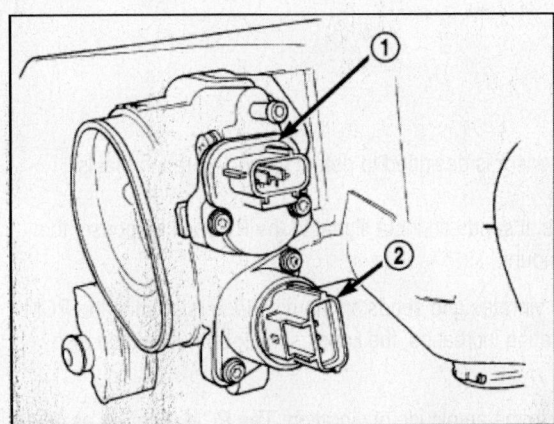

**Showing the location of the Idle Air Control Motor on 3.2L engine**

**To install:**

8. Install the IAC to the throttle body.
9. Tighten mounting screws to 5.1 Nm (45 inch lbs.) torque.
10. Attach electrical connector to the IAC.
11. Connect the negative battery cable.

**Connector Pinouts**

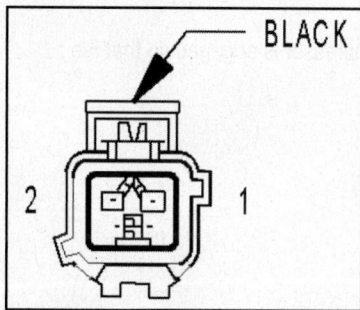

| PIN # | CIRCUIT | FUNCTION |
|-------|-----------|------------------|
| 1 | K39 18GY/RD | IAC MOTOR CONTROL |
| 2 | K60 18YL/BK | IAC RETURN |

**IAC Motor Connector Pinout**

| PIN # | CIRCUIT | FUNCTION |
|-------|-----------|------------------|
| 1 | K21 20BK/RD | IAT SIGNAL |
| 2 | K4 20BK/LB | SENSOR GROUND |

**IAT Sensor Connector Pinout**

## KNOCK SENSOR

### Description & Operation

The knock sensor threads into the cylinder block. The knock sensor is designed to detect engine vibration that is caused by detonation.

When the knock sensor detects a knock in one of the cylinders, it sends an input signal to the PCM. In response, the PCM retards ignition timing for all cylinders by a scheduled amount.

Knock sensors contain a piezoelectric material that constantly vibrates and sends an input voltage (signal) to the PCM while the engine operates. As the intensity of the crystal's vibration increases, the knock sensor output voltage also increases.

The voltage signal produced by the knock sensor increases with the amplitude of vibration. The PCM receives as an input the knock sensor voltage signal. If the signal rises above a predetermined level, the PCM will store that value in memory and retard ignition timing to reduce engine knock. If the knock sensor voltage exceeds a preset value, the PCM retards ignition timing for all cylinders. It is not a selective cylinder retard. The PCM ignores knock sensor input during engine idle conditions. Once the engine speed exceeds a specified value, knock retard is allowed.

Knock retard uses its own short term and long-term memory program. Long-term memory stores previous detonation information in its battery-backed RAM. The maximum authority that long term memory has over timing retard can be calibrated. Short-term memory is allowed to retard timing up to a preset amount under all operating conditions (as long as rpm is above the minimum rpm) except WOT. The PCM, using short-term memory, can respond quickly to retard timing when engine knock is detected. Short-term memory is lost any time the ignition key is turned off.

***NOTE: Over or under tightening affects knock sensor performance, possibly causing improper spark control.***

### Removal & Installation

The sensors screw into the cylinder block, directly below the intake manifold.

1. Remove intake manifold plenum
2. Disconnect electrical connector from knock sensor.
3. Use a crowfoot socket to remove the knock sensors.

**To install:**

4. Install knock sensor. Tighten knock sensor to 10 Nm (7 ft. lbs.) torque. Over or under tightening effects knock sensor performance resulting in possible improper spark control.

5. Attach electrical connector to knock sensor.

6. Install intake manifold plenum.

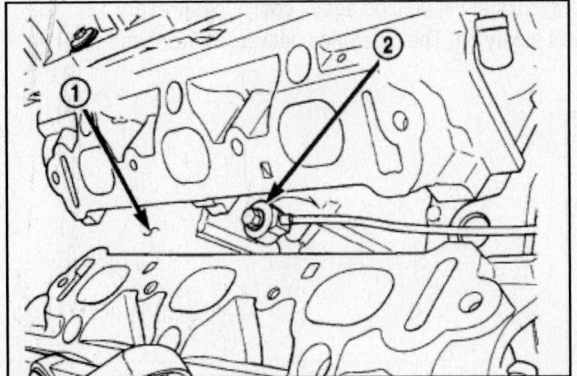

**Identifying the Knock Sensor (2) location on V6 engines (1)**

**Connector Pinouts**

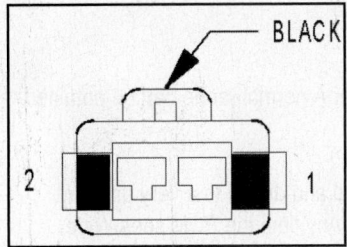

| PIN # | CIRCUIT | FUNCTION |
|-------|---------|----------|
| 1 | K45 20BK/VT | KNOCK SENSOR RETURN |
| 2 | K42 20DB/LG | KNOCK SENSOR SIGNAL |

**Knock Sensor Connector Pinout**

## MANIFOLD ABSOLUTE PRESSURE (MAP) SENSOR

### Description & Operation

The MAP sensor mounts to the driver side of the intake manifold plenum.

The MAP serves as a PCM input, using a silicon based sensing unit, to provide data on the manifold vacuum that draws the air/fuel mixture into the combustion chamber. The PCM requires this information to determine injector pulse width and spark advance. When MAP equals Barometric pressure, the pulse width will be at maximum.

Also like the cam and crank sensors, a 5-volt reference is supplied from the PCM and returns a voltage signal to the PCM that reflects manifold pressure. The zero pressure reading is 0.5 volts and full scale is 4.5 volts. For a pressure swing of 0 - 15 psi the voltage changes 4.0 volts. The sensor is supplied a regulated 4.8 - 5.1 volts to operate the sensor. Like the cam and crank sensors ground is provided through the sensor return circuit.

The MAP sensor input is the number one contributor to pulse width. The most important function of the MAP sensor is to determine barometric pressure. The PCM needs to know if the vehicle is at sea level or is it in Denver at 5000 feet above sea level, because the air density changes with altitude. It will also help to correct for varying weather conditions. If a hurricane was coming through the pressure would be very, very low or there could be a real fair weather, high-pressure area. This is important because as air pressure changes the barometric pressure changes. Barometric pressure and altitude have a direct inverse correlation, as altitude goes up barometric goes down. The first thing that happens as the key is rolled on, before reaching the crank position, the PCM powers up, comes around and looks at the MAP voltage, and based upon the voltage it sees, it knows the current barometric pressure relative to altitude. Once the engine starts, the PCM looks at the voltage again, continuously every 12 milliseconds, and compares the current voltage to what it was at key on. The difference between current and what it was at key on is manifold vacuum.

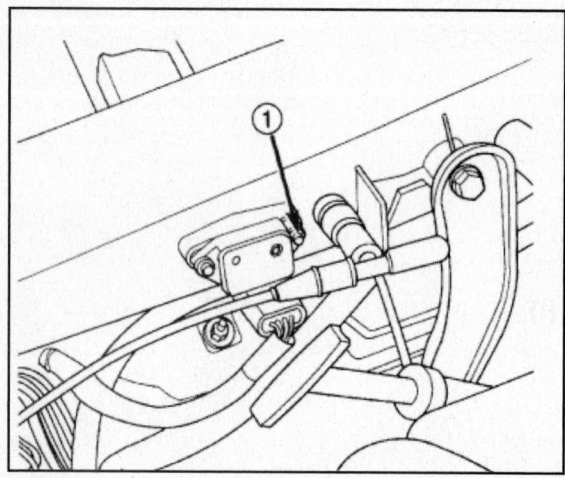

**Location of Manifold Absolute Pressure Sensor (1) on 3.2L**

During key ON (engine not running) the sensor reads (updates) barometric pressure. A normal range can be obtained by monitoring known good sensor in you work area.

As the altitude increases the air becomes thinner (less oxygen). If a vehicle is started and driven to a very different altitude than where it was at key ON, the barometric pressure needs to be updated. Any time the PCM sees Wide Open throttle, based upon TPS angle and RPM it will update barometric pressure in the MAP memory cell. With periodic updates, the PCM can make its calculations more effectively. The PCM uses the MAP sensor to aid in calculating the following:

- Barometric pressure
- Engine load
- Manifold pressure
- Injector pulse-width
- Spark-advance programs
- Shift-point strategies (F4AC1 transmissions only, via the CCD bus)
- Idle speed
- Decel fuel shutoff

The MAP sensor signal is provided from a single piezo-resistive element located in the center of a diaphragm. The element and diaphragm are both made of silicone. As the pressures changes the diaphragm moves causing the element to deflect which stresses the silicone. When silicone is exposed to stress its resistance changes. As manifold vacuum increases, the MAP sensor input voltage decreases proportionally. The sensor also contains electronics that condition the signal and provide temperature compensation.

The PCM recognizes a decrease in manifold pressure by monitoring a decrease in voltage from the reading stored in the barometric pressure memory cell. The MAP sensor is a linear sensor; as pressure changes, voltage changes proportionately. The range of voltage output from the sensor is usually between 4.6 volts at sea level to as low as 0.3 volts at 26 in. Hg. Barometric pressure is the pressure exerted by the atmosphere upon an object. At sea level on a standard day, no storm, barometric pressure is 29.92 in. Hg. For every 100 feet of altitude the barometric pressure drops 10 in. Hg. If a storm goes through it can either add, high-pressure, or decrease, low pressure, from what should be present for that altitude. Always know the average pressure and corresponding barometric pressure is for your area. The MAP sensor mounts to the driver side of the intake manifold plenum.

### Removal & Installation

1. Remove the negative battery cable.
2. Disconnect the electrical connector from the MAP sensor.
3. Remove bolts from sensor.
4. Remove sensor.

### To install:

5. Bolt sensor to intake.
6. Attach electrical connector to sensor.
7. Install the negative battery cable.

### Connector Pinouts

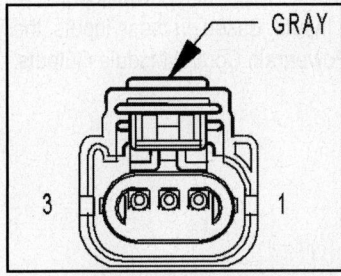

| PIN # | CIRCUIT | FUNCTION |
|-------|-----------|--------------|
| 1 | K1 20DG/RD | MAP SIGNAL |
| 2 | K4 18BK/LB | SENSOR GROUND |
| 3 | K6 20VT/WT | 5 VOLT SUPPLY |

**MAP Sensor Connector Pinout**

## NATURAL VACUUM LEAK DETECTOR (NLVD)

**Connector Pinouts**

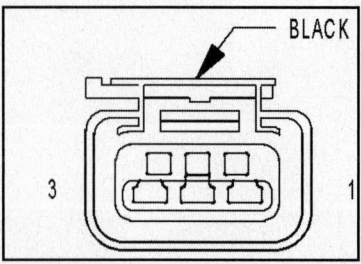

| PIN # | CIRCUIT | FUNCTION |
|-------|---------|----------|
| 1 | Z1 20BK | GROUND |
| 2 | K107 18OR | NVLD SIGNAL |
| 3 | K106 18WT/DG | NVLD SOLENOID CONTROL |

**NVLD Connector Pinout**

## POWERTRAIN CONTROL MODULE (PCM)

**Description & Operation**

The Powertrain Control Module (PCM) is a digital computer containing a microprocessor. The PCM receives input signals from various switches and sensors referred to as Powertrain Control Module Inputs. Based on these inputs, the PCM adjusts various engine and vehicle operations through devices referred to as Powertrain Control Module Outputs.

***PCM INPUTS***

- Air Conditioning Pressure Transducer
- ASD Relay
- Battery Voltage
- Brake Switch
- Camshaft Position Sensor
- Crankshaft Position Sensor
- Distance Sensor (from TCM)
- EGR Position Feedback
- Engine Coolant Temperature Sensor
- Heated Oxygen Sensors
- Ignition sense
- Intake Air Temperature Sensor
- Knock Sensor
- Leak Detection Pump Feedback
- Manifold Absolute Pressure (MAP) Sensor
- Park/Neutral (from TCM)
- PCI Bus
- Power Steering Pressure Switch
- Proportional Purge Sense
- SCI Receive
- Speed Control

- Throttle Position Sensor
- Torque Management Input (from TCM)
- Transmission Control Module (TCM)
- Transaxle Gear Engagement (from TCM)
- Vehicle Speed (from TCM)

### PCM OUTPUTS

- Air Conditioning Clutch Relay
- Automatic Shut Down (ASD) and Fuel Pump Relays
- Data Link Connector (PCI and SCI Transmit)
- Double Start Override
- EGR Solenoid
- Fuel Injectors
- Generator Field
- High Speed Fan Relay
- Idle Air Control Motor
- Ignition Coils
- Leak Detection Pump
- Low Speed Fan Relay
- MTV Actuator
- Proportional Purge Solenoid
- SRV Valve
- Speed Control Relay
- Speed Control Vent Relay
- Speed Control Vacuum Relay
- 8 Volt Output
- 5 Volt Output

Based on inputs it receives, the Powertrain Control Module (PCM) adjusts fuel injector pulse width, idle speed, ignition timing, and canister purge operation. The PCM regulates the cooling fans, air conditioning and speed control systems. The PCM changes generator charge rate by adjusting the generator field. The PCM adjusts injector pulse width (air-fuel ratio) based on the following inputs.

- Battery Voltage
- Intake Air Temperature Sensor
- Engine Coolant Temperature
- Engine Speed (Crankshaft Position Sensor)
- Exhaust Gas Oxygen Content (heated oxygen sensors)
- Manifold Absolute Pressure
- Throttle Position

The PCM adjusts engine idle speed through the idle air control motor based on the following inputs.

- Brake Switch
- Engine Coolant Temperature
- Engine Speed (Crankshaft Position Sensor)
- Park/Neutral (transmission gear selection)
- Transaxle Gear Engagement

- Throttle Position
- Vehicle Speed (from TCM)

The PCM adjusts ignition timing based on the following inputs.

- Intake Air Temperature
- Engine Coolant Temperature
- Engine Speed (Crankshaft Position Sensor)
- Knock Sensor
- Manifold Absolute Pressure
- Park/Neutral (transmission gear selection)
- Transaxle Gear Engagement
- Throttle Position

The Automatic Shut Down (ASD) and Fuel Pump relays are mounted externally, but are turned on and off by the PCM through the same circuit.

The camshaft and crankshaft signals are sent to the Powertrain Control Module. If the PCM does not receive both signals within approximately one second of engine cranking, it deactivates the ASD and fuel pump relays. When these relays are deactivated, power is shut off to the fuel injectors, ignition coils, fuel pump and the heating element in each oxygen sensor.

The PCM contains a voltage converter that changes battery voltage to a regulated 8.0 volts. The 8.0 volts power the Camshaft Position Sensor, Crankshaft Position Sensor and Vehicle Speed Sensor. The PCM also provides a 5.0-volt supply for the Engine Coolant Temperature (ECT) Sensor, Intake Air Temperature (IAT) Sensor, Manifold Absolute Pressure (MAP) Sensor and Throttle Position (TPS) Sensor.

The PCM engine control strategy prevents reduced idle speeds until after the engine operates for 200 miles. If the PCM is replaced after 200 miles of usage, update the mileage in the new PCM. Use the DRB scan tool to change the mileage in the PCM.

### PCM OPERATING MODES

As input signals to the Powertrain Control Module (PCM) change, the PCM adjusts its response to output devices. For example, the PCM must calculate a different injector pulse width and ignition timing for idle than it does for wide-open throttle. There are several different modes of operation that determine how the PCM responds to the various input signals.

There are two types of engine control operation: open loop and closed loop. In open loop operation, the PCM receives input signals and responds according to preset programming. Inputs from the heated oxygen sensors are not monitored. In closed loop operation, the PCM monitors the inputs from the heated oxygen sensors. This input indicates to the PCM whether or not the calculated injector pulse width results in the ideal air-fuel ratio of 14.7 parts air to 1 part fuel. By monitoring the exhaust oxygen content through the oxygen sensor, the PCM can fine-tune injector pulse width. Fine tuning injector pulse width allows the PCM to achieve the lowest emission levels while maintaining optimum fuel economy.

The engine start-up (crank), engine warm-up, and wide-open throttle modes are open loop modes. Under most operating conditions, closed loop modes occur with the engine at operating temperature.

### Ignition Switch On (Engine Off) Mode

When the ignition switch activates the fuel injection system, the following actions occur:

- The PCM determines atmospheric air pressure form the MAP sensor input to determine basic fuel strategy.
- The PCM monitors the engine coolant temperature sensor and throttle position sensor input. The PCM modifies fuel strategy based on this input. When the key is in the "ON" position and the engine is not running (zero rpm), the auto shutdown relay and fuel pump relay are not energized. Therefore, voltage is not supplied to the fuel pump, ignition coil, and fuel injectors.

### Engine Start-up Mode

This is an open loop mode. The following actions occur when the starter motor is engaged:

- The auto shutdown and fuel pump relays are energized. If the PCM does not receive the camshaft and crankshaft signal within approximately one second, these relays are de-energized.

- The PCM energizes all fuel injectors until it determines crankshaft position from the camshaft and crankshaft signals. The PCM determines crankshaft position within one engine revolution. After the crankshaft position has been determined, the PCM energizes the fuel injectors in sequence. The PCM adjusts the injector pulse width and synchronizes the fuel injectors by controlling the fuel injectors' ground paths.

Once the engine idles within 64 rpm of its target engine speed, the PCM compares the current MAP sensor value with the value received during the ignition switch ON (zero rpm) mode. A diagnostic trouble code is written to PCM memory if a minimum difference between the two values is not found. Once the auto shutdown and fuel pump relays have been energized, the PCM determines the fuel injector pulse width based on the following:

- Engine coolant temperature
- Manifold absolute pressure
- Intake air temperature
- Engine revolutions
- Throttle position

The PCM determines the spark advance based on the following:

- Engine coolant temperature
- Crankshaft position
- Camshaft position
- Intake air temperature
- Manifold absolute pressure
- Throttle position

### Engine Warm-Up Mode

This is an open loop mode. The PCM adjusts injector pulse width and controls injector synchronization by controlling the fuel injectors' ground paths. The PCM adjusts ignition timing and engine idle speed. The PCM adjusts the idle speed by controlling the idle air control motor and spark advance.

### Cruise or Idle Mode

When the engine is at normal operating temperature, this is a closed loop mode.

### Acceleration Mode

This is a closed loop mode. The PCM recognizes an increase in throttle position and a decrease in manifold vacuum as engine load increases. In response, the PCM increases the injector pulse width to meet the increased load. The A/C compressor may be de-energized for a short period of time.

### Deceleration Mode

This is a closed loop mode. The PCM recognizes a decrease in throttle position and an increase in manifold vacuum as engine load decreases. In response, the PCM decreases the injector pulse width to meet the decreased load. Full injector shut OFF may be obtained during high-speed deceleration.

### Wide Open Throttle Mode

This is an open loop mode. The throttle position sensor notifies the PCM of a wide-open throttle condition. Once a wide-open throttle is sensed, the PCM de-energizes the A/C compressor clutch relay for 15 seconds.

### PROGRAMMING THE POWERTRAIN CONTROL MODULE

**NOTE:** *Before replacing the PCM for a failed driver, control circuit or ground circuit, be sure to check the related component/circuit integrity for failures not detected due to a double fault in the circuit. Most PCM driver/control circuit failures are caused by internal failure to components (i.e. 12-volt pull-ups, drivers and ground sensors). These failures are difficult to detect when a double fault has occurred and only one DTC has set.*

**NOTE:** *if the PCM and the skim are replaced at the same time, program the VIN into the PCM first. All vehicle keys will then need to be replaced and programmed to the new skim.*

The SKIS "Secret Key" is an ID code that is unique to each SKIS. This code is programmed and stored in the SKIM, engine controller and transponder ship (ignition key). When replacing the PCM it is necessary to program the secret key into the PCM.

1. Turn the ignition ON (transmission in Park/Neutral).
2. Use the DRB and select "THEFT ALARM", "SKIM", and then "MISCELLANEOUS".
3. Select "PCM REPLACED".
4. Enter secured access mode by entering the vehicle four-digit PIN.
5. Press "ENTER" to transfer the secret key (the SKIM will send the secret key to the PCM).

**NOTE :** *If three attempts are made to enter the secure access mode using an incorrect pin, secured access mode will be locked out for one hour. To exit this lockout mode, turn the ignition to the run position for one hour then enter the correct pin. (Ensure all accessories are turned off. Also monitor the battery state and connect a battery charger if necessary).*

The PCM engine control strategy prevents reduced idle speeds until after the engine operates for 200 miles. If the PCM is replaced after 200 miles of usage, update the mileage and vehicle identification number (VIN) in the new PCM. Use the DRB scan tool to change the mileage and VIN in the PCM. If this step is not done a diagnostic trouble code (DTC) may be set and SKIM must be done or car will not start if it is a SKIM equipped car. If a SKIM car you must do a secret key transfer also.

### Removal & Installation

To avoid possible voltage spike damage to PCM, ignition key must be off, and the negative battery cable must be disconnected before unplugging the PCM connectors. Note radio programs.

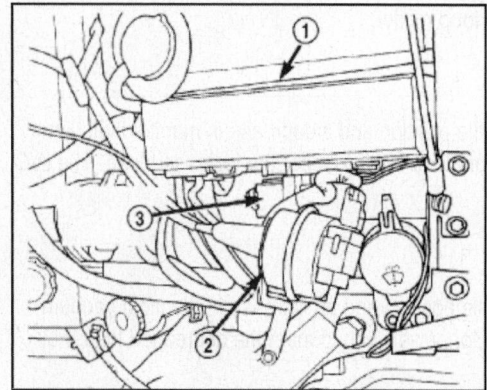

**Locating the Power Distribution Center (1), Speed Control Servo (2), and the Powertrain Control Module (3)**

1. Remove speed control servo and bracket and reposition out of the way.
2. Reposition wiring harness out of the way.

3. Remove washer bottle filler neck.

4. Remove the TCM and reposition it out of the way.

5. Remove the PCM.

6. Disconnect PCM 2 40-way connector.

**To install:**

7. Attach two 40-way connectors to PCM.

8. Install PCM. Tighten bolt to 4 Nm (35 inch lbs.) torque.

9. Install the TCM.

10. Install washer bottle filler neck.

11. Reposition wining harness.

12. Install speed control servo and bracket and tighten fasteners.

13. Connect negative battery cable and reprogram radio and clock.

14. Using a proper scan tool, program mileage and vehicle identification number (VIN) into PCM.

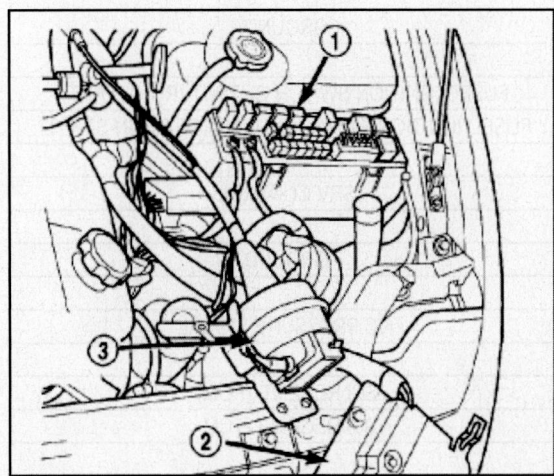

**Showing the Transmission Controller (2), in relation to the PDC (1) and the Speed Control Servo (3)**

### Connector Pinouts

*PCM C1 CONNECTOR*

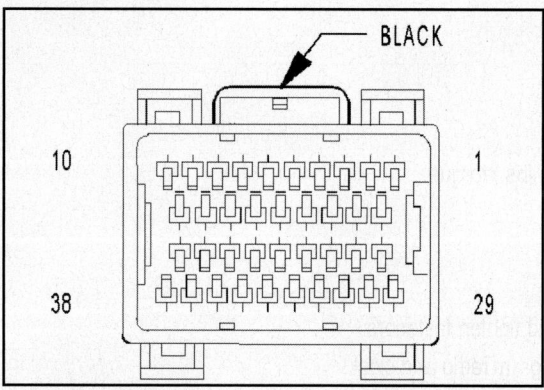

| PIN # | CIRCUIT | FUNCTION |
|---|---|---|
| 1-8 | - | - |
| 9 | Z12 16BK/TN | GROUND |
| 10 | - | - |
| 11 | F12 20DB/WT | FUSED IGNITION SWITCH OUTPUT (RUN-START) |
| 12 | F11 20RD/WT | FUSED IGNITION SWITCH OUTPUT (OFF-RUN-START) |
| 13-15 | - | - |
| 16 | K236 18GY/PK (3.5L HIGH OUTPUT) | SRV CONTROL |
| 17 | - | - |
| 18 | Z12 16BK/TN | GROUND |
| 19-20 | - | - |
| 21 | C18 20DB | A/C PRESSURE SIGNAL |
| 22-24 | - | - |
| 25 | D20 20LG | SCI RECEIVE (PCM) |
| 26 | D19 20VT/OR | SCI RECEIVE (TCM) |
| 27-28 | - | - |
| 29 | A209 20RD | FUSED B(+) |
| 30 | T751 20YL/BK | FUSED IGNITION SWITCH OUTPUT (START) |
| 31 | K141 20TN/WT | O2 1/2 SIGNAL |
| 32 | K904 18DB/DG | O2 RETURN (DOWN) |
| 33 | K341 20PK/WT | O2 2/2 SIGNAL |
| 34-35 | - | - |
| 36 | D21 20PK/TN | SCI TRANSMIT (PCM) |
| 37 | D15 20WT/DG | SCI TRANSMIT (TCM) |
| 38 | D25 18VT/YL | PCI BUS (PCM) |

**PCM C1 (Black) Connector Pinout**

**PCM C2 CONNECTOR**

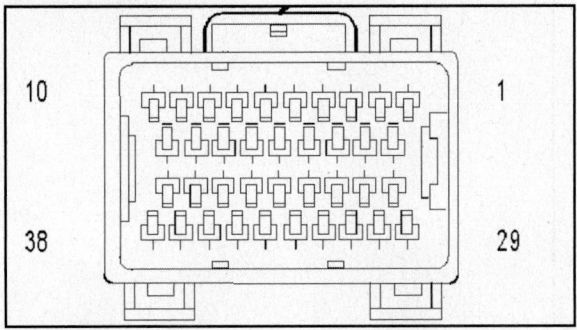

| PIN # | CIRCUIT | FUNCTION |
|:---:|:---:|:---:|
| 1 | K96 16TN/LB | COIL CONTROL NO. 6 |
| 2 | K95 16TN/DG | COIL CONTROL NO. 5 |
| 3 | K94 16TN/LG | COIL CONTROL NO. 4 |
| 4 | K58 18BR/DB | INJECTOR CONTROL NO. 6 |
| 5 | K38 18GY | INJECTOR CONTROL NO. 5 |
| 6 | - | - |
| 7 | K93 16 TN/OR | COIL CONTROL NO. 3 |
| 8 | - | - |
| 9 | K92 16TN/PK | COIL CONTROL NO. 2 |
| 10 | K91 16TN/RD | COIL CONTROL NO.1 |
| 11 | K14 18LB/BR | INJECTOR CONTROL NO. 4 |
| 12 | K13 18YL/WT | INJECTOR CONTROL NO. 3 |
| 13 | K12 18TN/WT | INJECTOR CONTROL NO. 2 |
| 14 | K11 18WT/DB | INJECTOR CONTROL NO. 1 |
| 15 | - | - |
| 16 | K36 18VT/RD | MTV CONTROL |
| 17 | K299 18BR/WT | O2 2/1 HEATER CONTROL |
| 18 | K99 18BR/OR | O2 1/1 HEATER CONTROL |
| 19 | K20 18DG | GEN FIELD CONTROL (+) |
| 20 | K2 20TN/BK | ECT SIGNAL |
| 21 | K22 20OR/DB | TP SIGNAL |
| 22 | - | - |
| 23 | K1 20DG/RD | MAP SIGNAL |
| 24 | K45 20BK/VT | KS RETURN |
| 25 | K42 20DB/LG | KS SIGNAL |
| 26 | - | - |
| 27 | K4 18BK/LB | SENSOR GROUND |
| 28 | K60 18YL/BK | IAC RETURN |
| 29 | K6 20VT/WT | 5 VOLT SUPPLY |
| 30 | K21 20BK/RD | IAT SIGNAL |
| 31 | K41 20BK/DG | O2 1/1 SIGNAL |
| 32 | K902 18BR/DG | O2 RETURN (UP) |
| 33 | K241 20LG/RD | 02 2/1 SIGNAL |
| 34 | K44 20TN/YL | CMP SIGNAL |
| 35 | K24 20GY/BK | CKP SIGNAL |
| 36 | - | - |
| 37 | - | - |
| 38 | K39 18GY/RD | IAC MOTOR CONTROL |

**PCM C2 Connector Pinout**

*C3 CONNECTOR*

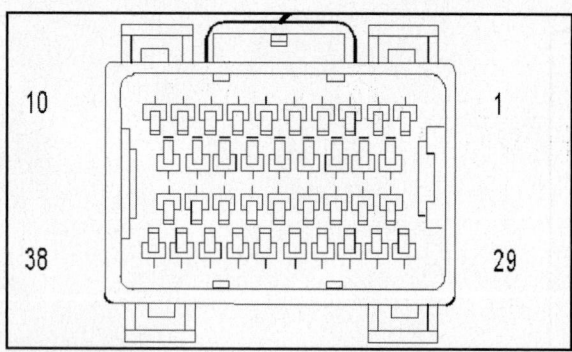

| PIN # | CIRCUIT | FUNCTION |
|---|---|---|
| 1-2 | - | - |
| 3 | K51 20DB/YL | ASD RELAY CONTROL |
| 4 | C27 20DB/PK | HIGH SPEED RAD FAN RELAY CONTROL |
| 5 | V35 20LG/RD | S/C VENT CONTROL |
| 6 | C24 20DB/PK | LOW RAD FAN RELAY CONTROL |
| 7 | V32 20YL/RD | S/C SUPPLY |
| 8 | K106 18WT/DG | NVLD SOLENOID CONTROL |
| 9 | K199 18BR/VT | O2 1/2 HEATER CONTROL |
| 10 | K399 18BR/GY | O2 2/2 HEATER CONTROL |
| 11 | C28 20DB/OR | A/C CLUTCH RELAY CONTROL |
| 12 | V36 18TN/RD | S/C VACUUM CONTROL |
| 13-17 | - | - |
| 18 | F142 16OR/DG | ASD RELAY OUTPUT |
| 19 | F142 16OR/DG | ASD RELAY OUTPUT |
| 20 | K52 18PK/BK | EVAP PURGE CONTROL |
| 21-22 | - | - |
| 23 | K29 20WT/PK | BRAKE SWITCH SIGNAL |
| 24-25 | - | - |
| 26 | T44 20YL (AUTOSTICK) | AUTOSTICK DOWNSHIFT SWITCH SENSE |
| 27 | T5 20LG/RD (AUTOSTICK) | AUTOSTICK UPSHIFT SWITCH SIGNAL |
| 28 | F142 16OR/DG | ASD RELAY OUTPUT |
| 29 | K108 18DG/LG | EVAP PURGE RETURN |
| 30-31 | - | - |
| 32 | K25 20VT/LG | AAT SIGNAL |
| 33 | - | - |
| 34 | V37 20RD/LG | S/C SWITCH SIGNAL |
| 35 | K107 18OR/RD | NVLD SWITCH SIGNAL |
| 36 | - | - |
| 37 | K31 20BR | FUEL PUMP RELAY CONTROL |
| 38 | K90 20TN | STARTER RELAY CONTROL |

**PCM C3 (white) Connector Pinout**

*C4 CONNECTOR*

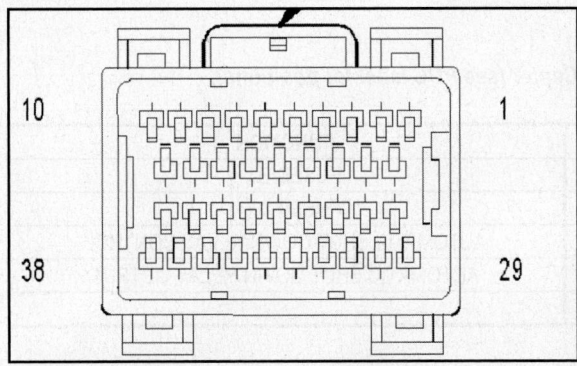

| PIN # | CIRCUIT | FUNCTION |
|-------|---------|----------|
| 1 | T60 16BR | OVERDRIVE SOLENOID CONTROL |
| 2 | T59 16PK | UNDERDRIVE SOLENOID CONTROL |
| 3-5 | - | - |
| 6 | T19 16WT | 2-4 SOLENOID CONTROL |
| 7-9 | - | - |
| 10 | T20 16LB | LOW/REVERSE SOLENOID CONTROL |
| 11 | - | - |
| 12 | Z14 16BK/YL | GROUND |
| 13 | Z13 16BK/RD | GROUND |
| 14 | Z13 16BK/RD | GROUND |
| 15 | T1 20LG/BK | TRS T1 SENSE |
| 16 | T3 20VT | TRS T3 SENSE |
| 17 | - | - |
| 18 | T15 20LG | TRANSMISSION CONTROL RELAY CONTROL |
| 19 | T16 16RD | TRANSMISSION CONTROL RELAY OUTPUT |
| 20-21 | - | - |
| 22 | T9 16OR/BK | OVERDRIVE PRESSURE SWITCH SENSE |
| 23-26 | - | - |
| 27 | T41 20BK/WT | TRS T41 SENSE |
| 28 | T16 16RD | TRANSMISSION CONTROL RELAY OUTPUT |
| 29 | T50 16DG | LOW/REVERSE PRESSURE SWITCH SENSE |
| 30 | T47 16YL/BK | 2-4 PRESSURE SWITCH SENSE |
| 31 | - | - |
| 32 | T14 20LG/WT | OUTPUT SPEED SENSOR SIGNAL |
| 33 | T52 20RD/BK | INPUT SPEED SENSOR SIGNAL |
| 34 | T13 20DB/BK | SPEED SENSOR GROUND |
| 35 | T54 20VT/PK | TRANSMISSION TEMPERATURE SENSOR SIGNAL |
| 36 | - | - |
| 37 | T42 20VT/WT | TRS T42 SENSE |
| 38 | T16 16RD | TRANSMISSION CONTROL RELAY OUTPUT |

**PCM C4 (Green) Connector Pinout**

POWER DISTRIBUTION CENTER RELAYS

**Connector Pinouts**

*NOTE: All relays are located in the Power Distribution Center (see PDC label for positions).*

| PIN # | CIRCUIT | FUNCTION |
|-------|---------|----------|
| A | A209 20RD | FUSED B (+) |
| B | A14 14RD/WT | FUSED B (+) |
| C | K51 20DB/YL | AUTOMATIC SHUT DOWN RELAY CONTROL |
| D | A142 14DG/OR | AUTOMATIC SHUT DOWN RELAY OUTPUT |
| E | - | - |

**Auto Shutdown Relay Connector Pinout**

| PIN # | CIRCUIT | FUNCTION |
|-------|---------|----------|
| A | F12 20DB/WT | FUSED IGNITION SWITCH OUTPUT (RUN-START) |
| B | A1 12RD | FUSED B(+) |
| C | K31 20BR | FUEL PUMP RELAY CONTROL |
| D | A141 12DG/WT | FUEL PUMP RELAY OUTPUT |
| E | - | - |

**Fuel Pump Relay Connector Pinout**

| PIN # | CIRCUIT | FUNCTION |
|-------|---------|----------|
| A | F18 20LG/BK | FUSED IGNITION SWITCH OUTPUT (RUN-START) |
| B | A17 12RD/BR | FUSED B (+) |
| C | C27 20DB/PK | HIGH SPEED RAD FAN RELAY CONTROL |
| D | C25 12YL | HIGH SPEED RAD FAN RELAY OUTPUT |
| E | - | - |

**Hi-Speed Radiator Fan Relay Connector Pinout**

| PIN # | CIRCUIT | FUNCTION |
|-------|---------|----------|
| A | F18 20LG/BK | FUSED IGNITION SWITCH OUTPUT (RUN-START) |
| B | A16 12GY | FUSED B (+) |
| C | C24 20DB/PK | LOW SPEED RAD FAN RELAY CONTROL |
| D | C23 12DG | LOW SPEED RAD FAN RELAY OUTPUT |
| E | - | - |

**Lo-Speed Radiator Fan Relay Connector Pinout**

RADIATOR FAN MOTOR

**Connector Pinouts**

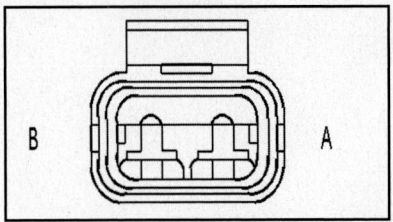

| PIN # | CIRCUIT | FUNCTION |
|-------|---------|----------|
| A | Z1 12BK | GROUND |
| B | 12RD | RADIATOR FAN MOTOR NO. 1 CONTROL |

**Radiator Fan Motor No. 1 Connector Pinout**

| PIN # | CIRCUIT | FUNCTION |
|-------|---------|----------|
| A | 12BK | GROUND |
| B | C23 12RD | LOW SPEED RADIATOR FAN RELAY OUTPUT |

**Radiator Fan Motor No. 2 Connector Pinout**

SENTRY KEY IMMOBILIZER (SKI) MODULE

**Connector Pinouts**

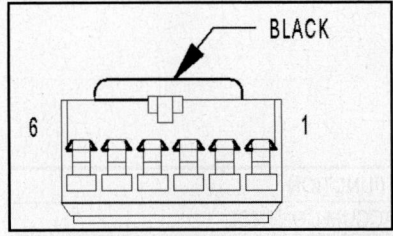

| PIN # | CIRCUIT | FUNCTION |
|-------|---------|----------|
| 1 | - | - |
| 2 | D25 20VT/YL | PCI BUS |
| 3 | - | - |
| 4 | G5 20DB/WT | FUSED IGNITION SWITCH OUTPUT (RUN-START) |
| 5 | Z2 20BK/LG | GROUND |
| 6 | M1 20PK | FUSED B (+) |

**SKI Module Connector Pinout**

## SHORT RUNNER VALVE SOLENOID

**Connector Pinouts**

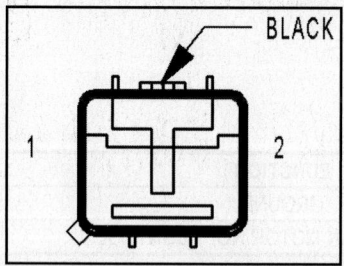

| PIN # | CIRCUIT | FUNCTION |
|-------|---------|----------|
| 1 | F42 18DG/LG | AUTOMATIC SHUT DOWN RELAY OUTPUT |
| 2 | K236 18GY/PK | SRV CONTROL |

**3.5L High Output Engine Short Runner Valve Solenoid Connector Pinout**

## SPEED CONTROL SERVO CONNECTOR

**Connector Pinouts**

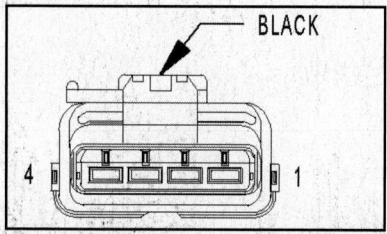

| PIN # | CIRCUIT | FUNCTION |
|-------|---------|----------|
| 1 | V36 18TN/RD | S/C VACCUM CONTROL |
| 2 | V35 20LG/RD | S/C VENT CONTROL |
| 3 | V30 20DB/RD | SPEED CONTROL BRAKE SWITCH OUTPUT |
| 4 | Z1 20BK | GROUND |

**Speed Control Servo Connector Pinout**

SPEED CONTROL SWITCH CONNECTORS

**Connector Pinouts**

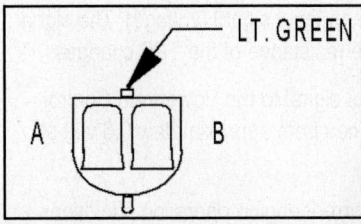

| PIN # | CIRCUIT | FUNCTION |
|---|---|---|
| A | V37 22RD/LG | S/C SWITCH SIGNAL |
| B | K4 22BK/LB | SENSOR GROUND |

**Speed Control (Left) Switch Connector Pinout**

| PIN # | CIRCUIT | FUNCTION |
|---|---|---|
| A | V37 22RD/LG | S/C SWITCH SIGNAL |
| B | K4 22BK/LB | SENSOR GROUND |

**Speed Control (Right) Switch Connector Pinout**

SPEED PROPORTIONAL STEERING SOLENOID CONNECTOR

**Connector Pinouts**

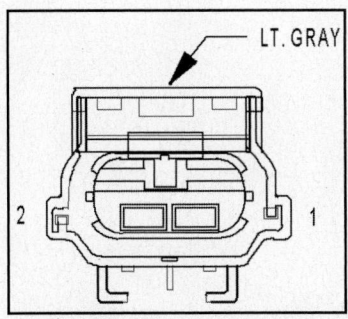

| PIN # | CIRCUIT | FUNCTION |
|---|---|---|
| 1 | S76 18LG/PK | SPEED PROPORTIONAL STEERING SOLENOID (-) |
| 2 | S77 18VT/OR | SPEED PROPORTIONAL STEERING SOLENOID (+) |

**Speed Proportional Steering Solenoid Connector Pinout**

THROTTLE POSITION (TP) SENSOR

### Description & Operation

The throttle position sensor mounts to the side of the throttle body. The sensor connects to the throttle blade shaft. The TPS is a variable resistor that provides the Powertrain Control Module (PCM) with an input signal (voltage). The signal represents throttle blade position. As the position of the throttle blade changes, the resistance of the TPS changes.

The PCM supplies approximately 5 volts to the TPS. The TPS output voltage (input signal to the Powertrain Control Module) represents throttle blade position. The TPS output voltage to the PCM varies from approximately 0.6 volt at minimum throttle opening (idle) to a maximum of 4.5 volts at wide-open throttle.

Along with inputs from other sensors, the PCM uses the TPS input to determine current engine operating conditions. The PCM also adjusts fuel injector pulse width and ignition timing based on these inputs.

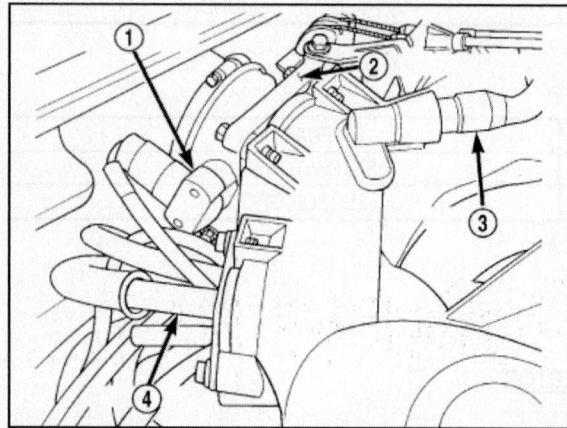

Locating the Throttle Position Sensor (1) on the throttle body (2), near the PCV valve (3) and EGR tube (4) on 2.7L engines

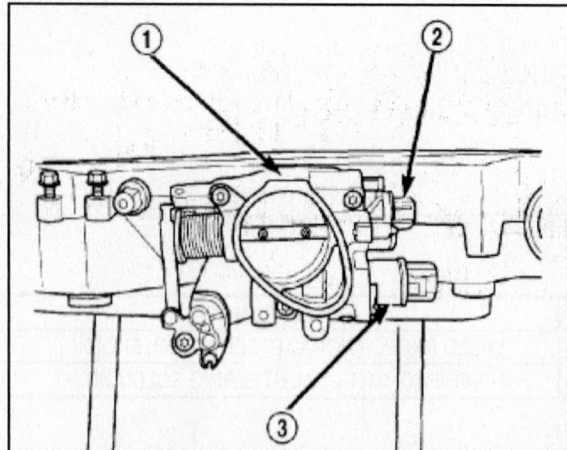

Locating the Throttle Position Sensor (2) on the throttle body, near the IAC motor (3) on 3.2L and 3.5L engines

### Removal & Installation

1. Disconnect the negative battery cable.
2. Disconnect the air plenum.
3. Disconnect the TPS electrical connector.
4. Remove the TPS mounting screws.
5. Remove the TPS.

**To install:**

*NOTE: The throttle shaft end of the throttle body slides into a socket in the TPS. The socket has two tabs inside it. The throttle shaft rests against the tabs. When indexed correctly the TPS can rotate clockwise a few degrees to line up the mounting screw holes with the screw holes in the throttle body. The TPS has slight tension when rotated into position. If it is difficult to rotate the TPS into position, install the sensor with the throttle shaft on the other side of the tabs in the socket. Tighten mounting screws to 3 Nm (25 inch lbs.) torque.*

6. After installing the TPS, the throttle plate should be closed. If the throttle plate is open, install the sensor on the other side of the tabs in the socket.
7. Attach electrical connector to the TPS.
8. Install air plenum and tighten the clamp.
9. Connect the negative battery cable.

**Testing**

The Throttle Position Sensor (TPS) can be tested with a digital voltmeter. The center terminal of the sensor is the output terminal.

With the ignition switch in the ON position, check the output voltage at the center terminal wire of the connector. Check the output voltage at idle and at wide open throttle (WOT). At idle, TPS output voltage should be greater then 0.6 volts. At wide-open throttle, TPS output voltage should be less than 4.5 volts. The output voltage should gradually increase as the throttle plate moves slowly from idle to WOT. Check for spread terminals at the sensor and PCM connections before replacing the TPS.

**Connector Pinouts**

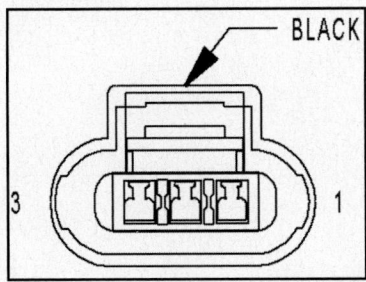

| PIN # | CIRCUIT | FUNCTION |
|-------|---------|----------|
| 1 | K6 20VT/WT | 5 VOLT SUPPLY |
| 2 | K22 20OR/DB | TP SIGNAL |
| 3 | K4 20BK/LB | SENSOR GROUND |

**TP Sensor Connector Pinout**

## TRANSMISSION SOLENOID/PRESSURE SWITCH

**Connector Pinouts**

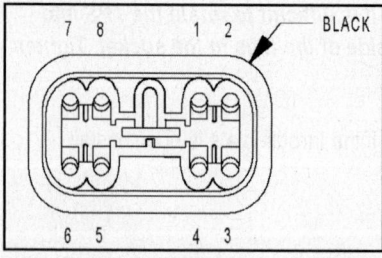

| PIN # | CIRCUIT | FUNCTION |
|-------|---------|----------|
| 1 | T9 16OR/BK | OVERDRIVE PRESSURE SWITCH SENSE |
| 2 | T50 16DG | LOW/REVERSE PRESSURE SWITCH SENSE |
| 3 | T47 16YL/BK | 2-4 PRESSURE SWITCH SENSE |
| 4 | T16 16RD | TRANSMISSION CONTROL RELAY OUTPUT |
| 5 | T19 16WT | 2-4 SOLENOID CONTROL |
| 6 | T20 16LB | LOW/REVERSE SOLENOID CONTROL |
| 7 | T60 16BR | OVERDRIVE SOLENOID CONTROL |
| 8 | T59 16PK | UNDERDRIVE SOLENOID CONTROL |

**Transmission Solenoid/Pressure Switch Connector Pinout**

## DAIMLERCHRYSLER CARS: PT CRUISER

ACCELERATOR PEDAL POSITION SENSOR (APPS)

### Description

The Accelerator Pedal Position Sensor (APPS) is a variable resistor that provides the PCM with an input signal (voltage). The signal represents throttle blade position. As the position of the accelerator pedal changes, the resistance of the APPS changes.

### Removal & Installation

1. Remove the air cleaner cover.
2. Disconnect the negative battery cable.
3. Unlock the electrical connector and then disconnect the electrical connector from the module.
4. Remove the mounting bolt.
5. Remove assembly from the mounting bracket.
6. Open the APPS module cover and disconnect the cable from the cam and module.

### To install:

7. Open APPS module and connect the cable.
8. Install APPS module.
9. Tighten the mounting bolt.
10. Connect the negative battery cable.
11. Install the air cleaner cover.

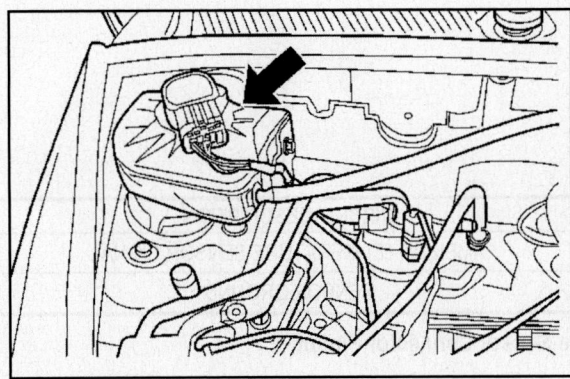

**Showing location of Accelerator Pedal Position Sensor**

## AMBIENT TEMPERATURE SENSOR

**Connector Pinouts**

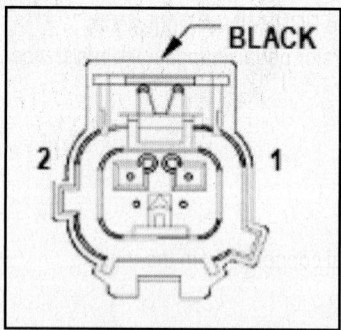

| PIN # | WIRE COLOR | CIRCUIT DESCRIPTION |
|-------|------------|---------------------|
| 1 | BR/OR | AAT SIGNAL |
| 2 | BR/YL | SENSOR GROUND 2 |

**PT Cruiser 2.4L (w/NGC) Ambient Temperature Sensor Connector Pinout**

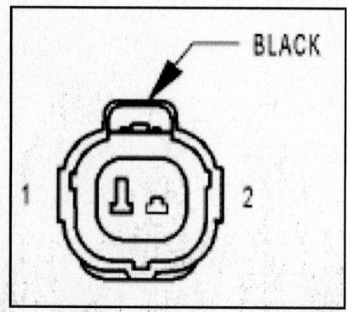

| PIN # | WIRE COLOR | CIRCUIT DESCRIPTION |
|-------|------------|---------------------|
| 1 | VT/LG | AMBIENT TEMPERATURE SENSOR SIGNAL |
| 2 | BK/LB | SENSOR GROUND |

**PT Cruiser all except 2.4L (w/NGC) Ambient Temperature Sensor Connector Pinout**

## AUTOMATIC SHUTDOWN (ASD)

**Description & Operation**

The Automatic Shutdown (ASD) is an input to the Powertrain Control Module from the relay in the Power Distribution Center. The ASD sense circuit informs the PCM when the ASD relay energizes. A 12-volt signal at this input indicates to the PCM that the ASD has been activated. This input is used only to sense that the ASD relay is energized.

When energized, the ASD relay provides power to operate the injectors, ignition coil, generator field, O2 sensor heaters (both upstream and downstream), and also provides a sense circuit to the PCM for diagnostic purposes. The PCM energizes the ASD any time there is a Crankshaft Position sensor signal that exceeds a predetermined value. The ASD relay can also be energized after the engine has been turned off to perform an O2 sensor heater test, if vehicle is equipped with OBD II diagnostics.

With SBEC III, the ASD relay's electromagnet is fed battery voltage, not ignition voltage. The PCM still provides the ground. As mentioned earlier, the PCM energizes the ASD relay during an O2 sensor heater test. This test is performed only after the engine has been shut off. The PCM still operates internally to perform several checks, including monitoring the O2 sensor heaters.

### AUTOMATIC SHUTDOWN RELAY

The ASD relay and fuel pump relay are located in the Power Distribution Center (PDC) near the Air Cleaner. The inside top of the PDC cover has a label showing relay and fuse location. They are ISO relays.

The PCM operates the Automatic Shut Down (ASD) relay and fuel pump relay. The PCM operates them by switching the ground path for the relays on and off. The ASD relay connects battery voltage to the fuel injectors, ignition coil, generator field, and the heating elements in the oxygen sensors. The fuel pump relay connects battery voltage to the fuel pump.

The PCM turns the ground path off when the ignition switch is in the OFF position. Both relays are off. When the ignition switch is in the ON or Crank position, the PCM monitors the crankshaft position sensor and camshaft position sensor signals to determine engine speed. If the PCM does not receive a crankshaft position sensor signal and camshaft position sensor signal when the ignition switch is in the Run position, it de-energizes both relays. When the relays are de-energized, battery voltage is not supplied to the fuel injectors, ignition coil, generator field, and the heating elements in the oxygen sensors, and fuel pump.

## CAMSHAFT POSITION (CMP) SENSOR

### Description & Operation

The camshaft position sensor is mounted to the rear of the cylinder head. The sensor also acts as a thrust plate to control camshaft endplay.

The PCM sends approximately 5 volts to the Hall-effect sensor. This voltage is required to operate the Hall-effect chip and the electronics inside the sensor. The input to the PCM occurs on a 5-volt output reference circuit. A ground for the sensor is provided through the sensor return circuit. The PCM identifies camshaft position by registering the change from 5 to 0 volts, as signaled from the Camshaft Position sensor.

On 2.0/2.4L engines a target magnet attaches to the rear of the camshaft and indexes to the correct position. The target magnet has fourteen different poles arranged in an asymmetrical pattern. As the target magnet rotates, the camshaft position sensor senses the change in polarity.

The PCM determines fuel injection synchronization and cylinder identification from inputs provided by the camshaft position sensor (2.0L and 2.4L) and crankshaft position sensor. From the two inputs, the PCM determines crankshaft position.

The sensor input switches from high (5 volts) to low (0.30 volts) as the target magnet rotates. When the north pole of the target magnet passes under the sensor, the output switches high. The sensor output switches low when the south pole of the target magnet passes underneath.

On 1.6L a raised platform on the cam sprocket serves as a target. When the sensor detects the step, the input voltage from the sensor to the PCM switches from high (5 volts) to low (0.3 volts). As the step returns away from the sensor, the input voltage switches back to high (5 volts).

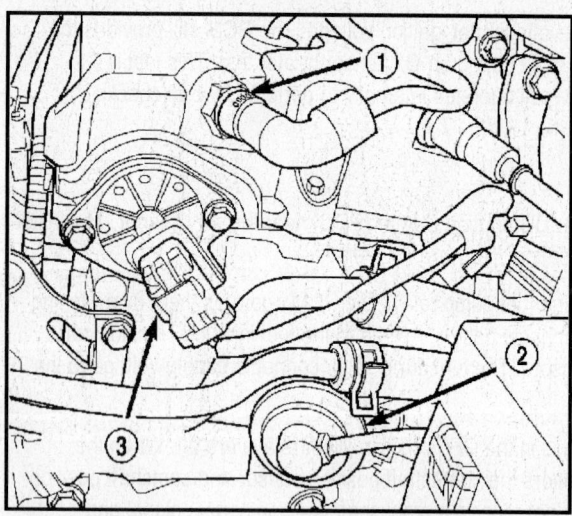

**Locating CMP Sensor (3), PCV Valve (1) & EGR Valve (3) on DOHC engines**

## Removal & Installation

### 2.0L & 2.4L

1. Remove the air cleaner lid, disconnect the inlet air temperature sensor and makeup air hose.
2. Remove the negative battery cable.
3. Disconnect electrical connector from camshaft position sensor.
4. Remove camshaft position sensor mounting screws. Remove sensor.
5. Loosen screw attaching target magnet to rear of camshaft.

**To install:**

The target magnet has locating dowels that fit into machined locating holes in the end of the camshaft.

***CAUTION: Installation of the electrical connector or sensor at an angle may damage the sensor pins.***

6. Install target magnet in end of camshaft. Tighten mounting screw to 3.6 Nm (32 ±5 inch lbs.) torque. Over-torquing could cause cracks in magnet. If magnet cracks replace it.
7. Install camshaft position sensor. Tighten sensor mounting screws to 13 Nm (115 ±15 inch lbs.) torque.
8. Carefully attach electrical connector to camshaft position sensor.
9. Install the negative battery cable.
10. Install the air cleaner lid; connect the inlet air temperature sensor and makeup air hose.

**Connector Pinouts**

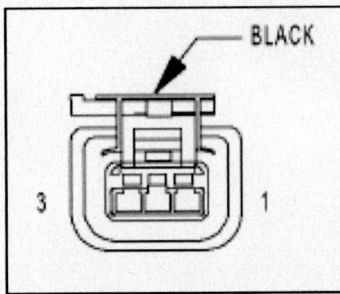

| PIN # | WIRE COLOR | CIRCUIT DESCRIPTION |
|-------|-----------|---------------------|
| 1 | GY/PK (1.6L) | CAMSHAFT POSITION SENSOR 5 VOLT SUPPLY |
| 1 | OR (EXCEPT 1.6L) | 5 VOLT SUPPLY |
| 2 | TN/YL (1.6L) | CAMSHAFT POSITION SENSOR SIGNAL |
| 2 | BK/LB (EXCEPT 1.6L) | SENSOR GROUND 1 |
| 3 | BR/GY (1.6L) | CAMSHAFT POSITION SENSOR GROUND |
| 3 | TN/YL (EXCEPT 1.6L) | CMP SIGNAL |

**PT Cruiser CMP Sensor Connector Pinout**

CRANKSHAFT POSITION (CKP) SENSOR

**Description & Operation**

The crankshaft position sensor mounts to the engine block below the generator and near the oil filter.

The PCM sends approximately 9 volts to the Hall-effect sensor. This voltage is required to operate the Hall-effect chip and the electronics inside the sensor. A ground for the sensor is provided through the sensor return circuit. The input to the PCM occurs on a 5-volt output reference circuit that operates as follows: The Hall-effect sensor contains a powerful magnet. As the magnetic field passes over the dense portion of the counterweight, the 5-volt signal is pulled to ground (.3 volts) through a transistor in the sensor. When the magnetic field passes over the notches in the crankshaft counterweight, the magnetic field turns off the transistor in the sensor, causing the PCM to register the 5-volt signal. The PCM identifies crankshaft position by registering the change from 5 to 0 volts, as signaled from the Crankshaft Position sensor.

The PCM determines what cylinder to fire from the crankshaft position sensor input and the camshaft position sensor input. The second crankshaft counterweight has machined into it two sets of four timing reference notches including a 60 degree signature notch. From the crankshaft position sensor input the PCM determines engine speed and crankshaft angle (position).

The notches generate pulses from high to low in the crankshaft position sensor output voltage. When a metal portion of the counterweight aligns with the crankshaft position sensor, the sensor output voltage goes low (less than 0.5 volts) When a notch aligns with the sensor, voltage goes high (5.0 volts) As a group of notches pass under the sensor, the output voltage switches from low (metal) to high (notch) then back to low.

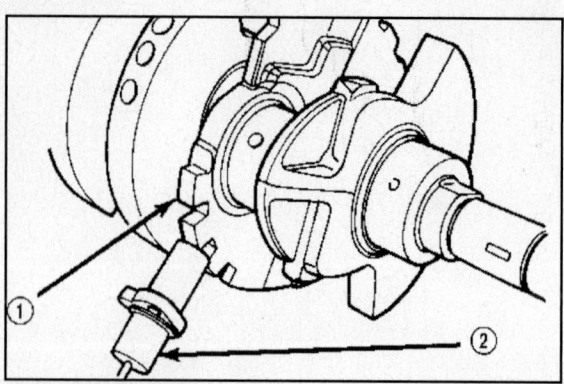

Showing typical reference slots (1) for CKP Sensor (2)

### Removal & Installation

1. Remove the air cleaner lid, disconnect the inlet air temperature sensor and makeup air hose.
2. Remove the negative battery cable.
3. Raise the vehicle and support.
4. Disconnect electrical connector from crankshaft position sensor.
5. Remove sensor mounting screw. Remove sensor.

### To install:

6. Install sensor and tighten the screw to 13 Nm (115 inch lbs.)
7. Connect electrical connector to crankshaft position sensor.
8. Lower vehicle.
9. Install the negative battery cable.
10. Install the air cleaner lid, connect the inlet air temperature sensor and makeup air hose.

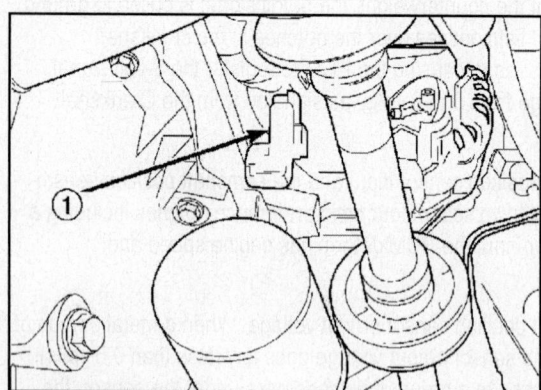

Location of CKP Sensor (1) on lower front of engine (2.4L engine shown)

**Connector Pinouts**

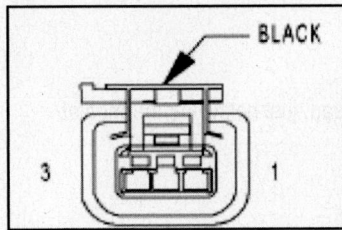

| PIN # | WIRE COLOR | CIRCUIT DESCRIPTION |
|---|---|---|
| 1 | BR/WT (1.6L) | CRANKSHAFT POSITION SENSOR GROUND |
| 1 | OR (EXCEPT 1.6L) | 5 VOLT SUPPLY |
| 2 | GY/BK (1.6L) | CRANKSHAFT POSITION SENSOR SIGNAL |
| 2 | BK/LB (EXCEPT 1.6L) | SENSOR GROUND 1 |
| 3 | OR (1.6L) | CRANKSHAFT POSITION SENSOR 5 VOLT SUPPLY |
| 3 | GY/BK (EXCEPT 1.6L) | CKP SIGNAL |

**CKP Sensor**

## DATA LINK CONNECTOR

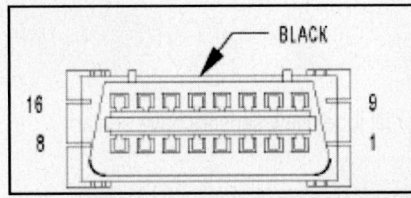

| PIN # | WIRE COLOR | CIRCUIT DESCRIPTION |
|---|---|---|
| 1 | - | NOT USED |
| 2 | VT/YL | PCI BUS |
| 3 | WT/VT | NOT USED |
| 4 | BK/TN | GROUND |
| 5 | BK/TN | GROUND |
| 6 | - | NOT USED |
| 7 | PK (DIESEL) | SCI TRANSMIT (ECM) |
| 7 | PK (GAS) | SCI TRANSMIT (PCM) |
| 8 | - | NOT USED |
| 9 | WT/DG | FLASH PROGRAM ENABLE |
| 10 | - | NOT USED |
| 11 | - | NOT USED |
| 12 | LG (GAS) | SCI RECEIVE |
| 13 | - | NOT USED |
| 14 | - | NOT USED |
| 15 | WT/DG (EATX) | SCI TRANSMIT (TCM) |
| 16 | 8RD/WT | FUSED B (+) |

**Data Link Connector Pinout**

## ENGINE COOLANT TEMPERATURE (ECT) SENSOR

### Description

*1.6L*

The engine coolant temperature (ECT) sensor threads into the rear of the cylinder head, just below the thermostat housing. The ECT Sensor is a Negative Thermal Coefficient (NTC) sensor.

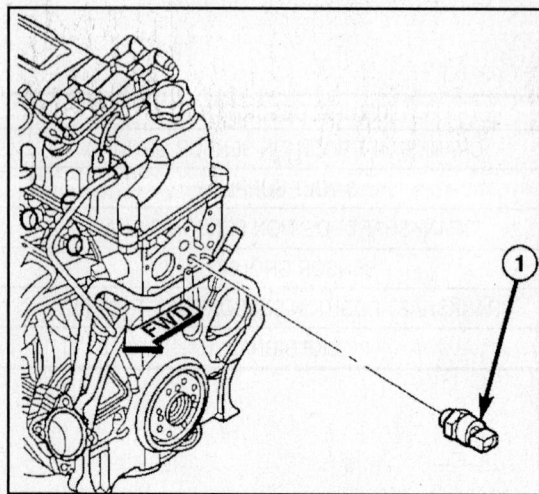

**Location of ECT Sensor (1) on 1.6L engine**

*2.0L & 2.4L*

The coolant sensor threads into the front of the cylinder head near the radiator fill tube. New sensors have sealant applied to the threads.

The ECT Sensor is a Negative Thermal Coefficient (NTC), dual range Sensor. The resistance of the ECT Sensor changes as coolant temperature changes. This results in different input voltages to the PCM. The PCM also uses the ECT Sensor input to operate the low and high-speed radiator cooling fans.

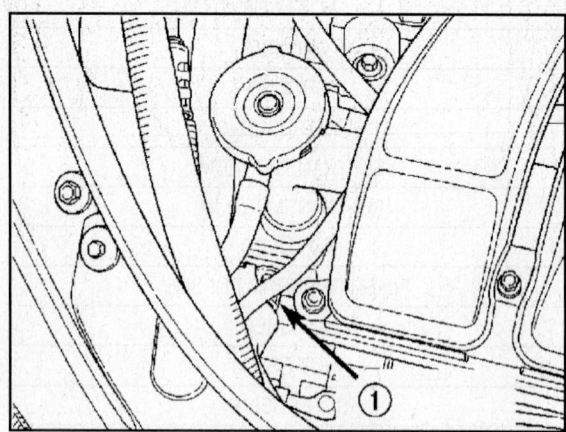

**Showing location of ECT Sensor (1) on 2.4L engine**

## Operation

*1.6L*

The ECT sensor provides an input to the PCM. As temperature increases, resistance of the sensor decreases. As coolant temperature varies, the ECT sensor resistance changes resulting in a different voltage value at the PCM ECT sensor signal circuit. The ECT sensor provides input for various PCM operations. The PCM uses the input to control air-fuel mixture, timing, and radiator fan on/off times. The ECT sensor input is also used for temperature gauge operation.

*2.0L & 2.4L*

The PCM sends 5 volts to the sensor and is grounded through the sensor return line. As temperature increases, resistance in the sensor decreases. As coolant temperature varies, the coolant temperature sensor resistance changes resulting in a different voltage value at the PCM engine coolant sense circuit.

When the engine is cold, the PCM will provide slightly richer air-fuel mixtures and higher idle speeds until normal operating temperatures are reached.

The combination coolant temperature sensor has two elements. One element supplies coolant temperature signal to the PCM. The other element supplies coolant temperature signal to the instrument panel gauge cluster. The PCM determines engine coolant temperature from the coolant temperature sensor.

As coolant temperature varies the coolant temperature sensors resistance changes resulting in a different input voltage to the PCM and the instrument panel gauge cluster. When the engine is cold, the PCM will provide slightly richer air-fuel mixtures and higher idle speeds until normal operating temperatures are reached.

The PCM has a dual temperature range program for better sensor accuracy at cold temperatures. At key-ON the PCM sends a regulated 5-volt signal through a 10,000-ohm resistor to the sensor. When the sensed voltage reaches approximately 1.25 volts the PCM turns on the transistor. The transistor connects a 1,000-ohm resistor in parallel with the 10,000-ohm resistor. With this drop in resistance the PCM recognizes an increase in voltage on the input circuit.

## Removal & Installation

*1.6L*

1. Disconnect negative battery cable.
2. Disconnect positive battery cable.
3. Remove battery.
4. Partially drain cooling system.
5. Disconnect coolant temperature sensor electrical connector.
6. Remove coolant temperature sensor.

## To install:

7. Install coolant temperature sensor. Tighten sensor to 17 Nm (150 inch lbs.).
8. Reconnect coolant temperature sensor connector.
9. Install battery.
10. Connect positive and negative battery cables.
11. Fill cooling system.

*2.0L & 2.4L*

1. Remove the air cleaner lid and makeup air hose.
2. Remove the negative battery cable.
3. With the engine cold, drain coolant until level drops below sensor level.
4. Disconnect coolant sensor electrical connector.
5. Remove coolant sensor.

**To Install:**

6. Install coolant sensor. Tighten sensor to 17-21 Nm (12-16 ft. lbs.) torque.
7. Attach electrical connector to sensor.
8. Fill cooling system.
9. Install the negative battery cable.
10. Install the air cleaner lid and makeup air hose.

**Connector Pinouts**

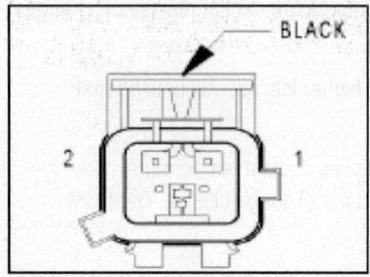

| PIN # | WIRE COLOR | CIRCUIT DESCRIPTION |
|---|---|---|
| 1 | BK/LB (1.6L) | ENGINE COOLANT TEMPERATURE SENSOR GROUND |
| 1 | BK/LB (EXCEPT 1.6L) | SENSOR GROUND 1 |
| 2 | TN/BK (1.6L) | ENGINE COOLANT TEMPERATURE SENSOR SIGNAL |
| 2 | TN/BK (EXCEPT 1.6L) | ECT SIGNAL |

**ECT Sensor Connector Pinout**

ENGINE OIL PRESSURE (EOP) SWITCH

**Connector Pinouts**

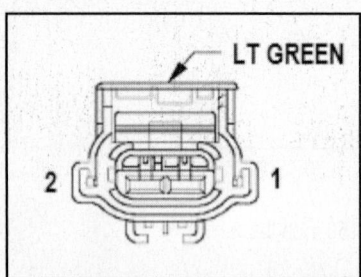

| PIN # | WIRE COLOR | CIRCUIT DESCRIPTION |
|---|---|---|
| 1 | GY (1.6L) | ENGINE OIL PRESSURE SWITCH SIGNAL |
| 1 | VT/LG (DIESEL) | ENGINE OIL PRESSURE SWITCH SIGNAL |
| 1 | GY (EXCEPT 1.6L/DIESEL) | OIL PRESSURE SIGNAL |
| 2 | - | NOT USED |

**EOP Switch Connector Pinout**

## EXHAUST GAS RECIRCULATION (EGR) VALVE

### Description & Operation

*2.4L*

The EGR system consists of:

- EGR tube
- EGR valve
- Electric EGR Transducer
- Connecting hoses

The EGR system reduces oxides of nitrogen (NOx) in engine exhaust and helps prevent detonation (engine knock) Under normal operating conditions, engine cylinder temperature can reach more than 3000°F. Formation of NOx increases proportionally with combustion temperature. To reduce the emission of these oxides, the cylinder temperature must be lowered. The system allows a predetermined amount of hot exhaust gas to recirculate and dilute the incoming air/fuel mixture. The diluted air/fuel mixture reduces peak flame temperature during combustion.

The electric EGR transducer contains an electrically operated solenoid and a back-pressure transducer Electric EGR Transducer The Powertrain Control Module (PCM) operates the solenoid. The PCM determines when to energize the solenoid. Exhaust system back-pressure controls the transducer

When the PCM energizes the solenoid, vacuum does not reach the transducer. Vacuum flows to the transducer when the PCM de-energizes the solenoid.

When exhaust system back-pressure becomes high enough, it fully closes a bleed valve in the transducer. When the PCM de-energizes the solenoid and back-pressure closes the transducer bleed valve, vacuum flows through the transducer to operate the EGR valve.

De-energizing the solenoid, but not fully closing the transducer bleed hole (because of low back-pressure), varies the strength of vacuum applied to the EGR valve. Varying the strength of the vacuum changes the amount of EGR supplied to the engine. This provides the correct amount of exhaust gas recirculation for different operating conditions.

This system does not allow EGR at idle. A failed or malfunctioning EGR system can cause engine spark knock, sags or hesitation, rough idle, engine stalling and increased emissions.

### Removal & Installation

*2.4L*

1. Remove the air cleaner lid, disconnect the inlet air temperature sensor and makeup air hose.
2. Remove the negative battery cable.
3. Loosen the bolts at the intake manifold.
4. Remove EGR tube bolts at EGR Valve.
5. Unclip EGR transducer bracket and remove.
6. Disconnect the vacuum supply hose to EGR transducer solenoid.
7. Unlock the connector then disconnect the electrical connector from solenoid.
8. Remove the 2 bolts at EGR valve to cylinder head. Remove EGR valve and transducer.
9. Clean gasket surfaces. Discard old gaskets. If necessary, clean EGR passages.

**To install:**

10. Connect vacuum supply tube to solenoid.
11. Attach electrical connector to solenoid.
12. Install EGR transducer into the bracket and snap closed.
13. Loosely install EGR valve with new gaskets.
14. Finger-tighten EGR tube fasteners.
15. Tighten EGR valve mounting screws to 22 Nm (195 inch lbs.) torque.

16. Tighten EGR tube fasteners to 11 Nm (97 inch lbs.) torque.
17. Be sure to route hoses and wiring away from hot-contact surfaces.
18. Install the negative battery cable.
19. Install the air cleaner lid; connect the inlet air temperature sensor and makeup air hose.

**Diagnosis**

*2.4L*

If the EGR system operates incorrectly, replace the entire EGR valve and transducer together. The EGR valve and electrical transducer are calibrated together.

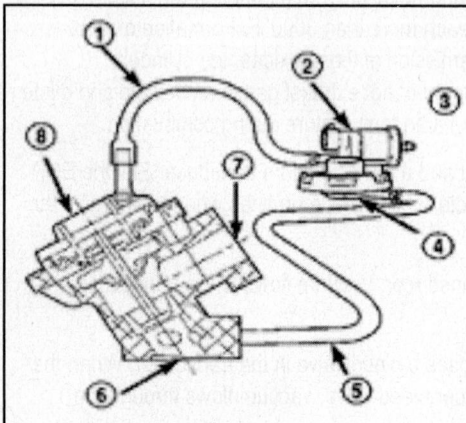

1. Vacuum Hose
2. EGR Valve Control Assembly
3. Intake Manifold Vacuum
4. Transducer Portion of Control Valve
5. Back-Pressure Hose
6. To Cylinder Head
7. To Intake Manifold
8. EGR Valve Assembly

**EGR Valve (2) and Transducer (4) shown with system components for 2.4L engine**

EVAP LEAK DETECTION PUMP

**Description & Operation**

The leak detection pump is a device used to detect a leak in the evaporative system. The pump contains a 3-port solenoid, a pump that contains a switch, a spring-loaded canister vent valve seal, 2 check valves and a spring/diaphragm.

Immediately after a cold start, when the engine temperature is between 40°F and 86°F, the 3-port solenoid is briefly energized. This initializes the pump by drawing air into the pump cavity and also closes the vent seal. During non-test test conditions, the vent seal is held open by the pump diaphragm assembly, which pushes it open at the full travel position. The vent seal will remain closed while the pump is cycling. This is due to the operation of the 3-port solenoid, which prevents the diaphragm assembly from reaching full travel. After the brief initialization period, the solenoid is de-energized, allowing atmospheric pressure to enter the pump cavity. This permits the spring to drive the diaphragm which forces air out of the pump cavity and into the vent system. When the solenoid is energized and de-energized, the cycle is repeated creating flow in typical diaphragm pump fashion. The pump is controlled in 2 modes:

*PUMP MODE*

The pump is cycled at a fixed rate to achieve a rapid pressure build in order to shorten the overall test time.

*TEST MODE*

The solenoid is energized with a fixed duration pulse. Subsequent fixed pulses occur when the diaphragm reaches the switch closure point. The spring in the pump is set so that the system will achieve an equalized pressure of about 7.5 inches of water.

When the pump starts, the cycle rate is quite high. As the system becomes pressurized, pump rate drops. If there is no leak, the pump will quit. If there is a leak, the test is terminated at the end of the test mode. If there is no leak, the purge monitor is run. If the cycle rate increases due to the flow through the purge system, the test is passed and the diagnostic is complete. The canister vent valve will unseal the system after completion of the test sequence as the pump diaphragm assembly moves to the full travel position.

## Removal & Installation

1. Properly release the fuel pressure.
2. Remove the air cleaner lid; disconnect the inlet air temperature sensor and makeup air hose.
3. Remove the negative battery cable.
4. Raise vehicle and support.
5. Drain fuel tank.
6. Remove fuel tank and EVAP system.
7. Remove the 3 mounting screws.
8. Remove LDP from bracket.
9. Remove hoses from EVAP canister.
10. Disconnect electrical connector for Leak Detection Pump (LDP).

## To install:

11. Connect electrical connector to the Leak Detection Pump (LDP)
12. Install hoses and lines.
13. Install LDP onto bracket and tighten to 3.9 Nm (35 inch lbs.)
14. Install the fuel tank and EVAP system.
15. Lower vehicle.
16. Install the negative battery cable.
17. Install the air cleaner lid, connect the inlet air temperature sensor and makeup air hose.
18. Fill fuel tank. Use the scan tool to pressurize the fuel system. Check for leaks.

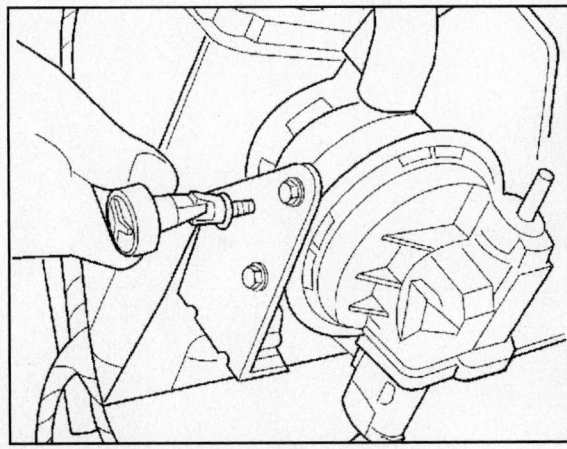

**Removing Leak Detection Pump (LDP)**

EVAP PROPORTIONAL PURGE SOLENOID

## Description & Operation

All vehicles use a proportional purge solenoid. The solenoid regulates the rate of vapor flow from the EVAP canister to the throttle body. The PCM operates the solenoid.

During the cold start warm-up period and the hot start time delay, the PCM does not energize the solenoid. When de-energized, no vapors are purged. The proportional purge solenoid operates at a frequency of 200 Hz and is controlled by an engine controller circuit that senses the current being applied to the proportional purge solenoid and then adjusts that current to achieve the desired purge flow. The proportional purge solenoid controls the purge rate of fuel vapors from the vapor canister and fuel tank to the engine intake manifold.

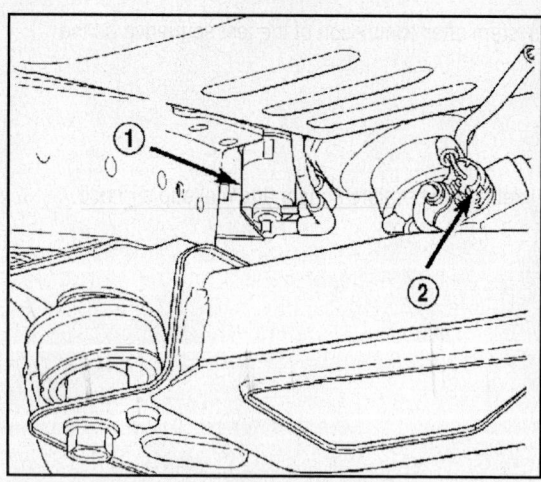

Showing location of EVAP Proportional Purge Solenoid (1) & Power Steering Switch (2)

### Removal & Installation

1. Remove the negative battery cable.
2. Remove solenoid from bracket by pulling up on solenoid.
3. Disconnect electrical connector from solenoid.
4. Disconnect vacuum tubes from solenoid.

### To install:

*NOTE: The top of the solenoid has TOP printed on it. The solenoid will not operate unless it is installed correctly.*

5. Connect vacuum tube to solenoid.
6. Connect electrical connector to solenoid.
7. Install solenoid on bracket.
8. Install the negative battery cable.

### Connector Pinouts

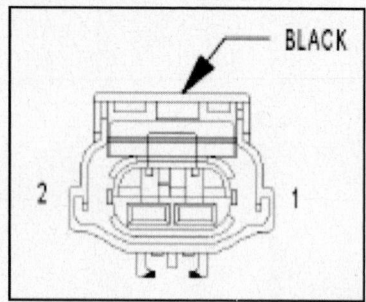

| PIN # | WIRE COLOR | CIRCUIT DESCRIPTION |
|-------|-----------|---------------------|
| 1 | DB/WT (1.6L) | FUSED IGNITION SWITCH OUTPUT (RUN-START) |
| 1 | WT/TN (EXCEPT 1.6L) | EVAP PURGE RETURN |
| 2 | PK/BK | EVAP PURGE CONTROL |

EVAP Purge Solenoid Valve Connector Pinout

## FUEL INJECTOR CONNECTOR

**Connector Pinouts**

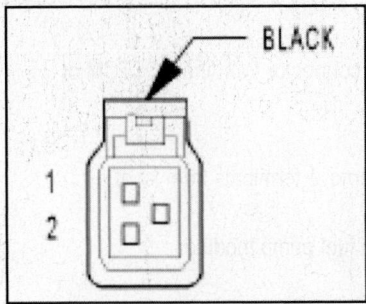

| PIN # | WIRE COLOR | CIRCUIT DESCRIPTION |
|-------|-----------|---------------------|
| 1 | WT/DB | FUEL INJECTOR NO. 1 DRIVER |
| 1 | TN | FUEL INJECTOR NO. 2 DRIVER |
| 1 | YL/WT | FUEL INJECTOR NO. 3 DRIVER |
| 1 | LB/BR | FUEL INJECTOR NO. 4 DRIVER |
| 2 | DG/OR | AUTOMATIC SHUT DOWN RELAY OUTPUT |

**Fuel Injector Connector Pinout**

## FUEL LEVEL SENSOR/FUEL PUMP MODULE

**Description & Operation**

The level sensor is attached to the side of the fuel pump module. The level sensor consists of a float, an arm, and a variable resistor.

As the fuel level increases, the float and arm move up. This decreases the sending unit resistance, causing the fuel gauge on the instrument panel to read full. The fuel level sensor (fuel gauge sending unit) sends a signal to the BCM and the BCM sends the signal over the PCI bus circuit to the PCM to indicate fuel level. The purpose of this feature is to prevent a false setting of misfire and fuel system monitor trouble codes if the fuel level is less than approximately 15 percent of its rated capacity. It is also used to send a signal for fuel gauge operation via the PCI bus circuits.

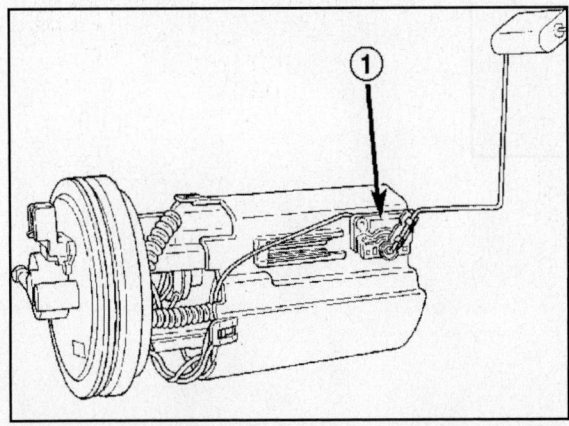

**Location of Fuel Level Sensor (1) on fuel pump module**

## Removal & Installation

*NOTE: The pump module harness on TOP of flange is not serviceable or removable*

1. Remove fuel pump module.
2. Depress retaining tab and remove the fuel pump/level sensor connector from the BOTTOM of the fuel pump module electrical connector.
3. Remove the wedge lock from the electrical connector.
4. Using Special Tool C-4334 terminal remover, or equivalent, remove terminals from level sensor connector.
5. Depress the tab and slide the sensor toward the bottom of the fuel pump module.
6. Slide level sensor out of channel in module.

## To install:

7. Wrap wires into groove in back of level sensor.
8. While feeding wires into guide grooves, slide level sensor up into channel until it snaps into place. Ensure tab at bottom of sensor locks in place.
9. Install level sensor wires in connector. Push the wires up through the connector and then pull them down until they lock in place. Ensure signal and ground wires are installed in the correct position.
10. Install locking wedge on connector.
11. Push connector up into bottom of fuel pump module electrical connector.
12. Install fuel pump module.

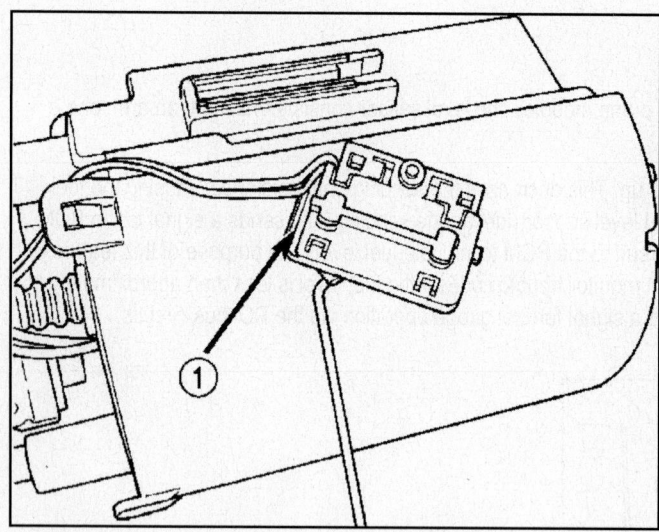

**Identifying wiring groove (1) on Fuel Level Sensor**

**Connector Pinouts**

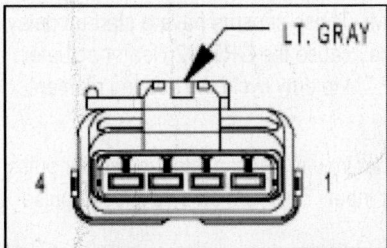

| PIN # | WIRE COLOR | CIRCUIT DESCRIPTION |
|-------|------------|---------------------|
| 1 | BK | GROUND |
| 2 | BK/LG | GROUND |
| 3 | DB | FUEL LEVEL SENSOR SIGNAL |
| 4 | DG/WT | FUEL PUMP RELAY OUTPUT |

**Fuel Pump Module Connector Pinout**

## GENERATOR

**Connector Pinouts**

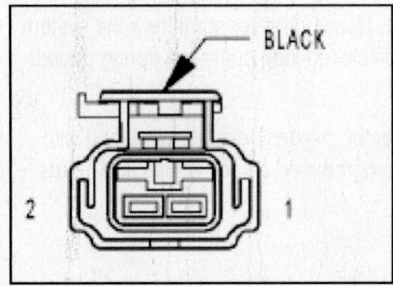

| PIN # | WIRE COLOR | CIRCUIT DESCRIPTION |
|-------|------------|---------------------|
| 1 | DG (1.6L) | GENERATOR FIELD CONTROL |
| 1 | BK (2.4L) | GROUND |
| 2 | DG/OR (1.6L) | AUTOMATIC SHUT DOWN RELAY OUTPUT |
| 2 | DG (2.4L) | GEN FIELD CONTROL |

**Generator Connector Pinout**

## HEATED OXYGEN SENSORS (HO2S)

**Description & Operation**

The upstream oxygen sensor is threaded into the outlet flange of the exhaust manifold. The downstream oxygen sensor is threaded into the side of the catalytic converter

The O2 sensors produce voltages from 0 to 1 volt, depending upon the oxygen content of the exhaust gas in the exhaust manifold. When a large amount of oxygen is present (caused by a lean air/fuel mixture), the sensors produce a voltage below 0.45 volts. When there is a lesser amount present (rich air/fuel mixture) it produces a voltage above 0.45 volts. By monitoring the oxygen content and converting it to electrical voltage, the sensors act as a rich- lean switch.

The oxygen sensors are equipped with a heating element that keeps the sensors at proper operating temperature during all operating modes. Maintaining correct sensor temperature at all times allows the system to enter into closed loop operation sooner. Also, it allows the system to remain in closed loop operation during periods of extended idle. Upstream O2s (California emission equipped) are Pulse Width Modulated (PWM). These sensors have a start-up delay to remove moisture from the heater element on cold start. This start-up delay may cause the DRB O2 Heater actuator test to be denied. The ground circuit is controlled by the PCM. This allows the PCM to duty cycle the heating element for the upstream O2s

In Closed Loop operation the PCM monitors the O2 sensor input (along with other inputs) and adjusts the injector pulse width accordingly. During Open Loop operation the PCM ignores the O2 sensor input. The PCM adjusts injector pulse width based on preprogrammed (fixed) values and inputs from other sensors.

The Automatic Shutdown (ASD) relay supplies battery voltage to both the upstream and downstream heated oxygen sensors. The oxygen sensors are equipped with a heating element. The heating elements reduce the time required for the sensors to reach operating temperature.

### UPSTREAM OXYGEN SENSOR 1/1

The input from the upstream heated oxygen sensor tells the PCM the oxygen content of the exhaust gas. Based on this input, the PCM fine tunes the air-fuel ratio by adjusting injector pulse width.

The sensor input switches from 0 to 1 volt, depending upon the oxygen content of the exhaust gas in the exhaust manifold. When a large amount of oxygen is present (caused by a lean air-fuel mixture), the sensor produces voltage as low as 0.1 volt. When there is a lesser amount of oxygen present (rich air-fuel mixture) the sensor produces a voltage as high as 1.0 volt. By monitoring the oxygen content and converting it to electrical voltage, the sensor acts as a rich-lean switch.

The heating element in the sensor provides heat to the sensor ceramic element. Heating the sensor allows the system to enter into closed loop operation sooner. Also, it allows the system to remain in closed loop operation during periods of extended idle.

In Closed Loop, the PCM adjusts injector pulse width based on the upstream heated oxygen sensor input along with other inputs. In Open Loop, the PCM adjusts injector pulse width based on preprogrammed (fixed) values and inputs from other sensors.

### DOWNSTREAM OXYGEN SENSOR 1/2

The Downstream O2 Sensor has two functions. One function is measuring catalyst efficiency. This is an OBD II requirement. The oxygen content of the exhaust gasses has significantly less fluctuation than at the inlet if the converter is working properly. The PCM compares upstream and Downstream O2 Sensor switch rates under specific operating conditions to determine if the catalyst is functioning properly.

The other function is a downstream fuel control. The upstream O2 goal varies within the window of operation of the O2 Sensor. In the past the goal was a preprogrammed fixed value based upon where it believed the catalyst operated most efficiently.

While the Upstream O2 Sensor input is used to maintain the 14.7:1 air/fuel ratio, variations in engines, exhaust systems and catalytic converters may cause this ratio to not be the most ideal for a particular catalyst and engine. To help maintain the catalyst operating at maximum efficiency, the PCM will fine tune the air/fuel ratio entering the catalyst based upon the oxygen content leaving the catalyst. This is accomplished by modifying the Upstream O2 Sensor voltage goal.

If the exhaust leaving the catalyst has too much oxygen (lean) the PCM increases the upstream O2 goal, which increases fuel in the mixture causing less oxygen to be left over. Conversely, if the oxygen content leaving the catalyst has is too little oxygen (rich) the PCM decreases the upstream O2 goal down which removes fuel from the mixture causing more oxygen to be left over. This function only occurs during downstream closed loop mode operation.

## Removal & Installation

### 1/1 UPSTREAM

1. Remove the air cleaner lid and makeup air hose.
2. Remove the negative battery cable.
3. Disconnect electrical connector from sensor.
4. Remove sensor using an oxygen sensor special tool C-4907.

**To install:**

5. After removing the sensor, the exhaust manifold threads must be cleaned with an 18 mm X 1.5 + 6E tap. If reusing the original sensor, coat the sensor threads with an anti-seize compound such as Loctite® 771-64, or equivalent. New sensors have compound on the threads and do not require an additional coating. Tighten the sensor to 28 Nm (20 ft. lbs.) torque.
6. Connect electrical connector to sensor. Install the negative battery cable.
7. Install the air cleaner lid and makeup air hose.

### 1/2 DOWNSTREAM

The downstream heated oxygen sensor threads into the exhaust pipe behind the catalytic converter.

1. Remove the air cleaner lid and makeup air hose.
2. Remove the negative battery cable.
3. Raise vehicle and support.
4. Disconnect electrical connector from sensor.
5. Disconnect sensor electrical harness from clips along body.
6. Remove sensor using an oxygen sensor crowfoot wrench.

**To install:**

7. After removing the sensor, the exhaust manifold threads must be cleaned with an 18 mm X 1.5 + 6E tap. If reusing the original sensor, coat the sensor threads with an anti-seize compound such as Loctite® 771-64, or equivalent. New sensors have compound on the threads and do not require an additional coating. Tighten the sensor to 28 Nm (20 ft. lbs.) torque.
8. Connect sensor electrical harness to clips along body.
9. Connect electrical connector to sensor.
10. Lower vehicle.
11. Install the negative battery cable.
12. Install the air cleaner lid and makeup air hose.

**Connector Pinouts**

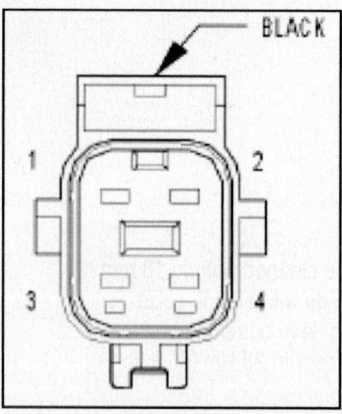

| PIN # | WIRE COLOR | CIRCUIT DESCRIPTION |
|---|---|---|
| 1 | BR/VT (1.6L) | Oxygen Sensor 1/2 Control |
| 1 | BK (Except 1.6L) | Ground |
| 2 | DG/OR (1.6L) | Automatic Shut Down Relay Output |
| 2 | BR/VT (Except 1.6L) | O2 1/2 Heater Control |
| 3 | DB/DG (1.6L) | Oxygen Sensor 1/2 Ground |
| 3 | DB/DG (Except 1.6L) | O2 Return |
| 4 | TN/WT (1.6L) | Oxygen Sensor 1/2 Signal |
| 4 | TN/WT (Except 1.6L) | O2 1/2 Signal |

**Downstream HO2S Connector Pinout**

| PIN # | WIRE COLOR | CIRCUIT DESCRIPTION |
|---|---|---|
| 1 | DG/OR (1.6L) | AUTOMATIC SHUT DOWN RELAY OUTPUT |
| 1 | BK (EXCEPT 1.6L) | GROUND |
| 2 | OR/RD (1.6L) | OXYGEN SENSOR 1/1 CONTROL |
| 2 | BR/OR (EXCEPT 1.6L) | O2 1/1 HEATER CONTROL |
| 3 | BR/DG (1.6L) | OXYGEN SENSOR 1/1 GROUND |
| 3 | DB/DG (EXCEPT 1.6L) | O2 RETURN |
| 4 | BK/DG (1.6L) | OXYGEN SENSOR 1/1 SIGNAL |
| 4 | BK/DG (EXCEPT 1.6L) | O2 1/1 SIGNAL |

**Upstream HO2S Connector Pinout**

## IDLE AIR CONTROL (IAC) MOTOR

### Description & Operation

The Idle Air Control (IAC) motor is mounted on the throttle body. The PCM operates the idle air control motor.

The PCM adjusts engine idle speed through the idle air control motor to compensate for engine load, coolant temperature or barometric pressure changes. The throttle body has an air bypass passage that provides air for the engine during closed throttle idle. The idle air control motor pintle protrudes into the air bypass passage and regulates the airflow through it.

The PCM adjusts engine idle speed by moving the IAC motor pintle in and out of the bypass passage. The adjustments are based on received PCM inputs. The inputs are from the throttle position sensor, crankshaft position sensor, coolant temperature sensor, MAP sensor, vehicle speed sensor and various switch operations (brake, park/neutral, A/C).

When engine rpm is above idle speed, the IAC is used for the following functions:

- Off-idle dashpot
- Deceleration air flow control
- A/C compressor load control (also opens the passage slightly before the compressor is engaged so that the engine rpm does not dip down when the compressor engages)

Target idle is determined by the following inputs:

- Gear position
- ECT Sensor
- Battery voltage
- Ambient/Battery Temperature Sensor
- VSS
- TPS
- MAP Sensor

### Removal & Installation

1. Remove the air cleaner lid, disconnect the inlet air temperature sensor and makeup air hose.
2. Remove the negative battery cable.
3. Disconnect electrical connector from idle air control motor.
4. Remove idle air control motor mounting screws.
5. Remove idle air control motor. Ensure O-ring is removed with the motor.

### To install:

6. The new idle air control motor requires a new O-ring. If pintle measures more than 1 inch (25 mm) it must be retracted. Use the scan tool "AIS Motor Open/Close Test" to retract the pintle (battery must be connected).
7. Carefully place idle air control motor into throttle body.
8. Install mounting screws. Tighten screws to: 6.2 Nm (55 inch lbs.) torque.
9. Attach electrical connectors to idle air control motor.
10. Install the negative battery cable.
11. Install the air cleaner lid, connect the inlet air temperature sensor and makeup air hose.

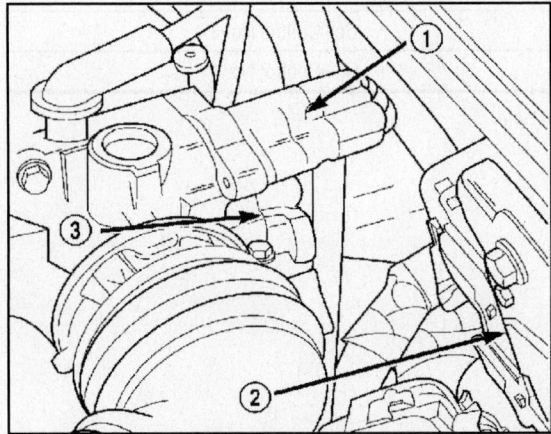

**Locating the IAC Motor (1), TP Sensor (2) and PCM (3) on PT Cruiser 2.4L engine**

**Connector Pinouts**

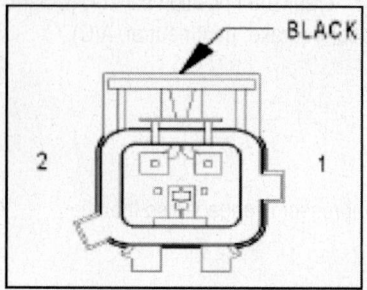

| PIN # | WIRE COLOR | CIRCUIT DESCRIPTION |
|-------|-----------|---------------------|
| 1 | VT/GY | IAC Motor Control |
| 2 | BR/VT | IAC Return |

**2.4L IAC Motor Connector Pinout**

## IGNITION COIL

**Connector Pinouts**

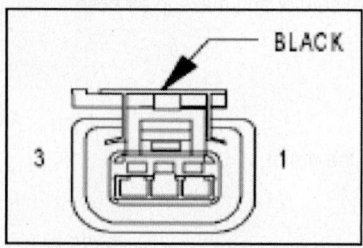

| PIN # | WIRE COLOR | CIRCUIT DESCRIPTION |
|-------|-----------|---------------------|
| 1 | DB/TN (1.6L) | Ignition Coil No. 2 Driver |
| 1 | DB/TN (2.4L Except Turbo) | Coil Control No. 2 |
| 1 | DB/TN (2.4L Turbo) | Coil Control No. 2 |
| 2 | DG/OR (1.6L/2.4l Turbo) | Automatic Shut Down Relay Output |
| 2 | DG/OR (2.4L Except Turbo) | Automatic Shut Down Relay Output |
| 3 | BK/GY (1.6L) | Ignition Coil No. 1 Driver |
| 3 | BK/GY (2.4L Except Turbo) | Coil Control No. 1 |
| 3 | BK/GY (2.4L Turbo) | Coil Control No. 1 |

**Ignition Coil Connector Pinout**

<u>INPUT SPEED SENSOR</u>

## Description & Operation

The Input Speed Sensor is a two-wire magnetic pickup device that generates AC signals as rotation occurs. It is threaded into the transaxle case, sealed with an O-ring, and is considered a primary input to the Powertrain/Transmission Control Module.

The Input Speed Sensor provides information on how fast the input shaft is rotating. As the teeth of the input clutch hub pass by the sensor coil, an AC voltage is generated and sent to the PCM/TCM. The PCM/TCM interprets this information as input shaft rpm.

The PCM/TCM compares the input speed signal with output speed signal to determine the following:

- Transmission gear ratio
- Speed ratio error detection
- CVI calculation

The PCM/TCM also compares the input speed signal and the engine speed signal to determine the following:

- Torque converter clutch slippage
- Torque converter element speed ratio

## Removal & Installation

*CAUTION: When disconnecting speed sensor connector, be sure that the connector weather seal does not fall off or remain in old sensor.*

   1. Disconnect the battery cables.
   2. Remove air cleaner assembly.
   3. Remove the battery hold down clamp and remove the battery.
   4. Remove the battery tray.
   5. Disconnect input speed sensor connector.
   6. Unscrew and remove input speed sensor.
   7. Inspect speed sensor O-ring and replace if necessary.

**To install:**

*CAUTION: When disconnecting speed sensor connector, be sure that the connector weather seal does not fall off or remain in old sensor.*

   8. Verify O-ring is installed into position.
   9. Install and tighten input speed sensor to 27 Nm (20 ft. lbs.).
  10. Connect speed sensor connector.
  11. Install the battery tray.
  12. Install the battery and hold down clamp.
  13. Install air cleaner assembly.
  14. Connect battery cables.

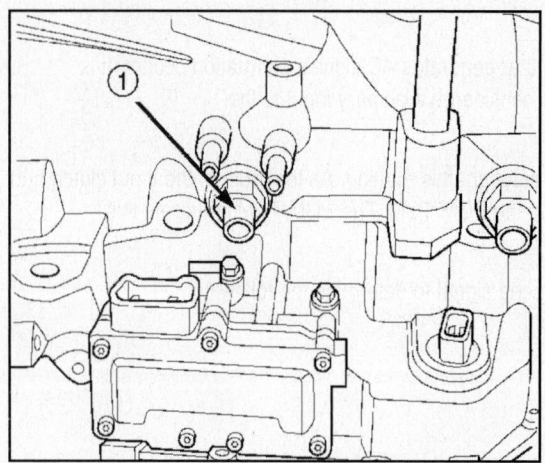

**Showing location of Input Speed Sensor (1) on PT Cruiser engines**

INTAKE AIR TEMPERATURE (IAT) SENSOR

### Description & Operation

The IAT sensor attaches to the intake air duct or is in the air tube. The IAT Sensor is a Negative Temperature Coefficient (NTC) Sensor that provides information to the PCM regarding the temperature of the air entering the intake manifold. The PCM sends 5 volts to the sensor and is grounded through the sensor return line. As temperature increases, resistance in the sensor decreases.

*INTAKE AIR TEMPERATURE*

The inlet air temperature sensor replaces the intake air temperature sensor and the battery temperature sensor. The PCM uses the information from the inlet air temperature sensor to determine values to use as an intake air temperature sensor and a battery temperature sensor. The Intake Air Temperature (IAT) sensor value is used by the PCM to determine air density. The PCM uses this information to calculate:

- Injector pulse width
- Adjustment of ignition timing (to prevent spark knock at high intake air temperatures)

*BATTERY TEMPERATURE*

The intake air temperature sensor replaces the intake air temperature sensor and the battery temperature sensor. The PCM uses the information from the inlet air temperature sensor to determine values for the PCM to use as an intake air temperature sensor and a battery temperature sensor.

The battery temperature information along with data from monitored line voltage (B+), is used by the PCM to vary the battery charging rate. System voltage will be higher at colder temperatures and is gradually reduced at warmer temperatures.

The battery temperature information is also used for OBD II diagnostics. Certain faults and OBD II monitors are either enabled or disabled depending upon the battery temperature sensor input (example: disable purge, enable LDP) Most OBD II monitors are disabled below 20°F.

### Removal & Installation

1. Remove the air cleaner lid; disconnect the inlet air temperature sensor and makeup air hose.
2. Disconnect the negative battery cable.
3. Disconnect electrical connector from the sensor
4. Remove the sensors.

**To install:**

5. Install sensor.

6. Attach electrical connector to sensor.

7. Connect the negative battery cable.

8. Install the air cleaner lid; connect the inlet air temperature sensor and makeup air hose.

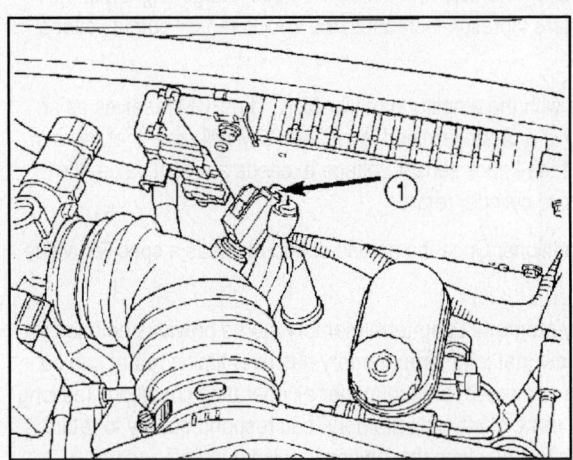

**Showing location of Intake Air Temperature Sensor (1)**

**Connector Pinouts**

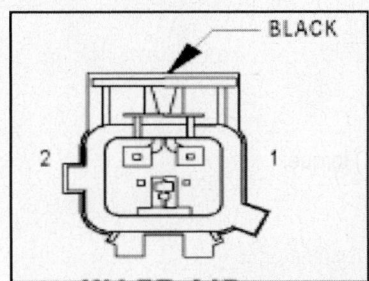

| PIN # | WIRE COLOR | CIRCUIT DESCRIPTION |
|-------|-----------|---------------------|
| 1 | BK/RD (1.6L) | Inlet Air Temperature Sensor Signal |
| 1 | BK/RD (Except 1.6L) | IAT Signal |
| 2 | RD (1.6L) | Inlet Air Temperature Sensor Ground |
| 2 | BK/LB (Except 1.6L) | Sensor Ground 1 |

**IAT Sensor Connector Pinout**

KNOCK SENSOR

### Description

The knock sensor threads into the cylinder block. The knock sensor is designed to detect engine vibration that is caused by detonation. When the knock sensor detects a knock in one of the cylinders, it sends an input signal to the PCM. In response, the PCM retards ignition timing for all cylinders by a scheduled amount.

Knock sensors contain a piezoelectric material, which constantly vibrates and sends an input voltage (signal) to the PCM while the engine operates. As the intensity of the crystal's vibration increases, the knock sensor output voltage also increases.

The voltage signal produced by the knock sensor increases with the amplitude of vibration. The PCM receives as an input the knock sensor voltage signal. If the signal rises above a predetermined level, the PCM will store that value in memory and retard ignition timing to reduce engine knock. If the knock sensor voltage exceeds a preset value, the PCM retards ignition timing for all cylinders. It is not a selective cylinder retard.

The PCM ignores knock sensor input during engine idle conditions. Once the engine speed exceeds a specified value, knock retard is allowed.

Knock retard uses its own short term and long-term memory program. Long-term memory stores previous detonation information in its battery-backed RAM. The maximum authority that long term memory has over timing retard can be calibrated. Short-term memory is allowed to retard timing up to a preset amount under all operating conditions (as long as rpm is above the minimum rpm) except WOT. The PCM, using short-term memory, can respond quickly to retard timing when engine knock is detected. Short-term memory is lost any time the ignition key is turned off.

***NOTE: Over or under tightening affects knock sensor performance, possibly causing improper spark control.***

### Removal & Installation

1. Disconnect electrical connector from knock sensor.
2. Use a crowfoot socket to remove the knock sensors.

**To install:**

3. Install knock sensor. Tighten knock sensor to 10 Nm (7 ft. lbs.) torque.
4. Attach electrical connector to knock sensor.

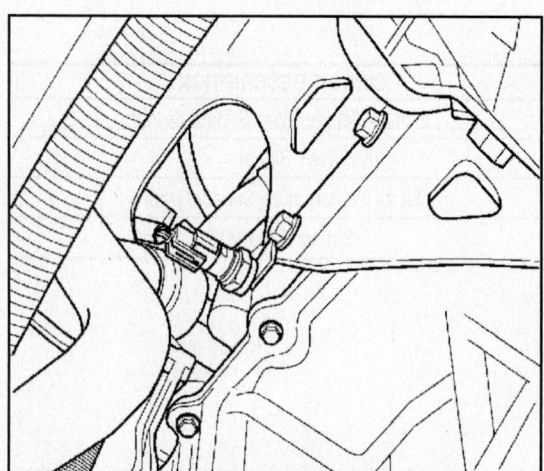

**Location of Knock Sensor on 2.4L engine**

**Connector Pinouts**

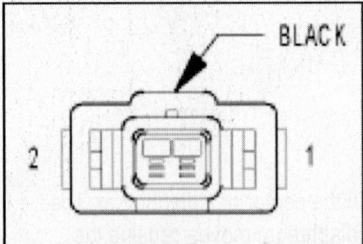

| PIN # | WIRE COLOR | CIRCUIT DESCRIPTION |
|:---:|:---:|:---:|
| 1 | DB/LG (1.6L) | Knock Sensor Signal |
| 1 | DB/LG (EXCEPT 1.6L) | KS Signal |
| 2 | BK/VT (1.6L) | Knock Sensor Ground |
| 2 | BK/VT (EXCEPT 1.6L) | KS Return |

**Knock Sensor Connector Pinout**

## MANIFOLD ABSOLUTE PRESSURE (MAP) SENSOR

### Description & Operation

The MAP sensor mounts to the intake manifold. The MAP serves as a PCM input, using a silicon based sensing unit, to provide data on the manifold vacuum that draws the air/fuel mixture into the combustion chamber. The PCM requires this information to determine injector pulse width and spark advance. When MAP equals Barometric pressure, the pulse width will be at maximum.

Also like the cam and crank sensors, a 5-volt reference is supplied from the PCM and returns a voltage signal to the PCM that reflects manifold pressure. The zero pressure reading is 0.5V and full scale is 4.5v. For a pressure swing of 0 - 15 psi the voltage changes 4.0v. The sensor is supplied a regulated 4.8 to 5.1 volts to operate the sensor. Like the cam and crank sensors ground is provided through the sensor return circuit.

The MAP sensor input is the number one contributor to pulse width. The most important function of the MAP sensor is to determine barometric pressure. The PCM needs to know if the vehicle is at sea level or is it in Denver at 5000 feet above sea level, because the air density changes with altitude. It will also help to correct for varying weather conditions. If a hurricane was coming through the pressure would be very, very low or there could be a real fair weather, high-pressure area. This is important because as air pressure changes the barometric pressure changes. Barometric pressure and altitude have a direct inverse correlation, as altitude goes up barometric goes down. The first thing that happens as the key is rolled on, before reaching the crank position, the PCM powers up, comes around and looks at the MAP voltage, and based upon the voltage it sees, it knows the current barometric pressure relative to altitude. Once the engine starts, the PCM looks at the voltage again, continuously every 12 milliseconds, and compares the current voltage to what it was at key on. The difference between current and what it was at key on is manifold vacuum.

During key ON (engine not running) the sensor reads (updates) barometric pressure. A normal range can be obtained by monitoring known good sensor in you work area.

As the altitude increases the air becomes thinner (less oxygen) If a vehicle is started and driven to a very different altitude than where it was at key ON the barometric pressure needs to be updated. Any time the PCM sees Wide Open throttle, based upon TPS angle and RPM it will update barometric pressure in the MAP memory cell. With periodic updates, the PCM can make its calculations more effectively.

The PCM uses the MAP sensor to aid in calculating the following:

- Barometric pressure

- Engine load
- Manifold pressure
- Injector pulse-width
- Spark-advance programs
- Shift-point strategies (F4AC1 transmissions only, via the CCD bus)
- Idle speed
- Decel fuel shutoff

The MAP sensor signal is provided from a single piezo-resistive element located in the center of a diaphragm. The element and diaphragm are both made of silicone. As the pressures changes the diaphragm moves causing the element to deflect which stresses the silicone. When silicone is exposed to stress its resistance changes. As manifold vacuum increases, the MAP sensor input voltage decreases proportionally. The sensor also contains electronics that condition the signal and provide temperature compensation.

The PCM recognizes a decrease in manifold pressure by monitoring a decrease in voltage from the reading stored in the barometric pressure memory cell. The MAP sensor is a linear sensor; as pressure changes, voltage changes proportionately. The range of voltage output from the sensor is usually between 4.6 volts at sea level to as low as 0.3 volts at 26 in. of Hg. Barometric pressure is the pressure exerted by the atmosphere upon an object. At sea level on a standard day, no storm, barometric pressure is 29.92 in Hg. For every 100 feet of altitude barometric pressure drops 10 in. Hg. You should make a habit of knowing what the average pressure and corresponding barometric pressure is for your area. Always use Diagnostic Test Procedures for MAP sensor Testing.

### Removal & Installation

*1.6L*

1. Remove the air cleaner cover.
2. Disconnect the negative battery cable.
3. Disconnect the electrical connector from the MAP sensor.
4. Remove the screws from the MAP sensor.
5. Remove the MAP sensor.

**To install:**

6. Make sure that the manifold is clean.
7. Install sensor to manifold.
8. Tighten screws.
9. Connect the electrical connector to the sensor.
10. Connect the negative battery cable
11. Install the air cleaner cover.

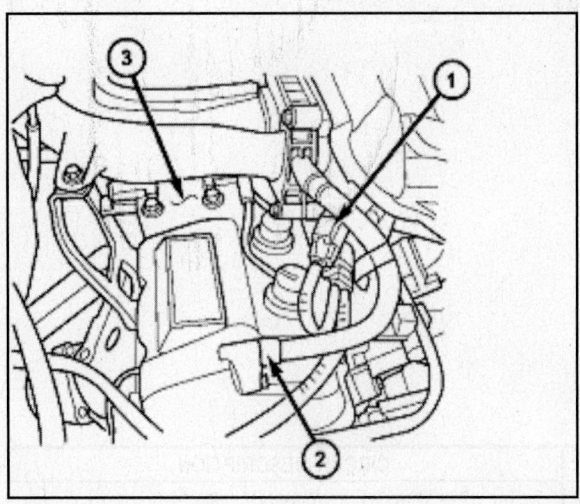

**MAP sensor (1) location on 1.6L engine; also showing PCV valve (2) and ignition coil (3)**

*2.0L & 2.4L*

1. Remove the air cleaner lid and makeup air hose.
2. Remove the negative battery cable.
3. Disconnect the electrical connector from the MAP sensor.
4. Remove sensor mounting screws.
5. Remove sensor.

**To install:**

6. Insert sensor into intake manifold while making sure not to damage O-ring seal.
7. Tighten mounting screws to 2 Nm (20 inch lbs.) torque for plastic manifold.
8. Attach electrical connector to sensor.
9. Install the negative battery cable.
10. Install the air cleaner lid and makeup air hose.

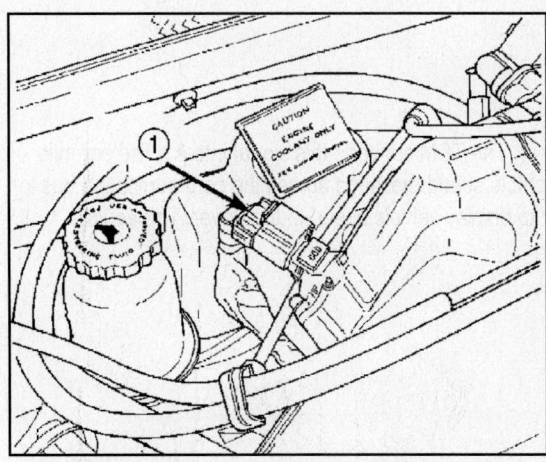

**Locating the MAP Sensor (1) on PT Cruiser 2.0L & 2.4L engines**

## NATURAL VACUUM LEAK DETECTION (NVLD)

**Connector Pinouts**

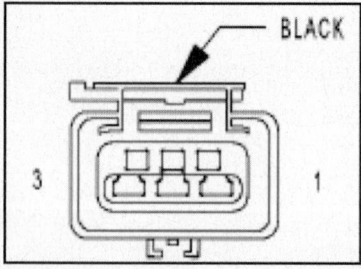

| PIN # | WIRE COLOR | CIRCUIT DESCRIPTION |
|-------|-----------|---------------------|
| 1 | BK | Ground |
| 2 | OR/YL | NVLD Switch Signal |
| 3 | WT/DG | NVLD Solenoid Control |

**NVLD Connector Pinout**

## OUTPUT SPEED SENSOR

**Description & Operation**

The Output Speed Sensor is a two-wire magnetic pickup device that generates an AC signal as rotation occurs. It is threaded into the transaxle case, sealed with an O-ring, and is considered a primary input to the Transmission Control Module (TCM)

The Output Speed Sensor provides information on how fast the output shaft is rotating. As the rear planetary carrier park pawl lugs pass by the sensor coil, an AC voltage is generated and sent to the TCM. The TCM interprets this information as output shaft rpm

The TCM compares the input and output speed signals to determine the following:

- Transmission gear ratio
- Speed ratio error detection
- CVI calculation

**VEHICLE SPEED SIGNAL**

The vehicle speed signal is taken from the Output Speed Sensor. The TCM converts this signal into a pulse per mile signal and sends it to the PCM. The PCM, in turn, sends the vehicle speed message across the communication bus to the BCM. The BCM sends this signal to the Instrument Cluster to display vehicle speed to the driver. The vehicle speed signal pulse is roughly 8000 pulses per mile.

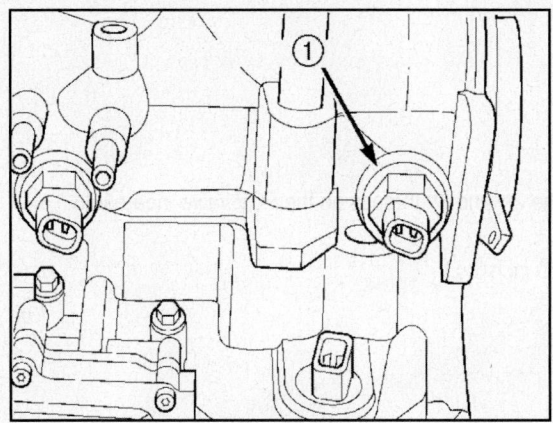

**Location of Output Speed Sensor (1) on PT Cruiser engines**

## Removal & Installation

*CAUTION: When disconnecting speed sensor connector, be sure that the connector weather seal does not fall off or remain in old sensor.*

1.  Disconnect the battery cables.
2.  Remove air cleaner assembly.
3.  Remove the battery hold down clamp and remove the battery.
4.  Remove the battery tray.
5.  Disconnect output speed sensor connector.
6.  Unscrew and remove output speed sensor.
7.  Inspect speed sensor O-ring and replace if necessary.

### To install:

8.  Verify O-ring is installed into position.
9.  Install and tighten output speed sensor to 27 Nm (20 ft. lbs.)
10.  Connect speed sensor connector.
11.  Install the battery tray.
12.  Install the battery and hold down clamp.
13.  Install air cleaner assembly.

## POSITIVE CRANKCASE VENTILATION (PCV) VALVE

### Description & Operation

The PCV valve contains a spring-loaded plunger. The plunger meters the amount of crankcase vapors routed into the combustion chamber based on intake manifold vacuum. When the engine is not operating or during an engine backfire, the spring forces the plunger back against the seat. This prevents vapors from flowing through the valve.

When the engine is at idle or cruising, high manifold vacuum is present. At these times manifold vacuum is able to completely compress the spring and pull the plunger to the top of the valve. In this position there is minimal vapor flow through the valve

During periods of moderate intake manifold vacuum the plunger is only pulled part way back from the inlet. This results in maximum vapor flow through the valve.

**Removal & Installation**

1. Remove hose to PCV valve.
2. Remove PCV valve.

**To install:**

3. Apply Mopar® Teflon Thread Sealant to the 2nd and 3rd threads on the PCV valve, nearest to the chamfered end.
4. Install PCV valve and tighten to 7.9 Nm (70 inch lbs.)
5. Connect hose.

## POWER STEERING FLUID PRESSURE SWITCH

**Description & Operation**

A power steering pressure switch is used to improve the vehicle's idle quality. The pressure switch improves vehicle idle quality by causing a readjustment of the engine idle speed as necessary when increased fluid pressure is sensed in the power steering system. The power steering pressure switch is mounted directly to the power steering gear.

The pressure switch functions by signaling the powertrain control module that an increase in pressure of the power steering system is putting additional load on the engine. This type of condition exists when the front tires of the vehicle are turned while the vehicle is stationary and the engine is at idle speed. When the powertrain control module receives the signal from the power steering pressure switch, it directs the engine to increase its idle speed. This increase in engine idle speed compensates for the additional load, thus maintaining the required engine idle speed and idle quality.

**Removal & Installation**

1. Disconnect negative battery cable from the negative post of the battery. Be sure cable is isolated from negative post on battery.
2. Raise the vehicle.
3. Locate the power steering fluid pressure switch on the backside of the power steering gear.
4. Remove the vehicle wiring harness connector from the power steering fluid pressure switch.
5. Using a 7/8 in. deep-well socket unscrew and remove the power steering fluid pressure switch from the power steering gear.

**To install:**

6. By hand, screw the power steering pressure switch into the power steering gear until it is fully seated.
7. Using a 7/8 in. deep-well socket, tighten the power steering pressure switch to a maximum torque of 8 Nm (70 inch lbs.) Over-torquing will result in stripping the threads out of the power steering pressure switch port in the steering gear.
8. Install the vehicle wiring harness connector. Be sure the latch on the wiring harness connector is fully engaged with the locking tab on the power steering pressure switch.
9. Lower the vehicle.
10. Fill the power steering fluid reservoir to the correct fluid level. Use only Mopar® Power Steering Fluid, or approved equivalent.
11. Connect the negative cable to the negative post of the battery.
12. Start the engine and turn the steering wheel several times stop-to-stop to bleed any air from the fluid in the power steering system. Stop the engine, check the fluid level, and inspect the system for leaks.

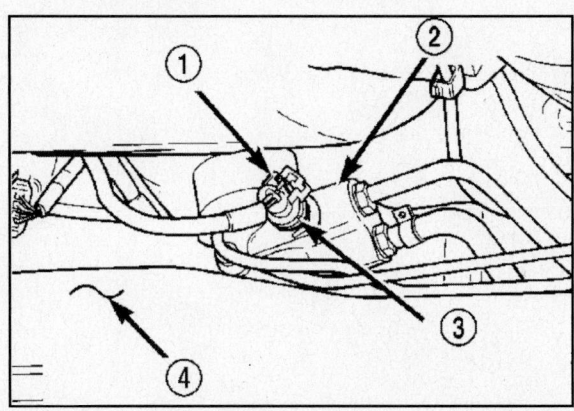

**Showing location of PSP switch wiring harness connector (1), power steering gear (2) and PSP Switch (3) near rear of front crossmember (4) on PT Cruiser**

POWERTRAIN CONTROL MODULE (PCM)

### Description & Operation

The Powertrain Control Module (PCM) is a digital computer containing a microprocessor.

The PCM receives input signals from various switches and sensors that are referred to as PCM Inputs. Based on these inputs, the PCM adjusts various engine and vehicle operations through devices that are referred to as PCM Outputs.

*PCM INPUTS:*

- Air Conditioning Controls
- Battery Voltage
- Battery Temperature Sensor
- Brake Switch
- Camshaft Position Sensor
- Crankshaft Position Sensor
- CCD Bus
- Engine Coolant Temperature Sensor
- Fuel Level Sensor
- Ignition Switch
- Intake Air Temperature Sensor
- Knock Sensor (2.0/2.4L only)
- Leak Detection Pump
- Manifold Absolute Pressure (MAP) Sensor
- Oxygen Sensors
- Power Steering Pressure Switch
- SCI Receive
- Speed Control Switches
- Throttle Position Sensor
- Transmission Park/Neutral Switch (automatic transmission)
- Vehicle Speed Sensor

*PCM OUTPUTS:*

- Air Conditioning Clutch Relay
- Auto Shutdown (ASD) Relay
- Charging Indicator Lamp

- CCD Bus
- SCI Transmit
- Proportional Purge Solenoid
- EGR Solenoid
- Fuel Injectors
- Fuel Pump Relay
- Generator Field
- Idle Air Control Motor
- Ignition Coils
- Malfunction Indicator (Check Engine) Lamp
- Radiator Fan Relays
- Speed Control Solenoids

Based on inputs it receives, the PCM adjusts fuel injector pulse width, idle speed, ignition spark advance, ignition coil dwell and EVAP canister purge operation. The PCM regulates the cooling fan, air conditioning and speed control systems. The PCM changes generator charge rate by adjusting the generator field. The PCM also performs diagnostics.

The PCM adjusts injector pulse width (air-fuel ratio) based on the following inputs:

- Battery voltage
- Coolant temperature
- Exhaust gas content (oxygen sensor)
- Engine speed (crankshaft position sensor)
- Intake air temperature
- Manifold absolute pressure
- Throttle position

The PCM adjusts ignition timing based on the following inputs:

- Coolant temperature
- Engine speed (crankshaft position sensor)
- Knock sensor
- Manifold absolute pressure
- Throttle position
- Transmission gear selection (park/neutral switch)
- Intake air temperature

The PCM also adjusts engine idle speed through the idle air control motor based on the following inputs:

- Air conditioning sense
- Battery voltage
- Battery temperature
- Brake switch
- Coolant temperature
- Engine speed (crankshaft position sensor)
- Engine run time
- Manifold absolute pressure
- Power steering pressure switch
- Throttle position

- Transmission gear selection (park/neutral switch)
- Vehicle distance (speed)

The Auto Shutdown (ASD) and fuel pump relays are located in the Power Distribution Center (PDC).

The camshaft position sensor and crankshaft position sensor signals are sent to the PCM. If the PCM does not receive the signal within approximately 1 second of engine cranking, it deactivates the ASD relay and fuel pump relay. When these relays are deactivated, power is shut off from the fuel injectors, ignition coils, oxygen sensor heating elements and fuel pump.

The PCM contains a voltage converter that changes battery voltage to a regulated 9 volts direct current to power the camshaft position sensor, crankshaft position sensor and vehicle speed sensor. The PCM also provides a 5-volt direct current supply for the manifold absolute pressure sensor, throttle position sensor, and A/C pressure switch.

### PCM GROUND

Ground is provided through multiple pins of the PCM connector. Depending on the vehicle there may be as many as three different ground pins. There are power grounds and sensor grounds. The power grounds are used to control the ground side of any relay, solenoid, ignition coil or injector. The signal ground is used for any input that uses sensor return for ground, and the ground side of any internal processing component.

The SBEC III case is shielded to prevent RFI and EMI. The PCM case is grounded and must be firmly attached to a good, clean body ground. Internally all grounds are connected together, however there is noise suppression on the sensor ground. For EMI and RFI protection the case is also grounded separately from the ground pins.

**NOTE: Use the proper scan tool to reprogram the new PCM with the vehicle's original identification number (VIN) and the vehicle's original mileage. If this step is not done a diagnostic trouble code (DTC) may be set.**

The PCM engine control strategy prevents reduced idle speeds until after the engine operates for 200 miles If the PCM is replaced after 200 miles of usage, update the mileage and vehicle identification number (VIN) in the new PCM. Use the DRB scan tool to change the mileage and VIN in the PCM. If this step is not done a Diagnostic Trouble Code (DTC) may be set. The PCM attaches to a bracket welded on the dash panel.

### Removal & Installation

1. Remove the air cleaner lid, disconnect the inlet air temperature sensor and makeup air hose.
2. Remove the negative battery cable.
3. Remove screws attaching PCM to dash panel bracket.
4. Unlock the connector then disconnect both 40-way connectors from the PCM.
5. Lift PCM up to remove it from vehicle.

### To install:

6. Install PCM. Tighten mounting screws to 10.7 Nm (95 inch lbs.).
7. Attach both 40-way connectors to PCM. Lock the tabs.
8. Install the negative battery cable.
9. Install the air cleaner lid; connect the inlet air temperature sensor and makeup air hose.

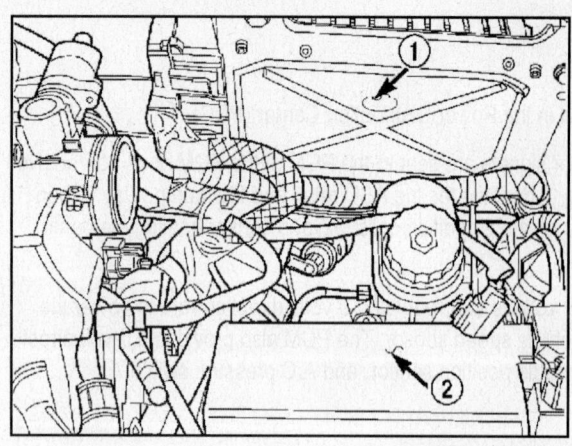

**Identifying location of PCM (1) and Power Distribution Center (2) on PT Cruiser**

## POWER DISTRIBUTION CENTER – RELAY LOCATIONS

**Connector Pinouts**

*NOTE: For location of specific relay in Power Distribution Center (PDC), see label on PDC lid.*

| PIN # | CIRCUIT | FUNCTION |
|-------|---------|----------|
| 80 | K51 20DB/VT | AUTOMATIC SHUT DOWN RELAY CONTROL |
| 81 | A14 14RD/TN | FUSED B (+) |
| 82 | A142 14DG/OR | AUTOMATIC SHUT DOWN RELAY OUTPUT |
| 83 | - | - |
| 84 | A14 14RD/TN | FUSED B (+) |

**Auto Shutdown Relay Connector Pinout**

| PIN # | CIRCUIT | FUNCTION |
|-------|---------|----------|
| 12 | K31 20BR/LG | FUEL PUMP RELAY CONTROL |
| 13 | - | - |
| 14 | F12 18DB/WT | FUSED IGNITION SWITCH OUTPUT (RUN-START) |
| 15 | A141 14DG/WT | FUEL PUMP RELAY OUTPUT |
| 16 | A1 18RD | FUSED B (+) |

**Fuel Pump Relay Connector Pinout**

SPEED CONTROL SWITCH

**Removal & Installation**

The speed control switches is mounted in the steering wheel and wired through the clock spring device under the airbag module.

*WARNING: If removal & installation of airbag module is necessary, follow proper instructions and precautions.*

1. Remove the air cleaner lid; disconnect the inlet air temperature sensor and makeup air hose.
2. Remove the negative battery cable.
3. Turn off ignition.
4. Remove air bag.
5. Remove the top mounting screw.
6. Rotate steering wheel so that the switch is in the 6 o'clock position. Remove 2 screws from the backside of the speed control switch.
7. Disconnect the electrical connector.
8. Remove switch.

**To install:**

9. The speed control switch is mounted in the steering wheel and wired through the clock spring device under the airbag module.
10. Connect the electrical connector.
11. Install switch and tighten the screws to 1.6 Nm (15 inch lbs.). Make sure rubber seal is in place around switch.
12. Install airbag.
13. Install the negative battery cable.
14. Install the air cleaner lid; connect the inlet air temperature sensor and makeup air hose.

SENTRY KEY IMMOBILIZER MODULE (SKIM)

**Description & Operation**

The Sentry Key Immobilizer Module (SKIM) contains a Radio Frequency (RF) transceiver and a central processing unit, which includes the Sentry Key Immobilizer System (SKIS) program logic. The SKIS programming enables the SKIM to program and retain in memory the codes of at least two, but no more than eight electronically coded Sentry Key transponders. The SKIS programming also enables the SKIM to communicate over the Programmable Communication Interface (PCI) bus network with the Powertrain Control Module (PCM), and/or the DRB III® scan tool.

The SKIM transmits and receives RF signals through a tuned antenna enclosed within a molded plastic ring formation that is integral to the SKIM housing. When the SKIM is properly installed on the steering column, the antenna ring is oriented around the circumference of the ignition lock cylinder housing. This antenna ring must be located within eight millimeters (0.31 in.) of the Sentry Key in order to ensure proper RF communication between the SKIM and the Sentry Key transponder.

For added system security, each SKIM is programmed with a unique "Secret Key" code and a security code. The SKIM keeps the "Secret Key" code in memory. The SKIM also sends the "Secret Key" code to each of the programmed Sentry Key transponders. The security code is used by the assembly plant to access the SKIS for initialization, or by the dealer technician to access the system for service. The SKIM also stores in its memory the Vehicle Identification Number (VIN), which it learns through a PCI bus message from the PCM during initialization.

The SKIM and the PCM both use software that includes a rolling code algorithm strategy, which helps to reduce the possibility of unauthorized SKIS disarming. The rolling code algorithm ensures security by preventing an override of the

SKIS through the unauthorized substitution of the SKIM or the PCM. However, the use of this strategy also means that replacement of either the SKIM or the PCM units will require a system initialization procedure to restore system operation.

When the ignition switch is turned to the ON or START positions, the SKIM transmits an RF signal to excite the Sentry Key transponder. The SKIM then listens for a return RF signal from the transponder of the Sentry Key that is inserted in the ignition lock cylinder. If the SKIM receives an RF signal with valid "Secret Key" and transponder identification codes, the SKIM sends a "valid key" message to the PCM over the PCI bus. If the SKIM receives an invalid RF signal or no response, it sends "invalid key" messages to the PCM. The PCM will enable or disable engine operation based upon the status of the SKIM messages.

The SKIM also sends messages to the Instrument Cluster that controls the VTSS indicator LED. The SKIM sends messages to the Instrument Cluster to turn the LED on for about three seconds when the ignition switch is turned to the ON position as a bulb test. After completion of the bulb test, the SKIM sends bus messages to keep the LED off for a duration of about one second. Then the SKIM sends messages to turn the LED on or off based upon the results of the SKIS self-tests. If the VTSS indicator LED comes on and stays on after the bulb test, it indicates that the SKIM has detected a system malfunction and/or that the SKIS has become inoperative.

If the SKIM detects an invalid key when the ignition switch is turned to the ON position, it sends messages to flash the VTSS indicator LED. The SKIM can also send messages to flash the LED and to generate a single audible chime tone. These functions serve as an indication to the customer that the SKIS has been placed in its "Customer Learn" programming mode. See Sentry Key Immobilizer System Transponder Programming in this section for more information on the "Customer Learn" programming mode.

*NOTE: For diagnosis or initialization of the SKIM and the PCM, a proper scan tool and the proper Powertrain Diagnostic Procedures are required. The SKIM cannot be repaired and, if faulty or damaged, the unit must be replaced.*

*WARNING: On vehicles equipped with airbags, check all information and precautions on restraint systems before attempting any steering wheel, steering column, or instrument panel component diagnosis or service. Failure to take the proper precautions could result in accidental airbag deployment and possible personal injury.*

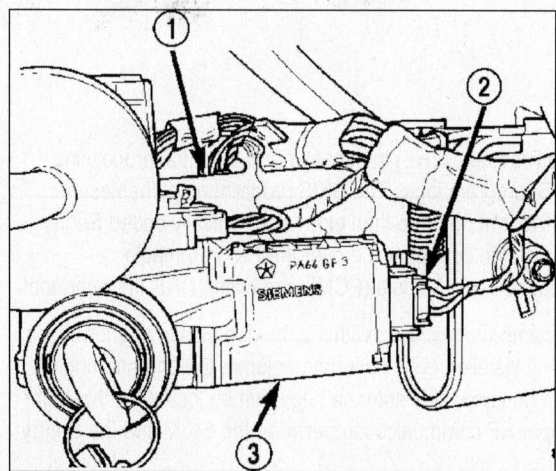

**Showing steering column (1) location of SKIM electrical connector (2) & SKIM (3)**

### Removal & Installation

1. Open hood.
2. Disconnect and isolate the battery negative cable.
3. Remove the steering column upper and lower shrouds.
4. Disconnect the steering column wire harness connector from the SKIM.

5. Remove the two screws securing the SKIM module to the top of the steering column.

6. Rotate the SKIM and its mounting bracket upwards and then to the side away from the steering column to slide the SKIM antenna ring from around the ignition switch lock cylinder housing.

7. Remove the SKIM from the vehicle.

**To install:**

**NOTE: If the SKIM is replaced with a new unit, a proper scan tool MUST be used to initialize the new SKIM and to program at least 2 Sentry Key transponders.**

8. Rotate the SKIM and its mounting bracket to the side towards the steering column to slide the SKIM antenna ring around the ignition switch lock cylinder housing and then downwards.

9. Install the two screws securing the SKIM module to the top of the steering column.

10. Connect the steering column wire harness connector to the SKIM.

11. Install the steering column upper and lower shrouds.

12. Connect the battery negative cable.

13. Verify vehicle and system operation.

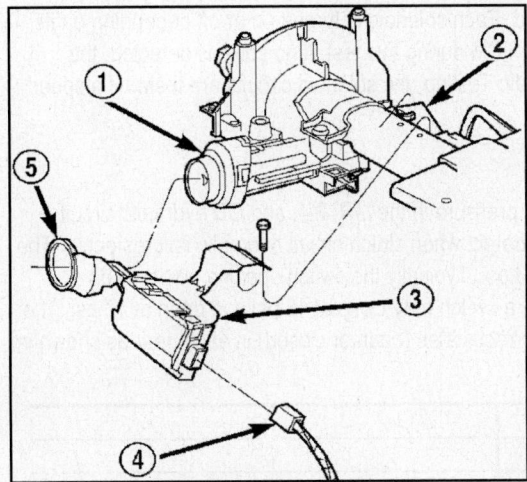

**Removing SKIM (3) from ignition switch (1) and steering column (2) after disconnecting electrical connector (4), removing mounting bolt and sliding SKIM antenna (5) from ignition switch**

## SOLENOID/PRESSURE SWITCH ASSEMBLY

### Description

The Solenoid/Pressure Switch Assembly is external to the transaxle and mounted to the transaxle case. The assembly consists of four solenoids that control hydraulic pressure to the LR/CC, 2/4, OD, and UD friction elements. The reverse clutch is controlled by line pressure from the manual valve in the valve body. The solenoids are contained within the Solenoid/Pressure Switch Assembly, and can only be serviced by replacing the assembly.

The solenoid assembly also contains pressure switches that monitor and send hydraulic circuit information to the PCM/TCM. Likewise, the pressure switches can only be service by replacing the assembly.

### Operation

#### SOLENOIDS

The solenoids receive electrical power from the Transmission Control Relay through a single wire. The PCM/TCM energizes or operates the solenoids individually by grounding the return wire of the solenoid needed. When a solenoid is energized, the solenoid valve shifts, and a fluid passage is opened or closed (vented or applied), depending on its default operating state. The result is application or release of a frictional element.

The 2/4 and UD solenoids are normally applied, which by design allow fluid to pass through in their relaxed or "off" state. This allows transaxle limp-in (P,R,N,2) in the event of an electrical failure.

The continuity of the solenoids and circuits are periodically tested. Each solenoid is turned on or off depending on its current state. An inductive spike should be detected by the PCM/TCM during this test. It no spike is detected, the circuit is tested again to verify the failure. In addition to the periodic Testing, the solenoid circuits are tested if a speed ratio or pressure switch error occurs.

#### PRESSURE SWITCHES

The PCM/TCM relies on three pressure switches to monitor fluid pressure in the L/R, 2/4, and OD hydraulic circuits. The primary purpose of these switches is to help the PCM/TCM detect when clutch circuit hydraulic failures occur. The range for the pressure switch closing and opening points is 11-23 psi. Typically the switch opening point will be approximately one psi lower than the closing point. For example, a switch may close at 18 psi and open at 17 psi. The switches are continuously monitored by the PCM/TCM for the correct states (open or closed) in each gear as shown in the following chart:

| GEAR | L/R | 2/4 | OD |
|------|-----|-----|-----|
| R | OP | OP | OP |
| P/N | CL | OP | OP |
| 1st | CL | OP | OP |
| 2nd | OP | CL | OP |
| D | OP | OP | CL |
| OD | OP | CL | CL |
| OP = OPEN | | | |
| CL = CLOSED | | | |

### Indicating pressure switch states in various operating modes

A Diagnostic Trouble Code (DTC) will set if the PCM/TCM senses any switch open or closed at the wrong time in a given gear.

The PCM/TCM also tests the 2/4 and OD pressure switches when they are normally off (OD and 2/4 are tested in 1st gear, OD in 2nd gear, and 2/4 in 3rd gear). The test simply verifies that they are operational, by looking for a closed state when the corresponding element is applied. Immediately after a shift into 1st, 2nd, or 3rd gear with the engine speed above 1000 rpm, the PCM/TCM momentarily turns on element pressure to the 2/4 and/or OD clutch circuits to identify that the appropriate switch has closed. If it doesn't close, it is tested again. If the switch fails to close the second time, the appropriate Diagnostic Trouble Code (DTC) will set.

### Removal & Installation

1. Disconnect the battery cables.
2. Remove air cleaner assembly.
3. Remove the battery hold down clamp and remove the battery.
4. Remove the battery tray.
5. Disconnect and remove the input speed sensor.
6. Disconnect the transmission oil cooler lines. Cap off hoses and fittings to prevent foreign matter intrusion.
7. Disconnect the solenoid/pressure switch assembly connector.
8. Remove the 3 solenoid/pressure switch assembly-to-transaxle case bolts.
9. Remove solenoid/pressure switch assembly and gasket.

*CAUTION: Be sure to keep foreign material from entering ports in transaxle case. Erratic transaxle operation and/or failure can result.*

### To install:

10. Install solenoid/pressure switch assembly to case using a new gasket.
11. Install bolts and torque to 13 Nm (110 inch lbs.).
12. Install 8-way connector and torque screw to 4 Nm (35 inch lbs.).
13. Uncap and install transmission oil cooler lines.
14. Install input speed sensor and torque to 27 Nm (20 ft. lbs.).
15. Install the battery tray.
16. Install the battery and hold down clamp.
17. Install air cleaner assembly.
18. Connect battery cables.
19. Perform Transaxle Quick Learn Procedure.

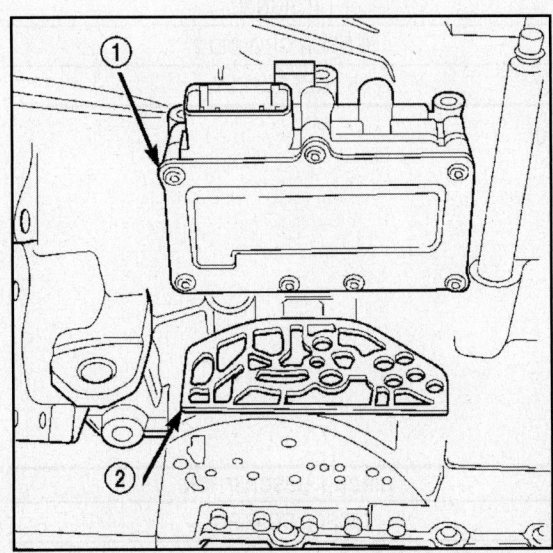

**Showing the Solenoid/Pressure Switch (1) and gasket (2)**

## SURGE SOLENOID (TURBO)

**Connector Pinouts**

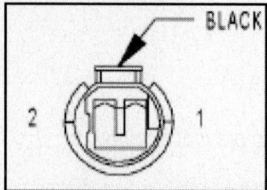

| PIN # | WIRE COLOR | CIRCUIT DESCRIPTION |
|-------|------------|---------------------|
| 1 | DB/YL | SURGE SOL CONTROL |
| 2 | DG/OR | AUTOMATIC SHUT DOWN RELAY OUTPUT |

**2.4L Turbo Surge Solenoid Connector Pinout**

## THROTTLE INLET PRESSURE SENSOR & SOLENOID

**Connector Pinouts**

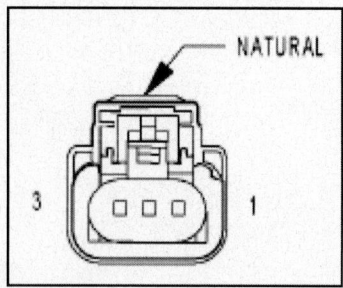

| PIN # | WIRE COLOR | CIRCUIT DESCRIPTION |
|-------|------------|---------------------|
| 1 | DB/LG | TIP SIGNAL |
| 2 | BR/YL | SENSOR GROUND 2 |
| 3 | VT/WT | 5 VOLT SUPPLY |

**2.4L Turbo Throttle Inlet Pressure Sensor Connector Pinout**

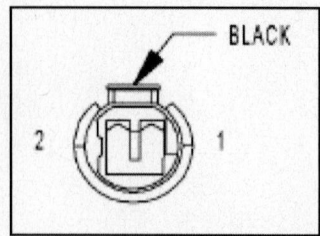

| PIN # | WIRE COLOR | CIRCUIT DESCRIPTION |
|-------|------------|---------------------|
| 1 | LB | TIP SOL CONTROL |
| 2 | DB/WT | FUSED IGNITION SWITCH OUTPUT (RUN-START) |

**2.4L Turbo Throttle Inlet Pressure Solenoid Connector Pinout**

## THROTTLE POSITION (TP) SENSOR

### Description & Operation

The throttle position sensor mounts to the side of the throttle body. The Throttle Position Sensor (TPS) connects to the throttle blade shaft. The TPS is a variable resistor that provides the PCM with an input signal (voltage) The signal represents throttle blade position. As the position of the throttle blade changes, the resistance of the TPS changes.

A single sensor ground is used for all O2 sensors (2 sensors on 4-cyl. vehicles and 4 sensors on 6-cyl. vehicles).

As vehicles accumulate mileage, the catalytic converter deteriorates. The deterioration results in a less efficient catalyst. To monitor catalytic converter deterioration, the fuel injection system uses 2 heated oxygen sensors: one sensor upstream of the catalytic converter, one downstream of the converter. The PCM compares the reading from the sensors to calculate the catalytic converter oxygen storage capacity and converter efficiency. Also, the PCM uses the upstream heated oxygen sensor input when adjusting injector pulse width.

When the catalytic converter efficiency drops below emission standards, the PCM stores a diagnostic trouble code and illuminates the malfunction indicator lamp (MIL).

The O2 sensors produce voltages from 0 to 1 volt, depending upon the oxygen content of the exhaust gas. When a large amount of oxygen is present (caused by a lean air/fuel mixture, can be caused by misfire and exhaust leaks), the sensors produce a low voltage. When there is a lesser amount of oxygen present (caused by a rich air/fuel mixture, can be caused by internal engine problems) it produces a higher voltage. By monitoring the oxygen content and converting it to electrical voltage, the sensors act as a rich-lean switch.

The oxygen sensors are equipped with a heating element that keeps the sensors at proper operating temperature during all operating modes. Maintaining correct sensor temperature at all times allows the system to enter into closed loop operation sooner. Also, it allows the system to remain in closed loop operation during periods of extended idle.

In Closed Loop operation the PCM monitors the O2 sensors input (along with other inputs) and adjusts the injector pulse width accordingly. During Open Loop operation the PCM ignores the O2 sensor input. The PCM adjusts injector pulse width based on preprogrammed (fixed) values and inputs from other sensors.

The Automatic Shutdown (ASD) relay supplies battery voltage to both the upstream and downstream heated oxygen sensors. The oxygen sensors are equipped with a heating element. The heating elements reduce the time required for the sensors to reach operating temperature. The PCM uses pulse width modulation to control the ground side of the heater to regulate the temperature on 4-cyl. upstream O2 heater only. All other 4-cyl. and 6-cyl. O2 heaters do not use pulse width modulation.

### *UPSTREAM OXYGEN SENSOR*

The input from the upstream heated oxygen sensor tells the PCM the oxygen content of the exhaust gas. Based on this input, the PCM fine tunes the air-fuel ratio by adjusting injector pulse width.

The sensor input switches from 0 to 1 volt, depending upon the oxygen content of the exhaust gas in the exhaust manifold. When a large amount of oxygen is present (caused by a lean air-fuel mixture), the sensor produces voltage as low as 0.1 volt. When there is a lesser amount of oxygen present (rich air-fuel mixture) the sensor produces a voltage as high as 1.0 volt. By monitoring the oxygen content and converting it to electrical voltage, the sensor acts as a rich-lean switch.

The heating element in the sensor provides heat to the sensor ceramic element. Heating the sensor allows the system to enter into closed loop operation sooner. Also, it allows the system to remain in closed loop operation during periods of extended idle.

In Closed Loop, the PCM adjusts injector pulse width based on the upstream heated oxygen sensor input along with other inputs. In Open Loop, the PCM adjusts injector pulse width based on preprogrammed (fixed) values and inputs from other sensors.

### *DOWNSTREAM OXYGEN SENSOR*

The downstream heated oxygen sensor input is used to detect catalytic converter deterioration. As the converter deteriorates, the input from the downstream sensor begins to match the upstream sensor input except for a slight time delay. By comparing the downstream heated oxygen sensor input to the input from the upstream sensor, the PCM calculates catalytic converter efficiency. Also used to establish the upstream O2 goal voltage (switching point).

The PCM supplies approximately 5 volts DC to the TPS. The TPS output voltage (input signal to the PCM) represents throttle blade position. The TPS output voltage to the PCM varies from approximately 0.5 volt at minimum throttle opening (idle) to a maximum of 4.5 volts at wide-open throttle.

Along with inputs from other sensors, the PCM uses TPS input to determine current engine operating conditions. The PCM also adjusts fuel injector pulse width and ignition timing based on these inputs.

### Removal & Installation

1. Remove the air cleaner lid and makeup air hose. Loosen the clamp and relocate assembly.
2. Remove the negative battery cable.
3. Disconnect electrical connector from the throttle position sensor.
4. Remove throttle position sensor mounting screws.
5. Remove throttle position sensor.

### To install:

6. The throttle shaft end of the throttle body slides into a socket in the TPS. The socket has 2 tabs inside it. The throttle shaft rests against the tabs. When indexed correctly, the TPS can rotate clockwise a few degrees to line up the mounting screw holes with the screw holes in the throttle body. The TPS has slight tension when rotated into position. If it is difficult to rotate the TPS into position, reinstall the sensor with the throttle shaft on the other side of the tabs in the socket of the TPS. Tighten mounting screws to 6.2 Nm (55 inch lbs.) torque.
7. After installing the TPS, the throttle plate should be closed. If the throttle plate is open, install the sensor on the other side of the tabs in the socket.
8. Attach electrical connectors to the throttle position sensor.
9. Install the negative battery cable.
10. Install the air cleaner lid and makeup air hose and tighten clamp.

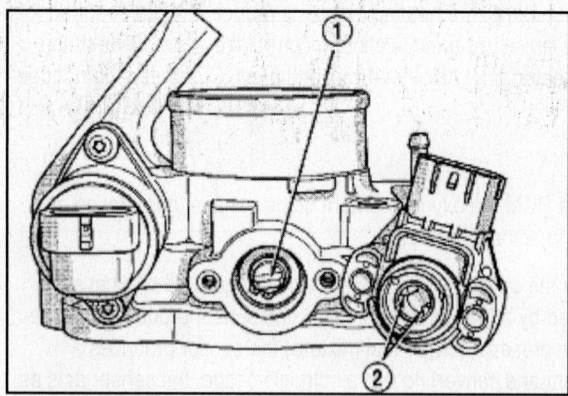

**Installing the TPS, noting the position of the locating tabs (2) on the throttle shaft (1)**

VEHICLE SPEED CONTROL

**Description**

Interactive Speed Control (used on A/T only) means that communication between the PCM and the TCM is taking place. Interactive speed control avoids unnecessary shifting for smoother, quieter operation and when downshifts are required, makes the shifts smoother.

When climbing a grade the interactive speed control tries to maintain the set speed by increasing the throttle opening, while inability/delaying downshifts.

If opening the throttle alone cannot maintain the set speed and the vehicle speed drops more than 3 mph below the set speed, the transmission will downshift to third gear. If the vehicle continues to lose speed, by more than 6 mph, the transmission will downshift again maintain the set speed. After the vehicle encounters a less-steep grade, or has crested the grade (reduced the load on the powertrain) and can maintain the set speed at a reduced throttle position, the transmission will upshift, as appropriate, until the set speed can be maintained in Overdrive.

Downshift delay features have been added to reduce the number and frequency of downshifts when operating in hilly or mountainous country.

While operating, interactive speed control delays or avoids downshifts by allowing up to nearly wide open throttle without the TCM scheduling a downshift. If the interactive speed control is not engaged or if the throttle is manually overridden by the driver while interactive speed control is engaged, the downshift delay feature is not activated.

Torque converter lock and unlock shifts are not affected by the downshift delay feature and will occur at the same throttle angle at a given speed regardless of whether interactive speed control operates or not.

All vehicles equipped with a four speed automatic transmission have a grade hunting feature for the 2nd to 3rd gear upshift and the 3rd to Overdrive upshift.

The TCM identifies the powertrain loading conditions and selects the proper gear to maintain the current vehicle speed. Under moderate loading conditions the transaxle will stay in 3rd gear until the top of the grade is reached or the powertrain loading is reduced.

If powertrain loading is severe, the transaxle may shift into 2nd gear and remain there until powertrain loading is reduced, then a 2nd to 3rd gear upshift will be scheduled. Grade hunting features always operate regardless of whether or not the interactive speed control is engaged.

***NOTE: If the interactive speed control is not engaged and powertrain loading is not reduced, the driver may have to completely lift off of the throttle before an upshift will occur.***

If the driver does lift off the throttle to induce an upshift under these conditions, vehicle speed will reduce and the Overdrive to 3rd and 3rd to 2nd gear downshifts will reoccur when the throttle is reapplied. If grade-hunting is repeatedly induced by the driver, transaxle damage may result.

Transmission control software includes an automatic speed control overspeed reduction feature. This maintains vehicle speed at the selected set point when descending a grade.

The Transmission Control Module (TCM) first senses that the speed control is set. If the set speed is exceeded by more than 4 mph (6.5 km/hr) and the throttle is closed, the TCM causes the transaxle to downshift to THIRD gear. After downshifting, the automatic speed control resumes normal operation. To ensure that an upshift is appropriate after the set speed is reached, the TCM waits until the speed control system opens the throttle at least 6 degrees before upshifting to OVERDRIVE again.

If the driver applies the brakes, canceling automatic speed control operation with the transaxle still in THIRD gear, the TCM maintains this gear until the driver opens the throttle at least 6 degrees to avoid an inappropriate upshift. The upshift is also delayed for 2.5 seconds after reaching the 6 degrees throttle opening in anticipation that the driver might open the throttle enough to require THIRD gear. This will avoid unnecessary and disturbing transmission cycling. If the automatic speed control RESUME feature is used after braking, the upshift is delayed until the set speed is achieved to reduce cycling and provide better response.

The speed control system is electronically controlled and vacuum operated. The electronic control is integrated into the Powertrain Control Module, located on the left side of the engine compartment next to the air cleaner. The controls are located on the steering wheel and consist of a single switch. The ON, OFF, RESUME, ACCEL, SET, COAST, and CANCEL, lever is located on the right of the steering wheel. The system is designed to operate at speeds above 25 mph (40 km/h)

**WARNING: THE USE OF SPEED CONTROL IS NOT RECOMMENDED WHEN DRIVING CONDITIONS DO NOT PERMIT MAINTAINING A CONSTANT SPEED, SUCH AS IN HEAVY TRAFFIC OR ON ROADS THAT ARE WINDING, ICY, SNOW COVERED, OR SLIPPERY.**

## Operation

When speed control is activated by depressing the ON switch, the PCM allows a set speed to be stored in RAM for speed control. To store a set speed, depress the SET switch while the vehicle is moving at a speed between 25 and 85 mph. In order for the speed control to engage, the brakes cannot be applied, nor can the gear selector be indicating the transmission is in Park or Neutral.

The speed control can be disengaged manually by:

- Stepping on the brake pedal
- Depressing the OFF switch
- Depressing the CANCEL switch.
- Depressing the clutch pedal

**NOTE: Depressing the OFF switch or turning off the ignition switch will erase the set speed stored in the PCM.**

For added safety, the speed control system is programmed to disengaged for any of the following conditions:

- An indication of Park or Neutral
- A rapid increase rpm (indicates that the clutch has been disengaged)
- Excessive engine rpm (indicates that the transmission may be in a low gear)
- The speed signal increases at a rate of 10 mph per second (indicates that the co-efficient of friction between the road surface and tires is extremely low)
- The speed signal decreases at a rate of 10 mph per second (indicates that the vehicle may have decelerated at an extremely high rate)

Once the speed control has been disengaged, depressing the RESUME switch when speed is greater than 25 mph allows the vehicle to resume control to the target speed that was stored in the PCM.

While the speed control is engaged, the driver can increase the vehicle speed by depressing the ACCEL switch. The new target speed is stored in the PCM when the ACCEL switch is released. The PCM also has a "tap-up" feature in which target speed increases by 2 mph for each momentary switch activation of the ACCEL switch. The PCM also provides a means to decelerate to a new lower target speed without disengaging speed control. Depress and hold the COAST switch until the desired speed is reached, then release the switch.

### SPEED CONTROL SWITCHES

The switches are mounted on a stalk extending from the right side of the steering wheel hub. It rotates with the steering wheel. The switch cannot be repaired. If the switch fails, the entire switch module must be replaced.

The speed control system has five separate voltage inputs that are provided from a single multiplexed (MUX) switch that supply's different voltage inputs to the PCM. The switch names are: ON, OFF, SET, COAST, RESUME, ACCEL, and CANCEL. Based on conditions when the buttons or lever is pushed or pulled (and released), the five voltages ranges provided to the PCM result in the following functions: ON, OFF, SET, COAST, RESUME, ACCEL, and CANCEL.

All the functions are included, but they operation of the switch is different then most other applications. A push button in the end of the stalk toggles the system voltage between turning the speed control system on (for the first press) and off (for the second press) The word "Cruise" will then illuminate on the cluster. Pushing the stalk down sets the cruise to maintain the current vehicle speed and is set into memory. Holding the stalk down allows the vehicle to coast to a lower speed. Pushing the stalk up causes the vehicle to resume the previously set speed that is set in memory. Holding it in the up position causes the vehicle speed to increase until the switch is released or the maximum allowable set speed is reached. Pulling the stalk toward the steering wheel cancels the operation of the speed control, allowing the vehicle to coast, but retaining the set speed in memory. Moving the stalk up momentarily causes the set speed to in crease by 2 mph. Repeated movement of the stalk in the same direction has a cumulative effect on the set speed. The minimum set speed is reduced to 25 mph and a maximum set speed is 85 mph.

### VEHICLE SPEED SIGNAL – AUTOMATIC TRANSAXLES

On A/T vehicles, the transaxle control module (TCM) supplies the vehicle speed signal to the PCM based on the output shaft speed. The PCM sends a 5-volt signal to the TCM. The TCM switches this signal to a ground, and then opens the circuit at a rate of 8000 pulses per mile. When the PCM counts 8000 pulses, the PCM assumes the vehicle has traveled one mile. The output speed sensor is located on the side of the transaxle.

The speed and distance signals, along with a closed throttle signal from the TPS, determine if a closed throttle deceleration or normal idle condition (vehicle stopped) exists. Under deceleration conditions, the PCM adjusts the idle air control motor to maintain a desired MAP value. Under idle conditions, the PCM adjusts the idle air control motor to maintain a desired engine speed.

### VEHICLE SPEED SIGNAL – MANUAL TRANSAXLES

The vehicle speed sensor is located in the transmission extension housing on M/T vehicles. The sensor input is used by the PCM to determine vehicle speed and distance traveled and is a Hall-effect sensor.

The Hall-effect sensor generates 8 pulses per sensor revolution. These signals, in conjunction with a closed throttle signal from the throttle position sensor, indicate a closed throttle deceleration to the PCM. Under deceleration conditions, the PCM adjusts the Idle Air Control (IAC) motor to maintain a desired MAP value.

Like all Hall-effect sensors, the electronics of the sensor needs a power source. This power source is provided by the PCM. The sensor switches a 5-volt signal sent from the PCM from a ground to an open circuit. It is the same 9-volt power supply that is used by the CKP and CMP sensors.

When the vehicle is stopped at idle, a closed throttle signal is received by the PCM (but a speed sensor signal is not received) Under idle conditions, the PCM adjusts the IAC motor to maintain a desired engine speed.

The vehicle speed sensor signal is also used to operate the following functions or systems:

- Speedometer
- Speed control
- Daytime Running Lights (Canadian Vehicles only)

The VSS used on 3-speed automatic and manual transaxle vehicles. A pinion gear that is meshed with the right axle drive shaft mechanically drives this sensor. When the PCM counts 4000 pulses, the PCM assumes the vehicle has traveled one mile.

### Removal & Installation

**WARNING: IF Removal & Installation OF AIRBAG MODULE IS NECESSARY, REFER TO THE RESTRAINT SYSTEMS.**

1. Remove the air cleaner lid; disconnect the inlet air temperature sensor and makeup air hose.
2. Remove the negative battery cable.
3. Turn off ignition.
4. Remove air bag.
5. Remove the top mounting screw.
6. Rotate steering wheel so that the switch is in the 6 o'clock position. Remove 2 screws from the backside of the speed control switch.

7. Disconnect the electrical connector.

8. Remove switch.

**To install:**

9. Connect the electrical connector.

10. Install switch and tighten the screws to 1.6 Nm (15 inch lbs.) Make sure rubber seal is in place around switch.

11. Install airbag.

12. Install the negative battery cable.

13. Install the air cleaner lid, connect the inlet air temperature sensor and makeup air hose.

## Testing

### ROAD TEST

Perform a vehicle road test to verify reports of speed control system malfunction. The road test should include attention to the speedometer. Speedometer operation should be smooth and without flutter at all speeds.

Flutter in the speedometer indicates a problem that might cause surging in the speed control system. The cause of any speedometer problems should be corrected before proceeding.

If a road test verifies a surge following a set, or an inoperative system, and the speedometer operates properly, check for:

- A Diagnostic Trouble Code (DTC) If a DTC exists, conduct tests per the Powertrain Diagnostic Procedures service manual.

- A misadjusted brake (stop) lamp switch. This could also cause an intermittent problem.

- Loose or corroded electrical connections at the servo. Corrosion should be removed from electrical terminals and a light coating of multipurpose grease applied.

- Leaking vacuum reservoir.

- Loose or leaking vacuum hoses or connections.

- Defective one-way vacuum check valve.

- Secure attachment at both ends of the speed control servo cable.

- Smooth operation of throttle linkage and throttle body air valve.

- Conduct electrical test at PCM.

- Failed speed control servo. Do the servo vacuum test.

**CAUTION: When probing for voltage or continuity at electrical connectors, care must be taken not to damage connector, terminals or seals. If these components are damaged, intermittent or complete system failure may occur.**

### CHECKING FOR DIAGNOSTIC CODES

When trying to verify a speed control system electronic malfunction: Connect a DRB scan tool if available to the data link connector. The connector is located at left side of the steering column, and at lower edge of the panel.

A speed control malfunction may occur without a diagnostic code being indicated.

**NOTE: For further information and usage of the scan tool and a more complete list of Diagnostic Trouble Code and No Trouble Codes, refer to the section on DIAGNOSTIC TROUBLE CODES.**

## VEHICLE SPEED SENSOR – MANUAL TRANSMISSION

### Description

The Vehicle Speed Sensor (VSS) is a pulse generator mounted to an adapter near the transmission output shaft. The sensor is driven through the adapter by a speedometer pinion gear. The VSS pulse signal to the speedometer/odometer is monitored by the PCM speed control circuitry to determine vehicle speed and to maintain speed control set speed.

### Removal & Installation

1. Raise vehicle on hoist.
2. Disconnect the speed sensor connector.

*CAUTION: Clean area around speed sensor before removing to prevent dirt from entering the transaxle during speed sensor Removal & Installation.*

3. Remove speed sensor retaining bolt.
4. Remove speed sensor from transaxle.

*CAUTION: Carefully remove vehicle speed sensor so sensor drive gear does not fall into transaxle. Should it fall into the transaxle during sensor Removal & Installation, drive gear must be reattached to sensor.*

5. Remove speed sensor drive gear from speed sensor.

### To install:

6. Install pinion gear to speed sensor.
7. Using a NEW O-ring, install the speed sensor to the transaxle.
8. Install the bolt and torque to 7 Nm (60 inch lbs.).
9. Connect speed sensor connector.
10. Lower vehicle and road test to verify proper speedometer operation.

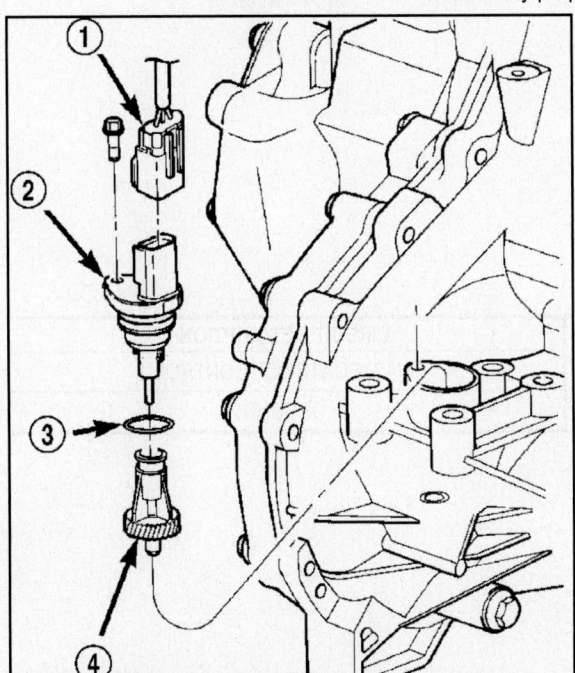

**Exploded view of removing the Vehicle Speed Sensor connector (1), VSS (2), O-ring (3) and pinion (4) from the manual transmission-equipped PT Cruiser**

**Connector Pinouts**

*GAS W/MTX*

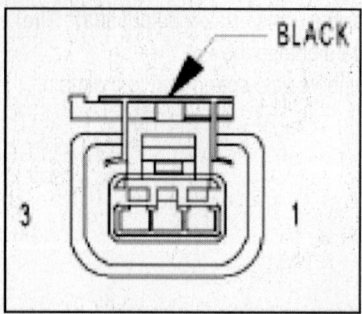

| PIN # | WIRE COLOR | CIRCUIT DESCRIPTION |
|---|---|---|
| 1 | OR/PK (1.6L) | VEHICLE SPEED SENSOR 5V SUPPLY |
| 1 | OR (EXCEPT 1.6L) | 5 VOLT SUPPLY |
| 2 | BR/OR (1.6L) | VEHICLE SPEED SENSOR GROUND |
| 2 | BK/LB (EXCEPT 1.6L) | SENSOR GROUND 1 |
| 3 | WT/OR (1.6L) | VEHICLE SPEED SENSOR SIGNAL |
| 3 | WT/OR (EXCEPT 1.6L) | VEHICLE SPEED SIGNAL |

**Vehicle Speed Sensor (with M/T)**

## WASTE GATE SOLENOID CONNECTOR

**Connector Pinouts**

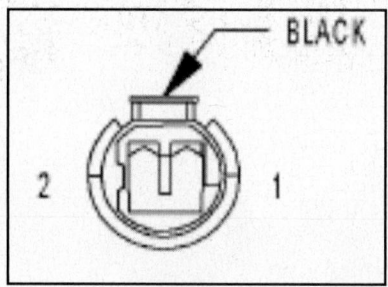

| PIN # | WIRE COLOR | CIRCUIT DESCRIPTION |
|---|---|---|
| 1 | DB/GY | WASTEGATE SOL CONTROL |
| 2 | BK | GROUND |

**2.4L Turbo Wastegate Solenoid Connector Pinout**

## DAIMLERCHRYSLER CARS: NEON

ACCELERATOR PEDAL POSITION (APP) SENSOR

**Connector Pinouts**

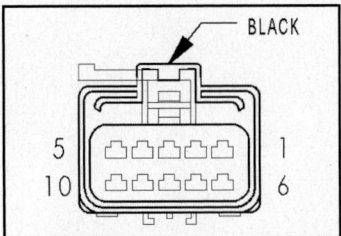

| PIN # | CIRCUIT | FUNCTION |
|---|---|---|
| 1 | - | - |
| 2 | K167 20BR/YL | ACCELERATOR PEDAL POSITION SENSOR GROUND NO. 1 |
| 3 | F858 20PK | ACCELERATOR PEDAL POSITION SENSOR 5 VOLT SUPPLY NO. 1 |
| 4 | K80 20DG/LG | ACCELERATOR PEDAL POSITION SENSOR SIGNAL NO. 1 |
| 5 | F859 20DG/PK | ACCELERATOR PEDAL POSITION SENSOR 5 VOLT SUPPLY NO. 2 |
| 6-7 | - | - |
| 8 | K981 20BR/DG | ACCELERATOR PEDAL POSITION SENSOR GROUND NO. 2 |
| 9 | - | - |
| 10 | K81 20DB/DG | ACCELERATOR PEDAL POSITION SENSOR SIGNAL NO. 2 |

**Neon 1.6L APP Sensor Connector Pinout**

AUTOMATIC SHUTDOWN (ASD) SYSTEM

**Description & Operation**

The ASD sense circuit informs the PCM when the ASD relay energizes. A 12-volt signal at this input indicates to the PCM that the ASD has been activated. This input is used only to sense that the ASD relay is energized.
When energized, the ASD relay supplies battery voltage to the fuel injectors, ignition coils and the heating element in each oxygen sensor. If the PCM does not receive 12 volts from this input after grounding the ASD relay, it sets a diagnostic trouble code (DTC).

*AUTOMATIC SHUTDOWN RELAY*

The ASD relay is located in the PDC. The inside top of the PDC cover has a label showing relay and fuse location. The Automatic Shutdown (ASD) relay supplies battery voltage to the fuel injectors, electronic ignition coil and the heating elements in the oxygen sensors generator field and PCM sense circuit.
A bus bar in the Power Distribution Center (PDC) supplies voltage to the solenoid side and contact side of the relay. The ASD relay power circuit contains a fuse between the buss bar in the PDC and the relay. The fuse also protects the power circuit for the fuel pump relay and pump. The fuse is located in the PDC.

The PCM controls the relay by switching the ground path for the solenoid side of the relay on and off. The PCM turns the ground path off when the ignition switch is in the OFF position unless the O2 Heater Monitor test is being run. When the ignition switch is in the ON or CRANK position, the PCM monitors the Crankshaft Position Sensor and Camshaft Position Sensor signals to determine engine speed and ignition timing (coil dwell). If the PCM does not receive the Crankshaft Position Sensor and Camshaft Position Sensor signals when the ignition switch is in the RUN position, it will de-energize the ASD relay.

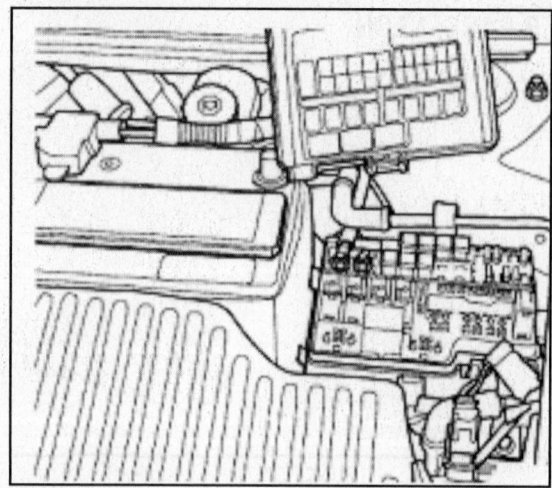

**Showing location of Power Distribution Center (PDC)**

| AUTO SHUTDOWN RELAY PINS, COLORS & FUNCTION TABLE | | |
|---|---|---|
| CAV | CIRCUIT | FUNCTION |
| 34 | A14 18RD/WT | FUSED B+ |
| 35 | A15 18RD/WT | FUSED B+ |
| 36 | A142 18DG/OR | AUTOMATIC SHUTDOWN RELAY |
| 37 | --- | --- |
| 38 | K51 18DB/YL | AUTOMATIC SHUTDOWN RELAY |

**Auto Shutdown relay pins, colors and functions chart**

## BATTERY TEMPERATURE SENSOR

**Connector Pinouts**

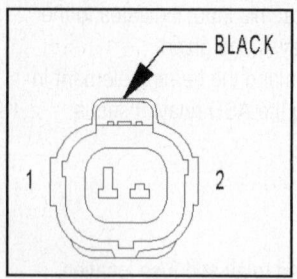

| PIN # | CIRCUIT | FUNCTION |
|---|---|---|
| 1 | K118 20PK/YL | BATTERY TEMP SIGNAL |
| 2 | K167 20BR/YL (2.4L TURBO) | SENSOR GROUND 2 |
| 2 | K167 20BR/YL (EXCEPT 2.4L TURBO) | SENSOR GROUND |

**Neon Battery Temperature Sensor**

## CAMSHAFT POSITION (CMP) SENSOR

**Description & Operation**

The camshaft position sensor (along with the crankshaft position sensor) provides inputs to the PCM to determine fuel injection synchronization and cylinder identification. From these inputs, the PCM determines crankshaft position. The camshaft position sensor is attached to the rear of the cylinder head.

A target magnet attaches to the rear of the camshaft and indexes to the proper position. The target magnet has four different poles arranges in an asymmetrical pattern. As the target magnet rotates, the camshaft position sensor senses the change in polarity. The sensor output switches from high (5.0 volts) to low (0.5 volts) as the target magnet rotates. When the north pole of the target magnet passes under the sensor, the output switches high. The sensor output switches low when the south pole of the target magnet passes underneath. The sensor also acts as a thrust plate to control camshaft endplay.

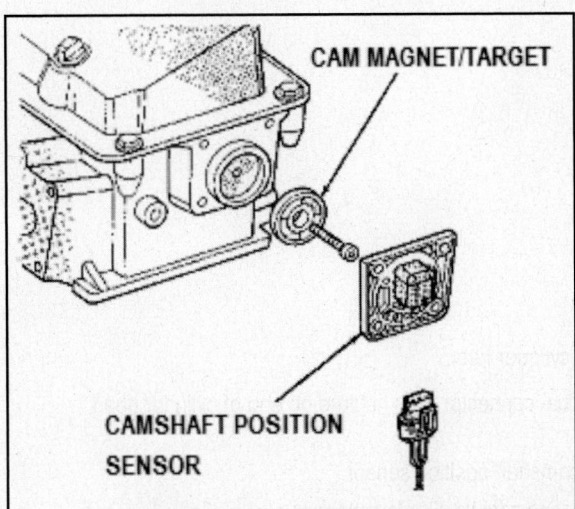

**The camshaft position sensor is mounted to the rear of the cylinder head-SOHC engine shown – 1995-99 Neon**

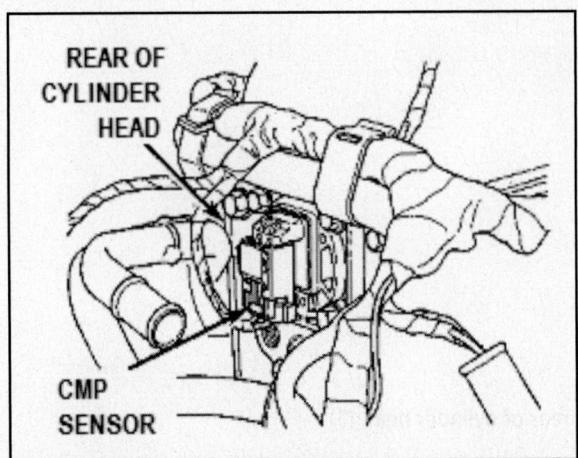

**Location of the camshaft position sensor-DOHC engine shown – 1995-99 Neon**

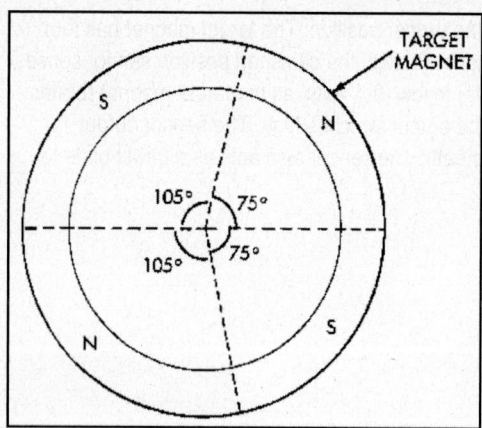

**View of the target magnet polarity – 1995-99 Neon**

## Removal & Installation

The camshaft position sensor is mounted to the rear of the cylinder head.

1. Remove brake booster hose and electrical connector from holders on end of cylinder head cover and reposition.
2. Disconnect electrical connectors from camshaft position sensor.
3. Remove camshaft position sensor mounting screws. Remove sensor.

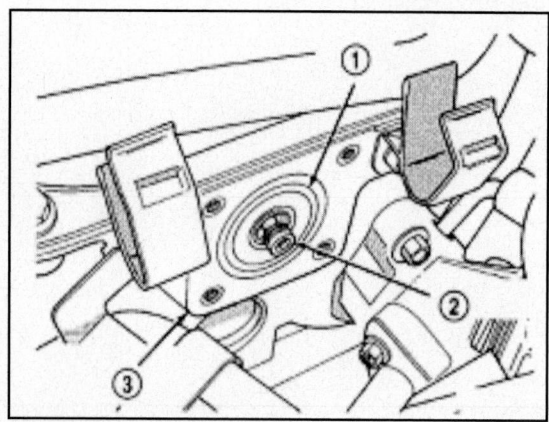

**Showing the target magnet (1) and mounting bolt (2) at rear of cylinder head (3)**

**To install:**

The camshaft position sensor is mounted to the rear of the cylinder head. The target magnet has two locating dowels that fit into machined locating holes in end of the camshaft.

4. Install target magnet in end of camshaft. Tighten mounting screw to 3.4 Nm (30 inch lbs.) torque.
5. Install camshaft position sensor. Tighten sensor mounting screws to 9 Nm (80 inch lbs.) torque.
6. Place brake booster hose and electrical harness in holders on end of valve cover.
7. Attach electrical connectors to camshaft position sensor.

**Testing**

*WITH AN OSCILLOSCOPE*

To test this sensor, you will need the use of an oscilloscope.

Visually check the connector, making sure it is attached properly and all of the terminals are straight, tight and free of corrosion. The output voltage of a proper operating camshaft or crankshaft position sensor switches from high (5.0 volts) to low (0.3 volts). By connecting an oscilloscope to the sensor output circuit, you can view the square wave pattern produced by the voltage swing.

*WITH A TEST LIGHT*

1. Check battery voltage. Voltage should be approximately 12.66 volts or higher to perform failure to start test.

2. Disconnect the harness connector from the coil pack.

3. Connect a test light to the B+ (battery voltage) terminal of the coil electrical connector and ground. The B+ wire for the DIS coil is the center terminal. Do not spread the terminal with the test light probe.

4. Turn the ignition key to the ON position. The test light should flash ON and then OFF. Do not turn the Key to OFF position, leave it in the ON position.

5. If the test light flashes momentarily, the PCM grounded the ASD relay. Proceed to step 5.

6. If the test light did not flash, the ASD relay did not energize. The cause is either the relay or one of the relay circuits. Use the DRB or scan tool to test the ASD relay and circuits. Check for codes and follow procedures.

7. Crank the engine. (If the key was placed in the OFF position after step 4, place the key in the ON position before cranking. Wait for the test light to flash once, then crank the engine.)

8. If the test light momentarily flashes during cranking, the PCM is not receiving a crankshaft position sensor signal.

9. If the test light did not flash during cranking, unplug the crankshaft position sensor connector. Turn the ignition key to the off position. Turn the key to the ON position, wait for the test light to momentarily flash once, then crank the engine. If the test light momentarily flashes, the crankshaft position sensor is shorted and must be replaced. If the light did not flash, the cause of the no-start is in either the crankshaft position sensor/camshaft position sensor 8 volt supply circuit, or the camshaft position sensor output or ground circuits.

## CAMSHAFT POSITION (CMP) SENSOR CONNECTOR

**Connector Pinouts**

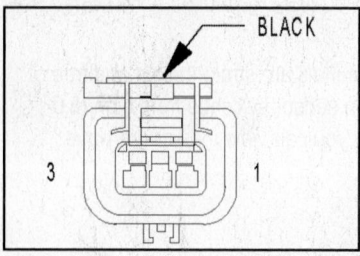

| PIN # | CIRCUIT | FUNCTION |
|-------|---------|----------|
| 1 | F854 20GY/PK (1.6L) | CAMSHAFT POSITION 5 VOLT SUPPLY |
| 1 | K7 20OR (2.0L) | 5 VOLT SUPPLY |
| 1 | K6 20VT/WT (2.4L TURBO) | 5 VOLT SUPPLY |
| 2 | K44 20TN/YL (1.6L) | CAMSHAFT POSITION SENSOR SIGNAL |
| 2 | K4 20BK/LB (2.0L) | SENSOR GROUND |
| 2 | K167 20BR/YL (2.4L TURBO) | SENSOR GROUND 2 |
| 3 | K944 20BR/GY (1.6L) | CAMSHAFT POSITION SENSOR GROUND |
| 3 | K44 20TN/YL (2.0L/2.4L TURBO) | CMP SIGNAL |

**CMP Sensor Connector Pinout**

## CLUTCH OPERATION SWITCH

**Description & Operation**

The clutch interlock/upstop switch is an assembly consisting of two switches: an engine starter inhibit switch (interlock) and a clutch pedal upstop switch. The switch assembly is located in the clutch/brake pedal bracket assembly, each switch being fastened by four plastic wing tabs.

### CLUTCH INTERLOCK SWITCH

The clutch interlock switch prevents engine starter operation and inadvertent vehicle movement with the clutch engaged and the transaxle in gear. The switch is open while the clutch pedal is at rest. When the clutch pedal is fully depressed, the pedal blade contacts and closes the switch, sending signal to the PCM, allowing engine starter operation. The interlock switch is not adjustable.

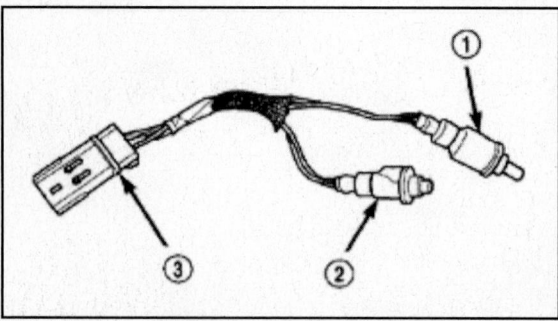

**Showing Upstop Switch (1), Interlock Switch (2) and connector (3) removed from vehicle**

### CLUTCH PEDAL UPSTOP SWITCH

With the clutch pedal at rest, the clutch pedal upstop switch is closed, allowing speed control operation. When the clutch pedal is depressed, the upstop switch opens and signals the PCM to cancel speed control operation, and enter a modified engine calibration schedule to improve driveability during gear-to-gear shifts. The upstop switch is not adjustable.

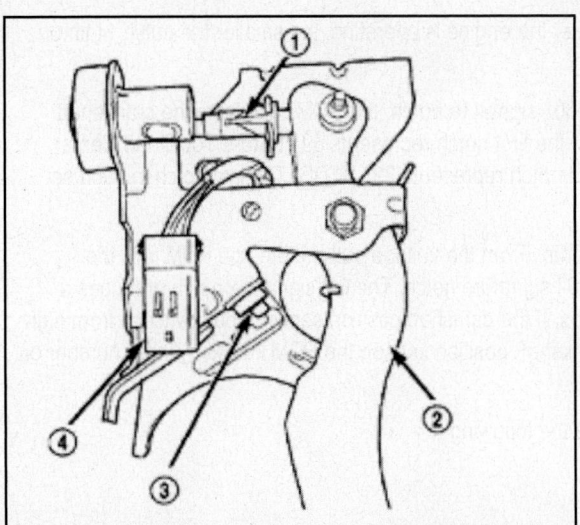

**Identifying the Upstop Switch (1), clutch pedal (2), interlock switch (3), and switch connector (4)**

## CRANKSHAFT POSITION (CKP) SENSOR

### Description & Operation

The PCM determines what cylinder to fire from the crankshaft position sensor input and the camshaft position sensor input. The second crankshaft counterweight has two sets of four timing reference notches, including a 60° signature notch. From the crankshaft position sensor input, the PCM determines engine speed and crankshaft angle (position).

The notches generate pulses from high to low in the crankshaft position sensor output voltage. When a metal portion of the counterweight aligns with the crankshaft position sensor, the sensor output voltage goes low (less than 0.5 volts). when a notch aligns with the sensor, voltage goes high (5.0 volts). As a group of notches pass under the sensor, the output voltage switches from low (metal) to high (notch) then back to low.

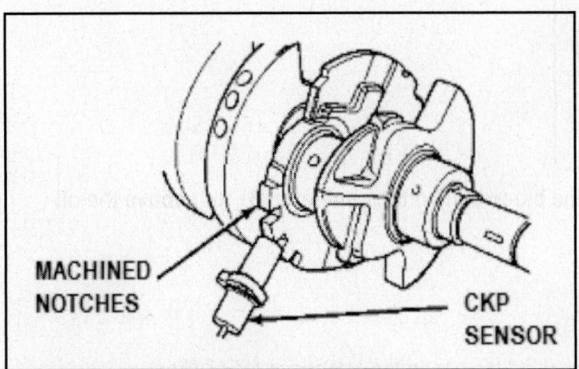

**Timing reference notches on crankshaft**

The Crankshaft Position Sensor (CKP) is a Hall-effect sensor. The PCM determines what cylinder to fire from the Crankshaft Position Sensor input and the Camshaft Position Sensor input. The second crankshaft counterweight has two sets of four timing reference notches including a 60-degree signature notch. From the Crankshaft Position Sensor input, the PCM determines engine speed and crankshaft angle (position). The PCM sends approximately 8 volts to the Hall-effect sensor. This voltage is required to operate the Hall-effect chip and the electronics inside the sensor. A ground for the sensor is provided through the sensor return circuit. The input to the PCM occurs on a 5-volt output reference circuit.

If available, an oscilloscope can display the square wave patterns of each voltage pulse. From the width of the output voltage pulses, the PCM calculates engine speed, The width of the pulses represent the amount of time the output voltage stays high before switching back to low. The period of time the sensor output voltage stays high before switching back to low is referred to as pulse width. The faster the engine is operating, the smaller the pulse width on the oscilloscope.

By counting the pulses and referencing the pulse from the 60° signature notch, the PCM calculates the crankshaft angle (position). In each group of timing reference notches, the first notch represents 69° Before Top Dead Center (BTDC). The second notch represents 49° BTDC. The third notch represents 29° BTDC. The last notch in each set represents 9° BTDC.

The timing reference notches are machined at 20° increments. From the voltage pulse width, the PCM tells the difference between the timing reference notches and the 60° signature notch. The 60° signature notch produces a longer pulse width than the smaller timing reference notches. If the camshaft position sensor input switches from high to low when the 60° signature notch passes under the crankshaft position sensor, the PCM knows cylinder number on is the next cylinder at TDC.

The PCM uses the Crankshaft Position Sensor to calculate the following:

- Engine RPM

- TDC number 1 and 4

- Ignition coil synchronization

- Injector synchronization

- Camshaft-to-crankshaft misalignment (Timing belt skipped 1 tooth or more diagnostic trouble code).

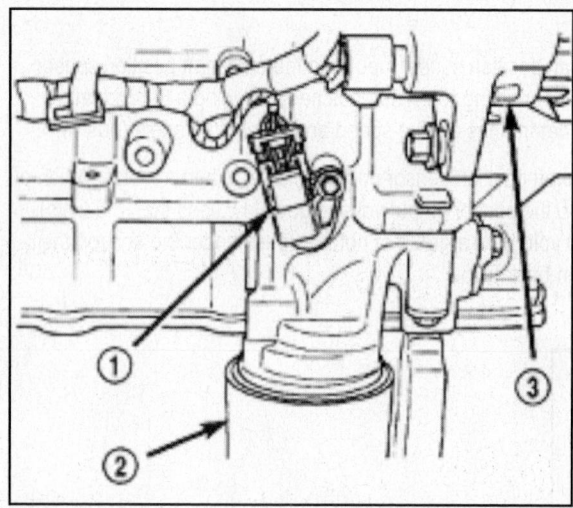

**The Crankshaft Position Sensor (1) mounts to the engine block behind the generator (3), just above the oil filter (2)**

### Removal & Installation

The crankshaft position sensor mounts to the engine block behind the generator, just above the oil filter.

   10. Disconnect electrical connector from crankshaft position sensor.

   11. Remove sensor mounting screw. Remove sensor.

   12. Disconnect the negative battery cable.

   13. Raise vehicle and support.

   14. Disconnect the electrical connector.

   15. Remove bolt from Crankshaft sensor.

   16. Remove sensor.

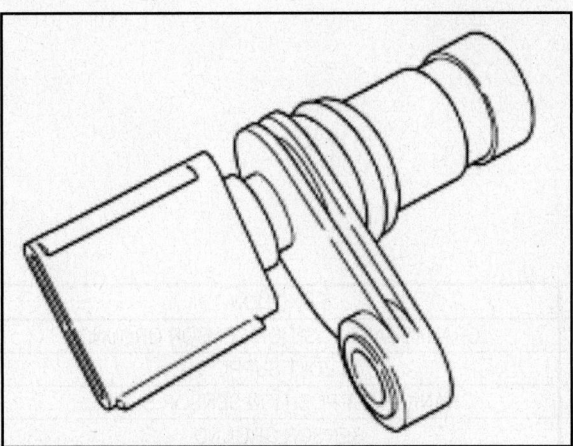

**Showing the Crankshaft Position (CKP) sensor**

**To install:**

The crankshaft position sensor mounts to the engine block behind the generator, just above the oil filter

    17. Install sensor. Install sensor mounting screw and tighten.

    18. Connect electrical connector to crankshaft position sensor.

    19. Install bolt and tighten to 10 Nm (90 inch lbs.).

    20. Connect the electrical connector.

    21. Lower vehicle. Connect the negative battery cable.

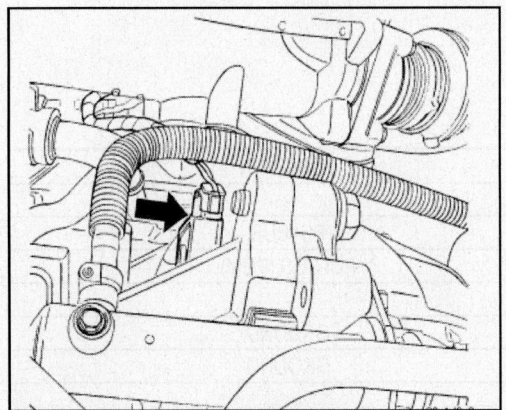

**Typical CKP Sensor location on Neon**

**Testing**

To test this sensor, you will need the use of an oscilloscope.

    1. Visually check the connector, making sure it is attached properly and all of the terminals are straight, tight and free of corrosion.

    2. The output voltage of a proper operating camshaft or crankshaft position sensor switches from high (5.0 volts) to low (0.3 volts). By connecting an oscilloscope to the sensor output circuit, you can view the square wave pattern produced by the voltage swing.

    3. Compare the scope pattern readings and results and repair or replace sensor as necessary.

**Connector Pinouts**

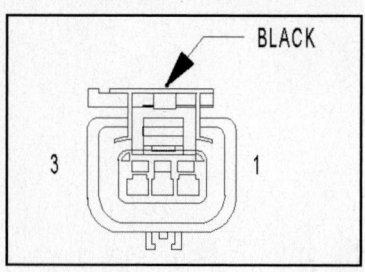

| PIN # | CIRCUIT | FUNCTION |
|-------|---------|----------|
| 1 | K924 20BR/WT (1.6L) | CRANKSHAFT POSITION SENSOR GROUND |
| 1 | K7 20OR (2.0L/2.4L TURBO) | 5 VOLT SUPPLY |
| 2 | K24 20GY/BK (1.6L) | CRANKSHAFT POSITION SENSOR SIGNAL |
| 2 | K4 20BK/LB (2.0L/2.4L TURBO) | SENSOR GROUND |
| 3 | K7 20OR (1.6L) | CRANKSHAFT POSITION 5 VOLT SUPPLY |
| 3 | K24 20GY/BK (2.0L/2.4L TURBO) | CKP SIGNAL |

**CKP Sensor Connector Pinout**

## DATA LINK CONNECTOR

**Connector Pinouts**

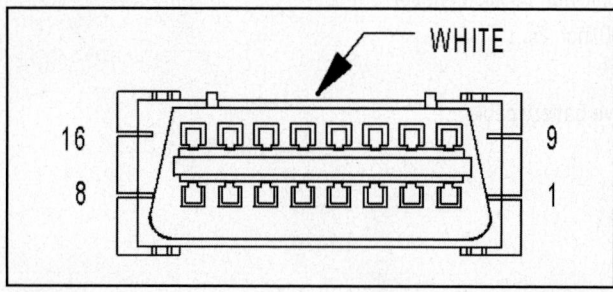

| PIN # | CIRCUIT | FUNCTION |
|-------|---------|----------|
| 1 | - | - |
| 2 | D25 20VT/YL (1.6L) | PCI BUS |
| 2 | D25 20VT/YL (2.0L/2.4L TURBO) | PCI BUS (PCM) |
| 3 | - | - |
| 4 | Z12 20BK/TN | GROUND |
| 5 | Z12 20BK/TN | GROUND |
| 6 | - | - |
| 7 | D21 20PK | SCI TRANSMIT (PCM) |
| 8 | - | - |
| 9 | D6 20PK/LB (2.0L) | SCI RECEIVE (TCM) |
| 10 | - | - |
| 11 | - | - |
| 12 | D20 20LG | SCI RECEIVE (PCM) |
| 13 | - | - |
| 14 | - | - |
| 15 | D15 20WT/DG (2.0L) | SCI TRANSMIT (TCM) |
| 16 | A14 18RD/WT | FUSED B(+) |

**Neon Data Link Connector Pinout**

## ELECTRONIC THROTTLE CONTROL MODULE

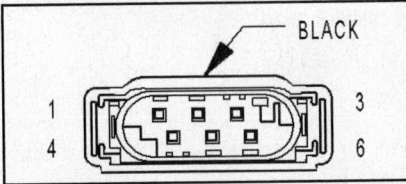

| PIN # | CIRCUIT | FUNCTION |
|---|---|---|
| 1 | K22 20BR/OR | THROTTLE POSITION SENSOR SIGNAL NO. 1 |
| 2 | K122 20DB/GY | THROTTLE POSITION SENSOR SIGNAL NO. 2 |
| 3 | F855 20OR/PK | THROTTLE POSITION SENSOR 5 VOLT SUPPLY |
| 4 | V50 20VT/OR | ELECTRONIC THROTTLE CONTROL POSITIVE MOTOR CONTROL |
| 5 | K922 20DB/OR | THROTTLE POSITION SENSOR GROUND |
| 6 | V51 20VT/BR | ELECTRONIC THROTTLE CONTROL NEGATIVE MOTOR CONTROL |

**Neon 1.6L Electronic Throttle Control Module Connector Pinout**

## ENGINE COOLANT TEMPERATURE (ECT) SENSOR

### Description & Operation

The coolant sensor threads into the rear of the cylinder head, next to the camshaft position sensor. New sensors have sealant applied to the threads.

The ECT Sensor is a Negative Thermal Coefficient (NTC), dual range Sensor. The resistance of the ECT Sensor changes as coolant temperature changes. This results in different input voltages to the PCM. The PCM also uses the ECT Sensor input to operate the low and high-speed radiator cooling fans.

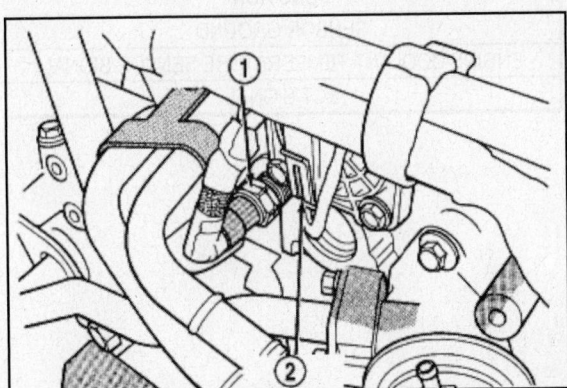

**Locating the ECT Sensor (1) and the CMP Sensor (2) on Neon**

The combination coolant temperature sensor has two elements. One element supplies coolant temperature signal to the PCM. The other element supplies coolant temperature signal to the instrument panel gauge cluster. The PCM determines engine coolant temperature from the coolant temperature sensor.

As coolant temperature varies the coolant temperature sensors resistance changes resulting in a different input voltage to the PCM and the instrument panel gauge cluster. When the engine is cold, the PCM will provide slightly richer air-fuel mixtures and higher idle speeds until normal operating temperatures are reached.

The PCM has a dual temperature range program for better sensor accuracy at cold temperatures. At key-ON the PCM sends a regulated 5-volt signal through a 10,000-ohm resistor to the sensor. When the sensed voltage reaches approximately 1.25 volts the PCM turns on the transistor. The transistor connects a 1,000-ohm resistor in parallel with the 10,000-ohm resistor. With this drop in resistance the PCM recognizes an increase in voltage on the input circuit.

## Removal & Installation

1. With the engine cold, drain coolant until level drops below cylinder head.
2. Disconnect coolant sensor electrical connector.
3. Remove coolant sensor.

**To install:**

4. Install coolant sensor. Tighten sensor to 18 Nm (165 inch lbs.) torque.
5. Attach electrical connector to sensor.
6. Fill cooling system.

## Connector Pinouts

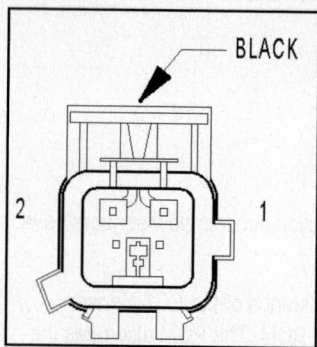

| PIN # | CIRCUIT | FUNCTION |
|-------|---------|----------|
| 1 | K4 20BK/LB | SENSOR GROUND |
| 2 | K2 20TN/BK (1.6L) | ENGINE COOLANT TEMPERATURE SENSOR SIGNAL |
| 2 | K2 20TN/BK (2.4L TURBO) | ECT SIGNAL |

**Neon 1.6L & 2.4L Turbo ECT Sensor**

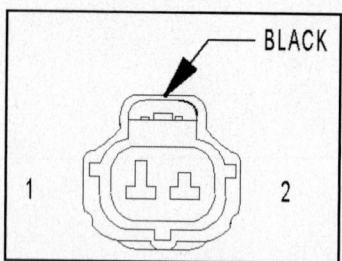

| PIN # | CIRCUIT | FUNCTION |
|-------|---------|----------|
| 1 | K4 20BK/LB | SENSOR GROUND |
| 2 | K2 20VT/LG | ECT SIGNAL |

**Neon 2.0L ECT Sensor**

EVAP PURGE SOLENOID

**Connector Pinouts**

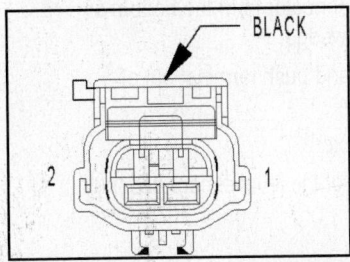

| PIN # | CIRCUIT | FUNCTION |
|-------|---------|----------|
| 1 | F12 18DB/WT (1.6L) | FUSED IGNITION SWITCH OUTPUT (RUN-START) |
| 1 | K108 20WT/TN (2.0L/2.4L TURBO) | EVAP/PURGE RETURN |
| 2 | K52 20PK/BK (1.6L) | EVAP/PURGE SOLENOID CONTROL |
| 2 | K52 20PK/BK (2.0L/2.4L TURBO) | EVAP/PURGE CONTROL |

**Neon EVAP Purge Solenoid Connector Pinout**

FUEL LEVEL SENSOR/FUEL PUMP MODULE

**Description & Operation**

The fuel level sensor (fuel gauge sending unit) is attached to the side of the fuel pump module. The level sensor consists of a float, an arm, and a variable resistor. As the fuel level increases and the float and arm moves up, a decrease results in the sending unit resistance. This causes the fuel gauge on the instrument panel to read FULL. The Fuel Level Sensor (fuel gauge sending unit) sends a signal to the BCM. In turn, the BCM provides the signal over the PCI bus circuit to the PCM to indicate fuel level.

The purpose of this feature is to prevent a false setting of misfire and fuel system monitor trouble codes, if the fuel level is less than 15% of its rated capacity. It is also used to send a signal for fuel gauge operation via the PCI bus circuits.

The fuel gauge level sending unit is attached to the fuel pump module. The fuel level sensor (fuel gauge sending unit) sends a signal to the PCM to indicate fuel level.

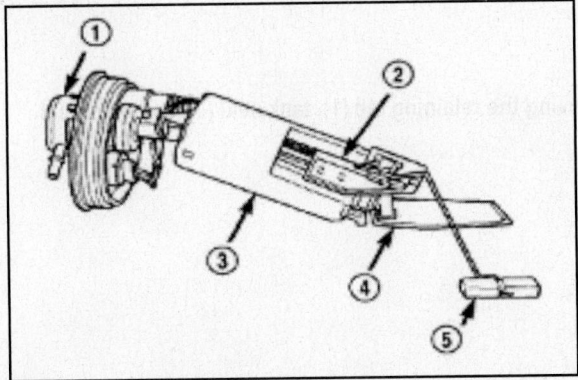

**The fuel sending unit assembly, showing the fuel filter/pressure regulator (1), fuel level sensor (2), fuel reservoir (3), inlet strainer (4), and float (5)**

**Removal & Installation**

1. Remove fuel pump module.
2. Depress retaining tab and remove the fuel pump/level sensor connector from the bottom of the fuel pump module electrical connector. Pull off blue locking wedge.
3. Using a small screwdriver lift locking finger away from terminal and push terminal out of connector
4. Push level sensor signal and ground terminals out of the connector.
5. Insert a screwdriver between the fuel pump module and the top of the level sensor housing. Push level sensor down slightly.
6. Slide level sensor wires through opening fuel pump module.
7. Slide level sensor out of installation channel in module.

**To install:**

8. Insert level sensor wires into bottom of opening in module.
9. Wrap wires into groove in back of level sensor.
10. While feeding wires into guide grooves, slide level sensor up into channel until it snaps into place. Ensure tab at bottom of sensor locks in place.
11. Install level sensor wires in connector. Push the wires up through the connector and then pull them down until they lock in place. Ensure signal and ground wires are installed in the correct position.
12. Install locking wedge on connector. Push connector up into bottom of fuel pump module electrical connector.
13. Install fuel pump module.

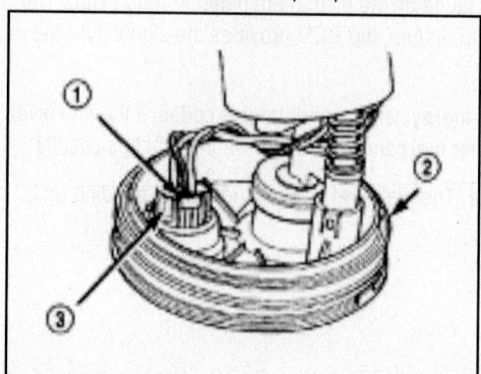

**Disassembling lower end of the fuel sending unit, removing the retaining tab (1), tank seal (2), and electrical connector (3)**

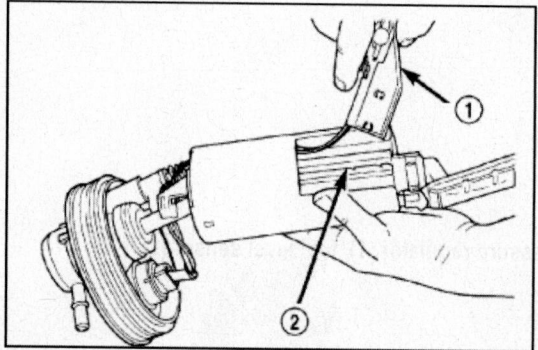

**Detaching electrical connector (1) from blue locking wedge (2)**

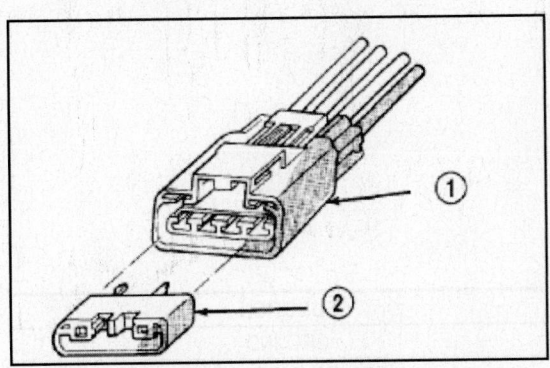

Sliding fuel level sensor (1) out of channel (2)

**Connector Pinouts**

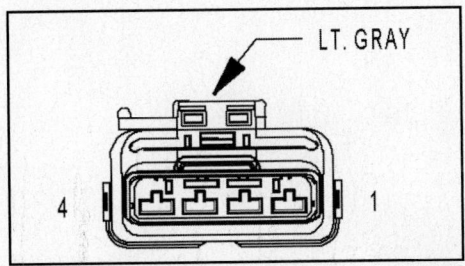

| PIN # | CIRCUIT | FUNCTION |
|---|---|---|
| 1 | Z1 18BK | GROUND |
| 2 | Z2 20BK/LG | GROUND |
| 3 | G4 20DB | FUEL LEVEL SENSOR SIGNAL |
| 4 | A141 18DG/WT | FUEL PUMP RELAY OUTPUT |

**Neon Fuel Pump Module Connector Pinout**

GENERATOR

**Connector Pinouts**

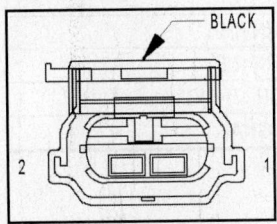

| PIN # | CIRCUIT | FUNCTION |
|---|---|---|
| 1 | K20 18DG (1.6L) | GENERATOR FIELD DRIVER (+) |
| 1 | Z1 18BK (2.0L) | GROUND |
| 2 | A142 18DG/OR (1.6L) | AUTOMATIC SHUT DOWN RELAY OUTPUT |
| 2 | K20 18DG (2.0L) | GEN FIELD CONTROL |

**Neon 1.6L & 2.0L Generator Connector Pinout**

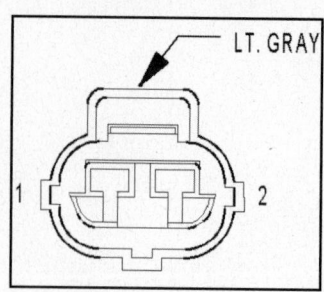

| PIN # | CIRCUIT | FUNCTION |
|-------|---------|----------|
| 1 | Z1 18BK | GROUND |
| 2 | K20 18DG | GEN FIELD CONTROL |

**Neon 2.4L Turbo Generator Connector Pinout**

HEATED OXYGEN SENSORS (HO2S)

**Connector Pinouts**

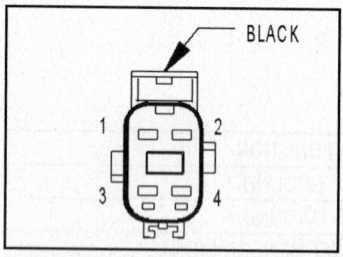

| PIN # | CIRCUIT | FUNCTION |
|-------|---------|----------|
| 1 | A142 18DG/OR (1.6L) | AUTOMATIC SHUT DOWN RELAY OUTPUT |
| 1 | Z1 18BK (2.0L/2.4L TURBO) | GROUND |
| 2 | K79 18OR/RD (1.6L) | OXYGEN SENSOR 1/1 HEATER CONTROL |
| 2 | K99 18BR/OR (2.0L/2.4L TURBO) | O2 1/1 HEATER CONTROL |
| 3 | K902 20BR (1.6L) | OXYGEN SENSOR GROUND |
| 3 | K904 18BR/OR (2.0L RT) | O2 RETURN (UP) |
| 3 | K904 18DB/DG (2.0L) | O2 RETURN (UP) |
| 3 | K904 20DB/DG (2.4L TURBO) | O2 RETURN (UP) |
| 4 | K41 20BK/DG (1.6L) | OXYGEN SENSOR 1/1 SIGNAL |
| 4 | K41 20BK/DG (2.0L/2.4L TURBO) | O2 1/1 SIGNAL |

**Neon O2 1/1 Upstream Sensor**

| PIN # | CIRCUIT | FUNCTION |
|-------|---------|----------|
| 1 | K199 18BR/VT (1.6L) | OXYGEN SENSOR 1/2 HEATER CONTROL |
| 1 | Z1 18BK (2.0L) | GROUND |
| 1 | Z1 20BK (2.4L TURBO) | GROUND |
| 2 | A142 18DG/OR (1.6L) | AUTOMATIC SHUT DOWN RELAY OUTPUT |
| 2 | K199 18BR/VT (2.0L/2.4L TURBO) | O2 1/2 HEATER CONTROL |
| 3 | K904 20DB/DG (1.6L) | OXYGEN SENSOR GROUND |
| 3 | K904 20DB/DG (2.0L/2.4L TURBO) | O2 RETURN (UP) |
| 4 | K141 20TN/WT (1.6L) | OXYGEN SENSOR 1/2 SIGNAL |
| 4 | K141 20TN/WT (2.0L/2.4L TURBO) | O2 1/2 SIGNAL |

**Neon O2 1/2 Downstream Sensor**

<u>IDLE AIR CONTROL (IAC) MOTOR</u>

### Description & Operation

The Idle Air Control (IAC) motor is mounted on the throttle body. The PCM operates the idle air control motor. The PCM adjusts engine idle speed through the idle air control motor to compensate for engine load, coolant temperature or barometric pressure changes. The throttle body has an air bypass passage that provides air for the engine during closed throttle idle. The IAC motor pintle protrudes into the air bypass passage and regulates airflow through it.

The PCM adjusts engine idle speed by moving the IAC motor pintle in and out of the bypass passage. The adjustments are based on inputs the PCM receives. The inputs are from the TP sensor, CKP sensor, ECT sensor, MAP sensor, VSS, and various switch operations (brake, park/neutral, air conditioning).

When engine rpm is above idle speed, the IAC is used for the following functions:

- Off-idle dashpot
- Deceleration air flow control
- A/C compressor load control (also opens the passage slightly before the compressor is engaged so that the engine rpm does not dip down when the compressor engages)

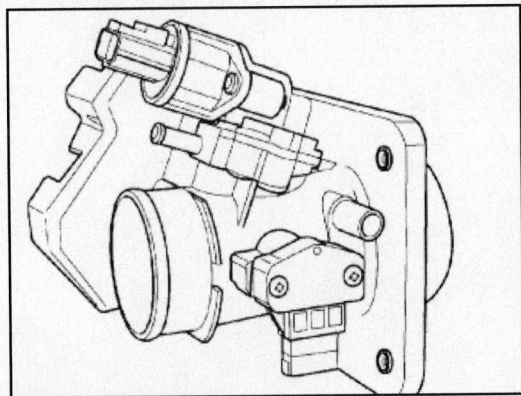

**Showing the Idle Air Control Motor**

*TARGET IDLE*

Target idle is determined by the following inputs:

- Gear position
- ECT Sensor
- Battery voltage
- Ambient/Battery Temperature Sensor
- VSS
- TPS
- MAP Sensor

***NOTE:*** *When servicing throttle body components, always reassemble components with new O-rings and seals where applicable.*

***CAUTION:*** *NEVER use lubricants on O-rings or seals, damage may result. If assembly of component is difficult, use water to aid assembly. Use care when removing hoses to prevent damage to hose or hose nipple.*

## Removal & Installation

1. Disconnect negative cable from battery.
2. Remove electrical connector from idle air control motor.
3. Remove idle air control motor mounting screws.
4. Remove motor from throttle body. Ensure the O-ring is removed with the motor.

## To install:

The new idle air control motor has a new O-ring installed on it. If pintle measures more than 1 inch (25 mm) it must be retracted. Use the scan tool "Idle Air Control Motor Open/Close" test to retract the pintle (battery must be connected.).

5. Carefully place idle air control motor into throttle body.
6. Install mounting screws. Tighten screws to 4.5 Nm (40 inch lbs.) torque.
7. Connect electrical connector to idle air control motor.
8. Connect negative cable to battery

## Connector Pinouts

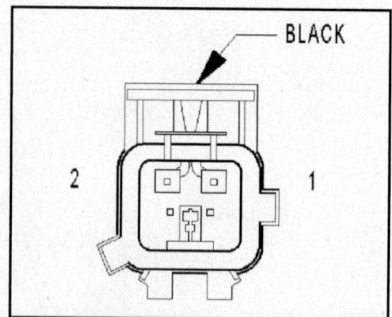

| PIN # | CIRCUIT | FUNCTION |
|-------|---------|----------|
| 1 | K610 20VT/GY (2.0L) | IAC MOTOR CONTROL |
| 1 | K610 18VT/GY (2.4L TURBO) | IAC MOTOR CONTROL |
| 2 | K961 20BR/VT (2.0L) | IAC RETURN |
| 2 | K961 18BR/VT (2.4L TURBO) | IAC RETURN |

**Neon 2.0L & 2.4L Turbo IAC Motor Connector Pinout**

## IGNITION COIL

### Description & Operation

High tension leads route to each cylinder from the coil. The coil fires 2 spark plugs at every power-stroke: one at compression; the other on the exhaust stroke. Coil No. 1 fires cylinders 1 and 4. Coil No. 2 fires cylinders Nos. 2 and 3. The PCM determines which of the coils to charge and fire at the correct time.

The Auto Shutdown (ASD) relay provides battery voltage to the ignition coil. The PCM provides a ground contact (circuit) for energizing the coil. When the PCM breaks the contact, the energy in the coil primary transfers to the secondary causing the spark. The PCM will de-energize the ASD relay if it does not receive the crankshaft position sensor and camshaft position sensor inputs. Refer to Auto Shutdown (ASD) Relay.

Base timing is non-adjustable, but is set from the factory at approximately 10° BTDC when the engine is warm and idling.

There is an adaptive dwell strategy that runs dwell from 4-6 msec when engine speed is below 3000 rpm and battery voltage is 12-14 volts. During cranking, dwell can be as much as 200 msec. The adaptive dwell is driven by the sensed current flow through the injector drivers. Current flow is limited to 8 amps. The coil pack assembly consists of 2 coils molded together. The coil pack is mounted on the valve cover.

*WARNING: The direct ignition system generates approximately 40,000 volts. Personal injury could result from contact with this system.*

The low resistance of the primary coils can allow current flow in excess of 15 amps. The PCM has a current sensing device in the coil output circuit. As dwell time starts, the PCM allows current to flow. When the sensing device registers 8 amps, the PCM begins to regulate current flow to maintain and not exceed 8 amps through the remainder of the dwell time. This prevents the PCM from being damaged by excess current flow.

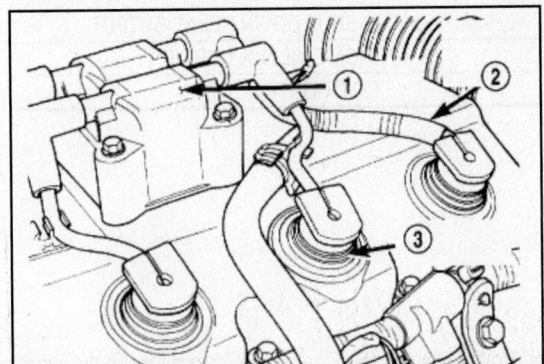

Showing the location of the ignition coil (1), spark plug cable (2) and insulator (3) on 2.0L engines

### Removal & Installation

1. Disconnect electrical connector from coil pack.
2. Remove coil pack mounting bolts.
3. Remove coil pack.

To install:

4. Install coil pack on valve cover.
5. Transfer spark plug cables to new coil pack. The coil pack towers are numbered with the cylinder identification. Be sure the ignition cables snap onto the towers.

| COIL MANUFACTURER | PRIMARY RESISTANCE | SECONDARY RESISTANCE |
|---|---|---|
| WEASTEC (STEEL TOWERS) | 0.45-0.65 OHMS @ 21-27°C | 11.5-13.5K OHMS @ 21-27°C |
| DIAMOND (COPPER TOWERS) | 0.53-0.65 OHMS @ 21-27°C | 10.9-14.7K OHMS @ 21-27°C |

**Identifying ignition coil resistances**

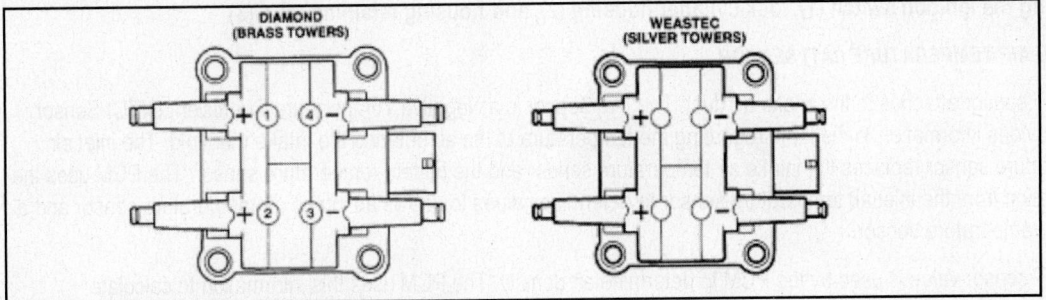

**Identifying the types of ignition coils used on Neon**

**Connector Pinouts**

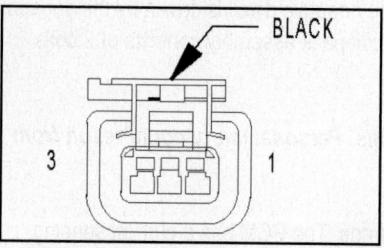

| PIN # | CIRCUIT | FUNCTION |
|-------|---------|----------|
| 1 | K17 18DB/TN (1.6L) | IGNITION COIL NO. 2 DRIVER |
| 1 | K17 16DB/TN (2.0L/2.4L TURBO) | COIL CONTROL NO. 2 |
| 2 | A142 18DG/OR | AUTOMATIC SHUT DOWN RELAY OUTPUT |
| 3 | K19 18BK/GY (1.6L) | IGNITION COIL NO. 1 DRIVER |
| 3 | K19 16BK/GY (2.0L/2.4L TURBO) | COIL CONTROL NO. 1 |

**Neon Ignition Coil Module Connector Pinout**

## IGNITION SWITCH

### Description & Operation

In the RUN position, the ignition switch connects power from the Power Distribution Center (PDC) to a fuse in the fuse block, back to a bus bar in the PDC. The bus bar feeds circuits for the Powertrain Control Module (PCM), Proportional purge solenoid, EGR solenoid, and ABS system. The bus bar in the PDC feeds the coil side of the radiator fan relay, A/C compressor clutch relay, and the fuel pump relay. It also feeds the Airbag Control Module (ACM). The ignition switch attaches to the lock cylinder housing on the end opposite the lock cylinder.

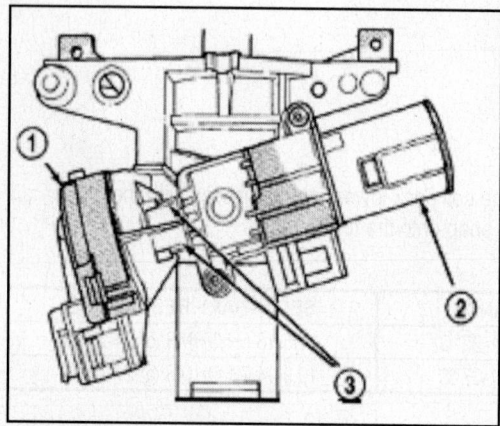

**Showing the ignition switch (1), lock cylinder housing (2), and housing retaining tabs (3)**

### INTAKE AIR TEMPERATURE (IAT) SENSOR

The IAT sensor attaches to the intake air duct. The IAT Sensor is a Negative Temperature Coefficient (NTC) Sensor that provides information to the PCM regarding the temperature of the air entering the intake manifold. The inlet air temperature sensor replaces the intake air temperature sensor and the battery temperature sensor. The PCM uses the information from the inlet air temperature sensor to determine values to use as an intake air temperature sensor and a battery temperature sensor.

The IAT sensor value is used by the PCM to determine air density. The PCM uses this information to calculate:

- Injector pulse width

- Adjustment of ignition timing (to prevent spark knock at high intake air temperatures)

### Battery Temperature

The inlet air temperature sensor replaces the intake air temperature sensor and the battery temperature sensor. The PCM uses the information from the inlet air temperature sensor to determine values for the PCM to use as an intake air temperature sensor and a battery temperature sensor.

The battery temperature information along with data from monitored line voltage (B+), is used by the PCM to vary the battery charging rate. System voltage will be higher at colder temperatures and is gradually reduced at warmer temperatures.

The battery temperature information is also used for OBD II diagnostics. Certain faults and OBD II monitors are either enabled or disabled depending upon the battery temperature sensor input (example: disable purge and EGR, enable LDP). Most OBD II monitors are disabled below 20°F.

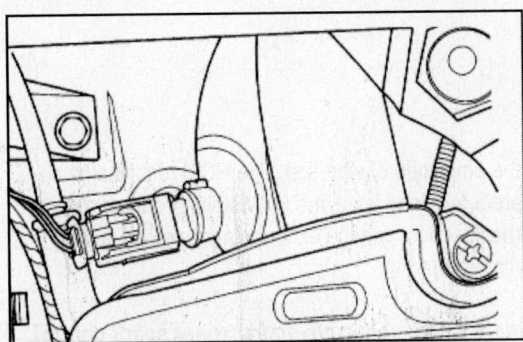

**Location of Intake Air Temperature (IAT) sensor**

### Connector Pinouts

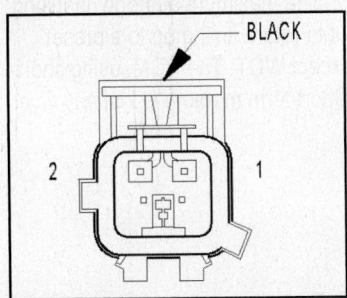

| PIN # | CIRCUIT | FUNCTION |
|---|---|---|
| 1 | K21 20BK/RD (1.6L) | INLET AIR TEMPERATURE SENSOR SIGNAL |
| 1 | K21 20BK/RD (2.0L/2.4L TURBO) | IAT SIGNAL |
| 2 | K921 20BR/LG (1.6L) | INLET AIR TEMPERATURE SENSOR GROUND |
| 2 | K167 20BR/YL (2.0L) | SENSOR GROUND |
| 2 | K4 20BK/LB (2.4L TURBO) | SENSOR GROUND |

**Neon IAT Sensor**

### KNOCK SENSOR

### Description & Operation

The knock sensor threads into the side of the cylinder block in front of the starter. The knock sensor is designed to detect engine vibration that is caused by detonation.

When the knock sensor detects a knock in one of the cylinders, it sends an input signal to the PCM. In response, the PCM retards ignition timing for all cylinders by a scheduled amount. Knock sensors contain a piezoelectric material, which sends an input voltage (signal) to the PCM. As the intensity of the engine knock vibration increases, the knock sensor output voltage also increases.

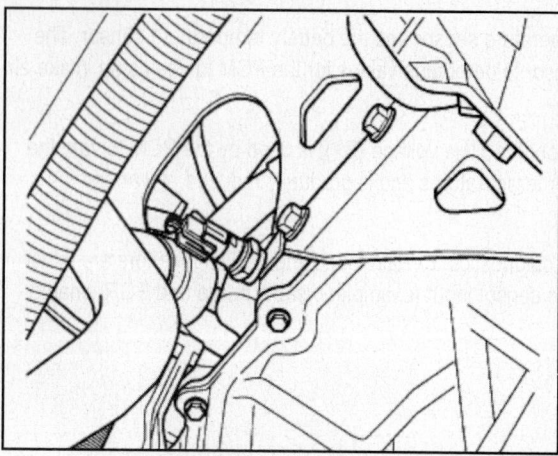

**Location of Knock Sensor on 2.0L engines**

The voltage signal produced by the knock sensor increases with the amplitude of vibration. The PCM receives as an input the knock sensor voltage signal. If the signal rises above a predetermined level, the PCM will store that value in memory and retard ignition timing to reduce engine knock. If the knock sensor voltage exceeds a preset value, the PCM retards ignition timing for all cylinders. It is not a selective cylinder retard.

***NOTE: Over or under-tightening affects knock sensor performance, possibly causing improper spark control.***

The PCM ignores knock sensor input during engine idle conditions. Once the engine speed exceeds a specified value, knock retard is allowed. Knock retard uses its own short term and long-term memory program.

Long-term memory stores previous detonation information in its battery-backed RAM. The maximum authority that long term memory has over timing retard can be calibrated. Short-term memory is allowed to retard timing up to a preset amount under all operating conditions (as long as rpm is above the minimum rpm) except WOT. The PCM, using short-term memory, can respond quickly to retard timing when engine knock is detected. Short-term memory is lost any time the ignition key is turned off.

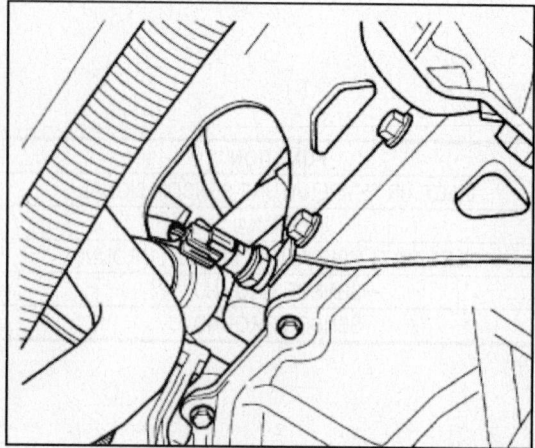

**The knock sensor threads into the side of the cylinder block in front of the starter**

### Removal & Installation

1. Disconnect electrical connector from knock sensor.
2. Use a crowfoot socket to remove the knock sensors.

**To install:**

3. Install knock sensor. Tighten knock sensor to 10 Nm (7 ft. lbs.) torque.
4. Attach electrical connector to knock sensor.

**Connector Pinouts**

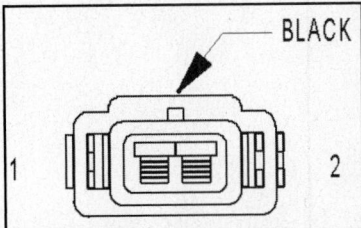

| PIN # | CIRCUIT | FUNCTION |
|-------|---------|----------|
| 1 | K42 20DB/LG (1.6L) | KNOCK SENSOR SIGNAL |
| 1 | K42 20DB/LG (2.0L/2.4L TURBO) | KS SIGNAL |
| 2 | K942 20BR/DB (1.6L) | KNOCK SENSOR GROUND |
| 2 | K45 20BK/VT (2.0L/2.4L TURBO) | KS RETURN |

**Neon Knock Sensor Connector Pinout**

LEAK DETECTION PUMP

**Description & Operation**

The evaporative emission system is designed to prevent the escape of fuel vapors from the fuel system. Leaks in the system, even small ones, can allow fuel vapors to escape into the atmosphere. Government regulations require on-board Testing to make sure that the evaporative (EVAP) system is functioning properly. The leak detection system tests for EVAP system leaks and blockage. It also performs self-diagnostics.

During self-diagnostics, the Powertrain Control Module (PCM) first checks the Leak Detection Pump (LDP) for electrical and mechanical faults. If the first checks pass, the PCM then uses the LDP to seal the vent valve and pump air into the system to pressurize it. If a leak is present, the PCM will continue pumping the LDP to replace the air that leaks out. The PCM determines the size of the leak based on how fast/long it must pump the LDP as it tries to maintain pressure in the system.

***EVAP LEAK DETECTION SYSTEM COMPONENTS***

**Service Port**: Used with special tools like the Miller Evaporative Emissions Leak Detector (EELD) to test for leaks in the system.
**EVAP Purge Solenoid**: The PCM uses the EVAP purge solenoid to control purging of excess fuel vapors stored in the EVAP canister. It remains closed during leak Testing to prevent loss of pressure.
**EVAP Canister**: The EVAP canister stores fuel vapors from the fuel tank for purging. EVAP Purge Orifice: Limits purge volume.
**EVAP System Air Filter**: Provides air to the LDP for pressurizing the system. It filters out dirt while allowing a vent to atmosphere for the EVAP system.

The main purpose of the LDP is to pressurize the fuel system for leak checking. It closes the EVAP system vent to atmospheric pressure so the system can be pressurized for leak Testing. The diaphragm is powered by engine vacuum. It pumps air into the EVAP system to develop a pressure of about 7.5' H2O (1/4) psi. A reed switch in the LDP allows the PCM to monitor the position of the LDP diaphragm. The PCM uses the reed switch input to monitor how fast the LDP is pumping air into the EVAP system. This allows detection of leaks and blockage.

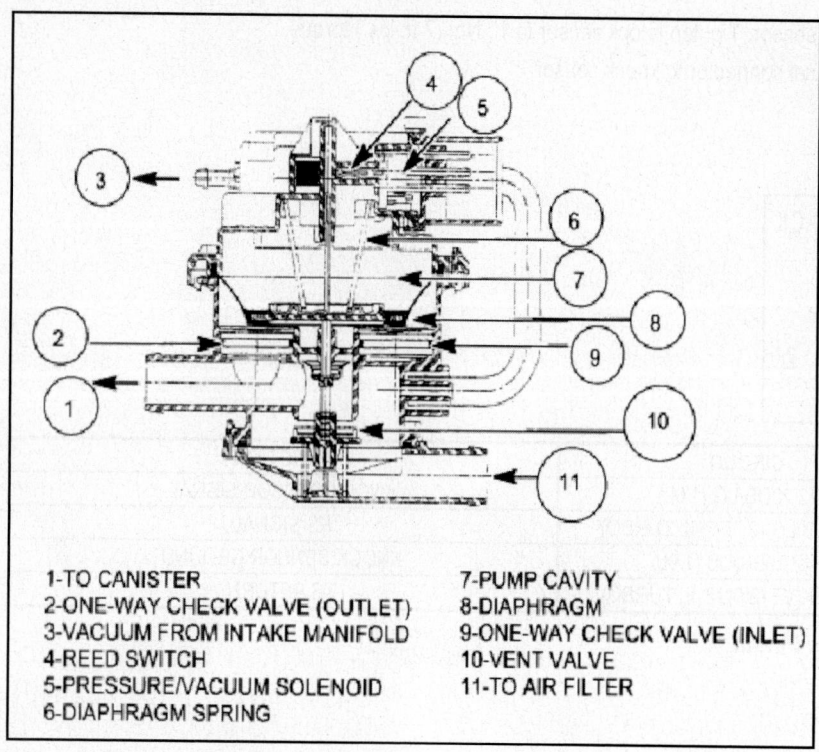

1-TO CANISTER
2-ONE-WAY CHECK VALVE (OUTLET)
3-VACUUM FROM INTAKE MANIFOLD
4-REED SWITCH
5-PRESSURE/VACUUM SOLENOID
6-DIAPHRAGM SPRING

7-PUMP CAVITY
8-DIAPHRAGM
9-ONE-WAY CHECK VALVE (INLET)
10-VENT VALVE
11-TO AIR FILTER

**Leak Detection Pump (LDP) Components**

**Testing**

*ENABLING CONDITIONS TO RUN EVAP LEAK DETECTION TEST*

*NOTE: The following values are approximate and vehicle specific. Use the values seen in pretest/monitor test screen on the DRB III or suitable scan tool. See TSB 25-02-98 for more detail.*

- Cold start: with ambient temperature (obtained from modeling the inlet air temperature sensor on passenger vehicles and the battery temperature sensor on Jeep & truck vehicles) between 4C°(40°F) and 32°C (90°F) for 0.040 leak. Between 4C°(40°F) and 29°C (85°F) for 0.020 leak.
- Engine coolant temperature within: -12°to -8°C (10°to 18°F) of battery/ambient.
- Battery voltage between 10 and 15 volts.

*NOTE: If battery voltage drops below 10 volts for more than 5 seconds during engine cranking, the EVAP leak detection test will not run.*

- Low fuel warning light off (fuel level must be between 15% and 85% for 0.040 leak and 30% and 85% for 0.020 leak).
- MAP sensor reading 22 in Hg or above (This is the manifold absolute pressure, not vacuum).

*NO ENGINE STALL DURING TEST*

If the system does not pass the EVAP Leak Detection Test, the following DTCs may be set:

- P0442 - EVAP LEAK MONITOR 0.040" LEAK DETECTED
- P0455 - EVAP LEAK MONITOR LARGE LEAK DETECTED
- P0456 - EVAP LEAK MONITOR 0.020" LEAK DETECTED
- P1486 - EVAP LEAK MON PINCHED HOSE FOUND

- P1494 - LEAK DETECTION PUMP SW OR MECH FAULT
- P1495 - LEAK DETECTION PUMP SOLENOID CIRCUIT

A DTC will not be set if a one-trip fault is set or if MIL is illuminated for any of the following:

- Purge Solenoid Electrical Fault
- All Engine Controller Self Test Faults
- All Cam and/or Crank Sensor Fault
- All Map Sensor Faults
- Ambient/battery Temperature Sensor Electrical Faults
- All Coolant Sensor Faults
- All TPS Faults
- LDP Pressure Switch Fault
- EGR Solenoid Fault
- All Injector Faults
- BARO Out Of Range
- Vehicle Speed Faults
- LDP Solenoid Circuit

When the ignition key is turned to "ON" the LDP diaphragm should be in the down position and the LDP reed switch should be closed. If the EVAP system has residual pressure, the LDP diaphragm may be up. This could result in the LDP reed switch being open when the key is turned to "ON" and a P1494 fault could be set because the PCM is expecting the reed switch to be closed.

After the key is turned "ON", the PCM immediately tests the LDP solenoid circuit for electrical faults. If a fault is detected, DTC P1495 will set, the MIL will illuminate, and the remaining EVAP Leak Detection Test is canceled.

*NOTE: If battery temperature is not within range, or if the engine coolant temperature is not within a specified range of the battery temperature, the PCM will not run tests for DTC P1494, P1486, P0442, P0455 and P0441. These temperature calibrations may be different between models.*

### Removal & Installation

1. Raise and support vehicle on a hoist.
2. Push locking tab on electrical connector to unlock and remove connector.
3. Loosen the sway bar bracket to remove the pump bracket.
4. Remove pump and bracket as an assembly.
5. Disconnect lines from LDP.
6. Remove filter. Remove pump from bracket.

**To install:**

7. Install pump to bracket and tighten bolts to 1.2 Nm (10.6 inch lbs.).
8. Install filter and tighten to 2.8 Nm (25 inch lbs.).
9. Before installing hoses to LDP, make sure they are not cracked or split. If a hose leaks, it will cause the Check Engine Lamp to illuminate. Connect lines to the LDP.

*NOTE : The LDP bracket must be between the rail and sway bar bracket.*

10. Install pump and bracket assembly to body and tighten bolts to 5.0 Nm (45 inch lbs.).
11. Install sway bar bracket bolt and tighten bolts to 33.8 Nm (25 ft. lbs.).
12. Install electrical connector to pump and push locking tab to lock.
13. Use the scan tool, verify proper operation of LDP.

## MANIFOLD ABSOLUTE PRESSURE (MAP) SENSOR

### Description & Operation

The MAP sensor mounts to the intake manifold. The PCM supplies 5 volts direct current to the MAP sensor. The MAP sensor converts intake manifold pressure into voltage. The PCM monitors the MAP sensor output voltage. As vacuum increases MAP sensor voltage decreases proportionately. Also. as vacuum decreases, MAP sensor voltage increase proportionately.

At key on, before the engine is started, the PCM determines atmospheric air pressure from the MAP sensor voltage. While the engine operates, the PCM determines intake manifold pressure from the MAP sensor voltage. Based on MAP sensor voltage and inputs from other sensors, the PCM adjusts spark advance and the air/fuel mixture.

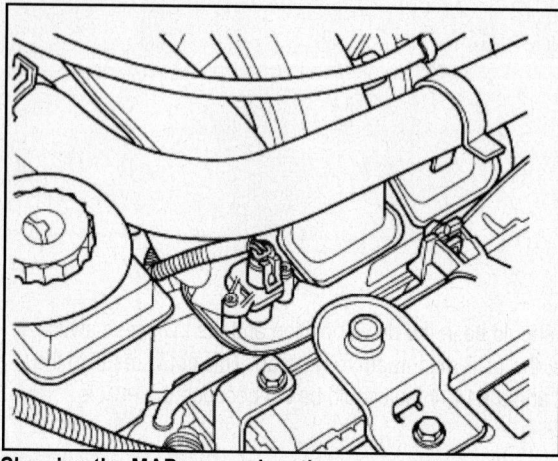

**Showing the MAP sensor location**

If the PCM considers the MAP Sensor information inaccurate, the PCM moves into "limp-in" mode. When the MAP Sensor is in limp-in, the PCM limits the engine speed as a function of the Throttle Position Sensor (TPS) to between 1500 and 4000 rpm. If the MAP Sensor sends realistic signals once again, the PCM moves out of limp-in and resumes using the MAP values. During limp-in a DTC is set and the MIL illuminates.

### Removal & Installation

1. Disconnect the electrical connector from the MAP sensor.
2. Remove sensor mounting screws.
3. Remove sensor.

**To install:**

4. Insert sensor into intake manifold while making sure not to damage O-ring seal.
5. Tighten mounting screws to 4.5 Nm (40 inch lbs.) torque for plastic manifold.
6. Attach electrical connector to sensor.

**Connector Pinouts**

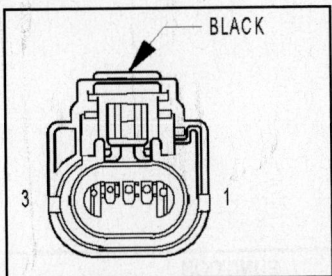

| PIN # | CIRCUIT | FUNCTION |
|---|---|---|
| 1 | K1 20DG/RD | MAP SENSOR SIGNAL |
| 2 | K901 20DB/VT | SENSOR GROUND |
| 3 | K6 18VT/WT | 5 VOLT SUPPLY |

**Neon 1.6L MAP Sensor**

| PIN # | CIRCUIT | FUNCTION |
|---|---|---|
| 1 | K1 20DG/RD | MAP SIGNAL |
| 2 | K4 20BK/LB | SENSOR GROUND |
| 3 | K7 20OR | 5 VOLT SUPPLY |

**Neon 2.0L MAP Sensor**

| PIN # | CIRCUIT | FUNCTION |
|---|---|---|
| 1 | K1 20DG/RD | MAP SIGNAL |
| 2 | K4 20BK/LB | SENSOR GROUND |
| 3 | K7 20OR | 5 VOLT SUPPLY |

**Neon 2.4L Turbo MAP Sensor**

MANIFOLD TUNING VALVE SOLENOID

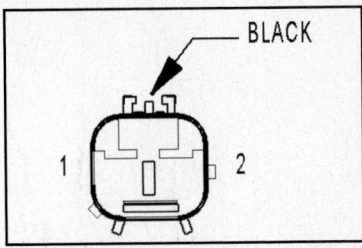

| PIN # | CIRCUIT | FUNCTION |
|---|---|---|
| 1 | K201 20BR/YL | MTV RELAY OUTPUT |
| 2 | Z1 18BK | GROUND |

**Neon Manifold Tuning Valve Solenoid (RT) Connector Pinout**

## NATURAL VACUUM LEAK DETECTOR (NVLD)

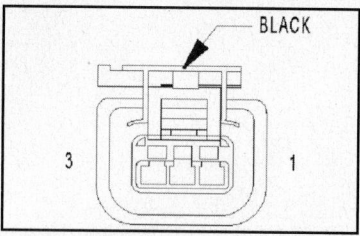

| PIN # | CIRCUIT | FUNCTION |
|-------|---------|----------|
| 1 | Z12 20BK (EARLY BUILD) | GROUND |
| 1 | Z1 20BK (NORMAL BUILD) | GROUND |
| 2 | K107 20OR | NVLD SWITCH SIGNAL |
| 3 | K106 20WT/DG | NVLD SOLENOID CONTROL |

**Neon 1.6L & 2.0L NVLD Connector Pinout**

## POWER STEERING PRESSURE (PSP) SWITCH

### Description & Operation

A power steering pressure switch is used to improve the vehicle's idle quality. The pressure switch improves vehicle idle quality by causing a readjustment of the engine idle speed, as necessary, when increased fluid pressure is sensed in the power steering system.

The pressure switch functions by signaling the Powertrain Control Module that an increase in pressure of the power steering system is putting additional load on the engine. This type of condition exists when the front tires of the vehicle are turned while the vehicle is stationary and the engine is at idle speed. These periods of high pump load and low engine rpm are most commonly found during parking maneuvers. When the Powertrain Control Module receives the signal from the power steering pressure switch, it directs the engine to increase its idle speed. This increase in engine idle speed compensates for the additional load, thus maintaining the required engine idle speed and idle quality.

When power steering pump pressure exceeds 2758 kPa (400 psi), the switch is open. The PCM increases idle airflow through the IAC motor to prevent engine stalling. The PCM sends 12 volts through a resister to the sensor circuit to ground. When pump pressure is low, the switch is closed.

**The power steering pressure switch (1) is located on the power steering gear**

**Connector Pinouts**

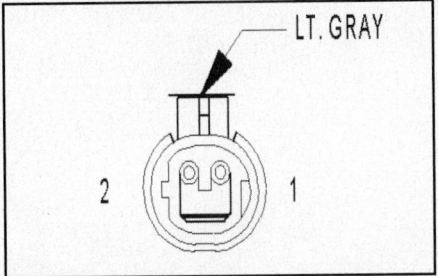

| PIN # | CIRCUIT | FUNCTION |
|---|---|---|
| 1 | K10 20DB/OR (1.6L) | POWER STEERING PRESSURE SWITCH SENSE |
| 1 | K10 20DB/OR (2.0L/2.4L TURBO) | PSP SWITCH SIGNAL |
| 2 | Z12 20BK/TN | GROUND |

**Neon PSP Switch Connector Pinout**

## POWERTRAIN CONTROL MODULE (PCM)

### Description & Operation

The Powertrain Control Module (PCM) is a digital computer containing a microprocessor. The PCM receives input signals from various switches and sensors that are referred to as PCM Inputs. Based on these inputs, the PCM adjusts various engine and vehicle operations through devices that are referred to as PCM Outputs.

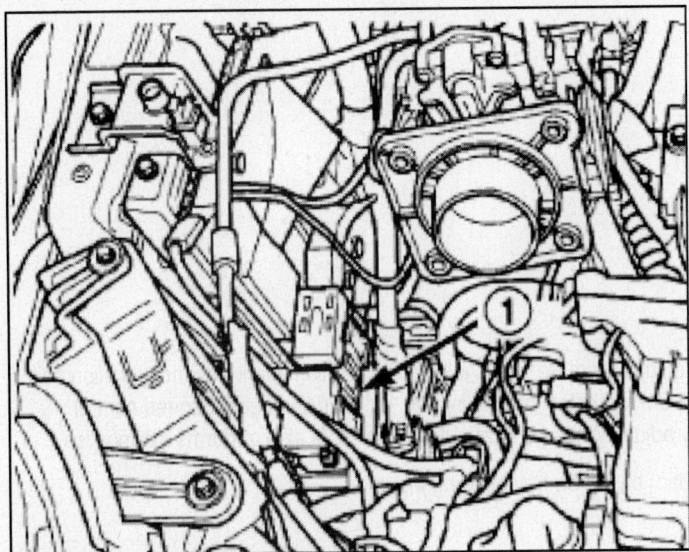

**Location of the PCM on 2.0L engines**

*PCM INPUTS:*

- Air Conditioning Controls
- Battery Voltage
- Inlet Air/Battery Temperature Sensor
- Brake Switch
- Camshaft Position Sensor
- Crankshaft Position Sensor
- Engine Coolant Temperature Sensor
- Fuel Level Sensor

- Ignition Switch
- Inlet Air/Intake Air Temperature Sensor
- Knock Sensor
- Manifold Absolute Pressure (MAP) Sensor
- Oxygen Sensors
- Power Steering Pressure Switch
- SCI Receive
- Speed Control Switches
- Throttle Position Sensor
- Transmission Park/Neutral Switch (automatic transmission)
- Vehicle Speed Sensor

### PCM OUTPUTS:

- Air Conditioning WOT Relay
- Auto Shutdown (ASD) Relay
- Charging Indicator Lamp
- Data Link Connector
- Proportional Purge Solenoid
- EGR Solenoid
- Fuel Injectors
- Fuel Pump Relay
- Generator Field
- Idle Air Control Motor
- Ignition Coils
- Malfunction Indicator (Check Engine) Lamp
- Radiator Fan Relay
- Speed Control Solenoids
- Tachometer
- Torque Converter Clutch Solenoid

Based on inputs it receives, the PCM adjusts fuel injector pulse width, idle speed, ignition spark advance, ignition coil dwell and EVAP canister purge operation. The PCM regulates the cooling fan, air conditioning and speed control systems, and changes generator charge rate by adjusting the generator field. The PCM also performs diagnostics.

The PCM adjusts injector pulse width (air-fuel ratio) based on the following inputs:

- Battery voltage
- Coolant temperature
- Inlet Air/Intake air temperature
- Exhaust gas content (oxygen sensor)
- Engine speed (crankshaft position sensor)
- Manifold absolute pressure
- Throttle position

The PCM adjusts ignition timing based on the following inputs:

- Coolant temperature
- Inlet Air/Intake air temperature
- Engine speed (crankshaft position sensor)

- Knock sensor
- Manifold absolute pressure
- Throttle position
- Transmission gear selection (park/neutral switch)

The PCM also adjusts engine idle speed through the idle air control motor based on the following inputs:

- Air conditioning sense
- Battery voltage
- Battery temperature
- Brake switch
- Coolant temperature
- Engine speed (crankshaft position sensor)
- Engine run time
- Manifold absolute pressure
- Power steering pressure switch
- Throttle position
- Transmission gear selection (park/neutral switch)
- Vehicle distance (speed)

The crankshaft position sensor signal is sent to the PCM. If the PCM does not receive the signal within approximately one second of engine cranking, it deactivates the ASD relay and fuel pump relay. When these relays deactivate, power is shut off from the fuel injectors, ignition coils, heating element in the oxygen sensors and the fuel pump.

The PCM contains a voltage converter that changes battery voltage to a regulated 8 volts direct current to power the camshaft position sensor, crankshaft position sensor and vehicle speed sensor. The PCM also provides a 5-volt direct current supply for the manifold absolute pressure sensor and throttle position sensor.

**PCM Ground**

Ground is provided through multiple pins of the PCM connector. Depending on the vehicle there may be as many as three different ground pins. There are power grounds and sensor grounds.

The power grounds are used to control the ground side of any relay, solenoid, ignition coil or injector. The signal ground is used for any input that uses sensor return for ground, and the ground side of any internal processing component.

The SBEC III case is shielded to prevent RFI and EMI. The PCM case is grounded and must be firmly attached to a good clean body ground.

Internally all grounds are connected together, however there is noise suppression on the sensor ground. For EMI and RFI protection the case is also grounded separately from the ground pins.

**Removal & Installation**

1. Disconnect the negative battery cable.
2. Remove the air cleaner box.
3. Remove the gray and black connector from the PCM.
4. Remove the harness clip bracket from PCM bracket.
5. Remove the nut from the upper bracket mount.
6. Raise vehicle and support on hoist.
7. Remove 2 lower bracket bolts.
8. Remove 4 screws from bracket and remove bracket from PCM.

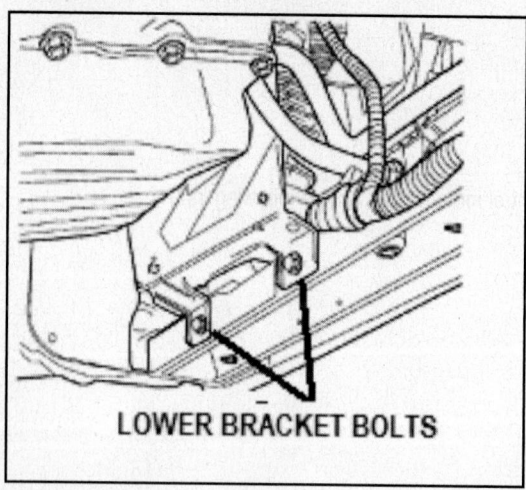

**LOWER BRACKET BOLTS**

**Locating 2 lower bracket bolts for PCM Removal & Installation**

**To install:**

9. Install bracket to PCM and tighten screws.
10. Install PCM and bracket to body and tighten the 2 lower bolts.
11. Lower vehicle.
12. Install upper bracket nut and tighten.
13. Clip in wiring harness bracket.
14. Install gray and black connectors to the PCM.
15. Install the air cleaner box.
16. Connect the negative battery cable.

**Connector Pinouts**

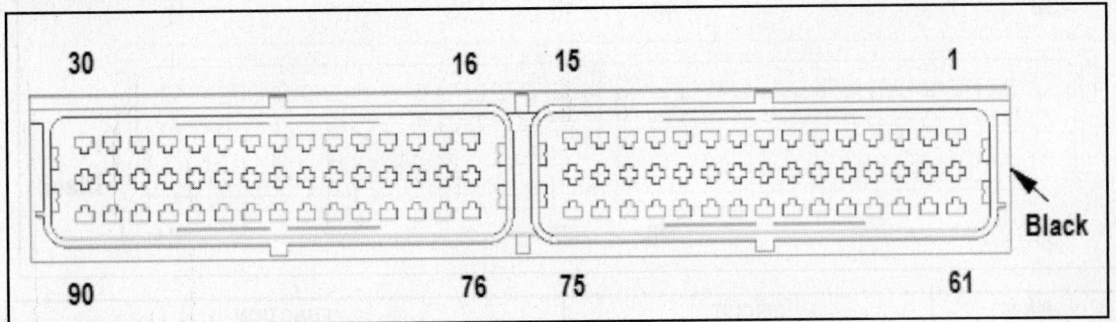

| PIN # | CIRCUIT | FUNCTION |
|---|---|---|
| 1 | A14 18RD/WT | FUSED B (+) |
| 2-4 | - | - |
| 5 | C28 18DB/OR | A/C COMPRESSOR CLUTCH RELAY CONTROL |
| 6 | D25 20VT/YL | PCI BUS |
| 7 | K81 20DB/DG | ACCELERATOR PEDAL POSITION SENSOR SIGNAL NO. 2 |
| 8 | C20 18BR/OR | A/C SWITCH SENSE |
| 9 | K921 20BR/LG | INLET AIR TEMPERATURE SENSOR GROUND |
| 10 | N907 20BR/OR | SENSOR GROUND |
| 11 | K901 20DB/VT | SENSOR GROUND |
| 12 | F858 20PK | ACCELERATOR PEDAL POSITION SENSOR 5 VOLT SUPPLY NO. 1 |
| 13 | F859 20DG/PK | ACCELERATOR PEDAL POSITION SENSOR 5 VOLT SUPPLY NO. 2 |
| 14 | F855 20OR/PK | THROTTLE POSITION SENSOR 5 VOLT SUPPLY |
| 15 | - | - |
| 16 | K1 20DG/RD | MAP SENSOR SIGNAL |
| 17 | K2 20TN/BK | ENGINE COOLANT TEMPERATURE SENSOR SIGNAL |
| 18 | K41 20BK/DG | OXYGEN SENSOR 1/1 SIGNAL |
| 19 | K167 20BR/YL | ACCELERATOR PEDAL POSITION SENSOR GROUND NO. 1 |
| 20 | K42 20DB/LG | KNOCK SENSOR SIGNAL |
| 21 | K24 20GY/BK | CRANKSHAFT POSITION SENSOR SIGNAL |
| 22 | K981 20BR/DG | ACCELERATOR PEDAL POSITION SENSOR GROUND NO. 2 |
| 23 | K922 20DB/OR | THROTTLE POSITION SENSOR GROUND |
| 24 | K902 20BR | OXYGEN SENSOR GROUND |
| 25 | K924 20BR/WT | CRANKSHAFT POSITION SENSOR GROUND |
| 26 | K942 20BR/DB | KNOCK SENSOR GROUND |
| 27 | - | - |
| 28 | Z12 16BK/TN | GROUND |
| 29 | Z12 16BK/TN | GROUND |
| 30 | K17 18DB/TN | IGNITION COIL NO. 2 DRIVER |
| 31 | V50 20VT/OR | ELECTRONIC THROTTLE CONTROL POSITIVE MOTOR CONTROL |
| 32 | V51 20VT/BR | ELECTRONIC THROTTLE CONTROL NEGATIVE MOTOR CONTROL |
| 33 | K4 20BK/LB | SENSOR GROUND |
| 34 | K51 18DB/YL | AUTOMATIC SHUT DOWN RELAY CONTROL |
| 35 | K90 20TN | ENGINE STARTER MOTOR RELAY CONTROL |
| 36-38 | - | - |
| 39 | K10 20WT | POWER STEERING PRESSURE SWITCH SENSE |
| 40 | G9 20GY/BK | BRAKE FLUID LEVEL SWITCH SENSE |
| 41 | D20 20LG | SCI RECEIVE |

**Neon 1.6L PCM 90-Pin Connector Pinout (1 of 3)**

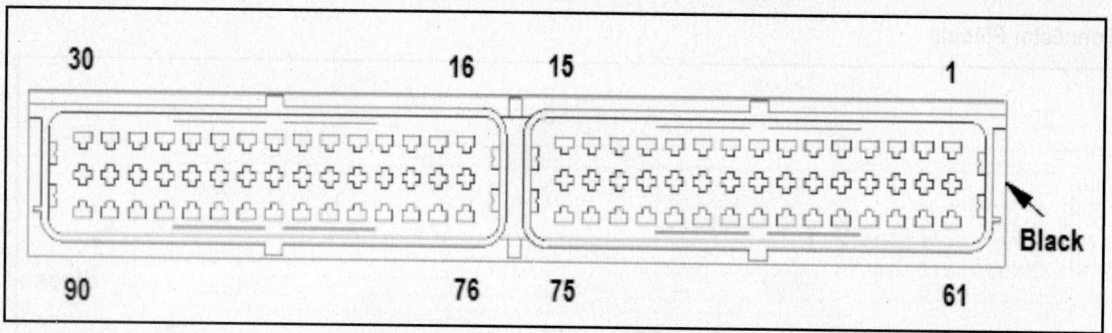

| PIN # | CIRCUIT | FUNCTION |
|---|---|---|
| 42 | K944 20BR/GY | CAMSHAFT POSITION SENSOR GROUND |
| 43 | F854 20GY/PK | CAMSHAFT POSITION 5 VOLT SUPPLY |
| 44 | K7 20OR | CRANKSHAFT POSITION 5 VOLT SUPPLY |
| 45 | F857 20OR/PK | VEHICLE SPEED SENSOR 5 VOLT SUPPLY |
| 46 | K22 20BR/OR | THROTTLE POSITION SENSOR SIGNAL NO. 1 |
| 47 | K141 20TN/WT | OXYGEN SENSOR ½ SIGNAL |
| 48 | - | - |
| 49 | K44 20TN/YL | CAMSHAFT POSITION SENSOR SIGNAL |
| 50 | G6 20GY | ENGINE OIL PRESSURE SWITCH SENSE |
| 51-52 | - | - |
| 53 | K904 20DB/DG | OXYGEN SENSOR GROUND |
| 54 | K11 18WT/DB | FUEL INJECTOR NO. 1 DRIVER |
| 55 | K52 20PK/BK | EVAP/PURGE SOLENOID CONTROL |
| 56-58 | - | - |
| 59 | Z12 16BK/TN | GROUND |
| 60 | K19 18BK/GY | IGNITION COIL NO. 1 DRIVER |
| 61 | F12 18DB/WT | FUSED IGNITION SWITCH OUTPUT (RUN-START) |
| 62 | - | - |
| 63 | K914 20BR/WT | GROUND |
| 64 | - | - |
| 65 | C27 18DB/PK | RADIATOR FAN HIGH RELAY CONTROL |
| 66 | K31 18BR | FUEL PUMP RELAY CONTROL |
| 67 | - | - |
| 68 | L50 20WT/TN | BRAKE LAMP SWITCH OUTPUT |
| 69 | K119 20LG/BK | CLUTCH UP SWITCH SIGNAL |
| 70 | T141 20YL/RD | CLUTCH INTERLOCK SWITCH SENSE |
| 71 | K29 20WT/PK | BRAKE LAMP SWITCH SENSE |
| 72 | K21 20BK/RD | INLET AIR TEMPERATURE SENSOR SIGNAL |
| 73 | D21 20PK | SCI TRANSMIT |
| 74 | K80 20DG/LG | ACCELERATOR PEDAL POSITION SENSOR SIGNAL NO. 1 |
| 75 | K6 18VT/WT | 5 VOLT SUPPLY |
| 76 | V37 20RD/LG | SPEED CONTROL SWITCH SIGNAL |
| 77 | K122 20DB/GY | THROTTLE POSITION SENSOR SIGNAL NO. 2 |
| 78 | - | - |
| 79 | G7 20WT/OR | VEHICLE SPEED SENSOR SIGNAL |
| 80 | - | - |
| 81 | K79 18OR/RD | OXYGEN SENSOR 1/1 HEATER CONTROL |
| 82 | K12 18TN | FUEL INJECTOR NO. 2 DRIVER |

**Neon 1.6L PCM 90-Pin Connector Pinout (2 of 3)**

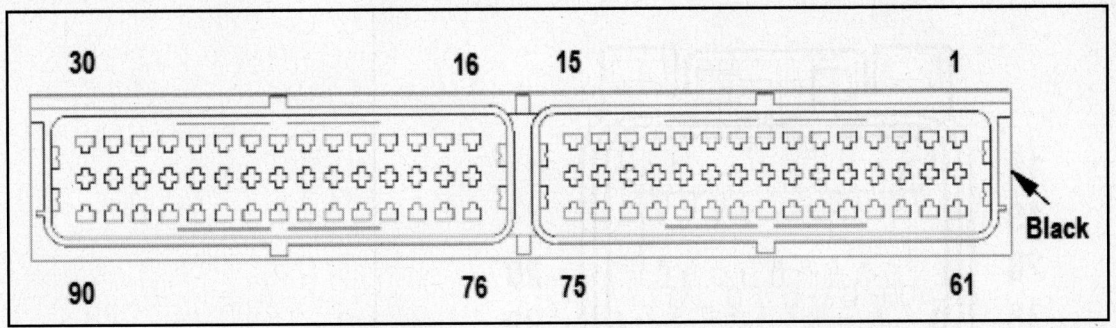

| PIN # | CIRCUIT | FUNCTION |
|-------|---------|----------|
| 83 | K13 18YL/WT | FUEL INJECTOR NO. 3 DRIVER |
| 84 | K14 18LB/BR | FUEL INJECTOR NO. 4 DRIVER |
| 85 | K199 18BR/VT | OXYGEN SENSOR 1/2 HEATER CONTROL |
| 86-87 | - | |
| 88 | A142 18DG/OR | AUTOMATIC SHUT DOWN RELAY OUTPUT |
| 89 | K20 18DG | GENERATOR FIELD DRIVER (+) |
| 90 | Z12 16BK/TN | GROUND |

**Neon 1.6L PCM 90-Pin Connector Pinout (3 of 3)**

*1.6L W/NEXT GENERATION CONTROLLER (NGC)*

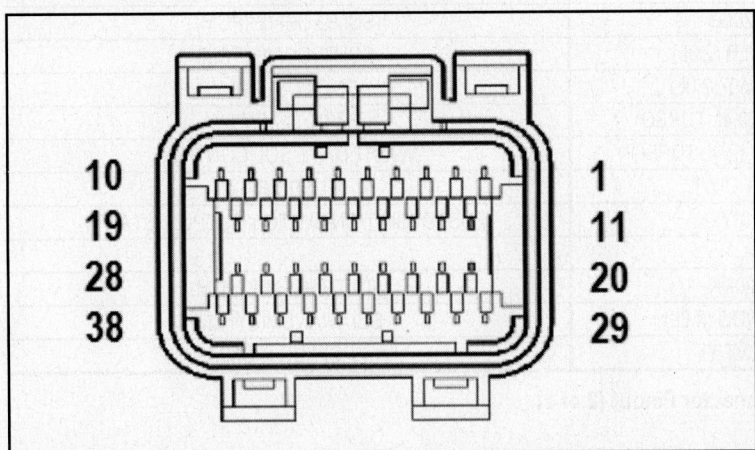

| PIN # | CIRCUIT | FUNCTION |
|-------|---------|----------|
| 1-8 | - | - |
| 9 | Z11 18BK/WT | GROUND |
| 10 | - | - |
| 11 | F12 18DB/WT (2.0L) | FUSED IGNITION SWITCH OUTPUT (RUN-START) |
| 11 | F12 18DB/RD (2.4L TURBO) | FUSED IGNITION SWITCH OUTPUT (RUN-START) |
| 12 | F11 20RD/WT (2.0L AUTOSTICK) | IGNITION SWITCH OUTPUT (OFF-RUN-START) |
| 12 | F11 20RD/WT (2.0L EXCEPT AUTOSTICK) | FUSED IGNITION SWITCH OUTPUT (RUN-START) |
| 13 | G7 20WT/OR | VEHICLE SPEED SIGNAL |
| 14 | G9 20GY/BK | BRAKE FLUID LEVEL SWITCH SENSE |
| 15 | K55 18LB (2.4L TURBO) | TIP SOL CONTROL |

**Neon 1.6L w/NGC PCM C1 38-pin Connector Pinout (1 of 2)**

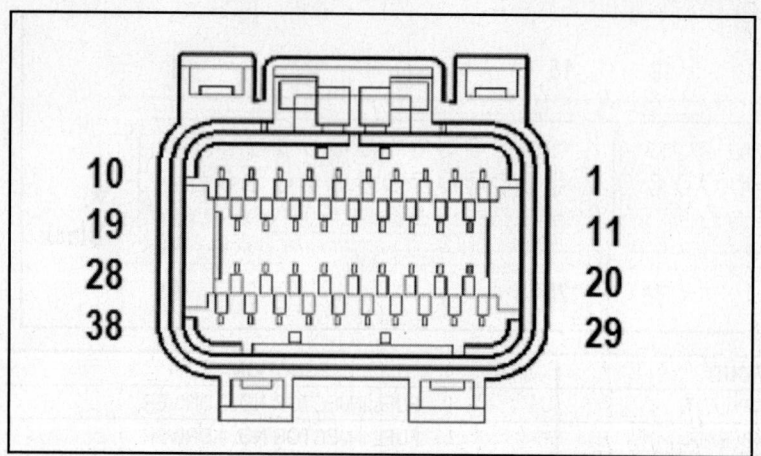

| PIN # | CIRCUIT | FUNCTION |
|---|---|---|
| 16 | - | - |
| 17 | K150 18DB/YL (2.4L TURBO) | SURGE SOL CONTROL |
| 18 | Z12 18BK/TN | GROUND |
| 19 | - | - |
| 20 | G6 20GY | OIL PRESSURE SIGNAL |
| 21 | - | - |
| 22 | K145 20BR/OR | AAT SIGNAL |
| 23 | K153 18LB (2.4L TURBO) | TIP SIGNAL |
| 24 | - | - |
| 25 | D20 20LG | SCI RECEIVE (PCM) |
| 26 | D6 20PK/LB (2.0L) | SCI RECEIVE (TCM) |
| 27 | K6 20VT/WT (2.0L) | 5 VOLT SUPPLY |
| 27 | K6 18VT/WT (2.4L TURBO) | 5 VOLT SUPPLY |
| 28 | K137 18DB/GY (2.4L TURBO) | WASTEGATE SOL CONTROL |
| 29 | A14 18RD/WT | FUSED B(+) |
| 30 | A41 16YL | FUSED IGNITION SWITCH OUTPUT (START) |
| 31-35 | - | - |
| 36 | D21 20PK | SCI TRANSMIT (PCM) |
| 37 | D15 20WT/DG (2.0L) | SCI TRANSMIT (TCM) |
| 38 | D25 20VT/YL | PCI BUS (PCM) |

**Neon 1.6L w/NGC PCM C1 38-pin Connector Pinout (2 of 2)**

**2.0L W/NEXT GENERATION CONTROLLER (NGC)**

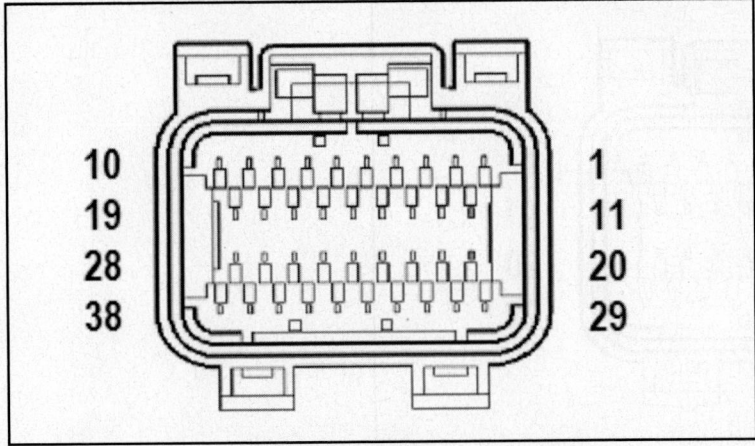

| PIN # | CIRCUIT | FUNCTION |
|-------|---------|----------|
| 1-8 | - | - |
| 9 | K17 16DB/TN | COIL CONTROL NO. 2 |
| 10 | K19 16BK/GY | COIL CONTROL NO. 1 |
| 11 | K14 18LG/BR | INJECTOR CONTROL NO. 4 |
| 12 | K13 18YL/WT | INJECTOR CONTROL NO. 3 |
| 13 | K12 18TN | INJECTOR CONTROL NO. 2 |
| 14 | K11 18WT/DB | INJECTOR CONTROL NO. 1 |
| 15 | - | - |
| 16 | K200 20BR/WT (RT) | MTV CONTROL |
| 17 | K199 18BR/VT | O2 1/2 HEATER CONTROL |
| 18 | K99 18BR/OR | O2 1/1 HEATER CONTROL |
| 19 | K20 18DG | GEN FIELD CONTROL |
| 20 | K2 20VT/LG | ECT SIGNAL |
| 21 | K22 20OR/DB | TP SIGNAL |
| 22 | - | - |
| 23 | K1 20DG/RD | MAP SIGNAL |
| 24 | K45 20BK/VT | KS RETURN |
| 25 | K42 20DB/LG | KS SIGNAL |
| 26 | - | - |
| 27 | K4 18BK/LB | SENSOR GROUND |
| 28 | K961 18BR/VT | IAC RETURN |
| 29 | K7 20OR | 5 VOLT SUPPLY |
| 30 | K21 20BK/RD | IAT SIGNAL |
| 31 | K41 20BK/DG | O2 1/1 SIGNAL |
| 32 | K904 18DB/DG (EXCEPT RT) | O2 RETURN (UP) |
| 32 | K904 18BR/OR (RT) | O2 RETURN (UP) |
| 33 | K141 20TN/WT | O2 1/2 SIGNAL |
| 34 | K44 20TN/YL | CMP SIGNAL |
| 35 | K24 20GY/BK | CKP SIGNAL |
| 36-37 | - | - |
| 38 | K610 18VT/GY | IAC MOTOR CONTROL |

**Neon 2.0L (w/NGC) PCM C2 30-pin Connector Pinout**

### 2.4L TURBO W/NEXT GENERATION CONTROLLER (NGC)

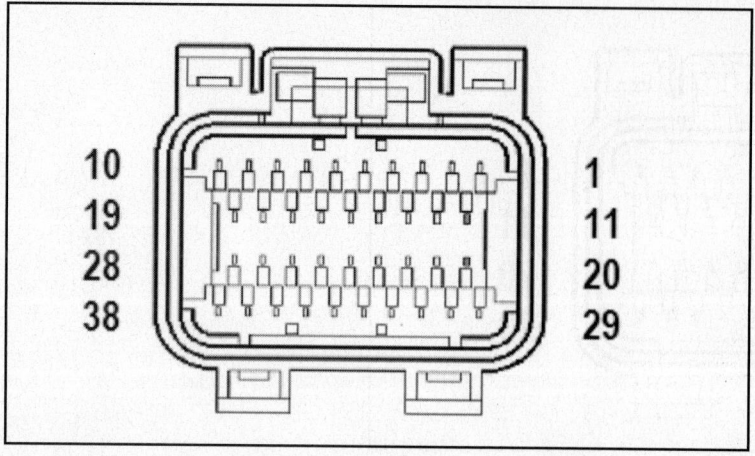

| PIN # | CIRCUIT | FUNCTION |
|-------|---------|----------|
| 1-8 | - | - |
| 9 | K17 16DB/TN | COIL CONTROL NO. 2 |
| 10 | K19 16BK/GY | COIL CONTROL NO. 1 |
| 11 | K14 18LB/BR | INJECTOR CONTROL NO. 4 |
| 12 | K13 18YL/WT | INJECTOR CONTROL NO. 3 |
| 13 | K12 18TN | INJECTOR CONTROL NO. 2 |
| 14 | K11 18WT/DB | INJECTOR CONTROL NO. 1 |
| 15-16 | - | - |
| 17 | K199 18BR/VT | O2 1/2 HEATER CONTROL |
| 18 | K99 18BR/OR | O2 1/1 HEATER CONTROL |
| 19 | K20 18DG | GEN FIELD CONTROL |
| 20 | K2 20TN/BK | ECT SIGNAL |
| 21 | K22 20OR/DB | TP SIGNAL |
| 22 | - | - |
| 23 | K1 20DG/RD | MAP SIGNAL |
| 24 | K45 20BK/VT | KS RETURN |
| 25 | K42 20DB/LG | KS SIGNAL |
| 26 | - | - |
| 27 | K4 18BK/LB | SENSOR GROUND |
| 28 | K961 18BR/VT | IAC RETURN |
| 29 | K7 18OR | 5 VOLT SUPPLY |
| 30 | K21 20BK/RD | IAT SIGNAL |
| 31 | K41 20BK/DG | O2 1/1 SIGNAL |
| 32 | K904 18DB/DG | O2 RETURN (UP) |
| 33 | K141 20TN/WT | O2 1/2 SIGNAL |
| 34 | K44 20TN/YL | CMP SIGNAL |
| 35 | K24 20GY/BK | CKP SIGNAL |
| 36-37 | - | - |
| 38 | K610 18VT/GY | IAC MOTOR CONTROL |

**Neon 2.4L Turbo (NGC) PCM C2 38-pin Connector Pinout**

### 2.0L & 2.4L TURBO W/NEXT GENERATION CONTROLLER (NGC)

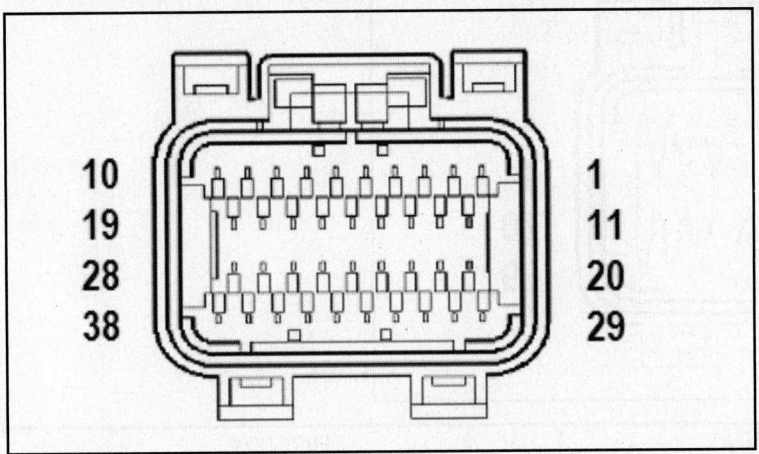

| PIN # | CIRCUIT | FUNCTION |
|---|---|---|
| 1-2 | - | - |
| 3 | K51 20DB/YL (2.0L) | AUTOMATIC SHUT DOWN RELAY CONTROL |
| 3 | K51 18DB/YL (2.4L TURBO) | AUTOMATIC SHUT DOWN RELAY CONTROL |
| 4 | C27 18DB/PK (2.4L TURBO) | HIGH SPEED RAD FAN RELAY CONTROL |
| 5 | V35 20LG/RD (2.0L) | S/C VENT CONTROL |
| 6 | C27 20DB/PK (2.0L) | RAD FAN HIGH RELAY CONTROL |
| 6 | C24 18DB/RD (2.4L TURBO) | RAD FAN LOW RELAY CONTROL |
| 7 | V32 20YL/RD (2.0L) | S/C SUPPLY |
| 8 | K106 20WT/DG (2.0L) | NVLD SOLENOID CONTROL |
| 8 | K106 18WT/DG (2.4L TURBO) | NVLD SOLENOID CONTROL |
| 9-10 | - | - |
| 11 | C28 20DB/OR (2.0L) | A/C CLUTCH RELAY CONTROL |
| 11 | C28 18DB/OR (2.4L TURBO) | A/C CLUTCH RELAY CONTROL |
| 12 | V36 20TN/RD (2.0L) | S/C VACUUM CONTROL |
| 13-16 | - | - |
| 17 | K167 20BR/YL (2.0L) | SENSOR GROUND |
| 17 | K167 18BR/YL (2.4L TURBO) | SENSOR GROUND 2 |
| 18 | | |
| 19 | A142 18DG/OR | AUTOMATIC SHUT DOWN RELAY OUTPUT |
| 20 | K52 20PK/BK | EVAP/PURGE CONTROL |
| 21 | T141 20YL/RD | CLUTCH INTERLOCK SWITCH SIGNAL |
| 22 | - | - |
| 23 | K29 20WT/PK | BRAKE SWITCH SIGNAL |
| 24 | C20 20BR/OR (2.0L) | A/C SWITCH SENSE |
| 24 | C20 20BR (2.4L TURBO) | A/C SWITCH SENSE |
| 25 | - | - |
| 26 | T44 20YL/LB (2.0L EATX EXCEPT EXPORT) | AUTOSTICK DOWNSHIFT SWITCH SIGNAL |
| 26 | T44 18YL/LB (2.0L EATX EXPORT) | AUTOSTICK DOWNSHIFT SWITCH SIGNAL |
| 26 | K119 20LG/BK (MTX EXC. 2.4L TURBO) | CLUTCH UP SWITCH SIGNAL |
| 26 | K119 20LG/OR (MTX LHD 2.0L EXPORT) | CLUTCH UP SWITCH SIGNAL |
| 27 | T5 20LG/LB (2.0L EXCEPT EXPORT) | AUTOSTICK UPSHIFT SWITCH SIGNAL |
| 27 | T5 18LG/LB (2.0L EXPORT) | AUTOSTICK UPSHIFT SWITCH SIGNAL |
| 28 | A142 18DG/OR | AUTOMATIC SHUT DOWN RELAY OUTPUT |
| 29 | K108 20WT/TN | EVAP/PURGE RETURN |

**Neon (w/NGC) PCM C3 38-pin Connector Pinout (1 of 2)**

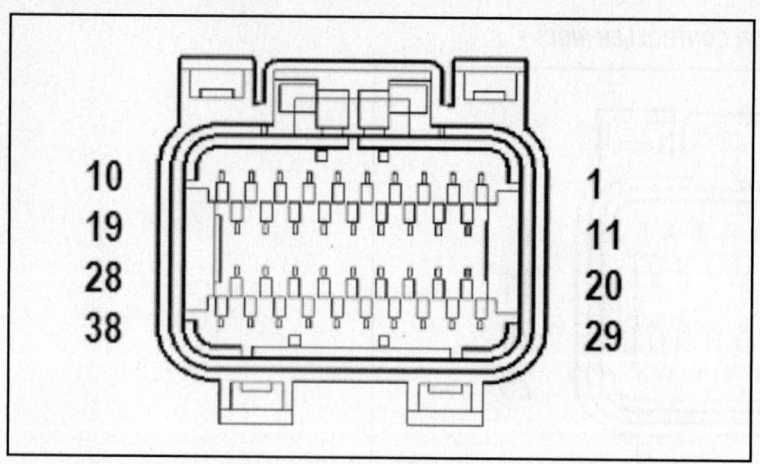

| PIN # | CIRCUIT | FUNCTION |
|---|---|---|
| 30 | K10 20DB/OR | PSP SWITCH SIGNAL |
| 31 | - | - |
| 32 | K118 20PK/YL | BATTERY TEMP SIGNAL |
| 33 | - | - |
| 34 | V37 20RD/LG (2.0L) | S/C SWITCH SIGNAL |
| 35 | K107 20OR (2.0L) | NVLD SWITCH SIGNAL |
| 35 | K107 18OR (2.4L TURBO) | NVLD SWITCH SIGNAL |
| 36 | - | - |
| 37 | K31 20BR (2.0L) | FUEL PUMP RELAY CONTROL |
| 37 | K31 18BR (2.4L TURBO) | FUEL PUMP RELAY CONTROL |
| 38 | K90 20TN | STARTER RELAY CONTROL |

Neon (w/NGC) PCM C3 38-pin Connector Pinout (2 of 2)

WITH EATX & NGC

| PIN # | CIRCUIT | FUNCTION |
|---|---|---|
| 1 | T60 18BR | OVERDRIVE SOLENOID CONTROL |
| 2 | T59 18PK/BK | UNDERDRIVE SOLENOID CONTROL |
| 3 | - | - |
| 4 | - | - |
| 5 | - | - |
| 6 | T19 18WT | 2-4 SOLENOID CONTROL |
| 7-9 | - | - |
| 10 | T20 18LB | LOW/REVERSE SOLENOID CONTROL |
| 11 | - | - |
| 12 | Z13 16BK/RD | GROUND |
| 13 | - | - |
| 14 | Z13 16BK/RD | GROUND |
| 15 | T1 20LG/BK | TRS T1 SENSE |
| 16 | T3 20VT | TRS T3 SENSE |
| 17 | - | - |

Neon w/EATX & NGC PCM C4 38-pin Connector Pinout (1 of 2)

WITH EATX & NGC (CONTINUED)

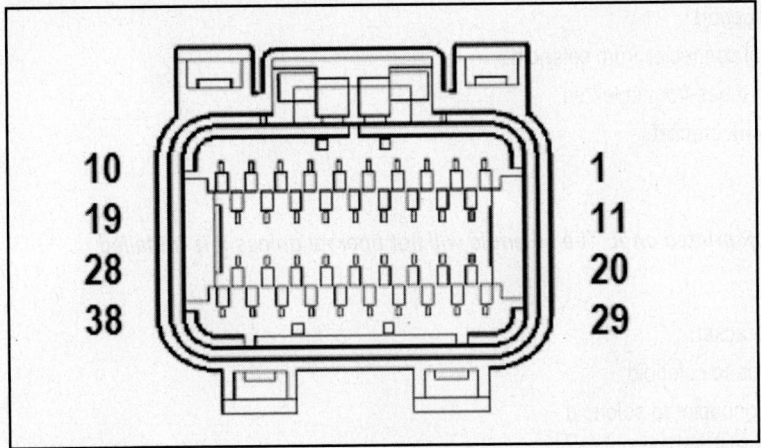

| PIN # | CIRCUIT | FUNCTION |
|---|---|---|
| 18 | T15 20LG | TRANSMISSION CONTROL RELAY CONTROL |
| 19 | T16 16RD | TRANSMISSION CONTROL RELAY OUTPUT |
| 20 | - | - |
| 21 | | |
| 22 | T9 18OR/BK | OVERDRIVE PRESSURE SWITCH SENSE |
| 23-26 | - | - |
| 27 | T41 20BK/WT | TRS T41 SENSE |
| 28 | T16 16RD | TRANSMISSION CONTROL RELAY OUTPUT |
| 29 | T50 18DG | LOW/REVERSE PRESSURE SWITCH SENSE |
| 30 | T47 18YL/BK | 2-4 PRESSURE SWITCH SENSE |
| 31 | - | - |
| 32 | T14 20LG/WT | OUTPUT SPEED SENSOR SIGNAL |
| 33 | T52 20RD/BK | INPUT SPEED SENSOR SIGNAL |
| 34 | T13 20DB/BK | SPEED SENSOR GROUND |
| 35 | T54 20VT/PK | TRANSMISSION TEMPERATURE SENSOR SIGNAL |
| 36 | - | - |
| 37 | T42 20VT/WT | TRS T42 SENSE |
| 38 | - | - |

**Neon w/EATX & NGC PCM C4 38-pin Connector Pinout (2 of 2)**

## PROPORTIONAL PURGE SOLENOID

### Description & Operation

All vehicles use a proportional purge solenoid. The solenoid regulates the rate of vapor flow from the EVAP canister to the throttle body. The PCM operates the solenoid.

During the cold start warm-up period and the hot start time delay, the PCM does not energize the solenoid. When de-energized, no vapors are purged.

The proportional purge solenoid operates at a frequency of 200 Hz and is controlled by an engine controller circuit that senses the current being applied to the proportional purge solenoid and then adjusts that current to achieve the desired purge flow. The proportional purge solenoid controls the purge rate of fuel vapors from the vapor canister and fuel tank to the engine intake manifold. The solenoid attaches to a bracket near the steering gear.

Removal & Installation

1. Raise vehicle and support.
2. Disconnect electrical connector from solenoid.
3. Disconnect vacuum tubes from solenoid.
4. Remove solenoid from bracket.

**To install:**

*Note: The top of the solenoid has TOP printed on it. The solenoid will not operate unless it is installed correctly.*

5. Install solenoid on bracket.
6. Connect vacuum tube to solenoid.
7. Connect electrical connector to solenoid.
8. Lower vehicle.

## THROTTLE POSITION (TP) SENSOR

### Description & Operation

The throttle position sensor mounts to the side of the throttle body. The Throttle Position Sensor (TPS) connects to the throttle blade shaft. The TPS is a variable resistor that provides the PCM with an input signal (voltage). The signal represents throttle blade position. As the position of the throttle blade changes, the resistance of the TPS changes.

The PCM supplies approximately 5 volts DC to the TPS. The TPS output voltage (input signal to the powertrain control module) represents throttle blade position. The TPS output voltage to the PCM varies from approximately 0.35 to 1.03 volts at minimum throttle opening (idle) to a maximum of 3.1 to 4.0 volts at wide-open throttle.

Along with inputs from other sensors, the PCM uses the TPS input to determine current engine operating conditions. The PCM also adjusts fuel injector pulse width and ignition timing based on these inputs. When the TPS indicates a voltage that is too high, too low or not believable, the PCM sets a DTC. When the DTC is set, the MIL is illuminated and the PCM moves into limp-in mode. Limp-in for the TPS is divided into three categories:

- Idle
- Part-throttle
- Wide-open throttle (WOT)

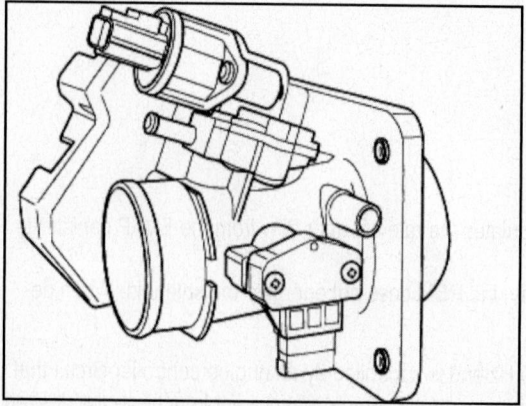

**Identifying the Throttle Position sensor on the throttle body**

**Removal & Installation**

1. Disconnect the negative battery cable.
2. Loosen the clamp for the air duct at the throttle body.
3. Remove the mounting bolt and nut for the air cleaner box.
4. Pull the air cleaner box and throttle body up to access the throttle position sensor.
5. Remove the throttle position sensor.

**To install:**

6. Install the throttle position
7. Locate the air cleaner box and throttle body and tighten the mounting bolt and nut.
8. Install the air duet hose and tighten the clamp.
9. Connect the negative battery cable.

## PARK/NEUTRAL POSITION (PNP) SWITCH

### Description & Operation

The park/neutral position switch is located on the automatic transaxle housing.

Manual transaxles do not use park/neutral switches. The switch provides an input to the PCM to indicate whether the automatic transaxle is in Park/Neutral, or a drive gear selection. This input is used to determine idle speed (varying with gear selection) and ignition timing advance. The park/neutral input is also used to cancel vehicle speed control. The park/neutral switch is sometimes referred to as the neutral safety switch.

The PCM delivers 8.5 volts to the center terminal of the Park/Neutral switch. When the gearshift lever is moved to either the Park or the Neutral position, the PCM receives a ground signal from the Park/Neutral switch. With the shift lever positioned in Drive or Reverse, the Park/Neutral switch contacts open, causing the signal to the PCM to go high.

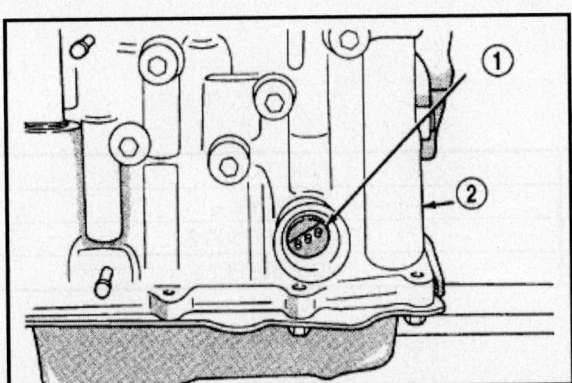

**Locating the Park/Neutral switch (1) on transaxle housing (2)**

## SENTRY KEY IMMOBILIZER MODULE (SKIM)

**Connector Pinouts**

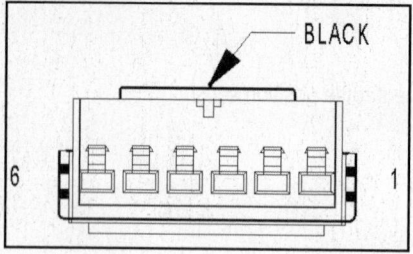

| PIN # | CIRCUIT | FUNCTION |
|-------|---------|----------|
| 1 | - | - |
| 2 | D25 22VT/YL | PCI BUS |
| 3 | - | - |
| 4 | G5 20DB/WT | FUSED IGNITION SWITCH OUTPUT (RUN-START) |
| 5 | Z2 20BK/LG | GROUND |
| 6 | M1 20PK | FUSED B (+) |

**Neon SKIM Connector Pinout**

## THROTTLE INTAKE AIR PRESSURE SENSOR

**Connector Pinouts**

***2.4L TURBO***

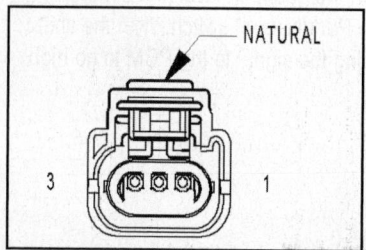

| PIN # | CIRCUIT | FUNCTION |
|-------|---------|----------|
| 1 | K153 18LB | TIP SIGNAL |
| 2 | K167 20BR/YL | SENSOR GROUND 2 |
| 3 | K6 18VT/WT | 5 VOLT SUPPLY |

**Neon 2.4L Turbo Throttle Intake Air Pressure Sensor**

## THROTTLE POSITION (TP) SENSOR CONNECTORS

**Connector Pinouts**

*1.6L & 2.0L*

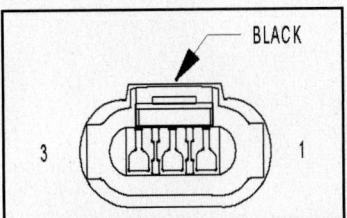

| PIN # | CIRCUIT | FUNCTION |
|-------|---------|----------|
| 1 | K167 20BR/YL | SENSOR GROUND |
| 2 | K22 20OR/DB (1.6L) | THROTTLE POSITION SENSOR SIGNAL NO. 1 |
| 2 | K22 20OR/DB (2.0L) | TP SIGNAL |
| 3 | K6 20VT/WT | 5 VOLT SUPPLY |

**Neon 1.6L & 2.0L TP Sensor Connector Pinout**

*2.4L TURBO*

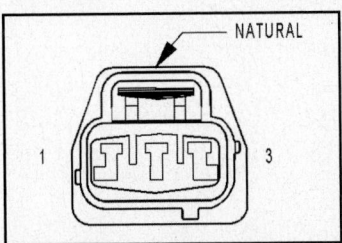

| PIN # | CIRCUIT | FUNCTION |
|-------|---------|----------|
| 1 | K4 20BK/LB | SENSOR GROUND |
| 2 | K22 20OR/DB | TP SIGNAL |
| 3 | K7 20OR | 5 VOLT SUPPLY |

**Neon 2.4L Turbo TP Sensor Connector Pinout**

## VEHICLE SPEED CONTROL SERVO

**Connector Pinouts**

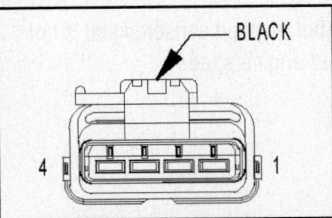

| PIN # | CIRCUIT | FUNCTION |
|-------|---------|----------|
| 1 | V36 20TN/RD | S/C VACUUM CONTROL |
| 2 | V35 20LG/RD | S/C VENT CONTROL |
| 3 | V30 20DB/RD | SPEED CONTROL BRAKE LAMP SWITCH OUTPUT |
| 4 | Z1 20BK | GROUND |

**Neon Vehicle Speed Control Servo Connector Pinout**

## VEHICLE SPEED SIGNAL (VSS)

### Description & Operation

The PCM requires the VSS to be able to control the following programs:

- Speed Control
- IAC motor (during deceleration)
- Injection pulse width (during deceleration)
- OBD II diagnostics
- PCM mileage EEPROM
- Road speed shutdown
- Speedometer/Odometer (bused message)

**NOTE: Road Speed Shutdown is the PCM shutting oft fuel injectors above a preset vehicle speed.**

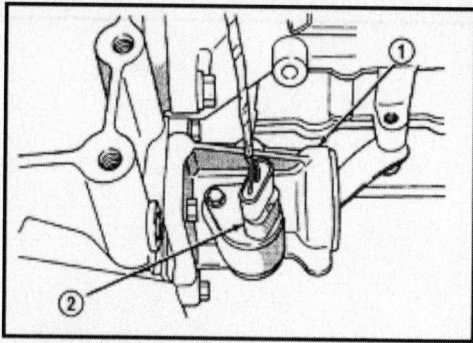

**The vehicle speed sensor (2) is located in the transmission extension housing (1)**

The vehicle speed sensor on 3-speed automatic and manual transaxle vehicles is a Hall-effect sensor. This sensor is mechanically driven by a pinion gear that is in mesh with the right axle drive shaft. The hall-effect sensor switches a 5-volt signal sent from the PCM from a ground to an open circuit.

Like all Hall-effect sensors, the electronics of the sensor needs a power source. This power source is provided by the PCM. It is the same 8-volt power supply that is used by the CKP and CMP sensors.

The vehicle speed sensor generates 8 pulses per sensor revolution. This signal, in conjunction with a closed throttle signal from the throttle position sensor, indicates a closed throttle deceleration to the PCM. Under deceleration conditions, the PCM adjusts the Idle Air Control (IAC) motor to maintain a desired MAP value.

When the vehicle is stopped at idle, a closed throttle signal is received by the PCM (but a speed sensor signal is not received). Under idle conditions, the PCM adjusts the IAC motor to maintain a desired engine speed

### Removal & Installation

1. Raise vehicle on hoist.
2. Disconnect the speed sensor connector.
3. Clean area around speed sensor before removing.
4. Remove speed sensor retaining bolt.
5. Remove speed sensor from transaxle, carefully so that sensor drive gear does not fall into transaxle. Should sensor drive gear fall into the transaxle during sensor removal, drive gear must be reattached to sensor.
6. Remove speed sensor drive gear from speed sensor.

**To install:**

7. Install pinion gear to speed sensor.

8. Using a NEW O-ring, install the speed sensor to the transaxle.

9. Install the bolt and torque to 7 Nm (60 inch lbs.).

10. Connect speed sensor connector.

11. Lower vehicle and road test to verify proper speedometer operation.

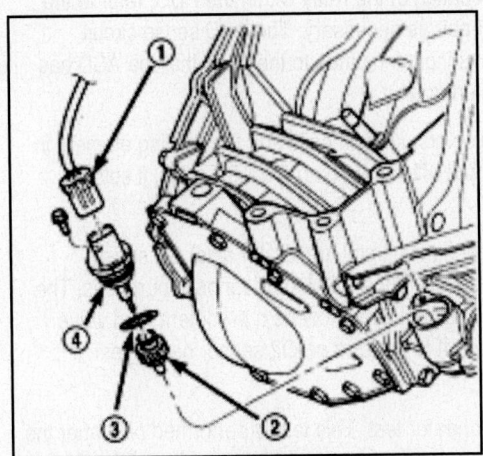

**An exploded view of the VSS (4), with electrical connector (1), speedometer pinion (2) and O-ring (3)**

# DAIMLERCHRYSLER CARS: SEBRING & STRATUS

## AUTO SHUT DOWN (ASD) RELAY

### Description & Operation

The relay is located in the Power Distribution Center (PDC). For the location of the relay within the PDC, refer to the PDC cover for location. Check electrical terminals for corrosion and repair as necessary. The ASD sense circuit informs the PCM when the ASD relay energizes. A 12-volt signal at this input indicates to the PCM that the ASD has been activated. This input is used only to sense that the ASD relay is energized.

When energized, the ASD relay supplies battery voltage to the fuel injectors, ignition coils and the heating element in each oxygen sensor. If the PCM does not receive 12 volts from this input after grounding the ASD relay, it sets a Diagnostic Trouble Code (DTC).

When energized, the ASD relay provides power to operate the injectors, ignition coil, generator field, O2 sensor heaters (both upstream and downstream), and also provides a sense circuit to the PCM for diagnostic purposes. The PCM energizes the ASD any time there is a Crankshaft Position sensor signal that exceeds a predetermined value. The ASD relay can also be energized after the engine has been turned off to perform an O2 sensor heater test, if vehicle is equipped with OBD II diagnostics.

As mentioned, the PCM energizes the ASD relay during an O2 sensor heater test. This test is performed only after the engine has been shut off. The PCM still operates internally to perform several checks, including monitoring the O2 sensor heaters.

*Note: For location of relays in Power Distribution Center, see label on lid of PDC>*

## CAMSHAFT POSITION (CMP) SENSOR

### Description & Operation

The CMP sensor contains a Hall-effect device that provides cylinder identification to the Powertrain Control Module (PCM). The sensor generates pulses as groups of notches on the camshaft sprocket pass underneath it. The PCM keeps track of crankshaft rotation and identifies each cylinder by the pulses generated by the notches on the camshaft sprocket. Crankshaft pulses follow each group of camshaft pulses.

### Removal & Installation

The camshaft position sensor is mounted in the front of the head.

1. Disconnect electrical connector from sensor.
2. Remove camshaft position sensor screw.
3. Without pulling on the connector, pull the sensor out of the chain case cover.

### To install:

4. Install sensor in the chain case cover and push sensor in until contact is made with the boss on the head.
5. While holding the sensor in this position, install and tighten the retaining bolt to 12 Nm (105 inch lbs.) torque.
6. Attach electrical connector to sensor.
7. Install the sensor.

### Testing

1. Disconnect the camshaft position sensor connector, and connect the test harness special tool (MB991709) in between. (All terminals should be connected.)
2. Connect the oscilloscope probe to camshaft position sensor connector terminal 2.
3. Disconnect the crankshaft position sensor connector, and connect the test harness special tool (MD998478) in between.

4. Connect the oscilloscope probe to crankshaft position sensor connector terminal 2 (black clip of special tool).

5. When metal aligns with the sensor, voltage goes low (less than 0.3 volts). When a notch aligns with the sensor, voltage spikes high (5.0 volts). As a group of notches pass under the sensor, the voltage switches from low (metal) to high (notch) then back to low. The number of notches determines the amount of pulses. If available, an oscilloscope or scan tool with special module can display the square wave patterns of each timing event.

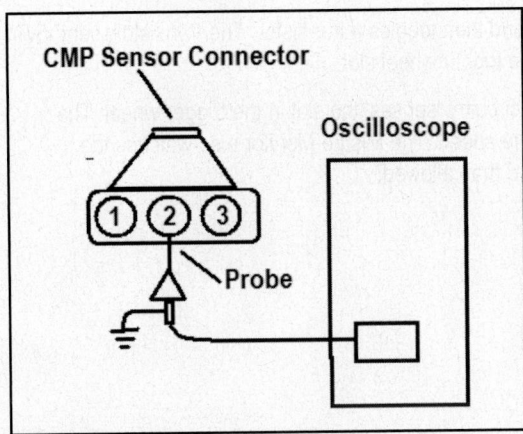

**Camshaft Position Sensor Testing**

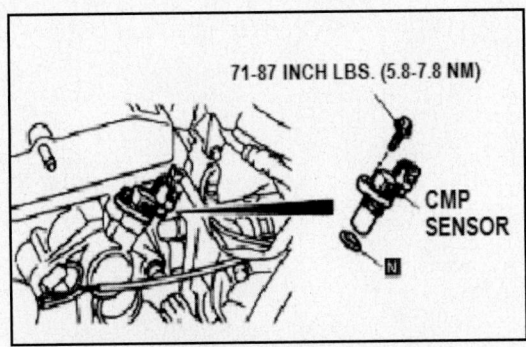

**Camshaft Position Sensor Location on 2.7L**

**Connector Pinouts**

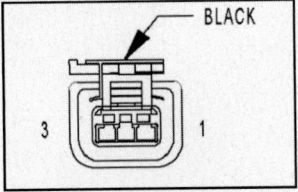

| PIN # | CIRCUIT | FUNCTION |
|-------|---------|----------|
| 1 | K6 20VT/WT (2.0L/2.4L) | 5 VOLT SUPPLY |
| 1 | K7 20OR (2.7L) | 8 VOLT SUPPLY |
| 2 | K4 20BK/LB | SENSOR GROUND 1 |
| 3 | K44 20TN/YL (2.0L/2.4L) | CMP SIGNAL |
| 3 | K44 20TN/YL (2.7L) | CAMSHAFT POSITION SENSOR SIGNAL |

**Sebring, Stratus CMP Sensor**

## CRANKSHAFT POSITION (CKP) & CAMSHAFT POSITION (CMP) SENSORS

### Description & Operation

#### 1995-99 SEBRING

The Crankshaft Position (CKP) and Camshaft Position (CMP) sensors used with this EI system provide engine position and speed data to the powertrain control module (PCM). Both sensors are Hall effect devices. The CKP sensor is mounted on the transaxle, adjacent to a trigger wheel on the torque converter drive plate. The CMP sensor is mounted to the Bank 2 cylinder head, adjacent to a trigger wheel on the cam gear.

Both trigger wheels have slots that the Hall effect circuit senses and then toggles a transistor. The transistors send 5v pulses to the PCM that represents the position and duration of the trigger wheel slot.

The signal to the PCM will remain high as long as the Hall effect circuitry 'senses' the slot in the trigger wheel. The PCM uses the pulsewidth from the CKP sensor to calculate engine speed. The Misfire Monitor also watches the pulsewidth to determine if crankshaft velocity has decreased more than allowed.

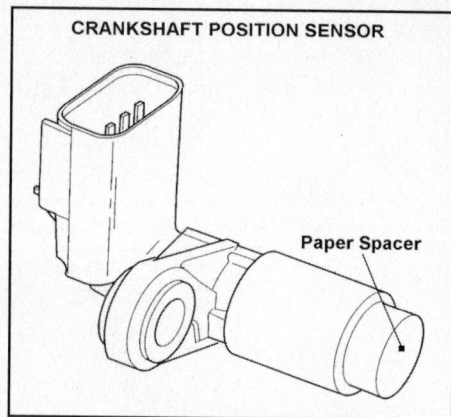

Crankshaft position sensor – 95-99 Sebring

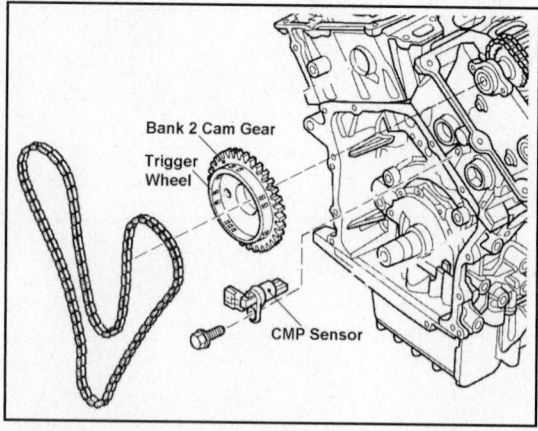

Camshaft position sensor location – 95-99 Sebring

#### SENSOR SIGNAL RELATIONSHIPS

The CMP trigger wheel contains slots grouped in the following order: 1, 2, 3, 1, 3, and 2. Since each number is repeated, the PCM must keep track of the sequence over time in order to properly calculate cylinder position. Also, these pulses do not occur at TDC. They only signal the PCM which cylinder will be arriving at TDC next. The PCM then uses CKP sensor pulses to "count down" to TDC for the designated cylinder.

There are 13 slots in a ring on the CKP trigger wheel (in two groups of 4 and one group of 5 slots). The PCM determines engine position at the falling edge of the fourth pulse in each group. The slots are located at 69°, 49°, 29°, and 9° before TDC for whichever pair of cylinders is approaching TDC. After the CMP pulse designates which cylinder is expected to arrive at TDC next, the falling edge of the fourth CKP pulse designates 9 before TDC.

The extra (thirteenth) slot occurs 11° after TDC, and helps the PCM to determine engine position at startup without having to wait for the complete CMP pulse sequence.

### CKP & CMP SENSOR CIRCUITS

The CKP and CMP sensors are both connected to the PCM by 3 wires. The sensors share an 8v reference voltage from PCM Pin 2-44. They also share the sensor ground with most other engine sensors (PCM Pin 2-43). The 0-5 volt return signal is sent to PCM Pin 1-32. The CMP 0-5 volt return signal is sent to PCM Pin 1-33.

### Testing

#### CKP SENSOR

#### Testing with DVOM

This test is used to monitor the DC voltage and frequency (HZ) in order to look for circuit or sensor faults. If the transistor switches, but cannot pull the signal to ground, the average DC voltage displayed will jump significantly. Also, the transistor may fail and toggle rapidly, producing a frequency that is not appropriate for the engine rpm.

It is important to monitor these values over time, as many failures occur intermittently.

1.   Place the shift selector in Park and block the drive wheels for safety.

2.   Connect the DVOM positive lead to the CKP sensor signal circuit (GY/BK wire) at Pin 1-32 and the negative lead to the battery negative post.

3.   If the CKP sensor signal is at Hot Idle, the DVOM would show a CKP signal of only 0.48v DC because the tester is averaging the signal voltage. The average value is low because the signal is low much longer than it is high (a duty cycle of 91.5 % may be shown on the DVOM).

4.   Use the DVOM MIN/MAX feature, if available, to verify that the CKP signal is really switching from 0-5v.

5.   Another way to use the DVOM to test the CKP sensor signal is to observe the frequency of the signal. At hot idle, the DVOM display might show 140.9 Hz. (140.9 cycles per second X 60 seconds / 13 slots per revolution = 650.3 rpm). Using this formula for a "no start", it is easy to calculate that at 200 rpm cranking speed, the DVOM should read about 43 Hz.

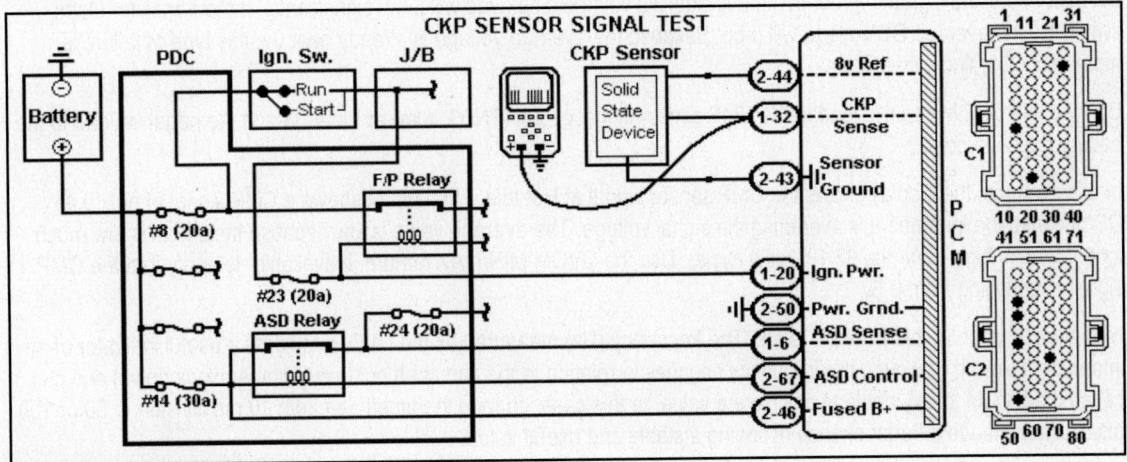

**CKP Sensor Test Schematic**

#### Testing the CKP Sensor with Lab Scope

The lab scope is used to monitor the CKP signal over time to verify that the transistor in the sensor continues to toggle the signal from 0-5 volts at the appropriate frequency. Watch for unwanted toggling, failure of the transistor to pull the signal to ground, or intermittently missing one or more pulses. All of these failures can cause driveability problems, trouble codes and no start conditions, depending on the severity of the failure. The Lab Scope can be used to test the CKP sensor as it provides a very accurate view of the sensor waveform and sensor relationships.

Place the shift selector in Park and block the drive wheels for safety. Connect the positive lead to the CKP sensor signal circuit (GY/BK wire) at Pin 1-32 and the negative lead to the battery negative post.

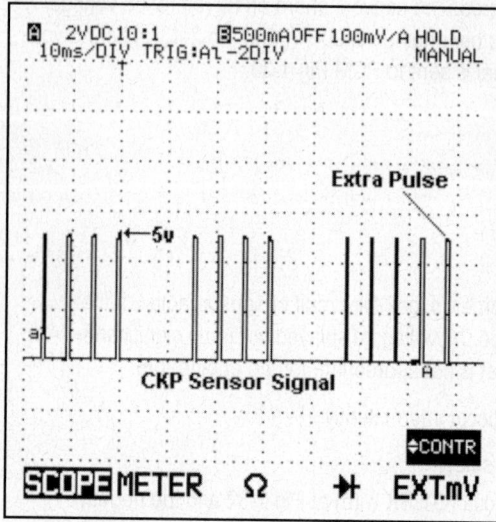

**To make the waveforms as clear as possible, set the scope settings to match the examples**

## CMP SENSOR

Prior to Testing, ensure the shift selector in Park and block the drive wheels for safety.

### Testing the CMP Sensor with a DVOM

This test is used to monitor the DC voltage to look for circuit or sensor faults. If the CMP sensor transistor switches, but cannot pull the signal to ground, the average DC voltage displayed will jump significantly. If the transistor stops switching, the average DC voltage will drop. Because the average voltage is already near 0v, this type of failure is more difficult to find with this test.

Connect the DVOM positive lead to the CMP sensor signal circuit (TN/YL wire) at Pin 1-33 and the negative lead to the battery negative post.

In this example, the display shows the CMP sensor signal at Hot Idle. The DVOM shows a CMP signal of only 0.46v DC because the instrument is averaging the signal voltage. The average value is low because the signal is low much longer than is high (note the 92.1 % duty cycle). Use the DVOM MIN/MAX feature, if available, to verify that the CMP signal is switching from 0-5v.

Note the HZ value in this example is OL. The frequency (Hz) measured by the DVOM could be a useful indicator of an intermittent signal. However, the frequency changes in relation to the camshaft position (number of windows) and can cause the DVOM to be unable to calculate a value. In this case, change the time base from 10 ms/division to 50 or 100 ms/division to have a better chance of seeing a stable and useful value.

### Using a Lab Scope to Test the CMP Sensor

The lab scope test is used to monitor this signal over time to verify that the transistor in the sensor continues to toggle the signal from 0-5 volts. Watch for unwanted toggling, failure of the transistor to pull the signal all the way to ground, or intermittently missing one or more pulses. All of these failures can cause drivability problems, trouble codes and no start conditions, depending on the severity of the failure.

The Lab Scope can be used to test the CMP sensor circuit as it provides a very accurate view of circuit activity and any signal glitches. Place the shift selector in Park and block the drive wheels for safety.

Connect the Channel 'A' positive probe to the CMP sensor circuit (TN/YL wire) at PCM Pin 1-33, and the negative probe to the battery negative post.

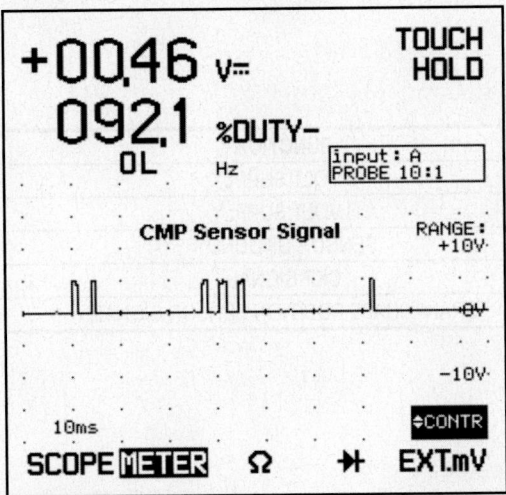

**Lab Scope Test Example**

In the example, the trace shows the CMP sensor signal at Hot Idle. The trace represents 720° of crankshaft rotation, or 360° of camshaft rotation. All six (6) CMP pulses can be seen, in order (1, 2, 3, 1, 3, and 2).

Each group of CMP pulses (1, 2, or 3 pulses) tells the PCM which cylinder is approaching TDC next. The PCM then starts a countdown to TDC using the CKP pulses.

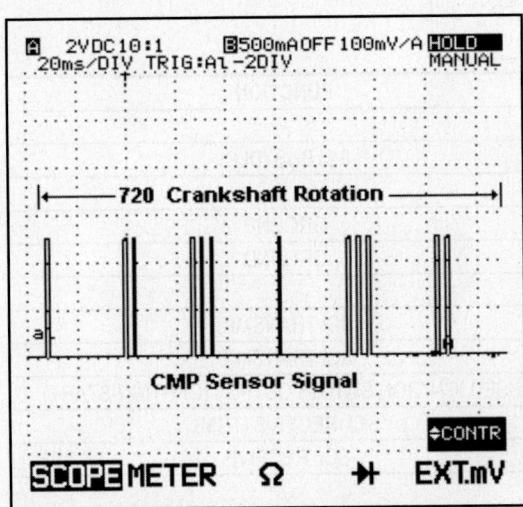

**Example scope pattern of CMP Sensor at hot idle**

**Connector Pinouts**

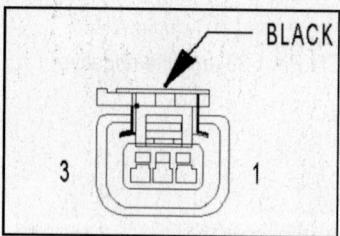

| PIN # | CIRCUIT | FUNCTION |
|-------|---------|----------|
| 1 | K6 20VT/WT (2.0L/2.4L) | 5 VOLT SUPPLY |
| 1 | K7 20OR (2.7L) | 8 VOLT SUPPLY |
| 2 | K4 20BK/LB | SENSOR GROUND 1 |
| 3 | K24 20GY/BK (2.0L/2.4L) | CKP SIGNAL |
| 3 | K24 20GY/BK (2.7L) | CRANKSHAFT POSITION SENSOR SIGNAL |

**Sebring, Stratus CKP Sensor**

DATA LINK CONNECTOR

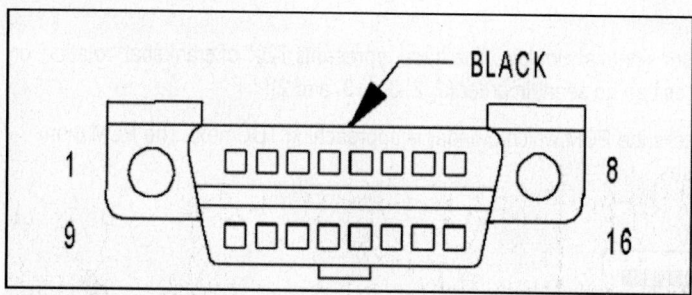

| PIN # | CIRCUIT | FUNCTION |
|-------|---------|----------|
| 1 | - | - |
| 2 | D25 20VT/YL | PCI BUS (DLC) |
| 3 | - | - |
| 4 | Z305 20BK/YL | GROUND |
| 5 | Z306 20BK/VT | GROUND |
| 6 | - | - |
| 7 | D21 20PK (2.0L/2.4L) | SCI TRANSMIT |
| 7 | D21 20PK (2.7L) | SCI TRANSMIT |
| 8 | F11 20RD/WT | FUSED IGNITION SWITCH OUTPUT (OFF-RUN-START) |
| 9 | D6 20PK/LB (2.0L/2.4L EATX) | SCI RECEIVE (TCM) |
| 9 | D6 20PK/LB (2.7L EATX) | SCI RECEIVE |
| 10 | - | - |
| 11 | - | - |
| 12 | D20 20LG (2.0L/2.4L) | SCI RECEIVE |
| 12 | D20 20LG (2.7L) | SCI RECEIVE |
| 13 | - | - |
| 14 | - | - |
| 15 | D15 20WT/DG | SCI TRANSMIT (TCM) |
| 16 | M1 20PK | FUSED B (+) |

**Sebring, Stratus Data Link Connector Pinout**

## ENGINE COOLANT TEMPERATURE (ECT) SENSOR

**Testing**

*CAUTION: Be careful not to touch the connector (resin section) with the tool when removing and installing.*

1. Drain engine coolant, and then remove the engine coolant temperature sensor.
2. With the temperature-sensing portion of engine coolant temperature sensor immersed in hot water, check the resistance. Standard value:

5.1 - 6.5 k/ohms at 0 °C (32 °F)

2.1 - 2.7 k/ohms at 20 °C (68 °F)

0.9 - 1.3 k/ohms at 40 °C (104 °F)

0.26 - 0.36 k/ohms at 80 °C (176 °F)

a. If resistance deviates from the standard value greatly, replace the sensor.
b. Apply 3M® AAD part number 8731, or equivalent, to threaded portion.
c. Install the engine coolant temperature sensor and tighten it to the specified torque. Tightening torque: 15-29 ft. lbs. (19-39 Nm).

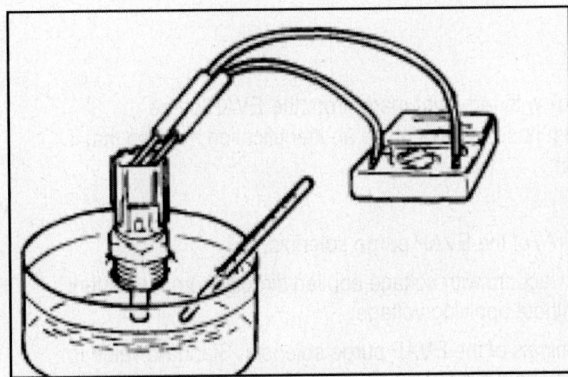

**Coolant Temperature Sensor Testing**

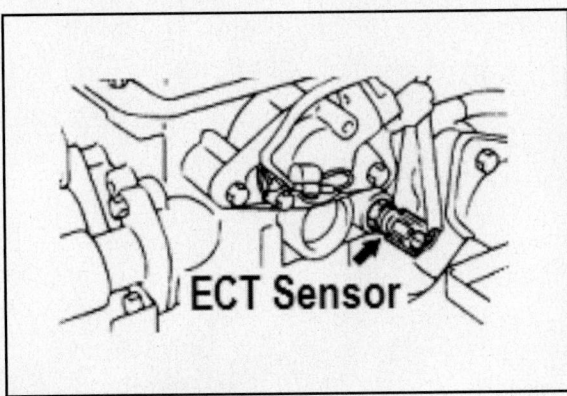

**ECT Sensor Location 2.4L**

## Connector Pinouts

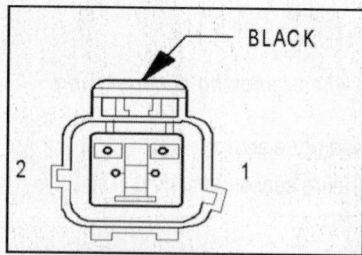

BLACK

2   1

| PIN # | CIRCUIT | FUNCTION |
|-------|---------|----------|
| 1 | K2 20TN/BK (2.0L/2.4L) | ECT SIGNAL |
| 1 | K2 20TN/BK (2.7L) | ENGINE COOLANT TEMPERATURE SENSOR SIGNAL |
| 2 | K4 20BK/LB | SENSOR GROUND 1 |

**Sebring, Stratus ECT Sensor**

## EVAP CANISTER PURGE SOLENOID

**Testing**

1. Remove the air intake hose.
2. Disconnect the vacuum hose (black, black with red paint mark) from the EVAP purge solenoid. When disconnecting the vacuum hose, always place an identification mark so that it can be reconnected at its original position.
3. Disconnect the harness connector.
4. Connect a hand vacuum pump to nipple (A) of the EVAP purge solenoid.
5. Check for air-tight integrity by applying a vacuum with voltage applied directly from the battery to the EVAP purge solenoid valve and without applying voltage.
6. Measure the resistance between the terminals of the EVAP purge solenoid. Standard value is 30 - 34 Ohms at 68°F (20°C). Replace solenoid if resistance is out of specification.

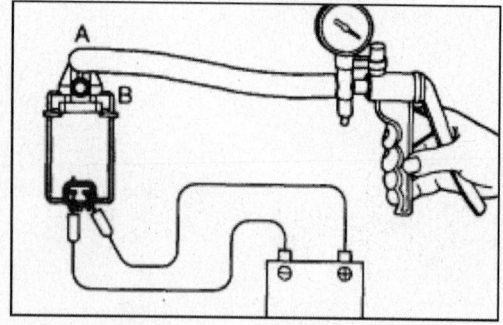

**Testing Canister Purge Solenoid**

| BATTERY POSITION VOLTAGE | NORMAL CONDITION |
|--------------------------|------------------|
| APPLIED | VACUUM LEAKS |
| NOT APPLIED | VACUUM MAINTAINED |

**Canister Purge Valve Testing Chart**

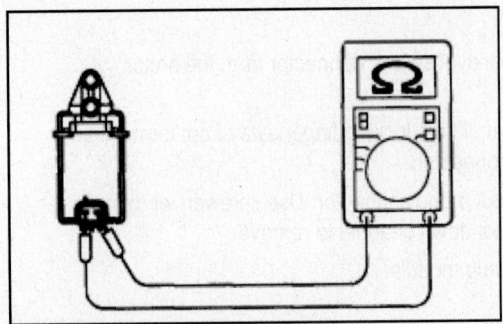

**EVAP Purge Solenoid Voltage Test**

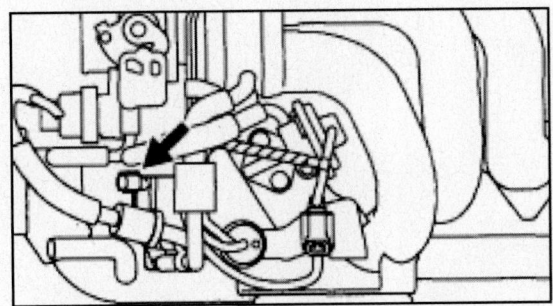

**Canister Purge Valve Location**

**Connector Pinouts**

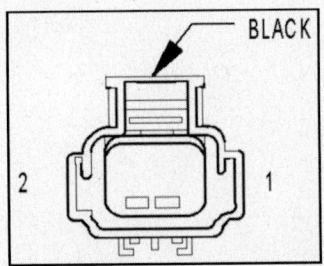

| PIN # | CIRCUIT | FUNCTION |
|-------|---------|----------|
| 1 | K52 20PK/BK (2.0L/2.4L) | EVAP PURGE CONTROL |
| 1 | K52 20PK/BK (2.7L) | EVAPORATIVE EMISSION SOLENOID CONTROL |
| 2 | K108 20WT/TN (2.0L/2.4L) | EVAP/PURGE RETURN |
| 2 | K108 20WT/TN (2.7L) | EVAPORATIVE SOLENOID SENSE |

**Sebring, Stratus EVAP Purge Solenoid Connector Pinout**

## FUEL LEVEL SENDING UNIT/FUEL PUMP MODULE

### Description & Operation

The fuel gauge level sending unit is attached to the side of fuel pump module. The level sensor is a variable resistor. Its resistance changes with the amount of fuel in the tank. The float arm attached to the sensor moves as the fuel level changes. The fuel level input is used as an input for OBD II. If the fuel level is below 15% or above 85% of total tank capacity several monitors are disabled. There are diagnostics for the level circuit open and shorted.

## Removal & Installation

1. Remove fuel pump module.
2. Depress retaining tab and remove the fuel pump/level sensor connector from the bottom of the fuel pump module electrical connector.
3. Pull off locking wedge. Using a small screwdriver, lift the locking finger away from terminal and push level sensor and ground terminals out of connector.
4. Push level sensor signal and ground terminals out of the connector. Use screwdriver to move locking tab on level sensor and move level sensor down channel to remove.
5. Slide level sensor wires through opening fuel pump module.
6. Slide level sensor out of channel in module.

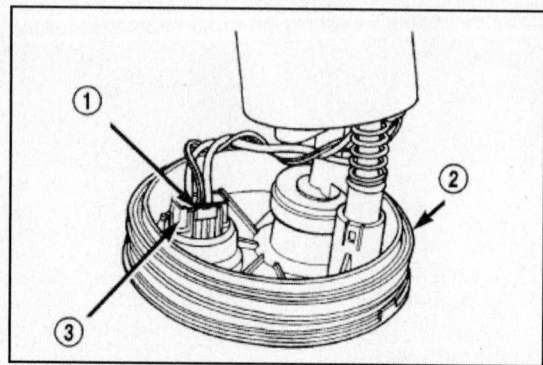

**Fuel Pump Level sensor connector (3), showing tank seal (2) and retaining tab (1)**

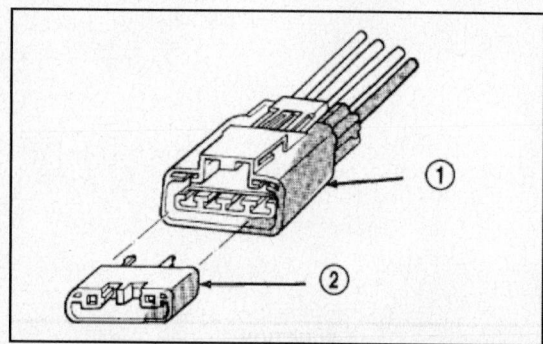

**Wire terminal locking wedge (2) and electrical connector (1)**

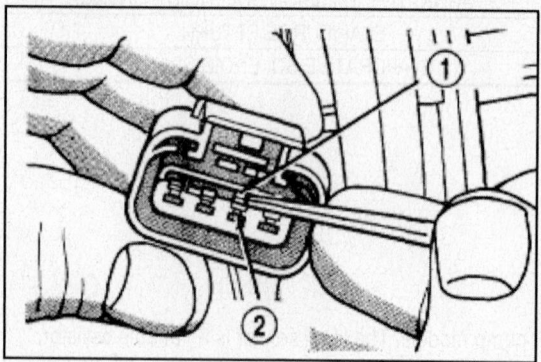

**Properly removing wires (1) from the connector (2)**

**To install:**

7. Insert level sensor wires into bottom of opening in module.

8. Wrap wires into groove in back of level sensor.

9. While feeding wires into guide grooves, slide level sensor up into channel until it snaps into place. Ensure tab at bottom of sensor locks in place.

10. Install level sensor wires in connector. Push the wires up through the connector and then pull them down until they lock in place. Ensure signal and ground wires are installed in the correct position.

11. Install locking wedge on connector.

12. Push connector up into bottom of fuel pump module electrical connector.

13. Install fuel pump module.

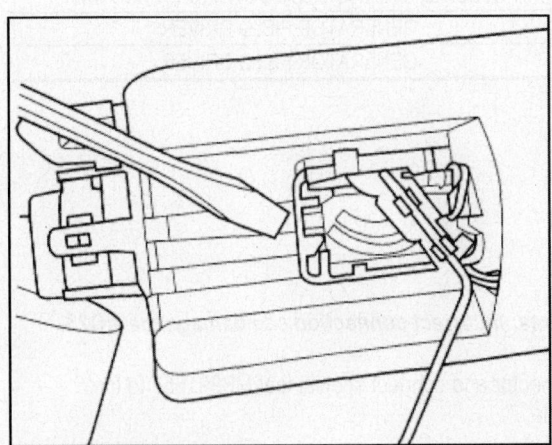

**Loosening Level Sensor**

**Connector Pinouts**

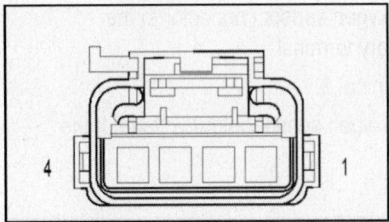

| PIN # | CIRCUIT | FUNCTION |
|-------|---------|----------|
| 1 | Z211 12BK | GROUND |
| 2 | Z211 18BK | GROUND |
| 3 | G4 18DB | FUEL LEVEL SENSOR SIGNAL |
| 4 | A141 14RD | FUEL PUMP RELAY OUTPUT |

**Sebring, Stratus Fuel Pump Module Connector Pinout**

## GENERATOR

**Connector Pinouts**

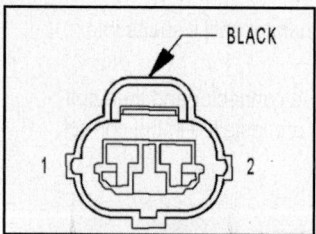

| PIN # | CIRCUIT | FUNCTION |
|:-----:|---------|----------|
| 1 | Z12 20BK/TN (2.0L/2.4L) | GROUND |
| 1 | F142 18OR/DG (2.7L) | FUSED AUTOMATIC SHUT DOWN RELAY OUTPUT |
| 2 | K20 18DG (2.0L/2.4L) | GENERATOR FIELD DRIVER |
| 2 | K20 18DG (2.7L) | GENERATOR FIELD DRIVER |

**Sebring, Stratus Generator Connector Pinout**

## HEATED OXYGEN SENSOR

**Testing**

*CAUTION: Be very careful when connecting the jumper wires; incorrect connection can damage the HO2S.*

1. Disconnect the heated oxygen sensor connector and connect special tool MB991658 to the connector on the heated oxygen sensor side.

2. Make sure that there is continuity of 11 - 18 Ohms at 20°C (68°F) between terminal **3** and terminal **4** on the heated oxygen sensor connector

3. If there is no continuity, replace the heated oxygen sensor.

4. Warm up the engine until engine coolant is 80°C (176°F).

5. Use the jumper wires to connect terminal **3** of the heated oxygen sensor connector to the positive battery terminal and terminal 4 to the negative battery terminal.

6. Connect a digital voltage meter between terminal **1** and terminal **2**.

7. While repeatedly revving the engine, measure the heated oxygen sensor output voltage. If the sensor is defective, replace the heated oxygen sensor.

| ENGINE CONDITION | O2 SENSOR OUTPUT VOLTAGE | REMARKS |
|------------------|:------------------------:|---------|
| WHEN REVVING ENGINE | 0.6-1.0V | NORMAL VOLTAGE RESPONSE TO RICHER A/F MIXTURE CAUSED BY REVVING ENGINE |

**Heated Oxygen Sensor Voltage Chart**

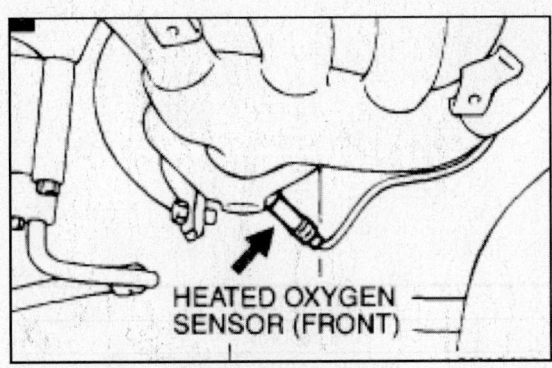

**Heated Oxygen Sensor Front**

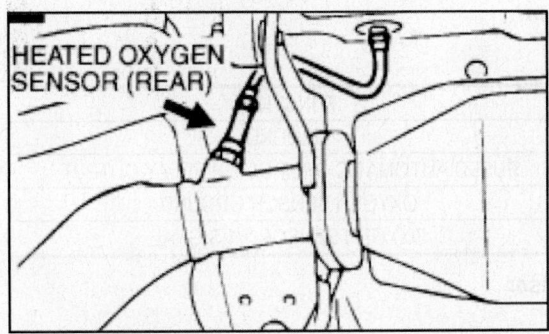

**Heated Oxygen Sensor Rear**

**Connector Pinouts**

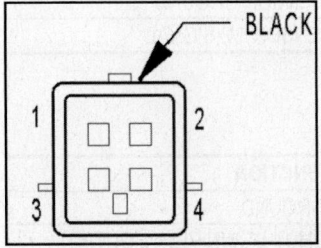

| PIN # | CIRCUIT | FUNCTION |
|-------|---------|----------|
| 1 | Z228 20BK | GROUND |
| 2 | K99 18BR/OR | O2 1/1 HEATER CONTROL |
| 3 | K902 18BR/DG | O2 RETURN (UP) |
| 4 | K41 20BK/LG | O2 1/1 SIGNAL |

**Sebring, Stratus 2.0L & 2.4L 1/1 (Upstream) HO2S Sensor**

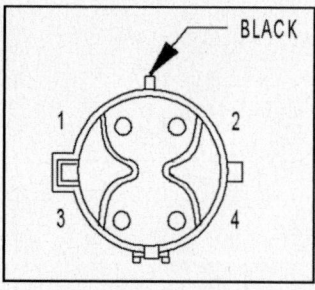

| PIN # | CIRCUIT | FUNCTION |
|---|---|---|
| 1 | Z186 20BK | GROUND |
| 2 | K199 18BR/VT | O2 1/2 HEATER CONTROL |
| 3 | K904 18DB/DG | O2 RETURN (DOWN) |
| 4 | K141 20TN/WT | O2 1/2 SIGNAL |

**Sebring, Stratus 2.0L & 2.4L 1/2 (Downstream) HO2S Sensor**

| PIN # | CIRCUIT | FUNCTION |
|---|---|---|
| 1 | Z192 20BK | GROUND |
| 2 | F142 18OR/DG | FUSED AUTOMATIC SHUT DOWN RELAY OUTPUT |
| 3 | K127 18DB/LG | OXYGEN SENSOR GROUND |
| 4 | K41 20BK/DG | OXYGEN SENSOR 1/1 SIGNAL |

**Sebring, Stratus 2.7L 1/1 (Right Bank Upstream) HO2S Sensor**

| PIN # | CIRCUIT | FUNCTION |
|---|---|---|
| 1 | Z188 20BK | GROUND |
| 2 | F142 18OR/DG | FUSED AUTOMATIC SHUT DOWN RELAY OUTPUT |
| 3 | K127 18DB/LG | OXYGEN SENSOR GROUND |
| 4 | K141 20TN/WT | OXYGEN SENSOR 1/2 SIGNAL |

**Sebring, Stratus 2.7L 1/2 (Right Bank Downstream) HO2S Sensor**

| PIN # | CIRCUIT | FUNCTION |
|---|---|---|
| 1 | Z193 20BK | GROUND |
| 2 | F142 18OR/DG | FUSED AUTOMATIC SHUT DOWN RELAY OUTPUT |
| 3 | K127 18DB/LG | OXYGEN SENSOR GROUND |
| 4 | K241 20LG/RD | OXYGEN SENSOR 2/1 SIGNAL |

**Sebring, Stratus 2.7L 2/1 (Left Bank Upstream) HO2S Sensor**

| PIN # | CIRCUIT | FUNCTION |
|---|---|---|
| 1 | Z186 20BK | GROUND |
| 2 | F142 18OR/DG | FUSED AUTOMATIC SHUT DOWN RELAY OUTPUT |
| 3 | K127 18DB/LG | OXYGEN SENSOR GROUND |
| 4 | K341 20PK/WT | OXYGEN SENSOR 2/2 SIGNAL |

**Sebring, Stratus 2.7L 2/2 (Left Bank Downstream) HO2S Sensor**

## IDLE AIR CONTROL (IAC) MOTOR

### Description & Operation

The idle air control motor (IAC) attaches to the throttle body. It is an electric stepper motor. The PCM adjusts engine idle speed through the idle air control motor to compensate for engine load, coolant temperature or barometric pressure changes. The throttle body has an air bypass passage that provides air for the engine during closed throttle idle. The idle air control motor pintle protrudes into the air bypass passage and regulates airflow through it.

The PCM adjusts engine idle speed by moving the IAC motor pintle in and out of the bypass passage. The adjustments are based on inputs the PCM receives. The inputs are from the throttle position sensor, crankshaft position sensor, coolant temperature sensor, MAP sensor, vehicle speed sensor and various switch operations (brake, park/neutral, air conditioning).

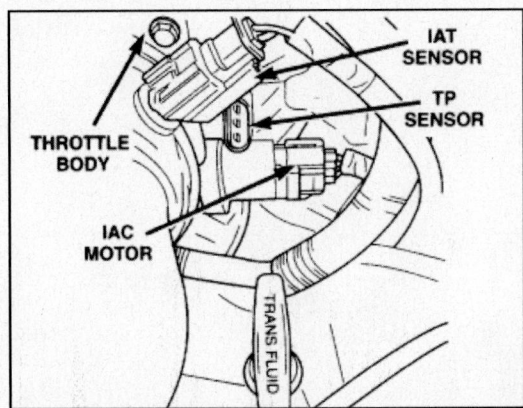

**Idle Air Control Motor (IAC)**

When engine rpm is above idle speed, the IAC is used for the following functions:

- Off-idle dashpot
- Deceleration air flow control
- A/C compressor load control (also opens the passage slightly before the compressor is engaged so that the engine rpm does not dip down when the compressor engages)

### Target Idle

**Target idle is determined by the following inputs:**

- Gear position ECT Sensor
- Battery voltage
- Ambient/Battery Temperature Sensor
- VSS
- TPS
- MAP Sensor

### Removal & Installation

1. Disconnect the negative battery cable.
2. Disconnect the IAC electrical connector.
3. Remove the IAC mounting screws.
4. Remove the IAC.

**To install:**

5. Install the IAC to the throttle body.
6. Tighten mounting screws to 5.1 Nm (45 inch lbs.) torque.
7. Attach electrical connector to the IAC.
8. Connect the negative battery cable.

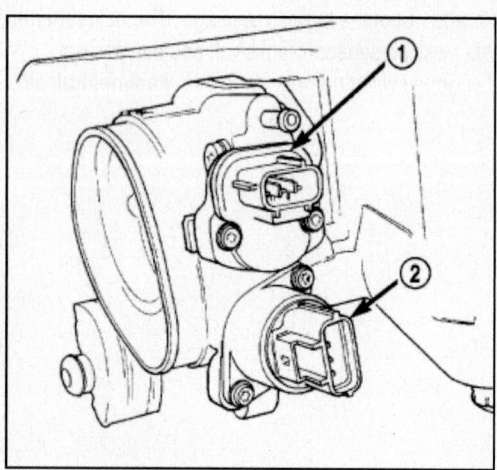

**Idle Air Control Motor (2) and TP sensor (1)**

**Connector Pinouts**

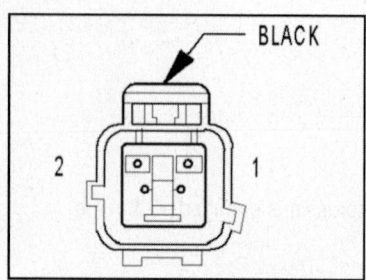

| PIN # | CIRCUIT | FUNCTION |
|-------|---------|----------|
| 1 | K39 18GY/RD | IAC MOTOR CONTROL |
| 2 | K60 18YL/BK | IAC RETURN |

**Sebring, Stratus 2.0 & 2.4L IAC Motor Connector Pinout**

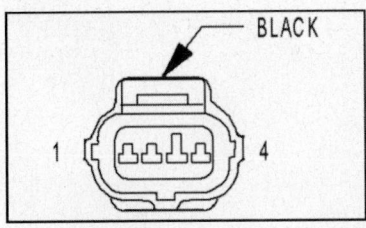

| PIN # | CIRCUIT | FUNCTION |
|---|---|---|
| 1 | K39 20GY/RD | IDLE AIR CONTROL MOTOR NO. 1 DRIVER |
| 2 | K60 20YL/BK | IDLE AIR CONTROL MOTOR NO. 2 DRIVER |
| 3 | K40 20BR/WT | IDLE AIR CONTROL MOTOR NO. 3 DRIVER |
| 4 | K59 20VT/BK | IDLE AIR CONTROL MOTOR NO. 4 DRIVER |

**Sebring, Stratus 2.7L IAC Motor Connector Pinout**

## INTAKE AIR TEMPERATURE (IAT) SENSOR

### Description & Operation

The IAT Sensor is a Negative Temperature Coefficient (NTC) Sensor that provides information to the PCM regarding the temperature of the air entering the intake manifold.

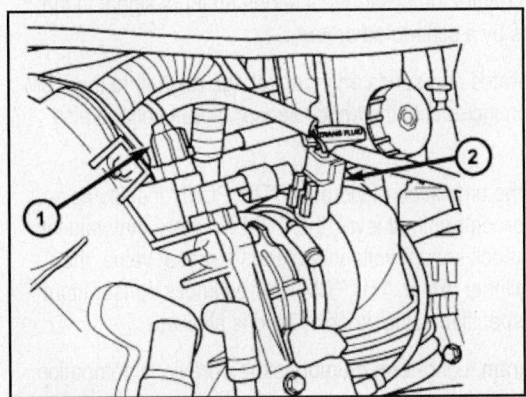

**Locating the TP sensor (1) and Intake Air Temperature (IAT) sensor (2)**

The inlet air temperature sensor replaces the intake air temperature sensor and the battery temperature sensor. The PCM uses the information from the inlet air temperature sensor to determine values to use as an intake air temperature sensor and a battery temperature sensor. The Intake Air Temperature **(IAT)** sensor value is used by the PCM to determine air density. The PCM uses this information to calculate:

- Injector pulse width
- Adjustment of ignition timing (to prevent spark knock at high intake air temperatures

### *BATTERY TEMPERATURE*

The inlet air temperature sensor replaces the intake air temperature sensor and the battery temperature sensor. The PCM uses the information from the inlet air temperature sensor to determine values for the PCM to use as an intake air temperature sensor and a battery temperature sensor.

The battery temperature information along with data from monitored line voltage (B+), is used by the PCM to vary the battery charging rate. System voltage will be higher at colder temperatures and is gradually reduced at warmer temperatures.

The battery temperature information is also used for OBD II diagnostics. Certain faults and OBD II monitors are either enabled or disabled depending upon the battery temperature sensor input (example: disable purge, enable LDP). Most OBD II monitors are disabled below 20°F.

**Connector Pinouts**

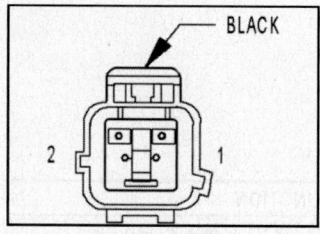

| PIN # | CIRCUIT | FUNCTION |
|-------|---------|----------|
| 1 | K21 20BK/RD (2.0L/2.4L) | IAT SIGNAL |
| 1 | K21 20BK/RD (2.7L) | INLET AIR TEMPERATURE SENSOR SIGNAL |
| 2 | K4 20BK/LB | SENSOR GROUND 1 |

**Sebring, Stratus IAT Sensor**

## KNOCK SENSOR

### Description & Operation

The knock sensor threads into the cylinder block. The knock sensor is designed to detect engine vibration that is caused by detonation. When the knock sensor detects a knock in one of the cylinders, it sends an input signal to the PCM. In response, the PCM retards ignition timing for all cylinders by a scheduled amount.

Knock sensors contain a piezoelectric material that constantly vibrates and sends an input voltage (signal) to the PCM while the engine operates. As the intensity of the crystal's vibration increases, the knock sensor output voltage also increases.

The voltage signal produced by the knock sensor increases with the amplitude of vibration. The PCM receives as an input the knock sensor voltage signal. If the signal rises above a predetermined level, the PCM will store that value in memory and retard ignition timing to reduce engine knock. If the knock sensor voltage exceeds a preset value, the PCM retards ignition timing for all cylinders. It is not a selective cylinder retard. The PCM ignores knock sensor input during engine idle conditions. Once the engine speed exceeds a specified value, knock retard is allowed.

Knock retard uses its own short-term and long-term memory program. Long-term memory stores previous detonation information in its battery-backed RAM. The maximum authority that long-term memory has over timing retard can be calibrated. Short-term memory is allowed to retard timing up to a preset amount under all operating conditions (as long as rpm is above the minimum rpm) except WOT. The PCM, using short-term memory, can respond quickly to retard timing when engine knock is detected. Short-term memory is lost any time the ignition key is turned OFF.

*NOTE: Over or under tightening affects knock sensor performance, possibly causing improper spark control.*

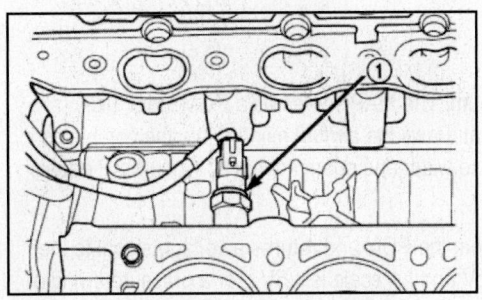

**Knock Sensor location on 2.7L engine**

**Removal & Installation**

1. Remove intake manifold plenum.
2. Remove the passenger side cylinder head.
3. Disconnect electrical connector from knock sensor. Use a crowfoot socket to remove the knock sensors.

**To install:**

*CAUTION: Over or under tightening affects knock sensor performance resulting in possible improper spark control.*

4. Install knock sensor. Tighten knock sensor to 10 Nm (7 ft. lbs.) torque.
5. Install the passenger side cylinder head.
6. Attach electrical connector to knock sensor.
7. Install intake manifold plenum.

**Connector Pinouts**

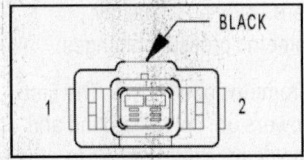

| PIN # | CIRCUIT | FUNCTION |
|-------|---------|----------|
| 1 | K42 20DB/LG | KS SIGNAL |
| 2 | K45 20BK/VT | KS RETURN |

**Sebring, Stratus 2.0L & 2.4L Knock Sensor**

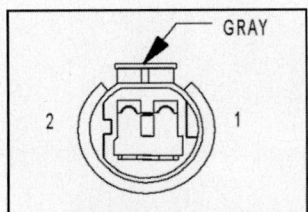

| PIN # | CIRCUIT | FUNCTION |
|-------|---------|----------|
| 1 | K4 20BK/LB | SENSOR GROUND 1 |
| 2 | K42 18BK/VT | KNOCK SENSOR SIGNAL |

**Sebring, Stratus 2.7L Knock Sensor**

## MANIFOLD ABSOLUTE PRESSURE (MAP) SENSOR

### Description & Operation

The MAP sensor mounts to the driver side of the intake manifold plenum. The MAP serves as a PCM input, using a silicon based sensing unit, to provide data on the manifold vacuum that draws the air/fuel mixture into the combustion chamber. The PCM requires this information to determine injector pulse width and spark advance. When MAP equals Barometric pressure, the pulse width will be at maximum.

Also, like the cam and crank sensors, a 5-volt reference is supplied from the PCM and returns a voltage signal to the PCM that reflects manifold pressure. The zero pressure reading is 0.5V and full scale is 4.5V. For a pressure swing of 0 - 15 psi the voltage changes 4.0V. The sensor is supplied a regulated 4.8 to 5.1 volts to operate the sensor. Like the cam and crank sensors ground is provided through the sensor return circuit.

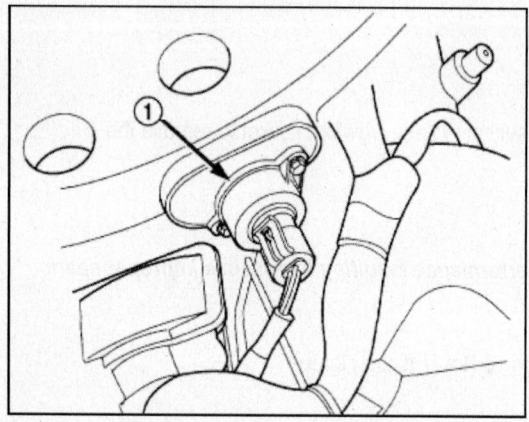

**Manifold Absolute Pressure (MAP) Sensor (1)**

The MAP sensor input is the number one contributor to pulse width. The most important function of the MAP sensor is to determine barometric pressure. The PCM needs to know if the vehicle is at sea level or is it in Denver at 5000 feet above sea level, because the air density changes with altitude. It will also help to correct for varying weather conditions. If a hurricane was coming through the pressure would be very, very low or there could be a real fair weather, high-pressure area. This is important because as air pressure changes the barometric pressure changes.

Barometric pressure and altitude have a direct inverse correlation, as altitude goes up barometric goes down. The first thing that happens as the key is rolled on, before reaching the crank position, the PCM powers up, comes around and looks at the MAP voltage, and based upon the voltage it sees, it knows the current barometric pressure relative to altitude. Once the engine starts, the PCM looks at the voltage again, continuously every 12 milliseconds, and compares the current voltage to what it was at key ON. The difference between current and what it was at key on is manifold vacuum.

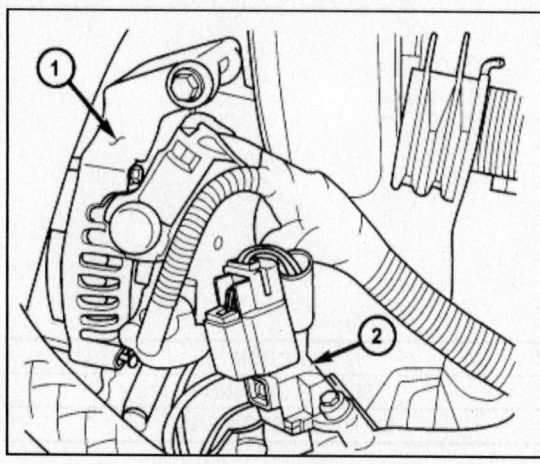

**Generator (1) and MAP Sensor (2) location on 2.7L**

During key ON (engine not running) the sensor reads (updates) barometric pressure. A normal range can be obtained by monitoring known good sensor in your work area.

As the altitude increases the air becomes thinner (less oxygen). If a vehicle is started and driven to a very different altitude than where it was at "Key On", the barometric pressure needs to be updated. Any time the PCM sees Wide Open throttle, based upon TPS angle and RPM it will update barometric pressure in the MAP memory cell. With periodic updates, the PCM can make its calculations more effectively.

The PCM uses the MAP sensor to aid in calculating the following:

- Barometric pressure
- Engine load
- Manifold pressure
- Injector pulse-width
- Spark-advance programs
- Shift-point strategies (F4AC1 transmissions only, via the CCD bus)
- Idle speed
- Decel fuel shutoff

The MAP sensor signal is provided from a single piezo-resistive element located in the center of a diaphragm. The element and diaphragm are both made of silicone. As the pressures changes the diaphragm moves causing the element to deflect which stresses the silicone. When silicone is exposed to stress its resistance changes. As manifold vacuum increases, the MAP sensor input voltage decreases proportionally. The sensor also contains electronics that condition the signal and provide temperature compensation.

The PCM recognizes a decrease in manifold pressure by monitoring a decrease in voltage from the reading stored in the barometric pressure memory cell. The MAP sensor is a linear sensor; as pressure changes, voltage changes proportionately. The range of voltage output from the sensor is usually between 4.6 volts at sea level to as low as 0.3 volts at 26 in. of Hg. Barometric pressure is the pressure exerted by the atmosphere upon an object. At sea level on a standard day, no storm, barometric pressure is 29.92 in Hg. For every 100 feet of altitude barometric pressure drops 0.10 in. Hg. If a storm goes through it can either add, high-pressure, or decrease, low pressure, from what should be present for that altitude. You should make a habit of knowing what the average pressure and corresponding barometric pressure is for your area. Always use the Diagnostic Test Procedures for MAP sensor Testing.

### Removal & Installation

1. Remove the negative battery cable.
2. Disconnect the electrical connector from the MAP sensor.
3. Remove sensor.

### To install:

4. The sensor mounts onto intake manifold plenum. Tighten screws to 4.5 Nm (40 inch lbs.) torque.
5. Attach electrical connector to sensor.
6. Install the negative battery cable.

## Connector Pinouts

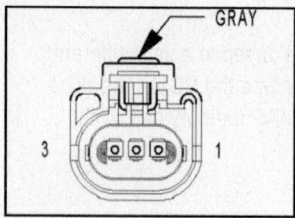

| PIN # | CIRCUIT | FUNCTION |
|-------|---------|----------|
| 1 | K1 20DG/RD | MAP SIGNAL |
| 2 | K4 20BK/LB | SENSOR GROUND 1 |
| 3 | K6 20VT/WT | 5 VOLT SUPPLY |

**Sebring, Stratus 2.0L & 2.4L MAP Sensor**

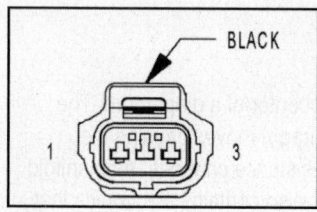

| PIN # | CIRCUIT | FUNCTION |
|-------|---------|----------|
| 1 | K6 20VT/WT | 5 VOLT SUPPLY |
| 2 | K4 20BK/LB | SENSOR GROUND 1 |
| 3 | K1 20DG/RD | MAP SENSOR SIGNAL |

**Sebring, Stratus 2.7L MAP Sensor**

POWER DISTRIBUTION CENTER – RELAY LOCATIONS

*Note: Illustration not available (see lid on power distribution center in vehicle for layout of relays.*

| PIN # | CIRCUIT | FUNCTION |
|---|---|---|
| 70 | C24 20DB/TN (2.0L/2.4L) | LOW RAD FAN RELAY CONTROL |
| 70 | C24 20DB/TN (2.7L) | LOW SPEED RADIATOR FAN RELAY CONTROL |
| 71 | F12 20DB/WT | FUSED IGNITION SWITCH OUTPUT (RUN-START) |
| 72 | C23 12DG | LOW SPEED RADIATOR FAN RELAY OUTPUT |
| 73 | - | - |
| 74 | A16 12RD/LG | FUSED B (+) |

**Lo-Speed Radiator Fan Relay pinout**

| PIN # | CIRCUIT | FUNCTION |
|---|---|---|
| 75 | C27 20DB/PK (2.7L) | HIGH SPEED RADIATOR FAN RELAY CONTROL |
| 75 | C27 20DB/PK (2.0L/2.4L) | HIGH SPEED RAD FAN RELAY CONTROL |
| 76 | F12 20DB/WT | FUSED IGNITION SWITCH OUTPUT (RUN-START) |
| 77 | C25 12YL | HIGH SPEED RADIATOR FAN RELAY OUTPUT |
| 78 | - | - |
| 79 | A16 12RD/LG | FUSED B (+) |

**Hi-Speed Radiator Fan Relay pinout**

| PIN # | CIRCUIT | FUNCTION |
|---|---|---|
| 7 | K90 20TN | ENGINE STARTER MOTOR RELAY CONTROL |
| 8 | - | - |
| 9 | A41 16YL | FUSED IGNITION SWITCH OUTPUT (START) |
| 9 | T141 16YL/RD (2.7L MTX) | FUSED IGNITION SWITCH OUTPUT (START) |
| 10 | T40 16BR | ENGINE STARTER MOTOR RELAY OUTPUT |
| 11 | A1 18RD | FUSED B(+) |

**Starter Relay pinout**

| PIN # | CIRCUIT | FUNCTION |
|---|---|---|
| 2 | Z246 20BK/RD | GROUND |
| 3 | - | - |
| 4 | T15 20LG | TRANSMISSION CONTROL RELAY CONTROL |
| 5 | T16 20RD | TRANSMISSION CONTROL RELAY OUTPUT |
| 6 | A24 18BK | FUSED B (+) |

**Transmission Control Relay Pinout**

## POWERTRAIN CONTROL MODULE (PCM)

### C1 CONNECTOR (W/NGC)

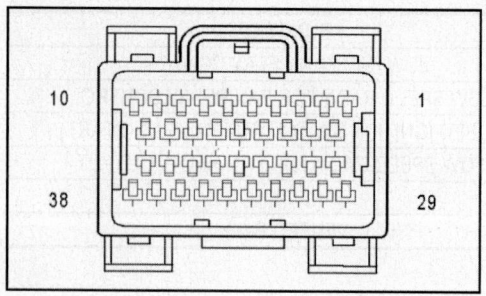

| PIN # | CIRCUIT | FUNCTION |
|---|---|---|
| 1-8 | - | - |
| 9 | Z12 16BK/TN | GROUND |
| 10 | - | - |
| 11 | F12 18DB/WT | FUSED IGNITION SWITCH OUTPUT (RUN-START) |
| 12 | F11 18RD/WT | FUSED IGNITION SWITCH OUTPUT (OFF-RUN-START) |
| 13 | G7 18WT/OR | VEHICLE SPEED SIGNAL |
| 14-17 | - | - |
| 18 | Z12 16BK/TN | GROUND |
| 19-20 | - | - |
| 21 | C18 20DB | A/C PRESSURE SIGNAL |
| 22-24 | - | - |
| 25 | D20 20LG | SCI RECEIVE |
| 26 | D6 20PK/LB (EATX) | SCI RECEIVE (TCM) |
| 27 | K7 18OR (MTX) | 5 VOLT SUPPLY |
| 28 | - | - |
| 29 | A14 16RD/TN | FUSED B(+) |
| 30 | A41 16YL | FUSED IGNITION SWITCH OUTPUT (START) |
| 31 | K141 20TN/WT | O2 1/2 SIGNAL |
| 32 | K904 18DB/DG | O2 RETURN (DOWN) |
| 33-35 | - | - |
| 36 | D21 20PK | SCI TRANSMIT |
| 37 | D15 20WT/DG | SCI TRANSMIT (TCM) |
| 38 | D25 20YL/VT (EATX) | PCI BUS (PCM) |
| 38 | D25 20OR (MTX) | PCI BUS (PCM) |

**Sebring, Stratus (w/NGC) PCM C1 Connector Pinout**

*C2 CONNECTOR (W/NGC)*

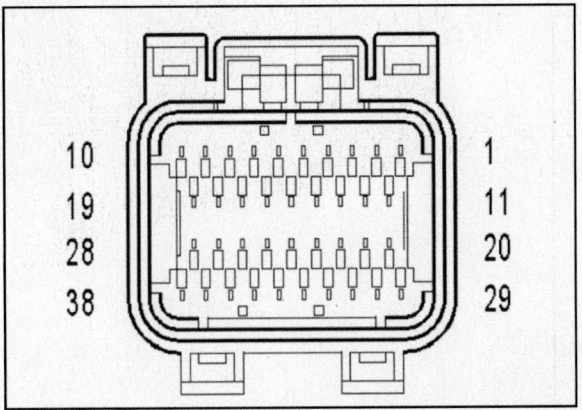

| PIN # | CIRCUIT | FUNCTION |
|---|---|---|
| 1-8 | - | - |
| 9 | K17 18DB/TN | IGNITION COIL NO. 2 DRIVER |
| 10 | K19 18BK/GY | IGNITION COIL NO. 1 DRIVER |
| 11 | K14 18LB/BR | FUEL INJECTOR NO. 4 DRIVER |
| 12 | K13 18YL/WT | FUEL INJECTOR NO. 3 DRIVER |
| 13 | K12 18TN | FUEL INJECTOR NO. 2 DRIVER |
| 14 | K11 18WT/DB | FUEL INJECTOR NO. 1 DRIVER |
| 15-16 | - | - |
| 17 | K199 18BR/VT | O2 1/2 HEATER CONTROL |
| 18 | K99 18BR/OR | O2 1/1 HEATER CONTROL |
| 19 | K20 18DG | GENERATOR FIELD DRIVER |
| 20 | K2 20TN/BK | ECT SIGNAL |
| 21 | K22 20OR/DB | TP SIGNAL |
| 22 | - | - |
| 23 | K1 20DG/RD | MAP SIGNAL |
| 24 | K45 20BK/VT | KS RETURN |
| 25 | K42 20DB/LG | KS SIGNAL |
| 26 | - | - |
| 27 | K4 20BK/LB | SENSOR GROUND 1 |
| 28 | K60 18YL/BK | IAC RETURN |
| 29 | K6 20VT/WT | 5 VOLT SUPPLY |
| 30 | K21 20BK/RD | IAT SIGNAL |
| 31 | K41 20BK/DG | O2 1/1 SIGNAL |
| 32 | K902 18BR/DG | O2 RETURN (UP) |
| 33 | K141 20TN/WT | O2 1/2 SIGNAL |
| 34 | K44 20TN/YL | CMP SIGNAL |
| 35 | K24 20GY/BK | CKP SIGNAL |
| 36-37 | - | - |
| 38 | K39 18GY/RD | IAC MOTOR CONTROL |

**Sebring, Stratus (w/NGC) PCM C2 Connector Pinout**

## C3 CONNECTOR (W/NGC)

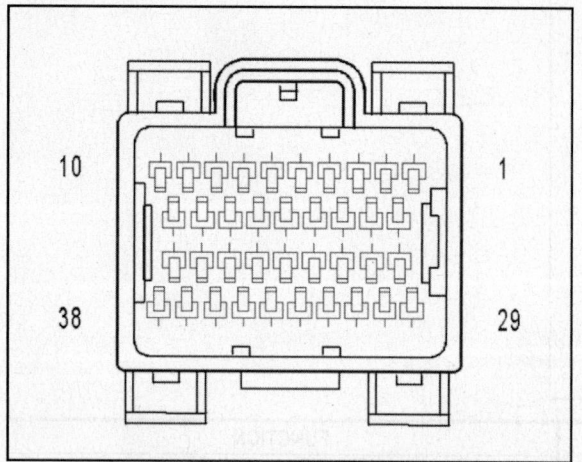

| PIN # | CIRCUIT | FUNCTION |
|---|---|---|
| 1-2 | - | - |
| 3 | K51 20DB/VT | AUTOMATIC SHUT DOWN RELAY CONTROL |
| 4 | C27 20DB/PK | HIGH SPEED RAD FAN RELAY CONTROL |
| 5 | V35 20LG/RD | S/C VENT CONTROL |
| 6 | C24 20DB/TN | LOW SPEED RAD FAN RELAY CONTROL |
| 7 | V32 20YL/RD | S/C SUPPLY |
| 8 | K106 20WT/DG | NVLD SOLENOID CONTROL |
| 9 | K199 18BR/VT | O2 1/2 HEATER CONTROL |
| 10 | - | - |
| 11 | C28 20DB/OR | A/C CLUTCH RELAY CONTROL |
| 12 | V36 20TN/RD | S/C VACUUM CONTROL |
| 13-16 | - | - |
| 17 | K4 18BK/LB | SENSOR GROUND 1 |
| 18 | - | - |
| 19 | F142 16OR/DG | FUSED AUTOMATIC SHUT DOWN RELAY OUTPUT |
| 20 | K52 20PK/BK | EVAP PURGE CONTROL |
| 21 | T141 20YL/RD (MTX) | FUSED IGNITION SWITCH OUTPUT (START) |
| 22 | - | - |
| 23 | K29 20WT/PK | BRAKE SWITCH SIGNAL |
| 24-25 | - | - |
| 26 | T44 20YL | AUTOSTICK DOWNSHIFT SWITCH SIGNAL |
| 27 | T5 20LG | AUTOSTICK UPSHIFT SWITCH SIGNAL |
| 28 | F142 16OR/DG | FUSED AUTOMATIC SHUT DOWN RELAY OUTPUT |
| 29 | K108 20WT/TN | EVAP/PURGE RETURN |
| 30 | K10 18DB/LG | PSP SWITCH SIGNAL |
| 31 | - | - |
| 32 | K25 18VT/LG | AAT SIGNAL |
| 33 | - | - |
| 34 | V37 20PK/LG | S/C SWITCH SIGNAL |
| 35 | K107 20OR | LEAK DETECTION PUMP SWITCH SENSE |
| 36 | - | - |
| 37 | K31 20BR/LG | FUEL PUMP RELAY CONTROL |
| 38 | K90 20TN | STARTER RELAY CONTROL |

**Sebring, Stratus (w/NGC) PCM C3 Connector Pinout**

## C4 CONNECTOR (W/NGC)

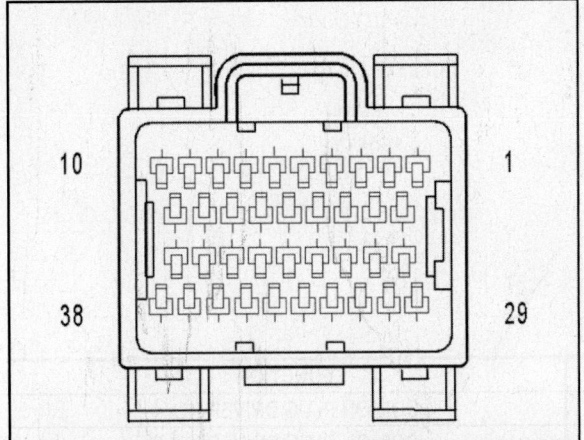

| PIN # | CIRCUIT | FUNCTION |
|-------|---------|----------|
| 1 | T60 20BR | OVERDRIVE SOLENOID CONTROL |
| 2 | T59 20PK | UNDERDRIVE SOLENOID CONTROL |
| 3-6 | - | |
| 6 | T19 20WT | 2-4 SOLENOID CONTROL |
| 7-9 | - | - |
| 10 | T20 20LB | LOW/REVERSE SOLENOID CONTROL |
| 10 | T20 16LB | LOW/REVERSE SOLENOID CONTROL |
| 11 | - | - |
| 12 | Z14 16BK/YL | GROUND |
| 13 | Z13 16BK/RD | GROUND |
| 14 | Z13 16BK/RD | GROUND |
| 15 | T1 20LG/BK | TRS T1 SENSE |
| 16 | T3 20VT | TRS T3 SENSE |
| 17 | - | |
| 18 | T15 20LG | TRANSMISSION CONTROL RELAY CONTROL |
| 19 | T16 20RD | TRANSMISSION CONTROL RELAY OUTPUT |
| 20-21 | - | - |
| 22 | T9 20OR/BK | OVERDRIVE PRESSURE SWITCH SENSE |
| 23-26 | - | - |
| 27 | T41 20BK/WT | TRS T41 SENSE |
| 28 | T16 20RD | TRANSMISSION CONTROL RELAY OUTPUT |
| 29 | T50 20DG | LOW/REVERSE PRESSURE SWITCH SENSE |
| 30 | T47 20YL/WT | 2-4 PRESSURE SWITCH SENSE |
| 31 | - | - |
| 32 | T14 20LG/WT | OUTPUT SPEED SENSOR SIGNAL |
| 33 | T52 20RD/BK | INPUT SPEED SENSOR SIGNAL |
| 34 | T13 20DB/BK | SPEED SENSOR GROUND |
| 35 | T54 20VT/YL | TRANSMISSION TEMPERATURE SENSOR SIGNAL |
| 36 | - | - |
| 37 | T42 20VT/WT | TRS T42 SENSE |
| 38 | T16 20RD | TRANSMISSION CONTROL RELAY OUTPUT |

**Sebring, Stratus (w/NGC) PCM C3 Connector Pinout**

*C1 CONNECTOR (W/SBEC)*

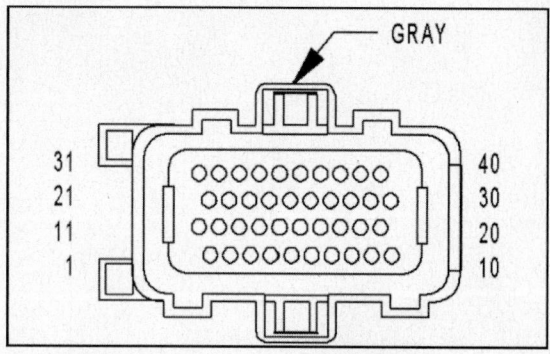

| PIN # | CIRCUIT | FUNCTION |
|---|---|---|
| 1 | K94 18TN/LG | COIL ON PLUG DRIVER NO. 4 |
| 2 | K93 18TN/OR | COIL ON PLUG DRIVER NO. 3 |
| 3 | K92 18TN/PK | COIL ON PLUG DRIVER NO. 2 |
| 4 | K96 18TN/LB | COIL ON PLUG DRIVER NO. 6 |
| 5 | V32 20YL/RD | SPEED CONTROL POWER SUPPLY |
| 6 | A142 14DG/OR | AUTOMATIC SHUT DOWN RELAY OUTPUT |
| 7 | K13 18YL/WT | FUEL INJECTOR NO. 3 DRIVER |
| 8 | K20 18DG | GENERATOR FIELD DRIVER |
| 9 | - | - |
| 10 | Z108 16BK/TN | GROUND |
| 11 | K91 18TN/RD | COIL ON PLUG DRIVER NO. 1 |
| 12 | - | - |
| 13 | K11 18WT/DB | FUEL INJECTOR NO. 1 DRIVER |
| 14 | K58 18BR/DB | FUEL INJECTOR NO. 6 DRIVER |
| 15 | K38 18GY | FUEL INJECTOR NO. 5 DRIVER |
| 16 | K14 18LB/BR | FUEL INJECTOR NO. 4 DRIVER |
| 17 | K12 18TN | FUEL INJECTOR NO. 2 DRIVER |
| 18-19 | - | - |
| 20 | F12 18DB/WT | FUSED IGNITION SWITCH OUTPUT (RUN-START) |
| 21 | K95 18TN/DG | COIL ON PLUG DRIVER NO. 5 |
| 22-24 | - | - |
| 25 | K42 18DB/LG | KNOCK SENSOR SIGNAL |
| 26 | K2 20TN/BK | ENGINE COOLANT TEMPERATURE SENSOR SIGNAL |
| 27 | K127 18DB/LG | OXYGEN SENSOR GROUND |
| 28 | - | - |
| 29 | K241 20LG/RD | OXYGEN SENSOR 2/1 SIGNAL |
| 30 | K41 20BK/DG | OXYGEN SENSOR 1/1 SIGNAL |
| 31 | K90 20TN | STARTER RELAY CONTROL |
| 32 | K24 20GY/BK | CRANKSHAFT POSITION SENSOR SIGNAL |
| 33 | K44 20TN/YL | CAMSHAFT POSITION SENSOR SIGNAL |
| 34 | - | - |
| 35 | K22 20OR/DB | THROTTLE POSITION SENSOR SIGNAL |
| 36 | K1 20DG/RD | MAP SENSOR SIGNAL |
| 37 | K21 20BK/RD | INLET AIR TEMPERATURE SENSOR SIGNAL |
| 38 | - | - |
| 39 | K36 18VT/RD | MANIFOLD SOLENOID CONTROL |
| 40 | - | - |

**Sebring, Stratus (w/SBEC) PCM C1 Connector Pinout**

## C2 CONNECTOR (W/SBEC)

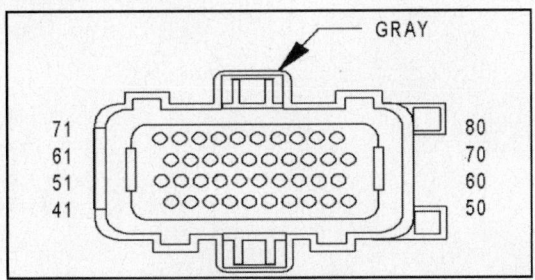

| PIN # | CIRCUIT | FUNCTION |
|---|---|---|
| 41 | V37 20PK/LG | SPEED CONTROL SWITCH SIGNAL |
| 42 | C18 20DB | A/C PRESSURE SIGNAL |
| 43 | K4 18BK/LB | SENSOR GROUND 1 |
| 44 | K7 18OR/WT | 8 VOLT SUPPLY |
| 45 | - | - |
| 46 | A14 14RD/TN | FUSED B (+) |
| 47 | Z109 16BK | GROUND |
| 48 | K40 20BR/WT | IDLE AIR CONTROL MOTOR NO. 3 DRIVER |
| 49 | K60 20YL/BK | IDLE AIR CONTROL MOTOR NO. 2 DRIVER |
| 50 | Z107 16BK/TN | GROUND |
| 51 | K141 20TN/WT | OXYGEN SENSOR 1/2 SIGNAL |
| 52 | K25 18VT/LG | BATTERY TEMPERATURE SENSOR SIGNAL |
| 53 | K341 20PK/WT (2.7L EATX) | OXYGEN SENSOR 2/2 SIGNAL |
| 54 | - | - |
| 55 | C24 20DB/TN | LOW SPEED RADIATOR FAN RELAY CONTROL |
| 56 | V36 20TN/RD | SPEED CONTROL VACUUM SOLENOID CONTROL |
| 57 | K39 20GY/RD | IDLE AIR CONTROL MOTOR NO. 1 DRIVER |
| 58 | K59 20VT/BK | IDLE AIR CONTROL MOTOR NO. 4 DRIVER |
| 59 | D25 20OR | PCI BUS |
| 60 | - | - |
| 61 | K6 18VT/WT | 5 VOLT SUPPLY |
| 62 | K29 20WT/PK | BRAKE SWITCH SENSE |
| 63 | T10 20YL/DG (EATX) | TORQUE MANAGEMENT REQUEST SENSE |
| 64 | C28 20DB/OR | A/C COMPRESSOR CLUTCH RELAY CONTROL |
| 65 | D21 20PK | SCI TRANSMIT |
| 66 | G7 18WT/OR (ABS) | VEHICLE SPEED SENSOR SIGNAL |
| 67 | K51 20DB/VT | AUTOMATIC SHUT DOWN RELAY CONTROL |
| 68 | K52 20PK/BK | EVAPORATIVE EMISSION SOLENOID CONTROL |
| 69 | C27 20DB/PK | HIGH SPEED RADIATOR FAN RELAY CONTROL |
| 70 | K108 20WT/TN | EVAPORATIVE SOLENOID SENSE |
| 71 | K71 20WT/RD (EATX) | EATX RPM SIGNAL |
| 72 | K107 20OR | LEAK DETECTION PUMP SWITCH SENSE |
| 73 | - | - |
| 74 | K31 20BR/LG | FUEL PUMP RELAY CONTROL |
| 75 | D20 20LG | SCI RECEIVE |
| 76 | T41 20BK/LB (EATX) | TRS T41 SENSE |
| 77 | K106 20WT/DG | LEAK DETECTION PUMP SOLENOID CONTROL |
| 78-79 | - | - |
| 80 | V35 20LG/RD | SPEED CONTROL VENT SOLENOID CONTROL |

**Sebring, Stratus (w/SBEC) PCM C2 Connector Pinout**

### POWER STEERING PRESSURE (PSP) SWITCH

**Connector Pinouts**

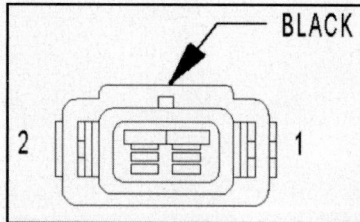

| PIN # | CIRCUIT | FUNCTION |
|-------|---------|----------|
| 1 | K10 20DB/LG | STEERING PRESSURE SWITCH SIGNAL |
| 2 | Z244 20BK | GROUND |

**Sebring, Stratus 2.0L PSP Switch Connector Pinout**

### SOLENOID/PRESSURE SWITCH

**Connector Pinouts**

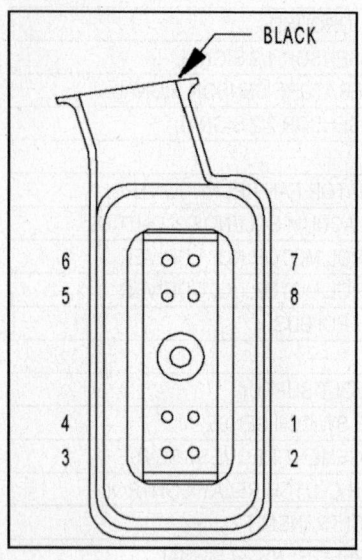

| PIN # | CIRCUIT | FUNCTION |
|-------|---------|----------|
| 1 | T47 18YL/BK | 2-4 PRESSURE SWITCH SENSE |
| 2 | T50 18DG | LOW/REVERSE PRESSURE SWITCH SENSE |
| 3 | T9 18OR/BK | OVERDRIVE PRESSURE SWITCH SENSE |
| 4 | T16 16RD | TRANSMISSION CONTROL RELAY OUTPUT |
| 5 | T59 18PK | UNDERDRIVE SOLENOID CONTROL |
| 6 | T60 18BR | OVERDRIVE SOLENOID CONTROL |
| 7 | T20 18LB | LOW/REVERSE SOLENOID CONTROL |
| 8 | T19 18WT | 2-4 SOLENOID CONTROL |

**Sebring, Stratus (41TE A/T) Solenoid/Pressure Switch Connector Pinout**

THROTTLE POSITION (TP) SENSOR

**Testing**

*TESTING USING A DVOM*

1. Disconnect the throttle position sensor connector.

2. Measure resistance between the throttle position sensor side connector terminal 1 and terminal 4. Standard value: 3.5 - 6.5 k/Ohms.

3. Measure resistance between the throttle position sensor side connector terminal 1 and terminal 3.

4. If resistance is outside the standard value, or if it doesn't change smoothly, replace the throttle position sensor.

5. Manually check operation by opening throttle valve slowly until it is fully open. Changes in ohm readings should be smooth, corresponding to movement of throttle valve.

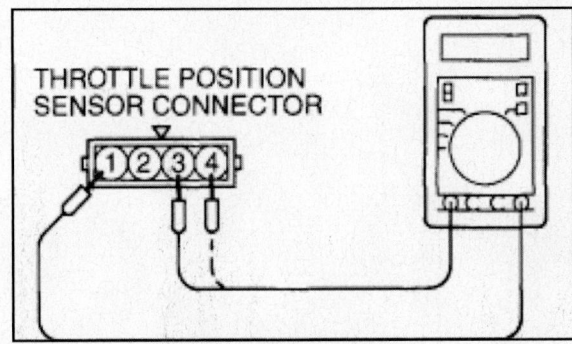

**Checking the Terminal Resistance**

*NOTE: After replacement, the throttle position sensor should be adjusted.*

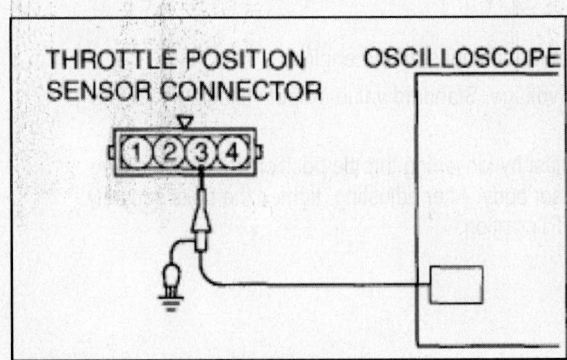

**Checking using an oscilloscope**

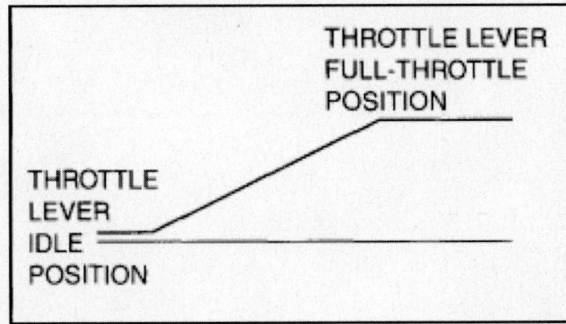

**Showing a normal waveform**

### TESTING USING OSCILLOSCOPE

1. Disconnect the throttle position sensor connector and connect the test harness special tool (MB991348) in between. (All terminals should be connected.)

2. Connect the oscilloscope probe to the throttle position sensor side connector terminal 3.

3. Turn the ignition switch ON position. Slowly move the throttle lever from the idle position to the full-throttle position and check then if the waveform is free from any noise.

4. If any noise is recognized, replace the throttle position sensor.

*NOTE: After replacement, the throttle position sensor should be adjusted.*

**Adjustment**

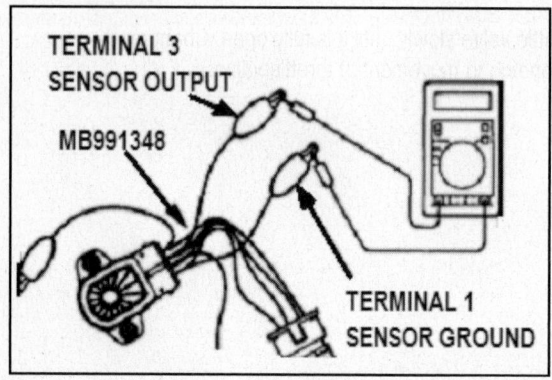

**Setup for testing and making adjustment to TP sensor**

1. Disconnect the throttle position sensor connector, and connector the special tool MB991348 in between. (All terminals should be connected.)

2. Connect a digital voltmeter between throttle position sensor terminal 3 (sensor output) and terminal 1 (sensor ground).

3. Turn the ignition switch to the ON position (but do not start the engine).

4. Check the throttle position sensor output voltage. Standard value output voltage is 535-735 mV.

5. If not within the standard value range, adjust by loosening throttle position sensor mounting bolts and turning the throttle position sensor body. After adjusting, tighten the bolts securely. Turn the ignition switch to the LOCK (OFF) position.

**Throttle Position Sensor Adjusting**

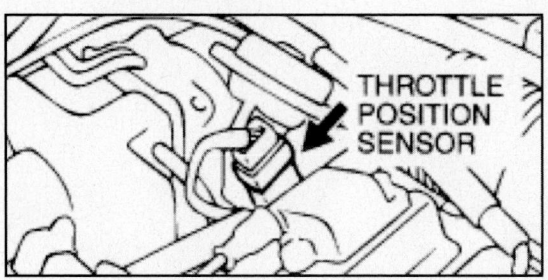

Showing on-engine location of the Throttle Position Sensor on 2.4L engines

**Connector Pinouts**

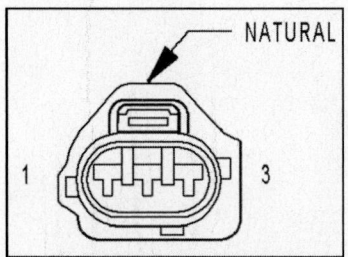

| PIN # | CIRCUIT | FUNCTION |
|-------|---------|----------|
| 1 | K6 20VT/WT | 5 VOLT SUPPLY |
| 2 | K22 20OR/DB | TP SIGNAL |
| 3 | K4 20BK/LB | SENSOR GROUND 1 |

**Sebring, Stratus 2.0L & 2.4L TP Sensor**

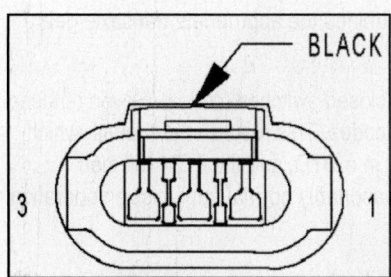

| PIN # | CIRCUIT | FUNCTION |
|-------|---------|----------|
| 1 | K6 20VT/WT | 5V SUPPLY |
| 2 | K22 20OR/DB | THROTTLE POSITION SENSOR SIGNAL |
| 3 | K4 20BK/LB | SENSOR GROUND 1 |

**Sebring, Stratus 2.7L TP Sensor**

## TRANSMISSION RANGE (TR) SENSOR

### Description

The Transmission Range Sensor (TRS) is mounted to the top of the valve body inside the transaxle and can only be serviced by removing the valve body. The electrical connector extends through the transaxle case. The Transmission Range Sensor (TRS) has four switch contacts that monitor shift lever position and send the information to the TCM. The TRS also has an integrated temperature sensor (thermistor) that communicates transaxle temperature to the TCM and PCM.

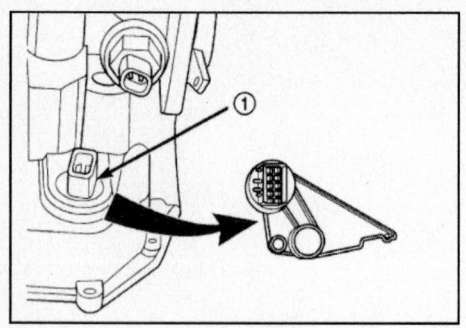

**Transmission Range Sensor (1) location**

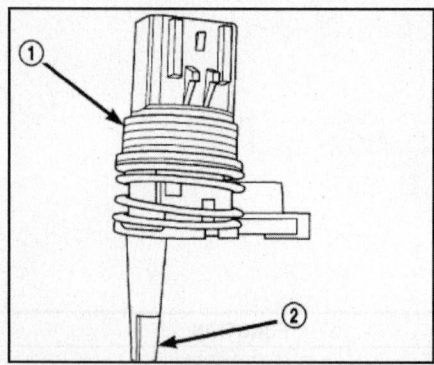

**Transmission Range Sensor (1) and Temperature Sensor (2)**

## Operation

The Transmission Range Sensor (TRS) communicates shift lever position to the TCM as a combination of open and closed switches. Each shift lever position has an assigned combination of switch states (open/closed) that the TCM receives from four sense circuits. The TCM interprets this information and determines the appropriate transaxle gear position and shift schedule.

Since there are four switches, there are 16 possible combinations of open and closed switches (codes). Seven of these codes are related to gear position and three are recognized as "between gear" codes. This results in six codes, which should never occur. These are called "invalid" codes. An invalid code will result in a DTC, and the TCM will then determine the shift lever position based on pressure switch data. This allows reasonably normal transmission operation with a TRS failure.

### TRANSMISSION TEMPERATURE SENSOR

The TRS has an integrated thermistor that the TCM uses to monitor the transmission's sump temperature. Since fluid temperature can affect transmission shift quality and converter lock up, the TCM requires this information to determine which shift schedule to operate in. The PCM also monitors this temperature data so it can energize the vehicle cooling fan(s) when a transmission "overheat" condition exists. If the thermistor circuit fails, the TCM will revert to calculated oil temperature usage.

| SELECTOR LEVER POSITION | T42 | T41 | T3 | T1 |
|---|---|---|---|---|
| PARK | CLOSED | CLOSED | CLOSED | OPEN |
| REVERSE | CLOSED | OPEN | OPEN | OPEN |
| NEUTRAL | CLOSED | CLOSED | OPEN | CLOSED |
| OVERDRIVE | OPEN | OPEN | OPEN | CLOSED |
| THIRD | OPEN | OPEN | CLOSED | OPEN |
| LOW | CLOSED | OPEN | CLOSED | CLOSED |

### *CALCULATED TEMPERATURE*

A failure in the temperature sensor or circuit will result in calculated temperature being substituted for actual temperature. Calculated temperature is a predicted fluid temperature, which is calculated from a combination of inputs:

- Battery (ambient) temperature
- Engine coolant temperature
- In-gear run time since start-up

### Removal & Installation

1. Remove valve body assembly from transaxle.
2. Remove transmission range sensor retaining screw and remove sensor from valve body.
3. Remove TRS from manual shaft.

### To install:

4. Install transmission range sensor to the valve body and torque retaining screw to 5 Nm (45 inch lbs.).
5. Install valve body to transaxle.

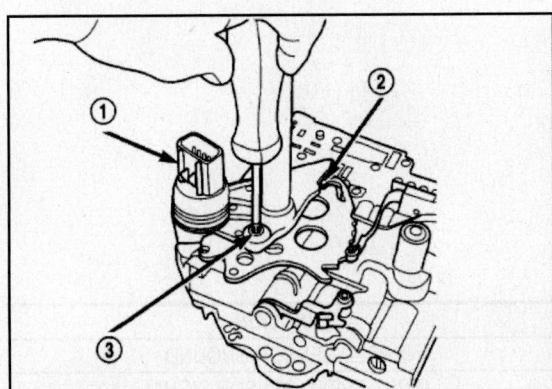

**Removing the Transmission Range Sensor (1), manual valve control pin (2) and retaining screw (3)**

### VEHICLE SPEED CONTROL (VSC) SERVO

**Connector Pinouts**

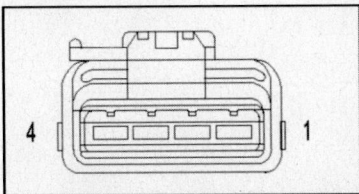

| PIN # | CIRCUIT | FUNCTION |
|:-----:|:--------|:---------|
| 1 | V36 20TN/RD (2.0L/2.4L) | S/C VACUUM CONTROL |
| 1 | V36 20TN/RD (2.7L) | SPEED CONTROL VACUUM SOLENOID CONTROL |
| 2 | V35 20LG/RD (2.0L/2.4L) | S/C VENT CONTROL |
| 2 | V35 20LG/RD (2.7L) | SPEED CONTROL VENT SOLENOID CONTROL |
| 3 | V30 20DB/RD | SPEED CONTROL BRAKE SWITCH OUTPUT |
| 4 | Z190 20BK | GROUND |

**Sebring, Stratus VSC Servo Connector Pinout**

## VEHICLE SPEED SENSOR (VSS)

**Connector Pinouts**

*M/T*

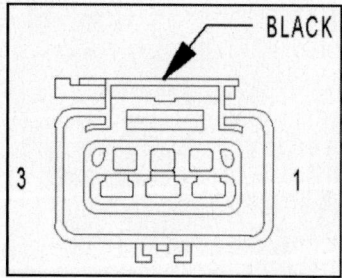

| PIN # | CIRCUIT | FUNCTION |
|-------|---------|----------|
| 1 | K7 18OR/WT | 8V SUPPLY |
| 2 | K4 20BK/LB | SENSOR GROUND 1 |
| 3 | G7 18WT/OR | VEHICLE SPEED SENSOR SIGNAL |

**Sebring, Stratus 2.0L (M/T) VSS Connector Pinout**

*A/T INPUT*

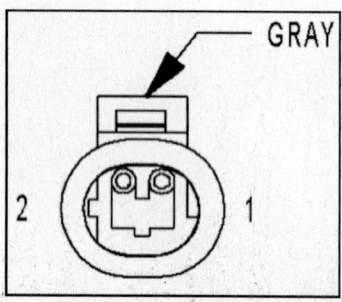

| PIN # | CIRCUIT | FUNCTION |
|-------|---------|----------|
| 1 | T13 18DB/BK | SPEED SENSOR GROUND |
| 2 | T52 18RD/BK | INPUT SPEED SENSOR SIGNAL |

**Sebring, Stratus (A/T) Input VSS Connector Pinout**

*A/T OUTPUT*

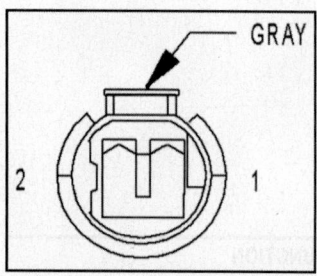

| PIN # | CIRCUIT | FUNCTION |
|-------|---------|----------|
| 1 | T13 18DB/BK | SPEED SENSOR GROUND |
| 2 | T14 18LG/WT | OUTPUT SPEED SENSOR SIGNAL |

**Sebring, Stratus (A/T) Output VSS Connector Pinout**

# DAIMLERCHRYSLER CARS: PACIFICA

CAMSHAFT POSITION (CMP) SENSOR

## Description & Operation

The camshaft position sensor contains a Hall-effect device that provides cylinder identification to the Powertrain Control Module (PCM). The sensor generates pulses as groups of notches on the camshaft sprocket pass underneath it. The PCM keeps track of crankshaft rotation and identifies each cylinder by the pulses generated by the notches on the camshaft sprocket.

When metal aligns with the sensor, voltage goes low (less than 0.3v). When a notch aligns with the sensor, voltage spikes high (5.0v). As a group of notches pass under the sensor, the voltage switches from low (metal) to high (notch) then back to low. The number of notches determines the amount of pulses. If available, an oscilloscope can display the square wave patterns of each timing event.

## Removal & Installation

*3.5L*

1. Disconnect the negative battery cable.
2. Remove air cleaner box.
3. Unlock the camshaft sensor electrical connector.
4. Disconnect the electrical connectors to the camshaft sensor.
5. Remove the mounting bolts.
6. Remove camshaft sensor.

**To install:**

7. Install camshaft sensor.
8. Install the mounting bolts and tighten, refer to the torque chart for value.
9. Install and lock the camshaft sensor electrical connector.
10. Install air cleaner box.
11. Connect the negative battery cable.

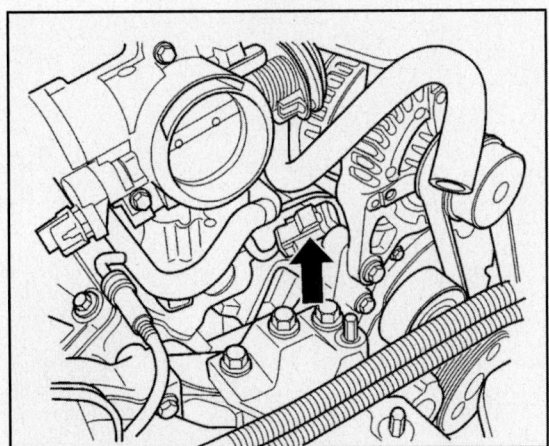

**Showing CMP Sensor Location (Pacifica 3.5L V6)**

*3.8L*

1. Disconnect the negative battery cable.
2. Unlock the camshaft sensor electrical connector.
3. Disconnect the electrical connectors to the camshaft sensor.
4. Loosen the mounting bolts.
5. Remove camshaft sensor.
6. Camshaft Position Sensor removed.

**To install:**

7. Install camshaft position sensor.
8. Tighten the mounting bolt to Nm 12 (106 inch lbs.).
9. Install and lock the camshaft sensor electrical connector.
10. Connect the negative battery cable.

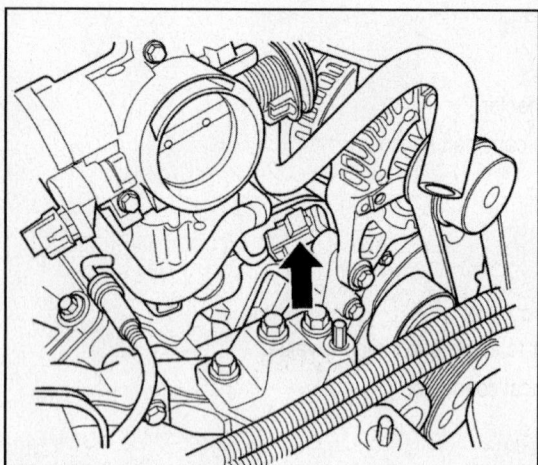

**CMP Sensor Location (Pacifica 3.8L V8)**

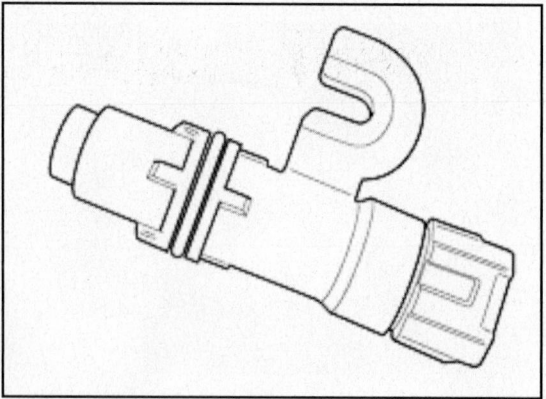

**Showing CMP Sensor Removed (Pacifica 3.8L V8)**

**Connector Pinouts**

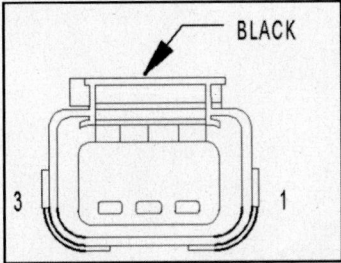

| CAV | CIRCUIT | FUNCTION |
|-----|---------|----------|
| 1 | F855 20PK/YL | 5 VOLT SUPPLY |
| 2 | K900 20DB/DG | SENSOR GROUND |
| 3 | K44 20DB/GY | CMP SIGNAL |

**3.5L & 3.8L CMP Sensor**

## CRANKSHAFT POSITION (CKP) SENSOR

**Removal & Installation**

The crankshaft sensor is located on the driver side of the vehicle, above the differential housing. The bottom of the sensor sits above the drive plate.

1. Disconnect the negative and positive battery cable.
2. Remove the battery.
3. Unlock and disconnect electrical connector from crankshaft position sensor.

**To install:**

4. Install sensor and push sensor down until contact is made with the transmission case. While holding the sensor in this position, install and tighten the retaining bolt to 12 Nm (105 inch lbs.) torque.
5. Connect electrical connector and lock to crankshaft position sensor.
6. Install the battery and connect the positive cable, then the negative battery cable.

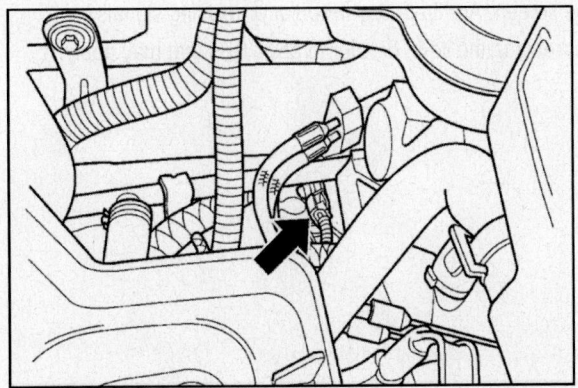

**Location of CKP Sensor on Pacifica 3.5L Engine**

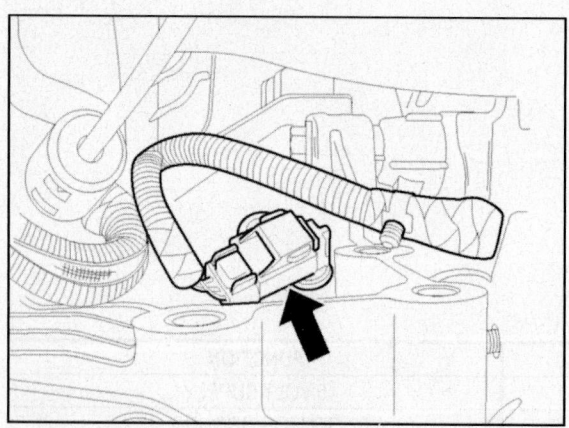

**Location of CKP Sensor on Pacifica 3.8L Engine**

## Testing

1. Visually inspect the related wire harness. Look for any chafed, pierced, pinched, or partially broken wires.

2. Visually inspect the related wire harness connectors. Look for broken, bent, pushed out, or corroded terminals.

3. Ensure the Crankshaft Position Sensor and the Camshaft Position Sensor are properly installed and the mounting bolt(s) tight.

4. Remove the Camshaft Position Sensor and inspect the Tone Wheel/Pulse Ring for damage, foreign material, or excessive movement.

5. If no problems were evident, start the engine.

6. Gently tap on the Cam Position Sensor and wiggle the Sensor.

7. Inspect the Sensor harness connector, PCM harness connector, Sensor connector, and PCM connector for loose, bent, corroded, or pushed out pins or terminals.

8. Inspect the related wire harness and the splices in the CMP circuits.

9. Repair any wiring and/or connector concerns, or replace the Camshaft Position Sensor.

10. With the proper lab scope probe and the Miller special tool #6801, backprobe the CKP Signal circuit at the CKP harness connector.

11. Start the engine. Observe the lab scope screen. Are there any irregular or missing signals?

12. If so, determine the cause and repair or replace the item, or refer to any TSBs that may apply.

## Connector Pinouts

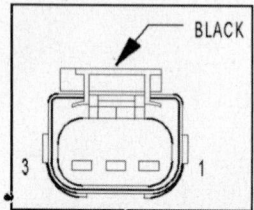

| CAV | CIRCUIT | FUNCTION |
|:---:|:---:|:---:|
| 1 | F855 20PK/YL | 5 VOLT SUPPLY |
| 2 | K900 20DB/DG | SENSOR GROUND |
| 3 | K24 20BR/LB | CKP SIGNAL |

**3.5L & 3.8L CKP Sensor**

<u>ENGINE COOLANT TEMPERATURE (ECT) SENSOR</u>

**Description**

*3.5L*

The engine coolant temperature sensor threads into a coolant passage on lower intake manifold near the thermostat. New sensors have sealant applied to the threads.

*3.8L*

The engine coolant temperature sensor threads into a coolant passage on lower intake manifold near the thermostat. New sensors have sealant applied to the threads.

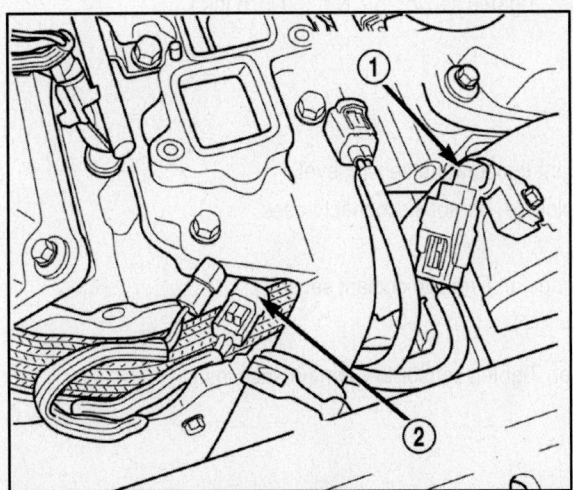

**Locating CMP Sensor (1) and ECT Sensor (2) on 3.5L Engine**

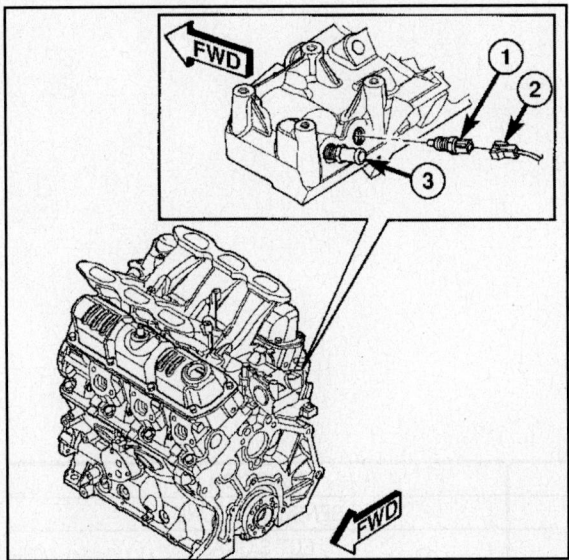

**Locating ECT Sensor (1) on Pacifica 3.8L engine with connector (2) and heater supply fitting (3)**

**Removal & Installation**

*3.5L*

**WARNING: Hot, pressurized coolant can cause injury by scalding. Cooling system must be partially drained before removing the coolant temperature sensor.**

1. Drain cooling system below engine coolant temperature sensor level.
2. Disconnect negative cable from remote jumper terminal.
3. With the engine cold, disconnect coolant sensor electrical connector and remove the sensor.

**To install:**

4. Install engine coolant temperature sensor. Tighten sensor to 7 Nm (60 inch lbs.).
5. Connect electrical connector to sensor.
6. Fill cooling system.

**3.8L**

1. Drain cooling system below engine coolant temperature sensor level.
2. Remove power steering reservoir and relocate. Do not disconnect hoses.
3. Remove ignition coil and bracket.
4. Disconnect coolant sensor electrical connector. Remove coolant sensor.

**To Install:**

5. Install engine coolant temperature sensor. Tighten sensor to 7 Nm (60 inch lbs.).
6. Connect electrical connector to sensor.
7. Install ignition coil bracket.
8. Install ignition coil.
9. Install power steering reservoir.
10. Fill cooling system.

**Connector Pinouts**

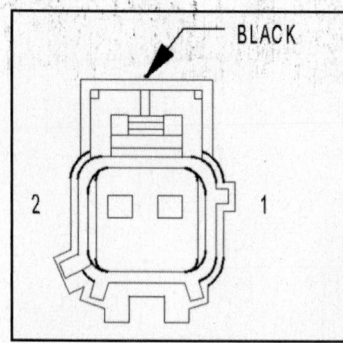

| CAV | CIRCUIT | FUNCTION |
|-----|---------|----------|
| 1 | K900 20DB/DG | SENSOR GROUND |
| 2 | K2 18VT/OR | ECT SIGNAL |

**3.5L & 3.8L ECT Sensor**

EXHAUST GAS RECIRCULATION (EGR) VALVE

**Description & Operation**

The engines use Exhaust Gas Recirculation (EGR) systems. The EGR system reduces oxides of nitrogen (NOx) in engine exhaust and helps prevent detonation (engine knock). Under normal operating conditions, engine cylinder temperature can reach more than 3000°F. Formation of NOx increases proportionally with combustion temperature. To reduce the emission of these oxides, the cylinder temperature must be lowered. The system allows a predetermined amount of hot exhaust gas to recirculate and dilute the incoming air/fuel mixture. The diluted air/fuel mixture reduces peak flame temperature during combustion.

The electric EGR transducer contains an electrically operated solenoid and a backpressure transducer. The Powertrain Control Module (PCM) operates the solenoid. The PCM determines when to energize the solenoid. Exhaust system backpressure controls the transducer. When the PCM energizes the solenoid, vacuum does not reach the transducer. Vacuum flows to the transducer when the PCM de-energizes the solenoid.

When exhaust system backpressure becomes high enough, it fully closes a bleed valve in the transducer. When the PCM de-energizes the solenoid and backpressure closes the transducer bleed valve, vacuum flows through the transducer to operate the EGR valve.

De-energizing the solenoid, but not fully closing the transducer bleed hole (because of low back-pressure), varies the strength of vacuum applied to the EGR valve. Varying the strength of the vacuum changes the amount of EGR supplied to the engine. This provides the correct amount of exhaust gas recirculation for different operating conditions.

This system does not allow EGR at idle. A failed or malfunctioning EGR system can cause engine spark knock, sags or hesitation, rough idle, engine stalling and increased emissions.

**Removal & Installation**

*3.5L*

1. Disconnect the negative and then the positive battery cable.
2. Remove battery.
3. Unlock and disconnect the electrical connector from the EGR valve.
4. Remove the 2 EGR tube bolts.
5. Remove the 2 mounting bolts for the EGR valve and gasket.

**To install:**

6. Install the 2 mounting bolts into the EGR valve and gasket.
7. Loose install the EGR valve to cylinder head.
8. Loose install the EGR tube to EGR valve.
9. Tighten the EGR upper tube to EGR valve bolts to 11.3 Nm (100 inch lbs.) torque.
10. Connect and lock the electrical connector for the EGR valve.
11. Install the battery. Connect the positive then the negative battery cable.

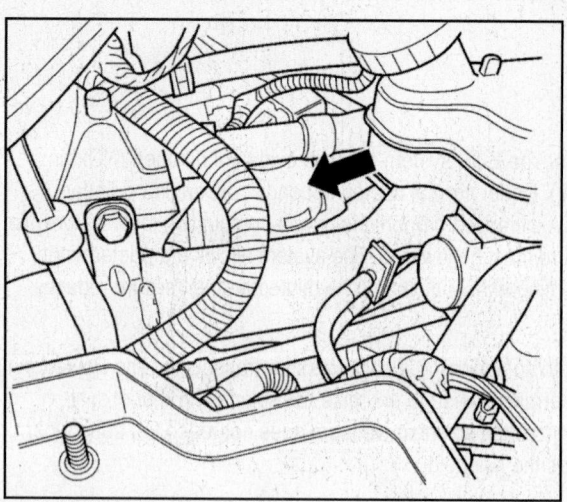

**EGR Valve location on Pacifica 3.5L engine**

*3.8L*

1. Disconnect negative battery cable.
2. Unlock and disconnect the electrical connector from the EGR valve.
3. Remove EGR tube bracket bolt. Remove the 2 EGR tube bolts.
4. Remove the 2 mounting bolts for the EGR valve and gasket.

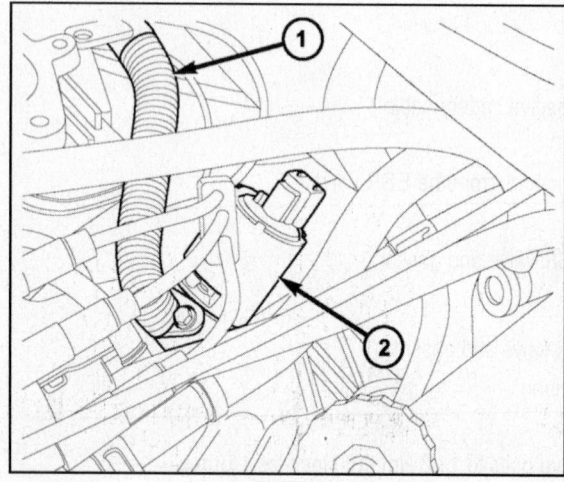

**EGR Valve (2) and hose (1) location on Pacifica 3.8L engine**

**To Install:**

5. Install the 2 mounting bolts into the EGR valve and gasket.
6. Loose install the EGR valve to cylinder head.
7. Loose install the EGR tube to EGR valve.
8. Install and tighten EGR tube bracket bolt.
9. Tighten the EGR tube to EGR valve bolts to 11.3 Nm (100 inch lbs.) torque.
10. Connect and lock the electrical connector for the EGR valve.
11. Install the battery. Connect the positive then the negative battery cable.

**Connector Pinouts**

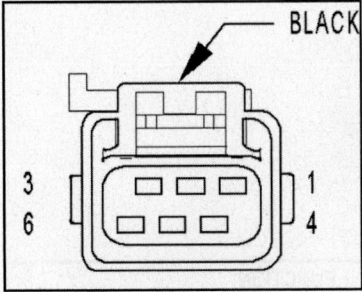

| CAV | CIRCUIT | FUNCTION |
|-----|---------|----------|
| 1 | K34 20DB/LG | EGR SIGNAL |
| 2 | F855 20PK/YL | 5 VOLT SUPPLY |
| 3 | K900 20DB/DG | SENSOR GROUND |
| 4 | Z335 18BK/WT | GROUND |
| 5 | - | - |
| 6 | K35 18DB/VT | EGR SOL CONTROL |

**3.5L & 3.8L EGR Solenoid Connector Pinout**

## EVAP PURGE SOLENOID

### Removal & Installation

1. Disconnect the negative battery cable.
2. Unlock and disconnect the electrical connector.
3. Release tab and pull solenoid from bracket.
4. Remove the vacuum lines from solenoid.

**To install:**

5. Connect the vacuum lines to solenoid.
6. Install solenoid to bracket.
7. Connect the electrical connector to the solenoid and lock.
8. Connect the negative battery cable.

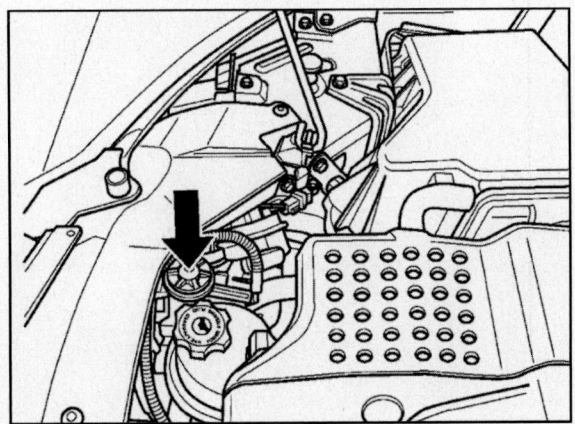

**Locating the EVAP Purge Solenoid**

**Connector Pinouts**

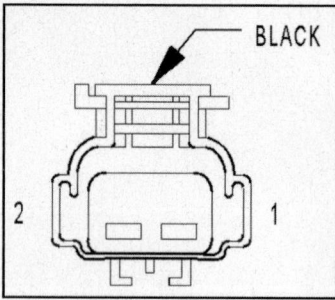

| CAV | CIRCUIT | FUNCTION |
|---|---|---|
| 1 | K52 20DB/WT | EVAP PURGE CONTROL |
| 2 | K70 20DB/BR | EVAP PURGE RETURN |

**3.5L & 3.8L EVAP Purge Solenoid Connector Pinout**

## FUEL LEVEL SENDING UNIT & SENSOR

**Removal & Installation**

1. Release the fuel pressure.
2. Disconnect negative battery cable.
3. Remove fuel tank.
4. Vacuum area before removing module lock ring.
5. Disconnect vapor line and electrical connector.
6. Module lock ring contact points.
7. Remove module lock ring, using a brass punch and hammer to loosen.
8. Remove module top.
9. Tip module onto its side to drain fuel from reservoir in module.
10. Drain fuel tank, use an approved gasoline draining station.
11. Disconnect electrical connector for fuel level sending card.
12. Remove fuel level sending card by prying on locking tab and pulling card down toward bottom of the pump module.

*NOTE: There are 2 locking tabs on sending card; one on the front and one on back.*

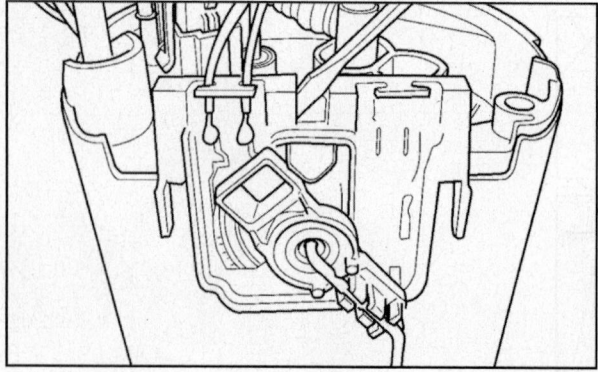

**Removing Fuel Level Sensor card**

**To install:**

1. Install fuel level sending card to module.
2. Connect electrical connector for fuel level sending card.
3. Install module top to module, ensure alignment pins are in place.
4. Install module into the tank with the level unit towards rear of tank.
5. Install a new seal onto the tank, making sure seal is properly seated in tank groove.
6. Install module lock ring, using brass punch and hammer to lock the lock ring.
7. Connect vapor line and electrical connector.
8. Install fuel tank.
9. Connect negative battery cable.
10. Fill fuel tank. Use the scan tool to pressurize the fuel system. Check for leaks.

**Connector Pinouts**

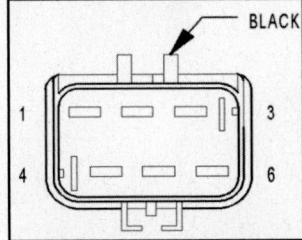

| CAV | CIRCUIT | FUNCTION |
|-----|---------|----------|
| 1 | N1 16DB/OR | FUEL PUMP RELAY OUTPUT |
| 2 | N4 20DB/YL | FUEL LEVEL SENSOR SIGNAL NO. 1 |
| 3 | N5 20DB/WT | FUEL LEVEL SENSOR SIGNAL NO. 2 |
| 4 | - | - |
| 5 | Z210 20BK/LB | GROUND |
| 6 | Z201 16BK | GROUND |

**3.5L & 3.8L Fuel Pump Module Connector Pinout**

HEATED OXYGEN SENSORS (HO2S)

**Removal & Installation**

*3.5L*

*CAUTION: When disconnecting the sensor electrical connector, do not pull directly on wire going into sensor.*

1. Disconnect the negative battery cable.
2. Disconnect the heated oxygen sensor electrical connector.
3. Use a socket such as a crowfoot wrench to remove oxygen sensor.

**To install:**

*NOTE: When replacing an O2 Sensor, the PCM RAM memory must be cleared, either by disconnecting the PCM C-1 connector or momentarily disconnecting the Battery negative terminal. The NGC learns the characteristics of each O2 heater element and these old values should be cleared when installing a new O2 sensor. The customer may experience driveability issues if this is not performed.*

4.   After removing the sensor, the threads must be cleaned with an 18 mm X 1.5 + 6E tap. If reusing the original sensor, coat the sensor threads with an anti-seize compound such as Loctite 771- 64 or equivalent. New sensors have compound on the threads and do not require an additional coating. Tighten the sensor to 28 Nm (20 ft. lbs.) torque.

5.   Connect the heated oxygen sensor electrical connector.

6.   Install the wiring clip to the heat shield.

7.   Connect the negative battery cable.

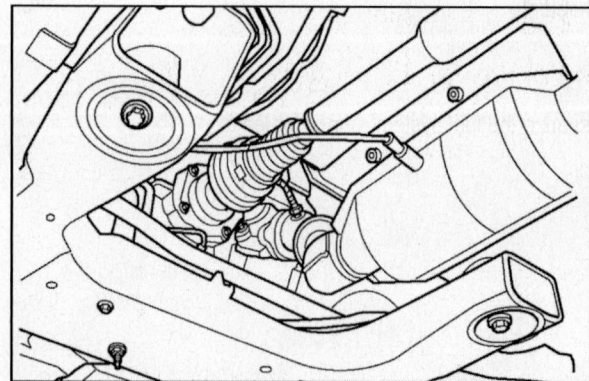

Showing location of O2 sensor on Pacifica 3.5L engine

### 3.8L – DOWNSTREAM O2 SENSOR

*CAUTION: When disconnecting the sensor electrical connector, do not pull directly on wire going into sensor.*

1.   Disconnect negative battery cable.

2.   Raise and support vehicle.

3.   Disconnect the electrical connector.

4.   Use a socket such as a crowfoot wrench to remove oxygen sensor.

### 3.8L – UPSTREAM O2 SENSOR

*CAUTION: When disconnecting the sensor electrical connector, do not pull directly on wire going into sensor.*

1.   Disconnect negative battery cable.

2.   Disconnect the upper O2 sensor connector.

3.   Use a socket such as a crowfoot wrench to remove oxygen sensor.

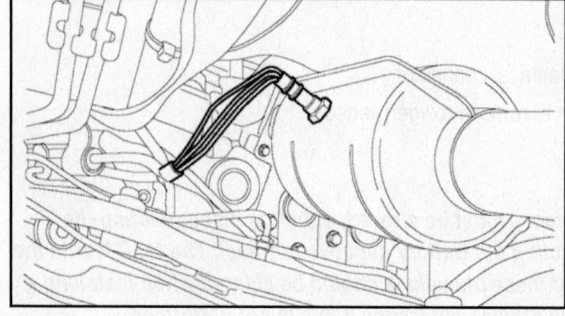

Location of downstream O2 sensor on Pacifica 3.8L engine

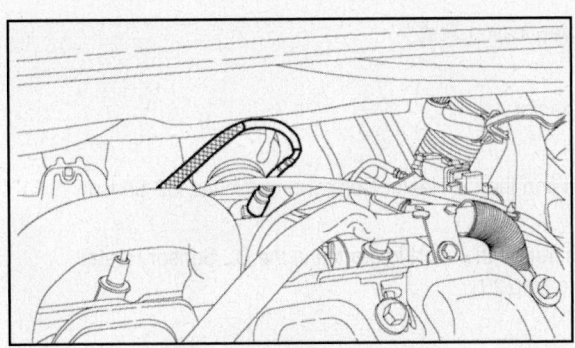

**Location of upstream O2 sensor on Pacifica 3.8L engine**

## To install:

### 3.8L – UPSTREAM & DOWNSTREAM O2 SENSORS

*NOTE: When replacing an O2 Sensor, the PCM RAM memory must be cleared, either by disconnecting the PCM C-1 connector or momentarily disconnecting the Battery negative terminal. The NGC learns the characteristics of each O2 heater element and these old values should be cleared when installing a new O2 sensor. The customer may experience driveability issues if this is not performed.*

1. After removing the sensor, the threads must be cleaned with an 18 mm X 1.5 + 6E tap. If reusing the original sensor, coat the sensor threads with an anti-seize compound such as Loctite 771- 64 or equivalent. New sensors have compound on the threads and do not require an additional coating. Tighten the sensor to 28 Nm (20 ft. lbs.) torque.

2. Connect the heated oxygen sensor electrical connector.

3. Connect the negative battery cable.

## Testing

*NOTE: Allow the O2 Sensor to cool down before conducting the test. The O2 Sensor voltage should stabilize at 5.0 volts. Raising the hood may help in reducing under hood temps quicker.*

1. Turn the ignition on, engine not running. With the scan tool, actuate the "O2 Heater Test."

2. With the scan tool, monitor O2 Sensor voltage for at least 2 minutes. Does the O2 Sensor voltage stay above 4.5 volts? If so, sensor is normal.

3. If voltage does not stay above 4.5v as indicated, check wiring and connectors. Check O2 sensor for contamination.

4. If wiring from the O2 sensor is damaged, DO NOT repair it; replace the O2 sensor.

5. Turn the ignition off. Allow the O2 sensor to cool down to room temperature.

6. Disconnect the O2 Sensor harness connector.

7. Measure the resistance across the O2 Sensor Heater element component side. Resistance for either O2 Sensor should be 2.1-2.7 ohms.

8. If the O2 Sensor is not within range, turn the ignition off.

9. Disconnect the O2 Sensor harness connector. Turn the ignition to ON; engine not running.

10. With the proper scan tool, actuate the "O2 Heater Test."

11. Using a 12-volt test light connected to ground, probe the O2 Heater Control circuit in the O2 Sensor harness connector.

12. If the test light illuminates brightly and flashes on and off, replace the O2 Sensor.

13. Confirm the repair with the appropriate Verification Test.

14. Turn the ignition off.

15. Disconnect the O2 Sensor harness connector.

16. Disconnect the PCM harness connector.

17. Measure the resistance between ground and the O2 Heater Control circuit in the O2 Sensor harness connector.

18. Is the resistance below 5.0 ohms? If yes, repair the short to ground in the O2 Sensor Heater Control circuit. If resistance is okay, go to step 20.

19. Confirm the repair.

20. If resistance was okay when measured in step 17, check the PCM harness connector terminals for corrosion, damage, or terminal push out. Repair as necessary.

21. If the resistance is still not within specified range, replace and program the Powertrain Control Module in accordance with the service information.

22. Confirm the repair.

**Connector Pinouts**

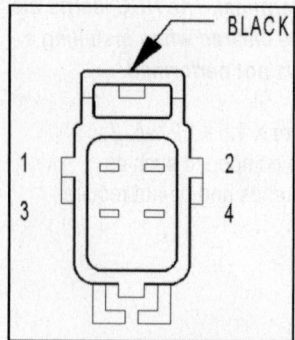

| CAV | CIRCUIT | FUNCTION |
|-----|---------|----------|
| 1 | Z42 18BK/LG | GROUND |
| 2 | K99 18BR/TN | O2 1/1 HEATER CONTROL |
| 3 | K902 18BR/DG | O2 RETURN (UP) |
| 4 | K41 18DB/LB | O2 1/1 SIGNAL |

**3.5L & 3.8L Upstream 1/1 HO2S Connector Pinout**

| CAV | CIRCUIT | FUNCTION |
|-----|---------|----------|
| 1 | Z43 18BK/LB | GROUND |
| 2 | K299 18BR/OR | O2 1/2 HEATER CONTROL |
| 3 | K904 20DB/DG | O2 RETURN (DOWN) |
| 4 | K141 20DB/YL | O2 1/2 SIGNAL |

**3.5L & 3.8L Downstream 1/2 HO2S Connector Pinout**

IDLE AIR CONTROL (IAC) MOTOR

**Removal & Installation**

1. Disconnect the negative battery cable.
2. Disconnect the IAC electrical connector.
3. Remove the IAC mounting screws.
4. Remove the IAC.

**To install:**

When servicing throttle body components, always reassemble components with new O-rings and seals where applicable. If assembly of component is difficult, a light coat of engine oil may be applied to the O-rings ONLY to aid assembly.

1. Install the IAC to the throttle body.
2. Tighten mounting screw to 5.5 Nm (49 inch lbs.) torque.
3. Attach electrical connector to the IAC.
4. Connect the negative battery cable.

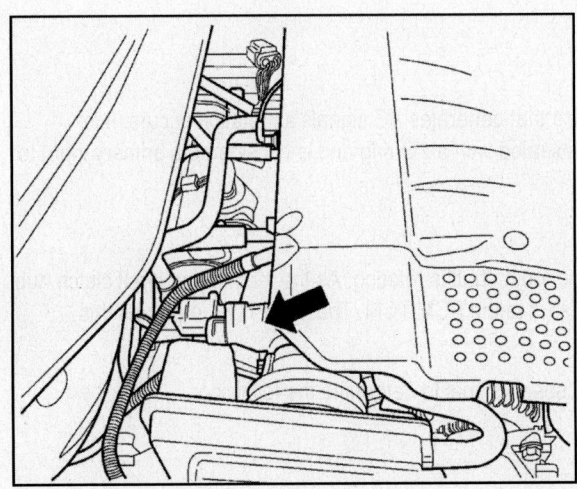

**Location of Idle Air Control Motor on Pacifica 3.5L engine**

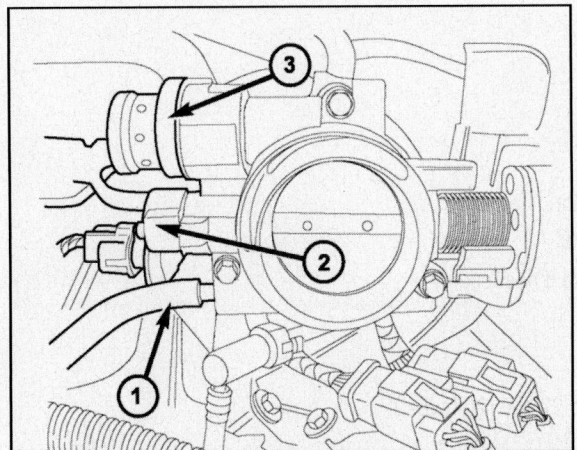

**Location of Idle Air Control Motor (3), throttle body (2) and fuel purge line (3) on Pacifica 3.8L engine**

**Connector Pinouts**

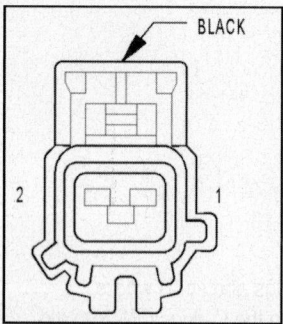

| CAV | CIRCUIT | FUNCTION |
|-----|---------|----------|
| 1 | K61 20VT/GY | IAC MOTOR CONTROL |
| 2 | K961 20BR/VT | IAC RETURN |

**3.5L & 3.8L IAC Motor Connector Pinout**

INPUT SPEED SENSOR

**Description**

The Input Speed Sensor is a two-wire magnetic pickup device that generates AC signals as rotation occurs. It is threaded into the transaxle case. The Input Speed Sensor is sealed with an O-ring and is considered a primary input to the Powertrain/Transmission Control Module.

**Operation**

The Input Speed Sensor provides information on how fast the input shaft is rotating. As the teeth of the input clutch hub (3) pass by the sensor coil, an AC voltage is generated and sent to the PCM/TCM. The PCM/TCM interprets this information as input shaft rpm.

The PCM/TCM compares the input speed signal with output speed signal to determine the following:

- Transmission gear ratio
- Speed ratio error detection
- CVI calculation

The PCM/TCM also compares the input speed signal and the engine speed signal to determine the following:

- Torque converter clutch slippage
- Torque converter element speed ratio

**Removal & Installation**

1. Disconnect battery negative cable.
2. Disconnect input speed sensor connector.
3. Unscrew and remove input speed sensor.
4. Inspect speed sensor O-ring and replace if necessary.

**To install:**

5. Verify O-ring is installed into position.
6. Install and tighten input speed sensor to 27 Nm (20 ft. lbs.).
7. Connect speed sensor connector.
8. Connect battery negative cable.

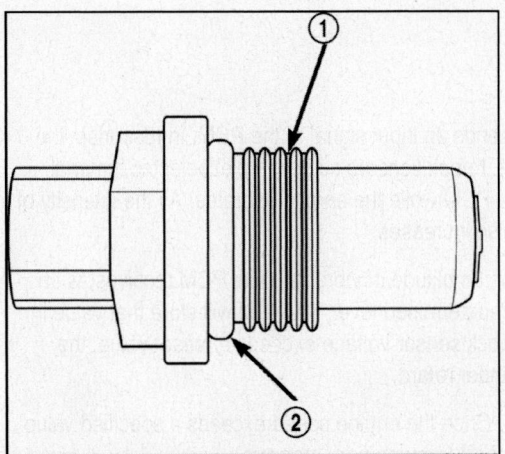

**Showing O-ring (2) location on Input Speed Sensor (1) on Pacifica 3.5L and 3.8L engines**

INTAKE AIR TEMPERATURE (IAT) SENSOR

**Removal & Installation**

1. Disconnect the negative battery cable.
2. Unlock and disconnect the electrical connector.
3. Remove sensor from hose.

**To install:**

4. Install sensor into inlet hose.
5. Connect and lock the electrical connector.
6. Connect the negative battery cable.

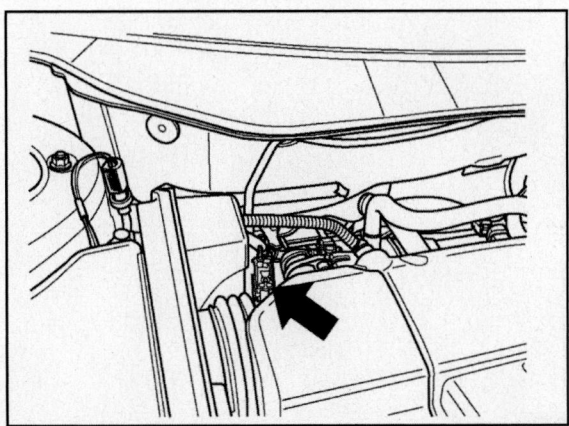

**Showing Intake Air Temperature Sensor location on Pacifica 3.5L and 3.8L engines**

## KNOCK SENSOR

### Description & Operation

When the knock sensor detects a knock in one of the cylinders, it sends an input signal to the PCM. In response, the PCM retards ignition timing for all cylinders by a scheduled amount. Knock sensors contain a piezoelectric material, which constantly vibrates and sends an input voltage (signal) to the PCM while the engine operates. As the intensity of the crystal's vibration increases, the knock sensor output voltage also increases.

The voltage signal produced by the knock sensor increases with the amplitude of vibration. The PCM receives as an input the knock sensor voltage signal. If the signal rises above a predetermined level, the PCM will store that value in memory and retard ignition timing to reduce engine knock. If the knock sensor voltage exceeds a preset value, the PCM retards ignition timing for all cylinders. It is not a selective cylinder retard.

The PCM ignores knock sensor input during engine idle conditions. Once the engine speed exceeds a specified value, knock retard is allowed. Knock retard uses its own long-term and short-term memory program.

Long-term memory stores previous detonation information in its battery-backed RAM. The maximum authority that long term memory has over timing retard can be calibrated. Short-term memory is allowed to retard timing up to a preset amount under all operating conditions (as long as rpm is above the minimum rpm) except WOT. The PCM, using short-term memory, can respond quickly to retard timing when engine knock is detected. Short-term memory is lost any time the ignition key is turned off.

*NOTE: Improper tightening affects sensor performance, possibly causing improper spark control.*

### Removal & Installation

**3.5L**

1. Disconnect the negative battery cable.
2. Remove the upper intake manifold.
3. Disconnect the electrical connector.
4. Remove the knock sensor.

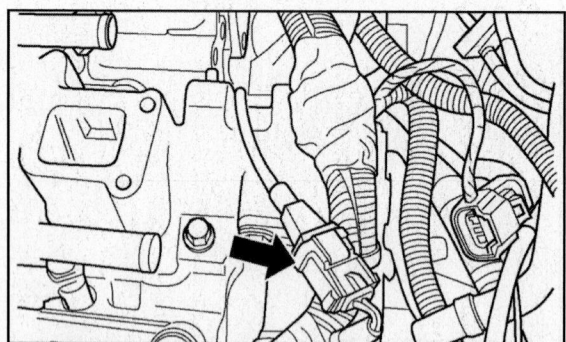

**Showing Knock Sensor Connector location on Pacifica 3.5L engine**

To install:

*NOTE: Improper tightening affects sensor performance, possibly causing improper spark control.*

5. Install knock sensor. Tighten knock sensor to 20 Nm (15.2 inch lbs.) torque.
6. Route the knock sensor wire in the proper location.
7. Install the upper intake manifold.
8. Connect the electrical connector.
9. Connect the negative battery cable.

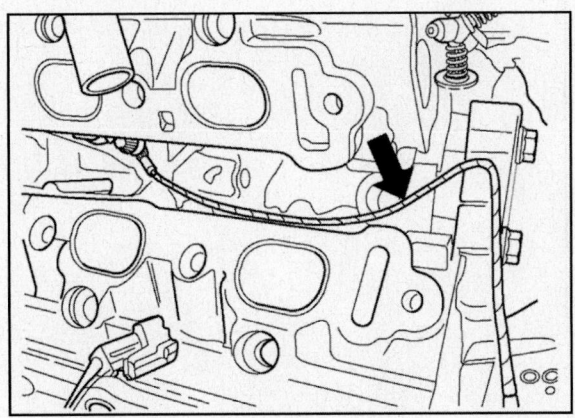

**Observe proper wire routing (Pacifica 3.5L engine shown)**

3.8L

10. Disconnect the negative battery cable.

11. Raise vehicle and support.

12. Disconnect the electrical connector.

13. Remove mounting bolt and sensor.

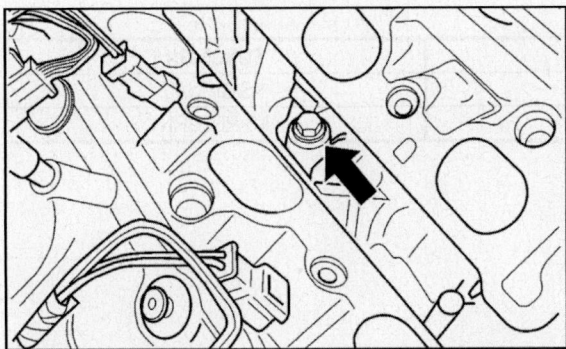

**Showing Knock Sensor location on Pacifica 3.5L engine**

**To install:**

1. Install knock sensor. Tighten knock sensor to 20 Nm (15.2 inch lbs.) torque.

2. Route the knock sensor wire in the proper location.

3. Connect the electrical connector.

4. Lower vehicle.

5. Connect the negative battery cable.

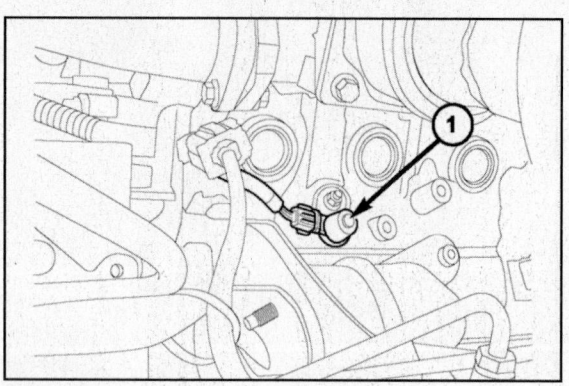

Showing KS location on Pacifica 3.8L engine

**Connector Pinouts**

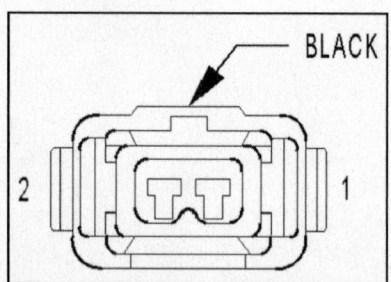

| CAV | CIRCUIT | FUNCTION |
|-----|---------|----------|
| 1 | K42 20DB/OR | KS SIGNAL |
| 2 | K942 20BR/LG | KS RETURN |

**3.5L & 3.8L Knock Sensor on 3.5L Engine**

## MANIFOLD ABSOLUTE PRESSURE (MAP) SENSOR

### Removal & Installation

1. Remove the negative battery cable.
2. Disconnect the electrical connector from the MAP sensor.
3. Remove bolt from sensor.
4. Remove sensor.

**To install:**

*3.5L*

1. Position the sensor onto intake manifold plenum. Tighten screws to 4.5 Nm (40 inch lbs.) torque.
2. Attach electrical connector to sensor.
3. Install the negative battery cable.

*3.8L*

NOTE: *When servicing components, always reassemble components with new O-rings and seals where applicable. If assembly of component is difficult, a light coat of engine oil may be applied to the O-RINGS ONLY to aid assembly.*

1. Install MAP sensor.
2. Install and tighten screws.

3. Connect and lock electrical connector.
4. Connect negative battery cable.

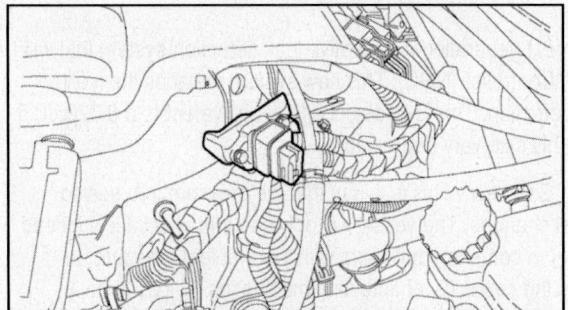

**Location of MAP Sensor on Pacifica 3.5L engine**

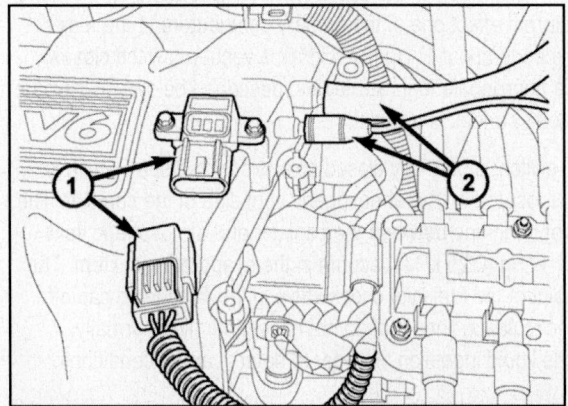

**Location of MAP sensor and connector (1) and vacuum lines (2) on Pacifica 3.8L engine**

**Connector Pinout**

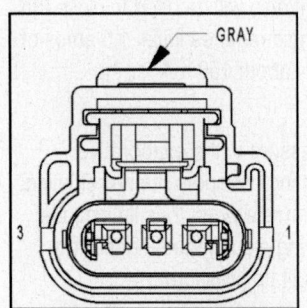

| CAV | CIRCUIT | FUNCTION |
|-----|---------|----------|
| 1 | K1 20VT/BR | MAP SIGNAL |
| 2 | K900 20DB/DG | SENSOR GROUND |
| 3 | F855 20PK/YL | 5 VOLT SUPPLY |

**3.5L & 3.8L MAP Sensor**

NATURAL VACUUM LEAK DETECTION (NVLD) SYSTEM

**Description & Operation**

The Natural Vacuum Leak Detection (NVLD) system is the next generation evaporative leak detection system that will first be used on vehicles equipped with the Next Generation Controller (NGC). This new system replaces the leak detection pump as the method of evaporative system leak detection. This is to detect a leak equivalent to a 0.020" (0.5 mm) hole. This system has the capability to detect holes of this size very dependably.

The basic leak detection theory employed with NVLD is the "Gas Law". This is to say that the pressure in a sealed vessel will change if the temperature of the gas in the vessel changes. The vessel will only see this effect if it is indeed sealed. Even small leaks will allow the pressure in the vessel to come to equilibrium with the ambient pressure. In addition to the detection of very small leaks, this system has the capability of detecting medium as well as large evaporative system leaks.

A vent valve seals the canister vent during engine off conditions. If the vapor system has a leak of less than the failure threshold, the evaporative system will be pulled into a vacuum, either due to the cool down from operating temperature or diurnal ambient temperature cycling. This considers the diurnal effect one of the primary contributors to the leak determination diagnosis. When the vacuum in the system exceeds about 1" Hg (0.25 kPa), a vacuum switch closes. The switch closure sends a signal to the NGC. The NGC, via appropriate logic strategies (described below), utilizes the switch signal, or lack thereof, to make a determination of whether a leak is present.

The NVLD device is designed with a normally open vacuum switch, a normally closed solenoid, and a seal, which is actuated by both the solenoid and a diaphragm. The NVLD is located on the atmospheric vent side of the canister. The NVLD assembly may be mounted on top of the canister outlet, or in-line between the canister and atmospheric vent filter. The normally open vacuum switch will close with about 1" Hg (0.25 kPa) vacuum in the evaporative system. The diaphragm actuates the switch. This is above the opening point of the fuel inlet check valve in the fill tube so cap off leaks can be detected. Submerged fill systems must have recirculation lines that do not have the in-line normally closed check valve that protects the system from failed nozzle liquid ingestion, in order to detect cap off conditions.

The normally closed valve in the NVLD is intended to maintain the seal on the evaporative system during the engine off condition. If vacuum in the evaporative system exceeds 3" to 6" Hg (0.75 to 1.5 kPa), the valve will be pulled off the seat, opening the seal. This will protect the system from excessive vacuum as well as allowing sufficient purge flow in the event that the solenoid was to become inoperative.

The solenoid actuates the valve to unseal the canister vent while the engine is running. It also will be used to close the vent during the medium and large leak tests and during the purge flow check. This solenoid requires initial 1.5 amps of current to pull the valve open but after 100 ms. will be duty cycled down to an average of about 150 mA for the remainder of the drive cycle.

Another feature in the device is a diaphragm that will open the seal in the NVLD with pressure in the evaporative system. The device will "blow off" at about 0.5" Hg (0.12 kPa) pressure to permit the venting of vapors during refueling. An added benefit to this is that it will also allow the tank to "breathe" during increasing temperatures, thus limiting the pressure in the tank to this low level. This is beneficial because the induced vacuum during a subsequent declining temperature will cause the switch to close sooner than if the tank had to decay from a built up pressure.

The device itself has 3 wires: Switch sense, solenoid driver and ground. The NGC utilizes a high-side driver to energize and duty-cycle the solenoid.

**Removal & Installation**

1. Disconnect the negative battery cable.
2. Raise vehicle and support.
3. Unlock and disconnect the electrical connector.
4. Remove clamps and remove hoses.
5. Press tab and release Natural Vacuum Leak Detection pump.

**To install:**

6. Slide the Natural Vacuum Leak Detection pump onto the bracket and make sure that the tab locks into the bracket.

7. Install hoses and clamps.

8. Connect the electrical connector and lock.

9. Lower vehicle. Connect the negative battery cable.

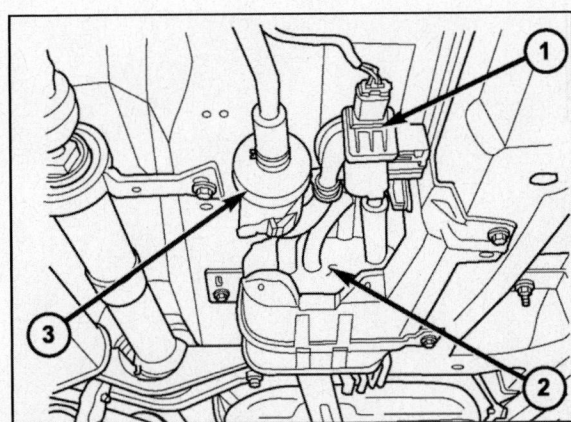

**Showing location of locking tab (1), NVLD Pump (2) and pump filter (3) on Pacifica**

## OUTPUT SPEED SENSOR

### Description

The Output Speed Sensor is a two-wire magnetic pickup device that generates an AC signal as rotation occurs. It is threaded into the transaxle case. The Output Speed Sensor is sealed with an O-ring, and is considered a primary input to the Powertrain/Transmission Control.

### Operation

The Output Speed Sensor provides information on how fast the output shaft is rotating. As the rear planetary carrier park pawl lugs pass by the sensor coil, an AC voltage is generated and sent to the PCM/TCM. The PCM/TCM interprets this information as output shaft rpm.

The PCM/TCM compares the input and output speed signals to determine the following:

- Transmission gear ratio
- Speed ratio error detection
- CVI calculation

### *VEHICLE SPEED SIGNAL*

The vehicle speed signal is taken from the Output Speed Sensor. The PCM converts this signal into a pulse per mile signal and sends the vehicle speed message across the communication bus to the BCM. The BCM sends this signal to the Instrument Cluster to display vehicle speed to the driver. The vehicle speed signal pulse is roughly 8000 pulses per mile.

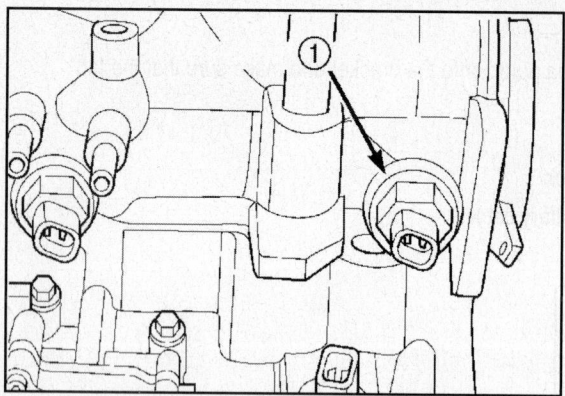

Location of Output Speed Sensor (1) on Pacifica engine

## Removal & Installation

1. Disconnect battery negative cable.
2. Raise vehicle on hoist.
3. Disconnect output speed sensor connector.
4. Unscrew and remove output speed sensor.
5. Inspect speed sensor O-ring and replace if necessary.

**To install:**

6. Verify O-ring is installed into position.
7. Install and tighten output speed sensor to 27 Nm (20 ft. lbs.).
8. Connect speed sensor connector.
9. Connect battery negative cable.

## ON-BOARD REFUELING VAPOR RECOVERY

### Description & Operation

The emission control principle in the On-Board Refueling Vapor Recovery (ORVR) system is that the fuel flowing into the filler tube (approx. 1 in. I.D.) creates an aspiration effect, which draws air into the fill tube. During refueling, the fuel tank is vented to the vapor canister to capture escaping vapors. With air flowing into the filler tube, there are no fuel vapors escaping to the atmosphere. Once the refueling vapors are captured by the canister, the vehicle's computer controlled purge system draws vapor out of the canister for the engine to burn. The vapors flow is metered by the purge solenoid so that there is no or minimal impact on driveability or tailpipe emissions.

As fuel starts to flow through the fill tube, it opens the normally closed check valve and enters the fuel tank. Vapor or air is expelled from the tank through the control valve to the vapor canister. Vapor is absorbed in the canister until vapor flow in the lines stops, either following shut-off or by having the fuel level in the tank rise high enough to close the control valve. The control valve contains a float that rises to seal the large diameter vent path to the canister. At this point in the fueling of the vehicle, the tank pressure increases, the check valve closes (preventing tank fuel from spitting back at the operator), and fuel then rises up the filler tube to shut-off the dispensing nozzle.

If the engine is shut-off while the On-Board diagnostics test is running, low level tank pressure can be trapped in the fuel tank and fuel can not be added to the tank until the pressure is relieved. This is due to the leak detection pump closing the vapor outlet from the top of the tank and the one-way check valve not allowing the tank to vent through the fill tube to atmosphere. Therefore, when fuel is added, it will back-up in the fill tube and shut off the dispensing nozzle. The pressure can be eliminated in two ways: 1. Vehicle purge must be activated and for a long enough period to eliminate the pressure. 2. Removing the fuel cap and allowing enough time for the system to vent thru the recirculation tube.

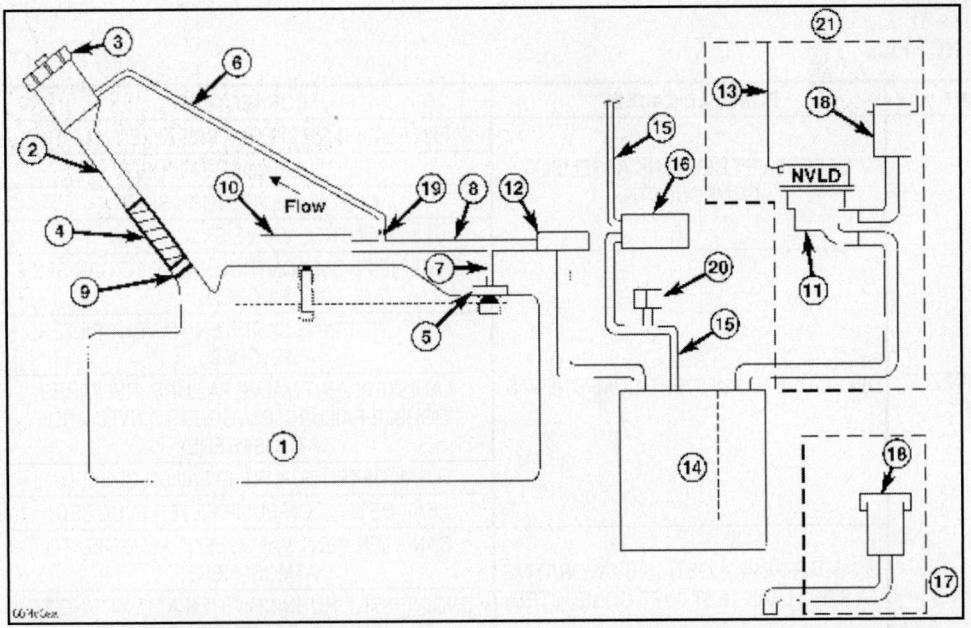

**Schematic of On-Board Refueling Vapor Recovery System on Pacifica 3.5L and 3.8L engines**

1 - Fuel Tank (Plastic)

2 - Fuel Filler Tube

3 - Fuel Cap (Pressure/Relief)

4 - Fill Tube To Fuel Tank Connector (Elastomeric)

5 - Tank Vent/Rollover Valve(s)

6 - Vapor Recirculation Line

7 - Tank Vapor Line

8 - Vapor Line To Canister

9 - Check Valve (N/C)

10 - Control Valve

11 - Natural Vacuum Lead Detection (NVLD)

12 - Liquid Separator (If Equipped)

13 - Engine Wiring Harness To NVLD

14 - Vapor Canister

15 - Purge Line

16 - Purge Device

17 - Without NVLD

18 - Breather Element

19 - Flow Control Orifice

20 - Service Port

21 - With NVLD

**Legend for ORVR system schematic on Pacifica**

**Diagnosis & Testing –**

**VEHICLE DOES NOT FILL**

| CONDITION | POSSIBLE CAUSES | CORRECTION |
|---|---|---|
| PRE-MATURE NOZZLE SHUT-OFF | DEFECTIVE FUEL TANK ASSEMBLY COMPONENTS. | FILL TUBE IMPROPERLY INSTALLED (SUMP) |
| | | FILL TUBE HOSE PINCHED. |
| | | CHECK VALVE STUCK SHUT. |
| | | CONTROL VALVE STUCK SHUT. |
| | DEFECTIVE VAPOR/VENT COMPONENTS. | VENT LINE FROM CONTROL VALVE TO CANISTER PINCHED. |
| | | VENT LINE FROM CANISTER TO VENT FILTER PINCHED. |
| | | CANISTER VENT VALVE FAILURE (REQUIRES DOUBLE FAILURE, PLUGGED TO NVLD AND ATMOSPHERE). |
| | | LEAK DETECTION PUMP FAILED CLOSED. |
| | | LEAK DETECTION PUMP FILTER PLUGGED. |
| | ON-BOARD DIAGNOSTICS EVAPORATIVE SYSTEM LEAK TEST JUST CONDUCTED. | CANISTER VENT VALVE VENT PLUGGED TO ATMOSPHERE. |
| | | ENGINE STILL RUNNING WHEN ATTEMPTING TO FILL (SYSTEM DESIGNED NOT TO FILL). |
| | DEFECTIVE FILL NOZZLE. | TRY ANOTHER NOZZLE. |
| FUEL SPITS OUT OF FILLER TUBE. | DURING FILL. | SEE PRE-MATURE SHUT-OFF. |
| | AT CONCLUSION OF FILL. | DEFECTIVE FUEL HANDLING COMPONENT. (CHECK VALVE STUCK OPEN). |
| | | DEFECTIVE VAPOR/VENT HANDLING COMPONENT. |
| | | DEFECTIVE FILL NOZZLE. |

POSITIVE CRANKCASE VENTILATION (PCV) VALVE & SYSTEM

**Description**

The PCV valve contains a spring loaded plunger. The plunger meters the amount of crankcase vapors routed into the combustion chamber based on intake manifold vacuum.

**Removal & Installation**

1. Remove hose from PCV valve. The valve is on the rear valve cover.
2. Unscrew the PCV valve.
3. Remove PCV valve.

**To install:**

4. Install PCV valve and tighten.
5. Install PCV hose.

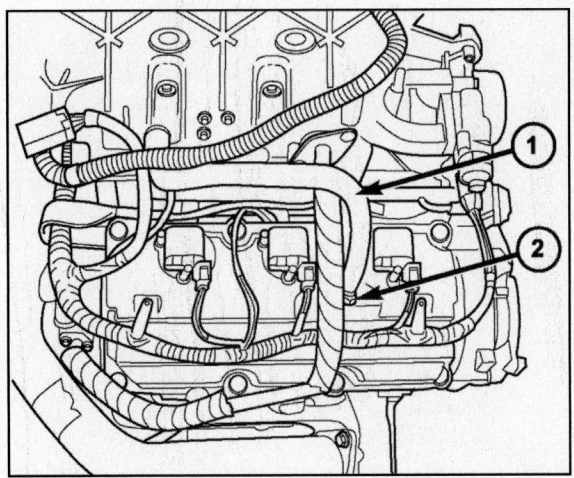

**Location of PCV hose (1) and PCV valve (2) on Pacifica engine**

**Diagnosis & Testing**

*WARNING: APPLY PARKING BRAKE AND/OR BLOCK WHEELS BEFORE PERFORMING ANY TEST OR ADJUSTMENT WITH THE ENGINE OPERATING.*

1. With engine idling, remove the hose from the PCV valve. If the valve is not plugged, a hissing noise will be heard as air passes through the valve. A strong vacuum should also be felt when a finger is placed over the valve inlet.

2. Install hose on PCV valve. Remove the make-up air hose from the air plenum at the rear of the engine. Hold a piece of stiff paper (parts tag) loosely over the end of the make-up air hose.

3. After allowing approximately one minute for crankcase pressure to reduce, the paper should draw up against the hose with noticeable force. If the engine does not draw the paper against the grommet after installing a new valve, replace the PCV valve hose.

4. Turn the engine off. Remove the PCV valve from intake manifold. The valve should rattle when shaken.

5. Replace the PCV valve and retest the system if it does not operate as described in the preceding tests. Do NOT attempt to clean the old PCV valve.

## POWERTRAIN CONTROL MODULE (PCM)

**Description**

*PCM INPUTS & OUTPUTS*

The Powertrain Control Module (PCM) is a digital computer containing a microprocessor. The PCM receives input signals from various switches and sensors referred to as Powertrain Control Module Inputs. Based on these inputs, the PCM adjusts various engine and vehicle operations through devices referred to as Powertrain Control Module Outputs.

*PCM INPUTS:*

- Air Conditioning Pressure Transducer
- Ambient temperature Sensor
- ASD Relay
- Battery Temperature Sensor (NGC)
- Battery Voltage
- Brake Switch

- Camshaft Position Sensor
- Crankshaft Position Sensor
- EGR Position Feedback
- Engine Coolant Temperature Sensor
- Heated Oxygen Sensors
- Ignition sense
- Intake Air Temperature Sensor
- Knock Sensor
- NVLD Assembly
- Manifold Absolute Pressure (MAP) Sensor
- Park/Neutral
- PCI Bus
- Power Steering Pressure Switch
- EVAP Purge Return
- SCI Receive
- Speed Control
- Throttle Position Sensor
- Transmission Control Relay (Switched B+)
- Transmission Pressure Switches
- Transmission Temperature Sensor
- Transmission Input Shaft Speed Sensor
- Transmission Output Shaft Speed Sensor
- Transaxle Gear Engagement
- Vehicle Speed

*PCM OUTPUTS:*

- Air Conditioning Clutch Relay
- Automatic Shut Down (ASD) and Fuel Pump Relays
- Data Link Connector (PCI and SCI Transmit)
- Double Start Override
- EGR Solenoid
- Fuel Injectors
- Generator Field
- High Speed Fan Relay
- Idle Air Control Motor
- Ignition Coils
- NVLD Assembly
- Low Speed Fan Relay
- MTV Actuator
- EVAP Purge
- SRV Valve
- Speed Control Vent Solenoid
- Speed Control Vacuum Solenoid
- 5-Volt Output

- Torque Reduction Request
- Transmission Control Relay
- Transmission Solenoids
- Vehicle Speed

Based on inputs it receives, the powertrain control module (PCM) adjusts fuel injector pulse width, idle speed, ignition timing, and canister purge operation. The PCM regulates the cooling fans, air conditioning and speed control systems. The PCM changes generator charge rate by adjusting the generator field.

The PCM adjusts injector pulse width (air-fuel ratio) based on the following inputs.

- Battery Voltage
- Intake Air Temperature Sensor
- Engine Coolant Temperature
- Engine Speed (crankshaft position sensor)
- Exhaust Gas Oxygen Content (heated oxygen sensors)
- Manifold Absolute Pressure
- Throttle Position

The PCM adjusts engine idle speed through the idle air control motor based on the following inputs:

- Brake Switch
- Engine Coolant Temperature
- Engine Speed (crankshaft position sensor)
- Park/Neutral
- Transaxle Gear Engagement
- Throttle Position
- Vehicle Speed

The PCM adjusts ignition timing based on the following inputs:

- Intake Air Temperature
- Engine Coolant Temperature
- Engine Speed (crankshaft position sensor)
- Knock Sensor
- Manifold Absolute Pressure
- Park/Neutral
- Transaxle Gear Engagement
- Throttle Position

The automatic shut down (ASD) and fuel pump relays are mounted externally, but turned on and off by the powertrain control module through the same circuit.

The camshaft and crankshaft signals are sent to the powertrain control module. If the PCM does not receive both signals within approximately one second of engine cranking, it deactivates the ASD and fuel pump relays. When these relays are deactivated, power is shut off to the fuel injectors, ignition coils, fuel pump and the heating element in each oxygen sensor.

The PCM engine control strategy prevents reduced idle speeds until after the engine operates for 320 km (200 miles). If the PCM is replaced after 320 km (200 miles) of usage, update the mileage in new PCM. Use a proper scan tool to change the mileage in the PCM.

*TRANSMISSION CONTROL*

### Clutch Volume Index (CVI)

An important function of the PCM is to monitor Clutch Volume Index (CVI). CVI represents the volume of fluid needed to compress a clutch pack. The PCM monitors gear ratio changes by monitoring the Input and Output Speed Sensors. The Input, or Turbine Speed Sensor sends an electrical signal to the PCM that represents input shaft rpm. The Output Speed Sensor provides the PCM with output shaft speed information.

By comparing the two inputs, the PCM can determine transaxle gear ratio. This is important to the CVI calculation because the PCM determines CVI by monitoring how long it takes for a gear change to occur. Using a proper scan tool and reading the Input/Output Speed Sensor values in the scan tool display can determine gear ratios. Gear ratio can be obtained by dividing the Input Speed Sensor value by the Output Speed Sensor value.

For example, if the input shaft is rotating at 1000 rpm and the output shaft is rotating at 500 rpm, then the PCM can determine that the gear ratio is 2:1. In direct drive (3rd gear), the gear ratio changes to 1:1. The gear ratio changes as clutches are applied and released. By monitoring the length of time it takes for the gear ratio to change following a shift request, the PCM can determine the volume of fluid used to apply or release a friction element.

The volume of transmission fluid needed to apply the friction elements are continuously updated for adaptive controls. As friction material wears, the volume of fluid need to apply the element increases.

Certain mechanical problems within the clutch assemblies (broken return springs, out of position snap rings, excessive clutch pack clearance, improper assembly, etc.) can cause inadequate or out-of-range clutch volumes. Also, defective Input/Output Speed Sensors and wiring can cause these conditions. The following chart identifies the appropriate clutch volumes and when they are monitored/updated:

| CLUTCH VOLUMES | | | | |
|---|---|---|---|---|
| CLUTCH | WHEN UPDATED | | | PROPER CLUTCH VOLUME |
| | SHIFT SEQUENCE | OIL TEMPERATURE | THROTTLE ANGLE | |
| L/R | 2-1 OR 3-1 COAST DOWNSHIFT | > 70° | < 5° | 35 TO 83 |
| 2/4 | 1-2 SHIFT | > 110° | 5 - 54° | 20 TO 77 |
| OD | 2-3 SHIFT | | | 48 TO 150 |
| UD | 4-3 OR 4-2 SHIFT | | > 5° | 24 TO 70 |

### Shift Schedules

As mentioned earlier, the PCM has programming that allows it to select a variety of shift schedules. Shift schedule selection is dependent on the following:

- Shift lever position
- Throttle position
- Engine load
- Fluid temperature
- Software level

As driving conditions change, the PCM appropriately adjusts the shift schedule. Refer to the SHIFT Operation chart to determine the appropriate operation expected, depending on driving conditions.

## Operation

### *PCM GROUND*

Ground is provided through multiple pins of the PCM connector. Depending on the vehicle there may be as many as two different ground pins. There are power grounds and sensor grounds.

The power grounds are used to control the ground side relays, solenoids, ignition coil or injectors. The signal ground is used for any input that uses sensor return for ground, and the ground side of any internal processing component.

The PCM case is shielded to prevent RFI and EMI. The PCM case is grounded and must be firmly attached to a good, clean body ground.

Internally all grounds are connected together, however there is noise suppression on the sensor ground. For EMI and RFI protection the housing and cover are also grounded separately from the ground pins.

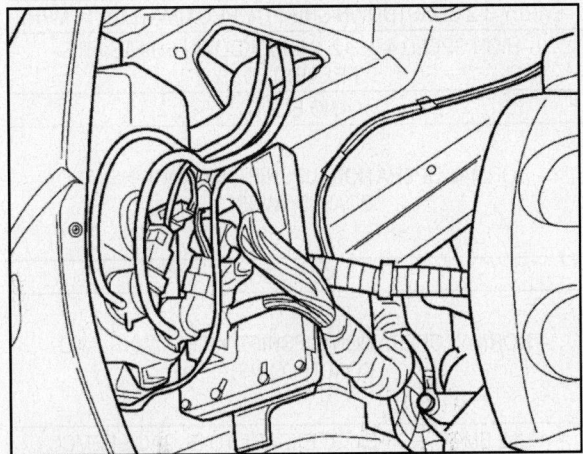

**PCM location shown for Pacifica**

### Removal & Installation

1. Disconnect the negative battery cable.
2. Raise vehicle and support.
3. Remove the left front wheel.
4. Remove the left front splash shield.
5. Unlock and remove the 4 connectors.
6. Remove the 3 mounting screws.
7. Remove the PCM.

**To install:**

*NOTE: The PCM connectors are color-coded.*

8. Install the PCM.
9. Install the 3 mounting screws.
10. Tighten screws.
11. Install and lock the 4 electrical connectors.
12. Install the left front splash shield.
13. Install the left front wheel.
14. Lower the vehicle. Connect the negative battery cable.

| SCHEDULE | CONDITION | EXPECTED OPERATION |
|---|---|---|
| EXTREME COLD | OIL TEMPERATURE AT START-UP BELOW -16° F | PARK, REVERSE, NEUTRAL AND 2ND GEAR ONLY (PREVENTS SHIFTING WHICH MAY FAIL A CLUTCH WITH FREQUENT SHIFTS) |
| COLD | OIL TEMPERATURE AT START-UP ABOVE -12° F AND BELOW 36° F | – DELAYED 2-3 UPSHIFT (APPROXIMATELY 22-31 MPH) |
| | | – DELAYED 3-4 UPSHIFT (45-53 MPH) |
| | | – EARLY 4-3 COASTDOWN SHIFT (APPROXIMATELY 30 MPH) |
| | | – EARLY 3-2 COASTDOWN SHIFT (APPROXIMATELY 17 MPH) |
| | | – HIGH SPEED 4-2, 3-2, 2-1 KICKDOWN SHIFTS ARE PREVENTED |
| | | – NO EMCC |
| WARM | OIL TEMPERATURE AT START-UP ABOVE 36° F AND BELOW 80 DEGREE F | – NORMAL OPERATION (UPSHIFT, KICKDOWNS, AND COASTDOWNS) |
| | | – NO EMCC |
| HOT | OIL TEMPERATURE AT START-UP ABOVE 80° F | – NORMAL OPERATION (UPSHIFT, KICKDOWNS, AND COASTDOWNS) |
| | | – FULL EMCC, NO PEMCC EXCEPT TO ENGAGE FEMCC (EXCEPT AT CLOSED THROTTLE AT SPEEDS ABOVE 70-83 MPH) |
| OVERHEAT | OIL TEMPERATURE ABOVE 240° F OR ENGINE COOLANT TEMPERATURE ABOVE 244° F | – DELAYED 2-3 UPSHIFT (25-32 MPH) |
| | | – DELAYED 3-4 UPSHIFT (41-48 MPH) |
| | | – 3RD GEAR FEMCC FROM 30-48 MPH |
| | | – 3RD GEAR PEMCC FROM 27-31 MPH |
| SUPER OVERHEAT | OIL TEMPERATURE ABOVE 260° F | – ALL "OVERHEAT" SHIFT SCHEDULE FEATURES APPLY |
| | | – 2ND GEAR PEMCC ABOVE 22 MPH |
| | | – ABOVE 22 MPH THE TORQUE CONVERTER WILL NOT UNLOCK UNLESS THE THROTTLE IS CLOSED OR IF A WIDE OPEN THROTTLE 2ND PEMCC TO 1 KICKDOWN IS MADE |

**Shift Operation Chart**

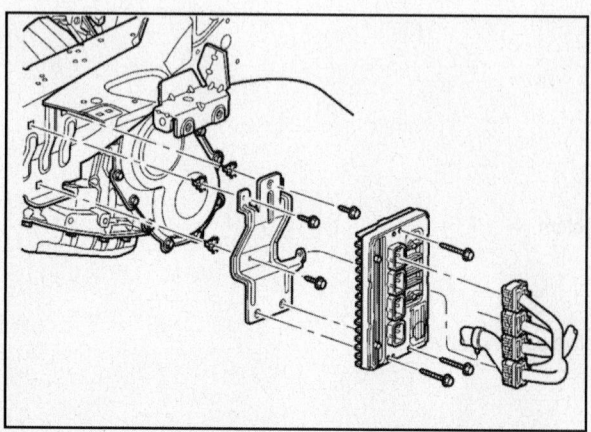

**Removing/installing the PCM on Pacifica**

**Connector Pinout**

*C1 CONNECTOR*

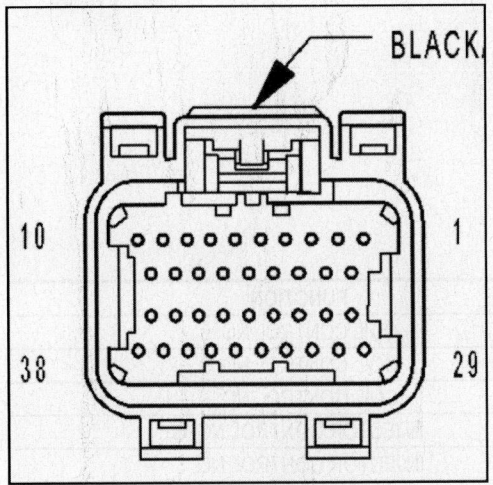

| CAV | CIRCUIT | FUNCTION |
|---|---|---|
| 1-8 | - | - |
| 9 | Z130 16BK/BR | GROUND |
| 10 | - | - |
| 11 | F202 20PK/GY | FUSED IGNITION SWITCH OUTPUT (RUN-START) |
| 12 | F1 20PK/WT | IGNITION UNLOCK-RUN-START |
| 13-15 | - | - |
| 16 | K236 16DB/LG | SRV SOL CONTROL |
| 17 | - | - |
| 18 | Z131 16BK/DG | GROUND |
| 19 | - | - |
| 20 | G6 16VT/GY | OIL PRESSURE SIGNAL |
| 21 | - | - |
| 22 | G31 20OR/VT | AAT SIGNAL |
| 23-24 | - | - |
| 25 | D20 20WT/LG | SCI RECEIVE (PCM) |
| 26 | D123 20WT/BR | FLASH PROGRAM ENABLE |
| 27-28 | - | - |
| 29 | A109 18OR/RD | FUSED B(+) |
| 30 | T751 20YL | IGNITION SWITCH OUTPUT (START) |
| 31 | K141 20DB/YL | O2 1/2 SIGNAL |
| 32 | K904 20DB/DG | O2 RETURN (DOWN) |
| 33-35 | - | - |
| 36 | D21 20WT/GY | SCI TRANSMIT (PCM) |
| 37 | D15 20BR/WT | SCI TRANSMIT (TCM) |
| 38 | D25 20WT/VT | PCI BUS |

**3.5L & 3.8L PCM C1 Connector Pinout**

*C2 CONNECTOR*

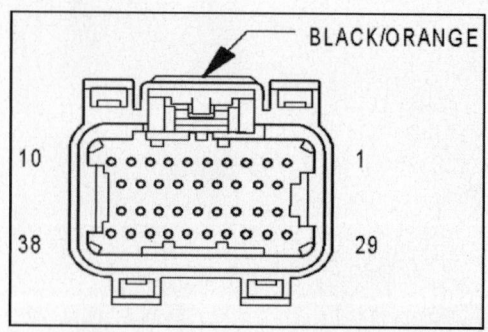

| CAV | CIRCUIT | FUNCTION |
|---|---|---|
| 1 | K10 16DB/OR | COIL CONTROL NO. 6 |
| 2 | K16 16DB/YL | COIL CONTROL NO. 5 |
| 3 | K15 16DB | COIL CONTROL NO. 4 |
| 4 | K58 16BR/VT | INJECTOR CONTROL NO. 6 |
| 5 | K38 16BR/OR | INJECTOR CONTROL NO. 5 |
| 6 | - | - |
| 7 | K18 16DB/OR | COIL CONTROL NO. 3 |
| 8 | K35 18DB/VT | EGR SOL CONTROL |
| 9 | K17 16DB/TN | COIL CONTROL NO. 2 |
| 10 | K19 16DB/DG | COIL CONTROL NO. 1 |
| 11 | K14 16BR/TN | INJECTOR CONTROL NO. 4 |
| 12 | K13 16BR/LB | INJECTOR CONTROL NO. 3 |
| 13 | K12 16BR/DB | INJECTOR CONTROL NO. 2 |
| 14 | K11 16BR/YL | INJECTOR CONTROL NO. 1 |
| 15 | - | - |
| 16 | K36 20DB/YL | MTV CONTROL |
| 17 | - | - |
| 18 | K99 18BR/TN | O2 1/1 HEATER CONTROL |
| 19 | K20 18BR/GY | GEN FIELD CONTROL |
| 20 | K2 18VT/OR | ECT SIGNAL |
| 21 | K22 18BR/OR | TP SIGNAL |
| 22 | K34 20DB/LG | EGR SIGNAL |
| 23 | K1 20VT/BR | MAP SIGNAL |
| 24 | K942 20BR/LG | KS RETURN |
| 25 | K42 20DB/OR | KS SIGNAL |
| 26 | - | - |
| 27 | K900 20DB/DG | SENSOR GROUND |
| 28 | K961 20BR/VT | IAC RETURN |
| 29 | F855 20PK/YL | 5 VOLT SUPPLY |
| 30 | K21 20BR/WT | IAT SIGNAL |
| 31 | K41 18DB/LB | O2 1/1 SIGNAL |
| 32 | K902 18BR/DG | O2 RETURN (UP) |
| 33 | - | - |
| 34 | K44 20DB/GY | CMP SIGNAL |
| 35 | K24 20BR/LB | CKP SIGNAL |
| 36-37 | - | - |
| 38 | K61 20VT/GY | IAC MOTOR CONTROL |

**3.5L & 3.8L PCM C2 Connector Pinout**

*C3 CONNECTOR*

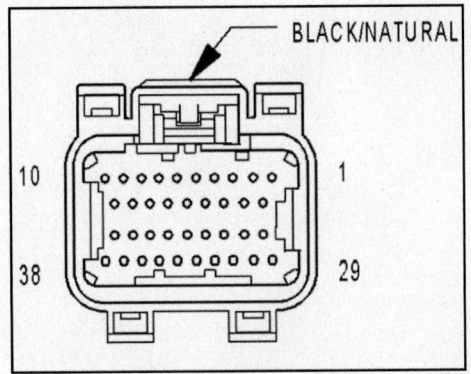

| CAV | CIRCUIT | FUNCTION |
|-----|---------|----------|
| 1-2 | - | - |
| 3 | K51 18BR/GY | AUTOMATIC SHUT DOWN RELAY CONTROL |
| 4 | - | - |
| 5 | V35 20VT/OR | S/C VENT CONTROL |
| 6 | K173 20BR/VT | RAD FAN RELAY CONTROL |
| 7 | V32 20VT/YL | S/C SUPPLY |
| 8 | K106 20VT/LB | NVLD SOL CONTROL |
| 9 | K299 18BR/OR | O2 1/2 HEATER CONTROL |
| 10 | - | - |
| 11 | C13 20LB/OR | A/C CLUTCH RELAY CONTROL |
| 12 | V36 20YL/VT | S/C VACUUM CONTROL |
| 13-18 | - | - |
| 19 | K342 16BR/WT | AUTOMATIC SHUT DOWN RELAY OUTPUT |
| 20 | K52 20DB/WT | EVAP PURGE CONTROL |
| 21-22 | - | - |
| 23 | B29 20DG/WT | BRAKE SWITCH SIGNAL |
| 24-25 | - | - |
| 26 | T44 20YL/DG | AUTOSTICK DOWNSHIFT SWITCH SIGNAL |
| 27 | T5 20DG/YL | AUTOSTICK UPSHIFT SWITCH SIGNAL |
| 28 | K342 16BR/WT | AUTOMATIC SHUT DOWN RELAY OUTPUT |
| 29 | K70 20DB/BR | EVAP PURGE RETURN |
| 30 | - | - |
| 31 | C18 20LB/BR | A/C PRESSURE TRANSDUCER SIGNAL |
| 32 | K91 20DB/YL | BATTERY TEMP SIGNAL |
| 33 | - | - |
| 34 | V37 20VT | S/C SWITCH SIGNAL |
| 35 | K107 20VT/WT | NVLD SWITCH SIGNAL |
| 36 | - | - |
| 37 | K31 20BR | FUEL PUMP RELAY CONTROL |
| 38 | T752 20DG/OR | STARTER RELAY CONTROL |

**3.5L & 3.8L PCM C3 Connector Pinout**

*C4 CONNECTOR*

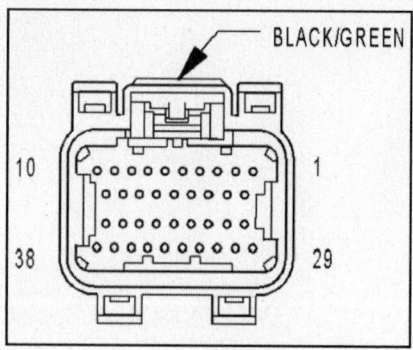

| CAV | CIRCUIT | FUNCTION |
|-----|---------|----------|
| 1 | T60 18YL/GY | OD SOLENOID CONTROL |
| 2 | T59 18YL/LB | UD SOLENOID CONTROL |
| 3-5 | - | - |
| 6 | T19 18DB/YL | 2-4 SOLENOID CONTROL |
| 7-9 | - | - |
| 10 | T20 18WT/DG | L/R SOLENOID CONTROL |
| 11-12 | - | - |
| 13 | Z133 16BK/LG | GROUND |
| 14 | Z133 16BK/LG | GROUND |
| 15 | T1 18LB/DG | TRS T1 SENSE |
| 16 | T3 18DG/DB | TRS T3 SENSE |
| 17 | - | - |
| 18 | T15 20YL/BR | TRANSMISSION CONTROL RELAY CONTROL |
| 19-21 | - | - |
| 22 | T9 18DG/TN | OD PRESSURE SWITCH SENSE |
| 23-26 | - | - |
| 27 | T41 18YL/DB | TRS T41 SENSE |
| 28 | T16 16YL/OR | TRANSMISSION CONTROL RELAY OUTPUT |
| 29 | T50 18YL/TN | L/R PRESSURE SWITCH SENSE |
| 30 | T47 18YL/DG | 2-4 PRESSURE SWITCH SENSE |
| 31 | - | - |
| 32 | T14 18DG/BR | OUTPUT SPEED SENSOR SIGNAL |
| 33 | T52 18DG/WT | INPUT SPEED SENSOR SIGNAL |
| 34 | T13 18DG/VT | SPEED SENSOR GROUND |
| 35 | T54 18DG/OR | TRANSMISSION TEMPERATURE SENSOR SIGNAL |
| 36 | - | - |
| 37 | T42 18DG/YL | TRS T42 SENSE |
| 38 | T16 16YL/OR | TRANSMISSION CONTROL RELAY OUTPUT |

**3.5L & 3.8L PCM C4 Connector Pinout**

SPEED CONTROL SWITCH

**Removal & Installation**

1. Disconnect the negative battery cable.
2. Remove the Driver side airbag.
3. Remove the speed control switch mounting screw.
4. Disconnect the electrical connector from switch.
5. Remove switch.

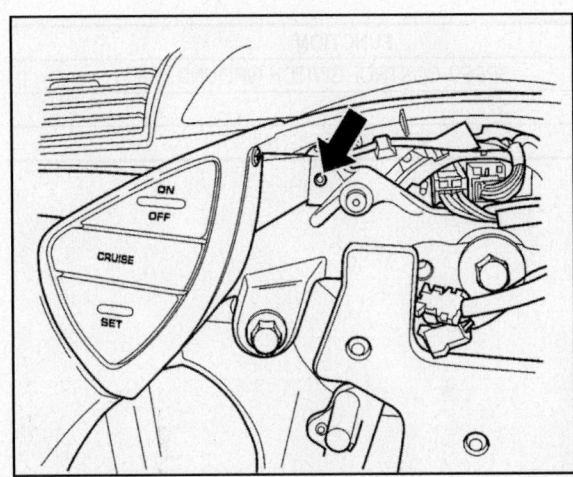

**Showing location of speed sensor switch on Pacifica 3.5L**

**To install:**

6. Connect the electrical connector to switch.
7. Install switch.
8. Install the speed control switch mounting screw.
9. Tighten Screw.
10. Install the Driver side Airbag.
11. Connect the negative battery cable.

**Connector Pinouts**

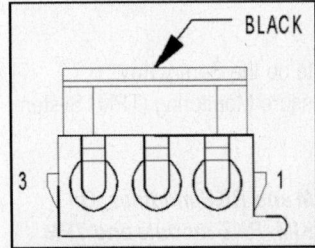

| CAV | CIRCUIT | FUNCTION |
|-----|-----------|-----------------------------------|
| 1 | Z23 20BK/VT | SPEED CONTROL SWITCH GROUND |
| 2 | - | - |
| 3 | V37 20VT | S/C SWITCH SIGNAL |

**Speed Control Switch (Left) Connector Pinout**

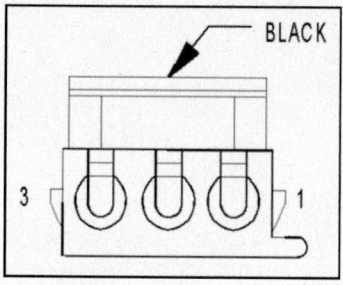

| CAV | CIRCUIT | FUNCTION |
|-----|---------|----------|
| 1 | Z23 20BK/VT | SPEED CONTROL SWITCH GROUND |
| 2 | - | - |
| 3 | V37 20VT | S/C SWITCH SIGNAL |

**Speed Control Switch (Right) Connector Pinout**

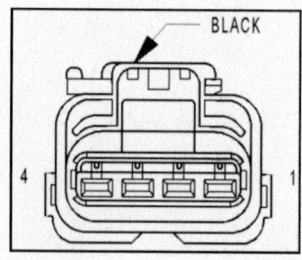

| CAV | CIRCUIT | FUNCTION |
|-----|---------|----------|
| 1 | V36 20YL/VT | S/C VACUUM CONTROL |
| 2 | V35 20VT/OR | S/C VENT CONTROL |
| 3 | V30 20VT/WT | SPEED CONTROL BRAKE SWITCH OUTPUT |
| 4 | Z155 18BK/LG | GROUND |

**Speed Control Servo Connector Pinout**

## SENTRY KEY REMOTE ENTRY MODULE (SKREEM)

### Description

The Sentry Key Remote Entry Module (SKREEM) performs the functions of what used to be the Sentry Key Immobilizer Module (SKIM), the Remote Keyless Entry (RKE) Module, and the Tire Pressure Monitoring (TPM) System (formerly located within the Compass Mini-Trip Computer (CMTC).

*NOTE: On EARLY BUILD vehicles, the SKREEM controls the functions of the SKIM and RKE modules. On NORMAL BUILD vehicles, the SKREEM controls the functions of the SKIM, RKE module and TPM system. If diagnosing the TPM, use the proper scan tool to determine if the SKREEM includes TPM system functionality.*

### SENTRY KEY IMMOBILIZER

The Sentry Key Immobilizer System (SKIS) authenticates an electronically coded Transponder Key placed into the ignition and sends a valid/invalid key message to the Powertrain Control Module (PCM) based upon the results. The "VALID/INVALID KEY" message communication is performed using a rolling code algorithm via the Programmable Communication Interface (PCI) data bus. A "VALID KEY" message must be sent to the Powertrain Control Module (PCM) within two seconds of ignition ON to free the engine from immobilization.

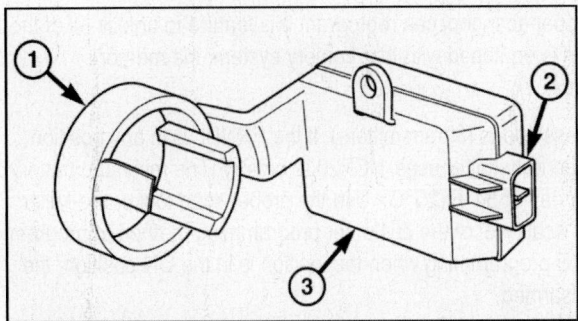

**Identifying SKREEM Components: antenna (1), connector (2), and SKREEM (3) on Pacifica models**

### REMOTE KEYLESS ENTRY (RKE)

The RKE transmitter uses radio frequency signals to communicate with the SKREEM module. The SKREEM is on the PCI bus. When the operator presses a button on the transmitter, it sends a specific request to the SKREEM. In turn the SKREEM sends the appropriate request over the PCI Bus to the:

- Driver Door Module (DDM) to control the driver front door lock and unlock functions, the arming and disarming of the Vehicle Theft Security System (VTSS) (if equipped), and the activation of illuminated entry.

- Integrated Power Module (IPM) to activate the park lamps, the headlamps, and the horn for horn chirp. If requested, the DDM sends a request over the PCI Bus to the:

- Passenger Door Module (PDM) to control the passenger front, rear driver, and rear passenger door lock and unlock functions.

- Power Liftgate Module (PLGM) to control the liftgate lock and unlock functions.

### TIRE PRESSURE MONITORING (TPM) - NORMAL BUILD VEHICLES

If equipped with the Tire Pressure Monitoring (TPM) System, each of the vehicles four wheels will have a valve stem with a pressure sensor and radio transmitter built in. Signals from the tire pressure sensor/transmitter are received and interpreted by the SKREEM.

A sensor/transmitter in a mounted wheel will broadcast its detected pressure once per minute when the vehicle is moving faster than 15 mph (24 km/h). Each sensor/transmitter's broadcast is uniquely coded so that the SKREEM can determine the location.

## Operation

### SENTRY KEY IMMOBILIZER

The Sentry Key Remote Entry Module (SKREEM) receives an encrypted Radio Frequency (RF) signal from the transponder key. The SKREEM then decrypts the signal and broadcasts the requested remote commands to the appropriate modules in the vehicle over the Programmable Communication Interface (PCI) data bus. A valid transponder key ID must be incorporated into the RF signal in order for the SKREEM to pass the message on to the appropriate modules.

Automatic transponder key synchronization is done by the SKREEM if a valid transponder key is inserted into the ignition cylinder, and the ignition is turned ON. This provides a maximum operation window for RKE functions.

Each Sentry Key Remote Entry System (SKREES) consists of a SKREEM and a transponder key. Each system has a secret key code unique to that system. The secret key is electronically coded in the SKREEM and in all programmed transponder keys. It is used for immobilization and RKE functions for data security. In addition, each transponder key will have a unique identification.

### REMOTE KEYLESS ENTRY

After pressing the lock button on the RKE transmitter, all of the door locks will lock, the illuminated entry will turn off (providing all doors are closed), and the VTSS (if equipped) will arm. After pressing the unlock button, on the RKE transmitter, one time, the driver door lock will unlock, the illuminated entry will turn on the courtesy lamps, and the VTSS (if equipped) will disarm. After pressing the unlock button a second time, the remaining door locks will unlock.

The Electronic Vehicle Information Center (EVIC) or the proper scan tool can reprogram this feature to unlock all of the door locks with one press of the unlock button. If the vehicle is equipped with the memory system, the memory message will identify which transmitter (1 or 2) sent the signal.

The SKREEM is capable of retaining up to 8 individual access codes (8 transmitters). If the PRNDL is in any position except park, the SKREEM will disable the RKE. The 4-button transmitter uses 1-CR2032 battery. The minimum battery life is approximately 4.7 years based on 20 transmissions a day at 84°F (25°C). Use the proper scan tool or the Miller Tool 9001 RF Detector to test the RKE transmitter. Use the scan tool or the customer programming method to program the RKE system. However, the SKREEM will only allow RKE programming when the ignition is in the ON position, the PRNDL is in park position, and the VTSS (if equipped) is disarmed.

### TIRE PRESSURE MONITORING (TPM) - NORMAL BUILD VEHICLES

The SKREEM monitors the signals from the tire pressure sensor/transmitters and determines if any tire has gone below the low-pressure threshold of 25 psi with system indicator ON or below 30 psi with system indicator OFF.

### CRITICAL AND NON-CRITICAL SYSTEM ALERTS

**CRITICAL:** A critical alert will be triggered when a tire pressure has gone below a set threshold pressure. The SKREEM will display "X TIRE(S) LOW PRESSURE". "X" will be the number of tires reporting low pressure. The message will display for the duration of the current ignition cycle or until an EVIC button is pressed. If the display is removed without correcting the condition, it will reappear 300 seconds to warn the driver of the low-pressure condition.

**NON-CRITICAL:** A non-critical alert will be triggered when no signal is received from a sensor/transmitter or when a sensor/transmitter low battery condition is detected. The EVIC will display "SERVICE TIRE PRESS. SYSTEM."

### DIAGNOSIS & Testing

*NOTE: On EARLY BUILD vehicles, the SKREEM controls the functions of the SKIM and RKE modules. On NORMAL BUILD vehicles, the SKREEM controls the functions of the SKIM, RKE module and TPM system. If diagnosing the TPM, use the proper scan tool to determine if the SKREEM includes TPM system functionality.*

For proper diagnosis and Testing of the Sentry Key Remote Entry Module (SKREEM), use a proper scan tool and refer to the proper Body Diagnostic Procedures information.

### Removal & Installation

1. Open hood, disconnect and isolate the battery negative cable.
2. Wait two minutes for the system reserve capacitor to discharge before beginning any system or component service.
3. Remove knee blocker airbag.
4. Unsnap instrument panel center stack left bezel from instrument panel assembly.
5. Reach up behind the left side of the center stack and disconnect the ignition switch and SKREEM electrical connectors.
6. From underneath the ignition switch/lock assembly, loosen the lower retaining screw.
7. Remove the two mounting nuts from the front of the instrument panel.
8. Push the ignition switch/lock assembly inward toward the bulkhead.
9. Remove the SKREEM mounting screw on the ignition lock assembly and separate the SKREEM from the ignition lock assembly.

### To install:

10. Position the SKREEM on the ignition lock housing and install the retaining screw.
11. Position the ignition switch/lock assembly up in behind the instrument panel and push rearward.
12. Install the two mounting nuts to the front of the instrument panel.
13. From underneath the ignition switch/lock assembly, tighten the lower retaining screw.

14. Reach up behind the left side of the center stack and connect the ignition switch and SKREEM electrical connectors.

15. Position the instrument panel center stack left bezel over retaining slots and firmly snap into place.

16. Install the knee blocker airbag.

17. Verify vehicle and system operation.

**WARNING: DO NOT connect the battery negative cable. Failure to do so could result in occupant personal injury or death.**

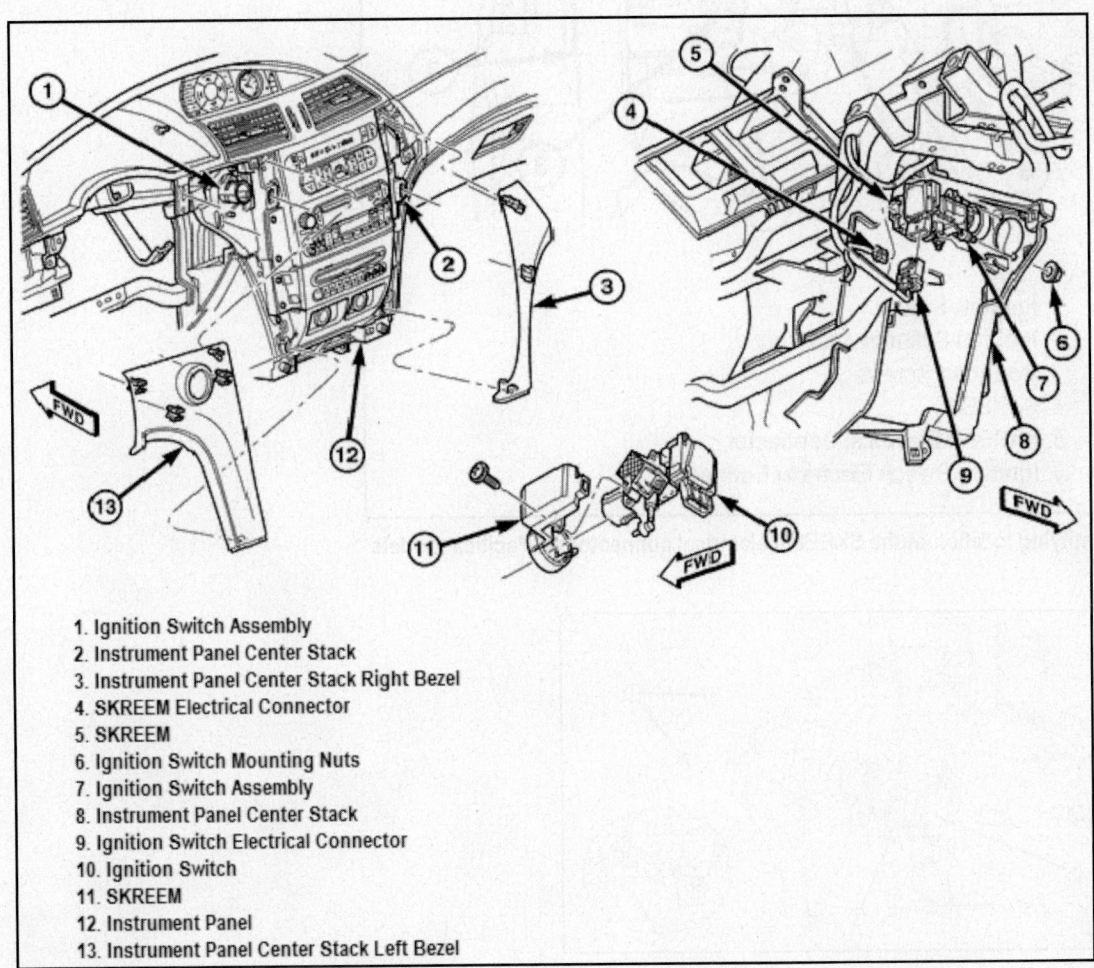

1. Ignition Switch Assembly
2. Instrument Panel Center Stack
3. Instrument Panel Center Stack Right Bezel
4. SKREEM Electrical Connector
5. SKREEM
6. Ignition Switch Mounting Nuts
7. Ignition Switch Assembly
8. Instrument Panel Center Stack
9. Ignition Switch Electrical Connector
10. Ignition Switch
11. SKREEM
12. Instrument Panel
13. Instrument Panel Center Stack Left Bezel

**Showing instrument panel disassembled for SKREEM removal and installation on Pacifica models**

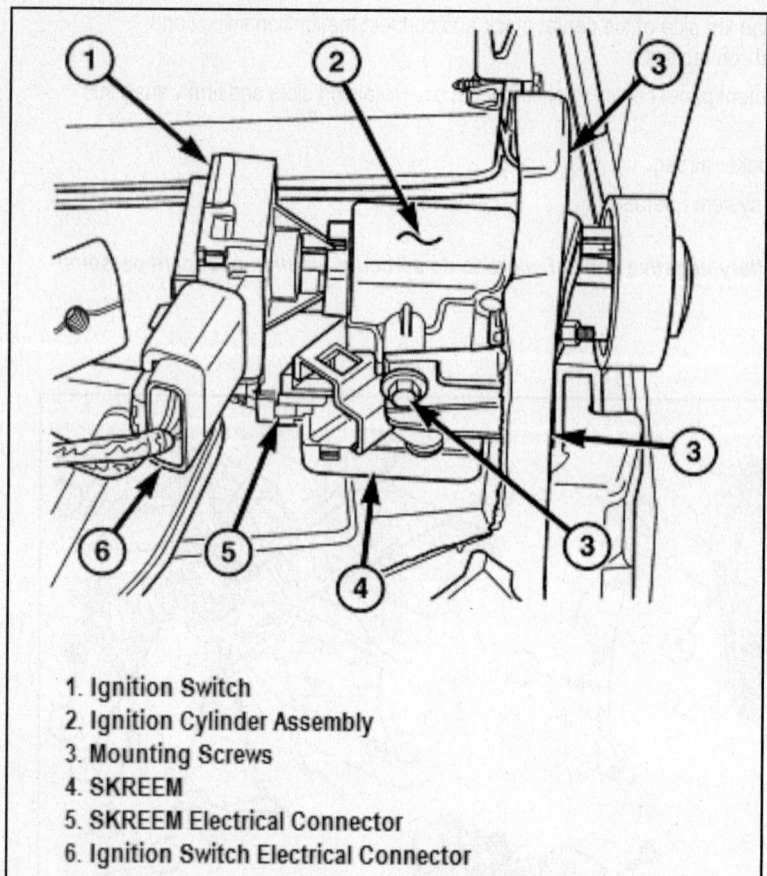

1. Ignition Switch
2. Ignition Cylinder Assembly
3. Mounting Screws
4. SKREEM
5. SKREEM Electrical Connector
6. Ignition Switch Electrical Connector

Identifying location of the SKREEM electrical connector on Pacifica models

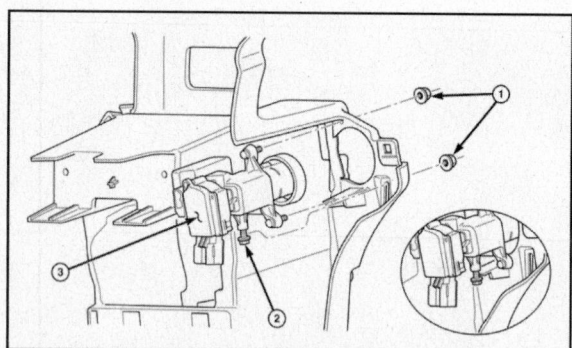

Locating ignition switch mounting nuts (1), screw (2) and switch assembly (3)

**Connector Pinouts**

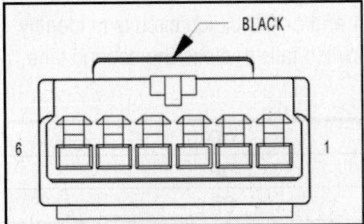

| CAV | CIRCUIT | FUNCTION |
|-----|---------|----------|
| 1 | - | - |
| 2 | D25 20WT/VT | PCI BUS |
| 3 | - | - |
| 4 | F20 20PK/GY | IGNITION SWITCH OUTPUT (RUN-START) |
| 5 | Z120 20BK/WT | GROUND |
| 6 | A118 20RD/OR | FUSED B(+) |

**SKREEM Module Connector Pinout**

## SOLENOID/PRESSURE SWITCH ASSEMBLY

### Description & Operation

The Solenoid/Pressure Switch Assembly is external to the transaxle and mounted to the transaxle case. The assembly consists of four solenoids that control hydraulic pressure to the LR/CC, 2/4, OD, and UD friction elements. The reverse clutch is controlled by line pressure from the manual valve in the valve body. The solenoids are contained within the Solenoid/Pressure Switch Assembly, and can only be serviced by replacing the assembly.

The solenoid assembly also contains pressure switches that monitor and send hydraulic circuit information to the PCM/TCM. Likewise, the pressure switches can only be service by replacing the assembly.

### SOLENOIDS

The solenoids receive electrical power from the Transmission Control Relay through a single wire. The PCM/TCM energizes or operates the solenoids individually by grounding the return wire of the solenoid needed. When a solenoid is energized, the solenoid valve shifts, and a fluid passage is opened or closed (vented or applied), depending on its default operating state. The result is an application or release of a frictional element.

The 2/4 and UD solenoids are normally applied, which by design allow fluid to pass through in their relaxed or "off" state. This allows transaxle limp-in (P,R,N,2) in the event of an electrical failure.

The continuity of the solenoids and circuits are periodically tested. Each solenoid is turned on or off depending on its current state. An inductive spike should be detected by the PCM/TCM during this test. It no spike is detected, the circuit is tested again to verify the failure. In addition to the periodic Testing, the solenoid circuits are tested if a speed ratio or pressure switch error occurs.

### PRESSURE SWITCHES

The PCM/TCM relies on three pressure switches to monitor fluid pressure in the L/R, 2/4, and OD hydraulic circuits. The primary purpose of these switches is to help the PCM/TCM detect when clutch circuit hydraulic failures occur. The range for the pressure switch closing and opening points is 11-23 psi. Typically the switch opening point will be approximately one psi lower than the closing point. For example, a switch may close at 18 psi and open at 17 psi. The switches are continuously monitored by the PCM/TCM for the correct states (open or closed) in each gear as shown in the following chart:

A Diagnostic Trouble Code (DTC) will set if the PCM/TCM senses any switch open or closed at the wrong time in a given gear.

The PCM/TCM also tests the 2/4 and OD pressure switches when they are normally off (OD and 2/4 are tested in 1st gear, OD in 2nd gear, and 2/4 in 3rd gear). The test simply verifies that they are operational, by looking for a closed state when the corresponding element is applied. Immediately after a shift into 1st, 2nd, or 3rd gear with the engine speed above 1000 rpm, the PCM/TCM momentarily turns on element pressure to the 2/4 and/or OD clutch circuits to identify that the appropriate switch has closed. If it doesn't close, it is tested again. If the switch fails to close the second time, the appropriate Diagnostic Trouble Code (DTC) will set.

| GEAR | L/R | 2/4 | OD |
|------|-----|-----|-----|
| R | OP | OP | OP |
| P/N | CL | OP | OP |
| 1ST | CL | OP | OP |
| 2ND | OP | CL | OP |
| D | OP | OP | CL |
| OD | OP | CL | CL |
| OP = OPEN | | | |
| CL = CLOSED | | | |

Indicating pressure switch states in solenoid/pressure switch assembly

### Removal & Installation

NOTE: *If solenoid/pressure switch assembly is being replaced, the "Quick-Learn" procedure must be performed.*

1. Disconnect battery negative cable.
2. Remove air cleaner assembly.
3. Disconnect solenoid/pressure switch assembly connector.
4. Disconnect input speed sensor connector.
5. Remove input speed sensor.
6. Remove 3 solenoid/pressure switch assembly-to-transaxle case bolts.
7. Remove solenoid/pressure switch assembly and gasket. Use care to prevent gasket material and foreign objects from become lodged in the transaxle case ports.

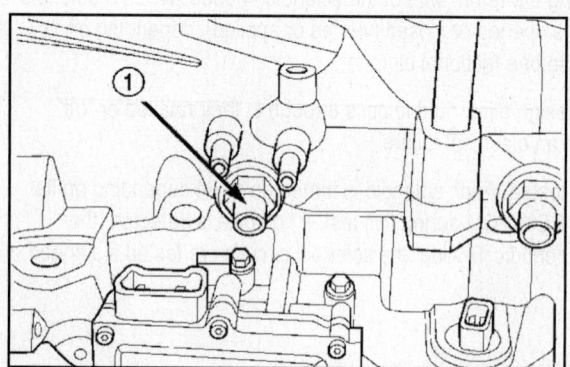

Location of Input Speed Sensor (1) with Solenoid/Pressure Switch Assembly just below it on Pacifica engines

**To install:**

*NOTE: If solenoid/pressure switch assembly is being replaced, it is necessary to perform the "Quick-Learn" procedure.*

1. Install solenoid/pressure switch assembly and new gasket to transaxle.
2. Install and torque three bolts to 13 Nm (110 inch lbs.).
3. Install input speed sensor and torque to 27 Nm (20 ft. lbs.).
4. Connect input speed sensor connector.
5. Install solenoid/pressure switch 8-way connector and torque to 4 Nm (35 inch lbs.).
6. Install air cleaner assembly.
7. Connect battery negative cable.

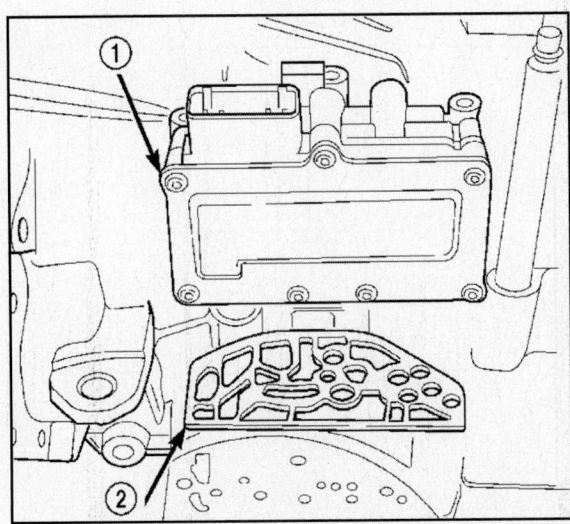

**Identifying the Solenoid/Pressure Switch Assembly (1) and gasket (2)**

THROTTLE POSITION (TP) SENSOR

**Removal & Installation**

*3.5L*

1. Disconnect the negative battery cable.
2. Disconnect the TPS electrical connector.
3. Remove the TPS mounting screws and remove the TPS.

*3.8L*

1. Disconnect negative battery cable.
2. Disconnect the TPS electrical connector.
3. Remove the TPS mounting screws.
4. Remove the TPS.

**To install:**

*3.5L & 3.8L*

*Note: The throttle shaft end of the throttle body slides into a socket in the TPS. The socket has two tabs inside it. The throttle shaft rests against the tabs. When indexed correctly, the TPS can rotate clockwise a few degrees to line up the mounting screw holes with the screw holes in the throttle body. The TPS has slight tension when rotated into position. If it is difficult to rotate the TPS into position, install the sensor with the throttle shaft on the other side of the tabs in the socket.*

1. Tighten TP sensor mounting screws to 5.1 Nm (45 inch lbs.) torque.
2. The throttle plate should be closed. If the throttle plate is open, install the sensor on the other side of the tabs in the socket.
3. Attach electrical connector to the TP sensor.
4. Connect the negative battery cable.

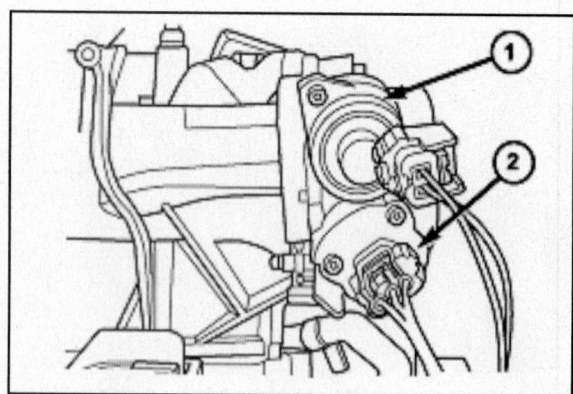

Locating TP sensor (2) and IAC motor (1) on the Pacifica 3.5L engine

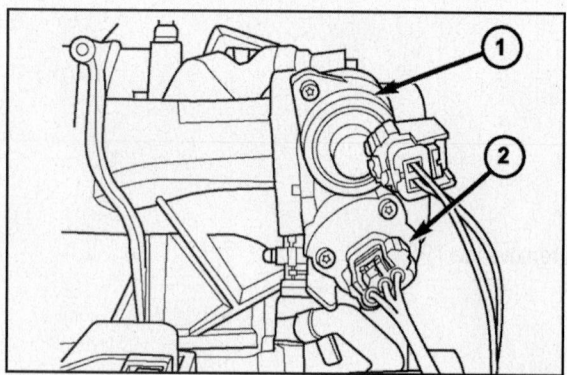

Locating TP sensor (2) and IAC motor (1) on the Pacifica 3.8L engine

**Connector Pinouts**

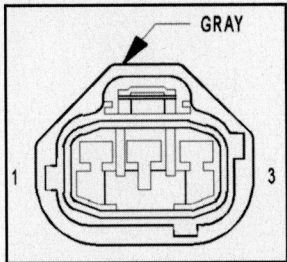

| CAV | CIRCUIT | FUNCTION |
|---|---|---|
| 1 | K900 20DB/DG | SENSOR GROUND |
| 2 | K22 18BR/OR | TP SIGNAL |
| 3 | F855 20PK/YL | 5 VOLT SUPPLY |

**TP Sensor Connector Pinout**

## DAIMLERCHRYSLER MINI-VANS: CARAVAN, TOWN & COUNTRY & VOYAGER

AMBIENT TEMPERATURE SENSOR

**Connector Pinouts**

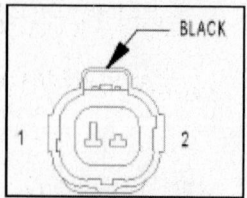

| PIN # | WIRE COLOR | CIRCUIT DESCRIPTION |
|-------|------------|---------------------|
| 1 | VT/LG (DIESEL) | AMBIENT TEMPERATURE SENSOR SIGNAL |
| 1 | VT/LG (GAS) | AAT SIGNAL |
| 2 | VT/BR (DIESEL) | AMBIENT TEMPERATURE SENSOR RETURN |
| 2 | DB/DG (GAS) | SENSOR GROUND |

**Ambient Temperature Sensor**

AUTOMATIC SHUTDOWN (ASD) RELAY

**Description & Operation**

The Powertrain Control Module (PCM) operates the ASD relay by switching the ground path on and off. The ASD relay supplies battery voltage to the fuel injectors, electronic ignition coil and the heating elements in the oxygen sensors.

The PCM controls the relay by switching the ground path for the solenoid side of the relay on and off. The PCM turns the ground path off when the ignition switch is in the OFF position unless the 02 Heater Monitor test is being run. When the ignition switch is in the ON or CRANK position, the PCM monitors the crankshaft position sensor and camshaft position sensor signals to determine engine speed and ignition timing (coil dwell). If the PCM does not receive the crankshaft position sensor and camshaft position sensor signals when the ignition switch is in the RUN position, it will de-energize the ASD relay.

### AUTOMATIC SHUTDOWN (ASD) SENSE

The ASD sense circuit informs the PCM when the ASD relay energizes. A 12-volt signal at this input indicates to the PCM that the ASD has been activated. This input is used only to sense that the ASD relay is energized. When energized, the ASD relay supplies battery voltage to the fuel injectors, ignition coils and the heating element in each oxygen sensor. If the PCM does not receive 12 volts from this input after grounding the ASD relay, it sets a Diagnostic Trouble Code (DTC).

### AUTOMATIC SHUTDOWN RELAY

The ASD relay and fuel pump relay are located in the Power Distribution Center (PDC) near the Air Cleaner. The inside top of the PDC cover has a label showing relay and fuse location.

The PCM operates the Automatic Shut Down (ASD) relay and fuel pump relays. The PCM operates them by switching the ground path for the solenoid side of the relays on and off. The ASD relay connects battery voltage to the fuel injectors and ignition coil. The fuel pump relay connects battery voltage to the fuel pump.

A Buss bar in the Power Distribution Center (PDC) supplies voltage to the coil side and contact side of the relay. The ASD relay power circuit contains a 25-amp fuse between the buss bar in the PDC and the relay. The fuses are located in the PDC.

| CAVITY | CIRCUIT | FUNCTION |
|--------|---------|----------|
| 30 | INTERNAL | FUSED B+ |
| 85 | K51 18DB/YL | AUTO. SHUTDOWN RELAY CONTROL |
| 86 | A0 GROUND | B+ |
| 87 | A142 18DG/OR | AUTO. SHUTDOWN RELAY OUTPUT |
| 87A | --- | NOT USED |

**ASD relay pinout information**

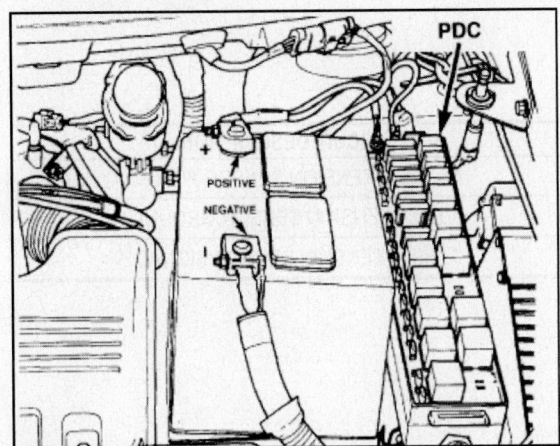

**Showing the location of the Power Distribution Center**

BATTERY TEMPERATURE SENSOR

**Connector Pinouts**

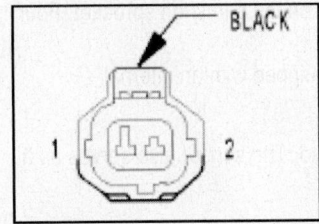

| PIN # | WIRE COLOR | CIRCUIT DESCRIPTION |
|-------|-----------|---------------------|
| 1 | BR/TN (DIESEL) | BATTERY TEMPERATURE SENSOR SIGNAL |
| 1 | DB/YL (GAS) | BATTERY TEMP SIGNAL |
| 2 | DB/DG (DIESEL) | SENSOR GROUND |
| 2 | DB/DG (GAS) | SENSOR GROUND |

**Caravan, Town & Country, Voyager Battery Temperature Sensor**

## BELT TENSION SENSOR CONNECTOR

**Connector Pinouts**

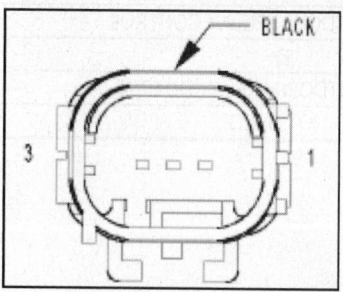

| PIN # | WIRE COLOR | CIRCUIT DESCRIPTION |
|-------|-----------|---------------------|
| 1 | LG/LB | BELT TENSION SENSOR FEED |
| 2 | LG/BR | BELT TENSION SENSOR GROUND |
| 3 | WT/OR | BELT TENSION SENSOR SIGNAL |

**Caravan, Town & Country, Voyager Belt Tension Sensor**

## CAMSHAFT POSITION (CMP) SENSOR

### Description & Operation

The camshaft position sensor (along with the crankshaft position sensor) provides inputs to the PCM to determine fuel injection synchronization and cylinder identification. From these inputs, the PCM determines crankshaft position.

The camshaft position sensor provides cylinder identification to the Powertrain Control Module (PCM). The sensor generates pulses as groups of notches on the camshaft sprocket pass underneath it. The PCM keeps track of crankshaft rotation and identifies each cylinder by the pulses generated by the notches on the camshaft sprocket. Four crankshaft pulses follow each group of camshaft pulses.
The 3.0L engine is equipped with a camshaft driven mechanical distributor, which is equipped with an internal camshaft position (fuel sync) sensor.

On the 2.4L engine, the camshaft position sensor mounts to the rear of the cylinder head. The sensor also serves as a thrust plate to control endplay of the camshaft.

On 3.3L and 3.8L engines, the camshaft position sensor mounts to the top of the timing case cover, in which the bottom of the sensor is positioned above the camshaft sprocket.

When the PCM receives 2 cam pulses followed by the long flat spot on the camshaft sprocket, it knows that the crankshaft timing marks for cylinder 1 are next (on driveplate). When the PCM receives one camshaft pulse after the long flat spot on the sprocket, cylinder number 2 crankshaft timing marks are next. After 3 camshaft pulses, the PCM knows cylinder 4 crankshaft timing marks follow. One camshaft pulse after the 3 pulses indicates cylinder 5. The 2 camshaft pulses which occur after cylinder No. 5 will signal cylinder No. 6. The PCM can synchronize on cylinders No. 1 or No. 4.

When metal aligns with the sensor, voltage goes low (less than 0.3 volts). When a notch aligns with the sensor, voltage switches high (5.0 volts). As a group of notches pass under the sensor, the voltage switches from low (metal) to high (notch) then back to low. The number of notches will determine the amount of pulses. If available, an oscilloscope can display the square wave patterns of each timing event.

Top Dead Center (TDC) does not occur when notches on the camshaft sprocket pass below the cylinder. TDC occurs after the camshaft pulse (or pulses) and after the 4 crankshaft pulses associated with the particular cylinder. The arrows and cylinder call outs represent which cylinder the flat spot and notches identify, they do not indicate TDC position.

The PCM determines fuel injection synchronization and cylinder identification from inputs provided by the camshaft position sensor and crankshaft position sensor. From the two inputs, the PCM determines crankshaft position.

The sensor generates pulses as groups of notches on the camshaft sprocket pass underneath it. The PCM keeps track of crankshaft rotation and identifies each cylinder by the pulses generated by the notches on the camshaft sprocket. Four crankshaft pulses follow each group of camshaft pulses.

The distance between the bottom of sensor and the camshaft sprocket is critical to the operation of the system.

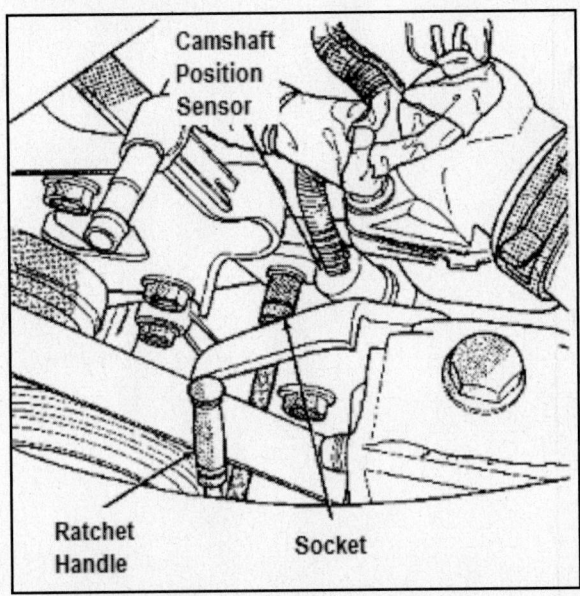

**CMP sensor location on 3.3L and 3.8L engines**

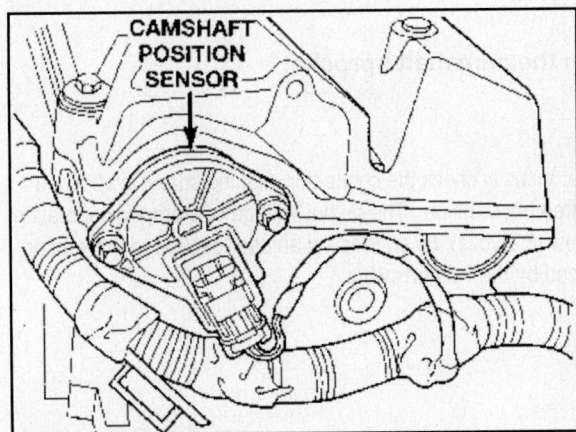

**Camshaft position sensor location-2.4L engine**

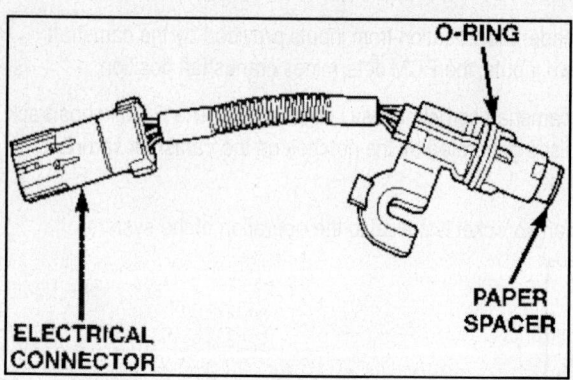

**Camshaft position sensor and spacer-3.3L and 3.8L engines**

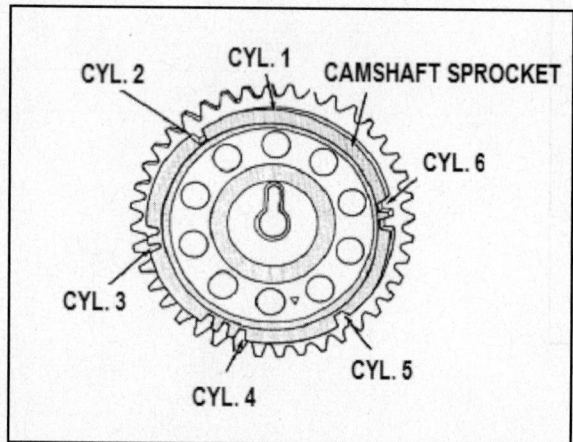

**Identifying notch and cylinder relationship on the camshaft sprocket**

### Testing

To test this sensor, you will need the use of an oscilloscope. Visually check the connector, making sure it is attached properly and that all of the terminals are straight, tight and free of corrosion. The output voltage of a properly operating camshaft position sensor switches from high (5.0 volts) to low (0.3 volts). By connecting an oscilloscope to the sensor output circuit, you can view the square wave pattern produced by the voltage swing.

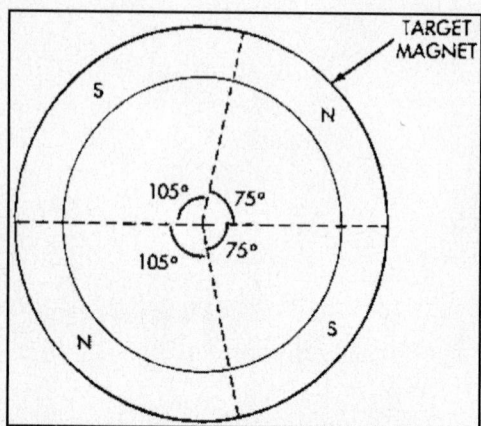

**Camshaft position sensor target magnet polarity-2.4L engine**

**Connector Pinouts**

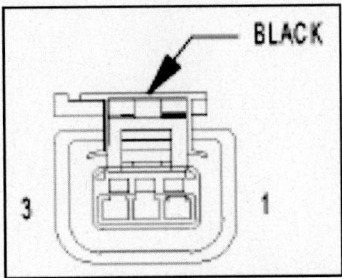

| PIN # | WIRE COLOR | CIRCUIT DESCRIPTION |
|-------|-----------|---------------------|
| 1 | PK/YL (2.4L) | 5 VOLT SUPPLY |
| 1 | PK/YL (3.3L/3.8L) | 5 VOLT SUPPLY |
| 2 | DB (3.3L/3.8L LATE BUILD EXPORT) | SENSOR GROUND |
| 2 | DB/DG (EXC. 3.3L/3.8L LATE BUILD EXPORT) | SENSOR GROUND |
| 3 | DB/GY | CMP SIGNAL |

**CMP Sensor**

## CRANKSHAFT POSITION (CKP) SENSOR

### Description & Operation

The PCM determines what cylinder to fire from the crankshaft position sensor input and the camshaft position sensor input. On 4-cylinder engines, the second crankshaft counterweight has two sets of four timing reference notches, including a 60° signature notch. From the crankshaft position sensor input, the PCM determines engine speed and crankshaft angle (position). On 6-cylinder engines, this sensor is a Hall effect device that detects notches in the flexplate.

The notches generate pulses from high to low in the crankshaft position sensor output voltage. When a metal portion of the notches line up with the crankshaft position sensor, the sensor output voltage goes low (less than 0.5 volts). When a notch aligns with the sensor, voltage goes high (5.0 volts). As a group of notches pass under the sensor, the output voltage switches from low (metal) to high (notch), then back to low.

If available, an oscilloscope can display the square wave patterns of each voltage pulse. From the width of the output voltage pulses, the PCM calculates engine speed. The width of the pulses represents the amount of time the output voltage stays high before switching back to low. The period of time the sensor output voltage stays high before switching back to low is referred to as pulse width. The faster the engine is operating, the smaller the pulse width on the oscilloscope.

On 4-cylinder engines, the crankshaft position sensor is mounted to the engine block behind the alternator, just above the oil filter. On 6-cylinder engines, the crankshaft position sensor is mounted on the transaxle housing, above the vehicle speed sensor.

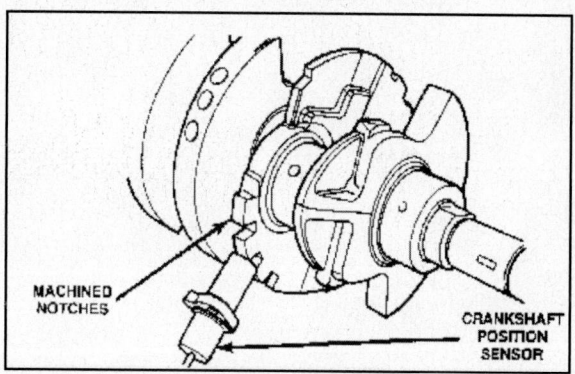

Timing reference notches on the 2.4L engine

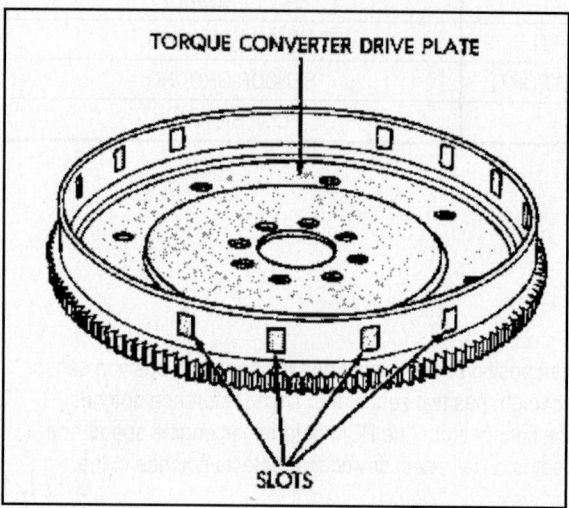

Timing reference notches on 3.0L, 3.3L and 3.8L engines

### Testing

To test this sensor, you will need the use of an oscilloscope. Visually check the connector, making sure it is attached properly and that all of the terminals are straight, tight and free of corrosion. Also inspect the notches in the crankshaft (4-cylinder) or flywheel (6-cylinder) for damage, and replace if necessary. The output voltage of a properly operating crankshaft position sensor switches from high (5.0 volts) to low (0.3 volts). By connecting an oscilloscope to the sensor output circuit, you can view the square wave pattern produced by the voltage swing.

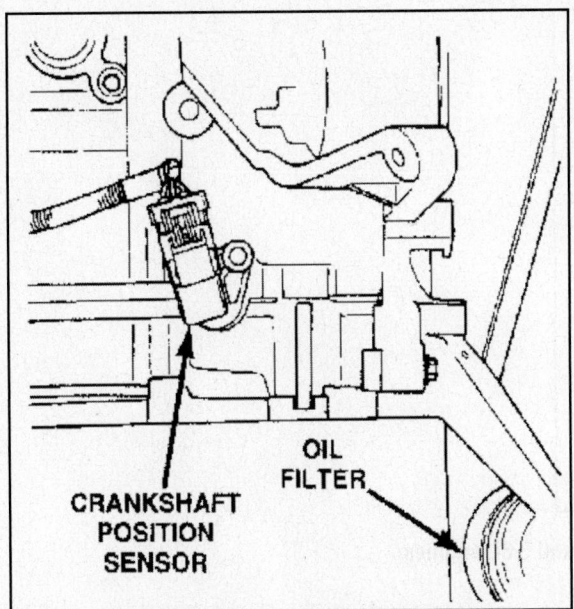

Location of the crankshaft position sensor on the 2.4L engine

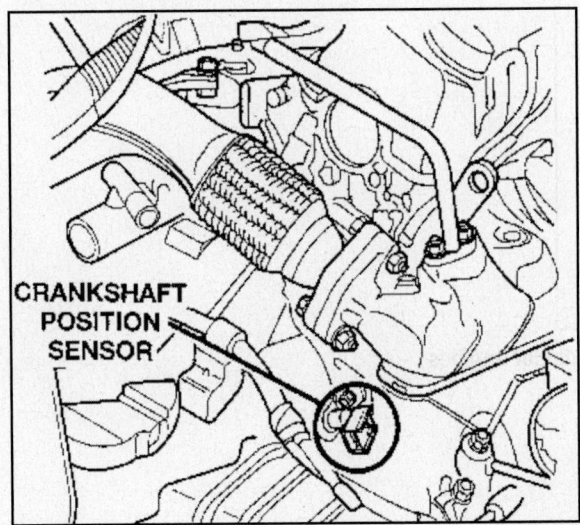

Location of the crankshaft position sensor on the 3.0L engine

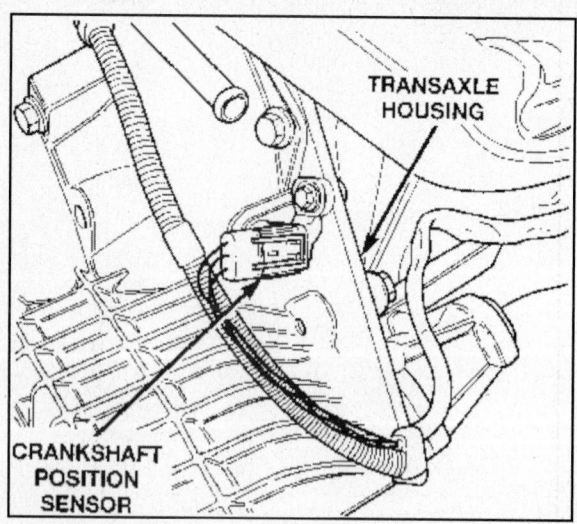

Location of the crankshaft position sensor on the 3.3L and 3.8L engines

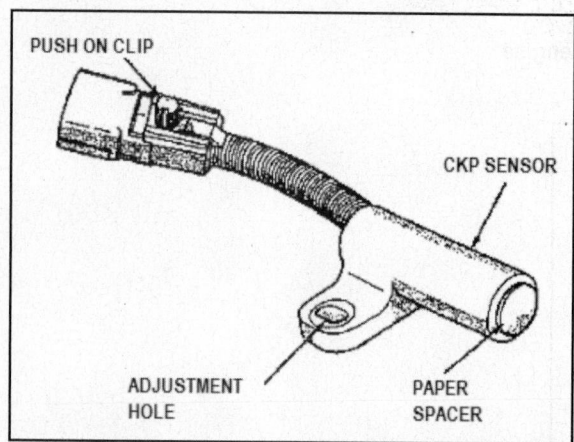

Crankshaft position sensor and spacer on 1996-97 6-cylinder engines

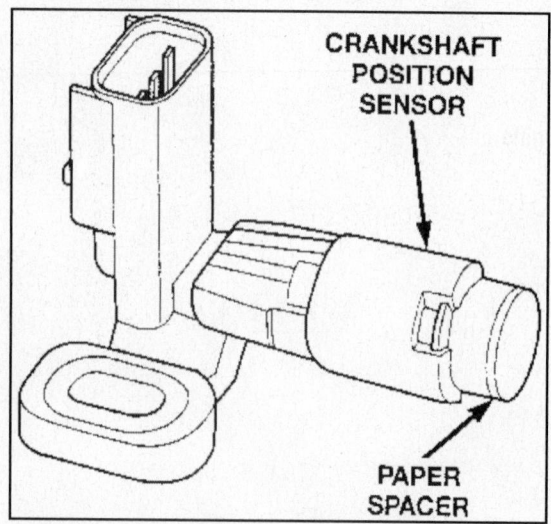

Crankshaft position sensor and spacer on 1998 and newer 6-cylinder engines

**Connector Pinouts**

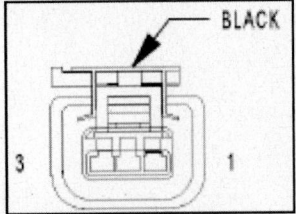

| PIN # | WIRE COLOR | CIRCUIT DESCRIPTION |
|-------|-----------|---------------------|
| 1 | PK/YL (2.4L) | 5 VOLT SUPPLY |
| 1 | PK/YL (3.3L/3.8L) | 5 VOLT SUPPLY |
| 2 | DB/DG | SENSOR GROUND |
| 3 | BR/LB | CKP SIGNAL |

**CKP Sensor**

## ENGINE COOLANT TEMPERATURE (ECT) SENSOR

### Description & Operation

The engine coolant temperature sensor is a variable resistor with a range of -40°F-265°F (-5°C-129°C).

The engine coolant temperature sensor provides an input voltage to the PCM. As the coolant temperature varies, the sensor resistance changes resulting in a different input voltage to the PCM.

When the engine is cold, the PCM will demand slightly richer air/fuel mixtures and higher idle speeds until normal operating temperatures are reached.

The engine coolant temperature sensor is also utilized for control of the cooling fan.

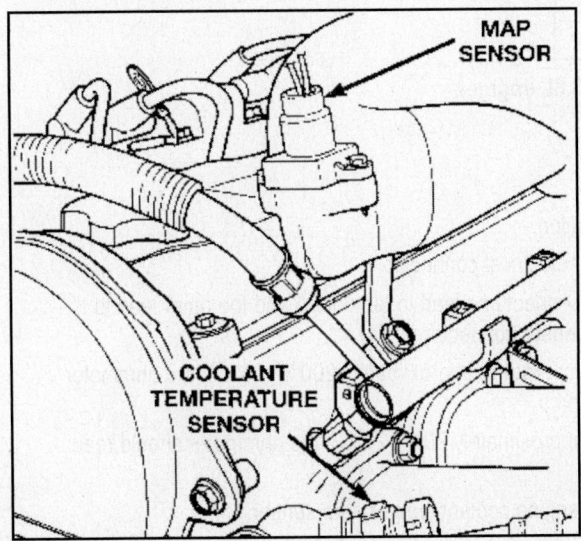

Engine coolant temperature sensor location-2.4L engine

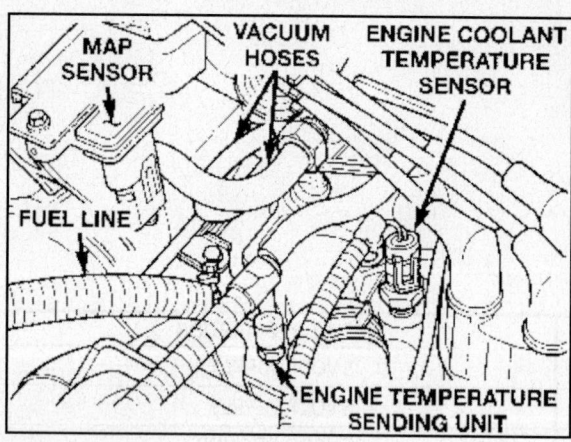

Engine coolant temperature sensor location-3.0L engine

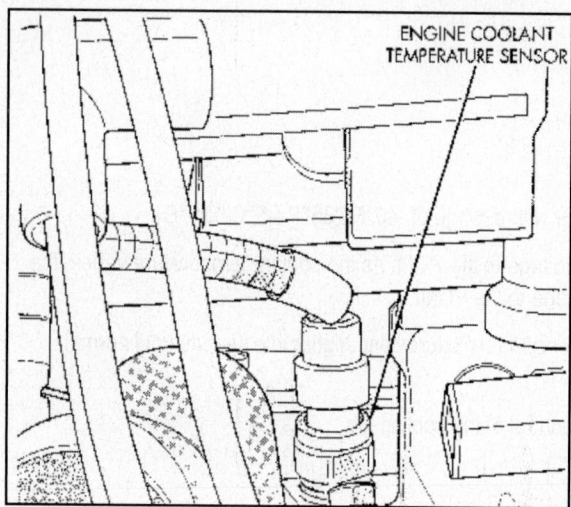

Engine coolant temperature sensor location-3.3L and 3.8L engines

### Testing

1. Turn the ignition switch to the OFF position.
2. Detach the coolant temperature sensor electrical connector.
3. Using a DVOM set to the ohms scale, connect one lead to terminal A and the other lead to terminal B of the coolant temperature sensor connector.
4. With the engine at normal operating temperature, approximately 200°F (93°C), the ohmmeter should read approximately 700-1000 ohms.
5. With the engine at room temperature, approximately 70°F (21°C), the ohmmeter should read approximately 7000-13,000 ohms.
6. If not within specifications, replace the engine coolant temperature sensor.

*NOTE: Test the resistance of the wiring harness between PCM terminal 26 and the sensor wiring harness connector. Also check for continuity between PCM connector terminal 43 and the sensor wiring harness connector. If the resistance measures greater than 1 ohm, repair the wiring harness as necessary.*

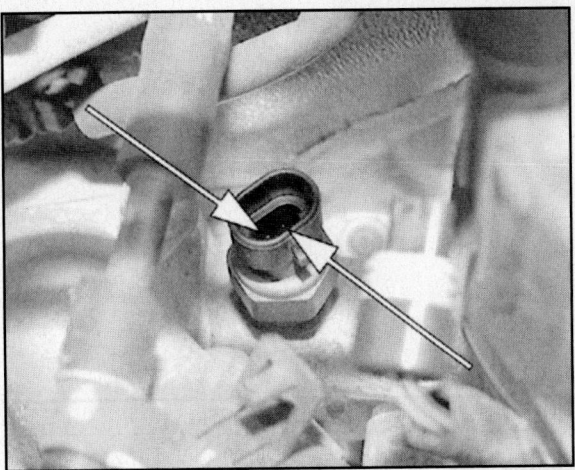

**Before testing the sensor, inspect the sensor terminals for damage or corrosion and replace if necessary**

**Connector Pinouts**

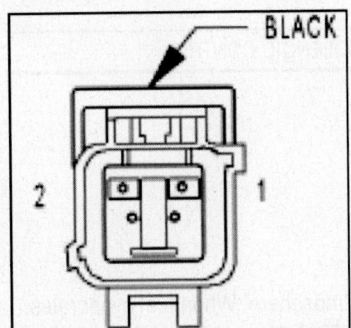

| PIN # | WIRE COLOR | CIRCUIT DESCRIPTION |
|-------|-----------|---------------------|
| 1 | DB/DG | SENSOR GROUND |
| 2 | VT/OR | ECT SIGNAL |

**ECT Sensor**

## EXHAUST GAS RECIRCULATION (EGR) SOLENOID

### Connector Pinouts

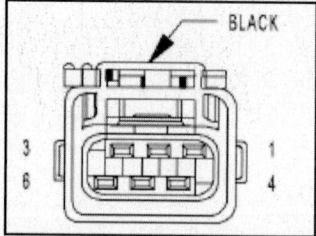

| PIN # | WIRE COLOR | CIRCUIT DESCRIPTION |
|-------|-----------|---------------------|
| 1 | DB/LG | EGR SOLENOID SIGNAL |
| 2 | PK/YL | 5 VOLT SUPPLY |
| 3 | DB/DG | SENSOR GROUND |
| 4 | BK/DB | GROUND |
| 5 | - | - |
| 6 | DB/VT | EGR SOLENOID CONTROL |

**3.3L, 3.8L (Late Build) EGR Solenoid Connector Pinout**

## EVAP PURGE SOLENOID

### Description & Operation

The evaporation control system prevents the emission of fuel tank vapors into the atmosphere. When fuel evaporates in the fuel tank, the vapors pass through vent hoses or tubes to an activated carbon filled evaporative canister. The canister temporarily holds the vapors. The Powertrain Control Module (PCM) allows intake manifold vacuum to draw vapors into the combustion chambers during certain operating conditions. All engines use a proportional purge solenoid system. The PCM controls vapor flow by operating the purge solenoid. Refer to Proportional Purge Solenoid in this section.

**NOTE: The evaporative system uses specially manufactured hoses. If they need replacement, only use fuel resistant hose. Also the hoses must be able to pass an Ozone compliance test.**

#### PROPORTION PURGE SOLENOID VALVE

All vehicles use a proportional purge solenoid. The solenoid regulates the rate of vapor flow from the EVAP canister to the throttle body. The PCM operates the solenoid.

During the cold start warm-up period and the hot start time delay, the PCM does not energize the solenoid. When de-energized, no vapors are purged.

The proportional purge solenoid operates at a frequency of 200 hz and is controlled by an engine controller circuit that senses the current being applied to the proportional purge solenoid and then adjusts that current to achieve the desired purge flow. The proportional purge solenoid controls the purge rate of fuel vapors from the vapor canister and fuel tank to the engine intake manifold.

### Removal & Installation

The solenoid attaches to a bracket near the radiator on the passenger side of vehicle. The solenoid will not operate unless it is installed correctly.

1. Disconnect electrical connector from solenoid.
2. Disconnect vacuum tubes from solenoid.

3. Remove solenoid from bracket.

**To install:**

The top of the solenoid has TOP printed on it. The solenoid will not operate unless it is installed correctly.

4. Install solenoid on bracket.

5. Connect vacuum tube to solenoid.

6. Connect electrical connector to solenoid.

**Connector Pinouts**

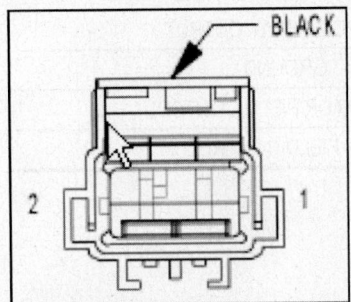

| PIN # | WIRE COLOR | CIRCUIT DESCRIPTION |
|-------|-----------|---------------------|
| 1 | DB/WT | EVAP PURGE CONTROL |
| 2 | DB/BR | EVAP PURGE RETURN |

**EVAP Purge Solenoid Connector Pinout**

## FUEL INJECTORS

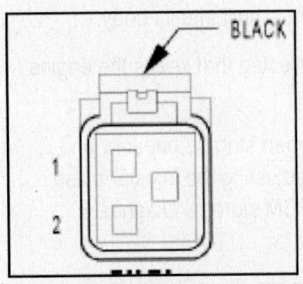

| PIN # | WIRE COLOR | CIRCUIT DESCRIPTION |
|-------|-----------|---------------------|
| 1 | BR/WT | ASD RELAY OUTPUT |
| 2 | BR/YL | INJECTOR CONTROL NO. 1 |
| 2 | BR/DB | INJECTOR CONTROL NO. 2 |
| 2 | BR/LB | INJECTOR CONTROL NO. 3 |
| 2 | BR/TN | INJECTOR CONTROL NO. 4 |
| 2 | BR/OR | INJECTOR CONTROL NO. 5 |
| 2 | BR/VT | INJECTOR CONTROL NO. 6 |

**Fuel Injector Connector Pinouts**

GENERATOR

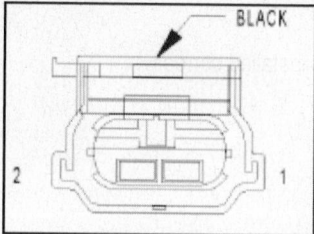

| PIN # | WIRE COLOR | CIRCUIT DESCRIPTION |
|---|---|---|
| 1 | BR/WT (DIESEL) | ECM/PCM RELAY OUTPUT |
| 1 | BK (GAS) | GROUND |
| 2 | BR/GY (DIESEL) | GENERATOR FIELD CONTROL |
| 2 | BR/GY (GAS) | GEN FIELD CONTROL |

**Generator Connector Pinout**


HEATED OXYGEN SENSORS (HO2S)

**Description & Operation**

As a vehicle accrues mileage, the catalytic converter deteriorates. The deterioration results in a less effective catalyst. To monitor catalytic converter deterioration, the fuel injection system uses two heated oxygen sensors: one is upstream of the catalytic converter and one downstream of the converter.

The heated oxygen sensor, or HO2S sensor is usually located near the catalytic converter. It produces a voltage signal of 0.1-1.0 volts based on the amount of oxygen in the exhaust gas. When a low amount of oxygen is present (caused by a rich air/fuel mixture), the sensor produces a high voltage. When a high amount of oxygen is present (caused by a lean air/fuel mixture), the sensor produces a low voltage. Because an accurate voltage signal is only produced if the sensor temperature is above approximately 600°F (315°C), a fast-acting heating element is built into its body.

The PCM uses the HO2S sensor voltage signal to constantly adjust the amount of fuel injected that keeps the engine at its peak efficiency.

The PCM compares the reading from the sensors to calculate the catalytic converter oxygen storage capacity and storage efficiency. The PCM also uses the upstream heated oxygen sensor input when adjusting the injector pulse width. When the catalytic converter efficiency drops below preset emission criteria, the PCM stores a Diagnostic Trouble Code (DTC) and illuminates the Malfunction Indicator Lamp (MIL).

The automatic shutdown relay supplies battery voltage to both of the heated oxygen sensors. The sensors have heating elements that reduce the amount of time it takes for the sensors to reach operating temperature.

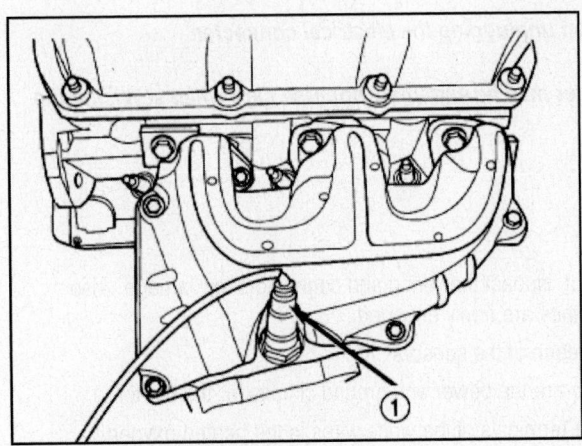

**Showing the upstream HO2S (1)**

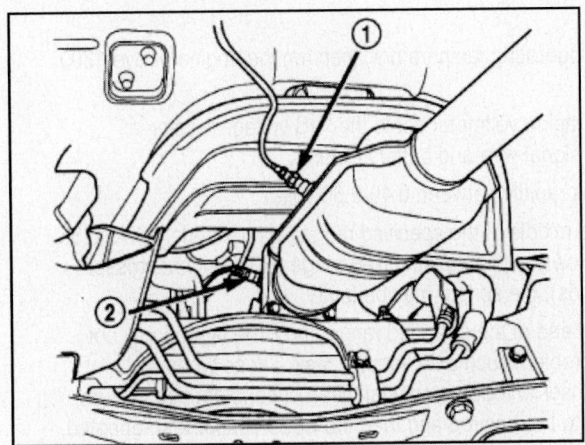

**Locating the downstream (1) HO2S and the upstream (2) HO2S**

### Removal & Installation

1. Disconnect the negative battery cable.
2. Raise and support the vehicle.
3. Disconnect the electrical connector.
4. Use a special socket or crows foot wrench to remove the sensor.

*When the sensor is removed, the exhaust manifold threads must be cleaned with an 18 mm X 1.5 + 6E tap. Note: If using the original sensor, coat the threads with Loctite 771-64 anti-seize compound or equivalent.*

**To install:**

*Note: Threads of new oxygen sensors are factory coated with anti-seize compound to aid in removal. DO NOT add any additional anti-seize compound to the threads of a new oxygen sensor.*

5. Install sensor and tighten to 27 Nm (20 ft. lbs.).
6. Connect the electrical connector.
7. Lower vehicle.
8. Install the negative battery cable.

*CAUTION: Do NOT pull on the oxygen sensor wire when unplugging the electrical connector.*

*WARNING: The exhaust manifold and catalytic converter may be extremely hot. Use care when servicing the oxygen sensor.*

**Testing**

### HEATING ELEMENT

1. Before Testing any electrical component, inspect the wiring and connectors for damage. Also wiggle the connectors to ensure a that they are firmly engaged.
2. Disconnect the electrical harness from each of the sensors.
3. The white wires in the sensor connector are the power and ground circuits for the heater.
4. Connect the ohmmeter test leads to the terminals of the white wires in the heated oxygen sensor connector.
5. Check the resistance of the sensor, if it is not within 4-7 ohms, replace the sensor.

### SENSOR

1. Start the engine and bring it to normal operating temperature, then run the engine above 1200 rpm for two minutes.
2. Backprobe with a high impedance averaging voltmeter set to the "DC voltage" scale. Backprobe between the HO2S sensor signal wire and battery ground.
3. Verify that the sensor voltage fluctuates rapidly between 0.40-0.60 volts.
4. If the sensor voltage is stabilized at the middle of the specified range (approximately 0.45-0.55 volts) or if the voltage fluctuates very slowly within the specified range (H02S signal crosses 0.5 volts less than 5 times in ten seconds), the sensor may be faulty.
5. If the sensor voltage stabilizes at either end of the specified range, the PCM is probably not able to compensate for a mechanical problem such as a vacuum leak. These types of mechanical problems will cause the sensor to report a constant lean or constant rich mixture. The mechanical problem will first have to be repaired and then the H02S sensor test repeated.
6. Pull a vacuum hose located after the throttle plate. Voltage should drop to approximately 0.12 volts (while still fluctuating rapidly). This tests the ability of the sensor to detect a lean mixture condition. Reattach the vacuum hose.
7. Richen the mixture using a propane enrichment tool. Sensor voltage should rise to approximately 0.90 volts (while still fluctuating rapidly). This tests the ability of the sensor to detect a rich mixture condition.
8. If the sensor voltage is above or below the specified range, the sensor and/or the sensor wiring may be faulty. Check the wiring for any breaks, repair as necessary and repeat the test.

*NOTE: Further sensor operational Testing requires the use of a special tester DRB scan tool or equivalent.*

**Connector Pinouts**

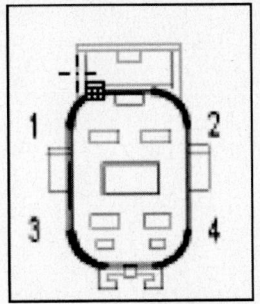

| PIN # | WIRE COLOR | CIRCUIT DESCRIPTION |
|-------|------------|---------------------|
| 1 | BK/LG | GROUND |
| 2 | BR/LG | O2 1/1 HEATER CONTROL |
| 3 | BR/DG | O2 RETURN (UP) |
| 4 | DB/LB | O2 1/1 SIGNAL |

**Upstream O2 Sensor**

| 1 | BK/DB | GROUND |
|---|-------|--------|
| 2 | BR/WT | O2 1/2 HEATER CONTROL |
| 3 | DB/DG (2.4L) | O2 RETURN (DOWN) |
| 3 | BR/DG (3.3L/3.8L) | O2 RETURN (DOWN) |
| 4 | DB/YL | O2 1/2 SIGNAL |

**Downstream O2 Sensor**

## IDLE AIR CONTROL (IAC) MOTOR

### Description & Operation

The Idle Air Control (IAC) motor, attached to the side of the throttle body, is operated by the PCM. The PCM adjusts engine idle speed through the idle air control motor to compensate for load on the engine, or ambient conditions.

The throttle body has an air bypass passage that provides air for the engine during closed throttle idle. The idle air control motor pintle protrudes into the air bypass passage and regulates the airflow through it. The PCM adjusts the idle speed by moving the IAC motor pintle in and out of the bypass passage. The speed is based on various sensor and switch inputs received by the PCM. The inputs are from the throttle position sensor, crankshaft position sensor, coolant temperature sensor, as well as various switch operations (brake, park/neutral, air conditioning). Increasing airflow when the throttle is closed quickly after a driving condition also prevents deceleration die out.

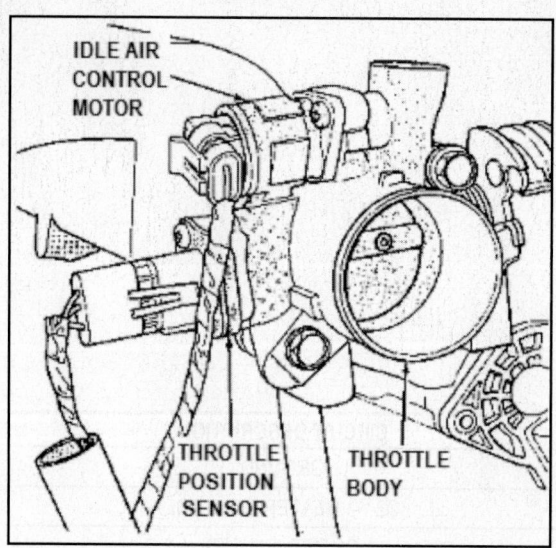

The IAC motor and TP Sensor location on the throttle body for 2.4L engine

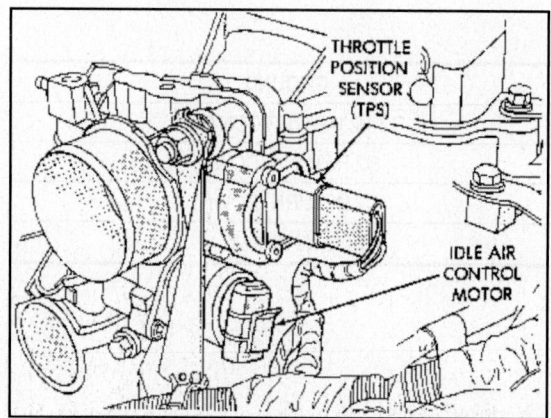

The IAC motor and TP Sensor location on the throttle body for 3.0L engine

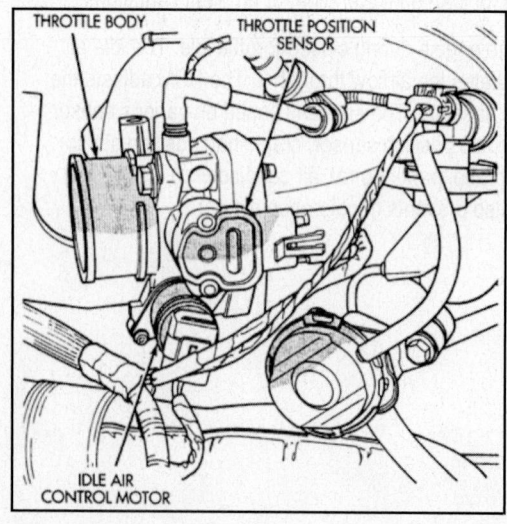

The IAC motor and TP Sensor location on the throttle body for 3.3L and 3.8L engines

**Removal & Installation**

When servicing throttle body components, always reassemble components with new O-rings and seals where applicable. Never use lubricants on O-rings or seals as damage may result. If assembly of component is difficult, use water to aid assembly. Use care when removing hoses to prevent damage to hose or hose nipple.

1. Disconnect negative cable from battery.
2. Remove electrical connector from idle air control motor.
3. Remove idle air control motor mounting screws.
4. Remove motor from throttle body. Ensure the O-ring is removed with the motor.

**To install:**

The new idle air control motor has a new O-ring installed on it. If pintle measures more than 1 inch (25 mm) it must be retracted. Use the scan tool "Idle Air Control Motor Open/Close" test function to retract the pintle (battery must be connected.)

5. Carefully place idle air control motor into throttle body.
6. Install mounting screws. Tighten screws to 2 Nm (17 inch lbs.) torque.
7. Connect electrical connector to idle air control motor.
8. Connect negative cable to battery.

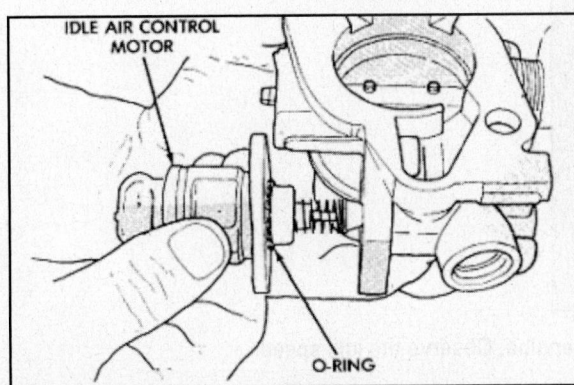

**Removing the Idle Air Control Motor**

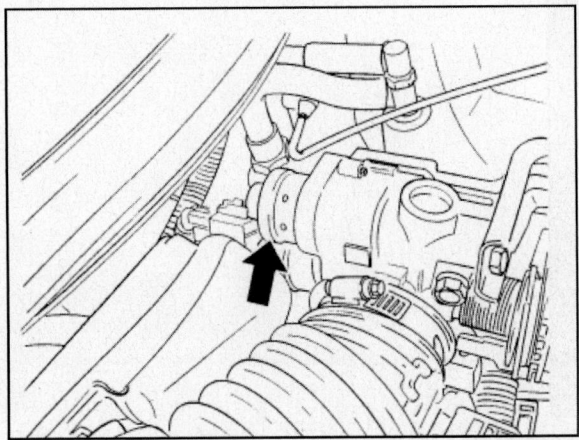

**Indicating the location of the Idle Air Control Motor on 3.3L VIN 3**

**Testing**

1. Visually check the connector, making sure it is properly attached and all of the terminals are straight, tight and free of corrosion.

2. You need to have access to a DRB® or equivalent scan tool to accurately test the Idle Air Control (IAC) motor and related circuits. Make sure to carefully follow all of the scan tool manufacturers directions when testing the IAC motor.

3. If you do not have access to a scan tool, this simple test should give you an indication if the circuit is working properly. First attach a tachometer to the engine, and then start the engine.

4. Observe the idle speed. Pull a vacuum hose (like the one leading from the brake booster to the intake manifold). The idle speed should rise, and then fall as the IAC motor tries to compensate for the vacuum leak.

5. Reattach the vacuum hose. The idle should drop, and then stabilize.

6. If the engine reacted as indicated, the circuit is probably OK.

Connect a tachometer to the engine, and then start the engine. Observe the idle speed...

... then, while observing the idle speed, disconnect a vacuum hose (such as brake booster-to-intake manifold). The IAC motor should compensate for the leak

**Connector Pinouts**

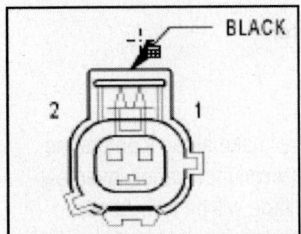

| PIN # | WIRE COLOR | CIRCUIT DESCRIPTION |
|-------|-----------|---------------------|
| 1 | VT/GY | IAC MOTOR CONTROL |
| 2 | BR/VT | IAC RETURN |

**IAC Motor Connector Pinout**

IGNITION COIL

**Connector Pinouts**

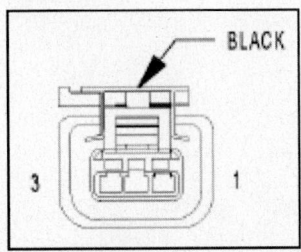

| PIN # | WIRE COLOR | CIRCUIT DESCRIPTION |
|-------|-----------|---------------------|
| 1 | DB/TN | COIL CONTROL NO. 2 |
| 2 | BR/WT | ASD RELAY OUTPUT |
| 3 | DB/DG | COIL CONTROL NO. 1 |

**2.4L Ignition Coil Connector Pinout**

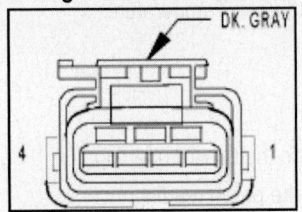

| PIN # | WIRE COLOR | CIRCUIT DESCRIPTION |
|-------|-----------|---------------------|
| 1 | BR/OR | COIL CONTROL NO. 3 |
| 2 | BR/WT | ASD RELAY OUTPUT |
| 3 | DB/DG | COIL CONTROL NO. 1 |
| 4 | DB/TN | COIL CONTROL NO. 2 |

**3.3L & 3.8L Ignition Coil Connector Pinout**

## INTAKE AIR TEMPERATURE (IAT) SENSOR

*NOTE: Only the 2.4L engine is equipped with the Intake Air Temperature (IAT) sensor.*

### Description & Operation

The IAT sensor threads into the intake manifold, where it measures the temperature of the intake air as it enters the engine. The sensor is a Negative Temperature Coefficient (NTC) thermistor-type sensor (resistance varies inversely with temperature). This means at high temperatures, resistance decreases and so the voltage will be low. At cold temperatures, the resistance is high and so the voltage will also be high. This allows the sensor to provide an analog voltage signal to the PCM. The PCM uses this signal to compensate for changes in air density due to temperature.

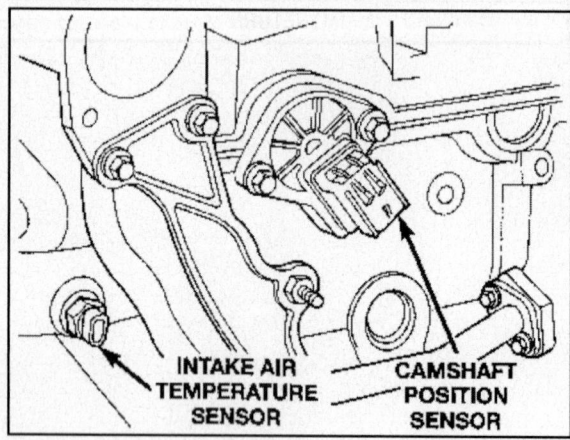

Intake Air Temperature (IAT) sensor location

### Testing

1. Visually check the connector, making sure it is attached properly and all of the terminals are straight, tight and free of corrosion.
2. With the engine OFF, turn the ignition key to the ON position.
3. Do not allow more than 5 minutes delay between the next 2 steps.
4. Using a DRB, or equivalent scan tool, read the information on the Intake Air Temperature (IAT) sensor and record the reading.
5. Turn the ignition switch OFF.
6. Remove the IAT sensor.
7. Using a temperature probe, quickly measure intake temperature inside the sensor opening.
8. Replace the IAT sensor if the scan tool reading is NOT within 10° of the probe reading.
9. Using a DRB, or equivalent scan tool, read the IAT sensor voltage.
10. If the voltage reading measures outside of the 0.5-4.5 volt range, disengage the IAT sensor connector.
11. Using the scan tool, read the IAT sensor voltage.
12. If the voltage reading measures greater than 4 volts, replace the IAT sensor.
13. Connect a jumper wire between the IAT signal and sensor ground circuits, then, along with the scan tool, read the sensor voltage.
14. If the voltage reading measures less than 1 volt, replace the IAT sensor.

**Connector Pinouts**

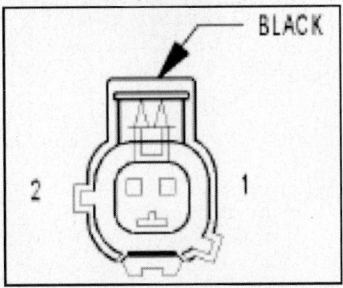

| PIN # | WIRE COLOR | CIRCUIT DESCRIPTION |
|-------|------------|---------------------|
| 1 | DB/LG | IAT SIGNAL |
| 2 | DB/DG | SENSOR GROUND |

**IAT Sensor**

## KNOCK SENSOR

### Description & Operation

The knock sensor is threaded into the side of the cylinder block, in front of the starter. When the knock sensor detects a knock in one of the cylinders, it sends an input signal to the PCM. In response, the PCM retards ignition timing for all cylinders by a specific amount. Knock sensors contain a piezoelectric material that sends an input signal (voltage) to the PCM. As the intensity of the engine knock vibration increases, the knock sensor output voltage also increases.

When the knock sensor detects a knock in one of the cylinders, it sends an input signal to the PCM. In response, the PCM retards ignition timing for all cylinders by a scheduled amount.

Knock sensors contain a piezoelectric material that constantly vibrates and sends an input voltage (signal) to the PCM while the engine operates. As the intensity of the crystal's vibration increases, the knock sensor output voltage also increases.

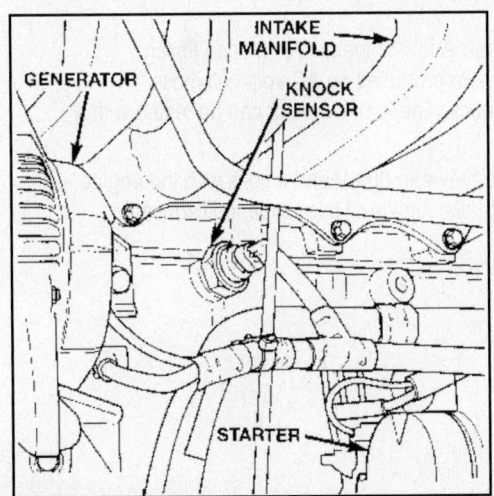

**Knock sensor location on the 2.4L engine**

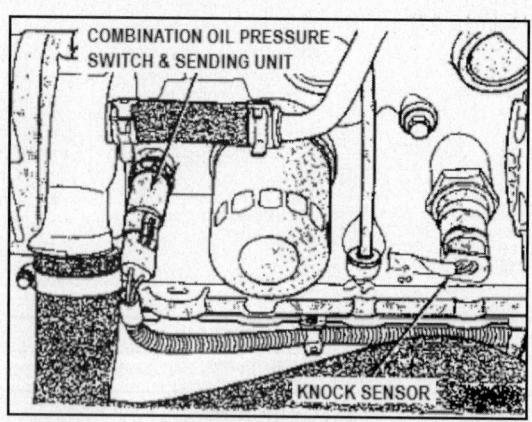

**Knock sensor location on the 3.3L and 3.8L engines**

The voltage signal produced by the knock sensor increases with the amplitude of vibration. The PCM receives as an input the knock sensor voltage signal. If the signal rises above a predetermined level, the PCM will store that value in memory and retard ignition timing to reduce engine knock. If the knock sensor voltage exceeds a preset value, the PCM retards ignition timing for all cylinders. It is not a selective cylinder retard. The PCM ignores knock sensor input during engine idle conditions. Once the engine speed exceeds a specified value, knock retard is allowed. Knock retard uses its own short-term and long-term memory program.

Long-term memory stores previous detonation information in its battery-backed RAM. The maximum authority that long term memory has over timing retard can be calibrated. Short-term memory is allowed to retard timing up to a preset amount under all operating conditions (as long as rpm is above the minimum rpm) except WOT. The PCM, using short-term memory, can respond quickly to retard timing when engine knock is detected. Short-term memory is lost any time the ignition key is turned OFF.

### Testing

1. Visually check the connector, making sure it is attached properly and that all of the terminals are straight, tight and free of corrosion.

2. A number of factors affect the engine knock sensor. A few of these are: ignition timing, cylinder pressure, fuel octane, etc. The knock sensor produces an AC voltage whose amplitude increases with the amount of engine knock. The knock sensor can be tested with a digital voltmeter.

3. The knock sensor output voltage should measure between 80mV and 4 volts with the engine running between 576 and 2208 rpm. If the output falls outside of this range, a Diagnostic Trouble Code (DTC) will set.

### Connector Pinouts

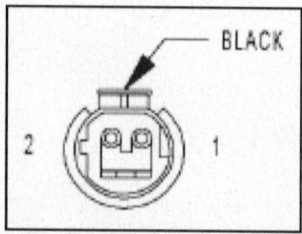

| PIN # | WIRE COLOR | CIRCUIT DESCRIPTION |
|:-----:|:----------:|:-------------------:|
| 1 | DB/YL | KS SIGNAL |
| 2 | BR/LG | KS RETURN |

**Knock Sensor Connector Pinout**

## MANIFOLD ABSOLUTE PRESSURE (MAP) SENSOR

### Description & Operation

The PCM supplies 5 volts of direct current to the Manifold Absolute Pressure (MAP) sensor. The MAP sensor then converts the intake manifold pressure into voltage. The PCM monitors the MAP sensor output voltage. As vacuum increases, the MAP sensor voltage decreases proportionately. Also, as vacuum decreases, the MAP sensor voltage increases proportionally.

With the ignition key ON, before the engine is started, the PCM determines atmospheric air pressure from the MAP sensor voltage. While the engine operates, the PCM figures out intake manifold pressure from the MAP sensor voltage. Based on the MAP sensor voltage and inputs from other sensors, the PCM adjusts spark advance and the air/fuel ratio. The MAP sensor is mounted to the intake manifold, near the throttle body inlet to the manifold. The sensor connects electrically to the PCM.

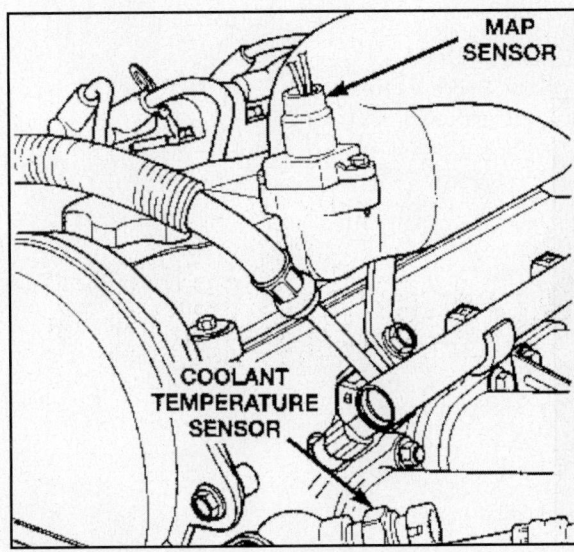

MAP sensor location on 2.4L engine

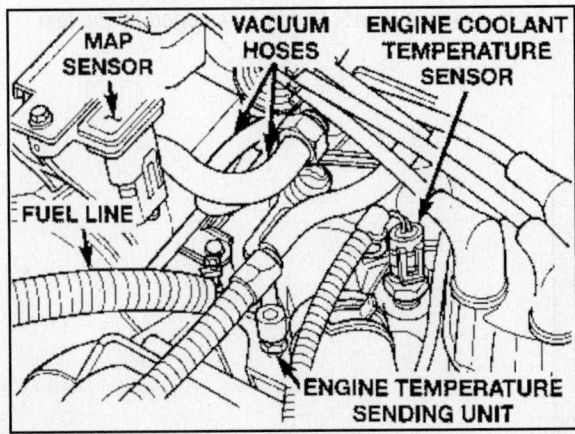

MAP sensor location on 3.0L engine

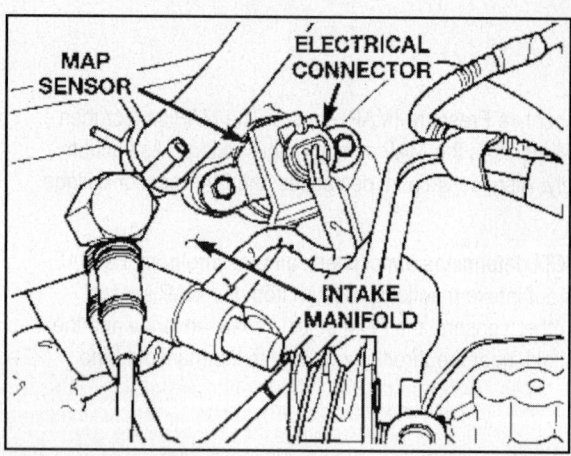

MAP sensor location on 3.3L and 3.8L engines

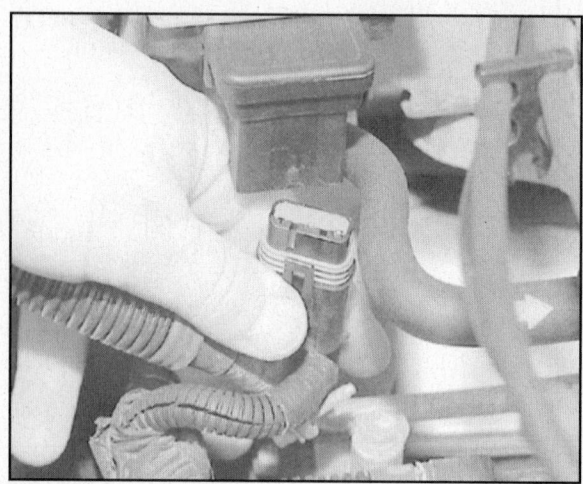

Disengage the MAP sensor connector (3.0L engine shown)

*WARNING: When Testing the MAP sensor, make sure the harness wires do not become damaged by the test meter probes.*

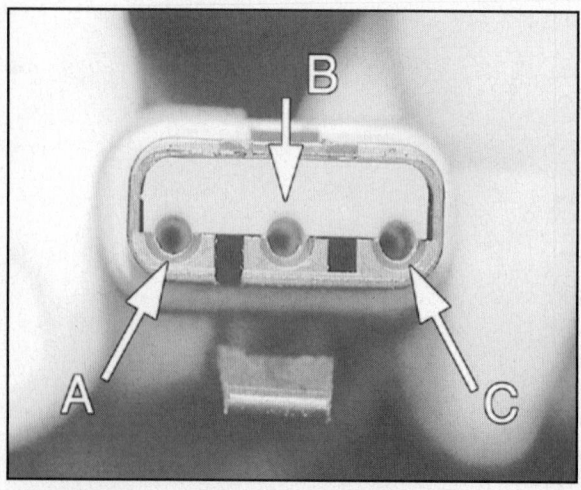

MAP sensor connector terminal identifications: (A) 5-volt supply, (B) sensor signal, (C) ground (3.0L engine application shown)

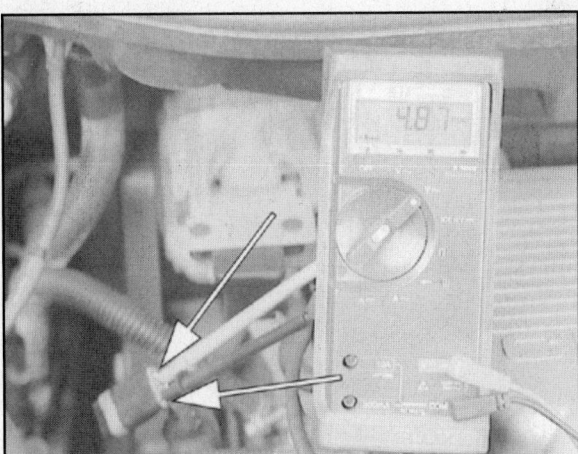

**Using a digital volt-ohmmeter, test the MAP sensor voltage-3.0L engine shown**

### Testing

1. Visually check the connector, making sure it is attached properly and that all of the terminals are straight, tight and free of corrosion.

2. Test the MAP sensor output voltage at the sensor connector between terminals B and C (2.4L, 3.3L and 3.8L engines), or A and B (3.0L engine).

3. With the ignition switch ON and the engine not running, the output voltage should be 4-5 volts. The voltage should fall to 1.5-2.1 volts with a hot, neutral idle speed condition. If OK, go to the next step. If not OK, go to Step 5.

4. Test the PCM terminal 36 for the same voltage described in the previous step to make sure the wire harness is OK. Repair as necessary.

5. Test the MAP sensor ground circuit at the sensor connector terminal A (2.4L, 3.3L and 3.8L engines) or C (3.0L engine) and PCM terminal 43. If OK, go to the next step. If not OK, repair as necessary.

6. Test the MAP sensor supply voltage between the sensor connector terminals A and B (2.4L, 3.3L and 3.8L engines), or A and C (3.0L engine) with the ignition key in the ON position. The voltage should be about 4.5-5.5 volts.

7. There should also be 4.5-5.5 volts at terminal 61 of the PCM. If OK, replace the MAP sensor.

8. If not, repair or replace the wire harness as required.

## MAP SENSOR

### Connector Pinouts

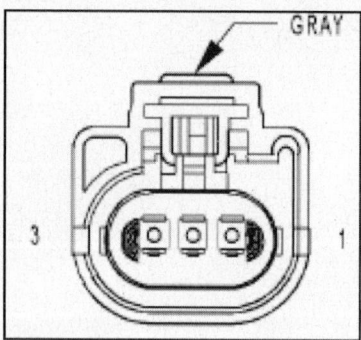

| PIN # | WIRE COLOR | CIRCUIT DESCRIPTION |
|-------|-----------|---------------------|
| 1 | VT/BR | MAP SIGNAL |
| 2 | DB/DG | SENSOR GROUND |
| 3 | PK/YL | 5 VOLT SUPPLY |

**MAP Sensor**

## NATURAL VACUUM LEAK DETECTION (NVLD) PUMP

### Removal & Installation

1. Disconnect the negative battery cable.
2. Raise and support the vehicle.
3. Remove 3 hoses.
4. Remove the electrical connector.
5. Remove the 3 screws and remove LDP pump.

**To install:**

6. Install LDP.
7. Install the 3 screws and tighten.
8. Install the electrical connector.
9. Install the 3 hoses.
10. Lower vehicle.
11. Connect the negative battery cable.

**Connector Pinouts**

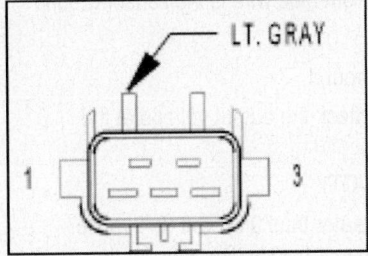

| PIN # | WIRE COLOR | CIRCUIT DESCRIPTION |
|-------|------------|---------------------|
| 1 | BK/WT | GROUND |
| 2 | VT/WT | NVLD SWITCH SIGNAL |
| 3 | VT/LB | NVLD SOL CONTROL |

**NVLD Connector Pinout**

## OIL PRESSURE SWITCH

**Connector Pinouts**

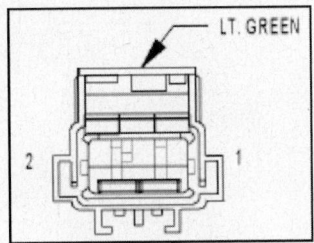

| PIN # | WIRE COLOR | CIRCUIT DESCRIPTION |
|-------|------------|---------------------|
| 1 | VT/GY | OIL PRESSURE SIGNAL |
| 2 | - | NOT USED |

**Oil Pressure Switch Connector Pinout**

## THROTTLE POSITION (TP) SENSOR

### Description & Operation

The Throttle Position Sensor (TPS) is mounted to the side of the throttle body and connects to the throttle blade shaft. The TPS is a variable resistor that provides the PCM with an input signal (voltage). The signal represents throttle blade position. As the position of the throttle blade changes, the resistance of the TPS changes.

The PCM supplies about 5 volts of DC current to the TPS. The TPS output voltage (input signal to the PCM) represents throttle blade position. The TPS output voltage to the PCM varies from about 0.5 volt at idle to a maximum of 4.0 volts at wide open throttle. The PCM uses the TPS input, and other sensor input, to determine current engine operating conditions. The PCM also adjusts fuel injector pulse width and ignition timing based on these inputs.

### Testing

In order to perform a complete test of the TPS and related circuits, you must use a DRB® or equivalent scan tool, and follow the manufacturer's directions. To check the Throttle Position Sensor (TPS) only, proceed with the following tests.

Visually check the connector, making sure it is attached properly and that all of the terminals are straight, tight and free of corrosion.

The TPS can be tested using a digital ohmmeter. The center terminal of the sensor supplies the output voltage. The outer terminal with the violet/white wire is the 5-volt supply terminal and the black/light blue wire is the sensor ground terminal.

1. Connect the DVOM between the center terminal and sensor ground.

2. With the ignition key to the ON position and the engine OFF, check the output voltage at the center terminal wire of the connector.

3. Check the output voltage at idle and at Wide Open Throttle (WOT):

4. For 1996 vehicles at idle, the TPS output voltage should be greater than 0.35 volt (0.4 volt for the 2.4L engine). At WOT, the output voltage should be less than 4.5 volts (3.8 volts for the 2.4L engine).

5. For 1997 and later vehicles at idle, the TPS output voltage should be about 0.38-1.20 volts. At WOT, the output voltage should be about 3.1-4.4 volts.

6. The output voltage should gradually increase as the throttle plate moves slowly from idle to WOT.

7. If voltage measures outside these values, replace the TPS.

8. Before replacing the TPS, check for spread terminals and also inspect the PCM connections.

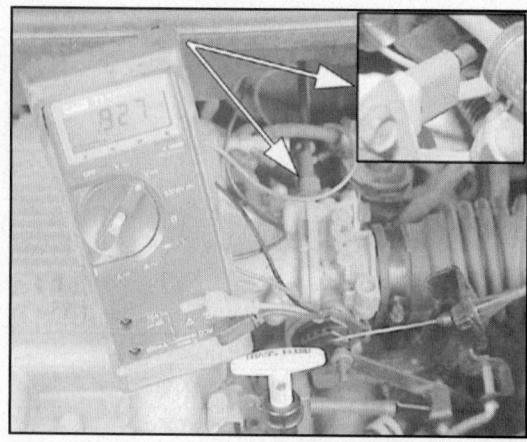

**TPS output voltage should be greater than 0.35 volt (0.4 volt for the 2.4L engine)**

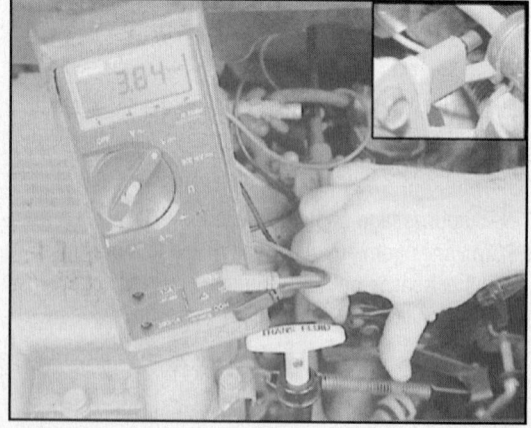

**At WOT (Wide Open Throttle), the output voltage should be less than 4.5 volts (3.8 volts for the 2.4L engine)**

**Connector Pinouts**

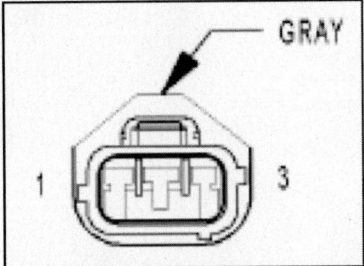

| PIN # | WIRE COLOR | CIRCUIT DESCRIPTION |
|-------|------------|---------------------|
| 1 | DB/DG | SENSOR GROUND |
| 2 | BR/OR | TP SIGNAL |
| 3 | PK/YL | 5 VOLT SUPPLY |

**TP Sensor Connector Pinout**

**VEHICLE SPEED SENSOR (VSS)**

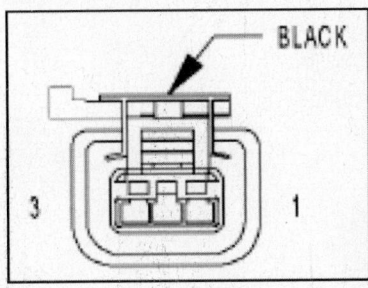

| 1 | PK/YL (2.4L) | 5 VOLT SUPPLY |
|---|--------------|----------------|
| 1 | PK/GY (DIESEL) | FUSED IGNITION SWITCH OUTPUT (RUN-START) |
| 2 | DB/DG (2.4L) | SENSOR GROUND |
| 2 | DB/DG (DIESEL) | SENSOR GROUND |
| 3 | DB/OR (2.4L) | VEHICLE SPEED SIGNAL |
| 3 | DB/OR (DIESEL) | VEHICLE SPEED SENSOR SIGNAL |

**Vehicle Speed Sensor (with M/T)**

## DAIMLERCHRYSLER SUV: DAKOTA

BATTERY TEMPERATURE SENSOR CONNECTORS

**Connector Pinouts**

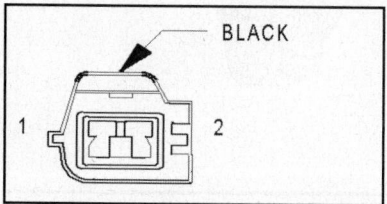

| PIN # | CIRCUIT | FUNCTION |
|-------|---------|----------|
| 1 | K4 18BK/LB | SENSOR GROUND |
| 2 | K118 20PK/YL 2.5L/3.9L/5.9L | BATTERY TEMPERATURE SENSOR SIGNAL |
| 2 | K118 20PK/YL 4.7L | AAT SIGNAL |

**3.9L Battery Temperature Sensor**

BODY CONTROL MODULE

**Connector Pinouts**

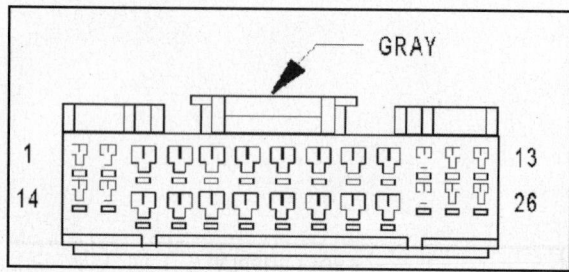

| PIN # | CIRCUIT | FUNCTION |
|-------|---------|----------|
| 1 | Z2 18BK/LG | GROUND |
| 2 | Z1 14BK | GROUND |
| 3 | M3 20PK/DB | CARGO LAMP SWITCH SENSE |
| 4 | L79 20RD/YL | PARK LAMP RELAY CONTROL |
| 5 | X3 20BK/RD | HORN RELAY CONTROL |
| 6 | G69 20BK/OR | VTSS INDICATOR DRIVER |
| 7 | L308 22LG/RD | PARK LAMP SWITCH SENSE |
| 8 | V52 18DG/RD | INTERMITTENT FRONT WIPER MODE SENSE |
| 9 | M4 18GY/BK | INTERIOR LAMP DEFEAT |
| 10 | D25 20VT/YL | PCI BUS |
| 11 | L4 18VT/WT | LOW BEAM SWITCH OUTPUT |
| 12 | X20 20RD/BK | RADIO CONTROL MUX |

**2003 Dakota 3.9L Body Control Module C1 Central Timer Module Connector Pinout (1 of 2)**

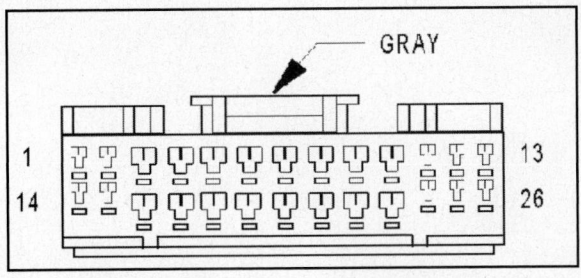

| PIN # | CIRCUIT | FUNCTION |
|-------|---------|----------|
| 13 | V51 20WT | INTERMITTENT FRONT WIPER SWITCH SIGNAL |
| 14 | Y158 20YL/VT | INTERIOR LAMP DRIVER |
| 15 | V10 18BR | FRONT WASHER PUMP/MOTOR CONTROL |
| 16 | Z2 18BK/LG | GROUND |
| 17 | - | NOT USED |
| 18 | G34 22RD/GY | HIGH BEAM INDICATOR DRIVER |
| 19 | G26 22LB | KEY-IN IGNITION SWITCH SENSE |
| 20 | - | NOT USED |
| 21 | M11 20PK/LB | COURTESY LAMP SWITCH SENSE |
| 22 | L27 20WT/TN | FOG LAMP SWITCH SENSE |
| 23 | L80 18WT/DG | HEADLAMP SWITCH OFF SENSE |
| 24 | L3 18RD/OR | HIGH BEAM SWITCH OUTPUT |
| 25 | F12 20DB/WT | FUSED IGNITION SWITCH OUTPUT (RUN-START) |
| 26 | V6 18DB | FUSED IGNITION SWITCH OUTPUT (RUN-ACC) |

**2003 Dakota 3.9L Body Control Module C1 Central Timer Module Connector Pinout (2 of 2)**

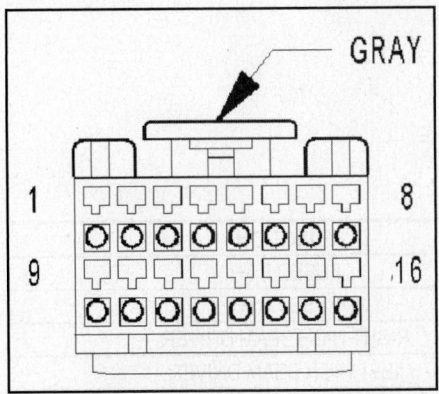

| PIN # | CIRCUIT | FUNCTION |
|-------|---------|----------|
| 1 | M2 20YL | COURTESY LAMP DRIVER |
| 2 | P96 20WT/LG | PASSENGER DOOR SWITCH MUX |
| 3 | Y192 20YL/RD | CARGO LAMP DRIVER |
| 4 | P33 18OR/BK | DOOR LOCK RELAY OUTPUT |
| 5 | P34 18PK/BK | DOOR UNLOCK RELAY OUTPUT |

**2003 Dakota 3.9L BCM C2 Central Timer Module Connector Pinout (1 of 2)**

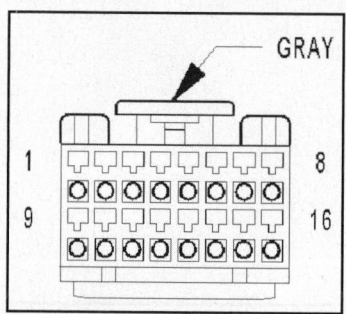

| PIN # | CIRCUIT | FUNCTION |
|---|---|---|
| 6 | P59 18DB | DRIVER DOOR UNLOCK RELAY OUTPUT |
| 7 | - | - |
| 8 | G11 20WT/BK | PARK BRAKE SWITCH SENSE |
| 9 | - | - |
| 10 | P97 18WT/DG (2 DOOR) | DRIVER DOOR SWITCH MUX |
| 10 | P97 20WT/DG (4 DOOR) | DRIVER DOOR SWITCH MUX |
| 11 | - | - |
| 12 | - | - |
| 13 | G74 20TN/RD | PASSENGER DOOR AJAR SWITCH SENSE |
| 14 | G73 22LG/OR | CYLINDER LOCK SWITCH MUX |
| 15 | G75 22TN | DRIVER DOOR AJAR SWITCH SENSE |
| 16 | Y158 20YL/VT | INTERIOR LAMP DRIVER |

**2003 Dakota 3.9L BCM C2 Central Timer Module Connector Pinout**

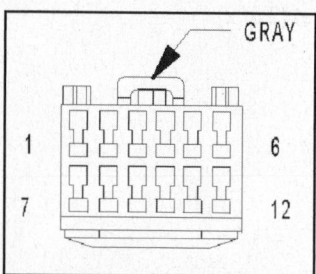

| PIN # | CIRCUIT | FUNCTION |
|---|---|---|
| 1 | A302 16RD/TN (3.9L/5.9L) | FUSED B (+) |
| 1 | A302 16RD/LG (4.7L) | FUSED B (+) |
| 2 | L34 18RD/OR | RIGHT HIGH BEAM DRIVER |
| 3 | L33 18LG/BR | LEFT HIGH BEAM DRIVER |
| 4 | L43 18VT | LEFT LOW BEAM DRIVER |
| 5 | L44 18VT/RD | RIGHT LOW BEAM DRIVER |
| 6 | A301 16RD/TN (4.7L) | FUSED B (+) |
| 6 | A301 16RD/LG (3.9L/5.9L) | FUSED B (+) |
| 7 | V18 18YL/DG | FRONT WIPER RELAY CONTROL |
| 8 | L26 20WT/VT | FOG LAMP RELAY CONTROL |
| 9 | - | NOT USED |
| 10 | V5 18DG | FRONT WIPER PARK SWITCH SENSE |
| 11 | V5 18DG | FRONT WIPER PARK SWITCH SENSE |
| 12 | V10 18BR | FRONT WASHER PUMP/MOTOR CONTROL |

**2003 Dakota 3.9L BCM C3 Central Timer Module Connector Pinout**

CAMSHAFT POSITION (CMP) SENSOR

## Description & Operation

### EXC. 4.7L

The Camshaft Position (CMP) sensor is located in the distributor. The sensor contains a Hall effect device called a sync signal generator to generate a fuel sync signal. This sync signal generator detects a rotating pulse ring (shutter) on the distributor shaft. The pulse ring rotates 180 degrees through the sync signal generator. Its signal is used in conjunction with the Crankshaft Position (CKP) sensor to differentiate between fuel injection and spark events. It is also used to synchronize the fuel injectors with their respective cylinders.

When the leading edge of the pulse ring (shutter) enters the sync signal generator, the following occurs: The interruption of magnetic field causes the voltage to switch high resulting in a sync signal of approximately 5 volts. When the trailing edge of the pulse ring (shutter) leaves the sync signal generator, the following occurs: The change of the magnetic field causes the sync signal voltage to switch low to 0 volts.

### 4.7L

The Camshaft Position Sensor (CMP) on the 4.7L V–8 engine is bolted to the front/top of the right cylinder head. The CMP sensor contains a hall effect device called a sync signal generator to generate a fuel sync signal. This sync signal generator detects notches located on a tone wheel. The tone wheel is located at the front of the camshaft for the right cylinder head. As the tone wheel rotates, the notches pass through the sync signal generator. The pattern of the notches (viewed counter-clockwise from front of engine) is: 1 notch, 2 notches, 3 notches, 3 notches, 2 notches 1 notch, 3 notches and 1 notch. The signal from the CMP sensor is used in conjunction with the crankshaft position sensor to differentiate between fuel injection and spark events. It is also used to synchronize the fuel injectors with their respective cylinders.

## Removal & Installation

### EXC. 4.7L

The camshaft position sensor is located in the distributor on all 2.5/3.9/5.2/5.9L engines.

*Note: Distributor removal is not necessary to remove camshaft position sensor.*

1. Remove air cleaner assembly.
2. Disconnect negative cable from battery.
3. Remove distributor cap from distributor (two screws).
4. Disconnect camshaft position sensor wiring harness from main engine wiring harness.
5. Remove distributor rotor from distributor shaft.
6. Lift the camshaft position sensor assembly from the distributor housing.

**To install:**

7. Install camshaft position sensor to distributor. Align sensor into notch on distributor housing.
8. Connect wiring harness.
9. Install rotor.
10. Install distributor cap. Tighten mounting screws.
11. Install air cleaner assembly.

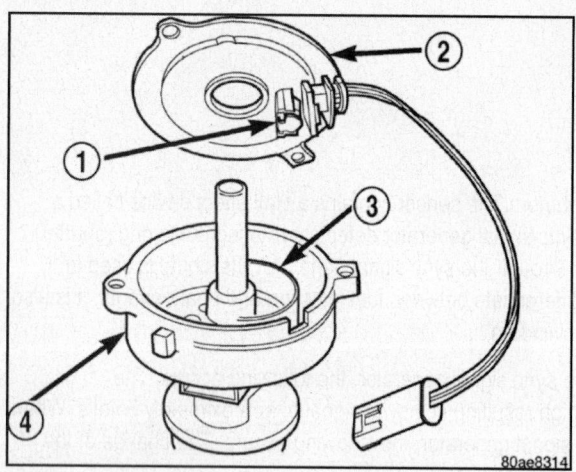

**Identifying the synch signal generator (1), CMP sensor (2), pulse ring (3) and distributor assembly (4) on 2.5L, 3.9L, 5.2L and 5.9L engines**

### 4.7L

The Camshaft Position Sensor (CMP) on the 4.7L V–8 engine is bolted to the front/top of the right cylinder head.

*Note: It is easier to remove/install sensor from under vehicle.*

1.  Raise and support vehicle.
2.  Disconnect electrical connector at CMP sensor.
3.  Remove sensor mounting bolt.
4.  Carefully twist sensor from cylinder head.
5.  Check condition of sensor O-ring.

**To install:**

The Camshaft Position Sensor (CMP) on the 4.7L V–8 engine is bolted to the front/top of the right cylinder head.

6.  Clean out machined hole in cylinder head.
7.  Apply a small amount of engine oil to sensor O-ring.
8.  Install sensor into cylinder head with a slight rocking action. Do not twist sensor into position as damage to O-ring may result.

*CAUTION: Before tightening sensor mounting bolt, be sure sensor is completely flush to cylinder head. If sensor is not flush, damage to sensor mounting tang may result.*

9.  Install mounting bolt and tighten to 12 Nm (106 inch lbs.) torque.
10.  Connect electrical connector to sensor.
11.  Lower vehicle.

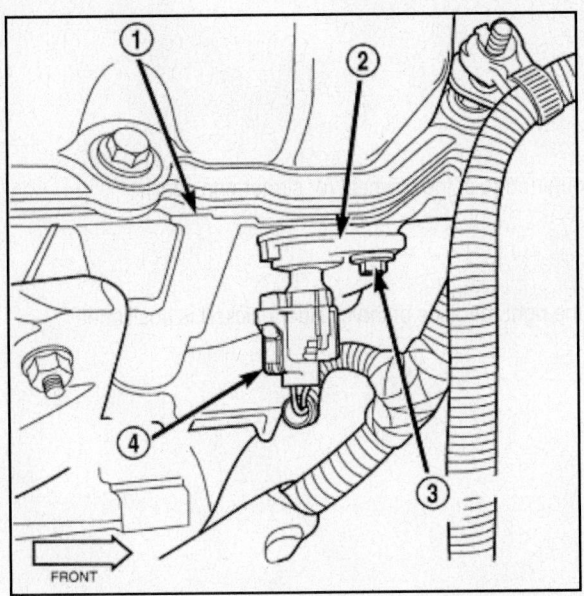

Identifying the CMP sensor (2) location on the right cylinder head (1); also showing the CMP sensor mounting bolt (3) and electrical connector (4) on 4.7L engine

**Connector Pinouts**

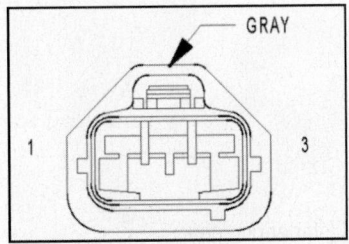

| PIN # | CIRCUIT | FUNCTION |
|-------|---------|----------|
| 1 | K44 20DB/GY (4.7L) | CMP SIGNAL |
| 1 | K44 20TN/YL (3.9L/5.9L) | CAMSHAFT POSITION SENSOR SIGNAL |
| 2 | K4 16BK/LB (4.7L) | SENSOR GROUND |
| 2 | K4 18BK/LB (3.9L/5.9L) | SENSOR GROUND |
| 3 | K6 16VT/WT (4.7L) | 5 VOLT SUPPLY |
| 3 | K7 18OR (3.9L/5.9L) | 5 VOLT SUPPLY |

**2003 Dakota 3.9L, 4.7L & 5.9L CMP Sensor**

| PIN # | CIRCUIT | FUNCTION |
|-------|---------|----------|
| 1 | K44 20DB/GY | CMP SIGNAL |
| 2 | K4 16BK/LB | SENSOR GROUND |
| 3 | K6 16VT/WT | 5 VOLT SUPPLY |

**2003-04 Dakota 3.7L CMP Sensor**

CKP SENSOR CONNECTOR

## DESCRIPTION

*2.5L*

The Crankshaft Position (CKP) sensor is located near the outer edge of the flywheel (or starter ring gear).

*3.7L V-6*

The Crankshaft Position (CKP) sensor (2) is mounted into the right rear side of the cylinder block. It is positioned and bolted into a machined hole.

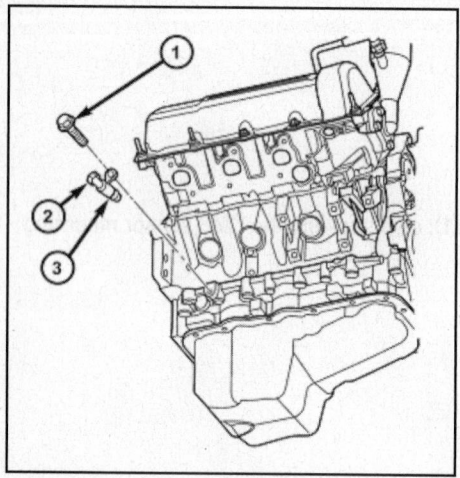

**Showing location of CKP Sensor on 3.7L engine**

*3.9L*

The Crankshaft Position (CKP) sensor is located near the outer edge of the flywheel (starter ring gear).

*4.7L*

The Crankshaft Position Sensor (CKP) is mounted into the right-rear side of the engine block.

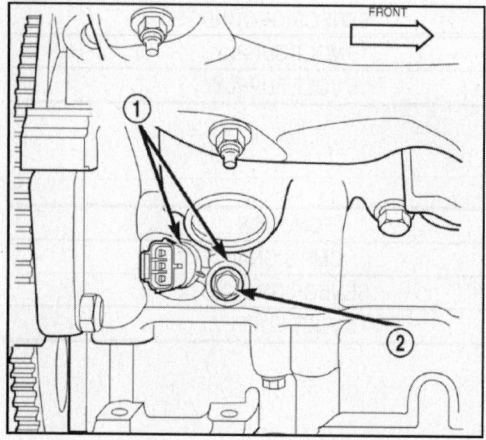

**Showing Location of CKP Sensor on 4.7L engine**

### 5.2L & 5.9L

The Crankshaft Position (CKP) sensor is located near the outer edge of the flywheel (starter ring gear).

### Operation

Engine speed and crankshaft position are provided through the CKP sensor. The sensor generates pulses that are the input sent to the Powertrain Control Module (PCM). The PCM interprets the sensor input to determine the crankshaft position. The PCM then uses this position, along with other inputs, to determine injector sequence and ignition timing.

The sensor is a hall effect device combined with an internal magnet. It is also sensitive to steel within a certain distance from it.

### 2.5L

The flywheel/drive plate has groups of four notches at its outer edge. On 2.5L 4-cylinder engines there are two sets of notches. The notches cause a pulse to be generated when they pass under the sensor. The pulses are the input to the PCM. For each engine revolution there are two groups of four pulses generated on 2.5L 4-cylinder engines.

The trailing edge of the fourth notch, which causes the pulse, is four degrees before top dead center (TDC) of the corresponding piston. The engine will not operate if the PCM does not receive a CKP sensor input.

### 3.7L V-6 & 4.7L V8

A tone wheel (target wheel) is bolted to the engine crankshaft (1). This tone wheel has sets of notches (2) at its outer edge. The notches cause a pulse to be generated when they pass under the sensor. The pulses are the input to the PCM.

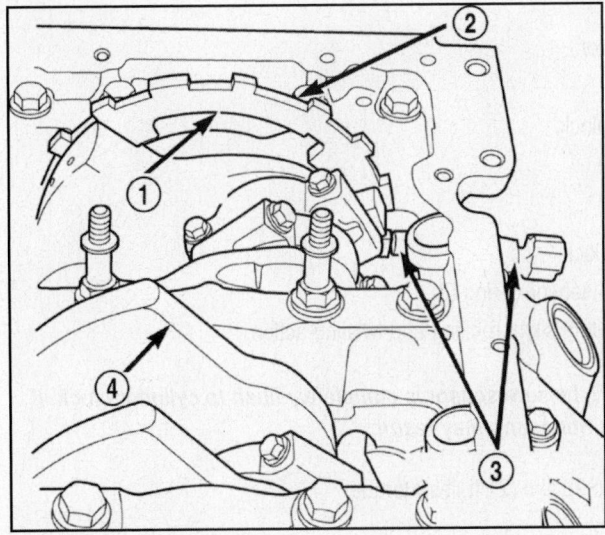

**Indicating CKP Related Components on 3.7L & 4.8L engines**

### 4.7L

On the 4.7L V–8 engine, a tone wheel is bolted to the engine crankshaft. This tone wheel has sets of notches at its outer edge. The notches cause a pulse to be generated when they pass under the sensor. The pulses are the input to the PCM.

### 5.2L & 5.9L

On 5.2/5.9L V-8 engines, the flywheel/drive plate has 8 single notches, spaced every 45 degrees, at its outer edge. The notches cause a pulse to be generated when they pass under the sensor. The pulses are the input to the PCM. For each engine revolution, there are 8 pulses generated on V-8 engines. The engine will not operate if the PCM does not receive a CKP sensor input.

### Removal & Installation

**2.5L**

1. Remove air tube between throttle body and air cleaner housing.
2. Near rear of intake manifold, disconnect pigtail harness (on the sensor) from main electrical harness.
3. Remove 2 sensor mounting bolts.
4. Remove sensor.
5. Remove clip from sensor wire harness.

**To install:**

6. Install sensor flush against opening in transmission housing.
7. Install and tighten two sensor mounting bolts to 12 Nm (9 ft. lbs.) torque.

**CAUTION: Two bolts are used to secure the sensor to transmission. These bolts are specially machined to correctly space the unit to flywheel. Do not attempt to install any other bolts.**

8. Connect electrical connector to sensor.
9. Install clip on sensor wire harness.
10. Install air tube between throttle body and air cleaner housing.

**3.7L**

1. Raise vehicle.
2. Disconnect sensor electrical connector.
3. Remove sensor mounting bolt (1).
4. Carefully twist sensor from cylinder block.
5. Check condition of sensor O-ring (3).

**To install:**

6. Clean out machined hole in engine block.
7. Apply a small amount of engine oil to sensor O-ring (3).
8. Install sensor (2) into engine block with a slight rocking and twisting action.

**CAUTION: Before tightening sensor mounting bolt (1), be sure sensor is completely flush to cylinder block. If sensor is not flush, damage to sensor mounting tang may result.**

9. Install mounting bolt (1) and tighten to 28 Nm (21 ft. lbs.) torque.
10. Connect electrical connector to sensor.
11. Lower vehicle.

**3.9L, 5.2L & 5.9L**

1. Remove right front tire and right front wheelhouse liner.
2. Disconnect crankshaft position sensor pigtail harness from main wiring harness.
3. Remove two sensor (recessed hex head) mounting bolts.
4. Remove sensor from engine.

**To install:**

5. Position crankshaft position sensor to engine.
6. Install mounting bolts and tighten to 8 Nm (70 inch lbs.) torque.
7. Connect main harness electrical connector to sensor.
8. Install right front tire and right front wheelhouse liner.

*4.7L*

1. Disconnect CKP electrical connector at sensor.

2. Remove CKP mounting bolt.

3. Carefully twist sensor from cylinder block.

4. Remove sensor from vehicle.

5. Check condition of sensor O-ring.

**To install:**

6. Clean out machined hole in engine block.

7. Apply a small amount of engine oil to sensor O-ring.

8. Install sensor into engine block with a slight rocking action. Do not twist sensor into position as damage to O-ring may result.

*CAUTION: Before tightening sensor mounting bolt, be sure sensor is completely flush to cylinder block. If sensor is not flush, damage to sensor mounting tang may result.*

9. Install mounting bolt and tighten to 28 Nm (21 ft. lbs.) torque.

10. Connect electrical connector to sensor.

**Connector Pinouts**

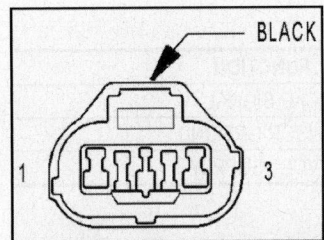

| CAV | CIRCUIT | FUNCTION |
|-----|---------|----------|
| 1 | K24 18GY/BK | CRANKSHAFT POSITION SENSOR SIGNAL |
| 2 | K4 18BK/LB | SENSOR GROUND |
| 3 | K7 18OR | 5V SUPPLY |

**2000-02 Dakota All Engines CKP Sensor**

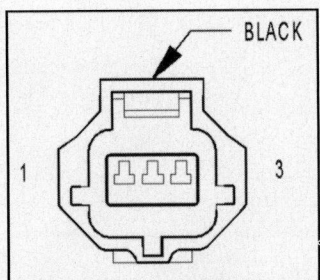

| PIN # | CIRCUIT | FUNCTION |
|-------|---------|----------|
| 1 | K24 20DB/WT | CKP SIGNAL |
| 2 | K4 18BK/LB | SENSOR GROUND |
| 3 | K7 18OR | 5 VOLT SUPPLY |

**2003-04 Dakota 3.7L CKP Sensor**

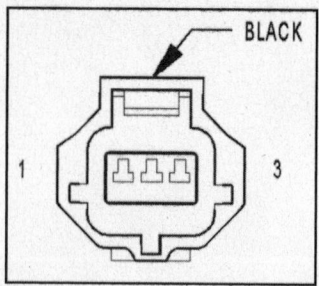

| CAV | CIRCUIT | FUNCTION |
|-----|---------|----------|
| 1 | K24 20BR/LB | CKP SIGNAL |
| 2 | K900 20DB/DG | SENSOR GROUND |
| 3 | F855 20PK/YL | 5 VOLT SUPPLY |

**2005-06 Dakota 3.7L CKP Sensor**

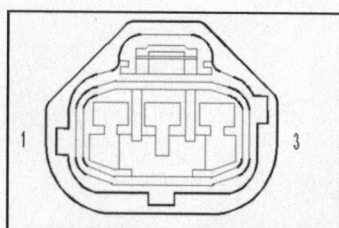

| CAV | CIRCUIT | FUNCTION |
|-----|---------|----------|
| 1 | K24 20BR/LB | CKP SIGNAL |
| 2 | K900 20DB/DG | SENSOR GROUND |
| 3 | F855 20PK/YL | 5 VOLT SUPPLY |

**2004-06 Dakota 4.7L CKP Sensor**

DATA LINK CONNECTORS

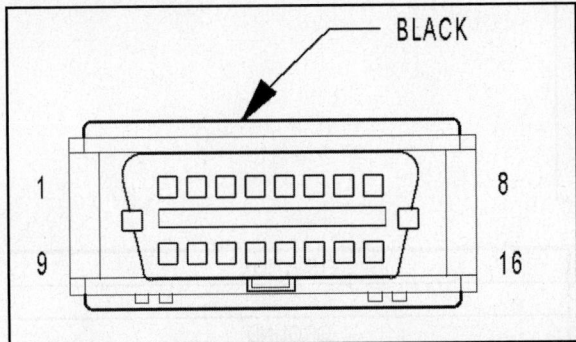

| PIN # | CIRCUIT | FUNCTION |
|-------|---------|----------|
| 1-2 | - | NOT USED |
| 3 | D1 20VT/BR | CCD BUS B(+) |
| 4 | Z11 18BK/WT | GROUND |
| 5 | Z12 16BK/TN | GROUND |
| 6 | D20 20LG | SCI RECEIVE |
| 7 | D21 20PK/DB | SCI TRANSMIT |
| 8-10 | - | NOT USED |
| 11 | D2 20WT/BK | CCD BUS(-) |
| 12-13 | - | NOT USED |
| 14 | D22 20PK/BK | SCI RECEIVE |
| 15 | - | NOT USED |
| 16 | M1 18PK | FUSED B(+) |

**2000-02 Dakota Data Link Connector Pinout**

| PIN # | CIRCUIT | FUNCTION |
|-------|---------|----------|
| 1 | - | NOT USED |
| 2 | D25 20VT/YL | PCI BUS |
| 3 | - | NOT USED |
| 4 | Z12 18BK/TN | GROUND |
| 5 | Z1 18BK | GROUND |
| 6 | - | NOT USED |
| 7 | D21 20PK/WT | SCI TRANSMIT (PCM) |
| 8 | - | NOT USED |
| 9 | D6 20PK/LB | SCI RECEIVE (TCM) |
| 10-11 | - | NOT USED |
| 12 | D20 20LG | SCI RECEIVE (PCM) |
| 13-14 | - | NOT USED |
| 15 | D15 18WT/DG | SCI TRANSMIT (TCM) |
| 16 | M1 20PK | FUSED B (+) |

**2003-04 Dakota Data Link Connector Pinout**

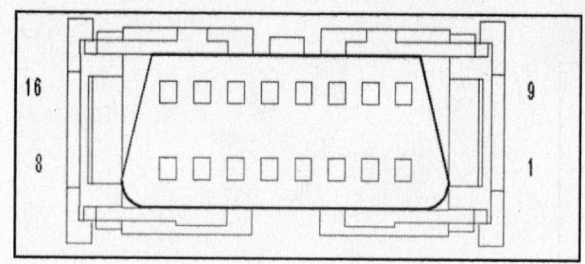

| CAV | CIRCUIT | FUNCTION |
|-----|---------|----------|
| 1-3 | - | - |
| 4 | Z11 18BK/LG | GROUND |
| 5 | Z111 18BK/WT | GROUND |
| 6 | D52 20WT/LB | CAN C DIAGNOSTIC (+) |
| 7 | D21 20WT/GY | SCI TRANSMIT (PCM) |
| 8 | - | - |
| 9 | D16 20WT/OR | SCI RECEIVE (TCM) |
| 10-11 | - | - |
| 12 | D20 20WT/LG | SCI RECEIVE (PCM) |
| 13 | - | - |
| 14 | D51 20PK/RD | CAN C DIAGNOSTIC (-) |
| 15 | D15 20BR/WT | SCI TRANSMIT (TCM) |
| 16 | A918 18RD/LB | FUSED B(+) |

**2005-06 Dakota Data Link Connector Pinout**

EVAP PURGE SOLENOID VALVE

**Connector Pinouts**

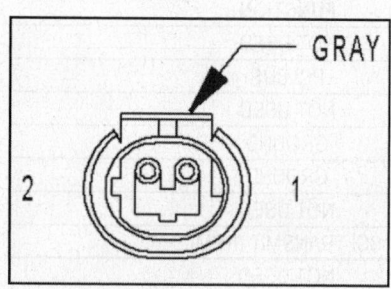

| CAV | CIRCUIT | FUNCTION |
|-----|---------|----------|
| 1 | K52 18PK/BK | DUTY CYCLE EVAP/PURGE SOLENOID CONTROL |
| 2 | F12 20DB/WT | FUSED IGNITION SWITCH OUTPUT (RUN-START) |

**2000-04 Dakota 2.5L, 3.9L, 4.7L (Exc. 03-06) & 5.9L EVAP Purge Solenoid Valve Connector Pinout**

**Connector Pinouts**

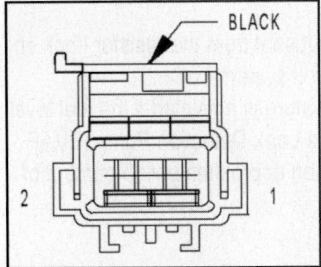

| PIN # | CIRCUIT | FUNCTION |
|-------|---------|----------|
| 1 | K70 18DB/BR | EVAP PURGE SIGNAL |
| 2 | K52 18PK/BK | EVAP/PURGE CONTROL |

**2003-04 Dakota 4.7L EVAP Purge Solenoid Valve Connector Pinout**

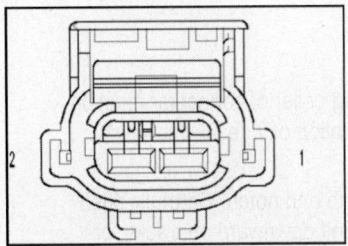

| CAV | CIRCUIT | FUNCTION |
|-----|---------|----------|
| 1 | K70 20DB/BR | EVAP PURGE SOL SIGNAL |
| 2 | K52 20DB/WT | EVAP PURGE SOL CONTROL |

**2005-06 Dakota 3.7L & 4.7L EVAP Purge Solenoid Valve Connector Pinout**

<u>FUEL LEVEL SENDING UNIT & SENSOR</u>

**Description & Operation**

The fuel gauge sending unit (fuel level sensor) is attached to the side of the fuel pump module. The sending unit consists of a float, an arm, and a variable resistor track (card).

The fuel pump module has 4 different circuits (wires). Two of these circuits are used for the fuel gauge sending unit for fuel gauge operation, and for certain OBD II emission requirements. The other 2 wires are used for electric fuel pump operation.

**For Fuel Gauge Operation:** A constant input voltage source of about 12 volts (battery voltage) is supplied to the resistor track on the fuel gauge sending unit. This is fed directly from the Powertrain Control Module (PCM).

*NOTE: For diagnostic purposes, this 12V power source can only be verified with the circuit opened (fuel pump module electrical connector unplugged). With the connectors plugged, output voltages will vary from about 0.6 volts at FULL, to about 8.6 volts at EMPTY (about 8.6 volts at EMPTY for Jeep models, and about 7.0 volts at EMPTY for Dodge Truck models).*

The resistor track is used to vary the voltage (resistance) depending on fuel tank float level. As fuel level increases, the float and arm move up, which decreases voltage. As fuel level decreases, the float and arm move down, which increases voltage. The varied voltage signal is returned back to the PCM through the sensor return circuit.

Both of the electrical circuits between the fuel gauge sending unit and the PCM are hard-wired (not multi-plexed). After the voltage signal is sent from the resistor track, and back to the PCM, the PCM will interpret the resistance (voltage) data and send a message across the multiplex bus circuits to the instrument panel cluster. Here it is translated into the appropriate fuel gauge level reading.

**For OBD II Emission Monitor Requirements:** The PCM will monitor the voltage output sent from the resistor track on the sending unit to indicate fuel level. The purpose of this feature is to prevent the OBD II system from recording/setting false misfire and fuel system monitor diagnostic trouble codes. The feature is activated if the fuel level in the tank is less than approximately 15 percent of its rated capacity. If equipped with a Leak Detection Pump (EVAP system monitor), this feature will also be activated if the fuel level in the tank is more than approximately 85 percent of its rated capacity.

### Removal & Installation

The fuel gauge sending unit (fuel level sensor) and float assembly is located on the side of fuel pump module. The fuel pump module is located inside of fuel tank.

#### ALL MODELS

1. Remove fuel tank.
2. Remove fuel pump module.

#### 2 DOOR MODELS

1. Unplug 4–way electrical connector.
2. Disconnect 2 sending unit wires at 4–way connector. The locking collar of connector must be removed before wires can be released from connector. Note location of wires within 4-way connector.
3. The sending unit is retained to pump module with a small lock tab and notch. Carefully push lock tab to the side and away from notch while sliding sending unit downward on tracks for Removal & Installation. Note wire routing while removing unit from module.

#### 4 DOOR MODELS

1. Remove electrical connector at sending unit terminals.
2. Press on release tab to remove sending unit from pump module.

### To install:

The fuel gauge sending unit (fuel level sensor) and float assembly is located on the side of fuel pump module. The fuel pump module is located inside of fuel tank.

### 2 Door Models

1. Position sending unit into tracks. Note wire routing.
2. Push unit on tracks until lock tab snaps into notch.
3. Connect 2 sending unit wires into 4–way connector and install locking collar.
4. Connect 4–way electrical connector to module.

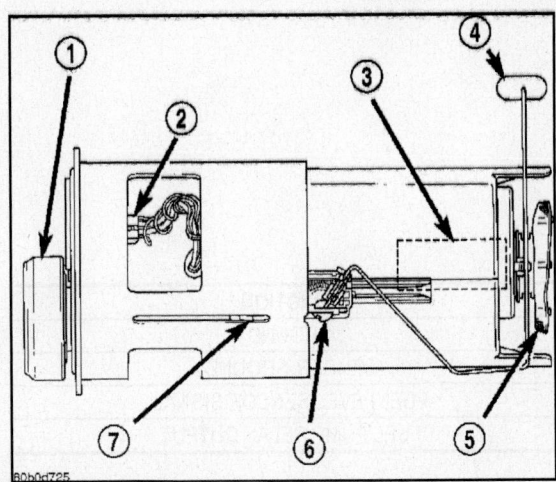

Indicating the 2-door model fuel pump module components: 1 – fuel filter/pressure regulator; 2 – electrical connector; 3 – electric fuel pump; 4 – fuel gauge float; 5 – fuel pump inlet filter; 6 – fuel gauge sending unit; 7 – module locking tabs

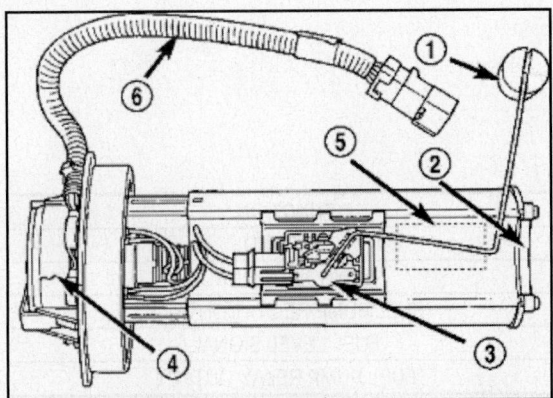

Indicating the 4-door model fuel pump module components: 1 – fuel gauge float; 2 – pickup filter; 3 – fuel gauge sending unit; 4 – fuel filter/pressure regulator; 5 – electrical fuel pump; 6 – wiring harness

### 4 DOOR MODELS

1. Position sending unit to pump module and snap into place.
2. Connect electrical connector to terminals.

### ALL MODELS

1. Install fuel pump module.
2. Install fuel tank.

**Connector Pinouts**

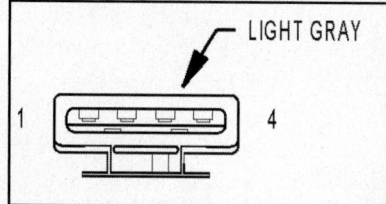

| PIN # | CIRCUIT | FUNCTION |
|-------|---------|----------|
| 1 | Z1 18BK | GROUND |
| 2 | K4 18BK/LB | SENSOR GROUND |
| 3 | G4 18DB | FUEL LEVEL SENSOR SIGNAL |
| 4 | A61 16DG/BK | FUEL PUMP RELAY OUTPUT |

**2000-02 Dakota Fuel Pump Module Connector Pinout**

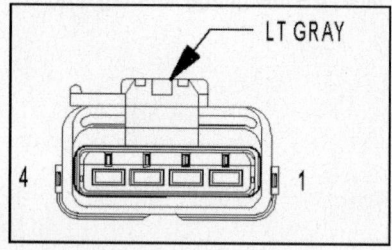

| PIN # | CIRCUIT | FUNCTION |
|-------|---------|----------|
| 1 | Z1 18BK | GROUND |
| 2 | K4 18BK/LB | SENSOR GROUND |
| 3 | K226 18DB/WT (3.9L/5.9L) | FUEL PUMP RELAY CONTROL |
| 3 | K226 18DB/WT (4.7L) | FUEL LEVEL SIGNAL |
| 4 | A61 16DG/BK | FUEL PUMP RELAY OUTPUT |

**2003-04 Dakota (Exc. 3.7L) Fuel Pump Module Connector Pinout**

| PIN # | CIRCUIT | FUNCTION |
|-------|---------|----------|
| 1 | Z1 18BK | GROUND |
| 2 | K4 18BK/LB | SENSOR GROUND 1 |
| 3 | K226 18DB/WT | FUEL LEVEL SIGNAL |
| 4 | A61 16DG/BK | FUEL PUMP RELAY OUTPUT |

**2003-04 Dakota 3.7L Fuel Pump Module Connector Pinout**

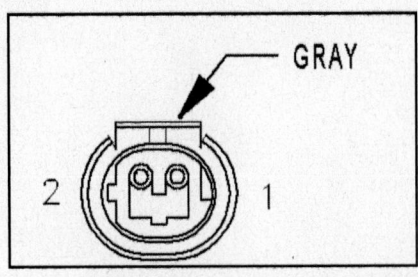

| PIN # | CIRCUIT | FUNCTION |
|---|---|---|
| 1 | F12 18DB/WT (2.5L) | FUSED IGNITION SWITCH OUTPUT (ST-RUN) |
| 1 | K52 18PK/BK (3.9L/4.7L/5.9L) | DUTY CYCLE EVAP/PURGE SOLENOID CONTROL |
| 2 | F12 18DB/WT (3.9L/4.7L/5.9L) | FUSED IGNITION SWITCH OUTPUT (ST-RUN) |
| 2 | K52 18PK/BK (2.5L) | DUTY CYCLE EVAP/PURGE SOLENOID CONTROL |

**2000 Dakota Fuel Pump Proportional Purge Solenoid Connector Pinout**

## GENERATOR

**Connector Pinouts**

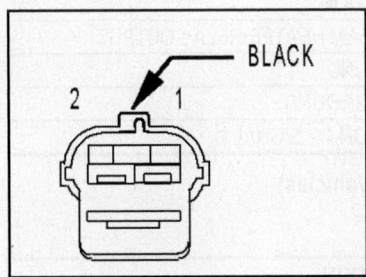

| PIN # | CIRCUIT | FUNCTION |
|---|---|---|
| 1 | K125 18WT/DB | GENERATOR SOURCE |
| 2 | K20 18DG | GENERATOR FIELD |

**2000 Dakota Generator Connector Pinout**

HEATED OXYGEN SENSOR (HO2S)

**Connector Pinouts**

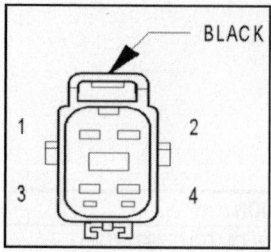

| CAV | CIRCUIT | FUNCTION |
|-----|---------|----------|
| 1 | F142 18OR/DG | FUSED AUTOMATIC SHUT DOWN RELAY OUTPUT |
| 2 | K100 18VT/WT | PWM 1/1 HEATER DRIVER |
| 3 | K4 18BK/LB | SENSOR GROUND |
| 4 | K41 18BK/DG | OXYGEN SENSOR 1/1 SIGNAL |

**2000-02 Dakota 2.5L, 3.7L, 4.7L, 5.9 1/1 Upstream O2 Sensor**

| CAV | CIRCUIT | FUNCTION |
|-----|---------|----------|
| 1 | F242 18DG/PK | OXYGEN SENSOR DOWNSTREAM HEATER RELAY OUTPUT |
| 2 | Z11 18BK/WT | GROUND |
| 3 | K4 18BK/LB | SENSOR GROUND |
| 4 | K141 18TN/WT | OXYGEN SENSOR 1/2 SIGNAL |

**2000-02 Dakota 2.5L, 3.7L, 4.7L 5.9L 1/2 Downstream O2 Sensor (California Vehicles)**

| CAV | CIRCUIT | FUNCTION |
|-----|---------|----------|
| 1 | F142 18OR/DG | FUSED AUTOMATIC SHUT DOWN RELAY OUTPUT |
| 2 | K200 18VT/OR | PWM 1/2 HEATER DRIVER |
| 3 | K4 18BK/LB | SENSOR GROUND |
| 4 | K141 18TN/WT | OXYGEN SENSOR 1/2 SIGNAL |

**2000-02 Dakota 2.5L, 3.7L, 4.7L, 5.9L 1/2 Downstream O2 Sensor (Except California Vehicles)**

| CAV | CIRCUIT | FUNCTION |
|-----|---------|----------|
| 1 | F142 18OR/DG | FUSED AUTOMATIC SHUT DOWN RELAY OUTPUT |
| 2 | K200 18VT/OR | PWM 2/1 HEATER |
| 3 | K4 18BK/LB | SENSOR GROUND |
| 4 | K241 18LG/RD | OXYGEN SENSOR 2/1 SIGNAL |

**2000-02 Dakota 2.5L, 3.7L, 4.7L, 5.9L 2/1 Upstream O2 Sensor (California Vehicles)**

| CAV | CIRCUIT | FUNCTION |
|-----|---------|----------|
| 1 | F242 18DG/PK | OXYGEN SENSOR DOWNSTREAM HEATER RELAY OUTPUT |
| 2 | Z11 18BK/WT | GROUND |
| 3 | K4 18BK/LB | SENSOR GROUND |
| 4 | K341 18TN/WT | OXYGEN SENSOR 2/2 SIGNAL |

**2000-02 Dakota 2.5L, 3.7L, 4.7L, 5.9L 2/2 Downstream O2 Sensor (California Vehicles)**

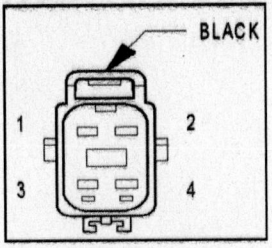

| PIN # | CIRCUIT | FUNCTION |
|---|---|---|
| 1 | K99 18BR/OR | O2 1/1 HEATER CONTROL |
| 2 | Z11 18BK/WT | GROUND |
| 3 | K902 18BK/DG | O2 RETURN (UP) |
| 4 | K41 18BK/DG | O2 1/1 SIGNAL |

**2003-04 Dakota 3.7L & 4.7L 1/1 Upstream O2 Sensor**

| PIN # | CIRCUIT | FUNCTION |
|---|---|---|
| 1 | K199 18BR/VT | O2 1/2 HEATER CONTROL |
| 2 | Z11 18BK/WT | GROUND |
| 3 | K904 18DB/DG | O2 RETURN (DOWN) |
| 4 | K141 18TN/WT | O2 1/2 SIGNAL |

**2003-04 Dakota 4.7L 1/2 Downstream O2 Sensor**

| PIN # | CIRCUIT | FUNCTION |
|---|---|---|
| 1 | K199 18BR/WT | O2 1/2 HEATER CONTROL |
| 2 | Z11 18BK/WT | GROUND |
| 3 | K904 18DB/DG | O2 RETURN (DOWN) |
| 4 | K141 18TN/WT | O2 1/2 SIGNAL |

**2003-04 Dakota 3.7L 1/2 Downstream O2 Sensor**

| PIN # | CIRCUIT | FUNCTION |
|---|---|---|
| 1 | K299 18BR/WT | O2 2/1 HEATER CONTROL |
| 2 | Z11 18BK/WT | GROUND |
| 3 | K902 18BR/DG | O2 RETURN (UP) |
| 4 | K241 18LG/RD | O2 2/1 SIGNAL |

**2003-04 Dakota 3.7L & 4.7L 2/1 Upstream O2 Sensor**

| PIN # | CIRCUIT | FUNCTION |
|---|---|---|
| 1 | K399 18BR/GY | O2 2/2 HEATER CONTROL |
| 2 | Z11 18BK/WT | GROUND |
| 3 | K904 18DB/DG | O2 RETURN (DOWN) |
| 4 | K341 18TN/WT | O2 2/2 SIGNAL |

**2003-04 Dakota 4.7L 2/2 Downstream O2 Sensor**

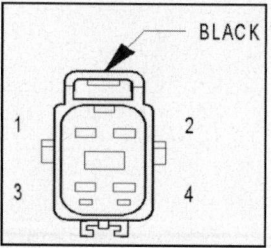

| PIN # | CIRCUIT | FUNCTION |
|---|---|---|
| 1 | K399 18BR/GY | O2 2/2 HEATER CONTROL |
| 2 | Z11 18BK/WT | GROUND |
| 3 | K904 18DB/DG | O2 RETURN (UP) |
| 4 | K341 18TN/WT | O2 2/2 SIGNAL |

**2003-04 Dakota 3.7L 2/2 Downstream O2 Sensor**

## IDLE AIR CONTROL (IAC) MOTOR

### Description & Operation

The IAC stepper motor is mounted to the throttle body, and regulates the amount of air bypassing the control of the throttle plate. As engine loads and ambient temperatures change, engine rpm changes. A pintle on the IAC stepper motor protrudes into a passage in the throttle body, controlling air flow through the passage. The IAC is controlled by the Powertrain Control Module (PCM) to maintain the target engine idle speed.

At idle, engine speed can be increased by retracting the IAC motor pintle and allowing more air to pass through the port, or it can be decreased by restricting the passage with the pintle and diminishing the amount of air bypassing the throttle plate.

The IAC is called a stepper motor because it is moved (rotated) in steps, or increments. Opening the IAC opens an air passage around the throttle blade which increases RPM. The PCM uses the IAC motor to control idle speed (along with timing) and to reach a desired MAP during decel (keep engine from stalling).

The IAC motor has 4 wires with 4 circuits. Two of the wires are for 12 volts and ground to supply electrical current to the motor windings to operate the stepper motor in one direction. The other 2 wires are also for 12 volts and ground to supply electrical current to operate the stepper motor in the opposite direction.

To make the IAC go in the opposite direction, the PCM just reverses polarity on both windings. If only 1 wire is open, the IAC can only be moved 1 step (increment) in either direction. To keep the IAC motor in position when no movement is needed, the PCM will energize both windings at the same time. This locks the IAC motor in place.

In the IAC motor system, the PCM will count every step that the motor is moved. This allows the PCM to determine the motor pintle position. If the memory is cleared, the PCM no longer knows the position of the pintle. So at the first key ON, the PCM drives the IAC motor closed, regardless of where it was before. This zeros the counter. From this point the PCM will back out the IAC motor and keep track of its position again.

When engine rpm is above idle speed, the IAC is used for the following:

- Off-idle dashpot (throttle blade will close quickly but idle speed will not stop quickly)
- Deceleration air flow control
- A/C compressor load control (also opens the passage slightly before the compressor is engaged so that the engine rpm does not dip down when the compressor engages)
- Power steering load control
- The PCM can control polarity of the circuit to control direction of the stepper motor.

**IAC Stepper Motor Program:** The PCM is also equipped with a memory program that records the number of steps the IAC stepper motor most recently advanced to during a certain set of parameters. For example: The PCM was attempting to maintain a 1000 rpm target during a cold start-up cycle. The last recorded number of steps for that may have been 125. That value would be recorded in the memory cell so that the next time the PCM recognizes the identical conditions, the PCM recalls that 125 steps were required to maintain the target. This program allows for greater customer satisfaction due to greater control of engine idle.

Another function of the memory program, which occurs when the power steering switch (if equipped), or the A/C request circuit, requires that the IAC stepper motor control engine rpm, is the recording of the last targeted steps into the memory cell. The PCM can anticipate A/C compressor loads. This is accomplished by delaying compressor operation for approximately 0.5 seconds until the PCM moves the IAC stepper motor to the recorded steps that were loaded into the memory cell. Using this program helps eliminate idle-quality changes as loads change. Finally, the PCM incorporates a "No-Load" engine speed limiter of approximately 1800 - 2000 rpm, when it recognizes that the TPS is indicating an idle signal and IAC motor cannot maintain engine idle.

A (factory adjusted) set screw is used to mechanically limit the position of the throttle body throttle plate. **Never attempt to adjust the engine idle speed using this screw.** All idle speed functions are controlled by the IAC motor through the PCM.

*NOTE: A separate IAC motor is not used on 5.7L engines.*

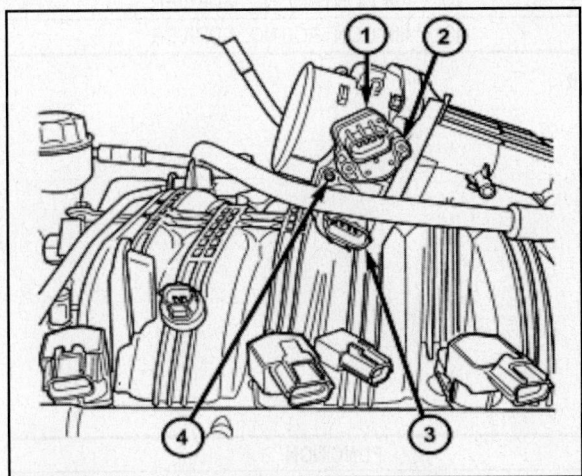

**Showing IAC Motor Location on 3.7L Engine**

### Removal & Installation

1. If necessary, remove the air duct and/or air resonator box at the throttle body.
2. Disconnect electrical connector from IAC motor.
3. Remove two mounting bolts (screws).
4. Remove IAC motor from throttle body.

### To install:

5. Install IAC motor to throttle body.
6. Install and tighten two mounting bolts (screws) to 7 Nm (60 inch lbs.) torque.
7. Install electrical connector.
8. If removed, install air duct/air box to throttle body.

## Connector Pinouts

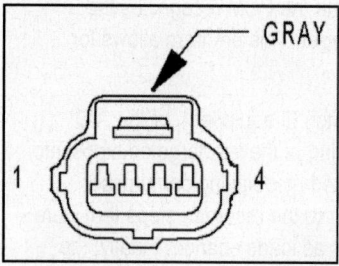

| CAV | CIRCUIT | FUNCTION |
|-----|---------|----------|
| 1 | K39 18GY/RD (2.5L) | IDLE AIR CONTROL NO. 1 DRIVER |
| 1 | K39 16GY/RD (4.7L) | IDLE AIR CONTROL NO. 1 DRIVER |
| 2 | K60 18YL/BK (2.5L) | IDLE AIR CONTROL NO. 2 DRIVER |
| 2 | K60 16YL/BK (4.7L) | IDLE AIR CONTROL NO. 2 DRIVER |
| 3 | K40 18BR/WT (2.5L) | IDLE AIR CONTROL NO. 3 DRIVER |
| 3 | K40 16BR/WT (4.7L) | IDLE AIR CONTROL NO. 3 DRIVER |
| 4 | K59 18VT/BK (2.5L) | IDLE AIR CONTROL NO. 4 DRIVER |
| 4 | K59 16VT/BK (4.7L) | IDLE AIR CONTROL NO. 4 DRIVER |

**2000-02 Dakota 2.5L & 4.7L IAC Motor Connector Pinout**

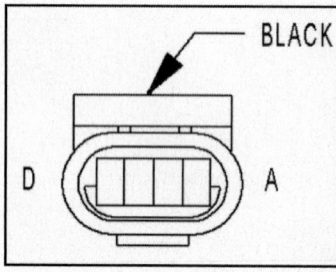

| CAV | CIRCUIT | FUNCTION |
|-----|---------|----------|
| D | K39 18GY/RD | IDLE AIR CONTROL NO. 1 DRIVER |
| C | K60 18YL/BK | IDLE AIR CONTROL NO. 2 DRIVER |
| B | K40 18BR/WT | IDLE AIR CONTROL NO. 3 DRIVER |
| A | K59 18VT/BK | IDLE AIR CONTROL NO. 4 DRIVER |

**2000-02 Dakota 3.9L & 5.9L IAC Motor Connector Pinout**

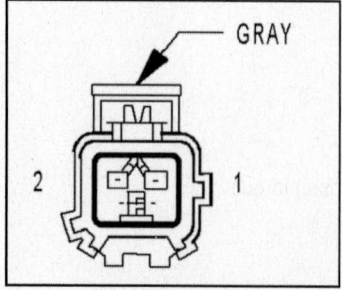

| PIN # | CIRCUIT | FUNCTION |
|-------|---------|----------|
| 1 | K60 16YL/BK | IAC RETURN |
| 2 | K39 16GY/RD | IAC MOTOR CONTROL |

**2003-04 Dakota 3.7L Idle Air Control Motor Connector Pinout**

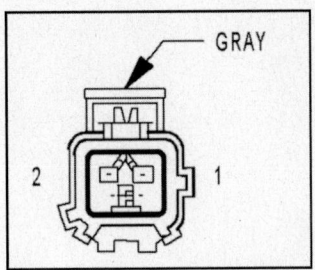

| CAV | CIRCUIT | FUNCTION |
|-----|---------|----------|
| 1 | K961 20BR/VT | IAC SIGNAL |
| 2 | K61 20VT/GY | IAC CONTROL |

**2005-06 Dakota 3.7L & 4.7L IAC Motor Connector Pinout**

IGNITION COIL CAPACITOR CONNECTORS

**Connector Pinouts**

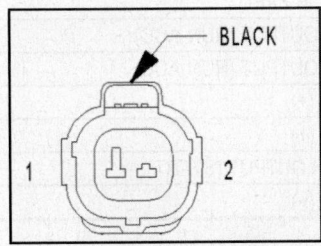

| PIN # | CIRCUIT | FUNCTION |
|-------|---------|----------|
| 1 | A142 16DG/OR | ASD RELAY OUTPUT |
| 2 | - | - |

**2003-04 Dakota 3.7L ignition Coil Capacitor Pinout**

IGNITION SWITCH CONNECTORS

**Connector Pinouts**

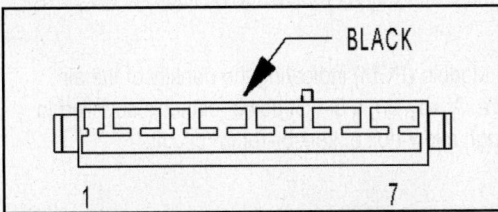

| PIN # | CIRCUIT | FUNCTION |
|-------|---------|----------|
| 1 | A41 14YL | IGNITION SWITCH OUTPUT (START) |
| 2 | A21 12DB | IGNITION SWITCH OUTPUT (ST-RUN) |
| 3 | - | NOT USED |
| 4 | A2 12PK/BK | FUSED B (+) |
| 5 | A22 12BK/OR | IGNITION SWITCH OUTPUT (RUN) |
| 6 | A31 12BK/WT | IGNITION SWITCH OUTPUT (RUN-ACC) |
| 7 | A1 12RD | FUSED (B+) |

**2000-02 Dakota 2.5L Ignition Switch Connector Pinout**

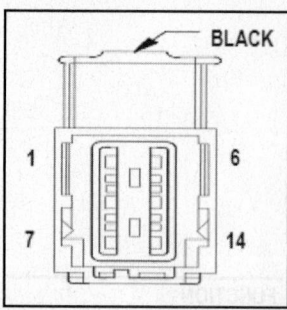

| PIN # | CIRCUIT | FUNCTION |
|-------|---------|----------|
| 1 | A111 10RD/LB | FUSED B (+) |
| 2 | A81 18DG/RD | FUSED IGNITION SWITCH OUTPUT (RUN-START) |
| 3 | A21 16DB | FUSED IGNITION SWITCH OUTPUT (RUN-START) |
| 4 | A19 16RD/YL | FUSED B (+) |
| 5 | G26 22LB | KEY-IN IGNITION SWITCH SENSE |
| 6 | Z1 20BK | GROUND |
| 7 | A22 12BK/OR | FUSED IGNITION SWITCH OUTPUT (RUN) |
| 8 | C1 12DG | BLOWER MOTOR FEED |
| 9 | A31 12BK/WT | FUSED IGNITION SWITCH OUTPUT (RUN-ACC) |
| 10 | A30 12RD/WT | FUSED IGNITION SWITCH OUTPUT (RUN-ACC) |
| 11 | A1 12RD | FUSED B (+) |
| 12 | A18 12RD/BK | FUSED B (+) |
| 13 | A41 16YL | FUSED IGNITION SWITCH OUTPUT (START) |
| 14 | A2 12PK/BK | FUSED B (+) |

**2003-04 Dakota 3.7L & 4.7L Ignition Switch Connector Pinout**

INTAKE AIR TEMPERATURE (IAT) SENSOR

## Description & Operation

The 2–wire Intake Manifold Air Temperature (IAT) sensor is installed in the intake manifold with the sensor element extending into the air stream.

The IAT sensor is a two-wire Negative Thermal Coefficient (NTC) sensor. Meaning, as intake manifold temperature increases, resistance (voltage) in the sensor decreases. As temperature decreases, resistance (voltage) in the sensor increases.

The IAT sensor provides an input voltage to the Powertrain Control Module (PCM) indicating the density of the air entering the intake manifold based upon intake manifold temperature. At key-on, a 5–volt power circuit is supplied to the sensor from the PCM. The sensor is grounded at the PCM through a low-noise, sensor-return circuit.

The PCM uses this input to calculate the following:

- Injector pulse-width
- Adjustment of spark timing (to help prevent spark knock with high intake manifold air-charge temperatures)
- The resistance values of the IAT sensor are the same as for the Engine Coolant Temperature (ECT) sensor.

**Removal & Installation**

*3.7L*

The intake manifold air temperature (IAT) sensor is installed into the left side of intake manifold plenum.

1. Disconnect electrical connector from IAT sensor.
2. Clean dirt from intake manifold at sensor base.
3. Gently lift on small plastic release tab and rotate sensor about 1/4 turn counter-clockwise for removal.
4. Check condition of sensor O-ring.

**To install:**

1. Check condition of sensor O-ring.
2. Clean sensor mounting hole in intake manifold.
3. Position sensor into intake manifold and rotate clockwise until past release tab.
4. Install electrical connector.

*4.7L*

The Intake Manifold Air Temperature (IAT) sensor is installed into the intake manifold plenum near the left side of the throttle body.

*Threaded Type Sensor*

1. Disconnect electrical connector from sensor .
2. Remove sensor from intake manifold.

*Snap-In Type Sensor*

1. Disconnect electrical connector from IAT sensor.
2. Clean dirt from intake manifold at sensor base.
3. Gently lift on small plastic release tab and rotate sensor about 1/4 turn counter-clockwise for removal.
4. Check condition of sensor O-ring.

**To install:**

*4.7L*

*Threaded Type Sensor*

5. Install sensor into intake manifold. Tighten sensor to 28 Nm (20 ft. lbs.) torque.
6. Connect electrical connector to sensor.

*Snap-In Type Sensor*

1. Check condition of sensor O-ring.
2. Clean sensor mounting hole in intake manifold.
3. Position sensor into intake manifold and rotate clockwise until past release tab.
4. Install electrical connector.

**Connector Pinouts**

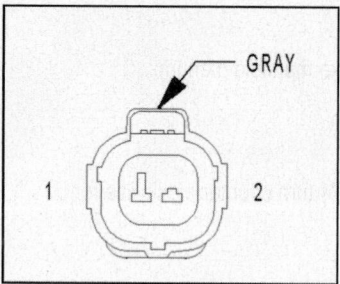

| CAV | CIRCUIT | FUNCTION |
|-----|---------|----------|
| 1 | K4 18BK/LB | SENSOR GROUND 1 |
| 2 | K21 20BK/RD | IAT SIGNAL |

**3.6L & 4.6L IAT Sensor Connector Pinout**

## KNOCK SENSOR CONNECTOR

**Connector Pinouts**

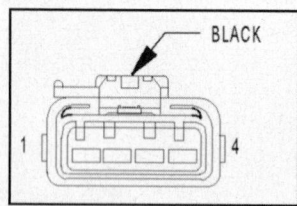

| PIN # | CIRCUIT | FUNCTION |
|-------|---------|----------|
| 1 | K42 18DB/YL | KNOCK SENSOR NO. 1 SIGNAL |
| 2 | K942 18BR/LG | KNOCK SENSOR NO. 1 RETURN |
| 3 | K242 18BR/WT | KNOCK SENSOR NO. 2 SIGNAL |
| 4 | K924 18WT/BR | KNOCK SENSOR NO. 2 RETURN |

**2004 3.7L Dakota Knock Sensor**

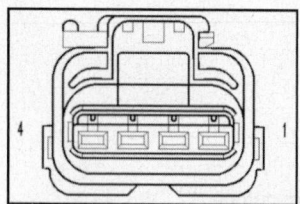

| CAV | CIRCUIT | FUNCTION |
|-----|---------|----------|
| 1 | K42 20DB/OR | KNOCK SENSOR NO. 1 SIGNAL |
| 2 | K942 20BR/LG | KNOCK SENSOR NO. 1 RETURN |
| 3 | K242 20BR/WT | KNOCK SENSOR NO. 2 SIGNAL |
| 4 | K924 20PK/RD | KNOCK SENSOR NO. 2 RETURN |

**2005 3.7L & 4.7L Dakota Knock Sensor**

LEAK DETECTION PUMP

*NOTE: On 2003 & later 4.7L engines, see NATURAL VACUUM LEAK DETECTION (NVLD).*

### Description & Operation

The Leak Detection Pump (LDP) is used only with certain emission packages. The LDP is a device used to detect a leak in the evaporative system. The pump contains a 3-port solenoid, a pump that contains a switch, a spring loaded canister vent valve seal, 2 check valves and a spring/diaphragm.

Immediately after a cold start, engine temperature between 40°F and 86°F, the 3 port solenoid is briefly energized. This initializes the pump by drawing air into the pump cavity and also closes the vent seal. During non-test test conditions, the vent seal is held open by the pump diaphragm assembly which pushes it open at the full travel position. The vent seal will remain closed while the pump is cycling. This is due to the operation of the 3 port solenoid which prevents the diaphragm assembly from reaching full travel. After the brief initialization period, the solenoid is de-energized, allowing atmospheric pressure to enter the pump cavity. This permits the spring to drive the diaphragm which forces air out of the pump cavity and into the vent system. When the solenoid is energized and de-energized, the cycle is repeated creating flow in typical diaphragm pump fashion. The pump is controlled in 2 modes:

### PUMP MODE

The pump is cycled at a fixed rate to achieve a rapid pressure build in order to shorten the overall test time.

### TEST MODE

The solenoid is energized with a fixed duration pulse. Subsequent fixed pulses occur when the diaphragm reaches the switch closure point.

The spring in the pump is set so that the system will achieve an equalized pressure of about 7.5 inches of water.

When the pump starts, the cycle rate is quite high. As the system becomes pressurized pump rate drops. If there is no leak the pump will quit. If there is a leak, the test is terminated at the end of the test mode.

If there is no leak, the purge monitor is run. If the cycle rate increases due to the flow through the purge system, the test is passed and the diagnostic is complete.

The canister vent valve will unseal the system after completion of the test sequence as the pump diaphragm assembly moves to the full travel position.

### Removal & Installation

The LDP is located in the engine compartment under the battery tray and Power Distribution Center (PDC). The LDP filter is attached to the outside of battery tray. The LDP and LDP filter are replaced (serviced) as one unit.

1. Disconnect negative battery cable at battery.
2. Remove battery.
3. Carefully disconnect rubber hose from bottom of LDP filter.
4. Remove clip retaining LDP filter to battery tray and remove filter from tray.
5. Disconnect battery temperature sensor pigtail wiring harness at bottom of battery tray.
6. To gain access to LDP, the PDC must be partially removed. Remove PDC-to-fender mounting screw at rear of PDC. Unsnap PDC from battery tray.
7. To prevent damage to PDC wiring, carefully position PDC to gain access to LDP.
8. Remove battery tray.
9. Carefully remove vapor/vacuum lines at LDP.
10. Disconnect electrical connector at LDP.
11. Remove 3 LDP mounting screws and remove LDP from vehicle.

**To install:**

12. Install LDP to bottom of battery tray. Tighten screws to 1 Nm (11 inch lbs..) torque.

13. Carefully install vapor/vacuum lines to LDP.

*CAUTION: The vapor/vacuum lines and hoses must be firmly connected. Check the vapor/vacuum lines at the LDP. LDP filter and EVAP canister purge solenoid for damage or leaks. If a leak is present, a Diagnostic Trouble Code (DTC) may be set.*

14. Connect electrical connector to LDP.
15. Install battery tray.
16. Install PDC to fender and battery tray (snaps on to battery tray).
17. Install LDP filter to battery tray (one clip).
18. Install connecting hose to bottom of LDP filter.
19. Connect battery temperature sensor pigtail wiring harness.
20. Install battery.
21. Connect negative battery cable to battery.

## Connector Pinouts

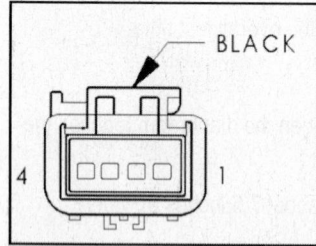

| CAV | CIRCUIT | FUNCTION |
|-----|---------|----------|
| 1 | - | - |
| 2 | K125 18WT/DB | GENERATOR SOURCE |
| 3 | K106 18WT/DG | LEAK DETECTION PUMP SOLENOID CONTROL |
| 4 | K107 18OR | LEAK DETECTION PUMP SWITCH SENSE |

**2000-04 Dakota (Exc. 2003 & Later 4.7L) EVAP Leak Detection Pump Connector Pinout**

## MAP SENSOR CONNECTORS

### Description

**EXCEPT 4.7L**

The Manifold Absolute Pressure (MAP) sensor is attached to the side of the engine throttle body with 2 screws. The sensor is connected to the throttle body with a rubber L-shaped fitting.

**4.7L**

The MAP sensor is located on the front of the intake manifold. An O-ring seals the sensor to the intake manifold.

### Operation

The MAP sensor is used as an input to the Powertrain Control Module (PCM). It contains a silicon based sensing unit to provide data on the manifold vacuum that draws the air/fuel mixture into the combustion chamber. The PCM requires this information to determine injector pulse width and spark advance. When manifold absolute pressure (MAP) equals Barometric pressure, the pulse width will be at maximum.

A 5-volt reference is supplied from the PCM and returns a voltage signal to the PCM that reflects manifold pressure. The zero pressure reading is 0.5V and full scale is 4.5V. For a pressure swing of 0–15 psi, the voltage changes 4.0V.

To operate the sensor, it is supplied a regulated 4.8 to 5.1 volts. Ground is provided through the low-noise, sensor return circuit at the PCM.

The MAP sensor input is the number one contributor to fuel injector pulse width. The most important function of the MAP sensor is to determine barometric pressure. The PCM needs to know if the vehicle is at sea level or at a higher altitude, because the air density changes with altitude. It will also help to correct for varying barometric pressure. Barometric pressure and altitude have a direct inverse correlation; as altitude goes up, barometric goes down. At key-on, the PCM powers up and looks at MAP voltage, and based upon the voltage it sees, it knows the current barometric pressure (relative to altitude). Once the engine starts, the PCM looks at the voltage again, continuously every 12 milliseconds, and compares the current voltage to what it was at key-on. The difference between current voltage and what it was at key-on, is manifold vacuum.

During key-on (engine not running) the sensor reads (updates) barometric pressure. A normal range can be obtained by monitoring a known good sensor.

As the altitude increases, the air becomes thinner (less oxygen). If a vehicle is started and driven to a very different altitude than where it was at key-on, the barometric pressure needs to be updated. Any time the PCM sees Wide Open Throttle (WOT), based upon Throttle Position Sensor (TPS) angle and RPM, it will update barometric pressure in the MAP memory cell. With periodic updates, the PCM can make its calculations more effectively.

The PCM uses the MAP sensor input to aid in calculating the following:

- Manifold pressure
- Barometric pressure
- Engine load
- Injector pulse-width
- Spark-advance programs
- Shift-point strategies (certain automatic transmissions only)
- Idle speed
- Decel fuel shutoff

The MAP sensor signal is provided from a single piezoresistive element located in the center of a diaphragm. The element and diaphragm are both made of silicone. As manifold pressure changes, the diaphragm moves causing the element to deflect, which stresses the silicone. When silicone is exposed to stress, its resistance changes. As manifold vacuum increases, the MAP sensor input voltage decreases proportionally. The sensor also contains electronics that condition the signal and provide temperature compensation.

The PCM recognizes a decrease in manifold pressure by monitoring a decrease in voltage from the reading stored in the barometric pressure memory cell. The MAP sensor is a linear sensor; meaning as pressure changes, voltage changes proportionately. The range of voltage output from the sensor is usually between 4.6 volts at sea level to as low as 0.3 volts at 26 in. of Hg. Barometric pressure is the pressure exerted by the atmosphere upon an object. At sea level on a standard day, no storm, barometric pressure is approximately 29.92 in. Hg. For every 100 feet of altitude, barometric pressure drops .10 in. Hg. If a storm goes through it can change barometric pressure from what should be present for that altitude. You should know what the average pressure and corresponding barometric pressure is for your area.

### Removal & Installation

#### EXCEPT 4.7L

The MAP sensor is mounted to the side of the throttle body. An L-shaped rubber fitting is used to connect the MAP sensor to throttle body.

1. Remove air duct at throttle body.
2. Remove electrical connector at sensor.
3. Remove two MAP sensor mounting bolts (screws).
4. While removing MAP sensor, slide the rubber L-shaped fitting from the throttle body.
5. Remove rubber L-shaped fitting from MAP sensor.

**To install:**

6. Install rubber L-shaped fitting to MAP sensor.

7. Position sensor to throttle body while guiding rubber fitting over throttle body vacuum nipple.

8. Install MAP sensor mounting bolts (screws). Tighten screws to 3 Nm (25 inch lbs..) torque.

9. Install electrical connector at sensor.

10. Install air duct at throttle body.

## 4.7L

The MAP sensor is located on the front of the intake manifold. An O-ring seals the sensor to the intake manifold.

1. Disconnect electrical connector at sensor.

2. Clean area around MAP sensor.

3. Remove 2 sensor mounting bolts.

4. Remove MAP sensor from intake manifold.

**To install:**

5. Clean MAP sensor mounting hole at intake manifold.

6. Check MAP sensor O-ring seal for cuts or tears.

7. Position sensor into manifold.

8. Install MAP sensor mounting bolts (screws). Tighten screws to 3 Nm (25 inch lbs..) torque.

9. Connect electrical connector.

### Connector Pinouts

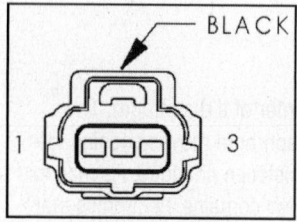

| CAV | CIRCUIT | FUNCTION |
|---|---|---|
| 1 | K7 18OR | 5V SUPPLY |
| 2 | K4 18BK/LB | SENSOR GROUND |
| 3 | K1 18DG/RD | MAP SENSOR SIGNAL |

**2000-04 Dakota 4.7L MAP Sensor**

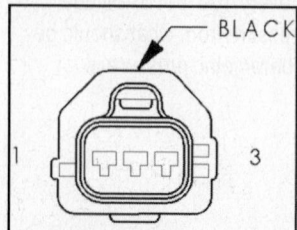

| CAV | CIRCUIT | FUNCTION |
|---|---|---|
| 1 | K4 18BK/LB | SENSOR GROUND |
| 2 | K1 18DG/RD | MAP SENSOR SIGNAL |
| 3 | K7 18OR | 5V SUPPLY |

**2000-04 Dakota Except 4.7L MAP Sensor**

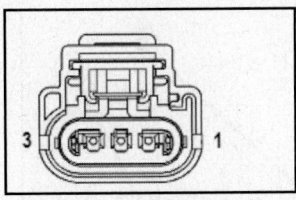

| CAV | CIRCUIT | FUNCTION |
|-----|---------|----------|
| 1 | K1 20VT/BR | MAP SIGNAL |
| 2 | K900 20DB/DG | SENSOR GROUND |
| 3 | F856 20YL/PK | 5 VOLT SUPPLY |

**2005-06 Dakota 3.7L & 4.7L MAP Sensor**


NATURAL VACUUM LEAK DETECTION (NVLD)

*NOTE: The NVLD system is used on 4.7L engines with Next Generation Controller (NGC).*

**Description & Operation**

Vehicles equipped with NGC engine control modules use an NVLD pump and system. Vehicles equipped with JTEC engine control modules use a leak detection pump. Refer to Leak Detection Pump (LDP) for additional information.

The Natural Vacuum Leak Detection (NVLD) system is the next generation evaporative leak detection system that will first be used on vehicles equipped with the Next Generation Controller (NGC). This new system replaces the leak detection pump as the method of evaporative system leak detection. This is to detect a leak equivalent to a 0.020" (0.5 mm) hole. This system has the capability to detect holes of this size very dependably.

The basic leak detection theory employed with NVLD is the "Gas Law". This is to say that the pressure in a sealed vessel will change if the temperature of the gas in the vessel changes. The vessel will only see this effect if it is indeed sealed. Even small leaks will allow the pressure in the vessel to come to equilibrium with the ambient pressure. In addition to the detection of very small leaks, this system has the capability of detecting medium as well as large evaporative system leaks.

A vent valve seals the canister vent during engine off conditions. If the vapor system has a leak of less than the failure threshold, the evaporative system will be pulled into a vacuum, either due to the cool down from operating temperature or diurnal ambient temperature cycling. This considers the diurnal effect one of the primary contributors to the leak determination diagnostic. When the vacuum in the system exceeds about 1" H2O (0.25 KPA), a vacuum switch closes. The switch closure sends a signal to the NGC. The NGC, via appropriate logic strategies, utilizes the switch signal, or lack thereof, to make a determination of whether a leak is present.

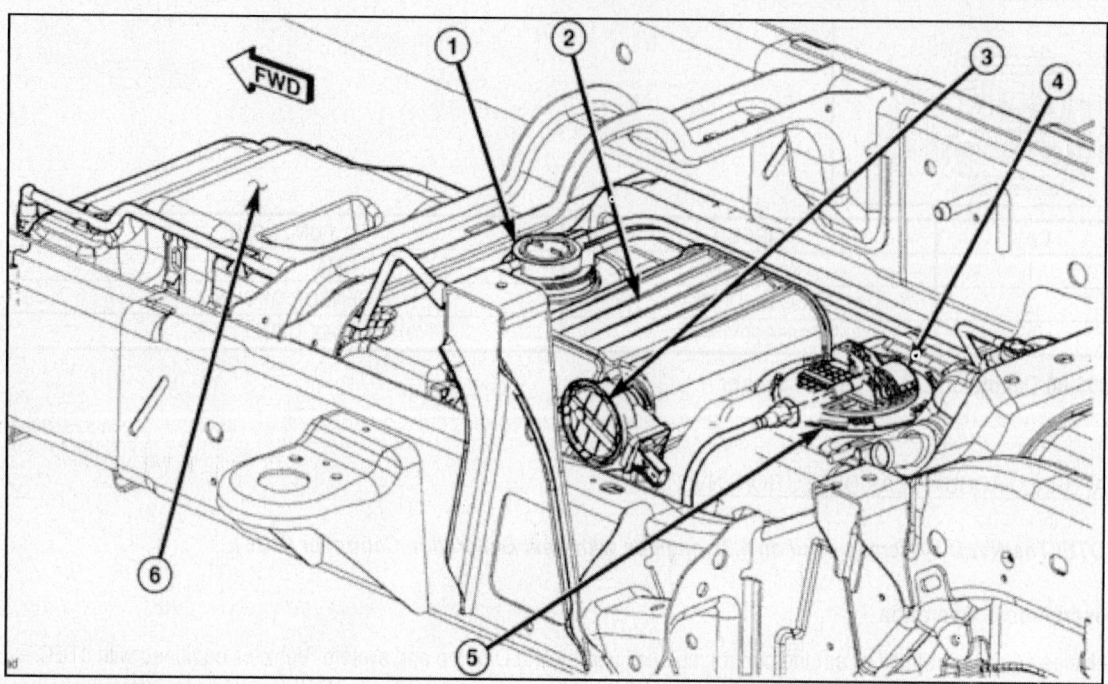

**The NVLD pump (3) is attached to the end of the EVAP canister (2) in front of the fuel tank (6)**

The NVLD device is designed with a normally open vacuum switch, a normally closed solenoid, and a seal, which is actuated by both the solenoid and a diaphragm. The NVLD is located on the atmospheric vent side of the canister. The NVLD assembly may be mounted on top of the canister outlet, or in-line between the canister and atmospheric vent filter. The normally open vacuum switch will close with about 1" H2O (0.25 KPA) vacuum in the evaporative system. The diaphragm actuates the switch. This is above the opening point of the fuel inlet check valve in the fill tube so cap off leaks can be detected. Submerged fill systems must have recirculation lines that do not have the in-line normally closed check valve that protects the system from failed nozzle liquid ingestion, in order to detect cap off conditions.

The normally closed valve in the NVLD is intended to maintain the seal on the evaporative system during the engine off condition. If vacuum in the evaporative system exceeds 3" to 6" H2O (0.75 to 1.5 KPA), the valve will be pulled off the seat, opening the seal. This will protect the system from excessive vacuum as well as allowing sufficient purge flow in the event that the solenoid was to become inoperative.

The solenoid actuates the valve to unseal the canister vent while the engine is running. It also will be used to close the vent during the medium and large leak tests and during the purge flow check. This solenoid requires initial 1.5 amps of current to pull the valve open but after 100 ms. will be duty cycled down to an average of about 150 mA for the remainder of the drive cycle.

Another feature in the device is a diaphragm that will open the seal in the NVLD with pressure in the evaporative system. The device will "blow off" at about 0.5" H2O (0.12 KPA) pressure to permit the venting of vapors during refueling. An added benefit to this is that it will also allow the tank to "breathe" during increasing temperatures, thus limiting the pressure in the tank to this low level. This is beneficial because the induced vacuum during a subsequent declining temperature will achieve the switch closed (pass threshold) sooner than if the tank had to decay from a built up pressure.

The device itself has 3 wires: Switch sense, solenoid driver and ground. It also includes a resistor to protect the switch from a short to battery or a short to ground. The NGC utilizes a high-side driver to energize and duty-cycle the solenoid.

## Removal & Installation

1.  Raise and support vehicle.
2.  Remove left-rear tire.
3.  Remove plastic shield (4) in front of left-rear tire. Access to both the EVAP canister (1) and NVLD pump (6) is from the area in front of the removed tire.
4.  Disconnect electrical connector at NVLD pump.
5.  Remove vapor line at NVLD pump. Pry outward on tab (3) and rotate pump (6) clockwise about 70° for removal.
6.  Remove NVLD pump O-ring (5) from EVAP canister (1).

**To install:**

7.  Install new NVLD pump O-ring (5) to EVAP canister (1).
8.  Position NVLD pump (6) into EVAP canister (1).
9.  Rotate pump (6) until tab aligns with notch in EVAP canister (1).
10. Carefully install vapor/vacuum lines to NVLD pump.

*CAUTION: The vapor/vacuum lines and hoses must be firmly connected. Check the vapor/vacuum lines at the NVLD pump, filter and EVAP canister purge solenoid for damage or leaks. If a leak is present, a Diagnostic Trouble Code (DTC) may be set.*

11. Connect electrical connector to pump.
12. Install plastic shield in front of left-rear tire.
13. Install left-rear tire.

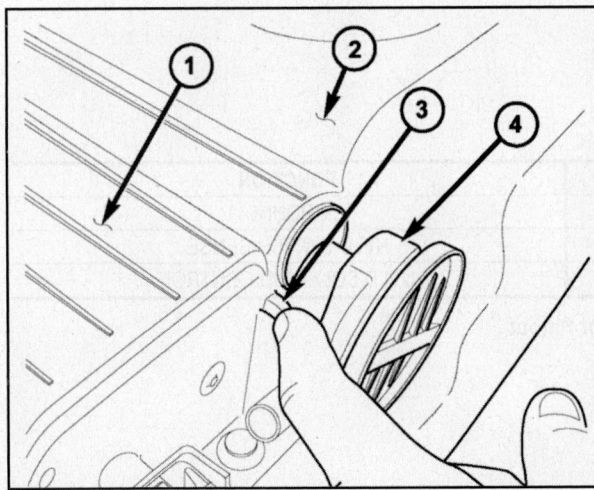

Removing the plastic shield (4) to access the NVLD pump.

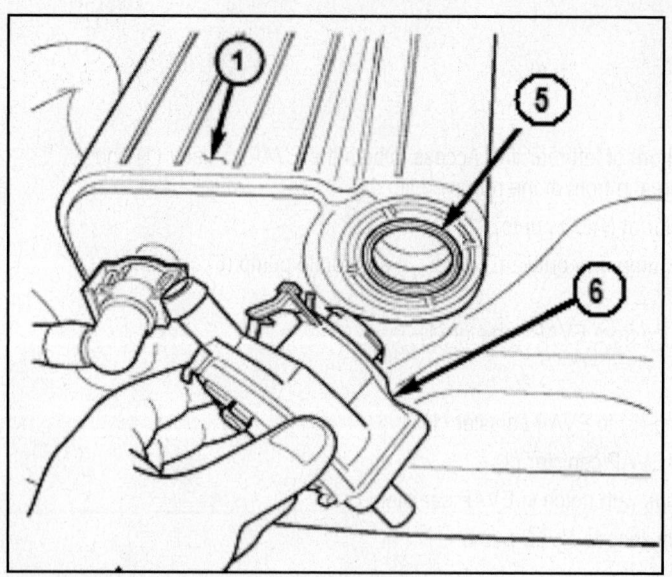

**Removing NVLD pump (6) from EVAP canister (1). Remove O-ring (5).**

**Connector Pinouts**

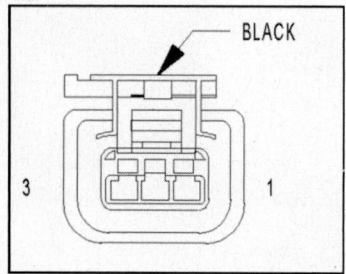

| PIN # | CIRCUIT | FUNCTION |
|-------|---------|----------|
| 1 | Z1 20BK | GROUND |
| 2 | K107 20OR | NVLD SWITCH SENSE |
| 3 | K106 20WT/DG | NVLD SOLENOID CONTROL |

**2003-06 Dakota 4.7L EVAP NVLD Assembly Connector Pinout**

OIL PRESSURE SENSOR CONNECTORS

**Connector Pinouts**

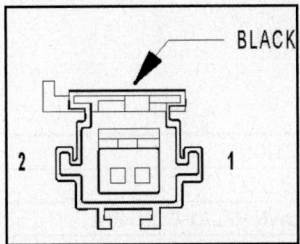

| PIN # | CIRCUIT | FUNCTION |
|-------|---------|----------|
| 1 | K4 18BK/LB | SENSOR GROUND |
| 2 | G60 16GY/YL (3.9L/5.9L) | ENGINE OIL PRESSURE SENSOR SIGNAL |
| 2 | G60 18GY/YL (2.5L) | ENGINE OIL PRESSURE SENSOR SIGNAL |

**2000 Dakota 2.5L, 3.9L, 5.9L Engine Oil Pressure Sensor**

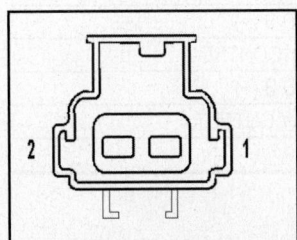

| PIN # | CIRCUIT | FUNCTION |
|-------|---------|----------|
| 1 | K4 18BK/LB | SENSOR GROUND |
| 2 | G60 18GY/YL | ENGINE OIL PRESSURE SENSOR SIGNAL |

**2000 Dakota 4.7L Engine Oil Pressure Sensor**

## POWER DISTRIBUTION CENTER RELAYS

**Connector Pinouts**

*Note: For location of relays in Power Distribution Center (PDC), check label on PDC lid.*

*AUTO SHUTDOWN RELAYS*

| PIN # | CIRCUIT | FUNCTION |
|-------|---------|----------|
| 30 | A16 14GY | FUSED B (+) |
| 85 | K51 20DB/YL | AUTOMATIC SHUT DOWN RELAY CONTROL |
| 86 | A14 16RD/WT (3.9L/5.9L) | FUSED B (+) |
| 86 | A16 14GY (4.7L) | FUSED B (+) |
| 87A | - | NOT USED |
| 87 | A142 14DG/OR | AUTOMATIC SHUT DOWN RELAY OUTPUT |

**2003 Dakota 3.9L, 4.7L & 5.9L ASD Relay**

| PIN # | CIRCUIT | FUNCTION |
|-------|---------|----------|
| 30 | A16 14GY | FUSED B (+) |
| 85 | K51 20DB/YL | ASD RELAY CONTROL |
| 86 | A16 14GY | FUSED B (+) |
| 87 | A142 14DG/OR | ASD RELAY OUTPUT |
| 87A | - | - |

**2003-04 Dakota 3.7L ASD Relay**

*FUEL PUMP RELAYS*

| PIN # | CIRCUIT | FUNCTION |
|-------|---------|----------|
| 30 | A14 16RD/WT | FUSED B (+) |
| 85 | K151 18LB/OR | FUEL PUMP RELAY CONTROL |
| 86 | F18 20LG/BK | FUSED IGNITION SWITCH OUTPUT (ST-RUN) |
| 87A | - | NOT USED |
| 87 | A61 16DG/BK | FUEL PUMP RELAY OUTPUT |

**2000 Dakota Fuel Pump Relay**

| PIN # | CIRCUIT | FUNCTION |
|-------|---------|----------|
| 30 | A14 16RD/WT | FUSED B (+) |
| 85 | K31 20BR | FUEL PUMP RELAY CONTROL |
| 86 | F18 20LG/BK | FUSED IGNITION SWITCH OUTPUT (RUN-START) |
| 87 | A61 16DG/BK | FUEL PUMP RELAY OUTPUT |
| 87A | - | - |

**2003-04 Dakota 3.7L Fuel Pump Relay**

*STARTER MOTOR RELAYS*

| PIN # | CIRCUIT | FUNCTION |
|---|---|---|
| 30 | A2 12PK/BK | FUSED B (+) |
| 85 | T41 18BK/WT | PARK/NEUTRAL POSITION SWITCH SENSE |
| 86 | A41 14DB/YL | IGNITION SWITCH OUTPUT (START) |
| 87A | - | NOT USED |
| 87 | T40 12BR | ENGINE STARTER MOTOR RELAY OUTPUT |

**2000 Dakota 2.5L Starter Motor Relay Pinout**

| PIN # | CIRCUIT | FUNCTION |
|---|---|---|
| 30 | A149 12RD/TN | FUSED B (+) |
| 85 | T752 18DG/OR | STARTER RELAY CONTROL |
| 86 | A169 18RD/YL | FUSED IGNITION SWITCH OUTPUT (START) |
| 87 | T40 12BR | STARTER MOTOR RELAY OUTPUT |
| 87A | - | - |

**2003-04 Dakota 3.7L Starter Motor Relay Pinout**

*TRANSMISSION CONTROL RELAYS*

| PIN # | CIRCUIT | FUNCTION |
|---|---|---|
| 30 | F84 16YL/WT | FUSED B (+) |
| 85 | Z1 20BK | GROUND |
| 86 | T15 18LG | TRANSMISSION CONTROL RELAY CONTROL |
| 87 | T16 16RD | TRANSMISSION CONTROL RELAY OUTPUT |
| 87A | - | - |

**2003-04 Dakota Transmission Control (A/T) Relay Pinout**

| PIN # | CIRCUIT | FUNCTION |
|---|---|---|
| 30 | F84 16YL/WT | FUSED B (+) |
| 85 | K30 18PK (3.9L/5.9L) | TRANSMISSION CONTROL RELAY CONTROL |
| 85 | Z1 20BK (4.7L) | GROUND |
| 86 | T15 18LG (4.7L) | TRANSMISSION CONTROL RELAY CONTROL |
| 86 | K125 18WT/DB (3.9L/5.9L) | GENERATOR SOURCE |
| 87 | T16 16RD | TRANSMISSION CONTROL RELAY OUTPUT |
| 87A | - | NOT USED |

**2003 Dakota Transmission Control (A/T) Relay Pinout**

## POWERTRAIN CONTROL MODULE (PCM)

### Description

The Powertrain Control Module (PCM) is located in the engine compartment. The PCM is referred to as JTEC.

### *5-VOLT SUPPLIES*

Two different Powertrain Control Module (PCM) 5-volt supply circuits are used; primary and secondary.

### Ignition Circuit Sense

This circuit ties the ignition switch to the Powertrain Control Module (PCM). Battery voltage is supplied to the PCM through the ignition switch when the ignition is in the Run or Start position. This is referred to as the "ignition sense" circuit and is used to "wake up" the PCM.

### *MODES OF OPERATION*

As input signals to the Powertrain Control Module (PCM) change, the PCM adjusts its response to the output devices. The PCM will operate in two different modes: Open Loop and Closed Loop.

During Open Loop modes, the PCM receives input signals and responds only according to preset PCM programming. Input from the oxygen (O2S) sensors is not monitored during Open Loop modes.

During Closed Loop modes, the PCM will monitor the oxygen (O2S) sensors input. This input indicates to the PCM whether or not the calculated injector pulse width results in the ideal air-fuel ratio. This ratio is 14.7 parts air-to-1 part fuel. By monitoring the exhaust oxygen content through the O2S sensor, the PCM can fine-tune the injector pulse width. This is done to achieve optimum fuel economy combined with low emission engine performance.

The fuel injection system has the following modes of operation:

- Ignition switch ON
- Engine start-up (crank)
- Engine warm-up
- Idle
- Cruise
- Acceleration
- Deceleration
- Wide open throttle (WOT)
- Ignition switch OFF

The ignition switch On, engine start-up (crank), engine warm-up, acceleration, deceleration and wide open throttle modes are Open Loop modes. The idle and cruise modes, (with the engine at operating temperature) are Closed Loop modes.

### Ignition Switch (Key-On) Mode

This is an Open Loop mode. When the fuel system is activated by the ignition switch, the following actions occur:

- The PCM pre-positions the idle air control (IAC) motor.
- The PCM determines atmospheric air pressure from the MAP sensor input to determine basic fuel strategy.
- The PCM monitors the engine coolant temperature sensor input. The PCM modifies fuel strategy based on this input.
- Intake manifold air temperature sensor input is monitored.
- Throttle position sensor (TPS) is monitored.
- The auto shutdown (ASD) relay is energized by the PCM for approximately three seconds.
- The fuel pump is energized through the fuel pump relay by the PCM. The fuel pump will operate for approximately three seconds unless the engine is operating or the starter motor is engaged.

- The O2S sensor heater element is energized via the ASD relay. The O2S sensor input is not used by the PCM to calibrate air-fuel ratio during this mode of operation.

### Engine Start-Up Mode

This is an Open Loop mode. The following actions occur when the starter motor is engaged.

The PCM receives inputs from:

- Battery voltage
- Engine coolant temperature sensor
- Crankshaft position sensor
- Intake manifold air temperature sensor
- Manifold absolute pressure (MAP) sensor
- Throttle position sensor (TPS)
- Starter motor relay
- Camshaft position sensor signal

The PCM monitors the crankshaft position sensor. If the PCM does not receive a crankshaft position sensor signal within 3 seconds of cranking the engine, it will shut down the fuel injection system. The fuel pump is activated by the PCM through the fuel pump relay.

Voltage is applied to the fuel injectors with the ASD relay via the PCM. The PCM will then control the injection sequence and injector pulse width by turning the ground circuit to each individual injector on and off. The PCM determines the proper ignition timing according to input received from the crankshaft position sensor.

### Engine Warm-Up Mode

This is an Open Loop mode. During engine warm-up, the PCM receives inputs from:

- Battery voltage
- Crankshaft position sensor
- Engine coolant temperature sensor
- Intake manifold air temperature sensor
- Manifold absolute pressure (MAP) sensor
- Throttle position sensor (TPS)
- Camshaft position sensor signal (in the distributor)
- Park/neutral switch (gear indicator signal—auto. trans. only)
- Air conditioning select signal (if equipped)
- Air conditioning request signal (if equipped)

Based on these inputs the following occurs:

- Voltage is applied to the fuel injectors with the ASD relay via the PCM. The PCM will then control the injection sequence and injector pulse width by turning the ground circuit to each individual injector on and off.
- The PCM adjusts engine idle speed through the idle air control (IAC) motor and adjusts ignition timing.
- The PCM operates the A/C compressor clutch through the clutch relay. This is done if the vehicle operator has selected A/C and requested by the A/C thermostat.
- When engine has reached operating temperature, the PCM will begin monitoring O2S sensor input. The system will then leave the warm-up mode and go into closed loop operation.

### Idle Mode

When the engine is at operating temperature, this is a Closed Loop mode. At idle speed, the PCM receives inputs from:

- Air conditioning select signal (if equipped)
- Air conditioning request signal (if equipped)
- Battery voltage
- Crankshaft position sensor
- Engine coolant temperature sensor
- Intake manifold air temperature sensor
- Manifold absolute pressure (MAP) sensor
- Throttle position sensor (TPS)
- Camshaft position sensor signal (in the distributor)
- Battery voltage
- Park/neutral switch (gear indicator signal—auto. trans. only)
- Oxygen sensors

Based on these inputs, the following occurs:

- Voltage is applied to the fuel injectors with the ASD relay via the PCM. The PCM will then control injection sequence and injector pulse width by turning the ground circuit to each individual injector on and off.
- The PCM monitors the O2S sensor input and adjusts air-fuel ratio by varying injector pulse width. It also adjusts engine idle speed through the idle air control (IAC) motor.
- The PCM adjusts ignition timing by increasing and decreasing spark advance.
- The PCM operates the A/C compressor clutch through the clutch relay. This happens if A/C has been selected by the vehicle operator and requested by the A/C thermostat.

### Cruise Mode

When the engine is at operating temperature, this is a Closed Loop mode. At cruising speed, the PCM receives inputs from:

- Air conditioning select signal (if equipped)
- Air conditioning request signal (if equipped)
- Battery voltage
- Engine coolant temperature sensor
- Crankshaft position sensor
- Intake manifold air temperature sensor
- Manifold absolute pressure (MAP) sensor
- Throttle position sensor (TPS)
- Camshaft position sensor signal (in the distributor)
- Park/neutral switch (gear indicator signal—auto. trans. only)
- Oxygen (O2S) sensors

Based on these inputs, the following occurs:

- Voltage is applied to the fuel injectors with the ASD relay via the PCM. The PCM will then adjust the injector pulse width by turning the ground circuit to each individual injector on and off.
- The PCM monitors the O2S sensor input and adjusts air-fuel ratio. It also adjusts engine idle speed through the idle air control (IAC) motor.
- The PCM adjusts ignition timing by turning the ground path to the coil on and off.

● The PCM operates the A/C compressor clutch through the clutch relay. This happens if A/C has been selected by the vehicle operator and requested by the A/C thermostat.

### Acceleration Mode

This is an Open Loop mode. The PCM recognizes an abrupt increase in throttle position or MAP pressure as a demand for increased engine output and vehicle acceleration. The PCM increases injector pulse width in response to increased throttle opening.

### Deceleration Mode

When the engine is at operating temperature, this is an Open Loop mode. During hard deceleration, the PCM receives the following inputs:

● Air conditioning select signal (if equipped)
● Air conditioning request signal (if equipped)
● Battery voltage
● Engine coolant temperature sensor
● Crankshaft position sensor
● Intake manifold air temperature sensor
● Manifold absolute pressure (MAP) sensor
● Throttle position sensor (TPS)
● Camshaft position sensor signal (in the distributor)
● Park/neutral switch (gear indicator signal—auto. trans. only)
● Vehicle speed sensor

If the vehicle is under hard deceleration with the proper rpm and closed throttle conditions, the PCM will ignore the oxygen sensor input signal. The PCM will enter a fuel cut-off strategy in which it will not supply a ground to the injectors. If a hard deceleration does not exist, the PCM will determine the proper injector pulse width and continue injection.

Based on the above inputs, the PCM will adjust engine idle speed through the idle air control (IAC) motor.

The PCM adjusts ignition timing by turning the ground path to the coil on and off.

### Wide Open Throttle Mode

This is an Open Loop mode. During wide open throttle operation, the PCM receives the following inputs:

● Battery voltage
● Crankshaft position sensor
● Engine coolant temperature sensor
● Intake manifold air temperature sensor
● Manifold absolute pressure (MAP) sensor
● Throttle position sensor (TPS)
● Camshaft position sensor signal (in the distributor)

During wide open throttle conditions, the following occurs:

● Voltage is applied to the fuel injectors with the ASD relay via the PCM. The PCM will then control the injection sequence and injector pulse width by turning the ground circuit to each individual injector on and off. The PCM ignores the oxygen sensor input signal and provides a predetermined amount of additional fuel. This is done by adjusting injector pulse width.

- The PCM adjusts ignition timing by turning the ground path to the coil on and off.

### Ignition Switch Off Mode

When ignition switch is turned to OFF position, the PCM stops operating the injectors, ignition coil, ASD relay and fuel pump relay.

### POWER GROUNDS

The Powertrain Control Module (PCM) has 2 main grounds. Both of these grounds are referred to as power grounds. All of the high-current, noisy, electrical devices are connected to these grounds as well as all of the sensor returns. The sensor return comes into the sensor return circuit, passes through noise suppression, and is then connected to the power ground.

The power ground is used to control ground circuits for the following PCM loads:

- Generator field winding
- Fuel injectors
- Ignition coil(s)
- Certain relays/solenoids
- Certain sensors

### SENSOR RETURN

The Sensor Return circuits are internal to the Powertrain Control Module (PCM).

Sensor Return provides a low–noise ground reference for all engine control system sensors.

### SIGNAL GROUND

Signal ground provides a low noise ground to the data link connector.

### Operation

The PCM operates the fuel system. The PCM is a pre-programmed, triple microprocessor digital computer. It regulates ignition timing, air-fuel ratio, emission control devices, charging system, certain transmission features, speed control, air conditioning compressor clutch engagement and idle speed. The PCM can adapt its programming to meet changing operating conditions.

The PCM receives input signals from various switches and sensors. Based on these inputs, the PCM regulates various engine and vehicle operations through different system components. These components are referred to as Powertrain Control Module (PCM) Outputs. The sensors and switches that provide inputs to the PCM are considered Powertrain Control Module (PCM) Inputs.

The PCM adjusts ignition timing based upon inputs it receives from sensors that react to: engine rpm, manifold absolute pressure, engine coolant temperature, throttle position, transmission gear selection (automatic transmission), vehicle speed and the brake switch.

The PCM adjusts idle speed based on inputs it receives from sensors that react to: throttle position, vehicle speed, transmission gear selection, engine coolant temperature and from inputs it receives from the air conditioning clutch switch and brake switch.

Based on inputs that it receives, the PCM adjusts ignition coil dwell. The PCM also adjusts the generator charge rate through control of the generator field and provides speed control operation.

### PCM INPUTS:

- A/C request (if equipped with factory A/C)
- A/C select (if equipped with factory A/C)
- Auto shutdown (ASD) sense
- Battery temperature

- Battery voltage
- Brake switch
- CCD bus (+) circuits
- CCD bus (-) circuits
- Camshaft position sensor signal
- Crankshaft position sensor
- Data link connection for DRB scan tool
- Engine coolant temperature sensor
- Fuel level
- Generator (battery voltage) output
- Ignition circuit sense (ignition switch in on/off/crank/run position)
- Intake manifold air temperature sensor
- Leak detection pump (switch) sense (if equipped)
- Manifold absolute pressure (MAP) sensor
- Oil pressure
- Output shaft speed sensor
- Overdrive/override switch
- Oxygen sensors
- Park/neutral switch (auto. trans. only)
- Power ground
- Sensor return
- Signal ground
- Speed control multiplexed single wire input
- Throttle position sensor
- Transmission governor pressure sensor
- Transmission temperature sensor
- Vehicle speed inputs from ABS or RWAL system

*PCM OUTPUTS:*

- A/C clutch relay
- Auto shutdown (ASD) relay
- CCD bus (+/-) circuits for: speedometer, voltmeter, fuel gauge, oil pressure gauge/lamp, engine temperature gauge and speed control warning lamp
- Data link connection for DRB scan tool
- EGR valve control solenoid (if equipped)
- EVAP canister purge solenoid
- Five volt sensor supply (primary)
- Five volt sensor supply (secondary)
- Fuel injectors
- Fuel pump relay
- Generator field driver (-)
- Generator field driver (+)
- Generator lamp (if equipped)
- Idle air control (IAC) motor
- Ignition coil

- Leak detection pump (if equipped)
- Malfunction indicator lamp (Check engine lamp); driven through CCD circuits
- Overdrive indicator lamp (if equipped)
- Radiator cooling fan (2.5L engine only)
- Speed control vacuum solenoid
- Speed control vent solenoid
- Tachometer (if equipped). Driven through CCD circuits.
- Transmission converter clutch circuit
- Transmission 3–4 shift solenoid
- Transmission relay
- Transmission temperature lamp (if equipped)
- Transmission variable force solenoid

### 5-VOLT SUPPLIES

### Primary 5–Volt Supplies:

- The required 5-volt power source to the Crankshaft Position (CKP) sensor.
- The required 5-volt power source to the Camshaft Position (CMP) sensor.
- A reference voltage for the Manifold Absolute Pressure (MAP) sensor.
- A reference voltage for the Throttle Position Sensor (TPS) sensor.

### Secondary 5–Volt Supplies:

- The required 5-volt power source to the oil pressure sensor.
- The required 5-volt power source for the Vehicle Speed Sensor (VSS) (if equipped).
- The 5-volt power source to the transmission pressure sensor (if equipped with an RE automatic transmission).

### IGNITION CIRCUIT SENSE

The ignition circuit sense input tells the PCM the ignition switch has energized the ignition circuit.

Battery voltage is also supplied to the PCM through the ignition switch when the ignition is in the RUN or START position. This is referred to as the "ignition sense" circuit and is used to "wake up" the PCM.

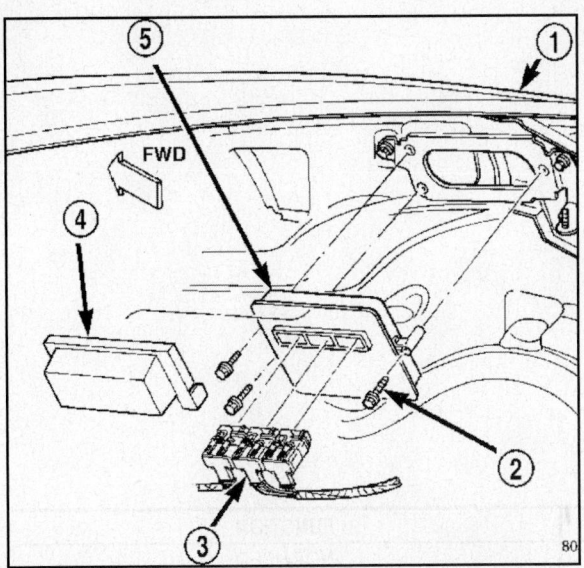

Showing location of the PCM (5) on the RF fender (1). Remove the mounting bolts (2), multi-pin connectors (3) and cover (4) before removing the PCM

### Removal & Installation

*NOTE: Use the proper scan tool to reprogram the new powertrain control module (PCM) with the vehicles original identification number (VIN) and the vehicles original mileage. If this step is not done, a diagnostic trouble code (DTC) may be set.*

The PCM is located in the engine compartment.

*WARNING: To avoid possible voltage spike damage to the PCM, ignition key must be off, and negative battery cable must be disconnected before unplugging PCM connectors.*

1. Disconnect and isolate the negative battery cable.
2. Remove the cover over the electrical connectors.
3. Carefully unplug the three 32–way connectors (four 38–way connectors if equipped with NGC) from PCM.
4. Remove the PCM mounting bolts and remove the PCM from vehicle.

### To install:

*NOTE: Use the proper scan tool to reprogram the new powertrain control module (PCM) with the vehicles original identification number (VIN) and the vehicles original mileage. If this step is not done, a diagnostic trouble code (DTC) may be set.*

5. Install the PCM and mounting bolts to vehicle.
6. Tighten bolts to 3–5 Nm (30–40 inch lbs.).
7. Check pin connectors in the PCM. Also check the three 32–way connectors (four 38–way connectors if equipped with NGC) for corrosion or damage. Repair as necessary.
8. Install the three 32–way connectors (four 38–way connectors if equipped with NGC) to PCM.
9. Install the cover over electrical connectors.
10. Reconnect the negative battery cable.

## Connector Pinouts

### 2.5L C1 CONNECTOR

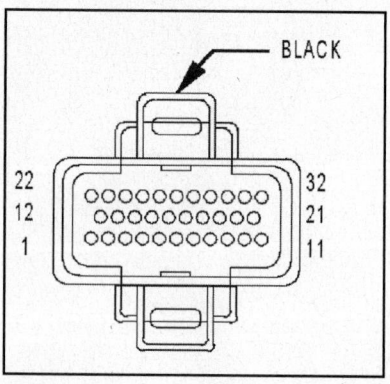

| PIN # | CIRCUIT | FUNCTION |
|---|---|---|
| 1 | - | NOT USED |
| 2 | F18 18LG/BK | FUSED IGNITION SWITCH OUTPUT (ST-RUN) |
| 3 | - | NOT USED |
| 4 | K4 18BK/LB | SENSOR GROUND |
| 5-6 | - | NOT USED |
| 7 | K19 16BK/GY | IGNITION COIL NO.1 DRIVER |
| 8 | K24 18GY/BK | CRANKSHAFT POSITION SENSOR SIGNAL |
| 9 | - | NOT USED |
| 10 | K60 18YL/BK | IDLE AIR CONTROL NO. 2 DRIVER |
| 11 | K40 18BR/WT | IDLE AIR CONTROL NO. 3 DRIVER |
| 12 | K10 18DB/OR | POWER STEERING PRESSURE SWITCH SENSE |
| 13-14 | - | NOT USED |
| 15 | K21 18BK/RD | INTAKE AIR TEMPERATURE SIGNAL |
| 16 | K2 18TN/BK | ENGINE COOLANT TEMPERATURE SENSOR SIGNAL |
| 17 | K7 18OR | 5V SUPPLY |
| 18 | K44 16TN/YL | CAMSHAFT POSITION SENSOR SIGNAL |
| 19 | K39 18GY/RD | IDLE AIR CONTROL NO. 1 DRIVER |
| 20 | K59 18VT/BK | IDLE AIR CONTROL NO. 4 DRIVER |
| 21 | - | NOT USED |
| 22 | A14 16RD/WT | FUSED B (+) |
| 23 | K22 18OR/DB | THROTTLE POSITION SENSOR SIGNAL |
| 24 | K141 18TN/WT | OXYGEN SENSOR 1/1 SIGNAL |
| 25 | K341 18OR/BK | OXYGEN SENSOR 1/2 SIGNAL |
| 26 | - | NOT USED |
| 27 | K1 18DG/RD | MAP SENSOR SIGNAL |
| 28-30 | - | NOT USED |
| 31 | Z12 14BK/TN | GROUND |
| 32 | Z12 14BK/TN | GROUND |

**2000-02 Dakota 2.5L PCM C1 (Black) Connector Pinout**

*3.9L & 5.9L C1 CONNECTOR*

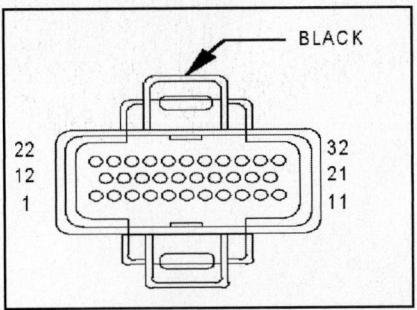

| PIN # | CIRCUIT | FUNCTION |
|-------|---------|----------|
| 1 | K93 16TN/OR (4.7L) | COIL ON PLUG NO. 3 DRIVER |
| 2 | F18 18LG/BK | FUSED IGNITION SWITCH OUTPUT (ST-RUN) |
| 3 | K94 16TN/LG (4.7L) | COIL ON PLUG NO. 4 DRIVER |
| 4 | K4 18BK/LB | SENSOR GROUND |
| 5 | K96 16TN/LB (4.7L) | COIL ON PLUG NO. 6 DRIVER |
| 6 | T41 18BK/WT (A/T) | PARK/NEUTRAL POSITION SWITCH SENSE |
| 7 | K19 16BK/GY | COIL ON PLUG NO. 1 DRIVER |
| 8 | K24 18GY/BK | CRANKSHAFT POSITION SENSOR SIGNAL |
| 9 | K18 16DB/GY (4.7L) | COIL ON PLUG NO. 8 DRIVER |
| 10 | K60 18YL/BK | IDLE AIR CONTROL NO. 2 DRIVER |
| 11 | K40 18BR/WT | IDLE AIR CONTORL NO. 3 DRIVER |
| 12 | K10 18DB/OR (4.7L) | POWER STEERING PRESSURE SWITCH SENSE |
| 13-14 | - | NOT USED |
| 15 | K21 18BK/RD | INTAKE AIR TEMPERATURE SIGNAL |
| 16 | K2 18TN/BK | ENGINE COOLANT TEMPERATURE SENSOR SIGNAL |
| 17 | K7 18OR | 5V SUPPLY |
| 18 | K44 16TN/YL | CAMSHAFT POSITION SENSOR SIGNAL |
| 19 | K39 18GY/RD | IDLE AIR CONTROL NO. 1 DRIVER |
| 20 | K59 18VT/BK | IDLE AIR CONTROL NO. 4 DRIVER |
| 21 | K95 16TN/DG (4.7L) | COIL ON PLUG NO. 5 DRIVER |
| 22 | A14 16RD/WT | FUSED B (+) |
| 23 | K22 18OR/DB | THROTTLE POSITION SENSOR SIGNAL |
| 24 | K141 18TN/WT (3.9L/5.9L) | OXYGEN SENSOR 1/1 SIGNAL |
| 25 | K341 18OR/BK (4.7L) | OXYGEN SENSOR 1/1 SIGNAL |
| 26 | K441 18OR/TN (4.7L CA.) | OXYGEN SENSOR 2/1 SIGNAL |
| 27 | K1 18DG/RD | MAP SENSOR SIGNAL |
| 28 | - | NOT USED |
| 29 | K341 18PK/WT (4.7L CA.) | OXYGEN SENSOR 2/2 SIGNAL |
| 30 | - | NOT USED |
| 31 | Z12 14BK/TN | GROUND |
| 32 | Z12 14BK/TN | GROUND |

**2000-02 Dakota 3.9L & 5.9L PCM C1 Connector Pinout**

**3.9L & 5.9L (JTEC) C1 CONNECTOR**

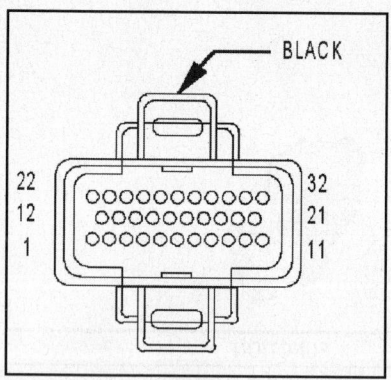

| PIN # | CIRCUIT | FUNCTION |
|-------|---------|----------|
| 1 | - | NOT USED |
| 2 | F18 18LG/BK | FUSED IGNITION SWITCH OUTPUT (RUN-START) |
| 3 | - | NOT USED |
| 4 | K4 18BK/LB | SENSOR GROUND |
| 5 | - | NOT USED |
| 6 | T41 18BK/WT | PARK/NEUTRAL POSITION SWITCH SENSE (T41) |
| 7 | K19 16BK/GY | COIL DRIVER NO. 1 |
| 8 | K24 20GY/BK | CRANKSHAFT POSITION SENSOR SIGNAL |
| 9 | - | NOT USED |
| 10 | K60 18YL/BK | IDLE AIR CONTROL NO. 2 DRIVER |
| 11 | K40 18BR/WT | IDLE AIR CONTROL NO. 3 DRIVER |
| 12-14 | - | NOT USED |
| 15 | K21 20BK/RD | INTAKE AIR TEMPERATURE SIGNAL |
| 16 | K2 18TN/BK | ENGINE COOLANT TEMPERATURE SENSOR SIGNAL |
| 17 | K7 18OR | 5 VOLT SUPPLY |
| 18 | K44 20TN/YL | CAMSHAFT POSITION SENSOR SIGNAL |
| 19 | K39 18GY/RD | IDLE AIR CONTROL NO. 1 DRIVER |
| 20 | K59 18VT/BK | IDLE AIR CONTROL NO. 4 DRIVER |
| 21 | - | NOT USED |
| 22 | A14 16RD/WT | FUSED B (+) |
| 23 | K22 20OR/DB | THROTTLE POSITION SENSOR SIGNAL |
| 24 | K41 18BK/DG | OXYGEN SENSOR 1/1 SIGNAL |
| 25 | K141 18TN/WT | OXYGEN SENSOR 1/2 SIGNAL |
| 26 | K241 18LG/RD | OXYGEN SENSOR 2/1 SIGNAL |
| 27 | K1 20DG/RD | MAP SENSOR SIGNAL |
| 28 | - | NOT USED |
| 29 | K341 18TN/WT | OXYGEN SENSOR 2/2 SIGNAL |
| 30 | - | NOT USED |
| 31 | Z12 14BK/TN | GROUND |
| 32 | Z12 14BK/TN | GROUND |

**2003-04 Dakota 3.9L & 5.9L (JTEC) PCM C1 Connector Pinout**

*2000-02 C2 CONNECTOR*

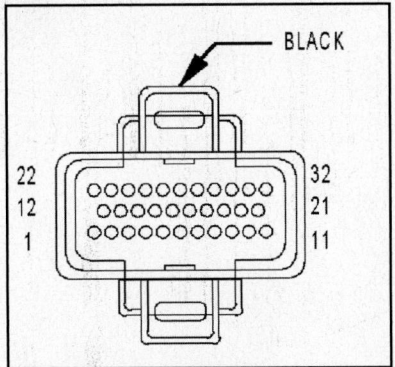

| PIN # | CIRCUIT | FUNCTION |
|-------|---------|----------|
| 1-3 | - | NOT USED |
| 4 | K11 18WT/DB | FUEL INJECTOR NO. 1 DRIVER |
| 5 | K13 18YL/WT | FUEL INJECTOR NO. 3 DRIVER |
| 6-9 | - | NOT USED |
| 10 | K20 18DG | GENERATOR FIELD |
| 11-14 | - | NOT USED |
| 15 | K12 18TN | FUEL INJECTOR NO. 2 DRIVER |
| 16 | K14 18LB/BR | FUEL INJECTOR NO. 4 DRIVER |
| 17-22 | - | NOT USED |
| 23 | G60 18GY/YL | ENGINE OIL PRESSURE SENSOR SIGNAL |
| 24-26 | - | NOT USED |
| 27 | G7 18WT/OR | VEHICLE SPEED SENSOR SIGNAL |
| 28-32 | - | NOT USED |

**2000-02 Dakota PCM C2 (White) Connector Pinout**

**2003-04 3.9L & 5.9L (JTEC) C2 CONNECTOR**

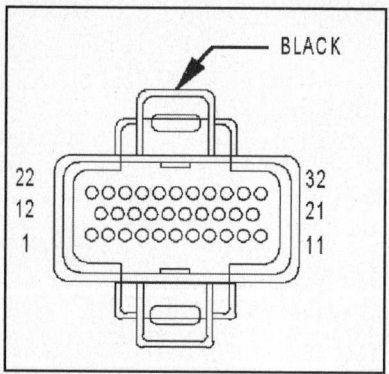

| PIN # | CIRCUIT | FUNCTION |
|---|---|---|
| 1 | T34 18GY/BK | TRANSMISSION TEMPERATURE SENSOR SIGNAL |
| 2 | K26 18VT | FUEL INJECTOR NO. 7 DRIVER |
| 3 | - | NOT USED |
| 4 | K11 18WT/DB | FUEL INJECTOR NO. 1 DRIVER |
| 5 | K13 18YL/WT | FUEL INJECTOR NO. 3 DRIVER |
| 6 | K38 18GY | FUEL INJECTOR NO. 5 DRIVER |
| 7 | - | NOT USED |
| 8 | K88 18VT/WT | GOVERNOR PRESSURE SOLENOID CONTROL |
| 9 | - | NOT USED |
| 10 | K20 18DG | GENERATOR FIELD DRIVER |
| 11 | K54 18OR/BK | TORQUE CONVERTER CLUTCH SOLENOID CONTROL |
| 12 | K58 18BR/DB | FUEL INJECTOR NO. 6 DRIVER |
| 13 | K28 18GY/LB | FUEL INJECTOR NO. 8 DRIVER |
| 14 | - | NOT USED |
| 15 | K12 18TN | FUEL INJECTOR NO. 2 DRIVER |
| 16 | K14 18LB/BR | FUEL INJECTOR NO. 4 DRIVER |
| 17 | C24 20DB/PK | RADIATOR FAN RELAY CONTROL |
| 18 | - | NOT USED |
| 19 | C18 20DB | A/C PRESSURE SIGNAL |
| 20 | - | NOT USED |
| 21 | T60 18BR | 3-4 SHIFT SOLENOID CONTROL |
| 22 | - | NOT USED |
| 23 | G60 16GY/YL | ENGINE OIL PRESSURE SWITCH SIGNAL |
| 24 | - | NOT USED |
| 25 | T13 18DB/BK | SPEED SENSOR GROUND |
| 26 | - | NOT USED |
| 27 | G7 18WT/OR | VEHICLE SPEED SENSOR SIGNAL |
| 28 | T14 18LG/WT | OUTPUT SPEED SENSOR SIGNAL |
| 29 | T25 18LG/RD | GOVERNOR PRESSURE SIGNAL |
| 30 | K30 18PK | TRANSMISSION CONTROL RELAY CONTROL |
| 31 | K6 18VT/WT | 5 VOLT SUPPLY |
| 32 | - | NOT USED |

**2003-04 Dakota 3.9L & 5.9L (JTEC) PCM C2 Connector Pinout (2 of 2)**

**2003-04 3.9L & 5.9L (JTEC) C3 CONNECTOR**

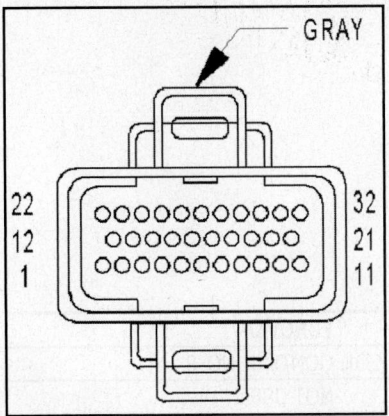

| PIN # | CIRCUIT | FUNCTION |
|---|---|---|
| 1 | C13 20DB/OR | A/C CLUTCH RELAY CONTROL |
| 2 | - | NOT USED |
| 3 | K51 20DB/YL | AUTOMATIC SHUT DOWN RELAY CONTROL |
| 4 | V36 18TN/RD | SPEED CONTROL VACUUM SOLENOID CONTROL |
| 5 | V35 20LG/RD | SPEED CONTROL VENT SOLENOID CONTROL |
| 6-7 | - | NOT USED |
| 8 | K100 20VT/WT | PWM 1/1 HEATER CONTROL |
| 9 | K512 18DG/BK | OXYGEN SENSOR DOWNSTREAM HEATER RELAY CONTROL |
| 10 | K106 18WT/DG | LEAK DETECTION PUMP SOLENOID CONTROL |
| 11 | V32 18YL/RD | SPEED CONTROL SUPPLY |
| 12 | A142 14DG/OR | AUTOMATIC SHUT DOWN RELAY OUTPUT |
| 13 | T6 18OR/WT | OVERDRIVE OFF SWITCH SENSE |
| 14 | K107 20OR | LEAK DETECTION PUMP SWITCH SENSE |
| 15 | K118 20PK/YL | BATTERY TEMPERATURE SENSOR SIGNAL |
| 16 | K200 18VT/OR | PWM 1/2 HEATER DRIVER |
| 17-18 | - | NOT USED |
| 19 | K31 20BR | FUEL PUMP RELAY CONTROL |
| 20 | K52 18PK/BK | DUTY CYCLE EVAP/PURGE SOLENOID CONTROL |
| 21 | - | NOT USED |
| 22 | C20 18BR | A/C SWITCH SENSE |
| 23 | - | NOT USED |
| 24 | V40 20WT/PK | BRAKE SWITCH SENSE |
| 25 | K125 18WT/DB | GENERATOR SOURCE |
| 26 | K226 18DB/WT | FUEL PUMP RELAY CONTROL |
| 27 | D21 18PK | SCI TRANSMIT |
| 28 | - | NOT USED |
| 29 | D20 20LG | SCI RECEIVE |
| 30 | D25 18VT/YL | PCI BUS |
| 31 | - | NOT USED |
| 32 | V37 18RD/LG | SPEED CONTROL SWITCH SIGNAL |

**2003-04 Dakota 3.9 & 5.9L (JTEC) PCM C3 Connector Pinout**

## 2003-06 4.7L (NGC) C1 CONNECTOR

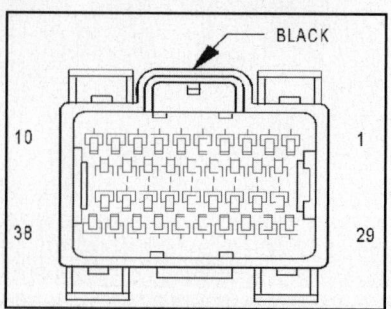

| PIN # | CIRCUIT | FUNCTION |
|---|---|---|
| 1 | K98 16LB/RD | COIL CONTROL NO. 8 |
| 2 | - | NOT USED |
| 3 | K97 16BR | COIL CONTROL NO. 7 |
| 4 | K28 18GY/LB | INJECTOR CONTROL NO. 8 |
| 5 | K26 18VT | INJECTOR CONTROL NO. 7 |
| 6-8 | - | NOT USED |
| 9 | Z12 16BK/TN | GROUND |
| 10 | | NOT USED |
| 11 | F18 18LG/BK | FUSED IGNITION SWITCH OUTPUT (RUN-START) |
| 12 | F11 20RD/WT | FUSED IGNITION SWITCH OUTPUT (OFF-RUN-START) |
| 13 | G7 18WT/OR | VEHICLE SPEED SIGNAL |
| 14-17 | - | NOT USED |
| 18 | Z12 16BK/TN | GROUND |
| 19 | - | NOT USED |
| 20 | G60 18GY/YL | OIL PRESSURE SIGNAL |
| 21 | C18 20LB/BR | A/C PRESSURE SIGNAL |
| 22 | G31 18VT/LG | AMBIENT TEMPERATURE SENSOR SIGNAL |
| 23-24 | - | NOT USED |
| 25 | D20 18LG | SCI RECEIVE (PCM) |
| 26 | D6 18PK/LB | SCI RECEIVE (TCM) |
| 27 | K6 18VT/WT | 5 VOLT SUPPLY |
| 28 | - | NOT USED |
| 29 | A14 16RD/WT | FUSED B (+) |
| 30 | A169 18RD/YL | FUSED IGNITION SWITCH OUTPUT (START) |
| 31 | K141 18TN/WT | O2 1/2 SIGNAL |
| 32 | K902 18BR/DG | O2 RETURN (DOWN) |
| 33 | K341 18TN/WT | O2 2/2 SIGNAL |
| 34-35 | - | NOT USED |
| 36 | D21 18PK | SCI TRANSMIT (PCM) |
| 37 | D15 18WT/DG | SCI TRANSMIT (TCM) |
| 38 | D25 18WT/VT | PCI BUS (PCM) |

**2003-06 Dakota 4.7L (NGC) PCM C1 Connector Pinout**

**2003-06 4.7L (NGC) C2 CONNECTOR**

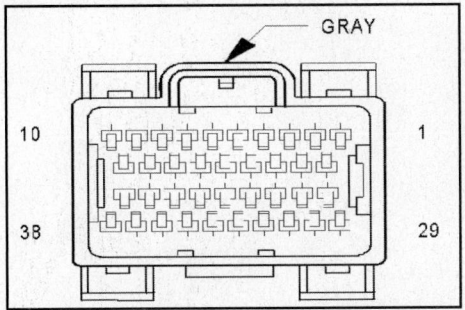

| PIN # | CIRCUIT | FUNCTION |
|---|---|---|
| 1 | K96 16TN/LB | COIL CONTROL NO. 6 |
| 2 | K95 16TN/DG | COIL CONTROL NO. 5 |
| 3 | K94 16TN/LG | COIL CONTROL NO. 4 |
| 4 | K58 18BR/DB | INJECTOR CONTROL NO. 6 |
| 5 | K38 18GY | INJECTOR CONTROL NO. 5 |
| 6 | - | NOT USED |
| 7 | K93 16TN/OR | COIL CONTROL NO. 3 |
| 8 | - | NOT USED |
| 9 | K92 16TN/PK | COIL CONTROL NO. 2 |
| 10 | K91 16TN/RD | COIL CONTROL NO. 1 |
| 11 | K14 18LB/BR | INJECTOR CONTROL NO. 4 |
| 12 | K13 18YL/WT | INJECTOR CONTROL NO. 3 |
| 13 | K12 18TN | INJECTOR CONTROL NO. 2 |
| 14 | K11 18WT/DB | INJECTOR CONTROL NO. 1 |
| 15-16 | - | NOT USED |
| 17 | K199 18BR/VT | O2 2/1 HEATER CONTROL |
| 18 | K99 18BR/OR | O2 1/1 HEATER CONTROL |
| 19 | K20 18DG | GEN FIELD CONTROL |
| 20 | K2 18VT/OR | ECT SENSOR |
| 21 | K22 20BR/OR | TP SENSOR NO. 1 SIGNAL |
| 22 | - | NOT USED |
| 23 | K1 20VT/BR | MAP SIGNAL |
| 24-26 | - | NOT USED |
| 27 | K4 18BK/LB | SENSOR GROUND |
| 28 | K60 16YL/BK | IAC RETURN |
| 29 | K7 18OR | 5 VOLT SUPPLY |
| 30 | K21 20BK/RD | IAT SIGNAL |
| 31 | K41 18BK/DG | O2 1/1 SIGNAL |
| 32 | K904 18DB/DG | O2 RETURN (UP) |
| 33 | K241 18LG/RD | O2 2/1 SIGNAL |
| 34 | K44 20DB/GY | CMP SIGNAL |
| 35 | K24 20DB/WT | CKP SIGNAL |
| 36-37 | - | NOT USED |
| 38 | K39 16GY/RD | IAC MOTOR CONTROL |

**2003-06 Dakota 4.7L (NGC) PCM C2 Connector Pinout**

*2003-06 4.7L (NGC) C3 CONNECTOR*

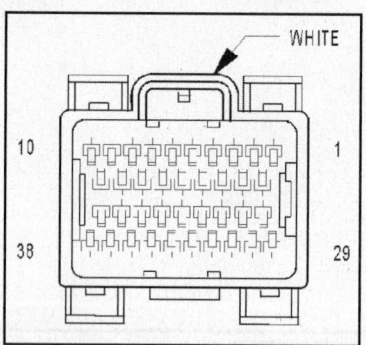

| PIN # | CIRCUIT | FUNCTION |
|-------|---------|----------|
| 1-2 | - | NOT USED |
| 3 | K51 20DB/YL | AUTOMATIC SHUT DOWN RELAY CONTROL |
| 4 | - | NOT USED |
| 5 | V35 20LG/RD | S/C VENT CONTROL |
| 6 | C24 20DB/PK | LOW RAD FAN RELAY CONTROL |
| 7 | V32 18VT/YL | S/C SUPPLY |
| 8 | K106 20WT/DG | NVLD SOLENOID CONTROL |
| 9 | K299 18BR/WT | O2 1/2 HEATER CONTROL |
| 10 | K399 18BR/GY | O2 2/2 HEATER CONTROL |
| 11 | C13 20LB/OR | A/C CLUTCH RELAY CONTROL |
| 12 | V36 20VT/DG | S/C VACUUM CONTROL |
| 13-18 | - | NOT USED |
| 19 | A142 16DG/OR | AUTOMATIC SHUT DOWN RELAY OUTPUT |
| 20 | K52 18PK/BK | EVAP PURGE CONTROL |
| 21 | T41 18BK/WT | TRS T41 SENSE |
| 22 | - | NOT USED |
| 23 | V40 20WT/PK | BRAKE SWITCH SIGNAL |
| 24-27 | - | NOT USED |
| 28 | A142 16DG/OR | AUTOMATIC SHUT DOWN RELAY OUTPUT |
| 29 | K70 18DB/BR | EVAP PURGE SIGNAL |
| 30 | K10 18DB/OR | STEERING PRESSURE SWITCH SIGNAL |
| 31 | - | NOT USED |
| 32 | K118 20PK/YL | AAT SIGNAL |
| 33 | K226 18DB/WT | FUEL LEVEL SIGNAL |
| 34 | V37 18VT/TN | S/C SWITCH SIGNAL |
| 35 | K107 20OR | NVLD SWITCH SIGNAL |
| 36 | - | NOT USED |
| 37 | K31 20BR | FUEL PUMP RELAY CONTROL |
| 38 | T752 18DG/OR | STARTER RELAY CONTROL |

**2003-06 Dakota 4.7L (NGC) PCM C3 Connector Pinout**

**2003-06 4.7L (NGC) C4 CONNECTOR**

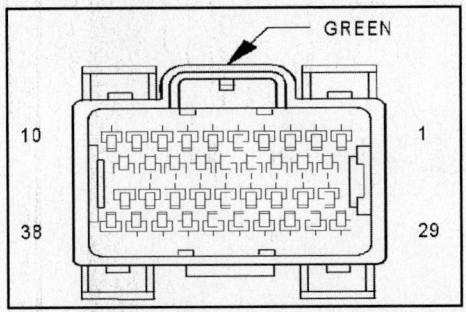

| PIN # | CIRCUIT | FUNCTION |
|---|---|---|
| 1 | T60 18YL/GY | OVERDRIVE SOLENOID CONTROL |
| 2 | T159 18DG/WT | 4C SOLENOID CONTROL |
| 3, 7, 9 | - | NOT USED |
| 4 | T118 18YL/DB | MS SOLENOID CONTROL |
| 5 | - | NOT USED |
| 6 | T119 18WT/DB | 2C SOLENOID CONTROL |
| 8 | T59 18YL/LB | UNDERDRIVE SOLENOID CONTROL |
| 10 | T120 18LG | LR SOLENOID CONTROL |
| 11 | T140 18VT/LG | PRESSURE CONTROL SOLENOID CONTROL |
| 12 | Z13 16BK/RD | GROUND |
| 13 | Z13 16BK/RD | GROUND |
| 14 | Z13 16BK/RD | GROUND |
| 15 | T1 18DG/LB | TRS T1 SENSE |
| 16 | T3 18DG/DB | TRS T3 SENSE |
| 17 | T6 20OR/WT | OVERDRIVE OFF SWITCH SENSE |
| 18 | T15 18YL/BR | TRANSMISSION CONTROL RELAY CONTROL |
| 19 | T16 18RD | TRANSMISSION CONTROL RELAY OUTPUT |
| 20 | T48 18DB | 4C PRESSURE SWITCH SENSE |
| 21 | T29 18GY | UNDERDRIVE PRESSURE SWITCH SENSE |
| 22 | T9 18YL/LG | OVERDRIVE PRESSURE SWITCH SENSE |
| 23-25 | - | NOT USED |
| 26 | T4 18PK/OR | TRS T2 SENSE |
| 27 | - | NOT USED |
| 28 | T16 18RD | TRANSMISSION CONTROL RELAY OUTPUT |
| 29 | T50 18YL/TN | LOW/REVERSE PRESSURE SWITCH SENSE |
| 30 | T47 18YL/DG | 2C PRESSURE SWITCH SENSE |
| 31 | T38 18VT/TN | LINE PRESSURE SENSOR SIGNAL |
| 32 | T14 18LG/WT | OUTPUT SPEED SENSOR SIGNAL |
| 33 | T52 18RD/BK | INPUT SPEED SENSOR SIGNAL |
| 34 | T13 18DB/BK | SPEED SENSOR GROUND |
| 35 | T54 18DG/OR | TRANSMISSION TEMPERATURE SENSOR SIGNAL |
| 36 | - | NOT USED |
| 37 | T42 18DG/YL | TRS T42 SENSE |
| 38 | T16 18RD | TRANSMISSION CONTROL RELAY OUTPUT |

**2003-06 Dakota 4.7 (NGC) PCM C4 Connector Pinout**

## POWER STEERING PRESSURE (PSP) SWITCH

**Connector Pinouts**

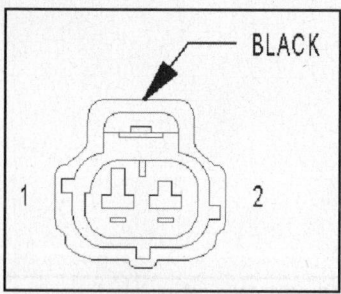

| PIN # | CIRCUIT | FUNCTION |
|-------|---------|----------|
| 1 | Z1 18BK | GROUND |
| 2 | K10 18DB/OR | STEERING PRESSURE SWITCH SIGNAL |

**2003-05 Dakota Power Steering Pressure Switch Connector**

## SENTRY KEY IMMOBILIZER (SKI) SYSTEM

**Description**

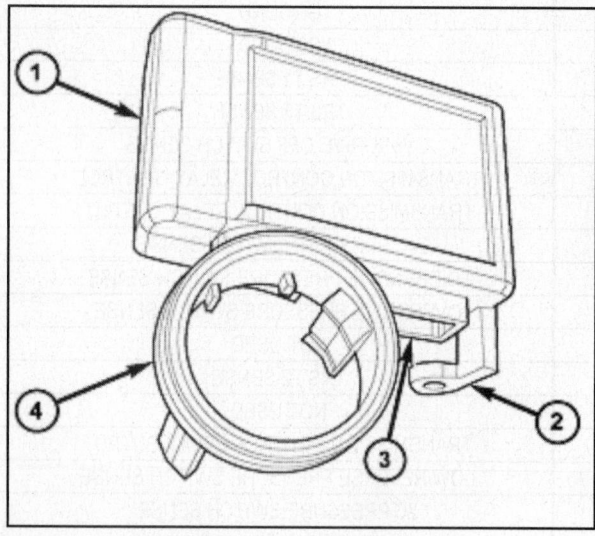

**Showing components of the SKREEM (1): plastic mounting tab (2); integral connector receptacle (3); antenna ring (4)**

The Sentry Key Remote Entry Module (SKREEM) (1) is sometimes referred to as the Wireless Control Module (WCM). The SKREEM is the primary component of the Sentry Key Immobilizer System (SKIS) and is also the receiver for the Remote Keyless Entry (RKE) system. The SKREEM is located on the right side of the steering column, near the ignition lock cylinder housing and is concealed beneath the steering column shrouds. The molded black plastic housing for the SKREEM has an integral molded plastic halo-like antenna ring (4) that extends from the bottom. When the SKREEM is properly installed on the steering column, the antenna ring is oriented around the circumference of the ignition lock cylinder housing.

A single integral connector receptacle (3) is located just behind the antenna ring on the bottom of the SKREEM housing. An integral molded plastic mounting tab (2) on the rear corner of the SKREEM housing has a hole in the center through which a screw passes to secure the unit to the steering column. The SKREEM is connected to the vehicle electrical system through a single take out and connector of the instrument panel wire harness.

Two SKREEM modules are used: one for vehicles equipped with RKE only, and one for vehicles equipped with RKE and SKIS. The SKREEM cannot be adjusted or repaired. If inoperative or damaged, the entire SKREEM unit must be replaced.

## Operation

The Sentry Key Remote Entry Module (SKREEM) contains a Radio Frequency (RF) transceiver and a microprocessor. The SKREEM transmits RF signals to, and receives RF signals from the Sentry Key transponder through a tuned antenna enclosed within the molded plastic antenna ring integral to the SKREEM housing. If this antenna ring is not mounted properly around the ignition lock cylinder housing, communication problems between the SKREEM and the transponder may arise. These communication problems will result in Sentry Key transponder-related faults. The SKREEM also serves as the Remote Keyless Entry (RKE) RF receiver. The SKREEM communicates over the Controller Area Network (CAN) data

The SKREEM retains in memory the ID numbers of any Sentry Key transponder that is programmed into it. A maximum of eight Sentry Key transponders can be programmed into the SKREEM. For added system security, each SKREEM is programmed with a unique Secret Key code. This code is stored in memory, sent over the CAN data bus to the PCM, and is encoded to the transponder of every Sentry Key that is programmed into the SKREEM. Therefore, the Secret Key code is a common element that is found in every component of the Sentry Key Immobilizer System (SKIS). Another security code, called a PIN, is used to gain access to the SKREEM Secured Access Mode. The Secured Access Mode is required during service to perform the SKIS initialization and Sentry Key transponder programming procedures. The SKREEM also stores the Vehicle Identification Number (VIN) in its memory, which it learns through a CAN data bus message from the PCM during SKIS initialization.

In the event that a SKREEM replacement is required, the Secret Key code can be transferred to the new SKREEM from the PCM using the diagnostic scan tool and the SKIS initialization procedure. Proper completion of the SKIS initialization will allow the existing Sentry Keys to be programmed into the new SKREEM so that new keys will not be required. In the event that the original Secret Key code cannot be recovered, SKREEM replacement will also require new Sentry Keys. The diagnostic scan tool will alert the technician during the SKIS initialization procedure if new Sentry Keys are required.

When the ignition switch is turned to the ON position, the SKREEM transmits an RF signal to the transponder in the ignition key. The SKREEM then waits for an RF signal response from the transponder. If the response received identifies the key as valid, the SKREEM sends a valid key message to the PCM over the CAN data bus. If the response received identifies the key as invalid or if no response is received from the key transponder, the SKREEM sends an invalid key message to the PCM. The PCM will enable or disable engine operation based upon the status of the SKREEM messages. It is important to note that the default condition in the PCM is an invalid key; therefore, if no message is received from the SKREEM by the PCM, the engine will be disabled and the vehicle immobilized after two seconds of running.

The SKREEM also sends security indicator status messages to the EMIC over the CAN data bus to tell the EMIC how to operate the security indicator. The security indicator status message from the SKREEM tells the EMIC to turn the indicator on for about three seconds each time the ignition switch is turned to the ON position as a bulb test. After completion of the bulb test, the SKREEM sends security indicator status messages to the EMIC to turn the indicator off, turn the indicator on, or to flash the indicator on and off. If the security indicator flashes or stays on solid after the bulb test, it signifies a SKIS fault. If the SKREEM detects a system malfunction or if the SKIS has become inoperative, the security indicator will stay on solid. If the SKREEM detects an invalid key or if a key transponder-related fault exists, the security indicator will flash. If the vehicle is equipped with the Customer Learn transponder programming feature, the SKREEM will also send messages to the EMIC to flash the security indicator whenever the Customer Learn programming mode is being utilized.

## Removal & Installation

**WARNING: To avoid personal injury or death, on vehicles equipped with airbags, disable the supplemental restraint system before attempting any steering wheel, steering column, airbag, occupant classification system, seat belt tensioner, impact sensor, or instrument panel component diagnosis or service. Disconnect and isolate the battery negative (ground) cable, then wait two minutes for the system capacitor to discharge before performing further diagnosis or service. This is the only sure**

*way to disable the supplemental restraint system. Failure to take the proper precautions could result in accidental airbag deployment.*

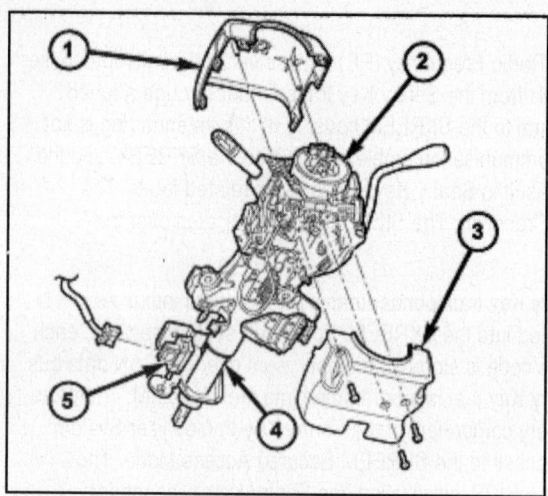

**Removing steering column (2) shrouds (1 and 3) and detaching wiring from electrical connector (5) on lower steering column (4)**

1. Disconnect and isolate the battery negative cable.
2. Grasp the tilt steering column knob firmly and pull it straight rearward to remove it from the tilt adjuster lever on the left side of the column.
3. From below the steering column, remove the two outboard screws that secure the upper column shroud (1) to the lower shroud (3).
4. Using hand pressure, push gently inward on both sides of the upper shroud above the parting line of the lower shroud to release the snap features that secure the two shroud halves to each other.
5. Remove the upper shroud from the lower shroud and the steering column.
6. Remove the one center screw that secures the lower shroud to the steering column (4).
7. Remove the lower shroud from the steering column.
8. Remove the screw (2) that secures the SKREEM (1) to the steering column housing.
9. Disengage the antenna ring (3) from around the ignition lock cylinder housing (4) and remove the SKREEM.

**To install:**

10. Position the Sentry Key Remote Entry Module (SKREEM) (1) onto the steering column with the antenna ring (3) oriented around the ignition lock cylinder housing (4).
11. Install and tighten the screw (2) that secures the SKREEM to the steering column housing. Tighten the screw to 2 Nm (20 inch lbs.).
12. Reconnect the wire harness connector (2) to the SKREEM (1).
13. Position the lower shroud (3) onto the steering column (4).
14. From below the steering column, install and tighten the one center screw that secures the lower shroud to the steering column. Tighten the screw to 2 Nm (20 inch lbs.).
15. Position the upper shroud (1) onto the steering column over the lower shroud. On vehicles equipped with an automatic transmission, be certain to engage the gearshift lever gap hider into the opening in the right side of both shroud halves.
16. Align the snap features on the upper shroud with the receptacles in the lower shroud and apply hand pressure to snap them together.

17. Install and tighten the two outboard screws that secure the upper shroud to the lower shroud. Tighten the screws to 2 Nm (20 inch lbs.).

18. Align the tilt steering column knob with the tilt adjuster lever on the left side of the steering column and use hand pressure to snap it back into place.

19. Reconnect the battery negative cable.

*NOTE: On vehicles equipped with the optional Sentry Key Immobilizer System (SKIS) if the SKREEM is replaced with a new unit, a diagnostic scan tool MUST be used to initialize the new SKREEM and to program at least two Sentry Key transponders before the vehicle can be operated.*

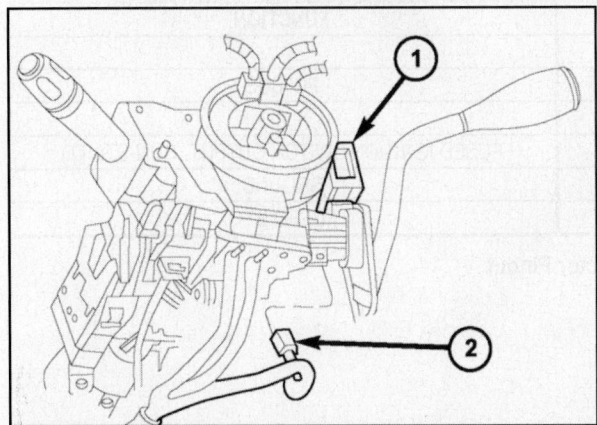

**Disconnect the wire harness connector (2) from the Sentry Key Remote Entry Module (SKREEM) (1).**

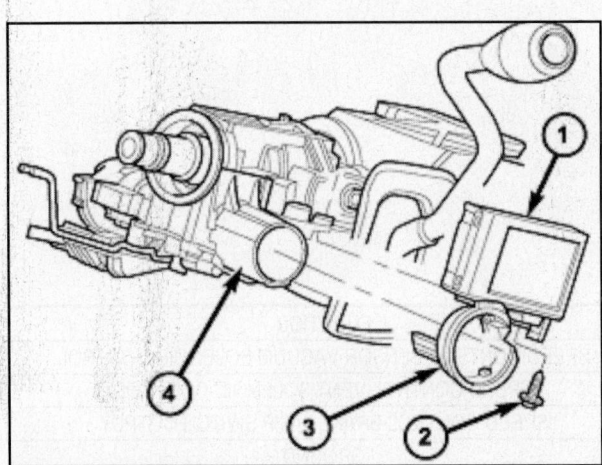

**Removing SKREEM (1) and halo antenna (3) from the lock cylinder (4), after removing screw (2)**

**Connector Pinouts**

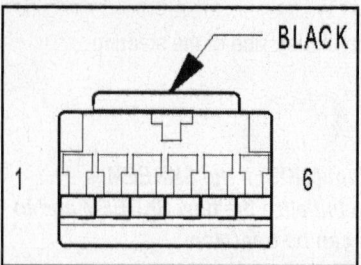

| PIN # | CIRCUIT | FUNCTION |
|-------|---------|----------|
| 1 | - | - |
| 2 | D25 20VT/YL | PCI BUS |
| 3 | - | - |
| 4 | F18 20LG/BK | FUSED IGNITION SWITCH OUTPUT (RUN-START) |
| 5 | Z11 20BK/WT | GROUND |
| 6 | M1 20PK | FUSED B (+) |

**2005-06 Dakota Sentry Key Immobilizer (SKI) Connector Pinout**


## SPEED CONTROL SERVO & SWITCH

**Connector Pinouts**

*SERVO*

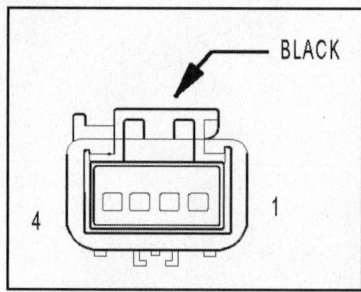

| PIN # | CIRCUIT | FUNCTION |
|-------|---------|----------|
| 1 | V36 18TN/RD | SPEED CONTROL SENSOR VACUUM SOLENOID CONTROL |
| 2 | V35 18LG/RD | SPEED CONTROL VENT SOLENOID CONTROL |
| 3 | V30 20DB/RD | SPEED CONTROL BRAKE LAMP SWITCH OUTPUT |
| 4 | Z1 20BK | GROUND |

**2000-02 Dakota 3.9L, 4.7L, 5.9L Vehicle Speed Control Servo Connector Pinout**

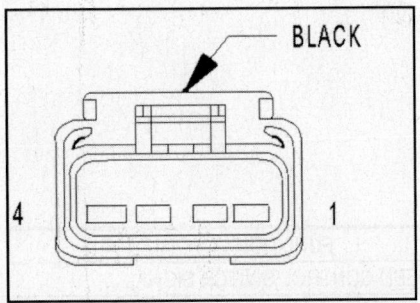

| PIN # | CIRCUIT | FUNCTION |
|-------|---------|----------|
| 1 | V36 20TN/RD | SPEED CONTROL VACUUM SOLENOID CONTROL |
| 2 | V35 20LG/RD | SPEED CONTROL VENT SOLENOID CONTROL |
| 3 | V30 20DB/RD | BRAKE LAMP SWITCH OUTPUT |
| 4 | Z1 18BK | GROUND |

**2003-05 Dakota Speed Control Servo**

### SPEED CONTROL COLUMN SWITCHES

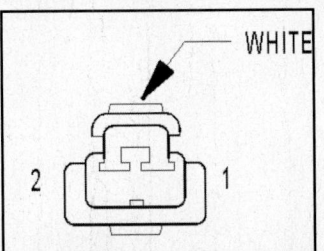

| PIN # | CIRCUIT | FUNCTION |
|-------|---------|----------|
| 1 | K4 22WT | SENSOR GROUND |
| 2 | V37 22DG/RD | SPEED CONTROL SWITCH SIGNAL |

**2000-02 Dakota Left Speed Control Switch Connector Pinout**

| PIN # | CIRCUIT | FUNCTION |
|-------|---------|----------|
| 1 | K4 22WT | SENSOR GROUND |
| 2 | V37 22DG/RD | SPEED CONTROL SWITCH SIGNAL |

**2000-02 Dakota Right Speed Control Switch Connector Pinout**

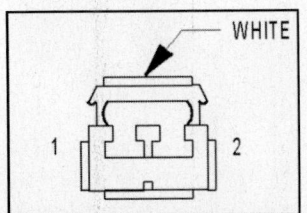

| PIN # | CIRCUIT | FUNCTION |
|-------|---------|----------|
| 1 | V37 22RD/LG | SPEED CONTROL SWITCH SIGNAL |
| 2 | K4 22BK/LB | SENSOR GROUND |

**2003-05 Dakota Left Speed Control Switch Connector Pinout**

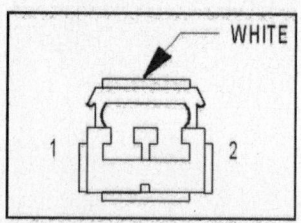

| PIN # | CIRCUIT | FUNCTION |
|---|---|---|
| 1 | V37 22RD/LG | SPEED CONTROL SWITCH SIGNAL |
| 2 | K4 22BK/LB | SENSOR GROUND |

**2003-05 Dakota Right Speed Control Switch Connector Pinout**

## THROTTLE POSITION (TP) SENSOR

**Connector Pinouts**

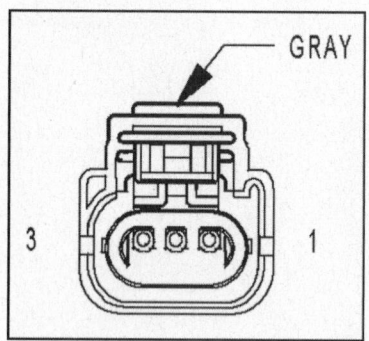

| PIN # | CIRCUIT | FUNCTION |
|---|---|---|
| 1 | K7 18OR | 5 VOLT SUPPLY |
| 2 | K4 18BK/LB (3.7L) | SENSOR GROUND 1 |
| 2 | K22 20BR/OR (4.7L) | TP NO. 1 SIGNAL |
| 3 | K22 20BR/OR (3.7L) | TP NO. 1 SIGNAL |
| 3 | K4 18BK/LB (4.7L) | SENSOR GROUND |

**2003-05 Dakota 3.7L & 4.7L TP Sensor**

## DAIMLERCHRYSLER SUV: CHEROKEE, GRAND CHEROKEE, LIBERTY, WRANGLER

AUTO SHUTDOWN (ASD) RELAY

### Description & Operation

The 5-pin, 12-volt, Automatic Shutdown (ASD) relay is located in the Power Distribution Center (PDC). Refer to label on PDC cover for relay location.

The ASD relay supplies battery voltage (12+ volts) to the fuel injectors and ignition coil(s). With certain emissions packages it also supplies 12 volts to the oxygen sensor heating elements and the oxygen sensor heater relay. The ground circuit for the coil, within the ASD relay, is controlled by the Powertrain Control Module (PCM). The PCM operates the ASD relay by switching its ground circuit ON and OFF. The ASD relay will be shut-down, meaning the 12-volt power supply to the ASD relay will be de-activated by the PCM if the ignition key is left in the ON position. This is if the engine has not been running for approximately 1.8 seconds.

*ASD SENSE*

A 12-volt signal at this input indicates to the PCM that the ASD has been activated. The relay is used to connect the oxygen sensor heater elements, oxygen sensor heater relay ignition coil and fuel injectors to 12-volt power supply.

This input is used only to sense that the ASD relay is energized. If the Powertrain Control Module (PCM) does not see 12 volts at this input when the ASD should be activated, it will set a Diagnostic Trouble Code (DTC).

### Removal & Installation

The ASD relay is located in the Power Distribution Center (PDC). Refer to label on PDC cover for relay location.

1. Remove PDC cover.
2. Remove relay from PDC.
3. Check condition of relay terminals and PDC connector terminals for damage or corrosion. Repair if necessary before installing relay.
4. Check for pin height (pin height should be the same for all terminals within the PDC connector). Repair if necessary before installing relay.

### Testing

- Terminal number **30** is connected to battery voltage. For both the ASD and fuel pump relays, terminal **30** is connected to battery voltage at all times.
- The PCM grounds the coil side of the relay through terminal number **85**.
- Terminal number **86** supplies voltage to the coil side of the relay.
- When the PCM de-energizes the ASD and fuel pump relays, terminal number **87A** connects to terminal **30**. This is the OFF position. In the OFF position,
- Voltage is not supplied to the rest of the circuit. Terminal **87A** is the center terminal on the relay.
- When the PCM energizes the ASD and fuel pump relays, terminal **87** connects to terminal **30**. This is the ON position. Terminal **87** supplies voltage to the rest of the circuit.

    1. Remove relay from connector before Testing. With the relay removed from the vehicle, use an ohmmeter to check the resistance between terminals 85 and 86. The resistance should be 75 ohms ±5 ohms.
    2. Connect the ohmmeter between terminals 30 and 87A. The ohmmeter should show continuity between terminals 30 and 87A.
    3. Connect the ohmmeter between terminals 87 and 30. The ohmmeter should not show continuity at this time.
    4. Connect one end of a jumper wire (16 gauge or smaller) to relay terminal 85. Connect the other end of the jumper wire to the ground side of a 12-volt power source.
    5. Connect one end of another jumper wire (16 gauge or smaller) to the power side of the 12-volt power source. Do not attach the other end of the jumper wire to the relay at this time.

**WARNING: Do NOT allow ohmmeter to contact terminals 85 or 86 during this test. Damage to ohmmeter may result.**

6. Attach the other end of the jumper wire to relay terminal 86. This activates the relay. The ohmmeter should now show continuity between relay terminals 87 and 30. The ohmmeter should not show continuity between relay terminals 87A and 30.

7. Disconnect jumper wires. Replace the relay if it did not pass the continuity and resistance tests. If the relay passed the tests, it operates properly. Check the remainder of the ASD and fuel pump relay circuits.

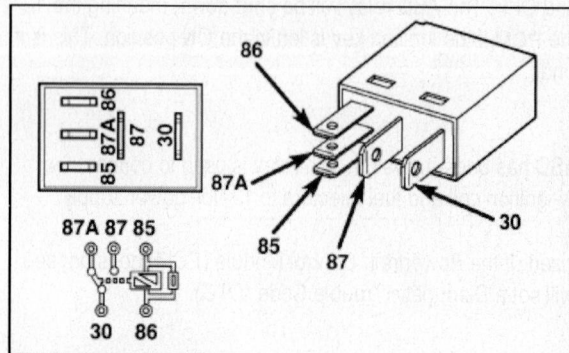

**Type 1 ISO Micro relay**

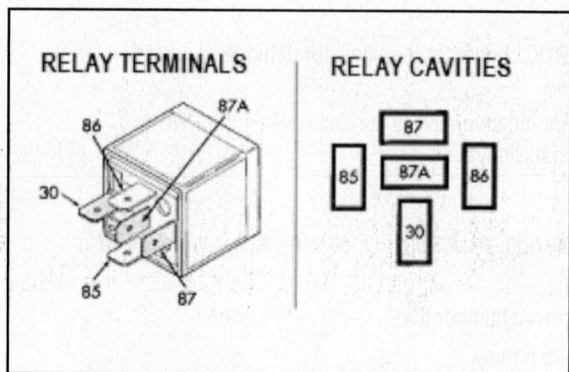

- #30    Common feed
- #85    Coil ground
- #86    Coil battery
- #87    Normally open
- #87A   Normally closed

**Type 2 ASD and fuel pump relay**

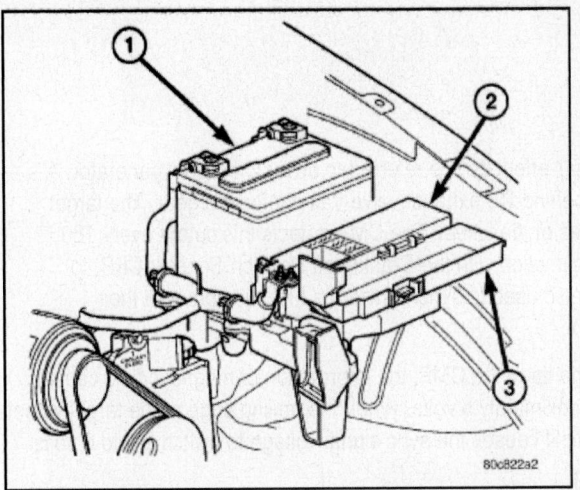

80c822a2

**Power Distribution Center (PDC) Battery (1) PDC (2) PDC cover (3)**

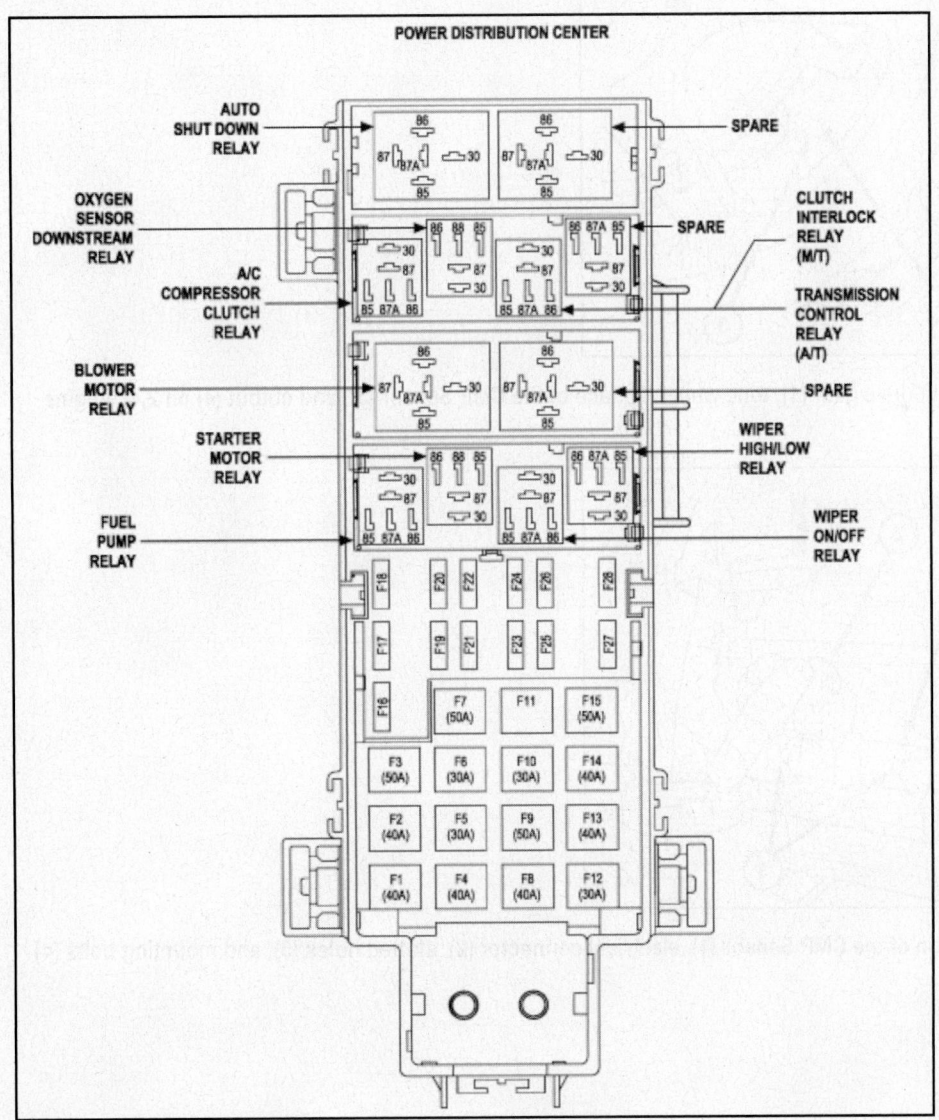

**Power Distribution Center**

CAMSHAFT POSITION (CMP) SENSOR

### Description & Operation

*2.4L*

The Camshaft Position Sensor (CMP) sensor contains a hall effect device referred to as a sync signal generator. A rotating target wheel (tone wheel) for the CMP is located behind the exhaust valve-camshaft drive gear. The target wheel is equipped with a cutout (notch) around 180 degrees of the wheel. The CMP detects this cutout every 180 degrees of camshaft gear rotation. Its signal is used in conjunction with the Crankshaft Position Sensor (CKP) to differentiate between fuel injection and spark events. It is also used to synchronize the fuel injectors with their respective cylinders.

When the leading edge of the target wheel cutout enters the tip of the CMP, the interruption of magnetic field causes the voltage to switch high, resulting in a sync signal of approximately 5 volts. When the trailing edge of the target wheel cutout leaves the tip of the CMP, the change of magnetic field causes the sync signal voltage to switch low to 0 volts.

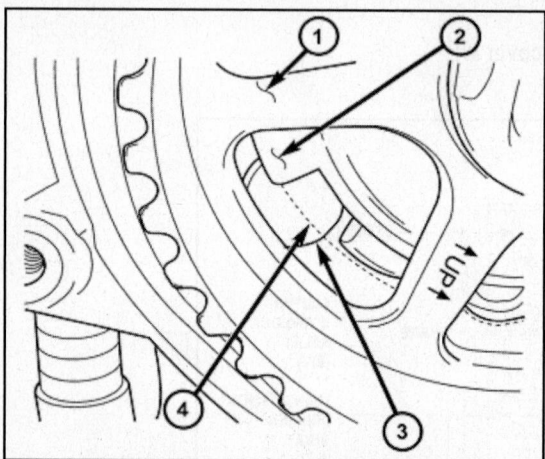

Identifying camshaft drive gear (1), tone wheel (2), face of the CMP Sensor (3), and cutout (4) on 2.4L engine

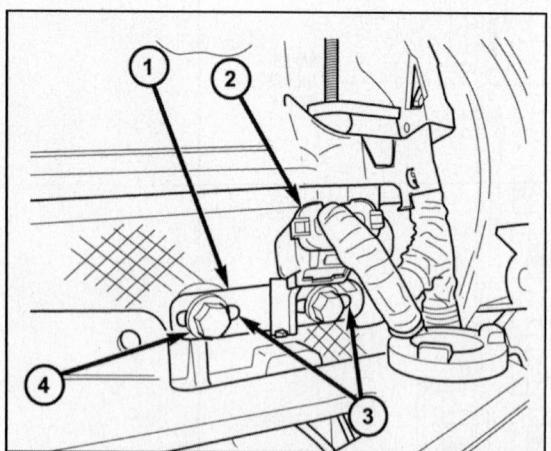

Showing the location of the CMP Sensor (1), electrical connector (2), slotted holes (3), and mounting bolts (4) on 2.4L engines

**2.5L**

On the 2.5L 4–cylinder engine the Camshaft Position (CMP) sensor is located in the distributor.

The sensor contains a Hall effect device called a sync signal generator to generate a fuel sync signal. This sync signal generator detects a rotating pulse ring (shutter) on the distributor shaft. The pulse ring rotates 180 degrees through the sync signal generator. Its signal is used in conjunction with the Crankshaft Position (CKP) sensor to differentiate between fuel injection and spark events. It is also used to synchronize the fuel injectors with their respective cylinders.

When the leading edge of the pulse ring (shutter) enters the sync signal generator, the following occurs: The interruption of magnetic field causes the voltage to switch high resulting in a sync signal of approximately 5 volts. When the trailing edge of the pulse ring (shutter) leaves the sync signal generator, the following occurs: The change of the magnetic field causes the sync signal voltage to switch low to 0 volts.

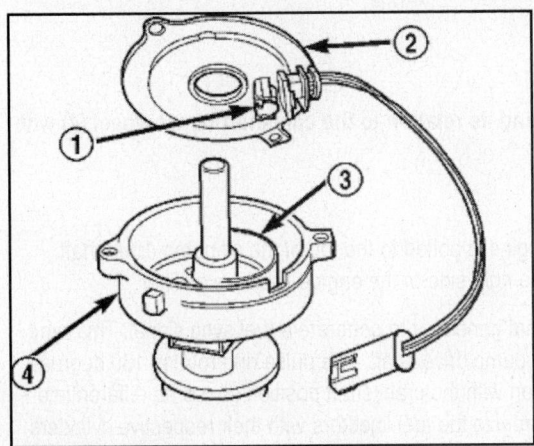

Showing the CMP Sensor (2) in the distributor on 2.5L engines: 1 – synch signal generator; 2 – CMP sensor; 3 – pulse ring; 4 – distributor assembly

**3.7L**

The Camshaft Position Sensor (CMP) sensor contains a hall effect device referred to as a sync signal generator. A rotating target wheel (tone wheel) for the CMP is located at the front of the camshaft for the right cylinder head. This sync signal generator detects notches located on a tone wheel. As the tone wheel rotates, the notches pass through the sync signal generator. The signal from the CMP sensor is used in conjunction with the Crankshaft Position Sensor (CKP) to differentiate between fuel injection and spark events. It is also used to synchronize the fuel injectors with their respective cylinders.

When the leading edge of the target wheel notch enters the tip of the CMP, the interruption of magnetic field causes the voltage to switch high, resulting in a sync signal of approximately 5 volts. When the trailing edge of the target wheel notch leaves then tip of the CMP the change of the magnetic field causes the sync signal voltage to switch to 0 volts.

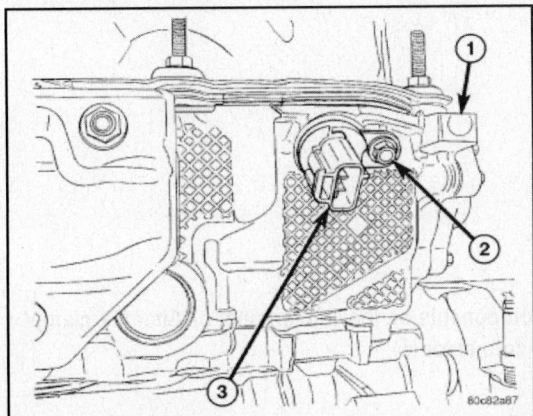

Showing the CMP sensor (3) and mounting bolt (2) on the right front cylinder head (1) on the 3.7L engine

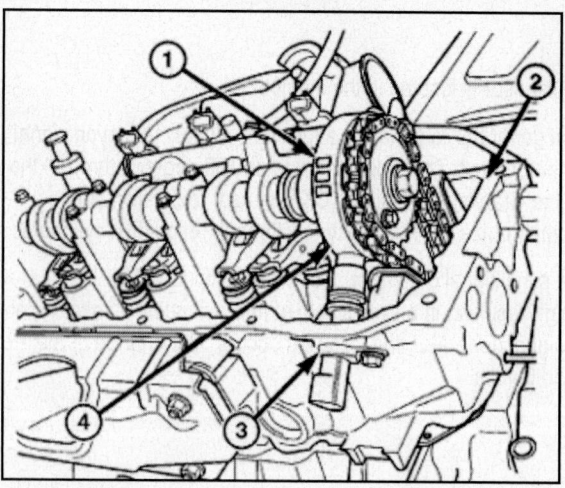

**Showing the CMP sensor (3) on the right cylinder head (2) and its relation to the camshaft target wheel (4) with the indicating notches (1) on the 3.7L engine**

*4.0L*

The Camshaft Position Sensor (CMP) on the 4.0L 6–cylinder engine is bolted to the top of the oil pump drive shaft assembly. The sensor and drive shaft assembly is located on the right side of the engine near the oil filter.

The CMP sensor contains a Hall effect device called a sync signal generator to generate a fuel sync signal. This sync signal generator detects a rotating pulse ring (shutter) on the oil pump drive shaft. The pulse ring rotates 180 degrees through the sync signal generator. Its signal is used in conjunction with the crankshaft position sensor to differentiate between fuel injection and spark events. It is also used to synchronize the fuel injectors with their respective cylinders.

When the leading edge of the pulse ring (shutter) enters the sync signal generator, the following occurs: The interruption of magnetic field causes the voltage to switch high resulting in a sync signal of approximately 5 volts.

When the trailing edge of the pulse ring (shutter) leaves the sync signal generator, the following occurs: The change of the magnetic field causes the sync signal voltage to switch low to 0 volts.

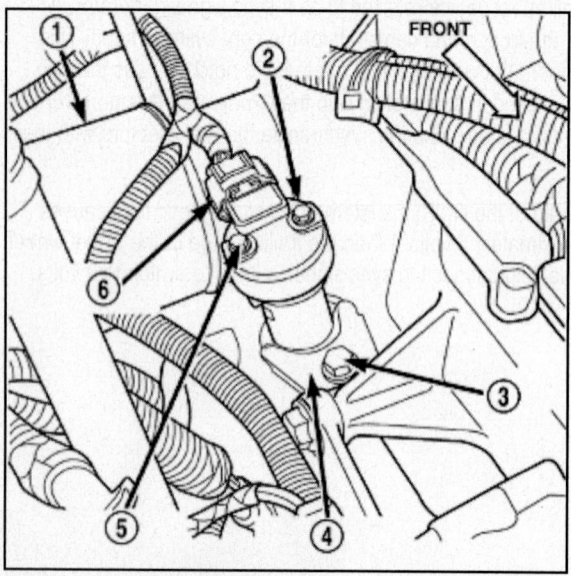

**Identifying the location of the CMP sensor (2) and related components on the 4.0L engine: oil filter (1); clamp bolt (3); hold-down clamp (4); mounting bolts (5); electrical connector (6)**

### 4.7L

The Camshaft Position Sensor (CMP) on the 4.7L V–8 engine is bolted to the front/top of the right cylinder head.

The CMP sensor contains a Hall effect device called a sync signal generator to generate a fuel sync signal. This sync signal generator detects notches located on a tone wheel. The tone wheel is located at the front of the camshaft for the right cylinder head. As the tone wheel rotates, the notches pass through the sync signal generator. The pattern of the notches (viewed counter-clockwise from front of engine) is: 1 notch, 2 notches, 3 notches, 3 notches, 2 notches 1 notch, 3 notches and 1 notch. The signal from the CMP sensor is used in conjunction with the crankshaft position sensor to differentiate between fuel injection and spark events. It is also used to synchronize the fuel injectors with their respective cylinders.

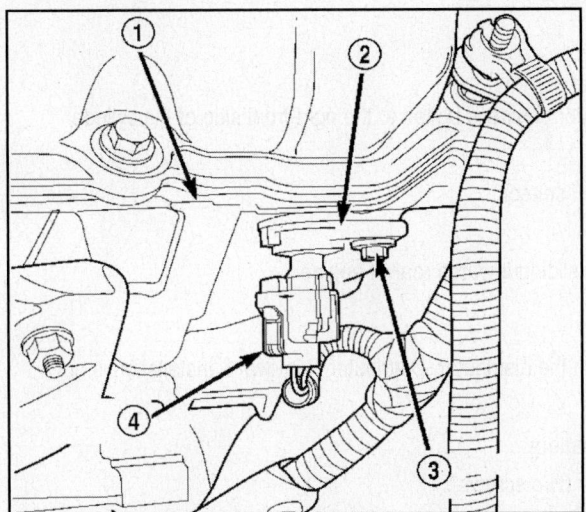

**Showing the CMP Sensor location on 4.7L engine: RH cylinder head (1); CMP sensor (2); mounting bolt (3); electrical connector (4)**

### 5.7L V-8

The Camshaft Position Sensor (CMP) on the 5.7L V-8 engine is located below the generator on the timing chain / case cover on the right/front side of engine.

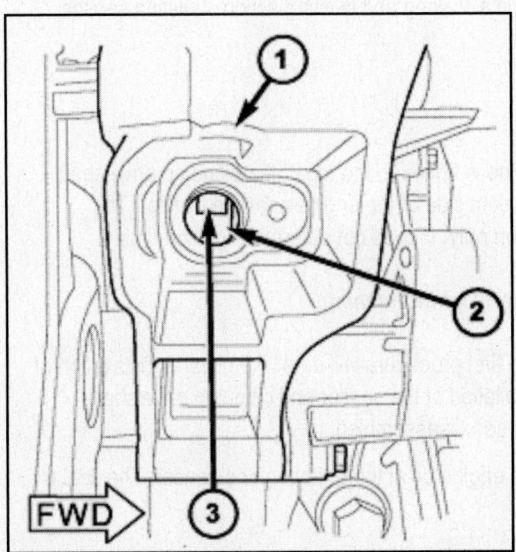

**Showing the location of the camshaft (2) and the housing notch (3) on 5.7L engine**

The CMP sensor is used in conjunction with the crankshaft position sensor to differentiate between fuel injection and spark events. It is also used to synchronize the fuel injectors with their respective cylinders. The sensor generates electrical pulses. These pulses (signals) are sent to the Powertrain Control Module (PCM). The PCM will then determine crankshaft position from both the camshaft position sensor and crankshaft position sensor.

The tone wheel is located at the front of the camshaft (2). As the tone wheel rotates, notches (3) pass through the sync signal generator.

When the cam gear is rotating, the sensor will detect the notches. Input voltage from the sensor to the PCM will then switch from a low (approximately 0.3 volts) to a high (approximately 5 volts). When the sensor detects a notch has passed, the input voltage switches back low to approximately 0.3 volts.

### Removal & Installation

#### 2.4L

The Camshaft Position Sensor (CMP) on the 2.4L 4–cylinder engine is bolted to the right-front side of the cylinder head. Sensor position (depth) is adjustable.

1. Disconnect electrical connector at CMP sensor.
2. Remove 2 sensor mounting bolts.
3. Remove sensor from cylinder head by sliding towards rear of engine.

#### 2.5L

On 2.5L engines, the camshaft position sensor is located in the distributor. Distributor Removal & Installation is not necessary to remove camshaft position sensor.

1. Disconnect negative battery cable at battery.
2. Remove distributor cap from distributor (two screws).
3. Disconnect camshaft position sensor wiring harness from main engine wiring harness.
4. Remove distributor rotor from distributor shaft.
5. Lift camshaft position sensor assembly from distributor housing.

#### 3.7L

The Camshaft Position Sensor (CMP) on this engine is bolted to the front/top of the right cylinder head.

1. Disconnect electrical connector at CMP sensor.
2. Remove sensor mounting bolt.
3. Carefully remove sensor from cylinder head in a rocking and twisting action. Twisting sensor eases Removal & Installation.
4. Check condition of sensor O-ring.

#### 4.0L

The Camshaft Position Sensor (CMP) on the 4.0L 6–cylinder engine is bolted to the top of the oil pump drive shaft assembly. The sensor and drive shaft assembly is located on the right side of the engine near the oil filter. The rotational position of oil pump drive determines fuel synchronization only. It does not determine ignition timing.

*CAUTION: Do NOT attempt to rotate the oil pump drive to modify ignition timing.*

Two different procedures are used for Removal & Installation. The first procedure will detail Removal & Installation of the sensor only. The second procedure will detail Removal & Installation of the sensor and oil pump drive shaft assembly. The second procedure is to be used if the engine has been disassembled.

An internal oil seal is used in the drive shaft housing that prevents engine oil at the bottom of the sensor. The seal is not serviceable.

### Sensor Only

1. Disconnect electrical connector at CMP sensor.
2. Remove 2 sensor mounting bolts.
3. Remove sensor from oil pump drive.

### Oil Pump Drive & Sensor

**CAUTION: If the CMP and oil pump drive are to be removed and installed, do NOT allow engine crankshaft or camshaft to rotate. CMP sensor relationship will be lost.**

1. Disconnect electrical connector at CMP sensor.
2. Remove 2 sensor mounting bolts
3. Remove sensor from oil pump drive.
4. Before proceeding to next step, mark and note rotational position of oil pump drive in relationship to engine block. After installation:, the CMP sensor should face rear of engine 0°.
5. Remove hold-down bolt and clamp.
6. While pulling assembly from engine, note direction and position of pulse ring. After Removal & Installation, look down into top of oil pump and note direction and position of slot at top of oil pump gear.
7. Remove and discard old oil pump drive-to-engine block gasket.

### 4.7L

The Camshaft Position Sensor (CMP) on the 4.7L V–8 engine is bolted to the front/top of the right cylinder head.

**NOTE: It is easier to remove/install sensor from under vehicle.**

1. Raise and support vehicle.
2. Disconnect electrical connector at CMP sensor.
3. Remove sensor mounting bolt.
4. Carefully twist sensor from cylinder head. Check condition of sensor O-ring.

**NOTE: Some 4.7L are equipped with a sensor spacer shim. If equipped, this shim will be located at sensor bolt hole, between the cylinder head and sensor mounting tang. Save this shim for installation.**

### 5.7L V-8

1. Disconnect electrical connector from the CMP sensor.
2. Remove sensor mounting bolt.
3. Carefully twist sensor from cylinder head. Check condition of sensor O-ring.

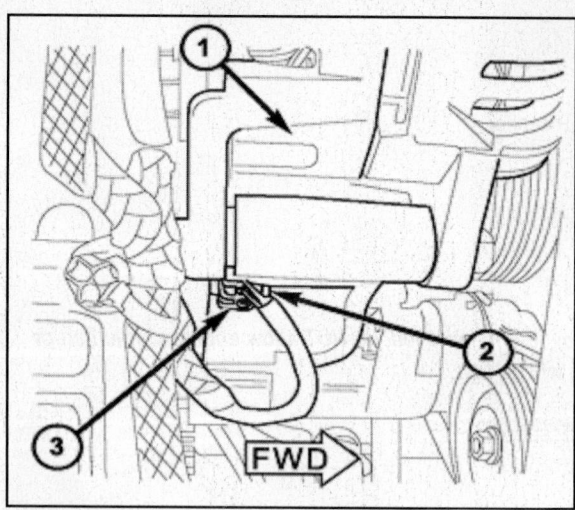

The Camshaft Position Sensor (CMP) (2) and electrical connector (3) on the 5.7L V-8 engine are located on right side of timing chain cover below generator (1)

## To install:

### 2.4L

The Camshaft Position Sensor (CMP) on the 2.4L 4–cylinder engine is bolted to the right-front side of the cylinder head. Sensor position (depth) is adjustable.

1. Remove plastic, upper timing belt cover (timing gear cover) by removing 3 bolts. Before attempting to remove cover, remove electrical connector from Engine Coolant Temperature (ECT) sensor. This will prevent damage to sensor.

2. Rotate (bump over) engine until camshaft timing gear and target wheel (tone wheel) are positioned and aligned to face of sensor.

**CAUTION: If gear and tone wheel are not properly positioned, damage to both sensor and target wheel will occur when attempting to start engine. Face of sensor MUST be behind target wheel while adjusting.**

3. Position sensor to cylinder head and install 2 sensor mounting bolts finger tight.

4. Set air gap between rear of target wheel and face of sensor to .030". This can best be accomplished using an L-shaped, wire-type spark plug gapping gauge. A piece of .030" brass shim stock may also be used.

5. Gently push sensor forward until it contacts gapping gauge. Do not push hard on sensor. Tighten 2 sensor mounting bolts.

**CAUTION: After tightening sensor mounting bolts, recheck air gap and adjust as necessary. Retorque bolts.**

6. Install upper timing belt cover and 3 bolts.

7. Connect electrical connector to ECT sensor.

8. Connect electrical connector to CMP sensor.

### 2.5L

On 2.5L engines, the camshaft position sensor is located in the distributor.

1. Install camshaft position sensor to distributor. Align sensor into notch on distributor housing.

2. Connect wiring harness.

3. Install rotor.

4.  Install distributor cap. Tighten mounting screws.

### 3.7L

The Camshaft Position Sensor (CMP) on this engine is bolted to the front/top of the right cylinder head.

1.  Clean out machined hole in cylinder head.
2.  Apply a small amount of engine oil to sensor O-ring.
3.  Install sensor into cylinder head with a slight rocking and twisting action.

**CAUTION: Before tightening sensor mounting bolt, be sure sensor is completely flush to cylinder head. If sensor is not flush, damage to sensor mounting tang may result.**

4.  Install mounting bolt and tighten.
5.  Connect electrical connector to sensor.

### 4.0L

#### Sensor Only

The Camshaft Position Sensor (CMP) on the 4.0L 6–cylinder engine is bolted to the top of the oil pump drive shaft assembly. The sensor and drive shaft assembly is located on the right side of the engine near the oil filter.

1.  Install sensor to oil pump drive.
2.  Install 2 sensor mounting bolts and tighten to 2 Nm (15 inch lbs.) torque.
3.  Connect electrical connector to CMP sensor.

#### Oil Pump Drive & Sensor

1.  Clean oil pump drive mounting hole area of engine block.
2.  Install new oil pump drive-to-engine block gasket.
3.  Temporarily install a toothpick or similar tool through access hole at side of oil pump drive housing. Align toothpick into mating hole on pulse ring.
4.  Install oil pump drive into engine while aligning into slot on oil pump. Rotate oil pump drive back to its original position and install hold-down clamp and bolt. Finger-tighten bolt. Do NOT do a final tightening of bolt at this time.
5.  If engine crankshaft or camshaft has been rotated, such as during engine tear-down, CMP sensor relationship must be reestablished.
6.  Remove ignition coil rail assembly.
7.  Remove cylinder number 1 spark plug.
8.  Hold a finger over the open spark plug hole. Rotate engine at vibration dampener bolt until compression (pressure) is felt.
9.  Slowly continue to rotate engine.

**NOTE: Do this until timing index mark on vibration damper pulley aligns with top dead center (TDC) mark (0 degree) on timing degree scale. Always rotate engine in direction of normal rotation. Do NOT rotate engine backward to align timing marks.**

10. Install oil pump drive into engine while aligning into slot on oil pump. If pump drive will not drop down flush to engine block, the oil pump slot is not aligned. Remove oil pump drive and align slot in oil pump to shaft at bottom of drive. Install into engine. Rotate oil pump drive back to its original position and install hold-down clamp and bolt. Finger-tighten the bolt. Do NOT do a final tightening of bolt at this time.
11. Remove toothpick from housing.

12.    Install sensor to oil pump drive. After installation:, the CMP sensor should face rear of engine: 0°.

13.    Install 2 sensor mounting bolts and tighten to 2 Nm (15 inch lbs.) torque.

14.    Connect electrical connector to CMP sensor.

15.    If removed, install spark plug and ignition coil rail.

16.    To verify correct rotational position of oil pump drive, the DRB scan tool must be used.

*WARNING: When performing the following test, the engine will be running. Be careful not to stand in line with the fan blades or fan belt. Do NOT wear loose clothing.*

17.    Connect the scan tool to data link connector. The data link connector is located in passenger compartment, below and to left of steering column.

18.    Gain access to "SET SYNC" screen on the scan tool.

19.    Follow the directions on the scan tool screen and start engine. Bring to operating temperature (engine must be in "closed loop" mode).

20.    With engine running at idle speed, the words "IN RANGE" should appear on the scan tool screen along with "0°". This indicates correct position of oil pump drive.

21.    If a plus (+) or a minus (-) is displayed next to degree number, and/or the degree displayed is not zero, loosen, but do NOT remove hold-down clamp bolt. Rotate oil pump drive until "IN RANGE" appears on screen. Continue to rotate oil pump drive until achieving as close to 0° as possible.

*NOTE: The degree scale on "SET SYNC" screen of the scan tool is referring to fuel synchronization only. It is not referring to ignition timing. Because of this, do NOT attempt to adjust ignition timing using this method. Rotating oil pump drive will have no effect on ignition timing. All ignition timing values are controlled by powertrain control module (PCM).*

22.    Tighten hold-down clamp bolt to 23 Nm (17 ft. lbs.) torque.

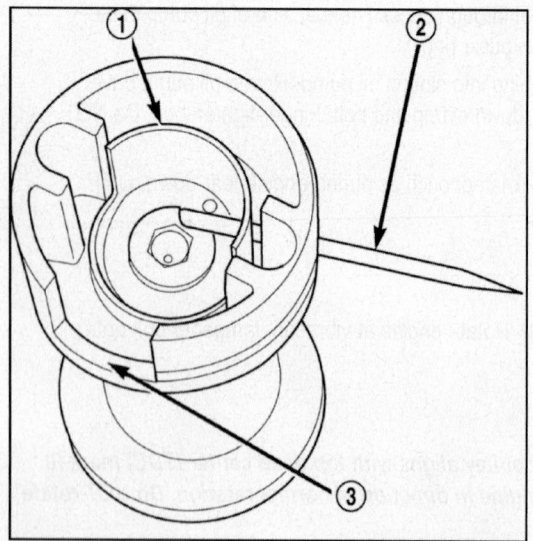

Setting the CMP sensor pulse ring alignment on 4.0L engine: 1 – pulse ring (shutter); 2 – toothpick for holding position; 3 – sensor base (oil pump driveshaft assembly)

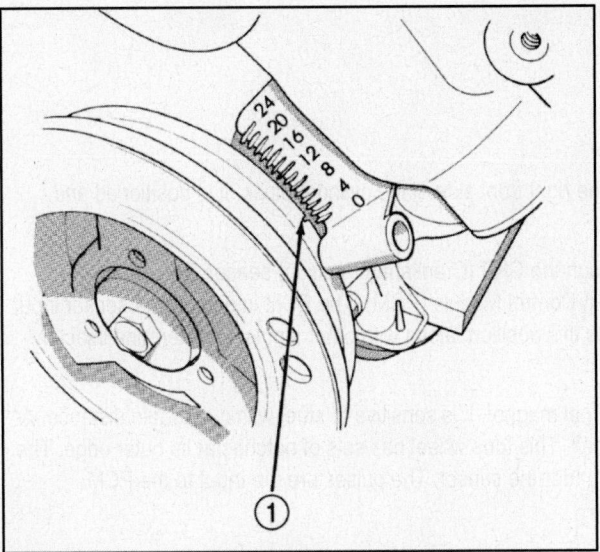

**Setting timing mark alignment on 4.0L engine**

### 4.7L

The Camshaft Position Sensor (CMP) on the 4.7L V–8 engine is bolted to the front/top of the right cylinder head.

1. Clean out machined hole in cylinder head.
2. Apply a small amount of engine oil to sensor O-ring.
3. Install sensor into cylinder head with a slight rocking action. Do NOT twist sensor into position as damage to O-ring may result.

*CAUTION: Before tightening sensor mounting bolt, be sure sensor is completely flush to cylinder head. If sensor is not flush, damage to sensor mounting tang may result.*

4. Install mounting bolt and tighten to 12 Nm (106 inch lbs.) torque.
5. Connect electrical connector to sensor.
6. Lower vehicle.

### 5.7L V-8

1. Clean out machined hole in cylinder head.
2. Apply a small amount of engine oil to sensor O-ring.
3. Install sensor into cylinder head with a slight rocking action. Do not twist sensor into position as damage to O-ring may result.

*CAUTION: Before tightening sensor mounting bolt, be sure sensor is completely flush to timing chain cover. If sensor is not flush, damage to sensor mounting tang may result.*

4. Install mounting bolt and tighten.
5. Connect electrical connector to sensor.

CRANKSHAFT POSITION (CKP) SENSOR

**Description & Operation**

*2.4L*

The Crankshaft Position (CKP) sensor is mounted into the right front side of the cylinder block. It is positioned and bolted into a machined hole.

Engine speed and crankshaft position are provided through the CKP (Crankshaft Position) sensor. The sensor generates pulses that are the input sent to the Powertrain Control Module (PCM). The PCM interprets the sensor input to determine the crankshaft position. The PCM then uses this position, along with other inputs, to determine injector sequence and ignition timing.

The sensor is a Hall effect device combined with an internal magnet. It is sensitive to steel within a certain distance. A tone wheel (target wheel) is a part of the engine crankshaft. This tone wheel has sets of notches at its outer edge. The notches cause a pulse to be generated when they pass under the sensor. The pulses are the input to the PCM.

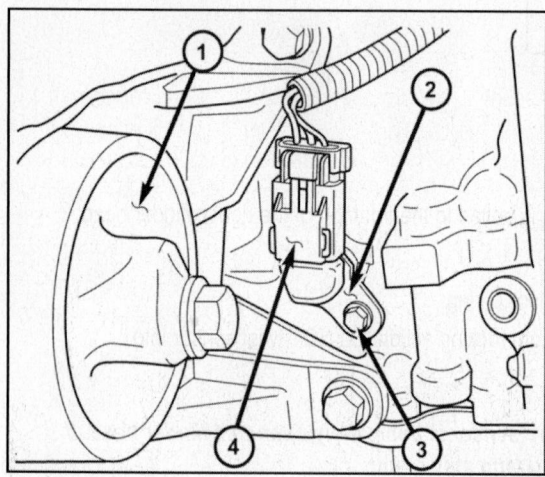

**Identifying the RF engine mount (1), CKP sensor (2), mounting bolt (3), and electrical connector (4) on 2.4L engines**

*3.7L*

The Crankshaft Position (CKP) sensor is mounted into the right rear side of the cylinder block. It is positioned and bolted into a machined hole.

Engine speed and crankshaft position are provided through the CKP (Crankshaft Position) sensor. The sensor generates pulses that are the input sent to the Powertrain Control Module (PCM). The PCM interprets the sensor input to determine the crankshaft position. The PCM then uses this position, along with other inputs, to determine injector sequence and ignition timing.

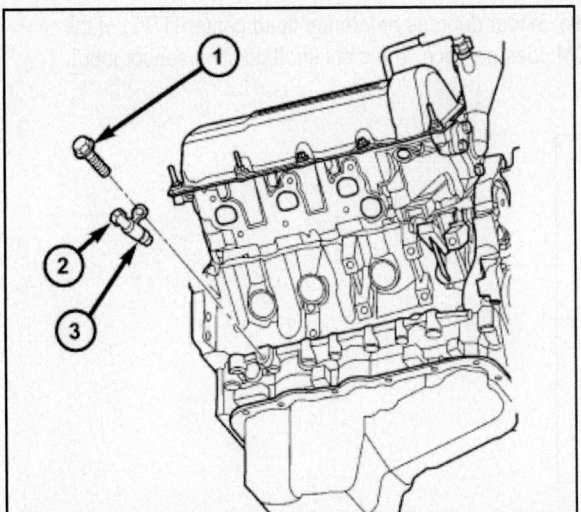

**Mounting bolt (1) CKP sensor (2) O-ring (3) on the 3.7L engine**

The sensor is a Hall effect device combined with an internal magnet. It is also sensitive to steel within a certain distance from it. A tone wheel (target wheel) is bolted to the engine crankshaft. This tone wheel has sets of notches at its outer edge. The notches cause a pulse to be generated when they pass under the sensor. The pulses are the input to the PCM.

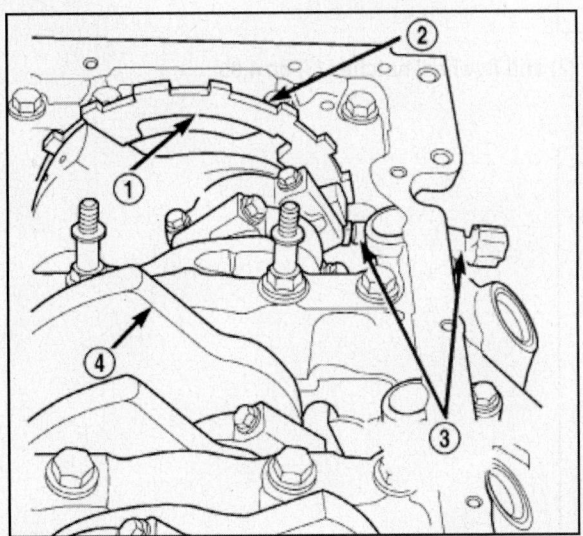

**Illustrating the tone wheel (1), notches (2), CKP sensor (3), and crankshaft (4) on the 3.7L engine**

### 4.0L

The Crankshaft Position Sensor (CKP) is mounted to the transmission bellhousing at the left/rear side of the engine block. Engine speed and crankshaft position are provided through the crankshaft position sensor. The sensor generates pulses that are the input sent to the powertrain control module (PCM). The PCM interprets the sensor input to determine the crankshaft position. The PCM then uses this position, along with other inputs, to determine injector sequence and ignition timing. The sensor is a Hall effect device combined with an internal magnet. It is also sensitive to steel within a certain distance from it.

On 4.0L 6-cylinder engines, the flywheel/drive plate has 3 sets of four notches at its outer edge. The notches cause a pulse to be generated when they pass under the sensor. The pulses are the input to the PCM. For each engine revolution there are 3 sets of four pulses generated.

The trailing edge of the fourth notch, which causes the pulse, is four degrees before top dead center (TDC) of the corresponding piston. The engine will not operate if the PCM does not receive a crankshaft position sensor input.

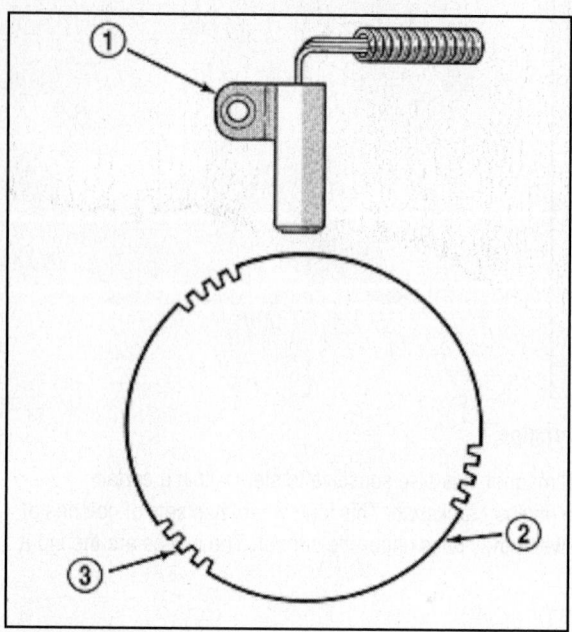

**Illustrating CKP sensor (1) relationship to the flywheel (2) and flywheel notches (3) on 4.0L**

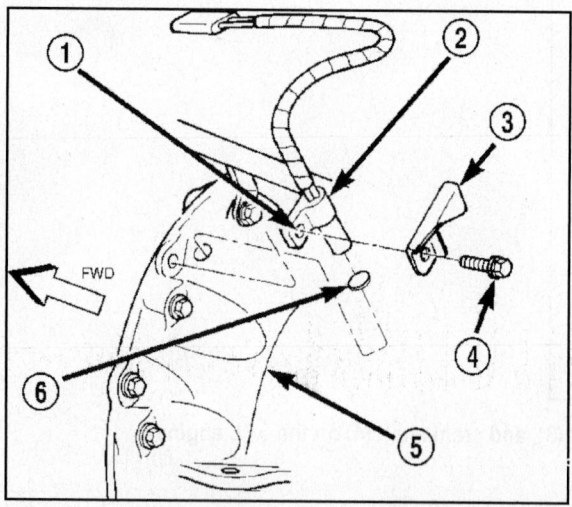

**Showing the Crankshaft Position (CKP) sensor (2) and related components – slotted hole (1), wire shield (3), mounting bolt (4), transmission housing (5), and paper spacer (6) for installation reference**

*4.7L*

The Crankshaft Position Sensor (CKP) is mounted into the engine block above the starter motor. Engine speed and crankshaft position are provided through the crankshaft position sensor. The sensor generates pulses that are the input sent to the powertrain control module (PCM). The PCM interprets the sensor input to determine the crankshaft position. The PCM then uses this position, along with other inputs, to determine injector sequence and ignition timing. The sensor is a Hall effect device combined with an internal magnet. It is also sensitive to steel within a certain distance from it.

On the 4.7L V–8 engine, a tone wheel is bolted to the engine crankshaft. This tone wheel has sets of notches at its outer edge. The notches cause a pulse to be generated when they pass under the sensor. The pulses are the input to the PCM.

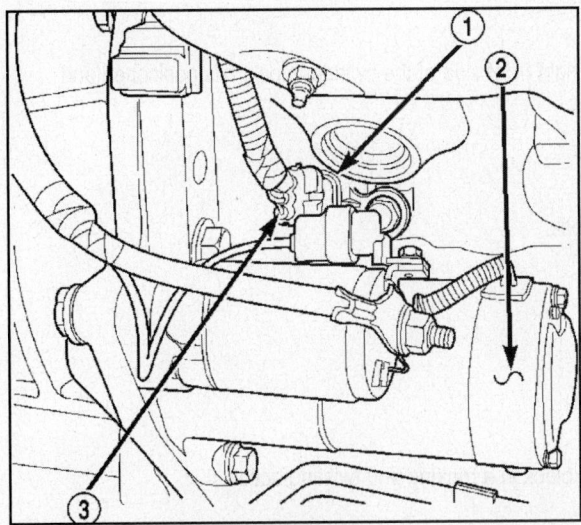

**Identifying location of Crankshaft Position (CKP) sensor (1), near starter (2) and electrical connector (3)**

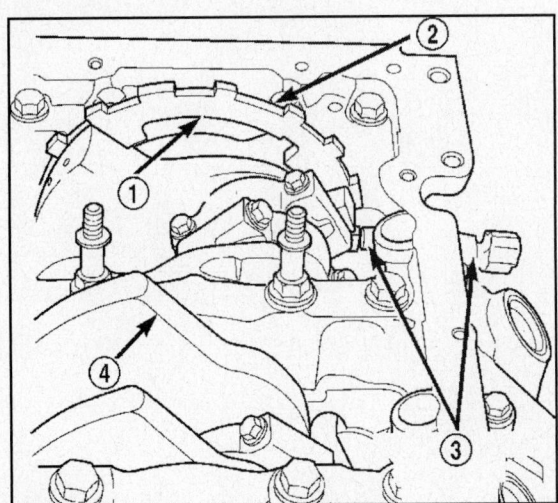

**Showing the tone wheel (1) in the engine, indicating its notches (2), the CKP sensor (3) and the crankshaft (4)**

*5.7L V8*

The crankshaft position sensor is located at the left rear of the engine just above the starter motor. The sensor detects the crankshaft position without contact (Hall effect) by means of missing segments on the tone wheel behind flex plate. The electronic control module (ECM) detects TDC position of cylinder 1 by means of the signal supplied by the sensor. Injection timing is synchronized by means of the camshaft signal and the crankshaft signal. This sensor is used to detect engine speed.

When the crankshaft rotates, an alternating voltage is generated in the crankshaft position sensor by the gaps of the tone wheel located behind the flex plate. In this case, the metal portion of the tone wheel generates a positive voltage pulse and the gap in the tone wheel a negative voltage pulse. The distance from the positive to the negative voltage peak equals the length of the gap.

The gap created by 3 missing teeth has the effect that no voltage is generated in the crankshaft position sensor. The ECM analyzes this gap, or time, without a signal from the crankshaft sensor, in order to detect the TDC position of cylinder 1.

## Removal & Installation

### 2.4L

The Crankshaft Position (CKP) sensor is mounted into the right front side of the cylinder block. It is positioned and bolted into a machined hole.

1. Disconnect sensor electrical connector.
2. Remove sensor bolt.
3. Carefully twist sensor from cylinder block.
4. Check condition of sensor O-ring.

### 3.7L

1. Raise vehicle.
2. Disconnect sensor electrical connector.
3. Remove sensor mounting bolt.
4. Carefully remove sensor from cylinder block in a rocking and twisting action.
5. Check condition of sensor O-ring.

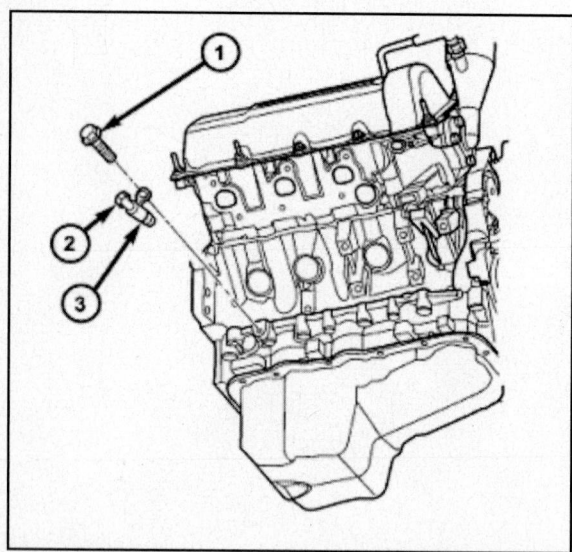

**Showing the sensor mounting bolt (1), CKP sensor (2) and O-ring (3) on the 3.7L engine**

### 4.0L

The Crankshaft Position (CKP) sensor is mounted to the transmission bellhousing at the left/rear side of the engine block. The sensor is adjustable and is attached with one bolt. A wire shield/router is attached to the sensor.

1. Disconnect sensor pigtail harness (3–way connector) from main engine wiring harness.
2. Remove sensor mounting bolt.
3. Remove wire shield and sensor.

*4.7L*

The Crankshaft Position (CKP) sensor is bolted to the side of the engine cylinder block above the starter motor. It is positioned into a machined hole at the side of the engine block.

1. Remove starter motor.
2. Disconnect CKP electrical connector at sensor.
3. Remove CKP mounting bolt.
4. Carefully twist sensor from cylinder block.
5. Remove sensor from vehicle.
6. Check condition of sensor O-ring.

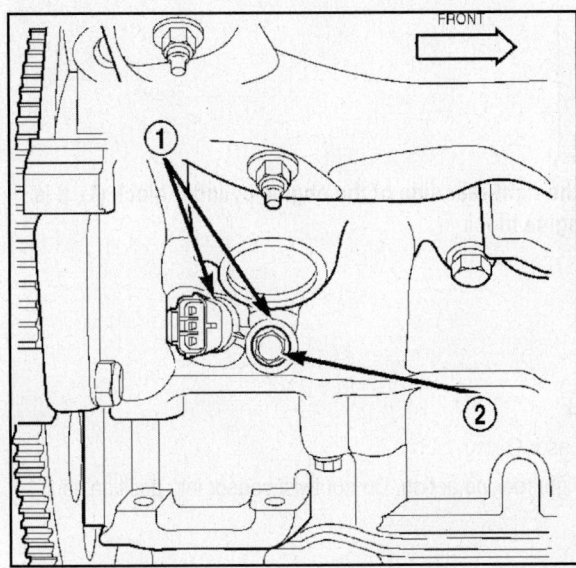

**Showing the CKP sensor (1) and its mounting bolt (2) for Removal & Installation on 4.7L engine**

*5.7L V8*

1. Raise vehicle.
2. Disconnect CKP electrical connector at sensor.
3. Remove CKP mounting bolt.
4. Carefully twist sensor from cylinder block.
5. Remove sensor from vehicle.
6. Check condition of sensor O-ring.

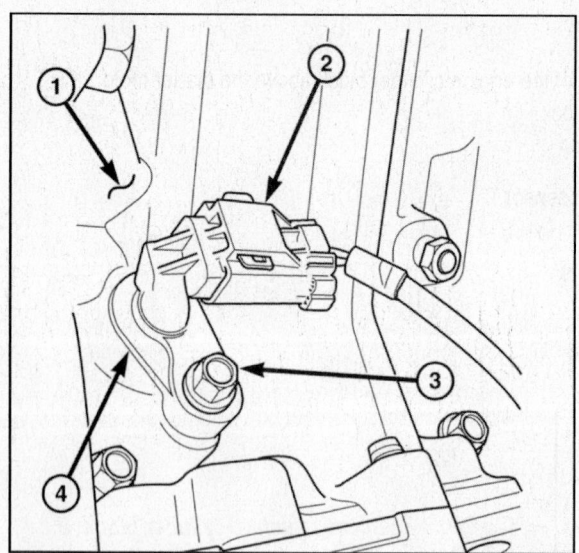

The Crankshaft Position (CKP) sensor (2) is located at the right-rear side of the engine cylinder block (1). It is positioned and bolted into a machined hole (4) in the engine block

**To install:**

*2.4L*

1.  Clean out machined hole in engine block.
2.  Apply a small amount of engine oil to sensor O-ring.
3.  Install sensor into engine block with a slight rocking action. Do not twist sensor into position as damage to O-ring may result.

*CAUTION: Before tightening sensor mounting bolt, be sure sensor is completely flush to cylinder block. If sensor is not flush, damage to sensor mounting tang may result.*

4.  Install mounting bolt and tighten to 28 Nm (21 ft. lbs.) torque.
5.  Connect electrical connector to sensor.

*3.7L*

1.  Clean out machined hole in engine block.
2.  Apply a small amount of engine oil to sensor O-ring.
3.  Install sensor into engine block with a slight rocking and twisting action.

*CAUTION: Before tightening sensor mounting bolt, be sure sensor is completely flush to cylinder block. If sensor is not flush, damage to sensor mounting tang may result.*

4.  Install mounting bolt and tighten to 28 Nm (21 ft. lbs.) torque.
5.  Connect electrical connector to sensor.

*4.0L*

New replacement sensors will be equipped with a paper spacer glued to bottom of sensor. If installing (returning) a used sensor to vehicle, a new paper spacer must be installed to bottom of sensor. This spacer will be ground off the first time engine is started. If spacer is not used, sensor will be broken the first time engine is started.

**New Sensors**: Be sure paper spacer is installed to bottom of sensor. If not, obtain spacer PN05252229.

**Used Sensors**: Clean bottom of sensor and install spacer PN05252229.

1. Install sensor into transmission bellhousing hole.
2. Position sensor wire shield to sensor.
3. Push sensor against flywheel/drive plate. With sensor pushed against flywheel/drive plate, tighten mounting bolt to 7 Nm (60 inch lbs.) torque.
4. Route sensor wiring harness into wire shield.
5. Connect sensor pigtail harness electrical connector to main wiring harness.

*4.7L*

1. Clean out machined hole in engine block.
2. Apply a small amount of engine oil to sensor O-ring.
3. Install sensor into engine block with a slight rocking action. Do not twist sensor into position as damage to O-ring may result.

**CAUTION: *Before tightening sensor mounting bolt, be sure sensor is completely flush to cylinder block. If sensor is not flush, damage to sensor mounting tang may result.***

4. Install mounting bolt and tighten to 28 Nm (21 ft. lbs.) torque.
5. Connect electrical connector to sensor.
6. Install starter motor.

*5.7L V8*

**CAUTION: *Before tightening sensor mounting bolt, be sure sensor is completely flush to cylinder block. If sensor is not flush, damage to sensor mounting tang may result.***

1. Clean out machined hole in engine block.
2. Apply a small amount of engine oil to sensor O-ring.
3. Install sensor (4) into engine block with a slight rocking and twisting action.
4. Install mounting bolt and tighten to 28 Nm (21 ft. lbs.) torque.
5. Connect electrical connector (2) to sensor.
6. Lower vehicle.

## CRANKCASE VENTILATION SYSTEM

**NOTE: *On 4.7L engines, see POSITIVE CRANKCASE VENTILATION SYSTEM***

### Description

The 4.0L 6-cylinder engine is equipped with a Crankcase Ventilation (CCV) system. The system consists of a fixed orifice fitting of a calibrated size. This fitting is pressed into a rubber grommet located on the top/rear of cylinder head (valve) cover, a pair of breather tubes (lines) to connect the system components, the air cleaner housing, and an air inlet fitting.

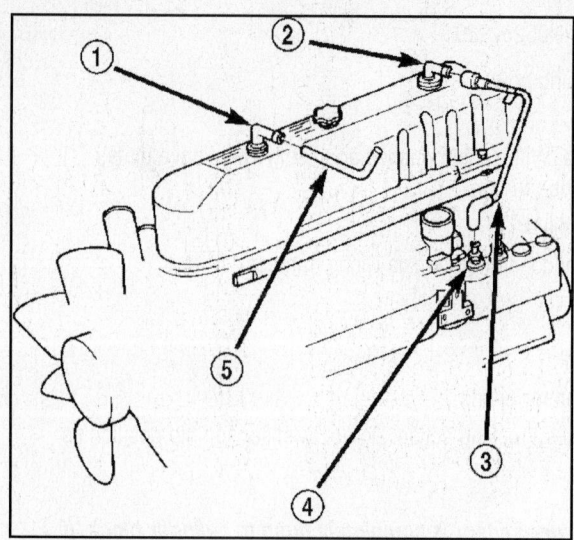

**Identifying the layout of the Crankcase Ventilation System components: air inlet fitting (1), fixed orifice fitting (2), rear CCV breather tube (3), intake manifold fitting (4), and front CCV breather tube (5)**

### Operation

The CCV system performs the same function as a conventional PCV system, but does not use a vacuum controlled PCV valve. The fixed orifice fitting meters the amount of crankcase vapors drawn out of the engine. When the engine is operating, fresh air enters the engine and mixes with crankcase vapors. Engine vacuum draws the vapor/air mixture through the fixed orifice and into the intake manifold. The vapors are then consumed during engine combustion.

### Removal & Installation

1. Pull fixed orifice fitting from valve cover grommet.
2. Separate fitting from CCV breather tube.

**To install:**

*NOTE: When installing fixed orifice fitting, be sure locations of fixed orifice fitting and air inlet fitting have not been inadvertently exchanged. The fixed orifice fitting is Light Grey in color and is located at rear of valve cover. The air inlet fitting is Black in color and is located at front of valve cover.*

3. Connect fitting to CCV breather tube.
4. Return fixed orifice fitting to valve cover grommet.

### Diagnosis & Testing

*Note: Before attempting diagnosis, be sure locations of fixed orifice fitting and air inlet fitting have not been inadvertently exchanged. The fixed orifice fitting is Light Grey in color and is located at rear of valve cover. The air inlet fitting is Black in color and is located at of valve cover.*

1. Pull fixed orifice fitting from valve cover and leave tube attached.
2. Start engine and bring to idle speed.
3. If fitting is not plugged, a hissing noise will be heard as air passes through fitting orifice. Also, a strong vacuum should be felt with a finger placed at fitting inlet.
4. If vacuum is not present, remove fitting orifice fitting from tube. Start engine. If vacuum can now be felt, replace fixed orifice fitting. Do not attempt to clean plastic fitting.
5. If vacuum is still not felt at hose, check line/hose for kinks or for obstruction. If necessary, clean out intake manifold fitting at intake manifold. Do this by turning a 1/4-inch drill (by hand)

through the fitting to dislodge any solid particles. Blow out the fitting with shop air. If necessary, use a smaller drill to avoid removing any metal from the fitting.

6. Return fixed orifice fitting to valve cover and leave tube attached.

7. Disconnect air inlet fitting and its attached hose at front of valve cover. Start engine and bring to idle speed. Hold a piece of stiff paper (such as a parts tag) loosely over the rubber grommet (opening) of the disconnected air inlet fitting.

8. The paper should be drawn against the rubber grommet with noticeable force. This will be after allowing approximately one minute for crankcase pressure to reduce.

9. If vacuum is not present, check breather hoses/tubes/lines for obstructions or restrictions.

10. After Testing, reconnect all system hoses/tubes/lines.

EVAP CANISTER PURGE SOLENOID

### Description & Operation

The duty cycle EVAP canister purge solenoid (DCP) regulates the rate of vapor flow from the EVAP canister to the intake manifold. The Powertrain Control Module (PCM) operates the solenoid.

During the cold start warm-up period and the hot start time delay, the PCM does not energize the solenoid. When de-energized, no vapors are purged. The PCM de-energizes the solenoid during open loop operation.

The engine enters closed loop operation after it reaches a specified temperature and the time delay ends. During closed loop operation, the PCM cycles (energizes and de-energizes) the solenoid 5 or 10 times per second, depending upon operating conditions. The PCM varies the vapor flow rate by changing solenoid pulse width. Pulse width is the amount of time that the solenoid is energized. The PCM adjusts solenoid pulse width based on engine operating condition.

### Removal & Installation

The duty cycle evaporative (EVAP) canister purge solenoid is located in the engine compartment near the brake master cylinder.

1. Disconnect electrical connector at solenoid.

2. Disconnect vacuum lines at solenoid.

3. Lift solenoid slot from mounting bracket for Removal & Installation.

### To install:

4. Position solenoid slot to mounting bracket.

5. Connect vacuum lines to solenoid. Be sure vacuum lines are firmly connected and not leaking or damaged. If leaking, a Diagnostic Trouble Code (DTC) may be set with certain emission packages.

6. Connect electrical connector to solenoid.

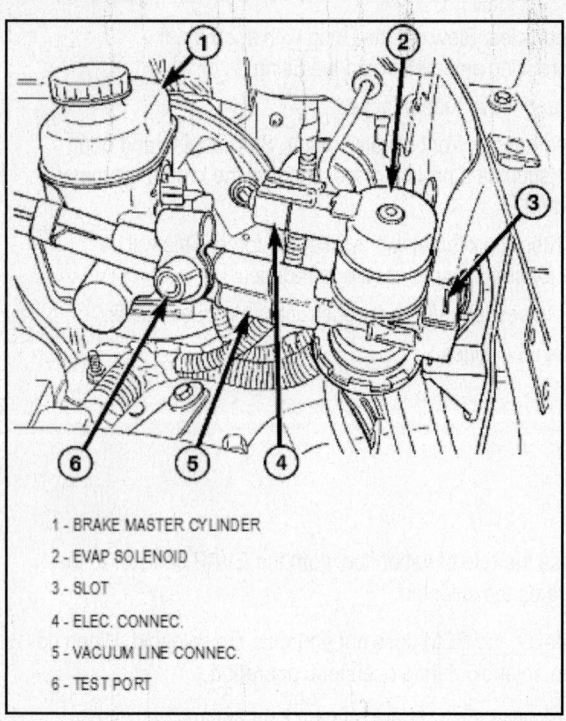

1 - BRAKE MASTER CYLINDER
2 - EVAP SOLENOID
3 - SLOT
4 - ELEC. CONNEC.
5 - VACUUM LINE CONNEC.
6 - TEST PORT

**Indicating the location of the EVAP Solenoid (2) and related components**

### EVAP LEAK DETECTION PUMP

### Description

The evaporative emission system is designed to prevent the escape of fuel vapors from the fuel system. Leaks in the system, even small ones, can allow fuel vapors to escape into the atmosphere. Government regulations require onboard Testing to make sure that the evaporative (EVAP) system is functioning properly. The leak detection system tests for EVAP system leaks and blockage. It also performs self-diagnostics. During self-diagnostics, the Powertrain Control Module (PCM) first checks the Leak Detection Pump (LDP) for electrical and mechanical faults. If the first checks pass, the PCM then uses the LDP to seal the vent valve and pump air into the system to pressurize it. If a leak is present, the PCM will continue pumping the LDP to replace the air that leaks out. The PCM determines the size of the leak based on how fast/long it must pump the LDP as it tries to maintain pressure in the system.

#### *EVAP LEAK DETECTION SYSTEM COMPONENTS*

- Service Port: Used with special tools like the Miller Evaporative Emissions Leak Detector (EELD) to test for leaks in the system.

- EVAP Purge Solenoid: The PCM uses the EVAP purge solenoid to control purging of excess fuel vapors stored in the EVAP canister. It remains closed during leak testing to prevent loss of pressure.

- EVAP Canister: The EVAP canister stores fuel vapors from the fuel tank for purging.

- EVAP Purge Orifice: Limits purge volume.

- EVAP System Air Filter: Provides air to the LDP for pressurizing the system. It filters out dirt while allowing a vent to atmosphere for the EVAP system.

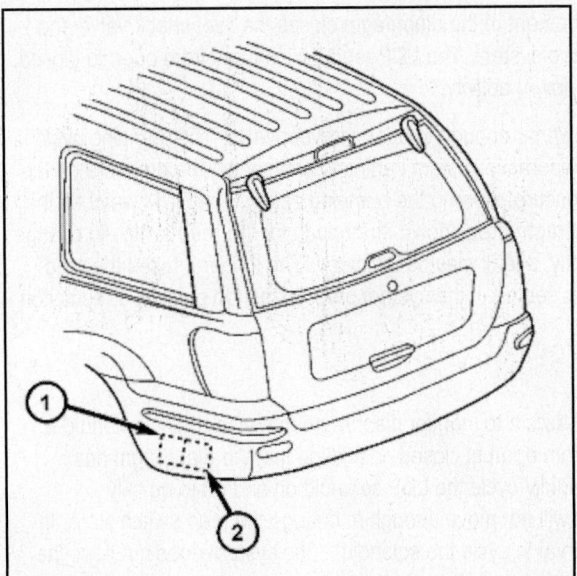

**Showing the location of the Leak Detection Pump (1) and EVAP Canister (2) at left rear of body**

## Operation

The main purpose of the LDP is to pressurize the fuel system for leak checking. It closes the EVAP system vent to atmospheric pressure so the system can be pressurized for leak Testing. The diaphragm is powered by engine vacuum. It pumps air into the EVAP system to develop a pressure of about 0.25 psi. A reed switch in the LDP allows the PCM to monitor the position of the LDP diaphragm. The PCM uses the reed switch input to monitor how fast the LDP is pumping air into the EVAP system. This allows detection of leaks and blockage.

The LDP assembly consists of several parts. The solenoid is controlled by the PCM, and it connects the upper pump cavity to either engine vacuum or atmospheric pressure. A vent valve closes the EVAP system to atmosphere, sealing the system during leak Testing. The pump section of the LDP consists of a diaphragm that moves up and down to bring air in through the air filter and inlet check valve, and pump it out through an outlet check valve into the EVAP system. The diaphragm is pulled up by engine vacuum, and pushed down by spring pressure, as the LDP solenoid turns on and off.

The LDP also has a magnetic reed switch to signal diaphragm position to the PCM. When the diaphragm is down, the switch is closed, which sends a 12v (system voltage) signal to the PCM. When the diaphragm is up, the switch is open, and there is no voltage sent to the PCM. This allows the PCM to monitor LDP pumping action as it turns the LDP solenoid on and off.

### LDP AT REST (NOT POWERED)

When the LDP is at rest (no electrical/vacuum) the diaphragm is allowed to drop down if the internal (EVAP system) pressure is not greater than the return spring. The LDP solenoid blocks the engine vacuum port and opens the atmospheric pressure port connected through the EVAP system air filter. The vent valve is held open by the diaphragm. This allows the canister to see atmospheric pressure.

### DIAPHRAGM UPWARD MOVEMENT

When the PCM energizes the LDP solenoid, the solenoid blocks the atmospheric port leading through the EVAP air filter and at the same time opens the engine vacuum port to the pump cavity above the diaphragm. The diaphragm moves upward when vacuum above the diaphragm exceeds spring force. This upward movement closes the vent valve. It also causes low pressure below the diaphragm, unseating the inlet check valve and allowing air in from the EVAP air filter. When the diaphragm completes its upward movement, the LDP reed switch turns from closed to open.

### DIAPHRAGM DOWNWARD MOVEMENT

Based on reed switch input, the PCM de-energizes the LDP solenoid, causing it to block the vacuum port, and open the atmospheric port. This connects the upper pump cavity to atmosphere through the EVAP air filter. The spring is now able to push the diaphragm down. The downward movement of the diaphragm closes the inlet check valve and opens the outlet check valve pumping air into the evaporative system. The LDP reed switch turns from open to closed, allowing the PCM to monitor LDP pumping (diaphragm up/down) activity.

During the pumping mode, the diaphragm will not move down far enough to open the vent valve. The pumping cycle is repeated as the solenoid is turned on and off. When the evaporative system begins to pressurize, the pressure on the bottom of the diaphragm will begin to oppose the spring pressure, slowing the pumping action. The PCM watches the time from when the solenoid is de-energized, until the diaphragm drops down far enough for the reed switch to change from opened to closed. If the reed switch changes too quickly, a leak may be indicated. The longer it takes the reed switch to change state, the tighter the evaporative system is sealed. If the system pressurizes too quickly, a restriction somewhere in the EVAP system may be indicated.

### PUMPING ACTION

Action : During portions of this test, the PCM uses the reed switch to monitor diaphragm movement. The solenoid is only turned on by the PCM after the reed switch changes from open to closed, indicating that the diaphragm has moved down. At other times during the test, the PCM will rapidly cycle the LDP solenoid on and off to quickly pressurize the system. During rapid cycling, the diaphragm will not move enough to change the reed switch state. In the state of rapid cycling, the PCM will use a fixed time interval to cycle the solenoid. If the system does not pass the EVAP Leak Detection Test, the following DTCs may be set:

- P0442 - EVAP LEAK MONITOR 0.040" LEAK DETECTED
- P0455 - EVAP LEAK MONITOR LARGE LEAK DETECTED
- P0456 - EVAP LEAK MONITOR 0.020" LEAK DETECTED
- P1486 - EVAP LEAK MON PINCHED HOSE FOUND
- P1494 - LEAK DETECTION PUMP SW OR MECH FAULT
- P1495 - LEAK DETECTION PUMP SOLENOID CIRCUIT

## Removal & Installation

The Leak Detection Pump (LDP) is located under the left quarter panel behind the left/rear wheel. It is attached to a two-piece support bracket. The LDP and LDP filter are replaced (serviced) as one unit.

1. Remove stone shield behind left/rear wheel. Drill out plastic rivets for removal.
2. Remove 3 LDP mounting bolts.
3. Remove support bracket brace bolt.
4. Loosen, but do not remove 2 support bracket nuts at frame rail.
5. To separate and lower front section of two-piece support bracket, remove 3 attaching bolts on bottom of support bracket. While lowering support bracket, disconnect LDP wiring clip.
6. Disconnect electrical connector at LDP.
7. Carefully remove vapor/vacuum lines at LDP.
8. Remove LDP.

**To install:**

9. Position LDP and carefully install vapor/vacuum lines to LDP and LDP filter.

**CAUTION: The vapor/vacuum lines and hoses must be firmly connected. Check the vapor/vacuum lines at the LDP; LDP filter and EVAP canister purge solenoid for damage or leaks. If a leak is present, a Diagnostic Trouble Code (DTC) may be set.**

10. Connect electrical connector to LDP.
11. While raising front section of support bracket, connect LDP wiring clip.
12. Install 3 LDP mounting bolts.
13. Join front and rear sections of two-piece support bracket by installing 3 bolts on bottom of support bracket. Do not tighten bolts at this time.

14. Install support bracket brace bolt. Do not tighten bolt at this time.

15. Tighten 2 support bracket nuts at frame rail and 3 support bracket bolts and brace bolt.

16. Position stone shield behind left/rear wheel. Install new plastic rivets.

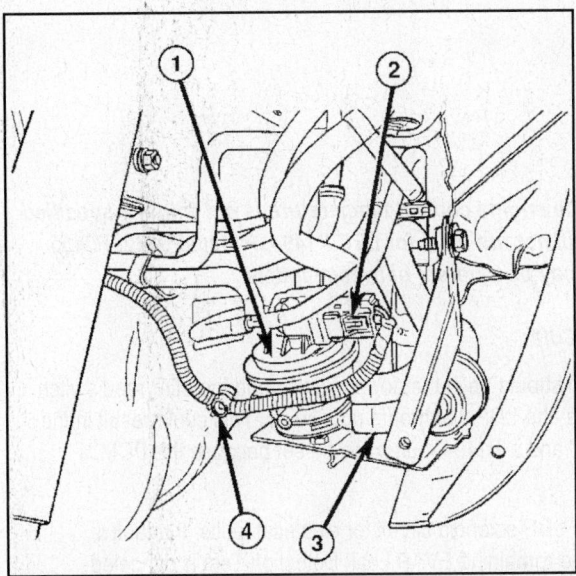

**Identifying the location of the Leak Detection Pump (1), electrical connector (2), vapor/vacuum lines (3), and wiring clip (4)**

### Diagnosis & Testing

#### *ENABLING CONDITIONS TO RUN EVAP LEAK DETECTION TEST*

Cold start: with ambient temperature (obtained from modeling the inlet air temperature sensor on passenger vehicles and the battery temperature sensor on Jeep & Dodge Truck vehicles) between 4° C (40° F) and 32° C (90° F) for 0.040 leaks. Between 4° C (40° F) and 29° C (85° F) for 0.020 leak. Conditions should be:

● Engine coolant temperature within: -12° to -8° C (10° to 18° F) of battery/ambient.

● Battery voltage between 10 and 15 volts.

● Low fuel warning light off (fuel level must be between 15% and 85%.

● MAP sensor reading 22 in Hg or above (This is the manifold absolute pressure, not vacuum).

● No engine stall during test.

*NOTE: If battery voltage drops below 10 volts for more than 5 seconds during engine cranking, the EVAP leak detection test will not run.*

*NOTE: The following values are approximate and vehicle specific. Use the values seen in pre test/monitor test screen on the scan tool. Also see TSB 25-02-98 for more detail.*

A DTC will not be set if a one-trip fault is set or if the MIL is illuminated for any of the following:

● Purge Solenoid Electrical Fault

● All TPS Faults

● All Engine Controller Self Test Faults

● LDP Pressure Switch Fault

● All Cam and/or Crank Sensor Fault

- EGR Solenoid Electrical Fault
- All MAP Sensor Faults
- All Injector Faults
- Ambient/Battery Temperature Sensor Electrical Faults
- BARO Out of Range
- Vehicle Speed Faults
- All Coolant Sensor Faults
- LDP Solenoid Circuit

**NOTE: If battery temperature is not within range, or if the engine coolant temperature is not within a specified range of the battery temperature, the PCM will not run tests for DTC P1494, P1486, P0442, P0455 and P0441. These temperature calibrations may be different between models.**

### SECTION 1 - P1495 LEAK DETECTION PUMP SOLENOID CIRCUIT

When the ignition key is turned to "ON", the LDP diaphragm should be in the down position and the LDP reed switch should be closed. If the EVAP system has residual pressure, the LDP diaphragm may be up. This could result in the LDP reed switch being open when the key is turned to "ON" and a P1494 fault could be set because the PCM is expecting the reed switch to be closed.

After the key is turned "ON", the PCM immediately tests the LDP solenoid circuit for electrical faults. If a fault is detected, DTC P1495 will set, the MIL will illuminate, and the remaining EVAP Leak Detection Test is canceled.

### SECTION 2 - P1494 LEAK DETECTION PUMP SWITCH OR MECHANICAL FAULT

If DTC P1495 is not set, the PCM will check for DTC P1494. If the LDP reed switch was closed when the key was turned to "ON", the PCM energizes the LDP solenoid for up to 8 seconds and monitors the LDP switch. As the LDP diaphragm is pulled up by engine vacuum, the LDP reed switch should change from closed to open. If it does not, the PCM sets a temporary fault (P1494) in memory, and waits until the next time the Enabling Conditions are met to run the test again. If this is again detected, P1494 is stored and the MIL is illuminated. If the problem is not detected during the next enabling cycle, the temporary fault will be cleared.

However, if the PCM detects the reed switch open when the key is turned to "ON", the PCM must determine if this condition is due to residual pressure in the EVAP system, or an actual fault. The PCM stores information in memory on EVAP system purging from previous engine run or drive cycles.

If little or no purging took place, residual pressure could be holding the LDP diaphragm up, causing the LDP switch to be open. Since this is not a malfunction, the PCM cancels the EVAP Leak Detection Test without setting the temporary fault.

If there was sufficient purging during the previous cycle to eliminate EVAP system pressure, the PCM judges that this is a malfunction and sets a temporary fault in memory. The next time that the Enabling Conditions are met, the test will run again. If the fault is again detected, the MIL will illuminate and DTC P1494 will be stored. If the fault is not detected, the temporary fault will be cleared.

### SECTION 3 - P1486 EVAP LEAK MONITOR PINCHED HOSE FOUND

If no fault has been detected so far, the PCM begins Testing for possible blockage in the EVAP system between the LDP and the fuel tank. This is done by monitoring the time required for the LDP to pump air into the EVAP system during two to three pump cycles. If no blockage is present, the LDP diaphragm is able to quickly pump air out of the LDP each time the PCM turns off the LDP solenoid. If a blockage is present, the PCM detects that the LDP takes longer to complete each pump cycle. If the pump cycles take longer than expected (approximately 6 to 10 seconds) the PCM will suspect a blockage. On the next drive when Enabling Conditions are met, the test will run again. If blockage is again detected, P1486 is stored, and the MIL is illuminated.

### *SECTION 4 - NO DTC CAN BE SET DURING THIS TIME*

After the LDP blockage tests are completed, the PCM then tests for EVAP system leakage. First, the PCM commands the LDP to rapidly pump for 20 to 50 seconds (depending on fuel level) to build pressure in the EVAP system. This evaluates the system to see if it can be sufficiently pressurized. This evaluation (rapid pump cycling) may occur several times prior to leak checking. The LDP reed switch does not close and open during rapid pumping because the diaphragm does not travel through its full range during this part of the test.

### *SECTION 5 - P0456, P0442, P0455 EVAP LEAK MONITOR AND LEAK DETECTED*

Next, the PCM performs one or more test cycles by monitoring the time required for the LDP reed switch to close (diaphragm to drop) after the LDP solenoid is turned off.

If the switch does not close, or closes after a long delay, it means that the system does not have any significant leakage and the EVAP Leak Detection Test is complete.

However, if the LDP reed switch closes quickly, there may be a leak or the fuel level may be low enough that the LDP must pump more to finish pressurizing the EVAP system. In this case, the PCM will rapidly pump the LDP again to build pressure in the EVAP system, and follow that by monitoring the time needed for several LDP test cycles. This process of rapid pumping followed by several LDP test cycles may repeat several times before the PCM judges that a leak is present.

When leaks are present, the LDP test cycle time will be inversely proportional to the size of the leak. The larger the leak, the shorter the test cycle time. The smaller the leak, the longer the test cycle time. DTC's may be set when a leak as small as 0.5 mm (0.020") diameter is present.

If the system detects a leak, a temporary fault will be stored in PCM memory. The time it takes to detect a .020, .040, or large leak is based on calibrations that vary from model to model. The important point to remember is if a leak is again detected on the next EVAP Leak Detection Test, the MIL will illuminate and a DTC will be stored based on the size of leak detected. If no leak is detected during the next test, the temporary fault will be cleared.

### Diagnostic Tips

During diagnosis, you can compare the LDP solenoid activity with the monitor sequence in Figure 6. If the PCM detects a problem that could set a DTC, the Testing is halted and LDP solenoid activity will stop. As each section of the test begins, it indicates that the previous section passed successfully. By watching to see which tests complete, you can see if any conditions are present that the PCM considers abnormal.

For example, if the LDP solenoid is energized for the test cycles to test for blockage (P1486), it means that the LDP has already passed its test for P1494. Then, if the PCM detects a possible blockage, it will set a temporary fault without turning on the MIL and continue the leak portion of the test. However, the PCM will assume that the system is already pressurized and skip the rapid pump cycles.

Always diagnose leaks, if possible, before disconnecting connections. Disconnecting connections may mask a leak condition. Keep in mind that if the purge solenoid seat is leaking, it could go undetected since the leak would end up in the intake manifold. Disconnect the purge solenoid at the manifold when leak checking. In addition, a pinched hose fault (P1486) could set if the purge solenoid does not purge the fuel system properly (blocked seat). The purge solenoid must vent the fuel system prior to the LDP system test. If the purge solenoid cannot properly vent the system the LDP cannot properly complete the test for P1486 and this fault can set due to pressure being in the EVAP system during the test sequence.

Multiple actuations of the scan tool's Leak Detection Pump (LDP) Monitor Test can hide a 0.020" leak because of excess vapor generation. Additionally, any source for additional vapor generation can hide a small leak in the EVAP system. Excess vapor generation can delay the fall of the LDP diaphragm thus hiding the small leak. An example of this condition could be bringing a cold vehicle into a warm shop for Testing or high ambient temperatures.

Fully plugged and partially plugged underhood vacuum lines have been known to set MIL conditions. P1494 and P0456 can be set for this reason. Always, thoroughly, check plumbing for pinches or blockage before condemning components.

## Test Equipment

The Evaporative Emission Leak Detector (EELD), Miller Special Tool 8404, is capable of visually detecting leaks in the evaporative system and will take the place of the ultrasonic leak detector 6917A. The EELD utilizes shop air and a smoke generator to visually detect leaks down to 0.020 or smaller. The food grade oil used to make the smoke includes an UV trace dye that will leave telltale signs of the leak under a black light. This is helpful when components have to be removed to determine the exact leak location. For detailed test instructions, follow the operator's manual packaged with the EELD.

*NOTE: Be sure that the PCM has the latest software update. Reprogram as indicated by any applicable Technical Service Bulletin. After LDP repairs are completed, verify the repair by running the DRB III® Leak Detection Pump (LDP) Monitor Test as described in Technical Service Bulletin 18-12-99.*

## FUEL LEVEL SENDING UNIT/SENSOR

### Description & Operation

The fuel gauge sending unit (fuel level sensor) is attached to the side of the fuel pump module. The sending unit consists of a float, an arm, and a variable resistor track (card).

The fuel pump module has 4 different circuits (wires). Two of these circuits are used for the fuel gauge sending unit for fuel gauge operation, and for certain OBD II emission requirements. The other 2 wires are used for electric fuel pump operation.

**For Fuel Gauge Operation:** A constant input voltage source of about 12 volts (battery voltage) is supplied to the resistor track on the fuel gauge sending unit. This is fed directly from the Powertrain Control Module (PCM).

*NOTE: For diagnostic purposes, this 12V power source can only be verified with the circuit opened (fuel pump module electrical connector unplugged). With the connectors plugged, output voltages will vary from about 0.6 volts at FULL, to about 8.6 volts at EMPTY (about 8.6 volts at EMPTY for Jeep models, and about 7.0 volts at EMPTY for Dodge Truck models).*

The resistor track is used to vary the voltage (resistance) depending on fuel tank float level. As fuel level increases, the float and arm move up, which decreases voltage. As fuel level decreases, the float and arm move down, which increases voltage. The varied voltage signal is returned back to the PCM through the sensor return circuit.

Both of the electrical circuits between the fuel gauge sending unit and the PCM are hard-wired (not multiplexed). After the voltage signal is sent from the resistor track, and back to the PCM, the PCM will interpret the resistance (voltage) data and send a message across the multiplex bus circuits to the instrument panel cluster. Here it is translated into the appropriate fuel gauge level reading.

**For OBD II Emission Monitor Requirements:** The PCM will monitor the voltage output sent from the resistor track on the sending unit to indicate fuel level. The purpose of this feature is to prevent the OBD II system from recording/setting false misfire and fuel system monitor diagnostic trouble codes. The feature is activated if the fuel level in the tank is less than approximately 15 percent of its rated capacity. If equipped with a Leak Detection Pump (EVAP system monitor), this feature will also be activated if the fuel level in the tank is more than approximately 85 percent of its rated capacity.

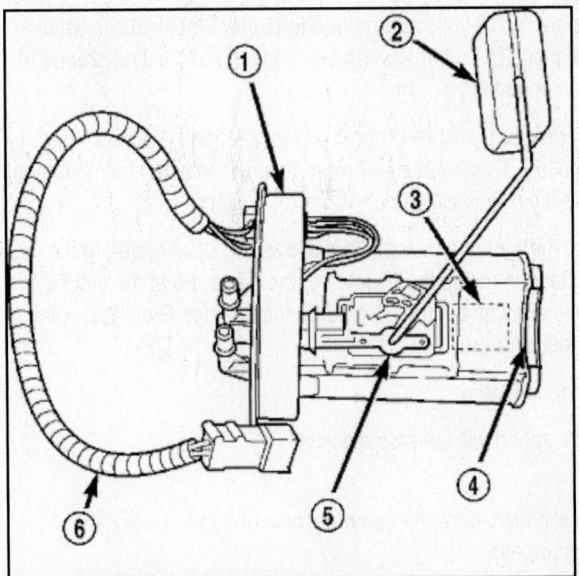

**Identifying the components of the fuel pump module assembly: 1 – fuel pump module; 2 – fuel gauge float; 3 – electrical fuel pump; 4 – inlet filter; 5 – fuel gauge sending unit; 6 – pigtail harness**

### Removal & Installation

The fuel gauge sending unit (fuel level sensor) and float assembly is located on the side of fuel pump module. The fuel pump module is located within the fuel tank.

1. Remove fuel tank.
2. Remove fuel pump module.
3. Remove electrical wire connector at sending unit terminals.
4. Press upward on release tab to remove sending unit from pump module.

**To install:**

5. Position sending unit to pump module and snap into place.
6. Connect electrical connector to terminals.
7. Install fuel pump module.
8. Install fuel tank.

## IDLE AIR CONTROL (IAC) MOTOR

### Description & Operation

The IAC stepper motor is mounted to the throttle body, and regulates the amount of air bypassing the control of the throttle plate. As engine loads and ambient temperatures change, engine rpm changes. A pintle on the IAC stepper motor protrudes into a passage in the throttle body, controlling airflow through the passage. The IAC is controlled by the Powertrain Control Module (PCM) to maintain the target engine idle speed.

At idle, engine speed can be increased by retracting the IAC motor pintle and allowing more air to pass through the port, or it can be decreased by restricting the passage with the pintle and diminishing the amount of air bypassing the throttle plate. The IAC is called a stepper motor because it is moved (rotated) in steps, or increments. Opening the IAC opens an air passage around the throttle blade, which increases RPM.

The PCM uses the IAC motor to control idle speed (along with timing) and to reach a desired MAP during decel (keep engine from stalling).

The IAC motor has 4 wires with 4 circuits. Two of the wires are for 12 volts and ground to supply electrical current to the motor windings to operate the stepper motor in one direction. The other 2 wires are also for 12 volts and ground to supply electrical current to operate the stepper motor in the opposite direction.

To make the IAC go in the opposite direction, the PCM just reverses polarity on both windings. If only 1 wire is open, the IAC can only be moved 1 step (increment) in either direction. To keep the IAC motor in position when no movement is needed, the PCM will energize both windings at the same time. This locks the IAC motor in place.

In the IAC motor system, the PCM will count every step that the motor is moved. This allows the PCM to determine the motor pintle position. If the memory is cleared, the PCM no longer knows the position of the pintle. So at the first key ON, the PCM drives the IAC motor closed, regardless of where it was before. This zeros the counter. From this point the PCM will back out the IAC motor and keep track of its position again.

When engine rpm is above idle speed, the IAC is used for the following:

- Off-idle dashpot (throttle blade will close quickly but idle speed will not stop quickly)
- Deceleration airflow control
- A/C compressor load control (also opens the passage slightly before the compressor is engaged so that the engine rpm does not dip down when the compressor engages)
- Power steering load control
- The PCM can control polarity of the circuit to control direction of the stepper motor.

### IAC STEPPER MOTOR PROGRAM

The PCM is also equipped with a memory program that records the number of steps the IAC stepper motor most recently advanced to during a certain set of parameters. For example: The PCM was attempting to maintain a 1000 rpm target during a cold start-up cycle. The last recorded number of steps for that may have been 125. That value would be recorded in the memory cell so that the next time the PCM recognizes the identical conditions; the PCM recalls that 125 steps were required to maintain the target. This program allows for greater customer satisfaction due to greater control of engine idle.

Another function of the memory program, which occurs when the power steering switch (if equipped), or the A/C request circuit, requires that the IAC stepper motor control engine rpm, is the recording of the last targeted steps into the memory cell. The PCM can anticipate A/C compressor loads. This is accomplished by delaying compressor operation for approximately 0.5 seconds until the PCM moves the IAC stepper motor to the recorded steps that were loaded into the memory cell. Using this program helps eliminate idle-quality changes as loads change. Finally, the PCM incorporates a "No-Load" engine speed limiter of approximately 1800 - 2000 rpm, when it recognizes that the TPS is indicating an idle signal and IAC motor cannot maintain engine idle. A (factory adjusted) set screw is used to mechanically limit the position of the throttle body throttle plate. The IAC motor through the PCM controls all idle speed functions.

*CAUTION: Never attempt to adjust the engine idle speed using this screw.*

### Removal & Installation

*3.7L*

The Idle Air Control (IAC) motor is located on the side of the throttle body.

1. Disconnect electrical connector from IAC motor.
2. Remove two mounting bolts (screws).
3. Remove IAC motor from throttle body.

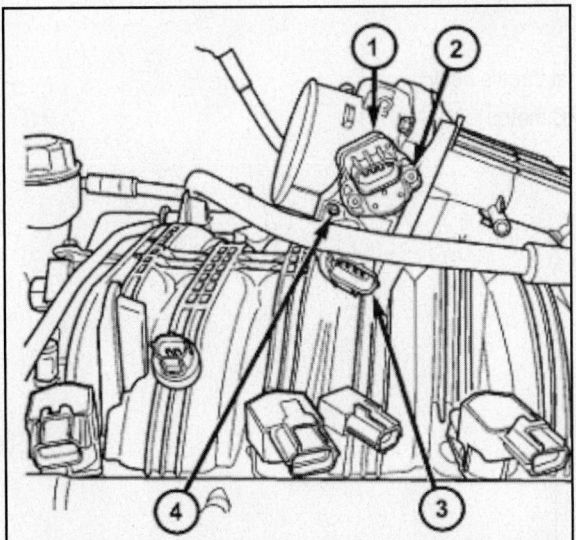

**Throttle position sensor (1), mounting screws (2), idle air control motor (3), and mounting screws (4) on the 3.7L engine**

*4.0L*

The IAC motor is located on the throttle body.

1. Remove air duct and air resonator box at throttle body.
2. Disconnect electrical connector from IAC motor.
3. Remove two mounting bolts (screws).
4. Remove IAC motor from throttle body.

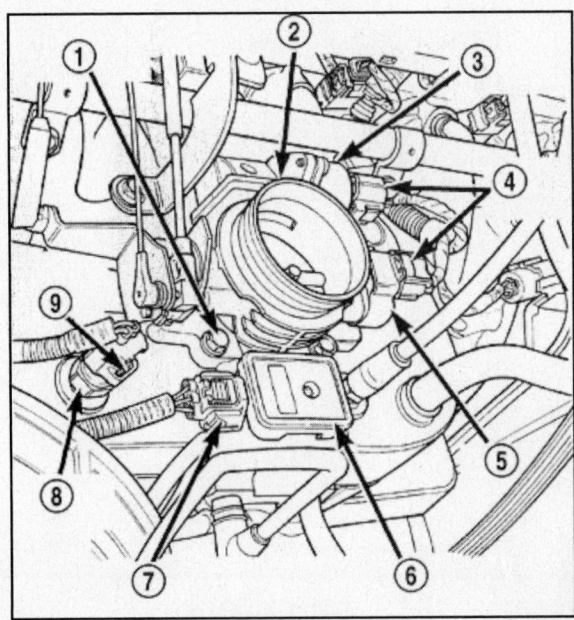

**Identifying 4.0L engine components on or near throttle body: 1 – mounting bolts; 2 – throttle body; 3 – IAC motor; 4 – electrical connector; 5 – TP sensor; 6 – MAP sensor; 7 – electrical connector; 8 – IAT sensor; 9 – electrical connector**

*4.7L*

1. Remove air duct and air resonator box at throttle body.
2. Disconnect electrical connector from IAC motor.
3. Remove two mounting bolts (screws).
4. Remove IAC motor from throttle body.

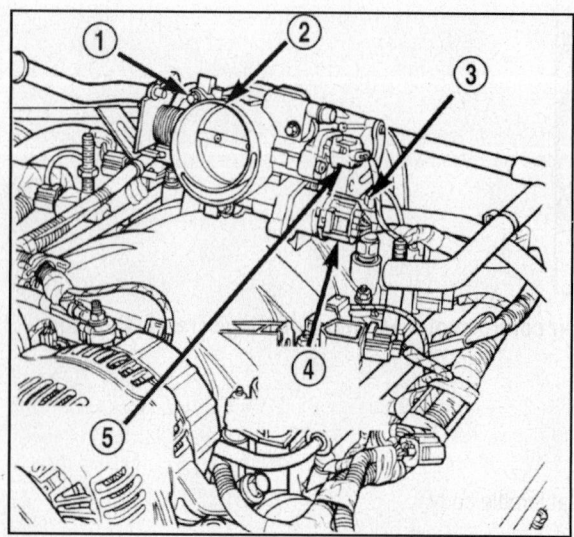

Identifying 4.7L engine components on or near throttle body: 1 – mounting bolts; 2 – throttle body; 3 – IAT sensor connector; 4 – IAC motor connector; 5 – TP sensor connector

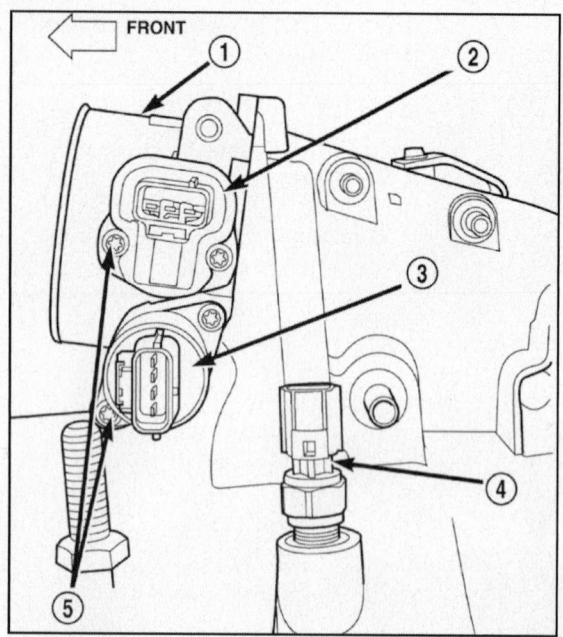

Closer view of the IAC motor (3), throttle body (1), TP sensor (2), IAT sensor (4), and mounting screws (5) on 4.7L engine

**To install:**

*3.7L*

1. Install IAC motor to throttle body.
2. Install and tighten two mounting bolts (screws) to 7 Nm (60 in lbs) torque.
3. Install electrical connector.

*4.0L & 4.7L*

The IAC motor is located on the throttle body.

1. Install IAC motor to throttle body.
2. Install and tighten two mounting bolts (screws) to 7 Nm (60 inch lbs.) torque.
3. Install electrical connector.
4. Install air cleaner duct/air box to throttle body.

## INTAKE AIR TEMPERATURE (IAT) SENSOR

### Description & Operation

The 2–wire Intake Manifold Air Temperature (IAT) sensor is installed in the intake manifold with the sensor element extending into the air stream.

The IAT sensor is a two-wire Negative Thermal Coefficient (NTC) sensor. Meaning, as intake manifold temperature increases, resistance (voltage) in the sensor decreases. As temperature decreases, resistance (voltage) in the sensor increases.

The IAT sensor provides an input voltage to the Powertrain Control Module (PCM) indicating the density of the air entering the intake manifold based upon intake manifold temperature. At key-on, a 5–volt power circuit is supplied to the sensor from the PCM. The sensor is grounded at the PCM through a low-noise, sensor-return circuit.

The PCM uses this input to calculate the following:

- Injector pulse-width
- Adjustment of spark timing (to help prevent spark knock with high intake manifold air-charge temperatures)

The resistance values of the IAT sensor are the same as for the Engine Coolant Temperature (ECT) sensor.

### Removal & Installation

*3.7L*

The Intake Manifold Air Temperature (IAT) sensor is installed into the left side of intake manifold plenum.

1. Disconnect electrical connector from IAT sensor. Clean dirt from intake manifold at sensor base.
2. Gently lift on small plastic release tab or and rotate sensor about 1/4 turn counterclockwise for Removal & Installation. Check condition of sensor O-ring.

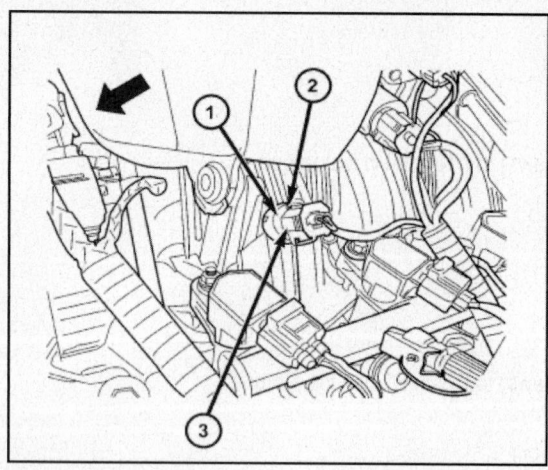

**IAT sensor (1), release tab (2), and electrical connector (3) for sensor Removal & Installation on 3.7L engine**

### 4.0L

The Intake Manifold Air Temperature (IAT) sensor is installed into the intake manifold plenum near the front of the throttle body.

1. Disconnect electrical connector from sensor.
2. Remove sensor from intake manifold.

### 4.7L

The Intake Manifold Air Temperature (IAT) sensor is located on the left side of the intake manifold.

#### Threaded Type Sensor

1. Disconnect electrical connector from sensor.
2. Remove sensor from intake manifold.

#### Snap-In Type Sensor

1. Disconnect electrical connector from IAT sensor.
2. Clean dirt from intake manifold at sensor base.
3. Gently lift on small plastic release tab and rotate sensor about 1/4 turn counter-clockwise for Removal & Installation.
4. Check condition of sensor O-ring.

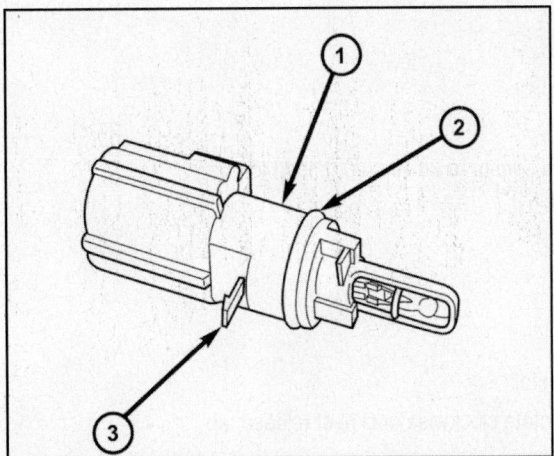

**Showing the IAT sensor (1), sensor O-ring (2), and the release tab (3)**

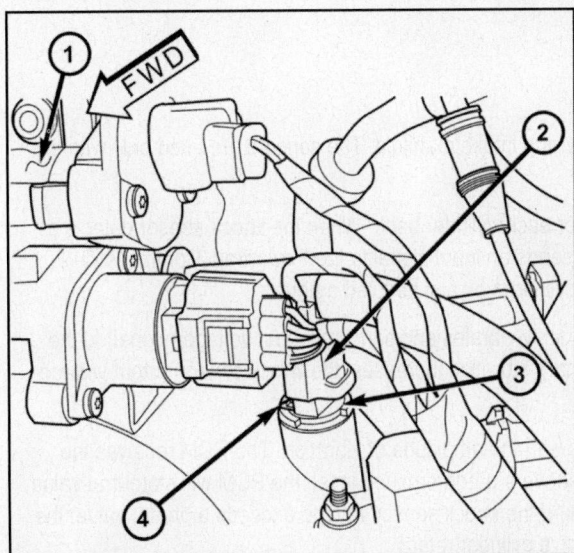

**Showing detail of the IAT sensor (3), release tab (4), and connector (2) on the left side of the throttle body (1) on 4.7L engine**

## To install:

### 3.7L

The intake manifold air temperature (IAT) sensor is installed into the left side of intake manifold plenum.

1. Check condition of sensor O-ring.
2. Clean the sensor mounting hole in the intake manifold.
3. Position the sensor into intake manifold and rotate clockwise until past release tab.
4. Install electrical connector.

### 4.0L

The Intake Manifold Air Temperature (IAT) sensor is installed into the intake manifold plenum near the front of the throttle body.

1. Install sensor into intake manifold. Tighten sensor to 28 Nm (20 ft. lbs.) torque.
2. Connect electrical connector to sensor.

*4.7L*

### Threaded Type Sensor

1. Install sensor into intake manifold. Tighten sensor to 28 Nm (20 ft. lbs.) torque.
2. Connect electrical connector to sensor.

### Snap-In Type Sensor

1. Check condition of sensor O-ring.
2. Clean sensor mounting hole in intake manifold.
3. Position sensor into intake manifold and rotate clockwise until past release tab.
4. Install electrical connector.

## KNOCK SENSOR

### Description & Operation

The 2 knock sensors are bolted into the cylinder block under the intake manifold. The sensors are used only with the 3.7L engine.

Two knock sensors are used on the 3.7L V-6 engine; one for each cylinder bank. When the knock sensor detects a knock in one of the cylinders on the corresponding bank, it sends an input signal to the Powertrain Control Module (PCM). In response, the PCM retards ignition timing for all cylinders by a scheduled amount.

Knock sensors contain a piezoelectric material, which constantly vibrates and sends an input voltage (signal) to the PCM while the engine operates. As the intensity of the crystal's vibration increases, the knock sensor output voltage also increases.

The voltage signal produced by the knock sensor increases with the amplitude of vibration. The PCM receives the knock sensor voltage signal as an input. If the signal rises above a predetermined level, the PCM will store that value in memory and retard ignition timing to reduce engine knock. If the knock sensor voltage exceeds a preset value, the PCM retards ignition timing for all cylinders. It is not a selective cylinder retard.

The PCM ignores knock sensor input during engine idle conditions. Once the engine speed exceeds a specified value, knock retard is allowed.

Knock retard uses its own short-term and long-term memory program. Long-term memory stores previous detonation information in its battery-backed RAM. The maximum authority that long term memory has over timing retard can be calibrated. Short-term memory is allowed to retard timing up to a preset amount under all operating conditions (as long as rpm is above the minimum rpm) except at Wide Open Throttle (WOT). The PCM, using short-term memory, can respond quickly to retard timing when engine knock is detected. Short-term memory is lost any time the ignition key is turned OFF.

*NOTE: Over or under tightening the sensor mounting bolts will affect knock sensor performance, possibly causing improper spark control. Always use the specified torque when installing the knock sensors.*

### Removal & Installation

The 2 knock sensors are bolted into the cylinder block under the intake manifold.

*NOTE: The left sensor is identified by an identification tag (LEFT). It is also identified by a larger bolt head. The Powertrain Control Module (PCM) must have and know the correct sensor left/right positions. Do not mix the sensor locations.*

1. Disconnect knock sensor dual pigtail harness from engine wiring harness. This connection is made near rear of left valve cover.
2. Remove intake manifold.
3. Remove sensor mounting bolts. Remove sensors from engine.

*NOTE: Note the foam strip on bolt threads. This foam is used only to retain the bolts to sensors for plant assembly. It is not used as a sealant. Do not apply any adhesive, sealant or thread locking compound to these bolts.*

To install:

*WARNING: Over or under tightening the sensor mounting bolts will affect knock sensor performance, possibly causing improper spark control. Always use the specified torque when installing the knock sensors. The torque for the knock senor bolt is relatively light for an 8mm bolt.*

*NOTE: Note foam strip on bolt threads. This foam is used only to retain the bolts to sensors for plant assembly. It is not used as a sealant. Do not apply any adhesive, sealant or thread locking compound to these bolts.*

4. Thoroughly clean knock sensor mounting holes.
5. Install sensors into cylinder block.
6. Install and tighten mounting bolts.
7. Install intake manifold.
8. Connect knock sensor wiring harness to engine harness at rear of intake manifold.

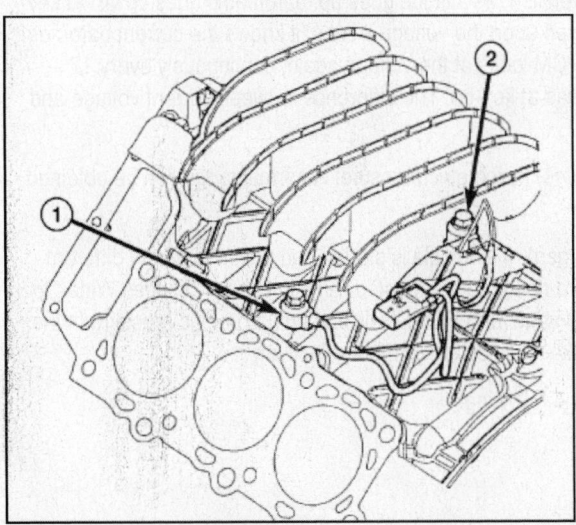

**Knock sensor locations (1) and mounting bolts (2) – 3.7L engine**

## MANIFOLD ABSOLUTE PRESSURE (MAP) SENSOR

### Description

*3.7L*

The Manifold Absolute Pressure (MAP) sensor is mounted into the front of the intake manifold with 2 screws.

*4.0L*

On the 4.0L six-cylinder engine the MAP sensor is mounted to the engine throttle body. On the 4.7L V-8 engine the MAP sensor is mounted to front of the intake manifold.

*4.7L*

The MAP sensor is located on the front of the intake manifold. An O-ring seals the sensor to the intake manifold.

### Operation

The MAP sensor is used as an input to the Powertrain Control Module (PCM). It contains a silicon based sensing unit to provide data on the manifold vacuum that draws the air/fuel mixture into the combustion chamber. The PCM requires this information to determine injector pulse width and spark advance. When manifold absolute pressure (MAP) equals Barometric pressure, the pulse width will be at maximum.

A 5-volt reference is supplied from the PCM and returns a voltage signal to the PCM that reflects manifold pressure. The zero pressure reading is 0.5V and full scale is 4.5V. For a pressure swing of 0–15 psi, the voltage changes 4.0V. To operate the sensor, it is supplied a regulated 4.8 to 5.1 volts. Ground is provided through the low-noise, sensor return circuit at the PCM.

The MAP sensor input is the number one contributor to fuel injector pulse width. The most important function of the MAP sensor is to determine barometric pressure. The PCM needs to know if the vehicle is at sea level or at a higher altitude, because the air density changes with altitude. It will also help to correct for varying barometric pressure. Barometric pressure and altitude have a direct inverse correlation; as altitude goes up, barometric goes down. At key-on, the PCM powers up and looks at MAP voltage, and based upon the voltage it sees, it knows the current barometric pressure (relative to altitude). Once the engine starts, the PCM looks at the voltage again, continuously every 12 milliseconds, and compares the current voltage to what it was at key-on. The difference between current voltage and what it was at Key On is the manifold vacuum.

During key-on (engine not running) the sensor reads (updates) barometric pressure. A normal range can be obtained by monitoring a known good sensor.

As the altitude increases, the air becomes thinner (less oxygen). If a vehicle is started and driven to a very different altitude than where it was at key-on, the barometric pressure needs to be updated. Any time the PCM sees Wide Open Throttle (WOT), based upon Throttle Position Sensor (TPS) angle and RPM, it will update barometric pressure in the MAP memory cell. With periodic updates, the PCM can make its calculations more effectively.

The PCM uses the MAP sensor input to aid in calculating the following:

- Manifold pressure
- Barometric pressure
- Engine load
- Injector pulse-width
- Spark-advance programs
- Shift-point strategies (certain automatic transmissions only)
- Idle speed
- Decel fuel shutoff

The MAP sensor signal is provided from a single piezo-resistive element located in the center of a diaphragm. The element and diaphragm are both made of silicone. As manifold pressure changes, the diaphragm moves causing the element to deflect, which stresses the silicone. When silicone is exposed to stress, its resistance changes. As manifold vacuum increases, the MAP sensor input voltage decreases proportionally. The sensor also contains electronics that condition the signal and provide temperature compensation.

The PCM recognizes a decrease in manifold pressure by monitoring a decrease in voltage from the reading stored in the barometric pressure memory cell. The MAP sensor is a linear sensor; meaning as pressure changes, voltage changes proportionately. The range of voltage output from the sensor is usually between 4.6 volts at sea level to as low as 0.3 volts at 26 in. of Hg. Barometric pressure is the pressure exerted by the atmosphere upon an object. At sea level on a standard day, no storm, barometric pressure is approximately 29.92 in Hg. For every 100 feet of altitude, barometric pressure drops .10 in. Hg. If a storm goes through it can change barometric pressure from what should be present for that altitude. You should know what the average pressure and corresponding barometric pressure is for your area.

### Removal & Installation

*3.7L*

The Manifold Absolute Pressure (MAP) sensor is mounted into the front of the intake manifold. An O-ring is used to seal the sensor to the intake manifold.

1. Disconnect electrical connector at sensor.
2. Clean area around MAP sensor.
3. Remove 2 sensor mounting screws.
4. Remove MAP sensor from intake manifold.
5. Check condition of sensor O-ring.

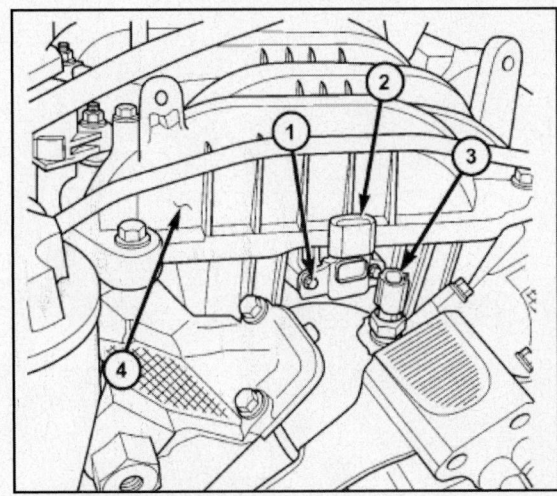

**Indicating the location of the MAP Sensor mounting screws: MAP sensor screws (1), MAP sensor (2), ECT sensor (3), and intake manifold (4)**

*4.0L*

The MAP sensor is mounted to the side of the throttle body. An L-shaped rubber fitting is used to connect the MAP sensor to throttle body.

1. Remove air cleaner duct and air resonator box at throttle body.
2. Remove MAP sensor mounting bolt (screw).
3. While removing MAP sensor, slide the rubber L-shaped fitting from the throttle body.
4. Remove rubber L-shaped fitting from MAP sensor.

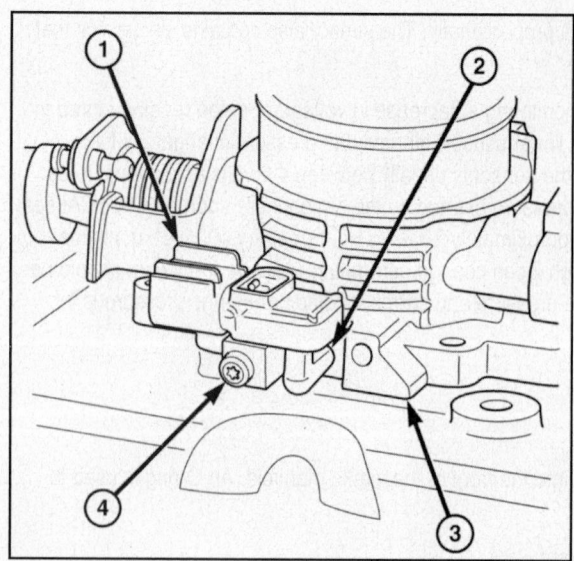

**Identifying the MAP sensor (2), rubber fitting (2), throttle body (3) and mounting screw (4) on 4.0L engine**

### 4.7L

The MAP sensor is located on the front of the intake manifold. An O-ring seals the sensor to the intake manifold.

1. Disconnect electrical connector at sensor.
2. Clean area around MAP sensor.
3. Remove 2 sensor mounting bolts.
4. Remove MAP sensor from intake manifold.

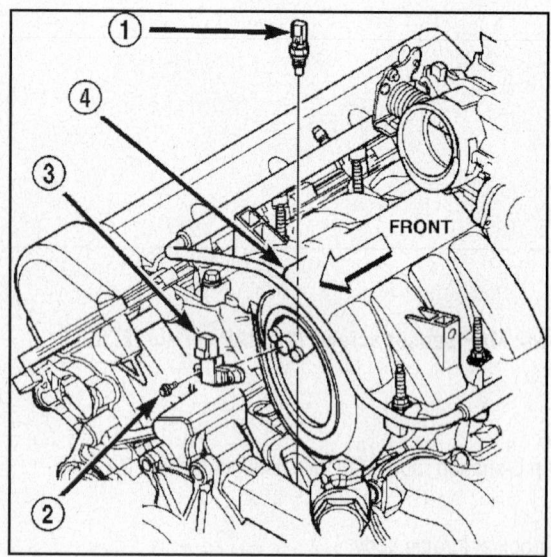

**Showing Removal & Installation of the MAP sensor (3) from the intake manifold (4); also showing the ECT sensor (1) and its mounting bolts (2) on 4.7L engine**

**To install:**

*3.7L*

1. Clean MAP sensor mounting hole at intake manifold.
2. Check MAP sensor O-ring seal for cuts or tears.
3. Position sensor into manifold.
4. Install MAP sensor mounting bolts (screws). Tighten screws to 3 Nm (25 inch lbs.) torque.
5. Connect electrical connector.

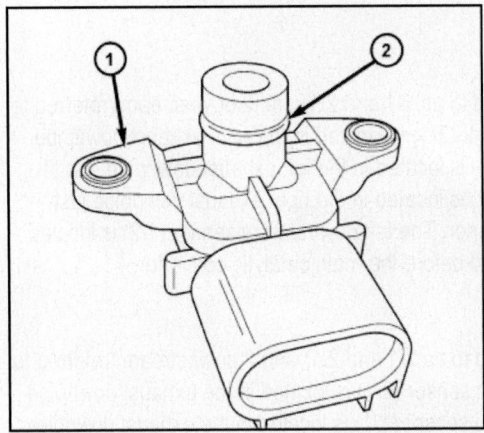

**MAP Sensor Illustration: MAP sensor (1) O-ring (2)**

*4.0L*

The MAP sensor is mounted to the side of the throttle body. An L-shaped rubber fitting is used to connect the MAP sensor to throttle body.

1. Install rubber L-shaped fitting to MAP sensor.
2. Position sensor to throttle body while guiding rubber fitting over throttle body vacuum nipple.
3. Install MAP sensor mounting bolt (screw). Tighten screw to 3 Nm (25 inch lbs.) torque.
4. Install air cleaner duct/air box.

*4.7L*

The MAP sensor is located on the front of the intake manifold. An O-ring seals the sensor to the intake manifold.

1. Clean MAP sensor mounting hole at intake manifold.
2. Check MAP sensor O-ring seal for cuts or tears.
3. Position sensor into manifold.
4. Install MAP sensor mounting bolts (screws). Tighten screws to 3 Nm (25 inch lbs.) torque.
5. Connect electrical connector.

## HEATED OXYGEN SENSOR (HO2S)

### Description

The Oxygen Sensors (O2S) are attached to, and protrude into the vehicle exhaust system. Depending on the emission package, the vehicle may use a total of either 2 or 4 sensors.

#### FEDERAL EMISSIONS PACKAGE

Two sensors are used: upstream (referred to as 1/1) and downstream (referred to as 1/2). With this emission package, the upstream sensor (1/1) is located just before the main catalytic converter. The downstream sensor (1/2) is located just after the main catalytic converter.

#### 4.7L V-8 WITH CALIFORNIA EMISSIONS PACKAGE

On this emissions package, 4 sensors are used: 2 upstream (referred to as 1/1 and 2/1) and 2 downstream (referred to as 1/2 and 2/2). With this emission package, the right upstream sensor (2/1) is located in the right exhaust downpipe just before the mini-catalytic converter. The left upstream sensor (1/1) is located in the left exhaust downpipe just before the mini-catalytic converter. The right downstream sensor (2/2) is located in the right exhaust downpipe just after the mini-catalytic converter, and before the main catalytic converter. The left downstream sensor (1/2) is located in the left exhaust downpipe just after the mini-catalytic converter, and before the main catalytic converter.

#### 4.0L 6-CYLINDER WITH CALIFORNIA EMISSIONS PACKAGE

On this emissions package, 4 sensors are used: 2 upstream (referred to as 1/1 and 2/1) and 2 downstream (referred to as 1/2 and 2/2). With this emission package, the rear/upper upstream sensor (2/1) is located in the exhaust downpipe just before the rear mini-catalytic converter. The front/upper upstream sensor (1/1) is located in the exhaust downpipe just before the front mini-catalytic converter. The rear/lower downstream sensor (2/2) is located in the exhaust downpipe just after the rear mini-catalytic converter, and before the main catalytic converter. The front/lower downstream sensor (1/2) is located in the exhaust downpipe just after the front mini-catalytic converter, and before the main catalytic converter.

### Operation

An O2 sensor is a galvanic battery that provides the PCM with a voltage signal (0-1 volt) inversely proportional to the amount of oxygen in the exhaust. In other words, if the oxygen content is low, the voltage output is high; if the oxygen content is high the output voltage is low. The PCM uses this information to adjust injector pulse-width to achieve the 14.7-to-1 air/fuel ratio necessary for proper engine operation and to control emissions.

The O2 sensor must have a source of oxygen from outside of the exhaust stream for comparison. Current O2 sensors receive their fresh oxygen (outside air) supply through the O2 sensor case housing.

Four wires (circuits) are used on each O2 sensor: a 12-volt feed circuit for the sensor heating element; a ground circuit for the heater element; a low-noise sensor return circuit to the PCM, and an input circuit from the sensor back to the PCM to detect sensor operation.

#### OXYGEN SENSOR HEATERS/HEATER RELAYS

Depending on the emissions package, the heating elements within the sensors will be supplied voltage from either the ASD relay, or 2 separate oxygen sensor relays.

The O2 sensor uses a Positive Thermal Co-efficient (PTC) heater element. As temperature increases, resistance increases. At ambient temperatures around 70°F, the resistance of the heating element is approximately 4.5 ohms on 4.0L engines. It is approximately 13.5 ohms on the 4.7L engine. As the sensor's temperature increases, resistance in the heater element increases. This allows the heater to maintain the optimum operating temperature of approximately 930°-1100°F (500°-600° C). Although the sensors operate the same, there are physical differences, due to the environment that they operate in, that keep them from being interchangeable.

Maintaining correct sensor temperature at all times allows the system to enter into closed loop operation sooner. Also, it allows the system to remain in closed loop operation during periods of extended idle.

In Closed Loop operation, the PCM monitors certain O2 sensor input(s) along with other inputs, and adjusts the injector pulse width accordingly. During Open Loop operation, the PCM ignores the O2 sensor input. The PCM adjusts injector pulse width based on preprogrammed (fixed) values and inputs from other sensors.

### UPSTREAM SENSOR (NON-CALIFORNIA EMISSIONS)

The upstream sensor (1/1) provides an input voltage to the PCM. The input tells the PCM the oxygen content of the exhaust gas. The PCM uses this information to fine tune fuel delivery to maintain the correct oxygen content at the downstream oxygen sensor. The PCM will change the air/fuel ratio until the upstream sensor inputs a voltage that the PCM has determined will make the downstream sensor output (oxygen content) correct. The upstream oxygen sensor also provides an input to determine catalytic converter efficiency.

### DOWNSTREAM SENSOR (NON-CALIFORNIA EMISSIONS)

The downstream oxygen sensor (1/2) is also used to determine the correct air-fuel ratio. As the oxygen content changes at the downstream sensor, the PCM calculates how much air-fuel ratio change is required. The PCM then looks at the upstream oxygen sensor voltage and changes fuel delivery until the upstream sensor voltage changes enough to correct the downstream sensor voltage (oxygen content). The downstream oxygen sensor also provides an input to determine catalytic converter efficiency.

### UPSTREAM SENSORS (CALIFORNIA ENGINES)

Two upstream sensors are used (1/1 and 2/1). The 1/1 sensor is the first sensor to receive exhaust gases from the #1 cylinder. They provide an input voltage to the PCM. The input tells the PCM the oxygen content of the exhaust gas. The PCM uses this information to fine tune fuel delivery to maintain the correct oxygen content at the downstream oxygen sensors. The PCM will change the air/fuel ratio until the upstream sensors input a voltage that the PCM has determined will make the downstream sensors output (oxygen content) correct. The upstream oxygen sensors also provide an input to determine mini-catalyst efficiency. Main catalytic converter efficiency is not calculated with this package.

### DOWNSTREAM SENSORS (CALIFORNIA ENGINES)

Two downstream sensors are used (1/2 and 2/2). The downstream sensors are used to determine the correct air-fuel ratio. As the oxygen content changes at the downstream sensor, the PCM calculates how much air-fuel ratio change is required. The PCM then looks at the upstream oxygen sensor voltage, and changes fuel delivery until the upstream sensor voltage changes enough to correct the downstream sensor voltage (oxygen content).

The downstream oxygen sensors also provide an input to determine mini-catalyst efficiency. Main catalytic converter efficiency is not calculated with this package.

Engines equipped with either a downstream sensor(s), or a post-catalytic sensor, will monitor catalytic converter efficiency. If efficiency is below emission standards, the Malfunction Indicator Lamp (MIL) will be illuminated and a Diagnostic Trouble Code (DTC) will be set.

**Removal & Installation**

*CAUTION: Never apply any type of grease to the oxygen sensor electrical connector, or attempt any soldering of the sensor wiring harness.*

*WARNING: The exhaust manifold, exhaust pipes and catalytic converter(s) become very hot during engine operation. Allow engine to cool before removing oxygen sensor.*

1. Raise and support vehicle.
2. Disconnect O2S pigtail harness from main wiring harness.
3. If equipped, disconnect sensor wire harness mounting clips from engine or body.

*CAUTION: When disconnecting sensor electrical connector, do not pull directly on wire going into sensor.*

4. Remove O2S sensor with an oxygen sensor.

**To install:**

*NOTE: Threads of new oxygen sensors are factory coated with anti-seize compound to aid in installation.*

*CAUTION: DO NOT add any additional anti-seize compound to threads of a new oxygen sensor.*

5. Install O2S. Tighten to 30 Nm (22 ft. lbs.) torque.
6. Connect O2S wire connector to main wiring harness.
7. If equipped, connect sensor wire harness mounting clips to engine or body. When equipped, the O2S pigtail harness must be clipped and/or bolted back to their original positions on engine or body to prevent mechanical damage to wiring.
8. Lower vehicle.

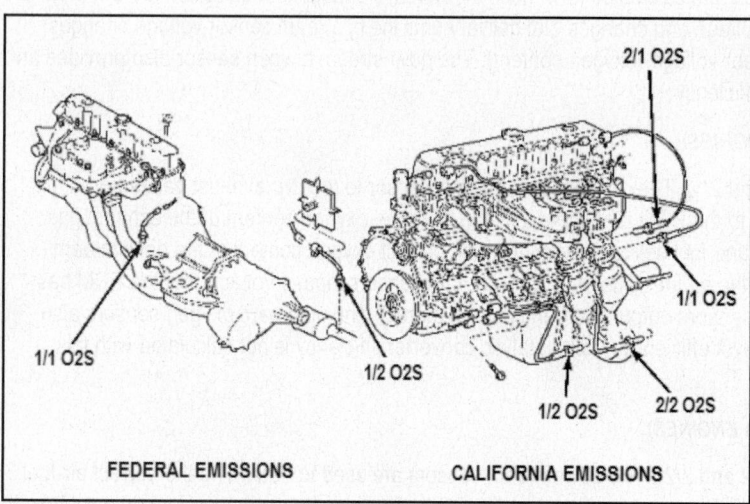

**4.0L engine HO2S locations**

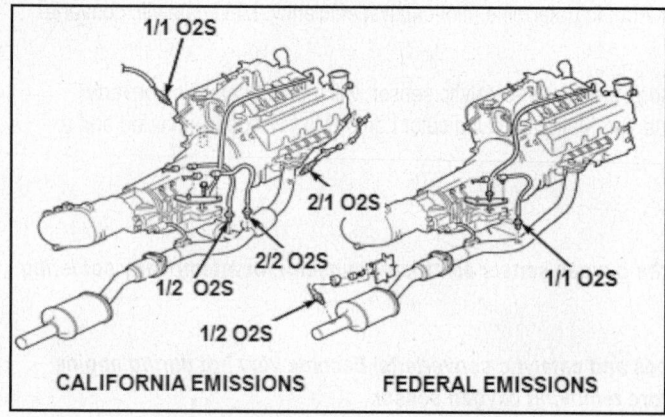

**4.7L engine HO2S locations**

## POSITIVE CRANKCASE VENTILATION (PCV) SYSTEM

### Description & Operation

The 4.7L V-8 engine is equipped with a closed crankcase ventilation system and a Positive Crankcase Ventilation (PCV) valve. This system consists of:

- PCV valve mounted to the oil filler housing. The PCV valve is sealed to the oil filler housing with an O-ring.
- Air cleaner housing
- Interconnected breathers (two) threaded into the rear of each cylinder head.
- Tubes and hose to connect the system components.

The PCV system operates by engine intake manifold vacuum. Filtered air is routed into the crankcase through the air cleaner hose and crankcase breathers. The metered air, along with crankcase vapors, is drawn through the PCV valve and into a passage in the intake manifold. The PCV system manages crankcase pressure and meters blow-by gases to the intake system, reducing engine sludge formation.

The PCV valve contains a spring-loaded plunger. This plunger meters the amount of crankcase vapors routed into the combustion chamber based on intake manifold vacuum.

When the engine is not operating, or during an engine pop-back, the spring forces the plunger back against the seat. This will prevent vapors from flowing through the valve. During periods of high manifold vacuum, such as idle or cruising speeds, vacuum is sufficient to completely compress spring. It will then pull the plunger to the top of the valve. In this position there is minimal vapor flow through the valve. During periods of moderate manifold vacuum, the plunger is only pulled part way back from inlet. This results in maximum vapor flow through the valve.

### Removal & Installation

The PCV valve is located on the oil filler tube. Two locating tabs are located on the side of the valve. These 2 tabs fit into a cam lock in the oil filler tube. An O-ring seals the valve to the filler tube.

1. Disconnect PCV line/hose by disconnecting rubber hose at PCV valve fitting.
2. Remove PCV valve at oil filler tube by rotating PCV valve downward (counter-clockwise) until locating tabs have been freed at cam lock. After tabs have cleared, pull valve straight out from filler tube.

**CAUTION: To prevent damage to PCV valve locating tabs, valve must be pointed downward for Removal & Installation. Do not force valve from oil filler tube.**

3. After valve is removed, check condition of valve O-ring.

### To install:

4. Return the PCV valve back to oil filler tube by placing valve locating tabs into cam lock.
5. Press the PCV valve in and rotate valve upward. A slight click will be felt when tabs have engaged cam lock. The valve should be pointed towards rear of vehicle.
6. Connect PCV line/hose and rubber hose to PCV valve.

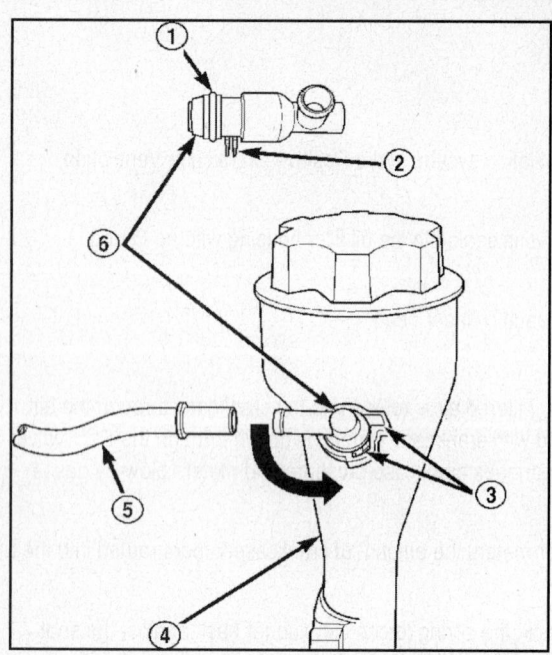

**Identifying the PCV components: 1 – O-ring; 2 – locating tabs; 3 – cam lock; 4 – oil filler tube; 5 – PCV line/hose; 6 – PCV valve**

### Diagnosis & Testing

1. Disconnect PCV line/hose by disconnecting rubber connecting hose at PCV valve fitting.
2. Remove PCV valve at oil filler tube by rotating PCV valve downward until locating tabs have been freed at cam lock. After tabs have cleared, pull valve straight out from filler tube.

*CAUTION: To prevent damage to PCV valve locating tabs, valve must be pointed downward for installation. Do not force valve from oil filler tube.*

3. After valve is removed, check condition of valve O-ring. Also, PCV valve should rattle when shaken.
4. Reconnect PCV valve to its connecting line/hose.
5. Start engine and bring to idle speed.
6. If valve is not plugged, a hissing noise will be heard as air passes through valve. Also, a strong vacuum should be felt with a finger placed at valve inlet.
7. If vacuum is not felt at valve inlet, check line/hose for kinks or for obstruction. If necessary, clean out intake manifold fitting at rear of manifold. Do this by turning a 1/4-inch drill (by hand) through the fitting to dislodge any solid particles. Blow out the fitting with shop air. If necessary, use a smaller drill to avoid removing any metal from the fitting.

*WARNING: Do not attempt to clean the old PCV valve.*

8. Return PCV valve back to oil filler tube by placing valve locating tabs into cam lock. Press PCV valve in and rotate valve upward. A slight click will be felt when tabs have engaged cam lock. Valve should be pointed towards rear of vehicle.
9. Connect PCV line/hose and connecting rubber hose to PCV valve.
10. Disconnect rubber hose from fresh air fitting at left side of air cleaner resonator box. Start engine and bring to idle speed. Hold a piece of stiff paper (such as a parts tag) loosely over the opening of the disconnected rubber hose.

11. The paper should be drawn against the hose opening with noticeable force. This will be after allowing approximately one minute for crankcase pressure to reduce.

12. If vacuum is not present, disconnect each PCV system hose at top of each breather. Check for obstructions or restrictions.

13. If vacuum is still not present, remove each PCV system breather from each cylinder head. Check for obstructions or restrictions. If plugged, replace breather. Tighten breather to 12 Nm (106 inch lbs.) torque. Do not attempt to clean breather

14. If vacuum is still not present, disconnect each PCV system hose at each fitting and check for obstructions or restrictions.

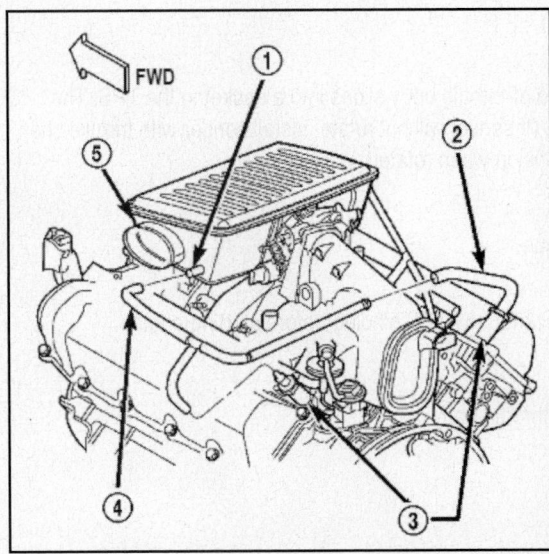

**Identifying the fresh air fitting (1), connecting tubes/hoses (2), crankcase breathers (3), rubber hose (4), and air cleaner resonator (5) for PCV system installation on 4.7L engine**

## THROTTLE POSITION (TP) SENSOR

### Description & Operation

The 3–wire Throttle Position Sensor (TPS) is mounted on the throttle body and is connected to the throttle blade.

The TPS is a 3–wire variable resistor that provides the Powertrain Control Module (PCM) with an input signal (voltage) that represents the throttle blade position of the throttle body. The sensor is connected to the throttle blade shaft. As the position of the throttle blade changes, the resistance (output voltage) of the TPS changes.

The PCM supplies approximately 5 volts to the TPS. The TPS output voltage (input signal to the PCM) represents the throttle blade position. The PCM receives an input signal voltage from the TPS. This will vary in an approximate range of from .26 volts at minimum throttle opening (idle), to 4.49 volts at wide-open throttle. Along with inputs from other sensors, the PCM uses the TPS input to determine current engine operating conditions. In response to engine operating conditions, the PCM will adjust fuel injector pulse width and ignition timing.

The PCM needs to identify the actions and position of the throttle blade at all times. This information is needed to assist in performing the following calculations:

- Ignition timing advance
- Fuel injection pulse-width
- Idle (learned value or minimum TPS)
- Off-idle (0.06 volt)

- Wide Open Throttle (WOT) open loop (2.608 volts above learned idle voltage)
- Deceleration fuel lean out
- Fuel cutoff during cranking at WOT (2.608 volts above learned idle voltage)
- A/C WOT cutoff (certain automatic transmissions only)

## Removal & Installation

### 3.7L & 4.0L

1. Disconnect TPS electrical connector.
2. Remove TPS mounting screws.
3. Remove TPS.

## To install:

The TPS is mounted to the throttle body. The throttle shaft end of throttle body slides into a socket in the TPS. The TPS must be installed so that it can be rotated a few degrees. (If sensor will not rotate, install sensor with throttle shaft on other side of socket tangs). The TPS will be under slight tension when rotated.

4. Install TPS and retaining screws.
5. Tighten screws to 7 Nm (60 inch lbs.) torque.
6. Connect TPS electrical connector to TPS.
7. Manually operate throttle (by hand) to check for any TPS binding before starting engine.

### 4.7L

1. Remove air duct and air resonator box at throttle body.
2. Disconnect TPS electrical connector.
3. Remove two TPS mounting bolts (screws).
4. Remove TPS from throttle body.

## To install:

The throttle shaft end of throttle body slides into a socket in TPS. The TPS must be installed so that it can be rotated a few degrees. If sensor will not rotate, install sensor with throttle shaft on other side of socket tangs. The TPS will be under slight tension when rotated.

5. Install TPS and two retaining bolts.
6. Tighten bolts to 7 Nm (60 inch lbs.) torque.
7. Manually operate throttle control lever by hand to check for any binding of TPS.
8. Connect TPS electrical connector to TPS.
9. Install air duct/air box to throttle body.

## TRANSMISSION RANGE SENSOR/SOLENOID ASSEMBLY

### Description & Operation

The transmission solenoid/TRS assembly, used on 45RFE/545RFE transmissions, is internal to the transmission and mounted on the valve body assembly. The assembly consists of six solenoids that control hydraulic pressure to the six friction elements (transmission clutches), and the torque converter clutch. The pressure control solenoid is located on the side of the solenoid/TRS assembly. The solenoid/TRS assembly also contains five pressure switches that feed information to the TCM.

### SOLENOIDS

Solenoids are used to control the L/R, 2C, 4C, OD, and UD friction elements. The reverse clutch is controlled by line pressure and the position of the manual valve in the valve body. All the solenoids are contained within the Solenoid and Pressure Switch Assembly. The solenoid and pressure switch assembly contains one additional solenoid, Multi-Select (MS), which serves primarily to provide 2nd and 3rd gear limp-in operation.

The solenoids receive electrical power from the Transmission Control Relay through a single wire. The TCM energizes or operates the solenoids individually by grounding the return wire of the solenoid as necessary. When a solenoid is energized, the solenoid valve shifts, and a fluid passage is opened or closed (vented or applied), depending on its default operating state. The result is an apply or release of a frictional element.

The MS and UD solenoids are normally applied to allow transmission limp-in in the event of an electrical failure.

The continuity of the solenoids and circuits are periodically tested. Each solenoid is turned on or off depending on its current state. An inductive spike should be detected by the TCM during this test. If no spike is detected, the circuit is tested again to verify the failure. In addition to the periodic Testing, the solenoid circuits are tested if a speed ratio or pressure switch error occurs.

### PRESSURE SWITCHES

The TCM relies on five pressure switches to monitor fluid pressure in the L/R, 2C, 4C, UD, and OD hydraulic circuits. The primary purpose of these switches is to help the TCM detect when clutch circuit hydraulic failures occur. The switches close at 23 psi and open at 11 psi, and simply indicate whether or not pressure exists. The switches are continuously monitored by the TCM for the correct states (open or closed) in each gear as shown in the following chart:

| GEAR | L/R | 2C | 4C | UD | OD |
|------|-----|-----|-----|-----|-----|
| R | OP | OP | OP | OP | OP |
| P/N | CL | OP | OP | OP | OP |
| 1ST | CL* | OP | OP | CL | OP |
| 2ND | OP | CL | OP | CL | OP |
| 2ND PRIME | OP | OP | CL | CL | OP |
| D | OP | OP | OP | CL | CL |
| 4TH | OP | OP | CL | OP | CL |
| 5TH | OP | CL | OP | OP | CL |

*L/R is closed if output speed is below 100 rpm in Drive and Manual 2. L/R is open in Manual 1.

**NOTE: A Diagnostic Trouble Code (DTC) will set if the TCM senses any switch open or closed at the wrong time in a given gear.**

### Removal & Installation

1. Remove the valve body from the transmission.
2. Remove the screws holding the transmission solenoid/TRS assembly onto the valve body.
3. Separate the transmission solenoid/TRS assembly from the valve body.

**To install:**

4. Place TRS selector plate in the PARK position.
5. Position the transmission solenoid/TRS assembly onto the valve body. Be sure that both alignment dowels are fully seated in the valve body and that the TRS switch contacts are properly positioned in the selector plate
6. Install the screws to hold the transmission solenoid/TRS assembly onto the valve body.
7. Tighten the solenoid assembly screws adjacent to the arrows cast into the bottom of the valve body first. Tighten the screws to 5.7 Nm (50 inch lbs.).
8. Tighten the remainder of the solenoid assembly screws to 5.7 Nm (50 inch lbs.).
9. Install the valve body into the transmission.

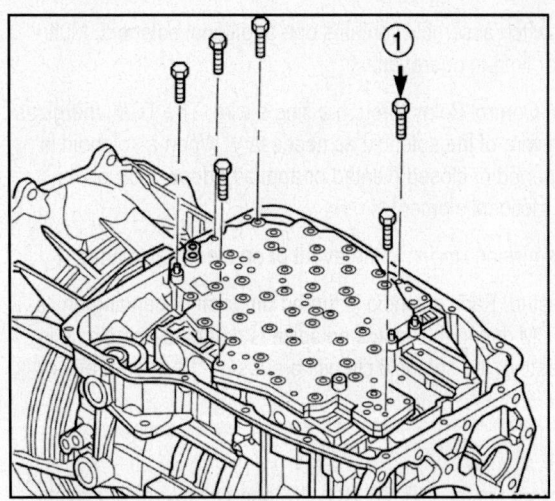

**Valve body mounting bolts**

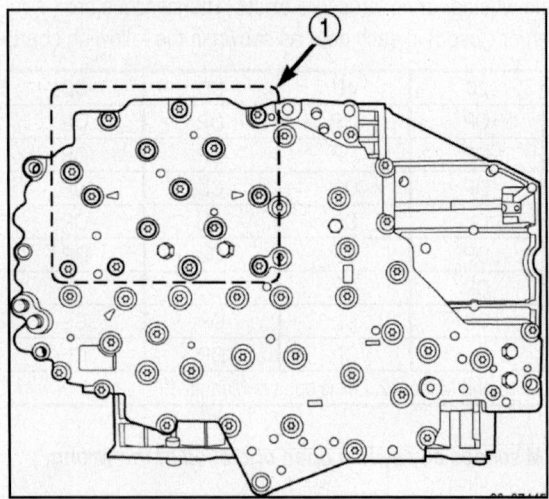

**Solenoid pack mounting bolts**

## VEHICLE SPEED CONTROL

### Description & Operation

The speed control system is electronically controlled and vacuum operated. Electronic control of the speed control system is integrated into the Powertrain Control Module (PCM). The controls consist of two steering wheel mounted switches. The switches are labeled: ON/OFF, RES/ACCEL, SET, COAST, and CANCEL.

The system is designed to operate at speeds above 30 mph (50 km/h). When speed control is selected by depressing the ON switch, the PCM allows a set speed to be stored in PCM RAM for speed control. To store a set speed, depress the SET switch while the vehicle is moving at a speed between 35 and 85 mph. In order for the speed control to engage, the brakes cannot be applied, nor can the gear selector be indicating the transmission is in Park or Neutral.

The speed control can be disengaged manually by:

- Stepping on the brake pedal
- Depressing the OFF switch
- Depressing the CANCEL switch.
- Depressing the clutch pedal (if equipped).

*NOTE: Depressing the OFF switch or turning off the ignition switch will erase the set speed stored in the PCM.*

For added safety, the speed control system is programmed to disengage for any of the following conditions:

- An indication of Park or Neutral
- A rapid increase rpm (indicates that the clutch has been disengaged)
- Excessive engine rpm (indicates that the transmission may be in a low gear)

The speed signal increases at a rate of 10 mph per second (indicates that the coefficient of friction between the road surface and tires is extremely low). The speed signal decreases at a rate of 10 mph per second (indicates that the vehicle may have decelerated at an extremely high rate).

Once the speed control has been disengaged, depressing the RES/ACCEL switch (when speed is greater than 30 mph) restores the vehicle to the target speed that was stored in the PCM.

While the speed control is engaged, the driver can increase the vehicle speed by depressing the RES/ACCEL switch. The new target speed is stored in the PCM when the RES/ACCEL is released. The PCM also has a "tap-up" feature in which vehicle speed increases at a rate of approximately 2 mph for each momentary switch activation of the RES/ACCEL switch.

A "tap down" feature is used to decelerate without disengaging the speed control system. To decelerate from an existing recorded target speed, momentarily depress the COAST switch. For each switch activation, speed will be lowered approximately 1 mph.

### OVERSHOOT/UNDERSHOOT

If the vehicle operator repeatedly presses and releases the SET button with their foot off of the accelerator (referred to as a "lift foot set"), the vehicle may accelerate and exceed the desired set speed by up to 5 mph (8 km/h). It may also decelerate to less than the desired set speed, before finally achieving the desired set speed.

The Speed Control System has an adaptive strategy that compensates for vehicle-to-vehicle variations in speed control cable lengths. When the speed control is set with the vehicle operator's foot off of the accelerator pedal, the speed control thinks there is excessive speed control cable slack and adapts accordingly. If the "lift foot sets" are continually used, a speed control overshoot/undershoot condition will develop.

To "unlearn" the overshoot/undershoot condition, the vehicle operator has to press and release the set button while maintaining the desired set speed using the accelerator pedal (not decelerating or accelerating), and then turning the cruise control switch to the OFF position (or press the CANCEL button if equipped) after waiting 10 seconds. This procedure must be performed approximately 10–15 times to completely unlearn the overshoot/undershoot condition.

### Removal & Installation

## SERVO CABLE

The speed control servo is attached to a bracket. The bracket and servo assembly are located below the battery tray.

1. Disconnect negative battery cable at battery.
2. Disconnect positive battery cable at battery.
3. Remove air cleaner housing at top of throttle body and disconnect servo cable at throttle body.
4. Remove battery from battery tray.
5. Disconnect wiring at battery tray.
6. Disconnect positive battery cable at Power Distribution Center (PDC).
7. Loosen PDC at battery tray.
8. Remove 4 battery tray bolts. One of these bolts attaches to speed control bracket flange that supports battery tray. While removing battery tray, disconnect battery temperature sensor electrical connector at sensor.
9. Disconnect vacuum line at servo vacuum hose fitting.
10. Disconnect electrical connector at servo.
11. If servo and mounting bracket are being removed as one assembly, remove two mounting nuts. These are located above right-front tire. Remove inner fender clips and pry inner fender back slightly to gain access to mounting nuts.
12. If servo is being removed from its mounting bracket, remove 2 mounting nuts holding servo cable sleeve to bracket.
13. Pull speed control cable sleeve and servo away from servo mounting bracket to expose cable retaining clip and remove clip.
14. Remove servo from mounting bracket or, remove servo and mounting bracket as one assembly.

### To install:

15. Position servo to mounting bracket.
16. Align hole in cable connector with hole in servo pin. Install cable-to-servo retaining clip.
17. Insert servo mounting studs through holes in servo mounting bracket.
18. Install servo cable mounting nuts and tighten to 8.5 Nm (75 inch lbs.) torque. If servo and bracket is being installed as one assembly, install 2 mounting nuts and tighten to 28 Nm ±6 Nm (250 inch lbs. ±50 inch lbs.) torque.
19. Connect vacuum line at servo.
20. Connect electrical connector at servo.
21. Connect servo cable to throttle body.
22. Install battery tray and battery temperature sensor.
23. Connect wiring to battery tray.
24. Install battery to battery tray.
25. Connect positive battery cable to Power Distribution Center (PDC).
26. Connect positive battery cable to battery.
27. Connect negative battery cable to battery.
28. Before starting engine, operate accelerator pedal to check for any binding.

## SERVO SWITCH

**WARNING: Before beginning any airbag system component Removal & Installation, remove and isolate the negative (-) cable from the battery. This is the only sure way to disable the airbag system. Then wait two minutes for system capacitor to discharge before further system service. Failure to do this could result in accidental airbag deployment and possible injury.**

1. Disconnect and isolate negative battery cable.
2. Remove airbag module.
3. Remove electrical connector at switch.
4. Remove switch-to-steering wheel mounting screw.
5. Remove switch.

**To install:**

6. Install switch and mounting screw.
7. Tighten screw to 1.5 Nm (15 inch lbs.) torque.
8. Install electrical connector to switch.
9. Install airbag module.
10. Connect negative battery cable.

### VACUUM RESERVOIR

The vacuum reservoir is located in the right/front corner of the vehicle behind the front bumper fascia.

1. Remove front bumper and grill assembly.
2. Remove 1 support bolt near front of reservoir.
3. Remove 2 reservoir mounting bolts.
4. Remove reservoir from vehicle to gain access to vacuum hose. Disconnect vacuum hose from reservoir fitting at rear of reservoir.

**To install:**

5. Connect vacuum hose to reservoir.
6. Install reservoir and tighten 2 bolts to 3 Nm (25 inch lbs.) torque.
7. Install front bumper and grill assembly.

## Diagnosis & Testing

### VACUUM RESERVOIR

1. Disconnect vacuum hose at speed control servo and install a vacuum gauge into the disconnected hose.
2. Start engine and observe gauge at idle. Vacuum gauge should read at least ten inches of mercury.
3. If vacuum is less than ten inches of mercury, determine source of leak. Check vacuum line to engine for leaks. Also check actual engine intake manifold vacuum. If manifold vacuum does not meet this requirement, check for poor engine performance and repair as necessary.
4. If vacuum line to engine is not leaking, check for leak at vacuum reservoir. Remove vacuum reservoir.
5. Disconnect vacuum line at reservoir and connect a hand-operated vacuum pump to reservoir fitting. Apply vacuum. Reservoir vacuum should not bleed off. If vacuum is being lost, replace reservoir.
6. Verify operation of one-way check valve and check it for leaks.
7. Locate one-way check valve. The valve is located in vacuum line between vacuum reservoir and engine vacuum source. Disconnect vacuum hoses (lines) at each end of valve.
8. Connect a hand-operated vacuum pump to reservoir end of check valve. Apply vacuum. Vacuum should not bleed off. If vacuum is being lost, replace one-way check valve.
9. Connect a hand-operated vacuum pump to vacuum source end of check valve. Apply vacuum. Vacuum should flow through valve. If vacuum is not flowing, replace one-way check

valve. Seal the fitting at opposite end of valve with a finger and apply vacuum. If vacuum will not hold, diaphragm within check valve has ruptured. Replace valve.

### Road Test

Perform a vehicle road test to verify reports of speed control system malfunction. The road test should include attention to the speedometer. Speedometer operation should be smooth and without flutter at all speeds.

Flutter in the speedometer indicates a problem that might cause surging in the speed control system. The cause of any speedometer problems should be corrected before proceeding.

If a road test verifies a system problem and the speedometer operates properly, check for:

- A Diagnostic Trouble Code (DTC). If a DTC exists, conduct appropriate test.
- A misadjusted brake (stop) lamp switch. This could also cause an intermittent problem.
- Loose, damaged or corroded electrical connections at the servo. Corrosion should be removed from electrical terminals and a light coating of Mopar® Multipurpose Grease, or equivalent, applied.
- Leaking vacuum reservoir.
- Loose or leaking vacuum hoses or connections.
- Defective one-way vacuum check valve.
- Secure attachment of both ends of the speed control servo cable.
- Smooth operation of throttle linkage and throttle body air valve.
- Failed speed control servo. Do the servo vacuum test.

*CAUTION: When test probing for voltage or continuity at electrical connectors, care must be taken not to damage connector, terminals or seals. If these components are damaged, intermittent or complete system failure may occur.*

## DODGE TRUCKS & VANS

ACCELERATOR PEDAL POSITION (APP) SENSOR

### Description & Operation

The Accelerator Pedal Position Sensor (APPS) assembly is located under the vehicle battery tray. A cable connects the assembly to the accelerator pedal. A plastic cover with a movable door is used to cover the assembly.

*NOTE: The APPS is used only with the 5.7L V-8 engine.*

The Accelerator Pedal Position Sensor (APPS) is a linear potentiometer. It provides the Powertrain Control Module (PCM) with a DC voltage signal proportional to the angle, or position of the accelerator pedal. The APPS signal is translated (along with other sensors) to place the throttle plate (within the throttle body) to a pre-determined position.

A mechanical cable is used between the accelerator pedal and the APPS assembly. Although a cable is used between the pedal and APPS, a mechanical cable is not used at the throttle body. Throttle plate position is electrically determined.

### Removal & Installation

The APPS is serviced (replaced) as one assembly including the sensor, plastic housing and cable. The APPS assembly is located under the vehicle battery tray. Access to APPS is gained from over top of left / front tire.

1. Disconnect negative battery cable at battery.
2. Disconnect APPS cable at accelerator pedal.
3. Remove wheel house liner at left / front wheel.
4. Gain access to APPS electrical connector by opening swing-down door.
5. Disconnect electrical connector.
6. Remove 3 mounting bolts.
7. Remove APPS assembly from battery tray.
8. If cable is to be separated at APPS, unsnap cable clip from ball socket. Release cable from plastic housing by pressing on small cable release tab.

**To install:**

9. Install Accelerator Pedal Position Sensor (APPS) cable to accelerator pedal. Refer to Accelerator Pedal Removal / Installation.
10. Connect electrical connector to APPS.
11. If necessary, connect cable to APPS lever ball socket (snaps on).
12. Snap APPS cable cover closed.
13. Position APPS assembly to bottom of battery tray and install 3 bolts.
14. Install wheelhouse liner.
15. The 5.7L V-8 engine is equipped with a fully electronic accelerator pedal position sensor. If equipped with a 5.7L, also perform the following 3 steps:

Connect negative battery cable to battery.

Turn ignition switch ON, but do not crank engine.

Leave ignition switch ON for a minimum of 10 seconds. This will allow PCM to learn electrical parameters.

*NOTE: The DRB III® Scan Tool may also be used to learn electrical parameters. Go to the Miscellaneous menu, and then select ETC Learn.*

16. If the previous step is not performed, a Diagnostic Trouble Code (DTC) will be set.
17. If necessary, use a proper scan tool to erase any Diagnostic Trouble Codes (DTC's) from PCM.

**Connector Pinouts**

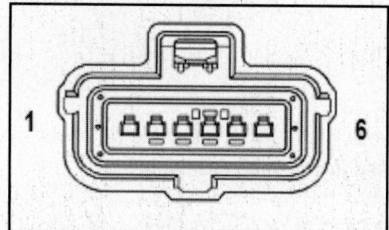

| PIN # | CIRCUIT | FUNCTION |
|-------|---------|----------|
| 1 | YL/PK | 5 VOLT SUPPLY |
| 2 | WT/BR | APPS NO. 2 SIGNAL |
| 3 | BR/VT | APPS NO. 2 RETURN |
| 4 | BR/YL | APPS NO. 1 RETURN |
| 5 | BR/WT | APPS NO. 1 SIGNAL |
| 6 | PK/YL | 5 VOLT SUPPLY |

**5.7L HEMI APP Sensor**

AMBIENT AIR TEMPERATURE (AAT) SENSOR

**Connector Pinouts**

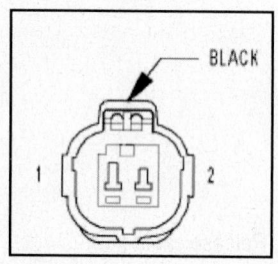

| PIN # | CIRCUIT | FUNCTION |
|-------|---------|----------|
| 1 | VT/LG | AAT SIGNAL |
| 2 | VT/YL (EXCEPT NGC) | SENSOR GROUND |
| 2 | DB/DG (NGC) | SENSOR GROUND |

**AAT Sensor**

BATTERY TEMPERATURE SENSOR CONNECTOR

### Description & Operation

The Battery Temperature Sensor (BTS) is attached to the battery tray located under the battery.

The BTS is used to determine the battery temperature and control battery charging rate. This temperature data, along with data from monitored line voltage, is used by the PCM (ECM Diesel) to vary the battery charging rate. System voltage will be higher at colder temperatures and is gradually reduced at warmer temperatures.

The PCM sends 5 volts to the sensor and is grounded through the sensor return line. As temperature increases, resistance in the sensor decreases and the detection voltage at the PCM increases.

The BTS is also used for OBD II diagnostics. Certain faults and OBD II monitors are either enabled or disabled, depending upon BTS input (for example, disable purge and enable Leak Detection Pump (LDP) and O2 sensor heater tests). Most OBD II monitors are disabled below 20°F.

### Removal & Installation

1. Remove battery.
2. Pry sensor straight up from battery tray mounting hole to gain access to electrical connector.
3. Disconnect sensor from engine wire harness electrical connector.

### To install:

4. Pull electrical connector up through mounting hole in top of battery tray.
5. Connect sensor.
6. Snap sensor into battery tray.
7. Install battery.

### Connector Pinouts

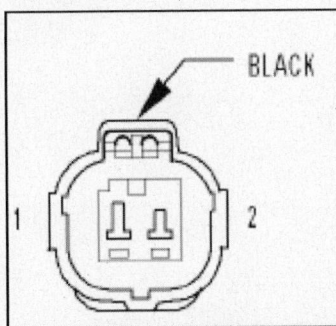

| PIN # | CIRCUIT | FUNCTION |
|---|---|---|
| 1 | DB/VT | BATT TEMP SIGNAL |
| 2 | DB/DG (EXCEPT SRT) | SENSOR GROUND |
| 2 | DB/DG (SRT) | SENSOR GROUND |

### Battery Temperature Sensor

## BRAKE SPEED SENSOR

### Testing

#### RESISTANCE TEST

1. Raise and safely support the vehicle securely on jackstands.
2. Ensure the ignition key is in the OFF position.
3. Disconnect the speed sensor electrical harness.
4. Measure resistance between the terminals on the speed sensor
5. Resistance should be 1000-2500 ohms.
6. If resistance is not within specification, check the harness for continuity.
7. If continuity exists, the sensor is faulty.

#### AIR GAP

1. Raise and safely support the vehicle securely on jackstands.
2. Remove the sensor from the differential.
3. Measure and record the distance from the underside of the sensor flange to the end of the sensor pole piece. This distance represents dimension "B". This dimension should be 1.07-1.08 in. (27.18-27.43mm). If dimension is not within specification, replace the speed sensor.
4. Measure and record the distance between the sensor mounting surface of the differential case and the teeth at the top of the exciter ring. This distance represents dimension "A". This dimension should be 1.085-1.120 in. (27.56-28.45mm). If dimension is not within specification, replace the exciter ring.
5. Subtract dimension "B" from dimension "A" to determine sensor air gap. The gap should be 0.005-0.050 in. (0.127-1.27mm).

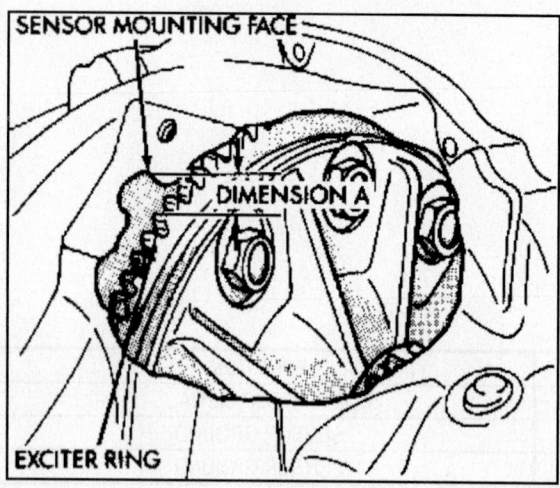

**Measure and record the distance ("A") between the sensor mounting surface of the differential case and the teeth at the top of the exciter ring**

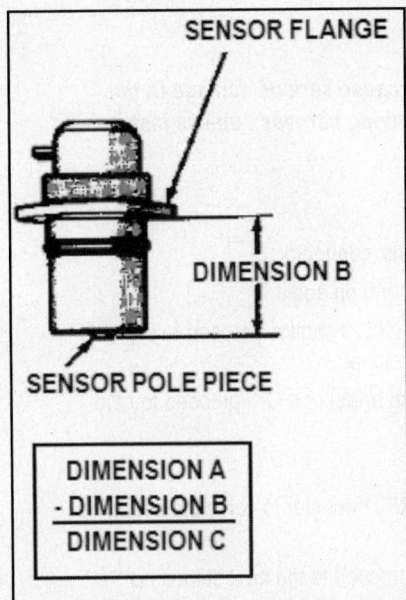

**Measure and record the distance from the underside of the sensor flange to the end of the sensor pole piece**

CAMSHAFT POSITION (CMP) SENSOR

**Description & Operation**

The Camshaft Position (CMP) sensor provides camshaft position information that is used by the ECM for fuel and ignition system synchronization. The sensor is a Hall effect digital sensor. It has a metal pulse ring and a pickup assembly located inside the distributor. The CMP signal is a digital on/off type signal. When the pulse ring travels through the pickup, a permanent magnet inside the pickup creates magnetism, which induces voltage. The pulse ring has slots, or one large slot; as the slot(s) pass, the pickup loses its magnetism and voltage is lost, thereby generating the on/off signal. The CMP and Crankshaft Position (CKP) sensor let the ECM know the position of the camshaft and crankshaft, so the engine can be properly timed.

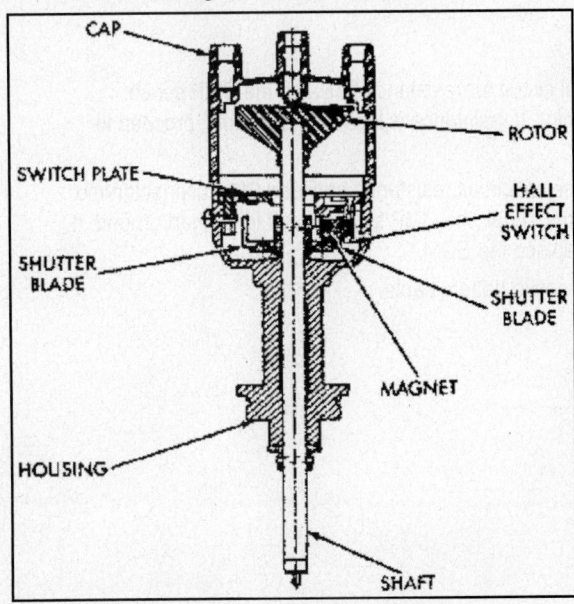

**Distributor with an integral Hall effect switch (CMP sensor)**

Testing

*WARNING: Testing the wiring harness with the ECM still connected can cause serious damage to the processor. ALWAYS disconnect the ECM before Testing the wiring harness., unless instructed otherwise.*

1. Remove the engine cover.

2. With the DRB-II scan tool or equivalent, plug in to diagnostic connector.

3. Using the DRB-II, erase all DTC's and turn the ignition off and on again.

4. Crank the engine and observe CAM SYNC; if NO CAM SYNC is shown, proceed to the next step. If CAM SYNC was present, the CMP sensor is operational.

5. Inspect the wiring and connectors, repair as necessary and retest or if OK, proceed to next step.

6. Unplug the CMP sensor connector.

7. Turn the ignition ON and, using a voltmeter, probe the CMP connector to verify the supply voltage using supplied wiring schematic.

8. On 1995 and earlier vehicles, if voltage is 7.0v or above, proceed to the next step; if no voltage or insufficient voltage is detected, repair the circuit and retest.

9. On 1996-98 vehicles, the voltage should be 4.5v or above. If voltage does not met specifications, repair circuit and retest, if OK, proceed to next step.

10. Connect a jumper between the CMP signal circuit and CMP ground circuit (use supplied schematic). Without turning key off, attempt to start vehicle while making and breaking connection. If vehicle starts, replace CMP sensor. If vehicle does not start proceed to next step.

11. Remove the distributor cap and ensure that distributor turns while cranking engine. If distributor turns, proceed to next step, if it does not turn, repair as necessary and retest.

12. Disconnect the negative battery cable.

13. Using an ohmmeter, probe the ground circuit between the CMP sensor connector and a engine ground. If resistance is less than 5.0 ohms, proceed to next step, if not, repair circuit and retest.

14. Disconnect the ECM wiring harness.

15. Using an ohmmeter, test the CMP signal circuit for resistance between the CMP sensor connector and the ECM harness connector. If resistance is less than 5.0 ohms, proceed to next step, if not, repair circuit and retest.

16. Using an ohmmeter measure the CMP signal circuit resistance between CMP connector and ground, if resistance is less than 5.0 ohms, repair the CMP signal circuit for a short ground. If the resistance is more than 5.0 ohms, replace the ECM.

17. Install the engine cover. Connect the negative battery cable.

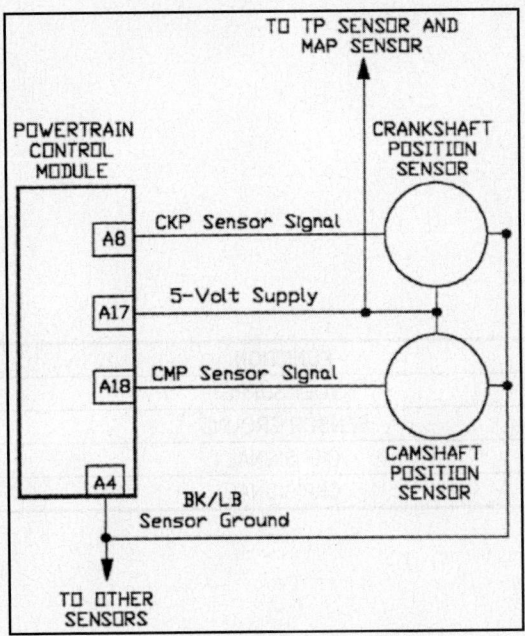

The CMP sensor circuit schematic

**Connector Pinouts**

*3.7L ENGINE*

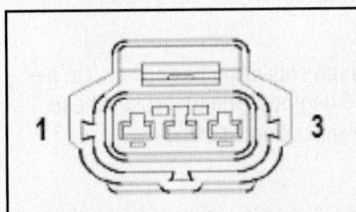

| PIN # | CIRCUIT | FUNCTION |
|-------|---------|----------|
| 1 | BR/LB | CKP SIGNAL |
| 2 | DB/DG | SENSOR GROUND |
| 3 | PK/YL | 5 VOLT SUPPLY |

**Dakota 3.7L CMP Sensor**

*4.7L ENGINE*

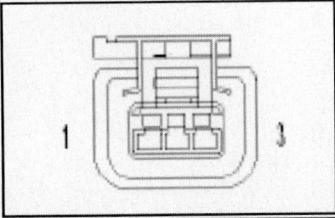

| PIN # | CIRCUIT | FUNCTION |
|-------|---------|----------|
| 1 | BR/LB | CKP SIGNAL |
| 2 | DB/DG | SENSOR GROUND |
| 3 | PK/YL | 5 VOLT SUPPLY |

**Dakota 4.7L CMP Sensor**

*5.7L ENGINE*

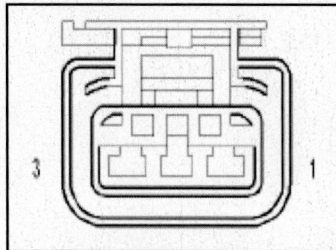

| PIN # | CIRCUIT | FUNCTION |
|-------|---------|----------|
| 1 | PK/YL | 5 VOLT SUPPLY |
| 2 | DB/DG | SENSOR GROUND |
| 3 | BR/LB (5.7L) | CKP SIGNAL |
| 3 | BR/LB (SRT) | CKP SIGNAL |

**Dakota 5.7L CMP Sensor**

CRANKSHAFT POSITION (CKP) SENSOR

**Description & Operation**

The CKP sensor, located on the back of the engine, runs off the flywheel/driveplate and provides the ECM with crankshaft position and engine speed. The sensor is a Hall effect type that senses the passing of teeth on a flywheel/driveplate, because the teeth disrupt the magnetic field of the sensor. This disruption creates a pulse generation, which is monitored by the ECM.

On the 3.9L engine, the flywheel/driveplate has three sets of double notches and three sets of single notches. On the 5.2L and the 5.9L, the flywheel/driveplate has eight single notches spaced every 45 degrees. The ECM uses these pulses to properly set the ignition timing. With out these signals, the vehicles will crank but not start.

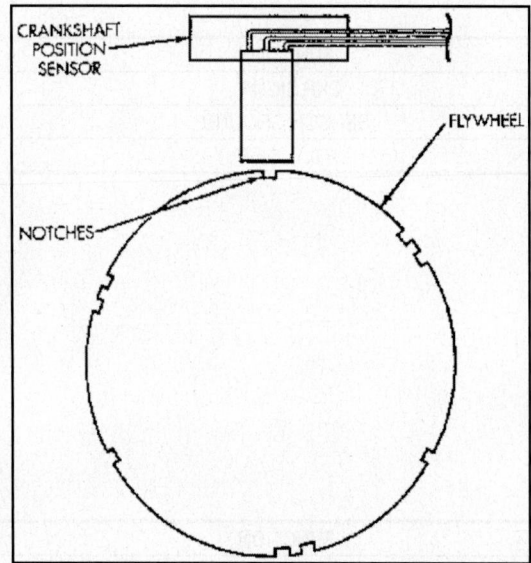

**CKP sensor operation-3.9L engine**

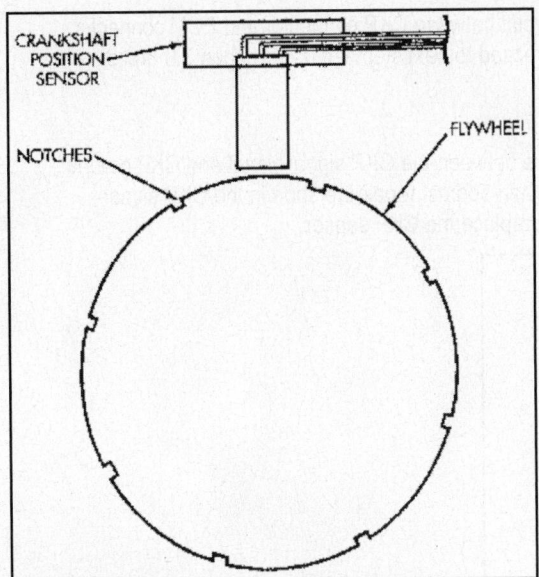

**CKP sensor operation-5.2L and 5.9L engines**

Testing

*WARNING: Testing the wiring harness with the ECM still connected can cause serious damage to the processor. ALWAYS disconnect the ECM before Testing the wiring harness., unless instructed otherwise.*

1. Remove the engine cover.
2. Using the DRB-II scan tool or equivalent, plug into diagnostic connector and erase DTC's.
3. Turn ignition OFF and then ON again.
4. Crank engine until starts or for 10 seconds, whichever first. Monitor Crank Signal on scan tool. If Crank Signal is detected, CKP sensor is OK, if No Crank Signal is detected, proceed with diagnostics.
5. Inspect wiring and connectors and repair as necessary and retest. If the wiring and connectors are OK, proceed to the next step.
6. With the ignition OFF, unplug the CKP sensor signal. Using a voltmeter, check the supply voltage using supplied schematics.
7. On 1995 and earlier vehicles, it should be 7.0v or more; if not, repair circuit and retest. If voltage is OK, proceed to next step.
8. On 1996-98 vehicles, it should be 4.5v or more; if not, repair circuit and retest. If voltage is OK, proceed to the next step.
9. Using the scan tool, erase DTC's in the ECM, then turn the ignition OFF and install a jumper wire on the CKP signal circuit. Turn the ignition ON.
10. While observing the display on scan tool, tap the other end of the jumper wire to the sensor ground. If the scan tool shows No Cam Sync, replace the CKP sensor. If the scan tool shows a Cam Sync, proceed to next step.
11. With the ignition OFF, using an ohmmeter probe the ground signal on the CKP sensor connector, measure the resistance between the connector and ground. If it is less than 5.0 ohms, proceed to the next step; if it is more than 5.0v, repair the circuit and retest.
12. Disconnect the negative battery cable.
13. Disconnect the ECM wiring harness. Inspect for any damaged pins or connector terminals and repair and retest if necessary.

14. Using an ohmmeter, test the CKP signal circuit between CKP connector and ECM connector for resistance. If it is less than 5.0 ohms, proceed to next step, if it is more than 5.0 ohms, repair the circuit and retest.

15. Plug in the CKP sensor connector.

16. Using an ohmmeter, measure the resistance between the CKP signal circuit and CKP ground at the ECM connector. If resistance is less than 5ohms, repair the short in the CKP signal wire. If the resistance is more than 5 ohms, replace the CKP sensor.

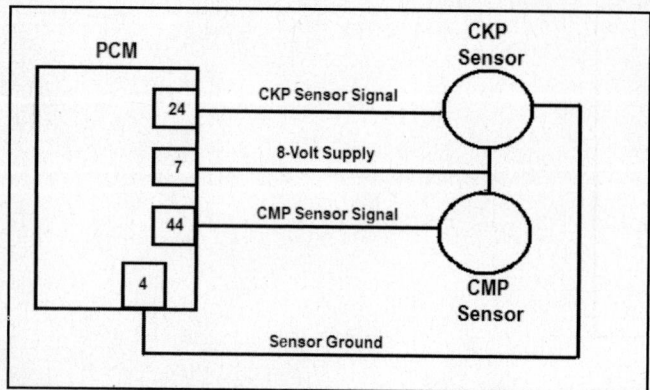

**The CKP sensor circuit schematic**

DATA LINK CONNECTOR

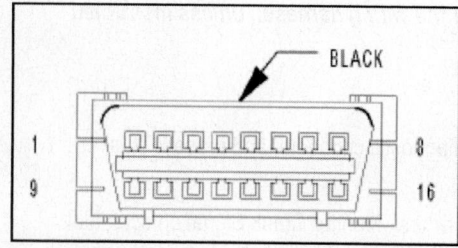

| PIN # | CIRCUIT | FUNCTION |
|---|---|---|
| 1 | - | - |
| 2 | WT/VT | PCI BUS |
| 3 | - | - |
| 4 | BK/LG | GROUND |
| 5 | BK/WT | GROUND |
| 6 | - | - |
| 7 | WT/BR (DIESEL) | SCI TRANSMIT (ECM) |
| 7 | WT/BR (GAS) | SCI TRANSMIT (PCM) |
| 8 | - | - |
| 9 | WT/OR (EXCEPT SRT) | SCI RECEIVE (TCM) |
| 10-11 | - | - |
| 12 | WT/LG (DIESEL) | SCI RECEIVE (ECM) |
| 12 | WT/LG (GAS) | SCI RECEIVE (PCM) |
| 13-14 | - | - |
| 15 | WT/DG (EXCEPT SRT) | SCI TRANSMIT (TCM) |
| 16 | GY/RD | FUSED B (+) |

**Ram Data Link Connector Pinout**

## ENGINE CONTROL MODULE (ECM)

### Description & Operation

The Engine Control Module (ECM) performs many functions on your vehicle. The module accepts information from various engine sensors and computes the required fuel flow rate necessary to maintain the correct amount of air/fuel ratio throughout the entire engine operational range.

*NOTE: This component may also be referred to as the Powertrain Control Module (PCM).*

Based on the information that is received and programmed into the ECM memory, the ECM generates output signals to control relays, actuators and solenoids. The ECM also sends out a command to the fuel injectors that meter the appropriate quantity of fuel. The module automatically senses and compensates for any changes in altitude when driving your vehicle.

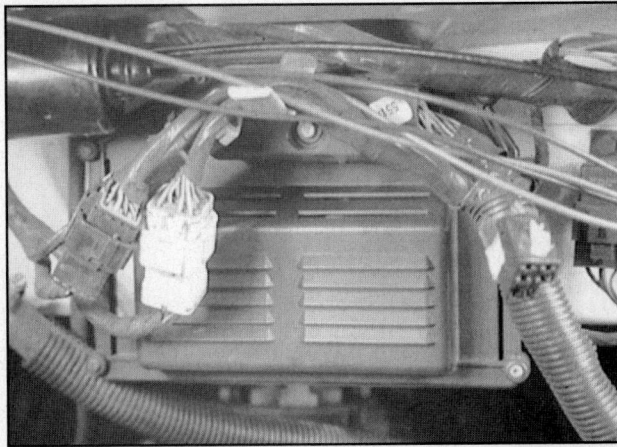

**The ECM is located on the cowl in the engine compartment**

### Removal & Installation

*WARNING: The negative battery cable MUST BE DISCONNECTED prior to servicing the ECM. Voltage spikes could damage the ECM while handling it.*

1. Disconnect the negative battery cable.
2. Remove the retaining tabs on the cover over the ECM and remove the ECM cover.
3. Remove the ECM wiring harness connector(s).
4. Remove the ECM retaining bolts and remove the ECM.

**To install:**

5. Install the ECM and tighten mounting bolts to 35 inch lbs. (4 Nm).
6. Inspect the ECM pins and the terminals on the harness connector(s) for damage and corrosion before installing. Install wiring harness connector(s) onto ECM.
7. Install the ECM cover.
8. Connect the negative battery cable.
9. The scan tool must be used to program the new processor with the vehicle's VIN number and original mileage. If this step is not done, a DTC can be set.

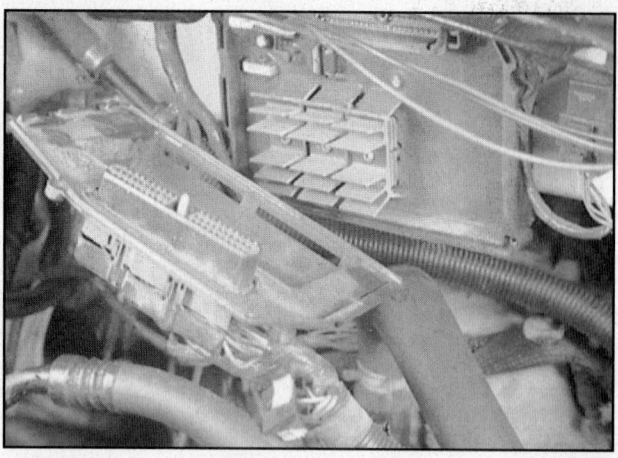

The ECM with its connector and cover removed

Handle the ECM with care after Removal & Installation from the vehicle; dropping it could destroy the circuits inside

## Connector Pinouts

### C1 CONNECTOR

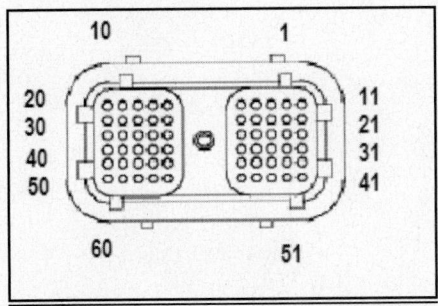

| PIN # | CIRCUIT | FUNCTION |
|---|---|---|
| 1 | DB | WASTEGATE SOLENOID CONTROL |
| 2-3 | - | |
| 4 | WT/LB | CCD BUS (+) |
| 5 | WT/DB | CCD BUS (-) |
| 6 | BR/LB | FUEL PUMP RELAY RETURN |
| 7 | BR/GY | GEN FIELD CONTROL |
| 8 | TN/YL (A/T) | TTVA MOTOR (+) |
| 9-10 | - | |
| 11 | TN/BK | ECT SIGNAL |
| 12 | BR/YL | FUEL RAIL SENSOR SIGNAL |
| 13-14 | - | - |
| 15 | DB/LG | INTAKE AIR TEMPERATURE SENSOR SIGNAL |
| 16-17 | - | - |
| 18 | BR/WT | APPS NO. 1 SIGNAL |
| 19 | BR | FUEL PUMP RELAY CONTROL |
| 20 | TN/OR (A/T) | TTVA MOTOR (-) |
| 21 | - | - |
| 22 | LB/BR | CKP SIGNAL |
| 23 | RD/WT | SENSOR GROUND |
| 24 | DB/GY | CMP SIGNAL |
| 25 | LG | 5 VOLT SUPPLY |
| 26 | BK/LB | A/C PRESSURE SIGNAL |
| 27 | DB/BR | 5 VOLT SUPPLY |
| 28 | WT/BR | APPS NO. 2 SIGNAL |
| 29 | DG/YL | SENSOR GROUND |
| 30 | WT/DB | INJECTOR CONTROL NO. 1 |
| 31 | VT/LG | WATER IN FUEL SENSOR SIGNAL |
| 32 | BR/OR | SENSOR GROUND |
| 33 | BR/YL | APPS NO. 1 RETURN |
| 34 | - | - |
| 35 | OR (A/T) | TTVA POSITION SENSOR SIGNAL |
| 36 | BR/TN | INJECTOR CONTROL NO. 4 |
| 37-39 | - | - |
| 40 | TN | INJECTOR CONTROL NO. 3 |
| 41 | - | |

**Dodge Ram ECM C1 Connector Pinout (1 of 2)**

## C1 CONNECTOR (CONTINUED)

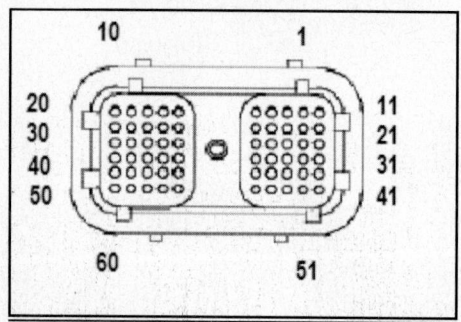

| PIN # | CIRCUIT | FUNCTION |
|-------|---------|----------|
| 42 | BR/VT | 5 VOLT SUPPLY |
| 43 | LB | BOOST PRESSURE SENSOR SIGNAL |
| 44 | - | - |
| 45 | - | - |
| 46 | BR/VT | INJECTOR CONTROL NO. 6 |
| 47 | - | - |
| 48 | - | - |
| 49 | - | - |
| 50 | BR/DB | INJECTOR CONTROL NO. 2 |
| 51 | - | - |
| 52 | BR/OR | SENSOR GROUND |
| 53 | VT/RD | FUEL RAIL SENSOR RETURN |
| 54 | PK/YL | 5 VOLT SUPPLY |
| 55 | - | - |
| 56 | GY | INJECTOR CONTROL NO. 5 |
| 57 | TN/PK | INJECTOR HIGH SIDE DRIVER-BANK 2 |
| 58 | - | - |
| 59 | BR/LG | INJECTOR HIGH SIDE DRIVER-BANK 1 |
| 60 | - | - |

**Dodge Ram ECM C1 Connector Pinout (2 of 2)**

*C2 CONNECTOR*

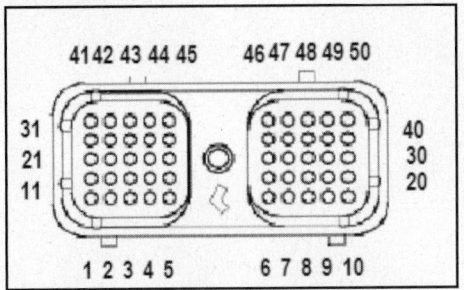

| PIN # | CIRCUIT | FUNCTION |
|---|---|---|
| 1-2 | - | - |
| 3 | VT/WT | INLET AIR TEMPERATURE SENSE |
| 4 | - | - |
| 5 | BR/OR | INTAKE AIR HEATER NO. 2 RELAY CONTROL |
| 6 | BR/YL | INTAKE AIR HEATER NO. 1 RELAY CONTROL |
| 7 | BR | FUEL PUMP RELAY CONTROL |
| 8 | - | - |
| 9 | YL/DB (A/T) | PARK/NEUTRAL POSITION SWITCH SENSE (T41) |
| 10 | BR/LB | FAN SPEED SENSOR |
| 11 | DG/YL | VEHICLE SPEED SIGNAL NO. 1 |
| 12 | VT/GY | OIL PRESSURE SIGNAL |
| 13 | DG (A/T) | TOW/HAUL OVERDRIVE OFF SWITCH SENSE |
| 14 | DG (A/T) | GOVERNOR PRESSURE SOLENOID CONTROL |
| 15 | DG/TN (A/T) | 3-4 SOLENOID CONTROL |
| 16 | WT/BR | SCI TRANSMIT (ECM) |
| 17 | - | - |
| 18 | YL/BR (A/T) | GOVERNOR PRESSURE SENSOR SIGNAL |
| 19 | WT/LG | SCI RECEIVE (ECM) |
| 20 | RD | FUSED B(+) |
| 21 | BK | GROUND |
| 22 | - | - |
| 23 | YL/PK (A/T) | 5 VOLT SUPPLY |
| 24 | DB/DG | SENSOR GROUND |
| 25 | YL/LB (A/T) | TORQUE CONVERTER CLUTCH SOLENOID CONTROL |
| 26 | DB/WT | FUEL LEVEL SENSOR SIGNAL |
| 27 | - | - |
| 28 | WT/VT | PCI BUS |
| 29 | DG/OR (A/T) | TRANSMISSION TEMPERATURE SENSOR SIGNAL |
| 30 | RD | FUSED B(+) |
| 31 | YL/DB (A/T) | TRANSMISSION CONTROL RELAY CONTROL |
| 32 | PK/GY | FUSED IGNITION SWITCH OUTPUT (RUN-START) |
| 33 | VT/BR | 5 VOLT SUPPLY |
| 34 | - | - |
| 35 | BR/YL | INLET AIR PRESSURE SENSE |
| 36 | VT/YL | BRAKE SWITCH NO. 2 SIGNAL |
| 37 | DG/WT | BRAKE SWITCH NO. 1 SIGNAL |
| 38-39 | - | - |

**Dodge Ram ECM C2 Connector Pinout (1 of 2)**

**C2 CONNECTOR - CONTINUED**

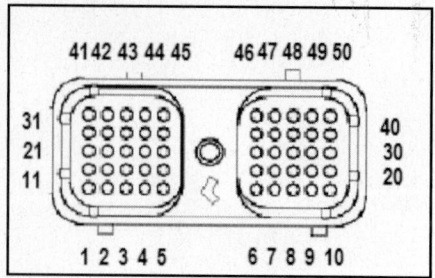

| PIN # | CIRCUIT | FUNCTION |
|---|---|---|
| 40 | RD | FUSED B(+) |
| 41 | LB/OR | A/C CLUTCH RELAY CONTROL |
| 42 | - | - |
| 43 | BR/OR | FAN CLUTCH CONTROL |
| 44 | DG/BR (A/T) | OUTPUT SPEED SENSOR SIGNAL |
| 45 | DG/VT (A/T) | SPEED SENSOR GROUND |
| 46 | VT | S/C SWITCH NO. 1 SIGNAL |
| 47 | DB/VT | BATT TEMP SIGNAL |
| 48 | BR/VT | APPS NO. 2 RETURN |
| 49 | BK | GROUND |
| 50 | BK | GROUND |

**Ram ECM C2 Connector Pinout (2 of 2)**

## ENGINE COOLANT TEMPERATURE (ECT) SENDER

### Testing

A quick way to determine if the gauge, the MIL, or the sending unit is faulty is to disconnect the sending unit electrical harness and ground it (if two-terminal, jumper between the terminals) with the ignition ON. If the gauge responds or the light illuminates, the sending unit may be faulty. Proceed with the following sending unit test.

1. Disconnect the sending unit electrical harness.
2. Remove the radiator cap and place a mechanic's thermometer in the coolant.
3. Using an ohmmeter, check the resistance between the sending unit terminals.
4. Resistance should be high with the engine coolant cold and low with the engine coolant hot.
5. It is best to check resistance with the engine cool, then start the engine and watch the resistance change as the engine warms.
6. If resistance does not drop as engine temperature rises, the sending unit is faulty.

The temperature sender is located at the front of the engine, adjacent to the thermostat

**Connector Pinouts**

*3.7L & 4.7L ENGINES*

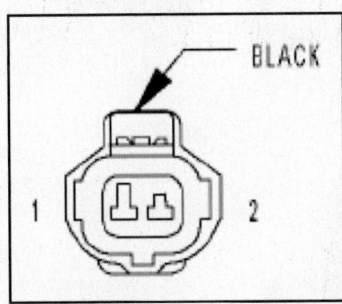

| PIN # | CIRCUIT | FUNCTION |
|-------|---------|----------|
| 1 | DB/DG | SENSOR GROUND |
| 2 | VT/OR | ECT SIGNAL |

**Ram 3.7L & 4.7L ECT Sensor Connector Pinout**

*5.7L/SRT ENGINES*

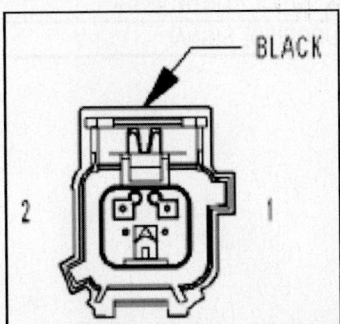

| PIN # | CIRCUIT | FUNCTION |
|-------|---------|----------|
| 1 | DB/DG (5.7L) | SENSOR GROUND |
| 1 | VT/OR (SRT) | ECT SIGNAL |
| 2 | VT/OR (5.7L) | ECT SIGNAL |
| 2 | DB/DG (SRT) | SENSOR GROUND |

**Ram 5.7L/SRT ECT Sensor**

## ENGINE OIL PRESSURE (EOP) SENSOR

**Connector Pinouts**

*SRT MODELS ONLY*

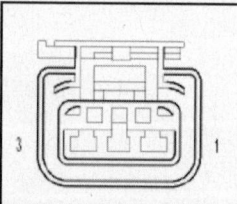

| PIN # | CIRCUIT | FUNCTION |
|-------|---------|----------|
| 1 | PK/YL | 5 VOLT SUPPLY |
| 2 | VT/GY | OIL PRESSURE SIGNAL |
| 3 | DB/DG | SENSOR GROUND |

**Ram SRT Models Engine Oil Pressure Sensor**

## EVAP PURGE SOLENOID VALVE

**Connector Pinouts**

*NEXT-GENERATION CONTROLLER (NGC) EQUIPPED*

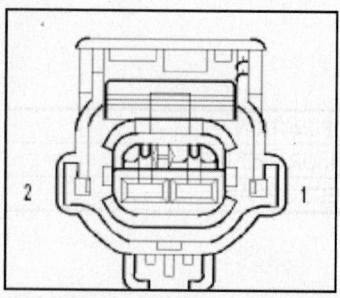

| PIN # | CIRCUIT | FUNCTION |
|-------|---------|----------|
| 1 | K52 18DB/WT | EVAP PURGE SOL CONTROL |
| 2 | K70 18DB/BR | EVAP PURGE SOL SIGNAL |

**Ram w/NGC EVAP Purge Solenoid Valve Connector Pinout**

*SRT MODELS*

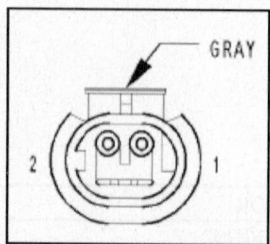

| PIN # | CIRCUIT | FUNCTION |
|-------|---------|----------|
| 1 | F202 20PK/GY | FUSED IGNITION SWITCH OUTPUT (RUN-START) |
| 2 | K70 18DB/BR | EVAP PURGE RETURN |

**Ram SRT models EVAP Purge Solenoid Valve Connector Pinout**

## EXHAUST GAS RECIRCULATION (EGR) VALVE

**Connector Pinouts**

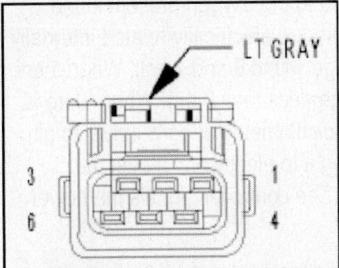

| PIN # | CIRCUIT | FUNCTION |
|-------|---------|----------|
| 1 | DB/LG | EGR SIGNAL |
| 2 | YL/PK | 5 VOLT SUPPLY |
| 3 | DB/DG | SENSOR GROUND |
| 4 | BK/WT | GROUND |
| 5 | - | NOT USED |
| 6 | DB/VT | EGR SOL CONTROL |

**Ram 4.7L & 5.7L EGR Valve Connector Pinout**

## FUEL PUMP MODULE

**Connector Pinouts**

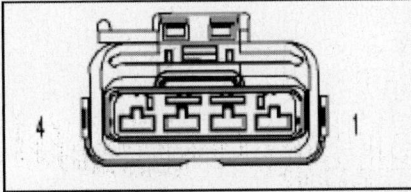

| PIN # | CIRCUIT | FUNCTION |
|-------|---------|----------|
| 1 | BK/OR | GROUND |
| 2 | DB/DG | SENSOR GROUND |
| 3 | DB/WT | FUEL LEVEL SENSOR SIGNAL |
| 4 | OR/RD | FUEL PUMP RELAY OUTPUT |

**Ram Fuel Pump Module Connector Pinout**

<u>HEATED OXYGEN SENSOR (HO2S)</u>

### Description & Operation

The oxygen sensor (O2) is a device that produces an electrical voltage when exposed to the oxygen present in the exhaust gases. The sensor is mounted in the exhaust manifold. Some oxygen sensors are electrically heated internally for faster switching when the engine is running. The oxygen sensor produces a voltage within 0 and 1 volt. When there is a large amount of oxygen present (lean mixture), the sensor produces a low voltage (less than 0.4v). When there is a lesser amount present (rich mixture) it produces a higher voltage (0.6-1.0v). The stoichiometric or correct fuel to air ratio will read between 0.4 and 0.6v. By monitoring the oxygen content and converting it to electrical voltage, the sensor acts as a rich-lean switch. The voltage is transmitted to the engine controller. The controller signals the power module to trigger the fuel injector.

Later models, have two sensors, one before the catalytic converter and one after. This is done for a catalyst efficiency monitor that is a part of the OBD-II engine controls that are on these year vehicles. The one before the catalyst measures the exhaust emissions right out of the engine, and sends the signal to the ECM about the state of the mixture as previously talked about. The second sensor reports the difference in the emissions after the exhaust gases have gone through the catalyst. This sensor reports to the ECM the amount of emissions reduction the catalyst is performing.

The oxygen sensor will not work until a predetermined temperature is reached, until this time the engine controller is running in what as known as OPEN LOOP operation. OPEN LOOP means that the engine controller has not yet begun to correct the air-to-fuel ratio by reading the oxygen sensor. After the engine comes to operating temperature, the engine controller will monitor the oxygen sensor and correct the air/fuel ratio from the sensor's readings. This is what is known as CLOSED LOOP operation.

A heated oxygen sensor has a heating element that keeps the sensor at proper operating temperature during all operating modes. Maintaining correct sensor temperature at all times allows the system to enter into CLOSED LOOP operation sooner.

In CLOSED LOOP operation the engine controller monitors the sensor input (along with other inputs) and adjusts the injector pulse width accordingly. During OPEN LOOP operation the engine controller ignores the sensor input and adjusts the injector pulse to a preprogrammed value based on other inputs.

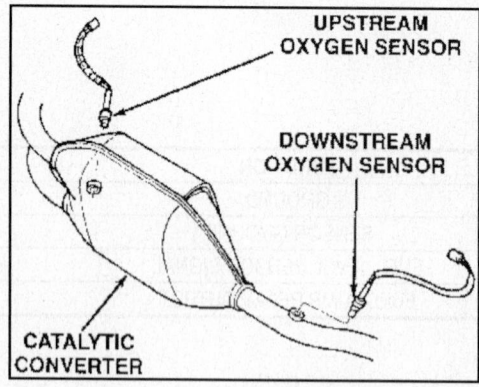

**Dual O₂ sensors found on 1996 and later vehicles, as part of the catalyst efficiency monitor**

### Testing

*CAUTION: The exhaust pipe gets extremely hot during engine operation, and if touched, severe burns can occur. If servicing the oxygen sensor, avoid contacting the exhaust system.*

1. Start the engine and bring it up to operating temperature.
2. Raise and support the vehicle.

3. Backprobe the O2 sensor between the O2 sensor output wire and ground with a suitable high impedance voltmeter.

4. The O2 sensor should be rapidly switching between 0 and 1v. If working properly, it should be switching from a lean mixture (less than 0.4v) to a rich mixture (0.6-1.0v), and back. The average voltage should fall between 0.4-0.6v.

5. If the sensor switches slowly, or is stuck in the middle of the range, the O2 may be faulty.

6. If the sensor is stuck rich or lean, it most likely indicates a problem with the engine; for example, a vacuum leak would cause the O2 to read a lean mixture, and a malfunctioning fuel pressure regulator would cause a rich mixture.

7. If the O2 sensor is above or below the specified range (0-1v), a wiring or computer problem is most likely the cause.

8. Lower the vehicle. Turn the engine off.

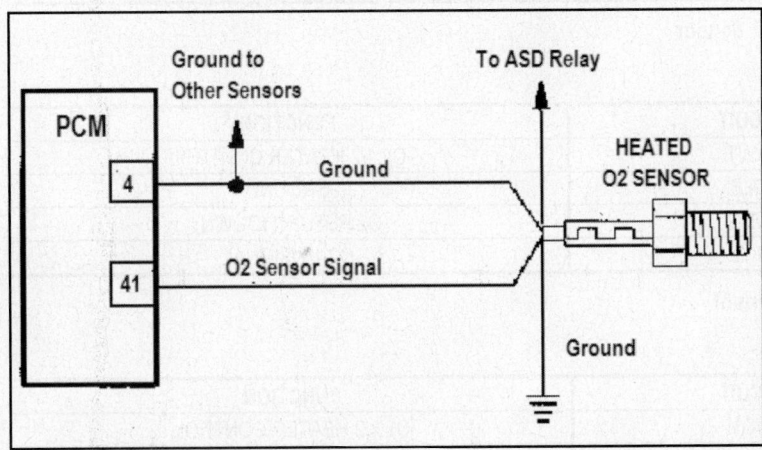

**Heated oxygen sensor circuit schematic**

**Connector Pinouts**

*3.7L & 5.7L ENGINES*

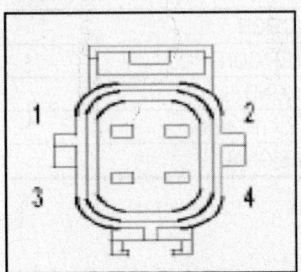

| PIN # | CIRCUIT | FUNCTION |
|---|---|---|
| 1 | BR/LG | O2 1/1 HEATER CONTROL |
| 2 | BK/LG | GROUND |
| 3 | BR/DG | O2 RETURN (UP) |
| 4 | DB/LB | O2 1/1 SIGNAL |

**Ram 3.7L & 5.7L (L.D.) Left Front O2 Sensor**

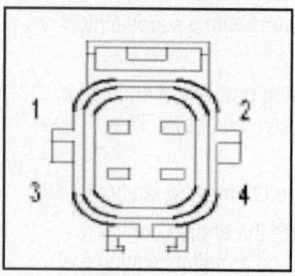

| PIN # | CIRCUIT | FUNCTION |
|-------|---------|----------|
| 1 | BR/VT | O2 2/1 HEATER CONTROL |
| 2 | BK | GROUND |
| 3 | BR/DG | O2 RETURN (UP) |
| 4 | DB/LG | O2 2/1 SIGNAL |

**Ram 3.7L & 5.7L (L.D.) Right Front O2 Sensor**

| PIN # | CIRCUIT | FUNCTION |
|-------|---------|----------|
| 1 | BR/WT | O2 1/2 HEATER CONTROL |
| 2 | BK/LB | GROUND |
| 3 | DB/DG | O2 RETURN (DOWN) |
| 4 | DB/YL | O2 1/2 SIGNAL |

**Ram 3.7 & 5.7L (L.D.) Left Rear O2 Sensor**

| PIN # | CIRCUIT | FUNCTION |
|-------|---------|----------|
| 1 | BR/GY | O2 2/2 HEATER CONTROL |
| 2 | BK | GROUND |
| 3 | DB/DG | O2 RETURN (DOWN) |
| 4 | BR | O2 2/2 SIGNAL |

**Ram 3.7 & 5.7L (L.D.) Right Rear O2 Sensor**

| PIN # | CIRCUIT | FUNCTION |
|-------|---------|----------|
| 1 | BR/WT | O2 1/2 HEATER CONTROL |
| 2 | BK/LB | GROUND |
| 3 | DB/DG | O2 RETURN (DOWN) |
| 4 | DB/YL | O2 1/2 SIGNAL |

**Ram 5.7L (H.D.) Left Rear O2 Sensor**

| PIN # | CIRCUIT | FUNCTION |
|-------|---------|----------|
| 1 | BR/GY | O2 2/2 HEATER CONTROL |
| 2 | BK | GROUND |
| 3 | DB/DG | O2 RETURN (DOWN) |
| 4 | BR | O2 2/2 SIGNAL |

**Ram 5.7L (H.D.) Right Rear O2 Sensor**

**4.7L ENGINES**

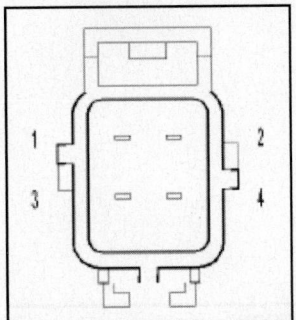

| PIN # | CIRCUIT | FUNCTION |
|-------|---------|----------|
| 1 | BR/LG | O2 1/1 HEATER CONTROL |
| 2 | BK/LG | GROUND |
| 3 | BR/DG | O2 RETURN (UP) |
| 4 | DB/LB | O2 1/1 SIGNAL |

**Ram 4.7L Left Front O2 Sensor**

| PIN # | CIRCUIT | FUNCTION |
|-------|---------|----------|
| 1 | BR/VT | O2 2/1 HEATER CONTROL |
| 2 | BK | GROUND |
| 3 | BR/DG | O2 RETURN (UP) |
| 4 | DB/LG | O2 2/1 SIGNAL |

**Ram 4.7L Right Front O2 Sensor**

| PIN # | CIRCUIT | FUNCTION |
|-------|---------|----------|
| 1 | BR/WT | O2 1/2 HEATER CONTROL |
| 2 | BK/LB | GROUND |
| 3 | DB/DG | O2 RETURN (DOWN) |
| 4 | DB/YL | O2 1/2 SIGNAL |

**Ram 4.7L Left Rear O2 Sensor**

| PIN # | CIRCUIT | FUNCTION |
|-------|---------|----------|
| 1 | BR/GY | O2 2/2 HEATER CONTROL |
| 2 | BK | GROUND |
| 3 | DB/DG | O2 RETURN (DOWN) |
| 4 | BR | O2 2/2 SIGNAL |

**Ram 4.7L Right Rear O2 Sensor**

**SRT MODELS**

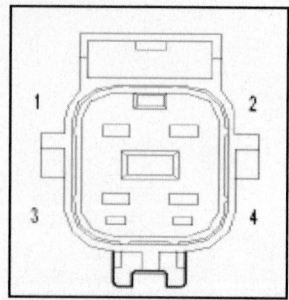

| PIN # | CIRCUIT | FUNCTION |
|-------|---------|----------|
| 1 | BR/YL | FUSED ASD RELAY OUTPUT |
| 2 | BR/LG | O2 1/1 HEATER CONTROL |
| 3 | DB/DG | SENSOR GROUND |
| 4 | DB/LB | O2 1/1 SIGNAL |

**Ram SRT Left Front O2 Sensor**

| PIN # | CIRCUIT | FUNCTION |
|-------|---------|----------|
| 1 | BR/YL | FUSED ASD RELAY OUTPUT |
| 2 | BR/VT | O2 2/1 HEATER CONTROL |
| 3 | DB/DG | SENSOR GROUND |
| 4 | DB/LG | O2 2/1 SIGNAL |

**Ram SRT Right Front O2 Sensor**

| PIN # | CIRCUIT | FUNCTION |
|-------|---------|----------|
| 1 | BR/GY | O2 SENSOR DOWNSTREAM RELAY OUTPUT |
| 2 | BK | GROUND |
| 3 | DB/DG | SENSOR GROUND |
| 4 | DB/YL | O2 1/2 SIGNAL |

**Ram SRT Left Rear O2 Sensor**

| PIN # | CIRCUIT | FUNCTION |
|-------|---------|----------|
| 1 | BR/GY | O2 SENSOR DOWNSTREAM RELAY OUTPUT |
| 2 | BK | GROUND |
| 3 | DB/DG | SENSOR GROUND |
| 4 | BR | O2 2/2 SIGNAL |

**Ram SRT Right Rear O2 Sensor**

<u>IDLE AIR CONTROL (IAC) MOTOR</u>

## Description & Operation

The Idle Air Control (IAC) motor is mounted to the throttle body and is operated by the engine controller. The throttle body has an air control passage that provides air for the engine at idle (when the throttle plate is closed). The IAC motor pintle protrudes into the air control passage and regulates airflow through it. Based on various sensor inputs, the engine controller adjusts engine speed by moving the pintle in and out of the air control passage. The IAC motor is positioned when the ignition is turned to the ON position.

## Testing

**WARNING: When the IAC motor is removed from the throttle body, do not extend the pintle more than 0.250 inch (6.35mm). If the pintle is extended more than this amount, it may separate from the IAC motor and the motor will have to be replaced.**

1. To perform a complete test of the IAC motor, you will need the DRB-II scan tool, or equivalent. This test is a test of the IAC motor only. You will need access to the special factory IAC motor exerciser tool No. 7558, or equivalent.
2. Set the parking brake and block the drive wheels.
3. Route all tester cables away from the cooling fan, drive belt(s), pulleys and exhaust system.
4. Do NOT operate the engine indoors, but provide proper ventilation.
5. Always return the engine idle speed to normal before disconnecting the exerciser tool.
6. Remove the engine cover.
7. With the engine OFF, unplug the IAC motor wire connector.
8. Plug the exerciser tool No. 7558 harness connector onto the IAC motor.
9. Connect the red clip of the exerciser tool to the positive battery terminal, and the black to the negative battery terminal. The red lamp will illuminate when the tool is properly connected.
10. Start the engine. When the switch is in the high or low position, the lamp on the exerciser tool will flash. This indicates that the voltage pulses are being sent to the IAC stepper motor.
11. Move the switch to the HIGH position. The engine speed should increase. Move the switch to the LOW position. The engine speed should decrease.
12. If the engine speed changed predictably while using the exerciser tool, the IAC motor is working correctly. Disconnect the exerciser tool and install the IAC motor wiring connector on IAC motor.
13. If the engine speed does not change, turn the ignition OFF and proceed to Step 6.
14. Remove the IAC motor from the throttle body.
15. With the ignition OFF, cycle the exerciser tool switch between the HIGH and LOW positions. Keep your attention on the pintle. It should move in-and-out of the motor.
16. If the pintle does not move, replace the IAC motor. Start the engine and test the replacement motor operation as described in Step 5.
17. If the pintle operates properly, check the IAC motor bore in the throttle body for blockage and clean as needed. Install the IAC motor and retest. If blockage is not found, more complete testing will be required using the DRB-II scan tool or equivalent.

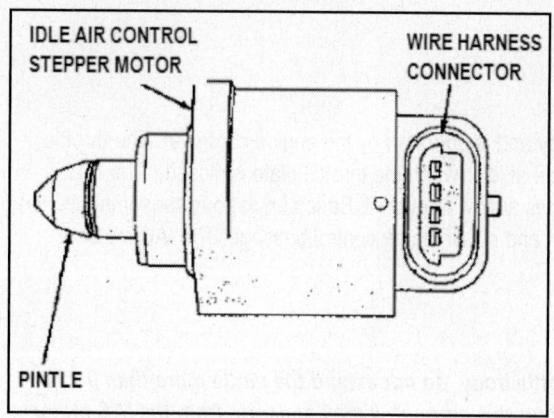

**IAC motor assembly**

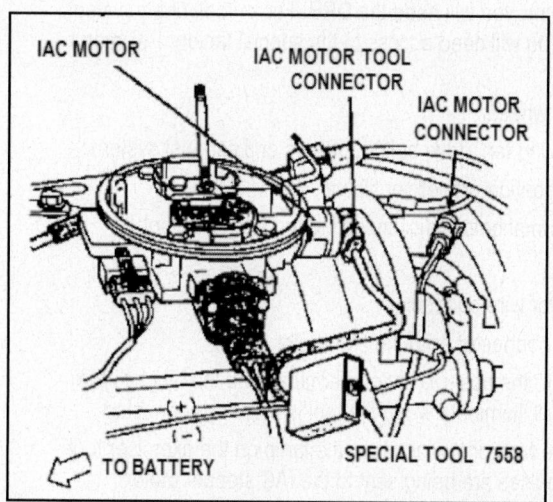

**Connect special tool No. 7558, or equivalent, to the IAC motor**

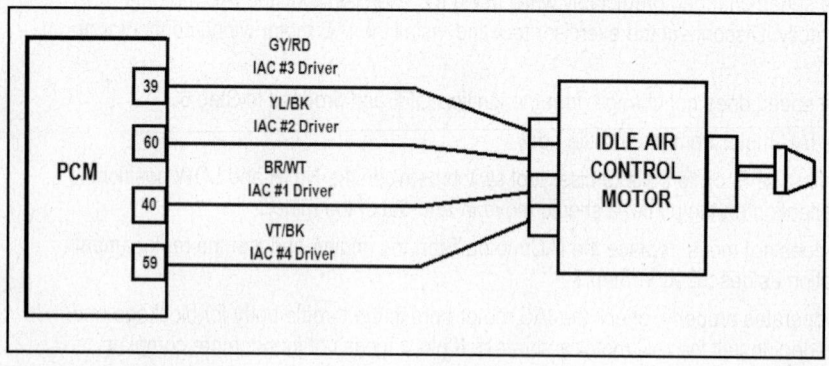

**The IAC motor and driver circuits**

## IDLE SPEED CONTROL ACTUATOR

### Description & Operation

The Idle Speed Control (ISC) actuator is mounted to the throttle body and is controlled by the ECM. The ISC contains the idle contact switch and provides an input signal to the ECM. Various other sensors on the car give input to the ECM that helps to control the ISC. The ECM supplies current and a ground path to the ISC actuator. This enables the ECM to increase or decrease the throttle stop angle by extending or retracting the ISC actuator. The throttle lever rests against an adjustment screw at the end of the actuator. The actuator extends or retracts to control engine speed and to set throttle stop angle during deceleration.

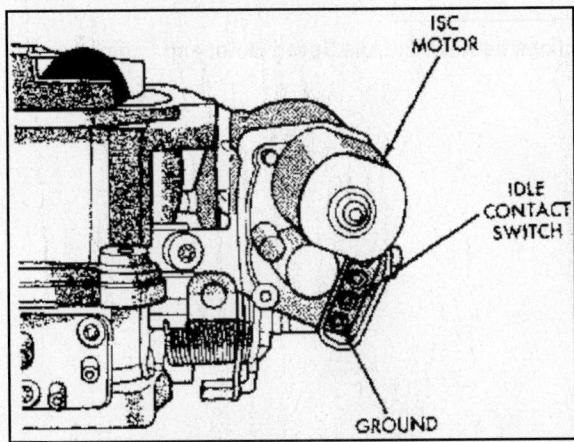

**ISC actuator with its electrical connector removed-note the Idle Contact Switch circuit**

### Testing

*WARNING: Testing the wiring harness with the ECM still connected can cause serious damage to the processor. ALWAYS disconnect the ECM before Testing the wiring harness., unless instructed otherwise.*

1. Remove the engine cover.
2. Remove the air cleaner assembly.
3. While watching the ISC actuator, have an assistant start the engine, shut it off and turn the ignition OFF. The ISC actuator should move outward to preset a fast idle after the next start, after the ignition is turned OFF. If it does not move outward, continue diagnostics. If the actuator moves outward, start engine and let idle to verify proper idle speed, if idle speed is out of specifications, set the idle speed using the procedure on reinstallation of the ISC.
4. Check the ISC actuator to see if it is frozen or sticking, if it is, replace it and retest. If it is OK, continue diagnostics.
5. Start the engine and backprobe the ISC connector to voltage is getting to ISC actuator. Voltage should be fluctuating between 2.0 and 6.0 volts. If voltage is OK, continue diagnostics. If voltage is not OK, repair circuit and retest.
6. Shut the engine off.
7. Disconnect the negative battery cable.
8. Disconnect the ECM wiring harness.
9. Using an ohmmeter, measure the resistance on the ground circuit, using the supplied wiring diagrams, between the ISC connector and ECM harness connector. Resistance should be less than 5.0 ohms, if not repair ground circuit and retest. If resistance is OK, replace ISC motor.

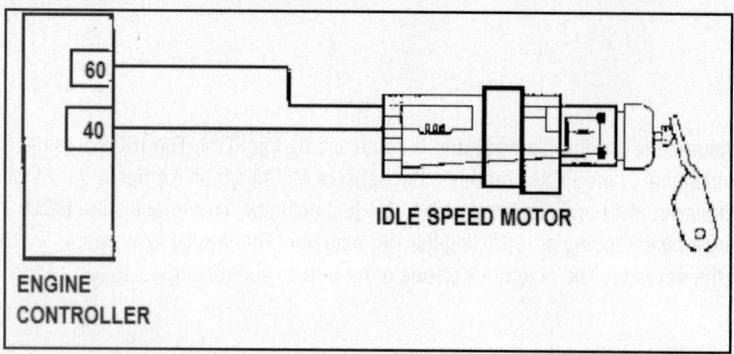

ISC actuator (motor) circuit schematic, showing connections between the Idle Speed Motor and Engine Controller

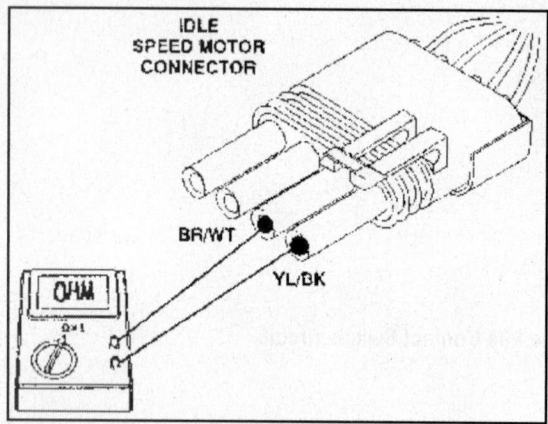

Testing ISC circuit resistance with an ohmmeter

## INTAKE AIR TEMPERATURE (IAT) SENSOR

### Description & Operation

The Intake Air Temperature (IAT) sensor determines the air temperature inside the intake manifold. Resistance changes in response to the ambient air temperature. The sensor resistance decreases as the air temperature increases. This provides a signal to the ECM indicating the temperature of the incoming air charge. This sensor helps the ECM to determine spark timing and air/fuel ratio. The IAT is threaded and screws into the intake manifold.

*NOTE: Only Multi-port Fuel Injected (MFI) vehicles are equipped with an IAT sensor.*

### Testing

*WARNING: Testing the wiring harness with the ECM still connected can cause serious damage to the processor. ALWAYS disconnect the ECM before testing the wiring harness, unless instructed otherwise.*

1. Disconnect the negative battery cable.
2. Remove the engine cover.
3. Unplug the electrical connector at the sensor.
4. Using an ohmmeter, measure the resistance across the two terminals of the sensor. The sensor resistance should be less than 1340 ohms with the engine at normal operating temperature. Replace the sensor if not within specifications.

5. To test the wiring harness, unplug the ECM; test the harness between the proper ECM connector terminal and the sensor connector. Also check the harness between the proper ECM connector terminal and the sensor connector. If resistance is more than 1 ohm, repair the wiring harness.

6. Install the engine cover.

7. Connect the negative battery cable.

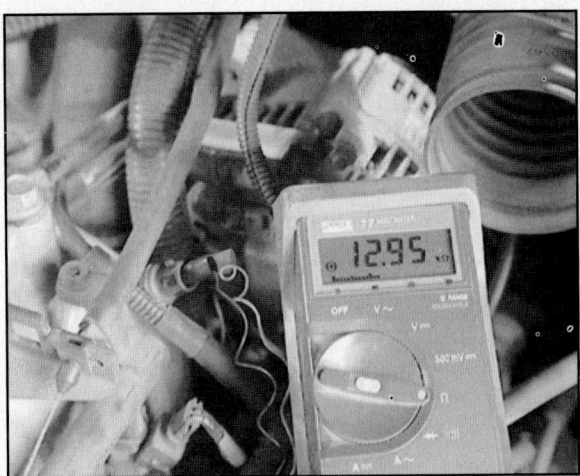

**Using an ohmmeter, test the intake air temperature sensor for resistance across its two terminals**

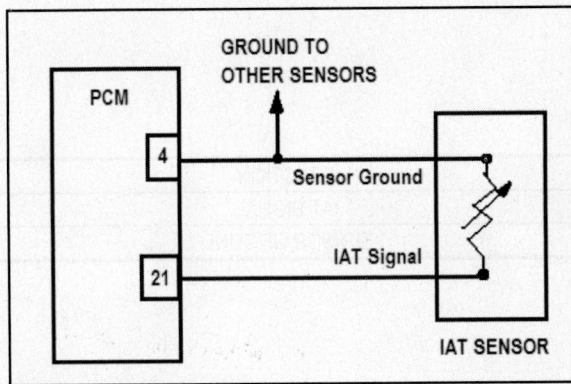

**The IAT sensor circuit schematic**

### Connector Pinouts

*3.7L & 4.7L ENGINES*

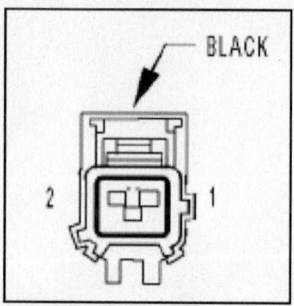

| PIN # | CIRCUIT | FUNCTION |
|-------|---------|----------|
| 1 | DB/DG | SENSOR GROUND |
| 2 | DB/LG | IAT SIGNAL |

**Ram 3.7L & 4.7L IAT Sensor**

*5.7L/SRT*

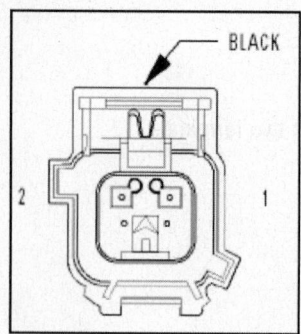

| PIN # | CIRCUIT | FUNCTION |
|-------|---------|----------|
| 1 | DB/LG | IAT SIGNAL |
| 2 | DB/DG | SENSOR GROUND |

**Ram 5.7L/SRT models IAT Sensor**

## KNOCK SENSOR

**Connector Pinouts**

**3.7L & 4.7L**

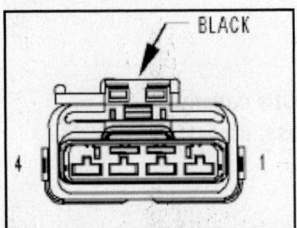

| PIN # | CIRCUIT | FUNCTION |
|-------|---------|----------|
| 1 | DB/YL (3.7L) | KNOCK SENSOR NO. 1 SIGNAL |
| 1 | BR/LG (4.7L) | KNOCK SENSOR NO. 1 RETURN |
| 2 | BR/LG (3.7L) | KNOCK SENSOR NO. 1 RETURN |
| 2 | DB/YL (4.7L) | KNOCK SENSOR NO. 1 SIGNAL |
| 3 | BR/WT (3.7L) | KNOCK SENSOR NO. 2 SIGNAL |
| 3 | WT/BR (4.7L) | KNOCK SENSOR NO. 2 RETURN |
| 4 | WT/BR (3.7L) | KNOCK SENSOR NO. 2 RETURN |
| 4 | BR/WT (4.7L) | KNOCK SENSOR NO. 2 SIGNAL |

**Ram 3.7L & 4.7L Knock Sensor**

**5.7L/SRT**

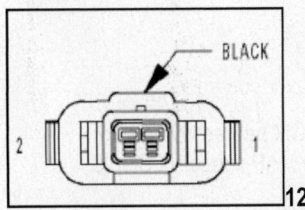

| PIN # | CIRCUIT | FUNCTION |
|-------|---------|----------|
| 1 | DB/YL | KNOCK SENSOR NO. 1 SIGNAL |
| 2 | BR/LG (5.7L) | KNOCK SENSOR NO. 1 RETURN |
| 2 | DB/DG (SRT) | SENSOR GROUND |

**Ram 5.7L/SRT models Knock Sensor**

## MANIFOLD ABSOLUTE PRESSURE (MAP) SENSOR

### Description & Operation

The Manifold Absolute Pressure (MAP) sensor measures the pressure inside the intake manifold, by measuring the vacuum level. It sends a voltage signal to the ECM in relation to the pressure inside the manifold, which varies according to engine load and altitude. The ECM uses this signal to adjust the air/fuel ratio.

### Testing

***WARNING: Testing the wiring harness with the ECM still connected can cause serious damage to the processor. ALWAYS disconnect the ECM before testing the wiring harness, unless instructed otherwise.***

1. Remove the engine cover.
2. Remove the air cleaner assembly.
3. Inspect the L-shaped tube from the throttle body to the MAP for cracks, blockage, and damage. Repair as necessary.
4. Unplug the MAP sensor connector.
5. Test the MAP sensor output voltage at the MAP sensor connector between terminals A and B. With ignition
6. ON and the engine OFF, the voltage should be between 4-5 volts. If no voltage is present, proceed to Step 8.
7. Test MAP sensor supply voltage at sensor connector between terminals A and C with the ignition ON and engine off, voltage should be 4.5-5.0 volts. If no voltage is present, proceed to Step 8.
8. Test the MAP sensor ground circuit at terminal A of the MAP connector; voltage should be less than 0.2 volts. If not inspect for open harness from pin 4 of ECM harness and terminal A. If no voltage is present, proceed to the next step.
9. Turn the ignition OFF.
10. Disconnect the ECM harness from the ECM.
11. Test the MAP sensor output voltage at the ECM connector. At the ECM harness connector test pin 1 of the connector, the voltage should be the same as at the MAP sensor in Step 5.
12. Check ECM harness connector pin 6 for the same voltage as in Step 6.
13. If all of the above tests pass, plug in ECM wiring harness and MAP sensor connector.
14. With the ignition in the ON position, and the engine OFF, remove the tube from the throttle body to the MAP sensor.
15. Connect a vacuum pump to the nipple on the MAP sensor, and pump the sensor to 20-27 in. Hg. of vacuum. Check the sensor output voltage, it should be below 1.8 volts. If not, replace the MAP sensor. If the voltage is OK, proceed to next step.
16. Relieve vacuum pressure on the sensor, and then check the output voltage. The voltage should be 4-5 volts; if not, replace the MAP sensor.
17. Install the air cleaner assembly.
18. Install the engine cover.

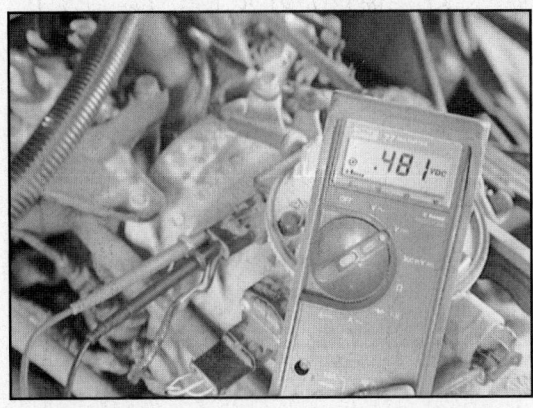

**Use a voltmeter to check the MAP sensor for signal**

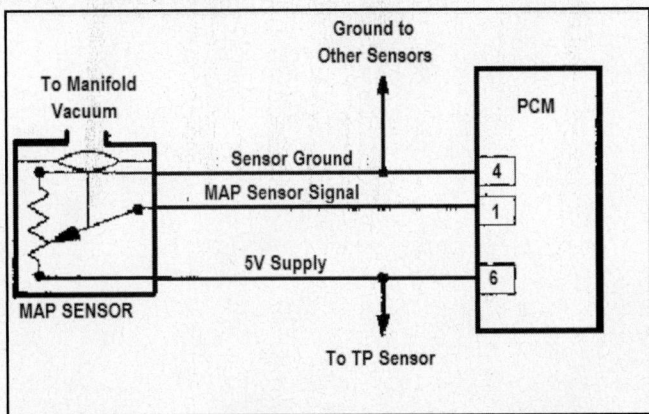

**The MAP sensor circuit schematic for MFI equipped vehicles**

**Connector Pinouts**

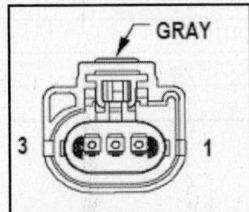

| PIN # | CIRCUIT | FUNCTION |
|---|---|---|
| 1 | VT/BR | MAP SIGNAL |
| 2 | DB/DG (3.7L/4.7L) | SENSOR GROUND |
| 2 | DB/DG (5.7L) | SENSOR GROUND |

**Ram (NGC) MAP Sensor**

## NATURAL VACUUM LEAK DETECTION (NVLD)

**Connector Pinouts**

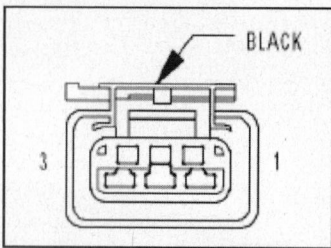

| PIN # | CIRCUIT | FUNCTION |
|-------|---------|----------|
| 1 | - | - |
| 2 | BR/DG | GENERATOR SOURCE |
| 3 | VT/LB | LEAK DETECTION PUMP SOLENOID CONTROL |
| 4 | VT/WT | LEAK DETECTION PUMP SWITCH SENSE |

**Ram (NGC) NVLD Connector Pinout**

## SPEED CONTROL SERVO

**Connector Pinouts**

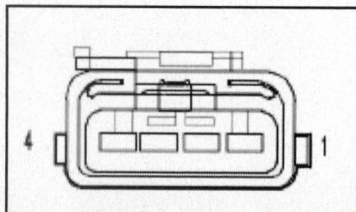

| PIN # | CIRCUIT | FUNCTION |
|-------|---------|----------|
| 1 | V36 20VT/YL (EXCEPT SRT) | S/C VACUUM SOL CONTROL |
| 1 | V36 18VT/YL (SRT) | S/C VACUUM SOL CONTROL |
| 2 | V35 20VT/OR (EXCEPT SRT) | S/C VENT SOL CONTROL |
| 2 | V35 18VT/OR (SRT) | S/C VENT SOL CONTROL |
| 3 | V30 20VT/WT | S/C BRAKE SWITCH OUTPUT |
| 4 | Z913 20BK | GROUND |

**Ram 3.7L, 4.7L & SRT Speed Control Servo Connector Pinout**

## THROTTLE BODY TEMPERATURE SENSOR

### Description & Operation

Throttle Body Injected (TBI) vehicles equipped with a 5.2L or 5.9L engine have a Throttle Body Temperature (TBT) sensor. The sensor monitors throttle body temperature, which is the same as fuel temperature. It is mounted on the throttle body. This sensor provides the ECM with information on fuel temperature, which allows the ECM to richen the air/fuel mixture for a hot restart condition.

### Testing

**WARNING: Testing the wiring harness with the ECM still connected can cause serious damage to the processor. ALWAYS disconnect the ECM before Testing the wiring harness, unless instructed otherwise.**

1. Disconnect the negative battery cable. Remove the engine cover.

2. Unplug the electrical connector at the sensor.

3. Using an ohmmeter, measure the resistance across the two terminals of the sensor. The sensor resistance should be less than 1340 ohms with the engine at normal operating temperature. Replace the sensor if not within specifications.

4. To test the wiring harness, unplug the ECM; test the harness between the proper ECM connector terminal and the sensor connector. If resistance is more than 1 ohm, repair the wiring harness.

5. Install the engine cover. Connect the negative battery cable.

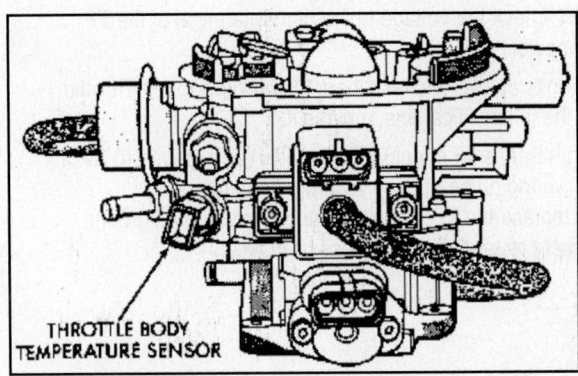

**Location of the throttle body temperature sensor on 5.2L and 5.9L TBI engines**

**Connector Pinouts**

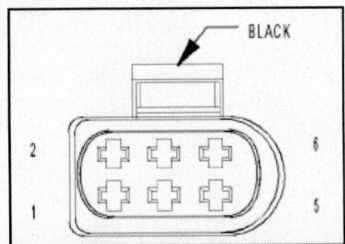

| PIN # | CIRCUIT | FUNCTION |
|-------|---------|----------|
| 1 | K22 18BR/OR | TP NO. 1 SIGNAL |
| 2 | F855 18PK/YL | 5 VOLT SUPPLY |
| 3 | K447 18TN/YL | ETC MOTOR (+) |
| 4 | K122 18BR/DG | TP NO. 2 SIGNAL |
| 5 | K448 18TN/OR | ETC MOTOR (-) |
| 6 | K922 18BR/DB | TP SENSOR RETURN |

**Ram 5.7L Throttle Body Connector Pinout**

## THROTTLE POSITION SENSOR

### Description & Operation

The Throttle Position (TP) sensor is a potentiometer that provides a signal to the ECM that is directly proportional to the throttle plate position. The TP sensor is mounted on the side of the throttle body and is connected to the throttle plate shaft. The TP sensor monitors throttle plate movement and position, and transmits an appropriate electrical signal to the ECM. These signals are used by the ECM to adjust the air/fuel mixture, spark timing and EGR operation according to engine load at idle, part throttle, or full throttle. The TP sensor is not adjustable.

### Testing

1. Remove the engine cover.
2. Remove the air cleaner assembly.
3. With the engine OFF and the ignition ON, check the voltage at the center terminal of the TP sensor by carefully backprobing the connector.
4. Voltage should be between 0.2 and 1.4 volts at idle, and less than 4.8v at Wide Open Throttle (WOT). If the TP sensor does not meet these specifications, replace it.
5. If no voltage is present, check the wiring harness for supply voltage (5.0v) and ground (0.3v or less), by referring to your corresponding wiring guide. If supply voltage and ground are present, but no output voltage from TP, replace the TP sensor. If supply voltage and ground do not meet specifications, make necessary repairs to the harness or ECM.
6. Replace the air cleaner assembly.
7. Replace the engine cover.

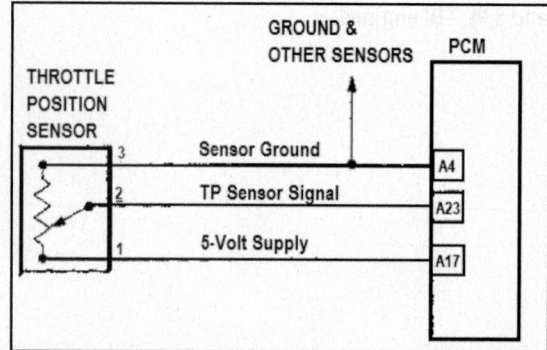

**The TP sensor circuit schematic**

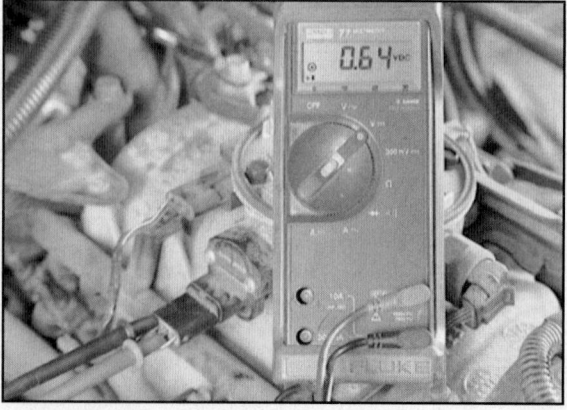

**Using a voltmeter to check the TP sensor for signal voltage**

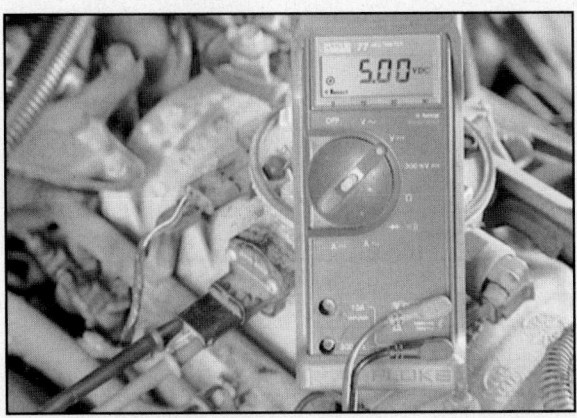

Using a voltmeter to check the TP sensor for reference (supply) voltage

You can use the data display function of the scan tool to get the TP sensor readings

**Connector Pinouts**

*3.7L ENGINES*

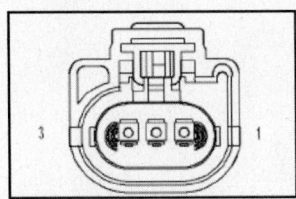

| PIN # | CIRCUIT | FUNCTION |
|-------|---------|----------|
| 1 | F855 18PK/YL | 5 VOLT SUPPLY |
| 2 | K900 18DB/DG | SENSOR GROUND |
| 3 | K22 18BR/OR | TP NO. 1 SIGNAL |

**Ram 3.7L TP Sensor**

## 4.7L ENGINES

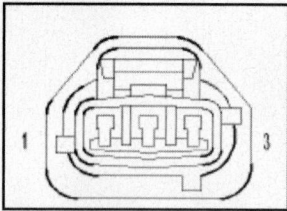

| PIN # | CIRCUIT | FUNCTION |
|-------|---------|----------|
| 1 | F855 18PK/YL | 5 VOLT SUPPLY |
| 2 | K22 18BR/OR | TP NO. 1 SIGNAL |
| 3 | K900 18DB/DG | SENSOR GROUND |

**Ram 4.7L TP Sensor**

## SRT MODELS

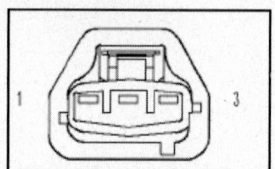

| PIN # | CIRCUIT | FUNCTION |
|-------|---------|----------|
| 1 | K900 20DB/DG | SENSOR GROUND |
| 2 | K22 18BR/OR | THROTTLE POSITION SENSOR SIGNAL |
| 3 | F855 20PK/YL | 5 VOLT SUPPLY |

**Ram SRT TP Sensor**

## Contents

## WHAT TO DO WHEN THERE ARE NO DTCS

Do not attempt to diagnose a Drivability Symptoms without having a logical plan to use to determine which Engine Control system is the cause of the symptom - this plan should include a way to determine which systems do not have a problem! *Drivability symptom diagnosis is a part of an organized approach to problem solving and repair.*

### Drivability Symptom Index Table
To use this list, locate the symptom that matches a particular problem and refer to the areas to test. The items listed under each symptom may not apply to all models, engines or vehicle systems. The repair steps indicate what vehicle component or system to test.

Note:    *The Drivability Symptoms in this list are intended to be generic. While they apply to most vehicles, some vehicles may not have all of the components listed. Refer to other Chilton repair manuals and electronic media for specific tests.*

### Symptom Test Table

| Symptom Description | Suggested Areas to Test |
|---|---|
| Test 1 - No Start, Hard Start Condition<br>● No Crank<br>● Hard Start, Long Crank, Erratic Crank<br>● Stall After Start<br>● No Start, Normal Crank<br>● No Start, MIL is off (if the VREF shorts to ground) | - Check battery, battery circuits to starter<br>- Check for a damaged flywheel, engine compression, base timing and minimum air rate<br>- Check for a failed fuel pump relay<br>- Check for distributor rotor "punch-through"<br>- Check for a faulty ignition control module (ICM)<br>- Check for a VREF circuit shorted to ground<br>- Check SKIM (security system) with a Scan Tool |
| Test 2 - Rough Idle or Stalls Condition<br>● Low or slow idle speed<br>● Fast idle speed<br>● Hunting or rolling idle speed<br>● Slow return to idle speed<br>● Stalls or almost stalls | - Check for engine vacuum leaks<br>- Check the condition of the PCV valve and lines<br>- Check for excessive carbon buildup<br>- Check for a restricted exhaust (in Section 2)<br>- Check base idle speed, check for low fuel pressure<br>- Check the throttle linkage for sticking or binding |
| Test 3 - Runs Rough Condition<br>● At idle speed<br>● During acceleration<br>● At cruise speed<br>● During deceleration | - Check for engine vacuum leaks at intake manifold<br>- Check condition of ignition secondary components<br>- Check base timing and idle speed settings<br>- Check for low or high fuel pressure<br>- Check for dirty, leaking or shorted fuel injectors<br>- Check for excessive carbon buildup on valves |
| Test 4 - Cuts-out, Misses Condition<br>● At idle speed<br>● During acceleration<br>● At cruise speed<br>● During deceleration | - Check for engine vacuum leaks at intake manifold<br>- Check condition of ignition secondary components<br>- Check that spark timing advance is available<br>- Check for low or high fuel pressure<br>- Check for dirty, leaking or shorted fuel injectors<br>- Check for excessive carbon buildup on valves |
| Test 5 - Bucks, Jerks Condition<br>● During acceleration<br>● At cruise speed<br>● During deceleration | - Check for engine vacuum leaks at intake manifold<br>- Check condition of ignition secondary components<br>- Check that spark timing advance is available<br>- Check for low or high fuel pressure<br>- Check for dirty, leaking or shorted fuel injectors<br>- Check operation of the TCC solenoid, brake switch |

SYMPTOM DIAGNOSIS TESTS

### Test 1: No Start, Hard Start Condition

**Note:** *If there is no spark output or fuel pressure available, check for a failed fuel pump relay, no power to the PCM, or loss of the ignition reference signal to the PCM.*

### PRELIMINARY CHECKS

Prior to starting this symptom test routine, inspect these underhood items:

1. Check battery charge and condition, starter current draw.
2. Verify the starter relay operation and that the engine cranks (turns over).
3. Verify the check engine light (MIL) operation - if it does not activate, check the PCM power and ground circuits, and check for 5v supply at the MAP or TP sensor.
4. Check Air Intake system for restrictions (inspect air inlet tubes, air filter for dirt, etc.).
5. Check the status of the Smart Key Immobilizer System (SKIM) with the Scan Tool.

### Test 1 Chart

| Step | Action | Yes | No |
|---|---|---|---|
| 1 | Step Description: No Start Condition Only<br>» Check battery cables, state of charge.<br>» If the engine does not rotate, inspect for a locked engine (hydrostatic lockup condition).<br>» Does the engine crank normally? | Go to Step 2. | Repair the fault in the battery, starter, or Base Engine. Retest for the symptom when all repairs are done. |
| 2 | Step Description: Check the Fuel System<br>» Verify that the pump operates at key on.<br>» Check the fuel pump relay operation. If the relay does not operate, check for blown fuse.<br>» Inspect pump for a leak-down condition<br>» Test fuel pressure, volume and quality.<br>» Test the operation of the fuel regulator.<br>» Are there any faults in the Fuel system? | Make needed repairs. | Go to Step 3. |
| 3 | Step Description: Check the Ignition System<br>» Inspect ignition secondary components for damage (look for rotor "punch-through").<br>» Inspect the coils for signs of spark leakage at coil towers or primary connections.<br>» Check the spark output with a spark tester.<br>» Test Ignition system with an engine analyzer.<br>» Are there any faults in the Ignition system? | Make repairs to the Ignition system. Then retest the symptom. | Go to Step 4. |
| 4 | Step Description: Check the Exhaust System<br>» Check Exhaust system for leaks or damage.<br>» Check the Exhaust system for a restriction using the Vacuum or Pressure Gauge Test (e.g., exhaust backpressure reading should not exceed 1.5 psi at cruise speeds).<br>» Are there any faults in the Exhaust system? | Make repairs to the Exhaust system. Then retest the symptom. | Go to Step 5. |
| 5 | Step Description: Check the MAP Sensor<br>» Disconnect the MAP sensor and attempt to start the engine.<br>» Does the engine start and run normally? | Replace the MAP sensor. Retest for the symptom when repairs are completed. | Go to Step 6. |

## Test 1: No Start, Hard Start Condition (Continued)

### Test 1 Chart (Continued)

| Step | Action (Hard Start Only) | Yes | No |
|------|--------------------------|-----|-----|
| 6 | Step Description: Check for a Hot Engine<br>» Check for signs of an engine overheating condition related to a Hard Start Symptom.<br>» Does the engine appear to be overheated? | Make the repairs to correct the hot engine and then retest for the symptom when done. | Go to Step 7. |
| 7 | Step Description: Check ECT Sensor PID<br>» Connect a Scan Tool and turn the key to on.<br>» Read the ECT sensor (compare to chart).<br>» Has the ECT sensor shifted out of range? | Replace the ECT sensor. Then retest for the symptom when all repairs are completed. | Go to Step 8. |
| 8 | Step Description: Check the PCV System<br>» Inspect the PCV system components for broken parts or loose connections.<br>» Test the operation of the PCV valve.<br>» Are there any faults in the PCV system? | Repair the PCV system. Refer to the PCV system tests in this manual. Retest the symptom when all repairs are done. | Go to Step 9. |
| 9 | Step Description: Check the EVAP System<br>» Inspect for damaged or disconnected EVAP system components.<br>» Inspect for a fuel saturated charcoal canister.<br>» Are there any faults in the EVAP system? | Refer to the EVAP system tests in this manual. Retest for the symptom when all repairs are completed. | Go to Step 10. |
| 10 | Step Description: Test the Base Engine<br>» Check the engine compression.<br>» Test valve timing and timing chain condition.<br>» Check for a worn camshaft or valve train.<br>» Check for any large intake manifold leaks.<br>» Are there any faults in the Base Engine? | Repair the Base Engine. Refer to the Base Engine Tests in this manual. Retest symptom when done. | Return to Step 2 to repeat the test steps in this series to locate and repair the "No Start, Hard Start" condition. |

## Test 2: Rough, Low or High Idle Speed Condition

**Note:**    *If the vehicle has a rough idle and the base timing, idle speed and the IAC (or AIS) motor operates properly, check the engine for excessive carbon buildup.*

### PRELIMINARY CHECKS

Prior to starting this symptom test routine, inspect these underhood items:

1. All related vacuum lines for proper routing and integrity.
2. All related electrical connectors and wiring harnesses for faults (Wiggle Test).
3. Check the throttle linkage for a sticking or binding condition.
4. Air Intake system for restrictions (air inlet tubes, dirty air filter, etc.).
5. Search for any technical service bulletins related to this symptom.
6. Turn the key to off. Unplug the MAP sensor connection and restart the engine to recheck for the idle concern. If the condition is gone, replace the MAP sensor.

**Test 2: Rough, Low or High Idle Speed Condition (Continued)**

**Test 2 Chart**

| Step | Action | Yes | No |
|---|---|---|---|
| 1 | Step Description: Verify the rough idle or stall<br>» Does the engine have a warm engine rough idle, low idle or high idle condition in P or N? | Go to Step 2. | Fault is intermittent. Return to the Symptom List and select another fault. |
| 2 | Step Description: Verify idle speed & timing<br>» Verify the base timing is within specifications<br>» Verify that the base idle speed is set properly » Are the timing and idle speed set properly? | Go to Step 3. | Set the base idle speed and timing to the specifications and then retest for the symptom. |
| 3 | Step Description: Check AIS / IAC Operation<br>» Check the AIS or IAC motor operation<br>» Inspect the AIS/IAC housing in throttle body for restricted passages. Clean as needed.<br>» Set the parking brake, block the drive wheels and turn the A/C off. Install the Scan Tool.<br>» IAC Motor Tester - Turn the key off and then connect the IAC tester to the IAC valve.<br>» Start the engine and use the IAC tester to extend and retract the IAC valve.<br>» ATM Test - Start the engine. Use the tool to change the speed from min-idle to 1500 rpm.<br>» Did the idle speed change as commanded? | Install an Aftermarket Noid light and check the operation of the PCM and AIS or IAC motor circuits. Check the motor for signs of open or shorted circuits. Replace the IAC motor or PCM as needed or make repairs to the IAC motor wiring.<br>If all are okay, go to Step 4. | If the AIS/IAC motor passages are clean and engine speed did not change as described when the AIS/IAC motor was extended and retracted, replace the AIS/IAC motor.<br>Then retest for the condition. |
| 4 | Step Description: Check/compare PID values<br>» Connect Scan Tool & turn off all accessories.<br>» Start the engine and allow it to fully warmup.<br>» Monitor all related PIDs on the Scan Tool.<br>» Verify the P/N switch input in gear and Park.<br>» Check the O2S operation with a Lab Scope.<br>» Are all PIDs within normal range? | Go to Step 5.<br>Note: An IAC motor count of over 80 indicates the pintle is extended and an IAC count of (0) indicates the pintle is retracted. | One or more of the PIDs are out of range when compared to "known good" values. Make repairs to the system that is out of range, then retest for the symptom. |
| 5 | Step Description: Check the Ignition System<br>» Inspect the coils for signs of spark leakage at coil towers or primary connections.<br>» Check the spark output with a spark tester.<br>» Test Ignition system with an engine analyzer.<br>» Were any faults found in the Ignition system? | Make repairs as needed<br> | Go to Step 6. |
| 6 | Step Description: Check the Fuel System<br>» Inspect the Fuel delivery system for leaks.<br>» Test the fuel pressure, quality and volume.<br>» Test the operation of the pressure regulator.<br>» Were any faults found in the Fuel system? | Make repairs as needed<br>Fuel Pressure Gauge<br>Fuel Rail Test Port | Go to Step 7. |
| 7 | Step Description: Check the Exhaust System<br>» Check Exhaust system for leaks or damage.<br>» Check the Exhaust system for a restriction using the Vacuum or Pressure Gauge Test (e.g., exhaust backpressure reading should not exceed 1.5 psi at cruise speeds).<br>» Were any faults found in Exhaust System? | Make repairs to the Exhaust system. Then retest the symptom.<br>**Inspect for Damage**<br> | Go to Step 8. |

## Rough, Low or High Idle Speed Condition (Continued)

### Test 2 Chart (Continued)

| Step | Action | Yes | No |
|---|---|---|---|
| 8 | Step Description: Check the PCV System<br>» Inspect the PCV system components for broken parts or loose connections.<br>» Test the operation of the PCV valve.<br>» Were any faults found in the PCV system? | Make repairs to the PCV system. Refer to the PCV system tests in this manual. Then retest for the condition. | Go to Step 9. |
| 9 | Step Description: Check the EVAP System<br>» Inspect for damaged or disconnected EVAP system components or a saturated canister.<br>» Were any faults found in the EVAP system? | Make repairs to EVAP system (use the EVAP tests in this manual). Retest for the condition. | Go to Step 10. |
| 10 | Step Description: Check the Base Engine<br>» Test the engine compression.<br>» Test valve timing and timing chain condition.<br>» Check for a worn camshaft or valve train.<br>» Check for any large intake manifold leaks.<br>» Were any faults found in the Base Engine? | Make repairs as needed to the Base Engine. Refer to the Base Engine tests in this manual. Then retest for the condition when repairs are completed. | Go to Step 2 and repeat the tests from the beginning to locate and repair the cause of the "Rough, Low or High Idle Speed" condition. |

## Test 3: Runs Rough Condition

### PRELIMINARY CHECKS

Prior to starting this symptom test routine, inspect these underhood items:

1. All related vacuum lines for proper routing and integrity
2. Air Intake system for restrictions (air inlet tubes, dirty air filter, etc.)
3. Search for any technical service bulletins related to this symptom.

### Test 3 Chart

| Step | Action | Yes | No |
|---|---|---|---|
| 1 | Step Description: Verify engine runs rough<br>» Start the engine and allow it to idle in P or N.<br>» Does the engine run rough when warm in Park or Neutral position? | Check for any stored codes. If codes are set, repair codes and retest. If no codes are set, go to Step 3. | Go to Step 2. |
| 2 | Step Description: Condition does not exist!<br>» Inspect various underhood items that could cause an intermittent Runs Rough condition (i.e., dirt in the throttle body, vacuum leaks, IAC motor connections, etc.).<br>» Were any problems located in this step? | Correct the problems. Do a PCM reset and engine "idle relearn" procedure. Then verify the "runs rough" condition is repaired. | The problem is not present at this time. It may be an intermittent problem. |
| 3 | Step Description: Check/compare PID values<br>» Connect a Scan Tool to the test connector.<br>» Turn off all accessories.<br>» Start the engine and allow it to fully warmup.<br>» Monitor all related PIDs on the Scan Tool.<br>» Were all PIDs within their normal range? | Go to Step 4.<br>Note: The IAC motor should read from 5-50 counts. Check the LONGFT reading for a large shift into the negative range (due to a rich condition). | One or more of the PIDs are out of range when compared to "known good" values. Make repairs to the system that is out of range, then retest for the symptom. |

## Test 3: Runs Rough Condition (Continued)

### Test 3 Chart (Continued)

| Step | Action | Yes | No |
|------|--------|-----|-----|
| 4 | Step Description: Check the Ignition System<br>» Inspect the coils for signs of spark leakage at coil towers or primary connections.<br>» Check the spark output with a spark tester.<br>» Test Ignition system with an engine analyzer.<br>» Were any faults found in the Ignition system? | Make repairs as needed | Go to Step 5. |
| 5 | Step Description: Check the Fuel System<br>» Inspect the Fuel delivery system for leaks.<br>» Test the fuel pressure, quality and volume.<br>» Test the operation of the pressure regulator.<br>» Were any faults found in the Fuel system? | Make repairs as needed | Go to Step 6. |
| 6 | Step Description: Check the Exhaust System<br>» Check Exhaust system for leaks or damage.<br>» Check the Exhaust system for a restriction using the Vacuum or Pressure Gauge Test (e.g., exhaust backpressure reading should not exceed 1.5 psi at cruise speeds).<br>» Were any faults found in Exhaust System? | Make repairs to the Exhaust system. Then retest the symptom.<br>**Inspect for Damage** | Go to Step 7. |
| 7 | Step Description: Check the PCV System<br>» Inspect the PCV system components for broken parts or loose connections.<br>» Test the operation of the PCV valve.<br>» Were any faults found in the PCV system? | Make repairs to the PCV system. Refer to the PCV system tests in this manual. Then retest for the condition. | Go to Step 9. |
| 8 | Step Description: Check the EVAP System<br>» Inspect for damaged or disconnected EVAP system components or a saturated canister.<br>» Were any faults found in the EVAP system? | Make repairs to EVAP system (use the EVAP tests in this manual). Retest for the condition. | Go to Step 10. |
| 9 | Step Description: Check Engine Condition<br>» Test the engine compression.<br>» Test valve timing and timing chain condition.<br>» Check for a worn camshaft or valve train.<br>» Check for any large intake manifold leaks.<br>» Were any faults found in the Base Engine? | Make repairs as needed to the Base Engine. Refer to the Base Engine tests in this manual. Then retest for the condition when repairs are completed. | Return to Step 2 and repeat the tests from the beginning to locate and repair the cause of the "Runs Rough" condition. |

## Example EVAP System Graphic

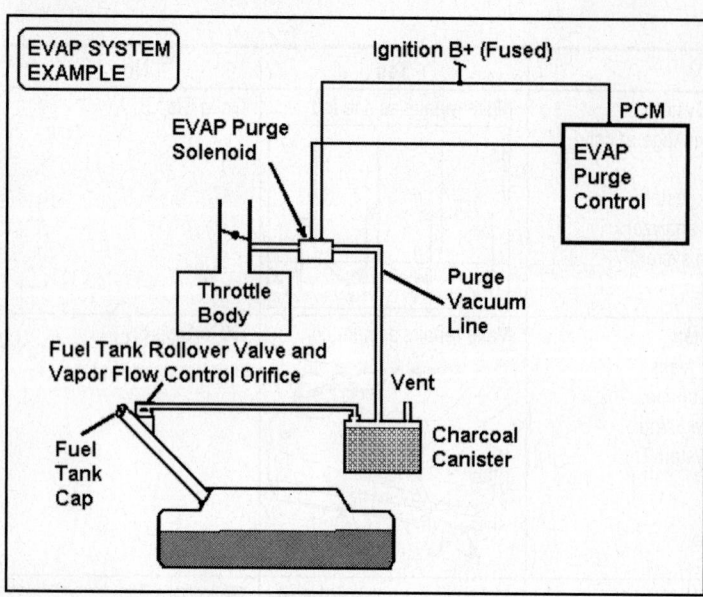

## Test 4: Cuts-out or Misses Condition

### PRELIMINARY CHECKS

Prior to starting this symptom test routine, inspect these underhood items:

1. All related vacuum lines for proper routing and integrity
2. Search for any technical service bulletins related to this symptom.

### Test 4 Chart

| Step | Action | Yes | No |
|------|--------|-----|-----|
| 1 | Step Description: Verify Cuts-out condition<br>» Start the engine and attempt to verify the Cuts-out or misses condition.<br>» Does the engine have a cuts-out condition? | Check for any stored codes. If codes are set, repair codes and retest. If no codes are set, go to Step 3. | Go to Step 2. |
| 2 | Step Description: Condition does not exist!<br>» Inspect various underhood items that could cause an intermittent Cuts-out condition (i.e., EVAP, Fuel or Ignition system components).<br>» Were any problems located in this step? | Correct the problems. Do a PCM reset and "Fuel Trim Relearn" procedure. Then verify condition is repaired. | The problem is not present at this time. It may be an intermittent problem. |
| 3 | Step Description: Check/compare PID values<br>» Connect a Scan Tool to the test connector.<br>» Turn off all accessories.<br>» Start the engine and allow it to fully warmup.<br>» Monitor all related PIDs on the Scan Tool (i.e., ECT IAC Counts and LONGFT at idle).<br>» Were all PIDs within their normal range? | Go to Step 4.<br>Note: The IAC motor should be from 5-50 counts. Watch fuel trim (%) for a large shift into the negative (-) range (due to a rich condition). | One or more of the PIDs are out of range when compared to "known good" values. Make repairs to the system that is out of range, then retest for the symptom. |

**Test 4: Cuts Out or Misses Condition (Continued)**

**Test 4 Chart (Continued)**

| Step | Action | Yes | No |
|---|---|---|---|
| 4 | Step Description: Check the Ignition System<br>» Inspect the coils for signs of spark leakage at coil towers or primary connections.<br>» Check the spark output with a spark tester.<br>» Test Ignition system with an engine analyzer.<br>» Were any faults found in the Ignition system? | Make repairs as needed | Go to Step 5. |
| 5 | Step Description: Check the Fuel System<br>» Inspect the Fuel delivery system for leaks.<br>» Test the fuel pressure, quality and volume.<br>» Test the operation of the pressure regulator.<br>» Were any faults found in the Fuel system? | Make repairs as needed | Go to Step 6. |
| 6 | Step Description: Check the Exhaust System<br>» Check Exhaust system for leaks or damage.<br>» Check the Exhaust system for a restriction using the Vacuum or Pressure Gauge Test (e.g., exhaust backpressure reading should not exceed 1.5 psi at cruise speeds).<br>» Were any faults found in Exhaust System? | Make repairs to the Exhaust system. Then retest the symptom. | Go to Step 7. |
| 7 | Step Description: Check the PCV System<br>» Inspect the PCV system components for broken parts or loose connections.<br>» Test the operation of the PCV valve.<br>» Were any faults found in the PCV system? | Make repairs to the PCV system. Refer to the PCV system tests in this manual. Then retest for the condition. | Go to Step 8. |
| 8 | Step Description: Check the EVAP System<br>» Inspect for damaged or disconnected EVAP system components<br>» Check for a saturated EVAP canister.<br>» Were any faults found in the EVAP system? | Make repairs to EVAP system (use the EVAP tests in this manual). Retest for the condition. | Go to Step 9. |
| 9 | Step Description: Check the AIR system<br>» Inspect AIR system for broken parts, leaking valves or disconnected hoses (see graphic).<br>» Test the operation of Secondary AIR system.<br>» Were any faults found in the AIR system? | Make repairs as needed. Refer to the Secondary AIR system tests in this manual. Retest for the condition. | Go to Step 10. |
| 10 | Step Description: Check Engine Condition<br>» Test the engine compression.<br>» Test valve timing and timing chain condition.<br>» Check for a worn camshaft or valve train.<br>» Check for any large intake manifold leaks.<br>» Were any faults found in the Base Engine? | Make repairs as needed to the Base Engine. Refer to the Base Engine tests in this manual. Then retest for the condition when repairs are completed. | Go to Step 2 and repeat the tests from the beginning to locate and repair the cause of the "Cuts Out or Misses" condition. |

**Typical Secondary Air System Graphic**

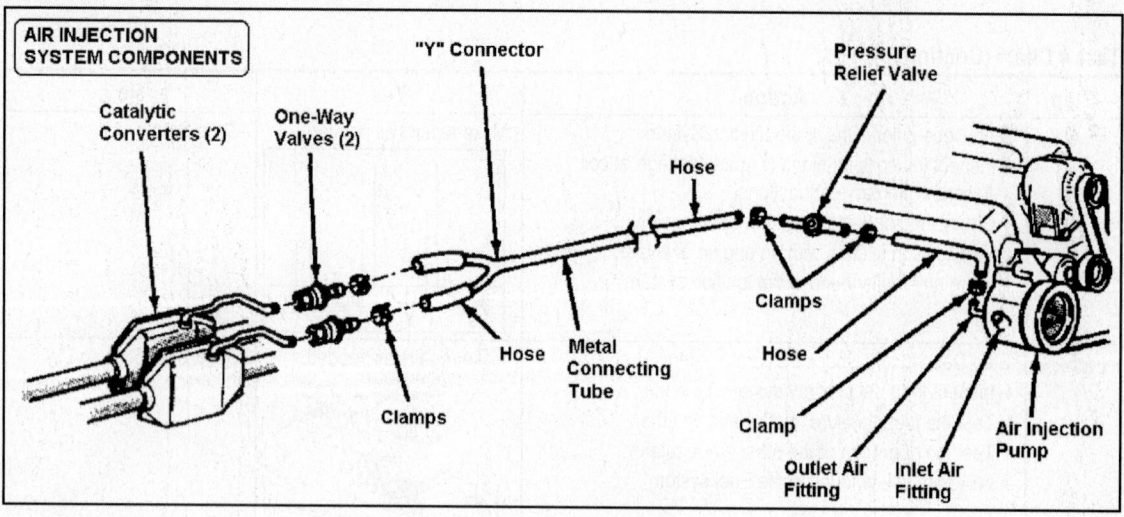

**Test 5: Surge Condition**

*PRELIMINARY CHECKS*
1. Discuss how the operation of the torque converter clutch (TCC) or air conditioning compressor can affect the "feel" of the vehicle during normal operation. Refer to the information in the Owner's Manual to explain how these devices normally operate.
2. Search for any technical service bulletins related to this symptom.

**Test 5 Chart**

| Step | Action | Yes | No |
|---|---|---|---|
| 1 | Step Description: Verify the surge condition<br>» Drive the vehicle and attempt to verify that the vehicle surges at cruise speeds.<br>» Does the engine have a surge condition? | Check for any stored codes. If codes are set, repair codes and retest. If no codes are set, go to Step 3. | Go to Step 2. |
| 2 | Step Description: Condition does not exist!<br>» Inspect various underhood items that could cause an intermittent surge condition (check for leaks in the MAP sensor vacuum lines).<br>» Were any problems located in this step? | Correct the problems. Do a PCM reset and "Fuel Trim Relearn" procedure. Then verify condition is repaired. | The problem is not present at this time. It may be an intermittent problem. |
| 3 | Step Description: Check/compare PID values<br>» Connect a Scan Tool to the test connector.<br>» Start the engine and allow it to fully warmup.<br>» Monitor all related PIDs on Scan Tool (HO2S switching, LONGFT, and the TCC operation)<br>» Compare VSS PID reading to speedometer.<br>» Were all PIDs within their normal range? | Go to Step 4.<br>Note: Verify that the front HO2S responds quickly to throttle changes. Check for silicon contamination on the front HO2S (this can cause a rich A/F signal). | One or more of the PIDs are out of range when compared to "known good" values. Make repairs to the system that is out of range, then retest for the symptom. |

**Test 5: Surge Condition (Continued)**

**Test 5 Chart (Continued)**

| Step | Action | Yes | No |
|------|--------|-----|-----|
| 4 | Step Description: Check the Ignition System<br>» Inspect the coils for signs of spark leakage at coil towers or primary connections.<br>» Check the spark output with a spark tester.<br>» Test Ignition system with an engine analyzer.<br>» Were any faults found in the Ignition system? | Make repairs as needed<br> | Go to Step 5. |
| 5 | Step Description: Check the Fuel System<br>» Inspect the Fuel delivery system for leaks.<br>» Test the fuel pressure, quality and volume.<br>» Test the operation of the pressure regulator.<br>» Were any faults found in the Fuel system? | Make repairs as needed<br> | Go to Step 6. |
| 6 | Step Description: Check the Exhaust System<br>» Check Exhaust system for leaks or damage.<br>» Check the Exhaust system for a restriction using the Vacuum or Pressure Gauge Test (e.g., exhaust backpressure reading should not exceed 1.5 psi at cruise speeds).<br>» Were any faults found in Exhaust System? | Make repairs to the Exhaust system. Then retest the symptom.<br>**Inspect for Damage**<br> | Return to Step 2 and repeat the tests from the beginning to locate and repair the cause of the "Surge" condition. |

## INTERMITTENT FAULT TESTS

Many trouble code repair charts end with a result that reads "Fault Not Present at this Time." What this expression means is that the conditions that were present when a code set or drivability symptom occurred are no longer there or were not met. In effect, the problem was present at least once, but is not present at this time. However, it is likely to return in the future, so it should be diagnosed and repaired if at all possible.

One way to find an intermittent problem is to gather the information that was present when the problem occurred. In the case of a Code Fault, this can be done in two ways: by capturing the data in Snapshot or Movie mode or by driver observations.

The PCM has to detect the fault for a specific period of time before a trouble code will set. While intermittent problems may appear to be occasional in nature, they usually occur under specific conditions. Therefore, you should identify and duplicate these conditions. Since intermittent faults are difficult to duplicate, a logical routine (checklist) must be followed when attempting to find the faulty component, system or circuit. The tests on the next page can be used to help find the cause of an intermittent fault.

Some intermittent faults occur due to a loose connection, wiring problem or warped circuit board. An intermittent fault can also be caused by poor test techniques that cause damage to the male or female ends of a connector.

**Test for Loose Connectors**
To test for a loose or damaged connection, take the male end of a connector from another wiring harness and carefully push it into the "suspect" female terminal to verify that the opening is tight. There should be some resistance felt as the male connector is inserted in the terminal connection.

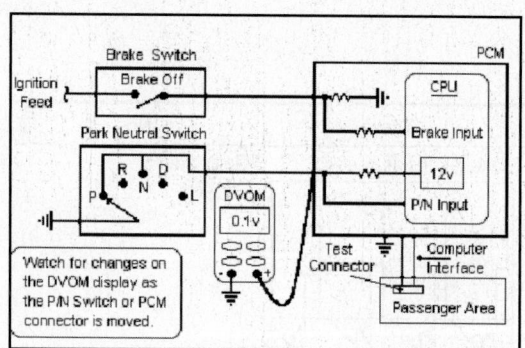

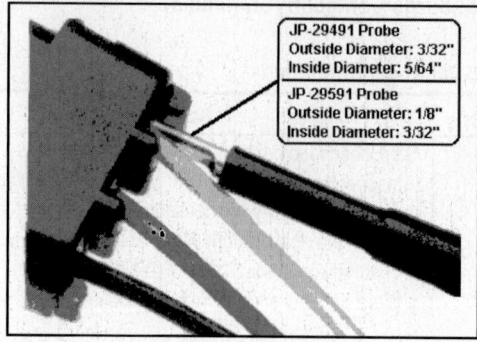

### The Wiggle Test

A wiggle test can be used to locate the cause of some intermittent faults. The sensor, switch or the PCM wiring can be back-probed, as shown, while the test is done.

During testing, move or wiggle the suspect device, connector or wiring while watching for a change.
If the DVOM has a Min/Max record mode, use this mode during the test.

## OTHER DIAGNOSIS & TESTING

### Vehicle Does Not Fill

### Test Chart

| CONDITION | POSSIBLE CAUSES | CORRECTION |
|---|---|---|
| Pre-Mature Nozzle Shut-Off | Defective fuel tank assembly components. | Fill tube improperly installed (sump) |
| | | Fill tube hose pinched. |
| | | Check valve stuck shut. |
| | | Control valve stuck shut. |
| | Defective vapor/vent components. | Vent line from control valve to canister pinched. |
| | | Vent line from canister to vent filter pinched. |
| | | Canister vent valve failure (requires double failure, plugged to NVLD and atmosphere). |
| | | Leak detection pump failed closed. |
| | | Leak detection pump filter plugged. |
| | On-Board diagnostics evaporative system leak test just conducted. | Canister vent valve vent plugged to atmosphere. |
| | | Engine still running when attempting to fill (System designed not to fill). |
| | Defective fill nozzle. | Try another nozzle. |
| Fuel Spits Out Of Filler Tube. | During fill. | See Pre-Mature Shut-Off. |
| | At conclusion of fill. | Defective fuel handling component. (Check valve stuck open). |
| | | Defective vapor/vent handling component. |
| | | Defective fill nozzle. |

# Contents

## DAIMLERCHRYSLER MINI-VANS

## DODGE TRUCKS & VANS

## DODGE & JEEP SUVS

Cherokee, Durango, Grand Cherokee, Liberty, Wrangler

## About This Section

### Introduction

This section contains Pin Tables for DaimlerChrysler vehicles from 1995-2005, and slightly older, in some applications. It can be used to help you repair Trouble Code and No Code problems related to the PCM.

VEHICLE COVERAGE

PIN Charts for DaimlerChrysler vehicles, from 2005 and back as far as 1993, are covered in this material.

HOW TO USE THIS SECTION

This section can be used to look up the location of a particular pin, a wire color, or to find a "known good" value of a circuit. To locate the PCM information for a particular vehicle, find the model, correct engine size (with VIN Code) and finally the year of the vehicle.

For example, to look up the PCM terminals for a 1999 Cirrus 2.5L 4v V6 VIN H, go to the Contents Page to find the appropriate table location containing the year, model and engine you need. Then go to that page to find:

**1998-2000 Cirrus 2.5L 4v VIN H (A/T) 'C1' Connector**

| PCM Pin # | Wire Color | Circuit Description (40-Pin) | Value at Hot Idle |
|-----------|------------|------------------------------|-------------------|
| 4 | BK/GY | Coil 1 Driver Control | 5°, at 55 mph: 8° dwell |
| 6 | DG/OR | ASD Relay Output (B+) | 12-14v |
| 7 | YL/WT | Injector 3 Driver | 1.0-4.0 ms |
| 8 | DG | Generator Field Driver | Digital Signal: 0-12-0v |

In this example, the Coil Driver circuit is connected to Pin 4 of the 40-Pin connector with a BK/GY wire. The Hot Idle value is 5° while the 55 mph value is 8°. Note the change in dwell as the mph changed.

The ASD relay output signal is connected to Pin 6 of the 40-Pin connector (DG/OR wire). This signal indicates the voltage output of the ASD relay during vehicle operation. This signal should always read near battery voltage with the engine running.

*NOTE: Some tables in this section include pin values at hot idle, give pin wire colors and include some connector pin views. There are some tables that do not contain all of these elements. The information was prepared using the best available data from the manufacturer.*

**DaimlerChrysler Cars**

300M & MAGNUM

**1999 300M 3.5L VIN G 'C1' Connector**

| PCM Pin # | Wire Color | Circuit Description (40-Pin) | Value at Hot Idle |
|---|---|---|---|
| 1 | TN/LG | Coil 4 Driver | 5°, at 55 mph: 8° dwell |
| 2 | TN/OR | Coil 3 Driver | 5°, at 55 mph: 8° dwell |
| 3 | TN/PK | Coil 2 Driver | 5°, at 55 mph: 8° dwell |
| 4 | TN/LB | Coil 6 Driver | 5°, at 55 mph: 8° dwell |
| 5 | YL/RD | S/C Power Supply | 12-14v |
| 6 | DG/OR | ASD Relay Output | 12-14v |
| 7 | YL/WT | Injector 3 Driver | 1.0-4.0 ms |
| 8 | DG | Generator Field Driver | Digital Signal: 0-12-0v |
| 9, 12 | --- | Not Used | --- |
| 10 | BK/TN | Power Ground | <0.1v |
| 11 | TN/RD | Coil 1 Driver | 5°, at 55 mph: 8° dwell |
| 13 | WT/DB | Injector 1 Driver | 1.0-4.0 ms |
| 14 | BR/DB | Injector 6 Driver | 1.0-4.0 ms |
| 15 | GY | Injector 5 Driver | 1.0-4.0 ms |
| 16 | LB/BR | Injector 4 Driver | 1.0-4.0 ms |
| 17 | TN | Injector 2 Driver | 1.0-4.0 ms |
| 18 | GY/PK | Short Runner Valve Solenoid | Valve On: 1v, Off: 12v |
| 19 | --- | Not Used | --- |
| 20 | DB/WT | Ignition Switch Output | 12-14v |
| 21 | TN/DG | Coil 5 Driver | 5°, at 55 mph: 8° dwell |
| 22-23 | --- | Not Used | --- |
| 24 | DB/LG | Knock Sensor Signal | 0.080v AC |
| 25 | BK/PK | Knock Sensor Ground | <0.050v |
| 26 | TN/BK | ECT Sensor Signal | At 180°F: 2.80v |
| 27 | BK/OR | Oxygen Sensor Ground | <0.050v |
| 28 | --- | Not Used | --- |
| 29 | LG/RD | HO2S-21 (B2 S1) Signal | 0.1-1.1v |
| 30 | BK/DG | HO2S-11 (B1 S1) Signal | 0.1-1.1v |
| 31 | TN/WT | Starter Relay Control | KOEC: 9-11v |
| 32 | GY/BK | CKP Sensor Signal | Digital Signal: 0-5-0v |
| 33 | TN/YL | CMP Sensor Signal | Digital Signal: 0-5-0v |
| 34 | LG/PK | EGR Sensor Signal | 0.6-0.8v |
| 35 | OR/DB | TP Sensor Signal | 0.6-1.0v |
| 36 | DG/RD | MAP Sensor Signal | 1.5-1.7v |
| 37 | BK/RD | IAT Sensor Signal | At 100°F: 1.83v |
| 38 | --- | Not Used | --- |
| 39 | PK/RD | Manifold Tuning Valve Control | Valve Off: 12v, On: 1v |
| 40 | GY/YL | EGR Solenoid Control | 12v, at 55 mph: 1v |

**Standard Colors and Abbreviations**

| Abbreviation | Color | Abbreviation | Color | Abbreviation | Color |
|---|---|---|---|---|---|
| BK | Black | GY | Gray | RD | Red |
| BL | Blue | GN | Green | TN | Tan |
| BR | Brown | LG | Light Green | VT | Violet |
| DB | Dark Blue | OR | Orange | WT | White |
| DG | Dark Green | PK | Pink | YL | Yellow |

## 1999 300M 3.5L VIN G 'C2' Connector

| PCM Pin # | Wire Color | Circuit Description (40-Pin) | Value at Hot Idle |
|---|---|---|---|
| 41 | RD/LG | S/C Set Switch Signal | S/C & Set Switch On: 3.8v |
| 42 | DB | A/C Pressure Switch Signal | A/C On: 0.45-4.85v |
| 43 | BK/LB | Sensor Ground | <0.050v |
| 44 | OR | 8-Volt Supply | 7.9-8.1v |
| 45 | DB/LG | PSP Switch Signal | Straight: 0v, Turning: 5v |
| 46 | RD/WT | Battery Power (Fused B+) | 12-14v |
| 47 | BK/LB | Sensor Ground | <0.050v |
| 48 | BR/WT | IAC 3 Driver | DC pulse signals: 0.8-11v |
| 49 | YL/BK | IAC 2 Driver | DC pulse signals: 0.8-11v |
| 50 | BK/TN | Power Ground | <0.1v |
| 51 | TN/WT | HO2S-12 (B1 S2) Signal | 0.1-1.1v |
| 53 | PK/WT | HO2S-22 (B2 S2) Signal | 0.1-1.1v |
| 55 | DB/PK | Low Speed Fan Relay | Relay Off: 12v, On: 1v |
| 56 | TN/RD | S/C Vacuum Solenoid | Vacuum Increasing: 1v |
| 57 | GY/RD | IAC 1 Driver | DC pulse signals: 0.8-11v |
| 58 | PK/BK | IAC 4 Driver | DC pulse signals: 0.8-11v |
| 59 | YL/PK | PCI Data Bus (J1850) | Digital Signals: 0-7-0v |
| 61 | PK/WT | 5-Volt Supply | 4.9-5.1v |
| 62 | WT/PK | Brake Switch Signal | Brake Off: 0v, On: 12v |
| 63 | YL/DG | Torque Management Request | Digital Signals |
| 64 | DB/OR | A/C Clutch Relay Control | Relay Off: 12v, On: 1v |
| 65 | PK | SCI Transmit | 0v |
| 66 | WT/OR | Vehicle Speed Signal | Digital Signal |
| 67 | DB/YL | ASD Relay Control | Relay Off: 12v, On: 1v |
| 68 | PK/BK | EVAP Purge Solenoid Control | PWM Signal: 0-12-0v |
| 69 | DB/LG | High Speed Fan Relay | Relay Off: 12v, On: 1v |
| 70 | DG/LG | EVAP Purge Solenoid Sense | 0-1v |
| 71 | WT/RD | EATX RPM Signal | Digital Signals |
| 72 | OR/RD | LDP Switch Sense | Open: 12v, Closed: 0v |
| 73, 78-79 | --- | Not Used | --- |
| 74 | BR | Fuel Pump Relay Control | Relay Off: 12v, On: 1v |
| 75 | LG | SCI Receive | 0v |
| 76 | BK/PK | PNP Switch Signal | In P/N: 0v, Others: 5v |
| 77 | WT/DG | LDP Solenoid Control | PWM Signal: 0-12-0v |
| 80 | LG/RD | S/C Vent Solenoid | Vacuum Decreasing: 1v |

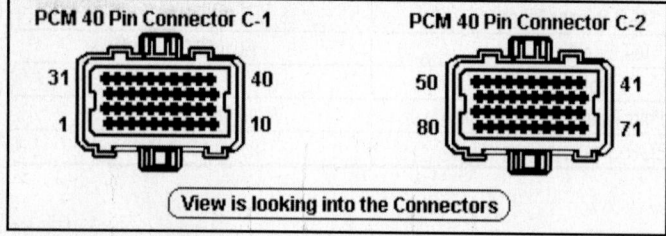

**Pin Connector Graphic**

## 2000-01 300M 3.5L VIN G 'C1' Connector

| PCM Pin # | Wire Color | Circuit Description (40-Pin) | Value at Hot Idle |
|-----------|-----------|------------------------------|-------------------|
| 1 | TN/LG | Coil 4 Driver | 5°, at 55 mph: 8° dwell |
| 2 | TN/OR | Coil 3 Driver | 5°, at 55 mph: 8° dwell |
| 3 | TN/PK | Coil 2 Driver | 5°, at 55 mph: 8° dwell |
| 4 | TN/LB | Coil 6 Driver | 5°, at 55 mph: 8° dwell |
| 5 | YL/RD | S/C Power Supply | 12-14v |
| 6 | DG/OR | ASD Relay Output | 12-14v |
| 7 | YL/WT | Injector 3 Driver | 1.0-4.0 ms |
| 8 | DG | Generator Field Driver | Digital Signal: 0-12-0v |
| 9 | --- | Not Used | --- |
| 10 | BK/TN | Power Ground | <0.1v |
| 11 | TN/RD | Coil 1 Driver | 5°, at 55 mph: 8° dwell |
| 12 | --- | Not Used | --- |
| 13 | WT/DB | Injector 1 Driver | 1.0-4.0 ms |
| 14 | BR/DB | Injector 6 Driver | 1.0-4.0 ms |
| 15 | GY | Injector 5 Driver | 1.0-4.0 ms |
| 16 | LB/BR | Injector 4 Driver | 1.0-4.0 ms |
| 17 | TN | Injector 2 Driver | 1.0-4.0 ms |
| 18 | GY/PK | Short Runner Valve Solenoid | Valve On: 1v, Off: 12v |
| 19 | --- | Not Used | --- |
| 20 | DB/WT | Ignition Switch Output | 12-14v |
| 21 | TN/DG | Coil 5 Driver | 5°, at 55 mph: 8° dwell |
| 22-23 | --- | Not Used | --- |
| 24 | DB/LG | Knock Sensor Signal | 0.080v AC |
| 25 | BK/VT | Knock Sensor Return | <0.050v |
| 26 | TN/BK | ECT Sensor Signal | At 180°F: 2.80v |
| 27 | BK/OR | Oxygen Sensor Ground | <0.050v |
| 28 | --- | Not Used | --- |
| 29 | LG/RD | HO2S-21 (B2 S1) Signal | 0.1-1.1v |
| 30 | BK/DG | HO2S-11 (B1 S1) Signal | 0.1-1.1v |
| 31 | TN/WT | Starter Relay Control | KOEC: 9-11v |
| 32 | GY/BK | CKP Sensor Signal | Digital Signal: 0-5-0v |
| 33 | TN/YL | CMP Sensor Signal | Digital Signal: 0-5-0v |
| 34 | LG/PK | EGR Sensor Signal | 0.6-0.8v |
| 35 | OR/DB | TP Sensor Signal | 0.6-1.0v |
| 36 | DG/RD | MAP Sensor Signal | 1.5-1.7v |
| 37 | BK/RD | IAT Sensor Signal | At 100°F: 1.83v |
| 38 | --- | Not Used | --- |
| 39 | VT/RD | Manifold Tuning Valve Control | Valve Off: 12v, On: 1v |
| 40 | GY/YL | EGR Solenoid Control | 12v, at 55 mph: 1v |

## Standard Colors and Abbreviations

| Abbreviation | Color | Abbreviation | Color | Abbreviation | Color |
|--------------|-------|--------------|-------|--------------|-------|
| BK | Black | GY | Gray | RD | Red |
| BL | Blue | GN | Green | TN | Tan |
| BR | Brown | LG | Light Green | VT | Violet |
| DB | Dark Blue | OR | Orange | WT | White |
| DG | Dark Green | PK | Pink | YL | Yellow |

## 2000-01 300M 3.5L VIN G 'C2' Connector

| PCM Pin # | Wire Color | Circuit Description (40-Pin) | Value at Hot Idle |
|---|---|---|---|
| 41 | RD/LG | S/C Set Switch Signal | S/C & Set Switch On: 3.8v |
| 42 | DB | A/C Pressure Switch Signal | A/C On: 0.45-4.85v |
| 43 | BK/LB | Sensor Ground | <0.050v |
| 44 | OR | 8-Volt Supply | 7.9-8.1v |
| 45 | DB/LG | PSP Switch Signal | Straight: 0v, Turning: 5v |
| 46 | RD/WT | Battery Power (Fused B+) | 12-14v |
| 47 | BK/LB | Sensor Ground | <0.050v |
| 48 | BR/WT | IAC 3 Driver | DC pulse signals: 0.8-11v |
| 49 | YL/BK | IAC 2 Driver | DC pulse signals: 0.8-11v |
| 50 | BK/TN | Power Ground | <0.1v |
| 51 | TN/WT | HO2S-12 (B1 S2) Signal | 0.1-1.1v |
| 53 | PK/WT | HO2S-22 (B2 S2) Signal | 0.1-1.1v |
| 55 | DB/PK | Low Speed Fan Relay | Relay Off: 12v, On: 1v |
| 56 | TN/RD | S/C Vacuum Solenoid | Vacuum Increasing: 1v |
| 57 | GY/RD | IAC 1 Driver | DC pulse signals: 0.8-11v |
| 58 | PK/BK | IAC 4 Driver | DC pulse signals: 0.8-11v |
| 59 | YL/PK | PCI Data Bus (J1850) | Digital Signals: 0-7-0v |
| 61 | PK/WT | 5-Volt Supply | 4.9-5.1v |
| 62 | WT/PK | Brake Switch Signal | Brake Off: 0v, On: 12v |
| 63 | YL/DG | Torque Management Request | Digital Signals |
| 64 | DB/OR | A/C Clutch Relay Control | Relay Off: 12v, On: 1v |
| 65 | LG | SCI Transmit | 0v |
| 66 | WT/OR | Vehicle Speed Signal | Digital Signal |
| 67 | DB/YL | ASD Relay Control | Relay Off: 12v, On: 1v |
| 68 | PK/BK | EVAP Purge Solenoid Control | PWM Signal: 0-12-0v |
| 69 | DB/LG | High Speed Fan Relay | Relay Off: 12v, On: 1v |
| 70 | DG/LG | EVAP Purge Solenoid Sense | 0-1v |
| 71 | WT/RD | EATX RPM | Digital Signals |
| 72 | OR/RD | LDP Switch Sense | Open: 12v, Closed: 0v |
| 73, 78-79 | --- | Not Used | --- |
| 74 | BR | Fuel Pump Relay Control | Relay Off: 12v, On: 1v |
| 75 | LG | SCI Receive | 5v |
| 76 | BK/PK | PNP Switch Signal | In P/N: 0v, Others: 5v |
| 77 | WT/DG | LDP Solenoid Control | PWM Signal: 0-12-0v |
| 80 | LG/RD | S/C Vent Solenoid | Vacuum Decreasing: 1v |

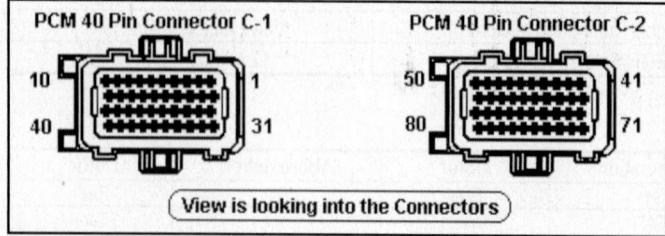

**Pin Connector Graphic**

## 2002-04 300M 3.5L VIN G & K 'C1' Black Connector

| PCM Pin # | Wire Color | Circuit Description (38-Pin) | Value at Hot Idle |
|---|---|---|---|
| 1-8 | --- | Not Used | --- |
| 9 | TN | Power Ground | <0.1v |
| 11 | DB/WT | Ignition Switch Output (Run-Start) | 12-14v |
| 12 | RD/WT | Ignition Switch Output (Off-Run-Start) | 12-14v |
| 13-15 | --- | Not Used | --- |
| 16 | GY/PK | Short Runner Valve Solenoid | Valve On: 1v, Off: 12v |
| 17, 19-20 | --- | Not Used | --- |
| 18 | BK/TN | Power Ground | <0.1v |
| 21 | DB | A/C Pressure Switch Signal | A/C On: 0.45-4.85v |
| 22-24 | --- | Not Used | --- |
| 25 | LG | SCI Receive (PCM) | 5v |
| 26 | VT/OR | SCI Receive (TCM) | 5v |
| 27-28 | --- | Not Used | --- |
| 29 | RD | Battery Power (Fused B+) | 12-14v |
| 30 | YL/BK | Ignition Switch Output (start) | 8-11v (cranking) |
| 31 | TN/WT | HO2S-12 (B1 S2) Signal | 0.1-1.1v |
| 32 | DB/LG | Oxygen Sensor Return (down) | <0.050v |
| 33 | PK/WT | HO2S-22 (B2 S2) Signal | 0.1-1.1v |
| 34-35 | --- | Not Used | --- |
| 36 | PK/TN | SCI Transmit (PCM) | 0v |
| 37 | WT/DG | SCI Transmit (TCM) | 0v |
| 38 | VT/YL | PCI Data Bus (J1850) | Digital Signals: 0-7-0v |

## Standard Colors and Abbreviations

| Abbreviation | Color | Abbreviation | Color | Abbreviation | Color |
|---|---|---|---|---|---|
| BK | Black | GY | Gray | RD | Red |
| BL | Blue | GN | Green | TN | Tan |
| BR | Brown | LG | Light Green | VT | Violet |
| DB | Dark Blue | OR | Orange | WT | White |
| DG | Dark Green | PK | Pink | YL | Yellow |

## 2002-04 300M 3.5L G & K 'C2' Gray Connector

| PCM Pin # | Wire Color | Circuit Description (38-Pin) | Value at Hot Idle |
|---|---|---|---|
| 1 | TN/LB | Coil 6 Driver | 5°, at 55 mph: 8° dwell |
| 2 | TN/DG | Coil 5 Driver | 5°, at 55 mph: 8° dwell |
| 3 | TN/LG | Coil 4 Driver | 5°, at 55 mph: 8° dwell |
| 4 | BR/DB | Injector 6 Driver | 1.0-4.0 ms |
| 5 | GY | Injector 5 Driver | 1.0-4.0 ms |
| 6, 15, 26 | --- | Not Used | --- |
| 7 | TN/OR | Coil 3 Driver | 5°, at 55 mph: 8° dwell |
| 8 | GY/YL | EGR Solenoid Control (Not used 04) | 12v, at 55 mph: 1v |
| 9 | TN/PK | Coil 2 Driver | 5°, at 55 mph: 8° dwell |
| 10 | TN/RD | Coil 1 Driver | 5°, at 55 mph: 8° dwell |
| 11 | LB/BR | Injector 4 Driver | 1.0-4.0 ms |
| 12 | YL/WT | Injector 3 Driver | 1.0-4.0 ms |
| 13 | TN | Injector 2 Driver | 1.0-4.0 ms |
| 14 | WT/DB | Injector 1 Driver | 1.0-4.0 ms |
| 16 | VT/RD | Manifold Tuning Solenoid Control | Valve Off: 12v, On: 1v |
| 17 | BR/WT | HO2S-21 (B2 S1) Heater Control | Heater On: 1v, Off: 12v |
| 18 | BR/OR | HO2S-11 (B1 S1) Heater Control | Heater On: 1v, Off: 12v |
| 19 | DG | Generator Field Driver | Digital Signal: 0-12-0v |
| 20 | TN/BK | ECT Sensor Signal | At 180°F: 2.80v |
| 21 | OR/DB | TP Sensor Signal | 0.6-1.0v |
| 22 | LG/PK | EGR Sensor Signal (Not used 04) | 0.6-0.8v |
| 23 | DG/RD | MAP Sensor Signal | 1.5-1.7v |
| 24 | BK/VT | Knock Sensor Return | <0.050v |
| 25 | DB/LG | Knock Sensor Signal | 0.080v AC |
| 27 | BK/LB | Sensor Ground | <0.1v |
| 28 | YL/BK | IAC Motor Return | 12-14v |
| 29 | VT/WT | 5-Volt Supply | 4.9-5.1v |
| 30 | BK/RD | IAT Sensor Signal | At 100°F: 1.83v |
| 31 | BK/DG | HO2S-11 (B1 S1) Signal | 0.1-1.1v |
| 32 | BR/DG | HO2S-11 (B1 S1) Ground (Up) | <0.1v |
| 33 | LG/RD | HO2S-21 (B2 S1) Signal | 0.1-1.1v |
| 34 | TN/YL | CMP Sensor Signal | Digital Signal: 0-5-0v |
| 35 | GY/BK | CKP Sensor Signal | Digital Signal: 0-5-0v |
| 36-37 | --- | Not Used | --- |
| 38 | GY/RD | IAC Motor Driver | DC pulse signals: 0.8-11v |

## Standard Colors and Abbreviations

| Abbreviation | Color | Abbreviation | Color | Abbreviation | Color |
|---|---|---|---|---|---|
| BK | Black | GY | Gray | RD | Red |
| BL | Blue | GN | Green | TN | Tan |
| BR | Brown | LG | Light Green | VT | Violet |
| DB | Dark Blue | OR | Orange | WT | White |
| DG | Dark Green | PK | Pink | YL | Yellow |

**2002-04 300M 3.5L VIN G & K 'C3' White Connector**

| PCM Pin # | Wire Color | Circuit Description (38-Pin) | Value at Hot Idle |
|---|---|---|---|
| 1-2, 13-17 | --- | Not Used | --- |
| 3 | DB/YL | ASD Relay Control | Relay Off: 12v, On: 1v |
| 4 | DB/PK | High Speed Radiator Fan Relay | Relay Off: 12v, On: 1v |
| 5 | LG/RD | S/C Vent Solenoid | Vacuum Decreasing: 1v |
| 6 | DB/PK | Low Speed Radiator Fan Relay | Relay Off: 12v, On: 1v |
| 7 | YL/RD | S/C Power Supply | 12-14v |
| 8 | WT/DG | Natural Vacuum Leak Detection Solenoid | Solenoid Off: 12v, On: 1v |
| 9 | BR/VT | HO2S-21 (B1 S2) Heater Control | Heater On: 1v, Off: 12v |
| 10 | BR/GY | HO2S-11 (B2 S2) Heater Control | Heater On: 1v, Off: 12v |
| 11 | DB/OR | A/C Clutch Relay Control | Relay Off: 12v, On: 1v |
| 12 | TN/RD | S/C Vacuum Solenoid | Vacuum Increasing: 1v |
| 18, 19 | OR/DG | Automatic Shutdown Relay Output | 12-14v |
| 20 | PK/BK | EVAP Purge Solenoid Control | PWM Signal: 0-12-0v |
| 21-22, 24-25 | --- | Not Used | --- |
| 23 | WT/PK | Brake Switch Signal | Brake Off: 0v, On: 12v |
| 26 | YL | AutoStick Downshift Switch | Digital Signal: 0v or 12v |
| 27 | LG/RD | AutoStick Upshift Switch | Digital Signal: 0v or 12v |
| 28 | OR/DG | Automatic Shutdown Relay Output | 12-14v |
| 29 | DG/LG | EVAP Purge Solenoid Return | 0-1v |
| 30-31, 33, 36 | --- | Not Used | --- |
| 32 | VT/LG | Ambient Air Temperature Sensor | At 100°F: 1.83v |
| 34 | RD/LG | S/C Set Switch Signal | S/C & Set Switch On: 3.8v |
| 35 | OR/RD | Natural Vacuum Leak Detection Switch Sense | 0.1v |
| 37 | BR | Fuel Pump Relay Control | Relay Off: 12v, On: 1v |
| 38 | TN | Starter Relay Control | KOEC: 9-11v |

## Standard Colors and Abbreviations

| Abbreviation | Color | Abbreviation | Color | Abbreviation | Color |
|---|---|---|---|---|---|
| BK | Black | GY | Gray | RD | Red |
| BL | Blue | GN | Green | TN | Tan |
| BR | Brown | LG | Light Green | VT | Violet |
| DB | Dark Blue | OR | Orange | WT | White |
| DG | Dark Green | PK | Pink | YL | Yellow |

## 2002-04 300M 3.5L VIN G, K 'C4' Green Connector

| PCM Pin # | Wire Color | Circuit Description (38-Pin) | Value at Hot Idle |
|---|---|---|---|
| 1, 3-5 | --- | Not Used | --- |
| 2 | BR | Overdrive Solenoid Control | Solenoid Off: 12v, On: 1v |
| 3 | PK | Underdrive Solenoid Control | Solenoid Off: 12v, On: 1v |
| 6 | WT | 2-4 Solenoid Control | Solenoid Off: 12v, On: 1v |
| 7-9, 11 | --- | Not Used | --- |
| 10 | LB | Low/Reverse Solenoid Control | Solenoid Off: 12v, On: 1v |
| 12 | BK/YL | Ground | <0.050v |
| 13, 14 | BK/RD | Ground | <0.050v |
| 15 | LG/BK | TRS T1 Sense | <0.050v |
| 16 | VT | TRS T3 Sense | <0.050v |
| 17, 20-21 | --- | Not Used | --- |
| 18 | LG, RD | Transmission Control Relay Control | Relay Off: 12v, On: 1v |
| 19 | RD | Transmission Control Relay Output | Relay Off: 12v, On: 1v |
| 22 | R/BK | Overdrive Pressure Switch Sense | |
| 23-26, 31, 36 | --- | Not Used | --- |
| 27 | BK/WT | TRS T41 Sense | <0.050v |
| 28, 38 | RD | Transmission Control Relay Output | Relay Off: 12v, On: 1v |
| 29 | LG | Low/Reverse Pressure Switch Sense | 12-14v |
| 30 | YL/BK | 2-4 Pressure Switch Sense | In Low/Reverse: 2-4v |
| 32 | LG/WT | Output Speed Sensor Signal | In 2-4 Position: 2-4v |
| 33 | RD/BK | Input Speed Sensor Signal | Moving: AC voltage |
| 34 | DB/BK | Speed Sensor Ground | Moving: AC voltage |
| 35 | VT/PK | Transmission Temperature Sensor Signal | <0.050v |
| 37 | VT/WT | TRS T42 Sense | In PRNL: 0v, Others 5v |

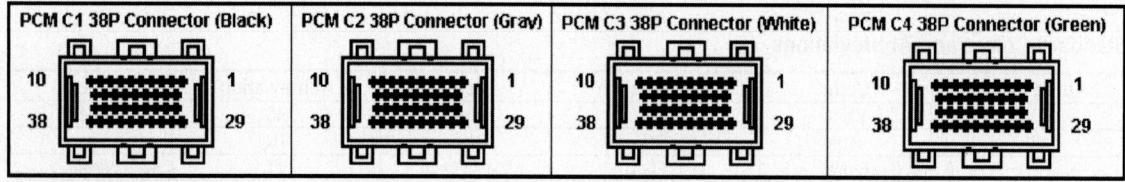

**Pin Connector Graphic**

## Standard Colors and Abbreviations

| Abbreviation | Color | Abbreviation | Color | Abbreviation | Color |
|---|---|---|---|---|---|
| BK | Black | GY | Gray | RD | Red |
| BL | Blue | GN | Green | TN | Tan |
| BR | Brown | LG | Light Green | VT | Violet |
| DB | Dark Blue | OR | Orange | WT | White |
| DG | Dark Green | PK | Pink | YL | Yellow |

**2005 300M 2.7L, 3.5L, 5.7L & Magnum 5.7L 'C1' Black/Black Connector**

| PCM Pin # | Wire Color | Circuit Description (38-Pin) | Value at Hot Idle |
|---|---|---|---|
| 1 | DB/YL (5.7L) | Coil Control No. 8 | |
| 2 | --- | Not Used | --- |
| 3 | DB/YL (5.7L) | Coil Control No. 7 | |
| 4 | BR/LB (5.7L) | Injector Control No. 8 | |
| 5 | BR/YL (5.7L) | Injector Control No. 7 | |
| 6 | --- | Not Used | --- |
| 7 | --- | Not Used | --- |
| 8 | --- | Not Used | --- |
| 9 | BK/BR | Ground | <0.050v |
| 10 | --- | Not Used | --- |
| 11 | PK/GY | Fused Ignition Switch Output (Run-Start) | |
| 12 | PK (2.7L/3.5L) | Ignition Unlock-Run-Start | |
| 13 | --- | Not Used | --- |
| 14 | --- | Not Used | --- |
| 15 | --- | Not Used | --- |
| 16 | DB/LG (3.5L) | SRV Control | |
| 17 | --- | Not Used | --- |
| 18 | BK/BR | Ground | <0.050v |
| 19 | --- | Not Used | --- |
| 20 | VT/GY | Engine Oil Pressure Signal | |
| 21 | --- | Not Used | --- |
| 22 | --- | Not Used | --- |
| 23 | --- | Not Used | --- |
| 24 | --- | Not Used | --- |
| 25 | WT/LG | SCI Receive (PCM) | |
| 26 | WT/OR | SCI Receive (TCM) | |
| 27 | YL/PK | 5 Volt Supply | |
| 28 | BR/LB (5.7L MDS) | MDS Sol Control No. 4 | |
| 29 | RD | Fused B (+) | |
| 30 | YL | Fused Ignition Switch Output (Start) | |
| 31 | DB/YL | O2 1/2 Signal | |
| 32 | DB/DG | O2 Return (Down) | |
| 33 | BR | O2 2/2 Signal | |
| 34 | WT/LG | CAN C Bus (+) | |
| 35 | WT/LB | CAN C Bus (-) | |
| 36 | WT/BR | SCI Transmit (PCM) | |
| 37 | WT/DG | SCI Transmit (TCM) | |
| 38 | --- | Not Used | --- |

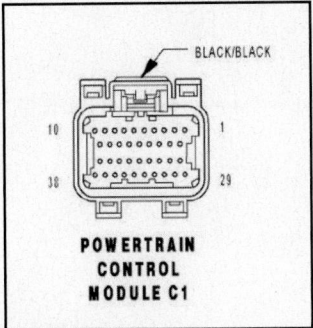

BLACK/BLACK

**POWERTRAIN CONTROL MODULE C1**

## 2005 300M 2.7L, 3.5L, 5.7L & Magnum 5.7L 'C1' Black/Orange Connector

| PCM Pin # | Wire Color | Circuit Description (38-Pin) | Value at Hot Idle |
|---|---|---|---|
| 1 | DB/OR | Coil Control No. 6 | |
| 2 | DB/YL | Coil Control No. 5 | |
| 3 | DB/GY | Coil Control No. 4 | |
| 4 | BR/VT | Injector Control No. 6 | |
| 5 | BR/OR | Injector Control No. 5 | |
| 6 | DB/GY | ETC Motor (+) | |
| 7 | DB/LG | Coil Control No. 3 | |
| 8 | DB/VT | EGR Sol Control | |
| 9 | DB/TN | Coil Control No. 2 | |
| 10 | DB/DG | Coil Control No. 1 | |
| 11 | BR/TN | Injector Control No. 4 | |
| 12 | BR/LB | Injector Control No. 3 | |
| 13 | BR/DB | Injector Control No. 2 | |
| 14 | BR/YL | Injector Control No. 1 | |
| 15 | BR/DB | TP Sensor Return | |
| 16 | BR (2.7L/3.5L) | MTV Control | |
| 16 | BR/DG (5.7L MDS) | MDS Sol Control No. 6 | |
| 17 | BR/VT | O2 2/1 Heater Control | |
| 18 | BR/LG | O2 1/1 Heater Control | |
| 19 | BR/GY | Gen Field Control | |
| 20 | VT/OR | ECT Signal | |
| 21 | BR/OR | TP Sensor No. 1 Signal | |
| 22 | DB/LG | EGR Signal | |
| 23 | VT/BR | MAP Signal | |
| 24 | BR/LG | Knock Sensor No. 1 Return | |
| 25 | DB/YL | Knock Sensor No. 1 Signal | |
| 26 | --- | Not Used | --- |
| 27 | DB/DG | Sensor Ground | |
| 28 | BR/DG | TP Sensor No. 2 Signal | |
| 29 | PK/YL | 5 Volt Supply | |
| 30 | DB/LG | IAT Signal | |
| 31 | DB/LB | O2 1/1 Signal | |
| 32 | BR/DG | O2 Return (Up) | |
| 33 | DB/LG | O2 2/1 Signal | |
| 34 | DB/GY | CMP Signal | |
| 35 | DB/WT | CKP Signal | |
| 36 | BR/WT (5.7L) | Knock Sensor No. 2 Signal | |
| 37 | WT/BR (5.7L) | Knock Sensor No. 2 Return | |
| 38 | DB/LG | ETC Motor (-) | |

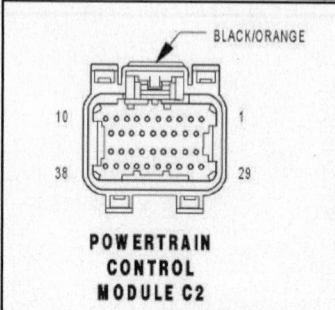

BLACK/ORANGE

10    1
38    29

**POWERTRAIN
CONTROL
MODULE C2**

**2005 300M 2.7L, 3.5L, 5.7L & Magnum 5.7L 'C3' Black/Natural Connector**

| PCM Pin # | Wire Color | Circuit Description (38-Pin) | Value at Hot Idle |
|---|---|---|---|
| 1 | --- | Not Used | --- |
| 2 | --- | Not Used | --- |
| 3 | BR/WT | Auto Shut Down Relay Control | |
| 4 | --- | Not Used | --- |
| 5 | BR/OR (5.7L MDS) | MDS Sol Control No. 7 | |
| 6 | BR/WT (5.7L MDS) | MDS Sol Control No. 1 | |
| 7 | --- | Not Used | --- |
| 8 | VT/GY | NVLD Sol Control | |
| 9 | BR/WT | O2 ½ Heater Control | |
| 10 | BR/GY | O2 2/2 Heater Control | |
| 11 | LB/OR | A/C Clutch Relay Control | |
| 12 | --- | Not Used | --- |
| 13 | --- | Not Used | --- |
| 14 | LB/DG | Brake Switch No. 2 Signal | |
| 15 | --- | Not Used | --- |
| 16 | BR/YL | APPS No. 1 Return | |
| 17 | BR/VT | APPS No. 2 Return | |
| 18 | --- | Not Used | --- |
| 19 | BR/GY | Fused Auto Shut Down Relay Output 3 | |
| 20 | DB/OR | EVAP Purge Control | |
| 21 | --- | Not Used | --- |
| 22 | --- | Not Used | --- |
| 23 | DG/WT | Brake Switch No. 1 Signal | |
| 24 | --- | Not Used | --- |
| 25 | BR/WT | APPS No. 1 Signal | |
| 26 | --- | Not Used | --- |
| 27 | --- | Not Used | --- |
| 28 | BR/WT | Fused Auto Shut Down Relay Output 1 | |
| 29 | DB/BR | EVAP Purge Return | |
| 30 | --- | Not Used | --- |
| 31 | --- | Not Used | --- |
| 32 | --- | Not Used | --- |
| 33 | VT/YL (5.7L) | Engine Oil Temperature Signal | |
| 34 | --- | Not Used | --- |
| 35 | VT/WT | NVLD Switch Signal | |
| 36 | WT/BR | APPS No. 2 Signal | |
| 37 | BR | Fuel Pump Relay Control | |
| 38 | DG/OR | Starter Relay Control | |

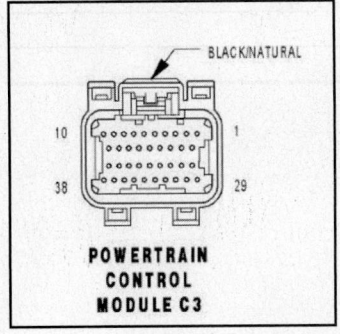

BLACK/NATURAL

10   1
38   29

**POWERTRAIN
CONTROL
MODULE C3**

**2005 300M 2.7L, 3.5L, 5.7L & Magnum 5.7L 'C4' Black/Green Connector**

| PCM Pin # | Wire Color | Circuit Description (38-Pin) | Value at Hot Idle |
|---|---|---|---|
| 1 | YL/GY | OD Solenoid Control | |
| 2 | YL/LB | UD Solenoid Control | |
| 3 | --- | Not Used | --- |
| 4 | --- | Not Used | --- |
| 5 | --- | Not Used | --- |
| 6 | YL/DB | 2-4 Solenoid Control | |
| 7 | --- | Not Used | --- |
| 8 | --- | Not Used | --- |
| 9 | --- | Not Used | --- |
| 10 | DG/WT | L/R Solenoid Control | |
| 11 | --- | Not Used | --- |
| 12 | BK/BR | Ground | |
| 13 | BK/BR | Ground | |
| 14 | BK/BR | Ground | |
| 15 | DG/LB | TRS T1 Sense | |
| 16 | DG/DB | TRS T3 Sense | |
| 17 | --- | Not Used | --- |
| 18 | YL/BR | Transmission Control Relay Control | |
| 19 | YL/OR | Transmission Control Relay Output | |
| 20 | --- | Not Used | --- |
| 21 | --- | Not Used | --- |
| 22 | DG/TN | OD Pressure Switch Sense | |
| 23 | --- | Not Used | --- |
| 24 | --- | Not Used | --- |
| 25 | --- | Not Used | --- |
| 26 | --- | Not Used | --- |
| 27 | YL/DB | TRS T41 Sense (P/N Sense) | |
| 28 | YL/OR | Transmission Control Relay Output | |
| 29 | YL/TN | L/R Pressure Switch Sense | |
| 30 | YL/DG | 2-4 Pressure Switch Sense | |
| 31 | --- | Not Used | --- |
| 32 | DG/BR | Output Speed Sensor Signal | |
| 33 | DG/OR | Input Speed Sensor Signal | |
| 34 | DG/VT | Sensor Ground | |
| 35 | DG/OR | Transmission Temperature Sensor Signal | |
| 36 | --- | Not Used | --- |
| 37 | DG/YL | TRS T42 Sense | |
| 38 | YL/OR | Transmission Control Relay Output | |

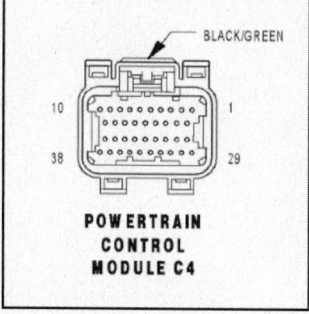

BLACK/GREEN

**POWERTRAIN
CONTROL
MODULE C4**

BREEZE, CIRRUS & STRATUS

### 1995-97 Stratus 2.0L VIN C (M/T) 'C2' White Connector

| PCM Pin # | Wire Color | Circuit Description (40-Pin) | Value at Hot Idle |
|---|---|---|---|
| 41 | PK/LG | S/C Set Switch Signal | S/C & Set Switch On: 3.8V |
| 42 | DB/YL | A/C Damped Pressure Switch | A/C On: 1v, Off: 12v |
| 43 | BK/LB | Sensor Ground | <0.050v |
| 44 | OR/WT | 8-Volt Supply | 7.9-8.1v |
| 45 | DB/LG | PSP Switch Signal | Straight: 0v, Turning: 5v |
| 46 | RD/TN | Battery Power (Fused B+) | 12-14v |
| 47 | --- | Not Used | --- |
| 48 | BR/GY | IAC 3 Driver | DC pulse signals: 0.8-11v |
| 49 | YL/BK | IAC 2 Driver | DC pulse signals: 0.8-11v |
| 50 | BK/TN | Power Ground | <0.1v |
| 51 | TN/WT | HO2S-12 (B1 S2) Signal | 0.1-1.1v |
| 52 | PK/LG | Battery Temperature Sensor | At 86°F: 1.96v |
| 53-54 | --- | Not Used | --- |
| 55 | DB/TN | Low Speed Fan Relay | Relay Off: 12v, On: 1v |
| 56 | --- | Not Used | --- |
| 57 | GY/RD | IAC 1 Driver | DC pulse signals: 0.8-11v |
| 58 | PK/GY | IAC 4 Driver | DC pulse signals: 0.8-11v |
| 59 | PK/BR | CCD Bus (+) | Digital Signal: 0-5-0v |
| 60 | WT/BK | CCD Bus (-) | <0.050v |
| 61 | PK/WT | 5-Volt Supply | 4.9-5.1v |
| 62 | WT/RD | Brake Switch Signal | Brake Off: 0v, On: 12v |
| 63 | YL/DG | Torque Management Request | Digital Signals |
| 64 | DB/OR | A/C Clutch Relay Control | Relay Off: 12v, On: 1v |
| 65 | PK/LB | SCI Transmit | 0v |
| 66 | WT/OR | Vehicle Speed Signal | Digital Signal |
| 67 | DB/PK | ASD Relay Control | Relay Off: 12v, On: 1v |
| 68 | PK/GY | EVAP Purge Solenoid Control | PWM signal: 0-12-0v |
| 69 | DB/PK | High Speed Fan Relay | Relay Off: 12v, On: 1v |
| 72 ('96-'97) | OR/DG | LDP Switch Signal | Switch Closed: 0v, open: 12v |
| 73 | --- | Not Used | --- |
| 74 | BR/LG | Fuel Pump Relay Control | Relay Off: 12v, On: 1v |
| 75 | LG/WT | SCI Receive | 0v |
| 76 | BR/WT | PNP Switch Signal | In P/N: 0v, Others: 5v |
| 77 ('96-'97) | WT/DG | LDP Solenoid Control | PWM signal: 0-12-0v |
| 78 | WT/PK | S/C Vacuum Solenoid | Vacuum Increasing: 1v |
| 80 | LG/RD | S/C Vent Solenoid | Vacuum Decreasing: 1v |

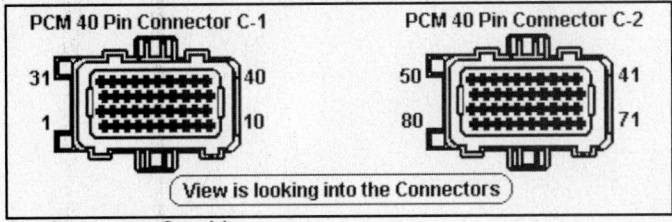

**Pin Connector Graphic**

## 1998-2000 Breeze, Cirrus & Stratus 2.0L VIN C (M/T) 'C1' Black Connector

| PCM Pin # | Wire Color | Circuit Description (40-Pin) | Value at Hot Idle |
|---|---|---|---|
| 1-2 | --- | Not Used | --- |
| 3 | DB/TN | Coil 2 Driver | 5°, at 55 mph: 8° dwell |
| 4 | --- | Not Used | --- |
| 5 | YL/RD | S/C Power Supply | 12-14v |
| 6 | DG/OR | ASD Relay Output | 12-14v |
| 7 | YL/WT | Injector 3 Driver | 1-4 ms |
| 8 | DG | Generator Field Control | Digital Signal: 0-12-0v |
| 9 | --- | Not Used | --- |
| 10 | BK/TN | Power Ground | <0.1v |
| 11 | BK/GY | Coil 1 Driver | 5°, at 55 mph: 8° dwell |
| 12 | --- | Not Used | --- |
| 13 | WT/DB | Injector 1 Driver | 1-4 ms |
| 14-15 | --- | Not Used | --- |
| 16 | LB/BR | Injector 4 Driver | 1-4 ms |
| 17 | TN | Injector 2 Driver | 1-4 ms |
| 18-19 | --- | Not Used | --- |
| 20 | DB/WT | Ignition Switch Output | 12-14v |
| 21-24 | --- | Not Used | --- |
| 25 | DB/LG | Knock Sensor Signal | 0.080v AC |
| 26 | TN/BK | ECT Sensor Signal | At 180°F: 2.80v |
| 27 | BK/OR | HO2S Signal Ground | <0.050v |
| 28-29 | --- | Not Used | --- |
| 30 | BK/DG | HO2S-11 (B1 S1) Signal | 0.1-1.1v |
| 31 | TN | Starter Relay Control | KOEC: 9-11v |
| 32 | GY/BK | CKP Sensor Signal | Digital Signal: 0-5-0v |
| 33 | TN/YL | CMP Sensor Signal | Digital Signal: 0-5-0v |
| 34 | --- | Not Used | --- |
| 35 | OR/LB | TP Sensor Signal | 0.6-1.0v |
| 36 | DG/RD | MAP Sensor Signal | 1.5-1.7v |
| 37 | BK/RD | IAT Sensor Signal | At 100°F: 1.83v |
| 38-39 | --- | Not Used | --- |
| 40 | GY/YL | EGR Solenoid Control | 12v, 55 mph: 1v |

### Standard Colors and Abbreviations

| Abbreviation | Color | Abbreviation | Color | Abbreviation | Color |
|---|---|---|---|---|---|
| BK | Black | GY | Gray | RD | Red |
| BL | Blue | GN | Green | TN | Tan |
| BR | Brown | LG | Light Green | VT | Violet |
| DB | Dark Blue | OR | Orange | WT | White |
| DG | Dark Green | PK | Pink | YL | Yellow |

**1998-2000 Breeze, Cirrus & Stratus 2.0L VIN C (M/T) 'C2' White Connector**

| PCM Pin # | Wire Color | Circuit Description (40-Pin) | Value at Hot Idle |
|---|---|---|---|
| 41 | RD/LG | S/C Set Switch Signal | S/C & Set Switch On: 3.8V |
| 42 | DB/YL | A/C Damped Pressure Switch | A/C On: 1v, Off: 12v |
| 43 | BK/LB | Sensor Ground | <0.050v |
| 44 | OR/WT | 8-Volt Supply | 7.9-8.1v |
| 45 | DB/LG | PSP Switch Signal | Straight: 0v, Turning: 5v |
| 46 | RD/TN | Battery Power (Fused B+) | 12-14v |
| 47 | BK | Power Ground | <0.1v |
| 48 | BR/WT | IAC 3 Driver | DC pulse signals: 0.8-11v |
| 49 | YL/BK | IAC 2 Driver | DC pulse signals: 0.8-11v |
| 50 | BK/TN | Power Ground | <0.1v |
| 51 | TN/WT | HO2S-12 (B1 S2) Signal | 0.1-1.1v |
| 52 | VT/LG | Battery Temperature Sensor | At 86°F: 1.96v |
| 53-54, 71 | --- | Not Used | --- |
| 55 | DB/TN | Low Speed Fan Relay | Relay Off: 12v, On: 1v |
| 56 | WT/VT | S/C Vacuum Solenoid | Vacuum Increasing: 1v |
| 57 | GY/RD | IAC 1 Driver | DC pulse signals: 0.8-11v |
| 58 | VT/BK | IAC 4 Driver | DC pulse signals: 0.8-11v |
| 59 | VT/BR | CCD Bus (+) | Digital Signal: 0-5-0v |
| 60 | WT/BK | CCD Bus (-) | <0.050v |
| 61 | VT/WT | 5-Volt Supply | 4.9-5.1v |
| 62 | WT/RD | Brake Switch Signal | Brake Off: 0v, On: 12v |
| 63 | YL/DG | Torque Management Request | Digital Signals |
| 64 | DB/OR | A/C Clutch Relay Control | Relay Off: 12v, On: 1v |
| 65, 75 | PK, LG | SCI Transmit, SCI Receive | 0v |
| 66 | WT/OR | Vehicle Speed Signal | Digital Signal |
| 67 | DB/VT | ASD Relay Control | Relay Off: 12v, On: 1v |
| 68 | PK/GY | EVAP Purge Solenoid Control | PWM signal: 0-12-0v |
| 69 | DB/PK | High Speed Fan Relay | Relay Off: 12v, On: 1v |
| 70 | WT/TN | EVAP Purge Solenoid Sense | 0-1v |
| 72 | OR/DG | LDP Switch Signal | Switch Closed: 0v, open: 12v |
| 73, 78-79 | --- | Not Used | --- |
| 74 | BR/LG | Fuel Pump Relay Control | Relay Off: 12v, On: 1v |
| 76 | BK/WT | PNP Switch Signal | In P/N: 0v, Others: 5v |
| 77 | WT/DG | LDP Solenoid Control | PWM signal: 0-12-0v |
| 80 | LG/RD | S/C Vent Solenoid | Vacuum Decreasing: 1v |

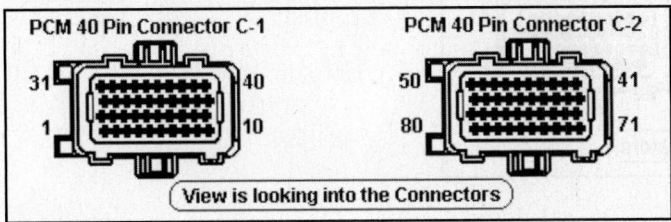

**Pin Connector Graphic**

**1995-97 Breeze, Cirrus & Stratus 2.4L VIN X (A/T) 'C1' Connector**

| PCM Pin # | Wire Color | Circuit Description (40-Pin) | Value at Hot Idle |
|---|---|---|---|
| 1, 8-9 | --- | Not Used | --- |
| 2 | BK/GY | Coil 1 Driver | 5°, at 55 mph: 8° dwell |
| 3 | DB/DG | Coil 2 Driver | 5°, at 55 mph: 8° dwell |
| 4 | DG | Generator Field Driver | Digital Signal: 0-12-0v |
| 5 | YL/PK | S/C Power Supply | 12-14v |
| 6 | DG/OR | ASD Relay Output | 12-14v |
| 7 | YL/WT | Injector 3 Driver | 1.0-4.0 ms |
| 8-9 | --- | Not Used | --- |
| 10 | BK/TN | Power Ground | <0.1v |
| 11-12 | --- | Not Used | --- |
| 13 | WT/LB | Injector 1 Driver | 1.0-4.0 ms |
| 14-15 | --- | Not Used | --- |
| 16 | LB/BR | Injector 4 Driver | 1.0-4.0 ms |
| 17 | TN | Injector 2 Driver | 1.0-4.0 ms |
| 18-19 | --- | Not Used | --- |
| 20 | DB/WT | Ignition Switch Output | 12-14v |
| 21-23 | --- | Not Used | --- |
| 24 ('95) | BK/LG | Knock Sensor Signal | 0.080v AC |
| 24 ('96-'97) | GY/BK | Knock Sensor Signal | 0.080v AC |
| 25 | --- | Not Used | --- |
| 26 | TN/BK | ECT Sensor Signal | At 180°F: 2.80v |
| 27-29 | --- | Not Used | --- |
| 30 | BK/DG | HO2S-11 (B1 S1) Signal | 0.1-1.1v |
| 31 | --- | Not Used | --- |
| 32 | GY/BK | CKP Sensor Signal | Digital Signal: 0-5-0v |
| 33 | TN/YL | CMP Sensor Signal | Digital Signal: 0-5-0v |
| 34 | --- | Not Used | --- |
| 35 | OR/LB | TP Sensor Signal | 0.6-1.0v |
| 36 | DG/RD | MAP Sensor Signal | 1.5-1.7v |
| 37 | BK/RD | IAT Sensor Signal | At 100°F: 1.83v |
| 38-39 | --- | Not Used | --- |
| 40 | GY/YL | EGR Solenoid Control | 12v, at 55 mph: 1v |

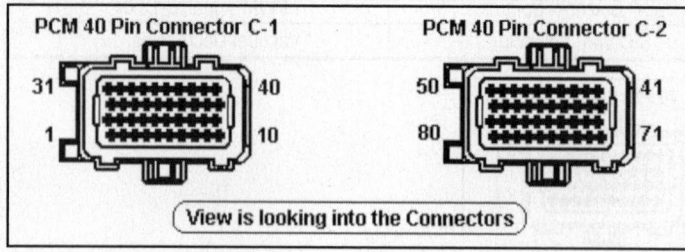

**PCM 40 Pin Connector C-1**

31    40
1    10

**PCM 40 Pin Connector C-2**

50    41
80    71

( View is looking into the Connectors )

**Pin Connector Graphic**

**1995-97 Breeze, Cirrus & Stratus 2.4L VIN X (A/T) 'C2' Connector**

| PCM Pin # | Wire Color | Circuit Description (40-Pin) | Value at Hot Idle |
|---|---|---|---|
| 41 | PK/LG | S/C Set Switch Signal | S/C & Set Switch On: 3.8v |
| 42 | DB/YL | A/C Pressure Switch Signal | A/C On: 0.45-4.85v |
| 43 | BK/LB | Sensor Ground | <0.050v |
| 44 | OR/WT | 8-Volt Supply | 7.9-8.1v |
| 45 | DB/LG | PSP Switch Signal | Straight: 0v, Turning: 5v |
| 46 ('95) | RD/WT | Battery Power (Fused B+) | 12-14v |
| 46 ('96-'97) | RD/TN | Battery Power (Fused B+) | 12-14v |
| 47 | BK | Power Ground | <0.1v |
| 48 | BR/GY | IAC 3 Driver | DC pulse signals: 0.8-11v |
| 49 | YL/BK | IAC 2 Driver | DC pulse signals: 0.8-11v |
| 50 | BK/TN | Power Ground | <0.1v |
| 51 ('96-'97) | TN/WT | HO2S-12 (B1 S2) Signal | 0.1-1.1v |
| 52 | PK/LG | Battery Temperature Sensor | At 86°F: 1.96v |
| 55 | DB/TN | Low Speed Fan Relay | Relay Off: 12v, On: 1v |
| 57 | GY/RD | IAC 1 Driver | DC pulse signals: 0.8-11v |
| 58 | PK/GY | IAC 4 Driver | DC pulse signals: 0.8-11v |
| 59 ('95) | YL/PK | CCD Bus (+) | Digital Signal: 0-5-0v |
| 59 ('96-'97) | PK/BR | CCD Bus (+) | Digital Signal: 0-5-0v |
| 60 | WT/BK | CCD Bus (-) | <0.050v |
| 61 | PK/WT | 5-Volt Supply | 4.9-5.1v |
| 62 | WT/RD | Brake Switch Signal | Brake Off: 0v, On: 12v |
| 63 ('96-'97) | YL/DG | Torque Management Request | Digital Signals |
| 64 | DB/OR | A/C Clutch Relay Control | Relay Off: 12v, On: 1v |
| 65 | PK/LB | SCI Transmit | 0v |
| 66 | WT/OR | Vehicle Speed Signal | Digital: 0-8-0-8v |
| 67 | DB/PK | ASD Relay Control | Relay Off: 12v, On: 1v |
| 68 ('95) | DG/LG | EVAP Purge Solenoid Control | PWM Signal: 0-12-0v |
| 68 ('96-'97) | PK/GY | EVAP Purge Solenoid Control | PWM Signal: 0-12-0v |
| 69 | DB/PK | High Speed Fan Relay | Relay Off: 12v, On: 1v |
| 72 ('96-'97) | OR/DG | LDP Switch Sense | Open: 12v, Closed: 0v |
| 74 | BR/LG | Fuel Pump Relay Control | Relay Off: 12v, On: 1v |
| 75 | LG/WT | SCI Receive | 0v |
| 76 ('96-'97) | BK/PK | PNP Switch Signal | In P/N: 0v, Others: 5v |
| 77 ('96-'97) | WT/DG | LDP Solenoid Control | PWM Signal: 0-12-0v |
| 78 | WT/PK | S/C Vacuum Solenoid | Vacuum Increasing: 1v |
| 80 | LG/RD | S/C Vent Solenoid | Vacuum Decreasing: 1v |

**Standard Colors and Abbreviations**

| Abbreviation | Color | Abbreviation | Color | Abbreviation | Color |
|---|---|---|---|---|---|
| BK | Black | GY | Gray | RD | Red |
| BL | Blue | GN | Green | TN | Tan |
| BR | Brown | LG | Light Green | VT | Violet |
| DB | Dark Blue | OR | Orange | WT | White |
| DG | Dark Green | PK | Pink | YL | Yellow |

## 1998-2000 Breeze, Cirrus & Stratus 2.4L VIN X (A/T) 'C1' Connector

| PCM Pin # | Wire Color | Circuit Description (40-Pin) | Value at Hot Idle |
|---|---|---|---|
| 1-2 | --- | Not Used | --- |
| 3 | DB/TN | Coil 2 Driver | 5°, at 55 mph: 8° dwell |
| 4 | --- | Not Used | |
| 5 | YL/RD | S/C On/Off Switch Signal | 0v or 6.7v |
| 6 | DG/OR | ASD Relay Output | 12-14v |
| 7 | YL/WT | Injector 3 Driver | 1.0-4.0 ms |
| 8 | DG | Generator Field Driver | Digital Signal: 0-12-0v |
| 9 | --- | Not Used | --- |
| 10 | BK/TN | Power Ground | <0.1v |
| 11 | BK/GY | Coil 2 Driver | 5°, at 55 mph: 8° dwell |
| 12 | --- | Not Used | --- |
| 13 | WT/LB | Injector 1 Driver | 1.0-4.0 ms |
| 14-15 | --- | Not Used | --- |
| 16 | LB/BR | Injector 4 Driver | 1.0-4.0 ms |
| 17 | TN | Injector 2 Driver | 1.0-4.0 ms |
| 18-19 | --- | Not Used | --- |
| 20 | DB/WT | Ignition Switch Output | 12-14v |
| 21-24 | --- | Not Used | --- |
| 25 | DB/LG | Knock Sensor Signal | 0.080v AC |
| 26 | TN/BK | ECT Sensor Signal | At 180°F: 2.80v |
| 27 | BK/OR | HO2S-11 (B1 S1) Ground | <0.050v |
| 28-29 | --- | Not Used | --- |
| 30 | BK/DG | HO2S-11 (B1 S1) Signal | 0.1-1.1v |
| 31 | --- | Not Used | --- |
| 32 | GY/BK | CKP Sensor Signal | Digital Signal: 0-5-0v |
| 33 | TN/YL | CMP Sensor Signal | Digital Signal: 0-5-0v |
| 34 | --- | Not Used | --- |
| 35 | OL/LB | TP Sensor Signal | 0.6-1.0v |
| 36 | DG/RD | MAP Sensor Signal | 1.5-1.7v |
| 37 | BK/RD | IAT Sensor Signal | At 100°F: 1.83v |
| 38-39 | --- | Not Used | --- |
| 40 | GY/YL | EGR Solenoid Control | 12v, at 55 mph: 1v |

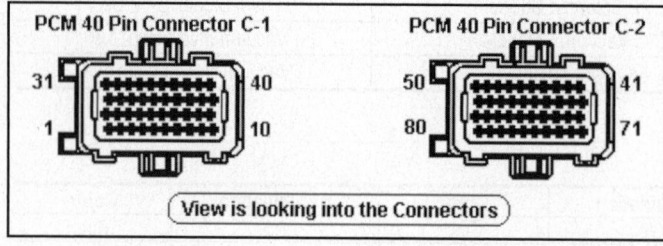

**Pin Connector Graphic**

**1998-2000 Breeze, Cirrus & Stratus 2.4L VIN X (A/T) 'C2' Connector**

| PCM Pin # | Wire Color | Circuit Description (40-Pin) | Value at Hot Idle |
|---|---|---|---|
| 41 | RD/LG | S/C Set Switch Signal | S/C & Set Switch On: 3.8v |
| 42 | DB/YL | A/C Pressure Switch Signal | A/C On: 0.45-4.85v |
| 43 | BK/LB | Sensor Ground | <0.050v |
| 44 | OR/WT | 8-Volt Supply | 7.9-8.1v |
| 45 | DB/LG | PSP Switch Signal | Straight: 0v, Turning: 5v |
| 46 | RD/TN | Battery Power (Fused B+) | 12-14v |
| 47 | BK | Power Ground | <0.1v |
| 48 | BR/WT | IAC 3 Driver | DC pulse signals: 0.8-11v |
| 49 | YL/BK | IAC 2 Driver | DC pulse signals: 0.8-11v |
| 50 | BK/TN | Power Ground | <0.1v |
| 51 | TN/WT | HO2S-12 (B1 S2) Signal | 0.1-1.1v |
| 52 | VT/LG | Battery Temperature Sensor | At 86°F: 1.96v |
| 53-54 | --- | Not Used | --- |
| 55 | DB/TN | Low Speed Fan Relay | Relay Off: 12v, On: 1v |
| 56 | WT/VT | S/C Vacuum Solenoid | Vacuum Increasing: 1v |
| 57 | GY/RD | IAC 1 Driver | DC pulse signals: 0.8-11v |
| 58 | VT/BK | IAC 4 Driver | DC pulse signals: 0.8-11v |
| 59 | VT/BR | CCD Bus (+) | Digital Signal: 0-5-0v |
| 60 | WT/BK | CCD Bus (-) | <0.050v |
| 61 | VT/WT | 5-Volt Supply | 4.9-5.1v |
| 62 | WT/RD | Brake Switch Signal | Brake Off: 0v, On: 12v |
| 63 | YL/DG | Torque Management Request | Digital Signals |
| 64 | DB/OR | A/C Clutch Relay Control | Relay Off: 12v, On: 1v |
| 65 | PK/LB | SCI Transmit | 0v |
| 66 | WT/OR | Vehicle Speed Signal | Digital: 0-8-0-8v |
| 67 | DB/VT | ASD Relay Control | Relay Off: 12v, On: 1v |
| 68 | PK/GY | EVAP Purge Solenoid Control | PWM Signal: 0-12-0v |
| 69 | DB/PK | High Speed Fan Relay | Relay Off: 12v, On: 1v |
| 70 | WT/TN | EVAP Purge Solenoid Sense | 0-1v |
| 71, 73 | --- | Not Used | --- |
| 72 | OR/DG | LDP Switch Sense | Open: 12v, Closed: 0v |
| 74 | BR/LG | Fuel Pump Relay Control | Relay Off: 12v, On: 1v |
| 75 | LG | SCI Receive | 0v |
| 76 | BK/WT | PNP Switch Signal | In P/N: 0v, Others: 5v |
| 77 | WT/DG | LDP Solenoid Control | PWM Signal: 0-12-0v |
| 78-79 | --- | Not Used | --- |
| 80 | LG/RD | S/C Vent Solenoid | Vacuum Decreasing: 1v |

**Standard Colors and Abbreviations**

| Abbreviation | Color | Abbreviation | Color | Abbreviation | Color |
|---|---|---|---|---|---|
| BK | Black | GY | Gray | RD | Red |
| BL | Blue | GN | Green | TN | Tan |
| BR | Brown | LG | Light Green | VT | Violet |
| DB | Dark Blue | OR | Orange | WT | White |
| DG | Dark Green | PK | Pink | YL | Yellow |

## 1995-97 Breeze, Cirrus & Stratus 2.5L VIN H (A/T) 'C1' Connector

| PCM Pin # | Wire Color | Circuit Description (40-Pin) | Value at Hot Idle |
|---|---|---|---|
| 1-3 | --- | Not Used | --- |
| 4 | DG | Generator Field Driver | Digital Signal: 0-12-0v |
| 5 | YL/PK | S/C Power Supply | 12-14v |
| 6 | DG/OR | ASD Relay Output | 12-14v |
| 7 | YL/WT | Injector 3 Driver | 1.0-4.0 ms |
| 8-9 | --- | Not Used | --- |
| 10 | BK/TN | Power Ground | <0.1v |
| 11 | BK/GY | Coil Driver | 5°, at 55 mph: 8° dwell |
| 12 | --- | Not Used | --- |
| 13 | WT/LB | Injector 1 Driver | 1.0-4.0 ms |
| 14 | BR/DG | Injector 6 Driver | 1.0-4.0 ms |
| 15 | GY | Injector 5 Driver | 1.0-4.0 ms |
| 16 | LB/BR | Injector 4 Driver | 1.0-4.0 ms |
| 17 | TN | Injector 2 Driver | 1.0-4.0 ms |
| 18-19 | --- | Not Used | --- |
| 20 | DB/WT | Ignition Switch Output | 12-14v |
| 21-23 | --- | Not Used | --- |
| 24 ('96-'97) | GY/BK | Knock Sensor Signal | 0.80v AC |
| 25 | --- | Not Used | --- |
| 26 | TN/BK | ECT Sensor Signal | At 180°F: 2.80v |
| 27-28 | --- | Not Used | --- |
| 29 ('95) | BK/DG | HO2S-11 (B1 S1) Signal | 0.1-1.1v |
| 30 | BK/DG | HO2S-11 (B1 S1) Signal | 0.1-1.1v |
| 30 ('95) | TN/WT | HO2S-12 (B1 S2) Signal | 0.1-1.1v |
| 31 | --- | Not Used | --- |
| 32 | GY/BK | CKP Sensor Signal | Digital Signal: 0-5-0v |
| 33 | TN/YL | CMP Sensor Signal | Digital Signal: 0-5-0v |
| 34 | --- | Not Used | --- |
| 35 | OR/LB | TP Sensor Signal | 0.6-1.0v |
| 36 | DG/RD | MAP Sensor Signal | 1.5-1.7v |
| 37 | BK/RD | IAT Sensor Signal | At 100°F: 1.83v |
| 38-39 | --- | Not Used | --- |
| 40 | GY/YL | EGR Solenoid Control | 12v, at 55 mph: 1v |

## Standard Colors and Abbreviations

| Abbreviation | Color | Abbreviation | Color | Abbreviation | Color |
|---|---|---|---|---|---|
| BK | Black | GY | Gray | RD | Red |
| BL | Blue | GN | Green | TN | Tan |
| BR | Brown | LG | Light Green | VT | Violet |
| DB | Dark Blue | OR | Orange | WT | White |
| DG | Dark Green | PK | Pink | YL | Yellow |

**1995-97 Breeze, Cirrus & Stratus 2.5L VIN H (A/T) 'C2' Connector**

| PCM Pin # | Wire Color | Circuit Description (40-Pin) | Value at Hot Idle |
|---|---|---|---|
| 41 | PK/LG | S/C Set Switch Signal | S/C & Set Switch On: 3.8v |
| 42 | DB/YL | A/C Pressure Switch Signal | A/C On: 0.45-4.85v |
| 43 | BK/LB | Sensor Ground | <0.050v |
| 44 | OR/WT | 8-Volt Supply | 7.9-8.1v |
| 45 | DB/LG | PSP Switch Signal | Straight: 0v, Turning: 5v |
| 46 | RD/TN | Battery Power (Fused B+) | 12-14v |
| 47 | --- | Not Used | --- |
| 48 | BR/GY | IAC 1 Driver | DC pulse signals: 0.8-11v |
| 49 | YL/BK | IAC 2 Driver | DC pulse signals: 0.8-11v |
| 50 | BK/TN | Power Ground | <0.1v |
| 51 ('96-'97) | TN/WT | HO2S-12 (B1 S2) Signal | 0.1-1.1v |
| 52 | PK/LG | Battery Temperature Sensor | At 86°F: 1.96v |
| 53-54 | --- | Not Used | --- |
| 55 | DB/TN | Low Speed Fan Relay | Relay Off: 12v, On: 1v |
| 56 | --- | Not Used | --- |
| 57 | GY/RD | IAC 3 Driver | DC pulse signals: 0.8-11v |
| 58 | PK/GY | IAC 4 Driver | DC pulse signals: 0.8-11v |
| 59 | PK/BR | CCD Bus (+) | Digital Signal: 0-5-0v |
| 60 | WT/BK | CCD Bus (-) | <0.050v |
| 61 | PK/WT | 5-Volt Supply | 4.9-5.1v |
| 62 | WT/RD | Brake Switch Signal | Brake Off: 0v, On: 12v |
| 63 | YL/DG | Torque Management Request | Digital Signals |
| 64 | DB/OR | A/C Clutch Relay Control | Relay Off: 12v, On: 1v |
| 65 | PK, LG | SCI Transmit, SCI Receive | 0v |
| 66 | WT/OR | Vehicle Speed Signal | Digital: 0-8-0-8v |
| 67 | DB/PK | ASD Relay Control | Relay Off: 12v, On: 1v |
| 68 | PK/GY | EVAP Purge Solenoid Control | PWM Signal: 0-12-0v |
| 69 | DB/PK | High Speed Fan Relay | Relay Off: 12v, On: 1v |
| 70-71 | --- | Not Used | --- |
| 72 ('96-'97) | OR/DG | LDP Switch Sense | Open: 12v, Closed: 0v |
| 73 | --- | Not Used | --- |
| 74 | BR/LG | Fuel Pump Relay Control | Relay Off: 12v, On: 1v |
| 75 | LG/WT | SCI Receive | 0v |
| 76 | BK/WT | PNP Switch Signal | In P/N: 0v, Others: 5v |
| 77 ('96-'97) | WT/DG | LDP Solenoid Control | PWM Signal: 0-12-0v |
| 78 | WT/PK | S/C Vacuum Solenoid | Vacuum Increasing: 1v |
| 79 | --- | Not Used | --- |
| 80 | LG/RD | S/C Vent Solenoid | Vacuum Decreasing: 1v |

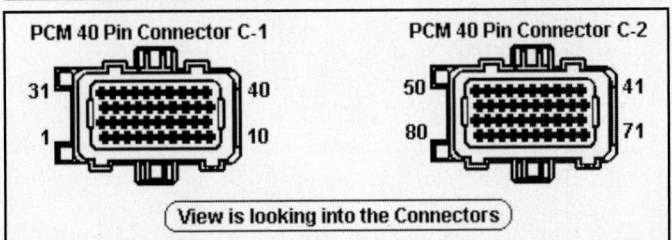

**Pin Connector Graphic**

## 1998-2000 Breeze, Cirrus & Stratus 2.5L VIN H (A/T) 'C1' Connector

| PCM Pin # | Wire Color | Circuit Description (40-Pin) | Value at Hot Idle |
|---|---|---|---|
| 1-3 | --- | Not Used | --- |
| 4 | BK/GY | Coil 1 Driver | 5°, at 55 mph: 8° dwell |
| 5 | YL/RD | S/C Power Supply | 12-14v |
| 6 | DG/OR | ASD Relay Output | 12-14v |
| 7 | YL/WT | Injector 3 Driver | 1.0-4.0 ms |
| 8 | DG | Generator Field Driver | Digital Signal: 0-12-0v |
| 9 | --- | Not Used | --- |
| 10 | BK/TN | Power Ground | <0.1v |
| 11-12 | --- | Not Used | --- |
| 13 | WT/DB | Injector 1 Driver | 1.0-4.0 ms |
| 14 | BR/DG | Injector 6 Driver | 1.0-4.0 ms |
| 15 | GY | Injector 5 Driver | 1.0-4.0 ms |
| 16 | LB/BR | Injector 4 Driver | 1.0-4.0 ms |
| 17 | TN | Injector 2 Driver | 1.0-4.0 ms |
| 18-19 | --- | Not Used | --- |
| 20 | DB/WT | Ignition Switch Output | 12-14v |
| 21-25 | --- | Not Used | --- |
| 26 | TN/BK | ECT Sensor Signal | At 180°F: 2.80v |
| 27 | BK/OR | Oxygen Sensor Ground | <0.050v |
| 28-29 | --- | Not Used | --- |
| 30 | BK/DG | HO2S-11 (B1 S1) Signal | 0.1-1.1v |
| 31 | --- | Not Used | --- |
| 32 | GY/BK | CKP Sensor Signal | Digital Signal: 0-5-0v |
| 33 | TN/YL | CMP Sensor Signal | Digital Signal: 0-5-0v |
| 34 | --- | Not Used | --- |
| 35 | OR/DB | TP Sensor Signal | 0.6-1.0v |
| 36 | DG/RD | MAP Sensor Signal | 1.5-1.7v |
| 37 | BK/RD | IAT Sensor Signal | At 100°F: 1.83v |
| 38-39 | --- | Not Used | --- |
| 40 | GY/YL | EGR Solenoid Control | 12v, at 55 mph: 1v |

## Standard Colors and Abbreviations

| Abbreviation | Color | Abbreviation | Color | Abbreviation | Color |
|---|---|---|---|---|---|
| BK | Black | GY | Gray | RD | Red |
| BL | Blue | GN | Green | TN | Tan |
| BR | Brown | LG | Light Green | VT | Violet |
| DB | Dark Blue | OR | Orange | WT | White |
| DG | Dark Green | PK | Pink | YL | Yellow |

**1998-2000 Breeze, Cirrus & Stratus 2.5L VIN H (A/T) 'C2' Connector**

| PCM Pin # | Wire Color | Circuit Description (40-Pin) | Value at Hot Idle |
|---|---|---|---|
| 41 | RD/LG | S/C Set Switch Signal | S/C & Set Switch On: 3.8v |
| 42 | DB/YL | A/C Pressure Switch Signal | A/C On: 0.45-4.85v |
| 43 | BK/LB | Sensor Ground | <0.050v |
| 44 | OR/WT | 8-Volt Supply | 7.9-8.1v |
| 45 | DB/LG | PSP Switch Signal | Straight: 0v, Turning: 5v |
| 46 | RD/TN | Battery Power (Fused B+) | 12-14v |
| 47, 50 | BK/TN | Power Ground | <0.1v |
| 48 | BR/WT | IAC 3 Driver | DC pulse signals: 0.8-11v |
| 49 | YL/BK | IAC 2 Driver | DC pulse signals: 0.8-11v |
| 51 | TN/WT | HO2S-12 (B1 S2) Signal | 0.1-1.1v |
| 52 | VT/LG | Battery Temperature Sensor | At 86ºF: 1.96v |
| 53-54 | --- | Not Used | --- |
| 55 | DB/TN | Low Speed Fan Relay | Relay Off: 12v, On: 1v |
| 56 | WT/PK | S/C Vacuum Solenoid | Vacuum Increasing: 1v |
| 57 | GY/RD | IAC 1 Driver | DC pulse signals: 0.8-11v |
| 58 | VT/BK | IAC 4 Driver | DC pulse signals: 0.8-11v |
| 59 | VT/BR | CCD Bus (+) | Digital Signal: 0-5-0v |
| 60 | WT/BK | CCD Bus (-) | <0.050v |
| 61 | VT/WH | 5-Volt Supply | 4.9-5.1v |
| 62 | WT/RD | Brake Switch Signal | Brake Off: 0v, On: 12v |
| 63 | YL/DG | Torque Management Request | Digital Signals |
| 64 | DB/OR | A/C Clutch Relay Control | Relay Off: 12v, On: 1v |
| 65, 75 | PK, LG | SCI Transmit, SCI Receive | 0v |
| 66 | WT/OR | Vehicle Speed Signal | Digital: 0-8-0-8v |
| 67 | DB/VT | ASD Relay Control | Relay Off: 12v, On: 1v |
| 68 | PK/GY | EVAP Purge Solenoid Control | PWM Signal: 0-12-0v |
| 69 | DB/PK | High Speed Fan Relay | Relay Off: 12v, On: 1v |
| 70 | WT/TN | EVAP Purge Solenoid Sense | 0-1v |
| 71, 73 | --- | Not Used | --- |
| 72 | OR/DG | LDP Switch Sense | Open: 12v, Closed: 0v |
| 74 | BR/LG | Fuel Pump Relay Control | Relay Off: 12v, On: 1v |
| 76 | BK/WT | PNP Switch Signal | In P/N: 0v, Others: 5v |
| 77 | WT/DG | LDP Solenoid Control | PWM Signal: 0-12-0v |
| 78-79 | --- | Not Used | --- |
| 80 | LG/RD | S/C Vent Solenoid | Vacuum Decreasing: 1v |

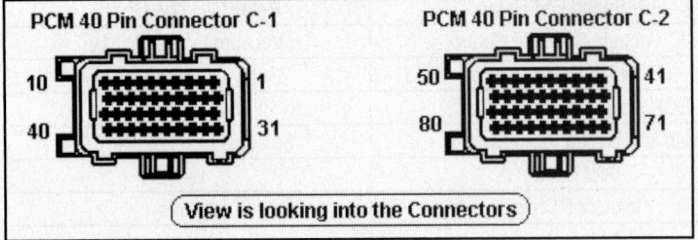

**PCM 40 Pin Connector C-1**     **PCM 40 Pin Connector C-2**

10   1    50   41

40   31    80   71

**View is looking into the Connectors**

**Pin Connector Graphic**

CONCORDE, INTREPID, LHS, NEW YORKER & VISION

### 1993-95 Vision 3.3L VIN T (A/T) 60-Pin Connector

| PCM Pin # | Wire Color | Circuit Description (60-Pin) | Value at Hot Idle |
|---|---|---|---|
| 1 | DG/RD | MAP Sensor Signal | 1.5-1.7v |
| 2 | TN/BK | ECT Sensor Signal | At 180°F: 2.80v |
| 3 | RD/WT | Fused Battery Power (B+) | 12-14v |
| 4 | BK/LB | Sensor Ground | <0.050v |
| 5 | BK/WT | Sensor Ground | <0.050v |
| 6 | PK/WT | 5-Volt Supply | 4.9-5.1v |
| 7 | OR | 8-Volt Supply | 7.9-8.1v |
| 8 | YL/DG | Torque Management Request | Digital Signals |
| 9 | DB/WT | Ignition Switch Power (B+) | 12-14v |
| 10 | GY/BK | Knock Sensor 2 Signal | 0.080v AC |
| 11 | BK/TN | Power Ground | <0.1v |
| 12 | BK/TN | Power Ground | <0.1v |
| 13 | LB/BR | Injector 4 Driver | 1-4 ms |
| 14 | YL/WT | Injector 3 Driver | 1-4 ms |
| 15 | TN | Injector 2 Driver | 1-4 ms |
| 16 | WT/DB | Injector 1 Driver | 1-4 ms |
| 17 | DB/YL | Coil 2 Driver | 5°, at 55 mph: 8° dwell |
| 17 ('94-'95) | WT | Coil 2 Driver | 5°, at 55 mph: 8° dwell |
| 18 | RD/YL | Coil 3 Driver | 5°, at 55 mph: 8° dwell |
| 18 ('94-'95) | RD | Coil 3 Driver | 5°, at 55 mph: 8° dwell |
| 19 | GY | Coil 1 Driver | 5°, at 55 mph: 8° dwell |
| 19 ('94-'95) | BK | Coil 1 Driver | 5°, at 55 mph: 8° dwell |
| 20 | DG | Alternator Field Control | Digital Signal: 0-12-0v |
| 21 | BK/RD | IAT Sensor Signal | At 100°F: 1.83v |
| 22 | OR/DB | TP Sensor Signal | 0.6-1.0v |
| 23 | RD/LG | S/C Set Switch Signal | S/C & Set Switch On: 3.8v |
| 24 | LB/DB | CKP Sensor Signal | Digital Signal: 0-5-0v |
| 25 | PK | SCI Transmit | 0v |
| 26 | VT/BR | CCD Bus (+) | Digital Signal: 0-5-0v |
| 27-28 | --- | Not Used | --- |
| 29 | WT/PK | Brake Pedal Position Switch | Brake Off: 0v, On: 12v |
| 30 | BK/LG | PNP Switch Signal | In P/N: 0v, Others: 5v |
| 31 | DB/PK | High Speed Fan Control | Relay Off: 12v, On: 1v |
| 32 | WT | Low Speed Fan Relay | Relay Off: 12v, On: 1v |
| 33 | TN/RD | S/C Vacuum Solenoid Control | Vacuum Increasing: 1v |
| 34 | DB/OR | A/C Clutch Relay Control | Relay Off: 12v, On: 1v |
| 35 | GY/YL | EGR Solenoid Control (Cal) | 12v, at 55 mph: 1v |
| 36 | VT | Manifold Tuning Valve Control | Solenoid Off: 12v, On: 1v |
| 37 | --- | Not Used | --- |
| 38 | BR/RD | Injector 5 Driver | 1-4 ms |
| 39 | GY/RD | AIS 1 Motor Control | DC pulse signals |
| 40 | BR/WT | AIS 3 Motor Control | DC pulse signals |

**1993-95 Vision 3.3L VIN T (A/T) 60-Pin Connector - Continued**

| PCM Pin # | Wire Color | Circuit Description (60-Pin) | Value at Hot Idle |
|---|---|---|---|
| 41 | BK/DG | HO2S-11 (B1 S1) Signal | 0.1-1.1v |
| 42 | BK/LG | Knock Sensor 1 Signal | 0.080v AC |
| 43 | --- | Not Used | --- |
| 44 | TN/YL | CMP Sensor Signal | Digital Signal: 0-5-0v |
| 45 | LG | SCI Receive | 5v |
| 46 | WT/BK | CCD Bus (-) | <0.050v |
| 47 | WT/OR | Vehicle Speed Signal | Digital Signal |
| 48 | DB | A/C Pressure Switch Signal | A/C On: 0.45-4.85v |
| 49 | TN/WT | HO2S-21 (B2 S1) Signal | 0.1-1.1v |
| 50 | --- | Not Used | --- |
| 51 | DB/YL | ASD Relay Control | Relay Off: 12v, On: 1v |
| 52 | PK/BK | EVAP Purge Solenoid Control | Solenoid Off: 12v, On: 1v |
| 53 | LG/RD | S/C Vent Solenoid Control | Vacuum Decreasing: 1v |
| 54 | --- | Not Used | --- |
| 55 | TN/RD | S/C Relay Control | Relay Off: 12v, On: 1v |
| 56 | --- | Not Used | --- |
| 57 | DG/OR | ASD Relay Output | 12-14v |
| 58 | BR/BK | Injector 6 Driver | 1-4 ms |
| 59 | PK/BK | AIS 4 Motor Control | DC pulse signals |
| 60 | YL/BK | AIS 2 Motor Control | DC pulse signals |

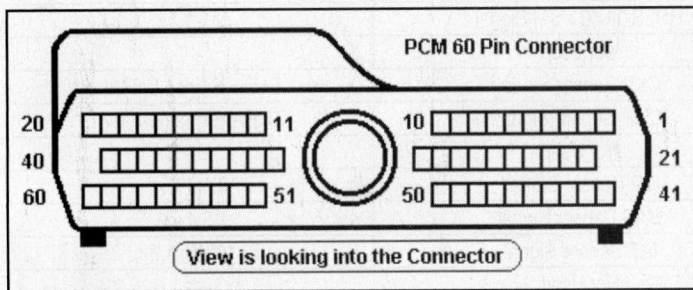

**Pin Connector Graphic**

**Standard Colors and Abbreviations**

| Abbreviation | Color | Abbreviation | Color | Abbreviation | Color |
|---|---|---|---|---|---|
| BK | Black | GY | Gray | RD | Red |
| BL | Blue | GN | Green | TN | Tan |
| BR | Brown | LG | Light Green | VT | Violet |
| DB | Dark Blue | OR | Orange | WT | White |
| DG | Dark Green | PK | Pink | YL | Yellow |

## 1996 Vision 3.3L VIN T (A/T) 'C1' Black Connector

| PCM Pin # | Wire Color | Circuit Description (40-Pin) | Value at Hot Idle |
|---|---|---|---|
| 1 | LG/RD | HO2S-22 (B2 S2) Signal | 0.1-1.1v |
| 2 | RD | Coil 3 Driver | 5º, at 55 mph: 8º dwell |
| 3 | WT | Coil 2 Driver | 5º, at 55 mph: 8º dwell |
| 4 | DG | Generator Field Driver | Digital Signal: 0-12-0v |
| 5 | YL/RD | S/C Power Supply (B+) | 12-14v |
| 6 | DG/OR | ASD Relay Output | 12-14v |
| 7 | YL/WT | Injector 3 Driver | 1-4 ms |
| 8-9 | --- | Not Used | --- |
| 10 | BK/TN | Power Ground | <0.1v |
| 11 | BK | Coil 1 Driver | 5º, at 55 mph: 8º dwell |
| 13 | WT/LB | Injector 1 Driver | 1-4 ms |
| 14 | BR/BK | Injector 6 Driver | 1-4 ms |
| 15 | BR/RD | Injector 5 Driver | 1-4 ms |
| 16 | LB/BR | Injector 4 Driver | 1-4 ms |
| 17 | TN | Injector 2 Driver | 1-4 ms |
| 18-19 | --- | Not Used | --- |
| 20 | DB/WT | Ignition Switch Power (B+) | 12-14v |
| 21-25 | --- | Not Used | --- |
| 26 | TN/BK | ECT Sensor Signal | At 180ºF: 2.80v |
| 27-28 | --- | Not Used | --- |
| 29 | TN/WT | HO2S-21 (B2 S1) Signal | 0.1-1.1v |
| 30 | BK/DG | HO2S-11 (B1 S1) Signal | 0.1-1.1v |
| 31 | --- | Not Used | --- |
| 32 | LB/DB | CKP Sensor Signal | Digital Signal: 0-5-0v |
| 33 | TN/YL | CMP Sensor Signal | Digital Signal: 0-5-0v |
| 34 | --- | Not Used | --- |
| 35 | OR/DB | TP Sensor Signal | 0.6-1.0v |
| 36 | DG/RD | MAP Sensor Signal | 1.5-1.7v |
| 37 | BK/RD | IAT Sensor Signal | At 100ºF: 1.83v |
| 38 | --- | Not Used | --- |
| 39 | PK/RD | Manifold Tuning Valve Control | Valve On: 1v, Off: 12v |
| 40 | GY/YL | EGR Solenoid Control | 12v, at 55 mph: 1v |

## Standard Colors and Abbreviations

| Abbreviation | Color | Abbreviation | Color | Abbreviation | Color |
|---|---|---|---|---|---|
| BK | Black | GY | Gray | RD | Red |
| BL | Blue | GN | Green | TN | Tan |
| BR | Brown | LG | Light Green | VT | Violet |
| DB | Dark Blue | OR | Orange | WT | White |
| DG | Dark Green | PK | Pink | YL | Yellow |

**1996 Vision 3.3L VIN T (A/T) 'C2' White Connector**

| PCM Pin # | Wire Color | Circuit Description (40-Pin) | Value at Hot Idle |
|---|---|---|---|
| 41 | RD/LG | S/C Set Switch Signal | S/C & Set Switch On: 3.8v |
| 42 | DB | A/C Pressure Sensor | A/C On: 0.45-4.85v |
| 43 | BK/LB | Sensor Ground | <0.050v |
| 44 | OR | 8-Volt Supply | 7.9-8.1v |
| 45 | --- | Not Used | --- |
| 46 | RD/WT | Fused Battery Power (B+) | 12-14v |
| 47 | BK/WT | Power Ground | <0.1v |
| 48 | BR/WT | IAC 1 Driver | DC pulse signals |
| 49 | YL/BK | IAC 2 Driver | DC pulse signals |
| 50 | BK/TN | Power Ground | <0.1v |
| 51 | TN/WT | HO2S-12 (B1 S2) Signal | 0.1-1.1v |
| 52-54 | --- | Not Used | --- |
| 55 | WT | Low Speed Fan Relay | Relay Off: 12v, On: 1v |
| 56 | --- | Not Used | --- |
| 57 | GY/RD | IAC 3 Driver | DC pulse signals |
| 58 | PK/BK | IAC 4 Driver | DC pulse signals |
| 59 | PK/BR | CCD Bus (+) | Digital signals: 0-5-0-5v |
| 60 | WT/BK | CCD Bus (-) | <0.050v |
| 61 | PK/WT | 5-Volt Supply | 4.9-5.1v |
| 62 | WT/PK | Brake Pedal Position Switch | Brake Off: 0v, On: 12v |
| 63 | YL/DG | Torque Management Request | Digital Signals |
| 64 | DB/OR | A/C Clutch Relay Control | Relay Off: 12v, On: 1v |
| 65 | PK | SCI Transmit | 0v |
| 66 | WT/OR | Vehicle Speed Signal | Digital Signal |
| 67 | DB/YL | ASD Relay Control | Relay Off: 12v, On: 1v |
| 68 | PK/BK | EVAP Purge Solenoid Control | PWM Signal: 0-12-0v |
| 69 | DB/PK | High Speed Fan Control | Relay Off: 12v, On: 1v |
| 70-73 | --- | Not Used | --- |
| 74 | BR | Fuel Pump Relay Control | Relay Off: 12v, On: 1v |
| 75 | LG | SCI Receive | 5v |
| 76 | BK/DG | PNP Switch Signal | In P/N: 0v, Others: 5v |
| 77 | --- | Not Used | --- |
| 78 | TN/RD | S/C Vacuum Solenoid Control | Vacuum Increasing: 1v |
| 79 | --- | Not Used | --- |
| 80 | LG/RD | S/C Vent Solenoid Control | Vacuum Decreasing: 1v |

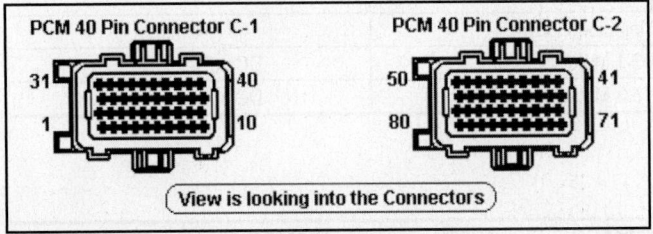

**Pin Connector Graphic**

### 1994 Vision 3.3L Flex Fuel VIN U (A/T) 60-Pin Connector

| PCM Pin # | Wire Color | Circuit Description (60-Pin) | Value at Hot Idle |
|---|---|---|---|
| 1 | DG/RD | MAP Sensor Signal | 1.5-1.7v |
| 2 | TN/BK | ECT Sensor Signal | At 180°F: 2.80v |
| 3 | RD/WT | Fused Battery Power (B+) | 12-14v |
| 4 | BK/LB | Sensor Ground | <0.050v |
| 5 | BK/WT | Sensor Ground | <0.050v |
| 6 | PK/WT | 5-Volt Supply | 4.9-5.1v |
| 7 | OR | 8-Volt Supply | 7.9-8.1v |
| 8 | YL/DG | Torque Management Request | Digital Signals |
| 9 | DB/WT | Ignition Switch Power (B+) | 12-14v |
| 10 | --- | Not Used | --- |
| 11 | BK/TN | Power Ground | <0.1v |
| 12 | BK/TN | Power Ground | <0.1v |
| 13 | LB/BR | Injector 4 Driver | 1-4 ms |
| 14 | YL/WT | Injector 3 Driver | 1-4 ms |
| 15 | TN | Injector 2 Driver | 1-4 ms |
| 16 | WT/DB | Injector 1 Driver | 1-4 ms |
| 17 | WT | Coil 2 Driver | 5°, at 55 mph: 8° dwell |
| 18 | RD/YL | Coil 3 Driver | 5°, at 55 mph: 8° dwell |
| 19 | BK | Coil 1 Driver | 5°, at 55 mph: 8° dwell |
| 20 | DG | Alternator Field Control | Digital Signal: 0-12-0v |
| 21 | BK/RD | IAT Sensor Signal | At 100°F: 1.83v |
| 22 | OR/DB | TP Sensor Signal | 0.6-1.0v |
| 23 | RD/LG | S/C Set Switch Signal | S/C & Set Switch On: 3.8v |
| 24 | GY/BK | CKP Sensor Signal | Digital Signal: 0-5-0v |
| 25 | PK | SCI Transmit | 0v |
| 26 | PK/BR | CCD Bus (+) | Digital Signal: 0-5-0v |
| 27-28 | --- | Not Used | --- |
| 29 | WT/PK | Brake Pedal Position Switch | Brake Off: 0v, On: 12v |
| 30 | BK/LG | PNP Switch Signal | In P/N: 0v, Others: 5v |
| 31 | DB/PK | High Speed Fan Control | Relay Off: 12v, On: 1v |
| 32 | WT | Low Speed Fan Relay | Relay Off: 12v, On: 1v |
| 33 | TN/RD | S/C Vacuum Solenoid Control | Vacuum Increasing: 1v |
| 34 | DB/OR | A/C Clutch Relay Control | Relay Off: 12v, On: 1v |
| 35 | GY/YL | EGR Solenoid Control (Cal) | 12v, at 55 mph: 1v |
| 36 | PK/RD | Manifold Tuning Valve Control | Solenoid Off: 12v, On: 1v |
| 37 | --- | Not Used | --- |
| 38 | BR/RD | Injector 5 Driver | 1-4 ms |
| 39 | GY/RD | AIS 1 Motor Control | DC pulse signals |
| 40 | BR/WT | AIS 3 Motor Control | DC pulse signals |

### Standard Colors and Abbreviations

| Abbreviation | Color | Abbreviation | Color | Abbreviation | Color |
|---|---|---|---|---|---|
| BK | Black | GY | Gray | RD | Red |
| BL | Blue | GN | Green | TN | Tan |
| BR | Brown | LG | Light Green | VT | Violet |
| DB | Dark Blue | OR | Orange | WT | White |
| DG | Dark Green | PK | Pink | YL | Yellow |

**1994 Vision 3.3L Flex Fuel VIN U (A/T) 60-Pin Connector - Continued**

| PCM Pin # | Wire Color | Circuit Description (60-Pin) | Value at Hot Idle |
|---|---|---|---|
| 41 | BK/DG | HO2S-11 (B1 S1) Signal | 0.1-1.1v |
| 42 | BK/LG | Knock Sensor Signal | 0.080v AC |
| 43 | --- | Not Used | --- |
| 44 | TN/YL | CMP Sensor Signal | Digital Signal: 0-5-0v |
| 45 | LG | SCI Receive | 5v |
| 46 | WT/BK | CCD Bus (-) | <0.050v |
| 47 | WT/OR | Vehicle Speed Signal | Digital Signal |
| 48 | DB | A/C Pressure Switch Signal | A/C On: 0.45-4.85v |
| 49 | TN/WT | HO2S-21 (B2 S1) Signal | 0.1-1.1v |
| 50 | YL/WT | Flex Fuel Sensor Signal | 0.5-4.5v |
| 51 | DB/YL | ASD Relay Control | Relay Off: 12v, On: 1v |
| 52 | PK/BK | EVAP Purge Solenoid Control | Solenoid Off: 12v, On: 1v |
| 53 | LG/RD | S/C Vent Solenoid Control | Vacuum Decreasing: 1v |
| 54 | --- | Not Used | --- |
| 55 | TN/RD | S/C Relay Output | 12-14v |
| 56 | --- | Not Used | --- |
| 57 | DG/OR | ASD Relay Output | 12-14v |
| 58 | BR/BK | Injector 6 Driver | 1-4 ms |
| 59 | PK/BK | AIS 4 Motor Control | DC pulse signals |
| 60 | YL/BK | AIS 2 Motor Control | DC pulse signals |

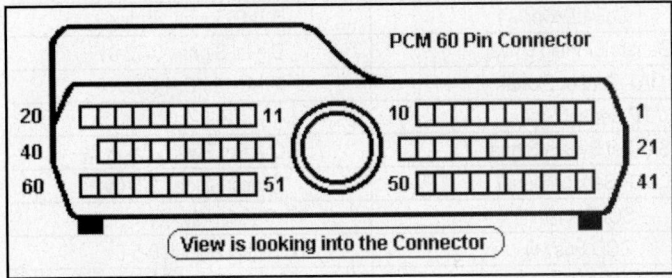

**Pin Connector Graphic**

**Standard Colors and Abbreviations**

| Abbreviation | Color | Abbreviation | Color | Abbreviation | Color |
|---|---|---|---|---|---|
| BK | Black | GY | Gray | RD | Red |
| BL | Blue | GN | Green | TN | Tan |
| BR | Brown | LG | Light Green | VT | Violet |
| DB | Dark Blue | OR | Orange | WT | White |
| DG | Dark Green | PK | Pink | YL | Yellow |

## 1993-95 Vision 3.5L VIN F (A/T) 60-Pin Connector

| PCM Pin # | Wire Color | Circuit Description (60-Pin) | Value at Hot Idle |
|---|---|---|---|
| 1 | DG/RD | MAP Sensor Signal | 1.5-1.7v |
| 2 | TN/BK | ECT Sensor Signal | At 180°F: 2.80v |
| 3 | RD/WT | Fused Battery Power (B+) | 12-14v |
| 4 | BK/LB | Sensor Ground | <0.050v |
| 5 | BK/WT | Sensor Ground | <0.050v |
| 6 | PK/WT | 5-Volt Supply | 4.9-5.1v |
| 7 | OR | 8-Volt Supply | 7.9-8.1v |
| 8 | YL/DG | Torque Management Request | Digital Signals |
| 9 | DB/WT | Ignition Switch Power (B+) | 12-14v |
| 10 | GY/BK | Knock Sensor 2 Signal | 0.080v AC |
| 11 | BK/TN | Power Ground | <0.1v |
| 12 | BK/TN | Power Ground | <0.1v |
| 13 | LB/BR | Injector 4 Driver | 1-4 ms |
| 14 | YL/WT | Injector 3 Driver | 1-4 ms |
| 15 | TN | Injector 2 Driver | 1-4 ms |
| 16 | WT/DB | Injector 1 Driver | 1-4 ms |
| 17 | DB/YL | Coil 2 Driver | 5°, at 55 mph: 8° dwell |
| 17 ('94-'95) | WT | Coil 2 Driver | 5°, at 55 mph: 8° dwell |
| 18 | RD/YL | Coil 3 Driver | 5°, at 55 mph: 8° dwell |
| 18 ('94-'95) | RD | Coil 3 Driver | 5°, at 55 mph: 8° dwell |
| 19 | GY | Coil 1 Driver | 5°, at 55 mph: 8° dwell |
| 19 ('94-'95) | BK | Coil 1 Driver | 5°, at 55 mph: 8° dwell |
| 20 | DG | Generator Field Driver | Digital Signal: 0-12-0v |
| 21 | BK/RD | IAT Sensor Signal | At 100°F: 1.83v |
| 22 | OR/DB | TP Sensor Signal | 0.6-1.0v |
| 23 | RD/LG | S/C Set Switch Signal | S/C & Set Switch On: 3.8v |
| 24 | LB/DB | CKP Sensor Signal | Digital Signal: 0-5-0v |
| 25 | PK | SCI Transmit | 0v |
| 26 | PK/BR | CCD Bus (+) | Digital Signal: 0-5-0v |
| 27-28 | --- | Not Used | --- |
| 29 | WT/PK | Brake Pedal Position Switch | Brake Off: 0v, On: 12v |
| 30 | BK/LG | PNP Switch Signal | In P/N: 0v, Others: 5v |
| 31 | DB/PK | High Speed Fan Control | Relay Off: 12v, On: 1v |
| 32 | WT | Low Speed Fan Relay | Relay Off: 12v, On: 1v |
| 33 | TN/RD | S/C Vacuum Solenoid Control | Vacuum Increasing: 1v |
| 34 | DB/OR | A/C Clutch Relay Control | Relay Off: 12v, On: 1v |
| 35 | GY/YL | EGR Solenoid Control | 12v, at 55 mph: 1v |
| 36 | PK | Manifold Tuning Valve Control | Solenoid Off: 12v, On: 1v |
| 37 | --- | Not Used | --- |
| 38 | BR/RD | Injector 5 Driver | 1-4 ms |
| 39 | GY/RD | IAC 1 Driver | DC pulse signals |
| 40 | BR/WT | IAC 3 Driver | DC pulse signals |

**1993-95 Vision 3.5L VIN F (A/T) 60-Pin Connector - Continued**

| PCM Pin # | Wire Color | Circuit Description (60-Pin) | Value at Hot Idle |
|---|---|---|---|
| 41 | BK/DG | HO2S-11 (B1 S1) Signal | 0.1-1.1v |
| 42 | BK/LG | Knock Sensor 1 Signal | 0.080v AC |
| 43 | --- | Not Used | --- |
| 44 | TN/YL | CMP Sensor Signal | Digital Signal: 0-5-0v |
| 45 | LG | SCI Receive | 5v |
| 46 | WT/BK | CCD Bus (-) | <0.050v |
| 47 | WT/OR | Vehicle Speed Signal | Digital Signal |
| 48 | DB | A/C Pressure Switch Signal | A/C On: 0.45-4.85v |
| 49 | TN/WT | HO2S-21 (B2 S1) Signal | 0.1-1.1v |
| 50 | --- | Not Used | --- |
| 51 | DB/YL | ASD Relay Control | Relay Off: 12v, On: 1v |
| 52 | PK/BK | EVAP Purge Solenoid Control | Solenoid Off: 12v, On: 1v |
| 53 | LG/RD | S/C Vent Solenoid Control | Vacuum Decreasing: 1v |
| 54 | --- | Not Used | --- |
| 55 | TN/RD | S/C Relay Power Output | Relay Off: 12v, On: 1v |
| 56 | --- | Not Used | --- |
| 57 | DG/OR | ASD Relay Output | 12-14v |
| 58 | BR/BK | Injector 6 Driver | 1-4 ms |
| 59 | PK/BK | IAC 4 Driver | DC pulse signals |
| 60 | YL/BK | IAC 2 Driver | DC pulse signals |

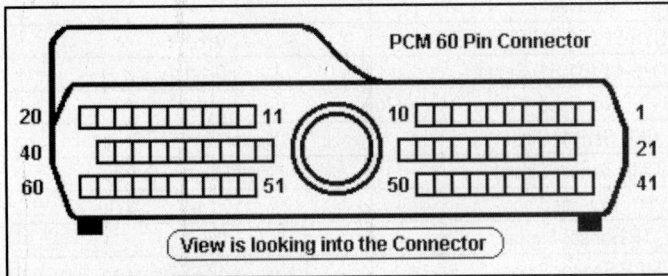

**Pin Connector Graphic**

**Standard Colors and Abbreviations**

| Abbreviation | Color | Abbreviation | Color | Abbreviation | Color |
|---|---|---|---|---|---|
| BK | Black | GY | Gray | RD | Red |
| BL | Blue | GN | Green | TN | Tan |
| BR | Brown | LG | Light Green | VT | Violet |
| DB | Dark Blue | OR | Orange | WT | White |
| DG | Dark Green | PK | Pink | YL | Yellow |

## 1996-97 Vision 3.5L VIN F (A/T) 'C1' Black Connector

| PCM Pin # | Wire Color | Circuit Description (40-Pin) | Value at Hot Idle |
|---|---|---|---|
| 1 | LG/RD | HO2S-22 (B2 S2) Signal | 0.1-1.1v |
| 2 | RD | Coil 3 Driver | 5°, at 55 mph: 8° dwell |
| 3 | WT | Coil 2 Driver | 5°, at 55 mph: 8° dwell |
| 4 | DG | Generator Field Driver | Digital Signal: 0-12-0v |
| 5 | YL/RD | S/C Power Supply (B+) | 12-14v |
| 6 | DG/OR | ASD Relay Output | 12-14v |
| 7 | YL/WT | Injector 3 Driver | 1-4 ms |
| 8-9 | --- | Not Used | --- |
| 10 | BK/TN | Power Ground | <0.1v |
| 11 | BK | Coil 1 Driver | 5°, at 55 mph: 8° dwell |
| 12 | --- | Not Used | --- |
| 13 | WT/LB | Injector 1 Driver | 1-4 ms |
| 14 | BR/BK | Injector 6 Driver | 1-4 ms |
| 15 | BR/RD | Injector 5 Driver | 1-4 ms |
| 16 | LB/BR | Injector 4 Driver | 1-4 ms |
| 17 | TN | Injector 2 Driver | 1-4 ms |
| 18-19 | --- | Not Used | --- |
| 20 | DB/WT | Ignition Switch Power (B+) | 12-14v |
| 21-23 | --- | Not Used | --- |
| 24 | BK/LG | Knock Sensor 1 Signal | 0.080v AC |
| 25 | GY/BK | Knock Sensor 2 Signal | 0.080v AC |
| 26 | TN/BK | ECT Sensor Signal | At 180°F: 2.80v |
| 27-28 | --- | Not Used | --- |
| 29 | TN/WT | HO2S-21 (B2 S1) Signal | 0.1-1.1v |
| 30 | BK/DG | HO2S-11 (B1 S1) Signal | 0.1-1.1v |
| 31 | --- | Not Used | --- |
| 32 | LB/DB | CKP Sensor Signal | Digital Signal: 0-5-0v |
| 33 | TN/YL | CMP Sensor Signal | Digital Signal: 0-5-0v |
| 34 | --- | Not Used | --- |
| 35 | OR/DB | TP Sensor Signal | 0.6-1.0v |
| 36 | DG/RD | MAP Sensor Signal | 1.5-1.7v |
| 37 | BK/RD | IAT Sensor Signal | At 100°F: 1.83v |
| 38 | --- | Not Used | --- |
| 39 | PK/RD | Manifold Tuning Valve Control | Solenoid Off: 12v, On: 1v |
| 40 | GY/YL | EGR Solenoid Control | 12v, 55 mph: 1v |

## Standard Colors and Abbreviations

| Abbreviation | Color | Abbreviation | Color | Abbreviation | Color |
|---|---|---|---|---|---|
| BK | Black | GY | Gray | RD | Red |
| BL | Blue | GN | Green | TN | Tan |
| BR | Brown | LG | Light Green | VT | Violet |
| DB | Dark Blue | OR | Orange | WT | White |
| DG | Dark Green | PK | Pink | YL | Yellow |

**1996-97 Vision 3.5L VIN F (A/T) 'C2' White Connector**

| PCM Pin # | Wire Color | Circuit Description (40-Pin) | Value at Hot Idle |
|---|---|---|---|
| 41 | RD/LG | S/C Set Switch Signal | S/C & Set Switch On: 3.8v |
| 42 | DB | A/C Pressure Switch Signal | A/C On: 0.45-4.85v |
| 43 | BK/LB | Sensor Ground | <0.050v |
| 44 | OR | 8-Volt Supply | 7.9-8.1v |
| 45 | --- | Not Used | --- |
| 46 | RD/WT | Fused Battery Power (B+) | 12-14v |
| 47 | BK/WT | Power Ground | <0.1v |
| 48 | BR/WT | IAC 1 Driver | DC pulse signals |
| 49 | YL/BK | IAC 2 Driver | DC pulse signals |
| 50 | BK/TN | Power Ground | <0.1v |
| 51 | TN/WT | HO2S-12 (B1 S2) Signal | 0.1-1.1v |
| 52-54 | --- | Not Used | --- |
| 55 | WT | Low Speed Fan Relay | Relay Off: 12v, On: 1v |
| 56 | --- | Not Used | --- |
| 57 | GY/RD | IAC 3 Driver | DC pulse signals |
| 58 | PK/BK | IAC 4 Driver | DC pulse signals |
| 59 | PK/BR | CCD Bus (+) | Digital Signal: 0-5-0v |
| 60 | WT/BK | CCD Bus (-) | <0.050v |
| 61 | PK/WT | 5-Volt Supply | 4.9-5.1v |
| 62 | WT/PK | Brake Pedal Position Switch | Brake Off: 0v, On: 12v |
| 63 | YL/DG | Torque Management Request | Digital Signals |
| 64 | DB/OR | A/C Clutch Relay Control | A/C Off: 12v, On: 1v |
| 65 | PK | SCI Transmit | 0v |
| 66 | WT/OR | Vehicle Speed Signal | Digital Signal |
| 67 | DB/YL | ASD Relay Control | Relay Off: 12v, On: 1v |
| 68 | PK/BK | EVAP Purge Solenoid Control | PWM Signal: 0-12-0v |
| 69 | DB/PK | High Speed Fan Control | Relay Off: 12v, On: 1v |
| 70-73 | --- | Not Used | --- |
| 74 | BR | Fuel Pump Relay Control | Relay Off: 12v, On: 1v |
| 75 | LG | SCI Receive | 5v |
| 76 | BK/DG | PNP Switch Signal | In P/N: 0v, Others: 5v |
| 77 | --- | Not Used | --- |
| 78 | TN/RD | S/C Vacuum Solenoid Control | Vacuum Increasing: 1v |
| 77 | --- | Not Used | --- |
| 80 | LG/RD | S/C Vent Solenoid Control | Vacuum Decreasing: 1v |

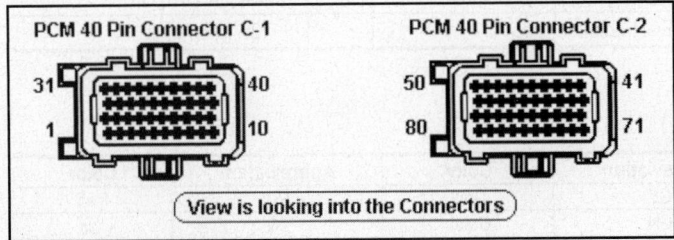

**Pin Connector Graphic**

## 1998-99 Concorde, Intrepid & Vision 2.7L VIN R (A/T) 'C1' Connector

| PCM Pin # | Wire Color | Circuit Description (40-Pin) | Value at Hot Idle |
|---|---|---|---|
| 1 | TN/LG | Coil 4 Driver | 5°, at 55 mph: 8° dwell |
| 2 | TN/OR | Coil 3 Driver | 5°, at 55 mph: 8° dwell |
| 3 | TN/PK | Coil 2 Driver | 5°, at 55 mph: 8° dwell |
| 4 | TN/LB | Coil 6 Driver | 5°, at 55 mph: 8° dwell |
| 5 | YL/RD | S/C Power Supply | 12-14v |
| 6 | DG/OR | ASD Relay Output | 12-14v |
| 7 | YL/WT | Injector 3 Driver | 1.0-4.0 ms |
| 8 | DG | Generator Field Driver | Digital Signal: 0-12-0v |
| 9 | --- | Not Used | --- |
| 10 | BK/TN | Power Ground | <0.1v |
| 11 | TN/RD | Coil 1 Driver | 5°, at 55 mph: 8° dwell |
| 12 | --- | Not Used | --- |
| 13 | WT/DB | Injector 1 Driver | 1.0-4.0 ms |
| 14 | BR/DB | Injector 6 Driver | 1.0-4.0 ms |
| 15 | GY | Injector 5 Driver | 1.0-4.0 ms |
| 16 | LB/BR | Injector 4 Driver | 1.0-4.0 ms |
| 17 | TN/WT | Injector 2 Driver | 1.0-4.0 ms |
| 18-19 | --- | Not Used | --- |
| 20 | DB/WT | Ignition Switch Output | 12-14v |
| 21 | TN/DG | Coil 5 Driver | 5°, at 55 mph: 8° dwell |
| 22-24 | --- | Not Used | --- |
| 25 | BK/PK | Knock Sensor Signal | 0.080v AC |
| 26 | TN/BK | ECT Sensor Signal | At 180°F: 2.80v |
| 27 | BK/OR | Oxygen Sensor Ground | <0.050v |
| 28 | --- | Not Used | --- |
| 29 | LG/RD | HO2S-21 (B2 S1) Signal | 0.1-1.1v |
| 30 | BK/DG | HO2S-11 (B1 S1) Signal | 0.1-1.1v |
| 31 | TN | Starter Relay Control | KOEC: 9-11v |
| 32 | GY/BK | CKP Sensor Signal | Digital Signal: 0-5-0v |
| 33 | TN/YL | CMP Sensor Signal | Digital Signal: 0-5-0v |
| 34 | LG/PK | EGR Sensor Signal | 0.6-0.8v |
| 35 | OR/DB | TP Sensor Signal | 0.6-1.0v |
| 36 | DG/RD | MAP Sensor Signal | 1.5-1.7v |
| 37 | BK/RD | IAT Sensor Signal | At 100°F: 1.83v |
| 38 | --- | Not Used | --- |
| 39 | VT/RD | Manifold Tuning Valve Control | Valve Off: 12v, On: 1v |
| 40 | GY/YL | EGR Solenoid Control | 12v, at 55 mph: 1v |

## Standard Colors and Abbreviations

| Abbreviation | Color | Abbreviation | Color | Abbreviation | Color |
|---|---|---|---|---|---|
| BK | Black | GY | Gray | RD | Red |
| BL | Blue | GN | Green | TN | Tan |
| BR | Brown | LG | Light Green | VT | Violet |
| DB | Dark Blue | OR | Orange | WT | White |
| DG | Dark Green | PK | Pink | YL | Yellow |

### 1998-99 Concorde, Intrepid & Vision 2.7L VIN R (A/T) 'C2' Connector

| PCM Pin # | Wire Color | Circuit Description (40-Pin) | Value at Hot Idle |
|---|---|---|---|
| 41 | RD/LG | S/C Set Switch Signal | S/C & Set Switch On: 3.8v |
| 42 | DB | A/C Pressure Switch Signal | A/C On: 0.45-4.85v |
| 43, 47 | BK/LB | Sensor Ground | <0.050v |
| 44 | OR | 8-Volt Supply | 7.9-8.1v |
| 45 | DB/LG | PSP Switch Signal | Straight: 0v, Turning: 5v |
| 46 | RD/WT | Battery Power (Fused B+) | 12-14v |
| 48 | BR/WT | IAC 3 Driver | DC pulse signals: 0.8-11v |
| 49 | YL/BK | IAC 2 Driver | DC pulse signals: 0.8-11v |
| 50 | BK/TN | Power Ground | <0.1v |
| 51 | TN/WT | HO2S-12 (B1 S2) Signal | 0.1-1.1v |
| 53 | PK/WT | HO2S-22 (B2 S2) Signal | 0.1-1.1v |
| 52, 54, 60 | --- | Not Used | --- |
| 55 | DB/PK | Low Speed Fan Relay | Relay Off: 12v, On: 1v |
| 56 | TN/RD | S/C Vacuum Solenoid | Vacuum Increasing: 1v |
| 57 | GY/RD | IAC 1 Driver | DC pulse signals: 0.8-11v |
| 58 | PK/BK | IAC 4 Driver | DC pulse signals: 0.8-11v |
| 59 | YL/PK | PCI Data Bus (J1850) | Digital Signals: 0-7-0v |
| 61 | PK/WT | 5-Volt Supply | 4.9-5.1v |
| 62 | WT/PK | Brake Switch Signal | Brake Off: 0v, On: 12v |
| 63 | YL/DG | Torque Management Request | Digital Signals |
| 64 | DB/OR | A/C Clutch Relay Control | Relay Off: 12v, On: 1v |
| 65, 75 | PK, LG | SCI Transmit, SCI Receive | 0v |
| 66 | WT/OR | Vehicle Speed Signal | Digital Signal |
| 67 | DB/YL | ASD Relay Control | Relay Off: 12v, On: 1v |
| 68 | PK/BK | EVAP Purge Solenoid Control | PWM Signal: 0-12-0v |
| 69 | DB/LG | High Speed Fan Relay | Relay Off: 12v, On: 1v |
| 70 | DG/LG | EVAP Purge Solenoid Sense | 0-1v |
| 71 | WT/RD | EATX RPM | Digital Signals |
| 72 | OR/RD | LDP Switch Sense | Open: 12v, Closed: 0v |
| 73, 78-79 | --- | Not Used | --- |
| 74 | BR | Fuel Pump Relay Control | Relay Off: 12v, On: 1v |
| 76 | BK/PK | PNP Switch Signal | In P/N: 0v, Others: 5v |
| 77 | WT/DG | LDP Solenoid Control | PWM Signal: 0-12-0v |
| 80 | LG/RD | S/C Vent Solenoid | Vacuum Decreasing: 1v |

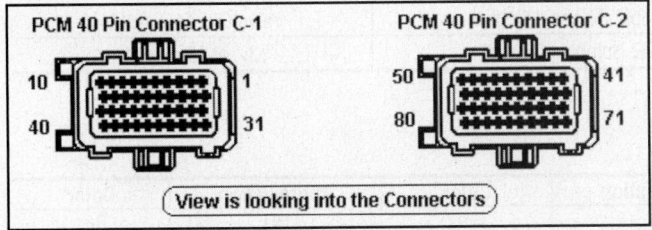

**Pin Connector Graphic**

### 2000-01 Concorde & Intrepid 2.7L VIN R, U & V 'C1' Connector

| PCM Pin # | Wire Color | Circuit Description (40-Pin) | Value at Hot Idle |
|---|---|---|---|
| 1 | TN/LG | Coil 4 Driver | 5°, at 55 mph: 8° dwell |
| 2 | TN/OR | Coil 3 Driver | 5°, at 55 mph: 8° dwell |
| 3 | TN/PK | Coil 2 Driver | 5°, at 55 mph: 8° dwell |
| 4 | TN/LB | Coil 6 Driver | 5°, at 55 mph: 8° dwell |
| 5 | YL/RD | S/C Power Supply | 12-14v |
| 6 | DG/OR | Auto Shutdown Relay Sense | 12-14v |
| 7 | YL/WT | Injector 3 Driver | 1.0-4.0 ms |
| 8 | DG | Generator Field Driver | Digital Signal: 0-12-0v |
| 9 | --- | Not Used | --- |
| 10 | BK/TN | Power Ground | <0.1v |
| 11 | TN/RD | Coil 1 Driver | 5°, at 55 mph: 8° dwell |
| 12 | --- | Not Used | --- |
| 13 | WT/DB | Injector 1 Driver | 1.0-4.0 ms |
| 14 | BR/DB | Injector 6 Driver | 1.0-4.0 ms |
| 15 | GY | Injector 5 Driver | 1.0-4.0 ms |
| 16 | LB/BR | Injector 4 Driver | 1.0-4.0 ms |
| 17 | TN/WT | Injector 2 Driver | 1.0-4.0 ms |
| 18-19 | --- | Not Used | --- |
| 20 | DB/WT | Ignition Switch Output | 12-14v |
| 21 | TN/DG | Coil 5 Driver | 5°, at 55 mph: 8° dwell |
| 22-24 | --- | Not Used | --- |
| 25 | BK/VT | Knock Sensor Signal | 0.080v AC |
| 26 | TN/BK | ECT Sensor Signal | At 180°F: 2.80v |
| 27 | BK/OR | Oxygen Sensor Ground | <0.050v |
| 28 | --- | Not Used | --- |
| 29 | LG/RD | HO2S-21 (B2 S1) Signal | 0.1-1.1v |
| 30 | BK/DG | HO2S-11 (B1 S1) Signal | 0.1-1.1v |
| 31 | TN | Starter Relay Control | KOEC: 9-11v |
| 32 | GY/BK | CKP Sensor Signal | Digital Signal: 0-5-0v |
| 33 | TN/YL | CMP Sensor Signal | Digital Signal: 0-5-0v |
| 34 | LG/PK | EGR Sensor Signal | 0.6-0.8v |
| 35 | OR/DB | TP Sensor Signal | 0.6-1.0v |
| 36 | DG/RD | MAP Sensor Signal | 1.5-1.7v |
| 37 | BK/RD | IAT Sensor Signal | At 100°F: 1.83v |
| 38 | --- | Not Used | --- |
| 39 | VT/RD | Manifold Solenoid Control | Valve On: 1v, Off: 12v |
| 40 | GY/YL | EGR Solenoid Control | 12v, at 55 mph: 1v |

### Standard Colors and Abbreviations

| Abbreviation | Color | Abbreviation | Color | Abbreviation | Color |
|---|---|---|---|---|---|
| BK | Black | GY | Gray | RD | Red |
| BL | Blue | GN | Green | TN | Tan |
| BR | Brown | LG | Light Green | VT | Violet |
| DB | Dark Blue | OR | Orange | WT | White |
| DG | Dark Green | PK | Pink | YL | Yellow |

**2000-01 Concorde & Intrepid 2.7L VIN R, U & V 'C2' Connector**

| PCM Pin # | Wire Color | Circuit Description (40-Pin) | Value at Hot Idle |
|---|---|---|---|
| 41 | RD/LG | S/C Set Switch Signal | S/C & Set Switch On: 3.8v |
| 42 | DB | A/C Pressure Switch Signal | A/C On: 0.45-4.85v |
| 43, 47 | BK/LB | Sensor Ground | <0.050v |
| 44 | OR | 8-Volt Supply | 7.9-8.1v |
| 45 | DB/LG | PSP Switch Signal | Straight: 0v, Turning: 5v |
| 46 | RD/WT | Battery Power (Fused B+) | 12-14v |
| 48 | BR/WT | IAC 3 Driver | DC pulse signals: 0.8-11v |
| 49 | YL/BK | IAC 2 Driver | DC pulse signals: 0.8-11v |
| 50 | BK/TN | Power Ground | <0.1v |
| 51 | TN/WT | HO2S-12 (B1 S2) Signal | 0.1-1.1v |
| 52 ('01) | VT/LG | Battery Temperature Sensor | At 86°F: 1.96v |
| 53 | PK/WT | HO2S-22 (B2 S2) Signal | 0.1-1.1v |
| 54, 60 | --- | Not Used | --- |
| 55 | DB/PK | Low Speed Fan Relay | Relay Off: 12v, On: 1v |
| 56 | TN/RD | S/C Vacuum Solenoid | Vacuum Increasing: 1v |
| 57 | GY/RD | IAC 1 Driver | DC pulse signals: 0.8-11v |
| 58 | PK/BK | IAC 4 Driver | DC pulse signals: 0.8-11v |
| 59 | YL/PK | PCI Data Bus (J1850) | Digital Signals: 0-7-0v |
| 61 | PK/WT | 5-Volt Supply | 4.9-5.1v |
| 62 | WT/PK | Brake Switch Signal | Brake Off: 0v, On: 12v |
| 63 | YL/DG | Torque Management Request | Digital Signals |
| 64 | DB/OR | A/C Clutch Relay Control | Relay Off: 12v, On: 1v |
| 65, 75 | PK, LG | SCI Transmit, SCI Receive | 0v |
| 66 | WT/OR | Vehicle Speed Signal | Digital Signal |
| 67 | DB/YL | ASD Relay Control | Relay Off: 12v, On: 1v |
| 68 | PK/BK | EVAP Purge Solenoid Control | PWM Signal: 0-12-0v |
| 69 | DB/LG | High Speed Fan Relay | Relay Off: 12v, On: 1v |
| 70 | DG/LG | EVAP Purge Solenoid Sense | 0-1v |
| 71 | WT/RD | EATX RPM | Digital Signals |
| 72 | OR/RD | LDP Switch Sense | Open: 12v, Closed: 0v |
| 73, 78-79 | --- | Not Used | --- |
| 74 | BR | Fuel Pump Relay Control | Relay Off: 12v, On: 1v |
| 76 | BK/PK | PNP Switch Signal | In P/N: 0v, Others: 5v |
| 77 | WT/DG | LDP Solenoid Control | PWM Signal: 0-12-0v |
| 80 | LG/RD | S/C Vent Solenoid | Vacuum Decreasing: 1v |

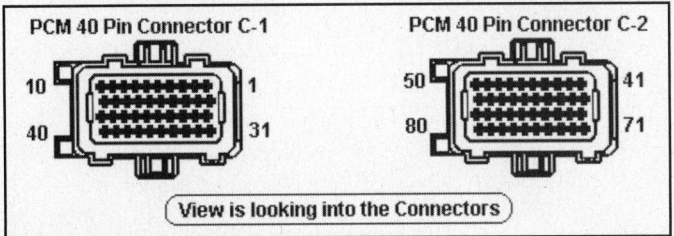

**Pin Connector Graphic**

## 2002-04 Concorde & Intrepid 2.7L VIN R 'C1' Black Connector

| PCM Pin # | Wire Color | Circuit Description (38-Pin) | Value at Hot Idle |
|-----------|------------|------------------------------|-------------------|
| 1-8, 10 | --- | Not Used | --- |
| 9 | BK/TN | Power Ground | <0.1v |
| 11 | DB/WT | Ignition Switch Output (Run-Start) | 12-14v |
| 12 | RD/WT | Ignition Switch Output (Off-Run-Start) | 12-14v |
| 13-16 | --- | Not Used | --- |
| 17 | --- | Not Used | --- |
| 18 | BK/TN | Power Ground | <0.1v |
| 19-20 | --- | Not Used | --- |
| 21 | DB | A/C Pressure Switch Signal | A/C On: 0.45-4.85v |
| 22-24 | --- | Not Used | --- |
| 25 | LG | SCI Receive (PCM) | 5v |
| 26 | VT/OR | SCI Receive (TCM) | 5v |
| 27-28 | --- | Not Used | --- |
| 29 | RD | Battery Power (Fused B+) | 12-14v |
| 30 | YL/BK | Ignition Switch Output (tart) | 8-11v (cranking) |
| 31 | TN/WT | HO2S-12 (B1 S2) Signal | 0.1-1.1v |
| 32 | DB/LG | Oxygen Sensor Return (down) | <0.050v |
| 33 | PK/WT | HO2S-22 (B2 S2) Signal | 0.1-1.1v |
| 34-35 | --- | Not Used | --- |
| 36 | PK/TN | SCI Transmit (PCM) | 0v |
| 37 | WT/DG | SCI Transmit (TCM) | 0v |
| 38 | VT/YL | PCI Data Bus (J1850) | Digital Signals: 0-7-0v |

## Standard Colors and Abbreviations

| Abbreviation | Color | Abbreviation | Color | Abbreviation | Color |
|--------------|-------|--------------|-------|--------------|-------|
| BK | Black | GY | Gray | RD | Red |
| BL | Blue | GN | Green | TN | Tan |
| BR | Brown | LG | Light Green | VT | Violet |
| DB | Dark Blue | OR | Orange | WT | White |
| DG | Dark Green | PK | Pink | YL | Yellow |

**2002-04 Concorde & Intrepid 2.7L VIN R 'C2' Gray Connector**

| PCM Pin # | Wire Color | Circuit Description (38-Pin) | Value at Hot Idle |
|---|---|---|---|
| 1 | TN/LB | Coil 6 Driver | 5°, at 55 mph: 8° dwell |
| 2 | TN/DG | Coil 5 Driver | 5°, at 55 mph: 8° dwell |
| 3 | TN/LG | Coil 4 Driver | 5°, at 55 mph: 8° dwell |
| 4 | BR/DB | Injector 6 Driver | 1.0-4.0 ms |
| 5 | GY | Injector 5 Driver | 1.0-4.0 ms |
| 6, 15, 26 | --- | Not Used | --- |
| 7 | TN/OR | Coil 3 Driver | 5°, at 55 mph: 8° dwell |
| 8 | GY/YL | EGR Solenoid Control | 12v, at 55 mph: 1v |
| 9 | TN/PK | Coil 2 Driver | 5°, at 55 mph: 8° dwell |
| 10 | TN/RD | Coil 1 Driver | 5°, at 55 mph: 8° dwell |
| 11 | LB/BR | Injector 4 Driver | 1.0-4.0 ms |
| 12 | YL/WT | Injector 3 Driver | 1.0-4.0 ms |
| 13 | TN | Injector 2 Driver | 1.0-4.0 ms |
| 14 | WT/DB | Injector 1 Driver | 1.0-4.0 ms |
| 16 | VT/RD | Manifold Tuning Solenoid Control | Valve Off: 12v, On: 1v |
| 17 | BR/WT | HO2S-21 (B2 S1) Heater Control | Heater On: 1v, Off: 12v |
| 18 | BR/OR | HO2S-11 (B1 S1) Heater Control | Heater On: 1v, Off: 12v |
| 19 | DG | Generator Field Driver | Digital Signal: 0-12-0v |
| 20 | TN/BK | ECT Sensor Signal | At 180°F: 2.80v |
| 21 | OR/DB | TP Sensor Signal | 0.6-1.0v |
| 22 | LG/PK | EGR Sensor Signal | 0.6-0.8v |
| 23 | DG/RD | MAP Sensor Signal | 1.5-1.7v |
| 24 | BK/VT | Knock Sensor Return | <0.050v |
| 25 | DB/LG | Knock Sensor Signal | 0.080v AC |
| 27 | BK/LB | Sensor Ground | <0.1v |
| 28 | YL/BK | IAC Motor Sense | 12-14v |
| 29 | VT/WT | 5-Volt Supply | 4.9-5.1v |
| 30 | BK/RD | IAT Sensor Signal | At 100°F: 1.83v |
| 31 | BK/DG | HO2S-11 (B1 S1) Signal | 0.1-1.1v |
| 32 | BR/DG | HO2S-11 (B1 S1) Ground (Up) | <0.1v |
| 33 | LG/RD | HO2S-21 (B2 S1) Signal | 0.1-1.1v |
| 34 | TN/YL | CMP Sensor Signal | Digital Signal: 0-5-0v |
| 35 | GY/BK | CKP Sensor Signal | Digital Signal: 0-5-0v |
| 36-37 | --- | Not Used | --- |
| 38 | GY/RD | IAC Motor Driver | DC pulse signals: 0.8-11v |

**Standard Colors and Abbreviations**

| Abbreviation | Color | Abbreviation | Color | Abbreviation | Color |
|---|---|---|---|---|---|
| BK | Black | GY | Gray | RD | Red |
| BL | Blue | GN | Green | TN | Tan |
| BR | Brown | LG | Light Green | VT | Violet |
| DB | Dark Blue | OR | Orange | WT | White |
| DG | Dark Green | PK | Pink | YL | Yellow |

## 2002-04 Concorde & Intrepid 2.7L VIN R 'C3' White Connector

| PCM Pin # | Wire Color | Circuit Description (38-Pin) | Value at Hot Idle |
|---|---|---|---|
| 1-2, 13-17 | --- | Not Used | --- |
| 3 | DB/YL | ASD Relay Control | Relay Off: 12v, On: 1v |
| 4 | DB/PK | High Speed Radiator Fan Relay | Relay Off: 12v, On: 1v |
| 5 | LG/RD | S/C Vent Solenoid | Vacuum Decreasing: 1v |
| 6 | DB/PK | Low Speed Radiator Fan Relay | Relay Off: 12v, On: 1v |
| 7 | YL/RD | S/C Power Supply | 12-14v |
| 8 | WT/DG | Natural Vacuum Leak Detection Solenoid | Solenoid Off: 12v, On: 1v |
| 9 | BR/VT | HO2S-21 (B2 S1) Heater Control | Heater On: 1v, Off: 12v |
| 10 | BR/GY | HO2S-11 (B1 S1) Heater Control | Heater On: 1v, Off: 12v |
| 11 | DB/OR | A/C Clutch Relay Control | Relay Off: 12v, On: 1v |
| 12 | TN/RD | S/C Vacuum Solenoid | Vacuum Increasing: 1v |
| 18, 19 | OR/DG | Automatic Shutdown Relay Output | 12-14v |
| 20 | PK/BK | EVAP Purge Solenoid Control | PWM Signal: 0-12-0v |
| 21-22, 24-25 | --- | Not Used | --- |
| 23 | WT/PK | Brake Switch Sense | Brake Off: 0v, On: 12v |
| 26 | YL | AutoStick Downshift Switch | Digital Signal: 0v or 12v |
| 27 | LG/RD | AutoStick Upshift Switch | Digital Signal: 0v or 12v |
| 28 | OR/DG | Automatic Shutdown Relay Output | 12-14v |
| 29 | DG/LG | EVAP Purge Solenoid Sense | 0-1v |
| 30-31, 33, 36 | --- | Not Used | --- |
| 32 | VT/LG | Ambient Air Temperature Sensor | At 100°F: 1.83v |
| 34 | RD/LG | S/C Set Switch Signal | S/C & Set Switch On: 3.8v |
| 35 | OR/RD | Natural Vacuum Leak Detection Switch Sense | 0.1v |
| 37 | BR | Fuel Pump Relay Control | Relay Off: 12v, On: 1v |
| 38 | TN | Starter Relay Control | KOEC: 9-11v |

## Standard Colors and Abbreviations

| Abbreviation | Color | Abbreviation | Color | Abbreviation | Color |
|---|---|---|---|---|---|
| BK | Black | GY | Gray | RD | Red |
| BL | Blue | GN | Green | TN | Tan |
| BR | Brown | LG | Light Green | VT | Violet |
| DB | Dark Blue | OR | Orange | WT | White |
| DG | Dark Green | PK | Pink | YL | Yellow |

**2002-04 Concorde & Intrepid 2.7L VIN R 'C4' Green Connector**

| PCM Pin # | Wire Color | Circuit Description (38-Pin) | Value at Hot Idle |
|---|---|---|---|
| 1, 3-5 | --- | Not Used | --- |
| 2 | BR | Overdrive Solenoid Control | Solenoid Off: 12v, On: 1v |
| 3 | PK | Underdrive Solenoid Control | Solenoid Off: 12v, On: 1v |
| 6 | WT | 2-4 Solenoid Control | Solenoid Off: 12v, On: 1v |
| 7-9, 11 | --- | Not Used | --- |
| 10 | LB | Low/Reverse Solenoid Control | Solenoid Off: 12v, On: 1v |
| 12 | BK/YL | Power Ground | <0.050v |
| 13, 14 | BK/RD | Power Ground | <0.050v |
| 15 | LG/BK | TRS T1 Sense | <0.050v |
| 16 | VT | TRS T3 Sense | <0.050v |
| 17, 20-21 | --- | Not Used | --- |
| 18, 19 | LG, RD | Transmission Control Relay Output | Relay Off: 12v, On: 1v |
| 19 | RD | Transmission Control Relay Output | Relay Off: 12v, On: 1v |
| 23-26, 31, 36 | --- | Not Used | --- |
| 27 | BK/WT | TRS T41 Sense | <0.050v |
| 28, 38 | RD | Transmission Control Relay Output | Relay Off: 12v, On: 1v |
| 29 | LG | Low/Reverse Pressure Switch Sense | 12-14v |
| 30 | YL/BK | 2-4 Pressure Switch Sense | In Low/Reverse: 2-4v |
| 32 | LG/WT | Output Speed Sensor Signal | In 2-4 Position: 2-4v |
| 33 | RD/BK | Input Speed Sensor Signal | Moving: AC voltage |
| 34 | DB/BK | Speed Sensor Ground | Moving: AC voltage |
| 35 | VT/PK | Transmission Temperature Sensor Signal | <0.050v |
| 37 | VT/WT | TRS T42 Sense | In PRNL: 0v, Others 5v |

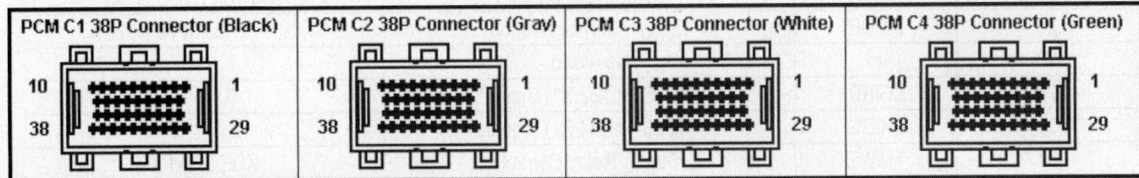

**Pin Connector Graphic**

**Standard Colors and Abbreviations**

| Abbreviation | Color | Abbreviation | Color | Abbreviation | Color |
|---|---|---|---|---|---|
| BK | Black | GY | Gray | RD | Red |
| BL | Blue | GN | Green | TN | Tan |
| BR | Brown | LG | Light Green | VT | Violet |
| DB | Dark Blue | OR | Orange | WT | White |
| DG | Dark Green | PK | Pink | YL | Yellow |

## 1998-99 Concorde, Intrepid & Vision 3.2L VIN J (A/T) 'C1' Connector

| PCM Pin # | Wire Color | Circuit Description (40-Pin) | Value at Hot Idle |
|---|---|---|---|
| 1 | TN/LG | Coil 4 Driver | 5º, at 55 mph: 8º dwell |
| 2 | TN/OR | Coil 3 Driver | 5º, at 55 mph: 8º dwell |
| 3 | TN/PK | Coil 2 Driver | 5º, at 55 mph: 8º dwell |
| 4 | TN/LB | Coil 6 Driver | 5º, at 55 mph: 8º dwell |
| 5 | YL/RD | S/C Power Supply | 12-14v |
| 6 | DG/OR | ASD Relay Output | 12-14v |
| 7 | YL/WT | Injector 3 Driver | 1.0-4.0 ms |
| 8 | DG | Generator Field Driver | Digital Signal: 0-12-0v |
| 9 | --- | Not Used | --- |
| 10 | BK/TN | Power Ground | <0.1v |
| 11 | TN/RD | Coil 1 Driver | 5º, at 55 mph: 8º dwell |
| 12 | --- | Not Used | --- |
| 13 | WT/DB | Injector 1 Driver | 1.0-4.0 ms |
| 14 | BR/DB | Injector 6 Driver | 1.0-4.0 ms |
| 15 | GY | Injector 5 Driver | 1.0-4.0 ms |
| 16 | LB/BR | Injector 4 Driver | 1.0-4.0 ms |
| 17 | TN | Injector 2 Driver | 1.0-4.0 ms |
| 18 | GY/PK | Short Runner Valve Control | Valve Off: 12v, On: 1v |
| 19 | --- | Not Used | --- |
| 20 | DB/WT | Ignition Switch Output | 12-14v |
| 21 | TN/DG | Coil 5 Driver | 5º, at 55 mph: 8º dwell |
| 22-23 | --- | Not Used | --- |
| 24 | DB/LG | Knock Sensor Signal | 0.080v AC |
| 25 | BK/VT | Knock Sensor Ground | <0.050v |
| 26 | TN/BK | ECT Sensor Signal | At 180ºF: 2.80v |
| 27 | BK/OR | Oxygen Sensor Ground | <0.050v |
| 28 | --- | Not Used | --- |
| 29 | LG/RD | HO2S-21 (B2 S1) Signal | 0.1-1.1v |
| 30 | BK/DG | HO2S-11 (B1 S1) Signal | 0.1-1.1v |
| 31 | TN/WT | Starter Relay Control | KOEC: 9-11v |
| 32 | GY/BK | CKP Sensor Signal | Digital Signal: 0-5-0v |
| 33 | TN/YL | CMP Sensor Signal | Digital Signal: 0-5-0v |
| 34 | LG/PK | EGR Sensor Signal | 0.6-0.8v |
| 35 | OR/DB | TP Sensor Signal | 0.6-1.0v |
| 36 | DG/RD | MAP Sensor Signal | 1.5-1.7v |
| 37 | BK/RD | IAT Sensor Signal | At 100ºF: 1.83v |
| 38 | --- | Not Used | --- |
| 39 | VT/RD | Manifold Tuning Valve Control | Valve Off: 12v, On: 1v |
| 40 | GY/YL | EGR Solenoid Control | 12v, at 55 mph: 1v |

## Standard Colors and Abbreviations

| Abbreviation | Color | Abbreviation | Color | Abbreviation | Color |
|---|---|---|---|---|---|
| BK | Black | GY | Gray | RD | Red |
| BL | Blue | GN | Green | TN | Tan |
| BR | Brown | LG | Light Green | VT | Violet |
| DB | Dark Blue | OR | Orange | WT | White |
| DG | Dark Green | PK | Pink | YL | Yellow |

**1998-99 Concorde, Intrepid & Vision 3.2L VIN J (A/T) 'C2' Connector**

| PCM Pin # | Wire Color | Circuit Description (40-Pin) | Value at Hot Idle |
|---|---|---|---|
| 41 | RD/LG | S/C Set Switch Signal | S/C & Set Switch On: 3.8v |
| 42 | DB | A/C Pressure Switch Signal | A/C On: 0.45-4.85v |
| 43 | BK/LB | Sensor Ground | <0.050v |
| 44 | OR | 8-Volt Supply | 7.9-8.1v |
| 45 | DB/LG | PSP Switch Signal | Straight: 0v, Turning: 5v |
| 46 | RD/WT | Battery Power (Fused B+) | 12-14v |
| 47, 50 | BK/WT | Power Ground | <0.1v |
| 48 | BR/WT | IAC 3 Driver | DC pulse signals: 0.8-11v |
| 49 | YL/BK | IAC 2 Driver | DC pulse signals: 0.8-11v |
| 51 | TN/WT | HO2S-12 (B1 S2) Signal | 0.1-1.1v |
| 52, 54, 60 | --- | Not Used | --- |
| 53 | PK/WT | HO2S-22 (B2 S2) Signal | 0.1-1.1v |
| 55 | DB/PK | Low Speed Fan Relay | Relay Off: 12v, On: 1v |
| 56 | TN/RD | S/C Vacuum Solenoid | Vacuum Increasing: 1v |
| 57 | GY/RD | IAC 1 Driver | DC pulse signals: 0.8-11v |
| 58 | VT/BK | IAC 4 Driver | DC pulse signals: 0.8-11v |
| 59 | YL/VT | PCI Data Bus (J1850) | Digital Signals: 0-7-0v |
| 61 | VT/WT | 5-Volt Supply | 4.9-5.1v |
| 62 | WT/PK | Brake Switch Signal | Brake Off: 0v, On: 12v |
| 63 | YL/DG | Torque Management Request | Digital Signals |
| 64 | DB/OR | A/C Clutch Relay Control | Relay Off: 12v, On: 1v |
| 65, 75 | PK, LG | SCI Transmit, SCI Receive | 0v |
| 66 | WT/OR | Vehicle Speed Signal | Digital Signal |
| 67 | DB/YL | ASD Relay Control | Relay Off: 12v, On: 1v |
| 68 | PK/BK | EVAP Purge Solenoid Control | PWM Signal: 0-12-0v |
| 69 | DB/LG | High Speed Fan Relay | Relay Off: 12v, On: 1v |
| 70 | DG/LG | EVAP Purge Solenoid Sense | 0-1v |
| 71 | WT/RD | EATX RPM | Digital Signals |
| 72 | OR/RD | LDP Switch Sense | Open: 12v, Closed: 0v |
| 73, 78-79 | --- | Not Used | --- |
| 74 | BR | Fuel Pump Relay Control | Relay Off: 12v, On: 1v |
| 76 | BK/PK | PNP Switch Signal | In P/N: 0v, Others: 5v |
| 77 | WT/DG | LDP Solenoid Control | PWM Signal: 0-12-0v |
| 80 | LG/RD | S/C Vent Solenoid | Vacuum Decreasing: 1v |

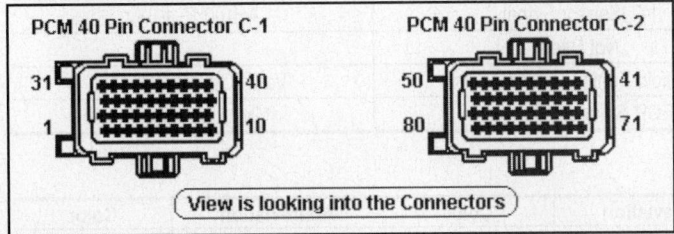

PCM 40 Pin Connector C-1     PCM 40 Pin Connector C-2

31    40     50    41
1    10     80    71

**View is looking into the Connectors**

**Pin Connector Graphic**

## 2000-01 Concorde & Intrepid 3.2L VIN J (A/T) 'C1' Connector

| PCM Pin # | Wire Color | Circuit Description (40-Pin) | Value at Hot Idle |
|---|---|---|---|
| 1 | TN/LG | Coil 4 Driver | 5°, at 55 mph: 8° dwell |
| 2 | TN/OR | Coil 3 Driver | 5°, at 55 mph: 8° dwell |
| 3 | TN/PK | Coil 2 Driver | 5°, at 55 mph: 8° dwell |
| 4 | TN/LB | Coil 6 Driver | 5°, at 55 mph: 8° dwell |
| 5 | YL/RD | S/C Power Supply | 12-14v |
| 6 | DG/OR | ASD Relay Output | 12-14v |
| 7 | YL/WT | Injector 3 Driver | 1.0-4.0 ms |
| 8 | DG | Generator Field Driver | Digital Signal: 0-12-0v |
| 9 | --- | Not Used | --- |
| 10 | BK/TN | Power Ground | <0.1v |
| 11 | TN/RD | Coil 1 Driver | 5°, at 55 mph: 8° dwell |
| 12 | --- | Not Used | --- |
| 13 | WT/DB | Injector 1 Driver | 1.0-4.0 ms |
| 14 | BR/DB | Injector 6 Driver | 1.0-4.0 ms |
| 15 | GY | Injector 5 Driver | 1.0-4.0 ms |
| 16 | LB/BR | Injector 4 Driver | 1.0-4.0 ms |
| 17 | TN | Injector 2 Driver | 1.0-4.0 ms |
| 18 | GY/PK | Short Runner Valve Control | Valve Off: 12v, On: 1v |
| 19 | --- | Not Used | --- |
| 20 | DB/WT | Ignition Switch Output | 12-14v |
| 21 | TN/DG | Coil 5 Driver | 5°, at 55 mph: 8° dwell |
| 22-23 | --- | Not Used | --- |
| 24 | DB/LG | Knock Sensor Signal | 0.080v AC |
| 25 | BK/VT | Knock Sensor Ground | <0.050v |
| 26 | TN/BK | ECT Sensor Signal | At 180°F: 2.80v |
| 27 | BK/OR | Oxygen Sensor Ground | <0.050v |
| 28 | --- | Not Used | --- |
| 29 | LG/RD | HO2S-21 (B2 S1) Signal | 0.1-1.1v |
| 30 | BK/DG | HO2S-11 (B1 S1) Signal | 0.1-1.1v |
| 31 | TN/WH | Starter Relay Control | KOEC: 9-11v |
| 32 | GY/BK | CKP Sensor Signal | Digital Signal: 0-5-0v |
| 33 | TN/YL | CMP Sensor Signal | Digital Signal: 0-5-0v |
| 34 | LG/PK | EGR Sensor Signal | 0.6-0.8v |
| 35 | OR/DB | TP Sensor Signal | 0.6-1.0v |
| 36 | DG/RD | MAP Sensor Signal | 1.5-1.7v |
| 37 | BK/RD | IAT Sensor Signal | At 100°F: 1.83v |
| 38 | --- | Not Used | --- |
| 39 | VT/RD | Manifold Tuning Valve Control | Valve Off: 12v, On: 1v |
| 40 | GY/YL | EGR Solenoid Control | 12v, at 55 mph: 1v |

## Standard Colors and Abbreviations

| Abbreviation | Color | Abbreviation | Color | Abbreviation | Color |
|---|---|---|---|---|---|
| BK | Black | GY | Gray | RD | Red |
| BL | Blue | GN | Green | TN | Tan |
| BR | Brown | LG | Light Green | VT | Violet |
| DB | Dark Blue | OR | Orange | WT | White |
| DG | Dark Green | PK | Pink | YL | Yellow |

**2000-01 Concorde & Intrepid 3.2L VIN J (A/T) 'C2' Connector**

| PCM Pin # | Wire Color | Circuit Description (40-Pin) | Value at Hot Idle |
|---|---|---|---|
| 41 | RD/LG | S/C Set Switch Signal | S/C & Set Switch On: 3.8v |
| 42 | DB | A/C Pressure Switch Signal | A/C On: 0.45-4.85v |
| 43 | BK/LB | Sensor Ground | <0.050v |
| 44 | OR | 8-Volt Supply | 7.9-8.1v |
| 45 | DB/LG | PSP Switch Signal | Straight: 0v, Turning: 5v |
| 46 | RD/WT | Battery Power (Fused B+) | 12-14v |
| 47, 50 | BK/WT | Power Ground | <0.1v |
| 48 | BR/WT | IAC 3 Driver | DC pulse signals: 0.8-11v |
| 49 | YL/BK | IAC 2 Driver | DC pulse signals: 0.8-11v |
| 51 | TN/WT | HO2S-12 (B1 S2) Signal | 0.1-1.1v |
| 52 ('01) | VT/LG | Battery Temperature Sensor | At 86°F: 1.96v |
| 53 | PK/WT | HO2S-22 (B2 S2) Signal | 0.1-1.1v |
| 54, 60 | --- | Not Used | --- |
| 55 | DB/PK | Low Speed Fan Relay | Relay Off: 12v, On: 1v |
| 56 | TN/RD | S/C Vacuum Solenoid | Vacuum Increasing: 1v |
| 57 | GY/RD | IAC 1 Driver | DC pulse signals: 0.8-11v |
| 58 | VT/BK | IAC 4 Driver | DC pulse signals: 0.8-11v |
| 59 | VT/YL | PCI Data Bus (J1850) | Digital Signals: 0-7-0v |
| 61 | VT/WT | 5-Volt Supply | 4.9-5.1v |
| 62 | WT/PK | Brake Switch Signal | Brake Off: 0v, On: 12v |
| 63 | YL/DG | Torque Management Request | Digital Signals |
| 64 | DB/OR | A/C Clutch Relay Control | Relay Off: 12v, On: 1v |
| 65, 75 | PK, LG | SCI Transmit, SCI Receive | 0v |
| 66 | WT/OR | Vehicle Speed Signal | Digital Signal |
| 67 | DB/YL | ASD Relay Control | Relay Off: 12v, On: 1v |
| 68 | PK/BK | EVAP Purge Solenoid Control | PWM Signal: 0-12-0v |
| 69 | DB/LG | High Speed Fan Relay | Relay Off: 12v, On: 1v |
| 70 | DG/LG | EVAP Purge Solenoid Sense | 0-1v |
| 71 | WT/RD | EATX RPM | Digital Signals |
| 72 | OR/RD | LDP Switch Sense | Open: 12v, Closed: 0v |
| 73, 78-79 | --- | Not Used | --- |
| 74 | BR | Fuel Pump Relay Control | Relay Off: 12v, On: 1v |
| 76 | BR/YL | TRS T41 Sense Signal | In P/N: 0v, Others: 5v |
| 77 | WT/DG | LDP Solenoid Control | PWM Signal: 0-12-0v |
| 80 | LG/RD | S/C Vent Solenoid | Vacuum Decreasing: 1v |

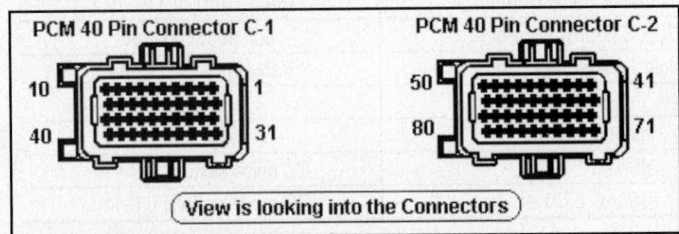

**Pin Connector Graphic**

## 1993-95 Concorde, Intrepid & Vision 3.3L VIN T & Flex-Fuel VIN U (A/T)

| PCM Pin # | Wire Color | Circuit Description (60-Pin) | Value at Hot Idle |
|---|---|---|---|
| 1 | DG/RD | MAP Sensor Signal | 1.5-1.7v |
| 2 | TN/BK | ECT Sensor Signal | At 180°F: 2.80v |
| 3 | RD/WT | Battery Power (Fused B+) | 12-14v |
| 4 | BK/LB | Sensor Ground | <0.050v |
| 5 | BK/WT | Sensor Ground | <0.050v |
| 6 | PK/WT | 5-Volt Supply | 4.9-5.1v |
| 7 | OR | 9-Volt Supply | 7.9-8.1v |
| 8 | YL/DG | Torque Management Request | Digital Signals |
| 9 | DB/WT | Ignition Switch Output | 12-14v |
| 10 | GY/BK | Knock Sensor 2 Signal | 0.080v AC |
| 11 | BK/TN | Power Ground | <0.1v |
| 12 | BK/TN | Power Ground | <0.1v |
| 13 | LB/BR | Injector 4 Driver | 1.0-4.0 ms |
| 14 | YL/WT | Injector 3 Driver | 1.0-4.0 ms |
| 15 | TN | Injector 2 Driver | 1.0-4.0 ms |
| 16 | WT/DB | Injector 1 Driver | 1.0-4.0 ms |
| 17 ('93) | DB/YL | Coil 2 Driver | 5°, at 55 mph: 8° dwell |
| 17 ('94-'95) | WT | Coil 2 Driver | 5°, at 55 mph: 8° dwell |
| 18 ('93) | RD/YL | Coil 3 Driver | 5°, at 55 mph: 8° dwell |
| 18 ('94-'95) | RD | Coil 3 Driver | 5°, at 55 mph: 8° dwell |
| 19 ('93) | GY | Coil 1 Driver | 5°, at 55 mph: 8° dwell |
| 19 ('94-'95) | BK | Coil 1 Driver | 5°, at 55 mph: 8° dwell |
| 20 | DG | Generator Field Driver | Digital Signal: 0-12-0v |
| 21 | BK/RD | IAT Sensor Signal | At 100°F: 1.83v |
| 22 | OR/DB | TP Sensor Signal | 0.6-1.0v |
| 23 | RD/LG | S/C Set Switch Signal | S/C & Set Switch On: 3.8v |
| 24 | LB/DB | CKP Sensor Signal | Digital Signal: 0-5-0v |
| 25 | PK | SCI Transmit | 0v |
| 26 | PK/BR | CCD Bus (+) | Digital Signal: 0-5-0v |
| 27-28 | --- | Not Used | --- |
| 29 | WT/PK | Brake Switch Signal | Brake Off: 0v, On: 12v |
| 30 | BK/LG | PNP Switch Signal | In P/N: 0v, Others: 5v |
| 31 | DB/PK | High Speed Fan Relay | Relay Off: 12v, On: 1v |
| 32 | WT | Low Speed Fan Relay | Relay Off: 12v, On: 1v |
| 33 | TN/RD | S/C Vacuum Solenoid | Vacuum Increasing: 1v |
| 34 | DB/OR | A/C Clutch Relay Control | Relay Off: 12v, On: 1v |
| 35 | GY/YL | EGR Solenoid Control | 12v, at 55 mph: 1v |
| 36 ('93) | PK | Manifold Tuning Valve Control | Valve Off: 12v, On: 1v |
| 37 | --- | Not Used | --- |
| 38 | BR/RD | Injector 5 Driver | 1.0-4.0 ms |
| 39 | GY/RD | AIS/IAC 1 Driver | DC pulse signals: 0.8-11v |
| 40 | BR/WT | AIS/IAC 3 Driver | DC pulse signals: 0.8-11v |

**1993-95 Concorde, Intrepid & Vision 3.3L VIN T & Flex Fuel VIN U (A/T) - Continued**

| PCM Pin # | Wire Color | Circuit Description (60-Pin) | Value at Hot Idle |
|---|---|---|---|
| 41 | BK/DG | HO2S-11 (B1 S1) Signal | 0.1-1.1v |
| 42 | BK/LG | Knock Sensor 1 Signal | 0.080v AC |
| 43, 54, 56 | --- | Not Used | --- |
| 44 | TN/YL | CMP Sensor Signal | Digital Signal: 0-5-0v |
| 45 | LG | SCI Receive | 0v |
| 46 | WT/BK | CCD Bus (-) | <0.050v |
| 47 | WT/OR | Vehicle Speed Signal | Digital Signal |
| 48 | DB | A/C Pressure Switch Signal | A/C On: 0.45-4.85v |
| 49 | TN/WT | HO2S-12 (B1 S2) Signal | 0.1-1.1v |
| 50 (Flex Fuel) | YL/WT | Flex Fuel Sensor Signal | 0.5-4.5v |
| 51 | DB/YL | ASD Relay Output | 12-14v |
| 52 | PK/BK | EVAP Purge Solenoid Control | Solenoid Off: 12v, On: 1v |
| 53 | LG/RD | S/C Vent Solenoid | Vacuum Decreasing: 1v |
| 54 | --- | Not Used | --- |
| 55 | TN/RD | S/C Power Supply | 12-14v |
| 56 | --- | Not Used | --- |
| 57 | DG/OR | ASD Relay Control | Relay Off: 12v, On: 1v |
| 58 | BR/BK | Injector 6 Driver | 1.0-4.0 ms |
| 59 | PK/BK | AIS/IAC 4 Driver | DC pulse signals: 0.8-11v |
| 60 | YL/BK | AIS/IAC 2 Driver | DC pulse signals: 0.8-11v |

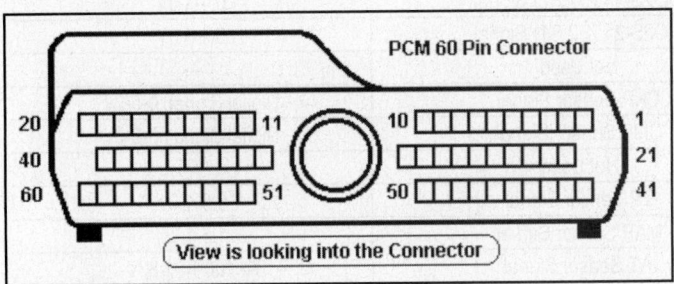

**Pin Connector Graphic**

## 1996-97 Concorde, Intrepid & Vision 3.3L VIN T (A/T) 'C1' Connector

| PCM Pin # | Wire Color | Circuit Description (40-Pin) | Value at Hot Idle |
|---|---|---|---|
| 1 | LG/RD | HO2S-12 (B1 S2) Signal | 0.1-1.1v |
| 2 | RD | Coil 3 Driver | 5°, at 55 mph: 8° dwell |
| 3 | WT | Coil 2 Driver | 5°, at 55 mph: 8° dwell |
| 4 | DG | Generator Field Driver | Digital Signal: 0-12-0v |
| 5 | YL/RD | S/C Power Supply | 12-14v |
| 6 | DG/OR | ASD Relay Output | 12-14v |
| 7 | YL/WT | Injector 3 Driver | 1.0-4.0 ms |
| 8-9 | --- | Not Used | --- |
| 10 | BK/TN | Power Ground | <0.1v |
| 11 | BK | Coil 1 Driver | 5°, at 55 mph: 8° dwell |
| 12 | --- | Not Used | --- |
| 13 | WT/LB | Injector 1 Driver | 1.0-4.0 ms |
| 14 | BR/BK | Injector 6 Driver | 1.0-4.0 ms |
| 15 | BR/RD | Injector 5 Driver | 1.0-4.0 ms |
| 16 | LB/BR | Injector 4 Driver | 1.0-4.0 ms |
| 17 | TN | Injector 2 Driver | 1.0-4.0 ms |
| 18-19 | --- | Not Used | --- |
| 20 | DB/WT | Ignition Switch Output | 12-14v |
| 21-25 | --- | Not Used | --- |
| 26 | TN/BK | ECT Sensor Signal | At 180°F: 2.80v |
| 27-28 | --- | Not Used | --- |
| 29 | TN/WT | HO2S-11 (B1 S1) Signal | 0.1-1.1v |
| 30 | BK/DG | HO2S-21 (B2 S1) Signal | 0.1-1.1v |
| 31 | --- | Not Used | --- |
| 32 | LB/DB | CKP Sensor Signal | Digital Signal: 0-5-0v |
| 33 | TN/YL | CMP Sensor Signal | Digital Signal: 0-5-0v |
| 34 | --- | Not Used | --- |
| 35 | OR/DB | TP Sensor Signal | 0.6-1.0v |
| 36 | DG/RD | MAP Sensor Signal | 1.5-1.7v |
| 37 | BK/RD | IAT Sensor Signal | At 100°F: 1.83v |
| 38 | --- | Not Used | --- |
| 39 | PK/RD | Manifold Tuning Valve Control | Valve Off: 12v, On: 1v |
| 40 | GY/YL | EGR Solenoid Control | 12v, at 55 mph: 1v |

## Standard Colors and Abbreviations

| Abbreviation | Color | Abbreviation | Color | Abbreviation | Color |
|---|---|---|---|---|---|
| BK | Black | GY | Gray | RD | Red |
| BL | Blue | GN | Green | TN | Tan |
| BR | Brown | LG | Light Green | VT | Violet |
| DB | Dark Blue | OR | Orange | WT | White |
| DG | Dark Green | PK | Pink | YL | Yellow |

**1996-97 Concorde, Intrepid & Vision 3.3L VIN T (A/T) 'C2' Connector**

| PCM Pin # | Wire Color | Circuit Description (40-Pin) | Value at Hot Idle |
|---|---|---|---|
| 41 | RD/LG | S/C Set Switch Signal | S/C & Set Switch On: 3.8v |
| 42 | DB | A/C Pressure Switch Signal | A/C On: 0.45-4.85v |
| 43 | BK/LB | Sensor Ground | <0.050v |
| 44 | OR | 8-Volt Supply | 7.9-8.1v |
| 45 | --- | Not Used | --- |
| 46 | RD/WT | Battery Power (Fused B+) | 12-14v |
| 47 | BK/WT | Power Ground | <0.1v |
| 48 | BR/WT | IAC 1 Driver | DC pulse signals: 0.8-11v |
| 49 | YL/BK | IAC 2 Driver | DC pulse signals: 0.8-11v |
| 50 | BK/TN | Power Ground | <0.1v |
| 51 | TN/WT | HO2S-22 (B2 S2) Signal | 0.1-1.1v |
| 52-54 | --- | Not Used | --- |
| 55 | WT | Low Speed Fan Relay | Relay Off: 12v, On: 1v |
| 56 | --- | Not Used | --- |
| 57 | GY/RD | IAC 3 Driver | DC pulse signals: 0.8-11v |
| 58 | PK/GY | IAC 4 Driver | DC pulse signals: 0.8-11v |
| 59 | PK/BR | CCD Bus (+) | Digital Signal: 0-5-0v |
| 60 | WT/BK | CCD Bus (-) | <0.050v |
| 61 | PK/WT | 5-Volt Supply | 4.9-5.1v |
| 62 | WT/PK | Brake Switch Signal | Brake Off: 0v, On: 12v |
| 63 | YL/DG | Torque Management Request | Digital Signals |
| 64 | DB/OR | A/C Clutch Relay Control | Relay Off: 12v, On: 1v |
| 65 | PK | SCI Transmit | 0v |
| 66 | WT/OR | Vehicle Speed Signal | Digital: 0-8-0-8v |
| 67 | DB/YL | ASD Relay Control | Relay Off: 12v, On: 1v |
| 68 | PK/BK | EVAP Purge Solenoid Control | PWM Signal: 0-12-0v |
| 69 | DB/PK | High Speed Fan Relay | Relay Off: 12v, On: 1v |
| 70-73 | --- | Not Used | --- |
| 74 | BR | Fuel Pump Relay Control | Relay Off: 12v, On: 1v |
| 75 | LG | SCI Receive | 0v |
| 76 | BK/DG | PNP Switch Signal | In P/N: 0v, Others: 5v |
| 77 | --- | Not Used | --- |
| 78 | TN/RD | S/C Vacuum Solenoid | Vacuum Increasing: 1v |
| 79 | --- | Not Used | --- |
| 80 | LG/RD | S/C Vent Solenoid | Vacuum Decreasing: 1v |

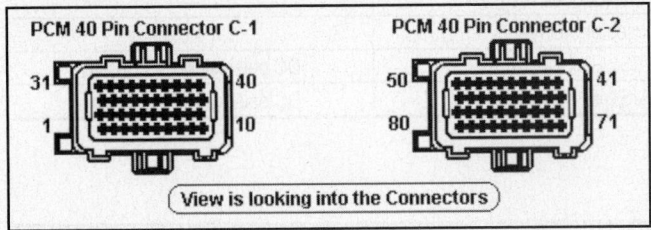

**PCM 40 Pin Connector C-1**     **PCM 40 Pin Connector C-2**

View is looking into the Connectors

**Pin Connector Graphic**

**1993-95 Concorde, Intrepid, LHS, New Yorker & Vision 3.5L VIN F (A/T)**

| PCM Pin # | Wire Color | Circuit Description (60-Pin) | Value at Hot Idle |
|-----------|-----------|-----------------------------|-------------------|
| 1 | DG/RD | MAP Sensor Signal | 1.5-1.7v |
| 2 | TN/BK | ECT Sensor Signal | At 180ºF: 2.80v |
| 3 | RD/WT | Battery Power (Fused B+) | 12-14v |
| 4 | BK/LB | Sensor Ground | <0.050v |
| 5 | BK/WT | Sensor Ground | <0.050v |
| 6 | PK/WT | 5-Volt Supply | 4.9-5.1v |
| 7 | OR | 8-Volt Supply | 7.9-8.1v |
| 8 | YL/DG | Torque Management Request | Digital Signals |
| 9 | DB/WT | Ignition Switch Output | 12-14v |
| 10 | GY/BK | Knock Sensor 2 Signal | 0.080v AC |
| 11 | BK/TN | Power Ground | <0.1v |
| 12 | BK/TN | Power Ground | <0.1v |
| 13 | LB/BR | Injector 4 Driver | 1.0-4.0 ms |
| 14 | YL/WT | Injector 3 Driver | 1.0-4.0 ms |
| 15 | TN | Injector 2 Driver | 1.0-4.0 ms |
| 16 | WT/DB | Injector 1 Driver | 1.0-4.0 ms |
| 17 | DB/YL | Coil 2 Driver | 5º, at 55 mph: 8º dwell |
| 18 | RD/YL | Coil 3 Driver | 5º, at 55 mph: 8º dwell |
| 19 | GY | Coil 1 Driver | 5º, at 55 mph: 8º dwell |
| 20 | DG | Generator Field Driver | Digital Signal: 0-12-0v |
| 21 | BK/RD | IAT Sensor Signal | At 100ºF: 1.83v |
| 22 | OR/DB | TP Sensor Signal | 0.6-1.0v |
| 23 | RD/LG | S/C Set Switch Signal | S/C & Set Switch On: 3.8v |
| 24 | LB/DB | CKP Sensor Signal | Digital Signal: 0-5-0v |
| 25 | PK | SCI Transmit | 0v |
| 26 | PK/BR | CCD Bus (+) | Digital Signal: 0-5-0v |
| 27-28 | --- | Not Used | --- |
| 29 | WT/PK | Brake Switch Signal | Brake Off: 0v, On: 12v |
| 30 | BK/LG | PNP Switch Signal | In P/N: 0v, Others: 5v |
| 31 | DB/PK | High Speed Fan Relay | Relay Off: 12v, On: 1v |
| 32 | WT | Low Speed Fan Relay | Relay Off: 12v, On: 1v |
| 33 | TN/RD | S/C Vacuum Solenoid | Vacuum Increasing: 1v |
| 34 | DB/OR | A/C Clutch Relay Control | Relay Off: 12v, On: 1v |
| 35 | GY/YL | EGR Solenoid Control | 12v, at 55 mph: 1v |
| 36 | PK | Manifold Tuning Valve Control | Valve Off: 12v, On: 1v |
| 37 | --- | Not Used | --- |
| 38 | BR/RD | Injector 5 Driver | 1.0-4.0 ms |
| 39 | GY/RD | AIS/IAC 1 Driver | DC pulse signals: 0.8-11v |
| 40 | BR/WT | AIS/IAC 3 Driver | DC pulse signals: 0.8-11v |

## Standard Colors and Abbreviations

| Abbreviation | Color | Abbreviation | Color | Abbreviation | Color |
|--------------|-------|--------------|-------|--------------|-------|
| BK | Black | GY | Gray | RD | Red |
| BL | Blue | GN | Green | TN | Tan |
| BR | Brown | LG | Light Green | VT | Violet |
| DB | Dark Blue | OR | Orange | WT | White |
| DG | Dark Green | PK | Pink | YL | Yellow |

**1993-95 Concorde, Intrepid, LHS, New Yorker & Vision 3.5L VIN F (A/T) - Continued**

| PCM Pin # | Wire Color | Circuit Description (60-Pin) | Value at Hot Idle |
|---|---|---|---|
| 41 | BK/DG | HO2S-11 (B1 S1) Signal | 0.1-1.1v |
| 42 | BK/LG | Knock Sensor 1 Signal | 0.080v AC |
| 43 | --- | Not Used | --- |
| 44 | TN/YL | CMP Sensor Signal | Digital Signal: 0-5-0v |
| 45 | LG | SCI Receive | 0v |
| 46 | WT/BK | CCD Bus (-) | <0.050v |
| 47 | WT/OR | Vehicle Speed Signal | Digital Signal |
| 48 | DB | A/C Pressure Switch Signal | A/C On: 0.45-4.85v |
| 49 | TN/WT | HO2S-12 (B1 S2) Signal | 0.1-1.1v |
| 50 | YL/WT | Flex Fuel Sensor Signal | 0.5-4.5v |
| 51 | DB/YL | ASD Relay Output | 12-14v |
| 52 | PK/BK | EVAP Purge Solenoid Control | Solenoid Off: 12v, On: 1v |
| 53 | LG/RD | S/C Vent Solenoid | Vacuum Decreasing: 1v |
| 54 | --- | Not Used | --- |
| 55 | TN/RD | S/C Power Supply | 12-14v |
| 56 | --- | Not Used | --- |
| 57 | DG/OR | ASD Relay Control | Relay Off: 12v, On: 1v |
| 58 | BR/BK | Injector 6 Driver | 1.0-4.0 ms |
| 59 | PK/BK | IAC 4 Driver | DC pulse signals: 0.8-11v |
| 60 | YL/BK | IAC 2 Driver | DC pulse signals: 0.8-11v |

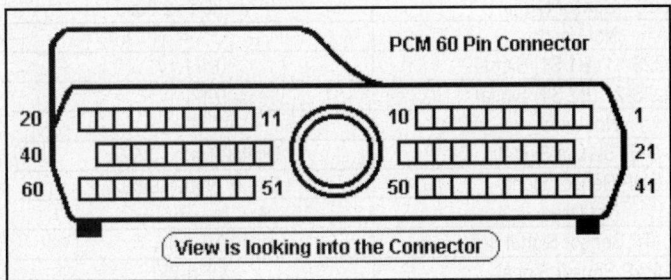

**Pin Connector Graphic**

## 1996-97 Concorde, Intrepid, LHS, New Yorker & Vision 3.5L VIN F (A/T) 'C1' Connector

| PCM Pin # | Wire Color | Circuit Description (40-Pin) | Value at Hot Idle |
|---|---|---|---|
| 1 | LG/RD | HO2S-12 (B1 S2) Signal | 0.1-1.1v |
| 2 | RD | Coil 3 Driver | 5°, at 55 mph: 8° dwell |
| 3 | WT | Coil 2 Driver | 5°, at 55 mph: 8° dwell |
| 4 | DG | Generator Field Driver | Digital Signal: 0-12-0v |
| 5 | YL/RD | S/C Power Supply | 12-14v |
| 6 | DG/OR | ASD Relay Output | 12-14v |
| 7 | YL/WT | Injector 3 Driver | 1.0-4.0 ms |
| 8-9 | --- | Not Used | --- |
| 10 | BK/TN | Power Ground | <0.1v |
| 11 | BK | Coil 1 Driver | 5°, at 55 mph: 8° dwell |
| 12 | --- | Not Used | --- |
| 13 | WT/LB | Injector 1 Driver | 1.0-4.0 ms |
| 14 | BR/BK | Injector 6 Driver | 1.0-4.0 ms |
| 15 | BR/RD | Injector 5 Driver | 1.0-4.0 ms |
| 16 | LB/BR | Injector 4 Driver | 1.0-4.0 ms |
| 17 | TN | Injector 2 Driver | 1.0-4.0 ms |
| 18-19 | --- | Not Used | --- |
| 20 | DB/WT | Ignition Switch Output | 12-14v |
| 21-23 | --- | Not Used | --- |
| 24 | BK/LG | Knock Sensor 1 Signal | 0.080v AC |
| 25 | GY/BK | Knock Sensor 2 Signal | 0.080v AC |
| 26 | TN/BK | ECT Sensor Signal | At 180°F: 2.80v |
| 27-28 | --- | Not Used | --- |
| 29 | TN/WT | HO2S-11 (B1 S1) Signal | 0.1-1.1v |
| 30 | BK/DG | HO2S-21 (B2 S1) Signal | 0.1-1.1v |
| 31 | --- | Not Used | --- |
| 32 | LB/DB | CKP Sensor Signal | Digital Signal: 0-5-0v |
| 33 | TN/YL | CMP Sensor Signal | Digital Signal: 0-5-0v |
| 34 | --- | Not Used | --- |
| 35 | OR/DB | TP Sensor Signal | 0.6-1.0v |
| 36 | DG/RD | MAP Sensor Signal | 1.5-1.7v |
| 37 | BK/RD | IAT Sensor Signal | At 100°F: 1.83v |
| 38 | --- | Not Used | --- |
| 39 | PK/RD | Manifold Tuning Valve Control | Valve Off: 12v, On: 1v |
| 40 | GY/YL | EGR Solenoid Control | 12v, at 55 mph: 1v |

## Standard Colors and Abbreviations

| Abbreviation | Color | Abbreviation | Color | Abbreviation | Color |
|---|---|---|---|---|---|
| BK | Black | GY | Gray | RD | Red |
| BL | Blue | GN | Green | TN | Tan |
| BR | Brown | LG | Light Green | VT | Violet |
| DB | Dark Blue | OR | Orange | WT | White |
| DG | Dark Green | PK | Pink | YL | Yellow |

**1996-97 Concorde, Intrepid, LHS, New Yorker & Vision 3.5L VIN F (A/T) 'C2' Connector**

| PCM Pin # | Wire Color | Circuit Description (40-Pin) | Value at Hot Idle |
|---|---|---|---|
| 41 | RD/LG | S/C Set Switch Signal | S/C & Set Switch On: 3.8v |
| 42 | DB | A/C Pressure Switch Signal | A/C On: 0.45-4.85v |
| 43 | BK/LB | Sensor Ground | <0.050v |
| 44 | OR | 8-Volt Supply | 7.9-8.1v |
| 45 | --- | Not Used | --- |
| 46 | RD/WT | Battery Power (Fused B+) | 12-14v |
| 47 | BK/WT | Power Ground | <0.1v |
| 48 | BR/WT | IAC 1 Driver | DC pulse signals: 0.8-11v |
| 49 | YL/BK | IAC 2 Driver | DC pulse signals: 0.8-11v |
| 50 | BK/TN | Power Ground | <0.1v |
| 51 | TN/WT | HO2S-22 (B2 S2) Signal | 0.1-1.1v |
| 52-54 | --- | Not Used | --- |
| 55 | WT | Low Speed Fan Relay | Relay Off: 12v, On: 1v |
| 56 | --- | Not Used | --- |
| 57 | GY/RD | IAC 3 Driver | DC pulse signals: 0.8-11v |
| 58 | PK/GY | IAC 4 Driver | DC pulse signals: 0.8-11v |
| 59 | PK/BR | CCD Bus (+) | Digital Signal: 0-5-0v |
| 60 | WT/BK | CCD Bus (-) | <0.050v |
| 61 | PK/WT | 5-Volt Supply | 4.9-5.1v |
| 62 | WT/PK | Brake Switch Signal | Brake Off: 0v, On: 12v |
| 63 | YL/DG | Torque Management Request | Digital Signals |
| 64 | DB/OR | A/C Clutch Relay Control | Relay Off: 12v, On: 1v |
| 65 | PK | SCI Transmit | 0v |
| 66 | WT/OR | Vehicle Speed Signal | Digital: 0-8-0-8v |
| 67 | DB/YL | ASD Relay Control | Relay Off: 12v, On: 1v |
| 68 | PK/BK | EVAP Purge Solenoid Control | PWM Signal: 0-12-0v |
| 69 | DB/PK | High Speed Fan Relay | Relay Off: 12v, On: 1v |
| 70-73 | --- | Not Used | --- |
| 74 | BR | Fuel Pump Relay Control | Relay Off: 12v, On: 1v |
| 75 | LG | SCI Receive | 0v |
| 76 | BK/DG | PNP Switch Signal | In P/N: 0v, Others: 5v |
| 77, 79 | --- | Not Used | --- |
| 78 | TN/RD | S/C Vacuum Solenoid | Vacuum Increasing: 1v |
| 79 | --- | Not Used | --- |
| 80 | LG/RD | S/C Vent Solenoid | Vacuum Decreasing: 1v |

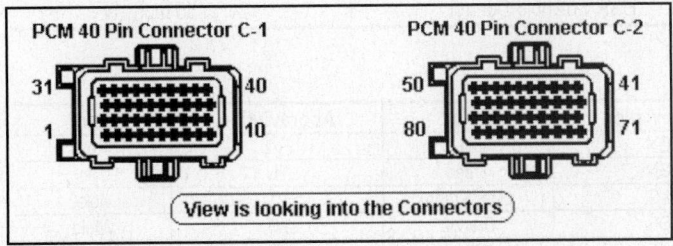

**Pin Connector Graphic**

## 1998-99 Concorde, Intrepid, LHS, New Yorker & Vision 3.5L VIN G (A/T) 'C1' Connector

| PCM Pin # | Wire Color | Circuit Description (40-Pin) | Value at Hot Idle |
|---|---|---|---|
| 1 | TN/LG | Coil 4 Driver | 5°, at 55 mph: 8° dwell |
| 2 | TN/OR | Coil 3 Driver | 5°, at 55 mph: 8° dwell |
| 3 | TN/PK | Coil 2 Driver | 5°, at 55 mph: 8° dwell |
| 4 | TN/LB | Coil 6 Driver | 5°, at 55 mph: 8° dwell |
| 5 | YL/RD | S/C Power Supply | 12-14v |
| 6 | DG/OR | ASD Relay Output | 12-14v |
| 7 | YL/WT | Injector 3 Driver | 1.0-4.0 ms |
| 8 | DG | Generator Field Driver | Digital Signal: 0-12-0v |
| 9, 12 | --- | Not Used | --- |
| 10 | BK/TN | Power Ground | <0.1v |
| 11 | TN/RD | Coil 1 Driver | 5°, at 55 mph: 8° dwell |
| 13 | WT/DB | Injector 1 Driver | 1.0-4.0 ms |
| 14 | BR/DB | Injector 6 Driver | 1.0-4.0 ms |
| 15 | GY | Injector 5 Driver | 1.0-4.0 ms |
| 16 | LB/BR | Injector 4 Driver | 1.0-4.0 ms |
| 17 | TN | Injector 2 Driver | 1.0-4.0 ms |
| 18 | GY/PK | Short Runner Valve Solenoid | Valve On: 1v, Off: 12v |
| 19 | --- | Not Used | --- |
| 20 | DB/WT | Ignition Switch Output | 12-14v |
| 21 | TN/DG | Coil 5 Driver | 5°, at 55 mph: 8° dwell |
| 22-23 | --- | Not Used | --- |
| 24 | DB/LG | Knock Sensor Signal | 0.080v AC |
| 25 | BK/PK | Knock Sensor Ground | <0.050v |
| 26 | TN/BK | ECT Sensor Signal | At 180°F: 2.80v |
| 27 | BK/OR | Oxygen Sensor Ground | <0.050v |
| 28, 38 | --- | Not Used | --- |
| 29 | LG/RD | HO2S-21 (B2 S1) Signal | 0.1-1.1v |
| 30 | BK/DG | HO2S-11 (B1 S1) Signal | 0.1-1.1v |
| 31 | TN/WT | Starter Relay Control | KOEC: 9-11v |
| 32 | GY/BK | CKP Sensor Signal | Digital Signal: 0-5-0v |
| 33 | TN/YL | CMP Sensor Signal | Digital Signal: 0-5-0v |
| 34 | LG/PK | EGR Sensor Signal | 0.6-0.8v |
| 35 | OR/DB | TP Sensor Signal | 0.6-1.0v |
| 36 | DG/RD | MAP Sensor Signal | 1.5-1.7v |
| 37 | BK/RD | IAT Sensor Signal | At 100°F: 1.83v |
| 39 | PK/RD | Manifold Tuning Valve Control | Valve Off: 12v, On: 1v |
| 40 | GY/YL | EGR Solenoid Control | 12v, at 55 mph: 1v |

### Standard Colors and Abbreviations

| Abbreviation | Color | Abbreviation | Color | Abbreviation | Color |
|---|---|---|---|---|---|
| BK | Black | GY | Gray | RD | Red |
| BL | Blue | GN | Green | TN | Tan |
| BR | Brown | LG | Light Green | VT | Violet |
| DB | Dark Blue | OR | Orange | WT | White |
| DG | Dark Green | PK | Pink | YL | Yellow |

**1998-99 Concorde, Intrepid, LHS, New Yorker & Vision 3.5L VIN G (A/T) 'C2' Connector**

| PCM Pin # | Wire Color | Circuit Description (40-Pin) | Value at Hot Idle |
|---|---|---|---|
| 41 | RD/LG | S/C Set Switch Signal | S/C & Set Switch On: 3.8v |
| 42 | DB | A/C Pressure Switch Signal | A/C On: 0.45-4.85v |
| 43 | BK/LB | Sensor Ground | <0.050v |
| 44 | OR | 8-Volt Supply | 7.9-8.1v |
| 45 | DB/LG | PSP Switch Signal | Straight: 0v, Turning: 5v |
| 46 | RD/WT | Battery Power (Fused B+) | 12-14v |
| 47 | BK/LB | Power Ground | <0.1v |
| 48 | BR/WT | IAC 3 Driver | DC pulse signals: 0.8-11v |
| 49 | YL/BK | IAC 2 Driver | DC pulse signals: 0.8-11v |
| 50 | BK/TN | Power Ground | <0.1v |
| 51 | TN/WT | HO2S-12 (B1 S2) Signal | 0.1-1.1v |
| 53 | PK/WT | HO2S-22 (B2 S2) Signal | 0.1-1.1v |
| 55 | DB/PK | Low Speed Fan Relay | Relay Off: 12v, On: 1v |
| 56 | TN/RD | S/C Vacuum Solenoid | Vacuum Increasing: 1v |
| 57 | GY/RD | IAC 1 Driver | DC pulse signals: 0.8-11v |
| 58 | PK/BK | IAC 4 Driver | DC pulse signals: 0.8-11v |
| 59 | YL/PK | PCI Data Bus (J1850) | Digital Signals: 0-7-0v |
| 61 | PK/WT | 5-Volt Supply | 4.9-5.1v |
| 62 | WT/PK | Brake Switch Signal | Brake Off: 0v, On: 12v |
| 63 | YL/DG | Torque Management Request | Digital Signals |
| 64 | DB/OR | A/C Clutch Relay Control | Relay Off: 12v, On: 1v |
| 65, 75 | PK, LG | SCI Transmit, SCI Receive | 0v |
| 66 | WT/OR | Vehicle Speed Signal | Digital Signal |
| 67 | DB/YL | ASD Relay Control | Relay Off: 12v, On: 1v |
| 68 | PK/BK | EVAP Purge Solenoid Control | PWM Signal: 0-12-0v |
| 69 | DB/LG | High Speed Fan Relay | Relay Off: 12v, On: 1v |
| 70 | DG/LG | EVAP Purge Solenoid Sense | 0-1v |
| 71 | WT/RD | EATX RPM | Digital Signals |
| 72 | OR/RD | LDP Switch Sense | Open: 12v, Closed: 0v |
| 73 | --- | Not Used | --- |
| 74 | BR | Fuel Pump Relay Control | Relay Off: 12v, On: 1v |
| 76 | BK/PK | PNP Switch Signal | In P/N: 0v, Others: 5v |
| 77 | WT/DG | LDP Solenoid Control | PWM Signal: 0-12-0v |
| 78-79 | --- | Not Used | --- |
| 80 | LG/RD | S/C Vent Solenoid | Vacuum Decreasing: 1v |

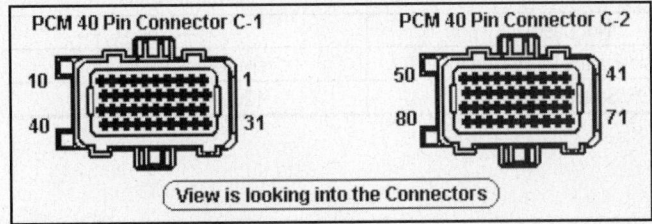

**Pin Connector Graphic**

### 2000-01 LHS 3.5L VIN G (A/T) 'C1' Connector

| PCM Pin # | Wire Color | Circuit Description (40-Pin) | Value at Hot Idle |
|---|---|---|---|
| 1 | TN/LG | Coil 4 Driver | 5°, at 55 mph: 8° dwell |
| 2 | TN/OR | Coil 3 Driver | 5°, at 55 mph: 8° dwell |
| 3 | TN/PK | Coil 2 Driver | 5°, at 55 mph: 8° dwell |
| 4 | TN/LB | Coil 6 Driver | 5°, at 55 mph: 8° dwell |
| 5 | YL/RD | S/C Power Supply | 12-14v |
| 6 | DG/OR | ASD Relay Output | 12-14v |
| 7 | YL/WT | Injector 3 Driver | 1.0-4.0 ms |
| 8 | DG | Generator Field Driver | Digital Signal: 0-12-0v |
| 9 | --- | Not Used | --- |
| 10 | BK/TN | Power Ground | <0.1v |
| 11 | TN/RD | Coil 1 Driver | 5°, at 55 mph: 8° dwell |
| 12 | --- | Not Used | --- |
| 13 | WT/DB | Injector 1 Driver | 1.0-4.0 ms |
| 14 | BR/DB | Injector 6 Driver | 1.0-4.0 ms |
| 15 | GY | Injector 5 Driver | 1.0-4.0 ms |
| 16 | LB/BR | Injector 4 Driver | 1.0-4.0 ms |
| 17 | TN | Injector 2 Driver | 1.0-4.0 ms |
| 18 | GY/PK | Short Runner Valve Solenoid | Valve On: 1v, Off: 12v |
| 19 | --- | Not Used | --- |
| 20 | DB/WT | Ignition Switch Output | 12-14v |
| 21 | TN/DG | Coil 5 Driver | 5°, at 55 mph: 8° dwell |
| 22-23 | --- | Not Used | --- |
| 24 | DB/LG | Knock Sensor Signal | 0.080v AC |
| 25 | BK/VT | Knock Sensor Ground | <0.050v |
| 26 | TN/BK | ECT Sensor Signal | At 180°F: 2.80v |
| 27 | BK/OR | Oxygen Sensor Ground | <0.050v |
| 28 | --- | Not Used | --- |
| 29 | LG/RD | HO2S-21 (B2 S1) Signal | 0.1-1.1v |
| 30 | BK/DG | HO2S-11 (B1 S1) Signal | 0.1-1.1v |
| 31 | TN/WT | Starter Relay Control | KOEC: 9-11v |
| 32 | GY/BK | CKP Sensor Signal | Digital Signal: 0-5-0v |
| 33 | TN/YL | CMP Sensor Signal | Digital Signal: 0-5-0v |
| 34 | LG/PK | EGR Sensor Signal | 0.6-0.8v |
| 35 | OR/DB | TP Sensor Signal | 0.6-1.0v |
| 36 | DG/RD | MAP Sensor Signal | 1.5-1.7v |
| 37 | BK/RD | IAT Sensor Signal | At 100°F: 1.83v |
| 38 | --- | Not Used | --- |
| 39 | VT/RD | Manifold Tuning Valve Control | Valve Off: 12v, On: 1v |
| 40 | GY/YL | EGR Solenoid Control | 12v, at 55 mph: 1v |

### Standard Colors and Abbreviations

| Abbreviation | Color | Abbreviation | Color | Abbreviation | Color |
|---|---|---|---|---|---|
| BK | Black | GY | Gray | RD | Red |
| BL | Blue | GN | Green | TN | Tan |
| BR | Brown | LG | Light Green | VT | Violet |
| DB | Dark Blue | OR | Orange | WT | White |
| DG | Dark Green | PK | Pink | YL | Yellow |

**2000-01 LHS 3.5L VIN G (A/T) 'C2' Connector**

| PCM Pin # | Wire Color | Circuit Description (40-Pin) | Value at Hot Idle |
|---|---|---|---|
| 41 | RD/LG | S/C Set Switch Signal | S/C & Set Switch On: 3.8v |
| 42 | DB | A/C Pressure Switch Signal | A/C On: 0.45-4.85v |
| 43 | BK/LB | Sensor Ground | <0.050v |
| 44 | OR | 8-Volt Supply | 7.9-8.1v |
| 45 | DB/LG | PSP Switch Signal | Straight: 0v, Turning: 5v |
| 46 | RD/WT | Battery Power (Fused B+) | 12-14v |
| 47 | BK | Power Ground | <0.1v |
| 48 | BR/WT | IAC 3 Driver | DC pulse signals: 0.8-11v |
| 49 | YL/BK | IAC 2 Driver | DC pulse signals: 0.8-11v |
| 50 | BK | Power Ground | <0.1v |
| 51 | TN/WT | HO2S-12 (B1 S2) Signal | 0.1-1.1v |
| 52 ('01) | VT/LG | Battery Temperature Sensor | At 86°F: 1.96v |
| 53 | PK/WT | HO2S-22 (B2 S2) Signal | 0.1-1.1v |
| 54 | --- | Not Used | --- |
| 55 | DB/PK | Low Speed Fan Relay | Relay Off: 12v, On: 1v |
| 56 | TN/RD | S/C Vacuum Solenoid | Vacuum Increasing: 1v |
| 57 | GY/RD | IAC 1 Driver | DC pulse signals: 0.8-11v |
| 58 | VT/BK | IAC 4 Driver | DC pulse signals: 0.8-11v |
| 59 | VT/YL | PCI Data Bus (J1850) | Digital Signals: 0-7-0v |
| 60 | --- | Not Used | --- |
| 61 | VT/WT | 5-Volt Supply | 4.9-5.1v |
| 62 | WT/PK | Brake Switch Signal | Brake Off: 0v, On: 12v |
| 63 | YL/DG | Torque Management Request | Digital Signals |
| 64 | DB/OR | A/C Clutch Relay Control | Relay Off: 12v, On: 1v |
| 65 | PK | SCI Transmit | 0v |
| 66 | WT/OR | Vehicle Speed Signal | Digital Signal |
| 67 | DB/YL | ASD Relay Control | Relay Off: 12v, On: 1v |
| 68 | PK/BK | EVAP Purge Solenoid Control | PWM Signal: 0-12-0v |
| 69 | DB/LG | High Speed Fan Relay | Relay Off: 12v, On: 1v |
| 70 | DG/LG | EVAP Purge Solenoid Sense | 0-1v |
| 71 | WT/RD | EATX RPM | Digital Signals |
| 72 | OR/RD | LDP Switch Sense | Open: 12v, Closed: 0v |
| 73 | --- | Not Used | --- |
| 74 | BR | Fuel Pump Relay Control | Relay Off: 12v, On: 1v |
| 75 | LG | SCI Receive | 0v |
| 76 | LG | TRS T41 Sense | In P/N: 0v, Others: 5v |
| 77 | WT/DG | LDP Solenoid Control | PWM Signal: 0-12-0v |
| 78-79 | --- | Not Used | --- |
| 80 | LG/RD | S/C Vent Solenoid | Vacuum Decreasing: 1v |

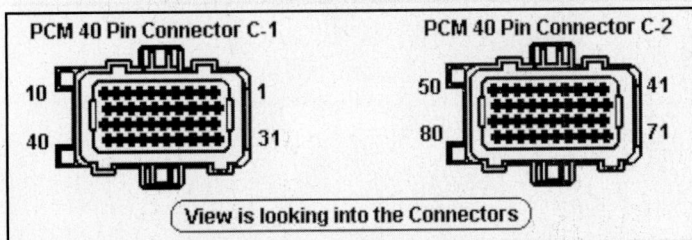

**Pin Connector Graphic**

## 2002-04 Concorde & Intrepid 3.5L VIN G & M (A/T) 'C1' Black Connector

| PCM Pin # | Wire Color | Circuit Description (38-Pin) | Value at Hot Idle |
|---|---|---|---|
| 1-8 | --- | Not Used | --- |
| 9 | BK/TN | Power Ground | <0.1v |
| 10 | --- | Not Used | --- |
| 11 | DB/WT | Fused Ignition Power (B+) | 12-14v |
| 12 | RD/WT | Fused Ignition Power (B+) | 12-14v |
| 13-15 | --- | Not Used | --- |
| 16 | GY/PK | Short Runner Valve Solenoid | Valve On: 1v, Off: 12v |
| 17 | --- | Not Used | --- |
| 18 | BK/TN | Power Ground | <0.1v |
| 19-20 | --- | Not Used | --- |
| 21 | DB | A/C Pressure Sensor | A/C On: 0.45-4.85v |
| 22-24 | --- | Not Used | --- |
| 25 | LG | SCI Receive | 5v |
| 26 | VT/OR | SCI Transmit | 0v |
| 27-28 | --- | Not Used | --- |
| 29 | RD | Fused Power (B+) | 12-14v |
| 30 | YL/BK | Fused Ignition Output (Start) | 10-12v |
| 31 | TN/WT | HO2S-12 (B1 S2) Signal | 0.1-1.1v |
| 32 | DB/DG | HO2S-12 (B1 S2) Ground | <0.050v |
| 33 | PK/WT | HO2S-22 (B2 S2) Signal | 0.1-1.1v |
| 34-35 | --- | Not Used | --- |
| 36 | PK/TN | SCI Transmit (PCM) | 0v |
| 37 | WT/DG | SCI Transmit (PCM) | 0v |
| 38 | VT/YL | PCI Data Bus (J1850) | Digital Signals: 0-7-0v |

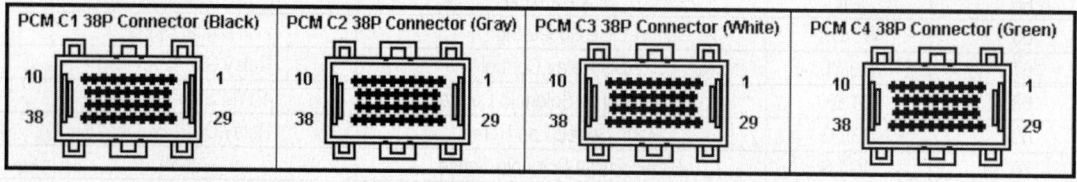

**Pin Connector Graphic**

## Standard Colors and Abbreviations

| Abbreviation | Color | Abbreviation | Color | Abbreviation | Color |
|---|---|---|---|---|---|
| BK | Black | GY | Gray | RD | Red |
| BL | Blue | GN | Green | TN | Tan |
| BR | Brown | LG | Light Green | VT | Violet |
| DB | Dark Blue | OR | Orange | WT | White |
| DG | Dark Green | PK | Pink | YL | Yellow |

**2002-04 Concorde & Intrepid 3.5L VIN G & M (A/T) 'C2' Gray Connector**

| PCM Pin # | Wire Color | Circuit Description (38-Pin) | Value at Hot Idle |
|---|---|---|---|
| 1 | TN/LB | Coil On Plug 6 Driver | 5°, at 55 mph: 8° dwell |
| 2 | TN/DG | Coil On Plug 5 Driver | 5°, at 55 mph: 8° dwell |
| 3 | TN/LG | Coil On Plug 4 Driver | 5°, at 55 mph: 8° dwell |
| 4 | BR/DB | Injector 6 Driver | 1.0-4.0 ms |
| 5 | GY | Injector 5 Driver | 1.0-4.0 ms |
| 6, 15, 26 | --- | Not Used | --- |
| 7 | TN/OR | Coil On Plug 3 Driver | 5°, at 55 mph: 8° dwell |
| 8 | GY/YL | EGR Solenoid Control | 12v, at 55 mph: 1v |
| 9 | TN/PK | Coil On Plug 2 Driver | 5°, at 55 mph: 8° dwell |
| 10 | TN/RD | Coil On Plug 1 Driver | 5°, at 55 mph: 8° dwell |
| 11 | LB/BR | Injector 4 Driver | 1.0-4.0 ms |
| 12 | YL/WT | Injector 3 Driver | 1.0-4.0 ms |
| 13 | TN/WT | Injector 2 Driver | 1.0-4.0 ms |
| 14 | WT/DB | Injector 1 Driver | 1.0-4.0 ms |
| 16 | VT/RD | Manifold Solenoid Control | Valve Off: 12v, On: 1v |
| 17 | BR/WT | HO2S-21 (B2 S1) Heater | Heater Off: 12v, On: 1v |
| 18 | BR/OR | HO2S-11 (B1 S1) Heater | Heater Off: 12v, On: 1v |
| 19 | DG | Generator Field Driver | Digital Signals: 0-12-0v |
| 20 | TN/BK | ECT Sensor Signal | At 180°F: 2.80v |
| 21 | DB | TP Sensor Signal | 0.6-1.0v |
| 22 | LG/PK | EGR Sensor Signal | 0.6-0.8v |
| 23 | DG/RD | MAP Sensor Signal | 1.5-1.7v |
| 24 | BK/VT | Knock Sensor Ground | <0.050v |
| 25 | DB/LG | Knock Sensor Signal | 0.080v AC |
| 27 | BK/LB | Sensor Ground | <0.050v |
| 28 | YL/BK | IAC Motor Sense | 12-14v |
| 29 | VT/WT | 5-Volt Supply | 4.9-5.1v |
| 30 | BK/RD | IAT Sensor Signal | At 100°F: 1.83v |
| 31 | BK/DG | HO2S-11 (B1 S1) Signal | 0.1-1.1v |
| 32 | BR/DG | HO2S-11 (B1 S2) Ground | <0.050v |
| 33 | LG/RD | HO2S-21 (B2 S1) Signal | 0.1-1.1v |
| 34 | TN/YL | CMP Sensor Signal | Digital Signal: 0-5-0v |
| 35 | GY/BK | CKP Sensor Signal | Digital Signal: 0-5-0v |
| 36-37 | --- | Not Used | --- |
| 38 | GY/RD | IAC Motor Driver Control | DC pulses: 0.8-11v |

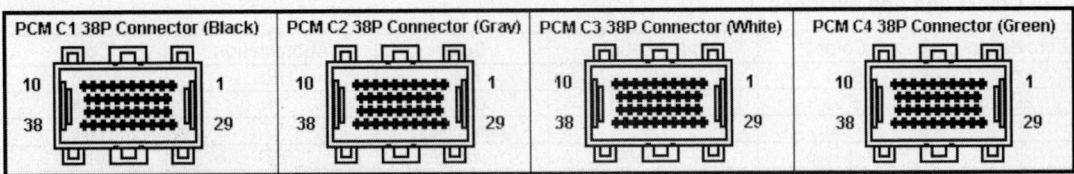

**Pin Connector Graphic**

## 2002-04 Concorde & Intrepid 3.5L VIN G & M (A/T) 'C3' White Connector

| PCM Pin # | Wire Color | Circuit Description (38-Pin) | Value at Hot Idle |
|---|---|---|---|
| 1-2, 13-17 | --- | Not Used | --- |
| 3 | DB/YL | ASD Relay Control | Relay Off: 12v, On: 1v |
| 4 | DB/PK | High Speed Fan Relay | Relay Off: 12v, On: 1v |
| 5 | DB/PK | S/C Vent Solenoid | Vacuum Decreasing: 1v |
| 6 | DB/PK | Low Speed Fan Relay | Relay Off: 12v, On: 1v |
| 7 | YL/RD | S/C Power Supply (B+) | 12-14v |
| 8 | WT/DG | Natural Vacuum Leak Detection Solenoid | Solenoid Off: 12v, On: 1v |
| 9 | BR/VT | HO2S-12 (B1 S2) Heater | Heater Off: 12v, On: 1v |
| 10 | BR/GY | HO2S-22 (B2 S2) Heater | Heater Off: 12v, On: 1v |
| 11 | DB/OR | A/C Clutch Relay Control | Relay Off: 12v, On: 1v |
| 12 | TN/RD | S/C Vacuum Solenoid | Vacuum Increasing: 1v |
| 18, 19 | OR/DG | Fused ASD Relay Power (B+) | 12-14v |
| 20 | PK/BK | EVAP Purge Solenoid Control | PWM Signal: 0-12-0v |
| 21-22, 24-25 | --- | Not Used | --- |
| 23 | WT/PK | Brake Switch Sense | Brake Off: 0v, On: 12v |
| 26 | YL | AutoStick Downshift Switch | Digital Signal: 0v or 12v |
| 27 | LG/RD | AutoStick Upshift Switch | Digital Signal: 0v or 12v |
| 28 | OR/DG | Fused ASD Relay Power (B+) | 12-14v |
| 29 | DG/LG | EVAP Purge Solenoid Sense | <0.1v |
| 30-31, 33, 36 | --- | Not Used | --- |
| 32 | VT/LG | Ambient Temperature Sensor | At 86°F: 1.96v |
| 34 | RD/LG | S/C Set Switch Signal | S/C & Set Switch On: 3.8v |
| 35 | BR | Natural Vacuum Leak Detection Switch | Open: 12v, Closed: 0v |
| 37 | BR | Fuel Pump Relay Control | Relay Off: 12v, On: 1v |
| 38 | TN | Starter Relay Control | Relay Off: 12v, On: 1v |

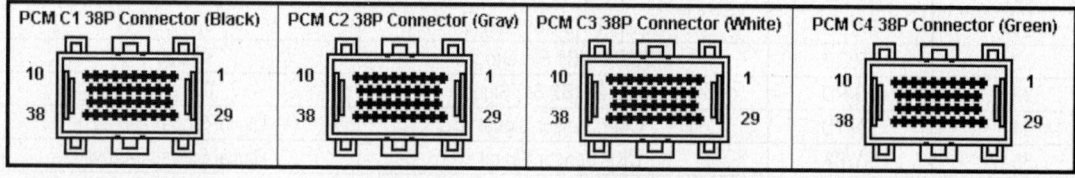

**Pin Connector Graphic**

## Standard Colors and Abbreviations

| Abbreviation | Color | Abbreviation | Color | Abbreviation | Color |
|---|---|---|---|---|---|
| BK | Black | GY | Gray | RD | Red |
| BL | Blue | GN | Green | TN | Tan |
| BR | Brown | LG | Light Green | VT | Violet |
| DB | Dark Blue | OR | Orange | WT | White |
| DG | Dark Green | PK | Pink | YL | Yellow |

**2002-04 Concorde & Intrepid 3.5L VIN G & M (A/T) 'C4' Green Connector**

| PCM Pin # | Wire Color | Circuit Description (38-Pin) | Value at Hot Idle |
|---|---|---|---|
| 1 | --- | Not Used | --- |
| 2 | BR | Overdrive Solenoid Control | Solenoid Off: 12v, On: 1v |
| 3 | PK | Underdrive Solenoid Control | Solenoid Off: 12v, On: 1v |
| 4-5 | --- | Not Used | --- |
| 6 | WT | A/T: 2-4 Solenoid Control | Solenoid Off: 12v, On: 1v |
| 7-9, 11 | --- | Not Used | --- |
| 10 | LB | Low/Reverse Solenoid control | Solenoid Off: 12v, On: 1v |
| 12 | BK/YL | Power Ground | <0.1v |
| 13, 14 | BK/RD | Power Ground | <0.1v |
| 15 | LG/BK | A/T: TRS T1 Sense | In NOL: 0v, Others: 5v |
| 16 | VT | A/T: TRS T3 Sense | In P3L: 0v, Others: 5v |
| 17 | --- | Not Used | --- |
| 18, 19 | LG, RD | Transmission Control Relay Control | Relay Off: 12v, On: 1v |
| 20-21 | --- | Not Used | --- |
| 22 | OR/BK | Overdrive Pressure Switch | In Overdrive: 2-4v |
| 23-26 | --- | Not Used | --- |
| 27 | BK/WT | A/T: TRS T41 Sense | In P/N: 0v, Others: 5v |
| 28 | RD | Trans. Control Relay Output | 12-14v |
| 29 | DG | Low/Reverse Pressure Switch | In Low/Reverse: 2-4v |
| 30 | YL/BK | A/T: 2-4 Pressure Switch | In 2-4 Position: 2-4v |
| 31 | --- | Not Used | --- |
| 32 | LG/WT | A/T: Output Speed Sensor | Moving: AC voltage |
| 33 | RD/BK | A/T: Input Speed Sensor | Moving: AC voltage |
| 34 | DB/BK | A/T: Speed Sensor Ground | <0.050v |
| 35 | VT/PK | Trans. Temperature Sensor | 3.2-3.4v at 104ºF |
| 36 | --- | Not Used | --- |
| 37 | VT/WT | A/T: TRS T42 Sense | In PRNL: 0v, Others 5v |
| 38 | RD | Trans. Control Relay Output | 12-14v |

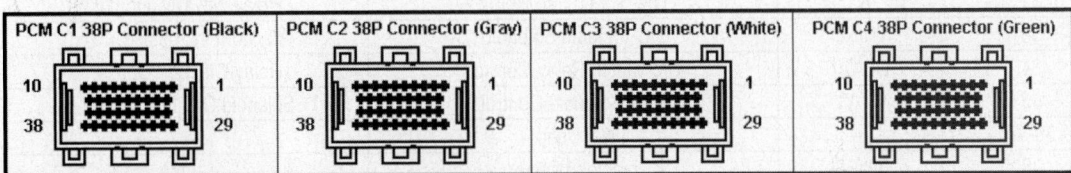

**Pin Connector Graphic**

**Standard Colors and Abbreviations**

| Abbreviation | Color | Abbreviation | Color | Abbreviation | Color |
|---|---|---|---|---|---|
| BK | Black | GY | Gray | RD | Red |
| BL | Blue | GN | Green | TN | Tan |
| BR | Brown | LG | Light Green | VT | Violet |
| DB | Dark Blue | OR | Orange | WT | White |
| DG | Dark Green | PK | Pink | YL | Yellow |

**1994-95 Intrepid 3.3L V6 Flex Fuel VIN U 60-Pin Connector**

| PCM Pin # | Wire Color | Circuit Description (60-Pin) | Value at Hot Idle |
|---|---|---|---|
| 1 | DG/RD | MAP Sensor Signal | 1.5-1.7v |
| 2 | TN/BK | ECT Sensor Signal | At 180ºF: 2.80v |
| 3 | RD/WT | Battery Power (Fused B+) | 12-14v |
| 4 | BK/LB | Sensor Ground | <0.050v |
| 5 | BK/WT | Sensor Ground | <0.050v |
| 6 | PK/WT | 5-Volt Supply | 4.9-5.1v |
| 7 | OR | 9-Volt Supply | 8.9-9.1v |
| 8 | YL/DG | Torque Management Request | Digital Signals |
| 9 | DB/WT | Ignition Switch Output | 12-14v |
| 10 | GY/BK | Knock Sensor 2 Signal | 0.080v AC |
| 11 | BK/TN | Power Ground | <0.1v |
| 12 | BK/TN | Power Ground | <0.1v |
| 13 | LB/BR | Injector 4 Driver | 1-4 ms |
| 14 | YL/WT | Injector 3 Driver | 1-4 ms |
| 15 | TN | Injector 2 Driver | 1-4 ms |
| 16 | WT/DB | Injector 1 Driver | 1-4 ms |
| 17 | DB/YL | Coil 2 Driver | 5º, at 55 mph: 8º dwell |
| 18 | RD/YL | Coil 3 Driver | 5º, at 55 mph: 8º dwell |
| 19 | GY | Coil 1 Driver | 5º, at 55 mph: 8º dwell |
| 20 | DG | Generator Field Control | Digital Signal: 0-12-0v |
| 21 | BK/RD | Air Temperature Sensor | At 100ºF: 2.51v |
| 22 | OR/DB | TP Sensor Signal | 0.6-1.0v |
| 23 | RD/LG | S/C Switch Signal | S/C & Set Switch On: 3.8V |
| 24 | LB/DB | CKP Sensor Signal | Digital Signal: 0-5-0v |
| 25 | PK | SCI Transmit | 0v |
| 26 | PK/BR | CCD Bus (+) | Digital Signal: 0-5-0v |
| 27-28 | --- | Not Used | --- |
| 29 | WT/PK | Brake Switch Signal | Brake Off: 0v, On: 12v |
| 30 | BK/LG | PNP Switch Signal | In P/N: 0v, Others: 5v |
| 31 | DB/PK | High Speed Fan Relay | Relay Off: 12v, On: 1v |
| 32 | WT | Low Speed Fan Relay | Relay Off: 12v, On: 1v |
| 33 | TN/RD | S/C Vacuum Solenoid | Vacuum Increasing: 1v |
| 34 | DB/OR | A/C Clutch Relay Control | Relay Off: 12v, On: 1v |
| 35 | GY/YL | EGR Solenoid Control | Solenoid Off: 12v, On: 1v |
| 36-37 | --- | Not Used | --- |
| 38 | BR/RD | Injector 5 Driver | 1-4 ms |
| 39 | GY/RD | AIS 1 Motor Control | DC pulse signals: 0.8-11v |
| 40 | BR/WT | AIS 3 Motor Control | DC pulse signals: 0.8-11v |

## Standard Colors and Abbreviations

| Abbreviation | Color | Abbreviation | Color | Abbreviation | Color |
|---|---|---|---|---|---|
| BK | Black | GY | Gray | RD | Red |
| BL | Blue | GN | Green | TN | Tan |
| BR | Brown | LG | Light Green | VT | Violet |
| DB | Dark Blue | OR | Orange | WT | White |
| DG | Dark Green | PK | Pink | YL | Yellow |

**1994-95 Intrepid 3.3L V6 Flex Fuel VIN U 60-Pin Connector - Continued**

| PCM Pin # | Wire Color | Circuit Description (60-Pin) | Value at Hot Idle |
|---|---|---|---|
| 41 | BK/DG | HO2S-11 (B1 S1) Signal | 0.1-1.1v |
| 42 | BK/LG | Knock Sensor 1 Signal | 0.080v AC |
| 43 | --- | Not Used | --- |
| 44 | TN/YL | CMP Sensor Signal | Digital Signal: 0-5-0v |
| 45 | LG | SCI Receive | 0v |
| 46 | WT/BK | CCD Bus (-) | <0.050v |
| 47 | WT/OR | Vehicle Speed Signal | Digital Signal |
| 48 | DB | A/C Pressure Switch Signal | A/C On: 0.45-4.85v |
| 49 | TN/WT | HO2S-21 (B2 S1) Signal | 0.1-1.1v |
| 50 | YL/WT | Flex Fuel Sensor Signal | 0.5-4.5v |
| 51 | DB/YL | ASD Relay Control | Relay Off: 12v, On: 1v |
| 52 | PK/BK | EVAP Purge Solenoid Control | Solenoid Off: 12v, On: 1v |
| 53 | LG/RD | S/C Vent Solenoid | Vacuum Decreasing: 1v |
| 54 | --- | Not Used | --- |
| 55 | TN/RD | S/C Power Supply | 12-14v |
| 56 | --- | Not Used | --- |
| 57 | DG/OR | ASD Relay Output | 12-14v |
| 58 | BR/BK | Injector 6 Driver | 1-4 ms |
| 59 | PK/BK | AIS 4 Motor Control | DC pulse signals: 0.8-11v |
| 60 | YL/BK | AIS 2 Motor Control | DC pulse signals: 0.8-11v |

CROSSFIRE

## 2004 Crossfire 3.2L VIN L 'C1' Black Connector

| PCM Pin # | Wire Color | Circuit Description (9-Pin) | Value at Hot Idle |
|---|---|---|---|
| 1 | --- | Not Used | --- |
| 2 | BK/WT | Coil On Plug Driver No. 1 | |
| 3 | BL | Sensor Ground | |
| 4 | GY/WT | Coil On Plug Driver No. 6 | |
| 5 | BL/WT | Coil On Plug Driver No. 4 | |
| 6 | BK | Throttle Control Motor 2 | |
| 7 | DG/WT | Coil On Plug Driver No. 3 | |
| 8 | --- | Not Used | --- |
| 9 | YL/WT | Coil On Plug Driver No. 2 | |

## 2004 Crossfire 3.2L VIN L 'C2' Black Connector

| PCM Pin # | Wire Color | Circuit Description (24-Pin) | Value at Hot Idle |
|---|---|---|---|
| 1-6 | --- | Not Used | --- |
| 7 | GY/WT | EGR Solenoid Control | |
| 8-11 | --- | Not Used | --- |
| 12 | VT/WT | Coil On Plug Driver No. 5 | |
| 13 | GY/BR | Air Pump Switch Over Solenoid Control | |
| 14 | DG/VT | Short Runner Valve Solenoid Control | |
| 15 | BR/GY | Oxygen Sensor 2/1 Heater Control | |
| 16 | --- | Not Used | --- |
| 17 | --- | Not Used | --- |
| 18 | GY/VT | Fuel Injector No. 6 Control | |
| 19 | GY/DG | Fuel Injector No. 3 Control | |
| 20 | --- | Not Used | --- |
| 21 | GY/YL | Fuel Injector No. 5 Control | |
| 22 | GY/RD | Fuel Injector No. 2 Control | |
| 23 | --- | Not Used | --- |
| 24 | BR/BK | Oxygen Sensor 1/1 Heater Control | |

## Standard Colors and Abbreviations

| Abbreviation | Color | Abbreviation | Color | Abbreviation | Color |
|---|---|---|---|---|---|
| BK | Black | GY | Gray | RD | Red |
| BL | Blue | GN | Green | TN | Tan |
| BR | Brown | LG | Light Green | VT | Violet |
| DB | Dark Blue | OR | Orange | WT | White |
| DG | Dark Green | PK | Pink | YL | Yellow |

**2004 Crossfire 3.2L VIN L 'C3' Black Connector**

| PCM Pin # | Wire Color | Circuit Description (52-Pin) | Value at Hot Idle |
|---|---|---|---|
| 1 | — | Not Used | |
| 2 | PK | Camshaft Position Sensor Signal | |
| 3 | BK | Sensor Ground | |
| 4 | VT | Right Knock Sensor Signal | |
| 5 | DG/RD | Engine Coolant Temperature Sensor Signal | |
| 6 | DG | Sensor Ground | |
| 7 | BK/BR | Oxygen Sensor 2/1 Signal | |
| 8 | YL/DG | IAT Sensor Signal | |
| 9 | BR/DG | Sensor Ground | |
| 10 | RD/DG | MAP Sensor Signal | |
| 11 | YL/WT | MAF Sensor Signal | |
| 12 | BR | Sensor Ground | |
| 13 | DG/WT | Camshaft Position Sensor Signal 1 | |
| 14 | --- | Not Used | --- |
| 15 | GY/BL | Oil Sensor Signal | |
| 16 | BK | Sensor Ground | |
| 17 | YL | Left Knock Sensor Signal | |
| 18 | --- | Not Used | --- |
| 19 | GY | Sensor Ground | |
| 20 | WT | Throttle Position Sensor Signal 2 | |
| 21 | YL | Throttle Position Sensor Signal 1 | |
| 22 | VT | 5 Volt Supply | |
| 23 | BK | Oxygen Sensor 1/1 Signal | |
| 24 | --- | Not Used | --- |
| 25 | BR/YL | 5 Volt Supply | |
| 26 | DG | Camshaft Position Sensor Signal 2 | |
| 27 | GY/BK | Fuel Injector No. 1 Control | |
| 28 | GY/BL | Fuel Injector No. 4 Control | |
| 29-38 | --- | Not Used | --- |
| 39 | RD/YL | 5 Volt Supply | |
| 40 | --- | Not Used | --- |
| 41 | --- | Not Used | --- |
| 42 | --- | Not Used | --- |
| 43 | --- | Not Used | --- |
| 44 | GY | Coil On Plug Driver No. 6 | |
| 45 | DG | Coil On Plug Driver No. 3 | |
| 46 | BL | Coil On Plug Driver No. 4 | |
| 47 | BK | Coil On Plug Driver No. 1 | |
| 48 | --- | Not Used | --- |
| 49 | VT | Coil On Plug Driver No. 5 | |
| 50 | YL | Coil On Plug Driver No. 2 | |
| 51 | --- | Not Used | --- |
| 52 | --- | Not Used | --- |

## 2004 Crossfire 3.2L VIN L 'C4' Black Connector

| PCM Pin # | Wire Color | Circuit Description (40-Pin) | Value at Hot Idle |
|:---:|:---:|:---:|:---:|
| 1 | WT | CAN C Bus (+) | |
| 2 | --- | Not Used | --- |
| 3 | GY/YL | Backup Lamp Switch Sense | |
| 4 | RD/BK | Clutch Pedal Position Switch Signal | |
| 5 | BR | Sensor Ground | |
| 6 | BK | Sensor Ground | |
| 7 | DG | Accelerator Pedal Position Sensor Signal No. 2 | |
| 8 | GY | Accelerator Pedal Position Sensor Signal No. 1 | |
| 9 | --- | Not Used | --- |
| 10 | BR/YL | Sensor Ground | |
| 11 | DG | CAN C Bus (-) | |
| 12 | DG/BK | Enhanced Accident Report Driver | |
| 13 | --- | Not Used | --- |
| 14 | BK | Kickdown Switch Signal | |
| 15 | YL/BK | Oxygen Sensor 2/2 Signal | |
| 16 | YL/DG | Oxygen Sensor 1/2 Signal | |
| 17 | DG | Fuel Tank Pressure Sensor Signal | |
| 18 | OR | Sensor Ground | |
| 19 | YL/WT | Accelerator Pedal Position Sensor 5 Volt Supply | |
| 20 | --- | Not Used | --- |
| 21 | --- | Not Used | --- |
| 22 | GY | Off Signal | |
| 23 | DG/RD | Accel/Set Signal | |
| 24 | --- | Not Used | --- |
| 25 | VT | Fused Ignition Switch Output | |
| 26 | DG/BL | On Signal | |
| 27 | --- | Not Used | --- |
| 28 | YL/DG | Decel/Set Signal | |
| 29 | BL | SCI Transmit | |
| 30 | BL/RD | Resume Signal | |
| 31 | --- | Not Used | --- |
| 32 | DG/YL | SCI Receive | |
| 33 | --- | Not Used | --- |
| 34 | --- | Not Used | --- |
| 35 | PK/RD | Sensor Ground | |
| 36 | BL/RD | Fuel Pump Relay Control | |
| 37 | BR/BL | Charcoal Canister Shut Off Valve Control | |
| 38 | BR/VT | Air Pump Relay Control | |
| 39 | GY | Radiator Fan Control | |
| 40 | BK/RD | 5 Volt Supply | |

## Standard Colors and Abbreviations

| Abbreviation | Color | Abbreviation | Color | Abbreviation | Color |
|:---:|:---:|:---:|:---:|:---:|:---:|
| BK | Black | GY | Gray | RD | Red |
| BL | Blue | GN | Green | TN | Tan |
| BR | Brown | LG | Light Green | VT | Violet |
| DB | Dark Blue | OR | Orange | WT | White |
| DG | Dark Green | PK | Pink | YL | Yellow |

## 2004 Crossfire 3.2L VIN L 'C5' White Connector

| PCM Pin # | Wire Color | Circuit Description (9-Pin) | Value at Hot Idle |
|---|---|---|---|
| 1 | BR/DG | Oxygen Sensor 1/2 Heater Control | |
| 2 | BR/RD | EVAP Purge Solenoid Control | |
| 3 | BK/DG | Fused B+ | |
| 4 | BR/WT | Oxygen Sensor 2/2 Heater Control | |
| 5 | BR | Ground | |
| 6 | BR | Ground | |
| 7 | BR | Ground | |
| 8 | BR | Ground | |
| 9 | RD/DG | Engine Control Relay Output | |

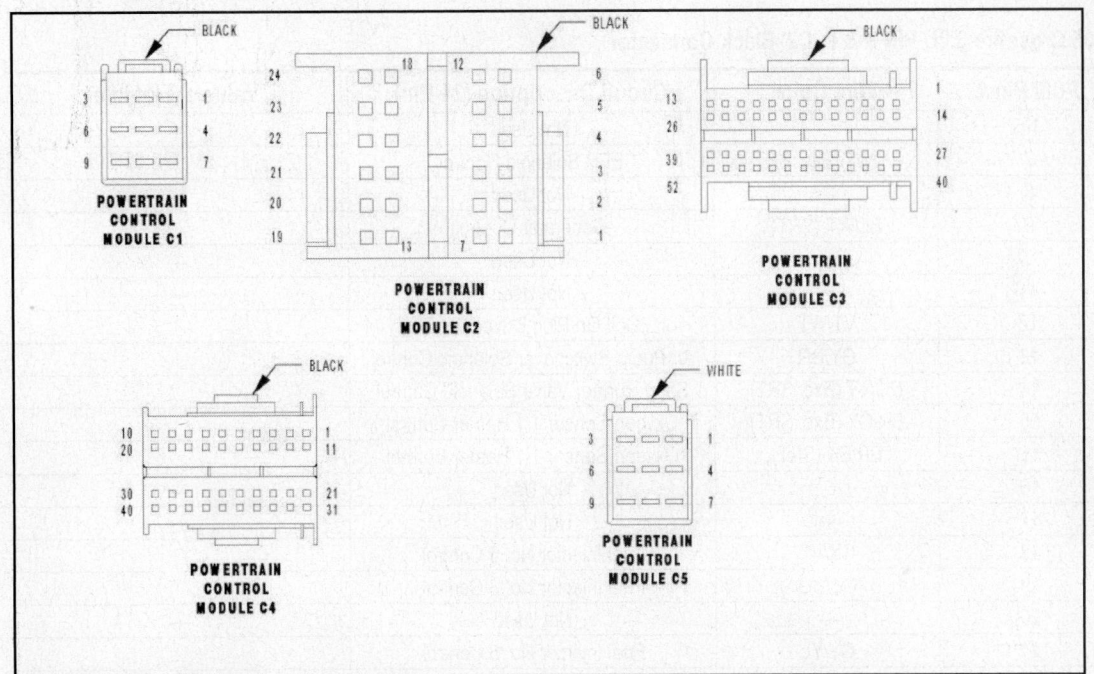

**Pin Connector Graphic**

## Standard Colors and Abbreviations

| Abbreviation | Color | Abbreviation | Color | Abbreviation | Color |
|---|---|---|---|---|---|
| BK | Black | GY | Gray | RD | Red |
| BL | Blue | GN | Green | TN | Tan |
| BR | Brown | LG | Light Green | VT | Violet |
| DB | Dark Blue | OR | Orange | WT | White |
| DG | Dark Green | PK | Pink | YL | Yellow |

## 2005 Crossfire 3.2L VIN N & L 'C1' Black Connector

| PCM Pin # | Wire Color | Circuit Description (9-Pin) | Value at Hot Idle |
|---|---|---|---|
| 1 | BK/PK (SRT) | Supercharger Clutch Control | |
| 2 | BK/WT | Coil On Plug Driver No. 1 | |
| 3 | BL | Throttle Control Motor 1 | |
| 4 | GY/WT | Coil On Plug Driver No. 6 | |
| 5 | BL/WT | Coil On Plug Driver No. 4 | |
| 6 | BK | Throttle Control Motor 2 | |
| 7 | DG/WT | Coil On Plug Driver No. 3 | |
| 8 | --- | Not Used | --- |
| 9 | YL/WT | Coil On Plug Driver No. 2 | |

## 2005 Crossfire 3.2L VIN N & L 'C2' Black Connector

| PCM Pin # | Wire Color | Circuit Description (24-Pin) | Value at Hot Idle |
|---|---|---|---|
| 1-6 | --- | Not Used | --- |
| 7 | GY/WT (Exc. SRT) | EFR Solenoid Control | |
| 8 | --- | Not Used | |
| 9 | BL/WT (SRT) | Generator Control | |
| 10 | --- | Not Used | --- |
| 11 | --- | Not Used | --- |
| 12 | VT/WT | Coil On Plug Driver No. 5 | |
| 13 | GY/BR | Air Pump Switchover Solenoid Control | |
| 14 | DG/VT (Exc. SRT) | Short Runner Valve Solenoid Control | |
| 15 | BR/GY (Exc. SRT) | Oxygen Sensor 1/1 Heater Control | |
| 15 | BR/BK (SRT) | Oxygen Sensor 1/1 Heater Control | |
| 16 | --- | Not Used | --- |
| 17 | --- | Not Used | --- |
| 18 | GY/VT | Fuel Injector No. 6 Control | |
| 19 | GY/DG | Fuel Injector No. 3 Control | |
| 20 | --- | Not Used | --- |
| 21 | GY/YL | Fuel Injector No. 5 Control | |
| 22 | GY/RD | Fuel Injector No. 2 Control | |
| 23 | --- | Not Used | --- |
| 24 | BR/BK (Exc. SRT) | Oxygen Sensor 2/1 Heater Control | |
| 24 | BR/GY (SRT) | Oxygen Sensor 2/1 Heater Control | |

## Standard Colors and Abbreviations

| Abbreviation | Color | Abbreviation | Color | Abbreviation | Color |
|---|---|---|---|---|---|
| BK | Black | GY | Gray | RD | Red |
| BL | Blue | GN | Green | TN | Tan |
| BR | Brown | LG | Light Green | VT | Violet |
| DB | Dark Blue | OR | Orange | WT | White |
| DG | Dark Green | PK | Pink | YL | Yellow |

**2005 Crossfire 3.2L VIN N & L 'C3' Black Connector**

| PCM Pin # | Wire Color | Circuit Description (52-Pin) | Value at Hot Idle |
|---|---|---|---|
| 1 | --- | Not Used | --- |
| 2 | PK | Camshaft Position Sensor Signal | |
| 3 | BK | Sensor Ground | |
| 4 | VT | Right Knock Sensor Signal | |
| 5 | DG/RD | Engine Coolant Temperature Sensor Signal | |
| 6 | DG | Sensor Ground | |
| 7 | BK | Oxygen Sensor 1/1 Signal | |
| 8 | YL/DG | IAT Sensor Signal | |
| 9 | BR/DG | Sensor Ground | |
| 10 | RD/DG (Exc. SRT) | MAP Sensor Signal | |
| 10 | RD/BK (SRT) | MAP Sensor Signal | |
| 11 | YL/WT (Exc. SRT) | MAF Sensor Signal | |
| 12 | BR (Exc. SRT) | Sensor Ground | |
| 13 | DG/WT | Crankshaft Position Sensor Signal 1 | |
| 14 | --- | Not Used | --- |
| 15 | GY/BL | Oil Sensor Signal | |
| 16 | BK | Sensor Ground | |
| 17 | YL | Left Knock Sensor Signal | |
| 18 | --- | Not Used | --- |
| 19 | GY | Sensor Ground | |
| 20 | WT | Throttle Position Sensor Signal 2 | |
| 21 | YL | Throttle Position Sensor Signal 1 | |
| 22 | VT | 5 Volt Supply | |
| 23 | BK (Exc. SRT) | Oxygen Sensor 2/1 Signal | |
| 23 | BK/BR (SRT) | Oxygen Sensor 2/1 Signal | |
| 24 | --- | Not Used | --- |
| 25 | BR/YL (Exc. SRT) | 5 Volt Supply | |
| 26 | DG | Crankshaft Position Sensor Signal 2 | |
| 27 | GY/BK | Fuel Injector No. 1 Control | |
| 28 | GY/BL | Fuel Injector No. 4 Control | |
| 29-38 | --- | Not Used | --- |
| 39 | RD/YL | 5 Volt Supply | |
| 40-43 | --- | Not Used | --- |
| 44 | GY | Coil On Plug Driver No. 6 | |
| 45 | DG | Coil On Plug Driver No. 3 | |
| 46 | BL | Coil On Plug Driver No. 4 | |
| 47 | BK | Coil On Plug Driver No. 1 | |
| 48 | --- | Not Used | --- |
| 49 | VT | Coil On Plug Driver No. 5 | |
| 50 | YL | Coil On Plug Driver No. 2 | |
| 51 | --- | Not Used | --- |
| 52 | --- | Not Used | --- |

## 2005 Crossfire 3.2L VIN N & L 'C4' Black Connector

| PCM Pin # | Wire Color | Circuit Description (40-Pin) | Value at Hot Idle |
|---|---|---|---|
| 1 | WT | CAN C Bus (+) | |
| 2 | --- | Not Used | --- |
| 3 | GY/YL (Exc. SRT) | Backup Lamp Switch Output | |
| 4 | RD/BK (M/T) | Clutch Pedal Position Switch Signal | |
| 5 | BR | Sensor Ground | |
| 6 | BK (MK) | Sensor Ground | |
| 6 | BK (MK25) | Sensor Ground | |
| 7 | DG (MK) | Accelerator Pedal Position Sensor Signal No. 2 | |
| 7 | DG (MK25) | Accelerator Pedal Position Sensor Signal No. 2 | |
| 8 | GY (MK) | Accelerator Pedal Position Sensor Signal No. 1 | |
| 8 | GY (MK25) | Accelerator Pedal Position Sensor Signal No. 1 | |
| 9 | --- | Not Used | --- |
| 10 | BR/YL | Sensor Ground | |
| 11 | DG | CAN C Bus (-) | |
| 12 | DG/BK | Enhanced Accident Report Driver | |
| 13 | --- | Not Used | --- |
| 14 | BK (A/T) | Kickdown Switch Signal | |
| 15 | YL/BK | Oxygen Sensor 1/2 Signal | |
| 16 | YL/DG | Oxygen Sensor 2/2 Signal | |
| 17 | DG | Fuel Tank Pressure Sensor Signal | |
| 18 | OR (MK) | Sensor Ground | |
| 18 | OR (MK25) | Sensor Ground | |
| 19 | YL/WT (MK) | Accelerator Pedal Position Sensor 5 Volt Supply | |
| 19 | YL/WT (MK25) | Accelerator Pedal Position Sensor 5 Volt Supply | |
| 20 | --- | Not Used | --- |
| 21 | --- | Not Used | --- |
| 22 | GY | Off Signal | |
| 23 | DG/RD | Accel/Set Signal | |
| 24 | --- | Not Used | --- |
| 25 | VT (A/T) | Fused Ignition Switch Output | |
| 25 | VT (M/T) | Clutch Interlock Switch Output | |
| 26 | DG/BL | On Signal | |
| 27 | --- | Not Used | --- |
| 28 | YL/DG | Decel/Set Signal | |
| 29 | BL | SCI Transmit | |
| 30 | BL/RD | Resume Signal | |
| 31 | --- | Not Used | --- |
| 32 | DG/YL | SCI Receive | |
| 33 | --- | Not Used | --- |
| 34 | GY/BL (SRT) | Charge Air Cooler Circulation Pump Control | |
| 35 | PK/RD | Sensor Ground | |
| 36 | BL/RD | Fuel Pump Relay Control | |
| 37 | BR/BL | Charcoal Canister Shutoff Valve Control | |
| 38 | BR/VT | Air Pump Relay Control | |
| 39 | GY | Radiator Fan Control | |
| 40 | BK/RD | 5 Volt Supply | |

**2005 Crossfire 3.2L VIN N & L 'C5' Black Connector**

| PCM Pin # | Wire Color | Circuit Description (9-Pin) | Value at Hot Idle |
|---|---|---|---|
| 1 | BR/DG | Oxygen Sensor 2/2 Heater Control | |
| 2 | BR/RD | EVAP Purge Solenoid Control | |
| 3 | BK/DG | Fused B (+) | |
| 4 | BR/WT | Oxygen Sensor 1/2 Heater Control | |
| 5 | BR | Ground | |
| 6 | BR | Ground | |
| 7 | BR | Ground | |
| 8 | BR | Ground | |
| 9 | RD/DG | Fused Engine Control Relay Output | |

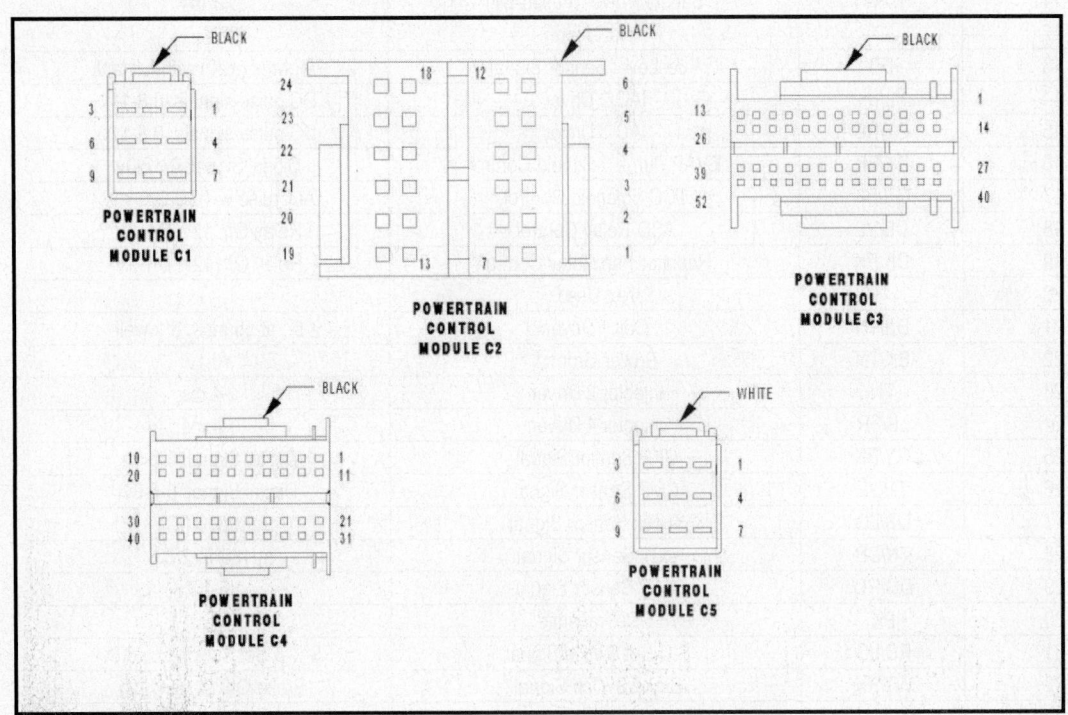

**Pin Connector Graphic**

**Standard Colors and Abbreviations**

| Abbreviation | Color | Abbreviation | Color | Abbreviation | Color |
|---|---|---|---|---|---|
| BK | Black | GY | Gray | RD | Red |
| BL | Blue | GN | Green | TN | Tan |
| BR | Brown | LG | Light Green | VT | Violet |
| DB | Dark Blue | OR | Orange | WT | White |
| DG | Dark Green | PK | Pink | YL | Yellow |

NEON

### 1995 Neon 2.0L VIN C 60-Pin Connector

| PCM Pin # | Wire Color | Circuit Description (60-Pin) | Value at Hot Idle |
|---|---|---|---|
| 1 | BK/GY | Coil 2 Driver | 5º, at 55 mph: 8º dwell |
| 2 | BK/TN | Power Ground | <0.1v |
| 3 | YL/WT | Injector 3 Driver | 1-4 ms |
| 4 | WT/DB | Injector 1 Driver | 1-4 ms |
| 5 | WT/OR | Vehicle Speed Signal | Digital Signal |
| 6 | BK/RD | IAT Sensor Signal | At 100ºF: 1.83v |
| 7 | TN/WT | HO2S-12 (B1 S2) Signal | 0.1-1.1v |
| 8 | BK/DG | HO2S-11 (B1 S1) Signal | 0.1-1.1v |
| 9 | LG | SCI Receive | 0v |
| 10 | OR/DB | TP Sensor Signal | 0.6-1.0v |
| 11 | RD/WT | Battery Power (Fused B+) | 12-14v |
| 12 | --- | Not Used | --- |
| 13 | DB | Fuel Level Sensor Signal | 70 ohms (±20) with full tank |
| 14 | YL/BK | IAC 2 Driver | DC pulse signals: 0.8-11v |
| 15 | GY/RD | IAC 3 Driver | DC pulse signals: 0.8-11v |
| 16 | PK/BK | EVAP Purge Solenoid Control | Digital Signal: 0-12-0v |
| 17 | OR/BK | TCC Solenoid Control | At Cruise w/TCC On: <1v |
| 18 | DB/YL | ASD Relay Control | Relay Off: 12v, On: 1v |
| 19 | DB/PK | Radiator Fan Relay Control | Relay Off: 12v, On: 1v |
| 20 | --- | Not Used | --- |
| 21 | DB/TN | Coil 1 Driver | 5º, at 55 mph: 8º dwell |
| 22 | BK/TN | Power Ground | <0.1v |
| 23 | TN | Injector 2 Driver | 1-4 ms |
| 24 | LB/BR | Injector 4 Driver | 1-4 ms |
| 25 | GY/BK | CKP Sensor Signal | Digital Signal: 0-5-0v |
| 26 | TN/YL | CMP Sensor Signal | Digital Signal: 0-5-0v |
| 27 | DB/LG | Knock Sensor Signal | 0.080v AC |
| 28 | TN/DB | ECT Sensor Signal | At 180ºF: 2.80v |
| 29 | DG/RD | MAP Sensor Signal | 1.5-1.7v |
| 30 | PK | SCI Transmit | 0v |
| 31 | RD/LG | S/C Set Switch Signal | S/C & Set Switch On: 3.8v |
| 32 | WT/PK | Brake Switch Signal | Brake Off: 0v, On: 12v |
| 33 | BR/OR | A/C Select Switch Sense | A/C On: 1v, Off: 12v |
| 34 | PK/BK | IAC 4 Driver | DC pulse signals: 0.8-11v |
| 35 | BR/WT | IAC 1 Driver | DC pulse signals: 0.8-11v |
| 36 | BK/PK | MIL (lamp) Control | Lamp On: 1v, Off: 12v |
| 37 | TN/BK | Generator Lamp Control | Lamp On: 1v, Off: 12v |
| 38 | BR | Fuel Pump Relay Control | Relay Off: 12v, On: 1v |
| 39 | GY/YL | EGR Solenoid Control | 12v, 55 mph: 1v |
| 40 | TN/RD | S/C Vacuum Solenoid | Vacuum Increasing: 1v |

**1995 Neon 2.0L VIN C 60-Pin Connector - Continued**

| PCM Pin # | Wire Color | Circuit Description (60-Pin) | Value at Hot Idle |
|-----------|-----------|-----------------------------|-------------------|
| 41 | DG | Generator Field Control | Digital Signal: 0-12-0v |
| 42 | DG/OR | ASD Relay Output | 12-14v |
| 43 | PK/WT | 5-Volt Supply | 4.9-5.1v |
| 44 | OR | 8-Volt Supply | 7.9-8.1v |
| 45-46 | --- | Not Used | --- |
| 47 | YL/RD | S/C Power Supply | 12-14v |
| 48 | --- | Not Used | --- |
| 49 | PK/LG | Battery Temperature Sensor | At 86°F: 1.96v |
| 50 | BR/YL | PNP Switch Signal | In P/N: 0v, Others: 5v |
| 51 | BK/LB | Sensor Ground | <0.050v |
| 52 | BK/WT | Power Ground | <0.1v |
| 53 | --- | Not Used | --- |
| 54 | LG/BK | Ignition Switch Output | 12-14v |
| 55 | PK/BK | Reverse Indicator Control | Lamp On: 1v, Off: 12v |
| 56 | WT | PSP Switch Signal | Straight: 0v, Turning: 5v |
| 57-58 | --- | Not Used | --- |
| 59 | DB/OR | A/C Clutch Relay Control | Relay Off: 12v, On: 1v |
| 60 | LG/RD | S/C Vent Solenoid | Vacuum Increasing: 1v |

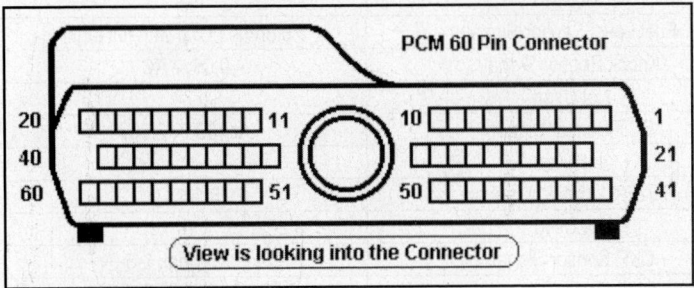

**Pin Connector Graphic**

**Standard Colors and Abbreviations**

| Abbreviation | Color | Abbreviation | Color | Abbreviation | Color |
|--------------|-------|--------------|-------|--------------|-------|
| BK | Black | GY | Gray | RD | Red |
| BL | Blue | GN | Green | TN | Tan |
| BR | Brown | LG | Light Green | VT | Violet |
| DB | Dark Blue | OR | Orange | WT | White |
| DG | Dark Green | PK | Pink | YL | Yellow |

## 1996-99 Neon 2.0L VIN C 'C1' Black Connector

| PCM Pin # | Wire Color | Circuit Description (40-Pin) | Value at Hot Idle |
|---|---|---|---|
| 1 | --- | Not Used | --- |
| 2 | BK/GY | Coil 1 Driver | 5º, at 55 mph: 8º dwell |
| 3 | DB/TN | Coil 2 Driver | 5º, at 55 mph: 8º dwell |
| 4 | DG | Generator Field Control | Digital Signal: 0-12-0v |
| 5 | YL/RD | S/C Power Supply | 12-14v |
| 6 | DG/OR | ASD Relay Output | 12-14v |
| 7 | YL/WT | Injector 3 Driver | 1-4 ms |
| 8 | BK/PK | MIL (lamp) Control | Lamp On: 1v, Off: 12v |
| 9 | --- | Not Used | --- |
| 10 | BK/TN | Power Ground | <0.1v |
| 11-12 | --- | Not Used | --- |
| 13 | WT/LB | Injector 1 Driver | 1-4 ms |
| 14-15 | --- | Not Used | --- |
| 16 | LB/BR | Injector 4 Driver | 1-4 ms |
| 17 | TN | Injector 2 Driver | 1-4 ms |
| 18 | LG | Radiator Fan Relay Control | Relay Off: 12v, On: 1v |
| 19 | --- | Not Used | --- |
| 20 | DB/WT | Ignition Switch Output | 12-14v |
| 21-22 | --- | Not Used | --- |
| 23 | DB | Fuel Level Sensor Signal | 70 ohms (±20) with full tank |
| 24 | DB/LG | Knock Sensor Signal | 0.080v AC |
| 25 | --- | Not Used | --- |
| 26 | TN/DB | ECT Sensor Signal | At 180ºF: 2.80v |
| 27-29 | --- | Not Used | --- |
| 30 | BK/DG | HO2S-11 (B1 S1) Signal | 0.1-1.1v |
| 31 | --- | Not Used | --- |
| 32 | GY/BK | CKP Sensor Signal | Digital Signal: 0-5-0v |
| 33 | TN/YL | CMP Sensor Signal | Digital Signal: 0-5-0v |
| 34 | --- | Not Used | --- |
| 35 | OR/LB | TP Sensor Signal | 0.6-1.0v |
| 36 | DG/RD | MAP Sensor Signal | 1.5-1.7v |
| 37 | BK/RD | IAT Sensor Signal | At 100ºF: 1.83v |
| 38 | BR/OR | A/C Select Switch Sense | A/C On: 1v, Off: 12v |
| 39 | --- | Not Used | --- |
| 40 | GY/YL | EGR Solenoid Control | 12v, 55 mph: 1v |

## Standard Colors and Abbreviations

| Abbreviation | Color | Abbreviation | Color | Abbreviation | Color |
|---|---|---|---|---|---|
| BK | Black | GY | Gray | RD | Red |
| BL | Blue | GN | Green | TN | Tan |
| BR | Brown | LG | Light Green | VT | Violet |
| DB | Dark Blue | OR | Orange | WT | White |
| DG | Dark Green | PK | Pink | YL | Yellow |

**1996-99 Neon 2.0L VIN C 'C2' White Connector**

| PCM Pin # | Wire Color | Circuit Description (40-Pin) | Value at Hot Idle |
|---|---|---|---|
| 41 | RD/LG | S/C Set Switch Signal | S/C & Set Switch On: 3.8v |
| 42 | --- | Not Used | --- |
| 43 | BK/LB | Sensor Ground | <0.050v |
| 44 | OR | 8-Volt Supply | 7.9-8.1v |
| 45 | WT | PSP Switch Signal | Straight: 0v, Turning: 5v |
| 46 | RD/WT | Battery Power (Fused B+) | 12-14v |
| 47 | BK/WT | Sensor Ground | <0.050v |
| 48 | BR/WT | IAC 3 Driver | DC pulse signals: 0.8-11v |
| 49 | YL/BK | IAC 2 Driver | DC pulse signals: 0.8-11v |
| 50 | BK/TN | Power Ground | <0.1v |
| 51 | TN/WT | HO2S-12 (B1 S2) Signal | 0.1-1.1v |
| 52 | PK/LG | Battery Temperature Sensor | At 86°F: 1.96v |
| 53-55 | --- | Not Used | --- |
| 56 | TN/BK | Generator Lamp Control | Lamp On: 1v, Off: 12v |
| 57 | GY/RD | IAC 1 Driver | DC pulse signals: 0.8-11v |
| 58 | PK | IAC 4 Driver | DC pulse signals: 0.8-11v |
| 59-60 | --- | Not Used | --- |
| 61 | PK/WT | 5-Volt Supply | 4.9-5.1v |
| 62 | WT/PK | Brake Switch Signal | Brake Off: 0v, On: 12v |
| 63 | --- | Not Used | --- |
| 64 | DB/OR | A/C Clutch Relay Control | Relay Off: 12v, On: 1v |
| 65 | PK | SCI Transmit | 0v |
| 66 | WT/OR | Vehicle Speed Signal | Digital Signal |
| 67 | DB/YL | ASD Relay Control | Relay Off: 12v, On: 1v |
| 68 | PK/BK | EVAP Purge Solenoid Control | PWM signal: 0-12-0v |
| 69-71 | --- | Not Used | --- |
| 72 | OR | LDP Switch Signal | Switch Closed: 0v, open: 12v |
| 73 | GY/LB | Tachometer Signal | Pulse Signals |
| 74 | BR | Fuel Pump Relay Control | Relay Off: 12v, On: 1v |
| 75 | LG | SCI Receive | 5v |
| 76 | BR/YL | PNP Switch Signal | In P/N: 0v, Others: 5v |
| 77 | WT/LG | LDP Solenoid Control | PWM signal: 0-12-0v |
| 78 | TN/RD | S/C Vacuum Solenoid | Vacuum Increasing: 1v |
| 79 | OR/BK | TCC Solenoid Control | At Cruise w/TCC On: <1v |
| 80 | LG/RD | S/C Vent Solenoid | Vacuum Decreasing: 1v |

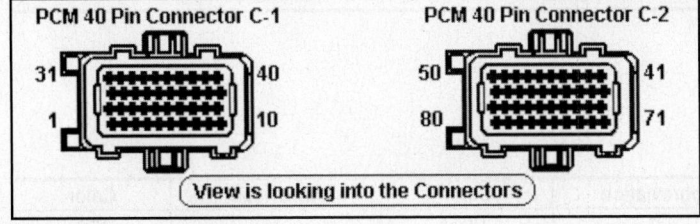

**Pin Connector Graphic**

## 2000-03 Neon 2.0L VIN C 'C1' Black Connector

| PCM Pin # | Wire Color | Circuit Description (40-Pin) | Value at Hot Idle |
|---|---|---|---|
| 1-2 | --- | Not Used | --- |
| 3 | DB/TN | Coil 2 Driver | 5°, at 55 mph: 8° dwell |
| 4 | --- | Not Used | --- |
| 5 | YL/RD | S/C Power Supply | 12-14v |
| 6 | DG/OR | ASD Relay Output | 12-14v |
| 7 | YL/WT | Injector 3 Driver | 1-4 ms |
| 8 | DG | Generator Field Control | Digital Signal: 0-12-0v |
| 9 | --- | Not Used | --- |
| 10 | BK/TN | Power Ground | <0.1v |
| 11 | BK/GY | Coil 1 Driver | 5°, at 55 mph: 8° dwell |
| 12 | GY | Engine Oil Pressure Sense | Switch open: 12v, closed: 0v |
| 13 | WT/LB | Injector 1 Driver | 1-4 ms |
| 14-15 | --- | Not Used | --- |
| 16 | LB/BR | Injector 4 Driver | 1-4 ms |
| 17 | TN | Injector 2 Driver | 1-4 ms |
| 18 | OR/RD | HO2S-11 (B1 S1) Heater | Heater On: 1v, Off: 12v |
| 19 | --- | Not Used | --- |
| 20 | DB/WT | Ignition Switch Output | 12-14v |
| 21-22 | --- | Not Used | --- |
| 23 | LG/BK | Clutch Switch Signal | Clutch Out: 5v, In: 0v |
| 24 | --- | Not Used | --- |
| 25 | DB/LG | Knock Sensor Signal | 0.080v AC |
| 26 | TN/DB | ECT Sensor Signal | At 180°F: 2.80v |
| 27 | BK/OR | HO2S-11 Ground | <0.050v |
| 28-29 | --- | Not Used | --- |
| 30 | BK/DG | HO2S-11 (B1 S1) Signal | 0.1-1.1v |
| 31 | TN | Starter Relay Control | KOEC: 9-11v |
| 32 | GY/BK | CKP Sensor Signal | Digital Signal: 0-5-0v |
| 33 | TN/YL | CMP Sensor Signal | Digital Signal: 0-5-0v |
| 34 | --- | Not Used | --- |
| 35 | OR/DB | TP Sensor Signal | 0.6-1.0v |
| 36 | DG/RD | MAP Sensor Signal | 1.5-1.7v |
| 37 | --- | Not Used | --- |
| 38 | BR/OR | A/C Select Switch Sense | A/C On: 1v, Off: 12v |
| 39 | BR/WT | Manifold Tuning Valve Relay | Relay Off: 12v, On: 1v |
| 40 | --- | Not Used | --- |

## Standard Colors and Abbreviations

| Abbreviation | Color | Abbreviation | Color | Abbreviation | Color |
|---|---|---|---|---|---|
| BK | Black | GY | Gray | RD | Red |
| BL | Blue | GN | Green | TN | Tan |
| BR | Brown | LG | Light Green | VT | Violet |
| DB | Dark Blue | OR | Orange | WT | White |
| DG | Dark Green | PK | Pink | YL | Yellow |

## 2000-03 Neon 2.0L VIN C 'C2' White Connector

| PCM Pin # | Wire Color | Circuit Description (40-Pin) | Value at Hot Idle |
|---|---|---|---|
| 41 | RD/LG | S/C Set Switch Signal | S/C & Set Switch On: 3.8v |
| 42, 48 | --- | Not Used | --- |
| 43 | BK/LB | Sensor Ground | <0.050v |
| 44 | OR | 8-Volt Supply | 7.9-8.1v |
| 45 | WT | PSP Switch Signal | Straight: 0v, Turning: 5v |
| 46 | RD/WT | Battery Power (Fused B+) | 12-14v |
| 47 | BK/WT | Power Ground | <0.1v |
| 49 | YL/BK | IAC 2 Driver | DC pulse signals: 0.8-11v |
| 50 | BK/TN | Power Ground | <0.1v |
| 51 | TN/WT | HO2S-12 (B1 S2) Signal | 0.1-1.1v |
| 52 | TN/LG | Inlet Air Temperature Sensor | At 86°F: 1.96v |
| 53-54 | --- | Not Used | --- |
| 55 | DB/PK | Radiator Fan Relay Control | Relay Off: 12v, On: 1v |
| 56 | TN/RD | S/C Vacuum Solenoid | Vacuum Increasing: 1v |
| 57 | GY/RD | IAC 1 Motor Sense (A/T) | 12-14v |
| 57 | BR/WT | IAC 1 Motor Sense (M/T) | 12-14v |
| 58, 60 | --- | Not Used | --- |
| 59 | VT/YL | PCM Data Bus (J1950) | Digital Signal: 0-5-0v |
| 61 | VT/WT | 5-Volt Supply | 4.9-5.1v |
| 62 | WT/PK | Brake Switch Signal | Brake Off: 0v, On: 12v |
| 63 | YL/DG | Torque Management Request | Digital Signals |
| 64 | DB/OR | A/C Clutch Relay Control | Relay Off: 12v, On: 1v |
| 65 | PK | SCI Transmit | 0v |
| 66 | WT/OR | Vehicle Speed Signal | Digital Signal |
| 67 | DB/YL | ASD Relay Control | Relay Off: 12v, On: 1v |
| 68 | PK/BK | EVAP Purge Solenoid Control | PWM signal: 0-12-0v |
| 69 | --- | Not Used | --- |
| 70 | DB | EVAP Purge Solenoid Sense | 0-1v |
| 71 | GY/RD | Brake Fluid Level Switch | Switch open: 12v, closed; 0v |
| 72 | YL | LDP Switch Signal | Switch Closed: 0v, open: 12v |
| 73, 79 | --- | Not Used | --- |
| 74 | BR | Fuel Pump Relay Control | Relay Off: 12v, On: 1v |
| 75 | LG | SCI Receive | 5v |
| 76 | BR/YL | TRS TR1 Sense / PNP Signal | In P/N: 0v, Others: 5v |
| 77 | WT/DB | LDP Solenoid Control | PWM signal: 0-12-0v |
| 78 | OR/BK | TCC Solenoid Control | Digital Signal: 0-12-0v |
| 80 | LG/RD | S/C Vent Solenoid | Vacuum Decreasing: 1v |

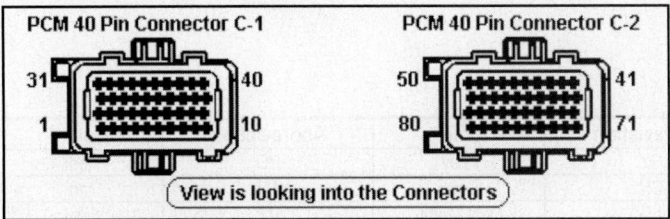

**Pin Connector Graphic**

## 2001-03 Neon ACR & R/T 2.0L VIN F 'C1' Black Connector

| PCM Pin # | Wire Color | Circuit Description (40-Pin) | Value at Hot Idle |
|---|---|---|---|
| 1-2 | --- | Not Used | --- |
| 3 | DB/TN | Coil 2 Driver | 5°, at 55 mph: 8° dwell |
| 4 | --- | Not Used | --- |
| 5 | YL/RD | S/C Power Supply | 12-14v |
| 6 | DG/OR | ASD Relay Output | 12-14v |
| 7 | YL/WT | Injector 3 Driver | 1-4 ms |
| 8 | DG | Generator Field Control | Digital Signal: 0-12-0v |
| 9 | --- | Not Used | --- |
| 10 | BK/TN | Power Ground | <0.1v |
| 11 | BK/GY | Coil 1 Driver | 5°, at 55 mph: 8° dwell |
| 12 | GY | Engine Oil Pressure Sense | Switch open: 12v, closed: 0v |
| 13 | WT/DB | Injector 1 Driver | 1-4 ms |
| 14-15 | --- | Not Used | --- |
| 16 | LB/BR | Injector 4 Driver | 1-4 ms |
| 17 | TN | Injector 2 Driver | 1-4 ms |
| 18 | OR/RD | HO2S-11 (B1 S1) Heater | Heater On: 1v, Off: 12v |
| 19 | --- | Not Used | --- |
| 20 | DB/WT | Ignition Switch Output | 12-14v |
| 21-22 | --- | Not Used | --- |
| 23 | LG/BK | Clutch Switch Signal | Clutch Out: 5v, In: 0v |
| 24 | --- | Not Used | --- |
| 25 | DB/LG | Knock Sensor Signal | 0.080v AC |
| 26 | TN/BK | ECT Sensor Signal | At 180°F: 2.80v |
| 27 | BK/OR | HO2S-11 Ground | <0.050v |
| 28-29 | --- | Not Used | --- |
| 30 | BK/DG | HO2S-11 (B1 S1) Signal | 0.1-1.1v |
| 31 | TN | Starter Relay Control | KOEC: 9-11v |
| 32 | GY/BK | CKP Sensor Signal | Digital Signal: 0-5-0v |
| 33 | TN/YL | CMP Sensor Signal | Digital Signal: 0-5-0v |
| 34 | --- | Not Used | --- |
| 35 | OR/DB | TP Sensor Signal | 0.6-1.0v |
| 36 | DG/RD | MAP Sensor Signal | 1.5-1.7v |
| 37 | --- | Not Used | --- |
| 38 | BR/OR | A/C Select Switch Sense | A/C On: 1v, Off: 12v |
| 39 | BR/WT | Manifold Tuning Valve Relay | Relay Off: 12v, On: 1v |
| 40 | --- | Not Used | --- |

## Standard Colors and Abbreviations

| Abbreviation | Color | Abbreviation | Color | Abbreviation | Color |
|---|---|---|---|---|---|
| BK | Black | GY | Gray | RD | Red |
| BL | Blue | GN | Green | TN | Tan |
| BR | Brown | LG | Light Green | VT | Violet |
| DB | Dark Blue | OR | Orange | WT | White |
| DG | Dark Green | PK | Pink | YL | Yellow |

**2001-03 Neon ACR & R/T 2.0L VIN F 'C2' White Connector**

| PCM Pin # | Wire Color | Circuit Description (40-Pin) | Value at Hot Idle |
|---|---|---|---|
| 41 | RD/LG | S/C Set Switch Signal | S/C & Set Switch On: 3.8v |
| 42, 48 | --- | Not Used | --- |
| 43 | BK/LB | Sensor Ground | <0.050v |
| 44 | OR | 8-Volt Supply | 7.9-8.1v |
| 45 | WT | PSP Switch Signal | Straight: 0v, Turning: 5v |
| 46 | RD/WT | Battery Power (Fused B+) | 12-14v |
| 47 | BK/WT | Power Ground | <0.1v |
| 49 | YL/BK | IAC 2 Driver | DC pulse signals: 0.8-11v |
| 50 | BK/TN | Power Ground | <0.1v |
| 51 | TN/WT | HO2S-12 (B1 S2) Signal | 0.1-1.1v |
| 52 | VT/LG | Inlet Air Temperature Sensor | At 86ºF: 1.96v |
| 53-54 | --- | Not Used | --- |
| 55 | DB/PK | Radiator Fan Relay Control | Relay Off: 12v, On: 1v |
| 56 | TN/RD | S/C Vacuum Solenoid | Vacuum Increasing: 1v |
| 57 | GY/RD | IAC Motor Sense (w/ EATX) | 12-14v |
| 57 | BR/WT | IAC Motor Sense (w/o EATX) | 12-14v |
| 58, 60 | --- | Not Used | --- |
| 59 | VT/YL | PCM Data Bus (J1950) | Digital Signal: 0-5-0v |
| 61 | VT/WT | 5-Volt Supply | 4.9-5.1v |
| 62 | WT/PK | Brake Switch Signal | Brake Off: 0v, On: 12v |
| 63 | YL/DG | Torque Management Request | Digital Signals |
| 64 | DB/OR | A/C Clutch Relay Control | Relay Off: 12v, On: 1v |
| 65 | PK | SCI Transmit | 0v |
| 66 | WT/OR | Vehicle Speed Signal | Digital Signal |
| 67 | DB/YL | ASD Relay Control | Relay Off: 12v, On: 1v |
| 68 | PK/BK | EVAP Purge Solenoid Control | PWM signal: 0-12-0v |
| 69, 73, 79 | --- | Not Used | --- |
| 70 | DB | EVAP Purge Solenoid Sense | 0-1v |
| 71 | GY/RD | Brake Fluid Level Sense | Switch open: 12v, closed; 0v |
| 72 | YL | LDP Switch Signal | Switch Closed: 0v, open: 12v |
| 74 | BR | Fuel Pump Relay Control | Relay Off: 12v, On: 1v |
| 75 | LG | SCI Receive | 0v |
| 76 | BR/YL | TRS TR1 Sense / PNP Signal | In P/N: 0v, Others: 5v |
| 76 | YL/RD | Clutch Interlock Switch Sense | Clutch Out: 12v, In: 0v |
| 77 | WT/DB | LDP Solenoid Control | PWM signal: 0-12-0v |
| 78 | OR/BK | TCC Solenoid Control | Digital Signal: 0-12-0v |
| 80 | LG/RD | S/C Vent Solenoid | Vacuum Decreasing: 1v |

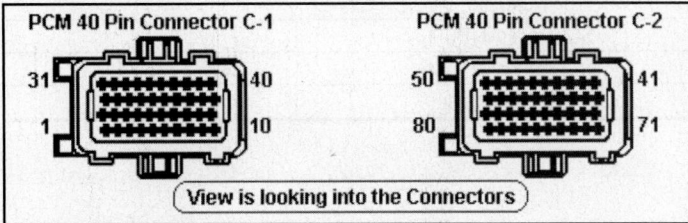

**Pin Connector Graphic**

**2004-05 Neon 2.0L, 2.4L Turbo 'C1' Black/Black Connector**

| PCM Pin # | Wire Color | Circuit Description (38-Pin) | Value at Hot Idle |
|---|---|---|---|
| 1 | --- | Not Used | --- |
| 2 | --- | Not Used | --- |
| 3 | --- | Not Used | --- |
| 4 | --- | Not Used | --- |
| 5 | --- | Not Used | --- |
| 6 | --- | Not Used | --- |
| 7 | --- | Not Used | --- |
| 8 | --- | Not Used | --- |
| 9 | BK/WT | Ground | |
| 10 | --- | Not Used | --- |
| 11 | DB/WT (2.0L) | Fused Ignition Switch Output (Run-Start) | |
| 11 | DB/RD (2.4L Turbo) | Fused Ignition Switch Output (Run-Start) | |
| 12 | RD/WT (2.0L AutoStick) | Ignition Switch Output (Off-Run-Start) | |
| 12 | RD/WT (2.0L Exc. AutoStick) | Fused Ignition Switch Output (Run-Start) | |
| 13 | WT/OR | Vehicle Speed Signal | |
| 14 | GY/BK | Brake Fluid Level Switch Sense | |
| 15 | LB (2.4L Turbo) | TP Sol Control | |
| 16 | --- | Not Used | --- |
| 17 | DB/YL (2.4L Turbo) | Surge Solenoid Control | |
| 18 | BK/TN | Ground | |
| 19 | --- | Not Used | --- |
| 20 | GY | Oil Pressure Signal | |
| 21 | --- | Not Used | --- |
| 22 | BR/OR | AAT Signal | |
| 23 | LB (2.4L Turbo) | TP Signal | |
| 24 | --- | Not Used | --- |
| 25 | LG | SCI Receive (PCM) | |
| 26 | PK/LB (2.0L) | SCI Receive (TCM) | |
| 27 | VT/WT (2.0L) | 5 Volt Supply | |
| 27 | VT/WT (2.4L Turbo) | 5 Volt Supply | |
| 28 | DB/GY (2.4L Turbo) | Wastegate Sol Control | |
| 29 | RD/WT | Fused B (+) | |
| 30 | YL | Fused Ignition Switch Output (Start) | |
| 31 | --- | Not Used | --- |
| 32 | --- | Not Used | --- |
| 33 | --- | Not Used | --- |
| 34 | --- | Not Used | --- |
| 35 | --- | Not Used | --- |
| 36 | PK | SCI Transmit (PCM) | |
| 37 | WT/DG (2.0L) | SCI Transmit (TCM) | |
| 38 | VT/YL | PCI Bus (PCM) | |

**2004-05 Neon 2.0L, 2.4L Turbo 'C2' Black/Orange Connector**

| PCM Pin # | Wire Color | Circuit Description (38-Pin) | Value at Hot Idle |
|---|---|---|---|
| 1 | --- | Not Used | --- |
| 2 | --- | Not Used | --- |
| 3 | --- | Not Used | --- |
| 4 | --- | Not Used | --- |
| 5 | --- | Not Used | --- |
| 6 | --- | Not Used | --- |
| 7 | --- | Not Used | --- |
| 8 | --- | Not Used | --- |
| 9 | DB/TN (2.0L) | Coil Control No. 2 | |
| 9 | DB/TN (2.4L Turbo) | Coil Control No. 2 | |
| 10 | BK/GY (2.0L) | Coil Control No. 1 | |
| 10 | BK/GY (2.4L Turbo) | Coil Control No. 1 | |
| 11 | LB/BR | Injector Control No. 4 | |
| 12 | YL/WT | Injector Control No. 3 | |
| 13 | TN | Injector Control No. 2 | |
| 14 | WT/DB | Injector Control No. 1 | |
| 15 | --- | Not Used | --- |
| 16 | VT/OR (RT) | MTV Control | |
| 17 | BR/VT | O2 1/2 Heater Control | |
| 18 | BR/OR | O2 1/1 Heater Control | |
| 19 | DG (2.0L) | Gen Field Control | |
| 19 | DG (2.4L Turbo) | Gen Field Control | |
| 20 | VT/LG (2.0L) | ECT Signal | |
| 20 | TN/BK (2.4L Turbo) | ECT Signal | |
| 21 | OR/DB | TP Signal | |
| 22 | --- | Not Used | --- |
| 23 | DG/RD | MAP Signal | |
| 24 | BK/VT | KS Return | |
| 25 | DB/LG | KS Signal | |
| 26 | --- | Not Used | --- |
| 27 | BK/LB (2.0L) | Sensor Ground | |
| 27 | BK/LB (2.4L Turbo) | Sensor Ground | |
| 28 | BR/WT (2.0L) | IAC Return | |
| 28 | BR/VT (2.4L Turbo) | IAC Return | |
| 29 | OR (2.0L) | 5 Volt Supply | |
| 29 | OR (2.4L Turbo) | 5 Volt Supply | |
| 30 | BK/RD | IAT Signal | |
| 31 | BK/DG | O2 1/1 Signal | |
| 32 | DB/DG (2.0L) | O2 Return | |
| 32 | DB/DG (2.4L Turbo) | O2 Return | |
| 33 | TN/WT | O2 1/2 Signal | |
| 34 | TN/YL | CMP Signal | |
| 35 | GY/BK | CKP Signal | |
| 36 | --- | Not Used | --- |
| 37 | --- | Not Used | --- |
| 38 | VT/GY (2.0L) | IAC Motor Control | |
| 38 | VT/GY (2.4L Turbo) | IAC Motor Control | |

## 2004-05 Neon 2.0L, 2.4L Turbo 'C3' Black/Natural Connector

| PCM Pin # | Wire Color | Circuit Description (38-Pin) | Value at Hot Idle |
|---|---|---|---|
| 1 | --- | Not Used | --- |
| 2 | --- | Not Used | --- |
| 3 | DB/YL (2.0L) | Automatic Shut Down Relay Control | |
| 3 | DB/YL (2.4L Turbo) | Automatic Shut Down Relay Control | |
| 4 | WT/DB (2.0L w/ETAX) | High Speed Rad Fan Relay Control | |
| 4 | DB/PK (2.4L Turbo) | High Speed Rad Fan Relay Control | |
| 5 | LG/RD (2.0L) | S/C Vent Control | |
| 6 | DB/PK (2.0L) | Rad Fan Relay Control | |
| 6 | DB/RD (2.4L Turbo) | Rad Fan Low Relay Control | |
| 7 | YL/RD (2.0L) | S/C Supply | |
| 8 | WT/DG (2.0L) | NVLD Solenoid Control | |
| 8 | WT/DG (2.4L Turbo) | NVLD Solenoid Control | |
| 9-10 | --- | Not Used | --- |
| 11 | DB/OR (2.0L) | A/C Clutch Relay Control | |
| 11 | DB/OR (2.4L Turbo) | A/C Clutch Relay Control | |
| 12 | TN/RD (2.0L) | S/C Vacuum Control | |
| 13-16 | --- | Not Used | --- |
| 17 | BR/YL (2.0L) | Sensor Ground 2 | |
| 17 | BR/YL (2.4L Turbo) | Sensor Ground 2 | |
| 18 | --- | Not Used | --- |
| 19 | DG/OR | Automatic Shut Down Relay Output | |
| 20 | PK/BK | EVAP/Purge Control | |
| 21 | YL/RD | Clutch Interlock Switch Signal | |
| 22 | --- | Not Used | --- |
| 23 | WT/PK | Brake Switch Signal | |
| 24 | BR/OR (2.0L) | A/C Switch Sense | |
| 24 | BR (2.4L Turbo) | A/C Switch Sense | |
| 25 | --- | Not Used | --- |
| 26 | YL/LB (2.0L EATX) | AutoStick Downshift Switch Signal | |
| 26 | LG/BK (2.0L MTX or 2.4L Turbo) | Clutch Up Switch Signal | |
| 27 | LG/LB (2.0L EATX) | AutoStick Upshift Switch Signal | |
| 28 | DG/OR | Automatic Shut Down Relay Output | |
| 29 | WT/TN | EVAP/Purge Return | |
| 30 | DB/OR | PSP Switch Signal | |
| 31 | --- | Not Used | --- |
| 32 | PK/YL (2.4L Turbo) | Battery Temp Signal | |
| 33 | --- | Not Used | --- |
| 34 | RD/LG (2.0L) | S/C Switch Signal | |
| 35 | OR (2.0L) | NVLD Switch Signal | |
| 35 | OR (2.4L Turbo) | NVLD Switch Signal | |
| 36 | --- | Not Used | --- |
| 37 | BR (2.0L) | Fuel Pump Relay Control | |
| 37 | BR (2.4L Turbo) | Fuel Pump Relay Control | |
| 38 | TN | Starter Relay Control | |

**2004-05 Neon 2.0L, 2.4L Turbo 'C4' Black/Green Connector**

| PCM Pin # | Wire Color | Circuit Description (38-Pin) | Value at Hot Idle |
|---|---|---|---|
| 1 | BR | OD Solenoid Control | |
| 2 | PK/BK | UD Solenoid Control | |
| 3 | --- | Not Used | --- |
| 4 | --- | Not Used | --- |
| 5 | --- | Not Used | --- |
| 6 | WT | 2-4 Solenoid Control | |
| 7 | --- | Not Used | --- |
| 8 | --- | Not Used | --- |
| 9 | --- | Not Used | --- |
| 10 | LB | L/R Solenoid Control | |
| 11 | --- | Not Used | --- |
| 12 | BK/RD | Ground | |
| 13 | --- | Not Used | --- |
| 14 | BK/RD | Ground | |
| 15 | LG/BK | TRS T1 Sense | |
| 16 | VT | TRS T3 Sense | |
| 17 | --- | Not Used | --- |
| 18 | LG | Transmission Control Relay Control | |
| 19 | RD | Transmission Control Relay Output | |
| 20 | --- | Not Used | --- |
| 21 | --- | Not Used | --- |
| 22 | OR/BK | OD Pressure Switch Sense | |
| 23 | --- | Not Used | --- |
| 24 | --- | Not Used | --- |
| 25 | --- | Not Used | --- |
| 26 | --- | Not Used | --- |
| 27 | BK/WT | TRS T41 Sense | |
| 28 | RD | Transmission Control Relay Output | |
| 29 | DG | L/R Pressure Switch Sense | |
| 30 | YL/BK | 2-4 Pressure Switch Sense | |
| 31 | --- | Not Used | --- |
| 32 | LG/WT | Output Speed Sensor Signal | |
| 33 | RD/BK | Input Speed Sensor Signal | |
| 34 | DB/BK | Speed Sensor Ground | |
| 35 | VT/PK | Transmission Temperature Sensor Signal | |
| 36 | --- | Not Used | --- |
| 37 | VT/WT | TRS T42 Sense | |
| 38 | --- | Not Used | --- |

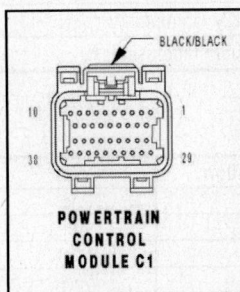

POWERTRAIN
CONTROL
MODULE C1

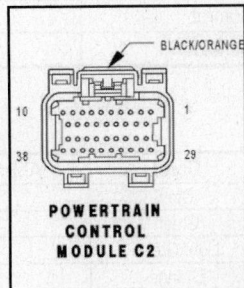

POWERTRAIN
CONTROL
MODULE C2

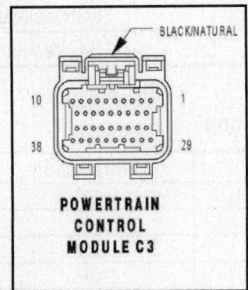

POWERTRAIN
CONTROL
MODULE C3

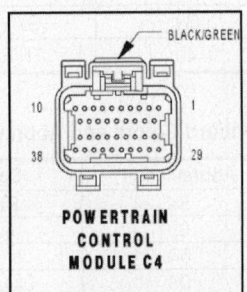

POWERTRAIN
CONTROL
MODULE C4

## 1995 Neon 2.0L VIN Y 60-Pin Connector

| PCM Pin # | Wire Color | Circuit Description (60-Pin) | Value at Hot Idle |
|-----------|------------|------------------------------|-------------------|
| 1 | BK/GY | Coil 2 Driver | 5°, at 55 mph: 8° dwell |
| 2 | BK/TN | Power Ground | <0.1v |
| 3 | YL/WT | Injector 3 Driver | 1-4 ms |
| 4 | WT/DB | Injector 1 Driver | 1-4 ms |
| 5 | WT/OR | Vehicle Speed Signal | Digital Signal |
| 6 | BK/RD | IAT Sensor Signal | At 100°F: 1.83v |
| 7 | DG/BK | HO2S-12 (B1 S2) Signal | 0.1-1.1v |
| 8 | BK/DG | HO2S-11 (B1 S1) Signal | 0.1-1.1v |
| 9 | LG | SCI Receive | 0v |
| 10 | OR/DB | TP Sensor Signal | 0.6-1.0v |
| 11 | RD/WT | Battery Power (Fused B+) | 12-14v |
| 12-13 | --- | Not Used | --- |
| 14 | YL/BK | IAC 2 Driver | DC pulse signals: 0.8-11v |
| 15 | GY/RD | IAC 3 Driver | DC pulse signals: 0.8-11v |
| 16 | PK/BK | EVAP Purge Solenoid Control | PWM signal: 0-12-0v |
| 17 | OR/BK | TCC Solenoid Control | At Cruise w/TCC On: <1v |
| 18 | DB/YL | ASD Relay Control | Relay Off: 12v, On: 1v |
| 19 | DB/PK | Radiator Fan Relay Control | Relay Off: 12v, On: 1v |
| 20 | --- | Not Used | --- |
| 21 | DB/YL | Coil 1 Driver | 5°, at 55 mph: 8° dwell |
| 22 | BK/TN | Power Ground | <0.1v |
| 23 | TN | Injector 2 Driver | 1-4 ms |
| 24 | LB/BR | Injector 4 Driver | 1-4 ms |
| 25 | GY/BK | CKP Sensor Signal | Digital Signal: 0-5-0v |
| 26 | TN/YL | CMP Sensor Signal | Digital Signal: 0-5-0v |
| 27 | BK/LG | Knock Sensor Signal | 0.080v AC |
| 28 | TN/BK | ECT Sensor Signal | At 180°F: 2.80v |
| 29 | DG/RD | MAP Sensor Signal | 1.5-1.7v |
| 30 | PK | SCI Transmit | 0v |
| 31 | RD/LG | S/C Set Switch Signal | S/C & Set Switch On: 3.8v |
| 32 | WT/PK | Brake Switch Signal | Brake Off: 0v, On: 12v |
| 33 | DG/RD | A/C Damped Pressure Switch | A/C On: 1v, Off: 12v |
| 34 | PK/BK | IAC 4 Driver | DC pulse signals: 0.8-11v |
| 35 | BR/WT | IAC 1 Driver | DC pulse signals: 0.8-11v |
| 36 | BK/PK | MIL (lamp) Control | Lamp On: 1v, Off: 12v |
| 37 | TN/BK | Generator Lamp Control | Lamp On: 1v, Off: 12v |
| 38 | BR | Fuel Pump Relay Control | Relay Off: 12v, On: 1v |
| 39 | GY/YL | EGR Solenoid Control | 12v, 55 mph: 1v |
| 40 | TN/RD | S/C Vacuum Solenoid | Vacuum Increasing: 1v |

## Standard Colors and Abbreviations

| Abbreviation | Color | Abbreviation | Color | Abbreviation | Color |
|--------------|-------|--------------|-------|--------------|-------|
| BK | Black | GY | Gray | RD | Red |
| BL | Blue | GN | Green | TN | Tan |
| BR | Brown | LG | Light Green | VT | Violet |
| DB | Dark Blue | OR | Orange | WT | White |
| DG | Dark Green | PK | Pink | YL | Yellow |

**1995 Neon 2.0L VIN Y 60-Pin Connector - Continued**

| PCM Pin # | Wire Color | Circuit Description (60-Pin) | Value at Hot Idle |
|---|---|---|---|
| 41 | DG | Generator Field Control | Digital Signal: 0-12-0v |
| 42 | DG/OR | ASD Relay Output | 12-14v |
| 43 | PK/WT | 5-Volt Supply | 4.9-5.1v |
| 44 | OR | 8-Volt Supply | 7.9-8.1v |
| 45-46 | --- | Not Used | --- |
| 47 | YL/RD | S/C Power Supply | 12-14v |
| 48 | GY/LB | Tachometer Signal | Pulse Signals |
| 49 | PK/LG | Battery Temperature Sensor | At 86°F: 1.96v |
| 50 | BR/YL | PNP Switch Signal | In P/N: 0v, Others: 5v |
| 51 | BK/LB | Sensor Ground | <0.050v |
| 52 | BK/WT | Sensor Ground | <0.050v |
| 53 | --- | Not Used | --- |
| 54 | LG/BK | Ignition Switch Output | 12-14v |
| 55 | --- | Not Used | --- |
| 56 | WT | PSP Switch Signal | Straight: 0v, Turning: 5v |
| 57-58 | --- | Not Used | --- |
| 59 | DB/OR | A/C Clutch Relay Control | Relay Off: 12v, On: 1v |
| 60 | LG/RD | S/C Vent Solenoid | Vacuum Decreasing: 1v |

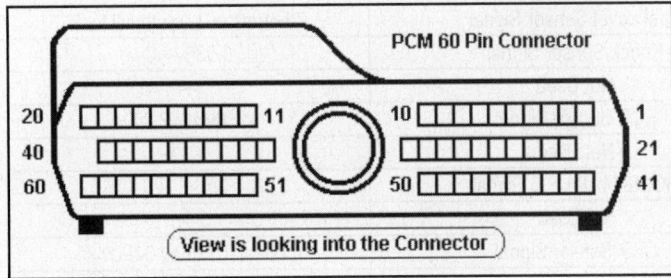

**Pin Connector Graphic**

**Standard Colors and Abbreviations**

| Abbreviation | Color | Abbreviation | Color | Abbreviation | Color |
|---|---|---|---|---|---|
| BK | Black | GY | Gray | RD | Red |
| BL | Blue | GN | Green | TN | Tan |
| BR | Brown | LG | Light Green | VT | Violet |
| DB | Dark Blue | OR | Orange | WT | White |
| DG | Dark Green | PK | Pink | YL | Yellow |

## 1996-99 Neon 2.0L VIN Y 'C1' Black Connector

| PCM Pin # | Wire Color | Circuit Description (40-Pin) | Value at Hot Idle |
|-----------|------------|------------------------------|-------------------|
| 1 | --- | Not Used | --- |
| 2 | BK/GY | Coil 1 Driver | 5°, at 55 mph: 8° dwell |
| 3 | DB/TN | Coil 2 Driver | 5°, at 55 mph: 8° dwell |
| 4 | DG | Generator Field Control | Digital Signal: 0-12-0v |
| 5 | YL/RD | S/C Power Supply | 12-14v |
| 6 | DG/OR | ASD Relay Output | 12-14v |
| 7 | YL/WT | Injector 3 Driver | 1-4 ms |
| 8 | BK/PK | MIL (lamp) Control | Lamp On: 1v, Off: 12v |
| 9 | --- | Not Used | --- |
| 10 | BK/TN | Power Ground | <0.1v |
| 11-12 | --- | Not Used | --- |
| 13 | WT/LB | Injector 1 Driver | 1-4 ms |
| 14-15 | --- | Not Used | --- |
| 16 | LB/BR | Injector 4 Driver | 1-4 ms |
| 17 | TN | Injector 2 Driver | 1-4 ms |
| 18 | LG | Radiator Fan Relay Control | Relay Off: 12v, On: 1v |
| 19 | --- | Not Used | --- |
| 20 | DB/WT | Ignition Switch Output | 12-14v |
| 21-22 | --- | Not Used | --- |
| 23 | DB | Fuel Level Sensor Signal | 70 ohms (±20) with full tank |
| 24 | BK/LG | Knock Sensor Signal | 0.080v AC |
| 25 | --- | Not Used | --- |
| 26 | TN/BK | ECT Sensor Signal | At 180°F: 2.80v |
| 27-29 | --- | Not Used | --- |
| 30 | BK/DG | HO2S-11 (B1 S1) Signal | 0.1-1.1v |
| 31 | --- | Not Used | --- |
| 32 | GY/BK | CKP Sensor Signal | Digital Signal: 0-5-0v |
| 33 | TN/YL | CMP Sensor Signal | Digital Signal: 0-5-0v |
| 34 | --- | Not Used | --- |
| 35 | OR/LB | TP Sensor Signal | 0.6-1.0v |
| 36 | DG/RD | MAP Sensor Signal | 1.5-1.7v |
| 37 | BK/RD | IAT Sensor Signal | At 100°F: 1.83v |
| 38 | BR/OR | A/C Select Switch Sense | A/C On: 1v, Off: 12v |
| 39 | --- | Not Used | --- |
| 40 | GY/YL | EGR Solenoid Control | 12v, 55 mph: 1v |

## Standard Colors and Abbreviations

| Abbreviation | Color | Abbreviation | Color | Abbreviation | Color |
|--------------|-------|--------------|-------|--------------|-------|
| BK | Black | GY | Gray | RD | Red |
| BL | Blue | GN | Green | TN | Tan |
| BR | Brown | LG | Light Green | VT | Violet |
| DB | Dark Blue | OR | Orange | WT | White |
| DG | Dark Green | PK | Pink | YL | Yellow |

**1996-99 Neon 2.0L VIN Y 'C2' White Connector**

| PCM Pin # | Wire Color | Circuit Description (40-Pin) | Value at Hot Idle |
|---|---|---|---|
| 41 | RD/LG | S/C Set Switch Signal | S/C & Set Switch On: 3.8v |
| 42 | --- | Not Used | --- |
| 43 | BK/LB | Sensor Ground | <0.050v |
| 44 | OR | 8-Volt Supply | 7.9-8.1v |
| 45 | WT | PSP Switch Signal | Straight: 0v, Turning: 5v |
| 46 | RD/WT | Battery Power (Fused B+) | 12-14v |
| 47 | BK/WT | Sensor Ground | <0.050v |
| 48 | BR/WT | IAC 3 Driver | DC pulse signals: 0.8-11v |
| 49 | YL/BK | IAC 2 Driver | DC pulse signals: 0.8-11v |
| 50 | BK/TN | Power Ground | <0.1v |
| 51 | TN/WT | HO2S-12 (B1 S2) Signal | 0.1-1.1v |
| 52 | PK/LG | Battery Temperature Sensor | At 86ºF: 1.96v |
| 53-55 | --- | Not Used | --- |
| 56 | TN/BK | Generator Lamp Control | Lamp On: 1v, Off: 12v |
| 57 | GY/RD | IAC 1 Driver | DC pulse signals: 0.8-11v |
| 58 | PK | IAC 4 Driver | DC pulse signals: 0.8-11v |
| 59-60 | --- | Not Used | --- |
| 61 | PK/WT | 5-Volt Supply | 4.9-5.1v |
| 62 | WT/PK | Brake Switch Signal | Brake Off: 0v, On: 12v |
| 63 | --- | Not Used | --- |
| 64 | DB/OR | A/C Clutch Relay Control | Relay Off: 12v, On: 1v |
| 65 | PK | SCI Transmit | 0v |
| 66 | WT/OR | Vehicle Speed Signal | Digital Signal |
| 67 | DB/YL | ASD Relay Control | Relay Off: 12v, On: 1v |
| 68 | PK/BK | EVAP Purge Solenoid Control | PWM signal: 0-12-0v |
| 69-71 | --- | Not Used | --- |
| 72 | OR | LDP Switch Signal | Switch Closed: 0v, open: 12v |
| 73 | GY/LB | Tachometer Signal | Pulse Signals |
| 74 | BR | Fuel Pump Relay Control | Relay Off: 12v, On: 1v |
| 75 | LG | SCI Receive | 5v |
| 76 | BR/YL | PNP Switch Signal | In P/N: 0v, Others: 5v |
| 77 | WT/LG | LDP Solenoid Control | PWM signal: 0-12-0v |
| 78 | TN/RD | S/C Vacuum Solenoid | Vacuum Increasing: 1v |
| 79 | OR/BK | TCC Solenoid Control | At Cruise w/TCC On: <1v |
| 80 | LG/RD | S/C Vent Solenoid | Vacuum Decreasing: 1v |

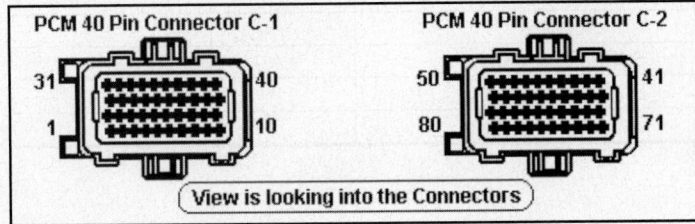

**Pin Connector Graphic**

<u>PACIFICA</u>

### 2004-05 Pacifica 3.5L, 3.8L 'C1' Black/Black Connector

| PCM Pin # | Wire Color | Circuit Description (38-Pin) | Value at Hot Idle |
|---|---|---|---|
| 1 | --- | Not Used | --- |
| 2 | --- | Not Used | --- |
| 3 | --- | Not Used | --- |
| 4 | --- | Not Used | --- |
| 5 | --- | Not Used | --- |
| 6 | --- | Not Used | --- |
| 7 | --- | Not Used | --- |
| 8 | --- | Not Used | --- |
| 9 | BK/BR | Ground | |
| 10 | --- | Not Used | --- |
| 11 | PK/GY | Fused Ignition Switch Output (Run-Start) | |
| 12 | PK/WT | Ignition Unlock-Run-Start | |
| 13 | --- | Not Used | --- |
| 14 | --- | Not Used | --- |
| 15 | --- | Not Used | --- |
| 16 | DB/LG | SRV Sol Control | |
| 17 | --- | Not Used | --- |
| 18 | BK/DG | Ground | |
| 19 | --- | Not Used | --- |
| 20 | VT/GY | Oil Pressure Signal | |
| 21 | --- | Not Used | --- |
| 22 | OR/VT | AAT Signal | |
| 23 | --- | Not Used | --- |
| 24 | --- | Not Used | --- |
| 25 | WT/LG | SCI Receive (PCM) | |
| 26 | WT/BR | Flash Program Enable | |
| 27 | --- | Not Used | --- |
| 28 | --- | Not Used | --- |
| 29 | OR/RD | Fused B (+) | |
| 30 | YL | Ignition Switch Output (Start) | |
| 31 | DB/YL | O2 1/2 Signal | |
| 32 | DB/DG | O2 Return (Down) | |
| 33 | --- | Not Used | --- |
| 34 | --- | Not Used | --- |
| 35 | --- | Not Used | --- |
| 36 | WT/GY | SCI Transmit (PCM) | |
| 37 | BR/WT | SCI Transmit (TCM) | |
| 38 | WT/VT | PCI Bus | |

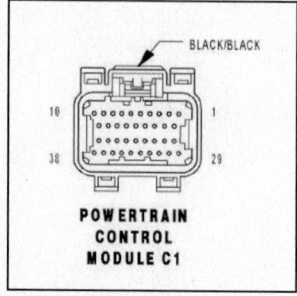

BLACK/BLACK

10    1

38    29

**POWERTRAIN
CONTROL
MODULE C1**

**2004-05 Pacifica 3.5L, 3.8L 'C2' Black/Orange Connector**

| PCM Pin # | Wire Color | Circuit Description (38-Pin) | Value at Hot Idle |
|---|---|---|---|
| 1 | DB/OR | Coil Control No. 6 | |
| 2 | DB/YL | Coil Control No. 5 | |
| 3 | DB | Coil Control No. 4 | |
| 4 | BR/VT | Injector Control No. 6 | |
| 5 | BR/OR | Injector Control No. 5 | |
| 6 | --- | Not Used | --- |
| 7 | DB/OR | Coil Control No. 3 | |
| 8 | DB/VT | EGR Sol Control | |
| 9 | DB/TN | Coil Control No. 2 | |
| 10 | DB/DG | Coil Control No. 1 | |
| 11 | BR/TN | Injector Control No. 4 | |
| 12 | BR/LB | Injector Control No. 3 | |
| 13 | BR/DB | Injector Control No. 2 | |
| 14 | BR/YL | Injector Control No. 1 | |
| 15 | --- | Not Used | --- |
| 16 | DB/YL | MTV Control | |
| 17 | --- | Not Used | --- |
| 18 | BR/TN | O2 1/1 Heater Control | |
| 19 | BR/GY | Gen Field Control | |
| 20 | VT/OR | ECT Signal | |
| 21 | BR/OR | TP Signal | |
| 22 | DB/LG | EGR Signal | |
| 23 | VT/BR | MAP Signal | |
| 24 | BR/LG | KS Return | |
| 25 | DB/OR | KS Signal | |
| 26 | --- | Not Used | --- |
| 27 | DB/DG | Sensor Ground | |
| 28 | BR/VT | IAC Return | |
| 29 | PK/YL | 5 Volt Supply | |
| 30 | BR/WT | IAT Signal | |
| 31 | DB/LB | O2 1/1 Signal | |
| 32 | BR/DG | O2 Return (Up) | |
| 33 | --- | Not Used | --- |
| 34 | DB/GY | CMP Signal | |
| 35 | BR/LB | CKP Signal | |
| 36 | --- | Not Used | --- |
| 37 | --- | Not Used | --- |
| 38 | VT/GY | IAC Motor Control | |

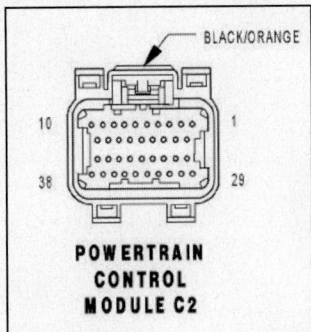

BLACK/ORANGE

10     1
38     29

**POWERTRAIN
CONTROL
MODULE C2**

## 2004-05 Pacifica 3.5L, 3.8L 'C3' Black/Natural Connector

| PCM Pin # | Wire Color | Circuit Description (38-Pin) | Value at Hot Idle |
|---|---|---|---|
| 1 | --- | Not Used | --- |
| 2 | --- | Not Used | --- |
| 3 | BR/GY | Automatic Shut Down Relay Control | |
| 4 | --- | Not Used | --- |
| 5 | VT/OR | S/C Vent Control | |
| 6 | BR/VT | Rad. Fan Relay Control | |
| 7 | VT/YL | S/C Supply | |
| 8 | VT/LB | NVLD Sol Control | |
| 9 | BR/OR | O2 1/2 Heater Control | |
| 10 | --- | Not Used | --- |
| 11 | LB/OR | A/C Clutch Relay Control | |
| 12 | YL/VT | S/C Vacuum Control | |
| 13 | --- | Not Used | --- |
| 14 | --- | Not Used | --- |
| 15 | --- | Not Used | --- |
| 16 | --- | Not Used | --- |
| 17 | --- | Not Used | --- |
| 18 | --- | Not Used | --- |
| 19 | BR/WT | Automatic Shut Down Relay Output | |
| 20 | DB/WT | EVAP Purge Control | |
| 21 | --- | Not Used | --- |
| 22 | --- | Not Used | --- |
| 23 | DG/WT | Brake Switch Signal | |
| 24 | --- | Not Used | --- |
| 25 | --- | Not Used | --- |
| 26 | YL/DG | AutoStick Downshift Switch Signal | |
| 27 | DG/YL | AutoStick Upshift Switch Signal | |
| 28 | BR/WT | Automatic Shut Down Relay Output | |
| 29 | DB/BR | EVAP Purge Return | |
| 30 | --- | Not Used | --- |
| 31 | LB/BR | A/C Pressure Transducer Signal | |
| 32 | DB/YL | Battery Temp Signal | |
| 33 | --- | Not Used | --- |
| 34 | VT | S/C Switch Signal | |
| 35 | VT/WT | NVLD Switch Signal | |
| 36 | --- | Not Used | --- |
| 37 | BR | Fuel Pump Relay Control | |
| 38 | DG/OR | Starter Relay Control | |

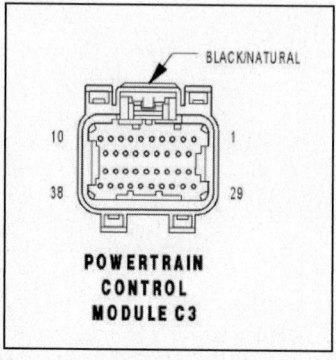

BLACK/NATURAL

10    1
38    29

**POWERTRAIN
CONTROL
MODULE C3**

## 2004-05 Pacifica 3.5L, 3.8L 'C3' Black/Green Connector

| PCM Pin # | Wire Color | Circuit Description (38-Pin) | Value at Hot Idle |
|---|---|---|---|
| 1 | YL/GY | OD Solenoid Control | |
| 2 | YL/LB | UD Solenoid Control | |
| 3 | --- | Not Used | --- |
| 4 | --- | Not Used | --- |
| 5 | --- | Not Used | --- |
| 6 | DB/YL | 2Not Used4 Solenoid Control | |
| 7 | --- | Not Used | --- |
| 8 | --- | Not Used | --- |
| 9 | --- | Not Used | --- |
| 10 | WT/DG | L/R Solenoid Control | |
| 11 | --- | Not Used | --- |
| 12 | --- | Not Used | --- |
| 13 | BK/LG | Ground | |
| 14 | BK/LG | Ground | |
| 15 | LB/DG | TRS T1 Sense | |
| 16 | DG/DB | TRS T3 Sense | |
| 17 | --- | Not Used | --- |
| 18 | YL/BR | Transmission Control Relay Control | |
| 19 | --- | Not Used | --- |
| 20 | --- | Not Used | --- |
| 21 | --- | Not Used | --- |
| 22 | DG/TN | OD Pressure Switch Sense | |
| 23 | --- | Not Used | --- |
| 24 | --- | Not Used | --- |
| 25 | --- | Not Used | --- |
| 26 | --- | Not Used | --- |
| 27 | YL/DB | TRS T41 Sense | |
| 28 | YL/OR | Transmission Control Relay Output | |
| 29 | YL/TN | L/R Pressure Switch Sense | |
| 30 | YL/DG | 2-4 Pressure Switch Sense | |
| 31 | --- | Not Used | --- |
| 32 | DG/BR | Output Speed Sensor Signal | |
| 33 | DG/WT | Input Speed Sensor Signal | |
| 34 | DG/VT | Speed Sensor Ground | |
| 35 | DG/OR | Transmission Temperature Sensor Signal | |
| 36 | --- | Not Used | --- |
| 37 | DG/YL | TRS T42 Sense | |
| 38 | YL/OR | Transmission Control Relay Output | |

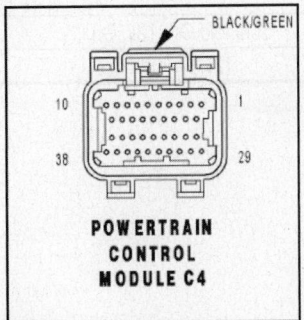

**POWERTRAIN CONTROL MODULE C4**

PROWLER

### 2001-02 Prowler 3.5L VIN G (A/T) 'C1' Connector

| PCM Pin # | Wire Color | Circuit Description (40-Pin) | Value at Hot Idle |
|---|---|---|---|
| 1 | TN/LG | COP 4 Driver | 6°, 55 mph: 9° dwell |
| 2 | RD/YL | COP 3 Driver | 6°, 55 mph: 9° dwell |
| 3 | DB/TN | COP 2 Driver | 6°, 55 mph: 9° dwell |
| 4 | TN/VT | COP 6 Driver | 6°, 55 mph: 9° dwell |
| 5 | YL/RD | S/C On/Off Switch Power | 12-14v |
| 6 | DG/OR | ASD Relay Output | 12-14v |
| 7 | YL/WT | Injector 3 Driver | 1-4 ms |
| 8 | DG | Generator Field Driver | Digital Signal: 0-12-0v |
| 9 | --- | Not Used | --- |
| 10 | BK/TN | Power Ground | <0.1v |
| 11 | GY/OR | COP 1 Driver | 6°, 55 mph: 9° dwell |
| 12 | --- | Not Used | --- |
| 13 | WT/DB | Injector 1 Driver | 1-4 ms |
| 14 | BR/DB | Injector 6 Driver | 1-4 ms |
| 15 | GY | Injector 5 Driver | 1-4 ms |
| 16 | LB/BR | Injector 4 Driver | 1-4 ms |
| 17 | TN | Injector 2 Driver | 1-4 ms |
| 18 | DB/BK | Short Runner Valve Solenoid | Valve On: 1v, Off: 12v |
| 19 | --- | Not Used | --- |
| 20 | DB/YL | Ignition Switch Output | 12-14v |
| 21 | TN/PK | COP 5 Driver | 6°, 55 mph: 9° dwell |
| 22-23 | --- | Not Used | --- |
| 24 | DB/LG | Knock Sensor Signal | 0.080v AC |
| 25 | GY/LG | Knock Sensor Return | <0.050v |
| 26 | TN/BK | ECT Sensor Signal | At 180°F: 2.80v |
| 27 | BK/YL | HO2S Ground | <0.050v |
| 28 | BK/LG | Transmission Fan Relay | Relay Off: 12v, On: 1v |
| 29 | LG/RD | HO2S-21 (B2 S1) Signal | 0.1-1.1v |
| 30 | BK/DG | HO2S-11 (B1 S1) Signal | 0.1-1.1v |
| 31 | TN/DG | Starter Relay Control | KOEC: 9-11v |
| 32 | GY/BK | CKP Sensor Signal | Digital Signal: 0-5-0v |
| 33 | TN/YL | CMP Sensor Signal | Digital Signal: 0-5-0v |
| 34 | LG/PK | EGR Sensor Signal | 0.6-0.8v |
| 35 | OR/DB | TP Sensor Signal | 0.6-1.0v |
| 36 | DG/RD | MAP Sensor Signal | 1.5-1.7v |
| 37 | BK/RD | IAT Sensor Signal | At 100°F: 1.83v |
| 38 | --- | Not Used | --- |
| 39 | VT/RD | Manifold Tuning Valve Control | Solenoid Off: 12v, On: 1v |
| 40 | GY/YL | EGR Solenoid Control | 12v, 55 mph: 1v |

### 2001-02 Prowler 3.5L VIN G (A/T) 'C2' Connector

| PCM Pin # | Wire Color | Circuit Description (40-Pin) | Value at Hot Idle |
|---|---|---|---|
| 41 | RD/LG | S/C Set Switch Signal | S/C & Set Switch On: 3.8v |
| 42 | DB | A/C Pressure Switch Signal | A/C On: 0.45-4.85v |
| 43, 47 | BK/LB | Sensor & Power Ground | <0.1v |
| 44 | OR | 8-Volt Supply | 7.9-8.1v |
| 45, 54 | --- | Not Used | --- |
| 46 | RD/WT | Battery Power (Fused B+) | 12-14v |
| 48 | BR/WT | IAC 3 Driver | DC pulse signals: 0.8-11v |
| 49 | YL/BK | IAC 2 Driver | DC pulse signals: 0.8-11v |
| 50 | BK/TN | Power Ground | <0.1v |
| 51 | TN/WT | HO2S-12 (B1 S2) Signal | 0.1-1.1v |
| 52 | PK/YL | Battery Temperature Sensor | At 86°F: 1.96v |
| 53 | BK/OR | HO2S-22 (B2 S2) Signal | 0.1-1.1v |
| 55 | DB/RD | Low Speed Fan Relay | Relay Off: 12v, On: 1v |
| 56 | TN/RD | S/C Vacuum Solenoid | Vacuum Increasing: 1v |
| 57 | GY/RD | IAC 1 Driver | DC pulse signals: 0.8-11v |
| 58 | VT/BK | IAC 4 Driver | DC pulse signals: 0.8-11v |
| 59, 60 | VT, WT | CCD Bus (+), CCD Bus (-) | Digital Signal: 0-5-0v |
| 61 | VT/WT | 5-Volt Supply | 4.9-5.1v |
| 62 | WT/PK | Brake Switch Signal | Brake Off: 0v, On: 12v |
| 63 | YL/DG | Torque Management Request | Digital Signals |
| 64 | DB/OR | A/C Clutch Relay Control | Relay Off: 12v, On: 1v |
| 65, 75 | PK, LG | SCI Transmit, SCI Receive | 0v |
| 66 | WT/OR | Vehicle Speed Signal | Digital Signal |
| 67 | DB/WT | ASD Relay Control | Relay Off: 12v, On: 1v |
| 68 | PK/BK | EVAP Purge Solenoid Control | PWM Signal: 0-12-0v |
| 69 | DB/LG | High Speed Fan Control | Relay Off: 12v, On: 1v |
| 70 | PK/LB | EVAP Purge Solenoid Sense | 0-1v |
| 71 | WT/RD | EATX RPM Signal | Digital Signals |
| 72 | LB | LDP Switch Signal | Closed: 0v, Open: 12v |
| 73 | GY/LB | Tachometer Signals | Pulse Signals |
| 74 | BR | Fuel Pump Relay Control | Relay Off: 12v, On: 1v |
| 76 | BK/VT | TRS T41 Sense Signal | In P/N: 0v, Others: 5v |
| 77 | WT/DG | LDP Solenoid Control | PWM Signal: 0-12-0v |
| 78-79 | --- | Not Used | --- |
| 80 | LG/WT | S/C Vent Solenoid | Vacuum Decreasing: 1v |

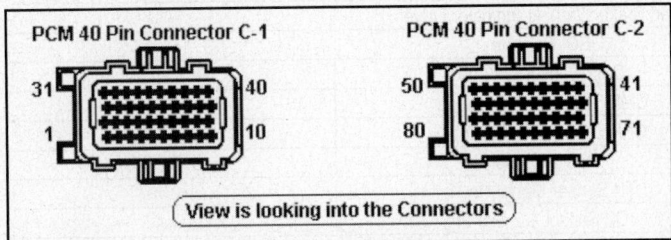

**Pin Connector Graphic**

PT CRUISER

**2004-05 PT Cruiser 1.6L 90-Pin Connector**

| PCM Pin # | Wire Color | Circuit Description (90-Pin) | Value at Hot Idle |
|---|---|---|---|
| 1 | GY/WT | Fused B (+) | |
| 2 | --- | Not Used | --- |
| 3 | --- | Not Used | --- |
| 4 | --- | Not Used | --- |
| 5 | DB/OR | A/C Compressor Clutch Relay Control | |
| 6 | VT/YL | PCI Bus | |
| 7 | DB/DG | APP Sensor No. 2 Signal | |
| 8 | BR/WT | A/C High Pressure Signal | |
| 9 | RD | IAT Sensor Ground | |
| 10 | BR/OR | VSS Ground | |
| 11 | DB/VT | Manifold Absolute Pressure Sensor Ground | |
| 12 | DG/LG | APP Sensor No. 1 5 Volt Supply | |
| 13 | DG/PK | APP Sensor No. 2 5 Volt Supply | |
| 14 | OR/PK | TP Sensor 5 Volt Supply | |
| 15 | --- | Not Used | --- |
| 16 | DG/RD | MAP Sensor Signal | |
| 17 | TN/BK | ECT Sensor Signal | |
| 18 | BK/DG | O2 Sensor 1/1 Signal | |
| 19 | BR/YL | APP Sensor No. 1 Ground | |
| 20 | DB/LG | KS Signal | |
| 21 | GY/BK | CKP Sensor Signal | |
| 22 | BR/DG | APP Sensor No. 2 Ground | |
| 23 | DB/OR | TP Sensor Ground | |
| 24 | BR/DG | O2 Sensor 1/1 Ground | |
| 25 | BR/WT | CKP Sensor Ground | |
| 26 | BK/VT | KS Ground | |
| 27 | --- | Not Used | --- |
| 28 | BK/TN | Ground | |
| 29 | BK/TN | Ground | |
| 30 | DB/TN | Ignition Coil No. 2 Driver | |
| 31 | VT/OR | ETC Positive Motor Control | |
| 32 | WT | ETC Negative Motor Control | |
| 33 | BK/LB | ECT Sensor Ground | |
| 34 | DB/YL | ASD Relay Control | |
| 35 | TN | Starter Motor Relay Control | |
| 36 | YL/RD | Low Speed Radiator Fan Relay Control | |
| 37 | --- | Not Used | --- |
| 38 | --- | Not Used | --- |
| 39 | DB/OR | PSP Switch Signal | |
| 40 | GY/BK | Red Brake Warning Indicator Driver | |
| 41 | LG | SCI Receive | |
| 42 | BR/GY | CMP Sensor Ground | |
| 43 | GY/PK | CMP Sensor 5 Volt Supply | |
| 44 | OR | CKP Sensor 5 Volt Supply | |
| 45 | OR/PK | VSS 5v Supply | |
| 46 | OR/DB | TP Sensor Signal No. 1 | |
| 47 | TN/WT | O2 Sensor 1/2 Signal | |
| 48 | --- | Not Used | --- |
| 49 | TN/YL | CMP Sensor Signal | |
| 50 | GY | Engine Oil Pressure Switch Signal | |
| 51 | --- | Not Used | --- |
| 52 | --- | Not Used | --- |
| 53 | DB/DG | O2 Sensor 1/2 Ground | |
| 54 | WT/DB | Fuel Injector No. 1 Driver | |
| 55 | PK/BK | EVAP/Purge Solenoid Control | |

**2004-05 PT Cruiser 1.6L 90-Pin Connector - Continued**

| PCM Pin # | Wire Color | Circuit Description (90-Pin) |
|-----------|-----------|------------------------------|
| 56 | --- | Not Used |
| 57 | --- | Not Used |
| 58 | --- | Not Used |
| 59 | BK/TN | Ground |
| 60 | BK/GY | Ignition Coil No. 1 Driver |
| 61 | DB/WT | Fused Ignition Switch Output (Run-Start) |
| 62 | --- | Not Used |
| 63 | --- | Not Used |
| 64 | --- | Not Used |
| 65 | DB/PK | High Speed Radiator Fan Relay Control |
| 66 | BR | Fuel Pump Relay Control |
| 67 | --- | Not Used |
| 68 | WT/TN | Brake Lamp Switch Output |
| 69 | LG/BK | Clutch Upstop Switch Signal |
| 70 | YL/RD | Clutch Interlock Switch Signal |
| 71 | WT/PK | Brake Lamp Switch Signal |
| 72 | BK/RD | Inlet Air Temperature Sensor Signal |
| 73 | PK | SCI Transmit |
| 74 | BK/LG | APP Sensor No. 1 Signal |
| 75 | VT/WT | MAP Sensor 5v Supply |
| 76 | --- | Not Used |
| 77 | DB/GY | TP Sensor Signal No. 2 |
| 78 | --- | Not Used |
| 79 | WT/OR | Vehicle Speed Sensor Signal |
| 80 | --- | Not Used |
| 81 | OR/RD | Oxygen Sensor 1/1 Control |
| 82 | TN | Fuel Injector No. 2 Driver |
| 83 | YL/WT | Fuel Injector No. 3 Driver |
| 84 | LB/BR | Fuel Injector No. 4 Driver |
| 85 | BR/VT | Oxygen Sensor 1/2 Control |
| 86 | --- | Not Used |
| 87 | --- | Not Used |
| 88 | DG/OR | Automatic Shut Down Relay Output |
| 89 | DG | Generator Field Control |
| 90 | BK/TN | Ground |

BLACK

POWERTRAIN
CONTROL
MODULE
(1.6L)

**Standard Colors and Abbreviations**

| Abbreviation | Color | Abbreviation | Color | Abbreviation | Color |
|--------------|-------|--------------|-------|--------------|-------|
| BK | Black | GY | Gray | RD | Red |
| BL | Blue | GN | Green | TN | Tan |
| BR | Brown | LG | Light Green | VT | Violet |
| DB | Dark Blue | OR | Orange | WT | White |
| DG | Dark Green | PK | Pink | YL | Yellow |

## 2001-02 PT Cruiser 2.4L VIN B A/T 'C1' Black Connector

| PCM Pin # | Wire Color | Circuit Description (40-Pin) | Value at Hot Idle |
|---|---|---|---|
| 1-2 | --- | Not Used | --- |
| 3 | DB/TN | Coil 2 Driver | 5°, 55 mph: 8° dwell |
| 4 | --- | Not Used | --- |
| 5 | YL/RD | Speed Control Power Supply | 0v or 6.7v |
| 6 | DG/OR | ASD Relay Output | 12-14v |
| 7 | YL/WT | Injector 3 Driver | 1.0-4.0 ms |
| 8 | DG | Generator Field Driver | Digital Signal: 0-12-0v |
| 9 | --- | Not Used | --- |
| 10 | BK/TN | Power Ground | <0.1v |
| 11 | BK/GY | Coil 2 Driver | 5°, 55 mph: 8° dwell |
| 12 | GY | Engine Oil Pressure Sensor | 1.6v at 24 psi |
| 13 | WT/DB | Injector 1 Driver | 1.0-4.0 ms |
| 14-15 | --- | Not Used | --- |
| 16 | LB/BR | Injector 4 Driver | 1.0-4.0 ms |
| 17 | TN | Injector 2 Driver | 1.0-4.0 ms |
| 18 | OR/RD | HO2S-11 (B1 S1) Heater | PWM Signal: 0-12-0v |
| 19 | --- | Not Used | --- |
| 20 | DB/WT | Ignition Switch Output | 12-14v |
| 21-22 | --- | Not Used | --- |
| 23 | LG/BK | Clutch Upstop Switch Signal | Pedal Up: 0v, Down: 1v |
| 24 | --- | Not Used | --- |
| 25 | DB/LG | Knock Sensor Signal | 0.080v AC |
| 26 | TN/BK | ECT Sensor Signal | At 180°F: 2.80v |
| 27 | BK/OR | Oxygen Sensor Ground | <0.050v |
| 28-29 | --- | Not Used | --- |
| 30 | BK/DG | HO2S-11 (B1 S1) Signal | 0.1-1.1v |
| 31 | TN | Engine Starter Motor Relay | KOEC: 9-11v |
| 32 | GY/BK | CKP Sensor Signal | Digital Signal: 0-5-0v |
| 33 | TN/YL | CMP Sensor Signal | Digital Signal: 0-5-0v |
| 34 | --- | Not Used | --- |
| 35 | OR/DB | TP Sensor Signal | 0.6-1.0v |
| 36 | DG/RD | MAP Sensor Signal | 1.5-1.7v |
| 37 | --- | Not Used | --- |
| 38 | BR | A/C Switch Sense | A/C Off: 12v, On: 1v |
| 37 | --- | Not Used | --- |
| 40 | GY/YL | EGR Solenoid Control | 12v, at 55 mph: 1v |

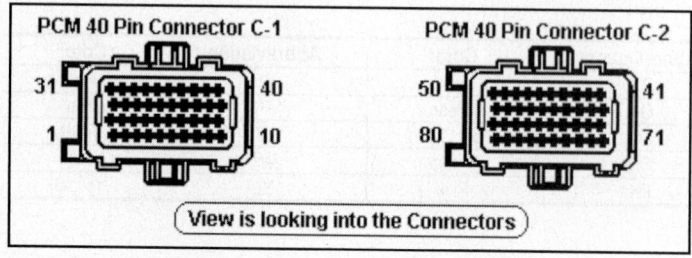

**Pin Connector Graphic**

**2001-02 PT Cruiser 2.4L VIN B A/T 'C2' Gray Connector**

| PCM Pin # | Wire Color | Circuit Description (40-Pin) | Value at Hot Idle |
|---|---|---|---|
| 41 | OR/DG | Speed Control Set Switch Signal | S/C & Set Switch On: 3.8v |
| 42 | --- | Not Used | --- |
| 43 | BK/LB | Sensor Ground | <0.050v |
| 44 | OR | 8-Volt Supply | 7.9-8.1v |
| 45 | DB/OR | PSP Switch Signal | Straight: 0v, Turning: 5v |
| 46 | RD/WT | Battery Power (Fused B+) | 12-14v |
| 47 | BK/WT | Power Ground | <0.1v |
| 48 | BR/WT | IAC 3 Driver | Pulse Signals: 0.8-11v |
| 49 | YL/BK | IAC 2 Driver | Pulse Signals: 0.8-11v |
| 50 | BK/TN | Power Ground | <0.1v |
| 51 | TN/WT | HO2S-12 (B1 S2) Signal | 0.1-1.1v |
| 52 | BK/RD | Inlet Air Temperature Sensor | At 86°F: 1.96v |
| 53-54 | --- | Not Used | --- |
| 55 | YL/RD | Low Speed Fan Relay | Relay Off: 12v, On: 1v |
| 56 | TN/RD | Speed Control Vacuum Solenoid | Vacuum Increasing: 1v |
| 57 | GY/RD | IAC 1 Driver | Pulse Signals: 0.8-11v |
| 58 | VT/BK | IAC 4 Driver | Pulse Signals: 0.8-11v |
| 59 | VT/YL | PCI Data Bus (J1850) | Digital Signal: 0-7-0v |
| 60-61 | --- | Not Used | --- |
| 61 | WT/PK | 5-Volt Supply | 4.9-5.1v |
| 62 | WT/PK | Brake Switch Signal | Brake Off: 0v, On: 12v |
| 63 | YL/DG | Torque Management Request | Digital Signals |
| 64 | DB/OR | A/C Clutch Relay Control | Relay Off: 12v, On: 1v |
| 65 | PK | SCI Transmit | 0v |
| 66 | WT/OR | Vehicle Speed Signal | Digital Signal |
| 67 | DB/YL | ASD Relay Control | Relay Off: 12v, On: 1v |
| 68 | PK/BK | EVAP Purge Solenoid Control | PWM Signal: 0-12-0v |
| 69 | DB/PK | High Speed Fan Relay | Relay Off: 12v, On: 1v |
| 70 | VT/RD | EVAP Purge Solenoid Sense | 0-1v |
| 71 | GY/BK | Red Brake Warning Indicator | Lamp Off: 12v, On: 1v |
| 72 | OR/YL | LDP Switch Sense | Open: 12v, Closed: 0v |
| 73 | --- | Not Used | --- |
| 74 | BR | Fuel Pump Relay Control | Relay Off: 12v, On: 1v |
| 75 | LG | SCI Receive | 5v |
| 76 | BK/WT | TRS T41 Signal | In P/N: 0v, Others: 5v |
| 77 | WT/DG | LDP Solenoid Control | PWM Signal: 0-12-0v |
| 78 | OR/BK | TCC Solenoid Control | Solenoid Off: 12v, On: 1v |
| 79 | --- | Not Used | --- |
| 80 | LG/RD | Speed Control Vent Solenoid | Vacuum Decreasing: 1v |

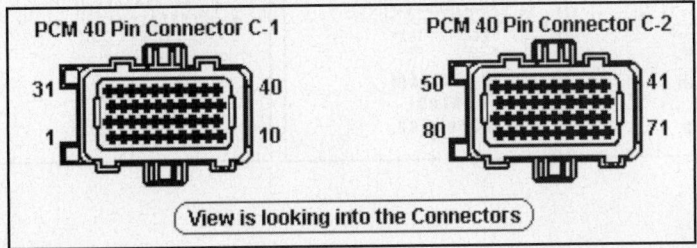

**Pin Connector Graphic**

## 2003-05 PT Cruiser 2.4L L4 'C1' Connector

| PCM Pin # | Wire Color | Circuit Description (38-Pin) | Value at Hot Idle |
|-----------|------------|-----------------------------|-------------------|
| 1-8 | --- | Not Used | --- |
| 9 | BK/TN | Ground | |
| 10 | --- | Not Used | --- |
| 11 | DB/WT | Fused Ignition Switch Output (Run-Start) | |
| 12 | RD/WT (EATX) | Fused Ignition Switch Output (Unlock-Run-Start) | |
| 13 | WT/OR (MTX) | Vehicle Speed Signal | |
| 14 | GY/BK | Red Brake Warning Indicator Driver | |
| 15 | LB (2.4L Turbo) | TP Solenoid Control | |
| 16 | --- | Not Used | --- |
| 17 | DB/YL (2.4L Turbo) | Surge Solenoid Control | |
| 18 | BK/TN | Ground | |
| 19 | --- | Not Used | --- |
| 20 | GY | Oil Pressure Signal | |
| 21 | --- | Not Used | --- |
| 22 | BR/OR | AAT Signal | |
| 23 | DB/LG (2.4L Turbo) | Tip Signal | |
| 24 | --- | Not Used | --- |
| 25 | LG | SCI Receive | |
| 26 | PK/LB (EATX) | SCI Receive (TCM) | |
| 27 | VT/WT (2.4L Turbo) | 5 Volt Supply | |
| 28 | DB/GY (2.4L Turbo) | Wastegate Sol Control | |
| 29 | RD/WT | Fused B (+) | |
| 30 | YL | Fused Ignition Switch Output (Start) | |
| 31-35 | --- | Not Used | --- |
| 36 | PK | SCI Transmit | |
| 37 | WT/DG (EATX) | SCI Transmit (TCM) | |
| 38 | VT/YL | PCI Bus | |

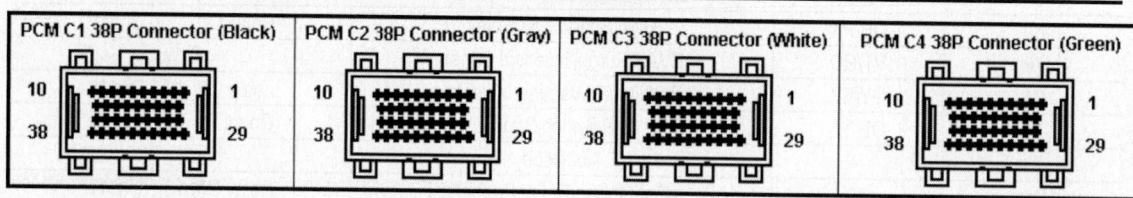

Pin Connector Graphic 2003

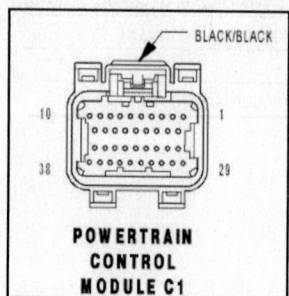
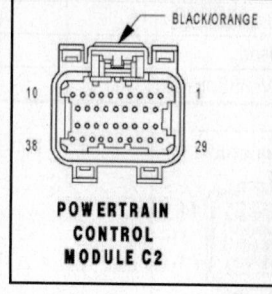
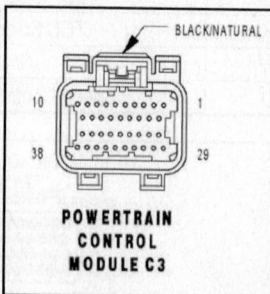
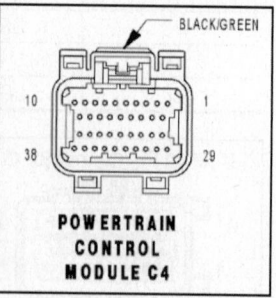

Pin Connector Graphic 2004-05

**2003-05 PT Cruiser 2.4L L4 'C2' Connector**

| PCM Pin # | Wire Color | Circuit Description (38-Pin) | Value at Hot Idle |
|---|---|---|---|
| 1-8 | --- | Not Used | --- |
| 9 | DB/TN | Coil Control No. 2 | |
| 10 | BK/GY | Coil Control No. 1 | |
| 11 | LB/BR | Injector Control No. 4 | |
| 12 | YL/WT | Injector Control No. 3 | |
| 13 | TN | Injector Control No. 2 | |
| 14 | WT/DB | Injector Control No. 1 | |
| 15-16 | --- | Not Used | --- |
| 17 | BR/VT | O2 1/2 Heater Control | |
| 18 | BR/OR | O2 1/1 Heater Control | |
| 19 | DG | Gen Field Control | |
| 20 | TN/BK | ECT Signal | |
| 21 | OR/DB | TP Signal | |
| 22 | --- | Not Used | --- |
| 23 | DG/RD | MAP Signal | |
| 24 | BK/VT | KS Return | |
| 25 | DB/LG | KS Signal | |
| 26 | --- | Not Used | --- |
| 27 | BK/LB | Sensor Ground 1 | |
| 28 | BR/VT | IAC Return | |
| 29 | OR | 5 Volt Supply | |
| 30 | BK/RD | IAT Signal | |
| 31 | BK/DG | O2 1/1 Signal | |
| 32 | DB/DG | O2 Return | |
| 33 | TN/WT | O2 1/2 Signal | |
| 34 | TN/YL | CMP Signal | |
| 35 | GY/BK | CKP Signal | |
| 36 | --- | Not Used | --- |
| 37 | --- | Not Used | --- |
| 38 | --- | Not Used | --- |

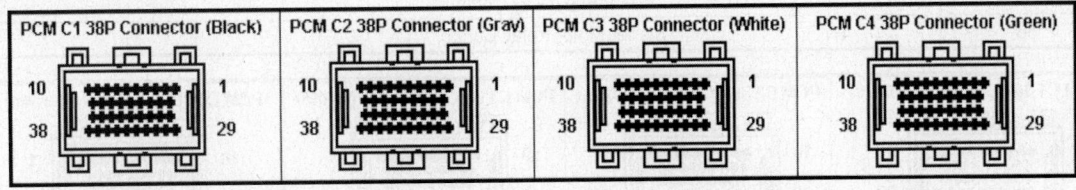

**Pin Connector Graphic 2003**

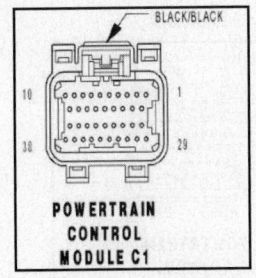 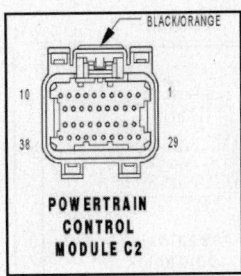 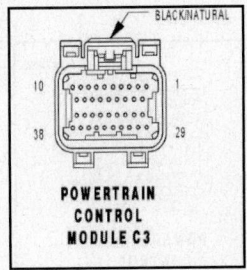 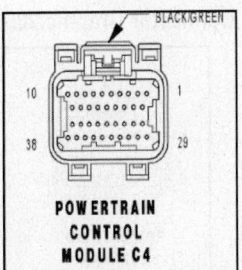

**Pin Connector Graphic 2004-05**

**2003-05 PT Cruiser 2.4L L4 'C3' White Connector**

| PCM Pin # | Wire Color | Circuit Description (38-Pin) | Value at Hot Idle |
|---|---|---|---|
| 1-2 | --- | Not Used | --- |
| 3 | DB/YL | Automatic Shut Down Relay Control | |
| 4 | DB/PK | High Speed Rad Fan Relay Control | |
| 5 | LG/RD | S/C Vent Control | |
| 6 | YL/RD | Low Speed Rad Fan Relay Control | |
| 6 | DB/WT (2.4L Turbo) | Low Speed Rad Fan Relay Control | |
| 7 | YL/RD | S/C Supply | |
| 8 | WT/DG | NVLD Solenoid Control | |
| 9-10 | --- | Not Used | --- |
| 11 | DB/OR | A/C Clutch Relay Control | |
| 12 | TN/RD | S/C Vacuum Control | |
| 13-16 | --- | Not Used | --- |
| 17 | BR/YL | Sensor Ground 2 | |
| 18 | --- | Not Used | --- |
| 19 | DG/OR | Automatic Shut Down Relay Output | |
| 20 | PK/BK | EVAP Purge Control | |
| 21 | YL/RD | Clutch Interlock Switch Signal | |
| 22 | --- | Not Used | --- |
| 23 | WT/PK | Brake Switch Signal | |
| 24 | DB/WT | A/C High Pressure Signal | |
| 25 | --- | Not Used | --- |
| 26 | YL/LB (Exc. Turbo) | AutoStick Downshift Switch Signal | |
| 26 | YL (Turbo) | AutoStick Downshift Switch Signal | |
| 27 | LG/LB (Exc. Turbo) | AutoStick Upshift Switch Signal | |
| 27 | LG (Turbo) | AutoStick Upshift Switch Signal | |
| 28 | DG/OR | Automatic Shutdown Relay Output | |
| 29 | WT/TN | EVAP Purge Return | |
| 30 | DB/OR | PSP Switch Signal | |
| 31-33 | --- | Not Used | --- |
| 34 | RD/LG | S/C Switch Signal | |
| 35 | OR | NVLD Switch Signal | |
| 36 | --- | Not Used | --- |
| 37 | BR | Fuel Pump Relay Control | |
| 38 | TN | Starter Motor Relay Control | |

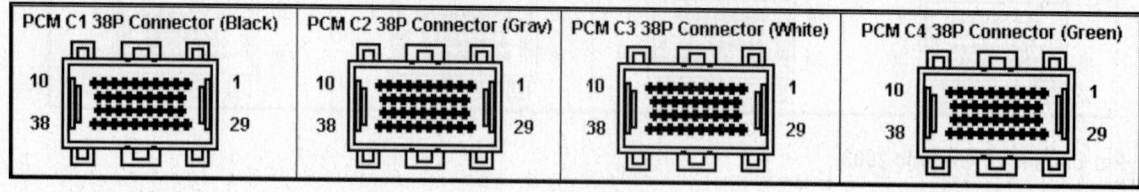

**Pin Connector Graphic 2003**

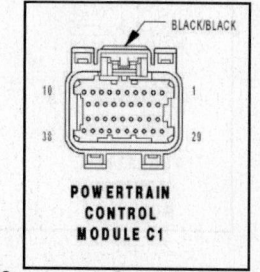

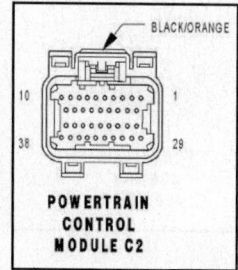

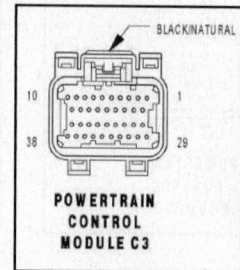

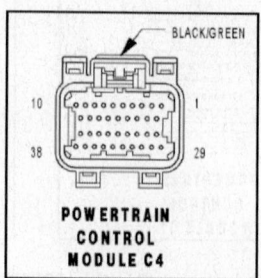

**Pin Connector Graphic 2004-05**

**2003-05 PT Cruiser 2.4L L4 'C4' Green Connector**

| PCM Pin # | Wire Color | Circuit Description (38-Pin) | Value at Hot Idle |
|---|---|---|---|
| 1 | BR | OD Solenoid Control | |
| 2 | PK (2.4L Exc. Turbo) | UD Solenoid Control | |
| 2 | PK/BK (2.4L Turbo) | UD Solenoid Control | |
| 3-5 | --- | Not Used | --- |
| 6 | WT | 2-4 Solenoid Control | |
| 7-9 | --- | Not Used | --- |
| 10 | LB | L/R Solenoid Control | |
| 11 | --- | Not Used | --- |
| 12 | BK/YL | Ground | |
| 13 | --- | Not Used | --- |
| 14 | BK/RD | Ground | |
| 15 | LG/BK | TRS T1 Sense | |
| 16 | VT | TRS T3 Sense | |
| 17 | --- | Not Used | --- |
| 18 | LG | Transmission Control Relay Control | |
| 19 | RD | Transmission Control Relay Output | |
| 20-21 | --- | Not Used | --- |
| 22 | OR/BK | OD Pressure Switch Sense | |
| 23-26 | --- | Not Used | --- |
| 27 | BK/WT | TRS T41 Sense | |
| 28 | RD | Transmission Control Relay Output | |
| 29 | DG | L/R Pressure Switch Sense | |
| 30 | YL/BK | 2-4 Pressure Switch Sense | |
| 31 | --- | Not Used | --- |
| 32 | LG/WT | Output Speed Sensor Signal | |
| 33 | RD/BK | Input Speed Sensor Signal | |
| 34 | DB/BK | Speed Sensor Ground | |
| 35 | VT/PK | Transmission Temperature Sensor Signal | |
| 36 | --- | Not Used | --- |
| 37 | VT/WT | TRS T42 Sense | |
| 38 | --- | Not Used | --- |

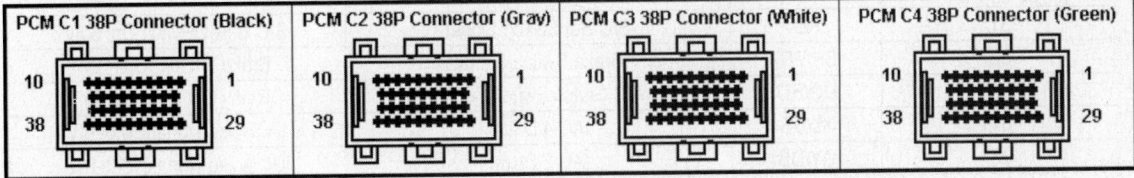

**Pin Connector Graphic 2003**

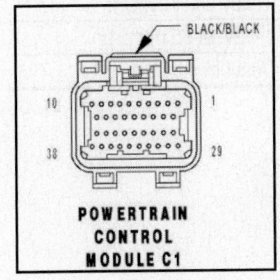
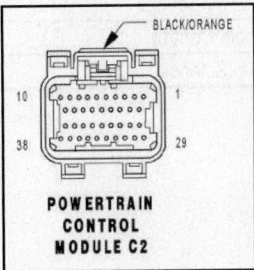
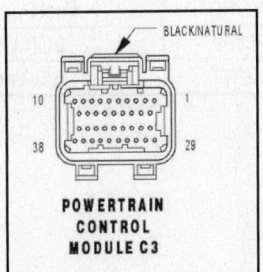
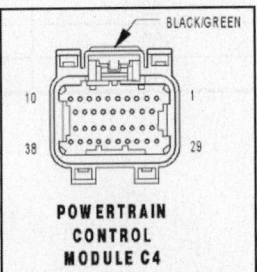

**Pin Connector Graphic 2004-05**

SEBRING SEDAN & CONVERTIBLE, STRATUS SEDAN & CONVERTIBLE, & AVENGER

**1995 Sebring & Avenger 2.0L VIN Y 60-Pin Connector**

| PCM Pin # | Wire Color | Circuit Description (60-Pin) | Value at Hot Idle |
|---|---|---|---|
| 1 | BK/GY | Coil 2 Driver | 5°, at 55 mph: 8° dwell |
| 2 | BK | Power Ground | <0.1v |
| 3 | YL/DG | Injector 3 Driver | 1.0-4.0 ms |
| 4 | LG/BK | Injector 1 Driver | 1.0-4.0 ms |
| 5 | YL/WT | Vehicle Speed Signal | Digital Signal |
| 6 | BR/DB | IAT Sensor Signal | At 100°F: 1.83v |
| 7 | WT/DG | HO2S-12 (B1 S2) Signal | 0.1-1.1v |
| 8 | WT/BK | HO2S-11 (B1 S1) Signal | 0.1-1.1v |
| 9 | DB/BK | SCI Receive | 0v |
| 10 | BR/RD | TP Sensor Signal | 0.6-1.0v |
| 11 | RD/BK | Battery Power (Fused B+) | 12-14v |
| 12 | --- | Not Used | --- |
| 13 | YL/DB | Fuel Level Sensor Signal | Digital Signal |
| 14 | OR | IAC 2 Driver | DC pulse signals: 0.8-11v |
| 15 | GY | IAC 3 Driver | DC pulse signals: 0.8-11v |
| 16 | LG/BK | EVAP Purge Solenoid Control | PWM Signal: 0-12-0v |
| 17 | LG | Speed Control Lamp Driver | S/C On: 1v, S/C Off: 12v |
| 18 | RD/WT | ASD Relay Control | Relay Off: 12v, On: 1v |
| 19 | DG/BK | Low Speed Fan Relay | Relay Off: 12v, On: 1v |
| 20 | RD/YL | Secondary Air Solenoid | Valve Off: 12v, On: 1v |
| 21 | BK/DB | Coil 1 Driver | 5°, at 55 mph: 8° dwell |
| 22 | BK | Power Ground | <0.1v |
| 23 | YL/RD | Injector 2 Driver | 1.0-4.0 ms |
| 24 | LG/RD | Injector 4 Driver | 1.0-4.0 ms |
| 25 | DB/WT | CKP Sensor Signal | Digital Signal: 0-5-0v |
| 26 | DB/RD | CMP Sensor Signal | Digital Signal: 0-5-0v |
| 27 | WT/YL | Knock Sensor Signal | 0.080v AC |
| 28 | DG/WT | ECT Sensor Signal | At 180°F: 2.80v |
| 29 | YL/BK | MAP Sensor Signal | 1.5-1.7v |
| 30 | PK | SCI Transmit | 0v |
| 31 | RD | S/C Set Switch Signal | S/C & Set Switch On: 3.8v |
| 32 | BR/WT | Brake Switch Signal | Brake Off: 0v, On: 12v |
| 33 | DG/RD | A/C Clutch Switch Sense | Relay Off: 12v, On: 1v |
| 34 | YL/DB | IAC 4 Driver | DC pulse signals: 0.8-11v |
| 35 | GY/DB | IAC 2 Driver | DC pulse signals: 0.8-11v |
| 36 | DG/RD | Amber MIL Control | MIL On: 1v, MIL Off: 12v |
| 37 | DB | Generator Lamp Control | Lamp On: 1v, Off: 12v |
| 38 | WT/RD | Fuel Pump Relay Control | Relay Off: 12v, On: 1v |
| 39 | RD/DB | EGR Solenoid Control | 12v, at 55 mph: 1v |
| 40 | LG/WT | S/C Vacuum Solenoid | Vacuum Increasing: 1v |

**1995 Sebring & Avenger 2.0L VIN Y 60-Pin Connector - Continued**

| PCM Pin # | Wire Color | Circuit Description (60-Pin) | Value at Hot Idle |
|---|---|---|---|
| 41 | DB | Alternator Field Control | Digital Signal: 0-12-0v |
| 42 | BK/RD | ASD Relay Output | 12-14v |
| 43 | DG/YL | 5-Volt Supply | 4.9-5.1v |
| 44 | YL | 8-Volt Supply | 7.9-8.1v |
| 45 | WT/DB | CCD Bus (-) | <0.050v |
| 46 | BK/DB | CCD Bus (+) | Digital Signal: 0-5-0v |
| 47 | --- | Not Used | --- |
| 48 | WT | Tachometer Signal | Pulse Signals |
| 49 | --- | Not Used | --- |
| 50 | BK/YL | PNP Switch Signal | In P/N: 0v, Others: 5v |
| 51 | BK/DG | Sensor Ground | <0.050v |
| 52 | BK | Power Ground | <0.1v |
| 53 | --- | Not Used | --- |
| 54 | BK/WT | Ignition Switch Output | 12-14v |
| 55 | RD/BK | Speed Control Mode Signal | S/C On: 1v, Off: 12v |
| 56 | DB/YL | PSP Switch Signal | Straight: 0v, Turning: 5v |
| 57 | DG/OR | High Speed Fan Relay | Relay Off: 12v, On: 1v |
| 58 | DB/YL | Bulb (lamp) Check Control | B/C On: 1v, B/C Off: 12v |
| 59 | DG | A/C Clutch Relay Control | Relay Off: 12v, On: 1v |
| 60 | BK/YL | S/C Vent Solenoid | Vacuum Decreasing: 1v |

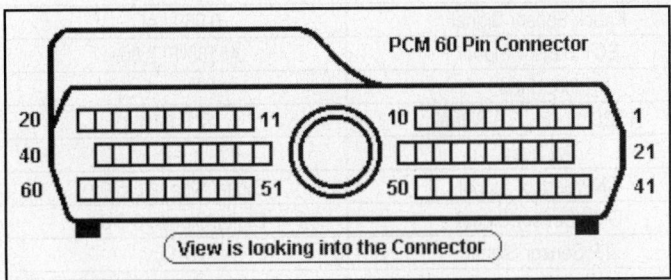

**Pin Connector Graphic**

**Standard Colors and Abbreviations**

| Abbreviation | Color | Abbreviation | Color | Abbreviation | Color |
|---|---|---|---|---|---|
| BK | Black | GY | Gray | RD | Red |
| BL | Blue | GN | Green | TN | Tan |
| BR | Brown | LG | Light Green | VT | Violet |
| DB | Dark Blue | OR | Orange | WT | White |
| DG | Dark Green | PK | Pink | YL | Yellow |

**1996-97 Sebring & Avenger 2.0L VIN Y 'C1' Connector**

| PCM Pin # | Wire Color | Circuit Description (40-Pin) | Value at Hot Idle |
|---|---|---|---|
| 1, 5 | --- | Not Used | --- |
| 2 | BK/DB | Coil 1 Driver | 5°, at 55 mph: 8° dwell |
| 3 | BR | Coil 2 Driver | 5°, at 55 mph: 8° dwell |
| 4 | DB | Alternator Field Control | Digital Signal: 0-12-0v |
| 6 | BK/RD | ASD Relay Output | 12-14v |
| 7 | YL/DG | Injector 3 Driver | 1.0-4.0 ms |
| 8 | DG/RD | Amber MIL Control | MIL On: 1v, MIL Off: 12v |
| 9 | DB/YL | Bulb (lamp) Check Control | B/C On: 1v, B/C Off: 12v |
| 10 | BK | Power Ground | <0.1v |
| 11 | --- | Not Used | --- |
| 12 | RD/BK | S/C On Switch Signal | S/C On: 1v, Off: 12v |
| 13 | LG/BK | Injector 1 Driver | 1.0-4.0 ms |
| 14-15 | --- | Not Used | --- |
| 16 | LG/RD | Injector 4 Driver | 1.0-4.0 ms |
| 17 | YL/RD | Injector 2 Driver | 1.0-4.0 ms |
| 18 | --- | Not Used | --- |
| 19 | DG/OR | High Speed Fan Relay | Relay Off: 12v, On: 1v |
| 20 | BK/WT | Ignition Switch Output | 12-14v |
| 21, 25 | --- | Not Used | --- |
| 22 | LG | Speed Control Lamp Driver | S/C On: 1v, S/C Off: 12v |
| 23 | YL/DB | Fuel Level Sensor Signal | 70 ohms (±20) w/full tank |
| 24 | WT/YL | Knock Sensor Signal | 0.080v AC |
| 26 | DG/WT | ECT Sensor Signal | At 180°F: 2.80v |
| 27-29 | --- | Not Used | --- |
| 30 | WT/BK | HO2S-11 (B1 S1) Signal | 0.1-1.1v |
| 31, 34 | --- | Not Used | --- |
| 32 | DB/WT | CKP Sensor Signal | Digital Signal: 0-5-0v |
| 33 | DB/RD | CMP Sensor Signal | Digital Signal: 0-5-0v |
| 35 | BR/RD | TP Sensor Signal | 0.6-1.0v |
| 36 | YL/BK | MAP Sensor Signal | 1.5-1.7v |
| 37 | BR/DB | IAT Sensor Signal | At 100°F: 1.83v |
| 38 | DG/RD | A/C Select Switch Sense | Relay Off: 12v, On: 1v |
| 39 | --- | Not Used | --- |
| 40 | RD/DB | EGR Solenoid Control | 12v, at 55 mph: 1v |

## Standard Colors and Abbreviations

| Abbreviation | Color | Abbreviation | Color | Abbreviation | Color |
|---|---|---|---|---|---|
| BK | Black | GY | Gray | RD | Red |
| BL | Blue | GN | Green | TN | Tan |
| BR | Brown | LG | Light Green | VT | Violet |
| DB | Dark Blue | OR | Orange | WT | White |
| DG | Dark Green | PK | Pink | YL | Yellow |

**1996-97 Sebring & Avenger 2.0L VIN Y 'C2' Connector**

| PCM Pin # | Wire Color | Circuit Description (40-Pin) | Value at Hot Idle |
|---|---|---|---|
| 41 | RD | S/C Set Switch Signal | S/C & Set Switch On: 3.8v |
| 42 | --- | Not Used | --- |
| 43 | BK/DG | Sensor Ground | <0.050v |
| 44 | YL | 8-Volt Supply | 7.9-8.1v |
| 45 | DB/YL | PSP Switch Signal | Straight: 0v, Turning: 5v |
| 46 | RD/BK | Battery Power (Fused B+) | 12-14v |
| 47 | BK/DG | Sensor Ground | <0.050v |
| 48 | GY/DB | IAC 3 Driver | DC pulse signals: 0.8-11v |
| 49 | OR | IAC 2 Driver | DC pulse signals: 0.8-11v |
| 50 | BK | Power Ground | <0.1v |
| 51 | WT/DG | HO2S-12 (B1 S2) Signal | 0.1-1.1v |
| 52-54 | --- | Not Used | --- |
| 55 | DG/BK | Low Speed Fan Relay | Relay Off: 12v, On: 1v |
| 56 | DB | Generator Lamp Control | Lamp On: 1v, Off: 12v |
| 57 | GY | IAC 1 Driver | DC pulse signals: 0.8-11v |
| 58 | YL/DB | IAC 4 Driver | DC pulse signals: 0.8-11v |
| 59 | BK/DB | CCD Bus (+) | Digital Signal: 0-5-0v |
| 60 | WT/DB | CCD Bus (-) | <0.050v |
| 61 | DG/YL | 5-Volt Supply | 4.9-5.1v |
| 62 | BR/WT | Brake Switch Signal | Brake Off: 0v, On: 12v |
| 63 | OR/BK | Torque Management Request | Digital Signals |
| 64 | DG | A/C Clutch Relay Control | Relay Off: 12v, On: 1v |
| 65 | PK | SCI Transmit | 0v |
| 66 | YL/WT | Vehicle Speed Signal | Digital Signal |
| 67 | RD/WT | ASD Relay Control | Relay Off: 12v, On: 1v |
| 68 | LG/BK | EVAP Purge Solenoid Control | PWM Signal: 0-12-0v |
| 69 | DG/OR | High Speed Fan Relay | Relay Off: 12v, On: 1v |
| 70-71 | --- | Not Used | --- |
| 72 | RD/DB | LDP Switch Sense | Open: 12v, Closed: 0v |
| 73 | WT | Tachometer Signal | Pulse Signals |
| 74 | BK/DB | Fuel Pump Relay Control | Relay Off: 12v, On: 1v |
| 75 | DB/BK | SCI Receive | 0v |
| 76 | BK/YL | PNP Switch Signal | In P/N: 0v, Others: 5v |
| 77 | RD/YL | LDP Solenoid Control | PWM Signal: 0-12-0v |
| 78 | LG/WT | S/C Vacuum Solenoid | Vacuum Increasing: 1v |
| 79 | RD/YL | M/T: Aspirator Solenoid | Valve Off: 12v, On: 1v |
| 80 | BK/YL | S/C Vent Solenoid | Vacuum Decreasing: 1v |

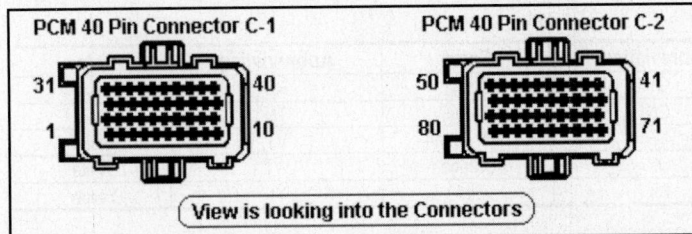

**Pin Connector Graphic**

## 1998-99 Sebring & Avenger 2.0L VIN Y 'C1' Connector

| PCM Pin # | Wire Color | Circuit Description (40-Pin) | Value at Hot Idle |
|---|---|---|---|
| 1-2 | --- | Not Used | --- |
| 3 | BR | Coil 2 Driver | 5°, at 55 mph: 8° dwell |
| 4-5 | --- | Not Used | --- |
| 6 | BK/RD | ASD Relay Output | 12-14v |
| 7 | YL/DG | Injector 3 Driver | 1.0-4.0 ms |
| 8 | DB | Alternator Field Control | Digital Signal: 0-12-0v |
| 9 | DB/YL | Bulb (lamp) Check Control | B/C On: 1v, B/C Off: 12v |
| 10 | BK | Power Ground | <0.1v |
| 11 | BK/DB | Coil 1 Driver | 5°, at 55 mph: 8° dwell |
| 12 | RD/BK | S/C On Switch Signal | S/C On: 1v, Off: 12v |
| 13 | LG/BK | Injector 1 Driver | 1.0-4.0 ms |
| 14-15 | --- | Not Used | --- |
| 16 | LG/RD | Injector 4 Driver | 1.0-4.0 ms |
| 17 | YL/RD | Injector 2 Driver | 1.0-4.0 ms |
| 18 | LG | Speed Control Lamp Driver | S/C On: 1v, S/C Off: 12v |
| 19 | DG/OR | High Speed Fan Relay | Relay Off: 12v, On: 1v |
| 20 | BK/WT | Ignition Switch Output | 12-14v |
| 21 | --- | Not Used | --- |
| 22 | DG/RD | Amber MIL Control | MIL On: 1v, MIL Off: 12v |
| 23 | YL/DB | Fuel Level Sensor Signal | 70 ohms (±20) w/full tank |
| 24 | --- | Not Used | --- |
| 25 | WT/YL | Knock Sensor Signal | 0.080v AC |
| 26 | DG/WT | ECT Sensor Signal | At 180°F: 2.80v |
| 27-29 | --- | Not Used | --- |
| 30 | WT/BK | HO2S-11 (B1 S1) Signal | 0.1-1.1v |
| 31 | --- | Not Used | --- |
| 32 | DB/WT | CKP Sensor Signal | Digital Signal: 0-5-0v |
| 33 | DB/RD | CMP Sensor Signal | Digital Signal: 0-5-0v |
| 34 | --- | Not Used | --- |
| 35 | BR/RD | TP Sensor Signal | 0.6-1.0v |
| 36 | YL/BK | MAP Sensor Signal | 1.5-1.7v |
| 37 | BR/DB | IAT Sensor Signal | At 100°F: 1.83v |
| 38 | DG/RD | A/C Select Switch Sense | Relay Off: 12v, On: 1v |
| 39 | DB | Generator Lamp Control | Lamp On: 1v, Off: 12v |
| 40 | RD/DB | EGR Solenoid Control | 12v, at 55 mph: 1v |

## Standard Colors and Abbreviations

| Abbreviation | Color | Abbreviation | Color | Abbreviation | Color |
|---|---|---|---|---|---|
| BK | Black | GY | Gray | RD | Red |
| BL | Blue | GN | Green | TN | Tan |
| BR | Brown | LG | Light Green | VT | Violet |
| DB | Dark Blue | OR | Orange | WT | White |
| DG | Dark Green | PK | Pink | YL | Yellow |

**1998-99 Sebring & Avenger 2.0L VIN Y 'C2' Connector**

| PCM Pin # | Wire Color | Circuit Description (40-Pin) | Value at Hot Idle |
|---|---|---|---|
| 41 | RD | S/C Set Switch Signal | S/C & Set Switch On: 3.8v |
| 42-43 | --- | Not Used | --- |
| 44 | YL | 8-Volt Supply | 7.9-8.1v |
| 45 | DB/YL | PSP Switch Signal | Straight: 0v, Turning: 5v |
| 46 | RD/BK | Battery Power (Fused B+) | 12-14v |
| 47 | BK | Sensor Ground | <0.050v |
| 48 | GY/DB | IAC 3 Driver | DC pulse signals: 0.8-11v |
| 49 | OR | IAC 2 Driver | DC pulse signals: 0.8-11v |
| 50 | BK | Power Ground | <0.1v |
| 51 | WT/DG | HO2S-12 (B1 S2) Signal | 0.1-1.1v |
| 52-54 | --- | Not Used | --- |
| 55 | DG/BK | Low Speed Fan Relay | Relay Off: 12v, On: 1v |
| 56 | LG/WT | S/C Vacuum Solenoid | Vacuum Increasing: 1v |
| 57 | GY | IAC 1 Driver | DC pulse signals: 0.8-11v |
| 58 | YL/DB | IAC 4 Driver | DC pulse signals: 0.8-11v |
| 59 | BK/DB | CCD Bus (+) | Digital Signal: 0-5-0v |
| 60 | WT/DB | CCD Bus (-) | <0.050v |
| 61 | DG/YL | 5-Volt Supply | 4.9-5.1v |
| 62 | BR/WT | Brake Switch Signal | Brake Off: 0v, On: 12v |
| 63 | OR/BK | Torque Management Request | Digital Signals |
| 64 | DG | A/C Clutch Relay Control | Relay Off: 12v, On: 1v |
| 65 | PK | SCI Transmit | 0v |
| 66 | YL/WT | Vehicle Speed Signal | Digital Signal |
| 67 | RD/WT | ASD Relay Control | Relay Off: 12v, On: 1v |
| 68 | LG/BK | EVAP Purge Solenoid Control | PWM Signal: 0-12-0v |
| 69 | DG/OR | High Speed Fan Relay | Relay Off: 12v, On: 1v |
| 70 | BK/WT | EVAP Purge Solenoid Sense | 0-1v |
| 72 | RD/DB | LDP Switch Sense | Open: 12v, Closed: 0v |
| 73 | WT | Tachometer Signal | Pulse Signals |
| 74 | BK/DB | Fuel Pump Relay Control | Relay Off: 12v, On: 1v |
| 75 | DB/BK | SCI Receive | 0v |
| 76 | BK/YL | PNP Switch Signal | In P/N: 0v, Others: 5v |
| 77 | RD/YL | LDP Solenoid Control | PWM Signal: 0-12-0v |
| 78-79 | --- | Not Used | --- |
| 80 | BK/YL | S/C Vent Solenoid | Vacuum Decreasing: 1v |

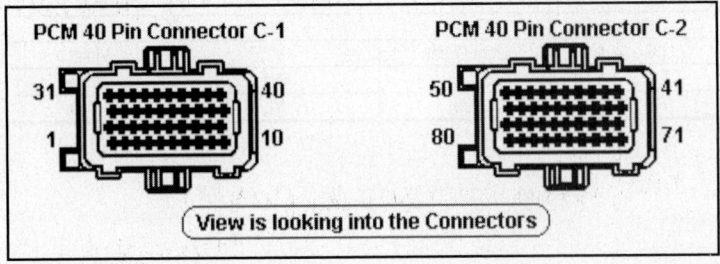

**Pin Connector Graphic**

## 1996-97 Sebring & Stratus 2.4L VIN X (A/T) 'C1' Connector

| PCM Pin # | Wire Color | Circuit Description (40-Pin) | Value at Hot Idle |
|---|---|---|---|
| 1 | --- | Not Used | --- |
| 2 | BK/GY | Coil 1 Driver | 5°, at 55 mph: 8° dwell |
| 3 | DB/DG | Coil 2 Driver | 5°, at 55 mph: 8° dwell |
| 4 | DB | Alternator Field Control | Digital Signal: 0-12-0v |
| 4 ('97) | DG | Alternator Field Control | Digital Signal: 0-12-0v |
| 5 | YL/PK | S/C Power Supply | 12-14v |
| 6 | DG/OR | ASD Relay Output | 12-14v |
| 7 | YL/WT | Injector 3 Driver | 1.0-4.0 ms |
| 8-9 | --- | Not Used | --- |
| 10 | BK/TN | Power Ground | <0.1v |
| 11-12 | --- | Not Used | --- |
| 13 | WT/LB | Injector 1 Driver | 1.0-4.0 ms |
| 14-15 | --- | Not Used | --- |
| 16 | LB/BR | Injector 4 Driver | 1.0-4.0 ms |
| 17 | TN | Injector 2 Driver | 1.0-4.0 ms |
| 18-19 | --- | Not Used | --- |
| 20 | DB/WT | Ignition Switch Output | 12-14v |
| 21-23 | --- | Not Used | --- |
| 24 | GY/BK | Knock Sensor Signal | 0.080v AC |
| 25 | --- | Not Used | --- |
| 26 | TN/BK | ECT Sensor Signal | At 180°F: 2.80v |
| 27-29 | --- | Not Used | --- |
| 30 | BK/DG | HO2S-11 (B1 S1) Signal | 0.1-1.1v |
| 31 | --- | Not Used | --- |
| 32 | GY/BK | CKP Sensor Signal | Digital Signal: 0-5-0v |
| 33 | TN/YL | CMP Sensor Signal | Digital Signal: 0-5-0v |
| 34 | --- | Not Used | --- |
| 35 | OR/LB | TP Sensor Signal | 0.6-1.0v |
| 36 | DG/RD | MAP Sensor Signal | 1.5-1.7v |
| 37 | BK/RD | IAT Sensor Signal | At 100°F: 1.83v |
| 38-39 | --- | Not Used | --- |
| 40 | GY/YL | EGR Solenoid Control | 12v, at 55 mph: 1v |

## Standard Colors and Abbreviations

| Abbreviation | Color | Abbreviation | Color | Abbreviation | Color |
|---|---|---|---|---|---|
| BK | Black | GY | Gray | RD | Red |
| BL | Blue | GN | Green | TN | Tan |
| BR | Brown | LG | Light Green | VT | Violet |
| DB | Dark Blue | OR | Orange | WT | White |
| DG | Dark Green | PK | Pink | YL | Yellow |

**1996-97 Sebring & Stratus 2.4L VIN X (A/T) 'C2' Connector**

| PCM Pin # | Wire Color | Circuit Description (40-Pin) | Value at Hot Idle |
|---|---|---|---|
| 41 | PK/LG | S/C Set Switch Signal | S/C & Set Switch On: 3.8v |
| 42 | DB/YL | A/C Pressure Switch Signal | A/C On: 0.45-4.85v |
| 43 | BK/LB | Sensor Ground | <0.050v |
| 44 | OR/WT | 8-Volt Supply | 7.9-8.1v |
| 45 | DB/LG | PSP Switch Signal | Straight: 0v, Turning: 5v |
| 46 | RD/TN | Battery Power (Fused B+) | 12-14v |
| 47, 56 | --- | Not Used | --- |
| 48 | BR/GY | IAC 1 Driver | DC pulse signals: 0.8-11v |
| 49 | YL/BK | IAC 2 Driver | DC pulse signals: 0.8-11v |
| 50 | BK/TN | Power Ground | <0.1v |
| 51 | TN/WT | HO2S-12 (B1 S2) Signal | 0.1-1.1v |
| 52 | PK/LG | Battery Temperature Sensor | At 86°F: 1.96v |
| 55 | DB/TN | Low Speed Fan Relay | Relay Off: 12v, On: 1v |
| 57 | GY/RD | IAC 3 Driver | DC pulse signals: 0.8-11v |
| 58 | PK/GY | IAC 4 Driver | DC pulse signals: 0.8-11v |
| 59 | PK/BR | CCD Bus (+) | Digital Signal: 0-5-0v |
| 60 | WT/BK | CCD Bus (-) | <0.050v |
| 61 | PK/WT | 5-Volt Supply | 4.9-5.1v |
| 62 | WT/RD | Brake Switch Signal | Brake Off: 0v, On: 12v |
| 63 | YL/DG | Torque Management Request | Digital Signals |
| 64 | DB/OR | A/C Clutch Relay Control | Relay Off: 12v, On: 1v |
| 65 | PK/LB | SCI Transmit | 0v |
| 66 | WT/OR | Vehicle Speed Signal | Digital Signal |
| 67 | DB/PK | ASD Relay Control | Relay Off: 12v, On: 1v |
| 68 | PK/GY | EVAP Purge Solenoid Control | PWM Signal: 0-12-0v |
| 69 | DB/PK | High Speed Fan Relay | Relay Off: 12v, On: 1v |
| 70-71, 79 | --- | Not Used | --- |
| 72 | OR/DG | LDP Switch Sense | Open: 12v, Closed: 0v |
| 74 | BR/LG | Fuel Pump Relay Control | Relay Off: 12v, On: 1v |
| 75 | LG/WT | SCI Receive | 0v |
| 76 | BK/WT | PNP Switch Signal | In P/N: 0v, Others: 5v |
| 77 | WT/DG | LDP Solenoid Control | PWM Signal: 0-12-0v |
| 78 | WT/PK | S/C Vacuum Solenoid | Vacuum Increasing: 1v |
| 80 | LG/RD | S/C Vent Solenoid | Vacuum Decreasing: 1v |

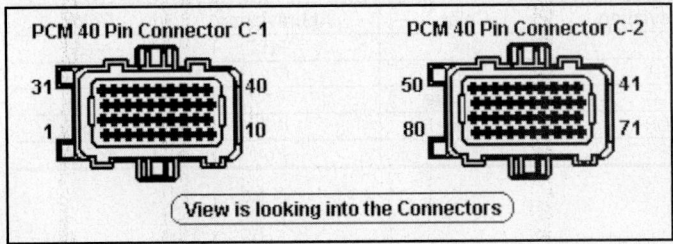

**Pin Connector Graphic**

**1998-2000 Sebring & Stratus 2.4L VIN X (A/T) 'C1' Connector**

| PCM Pin # | Wire Color | Circuit Description (40-Pin) | Value at Hot Idle |
|-----------|-----------|------------------------------|-------------------|
| 1-2 | --- | Not Used | --- |
| 3 | DB/TN | Coil 2 Driver | 5°, at 55 mph: 8° dwell |
| 4 | --- | Not Used | --- |
| 5 | YL/RD | S/C Power Supply | 12-14v |
| 6 | DG/OR | ASD Relay Output | 12-14v |
| 7 | YL/WT | Injector 3 Driver | 1.0-4.0 ms |
| 8 | DG | Alternator Field Control | Digital Signal: 0-12-0v |
| 9 | --- | Not Used | --- |
| 10 | BK/TN | Power Ground | <0.1v |
| 11 | BK/GY | Coil 1 Driver | 5°, at 55 mph: 8° dwell |
| 12 | --- | Not Used | --- |
| 13 | WT/DB | Injector 1 Driver | 1.0-4.0 ms |
| 14-15 | --- | Not Used | --- |
| 16 | LB/BR | Injector 4 Driver | 1.0-4.0 ms |
| 17 | TN | Injector 2 Driver | 1.0-4.0 ms |
| 18-19 | --- | Not Used | --- |
| 20 | DB/WT | Ignition Switch Output | 12-14v |
| 21-24 | --- | Not Used | --- |
| 25 | DB/LG | Knock Sensor Signal | 0.080v AC |
| 26 | TN/BK | ECT Sensor Signal | At 180°F: 2.80v |
| 27 | BK/OR | Oxygen Sensor Ground | <0.050v |
| 28-29 | --- | Not Used | --- |
| 30 | BK/DG | HO2S-11 (B1 S1) Signal | 0.1-1.1v |
| 31 | TN | Starter Relay Control | KOEC: 9-11v |
| 32 | GY/BK | CKP Sensor Signal | Digital Signal: 0-5-0v |
| 33 | TN/YL | CMP Sensor Signal | Digital Signal: 0-5-0v |
| 34 | --- | Not Used | --- |
| 35 | OR/DB | TP Sensor Signal | 0.6-1.0v |
| 36 | DG/RD | MAP Sensor Signal | 1.5-1.7v |
| 37 | BK/RD | IAT Sensor Signal | At 100°F: 1.83v |
| 38-39 | --- | Not Used | --- |
| 40 | GY/YL | EGR Solenoid Control | 12v, at 55 mph: 1v |

## Standard Colors and Abbreviations

| Abbreviation | Color | Abbreviation | Color | Abbreviation | Color |
|--------------|-------|--------------|-------|--------------|-------|
| BK | Black | GY | Gray | RD | Red |
| BL | Blue | GN | Green | TN | Tan |
| BR | Brown | LG | Light Green | VT | Violet |
| DB | Dark Blue | OR | Orange | WT | White |
| DG | Dark Green | PK | Pink | YL | Yellow |

**1998-2000 Sebring & Stratus 2.4L VIN X (A/T) 'C2' Connector**

| PCM Pin # | Wire Color | Circuit Description (40-Pin) | Value at Hot Idle |
|---|---|---|---|
| 41 | RD/LG | S/C Set Switch Signal | S/C & Set Switch On: 3.8v |
| 42 | DB/YL | A/C Pressure Switch Signal | A/C On: 0.45-4.85v |
| 43 | BK/LB | Sensor Ground | <0.050v |
| 44 | OR/WT | 8-Volt Supply | 7.9-8.1v |
| 45 | DB/LG | PSP Switch Signal | Straight: 0v, Turning: 5v |
| 46 | RD/TN | Battery Power (Fused B+) | 12-14v |
| 47 | BK/WT | Sensor Ground | <0.050v |
| 48 | BR/WT | IAC 3 Driver | DC pulse signals: 0.8-11v |
| 49 | YL/BK | IAC 2 Driver | DC pulse signals: 0.8-11v |
| 50 | BK/TN | Power Ground | <0.1v |
| 51 | TN/WT | HO2S-12 (B1 S2) Signal | 0.1-1.1v |
| 52 | PK/YL | Battery Temperature Sensor | At 86ºF: 1.96v |
| 53-54 | --- | Not Used | --- |
| 55 | DB/TN | Low Speed Fan Relay | Relay Off: 12v, On: 1v |
| 56 | WT/PK | S/C Vacuum Solenoid | Vacuum Increasing: 1v |
| 57 | GY/RD | IAC 1 Driver | DC pulse signals: 0.8-11v |
| 58 | PK/BK | IAC 4 Driver | DC pulse signals: 0.8-11v |
| 59 | PK/BR | CCD Bus (+) | Digital Signal: 0-5-0v |
| 60 | WT/BK | CCD Bus (-) | <0.050v |
| 61 | PK/WT | 5-Volt Supply | 4.9-5.1v |
| 62 | WT/RD | Brake Switch Signal | Brake Off: 0v, On: 12v |
| 63 | YL/DG | Torque Management Request | Digital Signals |
| 64 | DB/OR | A/C Clutch Relay Control | Relay Off: 12v, On: 1v |
| 65 | PK/LB | SCI Transmit | 0v |
| 66 | WT/OR | Vehicle Speed Signal | Digital Signal |
| 67 | DB/PK | ASD Relay Control | Relay Off: 12v, On: 1v |
| 68 | PK/GY | EVAP Purge Solenoid Control | PWM Signal: 0-12-0v |
| 69 | DB/PK | High Speed Fan Relay | Relay Off: 12v, On: 1v |
| 70 | WT/TN | EVAP Purge Solenoid Sense | 0-1v |
| 71-73 | --- | Not Used | --- |
| 74 | BR/LG | Fuel Pump Relay Control | Relay Off: 12v, On: 1v |
| 75 | LG | SCI Receive | 0v |
| 76 | BK/LB | PNP Switch Signal | In P/N: 0v, Others: 5v |
| 77-79 | --- | Not Used | --- |
| 80 | LG/RD | S/C Vent Solenoid | Vacuum Decreasing: 1v |

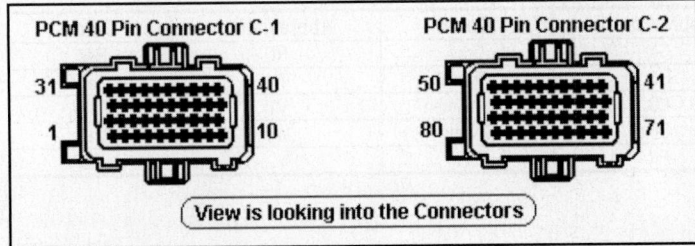

**Pin Connector Graphic**

## 2001-02 Sebring & Stratus 2.4L VIN X (A/T) 'C1' Connector

| PCM Pin # | Wire Color | Circuit Description (40-Pin) | Value at Hot Idle |
|---|---|---|---|
| 1-2 | --- | Not Used | --- |
| 3 | DB/TN | Coil 2 Driver | 5°, at 55 mph: 8° dwell |
| 4 | --- | Not Used | --- |
| 5 | YL/RD | S/C Power Supply | 12-14v |
| 6 | DG/OR | ASD Relay Output | 12-14v |
| 7 | YL/WT | Injector 3 Driver | 1.0-4.0 ms |
| 8 | DG | Alternator Field Control | Digital Signal: 0-12-0v |
| 9 | --- | Not Used | --- |
| 10 | BK/TN | Power Ground | <0.1v |
| 11 | BK/GY | Coil 1 Driver | 5°, at 55 mph: 8° dwell |
| 12 | --- | Not Used | --- |
| 13 | WT/DB | Injector 1 Driver | 1.0-4.0 ms |
| 14-15 | --- | Not Used | --- |
| 16 | LB/BR | Injector 4 Driver | 1.0-4.0 ms |
| 17 | TN | Injector 2 Driver | 1.0-4.0 ms |
| 18 | OR/RD | HO2S-11Heater Control | Heater On: 1v, Off: 12v |
| 19 | --- | Not Used | --- |
| 20 | DB/WT | Ignition Switch Output | 12-14v |
| 21-24 | --- | Not Used | --- |
| 25 | DB/LG | Knock Sensor Signal | 0.080v AC |
| 26 | TN/BK | ECT Sensor Signal | At 180°F: 2.80v |
| 27 | DB/OR | Oxygen Sensor Ground | <0.050v |
| 28-29 | --- | Not Used | --- |
| 30 | BK/DG | HO2S-11 (B1 S1) Signal | 0.1-1.1v |
| 31 | TN | Starter Relay Control | KOEC: 9-11v |
| 32 | GY/BK | CKP Sensor Signal | Digital Signal: 0-5-0v |
| 33 | TN/YL | CMP Sensor Signal | Digital Signal: 0-5-0v |
| 34 | --- | Not Used | --- |
| 35 | OR/DB | TP Sensor Signal | 0.6-1.0v |
| 36 | DG/RD | MAP Sensor Signal | 1.5-1.7v |
| 37 | BK/RD | IAT Sensor Signal | At 100°F: 1.83v |
| 38-39 | --- | Not Used | --- |
| 40 | GY/YL | EGR Solenoid Control | 12v, at 55 mph: 1v |

### Standard Colors and Abbreviations

| Abbreviation | Color | Abbreviation | Color | Abbreviation | Color |
|---|---|---|---|---|---|
| BK | Black | GY | Gray | RD | Red |
| BL | Blue | GN | Green | TN | Tan |
| BR | Brown | LG | Light Green | VT | Violet |
| DB | Dark Blue | OR | Orange | WT | White |
| DG | Dark Green | PK | Pink | YL | Yellow |

## 2001-02 Sebring & Stratus 2.4L VIN X (A/T) 'C2' Connector

| PCM Pin # | Wire Color | Circuit Description (40-Pin) | Value at Hot Idle |
|---|---|---|---|
| 41 | PK/LG | S/C Set Switch Signal | S/C & Set Switch On: 3.8v |
| 42 | DB | A/C Pressure Switch Signal | A/C On: 0.45-4.85v |
| 43 | BK/LB | Sensor Ground | <0.050v |
| 44 | OR/WT | 8-Volt Supply | 7.9-8.1v |
| 45 | DB/OR | PSP Switch Signal | Straight: 0v, Turning: 5v |
| 46 | RD/TN | Battery Power (Fused B+) | 12-14v |
| 47 | BK | Power Ground | <0.1v |
| 48 | BR/WT | IAC 3 Driver | DC pulse signals: 0.8-11v |
| 49 | YL/BK | IAC 2 Driver | DC pulse signals: 0.8-11v |
| 50 | BK/TN | Power Ground | <0.1v |
| 51 | TN/WT | HO2S-12 (B1 S2) Signal | 0.1-1.1v |
| 52 | VT/LG | Ambient Temperature Sensor | At 86°F: 1.96v |
| 53-54 | --- | Not Used | --- |
| 55 | DB/TN | Low Speed Fan Relay | Relay Off: 12v, On: 1v |
| 56 | TN/RD | S/C Vacuum Solenoid | Vacuum Increasing: 1v |
| 57 | GY/RD | IAC 1 Driver | DC pulse signals: 0.8-11v |
| 58 | VT/BK | IAC 4 Driver | DC pulse signals: 0.8-11v |
| 59 | OR | PCI Data Bus (J1850) | Digital Signals: 0-7-0v |
| 60 | --- | Not Used | --- |
| 61 | VT/WT | 5-Volt Supply | 4.9-5.1v |
| 62 | WT/PK | Brake Switch Signal | Brake Off: 0v, On: 12v |
| 63 | YL/DG | Torque Management Request | Digital Signals |
| 64 | DB/OR | A/C Clutch Relay Control | Relay Off: 12v, On: 1v |
| 65 | PK | SCI Transmit | 0v |
| 66 | WT/OR | Vehicle Speed Signal | Digital Signal |
| 67 | DB/VT | ASD Relay Control | Relay Off: 12v, On: 1v |
| 68 | PK/BK | EVAP Purge Solenoid Control | PWM Signal: 0-12-0v |
| 69 | DB/PK | High Speed Fan Relay | Relay Off: 12v, On: 1v |
| 70 | WT/TN | EVAP Purge Solenoid Sense | 0-1v |
| 71 | WT/RD | EATX RPM Signal | Digital Signals |
| 72 | OR | LDP Switch Sense | Open: 12v, Closed: 0v |
| 73 | --- | Not Used | --- |
| 74 | BR/LG | Fuel Pump Relay Control | Relay Off: 12v, On: 1v |
| 75 | LG | SCI Receive | 0v |
| 76 | BK/LB | TRS T41 Sense | In P/N: 0v, Others: 5v |
| 77 | WT/DG | LDP Solenoid Control | PWM Signal: 0-12-0v |
| 78-79 | --- | Not Used | --- |
| 80 | LG/RD | S/C Vent Solenoid | Vacuum Decreasing: 1v |

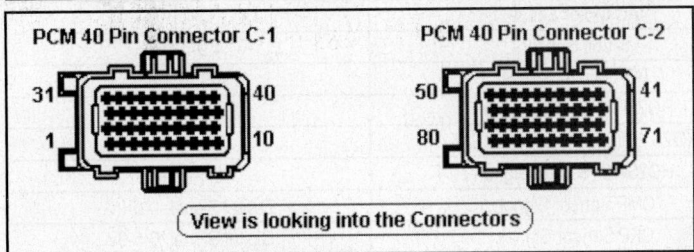

**Pin Connector Graphic**

## 2003 Sebring & Stratus 2.4L VIN X 'C1' Black Connector

| PCM Pin # | Wire Color | Circuit Description (38-Pin) | Value at Hot Idle |
|---|---|---|---|
| 1-8, 10 | --- | Not Used | --- |
| 9 | BK/TN | Power Ground | <0.1v |
| 11 | DB/WT | Ignition Switch Output (Run-Start) | 12-14v |
| 12 | RD/WT | Ignition Switch Output (Off-Run-Start) | 12-14v |
| 13 | WT/OR | Vehicle Speed Signal | Digital Signal |
| 14-17 | --- | Not Used | --- |
| 18 | BK/TN | Power Ground | <0.1v |
| 19-20 | --- | Not Used | --- |
| 21 | DB | A/C Pressure Switch Signal | A/C On: 0.45-4.85v |
| 22-24 | --- | Not Used | --- |
| 25 | LG | SCI Receive (PCM) | 5v |
| 26 | PK/LB | SCI Receive (TCM) | 5v |
| 27-28 | --- | Not Used | --- |
| 29 | RD/TN | Battery Power (Fused B+) | 12-14v |
| 30 | YL | Ignition Switch Output (tart) | 8-11v (cranking) |
| 31 | TN/WT | HO2S-12 (B1 S2) Signal | 0.1-1.1v |
| 32 | DB/LG | Oxygen Sensor Return (down) | <0.050v |
| 33-35 | --- | Not Used | --- |
| 36 | PK | SCI Transmit (PCM) | 0v |
| 37 | WT/DG | SCI Transmit (TCM) | 0v |
| 38 | YL/VT | PCI Data Bus (J1850) | Digital Signals: 0-7-0v |

## 2003 Sebring & Stratus 2.4L VIN X 'C2' Gray Connector

| PCM Pin # | Wire Color | Circuit Description (38-Pin) | Value at Hot Idle |
|---|---|---|---|
| 1-8 | --- | Not Used | --- |
| 9 | DB/TN | Coil 2 Driver | 5°, at 55 mph: 8° dwell |
| 10 | BK/GY | Coil 1 Driver | 5°, at 55 mph: 8° dwell |
| 11 | LB/BR | Injector 4 Driver | 1.0-4.0 ms |
| 12 | YL/WT | Injector 3 Driver | 1.0-4.0 ms |
| 13 | TN | Injector 2 Driver | 1.0-4.0 ms |
| 14 | WT/DB | Injector 1 Driver | 1.0-4.0 ms |
| 15-16, 22, 25 | --- | Not Used | --- |
| 17 | BR/VT | HO2S-21 (B2 S1) Heater Control | Heater On: 1v, Off: 12v |
| 18 | BR/OR | HO2S-11 (B1 S1) Heater Control | Heater On: 1v, Off: 12v |
| 19 | DG | Generator Field Driver | Digital Signal: 0-12-0v |
| 20 | TN/BK | ECT Sensor Signal | At 180°F: 2.80v |
| 21 | OR/DB | TP Sensor Signal | 0.6-1.0v |
| 23 | OR/RD | MAP Sensor Signal | 1.5-1.7v |
| 24 | BK/VT | Knock Sensor Return | <0.050v |
| 27 | BK/LB | Sensor Ground | <0.1v |
| 28 | YL/BK | IAC Motor Sense | 12-14v |
| 29 | VT/WT | 5-Volt Supply | 4.9-5.1v |
| 30 | BK/RD | IAT Sensor Signal | At 100°F: 1.83v |
| 31 | BK/DG | HO2S-11 (B1 S1) Signal | 0.1-1.1v |
| 32 | BR/DG | HO2S-11 (B1 S1) Ground (Up) | <0.1v |
| 33 | TN/WT | HO2S-12 (B1 S2) Signal | 0.1-1.1v |
| 34 | TN/YL | CMP Sensor Signal | Digital Signal: 0-5-0v |
| 35 | GY/BK | CKP Sensor Signal | Digital Signal: 0-5-0v |
| 36-37 | --- | Not Used | --- |
| 38 | GY/RD | IAC Motor Driver | DC pulse signals: 0.8-11v |

**2003 Sebring & Stratus 2.4L VIN X 'C3' White Connector**

| PCM Pin # | Wire Color | Circuit Description (38-Pin) | Value at Hot Idle |
|---|---|---|---|
| 1-2, 10, 13-16, 18 | --- | Not Used | --- |
| 3 | DB/VT | ASD Relay Control | Relay Off: 12v, On: 1v |
| 4 | DB/PK | High Speed Radiator Fan Relay | Relay Off: 12v, On: 1v |
| 5 | LG/RD | S/C Vent Solenoid | Vacuum Decreasing: 1v |
| 6 | DB/TN | Low Speed Radiator Fan Relay | Relay Off: 12v, On: 1v |
| 7 | YL/RD | S/C Power Supply | 12-14v |
| 8 | WT/DG | Natural Vacuum Leak Detection Solenoid | Solenoid Off: 12v, On: 1v |
| 9 | BR/VT | HO2S-12 (B1 S2) Heater Control | Heater On: 1v, Off: 12v |
| 11 | DB/OR | A/C Clutch Relay Control | Relay Off: 12v, On: 1v |
| 12 | TN/RD | S/C Vacuum Solenoid | Vacuum Increasing: 1v |
| 17 | LB | Sensor Ground No. 2 | <0.050v |
| 19 | OR/DG | Automatic Shutdown Relay Output | 12-14v |
| 20 | PK/BK | EVAP Purge Solenoid Control | PWM Signal: 0-12-0v |
| 21-22, 24-25 | --- | Not Used | --- |
| 23 | WT/PK | Brake Switch Sense | Brake Off: 0v, On: 12v |
| 26 | YL | AutoStick Downshift Switch | Digital Signal: 0v or 12v |
| 27 | LG | AutoStick Upshift Switch | Digital Signal: 0v or 12v |
| 28 | OR/DG | Automatic Shutdown Relay Output | 12-14v |
| 29 | WT/TN | EVAP Purge Solenoid Return | 0-1v |
| 30 | DB/LG | Power Steering Pressure Switch Signal | Straight: 0v, Turning: 5v |
| 31, 33, 36 | --- | Not Used | --- |
| 32 | VT/LG | Ambient Air Temperature Sensor | At 100°F: 1.83v |
| 34 | PK/LG | Speed Control Set Switch Signal | S/C & Set Switch On: 3.8v |
| 35 | OR | Natural Vacuum Leak Detection Switch Sense | 0.1v |
| 37 | BR/LG | Fuel Pump Relay Control | Relay Off: 12v, On: 1v |
| 38 | TN | Starter Relay Control | KOEC: 9-11v |

**2003 Sebring & Stratus 2.4L VIN X 'C4' Green Connector**

| PCM Pin # | Wire Color | Circuit Description (38-Pin) | Value at Hot Idle |
|---|---|---|---|
| 1 | BR | Overdrive Solenoid Control | Solenoid Off: 12v, On: 1v |
| 2 | PK | Underdrive Solenoid Control | Solenoid Off: 12v, On: 1v |
| 3-5, 7-9, 11 | --- | Not Used | --- |
| 6 | WT | 2-4 Solenoid Control | Solenoid Off: 12v, On: 1v |
| 10 | LB | Low/Reverse Solenoid Control | Solenoid Off: 12v, On: 1v |
| 12 | BK/YL | Power Ground | <0.050v |
| 13, 14 | BK/RD | Power Ground | <0.050v |
| 15 | LG/BK | TRS T1 Sense | <0.050v |
| 16 | VT | TRS T3 Sense | <0.050v |
| 17, 20-21 | --- | Not Used | --- |
| 18, 19, 31, 36 | LG, RD | Transmission Control Relay Output | Relay Off: 12v, On: 1v |
| 23-26 | --- | Not Used | --- |
| 27 | BK/WT | TRS T41 Sense | <0.050v |
| 28, 38 | RD | Transmission Control Relay Output | Relay Off: 12v, On: 1v |
| 29 | DG | Low/Reverse Pressure Switch Sense | 12-14v |
| 30 | YL/WT | 2-4 Pressure Switch Sense | In Low/Reverse: 2-4v |
| 32 | LG/WT | Output Speed Sensor Signal | In 2-4 Position: 2-4v |
| 33 | RD/BK | Input Speed Sensor Signal | Moving: AC voltage |
| 34 | DB/BK | Speed Sensor Ground | Moving: AC voltage |
| 35 | VT/YL | Transmission Temperature Sensor Signal | <0.050v |
| 37 | VT/WT | TRS T42 Sense | In PRNL: 0v, Others 5v |

**2004 Sebring & Stratus 2.4L VIN G, J, X (A/T) 'C1' Connector**

| PCM Pin # | Wire Color | Circuit Description (38-Pin) | Value at Hot Idle |
|---|---|---|---|
| 1 | --- | Not Used | --- |
| 2 | --- | Not Used | --- |
| 3 | --- | Not Used | --- |
| 4 | --- | Not Used | --- |
| 5 | --- | Not Used | --- |
| 6 | --- | Not Used | --- |
| 7 | --- | Not Used | --- |
| 8 | --- | Not Used | --- |
| 9 | BK/TN | Ground | |
| 10 | --- | Not Used | --- |
| 11 | DB/WT | Fused Ignition Switch Output (Run-Start) | |
| 12 | RD/WT | Fused Ignition Switch Output (Off-Start) | |
| 13 | WT/OR (MTX) | Vehicle Speed Signal | |
| 14 | --- | Not Used | --- |
| 15 | --- | Not Used | --- |
| 16 | --- | Not Used | --- |
| 17 | --- | Not Used | --- |
| 18 | BK/TN | Ground | |
| 19 | --- | Not Used | --- |
| 20 | --- | Not Used | --- |
| 21 | DB | A/C Pressure Signal | |
| 22 | --- | Not Used | --- |
| 23 | --- | Not Used | --- |
| 24 | --- | Not Used | --- |
| 25 | LG | SCI Receive (PCM) | |
| 26 | PK/LB (EATX) | SCI Receive (TCM) | |
| 27 | OR (MTX) | 5 Volt Supply | |
| 28 | --- | Not Used | --- |
| 29 | RD/TN | Fused B (+) | |
| 30 | YL | Fused Ignition Switch Output (Start) | |
| 31 | --- | Not Used | --- |
| 32 | DB/DG | O2 Return (Down) | |
| 33 | --- | Not Used | --- |
| 34 | --- | Not Used | --- |
| 35 | --- | Not Used | --- |
| 36 | PK | SCI Transmit (PCM) | |
| 37 | WT/DG (EATX) | SCI Transmit (TCM) | |
| 38 | YL/VT (EATX) | PCI Bus (PCM) | |
| 38 | OR (MTX) | PCI Bus (PCM) | |

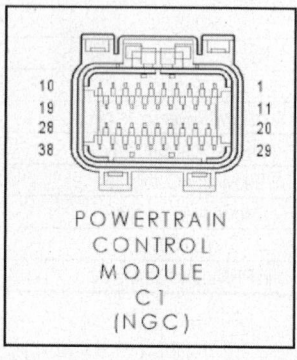

POWERTRAIN
CONTROL
MODULE
C1
(NGC)

**2004 Sebring & Stratus 2.4L VIN G, J, X (A/T) 'C2' Connector**

| PCM Pin # | Wire Color | Circuit Description (38-Pin) | Value at Hot Idle |
|---|---|---|---|
| 1 | --- | Not Used | --- |
| 2 | --- | Not Used | --- |
| 3 | --- | Not Used | --- |
| 4 | --- | Not Used | --- |
| 5 | --- | Not Used | --- |
| 6 | --- | Not Used | --- |
| 7 | --- | Not Used | --- |
| 8 | --- | Not Used | --- |
| 9 | DB/TN | Coil Control No. 2 | |
| 10 | BK/GY | Coil Control No. 1 | |
| 11 | LB/BR | Injector Control No. 4 | |
| 12 | YL/WT | Injector Control No. 3 | |
| 13 | TN | Injector Control No. 2 | |
| 14 | WT/DB | Injector Control No. 1 | |
| 15 | --- | Not Used | --- |
| 16 | --- | Not Used | --- |
| 17 | BR/VT | O2 1/2 Heater Control | |
| 18 | BR/OR | O2 1/1 Heater Control | |
| 19 | DG | Gen Field Control | |
| 20 | TN/BK | ECT Signal | |
| 21 | OR/DB | TP Signal | |
| 22 | --- | Not Used | --- |
| 23 | DG/RD | MAP Signal | |
| 24 | BK/VT | KS Return | |
| 25 | DB/LG | KS Signal | |
| 26 | --- | Not Used | --- |
| 27 | BK/LB | Sensor Ground 1 | |
| 28 | YL/BK | IAC Return | |
| 29 | VT/WT | 5 Volt Supply | |
| 30 | BK/RD | IAT Signal | |
| 31 | BK/DG | O2 1/1 Signal | |
| 32 | BR/DG | O2 Return (Up) | |
| 33 | TN/WT | O2 1/2 Signal | |
| 34 | TN/YL | CMP Signal | |
| 35 | GY/BK | CKP Signal | |
| 36 | --- | Not Used | --- |
| 37 | --- | Not Used | --- |
| 38 | GY/RD | IAC Motor Control | |

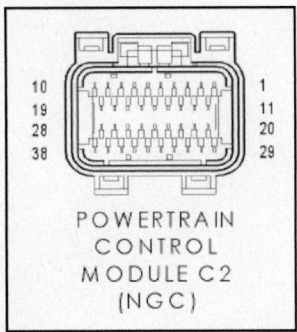

POWERTRAIN
CONTROL
MODULE C2
(NGC)

## 2004 Sebring & Stratus 2.4L VIN G, J, X (A/T) 'C3' Connector

| PCM Pin # | Wire Color | Circuit Description (38-Pin) | Value at Hot Idle |
|---|---|---|---|
| 1 | --- | Not Used | --- |
| 2 | --- | Not Used | --- |
| 3 | DB/VT | Automatic Shut Down Relay Control | |
| 4 | DB/PK | High Rad Fan Relay Control | |
| 5 | LG/RD | S/C Vent Control | |
| 6 | DB/TN | Low Rad Fan Relay Control | |
| 7 | YL/RD | S/C Supply | |
| 8 | WT/DG | NVLD Solenoid Control | |
| 9 | --- | Not Used | --- |
| 10 | --- | Not Used | --- |
| 11 | DB/OR | A/C Clutch Relay Control | |
| 12 | TN/RD | S/C Vacuum Control | |
| 13 | --- | Not Used | --- |
| 14 | --- | Not Used | --- |
| 15 | --- | Not Used | --- |
| 16 | --- | Not Used | --- |
| 17 | BK/LB | Sensor Ground 2 | |
| 18 | --- | Not Used | --- |
| 19 | OR/DG | Fused Automatic Shut Down Relay Output | |
| 20 | PK/BK | EVAP Purge Control | |
| 21 | YL/RD (MTX) | Fused Ignition Switch Output (Start) | |
| 22 | --- | Not Used | --- |
| 23 | WT/PK | Brake Switch Signal | |
| 24 | --- | Not Used | --- |
| 25 | --- | Not Used | --- |
| 26 | YL | AutoStick Downshift Switch Signal | |
| 27 | LG | AutoStick Upshift Switch Signal | |
| 28 | OR/DG | Fused Automatic Shut Down Relay Output | |
| 29 | WT/TN | EVAP Purge Return | |
| 30 | DB/LG | PSP Switch Signal | |
| 31 | --- | Not Used | --- |
| 32 | VT/LG | AAT Signal | |
| 33 | --- | Not Used | --- |
| 34 | PK/LG | S/C Switch Signal | |
| 35 | OR | NVLD Switch Signal | |
| 36 | --- | Not Used | --- |
| 37 | BR/LG | Fuel Pump Relay Control | |
| 38 | TN | Starter Relay Control | |

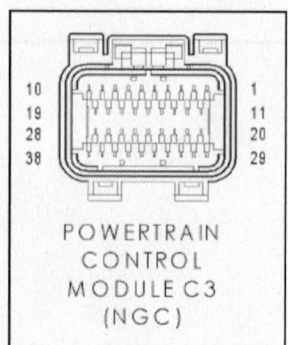

POWERTRAIN
CONTROL
MODULE C3
(NGC)

**2004 Sebring & Stratus 2.4L VIN G, J, X (A/T) 'C4' Connector**

| PCM Pin # | Wire Color | Circuit Description (38-Pin) | Value at Hot Idle |
|---|---|---|---|
| 1 | BR | Overdrive Solenoid Control | |
| 2 | PK | Underdrive Solenoid Control | |
| 3 | --- | Not Used | --- |
| 4 | --- | Not Used | --- |
| 5 | --- | Not Used | --- |
| 6 | WT | 2Not Used4 Solenoid Control | |
| 7 | --- | Not Used | --- |
| 8 | --- | Not Used | --- |
| 9 | --- | Not Used | --- |
| 10 | LB | Low/Reverse Solenoid Control | |
| 11 | --- | Not Used | --- |
| 12 | BK/YL | Ground | |
| 13 | BK/RD | Ground | |
| 14 | BK/RD | Ground | |
| 15 | LG/BK | TRS T1 Sense | |
| 16 | VT | TRS T3 Sense | |
| 17 | --- | Not Used | --- |
| 18 | LG | Transmission Control Relay Control | |
| 19 | RD | Transmission Control Relay Output | |
| 20 | --- | Not Used | --- |
| 21 | --- | Not Used | --- |
| 22 | OR/BK | Overdrive Pressure Switch Sense | |
| 23 | --- | Not Used | --- |
| 24 | --- | Not Used | --- |
| 25 | --- | Not Used | --- |
| 26 | --- | Not Used | --- |
| 27 | BK/WT | TRS T41 Sense | |
| 28 | RD | Transmission Control Relay Output | |
| 29 | DG | Low/Reverse Pressure Switch Sense | |
| 30 | YL/BK | 2-4 Pressure Switch Sense | |
| 31 | --- | Not Used | --- |
| 32 | LG/WT | Output Speed Sensor Signal | |
| 33 | RD/BK | Input Speed Sensor Signal | |
| 34 | DB/BK | Speed Sensor Ground | |
| 35 | VT/YL | Transmission Temperature Sensor Signal | |
| 36 | --- | Not Used | --- |
| 37 | VT/WT | TRS T42 Sense | |
| 38 | RD | Transmission Control Relay Output | |

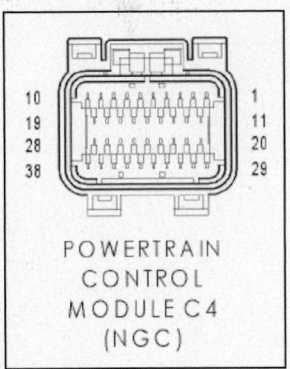

10
19
28
38

1
11
20
29

POWERTRAIN
CONTROL
MODULE C4
(NGC)

## 1996-97 Sebring 2.5L VIN H (A/T) 'C1' Connector

| PCM Pin # | Wire Color | Circuit Description (40-Pin) | Value at Hot Idle |
|---|---|---|---|
| 1-3 | --- | Not Used | --- |
| 4 | DG | Alternator Field Control | Digital Signal: 0-12-0v |
| 5 | YL/PK | S/C On Switch Signal | S/C On: 1v, Off: 12v |
| 6 | DG/OR | ASD Relay Output | 12-14v |
| 7 | YL/WT | Injector 3 Driver | 1.0-4.0 ms |
| 8-9 | --- | Not Used | --- |
| 10 | BK/TN | Power Ground | <0.1v |
| 11 | BK/GY | Coil 1 Driver | 5°, at 55 mph: 8° dwell |
| 12 | --- | Not Used | --- |
| 13 | WT/LB | Injector 1 Driver | 1.0-4.0 ms |
| 14 | BR/DG | Injector 6 Driver | 1.0-4.0 ms |
| 15 | GY | Injector 5 Driver | 1.0-4.0 ms |
| 16 | LB/BR | Injector 4 Driver | 1.0-4.0 ms |
| 17 | TN | Injector 2 Driver | 1.0-4.0 ms |
| 18-19 | --- | Not Used | --- |
| 20 | DB/WT | Ignition Switch Output | 12-14v |
| 21-25 | --- | Not Used | --- |
| 26 | TN/BK | ECT Sensor Signal | At 180°F: 2.80v |
| 27-29 | --- | Not Used | --- |
| 30 | BK/DG | HO2S-11 (B1 S1) Signal | 0.1-1.1v |
| 31 | --- | Not Used | --- |
| 32 | GY/BK | CKP Sensor Signal | Digital Signal: 0-5-0v |
| 33 | TN/YL | CMP Sensor Signal | Digital Signal: 0-5-0v |
| 34 | --- | Not Used | --- |
| 35 | OR/LB | TP Sensor Signal | 0.6-1.0v |
| 36 | DG/RD | MAP Sensor Signal | 1.5-1.7v |
| 37 | BK/RD | IAT Sensor Signal | At 100°F: 1.83v |
| 38-39 | --- | Not Used | --- |
| 40 | GY/YL | EGR Solenoid Control | 12v, at 55 mph: 1v |

## Standard Colors and Abbreviations

| Abbreviation | Color | Abbreviation | Color | Abbreviation | Color |
|---|---|---|---|---|---|
| BK | Black | GY | Gray | RD | Red |
| BL | Blue | GN | Green | TN | Tan |
| BR | Brown | LG | Light Green | VT | Violet |
| DB | Dark Blue | OR | Orange | WT | White |
| DG | Dark Green | PK | Pink | YL | Yellow |

**1996-97 Sebring 2.5L VIN H (A/T) 'C2' Connector**

| PCM Pin # | Wire Color | Circuit Description (40-Pin) | Value at Hot Idle |
|---|---|---|---|
| 41 | RD/LG | S/C Set Switch Signal | S/C & Set Switch On: 3.8v |
| 42 | DB/YL | A/C Pressure Switch Signal | A/C On: 0.45-4.85v |
| 43 | BK/LB | Sensor Ground | <0.050v |
| 44 | OR/WT | 8-Volt Supply | 7.9-8.1v |
| 45 | DB/LG | PSP Switch Signal | Straight: 0v, Turning: 5v |
| 46 | RD/TN | Battery Power (Fused B+) | 12-14v |
| 47, 53-54 | --- | Not Used | --- |
| 48 | BR/GY | IAC 3 Driver | DC pulse signals: 0.8-11v |
| 49 | YL/BK | IAC 2 Driver | DC pulse signals: 0.8-11v |
| 50 | BK/TN | Power Ground | <0.1v |
| 51 | WT/DG | HO2S-12 (B1 S2) Signal | 0.1-1.1v |
| 52 | PK/YL | Battery Temperature Sensor | At 86°F: 1.96v |
| 55 | DB/TN | Low Speed Fan Relay | Relay Off: 12v, On: 1v |
| 57 | GY/RD | IAC 1 Driver | DC pulse signals: 0.8-11v |
| 58 | PK/GY | IAC 4 Driver | DC pulse signals: 0.8-11v |
| 59 | PK/BR | CCD Bus (+) | Digital Signal: 0-5-0v |
| 60 | WT/BK | CCD Bus (-) | <0.050v |
| 61 | PK/WT | 5-Volt Supply | 4.9-5.1v |
| 62 | WT/RD | Brake Switch Signal | Brake Off: 0v, On: 12v |
| 63 | YL/DG | Torque Management Request | Digital Signals |
| 64 | DB/OR | A/C Clutch Relay Control | Relay Off: 12v, On: 1v |
| 65 | PK/LG | SCI Transmit | 0v |
| 66 | WT/OR | Vehicle Speed Signal | Digital Signal |
| 67 | DB/PK | ASD Relay Control | Relay Off: 12v, On: 1v |
| 68 | PK/GY | EVAP Purge Solenoid Control | PWM Signal: 0-12-0v |
| 69 | DB/PK | High Speed Fan Relay | Relay Off: 12v, On: 1v |
| 70-71 | --- | Not Used | --- |
| 72 | OR | LDP Switch Sense | Open: 12v, Closed: 0v |
| 74 | BR/LG | Fuel Pump Relay Control | Relay Off: 12v, On: 1v |
| 75 | LG | SCI Receive | 0v |
| 76 | BK/WT | PNP Switch Signal | In P/N: 0v, Others: 5v |
| 77 | WT/DG | LDP Solenoid Control | PWM Signal: 0-12-0v |
| 78 | WT/PK | S/C Vacuum Solenoid | Vacuum Increasing: 1v |
| 79 | --- | Not Used | --- |
| 80 | BK/YL | S/C Vent Solenoid | Vacuum Decreasing: 1v |

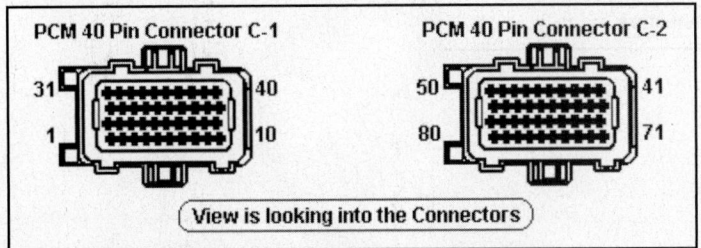

**Pin Connector Graphic**

## 1998 Sebring 2.5L VIN H (A/T) 'C1' Connector

| PCM Pin # | Wire Color | Circuit Description (40-Pin) | Value at Hot Idle |
|---|---|---|---|
| 1-3 | --- | Not Used | --- |
| 4 | BK/GY | Coil 1 Driver | 5°, at 55 mph: 8° dwell |
| 5 | YL/RD | S/C On Switch Signal | S/C On: 1v, Off: 12v |
| 6 | DG/OR | ASD Relay Output | 12-14v |
| 7 | YL/WT | Injector 3 Driver | 1.0-4.0 ms |
| 8 | DG | Alternator Field Control | Digital Signal: 0-12-0v |
| 9 | --- | Not Used | --- |
| 10 | BK/TN | Power Ground | <0.1v |
| 11-12 | --- | Not Used | --- |
| 13 | WT/DB | Injector 1 Driver | 1.0-4.0 ms |
| 14 | BR/DB | Injector 6 Driver | 1.0-4.0 ms |
| 15 | GY | Injector 5 Driver | 1.0-4.0 ms |
| 16 | LB/BR | Injector 4 Driver | 1.0-4.0 ms |
| 17 | TN | Injector 2 Driver | 1.0-4.0 ms |
| 18-19 | --- | Not Used | --- |
| 20 | DB/WT | Ignition Switch Output | 12-14v |
| 21-25 | --- | Not Used | --- |
| 26 | TN/BK | ECT Sensor Signal | At 180°F: 2.80v |
| 27 | BK/OR | Oxygen Sensor Ground | <0.050v |
| 28-29 | --- | Not Used | --- |
| 30 | BK/DG | HO2S-11 (B1 S1) Signal | 0.1-1.1v |
| 31 | TN | Starter Relay Control | KOEC: 9-11v |
| 32 | GY/BK | CKP Sensor Signal | Digital Signal: 0-5-0v |
| 33 | TN/YL | CMP Sensor Signal | Digital Signal: 0-5-0v |
| 34 | --- | Not Used | --- |
| 35 | OR/DB | TP Sensor Signal | 0.6-1.0v |
| 36 | DG/RD | MAP Sensor Signal | 1.5-1.7v |
| 37 | BK/RD | IAT Sensor Signal | At 100°F: 1.83v |
| 38-39 | --- | Not Used | --- |
| 40 | GY/YL | EGR Solenoid Control | 12v, at 55 mph: 1v |

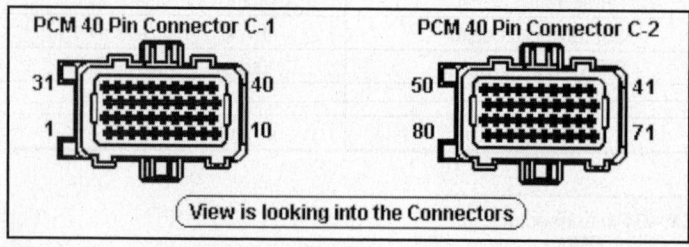

**Pin Connector Graphic**

## 1998 Sebring 2.5L VIN H (A/T) 'C2' Connector

| PCM Pin # | Wire Color | Circuit Description (40-Pin) | Value at Hot Idle |
|---|---|---|---|
| 41 | RD/LG | S/C Set Switch Signal | S/C & Set Switch On: 3.8v |
| 42 | DB/YL | A/C Pressure Switch Signal | A/C On: 0.45-4.85v |
| 43 | BK/LB | Sensor Ground | <0.050v |
| 44 | OR/WT | 8-Volt Supply | 7.9-8.1v |
| 45 | DB/LG | PSP Switch Signal | Straight: 0v, Turning: 5v |
| 46 | RD/TN | Battery Power (Fused B+) | 12-14v |
| 47 | BK/WT | Sensor Ground | <0.050v |
| 48 | BR/WT | IAC 1 Driver | DC pulse signals: 0.8-11v |
| 49 | YL/BK | IAC 2 Driver | DC pulse signals: 0.8-11v |
| 50 | BK/TN | Power Ground | <0.1v |
| 51 | TN/WT | HO2S-12 (B1 S2) Signal | 0.1-1.1v |
| 52 | PK/YL | Battery Temperature Sensor | At 86°F: 1.96v |
| 53-54 | --- | Not Used | --- |
| 55 | DB/TN | Low Speed Fan Relay | Relay Off: 12v, On: 1v |
| 56 | WT/VT | S/C Vacuum Solenoid | Vacuum Increasing: 1v |
| 57 | GY/RD | IAC 3 Driver | DC pulse signals: 0.8-11v |
| 58 | VT/GY | IAC 4 Driver | DC pulse signals: 0.8-11v |
| 59 | VT/BR | CCD Bus (+) | Digital Signal: 0-5-0v |
| 60 | WT/BK | CCD Bus (-) | <0.050v |
| 61 | VT/WT | 5-Volt Supply | 4.9-5.1v |
| 62 | WT/RD | Brake Switch Signal | Brake Off: 0v, On: 12v |
| 63 | YL/DG | Torque Management Request | Digital Signals |
| 64 | DB/OR | A/C Clutch Relay Control | Relay Off: 12v, On: 1v |
| 65 | PK/LB | SCI Transmit | 0v |
| 66 | WT/OR | Vehicle Speed Signal | Digital Signal |
| 67 | DB/VT | ASD Relay Control | Relay Off: 12v, On: 1v |
| 68 | PK/GY | EVAP Purge Solenoid Control | PWM Signal: 0-12-0v |
| 69 | DB/PK | High Speed Fan Relay | Relay Off: 12v, On: 1v |
| 70 | WT/TN | EVAP Purge Solenoid Sense | 0-1v |
| 71, 73 | --- | Not Used | --- |
| 72 | OR | LDP Switch Sense | Open: 12v, Closed: 0v |
| 74 | BR/LG | Fuel Pump Relay Control | Relay Off: 12v, On: 1v |
| 75 | LG | SCI Receive | 0v |
| 76 | BK/LB | PNP Switch Signal | In P/N: 0v, Others: 5v |
| 77 | WT/DG | LDP Solenoid Control | PWM Signal: 0-12-0v |
| 78-79 | --- | Not Used | --- |
| 80 | LG/RD | S/C Vent Solenoid | Vacuum Decreasing: 1v |

## Standard Colors and Abbreviations

| Abbreviation | Color | Abbreviation | Color | Abbreviation | Color |
|---|---|---|---|---|---|
| BK | Black | GY | Gray | RD | Red |
| BL | Blue | GN | Green | TN | Tan |
| BR | Brown | LG | Light Green | VT | Violet |
| DB | Dark Blue | OR | Orange | WT | White |
| DG | Dark Green | PK | Pink | YL | Yellow |

## 1999-2000 Sebring 2.5L VIN H (A/T) 'C1' Connector

| PCM Pin # | Wire Color | Circuit Description (40-Pin) | Value at Hot Idle |
|---|---|---|---|
| 1-3 | --- | Not Used | --- |
| 4 | BK/GY | Coil 1 Driver | 5º, at 55 mph: 8º dwell |
| 5 | YL/RD | S/C On Switch Signal | S/C On: 1v, Off: 12v |
| 6 | DG/OR | ASD Relay Output | 12-14v |
| 7 | YL/WT | Injector 3 Driver | 1.0-4.0 ms |
| 8 | DG | Alternator Field Control | Digital Signal: 0-12-0v |
| 9 | --- | Not Used | --- |
| 10 | BK/TN | Power Ground | <0.1v |
| 11-12 | --- | Not Used | --- |
| 13 | WT/DB | Injector 1 Driver | 1.0-4.0 ms |
| 14 | BR/DB | Injector 6 Driver | 1.0-4.0 ms |
| 15 | GY | Injector 5 Driver | 1.0-4.0 ms |
| 16 | LB/BR | Injector 4 Driver | 1.0-4.0 ms |
| 17 | TN | Injector 2 Driver | 1.0-4.0 ms |
| 18-19 | --- | Not Used | --- |
| 20 | DB/WT | Ignition Switch Output | 12-14v |
| 21-25 | --- | Not Used | --- |
| 26 | TN/BK | ECT Sensor Signal | At 180ºF: 2.80v |
| 27 | BK/OR | Oxygen Sensor Ground | <0.050v |
| 28-29 | --- | Not Used | --- |
| 30 | BK/DG | HO2S-11 (B1 S1) Signal | 0.1-1.1v |
| 31 | TN | Starter Relay Control | KOEC: 9-11v |
| 32 | GY/BK | CKP Sensor Signal | Digital Signal: 0-5-0v |
| 33 | TN/YL | CMP Sensor Signal | Digital Signal: 0-5-0v |
| 34 | --- | Not Used | --- |
| 35 | OR/LB | TP Sensor Signal | 0.6-1.0v |
| 36 | DG/RD | MAP Sensor Signal | 1.5-1.7v |
| 37 | BK/RD | IAT Sensor Signal | At 100ºF: 1.83v |
| 38-39 | --- | Not Used | --- |
| 40 | GY/YL | EGR Solenoid Control | 12v, at 55 mph: 1v |

## Standard Colors and Abbreviations

| Abbreviation | Color | Abbreviation | Color | Abbreviation | Color |
|---|---|---|---|---|---|
| BK | Black | GY | Gray | RD | Red |
| BL | Blue | GN | Green | TN | Tan |
| BR | Brown | LG | Light Green | VT | Violet |
| DB | Dark Blue | OR | Orange | WT | White |
| DG | Dark Green | PK | Pink | YL | Yellow |

**1999-2000 Sebring 2.5L VIN H (A/T) 'C2' Connector**

| PCM Pin # | Wire Color | Circuit Description (40-Pin) | Value at Hot Idle |
|---|---|---|---|
| 41 | RD/LG | S/C Set Switch Signal | S/C & Set Switch On: 3.8v |
| 42 | DB/YL | A/C Pressure Switch Signal | A/C On: 0.45-4.85v |
| 43 | BK/LB | Sensor Ground | <0.050v |
| 44 | OR/WT | 8-Volt Supply | 7.9-8.1v |
| 45 | DB/LG | PSP Switch Signal | Straight: 0v, Turning: 5v |
| 46 | RD/TN | Battery Power (Fused B+) | 12-14v |
| 47 | BK/WT | Sensor Ground | <0.050v |
| 48 | BR/GY | IAC 1 Driver | DC pulse signals: 0.8-11v |
| 49 | YL/BK | IAC 2 Driver | DC pulse signals: 0.8-11v |
| 50 | BK/TN | Power Ground | <0.1v |
| 51 | TN/WT | HO2S-12 (B1 S2) Signal | 0.1-1.1v |
| 52 | PK/YL | Battery Temperature Sensor | At 86°F: 1.96v |
| 53-54 | --- | Not Used | --- |
| 55 | DB/TN | Low Speed Fan Relay | Relay Off: 12v, On: 1v |
| 56 | WT/VT | S/C Vacuum Solenoid | Vacuum Increasing: 1v |
| 57 | GY/RD | IAC 3 Driver | DC pulse signals: 0.8-11v |
| 58 | VT/BK | IAC 4 Driver | DC pulse signals: 0.8-11v |
| 59 | VT/BR | CCD Bus (+) | Digital Signal: 0-5-0v |
| 60 | WT/BK | CCD Bus (-) | <0.050v |
| 61 | VT/WT | 5-Volt Supply | 4.9-5.1v |
| 62 | WT/RD | Brake Switch Signal | Brake Off: 0v, On: 12v |
| 63 | YL/DG | Torque Management Request | Digital Signals |
| 64 | DB/OR | A/C Clutch Relay Control | Relay Off: 12v, On: 1v |
| 65 | PK/LG | SCI Transmit | 0v |
| 66 | WT/OR | Vehicle Speed Signal | Digital Signal |
| 67 | DB/VT | ASD Relay Control | Relay Off: 12v, On: 1v |
| 68 | PK/GY | EVAP Purge Solenoid Control | PWM Signal: 0-12-0v |
| 69 | DB/PK | High Speed Fan Relay | Relay Off: 12v, On: 1v |
| 70 | WT/TN | EVAP Purge Solenoid Sense | 0-1v |
| 71-73 | --- | Not Used | --- |
| 74 | BR/LG | Fuel Pump Relay Control | Relay Off: 12v, On: 1v |
| 75 | LG | SCI Receive | 0v |
| 76 | BK/LB | TRS T41 Sense | In P/N: 0v, Others: 5v |
| 77-79 | --- | Not Used | --- |
| 80 | LG/RD | S/C Vent Solenoid | Vacuum Decreasing: 1v |

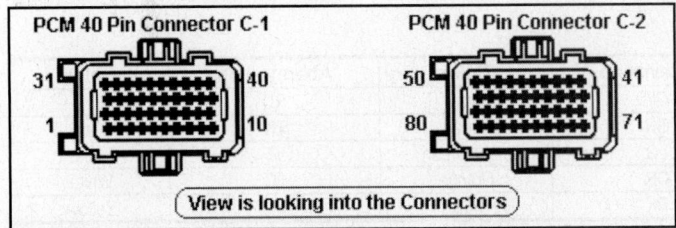

Pin Connector Graphic

### 1995 Sebring & Avenger 2.5L VIN N (A/T) 'C1' Connector

| PCM Pin # | Wire Color | Circuit Description (40-Pin) | Value at Hot Idle |
|---|---|---|---|
| 1-3 | --- | Not Used | --- |
| 4 | DB | Alternator Field Control | Digital Signal: 0-12-0v |
| 5 | --- | Not Used | --- |
| 6 | BK/RD | ASD Relay Output | 12-14v |
| 7 | YL/DG | Injector 3 Driver | 1.0-4.0 ms |
| 8 | DG/RD | Amber MIL Control | MIL On: 1v, MIL Off: 12v |
| 9 | DB/YL | Bulb Check Control | Lamp On: 1v, Off: 12v |
| 10 | BK | Power Ground | <0.1v |
| 11 | RD/DG | Coil Driver | 5°, at 55 mph: 8° dwell |
| 12 | RD/BK | S/C On Switch Signal | S/C On: 1v, Off: 12v |
| 13 | LG/BK | Injector 1 Driver | 1.0-4.0 ms |
| 14 | BR/RD | Injector 6 Driver | 1.0-4.0 ms |
| 15 | RD/WT | Injector 5 Driver | 1.0-4.0 ms |
| 16 | LG/RD | Injector 4 Driver | 1.0-4.0 ms |
| 17 | YL/RD | Injector 2 Driver | 1.0-4.0 ms |
| 18 | --- | Not Used | --- |
| 19 | DG/OR | High Speed Fan Relay | Relay Off: 12v, On: 1v |
| 20 | BK/WT | Ignition Switch Output | 12-14v |
| 21 | --- | Not Used | --- |
| 22 | LG | Speed Control Lamp Driver | Lamp On: 1v, Off: 12v |
| 23-25 | --- | Not Used | --- |
| 26 | DG/WT | ECT Sensor Signal | At 180°F: 2.80v |
| 27-28 | --- | Not Used | --- |
| 29 | DG/BK | HO2S-12 (B1 S2) Signal | 0.1-1.1v |
| 30 | WT/BK | HO2S-11 (B1 S1) Signal | 0.1-1.1v |
| 31 | --- | Not Used | --- |
| 32 | LB/WT | CKP Sensor Signal | Digital Signal: 0-5-0v |
| 33 | BR | CMP Sensor Signal | Digital Signal: 0-5-0v |
| 34 | --- | Not Used | --- |
| 35 | BR/RD | TP Sensor Signal | 0.6-1.0v |
| 36 | YL/BK | MAP Sensor Signal | 1.5-1.7v |
| 37 | BR/LB | IAT Sensor Signal | At 100°F: 1.83v |
| 38 | DG/RD | A/C Clutch Switch Sense | Relay Off: 12v, On: 1v |
| 39 | --- | Not Used | --- |
| 40 | RD/LB | EGR Solenoid Control | 12v, at 55 mph: 1v |

### Standard Colors and Abbreviations

| Abbreviation | Color | Abbreviation | Color | Abbreviation | Color |
|---|---|---|---|---|---|
| BK | Black | GY | Gray | RD | Red |
| BL | Blue | GN | Green | TN | Tan |
| BR | Brown | LG | Light Green | VT | Violet |
| DB | Dark Blue | OR | Orange | WT | White |
| DG | Dark Green | PK | Pink | YL | Yellow |

**1995 Sebring & Avenger 2.5L VIN N (A/T) 'C2' Connector**

| PCM Pin # | Wire Color | Circuit Description (40-Pin) | Value at Hot Idle |
|---|---|---|---|
| 41 | RD | S/C Set Switch Signal | S/C & Set Switch On: 3.8v |
| 42 | --- | Not Used | --- |
| 43 | BK/DG | Sensor Ground | <0.050v |
| 44 | YL | 8-Volt Supply | 7.9-8.1v |
| 45 | LB/YL | PSP Switch Signal | Straight: 0v, Turning: 5v |
| 46 | RD/BK | Battery Power (Fused B+) | 12-14v |
| 47 | BK | Sensor Ground | <0.050v |
| 48 | DG/LB | IAC 1 Driver | DC pulse signals: 0.8-11v |
| 49 | OR | IAC 2 Driver | DC pulse signals: 0.8-11v |
| 50 | BK | Power Ground | <0.1v |
| 51-54 | --- | Not Used | --- |
| 55 | DG/BK | Low Speed Fan Relay | Relay Off: 12v, On: 1v |
| 56 | LB | Generator Lamp Control | Lamp On: 1v, Off: 12v |
| 57 | DG | IAC 3 Driver | DC pulse signals: 0.8-11v |
| 58 | YL/LB | IAC 4 Driver | DC pulse signals: 0.8-11v |
| 59 | BK/DB | CCD Bus (+) | Digital Signal: 0-5-0v |
| 60 | WT/LB | CCD Bus (-) | <0.050v |
| 61 | DG/YL | 5-Volt Supply | 4.9-5.1v |
| 62 | BR/WT | Brake Switch Signal | Brake Off: 0v, On: 12v |
| 63 | OR/BK | Torque Management Request | Digital Signals |
| 64 | DG | A/C Clutch Relay Control | Relay Off: 12v, On: 1v |
| 65 | PK | SCI Transmit | 0v |
| 66 | YL/WT | Vehicle Speed Signal | Digital Signal |
| 67 | RD/WT | ASD Relay Control | Relay Off: 12v, On: 1v |
| 68 | LG/BK | EVAP Purge Solenoid Control | PWM Signal: 0-12-0v |
| 69 | DG/WT | High Speed Fan Relay | Relay Off: 12v, On: 1v |
| 70-72 | --- | Not Used | --- |
| 73 | WT | Tachometer Signal | Pulse Signals |
| 74 | WT/RD | Fuel Pump Relay Control | Relay Off: 12v, On: 1v |
| 75 | DB/BK | SCI Receive | 0v |
| 76 | BK/YL | PNP Switch Signal | In P/N: 0v, Others: 5v |
| 77, 79 | --- | Not Used | --- |
| 78 | LG/WT | S/C Vacuum Solenoid | Vacuum Increasing: 1v |
| 80 | BK/YL | S/C Vent Solenoid | Vacuum Decreasing: 1v |

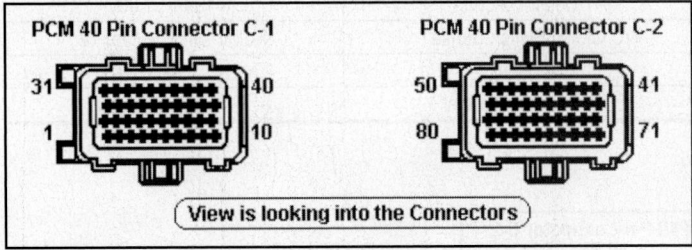

**Pin Connector Graphic**

**1996-97 Sebring & Avenger 2.5L VIN N (A/T) 'C1' Connector**

| PCM Pin # | Wire Color | Circuit Description (40-Pin) | Value at Hot Idle |
|---|---|---|---|
| 1 | WT/RD | HO2S-12 (B1 S2) Signal | 0.1-1.1v |
| 2-3, 5 | --- | Not Used | --- |
| 4 | DB | Alternator Field Control | Digital Signal: 0-12-0v |
| 6 | BK/RD | ASD Relay Output | 12-14v |
| 7 | YL/DG | Injector 3 Driver | 1.0-4.0 ms |
| 8 | DG/RD | Amber MIL Control | MIL On: 1v, MIL Off: 12v |
| 9 | DB/YL | Bulb (lamp) Check Control | B/C On: 1v, B/C Off: 12v |
| 10 | BK | Power Ground | <0.1v |
| 11 | RD/DG | Coil Driver | 5°, at 55 mph: 8° dwell |
| 12 | RD/BK | S/C On Switch Signal | S/C On: 1v, Off: 12v |
| 13 | LG/BK | Injector 1 Driver | 1.0-4.0 ms |
| 14 | BR/RD | Injector 6 Driver | 1.0-4.0 ms |
| 15 | RD/WT | Injector 5 Driver | 1.0-4.0 ms |
| 16 | LG/RD | Injector 4 Driver | 1.0-4.0 ms |
| 17 | YL/RD | Injector 2 Driver | 1.0-4.0 ms |
| 19 | DG/OR | High Speed Fan 2 Control | Relay Off: 12v, On: 1v |
| 20 | BK/WT | Ignition Switch Output | 12-14v |
| 21 | --- | Not Used | --- |
| 22 | LG | Speed Control Lamp Driver | S/C On: 1v, S/C Off: 12v |
| 23 | YL/DB | Fuel Level Sensor Signal | 70 ohms (±20) w/full tank |
| 24-25 | --- | Not Used | --- |
| 26 | DG/WT | ECT Sensor Signal | At 180°F: 2.80v |
| 27 | --- | Not Used | --- |
| 28 | DG/YL | High Speed Fan 1 Control | Relay Off: 12v, On: 1v |
| 29 | DG/BK | HO2S-11 (B1 S1) Signal | 0.1-1.1v |
| 30 | WT/BK | HO2S-21 (B2 S1) Signal | 0.1-1.1v |
| 32 | LB/WT | CKP Sensor Signal | Digital Signal: 0-5-0v |
| 33 | BR | CMP Sensor Signal | Digital Signal: 0-5-0v |
| 34 | --- | Not Used | --- |
| 35 | BR/RD | TP Sensor Signal | 0.6-1.0v |
| 36 | YL/BK | MAP Sensor Signal | 1.5-1.7v |
| 37 | BR/DB | IAT Sensor Signal | At 100°F: 1.83v |
| 38 | DG/RD | A/C Clutch Switch Sense | Relay Off: 12v, On: 1v |
| 39 | --- | Not Used | --- |
| 40 | RD/DB | EGR Solenoid Control | 12v, at 55 mph: 1v |

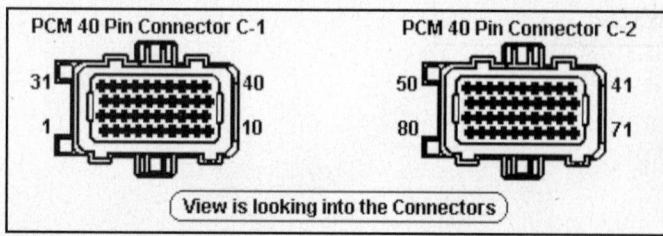

**PCM 40 Pin Connector C-1**    **PCM 40 Pin Connector C-2**

31     40     50     41

1     10     80     71

**View is looking into the Connectors**

**Pin Connector Graphic**

### 1996-97 Sebring & Avenger 2.5L VIN N (A/T) 'C2' Connector

| PCM Pin # | Wire Color | Circuit Description (40-Pin) | Value at Hot Idle |
|---|---|---|---|
| 41 | RD | S/C Set Switch Signal | S/C & Set Switch On: 3.8v |
| 42 | --- | Not Used | --- |
| 43 | BK/DG | Sensor Ground | <0.050v |
| 44 | YL | 8-Volt Supply | 7.9-8.1v |
| 45 | DB/YL | PSP Switch Signal | Straight: 0v, Turning: 5v |
| 46 | RD/BK | Battery Power (Fused B+) | 12-14v |
| 47 | BK | Sensor Ground | <0.050v |
| 48 | DG/DB | IAC 3 Driver | DC pulse signals: 0.8-11v |
| 49 | OR | IAC 2 Driver | DC pulse signals: 0.8-11v |
| 50 | BK | Power Ground | <0.1v |
| 51 | WT/DG | HO2S-22 (B2 S2) Signal | 0.1-1.1v |
| 52-54 | --- | Not Used | --- |
| 55 | DG/BK | Low Speed Fan Relay | Relay Off: 12v, On: 1v |
| 56 | DB | Generator Lamp Control | Lamp On: 1v, Off: 12v |
| 57 | DG | IAC 1 Driver | DC pulse signals: 0.8-11v |
| 58 | YL/DB | IAC 4 Driver | DC pulse signals: 0.8-11v |
| 59 | BK/DB | CCD Bus (+) | Digital Signal: 0-5-0v |
| 60 | WT/DB | CCD Bus (-) | <0.050v |
| 61 | DG/YL | 5-Volt Supply | 4.9-5.1v |
| 62 | BR/WT | Brake Switch Signal | Brake Off: 0v, On: 12v |
| 63 | OR/BK | Torque Management Request | Digital Signals |
| 64 | DG | A/C Clutch Relay Control | Relay Off: 12v, On: 1v |
| 65 | PK | SCI Transmit | 0v |
| 66 | YL/WT | Vehicle Speed Signal | Digital Signal |
| 67 | RD/WT | ASD Relay Control | Relay Off: 12v, On: 1v |
| 68 | LG/BK | EVAP Purge Solenoid Control | PWM Signal: 0-12-0v |
| 69 | DG/WT | High Condenser Fan Relay | Relay Off: 12v, On: 1v |
| 70-71, 79 | --- | Not Used | --- |
| 72 | RD/DB | LDP Switch Sense | Open: 12v, Closed: 0v |
| 73 | WT | Tachometer Signal | Pulse Signals |
| 74 | WT/RD | Fuel Pump Relay Control | Relay Off: 12v, On: 1v |
| 75 | DB/BK | SCI Receive | 0v |
| 76 | BK/YL | PNP Switch Signal | In P/N: 0v, Others: 5v |
| 77 | RD/YL | LDP Solenoid Control | PWM Signal: 0-12-0v |
| 78 | LG/WT | S/C Vacuum Solenoid | Vacuum Increasing: 1v |
| 80 | BK/YL | S/C Vent Solenoid | Vacuum Decreasing: 1v |

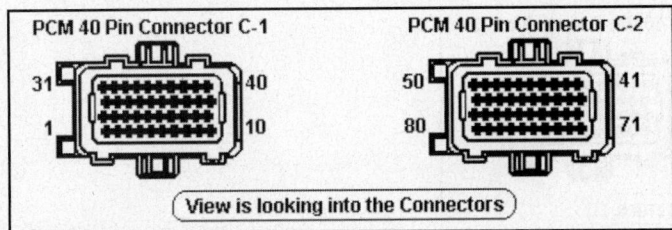

**Pin Connector Graphic**

**1998-2000 Sebring & Avenger 2.5L VIN N (A/T) 'C1' Connector**

| PCM Pin # | Wire Color | Circuit Description (40-Pin) | Value at Hot Idle |
|---|---|---|---|
| 1-3 | --- | Not Used | --- |
| 4 | RD/DG | Coil Driver | 5°, at 55 mph: 8° dwell |
| 5 | --- | Not Used | --- |
| 6 | BK/RD | ASD Relay Output | 12-14v |
| 7 | YL/DG | Injector 3 Driver | 1.0-4.0 ms |
| 8 | DB | Alternator Field Control | Digital Signal: 0-12-0v |
| 9 | DB/YL | Bulb (lamp) Check Control | B/C On: 1v, B/C Off: 12v |
| 10 | BK | Power Ground | <0.1v |
| 11 | --- | Not Used | --- |
| 12 | RD/BK | S/C On Switch Signal | S/C On: 1v, Off: 12v |
| 13 | LG/BK | Injector 1 Driver | 1.0-4.0 ms |
| 14 | BR/RD | Injector 6 Driver | 1.0-4.0 ms |
| 15 | RD/WT | Injector 5 Driver | 1.0-4.0 ms |
| 16 | LG/RD | Injector 4 Driver | 1.0-4.0 ms |
| 17 | YL/RD | Injector 2 Driver | 1.0-4.0 ms |
| 18 | LG | Speed Control Lamp Driver | S/C On: 1v, S/C Off: 12v |
| 19 | DG/OR | High Speed Fan 2 Control | Relay Off: 12v, On: 1v |
| 20 | BK/WT | Ignition Switch Output | 12-14v |
| 21 | --- | Not Used | --- |
| 22 | DG/RD | Amber MIL Control | MIL On: 1v, MIL Off: 12v |
| 23 | YL/DB | Fuel Level Sensor Signal | 70 ohms (±20) w/full tank |
| 24-25 | --- | Not Used | --- |
| 26 | DG/WT | ECT Sensor Signal | At 180°F: 2.80v |
| 27 | --- | Not Used | --- |
| 28 | DG/YL | High Speed Fan 1 Control | Relay Off: 12v, On: 1v |
| 29 | DG/BK | HO2S-11 (B1 S1) Signal | 0.1-1.1v |
| 30 | WT/BK | HO2S-12 (B1 S2) Signal | 0.1-1.1v |
| 31 | --- | Not Used | --- |
| 32 | DB/WT | CKP Sensor Signal | Digital Signal: 0-5-0v |
| 33 | BR | CMP Sensor Signal | Digital Signal: 0-5-0v |
| 34 | --- | Not Used | --- |
| 35 | BR/RD | TP Sensor Signal | 0.6-1.0v |
| 36 | YL/BK | MAP Sensor Signal | 1.5-1.7v |
| 37 | BR/DB | IAT Sensor Signal | At 100°F: 1.83v |
| 38 | DG/RD | A/C Clutch Switch Sense | Relay Off: 12v, On: 1v |
| 39 | DB | Generator Lamp Control | Lamp On: 1v, Off: 12v |
| 40 | RD/DB | EGR Solenoid Control | 12v, at 55 mph: 1v |

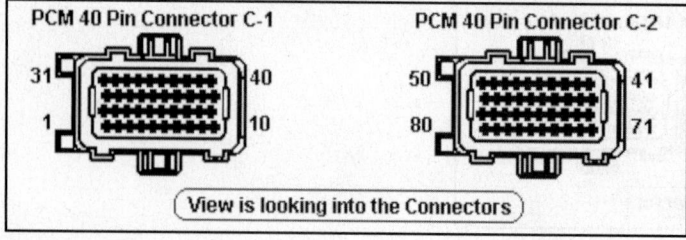

**Pin Connector Graphic**

## 1998-2000 Sebring & Avenger 2.5L VIN N (A/T) 'C2' Connector

| PCM Pin # | Wire Color | Circuit Description (40-Pin) | Value at Hot Idle |
|---|---|---|---|
| 41 | RD | S/C Set Switch Signal | S/C & Set Switch On: 3.8v |
| 42-43 | --- | Not Used | --- |
| 44 | YL | 8-Volt Supply | 7.9-8.1v |
| 45 | DB/YL | PSP Switch Signal | Straight: 0v, Turning: 5v |
| 46 | RD/BK | Battery Power (Fused B+) | 12-14v |
| 47 | BK | Sensor Ground | <0.050v |
| 48 | GY/DB | IAC 3 Driver | DC pulse signals: 0.8-11v |
| 49 | OR | IAC 2 Driver | DC pulse signals: 0.8-11v |
| 50 | BK | Power Ground | <0.1v |
| 51 | WT/DG | HO2S-22 (B2 S2) Signal | 0.1-1.1v |
| 52, 54 | --- | Not Used | --- |
| 53 | WT/RD | HO2S-21 (B2 S1) Signal | 0.1-1.1v |
| 55 | DG/BK | Low Speed Fan Relay | Relay Off: 12v, On: 1v |
| 56 | LG/WT | S/C Vacuum Solenoid | Vacuum Increasing: 1v |
| 57 | GY | IAC 1 Driver | DC pulse signals: 0.8-11v |
| 58 | YL/DB | IAC 4 Driver | DC pulse signals: 0.8-11v |
| 59 | BK/DB | CCD Bus (+) | Digital Signal: 0-5-0v |
| 60 | WT/DB | CCD Bus (-) | <0.050v |
| 61 | DG/YL | 5-Volt Supply | 4.9-5.1v |
| 62 | BR/WT | Brake Switch Signal | Brake Off: 0v, On: 12v |
| 63 | OR/BK | Torque Management Request | Digital Signals |
| 64 | DG | A/C Clutch Relay Control | Relay Off: 12v, On: 1v |
| 65 | PK | SCI Transmit | 0v |
| 66 | YL/WT | Vehicle Speed Signal | Digital Signal |
| 67 | RD/WT | ASD Relay Control | Relay Off: 12v, On: 1v |
| 68 | LG/BK | EVAP Purge Solenoid Control | PWM Signal: 0-12-0v |
| 69 | DG/WT | High Condenser Fan Relay | Relay Off: 12v, On: 1v |
| 70 | B/WT | EVAP Purge Solenoid Sense | 0-1v |
| 71 | --- | Not Used | --- |
| 72 | RD/DB | LDP Switch Sense | Open: 12v, Closed: 0v |
| 73 | WT | Tachometer Signal | Pulse Signals |
| 74 | WT/RD | Fuel Pump Relay Control | Relay Off: 12v, On: 1v |
| 75 | DB/BK | SCI Receive | 0v |
| 76 | BK/YL | PNP Switch Signal | PWM Signal: 0-12-0v |
| 77 | RD/YL | LDP Solenoid Control | PWM Signal: 0-12-0v |
| 78-79 | --- | Not Used | --- |
| 80 | BK/YL | S/C Vent Solenoid | Vacuum Decreasing: 1v |

## Standard Colors and Abbreviations

| Abbreviation | Color | Abbreviation | Color | Abbreviation | Color |
|---|---|---|---|---|---|
| BK | Black | GY | Gray | RD | Red |
| BL | Blue | GN | Green | TN | Tan |
| BR | Brown | LG | Light Green | VT | Violet |
| DB | Dark Blue | OR | Orange | WT | White |
| DG | Dark Green | PK | Pink | YL | Yellow |

## 2001-04 Sebring & Stratus 2.7L VIN R (A/T) 'C1' Connector

| PCM Pin # | Wire Color | Circuit Description (40-Pin) | Value at Hot Idle |
|---|---|---|---|
| 1 | TN/LG | COP 4 Driver Control | 5°, at 55 mph: 8° dwell |
| 2 | TN/OR | COP 3 Driver Control | 5°, at 55 mph: 8° dwell |
| 3 | TN/PK | COP 2 Driver Control | 5°, at 55 mph: 8° dwell |
| 4 | TN/LG | COP 6 Driver Control | 5°, at 55 mph: 8° dwell |
| 5 | YL/RD | S/C Power Supply | 12-14v |
| 6 | DG/OR | ASD Relay Output | 12-14v |
| 7 | YL/WT | Injector 3 Driver | 1.0-4.0 ms |
| 8 | DG | Alternator Field Control | Digital Signal: 0-12-0v |
| 9 | --- | Not Used | --- |
| 10 | BK/TN | Power Ground | <0.1v |
| 11 | TN/RD | COP 1 Driver Control | 5°, at 55 mph: 8° dwell |
| 12 | --- | Not Used | --- |
| 13 | WT/DB | Injector 1 Driver | 1.0-4.0 ms |
| 14 | BR/DB | Injector 6 Driver | 1.0-4.0 ms |
| 15 | GY | Injector 5 Driver | 1.0-4.0 ms |
| 16 | LB/BR | Injector 4 Driver | 1.0-4.0 ms |
| 17 | TN | Injector 2 Driver | 1.0-4.0 ms |
| 18-19 | --- | Not Used | --- |
| 20 | DB/WT | Ignition Switch Output | 12-14v |
| 21 | TN/DG | COP 5 Driver Control | 5°, at 55 mph: 8° dwell |
| 22-24 | --- | Not Used | --- |
| 25 | DB/LG | Knock Sensor Signal | 0.080v AC |
| 26 | TN/BK | ECT Sensor Signal | At 180°F: 2.80v |
| 27 | DB/LG | HO2S Sensor Ground | <0.050v |
| 28 | --- | Not Used | --- |
| 29 | LG/RD | HO2S-21 (B2 S1) Signal | 0.1-1.1v |
| 30 | BK/DG | HO2S-11 (B1 S1) Signal | 0.1-1.1v |
| 31 | TN | Starter Relay Control | KOEC: 9-11v |
| 32 | GY/BK | CKP Sensor Signal | Digital Signal: 0-5-0v |
| 33 | TN/YL | CMP Sensor Signal | Digital Signal: 0-5-0v |
| 34 | LG/PK | EGR Sensor Signal | 0.6-0.8v |
| 35 | OR/DB | TP Sensor Signal | 0.6-1.0v |
| 36 | DG/RD | MAP Sensor Signal | 1.5-1.7v |
| 37 | BK/RD | IAT Sensor Signal | At 100°F: 1.83v |
| 38 | --- | Not Used | --- |
| 39 | VT/RD | Manifold Solenoid Control | Solenoid Off: 12v, On: 1v |
| 40 | GY/YL | EGR Solenoid Control | 12v, at 55 mph: 1v |

### Standard Colors and Abbreviations

| Abbreviation | Color | Abbreviation | Color | Abbreviation | Color |
|---|---|---|---|---|---|
| BK | Black | GY | Gray | RD | Red |
| BL | Blue | GN | Green | TN | Tan |
| BR | Brown | LG | Light Green | VT | Violet |
| DB | Dark Blue | OR | Orange | WT | White |
| DG | Dark Green | PK | Pink | YL | Yellow |

**2001-04 Sebring & Stratus 2.7L VIN R (A/T) 'C2' Connector**

| PCM Pin # | Wire Color | Circuit Description (40-Pin) | Value at Hot Idle |
|---|---|---|---|
| 41 | PK/LG | S/C Set Switch Signal | S/C & Set Switch On: 3.8v |
| 42 | DB | A/C Pressure Switch Signal | A/C On: 0.45-4.85v |
| 43 | BK/LB | Sensor Ground | <0.050v |
| 44 | OR/WT | 8-Volt Supply | 7.9-8.1v |
| 45 | --- | Not Used | --- |
| 46 | RD/TN | Battery Power (Fused B+) | 12-14v |
| 47 | BK | Power Ground | <0.1v |
| 48 | BR/WT | IAC 3 Driver | DC pulse signals: 0.8-11v |
| 49 | YL/BK | IAC 2 Driver | DC pulse signals: 0.8-11v |
| 50 | BK/TN | Power Ground | <0.1v |
| 51 | TN/WT | HO2S-12 (B1 S2) Signal | 0.1-1.1v |
| 52 | VT/LG | Ambient Temperature Sensor | At 86°F: 1.96v |
| 53 | PK/WT | HO2S-22 (B2 S2) Signal | 0.1-1.1v |
| 54 | --- | Not Used | --- |
| 55 | DB/TN | Low Speed Fan Relay | Relay Off: 12v, On: 1v |
| 56 | TN/RD | S/C Vacuum Solenoid | Vacuum Increasing: 1v |
| 57 | GY/RD | IAC 1 Driver | DC pulse signals: 0.8-11v |
| 58 | VT/BK | IAC 4 Driver | DC pulse signals: 0.8-11v |
| 59 | OR | PCI Data Bus (J1850) | Digital Signals: 0-7-0v |
| 60 | --- | Not Used | --- |
| 61 | VT/WT | 5-Volt Supply | 4.9-5.1v |
| 62 | WT/PK | Brake Switch Signal | Brake Off: 0v, On: 12v |
| 63 | YL/DG | Torque Management Request | Digital Signals |
| 64 | DB/OR | A/C Clutch Relay Control | Relay Off: 12v, On: 1v |
| 65 | PK | SCI Transmit | Digital Signal: 0-5-0v |
| 66 | WT/OR | Vehicle Speed Signal | Digital Signal |
| 67 | DB/VT | ASD Relay Control | Relay Off: 12v, On: 1v |
| 68 | PK/BK | EVAP Purge Solenoid Control | PWM Signal: 0-12-0v |
| 69 | DB/PK | High Speed Fan Relay | Relay Off: 12v, On: 1v |
| 70 | WT/TN | EVAP Purge Solenoid Sense | 0-1v |
| 71 | WT/RD | EATX RPM Signal | Digital Signals |
| 72 | OR | LDP Switch Sense | Open: 12v, Closed: 0v |
| 73 | --- | Not Used | --- |
| 74 | BR/LG | Fuel Pump Relay Control | Relay Off: 12v, On: 1v |
| 75 | LG | SCI Receive | 0v |
| 76 | BK/LB | TRS T41 Sense | In P/N: 0v, Others: 5v |
| 77 | WT/DG | LDP Solenoid Control | PWM Signal: 0-12-0v |
| 78-79 | --- | Not Used | --- |
| 80 | LG/RD | S/C Vent Solenoid | Vacuum Decreasing: 1v |

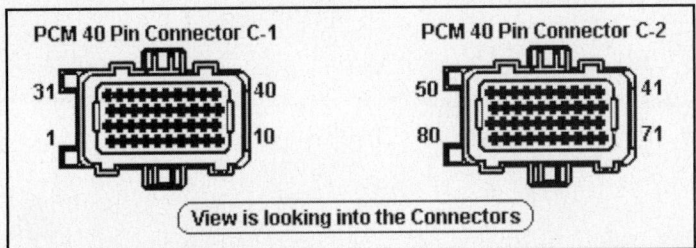

**Pin Connector Graphic**

**2004 Sebring 3.0L VIN H (M/T) 'C1' Connector**

| PCM Pin # | Wire Color | Circuit Description (40-Pin) | Value at Hot Idle |
|-----------|------------|------------------------------|-------------------|
| 1 | TN/LG | Coil On Plug Driver No. 4 | |
| 2 | TN/OR | Coil On Plug Driver No. 3 | |
| 3 | TN/PK | Coil On Plug Driver No. 2 | |
| 4 | TN/LB | Coil On Plug Driver No. 6 | |
| 5 | YL/RD | Speed Control Power Supply | |
| 6 | DG/OR | Automatic Shut Down Relay Output | |
| 7 | YL/WT | Fuel Injector No. 3 Driver | |
| 8 | DG | Generator Field Driver | |
| 9 | --- | Not Used | --- |
| 10 | BK/TN | Ground | |
| 11 | TN/RD | Coil On Plug Driver No. 1 | |
| 12 | --- | Not Used | --- |
| 13 | WT/DB | Fuel Injector No. 1 Driver | |
| 14 | BR/DB | Fuel Injector No. 6 Driver | |
| 15 | GY | Fuel Injector No. 5 Driver | |
| 16 | LB/BR | Fuel Injector No. 4 Driver | |
| 17 | TN | Fuel Injector No. 2 Driver | |
| 18 | --- | Not Used | --- |
| 19 | --- | Not Used | --- |
| 20 | DB/WT | Fused Ignition Switch Output (Run-Start) | |
| 21 | TN/DG | Coil On Plug Driver No. 5 | |
| 22 | --- | Not Used | --- |
| 23 | --- | Not Used | --- |
| 24 | --- | Not Used | --- |
| 25 | DB/LG | Knock Sensor Signal | |
| 26 | TN/BK | Engine Coolant Temperature Sensor Signal | |
| 27 | DB/LG | Oxygen Sensor Ground | |
| 28 | --- | Not Used | --- |
| 29 | LG/RD | Oxygen Sensor 2/1 Signal | |
| 30 | BK/DG | Oxygen Sensor 1/1 Signal | |
| 31 | TN | Starter Relay Control | |
| 32 | GY/BK | Crankshaft Position Sensor Signal | |
| 33 | TN/YL | Camshaft Position Sensor Signal | |
| 34 | --- | Not Used | --- |
| 35 | OR/DB | Throttle Position Sensor Signal | |
| 36 | DG/RD | MAP Sensor Signal | |
| 37 | BK/RD | Inlet Air Temperature Sensor Signal | |
| 38 | --- | Not Used | --- |
| 39 | --- | Not Used | --- |
| 40 | --- | Not Used | --- |

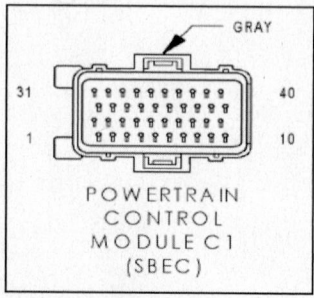

GRAY

31   40
1   10

POWERTRAIN
CONTROL
MODULE C1
(SBEC)

**2004 Sebring 3.0L VIN H (M/T) 'C2' Connector (Continued)**

| PCM Pin # | Wire Color | Circuit Description (40-Pin) | Value at Hot Idle |
|---|---|---|---|
| 41 | PK/LG | Speed Control Switch Signal | |
| 42 | DB | A/C Pressure Signal | |
| 43 | BK/LB | Sensor Ground 1 | |
| 44 | OR/WT | 8 Volt Supply | |
| 45 | --- | Not Used | --- |
| 46 | RD/TN | Fused B (+) | |
| 47 | BK | Ground | |
| 48 | BR/WT | Idle Air Control Motor No. 3 Driver | |
| 49 | YL/BK | Idle Air Control Motor No. 2 Driver | |
| 50 | BK/TN | Ground | |
| 51 | TN/WT | Oxygen Sensor 1/2 Signal | |
| 52 | VT/LG | Ambient Temperature Sensor Signal | |
| 53 | PK/WT (EATX) | Oxygen Sensor 2/2 Signal | |
| 54 | --- | Not Used | --- |
| 55 | DB/TN | Low Speed Radiator Fan Relay Control | |
| 56 | TN/RD | Speed Control Vacuum Solenoid Control | |
| 57 | GY/RD | Idle Air Control Motor No. 1 Driver | |
| 58 | VT/BK | Idle Air Control Motor No. 4 Driver | |
| 59 | OR | PCI Bus | |
| 60 | --- | Not Used | --- |
| 61 | VT/WT | 5 Volt Supply | |
| 62 | WT/PK | Brake Switch Sense | |
| 63 | YL/DG (EATX) | Torque Management Request Sense | |
| 64 | DB/OR | A/C Compressor Clutch Relay Control | |
| 65 | PK | SCI Transmit | |
| 66 | WT/OR (ABS) | Vehicle Speed Sensor Signal | |
| 67 | DB/VT | Automatic Shut Down Relay Control | |
| 68 | PK/BK | Evaporative Emission Solenoid Control | |
| 69 | DB/PK | High Speed Radiator Fan Relay Control | |
| 70 | WT/TN | Evaporative Solenoid Sense | |
| 71 | WT/RD (EATX) | EATX Rpm Signal | |
| 72 | OR | Leak Detection Pump Switch Sense | |
| 73 | --- | Not Used | --- |
| 74 | BR/LG | Fuel Pump Relay Control | |
| 75 | LG | SCI Receive | |
| 76 | BK/LB (EATX) | TRS T41 Sense | |
| 77 | WT/DG | Leak Detection Pump Solenoid Control | |
| 78 | --- | Not Used | --- |
| 79 | --- | Not Used | --- |
| 80 | LG/RD | Speed Control Vent Solenoid Control | |

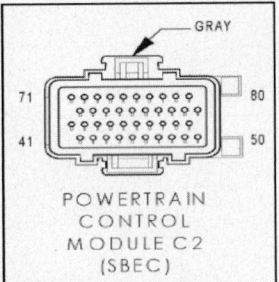

POWERTRAIN
CONTROL
MODULE C2
(SBEC)

**2005 Sebring 2.4L, 2.7L, 3.0L 'C1' Black/Black Connector**

| PCM Pin # | Wire Color | Circuit Description (38-Pin) | Value at Hot Idle |
|---|---|---|---|
| 1 | --- | Not Used | --- |
| 2 | --- | Not Used | --- |
| 3 | --- | Not Used | --- |
| 4 | --- | Not Used | --- |
| 5 | --- | Not Used | --- |
| 6 | --- | Not Used | --- |
| 7 | --- | Not Used | --- |
| 8 | --- | Not Used | --- |
| 9 | BK/TN | Ground | |
| 10 | --- | Not Used | --- |
| 11 | DB/WT | Fused Ignition Switch Output (Run-Start) | |
| 12 | RD/WT | Fused Ignition Switch Output (Off-Run-Start) | |
| 13 | WT/OR (MTX) | Vehicle Speed Signal | |
| 14 | --- | Not Used | --- |
| 15 | --- | Not Used | --- |
| 16 | --- | Not Used | --- |
| 17 | --- | Not Used | --- |
| 18 | BK/TN | Ground | |
| 19 | --- | Not Used | --- |
| 20 | --- | Not Used | --- |
| 21 | DB | A/C Pressure Signal | |
| 22 | --- | Not Used | --- |
| 23 | --- | Not Used | --- |
| 24 | --- | Not Used | --- |
| 25 | LG | SCI Receive (PCM) | |
| 26 | PK/LB (EATX) | SCI Receive (TCM) | |
| 27 | OR (MTX EXPORT) | 5 Volt Supply | |
| 28 | BK/OR (2.4L PZEV) | Air Pump Motor Relay Control | |
| 29 | RD/TN | Fused B (+) | |
| 30 | YL | Fused Ignition Switch Output (Start) | |
| 31 | TN/WT (2.7L) | O2 Sensor 1/2 Signal | |
| 32 | DB/DG | O2 Return (Down) | |
| 33 | PK/WT (2.7L) | O2 Sensor 2/2 Signal | |
| 34 | --- | Not Used | --- |
| 35 | --- | Not Used | --- |
| 36 | PK | SCI Transmit (PCM) | |
| 37 | WT/DG (EATX) | SCI Transmit (TCM) | |
| 38 | VT/YL (EATX) | PCI Bus (PCM) | |
| 38 | OR (MTX) | PCI Bus (PCM) | |

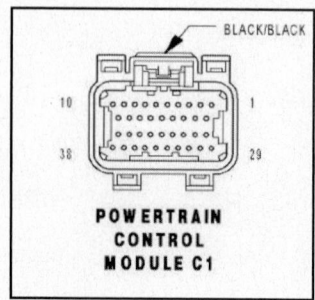

BLACK/BLACK

POWERTRAIN
CONTROL
MODULE C1

## 2005 Sebring 2.4L, 2.7L, 3.0L 'C2' Black/Orange Connector

| PCM Pin # | Wire Color | Circuit Description (38-Pin) | Value at Hot Idle |
|---|---|---|---|
| 1 | TN/LB (2.7L) | Coil Control No. 6 | |
| 2 | TN/DG (2.7L) | Coil Control No. 5 | |
| 3 | TN/LG (2.7L) | Coil Control No. 4 | |
| 4 | BR/DB (2.7L) | Injector Control No. 6 | |
| 5 | GY (2.7L) | Injector Control No. 5 | |
| 6 | --- | Not Used | --- |
| 7 | TN/OR (2.7L) | Coil Control No. 3 | |
| 8 | GY/YL (2.4L PZEV/ 2.7L) | EGR Solenoid Control | |
| 9 | DB/TN (2.0L/2.4L) | Coil Control No. 2 | |
| 9 | TN/PK (2.7L) | Coil Control No. 2 | |
| 10 | BK/GY (2.0L/2.4L) | Coil Control No. 1 | |
| 10 | TN/RD (2.7L) | Coil Control No. 1 | |
| 11 | LB/BR | Injector Control No. 4 | |
| 12 | YL/WT | Injector Control No. 3 | |
| 13 | TN | Injector Control No. 2 | |
| 14 | WT/DB | Injector Control No. 1 | |
| 15-16 | --- | Not Used | --- |
| 17 | BR/VT (2.0L/2.4L) | O2 1/2 Heater Control | |
| 17 | BR/WT (2.7L) | O2 2/1 Heater Control | |
| 18 | BR/OR | O2 1/1 Heater Control | |
| 19 | DG | Gen Field Control | |
| 20 | TN/BK | ECT Signal | |
| 21 | OR/DB | TP Signal | |
| 22 | LG/PK (2.4L PZEV/ 2.7L) | EGR Sensor Signal | |
| 23 | DG/RD | MAP Signal | |
| 24 | BK/VT | KS Return | |
| 25 | DB/LG | KS Signal | |
| 26 | DB (2.4L PZEV) | MAF Sensor Signal | |
| 27 | BK/LB | Sensor Ground 1 | |
| 28 | YL/BK | IAC Return | |
| 29 | VT/WT | 5 Volt Supply | |
| 30 | BK/RD | IAT Signal | |
| 31 | BK/DG | O2 1/1 Signal | |
| 32 | BR/DG | O2 Return (Up) | |
| 33 | TN/WT (2.0L/2.4L) | O2 1/2 Signal | |
| 33 | LG/RD (2.7L) | O2 2/1 Signal | |
| 34 | TN/YL | CMP Signal | |
| 35 | GY/BK | CKP Signal | |
| 36-37 | --- | Not Used | --- |
| 38 | GY/RD | IAC Motor Control | |

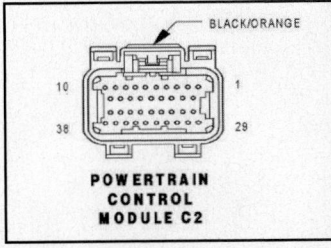

POWERTRAIN
CONTROL
MODULE C2

## 2005 Sebring 2.4L, 2.7L, 3.0L 'C3' Black/Natural Connector

| PCM Pin # | Wire Color | Circuit Description (38-Pin) | Value at Hot Idle |
|---|---|---|---|
| 1 | --- | Not Used | --- |
| 2 | --- | Not Used | --- |
| 3 | DB/VT | Automatic Shut Down Relay Control | |
| 4 | DB/PK | High Speed Rad Fan Relay Control | |
| 5 | LG/RD | S/C Vent Control | |
| 6 | DB/TN | Low Rad Fan Relay Control | |
| 7 | YL/RD | S/C Supply | |
| 8 | WT/DG | NVLD Solenoid Control | |
| 9 | BR/VT (2.7L) | O2 1/2 Heater Control | |
| 10 | BR/GY (2.7L) | O2 2/2 Heater Control | |
| 11 | DB/OR | A/C Clutch Relay Control | |
| 12 | TN/RD | S/C Vacuum Control | |
| 13 | --- | Not Used | --- |
| 14 | --- | Not Used | --- |
| 15 | --- | Not Used | --- |
| 16 | --- | Not Used | --- |
| 17 | BK/LB | Sensor Ground 1 | |
| 18 | --- | Not Used | --- |
| 19 | OR/DG | Fused Automatic Shut Down Relay Output | |
| 20 | PK/BK | EVAP Purge Control | |
| 21 | DG/OR | Clutch Interlock/ Upstop Switch Output | |
| 22 | --- | Not Used | --- |
| 23 | WT/PK | Brake Switch Signal | |
| 24 | --- | Not Used | --- |
| 25 | --- | Not Used | --- |
| 26 | YL | AutoStick Downshift Switch Signal | |
| 27 | LG | AutoStick Upshift Switch Signal | |
| 28 | OR/DG | Fused Automatic Shut Down Relay Output | |
| 29 | WT/TN | EVAP Purge Return | |
| 30 | DB/LG | PSP Switch Signal | |
| 31 | --- | Not Used | --- |
| 32 | VT/LG | AAT Signal | |
| 33 | --- | Not Used | --- |
| 34 | RD/LG | S/C Switch Signal | |
| 35 | OR | NVLD Switch Signal | |
| 36 | --- | Not Used | --- |
| 37 | BR/LG | Fuel Pump Relay Control | |
| 38 | TN | Starter Relay Control | |

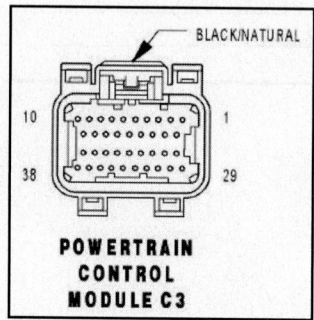

BLACK/NATURAL

POWERTRAIN
CONTROL
MODULE C3

## 2005 Sebring 2.4L, 2.7L, 3.0L 'C4' Black/Green Connector

| PCM Pin # | Wire Color | Circuit Description (38-Pin) | Value at Hot Idle |
|---|---|---|---|
| 1 | BR | OD Solenoid Control | |
| 2 | PK | UD Solenoid Control | |
| 3 | --- | Not Used | --- |
| 4 | --- | Not Used | --- |
| 5 | --- | Not Used | --- |
| 6 | WT | 2Not Used4 Solenoid Control | |
| 7 | --- | Not Used | --- |
| 8 | --- | Not Used | --- |
| 9 | --- | Not Used | --- |
| 10 | LB | L/R Solenoid Control | |
| 11 | --- | Not Used | --- |
| 12 | BK/YL | Ground | |
| 13 | BK/RD | Ground | |
| 14 | BK/RD | Ground | |
| 15 | LG/BK | TRS T1 Sense | |
| 16 | VT | TRS T3 Sense | |
| 17 | --- | Not Used | --- |
| 18 | LG | Transmission Control Relay Control | |
| 19 | RD | Transmission Control Relay Output | |
| 20 | --- | Not Used | --- |
| 21 | --- | Not Used | --- |
| 22 | OR/BK | OD Pressure Switch Sense | |
| 23 | --- | Not Used | --- |
| 24 | --- | Not Used | --- |
| 25 | --- | Not Used | --- |
| 26 | --- | Not Used | --- |
| 27 | BK/WT | TRS T41 Sense | |
| 28 | RD | Transmission Control Relay Output | |
| 29 | DG | L/R Pressure Switch Sense | |
| 30 | YL/BK | 2Not Used4 Pressure Switch Sense | |
| 31 | --- | Not Used | --- |
| 32 | LG/WT | Output Speed Sensor Signal | |
| 33 | RD/BK | Input Speed Sensor Signal | |
| 34 | DB/BK | Speed Sensor Ground | |
| 35 | VT/YL | Transmission Temperature Sensor Signal | |
| 36 | --- | Not Used | --- |
| 37 | VT/WT | TRS T42 Sense | |
| 38 | RD | Transmission Control Relay Output | |

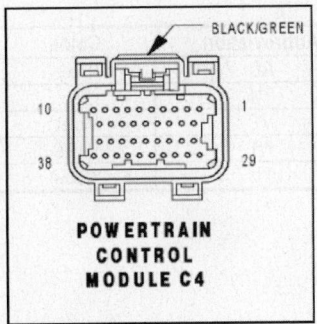

BLACK/GREEN

10        1

38        29

**POWERTRAIN CONTROL MODULE C4**

## SEBRING COUPE & STRATUS COUPE

### 2001-03 Sebring Coupe & Stratus Coupe 2.4L VIN G (M/T) C109 Connector

| PCM Pin # | Wire Color | Circuit Description (26-Pin) | Value at Hot Idle |
|:---:|:---:|:---:|:---:|
| 1 | YL/BL | Injector 1 Driver | 1.0-4.0 ms |
| 2 | YL/BK | Injector 3 Driver | 1.0-4.0 ms |
| 3 | --- | Not Used | --- |
| 4 | GN/BK | IAC Stepper Motor 'A' Signal | Pulse Signals |
| 5 | GN/WT | IAC Stepper Motor 'B' Signal | Pulse Signals |
| 6 | BL/RD | EGR Solenoid Control | Solenoid Off: 12v, On: 1v |
| 7 | --- | Not Used | --- |
| 8 | GN | A/C Clutch Relay Control | Relay Off: 12v, On: 1v |
| 9 | BL | EVAP Purge Solenoid Control | Solenoid Off: 12v, On: 1v |
| 10 | BK/BL | Ignition Coil 1 Control | Digital Signal: 0-12-0v |
| 11 | --- | Not Used | --- |
| 12 | RD | Ignition Power (MPI Relay) | 12-14v |
| 13 | BK | Power Ground | <0.1v |
| 14 | YL/BL | Injector 2 Driver | 1.0-4.0 ms |
| 15 | GN/YL | Injector 4 Driver | 1.0-4.0 ms |
| 16 | --- | Not Used | --- |
| 17 | GN/RD | IAC Stepper Motor 'A' Signal | Pulse Signals |
| 18 | BK/YL | IAC Stepper Motor 'B' Signal | Pulse Signals |
| 19 | GN/BL | Volume Airflow Sensor Reset | 1-3v, 3000 rpm: 6-9v |
| 20 | --- | Not Used | --- |
| 21 | BL/OR | Radiator Cooling Fan Control | Fan Off: 0.1v, On: 0.7v |
| 22 | BK | Fuel Pump Relay Control | Pump Off: 12v, On: 1v |
| 23 | WT/GN | Ignition Coil 2 Control | Digital Signal: 0-12-0v |
| 24 | --- | Not Used | --- |
| 25 | RD | Ignition Power (MPI Relay) | 12-14v |
| 26 | BK | Power Ground | <0.1v |

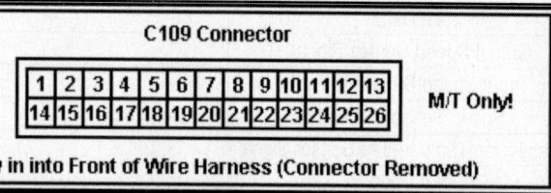

**PCM C109 26-Pin Connector Graphic**

### Standard Colors and Abbreviations

| Abbreviation | Color | Abbreviation | Color | Abbreviation | Color |
|:---:|:---:|:---:|:---:|:---:|:---:|
| BK | Black | GY | Gray | RD | Red |
| BL | Blue | GN | Green | TN | Tan |
| BR | Brown | LG | Light Green | VT | Violet |
| DB | Dark Blue | OR | Orange | WT | White |
| DG | Dark Green | PK | Pink | YL | Yellow |

**2001-03 Sebring Coupe & Stratus Coupe 2.4L VIN G (A/T) 'C110' Connector**

| PCM Pin # | Wire Color | Circuit Description (35-Pin) | Value at Hot Idle |
|---|---|---|---|
| 1 | YL/BL | Injector 1 Driver | 1.0-4.0 ms |
| 2 | GN/YL | Injector 4 Driver | 1.0-4.0 ms |
| 3 | BR/WT | HO2S-11 Heater Control | Digital Signal: 0-12-0v |
| 4-5 | --- | Not Used | --- |
| 6 | BL/RD | EGR Solenoid Control | Solenoid Off: 12v, On: 1v |
| 7 | --- | Not Used | --- |
| 8 | BK/RD | Generator Signal (lights on) | 0.2-3.5v |
| 9 | YL/RD | Injector 2 Driver | 1.0-4.0 ms |
| 10 | --- | Not Used | --- |
| 11 | BL/YL | Ignition Coil 1 Control | Digital Signal: 0-12-0v |
| 12 | WT/GN | Ignition Coil 2 Control | Digital Signal: 0-12-0v |
| 12-13 | --- | Not Used | --- |
| 14 | GN/BK | IAC Stepper Motor 'A' Signal | Pulse Signals |
| 15 | GN/WT | IAC Stepper Motor 'B' Signal | Pulse Signals |
| 16-17 | --- | Not Used | --- |
| 18 | BL/OR | Radiator Fan Relay Control | Fan Off: 0.1v, On: 0.7v |
| 19 | GN/BL | Volume Airflow Sensor Reset | 1-3v, 3000 rpm: 6-9v |
| 20 | GN | A/C Clutch Relay Control | Relay Off: 12v, On: 1v |
| 21 | BK | Fuel Pump Relay Control | Pump Off: 12v, On: 1v |
| 22 | RD/YL | MIL (lamp) Control | Lamp Off: 12v, On: 1v |
| 23 | --- | Not Used | --- |
| 24 | YL/BK | Injector 3 Driver | 1.0-4.0 ms |
| 25 | --- | Not Used | --- |
| 26 | BL/WT | HO2S-12 Heater Control | Digital Signal: 0-12-0v |
| 27 | --- | Not Used | --- |
| 28 | GN/RD | IAC Stepper Motor 'A' Signal | Pulse Signals |
| 29 | BK/YL | IAC Stepper Motor 'B' Signal | Pulse Signals |
| 30-33 | --- | Not Used | --- |
| 34 | BL | EVAP Purge Solenoid Control | PWM Signal: 0-12-0v |
| 35 | YL | EVAP Vent Solenoid Control | Solenoid Off: 12v, On: 1v |

C110 Connector

| 1 | 2 | | 3 | 4 | | | | | 5 | 6 | | 7 | 8 |
9 10 11 12 13 14 15 16 17 18 19 20 21 22 23
24 25 | 26 27 28 29 | 30 31 32 33 | 34 35

View in into Front of Wire Harness (Connector Removed)

**PCM C110 35-Pin Connector Graphic**

**Standard Colors and Abbreviations**

| Abbreviation | Color | Abbreviation | Color | Abbreviation | Color |
|---|---|---|---|---|---|
| BK | Black | GY | Gray | RD | Red |
| BL | Blue | GN | Green | TN | Tan |
| BR | Brown | LG | Light Green | VT | Violet |
| DB | Dark Blue | OR | Orange | WT | White |
| DG | Dark Green | PK | Pink | YL | Yellow |

### 2001-03 Sebring Coupe & Stratus Coupe 2.4L VIN G (M/T) C113 Connector

| PCM Pin # | Wire Color | Circuit Description (16 Pin) | Value at Hot Idle |
|---|---|---|---|
| 31-32 | --- | Not Used | --- |
| 33 | BR/RD | Generator Signal (Lights "on") | 0.2-3.5v |
| 34-35 | --- | Not Used | --- |
| 36 | RD/YL | MIL (lamp) Control | Lamp Off: 12v, On: 1v |
| 37 | YL | PSP Switch Signal | Straight: 5v, Turning: 0v |
| 38 | WT/VT | MPI (Power) Relay Control | Relay Off: 12v, On: 1v |
| 39-41 | --- | Not Used | --- |
| 41 | YL/BK | Generator 'FR' Terminal | 0.5-4.5v |
| 42-44 | --- | Not Used | --- |
| 45 | GN/RD | A/C Switch On/Off Signal | A/C Off: 0v, On: 12v |
| 46 | --- | Not Used | --- |

### 2001-03 Sebring Coupe & Stratus Coupe 2.4L VIN G (M/T) C116 Connector

| PCM Pin # | Wire Color | Circuit Description (12 Pin) | Value at Hot Idle |
|---|---|---|---|
| 51 | WT/BL | Immobilizer System Signal | Digital Signals |
| 52-53 | --- | Not Used | --- |
| 54 | BL/WT | HO2S-12 Heater Control | Digital Signal: 0-12-0v |
| 55 | YL | EVAP Vent Solenoid Control | Solenoid Off: 12v, On: 1v |
| 56 | GY/RD | Diagnosis Control (DLC #1) | 0v |
| 57 | --- | Not Used | --- |
| 58 | WT/RD | Tachometer Signals | Pulse Signals |
| 59 | --- | Not Used | --- |
| 60 | BR/WT | HO2S-11 Heater Control | Digital Signal: 0-12-0v |
| 61 | BR/WT | FTP Sensor Signal | 2.5v (fuel cap off) |
| 62 | RD/WT | ISO 9141 Bus (DLC #7) | 12v (no Scan Tool) |

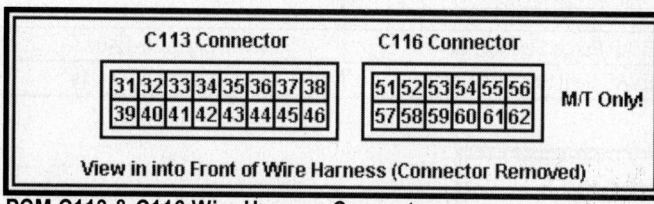

**PCM C113 & C116 Wire Harness Connector**

### Standard Colors and Abbreviations

| Abbreviation | Color | Abbreviation | Color | Abbreviation | Color |
|---|---|---|---|---|---|
| BK | Black | GY | Gray | RD | Red |
| BL | Blue | GN | Green | TN | Tan |
| BR | Brown | LG | Light Green | VT | Violet |
| DB | Dark Blue | OR | Orange | WT | White |
| DG | Dark Green | PK | Pink | YL | Yellow |

## 2001-03 Sebring Coupe & Stratus Coupe 2.4L VIN G (A/T) 'C114' Connector

| PCM Pin # | Wire Color | Circuit Description (26-Pin) | Value at Hot Idle |
|---|---|---|---|
| 41 | RD | Ignition Power (MPI Relay) | 12-14v |
| 42 | BK | Power Ground | <0.1v |
| 43 | WT/RD | Tachometer Signal | Pulse Signals |
| 44 | YL/GN | ECT Sensor Signal | 0.3-0.9v at 176ºF |
| 45 | GN/RD | Distributor CKP Sensor Signal | Digital Signal: 0-5-0v |
| 46 | GN/YL | Sensor Voltage Reference | 4.9-5.1v |
| 47 | RD | Ignition Power (MPI Relay) | 12-14v |
| 48 | BK | Power Ground | <0.1v |
| 49 | WT/VT | MPI (Power) Relay Control | Relay Off: 12v, On: 1v |
| 50 | WT/BL | A/T Control Relay | Relay Off: 12v, Off: 1v |
| 51 | --- | Not Used | --- |
| 52 | YL | PSP Switch Signal | Straight: 5v, Turning: 0v |
| 53 | --- | Not Used | --- |
| 54 | YL/BK | Generator 'FR' Terminal | 0.5-4.5v |
| 55 | GN/WT | BARO Sensor Signal | 3.7-4.3 at Sea Level |
| 56 | BL/YL | CMP Sensor Signal | Digital Signal: 0-5-0v |
| 57 | BK | Sensor Ground | <0.050v |
| 58 | BK/RD | Park Neutral Switch | In P/N: 12v, Others: 0v |
| 59 | BK/RD | Starter (Cranking) Signal | 9-11v (cranking) |
| 60 | --- | Not Used | --- |
| 61 | BL/WT | A/C Switch 2 Signal (Hi Blow) | 12v |
| 62-63 | --- | Not Used | --- |
| 64 | RD/BL | IAT Sensor Signal | 1.5-2.1v at 104ºF |
| 65 | WT/GN | Volume Airflow Sensor | 2.2-3.2v |
| 66 | OR/BL | Keep Alive Power | 12-14v |

```
                    C114 Connector

        41 42 43           44 45 46
        47 48 49 50 51 52 53 54 55 56 57
        58 59    60 61 62 63    64 65 66

    View in into Front of Wire Harness (Connector Removed)
```

**PCM C114 Wire Harness Connector Graphic**

## Standard Colors and Abbreviations

| Abbreviation | Color | Abbreviation | Color | Abbreviation | Color |
|---|---|---|---|---|---|
| BK | Black | GY | Gray | RD | Red |
| BL | Blue | GN | Green | TN | Tan |
| BR | Brown | LG | Light Green | VT | Violet |
| DB | Dark Blue | OR | Orange | WT | White |
| DG | Dark Green | PK | Pink | YL | Yellow |

**2001-03 Sebring Coupe & Stratus Coupe 2.4L VIN G (A/T) 'C117' Connector**

| PCM Pin # | Wire Color | Circuit Description (28-Pin) | Value at Hot Idle |
|---|---|---|---|
| 71 | WT | HO2S-11 (B1 S1) Signal | 0.1-1.1v |
| 72 | --- | Not Used | --- |
| 73 | GN | HO2S-12 (B1 S2) Signal | 0.1-1.1v |
| 74 | --- | Not Used | --- |
| 75 | GY/BL | Cruise Control Switch Signal | 12v or 0v |
| 76 | BK | Power Ground | <0.1v |
| 77 | RD/BL | A/T Control Relay Output | 12-14v |
| 78 | YL | TP Sensor Signal | 0.53-0.73v |
| 79 | YL/RD | Idle Position Switch Signal | 0v, Switch Open: 4-5v |
| 80 | WT/BL | Vehicle Speed Signal | Digital Signal |
| 81-82 | --- | Not Used | --- |
| 83 | GN/RD | A/C Switch On/Off Signal | A/C Off: 0v, On: 12v |
| 84 | GY/RD | DLC Diagnosis Control (#1) | 0v |
| 85 | RD/WT | DLC ISO 9141 Bus (#7) | 12v |
| 86-87 | --- | Not Used | --- |
| 88 | BK | Power Ground | <0.1v |
| 89 | RD/BL | A/T Control Relay Output | 12-14v |
| 90 | WT | Knock Sensor Signal | 0.080v AC |
| 91 | BL/RD | MAP Sensor Signal | 0.8-1.1v |
| 92 | BR/WT | FTP Sensor Signal | 2.5v (fuel cap off) |
| 93-96 | --- | Not Used | --- |
| 97 | BK | Power Ground | <0.1v |
| 98 | BK/WT | Ignition Switch Power | 12-14v |

```
                    C117 Connector

        71 72 73 74          75 76 77
        78 79 80 81 82 83 84 85 86 87 88 89
        90 91 92    93 94    95 96    97 98

        View in into Front of Wire Harness (Connector Removed)
```

**PCM C117 Wire Harness Connector Graphic**

**Standard Colors and Abbreviations**

| Abbreviation | Color | Abbreviation | Color | Abbreviation | Color |
|---|---|---|---|---|---|
| BK | Black | GY | Gray | RD | Red |
| BL | Blue | GN | Green | TN | Tan |
| BR | Brown | LG | Light Green | VT | Violet |
| DB | Dark Blue | OR | Orange | WT | White |
| DG | Dark Green | PK | Pink | YL | Yellow |

**2001-03 Sebring Coupe & Stratus Coupe 2.4L VIN G (M/T) C120 Connector**

| PCM Pin # | Wire Color | Circuit Description (22 Pin) | Value at Hot Idle |
|---|---|---|---|
| 71 | BR/RD | Starter (Cranking) Signal | 9-11v (cranking) |
| 72 | RD/BL | IAT Sensor Signal | 1.5-2.1v at 104°F |
| 73 | BL/RD | MAP Sensor Signal | 0.8-1.1v |
| 75 | GN | HO2S-12 (B1 S2) Signal | 0.1-1.1v |
| 76 | WT | HO2S-11 (B1 S1) Signal | 0.1-1.1v |
| 78 | WT | Knock Sensor Signal | 0.080v AC |
| 79 | WT/VT | DLC No. 2 Signal | N/A |
| 80 | OR/BK | Keep Alive Power | 12-14v |
| 81 | GN/YL | Sensor Voltage Reference | 4.9-5.1v |
| 82 | BK/WT | Ignition Switch Power | 12-14v |
| 83 | YL/GN | ECT Sensor Signal | 0.3-0.9v at 176°F |
| 84 | YL | TP Sensor Signal | 0.53-0.73v |
| 85 | GN/WT | BARO Sensor Signal | 3.7-4.3 at Sea Level |
| 86 | WT/BL | Vehicle Speed Signal | Digital Signal |
| 87 | YL/RD | Idle Position Switch Signal | 0v, Off-Idle: 4v |
| 88 | BL/YL | CMP (TDC) Sensor Signal | Digital Signal: 0-5-0v |
| 89 | GN/RD | CKP Sensor Signal | Digital Signal: 0-5-0v |
| 90 | WT/GN | Volume Airflow Sensor | 2.2-3.2v |
| 91 | BK | Power Ground | <0.1v |
| 92 | BK | Sensor Ground | <0.050v |

```
                    C120 Connector

          71 72 73 74 75 76 77 78 79 80 81      M/T Only!
          82 83 84 85 86 87 88 89 90 91 92

        View in into Front of Wire Harness (Connector Removed)
```

**PCM C120 Wire Harness Connector Graphic**

**Standard Colors and Abbreviations**

| Abbreviation | Color | Abbreviation | Color | Abbreviation | Color |
|---|---|---|---|---|---|
| BK | Black | GY | Gray | RD | Red |
| BL | Blue | GN | Green | TN | Tan |
| BR | Brown | LG | Light Green | VT | Violet |
| DB | Dark Blue | OR | Orange | WT | White |
| DG | Dark Green | PK | Pink | YL | Yellow |

**2001-03 Sebring Coupe & Stratus Coupe 2.4L VIN G (A/T) 'C121' Connector**

| PCM Pin # | Wire Color | Circuit Description (30-Pin) | Value at Hot Idle |
|---|---|---|---|
| 101 | BK/BL | PNP Switch 'P' Signal | In Park: 12v, Or 0v |
| 102 | YL | PNP Switch 'D' Signal | In Drive: 12v, Or 0v |
| 103 | WT | Input Shaft Speed Sensor | Moving: 0-12-0v |
| 104 | GN/YL | Output Shaft Speed Sensor | Moving: 0-12-0v |
| 105 | BR/RD | A/T Shift Mode 1st Indicator | 0v |
| 106 | RD/YL | A/T Second Solenoid | 12-14v |
| 107 | YL/RD | A/T TCC Solenoid | 12-14v |
| 108 | RD/BL | PNP Switch 'R' Signal | In Reverse: 12v, Or: 0v |
| 109 | WT | PNP Switch 'D3' Signal | In Drive 3: 12v, Or 0v |
| 110 | GY | PNP Switch 'Low' Signal | In Low: 12v, Or 0v |
| 111 | WT/GN | Immobilizer System Signal | Digital Signals |
| 112 | --- | Not Used | --- |
| 113 | WT/VT | DLC No. 2 Signal | 0v |
| 114 | WT/RD | DLC No. 2 Signal | 0v |
| 115 | --- | Not Used | --- |
| 116 | --- | Not Used | --- |
| 117 | WT/BL | A/T Shift Mode 3rd Indicator | 0v |
| 118 | YL/BL | A/T Shift Mode 2nd Indicator | 0v |
| 120 | RD | A/T Underdrive Solenoid | 12-14v |
| 121 | BR | PNP Switch 'N' Signal | In Neutral: 12v, Or 0v |
| 122 | YL/BL | PNP Switch 'D2' Signal | In Drive 2: 12v, Or 0v |
| 123 | GN/OR | Stop Light Switch Signal | Brake Off: 0v, On: 12v |
| 124 | BK/PK | TFT Sensor Signal | 3.2-3.4v at 104°F |
| 128 | YL/BL | A/T Shift Mode 4th Indicator | 0v |
| 129 | RD/WT | A/T Low/Reverse Solenoid | 12-14v |
| 130 | BL | A/T Overdrive Solenoid | 12-14v |

```
                    C121 Connector

   101 102    103 104              105 106 107
   108 109 110 111 112 113 114 115 116 117 118 119 120
   121 122 123    124 125    126 127 128    129 130

      View in into Front of Wire Harness (Connector Removed)
```

**PCM C121 Wire Harness Connector Graphic**

## Standard Colors and Abbreviations

| Abbreviation | Color | Abbreviation | Color | Abbreviation | Color |
|---|---|---|---|---|---|
| BK | Black | GY | Gray | RD | Red |
| BL | Blue | GN | Green | TN | Tan |
| BR | Brown | LG | Light Green | VT | Violet |
| DB | Dark Blue | OR | Orange | WT | White |
| DG | Dark Green | PK | Pink | YL | Yellow |

**2001-03 Sebring Coupe & Stratus Coupe 3.0L VIN H (M/T) 'C111' Connector**

| PCM Pin # | Wire Color | Circuit Description (35-Pin) | Value at Hot Idle |
|---|---|---|---|
| 1 | YL/BL | Injector 1 Driver | 1.0-4.0 ms |
| 2 | GN/YL | Injector 4 Driver | 1.0-4.0 ms |
| 3 | BR/WT | HO2S-21 Heater Control | Digital Signal: 0-12-0v |
| 4 | BL/WT | HO2S-11 Heater Control | Digital Signal: 0-12-0v |
| 5, 7 | --- | Not Used | --- |
| 6 | BL/RD | EGR Solenoid Control | Solenoid Off: 12v, On: 1v |
| 8 | BR/RD | Generator Signal (Lights "on") | 0.2-3.5v |
| 9 | YL/RD | Injector 2 Driver | 1.0-4.0 ms |
| 10 | GN/RD | Injector 5 Driver | 1.0-4.0 ms |
| 11 | BK/BL | Power Transistor Control | Digital Signal: 0-12-0v |
| 12-13 | --- | Not Used | --- |
| 14 | GN/BK | IAC Stepper Motor 'A' Signal | Pulse Signals |
| 15 | GN/WT | IAC Stepper Motor 'B' Signal | Pulse Signals |
| 16 | BL | EVAP Purge Solenoid Control | PWM Signal: 0-12-0v |
| 17 | --- | Not Used | --- |
| 18 | BL/OR | Radiator Cooling Fan Control | Fan Off: 0.1v, On: 0.7v |
| 19 | GN/BL | Volume Airflow Sensor Reset | 1-3v, 3000 rpm: 6-9v |
| 20 | GN | A/C Clutch Relay Control | Relay Off: 12v, On: 1v |
| 21 | BK | Fuel Pump Relay Control | Pump Off: 12v, On: 1v |
| 22 | RD/YL | MIL (lamp) Control | Lamp Off: 12v, On: 1v |
| 23 | --- | Not Used | --- |
| 24 | YL/BL | Injector 3 Driver | 1.0-4.0 ms |
| 25 | LG | Injector 6 Driver | 1.0-4.0 ms |
| 26 | BL/WT | HO2S-22 Heater Control | Digital Signal: 0-12-0v |
| 27 | BR | HO2S-12 Heater Control | Digital Signal: 0-12-0v |
| 28 | GN/RD | IAC Stepper Motor 'A' Signal | Pulse Signals |
| 29 | BK/YL | IAC Stepper Motor 'B' Signal | Pulse Signals |
| 30-34 | --- | Not Used | --- |
| 35 | YL | EVAP Vent Solenoid Control | Solenoid Off: 12v, On: 1v |

```
                    C111 Connector
  ┌──┬──┐  ┌──┬──┐  ┌──┐  ┌──┬──┐  ┌──┬──┐
  │1 │2 │  │3 │4 │  │  │  │5 │6 │  │7 │8 │
  ├──┼──┼──┼──┼──┼──┼──┼──┼──┼──┼──┼──┼──┤      M/T Only!
  │9 │10│11│12│13│14│15│16│17│18│19│20│21│22│23│
  ├──┼──┼──┼──┼──┼──┼──┼──┼──┼──┼──┼──┼──┤
  │24│25│  │26│27│28│29│  │30│31│32│33│  │34│35│
  └──┴──┘  └──┴──┴──┴──┘  └──┴──┴──┴──┘  └──┴──┘
      View in into Front of Wire Harness (Connector Removed)
```

**PCM C111 Wire Harness Connector Graphic**

**Standard Colors and Abbreviations**

| Abbreviation | Color | Abbreviation | Color | Abbreviation | Color |
|---|---|---|---|---|---|
| BK | Black | GY | Gray | RD | Red |
| BL | Blue | GN | Green | TN | Tan |
| BR | Brown | LG | Light Green | VT | Violet |
| DB | Dark Blue | OR | Orange | WT | White |
| DG | Dark Green | PK | Pink | YL | Yellow |

**2001-03 Sebring Coupe & Stratus Coupe 3.0L VIN H (A/T) 'C112' Connector**

| PCM Pin # | Wire Color | Circuit Description (35-Pin) | Value at Hot Idle |
|---|---|---|---|
| 1 | YL/BL | Injector 1 Driver | 1.0-4.0 ms |
| 2 | GN/YL | Injector 4 Driver | 1.0-4.0 ms |
| 3 | BR/WT | HO2S-21 Heater Control | Digital Signal: 0-12-0v |
| 4 | BL/WT | HO2S-11 Heater Control | Digital Signal: 0-12-0v |
| 5, 7 | --- | Not Used | --- |
| 6 | BL/RD | EGR Solenoid Control | Solenoid Off: 12v, On: 1v |
| 8 | BK/RD | Generator Signal (lights on) | 0.2-3.5v |
| 9 | YL/RD | Injector 2 Driver | 1.0-4.0 ms |
| 10 | RD/BL | Injector 5 Driver | 1.0-4.0 ms |
| 11 | BK/BL | Power Transistor Control | Digital Signal: 0-12-0v |
| 12-13 | --- | Not Used | --- |
| 14 | GN/BK | IAC Stepper Motor 'A' Signal | Pulse Signals |
| 15 | GN/WT | IAC Stepper Motor 'B' Signal | Pulse Signals |
| 16-17 | --- | Not Used | --- |
| 18 | BL/OR | Radiator Fan Relay Control | Fan Off: 0.1v, On: 0.7v |
| 19 | GN/BL | Volume Airflow Sensor Reset | 1-3v, 3000 rpm: 6-9v |
| 20 | GN | A/C Clutch Relay Control | Relay Off: 12v, On: 1v |
| 21 | BK | Fuel Pump Relay Control | Pump Off: 12v, On: 1v |
| 22 | RD/YL | MIL (lamp) Control | Lamp Off: 12v, On: 1v |
| 23 | --- | Not Used | --- |
| 24 | YL/BK | Injector 3 Driver | 1.0-4.0 ms |
| 25 | LG | Injector 6 Driver | 1.0-4.0 ms |
| 26 | BL/WT | HO2S-22 Heater Control | Digital Signal: 0-12-0v |
| 27 | BR | HO2S-12 Heater Control | Digital Signal: 0-12-0v |
| 28 | GN/RD | IAC Stepper Motor 'A' Signal | Pulse Signals |
| 29 | BL/YL | IAC Stepper Motor 'B' Signal | Pulse Signals |
| 30-33 | --- | Not Used | --- |
| 34 | BL | EVAP Purge Solenoid Control | PWM Signal: 0-12-0v |
| 35 | YL | EVAP Vent Solenoid Control | Solenoid Off: 12v, On: 1v |

```
                    C112 Connector
      ┌───┬───┬───┬───┬───────┬───┬───┬───┬───┐
      │ 1 │ 2 │ 3 │ 4 │       │ 5 │ 6 │ 7 │ 8 │
      ├──┬┴─┬─┴┬──┼┬──┼┬─┬────┼┬──┼┬──┼┬─┬┴──┤
      │9 │10│11│12│13│14│15│16│17│18│19│20│21│22│23│
      ├──┴┬─┴──┼──┼──┼──┼─────┼──┼──┼──┼──┴┬──┤
      │24 │25  │26│27│28│29    │30│31│32│33│34│35│
      └───┴────┴──┴──┴──┴──────┴──┴──┴──┴───┴──┘
      View in into Front of Wire Harness (Connector Removed)
```

**PCM C112 Wire Harness Connector Graphic**

**Standard Colors and Abbreviations**

| Abbreviation | Color | Abbreviation | Color | Abbreviation | Color |
|---|---|---|---|---|---|
| BK | Black | GY | Gray | RD | Red |
| BL | Blue | GN | Green | TN | Tan |
| BR | Brown | LG | Light Green | VT | Violet |
| DB | Dark Blue | OR | Orange | WT | White |
| DG | Dark Green | PK | Pink | YL | Yellow |

**2001-03 Sebring Coupe & Stratus Coupe 3.0L VIN H (A/T) 'C115' Connector**

| PCM Pin # | Wire Color | Circuit Description (26-Pin) | Value at Hot Idle |
|-----------|------------|------------------------------|-------------------|
| 41 | RD | Ignition Power (MPI Relay) | 12-14v |
| 42 | BK | Power Ground | <0.1v |
| 43 | WT | Tachometer Signal | Pulse Signals |
| 44 | YL/GN | ECT Sensor Signal | 0.3-0.9v at 176°F |
| 45 | GN/RD | Distributor CKP Sensor Signal | Digital Signal: 0-5-0v |
| 46 | GN | Sensor Voltage Reference | 4.9-5.1v |
| 47 | RD | Ignition Power (MPI Relay) | 12-14v |
| 48 | BK | Power Ground | <0.1v |
| 49 | WT/VT | MPI (Power) Relay Control | Relay Off: 12v, On: 1v |
| 50 | WT/BL | A/T Control Relay | Relay Off: 12v, On: 1v |
| 51 | --- | Not Used | --- |
| 52 | YL | PSP Switch Signal | Straight: 5v, Turning: 0v |
| 53 | --- | Not Used | --- |
| 54 | YL/BK | Generator 'FR' Terminal | 0.5-4.5v |
| 55 | GN/WT | BARO Sensor Signal | 3.7-4.3 at Sea Level |
| 56 | RD/WT | CMP (TDC) Sensor Signal | Digital Signal: 0-5-0v |
| 57 | BK | Sensor Ground | <0.050v |
| 58 | BR/RD | Starter (Cranking) Signal | 9-11v (cranking) |
| 59-60 | --- | Not Used | --- |
| 61 | BL/WT | A/C Switch 2 Signal (Hi Blow) | 12v |
| 62-63 | --- | Not Used | --- |
| 64 | RD/BL | IAT Sensor Signal | 1.5-2.1v at 104°F |
| 65 | WT/GN | Volume Airflow Sensor | 2.2-3.2v |
| 66 | OR/BL | Keep Alive Power | 12-14v |

**C115 Connector**

```
41 42 43        44 45 46
47 48 49 50 51 52 53 54 55 56 57
58 59    60 61 62 63    64 65 66
```

**View in into Front of Wire Harness (Connector Removed)**

**PCM C115 Wire Harness Connector Graphic**

**Standard Colors and Abbreviations**

| Abbreviation | Color | Abbreviation | Color | Abbreviation | Color |
|--------------|-------|--------------|-------|--------------|-------|
| BK | Black | GY | Gray | RD | Red |
| BL | Blue | GN | Green | TN | Tan |
| BR | Brown | LG | Light Green | VT | Violet |
| DB | Dark Blue | OR | Orange | WT | White |
| DG | Dark Green | PK | Pink | YL | Yellow |

## 2001-03 Sebring Coupe & Stratus Coupe 3.0L VIN H (M/T) 'C118' Connector

| PCM Pin # | Wire Color | Circuit Description (28-Pin) | Value at Hot Idle |
|---|---|---|---|
| 41 | --- | Not Used | --- |
| 42 | GN/YL | Sensor Voltage Reference | 4.9-5.1v |
| 43 | GN/RD | Distributor CKP Sensor Signal | Digital Signal: 0-5-0v |
| 44 | YL/GN | ECT Sensor Signal | 0.3-0.9v at 176°F |
| 45 | WT | Tachometer Signals | Pulse Signals |
| 46 | BK | Power Ground | <0.1v |
| 47 | RD | Ignition Power (MPI Relay) | 12-14v |
| 48 | --- | Not Used | --- |
| 49 | BK | Sensor Ground | <0.050v |
| 50 | RD/WT | CMP (TDC) Sensor Signal | Digital Signal: 0-5-0v |
| 51 | GN/WT | BARO Sensor Signal | 3.7-4.3 at Sea Level |
| 52 | YL/BK | Generator 'FR' Terminal | 0.5-4.5v |
| 53 | --- | Not Used | --- |
| 54 | YL | PSP Switch Signal | Straight: 5v, Turning: 0v |
| 55-56 | --- | Not Used | --- |
| 57 | WT/VT | MPI (Power) Relay Control | Relay Off: 12v, On: 1v |
| 58 | BK | Power Ground | <0.1v |
| 59 | RD | Ignition Power (MPI Relay) | 12-14v |
| 60 | OR/BK | Keep Alive Power | 12-14v |
| 61 | WT/GN | Volume Airflow Sensor | 2.2-3.2v |
| 62 | RD/BL | IAT Sensor Signal | 1.5-2.1v at 104°F |
| 63-64 | --- | Not Used | --- |
| 65 | BL/WT | A/C Switch 2 Signal (Hi Blow) | 12v |
| 66-67 | --- | Not Used | --- |
| 68 | BR/RD | Starter (Cranking) Signal | 9-11v (cranking) |

```
              C118 Connector
     ┌──┬──┬──┬──┐     ┌──┬──┬──┐
     │41│42│43│44│     │45│46│47│
     ├──┼──┼──┼──┼──┬──┼──┼──┼──┤     M/T Only!
     │48│49│50│51│52│53│54│55│56│57│58│59│
     ├──┼──┼──┼──┼──┬──┼──┼──┼──┤
     │60│61│  │62│63│64│  │65│66│  │67│68│
     └──┴──┴──┴──┴──┴──┴──┴──┴──┘

   View in into Front of Wire Harness (Connector Removed)
```

**PCM C118 Wire Harness Connector Graphic**

## Standard Colors and Abbreviations

| Abbreviation | Color | Abbreviation | Color | Abbreviation | Color |
|---|---|---|---|---|---|
| BK | Black | GY | Gray | RD | Red |
| BL | Blue | GN | Green | TN | Tan |
| BR | Brown | LG | Light Green | VT | Violet |
| DB | Dark Blue | OR | Orange | WT | White |
| DG | Dark Green | PK | Pink | YL | Yellow |

**2001-03 Sebring Coupe & Stratus Coupe 3.0L V6 VIN H (A/T) 'C119' Connector**

| PCM Pin # | Wire Color | Circuit Description (28-Pin) | Value at Hot Idle |
|---|---|---|---|
| 71 | WT | HO2S-21 (B2 S1) Signal | 0.1-1.1v |
| 72 | BL | HO2S-11 (B1 S1) Signal | 0.1-1.1v |
| 73 | GN | HO2S-22 (B2 S2) Signal | 0.1-1.1v |
| 74 | BR | HO2S-12 (B1 S2) Signal | 0.1-1.1v |
| 75 | GY/BL | Cruise Control Switch Signal | N/A |
| 76 | BK | Power Ground | <0.1v |
| 77 | RD/BL | A/T Control Relay Output | 12-14v |
| 78 | YL | TP Sensor Signal | 0.53-0.73v |
| 79 | YL/RD | Idle Position Switch Signal | 0v, Switch Open: 4-5v |
| 80 | WT/YL | Vehicle Speed Signal | Digital Signal |
| 81-82 | --- | Not Used | --- |
| 83 | GN/RD | A/C Switch On/Off Signal | A/C Off: 0v, On: 12v |
| 84 | GY/RD | Diagnosis Control (DLC #1) | 0v |
| 85 | RD/WT | ISO 9141 Bus (DLC #7) | 12v |
| 86-87 | --- | Not Used | --- |
| 88 | BK | Power Ground | <0.1v |
| 89 | RD/BL | A/T Control Relay Output | 12-14v |
| 90 | WT | Knock Sensor Signal | 0.080v AC |
| 91 | BL/RD | MAP Sensor Signal | 0.8-1.1v |
| 92 | BR/WT | FTP Sensor Signal | 2.5v (fuel cap off) |
| 93-96 | --- | Not Used | --- |
| 97 | BK | Power Ground | <0.1v |
| 98 | BK/WT | Ignition Switch Power | 12-14v |

```
                    C119 Connector
         71 72 73 74          75 76 77
         78 79 80 81 82 83 84 85 86 87 88 89
         90 91 92    93 94    95 96    97 98
      View in into Front of Wire Harness (Connector Removed)
```

**PCM C119 Wire Harness Connector Graphic**

**Standard Colors and Abbreviations**

| Abbreviation | Color | Abbreviation | Color | Abbreviation | Color |
|---|---|---|---|---|---|
| BK | Black | GY | Gray | RD | Red |
| BL | Blue | GN | Green | TN | Tan |
| BR | Brown | LG | Light Green | VT | Violet |
| DB | Dark Blue | OR | Orange | WT | White |
| DG | Dark Green | PK | Pink | YL | Yellow |

**2001-03 Sebring Coupe & Stratus Coupe 3.0L V6 VIN H (M/T) 'C122' Connector**

| PCM Pin # | Wire Color | Circuit Description (30-Pin) | Value at Hot Idle |
|---|---|---|---|
| 71 | WT | HO2S-21 (B2 S1) Signal | 0.1-1.1v |
| 72 | BL | HO2S-11 (B1 S1) Signal | 0.1-1.1v |
| 73 | GN | HO2S-22 (B2 S2) Signal | 0.1-1.1v |
| 74 | YL | HO2S-12 (B1 S2) Signal | 0.1-1.1v |
| 75 | --- | Not Used | --- |
| 76 | BK | Power Ground | <0.1v |
| 77 | --- | Not Used | --- |
| 78 | YL | TP Sensor Signal | 0.53-0.73v |
| 79 | YL/RD | Idle Position Switch Signal | 0v, Off-Idle: 4v |
| 80 | WT/BL | Vehicle Speed Signal | Digital Signal |
| 81-82 | --- | Not Used | --- |
| 83 | GN/RD | A/C Switch On/Off Signal | A/C Off: 0v, On: 12v |
| 84 | GY/RD | Diagnosis Control (DLC #1) | 0v |
| 85 | RD/WT | ISO 9141 Bus (DLC #7) | 12v |
| 86-87 | --- | Not Used | --- |
| 88 | BK | Power Ground | <0.1v |
| 89-90 | --- | Not Used | --- |
| 91 | WT | Knock Sensor Signal | 0.080v AC |
| 92 | BL/RD | MAP Sensor Signal | 0.8-1.1v |
| 93 | BR/WT | FTP Sensor Signal | 2.5v (fuel cap off) |
| 94-96 | --- | Not Used | --- |
| 97 | BK | Power Ground | <0.1v |
| 98 | WT/GN | Immobilizer System Signal | Digital Signals |
| 99 | BK/WT | Ignition Switch Power | 12-14v |
| 100 | WT/VT | DLC No. 2 Signal | 0v |

```
          C122 Connector
  ┌──────────────────────────────────┐
  │ 71 72  73 74        75 76 77      │
  │ 78 79 80 81 82 83 84 85 86 87 88 89 90 │   M/T Only!
  │ 91 92 93  94 95  96 97 98  99 100 │
  └──────────────────────────────────┘
   View in into Front of Wire Harness (Connector Removed)
```

**PCM C122 Wire Harness Connector Graphic**

**Standard Colors and Abbreviations**

| Abbreviation | Color | Abbreviation | Color | Abbreviation | Color |
|---|---|---|---|---|---|
| BK | Black | GY | Gray | RD | Red |
| BL | Blue | GN | Green | TN | Tan |
| BR | Brown | LG | Light Green | VT | Violet |
| DB | Dark Blue | OR | Orange | WT | White |
| DG | Dark Green | PK | Pink | YL | Yellow |

**2001-03 Sebring Coupe & Stratus Coupe 3.0L V6 VIN H (A/T) 'C123' Connector**

| PCM Pin # | Wire Color | Circuit Description (30-Pin) | Value at Hot Idle |
|---|---|---|---|
| 101 | BK/BL | PNP Switch 'P' Signal | In Park: 12v, Or 0v |
| 102 | YL | PNP Switch 'D' Signal | In Drive: 12v, Or 0v |
| 103 | WT | Input Shaft Speed Sensor | Moving: 0-12-0v |
| 104 | GN/YL | Output Shaft Speed Sensor | Moving: 0-12-0v |
| 105 | BR/RD | A/T Shift Mode 1st Indicator | 0v |
| 106 | RD/YL | A/T Second Solenoid | 12-14v |
| 107 | YL/RD | A/T TCC Solenoid | 12-14v |
| 108 | RD/BL | PNP Switch 'R' Signal | In Reverse: 12v, Or: 0v |
| 109 | WT | PNP Switch 'D3' Signal | In Drive 3: 12v, Or 0v |
| 110 | GN | PNP Switch 'Low' Signal | In Low: 12v, Or 0v |
| 111 | WT/GN | Immobilizer System Signal | Digital Signals |
| 112 | --- | Not Used | --- |
| 113 | WT/BL | DLC No. 2 Signal | N/A |
| 114-116 | --- | Not Used | --- |
| 117 | WT/BL | A/T Shift Mode 3rd Indicator | 0v |
| 118 | YL/BL | A/T Shift Mode 2nd Indicator | 0v |
| 120 | RD | A/T Underdrive Solenoid | 12-14v |
| 121 | BR | PNP Switch 'N' Signal | In Neutral: 12v, Or 0v |
| 122 | YL/BL | PNP Switch 'D2' Signal | In Drive 2: 12v, Or 0v |
| 123 | GN/OR | Stop Light Switch Signal | Brake Off: 0v, On: 12v |
| 124 | BL/PK | TFT Sensor Signal | 3.2-3.4v at 104°F |
| 125-127 | --- | Not Used | --- |
| 128 | YL/BL | A/T Shift Mode 4th Indicator | 0v |
| 129 | RD/WT | A/T Low/Reverse Solenoid | 12-14v |
| 130 | BL | A/T Overdrive Solenoid | 12-14v |

```
                    C123 Connector
  101 102    103 104           105 106 107
  108 109 110 111 112 113 114 115 116 117 118 119 120
  121 122 123    124 125      126 127 128    129 130
     View in into Front of Wire Harness (Connector Removed)
```

**PCM C123 Wire Harness Connector Graphic**

**Standard Colors and Abbreviations**

| Abbreviation | Color | Abbreviation | Color | Abbreviation | Color |
|---|---|---|---|---|---|
| BK | Black | GY | Gray | RD | Red |
| BL | Blue | GN | Green | TN | Tan |
| BR | Brown | LG | Light Green | VT | Violet |
| DB | Dark Blue | OR | Orange | WT | White |
| DG | Dark Green | PK | Pink | YL | Yellow |

## 2004 Sebring Coupe & Stratus Coupe 3.0L V6 VIN H (M/T) 'C1' Connector

| PCM Pin # | Wire Color | Circuit Description (30-Pin) | Value at Hot Idle |
|---|---|---|---|
| 1 | TN/LG | Coil On Plug Driver No. 4 | |
| 2 | TN/OR | Coil On Plug Driver No. 3 | |
| 3 | TN/PK | Coil On Plug Driver No. 2 | |
| 4 | TN/LB | Coil On Plug Driver No. 6 | |
| 5 | YL/RD | Speed Control Power Supply | |
| 6 | DG/OR | Automatic Shut Down Relay Output | |
| 7 | YL/WT | Fuel Injector No. 3 Driver | |
| 8 | DG | Generator Field Driver | |
| 9 | --- | Not Used | --- |
| 10 | BK/TN | Ground | |
| 11 | TN/RD | Coil On Plug Driver No. 1 | |
| 12 | --- | Not Used | --- |
| 13 | WT/DB | Fuel Injector No. 1 Driver | |
| 14 | BR/DB | Fuel Injector No. 6 Driver | |
| 15 | GY | Fuel Injector No. 5 Driver | |
| 16 | LB/BR | Fuel Injector No. 4 Driver | |
| 17 | TN | Fuel Injector No. 2 Driver | |
| 18 | --- | Not Used | --- |
| 19 | --- | Not Used | --- |
| 20 | DB/WT | Fused Ignition Switch Output (Run-Start) | |
| 21 | TN/DG | Coil On Plug Driver No. 5 | |
| 22 | --- | Not Used | --- |
| 23 | --- | Not Used | --- |
| 24 | --- | Not Used | --- |
| 25 | DB/LG | Knock Sensor Signal | |
| 26 | TN/BK | Engine Coolant Temperature Sensor Signal | |
| 27 | DB/LG | Oxygen Sensor Ground | |
| 28 | --- | Not Used | --- |
| 29 | LG/RD | Oxygen Sensor 2/1 Signal | |
| 30 | BK/DG | Oxygen Sensor 1/1 Signal | |
| 31 | TN | Starter Relay Control | |
| 32 | GY/BK | Crankshaft Position Sensor Signal | |
| 33 | TN/YL | Camshaft Position Sensor Signal | |
| 34 | --- | Not Used | --- |
| 35 | OR/DB | Throttle Position Sensor Signal | |
| 36 | DG/RD | MAP Sensor Signal | |
| 37 | BK/RD | Inlet Air Temperature Sensor Signal | |
| 38 | --- | Not Used | --- |
| 39 | --- | Not Used | --- |
| 40 | --- | Not Used | --- |

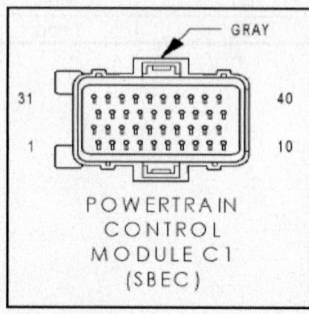

POWERTRAIN
CONTROL
MODULE C1
(SBEC)

**2004 Stratus 3.0L V6 VIN H (M/T) 'C2' Connector (Continued)**

| PCM Pin # | Wire Color | Circuit Description (40-Pin) | Value at Hot Idle |
|:---:|:---:|:---:|:---:|
| 41 | PK/LG | Speed Control Switch Signal | |
| 42 | DB | A/C Pressure Signal | |
| 43 | BK/LB | Sensor Ground 1 | |
| 44 | OR/WT | 8 Volt Supply | |
| 45 | --- | Not Used | --- |
| 46 | RD/TN | Fused B (+) | |
| 47 | BK | Ground | |
| 48 | BR/WT | Idle Air Control Motor No. 3 Driver | |
| 49 | YL/BK | Idle Air Control Motor No. 2 Driver | |
| 50 | BK/TN | Ground | |
| 51 | TN/WT | Oxygen Sensor 1/2 Signal | |
| 52 | VT/LG | Ambient Temperature Sensor Signal | |
| 53 | PK/WT (EATX) | Oxygen Sensor 2/2 Signal | |
| 54 | --- | Not Used | --- |
| 55 | DB/TN | Low Speed Radiator Fan Relay Control | |
| 56 | TN/RD | Speed Control Vacuum Solenoid Control | |
| 57 | GY/RD | Idle Air Control Motor No. 1 Driver | |
| 58 | VT/BK | Idle Air Control Motor No. 4 Driver | |
| 59 | OR | PCI Bus | |
| 60 | --- | Not Used | --- |
| 61 | VT/WT | 5 Volt Supply | |
| 62 | WT/PK | Brake Switch Sense | |
| 63 | YL/DG (EATX) | Torque Management Request Sense | |
| 64 | DB/OR | A/C Compressor Clutch Relay Control | |
| 65 | PK | SCI Transmit | |
| 66 | WT/OR (ABS) | Vehicle Speed Sensor Signal | |
| 67 | DB/VT | Automatic Shut Down Relay Control | |
| 68 | PK/BK | Evaporative Emission Solenoid Control | |
| 69 | DB/PK | High Speed Radiator Fan Relay Control | |
| 70 | WT/TN | Evaporative Solenoid Sense | |
| 71 | WT/RD (EATX) | EATX Rpm Signal | |
| 72 | OR | Leak Detection Pump Switch Sense | |
| 73 | --- | Not Used | --- |
| 74 | BR/LG | Fuel Pump Relay Control | |
| 75 | LG | SCI Receive | |
| 76 | BK/LB (EATX) | TRS T41 Sense | |
| 77 | WT/DG | Leak Detection Pump Solenoid Control | |
| 78 | --- | Not Used | --- |
| 79 | --- | Not Used | --- |
| 80 | LG/RD | Speed Control Vent Solenoid Control | |

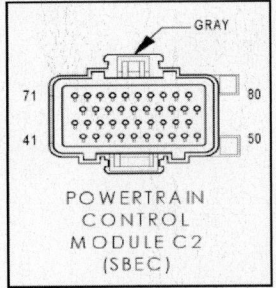

POWERTRAIN
CONTROL
MODULE C2
(SBEC)

**2005 Sebring Coupe & Stratus Coupe 2.4L, 2.7L, 3.0L 'C1' Black/Black Connector**

| PCM Pin # | Wire Color | Circuit Description (38-Pin) | Value at Hot Idle |
|---|---|---|---|
| 1 | --- | Not Used | --- |
| 2 | --- | Not Used | --- |
| 3 | --- | Not Used | --- |
| 4 | --- | Not Used | --- |
| 5 | --- | Not Used | --- |
| 6 | --- | Not Used | --- |
| 7 | --- | Not Used | --- |
| 8 | --- | Not Used | --- |
| 9 | BK/TN | Ground | |
| 10 | --- | Not Used | --- |
| 11 | DB/WT | Fused Ignition Switch Output (Run-Start) | |
| 12 | RD/WT | Fused Ignition Switch Output (Off-Run-Start) | |
| 13 | WT/OR (MTX) | Vehicle Speed Signal | |
| 14 | --- | Not Used | --- |
| 15 | --- | Not Used | --- |
| 16 | --- | Not Used | --- |
| 17 | --- | Not Used | --- |
| 18 | BK/TN | Ground | |
| 19 | --- | Not Used | --- |
| 20 | --- | Not Used | --- |
| 21 | DB | A/C Pressure Signal | |
| 22 | --- | Not Used | --- |
| 23 | --- | Not Used | --- |
| 24 | --- | Not Used | --- |
| 25 | LG | SCI Receive (PCM) | |
| 26 | PK/LB (EATX) | SCI Receive (TCM) | |
| 27 | OR (MTX EXPORT) | 5 Volt Supply | |
| 28 | BK/OR (2.4L PZEV) | Air Pump Motor Relay Control | |
| 29 | RD/TN | Fused B (+) | |
| 30 | YL | Fused Ignition Switch Output (Start) | |
| 31 | TN/WT (2.7L) | O2 Sensor 1/2 Signal | |
| 32 | DB/DG | O2 Return (Down) | |
| 33 | PK/WT (2.7L) | O2 Sensor 2/2 Signal | |
| 34 | --- | Not Used | --- |
| 35 | --- | Not Used | --- |
| 36 | PK | SCI Transmit (PCM) | |
| 37 | WT/DG (EATX) | SCI Transmit (TCM) | |
| 38 | VT/YL (EATX) | PCI Bus (PCM) | |
| 38 | OR (MTX) | PCI Bus (PCM) | |

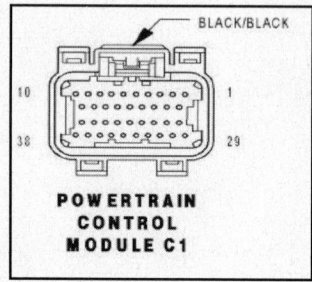

**BLACK/BLACK**

**POWERTRAIN CONTROL MODULE C1**

**2005 Sebring Coupe & Stratus Coupe 2.4L, 2.7L, 3.0L 'C2' Black/Orange Connector**

| PCM Pin # | Wire Color | Circuit Description (38-Pin) | Value at Hot Idle |
|---|---|---|---|
| 1 | TN/LB (2.7L) | Coil Control No. 6 | |
| 2 | TN/DG (2.7L) | Coil Control No. 5 | |
| 3 | TN/LG (2.7L) | Coil Control No. 4 | |
| 4 | BR/DB (2.7L) | Injector Control No. 6 | |
| 5 | GY (2.7L) | Injector Control No. 5 | |
| 6 | --- | Not Used | --- |
| 7 | TN/OR (2.7L) | Coil Control No. 3 | |
| 8 | GY/YL (2.4L PZEV/ 2.7L) | EGR Solenoid Control | |
| 9 | DB/TN (2.0L/2.4L) | Coil Control No. 2 | |
| 9 | TN/PK (2.7L) | Coil Control No. 2 | |
| 10 | BK/GY (2.0L/2.4L) | Coil Control No. 1 | |
| 10 | TN/RD (2.7L) | Coil Control No. 1 | |
| 11 | LB/BR | Injector Control No. 4 | |
| 12 | YL/WT | Injector Control No. 3 | |
| 13 | TN | Injector Control No. 2 | |
| 14 | WT/DB | Injector Control No. 1 | |
| 15-16 | --- | Not Used | --- |
| 17 | BR/VT (2.0L/2.4L) | O2 1/2 Heater Control | |
| 17 | BR/WT (2.7L) | O2 2/1 Heater Control | |
| 18 | BR/OR | O2 1/1 Heater Control | |
| 19 | DG | Gen Field Control | |
| 20 | TN/BK | ECT Signal | |
| 21 | OR/DB | TP Signal | |
| 22 | LG/PK (2.4L PZEV/ 2.7L) | EGR Sensor Signal | |
| 23 | DG/RD | MAP Signal | |
| 24 | BK/VT | KS Return | |
| 25 | DB/LG | KS Signal | |
| 26 | DB (2.4L PZEV) | MAF Sensor Signal | |
| 27 | BK/LB | Sensor Ground 1 | |
| 28 | YL/BK | IAC Return | |
| 29 | VT/WT | 5 Volt Supply | |
| 30 | BK/RD | IAT Signal | |
| 31 | BK/DG | O2 1/1 Signal | |
| 32 | BR/DG | O2 Return (Up) | |
| 33 | TN/WT (2.0L/2.4L) | O2 1/2 Signal | |
| 33 | LG/RD (2.7L) | O2 2/1 Signal | |
| 34 | TN/YL | CMP Signal | |
| 35 | GY/BK | CKP Signal | |
| 36-37 | --- | Not Used | --- |
| 38 | GY/RD | IAC Motor Control | |

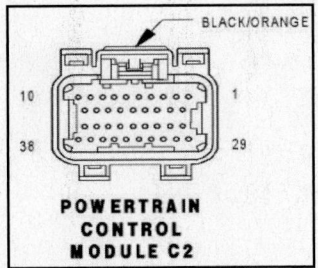

POWERTRAIN
CONTROL
MODULE C2

## 2005 Sebring Coupe & Stratus Coupe 2.4L, 2.7L, 3.0L 'C3' Black/Natural Connector

| PCM Pin # | Wire Color | Circuit Description (38-Pin) | Value at Hot Idle |
|---|---|---|---|
| 1 | --- | Not Used | --- |
| 2 | --- | Not Used | --- |
| 3 | DB/VT | Automatic Shut Down Relay Control | |
| 4 | DB/PK | High Speed Rad Fan Relay Control | |
| 5 | LG/RD | S/C Vent Control | |
| 6 | DB/TN | Low Rad Fan Relay Control | |
| 7 | YL/RD | S/C Supply | |
| 8 | WT/DG | NVLD Solenoid Control | |
| 9 | BR/VT (2.7L) | O2 1/2 Heater Control | |
| 10 | BR/GY (2.7L) | O2 2/2 Heater Control | |
| 11 | DB/OR | A/C Clutch Relay Control | |
| 12 | TN/RD | S/C Vacuum Control | |
| 13 | --- | Not Used | --- |
| 14 | --- | Not Used | --- |
| 15 | --- | Not Used | --- |
| 16 | --- | Not Used | --- |
| 17 | BK/LB | Sensor Ground 1 | |
| 18 | --- | Not Used | --- |
| 19 | OR/DG | Fused Automatic Shut Down Relay Output | |
| 20 | PK/BK | EVAP Purge Control | |
| 21 | DG/OR | Clutch Interlock/ Upstop Switch Output | |
| 22 | --- | Not Used | --- |
| 23 | WT/PK | Brake Switch Signal | |
| 24 | --- | Not Used | --- |
| 25 | --- | Not Used | --- |
| 26 | YL | AutoStick Downshift Switch Signal | |
| 27 | LG | AutoStick Upshift Switch Signal | |
| 28 | OR/DG | Fused Automatic Shut Down Relay Output | |
| 29 | WT/TN | EVAP Purge Return | |
| 30 | DB/LG | PSP Switch Signal | |
| 31 | --- | Not Used | --- |
| 32 | VT/LG | AAT Signal | |
| 33 | --- | Not Used | --- |
| 34 | RD/LG | S/C Switch Signal | |
| 35 | OR | NVLD Switch Signal | |
| 36 | --- | Not Used | --- |
| 37 | BR/LG | Fuel Pump Relay Control | |
| 38 | TN | Starter Relay Control | |

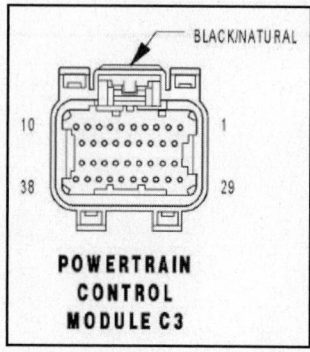

BLACK/NATURAL

10   1
38   29

**POWERTRAIN
CONTROL
MODULE C3**

**2005 Sebring Coupe & Stratus Coupe 2.4L, 2.7L, 3.0L 'C4' Black/Green Connector**

| PCM Pin # | Wire Color | Circuit Description (38-Pin) | Value at Hot Idle |
|---|---|---|---|
| 1 | BR | OD Solenoid Control | |
| 2 | PK | UD Solenoid Control | |
| 3 | --- | Not Used | --- |
| 4 | --- | Not Used | --- |
| 5 | --- | Not Used | --- |
| 6 | WT | 2-4 Solenoid Control | |
| 7 | --- | Not Used | --- |
| 8 | --- | Not Used | --- |
| 9 | --- | Not Used | --- |
| 10 | LB | L/R Solenoid Control | |
| 11 | --- | Not Used | --- |
| 12 | BK/YL | Ground | |
| 13 | BK/RD | Ground | |
| 14 | BK/RD | Ground | |
| 15 | LG/BK | TRS T1 Sense | |
| 16 | VT | TRS T3 Sense | |
| 17 | --- | Not Used | --- |
| 18 | LG | Transmission Control Relay Control | |
| 19 | RD | Transmission Control Relay Output | |
| 20 | --- | Not Used | --- |
| 21 | --- | Not Used | --- |
| 22 | OR/BK | OD Pressure Switch Sense | |
| 23 | --- | Not Used | --- |
| 24 | --- | Not Used | --- |
| 25 | --- | Not Used | --- |
| 26 | --- | Not Used | --- |
| 27 | BK/WT | TRS T41 Sense | |
| 28 | RD | Transmission Control Relay Output | |
| 29 | DG | L/R Pressure Switch Sense | |
| 30 | YL/BK | 2-4 Pressure Switch Sense | |
| 31 | --- | Not Used | --- |
| 32 | LG/WT | Output Speed Sensor Signal | |
| 33 | RD/BK | Input Speed Sensor Signal | |
| 34 | DB/BK | Speed Sensor Ground | |
| 35 | VT/YL | Transmission Temperature Sensor Signal | |
| 36 | --- | Not Used | --- |
| 37 | VT/WT | TRS T42 Sense | |
| 38 | RD | Transmission Control Relay Output | |

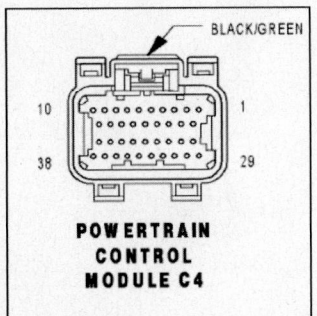

BLACK/GREEN

10

1

38

29

**POWERTRAIN
CONTROL
MODULE C4**

SPIRIT

**1993-95 Spirit 2.5L VIN K 60-Pin Connector**

| PCM Pin # | Wire Color | Circuit Description (60-Pin) | Value at Hot Idle |
|---|---|---|---|
| 1 | DG/RD | MAP Sensor Signal | 1.5-1.7v |
| 2 | TN/BK | ECT Sensor Signal | At 180°F: 2.80v |
| 3 | RD/WT | Battery Power (Fused B+) | 12-14v |
| 4 | BK/LB | Sensor Ground | <0.050v |
| 5 | BK/WT | Sensor Ground | <0.050v |
| 6 | PK/WT | 5-Volt Supply (MAP Sensor) | 4.9-5.1v |
| 7 | OR | 8-Volt Supply | 7.9-8.1v |
| 8 | WT | Ignition Switch Start Signal | 12-14v |
| 9 | DB | Ignition Switch Output | 12-14v |
| 10 | --- | Not Used | --- |
| 11 | BK/TN | Power Ground | <0.1v |
| 12 | BK/TN | Power Ground | <0.1v |
| 13-15 | --- | Not Used | --- |
| 16 | WT/DB | Injector 1 Driver | 1-4 ms |
| 17-18 | --- | Not Used | --- |
| 19 | BK/GY | Coil Driver | 5°, at 55 mph: 8° dwell |
| 20 | DG | Generator Field Driver | Digital Signal: 0-12-0v |
| 21 | --- | Not Used | --- |
| 22 | OR/DB | TP Sensor Signal | 0.6-1.0v |
| 23 | RD/LG | S/C Set Switch Signal | S/C & Set Switch On: 3.8V |
| 24 | GY/BK | Distributor Reference Signal | Digital Signal: 0-5-0v |
| 25 | PK | SCI Transmit | 0v |
| 26 | PK/BR | CCD Bus (+) | Digital Signal: 0-5-0v |
| 27 | BR | A/C Damped Pressure Switch | A/C On: 1v, Off: 12v |
| 28 | --- | Not Used | --- |
| 29 | WT/PK | Brake Switch Signal | Brake Off: 0v, On: 12v |
| 30 | BR/YL | PNP Switch Signal | In P/N: 0v, Others: 5v |
| 31 | DB/PK | Radiator Fan Relay Control | Relay Off: 12v, On: 1v |
| 32 | BK/PK | MIL (lamp) Control | Lamp On: 1v, Off: 12v |
| 33 | TN/RD | S/C Vacuum Solenoid | Vacuum Increasing: 1v |
| 34 | DB/OR | A/C Clutch Relay Control | Relay Off: 12v, On: 1v |
| 35 | GY/YL | EGR Solenoid Control (Calif.) | 12v, 55 mph: 1v |
| 36-38 | --- | Not Used | --- |
| 39 | GY/RD | AIS Motor Control | DC pulse signals: 0.8-11v |
| 40 | BR/WT | AIS Motor Control | DC pulse signals: 0.8-11v |

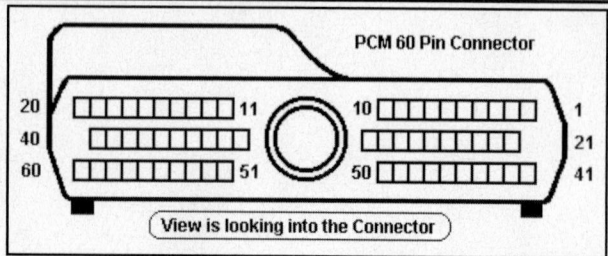

**Pin Connector Graphic**

**1993-95 Spirit 2.5L TBI VIN K 60-Pin Connector - Continued**

| PCM Pin # | Wire Color | Circuit Description (60-Pin) | Value at Hot Idle |
|-----------|-----------|------------------------------|-------------------|
| 41 | BK/DG | HO2S-11 (B1 S1) Signal | 0.1-1.1v |
| 42 | --- | Not Used | --- |
| 43 | GY/LB | Tachometer Signal | Pulse Signals |
| 44 | --- | Not Used | --- |
| 45 | LG | SCI Receive | 0v |
| 46 | WT/BK | CCD Bus (-) | <0.050v |
| 47 | WT/OR | Vehicle Speed Signal | Digital Signal |
| 48-50 | --- | Not Used | --- |
| 51 | DB/YL | ASD Relay Control | Relay Off: 12v, On: 1v |
| 52 | PK/BK | EVAP Purge Solenoid Control | Solenoid Off: 12v, On: 1v |
| 53 | LG/RD | S/C Vent Solenoid | Vacuum Decreasing: 1v |
| 54 | OR/BK | EMCC Solenoid | Solenoid Off: 12v, On: 1v |
| 55 | --- | Not Used | --- |
| 57 | DG/OR | Generator Field Source | 12-14v |
| 58 | --- | Not Used | --- |
| 59 | PK/BK | AIS Motor Control | DC pulse signals: 0.8-11v |
| 60 | YL/BK | AIS Motor Control | DC pulse signals: 0.8-11v |

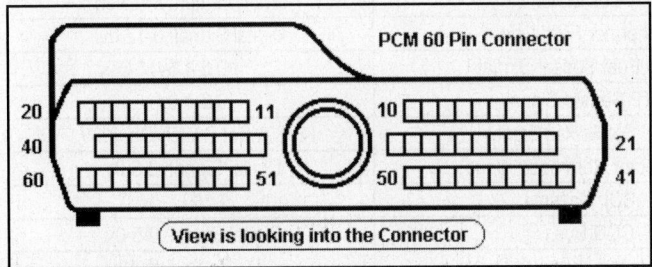

**Pin Connector Graphic**

**Standard Colors and Abbreviations**

| Abbreviation | Color | Abbreviation | Color | Abbreviation | Color |
|--------------|-------|--------------|-------|--------------|-------|
| BK | Black | GY | Gray | RD | Red |
| BL | Blue | GN | Green | TN | Tan |
| BR | Brown | LG | Light Green | VT | Violet |
| DB | Dark Blue | OR | Orange | WT | White |
| DG | Dark Green | PK | Pink | YL | Yellow |

## 1994-95 Spirit 2.5L Flex Fuel VIN V 60-Pin Connector

| PCM Pin # | Wire Color | Circuit Description (60-Pin) | Value at Hot Idle |
|---|---|---|---|
| 1 | DG/RD | MAP Sensor Signal | 1.5-1.7v |
| 2 | TN/BK | ECT Sensor Signal | At 180ºF: 2.80v |
| 3 | RD/WT | Battery Power (Fused B+) | 12-14v |
| 4 | BK/LB | Sensor Ground | <0.050v |
| 5 | BK/LB | Sensor Ground | <0.050v |
| 6 | PK/WT | 5-Volt Supply (MAP Sensor) | 4.9-5.1v |
| 7 | OR | 8-Volt Supply | 7.9-8.1v |
| 8 | WT | Ignition Switch Start Signal | 12-14v |
| 9 | DB | Ignition Switch Output | 12-14v |
| 10 | --- | Not Used | --- |
| 11 | BK/TN | Power Ground | <0.1v |
| 12 | BK/TN | Power Ground | <0.1v |
| 13 | LB/BR | Injector 4 Driver | 1-4 ms |
| 14 | YL/WT | Injector 3 Driver | 1-4 ms |
| 15 | TN | Injector 2 Driver | 1-4 ms |
| 16 | WT/DB | Injector 1 Driver | 1-4 ms |
| 17-18 | --- | Not Used | --- |
| 19 | BK/GY | Coil Driver | 5º, at 55 mph: 8º dwell |
| 20 | DG | Generator Field Driver | Digital Signal: 0-12-0v |
| 21 | BK/RD | Flex Fuel Sensor Signal | 0.5-4.5v |
| 22 | OR/DB | TP Sensor Signal | 0.6-1.0v |
| 23 | RD/LG | S/C Set Switch Signal | S/C & Set Switch On: 3.8V |
| 24 | GY/BK | Distributor Reference Signal | Digital Signal: 0-5-0v |
| 25 | PK | SCI Transmit | 0v |
| 26 | PK/BR | CCD Bus (+) | Digital Signal: 0-5-0v |
| 27 | BR | A/C Damped Pressure Switch | A/C On: 1v, Off: 12v |
| 28 | --- | Not Used | --- |
| 29 | WT/PK | Brake Switch Signal | Brake Off: 0v, On: 12v |
| 30 | BR/YL | PNP Switch Signal | In P/N: 0v, Others: 5v |
| 31 | DB/PK | Radiator Fan Relay Control | Relay Off: 12v, On: 1v |
| 32 | BK/PK | MIL (lamp) Control | Lamp On: 1v, Off: 12v |
| 33 | TN/RD | S/C Vacuum Solenoid | Vacuum Increasing: 1v |
| 34 | DB/OR | A/C Clutch Relay Control | Relay Off: 12v, On: 1v |
| 35 | GY/YL | EGR Solenoid Control (Calif.) | 12v, 55 mph: 1v |
| 36-38 | --- | Not Used | --- |
| 39 | GY/RD | AIS Motor Control | DC pulse signals: 0.8-11v |
| 40 | BR/WT | AIS Motor Control | DC pulse signals: 0.8-11v |

## Standard Colors and Abbreviations

| Abbreviation | Color | Abbreviation | Color | Abbreviation | Color |
|---|---|---|---|---|---|
| BK | Black | GY | Gray | RD | Red |
| BL | Blue | GN | Green | TN | Tan |
| BR | Brown | LG | Light Green | VT | Violet |
| DB | Dark Blue | OR | Orange | WT | White |
| DG | Dark Green | PK | Pink | YL | Yellow |

**1994-95 Spirit 2.5L Flex Fuel VIN V 60-Pin Connector - Continued**

| PCM Pin # | Wire Color | Circuit Description (60-Pin) | Value at Hot Idle |
|-----------|------------|------------------------------|-------------------|
| 41 | BK/DG | HO2S-11 (B1 S1) Signal | 0.1-1.1v |
| 42 | BK/RD | Knock Sensor Signal | 0.080v AC |
| 43 | GY/LB | Tachometer Signal | Pulse Signals |
| 44 | TN/YL | CMP Sensor Signal | Digital Signal: 0-5-0v |
| 45 | LG | SCI Receive | 0v |
| 46 | WT/BK | CCD Bus (-) | <0.050v |
| 47 | WT/OR | Vehicle Speed Signal | Digital Signal |
| 48-50 | --- | Not Used | --- |
| 51 | DB/YL | ASD Relay Control | Relay Off: 12v, On: 1v |
| 52 | PK/BK | EVAP Purge Solenoid Control | Solenoid Off: 12v, On: 1v |
| 53 | LG/RD | S/C Vent Solenoid | Vacuum Decreasing: 1v |
| 54 | OR/BK | EMCC Solenoid | Solenoid Off: 12v, On: 1v |
| 56 | --- | Not Used | --- |
| 57 | DG/OR | Generator Field Source | 12-14v |
| 58 | --- | Not Used | --- |
| 59 | PK/BK | AIS Motor Control | DC pulse signals: 0.8-11v |
| 60 | YL/BK | AIS Motor Control | DC pulse signals: 0.8-11v |

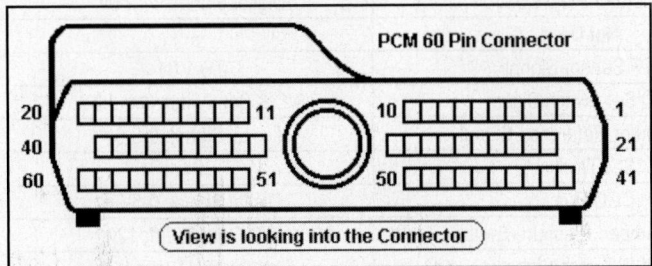

**Pin Connector Graphic**

**Standard Colors and Abbreviations**

| Abbreviation | Color | Abbreviation | Color | Abbreviation | Color |
|--------------|-------|--------------|-------|--------------|-------|
| BK | Black | GY | Gray | RD | Red |
| BL | Blue | GN | Green | TN | Tan |
| BR | Brown | LG | Light Green | VT | Violet |
| DB | Dark Blue | OR | Orange | WT | White |
| DG | Dark Green | PK | Pink | YL | Yellow |

## 1992-95 Spirit 3.0L VIN 3 (A/T) 60-Pin Connector

| PCM Pin # | Wire Color | Circuit Description (60-Pin) | Value at Hot Idle |
|---|---|---|---|
| 1 | DG/RD | MAP Sensor Signal | 1.5-1.7v |
| 2 | TN/BK | ECT Sensor Signal | At 180°F: 2.80v |
| 3 | RD/WT | Battery Power (Fused B+) | 12-14v |
| 4 | BK/LB | Sensor Ground | <0.050v |
| 5 | BK/LB | Sensor Ground | <0.050v |
| 6 | PK/WT | 5-Volt Supply | 4.9-5.1v |
| 7 | OR | 8-Volt Supply | 7.9-8.1v |
| 8 | --- | Not Used | --- |
| 9 | DB | ASD Relay Output | 12-14v |
| 10 | --- | Not Used | --- |
| 11-12 | BK/TN | Power Ground | <0.1v |
| 13 | LB/BR | Injector 4 Driver | 1-4 ms |
| 14 | YL/WT | Injector 3 Driver | 1-4 ms |
| 15 | TN | Injector 2 Driver | 1-4 ms |
| 16 | WT/DB | Injector 1 Driver | 1-4 ms |
| 17-18 | --- | Not Used | --- |
| 19 | BK/GY | Coil 1 Driver | 5°, at 55 mph: 8° dwell |
| 20 | DG | Generator Field Control | Digital Signal: 0-12-0v |
| 21 | --- | Not Used | --- |
| 22 | OR/DB | TP Sensor Signal | 0.6-1.0v |
| 23 | RD/LG | S/C Set Switch Signal | S/C & Set Switch On: 3.8V |
| 24 | GY/WT | Distributor Reference Signal | Digital Signal: 0-5-0v |
| 25 | PK | SCI Transmit | 0v |
| 26 | PK/BR | CCD Bus (+) | Digital Signal: 0-5-0v |
| 27 | BR | A/C Damped Pressure Switch | A/C On: 1v, Off: 12v |
| 28 | DB/OR | PSP Switch Signal | Straight: 0v, Turning: 5v |
| 29 | WT/PK | Brake Switch Signal | Brake Off: 0v, On: 12v |
| 30 | BR/YL | PNP Switch Signal | In P/N: 0v, Others: 5v |
| 31 | DB/PK | Radiator Fan Relay Control | Relay Off: 12v, On: 1v |
| 32 | BK/PK | MIL (lamp) Control | Lamp On: 1v, Off: 12v |
| 33 | TN/RD | S/C Vacuum Solenoid | Vacuum Increasing: 1v |
| 34 | DB/OR | A/C Clutch Relay Control | Relay Off: 12v, On: 1v |
| 35 | GY/YL | EGR Solenoid Control (Calif.) | 12v, 55 mph: 1v |
| 36-37 | --- | Not Used | --- |
| 38 | GY | Injector 5 Driver | 1-4 ms |
| 39 | GY/RD | AIS 3 Motor Control | DC pulse signals: 0.8-11v |
| 40 | BR/WT | AIS 1 Motor Control | DC pulse signals: 0.8-11v |

## Standard Colors and Abbreviations

| Abbreviation | Color | Abbreviation | Color | Abbreviation | Color |
|---|---|---|---|---|---|
| BK | Black | GY | Gray | RD | Red |
| BL | Blue | GN | Green | TN | Tan |
| BR | Brown | LG | Light Green | VT | Violet |
| DB | Dark Blue | OR | Orange | WT | White |
| DG | Dark Green | PK | Pink | YL | Yellow |

**1992-95 Spirit 3.0L VIN 3 (A/T) 60-Pin Connector - Continued**

| PCM Pin # | Wire Color | Circuit Description (60-Pin) | Value at Hot Idle |
|---|---|---|---|
| 41 | BK/DG | HO2S-11 (B1 S1) Signal | 0.1-1.1v |
| 42 | --- | Not Used | --- |
| 43 | GY/LB | Tachometer Signal | Pulse Signals |
| 44 | TN/YL | Distributor Sync Signal | Digital Signal: 0-5-0v |
| 45 | LG | SCI Receive | 0v |
| 46 | WT/BK | CCD Bus (-) | <0.050v |
| 47 | WT/OR | Vehicle Speed Signal | Digital Signal |
| 51 | DB/YL | ASD Relay Control | Relay Off: 12v, On: 1v |
| 52 | PK/BK | EVAP Purge Solenoid Control | Solenoid Off: 12v, On: 1v |
| 53 | LG/RD | S/C Vent Solenoid | Vacuum Decreasing: 1v |
| 54 | OR/BK | EMCC Solenoid | Solenoid Off: 12v, On: 1v |
| 55-56 | --- | Not Used | --- |
| 57 | DG/OR | Generator Field Source | 12-14v |
| 58 | BR/DB | Injector 5 Driver | 1-4 ms |
| 59 | PK/BK | AIS 4 Motor Control | DC pulse signals: 0.8-11v |
| 60 | YL/BK | AIS 2 Motor Control | DC pulse signals: 0.8-11v |

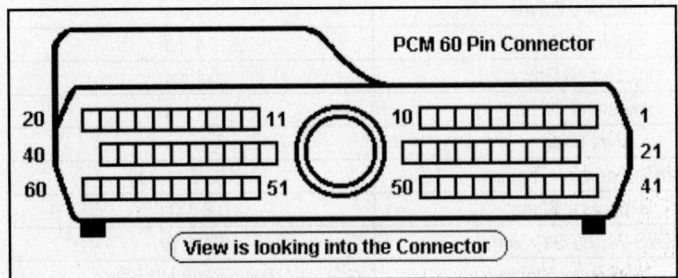

**Pin Connector Graphic**

**Standard Colors and Abbreviations**

| Abbreviation | Color | Abbreviation | Color | Abbreviation | Color |
|---|---|---|---|---|---|
| BK | Black | GY | Gray | RD | Red |
| BL | Blue | GN | Green | TN | Tan |
| BR | Brown | LG | Light Green | VT | Violet |
| DB | Dark Blue | OR | Orange | WT | White |
| DG | Dark Green | PK | Pink | YL | Yellow |

VIPER

## 1992-95 Viper 8.0L VIN E (M/T) 60-Pin Connector

| PCM Pin # | Wire Color | Circuit Description (60-Pin) | Value at Hot Idle |
|-----------|-----------|------------------------------|-------------------|
| 1 | DG/RD | MAP Sensor Signal | 1.5-1.7v |
| 2 | TN/BK | ECT Sensor Signal | At 180ºF: 2.80v |
| 3 | RD/WT | ASD Relay Output (B+) | 12-14v |
| 4 | BK/LB | Sensor Ground | <0.050v |
| 5 | BK/WT | Sensor Ground | <0.050v |
| 6 | VT/WT | 5-Volt Supply | 4.9-5.1v |
| 7 | OR | 8-Volt Supply | 7.9-8.1v |
| 8 | --- | Not Used | --- |
| 9 | DB | Ignition Switch Output | 12-14v |
| 10 | --- | Not Used | --- |
| 11 | BK/TN | Power Ground | <0.1v |
| 12 | BK/TN | Power Ground | <0.1v |
| 13 | BR/BK | Injector 6 Driver | 1-4 ms |
| 14 | YL/WT | Injector 3 Driver | 1-4 ms |
| 15 | WT/LG | Injector 9 Driver | 1-4 ms |
| 16 | TN | Injector 2 Driver | 1-4 ms |
| 17 | BK/LG | Injector 10 Driver | 1-4 ms |
| 18 | LB/BR | Injector 4 Driver | 1-4 ms |
| 19-20 | --- | Not Used | --- |
| 21 | BK/RD | Air Temperature Sensor | At 100ºF: 1.83v |
| 22 | OR/DB | TP Sensor Signal | 0.6-1.0v |
| 23 | LG/RD | HO2S-21 (B2 S1) Signal | 0.1-1.1v |
| 24 | GY/BK | CKP Sensor Signal | Digital Signal: 0-5-0v |
| 25 | PK | SCI Transmit | 0v |
| 26 | VT/BR | CCD Bus (+) | Digital Signal: 0-5-0v |
| 27 | BR | A/C Damped Pressure Switch | A/C On: 1v, Off: 12v |
| 28 | --- | Not Used | --- |
| 29 | WT/PK | Brake Switch Signal | Brake Off: 0v, On: 12v |
| 30 | --- | Not Used | --- |
| 31 | DB/PK | Radiator Fan Relay Control | Relay Off: 12v, On: 1v |
| 32 | BK/PK | MIL (lamp) Control | Lamp On: 1v, Off: 12v |
| 33 | --- | Not Used | --- |
| 34 | DB/OR | A/C Clutch Relay Control | Relay Off: 12v, On: 1v |
| 35 ('94-'95) | YL | Low / High Speed Fan Relay | Relay Off: 12v, On: 1v |
| 36 ('94-'95) | LB/BK | Reverse Lockout Solenoid | Solenoid Off: 12v, On: 1v |
| 37 | PK/VT | Serial COMM (multiplex) | Digital Signals |
| 38 | GY/LG | Injector 8 Driver | 1-4 ms |
| 39 | GY/RD | IAC 1 Driver (open) | DC pulse signals: 0.8-11v |
| 40 | BR/WT | IAC 3 Driver (open) | DC pulse signals: 0.8-11v |

**1992-95 Viper 8.0L VIN E (M/T) 60-Pin Connector - Continued**

| PCM Pin # | Wire Color | Circuit Description (60-Pin) | Value at Hot Idle |
|-----------|------------|------------------------------|-------------------|
| 41 | BK/DG | HO2S-11 (B1 S1) Signal | 0.1-1.1v |
| 42 | --- | Not Used | --- |
| 43 | GY/LB | Tachometer Signal | Pulse Signals |
| 44 | TN/YL | CMP Sensor Signal | Digital Signal: 0-5-0v |
| 45 | LG | SCI Receive | 0v |
| 46 | WT/BK | CCD Bus (-) | <0.050v |
| 47 | WT/TN | Vehicle Speed Signal | Digital Signal |
| 48-50 | --- | Not Used | --- |
| 51 | DB/YL | ASD Relay Control | Relay Off: 12v, On: 1v |
| 52 | PK/BK | EVAP Purge Solenoid Control | Solenoid Off: 12v, On: 1v |
| 53 | --- | Not Used | --- |
| 54 | OR/BK | Shift Indicator Lamp Control | Lamp On: 1v, Off: 12v |
| 55 | GY/OR | 2-3 Skip Shift Solenoid | Solenoid Off: 12v, On: 1v |
| 56 | --- | Not Used | --- |
| 57 | DG/OR | Voltage Sensor to PCM | 12-14v |
| 58 | VT | Injector 7 Driver | 1-4 ms |
| 59 | VT/BK | IAC 4 Driver (Close) | DC pulse signals: 0.8-11v |
| 60 | YL/BK | IAC 2 Driver (Close) | DC pulse signals: 0.8-11v |

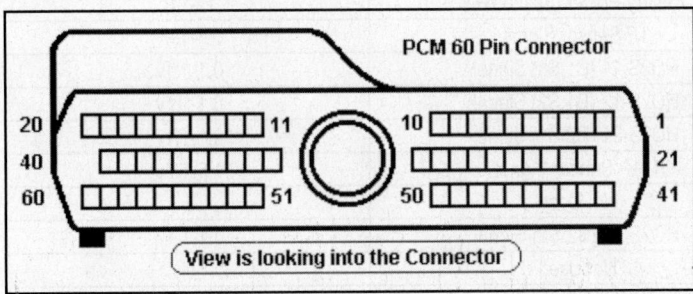

**Pin Connector Graphic**

**Standard Colors and Abbreviations**

| Abbreviation | Color | Abbreviation | Color | Abbreviation | Color |
|--------------|-------|--------------|-------|--------------|-------|
| BK | Black | GY | Gray | RD | Red |
| BL | Blue | GN | Green | TN | Tan |
| BR | Brown | LG | Light Green | VT | Violet |
| DB | Dark Blue | OR | Orange | WT | White |
| DG | Dark Green | PK | Pink | YL | Yellow |

**1996-97 Viper 8.0L VIN E (M/T) Black 'A' Connector**

| PCM Pin # | Wire Color | Circuit Description (32-Pin) | Value at Hot Idle |
|---|---|---|---|
| 1 | YL/GY | Coil 4 Driver | 5°, at 55 mph: 8° dwell |
| 2 | DB | Ignition Switch Output | 12-14v |
| 3 | RD/YL | Coil 3 Driver | 5°, at 55 mph: 8° dwell |
| 4 | BK/LB | Sensor Ground | <0.050v |
| 5 | DG/GY | Coil 5 Driver | 5°, at 55 mph: 8° dwell |
| 6 | --- | Not Used | --- |
| 7 | BK/GY | Coil 1 Driver | 5°, at 55 mph: 8° dwell |
| 8 | GY/BK | CKP Sensor Signal | Digital Signal: 0-5-0v |
| 9 | DB/WT | Coil 2 Driver | 5°, at 55 mph: 8° dwell |
| 10 | YL/BK | IAC 2 Driver | DC pulse signals: 0.8-11v |
| 11 | BR/WT | IAC 3 Driver | DC pulse signals: 0.8-11v |
| 12-14 | --- | Not Used | --- |
| 15 | BK/RD | IAT Sensor Signal | At 100°F: 1.83v |
| 16 | TN/BK | ECT Sensor Signal | At 180°F: 2.80v |
| 17 | OR | 5-Volt Supply | 4.9-5.1v |
| 18 | TN/YL | CMP Sensor Signal | Digital Signal: 0-5-0v |
| 19 | GY/RD | IAC 1 Driver | DC pulse signals: 0.8-11v |
| 20 | PK/BK | IAC 4 Driver | DC pulse signals: 0.8-11v |
| 21 | --- | Not Used | --- |
| 22 | RD/WT | Battery Power (Fused B+) | 12-14v |
| 23 | OR/DB | TP Sensor Signal | 0.6-1.0v |
| 24 | LG/RD | HO2S-11 (B1 S1) Signal | 0.1-1.1v |
| 25 | PK/WT | HO2S-12 (B1 S2) Signal | 0.1-1.1v |
| 26 | BK/DG | HO2S-21 (B2 S1) Signal | 0.1-1.1v |
| 27 | DG/RD | MAP Sensor Signal | 1.5-1.7v |
| 28 | --- | Not Used | --- |
| 29 | TN/WT | HO2S-22 (B2 S2) Signal | 0.1-1.1v |
| 30 | --- | Not Used | --- |
| 31 | DB | Power Ground | <0.1v |
| 32 | DB | Power Ground | <0.1v |

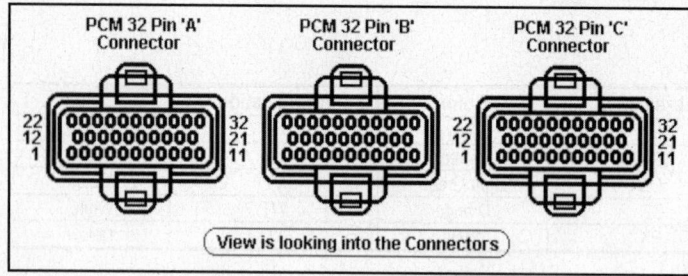

**Pin Connector Graphic**

## 1996-97 Viper 8.0L VIN E (M/T) White 'B' Connector

| PCM Pin # | Wire Color | Circuit Description (32-Pin) | Value at Hot Idle |
|---|---|---|---|
| 1 | --- | Not Used | --- |
| 2 | PK | Injector 7 Driver | 1-4 ms |
| 3 | WT/LG | Injector 9 Driver | 1-4 ms |
| 4 | WT/DB | Injector 1 Driver | 1-4 ms |
| 5 | YL/WT | Injector 3 Driver | 1-4 ms |
| 6 | GY/OR | Injector 5 Driver | 1-4 ms |
| 7-10 | --- | Not Used | --- |
| 11 | OR/BK | Shift Indicator Lamp Control | Lamp On: 1v, Off: 12v |
| 12 | BR/DB | Injector 6 Driver | 1-4 ms |
| 13 | GY/LB | Injector 8 Driver | 1-4 ms |
| 14 | BK/LG | Injector 10 Driver | 1-4 ms |
| 15 | TN | Injector 2 Driver | 1-4 ms |
| 16 | LB/BR | Injector 4 Driver | 1-4 ms |
| 17-20 | --- | Not Used | --- |
| 21 | LB/BK | Reverse Lockout Solenoid | Solenoid Off: 12v, On: 1v |
| 22-26 | --- | Not Used | --- |
| 27 | WT/TN | OSS Sensor Signal (+) | Moving: AC Pulse Signals |
| 28 | WT/OR | Vehicle Speed Signal | Digital Signal |
| 29-30 | --- | Not Used | --- |
| 31 | PK/WT | 5-Volt Supply | 4.9-5.1v |
| 32 | --- | Not Used | --- |

## Standard Colors and Abbreviations

| Abbreviation | Color | Abbreviation | Color | Abbreviation | Color |
|---|---|---|---|---|---|
| BK | Black | GY | Gray | RD | Red |
| BL | Blue | GN | Green | TN | Tan |
| BR | Brown | LG | Light Green | VT | Violet |
| DB | Dark Blue | OR | Orange | WT | White |
| DG | Dark Green | PK | Pink | YL | Yellow |

## 1996-97 Viper 8.0L VIN E (M/T) Grey 'C' Connector

| PCM Pin # | Wire Color | Circuit Description (32-Pin) | Value at Hot Idle |
|---|---|---|---|
| 1 | DB/OR | A/C Clutch Relay Control | Relay Off: 12v, On: 1v |
| 2 | DB/PK | Low Speed Fan Relay | Relay Off: 12v, On: 1v |
| 3 | DB/YL | ASD Relay Control | Relay Off: 12v, On: 1v |
| 4-5 | --- | Not Used | --- |
| 6 | GY/OR | 2-3 Skip Shift Solenoid | Solenoid Off: 12v, On: 1v |
| 7-11 | --- | Not Used | --- |
| 12 | DG/OR | ASD Relay Output | 12-14v |
| 13-14 | --- | Not Used | --- |
| 15 | PK/LG | Battery Temperature Sensor | At 86°F: 1.96v |
| 16 | --- | Not Used | --- |
| 17 | BK/PK | MIL (lamp) Control | Lamp On: 1v, Off: 12v |
| 18 | --- | Not Used | --- |
| 19 | BR/PK | Fuel Pump Relay Control | Relay Off: 12v, On: 1v |
| 20 | PK/BK | EVAP Purge Solenoid Control | PWM signal: 0-12-0v |
| 21 | YL | High Speed Fan Relay | Relay Off: 12v, On: 1v |
| 22 | BR | A/C Request Signal | A/C On: 1v, Off: 12v |
| 23 | LG | A/C Select Signal | A/C On: 1v, Off: 12v |
| 24 | WT/PK | Brake Switch Signal | Brake Off: 0v, On: 12v |
| 25 | --- | Not Used | --- |
| 26 | DB/WT | Fuel Level Sensor Signal | 70 ohms (±20) with full tank |
| 27 | PK | SCI Transmit | 0v |
| 28 | WT/BK | CCD Bus (-) | <0.050v |
| 29 | LG | SCI Receive | 0v |
| 30 | PK/BR | CCD Bus (+) | Digital Signal: 0-5-0v |
| 31 | GY/LB | Tachometer Signal | Pulse Signals |
| 32 | --- | Not Used | --- |

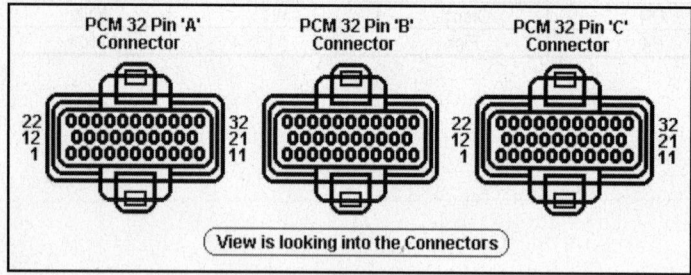

**Pin Connector Graphic**

**1998-99 Viper 8.0L VIN E (M/T) Black 'A' Connector**

| PCM Pin # | Wire Color | Circuit Description (32-Pin) | Value at Hot Idle |
|---|---|---|---|
| 1 | YL/GY | Coil 4 Driver | 6°, 55 mph: 9° dwell |
| 2 | DB/BK | Ignition Switch Output | 12-14v |
| 3 | RD/YL | Coil 3 Driver | 6°, 55 mph: 9° dwell |
| 4 | BK/LB | Sensor Ground | <0.050v |
| 5 | DG/GY | Coil 5 Driver | 6°, 55 mph: 9° dwell |
| 6 | --- | Not Used | --- |
| 7 | BK/GY | Coil 1 Driver | 6°, 55 mph: 9° dwell |
| 8 | GY/BK | CKP Sensor Signal | Digital Signal: 0-5-0v |
| 9 | DB/TN | Coil 2 Driver | 6°, 55 mph: 9° dwell |
| 10 | YL/BK | IAC 2 Driver | DC pulse signals: 0.8-11v |
| 11 | BR/WT | IAC 3 Driver | DC pulse signals: 0.8-11v |
| 12-14 | --- | Not Used | --- |
| 15 | BK/RD | IAT Sensor Signal | At 100°F: 1.83v |
| 16 | TN/BK | ECT Sensor Signal | At 180°F: 2.80v |
| 17 | OR | 5-Volt Supply | 4.9-5.1v |
| 18 | TN/YL | CMP Sensor Signal | Digital Signal: 0-5-0v |
| 19 | GY/RD | IAC 1 Driver | DC pulse signals: 0.8-11v |
| 20 | PK/BK | IAC 4 Driver | DC pulse signals: 0.8-11v |
| 21 | --- | Not Used | --- |
| 22 | RD/WT | Battery Power (Fused B+) | 12-14v |
| 23 | OR/DB | TP Sensor Signal | 0.6-1.0v |
| 24 | LG/RD | HO2S-11 (B1 S1) Signal | 0.1-1.1v |
| 25 | PK/WT | HO2S-12 (B1 S2) Signal | 0.1-1.1v |
| 26 | BK/DG | HO2S-21 (B2 S1) Signal | 0.1-1.1v |
| 27 | DG/RD | MAP Sensor Signal | 1.5-1.7v |
| 28 | --- | Not Used | --- |
| 29 | TN/WT | HO2S-22 (B2 S2) Signal | 0.1-1.1v |
| 30 | --- | Not Used | --- |
| 31 | BK/TN | Power Ground | <0.1v |
| 32 | BK/TN | Power Ground | <0.1v |

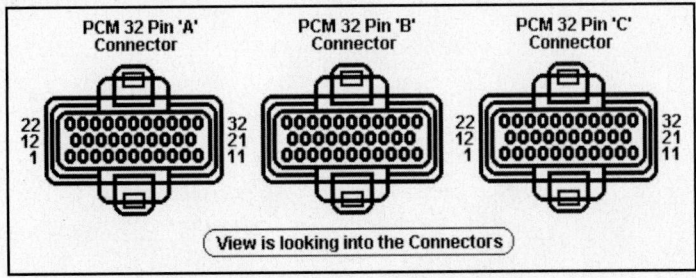

**Pin Connector Graphic**

## 1998-99 Viper 8.0L VIN E (M/T) White 'B' Connector

| PCM Pin # | Wire Color | Circuit Description (32-Pin) | Value at Hot Idle |
|---|---|---|---|
| 1 | --- | Not Used | --- |
| 2 | PK | Injector 7 Driver | 1-4 ms |
| 3 | WT/LG | Injector 9 Driver | 1-4 ms |
| 4 | WT/DB | Injector 1 Driver | 1-4 ms |
| 5 | YL/WT | Injector 3 Driver | 1-4 ms |
| 6 | GY | Injector 5 Driver | 1-4 ms |
| 7-9 | --- | Not Used | --- |
| 10 | DG | Generator Field Control | Digital Signal: 0-12-0v |
| 11 | OR/BK | Shift Indicator Lamp Control | Lamp On: 1v, Off: 12v |
| 12 | BR/BK | Injector 6 Driver | 1-4 ms |
| 13 | GY/LG | Injector 8 Driver | 1-4 ms |
| 14 | BK/LG | Injector 10 Driver | 1-4 ms |
| 15 | TN | Injector 2 Driver | 1-4 ms |
| 16 | LB/BR | Injector 4 Driver | 1-4 ms |
| 17-20 | --- | Not Used | --- |
| 21 | LB/BK | Reverse Lockout Solenoid | Solenoid Off: 12v, On: 1v |
| 22-26 | --- | Not Used | --- |
| 27 | WT/TN | OSS Sensor Signal (+) | Moving: AC Pulse Signals |
| 28 | WT/OR | Vehicle Speed Signal | Digital Signal |
| 29-30 | --- | Not Used | --- |
| 31 | PK/WT | 5-Volt Supply | 4.9-5.1v |
| 32 | --- | Not Used | --- |

## Standard Colors and Abbreviations

| Abbreviation | Color | Abbreviation | Color | Abbreviation | Color |
|---|---|---|---|---|---|
| BK | Black | GY | Gray | RD | Red |
| BL | Blue | GN | Green | TN | Tan |
| BR | Brown | LG | Light Green | VT | Violet |
| DB | Dark Blue | OR | Orange | WT | White |
| DG | Dark Green | PK | Pink | YL | Yellow |

**1998-99 Viper 8.0L VIN E (M/T) Grey 'C' Connector**

| PCM Pin # | Wire Color | Circuit Description (32-Pin) | Value at Hot Idle |
|:---:|:---:|:---:|:---:|
| 1 | DB/OR | A/C Clutch Relay Control | Relay Off: 12v, On: 1v |
| 2 | DB/PK | Low Speed Fan Relay | Relay Off: 12v, On: 1v |
| 3 | DB/YL | ASD Relay Control | Relay Off: 12v, On: 1v |
| 4-5 | --- | Not Used | --- |
| 6 | GY/OR | 2-3 Skip Shift Solenoid | Solenoid Off: 12v, On: 1v |
| 7-9 | --- | Not Used | --- |
| 10 | WT/DG | LDP Solenoid Control | PWM signal: 0-12-0v |
| 11 | --- | Not Used | --- |
| 12 | DG/OR | ASD Relay Output | 12-14v |
| 13 | --- | Not Used | --- |
| 14 | OR | LDP Switch Signal | Switch Closed: 0v, open: 12v |
| 15 | PK/LG | Battery Temperature Sensor | At 86ºF: 1.96v |
| 16 | DG/YL | Generator Lamp Control | Lamp On: 1v, Off: 12v |
| 17 | BK/PK | MIL (lamp) Control | Lamp On: 1v, Off: 12v |
| 18 | --- | Not Used | --- |
| 19 | BR/PK | Fuel Pump Relay Control | Relay Off: 12v, On: 1v |
| 20 | PK/BK | EVAP Purge Solenoid Control | Digital Signal: 0-12-0v |
| 21 | YL | High Speed Fan Relay | Relay Off: 12v, On: 1v |
| 22 | BR | A/C Request Signal | A/C On: 1v, Off: 12v |
| 23 | BR | A/C Select Signal | A/C On: 1v, Off: 12v |
| 24 | WT/PK | Brake Switch Signal | Brake Off: 0v, On: 12v |
| 25 | --- | Not Used | --- |
| 26 | DB | Fuel Level Sensor Signal | 70 ohms (±20) with full tank |
| 27 | PK | SCI Transmit | 0v |
| 28 | WT/BK | CCD Bus (-) | <0.050v |
| 29 | LG | SCI Receive | 0v |
| 30 | PK/BR | CCD Bus (+) | Digital Signal: 0-5-0v |
| 31 | GY/LB | Tachometer Signal | Pulse Signals |
| 32 | --- | Not Used | --- |

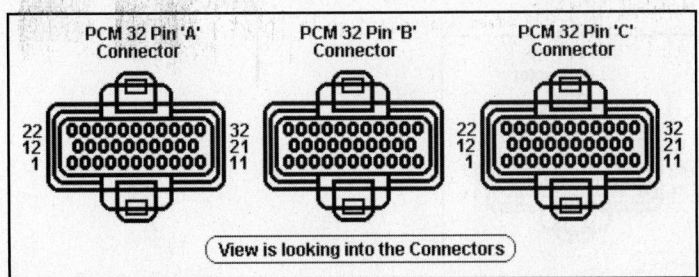

**Pin Connector Graphic**

**2000-02 Viper 8.0L VIN E (M/T) 32-Pin C1 Black Connector**

| PCM Pin # | Wire Color | Circuit Description (32-Pin) | Value at Hot Idle |
|-----------|-----------|------------------------------|-------------------|
| 1 | YL/GY | Coil 4 Driver | 6°, 55 mph: 9° dwell |
| 2 | DB/GY | Ignition Switch Output | 12-14v |
| 3 | RD/YL | Coil 3 Driver | 6°, 55 mph: 9° dwell |
| 4 | BK/LB | Sensor Ground | <0.050v |
| 5 | DG/GY | Coil 5 Driver | 6°, 55 mph: 9° dwell |
| 6 | --- | Not Used | --- |
| 7 | BK/GY | Coil 1 Driver | 6°, 55 mph: 9° dwell |
| 8 | GY/BK | CKP Sensor Signal | Digital Signal: 0-5-0v |
| 9 | DB/TN | Coil 2 Driver | 6°, 55 mph: 9° dwell |
| 10 | YL/BK | IAC 2 Driver | DC pulse signals: 0.8-11v |
| 11 | BR/WT | IAC 3 Driver | DC pulse signals: 0.8-11v |
| 12-14 | --- | Not Used | --- |
| 15 | BK/RD | IAT Sensor Signal | At 100°F: 1.83v |
| 16 | TN/BK | ECT Sensor Signal | At 180°F: 2.80v |
| 17 | OR | 5-Volt Supply | 4.9-5.1v |
| 18 | TN/YL | CMP Sensor Signal | Digital Signal: 0-5-0v |
| 19 | GY/RD | IAC 1 Driver | DC pulse signals: 0.8-11v |
| 20 | VT/BK | IAC 4 Driver | DC pulse signals: 0.8-11v |
| 21 | --- | Not Used | --- |
| 22 | PK/WT | Battery Power (Fused B+) | 12-14v |
| 23 | OR/DB | TP Sensor Signal | 0.6-1.0v |
| 24 | LG/RD | HO2S-11 (B1 S1) Signal | 0.1-1.1v |
| 25 | PK/WT | HO2S-12 (B1 S2) Signal | 0.1-1.1v |
| 26 | BK/DG | HO2S-21 (B2 S1) Signal | 0.1-1.1v |
| 27 | DG/RD | MAP Sensor Signal | 1.5-1.7v |
| 28 | --- | Not Used | --- |
| 29 | TN/WT | HO2S-22 (B2 S2) Signal | 0.1-1.1v |
| 30 | --- | Not Used | --- |
| 31 | BK/TN | Power Ground | <0.1v |
| 32 | BK/TN | Power Ground | <0.1v |

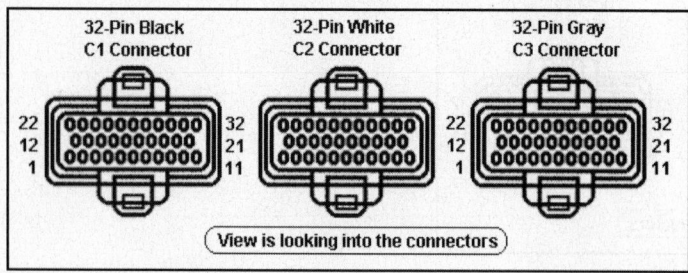

**Pin Connector Graphic**

**2000-02 Viper 8.0L VIN E (M/T) 32-Pin C2 White Connector**

| PCM Pin # | Wire Color | Circuit Description (32-Pin) | Value at Hot Idle |
|---|---|---|---|
| 1 | --- | Not Used | --- |
| 2 | VT | Injector 7 Driver | 1-4 ms |
| 3 | WT/LG | Injector 9 Driver | 1-4 ms |
| 4 | WT/DB | Injector 1 Driver | 1-4 ms |
| 5 | YL/WT | Injector 3 Driver | 1-4 ms |
| 6 | GY | Injector 5 Driver | 1-4 ms |
| 7-9 | --- | Not Used | --- |
| 10 | DG | Generator Field Control | Digital Signal: 0-12-0v |
| 11 | OR/BK | Shift Indicator Lamp Control | Lamp On: 1v, Off: 12v |
| 12 | BR/BK | Injector 6 Driver | 1-4 ms |
| 13 | GY/LG | Injector 8 Driver | 1-4 ms |
| 14 | BK/LG | Injector 10 Driver | 1-4 ms |
| 15 | TN | Injector 2 Driver | 1-4 ms |
| 16 | LB/BR | Injector 4 Driver | 1-4 ms |
| 17-20 | --- | Not Used | --- |
| 21 | LB/BK | Reverse Lockout Solenoid | Solenoid Off: 12v, On: 1v |
| 22-26 | --- | Not Used | --- |
| 27 | WT/TN | OSS Sensor Signal (+) | Moving: AC Pulse Signals |
| 28 | WT/OR | Vehicle Speed Signal | Digital Signal |
| 29-30 | --- | Not Used | --- |
| 31 | VT/WT | 5-Volt Supply | 4.9-5.1v |
| 32 | --- | Not Used | --- |

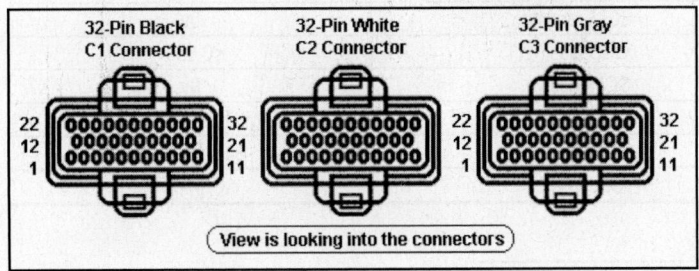

**Pin Connector Graphic**

**Standard Colors and Abbreviations**

| Abbreviation | Color | Abbreviation | Color | Abbreviation | Color |
|---|---|---|---|---|---|
| BK | Black | GY | Gray | RD | Red |
| BL | Blue | GN | Green | TN | Tan |
| BR | Brown | LG | Light Green | VT | Violet |
| DB | Dark Blue | OR | Orange | WT | White |
| DG | Dark Green | PK | Pink | YL | Yellow |

## 2000-02 Viper 8.0L VIN E (M/T) 32-Pin C3 Gray Connector

| PCM Pin # | Wire Color | Circuit Description (32-Pin) | Value at Hot Idle |
|---|---|---|---|
| 1 | DB/OR | A/C Clutch Relay Control | Relay Off: 12v, On: 1v |
| 2 | DB/PK | Low Speed Fan Relay | Relay Off: 12v, On: 1v |
| 3 | DB/YL | ASD Relay Control | Relay Off: 12v, On: 1v |
| 4-5 | --- | Not Used | --- |
| 6 | GY/OR | 2-3 Skip Shift Solenoid | Solenoid Off: 12v, On: 1v |
| 7-9 | --- | Not Used | --- |
| 10 | WT/DG | LDP Solenoid Control | PWM signal: 0-12-0v |
| 11 | --- | Not Used | --- |
| 12 | DG/OR | ASD Relay Output | 12-14v |
| 13 | --- | Not Used | --- |
| 14 | OR | LDP Switch Signal | Switch Closed: 0v, open: 12v |
| 15 | VT/LG | Battery Temperature Sensor | At 86°F: 1.96v |
| 16 | DG/YL | Generator Lamp Control | Lamp On: 1v, Off: 12v |
| 17 | BK/PK | MIL (lamp) Control | Lamp On: 1v, Off: 12v |
| 18 | --- | Not Used | --- |
| 19 | BR/VT | Fuel Pump Relay Control | Relay Off: 12v, On: 1v |
| 20 | PK/BK | EVAP Purge Solenoid Control | PWM signal: 0-12-0v |
| 21 | YL | High Speed Fan Relay | Relay Off: 12v, On: 1v |
| 22 | BR | A/C Switch Sense | A/C On: 1v, Off: 12v |
| 23 | BR | A/C Switch Sense | A/C On: 1v, Off: 12v |
| 24 | WT/PK | Brake Switch Signal | Brake Off: 0v, On: 12v |
| 25 | --- | Not Used | --- |
| 26 | DB | Fuel Level Sensor Signal | 70 ohms (±20) with full tank |
| 27 | PK | SCI Transmit | 0v |
| 28 | WT/BK | CCD Bus (-) | <0.050v |
| 29 | LG | SCI Receive | 0v |
| 30 | VT/BR | CCD Bus (+) | Digital Signal: 0-5-0v |
| 31 | GY/LB | Tachometer Signal | Pulse Signals |
| 32 | --- | Not Used | --- |

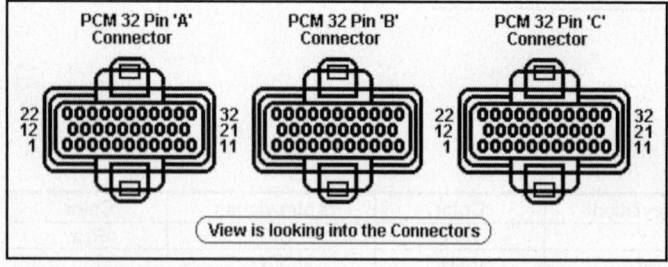

**Pin Connector Graphic**

### 2003 Viper 8.3L VIN Z (M/T) 32-Pin C1 Black Connector

| PCM Pin # | Wire Color | Circuit Description (32-Pin) | Value at Hot Idle |
|---|---|---|---|
| 1 | YL/GY | Coil 4 Driver | 6°, 55 mph: 9° dwell |
| 2 | DB/GY | Ignition Switch Output | 12-14v |
| 3 | RD/YL | Coil 3 Driver | 6°, 55 mph: 9° dwell |
| 4 | BK/LB | Sensor Ground | <0.050v |
| 5 | DG/GY | Coil 5 Driver | 6°, 55 mph: 9° dwell |
| 6 | --- | Not Used | --- |
| 7 | BK/GY | Coil 1 Driver | 6°, 55 mph: 9° dwell |
| 8 | GY/BK | CKP Sensor Signal | Digital Signal: 0-5-0v |
| 9 | DB/TN | Coil 2 Driver | 6°, 55 mph: 9° dwell |
| 10 | YL/BK | IAC 2 Driver | DC pulse signals: 0.8-11v |
| 11 | BR/WT | IAC 3 Driver | DC pulse signals: 0.8-11v |
| 12-14 | --- | Not Used | --- |
| 15 | BK/RD | IAT Sensor Signal | At 100°F: 1.83v |
| 16 | TN/BK | ECT Sensor Signal | At 180°F: 2.80v |
| 17 | OR | 5-Volt Supply | 4.9-5.1v |
| 18 | TN/YL | CMP Sensor Signal | Digital Signal: 0-5-0v |
| 19 | GY/RD | IAC 1 Driver | DC pulse signals: 0.8-11v |
| 20 | VT/BK | IAC 4 Driver | DC pulse signals: 0.8-11v |
| 21 | --- | Not Used | --- |
| 22 | PK/WT | Battery Power (Fused B+) | 12-14v |
| 23 | OR/DB | TP Sensor Signal | 0.6-1.0v |
| 24 | LG/RD | HO2S-11 (B1 S1) Signal | 0.1-1.1v |
| 25 | PK/WT | HO2S-12 (B1 S2) Signal | 0.1-1.1v |
| 26 | BK/DG | HO2S-21 (B2 S1) Signal | 0.1-1.1v |
| 27 | DG/RD | MAP Sensor Signal | 1.5-1.7v |
| 28 | --- | Not Used | --- |
| 29 | TN/WT | HO2S-22 (B2 S2) Signal | 0.1-1.1v |
| 30 | --- | Not Used | --- |
| 31 | BK/TN | Power Ground | <0.1v |
| 32 | BK/TN | Power Ground | <0.1v |

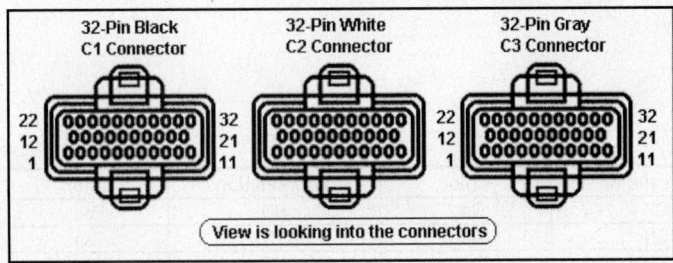

**Pin Connector Graphic**

## 2003 Viper 8.3L VIN Z (M/T) 32-Pin C2 White Connector

| PCM Pin # | Wire Color | Circuit Description (32-Pin) | Value at Hot Idle |
|---|---|---|---|
| 1 | --- | Not Used | --- |
| 2 | VT | Injector 7 Driver | 1-4 ms |
| 3 | WT/LG | Injector 9 Driver | 1-4 ms |
| 4 | WT/DB | Injector 1 Driver | 1-4 ms |
| 5 | YL/WT | Injector 3 Driver | 1-4 ms |
| 6 | GY | Injector 5 Driver | 1-4 ms |
| 7-9 | --- | Not Used | --- |
| 10 | DG | Generator Field Control | Digital Signal: 0-12-0v |
| 11 | OR/BK | Shift Indicator Lamp Control | Lamp On: 1v, Off: 12v |
| 12 | BR/BK | Injector 6 Driver | 1-4 ms |
| 13 | GY/LG | Injector 8 Driver | 1-4 ms |
| 14 | BK/LG | Injector 10 Driver | 1-4 ms |
| 15 | TN | Injector 2 Driver | 1-4 ms |
| 16 | LB/BR | Injector 4 Driver | 1-4 ms |
| 17-20 | --- | Not Used | --- |
| 21 | LB/BK | Reverse Lockout Solenoid | Solenoid Off: 12v, On: 1v |
| 22-26 | --- | Not Used | --- |
| 27 | WT/TN | OSS Sensor Signal (+) | Moving: AC Pulse Signals |
| 28 | WT/OR | Vehicle Speed Signal | Digital Signal |
| 29-30 | --- | Not Used | --- |
| 31 | VT/WT | 5-Volt Supply | 4.9-5.1v |
| 32 | --- | Not Used | --- |

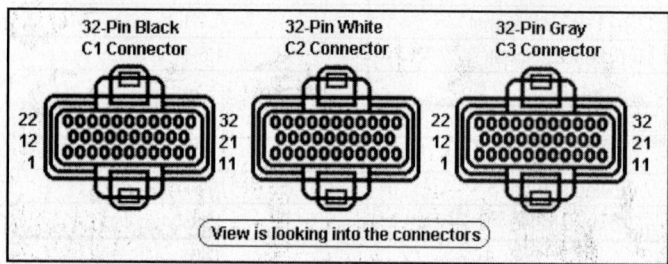

**Pin Connector Graphic**

## Standard Colors and Abbreviations

| Abbreviation | Color | Abbreviation | Color | Abbreviation | Color |
|---|---|---|---|---|---|
| BK | Black | GY | Gray | RD | Red |
| BL | Blue | GN | Green | TN | Tan |
| BR | Brown | LG | Light Green | VT | Violet |
| DB | Dark Blue | OR | Orange | WT | White |
| DG | Dark Green | PK | Pink | YL | Yellow |

## 2003 Viper 8.3L VIN Z (M/T) 32-Pin C3 Gray Connector

| PCM Pin # | Wire Color | Circuit Description (32-Pin) | Value at Hot Idle |
|---|---|---|---|
| 1 | DB/OR | A/C Clutch Relay Control | Relay Off: 12v, On: 1v |
| 2 | DB/PK | Low Speed Fan Relay | Relay Off: 12v, On: 1v |
| 3 | DB/YL | ASD Relay Control | Relay Off: 12v, On: 1v |
| 4-5 | --- | Not Used | --- |
| 6 | GY/OR | 2-3 Skip Shift Solenoid | Solenoid Off: 12v, On: 1v |
| 7-9 | --- | Not Used | --- |
| 10 | WT/DG | LDP Solenoid Control | PWM signal: 0-12-0v |
| 11 | --- | Not Used | --- |
| 12 | DG/OR | ASD Relay Output | 12-14v |
| 13 | --- | Not Used | --- |
| 14 | OR | LDP Switch Signal | Switch Closed: 0v, open: 12v |
| 15 | VT/LG | Battery Temperature Sensor | At 86°F: 1.96v |
| 16 | DG/YL | Generator Lamp Control | Lamp On: 1v, Off: 12v |
| 17 | BK/PK | MIL (lamp) Control | Lamp On: 1v, Off: 12v |
| 18 | --- | Not Used | --- |
| 19 | BR/VT | Fuel Pump Relay Control | Relay Off: 12v, On: 1v |
| 20 | PK/BK | EVAP Purge Solenoid Control | PWM signal: 0-12-0v |
| 21 | YL | High Speed Fan Relay | Relay Off: 12v, On: 1v |
| 22 | BR | A/C Switch Sense | A/C On: 1v, Off: 12v |
| 23 | BR | A/C Switch Sense | A/C On: 1v, Off: 12v |
| 24 | WT/PK | Brake Switch Signal | Brake Off: 0v, On: 12v |
| 25 | --- | Not Used | --- |
| 26 | DB | Fuel Level Sensor Signal | 70 ohms (±20) with full tank |
| 27 | PK | SCI Transmit | 0v |
| 28 | WT/BK | CCD Bus (-) | <0.050v |
| 29 | LG | SCI Receive | 0v |
| 30 | VT/BR | CCD Bus (+) | Digital Signal: 0-5-0v |
| 31 | GY/LB | Tachometer Signal | Pulse Signals |
| 32 | --- | Not Used | --- |

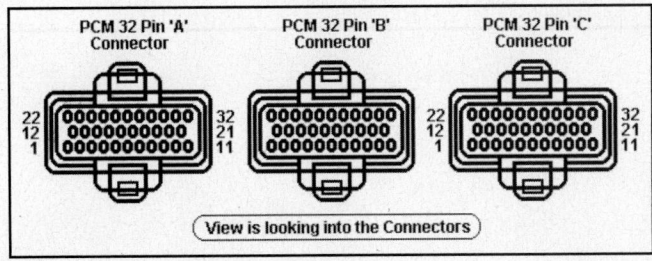

**Pin Connector Graphic**

**2004-05 Viper 8.3L VIN Z C1 Black Connector**

| PCM Pin # | Wire Color | Circuit Description (32-Pin) | Value at Hot Idle |
|---|---|---|---|
| 1 | YL/GY | Ignition Coil No. 4 Driver | |
| 2 | PK/GY | Fused Ignition Switch Output (Run-Start) | |
| 3 | RD/YL | Ignition Coil No. 3 Driver | |
| 4 | DB/DG | Sensor Ground | |
| 5 | DG/GY | Ignition Coil No. 5 Driver | |
| 6 | --- | Not Used | --- |
| 7 | BK/GY | Ignition Coil No. 1 Driver | |
| 8 | GY/BK | CKP Signal | |
| 9 | DB/TN | Ignition Coil No. 2 Driver | |
| 10 | GY/RD | IAC Motor No. 3 Driver | |
| 11 | YL/BK | IAC Motor No. 2 Driver | |
| 12 | --- | Not Used | --- |
| 13 | --- | Not Used | --- |
| 14 | --- | Not Used | --- |
| 15 | BK/RD | IAT Signal | |
| 16 | TN/BK | ECT Signal | |
| 17 | OR | 5v Supply | |
| 18 | TN/YL | CMP Signal | |
| 19 | VT/BK | IAC Motor No. 4 Driver | |
| 20 | BR/WT | IAC Motor No. 1 Driver | |
| 21 | --- | Not Used | --- |
| 22 | RD/WT | Fused B (+) | |
| 23 | OR/DB | TP Signal | |
| 24 | BK/DG | O2 1/1 Signal | |
| 25 | TN/WT | O2 1/2 Signal | |
| 26 | LG/RD | O2 2/1 Signal | |
| 27 | DG/RD | MAP Signal | |
| 28 | --- | Not Used | --- |
| 29 | TN/WT | O2 2/2 Signal | |
| 30 | --- | Not Used | --- |
| 31 | BK/TN | Ground | |
| 32 | BK/TN | Ground | |

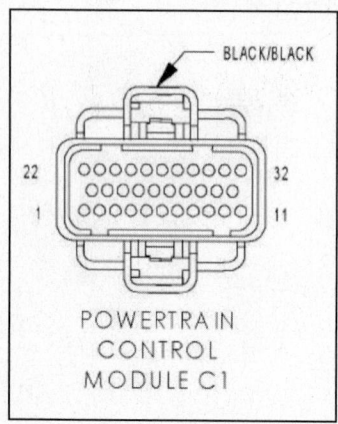

BLACK/BLACK

22    32

1    11

POWERTRAIN
CONTROL
MODULE C1

## 2004-05 Viper 8.3L VIN Z C2 White Connector

| PCM Pin # | Wire Color | Circuit Description (32-Pin) | Value at Hot Idle |
|---|---|---|---|
| 1 | --- | Not Used | --- |
| 2 | VT | Fuel Injector No. 7 Driver | |
| 3 | WT/LG | Fuel Injector No. 9 Driver | |
| 4 | WT/DB | Fuel Injector No. 1 Driver | |
| 5 | YL/WT | Fuel Injector No. 3 Driver | |
| 6 | GY | Fuel Injector No. 5 Driver | |
| 7 | --- | Not Used | --- |
| 8 | --- | Not Used | --- |
| 9 | --- | Not Used | --- |
| 10 | DG | Generator Field Driver | |
| 11 | --- | Not Used | --- |
| 12 | BR/DB | Fuel Injector No. 6 Driver | |
| 13 | GY/LB | Fuel Injector No. 8 Driver | |
| 14 | BK/LG | Fuel Injector No. 10 Driver | |
| 15 | TN | Fuel Injector No. 2 Driver | |
| 16 | LB/BR | Fuel Injector No. 4 Driver | |
| 17 | OR/RD | Hydraulic Cooling Fan Control | |
| 18 | --- | Not Used | --- |
| 19 | DB | A/C Pressure Signal | |
| 20 | --- | Not Used | --- |
| 21 | LB/BK | Reverse Lock-Out Solenoid Control | |
| 22 | --- | Not Used | --- |
| 23 | GY/YL | Oil Pressure Sensor Signal | |
| 24 | --- | Not Used | --- |
| 25 | --- | Not Used | --- |
| 26 | --- | Not Used | --- |
| 27 | WT/OR | Vehicle Speed Sensor Signal | |
| 28 | --- | Not Used | --- |
| 29 | VT/WT | Oil Temperature Sensor Signal | |
| 30 | --- | Not Used | --- |
| 31 | BR/PK | 5v Supply | |
| 32 | --- | Not Used | --- |

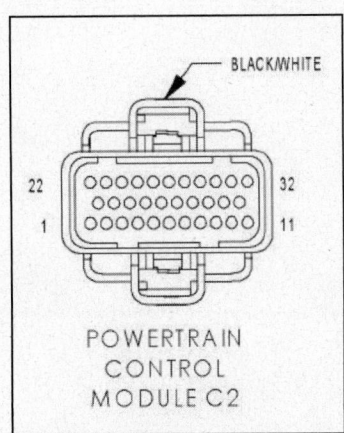

BLACK/WHITE

22  32
1  11

POWERTRAIN
CONTROL
MODULE C2

**2004-05 Viper 8.3L VIN Z C3 Gray Connector**

| PCM Pin # | Wire Color | Circuit Description (32-Pin) | Value at Hot Idle |
|---|---|---|---|
| 1 | DB/OR | A/C Clutch Relay Control | |
| 2 | --- | Not Used | --- |
| 3 | DB/YL | Automatic Shut Down Relay Control | |
| 4 | --- | Not Used | --- |
| 5 | --- | Not Used | --- |
| 6 | GY/OR | Skip Shift Solenoid Control | |
| 7 | --- | Not Used | --- |
| 8 | OR/RD | O2 1/1 Heater Control | |
| 9 | BR/YL | Oxygen Sensor Heater Relay Control | |
| 10 | WT/DG | Leak Detection Pump Solenoid Control | |
| 11 | --- | Not Used | --- |
| 12 | BR/WT | Automatic Shut Down Relay Output | |
| 13 | --- | Not Used | --- |
| 14 | OR | Leak Detection Pump Switch Sense | |
| 15 | VT/LG | Battery Temperature Sensor Signal | |
| 16 | BK/LG | O2 2/1 Heater Control | |
| 17 | TN | Starter Motor Relay Control | |
| 18 | --- | Not Used | --- |
| 19 | BR | Fuel Pump Relay Control | |
| 20 | PK/BK | EVAP/Purge Solenoid Control | |
| 21 | --- | Not Used | --- |
| 22 | BR | A/C Switch Sense | |
| 23 | --- | Not Used | --- |
| 24 | WT/PK | Brake Lamp Switch Sense | |
| 25 | WT/DB | Generator Source | |
| 26 | --- | Not Used | --- |
| 27 | PK | SCI Transmit | |
| 28 | - | Not Used | |
| 29 | LG | SCI Receive | |
| 30 | VT/YL | PCI Bus | |
| 31 | --- | Not Used | --- |
| 32 | --- | Not Used | --- |

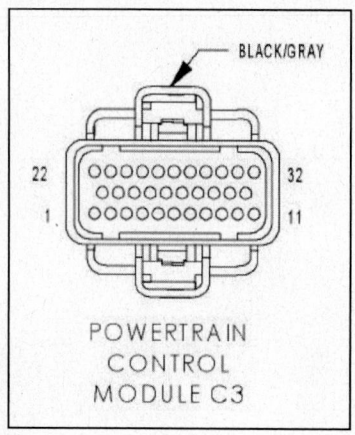

BLACK/GRAY

22        32
1         11

POWERTRAIN
CONTROL
MODULE C3

**DaimlerChrysler Mini-Vans**

CARAVAN, TOWN & COUNTRY, VOYAGER

**1996 Caravan, Grand Caravan, Voyager 2.4L VIN B (A/T) 'C1' Black Connector**

| PCM Pin # | Wire Color | Circuit Description (40-Pin) | Value at Hot Idle |
|---|---|---|---|
| 1-2 | --- | Not Used | --- |
| 3 | DB/TN | Coil 2 Driver Control | 5°, 55 mph: 8° dwell |
| 4 | DG | Generator Field Driver | Digital Signals: 0-12-0v |
| 5 | YL/RD | Speed Control Power Supply | 12-14v |
| 6 | DG/OR | ASD Relay Power (B+) | 12-14v |
| 7 | YL/WT | Injector 3 Driver | 1-4 ms |
| 8 | TN | Starter Relay Control | KOEC: 9-11v |
| 9 | --- | Not Used | --- |
| 10 | BK/TN | Power Ground | <0.1v |
| 11 | GY | Coil 1 Driver Control | 5°, 55 mph: 8° dwell |
| 12 | --- | Not Used | --- |
| 13 | WT/DB | Injector 1 Driver | 1-4 ms |
| 14-15 | --- | Not Used | --- |
| 16 | LB/BR | Injector 4 Driver | 1-4 ms |
| 17 | TN | Injector 2 Driver | 1-4 ms |
| 18-19 | --- | Not Used | --- |
| 20 | WT/BK | Ignition Switch Power (B+) | 12-14v |
| 21-23 | --- | Not Used | --- |
| 24 | DB/LG | Knock Sensor Signal | 0.080v AC |
| 25 | --- | Not Used | --- |
| 26 | TN/BK | ECT Sensor Signal | At 180°F: 2.80v |
| 27-29 | --- | Not Used | --- |
| 30 | BK/DG | HO2S-11 (B1 S1) Signal | 0.1-1.1v |
| 31 | --- | Not Used | --- |
| 32 | GY/BK | CKP Sensor Signal | Digital Signals: 0-5-0v |
| 33 | TN/YL | CMP Sensor Signal | Digital Signals: 0-5-0v |
| 34 | --- | Not Used | --- |
| 35 | OR/DB | TP Sensor Signal | 0.6-1.0v |
| 36 | DG/RD | MAP Sensor Signal | 1.5-1.6v |
| 37 | BK/RD | IAT Sensor Signal | At 100°F: 1.83v |
| 38-40 | --- | Not Used | --- |

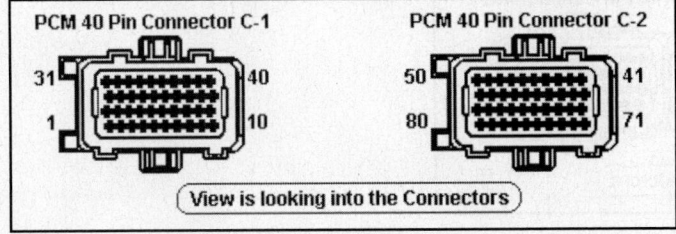

**Pin Connector Graphic**

**1996 Caravan, Grand Caravan, Voyager 2.4L VIN B (A/T) 'C2' White Connector**

| PCM Pin # | Wire Color | Circuit Description (40-Pin) | Value at Hot Idle |
|---|---|---|---|
| 41 | RD/LG | Speed Control Set Switch Signal | S/C & Set Switch On: 3.8v |
| 42 | DB | A/C Pressure Switch Signal | A/C On: 0.451-4.850v |
| 43 | BK/LB | Sensor Ground | <0.050v |
| 44 | OR | 8-Volt Supply | 7.9-8.1v |
| 45 | DB | A/C Switch Sense Signal | A/C Off: 12v, On: 1v |
| 46 | RD/WT | Keep Alive Power (B+) | 12-14v |
| 47 | --- | Not Used | --- |
| 48 | BR/WT | IAC 1 Driver Control | Pulse Signals |
| 49 | YL/BK | IAC 2 Driver Control | Pulse Signals |
| 50 | BK/TN | Power Ground | <0.1v |
| 51 | TN/WT | HO2S-12 (B1 S2) Signal | 0.1-1.1v |
| 52-55 | --- | Not Used | --- |
| 56 | OR/BK | Lockup Torque Converter | At Cruise w/TCC On: 1v |
| 57 | GY/RD | IAC 3 Driver Control | Pulse Signals |
| 58 | PK/BK | IAC 4 Driver Control | Pulse Signals |
| 59 | PK/BR | CCD Bus (+) | Digital Signals: 0-5-0v |
| 60 | WT/BK | CCD Bus (-) | <0.050v |
| 61 | PK/WT | 5-Volt Supply | 4.9-5.1v |
| 62 | WT/PK | Brake Switch Sense Signal | Brake Off: 0v, On: 12v |
| 63 | YL/DG | Torque Management Request | Digital Signals |
| 64 | DB/OR | A/C Clutch Relay Control | A/C Off: 12v, On: 1v |
| 65 | PK | SCI Transmit | 0v |
| 66 | WT/OR | Vehicle Speed Signal | Digital Signal |
| 67 | DB/YL | ASD Relay Control | Relay Off: 12v, On: 1v |
| 68 | PK/BK | EVAP Purge Solenoid Control | PWM Signal: 0-12-0v |
| 69-72 | --- | Not Used | --- |
| 73 | LG/DG | Radiator Fan Control Relay | Relay Off: 12v, On: 1v |
| 74 | BR | Fuel Pump Relay Control | F/P On: 1v, Off: 12v |
| 75 | LG | SCI Receive | 5v |
| 76 | BK/WT | PNP Switch Sense (EATX) | In P/N: 0v, Others: 5v |
| 77 | --- | Not Used | --- |
| 78 | TN/RD | Speed Control Vacuum Solenoid | Vacuum Increasing: 1v |
| 79 | --- | Not Used | --- |
| 80 | LG/RD | Speed Control Vent Solenoid | Vacuum Decreasing: 1v |

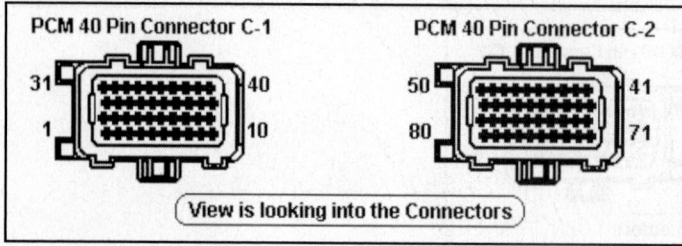

**Pin Connector Graphic**

**1997-98 Caravan, Grand Caravan, Voyager 2.4L VIN B (A/T) 'C1' Black Connector**

| PCM Pin # | Wire Color | Circuit Description (40-Pin) | Value at Hot Idle |
|---|---|---|---|
| 1-2 | --- | Not Used | --- |
| 3 | DB/TN | Coil 2 Driver Control | 5°, 55 mph: 8° dwell |
| 4 | DG | Generator Field Driver | Digital Signals: 0-12-0v |
| 5 | YL/RD | Speed Control Power Supply | 12-14v |
| 6 | DG/OR | ASD Relay Power (B+) | 12-14v |
| 7 | YL/WT | Injector 3 Driver | 1-4 ms |
| 8 | DB/OR | Starter Relay Control | KOEC: 9-11v |
| 9 | --- | Not Used | --- |
| 10 | BK/TN | Power Ground | <0.1v |
| 11 | GY | Coil 1 Driver Control | 5°, 55 mph: 8° dwell |
| 12 | --- | Not Used | --- |
| 13 | WT/DB | Injector 1 Driver | 1-4 ms |
| 14-15 | --- | Not Used | --- |
| 16 | LB/BR | Injector 4 Driver | 1-4 ms |
| 17 | T | Injector 2 Driver | 1-4 ms |
| 18-19 | --- | Not Used | --- |
| 20 | WT/BK | Ignition Switch Power (B+) | 12-14v |
| 21 | --- | Not Used | --- |
| 22 | BK/PK | MIL (lamp) Control | MIL Off: 12v, On: 1v |
| 23 | --- | Not Used | --- |
| 24 | DB/LG | Knock Sensor Signal | 0.080v AC |
| 25 | --- | Not Used | --- |
| 26 | TN/BK | ECT Sensor Signal | At 180°F: 2.80v |
| 27-29 | --- | Not Used | --- |
| 30 | BK/DG | HO2S-11 (B1 S1) Signal | 0.1-1.1v |
| 31 | --- | Not Used | --- |
| 32 | GY/BK | CKP Sensor Signal | Digital Signals: 0-5-0v |
| 33 | TN/YL | CMP Sensor Signal | Digital Signals: 0-5-0v |
| 34 | --- | Not Used | --- |
| 35 | OR/DB | TP Sensor Signal | 0.6-1.0v |
| 36 | DG/RD | MAP Sensor Signal | 1.5-1.6v |
| 37 | BK/RD | IAT Sensor Signal | At 100°F: 1.83v |
| 38-40 | --- | Not Used | --- |

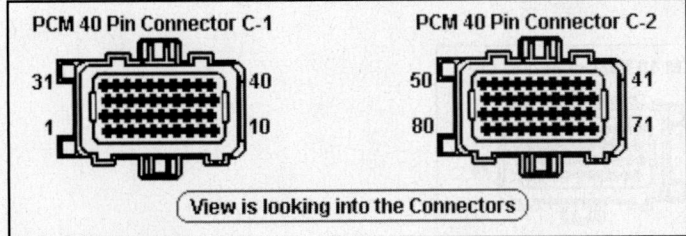

**Pin Connector Graphic**

**1997-98 Caravan, Grand Caravan, Voyager 2.4L VIN B (A/T) 'C2' White Connector**

| PCM Pin # | Wire Color | Circuit Description (40-Pin) | Value at Hot Idle |
|---|---|---|---|
| 41 | RD/LG | Speed Control Set Switch Signal | S/C & Set Switch On: 3.8v |
| 42 | DB | A/C Pressure Switch Signal | A/C On: 0.451-4.850v |
| 43 | BK/LB | Sensor Ground | <0.050v |
| 44 | OR | 8-Volt Supply | 7.9-8.1v |
| 45 | DG | A/C Switch Signal | A/C Off: 12v, On: 1v |
| 46 | RD/WT | Keep Alive Power (B+) | 12-14v |
| 47 | --- | Not Used | --- |
| 48 | BR/WT | IAC 3 Driver Control | Pulse Signals |
| 49 | YL/BK | IAC 2 Driver Control | Pulse Signals |
| 50 | BK/TN | Power Ground | <0.1v |
| 51 | TN/WT | HO2S-12 (B1 S2) Signal | 0.1-1.1v |
| 52-55 | --- | Not Used | --- |
| 56 | OR/BK | Lockup Torque Converter | At Cruise w/TCC On: 1v |
| 57 | GY/RD | IAC 1 Driver Control | Pulse Signals |
| 58 | PK/BK | IAC 4 Driver Control | Pulse Signals |
| 59 | PK/BR | CCD Bus (+) | Digital Signals: 0-5-0v |
| 60 | WT/BK | CCD Bus (-) | <0.050v |
| 61 | PK/WT | 5-Volt Supply | 4.9-5.1v |
| 62 | WT/PK | Brake Switch Sense Signal | Brake Off: 0v, On: 12v |
| 63 | YL/DG | Torque Management Request | Digital Signals |
| 64 | DB/OR | A/C Clutch Relay Control | A/C Off: 12v, On: 1v |
| 65, 75 | PK, LG | SCI Transmit, SCI Receive | 5v |
| 66 | WT/OR | Vehicle Speed Signal | Digital Signal |
| 67 | DB/YL | ASD Relay Control | Relay Off: 12v, On: 1v |
| 68 | PK/BK | EVAP Purge Solenoid Control | PWM Signal: 0-12-0v |
| 69-71 | --- | Not Used | --- |
| 72 | YL/BK | LDP Switch Sense Signal | LDP Switch Closed: 0v |
| 73 | LG/DB | Radiator Fan Control Relay | Relay Off: 12v, On: 1v |
| 74 | BR | Fuel Pump Relay Control | F/P On: 1v, Off: 12v |
| 76 | BK/WT | PNP Switch Sense (EATX) | In P/N: 0v, Others: 5v |
| 77 | WT/DG | LDP Solenoid Control | PWM Signal: 0-12-0v |
| 78 | TN/RD | Speed Control Vacuum Solenoid | Vacuum Increasing: 1v |
| 79 | --- | Not Used | --- |
| 80 | LG/RD | Speed Control Vent Solenoid | Vacuum Decreasing: 1v |

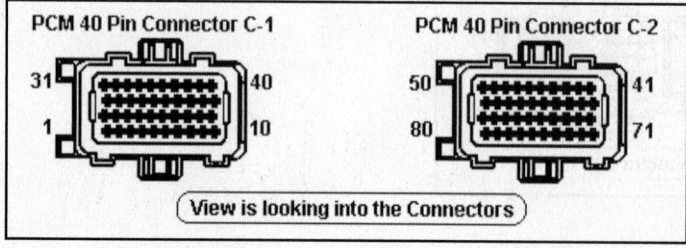

**Pin Connector Graphic**

**1999-2000 Caravan, Grand Caravan, Voyager 2.4L VIN B (A/T) 'C1' Black Connector**

| PCM Pin # | Wire Color | Circuit Description (40-Pin) | Value at Hot Idle |
|---|---|---|---|
| 1-2 | --- | Not Used | --- |
| 3 | DB/TN | Coil 2 Driver Control | 5°, 55 mph: 8° dwell |
| 4 | --- | Not Used | --- |
| 5 | YL/RD | Speed Control Power Supply | 12-14v |
| 6 | DG/OR | ASD Relay Power (B+) | 12-14v |
| 7 | YL/WT | Injector 3 Driver | 1-4 ms |
| 8 | DG | Generator Field Driver | Digital Signals: 0-12-0v |
| 9 | --- | Not Used | --- |
| 10 | BK/TN | Power Ground | <0.1v |
| 11 | GY/RD | Coil 1 Driver Control | 5°, 55 mph: 8° dwell |
| 12 | --- | Not Used | --- |
| 13 | WT/DB | Injector 1 Driver | 1-4 ms |
| 14-15 | --- | Not Used | --- |
| 16 | LB/BR | Injector 4 Driver | 1-4 ms |
| 17 | TN/WT | Injector 2 Driver | 1-4 ms |
| 18-19 | --- | Not Used | --- |
| 20 | WT/BK | Ignition Switch Power (B+) | 12-14v |
| 21 | --- | Not Used | --- |
| 22 | BK/PK | MIL (lamp) Control | MIL Off: 12v, On: 1v |
| 23-24 | --- | Not Used | --- |
| 25 | DB/LG | Knock Sensor Signal | 0.080v AC |
| 26 | TN/BK | ECT Sensor Signal | At 180°F: 2.80v |
| 27 | BK/OR | HO2S Ground | <0.050v |
| 28-29 | --- | Not Used | --- |
| 30 | BK/DG | HO2S-11 (B1 S1) Signal | 0.1-1.1v |
| 31 | TN | Smart Start Relay Control | KOEC: 9-11v |
| 32 | GY/BK | CKP Sensor Signal | Digital Signals: 0-5-0v |
| 33 | TN/YL | CMP Sensor Signal | Digital Signals: 0-5-0v |
| 34, 39 | --- | Not Used | --- |
| 35 | OR/DB | TP Sensor Signal | 0.6-1.0v |
| 36 | DG/RD | MAP Sensor Signal | 1.5-1.6v |
| 37 | BK/RD | IAT Sensor Signal | At 100°F: 1.83v |
| 38 | DG/LB | A/C Switch Sense Signal | A/C Off: 12v, On: 1v |
| 40 | GY/YL | EGR Solenoid Control | 12v, at 55 mph: 1v |

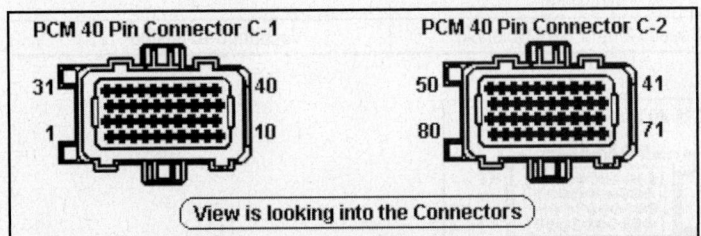

**Pin Connector Graphic**

## 1999-2000 Caravan, Grand Caravan, Voyager 2.4L VIN B (A/T) 'C2' White Connector

| PCM Pin # | Wire Color | Circuit Description (40-Pin) | Value at Hot Idle |
|---|---|---|---|
| 41 | RD/LG | Speed Control Set Switch Signal | S/C & Set Switch On: 3.8v |
| 42 | DB | A/C Pressure Switch Signal | A/C On: 0.451-4.850v |
| 43 | BK/LB | Sensor Ground | <0.050v |
| 44 | OR | 8-Volt Supply | 7.9-8.1v |
| 45 | --- | Not Used | --- |
| 46 | RD/WT | Keep Alive Power (B+) | 12-14v |
| 47 | --- | Not Used | --- |
| 48 | BR/WT | IAC 3 Driver Control | Pulse Signals |
| 49 | YL/BK | IAC 2 Driver Control | Pulse Signals |
| 50 | BK/TN | Power Ground | <0.1v |
| 51 | TN/WT | HO2S-12 (B1 S2) Signal | 0.1-1.1v |
| 52-55 | --- | Not Used | --- |
| 56 | TN/RD | Speed Control Vacuum Solenoid | Vacuum Increasing: 1v |
| 57 | GY/RD | IAC 1 Driver Control | Pulse Signals |
| 58 | PK/BK | IAC 4 Driver Control | Pulse Signals |
| 59 | PK/BR | CCD Bus (+) | Digital Signals: 0-5-0v |
| 60 | WT/BK | CCD Bus (-) | <0.050v |
| 61 | PK/WT | 5-Volt Supply | 4.9-5.1v |
| 62 | WT/PK | Brake Switch Sense Signal | Brake Off: 0v, On: 12v |
| 63 | YL/DG | Torque Management Request | Digital Signals |
| 64 | DB/OR | A/C Clutch Relay Control | Relay Off: 12v, On: 1v |
| 65 | PK | SCI Transmit | 0v |
| 66 | WT/OR | Vehicle Speed Signal | Digital Signal |
| 67 | DB/YL | ASD Relay Control | Relay Off: 12v, On: 1v |
| 68 | PK/BK | EVAP Purge Solenoid Control | PWM Signal: 0-12-0v |
| 69 | --- | Not Used | --- |
| 70 | P/R | EVAP Purge Solenoid Sense | 0-1v |
| 71 | --- | Not Used | --- |
| 72 | YL/BK | LDP Switch Sense Signal | LDP Switch Closed: 0v |
| 73 | LG/DB | Radiator Fan Control Relay | Relay Off: 12v, On: 1v |
| 74 | BR | Fuel Pump Relay Control | Relay Off: 12v, On: 1v |
| 75 | LG | SCI Receive | 5v |
| 76 | BK/WT | PNP Switch Sense (EATX) | In P/N: 0v, Others: 5v |
| 77 | WT/DG | LDP Solenoid Control | PWM Signal: 0-12-0v |
| 78 | OR/BK | Lockup Torque Converter | At Cruise w/TCC On: 1v |
| 79 | --- | Not Used | --- |
| 80 | LG/RD | Speed Control Vent Solenoid | Vacuum Decreasing: 1v |

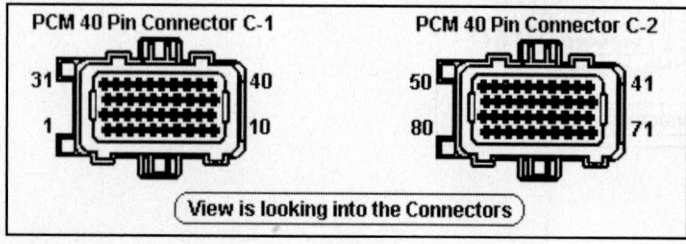

**Pin Connector Graphic**

**2001-02 Caravan, Grand Caravan, Voyager 2.4L VIN B (A/T) 'C1' Black Connector**

| PCM Pin # | Wire Color | Circuit Description (40-Pin) | Value at Hot Idle |
|---|---|---|---|
| 1-2 | --- | Not Used | --- |
| 3 | DB/TN | Coil 2 Driver Control | 5°, 55 mph: 8° dwell |
| 4 | --- | Not Used | --- |
| 5 | VT/YL | S/C On/Off Switch Sense | 12-14v |
| 6 | BR/WT | ASD Relay Power (B+) | 12-14v |
| 7 | BR/LB | Injector 3 Driver | 1-4 ms |
| 8 | BR/GY | Generator Field Driver | Digital Signals: 0-12-0v |
| 9 | --- | Not Used | --- |
| 10 | BK/BR | Power Ground | <0.1v |
| 11 | DB/GN | Coil 1 Driver Control | 5°, 55 mph: 8° dwell |
| 12 | VT/GY | Engine Oil Pressure Sensor | 1.6v at 24 psi |
| 13 | BR/YL | Injector 1 Driver | 1-4 ms |
| 14-15 | --- | Not Used | --- |
| 16 | BR/TN | Injector 4 Driver | 1-4 ms |
| 17 | BR/DB | Injector 2 Driver | 1-4 ms |
| 18 | BR/LG | HO2S-11 (B1 S1) Heater | PWM Signal: 0-12-0v |
| 19 | --- | Not Used | --- |
| 20 | PK/GY | Ignition Switch Power (B+) | 12-14v |
| 21-24 | --- | Not Used | --- |
| 25 | DB/YL | Knock Sensor Signal | 0.080v AC |
| 26 | VT/OR | ECT Sensor Signal | At 180°F: 2.80v |
| 27 | BR/LG | HO2S Ground | <0.050v |
| 28-29 | --- | Not Used | --- |
| 30 | DB/LB | HO2S-11 (B1 S1) Signal | 0.1-1.1v |
| 31 | DG/OR | Double Start Override | KOEC: 9-11v |
| 32 | BR/LB | CKP Sensor Signal | Digital Signals: 0-5-0v |
| 33 | DB/GY | CMP Sensor Signal | Digital Signals: 0-5-0v |
| 34 | --- | Not Used | --- |
| 35 | BR/OR | TP Sensor Signal | 0.6-1.0v |
| 36 | VT/BR | MAP Sensor Signal | 1.5-1.6v |
| 37 | DB/LG | IAT Sensor Signal | At 100°F: 1.83v |
| 38-39 | --- | Not Used | --- |
| 40 | DB/VT | EGR Solenoid Control | 12v, at 55 mph: 1v |

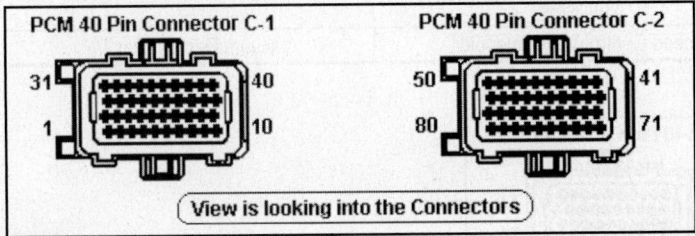

**Pin Connector Graphic**

**2001-02 Caravan, Grand Caravan, Voyager 2.4L VIN B (A/T) 'C2' White Connector**

| PCM Pin # | Wire Color | Circuit Description (40-Pin) | Value at Hot Idle |
|---|---|---|---|
| 41 | VT | Speed Control Set Switch Signal | S/C & Set Switch On: 3.8v |
| 42 | LB/BR | A/C Pressure Switch Signal | A/C On: 0.451-4.850v |
| 43 | DB/DG | Sensor Ground | <0.050v |
| 44 | BR/PK | 8-Volt Supply | 7.9-8.1v |
| 45 | --- | Not Used | --- |
| 46 | OR/RD | Keep Alive Power (B+) | 12-14v |
| 47 | --- | Not Used | --- |
| 48 | BR/LG | IAC 3 Driver Control | Pulse Signals |
| 49 | VT/LG | IAC 2 Driver Control | Pulse Signals |
| 50 | BK/DG | Power Ground | <0.1v |
| 51 | DB/YL | HO2S-12 (B1 S2) Signal | 0.1-1.1v |
| 52-55 | --- | Not Used | --- |
| 56 | VT/YL | Speed Control Vacuum Solenoid | Vacuum Increasing: 1v |
| 57 | VT/DG | IAC 1 Driver Control | Pulse Signals |
| 58 | BR/DG | IAC 4 Driver Control | Pulse Signals |
| 59 | WT/VT | PCI Data Bus (J1850) | Digital Signals: 0-7-0v |
| 60 | --- | Not Used | --- |
| 61 | PK/YL | 5-Volt Supply | 4.9-5.1v |
| 62 | DG/WT | Brake Switch Sense Signal | Brake Off: 0v, On: 12v |
| 63 | --- | Not Used | --- |
| 64 | LB/OR | A/C Clutch Relay Control | Relay Off: 12v, On: 1v |
| 65 | WT/BR | SCI Transmit Signal | 0v |
| 66 | DB/OR | Vehicle Speed Signal | Digital Signal |
| 67 | BR/WT | ASD Relay Control | Relay Off: 12v, On: 1v |
| 68 | DB/WT | EVAP Purge Solenoid Control | PWM Signal: 0-12-0v |
| 69, 71 | --- | Not Used | --- |
| 70 | DB/BR | EVAP Purge Solenoid Sense | 0-1v |
| 72 | VT/WT | LDP Switch Sense Signal | LDP Switch Closed: 0v |
| 73 | BR/VT | Radiator Fan Control Relay | Relay Off: 12v, On: 1v |
| 74 | BR | Fuel Pump Relay Control | Relay Off: 12v, On: 1v |
| 75 | WT/LG | SCI Receive Signal | 5v |
| 76 | YL/DB | PNP Switch Sense (EATX) | In P/N: 0v, Others: 5v |
| 77 | VT/LB | LDP Solenoid Control | PWM Signal: 0-12-0v |
| 78 | DB/WT | A/T: Lockup Torque Converter | At Cruise w/TCC On: 1v |
| 79 | --- | Not Used | --- |
| 80 | VT/OR | Speed Control Vent Solenoid | Vacuum Decreasing: 1v |

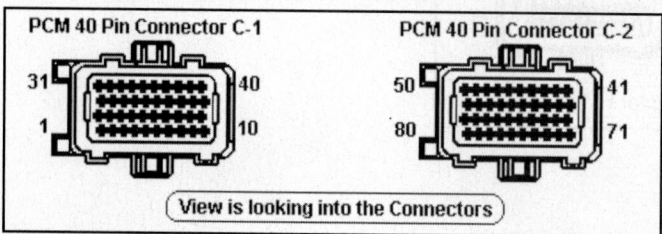

**Pin Connector Graphic**

## 2003 Caravan, Grand Caravan, Voyager 2.4L VIN B 'C1' Black Connector

| PCM Pin # | Wire Color | Circuit Description (38-Pin) | Value at Hot Idle |
|---|---|---|---|
| 1-8 | --- | Not Used | --- |
| 9 | BK/BR | Power Ground | <0.1v |
| 10 | --- | Not Used | --- |
| 11 | PK/GY | Ignition Switch Output (Run-Start) | 12-14v |
| 12 | PK/OR | Ignition Switch Output (Run-Start) | 12-14v |
| 13 | DB/OR | Vehicle Speed Signal | Digital Signal |
| 14 | GY/BK | Brake Fluid Level Switch Signal | Switch Open: 12v, Closed: 0v |
| 15-17 | --- | Not Used | --- |
| 18 | BK/DG | Power Ground | <0.1v |
| 19 | --- | Not Used | --- |
| 20 | VT/GY | Oil Pressure Signal | 1.6v at 24 psi |
| 21-24 | --- | Not Used | --- |
| 25 | WT/LG | SCI Receive (PCM) - DLC Pin 12 | 5v |
| 26 | WT/BR | Flash Program Enable | 5v |
| 27-28 | --- | Not Used | --- |
| 29 | RD/WT | Battery Power (B+) | 12-14v |
| 30 | YL | Ignition Switch Output (Start) | 8-11v (cranking) |
| 31 | DB/YL | HO2S-12 (B1 S2) Signal | 0.1-1.1v |
| 32 | DB/DG | Oxygen Sensor Return (Downstream) | <0.1v |
| 33-35 | --- | Not Used | --- |
| 36 | WT/BR | SCI Transmit (PCM) - DLC Pin 7 | 0v |
| 37 | DG/YL | SCI Transmit (TCM) | 0v |
| 38 | WT/VT | PCI Data Bus (J1850) - DLC Pin 2 | Digital Signals: 0-7-0v |

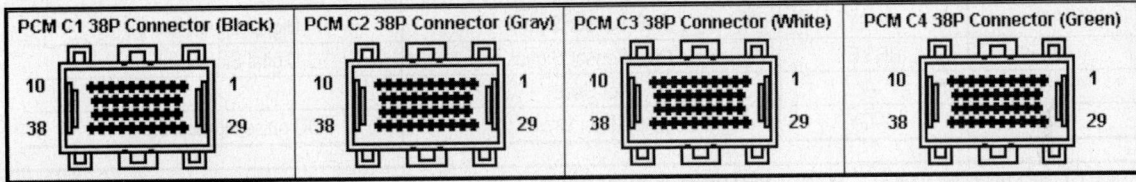

**Pin Connector Graphic**

## 2003 Caravan, Grand Caravan, Voyager 2.4L VIN B 'C2' Gray Connector

| PCM Pin # | Wire Color | Circuit Description (38-Pin) | Value at Hot Idle |
|---|---|---|---|
| 1-8 | --- | Not Used | --- |
| 9 | DB/TN | Coil 2 Driver | 5°, at 55 mph: 8° dwell |
| 10 | DB/DG | Coil 1 Driver | 5°, at 55 mph: 8° dwell |
| 11 | BR/TN | Injector 4 Driver | 1.0-4.0 ms |
| 12 | BR/LB | Injector 3 Driver | 1.0-4.0 ms |
| 13 | BR/DB | Injector 2 Driver | 1.0-4.0 ms |
| 14 | BR/YL | Injector 1 Driver | 1.0-4.0 ms |
| 15-17 | --- | Not Used | --- |
| 18 | BR/LG | HO2S-11 (B1 S1) Heater Control | Heater Off: 12v, On: 1v |
| 19 | BR/GY | Generator Field Driver | Digital Signal: 0-12-0v |
| 20 | VT/OR | ECT Sensor Signal | At 180°F: 2.80v |
| 21 | BR/OR | TP Sensor Signal | 0.6-1.0v |
| 22 | --- | Not Used | --- |
| 23 | VT/BR | MAP Sensor Signal | 1.5-1.7v |
| 24 | BR/LG | Knock Sensor Return | <0.050v |
| 25 | DB/YL | Knock Sensor Signal | 0.080v AC |
| 26 | --- | Not Used | --- |
| 27 | DB/DG | Sensor Ground | <0.1v |
| 28 | VR/VT | IAC Motor Return | 12-14v |
| 29 | PK/YL | 5-Volt Supply | 4.9-5.1v |
| 30 | DB/LG | IAT Sensor Signal | At 100°F: 1.83v |
| 31 | DB/LB | HO2S-11 (B1 S1) Signal | 0.1-1.1v |
| 32 | BR/DG | Oxygen Sensor Return (Upstream) | <0.1v |
| 33 | --- | Not Used | --- |
| 34 | DB/GY | CMP Sensor Signal | Digital Signal: 0-5-0v |
| 35 | BR/LB | CKP Sensor Signal | Digital Signal: 0-5-0v |
| 36-37 | --- | Not Used | --- |
| 38 | VT/GY | IAC Motor Driver | DC pulse signals: 0.8-11v |

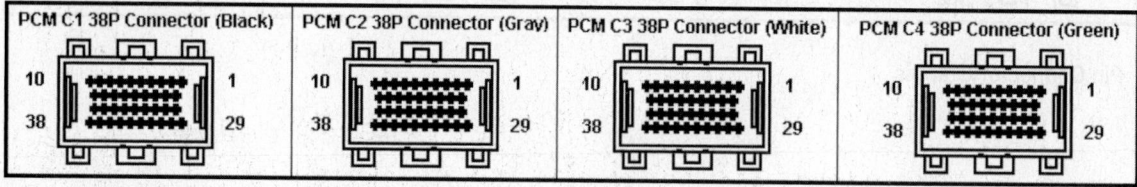

**Pin Connector Graphic**

**2003 Caravan, Grand Caravan, Voyager 2.4L VIN B 'C3' White Connector**

| PCM Pin # | Wire Color | Circuit Description (38-Pin) | Value at Hot Idle |
|---|---|---|---|
| 1-2, 4 | --- | Not Used | --- |
| 3 | BR/WT | ASD Relay Control | Relay Off: 12v, On: 1v |
| 5 | VT/OR | Speed Control Vent Solenoid | Vacuum Decreasing: 1v |
| 6 | BR/VT | Radiator Fan Relay Control | Relay Off: 12v, On: 1v |
| 7 | VT/YL | Speed Control Power Supply | 12-14v |
| 8 | VT/LG | Natural Vacuum Leak Detection Pump | Solenoid Off: 12v, On: 1v |
| 9 | BR/WT | HO2S-12 (B1 S2) Heater Control | --- |
| 10 | --- | Not Used | --- |
| 11 | LB/OR | A/C Clutch Relay Control | Relay Off: 12v, On: 1v |
| 12 | VT/YL | Speed Control Vacuum Solenoid | Vacuum Increasing: 1v |
| 13-18, 21-22 | --- | Not Used | --- |
| 19 | BR/WT | Automatic Shutdown Relay Output | 12-14v |
| 20 | DB/WT | EVAP Purge Solenoid Control | PWM Signal: 0-12-0v |
| 23 | DG/WT | Brake Switch Sense | Brake Off: 0v, On: 12v |
| 24-27 | --- | Not Used | --- |
| 28 | BR/WT | Automatic Shutdown Relay Output | 12-14v |
| 29 | DB/BR | EVAP Purge Sense | 0-1v |
| 30 | --- | Not Used | --- |
| 31 | LB/BR | A/C Pressure Sensor | A/C On: 0.45-4.85v |
| 32 | DB/YL | Battery Temperature Sensor | At 100°F: 1.83v |
| 33, 36 | --- | Not Used | --- |
| 34 | VT | Speed Control Set Switch Signal | S/C & Set Switch On: 3.8v |
| 35 | VT/WT | Natural Vacuum Leak Detection Switch Sense | 0.1v |
| 36 | --- | Not Used | --- |
| 37 | BR | Fuel Pump Relay Control | Relay Off: 12v, On: 1v |
| 38 | DG/OR | Starter Relay Control | KOEC: 9-11v |

## 2003 Caravan, Grand Caravan, Voyager 2.4L VIN B 'C4' Green Connector

| PCM Pin # | Wire Color | Circuit Description (38-Pin) | Value at Hot Idle |
|---|---|---|---|
| 1 | YL/GY | Overdrive Solenoid Control | Solenoid Off: 12v, On: 1v |
| 2 | YL/LB | Underdrive Solenoid Control | Solenoid Off: 12v, On: 1v |
| 3-5 | --- | Not Used | --- |
| 6 | YL/DB | 2-4 Solenoid Control | Solenoid Off: 12v, On: 1v |
| 7-9 | --- | Not Used | --- |
| 10 | DG/WT | Low/Reverse Solenoid Control | Solenoid Off: 12v, On: 1v |
| 11, 14 | --- | Not Used | --- |
| 12, 13 | BK/LG | Power Ground | <0.050v |
| 15 | DG/LB | TRS T1 Sense | <0.050v |
| 16 | DG/DB | TRS T3 Sense | <0.050v |
| 17, 20-21 | --- | Not Used | --- |
| 18, 38 | YL/BR | Transmission Control Relay Output | Relay Off: 12v, On: 1v |
| 19, 28 | YL/OR | Transmission Control Relay Output | Relay Off: 12v, On: 1v |
| 23-26 | --- | Not Used | --- |
| 27 | DG/GY | TRS T41 Sense | <0.050v |
| 29 | YL/TN | Low/Reverse Pressure Switch Sense | 12-14v |
| 30 | YL/DG | 2-4 Pressure Switch Sense | In Low/Reverse: 2-4v |
| 31, 36 | --- | Not Used | --- |
| 32 | DG/BR | Output Speed Sensor Signal | In 2-4 Position: 2-4v |
| 33 | DG/WT | Input Speed Sensor Signal | Moving: AC voltage |
| 34 | DG/VT | Speed Sensor Ground | Moving: AC voltage |
| 35 | DG/OR | Transmission Temperature Sensor Signal | <0.050v |
| 37 | DG/YL | TRS T42 Sense | In PRNL: 0v, Others 5v |

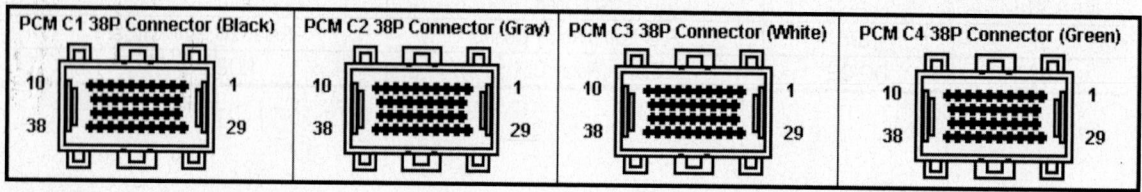

**Pin Connector Graphic**

**1994-95 Caravan, Voyager 2.5L VIN K 60-Pin Connector**

| PCM Pin # | Wire Color | Circuit Description (60-Pin) | Value at Hot Idle |
|---|---|---|---|
| 1 | DG/RD | MAP Sensor Signal | 1.5-1.6v |
| 2 | TN/BK | ECT Sensor Signal | At 180°F: 2.80v |
| 3 | RD/WT | Keep Alive Power (B+) | 12-14v |
| 4 | BK/LB | Sensor Ground | <0.050v |
| 5 | BK/WT | Sensor Ground | <0.050v |
| 6 | PK/WT | 5-Volt Supply | 4.9-5.1v |
| 7 | OR | 8-Volt Supply | 7.9-8.1v |
| 8 | WT | Ignition Switch (Start) Output | 12-14v |
| 9 | DB/WT | Ignition Switch Power (B+) | 12-14v |
| 10, 13-15 | --- | Not Used | --- |
| 11 | BK/TN | Power Ground | <0.1v |
| 12 | BK/TN | Power Ground | <0.1v |
| 16 | LB | Injector Control Driver | 1-4 ms |
| 17-18, 21, 28 | --- | Not Used | --- |
| 19 | BK/GY | Coil Driver Control | 5°, 55 mph: 8° dwell |
| 20 | DG | Alternator Field Control | Digital Signals: 0-12-0v |
| 22 | DB | TP Sensor Signal | 0.6-1.0v |
| 23 | RD | Speed Control Set Switch Signal | S/C & Set Switch On: 3.8v |
| 24 | GY/BK | Distributor Reference Signal | Digital Signals: 0-5-0v |
| 25 | PK | SCI Transmit | 0v |
| 26 | PK/BR | CCD Bus (+) | Digital Signals: 0-5-0v |
| 27 | BR | A/C Damped Pressure Switch | A/C Off: 12v, On: 1v |
| 29 | WT/PK | Brake Switch Sense Signal | Brake Off: 0v, On: 12v |
| 30 | BR/YL | PNP Switch Sense Signal | In P/N: 0v, Others: 5v |
| 31 | DB/PK | Low Speed Fan Control | Relay Off: 12v, On: 1v |
| 32 | --- | Not Used | --- |
| 33 | TN/RD | Speed Control Vacuum Solenoid | Vacuum Increasing: 1v |
| 34 | DB/OR | A/C Clutch Relay Control | Relay Off: 12v, On: 1v |
| 35-36 | --- | Not Used | --- |
| 39 | GY/RD | AIS Motor 3 Control | Pulse Signals |
| 40 | BR/WT | AIS Motor 1 Control | Pulse Signals |
| 41 | DG/WT | HO2S-11 (B1 S1) Signal | 0.1-1.1v |
| 42-44 | --- | Not Used | --- |
| 45 | LG | SCI Receive | 5v |
| 46 | WT/BK | CCD Bus (-) | <0.050v |
| 47 | WT/OR | Vehicle Speed Signal | Digital Signal |
| 48-50 | --- | Not Used | --- |
| 51 | DB/YL | ASD Relay Power (B+) | Relay Off: 12v, On: 1v |
| 52 | PK/BK | EVAP Purge Solenoid Control | Solenoid Off: 12v, On: 1v |
| 53 | LG/RD | Speed Control Vent Solenoid | Vacuum Decreasing: 1v |
| 54 | OR/BK | Lockup Torque Converter | At Cruise w/TCC On: 1v |
| 55 | YL | High Speed Fan Control | Relay Off: 12v, On: 1v |
| 56, 58 | --- | Not Used | --- |
| 57 | DG/OR | ASD Relay Control | Relay Off: 12v, On: 1v |
| 59 | PK/BK | AIS Motor 4 Control | Pulse Signals |
| 60 | YL/BK | AIS Motor 2 Control | Pulse Signals |

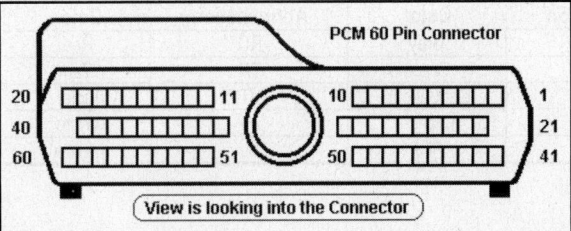

**Pin Connector Graphic**

## 1994-95 Caravan, Voyager 3.0L VIN 3 (A/T) 60-Pin Connector

| PCM Pin # | Wire Color | Circuit Description (60-Pin) | Value at Hot Idle |
|---|---|---|---|
| 1 | DG/RD | MAP Sensor Signal | 1.5-1.6v |
| 2 | TN/BK | ECT Sensor Signal | At 180°F: 2.80v |
| 3 | RD/WT | Keep Alive Power (B+) | 12-14v |
| 4 | BK/LB | Sensor Ground | <0.050v |
| 5 | BK/WT | Sensor Ground | <0.050v |
| 6 | PK/WT | 5-Volt Supply | 4.9-5.1v |
| 7 | OR | 8-Volt Supply | 7.9-8.1v |
| 9 | DB/WT | Ignition Switch Power (B+) | 12-14v |
| 10 | --- | Not Used | --- |
| 11 | BK/TN | Power Ground | <0.1v |
| 12 | BK/TN | Power Ground | <0.1v |
| 13 | LB/BR | Injector 4 Driver Control | 1-4 ms |
| 14 | YL/WT | Injector 3 Driver Control | 1-4 ms |
| 15 | TN | Injector 2 Driver Control | 1-4 ms |
| 16 | LB | Injector 1 Driver Control | 1-4 ms |
| 17-18 | --- | Not Used | --- |
| 19 | BK/GY | Coil Driver Control | 5°, 55 mph: 8° dwell |
| 20 | DG | Alternator Field Control | Digital Signals: 0-12-0v |
| 21 | --- | Not Used | --- |
| 22 | DB | TP Sensor Signal | 0.6-1.0v |
| 23 | RD | Speed Control Set Switch Signal | S/C & Set Switch On: 3.8v |
| 24 | GY/P | Distributor Reference Signal | Digital Signals: 0-5-0v |
| 25 | PK | SCI Transmit | 0v |
| 26 | PK/BR | CCD Bus (+) | Digital Signals: 0-5-0v |
| 27 | BR | A/C Damped Pressure Switch | A/C Off: 12v, On: 1v |
| 28 | --- | Not Used | --- |
| 29 | WT/PK | Brake Switch Sense Signal | Brake Off: 0v, On: 12v |
| 30 | BK/OR | PNP Switch Sense Signal | In P/N: 0v, Others: 5v |
| 31 | DB/PK | Low Speed Fan Control | Relay Off: 12v, On: 1v |
| 32 | --- | Not Used | --- |
| 33 | TN/RD | Speed Control Vacuum Solenoid | Vacuum Increasing: 1v |
| 34 | DB/OR | A/C Clutch Relay Control | Relay Off: 12v, On: 1v |
| 35 | GY/YL | EGR Solenoid Control | Solenoid Off: 12v, On: 1v |
| 36-37 | --- | Not Used | --- |
| 38 | GY | Injector 5 Driver Control | 1-4 ms |
| 39 | GY/RD | AIS Motor Control (open) | Pulse Signals |
| 40 | BR/WT | AIS Motor Control (close) | Pulse Signals |

## Standard Colors and Abbreviations

| Abbreviation | Color | Abbreviation | Color | Abbreviation | Color |
|---|---|---|---|---|---|
| BK | Black | GY | Gray | RD | Red |
| BL | Blue | GN | Green | TN | Tan |
| BR | Brown | LG | Light Green | VT | Violet |
| DB | Dark Blue | OR | Orange | WT | White |
| DG | Dark Green | PK | Pink | YL | Yellow |

**1994-95 Caravan, Voyager 3.0L VIN 3 (A/T) 60-Pin Connector - Continued**

| PCM Pin # | Wire Color | Circuit Description (60-Pin) | Value at Hot Idle |
|---|---|---|---|
| 41 | DG/WT | HO2S-11 (B1 S1) Signal | 0.1-1.1v |
| 42-43 | --- | Not Used | --- |
| 44 | TN/YL | Distributor Sync Signal | Digital Signals: 0-5-0v |
| 45 | LG | SCI Receive | 5v |
| 46 | WT/BK | CCD Bus (-) | <0.050v |
| 47 | WT/OR | Vehicle Speed Signal | Digital Signal |
| 48-50 | --- | Not Used | --- |
| 51 | DB/YL | ASD Relay Control | Relay Off: 12v, On: 1v |
| 52 | PK/BK | EVAP Purge Solenoid Control | Solenoid Off: 12v, On: 1v |
| 53 | LG/RD | Speed Control Vent Solenoid | Vacuum Decreasing: 1v |
| 54 | OR/BK | Lockup Torque Converter | At Cruise w/TCC On: 1v |
| 55 | YL | High Speed Fan Control | Relay Off: 12v, On: 1v |
| 56 | --- | Not Used | --- |
| 57 | DG/OR | ASD Relay Power (B+) | 12-14v |
| 58 | BR/DB | Injector 6 Driver Control | 1-4 ms |
| 59 | PK/BK | AIS Motor Control (close) | Pulse Signals |
| 60 | YL/BK | AIS Motor Control (close) | Pulse Signals |

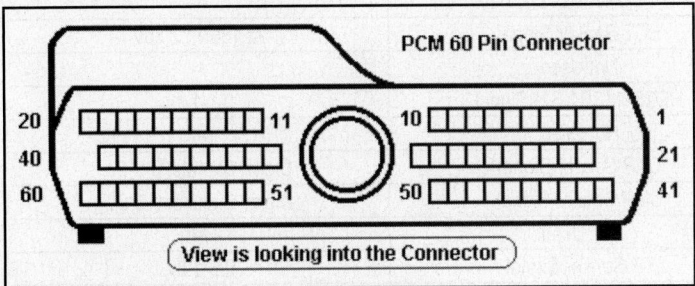

**Pin Connector Graphic**

**Standard Colors and Abbreviations**

| Abbreviation | Color | Abbreviation | Color | Abbreviation | Color |
|---|---|---|---|---|---|
| BK | Black | GY | Gray | RD | Red |
| BL | Blue | GN | Green | TN | Tan |
| BR | Brown | LG | Light Green | VT | Violet |
| DB | Dark Blue | OR | Orange | WT | White |
| DG | Dark Green | PK | Pink | YL | Yellow |

## 1996 Caravan, Voyager 3.0L VIN 3 (A/T) 'C1' Black Connector

| PCM Pin # | Wire Color | Circuit Description (40-Pin) | Value at Hot Idle |
|-----------|-----------|------------------------------|-------------------|
| 1-3 | --- | Not Used | --- |
| 4 | DG | Generator Field Driver | Digital Signals: 0-12-0v |
| 5 | YL/RD | Speed Control Power Supply | 12-14v |
| 6 | DG/OR | ASD Relay Power (B+) | 12-14v |
| 7 | YL/WT | Injector 3 Driver | 1-4 ms |
| 8 | TN | Starter Relay Control | KOEC: 9-11v |
| 9 | --- | Not Used | --- |
| 10 | BK/TN | Power Ground | <0.1v |
| 11 | GY | Coil 1 Driver Control | 5°, 55 mph: 8° dwell |
| 12 | --- | Not Used | --- |
| 13 | WT/DB | Injector 1 Driver | 1-4 ms |
| 14 | BR/DB | Injector 6 Driver | 1-4 ms |
| 15 | GY | Injector 5 Driver | 1-4 ms |
| 16 | LB/BR | Injector 4 Driver | 1-4 ms |
| 17 | TN | Injector 2 Driver | 1-4 ms |
| 18-19 | --- | Not Used | --- |
| 20 | WT/BK | Ignition Switch Power (B+) | 12-14v |
| 21-25 | --- | Not Used | --- |
| 26 | TN/BK | ECT Sensor Signal | At 180°F: 2.80v |
| 27-29 | --- | Not Used | --- |
| 30 | BK/DG | HO2S-11 (B1 S1) Signal | 0.1-1.1v |
| 31 | --- | Not Used | --- |
| 32 | GY/BK | CKP Sensor Signal | Digital Signals: 0-5-0v |
| 33 | TN/YL | CMP Sensor Signal | Digital Signals: 0-5-0v |
| 34 | --- | Not Used | --- |
| 35 | OR/DB | TP Sensor Signal | 0.6-1.0v |
| 36 | DG/RD | MAP Sensor Signal | 1.5-1.6v |
| 37-39 | --- | Not Used | --- |
| 40 | GY/YL | EGR Solenoid Control | 12v, at 55 mph: 1v |

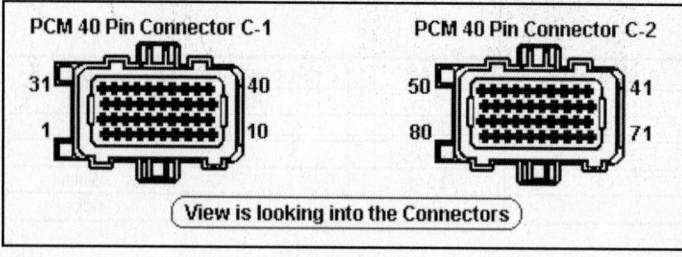

**Pin Connector Graphic**

### Standard Colors and Abbreviations

| Abbreviation | Color | Abbreviation | Color | Abbreviation | Color |
|--------------|-------|--------------|-------|--------------|-------|
| BK | Black | GY | Gray | RD | Red |
| BL | Blue | GN | Green | TN | Tan |
| BR | Brown | LG | Light Green | VT | Violet |
| DB | Dark Blue | OR | Orange | WT | White |
| DG | Dark Green | PK | Pink | YL | Yellow |

**1996 Caravan, Voyager 3.0L VIN 3 (A/T) 'C2' White Connector**

| PCM Pin # | Wire Color | Circuit Description (40-Pin) | Value at Hot Idle |
|---|---|---|---|
| 41 | RD/LG | Speed Control Set Switch Signal | S/C & Set Switch On: 3.8v |
| 42 | DB | A/C Pressure Switch Signal | A/C On: 0.451-4.850v |
| 43 | BK/LB | Sensor Ground | <0.050v |
| 44 | OR | 8-Volt Supply | 7.9-8.1v |
| 45 | DB | A/C Switch Sense Signal | A/C Off: 12v, On: 1v |
| 46 | RD/WT | Keep Alive Power (B+) | 12-14v |
| 47 | --- | Not Used | --- |
| 48 | BR/WT | IAC 1 Driver Control | Pulse Signals |
| 49 | YL/BK | IAC 2 Driver Control | Pulse Signals |
| 50 | BK/TN | Power Ground | <0.1v |
| 51 | TN/WT | HO2S-12 (B1 S2) Signal | 0.1-1.1v |
| 52-55 | --- | Not Used | --- |
| 56 | OR/BK | Lockup Torque Converter | At Cruise w/TCC On: 1v |
| 57 | GY/RD | IAC 3 Driver Control | Pulse Signals |
| 58 | PK/BK | IAC 4 Driver Control | Pulse Signals |
| 59 | PK/BR | CCD Bus (+) | Digital Signals: 0-5-0v |
| 60 | WT/BK | CCD Bus (-) | <0.050v |
| 61 | PK/WT | 5-Volt Supply | 4.9-5.1v |
| 62 | WT/PK | Brake Switch Sense Signal | Brake Off: 0v, On: 12v |
| 63 | YL/DG | Torque Management Request | Digital Signals |
| 64 | DB/OR | A/C Clutch Relay Control | Relay Off: 12v, On: 1v |
| 65 | PK | SCI Transmit Signal | 0v |
| 66 | WT/OR | Vehicle Speed Signal | Digital Signal |
| 67 | DB/YL | ASD Relay Control | Relay Off: 12v, On: 1v |
| 68 | PK/BK | EVAP Purge Solenoid Control | PWM Signal: 0-12-0v |
| 69-71 | --- | Not Used | --- |
| 72 | YL/BK | LDP Switch Sense Signal | LDP Switch Closed: 0v |
| 73 | LG/DG | Radiator Fan Control Relay | Relay Off: 12v, On: 1v |
| 74 | BR | Fuel Pump Relay Control | Relay Off: 12v, On: 1v |
| 75 | LG | SCI Receive Signal | 5v |
| 76 | BK/WT | PNP Switch Sense (EATX) | In P/N: 0v, Others: 5v |
| 77 | WT/DG | LDP Solenoid Control | PWM Signal: 0-12-0v |
| 78 | TN/RD | Speed Control Vacuum Solenoid | Vacuum Increasing: 1v |
| 79 | --- | Not Used | --- |
| 80 | LG/RD | Speed Control Vent Solenoid | Vacuum Decreasing: 1v |

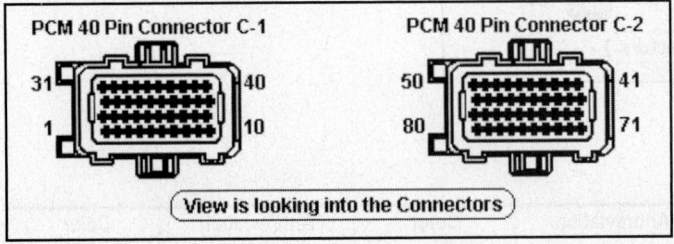

**Pin Connector Graphic**

## 1997-98 Caravan, Voyager 3.0L VIN 3 'C1' Black Connector

| PCM Pin # | Wire Color | Circuit Description (40-Pin) | Value at Hot Idle |
|---|---|---|---|
| 1-3 | --- | Not Used | --- |
| 4 | DG | Generator Field Driver | Digital Signals: 0-12-0v |
| 5 | YL/RD | Speed Control Power Supply | 12-14v |
| 6 | DG/OR | ASD Relay Power (B+) | 12-14v |
| 7 | YL/WT | Injector 3 Driver | 1-4 ms |
| 8 | TN | Starter Relay Control | KOEC: 9-11v |
| 9 | --- | Not Used | --- |
| 10 | BK/TN | Power Ground | <0.1v |
| 11 | GY | Coil 1 Driver Control | 5°, 55 mph: 8° dwell |
| 12 | --- | Not Used | --- |
| 13 | WT/DB | Injector 1 Driver | 1-4 ms |
| 14 | BR/DB | Injector 6 Driver | 1-4 ms |
| 15 | GY | Injector 5 Driver | 1-4 ms |
| 16 | LB/BR | Injector 4 Driver | 1-4 ms |
| 17 | TN | Injector 2 Driver | 1-4 ms |
| 18-19 | --- | Not Used | --- |
| 20 | WT/BK | Ignition Switch Power (B+) | 12-14v |
| 21 | --- | Not Used | --- |
| 22 | BK/PK | MIL (lamp) Control | MIL Off: 12v, On: 1v |
| 23-25 | --- | Not Used | --- |
| 26 | TN/BK | ECT Sensor Signal | At 180°F: 2.80v |
| 27-29 | --- | Not Used | --- |
| 30 | BK/DG | HO2S-11 (B1 S1) Signal | 0.1-1.1v |
| 31 | --- | Not Used | --- |
| 32 | GY/BK | CKP Sensor Signal | Digital Signals: 0-5-0v |
| 33 | TN/YL | CMP Sensor Signal | Digital Signals: 0-5-0v |
| 34 | --- | Not Used | --- |
| 35 | OR/DB | TP Sensor Signal | 0.6-1.0v |
| 36 | DG/RD | MAP Sensor Signal | 1.5-1.6v |
| 37-39 | --- | Not Used | --- |
| 40 | GY/YL | EGR Solenoid Control | 12v, at 55 mph: 1v |

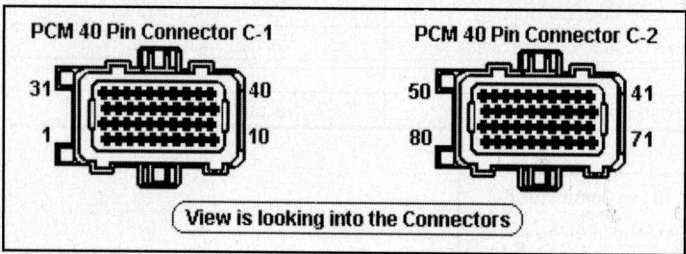

**Pin Connector Graphic**

## Standard Colors and Abbreviations

| Abbreviation | Color | Abbreviation | Color | Abbreviation | Color |
|---|---|---|---|---|---|
| BK | Black | GY | Gray | RD | Red |
| BL | Blue | GN | Green | TN | Tan |
| BR | Brown | LG | Light Green | VT | Violet |
| DB | Dark Blue | OR | Orange | WT | White |
| DG | Dark Green | PK | Pink | YL | Yellow |

**1997-98 Caravan, Voyager 3.0L VIN 3 'C2' White Connector**

| PCM Pin # | Wire Color | Circuit Description (40-Pin) | Value at Hot Idle |
|-----------|------------|------------------------------|-------------------|
| 41 | RD/LG | Speed Control Set Switch Signal | S/C & Set Switch On: 3.8v |
| 42 | DB | A/C Pressure Switch Signal | A/C On: 0.451-4.850v |
| 43 | BK/LB | Sensor Ground | <0.050v |
| 44 | OR | 8-Volt Supply | 7.9-8.1v |
| 45 | DG | A/C Switch Sense Signal | A/C Off: 12v, On: 1v |
| 46 | RD/WT | Keep Alive Power (B+) | 12-14v |
| 47 | --- | Not Used | --- |
| 48 | BR/WT | IAC 3 Driver Control | Pulse Signals |
| 49 | YL/BK | IAC 2 Driver Control | Pulse Signals |
| 50 | BK/TN | Power Ground | <0.1v |
| 51 | TN/WT | HO2S-12 (B1 S2) Signal | 0.1-1.1v |
| 52-55 | --- | Not Used | --- |
| 56 | OR/BK | Lockup Torque Converter | At Cruise w/TCC On: 1v |
| 57 | GY/RD | IAC 1 Driver Control | Pulse Signals |
| 58 | PK/BK | IAC 4 Driver Control | Pulse Signals |
| 59 | PK/BR | CCD Bus (+) | Digital Signals: 0-5-0v |
| 60 | WT/BK | CCD Bus (-) | <0.050v |
| 61 | PK/WT | 5-Volt Supply | 4.9-5.1v |
| 62 | WT/PK | Brake Switch Sense Signal | Brake Off: 0v, On: 12v |
| 63 | YL/DG | Torque Management Request | Digital Signals |
| 64 | DB/OR | A/C Clutch Relay Control | Relay Off: 12v, On: 1v |
| 65 | PK | SCI Transmit Signal | 0v |
| 66 | WT/OR | Vehicle Speed Signal | Digital Signal |
| 67 | DB/YL | ASD Relay Control | Relay Off: 12v, On: 1v |
| 68 | PK/BK | EVAP Purge Solenoid Control | PWM Signal: 0-12-0v |
| 69-71 | --- | Not Used | --- |
| 72 | YL/BK | LDP Switch Sense Signal | LDP Switch Closed: 0v |
| 73 | LG/DB | Radiator Fan Control Relay | Relay Off: 12v, On: 1v |
| 74 | BR | Fuel Pump Relay Control | Relay Off: 12v, On: 1v |
| 75 | LG | SCI Receive Signal | 5v |
| 76 | BK/WT | PNP Switch Sense (EATX) | In P/N: 0v, Others: 5v |
| 77 | WT/DG | LDP Solenoid Control | PWM Signal: 0-12-0v |
| 78 | TN/RD | Speed Control Vacuum Solenoid | Vacuum Increasing: 1v |
| 79 | --- | Not Used | --- |
| 80 | LG/RD | Speed Control Vent Solenoid | Vacuum Decreasing: 1v |

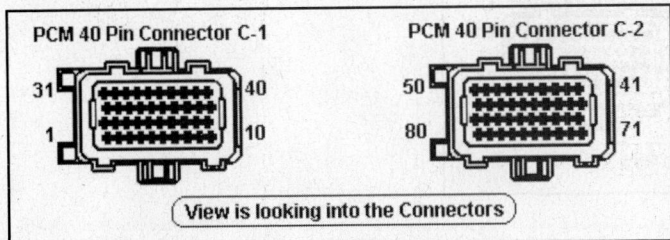

PCM 40 Pin Connector C-1

PCM 40 Pin Connector C-2

View is looking into the Connectors

**Pin Connector Graphic**

## 1999-2000 Caravan, Voyager 3.0L VIN 3 'C1' Black Connector

| PCM Pin # | Wire Color | Circuit Description (40-Pin) | Value at Hot Idle |
|---|---|---|---|
| 1 | --- | Not Used | --- |
| 2 | RD/YL | Coil 3 Driver Control | 5°, 55 mph: 8° dwell |
| 3 | DB/TN | Coil 2 Driver Control | 5°, 55 mph: 8° dwell |
| 4 ('99) | DG | Generator Field Driver | Digital Signals: 0-12-0v |
| 5 | YL/RD | Speed Control Power Supply | 12-14v |
| 6 | DG/OR | ASD Relay Power (B+) | 12-14v |
| 7 | YL/WT | Injector 3 Driver | 1-4 ms |
| 8 ('99) | TN | Smart Start Relay Control | KOEC: 9-11v |
| 8 | DG | Generator Field Driver | Digital Signals: 0-12-0v |
| 9, 12 | --- | Not Used | --- |
| 10 | BK/TN | Power Ground | <0.1v |
| 11 | GY/RD | Coil 1 Driver Control | 5°, 55 mph: 8° dwell |
| 13 | WT/DB | Injector 1 Driver | 1-4 ms |
| 14 | BR/DB | Injector 6 Driver | 1-4 ms |
| 15 | GY | Injector 5 Driver | 1-4 ms |
| 16 | LB/BR | Injector 4 Driver | 1-4 ms |
| 17 | TN/WT | Injector 2 Driver | 1-4 ms |
| 18-19 | --- | Not Used | --- |
| 20 | WT/BK | Ignition Switch Power (B+) | 12-14v |
| 21 | --- | Not Used | --- |
| 22 | BK/PK | MIL (lamp) Control | MIL Off: 12v, On: 1v |
| 23-24 | --- | Not Used | --- |
| 25 | DB/LG | Knock Sensor Signal | 0.080v AC |
| 26 | TN/BK | ECT Sensor Signal | At 180°F: 2.80v |
| 27 | BK/OR | HO2S Ground | <0.050v |
| 28-29 | --- | Not Used | --- |
| 30 | BK/DG | HO2S-11 (B1 S1) Signal | 0.1-1.1v |
| 31 | --- | Not Used | --- |
| 32 | GY/BK | CKP Sensor Signal | Digital Signals: 0-5-0v |
| 33 | TN/YL | CMP Sensor Signal | Digital Signals: 0-5-0v |
| 34 | --- | Not Used | --- |
| 35 | OR/DB | TP Sensor Signal | 0.6-1.0v |
| 36 | DG/RD | MAP Sensor Signal | 1.5-1.6v |
| 37-39 | --- | Not Used | --- |
| 40 | GY/YL | EGR Solenoid Control | 12v, at 55 mph: 1v |

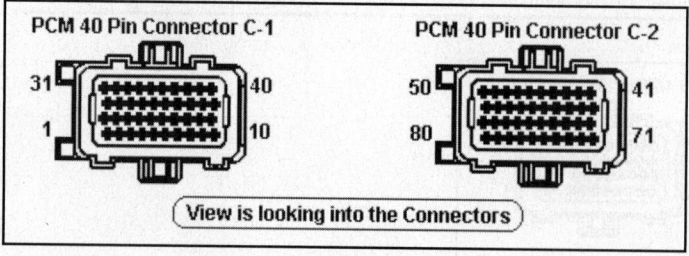

**Pin Connector Graphic**

**1999-2000 Caravan, Voyager 3.0L VIN 3 'C2' White Connector**

| PCM Pin # | Wire Color | Circuit Description (40-Pin) | Value at Hot Idle |
|---|---|---|---|
| 41 | RD/LG | Speed Control Set Switch Signal | S/C & Set Switch On: 3.8v |
| 42 | DB | A/C Pressure Switch Signal | A/C On: 0.451-4.850v |
| 43 | BK/LB | Sensor Ground | <0.050v |
| 44 | OR | 8-Volt Supply | 7.9-8.1v |
| 45 ('99) | DG/LB | A/C Switch Sense Signal | A/C Off: 12v, On: 1v |
| 46 | RD/WT | Keep Alive Power (B+) | 12-14v |
| 47 | --- | Not Used | --- |
| 48 | BR/WT | IAC 3 Driver Control | Pulse Signals |
| 49 | YL/BK | IAC 2 Driver Control | Pulse Signals |
| 50 | BK/TN | Power Ground | <0.1v |
| 51 | TN/WT | HO2S-12 (B1 S2) Signal | 0.1-1.1v |
| 52-56 | --- | Not Used | --- |
| 57 | GY/RD | IAC 1 Driver Control | Pulse Signals |
| 58 | PK/BK | IAC 4 Driver Control | Pulse Signals |
| 59 | PK/BR | CCD Bus (+) | Digital Signals: 0-5-0v |
| 60 | WT/BK | CCD Bus (-) | <0.050v |
| 61 | PK/WT | 5-Volt Supply | 4.9-5.1v |
| 62 | WT/PK | Brake Switch Sense Signal | Brake Off: 0v, On: 12v |
| 63 | YL/DG | Torque Management Request | Digital Signals |
| 64 | DB/OR | A/C Clutch Relay Control | Relay Off: 12v, On: 1v |
| 65 | PK | SCI Transmit Signal | 0v |
| 66 | WT/OR | Vehicle Speed Signal | Digital Signal |
| 67 | DB/YL | ASD Relay Control | Relay Off: 12v, On: 1v |
| 68 | PK/BK | EVAP Purge Solenoid Control | PWM Signal: 0-12-0v |
| 69-71 | --- | Not Used | --- |
| 72 ('99) | YL/BK | LDP Switch Sense Signal | LDP Switch Closed: 0v |
| 73 | LG/DB | Radiator Fan Control Relay | Relay Off: 12v, On: 1v |
| 74 | BR | Fuel Pump Relay Control | Relay Off: 12v, On: 1v |
| 75 | LG | SCI Receive Signal | 5v |
| 76 | BR/YL | PNP Switch Sense (EATX) | In P/N: 0v, Others: 5v |
| 77 ('99) | WT/DG | LDP Solenoid Control | PWM Signal: 0-12-0v |
| 78 | TN/RD | Speed Control Vacuum Solenoid | Vacuum Increasing: 1v |
| 79 | --- | Not Used | --- |
| 80 | LG/RD | Speed Control Vent Solenoid | Vacuum Decreasing: 1v |

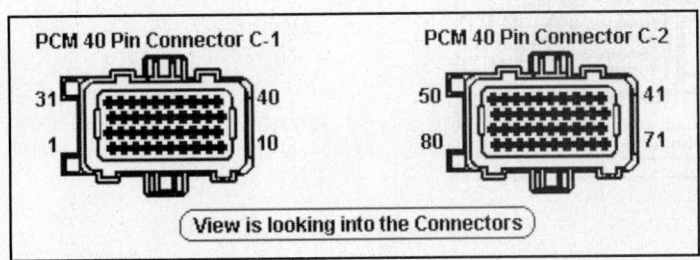

**Pin Connector Graphic**

## 2001 Caravan, Voyager 3.0L V6 OHC Flex Fuel VIN 3 'C1' Black Connector

| PCM Pin # | Wire Color | Circuit Description (40-Pin) | Value at Hot Idle |
|---|---|---|---|
| 1 | --- | Not Used | --- |
| 2 | DB/OR | Coil 3 Driver Control | 5°, 55 mph: 8° dwell |
| 3 | DB/TN | Coil 2 Driver Control | 5°, 55 mph: 8° dwell |
| 4 | --- | Not Used | --- |
| 5 | VT/YL | Speed Control Power Supply | 12-14v |
| 6 | BR/WT | ASD Relay Power (B+) | 12-14v |
| 7 | BR/LB | Injector 3 Driver | 1-4 ms |
| 8 | BR/GY | Generator Field Driver | Digital Signals: 0-12-0v |
| 9 | --- | Not Used | --- |
| 10 | BK/BR | Power Ground | <0.1v |
| 11 | DB/DG | Coil 1 Driver Control | 5°, 55 mph: 8° dwell |
| 12 | VT/GY | Engine Oil Pressure Sensor | 1.6v at 24 psi |
| 13 | BR/YL | Injector 1 Driver | 1-4 ms |
| 14 | BR/VT | Injector 6 Driver | 1-4 ms |
| 15 | BR/OR | Injector 5 Driver | 1-4 ms |
| 16 | BR/TN | Injector 4 Driver | 1-4 ms |
| 17 | BR/DB | Injector 2 Driver | 1-4 ms |
| 18 | BR/LG | HO2S-11 (B1 S1) Heater Control | Heater Off: 12v, On: 1v |
| 19 | --- | Not Used | --- |
| 20 | PK/GY | Ignition Switch Power (B+) | 12-14v |
| 21-24 | --- | Not Used | --- |
| 25 | DB/YL | Knock Sensor Signal | 0.080v AC |
| 26 | VT/OR | ECT Sensor Signal | At 180°F: 2.80v |
| 27 | BR/DG | Oxygen Sensor Ground (Both) | <0.050v |
| 28-29 | --- | Not Used | --- |
| 30 | DB/LB | HO2S-11 (B1 S1) Signal | 0.1-1.1v |
| 31 | DG/OR | Starter Motor Relay Control | Relay Off: 12v, On: 1v |
| 32 | BR/LB | CKP Sensor Signal | Digital Signals: 0-5-0v |
| 33 | DB/GY | CMP Sensor Signal | Digital Signals: 0-5-0v |
| 34 | --- | Not Used | --- |
| 35 | BR/OR | TP Sensor Signal | 0.6-1.0v |
| 36 | VT/BR | MAP Sensor Signal | 1.5-1.6v |
| 37 | DB/LG | Intake Air Temperature Sensor | At 100°F: 1.83v |
| 38-39 | --- | Not Used | --- |
| 40 | DB/VT | EGR Solenoid Control | 12v, at 55 mph: 1v |

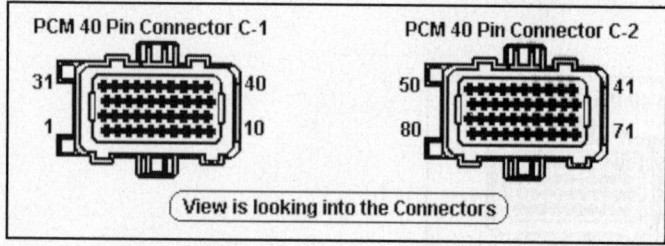

**Pin Connector Graphic**

**2002 Caravan, Voyager 3.0L V6 OHV Flex Fuel VIN 3 'C1' Black Connector**

| PCM Pin # | Wire Color | Circuit Description (40-Pin) | Value at Hot Idle |
|---|---|---|---|
| 1 | --- | Not Used | --- |
| 2 | DB/OR | Coil 3 Driver Control | 5°, 55 mph: 8° dwell |
| 3 | DB/TN | Coil 2 Driver Control | 5°, 55 mph: 8° dwell |
| 4 | --- | Not Used | --- |
| 5 | VT/YL | Speed Control Power Supply | 12-14v |
| 6 | BR/WT | ASD Relay Power (B+) | 12-14v |
| 7 | BR/LB | Injector 3 Driver | 1-4 ms |
| 8 | BR/GY | Generator Field Driver | Digital Signals: 0-12-0v |
| 9 | --- | Not Used | --- |
| 10 | BK/BR | Power Ground | <0.1v |
| 11 | DB/DG | Coil 1 Driver Control | 5°, 55 mph: 8° dwell |
| 12 | VT/GY | Engine Oil Pressure Sensor | 1.6v at 24 psi |
| 13 | BR/YL | Injector 1 Driver | 1-4 ms |
| 14 | BR/VT | Injector 6 Driver | 1-4 ms |
| 15 | BR/OR | Injector 5 Driver | 1-4 ms |
| 16 | BR/TN | Injector 4 Driver | 1-4 ms |
| 17 | BR/DB | Injector 2 Driver | 1-4 ms |
| 18 | BR/LG | HO2S-11 (B1 S1) Heater Control | Heater Off: 12v, On: 1v |
| 19 | --- | Not Used | --- |
| 20 | PK/GY | Ignition Switch Power (B+) | 12-14v |
| 21-24 | --- | Not Used | --- |
| 25 | DB/YL | Knock Sensor Signal | 0.080v AC |
| 26 | VT/OR | ECT Sensor Signal | At 180°F: 2.80v |
| 27 | BR/DG | Oxygen Sensor Ground (Both) | <0.050v |
| 28-29 | --- | Not Used | --- |
| 30 | DB/LB | HO2S-11 (B1 S1) Signal | 0.1-1.1v |
| 31 | DG/OR | Starter Motor Relay Control | Relay Off: 12v, On: 1v |
| 32 | BR/LB | CKP Sensor Signal | Digital Signals: 0-5-0v |
| 33 | DB/GY | CMP Sensor Signal | Digital Signals: 0-5-0v |
| 34 | --- | Not Used | --- |
| 35 | BR/OR | TP Sensor Signal | 0.6-1.0v |
| 36 | VT/BR | MAP Sensor Signal | 1.5-1.6v |
| 37 | DB/LG | Intake Air Temperature Sensor | At 100°F: 1.83v |
| 38-39 | --- | Not Used | --- |
| 40 | DB/VT | EGR Solenoid Control | 12v, at 55 mph: 1v |

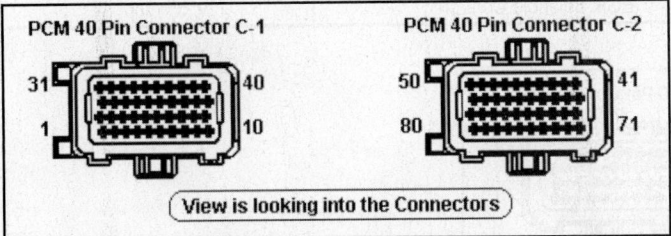

**Pin Connector Graphic**

## 1998-99 Caravan, Town & Country, Voyager 3.3L Flex Fuel VIN G 'C1' Connector

| PCM Pin # | Wire Color | Circuit Description (40-Pin) | Value at Hot Idle |
|---|---|---|---|
| 1 | --- | Not Used | --- |
| 2 | RD/YL | Coil Driver 3 (Cyl 3 & 6) | 5°, 55 mph: 8° dwell |
| 3 | DB/TN | Coil Driver 2 (Cyl 2 & 5) | 5°, 55 mph: 8° dwell |
| 4 | --- | Not Used | --- |
| 5 | YL/RD | Speed Control Power Supply | 12-14v |
| 6 | DG/OR | ASD Relay Output (B+) | 12-14v |
| 7 | WT/DB | Injector 3 Driver Control | 1-4 ms |
| 8 | DG | Generator Field Driver | Digital Signal: 0-12-0v |
| 9 | --- | Not Used | --- |
| 10 | BK/TN | Power Ground | <0.1v |
| 11 | GY/RD | Coil Driver 1 (Cyl 1 & 4) | 5°, 55 mph: 8° dwell |
| 12 | --- | Not Used | --- |
| 13 | YL/WT | Injector 1 Driver Control | 1-4 ms |
| 14 | BR/DB | Injector 6 Driver Control | 1-4 ms |
| 15 | GY | Injector 5 Driver Control | 1-4 ms |
| 16 | LB/BR | Injector 4 Driver Control | 1-4 ms |
| 17 | TN | Injector 2 Driver Control | 1-4 ms |
| 18-19 | --- | Not Used | --- |
| 20 | WT/BK | Ignition Switch Output (B+) | 12-14v |
| 21 | --- | Not Used | --- |
| 22 | BK/PK | Malfunction Indicator Lamp | MIL Off: 12v, On: 1v |
| 23-24 | --- | Not Used | --- |
| 25 | DB/LG | Knock Sensor Signal | 0.080v AC |
| 26 | TN/BK | ECT Sensor Signal | At 180°F: 2.80v |
| 27 | BK/OR | Oxygen Sensor Ground | <0.050v |
| 28-29 | --- | Not Used | --- |
| 30 | BK/DG | HO2S-11 (B1 S1) Signal | 0.1-1.1v |
| 31 | TN | Smart Start Relay Control | KOEC: 9-11v |
| 32 | GY/BK | CKP Sensor Signal | Digital Signal: 0-5-0v |
| 33 | TN/YL | CMP Sensor Signal | Digital Signal: 0-5-0v |
| 34 | --- | Not Used | --- |
| 35 | OR/DB | TP Sensor Signal | 0.6-1.0v |
| 36 | DG/RD | MAP Sensor Signal | 1.5-1.7v |
| 37 | --- | Not Used | --- |
| 38 | DG | A/C On/Off Switch Signal | A/C Off: 12v, On: 1v |
| 39 | --- | Not Used | --- |
| 40 | GY/YL | EGR Solenoid Control | 12v, 55 mph: 1v |

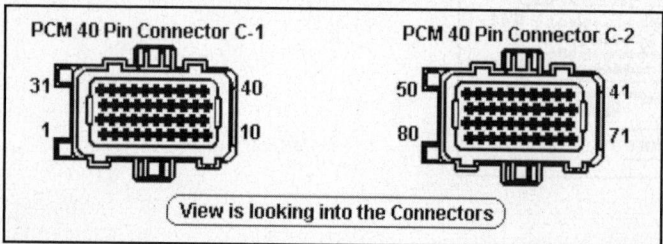

**Pin Connector Graphic**

**1998-99 Caravan, Town & Country, Voyager 3.3L Flex Fuel VIN G 'C2' Connector**

| PCM Pin # | Wire Color | Circuit Description (40-Pin) | Value at Hot Idle |
|---|---|---|---|
| 41 | RD/LG | Speed Control Set Switch Signal | S/C & Set Switch On: 3.8v |
| 42 | DB | A/C Pressure Switch Signal | A/C On: 0.451-4.850v |
| 43 | BK/LB | Sensor Ground | <0.050v |
| 44 | OR | 8-Volt Supply | 7.9-8.1v |
| 45 | DG | A/C Switch Sense | A/C Off: 12v, On: 1v |
| 46 | RD/WT | Fused Battery Power (B+) | 12-14v |
| 47 | --- | Not Used | --- |
| 48 | BR/WT | IAC 3 Driver Control | Pulse Signal: 0.8-11v |
| 49 | YL/BK | IAC 2 Driver Control | Pulse Signal: 0.8-11v |
| 50 | BK/TN | Power Ground | <0.1v |
| 51 | TN/WT | HO2S-12 (B1 S2) Signal | 0.1-1.1v |
| 52-53 | --- | Not Used | --- |
| 54 | DG/LB | Flex Fuel Sensor | Digital Signal |
| 55-56 | --- | Not Used | --- |
| 57 | GY/RD | IAC 1 Driver Control | Pulse Signal: 0.8-11v |
| 58 | VT/BK | IAC 4 Driver Control | Pulse Signal: 0.8-11v |
| 59 | VT/BR | CCD Bus (+) | Digital Signal: 0-5-0v |
| 60 | WT/BK | CCD Bus (-) | <0.050v |
| 61 | VT/WT | 5-Volt Supply | 4.9-5.1v |
| 62 | WT/PK | Brake Switch Sense Signal | Brake Off: 0v, On: 12v |
| 63 | YL/DG | Torque Management Request | Digital Signals |
| 64 | DB/OR | A/C Clutch Relay Control | Relay Off: 12v, On: 1v |
| 65 | PK | SCI Transmit | 0v |
| 66 | WT/OR | Vehicle Speed Signal | Digital Signal |
| 67 | DB/YL | Auto Shutdown Relay Control | Relay Off: 12v, On: 1v |
| 68 | PK/BK | EVAP Purge Solenoid Control | PWM Signal: 0-12-0v |
| 69 | --- | Not Used | --- |
| 70 | DG/LG | EVAP Purge Solenoid Sense | 0-1v |
| 71 | --- | Not Used | --- |
| 72 | DB/WT | LDP Switch Sense Signal | LDP Switch Closed: 0v |
| 73 | LG/DB | Radiator Fan Control Relay | Relay Off: 12v, On: 1v |
| 74 | BR | Fuel Pump Relay Control | Relay Off: 12v, On: 1v |
| 75 | LG | SCI Receive | 5v |
| 76 | BK/WT | PNP Switch Sense Signal | In P/N: 0v, Others: 5v |
| 77 | WT/DG | LDP Solenoid Control | PWM Signal: 0-12-0v |
| 78 | TN/RD | Speed Control Vacuum Solenoid Control | Vacuum Increasing: 1v |
| 79 | --- | Not Used | --- |
| 80 | LG/RD | Speed Control Vent Solenoid Control | Vacuum Decreasing: 1v |

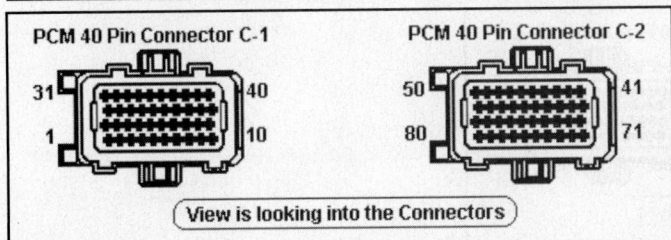

**Pin Connector Graphic**

**2000 Caravan, Town & Country, Voyager 3.3L Flex Fuel VIN G 'C1' Connector**

| PCM Pin # | Wire Color | Circuit Description (40-Pin) | Value at Hot Idle |
|---|---|---|---|
| 1 | --- | Not Used | --- |
| 2 | RD/YL | Coil Driver 3 (Cyl 3 & 6) | 5°, 55 mph: 8° dwell |
| 3 | DB/TN | Coil Driver 2 (Cyl 2 & 5) | 5°, 55 mph: 8° dwell |
| 4 | --- | Not Used | --- |
| 5 | YL/RD | Speed Control Power Supply | 12-14v |
| 6 | DG/OR | ASD Relay Output (B+) | 12-14v |
| 7 | YL/WT | Injector 3 Driver Control | 1-4 ms |
| 8 | DG | Generator Field Driver | Digital Signal: 0-12-0v |
| 9 | --- | Not Used | --- |
| 10 | BK/TN | Power Ground | <0.1v |
| 11 | GY/RD | Coil Driver 1 (Cyl 1 & 4) | 5°, 55 mph: 8° dwell |
| 12 | --- | Not Used | --- |
| 13 | WT/DB | Injector 1 Driver Control | 1-4 ms |
| 14 | BR/DB | Injector 6 Driver Control | 1-4 ms |
| 15 | GY | Injector 5 Driver Control | 1-4 ms |
| 16 | LB/BR | Injector 4 Driver Control | 1-4 ms |
| 17 | TN/WT | Injector 2 Driver Control | 1-4 ms |
| 18-19 | --- | Not Used | --- |
| 20 | WT/BK | Ignition Switch Output (B+) | 12-14v |
| 21 | --- | Not Used | --- |
| 22 | BK/PK | Malfunction Indicator Lamp | MIL Off: 12v, On: 1v |
| 23-24 | --- | Not Used | --- |
| 25 | DB/LG | Knock Sensor Signal | 0.080v AC |
| 26 | TN/BK | ECT Sensor Signal | At 180°F: 2.80v |
| 27 | BK/OR | Oxygen Sensor Ground | <0.050v |
| 28-29 | --- | Not Used | --- |
| 30 | BK/DG | HO2S-11 (B1 S1) Signal | 0.1-1.1v |
| 31 | TN | Smart Start Relay Control | KOEC: 9-11v |
| 32 | GY/BK | CKP Sensor Signal | Digital Signal: 0-5-0v |
| 33 | TN/YL | CMP Sensor Signal | Digital Signal: 0-5-0v |
| 34 | --- | Not Used | --- |
| 35 | OR/BK | TP Sensor Signal | 0.6-1.0v |
| 36 | DG/RD | MAP Sensor Signal | 1.5-1.7v |
| 37 | --- | Not Used | --- |
| 38 | DG/LB | A/C On/Off Switch Signal | A/C Off: 12v, On: 1v |
| 39 | --- | Not Used | --- |
| 40 | GY/YL | EGR Solenoid Control | 12v, 55 mph: 1v |

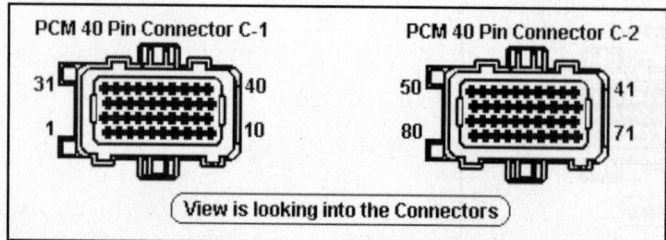

**Pin Connector Graphic**

**2000 Caravan, Town & Country, Voyager 3.3L Flex Fuel VIN G 'C2' Connector**

| PCM Pin # | Wire Color | Circuit Description (40-Pin) | Value at Hot Idle |
|---|---|---|---|
| 41 | RD/LG | Speed Control Set Switch Signal | S/C & Set Switch On: 3.8v |
| 42 | DB | A/C Pressure Switch Signal | A/C On: 0.451-4.850v |
| 43 | BK/LB | Sensor Ground | <0.050v |
| 44 | OR | 8-Volt Supply | 7.9-8.1v |
| 45, 47 | --- | Not Used | --- |
| 46 | RD/WT | Fused Battery Power (B+) | 12-14v |
| 48 | BR/WT | IAC 3 Driver Control | Pulse Signal: 0.8-11v |
| 49 | YL/BK | IAC 2 Driver Control | Pulse Signal: 0.8-11v |
| 50 | BK/TN | Power Ground | <0.1v |
| 51 | TN/WT | HO2S-12 (B1 S2) Signal | 0.1-1.1v |
| 52-53 | --- | Not Used | --- |
| 54 | YL/WT | Flex Fuel Sensor | Digital Signal |
| 55 | --- | Not Used | --- |
| 56 | TN/RD | Speed Control Vacuum Solenoid Control | Vacuum Increasing: 1v |
| 57 | GY/RD | IAC 1 Driver Control | Pulse Signal: 0.8-11v |
| 58 | PK/BK | IAC 4 Driver Control | Pulse Signal: 0.8-11v |
| 59 | PK/BR | CCD Bus (+) | Digital Signal: 0-5-0v |
| 60 | WT/BK | CCD Bus (-) | <0.050v |
| 61 | PK/WT | 5-Volt Supply | 4.9-5.1v |
| 62 | WT/PK | Brake Switch Sense Signal | Brake Off: 0v, On: 12v |
| 63 | YL/DG | Torque Management Request | Digital Signals |
| 64 | DB/OR | A/C Clutch Relay Control | Relay Off: 12v, On: 1v |
| 65 | PK | SCI Transmit | 0v |
| 66 | WT/OR | Vehicle Speed Signal | Digital Signal |
| 67 | DB/YL | Auto Shutdown Relay Control | Relay Off: 12v, On: 1v |
| 68 | PK/BK | EVAP Purge Solenoid Control | PWM Signal: 0-12-0v |
| 69 | --- | Not Used | --- |
| 70 | PK/RD | EVAP Purge Solenoid Sense | 0-1v |
| 71 | --- | Not Used | --- |
| 72 | DB/WT | LDP Switch Sense Signal | LDP Switch Closed: 0v |
| 73 | LG/DB | Radiator Fan Control Relay | Relay Off: 12v, On: 1v |
| 74 | BR | Fuel Pump Relay Control | Relay Off: 12v, On: 1v |
| 75 | LG | SCI Receive | 5V |
| 76 | BR/YL | PNP Switch Sense Signal | In P/N: 0v, Others: 5v |
| 77 | WT/DG | LDP Solenoid Control | PWM Signal: 0-12-0v |
| 78-79 | --- | Not Used | --- |
| 80 | LG/RD | Speed Control Vent Solenoid Control | Vacuum Decreasing: 1v |

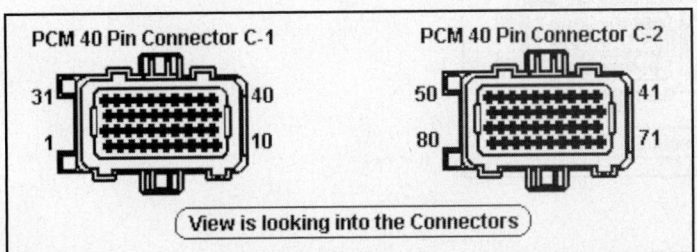

**Pin Connector Graphic**

## 2001-02 Caravan, Town & Country, Voyager 3.3L Flex Fuel VIN G A/T 'C1' Connector

| PCM Pin # | Wire Color | Circuit Description (40-Pin) | Value at Hot Idle |
|---|---|---|---|
| 1 | --- | Not Used | --- |
| 2 | DB/OR | Coil 3 Driver Control | 5°, 55 mph: 8° dwell |
| 3 | DB/TN | Coil 2 Driver Control | 5°, 55 mph: 8° dwell |
| 4 | --- | Not Used | --- |
| 5 | VT/YL | Speed Control On/Off Switch Sense | 12-14v |
| 6 | BR/WT | ASD Relay Power (B+) | 12-14v |
| 7 | BR/LB | Injector 3 Driver | 1-4 ms |
| 8 | BR/GY | Generator Field Driver | Digital Signal: 0-12-0v |
| 9 | --- | Not Used | --- |
| 10 | BK/TN | Power Ground | <0.1v |
| 11 | DB/DG | Coil 1 Driver Control | 5°, 55 mph: 8° dwell |
| 12 | VT/GY | Engine Oil Pressure Sensor | 1.6v at 24 psi |
| 13 | BR/YL | Injector 1 Driver Control | 1-4 ms |
| 14 | BR/VT | Injector 6 Driver Control | 1-4 ms |
| 15 | BR/OR | Injector 5 Driver Control | 1-4 ms |
| 16 | BR/TN | Injector 4 Driver Control | 1-4 ms |
| 17 | BR/DB | Injector 2 Driver Control | 1-4 ms |
| 18 | BR/LG | HO2S-11 (B1 S1) Heater | PWM Signal: 0-12-0v |
| 19 | --- | Not Used | --- |
| 20 | PK/GY | Ignition Switch Power (B+) | 12-14v |
| 21-24 | --- | Not Used | --- |
| 25 | DB/YL | Knock Sensor Signal | 0.080v AC |
| 26 | VT/OR | ECT Sensor Signal | At 180°F: 2.80v |
| 27 | BR/DG | Oxygen Sensor Ground | <0.050v |
| 28-29 | --- | Not Used | --- |
| 30 | DB/LB | HO2S-11 (B1 S1) Signal | 0.1-1.1v |
| 31 | DG/OR | Double Start Override Signal | KOEC: 9-11v |
| 32 | BR/LB | CKP Sensor Signal | Digital Signal: 0-5-0v |
| 33 | DB/GY | CMP Sensor Signal | Digital Signal: 0-5-0v |
| 34 | --- | Not Used | --- |
| 35 | BR/OR | TP Sensor Signal | 0.6-1.0v |
| 36 | VT/BR | MAP Sensor Signal | 1.5-1.6v |
| 37 | DB/LG | Intake Air Temp. Signal | At 100°F: 1.83v |
| 38-39 | --- | Not Used | --- |
| 40 | DB/VT | EGR Solenoid Control | 12v, at 55 mph: 1v |

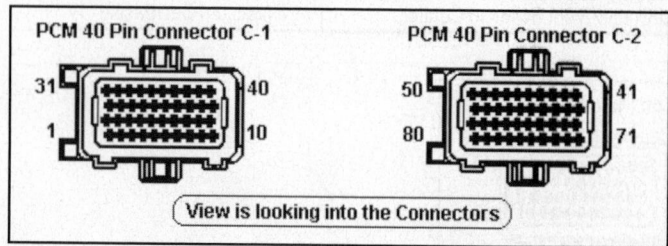

PCM 40 Pin Connector C-1    PCM 40 Pin Connector C-2

31 [ ] 40    50 [ ] 41
1 [ ] 10    80 [ ] 71

View is looking into the Connectors

**Pin Connector Graphic**

**2001-02 Caravan, Town & Country, Voyager 3.3L Flex Fuel VIN G A/T 'C2' Connector**

| PCM Pin # | Wire Color | Circuit Description (40-Pin) | Value at Hot Idle |
|---|---|---|---|
| 41 | VT | Speed Control Set Switch Signal | S/C & Set Switch On: 3.8v |
| 42 | LB/BR | A/C Pressure Switch Signal | A/C On: 0.451-4.850v |
| 43 | DB/DG | Sensor Ground | <0.050v |
| 44 | BR/PK | 8-Volt Supply | 7.9-8.1v |
| 45 | --- | Not Used | --- |
| 46 | OR/RD | Keep Alive Power (B+) | 12-14v |
| 47-48 | --- | Not Used | --- |
| 49 | VT/DG | IAC 1 Driver Control | Pulse Signals |
| 50 | BK/DG | Power Ground | <0.1v |
| 51 | DB/YL | HO2S-12 (B1 S2) Signal | 0.1-1.1v |
| 52-53 | --- | Not Used | --- |
| 54 | YL/WT | Flex Fuel Sensor | Digital Signals |
| 56 | VT/YL | Speed Control Vacuum Solenoid | Vacuum Increasing: 1v |
| 57 | VT/LG | IAC 2 Driver Control | Pulse Signals |
| 58 | --- | Not Used | --- |
| 59 | WT/VT | PCI Data Bus (J1850) | Digital Signals: 0-7-0v |
| 60 | --- | Not Used | --- |
| 61 | PK/YL | 5-Volt Supply | 4.9-5.1v |
| 62 | DG/WT | Brake Switch Sense Signal | Brake Off: 0v, On: 12v |
| 63 | DG/LG | Torque Management Request | Digital Signals |
| 64 | LB/OR | A/C Clutch Relay Control | Relay Off: 12v, On: 1v |
| 65 | WT/BR | SCI Transmit Signal | 0v |
| 66 | DB/OR | Vehicle Speed Signal | Digital Signal |
| 67 | BR/WT | ASD Relay Control | Relay Off: 12v, On: 1v |
| 68 | DB/WT | EVAP Purge Solenoid Control | PWM Signal: 0-12-0v |
| 69 | --- | Not Used | --- |
| 70 | DB/BR | EVAP Purge Solenoid Sense | 0-1v |
| 71 | --- | Not Used | --- |
| 72 | VT/WT | LDP Switch Sense Signal | LDP Switch Closed: 0v |
| 73 | BR/VT | Radiator Fan Control Relay | Relay Off: 12v, On: 1v |
| 74 | BR | Fuel Pump Relay Control | Relay Off: 12v, On: 1v |
| 75 | WT/LG | SCI Receive Signal | 5v |
| 76 | YL/LB | TRS T41 Signal | In P/N: 0v, Others: 5v |
| 77 | VT/LB | LDP Solenoid Control | PWM Signal: 0-12-0v |
| 78-79 | --- | Not Used | --- |
| 80 | LG/RD | Speed Control Vent Solenoid Control | Vacuum Decreasing: 1v |

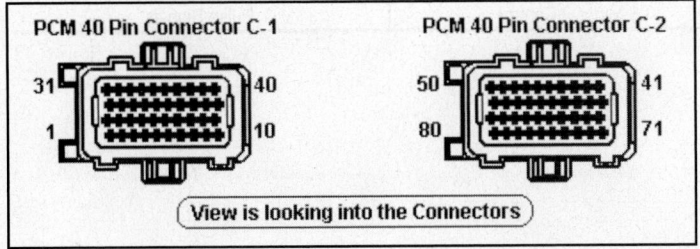

PCM 40 Pin Connector C-1
31 — 40
1 — 10

PCM 40 Pin Connector C-2
50 — 41
80 — 71

**View is looking into the Connectors**

**Pin Connector Graphic**

### 1994-95 Caravan, Town & Country, Voyager 3.3L CNG VIN J (A/T) 60-Pin Connector

| PCM Pin # | Wire Color | Circuit Description (60-Pin) | Value at Hot Idle |
|---|---|---|---|
| 1 | DG/RD | MAP Sensor Signal | 1.5-1.6v |
| 2 | TN/BK | ECT Sensor Signal | At 180°F: 2.80v |
| 3 | RD/WT | Keep Alive Power (B+) | 12-14v |
| 4, 5 | BK/LB, BK | Sensor Ground | <0.050v |
| 6 | PK/WT | 5-Volt Supply | 4.9-5.1v |
| 7 | OR | 8-Volt Supply | 7.9-8.1v |
| 8, 10 | --- | Not Used | --- |
| 9 | DB/WT | Ignition Switch Power (B+) | 12-14v |
| 11, 12 | BK/TN | Power Ground | <0.1v |
| 13 | PK or LB/BR | Injector 4 Driver Control | 1-4 ms |
| 14 | PK or YL/WT | Injector 3 Driver Control | 1-4 ms |
| 15 | PK/DB or Tan | Injector 2 Driver Control | 1-4 ms |
| 16 | PK or LB | Injector 1 Driver Control | 1-4 ms |
| 17 | DB/TN | Coil 2 Driver Control | 5°, 55 mph: 8° dwell |
| 18 | DB/GY | Coil 3 Driver Control | 5°, 55 mph: 8° dwell |
| 19 | BK/GY | Coil 1 Driver Control | 5°, 55 mph: 8° dwell |
| 20 | DG | Alternator Field Control | Digital Signals: 0-12-0v |
| 21 | TN/PK | Fuel Temperature Sensor | 0.5-4.5v |
| 22 | DB | TP Sensor Signal | 0.6-1.0v |
| 23 | RD | Speed Control Set Switch Signal | S/C & Set Switch On: 3.8v |
| 24 | GY/P | CKP Sensor Signal | Digital Signals: 0-5-0v |
| 25 | PK, LG | SCI Transmit, SCI Receive | 0v, 5v |
| 26, 46 | PK, WT/BK | CCD Bus (+), CCD Bus (-) Signal | Digital Signals: 0-5-0v, <0.050v |
| 27 | BR | A/C Switch Sense Signal | A/C Off: 12v, On: 1v |
| 28, 32, 35-37, 43 | --- | Not Used | --- |
| 29 | WT/PK | Brake Switch Sense Signal | Brake Off: 0v, On: 12v |
| 30 | BK/OR | PNP Switch Sense Signal | In P/N: 0v, Others: 5v |
| 31 | DB/PK | Low Speed Fan Control | Relay Off: 12v, On: 1v |
| 33 | TN/RD | Speed Control Vacuum Solenoid | Vacuum Increasing: 1v |
| 34 | DB/OR | A/C Clutch Relay Control | Relay Off: 12v, On: 1v |
| 38 | PK or GY | Injector 5 Driver Control | 1-4 ms |
| 39 | GY/RD | AIS Motor 3 Control | Pulse Signals |
| 40 | BR/WT | AIS Motor 1 Control | Pulse Signals |
| 41 | DG/WT | HO2S-11 (B1 S1) Signal | 0.1-1.1v |
| 42 | DB/LB | Fuel Low Pressure Sensor | 0.5-4.5v |
| 44 | TN/YL | CMP Sensor Signal | Digital Signals: 0-5-0v |
| 47 | WT/OR | Vehicle Speed Signal | Digital Signals |
| 48-50, 52, 54, 56 | --- | Not Used | --- |
| 51 | DB/YL | ASD Relay Control | Relay Off: 12v, On: 1v |
| 53 | LG/RD | Speed Control Vent Solenoid | Vacuum Decreasing: 1v |
| 55 | YL | High Speed Fan Control | Relay Off: 12v, On: 1v |
| 57 | DG/OR | ASD Relay Power (B+) | 12-14v |
| 58 | PK/TN | Injector 6 Driver Control | 1-4 ms |
| 58 ('95) | BR/DB | Injector 6 Driver Control | 1-4 ms |
| 59 | PK/BK | AIS Motor 4 Control | Pulse Signals |
| 60 | YL/BK | AIS Motor 2 Control | Pulse Signals |

## 1994-95 Caravan & Grand Caravan 3.3L VIN R (A/T) 60-Pin Connector

| PCM Pin # | Wire Color | Circuit Description (60-Pin) | Value at Hot Idle |
|---|---|---|---|
| 1 | DG/RD | MAP Sensor Signal | 1.5-1.6v |
| 2 | TN/BK | ECT Sensor Signal | At 180°F: 2.80v |
| 3 | RD/WT | Keep Alive Power (B+) | 12-14v |
| 4 | BK/LB | Sensor Ground | <0.050v |
| 5 | BK/WT | Sensor Ground | <0.050v |
| 6 | PK/WT | 5-Volt Supply | 4.9-5.1v |
| 7 | OR | 8-Volt Supply | 7.9-8.1v |
| 8 | --- | Not Used | --- |
| 9 | DB/WT | Ignition Switch Power (B+) | 12-14v |
| 10 | --- | Not Used | --- |
| 11 | BK/TN | Power Ground | <0.1v |
| 12 | BK/TN | Power Ground | <0.1v |
| 13 | LB/BR | Injector 4 Driver Control | 1-4 ms |
| 14 | YL/WT | Injector 3 Driver Control | 1-4 ms |
| 15 | TN | Injector 2 Driver Control | 1-4 ms |
| 16 | LB | Injector 1 Driver Control | 1-4 ms |
| 17 | DB/YL | Coil 2 Driver Control | 5°, 55 mph: 8° dwell |
| 18 | DB/GY | Coil 3 Driver Control | 5°, 55 mph: 8° dwell |
| 19 | BK/GY | Coil 1 Driver Control | 5°, 55 mph: 8° dwell |
| 20 | DG | Alternator Field Control | Digital Signals: 0-12-0v |
| 21 | --- | Not Used | --- |
| 22 | DB | TP Sensor Signal | 0.6-1.0v |
| 23 | RD | Speed Control Set Switch Signal | S/C & Set Switch On: 3.8v |
| 24 | GY/PK | CKP Sensor Signal | Digital Signals: 0-5-0v |
| 25 | PK | SCI Transmit | 0v |
| 26 | PK/BR | CCD Bus (+) | Digital Signals: 0-5-0v |
| 27 | BR | A/C Damped Pressure Switch | A/C Off: 12v, On: 1v |
| 28 | --- | Not Used | --- |
| 29 | WT/PK | Brake Switch Sense Signal | Brake Off: 0v, On: 12v |
| 30 | BK/OR | PNP Switch Sense Signal | In P/N: 0v, Others: 5v |
| 31 | DB/PK | Low Speed Fan Control | Relay Off: 12v, On: 1v |
| 32 | --- | Not Used | --- |
| 33 | TN/RD | Speed Control Vacuum Solenoid | Vacuum Increasing: 1v |
| 34 | DB/OR | A/C Clutch Relay Control | Relay Off: 12v, On: 1v |
| 35 | GY/YL | EGR Solenoid Control | Solenoid Off: 12v, On: 1v |
| 36-37 | --- | Not Used | --- |
| 38 | GY | Injector 5 Driver Control | 1-4 ms |
| 39 | GY/RD | AIS Motor Control (open) | Pulse Signals |
| 40 | BR/WT | AIS Motor Control (close) | Pulse Signals |

## Standard Colors and Abbreviations

| Abbreviation | Color | Abbreviation | Color | Abbreviation | Color |
|---|---|---|---|---|---|
| BK | Black | GY | Gray | RD | Red |
| BL | Blue | GN | Green | TN | Tan |
| BR | Brown | LG | Light Green | VT | Violet |
| DB | Dark Blue | OR | Orange | WT | White |
| DG | Dark Green | PK | Pink | YL | Yellow |

**1994-95 Caravan & Grand Caravan 3.3L VIN R (A/T) 60-Pin Connector - Continued**

| PCM Pin # | Wire Color | Circuit Description (60-Pin) | Value at Hot Idle |
|---|---|---|---|
| 41 | DG/WT | HO2S-11 (B1 S1) Signal | 0.1-1.1v |
| 42-43 | --- | Not Used | --- |
| 44 | TN/YL | CMP Sensor Signal | Digital Signals: 0-5-0v |
| 45 | LG | SCI Receive | 5v |
| 46 | WT/BK | CCD Bus (-) | <0.050v |
| 47 | WT/OR | Vehicle Speed Signal | Digital Signal |
| 48-50 | --- | Not Used | --- |
| 51 | DB/YL | ASD Relay Control | Relay Off: 12v, On: 1v |
| 52 | PK/BK | EVAP Purge Solenoid Control | Solenoid Off: 12v, On: 1v |
| 53 | LG/RD | Speed Control Vent Solenoid | Vacuum Decreasing: 1v |
| 54 | --- | Not Used | --- |
| 55 | YL | High Speed Fan Control | Relay Off: 12v, On: 1v |
| 56 | --- | Not Used | --- |
| 57 | DG/OR | ASD Relay Power (B+) | 12-14v |
| 58 | BR/DB | Injector 6 Driver Control | 1-4 ms |
| 59 | PK/BK | AIS Motor Control (close) | Pulse Signals |
| 60 | YL/BK | AIS Motor Control (close) | Pulse Signals |

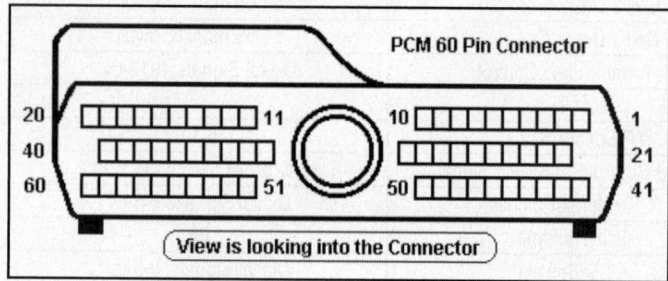

**Pin Connector Graphic**

**Standard Colors and Abbreviations**

| Abbreviation | Color | Abbreviation | Color | Abbreviation | Color |
|---|---|---|---|---|---|
| BK | Black | GY | Gray | RD | Red |
| BL | Blue | GN | Green | TN | Tan |
| BR | Brown | LG | Light Green | VT | Violet |
| DB | Dark Blue | OR | Orange | WT | White |
| DG | Dark Green | PK | Pink | YL | Yellow |

**1996 Caravan, Town & Country, Voyager 3.3L VIN R 'C1' Connector**

| PCM Pin # | Wire Color | Circuit Description (40-Pin) | Value at Hot Idle |
|---|---|---|---|
| 1 | --- | Not Used | --- |
| 2 | RD/YL | Coil Driver 3 (Cyl 3 & 6) | 5°, 55 mph: 8° dwell |
| 3 | DB/TN | Coil Driver 2 (Cyl 2 & 5) | 5°, 55 mph: 8° dwell |
| 4 | DG | Generator Field Driver | Digital Signal: 0-12-0v |
| 5 | YL/RD | Speed Control Power Supply | 12-14v |
| 6 | DG/OR | ASD Relay Output (B+) | 12-14v |
| 7 | YL/WT | Injector 3 Driver Control | 1-4 ms |
| 8 | TN | Smart Start Relay Control | KOEC: 9-11v |
| 9 | --- | Not Used | --- |
| 10 | BK/TN | Power Ground | <0.1v |
| 11 | GY/RD | Coil Driver 1 (Cyl 1 & 4) | 5°, 55 mph: 8° dwell |
| 12 | --- | Not Used | --- |
| 13 | WT/DB | Injector 1 Driver Control | 1-4 ms |
| 14 | BR/DB | Injector 6 Driver Control | 1-4 ms |
| 15 | GY | Injector 5 Driver Control | 1-4 ms |
| 16 | LB/BR | Injector 4 Driver Control | 1-4 ms |
| 17 | TN | Injector 2 Driver Control | 1-4 ms |
| 18-19 | --- | Not Used | --- |
| 20 | WT/BK | Ignition Switch Output (B+) | 12-14v |
| 21-25 | --- | Not Used | --- |
| 26 | TN/BK | ECT Sensor Signal | At 180°F: 2.80v |
| 27-29 | --- | Not Used | --- |
| 30 | BK/DG | HO2S-11 (B1 S1) Signal | 0.1-1.1v |
| 31 | --- | Not Used | --- |
| 32 | GY/BK | CKP Sensor Signal | Digital Signal: 0-5-0v |
| 33 | TN/YL | CMP Sensor Signal | Digital Signal: 0-5-0v |
| 34 | --- | Not Used | --- |
| 35 | OR/DB | TP Sensor Signal | 0.6-1.0v |
| 36 | DG/RD | MAP Sensor Signal | 1.5-1.7v |
| 37-39 | --- | Not Used | --- |
| 40 | GY/YL | EGR Solenoid Control | 12v, 55 mph: 1v |

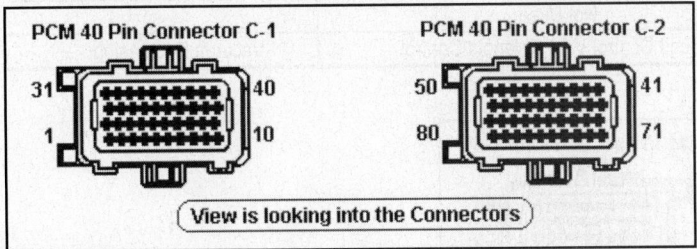

**Pin Connector Graphic**

## 1996 Caravan, Town & Country, Voyager 3.3L VIN R 'C2' Connector

| PCM Pin # | Wire Color | Circuit Description (40-Pin) | Value at Hot Idle |
|---|---|---|---|
| 41 | RD/LG | Speed Control Set Switch Signal | S/C & Set Switch On: 3.8v |
| 42 | DB | A/C Pressure Switch Signal | A/C On: 0.451-4.850v |
| 43 | BK/LB | Sensor Ground | <0.050v |
| 44 | OR | 8-Volt Supply | 7.9-8.1v |
| 45 | DB | A/C Switch Sense Signal | A/C Off: 12v, On: 1v |
| 46 | RD/WT | Fused Battery Power (B+) | 12-14v |
| 47 | --- | Not Used | --- |
| 48 | BR/WT | IAC 1 Driver Control | Pulse Signal: 0.8-11v |
| 49 | YL/BK | IAC 2 Driver Control | Pulse Signal: 0.8-11v |
| 50 | BK/TN | Power Ground | <0.1v |
| 52-56 | --- | Not Used | --- |
| 51 | TN/WT | HO2S-12 (B1 S2) Signal | 0.1-1.1v |
| 57 | GY/RD | IAC 3 Driver Control | Pulse Signal: 0.8-11v |
| 58 | PK/BK | IAC 4 Driver Control | Pulse Signal: 0.8-11v |
| 59 | PK/BR | CCD Bus (+) | Digital Signal: 0-5-0v |
| 60 | WT/BK | CCD Bus (-) | <0.050v |
| 61 | PK/WT | 5-Volt Supply | 4.9-5.1v |
| 62 | WT/PK | Brake Switch Sense Signal | Brake Off: 0v, On: 12v |
| 63 | YL/DG | Torque Management Request | Digital Signals |
| 64 | DB/OR | A/C Clutch Relay Control | Relay Off: 12v, On: 1v |
| 65 | PK | SCI Transmit | 0v |
| 66 | WT/OR | Vehicle Speed Signal | Digital Signal |
| 67 | DB/YL | Auto Shutdown Relay Control | Relay Off: 12v, On: 1v |
| 68 | PK/BK | EVAP Purge Solenoid Control | PWM Signal: 0-12-0v |
| 69-71 | --- | Not Used | --- |
| 72 | YL/BK | LDP Switch Sense Signal | LDP Switch Closed: 0v |
| 73 | LG/DB | Radiator Fan Control Relay | Relay Off: 12v, On: 1v |
| 74 | BR | Fuel Pump Relay Control | Relay Off: 12v, On: 1v |
| 75 | LG | SCI Receive | 5v |
| 76 | BK/WT | PNP Switch Sense Signal | In P/N: 0v, Others: 5v |
| 77 | WT/DG | LDP Solenoid Control | PWM Signal: 0-12-0v |
| 78 | TN/RD | Speed Control Vacuum Solenoid Control | Vacuum Increasing: 1v |
| 79 | --- | Not Used | --- |
| 80 | LG/RD | Speed Control Vent Solenoid Control | Vacuum Decreasing: 1v |

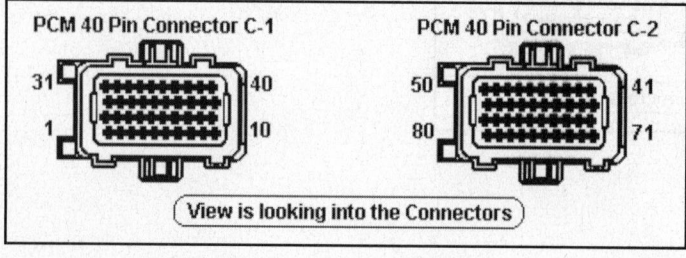

**Pin Connector Graphic**

**1997-98 Caravan, Town & Country, Voyager 3.3L VIN R 'C1' Connector**

| PCM Pin # | Wire Color | Circuit Description (40-Pin) | Value at Hot Idle |
|---|---|---|---|
| 1 | --- | Not Used | --- |
| 2 | RD/YL | Coil Driver 3 (Cyl 3 & 6) | 5°, 55 mph: 8° dwell |
| 3 | DB/TN | Coil Driver 2 (Cyl 2 & 5) | 5°, 55 mph: 8° dwell |
| 4 | DG | Generator Field Driver | Digital Signal: 0-12-0v |
| 5 | YL/RD | Speed Control Power Supply | 12-14v |
| 6 | DG/OR | ASD Relay Output (B+) | 12-14v |
| 7 | YL/WT | Injector 3 Driver Control | 1-4 ms |
| 8 | TN | Smart Start Relay Control | KOEC: 9-11v |
| 9 | --- | Not Used | --- |
| 10 | BK/TN | Power Ground | <0.1v |
| 11 | GY/RD | Coil Driver 1 (Cyl 1 & 4) | 5°, 55 mph: 8° dwell |
| 12 | --- | Not Used | --- |
| 13 | WT/DB | Injector 1 Driver Control | 1-4 ms |
| 14 | BR/DB | Injector 6 Driver Control | 1-4 ms |
| 15 | GY | Injector 5 Driver Control | 1-4 ms |
| 16 | LB/BR | Injector 4 Driver Control | 1-4 ms |
| 17 | TN | Injector 2 Driver Control | 1-4 ms |
| 18-19 | --- | Not Used | --- |
| 20 | WT/BK | Ignition Switch Output (B+) | 12-14v |
| 21, 23 | --- | Not Used | --- |
| 22 | BK/PK | Malfunction Indicator Lamp | MIL Off: 12v, On: 1v |
| 24 | DB/LG | Knock Sensor Signal | No knock: 2.5v DC |
| 25 | --- | Not Used | --- |
| 26 | TN/BK | ECT Sensor Signal | At 180°F: 2.80v |
| 27-29 | --- | Not Used | --- |
| 30 | BK/DG | HO2S-11 (B1 S1) Signal | 0.1-1.1v |
| 31 | --- | Not Used | --- |
| 32 | GY/BK | CKP Sensor Signal | Digital Signal: 0-5-0v |
| 33 | TN/YL | CMP Sensor Signal | Digital Signal: 0-5-0v |
| 34 | --- | Not Used | --- |
| 35 | OR/DB | TP Sensor Signal | 0.6-1.0v |
| 36 | DG/RD | MAP Sensor Signal | 1.5-1.7v |
| 37-39 | --- | Not Used | --- |
| 40 | GY/YL | EGR Solenoid Control | 12v, 55 mph: 1v |

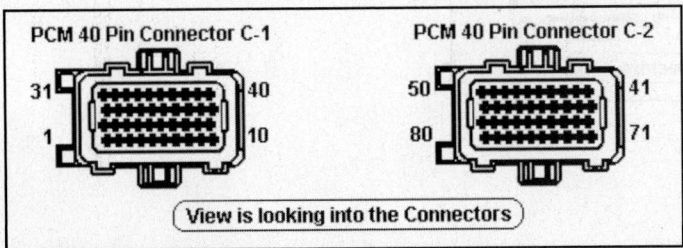

**Pin Connector Graphic**

**1997-98 Caravan, Town & Country, Voyager 3.3L VIN R 'C2' Connector**

| PCM Pin # | Wire Color | Circuit Description (40-Pin) | Value at Hot Idle |
|---|---|---|---|
| 41 | RD/LG | Speed Control Set Switch Signal | S/C & Set Switch On: 3.8v |
| 42 | DB | A/C Pressure Switch Signal | A/C On: 0.451-4.850v |
| 43 | BK/LB | Sensor Ground | <0.050v |
| 44 | OR | 8-Volt Supply | 7.9-8.1v |
| 45 | DG | A/C Switch Sense Signal | A/C Off: 12v, On: 1v |
| 46 | RD/WT | Fused Battery Power (B+) | 12-14v |
| 47 | --- | Not Used | --- |
| 48 | BR/WT | IAC 3 Driver Control | Pulse Signal: 0.8-11v |
| 49 | YL/BK | IAC 2 Driver Control | Pulse Signal: 0.8-11v |
| 50 | BK/TN | Power Ground | <0.1v |
| 51 | TN/WT | HO2S-12 (B1 S2) Signal | 0.1-1.1v |
| 52-56 | --- | Not Used | --- |
| 57 | GY/RD | IAC 1 Driver Control | Pulse Signal: 0.8-11v |
| 58 | PK/BK | IAC 4 Driver Control | Pulse Signal: 0.8-11v |
| 59 | PK/BR | CCD Bus (+) | Digital Signal: 0-5-0v |
| 60 | WT/BK | CCD Bus (-) | <0.050v |
| 61 | PK/WT | 5-Volt Supply | 4.9-5.1v |
| 62 | WT/PK | Brake Switch Sense Signal | Brake Off: 0v, On: 12v |
| 63 | YL/DG | Torque Management Request | Digital Signals |
| 64 | DB/OR | A/C Clutch Relay Control | Relay Off: 12v, On: 1v |
| 65 | PK | SCI Transmit | 0v |
| 66 | WT/OR | Vehicle Speed Signal | Digital Signal |
| 67 | DB/YL | Auto Shutdown Relay Control | Relay Off: 12v, On: 1v |
| 68 | PK/BK | EVAP Purge Solenoid Control | PWM Signal: 0-12-0v |
| 69-71 | --- | Not Used | --- |
| 72 | DB/WT | LDP Switch Sense Signal | LDP Switch Closed: 0v |
| 73 | LG/DG | Radiator Fan Control Relay | Relay Off: 12v, On: 1v |
| 74 | BR | Fuel Pump Relay Control | Relay Off: 12v, On: 1v |
| 75 | LG | SCI Receive | 5v |
| 76 | BK/Y | PNP Switch Sense Signal | In P/N: 0v, Others: 5v |
| 77 | WT/DG | LDP Solenoid Control | PWM Signal: 0-12-0v |
| 78 | TN/RD | Speed Control Vacuum Solenoid Control | Vacuum Increasing: 1v |
| 79 | --- | Not Used | --- |
| 80 | LG/RD | Speed Control Vent Solenoid Control | Vacuum Decreasing: 1v |

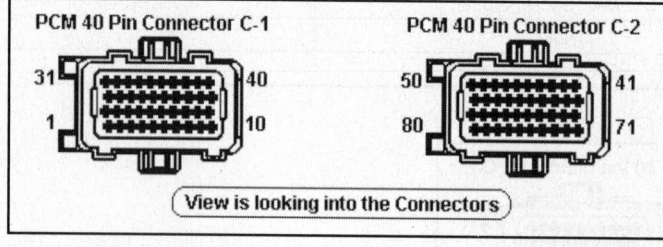

**Pin Connector Graphic**

**1999-2000 Caravan, Town & Country, Voyager 3.3L VIN R 'C1' Connector**

| PCM Pin # | Wire Color | Circuit Description (40-Pin) | Value at Hot Idle |
|---|---|---|---|
| 1, 4 | --- | Not Used | --- |
| 2 | RD/YL | Coil Driver 3 (Cyl 3 & 6) | 5°, 55 mph: 8° dwell |
| 3 | DB/TN | Coil Driver 2 (Cyl 2 & 5) | 5°, 55 mph: 8° dwell |
| 5 | YL/RD | Speed Control Power Supply | 12-14v |
| 6 | DG/OR | ASD Relay Output (B+) | 12-14v |
| 7 | YL/WT | Injector 3 Driver Control | 1-4 ms |
| 8 | DG | Generator Field Driver | Digital Signal: 0-12-0v |
| 9 | --- | Not Used | --- |
| 10 | BK/TN | Power Ground | <0.1v |
| 11 | GY/RD | Coil Driver 1 (Cyl 1 & 4) | 5°, 55 mph: 8° dwell |
| 12 | --- | Not Used | --- |
| 13 | WT/DB | Injector 1 Driver Control | 1-4 ms |
| 14 | BR/DB | Injector 6 Driver Control | 1-4 ms |
| 15 | GY | Injector 5 Driver Control | 1-4 ms |
| 16 | LB/BR | Injector 4 Driver Control | 1-4 ms |
| 17 | TN/WT | Injector 2 Driver Control | 1-4 ms |
| 18-19 | --- | Not Used | --- |
| 20 | WT/BK | Ignition Switch Output (B+) | 12-14v |
| 21, 23 | --- | Not Used | --- |
| 22 | BK/PK | Malfunction Indicator Lamp | MIL Off: 12v, On: 1v |
| 24 | --- | Not Used | --- |
| 25 | DB/LG | Knock Sensor Signal | No knock: 2.5v DC |
| 26 | TN/BK | ECT Sensor Signal | At 180°F: 2.80v |
| 27 | BK/OR | HO2S-11 Ground | <0.1v |
| 28-29 | --- | Not Used | --- |
| 30 | BK/DG | HO2S-11 (B1 S1) Signal | 0.1-1.1v |
| 31 | TN | Smart Start Relay Control | KOEC: 9-11v |
| 32 | GY/BK | CKP Sensor Signal | Digital Signal: 0-5-0v |
| 33 | TN/YL | CMP Sensor Signal | Digital Signal: 0-5-0v |
| 34 | --- | Not Used | --- |
| 35 | OR/DB | TP Sensor Signal | 0.6-1.0v |
| 36 | DG/RD | MAP Sensor Signal | 1.5-1.7v |
| 37, 39 | --- | Not Used | --- |
| 38 | DG/LB | A/C Switch Sense Signal | A/C Off: 12v, On: 1v |
| 40 | GY/YL | EGR Solenoid Control | 12v, 55 mph: 1v |

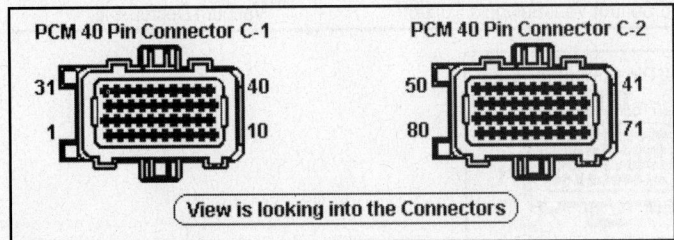

**Pin Connector Graphic**

**1999-2000 Caravan, Town & Country, Voyager 3.3L VIN R 'C2' Connector**

| PCM Pin # | Wire Color | Circuit Description (40-Pin) | Value at Hot Idle |
|---|---|---|---|
| 41 | RD/LG | Speed Control Set Switch Signal | S/C & Set Switch On: 3.8v |
| 42 | DB | A/C Pressure Switch Signal | A/C On: 0.451-4.850v |
| 43 | BK/LB | Sensor Ground | <0.050v |
| 44 | OR | 8-Volt Supply | 7.9-8.1v |
| 45 | --- | Not Used | --- |
| 46 | RD/WT | Fused Battery Power (B+) | 12-14v |
| 47 | --- | Not Used | --- |
| 48 | BR/WT | IAC 3 Driver Control | Pulse Signal: 0.8-11v |
| 49 | YL/BK | IAC 2 Driver Control | Pulse Signal: 0.8-11v |
| 50 | BK/TN | Power Ground | <0.1v |
| 51 | TN/WT | HO2S-12 (B1 S2) Signal | 0.1-1.1v |
| 52-55 | --- | Not Used | --- |
| 56 | TN/RD | Speed Control Vacuum Solenoid Control | Vacuum Increasing: 1v |
| 57 | GY/RD | IAC 1 Driver Control | Pulse Signal: 0.8-11v |
| 58 | PK/BK | IAC 4 Driver Control | Pulse Signal: 0.8-11v |
| 59 | PK/BR | CCD Bus (+) | Digital Signal: 0-5-0v |
| 60 | WT/BK | CCD Bus (-) | <0.050v |
| 61 | PK/WT | 5-Volt Supply | 4.9-5.1v |
| 62 | WT/PK | Brake Switch Sense Signal | Brake Off: 0v, On: 12v |
| 63 | YL/DG | Torque Management Request | Digital Signals |
| 64 | DB/OR | A/C Clutch Relay Control | Relay Off: 12v, On: 1v |
| 65 | PK | SCI Transmit Signal | 0v |
| 66 | WT/OR | Vehicle Speed Signal | Digital Signal |
| 67 | DB/YL | Auto Shutdown Relay Control | Relay Off: 12v, On: 1v |
| 68 | PK/BK | EVAP Purge Solenoid Control | PWM Signal: 0-12-0v |
| 69 | --- | Not Used | --- |
| 70 ('99) | DG/LG | EVAP Purge Solenoid Sense | 0-1v |
| 70 | PK/RD | EVAP Purge Solenoid Sense | 0-1v |
| 71 | --- | Not Used | --- |
| 72 | DB/WT | LDP Switch Sense Signal | LDP Switch Closed: 0v |
| 73 | LG/DG | Radiator Fan Control Relay | Relay Off: 12v, On: 1v |
| 74 | BR | Fuel Pump Relay Control | Relay Off: 12v, On: 1v |
| 75 | LG | SCI Receive Signal | 5v |
| 76 | BK/Y | PNP Switch Sense Signal | In P/N: 0v, Others: 5v |
| 77 | WT/DG | LDP Solenoid Control | PWM Signal: 0-12-0v |
| 78-79 | --- | Not used | --- |
| 80 | LG/RD | Speed Control Vent Solenoid Control | Vacuum Decreasing: 1v |

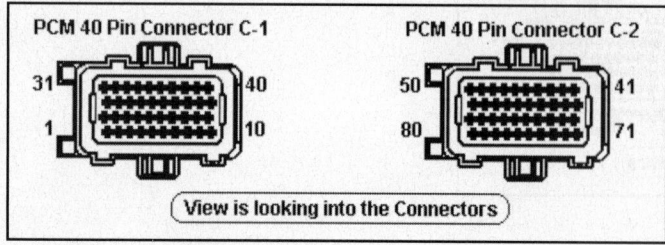

**Pin Connector Graphic**

## 2001-03 Caravan, Town & Country, Voyager 3.3L VIN R 'C1' Black Connector

| PCM Pin # | Wire Color | Circuit Description (38-Pin) | Value at Hot Idle |
|---|---|---|---|
| 1-8, 10 | --- | Not Used | --- |
| 9 | BK/BR | Power Ground | <0.1v |
| 11 | PK/GY | Ignition Switch Output (Run-Start) | 12-14v |
| 12 | PK/WT | Ignition Switch Output (Off-Run-Start) | 12-14v |
| 13-17, 19 | --- | Not Used | --- |
| 18 | BK/DG | Power Ground | <0.1v |
| 20 | VT/GY | Oil Pressure Sensor | A/C On: 0.45-4.85v |
| 21, 23-24 | --- | Not Used | --- |
| 22 | VT/LG | Ambient Air Temperature Sensor | At 100°F: 1.83v |
| 25 | WT/LG | SCI Receive (PCM) | 5v |
| 26 | WT/BR | Flash Program Enable | 5v |
| 27-28 | --- | Not Used | --- |
| 29 | OR/RD | Battery Power (B+) | 12-14v |
| 30 | YL | Ignition Switch Output (Start) | 8-11v (cranking) |
| 31 | DB/YL | HO2S-12 (B1 S2) Signal | 0.1-1.1v |
| 32 | DB/DG | Oxygen Sensor Return (down) | <0.050v |
| 33-35 | --- | Not Used | --- |
| 36 | WT/BR | SCI Transmit (PCM) | 0v |
| 37 | DG/YL | SCI Transmit (TCM) | 0v |
| 38 | WT/VT | PCI Data Bus (J1850) | Digital Signals: 0-7-0v |

## 2001-03 Caravan, Town & Country, Voyager 3.3L VIN R 'C2' Gray Connector

| PCM Pin # | Wire Color | Circuit Description (38-Pin) | Value at Hot Idle |
|---|---|---|---|
| 1-8 | --- | Not Used | --- |
| 9 | DB/TN | Coil 2 Driver | 5°, at 55 mph: 8° dwell |
| 10 | DB/DG | Coil 1 Driver | 5°, at 55 mph: 8° dwell |
| 11 | BR/TN | Injector 4 Driver Control | 1.0-4.0 ms |
| 12 | BR/LB | Injector 3 Driver Control | 1.0-4.0 ms |
| 13 | BR/DB | Injector 2 Driver Control | 1.0-4.0 ms |
| 14 | BR/YL | Injector 1 Driver Control | 1.0-4.0 ms |
| 15-17 | --- | Not Used | --- |
| 18 | BR/LG | HO2S-12 (B1 S2) Heater Control | Heater Off: 12v, On: 1v |
| 19 | BR/GY | Generator Field Driver | Digital Signal: 0-12-0v |
| 20 | VT/OR | ECT Sensor Signal | At 180°F: 2.80v |
| 21 | BR/OR | TP Sensor Signal | 0.6-1.0v |
| 22 | --- | Not Used | --- |
| 23 | VT/BR | MAP Sensor Signal | 1.5-1.7v |
| 24 | BR/LG | Knock Sensor Return | <0.050v |
| 25 | DB/YL | Knock Sensor Signal | 0.080v AC |
| 26 | --- | Not Used | --- |
| 27 | DB/DG | Sensor Ground | <0.1v |
| 28 | BR/VT | IAC Motor Sense | 12-14v |
| 29 | PK/YL | 5-Volt Supply | 4.9-5.1v |
| 30 | DB/LG | IAT Sensor Signal | At 100°F: 1.83v |
| 31 | DB/LG | HO2S-11 (B1 S1) Signal | 0.1-1.1v |
| 32 | BR/DG | HO2S-11 (B1 S1) Ground (Upstream) | <0.1v |
| 33 | --- | Not Used | --- |
| 34 | DB/GY | CMP Sensor Signal | Digital Signal: 0-5-0v |
| 35 | BR/LB | CKP Sensor Signal | Digital Signal: 0-5-0v |
| 36-37 | --- | Not Used | --- |
| 38 | GY/RD | IAC Motor Driver | DC pulse signals: 0.8-11v |

### 2001-03 Caravan, Town & Country, Voyager 3.3L VIN R 'C3' White Connector

| PCM Pin # | Wire Color | Circuit Description (38-Pin) | Value at Hot Idle |
|---|---|---|---|
| 1-2, 4 | --- | Not Used | --- |
| 3 | BR/WT | ASD Relay Control | Relay Off: 12v, On: 1v |
| 5 | VT/OR | Speed Control Vent Solenoid | Vacuum Decreasing: 1v |
| 6 | BR/VT | Radiator Fan Relay Control | Relay Off: 12v, On: 1v |
| 7 | VT/YL | Speed Control Power Supply | 12-14v |
| 8 | WT/DG | Natural Vacuum Leak Detection Solenoid | Solenoid Off: 12v, On: 1v |
| 9 | BR/WT | HO2S-12 (B1 S2) Heater Control | Heater On: 1v, Off: 12v |
| 10 | --- | Not Used | --- |
| 11 | LB/OR | A/C Clutch Relay Control | Relay Off: 12v, On: 1v |
| 12 | VT/YL | Speed Control Vacuum Solenoid | Vacuum Increasing: 1v |
| 13-18 | --- | Not Used | --- |
| 19 | BR/WT | Automatic Shutdown Relay Output | 12-14v |
| 20 | DB/WT | EVAP Purge Solenoid Control | PWM Signal: 0-12-0v |
| 21-22 | --- | Not Used | --- |
| 23 | DG/WT | Brake Switch Sense | Brake Off: 0v, On: 12v |
| 24-27 | --- | Not Used | --- |
| 28 | BR/WT | Automatic Shutdown Relay Output | 12-14v |
| 29 | DB/BR | EVAP Purge Solenoid Sense | 0-1v |
| 30, 33, 36 | --- | Not Used | --- |
| 31 | LB/BR | A/C Pressure Sensor | A/C On: 0.451-4.850v |
| 32 | DB/YL | Battery Temperature Sensor | At 100°F: 1.83v |
| 34 | VT | Speed Control Set Switch Signal | S/C & Set Switch On: 3.8v |
| 35 | VT/WT | Natural Vacuum Leak Detection Switch Sense | 0.1v |
| 37 | BR | Fuel Pump Relay Control | Relay Off: 12v, On: 1v |
| 38 | DG/OR | Starter Relay Control | KOEC: 9-11v |

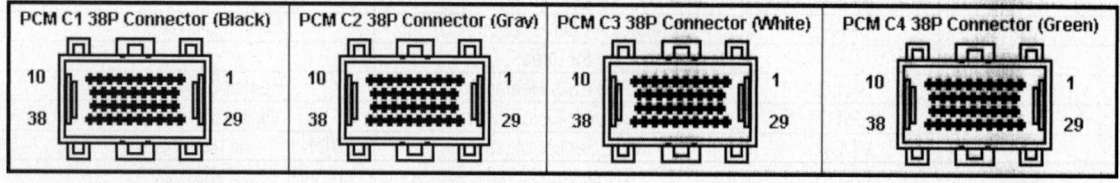

**Pin Connector Graphics**

**2001-03 Caravan, Town & Country, Voyager 3.3L VIN R 'C4' Green Connector**

| PCM Pin # | Wire Color | Circuit Description (38-Pin) | Value at Hot Idle |
|---|---|---|---|
| 1 | YL/GY | Overdrive Solenoid Control | Solenoid Off: 12v, On: 1v |
| 2 | YL/LB | Underdrive Solenoid Control | Solenoid Off: 12v, On: 1v |
| 3-5 | --- | Not Used | --- |
| 6 | YL/DB | 2-4 Solenoid Control | Solenoid Off: 12v, On: 1v |
| 7-9, 11 | --- | Not Used | --- |
| 10 | DG/WT | Low/Reverse Solenoid Control | Solenoid Off: 12v, On: 1v |
| 12, 13 | BK/LG | Power Ground | <0.050v |
| 14 | --- | Not Used | --- |
| 15 | DG/LB | TRS T1 Sense | <0.050v |
| 16 | DG/DB | TRS T3 Sense | <0.050v |
| 17, 20-21 | --- | Not Used | --- |
| 18 | YL/BR | Transmission Control Relay Control | Relay Off: 12v, On: 1v |
| 19, 38 | YL/OR | Transmission Control Relay Output | Relay Off: 12v, On: 1v |
| 23-26 | --- | Not Used | --- |
| 27 | DG/GY | TRS T41 Sense | <0.050v |
| 28, 38 | YL/OR | Transmission Control Relay Output | Relay Off: 12v, On: 1v |
| 29 | YL/TN | Low/Reverse Pressure Switch Sense | 12-14v |
| 30 | YL/DG | 2-4 Pressure Switch Sense | In Low/Reverse: 2-4v |
| 31, 36 | --- | Not Used | --- |
| 32 | DG/BR | Output Speed Sensor Signal | In 2-4 Position: 2-4v |
| 33 | DG/WT | Input Speed Sensor Signal | Moving: AC voltage |
| 34 | DG/VT | Speed Sensor Ground | Moving: AC voltage |
| 35 | DG/OR | Transmission Temperature Sensor Signal | <0.050v |
| 37 | DG/YL | TRS T42 Sense | In PRNL: 0v, Others 5v |

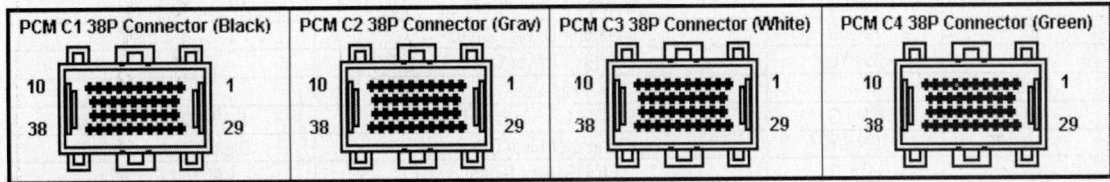

**Pin Connector Graphic**

**2001-03 Caravan, Town & Country, Voyager 3.3L CNG VIN 3 (A/T) 'C1' Black Connector**

| PCM Pin # | Wire Color | Circuit Description (40-Pin) | Value at Hot Idle |
|---|---|---|---|
| 1-8 | --- | Not Used | --- |
| 9 | BK/BR | Power Ground | <0.1v |
| 10 | --- | Not Used | --- |
| 11 | PK/GY | Ignition Switch Output (Start-Run) | 12-14v |
| 12 | PK/WT | Ignition Switch Output (Run/Start) | 12-14v |
| 13-17 | --- | Not Used | --- |
| 18 | BK/DG | Power Ground | <0.1v |
| 19, 21 | --- | Not Used | --- |
| 20 | VT/GY | Engine Oil Pressure Sensor | 1.6v at 24 psi |
| 22 | VT/LG | Ambient Air Temperature Sensor | At 100°F: 1.83v |
| 23-24 | --- | Not Used | --- |
| 25 | WT/LG | SCI Receive Signal | 5v |
| 26 | WT/BK | Flash Program Enable | --- |
| 27-28 | --- | Not Used | --- |
| 29 | OR/RD | Fused B+ | 12-14v |
| 30 | YL | Ignition Switch Output (Start) | 12-14v |
| 31 | DB/YL | HO2S-12 (B1 S2) Signal | 0.1-1.1v |
| 32 | DB/DG | Oxygen Sensor Ground | <0.050v |
| 33-35 | --- | Not Used | --- |
| 36 | WT/BR | SCI Transmit (PCM) | 0v |
| 37 | DG/YL | SCI Transmit (TCM) | 0v |
| 38 | WT/VT | PCI Data Bus (J1850) | Digital Signals: 0-7-0v |

**2001-03 Caravan, Town & Country, Voyager 3.3L CNG VIN 3 (A/T) 'C2' Black Connector**

| PCM Pin # | Wire Color | Circuit Description (40-Pin) | Value at Hot Idle |
|---|---|---|---|
| 1-8 | --- | Not Used | --- |
| 9 | DB/TN | Coil 2 Driver | 5°, 55 mph: 8° dwell |
| 10 | DB/DG | Coil 1 Driver | 5°, 55 mph: 8° dwell |
| 11 | BR/TN | Injector 4 Driver Control | 1-4 ms |
| 12 | BR/LB | Injector 3 Driver Control | 1-4 ms |
| 13 | BR/DB | Injector 2 Driver Control | 1-4 ms |
| 14 | BR/YL | Injector 1 Driver Control | 1-4 ms |
| 15-17 | --- | Not Used | --- |
| 18 | BR/LG | HO2S-11 (B1 S1) Heater Control | PWM Signal: 0-12-0v |
| 19 | BR/GY | Generator Field Driver | Digital Signal: 0-12-0v |
| 20 | VT/OR | ECT Sensor Signal | At 180°F: 2.80v |
| 21 | BR/OR | TP Sensor Signal | 0.6-1.0v |
| 22, 26 | --- | Not Used | --- |
| 23 | VT/BR | MAP Sensor Signal | 1.5-1.6v |
| 24 | BR/LG | Knock Sensor Return | <0.050v |
| 25 | DB/YL | Knock Sensor Signal | 0.080v AC |
| 27 | DB/DG | Sensor Ground | <0.1v |
| 28 | BR/VT | IAC Motor Return | <0.050v |
| 29 | PK/YL | 5-Volt Supply | 4.9-5.1v |
| 30 | DB/LG | IAT Sensor Signal | At 100°F: 1.83v |
| 31 | DB/LG | HO2S-11 (B1 S1) Sensor | 0.1-1.1v |
| 32 | BR/DG | HO2S-11 (B1 S1) Ground (Up) | <0.050v |
| 33, 36-37 | --- | Not Used | --- |
| 34 | BR/LB | CKP Sensor Signal | Digital Signal: 0-5-0v |
| 35 | DB/GY | CMP Sensor Signal | Digital Signal: 0-5-0v |
| 38 | VT/GY | IAC 1 Motor Control | Pulse Signals |

## 2001-03 Caravan, Town & Country, Voyager 3.3L CNG VIN 3 (A/T) 'C3' Black Connector

| PCM Pin # | Wire Color | Circuit Description (38-Pin) | Value at Hot Idle |
|---|---|---|---|
| 1-2 | --- | Not Used | --- |
| 3 | BR/WT | ASD Relay Control | Relay Off: 12v, On: 1v |
| 4 | --- | Not Used | --- |
| 5 | VT/OR | Speed Control Vent Solenoid | Vacuum Decreasing: 1v |
| 6 | BR/VT | Radiator Fan Relay Control | Relay Off: 12v, On: 1v |
| 7 | VT/YL | Speed Control Power Supply | 12-14v |
| 8 | WT/DG | Natural Vacuum Leak Detection Solenoid | Solenoid Off: 12v, On: 1v |
| 9 | BR/WT | HO2S-12 (Bank 1 Sensor 2) Heater Control | --- |
| 11 | LB/OR | A/C Clutch Relay Control | Relay Off: 12v, On: 1v |
| 12 | VT/YL | Speed Control Vacuum Solenoid | Vacuum Increasing: 1v |
| 13-18 | --- | Not Used | --- |
| 19 | BR/WT | Automatic Shutdown Relay Output | 12-14v |
| 20 | DB/WT | EVAP Purge Solenoid Control | PWM Signal: 0-12-0v |
| 21 | --- | Not Used | --- |
| 23 | DG/WT | Brake Switch Sense | Brake Off: 0v, On: 12v |
| 24-27 | --- | Not Used | --- |
| 28 | BR/WT | Automatic Shutdown Relay Output | 12-14v |
| 29 | DB/BR | EVAP Purge Solenoid Sense | 0-1v |
| 30 | --- | Not Used | --- |
| 31 | LB/BR | A/C Pressure Signal | A/C On: 0.451-4.850v |
| 32 | DB/YL | Battery Temperature Sensor | At 100°F: 1.83v |
| 33, 36 | --- | Not Used | --- |
| 34 | VT | Speed Control Set Switch Signal | S/C & Set Switch On: 3.8v |
| 35 | VT/WT | Natural Vacuum Leak Detection Switch Sense | 0.1v |
| 37 | BR | Fuel Pump Relay Control | Relay Off: 12v, On: 1v |
| 38 | DG/OR | Starter Relay Control | KOEC: 9-11v |

PCM C1 38P Connector (Black)  PCM C2 38P Connector (Gray)  PCM C3 38P Connector (White)  PCM C4 38P Connector (Green)

**Pin Connector Graphic**

### Standard Colors and Abbreviations

| Abbreviation | Color | Abbreviation | Color | Abbreviation | Color |
|---|---|---|---|---|---|
| BK | Black | GY | Gray | RD | Red |
| BL | Blue | GN | Green | TN | Tan |
| BR | Brown | LG | Light Green | VT | Violet |
| DB | Dark Blue | OR | Orange | WT | White |
| DG | Dark Green | PK | Pink | YL | Yellow |

## 2001-03 Caravan, Town & Country, Voyager 3.3L CNG VIN 3 (A/T) 'C4' Green Connector

| PCM Pin # | Wire Color | Circuit Description (38-Pin) | Value at Hot Idle |
|---|---|---|---|
| 1 | YL/LG | Overdrive Solenoid Control | Solenoid Off: 12v, On: 1v |
| 2 | YL/LB | Underdrive Solenoid Control | Solenoid Off: 12v, On: 1v |
| 3-5 | --- | Not Used | --- |
| 6 | YL/DB | 2-4 Solenoid Control | Solenoid Off: 12v, On: 1v |
| 7-9 | --- | Not Used | --- |
| 10 | DG/WT | Low/Reverse Solenoid Control | Solenoid Off: 12v, On: 1v |
| 11 | --- | Not Used | --- |
| 12 | BK/LG | Power Ground | <0.050v |
| 13 | --- | Not Used | --- |
| 14 | BK/LG | Power Ground | <0.050v |
| 15 | DG/LG | TRS T1 Sense | <0.050v |
| 16 | DG/DB | TRS T3 Sense | <0.050v |
| 17 | --- | Not Used | --- |
| 18 | YL/BR | Transmission Control Relay Control | Relay Off: 12v, On: 1v |
| 19 | YL/OR | Transmission Control Relay Output | 12-14v |
| 20-21 | --- | Not Used | --- |
| 23-26 | --- | Not Used | --- |
| 27 | DG/GY | TRS T41 Sense | <0.050v |
| 28 | YL/OR | Transmission Control Relay Output | 12-14v |
| 29 | YL/TN | Low/Reverse Pressure Switch Sense | 12-14v |
| 30 | YL/DG | 2-4 Pressure Switch Sense | In Low/Reverse: 2-4v |
| 31 | --- | Not Used | --- |
| 32 | DG/BR | Output Speed Sensor Signal | In 2-4 Position: 2-4v |
| 33 | DG/WT | Input Speed Sensor Signal | Moving: AC voltage |
| 34 | DG/VT | Speed Sensor Ground | Moving: AC voltage |
| 35 | DG/OR | Transmission Temperature Sensor Signal | 3.2-3.4v at 104ºF |
| 36 | --- | Not Used | --- |
| 37 | DG/YL | TRS T42 Sense | In PRNL: 0v, Others 5v |
| 38 | YL/OR | Transmission Control Relay Control | Relay Off: 12v; On: 1v |

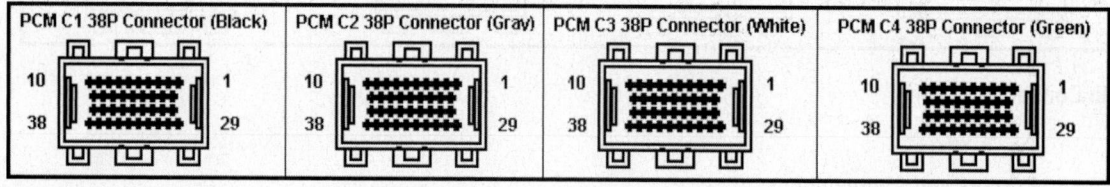

**Pin Connector Graphic**

## Standard Colors and Abbreviations

| Abbreviation | Color | Abbreviation | Color | Abbreviation | Color |
|---|---|---|---|---|---|
| BK | Black | GY | Gray | RD | Red |
| BL | Blue | GN | Green | TN | Tan |
| BR | Brown | LG | Light Green | VT | Violet |
| DB | Dark Blue | OR | Orange | WT | White |
| DG | Dark Green | PK | Pink | YL | Yellow |

## 2001-03 Caravan, Town & Country, Voyager 3.3L VIN 3 'C1' Black Connector

| PCM Pin # | Wire Color | Circuit Description (38-Pin) | Value at Hot Idle |
|---|---|---|---|
| 1-8, 10 | --- | Not Used | --- |
| 9 | BK/BR | Power Ground | <0.1v |
| 11 | PK/GY | Ignition Switch Output (Run-Start) | 12-14v |
| 12 | PK/WT | Ignition Switch Output (Off-Run-Start) | 12-14v |
| 13-17, 19 | --- | Not Used | --- |
| 18 | BK/DG | Power Ground | <0.1v |
| 20 | VT/GY | Oil Pressure Sensor | A/C On: 0.45-4.85v |
| 21, 23-24 | --- | Not Used | --- |
| 22 | VT/LG | Ambient Air Temperature Sensor | At 100°F: 1.83v |
| 25 | WT/LG | SCI Receive (PCM) | 5v |
| 26 | WT/BR | Flash Program Enable | 5v |
| 27-28 | --- | Not Used | --- |
| 29 | OR/RD | Battery Power (B+) | 12-14v |
| 30 | YL | Ignition Switch Output (Start) | 8-11v (cranking) |
| 31 | DB/YL | HO2S-12 (B1 S2) Signal | 0.1-1.1v |
| 32 | DB/DG | Oxygen Sensor Return (down) | <0.050v |
| 33-35 | --- | Not Used | --- |
| 36 | WT/BR | SCI Transmit (PCM) | 0v |
| 37 | DG/YL | SCI Transmit (TCM) | 0v |
| 38 | WT/VT | PCI Data Bus (J1850) | Digital Signals: 0-7-0v |

## 2001-03 Caravan, Town & Country, Voyager 3.3L VIN 3 'C2' Gray Connector

| PCM Pin # | Wire Color | Circuit Description (38-Pin) | Value at Hot Idle |
|---|---|---|---|
| 1-8 | --- | Not Used | --- |
| 9 | DB/TN | Coil 2 Driver | 5°, at 55 mph: 8° dwell |
| 10 | DB/DG | Coil 1 Driver | 5°, at 55 mph: 8° dwell |
| 11 | BR/TN | Injector 4 Driver Control | 1.0-4.0 ms |
| 12 | BR/LB | Injector 3 Driver Control | 1.0-4.0 ms |
| 13 | BR/DB | Injector 2 Driver Control | 1.0-4.0 ms |
| 14 | BR/YL | Injector 1 Driver Control | 1.0-4.0 ms |
| 15-17 | --- | Not Used | --- |
| 18 | BR/LG | HO2S-12 (B1 S2) Heater Control | Heater Off: 12v, On: 1v |
| 19 | BR/GY | Generator Field Driver | Digital Signal: 0-12-0v |
| 20 | VT/OR | ECT Sensor Signal | At 180°F: 2.80v |
| 21 | BR/OR | TP Sensor Signal | 0.6-1.0v |
| 22 | --- | Not Used | --- |
| 23 | VT/BR | MAP Sensor Signal | 1.5-1.7v |
| 24 | BR/LG | Knock Sensor Return | <0.050v |
| 25 | DB/YL | Knock Sensor Signal | 0.080v AC |
| 26 | --- | Not Used | --- |
| 27 | DB/DG | Sensor Ground | <0.1v |
| 28 | BR/VT | IAC Motor Sense | 12-14v |
| 29 | PK/YL | 5-Volt Supply | 4.9-5.1v |
| 30 | DB/LG | IAT Sensor Signal | At 100°F: 1.83v |
| 31 | DB/LB | HO2S-11 (B1 S1) Signal | 0.1-1.1v |
| 32 | BR/DG | HO2S-11 (B1 S1) Ground (Upstream) | <0.1v |
| 33 | --- | Not Used | --- |
| 34 | DB/GY | CMP Sensor Signal | Digital Signal: 0-5-0v |
| 35 | BR/LB | CKP Sensor Signal | Digital Signal: 0-5-0v |
| 36-37 | --- | Not Used | --- |
| 38 | VT/GY | IAC Motor Driver | DC pulse signals: 0.8-11v |

## 2001-03 Caravan, Town & Country, Voyager 3.3L VIN 3 'C3' White Connector

| PCM Pin # | Wire Color | Circuit Description (38-Pin) | Value at Hot Idle |
|---|---|---|---|
| 1-2, 4 | --- | Not Used | --- |
| 3 | BR/WT | ASD Relay Control | Relay Off: 12v, On: 1v |
| 5 | VT/OR | Speed Control Vent Solenoid | Vacuum Decreasing: 1v |
| 6 | BR/VT | Radiator Fan Relay Control | Relay Off: 12v, On: 1v |
| 7 | VT/YL | Speed Control Power Supply | 12-14v |
| 8 | WT/DG | Natural Vacuum Leak Detection Solenoid | Solenoid Off: 12v, On: 1v |
| 9 | BR/WT | HO2S-12 (B1 S2) Heater Control | Heater On: 1v, Off: 12v |
| 10 | --- | Not Used | --- |
| 11 | LB/OR | A/C Clutch Relay Control | Relay Off: 12v, On: 1v |
| 12 | VT/YL | Speed Control Vacuum Solenoid | Vacuum Increasing: 1v |
| 13-18 | --- | Not Used | --- |
| 18, 19 | BR/WT | Automatic Shutdown Relay Output | 12-14v |
| 20 | DB/WT | EVAP Purge Solenoid Control | PWM Signal: 0-12-0v |
| 21-22 | --- | Not Used | --- |
| 23 | DG/WT | Brake Switch Sense | Brake Off: 0v, On: 12v |
| 24-27 | --- | Not Used | --- |
| 28 | BR/WT | Automatic Shutdown Relay Output | 12-14v |
| 29 | DB/BR | EVAP Purge Solenoid Sense | 0-1v |
| 30-31, 33, 36 | --- | Not Used | --- |
| 31 | LB/BR | A/C Pressure Sensor | A/C On: 0.451-4.850v |
| 32 | DB/YL | Battery Temperature Sensor | At 100°F: 1.83v |
| 34 | VT | Speed Control Set Switch Signal | S/C & Set Switch On: 3.8v |
| 35 | VT/WT | Natural Vacuum Leak Detection Switch Sense | 0.1v |
| 37 | BR | Fuel Pump Relay Control | Relay Off: 12v, On: 1v |
| 38 | DG/OR | Starter Relay Control | KOEC: 9-11v |

## 2001-03 Caravan, Town & Country, Voyager 3.3L VIN 3 'C4' Green Connector

| PCM Pin # | Wire Color | Circuit Description (38-Pin) | Value at Hot Idle |
|---|---|---|---|
| 1 | YL/GY | Overdrive Solenoid Control | Solenoid Off: 12v, On: 1v |
| 2 | YL/LB | Underdrive Solenoid Control | Solenoid Off: 12v, On: 1v |
| 3-5 | --- | Not Used | --- |
| 6 | YL/LB | 2-4 Solenoid Control | Solenoid Off: 12v, On: 1v |
| 7-9, 11 | --- | Not Used | --- |
| 10 | DG/WT | Low/Reverse Solenoid Control | Solenoid Off: 12v, On: 1v |
| 12, 13 | BK/LG | Power Ground | <0.050v |
| 14 | --- | Not Used | --- |
| 15 | DG/LB | TRS T1 Sense | <0.050v |
| 16 | DG/DB | TRS T3 Sense | <0.050v |
| 17, 20-21 | --- | Not Used | --- |
| 18 | YL/BR | Transmission Control Relay Control | Relay Off: 12v, On: 1v |
| 19 | YL/OR, 38 | Transmission Control Relay Output | Relay Off: 12v, On: 1v |
| 23-26 | --- | Not Used | --- |
| 27 | DG/GY | TRS T41 Sense | <0.050v |
| 28, 38 | YL/OR | Transmission Control Relay Output | Relay Off: 12v, On: 1v |
| 29 | YL/TN | Low/Reverse Pressure Switch Sense | 12-14v |
| 30 | YL/DG | 2-4 Pressure Switch Sense | In Low/Reverse: 2-4v |
| 31, 36 | --- | Not Used | --- |
| 32 | DG/BR | Output Speed Sensor Signal | In 2-4 Position: 2-4v |
| 33 | DG/WT | Input Speed Sensor Signal | Moving: AC voltage |
| 34 | DG/VT | Speed Sensor Ground | Moving: AC voltage |
| 35 | DG/OR | Transmission Temperature Sensor Signal | <0.050v |
| 37 | DG/YL | TRS T42 Sense | In PRNL: 0v, Others 5v |

## 1994-95 Caravan, Town & Country, Voyager 3.8L VIN L 60-Pin Connector

| PCM Pin # | Wire Color | Circuit Description (60-Pin) | Value at Hot Idle |
|---|---|---|---|
| 1 | DG/RD | MAP Sensor Signal | 1.5-1.7v |
| 2 | TN/BK | ECT Sensor Signal | At 180°F: 2.80v |
| 3 | RD/WT | Fused Battery Power (B+) | 12-14v |
| 4 | BK/LB | Sensor Ground | <0.050v |
| 5 | BK/WT | Sensor Ground | <0.050v |
| 6 | PK/WT | 5-Volt Supply | 4.9-5.1v |
| 7 | OR | 8-Volt Supply | 7.9-8.1v |
| 9 | DB/WT | Ignition Switch Output (B+) | 12-14v |
| 11 | BK/TN | Power Ground | <0.1v |
| 12 | BK/TN | Power Ground | <0.1v |
| 13 | LB/BR | Injector Control 4 Driver | 1-4 ms |
| 14 | YL/WT | Injector Control 3 Driver | 1-4 ms |
| 15 | TN | Injector Control 2 Driver | 1-4 ms |
| 16 | LB | Injector Control 1 Driver | 1-4 ms |
| 17 | DB/YL | Coil Driver 2 Control | 5°, 55 mph: 8° dwell |
| 18 | DB/GY | Coil Driver 3 Control | 5°, 55 mph: 8° dwell |
| 19 | BK/GY | Coil Driver 1 Control | 5°, 55 mph: 8° dwell |
| 20 | DG | Alternator Field Control | Digital Signal: 0-12-0v |
| 22 | DB | TP Sensor Signal | 0.6-1.0v |
| 23 | RD | Speed Control Set Switch Signal | S/C & Set Switch On: 3.8v |
| 24 | GY/PK | CKP Sensor Signal | Digital Signal: 0-5-0v |
| 25 | PK | SCI Transmit | 0v |
| 26 | PK/BR | CCD Bus (+) | Digital Signal: 0-5-0v |
| 27 | BR | A/C Damped Pressure Switch | A/C Off: 12v, On: 1v |
| 29 | WT/PK | Brake Switch Sense Signal | Brake Off: 0v, On: 12v |
| 30 | BK/OR | PNP Switch Sense Signal | In P/N: 0v, Others: 5v |
| 31 | DB/PK | Low Speed Fan Control | Relay Off: 12v, On: 1v |
| 32 | --- | Not Used | --- |
| 33 | TN/RD | Speed Control Vacuum Solenoid Control | Vacuum Increasing: 1v |
| 34 | DB/OR | A/C Clutch Relay Control | Relay Off: 12v, On: 1v |
| 35 | GY/YL | EGR Solenoid Control | 12v, 55 mph: 1v |
| 36-37 | --- | Not Used | --- |
| 38 | GY | Injector Control 5 Driver | 1-4 ms |
| 39 | GY/RD | AIS Motor 3 Control | Pulse Signal: 0.8-11v |
| 40 | BR/WT | AIS Motor 1 Control | Pulse Signal: 0.8-11v |

**Pin Connector Graphic**

## Standard Colors and Abbreviations

| Abbreviation | Color | Abbreviation | Color | Abbreviation | Color |
|---|---|---|---|---|---|
| BK | Black | GY | Gray | RD | Red |
| BL | Blue | GN | Green | TN | Tan |
| BR | Brown | LG | Light Green | VT | Violet |
| DB | Dark Blue | OR | Orange | WT | White |
| DG | Dark Green | PK | Pink | YL | Yellow |

**1994-95 Caravan, Town & Country, Voyager 3.8L VIN L 60-Pin Connector - Continued**

| PCM Pin # | Wire Color | Circuit Description (60-Pin) | Value at Hot Idle |
|---|---|---|---|
| 41 | BK/DG | HO2S-11 (B1 S1) Signal | 0.1-1.1v |
| 42-43 | --- | Not Used | --- |
| 44 | TN/YL | CMP Sensor Signal | Digital Signal: 0-5-0v |
| 45 | LG | SCI Receive | 5v |
| 46 | WT/BK | CCD Bus (-) | <0.050v |
| 47 | WT/OR | Vehicle Speed Signal | Digital Signal |
| 50 | --- | Not Used | --- |
| 51 | DB/YL | ASD Relay Output (B+) | 12-14v |
| 52 | PK/BK | EVAP Purge Solenoid Control | Solenoid Off: 12v, On: 1v |
| 53 | LG/RD | Speed Control Vent Solenoid Control | Vacuum Decreasing: 1v |
| 55 | Y | High Speed Fan Control | Relay Off: 12v, On: 1v |
| 56 | --- | Not Used | --- |
| 57 | DG/OR | Fuel Injector Power Supply | 12-14v |
| 58 | BR/DB | Injector 6 Driver Control | 1-4 ms |
| 59 | PK/BK | AIS Motor 4 Control | Pulse Signal: 0.8-11v |
| 60 | YL/BK | AIS Motor 2 Control | Pulse Signal: 0.8-11v |

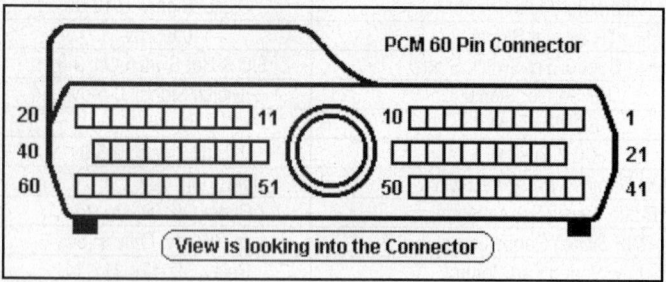

**Pin Connector Graphic**

**Standard Colors and Abbreviations**

| Abbreviation | Color | Abbreviation | Color | Abbreviation | Color |
|---|---|---|---|---|---|
| BK | Black | GY | Gray | RD | Red |
| BL | Blue | GN | Green | TN | Tan |
| BR | Brown | LG | Light Green | VT | Violet |
| DB | Dark Blue | OR | Orange | WT | White |
| DG | Dark Green | PK | Pink | YL | Yellow |

## 1996 Caravan, Town & Country, Voyager 3.8L VIN L 'C1' Connector

| PCM Pin # | Wire Color | Circuit Description (40-Pin) | Value at Hot Idle |
|---|---|---|---|
| 1 | --- | Not Used | --- |
| 2 | RD/YL | Coil Driver 3 (Cyl 3 & 6) | 5°, 55 mph: 8° dwell |
| 3 | DB/TN | Coil Driver 2 (Cyl 2 & 5) | 5°, 55 mph: 8° dwell |
| 4 | DG | Generator Field Driver | Digital Signal: 0-12-0v |
| 5 | YL/RD | Speed Control Power Supply | 12-14v |
| 6 | DG/OR | ASD Relay Output (B+) | 12-14v |
| 7 | YL/WT | Injector 3 Driver Control | 1-4 ms |
| 8 | TN | Smart Start Relay Control | KOEC: 9-11v |
| 9 | --- | Not Used | --- |
| 10 | BK/TN | Power Ground | <0.1v |
| 11 | GY | Coil Driver 1 (Cyl 1 & 4) | 5°, 55 mph: 8° dwell |
| 12 | --- | Not Used | --- |
| 13 | WT/DB | Injector 1 Driver Control | 1-4 ms |
| 14 | BR/DB | Injector 6 Driver Control | 1-4 ms |
| 15 | GY | Injector 5 Driver Control | 1-4 ms |
| 16 | LB/BR | Injector 4 Driver Control | 1-4 ms |
| 17 | TN | Injector 2 Driver Control | 1-4 ms |
| 18-19 | --- | Not Used | --- |
| 20 | WT/BK | Ignition Switch Output (B+) | 12-14v |
| 21-23 | --- | Not Used | --- |
| 24 | DB/LG | Knock Sensor Signal | No knock: 2.5v DC |
| 25 | --- | Not Used | --- |
| 26 | TN/BK | ECT Sensor Signal | At 180°F: 2.80v |
| 27-29 | --- | Not Used | --- |
| 30 | BK/DG | HO2S-11 (B1 S1) Signal | 0.1-1.1v |
| 31 | --- | Not Used | --- |
| 32 | GY/BK | CKP Sensor Signal | Digital Signal: 0-5-0v |
| 33 | TN/YL | CMP Sensor Signal | Digital Signal: 0-5-0v |
| 34 | --- | Not Used | --- |
| 35 | OR/DB | TP Sensor Signal | 0.6-1.0v |
| 36 | DG/RD | MAP Sensor Signal | 1.5-1.7v |
| 37-39 | --- | Not Used | --- |
| 40 | GY/YL | EGR Solenoid Control | 12v, 55 mph: 1v |

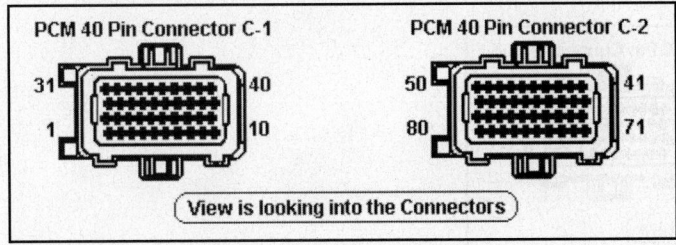

**Pin Connector Graphic**

## Standard Colors and Abbreviations

| Abbreviation | Color | Abbreviation | Color | Abbreviation | Color |
|---|---|---|---|---|---|
| BK | Black | GY | Gray | RD | Red |
| BL | Blue | GN | Green | TN | Tan |
| BR | Brown | LG | Light Green | VT | Violet |
| DB | Dark Blue | OR | Orange | WT | White |
| DG | Dark Green | PK | Pink | YL | Yellow |

**1996 Caravan, Town & Country, Voyager 3.8L VIN L 'C2' Connector**

| PCM Pin # | Wire Color | Circuit Description (40-Pin) | Value at Hot Idle |
|---|---|---|---|
| 41 | RD/LG | Speed Control Set Switch Signal | S/C & Set Switch On: 3.8v |
| 42 | DB | A/C Pressure Switch Signal | A/C On: 0.451-4.850v |
| 43 | BK/LB | Sensor Ground | <0.050v |
| 44 | OR | 8-Volt Supply | 7.9-8.1v |
| 45 | DB | A/C Switch Sense Signal | A/C Off: 12v, On: 1v |
| 46 | RD/WT | Fused Battery Power (B+) | 12-14v |
| 47 | --- | Not Used | --- |
| 48 | BR/WT | IAC 1 Driver Control | Pulse Signal: 0.8-11v |
| 49 | YL/BK | IAC 2 Driver Control | Pulse Signal: 0.8-11v |
| 50 | BK/TN | Power Ground | <0.1v |
| 51 | TN/WT | HO2S-12 (B1 S2) Signal | 0.1-1.1v |
| 52-56 | --- | Not Used | --- |
| 57 | GY/RD | IAC 3 Driver Control | Pulse Signal: 0.8-11v |
| 58 | PK/BK | IAC 4 Driver Control | Pulse Signal: 0.8-11v |
| 59 | PK/BR | CCD Bus (+) | Digital Signal: 0-5-0v |
| 60 | WT/BK | CCD Bus (-) | <0.050v |
| 61 | PK/WT | 5-Volt Supply | 4.9-5.1v |
| 62 | WT/PK | Brake Switch Sense Signal | Brake Off: 0v, On: 12v |
| 63 | YL/DG | Torque Management Request | Digital Signals |
| 64 | DB/OR | A/C Clutch Relay Control | Relay Off: 12v, On: 1v |
| 65 | PK | SCI Transmit | 0v |
| 66 | WT/OR | Vehicle Speed Signal | Digital Signal |
| 67 | DB/YL | Auto Shutdown Relay Control | Relay Off: 12v, On: 1v |
| 68 | PK/BK | EVAP Purge Solenoid Control | PWM Signal: 0-12-0v |
| 69-71 | --- | Not Used | --- |
| 72 | YL/BK | LDP Switch Sense Signal | LDP Switch Closed: 0v |
| 73 | LG/DB | Radiator Fan Control Relay | Relay Off: 12v, On: 1v |
| 74 | BR | Fuel Pump Relay Control | Relay Off: 12v, On: 1v |
| 75 | LG | SCI Receive | 5v |
| 76 | BK/WT | PNP Switch Sense Signal | In P/N: 0v, Others: 5v |
| 77 | WT/DG | LDP Solenoid Control | PWM Signal: 0-12-0v |
| 78 | TN/RD | Speed Control Vacuum Solenoid Control | Vacuum Increasing: 1v |
| 79 | --- | Not Used | --- |
| 80 | LG/RD | Speed Control Vent Solenoid Control | Vacuum Decreasing: 1v |

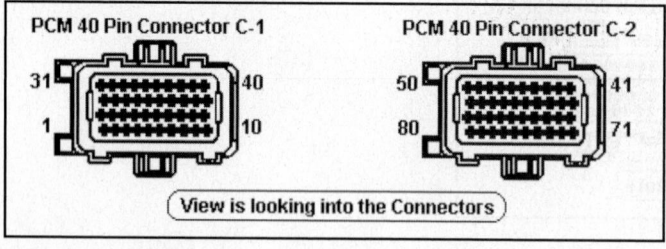

**PCM 40 Pin Connector C-1**     **PCM 40 Pin Connector C-2**

View is looking into the Connectors

**Pin Connector Graphic**

## 1997-98 Caravan, Town & Country, Voyager 3.8L VIN L 'C1' Connector

| PCM Pin # | Wire Color | Circuit Description (40-Pin) | Value at Hot Idle |
|---|---|---|---|
| 1 | --- | Not Used | --- |
| 2 | RD/YL | Coil Driver 3 (Cyl 3 & 6) | 5°, 55 mph: 8° dwell |
| 3 | DB/TN | Coil Driver 2 (Cyl 2 & 5) | 5°, 55 mph: 8° dwell |
| 4 | DG | Generator Field Driver | Digital Signal: 0-12-0v |
| 5 | YL/RD | Speed Control Power Supply | 12-14v |
| 6 | DG/OR | ASD Relay Output (B+) | 12-14v |
| 7 | YL/WT | Injector 3 Driver Control | 1-4 ms |
| 8 | TN | Smart Start Relay Control | KOEC: 9-11v |
| 9 | --- | Not Used | --- |
| 10 | BK/TN | Power Ground | <0.1v |
| 11 | GY/RD | Coil Driver 1 (Cyl 1 & 4) | 5°, 55 mph: 8° dwell |
| 12 | --- | --- | --- |
| 13 | WT/DB | Injector 1 Driver Control | 1-4 ms |
| 14 | BR/DB | Injector 6 Driver Control | 1-4 ms |
| 15 | GY | Injector 5 Driver Control | 1-4 ms |
| 16 | LB/BR | Injector 4 Driver Control | 1-4 ms |
| 17 | TN | Injector 2 Driver Control | 1-4 ms |
| 18-19 | --- | Not Used | --- |
| 20 | WT/BK | Ignition Switch Output (B+) | 12-14v |
| 21, 23 | --- | Not Used | --- |
| 22 | BK/PK | Malfunction Indicator Lamp | MIL Off: 12v, On: 1v |
| 24 | DB/LG | Knock Sensor Signal | No knock: 2.5v DC |
| 25 | --- | Not Used | --- |
| 26 | TN/BK | ECT Sensor Signal | At 180°F: 2.80v |
| 27-29 | --- | Not Used | --- |
| 30 | BK/DG | HO2S-11 (B1 S1) Signal | 0.1-1.1v |
| 31 | --- | Not Used | --- |
| 32 | GY/BK | CKP Sensor Signal | Digital Signal: 0-5-0v |
| 33 | TN/YL | CMP Sensor Signal | Digital Signal: 0-5-0v |
| 34 | --- | Not Used | --- |
| 35 | OR/DB | TP Sensor Signal | 0.6-1.0v |
| 36 | DG/RD | MAP Sensor Signal | 1.5-1.7v |
| 37-39 | --- | Not Used | --- |
| 40 | GY/YL | EGR Solenoid Control | 12v, 55 mph: 1v |

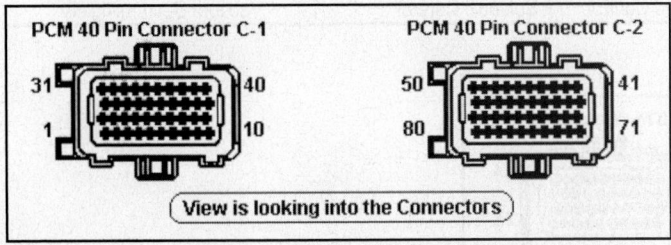

**Pin Connector Graphic**

**1997-98 Caravan, Town & Country, Voyager 3.8L VIN L 'C2' Connector**

| PCM Pin # | Wire Color | Circuit Description (40-Pin) | Value at Hot Idle |
|---|---|---|---|
| 41 | RD/LG | Speed Control Set Switch Signal | S/C & Set Switch On: 3.8v |
| 42 | DB | A/C Pressure Switch Signal | A/C On: 0.451-4.850v |
| 43 | BK/LB | Sensor Ground | <0.050v |
| 44 | OR | 8-Volt Supply | 7.9-8.1v |
| 45 | DG | A/C Switch Sense Signal | A/C Off: 12v, On: 1v |
| 46 | RD/WT | Fused Battery Power (B+) | 12-14v |
| 47 | --- | Not Used | --- |
| 48 | BR/WT | IAC 3 Driver Control | Pulse Signal: 0.8-11v |
| 49 | YL/BK | IAC 2 Driver Control | Pulse Signal: 0.8-11v |
| 50 | BK/TN | Power Ground | <0.1v |
| 51 | TN/WT | HO2S-12 (B1 S2) Signal | 0.1-1.1v |
| 52-56 | --- | Not Used | --- |
| 57 | GY/RD | IAC 1 Driver Control | Pulse Signal: 0.8-11v |
| 58 | PK/BK | IAC 4 Driver Control | Pulse Signal: 0.8-11v |
| 59 | PK/BR | CCD Bus (+) | Digital Signal: 0-5-0v |
| 60 | WT/BK | CCD Bus (-) | <0.050v |
| 61 | PK/WT | 5-Volt Supply | 4.9-5.1v |
| 62 | WT/PK | Brake Switch Sense Signal | Brake Off: 0v, On: 12v |
| 63 | YL/DG | Torque Management Request | Digital Signals |
| 64 | DB/OR | A/C Clutch Relay Control | Relay Off: 12v, On: 1v |
| 65 | PK | SCI Transmit | 0v |
| 66 | WT/OR | Vehicle Speed Signal | Digital Signal |
| 67 | DB/YL | Auto Shutdown Relay Control | Relay Off: 12v, On: 1v |
| 68 | PK/BK | EVAP Purge Solenoid Control | PWM Signal: 0-12-0v |
| 69-71 | --- | Not Used | --- |
| 72 | DB/WT | LDP Switch Sense Signal | LDP Switch Closed: 0v |
| 73 | LG/DG | Radiator Fan Control Relay | Relay Off: 12v, On: 1v |
| 74 | BR | Fuel Pump Relay Control | Relay Off: 12v, On: 1v |
| 75 | LG | SCI Receive | 5v |
| 76 | BR/YL | PNP Switch Sense Signal | In P/N: 0v, Others: 5v |
| 77 | WT/DG | LDP Solenoid Control | PWM Signal: 0-12-0v |
| 78 | TN/RD | Speed Control Vacuum Solenoid Control | Vacuum Increasing: 1v |
| 79 | --- | Not Used | --- |
| 80 | LG/RD | Speed Control Vent Solenoid Control | Vacuum Decreasing: 1v |

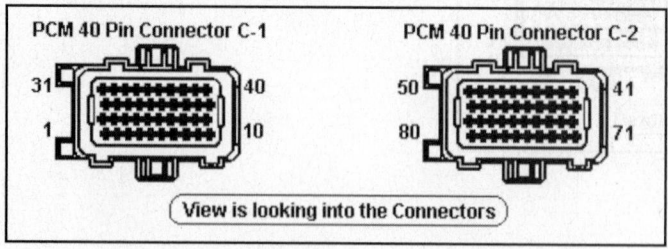

**Pin Connector Graphic**

**1999-2000 Caravan, Town & Country, Voyager 3.8L VIN L 'C1' Connector**

| PCM Pin # | Wire Color | Circuit Description (40-Pin) | Value at Hot Idle |
|---|---|---|---|
| 1 | --- | Not Used | --- |
| 2 | RD/YL | Coil Driver 3 (Cyl 3 & 6) | 5°, 55 mph: 8° dwell |
| 3 | DB/TN | Coil Driver 2 (Cyl 2 & 5) | 5°, 55 mph: 8° dwell |
| 4 | --- | Not Used | --- |
| 5 | YL/RD | Speed Control Power Supply | 12-14v |
| 6 | DG/OR | ASD Relay Output (B+) | 12-14v |
| 7 | YL/WT | Injector 3 Driver Control | 1-4 ms |
| 8 | DG | Generator Field Driver | Digital Signal: 0-12-0v |
| 9 | --- | Not Used | --- |
| 10 | BK/TN | Power Ground | <0.1v |
| 11 | GY/RD | Coil Driver 1 (Cyl 1 & 4) | 5°, 55 mph: 8° dwell |
| 12 | --- | Not Used | --- |
| 13 | WT/DB | Injector 1 Driver Control | 1-4 ms |
| 14 | BR/DB | Injector 6 Driver Control | 1-4 ms |
| 15 | GY | Injector 5 Driver Control | 1-4 ms |
| 16 | LB/BR | Injector 4 Driver Control | 1-4 ms |
| 17 | TN/WT | Injector 2 Driver Control | 1-4 ms |
| 18-19 | --- | Not Used | --- |
| 20 | WT/BK | Ignition Switch Output (B+) | 12-14v |
| 21 | --- | Not Used | --- |
| 22 | BK/PK | Malfunction Indicator Lamp | MIL Off: 12v, On: 1v |
| 23-24 | --- | Not Used | --- |
| 25 | DB/LG | Knock Sensor Signal | No knock: 2.5v DC |
| 26 | TN/BK | ECT Sensor Signal | At 180°F: 2.80v |
| 27 | BK/OR | HO2S-11 & HO2S-12 Ground | <0.1v |
| 28-29 | --- | Not Used | --- |
| 30 | BK/DG | HO2S-11 (B1 S1) Signal | 0.1-1.1v |
| 31 | TN | Smart Start Relay Control | KOEC: 9-11v |
| 32 | GY/BK | CKP Sensor Signal | Digital Signal: 0-5-0v |
| 33 | TN/YL | CMP Sensor Signal | Digital Signal: 0-5-0v |
| 34 | --- | Not Used | --- |
| 35 | OR/DB | TP Sensor Signal | 0.6-1.0v |
| 36 | DG/RD | MAP Sensor Signal | 1.5-1.7v |
| 37 | --- | Not Used | --- |
| 38 | DG/LB | A/C Switch Sense Signal | A/C Off: 12v, On: 1v |
| 39 | --- | Not Used | --- |
| 40 | GY/YL | EGR Solenoid Control | 12v, 55 mph: 1v |

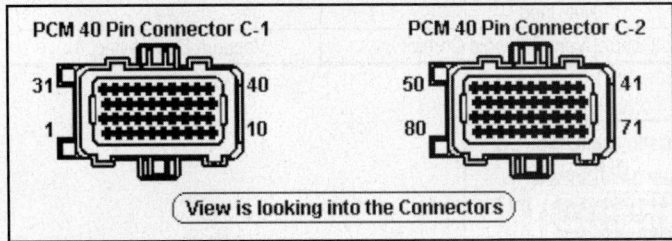

**Pin Connector Graphic**

## 1999-2000 Caravan, Town & Country, Voyager 3.8L VIN L 'C2' Connector

| PCM Pin # | Wire Color | Circuit Description (40-Pin) | Value at Hot Idle |
|---|---|---|---|
| 41 | RD/LG | Speed Control Set Switch Signal | S/C & Set Switch On: 3.8v |
| 42 | DB | A/C Pressure Switch Signal | A/C On: 0.451-4.850v |
| 43 | BK/LB | Sensor Ground | <0.050v |
| 44 | OR | 8-Volt Supply | 7.9-8.1v |
| 45 | --- | Not Used | --- |
| 46 | RD/WT | Fused Battery Power (B+) | 12-14v |
| 47 | --- | Not Used | --- |
| 48 | BR/WT | IAC 3 Driver Control | Pulse Signal: 0.8-11v |
| 49 | YL/BK | IAC 2 Driver Control | Pulse Signal: 0.8-11v |
| 50 | BK/TN | Power Ground | <0.1v |
| 51 | TN/WT | HO2S-12 (B1 S2) Signal | 0.1-1.1v |
| 52-55 | --- | Not Used | --- |
| 56 | TN/RD | Speed Control Vacuum Solenoid Control | Vacuum Increasing: 1v |
| 57 | GY/RD | IAC 1 Driver Control | Pulse Signal: 0.8-11v |
| 58 | PK/BK | IAC 4 Driver Control | Pulse Signal: 0.8-11v |
| 59 | PK/BR | CCD Bus (+) | Digital Signal: 0-5-0v |
| 60 | WT/BK | CCD Bus (-) | <0.050v |
| 61 | PK/WT | 5-Volt Supply | 4.9-5.1v |
| 62 | WT/PK | Brake Switch Sense Signal | Brake Off: 0v, On: 12v |
| 63 | YL/DG | Torque Management Request | Digital Signals |
| 64 | DB/OR | A/C Clutch Relay Control | Relay Off: 12v, On: 1v |
| 65 | PK | SCI Transmit | 0v |
| 66 | WT/OR | Vehicle Speed Signal | Digital Signal |
| 67 | DB/YL | Auto Shutdown Relay Control | Relay Off: 12v, On: 1v |
| 68 | PK/BK | EVAP Purge Solenoid Control | PWM Signal: 0-12-0v |
| 69 | --- | Not Used | --- |
| 70 | PK/RD | EVAP Purge Solenoid Sense | 0-1v |
| 71 | --- | Not Used | --- |
| 72 | DB/WT | LDP Switch Sense Signal | LDP Switch Closed: 0v |
| 73 | LG/DG | Radiator Fan Control Relay | Relay Off: 12v, On: 1v |
| 74 | BR | Fuel Pump Relay Control | Relay Off: 12v, On: 1v |
| 75 | LG | SCI Receive | 5v |
| 76 | BR/YL | PNP Switch Sense Signal | In P/N: 0v, Others: 5v |
| 77 | WT/DG | LDP Solenoid Control | PWM Signal: 0-12-0v |
| 78-79 | --- | Not Used | --- |
| 80 | LG/RD | Speed Control Vent Solenoid Control | Vacuum Decreasing: 1v |

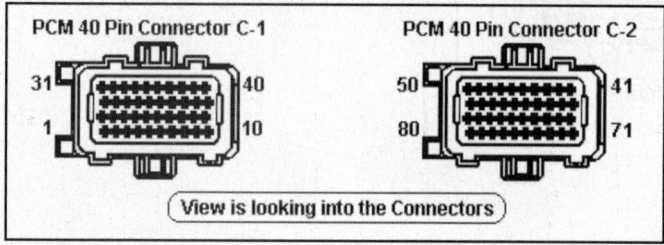

**PCM 40 Pin Connector C-1**    **PCM 40 Pin Connector C-2**

View is looking into the Connectors

**Pin Connector Graphic**

## 2001-02 Caravan, Town & Country, Voyager 3.8L VIN L A/T 'C1' Connector

| PCM Pin # | Wire Color | Circuit Description (40-Pin) | Value at Hot Idle |
|---|---|---|---|
| 1 | --- | Not Used | --- |
| 2 | DB/OR | Coil 3 Driver Control | 5°, 55 mph: 8° dwell |
| 3 | DB/TN | Coil 2 Driver Control | 5°, 55 mph: 8° dwell |
| 4 | --- | Not Used | --- |
| 5 | VT/YL | Speed Control On/Off Switch Sense | 12-14v |
| 6 | BR/WT | ASD Relay Power (B+) | 12-14v |
| 7 | BR/LB | Injector 3 Driver | 1-4 ms |
| 8 | BR/GY | Generator Field Driver | Digital Signal: 0-12-0v |
| 9 | --- | Not Used | --- |
| 10 | BK/BR | Power Ground | <0.1v |
| 11 | DB/GY | Coil 1 Driver Control | 5°, 55 mph: 8° dwell |
| 12 | VT/GY | Engine Oil Pressure Sensor | 1.6v at 24 psi |
| 13 | BR/YL | Injector 1 Driver Control | 1-4 ms |
| 14 | BR/VT | Injector 6 Driver Control | 1-4 ms |
| 15 | BR/OR | Injector 5 Driver Control | 1-4 ms |
| 16 | BR/TN | Injector 4 Driver Control | 1-4 ms |
| 17 | BR/DB | Injector 2 Driver Control | 1-4 ms |
| 18 | BR/LG | HO2S-11 (B1 S1) Heater | PWM Signal: 0-12-0v |
| 19 | --- | Not Used | --- |
| 20 | PK/GY | Ignition Switch Power (B+) | 12-14v |
| 21-24 | --- | Not Used | --- |
| 25 | DB/YL | Knock Sensor Signal | 0.080v AC |
| 26 | VT/OR | ECT Sensor Signal | At 180°F: 2.80v |
| 27 | BR/LG | HO2S Ground | <0.050v |
| 28-29 | --- | Not Used | --- |
| 30 | DB/LB | HO2S-11 (B1 S1) Signal | 0.1-1.1v |
| 31 | DG/OR | Double Start Override Signal | KOEC: 9-11v |
| 32 | BR/LB | CKP Sensor Signal | Digital Signal: 0-5-0v |
| 33 | DB/GY | CMP Sensor Signal | Digital Signal: 0-5-0v |
| 34 | --- | Not Used | --- |
| 35 | DB/OR | TP Sensor Signal | 0.6-1.0v |
| 36 | VT/BR | MAP Sensor Signal | 1.5-1.6v |
| 37 | DB/LG | IAT Sensor Signal | At 100°F: 1.83v |
| 38-39 | --- | Not Used | --- |
| 40 | DB/VT | EGR Solenoid Control | 12v, at 55 mph: 1v |

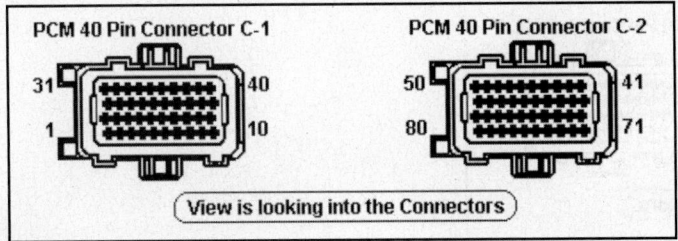

**Pin Connector Graphic**

**2001-02 Caravan, Town & Country, Voyager 3.8L VIN L A/T 'C2' Connector**

| PCM Pin # | Wire Color | Circuit Description (40-Pin) | Value at Hot Idle |
|---|---|---|---|
| 41 | VT | Speed Control Set Switch Signal | S/C & Set Switch On: 3.8v |
| 42 | LB/BR | A/C Pressure Switch Signal | A/C On: 0.451-4.850v |
| 43 | DB/DG | Sensor Ground | <0.050v |
| 44 | BR/PK | 8-Volt Supply | 7.9-8.1v |
| 45 | --- | Not Used | --- |
| 46 | OR/RD | Keep Alive Power (B+) | 12-14v |
| 47-48 | --- | Not Used | --- |
| 49 | VT/DG | IAC 1 Driver Control | Pulse Signals |
| 50 | BK/DG | Power Ground | <0.1v |
| 51 | DB/YL | HO2S-12 (B1 S2) Signal | 0.1-1.1v |
| 52-55 | --- | Not Used | --- |
| 56 | VT/YL | Speed Control Vacuum Solenoid | Vacuum Increasing: 1v |
| 57 | VT/LG | IAC 2 Driver Control | Pulse Signals |
| 58 | --- | Not Used | --- |
| 59 | WT/VT | PCI Data Bus (J1850) | Digital Signals: 0-7-0v |
| 60 | --- | Not Used | --- |
| 61 | PK/YL | 5-Volt Supply | 4.9-5.1v |
| 62 | DG/WT | Brake Switch Sense Signal | Brake Off: 0v, On: 12v |
| 63 | DG/LG | Torque Management Request | Digital Signals |
| 64 | LB/OR | A/C Clutch Relay Control | Relay Off: 12v, On: 1v |
| 65 | WT/BR | SCI Transmit Signal | 0v |
| 66 | DB/OR | Vehicle Speed Signal | Digital Signal |
| 67 | BR/WT | ASD Relay Control | Relay Off: 12v, On: 1v |
| 68 | DB/WT | EVAP Purge Solenoid Control | PWM Signal: 0-12-0v |
| 69 | --- | Not Used | --- |
| 70 | DB/BR | EVAP Purge Solenoid Sense | 0-1v |
| 71 | --- | Not Used | --- |
| 72 | VT/WT | LDP Switch Sense Signal | LDP Switch Closed: 0v |
| 73 | BR/VT | Radiator Fan Control Relay | Relay Off: 12v, On: 1v |
| 74 | BR | Fuel Pump Relay Control | Relay Off: 12v, On: 1v |
| 75 | WT/LG | SCI Receive Signal | 5v |
| 76 | YL/DB | TRS TR1 Signal | In P/N: 0v, Others: 5v |
| 77 | VT/LB | LDP Solenoid Control | PWM Signal: 0-12-0v |
| 78-79 | --- | Not Used | --- |
| 80 | VT/OR | Speed Control Vent Solenoid | Vacuum Decreasing: 1v |

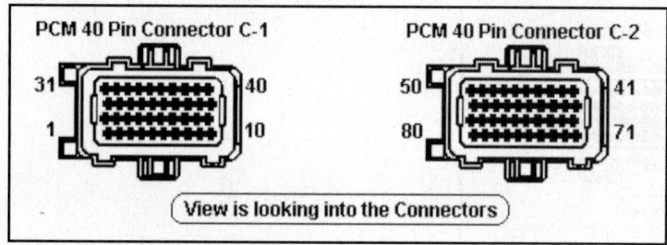

**Pin Connector Graphic**

## 2003 Caravan, Town & Country, Voyager 3.8L VIN L A/T 'C1' Black Connector

| PCM Pin # | Wire Color | Circuit Description (40-Pin) | Value at Hot Idle |
|---|---|---|---|
| 1-8 | --- | Not Used | --- |
| 9 | BK/BR | Power Ground | <0.1v |
| 10 | --- | Not Used | --- |
| 11 | PK/GY | Ignition Switch Output (Start-Run) | 12-14v |
| 12 | PK/WT | Ignition Switch Output (Run/Start) | 12-14v |
| 13-17 | --- | Not Used | --- |
| 18 | BK/DG | Power Ground | <0.1v |
| 19, 21 | --- | Not Used | --- |
| 20 | VT/GY | Engine Oil Pressure Sensor | 1.6v at 24 psi |
| 22 | VT/GY | Ambient Air Temperature Sensor | At 100ºF: 1.83v |
| 23-24 | --- | Not Used | --- |
| 25 | WT/LG | SCI Receive Signal - DLC Pin 12 | 5v |
| 26 | WT/BR | Flash Program Enable | 0v |
| 27-28 | --- | Not Used | --- |
| 29 | OR/RD | Fused B+ | 12-14v |
| 30 | YL | Ignition Switch Output (Start) | 12-14v |
| 31 | DB/YL | HO2S-12 (B1 S2) Signal | 0.1-1.1v |
| 32 | DB/DG | Oxygen Sensor Return (Downstream) | <0.050v |
| 33-35 | --- | Not Used | --- |
| 36 | WT/BR | SCI Transmit (PCM) - DLC Pin 7 | 0v |
| 37 | DG/YL | SCI Transmit (TCM) | 0v |
| 38 | WT/VT | PCI Data Bus (J1850) - DLC Pin 2 | Digital Signals: 0-7-0v |

## 2003 Caravan, Town & Country, Voyager 3.8L VIN L A/T 'C2' Gray Connector

| PCM Pin # | Wire Color | Circuit Description (40-Pin) | Value at Hot Idle |
|---|---|---|---|
| 1-8 | --- | Not Used | --- |
| 9 | DB/TN | Coil 2 Driver | 5º, 55 mph: 8º dwell |
| 10 | DB/DG | Coil 1 Driver | 5º, 55 mph: 8º dwell |
| 11 | BR/TN | Injector 4 Driver Control | 1-4 ms |
| 12 | BR/LB | Injector 3 Driver Control | 1-4 ms |
| 13 | BR/DB | Injector 2 Driver Control | 1-4 ms |
| 14 | BR/YL | Injector 1 Driver Control | 1-4 ms |
| 15-17 | --- | Not Used | --- |
| 18 | BR/LG | HO2S-11 (B1 S1) Heater Control | PWM Signal: 0-12-0v |
| 19 | BR/GY | Generator Field Driver | Digital Signal: 0-12-0v |
| 20 | VT/OR | ECT Sensor Signal | At 180ºF: 2.80v |
| 21 | BR/OR | TP Sensor Signal | 0.6-1.0v |
| 22, 26 | --- | Not Used | --- |
| 23 | VT/BR | MAP Sensor Signal | 1.5-1.6v |
| 24 | BR/LG | Knock Sensor Return | <0.050v |
| 25 | DB/YL | Knock Sensor Signal | 0.080v AC |
| 27 | DB/DG | Sensor Ground | <0.1v |
| 28 | BR/VT | IAC Motor Return | <0.050v |
| 29 | PK/YL | 5-Volt Supply | 4.9-5.1v |
| 30 | DB/LG | IAT Sensor Signal | At 100ºF: 1.83v |
| 31 | DB/LB | HO2S-11 (B1 S1) Sensor | 0.1-1.1v |
| 32 | BR/DG | Oxygen Sensor Return (Upstream) | <0.050v |
| 33 | --- | Not Used | --- |
| 34 | DB/GY | CKP Sensor Signal | Digital Signal: 0-5-0v |
| 35 | BR/L | CMP Sensor Signal | Digital Signal: 0-5-0v |
| 36-37 | --- | Not Used | --- |
| 38 | VT/GY | IAC 1 Motor Control | Pulse Signals |

## 2003 Caravan, Town & Country, Voyager 3.8L VIN L A/T 'C3' White Connector

| PCM Pin # | Wire Color | Circuit Description (38-Pin) | Value at Hot Idle |
|---|---|---|---|
| 1-2, 4 | --- | Not Used | --- |
| 3 | BR/WT | ASD Relay Control | Relay Off: 12v, On: 1v |
| 5 | VT/OR | Speed Control Vent Solenoid | Vacuum Decreasing: 1v |
| 6 | BR/VT | Radiator Fan Relay Control | Relay Off: 12v, On: 1v |
| 7 | VT/YL | Speed Control Power Supply | 12-14v |
| 8 | WT/DG | Natural Vacuum Leak Detection Solenoid | Solenoid Off: 12v, On: 1v |
| 9 | BR/WT | HO2S-12 (Bank 1 Sensor 2) Heater Control | --- |
| 11 | LB/OR | A/C Clutch Relay Control | Relay Off: 12v, On: 1v |
| 12 | VT/YL | Speed Control Vacuum Solenoid | Vacuum Increasing: 1v |
| 13-18 | --- | Not Used | --- |
| 19 | BR/WT | Automatic Shutdown Relay Output | 12-14v |
| 20 | DB/WT | EVAP Purge Solenoid Control | PWM Signal: 0-12-0v |
| 21 | YL | Ignition Switch Output (Start) on MTX | Cranking: 9-11v |
| 22, 24-27 | --- | Not Used | --- |
| 23 | DG/WT | Brake Switch Sense | Brake Off: 0v, On: 12v |
| 28 | BR/WT | Automatic Shutdown Relay Output | 12-14v |
| 29 | DB/BR | EVAP Purge Solenoid Sense | 0-1v |
| 30 | --- | Not Used | --- |
| 31 | LB/BR | A/C Pressure Signal | A/C On: 0.451-4.850v |
| 32 | DB/YL | Battery Temperature Sensor | At 100°F: 1.83v |
| 33, 36 | --- | Not Used | --- |
| 34 | VT | Speed Control Set Switch Signal | S/C & Set Switch On: 3.8v |
| 35 | VT/WT | Natural Vacuum Leak Detection Switch Sense | 0.1v |
| 37 | BR | Fuel Pump Relay Control | Relay Off: 12v, On: 1v |
| 38 | DG/OR | Starter Relay Control | KOEC: 9-11v |

## 2003 Caravan, Town & Country, Voyager 3.8L VIN L A/T 'C4' Green Connector

| PCM Pin # | Wire Color | Circuit Description (38-Pin) | Value at Hot Idle |
|---|---|---|---|
| 1 | YL/LG | Overdrive Solenoid Control | Solenoid Off: 12v, On: 1v |
| 2 | YL/LB | Underdrive Solenoid Control | Solenoid Off: 12v, On: 1v |
| 3-5, 7-9 | --- | Not Used | --- |
| 6 | YL/DB | 2-4 Solenoid Control | Solenoid Off: 12v, On: 1v |
| 10 | DG/WT | Low/Reverse Solenoid Control | Solenoid Off: 12v, On: 1v |
| 11, 13, 17 | --- | Not Used | --- |
| 12, 13 | BK/LG | Power Ground | <0.050v |
| 15 | DG/LB | TRS T1 Sense | <0.050v |
| 16 | DG/DB | TRS T3 Sense | <0.050v |
| 18 | YL/BR | Transmission Control Relay Control | Relay Off: 12v, On: 1v |
| 19, 28 | YL/OR | Transmission Control Relay Output | 12-14v |
| 20-21 | --- | Not Used | --- |
| 22 | DG/TN | Overdrive Pressure Switch Sense | 12-14v |
| 23-26 | --- | Not Used | --- |
| 27 | DG/GY | TRS T41 Sense | <0.050v |
| 29 | YL/TN | Low/Reverse Pressure Switch Sense | 12-14v |
| 30 | YL/DG | 2-4 Pressure Switch Sense | In Low/Reverse: 2-4v |
| 31, 36 | --- | Not Used | --- |
| 32 | DG/BR | Output Speed Sensor Signal | In 2-4 Position: 2-4v |
| 33 | DG/WT | Input Speed Sensor Signal | Moving: AC voltage |
| 34 | DG/VT | Speed Sensor Ground | Moving: AC voltage |
| 35 | DG/OR | Transmission Temperature Sensor Signal | 3.2-3.4v at 104°F |
| 37 | DG/YL | TRS T42 Sense | In PRNL: 0v, Others 5v |
| 38 | YL/OR | Transmission Control Relay Output | 12-14v |

## 2004-05 Caravan, Town & Country, Voyager 3.3L, 3.8L 'C1' Black/Black Connector

| PCM Pin # | Wire Color | Circuit Description (38-Pin) | Value at Hot Idle |
|:---:|:---:|:---|:---:|
| 1 | --- | Not Used | --- |
| 2 | --- | Not Used | --- |
| 3 | --- | Not Used | --- |
| 4 | --- | Not Used | --- |
| 5 | --- | Not Used | --- |
| 6 | --- | Not Used | --- |
| 7 | --- | Not Used | --- |
| 8 | --- | Not Used | --- |
| 9 | BK/BR | Ground | |
| 10 | --- | Not Used | --- |
| 11 | PK/GY | Fused Ignition Switch Output (Run-Start) | |
| 12 | PK/WT (EATX) | FCM Output (Unlock-Run-Start) | |
| 13 | DB/OR (MTX) | Vehicle Speed Signal | |
| 14 | --- | Not Used | --- |
| 15 | --- | Not Used | --- |
| 16 | --- | Not Used | --- |
| 17 | --- | Not Used | --- |
| 18 | BK/DG | Ground | |
| 19 | --- | Not Used | --- |
| 20 | VT/GY | Oil Pressure Signal | |
| 21 | --- | Not Used | --- |
| 22 | VT/LG | AAT Signal | |
| 23 | --- | Not Used | --- |
| 24 | --- | Not Used | --- |
| 25 | WT/LG | SCI Receive (PCM) | |
| 26 | WT/BR (2.4L EATX) | Flash Program Enable | |
| 26 | WT/OR (3.3L/3.8L) | Flash Program Enable | |
| 27 | --- | Not Used | --- |
| 28 | --- | Not Used | --- |
| 29 | OR/RD | Fused B (+) | |
| 30 | YL | Fused Ignition Switch Output (Start) | |
| 31 | DB/YL | O2 1/2 Signal | |
| 32 | DB/DG (2.4L) | O2 Return (Down) | |
| 32 | BR/DG (3.3L/3.8L) | O2 Return (Down) | |
| 33 | --- | Not Used | --- |
| 34 | --- | Not Used | --- |
| 35 | --- | Not Used | --- |
| 36 | WT/DG (2.4L) | SCI Transmit (PCM) | |
| 36 | WT/BR (3.3L/3.8L) | SCI Transmit (PCM) | |
| 37 | DG/YL (EATX) | SCI Transmit (TCM) | |
| 38 | WT/VT | PCI Bus | |

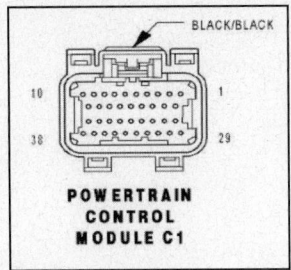

BLACK/BLACK

10    1

38    29

**POWERTRAIN
CONTROL
MODULE C1**

## 2004-05 Caravan, Town & Country, Voyager 3.3L, 3.8L 'C2' Black/Orange Connector

| PCM Pin # | Wire Color | Circuit Description (38-Pin) | Value at Hot Idle |
|---|---|---|---|
| 1 | --- | Not Used | --- |
| 2 | --- | Not Used | --- |
| 3 | --- | Not Used | --- |
| 4 | BR/VT (3.3L/3.8L) | Injector Control No. 6 | |
| 5 | BR/OR (3.3L/3.8L) | Injector Control No. 5 | |
| 6 | --- | Not Used | --- |
| 7 | BR/OR (3.3L/3.8L) | Coil Control No. 3 | |
| 8 | DB/VT (Exc. Early 3.8L) | EGR Solenoid Control | |
| 9 | DB/TN | Coil Control No. 2 | |
| 10 | DB/DG | Coil Control No. 1 | |
| 11 | BR/TN | Injector Control No. 4 | |
| 12 | BR/LB | Injector Control No. 3 | |
| 13 | BR/DB | Injector Control No. 2 | |
| 14 | BR/YL | Injector Control No. 1 | |
| 15 | --- | Not Used | --- |
| 16 | --- | Not Used | --- |
| 17 | --- | Not Used | --- |
| 18 | BR/LG | O2 1/1 Heater Control | |
| 19 | BR/GY | Gen Field Control | |
| 20 | VT/OR | ECT Signal | |
| 21 | BR/OR | TP Signal | |
| 22 | DB/LG (Exc. Early 3.8L) | EGR Solenoid Signal | |
| 23 | VT/BR | MAP Signal | |
| 24 | BR/LG (Exc. Early 3.8L) | KS Return | |
| 25 | DB/YL (Exc. Early 3.8L) | KS Signal | |
| 26 | --- | Not Used | --- |
| 27 | DB/DG | Sensor Ground | |
| 28 | BR/VT | IAC Return | |
| 29 | PK/YL | 5 Volt Supply | |
| 30 | DB/LG | IAT Signal | |
| 31 | DB/LB | O2 1/1 Signal | |
| 32 | BR/DG | O2 Return (Up) | |
| 33 | --- | Not Used | --- |
| 34 | DB/GY | CMP Signal | |
| 35 | BR/LB | CKP Signal | |
| 36 | --- | Not Used | --- |
| 37 | --- | Not Used | --- |
| 38 | VT/GY | IAC Motor Control | |

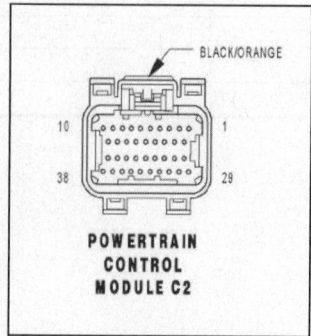

BLACK/ORANGE

10    1

38    29

**POWERTRAIN CONTROL MODULE C2**

### 2004-05 Caravan, Town & Country, Voyager 3.3L, 3.8L 'C3' Black/Natural Connector

| PCM Pin # | Wire Color | Circuit Description (38-Pin) | Value at Hot Idle |
|:---:|:---:|:---:|:---:|
| 1 | --- | Not Used | --- |
| 2 | --- | Not Used | --- |
| 3 | BR/WT | ASD Relay Control | |
| 4 | --- | Not Used | --- |
| 5 | VT/OR | S/C Vent Control | |
| 6 | BR/VT | Rad Fan Relay Control | |
| 7 | VT/YL | S/C Supply | |
| 8 | VT/LB (Exc. EXPORT) | NVLD Sol Control | |
| 9 | BR/WT | O2 1/2 Heater Control | |
| 10 | --- | Not Used | --- |
| 11 | LB/OR | A/C Clutch Relay Control | |
| 12 | VT/YL | S/C Vacuum Control | |
| 13 | --- | Not Used | --- |
| 14 | --- | Not Used | --- |
| 15 | --- | Not Used | --- |
| 16 | --- | Not Used | --- |
| 17 | --- | Not Used | --- |
| 18 | --- | Not Used | --- |
| 19 | BR/WT | ASD Relay Output | |
| 20 | DB/WT | EVAP Purge Control | |
| 21 | YL (MTX EXPORT) | Fused Ignition Switch Output (Start) | |
| 22 | --- | Not Used | --- |
| 23 | DG/WT | Brake Switch Signal | |
| 24 | --- | Not Used | --- |
| 25 | --- | Not Used | --- |
| 26 | --- | Not Used | --- |
| 27 | --- | Not Used | --- |
| 28 | BR/WT | ASD Relay Output | |
| 29 | DB/BR | EVAP Purge Return | |
| 30 | --- | Not Used | --- |
| 31 | LB/BR | A/C Pressure Signal | |
| 32 | DB/YL | Battery Temp Signal | |
| 33 | --- | Not Used | --- |
| 34 | VT | S/C Switch Signal | |
| 35 | VT/WT (Exc. EXPORT) | NVLD Switch Signal | |
| 36 | --- | Not Used | --- |
| 37 | BR | Fuel Pump Relay Control | |
| 38 | DG/OR | Starter Relay Control | |

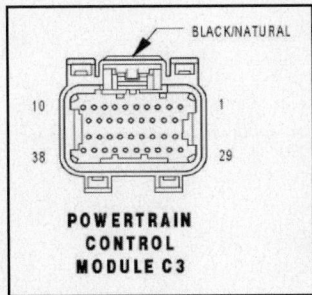

BLACK/NATURAL

10      1

38      29

**POWERTRAIN CONTROL MODULE C3**

**2004-05 Caravan, Town & Country, Voyager 3.3L, 3.8L 'C4' Black/Green Connector**

| PCM Pin # | Wire Color | Circuit Description (38-Pin) | Value at Hot Idle |
|---|---|---|---|
| 1 | YL/GY | OD Solenoid Control | |
| 2 | DB/LB | UD Solenoid Control | |
| 3 | --- | Not Used | --- |
| 4 | --- | Not Used | --- |
| 5 | --- | Not Used | --- |
| 6 | YL/DB | 2-4 Solenoid Control | |
| 7 | --- | Not Used | --- |
| 8 | --- | Not Used | --- |
| 9 | --- | Not Used | --- |
| 10 | DG/WT | L/R Solenoid Control | |
| 11 | --- | Not Used | --- |
| 12 | --- | Not Used | --- |
| 13 | BK/LG | Ground | |
| 14 | BK/LG | Ground | |
| 15 | DG/LB | TRS T1 Sense | |
| 16 | DG/DB | TRS T3 Sense | |
| 17 | --- | Not Used | --- |
| 18 | YL/BR | Transmission Control Relay Control | |
| 19 | YL/OR | Transmission Control Relay Output | |
| 20 | --- | Not Used | --- |
| 21 | --- | Not Used | --- |
| 22 | DG/TN | OD Pressure Switch Sense | |
| 23 | --- | Not Used | --- |
| 24 | --- | Not Used | --- |
| 25 | --- | Not Used | --- |
| 26 | --- | Not Used | --- |
| 27 | DG/GY | TRS T41 Sense | |
| 28 | YL/OR | Transmission Control Relay Output | |
| 29 | YL/TN | L/R Pressure Switch Sense | |
| 30 | YL/DG | 2-4 Pressure Switch Sense | |
| 31 | --- | Not Used | --- |
| 32 | DG/BR | Output Speed Sensor Signal | |
| 33 | DG/WT | Input Speed Sensor Signal | |
| 34 | DG/VT | Speed Sensor Ground | |
| 35 | DG/OR | Transmission Temperature Sensor Signal | |
| 36 | --- | Not Used | --- |
| 37 | DG/YL | TRS T42 Sense | |
| 38 | YL/OR | Transmission Control Relay Output | |

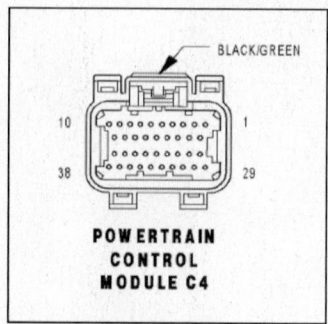

POWERTRAIN
CONTROL
MODULE C4

### Dodge Trucks & Vans

DAKOTA, RAM PICKUP & RAM VAN

**1993-95 Dakota 2.5L VIN G & K (M/T) Connector**

| PCM Pin # | Wire Color | Circuit Description (60-Pin) | Value at Hot Idle |
|-----------|------------|------------------------------|-------------------|
| 1 | DG/RD | MAP Sensor Signal | 1.5-1.6v |
| 2 | TN/BK | ECT Sensor Signal | At 180ºF: 2.80v |
| 3 | RD/WT | Keep Alive Power (B+) | 12-14v |
| 4 | BK/LB | Sensor Ground | <0.050v |
| 5 | BK/LB | Sensor Ground | <0.050v |
| 6 | PK/WT | 5-Volt Supply | 4.9-5.1v |
| 7 | OR | 8-Volt Supply | 7.9-8.1v |
| 8 | WT | Ignition Switch Power (B+) | 12-14v |
| 9 | DB | Ignition Switch Power (B+) | 12-14v |
| 10 | --- | Not Used | --- |
| 11 | BK/TN | Power Ground | <0.1v |
| 12 | BK/TN | Power Ground | <0.1v |
| 13-15 | --- | Not Used | --- |
| 16 | WT/DB | Injector 1 Driver Control | 1-4 ms |
| 17-18 | --- | Not Used | --- |
| 19 | GY | Coil Driver Control | 5º, 55 mph: 8º dwell |
| 20 | DG | Generator Field Control | Digital Signals: 0-12-0v |
| 21 | BK/RD | Throttle Body Temperature | At 100ºF: 2.51v |
| 22 | OR/DB | TP Sensor Signal | 0.6-1.0v |
| 23 | --- | Not Used | --- |
| 24 | GY/BK | Distributor Reference Signal | Digital Signals: 0-5-0v |
| 25 | PK | SCI Transmit | 0v |
| 27 | BR | A/C Clutch Signal | A/C Off: 12v, On: 1v |
| 28, 30 | --- | Not Used | --- |
| 29 | WT/PK | Brake Switch Sense Signal | Brake Off: 0v, On: 12v |
| 31 | DB/PK | Radiator Fan Control Relay | Relay Off: 12v, On: 1v |
| 32 | BK/PK | MIL (lamp) Control | MIL Off: 12v, On: 1v |
| 33 | --- | Not Used | --- |
| 34 | DB/OR | A/C WOT Relay Control | Relay Off: 12v, On: 1v |
| 35 | GY/YL | EGR Solenoid Control | Solenoid Off: 12v, On: 1v |
| 36-38 | --- | Not Used | --- |
| 39 | GY/RD | AIS Motor Control | Pulse Signals |
| 40 | BR/WT | AIS Motor Control | Pulse Signals |

### Standard Colors and Abbreviations

| Abbreviation | Color | Abbreviation | Color | Abbreviation | Color |
|--------------|-------|--------------|-------|--------------|-------|
| BK | Black | GY | Gray | RD | Red |
| BL | Blue | GN | Green | TN | Tan |
| BR | Brown | LG | Light Green | VT | Violet |
| DB | Dark Blue | OR | Orange | WT | White |
| DG | Dark Green | PK | Pink | YL | Yellow |

**1993-95 Dakota 2.5L VIN G & K (M/T) Connector - Continued**

| PCM Pin # | Wire Color | Circuit Description (60-Pin) | Value at Hot Idle |
|-----------|------------|------------------------------|-------------------|
| 41 | BK/DG | HO2S-11 (B1 S1) Signal | 0.1-1.1v |
| 42-45 | --- | Not Used | --- |
| 45 | LG | SCI Receive | 5v |
| 46 | --- | Not Used | --- |
| 47 | WT/OR | Vehicle Speed Signal | Digital Signal |
| 48-49 | --- | Not Used | --- |
| 51 | DB/YL | ASD Relay Control | Relay Off: 12v, On: 1v |
| 52 | PK/BK | EVAP Purge Solenoid Control | Solenoid Off: 12v, On: 1v |
| 53-55 | --- | Not Used | --- |
| 56 ('93) | GY/PK | Maintenance Indicator Lamp | Lamp Off: 12v, On: 1v |
| 57-58 | --- | Not Used | --- |
| 59 | PK/BK | AIS Motor Control | Pulse Signals |
| 60 | YL/BK | AIS Motor Control | Pulse Signals |

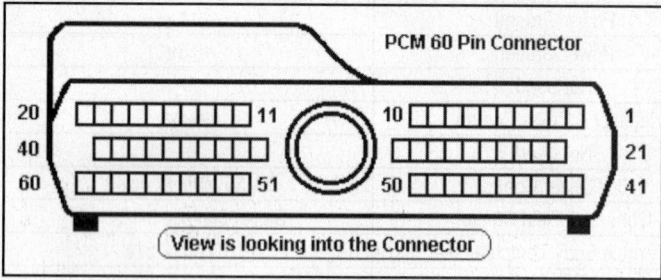

**Pin Connector Graphic**

**Standard Colors and Abbreviations**

| Abbreviation | Color | Abbreviation | Color | Abbreviation | Color |
|--------------|-------|--------------|-------|--------------|-------|
| BK | Black | GY | Gray | RD | Red |
| BL | Blue | GN | Green | TN | Tan |
| BR | Brown | LG | Light Green | VT | Violet |
| DB | Dark Blue | OR | Orange | WT | White |
| DG | Dark Green | PK | Pink | YL | Yellow |

## 1996 Dakota 2.5L VIN P (M/T) 'A' Black Connector

| PCM Pin # | Wire Color | Circuit Description (32-Pin) | Value at Hot Idle |
|-----------|------------|------------------------------|-------------------|
| 1, 3 | --- | Not Used | 12-14v |
| 2 | DB | Ignition Switch Power (B+) | 12-14v |
| 4 | BK/LB | Sensor Ground | <0.050v |
| 5-6 | --- | Not Used | 12-14v |
| 7 | GY | Coil 1 Driver Control | 5°, 55 mph: 8° dwell |
| 8 | GY/BK | CKP Sensor Signal | Digital Signals: 0-5-0v |
| 10 | YL/BK | IAC 2 Driver Control | Pulse Signals |
| 11 | BR/WT | IAC 3 Driver Control | Pulse Signals |
| 12 | OR | PSP Pressure Switch Sense | Straight: 0v, Turning: 5v |
| 13-14 | --- | Not Used | 12-14v |
| 15 | BK/RD | IAT Sensor Signal | At 100°F: 1.83v |
| 16 | TN/BK | ECT Sensor Signal | At 180°F: 2.80v |
| 17 | VT/WT | 5-Volt Supply | 4.9-5.1v |
| 18 | TN/YL | CMP Sensor Signal | Digital Signals: 0-5-0v |
| 19 | GY/RD | IAC 1 Driver Control | Pulse Signals |
| 20 | VT/BK | IAC 4 Driver Control | Pulse Signals |
| 21 | --- | Not Used | 12-14v |
| 22 | RD/WT | Keep Alive Power (B+) | 12-14v |
| 23 | OR/DB | TP Sensor Signal | 0.6-1.0v |
| 24 | TN/WT | HO2S-11 (B1 S1) Signal | 0.1-1.1v |
| 25 | TN/PK | HO2S-12 (B1 S2) Signal | 0.1-1.1v |
| 26 | --- | Not Used | 12-14v |
| 27 | DG/RD | MAP Sensor Signal | 1.5-1.6v |
| 28-30 | --- | Not Used | 12-14v |
| 31 | BK/TN | Power Ground | <0.1v |
| 32 | BK/TN | Power Ground | <0.1v |

## 1996 Dakota 2.5L VIN P (M/T) 'B' White Connector

| PCM Pin # | Wire Color | Circuit Description (32-Pin) | Value at Hot Idle |
|-----------|------------|------------------------------|-------------------|
| 1-3 | --- | Not Used | --- |
| 4 | WT/DB | Injector 1 Driver | 1-4 ms |
| 5 | YL/WT | Injector 3 Driver | 1-4 ms |
| 6-9 | --- | Not Used | --- |
| 10 | DG | Generator Field Control | Digital Signals: 0-12-0v |
| 11 | OR/BK | Upshift Lamp Driver | Lamp Off: 12v, On: 1v |
| 12-14 | --- | Not Used | --- |
| 15 | TN | Injector 2 Driver | 1-4 ms |
| 16 | LB/BR | Injector 4 Driver | 1-4 ms |
| 17-26 | --- | Not Used | --- |
| 27 | WT/OR | Vehicle Speed Signal | Digital Signal |
| 28-30 | --- | Not Used | --- |
| 31 | RD/YL | 5-Volt Supply | 4.9-5.1v |
| 32 | --- | Not Used | --- |

## 1996 Dakota 2.5L VIN P (M/T) 'C' Gray Connector

| PCM Pin # | Wire Color | Circuit Description (32-Pin) | Value at Hot Idle |
|---|---|---|---|
| 1 | DB/OR | A/C Clutch Relay Control | Relay Off: 12v, On: 1v |
| 2 | DB/PK | Radiator Fan Control Relay | Relay Off: 12v, On: 1v |
| 3 | DB/YL | ASD Relay Control | Relay Off: 12v, On: 1v |
| 4 | TN/RD | Speed Control Vacuum Solenoid | Vacuum Increasing: 1v |
| 5 | LG/RD | Speed Control Vent Solenoid | Vacuum Decreasing: 1v |
| 6 | LG/OR | Overdrive Off Lamp Control | Lamp Off: 12v, On: 1v |
| 7 | PK/BK | Transmission Temperature Lamp Control | Lamp Off: 12v, On: 1v |
| 8-10 | --- | Not Used | --- |
| 11 | YL/RD | Speed Control Power Supply | 12-14v |
| 12 | DG/OR | ASD Relay Power (B+) | 12-14v |
| 13 | OR/WT | Overdrive Off Switch Sense | Switch Off: 12v, On: 1v |
| 14 | --- | Not Used | --- |
| 15 | PK/YL | Battery Temperature Sensor | At 86°F: 1.96v |
| 16 | DG/YL | Generator Lamp Control | Lamp Off: 12v, On: 1v |
| 17 | BK/PK | MIL (lamp) Control | MIL Off: 12v, On: 1v |
| 18 | --- | Not Used | --- |
| 19 | BR/YL | Fuel Pump Relay Control | Relay Off: 12v, On: 1v |
| 20 | PK/BK | EVAP Purge Solenoid Control | PWM Signal: 0-12-0v |
| 21 | --- | Not Used | --- |
| 22 | BR | A/C Request Signal | A/C Off: 12v, On: 1v |
| 23 | LG/BK | A/C Select Signal | A/C Off: 12v, On: 1v |
| 24 | WT/PK | Brake Switch Sense Signal | Brake Off: 0v, On: 12v |
| 25 | --- | Not Used | --- |
| 26 | LB/BK | Fuel Level Sensor Signal | Full: 0.56v, 1/2 full: 2.5v |
| 27 | PK | SCI Transmit | 0v |
| 28 | WT/BK | CCD Bus (-) | <0.050v |
| 29 | LG | SCI Receive | 5v |
| 30 | VT/BR | CCD Bus (+) | Digital Signals: 0-5-0v |
| 31 | GY/LB | Tachometer Signal | Pulse Signals |
| 32 | WT/LG | Speed Control Set Switch Signal | S/C & Set Switch On: 3.8v |

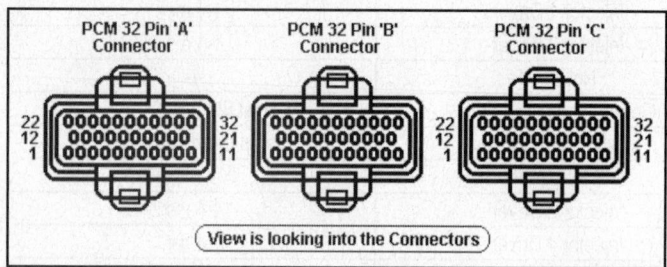

**Pin Connector Graphic**

### 1997 Dakota 2.5L VIN P (M/T) 'A' Black Connector

| PCM Pin # | Wire Color | Circuit Description (32-Pin) | Value at Hot Idle |
|---|---|---|---|
| 1 | --- | Not Used | --- |
| 2 | LG/BK | Ignition Switch Power (B+) | 12-14v |
| 3 | --- | Not Used | --- |
| 4 | BK/LB | Sensor Ground | <0.050v |
| 5 | --- | Not Used | --- |
| 6 | BK/WT | PNP Switch Sense Signal | In P/N: 0v, Others: 5v |
| 7 | BK/GY | Coil 1 Driver Control | 5°, 55 mph: 8° dwell |
| 8 | GY/BK | CKP Sensor Signal | Digital Signals: 0-5-0v |
| 9 | --- | Not Used | --- |
| 10 | YL/BK | IAC 2 Driver Control | Pulse Signals |
| 11 | BR/WT | IAC 3 Driver Control | Pulse Signals |
| 12 | GY/WT | PSP Switch Sense Signal | Straight: 0v, Turning: 5v |
| 13-14 | --- | Not Used | --- |
| 15 | BK/RD | IAT Sensor Signal | At 100°F: 1.83v |
| 16 | TN/BK | ECT Sensor Signal | At 180°F: 2.80v |
| 17 | VT/WT | 5-Volt Supply | 4.9-5.1v |
| 18 | TN/YL | CMP Sensor Signal | Digital Signals: 0-5-0v |
| 19 | GY/RD | IAC 1 Driver Control | Pulse Signals |
| 20 | VT/BK | IAC 4 Driver Control | Pulse Signals |
| 21 | --- | Not Used | --- |
| 22 | RD/WT | Keep Alive Power (B+) | 12-14v |
| 23 | OR/DB | TP Sensor Signal | 0.6-1.0v |
| 24 | TN/WT | HO2S-11 (B1 S1) Signal | 0.1-1.1v |
| 25 | TN/PK | HO2S-12 (B1 S2) Signal | 0.1-1.1v |
| 26 | --- | Not Used | --- |
| 27 | DG/RD | MAP Sensor Signal | 1.5-1.6v |
| 28-30 | --- | Not Used | --- |
| 31 | BK | Power Ground | <0.1v |
| 32 | BK | Power Ground | <0.1v |

### 1997 Dakota 2.5L VIN P (M/T) 'B' White Connector

| PCM Pin # | Wire Color | Circuit Description (32-Pin) | Value at Hot Idle |
|---|---|---|---|
| 1-3 | --- | Not Used | --- |
| 4 | WT/DB | Injector 1 Driver | 1-4 ms |
| 5 | YL/WT | Injector 3 Driver | 1-4 ms |
| 6-8 | --- | Not Used | --- |
| 10 | DG | Generator Field Control | Digital Signals: 0-12-0v |
| 11-14 | --- | Not Used | --- |
| 15 | TN | Injector 2 Driver | 1-4 ms |
| 16 | LB/BR | Injector 4 Driver | 1-4 ms |
| 17-21 | --- | Not Used | --- |
| 22 | --- | Not Used | --- |
| 23 | GY/YL | Engine Oil Pressure Sensor | 1.6v at 24 psi |
| 24-26 | --- | Not Used | --- |
| 27 | WT/OR | Vehicle Speed Signal | Digital Signal |
| 28-30 | --- | Not Used | --- |
| 31 | OR | 5-Volt Supply | 4.9-5.1v |
| 32 | --- | Not Used | --- |

## 1997 Dakota 2.5L VIN P (M/T) 'C' Gray Connector

| PCM Pin # | Wire Color | Circuit Description (32-Pin) | Value at Hot Idle |
|---|---|---|---|
| 1 | DB/OR | A/C Clutch Relay Control | Relay Off: 12v, On: 1v |
| 2 | DB/PK | Radiator Fan Control Relay | Relay Off: 12v, On: 1v |
| 3 | DB/YL | ASD Relay Control | Relay Off: 12v, On: 1v |
| 4 | TN/RD | Speed Control Vacuum Solenoid | Vacuum Increasing: 1v |
| 5 | LG/RD | Speed Control Vent Solenoid | Vacuum Decreasing: 1v |
| 6 | LG/OR | Overdrive 'Off' Lamp Control | Lamp Off: 12v, On: 1v |
| 7-10 | --- | Not Used | --- |
| 11 | YL/RD | Speed Control Power Supply | 12-14v |
| 12 | DG/OR | ASD Relay Power (B+) | 12-14v |
| 13 | OR/WT | Overdrive Off Switch Sense | Switch Off: 12v, On: 1v |
| 14 | --- | Not Used | --- |
| 15 | PK/YL | Battery Temperature Sensor | At 86°F: 1.96v |
| 16-18 | --- | Not Used | --- |
| 19 | LB/OR | Fuel Pump Relay Control | Relay Off: 12v, On: 1v |
| 20 | PK/BK | EVAP Purge Solenoid Control | PWM Signal: 0-12-0v |
| 21 | --- | Not Used | --- |
| 22 | BR | A/C Request Signal | A/C Off: 12v, On: 1v |
| 23 | LG/WT | A/C Select Signal | A/C Off: 12v, On: 1v |
| 24 | WT/PK | Brake Switch Sense Signal | Brake Off: 0v, On: 12v |
| 25 | DG/BK | Generator Field Source | 12-14v |
| 26 | DB | Fuel Level Sensor Signal | Full: 0.56v, 1/2 full: 2.5v |
| 27 | PK | SCI Transmit | 0v |
| 28 | WT/PK | CCD Bus (-) | <0.050v |
| 29 | LG | SCI Receive | 5v |
| 30 | VT/BR | CCD Bus (+) | Digital Signals: 0-5-0v |
| 31 | --- | Not Used | --- |
| 32 | GY/LB | Speed Control Set Switch Signal | S/C & Set Switch On: 3.8v |

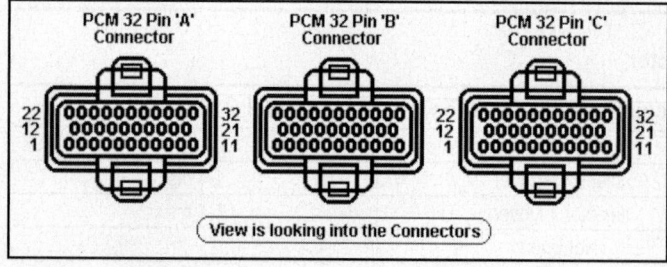

**Pin Connector Graphic**

## Standard Colors and Abbreviations

| Abbreviation | Color | Abbreviation | Color | Abbreviation | Color |
|---|---|---|---|---|---|
| BK | Black | GY | Gray | RD | Red |
| BL | Blue | GN | Green | TN | Tan |
| BR | Brown | LG | Light Green | VT | Violet |
| DB | Dark Blue | OR | Orange | WT | White |
| DG | Dark Green | PK | Pink | YL | Yellow |

**1998-99 Dakota 2.5L VIN P (M/T) 'C1' Black Connector**

| PCM Pin # | Wire Color | Circuit Description (32-Pin) | Value at Hot Idle |
|---|---|---|---|
| 1 | --- | Not Used | --- |
| 2 | LG/BK | Ignition Switch Power (B+) | 12-14v |
| 3 | --- | Not Used | --- |
| 4 | BK/LB | Sensor Ground | <0.050v |
| 5-6 | --- | Not Used | --- |
| 7 | BK/GY | Coil 1 Driver Control | 5°, 55 mph: 8° dwell |
| 8 | GY/BK | CKP Sensor Signal | Digital Signals: 0-5-0v |
| 9 | --- | Not Used | --- |
| 10 | YL/BK | IAC 2 Driver Control | Pulse Signals |
| 11 | BR/WT | IAC 3 Driver Control | Pulse Signals |
| 12 | GY/WT | PSP Switch Sense Signal | Straight: 0v, Turning: 5v |
| 13-14 | --- | Not Used | --- |
| 15 | BK/RD | IAT Sensor Signal | At 100°F: 1.83v |
| 16 | TN/BK | ECT Sensor Signal | At 180°F: 2.80v |
| 17 | PK/WT | 5-Volt Supply | 4.9-5.1v |
| 18 | TN/YL | CMP Sensor Signal | Digital Signals: 0-5-0v |
| 19 | GY/RD | IAC 1 Driver Control | Pulse Signals |
| 20 | VT/BK | IAC 4 Driver Control | Pulse Signals |
| 21 | --- | Not Used | --- |
| 22 | RD/WT | Keep Alive Power (B+) | 12-14v |
| 23 | OR/DB | TP Sensor Signal | 0.6-1.0v |
| 24 | TN/WT | HO2S-11 (B1 S1) Signal | 0.1-1.1v |
| 25 | OR/BK | HO2S-12 (B1 S2) Signal | 0.1-1.1v |
| 26 | --- | Not Used | --- |
| 27 | DG/RD | MAP Sensor Signal | 1.5-1.6v |
| 28-30 | --- | Not Used | --- |
| 31-32 | BK/TN | Power Ground | <0.1v |

**1998-99 Dakota 2.5L VIN P (M/T) 'C2' White Connector**

| PCM Pin # | Wire Color | Circuit Description (32-Pin) | Value at Hot Idle |
|---|---|---|---|
| 1-3 | --- | Not Used | --- |
| 4 | WT/DB | Injector 1 Driver | 1-4 ms |
| 5 | YL/WT | Injector 3 Driver | 1-4 ms |
| 6-9 | --- | Not Used | --- |
| 10 | DG | Generator Field Control | Digital Signals: 0-12-0v |
| 11-14 | --- | Not Used | --- |
| 15 | TN | Injector 2 Driver | 1-4 ms |
| 16 | LB/BR | Injector 4 Driver | 1-4 ms |
| 17-23 | --- | Not Used | --- |
| 23 | GY/YL | Engine Oil Pressure Sensor | 1.6v at 24 psi |
| 24-26 | --- | Not Used | --- |
| 27 | WT/OR | Vehicle Speed Signal | Digital Signal |
| 28-32 | --- | Not Used | --- |

**1998-99 Dakota 2.5L VIN P (M/T) 'C3' Gray Connector**

| PCM Pin # | Wire Color | Circuit Description (32-Pin) | Value at Hot Idle |
|---|---|---|---|
| 1 | DB/OR | A/C Clutch Relay Control | Relay Off: 12v, On: 1v |
| 2 | DB/PK | Radiator Fan Control Relay | Relay Off: 12v, On: 1v |
| 3 | DB/YL | ASD Relay Control | Relay Off: 12v, On: 1v |
| 4 | TN/RD | Speed Control Vacuum Solenoid | Vacuum Increasing: 1v |
| 5 | LG/RD | Speed Control Vent Solenoid | Vacuum Decreasing: 1v |
| 6 | LG/OR | Overdrive 'Off' Lamp Control | Lamp Off: 12v, On: 1v |
| 8-9 | --- | Not Used | --- |
| 10 | YL/DG | LDP Solenoid Control | PWM Signal: 0-12-0v |
| 11 | --- | Not Used | --- |
| 12 | DG/OR | ASD Relay Power (B+) | 12-14v |
| 13 | --- | Not Used | --- |
| 14 | YL/WT | LDP Switch Sense Signal | LDP Switch Closed: 0v |
| 15 | PK/YL | Battery Temperature Sensor | At 86°F: 1.96v |
| 16-18 | --- | Not Used | --- |
| 19 | LB/OR | Fuel Pump Relay Control | Relay Off: 12v, On: 1v |
| 20 | PK/BK | EVAP Purge Solenoid Control | PWM Signal: 0-12-0v |
| 21 | --- | Not Used | --- |
| 22 | BR | A/C Switch Sense | A/C Off: 12v, On: 1v |
| 23 | LG/WT | A/C Select Signal | A/C Off: 12v, On: 1v |
| 24 | WT/PK | Brake Switch Sense Signal | Brake Off: 0v, On: 12v |
| 25 | DG/BK | Generator Field Source | 12-14v |
| 26 | DB | Fuel Level Sensor Signal | Full: 0.56v, 1/2 full: 2.5v |
| 27 | PK | SCI Transmit | 0v |
| 28 | WT/PK | CCD Bus (-) | <0.050v |
| 29 | LG | SCI Receive | 5v |
| 30 | VT/BR | CCD Bus (+) | Digital Signals: 0-5-0v |
| 31 | --- | Not Used | --- |
| 32 | GY/LB | Speed Control Set Switch Signal | S/C & Set Switch On: 3.8v |

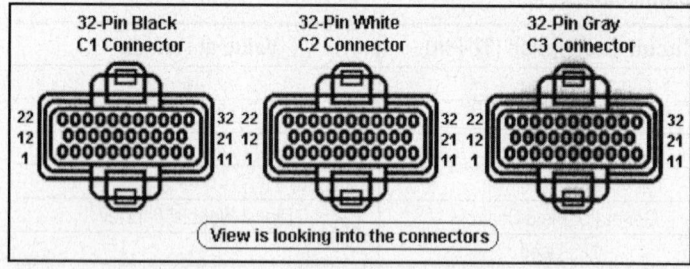

**Pin Connector Graphic**

**2000-01 Dakota 2.5L VIN P 'C1' Black Connector**

| PCM Pin # | Wire Color | Circuit Description (32-Pin) | Value at Hot Idle |
|---|---|---|---|
| 1 | --- | Not Used | --- |
| 2 | LG/BK | Ignition Switch Power (B+) | 12-14v |
| 3 | --- | Not Used | --- |
| 4 | BK/LB | Sensor Ground | <0.050v |
| 5-6 | --- | Not Used | --- |
| 7 | BK/GY | Coil 1 Driver Control | 5°, 55 mph: 8° dwell |
| 8 | GY/BK | CKP Sensor Signal | Digital Signals: 0-5-0v |
| 9 | --- | Not Used | --- |
| 10 | YL/BK | IAC 2 Driver Control | Pulse Signals |
| 11 | BR/WT | IAC 3 Driver Control | Pulse Signals |
| 12 | DB/OR | PSP Switch Sense Signal | Straight: 0v, Turning: 5v |
| 13-14 | --- | Not Used | --- |
| 15 | BK/RD | IAT Sensor Signal | At 100°F: 1.83v |
| 16 | TN/BK | ECT Sensor Signal | At 180°F: 2.80v |
| 17 | OR | 5-Volt Supply | 4.9-5.1v |
| 18 | TN/YL | CMP Sensor Signal | Digital Signals: 0-5-0v |
| 19 | GY/RD | IAC 1 Driver Control | Pulse Signals |
| 20 | VT/BK | IAC 4 Driver Control | Pulse Signals |
| 21 | --- | Not Used | --- |
| 22 | RD/WT | Keep Alive Power (B+) | 12-14v |
| 23 | OR/DB | TP Sensor Signal | 0.6-1.0v |
| 24 | TN/WT | HO2S-11 (B1 S1) Signal | 0.1-1.1v |
| 25 | OR/BK | HO2S-12 (B1 S2) Signal | 0.1-1.1v |
| 26 | --- | Not Used | --- |
| 27 | DG/RD | MAP Sensor Signal | 1.5-1.6v |
| 28-30 | --- | Not Used | --- |
| 31-32 | BK/TN | Power Ground | <0.1v |

**2000-01 Dakota 2.5L VIN P 'C2' White Connector**

| PCM Pin # | Wire Color | Circuit Description (32-Pin) | Value at Hot Idle |
|---|---|---|---|
| 1-3 | --- | Not Used | --- |
| 4 | WT/DB | Injector 1 Driver | 1-4 ms |
| 5 | YL/WT | Injector 3 Driver | 1-4 ms |
| 6-9 | --- | Not Used | --- |
| 10 | DG | Generator Field Control | Digital Signals: 0-12-0v |
| 11-14 | --- | Not Used | --- |
| 15 | TN | Injector 2 Driver | 1-4 ms |
| 16 | LB/BR | Injector 4 Driver | 1-4 ms |
| 17-18 | --- | Not Used | --- |
| 19 | DB | AC Pressure Sensor Signal | 0.90v at 79 psi |
| 20-22 | --- | Not Used | --- |
| 23 | GY/YL | Engine Oil Pressure Sensor | 1.6v at 24 psi |
| 24-26 | --- | Not Used | --- |
| 27 | WT/OR | Vehicle Speed Signal | Digital Signal |
| 28-32 | --- | Not Used | --- |

## 2000-01 Dakota 2.5L VIN P 'C3' Gray Connector

| PCM Pin # | Wire Color | Circuit Description (32-Pin) | Value at Hot Idle |
|---|---|---|---|
| 1 | DB/OR | A/C Clutch Relay Control | Relay Off: 12v, On: 1v |
| 2 | DB/PK | Radiator Fan Control Relay | Relay Off: 12v, On: 1v |
| 3 | DB/YL | ASD Relay Control | Relay Off: 12v, On: 1v |
| 4 | TN/RD | Speed Control Vacuum Solenoid | Vacuum Increasing: 1v |
| 5 | LG/RD | Speed Control Vent Solenoid | Vacuum Decreasing: 1v |
| 6-9 | --- | Not Used | --- |
| 10 | WT/DG | LDP Solenoid Control | PWM Signal: 0-12-0v |
| 11 | YL/RD | Speed Control Power Supply | 12-14v |
| 12 | DG/OR | ASD Relay Power (B+) | 12-14v |
| 13 | YL/DG | Torque Mgmt. Request Sense | Digital Signals |
| 14 | OR | LDP Switch Sense Signal | LDP Switch Closed: 0v |
| 15 | PK/YL | Battery Temperature Sensor | At 86°F: 1.96v |
| 16-18 | --- | Not Used | --- |
| 19 ('00) | LB/OR | Fuel Pump Relay Control | Relay Off: 12v, On: 1v |
| 19 ('01) | BR | Fuel Pump Relay Control | Relay Off: 12v, On: 1v |
| 20 | PK/BK | EVAP Purge Solenoid Control | PWM Signal: 0-12-0v |
| 21 | --- | Not Used | --- |
| 22 | BR | A/C Switch Sense | A/C Off: 12v, On: 1v |
| 23 | LG/WT | A/C Select Signal | A/C Off: 12v, On: 1v |
| 24 | WT/PK | Brake Switch Sense Signal | Brake Off: 0v, On: 12v |
| 25 | WT/DB | Generator Field Source | 12-14v |
| 26 | DB | Fuel Level Sensor Signal | Full: 0.56v, 1/2 full: 2.5v |
| 27 | PK | SCI Transmit | 0v |
| 27 ('00) | PK/DB | SCI Transmit | 0v |
| 28 ('00) | WT/BK | CCD Bus (-) | <0.050v |
| 28 ('01) | --- | Not Used | --- |
| 29 | LG | SCI Receive | 5v |
| 30 ('00) | VT/BR | CCD Bus (+) | Digital Signals: 0-5-0v |
| 30 ('01) | VT/BR | PCI Bus Signal (J1850) | Digital Signals: 0-5-0v |
| 31 | --- | Not Used | --- |
| 32 | RD/LG | Speed Control Set Switch Signal | S/C & Set Switch On: 3.8v |

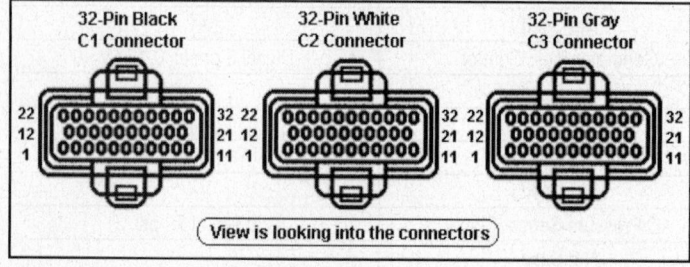

**Pin Connector Graphic**

## 2004 Dakota 3.7L, 4.7L 'C1' Black Connector

| PCM Pin # | Wire Color | Circuit Description (38-Pin) | Value at Hot Idle |
|---|---|---|---|
| 1 | LB/RD (4.7L) | Coil Control No. 8 | |
| 2 | --- | Not Used | --- |
| 3 | BR (4.7L) | Coil Control No. 7 | |
| 4 | GY/LB (4.7L) | Injector Control No. 8 | |
| 5 | VT (4.7L) | Injector Control No. 7 | |
| 6 | --- | Not Used | --- |
| 7 | --- | Not Used | --- |
| 8 | --- | Not Used | --- |
| 9 | BK/TN | Ground | |
| 10 | --- | Not Used | --- |
| 11 | LG/BK | Fused Ignition Switch Output (Run-Start) | |
| 12 | RD/WT | Fused Ignition Switch Output (Off-Run-Start) | |
| 13 | WT/OR | Vehicle Speed Signal | |
| 14 | --- | Not Used | --- |
| 15 | --- | Not Used | --- |
| 16 | --- | Not Used | --- |
| 17 | --- | Not Used | --- |
| 18 | BK/TN | Ground | |
| 19 | --- | Not Used | --- |
| 20 | GY/YL | Oil Pressure Signal | |
| 21 | LB/BR | A/C Pressure Signal | |
| 22 | VT/LG | Ambient Temperature Sensor Signal | |
| 23 | --- | Not Used | --- |
| 24 | --- | Not Used | --- |
| 25 | LG | SCI Receive | |
| 26 | PK/LB | SCI Receive (TCM) | |
| 27 | VT/WT | 5 Volt Supply | |
| 28 | --- | Not Used | --- |
| 29 | RD/WT | Fused B (+) | |
| 30 | RD/YL | Fused Ignition Switch Output (Start) | |
| 31 | TN/WT | O2 1/2 Signal | |
| 32 | BR/DG | O2 Return (Up) | |
| 33 | TN/WT | O2 2/2 Signal | |
| 34 | --- | Not Used | --- |
| 35 | --- | Not Used | --- |
| 36 | PK | SCI Transmit | |
| 37 | WT/DG | SCI Transmit (TCM) | |
| 38 | WT/VT | PCI Bus (PCM) | |

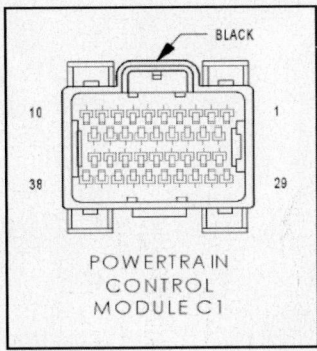

POWERTRAIN
CONTROL
MODULE C1

## 2004 Dakota 3.7L, 4.7L 'C2' Orange Connector

| PCM Pin # | Wire Color | Circuit Description (38-Pin) | Value at Hot Idle |
|---|---|---|---|
| 1 | TN/LB | Coil Control No. 6 | |
| 2 | TN/DG | Coil Control No. 5 | |
| 3 | TN/LG | Coil Control No. 4 | |
| 4 | BR/DB | Injector Control No. 6 | |
| 5 | GY | Fuel Injector No. 5 Driver | |
| 6 | --- | Not Used | --- |
| 7 | TN/OR | Coil Control No. 3 | |
| 8 | --- | Not Used | --- |
| 9 | TN/PK | Coil Control No. 2 | |
| 10 | TN/RD | Coil Control No. 1 | |
| 11 | LB/BR | Injector Control No. 4 | |
| 12 | YL/WT | Injector Control No. 3 | |
| 13 | TN | Injector Control No. 2 | |
| 14 | WT/DB | Injector Control No. 1 | |
| 15 | --- | Not Used | --- |
| 16 | --- | Not Used | --- |
| 17 | BR/WT | O2 2/1 Heater Control | |
| 18 | BR/OR | O2 1/1 Heater Control | |
| 19 | DG | Generator Field Driver | |
| 20 | VT/OR | ECT Sensor | |
| 21 | BR/OR | TP No.1 Signal | |
| 22 | --- | Not Used | --- |
| 23 | VT/BR | MAP Signal | |
| 24 | BR/LG (3.7L) | Knock Sensor No.1 Return | |
| 25 | DB/YL (3.7L) | Knock Sensor No.1 Signal | |
| 26 | --- | Not Used | --- |
| 27 | BK/LB | Sensor Ground 1 | |
| 28 | YL/BK | IAC Return | |
| 29 | OR | 5 Volt Supply | |
| 30 | BK/RD | IAT Signal | |
| 31 | BK/DG | O2 1/1 Signal | |
| 32 | DB/DG | O2 Return (Down) | |
| 33 | LG/RD | O2 2/1 Signal | |
| 34 | DB/GY | CMP Signal | |
| 35 | DB/WT | CKP Signal | |
| 36 | BR/WT (3.7L) | Knock Sensor No.2 Signal | |
| 37 | WT/BR (3.7L) | Knock Sensor No.2 Return | |
| 38 | GY/RD | IAC Motor Control | |

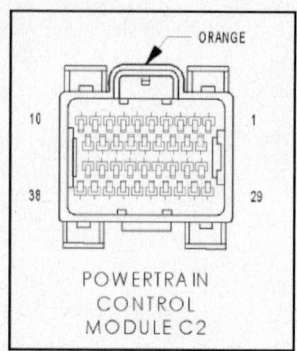

POWERTRAIN
CONTROL
MODULE C2

## 2004 Dakota 3.7L, 4.7L 'C3' White Connector

| PCM Pin # | Wire Color | Circuit Description (38-Pin) | Value at Hot Idle |
|---|---|---|---|
| 1 | --- | Not Used | --- |
| 2 | --- | Not Used | --- |
| 3 | DB/YL | ASD Relay Control | |
| 4 | --- | Not Used | --- |
| 5 | LG/RD | S/C Vent Control | |
| 6 | DB/PK | Radiator Fan Relay Control | |
| 7 | VT/YL | Speed Control Supply | |
| 8 | WT/DG | NVLD Solenoid Control | |
| 9 | BR/WT | O2 1/2 Heater Control | |
| 10 | BR/GY | O2 2/2 Heater Control | |
| 11 | LB/OR | A/C Clutch Relay Control | |
| 12 | VT/DG | S/C Vacuum Control | |
| 13 | --- | Not Used | --- |
| 14 | --- | Not Used | --- |
| 15 | --- | Not Used | --- |
| 16 | --- | Not Used | --- |
| 17 | --- | Not Used | --- |
| 18 | --- | Not Used | --- |
| 19 | DG/OR | ASD Relay Output | |
| 20 | PK/BK | EVAP/Purge Control | |
| 21 | DG/GY A/T | TRS T41 Sense | |
| 21 | BK/WT M/T | TRS T41 Sense | |
| 22 | --- | Not Used | --- |
| 23 | WT/PK | Brake Switch Signal | |
| 24 | --- | Not Used | --- |
| 25 | --- | Not Used | --- |
| 26 | --- | Not Used | --- |
| 27 | --- | Not Used | --- |
| 28 | DG/OR | ASD Relay Output | |
| 29 | DB/BR | EVAP Purge Signal | |
| 30 | --- | Not Used | --- |
| 31 | --- | Not Used | --- |
| 32 | PK/YL | AAT Signal | |
| 33 | DB/WT | Fuel Level Signal | |
| 34 | VT/TN | S/C Switch Signal | |
| 35 | OR | NVLD Switch Sense | |
| 36 | --- | Not Used | --- |
| 37 | BR | Fuel Pump Relay Control | |
| 38 | DG/OR | Starter Relay Control | |

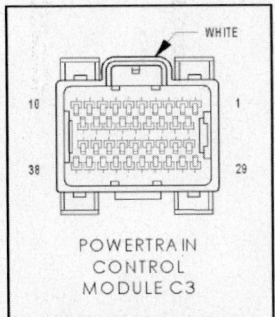

POWERTRAIN
CONTROL
MODULE C3

## 2004 Dakota 3.7L, 4.7L MPI 'C4' Green Connector

| PCM Pin # | Wire Color | Circuit Description (38-Pin) | Value at Hot Idle |
|---|---|---|---|
| 1 | YL/GY | Overdrive Solenoid Control | |
| 2 | DG/WT | 4C Solenoid Control | |
| 2 | YL/LB (3.7L) | Underdrive Solenoid Control | |
| 3 | --- | Not Used | --- |
| 4 | YL/DB | MS Solenoid Control | |
| 5 | --- | Not Used | --- |
| 6 | WT/DB | 2C Solenoid Control | |
| 6 | YL/DB (3.7L) | 2-4 Solenoid Control | |
| 7 | --- | Not Used | --- |
| 8 | YL/LB | Underdrive Solenoid Control | |
| 9 | --- | Not Used | --- |
| 10 | LG | L/R Solenoid Control | |
| 10 | DG/WT (3.7L) | Low/Reverse Solenoid Control | |
| 11 | VT/LG | Pressure Control Solenoid Control | |
| 12 | BK/RD | Ground | |
| 13 | BK/RD | Ground | |
| 14 | BK/RD | Ground | |
| 15 | DG/LB | TRS T1 Sense | |
| 16 | DG/DB | TRS T3 Sense | |
| 17 | OR/WT | Tow/Haul Overdrive Off Switch Sense | |
| 18 | YL/BR | Transmission Control Relay Control | |
| 19 | RD | Transmission Control Relay Output | |
| 20 | DB (4.7L) | 4C Pressure Switch Sense | |
| 21 | GY (4.7L) | Underdrive Pressure Switch Sense | |
| 22 | YL/LG | Overdrive Pressure Switch Sense | |
| 23-25 | --- | Not Used | --- |
| 26 | PK/OR (4.7L) | TRS T2 Sense | |
| 27 | --- | Not Used | --- |
| 28 | RD | Transmission Control Relay Output | |
| 29 | YL/TN | Low/Reverse Pressure Switch Sense | |
| 30 | YL | 2C Pressure Switch Sense | |
| 31 | VT/TN (4.7L) | Line Pressure Sensor Signal | |
| 32 | LG/WT | Output Speed Sensor Signal | |
| 33 | RD/BK | Input Speed Sensor Signal | |
| 34 | DB/BK | Speed Sensor Ground | |
| 35 | DG/OR | Transmission Temperature Sensor Signal | |
| 36 | --- | Not Used | --- |
| 37 | DG/YL | TRS T42 Sense | |
| 38 | RD (4.7L) | Transmission Control Relay Output | |

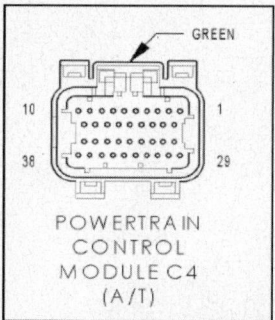

POWERTRAIN
CONTROL
MODULE C4
(A/T)

## 2005 Dakota 3.7L VIN K 'C1' Black/Black Connector

| PCM Pin # | Wire Color | Circuit Description (38-Pin) | Value at Hot Idle |
|---|---|---|---|
| 1 | --- | Not Used | --- |
| 2 | --- | Not Used | --- |
| 3 | --- | Not Used | --- |
| 4 | --- | Not Used | --- |
| 5 | --- | Not Used | --- |
| 6 | --- | Not Used | --- |
| 7 | --- | Not Used | --- |
| 8 | --- | Not Used | |
| 9 | BK | Ground | |
| 10 | --- | Not Used | --- |
| 11 | PK/GY | Fused Ignition Switch Output (Run-Start) | |
| 12 | PK/WT | Fused Ignition Switch Output (Run-Start) | |
| 13 | --- | Not Used | --- |
| 14 | --- | Not Used | --- |
| 15 | --- | Not Used | --- |
| 16 | --- | Not Used | --- |
| 17 | --- | Not Used | --- |
| 18 | BK | Ground | |
| 19 | --- | Not Used | --- |
| 20 | VT/GY | Engine Oil Pressure Signal | |
| 21 | --- | Not Used | --- |
| 22 | --- | Not Used | --- |
| 23 | --- | Not Used | --- |
| 24 | --- | Not Used | --- |
| 25 | WT/LG | SCI Receive (PCM) | |
| 26 | WT/OR | SCI Receive (TCM) | |
| 27 | YL/PK | 5 Volt Supply | |
| 28 | --- | Not Used | --- |
| 29 | RD | Fused B (+) | |
| 30 | PK/YL | Fused Ignition Switch Output (Run) | |
| 31 | DB/YL | O2 1/2 Signal | |
| 32 | BR/DG | O2 Return (Upstream) | |
| 33 | BR | O2 2/2 Signal | |
| 34 | WT/LG | CAN C Bus (+) | |
| 35 | WT/LB | CAN C Bus (-) | |
| 36 | WT/GY | SCI Transmit (PCM) | |
| 37 | BR/WT (A/T) | SCI Transmit (TCM) | |
| 38 | --- | Not Used | --- |

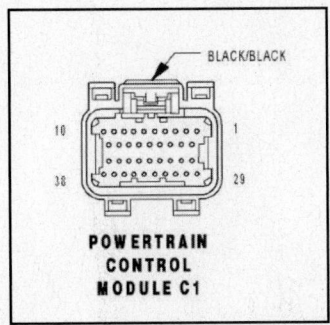

BLACK/BLACK

10    1

38    29

POWERTRAIN
CONTROL
MODULE C1

### 2005 Dakota 3.7L VIN K 'C2' Black/Orange Connector

| PCM Pin # | Wire Color | Circuit Description (38-Pin) | Value at Hot Idle |
|:---:|:---:|:---:|:---:|
| 1 | DB/OR | Coil Control No. | |
| 2 | DB/YL | Coil Control No. 5 | |
| 3 | DB | Coil Control No. 4 | |
| 4 | BR/VT | Injector Control No. | |
| 5 | BR/OR | Injector Control No. 5 | |
| 6 | --- | Not Used | --- |
| 7 | DB/OR | Coil Control No. 3 | |
| 8 | --- | Not Used | --- |
| 9 | DB/TN | Coil Control No. 2 | |
| 10 | DB/DG | Coil Control No. 1 | |
| 11 | BR/TN | Injector Control No. 4 | |
| 12 | BR/LB | Injector Control No. 3 | |
| 13 | BR/DB | Injector Control No. 2 | |
| 14 | BR/YL | Injector Control No. 1 | |
| 15 | --- | Not Used | --- |
| 16 | --- | Not Used | --- |
| 17 | BR/VT | O2 2/1 Heater Control | |
| 18 | BR/TN | O2 1/1 Heater Control | |
| 19 | BR/GY | Gen Field Control | |
| 20 | VT/OR | ECT Signal | |
| 21 | BR/OR | TP No. 1 Signal | |
| 22 | --- | Not Used | --- |
| 23 | VT/BR | MAP Signal | |
| 24 | BR/LG | Knock Sensor No. 1 Return | |
| 25 | DB/OR | Knock Sensor No. 1 Signal | |
| 26 | --- | Not Used | --- |
| 27 | DB/DG | Sensor Ground | |
| 28 | BR/VT | IAC Signal | |
| 29 | PK/YL | 5 Volt Supply | |
| 30 | BR/WT | IAT Signal | |
| 31 | DB/LB | O2 1/1 Signal | |
| 32 | DB/DG | O2 Return (Downstream) | |
| 33 | DB/LG | O2 2/1 Signal | |
| 34 | DB/GY | CMP Signal | |
| 35 | BR/LB | CKP Signal | |
| 36 | BR/WT | Knock Sensor No. 2 Signal | |
| 37 | PK/RD | Knock Sensor No. 2 Return | |
| 38 | VT/GY | IAC Control | |

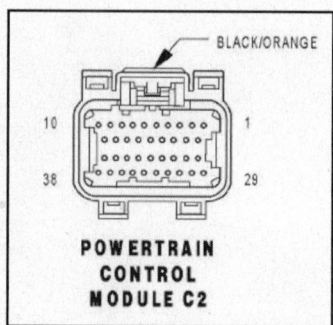

POWERTRAIN
CONTROL
MODULE C2

## 2005 Dakota 3.7L VIN K 'C3' Black/Natural

| PCM Pin # | Wire Color | Circuit Description (38-Pin) | Value at Hot Idle |
|---|---|---|---|
| 1 | --- | Not Used | ---- |
| 2 | --- | Not Used | ---- |
| 3 | BR/GY | ASD Relay Control | |
| 4 | --- | Not Used | ---- |
| 5 | VT/OR | S/C Vent Sol Control | |
| 6 | --- | Not Used | ---- |
| 7 | VT/YL | S/C Supply | |
| 8 | VT/LB | NVLD Sol Control | |
| 9 | BR/OR | O2 1/2 Heater Control | |
| 10 | BR/GY | O2 2/2 Heater Control | |
| 11 | LB/OR | A/C Clutch Relay Control | |
| 12 | VT/YL | S/C Vacuum Sol Control | |
| 13 | --- | Not Used | ---- |
| 14 | --- | Not Used | ---- |
| 15 | --- | Not Used | ---- |
| 16 | --- | Not Used | ---- |
| 17 | --- | Not Used | ---- |
| 18 | --- | Not Used | ---- |
| 19 | RD | ASD Relay Output | |
| 20 | DB/WT | EVAP Purge Sol Control | |
| 21 | YL/DB (A/T) | TRS T41 Sense (P/N Sense) | |
| 21 | YL/OR (M/T) | TRS T141 Sense (P/N Sense) | |
| 22 | --- | Not Used | ---- |
| 23 | DG/WT | Brake Switch Signal | |
| 24 | --- | Not Used | ---- |
| 25 | --- | Not Used | ---- |
| 26 | --- | Not Used | ---- |
| 27 | --- | Not Used | ---- |
| 28 | RD | ASD Relay Output | |
| 29 | DB/BR | EVAP Purge Sol Signal | |
| 30 | --- | Not Used | ---- |
| 31 | --- | Not Used | ---- |
| 32 | --- | Not Used | ---- |
| 33 | --- | Not Used | ---- |
| 34 | VT | S/C Switch No. 1 Signal | |
| 35 | VT/WT | NVLD Switch Signal | |
| 36 | --- | Not Used | ---- |
| 37 | BR | Fuel Pump Relay Control | |
| 38 | DG/OR | Starter Relay Control | |

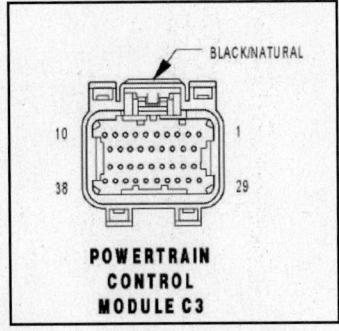

BLACK/NATURAL

10      1

38      29

**POWERTRAIN
CONTROL
MODULE C3**

## 2005 Dakota 3.7L VIN K (42RLE) 'C4' Black/Green

| PCM Pin # | Wire Color | Circuit Description (38-Pin) | Value at Hot Idle |
|:---:|:---:|:---:|:---:|
| 1 | YL/RD | OD Solenoid Control | |
| 2 | VT/PK | UD Solenoid Control | |
| 3 | --- | Not Used | ---- |
| 4 | --- | Not Used | ---- |
| 5 | --- | Not Used | ---- |
| 6 | YL/DB | 2-4 Solenoid Control | |
| 7 | --- | Not Used | ---- |
| 8 | --- | Not Used | ---- |
| 9 | --- | Not Used | ---- |
| 10 | DG/WT | L/R Solenoid Control | |
| 11 | --- | Not Used | ---- |
| 12 | BK | Ground | |
| 13 | BK | Ground | |
| 14 | BK | Ground | |
| 15 | YL/PK | TRS T1 Sense | |
| 16 | YL/PK | TRS T3 Sense | |
| 17 | DG | Tow/Haul Overdrive Off Switch Sense | |
| 18 | YL/BK | Transmission Control Relay Control | |
| 19 | YL/OR | Transmission Control Relay Output | |
| 20 | --- | Not Used | ---- |
| 21 | --- | Not Used | ---- |
| 22 | DG/TN | OD Pressure Switch Sense | |
| 23 | --- | Not Used | ---- |
| 24 | --- | Not Used | ---- |
| 25 | --- | Not Used | ---- |
| 26 | --- | Not Used | ---- |
| 27 | --- | Not Used | ---- |
| 28 | YL/OR | Transmission Control Relay Output | |
| 29 | VT/RD | L/R Pressure Switch Sense | |
| 30 | YL/DG | 2-4 Pressure Switch Sense | |
| 31 | --- | Not Used | ---- |
| 32 | YL/PK | Output Speed Sensor Signal | |
| 33 | YL/PK | Input Speed Sensor Signal | |
| 34 | DG/VT | Sensor Ground | |
| 35 | YL/PK | Transmission Temperature Sensor Signal | |
| 36 | --- | Not Used | ---- |
| 37 | YL/PK | TRS T42 Sense | |
| 38 | YL/OR | Transmission Control Relay Output | |

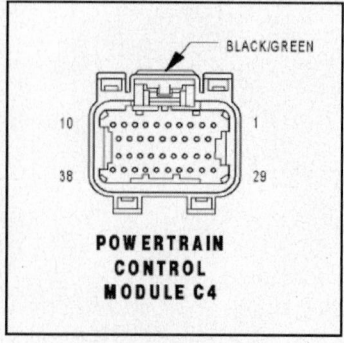

**POWERTRAIN CONTROL MODULE C4**

### 2005 Dakota 3.7L VIN K (545RFE) 'C4' Black/Green

| PCM Pin # | Wire Color | Circuit Description (38-Pin) | Value at Hot Idle |
|-----------|------------|-----------------------------|-------------------|
| 1 | YL/RD | OD Solenoid Control | |
| 2 | YL/DG | 4C Solenoid Control | |
| 3 | --- | Not Used | ---- |
| 4 | BR/YL | MS Solenoid Control | |
| 5 | --- | Not Used | ---- |
| 6 | DG/YL | 2C Solenoid Control | |
| 7 | --- | Not Used | ---- |
| 8 | VT/PK | UD Solenoid Control | |
| 9 | --- | Not Used | ---- |
| 10 | DG/WT | L/R Solenoid Control | |
| 11 | DG | Pressure Control Solenoid Control | |
| 12 | BK | Ground | |
| 13 | BK | Ground | |
| 14 | BK | Ground | |
| 15 | YL/PK | TRS T1 Sense | |
| 16 | YL/PK | TRS T3 Sense | |
| 17 | DG | Tow/Haul Overdrive Off Switch Sense | |
| 18 | YL/BK | Transmission Control Relay Control | |
| 19 | YL/OR | Transmission Control Relay Output | |
| 20 | BR/YL | 4C Pressure Switch Sense | |
| 21 | LB/RD | UD Pressure Switch Sense | |
| 22 | DG/TN | OD Pressure Switch Sense | |
| 23 | --- | Not Used | ---- |
| 24 | --- | Not Used | ---- |
| 25 | --- | Not Used | ---- |
| 26 | BR/YL | TRS T2 Sense | |
| 27 | --- | Not Used | ---- |
| 28 | YL/OR | Transmission Control Relay Output | |
| 29 | VT/RD | L/R Pressure Switch Sense | |
| 30 | BR/YL | 2C Pressure Switch Sense | |
| 31 | YL/BR | Line Pressure Sensor Signal | |
| 32 | YL/PK | Output Speed Sensor Signal | |
| 33 | YL/PK | Input Speed Sensor Signal | |
| 34 | DG/VT | Sensor Ground | |
| 35 | YL/PK | Transmission Temperature Sensor Signal | |
| 36 | --- | Not Used | ---- |
| 37 | YL/PK | TRS T42 Sense | |
| 38 | YL/OR | Transmission Control Relay Output | |

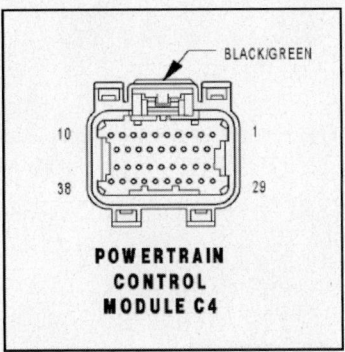

BLACK/GREEN

10
38
1
29

**POWERTRAIN
CONTROL
MODULE C4**

**1992-95 Dakota & Ram Pickup & Van 3.9L VIN X 60-Pin Connector**

| PCM Pin # | Wire Color | Circuit Description (60-Pin) | Value at Hot Idle |
|---|---|---|---|
| 1 | DG/RD | MAP Sensor Signal | 1.5-1.6v |
| 2 | TN/BK | ECT Sensor Signal | At 180°F: 2.80v |
| 3 | RD/WT | Keep Alive Power (B+) | 12-14v |
| 4 | BK/LB | Sensor Ground | <0.050v |
| 5 ('92-'93) | BK/WT | Sensor Ground | <0.050v |
| 5 ('94-'95) | --- | Not Used | --- |
| 6 | PK/WT | 5-Volt Supply | 4.9-5.1v |
| 7 | OR | 8-Volt Supply | 7.9-8.1v |
| 8 | --- | Not Used | |
| 9 | DB | Ignition Switch Power (B+) | 12-14v |
| 10 | OR/WT | Overdrive Override Switch | Switch Off: 12v, On: 1v |
| 11 | BK/TN | Power Ground | <0.1v |
| 12 | BK/TN | Power Ground | <0.1v |
| 13 | LB/BR | Injector 4 Driver Control | 1-4 ms |
| 14 | YL/WT | Injector 3 Driver Control | 1-4 ms |
| 15 | TN | Injector 2 Driver Control | 1-4 ms |
| 16 | WT/DG | Injector 1 Driver Control | 1-4 ms |
| 17-18 | --- | Not Used | --- |
| 19 | BK/GY | Coil Driver Control | 5°, 55 mph: 8° dwell |
| 20 | DG | Alternator Field Control | Digital Signals: 0-12-0v |
| 21 | BK/RD | Charge Temperature Sensor | At 100°F: 2.51v |
| 22 | OR/DB | TP Sensor Signal | 0.6-1.0v |
| 23 | --- | Not Used | --- |
| 24 | GY/BK | Distributor Reference Signal | Digital Signals: 0-5-0v |
| 25 | PK | SCI Transmit | 0v |
| 26 | --- | Not Used | --- |
| 27 | BR | A/C Clutch Signal | A/C Off: 12v, On: 1v |
| 28 | --- | Not Used | --- |
| 29 | WT/PK | Brake Switch Sense Signal | Brake Off: 0v, On: 12v |
| 30 | BR/YL | PNP Switch Sense Signal | In P/N: 0v, Others: 5v |
| 31 | --- | Not Used | --- |
| 32 | PK/BK | MIL (lamp) Control | MIL Off: 12v, On: 1v |
| 33 | TN/RD | Speed Control Vacuum Solenoid | Vacuum Increasing: 1v |
| 34 | DB/OR | A/C WOT Relay Control | Relay Off: 12v, On: 1v |
| 35 | GY/YL | EGR Solenoid Control | 12v, at 55 mph: 1v |

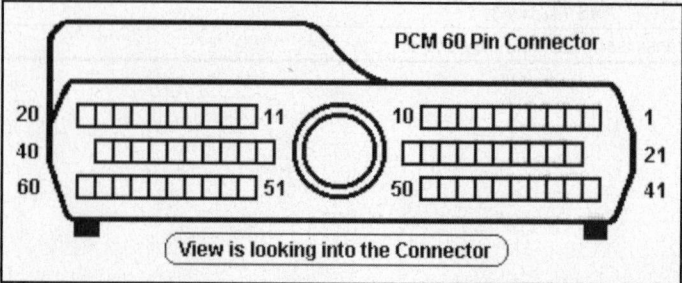

**PCM 60 Pin Connector**

View is looking into the Connector

**Pin Connector Graphic**

**1992-95 Dakota & Ram Pickup & Van 3.9L VIN X 60-Pin Connector - Continued**

| PCM Pin # | Wire Color | Circuit Description (60-Pin) | Value at Hot Idle |
|---|---|---|---|
| 36 | --- | Not Used | --- |
| 37 | BK/OR | Overdrive Lamp Control | Lamp Off: 12v, On: 1v |
| 38 | PK/BK | Injector 5 Driver Control | 1-4 ms |
| 39 | GY/RD | AIS Motor 4 Control | Pulse Signals |
| 40 | BR/WT | AIS Motor 2 Control | Pulse Signals |
| 41 | BK/DG | HO2S-11 (B1 S1) Signal | 0.1-1.1v |
| 42 | --- | Not Used | --- |
| 43 | GY/LB | Tachometer Signal | Pulse Signals |
| 44 | TN/YL | Distributor Sync Signal | Digital Signals: 0-5-0v |
| 45 | LG | SCI Receive | 5v |
| 46 | --- | Not Used | --- |
| 47 | WT/OR | Vehicle Speed Signal | Digital Signal |
| 48 | BR/RD | Speed Control Set Switch Signal | S/C & Set Switch On: 3.8v |
| 49 | YL/RD | Speed Control On/Off Switch | S/C On: 1v, Off: 12v |
| 50 | WT/LG | Speed Control Resume Switch | S/C On: 1v, Off: 12v |
| 51 | DB/YL | ASD Relay Control | Relay Off: 12v, On: 1v |
| 52 | PK/BK | EVAP Purge Solenoid Control | Solenoid Off: 12v, On: 1v |
| 53 | LG/RD | Speed Control Vent Solenoid | Vacuum Decreasing: 1v |
| 54 | OR/BK | Part Throttle Unlock Solenoid | Solenoid Off: 12v, On: 1v |
| 55 | OR/WT | Overdrive Lockout Solenoid | Solenoid Off: 12v, On: 1v |
| 56 ('92-'93) | GY/PK | Maintenance Indicator Lamp | Lamp Off: 12v, On: 1v |
| 57 | DG/BK | ASD Relay Power (B+) | 12-14v |
| 58 | LG/BK | Injector 6 Driver Control | 1-4 ms |
| 59 | PK/BK | AIS Motor 1 Control | Pulse Signals |
| 60 | GY/RD | AIS Motor 3 Control | Pulse Signals |

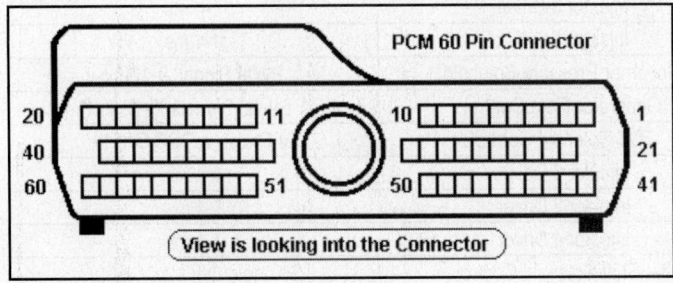

**Pin Connector Graphic**

**Standard Colors and Abbreviations**

| Abbreviation | Color | Abbreviation | Color | Abbreviation | Color |
|---|---|---|---|---|---|
| BK | Black | GY | Gray | RD | Red |
| BL | Blue | GN | Green | TN | Tan |
| BR | Brown | LG | Light Green | VT | Violet |
| DB | Dark Blue | OR | Orange | WT | White |
| DG | Dark Green | PK | Pink | YL | Yellow |

### 1996 Dakota 3.9L VIN X 'A' Black Connector

| PCM Pin # | Wire Color | Circuit Description (32-Pin) | Value at Hot Idle |
|---|---|---|---|
| 2 | DB | Ignition Switch Power (B+) | 12-14v |
| 4 | BK/LB | Sensor Ground | <0.050v |
| 6 | BR/YL | PNP Switch Sense Signal | In P/N: 0v, Others: 5v |
| 7 | GY | Coil 1 Driver Control | 5°, 55 mph: 8° dwell |
| 8 | GY/BK | CKP Sensor Signal | Digital Signals: 0-5-0v |
| 10 | YL/BK | IAC 2 Driver Control | Pulse Signals |
| 11 | BR/WT | IAC 3 Driver Control | Pulse Signals |
| 15 | BK/RD | IAT Sensor Signal | At 100°F: 1.83v |
| 16 | TN/BK | ECT Sensor Signal | At 180°F: 2.80v |
| 17 | PK/WT | 5-Volt Supply | 4.9-5.1v |
| 18 | TN/YL | CMP Sensor Signal | Digital Signals: 0-5-0v |
| 19 | GY/RD | IAC 1 Driver Control | Pulse Signals |
| 20 | PK/BK | IAC 4 Driver Control | Pulse Signals |
| 22 | RD/WT | Keep Alive Power (B+) | 12-14v |
| 23 | OR/DB | TP Sensor Signal | 0.6-1.0v |
| 24 | TN/WT | HO2S-11 (B1 S1) Signal | 0.1-1.1v |
| 25 | TN/PK | HO2S-12 (B1 S2) Signal | 0.1-1.1v |
| 27 | DG/RD | MAP Sensor Signal | 1.5-1.6v |
| 31-32 | BK/TN | Power Ground | <0.1v |

### 1996 Dakota 3.9L VIN X 'B' White Connector

| PCM Pin # | Wire Color | Circuit Description (32-Pin) | Value at Hot Idle |
|---|---|---|---|
| 1 | PK | Transmission Temperature Sensor Signal | At 200°F: 2.40v |
| 4 | WT/DB | Injector 1 Driver | 1-4 ms |
| 5 | YL/WT | Injector 3 Driver | 1-4 ms |
| 6 | GY | Injector 5 Driver | 1-4 ms |
| 8 | PK | Governor Pressure Solenoid | PWM Signal: 0-12-0v |
| 10 | DG | Generator Field Control | Digital Signals: 0-12-0v |
| 11 | OR/BK | TCC Solenoid Control | At Cruise w/TCC On: 1v |
| 12 | BR/DB | Injector 6 Driver | 1-4 ms |
| 15 | TN | Injector 2 Driver | 1-4 ms |
| 16 | LB/BR | Injector 4 Driver | 1-4 ms |
| 21 | BR | 3-4 Shift Solenoid Control | Solenoid Off: 12v, On: 1v |
| 25 | DB/BK | OSS Sensor (-) Signal | Moving: AC pulse signals |
| 27 | WT/OR | Vehicle Speed Signal | Digital Signal |
| 28 | LG/WT | OSS Sensor (+) Signal | Moving: AC pulse signals |
| 29 | LG/RD | Governor Pressure Sensor | 0.58v |
| 30 | PK/LB | Transmission Relay Control | Relay Off: 12v, On: 1v |
| 31 | RD/YL | 5-Volt Supply | 4.9-5.1v |

## 1996 Dakota 3.9L VIN X 'C' Gray Connector

| PCM Pin # | Wire Color | Circuit Description (32-Pin) | Value at Hot Idle |
|---|---|---|---|
| 1 | DB/OR | A/C Clutch Relay Control | Relay Off: 12v, On: 1v |
| 2 | DB/PK | Radiator Fan Control Relay | Relay Off: 12v, On: 1v |
| 3 | DB/YL | ASD Relay Control | Relay Off: 12v, On: 1v |
| 4 | TN/RD | Speed Control Vacuum Solenoid | Vacuum Increasing: 1v |
| 5 | LG/RD | Speed Control Vent Solenoid | Vacuum Decreasing: 1v |
| 6 | LG/OR | Overdrive 'Off' Lamp Control | Lamp Off: 12v, On: 1v |
| 7 | PK/BK | Transmission Temperature Lamp Control | Lamp Off: 12v, On: 1v |
| 8-10 | --- | Not Used | --- |
| 11 | YL/RD | Speed Control Power Supply | 12-14v |
| 12 | DG/OR | ASD Relay Power (B+) | 12-14v |
| 13 | OR/WT | Overdrive Off Switch Sense | Switch Off: 12v, On: 1v |
| 14 | --- | Not Used | |
| 15 | PK/YL | Battery Temperature Sensor | At 86°F: 1.96v |
| 16 | DG/Y | Generator Lamp Control | Lamp Off: 12v, On: 1v |
| 17 | BK/PK | MIL (lamp) Control | MIL Off: 12v, On: 1v |
| 18 | --- | Not Used | --- |
| 19 | BR | Fuel Pump Relay Control | Relay Off: 12v, On: 1v |
| 20 | PK/BK | EVAP Purge Solenoid Control | PWM Signal: 0-12-0v |
| 21 | --- | Not Used | --- |
| 22 | BR | A/C Request Signal | A/C Off: 12v, On: 1v |
| 23 | LG | A/C Select Signal | A/C Off: 12v, On: 1v |
| 24 | WT/PK | Brake Switch Sense Signal | Brake Off: 0v, On: 12v |
| 25 | --- | Not Used | --- |
| 26 | LB/BK | Fuel Level Sensor Signal | Full: 0.56v, 1/2 full: 2.5v |
| 27 | PK | SCI Transmit | 0v |
| 28 | WT/PK | CCD Bus (-) | <0.050v |
| 29 | LG | SCI Receive | 5v |
| 30 | PK/BR | CCD Bus (+) | Digital Signals: 0-5-0v |
| 31 | GY/LB | Tachometer Signal | Pulse Signals |
| 32 | WT/LG | Speed Control Set Switch Signal | S/C & Set Switch On: 3.8v |

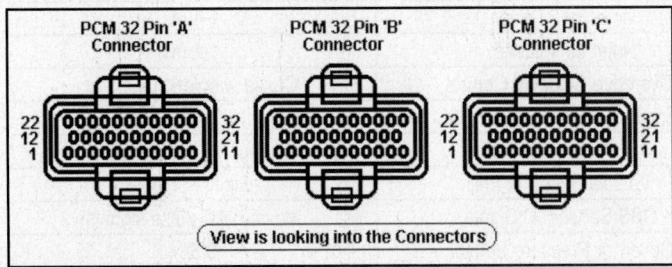

PCM 32 Pin 'A' Connector    PCM 32 Pin 'B' Connector    PCM 32 Pin 'C' Connector

View is looking into the Connectors

## Pin Connector Graphic

## Standard Colors and Abbreviations

| Abbreviation | Color | Abbreviation | Color | Abbreviation | Color |
|---|---|---|---|---|---|
| BK | Black | GY | Gray | RD | Red |
| BL | Blue | GN | Green | TN | Tan |
| BR | Brown | LG | Light Green | VT | Violet |
| DB | Dark Blue | OR | Orange | WT | White |
| DG | Dark Green | PK | Pink | YL | Yellow |

## 1997 Dakota 3.9L VIN X 'A' Black Connector

| PCM Pin # | Wire Color | Circuit Description (32-Pin) | Value at Hot Idle |
|---|---|---|---|
| 2 | LG/BK | Ignition Switch Power (B+) | 12-14v |
| 4 | BK/LB | Sensor Ground | <0.050v |
| 6 | BK/WT | PNP Switch Sense Signal | In P/N: 0v, Others: 5v |
| 7 | BK/GY | Coil 1 Driver Control | 5°, 55 mph: 8° dwell |
| 8 | GY/BK | CKP Sensor Signal | Digital Signals: 0-5-0v |
| 10 | YL/BK | IAC 2 Driver Control | Pulse Signals |
| 11 | BR/WT | IAC 3 Driver Control | Pulse Signals |
| 12 | GY/WT | PSP Switch Sense Signal | Straight: 0v, Turning: 5v |
| 15 | BK/RD | IAT Sensor Signal | At 100°F: 1.83v |
| 16 | TN/BK | ECT Sensor Signal | At 180°F: 2.80v |
| 17 | PK/WT | 5-Volt Supply | 4.9-5.1v |
| 18 | TN/YL | CMP Sensor Signal | Digital Signals: 0-5-0v |
| 19 | GY/RD | IAC 1 Driver Control | Pulse Signals |
| 20 | PK/BK | IAC 4 Driver Control | Pulse Signals |
| 22 | RD/WT | Keep Alive Power (B+) | 12-14v |
| 23 | OR/DB | TP Sensor Signal | 0.6-1.0v |
| 24 | OR/TN | HO2S-11 (B1 S1) Signal | 0.1-1.1v |
| 25 | OR/BK | HO2S-12 (B1 S2) Signal | 0.1-1.1v |
| 27 | DG/RD | MAP Sensor Signal | 1.5-1.6v |
| 31-32 | BK/TN | Power Ground | <0.1v |

## 1997 Dakota 3.9L VIN X 'B' White Connector

| PCM Pin # | Wire Color | Circuit Description (32-Pin) | Value at Hot Idle |
|---|---|---|---|
| 1 | PK | Transmission Temperature Sensor Signal | At 200°F: 2.40v |
| 4 | WT/DB | Injector 1 Driver | 1-4 ms |
| 5 | YL/WT | Injector 3 Driver | 1-4 ms |
| 6 | GY | Injector 5 Driver | 1-4 ms |
| 8 | PK | Governor Pressure Solenoid | PWM Signal: 0-12-0v |
| 10 | DG | Generator Field Control | Digital Signals: 0-12-0v |
| 11 | OR/BK | TCC Solenoid Control | At Cruise w/TCC On: 1v |
| 12 | BR/DB | Injector 6 Driver | 1-4 ms |
| 15 | TN | Injector 2 Driver | 1-4 ms |
| 16 | LB/BR | Injector 4 Driver | 1-4 ms |
| 21 | BR | Overdrive Solenoid Control | Cruise w/solenoid On: 1v |
| 23 | GY/YL | Engine Oil Pressure Sensor | 1.6v at 24 psi |
| 25 | DB/BK | OSS Sensor (-) Signal | Moving: AC pulse signals |
| 27 | WT/OR | Vehicle Speed Signal | Digital Signal |
| 28 | LG/WT | OSS Sensor (+) Signal | Moving: AC pulse signals |
| 29 | LG/RD | Governor Pressure Sensor | 0.58v |
| 30 | PK/BK | Transmission Relay Control | Relay Off: 12v, On: 1v |
| 31 | OR | 5-Volt Supply | 4.9-5.1v |
| 32 | --- | Not Used | --- |

**1997 Dakota 3.9L VIN X 'C' Gray Connector**

| PCM Pin # | Wire Color | Circuit Description (32-Pin) | Value at Hot Idle |
|---|---|---|---|
| 1 | DB/OR | A/C Clutch Relay Control | Relay Off: 12v, On: 1v |
| 2 | DB/PK | Radiator Fan Control Relay | Relay Off: 12v, On: 1v |
| 3 | DB/YL | ASD Relay Control | Relay Off: 12v, On: 1v |
| 4 | TN/RD | Speed Control Vacuum Solenoid | Vacuum Increasing: 1v |
| 5 | LG/RD | Speed Control Vent Solenoid | Vacuum Decreasing: 1v |
| 6 | LG/OR | Overdrive 'Off' Lamp Control | Lamp Off: 12v, On: 1v |
| 7-10 | --- | Not Used | --- |
| 11 | YL/RD | Speed Control Power Supply | 12-14v |
| 12 | DG/OR | ASD Relay Power (B+) | 12-14v |
| 13 | OR/WT | Overdrive Off Switch Sense | Switch Off: 12v, On: 1v |
| 14 | --- | Not Used | --- |
| 15 | PK/YL | Battery Temperature Sensor | At 86°F: 1.96v |
| 16-18 | --- | Not Used | --- |
| 19 | LB/OR | Fuel Pump Relay Control | Relay Off: 12v, On: 1v |
| 20 | PK/BK | EVAP Purge Solenoid Control | PWM Signal: 0-12-0v |
| 21 | --- | Not Used | --- |
| 22 | BR | A/C Request Signal | A/C Off: 12v, On: 1v |
| 23 | LG/WT | A/C Select Signal | A/C Off: 12v, On: 1v |
| 24 | WT/PK | Brake Switch Sense Signal | Brake Off: 0v, On: 12v |
| 25 | DG/BK | Generator Field Source | 12-14v |
| 26 | DB | Fuel Level Sensor Signal | Full: 0.56v, 1/2 full: 2.5v |
| 27 | PK | SCI Transmit | 0v |
| 28 | WT/BK | CCD Bus (-) | <0.050v |
| 29 | LG | SCI Receive | 5v |
| 30 | PK/BR | CCD Bus (+) | Digital Signals: 0-5-0v |
| 31 | --- | Not Used | --- |
| 32 | GY/LB | Speed Control Set Switch Signal | S/C & Set Switch On: 3.8v |

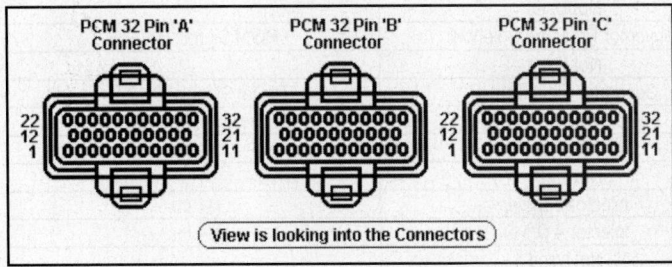

**Pin Connector Graphic**

## 1998-99 Dakota 3.9L VIN X 'C1' Black Connector

| PCM Pin # | Wire Color | Circuit Description (32-Pin) | Value at Hot Idle |
|---|---|---|---|
| 1, 3, 5 | --- | Not Used | --- |
| 2 | LG/BK | Ignition Switch Power (B+) | 12-14v |
| 4 | BK/LB | Sensor Ground | <0.050v |
| 6 | BK/WT | PNP Switch Sense Signal | In P/N: 0v, Others: 5v |
| 7 | BK/GY | Coil 1 Driver Control | 5°, 55 mph: 8° dwell |
| 8 | GY/BK | CKP Sensor Signal | Digital Signals: 0-5-0v |
| 9 | --- | Not Used | --- |
| 10 | YL/BK | IAC 2 Driver Control | Pulse Signals |
| 11 | BR/WT | IAC 3 Driver Control | Pulse Signals |
| 12-14 | --- | Not Used | --- |
| 15 | BK/RD | IAT Sensor Signal | At 100°F: 1.83v |
| 16 | TN/BK | ECT Sensor Signal | At 180°F: 2.80v |
| 17 | PK/WT | 5-Volt Supply | 4.9-5.1v |
| 18 | TN/YL | CMP Sensor Signal | Digital Signals: 0-5-0v |
| 19 | GY/RD | IAC 1 Driver Control | Pulse Signals |
| 20 | PK/BK | IAC 4 Driver Control | Pulse Signals |
| 21 | --- | Not Used | --- |
| 22 | RD/WT | Keep Alive Power (B+) | 12-14v |
| 23 | OR/DB | TP Sensor Signal | 0.6-1.0v |
| 24 | OR/TN | HO2S-11 (B1 S1) Signal | 0.1-1.1v |
| 25 | OR/BK | HO2S-12 (B1 S2) Signal | 0.1-1.1v |
| 26 | --- | Not Used | --- |
| 27 | DG/RD | MAP Sensor Signal | 1.5-1.6v |
| 28-30 | --- | Not Used | --- |
| 31 | BK/TN | Power Ground | <0.1v |
| 32 | BK/TN | Power Ground | <0.1v |

## 1998-99 Dakota 3.9L VIN X White 'C2' Connector

| PCM Pin # | Wire Color | Circuit Description (32-Pin) | Value at Hot Idle |
|---|---|---|---|
| 1 | PK | Transmission Temperature Sensor Signal | At 200°F: 2.40v |
| 3 | --- | Not Used | --- |
| 4 | WT/DB | Injector 1 Driver | 1-4 ms |
| 5 | YL/WT | Injector 3 Driver | 1-4 ms |
| 6 | GY | Injector 5 Driver | 1-4 ms |
| 7 | --- | Not Used | --- |
| 8 | PK | Governor Pressure Solenoid | PWM Signal: 0-12-0v |
| 9 | --- | Not Used | --- |
| 10 | DG | Generator Field Control | Digital Signals: 0-12-0v |
| 11 | OR/BK | TCC Solenoid Control | At Cruise w/TCC On: 1v |
| 12 | BR/DB | Injector 6 Driver | 1-4 ms |
| 13-14 | --- | Not Used | --- |
| 15 | TN | Injector 2 Driver | 1-4 ms |
| 16 | LB/BR | Injector 4 Driver | 1-4 ms |
| 17-20 | --- | Not Used | --- |
| 21 | BR | Overdrive Solenoid Control | Cruise w/solenoid On: 1v |
| 22 | --- | Not Used | --- |
| 23 | GY/YL | Engine Oil Pressure Sensor | 1.6v at 24 psi |
| 24 | --- | Not Used | --- |
| 25 | DB/BK | OSS Sensor (-) Signal | Moving: AC pulse signals |
| 26 | --- | Not Used | --- |
| 27 | WT/OR | Vehicle Speed Signal | Digital Signal |
| 28 | LG/WT | OSS Sensor (+) Signal | Moving: AC pulse signals |
| 29 | LG/RD | Governor Pressure Sensor | 0.58v |
| 30 | PK/BK | Transmission Relay Control | Relay Off: 12v, On: 1v |
| 31 | OR | 5-Volt Supply | 4.9-5.1v |
| 32 | --- | Not Used | --- |

## 1998-99 Dakota 3.9L VIN X 'C3' Gray Connector

| PCM Pin # | Wire Color | Circuit Description (32-Pin) | Value at Hot Idle |
|---|---|---|---|
| 1 | DB/OR | A/C Clutch Relay Control | Relay Off: 12v, On: 1v |
| 2 | DB/PK | Radiator Fan Control Relay | Relay Off: 12v, On: 1v |
| 3 | DB/YL | ASD Relay Control | Relay Off: 12v, On: 1v |
| 4 | TN/RD | Speed Control Vacuum Solenoid | Vacuum Increasing: 1v |
| 5 | LG/RD | Speed Control Vent Solenoid | Vacuum Decreasing: 1v |
| 6-9 | --- | Not Used | --- |
| 10 | YL/DG | LDP Solenoid Control | PWM Signal: 0-12-0v |
| 11 | YL/RD | Speed Control Power Supply | 12-14v |
| 12 | DG/OR | ASD Relay Power (B+) | 12-14v |
| 13 | OR/WT | Overdrive Off Switch Sense | Switch Off: 12v, On: 1v |
| 14 | YL/WT | LDP Switch Sense Signal | LDP Switch Closed: 0v |
| 15 | PK/YL | Battery Temperature Sensor | At 86°F: 1.96v |
| 16-18 | --- | Not Used | --- |
| 19 | LB/OR | Fuel Pump Relay Control | Relay Off: 12v, On: 1v |
| 20 | PK/BK | EVAP Purge Solenoid Control | PWM Signal: 0-12-0v |
| 21 | --- | Not Used | --- |
| 22 | BR | A/C Switch Signal | A/C Off: 12v, On: 1v |
| 23 | LG/WT | A/C Select Signal | A/C Off: 12v, On: 1v |
| 24 | WT/PK | Brake Switch Sense Signal | Brake Off: 0v, On: 12v |
| 25 | DG/BK | Generator Field Source | 12-14v |
| 26 | DB | Fuel Level Sensor Signal | Full: 0.56v, 1/2 full: 2.5v |
| 27 | PK | SCI Transmit | 0v |
| 28 | WT/BK | CCD Bus (-) | <0.050v |
| 29 | LG | SCI Receive | 5v |
| 30 | PK/BR | CCD Bus (+) | Digital Signals: 0-5-0v |
| 31 | --- | Not Used | --- |
| 32 | RD/LG | Speed Control Set Switch Signal | S/C & Set Switch On: 3.8v |

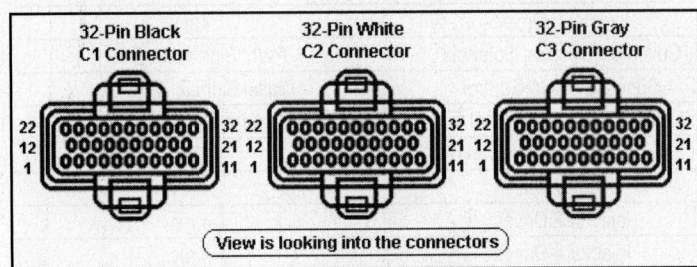

**Pin Connector Graphic**

## 2000 Dakota 3.9L VIN X 'C1' Black Connector

| PCM Pin # | Wire Color | Circuit Description (32-Pin) | Value at Hot Idle |
|---|---|---|---|
| 1, 3, 5, 9, 12-14 | --- | Not Used | --- |
| 2 | LG/BK | Ignition Switch Power (B+) | 12-14v |
| 4 | BK/LB | Sensor Ground | <0.050v |
| 6 | BK/WT | PNP Switch Sense Signal | In P/N: 0v, Others: 5v |
| 7 | BK/GY | Coil 1 Driver Control | 5°, 55 mph: 8° dwell |
| 8 | GY/BK | CKP Sensor Signal | Digital Signals: 0-5-0v |
| 10 | YL/BK | IAC 2 Driver Control | Pulse Signals |
| 11 | BR/WT | IAC 3 Driver Control | Pulse Signals |
| 15 | BK/RD | IAT Sensor Signal | At 100°F: 1.83v |
| 16 | TN/BK | ECT Sensor Signal | At 180°F: 2.80v |
| 17 | OR | 5-Volt Supply | 4.9-5.1v |
| 18 | TN/YL | CMP Sensor Signal | Digital Signals: 0-5-0v |
| 19 | GY/RD | IAC 1 Driver Control | Pulse Signals |
| 20 | VT/BK | IAC 4 Driver Control | Pulse Signals |
| 21 | --- | Not Used | --- |
| 22 | RD/WT | Keep Alive Power (B+) | 12-14v |
| 23 | OR/DB | TP Sensor Signal | 0.6-1.0v |
| 24 | TN/WT | HO2S-11 (B1 S1) Signal | 0.1-1.1v |
| 25 | OR/BK | HO2S-12 (B1 S2) Signal | 0.1-1.1v |
| 26, 28-30 | --- | Not Used | --- |
| 27 | DG/RD | MAP Sensor Signal | 1.5-1.6v |
| 31, 32 | BK/TN | Power Ground | <0.1v |

## 2000 Dakota 3.9L VIN X White 'C2' Connector

| PCM Pin # | Wire Color | Circuit Description (32-Pin) | Value at Hot Idle |
|---|---|---|---|
| 1 | PK | Transmission Temperature Sensor Signal | At 200°F: 2.40v |
| 2-3, 7, 9 | --- | Not Used | --- |
| 4 | WT/DB | Injector 1 Driver | 1-4 ms |
| 5 | YL/WT | Injector 3 Driver | 1-4 ms |
| 6 | GY | Injector 5 Driver | 1-4 ms |
| 8 | PK | Governor Pressure Solenoid | PWM Signal: 0-12-0v |
| 10 | DG | Generator Field Control | Digital Signals: 0-12-0v |
| 11 | OR/BK | TCC Solenoid Control | At Cruise w/TCC On: 1v |
| 12 | BR/DB | Injector 6 Driver | 1-4 ms |
| 13-14 | --- | Not Used | --- |
| 15 | TN | Injector 2 Driver | 1-4 ms |
| 16 | LB/BR | Injector 4 Driver | 1-4 ms |
| 17 | DB/PK | Radiator Fan Control Relay | Relay Off: 12v, On: 1v |
| 18-20, 22 | --- | Not Used | --- |
| 21 | BR | Overdrive Solenoid Control | Cruise w/solenoid On: 1v |
| 23 | GY/YL | Engine Oil Pressure Sensor | 1.6v at 24 psi |
| 24 | --- | Not Used | --- |
| 25 | DB/BK | OSS Sensor (-) Signal | Moving: AC pulse signals |
| 26, 32 | --- | Not Used | --- |
| 27 | WT/OR | Vehicle Speed Signal | Digital Signal |
| 28 | LG/WT | OSS Sensor (+) Signal | Moving: AC pulse signals |
| 29 | LG/RD | Governor Pressure Sensor | 0.58v |
| 30 | PK/BK | Transmission Relay Control | Relay Off: 12v, On: 1v |
| 31 | PK/WT | 5-Volt Supply | 4.9-5.1v |

## 2000 Dakota 3.9L VIN X 'C3' Gray Connector

| PCM Pin # | Wire Color | Circuit Description (32-Pin) | Value at Hot Idle |
|---|---|---|---|
| 1 | DB/OR | A/C Clutch Relay Control | Relay Off: 12v, On: 1v |
| 2 | --- | Not Used | --- |
| 3 | DB/YL | ASD Relay Control | Relay Off: 12v, On: 1v |
| 4 | TN/RD | Speed Control Vacuum Solenoid | Vacuum Increasing: 1v |
| 5 | LG/RD | Speed Control Vent Solenoid | Vacuum Decreasing: 1v |
| 6-9 | --- | Not Used | --- |
| 10 | WT/DG | LDP Solenoid Control | PWM Signal: 0-12-0v |
| 11 | YL/RD | Speed Control Power Supply | 12-14v |
| 12 | DG/OR | ASD Relay Power (B+) | 12-14v |
| 13 | OR/WT | Overdrive Off Switch Sense | Switch Off: 12v, On: 1v |
| 14 | OR | LDP Switch Sense Signal | LDP Switch Closed: 0v |
| 15 | PK/YL | Battery Temperature Sensor | At 86°F: 1.96v |
| 16-18 | --- | Not Used | --- |
| 19 | LB/OR | Fuel Pump Relay Control | Relay Off: 12v, On: 1v |
| 20 | PK/BK | EVAP Purge Solenoid Control | PWM Signal: 0-12-0v |
| 21 | --- | Not Used | --- |
| 22 | BR | A/C Switch Signal | A/C Off: 12v, On: 1v |
| 23 | LG/WT | A/C Select Signal | A/C Off: 12v, On: 1v |
| 24 | WT/PK | Brake Switch Sense Signal | Brake Off: 0v, On: 12v |
| 25 | WT/DB | Generator Field Source | 12-14v |
| 26 | DB | Fuel Level Sensor Signal | Full: 0.56v, 1/2 full: 2.5v |
| 27 | PK/DB | SCI Transmit | 0v |
| 28 | WT/BK | CCD Bus (-) | <0.050v |
| 29 | LG | SCI Receive | 5v |
| 30 | PK/BR | CCD Bus (+) | Digital Signals: 0-5-0v |
| 31 | --- | Not Used | --- |
| 32 | RD/LG | Speed Control Set Switch Signal | S/C & Set Switch On: 3.8v |

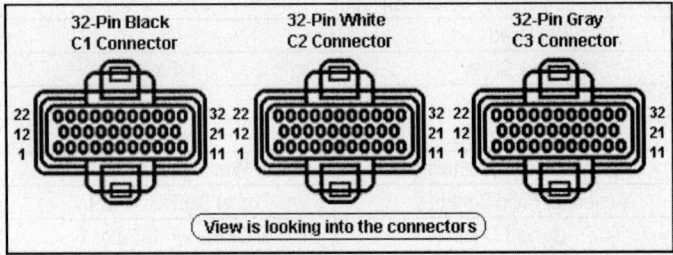

**Pin Connector Graphic**

## Standard Colors and Abbreviations

| Abbreviation | Color | Abbreviation | Color | Abbreviation | Color |
|---|---|---|---|---|---|
| BK | Black | GY | Gray | RD | Red |
| BL | Blue | GN | Green | TN | Tan |
| BR | Brown | LG | Light Green | VT | Violet |
| DB | Dark Blue | OR | Orange | WT | White |
| DG | Dark Green | PK | Pink | YL | Yellow |

## 2001 Dakota 3.9L VIN X 'C1' Black Connector

| PCM Pin # | Wire Color | Circuit Description (32-Pin) | Value at Hot Idle |
|---|---|---|---|
| 1, 3, 5, 9 | --- | Not Used | --- |
| 2 | LG/BK | Ignition Switch Power (B+) | 12-14v |
| 4 | BK/LB | Sensor Ground | <0.050v |
| 6 | BK/WT | PNP Switch Sense Signal | In P/N: 0v, Others: 5v |
| 7 | BK/GY | Coil 1 Driver Control | 5°, 55 mph: 8° dwell |
| 8 | GY/BK | CKP Sensor Signal | Digital Signals: 0-5-0v |
| 10 | YL/BK | IAC 2 Driver Control | Pulse Signals |
| 11 | BR/WT | IAC 3 Driver Control | Pulse Signals |
| 12-14 | --- | Not Used | --- |
| 15 | BK/RD | IAT Sensor Signal | At 100°F: 1.83v |
| 16 | TN/BK | ECT Sensor Signal | At 180°F: 2.80v |
| 17 | OR | 5-Volt Supply | 4.9-5.1v |
| 18 | TN/YL | CMP Sensor Signal | Digital Signals: 0-5-0v |
| 19 | GY/RD | IAC 1 Driver Control | Pulse Signals |
| 20 | VT/BK | IAC 4 Driver Control | Pulse Signals |
| 21 | --- | Not Used | --- |
| 22 | RD/WT | Keep Alive Power (B+) | 12-14v |
| 23 | OR/DB | TP Sensor Signal | 0.6-1.0v |
| 24 | BK/DG | HO2S-11 (B1 S1) Signal | 0.1-1.1v |
| 25 | TN/WT | HO2S-12 (B1 S2) Signal | 0.1-1.1v |
| 26 | LG/RD | HO2S-21 (B2 S1) Signal | 0.1-1.1v |
| 27 | DG/RD | MAP Sensor Signal | 1.5-1.6v |
| 28, 30 | --- | Not Used | --- |
| 29 | TN/WT | HO2S-22 (B2 S2) Signal | 0.1-1.1v |
| 31, 32 | BK/TN | Power Ground | <0.1v |

## 2001 Dakota 3.9L VIN X White 'C2' Connector

| PCM Pin # | Wire Color | Circuit Description (32-Pin) | Value at Hot Idle |
|---|---|---|---|
| 1 | GY/BK | Transmission Temperature Sensor Signal | At 200°F: 2.40v |
| 2-3, 7-9, 26, 32 | --- | Not Used | --- |
| 4 | WT/DB | Injector 1 Driver | 1-4 ms |
| 5 | YL/WT | Injector 3 Driver | 1-4 ms |
| 6 | GY | Injector 5 Driver | 1-4 ms |
| 8 | VT/WT | Variable Force Solenoid | PWM Signal: 0-12-0v |
| 10 | DB | Generator Field Control | Digital Signals: 0-12-0v |
| 11 | OR/BK | TCC Solenoid Control | At Cruise w/TCC On: 1v |
| 12 | BR/DB | Injector 6 Driver | 1-4 ms |
| 15 | TN | Injector 2 Driver | 1-4 ms |
| 16 | LB/BR | Injector 4 Driver | 1-4 ms |
| 17 | DB/PK | Radiator Fan Control Relay | Relay Off: 12v, On: 1v |
| 19 | DB | AC Pressure Sensor Signal | 0.90v at 79 psi |
| 21 | BR | 3-4 Shift Solenoid Control | Cruise w/solenoid on: 1v |
| 23 | GY/YL | Engine Oil Pressure Sensor | 1.6v at 24 psi |
| 25 | DB/BK | OSS Sensor (-) Signal | Moving: AC pulse signals |
| 27 | DB | Vehicle Speed Signal | Digital Signal |
| 28 | LG/WT | OSS Sensor (+) Signal | Moving: AC pulse signals |
| 29 | LG/RD | Governor Pressure Sensor | 0.58v |
| 30 | PK | Transmission Relay Control | Relay Off: 12v, On: 1v |
| 31 | VT/WT | 5-Volt Supply | 4.9-5.1v |

**2001 Dakota 3.9L VIN X 'C3' Gray Connector**

| PCM Pin # | Wire Color | Circuit Description (32-Pin) | Value at Hot Idle |
|-----------|-----------|------------------------------|-------------------|
| 1 | DB/OR | A/C Clutch Relay Control | Relay Off: 12v, On: 1v |
| 2 | --- | Not Used | --- |
| 3 | DB/YL | ASD Relay Control | Relay Off: 12v, On: 1v |
| 4 | TN/RD | Speed Control Vacuum Solenoid | Vacuum Increasing: 1v |
| 5 | LG/RD | Speed Control Vent Solenoid | Vacuum Decreasing: 1v |
| 6-7 | --- | Not Used | --- |
| 8 | VT/WT | HO2S-11 (B1 S1) Heater | PWM Signal: 0-12-0v |
| 9 | --- | Not Used | --- |
| 10 | WT/DG | LDP Solenoid Control | PWM Signal: 0-12-0v |
| 11 | YL/RD | Speed Control Power Supply | 12-14v |
| 12 | DG/OR | ASD Relay Power (B+) | 12-14v |
| 13 | OR/WT | Overdrive Off Switch Sense | Switch Off: 12v, On: 1v |
| 14 | OR | LDP Switch Sense Signal | LDP Switch Closed: 0v |
| 15 | PK/YL | Battery Temperature Sensor | At 86°F: 1.96v |
| 16 | VT/OR | HO2S-21 (B2 S1) Heater | PWM Signal: 0-12-0v |
| 17-18 | --- | Not Used | --- |
| 19 | BR | Fuel Pump Relay Control | Relay Off: 12v, On: 1v |
| 20 | PK/BK | EVAP Purge Solenoid Control | PWM Signal: 0-12-0v |
| 21 | --- | Not Used | --- |
| 22 | BR | A/C Switch Signal | A/C Off: 12v, On: 1v |
| 23 | --- | Not Used | --- |
| 24 | WT/PK | Brake Switch Sense Signal | Brake Off: 0v, On: 12v |
| 25 | WT/DB | Generator Field Source | 12-14v |
| 26 | DB/WT | Fuel Level Sensor Signal | Full: 0.56v, 1/2 full: 2.5v |
| 27 | PK | SCI Transmit | 0v |
| 28 | --- | Not Used | --- |
| 29 | LG | SCI Receive | 5v |
| 30 | VT/YL | PCI Bus Signal (J1850) | Digital Signals: 0-5-0v |
| 31 | --- | Not Used | --- |
| 32 | RD/LG | Speed Control Set Switch Signal | S/C & Set Switch On: 3.8v |

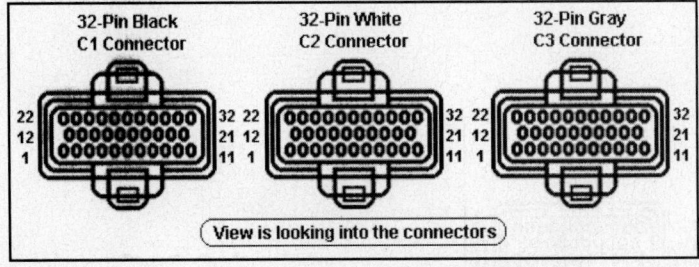

**Pin Connector Graphic**

**Standard Colors and Abbreviations**

| Abbreviation | Color | Abbreviation | Color | Abbreviation | Color |
|--------------|-------|--------------|-------|--------------|-------|
| BK | Black | GY | Gray | RD | Red |
| BL | Blue | GN | Green | TN | Tan |
| BR | Brown | LG | Light Green | VT | Violet |
| DB | Dark Blue | OR | Orange | WT | White |
| DG | Dark Green | PK | Pink | YL | Yellow |

## 2000 Dakota 4.7L VIN N 'C1' Black Connector

| PCM Pin # | Wire Color | Circuit Description (32-Pin) | Value at Hot Idle |
|---|---|---|---|
| 1 | TN/OR | Coil 3 Driver Control | 5°, 55 mph: 8° dwell |
| 2 | LG/BK | Ignition Switch Power (B+) | 12-14v |
| 3 | TN/LG | Coil 4 Driver Control | 5°, 55 mph: 8° dwell |
| 4 | BK/LB | Sensor Ground | <0.050v |
| 5 | TN/LB | Coil 6 Driver Control | 5°, 55 mph: 8° dwell |
| 6 | BK/WT | PNP Switch Sense Signal | In P/N: 0v, Others: 5v |
| 7 | BK/GY | Coil 1 Driver Control | 5°, 55 mph: 8° dwell |
| 8 | GY/BK | CKP Sensor Signal | Digital Signals: 0-5-0v |
| 9 | DB/GY | Coil 8 Driver Control | 5°, 55 mph: 8° dwell |
| 10 | YL/BK | IAC 2 Driver Control | Pulse Signals |
| 11 | BR/WT | IAC 3 Driver Control | Pulse Signals |
| 12 | DB/OR | PSP Switch Sense Signal | Straight: 0v, Turning: 5v |
| 13-14 | --- | Not Used | --- |
| 15 | BK/RD | IAT Sensor Signal | At 100°F: 1.83v |
| 16 | TN/BK | ECT Sensor Signal | At 180°F: 2.80v |
| 17 | OR | 5-Volt Supply | 4.9-5.1v |
| 18 | TN/YL | CMP Sensor Signal | Digital Signals: 0-5-0v |
| 19 | GY/RD | IAC 1 Driver Control | Pulse Signals |
| 20 | VT/BK | IAC 4 Driver Control | Pulse Signals |
| 21 | TN/DG | Coil 5 Driver Control | 5°, 55 mph: 8° dwell |
| 22 | RD/WT | Keep Alive Power (B+) | 12-14v |
| 23 | OR/DB | TP Sensor Signal | 0.6-1.0v |
| 24 | LG/RD | HO2S-11 (B1 S1) Signal (Cal) | 0.1-1.1v |
| 25 | TN/WT | HO2S-12 (B1 S2) Signal (Cal) | 0.1-1.1v |
| 26 | OR/TN | HO2S-21 (B2 S1) Signal (Cal) | 0.1-1.1v |
| 27 | DG/RD | MAP Sensor Signal | 1.5-1.6v |
| 28 | --- | Not Used | --- |
| 29 | PK/WT | HO2S-22 (B2 S2) Signal (Cal) | 0.1-1.1v |
| 30 | --- | Not Used | --- |
| 31 | BK/TN | Power Ground | <0.1v |
| 32 | BK/TN | Power Ground | <0.1v |

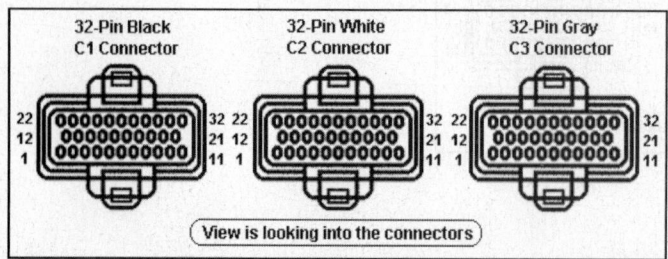

**Pin Connector Graphic**

**2000 Dakota 4.7L VIN N Gray 'C2' Connector**

| PCM Pin # | Wire Color | Circuit Description (32-Pin) | Value at Hot Idle |
|---|---|---|---|
| 1, 3, 8, 18-22 | --- | Not Used | --- |
| 2 | VT | Injector 7 Driver | 1-4 ms |
| 4 | WT/DB | Injector 1 Driver | 1-4 ms |
| 5 | YL/WT | Injector 3 Driver | 1-4 ms |
| 6 | GY | Injector 5 Driver | 1-4 ms |
| 7 | DB/TN | Coil 7 Driver Control | 5°, 55 mph: 8° dwell |
| 9 | TN/PK | Coil 2 Driver Control | 5°, 55 mph: 8° dwell |
| 10 | DG | Generator Field Control | Digital Signals: 0-12-0v |
| 11 | --- | Not Used | --- |
| 12 | BR/DB | Injector 6 Driver | 1-4 ms |
| 13 | GY/LB | Injector 8 Driver | 1-4 ms |
| 14 | --- | Not Used | --- |
| 15 | TN | Injector 2 Driver | 1-4 ms |
| 16 | LB/BR | Injector 4 Driver | 1-4 ms |
| 17 | DB/PK | Radiator Fan Relay Control | Relay Off: 12v, On: 1v |
| 23 | GY/YL | Engine Oil Pressure Sensor | 1.6v at 24 psi |
| 24-26 | --- | Not Used | --- |
| 27 | WT/OR | Vehicle Speed Signal | Digital Signal |
| 28-32 | --- | Not Used | --- |

**2000 Dakota 4.7L VIN N Black 'C3' Connector**

| PCM Pin # | Wire Color | Circuit Description (32-Pin) | Value at Hot Idle |
|---|---|---|---|
| 1 | DB/OR | A/C Clutch Relay Control | Relay Off: 12v, On: 1v |
| 2, 6-7, 16-18, 21 | --- | Not Used | --- |
| 3 | DB/YL | ASD Relay Control | Relay Off: 12v, On: 1v |
| 4 | TN/RD | Speed Control Vacuum Solenoid | Vacuum Increasing: 1v |
| 5 | LG/RD | Speed Control Vent Solenoid | Vacuum Decreasing: 1v |
| 8 | DG/WT | HO2S-11 HTR Control (Cal) | Relay Off: 12v, On: 1v |
| 9 | DG/BK | HO2S-12 HTR Control (Cal) | Relay Off: 12v, On: 1v |
| 10 | WT/DG | LDP Solenoid Control | PWM Signal: 0-12-0v |
| 11 | YL/RD | Speed Control Power Supply | 12-14v |
| 12 | DG/OR | ASD Relay Power (B+) | 12-14v |
| 13 | YL/DG | Torque Management Relay | Relay Off: 12v, On: 1v |
| 14 | OR | LDP Switch Sense Signal | LDP Switch Closed: 0v |
| 15 | PK/YL | Battery Temperature Sensor | At 86°F: 1.96v |
| 19 | LB/OR | Fuel Pump Relay Control | Relay Off: 12v, On: 1v |
| 20 | PK/BK | EVAP Purge Solenoid Control | PWM Signal: 0-12-0v |
| 22 | BR | A/C Switch Signal | A/C Off: 12v, On: 1v |
| 23 | LG/WT | A/C Select Signal | A/C Off: 12v, On: 1v |
| 24 | WT/PK | Brake Switch Sense Signal | Brake Off: 0v, On: 12v |
| 25 | WT/DB | Generator Field Source | 12-14v |
| 26 | DB | Fuel Level Sensor Signal | Full: 0.56v, 1/2 full: 2.5v |
| 27 | PK/DB | SCI Transmit | 0v |
| 28 | WT/BK | CCD Bus (-) | <0.050v |
| 29 | LG | SCI Receive | 5v |
| 30 | PK/BR | CCD Bus (+) | Digital Signals: 0-5-0v |
| 31 | --- | Not Used | --- |
| 32 | RD/LG | Speed Control Set Switch Signal | S/C & Set Switch On: 3.8v |

## 2001-02 Dakota 4.7L VIN N 'C1' Black Connector

| PCM Pin # | Wire Color | Circuit Description (32-Pin) | Value at Hot Idle |
|---|---|---|---|
| 1 | TN/OR | Coil 3 Driver Control | 5°, 55 mph: 8° dwell |
| 2 | LG/BK | Ignition Switch Power (B+) | 12-14v |
| 3 | TN/LG | Coil 4 Driver Control | 5°, 55 mph: 8° dwell |
| 4 | BK/LB | Sensor Ground | <0.050v |
| 5 | TN/LB | Coil 6 Driver Control | 5°, 55 mph: 8° dwell |
| 6 | BK/WT | PNP Switch Sense Signal | In P/N: 0v, Others: 5v |
| 7 | BK/GY | Coil 1 Driver Control | 5°, 55 mph: 8° dwell |
| 8 | GY/BK | CKP Sensor Signal | Digital Signals: 0-5-0v |
| 9 | LB/RD | Coil 8 Driver Control | 5°, 55 mph: 8° dwell |
| 10 | YL/BK | IAC 2 Driver Control | Pulse Signals |
| 11 | BR/WT | IAC 3 Driver Control | Pulse Signals |
| 12 | DB/OR | PSP Switch Sense Signal | Straight: 0v, Turning: 5v |
| 13-14 | --- | Not Used | --- |
| 15 | BK/RD | IAT Sensor Signal | At 100°F: 1.83v |
| 16 | TN/BK | ECT Sensor Signal | At 180°F: 2.80v |
| 17 | OR | 5-Volt Supply | 4.9-5.1v |
| 18 | TN/YL | CMP Sensor Signal | Digital Signals: 0-5-0v |
| 19 | GY/RD | IAC 1 Driver Control | Pulse Signals |
| 20 | VT/BK | IAC 4 Driver Control | Pulse Signals |
| 21 | TN/DG | Coil 5 Driver Control | 5°, 55 mph: 8° dwell |
| 22 | RD/WT | Keep Alive Power (B+) | 12-14v |
| 23 | OR/DB | TP Sensor Signal | 0.6-1.0v |
| 24 | BK/DG | HO2S-11 (B1 S1) Signal | 0.1-1.1v |
| 25 | TN/WT | HO2S-12 (B1 S2) Signal | 0.1-1.1v |
| 26 | LG/RD | HO2S-21 (B2 S1) Signal | 0.1-1.1v |
| 27 | DG/RD | MAP Sensor Signal | 1.5-1.6v |
| 28 | --- | Not Used | --- |
| 29 | TN/WT | HO2S-22 (B2 S2) Signal | 0.1-1.1v |
| 30 | --- | Not Used | --- |
| 31 | BK/TN | Power Ground | <0.1v |
| 32 | BK/TN | Power Ground | <0.1v |

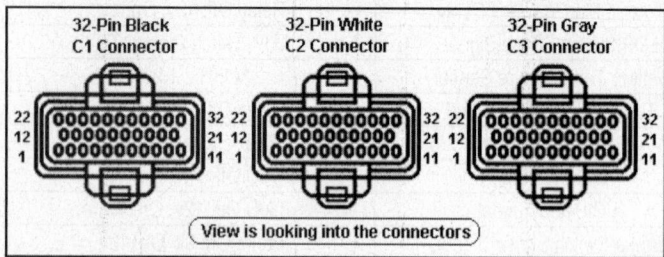

**Pin Connector Graphic**

## 2001-02 Dakota 4.7L VIN N Gray 'C2' Connector

| PCM Pin # | Wire Color | Circuit Description (32-Pin) | Value at Hot Idle |
|---|---|---|---|
| 1, 3, 8-11 | --- | Not Used | --- |
| 2 | VT | Injector 7 Driver | 1-4 ms |
| 4 | WT/DB | Injector 1 Driver | 1-4 ms |
| 5 | YL/WT | Injector 3 Driver | 1-4 ms |
| 6 | GY | Injector 5 Driver | 1-4 ms |
| 7 | DB/TN | Coil 7 Driver Control | 5°, 55 mph: 8° dwell |
| 9 | TN/PK | Coil 2 Driver Control | 5°, 55 mph: 8° dwell |
| 10 | DG | Generator Field Control | Digital Signals: 0-12-0v |
| 12 | BR/DB | Injector 6 Driver | 1-4 ms |
| 13 | GY/LB | Injector 8 Driver | 1-4 ms |
| 14, 18 | --- | Not Used | --- |
| 15 | TN | Injector 2 Driver | 1-4 ms |
| 16 | LB/BR | Injector 4 Driver | 1-4 ms |
| 17 | DB/PK | Radiator Fan Relay Control | Relay Off: 12v, On: 1v |
| 19 | DB | AC Pressure Sensor Signal | 0.90v at 79 psi |
| 20-22, 24-26 | --- | Not Used | --- |
| 23 | GY/YL | Engine Oil Pressure Sensor | 1.6v at 24 psi |
| 27 | WT/OR | Vehicle Speed Signal | Digital Signal |
| 28-30 | --- | Not Used | --- |
| 31 | VT/WT | 5-Volt Supply | 4.9-5.1v |
| 32 | --- | Not Used | --- |

## 2001-02 Dakota 4.7L VIN N Black 'C3' Connector

| PCM Pin # | Wire Color | Circuit Description (32-Pin) | Value at Hot Idle |
|---|---|---|---|
| 1 | DB/OR | A/C Clutch Relay Control | Relay Off: 12v, On: 1v |
| 3 | DB/YL | ASD Relay Control | Relay Off: 12v, On: 1v |
| 4 | TN/RD | Speed Control Vacuum Solenoid | Vacuum Increasing: 1v |
| 5 | LG/RD | Speed Control Vent Solenoid | Vacuum Decreasing: 1v |
| 8 | DG/WT | HO2S-11 HTR Control (Cal) | Relay Off: 12v, On: 1v |
| 9 | DG/BK | HO2S-12 HTR Control (Cal) | Relay Off: 12v, On: 1v |
| 10 | WT/DG | LDP Solenoid Control | PWM Signal: 0-12-0v |
| 11 | YL/RD | Speed Control Power Supply | 12-14v |
| 12 | DG/OR | ASD Relay Power (B+) | 12-14v |
| 13 | YL/DG | Torque Management Relay | Relay Off: 12v, On: 1v |
| 14 | OR | LDP Switch Sense Signal | LDP Switch Closed: 0v |
| 15 | PK/YL | Battery Temperature Sensor | At 86°F: 1.96v |
| 19 | LB/OR | Fuel Pump Relay Control | Relay Off: 12v, On: 1v |
| 20 | PK/BK | EVAP Purge Solenoid Control | PWM Signal: 0-12-0v |
| 22 | BR | A/C Switch Signal | A/C Off: 12v, On: 1v |
| 23 | LG/WT | A/C Select Signal | A/C Off: 12v, On: 1v |
| 24 | WT/PK | Brake Switch Sense Signal | Brake Off: 0v, On: 12v |
| 25 | WT/DB | Generator Field Source | 12-14v |
| 26 | DB | Fuel Level Sensor Signal | Full: 0.56v, 1/2 full: 2.5v |
| 27 | PK/DB | SCI Transmit | 0v |
| 28 | WT/BK | CCD Bus (-) | <0.050v |
| 29 | LG | SCI Receive | 5v |
| 30 | PK/BR | CCD Bus (+) | Digital Signals: 0-5-0v |
| 32 | RD/LG | Speed Control Set Switch Signal | S/C & Set Switch On: 3.8v |

## 2003 Dakota 4.7L VIN N 'C1' 38-Pin Black Connector

| PCM Pin # | Wire Color | Circuit Description (38-Pin) | Value at Hot Idle |
|---|---|---|---|
| 1 | LB/RD | Coil On Plug 8 Driver | 5°, at 55 mph: 8° dwell |
| 2 | --- | Not Used | --- |
| 3 | BR | Coil On Plug 7 Driver | 5°, at 55 mph: 8° dwell |
| 4 | GY/LG | Injector 8 Driver | 1.0-4.0 ms |
| 5 | VT | Injector 7 Driver | 1.0-4.0 ms |
| 6-9 | --- | Not Used | --- |
| 10 | BK/TN | Power Ground | <0.050v |
| 11 | LG/BK | Ignition Switch Output (Run-Start) | 12-14v |
| 12 | RD/WT | Ignition Switch Output (Off-Run-Start) | 12-14v |
| 13 | WT/OR | Vehicle Speed Signal | Digital Signals |
| 14-19 | --- | Not Used | --- |
| 20 | GY/YL | Oil Pressure Sensor | 1.6v at 24 psi |
| 21 | DB | A/C Pressure Sensor | A/C On: 0.45-4.85v |
| 22 | VT/LG | Ambient Temperature Sensor | At 86°F: 1.96v |
| 23-24 | --- | Not Used | --- |
| 25 | LG | SCI Receive (PCM) | 5v |
| 26 | PK/LB | SCI Receive (PCM) | 5v |
| 27 | VT/WT | 5-Volt Supply | 4.9-5.1v |
| 28 | --- | Not Used | --- |
| 29 | RD/WT | Fused Power (B+) | 12-14v |
| 30 | RD/YL | Fused Ignition Output (Start) | 9-11v |
| 31 | TN/WT | HO2S-12 (B1 S2) Signal | 0.1-1.1v |
| 32 | DB/DG | HO2S Return (Downstream) | <0.050v |
| 33 | TN/WT | HO2S-22 (B1 S2) Signal | 0.1-1.1v |
| 34-35 | --- | Not Used | --- |
| 36 | PK | SCI Transmit (PCM) | 0v |
| 37 | WT/DG | SCI Transmit (TCM) | 0v |
| 38 | WT/VT | PCI Data Bus (J1850) | Digital Signals: 0-7-0v |

## 2003 Dakota 4.7L VIN N 'C2' 38-Pin Gray Connector

| PCM Pin # | Wire Color | Circuit Description (38-Pin) | Value at Hot Idle |
|---|---|---|---|
| 1 | TN/LB | Coil On Plug 6 Driver | 5°, at 55 mph: 8° dwell |
| 2 | TN/DG | Coil On Plug 5 Driver | 5°, at 55 mph: 8° dwell |
| 3 | TN/LG | Coil On Plug 4 Driver | 5°, at 55 mph: 8° dwell |
| 4 | BR/DB | Injector 6 Driver | 1.0-4.0 ms |
| 5 | GY | Injector 5 Driver | 1.0-4.0 ms |
| 6, 8, 15-16 | --- | Not Used | --- |
| 7 | TN/OR | Coil On Plug 3 Driver | 5°, at 55 mph: 8° dwell |
| 9 | TN/PK | Coil On Plug 2 Driver | 5°, at 55 mph: 8° dwell |
| 10 | TN/RD | Coil On Plug 1 Driver | 5°, at 55 mph: 8° dwell |
| 11 | LB/BR | Injector 4 Driver | 1.0-4.0 ms |
| 12 | YL/WT | Injector 3 Driver | 1.0-4.0 ms |
| 13 | TN | Injector 2 Driver | 1.0-4.0 ms |
| 14 | WT/DG | Injector 1 Driver | 1.0-4.0 ms |
| 15-16 | --- | Not Used | --- |
| 17 | BR/VT | HO2S-21 (B2 S1) Heater | Heater Off: 12v, On: 1v |
| 18 | BR/OR | HO2S-11 (B1 S1) Heater | Heater Off: 12v, On: 1v |
| 19 | DG | Generator Field Driver | Digital Signals: 0-12-0v |
| 20 | VT/OR | ECT Sensor Signal | At 180°F: 2.80v |
| 21 | BR/OR | TP Sensor 1 Signal | 0.6-1.0v |
| 22 | --- | Not Used | --- |
| 23 | VT/BR | MAP Sensor Signal | 1.5-1.7v |
| 24-26 | --- | Not Used | --- |
| 27 | BK/LB | Sensor Ground | <0.050v |
| 28 | YL/BK | IAC Motor Return | 12-14v |
| 29 | OR | 5-Volt Supply | 4.9-5.1v |
| 30 | BK/RD | IAT Sensor Signal | At 100°F: 1.83v |
| 31 | BK/DG | HO2S-11 (B1 S1) Signal | 0.1-1.1v |
| 32 | DB/DG | HO2S Return (Upstream) | <0.050v |
| 33 | LG/RD | HO2S-21 (B2 S1) Signal | 0.1-1.1v |
| 34 | DB/GY | CMP Sensor Signal | Digital Signal: 0-5-0v |
| 35 | DB/WT | CKP Sensor Signal | Digital Signal: 0-5-0v |
| 36-37 | --- | Not Used | --- |
| 38 | GY/RD | IAC Motor Driver Control | DC pulses: 0.8-11v |

## 2003 Dakota 4.7L VIN N 'C3' 38-Pin White Connector

| PCM Pin # | Wire Color | Circuit Description (38-Pin) | Value at Hot Idle |
|---|---|---|---|
| 1-2, 4, 13-18 | --- | Not Used | --- |
| 3 | DB/YL | ASD Relay Control | Relay Off: 12v, On: 1v |
| 5 | LG/RD | S/C Vent Solenoid | Vacuum Decreasing: 1v |
| 6 | DB/PK | Low Speed Fan Relay | Relay Off: 12v, On: 1v |
| 7 | VT/YL | Speed Control Power Supply | 12-14v |
| 8 | WT/DG | Natural Vacuum Leak Detection Solenoid | Solenoid Off: 12v, On: 1v |
| 9 | BR/WT | HO2S-12 (B1 S2) Heater | Heater Off: 12v, On: 1v |
| 10 | BR/GY | HO2S-22 (B2 S2) Heater | Heater Off: 12v, On: 1v |
| 11 | LB/OR | A/C Clutch Relay Control | Relay Off: 12v, On: 1v |
| 12 | VT/DG | Speed Control Vacuum Solenoid | Vacuum Increasing: 1v |
| 19 | DG/OR | ASD Relay Output | 12-14v |
| 20 | PK/BK | EVAP Purge Solenoid Control | PWM Signal: 0-12-0v |
| 21-22, 24-27, 31, 36 | --- | Not Used | --- |
| 23 | WT/PK | Brake Switch Signal | Brake Off: 0v, On: 12v |
| 28 | DG/OR | ASD Relay Output | 12-14v |
| 29 | DB/BR | EVAP Purge Solenoid Sense | <0.1v |
| 30 | DB/OR | Power Steering Pressure Switch | Straight: 0v, Turning: 5v |
| 33 | DB/WT | Fuel Level Signal | Full: 0.56v, 1/2 full: 2.5v |
| 34 | VT/TN | Speed Control Set Switch Signal | S/C & Set Switch On: 3.8v |
| 35 | OR | Natural Vacuum Leak Detection Switch | Open: 12v, Closed: 0v |
| 37 | BR | Fuel Pump Relay Control | Relay Off: 12v, On: 1v |
| 38 | DG/OR | Starter Relay Control | Relay Off: 12v, On: 1v |

## 2003 Dakota 4.7L VIN N (A/T Only) 'C4' 38-Pin Green Connector

| PCM Pin # | Wire Color | Circuit Description (38-Pin) | Value at Hot Idle |
|---|---|---|---|
| 1 | YL/GY | Overdrive Solenoid Control | Solenoid Off: 12v, On: 1v |
| 2 | DG/WT | 4C Solenoid Control | Solenoid Off: 12v, On: 1v |
| 3, 5, 7, 9 | --- | Not Used | --- |
| 4 | YL/DB | MS Solenoid Control | Solenoid Off: 12v, On: 1v |
| 6 | WT/DB | 2C Solenoid Control | Solenoid Off: 12v, On: 1v |
| 8 | YL/LB | Underdrive Solenoid Control | Solenoid Off: 12v, On: 1v |
| 10 | LG | Low/Reverse Solenoid control | Solenoid Off: 12v, On: 1v |
| 11 | VT/LG | Pressure Control Solenoid | PWM Signals (0-12-0v) |
| 12-14, 19, 23-25 | BK/RD | Power Ground | <0.1v |
| 15 | DG/LB | TRS T1 Sense | In NOL: 0v, Others: 5v |
| 16 | DG/DB | TRS T3 Sense | In P3L: 0v, Others: 5v |
| 17 | DB/WT | Overdrive Off Switch Sense | In Overdrive: 2-4v |
| 18 | YL/BR | Transmission Control Relay Control | Relay Off: 12v, On: 1v |
| 20 | DB | 4C Pressure Switch Sense | In 4th Position: 2-4v |
| 21 | GY | Underdrive Pressure Switch Sense | In Underdrive Position: 2-4v |
| 22 | YL/LG | Overdrive Pressure Switch Sense | In Overdrive Position: 2-4v |
| 27, 28, 36, 38 | --- | Not Used | --- |
| 26 | PK/BK | TRS T2 Sense | In P/N: 0v, Others: 5v |
| 29 | DG | Low/Reverse Pressure Switch | In Low/Reverse: 2-4v |
| 30 | GY/DG | 2C Pressure Switch | In 2-4 Position: 2-4v |
| 31 | VT/TN | Line Pressure Switch | In 2-4 Position: 2-4v |
| 32 | LG/WT | Output Speed Sensor | Moving: AC voltage |
| 33 | RD/BK | Input Speed Sensor | Moving: AC voltage |
| 34 | DB/BK | Speed Sensor Ground | <0.050v |
| 35 | DG/OR | Transmission Temperature Sensor | 3.2-3.4v at 104°F |
| 37 | DG/YL | TRS T42 Sense | In PRNL: 0v, Others 5v |

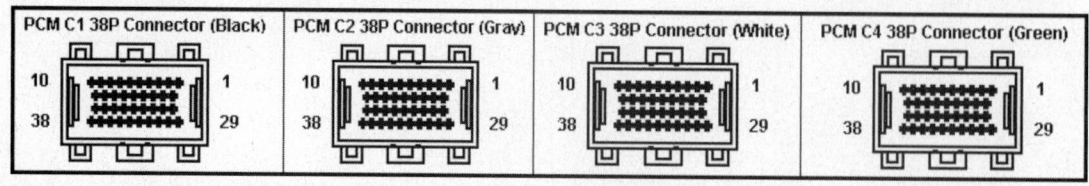

**2004-05 Dakota 4.7L VIN N 'C1' Black/Black Connector**

| PCM Pin # | Wire Color | Circuit Description (38-Pin) | Value at Hot Idle |
|---|---|---|---|
| 1 | DB/YL | Coil Control No. | |
| 2 | --- | Not Used | --- |
| 3 | BR | Coil Control No. | |
| 4 | BR/LB | Injector Control No. | |
| 5 | BR/YL | Injector Control No. | |
| 6 | --- | Not Used | --- |
| 7 | --- | Not Used | --- |
| 8 | --- | Not Used | --- |
| 9 | BK | Ground | |
| 10 | --- | Not Used | --- |
| 11 | PK/GY | Fused Ignition Switch Output (Run-Start) | |
| 12 | PK/WT | Fused Ignition Switch Output (Run-Start) | |
| 13 | --- | Not Used | --- |
| 14 | --- | Not Used | --- |
| 15 | --- | Not Used | --- |
| 16 | --- | Not Used | --- |
| 17 | --- | Not Used | --- |
| 18 | BK | Ground | |
| 19 | --- | Not Used | --- |
| 20 | VT/GY | Engine Oil Pressure Signal | |
| 21 | --- | Not Used | --- |
| 22 | --- | Not Used | --- |
| 23 | --- | Not Used | --- |
| 24 | --- | Not Used | --- |
| 25 | WT/LG | SCI Receive (PCM) | |
| 26 | WT/OR | SCI Receive (TCM) | |
| 27 | YL/PK | 5 Volt Supply | |
| 28 | --- | Not Used | --- |
| 29 | RD | Fused B (+) | |
| 30 | PK/YL | Fused Ignition Switch Output (Run) | |
| 31 | DB/YL | O2 1/2 Signal | |
| 32 | BR/DG | O2 Return (Upstream) | |
| 33 | BR | O2 2/2 Signal | |
| 34 | WT/LG | CAN C Bus (+) | |
| 35 | WT/LB | CAN C Bus (-) | |
| 36 | WT/GY | SCI Transmit (PCM) | |
| 37 | BR/WT (A/T) | SCI Transmit (TCM) | |
| 38 | --- | Not Used | --- |

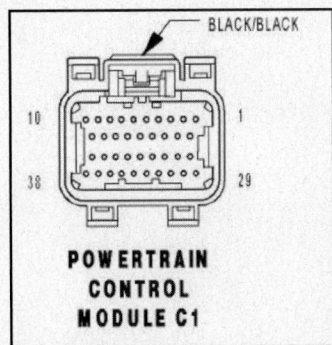

BLACK/BLACK

10   1

38   29

**POWERTRAIN
CONTROL
MODULE C1**

## 2004-05 Dakota 4.7L VIN N 'C2' Black/Orange Connector

| PCM Pin # | Wire Color | Circuit Description (38-Pin) | Value at Hot Idle |
|---|---|---|---|
| 1 | DB/OR | Coil Control No. | |
| 2 | DB/YL | Coil Control No. 5 | |
| 3 | DB | Coil Control No. 4 | |
| 4 | BR/VT | Injector Control No. | |
| 5 | BR/OR | Injector Control No. 5 | |
| 6 | --- | Not Used | --- |
| 7 | DB/OR | Coil Control No. 3 | |
| 8 | DB/VT | EGR Sol Control | |
| 9 | DB/TN | Coil Control No. 2 | |
| 10 | DB/DG | Coil Control No. 1 | |
| 11 | BR/TN | Injector Control No. 4 | |
| 12 | BR/LB | Injector Control No. 3 | |
| 13 | BR/DB | Injector Control No. 2 | |
| 14 | BR/YL | Injector Control No. 1 | |
| 15 | --- | Not Used | --- |
| 16 | --- | Not Used | --- |
| 17 | BR/VT | O2 2/1 Heater Control | |
| 18 | BR/TN | O2 1/1 Heater Control | |
| 19 | BR/GY | Gen Field Control | |
| 20 | VT/OR | ECT Signal | |
| 21 | BR/OR | TP No. 1 Signal | |
| 22 | DB/LG | EGR Signal | |
| 23 | VT/BR | MAP Signal | |
| 24 | BR/LG | Knock Sensor No. 1 Return | |
| 25 | DB/OR | Knock Sensor No. 1 Signal | |
| 26 | --- | Not Used | --- |
| 27 | DB/DG | Sensor Ground | |
| 28 | BR/VT | IAC Signal | |
| 29 | PK/YL | 5 Volt Supply | |
| 30 | BR/WT | IAT Signal | |
| 31 | DB/LB | O2 1/1 Signal | |
| 32 | DB/DG | O2 Return (Downstream) | |
| 33 | DB/LG | O2 2/1 Signal | |
| 34 | DB/GY | CMP Signal | |
| 35 | BR/LB | CKP Signal | |
| 36 | BR/WT | Knock Sensor No. 2 Signal | |
| 37 | PK/RD | Knock Sensor No. 2 Return | |
| 38 | VT/GY | IAC Control | |

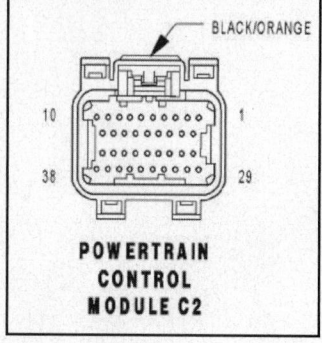

BLACK/ORANGE

10    1

38    29

**POWERTRAIN
CONTROL
MODULE C2**

### 2004-05 Dakota 4.7L VIN N 'C3' Black/Natural Connector

| PCM Pin # | Wire Color | Circuit Description (38-Pin) | Value at Hot Idle |
|---|---|---|---|
| 1 | --- | Not Used | --- |
| 2 | --- | Not Used | --- |
| 3 | BR/GY | ASD Relay Control | |
| 4 | --- | Not Used | --- |
| 5 | VT/OR | S/C Vent Sol Control | |
| 6 | --- | Not Used | --- |
| 7 | VT/YL | S/C Supply | |
| 8 | VT/LB | NVLD Sol Control | |
| 9 | BR/OR | O2 1/2 Heater Control | |
| 10 | BR/GY | O2 2/2 Heater Control | |
| 11 | LB/OR | A/C Clutch Relay Control | |
| 12 | VT/YL | S/C Vacuum Sol Control | |
| 13-18 | --- | Not Used | --- |
| 19 | RD | ASD Relay Output | |
| 20 | DB/WT | EVAP Purge Sol Control | |
| 21 | YL/DB (A/T) | TRS T41 Sense (P/N Sense) | |
| 21 | YL/OR (M/T) | TRS T141 Sense (P/N Sense) | |
| 22 | --- | Not Used | --- |
| 23 | DG/WT | Brake Switch Signal | |
| 24-27 | --- | Not Used | --- |
| 28 | RD | ASD Relay Output | |
| 29 | DB/BR | EVAP Purge Sol Signal | |
| 30-33 | --- | Not Used | --- |
| 34 | VT | S/C Switch No. 1 Signal | |
| 35 | VT/WT | NVLD Switch Signal | |
| 36 | --- | Not Used | --- |
| 37 | BR | Fuel Pump Relay Control | |
| 38 | DG/OR | Starter Relay Control | |

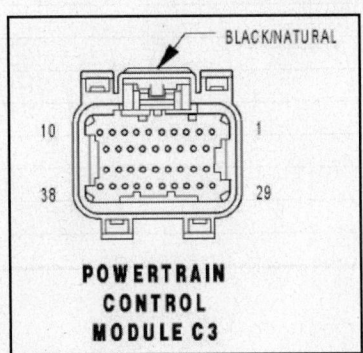

## 2004-05 Dakota 4.7L VIN N (42RLE) 'C4' Black/Green Connector

| PCM Pin # | Wire Color | Circuit Description (38-Pin) | Value at Hot Idle |
|---|---|---|---|
| 1 | YL/RD | OD Solenoid Control | |
| 2 | VT/PK | UD Solenoid Control | |
| 3 | --- | Not Used | --- |
| 4 | --- | Not Used | --- |
| 5 | --- | Not Used | --- |
| 6 | YL/DB | 2-4 Solenoid Control | |
| 7 | --- | Not Used | --- |
| 8 | --- | Not Used | --- |
| 9 | --- | Not Used | --- |
| 10 | DG/WT | L/R Solenoid Control | |
| 11 | --- | Not Used | --- |
| 12 | BK | Ground | |
| 13 | BK | Ground | |
| 14 | BK | Ground | |
| 15 | YL/PK | TRS T1 Sense | |
| 16 | YL/PK | TRS T3 Sense | |
| 17 | DG | Tow/Haul Overdrive Off Switch Sense | |
| 18 | YL/BK | Transmission Control Relay Control | |
| 19 | YL/OR | Transmission Control Relay Output | |
| 20 | --- | Not Used | --- |
| 21 | --- | Not Used | --- |
| 22 | DG/TN | OD Pressure Switch Sense | |
| 23 | --- | Not Used | --- |
| 24 | --- | Not Used | --- |
| 25 | --- | Not Used | --- |
| 26 | --- | Not Used | --- |
| 27 | --- | Not Used | --- |
| 28 | YL/OR | Transmission Control Relay Output | |
| 29 | VT/RD | L/R Pressure Switch Sense | |
| 30 | YL/DG | 2-4 Pressure Switch Sense | |
| 31 | --- | Not Used | --- |
| 32 | YL/PK | Output Speed Sensor Signal | |
| 33 | YL/PK | Input Speed Sensor Signal | |
| 34 | DG/VT | Sensor Ground | |
| 35 | YL/PK | Transmission Temperature Sensor Signal | |
| 36 | --- | Not Used | --- |
| 37 | YL/PK | TRS T42 Sense | |
| 38 | YL/OR | Transmission Control Relay Output | |

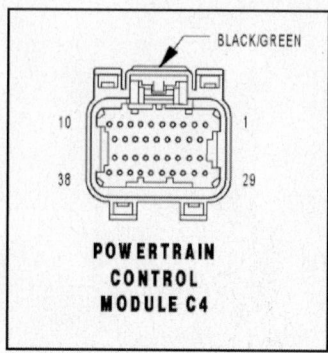

BLACK/GREEN

10      1
38      29

POWERTRAIN
CONTROL
MODULE C4

### 2004-05 Dakota 4.7L VIN N (545RFE) 'C4' Black/Green Connector

| PCM Pin # | Wire Color | Circuit Description (38-Pin) | Value at Hot Idle |
|---|---|---|---|
| 1 | YL/RD | OD Solenoid Control | |
| 2 | YL/DG | 4C Solenoid Control | |
| 3 | --- | Not Used | --- |
| 4 | BR/YL | MS Solenoid Control | |
| 5 | --- | Not Used | --- |
| 6 | DG/YL | 2C Solenoid Control | |
| 7 | --- | Not Used | --- |
| 8 | VT/PK | UD Solenoid Control | |
| 9 | --- | Not Used | --- |
| 10 | DG/WT | L/R Solenoid Control | |
| 11 | DG | Pressure Control Solenoid Control | |
| 12 | BK | Ground | |
| 13 | BK | Ground | |
| 14 | BK | Ground | |
| 15 | YL/PK | TRS T1 Sense | |
| 16 | YL/PK | TRS T3 Sense | |
| 17 | DG | Tow/Haul Overdrive Off Switch Sense | |
| 18 | YL/BK | Transmission Control Relay Control | |
| 19 | YL/OR | Transmission Control Relay Output | |
| 20 | BR/YL | 4C Pressure Switch Sense | |
| 21 | LB/RD | UD Pressure Switch Sense | |
| 22 | DG/TN | OD Pressure Switch Sense | |
| 23 | --- | Not Used | --- |
| 24 | --- | Not Used | --- |
| 25 | --- | Not Used | --- |
| 26 | BR/YL | TRS T2 Sense | |
| 27 | --- | Not Used | --- |
| 28 | YL/OR | Transmission Control Relay Output | |
| 29 | VT/RD | L/R Pressure Switch Sense | |
| 30 | BR/YL | 2C Pressure Switch Sense | |
| 31 | YL/BR | Line Pressure Sensor Signal | |
| 32 | YL/PK | Output Speed Sensor Signal | |
| 33 | YL/PK | Input Speed Sensor Signal | |
| 34 | DG/VT | Sensor Ground | |
| 35 | YL/PK | Transmission Temperature Sensor Signal | |
| 36 | --- | Not Used | --- |
| 37 | YL/PK | TRS T42 Sense | |
| 38 | YL/OR | Transmission Control Relay Output | |

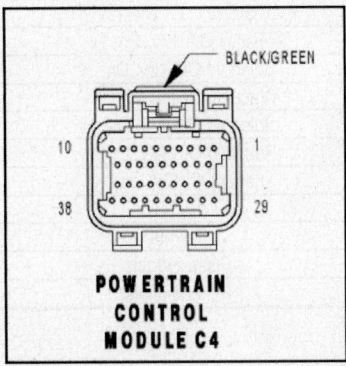

POWERTRAIN
CONTROL
MODULE C4

## 1992-95 Dakota 5.2L VIN Y (A/T) 60-Pin Connector

| PCM Pin # | Wire Color | Circuit Description (60-Pin) | Value at Hot Idle |
|---|---|---|---|
| 1 | DG/RD | MAP Sensor Signal | 1.5-1.6v |
| 2 | TN/BK | ECT Sensor Signal | At 180°F: 2.80v |
| 3 | RD/WT | Keep Alive Power (B+) | 12-14v |
| 4 | BK/LB | Sensor Ground | <0.050v |
| 5 ('92-'93) | BK/WT | Sensor Ground | <0.050v |
| 6 | PK/WT | 5-Volt Supply | 4.9-5.1v |
| 7 | OR | 8-Volt Supply | 7.9-8.1v |
| 8, 23, 26, 28, 31, 36 | --- | Not Used | --- |
| 9 | DB | Ignition Switch Power (B+) | 12-14v |
| 10 | OR/WT | Overdrive Override Switch | Switch Off: 12v, On: 1v |
| 11 | BK/TN | Power Ground | <0.1v |
| 12 | BK/TN | Power Ground | <0.1v |
| 13 | LB/BR | Injector 4 Driver | 1-4 ms |
| 14 | YL/WT | Injector 3 Driver | 1-4 ms |
| 15 | TN | Injector 2 Driver | 1-4 ms |
| 16 | WT/DG | Injector 1 Driver | 1-4 ms |
| 17 | DB/TN | Injector 7 Driver | 1-4 ms |
| 18 | DB/GY | Injector 8 Driver | 1-4 ms |
| 19 | BK/GY | Coil Driver Control | 5°, 55 mph: 8° dwell |
| 20 | DG | Alternator Field Control | Digital Signals: 0-12-0v |
| 21 | BK/RD | Charge Temperature Sensor | At 100°F: 2.51v |
| 22 | OR/DB | TP Sensor Signal | 0.6-1.0v |
| 24 | GY/BK | Distributor Reference Signal | Digital Signals: 0-5-0v |
| 25 | PK | SCI Transmit | 0v |
| 27 | BR | A/C Clutch Signal | A/C Off: 12v, On: 1v |
| 29 | WT/PK | Brake Switch Sense Signal | Brake Off: 0v, On: 12v |
| 30 | BR/YL | PNP Switch Sense Signal | In P/N: 0v, Others: 5v |
| 32 | PK/BK | MIL (lamp) Control | MIL Off: 12v, On: 1v |
| 33 | TN/RD | Speed Control Vacuum Solenoid | Vacuum Increasing: 1v |
| 34 | DB/OR | A/C WOT Relay Control | Relay Off: 12v, On: 1v |
| 35 ('92) | GY/YL | EGR Solenoid Control (Cal) | 12v, at 55 mph: 1v |
| 35 | GY/YL | EGR Solenoid Control | 12v, at 55 mph: 1v |
| 37 | BK/OR | Overdrive Lamp Control | Lamp Off: 12v, On: 1v |
| 38 | PK/BK | Injector 5 Driver | 1-4 ms |
| 39 | GY/RD | AIS Motor 4 Control | Pulse Signals |
| 40 | BR/WT | AIS Motor 2 Control | Pulse Signals |
| 41 | BK/DG | HO2S-11 (B1 S1) Signal | 0.1-1.1v |
| 42 | --- | Not Used | --- |
| 43 | GY/LB | Tachometer Signal | Pulse Signals |
| 44 | TN/YL | Distributor Sync Signal | Digital Signals: 0-5-0v |
| 45 | LG | SCI Receive | 5v |
| 46 | --- | Not Used | --- |
| 47 | WT/OR | Vehicle Speed Signal | Digital Signal |
| 48 | BR/RD | Speed Control Set Switch Signal | S/C & Set Switch On: 3.8v |
| 49 | YL/RD | Speed Control On/Off Switch | S/C On: 1v, Off: 12v |
| 50 | WT/LG | Speed Control Resume Switch | S/C On: 1v, Off: 12v |
| 51 | DB/YL | ASD Relay Control | Relay Off: 12v, On: 1v |
| 52 | PK/BK | EVAP Purge Solenoid Control | Solenoid Off: 12v, On: 1v |
| 53 | LG/RD | Speed Control Vent Solenoid | Vacuum Decreasing: 1v |
| 54 | OR/BK | Part Throttle Unlock Solenoid | Solenoid Off: 12v, On: 1v |
| 55 | OR/WT | Overdrive Lockout Solenoid | Solenoid Off: 12v, On: 1v |
| 56 ('92-'93) | GY/PK | Maintenance Indicator Lamp | Lamp Off: 12v, On: 1v |
| 57 | DG/BK | ASD Relay Power (B+) | 12-14v |
| 58 | LG/BK | Injector 6 Driver | 1-4 ms |
| 59 | PK/BK | AIS Motor 1 Control | Pulse Signals |
| 60 | GY/RD | AIS Motor 3 Control | Pulse Signals |

**1996 Dakota 5.2L VIN Y 'A' Black Connector**

| PCM Pin # | Wire Color | Circuit Description (32-Pin) | Value at Hot Idle |
|---|---|---|---|
| 1, 3, 5 | --- | Not Used | --- |
| 2 | DB | Ignition Switch Power (B+) | 12-14v |
| 4 | BK/LB | Sensor Ground | <0.050v |
| 6 | BR/YL | PNP Switch Sense Signal | In P/N: 0v, Others: 5v |
| 7 | GY | Coil 1 Driver Control | 5º, 55 mph: 8º dwell |
| 8 | GY/BK | CKP Sensor Signal | Digital Signals: 0-5-0v |
| 9 | --- | Not Used | --- |
| 10 | YL/BK | IAC 2 Driver Control | Pulse Signals |
| 11 | BR/WT | IAC 3 Driver Control | Pulse Signals |
| 12-14 | --- | Not Used | --- |
| 15 | BK/RD | IAT Sensor Signal | At 100ºF: 1.83v |
| 16 | TN/BK | ECT Sensor Signal | At 180ºF: 2.80v |
| 17 | PK/WT | 5-Volt Supply | 4.9-5.1v |
| 18 | TN/YL | CMP Sensor Signal | Digital Signals: 0-5-0v |
| 19 | GY/RD | IAC 1 Driver Control | Pulse Signals |
| 20 | PK/BK | IAC 4 Driver Control | Pulse Signals |
| 21 | --- | Not Used | --- |
| 22 | RD/WT | Keep Alive Power (B+) | 12-14v |
| 23 | OR/DB | TP Sensor Signal | 0.6-1.0v |
| 24 | TN/WT | HO2S-11 (B1 S1) Signal | 0.1-1.1v |
| 25 | TN/PK | HO2S-12 (B1 S2) Signal | 0.1-1.1v |
| 26 | --- | Not Used | --- |
| 27 | DG/RD | MAP Sensor Signal | 1.5-1.6v |
| 28-30 | --- | Not Used | --- |
| 31 | BK/TN | Power Ground | <0.1v |
| 32 | BK/TN | Power Ground | <0.1v |

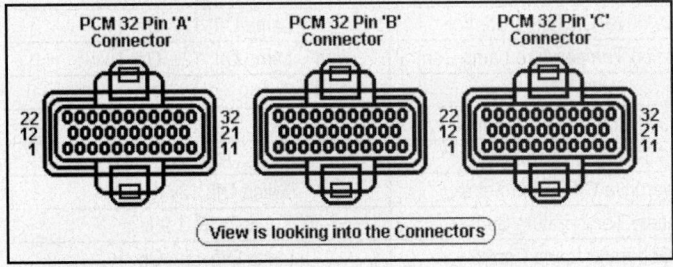

**Pin Connector Graphic**

**Standard Colors and Abbreviations**

| Abbreviation | Color | Abbreviation | Color | Abbreviation | Color |
|---|---|---|---|---|---|
| BK | Black | GY | Gray | RD | Red |
| BL | Blue | GN | Green | TN | Tan |
| BR | Brown | LG | Light Green | VT | Violet |
| DB | Dark Blue | OR | Orange | WT | White |
| DG | Dark Green | PK | Pink | YL | Yellow |

## 1996 Dakota 5.2L VIN Y 'B' White Connector

| PCM Pin # | Wire Color | Circuit Description (32-Pin) | Value at Hot Idle |
|---|---|---|---|
| 1 | PK | Transmission Temperature Sensor Signal | At 200°F: 2.40v |
| 2 | DB | Injector 7 Driver | 1-4 ms |
| 3, 7, 9, 14 | --- | Not Used | --- |
| 4 | WT/DB | Injector 1 Driver | 1-4 ms |
| 5 | YL/WT | Injector 3 Driver | 1-4 ms |
| 6 | GY | Injector 5 Driver | 1-4 ms |
| 8 | PK | Governor Pressure Solenoid | PWM Signal: 0-12-0v |
| 10 | DG | Generator Field Control | Digital Signals: 0-12-0v |
| 11 | OR/BK | TCC Solenoid Control | At Cruise w/TCC On: 1v |
| 12 | BR/DB | Injector 6 Driver | 1-4 ms |
| 13 | GY/LB | Injector 8 Driver | 1-4 ms |
| 15 | TN | Injector 2 Driver | 1-4 ms |
| 16 | LB/BR | Injector 4 Driver | 1-4 ms |
| 17-20, 22-24 | --- | Not Used | --- |
| 21 | BR | 3-4 Shift Solenoid Control | Solenoid Off: 12v, On: 1v |
| 25, 28 | DB, LG | OSS Sensor (-) Signal, (+) | Moving: AC pulse signals |
| 27 | WT/OR | Vehicle Speed Signal | Digital Signal |
| 29 | LG/RD | Governor Pressure Sensor | 0.58v |
| 30 | PK/LB | Transmission Relay Control | Relay Off: 12v, On: 1v |
| 31 | RD/YL | 5-Volt Supply | 4.9-5.1v |

## 1996 Dakota 5.2L VIN Y 'C' Gray Connector

| PCM Pin # | Wire Color | Circuit Description (32-Pin) | Value at Hot Idle |
|---|---|---|---|
| 1 | DB/OR | A/C Clutch Relay Control | Relay Off: 12v, On: 1v |
| 2 | DB/PK | Radiator Fan Control Relay | Relay Off: 12v, On: 1v |
| 3 | DB/YL | ASD Relay Control | Relay Off: 12v, On: 1v |
| 4 | TN/RD | Speed Control Vacuum Solenoid | Vacuum Increasing: 1v |
| 5 | LG/RD | Speed Control Vent Solenoid | Vacuum Decreasing: 1v |
| 6 | LG/OR | Overdrive 'Off' Lamp Control | Lamp Off: 12v, On: 1v |
| 7 | PK/BK | Transmission Temperature Lamp Control | Lamp Off: 12v, On: 1v |
| 8-10, 14, 18 | --- | Not Used | --- |
| 11 | YL/RD | Speed Control Power Supply | 12-14v |
| 12 | DG/OR | ASD Relay Power (B+) | 12-14v |
| 13 | OR/WT | Overdrive Off Switch Sense | Switch Off: 12v, On: 1v |
| 15 | PK/YL | Battery Temperature Sensor | At 86°F: 1.96v |
| 16 | DG/Y | Generator Lamp Control | Lamp Off: 12v, On: 1v |
| 17 | BK/PK | MIL (lamp) Control | MIL Off: 12v, On: 1v |
| 19 | BR | Fuel Pump Relay Control | Relay Off: 12v, On: 1v |
| 20 | PK/BK | EVAP Purge Solenoid Control | PWM Signal: 0-12-0v |
| 22 | BR | A/C Request Signal | A/C Off: 12v, On: 1v |
| 23 | LG | A/C Select Signal | A/C Off: 12v, On: 1v |
| 24 | WT/PK | Brake Switch Sense Signal | Brake Off: 0v, On: 12v |
| 26 | LB/BK | Fuel Level Sensor Signal | Full: 0.56v, 1/2 full: 2.5v |
| 27 | PK | SCI Transmit | 0v |
| 28 | WT | CCD Bus (-) Signal | <0.050v |
| 29 | LG | SCI Receive | 5v |
| 30 | PK/BR | CCD Bus (+) Signal | Digital Signals: 0-5-0v |
| 31 | GY/LB | Tachometer Signal | Pulse Signals |
| 32 | WT/LG | Speed Control Set Switch Signal | S/C & Set Switch On: 3.8v |

## 1997 Dakota 5.2L VIN Y 'A' Black Connector

| PCM Pin # | Wire Color | Circuit Description (32-Pin) | Value at Hot Idle |
|---|---|---|---|
| 1 | --- | Not Used | --- |
| 2 | LG/BK | Ignition Switch Power (B+) | 12-14v |
| 3 | --- | Not Used | --- |
| 4 | BK/LB | Sensor Ground | <0.050v |
| 5 | --- | Not Used | --- |
| 6 | BK/WT | PNP Switch Sense Signal | In P/N: 0v, Others: 5v |
| 7 | BK/GY | Coil 1 Driver Control | 5º, 55 mph: 8º dwell |
| 8 | GY/BK | CKP Sensor Signal | Digital Signals: 0-5-0v |
| 9 | --- | Not Used | --- |
| 10 | YL/BK | IAC 2 Driver Control | Pulse Signals |
| 11 | BR/WT | IAC 3 Driver Control | Pulse Signals |
| 12 | GY/WT | PSP Switch Sense Signal | Straight: 0v, Turning: 5v |
| 13-14 | --- | Not Used | --- |
| 15 | BK/RD | IAT Sensor Signal | At 100ºF: 1.83v |
| 16 | TN/BK | ECT Sensor Signal | At 180ºF: 2.80v |
| 17 | PK/WT | 5-Volt Supply | 4.9-5.1v |
| 18 | TN/YL | CMP Sensor Signal | Digital Signals: 0-5-0v |
| 19 | GY/RD | IAC 1 Driver Control | Pulse Signals |
| 20 | PK/BK | IAC 4 Driver Control | Pulse Signals |
| 21 | --- | Not Used | --- |
| 22 | RD/WT | Keep Alive Power (B+) | 12-14v |
| 23 | OR/DB | TP Sensor Signal | 0.6-1.0v |
| 24 | OR/TN | HO2S-11 (B1 S1) Signal | 0.1-1.1v |
| 25 | OR/BK | HO2S-12 (B1 S2) Signal | 0.1-1.1v |
| 26 | --- | Not Used | --- |
| 27 | DG/RD | MAP Sensor Signal | 1.5-1.6v |
| 28-30 | --- | Not Used | --- |
| 31 | BK/TN | Power Ground | <0.1v |
| 32 | BK/TN | Power Ground | <0.1v |

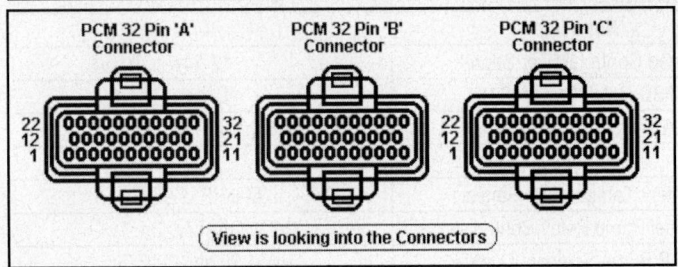

**Pin Connector Graphic**

## Standard Colors and Abbreviations

| Abbreviation | Color | Abbreviation | Color | Abbreviation | Color |
|---|---|---|---|---|---|
| BK | Black | GY | Gray | RD | Red |
| BL | Blue | GN | Green | TN | Tan |
| BR | Brown | LG | Light Green | VT | Violet |
| DB | Dark Blue | OR | Orange | WT | White |
| DG | Dark Green | PK | Pink | YL | Yellow |

## 1997 Dakota 5.2L VIN Y 'B' White Connector

| PCM Pin # | Wire Color | Circuit Description (32-Pin) | Value at Hot Idle |
|---|---|---|---|
| 1 | PK | Transmission Temperature Sensor Signal | At 200°F: 2.40v |
| 2 | PK/WT | Injector 7 Driver | 1-4 ms |
| 3, 7, 9, 14 | --- | Not Used | --- |
| 4 | WT/DB | Injector 1 Driver | 1-4 ms |
| 5 | YL/WT | Injector 3 Driver | 1-4 ms |
| 6 | GY | Injector 5 Driver | 1-4 ms |
| 8 | PK | Governor Pressure Solenoid | PWM Signal: 0-12-0v |
| 10 | DG | Generator Field Control | Digital Signals: 0-12-0v |
| 11 | OR/BK | TCC Solenoid Control | At Cruise w/TCC On: 1v |
| 12 | BR/DB | Injector 6 Driver | 1-4 ms |
| 13 | GY/LB | Injector 8 Driver | 1-4 ms |
| 15 | TN | Injector 2 Driver | 1-4 ms |
| 16 | LB/BR | Injector 4 Driver | 1-4 ms |
| 21 | BR | Overdrive Solenoid Control | Cruise w/solenoid On: 1v |
| 23 | GY/YL | Engine Oil Pressure Sensor | 1.6v at 24 psi |
| 25 | DB/BK | OSS Sensor (-) Signal | Moving: AC pulse signals |
| 27 | WT/OR | Vehicle Speed Signal | Digital Signal |
| 28 | LG/WT | OSS Sensor (+) Signal | Moving: AC pulse signals |
| 29 | LG/RD | Governor Pressure Sensor | 0.58v |
| 30 | PK/BK | Transmission Relay Control | Relay Off: 12v, On: 1v |
| 31 | OR | 5-Volt Supply | 4.9-5.1v |

## 1997 Dakota 5.2L VIN Y 'C' Gray Connector

| PCM Pin # | Wire Color | Circuit Description (32-Pin) | Value at Hot Idle |
|---|---|---|---|
| 1 | DB/OR | A/C Clutch Relay Control | Relay Off: 12v, On: 1v |
| 2 | DB/PK | Radiator Fan Control Relay | Relay Off: 12v, On: 1v |
| 3 | DB/YL | ASD Relay Control | Relay Off: 12v, On: 1v |
| 4 | TN/RD | Speed Control Vacuum Solenoid | Vacuum Increasing: 1v |
| 5 | LG/RD | Speed Control Vent Solenoid | Vacuum Decreasing: 1v |
| 6 | LG/OR | Overdrive Off Lamp Control | Lamp Off: 12v, On: 1v |
| 7-10, 16-18, 21, 31 | --- | Not Used | --- |
| 11 | YL/RD | Speed Control Power Supply | 12-14v |
| 12 | DG/OR | ASD Relay Power (B+) | 12-14v |
| 13 | OR/WT | Overdrive Off Switch Sense | Switch Off: 12v, On: 1v |
| 14 | --- | Not Used | --- |
| 15 | PK/YL | Battery Temperature Sensor | At 86°F: 1.96v |
| 19 | LB/OR | Fuel Pump Relay Control | Relay Off: 12v, On: 1v |
| 20 | PK/BK | EVAP Purge Solenoid Control | PWM Signal: 0-12-0v |
| 22 | BR | A/C Request Signal | A/C Off: 12v, On: 1v |
| 23 | LG/WT | A/C Select Signal | A/C Off: 12v, On: 1v |
| 24 | WT/PK | Brake Switch Sense Signal | Brake Off: 0v, On: 12v |
| 25 | DG/BK | Generator Field Source | 12-14v |
| 26 | DB | Fuel Level Sensor Signal | Full: 0.56v, 1/2 full: 2.5v |
| 27 | PK | SCI Transmit Signal | 0v |
| 28 | WT/BK | CCD Bus (-) | <0.050v |
| 29 | LG | SCI Receive Signal | 5v |
| 30 | PK/BR | CCD Bus (+) | Digital Signals: 0-5-0v |
| 32 | GY/LB | Speed Control Set Switch Signal | S/C & Set Switch On: 3.8v |

**1998-99 Dakota 5.2L VIN Y 'A' Black Connector**

| PCM Pin # | Wire Color | Circuit Description (32-Pin) | Value at Hot Idle |
|---|---|---|---|
| 1 | --- | Not Used | --- |
| 2 | LG/BK | Ignition Switch Power (B+) | 12-14v |
| 3 | --- | Not Used | --- |
| 4 | BK/LB | Sensor Ground | <0.050v |
| 5 | --- | Not Used | --- |
| 6 | BK/WT | PNP Switch Sense Signal | In P/N: 0v, Others: 5v |
| 7 | BK/GY | Coil 1 Driver Control | 5°, 55 mph: 8° dwell |
| 8 | GY/BK | CKP Sensor Signal | Digital Signals: 0-5-0v |
| 9 | --- | Not Used | --- |
| 10 | YL/BK | IAC 2 Driver Control | Pulse Signals |
| 11 | BR/WT | IAC 3 Driver Control | Pulse Signals |
| 12-14 | --- | Not Used | --- |
| 15 | BK/RD | IAT Sensor Signal | At 100°F: 1.83v |
| 16 | TN/BK | ECT Sensor Signal | At 180°F: 2.80v |
| 17 | PK/WT | 5-Volt Supply | 4.9-5.1v |
| 18 | TN/YL | CMP Sensor Signal | Digital Signals: 0-5-0v |
| 19 | GY/RD | IAC 1 Driver Control | Pulse Signals |
| 20 | PK/BK | IAC 4 Driver Control | Pulse Signals |
| 21 | --- | Not Used | --- |
| 22 | RD/WT | Keep Alive Power (B+) | 12-14v |
| 23 | OR/DB | TP Sensor Signal | 0.6-1.0v |
| 24 | OR/TN | HO2S-11 (B1 S1) Signal | 0.1-1.1v |
| 25 | OR/BK | HO2S-12 (B1 S2) Signal | 0.1-1.1v |
| 26 | --- | Not Used | --- |
| 27 | DG/RD | MAP Sensor Signal | 1.5-1.6v |
| 28-30 | --- | Not Used | --- |
| 31 | BK/TN | Power Ground | <0.1v |
| 32 | BK/TN | Power Ground | <0.1v |

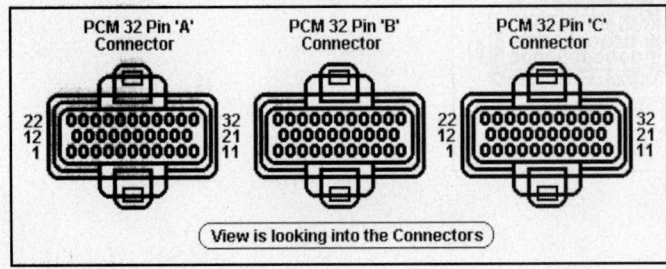

**Pin Connector Graphic**

**1998-99 Dakota 5.2L VIN Y 'B' White Connector**

| PCM Pin # | Wire Color | Circuit Description (32-Pin) | Value at Hot Idle |
|---|---|---|---|
| 1 | GY/BK | Transmission Temperature Sensor Signal | At 200°F: 2.40v |
| 2 | PK | Injector 7 Driver | 1-4 ms |
| 3 | --- | Not Used | --- |
| 4 | WT/DB | Injector 1 Driver | 1-4 ms |
| 5 | YL/WT | Injector 3 Driver | 1-4 ms |
| 6 | GY | Injector 5 Driver | 1-4 ms |
| 7 | --- | Not Used | --- |
| 8 | PK/WT | Variable Force Solenoid | PWM Signal: 0-12-0v |
| 9 | --- | Not Used | --- |
| 10 | DG | Generator Field Control | Digital Signals: 0-12-0v |
| 11 | OR/BK | TCC Solenoid Control | At Cruise w/TCC On: 1v |
| 12 | BR/DB | Injector 6 Driver | 1-4 ms |
| 13 | GY/LB | Injector 8 Driver | 1-4 ms |
| 14 | --- | Not Used | --- |
| 15 | TN | Injector 2 Driver | 1-4 ms |
| 16 | LB/BR | Injector 4 Driver | 1-4 ms |
| 17 ('00) | DB/P | Radiator Fan Relay Control | Relay Off: 12v, On: 1v |
| 21 | BR | Overdrive Solenoid Control | Cruise w/solenoid On: 1v |
| 23 | GY/YL | Engine Oil Pressure Sensor | 1.6v at 24 psi |
| 25 | DB/BK | OSS Sensor (-) Signal | Moving: AC pulse signals |
| 27 | WT/OR | Vehicle Speed Signal | Digital Signal |
| 28 | LG/WT | OSS Sensor (+) Signal | Moving: AC pulse signals |
| 29 | LG/RD | Governor Pressure Sensor | 0.58v |
| 30 | PK/BK | Transmission Relay Control | Relay Off: 12v, On: 1v |
| 31 | OR | 5-Volt Supply | 4.9-5.1v |

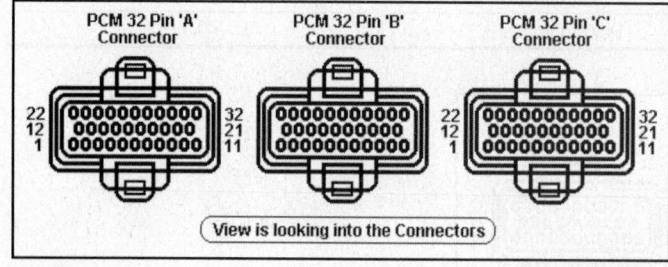

**Pin Connector Graphic**

## 1998-99 Dakota 5.2L VIN Y 'C' Gray Connector

| PCM Pin # | Wire Color | Circuit Description (32-Pin) | Value at Hot Idle |
|---|---|---|---|
| 1 | DB/OR | A/C Clutch Relay Control | Relay Off: 12v, On: 1v |
| 2 ('98-'99) | DB/PK | Radiator Fan Control Relay | Relay Off: 12v, On: 1v |
| 3 | DB/YL | ASD Relay Control | Relay Off: 12v, On: 1v |
| 4 | TN/RD | Speed Control Vacuum Solenoid | Vacuum Increasing: 1v |
| 5 | LG/RD | Speed Control Vent Solenoid | Vacuum Decreasing: 1v |
| 6-7 | --- | Not Used | --- |
| 8 | DG/BK | HO2S-11 HTR Control (California) | Relay Off: 12v, On: 1v |
| 9 | OR/RD | HO2S-12 HTR Control (California) | Relay Off: 12v, On: 1v |
| 10 | YL/DG | LDP Solenoid Control | PWM Signal: 0-12-0v |
| 11 | YL/RD | Speed Control Power Supply | 12-14v |
| 12 | DG/OR | ASD Relay Power (B+) | 12-14v |
| 13 | OR/WT | Overdrive Off Switch Sense | Switch Off: 12v, On: 1v |
| 14 | YL/WT | LDP Switch Sense Signal | LDP Switch Closed: 0v |
| 15 | PK/YL | Battery Temperature Sensor | At 86°F: 1.96v |
| 16-18 | --- | Not Used | --- |
| 19 | LB/OR | Fuel Pump Relay Control | Relay Off: 12v, On: 1v |
| 20 | PK/BK | EVAP Purge Solenoid Control | PWM Signal: 0-12-0v |
| 21 | --- | Not Used | --- |
| 22 | BR | A/C Request Signal | A/C Off: 12v, On: 1v |
| 23 | LG/WT | A/C Select Signal | A/C Off: 12v, On: 1v |
| 24 | WT/PK | Brake Switch Sense Signal | Brake Off: 0v, On: 12v |
| 25 | DG/BK | Generator Field Source | 12-14v |
| 26 | DB | Fuel Level Sensor Signal | Full: 0.56v, 1/2 full: 2.5v |
| 27 | PK | SCI Transmit Signal | 0v |
| 28 | WT | CCD Bus (-) Signal | <0.050v |
| 29 | LG | SCI Receive Signal | 5v |
| 30 | PK/BR | CCD Bus (+) Signal | Digital Signals: 0-5-0v |
| 31 | --- | Not Used | --- |
| 32 | RD/LG | Speed Control Set Switch Signal | S/C & Set Switch On: 3.8v |

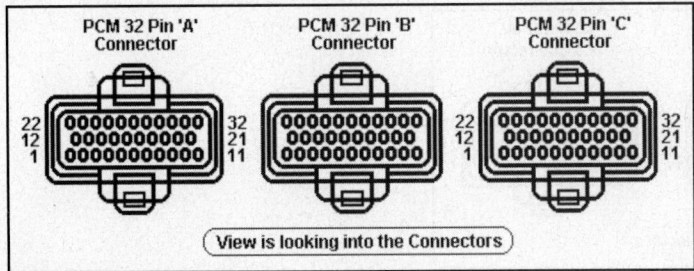

**Pin Connector Graphic**

### 1998-99 Dakota 5.9L VIN Z (A/T) 'C1' Black Connector

| PCM Pin # | Wire Color | Circuit Description (32-Pin) | Value at Hot Idle |
|---|---|---|---|
| 1 | --- | Not Used | --- |
| 2 | LG/BK | Ignition Switch Power (B+) | 12-14v |
| 3 | --- | Not Used | --- |
| 4 | BK/LB | Sensor Ground | <0.050v |
| 5 | --- | Not Used | --- |
| 6 | BK/WT | PNP Switch Sense Signal | In P/N: 0v, Others: 5v |
| 7 | BK/GY | Coil 1 Driver Control | 5°, 55 mph: 8° dwell |
| 8 | GY/BK | CKP Sensor Signal | Digital Signals: 0-5-0v |
| 9 | --- | Not Used | --- |
| 10 | YL/BK | IAC 2 Driver Control | Pulse Signals |
| 11 | BR/WT | IAC 3 Driver Control | Pulse Signals |
| 12-14 | --- | Not Used | --- |
| 15 | BK/RD | IAT Sensor Signal | At 100°F: 1.83v |
| 16 | TN/BK | ECT Sensor Signal | At 180°F: 2.80v |
| 17 | PK/WT | 5-Volt Supply | 4.9-5.1v |
| 18 | TN/YL | CMP Sensor Signal | Digital Signals: 0-5-0v |
| 19 | GY/RD | IAC 1 Driver Control | Pulse Signals |
| 20 | PK/BK | IAC 4 Driver Control | Pulse Signals |
| 21 | --- | Not Used | --- |
| 22 | RD/WT | Keep Alive Power (B+) | 12-14v |
| 23 | OR/DB | TP Sensor Signal | 0.6-1.0v |
| 24 | TN/WT | HO2S-11 (B1 S1) Signal | 0.1-1.1v |
| 25 | OR/BK | HO2S-12 (B1 S2) Signal | 0.1-1.1v |
| 26 | --- | Not Used | --- |
| 27 | DG/RD | MAP Sensor Signal | 1.5-1.6v |
| 28-29 | --- | Not Used | --- |
| 30 | --- | Not Used | --- |
| 31 | BK/TN | Power Ground | <0.1v |
| 32 | BK/TN | Power Ground | <0.1v |

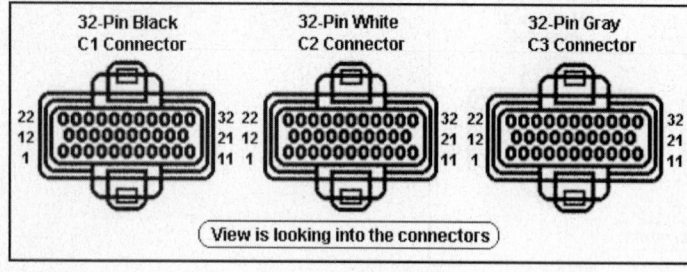

**Pin Connector Graphic**

**1998-99 Dakota 5.9L VIN Z (A/T) 'C2' White Connector**

| PCM Pin # | Wire Color | Circuit Description (32-Pin) | Value at Hot Idle |
|---|---|---|---|
| 1 | PK | Transmission Temperature Sensor Signal | At 200°F: 2.40v |
| 2 | VT | Injector 7 Driver | 1-4 ms |
| 4 | WT/DB | Injector 1 Driver | 1-4 ms |
| 5 | YL/WT | Injector 3 Driver | 1-4 ms |
| 6 | GY | Injector 5 Driver | 1-4 ms |
| 8 | PK | Governor Pressure Solenoid | PWM Signal: 0-12-0v |
| 10 | DG | Generator Field Control | Digital Signals: 0-12-0v |
| 11 | OR/BK | TCC Solenoid Control | At Cruise w/TCC On: 1v |
| 12 | BR/DB | Injector 6 Driver | 1-4 ms |
| 13 | GY/LB | Injector 8 Driver | 1-4 ms |
| 15 | TN | Injector 2 Driver | 1-4 ms |
| 16 | LB/BR | Injector 4 Driver | 1-4 ms |
| 17-20 | --- | Not Used | --- |
| 21 | BR | Overdrive Solenoid Control | Cruise w/solenoid On: 1v |
| 22, 32 | --- | Not Used | --- |
| 23 | GY/YL | Engine Oil Pressure Sensor | 1.6v at 24 psi |
| 25 | DB/BK | OSS Sensor (-) Signal | Moving: AC pulse signals |
| 27 | WT/OR | Vehicle Speed Signal | Digital Signal |
| 28 | LG/WT | OSS Sensor (+) Signal | Moving: AC pulse signals |
| 29 | LG/RD | Governor Pressure Sensor | 0.58v |
| 30 | PK/BK | Transmission Relay Control | Relay Off: 12v, On: 1v |
| 31 | VT/WT | 5-Volt Supply | 4.9-5.1v |

**1998-99 Dakota 5.9L VIN Z (A/T) 'C3' Gray Connector**

| PCM Pin # | Wire Color | Circuit Description (32-Pin) | Value at Hot Idle |
|---|---|---|---|
| 1 | DB/OR | A/C Clutch Relay Control | Relay Off: 12v, On: 1v |
| 3 | DB/YL | ASD Relay Control | Relay Off: 12v, On: 1v |
| 4 | TN/RD | Speed Control Vacuum Solenoid | Vacuum Increasing: 1v |
| 5 | LG/RD | Speed Control Vent Solenoid | Vacuum Decreasing: 1v |
| 6-9 | --- | Not Used | --- |
| 10 | YL/DG | LDP Solenoid Control | PWM Signal: 0-12-0v |
| 11 | YL/RD | Speed Control Power Supply | 12-14v |
| 12 | DG/OR | ASD Relay Power (B+) | 12-14v |
| 13 | OR/WT | Overdrive Off Switch Sense | Switch Off: 12v, On: 1v |
| 14 | YL/WT | LDP Switch Sense Signal | LDP Switch Closed: 0v |
| 15 | PK/YL | Battery Temperature Sensor | At 86°F: 1.96v |
| 19 | LB/OR | Fuel Pump Relay Control | Relay Off: 12v, On: 1v |
| 20 | PK/BK | EVAP Purge Solenoid Control | PWM Signal: 0-12-0v |
| 22 | BR | A/C Request Signal | A/C Off: 12v, On: 1v |
| 23 | LG/WT | A/C Select Signal | A/C Off: 12v, On: 1v |
| 24 | WT/PK | Brake Switch Sense Signal | Brake Off: 0v, On: 12v |
| 25 | DG, W | Generator Field Source | 12-14v |
| 26 | DB | Fuel Level Sensor Signal | Full: 0.56v, 1/2 full: 2.5v |
| 27 | PK/DB | SCI Transmit Signal | 0v |
| 28 | WT/BK | CCD Bus (-) Signal | <0.050v |
| 29 | LG | SCI Receive Signal | 5v |
| 30 | PK/BR | CCD Bus (+) Signal | Digital Signals: 0-5-0v |
| 32 | RD/LG | Speed Control Set Switch Signal | S/C & Set Switch On: 3.8v |

## 2000 Dakota 5.9L VIN Z (A/T) 'C1' Black Connector

| PCM Pin # | Wire Color | Circuit Description (32-Pin) | Value at Hot Idle |
|---|---|---|---|
| 1 | --- | Not Used | --- |
| 2 | LG/BK | Ignition Switch Power (B+) | 12-14v |
| 3 | --- | Not Used | --- |
| 4 | BK/LB | Sensor Ground | <0.050v |
| 5 | --- | Not Used | --- |
| 6 | BK/WT | PNP Switch Sense Signal | In P/N: 0v, Others: 5v |
| 7 | BK/GY | Coil 1 Driver Control | 5°, 55 mph: 8° dwell |
| 8 | GY/BK | CKP Sensor Signal | Digital Signals: 0-5-0v |
| 9 | --- | Not Used | --- |
| 10 | YL/BK | IAC 2 Driver Control | Pulse Signals |
| 11 | BR/WT | IAC 3 Driver Control | Pulse Signals |
| 12-14 | --- | Not Used | --- |
| 15 | BK/RD | IAT Sensor Signal | At 100°F: 1.83v |
| 16 | TN/BK | ECT Sensor Signal | At 180°F: 2.80v |
| 17 | VT/WT | 5-Volt Supply | 4.9-5.1v |
| 18 | TN/YL | CMP Sensor Signal | Digital Signals: 0-5-0v |
| 19 | GY/RD | IAC 1 Driver Control | Pulse Signals |
| 20 | VT/BK | IAC 4 Driver Control | Pulse Signals |
| 21 | --- | Not Used | --- |
| 22 | RD/WT | Keep Alive Power (B+) | 12-14v |
| 23 | OR/DB | TP Sensor Signal | 0.6-1.0v |
| 24 | TN/WT | HO2S-11 (B1 S1) Signal | 0.1-1.1v |
| 25 | OR/BK | HO2S-12 (B1 S2) Signal | 0.1-1.1v |
| 26 | --- | Not Used | --- |
| 27 | DG/RD | MAP Sensor Signal | 1.5-1.6v |
| 28-30 | --- | Not Used | --- |
| 31 | BK/TN | Power Ground | <0.1v |
| 32 | BK/TN | Power Ground | <0.1v |

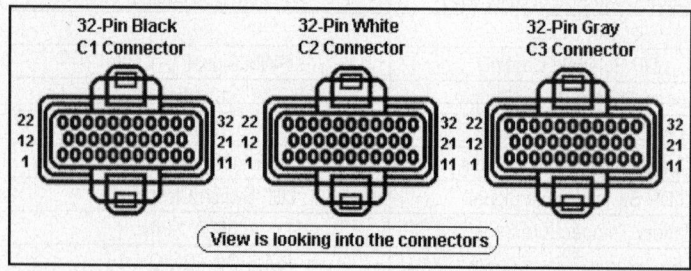

**Pin Connector Graphic**

## Standard Colors and Abbreviations

| Abbreviation | Color | Abbreviation | Color | Abbreviation | Color |
|---|---|---|---|---|---|
| BK | Black | GY | Gray | RD | Red |
| BL | Blue | GN | Green | TN | Tan |
| BR | Brown | LG | Light Green | VT | Violet |
| DB | Dark Blue | OR | Orange | WT | White |
| DG | Dark Green | PK | Pink | YL | Yellow |

## 2000 Dakota 5.9L VIN Z (A/T) White 'C2' Connector

| PCM Pin # | Wire Color | Circuit Description (32-Pin) | Value at Hot Idle |
|---|---|---|---|
| 1 | PK | Transmission Temperature Sensor Signal | At 200°F: 2.40v |
| 2 | VT | Injector 7 Driver | 1-4 ms |
| 4 | WT/DB | Injector 1 Driver | 1-4 ms |
| 5 | YL/WT | Injector 3 Driver | 1-4 ms |
| 6 | GY | Injector 5 Driver | 1-4 ms |
| 8 | PK | Governor Pressure Solenoid | PWM Signal: 0-12-0v |
| 10 | DG | Generator Field Control | Digital Signals: 0-12-0v |
| 11 | OR/BK | TCC Solenoid Control | At Cruise w/TCC On: 1v |
| 12 | BR/DB | Injector 6 Driver | 1-4 ms |
| 13 | GY/LB | Injector 8 Driver | 1-4 ms |
| 15 | TN | Injector 2 Driver | 1-4 ms |
| 16 | LB/BR | Injector 4 Driver | 1-4 ms |
| 17 | DB/PK | Radiator Fan Control Relay | Relay Off: 12v, On: 1v |
| 21 | BR | Overdrive Solenoid Control | Cruise w/solenoid On: 1v |
| 23 | GY/YL | Engine Oil Pressure Sensor | 1.6v at 24 psi |
| 25 | DB | OSS Sensor (-) Signal | Moving: AC pulse signals |
| 27 | WT/OR | Vehicle Speed Signal | Digital Signal |
| 28 | LG | OSS Sensor (+) Signal | Moving: AC pulse signals |
| 29 | LG/RD | Governor Pressure Sensor | 0.58v |
| 30 | PK | Transmission Relay Control | Relay Off: 12v, On: 1v |
| 31 | VT/WT | 5-Volt Supply | 4.9-5.1v |

## 2000 Dakota 5.9L VIN Z (A/T) 'C3' Gray Connector

| PCM Pin # | Wire Color | Circuit Description (32-Pin) | Value at Hot Idle |
|---|---|---|---|
| 1 | DB/OR | A/C Clutch Relay Control | Relay Off: 12v, On: 1v |
| 3 | DB/YL | ASD Relay Control | Relay Off: 12v, On: 1v |
| 4 | TN/RD | Speed Control Vacuum Solenoid | Vacuum Increasing: 1v |
| 5 | LG/RD | Speed Control Vent Solenoid | Vacuum Decreasing: 1v |
| 8 | DG/BK | HO2S-11 HTR Control (Cal) | Relay Off: 12v, On: 1v |
| 9 | OR/RD | HO2S-12 HTR Control (Cal) | Relay Off: 12v, On: 1v |
| 10 | YL/DG | LDP Solenoid Control | PWM Signal: 0-12-0v |
| 11 | YL/RD | Speed Control Power Supply | 12-14v |
| 12 | DG/OR | ASD Relay Power (B+) | 12-14v |
| 13 | OR/WT | Overdrive Off Switch Sense | Switch Off: 12v, On: 1v |
| 14 | YL/WT | LDP Switch Sense Signal | LDP Switch Closed: 0v |
| 15 | PK/YL | Battery Temperature Sensor | At 86°F: 1.96v |
| 19 | LB/OR | Fuel Pump Relay Control | Relay Off: 12v, On: 1v |
| 20 | PK/BK | EVAP Purge Solenoid Control | PWM Signal: 0-12-0v |
| 22 | BR | A/C Request Signal | A/C Off: 12v, On: 1v |
| 23 | LG/WT | A/C Select Signal | A/C Off: 12v, On: 1v |
| 24 | WT/PK | Brake Switch Sense Signal | Brake Off: 0v, On: 12v |
| 25 | DG, W | Generator Field Source | 12-14v |
| 26 | DB | Fuel Level Sensor Signal | Full: 0.56v, 1/2 full: 2.5v |
| 27 | PK/DB | SCI Transmit Signal | 0v |
| 28 | WT/BK | CCD Bus (-) Signal | <0.050v |
| 29 | LG | SCI Receive Signal | 5v |
| 30 | PK/BR | CCD Bus (+) Signal | Digital Signals: 0-5-0v |
| 32 | RD/LG | Speed Control Set Switch Signal | S/C & Set Switch On: 3.8v |

## 2001-03 Dakota 5.9L VIN Z (A/T) 'C1' Black Connector

| PCM Pin # | Wire Color | Circuit Description (32-Pin) | Value at Hot Idle |
|---|---|---|---|
| 1 | --- | Not Used | --- |
| 2 | LG/BK | Ignition Switch Power (B+) | 12-14v |
| 3 | --- | Not Used | --- |
| 4 | BK/LB | Sensor Ground | <0.050v |
| 5 | --- | Not Used | --- |
| 6 | BK/WT | PNP Switch Sense Signal | In P/N: 0v, Others: 5v |
| 7 | BK/GY | Coil 1 Driver Control | 5°, 55 mph: 8° dwell |
| 8 | GY/BK | CKP Sensor Signal | Digital Signals: 0-5-0v |
| 9 | --- | Not Used | --- |
| 10 | YL/BK | IAC 2 Driver Control | Pulse Signals |
| 11 | BR/WT | IAC 3 Driver Control | Pulse Signals |
| 12-14 | --- | Not Used | --- |
| 15 | BK/RD | IAT Sensor Signal | At 100°F: 1.83v |
| 16 | TN/BK | ECT Sensor Signal | At 180°F: 2.80v |
| 17 | OR | 5-Volt Supply | 4.9-5.1v |
| 18 | TN/YL | CMP Sensor Signal | Digital Signals: 0-5-0v |
| 19 | GY/RD | IAC 1 Driver Control | Pulse Signals |
| 20 | VT/BK | IAC 4 Driver Control | Pulse Signals |
| 21 | --- | Not Used | --- |
| 22 | RD/WT | Keep Alive Power (B+) | 12-14v |
| 23 | OR/DB | TP Sensor Signal | 0.6-1.0v |
| 24 | BK/DG | HO2S-11 (B1 S1) Signal | 0.1-1.1v |
| 25 | TN/WT | HO2S-12 (B1 S2) Signal | 0.1-1.1v |
| 26 | LG/RD | HO2S-21 (B2 S1) Signal | 0.1-1.1v |
| 27 | DG/RD | MAP Sensor Signal | 1.5-1.6v |
| 28 | --- | Not Used | --- |
| 29 | TN/WT | HO2S-22 (B2 S2) Signal | 0.1-1.1v |
| 30 | --- | Not Used | --- |
| 31 | BK/TN | Power Ground | <0.1v |
| 32 | BK/TN | Power Ground | <0.1v |

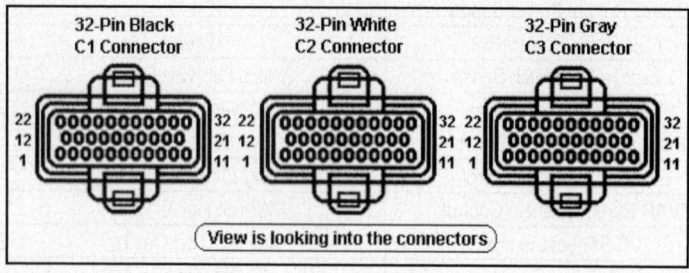

**Pin Connector Graphic**

**2001-03 Dakota 5.9L VIN Z (A/T) White 'C2' Connector**

| PCM Pin # | Wire Color | Circuit Description (32-Pin) | Value at Hot Idle |
|---|---|---|---|
| 1 | GY/BK | Transmission Temperature Sensor Signal | At 200ºF: 2.40v |
| 2 | VT | Injector 7 Driver | 1-4 ms |
| 3 | --- | Not Used | --- |
| 4 | WT/DB | Injector 1 Driver | 1-4 ms |
| 5 | YL/WT | Injector 3 Driver | 1-4 ms |
| 6 | GY | Injector 5 Driver | 1-4 ms |
| 8 | VT/WT | Variable Force Solenoid | PWM Signal: 0-12-0v |
| 9 | --- | Not Used | --- |
| 10 | DG | Generator Field Control | Digital Signals: 0-12-0v |
| 11 | OR/BK | TCC Solenoid Control | At Cruise w/TCC On: 1v |
| 12 | BR/DB | Injector 6 Driver | 1-4 ms |
| 13 | GY/LB | Injector 8 Driver | 1-4 ms |
| 14 | --- | Not Used | --- |
| 15 | TN | Injector 2 Driver | 1-4 ms |
| 16 | LB/BR | Injector 4 Driver | 1-4 ms |
| 17 | DB/PK | Radiator Fan Control Relay | Relay Off: 12v, On: 1v |
| 18 | --- | Not Used | --- |
| 19 | DB | AC Pressure Sensor Signal | 0.90v at 79 psi |
| 20 | --- | Not Used | --- |
| 21 | BR | 3-4 Shift Solenoid Control | Cruise w/solenoid on: 1v |
| 22 | --- | Not Used | --- |
| 23 | GY/YL | Engine Oil Pressure Sensor | 1.6v at 24 psi |
| 25 | DB/BK | OSS Sensor (-) Signal | Moving: AC pulse signals |
| 26 | --- | Not Used | --- |
| 27 | WT/OR | Vehicle Speed Signal | Digital Signals |
| 28 | LG/WT | OSS Sensor (+) Signal | Moving: AC pulse signals |
| 29 | LG/RD | Governor Pressure Sensor | 0.58v |
| 30 | PK | Transmission Relay Control | Relay Off: 12v, On: 1v |
| 31 | VT/WT | 5-Volt Supply | 4.9-5.1v |
| 32 | --- | Not Used | --- |

## 2001-03 Dakota 5.9L VIN Z (A/T) 'C3' Gray Connector

| PCM Pin # | Wire Color | Circuit Description (32-Pin) | Value at Hot Idle |
|---|---|---|---|
| 1 | DB/OR | A/C Clutch Relay Control | Relay Off: 12v, On: 1v |
| 2 | --- | Not Used | --- |
| 3 | DB/YL | ASD Relay Control | Relay Off: 12v, On: 1v |
| 4 | TN/RD | Speed Control Vacuum Solenoid | Vacuum Increasing: 1v |
| 5 | LG/RD | Speed Control Vent Solenoid | Vacuum Decreasing: 1v |
| 6-7 | --- | Not Used | --- |
| 8 | VT/WT | HO2S-11 (B1 S1) Heater | Heater Off: 12v, On: 1v |
| 9 | DG/BK | HO2S-12 (B1 S2) Heater | Heater Off: 12v, On: 1v |
| 10 | WT/DG | LDP Solenoid Control | PWM Signal: 0-12-0v |
| 11 | YL/RD | Speed Control Power Supply | 12-14v |
| 12 | DG/OR | ASD Relay Power (B+) | 12-14v |
| 13 | OR/WT | Overdrive Off Switch Sense | Switch Off: 12v, On: 1v |
| 14 | OR | LDP Switch Sense Signal | LDP Switch Closed: 0v |
| 15 | PK/YL | Battery Temperature Sensor | At 86°F: 1.96v |
| 16 | VT/OR | HO2S-12 (B1 S2) Heater Control | PWM Signal: 0-12-0v |
| 17-18 | --- | Not Used | --- |
| 19 | BR | Fuel Pump Relay Control | Relay Off: 12v, On: 1v |
| 20 | PK/BK | EVAP Purge Solenoid Control | PWM Signal: 0-12-0v |
| 21 | --- | Not Used | --- |
| 22 | BR | A/C Request Signal | A/C Off: 12v, On: 1v |
| 24 | WT/PK | Brake Switch Sense | Brake Off: 0v, On: 12v |
| 23 | --- | Not Used | --- |
| 25 | WT/DB | Generator Field Source | 12-14v |
| 26 | DB/WT | Fuel Level Sensor Signal | Full: 0.56v, 1/2 full: 2.5v |
| 27 | PK, LG | SCI Transmit, SCI Receive | 0v, 5v |
| 28 | --- | Not Used | --- |
| 29 | LG | SCI Receive | 5v |
| 30-31 | --- | Not Used | --- |
| 32 | RD/LG | Speed Control Switch Signal | S/C & Set Switch On: 3.8v |

**2001-03 Ram Pickup & Van 3.7L VIN K 'C1' Black Connector**

| PCM Pin # | Wire Color | Circuit Description (32-Pin) | Value at Hot Idle |
|-----------|------------|------------------------------|-------------------|
| 1 | TN/OR | Coil 3 Driver Control | 5°, 55 mph: 8° dwell |
| 2 | LG/BK | Ignition Switch Power (Start-Run) | 12-14v |
| 3 | TN/LG | Coil 4 Driver Control | 5°, 55 mph: 8° dwell |
| 4 | BK/LB | Sensor Ground | <0.050v |
| 5 | TN/LB | Coil 6 Driver Control | 5°, 55 mph: 8° dwell |
| 6 | BR/YL | PNP Switch Sense Signal | In P/N: 0v, Others: 5v |
| 7 | TN/RD | Coil 1 Driver Control | 5°, 55 mph: 8° dwell |
| 8 | GY/BK | CKP Sensor Signal | Digital Signals: 0-5-0v |
| 9 | --- | Not Used | --- |
| 10 | GY/RD | IAC 3 Driver Control | Pulse Signals |
| 11 | YL/BK | IAC 2 Driver Control | Pulse Signals |
| 12 | YL/BK | Power Steering Pressure Switch | Straight: 0v, Turning: 5v |
| 13 | YL | Ignition Switch Power (Start-Run) | 12-14v |
| 14 | --- | Not Used | --- |
| 15 | BK/RD | IAT Sensor Signal | At 100°F: 1.83v |
| 16 | TN/BK | ECT Sensor Signal | At 180°F: 2.80v |
| 17 | OR | 5-Volt Supply | 4.9-5.1v |
| 18 | TN/YL | CMP Sensor Signal | Digital Signals: 0-5-0v |
| 19 | VT/BK | IAC 4 Driver Control | Pulse Signals |
| 20 | BR/WT | IAC 1 Driver Control | Pulse Signals |
| 21 | TN/DG | Coil 3 Driver Control | 5°, 55 mph: 8° dwell |
| 22 | RD/WT | Keep Alive Power (B+) | 12-14v |
| 23 | OR/DB | TP Sensor Signal | 0.6-1.0v |
| 24 | BK/DG | HO2S-11 (B1 S1) Signal | 0.1-1.1v |
| 25 | TN/WT | HO2S-12 (B1 S2) Signal | 0.1-1.1v |
| 26 | --- | Not Used | --- |
| 27 | OR/RD | MAP Sensor Signal | 1.5-1.6v |
| 28-30 | --- | Not Used | --- |
| 31 | BK/TN | Power Ground | <0.1v |
| 32 | BK/TN | Power Ground | <0.1v |

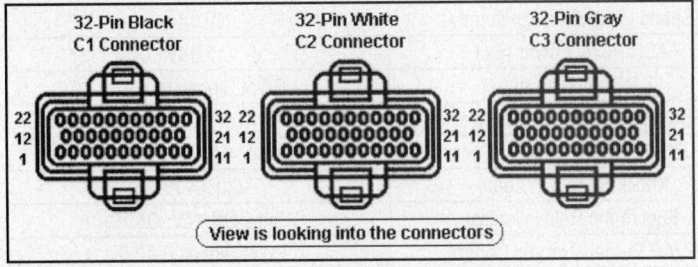

**Pin Connector Graphic**

**Standard Colors and Abbreviations**

| Abbreviation | Color | Abbreviation | Color | Abbreviation | Color |
|--------------|-------|--------------|-------|--------------|-------|
| BK | Black | GY | Gray | RD | Red |
| BL | Blue | GN | Green | TN | Tan |
| BR | Brown | LG | Light Green | VT | Violet |
| DB | Dark Blue | OR | Orange | WT | White |
| DG | Dark Green | PK | Pink | YL | Yellow |

## 2001-03 Ram Pickup & Van 3.7L VIN K (A/T) White 'C2' Connector

| PCM Pin # | Wire Color | Circuit Description (32-Pin) | Value at Hot Idle |
|---|---|---|---|
| 1-3 | --- | Not Used | --- |
| 4 | WT/DB | Injector 1 Driver | 1-4 ms |
| 5 | YL/WT | Injector 3 Driver | 1-4 ms |
| 6 | GY | Injector 5 Driver | 1-4 ms |
| 7-8 | --- | Not Used | --- |
| 9 | TN/PK | Coil 1 Driver Control | 5°, 55 mph: 8° dwell |
| 10 | DG | Generator Field Control | Digital Signal: 0-12-0v |
| 11-13 | --- | Not Used | --- |
| 15 | TN | Injector 2 Driver | 1-4 ms |
| 16 | LB/BR | Injector 4 Driver | 1-4 ms |
| 17 | DB/YL | Condenser Fan Control Relay | Relay Off: 12v, On: 1v |
| 18 | --- | Not Used | --- |
| 19 | DB | AC Pressure Sensor Signal | 0.90v at 79 psi |
| 20-22 | --- | Not Used | --- |
| 23 | GY/YL | Engine Oil Pressure Sensor | 1.6v at 24 psi |
| 24-26 | --- | Not Used | --- |
| 27 | WT/OR | Vehicle Speed Signal | Digital Signals |
| 28-30 | --- | Not Used | --- |
| 31 | VT/WT | 5-Volt Supply | 4.9-5.1v |
| 32 | --- | Not Used | --- |

## 2001-03 Ram Pickup & Van 3.7L VIN K (A/T) 'C3' Gray Connector

| PCM Pin # | Wire Color | Circuit Description (32-Pin) | Value at Hot Idle |
|---|---|---|---|
| 1 | DB/OR | A/C Clutch Relay Control | Relay Off: 12v, On: 1v |
| 2, 6, 13, 17 | --- | Not Used | --- |
| 3 | DB/YL | ASD Relay Control | Relay Off: 12v, On: 1v |
| 4 | TN/RD | Speed Control Vacuum Solenoid | Vacuum Increasing: 1v |
| 5 | LG/RD | Speed Control Vent Solenoid | Vacuum Decreasing: 1v |
| 7 | YL/BK | Knock Sensor 1 Signal | 0.080v AC |
| 8 | BR/VT | HO2S-11 (B1 S1) Heater Control | Heater Off: 12v, On: 1v |
| 9 | DB/OR | Heater Relay Control (Downstream) | Relay Off: 12v, On: 1v |
| 10 | WT/DG | LDP Switch Sense | LDP Switch Closed: 0v |
| 11 | YL/RD | Speed Control Power Supply | 12-14v |
| 12 | DG/OR | ASD Relay Power (B+) | 12-14v |
| 14 | OR | LDP Solenoid Control | PWM Signal: 0-12-0v |
| 15 | PK/YL | Battery Temperature Sensor | At 86°F: 1.96v |
| 16 | BR/WT | HO2S-12 (B1 S2) Heater Control | Heater Off: 12v, On: 1v |
| 18 | BR/VT | Knock Sensor 2 Signal | 0.080v AC |
| 19 | BR | Fuel Pump Relay Control | Relay Off: 12v, On: 1v |
| 20 | PK/BK | EVAP Purge Solenoid Control | PWM Signal: 0-12-0v |
| 21-23 | --- | Not Used | --- |
| 24 | WT/PK | Brake Switch Sense | Brake Off: 0v, On: 12v |
| 25 | WT/DB | Generator Field Source | 12-14v |
| 26 | LB/BK | Fuel Level Sensor Signal | Full: 0.56v, 1/2 full: 2.5v |
| 27 | PK | SCI Transmit | 0v |
| 28, 31 | --- | Not Used | --- |
| 29 | LG | SCI Receive | 5v |
| 30 | WT/BR | PCI Data Bus (J1850) | Digital Signals: 0-7-0v |
| 32 | RD/LG | Speed Control Switch Signal | S/C & Set Switch On: 3.8v |

**2004-05 Ram Pickup & Van 3.7L, 4.7L, 5.7L 'C1' Black/Black Connector**

| PCM Pin # | Wire Color | Circuit Description (38-Pin) | Value at Hot Idle |
|---|---|---|---|
| 1 | DB/YL (Exc. 3.7L) | Coil Control No. 8 | |
| 2 | --- | Not Used | --- |
| 3 | DB/YL (4.7L) | Coil Control No. 7 | |
| 3 | BR (5.7L) | Coil Control No. 7 | |
| 4 | BR/LB (Exc. 3.7L) | Injector Control No. 8 | |
| 5 | BR/YL (Exc. 3.7L) | Injector Control No. 7 | |
| 6 | --- | Not Used | --- |
| 7 | --- | Not Used | --- |
| 8 | DG/WT (5.7L) | Vehicle Speed Signal No. 2 | |
| 9 | BK/BR | Ground | |
| 10 | --- | Not Used | --- |
| 11 | PK/GY | Fused Ignition Switch Output (Run-Start) | |
| 12 | PK/WT | Fused Ignition Switch Output (Off-Run-Start) | |
| 13 | DG/YL | Vehicle Speed Signal No. 1 | |
| 14 | --- | Not Used | --- |
| 15 | --- | Not Used | --- |
| 16 | --- | Not Used | --- |
| 17 | --- | Not Used | --- |
| 18 | BK/DG | Ground | |
| 19 | --- | Not Used | --- |
| 20 | VT/GY | Oil Pressure Signal | |
| 21 | BK/LB | A/C Pressure Signal | |
| 22 | VT/LG | AAT Signal | |
| 23 | --- | Not Used | --- |
| 24 | --- | Not Used | --- |
| 25 | WT/LG (3.7L/4.7L) | SCI Receive (PCM) | |
| 25 | WT/LG (5.7L) | SCI Receive (PCM) | |
| 26 | WT/OR (Exc. 3.7L/4.7L A/T) | SCI Receive (TCM) | |
| 27 | YL/PK | 5 Volt Supply | |
| 28 | --- | Not Used | --- |
| 29 | RD | Fused B (+) | |
| 30 | YL | Ignition Switch Output (Start) | |
| 31 | DB/YL | O2 1/2 Signal | |
| 32 | BR/DG | O2 Return (Up) | |
| 33 | BR | O2 2/2 Signal | |
| 34 | --- | Not Used | --- |
| 35 | --- | Not Used | --- |
| 36 | WT/BR (3.7L/4.7L) | SCI Transmit (PCM) | |
| 36 | WT/BR (5.7L) | SCI Transmit (PCM) | |
| 37 | WT/DG (Exc. 3.7L/4.7L M/T) | SCI Transmit (TCM) | |
| 38 | WT/VT | PCI Bus | |

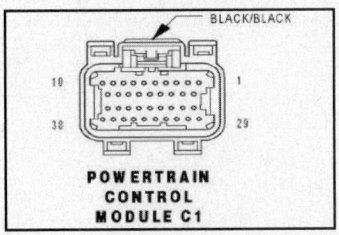

## 2004-05 Ram Pickup & Van 3.7L, 4.7L, 5.7L 'C2' Black/Orange Connector

| PCM Pin # | Wire Color | Circuit Description (38-Pin) | Value at Hot Idle |
|---|---|---|---|
| 1 | DB/OR | Coil Control No. 6 | |
| 2 | DB/YL | Coil Control No. 5 | |
| 3 | DB/GY | Coil Control No. 4 | |
| 4 | BR/VT | Injector Control No. 6 | |
| 5 | BR/OR | Injector Control No. 5 | |
| 6 | TN/YL (5.7L) | ETC Motor (+) | |
| 7 | DB/OR | Coil Control No. 3 | |
| 8 | DB/VT (4.7L/5.7L LD) | EGR Sol Control | |
| 9 | DB/TN | Coil Control No. 2 | |
| 10 | DB/DG | Coil Control No. 1 | |
| 11 | BR/TN | Injector Control No. 4 | |
| 12 | BR/LB | Injector Control No. 3 | |
| 13 | BR/DB | Injector Control No. 2 | |
| 14 | BR/YL | Injector Control No. 1 | |
| 15 | BR/DB (5.7L) | TP Sensor Return | |
| 16 | --- | Not Used | --- |
| 17 | BR/VT | O2 2/1 Heater Control | |
| 18 | BR/LG | O2 1/1 Heater Control | |
| 19 | BR/GY | Gen Field Control | |
| 20 | VT/OR | ECT Signal | |
| 21 | BR/OR | TP No. 1 Signal | |
| 22 | DB/LG (4.7L/5.7L LD) | EGR Signal | |
| 23 | VT/BR | MAP Signal | |
| 24 | BR/LG (Exc. 4.7L) | Knock Sensor No. 1 Return | |
| 25 | DB/YL (Exc. 4.7L) | Knock Sensor No. 1 Signal | |
| 26 | --- | Not Used | --- |
| 27 | DB/DG | Sensor Ground | |
| 28 | BR/DG (5.7L) | TP No. 2 Signal | |
| 28 | BR/VT (Exc. 5.7L) | IAC Signal | |
| 29 | PK/YL | 5 Volt Supply | |
| 30 | DB/LG | IAT Signal | |
| 31 | DB/LB | O2 1/1 Signal | |
| 32 | DB/DG | O2 Return (Down) | |
| 33 | DB/LG | O2 2/1 Signal | |
| 34 | DB/GY | CMP Signal | |
| 35 | BR/LB | CKP Signal | |
| 36 | BR/WT | Knock Sensor No. 2 Signal | |
| 37 | WT/BR | Knock Sensor No. 2 Return | |
| 38 | TN/OR (5.7L) | ETC Motor (-) | |
| 38 | VT/GY (Exc. 5.7L) | IAC Control | |

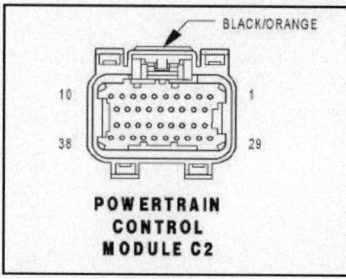

**POWERTRAIN
CONTROL
MODULE C2**

**2004-05 Ram Pickup & Van 3.7L, 4.7L, 5.7L 'C3' Black/Natural Connector**

| PCM Pin # | Wire Color | Circuit Description (38-Pin) | Value at Hot Idle |
|---|---|---|---|
| 1 | --- | Not Used | --- |
| 2 | --- | Not Used | --- |
| 3 | BR/WT | ASD Relay Control | |
| 4 | --- | Not Used | --- |
| 5 | VT/OR (3.7L/4.7L) | S/C Vent Sol Control | |
| 6 | BR/VT | Condenser Fan Relay Control | |
| 7 | VT/YL (3.7L/4.7L) | Speed Control Supply | |
| 8 | VT/LB | NVLD Sol Control | |
| 9 | BR/WT | O2 1/2 Heater Control | |
| 10 | BR/GY | O2 2/2 Heater Control | |
| 11 | LB/OR | A/C Clutch Relay Control | |
| 12 | VT/YL (3.7L/4.7L) | S/C Vacuum Sol Control | |
| 13 | --- | Not Used | --- |
| 14 | VT/YL (5.7L) | Brake Switch No. 2 Signal | |
| 15 | VT/BR (5.7L) | S/C Switch Return | |
| 16 | BR/YL (5.7L) | APPS No. 1 Return | |
| 17 | BR/VT (5.7L) | APPS No. 2 Return | |
| 18 | VT/OR (5.7L) | S/C Switch No. 2 Signal | |
| 19 | BR/WT | ASD Relay Output | |
| 20 | DB/WT | EVAP Purge Sol Control | |
| 21 | YL/DB | TRS T41 Sense | |
| 22 | VT (Exc. 3.7L/4.7L Base) | PTO Switch Sense | |
| 23 | DG/WT | Brake Switch No. 1 Signal | |
| 24 | --- | Not Used | --- |
| 25 | BR/WT (5.7L) | APPS No. 1 Signal | |
| 26 | --- | Not Used | --- |
| 27 | --- | Not Used | --- |
| 28 | BR/WT | ASD Relay Output | |
| 29 | DB/BR | EVAP Purge Sol Signal | |
| 30 | DB/WT (5.7L) | P/S Switch Signal | |
| 31 | --- | Not Used | --- |
| 32 | DB/VT | Battery Temp Signal | |
| 33 | DB/WT | Fuel Level Sensor Signal | |
| 34 | VT | S/C Switch No. 1 Signal | |
| 35 | VT/WT | NVLD Switch Signal | |
| 36 | WT/BR (5.7L) | APPS No. 2 Signal | |
| 37 | BR | Fuel Pump Relay Control | |
| 38 | DG/OR | Starter Relay Control | |

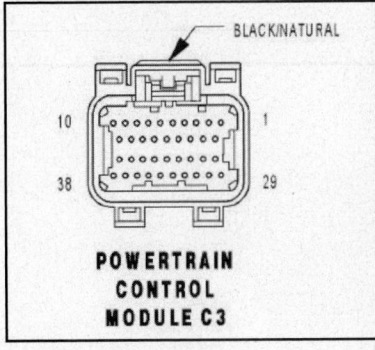

BLACK/NATURAL

POWERTRAIN
CONTROL
MODULE C3

## 2004-05 Ram Pickup & Van 3.7L, 4.7L, 5.7L 'C4' Black/Green Connector

| PCM Pin # | Wire Color | Circuit Description (38-Pin) | Value at Hot Idle |
|---|---|---|---|
| 1 | YL/GY | OD Solenoid Control | |
| 2 | YL/DG | 4C Solenoid Control | |
| 3 | --- | Not Used | --- |
| 4 | DG | MS Solenoid Control | |
| 5 | --- | Not Used | --- |
| 6 | YL/LG | 2C Solenoid Control | |
| 7 | --- | Not Used | --- |
| 8 | YL/LB | UD Solenoid Control | |
| 9 | --- | Not Used | --- |
| 10 | DG/WT | L/R Solenoid Control | |
| 11 | YL/VT (3.7L/4.7L) | Pressure Control Solenoid Control | |
| 11 | YL/GY (5.7L) | Pressure Control Solenoid Control | |
| 12 | BK | Ground | |
| 13 | BK | Ground | |
| 14 | BK | Ground | |
| 15 | DG/LB | TRS T1 Sense | |
| 16 | DG/DB | TRS T3 Sense | |
| 17 | DG | Tow/Haul Overdrive Off Switch Sense | |
| 18 | YL/DB | Transmission Control Relay Control | |
| 19 | YL/OR | Transmission Control Relay Output | |
| 20 | BR/YL | 4C Pressure Switch Sense | |
| 21 | YL/WT | UD Pressure Switch Sense | |
| 22 | DG/TN | OD Pressure Switch Sense | |
| 23 | --- | Not Used | --- |
| 24 | --- | Not Used | --- |
| 25 | --- | Not Used | --- |
| 26 | DG/VT (3.7L/4.7L) | TRS T2 Sense | |
| 26 | DG/LB (5.7L) | TRS T2 Sense | |
| 27 | --- | Not Used | --- |
| 28 | YL/OR | Transmission Control Relay Output | |
| 29 | YL/TN | L/R Pressure Switch Sense | |
| 30 | DG/LG (3.7L/4.7L) | 2C Pressure Switch Sense | |
| 30 | DG/YL (5.7L) | 2C Pressure Switch Sense | |
| 31 | YL/BR | Line Pressure Sensor Signal | |
| 32 | DG/BR | Output Speed Sensor Signal | |
| 33 | DG/OR | Input Speed Sensor Signal | |
| 34 | DG/VT | Speed Sensor Ground | |
| 35 | DG/OR | Transmission Temperature Sensor Signal | |
| 36 | --- | Not Used | --- |
| 37 | DG/YL | TRS T42 Sense | |
| 38 | YL/OR | Transmission Control Relay Output | |

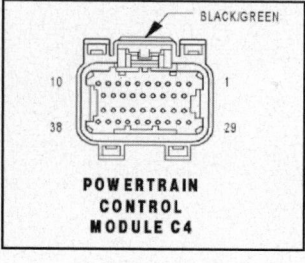

POWERTRAIN
CONTROL
MODULE C4

**1994-95 Ram Pickup & Van 3.9L VIN X 60-Pin Connector**

| PCM Pin # | Wire Color | Circuit Description (60-Pin) | Value at Hot Idle |
|---|---|---|---|
| 1 | DG/RD | MAP Sensor Signal | 1.5-1.6v |
| 2 | TN/BK | ECT Sensor Signal | At 180°F: 2.80v |
| 3 | RD/WT | Keep Alive Power (B+) | 12-14v |
| 4 | BK/LB | Sensor Ground | <0.050v |
| 5, 8, 17-18 | --- | Not Used | --- |
| 6 | PK/WT | 5-Volt Supply | 4.9-5.1v |
| 7 | OR | 8-Volt Supply | 7.9-8.1v |
| 9 | LG/BK | Ignition Switch Power (B+) | 12-14v |
| 10 | OR/WT | Overdrive Override Switch | Switch Off: 12v, On: 1v |
| 11 | BK/TN | Power Ground | <0.1v |
| 12 | BK/TN | Power Ground | <0.1v |
| 13 | LB/BR | Injector 4 Driver | 1-4 ms |
| 14 | YL/WT | Injector 3 Driver | 1-4 ms |
| 15 | TN | Injector 2 Driver | 1-4 ms |
| 16 | WT/DB | Injector 1 Driver | 1-4 ms |
| 19 | BK/GY | Coil Driver Control | 5°, 55 mph: 8° dwell |
| 20 | DG | Alternator Field Control | Digital Signals: 0-12-0v |
| 21 | BK/RD | Throttle Body Temperature | At 100°F: 2.51v |
| 22 | OR/DB | TP Sensor Signal | 0.6-1.0v |
| 23 | --- | Not Used | --- |
| 24 | GY/BK | Distributor Reference Signal | Digital Signals: 0-5-0v |
| 25 | PK | SCI Transmit Signal | 0v |
| 26, 28 | --- | Not Used | --- |
| 27 | BR | A/C Damped Pressure Switch | A/C Off: 12v, On: 1v |
| 29 | WT/PK | Brake Switch Sense Signal | Brake Off: 0v, On: 12v |
| 30 | BK/WT | PNP Switch Sense Signal | In P/N: 0v, Others: 5v |
| 31 | --- | Not Used | --- |
| 32 | BK/PK | MIL (lamp) Control | MIL Off: 12v, On: 1v |
| 33 | TN/RD | Speed Control Vacuum Solenoid | Vacuum Increasing: 1v |
| 34 | DB/OR | A/C Clutch Relay Control | Relay Off: 12v, On: 1v |
| 35 | GY/YL | EGR Solenoid Control | 12v, at 55 mph: 1v |
| 36 | --- | Not Used | --- |
| 37 | LG/OR | Overdrive Lamp Control | Lamp Off: 12v, On: 1v |
| 38 | GY | Injector 5 Driver | 1-4 ms |
| 39 | GY/RD | AIS Motor 4 Control | Pulse Signals |
| 40 | BR/WT | AIS Motor 2 Control | Pulse Signals |
| 41 | BK/DG | HO2S-11 (B1 S1) Signal | 0.1-1.1v |
| 42 | PK | Transmission Temperature Sensor Signal | At 200°F: 2.40v |
| 43 | GY/LB | Tachometer Signal | Pulse Signals |
| 44 | TN/YL | Distributor Sync Signal | Digital Signals: 0-5-0v |
| 45 | LG | SCI Receive Signal | 5v |
| 46 | --- | Not Used | --- |
| 47 | WT/OR | Vehicle Speed Signal | Digital Signal |
| 48 | BR/RD | Speed Control Set Switch Signal | S/C & Set Switch On: 3.8v |
| 49 | YL/RD | Speed Control On/Off Switch | S/C On: 1v, Off: 12v |
| 50 | WT/LG | Speed Control Resume Switch | S/C On: 1v, Off: 12v |
| 51 | DB/YL | ASD Relay Control | Relay Off: 12v, On: 1v |
| 52 | PK/BK | EVAP Purge Solenoid Control | Solenoid Off: 12v, On: 1v |
| 53 | LG/RD | Speed Control Vent Solenoid | Vacuum Decreasing: 1v |
| 54 | OR/BK | A/T: Overdrive Solenoid | Solenoid Off: 12v, On: 1v |
| 54 | OR/BK | M/T: SIL (lamp) Control | Lamp Off: 12v, On: 1v |
| 55 | OR/WT | Overdrive Lockout Solenoid | Solenoid Off: 12v, On: 1v |
| 56 | --- | Not Used | --- |
| 57 | DG/OR | ASD Relay Power (B+) | 12-14v |
| 58 | BR/DB | Injector 6 Driver | 1-4 ms |
| 59 | PK/BK | AIS Motor 1 Control | Pulse Signals |
| 60 | YL/BK | AIS Motor 3 Control | Pulse Signals |

**1996-97 Ram Pickup & Van 3.9L VIN X 'A' Black Connector**

| PCM Pin # | Wire Color | Circuit Description (32-Pin) | Value at Hot Idle |
|---|---|---|---|
| 2 | LG/BK | Ignition Switch Power (B+) | 12-14v |
| 4 | BK/LB | Sensor Ground | <0.050v |
| 6 | BK/WT | PNP Switch Sense Signal | In P/N: 0v, Others: 5v |
| 7 | BK/GY | Coil 1 Driver Control | 5°, 55 mph: 8° dwell |
| 8 | GY/BK | CKP Sensor Signal | Digital Signals: 0-5-0v |
| 10 | YL/BK | IAC 2 Driver Control | Pulse Signals |
| 11 | BR/WT | IAC 3 Driver Control | Pulse Signals |
| 15 | BK/RD | IAT Sensor Signal | At 100°F: 1.83v |
| 16 | TN/BK | ECT Sensor Signal | At 180°F: 2.80v |
| 17 | PK/WT | 5-Volt Supply | 4.9-5.1v |
| 18 | TN/YL | CMP Sensor Signal | Digital Signals: 0-5-0v |
| 19 | GY/RD | IAC 1 Driver Control | Pulse Signals |
| 20 | PK/BK | IAC 4 Driver Control | Pulse Signals |
| 22 | RD/WT | Keep Alive Power (B+) | 12-14v |
| 23 | OR/DB | TP Sensor Signal | 0.6-1.0v |
| 24 | TN/WT | HO2S-11 (B1 S1) Signal | 0.1-1.1v |
| 25 | OR/BK | HO2S-12 (B1 S2) Signal | 0.1-1.1v |
| 27 | DG/RD | MAP Sensor Signal | 1.5-1.6v |
| 31 | BK/TN | Power Ground | <0.1v |
| 32 | BK/TN | Power Ground | <0.1v |

**1996-97 Ram Pickup & Van 3.9L VIN X 'B' White Connector**

| PCM Pin # | Wire Color | Circuit Description (32-Pin) | Value at Hot Idle |
|---|---|---|---|
| 1 | PK | Transmission Temperature Sensor Signal | At 200°F: 2.40v |
| 4 | WT/DB | Injector 1 Driver | 1-4 ms |
| 5 | YL/WT | Injector 3 Driver | 1-4 ms |
| 6 | GY | Injector 5 Driver | 1-4 ms |
| 8 | PK/W | Governor Pressure Solenoid | PWM Signal: 0-12-0v |
| 10 | DG | Generator Field Control | Digital Signals: 0-12-0v |
| 11 | OR/BK | TCC Solenoid Control | At Cruise w/TCC On: 1v |
| 12 | BR/DB | Injector 6 Driver | 1-4 ms |
| 15 | TN | Injector 2 Driver | 1-4 ms |
| 16 | LB/BR | Injector 4 Driver | 1-4 ms |
| 21 | BR | 3-4 Shift Solenoid Control | Solenoid Off: 12v, On: 1v |
| 25 | DB/BK | OSS Sensor (-) Signal | Moving: AC pulse signals |
| 27 | WT/OR | Vehicle Speed Signal | Digital Signal |
| 28 | LG/BK | OSS Sensor (+) Signal | Moving: AC pulse signals |
| 29 | LG/WT | Governor Pressure Sensor | 0.58v |
| 30 | PK | Transmission Relay Control | Relay Off: 12v, On: 1v |
| 31 | OR | 5-Volt Supply | 4.9-5.1v |
| 32 | GY/YL | EGR Solenoid Control | 12v, at 55 mph: 1v |

## 1996-97 Ram Pickup & Van 3.9L VIN X 'C' Gray Connector

| PCM Pin # | Wire Color | Circuit Description (32-Pin) | Value at Hot Idle |
|-----------|-----------|----------------------------|-------------------|
| 1 | DB/OR | A/C Clutch Relay Control | Relay Off: 12v, On: 1v |
| 2 | --- | Not Used | --- |
| 3 | DB/YL | ASD Relay Control | Relay Off: 12v, On: 1v |
| 4 | TN/RD | Speed Control Vacuum Solenoid | Vacuum Increasing: 1v |
| 5 | LG/RD | Speed Control Vent Solenoid | Vacuum Decreasing: 1v |
| 6 | LG/OR | Overdrive Off Lamp Control | Lamp Off: 12v, On: 1v |
| 7 | PK/BK | Transmission Temperature Lamp Control | Lamp Off: 12v, On: 1v |
| 8-10 | --- | Not Used | --- |
| 11 | YL/RD | Speed Control Power Supply | 12-14v |
| 12 | DG/OR | ASD Relay Power (B+) | 12-14v |
| 13 | OR/WT | Overdrive Off Switch Sense | Switch Off: 12v, On: 1v |
| 14 | --- | Not Used | --- |
| 15 | PK/YL | Battery Temperature Sensor | At 86°F: 1.96v |
| 16 | TN/YL | Generator Lamp Control | Lamp Off: 12v, On: 1v |
| 17 | BK/PK | MIL (lamp) Control | MIL Off: 12v, On: 1v |
| 18 | GY/PK | Maintenance Indicator Lamp | Lamp Off: 12v, On: 1v |
| 19 | BR/WT | Fuel Pump Relay Control | Relay Off: 12v, On: 1v |
| 20 | PK | EVAP Purge Solenoid Control | PWM Signal: 0-12-0v |
| 21 | --- | Not Used | --- |
| 22 | BR | A/C Request Signal | A/C Off: 12v, On: 1v |
| 23 | LG | A/C Select Signal | A/C Off: 12v, On: 1v |
| 24 | WT/PK | Brake Switch Sense Signal | Brake Off: 0v, On: 12v |
| 25 | --- | Not Used | --- |
| 26 | DB/WT | Fuel Level Sensor Signal | Full: 0.56v, 1/2 full: 2.5v |
| 27 | PK/DB | SCI Transmit | 0v |
| 28 | --- | Not Used | --- |
| 29 | DG | SCI Receive | 5v |
| 30 | --- | Not Used | --- |
| 31 | GY/LB | Tachometer Signal | Pulse Signals |
| 32 | RD/LG | Speed Control Set Switch Signal | S/C & Set Switch On: 3.8v |

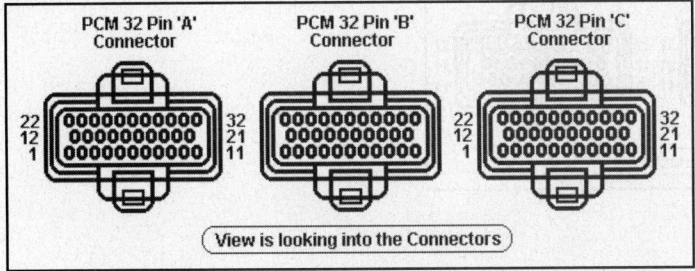

**Pin Connector Graphic**

### 1998-99 Ram Pickup & Van 3.9L VIN X 'C1' Black Connector

| PCM Pin # | Wire Color | Circuit Description (32-Pin) | Value at Hot Idle |
|---|---|---|---|
| 1, 3 | --- | Not Used | --- |
| 2 | LG/BK | Ignition Switch Power (B+) | 12-14v |
| 4 | BK/LB | Sensor Ground | <0.050v |
| 5 | --- | Not Used | --- |
| 6 | BK/WT | PNP Switch Sense Signal | In P/N: 0v, Others: 5v |
| 7 | BK/GY | Coil 1 Driver Control | 5°, 55 mph: 8° dwell |
| 8 | GY/BK | CKP Sensor Signal | Digital Signals: 0-5-0v |
| 9 | --- | Not Used | --- |
| 10 | YL/BK | IAC 2 Driver Control | Pulse Signals |
| 11 | BR/WT | IAC 3 Driver Control | Pulse Signals |
| 12 | --- | Not Used | --- |
| 13 | OR | Power Takeoff Switch Sense | Switch Off: 12v, On: 1v |
| 14 | --- | Not Used | --- |
| 15 | BK/RD | IAT Sensor Signal | At 100°F: 1.83v |
| 16 | TN/BK | ECT Sensor Signal | At 180°F: 2.80v |
| 17 | PK/WT | 5-Volt Supply | 4.9-5.1v |
| 18 | TN/YL | CMP Sensor Signal | Digital Signals: 0-5-0v |
| 19 | GY/RD | IAC 1 Driver Control | Pulse Signals |
| 20 | PK/BK | IAC 4 Driver Control | Pulse Signals |
| 21 | --- | Not Used | --- |
| 22 | RD/WT | Keep Alive Power (B+) | 12-14v |
| 23 | OR/DB | TP Sensor Signal | 0.6-1.0v |
| 24 | TN/WT | HO2S-11 (B1 S1) Signal | 0.1-1.1v |
| 25 | OR/BK | HO2S-12 (B1 S2) Signal | 0.1-1.1v |
| 26 | --- | Not Used | --- |
| 27 | DG/RD | MAP Sensor Signal | 1.5-1.6v |
| 28-30 | --- | Not Used | --- |
| 31 | BK/TN | Power Ground | <0.1v |
| 32 | BK/TN | Power Ground | <0.1v |

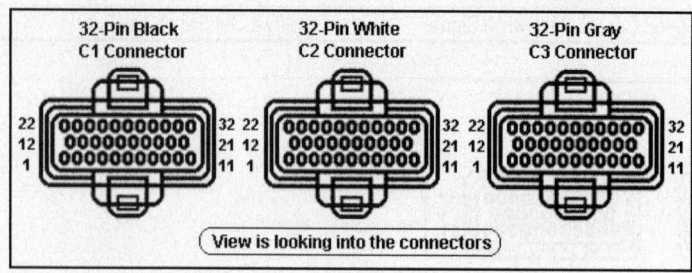

**Pin Connector Graphic**

### Standard Colors and Abbreviations

| Abbreviation | Color | Abbreviation | Color | Abbreviation | Color |
|---|---|---|---|---|---|
| BK | Black | GY | Gray | RD | Red |
| BL | Blue | GN | Green | TN | Tan |
| BR | Brown | LG | Light Green | VT | Violet |
| DB | Dark Blue | OR | Orange | WT | White |
| DG | Dark Green | PK | Pink | YL | Yellow |

**1998-99 Ram Pickup & Van 3.9L VIN X 'C2' White Connector**

| PCM Pin # | Wire Color | Circuit Description (32-Pin) | Value at Hot Idle |
|---|---|---|---|
| 1 | PK | Transmission Temperature Sensor Signal | At 200ºF: 2.40v |
| 2-3 | --- | Not Used | --- |
| 4 | WT/DB | Injector 1 Driver | 1-4 ms |
| 5 | YL/WT | Injector 3 Driver | 1-4 ms |
| 6 | GY | Injector 5 Driver | 1-4 ms |
| 7, 9 | --- | Not Used | --- |
| 8 | PK/W | Governor Pressure Solenoid | PWM Signal: 0-12-0v |
| 10 | DG | Generator Field Control | Digital Signals: 0-12-0v |
| 11 | OR/BK | TCC Solenoid Control | At Cruise w/TCC On: 1v |
| 12 | BR/DB | Injector 6 Driver | 1-4 ms |
| 15 | TN | Injector 2 Driver | 1-4 ms |
| 16 | LB/BR | Injector 4 Driver | 1-4 ms |
| 21 | BR | Overdrive Solenoid Control | Cruise w/solenoid On: 1v |
| 23 | GY/O | Engine Oil Pressure Sensor | 1.6v at 24 psi |
| 25 | DB/BK | OSS Sensor (-) Signal | Moving: AC pulse signals |
| 27 | WT/OR | Vehicle Speed Signal | Digital Signal |
| 28 | LG/BK | OSS Sensor (+) Signal | Moving: AC pulse signals |
| 29 | LG/WT | Governor Pressure Sensor | 0.58v |
| 30 | P | Transmission Relay Control | Relay Off: 12v, On: 1v |
| 31 | O | 5-Volt Supply | 4.9-5.1v |

**1998-99 Ram Pickup & Van 3.9L VIN X 'C3' Gray Connector**

| PCM Pin # | Wire Color | Circuit Description (32-Pin) | Value at Hot Idle |
|---|---|---|---|
| 1 | DB/OR | A/C Clutch Relay Control | Relay Off: 12v, On: 1v |
| 3 | DB/YL | ASD Relay Control | Relay Off: 12v, On: 1v |
| 4 | TN/RD | Speed Control Vacuum Solenoid | Vacuum Increasing: 1v |
| 5 | LG/RD | Speed Control Vent Solenoid | Vacuum Decreasing: 1v |
| 6 ('98) | LG/OR | Overdrive Off Lamp Control | Lamp Off: 12v, On: 1v |
| 8 | BR/VT | HO2S-11 (B1 S1) Heater | Heater Off: 12v, On: 1v |
| 9 | DG/PK | HO2S-12 (B1 S2) Heater | Heater Off: 12v, On: 1v |
| 10 | WT/DG | LDP Solenoid Control | PWM Signal: 0-12-0v |
| 11 | YL/RD | Speed Control Power Supply | 12-14v |
| 12 | DG/OR | ASD Relay Power (B+) | 12-14v |
| 13 | OR/WT | Overdrive Off Switch Sense | Switch Off: 12v, On: 1v |
| 14 | OR | LDP Switch Sense Signal | LDP Switch Closed: 0v |
| 15 | PK/YL | Battery Temperature Sensor | At 86ºF: 1.96v |
| 19 | BR/WT | Fuel Pump Relay Control | Relay Off: 12v, On: 1v |
| 20 | PK/WT | EVAP Purge Solenoid Control | PWM Signal: 0-12-0v |
| 22 | BR | A/C Switch Signal | A/C Off: 12v, On: 1v |
| 23 | LG/WT | A/C Select Signal | A/C Off: 12v, On: 1v |
| 24 | WT/PK | Brake Switch Sense Signal | Brake Off: 0v, On: 12v |
| 25 | DB | Generator Field Source | 12-14v |
| 26 | DB/WT | Fuel Level Sensor Signal | Full: 0.56v, 1/2 full: 2.5v |
| 27 | PK/DB | SCI Transmit Signal | 0v |
| 28 | WT/BK | CCD Bus (-) Signal | <0.050v |
| 29 | DG | SCI Receive Signal | 5v |
| 30 | VT/BR | CCD Bus (+) Signal | Digital Signals: 0-5-0v |
| 32 | RD/LG | Speed Control Set Switch Signal | S/C & Set Switch On: 3.8v |

**2000-01 Ram Pickup & Van 3.9L VIN X 'C1' Black Connector**

| PCM Pin # | Wire Color | Circuit Description (32-Pin) | Value at Hot Idle |
|---|---|---|---|
| 1 | --- | Not Used | --- |
| 2 | LG/BK | Ignition Switch Power (B+) | 12-14v |
| 3 | --- | Not Used | --- |
| 4 | BK/LB | Sensor Ground | <0.050v |
| 5 | --- | Not Used | --- |
| 6 | BK/WT | PNP Switch Sense Signal | In P/N: 0v, Others: 5v |
| 7 | BK/GY | Coil 1 Driver Control | 5°, 55 mph: 8° dwell |
| 8 | GY/BK | CKP Sensor Signal | Digital Signals: 0-5-0v |
| 9 | --- | Not Used | --- |
| 10 | YL/BK | IAC 2 Driver Control | Pulse Signals |
| 11 | BR/WT | IAC 3 Driver Control | Pulse Signals |
| 12 | --- | Not Used | --- |
| 13 | OR | Power Takeoff Switch Sense | Switch Off: 12v, On: 1v |
| 14 | --- | Not Used | --- |
| 15 | BK/RD | IAT Sensor Signal | At 100°F: 1.83v |
| 16 | TN/BK | ECT Sensor Signal | At 180°F: 2.80v |
| 17 | VT/WT | 5-Volt Supply | 4.9-5.1v |
| 18 | TN/YL | CMP Sensor Signal | Digital Signals: 0-5-0v |
| 19 | GY/RD | IAC 1 Driver Control | Pulse Signals |
| 20 | VT/BK | IAC 4 Driver Control | Pulse Signals |
| 21 | --- | Not Used | --- |
| 22 | RD/WT | Keep Alive Power (B+) | 12-14v |
| 23 | OR/DB | TP Sensor Signal | 0.6-1.0v |
| 24 | TN/WT | HO2S-11 (B1 S1) Signal | 0.1-1.1v |
| 25 | OR/BK | HO2S-12 (B1 S2) Signal | 0.1-1.1v |
| 26 | LG/RD | HO2S-21 (B2 S1) Signal (Cal) | 0.1-1.1v |
| 27 | DG/RD | MAP Sensor Signal | 1.5-1.6v |
| 28 | --- | Not Used | --- |
| 29 | OR/TN | HO2S-22 (B2 S2) Signal (Cal) | 0.1-1.1v |
| 30 | --- | Not Used | --- |
| 31 | BK/TN | Power Ground | <0.1v |
| 32 | BK/TN | Power Ground | <0.1v |

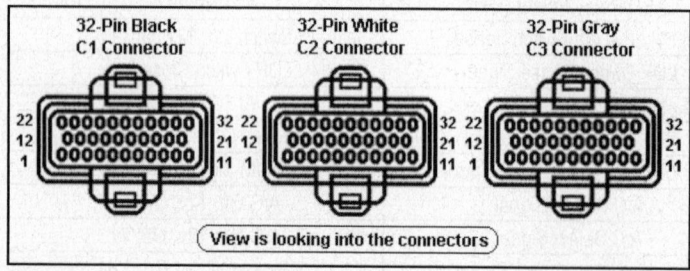

**Pin Connector Graphic**

**2000-01 Ram Pickup & Van 3.9L VIN X 'C2' White Connector**

| PCM Pin # | Wire Color | Circuit Description (32-Pin) | Value at Hot Idle |
|---|---|---|---|
| 1 | VT | Transmission Temperature Sensor Signal | At 200°F: 2.40v |
| 2-3, 7, 9 | --- | Not Used | --- |
| 4 | WT/DB | Injector 1 Driver | 1-4 ms |
| 5 | YL/WT | Injector 3 Driver | 1-4 ms |
| 6 | GY | Injector 5 Driver | 1-4 ms |
| 8 | VT/WT | Governor Pressure Solenoid | PWM Signal: 0-12-0v |
| 10 | DG | Generator Field Control | Digital Signals: 0-12-0v |
| 11 | OR/BK | TCC Solenoid Control | At Cruise w/TCC On: 1v |
| 12 | BR/DB | Injector 6 Driver | 1-4 ms |
| 15 | TN | Injector 2 Driver | 1-4 ms |
| 16 | LB/BR | Injector 4 Driver | 1-4 ms |
| 21 | BR | 3-4 Shift Solenoid Control | Cruise w/solenoid On: 1v |
| 23 | GY/OR | Engine Oil Pressure Sensor | 1.6v at 24 psi |
| 25 | DB/BK | OSS Sensor (-) Signal | Moving: AC pulse signals |
| 27 | WT/OR | Vehicle Speed Signal | Digital Signal |
| 28 | LG/BK | OSS Sensor (+) Signal | Moving: AC pulse signals |
| 29 | LG/WT | Governor Pressure Sensor | 0.58v |
| 30 | PK | Transmission Relay Control | Relay Off: 12v, On: 1v |
| 31 | OR | 5-Volt Supply | 4.9-5.1v |

**2000-01 Ram Pickup & Van 3.9L VIN X 'C3' Gray Connector**

| PCM Pin # | Wire Color | Circuit Description (32-Pin) | Value at Hot Idle |
|---|---|---|---|
| 1 | DB/OR | A/C Clutch Relay Control | Relay Off: 12v, On: 1v |
| 3 | DB/YL | ASD Relay Control | Relay Off: 12v, On: 1v |
| 4 | TN/RD | Speed Control Vacuum Solenoid | Vacuum Increasing: 1v |
| 5 | LG/RD | Speed Control Vent Solenoid | Vacuum Decreasing: 1v |
| 6-7 | --- | Not Used | --- |
| 8 | BR/VT | HO2S-11 Heater Control (Cal) | PWM Signal: 0-12-0v |
| 9 | DG/PK | HO2S-12 Heater Relay (Cal) | Relay Off: 12v, On: 1v |
| 10 | WT/DG | LDP Solenoid Control | PWM Signal: 0-12-0v |
| 11 | YL/RD | Speed Control Power Supply | 12-14v |
| 12 | DG/OR | ASD Relay Power (B+) | 12-14v |
| 13 | OR/WT | Overdrive Off Switch Sense | Switch Off: 12v, On: 1v |
| 14 | OR | LDP Switch Sense Signal | LDP Switch Closed: 0v |
| 15 | PK/YL | Battery Temperature Sensor | At 86°F: 1.96v |
| 16-18, 21 | --- | Not Used | --- |
| 19 | BR/WT | Fuel Pump Relay Control | Relay Off: 12v, On: 1v |
| 20 | PK/WT | EVAP Purge Solenoid Control | PWM Signal: 0-12-0v |
| 22 | BR | A/C Switch Signal | A/C Off: 12v, On: 1v |
| 23 | LG/WT | A/C Select Signal | A/C Off: 12v, On: 1v |
| 24 | WT/PK | Brake Switch Sense Signal | Brake Off: 0v, On: 12v |
| 25 | DB | Generator Field Source | 12-14v |
| 26 | DB/WT | Fuel Level Sensor Signal | Full: 0.56v, 1/2 full: 2.5v |
| 27 | PK/DB | SCI Transmit Signal | 0v |
| 28 | WT/BK | CCD Bus (-) Signal | <0.050v |
| 29 | DG | SCI Receive Signal | 5v |
| 30 | VT/BR | CCD Bus (+) Signal | Digital Signals: 0-5-0v |
| 32 | RD/LG | Speed Control Set Switch Signal | S/C & Set Switch On: 3.8v |

## 1993-95 Ram Pickup & Van 5.2L V8 CNG VIN T (A/T) 60-Pin Connector

| PCM Pin # | Wire Color | Circuit Description (60-Pin) | Value at Hot Idle |
|---|---|---|---|
| 1 | DG/RD | MAP Sensor Signal | 1.5-1.6v |
| 2 | TN/BK | ECT Sensor Signal | At 180ºF: 2.80v |
| 3 | RD | Keep Alive Power (B+) | 12-14v |
| 4 | BK/LB | Sensor Ground | <0.050v |
| 5 | BK/WT | Sensor Ground | <0.050v |
| 6 | PK/WT | 5-Volt Supply | 4.9-5.1v |
| 7 | OR | 8-Volt Supply | 7.9-8.1v |
| 8 | --- | Not Used | --- |
| 9 | DB | Ignition Switch Power (B+) | 12-14v |
| 10 | OR/WT | Overdrive Override Switch | Switch Off: 12v, On: 1v |
| 11 | BK/TN | Power Ground | <0.1v |
| 12 | BK/TN | Power Ground | <0.1v |
| 13 | W/LB | Injector 4 Driver | 1-4 ms |
| 14 | WT/YL | Injector 3 Driver | 1-4 ms |
| 15 | WT | Injector 2 Driver | 1-4 ms |
| 16 | WT/BR | Injector 1 Driver | 1-4 ms |
| 17 | WT/TN | Injector 7 Driver | 1-4 ms |
| 18 | W/GY | Injector 8 Driver | 1-4 ms |
| 19 | GY | Coil Driver Control | 5º, 55 mph: 8º dwell |
| 20 | DG/WT | Alternator Field Control | Digital Signals: 0-12-0v |
| 21 | BK/RD | Throttle Body Temperature | At 100ºF: 2.51v |
| 22 | OR/DB | TP Sensor Signal | 0.6-1.0v |
| 23 | DG/OR | Fuel Low Pressure Sensor | 0.5-4.5v |
| 24 | GY/BK | Distributor Reference Signal | Digital Signals: 0-5-0v |
| 25 | PK | SCI Transmit | 0v |
| 26 | --- | Not Used | --- |
| 27 | BR | A/C Clutch Signal | A/C Off: 12v, On: 1v |
| 28 | --- | Not Used | --- |
| 29 | WT/PK | Brake Switch Sense Signal | Brake Off: 0v, On: 12v |
| 30 | BR/YL | PNP Switch Sense Signal | In P/N: 0v, Others: 5v |
| 31 | --- | Not Used | --- |
| 32 | BK/PK | MIL (lamp) Control | MIL Off: 12v, On: 1v |
| 33 | TN/RD | Speed Control Vacuum Solenoid | Vacuum Increasing: 1v |
| 34 | DB/OR | A/C WOT Relay Control | Relay Off: 12v, On: 1v |
| 35 | GY/YL | EGR Solenoid Control | 12v, at 55 mph: 1v |
| 36 | --- | Not Used | --- |
| 37 | BK/OR | Overdrive Lamp Control | Lamp Off: 12v, On: 1v |
| 38 | WT/BK | Injector 5 Driver | 1-4 ms |
| 39 | GY/RD | AIS Motor Control | Pulse Signals |
| 40 | BR/WT | AIS Motor Control | Pulse Signals |

## Standard Colors and Abbreviations

| Abbreviation | Color | Abbreviation | Color | Abbreviation | Color |
|---|---|---|---|---|---|
| BK | Black | GY | Gray | RD | Red |
| BL | Blue | GN | Green | TN | Tan |
| BR | Brown | LG | Light Green | VT | Violet |
| DB | Dark Blue | OR | Orange | WT | White |
| DG | Dark Green | PK | Pink | YL | Yellow |

**1993-95 Ram Pickup & Van 5.2L V8 CNG VIN T (A/T) 60-Pin Connector - Continued**

| PCM Pin # | Wire Color | Circuit Description (60-Pin) | Value at Hot Idle |
|---|---|---|---|
| 41 | BK/DG | HO2S-11 (B1 S1) Signal | 0.1-1.1v |
| 42 | TN/PK | Fuel Temperature Sensor | 0.5-4.5v |
| 43 | --- | Not Used | --- |
| 44 | TN/YL | Distributor Sync Signal | Digital Signals: 0-5-0v |
| 45 | LG | SCI Receive | 5v |
| 46 | --- | Not Used | --- |
| 47 | WT/OR | Vehicle Speed Signal | Digital Signal |
| 48 | BR/RD | Speed Control Set Switch Signal | S/C & Set Switch On: 3.8v |
| 49 | YL/RD | Speed Control On/Off Switch | S/C On: 1v, Off: 12v |
| 50 | WT/LG | Speed Control Resume Switch | S/C On: 1v, Off: 12v |
| 51 | DB/YL | ASD Relay Control | Relay Off: 12v, On: 1v |
| 52 | --- | Not Used | --- |
| 53 | LG/RD | Speed Control Vent Solenoid | Vacuum Decreasing: 1v |
| 54 | OR/BK | Part Throttle Unlock Solenoid | Solenoid Off: 12v, On: 1v |
| 55 | OR/WT | Overdrive Lockout Solenoid | Solenoid Off: 12v, On: 1v |
| 56 | GY/PK | Maintenance Indicator Lamp | Lamp Off: 12v, On: 1v |
| 57 | DG/OR | ASD Relay Power (B+) | 12-14v |
| 58 | WT/RD | Injector 6 Driver | 1-4 ms |
| 59 | PK/BK | AIS Motor Control | Pulse Signals |
| 60 | YL/BK | AIS Motor Control | Pulse Signals |

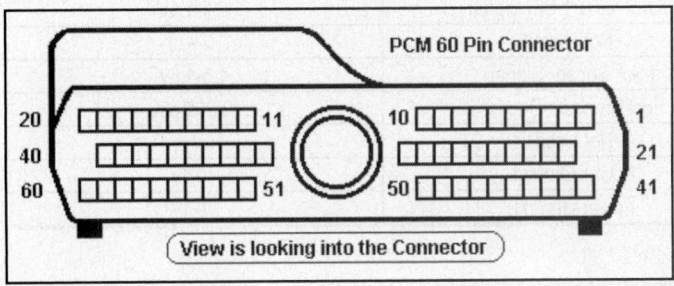

**Pin Connector Graphic**

**Standard Colors and Abbreviations**

| Abbreviation | Color | Abbreviation | Color | Abbreviation | Color |
|---|---|---|---|---|---|
| BK | Black | GY | Gray | RD | Red |
| BL | Blue | GN | Green | TN | Tan |
| BR | Brown | LG | Light Green | VT | Violet |
| DB | Dark Blue | OR | Orange | WT | White |
| DG | Dark Green | PK | Pink | YL | Yellow |

## 1996-97 Ram Pickup & Van 5.2L CNG VIN T 'A' Black Connector

| PCM Pin # | Wire Color | Circuit Description (32-Pin) | Value at Hot Idle |
|---|---|---|---|
| 1 | --- | Not Used | --- |
| 2 | LG/BK | Ignition Switch Power (B+) | 12-14v |
| 3 | --- | Not Used | --- |
| 4 | BK/LB | Sensor Ground | <0.050v |
| 5 | --- | Not Used | --- |
| 6 | BK/WT | PNP Switch Sense Signal | In P/N: 0v, Others: 5v |
| 7 | BK/GY | Coil 1 Driver Control | 5°, 55 mph: 8° dwell |
| 8 | GY/BK | CKP Sensor Signal | Digital Signals: 0-5-0v |
| 9 | --- | Not Used | --- |
| 10 | YL/BK | IAC 2 Driver Control | Pulse Signals |
| 11 | BR/WT | IAC 3 Driver Control | Pulse Signals |
| 12-14 | --- | Not Used | --- |
| 15 | BK/RD | IAT Sensor Signal | At 100°F: 1.83v |
| 16 | TN/BK | ECT Sensor Signal | At 180°F: 2.80v |
| 17 | PK/WT | 5-Volt Supply | 4.9-5.1v |
| 18 | TN/YL | CMP Sensor Signal | Digital Signals: 0-5-0v |
| 19 | GY/RD | IAC 1 Driver Control | Pulse Signals |
| 20 | PK/BK | IAC 4 Driver Control | Pulse Signals |
| 21 | --- | Not Used | --- |
| 22 | RD/WT | Keep Alive Power (B+) | 12-14v |
| 23 | OR/DB | TP Sensor Signal | 0.6-1.0v |
| 24 | TN/WT | HO2S-11 (B1 S1) Signal | 0.1-1.1v |
| 25 | OR/BK | HO2S-12 (B1 S2) Signal | 0.1-1.1v |
| 26 | --- | Not Used | --- |
| 27 | DG/RD | MAP Sensor Signal | 1.5-1.6v |
| 28 | DG/OR | Fuel Pressure Sensor | 0-255 psi |
| 29-30 | --- | Not Used | --- |
| 31 | BK/TN | Power Ground | <0.1v |
| 32 | BK/TN | Power Ground | <0.1v |

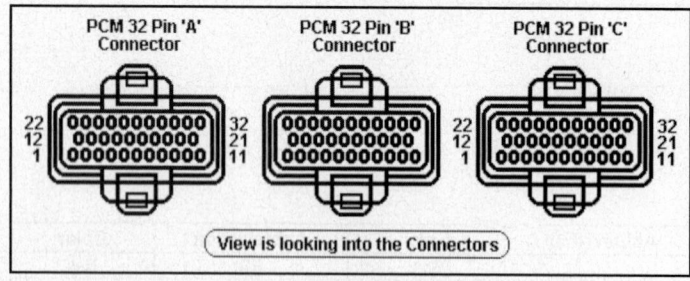

**Pin Connector Graphic**

**1996-97 Ram Pickup & Van 5.2L CNG VIN T 'B' White Connector**

| PCM Pin # | Wire Color | Circuit Description (32-Pin) | Value at Hot Idle |
|---|---|---|---|
| 1 | PK | Transmission Temperature Sensor Signal | At 200°F: 2.40v |
| 2 | PK/TN | Injector 7 Driver | 1-4 ms |
| 3, 7, 9, 14 | --- | Not Used | --- |
| 4 | WT/DB | Injector 1 Driver | 1-4 ms |
| 5 | YL/WT | Injector 3 Driver | 1-4 ms |
| 6 | GY | Injector 5 Driver | 1-4 ms |
| 8 | PK/WT | Governor Pressure Solenoid | PWM Signal: 0-12-0v |
| 10 | DG | Generator Field Control | Digital Signals: 0-12-0v |
| 11 | OR/BK | TCC Solenoid Control | At Cruise w/TCC On: 1v |
| 12 | BR/DB | Injector 6 Driver | 1-4 ms |
| 13 | GY/LB | Injector 8 Driver | 1-4 ms |
| 15 | TN | Injector 2 Driver | 1-4 ms |
| 16 | LB/BR | Injector 4 Driver | 1-4 ms |
| 17-20, 23-24 | --- | Not Used | --- |
| 21 | BR | 3-4 Shift Solenoid Control | Solenoid Off: 12v, On: 1v |
| 22 | TN/PK | Fuel Temperature Sensor | 0.5-4.5v |
| 25 | DB | OSS Sensor (+ | Moving: AC pulse signals |
| 27 | WT/OR | Vehicle Speed Signal | Digital Signal |
| 28 | LG | OSS Sensor (- Signals | Moving: AC pulse signals |
| 29 | LG/WT | Governor Pressure Sensor | 0.58v |
| 30 | PK | Transmission Relay Control | Relay Off: 12v, On: 1v |
| 31 | OR | 5-Volt Supply | 4.9-5.1v |
| 32 | GY/YL | EGR Solenoid Control | 12v, at 55 mph: 1v |

**1996-97 Ram Pickup & Van 5.2L CNG VIN T 'C' Gray Connector**

| PCM Pin # | Wire Color | Circuit Description (32-Pin) | Value at Hot Idle |
|---|---|---|---|
| 1 | DB/OR | A/C Clutch Relay Control | Relay Off: 12v, On: 1v |
| 3 | DB/YL | ASD Relay Control | Relay Off: 12v, On: 1v |
| 4 | TN/RD | Speed Control Vacuum Solenoid | Vacuum Increasing: 1v |
| 5 | LG/RD | Speed Control Vent Solenoid | Vacuum Decreasing: 1v |
| 6 | LG/OR | Overdrive Off Lamp Control | Lamp Off: 12v, On: 1v |
| 7 | PK/BK | Transmission Temperature Lamp Control | Lamp Off: 12v, On: 1v |
| 11 | YL/RD | Speed Control Power Supply | 12-14v |
| 12 | DG/OR | ASD Relay Power (B+) | 12-14v |
| 13 | OR/WT | Overdrive Off Switch Sense | Switch Off: 12v, On: 1v |
| 15 | PK/YL | Battery Temperature Sensor | At 86°F: 1.96v |
| 16 | TN/YL | Generator Lamp Control | Lamp Off: 12v, On: 1v |
| 17 | BK/PK | MIL (lamp) Control | MIL Off: 12v, On: 1v |
| 18 | GY/PK | Maintenance Indicator Lamp | Lamp Off: 12v, On: 1v |
| 19 | BR/WT | High Pressure Fuel Shutoff Relay | Relay Off: 12v, On: 1v |
| 22 | BR | A/C Request Signal | A/C Off: 12v, On: 1v |
| 23 | LG | A/C Select Signal | A/C Off: 12v, On: 1v |
| 24 | WT/PK | Brake Switch Sense Signal | Brake Off: 0v, On: 12v |
| 26 | DB/WT | Fuel Level Sensor Signal | Full: 0.56v, 1/2 full: 2.5v |
| 27 | PK | SCI Transmit | 0v |
| 28, 30 | --- | Not Used | --- |
| 29 | DG | SCI Receive | 5v |
| 31 | GY/LB | Tachometer Signal | Pulse Signals |
| 32 | RD/LG | Speed Control Set Switch Signal | S/C & Set Switch On: 3.8v |

## 1998-2003 Ram Pickup & Van 5.2L V8 OHV CNG VIN T 'A' Black Connector

| PCM Pin # | Wire Color | Circuit Description (32-Pin) | Value at Hot Idle |
|---|---|---|---|
| 1 | --- | Not Used | --- |
| 2 | LG/BK | Ignition Switch Power (B+) | 12-14v |
| 3 | --- | Not Used | --- |
| 4 | BK/LB | Sensor Ground | <0.050v |
| 5 | --- | Not Used | --- |
| 6 | BR/YL | PNP Switch Sense Signal | In P/N: 0v, Others: 5v |
| 7 | GY | Coil 1 Driver Control | 5°, 55 mph: 8° dwell |
| 8 | GY/BK | CKP Sensor Signal | Digital Signals: 0-5-0v |
| 9 | --- | Not Used | --- |
| 10 | YL/BK | IAC 2 Driver Control | Pulse Signals |
| 11 | BR/WT | IAC 3 Driver Control | Pulse Signals |
| 12-14 | --- | Not Used | --- |
| 15 | BK/RD | IAT Sensor Signal | At 100°F: 1.83v |
| 16 | TN/BK | ECT Sensor Signal | At 180°F: 2.80v |
| 17 | OR | 5-Volt Supply | 4.9-5.1v |
| 18 | TN/YL | CMP Sensor Signal | Digital Signals: 0-5-0v |
| 19 | GY/RD | IAC 1 Driver Control | Pulse Signals |
| 20 | VT/BK | IAC 4 Driver Control | Pulse Signals |
| 21 | --- | Not Used | --- |
| 22 | RD/WT | Keep Alive Power (B+) | 12-14v |
| 23 | OR/DB | TP Sensor Signal | 0.6-1.0v |
| 24 | BK/DG | HO2S-11 (B1 S1) Signal | 0.1-1.1v |
| 25 | TN/WT | HO2S-12 (B1 S2) Signal | 0.1-1.1v |
| 26 | --- | Not Used | --- |
| 27 | DG/RD | MAP Sensor Signal | 1.5-1.6v |
| 28 | DG/OR | Fuel Pressure Sensor | 0.5-4.5v |
| 29-30 | --- | Not Used | --- |
| 31 | BK/TN | Power Ground | <0.1v |
| 32 | BK/TN | Power Ground | <0.1v |

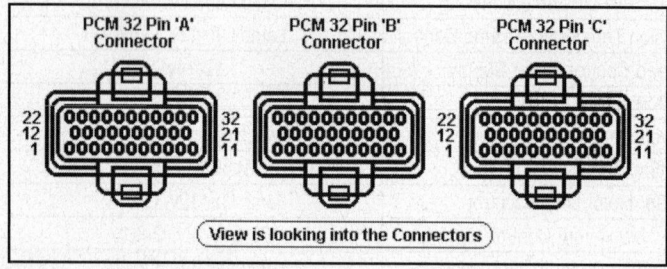

**Pin Connector Graphic**

## Standard Colors and Abbreviations

| Abbreviation | Color | Abbreviation | Color | Abbreviation | Color |
|---|---|---|---|---|---|
| BK | Black | GY | Gray | RD | Red |
| BL | Blue | GN | Green | TN | Tan |
| BR | Brown | LG | Light Green | VT | Violet |
| DB | Dark Blue | OR | Orange | WT | White |
| DG | Dark Green | PK | Pink | YL | Yellow |

**1998-2003 Ram Pickup & Van 5.2L CNG VIN T 'B' White Connector**

| PCM Pin # | Wire Color | Circuit Description (32-Pin) | Value at Hot Idle |
|---|---|---|---|
| 1 | GY/BK | Transmission Temperature Sensor Signal | At 200°F: 2.40v |
| 2 | VT | Injector 7 Driver | 1-4 ms |
| 3, 7 | --- | Not Used | --- |
| 4 | WT/DB | Injector 1 Driver | 1-4 ms |
| 5 | YL/WT | Injector 3 Driver | 1-4 ms |
| 6 | GY | Injector 5 Driver | 1-4 ms |
| 8 | VT/BK | Governor Pressure Solenoid | PWM Signal: 0-12-0v |
| 9, 14 | --- | Not Used | --- |
| 10 | DG/WT | Generator Field Control | Digital Signals: 0-12-0v |
| 11 | OR/BK | TCC Solenoid Control | At Cruise w/TCC On: 1v |
| 12 | BR/DB | Injector 6 Driver | 1-4 ms |
| 13 | GY/LB | Injector 8 Driver | 1-4 ms |
| 15 | TN | Injector 2 Driver | 1-4 ms |
| 16 | LB/BR | Injector 4 Driver | 1-4 ms |
| 17-20 | --- | Not Used | --- |
| 21 | BR | 3-4 Shift Solenoid Control | Solenoid Off: 12v, On: 1v |
| 22 | TN/PK | Fuel Temperature Sensor | 0.5-4.5v |
| 23 | GY/YL | Engine Oil Pressure Sensor | 1.6v at 24 psi |
| 24, 32 | --- | Not Used | --- |
| 25 | DB/BK | OSS Sensor (-) Signals | Moving: AC pulse signals |
| 27 | WT/OR | Vehicle Speed Signal | Digital Signals |
| 28 | LG/WT | OSS Sensor (+) Signals | Moving: AC pulse signals |
| 29 | LG/RD | Governor Pressure Sensor | 0.58v |
| 30 | PK | Transmission Relay Control | Relay Off: 12v, On: 1v |
| 31 | VT/WT | 5-Volt Supply | 4.9-5.1v |

**1998-2003 Ram Pickup & Van 5.2L CNG VIN T 'C' Gray Connector**

| PCM Pin # | Wire Color | Circuit Description (32-Pin) | Value at Hot Idle |
|---|---|---|---|
| 1 | DB/OR | A/C Clutch Relay Control | Relay Off: 12v, On: 1v |
| 2 | --- | Not Used | --- |
| 3 | DB/YL | ASD Relay Control | Relay Off: 12v, On: 1v |
| 4 | TN/RD | Speed Control Vacuum Solenoid | Vacuum Increasing: 1v |
| 5 | LG/RD | Speed Control Vent Solenoid | Vacuum Decreasing: 1v |
| 8 | OR/RD | HO2S-11 (B1 S1) Heater | PWM Signal: 0-12-0v |
| 9 | --- | Not Used | --- |
| 10 | WT/DG | LDP Solenoid Control | PWM Signal: 0-12-0v |
| 11 | YL/RD | Speed Control Power Supply | 12-14v |
| 12 | DG/OR | ASD Relay Power (B+) | 12-14v |
| 13 | OR/WT | Overdrive Off Switch Sense | Switch Off: 12v, On: 1v |
| 14 | OR | Leak Detection Pump Switch Sense | LDP Switch Closed: 0v |
| 15 | PK/YL | Battery Temperature Sensor | At 86°F: 1.96v |
| 16 | VT/TN | HO2S-12 (B1 S2) Heater | PWM Signal: 0-12-0v |
| 17-18, 20-21 | --- | Not Used | --- |
| 19 | BR | High Pressure Fuel Shutoff Relay | Relay Off: 12v, On: 1v |
| 22 | DB | A/C Pressure Switch Signal | A/C On: 0.451-4.850v |
| 23 | LG/GY | A/C Switch Sense | A/C Off: 12v, On: 1v |
| 24 | WT/PK | Brake Switch Sense | Brake Off: 0v, On: 12v |
| 25 | WT/DB | Generator Field Source | 12-14v |
| 26 | LB/BK | Fuel Level Sensor Signal | Full: 0.56v, 1/2 full: 2.5v |
| 27 | PK | SCI Transmit | 0v |
| 28 | WT/BK | CCD Bus (-) Signal | <0.050v |
| 29 | LG | SCI Receive | 5v |
| 30 | VT/BR | CCD Bus (+) Signal | Digital Signals: 0-5-0v |
| 31 | --- | Not Used | --- |
| 32 | WT/LG | Speed Control Set Switch Signal | S/C & Set Switch On: 3.8v |

### 1998-2000 Ram Pickup & Van 5.2L Propane VIN 2 'A' Black Connector

| PCM Pin # | Wire Color | Circuit Description (32-Pin) | Value at Hot Idle |
|---|---|---|---|
| 1 | --- | Not Used | --- |
| 2 | LG/BK | Ignition Switch Power (B+) | 12-14v |
| 3 | --- | Not Used | --- |
| 4 | BK/LB | Sensor Ground | <0.050v |
| 5 | --- | Not Used | --- |
| 6 | BR/YL | PNP Switch Sense Signal | In P/N: 0v, Others: 5v |
| 7 | GY | Coil 1 Driver Control | 5°, 55 mph: 8° dwell |
| 8 | GY/BK | CKP Sensor Signal | Digital Signals: 0-5-0v |
| 9 | --- | Not Used | --- |
| 10 | YL/BK | IAC 2 Driver Control | Pulse Signals |
| 11 | BR/WT | IAC 1 Driver Control | Pulse Signals |
| 12-14 | --- | Not Used | --- |
| 15 | BK/RD | IAT Sensor Signal | At 100°F: 1.83v |
| 16 | TN/BK | ECT Sensor Signal | At 180°F: 2.80v |
| 17 | OR | 5-Volt Supply | 4.9-5.1v |
| 18 | TN/YL | CMP Sensor Signal | Digital Signals: 0-5-0v |
| 19 | GY/RD | IAC 3 Driver Control | Pulse Signals |
| 20 | VT/BK | IAC 4 Driver Control | Pulse Signals |
| 21 | --- | Not Used | --- |
| 22 | RD/WT | Keep Alive Power (B+) | 12-14v |
| 22 ('01) | RD/WT | Keep Alive Power (B+) | 12-14v |
| 23 | OR/DB | TP Sensor Signal | 0.6-1.0v |
| 24 | BK/DG | HO2S-11 (B1 S1) Signal | 0.1-1.1v |
| 25 | TN/WT | HO2S-12 (B1 S2) Signal | 0.1-1.1v |
| 26 | --- | Not Used | --- |
| 27 | DG/RD | MAP Sensor Signal | 1.5-1.6v |
| 28 | DG/OR | Fuel Pressure Sensor | 0.5-4.5v |
| 29-30 | --- | Not Used | --- |
| 31 | BK/TN | Power Ground | <0.1v |
| 32 | BK/TN | Power Ground | <0.1v |

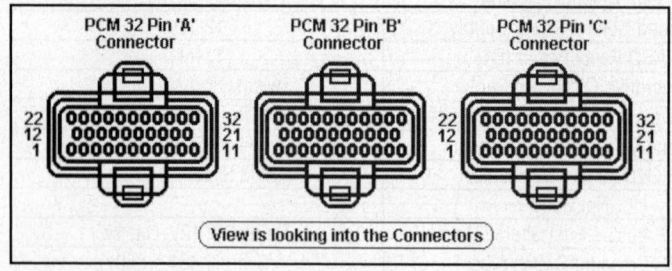

**Pin Connector Graphic**

### Standard Colors and Abbreviations

| Abbreviation | Color | Abbreviation | Color | Abbreviation | Color |
|---|---|---|---|---|---|
| BK | Black | GY | Gray | RD | Red |
| BL | Blue | GN | Green | TN | Tan |
| BR | Brown | LG | Light Green | VT | Violet |
| DB | Dark Blue | OR | Orange | WT | White |
| DG | Dark Green | PK | Pink | YL | Yellow |

## 1998-2000 Ram Pickup & Van 5.2L Propane VIN 2 'B' White Connector

| PCM Pin # | Wire Color | Circuit Description (32-Pin) | Value at Hot Idle |
|---|---|---|---|
| 1 | GY/BK | Transmission Temperature Sensor Signal | At 200°F: 2.40v |
| 2 | VT | Injector 7 Driver | 1-4 ms |
| 4 | WT/DB | Injector 1 Driver | 1-4 ms |
| 5 | YL/WT | Injector 3 Driver | 1-4 ms |
| 6 | GY | Injector 5 Driver | 1-4 ms |
| 8 | VT/WT | Governor Pressure Solenoid | PWM Signal: 0-12-0v |
| 10 | DG/WT | Generator Field Control | Digital Signals: 0-12-0v |
| 11 | OR/BK | TCC Solenoid Control | At Cruise w/TCC On: 1v |
| 12 | BR/DB | Injector 6 Driver | 1-4 ms |
| 13 | GY/LB | Injector 8 Driver | 1-4 ms |
| 15 | TN | Injector 2 Driver | 1-4 ms |
| 16 | LB/BR | Injector 4 Driver | 1-4 ms |
| 17-20, 24 | --- | Not Used | --- |
| 21 | BR | 3-4 Shift Solenoid Control | Solenoid Off: 12v, On: 1v |
| 22 | TN/PK | Fuel Temperature Sensor | 0.5-4.5v |
| 23 | GY/YL | Engine Oil Pressure Sensor | 1.6v at 24 psi |
| 25, 28 | DB, LG | OSS Sensor (-), (+) Signals | Moving: AC pulse signals |
| 27 | WT/OR | Vehicle Speed Signal | Digital Signal |
| 29 | LG/RD | Governor Pressure Sensor | 0.58v |
| 30 | PK | Transmission Relay Control | Relay Off: 12v, On: 1v |
| 31 | VT/WT | 5-Volt Supply | 4.9-5.1v |

## 1998-2000 Ram Pickup & Van 5.2L Propane VIN 2 'C' Gray Connector

| PCM Pin # | Wire Color | Circuit Description (32-Pin) | Value at Hot Idle |
|---|---|---|---|
| 1 | DB/OR | A/C Clutch Relay Control | Relay Off: 12v, On: 1v |
| 3 | DB/YL | ASD Relay Control | Relay Off: 12v, On: 1v |
| 4 | TN/RD | Speed Control Vacuum Solenoid | Vacuum Increasing: 1v |
| 5 | LG/RD | Speed Control Vent Solenoid | Vacuum Decreasing: 1v |
| 8 | OR/RD | HO2S-11 (B1 S1) Heater | PWM Signal: 0-12-0v |
| 9 | BK/OR | HO2S-11 Heater Relay (Cal) | Heater Off: 12v, On: 1v |
| 10 | WT/DG | LDP Solenoid Control | PWM Signal: 0-12-0v |
| 11 | YL/RD | Speed Control Power Supply | 12-14v |
| 12 | DG/OR | ASD Relay Power (B+) | 12-14v |
| 13 | OR/WT | Overdrive Off Switch Sense | Switch Off: 12v, On: 1v |
| 15 | PK/YL | Battery Temperature Sensor | At 86°F: 1.96v |
| 16 | VT/TN | HO2S-21 (B2 S1) Heater | PWM Signal: 0-12-0v |
| 19 | BR | High Pressure Fuel Shutoff Relay | Relay Off: 12v, On: 1v |
| 20 | PK/BK | EVAP Purge Solenoid Control | PWM Signal: 0-12-0v |
| 22 | DB | A/C Pressure Switch Signal | A/C On: 0.451-4.850v |
| 23 | LG/GY | A/C Select Signal | A/C Off: 12v, On: 1v |
| 24 | WT/PK | Brake Switch Sense Signal | Brake Off: 0v, On: 12v |
| 25 | WT/DB | Generator Field Source | 12-14v |
| 26 | DB | Fuel Level Sensor Signal | Full: 0.56v, 1/2 full: 2.5v |
| 27 | PK | SCI Transmit Signal | 0v |
| 28 | WT/BK | CCD Bus (-) Signal | <0.050v |
| 29 | LG | SCI Receive Signal | 5v |
| 30 | VT/BR | CCD Bus (+) Signal | Digital Signals: 0-5-0v |
| 32 | WT/LG | Speed Control Set Switch Signal | S/C & Set Switch On: 3.8v |

**1994-95 Ram Pickup & Van 5.2L VIN Y 60-Pin Connector**

| PCM Pin # | Wire Color | Circuit Description (60-Pin) | Value at Hot Idle |
|---|---|---|---|
| 1 | DG/RD | MAP Sensor Signal | 1.5-1.6v |
| 2 | TN/BK | ECT Sensor Signal | At 180°F: 2.80v |
| 3 | RD/WT | Keep Alive Power (B+) | 12-14v |
| 4 | BK/LB | Sensor Ground | <0.050v |
| 5, 8, 26, 36 | --- | Not Used | --- |
| 6 | PK/WT | 5-Volt Supply | 4.9-5.1v |
| 7 | OR | 8-Volt Supply | 7.9-8.1v |
| 9 | DB | Ignition Switch Power (B+) | 12-14v |
| 10 | OR/WT | Overdrive Override Switch | Switch Off: 12v, On: 1v |
| 11 | BK/TN | Power Ground | <0.1v |
| 12 | BK/TN | Power Ground | <0.1v |
| 13 | LB/BR | Injector 4 Driver | 1-4 ms |
| 14 | YL/WT | Injector 3 Driver | 1-4 ms |
| 15 | TN | Injector 2 Driver | 1-4 ms |
| 16 | WT/DB | Injector 1 Driver | 1-4 ms |
| 17 | DB/TN | Injector 7 Driver | 1-4 ms |
| 18 | RD/YL | Injector 8 Driver | 1-4 ms |
| 19 | BK/GY | Coil Driver Control | 5°, 55 mph: 8° dwell |
| 20 | DG | Alternator Field Control | Digital Signals: 0-12-0v |
| 21 | BK/RD | IAT Sensor Signal | At 100°F: 1.83v |
| 22 | OR/DB | TP Sensor Signal | 0.6-1.0v |
| 23 | TN/WT | Left HO2S-11 (B1 S1) Signal | 0.1-1.1v |
| 24 | GY/BK | Distributor Reference Signal | Digital Signals: 0-5-0v |
| 25 | PK | SCI Transmit | 0v |
| 27 | BR | A/C Damped Pressure Switch | A/C Off: 12v, On: 1v |
| 29 | WT/PK | Brake Switch Sense Signal | Brake Off: 0v, On: 12v |
| 30 | BK/WT | PNP Switch Sense Signal | In P/N: 0v, Others: 5v |
| 31 | PK/BK | Transmission Temperature Lamp Control | Lamp Off: 12v, On: 1v |
| 32 | BK/PK | MIL (lamp) Control | MIL Off: 12v, On: 1v |
| 33 | TN/RD | Speed Control Vacuum Solenoid | Vacuum Increasing: 1v |
| 34 | DB/OR | A/C WOT Relay Control | Relay Off: 12v, On: 1v |
| 35 | GY/YL | EGR Solenoid Control | 12v, at 55 mph: 1v |
| 37 | LG/OR | Overdrive Lamp Control | Lamp Off: 12v, On: 1v |
| 38 | GY | Injector 5 Driver | 1-4 ms |
| 39 | GY/RD | AIS Motor Control | Pulse Signals |
| 40 | BR/WT | AIS Motor Control | Pulse Signals |
| 41 | BK/DG | HO2S-11 (B1 S1) Signal | 0.1-1.1v |
| 42 | PK | Transmission Temperature Sensor Signal | At 200°F: 2.40v |
| 43 | GY/LB | Tachometer Signal | Pulse Signals |
| 44 | TN/YL | Distributor Sync Signal | Digital Signals: 0-5-0v |
| 45 | LG | SCI Receive | 5v |
| 47 | WT/OR | Vehicle Speed Signal | Digital Signal |
| 48 | BR/RD | Speed Control Set Switch Signal | S/C & Set Switch On: 3.8v |
| 49 | YL/RD | Speed Control On/Off Switch | S/C On: 1v, Off: 12v |
| 50 | WT/LG | Speed Control Resume Switch | S/C On: 1v, Off: 12v |
| 51 | DB/YL | ASD Relay Control | Relay Off: 12v, On: 1v |
| 52 | PK/BK | EVAP Purge Solenoid Control | Solenoid Off: 12v, On: 1v |
| 53 | LG/RD | Speed Control Vent Solenoid | Vacuum Decreasing: 1v |
| 54 | OR/BK | A/T: Overdrive Solenoid | Solenoid Off: 12v, On: 1v |
| 54 | OR/BK | M/T: Upshift Lamp Control | Lamp Off: 12v, On: 1v |
| 55 | BR | Overdrive Lockout Solenoid | Solenoid Off: 12v, On: 1v |
| 56 | GY/PK | Maintenance Indicator Lamp | Lamp Off: 12v, On: 1v |
| 57 | DG/OR | ASD Relay Power (B+) | 12-14v |
| 58 | BR/DB | Injector 6 Driver | 1-4 ms |
| 59 | PK/BK | AIS Motor Control | Pulse Signals |
| 60 | YL/BK | AIS Motor Control | Pulse Signals |

**1996-97 Ram Pickup & Van 5.2L VIN Y 'A' Black Connector**

| PCM Pin # | Wire Color | Circuit Description (32-Pin) | Value at Hot Idle |
|---|---|---|---|
| 1 | --- | Not Used | --- |
| 2 | LG/BK | Ignition Switch Power (B+) | 12-14v |
| 3 | --- | Not Used | --- |
| 4 | BK/LB | Sensor Ground | <0.050v |
| 5 | --- | Not Used | --- |
| 6 | BK/WT | PNP Switch Sense Signal | In P/N: 0v, Others: 5v |
| 7 | BK/GY | Coil 1 Driver Control | 5°, 55 mph: 8° dwell |
| 8 | GY/BK | CKP Sensor Signal | Digital Signals: 0-5-0v |
| 9 | --- | Not Used | --- |
| 10 | YL/BK | IAC 2 Driver Control | Pulse Signals |
| 11 | BR/WT | IAC 3 Driver Control | Pulse Signals |
| 12-14 | --- | Not Used | --- |
| 15 | BK/RD | IAT Sensor Signal | At 100°F: 1.83v |
| 16 | TN/BK | ECT Sensor Signal | At 180°F: 2.80v |
| 17 | PK/WT | 5-Volt Supply | 4.9-5.1v |
| 18 | TN/YL | CMP Sensor Signal | Digital Signals: 0-5-0v |
| 19 | GY/RD | IAC 1 Driver Control | Pulse Signals |
| 20 | OR/BK | IAC 4 Driver Control | Pulse Signals |
| 21 | --- | Not Used | --- |
| 22 | RD/WT | Keep Alive Power (B+) | 12-14v |
| 23 | OR/DB | TP Sensor Signal | 0.6-1.0v |
| 24 | TN/WT | HO2S-11 (B1 S1) Signal | 0.1-1.1v |
| 25 | OR/BK | HO2S-12 (B1 S2) Signal | 0.1-1.1v |
| 26 | --- | Not Used | --- |
| 27 | DG/RD | MAP Sensor Signal | 1.5-1.6v |
| 28-30 | --- | Not Used | --- |
| 31 | BK/TN | Power Ground | <0.1v |
| 32 | BK/TN | Power Ground | <0.1v |

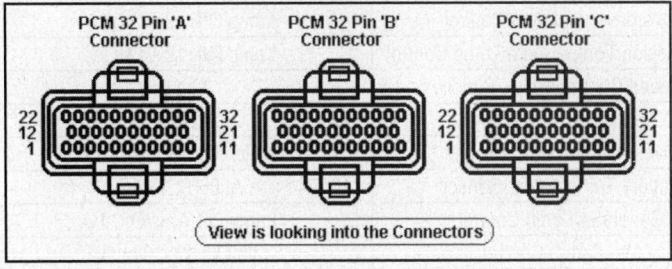

**Pin Connector Graphic**

**Standard Colors and Abbreviations**

| Abbreviation | Color | Abbreviation | Color | Abbreviation | Color |
|---|---|---|---|---|---|
| BK | Black | GY | Gray | RD | Red |
| BL | Blue | GN | Green | TN | Tan |
| BR | Brown | LG | Light Green | VT | Violet |
| DB | Dark Blue | OR | Orange | WT | White |
| DG | Dark Green | PK | Pink | YL | Yellow |

**1996-97 Ram Pickup & Van 5.2L VIN Y 'B' White Connector**

| PCM Pin # | Wire Color | Circuit Description (32-Pin) | Value at Hot Idle |
|---|---|---|---|
| 1 | VT | Transmission Temperature Sensor Signal | At 200°F: 2.40v |
| 2 | PK/TN | Injector 7 Driver | 1-4 ms |
| 3, 7, 9 | --- | Not Used | --- |
| 4 | WT/DB | Injector 1 Driver | 1-4 ms |
| 5 | YL/WT | Injector 3 Driver | 1-4 ms |
| 6 | GY | Injector 5 Driver | 1-4 ms |
| 8 | PK/WT | Governor Pressure Solenoid | PWM Signal: 0-12-0v |
| 10 | DG | Generator Field Control | Digital Signals: 0-12-0v |
| 11 | OR/BK | TCC Solenoid Control | At Cruise w/TCC On: 1v |
| 12 | BR/DB | Injector 6 Driver | 1-4 ms |
| 13 | GY/LB | Injector 8 Driver | 1-4 ms |
| 14 | --- | Not Used | --- |
| 15 | TN | Injector 2 Driver | 1-4 ms |
| 16 | LB/BR | Injector 4 Driver | 1-4 ms |
| 17-20, 22-24 | --- | Not Used | --- |
| 21 | BR | 3-4 Shift Solenoid Control | Solenoid Off: 12v, On: 1v |
| 25, 28 | DB, LG | OSS Sensor (-), (+) Signals | Moving: AC pulse signals |
| 26 | --- | Not Used | --- |
| 27 | WT/OR | Vehicle Speed Signal | Digital Signal |
| 29 | LG/WT | Governor Pressure Sensor | 0.58v |
| 30 | PK | Transmission Relay Control | Relay Off: 12v, On: 1v |
| 31 | OR | 5-Volt Supply | 4.9-5.1v |
| 32 | GY/YL | EGR Solenoid Control | 12v, at 55 mph: 1v |

**1996-97 Ram Pickup & Van 5.2L VIN Y 'C' Gray Connector**

| PCM Pin # | Wire Color | Circuit Description (32-Pin) | Value at Hot Idle |
|---|---|---|---|
| 1 | DB/OR | A/C Clutch Relay Control | Relay Off: 12v, On: 1v |
| 3 | DB/YL | ASD Relay Control | Relay Off: 12v, On: 1v |
| 4 | TN/RD | Speed Control Vacuum Solenoid | Vacuum Increasing: 1v |
| 5 | LG/RD | Speed Control Vent Solenoid | Vacuum Decreasing: 1v |
| 6 | LG/OR | Overdrive Off Lamp Control | Lamp Off: 12v, On: 1v |
| 7 | PK/BK | Transmission Temperature Lamp Control | Lamp Off: 12v, On: 1v |
| 11 | YL/RD | Speed Control Power Supply | 12-14v |
| 12 | DG/OR | ASD Relay Power (B+) | 12-14v |
| 13 | OR/WT | Overdrive Off Switch Sense | Switch Off: 12v, On: 1v |
| 15 | PK/YL | Battery Temperature Sensor | At 86°F: 1.96v |
| 16 | TN/YL | Generator Lamp Control | Lamp Off: 12v, On: 1v |
| 17 | BK/PK | MIL (lamp) Control | MIL Off: 12v, On: 1v |
| 18 | GY/PK | Maintenance Indicator Lamp | Lamp Off: 12v, On: 1v |
| 19 | BR/WT | Fuel Pump Relay Control | Relay Off: 12v, On: 1v |
| 20 | PK | EVAP Purge Solenoid Control | PWM Signal: 0-12-0v |
| 22 | BR | A/C Request Signal | A/C Off: 12v, On: 1v |
| 23 | LG | A/C Select Signal | A/C Off: 12v, On: 1v |
| 24 | WT/PK | Brake Switch Sense Signal | Brake Off: 0v, On: 12v |
| 26 | DB/WT | Fuel Level Sensor Signal | Full: 0.56v, 1/2 full: 2.5v |
| 27 | PK | SCI Transmit Signal | 0v |
| 29 | LG | SCI Receive Signal | 5v |
| 31 | GY/LB | Tachometer Signal | Pulse Signals |
| 32 | RD/LG | Speed Control Set Switch Signal | S/C & Set Switch On: 3.8v |

**1998-2001 Ram Pickup & Van 5.2L VIN Y 'C1' Black Connector**

| PCM Pin # | Wire Color | Circuit Description (32-Pin) | Value at Hot Idle |
|---|---|---|---|
| 1 | --- | Not Used | --- |
| 2 | LG/BK | Ignition Switch Power (B+) | 12-14v |
| 3 | --- | Not Used | --- |
| 4 | BK/LB | Sensor Ground | <0.050v |
| 5 | --- | Not Used | --- |
| 6 | BK/WT | PNP Switch Sense Signal | In P/N: 0v, Others: 5v |
| 7 | BK/GY | Coil 1 Driver Control | 5°, 55 mph: 8° dwell |
| 8 | GY/BK | CKP Sensor Signal | Digital Signals: 0-5-0v |
| 9 | --- | Not Used | --- |
| 10 | YL/BK | IAC 2 Driver Control | Pulse Signals |
| 11 | BR/WT | IAC 3 Driver Control | Pulse Signals |
| 12 | --- | Not Used | --- |
| 13 | OR | Power Takeoff Switch Sense | Switch Off: 12v, On: 1v |
| 14 | --- | Not Used | --- |
| 15 | BK/RD | IAT Sensor Signal | At 100°F: 1.83v |
| 16 | TN/BK | ECT Sensor Signal | At 180°F: 2.80v |
| 17 | VT/WT | 5-Volt Supply | 4.9-5.1v |
| 18 | TN/YL | CMP Sensor Signal | Digital Signals: 0-5-0v |
| 19 | GY/RD | IAC 1 Driver Control | Pulse Signals |
| 20 | PK/BK | IAC 4 Driver Control | Pulse Signals |
| 21 | --- | Not Used | --- |
| 22 | RD/WT | Keep Alive Power (B+) | 12-14v |
| 23 | OR/DB | TP Sensor Signal | 0.6-1.0v |
| 24 | TN/WT | HO2S-11 (B1 S1) Signal | 0.1-1.1v |
| 25 | OR/BK | HO2S-12 (B1 S2) Signal | 0.1-1.1v |
| 26 | LG/RD | HO2S-21 (B2 S1) Signal (Cal) | 0.1-1.1v |
| 27 | DG/RD | MAP Sensor Signal | 1.5-1.6v |
| 28 | --- | Not Used | --- |
| 29 | OR/TN | HO2S-22 (B2 S2) Signal (Cal) | 0.1-1.1v |
| 30 | --- | Not Used | --- |
| 31 | BK/TN | Power Ground | <0.1v |
| 32 | BK/TN | Power Ground | <0.1v |

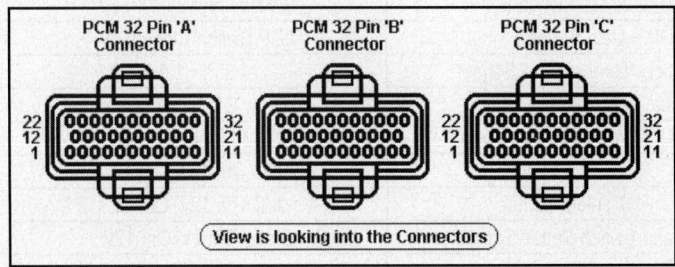

**Pin Connector Graphic**

### 1998-2001 Ram Pickup & Van 5.2L VIN Y 'C2' White Connector

| PCM Pin # | Wire Color | Circuit Description (32-Pin) | Value at Hot Idle |
|---|---|---|---|
| 1 | VT | Transmission Temperature Sensor Signal | At 200ºF: 2.40v |
| 2 | VT/TN | Injector 7 Driver | 1-4 ms |
| 4 | WT/DB | Injector 1 Driver | 1-4 ms |
| 5 | YL/WT | Injector 3 Driver | 1-4 ms |
| 6 | GY | Injector 5 Driver | 1-4 ms |
| 8 | VT/WT | Governor Pressure Solenoid | PWM Signal: 0-12-0v |
| 10 | DG | Generator Field Control | Digital Signals: 0-12-0v |
| 11 | OR/BK | TCC Solenoid Control | At Cruise w/TCC On: 1v |
| 12 | BR/DB | Injector 6 Driver | 1-4 ms |
| 13 | GY/LB | Injector 8 Driver | 1-4 ms |
| 15 | TN | Injector 2 Driver | 1-4 ms |
| 16 | LB/BR | Injector 4 Driver | 1-4 ms |
| 21 | BR | 3-4 Shift Solenoid Control | Cruise w/solenoid On: 1v |
| 23 | GY/OR | Engine Oil Pressure Sensor | 1.6v at 24 psi |
| 25 | DB/BK | OSS Sensor (-) Signal | Moving: AC pulse signals |
| 27 | WT/OR | Vehicle Speed Signal | Digital Signal |
| 28 | LG/BK | OSS Sensor (+) Signal | Moving: AC pulse signals |
| 29 | LG/WT | Governor Pressure Sensor | 0.58v |
| 30 | PK | Transmission Relay Control | Relay Off: 12v, On: 1v |
| 31 | OR | 5-Volt Supply | 4.9-5.1v |

### 1998-2001 Ram Pickup & Van 5.2L VIN Y 'C3' Gray Connector

| PCM Pin # | Wire Color | Circuit Description (32-Pin) | Value at Hot Idle |
|---|---|---|---|
| 1 | DB/OR | A/C Clutch Relay Control | Relay Off: 12v, On: 1v |
| 3 | DB/YL | ASD Relay Control | Relay Off: 12v, On: 1v |
| 4 | TN/RD | Speed Control Vacuum Solenoid | Vacuum Increasing: 1v |
| 5 | LG/RD | Speed Control Vent Solenoid | Vacuum Decreasing: 1v |
| 6 ('98) | LG/OR | Overdrive Off Lamp Control | Lamp Off: 12v, On: 1v |
| 8 | BR/VT | HO2S-11 Heater Control (Cal) | PWM Signal: 0-12-0v |
| 9 | DG/PK | HO2S-12 Heater Relay (Cal) | Relay Off: 12v, On: 1v |
| 10 | WT/DG | LDP Solenoid Control | PWM Signal: 0-12-0v |
| 11 | YL/RD | Speed Control Power Supply | 12-14v |
| 12 | DG/OR | ASD Relay Power (B+) | 12-14v |
| 13 | OR/WT | Overdrive Off Switch Sense | Switch Off: 12v, On: 1v |
| 14 | OR | LDP Switch Sense Signal | LDP Switch Closed: 0v |
| 15 | PK/YL | Battery Temperature Sensor | At 86ºF: 1.96v |
| 19 | BR/WT | Fuel Pump Relay Control | Relay Off: 12v, On: 1v |
| 20 | PK/WT | EVAP Purge Solenoid Control | PWM Signal: 0-12-0v |
| 22 | BR | A/C Switch Signal | A/C Off: 12v, On: 1v |
| 23 | LG/WT | A/C Select Signal | A/C Off: 12v, On: 1v |
| 24 | WT/PK | Brake Switch Sense Signal | Brake Off: 0v, On: 12v |
| 25 | DB | Generator Field Source | 12-14v |
| 26 | DB/WT | Fuel Level Sensor Signal | Full: 0.56v, 1/2 full: 2.5v |
| 27 | PK/DB | SCI Transmit Signal | 0v |
| 28 | WT/BK | CCD Bus (-) Signal | <0.050v |
| 29 | DG | SCI Receive Signal | 5v |
| 30 | VT/BR | CCD Bus (+) Signal | Digital Signals: 0-5-0v |
| 32 | RD/LG | Speed Control Set Switch Signal | S/C & Set Switch On: 3.8v |

**1994-95 Ram Pickup 5.9L VIN 5 & VIN Z 60-Pin Connector**

| PCM Pin # | Wire Color | Circuit Description (60-Pin) | Value at Hot Idle |
|---|---|---|---|
| 1 | DG/RD | MAP Sensor Signal | 1.5-1.6v |
| 2 | TN/BK | ECT Sensor Signal | At 180°F: 2.80v |
| 3 | RD/WT | Keep Alive Power (B+) | 12-14v |
| 4 | BK/LB | Sensor Ground | <0.050v |
| 5, 8, 26, 46 | --- | Not Used | --- |
| 6 | PK/WT | 5-Volt Supply | 4.9-5.1v |
| 7 | OR | 8-Volt Supply | 7.9-8.1v |
| 9 | LG/BK | Ignition Switch Power (B+) | 12-14v |
| 10 | OR/WT | Overdrive Override Switch | Switch Off: 12v, On: 1v |
| 11-12 | BK/TN | Power Ground | <0.1v |
| 13 | LB/BR | Injector 4 Driver | 1-4 ms |
| 14 | YL/WT | Injector 3 Driver | 1-4 ms |
| 15 | TN | Injector 2 Driver | 1-4 ms |
| 16 | WT/DB | Injector 1 Driver | 1-4 ms |
| 17 | DB/TN | Injector 7 Driver | 1-4 ms |
| 18 | RD/YL | Injector 8 Driver | 1-4 ms |
| 19 | BK/GY | Coil Driver Control | 5°, 55 mph: 8° dwell |
| 20 | DG | Alternator Field Control | Digital Signals: 0-12-0v |
| 21 | BK/RD | IAT Sensor Signal | At 100°F: 1.83v |
| 22 | OR/DB | TP Sensor Signal | 0.6-1.0v |
| 23 | TN/WT | HO2S-12 (B1 S2) Signal | 0.1-1.1v |
| 24 | GY/BK | Distributor Reference Signal | Digital Signals: 0-5-0v |
| 25 | PK | SCI Transmit | 0v |
| 27 | BR | A/C Damped Pressure Switch | A/C Off: 12v, On: 1v |
| 29 | WT/PK | Brake Switch Sense Signal | Brake Off: 0v, On: 12v |
| 30 | BK/WT | PNP Switch Sense Signal | In P/N: 0v, Others: 5v |
| 31 | PK/BK | Transmission Temperature Lamp Control | Lamp Off: 12v, On: 1v |
| 32 | BK/PK | MIL (lamp) Control | MIL Off: 12v, On: 1v |
| 33 | TN/RD | Speed Control Vacuum Solenoid | Vacuum Increasing: 1v |
| 34 | DB/OR | A/C WOT Relay Control | Relay Off: 12v, On: 1v |
| 35 | GY/YL | EGR Solenoid Control | 12v, at 55 mph: 1v |
| 37 | LG/OR | Overdrive Lamp Control | Lamp Off: 12v, On: 1v |
| 38 | GY | Injector 5 Driver | 1-4 ms |
| 39 | GY/RD | AIS Motor Control | Pulse Signals |
| 40 | BR/WT | AIS Motor Control | Pulse Signals |
| 41 | BK/DG | HO2S-11 (B1 S1) Signal | 0.1-1.1v |
| 42 | PK | Transmission Temperature Sensor Signal | At 200°F: 2.40v |
| 43 | GY/LB | Tachometer Signal | Pulse Signals |
| 44 | TN/YL | Distributor Sync Signal | Digital Signals: 0-5-0v |
| 45 | LG | SCI Receive | 5v |
| 47 | WT/OR | Vehicle Speed Signal | Digital Signal |
| 48 | BR/RD | Speed Control Set Switch Signal | S/C & Set Switch On: 3.8v |
| 49 | YL/RD | Speed Control On/Off Switch | S/C On: 1v, Off: 12v |
| 50 | WT/LG | Speed Control Resume Switch | S/C On: 1v, Off: 12v |
| 51 | DB/YL | ASD Relay Control | Relay Off: 12v, On: 1v |
| 52 | PK/BK | EVAP Purge Solenoid Control | Solenoid Off: 12v, On: 1v |
| 53 | LG/RD | Speed Control Vent Solenoid | Vacuum Decreasing: 1v |
| 54 | OR/BK | A/T: Overdrive Solenoid | Solenoid Off: 12v, On: 1v |
| 54 | OR/BK | M/T: Upshift Lamp Control | Lamp Off: 12v, On: 1v |
| 55 | BR | Overdrive Lockout Solenoid | Solenoid Off: 12v, On: 1v |
| 56 | GY/PK | Maintenance Indicator Lamp | Lamp Off: 12v, On: 1v |
| 57 | DG/OR | ASD Relay Power (B+) | 12-14v |
| 58 | BR/DB | Injector 6 Driver | 1-4 ms |
| 59 | PK/BK | AIS Motor Control | Pulse Signals |
| 60 | YL/BK | AIS Motor Control | Pulse Signals |

### 1996-97 Ram Pickup 5.9L VIN 5 & VIN Z 'A' Black Connector

| PCM Pin # | Wire Color | Circuit Description (32-Pin) | Value at Hot Idle |
|---|---|---|---|
| 1 | --- | Not Used | --- |
| 2 | LG/BK | Ignition Switch Power (B+) | 12-14v |
| 3 | --- | Not Used | --- |
| 4 | BK/LB | Sensor Ground | <0.050v |
| 5 | --- | Not Used | --- |
| 6 | BK/WT | PNP Switch Sense Signal | In P/N: 0v, Others: 5v |
| 7 | BK/GY | Coil 1 Driver Control | 5°, 55 mph: 8° dwell |
| 8 | GY/BK | CKP Sensor Signal | Digital Signals: 0-5-0v |
| 9 | --- | Not Used | --- |
| 10 | YL/BK | IAC 2 Driver Control | Pulse Signals |
| 11 | BR/WT | IAC 3 Driver Control | Pulse Signals |
| 12-14 | --- | Not Used | --- |
| 15 | BK/RD | IAT Sensor Signal | At 100°F: 1.83v |
| 16 | TN/BK | ECT Sensor Signal | At 180°F: 2.80v |
| 17 | PK/WT | 5-Volt Supply | 4.9-5.1v |
| 18 | TN/YL | CMP Sensor Signal | Digital Signals: 0-5-0v |
| 19 | GY/RD | IAC 1 Driver Control | Pulse Signals |
| 20 | PK/BK | IAC 4 Driver Control | Pulse Signals |
| 21 | --- | Not Used | --- |
| 22 | RD/WT | Keep Alive Power (B+) | 12-14v |
| 23 | OR/DB | TP Sensor Signal | 0.6-1.0v |
| 24 | TN/WT | HO2S-11 (B1 S1) Signal | 0.1-1.1v |
| 25 | OR/BK | L/D A/T: HO2S-12 Signal | 0.1-1.1v |
| 26 | BK/DG | H/D A/T: HO2S-12 Signal | 0.1-1.1v |
| 27 | DG/RD | MAP Sensor Signal | 1.5-1.6v |
| 28-30 | --- | Not Used | --- |
| 31 | BK/TN | Power Ground | <0.1v |
| 32 | BK/TN | Power Ground | <0.1v |

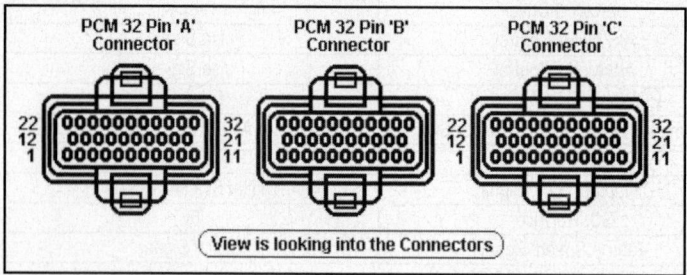

**Pin Connector Graphic**

### Standard Colors and Abbreviations

| Abbreviation | Color | Abbreviation | Color | Abbreviation | Color |
|---|---|---|---|---|---|
| BK | Black | GY | Gray | RD | Red |
| BL | Blue | GN | Green | TN | Tan |
| BR | Brown | LG | Light Green | VT | Violet |
| DB | Dark Blue | OR | Orange | WT | White |
| DG | Dark Green | PK | Pink | YL | Yellow |

**1996-97 Ram Pickup 5.9L VIN 5 & VIN Z 'B' White Connector**

| PCM Pin # | Wire Color | Circuit Description (32-Pin) | Value at Hot Idle |
|---|---|---|---|
| 1 | VT | Transmission Temperature Sensor Signal | At 200°F: 2.40v |
| 2 | PK/TN | Injector 7 Driver | 1-4 ms |
| 4 | WT/DB | Injector 1 Driver | 1-4 ms |
| 5 | YL/WT | Injector 3 Driver | 1-4 ms |
| 6 | GY | Injector 5 Driver | 1-4 ms |
| 8 | PK/WT | Governor Pressure Solenoid | PWM Signal: 0-12-0v |
| 10 | DG | Generator Field Control | Digital Signals: 0-12-0v |
| 11 | OR/BK | TCC Solenoid Control | At Cruise w/TCC On: 1v |
| 12 | BR/DB | Injector 6 Driver | 1-4 ms |
| 13 | GY/LB | Injector 8 Driver | 1-4 ms |
| 15 | TN | Injector 2 Driver | 1-4 ms |
| 16 | LB/BR | Injector 4 Driver | 1-4 ms |
| 21 | BR | 3-4 Shift Solenoid Control | Solenoid Off: 12v, On: 1v |
| 25 | DB/BK | OSS Sensor (-) Signal | Moving: AC pulse signals |
| 27 | WT/OR | Vehicle Speed Signal | Digital Signal |
| 28 | LG/BK | OSS Sensor (+) Signal | Moving: AC pulse signals |
| 29 | LG/WT | Governor Pressure Sensor | 0.58v |
| 30 | PK | Transmission Relay Control | Relay Off: 12v, On: 1v |
| 31 | OR | 5-Volt Supply | 4.9-5.1v |
| 32 | GY/YL | EGR Solenoid Control | 12v, at 55 mph: 1v |

**1996-97 Ram Pickup 5.9L VIN 5 & VIN Z 'C' Gray Connector**

| PCM Pin # | Wire Color | Circuit Description (32-Pin) | Value at Hot Idle |
|---|---|---|---|
| 1 | DB/OR | A/C Clutch Relay Control | Relay Off: 12v, On: 1v |
| 2, 8-10, 14 | --- | Not Used | --- |
| 3 | DB/YL | ASD Relay Control | Relay Off: 12v, On: 1v |
| 4 | TN/RD | Speed Control Vacuum Solenoid | Vacuum Increasing: 1v |
| 5 | LG/RD | Speed Control Vent Solenoid | Vacuum Decreasing: 1v |
| 6 | LG/OR | Overdrive Off Lamp Control | Lamp Off: 12v, On: 1v |
| 7 | PK/BK | Transmission Temperature Lamp Control | Lamp Off: 12v, On: 1v |
| 11 | YL/RD | Speed Control Power Supply | 12-14v |
| 12 | DG/OR | ASD Relay Power (B+) | 12-14v |
| 13 | OR/WT | Overdrive Off Switch Sense | Switch Off: 12v, On: 1v |
| 15 | PK/YL | Battery Temperature Sensor | At 86°F: 1.96v |
| 16 | TN/YL | Generator Lamp Control | Lamp Off: 12v, On: 1v |
| 17 | BK/PK | MIL (lamp) Control | MIL Off: 12v, On: 1v |
| 18 | GY/PK | Maintenance Indicator Lamp | Lamp Off: 12v, On: 1v |
| 19 | BR/WT | Fuel Pump Relay Control | Relay Off: 12v, On: 1v |
| 20 | PK | EVAP Purge Solenoid Control | PWM Signal: 0-12-0v |
| 21 | --- | Not Used | --- |
| 22 | BR | A/C Request Signal | A/C Off: 12v, On: 1v |
| 23 | LG | A/C Select Signal | A/C Off: 12v, On: 1v |
| 24 | WT/PK | Brake Switch Sense Signal | Brake Off: 0v, On: 12v |
| 21, 25 | --- | Not Used | --- |
| 26 | DB/WT | Fuel Level Sensor Signal | Full: 0.56v, 1/2 full: 2.5v |
| 27 | PK | SCI Transmit | 0v |
| 28 | WT/BK | CCD Bus (-) Signal | <0.050v |
| 29 | DG | SCI Receive | 5v |
| 30 | VT/BR | CCD Bus (+) Signal | Digital Signals: 0-5-0v |
| 31 | GY/LB | Tachometer Signal | Pulse Signals |
| 32 | RD/LG | Speed Control Set Switch Signal | S/C & Set Switch On: 3.8v |

## 1998-2001 Ram Pickup 5.9L VIN 5 & VIN Z 'C1' Black Connector

| PCM Pin # | Wire Color | Circuit Description (32-Pin) | Value at Hot Idle |
|---|---|---|---|
| 1 | --- | Not Used | --- |
| 2 | LG/BK | Ignition Switch Power (B+) | 12-14v |
| 3 | --- | Not Used | --- |
| 4 | BK/LB | Sensor Ground | <0.050v |
| 5 | --- | Not Used | --- |
| 6 | BK/WT | PNP Switch Sense Signal | In P/N: 0v, Others: 5v |
| 7 | BK/GY | Coil 1 Driver Control | 5°, 55 mph: 8° dwell |
| 8 | GY/BK | CKP Sensor Signal | Digital Signals: 0-5-0v |
| 9 | --- | Not Used | --- |
| 10 | YL/BK | IAC 2 Driver Control | Pulse Signals |
| 11 | BR/WT | IAC 3 Driver Control | Pulse Signals |
| 12, 14 | --- | Not Used | --- |
| 13 | OR | Power Takeoff Switch Sense | Switch Off: 12v, On: 1v |
| 15 | BK/RD | IAT Sensor Signal | At 100°F: 1.83v |
| 16 | TN/BK | ECT Sensor Signal | At 180°F: 2.80v |
| 17 | VT/WT | 5-Volt Supply | 4.9-5.1v |
| 18 | TN/YL | CMP Sensor Signal | Digital Signals: 0-5-0v |
| 19 | GY/RD | IAC 1 Driver Control | Pulse Signals |
| 20 | VT/BK | IAC 4 Driver Control | Pulse Signals |
| 21 | --- | Not Used | --- |
| 22 | RD/WT | Keep Alive Power (B+) | 12-14v |
| 23 | OR/DB | TP Sensor Signal | 0.6-1.0v |
| 24 | TN/WT | HO2S-11 (B1 S1) Signal | 0.1-1.1v |
| 24 (HD) | BK/DG | HO2S-11 (B1 S1) Signal | 0.1-1.1v |
| 25 | OR/BK | L/D Trans. - HO2S-12 | 0.1-1.1v |
| 25 ('00) | OR/BK | M/D A/T: HO2S-13 (Cal) | 0.1-1.1v |
| 26 | BK/DG | H/D A/T: HO2S-12 | 0.1-1.1v |
| 26 ('00-'01) | LG/RD | HO2S-21 (B2 S1) Signal (Cal) | 0.1-1.1v |
| 26 ('00) | LG/RD | M/D A/T: HO2S-12 (Cal) | 0.1-1.1v |
| 27 | DG/RD | MAP Sensor Signal | 1.5-1.6v |
| 28-30 | --- | Not Used | --- |
| 29 | OR/TN | HO2S-22 (B2 S2) Signal | 0.1-1.1v |
| 31 | BK/TN | Power Ground | <0.1v |
| 32 | BK/TN | Power Ground | <0.1v |

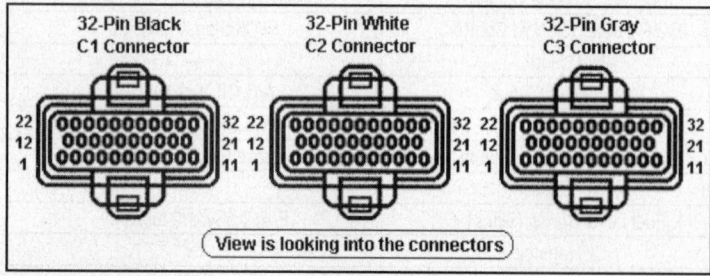

**Pin Connector Graphic**

**1998-2001 Ram Pickup 5.9L VIN 5 & VIN Z 'C2' White Connector**

| PCM Pin # | Wire Color | Circuit Description (32-Pin) | Value at Hot Idle |
|---|---|---|---|
| 1 | VT | Transmission Temperature Sensor Signal | At 200°F: 2.40v |
| 2 | VT/TN | Injector 7 Driver | 1-4 ms |
| 4 | WT/DB | Injector 1 Driver | 1-4 ms |
| 5 | YL/WT | Injector 3 Driver | 1-4 ms |
| 6 | GY | Injector 5 Driver | 1-4 ms |
| 8 | VT/WT | Governor Pressure Solenoid | PWM Signal: 0-12-0v |
| 10 | DG | Generator Field Control | Digital Signals: 0-12-0v |
| 11 | OR/BK | TCC Solenoid Control | At Cruise w/TCC On: 1v |
| 12 | BR/DB | Injector 6 Driver | 1-4 ms |
| 13 | GY/LB | Injector 8 Driver | 1-4 ms |
| 15 | TN | Injector 2 Driver | 1-4 ms |
| 16 | LB/BR | Injector 4 Driver | 1-4 ms |
| 21 | BR | 3-4 Shift Solenoid Control | Cruise w/solenoid On: 1v |
| 23 | GY/OR | Engine Oil Pressure Sensor | 1.6v at 24 psi |
| 25 | DB/BK | OSS Sensor (-) Signal | Moving: AC pulse signals |
| 27 | WT/OR | Vehicle Speed Signal | Digital Signal |
| 28 | LG/BK | OSS Sensor (+) Signal | Moving: AC pulse signals |
| 29 | LG/WT | Governor Pressure Sensor | 0.58v |
| 30 | PK | Transmission Relay Control | Relay Off: 12v, On: 1v |
| 31 | OR | 5-Volt Supply | 4.9-5.1v |

**1998-2001 Ram Pickup 5.9L VIN 5 & VIN Z 'C3' Gray Connector**

| PCM Pin # | Wire Color | Circuit Description (32-Pin) | Value at Hot Idle |
|---|---|---|---|
| 1 | DB/OR | A/C Clutch Relay Control | Relay Off: 12v, On: 1v |
| 3 | DB/YL | ASD Relay Control | Relay Off: 12v, On: 1v |
| 4 | TN/RD | Speed Control Vacuum Solenoid | Vacuum Increasing: 1v |
| 5 | LG/RD | Speed Control Vent Solenoid | Vacuum Decreasing: 1v |
| 6 ('98) | LG/OR | Overdrive Off Lamp Control | Lamp Off: 12v, On: 1v |
| 8 | BR/VT | HO2S-11 Heater Control (Cal) | Heater Off: 12v, On: 1v |
| 9 | DG/PK | HO2S-12 Heater Relay (Cal) | Relay Off: 12v, On: 1v |
| 10 | WT/DG | LDP Solenoid Control | PWM Signal: 0-12-0v |
| 11 | YL/RD | Speed Control Power Supply | 12-14v |
| 12 | DG/OR | ASD Relay Power (B+) | 12-14v |
| 13 | OR/WT | Overdrive Off Switch Sense | Switch Off: 12v, On: 1v |
| 14 | OR | LDP Switch Sense Signal | LDP Switch Closed: 0v |
| 15 | PK/YL | Battery Temperature Sensor | At 86°F: 1.96v |
| 19 | BR/WT | Fuel Pump Relay Control | Relay Off: 12v, On: 1v |
| 20 | PK/WT | EVAP Purge Solenoid Control | PWM Signal: 0-12-0v |
| 22 | BR | A/C Switch Signal | A/C Off: 12v, On: 1v |
| 23 | LG/WT | A/C Select Signal | A/C Off: 12v, On: 1v |
| 24 | WT/PK | Brake Switch Sense Signal | Brake Off: 0v, On: 12v |
| 25 | DB | Generator Field Source | 12-14v |
| 26 | DB/WT | Fuel Level Sensor Signal | Full: 0.56v, 1/2 full: 2.5v |
| 27, 29 | PK, LG | SCI Transmit, SCI Receive | 5v |
| 28 | WT/BK | CCD Bus (-) Signal | <0.050v |
| 30 | VT/BR | CCD Bus (+) Signal | Digital Signals: 0-5-0v |
| 32 | RD/LG | Speed Control Set Switch Signal | S/C & Set Switch On: 3.8v |

## 2002-03 Ram Pickup 5.9L VIN Z 'C1' Black Connector

| PCM Pin # | Wire Color | Circuit Description (32-Pin) | Value at Hot Idle |
|---|---|---|---|
| 1 | --- | Not Used | --- |
| 2 | LG/BK | Ignition Switch Power (B+) | 12-14v |
| 3 | --- | Not Used | --- |
| 4 | BK/LB | Sensor Ground | <0.050v |
| 5 | --- | Not Used | --- |
| 6 | BR/YL | PNP Switch Sense Signal | In P/N: 0v, Others: 5v |
| 7 | BK/GY | Coil 1 Driver Control | 5°, 55 mph: 8° dwell |
| 8 | GY/BK | CKP Sensor Signal | Digital Signals: 0-5-0v |
| 9 | --- | Not Used | --- |
| 10 | GY/RD | IAC 3 Driver Control | Pulse Signals |
| 11 | YL/BK | IAC 2 Driver Control | Pulse Signals |
| 12 | --- | Not Used | --- |
| 13 | OR | Power Takeoff Switch Sense | Switch Off: 12v, On: 1v |
| 14 | BR/WT | Manual Transmission Transfer Case | Switch Off: 12v, On: 1v |
| 15 | BR/RD | IAT Sensor Signal | At 100ºF: 1.83v |
| 16 | TN/BK | ECT Sensor Signal | At 180ºF: 2.80v |
| 17 | OR | 5-Volt Supply | 4.9-5.1v |
| 18 | TN/YL | CMP Sensor Signal | Digital Signals: 0-5-0v |
| 19 | VT/BK | IAC 4 Driver Control | Pulse Signals |
| 20 | BR/WT | IAC 1 Driver Control | Pulse Signals |
| 21 | --- | Not Used | --- |
| 22 | RD/WT | Keep Alive Power (B+) | 12-14v |
| 23 | OR/DB | TP Sensor Signal | 0.6-1.0v |
| 24 | BK/DG | HO2S-11 (B1 S1) Signal | 0.1-1.1v |
| 25 | TN/WT | HO2S-12 (B1 S2) Signal | 0.1-1.1v |
| 26 | BK/DG | H/D A/T: HO2S-12 | 0.1-1.1v |
| 26 | LG/RD | HO2S-21 (B2 S1) Signal (California) | 0.1-1.1v |
| 27 | OR/RD | MAP Sensor Signal | 1.5-1.6v |
| 28, 30 | --- | Not Used | --- |
| 29 | TN/WT | HO2S-22 (B2 S2) Signal | 0.1-1.1v |
| 31 | BK/TN | Power Ground | <0.1v |
| 32 | BK/TN | Power Ground | <0.1v |

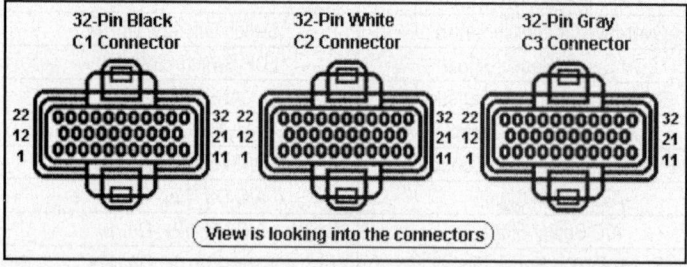

**Pin Connector Graphic**

**2002-03 Ram Pickup 5.9L VIN Z White 'C2' Connector**

| PCM Pin # | Wire Color | Circuit Description (32-Pin) | Value at Hot Idle |
|---|---|---|---|
| 1 | VT | Transmission Temperature Sensor Signal | At 200°F: 2.40v |
| 2 | VT/TN | Injector 7 Driver | 1-4 ms |
| 3, 7, 9, 14 | --- | Not Used | --- |
| 4 | WT/DB | Injector 1 Driver | 1-4 ms |
| 5 | YL/WT | Injector 3 Driver | 1-4 ms |
| 6 | GY | Injector 5 Driver | 1-4 ms |
| 8 | VT/WT | Governor Pressure Solenoid | PWM Signal: 0-12-0v |
| 10 | DG | Generator Field Control | Digital Signals: 0-12-0v |
| 11 | OR/BK | TCC Solenoid Control | At Cruise w/TCC On: 1v |
| 12 | BR/DB | Injector 6 Driver | 1-4 ms |
| 13 | GY/LB | Injector 8 Driver | 1-4 ms |
| 15 | TN | Injector 2 Driver | 1-4 ms |
| 16 | LB/BR | Injector 4 Driver | 1-4 ms |
| 17 | DB/YL | Condenser Fan Relay Control | Relay Off: 12v, On: 1v |
| 19 | DB | A/C Pressure Sensor | A/C On: 0.451-4.850v |
| 22, 24, 26, 32 | --- | Not Used | --- |
| 21 | BR | 3-4 Shift Solenoid Control | Cruise w/solenoid On: 1v |
| 23 | GY/YL | Engine Oil Pressure Sensor | 1.6v at 24 psi |
| 25 | DB/BK | OSS Sensor (-) Signal | Moving: AC pulse signals |
| 27 | WT/OR | Vehicle Speed Signal | Digital Signal |
| 28 | LG/WT | OSS Sensor (+) Signal | Moving: AC pulse signals |
| 29 | LG/RD | Governor Pressure Sensor | 0.58v |
| 30 | PK | Transmission Relay Control | Relay Off: 12v, On: 1v |
| 31 | VT/WT | 5-Volt Supply | 4.9-5.1v |

**2002-03 Ram Pickup 5.9L VIN Z 'C3' Gray Connector**

| PCM Pin # | Wire Color | Circuit Description (32-Pin) | Value at Hot Idle |
|---|---|---|---|
| 1 | DB/OR | A/C Clutch Relay Control | Relay Off: 12v, On: 1v |
| 2, 6-7, 17-18, 21-23 | --- | Not Used | --- |
| 3 | DB/YL | ASD Relay Control | Relay Off: 12v, On: 1v |
| 4 | TN/RD | Speed Control Vacuum Solenoid | Vacuum Increasing: 1v |
| 5 | LG/RD | Speed Control Vent Solenoid | Vacuum Decreasing: 1v |
| 8 | BR/VT | HO2S-11 Heater Control | Heater Off: 12v, On: 1v |
| 9 | DB/OR | HO2S-12 Heater Relay (Downstream) | Relay Off: 12v, On: 1v |
| 10 | WT/DG | LDP Solenoid Pump Switch Sense | LDP Switch Closed: 0v |
| 11 | YL/RD | Speed Control Power Supply | 12-14v |
| 12 | DG/OR | ASD Relay Power (B+) | 12-14v |
| 13 | YL/DG | Overdrive Off Switch Sense | Switch Off: 12v, On: 1v |
| 14 | OR | Leak Detection Pump Solenoid Control | PWM Signal: 0-12-0v |
| 15 | PK/YL | Battery Temperature Sensor | At 86°F: 1.96v |
| 16 | BR/WT | HO2S-21 (B2 S1) Heater Control | Heater Off: 12v, On: 1v |
| 19 | BR | Fuel Pump Relay Control | Relay Off: 12v, On: 1v |
| 20 | PK/BK | EVAP Purge Solenoid Control | PWM Signal: 0-12-0v |
| 24 | WT/PK | Brake Switch Sense Signal | Brake Off: 0v, On: 12v |
| 25 | WT/DB | Generator Field Source | 12-14v |
| 26 | LB/BK | Fuel Level Sensor Signal | Full: 0.56v, 1/2 full: 2.5v |
| 27 | PK | SCI Transmit | 0v |
| 28 | LG | SCI Receive | 5v |
| 30 | VT/BR | PCI Data Bus (J1850) | Digital Signals: 0-7-0v |
| 31 | --- | Not Used | --- |
| 32 | RD/LG | Speed Control Set Switch Signal | S/C & Set Switch On: 3.8v |

## 1993-95 Ram Van 5.9L VIN Z (A/T) 60-Pin Connector

| PCM Pin # | Wire Color | Circuit Description (60-Pin) | Value at Hot Idle |
|---|---|---|---|
| 1 | DG/RD | MAP Sensor Signal | 1.5-1.6v |
| 2 | TN/BK | ECT Sensor Signal | At 180°F: 2.80v |
| 3 | RD | Keep Alive Power (B+) | 12-14v |
| 4 | BK/LB | Sensor Ground | <0.050v |
| 5 ('93-'94) | BK/WT | Sensor Ground | <0.050v |
| 6 | PK/WT | 5-Volt Supply | 4.9-5.1v |
| 7 | OR | 8-Volt Supply | 7.9-8.1v |
| 8 | --- | Not Used | --- |
| 9 | DB | Ignition Switch Power (B+) | 12-14v |
| 10 | OR/WT | Overdrive Override Switch | Switch Off: 12v, On: 1v |
| 11 | BK/TN | Power Ground | <0.1v |
| 12 | BK/TN | Power Ground | <0.1v |
| 13 | LB/BR | Injector 4 Driver | 1-4 ms |
| 14 | YL/WT | Injector 3 Driver | 1-4 ms |
| 15 | TN | Injector 2 Driver | 1-4 ms |
| 16 | WT/DB | Injector 1 Driver | 1-4 ms |
| 17 | DB/TN | Injector 7 Driver | 1-4 ms |
| 18 | DB/GY | Injector 8 Driver | 1-4 ms |
| 19 ('93) | GY | Coil Driver Control | 5°, 55 mph: 8° dwell |
| 19 | BK/GY | Coil Driver Control | 5°, 55 mph: 8° dwell |
| 20 | DG/WT | Alternator Field Control | Digital Signals: 0-12-0v |
| 21 | BK/RD | IAT Sensor Signal | At 100°F: 1.83v |
| 22 | OR/DB | TP Sensor Signal | 0.6-1.0v |
| 23 | --- | Not Used | --- |
| 24 | GY/BK | Distributor Reference Signal | Digital Signals: 0-5-0v |
| 25 | PK | SCI Transmit | 0v |
| 26, 28 | --- | Not Used | --- |
| 27 ('93) | BR | A/C Clutch Signal | A/C Off: 12v, On: 1v |
| 27 | BR | A/C Damped Pressure Switch | A/C Off: 12v, On: 1v |
| 29 | WT/PK | Brake Switch Sense Signal | Brake Off: 0v, On: 12v |
| 30 | BR/O | PNP Switch Sense Signal | In P/N: 0v, Others: 5v |
| 31 | --- | Not Used | --- |
| 32 | BK/PK | MIL (lamp) Control | MIL Off: 12v, On: 1v |
| 33 | TN/RD | Speed Control Vacuum Solenoid | Vacuum Increasing: 1v |
| 34 | DB/OR | A/C WOT Relay Control | Relay Off: 12v, On: 1v |
| 35 | GY/YL | EGR Solenoid Control | 12v, at 55 mph: 1v |
| 36 | --- | Not Used | --- |
| 37 | BK/OR | Overdrive Lamp Control | Lamp Off: 12v, On: 1v |
| 38 | PK/BK | Injector 5 Driver | 1-4 ms |
| 39 | GY/RD | AIS Motor Control | Pulse Signals |
| 40 | BR/WT | AIS Motor Control | Pulse Signals |

**1993-95 Ram Van 5.9L VIN Z (A/T) 60-Pin Connector - Continued**

| PCM Pin # | Wire Color | Circuit Description (60-Pin) | Value at Hot Idle |
|---|---|---|---|
| 41 | BK/DG | HO2S-11 (B1 S1) Signal | 0.1-1.1v |
| 42-43 | --- | Not Used | --- |
| 44 | TN/YL | CMP Sensor Signal | Digital Signals: 0-5-0v |
| 45 | LG | SCI Receive | 5v |
| 46 | --- | Not Used | --- |
| 47 | WT/OR | Vehicle Speed Signal | Digital Signal |
| 48 | BR/RD | Speed Control Set Switch Signal | S/C & Set Switch On: 3.8v |
| 49 | YL/RD | Speed Control On/Off Switch | S/C On: 1v, Off: 12v |
| 50 | WT/LG | Speed Control Resume Switch | S/C On: 1v, Off: 12v |
| 51 | DB/YL | ASD Relay Control | Relay Off: 12v, On: 1v |
| 52 | PK/BK | EVAP Purge Solenoid Control | Solenoid Off: 12v, On: 1v |
| 53 | LG/RD | Speed Control Vent Solenoid | Vacuum Decreasing: 1v |
| 54 | OR/BK | Part Throttle Unlock Solenoid | Solenoid Off: 12v, On: 1v |
| 55 | OR/WT | Overdrive Lockout Solenoid | Solenoid Off: 12v, On: 1v |
| 56 | GY/PK | Maintenance Indicator Lamp | Lamp Off: 12v, On: 1v |
| 57 | DG/OR | ASD Relay Power (B+) | 12-14v |
| 58 | LG/BK | Injector 6 Driver | 1-4 ms |
| 59 | PK/BK | AIS Motor Control | Pulse Signals |
| 60 | YL/BK | AIS Motor Control | Pulse Signals |

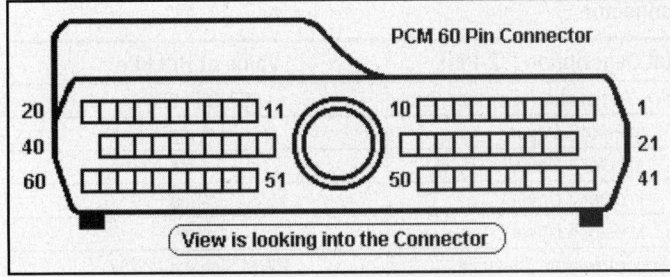

**Pin Connector Graphic**

**Standard Colors and Abbreviations**

| Abbreviation | Color | Abbreviation | Color | Abbreviation | Color |
|---|---|---|---|---|---|
| BK | Black | GY | Gray | RD | Red |
| BL | Blue | GN | Green | TN | Tan |
| BR | Brown | LG | Light Green | VT | Violet |
| DB | Dark Blue | OR | Orange | WT | White |
| DG | Dark Green | PK | Pink | YL | Yellow |

### 1996-97 Ram Van 5.9L VIN Z (A/T) 'A' Black Connector

| PCM Pin # | Wire Color | Circuit Description (32-Pin) | Value at Hot Idle |
|---|---|---|---|
| 2 | LG/BK | Ignition Switch Power (B+) | 12-14v |
| 4 | BK/LB | Sensor Ground | <0.050v |
| 6 | BR/O | PNP Switch Sense Signal | In P/N: 0v, Others: 5v |
| 7 | GY | Coil 1 Driver Control | 5°, 55 mph: 8° dwell |
| 8 | GY/BK | CKP Sensor Signal | Digital Signals: 0-5-0v |
| 10 | YL/BK | IAC 2 Driver Control | Pulse Signals |
| 11 | BR/WT | IAC 1 Driver Control | Pulse Signals |
| 15 | BK/RD | IAT Sensor Signal | At 100°F: 1.83v |
| 16 | TN/BK | ECT Sensor Signal | At 180°F: 2.80v |
| 17 | PK/WT | 5-Volt Supply | 4.9-5.1v |
| 18 | TN/YL | CMP Sensor Signal | Digital Signals: 0-5-0v |
| 19 | GY/RD | IAC 3 Driver Control | Pulse Signals |
| 20 | PK/BK | IAC 4 Driver Control | Pulse Signals |
| 22 | RD/WT | Keep Alive Power (B+) | 12-14v |
| 23 | OR/DB | TP Sensor Signal | 0.6-1.0v |
| 24 | BK/DG | HO2S-11 (B1 S1) Signal | 0.1-1.1v |
| 25 | TN/WT | HO2S-12 (B1 S2) Signal | 0.1-1.1v |
| 27 | DG/RD | MAP Sensor Signal | 1.5-1.6v |
| 31-32 | BK/TN | Power Ground | <0.1v |

### 1996-97 Ram Van 5.9L VIN Z (A/T) 'B' White Connector

| PCM Pin # | Wire Color | Circuit Description (32-Pin) | Value at Hot Idle |
|---|---|---|---|
| 1 | P | Transmission Temperature Sensor Signal | At 200°F: 2.40v |
| 2 | PK/TN | Injector 7 Driver | 1-4 ms |
| 4 | WT/DB | Injector 1 Driver | 1-4 ms |
| 5 | YL/WT | Injector 3 Driver | 1-4 ms |
| 6 | GY | Injector 5 Driver | 1-4 ms |
| 8 | PK/WT | Governor Pressure Solenoid | PWM Signal: 0-12-0v |
| 10 | DG | Generator Field Control | Digital Signals: 0-12-0v |
| 11 | OR/BK | TCC Solenoid Control | At Cruise w/TCC On: 1v |
| 12 | BR/DB | Injector 6 Driver | 1-4 ms |
| 13 | GY/LB | Injector 8 Driver | 1-4 ms |
| 15 | T | Injector 2 Driver | 1-4 ms |
| 16 | LB/BR | Injector 4 Driver | 1-4 ms |
| 21 | BR | 3-4 Shift Solenoid Control | Solenoid Off: 12v, On: 1v |
| 25 | DB/BK | OSS Sensor (-) Signal | Moving: AC pulse signals |
| 27 | WT/OR | Vehicle Speed Signal | Digital Signal |
| 28 | LG/WT | OSS Sensor (+) Signal | Moving: AC pulse signals |
| 29 | LG/RD | Governor Pressure Sensor | 0.58v |
| 30 | PK/LB | Transmission Relay Control | Relay Off: 12v, On: 1v |
| 31 | O | 5-Volt Supply | 4.9-5.1v |
| 32 | GY/YL | EGR Solenoid Control | 12v, at 55 mph: 1v |

**1996-97 Ram Van 5.9L VIN Z (A/T) 'C' Gray Connector**

| PCM Pin # | Wire Color | Circuit Description (32-Pin) | Value at Hot Idle |
|---|---|---|---|
| 1 | DB/OR | A/C Clutch Relay Control | Relay Off: 12v, On: 1v |
| 2 | --- | Not Used | --- |
| 3 | DB/YL | ASD Relay Control | Relay Off: 12v, On: 1v |
| 4 | TN/RD | Speed Control Vacuum Solenoid | Vacuum Increasing: 1v |
| 5 | LG/RD | Speed Control Vent Solenoid | Vacuum Decreasing: 1v |
| 6 | LG/OR | Overdrive Off Lamp Control | Lamp Off: 12v, On: 1v |
| 7-10 | --- | Not Used | --- |
| 11 | YL/RD | Speed Control Power Supply | 12-14v |
| 12 | DG/OR | ASD Relay Power (B+) | 12-14v |
| 13 | OR/WT | Overdrive Off Switch Sense | Switch Off: 12v, On: 1v |
| 14 | --- | Not Used | --- |
| 15 | PK/YL | Battery Temperature Sensor | At 86°F: 1.96v |
| 16 | TN/YL | Generator Lamp Control | Lamp Off: 12v, On: 1v |
| 17 | BK/PK | MIL (lamp) Control | MIL Off: 12v, On: 1v |
| 18 | --- | Not Used | --- |
| 19 | BR | Fuel Pump Relay Control | Relay Off: 12v, On: 1v |
| 20 | PK/BK | EVAP Purge Solenoid Control | PWM Signal: 0-12-0v |
| 21 | --- | Not Used | --- |
| 22 | BR/WT | A/C Request Signal | A/C Off: 12v, On: 1v |
| 23 | LG | A/C Select Signal | A/C Off: 12v, On: 1v |
| 24 | WT/PK | Brake Switch Sense Signal | Brake Off: 0v, On: 12v |
| 25 | --- | Not Used | --- |
| 26 | LB/BK | Fuel Level Sensor Signal | Full: 0.56v, 1/2 full: 2.5v |
| 27 | PK | SCI Transmit | 0v |
| 28 | --- | Not Used | --- |
| 29 | LG/WT | SCI Receive | 5v |
| 30-31 | --- | Not Used | --- |
| 32 | WT/LG | Speed Control Set Switch Signal | S/C & Set Switch On: 3.8v |

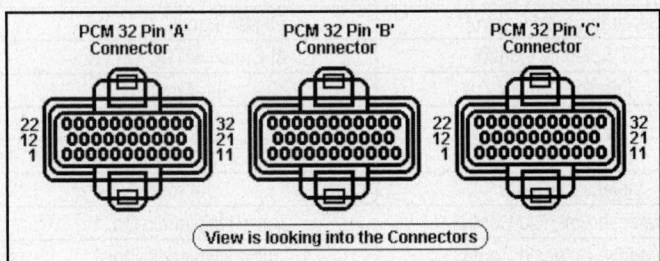

**Pin Connector Graphic**

### 1998-2000 Ram Van 5.9L VIN Z 'C1' Black Connector

| PCM Pin # | Wire Color | Circuit Description (32-Pin) | Value at Hot Idle |
|---|---|---|---|
| 2 | LG/BK | Ignition Switch Power (B+) | 12-14v |
| 4 | BK/LB | Sensor Ground | <0.050v |
| 6 | BK/WT | PNP Switch Sense Signal | In P/N: 0v, Others: 5v |
| 6 ('00) | BR/YL | PNP Switch Sense Signal | In P/N: 0v, Others: 5v |
| 7 | GY | Coil 1 Driver Control | 5°, 55 mph: 8° dwell |
| 8 | GY/BK | CKP Sensor Signal | Digital Signals: 0-5-0v |
| 10 | YL/BK | IAC 2 Driver Control | Pulse Signals |
| 11 | BR/WT | IAC 1 Driver Control | Pulse Signals |
| 15 | BK/RD | IAT Sensor Signal | At 100°F: 1.83v |
| 16 | TN/BK | ECT Sensor Signal | At 180°F: 2.80v |
| 17 | OR | 5-Volt Supply | 4.9-5.1v |
| 18 | TN/YL | CMP Sensor Signal | Digital Signals: 0-5-0v |
| 19 | GY/RD | IAC 3 Driver Control | Pulse Signals |
| 20 | PK/BK | IAC 4 Driver Control | Pulse Signals |
| 22 | RD/WT | Keep Alive Power (B+) | 12-14v |
| 23 | OR/DB | TP Sensor Signal | 0.6-1.0v |
| 24 | TN/WT | HO2S-11 (B1 S1) Signal | 0.1-1.1v |
| 25 | BK/DG | HO2S-12 (B1 S2) Signal | 0.1-1.1v |
| 27 | DG/RD | MAP Sensor Signal | 1.5-1.6v |
| 31-32 | BK/TN | Power Ground | <0.1v |

### 1998-2000 Ram Van 5.9L VIN Z White 'C2' Connector

| PCM Pin # | Wire Color | Circuit Description (32-Pin) | Value at Hot Idle |
|---|---|---|---|
| 1 | GY/BK | Transmission Temperature Sensor Signal | At 200°F: 2.40v |
| 2 | PK | Injector 7 Driver | 1-4 ms |
| 4 | WT/DB | Injector 1 Driver | 1-4 ms |
| 5 | YL/WT | Injector 3 Driver | 1-4 ms |
| 6 | GY | Injector 5 Driver | 1-4 ms |
| 8 | PK/WT | Governor Pressure Solenoid | PWM Signal: 0-12-0v |
| 10 | DG/WT | Generator Field Control | Digital Signals: 0-12-0v |
| 11 | OR/BK | TCC Solenoid Control | At Cruise w/TCC On: 1v |
| 12 | BR/DB | Injector 6 Driver | 1-4 ms |
| 13 | GY/LB | Injector 8 Driver | 1-4 ms |
| 15 | TN | Injector 2 Driver | 1-4 ms |
| 16 | LB/BR | Injector 4 Driver | 1-4 ms |
| 21 | OR/WT | Overdrive Solenoid Control | Cruise w/solenoid On: 1v |
| 21 ('00) | BR | Overdrive Solenoid Control | Cruise w/solenoid On: 1v |
| 23 | GY | Engine Oil Pressure Sensor | 1.6v at 24 psi |
| 25 | DB/BK | OSS Sensor (-) Signal | Moving: AC pulse signals |
| 27 | WT/OR | Vehicle Speed Signal | Digital Signal |
| 28 | LG/WT | OSS Sensor (+) Signal | Moving: AC pulse signals |
| 29 | LG/RD | Governor Pressure Sensor | 0.58v |
| 30 | LB/PK | Transmission Relay Control | Relay Off: 12v, On: 1v |
| 31 | PK/WT | 5-Volt Supply | 4.9-5.1v |

## 1998-2000 Ram Van 5.9L VIN Z 'C3' Gray Connector

| PCM Pin # | Wire Color | Circuit Description (32-Pin) | Value at Hot Idle |
|---|---|---|---|
| 1 | DB/OR | A/C Clutch Relay Control | Relay Off: 12v, On: 1v |
| 2 | --- | Not Used | --- |
| 3 | DB/YL | ASD Relay Control | Relay Off: 12v, On: 1v |
| 4 | TN/RD | Speed Control Vacuum Solenoid | Vacuum Increasing: 1v |
| 5 | LG/RD | Speed Control Vent Solenoid | Vacuum Decreasing: 1v |
| 6 ('98) | LG/OR | Overdrive Off Lamp Control | Lamp Off: 12v, On: 1v |
| 7-9 | --- | Not Used | --- |
| 10 | WT/DG | LDP Solenoid Control | PWM Signal: 0-12-0v |
| 11 | YL/RD | Speed Control Power Supply | 12-14v |
| 12 | DG/OR | ASD Relay Power (B+) | 12-14v |
| 13 | OR/WT | Overdrive Off Switch Sense | Switch Off: 12v, On: 1v |
| 14 | OR | LDP Switch Sense Signal | LDP Switch Closed: 0v |
| 15 | PK/YL | Battery Temperature Sensor | At 86°F: 1.96v |
| 16-18 | --- | Not Used | --- |
| 19 | BR | Fuel Pump Relay Control | Relay Off: 12v, On: 1v |
| 20 | PK/W | EVAP Purge Solenoid Control | PWM Signal: 0-12-0v |
| 20 ('00) | PK/BK | EVAP Purge Solenoid Control | PWM Signal: 0-12-0v |
| 21 | --- | Not Used | --- |
| 22 | DB | A/C Pressure Switch Signal | A/C On: 0.451-4.850v |
| 23 | LG/GY | A/C Select Signal | A/C Off: 12v, On: 1v |
| 24 | WT/PK | Brake Switch Sense Signal | Brake Off: 0v, On: 12v |
| 25 | OR/DB | Generator Field Source | 12-14v |
| 25 ('00) | WT/DB | Generator Field Source | 12-14v |
| 26 | LB/BK | Fuel Level Sensor Signal | Full: 0.56v, 1/2 full: 2.5v |
| 27 | PK | SCI Transmit | 0v |
| 28 | WT/BK | CCD Bus (-) | <0.050v |
| 29 | LG/WT | SCI Receive | 5v |
| 30 | PK/BR | CCD Bus (+) | Digital Signals: 0-5-0v |
| 31 | --- | Not Used | --- |
| 32 | WT/LG | Speed Control Set Switch Signal | S/C & Set Switch On: 3.8v |

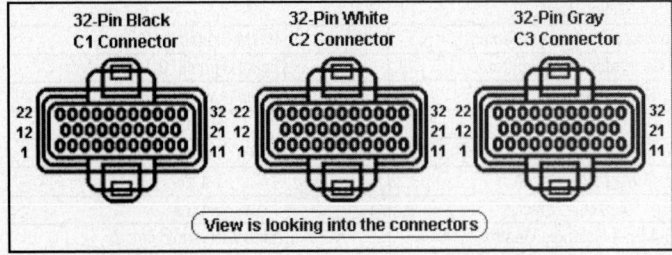

**Pin Connector Graphic**

### Standard Colors and Abbreviations

| Abbreviation | Color | Abbreviation | Color | Abbreviation | Color |
|---|---|---|---|---|---|
| BK | Black | GY | Gray | RD | Red |
| BL | Blue | GN | Green | TN | Tan |
| BR | Brown | LG | Light Green | VT | Violet |
| DB | Dark Blue | OR | Orange | WT | White |
| DG | Dark Green | PK | Pink | YL | Yellow |

### 2001-03 Ram Van 5.9L VIN Z (A/T) 'C1' Black Connector

| PCM Pin # | Wire Color | Circuit Description (32-Pin) | Value at Hot Idle |
|---|---|---|---|
| 1 | --- | Not Used | --- |
| 2 | LG/BK | Ignition Switch Power (B+) | 12-14v |
| 3 | --- | Not Used | --- |
| 4 | BK/LB | Sensor Ground | <0.050v |
| 5 | --- | Not Used | --- |
| 6 | BR/YL | PNP Switch Sense Signal | In P/N: 0v, Others: 5v |
| 7 | GY | Coil 1 Driver Control | 5°, 55 mph: 8° dwell |
| 8 | GY/BK | CKP Sensor Signal | Digital Signals: 0-5-0v |
| 9 | --- | Not Used | --- |
| 10 | YL/BK | IAC 2 Driver Control | Pulse Signals |
| 11 | BR/WT | IAC 1 Driver Control | Pulse Signals |
| 12-14 | --- | Not Used | --- |
| 15 | BK/RD | IAT Sensor Signal | At 100°F: 1.83v |
| 16 | TN/BK | ECT Sensor Signal | At 180°F: 2.80v |
| 17 | OR | 5-Volt Supply | 4.9-5.1v |
| 18 | TN/YL | CMP Sensor Signal | Digital Signals: 0-5-0v |
| 19 | GY/RD | IAC 3 Driver Control | Pulse Signals |
| 20 | VT/BK | IAC 4 Driver Control | Pulse Signals |
| 21 | --- | Not Used | --- |
| 22 | DG/BK | Keep Alive Power (B+) | 12-14v |
| 23 | OR/DB | TP Sensor Signal | 0.6-1.0v |
| 24 | DB/DG | HO2S-11 (B1 S1) Signal | 0.1-1.1v |
| 25 | TN/WT | HO2S-12 (B1 S2) Signal | 0.1-1.1v |
| 26 | --- | Not Used | --- |
| 27 | DG/RD | MAP Sensor Signal | 1.5-1.6v |
| 28-32 | --- | Not Used | --- |

### 2001-03 Ram Van 5.9L VIN Z (A/T) White 'C2' Connector

| PCM Pin # | Wire Color | Circuit Description (32-Pin) | Value at Hot Idle |
|---|---|---|---|
| 1 | GY/BK | Transmission Temperature Sensor Signal | At 200°F: 2.40v |
| 2 | VT | Injector 7 Driver | 1-4 ms |
| 3, 9, 14, 17-20 | --- | Not Used | --- |
| 4 | WT/DB | Injector 1 Driver | 1-4 ms |
| 5 | YL/WT | Injector 3 Driver | 1-4 ms |
| 6 | GY | Injector 5 Driver | 1-4 ms |
| 7 | --- | Not Used | --- |
| 8 | VT/BK | Governor Pressure Solenoid | PWM Signal: 0-12-0v |
| 10 | DG/WT | Generator Field Control | Digital Signals: 0-12-0v |
| 11 | OR/BK | TCC Solenoid Control | At Cruise w/TCC On: 1v |
| 12 | BR/DB | Injector 6 Driver | 1-4 ms |
| 13 | GY/LB | Injector 8 Driver | 1-4 ms |
| 15 | TN | Injector 2 Driver | 1-4 ms |
| 16 | LB/BR | Injector 4 Driver | 1-4 ms |
| 21 | BR | 3-4 Shift Solenoid Control | Cruise w/solenoid On: 1v |
| 22 | --- | Not Used | --- |
| 23 | GY/YL | Engine Oil Pressure Sensor | 1.6v at 24 psi |
| 24 | --- | Not Used | --- |
| 25 | DB | OSS Sensor (-) Signal | Moving: AC pulse signals |
| 26 | --- | Not Used | --- |
| 27 | WT/OR | Vehicle Speed Signal | Digital Signal |
| 28 | LG/RD | OSS Sensor (+) Signal | Moving: AC pulse signals |
| 29 | LG/RD | Governor Pressure Sensor | 0.58v |
| 30 | PK | Transmission Relay Control | Relay Off: 12v, On: 1v |
| 31 | VT/WT | 5-Volt Supply | 4.9-5.1v |
| 32 | --- | Not Used | --- |

**2001-03 Ram Van 5.9L VIN Z (A/T) 'C3' Gray Connector**

| PCM Pin # | Wire Color | Circuit Description (32-Pin) | Value at Hot Idle |
|---|---|---|---|
| 1 | DB/OR | A/C Clutch Relay Control | Relay Off: 12v, On: 1v |
| 2 | --- | Not Used | --- |
| 3 | DB/YL | ASD Relay Control | Relay Off: 12v, On: 1v |
| 4 | TN/RD | Speed Control Vacuum Solenoid | Vacuum Increasing: 1v |
| 5 | LG/RD | Speed Control Vent Solenoid | Vacuum Decreasing: 1v |
| 6-7 | --- | Not Used | --- |
| 8 | OR/RD | HO2S-11 (B1 S1) Heater | PWM Signal: 0-12-0v |
| 9 | --- | Not Used | --- |
| 10 | WT/DG | LDP Solenoid Control | PWM Signal: 0-12-0v |
| 11 | YL/RD | Speed Control Power Supply | 12-14v |
| 12 | DG/OR | ASD Relay Power (B+) | 12-14v |
| 13 | OR/WT | Overdrive Off Switch Sense | Switch Off: 12v, On: 1v |
| 14 | OR | LDP Switch Sense Signal | LDP Switch Closed: 0v |
| 15 | PK/YL | Battery Temperature Sensor | At 86°F: 1.96v |
| 16 | VT/TN | HO2S-21 (B2 S1) Heater | PWM Signal: 0-12-0v |
| 17-18 | --- | Not Used | --- |
| 19 | BR | Fuel Pump Relay Control | Relay Off: 12v, On: 1v |
| 20 | PK/BK | EVAP Purge Solenoid Control | PWM Signal: 0-12-0v |
| 21 | --- | Not Used | --- |
| 22 | DB | A/C Pressure Switch Signal | A/C On: 0.451-4.850v |
| 23 | LG/GY | A/C Select Signal | A/C Off: 12v, On: 1v |
| 24 | WT/PK | Brake Switch Sense Signal | Brake Off: 0v, On: 12v |
| 25 | WT/DB | Generator Field Source | 12-14v |
| 26 | LB/BK | Fuel Level Sensor Signal | Full: 0.56v, 1/2 full: 2.5v |
| 27 | PK | SCI Transmit | 0v |
| 28 | WT/BK | CCD Bus (-) | <0.050v |
| 29 | LG | SCI Receive | 5v |
| 30 | VT/BR | CCD Bus (+) | Digital Signals: 0-5-0v |
| 31 | --- | Not Used | --- |
| 32 | WT/LG | Speed Control Set Switch Signal | S/C & Set Switch On: 3.8v |

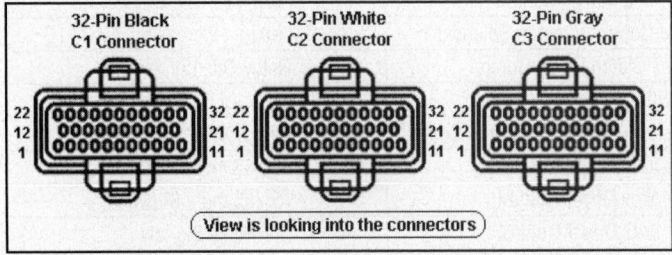

**Pin Connector Graphic**

**Standard Colors and Abbreviations**

| Abbreviation | Color | Abbreviation | Color | Abbreviation | Color |
|---|---|---|---|---|---|
| BK | Black | GY | Gray | RD | Red |
| BL | Blue | GN | Green | TN | Tan |
| BR | Brown | LG | Light Green | VT | Violet |
| DB | Dark Blue | OR | Orange | WT | White |
| DG | Dark Green | PK | Pink | YL | Yellow |

## 1994-95 Ram Pickup & Van 8.0L VIN W 60-Pin Connector

| PCM Pin # | Wire Color | Circuit Description (60-Pin) | Value at Hot Idle |
|-----------|-----------|------------------------------|-------------------|
| 1 | DG/RD | MAP Sensor Signal | 1.5-1.6v |
| 2 | TN/BK | ECT Sensor Signal | At 180°F: 2.80v |
| 3 | RD/WT | Keep Alive Power (B+) | 12-14v |
| 4 | BK/LB | Sensor Ground | <0.050v |
| 5 | --- | Not Used | --- |
| 6 | PK/WT | 5-Volt Supply | 4.9-5.1v |
| 7 | OR | 8-Volt Supply | 7.9-8.1v |
| 8 | --- | Not Used | --- |
| 9 | LG/BK | Ignition Switch Power (B+) | 12-14v |
| 10 | OR/WT | Overdrive Override Switch | Switch Off: 12v, On: 1v |
| 11 | BK/TN | Power Ground | <0.1v |
| 12 | BK/TN | Power Ground | <0.1v |
| 13 | LB/BR | Injector 5 & 8 Driver | 1-4 ms |
| 14 | YL/WT | Injector 3 & 6 Driver | 1-4 ms |
| 15 | TN | Injector 4 & 9 Driver | 1-4 ms |
| 16 | WT/DB | Injector 1 & 10 Driver | 1-4 ms |
| 17 | DB/TN | Coil Driver Control | 5°, 55 mph: 8° dwell |
| 18 | RD/YL | Coil Driver Control | 5°, 55 mph: 8° dwell |
| 19 | BK/GY | Coil Driver Control | 5°, 55 mph: 8° dwell |
| 20 | DG | Alternator Field Control | Digital Signals: 0-12-0v |
| 21 | BK/RD | IAT Sensor Signal | At 100°F: 1.83v |
| 22 | OR/DB | TP Sensor Signal | 0.6-1.0v |
| 23 | TN/WT | HO2S-11 (B1 S1) Signal | 0.1-1.1v |
| 24 | GY/BK | Distributor Reference Signal | Digital Signals: 0-5-0v |
| 25 | PK | SCI Transmit | 0v |
| 26 | --- | Not Used | --- |
| 27 | BR | A/C Damped Pressure Switch | A/C Off: 12v, On: 1v |
| 28 | --- | Not Used | --- |
| 29 | WT/PK | Brake Switch Sense Signal | Brake Off: 0v, On: 12v |
| 30 | BK/WT | Starter Relay Control | Relay Off: 12v, On: 1v |
| 31 | LG/OR | Overdrive Lamp Control | Lamp Off: 12v, On: 1v |
| 32 | BK/PK | MIL (lamp) Control | MIL Off: 12v, On: 1v |
| 33 | TN/RD | Speed Control Vacuum Solenoid | Vacuum Increasing: 1v |
| 34 | DB/OR | A/C Clutch Relay Control | Relay Off: 12v, On: 1v |
| 35 | GY/YL | EGR Solenoid Control | 12v, at 55 mph: 1v |
| 36 | PK/BK | Transmission Temperature Lamp Control | Lamp Off: 12v, On: 1v |
| 37 | WT/BK | Coil Driver Control | 5°, 55 mph: 8° dwell |
| 38 | BR/OR | Coil Driver Control | 5°, 55 mph: 8° dwell |
| 39 | GY/RD | IAC Driver Control | Pulse Signals |
| 40 | BR/WT | IAC Driver Control | Pulse Signals |

## Standard Colors and Abbreviations

| Abbreviation | Color | Abbreviation | Color | Abbreviation | Color |
|--------------|-------|--------------|-------|--------------|-------|
| BK | Black | GY | Gray | RD | Red |
| BL | Blue | GN | Green | TN | Tan |
| BR | Brown | LG | Light Green | VT | Violet |
| DB | Dark Blue | OR | Orange | WT | White |
| DG | Dark Green | PK | Pink | YL | Yellow |

**1994-95 Ram Pickup & Van 8.0L VIN W 60-Pin Connector - Continued**

| PCM Pin # | Wire Color | Circuit Description (60-Pin) | Value at Hot Idle |
|---|---|---|---|
| 41 | BK/DG | HO2S-12 (B1 S2) Signal | 0.1-1.1v |
| 42 | VT | Transmission Temperature Sensor Signal | At 200ºF: 2.40v |
| 43 | GY/LB | Tachometer Signal | Pulse Signals |
| 44 | TN/YL | CMP Sensor Signal | Digital Signals: 0-5-0v |
| 45 | LG | SCI Receive | 5v |
| 47 | WT/OR | Vehicle Speed Signal | Digital Signal |
| 48 | BR/RD | Speed Control Set Switch Signal | S/C & Set Switch On: 3.8v |
| 49 | YL/RD | Speed Control On/Off Switch | S/C On: 1v, Off: 12v |
| 50 | WT/LG | Speed Control Resume Switch | S/C On: 1v, Off: 12v |
| 51 | DB/YL | ASD Relay Control | Relay Off: 12v, On: 1v |
| 52 | PK/BK | EVAP Purge Solenoid Control | Solenoid Off: 12v, On: 1v |
| 53 | LG/RD | Speed Control Vent Solenoid | Vacuum Decreasing: 1v |
| 54 | OR/BK | A/T: Overdrive Solenoid | Solenoid Off: 12v, On: 1v |
| 54 | OR/BK | M/T: SIL (lamp) Control | Lamp Off: 12v, On: 1v |
| 55 | BR | Overdrive Lockout Solenoid | Solenoid Off: 12v, On: 1v |
| 56 | GY/PK | Maintenance Indicator Lamp | Lamp Off: 12v, On: 1v |
| 57 | DG/OR | ASD Relay Power (B+) | 12-14v |
| 58 | LG/BK | Injector 2 & 7 Driver | 1-4 ms |
| 59 | PK/BK | IAC Driver Control | Pulse Signals |
| 60 | YL/BK | IAC Driver Control | Pulse Signals |

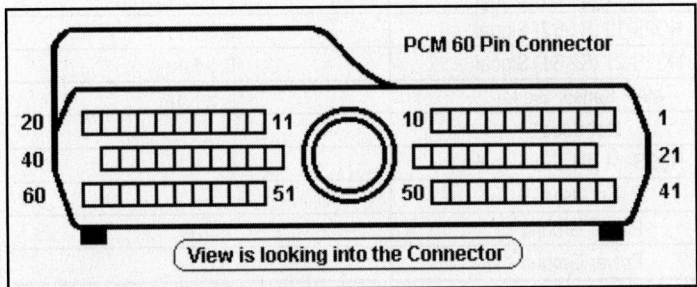

**Pin Connector Graphic**

**Standard Colors and Abbreviations**

| Abbreviation | Color | Abbreviation | Color | Abbreviation | Color |
|---|---|---|---|---|---|
| BK | Black | GY | Gray | RD | Red |
| BL | Blue | GN | Green | TN | Tan |
| BR | Brown | LG | Light Green | VT | Violet |
| DB | Dark Blue | OR | Orange | WT | White |
| DG | Dark Green | PK | Pink | YL | Yellow |

## 1996-97 Ram Pickup & Van 8.0L VIN W 'A' Black Connector

| PCM Pin # | Wire Color | Circuit Description (32-Pin) | Value at Hot Idle |
|---|---|---|---|
| 1 | YL/GY | Coil 4 Driver Control | 5°, 55 mph: 8° dwell |
| 2 | LG/BK | Ignition Switch Power (B+) | 12-14v |
| 3 | RD/BK | Coil 3 Driver Control | 5°, 55 mph: 8° dwell |
| 4 | BK/LB | Sensor Ground | <0.050v |
| 5 | DG/GY | Coil 5 Driver Control | 5°, 55 mph: 8° dwell |
| 6 | BK/WT | PNP Switch Sense Signal | In P/N: 0v, Others: 5v |
| 7 | BK/GY | Coil 1 Driver Control | 5°, 55 mph: 8° dwell |
| 8 | GY/BK | CKP Sensor Signal | Digital Signals: 0-5-0v |
| 9 | DB/WT | Coil 2 Driver Control | 5°, 55 mph: 8° dwell |
| 10 | YL/BK | IAC 2 Driver Control | Pulse Signals |
| 11 | BR/WT | IAC 3 Driver Control | Pulse Signals |
| 12-14 | --- | Not Used | --- |
| 15 | BK/RD | IAT Sensor Signal | At 100°F: 1.83v |
| 16 | TN/BK | ECT Sensor Signal | At 180°F: 2.80v |
| 17 | PK/WT | 5-Volt Supply | 4.9-5.1v |
| 18 | TN/YL | CMP Sensor Signal | Digital Signals: 0-5-0v |
| 19 | GY/RD | IAC 1 Driver Control | Pulse Signals |
| 20 | PK/BK | IAC 4 Driver Control | Pulse Signals |
| 21 | --- | Not Used | --- |
| 22 | RD/WT | Keep Alive Power (B+) | 12-14v |
| 23 | OR/DB | TP Sensor Signal | 0.6-1.0v |
| 24 | TN/WT | HO2S-11 (B1 S1) Signal | 0.1-1.1v |
| 25 | OR/BK | HO2S-12 (B1 S2) Signal | 0.1-1.1v |
| 26 | BK/DG | HO2S-21 (B2 S1) Signal | 0.1-1.1v |
| 27 | DG/RD | MAP Sensor Signal | 1.5-1.6v |
| 28 | --- | Not Used | --- |
| 29 | TN/RD | HO2S-13 (B1 S3) Signal | 0.1-1.1v |
| 30 | --- | Not Used | --- |
| 31 | BK/TN | Power Ground | <0.1v |
| 32 | BK/TN | Power Ground | <0.1v |

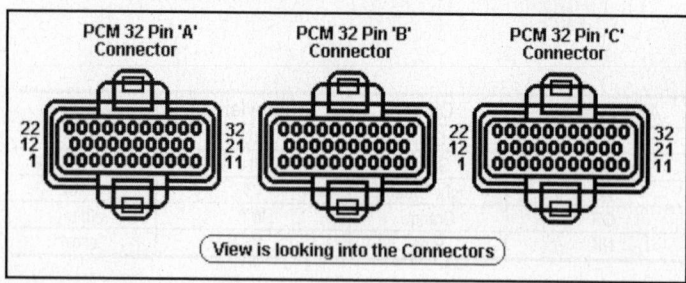

**Pin Connector Graphic**

## Standard Colors and Abbreviations

| Abbreviation | Color | Abbreviation | Color | Abbreviation | Color |
|---|---|---|---|---|---|
| BK | Black | GY | Gray | RD | Red |
| BL | Blue | GN | Green | TN | Tan |
| BR | Brown | LG | Light Green | VT | Violet |
| DB | Dark Blue | OR | Orange | WT | White |
| DG | Dark Green | PK | Pink | YL | Yellow |

## 1996-97 Ram Pickup & Van 8.0L VIN W 'B' White Connector

| PCM Pin # | Wire Color | Circuit Description (32-Pin) | Value at Hot Idle |
|---|---|---|---|
| 1 | VT | Transmission Temperature Sensor Signal | At 200°F: 2.40v |
| 2 | PK/TN | Injector 7 Driver | 1-4 ms |
| 3 | RD/BK | Injector 9 Driver | 1-4 ms |
| 4 | WT/DB | Injector 1 Driver | 1-4 ms |
| 5 | YL/WT | Injector 3 Driver | 1-4 ms |
| 6 | GY | Injector 5 Driver | 1-4 ms |
| 8 | PK/WT | Governor Pressure Solenoid | PWM Signal: 0-12-0v |
| 10 | DG | Generator Field Control | Digital Signals: 0-12-0v |
| 11 | OR/BK | TCC Solenoid Control | At Cruise w/TCC On: 1v |
| 12 | BR/DB | Injector 6 Driver | 1-4 ms |
| 13 | GY/LB | Injector 8 Driver | 1-4 ms |
| 14 | WT/DB | Injector 10 Driver | 1-4 ms |
| 15 | TN | Injector 2 Driver | 1-4 ms |
| 16 | LB/BR | Injector 4 Driver | 1-4 ms |
| 21 | BR | 3-4 Shift Solenoid Control | Solenoid Off: 12v, On: 1v |
| 25 | DB/BK | OSS Sensor (-) Signal | Moving: AC pulse signals |
| 27 | WT/OR | Vehicle Speed Signal | Digital Signal |
| 28 | LG/BK | OSS Sensor (+) Signal | Moving: AC pulse signals |
| 29 | LG/WT | Governor Pressure Sensor | 0.58v |
| 30 | PK | Transmission Relay Control | Relay Off: 12v, On: 1v |
| 31 | OR | 5-Volt Supply | 4.9-5.1v |
| 32 | GY/YL | EGR Solenoid Control | 12v, at 55 mph: 1v |

## 1996-97 Ram Pickup & Van 8.0L VIN W 'C' Gray Connector

| PCM Pin # | Wire Color | Circuit Description (32-Pin) | Value at Hot Idle |
|---|---|---|---|
| 1 | DB/OR | A/C Clutch Relay Control | Relay Off: 12v, On: 1v |
| 3 | DB/YL | ASD Relay Control | Relay Off: 12v, On: 1v |
| 4 | TN/RD | Speed Control Vacuum Solenoid | Vacuum Increasing: 1v |
| 5 | LG/RD | Speed Control Vent Solenoid | Vacuum Decreasing: 1v |
| 6 | LG/OR | Overdrive Off Lamp Control | Lamp Off: 12v, On: 1v |
| 7 | PK/BK | Transmission Temperature Lamp Control | Lamp Off: 12v, On: 1v |
| 11 | YL/RD | Speed Control Power Supply | 12-14v |
| 12 | DG/OR | ASD Relay Power (B+) | 12-14v |
| 13 | OR/WT | Overdrive Off Switch Sense | Switch Off: 12v, On: 1v |
| 15 | PK/YL | Battery Temperature Sensor | At 86°F: 1.96v |
| 16 | TN/YL | Generator Lamp Control | Lamp Off: 12v, On: 1v |
| 17 | BK/PK | MIL (lamp) Control | MIL Off: 12v, On: 1v |
| 18 | GY/PK | Maintenance Indicator Lamp | Lamp Off: 12v, On: 1v |
| 19 | BR/WT | Fuel Pump Relay Control | Relay Off: 12v, On: 1v |
| 20 | PK | EVAP Purge Solenoid Control | PWM Signal: 0-12-0v |
| 22 | BR | A/C Request Signal | A/C Off: 12v, On: 1v |
| 23 | LG | A/C Select Signal | A/C Off: 12v, On: 1v |
| 24 | WT/PK | Brake Switch Sense Signal | Brake Off: 0v, On: 12v |
| 26 | DB/WT | Fuel Level Sensor Signal | Full: 0.56v, 1/2 full: 2.5v |
| 27 | PK/DB | SCI Transmit Signal | 0v |
| 29 | DG | SCI Transmit Signal | 5v |
| 31 | GY/LB | Tachometer Signal | Pulse Signals |
| 32 | RD/LG | Speed Control Set Switch Signal | S/C & Set Switch On: 3.8v |

## 1998-2002 Ram Pickup & Van 8.0L VIN W 'C1' Black Connector

| PCM Pin # | Wire Color | Circuit Description (32-Pin) | Value at Hot Idle |
|---|---|---|---|
| 1 | YL/GY | Coil 4 Driver Control | 5°, 55 mph: 8° dwell |
| 2 | LG/BK | Ignition Switch Output (Run-Start) | 12-14v |
| 3 | RD/BK | Coil 3 Driver Control | 5°, 55 mph: 8° dwell |
| 4 | BK/LB | Sensor Ground | <0.1v |
| 5 | DG/GY | Coil 5 Driver Control | 5°, 55 mph: 8° dwell |
| 6 | BK/WT | PNP Switch Sense Signal | In P/N: 0v, Others: 5v |
| 7 | BK/GY | Coil 1 Driver Control | 5°, 55 mph: 8° dwell |
| 8 | GY/BK | CKP Sensor Signal | Digital Signals: 0-5-0v |
| 9 | DB/WT | Coil 2 Driver Control | 5°, 55 mph: 8° dwell |
| 10 | YL/BK | IAC 2 Driver Control | Pulse Signals |
| 11 | BR/WT | IAC 3 Driver Control | Pulse Signals |
| 12 | --- | Not Used | --- |
| 13 | OR | Power Takeoff Switch Sense | Switch Off: 12v, On: 1v |
| 14 | --- | Not Used | --- |
| 15 | BK/RD | IAT Sensor Signal | At 100°F: 1.83v |
| 16 | TN/BK | ECT Sensor Signal | At 180°F: 2.80v |
| 17 | VT/WT | 5-Volt Supply | 4.9-5.1v |
| 18 | TN/YL | CMP Sensor Signal | Digital Signals: 0-5-0v |
| 19 | GY/RD | IAC 1 Driver Control | Pulse Signals |
| 20 | VT/BK | IAC 4 Driver Control | Pulse Signals |
| 21 | --- | Not Used | --- |
| 22 | RD/WT | Keep Alive Power (B+) | 12-14v |
| 23 | OR/DB | TP Sensor Signal | 0.6-1.0v |
| 24 | BK/DG | HO2S-11 (B1 S1) Signal | 0.1-1.1v |
| 25 | OR/BK | HO2S-13 (B1 S3) Signal | 0.1-1.1v |
| 26 | LG/RD | HO2S-12 (B1 S2) Signal | 0.1-1.1v |
| 27 | DG/RD | MAP Sensor Signal | 1.5-1.6v |
| 28 | --- | Not Used | --- |
| 29 | TN/RD | HO2S-21 (B2 S1) Signal | 0.1-1.1v |
| 29 ('00-'01) | TN/WT | HO2S-21 (B2 S1) Signal | 0.1-1.1v |
| 30 | --- | Not Used | --- |
| 31 | BK/TN | Power Ground | <0.1v |
| 32 | BK/TN | Power Ground | <0.1v |

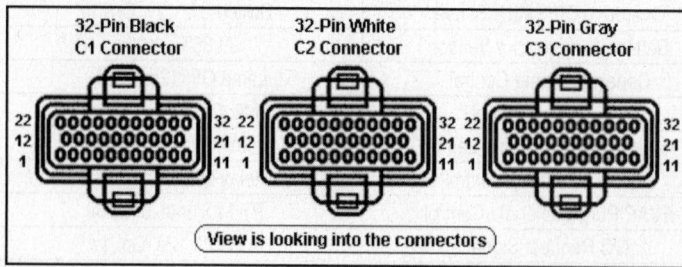

**Pin Connector Graphic**

**1998-2002 Ram Pickup & Van 8.0L VIN W 'C2' White Connector**

| PCM Pin # | Wire Color | Circuit Description (32-Pin) | Value at Hot Idle |
|---|---|---|---|
| 1 | VT | Transmission Temperature Sensor Signal | At 200°F: 2.40v |
| 2 | VT/TN | Injector 7 Driver | 1-4 ms |
| 3 | TN/BK | Injector 9 Driver | 1-4 ms |
| 4 | WT/DB | Injector 1 Driver | 1-4 ms |
| 5 | YL/WT | Injector 3 Driver | 1-4 ms |
| 6 | GY | Injector 5 Driver | 1-4 ms |
| 7 | --- | Not Used | --- |
| 8 | VT/WT | Governor Pressure Solenoid | PWM Signal: 0-12-0v |
| 9 | --- | Not Used | --- |
| 10 | DG | Generator Field Control | Digital Signals: 0-12-0v |
| 11 | OR/BK | TCC Solenoid Control | At Cruise w/TCC On: 1v |
| 12 | BR/DB | Injector 6 Driver | 1-4 ms |
| 13 | GY/LB | Injector 8 Driver | 1-4 ms |
| 14 | WT | Injector 10 Driver | 1-4 ms |
| 15 | TN | Injector 2 Driver | 1-4 ms |
| 16 | LB/BR | Injector 4 Driver | 1-4 ms |
| 17-20 | --- | Not Used | --- |
| 21 | BR | 3-4 Shift Solenoid Control | Solenoid Off: 12v, On: 1v |
| 22 | --- | Not Used | --- |
| 23 | GY/OR | Engine Oil Pressure Sensor | 1.6v at 24 psi |
| 24 | --- | Not Used | --- |
| 25 | DB/BK | OSS Sensor (-) Signal | Moving: AC pulse signals |
| 26 | --- | Not Used | --- |
| 27 | WT/OR | Vehicle Speed Signal | Digital Signal |
| 28 | LG/BK | OSS Sensor (+) Signal | Moving: AC pulse signals |
| 29 | LG/WT | Governor Pressure Sensor | 0.58v |
| 30 | PK | Transmission Relay Control | Relay Off: 12v, On: 1v |
| 31 | OR | 5-Volt Supply | 4.9-5.1v |
| 32 | --- | Not Used | --- |

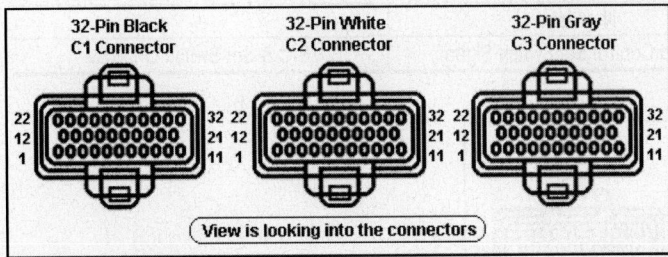

**Pin Connector Graphic**

## 1998-2002 Ram Pickup & Van 8.0L VIN W 'C3' Gray Connector

| PCM Pin # | Wire Color | Circuit Description (32-Pin) | Value at Hot Idle |
|---|---|---|---|
| 1 | DB/OR | A/C Clutch Relay Control | Relay Off: 12v, On: 1v |
| 2 | --- | Not Used | --- |
| 3 | DB/YL | ASD Relay Control | Relay Off: 12v, On: 1v |
| 4 | TN/RD | Speed Control Vacuum Solenoid | Vacuum Increasing: 1v |
| 5 | LG/RD | Speed Control Vent Solenoid | Vacuum Decreasing: 1v |
| 6 ('98) | LG/OR | Overdrive Off Lamp Control | Lamp Off: 12v, On: 1v |
| 7 | --- | Not Used | --- |
| 8 | BR/VT | HO2S-11 Heater Control (Cal) | Heater Off: 12v, On: 1v |
| 9 | DG/PK | HO2S-12 Heater Relay (Cal) | Relay Off: 12v, On: 1v |
| 10 | WT/OR | LDP Solenoid Control | PWM Signal: 0-12-0v |
| 10 ('00-'02) | WT/DG | LDP Solenoid Control | PWM Signal: 0-12-0v |
| 11 | YL/RD | Speed Control Power Supply | 12-14v |
| 12 | DG/OR | ASD Relay Power (B+) | 12-14v |
| 13 | OR/WT | Overdrive Off Switch Sense | Switch Off: 12v, On: 1v |
| 14 | OR | LDP Switch Sense Signal | LDP Switch Closed: 0v |
| 15 | PK/YL | Battery Temperature Sensor | At 86ºF: 1.96v |
| 16 | BR/WT | HO2S-21 Heater Control | Heater Off: 12v, On: 1v |
| 17-18 | --- | Not Used | --- |
| 19 | BR/WT | Fuel Pump Relay Control | Relay Off: 12v, On: 1v |
| 20 | PK/WT | EVAP Purge Solenoid Control | PWM Signal: 0-12-0v |
| 21 | --- | Not Used | --- |
| 22 | BR | A/C Switch Signal | A/C Off: 12v, On: 1v |
| 23 | LG/WT | A/C Select Signal | A/C Off: 12v, On: 1v |
| 24 | WT/PK | Brake Switch Sense Signal | Brake Off: 0v, On: 12v |
| 25 | DB | Generator Field Source | 12-14v |
| 26 | DB/WT | Fuel Level Sensor Signal | Full: 0.56v, 1/2 full: 2.5v |
| 27 | PK/DB | SCI Transmit Signal | 0v |
| 28 | WT/BK | CCD Bus (-) Signal | <0.050v |
| 29 | DG | SCI Receive Signal | 5v |
| 30 | VT/BR | CCD Bus (+) Signal | Digital Signals: 0-5-0v |
| 31 | --- | Not Used | --- |
| 32 | RD/LG | Speed Control Set Switch Signal | S/C & Set Switch On: 3.8v |

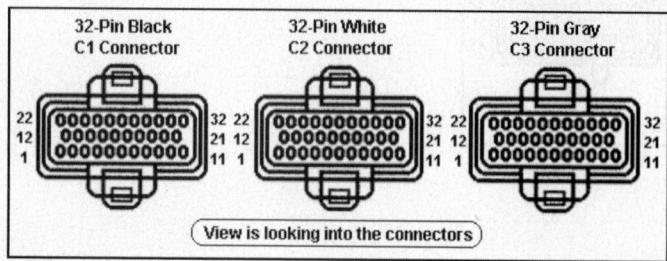

**Pin Connector Graphic**

**2003 Pickup & Van 8.0L VIN W 'C1' Black Connector**

| PCM Pin # | Wire Color | Circuit Description (32-Pin) | Value at Hot Idle |
|---|---|---|---|
| 1 | YL/GY | Coil 4 Driver Control | 5°, 55 mph: 8° dwell |
| 2 | LG/BK | Ignition Switch Output (Run-Start) | 12-14v |
| 3 | RD/BK | Coil 3 Driver Control | 5°, 55 mph: 8° dwell |
| 4 | BK/LB | Sensor Ground | <0.1v |
| 5 | DG/GY | Coil 5 Driver Control | 5°, 55 mph: 8° dwell |
| 6 | BR/YL | PNP Switch Sense Signal | In P/N: 0v, Others: 5v |
| 7 | BK/GY | Coil 1 Driver Control | 5°, 55 mph: 8° dwell |
| 8 | GY/BK | CKP Sensor Signal | Digital Signals: 0-5-0v |
| 9 | DB/TN | Coil 2 Driver Control | 5°, 55 mph: 8° dwell |
| 10 | GY/RD | IAC 3 Driver Control | Pulse Signals |
| 11 | YL/BK | IAC 2 Driver Control | Pulse Signals |
| 12-14 | --- | Not Used | --- |
| 15 | BK/RD | IAT Sensor Signal | At 100°F: 1.83v |
| 16 | TN/BK | ECT Sensor Signal | At 180°F: 2.80v |
| 17 | OR | 5-Volt Supply | 4.9-5.1v |
| 18 | TN/YL | CMP Sensor Signal | Digital Signals: 0-5-0v |
| 19 | VT/BK | IAC 4 Driver Control | Pulse Signals |
| 20 | BR/WT | IAC 1 Driver Control | Pulse Signals |
| 21 | --- | Not Used | --- |
| 22 | RD/WT | Keep Alive Power (B+) | 12-14v |
| 23 | OR/DB | TP Sensor Signal | 0.6-1.0v |
| 24 | BK/DG | HO2S-11 (B1 S1) Signal | 0.1-1.1v |
| 25 | TN/WT | HO2S-12 (B1 S2) Signal | 0.1-1.1v |
| 26 | LG/RD | HO2S-21 (B2 S1) Signal (California) | 0.1-1.1v |
| 27 | OR/RD | MAP Sensor Signal | 1.5-1.6v |
| 28 | --- | Not Used | --- |
| 29 | TN/WT | HO2S-22 (B2 S2) Signal (California) | 0.1-1.1v |
| 30 | --- | Not Used | --- |
| 31 | BK/TN | Power Ground | <0.1v |
| 32 | BK/TN | Power Ground | <0.1v |

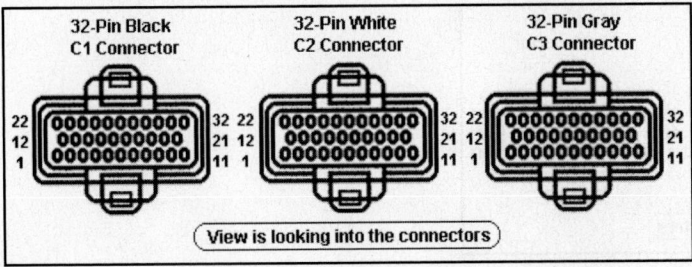

**Pin Connector Graphic**

## 2003 Ram Pickup & Van 8.0L VIN W 'C2' White Connector

| PCM Pin # | Wire Color | Circuit Description (32-Pin) | Value at Hot Idle |
|-----------|-----------|----------------------------|-------------------|
| 1 | VT | Transmission Temperature Sensor Signal | At 200ºF: 2.40v |
| 2 | VT/TN | Injector 7 Driver | 1-4 ms |
| 3 | TN | Injector 9 Driver | 1-4 ms |
| 4 | WT/DB | Injector 1 Driver | 1-4 ms |
| 5 | YL/WT | Injector 3 Driver | 1-4 ms |
| 6 | GY | Injector 5 Driver | 1-4 ms |
| 7 | --- | Not Used | --- |
| 8 | VT/WT | Governor Pressure Solenoid | PWM Signal: 0-12-0v |
| 9 | --- | Not Used | --- |
| 10 | DG | Generator Field Control | Digital Signals: 0-12-0v |
| 11 | OR/BK | TCC Solenoid Control | At Cruise w/TCC On: 1v |
| 12 | BR/DB | Injector 6 Driver | 1-4 ms |
| 13 | GY/LB | Injector 8 Driver | 1-4 ms |
| 14 | WT/DB | Injector 10 Driver | 1-4 ms |
| 15 | TN | Injector 2 Driver | 1-4 ms |
| 16 | LB/BR | Injector 4 Driver | 1-4 ms |
| 17 | DB/YL | Condenser Fan Relay Control | Relay Off: 12v, On: 1v |
| 18-20 | --- | Not Used | --- |
| 21 | LB/BR | 3-4 Shift Solenoid Control | Solenoid Off: 12v, On: 1v |
| 22 | --- | Not Used | --- |
| 23 | GY/YL | Oil Pressure Sensor | 1.6v at 24 psi |
| 24 | --- | Not Used | --- |
| 25 | DB/BK | OSS Sensor (-) Signal | Moving: AC pulse signals |
| 26 | --- | Not Used | --- |
| 27 | WT/OR | Vehicle Speed Signal | Digital Signals |
| 28 | LG/WT | OSS Sensor (+) Signal | Moving: AC pulse signals |
| 29 | LG/RD | Governor Pressure Sensor | 0.58v |
| 30 | PK | Transmission Relay Control | Relay Off: 12v, On: 1v |
| 31 | WT | 5-Volt Supply | 4.9-5.1v |
| 32 | --- | Not Used | --- |

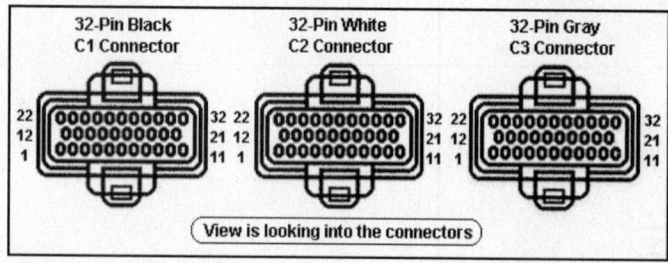

**Pin Connector Graphic**

## Standard Colors and Abbreviations

| Abbreviation | Color | Abbreviation | Color | Abbreviation | Color |
|--------------|-------|--------------|-------|--------------|-------|
| BK | Black | GY | Gray | RD | Red |
| BL | Blue | GN | Green | TN | Tan |
| BR | Brown | LG | Light Green | VT | Violet |
| DB | Dark Blue | OR | Orange | WT | White |
| DG | Dark Green | PK | Pink | YL | Yellow |

## 2003 Ram Pickup & Van 8.0L VIN W 'C3' Gray Connector

| PCM Pin # | Wire Color | Circuit Description (32-Pin) | Value at Hot Idle |
|---|---|---|---|
| 1 | OR | A/C Clutch Relay Control | Relay Off: 12v, On: 1v |
| 2 | --- | Not Used | --- |
| 3 | DB/YL | ASD Relay Control | Relay Off: 12v, On: 1v |
| 4 | TN/RD | Speed Control Vacuum Solenoid | Vacuum Increasing: 1v |
| 5 | LG/RD | Speed Control Vent Solenoid | Vacuum Decreasing: 1v |
| 6-7 | --- | Not Used | --- |
| 8 | BR/VT | HO2S-11 Heater Control (California) | Heater Off: 12v, On: 1v |
| 9 | DB/OR | HO2S-12 Heater Relay (California) | Relay Off: 12v, On: 1v |
| 10 | WT/OR | LDP Solenoid Control | PWM Signal: 0-12-0v |
| 11 | YL/RD | Speed Control Power Supply | 12-14v |
| 12 | DG/OR | ASD Relay Output (B+) | 12-14v |
| 13 | --- | Not Used | --- |
| 14 | OR | LDP Switch Sense Signal | LDP Switch Closed: 0v |
| 15 | PK/YL | Battery Temperature Sensor | At 86ºF: 1.96v |
| 16 | BR/WT | HO2S-21 Heater Control (Except California) | Heater Off: 12v, On: 1v |
| 17-18 | --- | Not Used | --- |
| 19 | BR | Fuel Pump Relay Control | Relay Off: 12v, On: 1v |
| 20 | PK/BK | EVAP Purge Solenoid Control | PWM Signal: 0-12-0v |
| 21-23 | --- | Not Used | --- |
| 24 | WT/PK | Brake Switch Sense | Brake Off: 0v, On: 12v |
| 25 | WT/OR | Generator Field Source | 12-14v |
| 26 | LB/BK | Fuel Level Sensor Signal | Full: 0.56v, 1/2 full: 2.5v |
| 27 | PK | SCI Transmit | 0v |
| 28 | --- | Not Used | --- |
| 29 | LG | SCI Receive | 5v |
| 30 | VT/BR | PCI Data Bus (J1850) | Digital Signals: 0-7-0v |
| 31 | --- | Not Used | --- |
| 32 | RD/LG | Speed Control Set Switch Signal | S/C & Set Switch On: 3.8v |

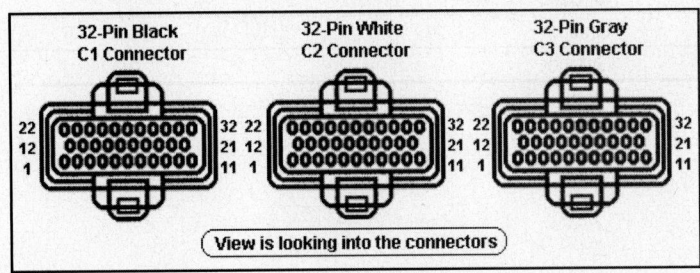

**Pin Connector Graphic**

## Standard Colors and Abbreviations

| Abbreviation | Color | Abbreviation | Color | Abbreviation | Color |
|---|---|---|---|---|---|
| BK | Black | GY | Gray | RD | Red |
| BL | Blue | GN | Green | TN | Tan |
| BR | Brown | LG | Light Green | VT | Violet |
| DB | Dark Blue | OR | Orange | WT | White |
| DG | Dark Green | PK | Pink | YL | Yellow |

## 2004-05 Ram Pickup 8.3L VIN H 'C1' Black Connector

| PCM Pin # | Wire Color | Circuit Description (32-Pin) | Value at Hot Idle |
|---|---|---|---|
| 1 | DB/WT | Coil Control No. 4 | |
| 2 | PK/GY | Fused Ignition Switch Output (Run-Start) | |
| 3 | DB/OR | Coil Control No. 3 | |
| 4 | DB/DG | Sensor Ground | |
| 5 | DB/YL | Coil Control No. 5 | |
| 6 | YL/DB (A/T) | Park Neutral Position Switch Sense (T41) | |
| 7 | DB/DG | Coil Control No. 1 | |
| 8 | BR/LB | CKP Signal | |
| 9 | DB/TN | Coil Control No. 2 | |
| 10 | VT/DG | Idle Air Control No. 2 Driver | |
| 11 | BR/LG | Idle Air Control No. 1 Driver | |
| 12 | --- | Not Used | --- |
| 13 | --- | Not Used | --- |
| 14 | --- | Not Used | --- |
| 15 | DB/LG | IAT Signal | |
| 16 | VT/OR | ECT Signal | |
| 17 | PK/YL | 5 Volt Supply | |
| 18 | DB/GY | CMP Signal | |
| 19 | VT/LG | Idle Air Control No. 3 Driver | |
| 20 | BR/DG | Idle Air Control No. 4 Driver | |
| 21 | --- | Not Used | --- |
| 22 | RD | Fused B (+) | |
| 23 | BR/OR | Throttle Position Sensor Signal | |
| 24 | DB/LB | O2 1/1 Signal | |
| 25 | DB/YL | O2 1/2 Signal | |
| 26 | DB/LG | O2 2/1 Signal | |
| 27 | VT/BR | MAP Sensor Signal | |
| 28 | --- | Not Used | --- |
| 29 | BR | O2 2/2 Signal | |
| 30 | --- | Not Used | --- |
| 31 | BK/BR | Ground | |
| 32 | BK/DG | Ground | |

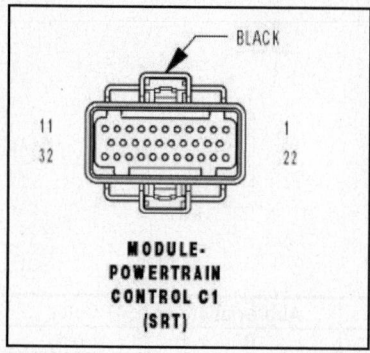

MODULE-
POWERTRAIN
CONTROL C1
(SRT)

**2004-05 Ram Pickup 8.3L VIN H 'C2' White Connector**

| PCM Pin # | Wire Color | Circuit Description (32-Pin) | Value at Hot Idle |
|:---:|:---:|:---:|:---:|
| 1 | DG/OR (A/T) | Transmission Temperature Sensor Signal | |
| 2 | BR/YL | Injector Control No. 7 | |
| 3 | BR | Injector Control No. 9 | |
| 4 | BR/YL | Injector Control No. 1 | |
| 5 | BR/LB | Injector Control No. 3 | |
| 6 | BR/OR | Injector Control No. 5 | |
| 7 | --- | Not Used | --- |
| 8 | DG (A/T) | Governor Pressure Solenoid Control | |
| 9 | --- | Not Used | --- |
| 10 | BR/GY | Gen Field Control | |
| 11 | YL/LB (A/T) | Torque Converter Clutch Solenoid Control | |
| 12 | BR/VT | Injector Control No. 6 | |
| 13 | BR/LB | Injector Control No. 8 | |
| 14 | BR/WT | Injector Control No. 10 | |
| 15 | BR/DB | Injector Control No. 2 | |
| 16 | BR/TN | Injector Control No. 4 | |
| 17 | LB/TN | Hydraulic Cooling Control Feed | |
| 18 | --- | Not Used | --- |
| 19 | BK/LB | A/C Pressure Signal | |
| 20 | --- | Not Used | --- |
| 21 | DG/TN (A/T) | 3-4 Solenoid Control | |
| 21 | DB/LB (M/T) | Solenoid Reverse Lockout Feed | |
| 22 | --- | Not Used | --- |
| 23 | VT/GY | Oil Pressure Signal | |
| 24 | --- | Not Used | --- |
| 25 | DG/VT (A/T) | Speed Sensor Ground | |
| 26 | --- | Not Used | --- |
| 27 | DG/YL | Vehicle Speed Signal No. 1 | |
| 28 | DG/BR (A/T) | Output Speed Sensor Signal | |
| 29 | YL/BR (A/T) | Governor Pressure Sensor Signal | |
| 30 | YL/DB (A/T) | Transmission Control Relay Control | |
| 31 | YL/PK | 5 Volt Supply | |
| 32 | --- | Not Used | --- |

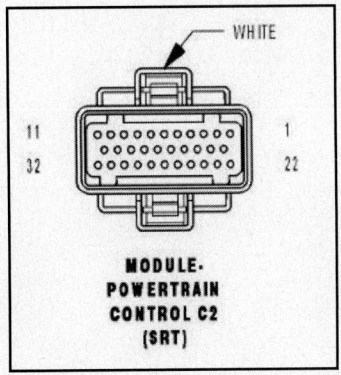

WHITE

11                     1
32                     22

**MODULE-
POWERTRAIN
CONTROL C2
(SRT)**

## 2004-05 Ram Pickup 8.3L VIN H 'C3' Gray Connector

| PCM Pin # | Wire Color | Circuit Description (32-Pin) | Value at Hot Idle |
|---|---|---|---|
| 1 | LB/OR | A/C Clutch Relay Control | |
| 2 | --- | Not Used | --- |
| 3 | BR/WT | ASD Relay Control | |
| 4 | VT/YL | S/C Vacuum Sol Control | |
| 5 | VT/OR | S/C Vent Sol Control | |
| 6 | --- | Not Used | --- |
| 7 | DB/YL | Knock Sensor No. 1 Signal | |
| 8 | BR/LG | O2 1/1 Heater Control | |
| 9 | BR/OR | Oxygen Sensor Downstream Relay Control | |
| 10 | VT/LB | Leak Detection Pump Solenoid Control | |
| 11 | VT/YL | Speed Control Supply | |
| 12 | BR/WT | ASD Relay Output | |
| 13 | DG | Tow/Haul Overdrive Off Switch Sense | |
| 14 | VT/WT | Leak Detection Pump Switch Sense | |
| 15 | DB/VT | Battery Temp Signal | |
| 16 | BR/VT | O2 2/1 Heater Control | |
| 17 | --- | Not Used | --- |
| 18 | BR/WT | Knock Sensor No. 2 Signal | |
| 19 | BR | Fuel Pump Relay Control | |
| 20 | DB/BR | EVAP Purge Return | |
| 21 | --- | Not Used | --- |
| 22 | --- | Not Used | --- |
| 23 | --- | Not Used | --- |
| 24 | DG/WT | Brake Switch No. 1 Signal | |
| 25 | BR/DG | Generator Source | |
| 26 | DB/WT | Fuel Level Sensor Signal | |
| 27 | WT/BR | SCI Transmit (PCM) | |
| 28 | --- | Not Used | --- |
| 29 | WT/LG | SCI Receive (PCM) | |
| 30 | WT/VT | PCI Bus | |
| 31 | --- | Not Used | --- |
| 32 | VT | S/C Switch No. 1 Signal | |

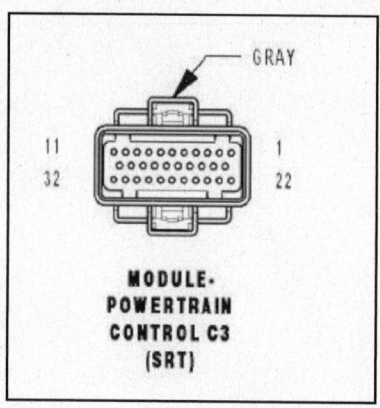

GRAY

11     1
32     22

**MODULE-
POWERTRAIN
CONTROL C3
(SRT)**

## Dodge Durango & Jeep SUVs

### 2004-05 Durango 3.7L, 4.7L, 5.7L 'C1' Black/Black Connector

| PCM Pin # | Wire Color | Circuit Description (38-Pin) | Value at Hot Idle |
|---|---|---|---|
| 1 | DB/YL (4.7L/5.7L) | Coil Control No. 8 | |
| 2 | --- | Not Used | --- |
| 3 | BR (4.7L/5.7L) | Coil Control No. 7 | |
| 4 | VT/LG (4.7L/5.7L) | Injector Control No. 8 | |
| 5 | YL/TN (4.7L/5.7L) | Injector Control No. 7 | |
| 6-8 | --- | Not Used | --- |
| 9 | BK | Ground | |
| 10 | --- | Not Used | --- |
| 11 | PK/GY | Fused Ignition Switch Output (Run-Start) | |
| 12 | PK/DG (2005) | Fused Ignition Switch Output (Off-Run-Start) | |
| 12 | PK/DG (2004) | Fused ASD Relay Output | |
| 13-17 | --- | Not Used | --- |
| 18 | BK | Ground | |
| 19 | --- | Not Used | --- |
| 20 | VT/GY | Engine Oil Pressure Signal | |
| 21-24 | --- | Not Used | --- |
| 25 | WT/BR | SCI Receive (PCM) | |
| 26 | WT/PK | SCI Receive (TCM) | |
| 27 | YL/PK | 5 Volt Supply | |
| 28 | --- | Not Used | --- |
| 29 | OR/GY | Fused B (+) | |
| 30 | PK/YL | Fused Ign. Switch Output (Start 2005; Run 2004) | |
| 31 | VT/DG | 02 1/2 Signal | |
| 32 | BR/DG | O2 Return (Upstream) | |
| 33 | BR | O2 2/2 Signal | |
| 34 | YL/RD | CAN C Bus (+) | |
| 35 | WT/LB | CAN C Bus (-) | |
| 36 | WT/GY | SCI Transmit (PCM) | |
| 37 | BR/WT | SCI Transmit (TCM) | |
| 38 | --- | Not Used | --- |

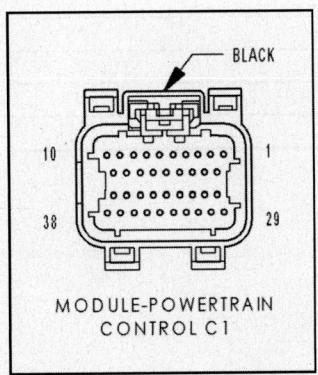

MODULE-POWERTRAIN
CONTROL C1

## 2004-05 Durango 3.7L, 4.7L, 5.7L 'C2' Black/Orange Connector

| PCM Pin # | Wire Color | Circuit Description (38-Pin) | Value at Hot Idle |
|---|---|---|---|
| 1 | VT/LB | Coil Control No. 6 | |
| 2 | DG/WT | Coil Control No. 5 | |
| 3 | DB | Coil Control No. 4 | |
| 4 | PK/DB | Injector Control No. 6 | |
| 5 | YL/RD | Injector Control No. 5 | |
| 6 | TN/YL (5.7L) | ETC Motor (+) | |
| 7 | VT/RD | Coil Control No. 3 | |
| 8 | LB/RD (4.7L/5.7L) | EGR Solenoid Control | |
| 9 | DB/TN | Coil Control No. 2 | |
| 10 | LB/RD | Coil Control No. 1 | |
| 11 | YL/LB | Injector Control No. 4 | |
| 12 | PK/DB | Injector Control No. 3 | |
| 13 | BR/DB | Injector Control No. 2 | |
| 14 | BR/YL | Injector Control No. 1 | |
| 15 | PK/GY (5.7L) | TP Sensor Return | |
| 16 | --- | Not Used | --- |
| 17 | GY/LB | O2 2/1 Heater Control | |
| 18 | LB/WT | O2 1/1 Heater Control | |
| 19 | BR/GY | Generator Field Control | |
| 20 | LB/RD | ECT Signal | |
| 21 | YL/DG | TP No. 1 Signal | |
| 22 | LB/GY | EGR Signal | |
| 23 | VT/BR | MAP Signal | |
| 24 | YL/DG | Knock Sensor No. 1 Return | |
| 25 | DB/OR | Knock Sensor No. 1 Signal | |
| 26 | --- | Not Used | --- |
| 27 | DB/DG | Sensor Ground | |
| 28 | BR/VT (3.7L/4.7L) | IAC Signal | |
| 28 | PK/LB (5.7L) | TP No. 2 Signal | |
| 29 | PK/YL | 5 Volt Supply | |
| 30 | BR/WT | IAT Signal | |
| 31 | DB/LB | O2 1/1 Signal | |
| 32 | DB/DG | O2 Return (Downstream) | |
| 33 | DB/LG | O2 2/1 Signal | |
| 34 | TN/LB | CMP Signal | |
| 35 | VT/DG | CKP Signal | |
| 36 | DB/OR | Knock Sensor No. 2 Signal | |
| 37 | DG/RD | Knock Sensor No. 2 Return | |
| 38 | VT/GY (3.7L/4.7L) | IAC Control | |
| 38 | TN/OR (5.7L) | ETC Motor (-) | |

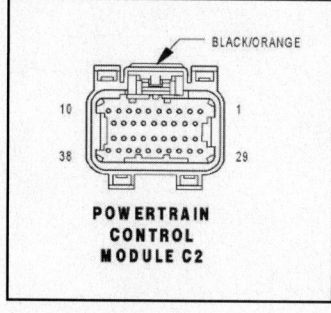

POWERTRAIN
CONTROL
MODULE C2

### 2004-05 Durango 3.7L, 4.7L, 5.7L 'C3' Black/Natural Connector

| PCM Pin # | Wire Color | Circuit Description (38-Pin) | Value at Hot Idle |
|---|---|---|---|
| 1 | --- | Not Used | --- |
| 2 | --- | Not Used | --- |
| 3 | BR/GY | ASD Relay Control | |
| 4 | --- | Not Used | --- |
| 5 | VT/OR (3.7L/4.7L) | S/C Vent Solenoid Control | |
| 6 | --- | Not Used | --- |
| 7 | VT/YL (3.7L/4.7L) | S/C Supply | |
| 8 | VT/LB | NVLD Sol Control | |
| 9 | DG/RD | O2 1/2 Heater Control | |
| 10 | BR/GY | O2 2/2 Heater Control | |
| 11 | LB/OR | A/C Clutch Relay Control | |
| 12 | VT/YL (3.7L/4.7L) | S/C Vacuum Solenoid Control | |
| 13 | --- | Not Used | --- |
| 14 | VT/YL (5.7L) | S/C Supply (2005) - Not Used (2004) | |
| 15 | VT/BR (5.7L) | S/C Switch Return | |
| 16 | BR/LB (5.7L) | APPS No. 1 Return | |
| 17 | BR/VT (5.7L) | APPS No. 2 Return | |
| 18 | VT/OR (5.7L) | S/C Switch No. 2 Signal | |
| 19 | RD | ASD Relay Output | |
| 20 | DB/WT | EVAP Purge Solenoid Signal | |
| 21 | YL/DB | TRS (T41) Sense (P/N Sense) | |
| 22 | - | - | |
| 23 | DG/WT | Brake Switch No. 1 Signal | |
| 24 | - | - | |
| 25 | BR/WT (5.7L) | APPS No. 1 Signal | |
| 26 | --- | Not Used | --- |
| 27 | --- | Not Used | --- |
| 28 | RD | ASD Relay Output | |
| 29 | DB/BR | EVAP Purge Solenoid Control | |
| 30 | DB/WT (5.7L) | P/S Pressure Switch Signal | |
| 31 | --- | Not Used | --- |
| 32 | --- | Not Used | --- |
| 33 | --- | Not Used | --- |
| 34 | VT | S/C Switch No. 1 Signal | |
| 35 | VT/WT | NVLD Switch Signal | |
| 36 | PK/RD (5.7L) | APPS No. 2 Signal | |
| 37 | BR | Fuel Pump Relay Control | |
| 38 | DG/OR | Engine Starter Motor Relay Control | |

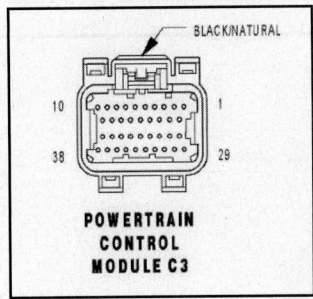

BLACK/NATURAL

10
38    1
29

**POWERTRAIN
CONTROL
MODULE C3**

**2004-05 Durango 3.7L, 4.7L, 5.7L 'C4' Black/Green Connector**

| PCM Pin # | Wire Color | Circuit Description (38-Pin) | Value at Hot Idle |
|---|---|---|---|
| 1 | VT/WT | OD Solenoid Control | |
| 2 | YL/LB (3.7L) | UD Solenoid Control | |
| 2 | VT/RD (4.7L/5.7L) | 4C Solenoid Control | |
| 3 | --- | Not Used | --- |
| 4 | VT/BR (4.7L/5.7L) | MS Solenoid Control | |
| 5 | --- | Not Used | --- |
| 6 | YL/DB (3.7L) | 2-4 Solenoid Control | |
| 6 | VT/DB (4.7L/5.7L) | 2C Solenoid Control | |
| 7 | --- | Not Used | --- |
| 8 | YL/LB (4.7L/5.7L) | UD Solenoid Control | |
| 9 | --- | Not Used | --- |
| 10 | VT/DG | L/R Solenoid Control | |
| 11 | DG | Pressure Control Solenoid Control | |
| 12 | BK | Ground | |
| 13 | BK | Ground | |
| 14 | BK | Ground (2005) - Not Used (2004) | |
| 15 | BR/DB | TRS T1 Sense | |
| 16 | DG/DB | TRS T3 Sense | |
| 17 | DG | Tow/Haul Overdrive Off Switch Sense | |
| 18 | VT/YL | Transmission Control Relay Control | |
| 19 | YL/OR | Transmission Control Relay Output | |
| 20 | BR/YL (4.7L/5.7L) | 4C Pressure Switch Sense | |
| 21 | YL/WT (4.7L/5.7L) | UD Pressure Switch Sense | |
| 22 | BR/VT | OD Pressure Switch Sense | |
| 23-25 | --- | Not Used | --- |
| 26 | BR/PK (4.7L/5.7L) | 2-4 Pressure Switch Sense (2005) | |
| 26 | BR/PK (4.7L/5.7L) | TRS T2 Sense (2004) | |
| 27 | --- | Not Used | --- |
| 28 | YL/OR | Transmission Control Relay Output | |
| 29 | VT/GY | Low/Reverse Pressure Switch Sense | |
| 30 | YL/DG (3.7L) | 2-4 Pressure Switch Sense | |
| 30 | DG/YL (4.7L/5.7L) | 2C Pressure Switch Sense | |
| 31 | YL/BR | Line Pressure Sensor Signal | |
| 32 | DG/BR | Output Speed Sensor Signal | |
| 33 | VT/OR | Input Speed Sensor Signal | |
| 34 | DG/VT | Speed Sensor Ground | |
| 35 | BR/RD | Transmission Temperature Sensor Signal | |
| 36 | --- | Not Used | --- |
| 37 | DG/YL | TRS T42 Sense | |
| 38 | YL/OR | Transmission Control Relay Output | |

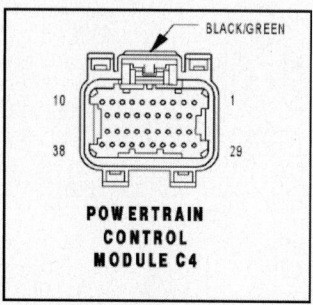

BLACK/GREEN

10    1

38    29

**POWERTRAIN
CONTROL
MODULE C4**

**1998-99 Durango 3.9L VIN X 'A' Black Connector**

| PCM Pin # | Wire Color | Circuit Description (32-Pin) | Value at Hot Idle |
|---|---|---|---|
| 1, 3, 5, 9 | --- | Not Used | --- |
| 2 | LG/BK | Ignition Switch Power (B+) | 12-14v |
| 4 | BK/LB | Sensor Ground | <0.1v |
| 6 | BK/WT | PNP Switch Sense Signal | In P/N: 0v, Others: 5v |
| 7 | BK/GY | Coil 1 Driver Control | 5°, 55 mph: 8° dwell |
| 8 | GY/BK | CKP Sensor Signal | Digital Signals: 0-5-0v |
| 10 | YL/BK | IAC 2 Driver Control | Pulse Signals |
| 11 | BR/WT | IAC 3 Driver Control | Pulse Signals |
| 12-14, 21 | --- | Not Used | --- |
| 15 | BK/RD | IAT Sensor Signal | At 100°F: 1.83v |
| 16 | TN/BK | ECT Sensor Signal | At 180°F: 2.80v |
| 17 | PK/WT | 5-Volt Supply | 4.9-5.1v |
| 18 | TN/YL | CMP Sensor Signal | Digital Signals: 0-5-0v |
| 19 | GY/RD | IAC 1 Driver Control | Pulse Signals |
| 20 | PK/BK | IAC 4 Driver Control | Pulse Signals |
| 22 | RD/WT | Fused Battery Power (B+) | 12-14v |
| 23 | OR/DB | TP Sensor Signal | 0.6-1.0v |
| 24 | TN/WT | HO2S-11 (B1 S1) Signal | 0.1-1.1v |
| 25 | PK/WT | HO2S-12 (B1 S2) Signal | 0.1-1.1v |
| 27 | DG/RD | MAP Sensor Signal | 1.5-1.6v |
| 26, 28-30 | --- | Not Used | --- |
| 31, 32 | BK/TN | Power Ground | <0.1v |

**1998-99 Durango 3.9L VIN X 'B' White Connector**

| PCM Pin # | Wire Color | Circuit Description (32-Pin) | Value at Hot Idle |
|---|---|---|---|
| 1 | GY/BK | Transmission Temperature Sensor | At 200°F: 2.40v |
| 2-3, 7, 9 | --- | Not Used | --- |
| 4 | WT/DB | Injector 1 Driver | 1-4 ms |
| 5 | YL/WT | Injector 3 Driver | 1-4 ms |
| 6 | GY | Injector 5 Driver | 1-4 ms |
| 7, 9, 13-14, 17-20 | --- | Not Used | --- |
| 8 | PK/WT | Governor Pressure Solenoid | PWM Signal: 0-12-0v |
| 10 | DG/WT | Generator Field Control | Digital Signals: 0-12-0v |
| 11 | OR/BK | TCC Solenoid Control | At Cruise w/TCC On: <1v |
| 12 | BR/DB | Injector 6 Driver | 1-4 ms |
| 15 | TN | Injector 2 Driver | 1-4 ms |
| 16 | LB/BR | Injector 4 Driver | 1-4 ms |
| 21 | BR | Overdrive Solenoid Control | Cruise w/solenoid on: <1v |
| 21 ('99) | BR | 3-4 Shift Solenoid Control | Solenoid Off: 12v, On: 1v |
| 23 | GY/YL | Engine Oil Pressure Sensor | 1.6v at 24 psi |
| 22, 24, 26, 32 | --- | Not Used | --- |
| 25 | DB | OSS Sensor (-) Signal | Moving: AC pulse signals |
| 27 | WT/OR | Vehicle Speed Signal | Digital Signal |
| 28 | LG | OSS Sensor (+) Signal | Moving: AC pulse signals |
| 29 | LG/RD | Governor Pressure Signal | 0.58v |
| 30 | PK | Transmission Relay Control | Relay Off: 12v, On: 1v |
| 31 | OR | 5-Volt Supply | 4.9-5.1v |

**1998-99 Durango 3.9L VIN X 'C' Gray Connector**

| PCM Pin # | Wire Color | Circuit Description (32-Pin) | Value at Hot Idle |
|---|---|---|---|
| 1 | DB/OR | A/C Clutch Relay Control | Relay Off: 12v, On: 1v |
| 2, 7-9 | --- | Not Used | --- |
| 3 | DB/YL | ASD Relay Control | Relay Off: 12v, On: 1v |
| 4 | TN/RD | Speed Control Vacuum Solenoid | Vacuum Increasing: 1v |
| 5 | LG/RD | Speed Control Vent Solenoid | Vacuum Decreasing: 1v |
| 6 ('98) | LG/OR | Overdrive 'Off' Lamp Control | O/D On: 1v, O/D Off: 12v |
| 10 | WT/DG | LDP Solenoid Control | PWM Signal: 0-12-0v |
| 11 | YL/RD | Speed Control Power Supply | 12-14v |
| 12 | DG/OR | ASD Relay Output | 12-14v |
| 13 | OR/WT | Overdrive 'Off' Switch Sense | Switch On: 1v, Off: 12v |
| 14 | OR | LDP Switch Sense Signal | Closed: 0v, Open: 12v |
| 15 | PK/YL | Battery Temperature Sensor | At 86°F: 1.96v |
| 16-18 | --- | Not Used | --- |
| 19 | DB/WT | Fuel Pump Relay Control | Relay Off: 12v, On: 1v |
| 20 | VT/BK | EVAP Purge Solenoid Control | PWM Signal: 0-12-0v |
| 21 | --- | Not Used | --- |
| 22 | BR | A/C Request Signal | A/C Off: 12v, On: 1v |
| 23 | LG/WT | A/C Select Signal | A/C Off: 12v, On: 1v |
| 24 | WT/PK | Brake Switch Sense Signal | Brake Off: 0v, On: 12v |
| 25 | WT/PK | Generator Field Source | 12-14v |
| 26 | DB | Fuel Level Sensor Signal | Full: 0.56v, 1/2 full: 2.5v |
| 27 | PK | SCI Transmit | 0v |
| 28 | WT/PK | CCD Bus (-) | <0.050v |
| 29 | PK/WT | SCI Receive | 5v |
| 30 | PK/BR | CCD Bus (+) | Digital Signals: 0-5-0v |
| 31 | --- | Not Used | --- |
| 32 | BR/YL | Speed Control Switch Signal | S/C & Set Switch On: 3.8v |

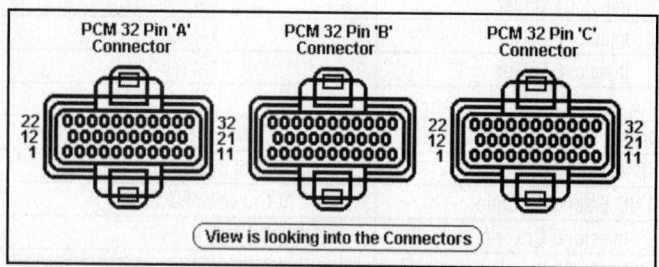

**Pin Connector Graphic**

**Standard Colors and Abbreviations**

| Abbreviation | Color | Abbreviation | Color | Abbreviation | Color |
|---|---|---|---|---|---|
| BK | Black | GY | Gray | RD | Red |
| BL | Blue | GN | Green | TN | Tan |
| BR | Brown | LG | Light Green | VT | Violet |
| DB | Dark Blue | OR | Orange | WT | White |
| DG | Dark Green | PK | Pink | YL | Yellow |

## 2000 Durango 4.7L VIN N (A/T) 'C1' Black Connector

| PCM Pin # | Wire Color | Circuit Description (32-Pin) | Value at Hot Idle |
|---|---|---|---|
| 1 | TN/OR | Coil 3 Driver Control | 5°, 55 mph: 8° dwell |
| 2 | LG/BK | Ignition Switch Power (B+) | 12-14v |
| 3 | TN/LG | Coil 4 Driver Control | 5°, 55 mph: 8° dwell |
| 4 | BK/LB | Sensor Ground | <0.1v |
| 5 | TN/LB | Coil 6 Driver Control | 5°, 55 mph: 8° dwell |
| 6 | BK/WT | PNP Switch Sense Signal | In P/N: 0v, Others: 5v |
| 7 | BK/GY | Coil 1 Driver Control | 5°, 55 mph: 8° dwell |
| 8 | GY/BK | CKP Sensor Signal | Digital Signals: 0-5-0v |
| 9 | RD/YL | Coil 8 Driver Control | 5°, 55 mph: 8° dwell |
| 10 | YL/BK | IAC 2 Driver Control | Pulse Signals |
| 11 | BR/WT | IAC 3 Driver Control | Pulse Signals |
| 12 | DB/OR | PSP Switch Sense Signal | Straight: 0v, Turning: 5v |
| 13-14 | --- | Not Used | --- |
| 15 | BK/RD | IAT Sensor Signal | At 100°F: 1.83v |
| 16 | TN/BK | ECT Sensor Signal | At 180°F: 2.80v |
| 17 | PK/WT | 5-Volt Supply | 4.9-5.1v |
| 18 | TN/YL | CMP Sensor Signal | Digital Signals: 0-5-0v |
| 19 | GY/RD | IAC 1 Driver Control | Pulse Signals |
| 20 | PK/BK | IAC 4 Driver Control | Pulse Signals |
| 21 | TN/RD | Coil 5 Driver Control | 5°, 55 mph: 8° dwell |
| 22 | RD/WT | Fused Battery Power (B+) | 12-14v |
| 23 | OR/DB | TP Sensor Signal | 0.6-1.0v |
| 24 | LG/RD | HO2S-11 (B1 S1) Signal | 0.1-1.1v |
| 25 | TN/WT | HO2S-12 (B1 S2) Signal | 0.1-1.1v |
| 26 | OR/TN | HO2S-21 (B2 S1) Signal (Cal) | 0.1-1.1v |
| 27 | DG/RD | MAP Sensor Signal | 1.5-1.6v |
| 28, 30 | --- | Not Used | --- |
| 29 | PK/WT | HO2S-22 (B2 S2) Signal (Cal) | 0.1-1.1v |
| 31, 32 | BK/TN | Power Ground | <0.1v |

## 2000 Durango 4.7L VIN N (A/T) 'C2' White Connector

| PCM Pin # | Wire Color | Circuit Description (32-Pin) | Value at Hot Idle |
|---|---|---|---|
| 2 | VT | Injector 7 Driver | 1-4 ms |
| 4 | WT/DB | Injector 1 Driver | 1-4 ms |
| 5 | YL/WT | Injector 3 Driver | 1-4 ms |
| 6 | GY | Injector 5 Driver | 1-4 ms |
| 7 | DB/TN | Coil 7 Driver Control | 5°, 55 mph: 8° dwell |
| 9 | TN/PK | Coil 2 Driver Control | 5°, 55 mph: 8° dwell |
| 10 | DG | Generator Field Control | Digital Signals: 0-12-0v |
| 12 | BR/DB | Injector 6 Driver | 1-4 ms |
| 13 | GY/LB | Injector 8 Driver | 1-4 ms |
| 15 | TN | Injector 2 Driver | 1-4 ms |
| 16 | LB/BR | Injector 4 Driver | 1-4 ms |
| 17 | DB/PK | Radiator Fan Relay Control | Relay Off: 12v, On: 1v |
| 23 | GY/YL | Engine Oil Pressure Sensor | 1.6v at 24 psi |
| 27 | WT/OR | Vehicle Speed Signal | Digital Signal |

## 2000 Durango 4.7L VIN N (A/T) 'C3' 32-Pin Black Connector

| PCM Pin # | Wire Color | Circuit Description (32-Pin) | Value at Hot Idle |
|---|---|---|---|
| 1 | DB/OR | A/C Clutch Relay Control | Relay Off: 12v, On: 1v |
| 3 | DB/YL | ASD Relay Control | Relay Off: 12v, On: 1v |
| 4 | TN/RD | Speed Control Vacuum Solenoid | Vacuum Increasing: 1v |
| 5 | LG/RD | Speed Control Vent Solenoid | Vacuum Decreasing: 1v |
| 8 | DG/BK | HO2S-11 Heater Relay (Cal) | Relay Off: 12v, On: 1v |
| 9 | OR/RD | HO2S-12 Heater Relay (Cal) | Relay Off: 12v, On: 1v |
| 10 | WT/DG | LDP Solenoid Control | PWM Signal: 0-12-0v |
| 11 | YL/RD | Speed Control Power Supply | 12-14v |
| 12 | DG/OR | ASD Relay Output | 12-14v |
| 13 | YL/DG | Torque Management Relay | Relay Off: 12v, On: 1v |
| 14 | OR | LDP Switch Sense Signal | Closed: 0v, Open: 12v |
| 15 | PK/YL | Battery Temperature Sensor | At 86°F: 1.96v |
| 19 | DB/WT | Fuel Pump Relay Control | Relay Off: 12v, On: 1v |
| 20 | PK/BK | EVAP Purge Solenoid Control | PWM Signal: 0-12-0v |
| 22 | BR | A/C Switch Signal | A/C Off: 12v, On: 1v |
| 23 | LG/WT | A/C Select Signal | A/C Off: 12v, On: 1v |
| 24 | WT/PK | Brake Switch Sense Signal | Brake Off: 0v, On: 12v |
| 25 | WL/DB | Generator Field Source | 12-14v |
| 26 | DB | Fuel Level Sensor Signal | Full: 0.56v, 1/2 full: 2.5v |
| 27 | PK/DB | SCI Transmit | 0v |
| 28 | WT/BK | CCD Bus (-) | <0.050v |
| 29 | PK/WT | SCI Receive | 5v |
| 30 | PT/BR | CCD Bus (+) | Digital Signals: 0-5-0v |
| 32 | RD/LG | Speed Control Switch Signal | S/C & Set Switch On: 3.8v |

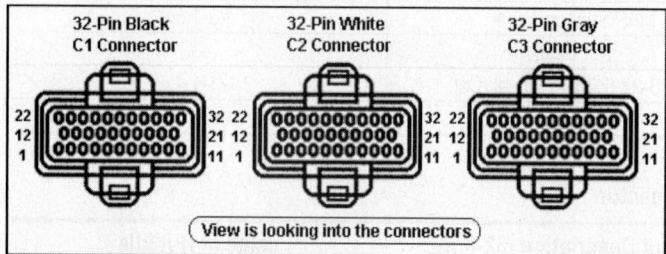

**Pin Connector Graphic**

## Standard Colors and Abbreviations

| Abbreviation | Color | Abbreviation | Color | Abbreviation | Color |
|---|---|---|---|---|---|
| BK | Black | GY | Gray | RD | Red |
| BL | Blue | GN | Green | TN | Tan |
| BR | Brown | LG | Light Green | VT | Violet |
| DB | Dark Blue | OR | Orange | WT | White |
| DG | Dark Green | PK | Pink | YL | Yellow |

### 2001-02 Durango 4.7L VIN N (A/T) 'C1' Black Connector

| PCM Pin # | Wire Color | Circuit Description (32-Pin) | Value at Hot Idle |
|---|---|---|---|
| 1 | TN/OR | Coil 3 Driver Control | 5°, 55 mph: 8° dwell |
| 2 | LG/BK | Ignition Switch Power (B+) | 12-14v |
| 3 | TN/LG | Coil 4 Driver Control | 5°, 55 mph: 8° dwell |
| 4 | BK/LB | Sensor Ground | <0.1v |
| 5 | TN/LB | Coil 6 Driver Control | 5°, 55 mph: 8° dwell |
| 6 | BK/WT | PNP Switch Sense Signal | In P/N: 0v, Others: 5v |
| 7 | BK/GY | Coil 1 Driver Control | 5°, 55 mph: 8° dwell |
| 8 | GY/BK | CKP Sensor Signal | Digital Signals: 0-5-0v |
| 9 | LB/RD | Coil 8 Driver Control | 5°, 55 mph: 8° dwell |
| 10 | YL/BK | IAC 2 Driver Control | Pulse Signals |
| 11 | BR/WT | IAC 3 Driver Control | Pulse Signals |
| 12 | DB/OR | PSP Switch Sense Signal | Straight: 0v, Turning: 5v |
| 13-14, 28, 30 | --- | Not Used | --- |
| 15 | BK/RD | IAT Sensor Signal | At 100°F: 1.83v |
| 16 | TN/BK | ECT Sensor Signal | At 180°F: 2.80v |
| 17 | OR | 5-Volt Supply | 4.9-5.1v |
| 18 | TN/YL | CMP Sensor Signal | Digital Signals: 0-5-0v |
| 19 | GY/RD | IAC 1 Driver Control | Pulse Signals |
| 20 | VT/BK | IAC 4 Driver Control | Pulse Signals |
| 21 | TN/DG | Coil 5 Driver Control | 5°, 55 mph: 8° dwell |
| 22 | RD/WT | Fused Battery Power (B+) | 12-14v |
| 23 | OR/DB | TP Sensor Signal | 0.6-1.0v |
| 24 | BK/DG | HO2S-11 (B1 S1) Signal | 0.1-1.1v |
| 25 | TN/WT | HO2S-12 (B1 S2) Signal | 0.1-1.1v |
| 26 | LG/RD | HO2S-21 (B2 S1) Signal (California) | 0.1-1.1v |
| 27 | DG/RD | MAP Sensor Signal | 1.5-1.6v |
| 29 | TN/WT | HO2S-22 (B2 S2) Signal (California) | 0.1-1.1v |
| 31, 32 | BK/TN | Power Ground | <0.1v |

### 2001-02 Durango 4.7L VIN N (A/T) 'C2' White Connector

| PCM Pin # | Wire Color | Circuit Description (32-Pin) | Value at Hot Idle |
|---|---|---|---|
| 1, 3, 8, 11 | --- | Not Used | --- |
| 2 | VT | Injector 7 Driver Control | 1-4 ms |
| 4 | WT/DB | Injector 1 Driver Control | 1-4 ms |
| 5 | YL/WT | Injector 3 Driver Control | 1-4 ms |
| 6 | GY | Injector 5 Driver | 1-4 ms |
| 7 | DB/TN | Coil 7 Driver Control | 5°, 55 mph: 8° dwell |
| 9 | TN/PK | Coil 2 Driver Control | 5°, 55 mph: 8° dwell |
| 10 | DG | Generator Field Control | Digital Signals: 0-12-0v |
| 12 | BR/DB | Injector 6 Driver Control | 1-4 ms |
| 13 | GY/LB | Injector 8 Driver Control | 1-4 ms |
| 14 | --- | Not Used | --- |
| 15 | TN | Injector 2 Driver Control | 1-4 ms |
| 16 | LB/BR | Injector 4 Driver Control | 1-4 ms |
| 17 | DB/PK | Radiator Fan Relay Control | Relay Off: 12v, On: 1v |
| 18-20, 24-26 | --- | Not Used | --- |
| 23 | GY/YL | Engine Oil Pressure Sensor | 1.6v at 24 psi |
| 27 | WT/OR | Vehicle Speed Signal | Digital Signals |
| 28-30, 32 | --- | Not Used | --- |
| 31 | VT/WT | 5-Volt Supply | 4.9-5.1v |

## 2001-02 Durango 4.7L VIN N (A/T) 'C3' 32-Pin Black Connector

| PCM Pin # | Wire Color | Circuit Description (32-Pin) | Value at Hot Idle |
|---|---|---|---|
| 1 | DB/OR | A/C Clutch Relay Control | Relay Off: 12v, On: 1v |
| 2 | --- | Not Used | --- |
| 3 | DB/YL | ASD Relay Control | Relay Off: 12v, On: 1v |
| 4 | TN/RD | Speed Control Vacuum Solenoid | Vacuum Increasing: 1v |
| 5 | LG/RD | Speed Control Vent Solenoid | Vacuum Decreasing: 1v |
| 6-7 | --- | Not Used | --- |
| 8 | VT/WT | HO2S-11 Heater Relay (California) | Relay Off: 12v, On: 1v |
| 9 | DG/BK | HO2S-12 Heater Relay (California) | Relay Off: 12v, On: 1v |
| 10 | WT/DG | LDP Solenoid Control | PWM Signal: 0-12-0v |
| 11 | YL/RD | Speed Control Power Supply | 12-14v |
| 12 | DG/OR | ASD Relay Output | 12-14v |
| 13 | YL/DG | Torque Management Relay | Relay Off: 12v, On: 1v |
| 14 | OR | LDP Switch Sense Signal | Closed: 0v, Open: 12v |
| 15 | PK/YL | Battery Temperature Sensor | At 86ºF: 1.96v |
| 16 | VT/OR | HO2S-12 Heater Control (California) | Digital Signals: 0-12-0v |
| 17-18 | --- | Not Used | --- |
| 19 | BR | Fuel Pump Relay Control | Relay Off: 12v, On: 1v |
| 20 | PK/BK | EVAP Purge Solenoid Control | PWM Signal: 0-12-0v |
| 21 | --- | Not Used | --- |
| 22 | BR | A/C Switch Signal | A/C Off: 12v, On: 1v |
| 23 | --- | Not Used | --- |
| 24 | WT/PK | Brake Switch Sense Signal | Brake Off: 0v, On: 12v |
| 25 | WT/DB | Generator Field Source | 12-14v |
| 26 | DB/WT | Fuel Pump Relay Control | Relay Off: 12v, On: 1v |
| 27 | PK | SCI Transmit | 0v |
| 28 | --- | Not Used | --- |
| 29 | LG | SCI Receive | 5v |
| 30 | VT/TN | PCI Data Bus (J1850) | Digital Signals: 0-7-0v |
| 31 | --- | Not Used | --- |
| 32 | RD/LG | Speed Control Switch Signal | S/C & Set Switch On: 3.8v |

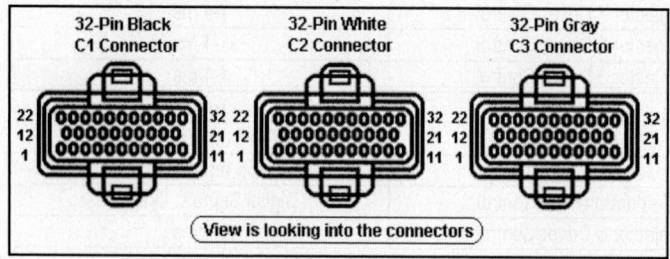

## Pin Connector Graphic

## Standard Colors and Abbreviations

| Abbreviation | Color | Abbreviation | Color | Abbreviation | Color |
|---|---|---|---|---|---|
| BK | Black | GY | Gray | RD | Red |
| BL | Blue | GN | Green | TN | Tan |
| BR | Brown | LG | Light Green | VT | Violet |
| DB | Dark Blue | OR | Orange | WT | White |
| DG | Dark Green | PK | Pink | YL | Yellow |

### 2003 Durango 4.7L VIN N 'C1' 38-Pin Black Connector

| PCM Pin # | Wire Color | Circuit Description (38-Pin) | Value at Hot Idle |
|---|---|---|---|
| 1 | LB/RD | Coil On Plug 8 Driver | 5°, at 55 mph: 8° dwell |
| 2, 6-8, 10 | --- | Not Used | --- |
| 3 | BR | Coil On Plug 7 Driver | 5°, at 55 mph: 8° dwell |
| 4 | GY/LG | Injector 8 Driver | 1.0-4.0 ms |
| 5 | VT | Injector 7 Driver | 1.0-4.0 ms |
| 9, 18 | BK/TN | Power Ground | <0.050v |
| 11 | LG/BK | Ignition Switch Output (Run-Start) | 12-14v |
| 12 | RD/WT | Ignition Switch Output (Off-Run-Start) | 12-14v |
| 13 | WT/OR | Vehicle Speed Signal | Digital Signals |
| 14-17, 19 | --- | Not Used | --- |
| 20 | GY/YL | Oil Pressure Sensor | 1.6v at 24 psi |
| 21 | LB/BR | A/C Pressure Sensor | A/C On: 0.45-4.85v |
| 22 | VT/OR | Ambient Temperature Sensor | At 86°F: 1.96v |
| 23-24, 28 | --- | Not Used | --- |
| 25 | LG | SCI Receive (PCM) | 5v |
| 26 | PK/LB | SCI Receive (PCM) | 5v |
| 27 | VT/WT | 5-Volt Supply | 4.9-5.1v |
| 29 | RD/WT | Fused Power (B+) | 12-14v |
| 30 | RD/YL | Fused Ignition Output (Start) | 9-11v |
| 31 | TN/WT | HO2S-12 (B1 S2) Signal | 0.1-1.1v |
| 32 | BR/DG | HO2S Return (Downstream) | <0.050v |
| 33 | TN/WT | HO2S-22 (B1 S2) Signal (California) | 0.1-1.1v |
| 34-35 | --- | Not Used | --- |
| 36 | PK | SCI Transmit (PCM) | 0v |
| 37 | WT/DG | SCI Transmit (PCM) | 0v |
| 38 | WT/VT | PCI Data Bus (J1850) | Digital Signals: 0-7-0v |

### 2003 Durango 4.7L VIN N 'C2' 38-Pin Gray Connector

| PCM Pin # | Wire Color | Circuit Description (38-Pin) | Value at Hot Idle |
|---|---|---|---|
| 1 | TN/LB | Coil On Plug 6 Driver | 5°, at 55 mph: 8° dwell |
| 2 | TN/DG | Coil On Plug 5 Driver | 5°, at 55 mph: 8° dwell |
| 3 | TN/LG | Coil On Plug 4 Driver | 5°, at 55 mph: 8° dwell |
| 4 | BR/DB | Injector 6 Driver | 1.0-4.0 ms |
| 5 | GY | Injector 5 Driver | 1.0-4.0 ms |
| 6, 8, 15-16 | --- | Not Used | --- |
| 7 | TN/OR | Coil On Plug 3 Driver | 5°, at 55 mph: 8° dwell |
| 9 | TN/PK | Coil On Plug 2 Driver | 5°, at 55 mph: 8° dwell |
| 10 | TN/RD | Coil On Plug 1 Driver | 5°, at 55 mph: 8° dwell |
| 11 | LB/BR | Injector 4 Driver | 1.0-4.0 ms |
| 12 | YL/WT | Injector 3 Driver | 1.0-4.0 ms |
| 13 | TN | Injector 2 Driver | 1.0-4.0 ms |
| 14 | WT/DG | Injector 1 Driver | 1.0-4.0 ms |
| 17 | BR/VT | HO2S-21 (B2 S1) Heater | Heater Off: 12v, On: 1v |
| 18 | BR/OR | HO2S-11 (B1 S1) Heater | Heater Off: 12v, On: 1v |
| 19 | DG | Generator Field Driver | Digital Signals: 0-12-0v |
| 20 | VT/OR | ECT Sensor Signal | At 180°F: 2.80v |
| 21 | BR/OR | TP Sensor 1 Signal | 0.6-1.0v |
| 22, 24-26 | --- | Not Used | --- |
| 23 | VT/BR | MAP Sensor Signal | 1.5-1.7v |
| 27 | BK/LB | Sensor Ground | <0.050v |
| 28 | YL/BK | IAC Motor Return | 12-14v |
| 29 | OR | 5-Volt Supply | 4.9-5.1v |
| 30 | DB/LG | IAT Sensor Signal | At 100°F: 1.83v |
| 31 | BK/DG | HO2S-11 (B1 S1) Signal | 0.1-1.1v |
| 32 | DB/DG | HO2S Return (Upstream) | <0.050v |
| 33 | LG/RD | HO2S-21 (B2 S1) Signal (California) | 0.1-1.1v |
| 34 | DB/GY | CMP Sensor Signal | Digital Signal: 0-5-0v |
| 35 | DB/WT | CKP Sensor Signal | Digital Signal: 0-5-0v |
| 36-37 | --- | Not Used | --- |
| 38 | GY/RD | IAC Motor Driver Control | DC pulses: 0.8-11v |

### 2003 Durango 4.7L VIN N 'C3' 38-Pin White Connector

| PCM Pin # | Wire Color | Circuit Description (38-Pin) | Value at Hot Idle |
|---|---|---|---|
| 1-2, 4, 13-18, 22 | --- | Not Used | --- |
| 3 | DB/YL | ASD Relay Control | Relay Off: 12v, On: 1v |
| 5 | LG/RD | Speed Control Vent Solenoid | Vacuum Decreasing: 1v |
| 6 | DB/PK | Low Speed Fan Relay | Relay Off: 12v, On: 1v |
| 7 | VT/YL | Speed Control Power Supply | 12-14v |
| 8 | WT/DG | Natural Vacuum Leak Detection Solenoid | Solenoid Off: 12v, On: 1v |
| 9 | BR/WT | HO2S-12 (B1 S2) Heater Control | Heater Off: 12v, On: 1v |
| 10 | BR/GY | HO2S-22 (B2 S2) Heater (California) | Heater Off: 12v, On: 1v |
| 11 | LB/OR | A/C Clutch Relay Control | Relay Off: 12v, On: 1v |
| 12 | VT/DG | Speed Control Vacuum Solenoid | Vacuum Increasing: 1v |
| 19, 28 | DG/OR | ASD Relay Output | 12-14v |
| 20 | PK/BK | EVAP Purge Solenoid Control | PWM Signal: 0-12-0v |
| 21 | BK/WT | Park Neutral Switch Sense (T41) | In P/N: 0v, Others: 5v |
| 24-27, 31, 36 | --- | Not Used | --- |
| 23 | WT/PK | Brake Switch Signal | Brake Off: 0v, On: 12v |
| 29 | DB/BR | EVAP Purge Solenoid Sense | <0.1v |
| 30 | DB/OR | Power Steering Pressure Switch | Straight: 0v, Turning: 5v |
| 32 | PK/YL | Battery Temperature Sensor | At 86°F: 1.96v |
| 33 | DB/WT | Fuel Level Signal | Full: 0.56v, 1/2 full: 2.5v |
| 34 | VT/TN | Speed Control Set Switch Signal | S/C & Set Switch On: 3.8v |
| 35 | OR | Natural Vacuum Leak Detection Switch | Open: 12v, Closed: 0v |
| 37 | BR | Fuel Pump Relay Control | Relay Off: 12v, On: 1v |
| 38 | DG/OR | Starter Relay Control | Relay Off: 12v, On: 1v |

### 2003 Durango 4.7L VIN N (A/T Only) 'C4' 38-Pin Green Connector

| | | | |
|---|---|---|---|
| 1 | YL/GY | Overdrive Solenoid Control | Solenoid Off: 12v, On: 1v |
| 2 | DG/WT | 4C Solenoid Control | Solenoid Off: 12v, On: 1v |
| 3, 5, 7, 9 | --- | Not Used | --- |
| 4 | YL/DB | MS Solenoid Control | Solenoid Off: 12v, On: 1v |
| 6 | WT/DB | 2C Solenoid Control | Solenoid Off: 12v, On: 1v |
| 8 | YL/LB | Underdrive Solenoid Control | Solenoid Off: 12v, On: 1v |
| 10 | LG | Low/Reverse Solenoid control | Solenoid Off: 12v, On: 1v |
| 11 | VT/LG | Pressure Control Solenoid | PWM Signals (0-12-0v) |
| 12-14 | BK/RD | Power Ground | <0.1v |
| 15 | DG/LB | TRS T1 Sense | In NOL: 0v, Others: 5v |
| 16 | DG/DB | TRS T3 Sense | In P3L: 0v, Others: 5v |
| 17 | OR/WT | Overdrive Off Switch Sense | In Overdrive: 2-4v |
| 18 | YL/BR | Transmission Control Relay Control | Relay Off: 12v, On: 1v |
| 19, 28, 38 | RD | Transmission Control Relay Output | 12-14v |
| 20 | DB | 4C Pressure Switch Sense | In 4th Position: 2-4v |
| 21 | GY | Underdrive Pressure Switch Sense | In Underdrive Position: 2-4v |
| 22 | YL/LG | Overdrive Pressure Switch Sense | In Overdrive Position: 2-4v |
| 23-25, 27, 36 | --- | Not Used | --- |
| 26 | PK/OR | TRS T2 Sense | In P/N: 0v, Others: 5v |
| 29 | YL/TN | Low/Reverse Pressure Switch | In Low/Reverse: 2-4v |
| 30 | YL/DG | 2C Pressure Switch | In 2-4 Position: 2-4v |
| 31 | VT/TN | Line Pressure Switch | In 2-4 Position: 2-4v |
| 32 | LG/WT | Output Speed Sensor | Moving: AC voltage |
| 33 | RD/BK | Input Speed Sensor | Moving: AC voltage |
| 34 | DB/BK | Speed Sensor Ground | <0.050v |
| 35 | DG/OR | Transmission Temperature Sensor | 3.2-3.4v at 104°F |
| 37 | DG/YL | TRS T42 Sense | In PRNL: 0v, Others 5v |

## 1998-2000 Durango 5.2L VIN Y 'C1' Black Connector

| PCM Pin # | Wire Color | Circuit Description (32-Pin) | Value at Hot Idle |
|---|---|---|---|
| 1 | --- | Not Used | --- |
| 2 | LG/BK | Ignition Switch Power (B+) | 12-14v |
| 3 | --- | Not Used | --- |
| 4 | BK/LB | Sensor Ground | <0.1v |
| 5 | --- | Not Used | --- |
| 6 | BK/WT | PNP Switch Sense Signal | In P/N: 0v, Others: 5v |
| 7 | BK/GY | Coil 1 Driver Control | 5°, 55 mph: 8° dwell |
| 8 | GY/BK | CKP Sensor Signal | Digital Signals: 0-5-0v |
| 9 | --- | Not Used | --- |
| 10 | YL/BK | IAC 2 Driver Control | Pulse Signals |
| 11 | BR/WT | IAC 3 Driver Control | Pulse Signals |
| 12-14 | --- | Not Used | --- |
| 15 | BK/RD | IAT Sensor Signal | At 100°F: 1.83v |
| 16 | TN/BK | ECT Sensor Signal | At 180°F: 2.80v |
| 17 | PK/WT | 5-Volt Supply | 4.9-5.1v |
| 18 | TN/YL | CMP Sensor Signal | Digital Signals: 0-5-0v |
| 19 | GY/RD | IAC 1 Driver Control | Pulse Signals |
| 20 | VT/BK | IAC 4 Driver Control | Pulse Signals |
| 21 | --- | Not Used | --- |
| 22 | RD/WT | Fused Battery Power (B+) | 12-14v |
| 23 | OR/DB | TP Sensor Signal | 0.6-1.0v |
| 24 | TN/WT | HO2S-11 (B1 S1) Signal | 0.1-1.1v |
| 25 | ON/BK | HO2S-12 (B1 S2) Signal | 0.1-1.1v |
| 26 | --- | Not Used | --- |
| 27 | DG/RD | MAP Sensor Signal | 1.5-1.6v |
| 31, 32 | BK/TN | Power Ground | <0.1v |

## 1998-2000 Durango 5.2L VIN Y 'C2' White Connector

| PCM Pin # | Wire Color | Circuit Description (32-Pin) | Value at Hot Idle |
|---|---|---|---|
| 1 | PK or GY | Transmission Temperature Sensor | At 200°F: 2.40v |
| 2 | PK/WT | Injector 7 Driver | 1-4 ms |
| 3, 7, 9, 14 | --- | Not Used | --- |
| 4 | WT/DB | Injector 1 Driver | 1-4 ms |
| 5 | YL/WT | Injector 3 Driver | 1-4 ms |
| 6 | GY | Injector 5 Driver | 1-4 ms |
| 8 (2000) | PK | Variable Force Solenoid | PWM Signal: 0-12-0v |
| 8 ('98-'99) | PK/WT | Governor Pressure Signal | 0.58v |
| 10 | DG | Generator Field Control | Digital Signals: 0-12-0v |
| 11 | OR/BK | TCC Solenoid Control | At Cruise w/TCC On: <1v |
| 12 | BR/DB | Injector 6 Driver | 1-4 ms |
| 13 | GY/LB | Injector 8 Driver | 1-4 ms |
| 15 | TN | Injector 2 Driver | 1-4 ms |
| 16 | LB/BR | Injector 4 Driver | 1-4 ms |
| 17 (2000) | DB/PK | Radiator Fan Relay Control | Relay Off: 12v, On: 1v |
| 18-20 | --- | Not Used | --- |
| 21 | BR | Overdrive Solenoid Control | Cruise w/solenoid on: <1v |
| 22 | --- | Not Used | --- |
| 23 | GY/YL | Engine Oil Pressure Sensor | 1.6v at 24 psi |
| 24 | --- | Not Used | --- |
| 25 | DB/BK | OSS Sensor (-) Signal | Moving: AC pulse signals |
| 26 | --- | Not Used | --- |
| 27 | WT/OR | Vehicle Speed Signal | Digital Signal |
| 28 | LG/WT | OSS Sensor (+) Signal | Moving: AC pulse signals |
| 29 (2000) | LG/RD | Governor Pressure Signal | 0.58v |
| 30 | PK | Transmission Relay Control | Relay Off: 12v, On: 1v |
| 31 | OR | 5-Volt Supply | 4.9-5.1v |
| 32 | --- | Not Used | --- |

### 1998-2000 Durango 5.2L VIN Y 'C3' Gray Connector

| PCM Pin # | Wire Color | Circuit Description (32-Pin) | Value at Hot Idle |
|---|---|---|---|
| 1 | DB/OR | A/C Clutch Relay Control | Relay Off: 12v, On: 1v |
| 2 ('98-'99) | DB/PK | Radiator Fan Control Relay | Relay Off: 12v, On: 1v |
| 3 | DB/YL | ASD Relay Control | Relay Off: 12v, On: 1v |
| 4 | TN/RD | Speed Control Vacuum Solenoid | Vacuum Increasing: 1v |
| 5 | LG/RD | Speed Control Vent Solenoid | Vacuum Decreasing: 1v |
| 8 | DG/BK | HO2S-11 HTR Control (CAL) | Relay Off: 12v, On: 1v |
| 9 | OR/RD | HO2S-12 HTR Control (CAL) | Relay Off: 12v, On: 1v |
| 10 ('98-'99) | YL/DG | LDP Solenoid Control | PWM Signal: 0-12-0v |
| 10 | WT/DG | LDP Solenoid Control | PWM Signal: 0-12-0v |
| 11 | YL/RD | Speed Control Power Supply | 12-14v |
| 12 | DG/OR | ASD Relay Output | 12-14v |
| 13 | OR/WT | Overdrive 'Off' Switch Sense | Switch On: 1v, Off: 12v |
| 14 ('98-'99) | YL/WT | LDP Switch Sense Signal | Closed: 0v, Open: 12v |
| 14 | OR | LDP Switch Sense Signal | Closed: 0v, Open: 12v |
| 15 | PK/YL | Battery Temperature Sensor | At 86°F: 1.96v |
| 19 ('98-'99) | LB/O | Fuel Pump Relay Control | Relay Off: 12v, On: 1v |
| 19 | DB/WT | Fuel Pump Relay Control | Relay Off: 12v, On: 1v |
| 20 | PK/BK | EVAP Purge Solenoid Control | PWM Signal: 0-12-0v |
| 22 | BR | A/C Switch Signal | A/C Off: 12v, On: 1v |
| 23 | LG/WT | A/C Select Signal | A/C Off: 12v, On: 1v |
| 24 | WT/PK | Brake Switch Sense Signal | Brake Off: 0v, On: 12v |
| 25 ('98-'99) | DG/BK | Generator Field Source | 12-14v |
| 25 | WT/DB | Generator Field Source | 12-14v |
| 26 | DB | Fuel Level Sensor Signal | Full: 0.56v, 1/2 full: 2.5v |
| 27 ('98-'99) | PK | SCI Transmit | 0v |
| 27 | PK/DB | SCI Transmit | 0v |
| 28 | WT/BK | CCD Bus (-) | <0.050v |
| 29 ('98-'99) | LG | SCI Receive | 5v |
| 29 | PK/WT | SCI Receive | 5v |
| 30 | PK/BR | CCD Bus (+) | Digital Signals: 0-5-0v |
| 32 | RD/LG | Speed Control Switch Signal | S/C & Set Switch On: 3.8v |

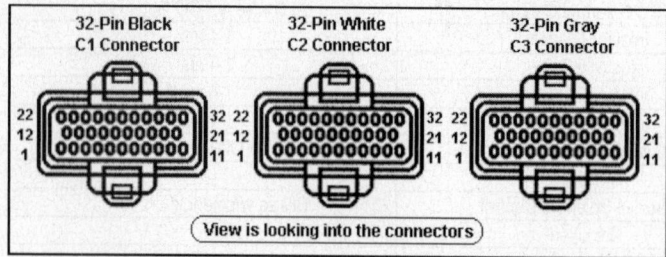

### Pin Connector Graphic

### Standard Colors and Abbreviations

| Abbreviation | Color | Abbreviation | Color | Abbreviation | Color |
|---|---|---|---|---|---|
| BK | Black | GY | Gray | RD | Red |
| BL | Blue | GN | Green | TN | Tan |
| BR | Brown | LG | Light Green | VT | Violet |
| DB | Dark Blue | OR | Orange | WT | White |
| DG | Dark Green | PK | Pink | YL | Yellow |

## 1998-2000 Durango 5.9L VIN Z 'A' Black Connector

| PCM Pin # | Wire Color | Circuit Description (32-Pin) | Value at Hot Idle |
|-----------|-----------|------------------------------|-------------------|
| 1, 5, 9 | --- | Not Used | --- |
| 2 | LG/BK | Ignition Switch Power (B+) | 12-14v |
| 3 | --- | Not Used | --- |
| 4 | BK/LB | Sensor Ground | <0.1v |
| 6 | BK/WT | PNP Switch Sense Signal | In P/N: 0v, Others: 5v |
| 7 | BK/GY | Coil 1 Driver Control | 5°, 55 mph: 8° dwell |
| 8 | GY/BK | CKP Sensor Signal | Digital Signals: 0-5-0v |
| 10 | YL/BK | IAC 2 Driver Control | Pulse Signals |
| 11 | BR/WT | IAC 3 Driver Control | Pulse Signals |
| 12-14 | --- | Not Used | --- |
| 15 | BK/RD | IAT Sensor Signal | At 100°F: 1.83v |
| 16 | TN/BK | ECT Sensor Signal | At 180°F: 2.80v |
| 17 | PK/WT | 5-Volt Supply | 4.9-5.1v |
| 18 | TN/YL | CMP Sensor Signal | Digital Signals: 0-5-0v |
| 19 | GY/RD | IAC 1 Driver Control | Pulse Signals |
| 20 | PK/BK | IAC 4 Driver Control | Pulse Signals |
| 21 | --- | Not Used | --- |
| 22 | RD/WT | Fused Battery Power (B+) | 12-14v |
| 23 | OR/DB | TP Sensor Signal | 0.6-1.0v |
| 24 | TN/WT | HO2S-11 (B1 S1) Signal | 0.1-1.1v |
| 25 ('98-'99) | OR/BK | HO2S-13 (Bank 1 Sensor 3) | 0.1-1.1v |
| 25 | OR/BK | HO2S-12 (B1 S2) Signal | 0.1-1.1v |
| 26 ('98-'99) | OR/TN | HO2S-21 (B2 S1) Signal | 0.1-1.1v |
| 27 | DG/RD | MAP Sensor Signal | 1.5-1.6v |
| 28 | --- | Not Used | --- |
| 29 ('98-'99) | TN/WT | HO2S-12 (B1 S2) Signal | 0.1-1.1v |
| 30 | --- | Not Used | --- |
| 31, 32 | BK/TN | Power Ground | <0.1v |

## 1998-2000 Durango 5.9L VIN Z 'B' White Connector

| PCM Pin # | Wire Color | Circuit Description (32-Pin) | Value at Hot Idle |
|-----------|-----------|------------------------------|-------------------|
| 1 | GY/BK | Transmission Temperature Sensor | At 200°F: 2.40v |
| 2 | PK | Injector 7 Driver | 1-4 ms |
| 3, 7, 9 | --- | Not Used | --- |
| 4 | WT/DB | Injector 1 Driver | 1-4 ms |
| 5 | YL/WT | Injector 3 Driver | 1-4 ms |
| 6 | GY | Injector 5 Driver | 1-4 ms |
| 8 | PK/WT | Variable Force Solenoid | PWM Signal: 0-12-0v |
| 10 | DG | Generator Field Control | Digital Signals: 0-12-0v |
| 11 | OR/BK | TCC Solenoid Control | At Cruise w/TCC On: <1v |
| 12 | BR/DB | Injector 6 Driver | 1-4 ms |
| 13 | GY/LB | Injector 8 Driver | 1-4 ms |
| 15 | TN | Injector 2 Driver | 1-4 ms |
| 16 | LB/BR | Injector 4 Driver | 1-4 ms |
| 17 (2000) | DB/PK | Radiator Fan Control Relay | Relay Off: 12v, On: 1v |
| 21 | BR | 3-4 Shift Solenoid Control | Cruise w/solenoid on: <1v |
| 23 | GY/YL | Engine Oil Pressure Sensor | 1.6v at 24 psi |
| 25 | DB | OSS Sensor (-) Signal | Moving: AC pulse signals |
| 27 | WT/OR | Vehicle Speed Signal | Digital Signal |
| 29 | LG/RD | Governor Pressure Signal | 0.58v |
| 28 | LG | OSS Sensor (+) Signal | Moving: AC pulse signals |
| 30 | PK | Transmission Relay Control | Relay Off: 12v, On: 1v |
| 31 | OR | 5-Volt Supply | 4.9-5.1v |

## 1998-2000 Durango 5.9L VIN Z 'C' Gray Connector

| PCM Pin # | Wire Color | Circuit Description (32-Pin) | Value at Hot Idle |
|---|---|---|---|
| 1 | DB/OR | A/C Clutch Relay Control | Relay Off: 12v, On: 1v |
| 2 | --- | Not Used | --- |
| 3 | DB/YL | ASD Relay Control | Relay Off: 12v, On: 1v |
| 4 | TN/RD | Speed Control Vacuum Solenoid | Vacuum Increasing: 1v |
| 5 | LG/RD | Speed Control Vent Solenoid | Vacuum Decreasing: 1v |
| 6-7 | --- | Not Used | --- |
| 8 | DG/BK | HO2S-11 HTR Control (California) | Relay Off: 12v, On: 1v |
| 9 | OR/RD | HO2S-12 HTR Control (California) | Relay Off: 12v, On: 1v |
| 10 ('98-'99) | YL/DG | LDP Solenoid Control | PWM Signal: 0-12-0v |
| 10 | WT/DG | LDP Solenoid Control | PWM Signal: 0-12-0v |
| 11 | YL/RD | Speed Control Power Supply | 12-14v |
| 12 | DG/OR | ASD Relay Output | 12-14v |
| 13 | OR/WT | Overdrive 'Off' Switch Sense | Switch On: 1v, Off: 12v |
| 14 | OR | LDP Switch Sense Signal | Closed: 0v, Open: 12v |
| 15 | PK/YL | Battery Temperature Sensor | At 86°F: 1.96v |
| 16-18, 21 | --- | Not Used | --- |
| 19 | DB/WT | Fuel Pump Relay Control | Relay Off: 12v, On: 1v |
| 20 | PK/BK | EVAP Purge Solenoid Control | PWM Signal: 0-12-0v |
| 22 | BR | A/C Switch Signal | A/C Off: 12v, On: 1v |
| 23 | LG/WT | A/C Select Signal | A/C Off: 12v, On: 1v |
| 24 | WT/PK | Brake Switch Sense Signal | Brake Off: 0v, On: 12v |
| 25 | WT/DB | Generator Field Source | 12-14v |
| 26 | DB | Fuel Level Sensor Signal | Full: 0.56v, 1/2 full: 2.5v |
| 27 | PK | SCI Transmit | 0v |
| 28 | WT | CCD Bus (-) | <0.050v |
| 29 | PK/WT | SCI Receive | 5v |
| 30 | PK/BR | CCD Bus (+) | Digital Signal: 0-5-0v |
| 31 | --- | Not Used | --- |
| 32 | RD/LG | Speed Control Switch Signal | S/C & Set Switch On: 3.8v |

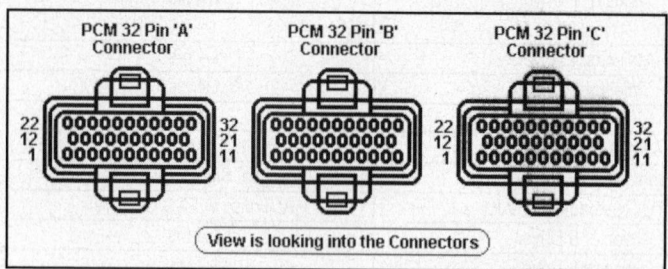

PCM 32 Pin 'A' Connector   PCM 32 Pin 'B' Connector   PCM 32 Pin 'C' Connector

View is looking into the Connectors

**Pin Connector Graphic**

## Standard Colors and Abbreviations

| Abbreviation | Color | Abbreviation | Color | Abbreviation | Color |
|---|---|---|---|---|---|
| BK | Black | GY | Gray | RD | Red |
| BL | Blue | GN | Green | TN | Tan |
| BR | Brown | LG | Light Green | VT | Violet |
| DB | Dark Blue | OR | Orange | WT | White |
| DG | Dark Green | PK | Pink | YL | Yellow |

**2001-03 Durango 5.9L VIN Z 'C1' Black Connector**

| PCM Pin # | Wire Color | Circuit Description (32-Pin) | Value at Hot Idle |
|---|---|---|---|
| 1, 3, 5, 9, 12-14 | --- | Not Used | --- |
| 2 | LG/BK | Ignition Switch Power (Run-Start) | 12-14v |
| 4 | BK/LB | Sensor Ground | <0.1v |
| 6 | BK/WT | PNP Switch Sense Signal | In P/N: 0v, Others: 5v |
| 7 | BK/GY | Coil 1 Driver Control | 5°, 55 mph: 8° dwell |
| 8 | GY/BK | CKP Sensor Signal | Digital Signals: 0-5-0v |
| 10 | YL/BK | IAC 2 Driver Control | Pulse Signals |
| 11 | BR/WT | IAC 3 Driver Control | Pulse Signals |
| 15 | BK/RD | IAT Sensor Signal | At 100°F: 1.83v |
| 16 | TN/BK | ECT Sensor Signal | At 180°F: 2.80v |
| 17 | OR | 5-Volt Supply | 4.9-5.1v |
| 18 | TN/YL | CMP Sensor Signal | Digital Signals: 0-5-0v |
| 19 | GY/RD | IAC 1 Driver Control | Pulse Signals |
| 20 | VT/BK | IAC 4 Driver Control | Pulse Signals |
| 21 | --- | Not Used | --- |
| 22 | RD/WT | Fused Battery Power (B+) | 12-14v |
| 23 | OR/DB | TP Sensor Signal | 0.6-1.0v |
| 24 | BK/DG | HO2S-11 (B1 S1) Signal | 0.1-1.1v |
| 25 | TN/WT | HO2S-12 (B1 S2) Signal (California) | 0.1-1.1v |
| 26 | LG/RD | HO2S-21 (B2 S1) Signal | 0.1-1.1v |
| 27 | DG/RD | MAP Sensor Signal | 1.5-1.6v |
| 28 | --- | Not Used | --- |
| 29 | TN/WT | HO2S-22 (B2 S2) Signal (California) | 0.1-1.1v |
| 30 | --- | Not Used | --- |
| 31 | BK/TN | Power Ground | <0.1v |
| 32 | BK/TN | Power Ground | <0.1v |

**2001-03 Durango 5.9L VIN Z 'C2' White Connector**

| PCM Pin # | Wire Color | Circuit Description (32-Pin) | Value at Hot Idle |
|---|---|---|---|
| 1 | GY/BK | Transmission Temperature Sensor | At 200°F: 2.40v |
| 2 | VT | Injector 7 Driver Control | 1-4 ms |
| 3, 7 | --- | Not Used | --- |
| 4 | WT/DB | Injector 1 Driver Control | 1-4 ms |
| 5 | YL/WT | Injector 3 Driver Control | 1-4 ms |
| 6 | GY | Injector 5 Driver Control | 1-4 ms |
| 8 | VT/WT | Governor Pressure Solenoid | PWM Signal: 0-12-0v |
| 9, 14 | --- | Not Used | --- |
| 10 | DB | Generator Field Control | Digital Signals: 0-12-0v |
| 11 | OR/BK | TCC Solenoid Control | At Cruise w/TCC On: 1v |
| 12 | BR/DB | Injector 6 Driver Control | 1-4 ms |
| 13 | GY/LB | Injector 8 Driver Control | 1-4 ms |
| 15 | TN | Injector 2 Driver Control | 1-4 ms |
| 16 | LB/BR | Injector 4 Driver Control | 1-4 ms |
| 17 | DB/PK | Radiator Fan Control Relay | Relay Off: 12v, On: 1v |
| 18, 22 | --- | Not Used | --- |
| 19 | DB | AC Pressure Sensor Signal | 0.90v at 79 psi |
| 21 | BR | Overdrive Solenoid Control | Cruise w/solenoid on: <1v |
| 23 | GY/YL | Engine Oil Pressure Sensor | 1.6v at 24 psi |
| 24, 26 | --- | Not Used | --- |
| 25 | DB/BK | OSS Sensor (-) Signal | Moving: AC pulse signals |
| 27 | WT/OR | Vehicle Speed Sensor | Digital Signals |
| 28 | LG/WT | OSS Sensor (+) | Moving: AC pulse signals |
| 29 | LG/RD | Governor Pressure Signal | 0.58v |
| 30 | PK | Transmission Relay Control | Relay Off: 12v, On: 1v |
| 31 | VT/WT | 5-Volt Supply | 4.9-5.1v |
| 32 | --- | Not Used | --- |

## 2001-03 Durango 5.9L VIN Z 'C3' Gray Connector

| PCM Pin # | Wire Color | Circuit Description (32-Pin) | Value at Hot Idle |
|---|---|---|---|
| 1 | DB/OR | A/C Clutch Relay Control | Relay Off: 12v, On: 1v |
| 2 | --- | Not Used | --- |
| 3 | DB/YL | ASD Relay Control | Relay Off: 12v, On: 1v |
| 4 | TN/RD | Speed Control Vacuum Solenoid | Vacuum Increasing: 1v |
| 5 | LG/RD | Speed Control Vent Solenoid | Vacuum Decreasing: 1v |
| 6-7 | --- | Not Used | --- |
| 8 | VT/WT | HO2S-11 (B1 S1) Heater Control | Relay Off: 12v, On: 1v |
| 9 | DG/BK | HO2S-12 (B1 S2) Heater Control | Relay Off: 12v, On: 1v |
| 10 | WT/DG | LDP Solenoid Control | PWM Signal: 0-12-0v |
| 11 | YL/RD | Speed Control Power Supply | 12-14v |
| 12 | DG/OR | ASD Relay Output | 12-14v |
| 13 | OR/WT | Overdrive 'Off' Switch Sense | Switch On: 1v, Off: 12v |
| 14 | OR | LDP Switch Sense Signal | Closed: 0v, Open: 12v |
| 15 | PK/YL | Battery Temperature Sensor | At 86ºF: 1.96v |
| 16 | VT/OR | HO2S-21 (B2 S1) Heater Control | Heater Off: 12v, On: PWM 0-12-0v |
| 17-18 | --- | Not Used | --- |
| 19 | BR | Fuel Pump Relay Control | Relay Off: 12v, On: 1v |
| 20 | PK/BK | EVAP Purge Solenoid Control | PWM Signal: 0-12-0v |
| 21 | --- | Not Used | --- |
| 22 | BR | A/C Switch Signal | A/C Off: 12v, On: 1v |
| 23 | --- | Not Used | --- |
| 24 | WT/PK | Brake Switch Sense | Brake Off: 0v, On: 12v |
| 25 | WT/DB | Generator Field Source | 12-14v |
| 26 | DB/WT | Fuel Level Sensor Signal | Full: 0.56v, 1/2 full: 2.5v |
| 27 | PK | SCI Transmit | 0v |
| 28 | --- | Not Used | --- |
| 29 | LG | SCI Receive | 5v |
| 30 | VT/YL | PCI Data Bus (J1850) | Digital Signals: 0-7-0v |
| 31 | --- | Not Used | --- |
| 32 | RD/LG | Speed Control Switch Signal | S/C & Set Switch On: 3.8v |

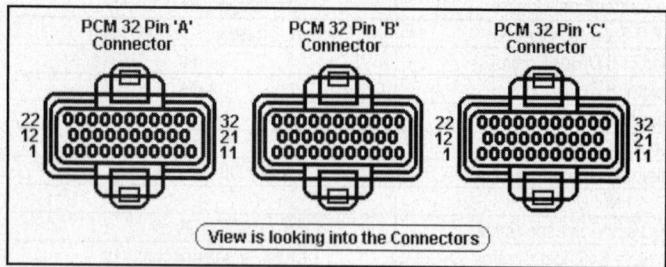

Pin Connector Graphic

**1993-95 Cherokee 2.5L VIN P 60-Pin Connector**

| PCM Pin # | Wire Color | Circuit Description (60-Pin) | Value at Hot Idle |
|---|---|---|---|
| 1 | DG/RD | MAP Sensor Signal | 1.5-1.6v |
| 2 | TN/BK | ECT Sensor Signal | At 180°F: 2.80v |
| 3 | RD | Fused Battery Power (B+) | 12-14v |
| 4 | BK/LB | Sensor Ground | <0.1v |
| 5 | BK/WT | Sensor Ground | <0.1v |
| 6 | PK/WT | 5-Volt Supply | 4.9-5.1v |
| 7 | OR | 8-Volt Supply | 7.9-8.1v |
| 8 | --- | Not Used | --- |
| 9 | DB | Ignition Switch Power (B+) | 12-14v |
| 10 | DB/WT | PSP Switch Sense Signal | Straight: 0v, Turning: 5v |
| 10 ('94-'95) | PK | PSP Switch Sense Signal | Straight: 0v, Turning: 5v |
| 11 | BK/TN | Power Ground | <0.1v |
| 12 | BK/TN | Power Ground | <0.1v |
| 13 | LB/BR | Injector 4 Driver | 1-4 ms |
| 14 | YL/WT | Injector 3 Driver | 1-4 ms |
| 15 | TN | Injector 2 Driver | 1-4 ms |
| 16 | WT/DB | Injector 1 Driver | 1-4 ms |
| 17-18 | --- | Not Used | --- |
| 19 | GY | Coil 1 Driver Control | 5°, 55 mph: 8° dwell |
| 20 | DG | Alternator Field Control | Digital Signals: 0-12-0v |
| 21 | BK/RD | IAT Sensor Signal | At 100°F: 1.83v |
| 22 | OR/DB | TP Sensor Signal | 0.6-1.0v |
| 23 | --- | Not Used | --- |
| 24 | GY/BK | CKP Sensor Signal | Digital Signals: 0-5-0v |
| 25 | PK | SCI Transmit | 0v |
| 26 | --- | Not Used | --- |
| 27 | LB | A/C Request Switch Sense | A/C Off: 12v, On: 1v |
| 28 | LG | A/C Select Switch Sense | A/C Off: 12v, On: 1v |
| 29 | WT/PK | Brake Switch Sense Signal | Brake Off: 0v, On: 12v |
| 30 | BR/YL | PNP Switch Sense Signal | In P/N: 0v, Others: 5v |
| 31 | --- | Not Used | --- |
| 32 | BK/PK | MIL (lamp) Control | MIL Off: 12v, On: 1v |
| 33 | TN/RD | Speed Control Vacuum Solenoid | Vacuum Increasing: 1v |
| 34 | DB/OR | A/C Clutch Relay Control | Relay Off: 12v, On: 1v |
| 35 | --- | Not Used | --- |
| 36 | DG/YL | Alternator Lamp Control | Lamp Off: 12v, On: 1v |
| 37 ('93) | RD/DB | Ballast Bypass Relay Control | Relay Off: 12v, On: 1v |
| 38 | --- | Not Used | --- |
| 39 | GY/RD | AIS Motor 1 Circuit Control | Pulse Signals |
| 40 | BR/WT | AIS Motor 3 Circuit Control | Pulse Signals |

## Standard Colors and Abbreviations

| Abbreviation | Color | Abbreviation | Color | Abbreviation | Color |
|---|---|---|---|---|---|
| BK | Black | GY | Gray | RD | Red |
| BL | Blue | GN | Green | TN | Tan |
| BR | Brown | LG | Light Green | VT | Violet |
| DB | Dark Blue | OR | Orange | WT | White |
| DG | Dark Green | PK | Pink | YL | Yellow |

### 1993-95 Cherokee 2.5L VIN P 60-Pin Connector

| PCM Pin # | Wire Color | Circuit Description (60-Pin) | Value at Hot Idle |
|---|---|---|---|
| 41 | BK/DG | HO2S-11 (B1 S1) Signal | 0.1-1.1v |
| 42 | --- | Not Used | --- |
| 43 | GY/LB | Tachometer Signal | Pulse Signals |
| 44 | TN/YL | CMP Sensor Signal | Digital Signals: 0-5-0v |
| 45 | LG | SCI Receive | 5v |
| 46 | --- | Not Used | --- |
| 47 | WT/OR | Vehicle Speed Signal | Digital Signals |
| 48 | BR/RD | Speed Control Set Switch | S/C & Set Switch On: 3.8v |
| 49 | YL/RD | Speed Control On/Off Switch | S/C Off: 12v, On: 1v |
| 50 | WT/LG | Speed Control Resume Switch | S/C Off: 12v, On: 1v |
| 51 | DB/YL | ASD Relay Control | Relay Off: 12v, On: 1v |
| 52 | --- | Not Used | --- |
| 53 | LG/RD | Speed Control Vent Solenoid | Vacuum Decreasing: 1v |
| 54 | OR/BK | A/T: Lockup Torque Converter | TCC on at Cruise: 1v |
| 54 | OR/BK | M/T: SIL (lamp) Control | Lamp Off: 12v, On: 1v |
| 55 | --- | Not Used | --- |
| 56 ('93) | GY/PK | Maintenance Indicator Lamp | Lamp Off: 12v, On: 1v |
| 57 | DG/OR | ASD Relay Output | 12-14v |
| 58 | --- | Not Used | --- |
| 59 | PK/BK | AIS Motor 4 Circuit Control | Pulse Signals |
| 60 | YL/BK | AIS Motor 2 Circuit Control | Pulse Signals |

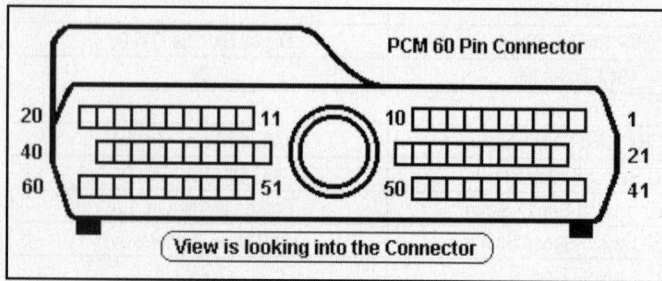

**Pin Connector Graphic**

### Standard Colors and Abbreviations

| Abbreviation | Color | Abbreviation | Color | Abbreviation | Color |
|---|---|---|---|---|---|
| BK | Black | GY | Gray | RD | Red |
| BL | Blue | GN | Green | TN | Tan |
| BR | Brown | LG | Light Green | VT | Violet |
| DB | Dark Blue | OR | Orange | WT | White |
| DG | Dark Green | PK | Pink | YL | Yellow |

### 1996 Cherokee 2.5L VIN P 'A' Black Connector

| PCM Pin # | Wire Color | Circuit Description (32-Pin) | Value at Hot Idle |
|---|---|---|---|
| 1 | --- | Not Used | --- |
| 2 | DB | Ignition Switch Power (B+) | 12-14v |
| 3 | --- | Not Used | --- |
| 4 | BK/LB | Sensor Ground | <0.1v |
| 5 | --- | Not Used | --- |
| 6 | BR/YL | PNP Switch Sense Signal | In P/N: 0v, Others: 5v |
| 6 | BK | M/T: Sensor Ground | <0.1v |
| 7 | GY | Coil 1 Driver Control | 5°, 55 mph: 8° dwell |
| 8 | GY/BK | CKP Sensor Signal | Digital Signals: 0-5-0v |
| 9 | --- | Not Used | --- |
| 10 | YL/BK | IAC 2 Driver Control | Pulse Signals |
| 11 | BR/WT | IAC 3 Driver Control | Pulse Signals |
| 12 | DB/OR | PSP Switch Sense Signal | Straight: 0v, Turning: 5v |
| 13-14 | --- | Not Used | --- |
| 15 | BK/RD | IAT Sensor Signal | At 100°F: 1.83v |
| 16 | TN/BK | ECT Sensor Signal | At 180°F: 2.80v |
| 17 | PK/WT | 5-Volt Supply | 4.9-5.1v |
| 18 | TN/YL | CMP Sensor Signal | Digital Signals: 0-5-0v |
| 19 | GY/RD | IAC 1 Driver Control | Pulse Signals |
| 20 | PK/BK | IAC 4 Driver Control | Pulse Signals |
| 21 | --- | Not Used | --- |
| 22 | RD | Fused Battery Power (B+) | 12-14v |
| 23 | OR/DB | TP Sensor Signal | 0.6-1.0v |
| 24 | BK/DG | HO2S-11 (B1 S1) Signal | 0.1-1.1v |
| 25 | TN/WT | HO2S-12 (B1 S2) Signal | 0.1-1.1v |
| 26 | --- | Not Used | --- |
| 27 | DG/RD | MAP Sensor Signal | 1.5-1.6v |
| 28-30 | --- | Not Used | --- |
| 31-32 | BK/TN | Power Ground | <0.1v |

### 1996 Cherokee 2.5L VIN P 'B' White Connector

| PCM Pin # | Wire Color | Circuit Description (32-Pin) | Value at Hot Idle |
|---|---|---|---|
| 1-3 | --- | Not Used | --- |
| 4 | WT/DB | Injector 1 Driver | 1-4 ms |
| 5 | YL/WT | Injector 3 Driver | 1-4 ms |
| 6-9 | --- | Not Used | --- |
| 10 | DG | Generator Field Control | Digital Signals: 0-12-0v |
| 11 | OR/BK | A/T: TCC Solenoid Control | TCC on at Cruise: 1v |
| 11 | OR/BK | M/T: SIL (lamp) Control | Lamp Off: 12v, On: 1v |
| 12-14 | --- | Not Used | --- |
| 15 | TN | Injector 2 Driver | 1-4 ms |
| 16 | LB/BR | Injector 4 Driver | 1-4 ms |
| 17-26 | --- | Not Used | --- |
| 27 | WT/OR | Vehicle Speed Signal | Digital Signal |
| 28-30 | --- | Not Used | --- |
| 31 | RD/YL | 5-Volt Supply | 4.9-5.1v |
| 32 | --- | Not Used | --- |

## 1996 Cherokee 2.5L VIN P 'C' Gray Connector

| PCM Pin # | Wire Color | Circuit Description (32-Pin) | Value at Hot Idle |
|---|---|---|---|
| 1 | DB/OR | A/C Clutch Relay Control | Relay Off: 12v, On: 1v |
| 2 | --- | Not Used | --- |
| 3 | DB/YL | ASD Relay Control | Relay Off: 12v, On: 1v |
| 4 | TN/RD | Speed Control Vacuum Solenoid | Vacuum Increasing: 1v |
| 5 | LG/RD | Speed Control Vent Solenoid | Vacuum Decreasing: 1v |
| 6-10 | --- | Not Used | --- |
| 11 | YL/RD | Speed Control Power Supply | 12-14v |
| 12 | DG/OR | ASD Relay Output | 12-14v |
| 13-14 | --- | Not Used | --- |
| 15 | PK/YL | Battery Temperature Sensor | At 86°F: 1.96v |
| 16 | PK/BK | Generator Lamp Control | Lamp Off: 12v, On: 1v |
| 17 | BK/PK | MIL (lamp) Control | MIL Off: 12v, On: 1v |
| 18 | --- | Not Used | --- |
| 19 | BR | Fuel Pump Relay Control | Relay Off: 12v, On: 1v |
| 20 | PK/BK | EVAP Purge Solenoid Control | PWM Signal: 0-12-0v |
| 21 | --- | Not Used | --- |
| 22 | DB/OR | A/C Request Signal | A/C Off: 12v, On: 1v |
| 23 | LG | A/C Select Signal | A/C Off: 12v, On: 1v |
| 24 | WT/PK | Brake Switch Sense Signal | Brake Off: 0v, On: 12v |
| 25 | --- | Not Used | --- |
| 26 | DB | Fuel Level Sensor Signal | Full: 0.5v, 1/2 full: 2.5v |
| 27 | PK | SCI Transmit | 0v |
| 28 | WT/BK | CCD Bus (-) | <0.050v |
| 29 | LG | SCI Receive | 5v |
| 30 | PK/BR | CCD Bus (+) | Digital Signals: 0-5-0v |
| 31 | GY/LB | Tachometer Signal | Pulse Signals |
| 32 | WT/LG | Speed Control Switch Signal | S/C & Set Switch On: 3.8v |

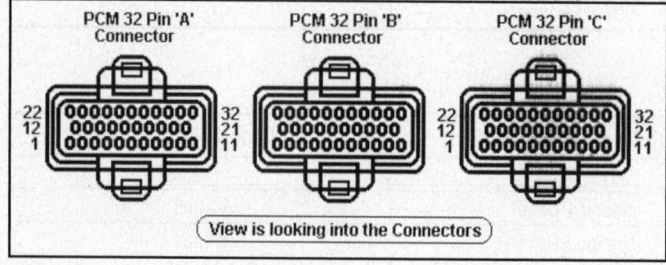

**Pin Connector Graphic**

## Standard Colors and Abbreviations

| Abbreviation | Color | Abbreviation | Color | Abbreviation | Color |
|---|---|---|---|---|---|
| BK | Black | GY | Gray | RD | Red |
| BL | Blue | GN | Green | TN | Tan |
| BR | Brown | LG | Light Green | VT | Violet |
| DB | Dark Blue | OR | Orange | WT | White |
| DG | Dark Green | PK | Pink | YL | Yellow |

## 1997 Cherokee 2.5L VIN P 'A' Black Connector

| PCM Pin # | Wire Color | Circuit Description (32-Pin) | Value at Hot Idle |
|---|---|---|---|
| 1 | --- | Not Used | --- |
| 2 | DB/WT | Ignition Switch Power (B+) | 12-14v |
| 3 | --- | Not Used | --- |
| 4 | BR/YL | Sensor Ground | <0.1v |
| 5 | --- | Not Used | --- |
| 6 | BK/WT | A/T: PNP Switch Signal | In P/N: 0v, Others: 5v |
| 6 | BK | M/T: Sensor Ground | <0.1v |
| 7 | GY | Coil 1 Driver Control | 5°, 55 mph: 8° dwell |
| 8 | GY/BK | CKP Sensor Signal | Digital Signals: 0-5-0v |
| 9 | --- | Not Used | --- |
| 10 | YL/BK | IAC 2 Driver Control | Pulse Signals |
| 11 | BR/WT | IAC 3 Driver Control | Pulse Signals |
| 12 | DB/BR | PSP Switch Sense Signal | Straight: 0v, Turning: 5v |
| 13-14 | --- | Not Used | --- |
| 15 | BK/RD | IAT Sensor Signal | At 100°F: 1.83v |
| 16 | TN/BK | ECT Sensor Signal | At 180°F: 2.80v |
| 17 | OR | 5-Volt Supply | 4.9-5.1v |
| 18 | TN/YL | CMP Sensor Signal | Digital Signals: 0-5-0v |
| 19 | GY/RD | IAC 1 Driver Control | Pulse Signals |
| 20 | PK/BK | IAC 4 Driver Control | Pulse Signals |
| 21 | --- | Not Used | --- |
| 22 | DG/BK | Fused Battery Power (B+) | 12-14v |
| 23 | OR/DB | TP Sensor Signal | 0.6-1.0v |
| 24 | BK/DG | HO2S-11 (B1 S1) Signal | 0.1-1.1v |
| 25 | TN/BK | HO2S-12 (B1 S2) Signal | 0.1-1.1v |
| 26 | --- | Not Used | --- |
| 27 | DG/RD | MAP Sensor Signal | 1.5-1.6v |
| 28-30 | --- | Not Used | --- |
| 31 | BK/TN | Power Ground | <0.1v |
| 32 | BK/TN | Power Ground | <0.1v |

## 1997 Cherokee 2.5L VIN P 'B' White Connector

| PCM Pin # | Wire Color | Circuit Description (32-Pin) | Value at Hot Idle |
|---|---|---|---|
| 1-3 | --- | Not Used | --- |
| 4 | WT/DB | Injector 1 Driver | 1-4 ms |
| 5 | YL/WT | Injector 3 Driver | 1-4 ms |
| 6-9 | --- | Not Used | --- |
| 10 | DG | Generator Field Control | Digital Signals: 0-12-0v |
| 11-14 | --- | Not Used | --- |
| 15 | T | Injector 2 Driver | 1-4 ms |
| 16 | LB/BR | Injector 4 Driver | 1-4 ms |
| 17-22 | --- | Not Used | --- |
| 23 | GY/YL | Engine Oil Pressure Sensor | 1.6v at 24 psi |
| 24-26, 28-30 | --- | Not Used | --- |
| 27 | WT/OR | Vehicle Speed Signal | Digital Signals |
| 31 | PK/OR | 5-Volt Supply | 4.9-5.1v |
| 32 | --- | Not Used | --- |

## 1997 Cherokee 2.5L VIN P 'C' Gray Connector

| PCM Pin # | Wire Color | Circuit Description (32-Pin) | Value at Hot Idle |
|---|---|---|---|
| 1 | DB/OR | A/C Clutch Relay Control | Relay Off: 12v, On: 1v |
| 2 | --- | Not Used | --- |
| 3 | DB/YL | ASD Relay Control | Relay Off: 12v, On: 1v |
| 4 | TN/RD | Speed Control Vacuum Solenoid | Vacuum Increasing: 1v |
| 5 | LG/RD | Speed Control Vent Solenoid | Vacuum Decreasing: 1v |
| 6-10 | --- | Not Used | --- |
| 11 | YL/RD | Speed Control Power Supply | 12-14v |
| 12 | DG/OR | ASD Relay Output | 12-14v |
| 13-14 | --- | Not Used | --- |
| 15 | PK/YL | Battery Temperature Sensor | At 86°F: 1.96v |
| 16-18 | --- | Not Used | --- |
| 19 | BR | Fuel Pump Relay Control | Relay Off: 12v, On: 1v |
| 20 | PK/BK | EVAP Purge Solenoid Control | PWM Signal: 0-12-0v |
| 21 | --- | Not Used | --- |
| 22 | DB/WT | A/C Request Signal | A/C Off: 12v, On: 1v |
| 23 | LG | A/C Select Signal | A/C Off: 12v, On: 1v |
| 24 | WT/PK | Brake Switch Sense Signal | Brake Off: 0v, On: 12v |
| 25 | --- | Not Used | --- |
| 26 | DB/LG | Fuel Level Sensor Signal | Full: 0.5v, 1/2 full: 2.5v |
| 27 | PK | SCI Transmit | 0v |
| 28 | WT/BK | CCD Bus (-) | <0.050v |
| 29 | LG | SCI Receive | 5v |
| 30 | PK/BR | CCD Bus (+) | Digital Signals: 0-5-0v |
| 31 | --- | Not Used | --- |
| 32 | RD/LG | Speed Control Switch Signal | S/C & Set Switch On: 3.8v |

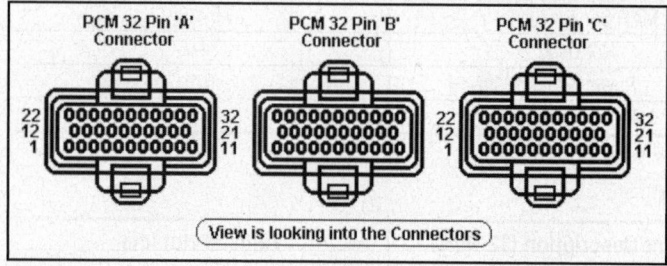

**Pin Connector Graphic**

## Standard Colors and Abbreviations

| Abbreviation | Color | Abbreviation | Color | Abbreviation | Color |
|---|---|---|---|---|---|
| BK | Black | GY | Gray | RD | Red |
| BL | Blue | GN | Green | TN | Tan |
| BR | Brown | LG | Light Green | VT | Violet |
| DB | Dark Blue | OR | Orange | WT | White |
| DG | Dark Green | PK | Pink | YL | Yellow |

**1998-2000 Cherokee 2.5L VIN P 'A' Black Connector**

| PCM Pin # | Wire Color | Circuit Description (32-Pin) | Value at Hot Idle |
|---|---|---|---|
| 1 | --- | Not Used | --- |
| 2 | DB/WT | Ignition Switch Power (B+) | 12-14v |
| 3 | --- | Not Used | --- |
| 4 | BR/YL | Sensor Ground | <0.1v |
| 5 | --- | Not Used | --- |
| 6 | BK/WT | A/T: PNP Switch Signal | In P/N: 0v, Others: 5v |
| 6 | BK | M/T: Sensor Ground | <0.1v |
| 7 | GY | Coil 1 Driver Control | 5º, 55 mph: 8º dwell |
| 8 | GY/BK | CKP Sensor Signal | Digital Signals: 0-5-0v |
| 9 | --- | Not Used | --- |
| 10 | YL/BK | IAC 2 Driver Control | Pulse Signals |
| 11 | BR/WT | IAC 3 Driver Control | Pulse Signals |
| 12 | DB/BR | PSP Switch Sense Signal | Straight: 0v, Turning: 5v |
| 13-14 | --- | Not Used | --- |
| 15 | BK/RD | IAT Sensor Signal | At 100ºF: 1.83v |
| 16 | TN/BK | ECT Sensor Signal | At 180ºF: 2.80v |
| 17 | OR | 5-Volt Supply | 4.9-5.1v |
| 18 | TN/YL | CMP Sensor Signal | Digital Signals: 0-5-0v |
| 19 | GY/RD | IAC 1 Driver Control | Pulse Signals |
| 20 | PK/BK | IAC 4 Driver Control | Pulse Signals |
| 21 | --- | Not Used | --- |
| 22 | DG/BK | Fused Battery Power (B+) | 12-14v |
| 23 | OR/DB | TP Sensor Signal | 0.6-1.0v |
| 24 | BK/DG | HO2S-11 (B1 S1) Signal | 0.1-1.1v |
| 25 ('98) | TN/BK | HO2S-12 (B1 S2) Signal | 0.1-1.1v |
| 25 | TN/WT | HO2S-12 (B1 S2) Signal | 0.1-1.1v |
| 26 | --- | Not Used | --- |
| 27 | DG/RD | MAP Sensor Signal | 1.5-1.6v |
| 28-30 | --- | Not Used | --- |
| 31-32 | BK/TN | Power Ground | <0.1v |

**1998-2000 Cherokee 2.5L VIN P 'B' White Connector**

| PCM Pin # | Wire Color | Circuit Description (32-Pin) | Value at Hot Idle |
|---|---|---|---|
| 1-3 | --- | Not Used | --- |
| 4 | WT/DB | Injector 1 Driver | 1-4 ms |
| 5 | YL/WT | Injector 3 Driver | 1-4 ms |
| 6-9 | --- | Not Used | --- |
| 10 | DG | Generator Field Control | Digital Signals: 0-12-0v |
| 11-12 | --- | Not Used | --- |
| 13 | OR/BK | TCC Solenoid Control | TCC on at Cruise: 1v |
| 14 | --- | Not Used | --- |
| 15 | TN | Injector 2 Driver | 1-4 ms |
| 16 | LB/BR | Injector 4 Driver | 1-4 ms |
| 17-22 | --- | Not Used | --- |
| 23 | GY/YL | Engine Oil Pressure Sensor | 1.6v at 24 psi |
| 24-26, 28-30 | --- | Not Used | --- |
| 27 | WT/OR | Vehicle Speed Signal | Digital Signal |
| 31 | PK/OR | 5-Volt Supply | 4.9-5.1v |
| 32 | --- | Not Used | --- |

## 1998-2000 Cherokee 2.5L VIN P 'C' Gray Connector

| PCM Pin # | Wire Color | Circuit Description (32-Pin) | Value at Hot Idle |
|---|---|---|---|
| 1 | DB/OR | A/C Clutch Relay Control | Relay Off: 12v, On: 1v |
| 2 | DB/PK | Radiator Fan Control Relay | Relay Off: 12v, On: 1v |
| 3 | DB/YL | ASD Relay Control | Relay Off: 12v, On: 1v |
| 4 | TN/RD | Speed Control Vacuum Solenoid | Vacuum Increasing: 1v |
| 5 | LG/RD | Speed Control Vent Solenoid | Vacuum Decreasing: 1v |
| 6-9 | --- | Not Used | --- |
| 10 | WT/DG | LDP Solenoid Control | PWM Signal: 0-12-0v |
| 11 | YL/RD | Speed Control Power Supply | 12-14v |
| 12 | DG/OR | ASD Relay Output | 12-14v |
| 13 | --- | Not Used | --- |
| 14 | WT/OR | LDP Switch Sense Signal | Closed: 0v, Open: 12v |
| 15 | PK/YL | Battery Temperature Sensor | At 86°F: 1.96v |
| 16-18 | --- | Not Used | --- |
| 19 | BR | Fuel Pump Relay Control | Relay Off: 12v, On: 1v |
| 20 | PK/BK | EVAP Purge Solenoid Control | PWM Signal: 0-12-0v |
| 21 | --- | Not Used | --- |
| 22 | DB/WT | A/C High Pressure Switch | Switch open: 12v, closed: 0v |
| 23 | LG | A/C Select Signal | A/C Off: 12v, On: 1v |
| 24 | WT/PK | Brake Switch Sense Signal | Brake Off: 0v, On: 12v |
| 25 | DG/OR | Generator Field Source | 12-14v |
| 26 | DB/LG | Fuel Level Sensor Signal | Full: 0.5v, 1/2 full: 2.5v |
| 27 | PK | SCI Transmit | 0v |
| 28 | WT/BK | CCD Bus (-) | <0.050v |
| 29 | LG/BK | SCI Receive | 5v |
| 30 | PK/BR | CCD Bus (+) | Digital Signals: 0-5-0v |
| 31 | --- | Not Used | --- |
| 32 | RD/LG | Speed Control Switch Signal | S/C & Set Switch On: 3.8v |

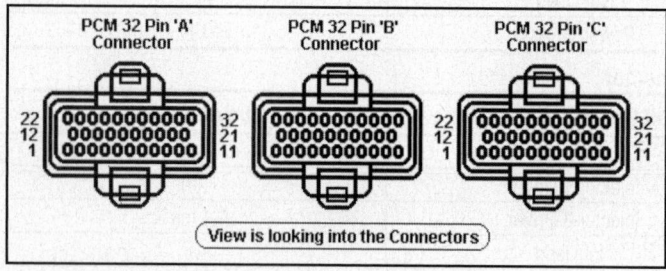

**Pin Connector Graphic**

## Standard Colors and Abbreviations

| Abbreviation | Color | Abbreviation | Color | Abbreviation | Color |
|---|---|---|---|---|---|
| BK | Black | GY | Gray | RD | Red |
| BL | Blue | GN | Green | TN | Tan |
| BR | Brown | LG | Light Green | VT | Violet |
| DB | Dark Blue | OR | Orange | WT | White |
| DG | Dark Green | PK | Pink | YL | Yellow |

## 1993-95 Cherokee 4.0L VIN S 60-Pin Connector

| PCM Pin # | Wire Color | Circuit Description (60-Pin) | Value at Hot Idle |
|---|---|---|---|
| 1 | DG/RD | MAP Sensor Signal | 1.5-1.6v |
| 2 | TN/BK | ECT Sensor Signal | At 180°F: 2.80v |
| 3 | RD | Fused Battery Power (B+) | 12-14v |
| 4 | BK/LB | Sensor Ground | <0.1v |
| 5 | BK/WT | Sensor Ground | <0.1v |
| 6 | PK/WT | 5-Volt Supply | 4.9-5.1v |
| 7 | OR | 8-Volt Supply | 7.9-8.1v |
| 8 | --- | Not Used | --- |
| 9 | DB | Ignition Switch Power (B+) | 12-14v |
| 10 | PK | Extended Idle Switch (Police) | Switch Off: 12v, On: 1v |
| 11 | BK/TN | Power Ground | <0.1v |
| 12 | BK/TN | Power Ground | <0.1v |
| 13 | LB/BR | Injector 4 Driver | 1-4 ms |
| 14 | YL/WT | Injector 3 Driver | 1-4 ms |
| 15 | TN | Injector 2 Driver | 1-4 ms |
| 16 | WT/DB | Injector 1 Driver | 1-4 ms |
| 17-18 | --- | Not Used | --- |
| 19 | GY | Coil 1 Driver Control | 5°, 55 mph: 8° dwell |
| 20 | DG | Alternator Field Control | Digital Signals: 0-12-0v |
| 21 | BK/RD | Air Temperature Sensor | At 100°F: 2.51v |
| 22 | OR/DB | TP Sensor Signal | 0.6-1.0v |
| 23 | --- | Not Used | --- |
| 24 | GY/BK | Distributor Reference Pickup | Digital Signals: 0-5-0v |
| 25 | PK | SCI Transmit | 0v |
| 26 | PK/BR | CCD Bus (+) | Digital Signals: 0-5-0v |
| 27 | LB | A/C Request Switch Sense | A/C Off: 12v, On: 1v |
| 28 | LG | A/C Select Switch Sense | A/C Off: 12v, On: 1v |
| 29 | WT/PK | Brake Switch Sense Signal | Brake Off: 0v, On: 12v |
| 30 | BR/YL | PNP Switch Sense Signal | In P/N: 0v, Others: 5v |
| 31 | DB/PK | Radiator Fan Control Relay | Relay Off: 12v, On: 1v |
| 32 | BK/PK | MIL (lamp) Control | MIL Off: 12v, On: 1v |
| 33 | TN/RD | Speed Control Vacuum Solenoid | Vacuum Increasing: 1v |
| 34 | DB/OR | A/C Clutch Relay Control | Relay Off: 12v, On: 1v |
| 35 | --- | Not Used | --- |
| 36 | DG/YL | Alternator Lamp Control | Lamp Off: 12v, On: 1v |
| 37 ('93) | RD/DB | Ballast Bypass Relay Control | Relay Off: 12v, On: 1v |
| 38 | PK/BK | Injector 5 Driver | 1-4 ms |
| 39 | GY/RD | AIS Motor 1 Circuit Control | Pulse Signals |
| 40 | BR/WT | AIS Motor 3 Circuit Control | Pulse Signals |

## Standard Colors and Abbreviations

| Abbreviation | Color | Abbreviation | Color | Abbreviation | Color |
|---|---|---|---|---|---|
| BK | Black | GY | Gray | RD | Red |
| BL | Blue | GN | Green | TN | Tan |
| BR | Brown | LG | Light Green | VT | Violet |
| DB | Dark Blue | OR | Orange | WT | White |
| DG | Dark Green | PK | Pink | YL | Yellow |

**1993-95 Cherokee 4.0L VIN S 60-Pin Connector - Continued**

| PCM Pin # | Wire Color | Circuit Description (60-Pin) | Value at Hot Idle |
|---|---|---|---|
| 41 | BK/DG | HO2S-11 (B1 S1) Signal | 0.1-1.1v |
| 42 | --- | Not Used | --- |
| 43 | GY/LB | Tachometer Signal | Pulse Signals |
| 44 | TN/YL | Distributor Sync Pickup | Digital Signals: 0-5-0v |
| 45 | LG | SCI Receive | 5v |
| 46 | WT/BK | CCD Bus (-) | <0.050v |
| 47 | WT/OR | Vehicle Speed Signal | Digital Signal |
| 48 | BR/RD | Speed Control Set Switch | S/C & Set Switch On: 3.8v |
| 49 | YL/RD | Speed Control On/Off Switch | S/C Off: 12v, On: 1v |
| 50 | WT/LG | Speed Control Resume Switch | S/C Off: 12v, On: 1v |
| 51 | DB/YL | ASD Relay Control | Relay Off: 12v, On: 1v |
| 52 | --- | Not Used | --- |
| 53 | LG/RD | Speed Control Vent Solenoid | Vacuum Decreasing: 1v |
| 54 | OR/BK | A/T: Lockup Torque Converter | TCC on at Cruise: 1v |
| 54 | OR/BK | M/T: SIL (lamp) Control | Lamp Off: 12v, On: 1v |
| 55 | --- | Not Used | --- |
| 56 ('93) | GY/PK | Maintenance Indicator Lamp | Lamp Off: 12v, On: 1v |
| 57 | DG/OR | ASD Relay Output | 12-14v |
| 58 | LG/BK | Injector 6 Driver | 1-4 ms |
| 59 | PK/BK | AIS Motor 4 Circuit Control | Pulse Signals |
| 60 | YL/BK | AIS Motor 2 Circuit Control | Pulse Signals |

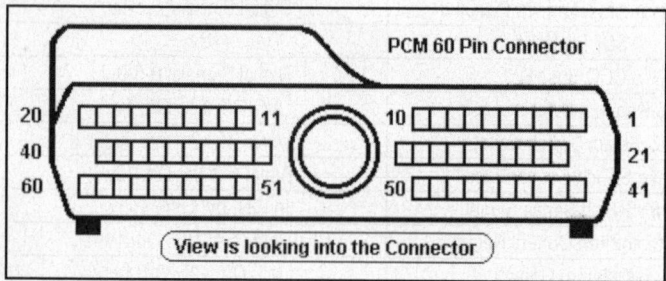

**Pin Connector Graphic**

## Standard Colors and Abbreviations

| Abbreviation | Color | Abbreviation | Color | Abbreviation | Color |
|---|---|---|---|---|---|
| BK | Black | GY | Gray | RD | Red |
| BL | Blue | GN | Green | TN | Tan |
| BR | Brown | LG | Light Green | VT | Violet |
| DB | Dark Blue | OR | Orange | WT | White |
| DG | Dark Green | PK | Pink | YL | Yellow |

### 1996 Cherokee 4.0L VIN S 'A' Black Connector

| PCM Pin # | Wire Color | Circuit Description (32-Pin) | Value at Hot Idle |
|---|---|---|---|
| 1, 3, 5, 9 | --- | Not Used | --- |
| 2 | DB | Ignition Switch Power (B+) | 12-14v |
| 4 | BK/LB | A/T: Analog Sensor Ground | <0.1v |
| 6 | BR/YL | PNP Switch Sense Signal | In P/N: 0v, Others: 5v |
| 6 | BK | M/T: Analog Sensor Ground | <0.1v |
| 7 | GY | Coil 1 Driver Control | 5º, 55 mph: 8º dwell |
| 8 | GY/BK | CKP Sensor Signal | Digital Signals: 0-5-0v |
| 10 | YL/BK | IAC 2 Driver Control | Pulse Signals |
| 11 | BR/WT | IAC 3 Driver Control | Pulse Signals |
| 12 | DB/OR | Extended Idle Switch (Police Special) | Switch Off: 12v, On: 1v |
| 13-14 | --- | Not Used | --- |
| 15 | BK/RD | IAT Sensor Signal | At 100ºF: 1.83v |
| 16 | TN/BK | ECT Sensor Signal | At 180ºF: 2.80v |
| 17 | PK/WT | 5-Volt Supply | 4.9-5.1v |
| 18 | TN/YL | CMP Sensor Signal | Digital Signals: 0-5-0v |
| 19 | GY/RD | IAC 1 Driver Control | Pulse Signals |
| 20 | PK/BK | IAC 4 Driver Control | Pulse Signals |
| 21 | --- | Not used | --- |
| 22 | RD | Fused Battery Power (B+) | 12-14v |
| 23 | OR/DB | TP Sensor Signal | 0.6-1.0v |
| 24 | BK/DG | HO2S-11 (B1 S1) Signal | 0.1-1.1v |
| 25 | TN/WT | HO2S-12 (B1 S2) Signal | 0.1-1.1v |
| 26 | --- | Not Used | --- |
| 27 | DG/RD | MAP Sensor Signal | 1.5-1.6v |
| 28-30 | --- | Not Used | --- |
| 31-32 | BK/TN | Power Ground | <0.1v |

### 1996 Cherokee 4.0L VIN S 'B' White Connector

| PCM Pin # | Wire Color | Circuit Description (32-Pin) | Value at Hot Idle |
|---|---|---|---|
| 1-3 | --- | Not Used | --- |
| 4 | WT/DB | Injector 1 Driver | 1-4 ms |
| 5 | YL/WT | Injector 3 Driver | 1-4 ms |
| 6 | GY | Injector 5 Driver | 1-4 ms |
| 7-9 | --- | Not Used | --- |
| 10 | DG | Generator Field Control | Digital Signals: 0-12-0v |
| 11 | OR/BK | A/T: TCC Solenoid Control | TCC on at Cruise: 1v |
| 11 | OR/BK | M/T: Upshift (lamp) Control | Lamp Off: 12v, On: 1v |
| 12 | BR/DB | Injector 6 Driver | 1-4 ms |
| 13-14 | --- | Not Used | --- |
| 15 | TN | Injector 2 Driver | 1-4 ms |
| 16 | LB/BR | Injector 4 Driver | 1-4 ms |
| 17-26 | --- | Not Used | --- |
| 27 | WT/OR | Vehicle Speed Signal | Digital Signal |
| 28-30 | --- | Not Used | --- |
| 31 | OR | 5-Volt Supply | 4.9-5.1v |
| 32 | --- | Not Used | --- |

### 1996 Cherokee 4.0L VIN S 'C' Gray Connector

| PCM Pin # | Wire Color | Circuit Description (32-Pin) | Value at Hot Idle |
|---|---|---|---|
| 1 | DB/OR | A/C Clutch Relay Control | Relay Off: 12v, On: 1v |
| 2 | DB/PK | Radiator Fan Control Relay | Relay Off: 12v, On: 1v |
| 3 | DB/YL | ASD Relay Control | Relay Off: 12v, On: 1v |
| 4 | TN/RD | Speed Control Vacuum Solenoid | Vacuum Increasing: 1v |
| 5 | LG/RD | Speed Control Vent Solenoid | Vacuum Decreasing: 1v |
| 6-10 | --- | Not Used | --- |
| 11 | YL/RD | Speed Control Power Supply | 12-14v |
| 12 | DG/OR | ASD Relay Output | 12-14v |
| 13-14 | --- | Not Used | --- |
| 15 | PK/YL | Battery Temperature Sensor | At 86°F: 1.96v |
| 16 | PK/BK | Generator Lamp Control | Lamp Off: 12v, On: 1v |
| 17 | BK/PK | MIL (lamp) Control | MIL Off: 12v, On: 1v |
| 18 | --- | Not Used | --- |
| 19 | BR | Fuel Pump Relay Control | Relay Off: 12v, On: 1v |
| 20 | PK/BK | EVAP Purge Solenoid Control | PWM Signal: 0-12-0v |
| 21 | --- | Not Used | --- |
| 22 | DB/OR | A/C Request Signal | A/C Off: 12v, On: 1v |
| 23 | LG | A/C Select Signal | A/C Off: 12v, On: 1v |
| 24 | WT/PK | Brake Switch Sense Signal | Brake Off: 0v, On: 12v |
| 25 | --- | Not Used | --- |
| 26 | DB | Fuel Level Sensor Signal | Full: 0.5v, 1/2 full: 2.5v |
| 27 | PK | SCI Transmit | 0v |
| 28 | WT/BK | CCD Bus (-) | <0.050v |
| 29 | LG | SCI Receive | 5v |
| 30 | PK/BR | CCD Bus (+) | Digital Signals: 0-5-0v |
| 31 | GY/LB | Tachometer Signal | Pulse Signals |
| 32 | WT/LG | Speed Control Switch Signal | S/C & Set Switch On: 3.8v |

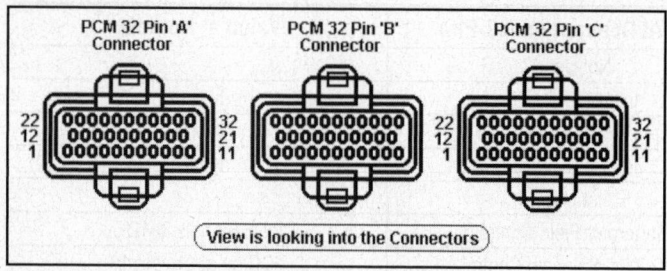

**Pin Connector Graphic**

### Standard Colors and Abbreviations

| Abbreviation | Color | Abbreviation | Color | Abbreviation | Color |
|---|---|---|---|---|---|
| BK | Black | GY | Gray | RD | Red |
| BL | Blue | GN | Green | TN | Tan |
| BR | Brown | LG | Light Green | VT | Violet |
| DB | Dark Blue | OR | Orange | WT | White |
| DG | Dark Green | PK | Pink | YL | Yellow |

## 1997 Cherokee 4.0L VIN S 'A' Black Connector

| PCM Pin # | Wire Color | Circuit Description (32-Pin) | Value at Hot Idle |
|---|---|---|---|
| 1, 3, 5 | --- | Not Used | --- |
| 2 | DB/WT | Ignition Switch Power (B+) | 12-14v |
| 4 | BR/YL | A/T: Analog Sensor Ground | <0.1v |
| 6 | BK/WT | PNP Switch Sense Signal | In P/N: 0v, Others: 5v |
| 6 | BK | M/T: Analog Sensor Ground | <0.1v |
| 7 | GY | Coil 1 Driver Control | 5°, 55 mph: 8° dwell |
| 8 | GY/BK | CKP Sensor Signal | Digital Signals: 0-5-0v |
| 9 | --- | Not Used | --- |
| 10 | YL/BK | IAC 2 Driver Control | Pulse Signals |
| 11 | BR/WT | IAC 3 Driver Control | Pulse Signals |
| 12 | GY | Extended Idle Switch (Police) | Switch Off: 12v, On: 1v |
| 13-14 | --- | Not Used | --- |
| 15 | BK/RD | IAT Sensor Signal | At 100°F: 1.83v |
| 16 | TN/BK | ECT Sensor Signal | At 180°F: 2.80v |
| 17 | OR | 5-Volt Supply | 4.9-5.1v |
| 18 | TN/YL | CMP Sensor Signal | Digital Signals: 0-5-0v |
| 19 | GY/RD | IAC 1 Driver Control | Pulse Signals |
| 20 | PK/BK | IAC 4 Driver Control | Pulse Signals |
| 21 | --- | Not Used | --- |
| 22 | DG/BK | Fused Battery Power (B+) | 12-14v |
| 23 | OR/DB | TP Sensor Signal | 0.6-1.0v |
| 24 | BK/DG | HO2S-11 (B1 S1) Signal | 0.1-1.1v |
| 25 | TN/BK | HO2S-12 (B1 S2) Signal | 0.1-1.1v |
| 26 | --- | Not Used | --- |
| 27 | DG/RD | MAP Sensor Signal | 1.5-1.6v |
| 28-30 | --- | Not Used | --- |
| 31 | BK/TN | Power Ground | <0.1v |
| 32 | BK/TN | Power Ground | <0.1v |

## 1997 Cherokee 4.0L VIN S 'B' White Connector

| PCM Pin # | Wire Color | Circuit Description (32-Pin) | Value at Hot Idle |
|---|---|---|---|
| 1-3 | --- | Not Used | --- |
| 4 | WT/DB | Injector 1 Driver | 1-4 ms |
| 5 | YL/WT | Injector 3 Driver | 1-4 ms |
| 6 | PK/BK | Injector 5 Driver | 1-4 ms |
| 7-9 | --- | Not Used | --- |
| 10 | DG | Generator Field Control | Digital Signals: 0-12-0v |
| 11-14 | --- | Not Used | --- |
| 15 | TN | Injector 2 Driver | 1-4 ms |
| 16 | LB/BR | Injector 4 Driver | 1-4 ms |
| 17-22 | --- | Not Used | --- |
| 23 | GY/YL | Engine Oil Pressure Sensor | 1.6v at 24 psi |
| 24-26 | --- | Not Used | --- |
| 27 | WT/OR | Vehicle Speed Signal | Digital Signal |
| 28-30 | --- | Not Used | --- |
| 31 | PK/OR | 5-Volt Supply | 4.9-5.1v |
| 32 | --- | Not Used | --- |

## 1997 Cherokee 4.0L VIN S 'C' Gray Connector

| PCM Pin # | Wire Color | Circuit Description (32-Pin) | Value at Hot Idle |
|---|---|---|---|
| 1 | DB/OR | A/C Clutch Relay Control | Relay Off: 12v, On: 1v |
| 2 | --- | Not Used | --- |
| 3 | DB/YL | ASD Relay Control | Relay Off: 12v, On: 1v |
| 4 | TN/RD | Speed Control Vacuum Solenoid | Vacuum Increasing: 1v |
| 5 | LG/RD | Speed Control Vent Solenoid | Vacuum Decreasing: 1v |
| 6-10 | --- | Not Used | --- |
| 11 | YL/RD | Speed Control Power Supply | 12-14v |
| 12 | DG/OR | ASD Relay Output | 12-14v |
| 13-14 | --- | Not Used | --- |
| 15 | PK/YL | Battery Temperature Sensor | At 86°F: 1.96v |
| 16-18 | --- | Not Used | --- |
| 19 | BR | Fuel Pump Relay Control | Relay Off: 12v, On: 1v |
| 20 | PK/BK | EVAP Purge Solenoid Control | PWM Signal: 0-12-0v |
| 21 | --- | Not Used | --- |
| 22 | DB/WT | A/C Request Signal | A/C Off: 12v, On: 1v |
| 23 | LG | A/C Select Signal | A/C Off: 12v, On: 1v |
| 24 | WT/PK | Brake Switch Sense Signal | Brake Off: 0v, On: 12v |
| 25 | --- | Not Used | --- |
| 26 | DB/LG | Fuel Level Sensor Signal | Full: 0.5v, 1/2 full: 2.5v |
| 27 | PK | SCI Transmit | 0v |
| 28 | WT/BK | CCD Bus (-) | <0.050v |
| 29 | LG | SCI Receive | 5v |
| 30 | PK/BR | CCD Bus (+) | Digital Signals: 0-5-0v |
| 31 | --- | Not Used | --- |
| 32 | RD/LG | Speed Control Switch Signal | S/C & Set Switch On: 3.8v |

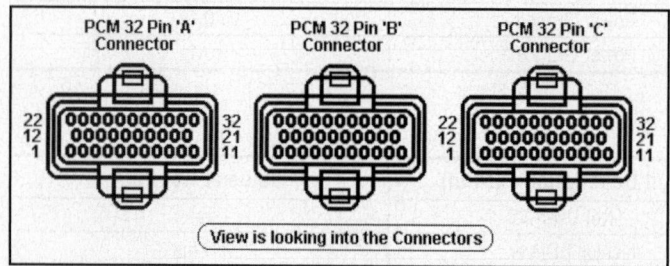

**Pin Connector Graphic**

## Standard Colors and Abbreviations

| Abbreviation | Color | Abbreviation | Color | Abbreviation | Color |
|---|---|---|---|---|---|
| BK | Black | GY | Gray | RD | Red |
| BL | Blue | GN | Green | TN | Tan |
| BR | Brown | LG | Light Green | VT | Violet |
| DB | Dark Blue | OR | Orange | WT | White |
| DG | Dark Green | PK | Pink | YL | Yellow |

**1998-99 Cherokee 4.0L V6 VIN S 'A' Black Connector**

| PCM Pin # | Wire Color | Circuit Description (32-Pin) | Value at Hot Idle |
|---|---|---|---|
| 1, 3, 5, 9 | --- | Not Used | --- |
| 2 | DB/WT | Ignition Switch Power (B+) | 12-14v |
| 4 | BR/YL | A/T: Analog Sensor Ground | <0.050v |
| 6 | BK/WT | PNP Switch Sense Signal | In P/N: 0v, Others: 5v |
| 6 | BK | M/T: Analog Sensor Ground | <0.050v |
| 7 | GY | Coil 1 Driver Control | 5°, 55 mph: 8° dwell |
| 8 | GY/BK | CKP Sensor Signal | Digital Signals: 0-5-0v |
| 10 | YL/BK | IAC 2 Driver Control | Pulse Signals |
| 11 | BR/WT | IAC 3 Driver Control | Pulse Signals |
| 12 | GY | Extended Idle Switch (Police) | Switch Off: 12v, On: 1v |
| 13-14 | --- | Not Used | --- |
| 15 | BK/RD | IAT Sensor Signal | At 100°F: 1.83v |
| 16 | TN/BK | ECT Sensor Signal | At 180°F: 2.80v |
| 17 | OR | 5-Volt Supply | 4.9-5.1v |
| 18 | TN/YL | CMP Sensor Signal | Digital Signals: 0-5-0v |
| 19 | GY/RD | IAC 1 Driver Control | Pulse Signals |
| 20 | PK/BK | IAC 4 Driver Control | Pulse Signals |
| 21 | --- | Not Used | --- |
| 22 | DG/BK | Fused Battery Power (B+) | 12-14v |
| 23 | OR/DB | TP Sensor Signal | 0.6-1.0v |
| 24 | BK/DG | HO2S-11 (B1 S1) Signal | 0.1-1.1v |
| 25 | TN/WT | HO2S-12 (B1 S2) Signal | 0.1-1.1v |
| 27 | DG/RD | MAP Sensor Signal | 1.5-1.6v |
| 28-30 | --- | Not Used | --- |
| 31 | BK/TN | Power Ground | <0.1v |
| 32 | BK/TN | Power Ground | <0.1v |

**1998-99 Cherokee 4.0L V6 VIN S 'B' White Connector**

| PCM Pin # | Wire Color | Circuit Description (32-Pin) | Value at Hot Idle |
|---|---|---|---|
| 1-3 | --- | Not Used | --- |
| 4 | WT/DB | Injector 1 Driver | 1-4 ms |
| 5 | YL/WT | Injector 3 Driver | 1-4 ms |
| 6 | PK/BK | Injector 5 Driver | 1-4 ms |
| 7-9 | --- | Not Used | --- |
| 10 | DG | Generator Field Control | Digital Signals: 0-12-0v |
| 11 ('98) | OR/BK | TCC Solenoid Control | TCC on at Cruise: 1v |
| 12 | LG/BK | Injector 6 Driver | 1-4 ms |
| 13-14 | --- | Not Used | --- |
| 15 | TN | Injector 2 Driver | 1-4 ms |
| 16 | LB/BR | Injector 4 Driver | 1-4 ms |
| 17-22 | --- | Not Used | --- |
| 23 | GY/YL | Engine Oil Pressure Sensor | 1.6v at 24 psi |
| 24-26 | --- | Not Used | --- |
| 27 | WT/OR | Vehicle Speed Signal | Digital Signal |
| 28-30 | --- | Not Used | --- |
| 31 | PK/OR | 5-Volt Supply | 4.9-5.1v |
| 32 | --- | Not Used | --- |

### 1998-99 Cherokee 4.0L V6 VIN S 'C' Gray Connector

| PCM Pin # | Wire Color | Circuit Description (32-Pin) | Value at Hot Idle |
|---|---|---|---|
| 1 | DB/OR | A/C Clutch Relay Control | Relay Off: 12v, On: 1v |
| 2 | DB/PK | Radiator Fan Control Relay | Relay Off: 12v, On: 1v |
| 3 | DB/YL | ASD Relay Control | Relay Off: 12v, On: 1v |
| 4 | TN/RD | Speed Control Vacuum Solenoid | Vacuum Increasing: 1v |
| 5 | LG/RD | Speed Control Vent Solenoid | Vacuum Decreasing: 1v |
| 6-9 | --- | Not Used | --- |
| 10 | WT/DG | LDP Solenoid Control | PWM Signal: 0-12-0v |
| 11 | YL/RD | Speed Control Power Supply | 12-14v |
| 12 | DG/OR | ASD Relay Output | 12-14v |
| 13 | --- | Not Used | --- |
| 14 ('98) | WT/OR | LDP Switch Sense Signal | Closed: 0v, Open: 12v |
| 14 | OR | Battery Temperature Sensor | At 86°F: 1.96v |
| 15 ('98) | PK/YL | Battery Temperature Sensor | At 86°F: 1.96v |
| 15 | PK/YL | LDP Switch Sense Signal | Closed: 0v, Open: 12v |
| 16-18 | --- | Not Used | --- |
| 19 | BR | Fuel Pump Relay Control | Relay Off: 12v, On: 1v |
| 20 | PK/BK | EVAP Purge Solenoid Control | PWM Signal: 0-12-0v |
| 21 | --- | Not Used | --- |
| 22 | DB/WT | A/C Request Signal | A/C Off: 12v, On: 1v |
| 23 | LG | A/C Select Signal | A/C Off: 12v, On: 1v |
| 24 | WT/PK | Brake Switch Sense Signal | Brake Off: 0v, On: 12v |
| 25 | DG/OR | Generator Field Source | 12-14v |
| 26 | DB/LG | Fuel Level Sensor Signal | Full: 0.5v, 1/2 full: 2.5v |
| 27 | PK | SCI Transmit | 0v |
| 28 | WT/BK | CCD Bus (-) | <0.050v |
| 29 | LG/BK | SCI Receive | 5v |
| 30 | PK/BR | CCD Bus (+) | Digital Signals: 0-5-0v |
| 31 | --- | Not Used | --- |
| 32 | RD/LG | Speed Control Switch Signal | S/C & Set Switch On: 3.8v |

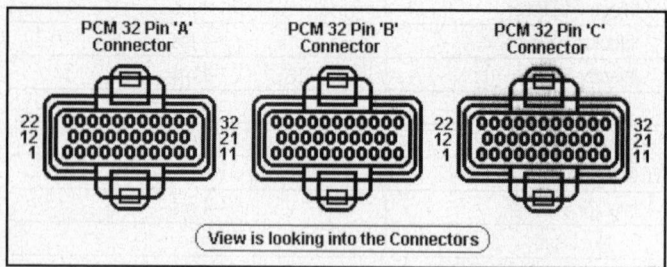

### Pin Connector Graphic

### Standard Colors and Abbreviations

| Abbreviation | Color | Abbreviation | Color | Abbreviation | Color |
|---|---|---|---|---|---|
| BK | Black | GY | Gray | RD | Red |
| BL | Blue | GN | Green | TN | Tan |
| BR | Brown | LG | Light Green | VT | Violet |
| DB | Dark Blue | OR | Orange | WT | White |
| DG | Dark Green | PK | Pink | YL | Yellow |

## 2000-01 Cherokee 4.0L V6 VIN S 'A' Black Connector

| PCM Pin # | Wire Color | Circuit Description (32-Pin) | Value at Hot Idle |
|---|---|---|---|
| 1 | RD/YL | Ignition Coil 3 Driver | 5°, 55 mph: 8° dwell |
| 2 | DB/WT | Ignition Switch Power (B+) | 12-14v |
| 3, 5, 9 | --- | Not Used | --- |
| 4 | BR/WT | Sensor Ground | <0.050v |
| 6 | BK/WT | PNP Switch Sense Signal | In P/N: 0v, Others: 5v |
| 6 | GY | Ignition Coil 1 Driver | 5°, 55 mph: 8° dwell |
| 8 | GY/BK | CKP Sensor Signal | Digital Signals: 0-5-0v |
| 10 | YL/BK | IAC 2 Driver Control | Pulse Signals |
| 11 | BR/WT | IAC 3 Driver Control | Pulse Signals |
| 12 | GY | Extended Idle Switch (California Police) | Switch Off: 12v, On: 1v |
| 13-14 | --- | Not Used | --- |
| 15 | BK/RD | IAT Sensor Signal | At 100°F: 1.83v |
| 16 | TN/BK | ECT Sensor Signal | At 180°F: 2.80v |
| 17 | OR | 5-Volt Supply | 4.9-5.1v |
| 18 | TN/YL | CMP Sensor Signal | Digital Signals: 0-5-0v |
| 19 | GY/RD | IAC 1 Driver Control | Pulse Signals |
| 20 | VT/BK | IAC 4 Driver Control | Pulse Signals |
| 21 | --- | Not Used | --- |
| 22 | DG/BK | Fused Battery Power (B+) | 12-14v |
| 23 | OR/DB | TP Sensor Signal | 0.6-1.0v |
| 24 | BK/DG | HO2S-11 (B1 S1) Signal | 0.1-1.1v |
| 25 | TN/WT | HO2S-12 (B1 S2) Signal | 0.1-1.1v |
| 26 | LG/RD | HO2S-21 (B2 S1) Signal | 0.1-1.1v |
| 27 | DG/RD | MAP Sensor Signal | 1.5-1.6v |
| 28-30 | --- | Not Used | --- |
| 31 | BK/TN | Power Ground | <0.1v |
| 32 | BK/TN | Power Ground | <0.1v |

## 2000-01 Cherokee 4.0L V6 VIN S 'B' White Connector

| PCM Pin # | Wire Color | Circuit Description (32-Pin) | Value at Hot Idle |
|---|---|---|---|
| 1-3 | --- | Not Used | --- |
| 4 | WT/DB | Injector 1 Driver Control | 1-4 ms |
| 5 | YL/WT | Injector 3 Driver Control | 1-4 ms |
| 6 | PK/BK | Injector 5 Driver Control | 1-4 ms |
| 7-8 | --- | Not Used | --- |
| 9 | DB/TN | Ignition Coil 2 Driver | 5°, 55 mph: 8° dwell |
| 10 | DG | Generator Field Control | Digital Signals: 0-12-0v |
| 11 | --- | Not Used | --- |
| 12 | LG/BK | Injector 6 Driver Control | 1-4 ms |
| 13-14 | --- | Not Used | --- |
| 15 | T | Injector 2 Driver Control | 1-4 ms |
| 16 | LB/BR | Injector 4 Driver Control | 1-4 ms |
| 17-22 | --- | Not Used | --- |
| 23 | GY/YL | Engine Oil Pressure Sensor | 1.6v at 24 psi |
| 24-26 | --- | Not Used | --- |
| 27 | WT/OR | Vehicle Speed Signal | Digital Signal |
| 28-30 | --- | Not Used | --- |
| 31 | VT/OR | 5-Volt Supply | 4.9-5.1v |
| 32 | --- | Not Used | --- |

### 2000-01 Cherokee 4.0L V6 VIN S 'C' Gray Connector

| PCM Pin # | Wire Color | Circuit Description (32-Pin) | Value at Hot Idle |
|---|---|---|---|
| 1 | DB/OR | A/C Clutch Relay Control | Relay Off: 12v, On: 1v |
| 2 | DB/PK | Radiator Fan Control Relay | Relay Off: 12v, On: 1v |
| 3 | DB/YL | ASD Relay Control | Relay Off: 12v, On: 1v |
| 4 | TN/RD | Speed Control Vacuum Solenoid | Vacuum Increasing: 1v |
| 5 | LG/RD | Speed Control Vent Solenoid | Vacuum Decreasing: 1v |
| 6-7 | --- | Not Used | --- |
| 8 | DB/OR | HO2S-11 Heater Relay | Relay Off: 12v, On: 1v |
| 9 | BR/VT | HO2S-12 Heater Relay | Relay Off: 12v, On: 1v |
| 10 | WT/DG | LDP Solenoid Control | PWM Signal: 0-12-0v |
| 11 | YL/RD | Speed Control Power Supply | 12-14v |
| 12 | DG/WT | ASD Relay Output | 12-14v |
| 13 | --- | Not Used | --- |
| 14 | WT/OR | LDP Switch Sense Signal | Closed: 0v, Open: 12v |
| 15 | PK/YL | Battery Temperature Sensor | At 86°F: 1.96v |
| 16-18 | --- | Not Used | --- |
| 19 | BR | Fuel Pump Relay Control | Relay Off: 12v, On: 1v |
| 20 | PK/BK | EVAP Purge Solenoid Control | PWM Signal: 0-12-0v |
| 21 | --- | Not Used | --- |
| 22 | DB/WT | A/C Request Signal | A/C Off: 12v, On: 1v |
| 23 | LG | A/C Select Signal | A/C Off: 12v, On: 1v |
| 24 | WT/PK | Brake Switch Sense Signal | Brake Off: 0v, On: 12v |
| 25 | DG/OR | Generator Field Source | 12-14v |
| 26 | DB/LG | Fuel Level Sensor Signal | Full: 0.5v, 1/2 full: 2.5v |
| 27 | PK | SCI Transmit | 0v |
| 28 | WT/BK | CCD Bus (-) | <0.050v |
| 29 | LG/BK | SCI Receive | 5v |
| 30 | VT/BR | CCD Bus (+) | Digital Signals: 0-5-0v |
| 31 | --- | Not Used | --- |
| 32 | RD/LG | Speed Control Switch Signal | S/C & Set Switch On: 3.8v |

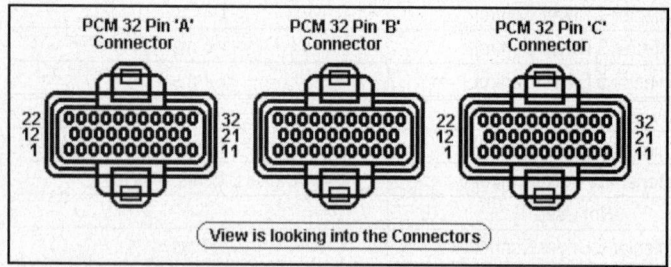

**Pin Connector Graphic**

### Standard Colors and Abbreviations

| Abbreviation | Color | Abbreviation | Color | Abbreviation | Color |
|---|---|---|---|---|---|
| BK | Black | GY | Gray | RD | Red |
| BL | Blue | GN | Green | TN | Tan |
| BR | Brown | LG | Light Green | VT | Violet |
| DB | Dark Blue | OR | Orange | WT | White |
| DG | Dark Green | PK | Pink | YL | Yellow |

## 1993-95 Grand Cherokee 4.0L V6 VIN S 60-Pin Connector

| PCM Pin # | Wire Color | Circuit Description (60-Pin) | Value at Hot Idle |
|---|---|---|---|
| 1 | RD/WT | MAP Sensor Signal | 1.5-1.6v |
| 2 | TN/BK | ECT Sensor Signal | At 180ºF: 2.80v |
| 3 | RD | Fused Battery Power (B+) | 12-14v |
| 4 | BK/LB | Sensor Ground | <0.1v |
| 5 | BK/TN | Sensor Ground | <0.1v |
| 6 | PK/WT | 5-Volt Supply | 4.9-5.1v |
| 7 | WT/BK | 8-Volt Supply | 7.9-8.1v |
| 8 | --- | Not Used | --- |
| 9 | LB/RD | Ignition Switch Power (B+) | 12-14v |
| 10 | --- | Not Used | --- |
| 11 | BK/TN | Power Ground | <0.1v |
| 12 | BK/TN | Power Ground | <0.1v |
| 13 | LB/BR | Injector 4 Driver | 1-4 ms |
| 14 | YL/WT | Injector 3 Driver | 1-4 ms |
| 15 | TN | Injector 2 Driver | 1-4 ms |
| 16 | W/LB | Injector 1 Driver | 1-4 ms |
| 17-18 | --- | Not Used | --- |
| 19 | GY/WT | Coil 1 Driver Control | 5º, 55 mph: 8º dwell |
| 20 | DG | Alternator Field Control | Digital Signals: 0-12-0v |
| 21 | BK/RD | IAT Sensor Signal | At 100ºF: 2.51v |
| 22 | OR/DB | TP Sensor Signal | 0.6-1.0v |
| 23, 31 | --- | Not Used | --- |
| 24 | RD/LG | CKP Sensor Signal | Digital Signals: 0-5-0v |
| 25 | BK | SCI Transmit | 0v |
| 26 | PK/BR | CCD Bus (+) | Digital Signals: 0-5-0v |
| 27 | DB/OR | A/C Pressure Switch Signal | A/C Off: 12v, On: 1v |
| 28 | LG | A/C Select Switch Sense | A/C Off: 12v, On: 1v |
| 29 | BR | Brake Switch Sense Signal | Brake Off: 0v, On: 12v |
| 30 | BR/WT | PNP Switch Sense Signal | In P/N: 0v, Others: 5v |
| 30 ('94-'95 | BK/WT | PNP Switch Sense Signal | In P/N: 0v, Others: 5v |
| 32 | BK/PK | MIL (lamp) Control | MIL Off: 12v, On: 1v |
| 33 | TN/RD | Speed Control Vacuum Solenoid | Vacuum Increasing: 1v |
| 34 | DB/OR | A/C Clutch Relay Control | Relay Off: 12v, On: 1v |
| 34 ('94-'95) | DB/RD | A/C Clutch Relay Control | Relay Off: 12v, On: 1v |
| 35-37 | --- | Not Used | --- |
| 38 | GY | Injector 5 Driver | 1-4 ms |
| 39 | YL/BK | AIS Motor 4 Circuit Control | Pulse Signals |
| 40 | BR/WT | AIS Motor 2 Circuit Control | Pulse Signals |

## Standard Colors and Abbreviations

| Abbreviation | Color | Abbreviation | Color | Abbreviation | Color |
|---|---|---|---|---|---|
| BK | Black | GY | Gray | RD | Red |
| BL | Blue | GN | Green | TN | Tan |
| BR | Brown | LG | Light Green | VT | Violet |
| DB | Dark Blue | OR | Orange | WT | White |
| DG | Dark Green | PK | Pink | YL | Yellow |

## 1993-95 Grand Cherokee 4.0L V6 VIN S 60-Pin Connector - Continued

| PCM Pin # | Wire Color | Circuit Description (60-Pin) | Value at Hot Idle |
|---|---|---|---|
| 41 | BK/OR | HO2S-11 (B1 S1) Signal | 0.1-1.1v |
| 42 | --- | Not Used | --- |
| 43 | GY/LB | Tachometer Signal | Pulse Signals |
| 44 | GY/BK | CMP Sensor Signal | Digital Signals: 0-5-0v |
| 45 | BK/YL | SCI Receive | 5v |
| 46 | WT/GY | CCD Bus (-) | <0.050v |
| 47 | WT/OR | Vehicle Speed Signal | Digital Signals |
| 48 | BR/RD | Speed Control Set Switch | S/C & Set Switch On: 3.8v |
| 49 | YL/RD | Speed Control On/Off Switch | S/C Off: 12v, On: 1v |
| 50 | WT/LG | Speed Control Resume Switch | S/C Off: 12v, On: 1v |
| 51 | PK | Fuel Pump Relay Control | Relay Off: 12v, On: 1v |
| 52 | --- | Not Used | --- |
| 53 | LG/RD | Speed Control Vent Solenoid | Vacuum Decreasing: 1v |
| 54 | OR/BK | M/T: SIL (lamp) Control | Lamp Off: 12v, On: 1v |
| 55 | --- | Not Used | --- |
| 56 ('93) | GY/PK | Maintenance Indicator Lamp | Lamp Off: 12v, On: 1v |
| 57 | DG/BK | ASD Relay Output | 12-14v |
| 58 | BR/YL | Injector 6 Driver | 1-4 ms |
| 59 | GY/RD | AIS Motor 1 Circuit Control | Pulse Signals |
| 60 | PK/BK | AIS Motor 3 Circuit Control | Pulse Signals |

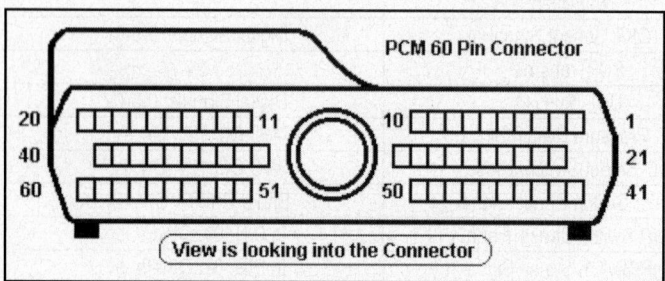

**Pin Connector Graphic**

## Standard Colors and Abbreviations

| Abbreviation | Color | Abbreviation | Color | Abbreviation | Color |
|---|---|---|---|---|---|
| BK | Black | GY | Gray | RD | Red |
| BL | Blue | GN | Green | TN | Tan |
| BR | Brown | LG | Light Green | VT | Violet |
| DB | Dark Blue | OR | Orange | WT | White |
| DG | Dark Green | PK | Pink | YL | Yellow |

## 1996-97 Grand Cherokee 4.0L V6 VIN S Black 'A' 32P Connector

| PCM Pin # | Wire Color | Circuit Description (32-Pin) | Value at Hot Idle |
|---|---|---|---|
| 1 | --- | Not Used | --- |
| 2 | OR | Ignition Switch Power (B+) | 12-14v |
| 3 | --- | Not Used | --- |
| 4 | BK/LB | Sensor Ground | <0.1v |
| 5 | --- | Not Used | --- |
| 6 | BK/WT | PNP Switch Sense Signal | In P/N: 0v, Others: 5v |
| 7 | GY/WT | Coil 1 Driver Control | 5º, 55 mph: 8º dwell |
| 8 | RD/LG | CKP Sensor Signal | Digital Signals: 0-5-0v |
| 9 | --- | Not Used | --- |
| 10 | PK/BK | IAC 4 Driver Control | Pulse Signals |
| 11 | BR/WT | IAC 3 Driver Control | Pulse Signals |
| 13 | --- | Not Used | --- |
| 14 ('96) | YL/RD | Fused Battery Power (B+) | 12-14v |
| 15 | BK/RD | IAT Sensor Signal | At 100ºF: 1.83v |
| 16 | TN/BK | ECT Sensor Signal | At 180ºF: 2.80v |
| 17 | WT/BK | 5-Volt Supply | 4.9-5.1v |
| 18 | GY/BK | CMP Sensor Signal | Digital Signals: 0-5-0v |
| 19 | YL/BK | IAC 2 Driver Control | Pulse Signals |
| 20 | GY/RD | IAC 1 Driver Control | Pulse Signals |
| 21 | --- | Not Used | --- |
| 22 | RD/YL | Fused Battery Power (B+) | 12-14v |
| 23 | OR/DB | TP Sensor Signal | 0.6-1.0v |
| 24 | BK/OR | HO2S-11 (B1 S1) Signal | 0.1-1.1v |
| 25 | BK/PK | HO2S-12 (B1 S2) Signal | 0.1-1.1v |
| 26 | --- | Not Used | --- |
| 27 | RD/WT | MAP Sensor Signal | 1.5-1.6v |
| 28-30 | --- | Not Used | --- |
| 31 | BK/TN | Power Ground | <0.1v |
| 32 | BK/TN | Power Ground | <0.1v |

## 1996-97 Grand Cherokee 4.0L V6 VIN S White 'B' 32P Connector

| PCM Pin # | Wire Color | Circuit Description (32-Pin) | Value at Hot Idle |
|---|---|---|---|
| 1 | PK | Transmission Temperature Sensor | At 200ºF: 2.40v |
| 2-3, 7, 13-14 | --- | Not Used | --- |
| 4 | WT/DB | Injector 1 Driver | 1-4 ms |
| 5 | YL/WT | Injector 3 Driver | 1-4 ms |
| 6 | GY | Injector 5 Driver | 1-4 ms |
| 8 | PK | Governor Pressure Solenoid | PWM Signal: 0-12-0v |
| 10 | DG | Generator Field Control | Digital Signals: 0-12-0v |
| 11 | DG/LB | TCC Solenoid Control | TCC on at Cruise: 1v |
| 12 | BR/YL | Injector 6 Driver | 1-4 ms |
| 15 | TN | Injector 2 Driver | 1-4 ms |
| 16 | LB/BR | Injector 4 Driver | 1-4 ms |
| 17-20 | --- | Not Used | --- |
| 21 | BR | Overdrive Solenoid Control | Solenoid on at Cruise: 1v |
| 22 | --- | Not Used | --- |
| 23 | GY/WT | Engine Oil Pressure Sensor | 1.6v at 24 psi |
| 24 | --- | Not Used | --- |
| 25 | DB | OSS Sensor (-) Signal | AC pulse signals |
| 26 | --- | Not Used | --- |
| 27 | WT/OR | Vehicle Speed Signal | Digital Signal |
| 28 | LG | OSS Sensor (+) Signal | AC pulse signals |
| 29 | LG | Governor Pressure Signal | 0.58v |
| 30 | BR/OR | Transmission Relay Control | Relay Off: 12v, On: 1v |
| 31 | PK/WT | 5-Volt Supply | 4.9-5.1v |

## 1996-97 Grand Cherokee 4.0L V6 VIN S Gray 'C' 32P Connector

| PCM Pin # | Wire Color | Circuit Description (32-Pin) | Value at Hot Idle |
|---|---|---|---|
| 1 | DB/RD | A/C Clutch Relay Control | Relay Off: 12v, On: 1v |
| 2 | --- | Not Used | --- |
| 3 | PK/WT | ASD Relay Control | Relay Off: 12v, On: 1v |
| 4 | TN/RD | Speed Control Vacuum Solenoid | Vacuum Increasing: 1v |
| 5 | LG/RD | Speed Control Vent Solenoid | Vacuum Decreasing: 1v |
| 6 | BR/YL | O/D Indicator (lamp) Control | O/D On: 1v, Off: 12v |
| 7-9 | --- | Not Used | --- |
| 10 | DG/RD | LDP Solenoid Control | PWM Signal: 0-12-0v |
| 11 | YL/RD | Speed Control Power Supply | 12-14v |
| 12 | DG/OR | ASD Relay Output | 12-14v |
| 13 | OR | Overdrive 'Off' Switch Sense | Switch On: 1v, Off: 12v |
| 14 | PK/RD | LDP Switch Sense Signal | Closed: 0v, Open: 12v |
| 15 | PK/YL | Battery Temperature Sensor | At 86°F: 1.96v |
| 16-18 | --- | Not Used | --- |
| 19 | DB | Fuel Pump Relay Control | Relay Off: 12v, On: 1v |
| 20 | PK/BK | EVAP Purge Solenoid Control | PWM Signal: 0-12-0v |
| 21 | --- | Not Used | --- |
| 22 | DB/BK | A/C Request Signal | A/C Off: 12v, On: 1v |
| 23 | --- | Not Used | --- |
| 24 | BR | Brake Switch Sense Signal | Brake Off: 0v, On: 12v |
| 25 | --- | Not Used | --- |
| 26 | LB/BK | Fuel Level Sensor Signal | Full: 0.5v, 1/2 full: 2.5v |
| 27 | BK/PK | SCI Transmit | 0v |
| 28 | WT/BK | CCD Bus (-) | <0.050v |
| 29 | BK/WT | SCI Receive | 5v |
| 30 | PK/BR | CCD Bus (+) | Digital Signals: 0-5-0v |
| 31 | --- | Not Used | --- |
| 32 | PK | Speed Control Switch Signal | S/C & Set Switch On: 3.8v |

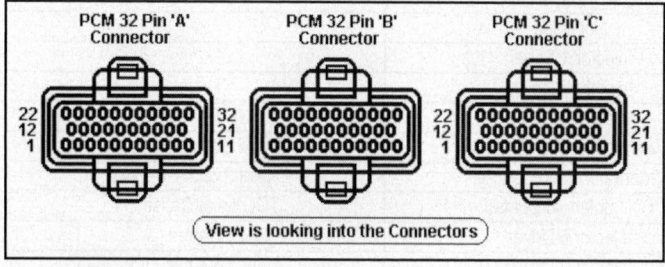

**Pin Connector Graphic**

## Standard Colors and Abbreviations

| Abbreviation | Color | Abbreviation | Color | Abbreviation | Color |
|---|---|---|---|---|---|
| BK | Black | GY | Gray | RD | Red |
| BL | Blue | GN | Green | TN | Tan |
| BR | Brown | LG | Light Green | VT | Violet |
| DB | Dark Blue | OR | Orange | WT | White |
| DG | Dark Green | PK | Pink | YL | Yellow |

## 1998 Grand Cherokee 4.0L VIN S 'A' Black Connector

| PCM Pin # | Wire Color | Circuit Description (32-Pin) | Value at Hot Idle |
|---|---|---|---|
| 1 | --- | Not Used | --- |
| 2 | OR | Ignition Switch Power (B+) | 12-14v |
| 3 | --- | Not Used | --- |
| 4 | BK/LB | Sensor Ground | <0.1v |
| 5 | --- | Not Used | --- |
| 6 | BK/WT | PNP Switch Sense Signal | In P/N: 0v, Others: 5v |
| 7 | GY/WT | Coil 1 Driver Control | 5°, 55 mph: 8° dwell |
| 8 | RD/LG | CKP Sensor Signal | Digital Signals: 0-5-0v |
| 9 | --- | Not Used | --- |
| 10 | PK/BK | IAC 4 Driver Control | Pulse Signals |
| 11 | BR/WT | IAC 3 Driver Control | Pulse Signals |
| 12-14 | --- | Not Used | --- |
| 15 | BK/RD | IAT Sensor Signal | At 100°F: 1.83v |
| 16 | TN/BK | ECT Sensor Signal | At 180°F: 2.80v |
| 17 | WT/BK | 5-Volt Supply | 4.9-5.1v |
| 18 | GY/BK | CMP Sensor Signal | Digital Signals: 0-5-0v |
| 19 | YL/BK | IAC 2 Driver Control | Pulse Signals |
| 20 | GY/RD | IAC 1 Driver Control | Pulse Signals |
| 21 | --- | Not Used | --- |
| 22 | RD/YL | Fused Battery Power (B+) | 12-14v |
| 23 | OR/DB | TP Sensor Signal | 0.6-1.0v |
| 24 | BK/OR | HO2S-11 (B1 S1) Signal | 0.1-1.1v |
| 25 | BK/PK | HO2S-12 (B1 S2) Signal | 0.1-1.1v |
| 26 | --- | Not Used | --- |
| 27 | RD/WT | MAP Sensor Signal | 1.5-1.6v |
| 28-30 | --- | Not Used | --- |
| 31 | BK/TN | Power Ground | <0.1v |
| 32 | BK/TN | Power Ground | <0.1v |

## 1998 Grand Cherokee 4.0L VIN S 'B' White Connector

| PCM Pin # | Wire Color | Circuit Description (32-Pin) | Value at Hot Idle |
|---|---|---|---|
| 1 | PK | Transmission Temperature Sensor | At 200°F: 2.40v |
| 2-3, 7, 9, 13-14 | --- | Not Used | --- |
| 4 | WT/DB | Injector 1 Driver | 1-4 ms |
| 5 | YL/WT | Injector 3 Driver | 1-4 ms |
| 6 | GY | Injector 5 Driver | 1-4 ms |
| 8 | PK | Governor Pressure Solenoid | PWM Signal: 0-12-0v |
| 10 | DG | Generator Field Control | Digital Signals: 0-12-0v |
| 11 | DG/LB | TCC Solenoid Control | TCC on at Cruise: 1v |
| 12 | BR/YL | Injector 6 Driver | 1-4 ms |
| 15 | TN | Injector 2 Driver | 1-4 ms |
| 16 | LB/BR | Injector 4 Driver | 1-4 ms |
| 17-20 | --- | Not Used | --- |
| 21 | BR | Overdrive Solenoid Control | Solenoid on at Cruise: 1v |
| 22 | --- | Not Used | --- |
| 23 | GY/WT | Engine Oil Pressure Sensor | 1.6v at 24 psi |
| 24 | --- | Not Used | --- |
| 25 | DB/BK | OSS Sensor Signal (-) | AC pulse signals |
| 26 | --- | Not Used | --- |
| 27 | WT/OR | Vehicle Speed Signal | Digital Signal |
| 28 | LG/WT | OSS Sensor Signal (+) | AC pulse signals |
| 29 | LG | Governor Pressure Signal | 0.58v |
| 30 | BR/OR | Transmission Relay Control | Relay Off: 12v, On: 1v |
| 31 | PK/OR | 5-Volt Supply | 4.9-5.1v |
| 32 | --- | Not Used | --- |

### 1998 Grand Cherokee 4.0L VIN S 'C' Gray Connector

| PCM Pin # | Wire Color | Circuit Description (32-Pin) | Value at Hot Idle |
|---|---|---|---|
| 1 | DB/RD | A/C Clutch Relay Control | Relay Off: 12v, On: 1v |
| 2 | --- | Not Used | --- |
| 3 | PK/WT | ASD Relay Control | Relay Off: 12v, On: 1v |
| 4 | TN/RD | Speed Control Vacuum Solenoid | Vacuum Increasing: 1v |
| 5 | LG/RD | Speed Control Vent Solenoid | Vacuum Decreasing: 1v |
| 6 | BR/YL | O/D Indicator (lamp) Control | O/D On: 1v, Off: 12v |
| 7-9 | --- | Not Used | --- |
| 10 | DG/RD | LDP Solenoid Control | PWM Signal: 0-12-0v |
| 11 | YL/RD | Speed Control Power Supply | 12-14v |
| 12 | DG/OR | ASD Relay Output | 12-14v |
| 13 | OR | Overdrive 'Off' Switch Sense | Switch On: 1v, Off: 12v |
| 14 | PK/RD | LDP Switch Sense Signal | Closed: 0v, Open: 12v |
| 15 | RD/YL | Battery Temperature Sensor | At 86°F: 1.96v |
| 16-18 | --- | Not Used | --- |
| 19 | DB | Fuel Pump Relay Control | Relay Off: 12v, On: 1v |
| 20 | PK/BK | EVAP Purge Solenoid Control | PWM Signal: 0-12-0v |
| 21 | --- | Not Used | --- |
| 22 | DB/BK | A/C Request Signal | A/C Off: 12v, On: 1v |
| 23 | --- | Not Used | --- |
| 24 | BR | Brake Switch Sense Signal | Brake Off: 0v, On: 12v |
| 25 | DG/PK | Generator Field Source | 12-14v |
| 26 | LB/BK | Fuel Level Sensor Signal | Full: 0.5v, 1/2 full: 2.5v |
| 27 | BK/PK | SCI Transmit | 0v |
| 28 | WT/BK | CCD Bus (-) | <0.050v |
| 29 | BK/WT | SCI Receive | 5v |
| 30 | PK/BR | CCD Bus (+) | Digital Signals: 0-5-0v |
| 31 | --- | Not Used | --- |
| 32 | PK | Speed Control Switch Signal | S/C & Set Switch On: 3.8v |

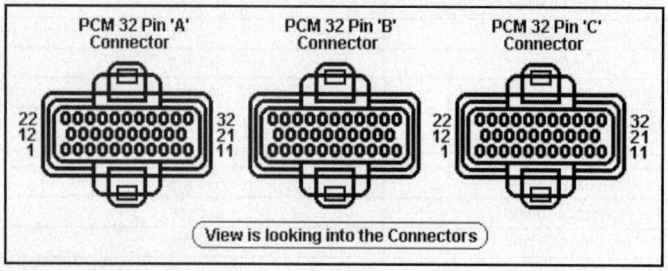

Pin Connector Graphic

### Standard Colors and Abbreviations

| Abbreviation | Color | Abbreviation | Color | Abbreviation | Color |
|---|---|---|---|---|---|
| BK | Black | GY | Gray | RD | Red |
| BL | Blue | GN | Green | TN | Tan |
| BR | Brown | LG | Light Green | VT | Violet |
| DB | Dark Blue | OR | Orange | WT | White |
| DG | Dark Green | PK | Pink | YL | Yellow |

## 1999 Grand Cherokee 4.0L VIN S 'A' Black Connector

| PCM Pin # | Wire Color | Circuit Description (32-Pin) | Value at Hot Idle |
|---|---|---|---|
| 1 | TN/OR | Ignition Coil 3 Driver Control | 5°, 55 mph: 8° dwell |
| 2 | OR/DB | Ignition Switch Power (B+) | 12-14v |
| 3 | --- | Not Used | --- |
| 4 | BK/LB | Sensor Ground | <0.050v |
| 5 | --- | Not Used | --- |
| 6 | BR/YL | PNP Switch Sense Signal | In P/N: 0v, Others: 5v |
| 7 | TN/RD | Ignition Coil 1 Driver Control | 5°, 55 mph: 8° dwell |
| 8 | GY/BK | CKP Sensor Signal | Digital Signals: 0-5-0v |
| 9 | --- | Not Used | --- |
| 10 | YL/BK | IAC 2 Driver Control | Pulse Signals |
| 11 | BR/WT | IAC 3 Driver Control | Pulse Signals |
| 12-14 | --- | Not Used | --- |
| 15 | BK/RD | IAT Sensor Signal | At 100°F: 1.83v |
| 16 | TN/BK | ECT Sensor Signal | At 180°F: 2.80v |
| 17 | OR | 5-Volt Supply | 4.9-5.1v |
| 18 | TN/YL | CMP Sensor Signal | Digital Signals: 0-5-0v |
| 19 | GY/RD | IAC 1 Driver Control | Pulse Signals |
| 20 | VT/BK | IAC 4 Driver Control | Pulse Signals |
| 21 | --- | Not Used | --- |
| 22 | RD/BK | Fused Battery Power (B+) | 12-14v |
| 23 | OR/RD | TP Sensor Signal | 0.6-1.0v |
| 24 | BK/DG | HO2S-11 (B1 S1) Signal | 0.1-1.1v |
| 25 | BK/DG | HO2S-12 (B1 S2) Signal | 0.1-1.1v |
| 26 | TN/WT | HO2S-21 (B2 S1) Signal (Cal) | 0.1-1.1v |
| 27 | DG/OR | MAP Sensor Signal | 1.5-1.6v |
| 28 | --- | Not Used | --- |
| 29 | PK/WT | HO2S-22 (B2 S2) Signal (Cal) | 0.1-1.1v |
| 30 | --- | Not Used | --- |
| 31 | BK/WT | Power Ground | <0.1v |
| 32 | BK/TN | Power Ground | <0.1v |

## 1999 Grand Cherokee 4.0L VIN S 'B' White Connector

| PCM Pin # | Wire Color | Circuit Description (32-Pin) | Value at Hot Idle |
|---|---|---|---|
| 1 | VT | Transmission Temperature Sensor | At 200°F: 2.40v |
| 2-3, 7, 13-14 | --- | Not Used | --- |
| 4 | WT/DB | Injector 1 Driver | 1-4 ms |
| 5 | YL/WT | Injector 3 Driver | 1-4 ms |
| 6 | GY | Injector 5 Driver | 1-4 ms |
| 8 | PK | Governor Pressure Solenoid | PWM Signal: 0-12-0v |
| 9 | TN/PK | Coil 2 Driver Control | 5°, 55 mph: 8° dwell |
| 10 | DG | Generator Field Control | Digital Signals: 0-12-0v |
| 11 | DG/LB | TCC Solenoid Control | TCC on at Cruise: 1v |
| 12 | BR/DB | Injector 6 Driver | 1-4 ms |
| 15 | TN | Injector 2 Driver | 1-4 ms |
| 16 | LB/BR | Injector 4 Driver | 1-4 ms |
| 17 | DB/PK | Radiator Fan Control Relay | Relay Off: 12v, On: 1v |
| 18-20 | --- | Not Used | --- |
| 21 | BR/WT | 3-4 Shift Solenoid Control | Solenoid on at Cruise: 1v |
| 22, 24 | --- | Not Used | --- |
| 23 | GY/YL | Engine Oil Pressure Sensor | 1.6v at 24 psi |
| 25 | DB/BK | OSS Sensor Signal (-) | AC pulse signals |
| 26 | --- | Not Used | --- |
| 27 | WT/OR | Vehicle Speed Signal | Digital Signal |
| 28 | LG/WT | OSS Sensor Signal (+) | AC pulse signals |
| 29 | LG/RD | Governor Pressure Signal | 0.58v |
| 30 | PK/YL | Transmission Relay Control | Relay Off: 12v, On: 1v |
| 31 | PK/OR | 5-Volt Supply | 4.9-5.1v |
| 32 | --- | Not Used | --- |

## 1999 Grand Cherokee 4.0L VIN S 'C' Gray Connector

| PCM Pin # | Wire Color | Circuit Description (32-Pin) | Value at Hot Idle |
|---|---|---|---|
| 1 | DB/OR | A/C Clutch Relay Control | Relay Off: 12v, On: 1v |
| 2 | --- | Not Used | --- |
| 3 | DB/YL | ASD Relay Control | Relay Off: 12v, On: 1v |
| 4 | TN/RD | Speed Control Vacuum Solenoid | Vacuum Increasing: 1v |
| 5 | LG/RD | Speed Control Vent Solenoid | Vacuum Decreasing: 1v |
| 6-7 | --- | Not Used | --- |
| 8 | GY | HO2S-12-22 Relay Signal | Relay Off: 12v, On: 1v |
| 9 | LB | HO2S-12-21 Relay Signal | Relay Off: 12v, On: 1v |
| 10 | WT/DG | LDP Solenoid Control | PWM Signal: 0-12-0v |
| 11 | OR/DG | Speed Control Power Supply | 12-14v |
| 12 | DG/LG | ASD Relay Output | 12-14v |
| 13 | OR/YL | Torque Management Relay | Relay Off: 12v, On: 1v |
| 14 | OR/PK | LDP Switch Sense Signal | Closed: 0v, Open: 12v |
| 15 | PK/LG | Battery Temperature Sensor | At 86°F: 1.96v |
| 16-18 | --- | Not Used | --- |
| 19 | BR | Fuel Pump Relay Control | Relay Off: 12v, On: 1v |
| 20 | PK/BK | EVAP Purge Solenoid Control | PWM Signal: 0-12-0v |
| 21 | --- | Not Used | --- |
| 22 | DB | A/C Pressure Signal | 0.90v at 79 psi |
| 23 | --- | Not Used | --- |
| 24 | WT/PK | Brake Switch Sense Signal | Brake Off: 0v, On: 12v |
| 25 | LB/RD | Generator Field Source | 12-14v |
| 26 | LB/YL | Fuel Level Sensor Signal | Full: 0.5v, 1/2 full: 2.5v |
| 27 | PK | SCI Transmit | 0v |
| 28 | WT/BK | CCD Bus (-) | <0.050v |
| 29 | LG/DG | SCI Receive | 5v |
| 30 | YL/PK | CCD Bus (+) | Digital Signals: 0-5-0v |
| 31 | --- | Not Used | --- |
| 32 | RD/LG | Speed Control Switch Signal | S/C & Set Switch On: 3.8v |

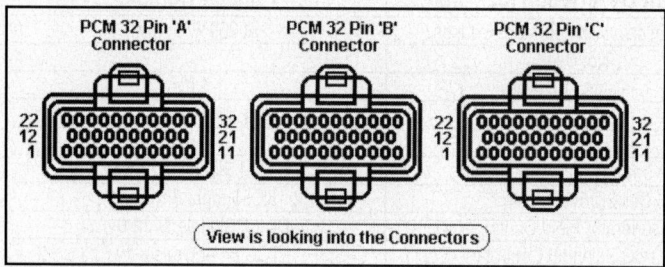

**Pin Connector Graphic**

## Standard Colors and Abbreviations

| Abbreviation | Color | Abbreviation | Color | Abbreviation | Color |
|---|---|---|---|---|---|
| BK | Black | GY | Gray | RD | Red |
| BL | Blue | GN | Green | TN | Tan |
| BR | Brown | LG | Light Green | VT | Violet |
| DB | Dark Blue | OR | Orange | WT | White |
| DG | Dark Green | PK | Pink | YL | Yellow |

### 2000-01 Grand Cherokee 4.0L VIN S 'A' Black Connector

| PCM Pin # | Wire Color | Circuit Description (32-Pin) | Value at Hot Idle |
|---|---|---|---|
| 1 | TN/OR | Ignition Coil 3 Driver Control | 5°, 55 mph: 8° dwell |
| 2 | OR/DB | Ignition Switch Power (B+) | 12-14v |
| 3 | --- | Not Used | --- |
| 4 | BK/LB | Sensor Ground | <0.050v |
| 5 | --- | Not Used | --- |
| 6 | BR/YL | PNP Switch Sense Signal | In P/N: 0v, Others: 5v |
| 7 | TN/RD | Ignition Coil 1 Driver Control | 5°, 55 mph: 8° dwell |
| 8 | GY/BK | CKP Sensor Signal | Digital Signals: 0-5-0v |
| 9 | --- | Not Used | --- |
| 10 | YL/BK | IAC 2 Driver Control | Pulse Signals |
| 11 | BR/WT | IAC 3 Driver Control | Pulse Signals |
| 12-14 | --- | Not Used | --- |
| 15 | BK/RD | IAT Sensor Signal | At 100°F: 1.83v |
| 16 | TN/BK | ECT Sensor Signal | At 180°F: 2.80v |
| 17 | OR | 5-Volt Supply | 4.9-5.1v |
| 18 | TN/YL | CMP Sensor Signal | Digital Signals: 0-5-0v |
| 19 | GY/RD | IAC 1 Driver Control | Pulse Signals |
| 20 | VT/BK | IAC 4 Driver Control | Pulse Signals |
| 21 | --- | Not Used | --- |
| 22 | RD/BK | Fused Battery Power (B+) | 12-14v |
| 23 | OR/RD | TP Sensor Signal | 0.6-1.0v |
| 24 | LG/RD | HO2S-11 (B1 S1) Signal | 0.1-1.1v |
| 25 | BK/DG | HO2S-12 (B1 S2) Signal | 0.1-1.1v |
| 26 | TN/WT | HO2S-21 (B2 S1) Signal | 0.1-1.1v |
| 27 | DG/RD | MAP Sensor Signal | 1.5-1.6v |
| 28 | --- | Not Used | --- |
| 29 | PK/WT | HO2S-22 (B2 S2) Signal | 0.1-1.1v |
| 30 | --- | Not Used | --- |
| 31 | BK/WT | Power Ground | <0.1v |
| 32 | BK/TN | Power Ground | <0.1v |

### 2000-01 Grand Cherokee 4.0L VIN S 'B' White Connector

| PCM Pin # | Wire Color | Circuit Description (32-Pin) | Value at Hot Idle |
|---|---|---|---|
| 1 | VT | Transmission Temperature Sensor | At 200°F: 2.40v |
| 2-3, 7, 13-14 | --- | Not Used | --- |
| 4 | WT/DB | Injector 1 Driver Control | 1-4 ms |
| 5 | YL/WT | Injector 3 Driver Control | 1-4 ms |
| 6 | GY | Injector 5 Driver Control | 1-4 ms |
| 8 | PK | Governor Pressure Solenoid | PWM Signal: 0-12-0v |
| 9 | TN/PK | Ignition Coil 2 Driver Control | 5°, 55 mph: 8° dwell |
| 10 | DG | Generator Field Control | Digital Signals: 0-12-0v |
| 11 | DG/LB | TCC Solenoid Control | TCC on at Cruise: 1v |
| 12 | BR/DB | Injector 6 Driver Control | 1-4 ms |
| 15 | TN | Injector 2 Driver Control | 1-4 ms |
| 16 | LB/BR | Injector 4 Driver Control | 1-4 ms |
| 17 | DB/PK | Radiator Fan Control Relay | Relay Off: 12v, On: 1v |
| 18, 20, 22 | --- | Not Used | --- |
| 19 | DB | AC Pressure Sensor Signal | 0.90v at 79 psi |
| 21 | BR/WT | 3-4 Shift Solenoid Control | Solenoid on at Cruise: 1v |
| 23 | GY/YL | Engine Oil Pressure Sensor | 1.6v at 24 psi |
| 24, 26 | --- | Not Used | --- |
| 25 | DB/BK | OSS Sensor Signal (-) | AC pulse signals |
| 27 | WT/OR | Vehicle Speed Signal | Digital Signal |
| 28 | LG/WT | OSS Sensor Signal (+) | AC pulse signals |
| 29 | LG/RD | Governor Pressure Signal | 0.58v |
| 30 | PK/YL | Transmission Relay Control | Relay Off: 12v, On: 1v |
| 31 | VT/BK | 5-Volt Supply | 4.9-5.1v |
| 32 | --- | Not Used | --- |

### 2000-01 Grand Cherokee 4.0L VIN S 'C' Gray Connector

| PCM Pin # | Wire Color | Circuit Description (32-Pin) | Value at Hot Idle |
|---|---|---|---|
| 1 | DB/OR | A/C Clutch Relay Control | Relay Off: 12v, On: 1v |
| 2 | --- | Not Used | --- |
| 3 | DB/YL | ASD Relay Control | Relay Off: 12v, On: 1v |
| 4 | TN/RD | Speed Control Vacuum Solenoid | Vacuum Increasing: 1v |
| 5 | LG/RD | Speed Control Vent Solenoid | Vacuum Decreasing: 1v |
| 6-7 | --- | Not Used | --- |
| 8 | GY | HO2S-11 Heater Relay | Relay Off: 12v, On: 1v |
| 9 | LB | HO2S-12 Heater Relay | Relay Off: 12v, On: 1v |
| 10 | WT/DG | LDP Solenoid Control | PWM Signal: 0-12-0v |
| 11 | OR/DG | Speed Control Power Supply | 12-14v |
| 12 | DG/LG | ASD Relay Output | 12-14v |
| 13 | OR/YL | Torque Management Request | Digital Signals |
| 14 | OR/PK | LDP Switch Sense Signal | Closed: 0v, Open: 12v |
| 15 | VT/LG | Battery Temperature Sensor | At 86°F: 1.96v |
| 16 | DB/WT | HO2S-21 Heater Relay | Relay Off: 12v, On: 1v |
| 17-18 | --- | Not Used | --- |
| 19 | BR | Fuel Pump Relay Control | Relay Off: 12v, On: 1v |
| 20 | PK/BK | EVAP Purge Solenoid Control | PWM Signal: 0-12-0v |
| 21 | --- | Not Used | --- |
| 22 | DB/YL | A/C Switch Sense | A/C Off: 12v, On: 1v |
| 23 | --- | Not Used | --- |
| 24 | WT/PK | Brake Switch Sense Signal | Brake Off: 0v, On: 12v |
| 25 | LB/RD | Generator Field Source | 12-14v |
| 26 | LB/YL | Fuel Level Sensor Signal | Full: 0.5v, 1/2 full: 2.5v |
| 27 | PK | SCI Transmit | 0v |
| 28 | WT/BK | CCD Bus (-) | <0.050v |
| 28 ('01) | --- | Not Used | --- |
| 29 | LG/DG | SCI Receive | 5v |
| 30 | YL/PK | CCD Bus (+) | Digital Signals: 0-5-0v |
| 30 ('01) | YL/VT | PCI Data Bus (J1850) | Digital Signals: 0-7-0v |
| 31 | --- | Not Used | --- |
| 32 | RD/LG | Speed Control Switch Signal | S/C & Set Switch On: 3.8v |

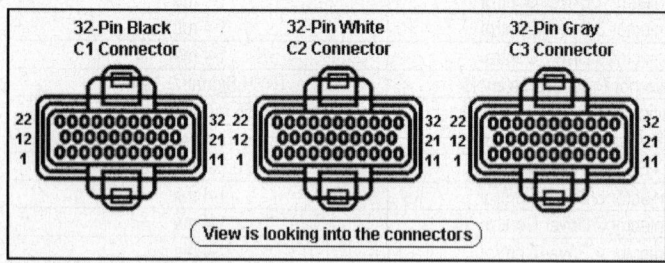

**Pin Connector Graphic**

### Standard Colors and Abbreviations

| Abbreviation | Color | Abbreviation | Color | Abbreviation | Color |
|---|---|---|---|---|---|
| BK | Black | GY | Gray | RD | Red |
| BL | Blue | GN | Green | TN | Tan |
| BR | Brown | LG | Light Green | VT | Violet |
| DB | Dark Blue | OR | Orange | WT | White |
| DG | Dark Green | PK | Pink | YL | Yellow |

## 2002-03 Grand Cherokee 4.0L VIN S 'A' Black Connector

| PCM Pin # | Wire Color | Circuit Description (32-Pin) | Value at Hot Idle |
|---|---|---|---|
| 1 | TN/OR | Ignition Coil 3 Driver Control | 5°, 55 mph: 8° dwell |
| 2 | OR/DB | Ignition Switch Power (B+) | 12-14v |
| 3, 5 | --- | Not Used | --- |
| 4 | BK/LB | Sensor Ground | <0.050v |
| 6 | BK/WT | Park Neutral Position Switch | In P/N: 0v, Others: 5v |
| 7 | TN/RD | Ignition Coil 1 Driver Control | 5°, 55 mph: 8° dwell |
| 8 | GY/BK | CKP Sensor Signal | Digital Signals: 0-5-0v |
| 9 | --- | Not Used | --- |
| 10 | YL/BK | IAC 2 Driver Control | Pulse Signals |
| 11 | BR/WT | IAC 3 Driver Control | Pulse Signals |
| 12-13 | --- | Not Used | --- |
| 14 | LG/BK | Transfer Case Position Switch | Switch Open: 12v, Closed: 0v |
| 15 | BK/RD | IAT Sensor Signal | At 100°F: 1.83v |
| 16 | TN/BK | ECT Sensor Signal | At 180°F: 2.80v |
| 17 | OR | 5-Volt Supply | 4.9-5.1v |
| 18 | TN/YL | CMP Sensor Signal | Digital Signals: 0-5-0v |
| 19 | GY/RD | IAC 1 Driver Control | Pulse Signals |
| 20 | VT/BK | IAC 4 Driver Control | Pulse Signals |
| 21 | --- | Not Used | --- |
| 22 | RD/BK | Fused Battery Power (B+) | 12-14v |
| 23 | OR/RD | TP Sensor Signal | 0.6-1.0v |
| 24 | BK/DG | HO2S-11 (B1 S1) Signal | 0.1-1.1v |
| 25 | TN/WT | HO2S-12 (B1 S2) Signal | 0.1-1.1v |
| 26 | LG/RD | HO2S-21 (B2 S1) Signal | 0.1-1.1v |
| 27 | DG/RD | MAP Sensor Signal | 1.5-1.6v |
| 28 | --- | Not Used | --- |
| 29 | TN/WT | HO2S-22 (B2 S2) Signal | 0.1-1.1v |
| 30 | --- | Not Used | --- |
| 31 | BK/WT | Power Ground | <0.1v |
| 32 | BK/TN | Power Ground | <0.1v |

## 2002-03 Grand Cherokee 4.0L VIN S 'B' White Connector

| PCM Pin # | Wire Color | Circuit Description (32-Pin) | Value at Hot Idle |
|---|---|---|---|
| 1 | VT | Transmission Temperature Sensor | At 200°F: 2.40v |
| 2-3, 7, 13-14 | --- | Not Used | --- |
| 4 | WT/DB | Injector 1 Driver Control | 1-4 ms |
| 5 | YL/WT | Injector 3 Driver Control | 1-4 ms |
| 6 | GY | Injector 5 Driver Control | 1-4 ms |
| 8 | PK | Governor Pressure Solenoid | PWM Signal: 0-12-0v |
| 9 | TN/PK | Ignition Coil 2 Driver Control | 5°, 55 mph: 8° dwell |
| 10 | DG | Generator Field Control | Digital Signals: 0-12-0v |
| 11 | LB | TCC Solenoid Control | TCC on at Cruise: 1v |
| 12 | BR/DB | Injector 6 Driver Control | 1-4 ms |
| 15 | TN | Injector 2 Driver Control | 1-4 ms |
| 16 | LB/BR | Injector 4 Driver Control | 1-4 ms |
| 17 | LG | Radiator Fan Motor Relay | Relay Off: 12v, On: 1v |
| 18, 20, 22 | --- | Not Used | --- |
| 19 | DB | AC Pressure Sensor Signal | 0.90v at 79 psi |
| 21 | BR | 3-4 Shift Solenoid Control | Solenoid on at Cruise: 1v |
| 23 | GY/YL | Engine Oil Pressure Sensor | 1.6v at 24 psi |
| 24, 26 | --- | Not Used | --- |
| 25 | DB/BK | Speed Sensor Ground | AC pulse signals |
| 27 | DG/YL | Vehicle Speed Signal | Digital Signals |
| 28 | LG/WT | Output Speed Sensor Signal (+) | AC pulse signals |
| 29 | LG/RD | Governor Pressure Signal | 0.58v |
| 30 | PK/YL | Transmission Relay Control | Relay Off: 12v, On: 1v |
| 31 | VT/BK | 5-Volt Supply | 4.9-5.1v |
| 32 | --- | Not Used | --- |

## 2002-03 Grand Cherokee 4.0L VIN S 'C' Gray Connector

| PCM Pin # | Wire Color | Circuit Description (32-Pin) | Value at Hot Idle |
|---|---|---|---|
| 1 | DB/OR | A/C Clutch Relay Control | Relay Off: 12v, On: 1v |
| 2 | --- | Not Used | --- |
| 3 | DB/YL | ASD Relay Control | Relay Off: 12v, On: 1v |
| 4 | TN/RD | Speed Control Vacuum Solenoid | Vacuum Increasing: 1v |
| 5 | LG/RD | Speed Control Vent Solenoid | Vacuum Decreasing: 1v |
| 6-7 | --- | Not Used | --- |
| 8 | BR/OR | HO2S-11 (B1 S1) Heater Control | Heater Off: 12v, On: 1v |
| 9 | RD/YL | Downstream Heater Control Relay | Relay Off: 12v, On: 1v |
| 10 | WT/DG | LDP Solenoid Control | PWM Signal: 0-12-0v |
| 11 | OR/DG | Speed Control Power Supply | 12-14v |
| 12 | DG/LG | ASD Relay Output | 12-14v |
| 13 | OR/WT | Overdrive Off Switch Sense | Switch Off: 12v, On: 1v |
| 14 | OR/PK | LDP Switch Sense Signal | Closed: 0v, Open: 12v |
| 15 | VT/LG | Battery Temperature Sensor | At 86°F: 1.96v |
| 16 | BR/WT | HO2S-12 (B1 S2) Heater Control | Heater Off: 12v, On: 1v |
| 17-18 | --- | Not Used | --- |
| 19 | BR | Fuel Pump Relay Control | Relay Off: 12v, On: 1v |
| 20 | PK/BK | EVAP Purge Solenoid Control | PWM Signal: 0-12-0v |
| 21-23 | --- | Not Used | --- |
| 24 | WT/PK | Brake Switch Sense Signal | Brake Off: 0v, On: 12v |
| 25 | WT/DB | Generator Field Source | 12-14v |
| 26 | LB/YL | Fuel Level Sensor Signal | Full: 0.5v, 1/2 full: 2.5v |
| 27 | PK | SCI Transmit | 0v |
| 28 | --- | Not Used | --- |
| 29 | LG | SCI Receive | 5v |
| 30 | WT/YL | PCI Data Bus (J1850) | Digital Signals: 0-7-0v |
| 31 | --- | Not Used | --- |
| 32 | DG/LG | Speed Control Switch Signal | S/C & Set Switch On: 3.8v |

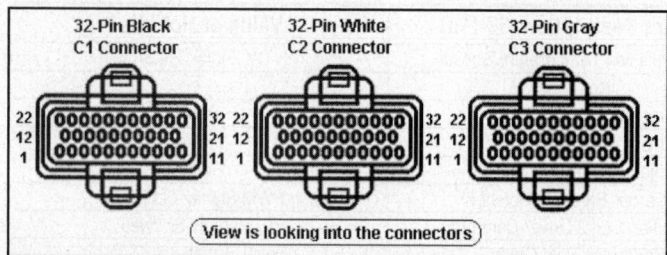

**Pin Connector Graphic**

## Standard Colors and Abbreviations

| Abbreviation | Color | Abbreviation | Color | Abbreviation | Color |
|---|---|---|---|---|---|
| BK | Black | GY | Gray | RD | Red |
| BL | Blue | GN | Green | TN | Tan |
| BR | Brown | LG | Light Green | VT | Violet |
| DB | Dark Blue | OR | Orange | WT | White |
| DG | Dark Green | PK | Pink | YL | Yellow |

**1999-2000 Grand Cherokee 4.7L VIN N 'C1' Black Connector**

| PCM Pin # | Wire Color | Circuit Description (32-Pin) | Value at Hot Idle |
|---|---|---|---|
| 1 | TN/OR | Ignition Coil 3 Driver Control | 7°, 55 mph: 9° dwell |
| 2 | OR/DB | Ignition Switch Power (B+) | 12-14v |
| 3 | TN/LG | Ignition Coil 4 Driver Control | 7°, 55 mph: 9° dwell |
| 4 | BK/LB | Sensor Ground | <0.1v |
| 5 | TN/LB | Ignition Coil 6 Driver Control | 7°, 55 mph: 9° dwell |
| 6 | BR/YL | PNP Switch Sense Signal | In P/N: 0v, Others: 5v |
| 7 | TN/RD | Ignition Coil 1 Driver Control | 7°, 55 mph: 9° dwell |
| 8 | GY/BK | CKP Sensor Signal | Digital Signals: 0-5-0v |
| 9 | LB/RD | Ignition Coil 8 Driver Control | 7°, 55 mph: 9° dwell |
| 10 | YL/BK | IAC 2 Driver Control | Pulse Signals |
| 11 | BR/WT | IAC 3 Driver Control | Pulse Signals |
| 12-14 | --- | --- | --- |
| 15 | BK/RD | IAT Sensor Signal | At 100°F: 1.83v |
| 16 | TN/BK | ECT Sensor Signal | At 180°F: 2.80v |
| 17 | OR | 5-Volt Supply | 4.9-5.1v |
| 18 | TN/YL | CMP Sensor Signal | Digital Signals: 0-5-0v |
| 19 | GY/BK | IAC 1 Driver Control | Pulse Signals |
| 20 | VT/BK | IAC 4 Driver Control | Pulse Signals |
| 21 | TN/DG | Ignition Coil 5 Driver Control | 7°, 55 mph: 9° dwell |
| 22 | RD/BK | Fused Battery Power (B+) | 12-14v |
| 23 | OR/RD | TP Sensor Signal | 0.6-1.0v |
| 24 | LG/RD | HO2S-11 (B1 S1) Signal | 0.1-1.1v |
| 25 | BK/DG | HO2S-12 (B1 S2) Signal | 0.1-1.1v |
| 26 | TN/WT | HO2S-21 (B2 S1) Signal | 0.1-1.1v |
| 27 | DG/RD | MAP Sensor Signal | 1.5-1.6v |
| 28, 30 | --- | Not Used | --- |
| 29 | PK/WT | HO2S-22 (B2 S2) Signal | 0.1-1.1v |
| 31 | BK/WT | Power Ground | <0.1v |
| 32 | BK/TN | Power Ground | <0.1v |

**1999-2000 Grand Cherokee 4.7L VIN N 'C2' White Connector**

| PCM Pin # | Wire Color | Circuit Description (32-Pin) | Value at Hot Idle |
|---|---|---|---|
| 1, 3, 8, 11, 14 | --- | Not Used | --- |
| 2 | DB/TN | Injector 7 Driver Control | 1-4 ms |
| 4 | WT/DB | Injector 1 Driver Control | 1-4 ms |
| 5 | YL/WT | Injector 3 Driver Control | 1-4 ms |
| 6 | GY | Injector 5 Driver Control | 1-4 ms |
| 7 | BR | Ignition Coil 7 Driver Control | 7°, 55 mph: 9° dwell |
| 9 | TN/PK | Ignition Coil 2 Driver Control | 7°, 55 mph: 9° dwell |
| 10 | DG | Generator Field Control | Digital Signals: 0-12-0v |
| 12 | BR/DB | Injector 6 Driver Control | 1-4 ms |
| 13 | DB/GY | Injector 8 Driver Control | 1-4 ms |
| 15 | TN | Injector 2 Driver Control | 1-4 ms |
| 16 | LB/BR | Injector 4 Driver Control | 1-4 ms |
| 17 | DB/RD | Radiator Fan Control Relay | Relay Off: 12v, On: 1v |
| 18, 20 | --- | Not Used | --- |
| 19 | DB | A/C Pressure Signal | 0.90v at 79 psi |
| 21 | BR/WT | 3-4 Shift Solenoid Control | Solenoid on at Cruise: 1v |
| 22 | --- | Not Used | --- |
| 23 | GY/YL | Engine Oil Pressure Sensor | 1.6v at 24 psi |
| 24-26 | --- | Not Used | --- |
| 27 | WT/OR | Vehicle Speed Signal | Digital Signal |
| 28-30 | --- | Not Used | --- |
| 31 | VT/BK | 5-Volt Supply | 4.9-5.1v |
| 32 | --- | Not Used | --- |

**1999-2000 Grand Cherokee 4.7L VIN N 'C3' Gray Connector**

| PCM Pin # | Wire Color | Circuit Description (32-Pin) | Value at Hot Idle |
|---|---|---|---|
| 1 | DB/OR | A/C Clutch Relay Control | Relay Off: 12v, On: 1v |
| 2 | --- | Not Used | --- |
| 3 | DB/YL | ASD Relay Control | Relay Off: 12v, On: 1v |
| 4 | TN/RD | Speed Control Vacuum Solenoid | Vacuum Increasing: 1v |
| 5 | LG/RD | Speed Control Vent Solenoid | Vacuum Decreasing: 1v |
| 6-7 | --- | Not Used | --- |
| 8 | GY | HO2S-12-22 Relay Signal | Relay Off: 12v, On: 1v |
| 9 | LB | HO2S-12-21 Relay Signal | Relay Off: 12v, On: 1v |
| 10 | WT/DG | LDP Solenoid Control | PWM Signal: 0-12-0v |
| 11 | OR/DG | Speed Control Power Supply | 12-14v |
| 12 | DG/LG | ASD Relay Output | 12-14v |
| 13 | OR/YL | Torque Management Relay | Relay Off: 12v, On: 1v |
| 14 | OR/PK | LDP Switch Sense Signal | Closed: 0v, Open: 12v |
| 15 | PK/LG | Battery Temperature Sensor | At 86ºF: 1.96v |
| 16-18 | --- | Not Used | --- |
| 19 | BR | Fuel Pump Relay Control | Relay Off: 12v, On: 1v |
| 20 | PK/BK | EVAP Purge Solenoid Control | PWM Signal: 0-12-0v |
| 21 | --- | Not Used | --- |
| 22 | DB | A/C Pressure Switch Signal | A/C On: 0.451-4.850v |
| 23 | --- | Not Used | --- |
| 24 | WT/PK | Brake Switch Sense Signal | Brake Off: 0v, On: 12v |
| 25 | LB/RD | Generator Field Source | 12-14v |
| 26 | LB/YL | Fuel Level Sensor Signal | Full: 0.5v, 1/2 full: 2.5v |
| 27 | PK | SCI Transmit | 0v |
| 28 | WT/BK | CCD Bus (-) | <0.050v |
| 29 | LG/DG | SCI Receive | 5v |
| 30 | YL/PK | CCD Bus (+) | Digital Signals: 0-5-0v |
| 31 | --- | Not Used | --- |
| 32 | RD/LG | Speed Control Switch Signal | S/C & Set Switch On: 3.8v |

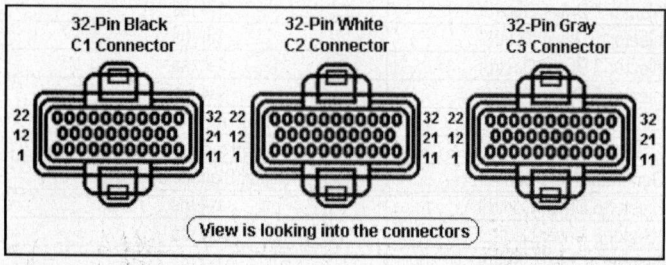

**Pin Connector Graphic**

### Standard Colors and Abbreviations

| Abbreviation | Color | Abbreviation | Color | Abbreviation | Color |
|---|---|---|---|---|---|
| BK | Black | GY | Gray | RD | Red |
| BL | Blue | GN | Green | TN | Tan |
| BR | Brown | LG | Light Green | VT | Violet |
| DB | Dark Blue | OR | Orange | WT | White |
| DG | Dark Green | PK | Pink | YL | Yellow |

## 2001 Grand Cherokee 4.7L VIN N 'C1' Black Connector

| PCM Pin # | Wire Color | Circuit Description (32-Pin) | Value at Hot Idle |
|---|---|---|---|
| 1 | TN/OR | Ignition Coil 3 Driver Control | 7°, 55 mph: 9° dwell |
| 2 | OR/DB | Ignition Switch Power (B+) | 12-14v |
| 3 | TN/LG | Ignition Coil 4 Driver Control | 7°, 55 mph: 9° dwell |
| 4 | BK/LB | Sensor Ground | <0.1v |
| 5 | TN/LB | Ignition Coil 6 Driver Control | 7°, 55 mph: 9° dwell |
| 6 | BR/YL | PNP Switch Sense Signal | In P/N: 0v, Others: 5v |
| 7 | TN/RD | Ignition Coil 1 Driver Control | 7°, 55 mph: 9° dwell |
| 8 | GY/BK | CKP Sensor Signal | Digital Signals: 0-5-0v |
| 9 | LB/RD | Ignition Coil 8 Driver Control | 7°, 55 mph: 9° dwell |
| 10 | YL/BK | IAC 2 Driver Control | Pulse Signals |
| 11 | BR/WT | IAC 3 Driver Control | Pulse Signals |
| 12-14 | --- | Not Used | --- |
| 15 | BK/RD | IAT Sensor Signal | At 100ºF: 1.83v |
| 16 | TN/BK | ECT Sensor Signal | At 180ºF: 2.80v |
| 17 | OR | 5-Volt Supply | 4.9-5.1v |
| 18 | TN/YL | CMP Sensor Signal | Digital Signals: 0-5-0v |
| 19 | GY/BK | IAC 1 Driver Control | Pulse Signals |
| 20 | VT/BK | IAC 4 Driver Control | Pulse Signals |
| 21 | TN/DG | Ignition Coil 5 Driver Control | 7°, 55 mph: 9° dwell |
| 22 | RD/BK | Fused Battery Power (B+) | 12-14v |
| 23 | OR/RD | TP Sensor Signal | 0.6-1.0v |
| 24 | LG/RD | HO2S-11 (B1 S1) Signal | 0.1-1.1v |
| 25 | BK/DG | HO2S-12 (B1 S2) Signal | 0.1-1.1v |
| 26 | TN/WT | HO2S-21 (B2 S1) Signal | 0.1-1.1v |
| 27 | DG/RD | MAP Sensor Signal | 1.5-1.6v |
| 28, 30 | --- | Not Used | --- |
| 29 | PK/WT | HO2S-22 (B2 S2) Signal | 0.1-1.1v |
| 31 | BK/WT | Power Ground | <0.1v |
| 32 | BK/TN | Power Ground | <0.1v |

## 2001 Grand Cherokee 4.7L VIN N 'C2' White Connector

| PCM Pin # | Wire Color | Circuit Description (32-Pin) | Value at Hot Idle |
|---|---|---|---|
| 1, 3, 8-11 | --- | Not Used | --- |
| 2 | DB/TN | Injector 7 Driver Control | 1-4 ms |
| 4 | WT/DB | Injector 1 Driver Control | 1-4 ms |
| 5 | YL/WT | Injector 3 Driver Control | 1-4 ms |
| 6 | GY | Injector 5 Driver Control | 1-4 ms |
| 7 | BR | Ignition Coil 7 Driver Control | 7°, 55 mph: 9° dwell |
| 9 | TN/PK | Ignition Coil 2 Driver Control | 7°, 55 mph: 9° dwell |
| 10 | DG | Generator Field Control | Digital Signals: 0-12-0v |
| 12 | BR/DB | Injector 6 Driver Control | 1-4 ms |
| 13 | DB/GY | Injector 8 Driver Control | 1-4 ms |
| 14 | --- | Not Used | --- |
| 15 | TN | Injector 2 Driver Control | 1-4 ms |
| 16 | LB/BR | Injector 4 Driver Control | 1-4 ms |
| 17 | DB/RD | Radiator Fan Control Relay | Relay Off: 12v, On: 1v |
| 18, 20, 22, 24 | --- | Not Used | --- |
| 19 | DB | A/C Pressure Signal | 0.90v at 79 psi |
| 21 | BR/WT | 3-4 Shift Solenoid Control | Solenoid on at Cruise: 1v |
| 23 | GY/YL | Engine Oil Pressure Sensor | 1.6v at 24 psi |
| 25 | DB/BK | Output Speed Sensor (N) | AC pulse signals |
| 26 | --- | Not Used | --- |
| 27 | WT/OR | Vehicle Speed Signal | Digital Signal |
| 28 | LG/WT | OSS Sensor Signal (P) | AC pulse signals |
| 29 | LG/RD | Governor Pressure Signal | 0.58v |
| 30 | PK/YL | Transmission Control Relay | Relay Off: 12v, On: 1v |
| 31 | VT/BK | 5-Volt Supply | 4.9-5.1v |
| 32 | --- | Not Used | --- |

**2001 Grand Cherokee 4.7L VIN N 'C3' Gray Connector**

| PCM Pin # | Wire Color | Circuit Description (32-Pin) | Value at Hot Idle |
|---|---|---|---|
| 1 | DB/OR | A/C Clutch Relay Control | Relay Off: 12v, On: 1v |
| 2 | --- | Not Used | --- |
| 3 | DB/YL | ASD Relay Control | Relay Off: 12v, On: 1v |
| 4 | TN/RD | Speed Control Vacuum Solenoid | Vacuum Increasing: 1v |
| 5 | LG/RD | Speed Control Vent Solenoid | Vacuum Decreasing: 1v |
| 6-7 | --- | Not Used | --- |
| 8 | GY | HO2S-12-22 Relay Signal | Relay Off: 12v, On: 1v |
| 9 | LB | HO2S-12-21 Relay Signal | Relay Off: 12v, On: 1v |
| 10 | WT/DG | LDP Solenoid Control | PWM Signal: 0-12-0v |
| 11 | OR/DG | Speed Control Power Supply | 12-14v |
| 12 | DG/LG | ASD Relay Output | 12-14v |
| 13 | OR/YL | Torque Management Relay | Relay Off: 12v, On: 1v |
| 14 | OR/PK | LDP Switch Sense Signal | Closed: 0v, Open: 12v |
| 15 | PK/LG | Battery Temperature Sensor | At 86°F: 1.96v |
| 16-18 | --- | Not Used | --- |
| 19 | BR | Fuel Pump Relay Control | Relay Off: 12v, On: 1v |
| 20 | PK/BK | EVAP Purge Solenoid Control | PWM Signal: 0-12-0v |
| 21 | --- | Not Used | --- |
| 22 | DB | A/C Pressure Switch Signal | A/C On: 0.451-4.850v |
| 23 | --- | Not Used | --- |
| 24 | WT/PK | Brake Switch Sense Signal | Brake Off: 0v, On: 12v |
| 25 | LB/RD | Generator Field Source | 12-14v |
| 26 | LB/YL | Fuel Level Sensor Signal | Full: 0.5v, 1/2 full: 2.5v |
| 27 | PK | SCI Transmit | 0v |
| 28 | WT/BK | CCD Bus (-) | <0.050v |
| 29 | LG/DG | SCI Receive | 5v |
| 30 | YL/PK | CCD Bus (+) | Digital Signals: 0-5-0v |
| 31 | --- | Not Used | --- |
| 32 | RD/LG | Speed Control Switch Signal | S/C & Set Switch On: 3.8v |

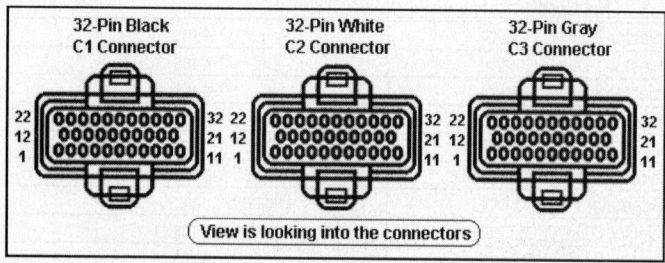

**Pin Connector Graphic**

**Standard Colors and Abbreviations**

| Abbreviation | Color | Abbreviation | Color | Abbreviation | Color |
|---|---|---|---|---|---|
| BK | Black | GY | Gray | RD | Red |
| BL | Blue | GN | Green | TN | Tan |
| BR | Brown | LG | Light Green | VT | Violet |
| DB | Dark Blue | OR | Orange | WT | White |
| DG | Dark Green | PK | Pink | YL | Yellow |

**2002-03 Grand Cherokee 4.7L VIN J, VIN N (A/T) 'C1' Black Connector**

| PCM Pin # | Wire Color | Circuit Description (32-Pin) | Value at Hot Idle |
|---|---|---|---|
| 1 | TN/OR | Ignition Coil 3 Driver Control | 7°, 55 mph: 9° dwell |
| 2 | OR/DB | Ignition Switch Power (B+) | 12-14v |
| 3 | TN/LG | Ignition Coil 4 Driver Control | 7°, 55 mph: 9° dwell |
| 4 | BK/LB | Sensor Ground | <0.1v |
| 5 | TN/LB | Ignition Coil 6 Driver Control | 7°, 55 mph: 9° dwell |
| 6 | BK/WT | PNP Switch Sense Signal | In P/N: 0v, Others: 5v |
| 7 | TN/RD | Ignition Coil 1 Driver Control | 7°, 55 mph: 9° dwell |
| 8 | GY/BK | CKP Sensor Signal | Digital Signals: 0-5-0v |
| 9 | LB/RD | Ignition Coil 8 Driver Control | 7°, 55 mph: 9° dwell |
| 10 | YL/BK | IAC 2 Driver Control | Pulse Signals |
| 11 | BR/WT | IAC 3 Driver Control | Pulse Signals |
| 12-13 | --- | --- | --- |
| 14 | LG/BK | Transfer Case Position Switch | Switch Open: 12v, Closed: 0v |
| 15 | BK/RD | IAT Sensor Signal | At 100°F: 1.83v |
| 16 | TN/BK | ECT Sensor Signal | At 180°F: 2.80v |
| 17 | OR | 5-Volt Supply | 4.9-5.1v |
| 18 | TN/YL | CMP Sensor Signal | Digital Signals: 0-5-0v |
| 19 | GY/BK | IAC 1 Driver Control | Pulse Signals |
| 20 | VT/BK | IAC 4 Driver Control | Pulse Signals |
| 21 | TN/DG | Ignition Coil 5 Driver Control | 7°, 55 mph: 9° dwell |
| 22 | RD/BK | Fused Battery Power (B+) | 12-14v |
| 23 | OR/RD | TP Sensor Signal | 0.6-1.0v |
| 24 | LG/RD | HO2S-11 (B1 S1) Signal | 0.1-1.1v |
| 25 | BK/DG | HO2S-12 (B1 S2) Signal | 0.1-1.1v |
| 26 | LG/RD | HO2S-21 (B2 S1) Signal | 0.1-1.1v |
| 27 | DG/RD | MAP Sensor Signal | 1.5-1.6v |
| 28, 30 | --- | Not Used | --- |
| 29 | PK/WT | HO2S-22 (B2 S2) Signal | 0.1-1.1v |
| 31 | BK/WT | Power Ground | <0.1v |
| 32 | BK/TN | Power Ground | <0.1v |

**2002-03 Grand Cherokee 4.7L VIN J, VIN N (A/T) 'C2' White Connector**

| PCM Pin # | Wire Color | Circuit Description (32-Pin) | Value at Hot Idle |
|---|---|---|---|
| 1, 3, 8, 11, 14 | --- | Not Used | --- |
| 2 | VT | Injector 7 Driver Control | 1-4 ms |
| 4 | WT/DB | Injector 1 Driver Control | 1-4 ms |
| 5 | YL/WT | Injector 3 Driver Control | 1-4 ms |
| 6 | GY | Injector 5 Driver Control | 1-4 ms |
| 7 | BR | Ignition Coil 7 Driver Control | 7°, 55 mph: 9° dwell |
| 9 | TN/PK | Ignition Coil 2 Driver Control | 7°, 55 mph: 9° dwell |
| 10 | DG | Generator Field Control | Digital Signals: 0-12-0v |
| 12 | BR/DB | Injector 6 Driver Control | 1-4 ms |
| 13 | GY/LB | Injector 8 Driver Control | 1-4 ms |
| 15 | TN | Injector 2 Driver Control | 1-4 ms |
| 16 | LB/BR | Injector 4 Driver Control | 1-4 ms |
| 17 | LG | Radiator Fan Control Relay | Relay Off: 12v, On: 1v |
| 18 | --- | Not Used | --- |
| 19 | DB | A/C Pressure Signal | 0.90v at 79 psi |
| 20-22 | --- | Not Used | --- |
| 23 | GY/YL | Engine Oil Pressure Sensor | 1.6v at 24 psi |
| 24-26 | --- | Not Used | --- |
| 27 | DG/YL | Vehicle Speed Signal | Digital Signal |
| 28-30 | --- | Not Used | --- |
| 31 | VT/BK | 5-Volt Supply | 4.9-5.1v |
| 32 | --- | Not Used | --- |

## 2002-03 Grand Cherokee 4.7L VIN J, VIN N (A/T) 'C3' Gray Connector

| PCM Pin # | Wire Color | Circuit Description (32-Pin) | Value at Hot Idle |
|---|---|---|---|
| 1 | DB/OR | A/C Clutch Relay Control | Relay Off: 12v, On: 1v |
| 2 | --- | Not Used | --- |
| 3 | DB/YL | ASD Relay Control | Relay Off: 12v, On: 1v |
| 4 | TN/RD | Speed Control Vacuum Solenoid | Vacuum Increasing: 1v |
| 5 | LG/RD | Speed Control Vent Solenoid | Vacuum Decreasing: 1v |
| 6 | --- | Not Used | --- |
| 7 | DB/LG | Knock Sensor 1 Signal | 0.080v AC |
| 8 | BR/OR | HO2S-11(B1 S1) Heater Control | Heater Off: 12v, On: 1v |
| 9 | RD/YL | Downstream Heater Relay Control | Relay Off: 12v, On: 1v |
| 10 | WT/DG | LDP Solenoid Control | PWM Signal: 0-12-0v |
| 11 | OR/DG | Speed Control Power Supply | 12-14v |
| 12 | DG/LG | ASD Relay Output | 12-14v |
| 13 | YL/DG | Torque Management Relay | Relay Off: 12v, On: 1v |
| 14 | OR/PK | LDP Switch Sense Signal | Closed: 0v, Open: 12v |
| 15 | VT/LG | Battery Temperature Sensor | At 86°F: 1.96v |
| 16 | BR/WT | HO2S-12 (B1 S1) Heater Control | Heater Off: 12v, On: 1v |
| 17 | --- | Not Used | --- |
| 18 | GY/BK | Knock Sensor 2 Signal | 0.080v AC |
| 19 | BR | Fuel Pump Relay Control | Relay Off: 12v, On: 1v |
| 20 | PK/BK | EVAP Purge Solenoid Control | PWM Signal: 0-12-0v |
| 21-23 | --- | Not Used | --- |
| 24 | WT/PK | Brake Switch Sense | Brake Off: 0v, On: 12v |
| 25 | WT/DB | Generator Field Source | 12-14v |
| 26 | LB/YL | Fuel Level Sensor Signal | Full: 0.5v, 1/2 full: 2.5v |
| 27 | PK | SCI Transmit | 0v |
| 28 | --- | Not Used | --- |
| 29 | LG/DG | SCI Receive | 5v |
| 30 | VT/YL | PCI Bus Signal (J1850) | Digital Signals: 0-7-0v |
| 31 | --- | Not Used | --- |
| 32 | RD/LG | Speed Control Switch Signal | S/C & Set Switch On: 3.8v |

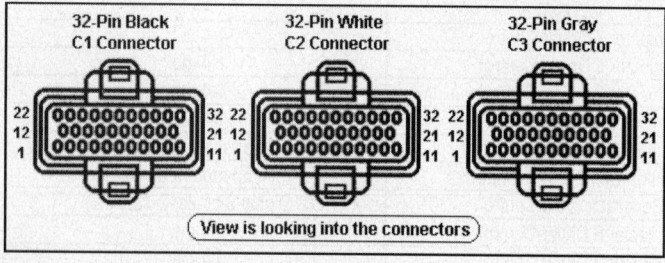

Pin Connector Graphic

## Standard Colors and Abbreviations

| Abbreviation | Color | Abbreviation | Color | Abbreviation | Color |
|---|---|---|---|---|---|
| BK | Black | GY | Gray | RD | Red |
| BL | Blue | GN | Green | TN | Tan |
| BR | Brown | LG | Light Green | VT | Violet |
| DB | Dark Blue | OR | Orange | WT | White |
| DG | Dark Green | PK | Pink | YL | Yellow |

**2004 Grand Cherokee 4.7L VIN N 'C1' Black Connector**

| PCM Pin # | Wire Color | Circuit Description (32-Pin) | Value at Hot Idle |
|---|---|---|---|
| 1 | TN/OR | Coil Driver No. 3 | |
| 2 | OR/DB | Fused Ignition Switch Output (Run-Start) | |
| 3 | TN/LG (4.7L) | Coil Driver No. 4 | |
| 4 | BK/LB | Sensor Ground | |
| 5 | TN/LB (4.7L) | Coil Driver No. 6 | |
| 6 | BK/WT | Park/Neutral Position Switch Sense (T41) | |
| 7 | TN/RD | Coil Driver No. 1 | |
| 8 | GY/BK | Crankshaft Position Sensor Signal | |
| 9 | LB/RD (4.7L) | Coil Driver No. 8 | |
| 10 | YL/BK | Idle Air Control No. 2 Driver | |
| 11 | BR/WT | Idle Air Control No. 3 Driver | |
| 12 | --- | Not Used | --- |
| 13 | --- | Not Used | --- |
| 14 | LG/BK | Transfer Case Position Sensor Input | |
| 15 | BK/RD | Intake Air Temperature Sensor Signal | |
| 16 | TN/BK | Engine Coolant Temperature Sensor Signal | |
| 17 | OR | 5 Volt Supply | |
| 18 | TN/YL | Camshaft Position Sensor Signal | |
| 19 | GY/BK | Idle Air Control No. 1 Driver | |
| 20 | VT/BK | Idle Air Control No. 4 Driver | |
| 21 | TN/DG (4.7L) | Coil Driver No. 5 | |
| 22 | RD/BK | Fused B (+) | |
| 23 | OR/RD | Throttle Position Sensor Signal | |
| 24 | BK/DG | Oxygen Sensor 1/1 Signal | |
| 25 | TN/WT | Oxygen Sensor 1/2 Signal | |
| 26 | LG/RD | Oxygen Sensor 2/1 Signal | |
| 27 | DG/RD | MAP Sensor Signal | |
| 28 | --- | Not Used | --- |
| 29 | TN/WT (4.0L) | Oxygen Sensor 2/2 Signal | |
| 30 | --- | Not Used | --- |
| 31 | BK/WT | Ground | |
| 32 | BK/TN | Ground | |

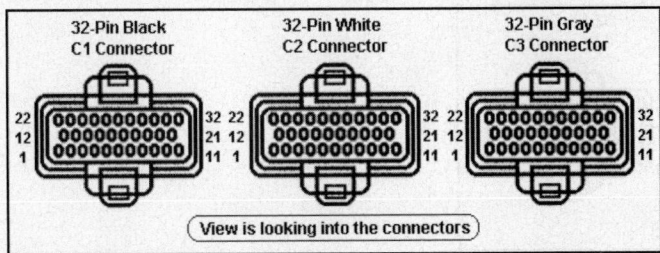

32-Pin Black C1 Connector    32-Pin White C2 Connector    32-Pin Gray C3 Connector

View is looking into the connectors

**Pin Connector Graphic**

## 2004 Grand Cherokee 4.7L VIN N 'C2' White Connector

| PCM Pin # | Wire Color | Circuit Description (32-Pin) | Value at Hot Idle |
|---|---|---|---|
| 1 | VT (4.0L) | Transmission Temperature Sensor Signal | |
| 2 | VT (4.7L) | Fuel Injector No. 7 Driver | |
| 3 | --- | Not Used | --- |
| 4 | WT/DB | Fuel Injector No. 1 Driver | |
| 5 | YL/WT | Fuel Injector No. 3 Driver | |
| 6 | GY | Fuel Injector No. 5 Driver | |
| 7 | BR (4.7L) | Coil Driver No. 7 | |
| 8 | PK (4.0L) | Governor Pressure Solenoid Control | |
| 9 | TN/PK | Coil Driver No. 2 | |
| 10 | DG | Generator Field Driver | |
| 11 | LB (4.0L) | Torque Converter Clutch Solenoid Control | |
| 12 | BR/DB | Fuel Injector No. 6 Driver | |
| 13 | GY/LB (4.7L) | Fuel Injector No. 8 Driver | |
| 14 | --- | Not Used | --- |
| 15 | TN | Fuel Injector No. 2 Driver | |
| 16 | LB/BR | Fuel Injector No. 4 Driver | |
| 17 | LG | Radiator Fan Relay Control | |
| 18 | --- | Not Used | --- |
| 19 | DB | A/C Pressure Signal | |
| 20 | --- | Not Used | --- |
| 21 | BR (4.0L) | 3-4 Shift Solenoid Control | |
| 22 | --- | Not Used | --- |
| 23 | GY/YL | Engine Oil Pressure Switch Signal | |
| 24 | --- | Not Used | --- |
| 25 | DB/BK (4.0L) | Speed Sensor Ground | |
| 26 | --- | Not Used | --- |
| 27 | DG/YL | Vehicle Speed Signal | |
| 28 | LG/WT (4.0L) | Output Speed Sensor Signal | |
| 29 | LG/RD (4.0L) | Governor Pressure Sensor Signal | |
| 30 | PK/YL (4.0L) | Transmission Control Relay Control | |
| 31 | VT/BK | 5 Volt Supply | |
| 32 | --- | Not Used | --- |

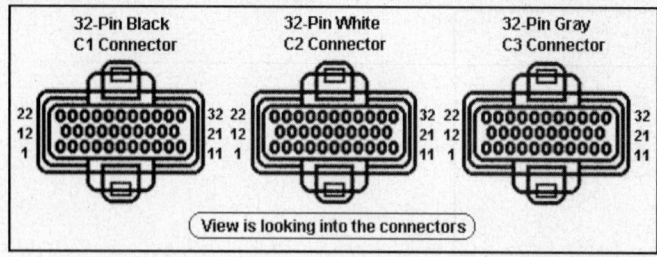

**Pin Connector Graphic**

**2004 Grand Cherokee 4.7L VIN N 'C3' Gray Connector**

| PCM Pin # | Wire Color | Circuit Description (32-Pin) | Value at Hot Idle |
|---|---|---|---|
| 1 | DB/OR | A/C Clutch Relay Control | |
| 2 | DB/PK (4.0L) | Low Speed Radiator Fan Relay Control | |
| 3 | DB/YL | Auto Shut Down Relay Control | |
| 4 | TN/RD | Speed Control Vacuum Solenoid Control | |
| 5 | LG/RD | Speed Control Vent Solenoid Control | |
| 6 | --- | Not Used | --- |
| 7 | DB/LG (4.7L H.O.) | Knock Sensor No. 1 Signal | |
| 8 | BR/OR | Oxygen Sensor 1/1 Heater Control | |
| 9 | RD/YL | Oxygen Sensor Downstream Relay Control | |
| 10 | WT/DG | Leak Detection Pump Solenoid Control | |
| 11 | OR/DG | Speed Control Power Supply | |
| 12 | DG/LG | Fused Auto Shut Down Relay Output | |
| 13 | OR/WT (4.0L LHD) | Overdrive Off Switch Sense | |
| 13 | OR/BK (4.0L RHD) | Overdrive Off Switch Sense | |
| 13 | DG/LG (4.7L LHD) | Torque Management Request Sense | |
| 13 | YL/DG (4.7L RHD) | Torque Management Request Sense | |
| 14 | OR/PK | Leak Detection Pump Switch Sense | |
| 15 | VT/LG | Battery Temperature Sensor Signal | |
| 16 | BR/WT | Oxygen Sensor 2/1 Heater Control | |
| 17 | --- | Not Used | --- |
| 18 | GY/BK (4.7L H.O.) | Knock Sensor No. 2 Signal | |
| 19 | BR | Fuel Pump Relay Control | |
| 20 | PK/BK | Duty Cycle EVAP/Purge Solenoid Control | |
| 21 | PK/DB (4.0L) | High Speed Radiator Fan Relay Control | |
| 22 | --- | Not Used | --- |
| 23 | --- | Not Used | --- |
| 24 | WT/PK | Brake Switch Sense | |
| 25 | WT/DB | Generator Source | |
| 26 | LB/YL | Fuel Level Sensor Signal | |
| 27 | PK | SCI Transmit | |
| 28 | --- | Not Used | --- |
| 29 | LG (LHD) | SCI Receive | |
| 29 | LG/DG (RHD) | SCI Receive | |
| 30 | VT/YL | PCI Bus | |
| 31 | --- | Not Used | --- |
| 32 | RD/LG | Speed Control Switch Signal | |

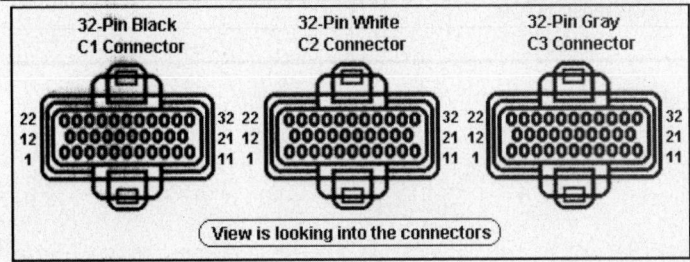

**Pin Connector Graphic**

## 2005 Grand Cherokee 4.7L VIN N 'C1' Black/Black Connector

| PCM Pin # | Wire Color | Circuit Description (32-Pin) | Value at Hot Idle |
|---|---|---|---|
| 1 | DB/BK (Exc. 3.7L) | Coil Control No. 8 | |
| 2 | --- | Not Used | --- |
| 3 | BR/LG (Exc. 3.7L) | Coil Control No. 7 | |
| 4 | BR/LB (Exc. 3.7L) | Injector Control No. 8 | |
| 5 | BR/YL (Exc. 3.7L) | Injector Control No. 7 | |
| 6 | --- | Not Used | --- |
| 7 | --- | Not Used | --- |
| 8 | --- | Not Used | --- |
| 9 | BK | Ground | |
| 10 | --- | Not Used | --- |
| 11 | PK/LG | Fused Ignition Switch Output (Run-Start) | |
| 12 | PK/WT (Exc. 3.7L) | Fused Ignition Switch Output (Run-Start) | |
| 13 | --- | Not Used | --- |
| 14 | --- | Not Used | --- |
| 15 | --- | Not Used | --- |
| 16 | --- | Not Used | --- |
| 17 | --- | Not Used | --- |
| 18 | BK/BR | Ground | |
| 19 | --- | Not Used | --- |
| 20 | VT/GY | Engine Oil Pressure Signal | |
| 21 | --- | Not Used | --- |
| 22 | --- | Not Used | --- |
| 23 | --- | Not Used | --- |
| 24 | --- | Not Used | --- |
| 25 | LG/WT | SCI Receive | |
| 26 | WT/OR (Exc. 3.7L) | SCI Receive (TCM) | |
| 27 | YL/PK | 5 Volt Supply | |
| 28 | BR/LB (5.7L) | Mds Sol Control No. 4 | |
| 29 | RD | Fused B (+) | |
| 30 | PK/OR | Fused Ignition Switch Output (Start) | |
| 31 | DB/YL | 02 1/2 Signal | |
| 32 | BR/DG | 02 Return (Upstream) | |
| 33 | BR | O2 2/2 Signal | |
| 34 | WT/LG | CAN C Bus (+) | |
| 35 | WT/LB | CAN C Bus (-) | |
| 36 | WT/VT | SCI Transmit | |
| 37 | BR/WT (Exc. 3.7L) | SCI Transmit (TCM) | |
| 38 | --- | Not Used | --- |

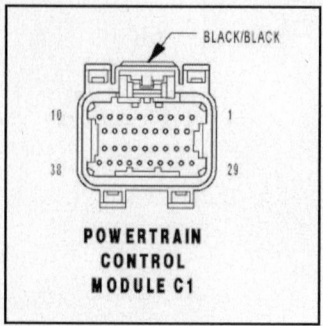

POWERTRAIN
CONTROL
MODULE C1

## 2005 Grand Cherokee 4.7L VIN N 'C2' Black/Orange Connector

| PCM Pin # | Wire Color | Circuit Description (32-Pin) | Value at Hot Idle |
|---|---|---|---|
| 1 | DB/OR | Coil Control No. 6 | |
| 2 | DB/YL | Coil Control No. 5 | |
| 3 | DB | Coil Control No. 4 | |
| 4 | BR/VT | Injector Control No. 6 | |
| 5 | BR/OR | Injector Control No. 5 | |
| 6 | --- | Not Used | --- |
| 7 | DB/BR | Coil Control No. 3 | |
| 8 | B/VT (4.7L) | EGR Sol Control | |
| 9 | DB/TN | Coil Control No. 2 | |
| 10 | YL/DB | Coil Control No. 1 | |
| 11 | BR/TN | Injector Control No. 4 | |
| 12 | BR/LB | Injector Control No. 3 | |
| 13 | BR/DB | Injector Control No. 2 | |
| 14 | BR/YL | Injector Control No. 1 | |
| 15 | --- | Not Used | --- |
| 16 | --- | Not Used | --- |
| 17 | BR/PK | O2 2/1 Heater Control | |
| 18 | OR/TN | O2 1/1 Heater Control | |
| 19 | BR/DG | Gen Field Control | |
| 20 | VT/OR | ECT Signal | |
| 21 | BR/OR | TP No. 1 Signal | |
| 22 | DB/LG (4.7L) | EGR Signal | |
| 23 | VT/BR | MAP Signal | |
| 24 | BR/LG | Knock Sensor No. 1 Return | |
| 25 | DB/OR | Knock Sensor No. 1 Signal | |
| 26 | --- | Not Used | --- |
| 27 | DB/DG | Sensor Ground | |
| 28 | BR/GY | IAC Signal | |
| 29 | PK/YL | 5 Volt Supply | |
| 30 | BR/WT | IAT Signal | |
| 31 | DB/LB | O2 1/1 Signal | |
| 32 | DB/WT | O2 Return (Downstream) | |
| 33 | DB/PK | O2 2/1 Signal | |
| 34 | GY/DB | CMP Signal | |
| 35 | BR | CKP Signal | |
| 36 | OR/WT | Knock Sensor No. 2 Signal | |
| 37 | WT/BR | Knock Sensor No. 2 Return | |
| 38 | VT/GY | IAC Control | |

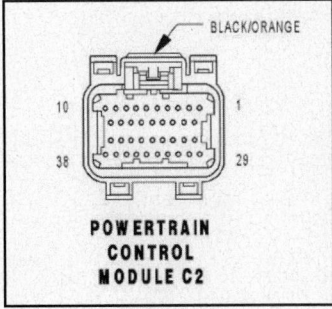

BLACK/ORANGE

10      1

38      29

**POWERTRAIN
CONTROL
MODULE C2**

## 2005 Grand Cherokee 5.7L 'C2' Black/Orange Connector

| PCM Pin # | Wire Color | Circuit Description (32-Pin) | Value at Hot Idle |
|---|---|---|---|
| 1 | DB/OR | Coil Control No. 6 | |
| 2 | DB/YL | Coil Control No. 5 | |
| 3 | DB | Coil Control No. 4 | |
| 4 | BR/VT | Injector Control No. 6 | |
| 5 | BR/OR | Injector Control No. 5 | |
| 6 | DB/GY | ETC Motor (+) | |
| 7 | DB/BR | Coil Control No. 3 | |
| 8 | DB/VT | EGR Sol Control | |
| 9 | DB/TN | Coil Control No. 2 | |
| 10 | YL/DB | Coil Control No. 1 | |
| 11 | BR/TN | Injector Control No. 4 | |
| 12 | BR/LB | Injector Control No. 3 | |
| 13 | BR/DB | Injector Control No. 2 | |
| 14 | BR/YL | Injector Control No. 1 | |
| 15 | BR/DB | TP Sensor Return | |
| 16 | BR/DG | Mds Sol Control No. 6 | |
| 17 | BR/GY | O2 2/1 Heater Control | |
| 18 | OR/TN | O2 1/1 Heater Control | |
| 19 | BR/DG | Gen Field Control | |
| 20 | VT/OR | ECT Signal | |
| 21 | BR/OR | TP No. 1 Signal | |
| 22 | DB/LG | EGR Signal | |
| 23 | VT/BR | MAP Signal | |
| 24 | BR/LG | Knock Sensor No. 1 Return | |
| 25 | DB/OR | Knock Sensor No. 1 Signal | |
| 26 | --- | Not Used | --- |
| 27 | DB/DG | Sensor Ground | |
| 28 | BR/DG | TP No. 2 Signal | |
| 29 | PK/YL | 5 Volt Supply | |
| 30 | BR/WT | IAT Signal | |
| 31 | DB/LB | O2 1/1 Signal | |
| 32 | DB/WT | O2 Return (Downstream) | |
| 33 | DB/PK | O2 2/1 Signal | |
| 34 | GY/DB | CMP Signal | |
| 35 | BR | CKP Signal | |
| 36 | OR/WT | Knock Sensor No. 2 Signal | |
| 37 | WT/BR | Knock Sensor No. 2 Return | |
| 38 | DB/LG | ETC Motor (-) | |

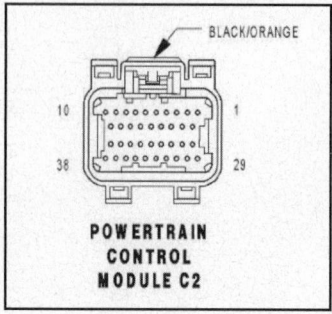

BLACK/ORANGE

POWERTRAIN
CONTROL
MODULE C2

**2005 Grand Cherokee 5.7L 'C3' Black/Natural Connector**

| PCM Pin # | Wire Color | Circuit Description (32-Pin) | Value at Hot Idle |
|---|---|---|---|
| 1 | --- | Not Used | --- |
| 2 | --- | Not Used | --- |
| 3 | BR/WT | ASD Relay Control | |
| 4 | --- | Not Used | --- |
| 5 | VT/OR (3.7L/4.7L) | S/C Vent Sol Control | |
| 5 | BR/OR (5.7L) | MDS Sol Control No. 7 | |
| 6 | BR/WT (5.7L) | MDS Sol Control No. 1 | |
| 7 | VT/YL (4.7L) | S/C Supply | |
| 8 | VT/LB | NVLD Sol Control | |
| 9 | BR/OR | O2 1/2 Heater Control | |
| 10 | BR/GY | O2 2/2 Heater Control | |
| 11 | LB/OR | A/C Clutch Relay Control | |
| 12 | VT/YL (3.7L/4.7L) | S/C Vacuum Sol Control | |
| 13 | --- | Not Used | --- |
| 14 | LB/DG (5.7L) | Brake Switch No. 2 Signal | |
| 15 | VT/BR | S/C Switch Return | |
| 16 | BR/YL (5.7L) | APPS No. 1 Return | |
| 17 | BR/VT (5.7L) | APPS No. 2 Return | |
| 18 | OR/VT (5.7L) | S/C Switch Sense 2 | |
| 19 | DG/LG | Fused ASD Relay Output | |
| 20 | DB/YL | EVAP Purge Sol Signal | |
| 21 | --- | Not Used | --- |
| 22 | --- | Not Used | --- |
| 23 | DG/WT | Brake Switch No. 1 Signal | |
| 24 | --- | Not Used | --- |
| 25 | BR/WT (5.7L) | APPS No. 1 Signal | |
| 26 | --- | Not Used | --- |
| 27 | DG/YL (4.7L/5.7L) | Autostick Up/Down (Ers) Sense | |
| 28 | DG/LG | Fused ASD Relay Output | |
| 29 | DB/BR | EVAP Purge Sol Control | |
| 30 | --- | Not Used | --- |
| 31 | --- | Not Used | --- |
| 32 | --- | Not Used | --- |
| 33 | VT/YL (5.7L) | Engine Oil Temperature Signal | |
| 34 | VT/OR | S/C Switch Sense 1 | |
| 35 | VT/WT | NVLD Switch Signal | |
| 36 | WT/BR (5.7L) | APPS No. 2 Signal | |
| 37 | BR | Fuel Pump Relay Control | |
| 38 | DG/OR | Engine Starter Motor Relay Control | |

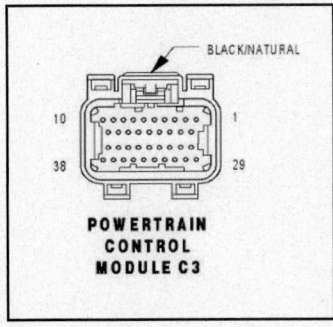

POWERTRAIN
CONTROL
MODULE C3

**2005 Grand Cherokee 5.7L 'C4' Black/Green Connector**

| PCM Pin # | Wire Color | Circuit Description (32-Pin) | Value at Hot Idle |
|---|---|---|---|
| 1 | YL/GY | OD Solenoid Control | |
| 2 | YL/DG | 4C Solenoid Control | |
| 3 | --- | Not Used | --- |
| 4 | DG | MS Solenoid Control | |
| 5 | --- | Not Used | --- |
| 6 | YL/LG | 2C Solenoid Control | |
| 7 | --- | Not Used | --- |
| 8 | YL/LB | UD Solenoid Control | |
| 9 | --- | Not Used | --- |
| 10 | DG/WT | L/R Solenoid Control | |
| 11 | YL/GY | Pressure Control Solenoid Control | |
| 12 | BK | Ground | |
| 13 | BK | Ground | |
| 14 | BK | Ground | |
| 15 | DG/LB | TRS T1 Sense | |
| 16 | DG/DB | TRS T3 Sense | |
| 17 | DG | Tow/Haul Overdrive Off Switch Sense | |
| 18 | YL/BR | Transmission Control Relay Control | |
| 19 | YL/OR | Transmission Control Relay Output | |
| 20 | BR/YL | 4C Pressure Switch Sense | |
| 21 | YL/WT | UD Pressure Switch Sense | |
| 22 | DG/TN | OD Pressure Switch Sense | |
| 23 | --- | Not Used | --- |
| 24 | --- | Not Used | --- |
| 25 | --- | Not Used | --- |
| 26 | DG/LB | 2-4 Pressure Switch Sense | |
| 27 | YL/DB | TRS T41 Sense (P/N) | |
| 28 | YL/OR | Transmission Control Relay Output | |
| 29 | YL/TN | L/R Pressure Switch Sense | |
| 30 | DG/YL | 2C Pressure Switch Sense | |
| 31 | YL/BR | Line Pressure Sensor Signal | |
| 32 | DG/BR | Output Speed Sensor Signal | |
| 33 | DG/WT | Input Speed Sensor Signal | |
| 34 | DG/VT | Sensor Ground | |
| 35 | DG/OR | Transmission Temperature Sensor Signal | |
| 36 | --- | Not Used | --- |
| 37 | DG/YL | TRS T42 Sense | |
| 38 | YL/OR | Transmission Control Relay Output | |

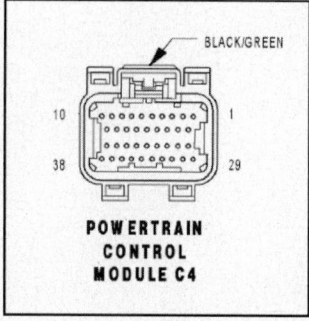

**POWERTRAIN CONTROL MODULE C4**

**1994-95 Grand Cherokee 5.2L VIN Y 60-Pin Connector**

| PCM Pin # | Wire Color | Circuit Description (60-Pin) | Value at Hot Idle |
|-----------|-----------|------------------------------|-------------------|
| 1 | RD/WT | MAP Sensor Signal | 1.5-1.6v |
| 2 | TN/BK | ECT Sensor Signal | At 180ºF: 2.80v |
| 3 | RD | Fused Battery Power (B+) | 12-14v |
| 4 | BK/LB | Sensor Ground | <0.1v |
| 5 | BK/TN | Power Ground | <0.1v |
| 6 | PK/WT | 5-Volt Supply | 4.9-5.1v |
| 7 | WT/BK | 8-Volt Supply | 7.9-8.1v |
| 8 | --- | Not Used | --- |
| 9 | LB | Ignition Switch Power (B+) | 12-14v |
| 10 | OR/BK | Overdrive Override Switch | Switch On: 1v, Off: 12v |
| 11 | BK/TN | Power Ground | <0.1v |
| 12 | BK/TN | Power Ground | <0.1v |
| 13 | LB/BR | Injector 4 Driver | 1-4 ms |
| 14 | YL/WT | Injector 3 Driver | 1-4 ms |
| 15 | TN | Injector 2 Driver | 1-4 ms |
| 16 | WT/DB | Injector 1 Driver | 1-4 ms |
| 17 | DB/WT | Injector 7 Driver | 1-4 ms |
| 18 | DB/YL | Injector 8 Driver | 1-4 ms |
| 19 | GY/WT | Coil Driver (Dwell) | 7º, 55 mph: 9º dwell |
| 20 | DG | Alternator Field Control | Digital Signals: 0-12-0v |
| 21 | BK/RD | IAT Sensor Signal | At 100ºF: 1.83v |
| 22 | OR/DB | TP Sensor Signal | 0.6-1.0v |
| 23 | --- | Not Used | --- |
| 24 | RD/LG | CKP Sensor Signal | Digital Signals: 0-5-0v |
| 25 | BK | SCI Transmit | 0v |
| 26 | PK/BR | CCD Bus (+) | Digital Signals: 0-5-0v |
| 27 | DB/OR | A/C Damped Pressure Switch | A/C Off: 12v, On: 1v |
| 28 | LG | A/C Select Signal | A/C Off: 12v, On: 1v |
| 29 | BR | Brake Switch Sense Signal | Brake Off: 0v, On: 12v |
| 30 | BK/WT | Starter Relay Control | Relay Off: 12v, On: 1v |
| 31 | --- | Not Used | --- |
| 32 | BK/PK | MIL (lamp) Control | MIL Off: 12v, On: 1v |
| 33 | TN/RD | Speed Control Vacuum Solenoid | Vacuum Increasing: 1v |
| 34 | DB/RD | A/C WOT Relay Control | Relay Off: 12v, On: 1v |
| 35 | GY/YL | EGR Solenoid Control | 12v, at 55 mph: 1v |
| 36 | --- | Not Used | --- |
| 37 | PK/OR | Overdrive (lamp) Control | O/D On: 1v, O/D Off: 12v |
| 38 | GY | Injector 5 Driver | 1-4 ms |
| 39 | YL/BK | AIS Motor 4 Circuit Control | Pulse Signals |
| 40 | BR/WT | AIS Motor 2 Circuit Control | Pulse Signals |

## Standard Colors and Abbreviations

| Abbreviation | Color | Abbreviation | Color | Abbreviation | Color |
|--------------|-------|--------------|-------|--------------|-------|
| BK | Black | GY | Gray | RD | Red |
| BL | Blue | GN | Green | TN | Tan |
| BR | Brown | LG | Light Green | VT | Violet |
| DB | Dark Blue | OR | Orange | WT | White |
| DG | Dark Green | PK | Pink | YL | Yellow |

## 1994-95 Grand Cherokee 5.2L VIN Y 60-Pin Connector - Connector

| PCM Pin # | Wire Color | Circuit Description (60-Pin) | Value at Hot Idle |
|-----------|-----------|------------------------------|-------------------|
| 41 | BK/OR | HO2S-11 (B1 S1) Signal | 0.1-1.1v |
| 42 | --- | Not Used | --- |
| 43 | GY/LB | Tachometer Signal | Pulse Signals |
| 44 | GY/BK | CMP Sensor Signal | Digital Signals: 0-5-0v |
| 45 | BK/YL | SCI Receive | 5v |
| 47 | WT/OR | Vehicle Speed Signal | Digital Signal |
| 48 | BR/RD | Speed Control Set Switch | S/C & Set Switch On: 3.8v |
| 49 | YL/RD | Speed Control On/Off Switch | S/C Off: 12v, On: 1v |
| 50 | WT/LG | Speed Control Resume Switch | S/C Off: 12v, On: 1v |
| 51 | PK | Fuel Pump Relay Control | Relay Off: 12v, On: 1v |
| 52 | PK/BK | EVAP Purge Solenoid Control | Solenoid Off: 12v, On: 1v |
| 53 | LG/RD | Speed Control Vent Solenoid | Vacuum Decreasing: 1v |
| 54 | PK/YL | TCC Solenoid Control | TCC on at Cruise: 1v |
| 55 | BR/LG | Overdrive Lockout Solenoid | Solenoid Off: 12v, On: 1v |
| 56 | --- | Not Used | --- |
| 57 | DG/BK | ASD Relay Output | 12-14v |
| 58 | BR/YL | Injector 6 Driver | 1-4 ms |
| 59 | PK/BK | AIS Motor 1 Circuit Control | Pulse Signals |
| 60 | YL/BK | AIS Motor 3 Circuit Control | Pulse Signals |

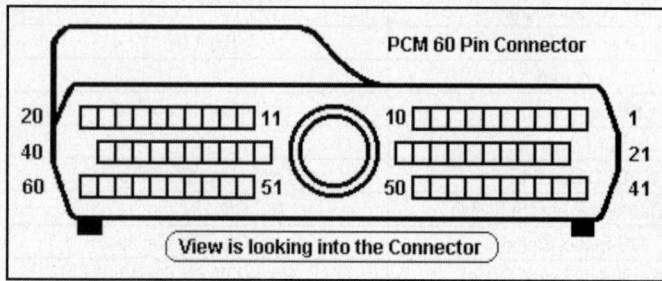

**Pin Connector Graphic**

## Standard Colors and Abbreviations

| Abbreviation | Color | Abbreviation | Color | Abbreviation | Color |
|--------------|-------|--------------|-------|--------------|-------|
| BK | Black | GY | Gray | RD | Red |
| BL | Blue | GN | Green | TN | Tan |
| BR | Brown | LG | Light Green | VT | Violet |
| DB | Dark Blue | OR | Orange | WT | White |
| DG | Dark Green | PK | Pink | YL | Yellow |

**1996-98 Grand Cherokee 5.2L VIN Y 'A' Black Connector**

| PCM Pin # | Wire Color | Circuit Description (32-Pin) | Value at Hot Idle |
|---|---|---|---|
| 1, 3, 5, 9 | --- | Not Used | --- |
| 2 | OR | Ignition Switch Power (B+) | 12-14v |
| 4 | BK/LB | Sensor Ground | <0.1v |
| 6 | BK/WT | PNP Switch Sense Signal | In P/N: 0v, Others: 5v |
| 7 | GY/WT | Coil 1 Driver Control | 7°, 55 mph: 9° dwell |
| 8 | RD/LG | CKP Sensor Signal | Digital Signals: 0-5-0v |
| 10 | PK/BK | IAC 4 Driver Control | Pulse Signals |
| 11 | BR/WT | IAC 3 Driver Control | Pulse Signals |
| 12-13 | --- | Not Used | --- |
| 14 | YL/RD | Fused Battery Power (B+) | 12-14v |
| 15 | BK/RD | IAT Sensor Signal | At 100°F: 1.83v |
| 16 | TN/BK | ECT Sensor Signal | At 180°F: 2.80v |
| 17 | WT/BK | 5-Volt Supply | 4.9-5.1v |
| 18 | GY/BK | CMP Sensor Signal | Digital Signals: 0-5-0v |
| 19 | YL/RD | IAC 2 Driver Control | Pulse Signals |
| 20 | GY/RD | IAC 1 Driver Control | Pulse Signals |
| 21 | --- | Not Used | --- |
| 22 | RD/YL | Fused Battery Power (B+) | 12-14v |
| 23 | OR/DB | TP Sensor Signal | 0.6-1.0v |
| 24 | BK/OR | HO2S-11 (B1 S1) Signal | 0.1-1.1v |
| 25 | BK/PK | HO2S-12 (B1 S2) Signal | 0.1-1.1v |
| 26 | --- | Not Used | --- |
| 27 | RD/WT | MAP Sensor Signal | 1.5-1.6v |
| 28-30 | --- | Not Used | --- |
| 31 | BK/TN | Power Ground | <0.1v |
| 32 | BK/TN | Power Ground | <0.1v |

**1996-98 Grand Cherokee 5.2L VIN Y 'B' White Connector**

| PCM Pin # | Wire Color | Circuit Description (32-Pin) | Value at Hot Idle |
|---|---|---|---|
| 1 | PK | Transmission Temperature Sensor | At 200°F: 2.40v |
| 2 | DB/WT | Injector 7 Driver | 1-4 ms |
| 3 | --- | Not Used | --- |
| 4 | WT/DB | Injector 1 Driver | 1-4 ms |
| 5 | YL/WT | Injector 3 Driver | 1-4 ms |
| 6 | GY | Injector 5 Driver | 1-4 ms |
| 7 | --- | Not Used | --- |
| 8 | PK | Governor Pressure Solenoid | PWM Signal: 0-12-0v |
| 10 | DG | Generator Field Control | Digital Signals: 0-12-0v |
| 11 | DG/LB | TCC Solenoid Control | TCC on at Cruise: 1v |
| 12 | BR/YL | Injector 6 Driver | 1-4 ms |
| 13 | DB/YL | Injector 8 Driver | 1-4 ms |
| 14 | --- | Not Used | --- |
| 15 | TN | Injector 2 Driver | 1-4 ms |
| 16 | LB/BR | Injector 4 Driver | 1-4 ms |
| 17-20 | --- | Not Used | --- |
| 21 | BR | Overdrive Solenoid Control | Solenoid Off: 12v, On: 1v |
| 22, 24 | --- | Not Used | --- |
| 23 | GY/WT | Engine Oil Pressure Sensor | 1.6v at 24 psi |
| 25 | DB/BK | OSS Sensor Signal (-) | AC pulse signals |
| 26 | --- | Not Used | --- |
| 27 | WT/OR | Vehicle Speed Signal | Digital Signal |
| 28 | LG/WT | OSS Sensor Signal (+) | AC pulse signals |
| 29 | LG | Governor Pressure Signal | 0.58v |
| 30 | BR/OR | Transmission Relay Control | Relay Off: 12v, On: 1v |
| 31 | PK/WT | 5-Volt Supply | 4.9-5.1v |
| 32 | --- | Not Used | --- |

### 1996-98 Grand Cherokee 5.2L VIN Y 'C' Gray Connector

| PCM Pin # | Wire Color | Circuit Description (32-Pin) | Value at Hot Idle |
|---|---|---|---|
| 1 | DB/RD | A/C Clutch Relay Control | Relay Off: 12v, On: 1v |
| 2 | --- | Not Used | --- |
| 3 | PK/WT | ASD Relay Control | Relay Off: 12v, On: 1v |
| 4 | TN/RD | Speed Control Vacuum Solenoid | Vacuum Increasing: 1v |
| 5 | LG/RD | Speed Control Vent Solenoid | Vacuum Decreasing: 1v |
| 6 | BR/YL | Overdrive 'Off' (lamp) Control | O/D On: 1v, O/D Off: 12v |
| 7-9 | --- | Not Used | --- |
| 10 | DG/RD | LDP Solenoid Control | PWM Signal: 0-12-0v |
| 11 | YL/RD | Speed Control Power Supply | 12-14v |
| 12 | DG/OR | ASD Relay Output | 12-14v |
| 13 | OR | Overdrive 'Off' Switch Sense | Switch On: 1v, Off: 12v |
| 14 | PK/RD | LDP Switch Sense Signal | Closed: 0v, Open: 12v |
| 15 | RD/YL | Battery Temperature Sensor | At 86°F: 1.96v |
| 16-18 | --- | Not Used | --- |
| 19 | DB | Fuel Pump Relay Control | Relay Off: 12v, On: 1v |
| 20 | PK/BK | EVAP Purge Solenoid Control | PWM Signal: 0-12-0v |
| 21 | --- | Not Used | --- |
| 22 | DB/BK | A/C Damped Pressure Switch | A/C Off: 12v, On: 1v |
| 23 | --- | Not Used | --- |
| 24 | BR | Brake Switch Sense Signal | Brake Off: 0v, On: 12v |
| 25 ('98) | DG/PK | Generator Field Source | 12-14v |
| 26 | LB/BK | Fuel Level Sensor Signal | Full: 0.5v, 1/2 full: 2.5v |
| 27 | BK/PK | SCI Transmit | 0v |
| 28 | WT/BK | CCD Bus (+) | Digital Signals: 0-5-0v |
| 29 | BK/WT | SCI Receive | 5v |
| 30 | PK/BR | CCD Bus (-) | <0.050v |
| 31 | --- | Not Used | --- |
| 32 | RD/LG | Speed Control Switch Signal | S/C & Set Switch On: 3.8v |
| 32 ('98) | PK | Speed Control Switch Signal | S/C & Set Switch On: 3.8v |

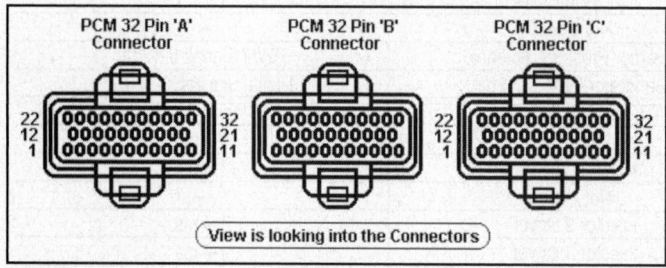

**Pin Connector Graphic**

### Standard Colors and Abbreviations

| Abbreviation | Color | Abbreviation | Color | Abbreviation | Color |
|---|---|---|---|---|---|
| BK | Black | GY | Gray | RD | Red |
| BL | Blue | GN | Green | TN | Tan |
| BR | Brown | LG | Light Green | VT | Violet |
| DB | Dark Blue | OR | Orange | WT | White |
| DG | Dark Green | PK | Pink | YL | Yellow |

## 1998 Grand Cherokee 5.9L VIN Z 'A' Black Connector

| PCM Pin # | Wire Color | Circuit Description (32-Pin) | Value at Hot Idle |
|---|---|---|---|
| 1 | --- | Not Used | --- |
| 2 | OR | Ignition Switch Power (B+) | 12-14v |
| 3 | --- | Not Used | --- |
| 4 | BK/LB | Sensor Ground | <0.1v |
| 5 | --- | Not Used | --- |
| 6 | BK/WT | PNP Switch Sense Signal | In P/N: 0v, Others: 5v |
| 7 | GY/WT | Coil 1 Driver Control | 7°, 55 mph: 9° dwell |
| 8 | RD/LG | CKP Sensor Signal | Digital Signals: 0-5-0v |
| 9 | --- | Not Used | --- |
| 10 | PK/BK | IAC 4 Driver Control | Pulse Signals |
| 11 | BR/WT | IAC 3 Driver Control | Pulse Signals |
| 12-14 | --- | Not Used | --- |
| 15 | BK/RD | IAT Sensor Signal | At 100°F: 1.83v |
| 16 | TN/BK | ECT Sensor Signal | At 180°F: 2.80v |
| 17 | WT/BK | 5-Volt Supply | 4.9-5.1v |
| 18 | GY/BK | CMP Sensor Signal | Digital Signals: 0-5-0v |
| 19 | YL/BK | IAC 2 Driver Control | Pulse Signals |
| 20 | GY/RD | IAC 1 Driver Control | Pulse Signals |
| 21 | --- | Not Used | --- |
| 22 | RD/YL | Fused Battery Power (B+) | 12-14v |
| 23 | OR/DB | TP Sensor Signal | 0.6-1.0v |
| 24 | BK/OR | HO2S-11 (B1 S1) Signal | 0.1-1.1v |
| 25 | BK/PK | HO2S-12 (B1 S2) Signal | 0.1-1.1v |
| 26 | --- | Not Used | --- |
| 27 | RD/WT | MAP Sensor Signal | 1.5-1.6v |
| 28-30 | --- | Not Used | --- |
| 31 | BK/TN | Power Ground | <0.1v |
| 32 | BK/TN | Power Ground | <0.1v |

## 1998 Grand Cherokee 5.9L VIN Z 'B' White Connector

| PCM Pin # | Wire Color | Circuit Description (32-Pin) | Value at Hot Idle |
|---|---|---|---|
| 1 | PK | Transmission Temperature Sensor | At 200°F: 2.40v |
| 2 | DB/WT | Injector 7 Driver | 1-4 ms |
| 3, 7, 9 | --- | Not Used | --- |
| 4 | WT/DB | Injector 1 Driver | 1-4 ms |
| 5 | YL/WT | Injector 3 Driver | 1-4 ms |
| 6 | GY | Injector 5 Driver | 1-4 ms |
| 8 | PK | Governor Pressure Solenoid | PWM Signal: 0-12-0v |
| 10 | DG | Generator Field Control | Digital Signals: 0-12-0v |
| 11 | DG/LB | TCC Solenoid Control | TCC on at Cruise: 1v |
| 12 | BR/YL | Injector 6 Driver | 1-4 ms |
| 13 | DB/YL | Injector 8 Driver | 1-4 ms |
| 14 | --- | Not Used | --- |
| 15 | TN | Injector 2 Driver | 1-4 ms |
| 16 | LB/BR | Injector 4 Driver | 1-4 ms |
| 17-20 | --- | Not Used | --- |
| 21 | BR | Overdrive Solenoid Control | Solenoid on at Cruise: 1v |
| 22 | --- | Not Used | --- |
| 23 | GY/WT | Engine Oil Pressure Sensor | 1.6v at 24 psi |
| 24, 26 | --- | Not Used | --- |
| 25 | DB/BK | OSS Sensor Signal (-) | AC pulse signals |
| 27 | WT/OR | Vehicle Speed Signal | Digital Signal |
| 28 | LG/WT | OSS Sensor Signal (+) | AC pulse signals |
| 29 | LG | Governor Pressure Signal | 0.58v |
| 30 | BR/OR | Transmission Relay Control | Relay Off: 12v, On: 1v |
| 31 | PK/WT | 5-Volt Supply | 4.9-5.1v |
| 32 | --- | Not Used | --- |

## 1998 Grand Cherokee 5.9L VIN Z 'C' Gray Connector

| PCM Pin # | Wire Color | Circuit Description (32-Pin) | Value at Hot Idle |
|---|---|---|---|
| 1 | DB/RD | A/C Clutch Relay Control | Relay Off: 12v, On: 1v |
| 2 | --- | Not Used | --- |
| 3 | PK/WT | ASD Relay Control | Relay Off: 12v, On: 1v |
| 4 | TN/RD | Speed Control Vacuum Solenoid | Vacuum Increasing: 1v |
| 5 | LG/RD | Speed Control Vent Solenoid | Vacuum Decreasing: 1v |
| 6 | BR/YL | Overdrive 'Off' (lamp) Control | O/D On: 1v, O/D Off: 12v |
| 7-9 | --- | Not Used | --- |
| 10 | DG/RD | LDP Solenoid Control | PWM Signal: 0-12-0v |
| 11 | YL/RD | Speed Control Power Supply | 12-14v |
| 12 | DG/OR | ASD Relay Output | 12-14v |
| 13 | OR | Overdrive 'Off' Switch Sense | Switch On: 1v, Off: 12v |
| 14 | PK/RD | LDP Switch Sense Signal | Closed: 0v, Open: 12v |
| 15 | RD/YL | Battery Temperature Sensor | At 86°F: 1.96v |
| 16-18 | --- | Not Used | --- |
| 19 | DB | Fuel Pump Relay Control | Relay Off: 12v, On: 1v |
| 20 | PK/BK | EVAP Purge Solenoid Control | PWM Signal: 0-12-0v |
| 21 | --- | Not Used | --- |
| 22 | DB/BK | A/C Switch Sense | A/C Off: 12v, On: 1v |
| 23 | --- | Not Used | --- |
| 24 | BR | Brake Switch Sense Signal | Brake Off: 0v, On: 12v |
| 25 | DG/PK | Generator Field Source | 12-14v |
| 26 | LB/BK | Fuel Level Sensor Signal | Full: 0.5v, 1/2 full: 2.5v |
| 27 | BK/PK | SCI Transmit | 0v |
| 28 | WT/BK | CCD Bus (-) | <0.050v |
| 29 | BK/WT | SCI Receive | 5v |
| 30 | PK/BR | CCD Bus (+) | Digital Signals: 0-5-0v |
| 31 | --- | Not Used | --- |
| 32 | PK | Speed Control Switch Signal | S/C & Set Switch On: 3.8v |

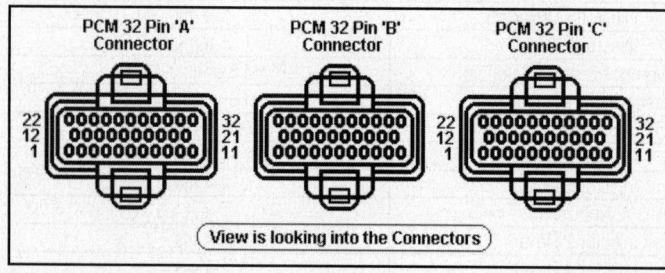

**Pin Connector Graphic**

## Standard Colors and Abbreviations

| Abbreviation | Color | Abbreviation | Color | Abbreviation | Color |
|---|---|---|---|---|---|
| BK | Black | GY | Gray | RD | Red |
| BL | Blue | GN | Green | TN | Tan |
| BR | Brown | LG | Light Green | VT | Violet |
| DB | Dark Blue | OR | Orange | WT | White |
| DG | Dark Green | PK | Pink | YL | Yellow |

**2002-03 Liberty 2.4L VIN 1 (M/T) 'C1' Black Connector**

| PCM Pin # | Wire Color | Circuit Description (32-Pin) | Value at Hot Idle |
|---|---|---|---|
| 1, 3, 5-6, 9 | --- | Not Used | --- |
| 2 | DB/WT | Ignition Switch Power (Start-Run) | 12-14v |
| 4 | BK/LB | Sensor Ground | <0.050v |
| 7 | BK/GY | Coil 1 Driver Control | 5º, 55 mph: 8º dwell |
| 8 | GY/BK | CKP Sensor Signal | Digital Signals: 0-5-0v |
| 10 | YL/BK | IAC 2 Driver Control | Pulse Signals |
| 11 | BR/WT | IAC 1 Driver Control | Pulse Signals |
| 12 | DB/OR | PSP Switch Sense Signal | Straight: 0v, Turning: 5v |
| 13 | YL/RD | Clutch Interlock Relay Control | Clutch Out: 12v, In: 0v |
| 14 | BR/WT | Transfer Case Position Switch | Switch Open: 12v, Closed: 0v |
| 15 | BK/RD | IAT Sensor Signal | At 100ºF: 1.83v |
| 16 | TN/BK | ECT Sensor Signal | At 180ºF: 2.80v |
| 17 | OR | 5-Volt Supply | 4.9-5.1v |
| 18 | TN/YL | CMP Sensor Signal | Digital Signals: 0-5-0v |
| 19 | GY/RD | IAC 3 Driver Control | DC Pulse Signals |
| 20 | VT/BK | IAC 4 Driver Control | DC Pulse Signals |
| 21 | --- | Not Used | --- |
| 22 | RD/WT | Battery Power (B+) | 12-14v |
| 23 | OR/DB | TP Sensor Signal | 0.6-1.0v |
| 24 | BK/DG | HO2S-11 (B1 S1) Signal | 0.1-1.1v |
| 25 | TN/WT | HO2S-12 (B1 S2) Signal | 0.1-1.1v |
| 26 | --- | Not Used | --- |
| 27 | DG/RD | MAP Sensor Signal | 1.5-1.6v |
| 28-30 | --- | Not Used | --- |
| 31 | BK/DB | Power Ground | <0.1v |
| 32 | BK/DB | Power Ground | <0.1v |

**2002-03 Liberty 2.4L VIN 1 (M/T) 'C2' White Connector**

| PCM Pin # | Wire Color | Circuit Description (32-Pin) | Value at Hot Idle |
|---|---|---|---|
| 1-3 | --- | Not Used | --- |
| 4 | WT/DB | Injector 1 Driver Control | 1-4 ms |
| 5 | YL/WT | Injector 3 Driver Control | 1-4 ms |
| 6-8 | --- | Not Used | --- |
| 9 | DB/TN | Coil 2 Driver Control | 5º, 55 mph: 8º dwell |
| 10 | DG | Generator Field Control | Digital Signals: 0-12-0v |
| 11-14 | --- | Not Used | --- |
| 15 | TN | Injector 2 Driver Control | 1-4 ms |
| 16 | LB/BR | Injector 4 Driver Control | 1-4 ms |
| 17 | LG | High Speed Radiator Fan Relay | Relay Off: 12v, On: 1v |
| 18 | --- | Not Used | --- |
| 19 | DB | A/C Pressure Sensor | A/C On: 0.451-4.850v |
| 20-22 | --- | Not Used | --- |
| 23 | GY/YL | Oil Pressure Sensor | 1.6v at 24 psi |
| 24-30 | --- | Not Used | --- |
| 31 | VT/WT | 5-Volt Supply | 4.9-5.1v |
| 32 | --- | Not Used | --- |

## 2002-03 Liberty 2.4L VIN 1 (M/T) 'C3' Gray Connector

| PCM Pin # | Wire Color | Circuit Description (32-Pin) | Value at Hot Idle |
|---|---|---|---|
| 1 | DG | A/C Clutch Relay Control | Relay Off: 12v, On: 1v |
| 2 | --- | Not Used | --- |
| 3 | DB/YL | ASD Relay Control | Relay Off: 12v, On: 1v |
| 4 | TN/RD | Speed Control Vacuum Solenoid | Vacuum Increasing: 1v |
| 5 | LG/RD | Speed Control Vent Solenoid | Vacuum Decreasing: 1v |
| 6 | TN | Clutch Switch Override Relay Control | Relay Off: 12v, On: 1v |
| 7 | --- | Not Used | --- |
| 8 | BR/OR | HO2S-11 (B1 S1) Heater Control | Heater Off: 12v, On: 1v |
| 9 | RD/YL | Oxygen Sensor Downstream Relay Control | Relay Off: 12v, On: 1v |
| 10 | WT/DG | LDP Solenoid Control | PWM Signal: 0-12-0v |
| 11 | YL/RD | Speed Control Power Supply | 12-14v |
| 12 | OR/DG | ASD Relay Output | 12-14v |
| 13 | YL/DG | Torque Management Request | Digital Signals |
| 14 | OR | LDP Switch Sense Signal | Closed: 0v, Open: 12v |
| 15 | PK/YL | Battery Temperature Sensor | At 86°F: 1.96v |
| 16 | BR/WT | HO2S-12 (B1 S2) Heater Control | Heater Off: 12v, On: 1v |
| 17 | DG/YL | Vehicle Speed Signal | Digital Signals |
| 18 | --- | Not Used | --- |
| 19 | BR | Fuel Pump Relay Control | Relay Off: 12v, On: 1v |
| 20 | PK/BK | EVAP Purge Solenoid Control | PWM Signal: 0-12-0v |
| 21 | --- | Not Used | --- |
| 22 | DB/OR | A/C Switch Sense | A/C Off: 12v, On: 1v |
| 23 | --- | Not Used | --- |
| 24 | WT/PK | Brake Switch Sense | Brake Off: 0v, On: 12v |
| 25 | WT/DB | Generator Field Source | 12-14v |
| 26 | DB/WT | Fuel Level Sensor Signal | Full: 0.5v, 1/2 full: 2.5v |
| 27 | PK | SCI Transmit | 0v |
| 28 | --- | Not Used | --- |
| 29 | LG | SCI Receive | 5v |
| 30 | VT/YL | PCI Bus Signal (J1850) | Digital Signals: 0-7-0v |
| 31 | --- | Not Used | --- |
| 32 | RD/LG | Speed Control Switch Signal | S/C & Set Switch On: 3.8v |

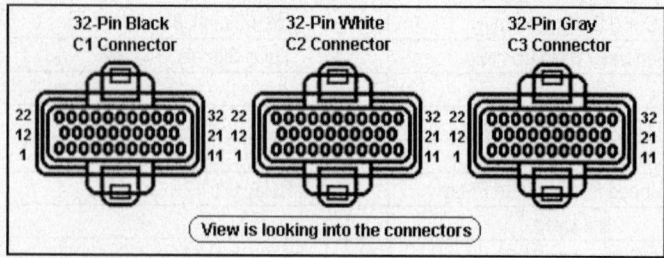

**Pin Connector Graphic**

**2004-05 Liberty 2.4L VIN 1 'C1' Black/Black Connector**

| PCM Pin # | Wire Color | Circuit Description (38-Pin) | Value at Hot Idle |
|:---:|:---:|:---:|:---:|
| 1 | --- | Not Used | --- |
| 2 | --- | Not Used | --- |
| 3 | --- | Not Used | --- |
| 4 | --- | Not Used | --- |
| 5 | --- | Not Used | --- |
| 6 | --- | Not Used | --- |
| 7 | --- | Not Used | --- |
| 8 | --- | Not Used | --- |
| 9 | BK/BR | Ground | |
| 10 | DB/YL | A/C Switch Sense | |
| 11 | PK/WT | Fused Ignition Switch Output (Run-Start) | |
| 12 | PK/WT | Fused Ignition Switch Output (Run-Start) | |
| 13 | DG/YL | Vehicle Speed Signal | |
| 14 | --- | Not Used | --- |
| 15 | --- | Not Used | --- |
| 16 | --- | Not Used | --- |
| 17 | --- | Not Used | --- |
| 18 | BK/DG | Ground | |
| 19 | --- | Not Used | --- |
| 20 | VT/GY | Engine Oil Pressure Signal | |
| 21 | LB/BR | A/C Pressure Signal | |
| 22 | VT/OR | AAT Signal | |
| 23 | --- | Not Used | --- |
| 24 | --- | Not Used | --- |
| 25 | WT/LG | SCI Receive (PCM) | |
| 26 | WT/OR (3.7L A/T) | SCI Receive (TCM) | |
| 27 | YL/PK | 5 Volt Supply | |
| 28 | --- | Not Used | --- |
| 29 | RD | Fused B (+) | |
| 30 | PK/OR | Fused Ignition Switch Output (Start) | |
| 31 | DB/YL | O2 1/2 Signal | |
| 32 | BR/DG | O2 Upstream Return | |
| 33 | BR (3.7L) | O2 2/2 Signal | |
| 34 | --- | Not Used | --- |
| 35 | --- | Not Used | --- |
| 36 | WT/GY | SCI Transmit (PCM) | |
| 37 | BR/WT (3.7L A/T) | SCI Transmit (TCM) | |
| 38 | WT/VT | PCI Bus | |

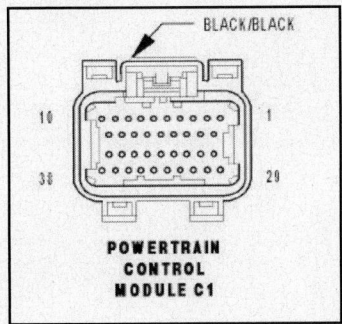

BLACK/BLACK

10    1

38    29

**POWERTRAIN
CONTROL
MODULE C1**

## 2004-05 Liberty 2.4L VIN 1 'C2' Black/Orange Connector

| PCM Pin # | Wire Color | Circuit Description (38-Pin) | Value at Hot Idle |
|---|---|---|---|
| 1 | DB/OR (3.7L) | Coil Control No. 6 | |
| 2 | DB/YL (3.7L) | Coil Control No. 5 | |
| 3 | DB (3.7L) | Coil Control No. 4 | |
| 4 | BR/VT (3.7L) | Injector Control No. 6 | |
| 5 | BR/OR (3.7L) | Injector Control No. 5 | |
| 6 | --- | Not Used | --- |
| 7 | DB (3.7L) | Coil Control No. 3 | |
| 8 | --- | Not Used | --- |
| 9 | DB/YL | Coil Control No. 2 | |
| 10 | YL/DB | Coil Control No. 1 | |
| 11 | BR/TN | Injector Control No. 4 | |
| 12 | BR/LB | Injector Control No. 3 | |
| 13 | BR/DB | Injector Control No. 2 | |
| 14 | BR/YL | Injector Control No. 1 | |
| 15 | --- | Not Used | --- |
| 16 | --- | Not Used | --- |
| 17 | BR/VT (3.7L) | O2 2/1 Heater Control | |
| 18 | BR/TN | O2 1/1 Heater Control | |
| 19 | BR/DG | Gen Field Control | |
| 20 | VT/OR | ECT Signal | |
| 21 | BR/OR | TP Signal | |
| 22 | --- | Not Used | --- |
| 23 | VT/BR | MAP Signal | |
| 24 | BR/LG (3.7L) | Knock Sensor No. 1 Return | |
| 25 | DB/OR (3.7L) | Knock Sensor No. 1 Signal | |
| 26 | BR/WT (4WD) | Transfer Case Position Sensor Input | |
| 27 | DB/DG | Sensor Ground | |
| 28 | BR/VT | IAC Signal | |
| 29 | PK/YL | 5 Volt Supply | |
| 30 | BR/WT | IAT Signal | |
| 31 | DB/LB | O2 1/1 Signal | |
| 32 | DB/DG | O2 Downstream Return | |
| 33 | DB/LG (3.7L) | O2 2/1 Signal | |
| 34 | DB/GY | CMP Signal | |
| 35 | BR/LB | CKP Signal | |
| 36 | BR/WT (3.7L) | Knock Sensor No. 2 Signal | |
| 37 | PK/RD (3.7L) | Knock Sensor No. 2 Return | |
| 38 | VT/GY | IAC Control | |

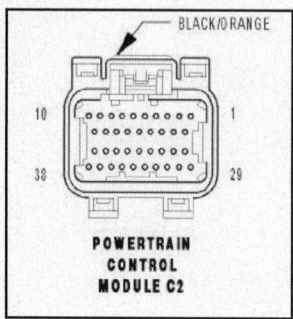

BLACK/ORANGE

POWERTRAIN
CONTROL
MODULE C2

## 2005 Liberty 2.8L Diesel VIN 5 'C1' Connector

| PCM Pin # | Wire Color | Circuit Description (96-Pin) | Value at Hot Idle |
|---|---|---|---|
| 1 | GY/BR | Fuel Injector No. 1 Low-Side Control | |
| 2 | YL/BR | Fuel Injector No. 2 Low-Side Control | |
| 3, 5-7, 9-10, 12-13 | --- | Not Used | --- |
| 4 | BR/OR | Fuel Pressure Solenoid Supply | |
| 8 | BR/YL | Fuel Temperature Sensor Signal | |
| 11 | DG/LG (A/T) | Torque Management Request Sense | |
| 14 | BR/GY | Camshaft Position Sensor Ground | |
| 15-24 | --- | Not Used | --- |
| 25 | OR/BR | Fuel Injector No. 1 Low-Side Control | |
| 26 | WT/BR | Fuel Injector No. 4 Low-Side Control | |
| 27 | --- | Not Used | --- |
| 28 | BR/LG | Fuel Quantity Solenoid Supply | |
| 29 | --- | Not Used | --- |
| 30 | VT | S/C Switch Signal No. 1 | |
| 31 | VT/OR | AAT Signal | |
| 32 | BR/LB | Inlet Pressure Sensor Ground | |
| 33 | --- | Not Used | --- |
| 34 | BR/WT | IAT Sensor Signal | |
| 35 | --- | Not Used | --- |
| 36 | BR | Inlet Pressure Sensor 5 Volt Supply | |
| 37 | YL/PK | Camshaft Position Sensor 5 Volt Supply | |
| 38 | DB/GY | Camshaft Position Sensor Signal | |
| 39-44 | --- | Not Used | --- |
| 45 | DB/LB | Cabin Heater Relay Control | |
| 46-48 | --- | Not Used | --- |
| 49 | BR/DB | Fuel Injector No. 2 High-Side Control | |
| 50 | --- | Not Used | --- |
| 51 | BR/TN | Fuel Injector No. 4 High-Side Control | |
| 52 | BR | Fuel Pressure Solenoid Control | |
| 53 | --- | Not Used | --- |
| 54 | VT/OR | S/C Switch Signal No. 2 | |
| 55-56 | --- | Not Used | --- |
| 57 | VT/OR | ECT Sensor Signal | |
| 58 | BR/LG | Inlet Pressure Sensor Signal | |
| 59 | BR/YL | Fuel Pressure Sensor Signal | |
| 60 | BR/YL | Boost Pressure Sensor 5 Volt Supply | |
| 61 | BR/OR | Mass Air Flow Sensor Ground | |
| 62 | --- | Not Used | --- |
| 63 | BR/OR | Boost Pressure Sensor Signal | |

**2005 Liberty 2.8L Diesel VIN 5 'C1' Connector (cont'd)**

| PCM Pin # | Wire color | Circuit Description (96-Pin) | Value at Hot Idle |
|---|---|---|---|
| 64 | BR/OR | Mass Air Flow Sensor 5 Volt Supply | |
| 65 | DB/DG | Sensor Ground | |
| 66 | --- | Not Used | --- |
| 67 | --- | Not Used | --- |
| 68 | BR/GY | EGR Air Flow Control Valve Control | |
| 69 | --- | Not Used | --- |
| 70 | --- | Not Used | --- |
| 71 | DB/OR | High Speed Rad Fan Relay Control | |
| 72 | --- | Not Used | --- |
| 73 | BR/LB | Fuel Injector No. 3 High-Side Control | |
| 74 | BR/YL | Fuel Injector No. 1 High-Side Control | |
| 75 | --- | Not Used | --- |
| 76 | BR/YL | Fuel Quantity Solenoid Control | |
| 77 | --- | Not Used | --- |
| 78 | --- | Not Used | --- |
| 79 | DG/OR | Clutch Upstop Switch Signal | |
| 80 | --- | Not Used | --- |
| 81 | --- | Not Used | --- |
| 82 | BR/WT | Transfer Case Position Sensor Input | |
| 83 | GY/BR | Sensor Ground | |
| 84 | YL/BR | Fuel Pressure Sensor Ground | |
| 85 | BR/OR | Mass Air Flow 5 Volt Supply | |
| 86 | BR/YL | Fuel Pressure Sensor 5 Volt Supply | |
| 87 | DB/BR | Crankshaft Position Sensor Signal No. 2 | |
| 88 | BR/LB | Crankshaft Position Sensor Signal No. 1 | |
| 89 | --- | Not Used | --- |
| 90 | DB/VT | EGR Solenoid Control | |
| 91 | DB/LG | Low Speed Rad Fan Relay Control | |
| 92 | DB/WT | Vacuum Reservoir Solenoid Control | |
| 93 | BR/WT | Boost Pressure Solenoid Control | |
| 94 | --- | Not Used | --- |
| 95 | --- | Not Used | --- |
| 96 | --- | Not Used | --- |

**Pin Connector Graphic**
**(Graphic Unavailable)**

**Standard Colors and Abbreviations**

| Abbreviation | Color | Abbreviation | Color | Abbreviation | Color |
|---|---|---|---|---|---|
| BK | Black | GY | Gray | RD | Red |
| BL | Blue | GN | Green | TN | Tan |
| BR | Brown | LG | Light Green | VT | Violet |
| DB | Dark Blue | OR | Orange | WT | White |
| DG | Dark Green | PK | Pink | YL | Yellow |

**2005 Liberty 2.8L Diesel VIN 5 'C2' Connector**

| PCM Pin # | Wire Color | Circuit Description (58-Pin) | Value at Hot Idle |
|---|---|---|---|
| 1 | BR/PK | Fused ASD Relay Output | |
| 2 | BK/DG | Ground | |
| 3 | BR/PK | Fused ASD Relay Output | |
| 4 | BK/DG | Ground | |
| 5 | BR/RD | Fused ASD Relay Output | |
| 6 | BK/DG | Ground | |
| 7 | --- | Not Used | --- |
| 8 | DB/YL | A/C Switch Sense | |
| 9-12 | --- | Not Used | --- |
| 13 | WT/BR | APP Sensor Signal 2 | |
| 14 | BR/VT | APP Sensor Ground 2 | |
| 15 | --- | Not Used | --- |
| 16 | --- | Not Used | --- |
| 17 | YL/DB (A/T) | TRS T41 Sense (P/N) | |
| 17 | YL/OR (M/T) | Clutch Interlock Switch Signal | |
| 18 | BR/WT (A/T) | Engine Rpm Signal | |
| 19 | PK/WT | Fused Ignition Switch Output (Run-Start) | |
| 20 | --- | Not Used | --- |
| 21 | --- | Not Used | --- |
| 22 | PK/OR | Fused Ignition Switch Output (Start) | |
| 23 | --- | Not Used | --- |
| 24 | BR/VT | APP Sensor 5 Volt Supply 1 | |
| 25 | BR/WT | APP Sensor Signal 1 | |
| 26 | BR/YL | APP Sensor Ground 1 | |
| 27 | --- | Not Used | --- |
| 28 | --- | Not Used | --- |
| 29 | VT/BR | APP Sensor 5 Volt Supply 2 | |
| 30 | --- | Not Used | --- |
| 31 | WT/GY | SCI Transmit (ECM) | |
| 32 | WT/TN | Primary Brake Switch Signal | |
| 33 | --- | Not Used | --- |
| 34 | DG/WT | Secondary Brake Switch Signal | |
| 35-39 | --- | Not Used | --- |
| 40 | LB/OR | A/C Clutch Relay Control | |
| 41-42 | --- | Not Used | --- |
| 43 | WT/BR | Glow Plug Module Signal | |
| 44 | BR/WT | ASD Relay Control | |
| 45 | BR/YL | Fuel Pump Relay Control | |
| 46 | LB/BR | Glow Plug Module Control | |
| 47-52 | --- | Not Used | --- |
| 53 | WT/LG | CAN C Bus (+) | |
| 54 | WT/LB | CAN C Bus (-) | |
| 55-57 | --- | Not Used | --- |
| 58 | DG/OR | Engine Starter Motor Relay Control | |

**Pin Connector Graphic**
**(Graphic Unavailable)**

**2005 Liberty 2.8L Diesel VIN 5 TCM Connector**

| PCM Pin # | Wire Color | Circuit Description (58-Pin) | Value at Hot Idle |
|---|---|---|---|
| 1 | DG/LB | TRS T1 Sense | |
| 2 | DG/LB | TRS T2 Sense | |
| 3 | DG/DB | TRS T3 Sense | |
| 4-5 | --- | Not Used | --- |
| 6 | BR/WT | Engine Rpm Signal | |
| 7 | WT/GY | SCI Transmit (ECM) | |
| 8 | PK/OR | Fused Ignition Switch Output (Start) | |
| 9 | DG/TN | OD Pressure Switch Sense | |
| 10 | DG/LG | Torque Management Request Sense | |
| 11 | PK/WT | Fused Ignition Switch Output (Run-Start) | |
| 12 | BR/WT | Accelerator Pedal Position Sensor Signal 1 | |
| 13 | DG/VT | Speed Sensor Ground | |
| 14 | DG/BR | Output Speed Sensor Signal | |
| 15 | YL/DB | Transmission Control Relay Control | |
| 16 | YL/OR | Transmission Control Relay Output | |
| 17 | YL/OR | Transmission Control Relay Output | |
| 18 | DG | Pressure Control Solenoid Control | |
| 19 | YL/LG | 2C Solenoid Control | |
| 20 | DG/WT | L/R Solenoid Control | |
| 21-28 | --- | Not Used | --- |
| 29 | YL/WT | UD Pressure Switch Sense | |
| 30 | YL/BR | Line Pressure Sensor Signal | |
| 31-35 | --- | Not Used | --- |
| 36 | YL/OR | Transmission Control Relay Output | |
| 37 | BK/LG | Ground | |
| 38 | BR/YL | 5 Volt Supply | |
| 39 | BK/LG | Ground | |
| 40 | YL/GY | MS Solenoid Control | |
| 41 | YL/DB | TRS T41 Sense (P/N) | |
| 42 | DG/YL | TRS T42 Sense | |
| 43 | WT/VT | PCI Bus | |
| 44-45 | --- | Not Used | --- |
| 46 | WT/OR | SCI Receive (ECM) | |
| 47 | DG/YL | 2C Pressure Switch Sense | |
| 48 | BR/YL | 4C Pressure Switch Sense | |
| 49 | DG | Tow/Haul Overdrive Off Switch Sense | |
| 50 | YL/TN | L/R Pressure Switch Sense | |
| 51 | BR/YL | Accelerator Pedal Position Sensor Ground 1 | |
| 52 | DG/WT | Input Speed Sensor Signal | |
| 53 | BK/LG | Ground | |
| 54 | DG/OR | Transmission Temperature Sensor Signal | |
| 55 | YL/LB | UD Solenoid Control | |
| 56 | RD | Fused B (+) | |
| 57 | BK/LG | Ground | |
| 58 | --- | Not Used | --- |
| 59 | YL/DG | 4C Solenoid Control | |
| 60 | YL/GY | OD Solenoid Control | |

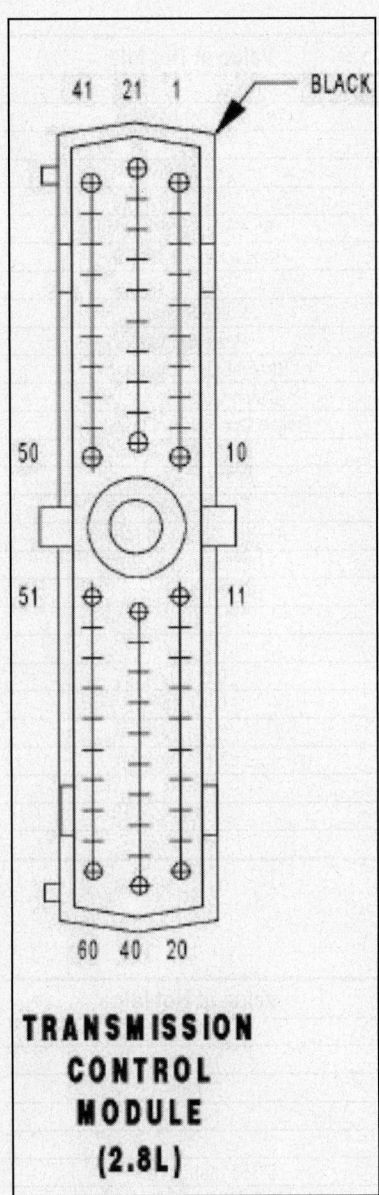

**2005 Liberty 2.8L Diesel VIN 5 TCM Connector**

**2002-03 Liberty 3.7L VIN K 'C1' Black Connector**

| PCM Pin # | Wire Color | Circuit Description (32-Pin) | Value at Hot Idle |
|---|---|---|---|
| 1 | --- | Not Used | --- |
| 2 | DB/WT | Ignition Switch Power (Start-Run) | 12-14v |
| 3 | --- | Not Used | --- |
| 4 | BK/LB | Sensor Ground | <0.050v |
| 5-6 | --- | Not Used | --- |
| 7 | BK/GY | Coil 1 Driver Control | 5°, 55 mph: 8° dwell |
| 8 | GY/BK | CKP Sensor Signal | Digital Signals: 0-5-0v |
| 9 | --- | Not Used | --- |
| 10 | YL/BK | IAC 2 Driver Control | Pulse Signals |
| 11 | BR/WT | IAC 1 Driver Control | Pulse Signals |
| 12 | DB/OR | PSP Switch Sense Signal | Straight: 0v, Turning: 5v |
| 13 | YL/RD | Clutch Interlock Relay Control | Clutch Out: 12v, In: 0v |
| 14 | BR/WT | Transfer Case Position Switch | Switch Open: 12v, Closed: 0v |
| 15 | BK/RD | IAT Sensor Signal | At 100°F: 1.83v |
| 16 | TN/BK | ECT Sensor Signal | At 180°F: 2.80v |
| 17 | OR | 5-Volt Supply | 4.9-5.1v |
| 18 | TN/YL | CMP Sensor Signal | Digital Signals: 0-5-0v |
| 19 | GY/RD | IAC 3 Driver Control | DC Pulse Signals |
| 20 | VT/BK | IAC 4 Driver Control | DC Pulse Signals |
| 21, 26 | --- | Not Used | --- |
| 22 | RD/WT | Battery Power (B+) | 12-14v |
| 23 | OR/DB | TP Sensor Signal | 0.6-1.0v |
| 24 | BK/DG | HO2S-11 (B1 S1) Signal | 0.1-1.1v |
| 25 | TN/WT | HO2S-12 (B1 S2) Signal | 0.1-1.1v |
| 27 | DG/RD | MAP Sensor Signal | 1.5-1.6v |
| 28-30 | --- | Not Used | --- |
| 31 | BK/DB | Power Ground | <0.1v |
| 32 | BK/DB | Power Ground | <0.1v |

**2002-03 Liberty 3.7L VIN K 'C2' White Connector**

| PCM Pin # | Wire Color | Circuit Description (32-Pin) | Value at Hot Idle |
|---|---|---|---|
| 1-3 | --- | Not Used | --- |
| 4 | WT/DB | Injector 1 Driver Control | 1-4 ms |
| 5 | YL/WT | Injector 3 Driver Control | 1-4 ms |
| 6 | GY | Injector 5 Driver Control | 1-4 ms |
| 7-8 | --- | Not Used | --- |
| 9 | DB/TN | Coil 2 Driver Control | 5°, 55 mph: 8° dwell |
| 10 | DG | Generator Field Control | Digital Signals: 0-12-0v |
| 11 | --- | Not Used | --- |
| 12 | BR/DB | Injector 6 Driver Control | 1-4 ms |
| 13-14 | --- | Not Used | --- |
| 15 | TN | Injector 2 Driver Control | 1-4 ms |
| 16 | LB/BR | Injector 4 Driver Control | 1-4 ms |
| 17 | LG | High Speed Radiator Fan Relay | Relay Off: 12v, On: 1v |
| 18 | --- | Not Used | --- |
| 19 | DB | A/C Pressure Sensor | A/C On: 0.451-4.850v |
| 20-22 | --- | Not Used | --- |
| 23 | GY/YL | Oil Pressure Sensor | 1.6v at 24 psi |
| 24-30 | --- | Not Used | --- |
| 31 | VT/WT | 5-Volt Supply | 4.9-5.1v |
| 32 | --- | Not Used | --- |

**2002-03 Liberty 3.7L VIN K 'C3' Gray Connector**

| PCM Pin # | Wire Color | Circuit Description (32-Pin) | Value at Hot Idle |
|---|---|---|---|
| 1 | DG | A/C Clutch Relay Control | Relay Off: 12v, On: 1v |
| 2 | --- | Not Used | --- |
| 3 | DB/YL | ASD Relay Control | Relay Off: 12v, On: 1v |
| 4 | TN/RD | Speed Control Vacuum Solenoid | Vacuum Increasing: 1v |
| 5 | LG/RD | Speed Control Vent Solenoid | Vacuum Decreasing: 1v |
| 6 | TN | Clutch Switch Override Relay Control | Relay Off: 12v, On: 1v |
| 7 | --- | Not Used | --- |
| 8 | BR/OR | HO2S-11 (B1 S1) Heater Control | Heater Off: 12v, On: 1v |
| 9 | RD/YL | Oxygen Sensor Downstream Relay Control | Relay Off: 12v, On: 1v |
| 10 | WT/DG | LDP Solenoid Control | PWM Signal: 0-12-0v |
| 11 | YL/RD | Speed Control Power Supply | 12-14v |
| 12 | OR/DG | ASD Relay Output | 12-14v |
| 13 | YL/DG | Torque Management Request | Digital Signals |
| 14 | OR | LDP Switch Sense Signal | Closed: 0v, Open: 12v |
| 15 | PK/YL | Battery Temperature Sensor | At 86°F: 1.96v |
| 16 | BR/WT | HO2S-12 (B1 S2) Heater Control | Heater Off: 12v, On: 1v |
| 17 | DG/YL | Vehicle Speed Signal | Digital Signals |
| 18 | --- | Not Used | --- |
| 19 | BR | Fuel Pump Relay Control | Relay Off: 12v, On: 1v |
| 20 | PK/BK | EVAP Purge Solenoid Control | PWM Signal: 0-12-0v |
| 21 | --- | Not Used | --- |
| 22 | DB/OR | A/C Switch Sense | A/C Off: 12v, On: 1v |
| 23 | --- | Not Used | --- |
| 24 | WT/PK | Brake Switch Sense | Brake Off: 0v, On: 12v |
| 25 | WT/DB | Generator Field Source | 12-14v |
| 26 | DB/WT | Fuel Level Sensor Signal | Full: 0.5v, 1/2 full: 2.5v |
| 27 | PK | SCI Transmit | 0v |
| 28 | --- | Not Used | --- |
| 29 | LG | SCI Receive | 5v |
| 30 | VT/YL | PCI Bus Signal (J1850) | Digital Signals: 0-7-0v |
| 31 | --- | Not Used | --- |
| 32 | RD/LG | Speed Control Switch Signal | S/C & Set Switch On: 3.8v |

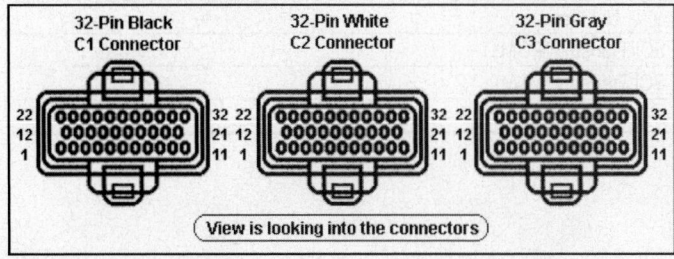

**Pin Connector Graphic**

**2004-05 Liberty 3.7L VIN K 'C1' Black/Black Connector**

| PCM Pin # | Wire Color | Circuit Description (38-Pin) | Value at Hot Idle |
|---|---|---|---|
| 1 | --- | Not Used | --- |
| 2 | --- | Not Used | --- |
| 3 | --- | Not Used | --- |
| 4 | --- | Not Used | --- |
| 5 | --- | Not Used | --- |
| 6 | --- | Not Used | --- |
| 7 | --- | Not Used | --- |
| 8 | --- | Not Used | --- |
| 9 | BK/BR | Ground | |
| 10 | DB/YL | A/C Switch Sense | |
| 11 | PK/WT | Fused Ignition Switch Output (Run-Start) | |
| 12 | PK/WT | Fused Ignition Switch Output (Run-Start) | |
| 13 | DG/YL | Vehicle Speed Signal | |
| 14 | --- | Not Used | --- |
| 15 | --- | Not Used | --- |
| 16 | --- | Not Used | --- |
| 17 | --- | Not Used | --- |
| 18 | BK/DG | Ground | |
| 19 | --- | Not Used | --- |
| 20 | VT/GY | Engine Oil Pressure Signal | |
| 21 | LB/BR | A/C Pressure Signal | |
| 22 | VT/OR | AAT Signal | |
| 23 | --- | Not Used | --- |
| 24 | --- | Not Used | --- |
| 25 | WT/LG | SCI Receive (PCM) | |
| 26 | WT/OR (3.7L A/T) | SCI Receive (TCM) | |
| 27 | YL/PK | 5 Volt Supply | |
| 28 | --- | Not Used | --- |
| 29 | RD | Fused B (+) | |
| 30 | PK/OR | Fused Ignition Switch Output (Start) | |
| 31 | DB/YL | O2 1/2 Signal | |
| 32 | BR/DG | O2 Upstream Return | |
| 33 | BR (3.7L) | O2 2/2 Signal | |
| 34 | --- | Not Used | --- |
| 35 | --- | Not Used | --- |
| 36 | WT/GY | SCI Transmit (PCM) | |
| 37 | BR/WT (3.7L A/T) | SCI Transmit (TCM) | |
| 38 | WT/VT | PCI Bus | |

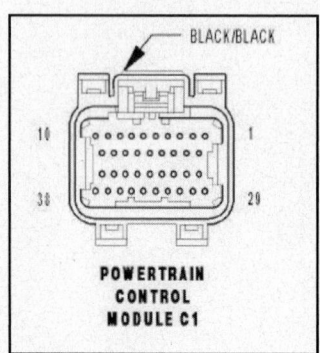

**2004-05 Liberty 3.7L VIN K 'C2' Black/Orange Connector**

| PCM Pin # | Wire Color | Circuit Description (38-Pin) | Value at Hot Idle |
|---|---|---|---|
| 1 | DB/OR (3.7L) | Coil Control No. 6 | |
| 2 | DB/YL (3.7L) | Coil Control No. 5 | |
| 3 | DB (3.7L) | Coil Control No. 4 | |
| 4 | BR/VT (3.7L) | Injector Control No. 6 | |
| 5 | BR/OR (3.7L) | Injector Control No. 5 | |
| 6 | --- | Not Used | --- |
| 7 | DB (3.7L) | Coil Control No. 3 | |
| 8 | --- | Not Used | --- |
| 9 | DB/YL | Coil Control No. 2 | |
| 10 | YL/DB | Coil Control No. 1 | |
| 11 | BR/TN | Injector Control No. 4 | |
| 12 | BR/LB | Injector Control No. 3 | |
| 13 | BR/DB | Injector Control No. 2 | |
| 14 | BR/YL | Injector Control No. 1 | |
| 15 | --- | Not Used | --- |
| 16 | --- | Not Used | --- |
| 17 | BR/VT (3.7L) | O2 2/1 Heater Control | |
| 18 | BR/TN | O2 1/1 Heater Control | |
| 19 | BR/DG | Gen Field Control | |
| 20 | VT/OR | ECT Signal | |
| 21 | BR/OR | TP Signal | |
| 22 | --- | Not Used | --- |
| 23 | VT/BR | MAP Signal | |
| 24 | BR/LG (3.7L) | Knock Sensor No. 1 Return | |
| 25 | DB/OR (3.7L) | Knock Sensor No. 1 Signal | |
| 26 | BR/WT (4WD) | Transfer Case Position Sensor Input | |
| 27 | DB/DG | Sensor Ground | |
| 28 | BR/VT | IAC Signal | |
| 29 | PK/YL | 5 Volt Supply | |
| 30 | BR/WT | IAT Signal | |
| 31 | DB/LB | O2 1/1 Signal | |
| 32 | DB/DG | O2 Downstream Return | |
| 33 | DB/LG (3.7L) | O2 2/1 Signal | |
| 34 | DB/GY | CMP Signal | |
| 35 | BR/LB | CKP Signal | |
| 36 | BR/WT (3.7L) | Knock Sensor No. 2 Signal | |
| 37 | PK/RD (3.7L) | Knock Sensor No. 2 Return | |
| 38 | VT/GY | IAC Control | |

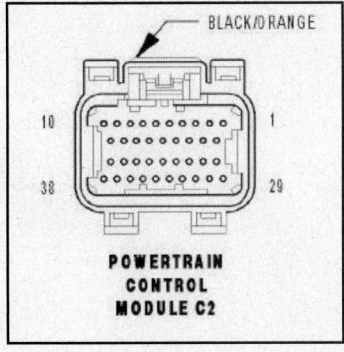

BLACK/ORANGE

10    1
38    29

POWERTRAIN
CONTROL
MODULE C2

**2004-05 Liberty 3.7L VIN K 'C3' Black/Natural Connector**

| PCM Pin # | Wire Color | Circuit Description (38-Pin) | Value at Hot Idle |
|---|---|---|---|
| 1 | --- | Not Used | --- |
| 2 | --- | Not Used | --- |
| 3 | BR/WT | ASD Relay Control | |
| 4 | DB/OR | High Speed Rad Fan Relay Control | |
| 5 | VT/OR (Exc. BASE) | S/C Vent Control | |
| 6 | DB/LG | Low Speed Rad Fan Relay Control | |
| 7 | VT/YL | S/C Power Supply | |
| 8 | VT/LB | NVLD Sol Control | |
| 9 | BR/OR | O2 1/2 Heater Control | |
| 10 | BR/GY | O2 2/2 Heater Control | |
| 11 | LB/OR | A/C Clutch Relay Control | |
| 12 | VT/YL (Exc. BASE) | S/C Vacuum Control | |
| 13 | --- | Not Used | --- |
| 14 | --- | Not Used | --- |
| 15 | --- | Not Used | --- |
| 16 | --- | Not Used | --- |
| 17 | --- | Not Used | --- |
| 18 | --- | Not Used | --- |
| 19 | PK/GY | Fused ASD Relay Output | |
| 20 | DB/WT | EVAP Purge Sol Control | |
| 21 | YL/OR (M/T) | Clutch Interlock Switch Signal | |
| 22 | --- | Not Used | --- |
| 23 | DG/WT (GAS) | Brake Switch No. 1 Signal | |
| 24 | --- | Not Used | --- |
| 25 | --- | Not Used | --- |
| 26 | --- | Not Used | --- |
| 27 | --- | Not Used | --- |
| 28 | PK/GY | Fused ASD Relay Output | |
| 29 | DB/BR | EVAP Purge Sol Signal | |
| 30 | DB/WT | P/S Pressure Signal | |
| 31 | --- | Not Used | --- |
| 32 | --- | Not Used | --- |
| 33 | DB/YL | Fuel Level Signal | |
| 34 | VT (Exc. BASE) | S/C Switch Signal No. 1 | |
| 35 | VT/WT | NVLD Switch Signal | |
| 36 | --- | Not Used | --- |
| 37 | BR | Fuel Pump Relay Control | |
| 38 | DG/OR | Engine Starter Motor Relay Control | |

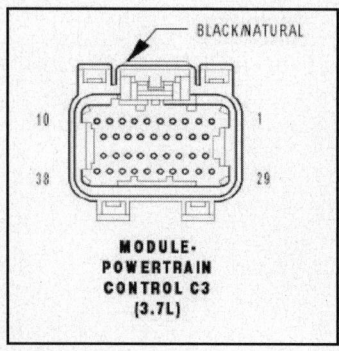

## 2004-05 Liberty 3.7L VIN K 'C4' Black/Green Connector

| PCM Pin # | Wire Color | Circuit Description (38-Pin) | Value at Hot Idle |
|---|---|---|---|
| 1 | YL/GY | OD Solenoid Control | |
| 2 | YL/LB | UD Solenoid Control | |
| 3 | --- | Not Used | --- |
| 4 | --- | Not Used | --- |
| 5 | --- | Not Used | --- |
| 6 | YL/DB | 2-4 Solenoid Control | |
| 7 | --- | Not Used | --- |
| 8 | --- | Not Used | --- |
| 9 | --- | Not Used | --- |
| 10 | DG/WT | L/R Solenoid Control | |
| 11 | --- | Not Used | |
| 12 | BK | Ground | |
| 13 | BK | Ground | |
| 14 | --- | Not Used | --- |
| 15 | DG/LB | TRS T1 Sense | |
| 16 | DG/DB | TRS T3 Sense | |
| 17 | DG | Tow/Haul Overdrive Off Switch Sense | |
| 18 | YL/DB | Transmission Control Relay Control | |
| 19 | YL/OR | Transmission Control Relay Output | |
| 20 | --- | Not Used | --- |
| 21 | --- | Not Used | --- |
| 22 | DG/TN | OD Pressure Switch Sense | |
| 23 | --- | Not Used | --- |
| 24 | --- | Not Used | --- |
| 25 | --- | Not Used | --- |
| 26 | --- | Not Used | --- |
| 27 | YL/DB | TRS T41 Sense (P/N) | |
| 28 | YL/OR | Transmission Control Relay Output | |
| 29 | YL/TN | L/R Pressure Switch Sense | |
| 30 | YL/DG | 2-4 Pressure Switch Sense | |
| 31 | YL/BR | Line Pressure Sensor Signal | |
| 32 | DG/BR | Output Speed Sensor Signal | |
| 33 | DG/WT | Input Speed Sensor Signal | |
| 34 | DG/VT | Speed Sensor Ground | |
| 35 | DG/OR | Transmission Temperature Sensor Signal | |
| 36 | --- | Not Used | --- |
| 37 | DG/YL | TRS T42 Sense | |
| 38 | --- | Not Used | --- |

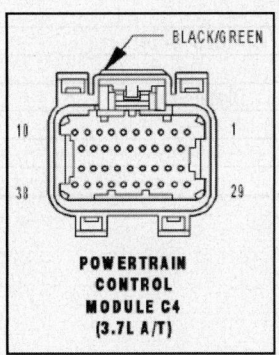

BLACK/GREEN

10    1

38    29

**POWERTRAIN CONTROL MODULE C4 (3.7L A/T)**

## 2003-04 Wrangler 2.4L VIN 1 'C1' Black Connector

| PCM Pin # | Wire Color | Circuit Description (32-Pin) | Value at Hot Idle |
|---|---|---|---|
| 1, 3 | --- | Not Used | --- |
| 2 | DB/WT | Ignition Switch Power (Start-Run) | 12-14v |
| 4 | BK/LB | Sensor Ground | <0.050v |
| 5 | --- | Not Used | --- |
| 6 | BK/WT | PNP Switch Sense | In P/N: 0v, Others: 5v |
| 7 | BK/GY | Coil 1 Driver Control | 5°, 55 mph: 8° dwell |
| 8 | GY/BK | CKP Sensor Signal | Digital Signals: 0-5-0v |
| 9 | --- | Not Used | --- |
| 10 | YL/BK | IAC 2 Driver Control | Pulse Signals |
| 11 | BR/WT | IAC 1 Driver Control | Pulse Signals |
| 12 | DB/OR | PSP Switch Sense Signal | Straight: 0v, Turning: 5v |
| 13 | YL/RD | Ignition Switch Power (Start) | Cranking: 9-11v |
| 14 | --- | Not Used | --- |
| 15 | BK/RD | IAT Sensor Signal | At 100°F: 1.83v |
| 16 | TN/BK | ECT Sensor Signal | At 180°F: 2.80v |
| 17 | OR | 5-Volt Supply | 4.9-5.1v |
| 18 | TN/YL | CMP Sensor Signal | Digital Signals: 0-5-0v |
| 19 | GY/RD | IAC 3 Driver Control | DC Pulse Signals |
| 20 | VT/BK | IAC 4 Driver Control | DC Pulse Signals |
| 21 | --- | Not Used | --- |
| 22 | RD/WT | Battery Power (B+) | 12-14v |
| 23 | OR/DB | TP Sensor Signal | 0.6-1.0v |
| 24 | BK/DG | HO2S-11 (B1 S1) Signal | 0.1-1.1v |
| 25 | TN/WT | HO2S-12 (B1 S2) Signal | 0.1-1.1v |
| 26 | --- | Not Used | --- |
| 27 | DG/RD | MAP Sensor Signal | 1.5-1.6v |
| 28-30 | --- | Not Used | --- |
| 31 | BK/TN | Power Ground | <0.1v |
| 32 | BK/TN | Power Ground | <0.1v |

## 2003-04 Wrangler 2.4L VIN 1 'C2' White Connector

| PCM Pin # | Wire Color | Circuit Description (32-Pin) | Value at Hot Idle |
|---|---|---|---|
| 1-3 | --- | Not Used | --- |
| 4 | WT/DB | Injector 1 Driver Control | 1-4 ms |
| 5 | YL/WT | Injector 3 Driver Control | 1-4 ms |
| 6-8 | --- | Not Used | --- |
| 9 | DB/TN | Coil 2 Driver Control | 5°, 55 mph: 8° dwell |
| 10 | DG | Generator Field Control | Digital Signals: 0-12-0v |
| 11-14 | --- | Not Used | --- |
| 15 | TN | Injector 2 Driver Control | 1-4 ms |
| 16 | LB/BR | Injector 4 Driver Control | 1-4 ms |
| 17 | LG | High Speed Radiator Fan Relay | Relay Off: 12v, On: 1v |
| 18 | --- | Not Used | --- |
| 19 | DB | A/C Pressure Sensor | A/C On: 0.451-4.850v |
| 20-22 | --- | Not Used | --- |
| 23 | GY/YL | Oil Pressure Sensor | 1.6v at 24 psi |
| 24-26 | --- | Not Used | --- |
| 27 | WT/OR | Vehicle Speed Signal | Digital Signals |
| 28-30 | --- | Not Used | --- |
| 31 | VT/WT | 5-Volt Supply | 4.9-5.1v |
| 32 | --- | Not Used | --- |

**2003-04 Wrangler 2.4L VIN 1 'C3' Gray Connector**

| PCM Pin # | Wire Color | Circuit Description (32-Pin) | Value at Hot Idle |
|---|---|---|---|
| 1 | OR | A/C Clutch Relay Control | Relay Off: 12v, On: 1v |
| 2 | --- | Not Used | --- |
| 3 | DB/YL | ASD Relay Control | Relay Off: 12v, On: 1v |
| 4 | TN/RD | Speed Control Vacuum Solenoid | Vacuum Increasing: 1v |
| 5 | LG/RD | Speed Control Vent Solenoid | Vacuum Decreasing: 1v |
| 6-9 | --- | Not Used | --- |
| 10 | WT/DG | LDP Solenoid Control | PWM Signal: 0-12-0v |
| 11 | YL/RD | Speed Control Power Supply | 12-14v |
| 12 | DG/PK | ASD Relay Output | 12-14v |
| 13 | YL/DG | Torque Management Request | Digital Signals |
| 14 | OR | LDP Switch Sense Signal | Closed: 0v, Open: 12v |
| 15 | PK/YL | Battery Temperature Sensor | At 86°F: 1.96v |
| 16-18 | --- | Not Used | --- |
| 19 | BR | Fuel Pump Relay Control | Relay Off: 12v, On: 1v |
| 20 | PK/BK | EVAP Purge Solenoid Control | PWM Signal: 0-12-0v |
| 21 | --- | Not Used | --- |
| 22 | DB/OR | A/C Switch Signal | A/C Off: 12v, On: 1v |
| 23 | LG | A/C Select Signal | A/C Off: 12v, On: 1v |
| 24 | WT/PK | Brake Switch Sense Signal | Brake Off: 0v, On: 12v |
| 25 | WT/DB | Generator Field Source | 12-14v |
| 26 | DB/LG | Fuel Level Sensor Signal | Full: 0.5v, 1/2 full: 2.5v |
| 27 | PK | SCI Transmit | 0v |
| 28 | --- | Not Used | --- |
| 29 | LG/WT | SCI Receive | 5v |
| 30 | VT/YL | PCI Bus Signal (J1850) | Digital Signals: 0-7-0v |
| 31 | --- | Not Used | --- |
| 32 | RD/LG | Speed Control Switch Signal | S/C & Set Switch On: 3.8v |

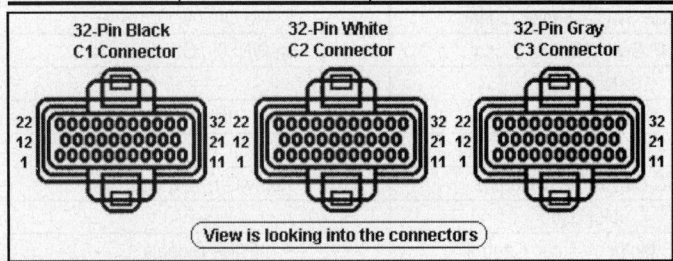

**Pin Connector Graphic**

**Standard Colors and Abbreviations**

| Abbreviation | Color | Abbreviation | Color | Abbreviation | Color |
|---|---|---|---|---|---|
| BK | Black | GY | Gray | RD | Red |
| BL | Blue | GN | Green | TN | Tan |
| BR | Brown | LG | Light Green | VT | Violet |
| DB | Dark Blue | OR | Orange | WT | White |
| DG | Dark Green | PK | Pink | YL | Yellow |

## 1993-96 Wrangler 2.5L VIN P 60-Pin Connector

| PCM Pin # | Wire Color | Circuit Description (60-Pin) | Value at Hot Idle |
|---|---|---|---|
| 1 | DG/RD | MAP Sensor Signal | 1.5-1.6v |
| 2 | TN/BK | ECT Sensor Signal | At 180°F: 2.80v |
| 3 | RD/WT | Fused Battery Power (B+) | 12-14v |
| 4 | BK/LB | Sensor Ground | <0.1v |
| 5 | BK/WT | Sensor Ground | <0.1v |
| 6 | PK/WT | 5-Volt Supply | 4.9-5.1v |
| 7 | OR | 8-Volt Supply | 7.9-8.1v |
| 8 | --- | Not Used | --- |
| 9 | DB/WT | Ignition Switch Power (B+) | 12-14v |
| 10 | --- | Not Used | --- |
| 11 | BK | Power Ground | <0.1v |
| 12 | BK | Power Ground | <0.1v |
| 13 | LB/BR | Injector 4 Driver | 1-4 ms |
| 14 | YL/WT | Injector 3 Driver | 1-4 ms |
| 15 | TN | Injector 2 Driver | 1-4 ms |
| 16 | WT/DB | Injector 1 Driver | 1-4 ms |
| 17-18 | --- | Not Used | --- |
| 19 | GY | Coil 1 Driver Control | 5°, 55 mph: 8° dwell |
| 20 | DG | Alternator Field Control | Digital Signals: 0-12-0v |
| 21 | BK/RD | IAT Sensor Signal | At 100°F: 1.83v |
| 22 | OR/DB | TP Sensor Signal | 0.6-1.0v |
| 23 | --- | Not Used | --- |
| 24 | GY/BK | CKP Sensor Signal | Digital Signals: 0-5-0v |
| 25 | BK | SCI Transmit | 0v |
| 26 | --- | Not Used | --- |
| 27 | LB | A/C Request Switch Sense | A/C Off: 12v, On: 1v |
| 28 | BR | A/C Select Switch Sense | A/C Off: 12v, On: 1v |
| 29 | WT/PK | Brake Switch Sense Signal | Brake Off: 0v, On: 12v |
| 30 | BR/YL | PNP Switch Sense Signal | In P/N: 0v, Others: 5v |
| 31 | --- | Not Used | --- |
| 32 | BK/PK | MIL (lamp) Control | MIL Off: 12v, On: 1v |
| 33 | --- | Not Used | --- |
| 34 | DB/OR | A/C Clutch Relay Control | Relay Off: 12v, On: 1v |
| 35-38 | --- | Not Used | --- |
| 39 | GY/RD | AIS Motor 1 Circuit Control | Pulse Signals |
| 40 | BR/WT | AIS Motor 3 Circuit Control | Pulse Signals |

## Standard Colors and Abbreviations

| Abbreviation | Color | Abbreviation | Color | Abbreviation | Color |
|---|---|---|---|---|---|
| BK | Black | GY | Gray | RD | Red |
| BL | Blue | GN | Green | TN | Tan |
| BR | Brown | LG | Light Green | VT | Violet |
| DB | Dark Blue | OR | Orange | WT | White |
| DG | Dark Green | PK | Pink | YL | Yellow |

**1993-96 Wrangler 2.5L VIN P 60-Pin Connector - Continued**

| PCM Pin # | Wire Color | Circuit Description (60-Pin) | Value at Hot Idle |
|---|---|---|---|
| 41 | BK/DG | HO2S-11 (B1 S1) Signal | 0.1-1.1v |
| 42 | --- | Not Used | --- |
| 43 | GY/LB | Tachometer Signal | Pulse Signals |
| 44 | TN/YL | CMP Sensor Signal | Digital Signals: 0-5-0v |
| 45 | LG | SCI Receive | 5v |
| 46 | --- | Not Used | --- |
| 47 | WT/OR | Vehicle Speed Signal | Digital Signal |
| 48-50 | --- | Not Used | --- |
| 51 | DB/YL | ASD Relay Control | Relay Off: 12v, On: 1v |
| 52-53 | --- | Not Used | --- |
| 54 | OR/BK | Shift Indicator Lamp Control | Lamp Off: 12v, On: 1v |
| 55 | --- | Not Used | --- |
| 56 ('93) | GY/PK | Maintenance Indicator Lamp | Lamp Off: 12v, On: 1v |
| 57 | DG/OR | ASD Relay Output | 12-14v |
| 58 | --- | Not Used | --- |
| 59 | PK/BK | AIS Motor 4 Circuit Control | Pulse Signals |
| 60 | YL/BK | AIS Motor 2 Circuit Control | Pulse Signals |

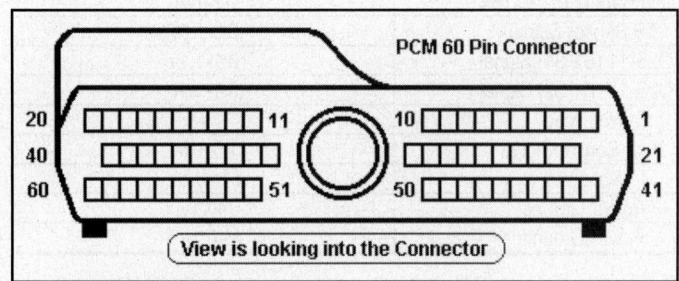

**Pin Connector Graphic**

**Standard Colors and Abbreviations**

| Abbreviation | Color | Abbreviation | Color | Abbreviation | Color |
|---|---|---|---|---|---|
| BK | Black | GY | Gray | RD | Red |
| BL | Blue | GN | Green | TN | Tan |
| BR | Brown | LG | Light Green | VT | Violet |
| DB | Dark Blue | OR | Orange | WT | White |
| DG | Dark Green | PK | Pink | YL | Yellow |

**1997 Wrangler 2.5L VIN P 'A' Black Connector**

| PCM Pin # | Wire Color | Circuit Description (32-Pin) | Value at Hot Idle |
|---|---|---|---|
| 1 | --- | Not Used | --- |
| 2 | RD/LG | Ignition Switch Power (B+) | 12-14v |
| 3 | --- | Not Used | --- |
| 4 | BR/YL | Sensor Ground | <0.1v |
| 5 | --- | Not Used | --- |
| 6 | BK/LB | PNP Switch Sense Signal | In P/N: 0v, Others: 5v |
| 7 | GY | Coil 1 Driver Control | 5°, 55 mph: 8° dwell |
| 8 | GY/BK | CKP Sensor Signal | Digital Signals: 0-5-0v |
| 9 | --- | Not Used | --- |
| 10 | YL/BK | IAC 2 Driver Control | Pulse Signals |
| 11 | BR/WT | IAC 3 Driver Control | Pulse Signals |
| 12 | DB/BR | PSP Switch Sense Signal | Straight: 0v, Turning: 5v |
| 13-14 | --- | Not Used | --- |
| 15 | BK/RD | IAT Sensor Signal | At 100°F: 1.83v |
| 16 | TN/BK | ECT Sensor Signal | At 180°F: 2.80v |
| 17 | OR | 5-Volt Supply | 4.9-5.1v |
| 18 | TN/YL | CMP Sensor Signal | Digital Signals: 0-5-0v |
| 19 | GY/RD | IAC 1 Driver Control | Pulse Signals |
| 20 | PK/BK | IAC 4 Driver Control | Pulse Signals |
| 21 | --- | Not Used | --- |
| 22 | RD/WT | Fused Battery Power (B+) | 12-14v |
| 23 | OR/DB | TP Sensor Signal | 0.6-1.0v |
| 24 | BK/DG | HO2S-11 (B1 S1) Signal | 0.1-1.1v |
| 25 | TN/WT | HO2S-12 (B1 S2) Signal | 0.1-1.1v |
| 26 | --- | Not Used | --- |
| 27 | DG/RD | MAP Sensor Signal | 1.5-1.6v |
| 28-30 | --- | Not Used | --- |
| 31 | BK/TN | Power Ground | <0.1v |
| 32 | BK/TN | Power Ground | <0.1v |

**1997 Wrangler 2.5L VIN P 'B' White Connector**

| PCM Pin # | Wire Color | Circuit Description (32-Pin) | Value at Hot Idle |
|---|---|---|---|
| 1-3 | --- | Not Used | --- |
| 4 | WT/DB | Injector 1 Driver | 1-4 ms |
| 5 | YL/WT | Injector 3 Driver | 1-4 ms |
| 6-9 | --- | Not Used | --- |
| 10 | DG | Generator Field Control | Digital Signals: 0-12-0v |
| 11 | OR/LG | TCC Solenoid Control (ATX) | TCC on at Cruise: 1v |
| 12-14 | --- | Not Used | --- |
| 15 | TN | Injector 2 Driver | 1-4 ms |
| 16 | LB/BR | Injector 4 Driver | 1-4 ms |
| 17-22 | --- | Not Used | --- |
| 23 | GY/YL | Engine Oil Pressure Sensor | 1.6v at 24 psi |
| 24-26 | --- | Not Used | --- |
| 27 | WT/OR | Vehicle Speed Signal | Digital Signal |
| 28-30 | --- | Not Used | --- |
| 31 | PK/OR | 5-Volt Supply | 4.9-5.1v |
| 32 | --- | Not Used | --- |

**1997 Wrangler 2.5L VIN P 'C' Gray Connector**

| PCM Pin # | Wire Color | Circuit Description (32-Pin) | Value at Hot Idle |
|---|---|---|---|
| 1 | DB/OR | A/C Clutch Relay Control | Relay Off: 12v, On: 1v |
| 2 | --- | Not Used | --- |
| 3 | DB/YL | ASD Relay Control | Relay Off: 12v, On: 1v |
| 4-11 | --- | Not Used | --- |
| 12 | DG/OR | ASD Relay Output | 12-14v |
| 13-14 | --- | Not Used | --- |
| 15 | PK/YL | Battery Temperature Sensor | At 86°F: 1.96v |
| 16-18 | --- | Not Used | --- |
| 19 | BR | Fuel Pump Relay Control | Relay Off: 12v, On: 1v |
| 20 | PK/BK | EVAP Purge Solenoid Control | PWM Signal: 0-12-0v |
| 21 | --- | Not Used | --- |
| 22 | DB/WT | A/C Request Signal | A/C Off: 12v, On: 1v |
| 23 | LG/DG | A/C Select Signal | A/C Off: 12v, On: 1v |
| 24 | WT/PK | Brake Switch Sense Signal | Brake Off: 0v, On: 12v |
| 25 | DG/OR | Generator Field Source | 12-14v |
| 26 | DB/LG | Fuel Level Sensor Signal | Full: 0.5v, 1/2 full: 2.5v |
| 27 | PK | SCI Transmit | 0v |
| 28 | WT/BK | CCD Bus (-) | <0.050v |
| 29 | LG | SCI Receive | 5v |
| 30 | PK/BR | CCD Bus (+) | Digital Signals: 0-5-0v |
| 31-32 | --- | --- | --- |

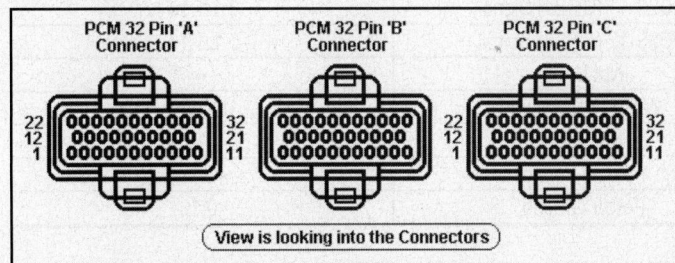

**Pin Connector Graphic**

**Standard Colors and Abbreviations**

| Abbreviation | Color | Abbreviation | Color | Abbreviation | Color |
|---|---|---|---|---|---|
| BK | Black | GY | Gray | RD | Red |
| BL | Blue | GN | Green | TN | Tan |
| BR | Brown | LG | Light Green | VT | Violet |
| DB | Dark Blue | OR | Orange | WT | White |
| DG | Dark Green | PK | Pink | YL | Yellow |

### 1998-2000 Wrangler 2.5L VIN P 'C1' Black Connector

| PCM Pin # | Wire Color | Circuit Description (32-Pin) | Value at Hot Idle |
|---|---|---|---|
| 1, 3 | --- | Not Used | --- |
| 2 ('98) | RD/LG | Ignition Switch Power (B+) | 12-14v |
| 2 | DB | Ignition Switch Power (B+) | 12-14v |
| 4 | BR/YL | Sensor Ground | <0.1v |
| 5 | --- | Not Used | --- |
| 6 | BR/LB | PNP Switch Sense Signal | In P/N: 0v, Others: 5v |
| 7 | GY | Coil 1 Driver Control | 5°, 55 mph: 8° dwell |
| 8 | GY/BK | CKP Sensor Signal | Digital Signals: 0-5-0v |
| 9 | --- | --- | --- |
| 10 | YL/BK | IAC 2 Driver Control | Pulse Signals |
| 11 | BR/WT | IAC 3 Driver Control | Pulse Signals |
| 12 | DB/BR | PSP Switch Sense Signal | Straight: 0v, Turning: 5v |
| 13-14 | --- | Not Used | --- |
| 15 | BK/RD | IAT Sensor Signal | At 100°F: 1.83v |
| 16 | TN/BK | ECT Sensor Signal | At 180°F: 2.80v |
| 17 | OR | 5-Volt Supply | 4.9-5.1v |
| 18 | TN/YL | CMP Sensor Signal | Digital Signals: 0-5-0v |
| 19 | GY/RD | IAC 1 Driver Control | Pulse Signals |
| 20 | PK/BK | IAC 4 Driver Control | Pulse Signals |
| 21 | --- | Not Used | --- |
| 22 | RD/WT | Fused Battery Power (B+) | 12-14v |
| 23 | OR/DB | TP Sensor Signal | 0.6-1.0v |
| 24 | BK/DG | HO2S-11 (B1 S1) Signal | 0.1-1.1v |
| 25 | TN/WT | HO2S-12 (B1 S2) Signal | 0.1-1.1v |
| 26 | --- | Not Used | --- |
| 27 | DG/RD | MAP Sensor Signal | 1.5-1.6v |
| 28-30 | --- | Not Used | --- |
| 31 | BK/TN | Power Ground | <0.1v |
| 32 | BK/TN | Power Ground | <0.1v |

### 1998-2000 Wrangler 2.5L VIN P 'C2' White Connector

| PCM Pin # | Wire Color | Circuit Description (32-Pin) | Value at Hot Idle |
|---|---|---|---|
| 1-3 | --- | Not Used | --- |
| 4 | WT/DB | Injector 1 Driver | 1-4 ms |
| 5 | YL/WT | Injector 3 Driver | 1-4 ms |
| 6-9 | --- | Not Used | --- |
| 10 | DG | Generator Field Control | Digital Signals: 0-12-0v |
| 11 | OR/LG | TCC Solenoid Control (ATX) | TCC on at Cruise: 1v |
| 12-14 | --- | Not Used | --- |
| 15 | TN | Injector 2 Driver | 1-4 ms |
| 16 | LB/BR | Injector 4 Driver | 1-4 ms |
| 17-22 | --- | Not Used | --- |
| 23 | GY/YL | Engine Oil Pressure Sensor | 1.6v at 24 psi |
| 24-26 | --- | Not Used | --- |
| 27 | WT/OR | Vehicle Speed Signal | Digital Signal |
| 28-30 | --- | Not Used | --- |
| 31 | PK/OR | 5-Volt Supply | 4.9-5.1v |
| 32 | --- | Not Used | --- |

### 1998-2000 Wrangler 2.5L VIN P 'C3' Gray Connector

| PCM Pin # | Wire Color | Circuit Description (32-Pin) | Value at Hot Idle |
|---|---|---|---|
| 1 | DB/OR | A/C Clutch Relay Control | Relay Off: 12v, On: 1v |
| 2 | --- | Not Used | --- |
| 3 | DB/YL | ASD Relay Control | Relay Off: 12v, On: 1v |
| 4 | TN/RD | Speed Control Vacuum Solenoid | Vacuum Increasing: 1v |
| 5 | LG/RD | Speed Control Vent Solenoid | Vacuum Decreasing: 1v |
| 6-9 | --- | Not Used | --- |
| 10 | WT/DG | LDP Solenoid Control | PWM Signal: 0-12-0v |
| 11 | YL/RD | Speed Control Power Supply | 12-14v |
| 12 | DG/PK | ASD Relay Output | 12-14v |
| 13 | --- | Not Used | --- |
| 14 | WT/OR | LDP Switch Sense Signal | Closed: 0v, Open: 12v |
| 15 | PK/YL | Battery Temperature Sensor | At 86ºF: 1.96v |
| 16-18 | --- | Not Used | --- |
| 19 | BR | Fuel Pump Relay Control | Relay Off: 12v, On: 1v |
| 20 | PK/BK | EVAP Purge Solenoid Control | PWM Signal: 0-12-0v |
| 21 | --- | Not Used | --- |
| 22 | DB/WT | A/C Switch Signal | A/C Off: 12v, On: 1v |
| 23 | LG | A/C Select Signal | A/C Off: 12v, On: 1v |
| 24 | WT/PK | Brake Switch Sense Signal | Brake Off: 0v, On: 12v |
| 25 | DG/OR | Generator Field Source | 12-14v |
| 26 | DB/LG | Fuel Level Sensor Signal | Full: 0.5v, 1/2 full: 2.5v |
| 27 | PK | SCI Transmit | 0v |
| 28 | WT/BK | CCD Bus (-) | <0.050v |
| 29 | LG | SCI Receive | 5v |
| 30 | PK/BR | CCD Bus (+) | Digital Signals: 0-5-0v |
| 31 | --- | Not Used | --- |
| 32 | RD/LG | Speed Control Switch Signal | S/C & Set Switch On: 3.8v |

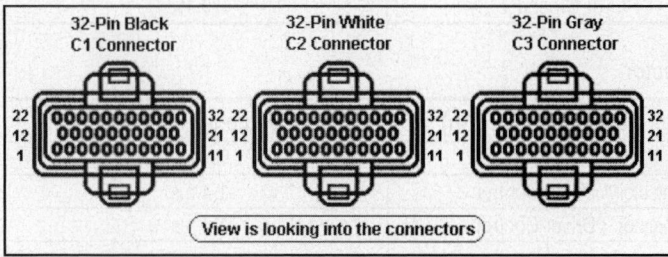

**Pin Connector Graphic**

### Standard Colors and Abbreviations

| Abbreviation | Color | Abbreviation | Color | Abbreviation | Color |
|---|---|---|---|---|---|
| BK | Black | GY | Gray | RD | Red |
| BL | Blue | GN | Green | TN | Tan |
| BR | Brown | LG | Light Green | VT | Violet |
| DB | Dark Blue | OR | Orange | WT | White |
| DG | Dark Green | PK | Pink | YL | Yellow |

**2001-02 Wrangler 2.5L VIN P 'C1' Black Connector**

| PCM Pin # | Wire Color | Circuit Description (32-Pin) | Value at Hot Idle |
|---|---|---|---|
| 1 | --- | Not Used | --- |
| 2 | DB | Ignition Switch Power (B+) | 12-14v |
| 3 | --- | Not Used | --- |
| 4 | BK/LB | Sensor Ground | <0.050v |
| 5 | --- | Not Used | --- |
| 6 | BR/LB | PNP Switch Sense Signal | In P/N: 0v, Others: 5v |
| 7 | GY | Coil 1 Driver Control | 5°, 55 mph: 8° dwell |
| 8 | GY/BK | CKP Sensor Signal | Digital Signals: 0-5-0v |
| 9 | --- | Not Used | --- |
| 10 | YL/BK | IAC 2 Driver Control | Pulse Signals |
| 11 | BR/WT | IAC 3 Driver Control | Pulse Signals |
| 12 | DB/BR | PSP Switch Sense Signal | Straight: 0v, Turning: 5v |
| 13-14 | --- | Not Used | --- |
| 15 | BK/RD | IAT Sensor Signal | At 100°F: 1.83v |
| 16 | TN/BK | ECT Sensor Signal | At 180°F: 2.80v |
| 17 | OR | 5-Volt Supply | 4.9-5.1v |
| 18 | TN/YL | CMP Sensor Signal | Digital Signals: 0-5-0v |
| 19 | GY/RD | IAC 1 Driver Control | Pulse Signals |
| 20 | VT/BK | IAC 4 Driver Control | Pulse Signals |
| 21 | --- | Not Used | --- |
| 22 | RD/WT | Fused Battery Power (B+) | 12-14v |
| 23 | OR/DB | TP Sensor Signal | 0.6-1.0v |
| 24 | BK/DG | HO2S-11 (B1 S1) Signal | 0.1-1.1v |
| 25 | TN/WT | HO2S-12 (B1 S2) Signal | 0.1-1.1v |
| 26 | --- | Not Used | --- |
| 27 | DG/RD | MAP Sensor Signal | 1.5-1.6v |
| 28-30 | --- | Not Used | --- |
| 31 | BK/TN | Power Ground | <0.1v |
| 32 | BK/TN | Power Ground | <0.1v |

**2001-02 Wrangler 2.5L VIN P 'C2' White Connector**

| PCM Pin # | Wire Color | Circuit Description (32-Pin) | Value at Hot Idle |
|---|---|---|---|
| 1-3 | --- | Not Used | --- |
| 4 | WT/DB | Injector 1 Driver Control | 1-4 ms |
| 5 | YL/WT | Injector 3 Driver Control | 1-4 ms |
| 6-9 | --- | Not Used | --- |
| 10 | DG | Generator Field Control | Digital Signals: 0-12-0v |
| 11 | OR/LG | TCC Solenoid Control (ATX) | TCC on at Cruise: 1v |
| 12-14 | --- | Not Used | --- |
| 15 | TN | Injector 2 Driver Control | 1-4 ms |
| 16 | LB/BR | Injector 4 Driver Control | 1-4 ms |
| 17-22 | --- | Not Used | --- |
| 23 | GY/YL | Engine Oil Pressure Sensor | 1.6v at 24 psi |
| 24-26 | --- | Not Used | --- |
| 27 | WT/OR | Vehicle Speed Signal | Digital Signal |
| 28-30 | --- | Not Used | --- |
| 31 | VT/OR | 5-Volt Supply | 4.9-5.1v |
| 32 | --- | Not Used | --- |

## 2001-02 Wrangler 2.5L VIN P 'C3' Gray Connector

| PCM Pin # | Wire Color | Circuit Description (32-Pin) | Value at Hot Idle |
|---|---|---|---|
| 1 | DB/OR | A/C Clutch Relay Control | Relay Off: 12v, On: 1v |
| 2 | --- | Not Used | --- |
| 3 | DB/YL | ASD Relay Control | Relay Off: 12v, On: 1v |
| 4 | TN/RD | Speed Control Vacuum Solenoid | Vacuum Increasing: 1v |
| 5 | LG/RD | Speed Control Vent Solenoid | Vacuum Decreasing: 1v |
| 6-9 | --- | Not Used | --- |
| 10 | WT/DG | LDP Solenoid Control | PWM Signal: 0-12-0v |
| 11 | YL/RD | Speed Control Power Supply | 12-14v |
| 12 | DG/PK | ASD Relay Output | 12-14v |
| 13 | --- | Not Used | --- |
| 14 | OR | LDP Switch Sense Signal | Closed: 0v, Open: 12v |
| 15 | PK/YL | Battery Temperature Sensor | At 86°F: 1.96v |
| 16-18 | --- | Not Used | --- |
| 19 | BR | Fuel Pump Relay Control | Relay Off: 12v, On: 1v |
| 20 | PK/BK | EVAP Purge Solenoid Control | PWM Signal: 0-12-0v |
| 21 | --- | Not Used | --- |
| 22 | DB/OR | A/C Switch Signal | A/C Off: 12v, On: 1v |
| 23 | LG | A/C Select Signal | A/C Off: 12v, On: 1v |
| 24 | WT/PK | Brake Switch Sense Signal | Brake Off: 0v, On: 12v |
| 25 | WT/DB | Generator Field Source | 12-14v |
| 26 | DB/LG | Fuel Level Sensor Signal | Full: 0.5v, 1/2 full: 2.5v |
| 27 | PK | SCI Transmit | 0v |
| 28 | --- | Not Used | --- |
| 29 | LG | SCI Receive | 5v |
| 30 | VT/YL | PCI Bus Signal (J1850) | Digital Signals: 0-7-0v |
| 31 | --- | Not Used | --- |
| 32 | RD/LG | Speed Control Switch Signal | S/C & Set Switch On: 3.8v |

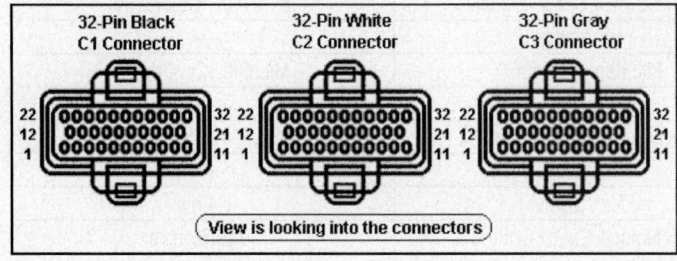

**Pin Connector Graphic**

## Standard Colors and Abbreviations

| Abbreviation | Color | Abbreviation | Color | Abbreviation | Color |
|---|---|---|---|---|---|
| BK | Black | GY | Gray | RD | Red |
| BL | Blue | GN | Green | TN | Tan |
| BR | Brown | LG | Light Green | VT | Violet |
| DB | Dark Blue | OR | Orange | WT | White |
| DG | Dark Green | PK | Pink | YL | Yellow |

**1993-95 Wrangler 4.0L VIN S 60-Pin Connector**

| PCM Pin # | Wire Color | Circuit Description (60-Pin) | Value at Hot Idle |
|---|---|---|---|
| 1 | DG/RD | MAP Sensor Signal | 1.5-1.6v |
| 2 | TN/BK | ECT Sensor Signal | At 180ºF: 2.80v |
| 3 | RD/WT | Fused Battery Power (B+) | 12-14v |
| 4 | BK/LB | Sensor Ground | <0.1v |
| 5 | BK/WT | Sensor Ground | <0.1v |
| 6 | PK/WT | 5-Volt Supply | 4.9-5.1v |
| 7 | OR | 8-Volt Supply | 7.9-8.1v |
| 8 | --- | Not Used | --- |
| 9 | DB/WT | Ignition Switch Power (B+) | 12-14v |
| 10 | --- | Not Used | --- |
| 11 | BK | Power Ground | <0.1v |
| 12 | BK | Power Ground | <0.1v |
| 13 | LB/BR | Injector 4 Driver | 1-4 ms |
| 14 | YL/WT | Injector 3 Driver | 1-4 ms |
| 15 | TN | Injector 2 Driver | 1-4 ms |
| 16 | WT/DB | Injector 1 Driver | 1-4 ms |
| 17-18 | --- | Not Used | --- |
| 19 | GY | Coil 1 Driver Control | 5º, 55 mph: 8º dwell |
| 20 | DG | Alternator Field Control | Digital Signals: 0-12-0v |
| 21 | BK/RD | Air Temperature Sensor | At 100ºF: 2.51v |
| 22 | OR/DB | TP Sensor Signal | 0.6-1.0v |
| 23 | --- | Not Used | --- |
| 24 | GY/BK | Distributor Reference Pickup | Digital Signals: 0-5-0v |
| 25 | BK | SCI Transmit | 0v |
| 26 | --- | Not Used | --- |
| 27 | LB | A/C Request Switch Sense | A/C Off: 12v, On: 1v |
| 28 | BR | A/C Select Switch Sense | A/C Off: 12v, On: 1v |
| 29 | WT/PK | Brake Switch Sense Signal | Brake Off: 0v, On: 12v |
| 30 | BR/YL | PNP Switch Sense Signal | In P/N: 0v, Others: 5v |
| 31 | --- | Not Used | --- |
| 32 | BK/PK | MIL (lamp) Control | MIL Off: 12v, On: 1v |
| 33 | --- | Not Used | --- |
| 34 | DB/OR | A/C Clutch Relay Control | Relay Off: 12v, On: 1v |
| 35-37 | --- | Not Used | --- |
| 38 | PK/BK | Injector 5 Driver | 1-4 ms |
| 39 | GY/RD | AIS Motor 1 Circuit Control | Pulse Signals |
| 40 | BR/WT | AIS Motor 3 Circuit Control | Pulse Signals |

## Standard Colors and Abbreviations

| Abbreviation | Color | Abbreviation | Color | Abbreviation | Color |
|---|---|---|---|---|---|
| BK | Black | GY | Gray | RD | Red |
| BL | Blue | GN | Green | TN | Tan |
| BR | Brown | LG | Light Green | VT | Violet |
| DB | Dark Blue | OR | Orange | WT | White |
| DG | Dark Green | PK | Pink | YL | Yellow |

**1993-95 Wrangler 4.0L VIN S 60-Pin Connector - Continued**

| PCM Pin # | Wire Color | Circuit Description (60-Pin) | Value at Hot Idle |
|---|---|---|---|
| 41 | BK/DG | HO2S-11 (B1 S1) Signal | 0.1-1.1v |
| 42 | --- | Not Used | --- |
| 43 | GY/LB | Tachometer Signal | Pulse Signals |
| 44 | TN/YL | Distributor Sync Pickup | Digital Signals: 0-5-0v |
| 45 | LG | SCI Receive | 5v |
| 46 | --- | Not Used | --- |
| 47 | WT/OR | Vehicle Speed Signal | Digital Signal |
| 48-50 | --- | Not Used | --- |
| 51 | DB/YL | ASD Relay Control | Relay Off: 12v, On: 1v |
| 52-53 | --- | Not Used | --- |
| 54 | OR/BK | M/T: SIL (lamp) Control | Lamp Off: 12v, On: 1v |
| 55 | --- | Not Used | --- |
| 56 ('93) | GY/PK | Maintenance Indicator Lamp | Lamp Off: 12v, On: 1v |
| 57 | DG/OR | ASD Relay Output | 12-14v |
| 58 | LG/BK | Injector 6 Driver | 1-4 ms |
| 59 | PK/BK | AIS Motor 4 Circuit Control | Pulse Signals |
| 60 | YL/BK | AIS Motor 2 Circuit Control | Pulse Signals |

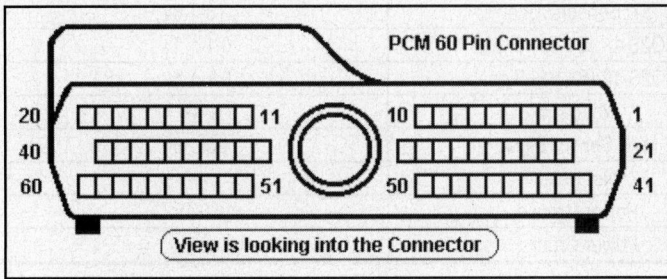

**Pin Connector Graphic**

**Standard Colors and Abbreviations**

| Abbreviation | Color | Abbreviation | Color | Abbreviation | Color |
|---|---|---|---|---|---|
| BK | Black | GY | Gray | RD | Red |
| BL | Blue | GN | Green | TN | Tan |
| BR | Brown | LG | Light Green | VT | Violet |
| DB | Dark Blue | OR | Orange | WT | White |
| DG | Dark Green | PK | Pink | YL | Yellow |

### 1997 Wrangler 4.0L VIN S 'A' Black Connector

| PCM Pin # | Wire Color | Circuit Description (32-Pin) | Value at Hot Idle |
|---|---|---|---|
| 1, 3 | --- | Not Used | --- |
| 2 | RD/LG | Ignition Switch Power (B+) | 12-14v |
| 4 | BR/YL | Sensor Ground | <0.1v |
| 5, 9 | --- | Not Used | --- |
| 6 | BR/LB | PNP Switch Sense Signal | In P/N: 0v, Others: 5v |
| 7 | GY | Coil 1 Driver Control | 5º, 55 mph: 8º dwell |
| 8 | GY/BK | CKP Sensor Signal | Digital Signals: 0-5-0v |
| 10 | YL/BK | IAC 2 Driver Control | Pulse Signals |
| 11 | BR/WT | IAC 3 Driver Control | Pulse Signals |
| 12-14 | --- | Not Used | --- |
| 15 | BK/RD | IAT Sensor Signal | At 100ºF: 1.83v |
| 16 | TN/BK | ECT Sensor Signal | At 180ºF: 2.80v |
| 17 | OR | 5-Volt Supply | 4.9-5.1v |
| 18 | TN/YL | CMP Sensor Signal | Digital Signals: 0-5-0v |
| 19 | GY/RD | IAC 1 Driver Control | Pulse Signals |
| 20 | PK/BK | IAC 4 Driver Control | Pulse Signals |
| 21 | --- | Not Used | --- |
| 22 | RD/WT | Fused Battery Power (B+) | 12-14v |
| 23 | OR/DB | TP Sensor Signal | 0.6-1.0v |
| 24 | BK/DG | HO2S-11 (B1 S1) Signal | 0.1-1.1v |
| 25 | TN/BK | HO2S-12 (B1 S2) Signal | 0.1-1.1v |
| 26 | --- | Not Used | --- |
| 27 | DG/RD | MAP Sensor Signal | 1.5-1.6v |
| 28-30 | --- | Not Used | --- |
| 31 | BK/TN | Power Ground | <0.1v |
| 32 | BK/TN | Power Ground | <0.1v |

### 1997 Wrangler 4.0L VIN S 'B' White Connector

| PCM Pin # | Wire Color | Circuit Description (32-Pin) | Value at Hot Idle |
|---|---|---|---|
| 1-3 | --- | Not Used | --- |
| 4 | WT/DB | Injector 1 Driver | 1-4 ms |
| 5 | YL/WT | Injector 3 Driver | 1-4 ms |
| 6 | PK/BK | Injector 5 Driver | 1-4 ms |
| 7-9 | --- | Not Used | --- |
| 10 | DG | Generator Field Control | Digital Signals: 0-12-0v |
| 11 | OR/LG | TCC Solenoid Control | TCC on at Cruise: 1v |
| 12 | LG/BK | Injector 6 Driver | 1-4 ms |
| 13-14 | --- | Not Used | --- |
| 15 | TN | Injector 2 Driver | 1-4 ms |
| 16 | LB/BR | Injector 4 Driver | 1-4 ms |
| 17-22 | --- | Not Used | --- |
| 23 | GY/YL | Engine Oil Pressure Sensor | 1.6v at 24 psi |
| 24-26 | --- | Not Used | --- |
| 27 | WT/OR | Vehicle Speed Signal | Digital Signal |
| 28-30 | --- | Not Used | --- |
| 31 | PK/OR | 5-Volt Supply | 4.9-5.1v |
| 32 | --- | Not Used | --- |

### 1997 Wrangler 4.0L VIN S 'C' Gray Connector

| PCM Pin # | Wire Color | Circuit Description (32-Pin) | Value at Hot Idle |
|---|---|---|---|
| 1 | DB/OR | A/C Clutch Relay Control | Relay Off: 12v, On: 1v |
| 2 | --- | Not Used | --- |
| 3 | DB/YL | ASD Relay Control | Relay Off: 12v, On: 1v |
| 4-11 | --- | Not Used | --- |
| 12 | DG/OR | ASD Relay Output | 12-14v |
| 13-14 | --- | Not Used | --- |
| 15 | PK/YL | Battery Temperature Sensor | At 86ºF: 1.96v |
| 16-18 | --- | Not Used | --- |
| 19 | BR | Fuel Pump Relay Control | Relay Off: 12v, On: 1v |
| 20 | PK/BK | EVAP Purge Solenoid Control | PWM Signal: 0-12-0v |
| 21 | --- | Not Used | |
| 22 | DB/WT | A/C Request Signal | A/C Off: 12v, On: 1v |
| 23 | LG/DG | A/C Select Signal | A/C Off: 12v, On: 1v |
| 24 | WT/PK | Brake Switch Sense Signal | Brake Off: 0v, On: 12v |
| 25 | DG/OR | Generator Field Source | 12-14v |
| 26 | DB/LG | Fuel Level Sensor Signal | Full: 0.5v, 1/2 full: 2.5v |
| 27 | PK | SCI Transmit | 0v |
| 28 | WT/BK | CCD Bus (-) | <0.050v |
| 29 | LG | SCI Receive | 5v |
| 30 | PK/BR | CCD Bus (+) | Digital Signals: 0-5-0v |
| 31-32 | --- | Not Used | --- |

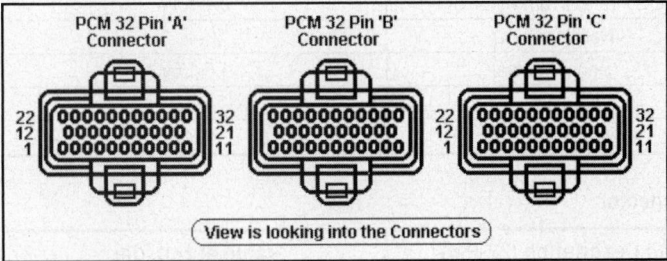

**Pin Connector Graphic**

### Standard Colors and Abbreviations

| Abbreviation | Color | Abbreviation | Color | Abbreviation | Color |
|---|---|---|---|---|---|
| BK | Black | GY | Gray | RD | Red |
| BL | Blue | GN | Green | TN | Tan |
| BR | Brown | LG | Light Green | VT | Violet |
| DB | Dark Blue | OR | Orange | WT | White |
| DG | Dark Green | PK | Pink | YL | Yellow |

**1998-2000 Wrangler 4.0L VIN S 'C1' Black Connector**

| PCM Pin # | Wire Color | Circuit Description (32-Pin) | Value at Hot Idle |
|---|---|---|---|
| 1, 3 | --- | Not Used | --- |
| 2 ('98) | RD/LG | Ignition Switch Power (B+) | 12-14v |
| 2 | DB | Ignition Switch Power (B+) | 12-14v |
| 4 | BR/YL | Sensor Ground | <0.1v |
| 5 | --- | Not Used | --- |
| 6 | BR/LB | PNP Switch Sense Signal | In P/N: 0v, Others: 5v |
| 7 | GY | Coil 1 Driver Control | 5°, 55 mph: 8° dwell |
| 8 | GY/BK | CKP Sensor Signal | Digital Signals: 0-5-0v |
| 9 | --- | Not Used | --- |
| 10 | YL/BK | IAC 2 Driver Control | Pulse Signals |
| 11 | BR/WT | IAC 3 Driver Control | Pulse Signals |
| 12-14 | --- | Not Used | --- |
| 15 | BK/RD | IAT Sensor Signal | At 100°F: 1.83v |
| 16 | TN/BK | ECT Sensor Signal | At 180°F: 2.80v |
| 17 | OR | 5-Volt Supply | 4.9-5.1v |
| 18 | TN/YL | CMP Sensor Signal | Digital Signals: 0-5-0v |
| 19 | GY/RD | IAC 1 Driver Control | Pulse Signals |
| 20 | PK/BK | IAC 4 Driver Control | Pulse Signals |
| 21 | --- | Not Used | --- |
| 22 | RD/WT | Fused Battery Power (B+) | 12-14v |
| 23 | OR/DB | TP Sensor Signal | 0.6-1.0v |
| 24 | BK/DG | HO2S-11 (B1 S1) Signal | 0.1-1.1v |
| 25 | TN/WT | HO2S-12 (B1 S2) Signal | 0.1-1.1v |
| 26 | --- | Not Used | --- |
| 27 | DG/RD | MAP Sensor Signal | 1.5-1.6v |
| 28-30 | --- | Not Used | --- |
| 31, 32 | BK/TN | Power Ground | <0.1v |

**1998-2000 Wrangler 4.0L VIN S 'C2' White Connector**

| PCM Pin # | Wire Color | Circuit Description (32-Pin) | Value at Hot Idle |
|---|---|---|---|
| 1-3 | --- | Not Used | --- |
| 4 | WT/DB | Injector 1 Driver | 1-4 ms |
| 5 | YL/WT | Injector 3 Driver | 1-4 ms |
| 6 | PK/BK | Injector 5 Driver | 1-4 ms |
| 7-9 | --- | Not Used | --- |
| 10 | DG | Generator Field Control | Digital Signals: 0-12-0v |
| 11 | OR/LG | TCC Solenoid Control (ATX) | TCC on at Cruise: 1v |
| 12 | LG/BK | Injector 6 Driver | 1-4 ms |
| 13-14 | --- | Not Used | --- |
| 15 | TN | Injector 2 Driver | 1-4 ms |
| 16 | LB/BR | Injector 4 Driver | 1-4 ms |
| 17-22 | --- | Not Used | --- |
| 23 | GY/YL | Engine Oil Pressure Sensor | 1.6v at 24 psi |
| 24-26 | --- | Not Used | --- |
| 27 | WT/OR | Vehicle Speed Signal | Digital Signal |
| 28-30 | --- | Not Used | --- |
| 31 | PK/OR | 5-Volt Supply | 4.9-5.1v |
| 32 | --- | Not Used | --- |

## 1998-2000 Wrangler 4.0L VIN S 'C3' Gray Connector

| PCM Pin # | Wire Color | Circuit Description (32-Pin) | Value at Hot Idle |
|---|---|---|---|
| 1 | DB/OR | A/C Clutch Relay Control | Relay Off: 12v, On: 1v |
| 2 | --- | Not Used | --- |
| 3 | DB/YL | ASD Relay Control | Relay Off: 12v, On: 1v |
| 4 | TN/RD | Speed Control Vacuum Solenoid | Vacuum Increasing: 1v |
| 5 | LG/RD | Speed Control Vent Solenoid | Vacuum Decreasing: 1v |
| 6-9 | --- | Not Used | --- |
| 10 | WT/DG | LDP Solenoid Control | PWM Signal: 0-12-0v |
| 11 | YL/RD | Speed Control Power Supply | 12-14v |
| 12 | DG/PK | ASD Relay Output | 12-14v |
| 13 | --- | Not Used | --- |
| 14 | WT/OR | LDP Switch Sense Signal | Closed: 0v, Open: 12v |
| 15 | PK/YL | Battery Temperature Sensor | At 86°F: 1.96v |
| 16-18 | --- | Not Used | --- |
| 19 | BR | Fuel Pump Relay Control | Relay Off: 12v, On: 1v |
| 20 | PK/BK | EVAP Purge Solenoid Control | PWM Signal: 0-12-0v |
| 21 | --- | Not Used | --- |
| 22 | DB/WT | A/C Switch Signal | A/C Off: 12v, On: 1v |
| 23 | LG | A/C Select Signal | A/C Off: 12v, On: 1v |
| 24 | WT/PK | Brake Switch Sense Signal | Brake Off: 0v, On: 12v |
| 25 | DG/OR | Generator Field Source | 12-14v |
| 26 | DB/LG | Fuel Level Sensor Signal | Full: 0.5v, 1/2 full: 2.5v |
| 27 | PK | SCI Transmit | 0v |
| 28 | WT/BK | CCD Bus (-) | <0.050v |
| 29 | LG | SCI Receive | 5v |
| 30 | PK/BR | CCD Bus (+) | Digital Signals: 0-5-0v |
| 31 | --- | Not Used | --- |
| 32 | RD/LG | Speed Control Switch Signal | S/C & Set Switch On: 3.8v |

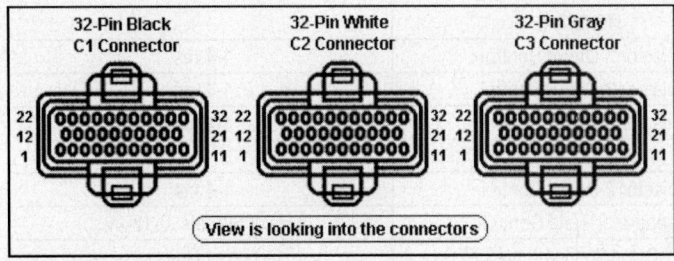

**Pin Connector Graphic**

## Standard Colors and Abbreviations

| Abbreviation | Color | Abbreviation | Color | Abbreviation | Color |
|---|---|---|---|---|---|
| BK | Black | GY | Gray | RD | Red |
| BL | Blue | GN | Green | TN | Tan |
| BR | Brown | LG | Light Green | VT | Violet |
| DB | Dark Blue | OR | Orange | WT | White |
| DG | Dark Green | PK | Pink | YL | Yellow |

## 2001 Wrangler 4.0L VIN S 'A' Black Connector

| PCM Pin # | Wire Color | Circuit Description (32-Pin) | Value at Hot Idle |
|-----------|-----------|------------------------------|-------------------|
| 1 | RD/YL | Ignition Coil 3 Driver Control | 5°, 55 mph: 8° dwell |
| 2 | DB | Ignition Switch Power (B+) | 12-14v |
| 3 | --- | Not Used | --- |
| 4 | BK/LB | Sensor Ground | <0.050v |
| 5, 9, 12-14, 21 | --- | Not Used | --- |
| 6 | BR/LB | PNP Switch Sense Signal | In P/N: 0v, Others: 5v |
| 7 | GY | Ignition Coil 1 Driver Control | 5°, 55 mph: 8° dwell |
| 8 | GY/BK | CKP Sensor Signal | Digital Signals: 0-5-0v |
| 10 | YL/BK | IAC 2 Driver Control | Pulse Signals |
| 11 | BR/WT | IAC 3 Driver Control | Pulse Signals |
| 15 | BK/RD | IAT Sensor Signal | At 100°F: 1.83v |
| 16 | TN/BK | ECT Sensor Signal | At 180°F: 2.80v |
| 17 | OR | 5-Volt Supply | 4.9-5.1v |
| 18 | TN/YL | CMP Sensor Signal | Digital Signals: 0-5-0v |
| 19 | GY/RD | IAC 1 Driver Control | Pulse Signals |
| 20 | VT/BK | IAC 4 Driver Control | Pulse Signals |
| 22 | RD/WT | Fused Battery Power (B+) | 12-14v |
| 23 | OR/DB | TP Sensor Signal | 0.6-1.0v |
| 24 | BK/DG | HO2S-11 (B1 S1) Signal | 0.1-1.1v |
| 25 | TN/WT | HO2S-12 (B1 S2) Signal | 0.1-1.1v |
| 26 | LG/RD | HO2S-21 (B2 S1) Signal (California) | 0.1-1.1v |
| 27 | DG/RD | MAP Sensor Signal | 1.5-1.6v |
| 28 | --- | Not Used | --- |
| 29 | TN | HO2S-22 (B2 S2) Signal (California) | 0.1-1.1v |
| 30 | --- | Not Used | --- |
| 31-32 | BK/TN | Power Ground | <0.1v |

## 2001 Wrangler 4.0L VIN S 'B' White Connector

| PCM Pin # | Wire Color | Circuit Description (32-Pin) | Value at Hot Idle |
|-----------|-----------|------------------------------|-------------------|
| 1-3 | --- | Not Used | --- |
| 4 | WT/DB | Injector 1 Driver Control | 1-4 ms |
| 5 | YL/WT | Injector 3 Driver Control | 1-4 ms |
| 6 | PK/BK | Injector 5 Driver Control | 1-4 ms |
| 7-8 | --- | Not Used | --- |
| 9 | DB/TN | Injector 2 Driver Control | 1-4 ms |
| 10 | DG | Generator Field Control | Digital Signals: 0-12-0v |
| 11 | OR/LG | TCC Solenoid Control (ATX) | TCC on at Cruise: 1v |
| 12 | LG/BK | Injector 6 Driver Control | 1-4 ms |
| 13-14 | --- | Not Used | --- |
| 15 | TN | Injector 2 Driver Control | 1-4 ms |
| 16 | LB/BR | Injector 4 Driver Control | 1-4 ms |
| 17-22 | --- | Not Used | --- |
| 23 | GY/YL | Engine Oil Pressure Sensor | 1.6v at 24 psi |
| 24-26 | --- | Not Used | --- |
| 27 | WT/OR | Vehicle Speed Signal | Digital Signal |
| 28-30 | --- | Not Used | --- |
| 31 | VT/OR | 5-Volt Supply | 4.9-5.1v |
| 32 | --- | Not Used | --- |

### 2001 Wrangler 4.0L MFI VIN S 'C' Gray Connector

| PCM Pin # | Wire Color | Circuit Description (32-Pin) | Value at Hot Idle |
|---|---|---|---|
| 1 | DB/OR | A/C Clutch Relay Control | Relay Off: 12v, On: 1v |
| 2 | --- | Not Used | --- |
| 3 | DB/YL | ASD Relay Control | Relay Off: 12v, On: 1v |
| 4 | TN/RD | Speed Control Vacuum Solenoid | Vacuum Increasing: 1v |
| 5 | LG/RD | Speed Control Vent Solenoid | Vacuum Decreasing: 1v |
| 6-7 | --- | Not Used | --- |
| 8 | BR/OR | HO2S-11 Heater Relay | Relay Off: 12v, On: 1v |
| 9 | RD/YL | HO2S-12 Heater Relay | Relay Off: 12v, On: 1v |
| 10 | WT/DG | LDP Solenoid Control | PWM Signal: 0-12-0v |
| 11 | YL/RD | Speed Control Power Supply | 12-14v |
| 12 | DG/PK | ASD Relay Output | 12-14v |
| 13 | --- | Not Used | --- |
| 14 | OR | LDP Switch Sense Signal | Closed: 0v, Open: 12v |
| 15 | PK/YL | Battery Temperature Sensor | At 86°F: 1.96v |
| 16 | BR/WT | HO2S-21 Heater Control | Relay Off: 12v, On: 1v |
| 17-18 | --- | Not Used | --- |
| 19 | BR | Fuel Pump Relay Control | Relay Off: 12v, On: 1v |
| 20 | PK/BK | EVAP Purge Solenoid Control | PWM Signal: 0-12-0v |
| 21 | --- | Not Used | --- |
| 22 | DB/OR | A/C Switch Signal | A/C Off: 12v, On: 1v |
| 23 | LG | A/C Select Signal | A/C Off: 12v, On: 1v |
| 24 | WT/PK | Brake Switch Sense Signal | Brake Off: 0v, On: 12v |
| 25 | WT/DB | Generator Field Source | 12-14v |
| 26 | DB/LG | Fuel Level Sensor Signal | Full: 0.5v, 1/2 full: 2.5v |
| 27 | PK | SCI Transmit | 0v |
| 28 | --- | Not Used | --- |
| 29 | LG | SCI Receive | 5v |
| 30 | VT/YL | PCI Bus Signal (J1850) | Digital Signals: 0-7-0v |
| 31 | --- | Not Used | --- |
| 32 | RD/LG | Speed Control Switch Signal | S/C & Set Switch On: 3.8v |

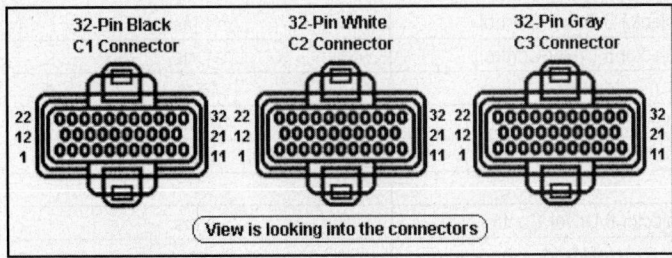

**Pin Connector Graphic**

### Standard Colors and Abbreviations

| Abbreviation | Color | Abbreviation | Color | Abbreviation | Color |
|---|---|---|---|---|---|
| BK | Black | GY | Gray | RD | Red |
| BL | Blue | GN | Green | TN | Tan |
| BR | Brown | LG | Light Green | VT | Violet |
| DB | Dark Blue | OR | Orange | WT | White |
| DG | Dark Green | PK | Pink | YL | Yellow |

## 2002-04 Wrangler 4.0L VIN S 'A' Black Connector

| PCM Pin # | Wire Color | Circuit Description (32-Pin) | Value at Hot Idle |
|---|---|---|---|
| 1 | RD/YL | Ignition Coil 3 Driver Control | 5°, 55 mph: 8° dwell |
| 2 | DB/WT | Ignition Switch Power (Start-Run) | 12-14v |
| 3, 5, 9, 12-14 | --- | Not Used | --- |
| 4 | BK/LB | Sensor Ground | <0.050v |
| 6 | BK/WT | PNP Switch Sense Signal | In P/N: 0v, Others: 5v |
| 7 | BR/GY | Ignition Coil 1 Driver Control | 5°, 55 mph: 8° dwell |
| 8 | GY/BK | CKP Sensor Signal | Digital Signals: 0-5-0v |
| 10 | YL/BK | IAC 2 Driver Control | DC Pulse Signals |
| 11 | BR/WT | IAC 3 Driver Control | DC Pulse Signals |
| 13 | YL/RD | Ignition Switch Power (Start) | 12-14v |
| 15 | BK/RD | IAT Sensor Signal | At 100°F: 1.83v |
| 16 | TN/BK | ECT Sensor Signal | At 180°F: 2.80v |
| 17 | OR | 5-Volt Supply | 4.9-5.1v |
| 18 | TN/YL | CMP Sensor Signal | Digital Signals: 0-5-0v |
| 19 | GY/RD | IAC 1 Driver Control | DC Pulse Signals |
| 20 | VT/BK | IAC 4 Driver Control | DC Pulse Signals |
| 21 | --- | Not Used | --- |
| 22 | RD/WT | Battery Power (B+) | 12-14v |
| 23 | OR/DB | TP Sensor Signal | 0.6-1.0v |
| 24 | BK/DG | HO2S-11 (B1 S1) Signal | 0.1-1.1v |
| 25 | TN/WT | HO2S-12 (B1 S2) Signal | 0.1-1.1v |
| 26 | LG/RD | HO2S-21 (B2 S1) Signal | 0.1-1.1v |
| 27 | DG/RD | MAP Sensor Signal | 1.5-1.6v |
| 28 | --- | Not Used | --- |
| 29 | TN/WT | HO2S-22 (B2 S2) Signal | 0.1-1.1v |
| 30 | --- | Not Used | --- |
| 31-32 | BK/TN | Power Ground | <0.1v |

## 2002-04 Wrangler 4.0L VIN S 'B' White Connector

| PCM Pin # | Wire Color | Circuit Description (32-Pin) | Value at Hot Idle |
|---|---|---|---|
| 1-3 | --- | Not Used | --- |
| 4 | WT/DB | Injector 1 Driver Control | 1-4 ms |
| 5 | YL/WT | Injector 3 Driver Control | 1-4 ms |
| 6 | GY | Injector 5 Driver Control | 1-4 ms |
| 7-8 | --- | Not Used | --- |
| 9 | DB/TN | Injector 2 Driver Control | 1-4 ms |
| 10 | DG | Generator Field Control | Digital Signals: 0-12-0v |
| 11 | --- | Not Used | --- |
| 12 | BR/DB | Injector 6 Driver Control | 1-4 ms |
| 13-14 | --- | Not Used | --- |
| 15 | TN | Injector 2 Driver Control | 1-4 ms |
| 16 | LB/BR | Injector 4 Driver Control | 1-4 ms |
| 17-22 | --- | Not Used | --- |
| 23 | GY/YL | Oil Pressure Sensor | 1.6v at 24 psi |
| 24-26 | --- | Not Used | --- |
| 27 | WT/OR | Vehicle Speed Signal | Digital Signals |
| 28-30 | --- | Not Used | --- |
| 31 | VT/WT | 5-Volt Supply | 4.9-5.1v |
| 32 | --- | Not Used | --- |

## 2002-04 Wrangler 4.0L MFI VIN S 'C' Gray Connector

| PCM Pin # | Wire Color | Circuit Description (32-Pin) | Value at Hot Idle |
|---|---|---|---|
| 1 | DB/OR | A/C Clutch Relay Control | Relay Off: 12v, On: 1v |
| 2 | --- | Not Used | --- |
| 3 | DB/YL | ASD Relay Control | Relay Off: 12v, On: 1v |
| 4 | TN/RD | Speed Control Vacuum Solenoid | Vacuum Increasing: 1v |
| 5 | LG/RD | Speed Control Vent Solenoid | Vacuum Decreasing: 1v |
| 6-7 | --- | Not Used | --- |
| 8 | BR/OR | Oxygen Sensor Upstream Relay Control | Relay Off: 12v, On: 1v |
| 9 | RD/YL | HO2S-12 Heater Control | Heater Off: 12v, On: 1v |
| 10 | WT/DG | LDP Solenoid Control | PWM Signal: 0-12-0v |
| 11 | YL/RD | Speed Control Power Supply | 12-14v |
| 12 | DG/PK | ASD Relay Output | 12-14v |
| 13 | YL/DG | Torque Management Request Signal | Digital Signals |
| 14 | OR | LDP Switch Sense Signal | Closed: 0v, Open: 12v |
| 15 | PK/YL | Battery Temperature Sensor | At 86°F: 1.96v |
| 16 | BR/WT | HO2S-21 Heater Control | Relay Off: 12v, On: 1v |
| 17-18 | --- | Not Used | --- |
| 19 | BR | Fuel Pump Relay Control | Relay Off: 12v, On: 1v |
| 20 | PK/BK | EVAP Purge Solenoid Control | PWM Signal: 0-12-0v |
| 21 | --- | Not Used | --- |
| 22 | DB/OR | A/C Switch Sense | A/C Off: 12v, On: 1v |
| 23 | LG | A/C Select Signal | A/C Off: 12v, On: 1v |
| 24 | WT/PK | Brake Switch Sense | Brake Off: 0v, On: 12v |
| 25 | WT/DB | Generator Field Source | 12-14v |
| 26 | DB/LG | Fuel Level Sensor Signal | Full: 0.5v, 1/2 full: 2.5v |
| 27 | PK | SCI Transmit | 0v |
| 28 | --- | Not Used | --- |
| 29 | LG/WT | SCI Receive | 5v |
| 30 | VT/YL | PCI Bus Signal (J1850) | Digital Signals: 0-7-0v |
| 31 | --- | Not Used | --- |
| 32 | RD/LB | Speed Control Switch Signal | S/C & Set Switch On: 3.8v |

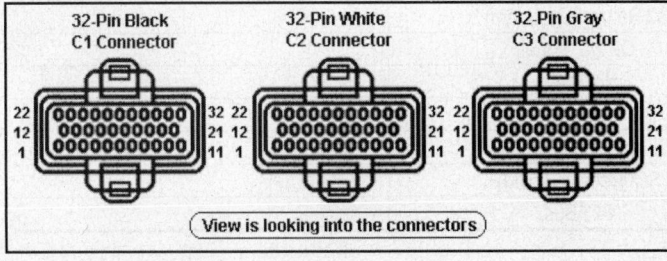

**Pin Connector Graphic**

### Standard Colors and Abbreviations

| Abbreviation | Color | Abbreviation | Color | Abbreviation | Color |
|---|---|---|---|---|---|
| BK | Black | GY | Gray | RD | Red |
| BL | Blue | GN | Green | TN | Tan |
| BR | Brown | LG | Light Green | VT | Violet |
| DB | Dark Blue | OR | Orange | WT | White |
| DG | Dark Green | PK | Pink | YL | Yellow |

**2005 Wrangler 4.0L VIN S 'C1' Black/Black Connector**

| PCM Pin # | Wire Color | Circuit Description (38-Pin) | Value at Hot Idle |
|---|---|---|---|
| 1 | --- | Not Used | --- |
| 2 | --- | Not Used | --- |
| 3 | --- | Not Used | --- |
| 4 | --- | Not Used | --- |
| 5 | --- | Not Used | --- |
| 6 | --- | Not Used | --- |
| 7 | --- | Not Used | --- |
| 8 | --- | Not Used | --- |
| 9 | BK/BR | Ground | |
| 10 | DB/YL (2.4L) | A/C Request Signal | |
| 10 | DB/YL (4.0L) | A/C Request Signal | |
| 11 | PK/WT | Fused Ignition Switch Output (Run-Start) | |
| 12 | PK/WT (A/T) | Fused Ignition Switch Output (Run-Start) | |
| 13 | DB/OR | Vehicle Speed Sensor Signal | |
| 14 | --- | Not Used | --- |
| 15 | --- | Not Used | --- |
| 16 | --- | Not Used | --- |
| 17 | --- | Not Used | --- |
| 18 | BK/DG | Ground | |
| 19 | --- | Not Used | --- |
| 20 | VT/GY | Oil Pressure Signal | |
| 21 | --- | Not Used | --- |
| 22 | VT/LG | AAT Signal | |
| 23 | --- | Not Used | --- |
| 24 | --- | Not Used | --- |
| 25 | WT/LG | SCI Receive (PCM) | |
| 26 | WT/OR | SCI Receive (TCM) | |
| 27 | YL/PK | 5 Volt Supply | |
| 28 | DG/LB | Overdrive Off Switch Indicator | |
| 29 | RD | Fused B (+) | |
| 30 | PK/OR | Ignition Switch Output (Start) | |
| 31 | DB/YL | O2 1/2 Signal | |
| 32 | BR/DG | O2 Return (Upstream) | |
| 33 | BR (4.0L) | O2 2/2 Signal | |
| 34 | --- | Not Used | --- |
| 35 | --- | Not Used | --- |
| 36 | WT/BR | SCI Transmit (PCM/Cab) | |
| 37 | WT/DG | SCI Receive (TCM) | |
| 38 | WT/VT | PCI Bus | |

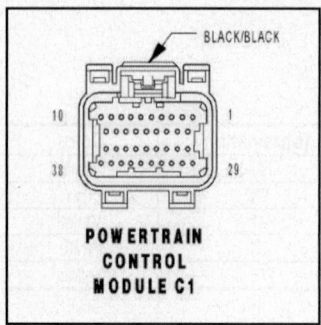

BLACK/BLACK

10    1

38    29

**POWERTRAIN CONTROL MODULE C1**

## 2005 Wrangler 4.0L VIN S 'C2' Black/Orange Connector

| PCM Pin # | Wire Color | Circuit Description (38-Pin) | Value at Hot Idle |
|---|---|---|---|
| 1 | --- | Not Used | --- |
| 2 | --- | Not Used | --- |
| 3 | --- | Not Used | --- |
| 4 | BR/VT (4.0L) | Injector Control No. 6 | |
| 5 | BR/OR (4.0L) | Injector Control No. 5 | |
| 6 | --- | Not Used | --- |
| 7 | DB/OR (4.0L) | Coil Control No. 3 | |
| 8 | --- | Not Used | --- |
| 9 | DB/TN | Coil Control No. 2 | |
| 10 | DB/DG | Coil Control No. 1 | |
| 11 | BR/TN | Injector Control No. 4 | |
| 12 | BR/LB | Injector Control No. 3 | |
| 13 | BR/DB | Injector Control No. 2 | |
| 14 | BR/YL | Injector Control No. 1 | |
| 15 | --- | Not Used | --- |
| 16 | --- | Not Used | --- |
| 17 | BR/VT | O2 2/1 Heater Control | |
| 18 | BR/LG | O2 1/1 Heater Control | |
| 19 | BR/DG | Gen Field Control | |
| 20 | VT/OR | ECT Signal | |
| 21 | BR/OR | TP No. 1 Signal | |
| 22 | --- | Not Used | --- |
| 23 | VT/BR | MAP Signal | |
| 24 | --- | Not Used | --- |
| 25 | --- | Not Used | --- |
| 26 | LG (Exc. Off-Rd. Pkg.) | Fused Sensor Ground | |
| 26 | BR/WT (Off-Rd. Pkg.) | Transfer Case Position Sensor Input | |
| 27 | DB/DG | Sensor Ground | |
| 28 | BR/VT | IAC Signal | |
| 29 | PK/YL | 5 Volt Supply | |
| 30 | DB/LG (2.4L) | IAT Signal | |
| 30 | BK/RD (4.0L) | IAT Signal | |
| 31 | DB/LB | O2 1/1 Signal | |
| 32 | DB/DG | O2 Return (Downstream) | |
| 33 | DB/LG (4.0L) | O2 2/1 Signal | |
| 34 | DB/GY | CMP Signal | |
| 35 | BR/LB | CKP Signal | |
| 36 | --- | Not Used | --- |
| 37 | --- | Not Used | --- |
| 38 | VT/GY | IAC Control | |

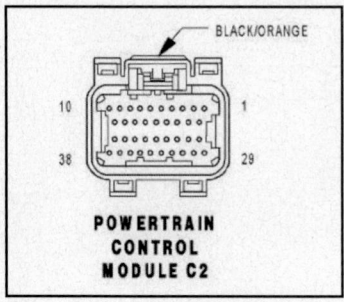

BLACK/ORANGE

10   1
38   29

**POWERTRAIN
CONTROL
MODULE C2**

### 2005 Wrangler 4.0L VIN S 'C3' Black/Natural Connector

| PCM Pin # | Wire Color | Circuit Description (38-Pin) | Value at Hot Idle |
|---|---|---|---|
| 1 | --- | Not Used | --- |
| 2 | --- | Not Used | --- |
| 3 | BR/WT | ASD Relay Control | |
| 4 | DB/OR | Radiator Fan Relay Control | |
| 5 | VT/LG (Speed Control) | S/C Vent Sol Control | |
| 6 | DB/LG | Low Speed Radiator Fan Relay Control | |
| 7 | VT/YL | S/C Supply | |
| 8 | VT/LB | NVLD Sol Control | |
| 9 | BR/WT | O2 1/2 Heater Control | |
| 10 | BR/GY (4.0L) | O2 2/2 Heater Control | |
| 11 | LB/OR | A/C Clutch Relay Control | |
| 12 | VT/YL | S/C Vacuum Sol Control | |
| 13 | --- | Not Used | --- |
| 14 | --- | Not Used | --- |
| 15 | --- | Not Used | --- |
| 16 | --- | Not Used | --- |
| 17 | --- | Not Used | --- |
| 18 | --- | Not Used | --- |
| 19 | BR | ASD Relay Output | |
| 20 | DB/WT | EVAP/Purge Sol Control | |
| 21 | YL/OR | Clutch Interlock Switch Sense | |
| 22 | --- | Not Used | --- |
| 23 | DG/WT | Brake Switch No. 1 Signal | |
| 24 | LB/OR | A/C Select Signal | |
| 25 | --- | Not Used | --- |
| 26 | --- | Not Used | --- |
| 27 | --- | Not Used | --- |
| 28 | BR | ASD Relay Output | |
| 29 | DB/BR | EVAP/Purge Sol Signal | |
| 30 | DB/WT | P/S Pressure Switch Signal | |
| 31 | LB/BR | A/C Pressure Signal | |
| 32 | DB/VT | Battery Temp Signal | |
| 33 | DB/WT | Fuel Level Sensor Signal | |
| 34 | VT | S/C Switch No. 1 Signal | |
| 35 | VT/WT | NVLD Switch Signal | |
| 36 | --- | Not Used | --- |
| 37 | BR | Fuel Pump Relay Control | |
| 38 | DG/OR | Starter Motor Relay Control | |

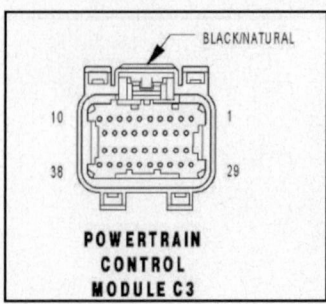

POWERTRAIN
CONTROL
MODULE C3

## 2005 Wrangler 4.0L VIN S 'C4' Black/Green Connector

| PCM Pin # | Wire Color | Circuit Description (38-Pin) | Value at Hot Idle |
|---|---|---|---|
| 1 | YL/GY | OD Solenoid Control | |
| 2 | YL/LB | UD Solenoid Control | |
| 3 | --- | Not Used | --- |
| 4 | --- | Not Used | --- |
| 5 | --- | Not Used | --- |
| 6 | YL/DB | 2-4 Solenoid Control | |
| 7 | --- | Not Used | --- |
| 8 | --- | Not Used | --- |
| 9 | --- | Not Used | --- |
| 10 | DG/WT | L/R Solenoid Control | |
| 11 | YL/GY | MS Solenoid Control | |
| 12 | BK | Ground | |
| 13 | BK | Ground | |
| 14 | --- | Not Used | --- |
| 15 | DG/LB | TRS T1 Sense | |
| 16 | DG/DB | TRS T3 Sense | |
| 17 | DG | Tow/Haul Overdrive Off Switch Sense | |
| 18 | YL/DB | Transmission Control Relay Control | |
| 19 | YL/OR | Transmission Control Relay Output | |
| 20 | --- | Not Used | --- |
| 21 | --- | Not Used | --- |
| 22 | DG/TN | OD Pressure Switch Sense | |
| 23 | --- | Not Used | --- |
| 24 | --- | Not Used | --- |
| 25 | --- | Not Used | --- |
| 26 | --- | Not Used | --- |
| 27 | YL/DB | TRS T41 Sense (P/N) | |
| 28 | YL/OR | Transmission Control Relay Output | |
| 29 | YL/TN | L/R Pressure Switch Sense | |
| 30 | YL/DG | 2-4 Pressure Switch Sense | |
| 31 | YL/BR | Line Pressure Sensor Signal | |
| 32 | DG/BR | Output Speed Sensor Signal | |
| 33 | DG/OR | Input Speed Sensor Signal | |
| 34 | DG/VT | Speed Sensor Ground | |
| 35 | DG/OR | Transmission Temperature Sensor Signal | |
| 36 | --- | Not Used | --- |
| 37 | DG/YL | TRS T42 Sense | |
| 38 | --- | Not Used | --- |

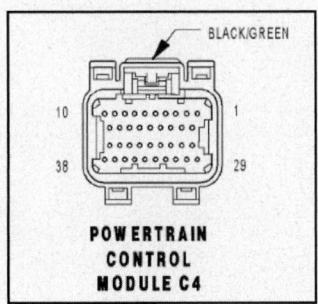

BLACK/GREEN

10    1
38    29

**POWERTRAIN
CONTROL
MODULE C4**